Chapters

Clinical Cases

1	An Introduction to Anatomy and Physiology	27	Using A&P to Save a Life	28
2	The Chemical Level of Organization	52	What is Wrong with my Baby?	53
3	The Cellular Level of Organization	90	When Your Heart is in the Wrong Place	91
4	The Tissue Level of Organization	139	The Rubber Girl	140
5	The Integumentary System	176	Skin Cells in Overdrive	177
6	Osseous Tissue and Bone Structure	204	A Case of Child Abuse?	205
7	The Axial Skeleton	232	Knocked Out	233
8	The Appendicular Skeleton	267	The Orthopedic Surgeon's Nightmare	268
9	Joints	289	What's Ailing the Birthday Girl?	290
10	Muscle Tissue	315	A Real Eye Opener	316
11	The Muscular System	358	The Weekend Warrior	359
12	Neural Tissue	411	Did FDR Really Have Polio?	412
13	The Spinal Cord, Spinal Nerves, and Spinal Reflexes	455	Prom Night	456
14	The Brain and Cranial Nerves	487	The Neuroanatomist's Stroke	488
15	Sensory Pathways and the Somatic Nervous System	534	Living with Cerebral Palsy	535
16	The Autonomic Nervous System and Higher-Order Functions	557	The First Day in Anatomy Lab	558
17	The Special Senses	589	A Chance to See	590
18	The Endocrine System	634	Stones, Bones, and Groans	635
19	Blood	678	A Mysterious Blood Disorder	679
20	The Heart	710	A Needle to the Chest	711
21	Blood Vessels and Circulation	749	Did Ancient Mummies Have Atherosclerosis?	750
22	The Lymphatic System and Immunity	807	Isn't There a Vaccine for That?	808
23	The Respiratory System	856	How Long Should a Cough Last?	857
24	The Digestive System	906	An Unusual Transplant	907
25	Metabolism and Energetics	961	The Miracle Supplement	962
26	The Urinary System	998	A Case of "Hidden" Bleeding	999
27	Fluid, Electrolyte, and Acid–Base Balance	1041	When Treatment Makes You Worse	1042
28	The Reproductive System	1076	A Post-Game Mystery	1077
29	Development and Inheritance	1121	The Twins That Looked Nothing Alike	1122

Spotlight Figures

1–1 Levels of Organization   1–10 Diagnostic Imaging Techniques

2–3 Chemical Notation

3–1 Anatomy of a Model Cell   3–7 Protein Synthesis, Processing, and Packaging   3–22 Membrane Transport
3–23 DNA replication   3–24 Stages of a Cell's Life Cycle

4–20 Inflammation and Regeneration

5–3 The Epidermis

6–11 Endochondral Ossification   6–16 Types of Fractures and Steps in Repair

7–4 Sectional Anatomy of the Skull

8–10 Sex Differences in the Human Skeleton

9–2 Joint Movement

10–9 Events at the Neuromuscular Junction   10–10 Excitation-Contraction Coupling
10–11 The Contraction Cycle & Cross-Bridge Formation

11–3 Muscle Action

12–9 Resting Membrane Potential   12–14 Generation of an Action Potential   12–15 Propagation of an Action Potential

13–8 Peripheral Distribution of Spinal Nerves   13–14 Reflexes

14–4 Formation & Circulation of Cerebrospinal Fluid

15–6 Somatic Sensory Pathways

16–2 Overview of the Autonomic Nervous System

17–2 Olfaction and Gustation   17–16 Refractive Problems   17–16 Photoreception

18–2 Structural Classification of Hormones   18–3 G Proteins and Second Messengers   18–18 Diabetes Mellitus
18–20 The General Adaptation Syndrome

19–1 The Composition of Whole Blood   19–9 Hemolytic Disease of the Newborn

20–10 Heart Disease and Heart Attacks   20–14 Cardiac Arrhythmias

21–33 Congenital Heart Problems

22–28 Cytokines of the Immune System

23–15 Respiratory Muscles and Pulmonary Ventilation   23–25 Control of Respiration

24–15 Regulation of Gastric Activity   24–27 Chemical Events in Digestion

25–11 Absorptive and Postabsorptive States

26–16 Summary of Renal Function

27–18 The Diagnosis of Acid-Base Disorders

28–12 Regulation of Male Reproduction   28–24 Regulation of Female Reproduction

29–5 Extraembryonic Membranes and Placenta Formation

FUNDAMENTALS OF
Anatomy&Physiology

Tenth Edition
Global Edition

Spotlight Figures

Spotlight Figures
are one- or two-page presentations that integrate text and art to communicate physiological, organizational, or clinical information in a visually effective format.

Clear steps use text and art to guide students through complex processes.

SPOTLIGHT
Figure 6–11
Endochondral Ossification

Endochondral ossification begins with the formation of a hyaline cartilage model. This model serves as the pattern for almost all bone development. Limb bone development is a good example of this process. By the time an embryo is six weeks old, the proximal bones of the limbs, the humerus (upper limb) and femur (lower limb), have formed, but they are composed entirely of cartilage. These cartilage models continue to grow by expansion of the cartilage matrix (interstitial growth) and the production of more cartilage at the outer surface (appositional growth).

Initiation of Ossification in the Developing Bone (Steps 1–4)

1 As the cartilage enlarges, chondrocytes near the center of the shaft increase greatly in size. The matrix is reduced to a series of small struts that soon begin to calcify. The enlarged chondrocytes then die and disintegrate, leaving cavities within the cartilage.

2 Blood vessels grow around the edges of the cartilage, and the cells of the perichondrium convert to osteoblasts. The shaft of the cartilage then becomes ensheathed in a superficial layer of bone.

3 Blood vessels penetrate the cartilage and invade the central region. Fibroblasts migrating with the blood vessels differentiate into osteoblasts and begin producing spongy bone at a primary ossification center. Bone formation then spreads along the shaft toward both ends of the former cartilage model.

4 Remodeling occurs as growth continues, creating a medullary cavity. The osseous tissue of the shaft becomes thicker, and the cartilage near each epiphysis is replaced by shafts of bone. Further growth involves increases in length and diameter.

Enlarging chondrocytes within calcifying matrix

Hyaline cartilage

Disintegrating chondrocytes of the cartilage model

Perichondrium

Bone collar

Epiphysis

Diaphysis

Blood vessel

Periosteum formed from perichondrium

Medullary cavity

Primary ossification center

Superficial bone

Spongy bone

Medullary cavity

Metaphysis

NEW The Tenth Edition has 18 new Spotlight Figures for a total of 50, including at least one in every chapter.

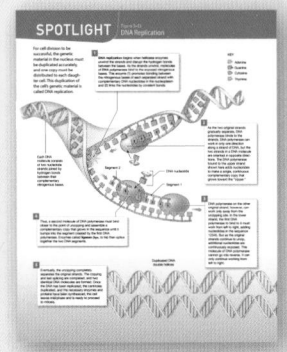

DNA Replication
Chapter 3, page 128

Joint Movement
Chapter 9, pages 296-297

Watch the MasteringA&P® > A&PFlix on Endochondral Ossification.

Increasing the Length of a Developing Long Bone (Steps 5–7)

During the initial stages of ossification, osteoblasts move away from the primary ossification center toward the epiphyses. But they do not complete the ossification of the model immediately, because the cartilages of the epiphyses continue to grow. The region where the cartilage is being replaced by bone lies at the metaphysis, the junction between the diaphysis and an epiphysis. On the diaphyseal (shaft) side of the metaphysis, osteoblasts continually invade the cartilage and replace it with bone, while on the epiphyseal side, new cartilage is produced at the same rate. The situation is like a pair of joggers, one in front of the other. As long as they are running at the same speed, they can run for miles without colliding. In this case, the osteoblasts and the epiphysis are both "running away" from the primary ossification center. As a result, the osteoblasts never catch up with the epiphysis, and the bone continues to grow longer and longer.

NEW There is one coordinating Spotlight Figure for every A&P Flix available in MasteringA&P.

The explanation is built directly into the illustration for efficient and effective learning.

5 Capillaries and osteoblasts migrate into the epiphyses, creating secondary ossification centers.

Hyaline cartilage

Epiphysis

Metaphysis

Periosteum

Compact bone

Secondary ossification center

6 The epiphyses eventually become filled with spongy bone. The metaphysis, a relatively narrow cartilaginous region called the **epiphyseal cartilage**, or *epiphyseal plate*, now separates the epiphysis from the diaphysis. On the shaft side of the metaphysis, osteoblasts continuously invade the cartilage and replace it with bone. New cartilage is produced at the same rate on the epiphyseal side.

Articular cartilage

Spongy bone

Epiphyseal cartilage

Diaphysis

Within the epiphyseal cartilage, the chondrocytes are organized into zones.

Chondrocytes at the epiphyseal side of the cartilage continue to divide and enlarge.

Chondrocytes degenerate at the diaphyseal side.

Osteoblasts migrate upward from the diaphysis and cartilage is gradually replaced by bone.

7 At puberty, the rate of epiphyseal cartilage production slows and the rate of osteoblast activity accelerates. As a result, the epiphyseal cartilage gets narrower and narrower, until it ultimately disappears. This event is called **epiphyseal closure**. The former location of the epiphyseal cartilage becomes a distinct **epiphyseal line** that remains after epiphyseal growth has ended.

Articular cartilage   Epiphyseal line

Spongy bone

Medullary cavity

A thin cap of the original cartilage model remains exposed to the joint cavity as the **articular cartilage**. This cartilage prevents damaging the joint from bone-to-bone contact.

The all-in-one-place presentation means no flipping back and forth between narrative and illustration to get the full story.

MasteringA&P®

NEW Spotlight Figure Coaching Activities

One Spotlight Figure in each chapter now has an assignable Coaching Activity in MasteringA&P.

The Epidermis
Chapter 5, page 180

Overview of the Autonomic Nervous System
Chapter 16, pages 560-561

Clinical Content

▌Clinical Notes appear within every chapter, expand upon topics just discussed, and present diseases and pathologies along with their relationship to normal function.

Clinical Note

Carpal Tunnel Syndrome The carpal bones articulate with one another at joints that permit limited sliding and twisting. Ligaments interconnect the carpal bones and help stabilize the wrist joint. The tendons of muscles that flex the fingers pass across the anterior surface of the wrist. These tendons are sandwiched between the intercarpal ligaments and a broad, superficial transverse ligament called the *flexor retinaculum*. Inflammation of the connective tissues between the flexor retinaculum and the carpal bones can compress the tendons and the adjacent *median nerve*. The result is pain, weakness, and reduced wrist mobility. This condition is called *carpal tunnel syndrome.*

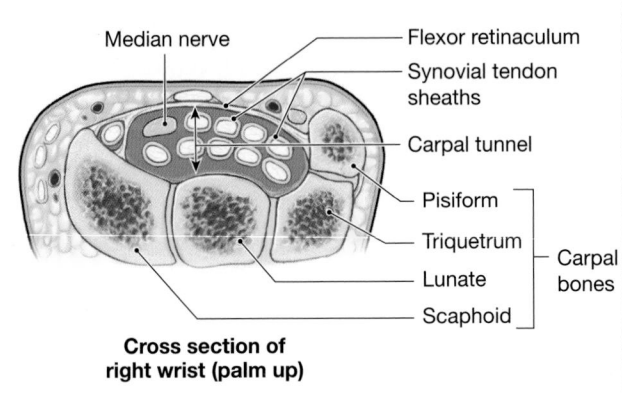

Cross section of
right wrist (palm up)

▌Related Clinical Terms end every chapter with a list of relevant clinical terms and definitions.

Related Clinical Terms

ankylosing spondylitis: A chronic, progressive inflammatory disease of the intervertebral spaces that causes abnormal fusion of the vertebrae.

arthroplasty: The surgical reconstruction or creation of an artificial joint.

arthroscopy: Insertion of a narrow tube containing optical fibers and a tiny camera (arthroscope) directly into the joint for visual examination.

joint mice: Small fibrous, cartilaginous, or bony loose bodies in the synovial cavity of a joint.

Lyme disease: An infectious disease transmitted to humans from the bite of a tick infected with *Borrelia burgdorferi,* causing flu-like symptoms and joint pain.

pannus: Granulation tissue (combination of fibrous connective tissue and capillaries) forming within a synovial membrane, which releases cartilage-destroying enzymes.

Examples of Clinical Notes:

Congenital Talipes Equinovarus

Chapter 8, page 284

Abnormal Bone Development

Chapter 6, page 222

AIDs

Chapter 22, page 845

NEW | Clinical Cases get students motivated for their future careers. Each chapter opens with a story-based Clinical Case related to the chapter content.

CLINICAL CASE

When Your Heart Is in the Wrong Place

Five-year-old Jackson came home from kindergarten with yet another ear infection. He is a happy child with a great appetite, but he can't seem to stay healthy. His nose is constantly plugged, and he always has a nasty cough. Sometimes he coughs so hard he vomits.

Jackson has been checked for cystic fibrosis, a genetic disease that causes thick, sticky, mucus secretions, to be sure this was not the cause of his earaches and chronic cough. There are no pets or smokers in the home.

Because this cough has gone on so long, Jackson is now getting a chest x-ray. The radiology technologist asked if she could take another film and carefully checked to be sure she had the "right" marker on Jackson's right side.

"Now that's funny," the tech said. "It looks like this x-ray is backward, but I know I took it correctly. Jackson's heart must be on the wrong side." **Could Jackson's heart really be on the wrong side? To find out, turn to the Clinical Case Wrap-Up on p. 112.**

| Each chapter ends with a Clinical Case Wrap-Up that incorporates the deeper content knowledge students will have gained from the chapter.

Examples of Clinical Cases:

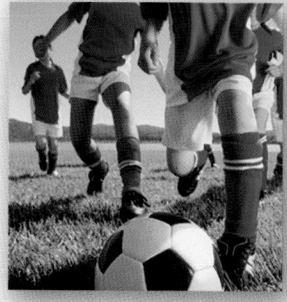

A Case of Child Abuse?
Chapter 6, page 205

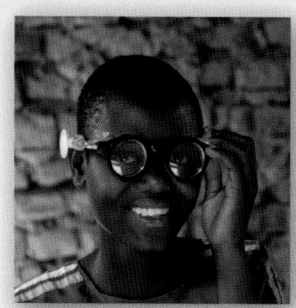

A Chance to See
Chapter 17, page 590

A Real Eye Opener
Chapter 10, page 316

MasteringA&P®

NEW Book-specific Clinical Case Activities

Every Clinical Case has a related assignable Activity in MasteringA&P.

MasteringA&P® Assignable Content

NEW ▌ Interactive and Adaptive Capabilities

- **Adaptive Follow-Up Assignments** allow instructors to easily assign personalized content for each individual student based on strengths and weaknesses identified by his or her performance on MasteringA&P parent assignments.

- **Dynamic Study Modules** help students acquire, retain, and recall information faster and more efficiently than ever before. The flashcard-style modules are available as a self-study tool or can be assigned by the instructor. They can be easily accessed with smartphones.

- **Learning Catalytics** is a "bring your own device" (laptop, smartphone, or tablet) student engagement, assessment, and classroom intelligence system. With Learning Catalytics, instructors can assess students in real time using open-ended tasks to probe student understanding.

NEW ▌ A&P Flix™ Coaching Activities

bring interactivity to these popular 3D movie-quality animations by asking students to manipulate the visuals.

▌ Video Tutor Coaching Activities

instruct and coach students on key A&P concepts from the book and are accompanied by questions with video hints and feedback specific to their misconceptions.

NEW ▌ Spotlight Figure Coaching Activities

bring interactivity to the Spotlight Figures from the book by asking students to manipulate the visuals.

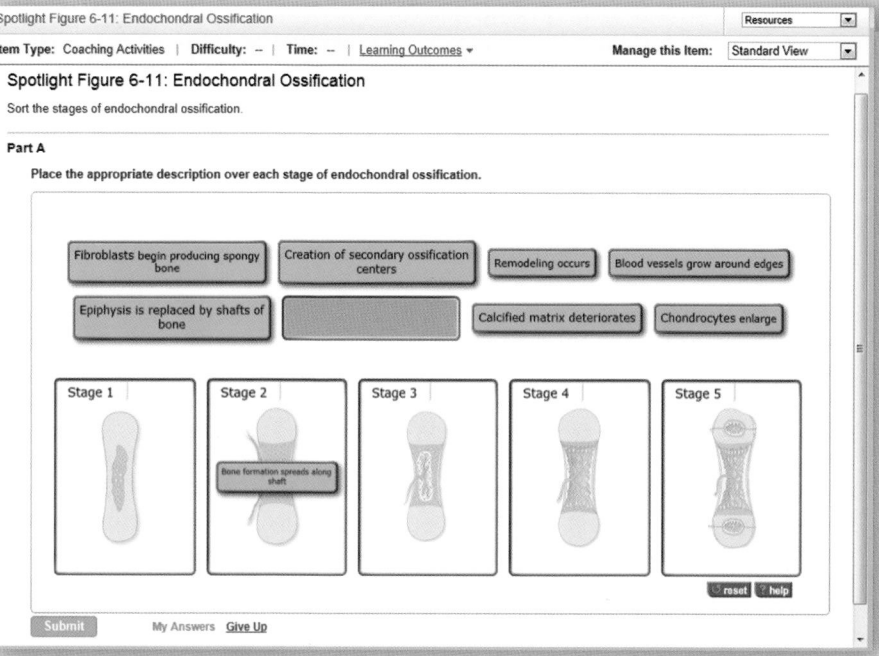

▌ Interactive Physiology® Coaching Activities

help students dive deeper into complex physiological processes using the Interactive Physiology tutorial program.

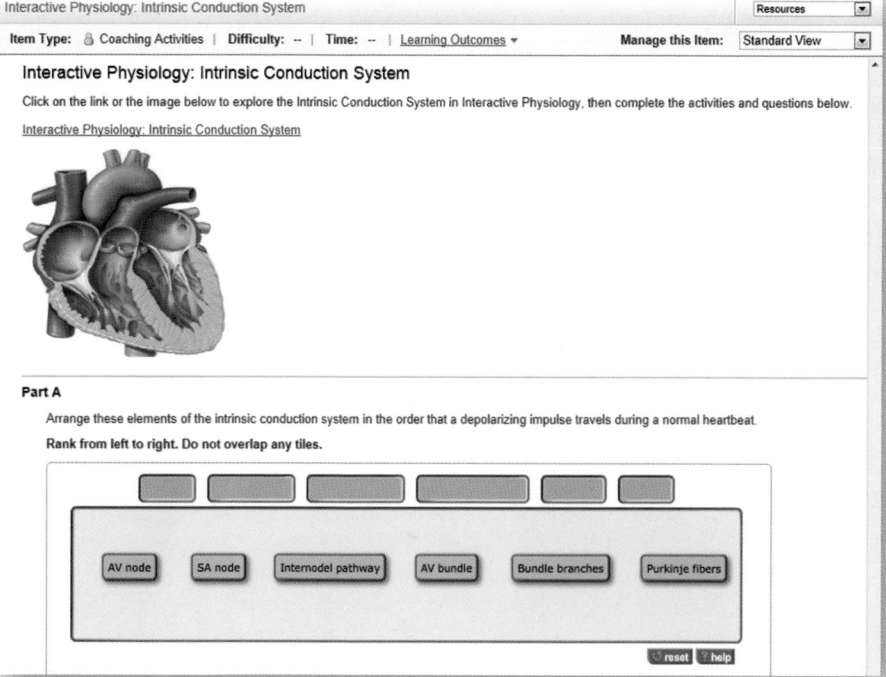

▌ Also Assignable in MasteringA&P®:

- **Art-labeling Activities** are drag and drop activities that allow students to assess their knowledge of terms and structures.
- **Art-based Questions** are conceptual questions related to art and instruct students with wrong-answer feedback.
- NEW **Book-specific Clinical Case Activities** stem from the story-based Clinical Cases in the book.
- NEW **Clinical Note Activities** prepare students for future careers in allied health fields.
- **Chemistry Review Activities** reinforce chemistry concepts necessary for an understanding of A&P.

- **PAL 3.0** and assessments
- **PhysioEx™ 9.1** and assessments
- **Reading Quiz Questions**
- **Chapter Test Questions**
- **Test Bank Questions**

MasteringA&P® Study Area

MasteringA&P® includes a Study Area that will help students get ready for tests with its simple three-step approach. Students can:

1. Take a pre-test and obtain a personalized study plan.
2. Learn and practice with animations, labeling activities, and interactive tutorials.
3. Self-test with quizzes and a chapter practice test.

❙ Practice Anatomy Lab™ (PAL™) 3.0

is a virtual anatomy study and practice tool that gives students 24/7 access to the most widely used lab specimens, including the human cadaver, anatomical models, histology, cat, and fetal pig. PAL 3.0 is easy to use and includes built-in audio pronunciations, rotatable bones, and simulated fill-in-the-blank lab practical exams.

NEW The PAL 3.0 App lets you access PAL 3.0 on your iPad or Android tablet. With the pinch-to-zoom feature, images can be instantly enlarged.

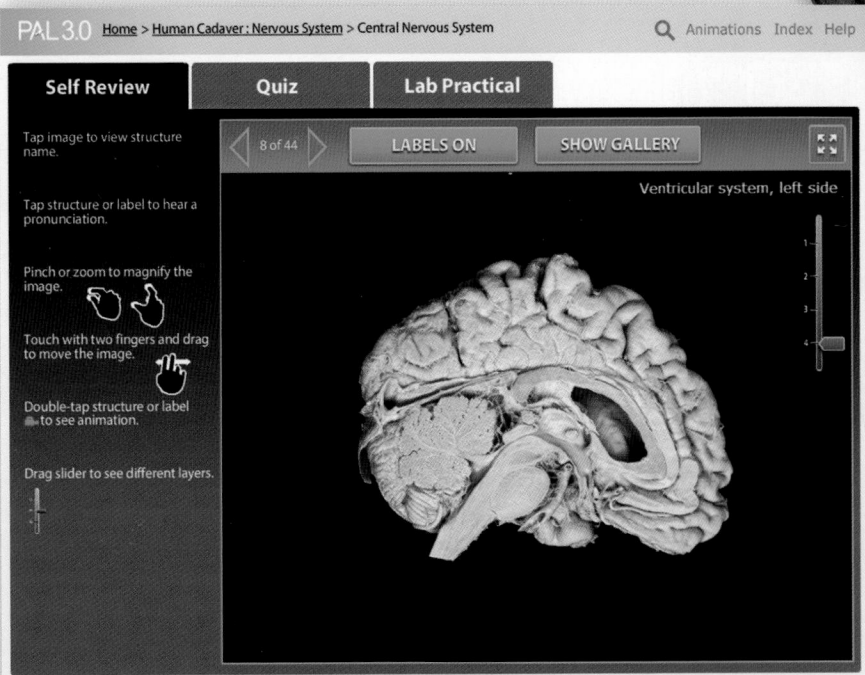

PAL 3.0   Home > Human Cadaver : Nervous System > Central Nervous System   🔍 Animations Index Help

| Self Review | Quiz | Lab Practical |

Tap image to view structure name.

Tap structure or label to hear a pronunciation.

Pinch or zoom to magnify the image.

Touch with two fingers and drag to move the image.

Double-tap structure or label to see animation.

Drag slider to see different layers.

◁ 8 of 44 ▷   LABELS ON   SHOW GALLERY

Ventricular system, left side

❙ Also Available in the Study Area: • eText   • PhysioEx™ 9.1   • Video Tutors

▌ A&P Flix™

are 3-D movie-quality animations
with self-paced tutorials and
gradable quizzes that help students
master the toughest topics in A&P:

- **NEW** Protein Synthesis
- Membrane Transport
- DNA Replication
- Mitosis
- **NEW** Endochondral Ossification
- Events at the Neuromuscular Junction
- The Cross Bridge Cycle
- Excitation-Contraction Coupling
- Resting Membrane Potential
- Generation of an Action Potential
- Propagation of an Action Potential
- **NEW** The Stretch Reflex
- **NEW** The Mechanism of Hormone Action: Second Messenger cAMP
- Origins, Insertions, Actions, and Innervations
  Over Over 60 animations on this topic
- Group Muscle Actions and Joints
  Over 50 animations on this topic

▌ Interactive Physiology® (IP)

helps students understand the hardest
part of A&P: physiology. Fun, interactive
tutorials, games, and quizzes give students
additional explanations to help them grasp
difficult concepts.

Modules:

- Muscular System
- Nervous System I
- Nervous System II
- Cardiovascular System
- Respiratory System
- Urinary System
- Fluids & Electrolytes
- Endocrine System
- Digestive System
- Immune System

Support for Instructors

Instructor's Manual
by Jeff Schinske
and Mary L. Puglia

MyReadinessTest™

NEW ▌ MyReadinessTest

by Lori K. Garrett

MyReadinessTest for A&P is a powerful online system that gets students prepared before their course starts. It assesses students' proficiency in study skills and foundational science and math concepts and provides coaching in core areas where students need additional practice and review.

▌ eText with Whiteboard Mode

The *Fundamentals of Anatomy & Physiology* eText comes with Whiteboard Mode, allowing instructors to use the eText for dynamic classroom presentations. Instructors can show one-page or two-page views from the book, zoom in or out to focus on select topics, and use the Whiteboard Mode to point to structures, circle parts of a process, trace pathways, and customize their presentations.

Instructors can also add notes to guide students, upload documents, and share their custom-enhanced eText with the whole class.

Instructors can find the eText with Whiteboard Mode on MasteringA&P.

FUNDAMENTALS OF

Anatomy & Physiology

Tenth Edition

Global Edition

Frederic H. Martini, Ph.D.
University of Hawaii at Manoa

Judi L. Nath, Ph.D.
Lourdes University, Sylvania, Ohio

Edwin F. Bartholomew, M.S.

William C. Ober, M.D.
Art Coordinator and Illustrator

Claire E. Ober, R.N.
Illustrator

Kathleen Welch, M.D.
Clinical Consultant

Ralph T. Hutchings
Biomedical Photographer

Clinical Cases by:
Ruth Anne O'Keefe, M.D.

PEARSON

Boston Columbus Indianapolis New York San Francisco Upper Saddle River
Amsterdam Cape Town Dubai London Madrid Milan Munich Paris Montreal Toronto
Delhi Mexico City São Paulo Sydney Hong Kong Seoul Singapore Taipei Tokyo

Executive Editor: *Leslie Berriman*
Head of Learning Asset Acquisition, Global Edition: *Laura Dent*
Assistant Editor: *Cady Owens*
Associate Project Editor: *Lisa Damerel*
Editorial Assistant: *Sharon Kim*
Director of Development: *Barbara Yien*
Development Editor: *Anne A. Reid*
Managing Editor: *Mike Early*
Assistant Managing Editor: *Nancy Tabor*
Project Manager: *Courtney Towson*
Acquisitions Editor, Global Edition: *Shona Mullen*
Associate Project Editor, Global Edition: *Binita Roy*
Senior Manufacturing Controller, Production,
Global Edition: *Trudy Kimber*
Director of Digital Product Development: *Lauren Fogel*

Executive Content Producer: *Liz Winer*
Content Producer: *Joe Mochnick*
Cover Photo Credit: © *Watthano/Shutterstock*
Production Management: *Norine Strang*
Compositor: *S4Carlisle Publishing Services, Inc.*
Copyeditor: *Michael Rossa*
Art Coordinator: *Kristina Seymour*
Design Manager: *Mark Ong*
Interior Designer: *tani hasegawa*
Cover Designer: *PreMedia Global*
Contributing Illustrators: *imagineeringart.com; Anita Impagliazzo*
Photo Researcher: *Maureen Spuhler*
Senior Procurement Specialist: *Stacey Weinberger*
Senior Anatomy & Physiology Specialist: *Derek Perrigo*
Senior Marketing Manager: *Allison Rona*

Notice: Our knowledge in clinical sciences is constantly changing. The authors and the publisher of this volume have taken care that the information contained herein is accurate and compatible with the standards generally accepted at the time of the publication. Nevertheless, it is difficult to ensure that all information given is entirely accurate for all circumstances. The authors and the publisher disclaim any liability, loss, or damage incurred as a consequence, directly or indirectly, of the use and application of any of the contents of this volume.

Pearson Education Limited
Edinburgh Gate
Harlow
Essex CM20 2JE
England

and Associated Companies throughout the world

Visit us on the World Wide Web at:
www.pearsonglobaleditions.com

ISBN 10: 1-292-05721-1
ISBN 13: 978-1-292-05721-7

British Library Cataloguing-in-Publication Data
A catalogue record for this book is available from the British Library

10 9 8 7 6 5 4 3 2 1
14

Typeset in 9/15 Giovanni Std by S4Carlisle Publishing Services, Inc.

Printed and bound by Courier Kendallville in The United States of America.

Text and Illustration Team

Frederic (Ric) H. Martini, Ph.D.
Author

Dr. Martini received his Ph.D. from Cornell University in comparative and functional anatomy for work on the pathophysiology of stress. In addition to professional publications that include journal articles and contributed chapters, technical reports, and magazine articles, he is the lead author of ten undergraduate texts on anatomy and physiology or anatomy. Dr. Martini is currently affiliated with the University of Hawaii at Manoa and has a long-standing bond with the Shoals Marine Laboratory, a joint venture between Cornell University and the University of New Hampshire. He has been active in the Human Anatomy and Physiology Society (HAPS) for over 20 years and was a member of the committee that established the course curriculum guidelines for A&P. He is now a President Emeritus of HAPS after serving as President-Elect, President, and Past-President over 2005–2007. Dr. Martini is also a member of the American Physiological Society, the American Association of Anatomists, the Society for Integrative and Comparative Biology, the Australia/New Zealand Association of Clinical Anatomists, the Hawaii Academy of Science, the American Association for the Advancement of Science, and the International Society of Vertebrate Morphologists.

Judi L. Nath, Ph.D.
Author

Dr. Judi Nath is a biology professor at Lourdes University, where she teaches anatomy and physiology, pathophysiology, and medical terminology. She received her Bachelor's and Master's degrees from Bowling Green State University and her Ph.D. from the University of Toledo. Dr. Nath is devoted to her students and strives to convey the intricacies of science in captivating ways that are meaningful, interactive, and exciting. She has won the Faculty Excellence Award—an accolade recognizing effective teaching, scholarship, and community service—multiple times. She is active in many professional organizations, notably the Human Anatomy and Physiology Society (HAPS), where she has served several terms on the board of directors. Dr. Nath is a coauthor of *Visual Anatomy & Physiology, Visual Essentials of Anatomy & Physiology,* and *Anatomy & Physiology* (all published by Pearson), and she is the sole author of *Using Medical Terminology.* Her favorite charities are those that have significantly affected her life, including the local Humane Society, the Cystic Fibrosis Foundation, and the ALS Association. On a personal note, Dr. Nath enjoys family life with her husband and their dogs.

Edwin F. Bartholomew, M.S.
Author

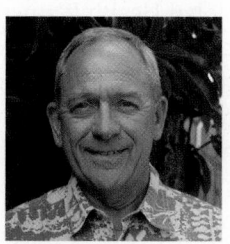

Edwin F. Bartholomew received his undergraduate degree from Bowling Green State University in Ohio and his M.S. from the University of Hawaii. Mr. Bartholomew has taught human anatomy and physiology at both the secondary and undergraduate levels and a wide variety of other science courses (from botany to zoology) at Maui Community College and at historic Lahainaluna High School, the oldest high school west of the Rockies. He is a coauthor of *Visual Anatomy & Physiology, Essentials of Anatomy & Physiology, Visual Essentials of Anatomy & Physiology, Structure and Function of the Human Body,* and *The Human Body in Health and Disease* (all published by Pearson). Mr. Bartholomew is a member of the Human Anatomy and Physiology Society (HAPS), the National Association of Biology Teachers, the National Science Teachers Association, the Hawaii Science Teachers Association, and the American Association for the Advancement of Science.

William C. Ober, M.D.
Art Coordinator and Illustrator

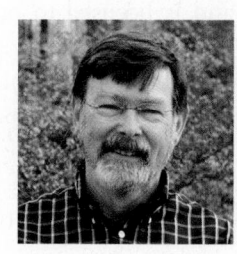

Dr. Ober received his undergraduate degree from Washington and Lee University and his M.D. from the University of Virginia. He also studied in the Department of Art as Applied to Medicine at Johns Hopkins University. After graduation, Dr. Ober completed a residency in Family Practice and later was on the faculty at the University of Virginia in the Department of Family Medicine and in the Department of Sports Medicine. He also served as Chief of Medicine of Martha Jefferson Hospital in Charlottesville, VA. He is currently a Visiting Professor of Biology at Washington and Lee University, where he has taught several courses and led student trips to the Galapagos Islands. He was on the Core Faculty at Shoals Marine Laboratory for 24 years, where he taught Biological Illustration every summer. Dr. Ober has collaborated with Dr. Martini on all of his textbooks in every edition.

Claire E. Ober, R.N.
Illustrator

Claire E. Ober, R.N., B.A., practiced family, pediatric, and obstetric nursing before turning to medical illustration as a full-time career. She returned to school at Mary Baldwin College, where she received her degree with distinction in studio art. Following a five-year apprenticeship, she has worked as Dr. Ober's partner in Medical & Scientific Illustration since 1986. She was on the Core Faculty at Shoals Marine Laboratory and co-taught the Biological Illustration course with Dr. Ober for 24 years. The textbooks illustrated by Medical & Scientific Illustration have won numerous design and illustration awards.

Kathleen Welch, M.D.
Clinical Consultant

Dr. Welch received her B.A. from the University of Wisconsin–Madison, her M.D. from the University of Washington in Seattle, and did her residency in Family Practice at the University of North Carolina in Chapel Hill. Participating in the Seattle WWAMI rural medical education program, she studied in Fairbanks, Anchorage, and Juneau, Alaska, with time in Boise, Idaho, and Anacortes, Washington, as well. For two years, she served as Director of Maternal and Child Health at the LBJ Tropical Medical Center in American Samoa and subsequently was a member of the Department of Family Practice at the Kaiser Permanente Clinic in Lahaina, Hawaii, and on the staff at Maui Memorial Hospital. She has been in private practice since 1987 and is licensed to practice in Hawaii and Washington State. Dr. Welch is a Fellow of the American Academy of Family Practice and a member of the Maui County Medical Society and the Human Anatomy and Physiology Society (HAPS). With Dr. Martini, she has coauthored both a textbook on anatomy and physiology and the *A&P Applications Manual*. She and Dr. Martini were married in 1979, and they have one son.

Ralph T. Hutchings
Biomedical Photographer

Mr. Hutchings was associated with the Royal College of Surgeons for 20 years. An engineer by training, he has focused for years on photographing the structure of the human body. The result has been a series of color atlases, including the *Color Atlas of Human Anatomy,* the *Color Atlas of Surface Anatomy,* and *The Human Skeleton* (all published by Mosby-Yearbook Publishing). For his anatomical portrayal of the human body, the International Photographers Association has chosen Mr. Hutchings as the best photographer of humans in the twentieth century. He lives in North London, where he tries to balance the demands of his photographic assignments with his hobbies of early motor cars and airplanes.

Ruth Anne O'Keefe, M.D.
Clinical Contributor

Dr. O'Keefe did her undergraduate studies at Marquette University, attended graduate school at the University of Wisconsin, and received her M.D. from George Washington University. She was the first woman to study orthopedics at The Ohio State University during her residency. She did fellowship training in trauma surgery at Loma Linda University in California. In addition to her private orthopedic practice, she has done orthopedic surgery around the world, taking her own surgical teams to places such as the Dominican Republic, Honduras, Peru, New Zealand, and Burkina Faso. She serves on the board of Global Health Partnerships, a group that partners with a clinic serving 35,000 people in remote Kenya. Dr. O'Keefe has always enjoyed teaching and now supervises medical students from the University of New Mexico doing ongoing research in Kenya. She lives in Albuquerque with her Sweet Ed. She is mother of four, grandmother of nine, and foster mother to many.

Preface

The Tenth Edition of *Fundamentals of Anatomy & Physiology* is a comprehensive textbook that fulfills the needs of today's students while addressing the concerns of their professors. We focused our attention on the question "How can we make this information meaningful, manageable, and comprehensible?" During the revision process, we drew upon our content knowledge, research skills, artistic talents, and years of classroom experience to make this edition the best yet.

The broad changes to this edition are presented in the **New to the Tenth Edition** section below, and the specific changes are presented in the **Chapter-by-Chapter Changes in the Tenth Edition** section that follows.

New to the Tenth Edition

In addition to the many technical changes in this edition, such as updated statistics and anatomy and physiology descriptions, we have made the following key changes:

- **NEW 50 Spotlight Figures** provide highly visual one- and two-page presentations of tough topics in the book, with a particular focus on physiology. In the Tenth Edition, 18 new Spotlight Figures have been added for a total of 50 across the chapters. There is now at least one Spotlight Figure in every chapter, as well as one Spotlight Figure corresponding to every A&P Flix.

- **NEW 29 Clinical Cases** get students motivated for their future careers. Each chapter opens with a story-based Clinical Case related to the chapter content and ends with a Clinical Case Wrap-Up that incorporates the deeper content knowledge students will have gained from the chapter.

- **NEW The repetition of the chapter-opening Learning Outcomes below the coordinated section headings within the chapters** underscores the connection between the HAPS-based Learning Outcomes and the associated teaching points. Author Judi Nath sat on the Human Anatomy and Physiology Society (HAPS) committee that developed the HAPS Learning Outcomes, recommended to A&P instructors, and the Learning Outcomes in this book are based on them. Additionally, the assessments in MasteringA&P are organized by these Learning Outcomes. As in the previous edition, full-sentence section headings, correlated with the Learning Outcomes, state a core fact or concept to help students readily see and learn the chapter content; and Checkpoints, located at the close of each section, ask students to pause and check their understanding of facts and concepts. If students cannot answer these questions within a matter of minutes, then they should reread the section before moving on. The Checkpoints reinforce the Learning Outcomes, resulting in a systematic integration of the Learning Outcomes over the course of the chapter. Answers to the Checkpoints are located in the blue Answers tab at the back of the book.

- **Easier narrative** uses simpler, shorter, more active sentences and a reading level that makes reading and studying easier for students.

- **Improved text-art integration** throughout the illustration program enhances the readability of figures. Several tables have been integrated directly into figures to help students make direct connections between tables and art.

- **Eponyms** are now included within the narrative, along with the anatomical terms used in *Terminologia Anatomica*.

- **NEW Assignable MasteringA&P activities** include the following:

  - **NEW Spotlight Figure Coaching Activities** are highly visual, assignable activities designed to bring interactivity to the Spotlight Figures in the book. Multi-part activities include the ranking and sorting types that ask students to manipulate the visuals.

  - **NEW Book-specific Clinical Case Activities** stem from the story-based Clinical Cases that appear at the beginning and end of each chapter in the book.

  - **NEW Adaptive Follow-up Assignments** allow instructors to easily assign personalized content for each individual student based on strengths and weaknesses identified by his or her performance on MasteringA&P parent assignments.

  - **NEW Dynamic Study Modules** help students acquire, retain, and recall information quickly and efficiently. The modules are available as a self-study tool or can be assigned by the instructor. They can be easily accessed with smartphones.

Chapter-by-Chapter Changes in the Tenth Edition

This annotated Table of Contents provides select examples of revision highlights in each chapter of the Tenth Edition. For a more complete list of changes, please contact the publisher.

Chapter 1: An Introduction to Anatomy and Physiology

- New Clinical Case: Using A&P to Save a Life
- New Spotlight Figure 1–10 Diagnostic Imaging Techniques
- New Clinical Note: Autopsies and Cadaver Dissection
- New Clinical Note: Auscultation
- Figure 1–7 Directional References revised
- Figure 1–8 Sectional Planes revised
- Figure 1–9 Relationships among the Subdivisions of the Body Cavities of the Trunk revised

Chapter 2: The Chemical Level of Organization

- New Clinical Case: What Is Wrong with My Baby?
- New Clinical Note: Radiation Sickness
- Clinical Note: Fatty Acids and Health revised
- Section 2-2 includes revised Molecular weight discussion
- Figure 2–4 The Formation of Ionic Bonds revised
- Figure 2–5 Covalent Bonds in Five Common Molecules revised
- Table 2–3 Important Functional Groups of Organic Compounds revised (to clarify structural group and R group)
- Protein Structure subsection includes new discussion of amino acids as zwitterions
- Figure 2–21 Protein Structure revised

Chapter 3: The Cellular Level of Organization

- New Clinical Case: When Your Heart Is in the Wrong Place
- New information added about cholesterol and other lipids
- New overview added about roles of microtubules
- Figure 3–5 The Endoplasmic Reticulum revised
- Clinical Note on DNA Fingerprinting revised
- Figure 3–13 The Process of Translation revised
- Figure 3–14 Diffusion revised
- Figure 3–17 Osmotic Flow across a Plasma Membrane revised
- New Spotlight Figure 3–22 Overview of Membrane Transport incorporates old Figures 3–18, 3–19, and 3–21 and old Table 3–2
- New Spotlight Figure 3–23 DNA Replication incorporates old Figure 3–23
- Spotlight Figure 3–24 Stages in a Cell's Life Cycle revised

Chapter 4: The Tissue Level of Organization

- New Clinical Case: The Rubber Girl
- Intercellular Connections subsection updated
- Figure 4–2 Cell Junctions revised
- Figure 4–8 The Cells and Fibers of Connective Tissue Proper revised
- Adipose Tissue subsection includes updated discussion of brown fat
- Figure 4–10 Loose Connective Tissues revised
- Spotlight Figure 4–20 Inflammation and Regeneration revised

Chapter 5: The Integumentary System

- New Clinical Case: Skin Cells in Overdrive
- Figure 5–1 The Components of the Integumentary System revised
- New Figure 5–2 The Cutaneous Membrane and Accessory Structures
- New Spotlight Figure 5–3 The Epidermis incorporates old Figures 5–2 and 5–3
- New Figure 5–5 Vitiligo
- New Figure 5–6 Sources of Vitamin D_3

- Clinical Note: Decubitus Ulcers revised with new photo
- New Figure 5–8 Reticular Layer of Dermis
- Figure 5–10 Dermal Circulation revised
- Figure 5–12 Hair Follicles and Hairs revised
- New Figure 5–11 Hypodermis

Chapter 6: Osseous Tissue and Bone Structure

- New Clinical Case: A Case of Child Abuse?
- Figure 6–1 A Classification of Bones by Shape revised
- New Figure 6–2 An Introduction to Bone Markings incorporates old Table 6–1
- New Spotlight Figure 6–11 Endochondral Ossification incorporates old Figure 6–10
- New Figure 6–12 Intramembranous Ossification
- Spotlight Figure 6–16 Types of Fractures and Steps in Repair revised
- Clinical Note: Abnormal Bone Development revised

Chapter 7: The Axial Skeleton

- New Clinical Case: Knocked Out
- New Clinical Note: Sinusitis
- Figure 7–2 Cranial and Facial Subdivisions of the Skull revised
- Figure 7–3 The Adult Skull revised to incorporate old Table 7–1
- New Spotlight Figure 7–4 Sectional Anatomy of the Skull incorporates old Figure 7–4 and parts of old Table 7–1
- Figure 7–6 The Frontal Bone revised
- Figure 7–14 The Nasal Complex revised
- Figure 7–22 The Thoracic Cage revised

Chapter 8: The Appendicular Skeleton

- New Clinical Case: The Orthopedic Surgeon's Nightmare
- New Clinical Note: Hip Fracture
- New Clinical Note: Runner's Knee
- New Clinical Note: Stress Fractures
- Carpal Bones subsection now lists the 8 carpal bones in two groups of 4 (proximal and distal carpal bones)
- Figure 8–6 Bones of the Right Wrist and Hand revised
- New Spotlight Figure 8–10 Sex Differences in the Human Skeleton incorporates old Figure 8–10, old Table 8–1, and old bulleted list in text
- Clinical Note: Carpal Tunnel Syndrome includes new illustration
- Figure 8–14 Bones of the Ankle and Foot revised
- Clinical Note: Congenital Talipes Equinovarus includes new photo

Chapter 9: Joints

- Chapter title changed from Articulations to Joints
- New Clinical Case: What's Ailing the Birthday Girl?
- New Clinical Note: Dislocation and Subluxation
- New Clinical Note: Damage to Intervertebral Discs
- Table 9–1 Functional and Structural Classifications of Articulations redesigned
- Spotlight Figure 9–2 Joint Movement incorporates old Figures 9–2 and 9–6 and subsection on Types of Synovial Joints
- Revised discussion of synovial fluid function in shock absorption
- Figure 9–6 Intervertebral Articulations expanded
- Figure 9–7 The Shoulder Joint revised
- Figure 9–10 The Right Knee Joint rearranged and revised
- Clinical Note: Knee Injuries revised

Chapter 10: Muscle Tissue

- New Clinical Case: A Real Eye Opener
- New subsection Electrical Impulses and Excitable Membranes added in Section 10-4
- New Spotlight Figure 10–10 Excitation–Contraction Coupling incorporates old Figures 10–9 and 10–10
- New Figure 10–13 Steps Involved in Skeletal Muscle Contraction and Relaxation incorporates old Table 10–1
- Treppe subsection includes new discussion of treppe in cardiac muscle
- Motor Units and Tension Production subsection includes new discussion of fasciculation
- Figure 10–20 Muscle Metabolism revised
- Table 10–2 Properties of Skeletal Muscle Fiber Types revised to make column sequences better parallel text discussion

Chapter 11: The Muscular System

- New Clinical Case: The Weekend Warrior
- Figure 11–1 Muscle Types Based on Pattern of Fascicle Organization revised
- Figure 11–2 The Three Classes of Levers revised
- New Spotlight Figure 11–3 Muscle Action
- Figure 11–14 An Overview of the Appendicular Muscles of the Trunk revised
- Figure 11–18 Muscles That Move the Hand and Fingers revised
- Figure 11–22 Extrinsic Muscles That Move the Foot and Toes revised

Chapter 12: Neural Tissue

- New Clinical Case: Did President Franklin D. Roosevelt Really Have Polio?
- New Figure 12–1 A Functional Overview of the Nervous System
- Figure 12–7 Schwann Cells, Peripheral Axons, and Formation of the Myelin Sheath revised and new part c step art added
- New Spotlight Figure 12–9 Resting Membrane Potential incorporates old Figure 12–9
- Figure 12–10 Electrochemical Gradients for Potassium and Sodium Ions revised
- Added ligand-gated channels as an alternative term for chemically gated channels
- New Spotlight Figure 12–15 Propagation of an Action Potential incorporates old Figures 12–6 and 12–15
- New Figure 12–16 Events in the Functioning of a Cholinergic Synapse incorporates old Figure 12–17 and old Table 12–4

Chapter 13: The Spinal Cord, Spinal Nerves, and Spinal Reflexes

- New Clinical Case: Prom Night
- New "Tips & Tricks" added to Cervical Plexus subsection
- Figure 13–7 Dermatomes revised
- New information on the Jendrassik maneuver added to Section 13-8
- New Figure 13–10 The Cervical Plexus incorporates old Table 13–1 and old Figure 13–11
- New Figure 13–11 The Brachial Plexus incorporates old Table 13–2 and old Figure 13–12
- New in-art Clinical Note: Sensory Innervation in the Hand added to Figure 13–11
- New Figure 13–12 The Lumbar and Sacral Plexuses incorporates old Table 13–3 and old Figure 13–13
- New in-art Clinical Note: Sensory Innervation in the Ankle and Foot added to Figure 13–12

- New Spotlight Figure 13–14 Spinal Reflexes incorporates old Figures 13–15, 13–17, 13–19, and 13–20

Chapter 14: The Brain and Cranial Nerves

- New Clinical Case: The Neuroanatomist's Stroke
- New Spotlight Figure 14–4 Formation and Circulation of Cerebrospinal Fluid incorporates old Figure 14–4
- Figure 14–5 The Diencephalon and Brain Stem revised
- New Figures 14–6 The Medulla Oblongata and 14–7 The Pons incorporate old Figure 14–6 and old Table 14–2
- New Figure 14–8 The Cerebellum incorporates old Figure 14–7 and old Table 14–3
- New Figure 14–9 The Midbrain incorporates old Figure 14–8, old Table 14–4, and a new cadaver photograph
- New Figure 14–11 The Hypothalamus in Sagittal Section incorporates old Figure 14–10 and old Table 14–6
- New Figure 14–12 The Limbic System incorporates old Figure 14–11 and old Table 14–7
- Figure 14–14 Fibers of the White Matter of the Cerebrum revised
- Figure 14–15 The Basal Nuclei revised
- Figure 14–16 Motor and Sensory Regions of the Cerebral Cortex revised
- New information on circumventricular organs added to Section 14-2

Chapter 15: Sensory Pathways and the Somatic Nervous System

- New Clinical Case: Living with Cerebral Palsy
- New Figure 15–1 An Overview of Events Occurring along the Sensory and Motor Pathways
- New Figure 15–3 Tonic and Phasic Sensory Receptors
- Spotlight Figure 15–6 Somatic Sensory Pathways revised
- Figure 15–8 Descending (Motor) Tracts in the Spinal Cord reorganized

Chapter 16: The Autonomic Nervous System and Higher-Order Functions

- New Clinical Case: The First Day in Anatomy Lab
- New Spotlight Figure 16–2 Overview of the Autonomic Nervous System incorporates old Figures 16–3 and 16–7
- Figure 16–3 Sites of Ganglia in Sympathetic Pathways revised
- Figure 16–4 The Distribution of Sympathetic Innervation revised

Chapter 17: The Special Senses

- New Clinical Case: A Chance to See
- Figure 17–1 The Olfactory Organs revised
- Spotlight Figure 17–2 Olfaction and Gustation revised
- Figure 17–3 Gustatory Receptors revised
- Figure 17–22 The Middle Ear revised
- Figures 17–23, 17–24, and 17–25 revised to indicate different orientations of maculae in the utricle and saccule
- Figure 17–32 Pathways for Auditory Sensations revised

Chapter 18: The Endocrine System

- New Clinical Case: Stones, Bones, and Groans
- New Spotlight Figure 18–3 G Proteins and Second Messengers incorporates old Figure 18–3
- Figure 18–7 The Hypophyseal Portal System and the Blood Supply to the Pituitary Gland revised
- Figure 18–11 The Thyroid Follicles revised
- New Figure 18–14 The Adrenal Gland incorporates old Figure 18–14 and old Table 18–5

Chapter 19: Blood

- New Clinical Case: A Mysterious Blood Disorder
- Figure 19–3 The Structure of Hemoglobin revised
- Table 19–4 includes revised names for Factors IX and XI and source of Factor X

Chapter 20: The Heart

- New Clinical Case: A Needle to the Chest
- Figure 20–3 The Superficial Anatomy of the Heart revised
- Figure 20–6 The Sectional Anatomy of the Heart revised
- Figure 20–12 Impulse Conduction through the Heart revised
- Figure 20–16 Phases of the Cardiac Cycle revised
- Figure 20–21 Autonomic Innervation of the Heart revised
- Figure 20–24 A Summary of the Factors Affecting Cardiac Output revised

Chapter 21: Blood Vessels and Circulation

- New Clinical Case: Did Ancient Mummies Have Atherosclerosis?
- Figure 21–2 Histological Structures of Blood Vessels revised
- Figure 21–8 Relationships among Vessel Diameter, Cross-Sectional Area, Blood Pressure, and Blood Velocity within the Systemic Circuit revised
- Figure 21–9 Pressures within the Systemic Circuit revised
- Figure 21–11 Forces Acting across Capillary Walls revised
- Figure 21–20 Arteries of the Chest and Upper Limb revised
- Figure 21–25 Arteries of the Lower Limb revised
- Figure 21–29 Flowcharts of Circulation to the Superior and Inferior Venae Cavae revised
- Figure 21–30 Venous Drainage from the Lower Limb revised

Chapter 22: The Lymphatic System and Immunity

- New Clinical Case: Isn't There a Vaccine for That?
- Figure 22–6 The Origin and Distribution of Lymphocytes revised
- Figure 22–11 Innate Defenses revised
- Complement System subsection – includes revised number of complement proteins in plasma (from 11 to more than 30)
- Figure 22–18 Antigens and MHC Proteins revised

Chapter 23: The Respiratory System

- New Clinical Case: How Long Should a Cough Last?
- Figure 23–1 The Structure of the Respiratory System reorganized
- Figure 23–3 The Structures of the Upper Respiratory System revised
- Figure 23–5 The Glottis and Surrounding Structures revised
- Figure 23–7 The Gross Anatomy of the Lungs revised
- Figure 23–9 The Bronchi, Lobules, and Alveoli of the Lung revised
- Figure 23–10 Alveolar Organization revised
- Figure 23–13 Mechanisms of Pulmonary Ventilation revised
- New Spotlight Figure 23–15 Respiratory Muscles and Pulmonary Ventilation incorporates old Figure 23–16
- Figure 23–16 Pulmonary Volumes and Capacities revised
- Spotlight Figure 23–25 Control of Respiration revised

Chapter 24: The Digestive System

- New Clinical Case: An Unusual Transplant
- Figure 24–10 The Esophagus revised
- Figure 24–12 The Stomach revised
- Figure 24–16 Segments of the Intestine revised
- Figure 24–21 The Anatomy and Physiology of the Gallbladder and Bile Ducts revised

Chapter 25: Metabolism and Energetics

- New Clinical Case: The Miracle Supplement
- Figure 25–9 Lipid Transport and Utilization revised
- Figure 25–12 MyPlate Plan revised
- Figure 25–14 Mechanisms of Heat Transfer revised

Chapter 26: The Urinary System

- New Clinical Case: A Case of "Hidden" Bleeding
- Revised all relevant figure labels by replacing "Renal lobe" with "Kidney lobe"
- Figure 26–6 The Functional Anatomy of a Representative Nephron and the Collecting System revised
- Spotlight Figure 26–16 Summary of Renal Function revised

Chapter 27: Fluid, Electrolyte, and Acid–Base Balance

- New Clinical Case: When Treatment Makes You Worse
- Figure 27–2 Cations and Anions in Body Fluids revised
- Figure 27–3 Fluid Gains and Losses revised
- Figure 27–11 The Role of Amino Acids in Protein Buffer Systems revised (to emphasize amino acids as zwitterions)
- Figure 27–13 Kidney Tubules and pH Regulation revised
- New Spotlight Figure 27–18 The Diagnosis of Acid–Base Disorders incorporates old Figure 27–18

Chapter 28: The Reproductive System

- New Clinical Case: A Post-Game Mystery
- Figure 28–1 The Male Reproductive System revised
- Figure 28–3 The Male Reproductive System in Anterior View revised and reorganized
- Figure 28–4 The Structure of the Testes revised
- Figure 28–7 Spermatogenesis revised
- Figure 28–13 The Female Reproductive System revised
- Figure 28–15 Oogenesis revised
- Figure 28–18 The Uterus revised

Chapter 29: Development and Inheritance

- New Clinical Case: The Twins That Looked Nothing Alike
- Revised all relevant chapter text by replacing "embryological" with "embryonic" for simplification
- New Spotlight Figure 29–5 Extraembryonic Membranes and Placenta Formation incorporates old Figure 29–5
- Table 29–2 An Overview of Prenatal Development includes revised sizes and weights at different gestational ages
- Figure 29–8 The Second and Third Trimesters revised
- Figure 29–9 Growth of the Uterus and Fetus revised
- Figure 29–13 Growth and Changes in Body Form and Proportion revised

Appendix

- New periodic table
- New codon chart

Acknowledgments

This textbook represents a group effort, and we would like to acknowledge the people who worked together with us to create this Tenth Edition.

Foremost on our thank-you list are the instructors who offered invaluable suggestions throughout the revision process. We thank them for their participation and list their names and affiliations below.

Lisa Conley, *Milwaukee Area Technical College*
Theresa G. D'Aversa, *Iona College*
Danielle Desroches, *William Paterson University*
Debra Galba-Machuca, *Portland Community College–Cascade*
Lauren Gollahon, *Texas Tech University*
Gigi Goochey, *Hawaii Community College*
Mark Haefele, *Community College of Denver*
Anthony Jones, *Tallahassee Community College*
William L' Amoreaux, *College of Staten Island*
J. Mitchell Lockhart, *Valdosta State University*
Scott Murdoch, *Moraine Valley Community College*
Louise Petroka, *Gateway Community College*
Cynthia Prentice-Craver, *Chemeketa Community College*
S. Michele Robichaux, *Nicholls State University*
Susan Rohde, *Triton College*
Yung Su, *Florida State University*
Bonnie Taylor, *Schoolcraft College*
Carol Veil, *Anne Arundel Community College*
Patricia Visser, *Jackson College*
Theresia Whelan, *State College of Florida – Manatee-Sarasota*
Samia Williams, *Santa Fe College*

The accuracy and currency of the clinical material in this edition and in the *A&P Applications Manual* in large part reflect the work of Kathleen Welch, M.D. Her professionalism and concern for practicality and common sense make the clinical information especially relevant for today's students. Additionally, our content expert on the Clinical Cases, Ruth Anne O'Keefe, M.D., provided constant, useful feedback on each chapter.

Virtually without exception, reviewers stressed the importance of accurate, integrated, and visually attractive illustrations in helping readers understand essential material. The revision of the art program was directed by Bill Ober, M.D. and Claire E. Ober, R.N. Their suggestions about presentation sequence, topics of clinical importance, and revisions to the proposed art were of incalculable value to us and to the project. The illustration program for this edition was further enhanced by the efforts of several other talented individuals. Jim Gibson designed most of the new Spotlight Figures in the art program and consulted on the design and layout of the individual figures. His talents have helped produce an illustration program that is dynamic, cohesive, and easy to understand. Anita Impagliazzo helped create the new photo/art combinations that have resulted in clearer presentations and a greater sense of realism in important anatomical figures. We are also grateful to the talented team at Imagineering (imagineeringart.com) for their dedicated and detailed illustrative work on key figures for this edition. The new color micrographs in this edition were provided by Dr. Robert Tallitsch, and his assistance is much appreciated. Many of the striking anatomy photos in the text and in *Martini's Atlas of the Human Body* are the work of biomedical photographer Ralph Hutchings; his images played a key role in the illustration program.

We also express our appreciation to the editors and support staff at Pearson Science.

We owe special thanks to Executive Editor, Leslie Berriman, for her creativity and dedication. Her vision helped shape this book in countless ways. Leslie's enthusiasm for publishing the highest quality material spills over onto the author/illustrator team. She is our biggest advocate and is always willing to champion our cause—despite the challenges of working with authors. We are appreciative of all her efforts on our behalf.

Assistant Editor, Cady Owens, and Associate Project Editor, Lisa Damerel, are unquestionably the very finest at what they do. While it is expected that editors pay attention to details and keep projects moving forward, Cady and Lisa are true professionals and extremely skilled at not only preparing our material for publication, but making sure it is the best it could possibly be. This past year could not have happened without them.

Annie Reid, our Development Editor, played a vital role in revising the Tenth Edition. Her unfailing attention to readability, consistency, and quality was invaluable to the authors in meeting our goal of delivering complex A&P content in a more student-friendly way.

We are grateful to Mike Rossa for his careful attention to detail and consistency in his copyedit of the text and art.

This book would not exist without the extraordinary dedication of the Production team, including Caroline Ayres, who solved many problems under pressure with unfailing good cheer. Norine Strang skillfully led her excellent team at S4Carlisle to move the book smoothly through composition.

The striking cover and clear, navigable interior design were created by tani hasegawa. Thanks also to Mark Ong, Design Manager, and Marilyn Perry, who devised innovative solutions for several complex design challenges.

Thanks to our photo researcher, Maureen Spuhler, and photo editor, Donna Kalal, for finding, obtaining, and coordinating all the photos in the photo program.

Thanks are also due to Sharon Kim, Editorial Assistant, who served as project editor for the print supplements for instructors and students and coordinated the administrative details of the entire textbook program. Dorothy Cox and Shannon Kong worked tirelessly to shepherd the print and media supplements through production. Thanks also to Stacey Weinberger for handling the physical manufacturing of the book.

We are also grateful to Joe Mochnick, Content Producer, and Liz Winer, Executive Content Producer, for their creative efforts on the media package, most especially MasteringA&P.

We would also like to express our gratitude to the following people at Pearson Science: Paul Corey, President, who continues to support all our texts; Barbara Yien, Director of Development, who kindly kept all phases moving forward under all circumstances; Allison Rona, Senior Marketing Manager; and the dedicated Pearson Science sales representatives for their continuing support of this project. Special thanks go to Frank Ruggirello, Vice President and Editorial Director, for working closely with Leslie in ensuring we have the resources necessary to publish what students need to succeed. And, a round of applause goes to Derek Perrigo, Senior A&P Specialist, our biggest cheerleader.

To help improve future editions, we encourage you to send any pertinent information, suggestions, or comments about the organization or content of this textbook to us directly, using the e-mail addresses below. We warmly welcome comments and suggestions and will carefully consider them in the preparation of the Eleventh Edition.

Frederic (Ric) H. Martini
Haiku, Hawaii
martini@pearson.com

Judi L. Nath
Sandusky, Ohio
nath@pearson.com

Edwin F. Bartholomew
Lahaina, Hawaii
bartholomew@pearson.com

Pearson wishes to thank and acknowledge the following people for their work on the Global Edition:

Contributors
Steven Fenby, Teesside University
Bernie Barnicoat, University of Hull

Reviewers
Martin Steggall, City University London
Harry Chummun, University of Greenwich

Contents

Preface   v

UNIT 1 LEVELS OF ORGANIZATION

1 An Introduction to Anatomy and Physiology   27

▌ An Introduction to Studying the Human Body   28

1-1 ▌ Anatomy and physiology directly affect your life   28

1-2 ▌ Anatomy is structure, and physiology is function   29

1-3 ▌ Anatomy and physiology are closely integrated   30
  Anatomy   30
  Physiology   31

1-4 ▌ Levels of organization progress from molecules to a complete organism   32

1-5 ▌ Homeostasis is the state of internal balance   33

1-6 ▌ Negative feedback opposes variations from normal, whereas positive feedback exaggerates them   36
  The Role of Negative Feedback in Homeostasis   36
  The Role of Positive Feedback in Homeostasis   38
  Systems Integration, Equilibrium, and Homeostasis   39

1-7 ▌ Anatomical terms describe body regions, anatomical positions and directions, and body sections   40
  Superficial Anatomy   40
  Sectional Anatomy   42

1-8 ▌ Body cavities of the trunk protect internal organs and allow them to change shape   44
  The Thoracic Cavity   48
  The Abdominopelvic Cavity   48

Chapter Review   49

Spotlights
  Levels of Organization   34
  Diagnostic Imaging Techniques   46

Clinical Case
  Using A&P to Save a Life   28

Clinical Notes
  Autopsies and Cadaver Dissection   31
  Auscultation   40

2 The Chemical Level of Organization   52

▌ An Introduction to the Chemical Level of Organization   53

2-1 ▌ Atoms are the basic particles of matter   53
  Atomic Structure   53
  Elements and Isotopes   54
  Atomic Weights   55
  Electrons and Energy Levels   56

2-2 ▌ Chemical bonds are forces formed by atom interactions   57
  Ionic Bonds   57
  Covalent Bonds   60
  Hydrogen Bonds   61
  States of Matter   61

2-3 ▌ Decomposition, synthesis, and exchange reactions are important chemical reactions in physiology   62
  Basic Energy Concepts   62
  Types of Chemical Reactions   63

2-4 ▌ Enzymes catalyze specific biochemical reactions by lowering the energy needed to start them   64

2-5 ▌ Inorganic compounds lack carbon, and organic compounds contain carbon   65

2-6 ▌ Physiological systems depend on water   65
  The Properties of Aqueous Solutions   66
  Colloids and Suspensions   67

2-7 ▌ Body fluid pH is vital for homeostasis   67

2-8 ▌ Acids, bases, and salts are inorganic compounds with important physiological roles   68
  Salts   69
  Buffers and pH Control   69

2-9 ▌ Carbohydrates contain carbon, hydrogen, and oxygen in a 1:2:1 ratio   69
  Monosaccharides   70
  Disaccharides and Polysaccharides   71

2-10 ▌ Lipids often contain a carbon-to-hydrogen ratio of 1:2   72
  Fatty Acids   72
  Eicosanoids   73
  Glycerides   74
  Steroids   74
  Phospholipids and Glycolipids   75

2-11 ▮ **Proteins contain carbon, hydrogen, oxygen, and nitrogen and are formed from amino acids** 77
Protein Structure 77
Protein Shape 78
Enzyme Function 80
Glycoproteins and Proteoglycans 82

2-12 ▮ **DNA and RNA are nucleic acids** 82
Structure of Nucleic Acids 82
RNA and DNA 82

2-13 ▮ **ATP is a high-energy compound used by cells** 84

2-14 ▮ **Chemicals and their interactions form functional units called cells** 85

Chapter Review 86
Spotlights
Chemical Notation 58
Clinical Case
What Is Wrong with My Baby? 53
Clinical Notes
Radiation Sickness 55
Fatty Acids and Health 74

3 The Cellular Level of Organization 90

▮ **An Introduction to Cells** 91

3-1 ▮ **The plasma membrane separates the cell from its surrounding environment and performs various functions** 91
Membrane Lipids 94
Membrane Proteins 95
Membrane Carbohydrates 96

3-2 ▮ **Organelles within the cytoplasm perform particular functions** 96
The Cytosol 96
The Organelles 97

3-3 ▮ **The nucleus contains DNA and enzymes essential for controlling cellular activities** 108
Contents of the Nucleus 109
Information Storage in the Nucleus 109

3-4 ▮ **DNA controls protein synthesis, cell structure, and cell function** 110
The Role of Gene Activation in Protein Synthesis 110
The Transcription of mRNA 110
Translation and Protein Synthesis 112
How the Nucleus Controls Cell Structure and Function 113

3-5 ▮ **Diffusion is a passive transport mechanism that assists membrane passage** 113
Diffusion 115
Diffusion across Plasma Membranes 117

3-6 ▮ **Carrier-mediated and vesicular transport assist membrane passage** 120
Carrier-Mediated Transport 120
Vesicular Transport 122

3-7 ▮ **The membrane potential results from the unequal distribution of positive and negative charges across the plasma membrane** 126

3-8 ▮ **Stages of a cell's life cycle include interphase, mitosis, and cytokinesis** 127
DNA Replication 127
Interphase, Mitosis, and Cytokinesis 127
The Mitotic Rate and Energy Use 129

3-9 ▮ **Several growth factors affect the cell life cycle** 129

3-10 ▮ **Tumors and cancers are characterized by abnormal cell growth and division** 132

3-11 ▮ **Differentiation is cellular specialization as a result of gene activation or repression** 134

Chapter Review 135
Spotlights
Anatomy of a Model Cell 92
Protein Synthesis, Processing, and Packaging 104
Overview of Membrane Transport 124
DNA Replication 128
Stages of a Cell's Life Cycle 130
Clinical Case
When Your Heart Is in the Wrong Place 91
Clinical Notes
Inheritable Mitochondrial Disorders 106
DNA Fingerprinting 110
Mutations 112
Drugs and the Plasma Membrane 116
Telomerase, Aging, and Cancer 133
Parkinson's Disease 134

4 The Tissue Level of Organization 139

▮ **An Introduction to the Tissue Level of Organization** 140

4-1 ▮ **The four tissue types are epithelial, connective, muscle, and neural** 140

4-2 ▮ **Epithelial tissue covers body surfaces, lines cavities and tubular structures, and serves essential functions** 140
Functions of Epithelial Tissue 141
Specializations of Epithelial Cells 141
Maintaining the Integrity of Epithelia 142

4-3 ▮ **Cell shape and number of layers determine the classification of epithelia** 144
Classification of Epithelia 145
Glandular Epithelia 149

4-4 ▌ **Connective tissue provides a protective structural framework for other tissue types** 152
    Classification of Connective Tissues 152
    Connective Tissue Proper 152

4-5 ▌ **Cartilage and bone provide a strong supporting framework** 159
    Cartilage 159
    Bone 162

4-6 ▌ **Tissue membranes are physical barriers of four types: mucous, serous, cutaneous, and synovial** 163
    Mucous Membranes 163
    Serous Membranes 163
    The Cutaneous Membrane 165
    Synovial Membranes 165

4-7 ▌ **Connective tissue creates the internal framework of the body** 165

4-8 ▌ **The three types of muscle tissue are skeletal, cardiac, and smooth** 166
    Skeletal Muscle Tissue 166
    Cardiac Muscle Tissue 168
    Smooth Muscle Tissue 168

4-9 ▌ **Neural tissue responds to stimuli and propagates electrical impulses throughout the body** 168

4-10 ▌ **The response to tissue injury involves inflammation and regeneration** 170
    Inflammation 170
    Regeneration 170

4-11 ▌ **With advancing age, tissue repair declines and cancer rates increase** 170
    Aging and Tissue Structure 170
    Aging and Cancer Incidence 172

Chapter Review 172
Spotlights
    Inflammation and Regeneration 171
Clinical Case
    The Rubber Girl 140
Clinical Notes
    Exfoliative Cytology 146
    Marfan's Syndrome 155

UNIT 2   SUPPORT AND MOVEMENT

5   The Integumentary System   176

▌ **An Introduction to the Integumentary System** 177

5-1 ▌ **The epidermis is composed of layers with various functions** 179
    Stratum Basale 179
    Stratum Spinosum 179
    Stratum Granulosum 181
    Stratum Lucidum 181
    Stratum Corneum 181

5-2 ▌ **Factors influencing skin color are epidermal pigmentation and dermal circulation** 181
    The Role of Epidermal Pigmentation 182
    The Role of Dermal Circulation 183

5-3 ▌ **Sunlight causes epidermal cells to convert a steroid into vitamin D$_3$** 184

5-4 ▌ **Epidermal growth factor has several effects on the epidermis and epithelia** 185

5-5 ▌ **The dermis is the tissue layer that supports the epidermis** 186
    Dermal Strength and Elasticity 186
    Cleavage Lines 187
    The Dermal Blood Supply 187
    Innervation of the Skin 187

5-6 ▌ **The hypodermis connects the dermis to underlying tissues** 188

5-7 ▌ **Hair is composed of keratinized dead cells that have been pushed to the surface** 189
    Hair Production 191
    The Hair Growth Cycle 191
    Types of Hairs 191
    Hair Color 191

5-8 ▌ **Sebaceous glands and sweat glands are exocrine glands found in the skin** 192
    Sebaceous Glands 192
    Sweat Glands 193
    Other Integumentary Glands 194
    Control of Glandular Secretions and the Homeostatic Role of the Integument 194

5-9 ▌ **Nails are keratinized epidermal cells that protect the tips of fingers and toes** 195

5-10 ▌ **Several phases are involved in repairing the integument following an injury** 195

5-11 ▌ **Effects of aging include skin thinning, wrinkling, and reduced melanocyte activity** 198

Chapter Review 201
Spotlights
    The Epidermis 180
Clinical Case
    Skin Cells in Overdrive 177
Clinical Notes
    Skin Cancer 184
    Decubitus Ulcers 187
    Liposuction 189
    Burns and Grafts 197
    Skin Abnormalities 198

6   Osseous Tissue and Bone Structure 204

▌ An Introduction to the Skeletal System 205

6-1   ▌ The skeletal system has five primary functions 205

6-2   ▌ Bones are classified according to shape and structure, and they have a variety of surface markings 206
Bone Shapes 206
Bone Markings 207
Bone Structure 208

6-3   ▌ Bone is composed of matrix and several types of cells: osteocytes, osteoblasts, osteogenic cells, and osteoclasts 208
Bone Matrix 209
Bone Cells 209

6-4   ▌ Compact bone contains parallel osteons, and spongy bone contains trabeculae 210
Compact Bone Structure 211
Spongy Bone Structure 212
The Periosteum and Endosteum 213

6-5   ▌ Bones form through ossification and enlarge through appositional growth and remodeling 215
Endochondral Ossification 215
Intramembranous Ossification 218
The Blood and Nerve Supplies to Bone 218

6-6   ▌ Bone growth and development depend on a balance between bone formation and bone resorption 218

6-7   ▌ Exercise, hormones, and nutrition affect bone development and the skeletal system 220
The Effects of Exercise on Bone 220
Nutritional and Hormonal Effects on Bone 220

6-8   ▌ Calcium plays a critical role in bone physiology 222
The Skeleton as a Calcium Reserve 222
Hormones and Calcium Balance 222

6-9   ▌ A fracture is a crack or break in a bone 224

6-10 ▌ Osteopenia has a widespread effect on aging skeletal tissue 225

Chapter Review 229
Spotlights
Endochondral Ossification 216
Types of Fractures and Steps in Repair 226
Clinical Case
A Case of Child Abuse? 205
Clinical Notes
Heterotopic Bone Formation 214
Abnormal Bone Development 222

7   The Axial Skeleton 232

▌ An Introduction to the Axial Skeleton 233

7-1   ▌ The 80 bones of the head and trunk make up the axial skeleton 233

7-2   ▌ The skull is composed of 8 cranial bones and 14 facial bones 233

7-3   ▌ Each orbital complex contains an eye, and the nasal complex encloses the nasal cavities 249
The Orbital Complexes 249
The Nasal Complex 249

7-4   ▌ Fontanelles are non-ossified areas between cranial bones that allow for brain growth in infants and small children 250

7-5   ▌ The vertebral column has four spinal curves 252
Spinal Curvature 252
Vertebral Anatomy 253

7-6   ▌ The five vertebral regions are the cervical, thoracic, lumbar, sacral, and coccygeal regions 254
Cervical Vertebrae 255
Thoracic Vertebrae 258
Lumbar Vertebrae 258
The Sacrum 258
The Coccyx 261

7-7   ▌ The thoracic cage protects organs in the chest and provides sites for muscle attachment 261
The Ribs 261
The Sternum 262

Chapter Review 264
Spotlights
Sectional Anatomy of the Skull 238
Clinical Case
Knocked Out 233
Clinical Notes
Temporomandibular Joint Syndrome 248
Sinusitis 251
Craniostenosis 252
Kyphosis, Lordosis, and Scoliosis 255

8   The Appendicular Skeleton 267

▌ An Introduction to the Appendicular Skeleton 268

8-1   ▌ The pectoral girdles—the clavicles and scapulae—attach the upper limbs to the axial skeleton 268
The Clavicles 270
The Scapulae 270

8-2 ▌ **The upper limbs are adapted for free movement** 271
The Humerus 271
The Ulna 273
The Radius 273
The Carpal Bones 273
The Metacarpal Bones and Phalanges 274

8-3 ▌ **The pelvic girdle—two hip bones—attaches the lower limbs to the axial skeleton** 276
The Pelvic Girdle 276
The Pelvis 276

8-4 ▌ **The lower limbs are adapted for movement and support** 280
The Femur 280
The Patella 280
The Tibia 282
The Fibula 282
The Tarsal Bones 283
The Metatarsal Bones and Phalanges 284

8-5 ▌ **Sex differences and age account for individual skeletal variation** 284
Chapter Review 286
Spotlights
Sex Differences in the Human Skeleton 279
Clinical Case
The Orthopedic Surgeon's Nightmare 268
Clinical Notes
Carpal Tunnel Syndrome 275
Hip Fracture 280
Runner's Knee 280
Stress Fractures 284
Congenital Talipes Equinovarus 284

9 **Joints** 289

▌ An Introduction to Joints 290

9-1 ▌ **Joints are categorized according to their range of motion or structure** 290

9-2 ▌ **Synovial joints are freely movable joints containing synovial fluid** 292
Articular Cartilage 292
Synovial Fluid 293
Accessory Structures 293
Factors That Stabilize Synovial Joints 293

9-3 ▌ **The structure and function of synovial joints enable various skeletal movements** 294
Types of Movements at Synovial Joints 295

9-4 ▌ **Intervertebral discs and ligaments are structural components of intervertebral joints** 300
Intervertebral Discs 300
Intervertebral Ligaments 301
Vertebral Movements 301

9-5 ▌ **The shoulder is a ball-and-socket joint, and the elbow is a hinge joint** 302
The Shoulder Joint 302
The Elbow Joint 304

9-6 ▌ **The hip is a ball-and-socket joint, and the knee is a hinge joint** 305
The Hip Joint 305
The Knee Joint 306

9-7 ▌ **With advancing age, arthritis and other degenerative changes impair joint mobility** 309

9-8 ▌ **The skeletal system supports and stores energy and minerals for other body systems** 310
Chapter Review 312
Spotlights
Joint Movement 296
Clinical Case
What's Ailing the Birthday Girl? 290
Clinical Notes
Bursitis and Bunions 294
Dislocation and Subluxation 294
Damage to Intervertebral Discs 301
Knee Injuries 309

10 **Muscle Tissue** 315

▌ An Introduction to Muscle Tissue 316

10-1 ▌ **Skeletal muscle performs six major functions** 316

10-2 ▌ **A skeletal muscle contains muscle tissue, connective tissues, blood vessels, and nerves** 317
Organization of Connective Tissues 318
Blood Vessels and Nerves 318

10-3 ▌ **Skeletal muscle fibers have distinctive features** 318
The Sarcolemma and Transverse Tubules 318
Myofibrils 319
The Sarcoplasmic Reticulum 320
Sarcomeres 321
Sliding Filaments and Muscle Contraction 324

10-4 ▌ **The nervous system communicates with skeletal muscles at the neuromuscular junction** 325
Electrical Impulses and Excitable Membranes 328
The Control of Skeletal Muscle Activity 328
Excitation–Contraction Coupling 328
Relaxation 334

10-5 ▌ **Sarcomere shortening and muscle fiber stimulation produce tension** 334
Tension Production by Muscle Fibers 334
Tension Production by Skeletal Muscles 338
Motor Units and Tension Production 338

10-6 | ATP provides energy for muscle contraction 342
    ATP and CP Reserves 342
    ATP Generation 343
    Energy Use and the Level of Muscular Activity 344
    Muscle Fatigue 344
    The Recovery Period 344
    Hormones and Muscle Metabolism 346

10-7 | Muscle performance capabilities depend on muscle fiber type and physical conditioning 346
    Types of Skeletal Muscle Fibers 346
    Muscle Performance and the Distribution of Muscle Fibers 347
    Muscle Hypertrophy and Atrophy 347
    Physical Conditioning 348

10-8 | Cardiac muscle tissue differs structurally and functionally from skeletal muscle tissue 350
    Structural Characteristics of Cardiac Muscle Tissue 350
    Functional Characteristics of Cardiac Muscle Tissue 351

10-9 | Smooth muscle tissue differs structurally and functionally from skeletal and cardiac muscle tissue 351
    Structural Characteristics of Smooth Muscle Tissue 352
    Functional Characteristics of Smooth Muscle Tissue 353

Chapter Review 354
Spotlights
    Events at the Neuromuscular Junction 326
    Excitation–Contraction Coupling 329
    The Contraction Cycle and Cross-Bridge Formation 330
Clinical Case
    A Real Eye Opener 316
Clinical Notes
    Tetanus 332
    Rigor Mortis 334
    Delayed-Onset Muscle Soreness 349

11 The Muscular System 358

| An Introduction to the Muscular System 359

11-1 | Fascicle arrangement is correlated with muscle power and range of motion 359
    Parallel Muscles 359
    Convergent Muscles 359
    Pennate Muscles 360
    Circular Muscles 361

11-2 | The three classes of levers increase muscle efficiency 361

11-3 | Muscle origins are at the fixed end of muscles, and insertions are at the movable end of muscles 362
    Origins and Insertions 362
    Actions 363

11-4 | Descriptive terms are used to name skeletal muscles 365
    Location in the Body 365
    Origin and Insertion 365
    Fascicle Organization 365
    Position 365
    Structural Characteristics 365
    Action 366
    Axial and Appendicular Muscles 367

11-5 | Axial muscles are muscles of the head and neck, vertebral column, trunk, and pelvic floor 367
    Muscles of the Head and Neck 367
    Muscles of the Vertebral Column 377
    Oblique and Rectus Muscles 380
    Muscles of the Pelvic Floor 382

11-6 | Appendicular muscles are muscles of the shoulders, upper limbs, pelvis, and lower limbs 384
    Muscles of the Shoulders and Upper Limbs 384
    Muscles of the Pelvis and Lower Limbs 395

11-7 | With advancing age, the size and power of muscle tissue decrease 405

11-8 | Exercise produces responses in multiple body systems 405

Chapter Review 407
Spotlights
    Muscle Action 364
Clinical Case
    The Weekend Warrior 359
Clinical Notes
    Intramuscular Injections 370
    Hernia 404

UNIT 3 CONTROL AND REGULATION

12 Neural Tissue 411

| An Introduction to Neural Tissue 412

12-1 | The nervous system has anatomical and functional divisions 412
    The Anatomical Divisions of the Nervous System 412
    The Functional Divisions of the Nervous System 413

12-2 | Neurons are nerve cells specialized for intercellular communication 414
    The Structure of Neurons 414
    The Classification of Neurons 416

12-3 | CNS and PNS neuroglia support and protect neurons 418
    Neuroglia of the Central Nervous System 418
    Neuroglia of the Peripheral Nervous System 423
    Neural Responses to Injuries 424

12-4 ▮ **The membrane potential is the electrical potential of the cell's interior relative to its surroundings** 424
The Membrane Potential 424
Changes in the Membrane Potential 428
Graded Potentials 430

12-5 ▮ **An action potential is an electrical event** 432
The All-or-None Principle 432
Generation of Action Potentials 432
Propagation of Action Potentials 433

12-6 ▮ **Axon diameter, in addition to myelin, affects propagation speed** 438

12-7 ▮ **At synapses, communication occurs among neurons or between neurons and other cells** 439
Synaptic Activity 439
General Properties of Synapses 439
Cholinergic Synapses 440

12-8 ▮ **Neurotransmitters and neuromodulators have various functions** 442
The Activities of Other Neurotransmitters 442
Neuromodulators 443
How Neurotransmitters and Neuromodulators Work 446

12-9 ▮ **Individual neurons process information by integrating excitatory and inhibitory stimuli** 447
Postsynaptic Potentials 447
Presynaptic Inhibition and Presynaptic Facilitation 449
The Rate of Generation of Action Potentials 449

Chapter Review 451
Spotlights
Resting Membrane Potential 426
Generation of an Action Potential 434
Propagation of an Action Potential 436
Clinical Case
Did Franklin D. Roosevelt Really Have Polio? 412
Clinical Notes
Rabies 416
Tumors 418
Demyelination 421

13 **The Spinal Cord, Spinal Nerves, and Spinal Reflexes** 455

▮ **An Introduction to the Spinal Cord, Spinal Nerves, and Spinal Reflexes** 456

13-1 ▮ **The brain and spinal cord make up the central nervous system (CNS), and the cranial nerves and spinal nerves make up the peripheral nervous system (PNS)** 456

13-2 ▮ **The spinal cord is surrounded by three meninges and carries sensory and motor information** 457
Gross Anatomy of the Spinal Cord 457
Spinal Meninges 459

13-3 ▮ **Gray matter integrates information and initiates commands, and white matter carries information from place to place** 461
Organization of Gray Matter 463
Organization of White Matter 463

13-4 ▮ **Spinal nerves form plexuses that are named according to their level of emergence from the vertebral canal** 463
Anatomy of Spinal Nerves 463
Peripheral Distribution of Spinal Nerves 464
Nerve Plexuses 464

13-5 ▮ **Interneurons are organized into functional groups called neuronal pools** 473

13-6 ▮ **Reflexes are rapid, automatic responses to stimuli** 475
The Reflex Arc 475
Classification of Reflexes 478

13-7 ▮ **Spinal reflexes vary in complexity** 479
Monosynaptic Reflexes 479
Polysynaptic Reflexes 480

13-8 ▮ **The brain can affect spinal cord–based reflexes** 481
Voluntary Movements and Reflex Motor Patterns 482
Reinforcement and Inhibition 482

Chapter Review 483
Spotlights
Peripheral Distribution of Spinal Nerves 466
Spinal Reflexes 476
Clinical Case
Prom Night 456
Clinical Notes
Anesthesia 461
Shingles 465
Sensory Innervation in the Hand 470
Sensory Innervation in the Ankle and Foot 473

14 **The Brain and Cranial Nerves** 487

▮ **An Introduction to the Brain and Cranial Nerves** 488

14-1 ▮ **The brain has several principal structures, each with specific functions** 488
Major Brain Regions and Landmarks 488
Embryology of the Brain 490
Ventricles of the Brain 490

14-2 ▮ **The brain is protected and supported by the cranial meninges, cerebrospinal fluid, and the blood–brain barrier** 491
The Cranial Meninges 491
Cerebrospinal Fluid 493
The Blood Supply to the Brain 495

14-3 | The medulla oblongata is continuous with the spinal cord and contains vital centers 496

14-4 | The pons contains nuclei and tracts that carry or relay sensory and motor information 498

14-5 | The cerebellum coordinates learned and reflexive patterns of muscular activity at the subconscious level 499

14-6 | The midbrain regulates auditory and visual reflexes and controls alertness 501

14-7 | The diencephalon integrates sensory information with motor output at the subconscious level 503
The Thalamus 503
The Hypothalamus 504

14-8 | The limbic system is a group of tracts and nuclei that function in emotion, motivation, and memory 506

14-9 | The cerebrum, the largest region of the brain, contains motor, sensory, and association areas 508
The Cerebral Cortex 508
The White Matter of the Cerebrum 508
The Basal Nuclei 508
Motor and Sensory Areas of the Cortex 512

14-10 | Cranial reflexes involve sensory and motor fibers of cranial nerves 529

Chapter Review 530
Spotlights
Formation and Circulation of Cerebrospinal Fluid 494
Clinical Case
The Neuroanatomist's Stroke 533
Clinical Notes
Epidural and Subdural Hemorrhages 493
Disconnection Syndrome 514
Aphasia and Dyslexia 515

15 Sensory Pathways and the Somatic Nervous System 534

An Introduction to Sensory Pathways and the Somatic Nervous System 535

15-1 | Sensory information from all parts of the body is routed to the somatosensory cortex 535

15-2 | Sensory receptors connect our internal and external environments with the nervous system 536
The Detection of Stimuli 536
The Interpretation of Sensory Information 537
Adaptation 538

15-3 | General sensory receptors are classified by the type of stimulus that excites them 539
Nociceptors 539
Thermoreceptors 540
Mechanoreceptors 540
Chemoreceptors 543

15-4 | Separate pathways carry somatic sensory and visceral sensory information 544
Somatic Sensory Pathways 544
Visceral Sensory Pathways 549

15-5 | The somatic nervous system is an efferent division that controls skeletal muscles 549
The Corticospinal Pathway 550
The Medial and Lateral Pathways 551
The Basal Nuclei and Cerebellum 552
Levels of Processing and Motor Control 553

Chapter Review 554
Spotlights
Somatic Sensory Pathways 546
Clinical Case
Living with Cerebral Palsy 535
Clinical Notes
Assessment of Tactile Sensitivities 543
Amyotrophic Lateral Sclerosis 552
Cerebral Palsy 553

16 The Autonomic Nervous System and Higher-Order Functions 557

An Introduction to the Autonomic Nervous System and Higher-Order Functions 558

16-1 | The autonomic nervous system is involved in the unconscious regulation of visceral functions and has sympathetic and parasympathetic divisions 558
Organization of the ANS 558
Divisions of the ANS 559

16-2 | The sympathetic division consists of preganglionic neurons and ganglionic neurons involved in using energy and increasing metabolic rate 562
Organization and Anatomy of the Sympathetic Division 563
Sympathetic Activation 566

16-3 | Stimulation of sympathetic neurons leads to the release of various neurotransmitters 566
Sympathetic Stimulation and the Release of NE and E 567
Sympathetic Stimulation and the Release of ACh and NO 567
Summary: The Sympathetic Division 568

16-4 | The parasympathetic division consists of preganglionic neurons and ganglionic neurons involved in conserving energy and lowering metabolic rate 568
Organization and Anatomy of the Parasympathetic Division 568
Parasympathetic Activation 568

16-5 ▌ **Stimulation of parasympathetic neurons leads to the release of the neurotransmitter ACh** 570

Neurotransmitter Release 570

Membrane Receptors and Responses 570

Summary: The Parasympathetic Division 570

16-6 ▌ **The sympathetic and parasympathetic divisions interact, creating dual innervation** 571

Anatomy of Dual Innervation 571

Autonomic Tone 572

16-7 ▌ **Visceral reflexes play a role in the integration and control of autonomic functions** 575

Visceral Reflexes 575

Higher Levels of Autonomic Control 576

The Integration of SNS and ANS Activities 577

16-8 ▌ **Higher-order functions include memory and states of consciousness** 578

Memory 578

States of Consciousness 580

16-9 ▌ **Neurotransmitters influence brain chemistry and behavior** 582

16-10 ▌ **Aging produces various structural and functional changes in the nervous system** 583

Chapter Review 585

Spotlight

Overview of the Autonomic Nervous System 560

Clinical Case

The First Day in Anatomy Lab 558

Clinical Notes

Amnesia 579

Categorizing Nervous System Disorders 581

Alzheimer's Disease 583

17 **The Special Senses** **589**

▌ An Introduction to the Special Senses 590

17-1 ▌ **Olfaction, the sense of smell, involves olfactory receptors responding to chemical stimuli** 590

Olfactory Receptors 590

Olfactory Pathways 591

Olfactory Discrimination 591

17-2 ▌ **Gustation, the sense of taste, involves taste receptors responding to chemical stimuli** 594

Taste Receptors 594

Gustatory Pathways 594

Gustatory Discrimination 595

17-3 ▌ **Internal eye structures contribute to vision, while accessory eye structures provide protection** 596

Accessory Structures of the Eye 596

The Eye 599

17-4 ▌ **Photoreceptors respond to light and change it into electrical signals essential to visual physiology** 607

Visual Physiology 607

The Visual Pathways 613

17-5 ▌ **Equilibrium sensations originate within the internal ear, while hearing involves the detection and interpretation of sound waves** 616

Anatomy of the Ear 616

Equilibrium 619

Hearing 622

Chapter Review 630

Spotlights

Olfaction and Gustation 592

Refractive Problems 608

Photoreception 610

Clinical Case

A Chance to See 590

Clinical Notes

Diabetic Retinopathy 600

Detached Retina 600

Glaucoma 603

Motion Sickness 619

18 **The Endocrine System** **634**

▌ An Introduction to the Endocrine System 635

18-1 ▌ **Homeostasis is preserved through intercellular communication** 635

18-2 ▌ **The endocrine system regulates physiological processes through the binding of hormones to receptors** 637

Classes of Hormones 637

Secretion and Distribution of Hormones 638

Mechanisms of Hormone Action 640

Control of Endocrine Activity by Endocrine Reflexes 642

18-3 ▌ **The bilobed pituitary gland is an endocrine organ that releases nine peptide hormones** 645

The Anterior Lobe of the Pituitary Gland 645

The Posterior Lobe of the Pituitary Gland 649

Summary: The Hormones of the Pituitary Gland 651

18-4 ▌ **The thyroid gland lies inferior to the larynx and requires iodine for hormone synthesis** 652

Thyroid Follicles and Thyroid Hormones 652

Functions of Thyroid Hormones 654

The C Cells of the Thyroid Gland and Calcitonin 655

18-5 ▌ **The four parathyroid glands, embedded in the posterior surface of the thyroid gland, secrete parathyroid hormone to elevate blood Ca^{2+}** 656

18-6 ▌ **The adrenal glands, consisting of a cortex and medulla, cap the kidneys and secrete several hormones** 657

The Adrenal Cortex 657

The Adrenal Medulla 660

18-7 ▌ **The pineal gland, attached to the roof of the third ventricle, secretes melatonin**   660

18-8 ▌ **The pancreas is both an exocrine organ and endocrine gland**   661
The Pancreatic Islets   662
Insulin   663
Glucagon   663

18-9 ▌ **Many organs have secondary endocrine functions**   665
The Intestines   665
The Kidneys   665
The Heart   666
The Thymus   666
The Gonads   667
Adipose Tissue   669

18-10 ▌ **Hormones interact to produce coordinated physiological responses**   669
Role of Hormones in Growth   670
Aging and Hormone Production   670

Chapter Review   674
Spotlights
Structural Classification of Hormones   639
G Proteins and Second Messengers   641
Diabetes Mellitus   664
The General Adaptation Syndrome   671
Clinical Case
Stones, Bones, and Groans   635
Clinical Notes
Diabetes Insipidus   649
Endocrine Disorders   668
Hormones and Athletic Performance   672

UNIT 4   FLUIDS AND TRANSPORT

19   Blood   678

▌ **An Introduction to Blood and the Cardiovascular System**   679

19-1 ▌ **Blood has several important functions and unique physical characteristics**   679

19-2 ▌ **Plasma, the fluid portion of blood, contains significant quantities of plasma proteins**   682
Plasma Proteins   682

19-3 ▌ **Red blood cells, formed by erythropoiesis, contain hemoglobin that can be recycled**   683
Abundance of RBCs   683
Structure of RBCs   684
Hemoglobin   685
RBC Formation and Turnover   686
RBC Production   687

19-4 ▌ **The ABO blood types and Rh system are based on antigen–antibody responses**   690
Cross-Reactions in Transfusions   692
Testing for Transfusion Compatibility   692

19-5 ▌ **The various types of white blood cells contribute to the body's defenses**   693
WBC Circulation and Movement   693
Types of WBCs   696
The Differential Count and Changes in WBC Profiles   697
WBC Production   698

19-6 ▌ **Platelets, disc-shaped structures formed from megakaryocytes, function in the clotting process**   700
Platelet Functions   701
Platelet Production   701

19-7 ▌ **Hemostasis involves vascular spasm, platelet plug formation, and blood coagulation**   701
The Vascular Phase   701
The Platelet Phase   702
The Coagulation Phase   703
Fibrinolysis   705

Chapter Review   706
Spotlights
The Composition of Whole Blood   680
Hemolytic Disease of the Newborn   694
Clinical Case
A Mysterious Blood Disorder   679
Clinical Notes
Collecting Blood for Analysis   682
Plasma Expanders   683
Abnormal Hemoglobin   686

20   The Heart   710

▌ **An Introduction to the Cardiovascular System**   711

20-1 ▌ **The heart is a four-chambered organ, supplied by the coronary circulation, that pumps oxygen-poor blood to the lungs and oxygen-rich blood to the rest of the body**   712
The Pericardium   712
Superficial Anatomy of the Heart   712
The Heart Wall   712
Cardiac Muscle Tissue   715
Internal Anatomy and Organization   715
Connective Tissues and the Cardiac Skeleton   721
The Blood Supply to the Heart   721

20-2 ▌ **The conducting system distributes electrical impulses through the heart, and an electrocardiogram records the associated electrical events**   723
Cardiac Physiology   723
The Conducting System   723

The Electrocardiogram   728

Contractile Cells   729

20-3 ▌ **Events during a complete heartbeat make up a cardiac cycle**   732

Phases of the Cardiac Cycle   733

Pressure and Volume Changes in the Cardiac Cycle   734

Heart Sounds   736

20-4 ▌ **Cardiodynamics examines the factors that affect cardiac output**   737

Overview: Factors Affecting Cardiac Output   737

Factors Affecting the Heart Rate   738

Factors Affecting the Stroke Volume   741

Summary: The Control of Cardiac Output   743

The Heart and the Cardiovascular System   744

Chapter Review   745

Spotlights

Heart Disease and Heart Attacks   724

Cardiac Arrhythmias   730

Clinical Case

A Needle to the Chest   711

Clinical Note

Abnormal Conditions Affecting Cardiac Output   734

21 Blood Vessels and Circulation   749

▌ **An Introduction to Blood Vessels and Circulation**   750

21-1 ▌ **Arteries, arterioles, capillaries, venules, and veins differ in size, structure, and functional properties**   750

The Structure of Vessel Walls   751

Differences between Arteries and Veins   752

Capillaries   755

Veins   758

The Distribution of Blood   759

21-2 ▌ **Pressure and resistance determine blood flow and affect rates of capillary exchange**   760

Pressure   760

Total Peripheral Resistance   760

An Overview of Cardiovascular Pressures   762

Capillary Pressures and Capillary Exchange   765

21-3 ▌ **Cardiovascular regulatory mechanisms involve autoregulation, neural mechanisms, and endocrine responses**   768

Autoregulation of Blood Flow within Tissues   768

Neural Mechanisms   769

Hormones and Cardiovascular Regulation   772

21-4 ▌ **The cardiovascular system adapts to physiological stress and maintains a special vascular supply to the brain, heart, and lungs**   775

The Cardiovascular Response to Exercise   775

The Cardiovascular Response to Hemorrhaging   777

Vascular Supply to Special Regions   778

21-5 ▌ **The pulmonary and systemic circuits of the cardiovascular system exhibit three general functional patterns**   779

21-6 ▌ **In the pulmonary circuit, deoxygenated blood enters the lungs in arteries, and oxygenated blood leaves the lungs by veins**   780

21-7 ▌ **The systemic circuit carries oxygenated blood from the left ventricle to tissues and organs other than the pulmonary exchange surfaces, and returns deoxygenated blood to the right atrium**   781

Systemic Arteries   781

Systemic Veins   789

21-8 ▌ **Modifications of fetal and maternal cardiovascular systems promote the exchange of materials, and independence occurs at birth**   798

Placental Blood Supply   798

Fetal Circulation in the Heart and Great Vessels   798

Cardiovascular Changes at Birth   799

21-9 ▌ **Aging affects the blood, heart, and blood vessels**   801

Chapter Review   803

Spotlight

Congenital Heart Problems   800

Clinical Case

Did Ancient Mummies Have Atherosclerosis?   750

Clinical Notes

Arteriosclerosis   754

Edema   767

22 The Lymphatic System and Immunity   807

▌ **An Introduction to the Lymphatic System and Immunity**   808

22-1 ▌ **Surface barriers and internal defenses make up innate defenses, and lymphocytes provide adaptive defenses**   808

22-2 ▌ **Lymphatic vessels, lymphocytes, lymphoid tissues, and lymphoid organs function in body defenses**   809

Functions of the Lymphatic System   810

Lymphatic Vessels   810

Lymphocytes   813

Lymphoid Tissues   816

Lymphoid Organs   816

The Lymphatic System and Body Defenses   820

22-3 ▌ **Innate (nonspecific) defenses do not discriminate between potential threats and respond the same regardless of the invader**   822

Physical Barriers   822

Phagocytes   822

Immune Surveillance 824

Interferons 825

Complement System 825

Inflammation 827

Fever 828

22-4 ▌ **Adaptive (specific) defenses respond to individual threats and are either cell-mediated or antibody-mediated** 828

Forms of Immunity 829

Properties of Adaptive Immunity 830

An Introduction to the Immune Response 830

22-5 ▌ **T cells play a role in initiating, maintaining, and controlling the immune response** 831

Antigen Presentation 831

Antigen Recognition 832

Activation of CD8 T Cells 834

Activation of CD4 T Cells 835

22-6 ▌ **B cells respond to antigens by producing specific antibodies** 836

B Cell Sensitization and Activation 836

Antibody Structure 837

Primary and Secondary Responses to Antigen
  Exposure 840

Summary of the Immune Response 841

22-7 ▌ **Immunocompetence enables a normal immune response; abnormal responses result in immune disorders** 844

The Development of Immunocompetence 844

Cytokines of the Immune System 844

Immune Disorders 844

Stress and the Immune Response 849

22-8 ▌ **The immune response diminishes as we age** 851

22-9 ▌ **The nervous and endocrine systems influence the immune response** 851

Chapter Review 852

Spotlight

Cytokines of the Immune System 846

Clinical Case

Isn't There a Vaccine for That? 808

Clinical Notes

Cancer and the Lymphatic System 814

Graft Rejection and Immunosuppression 832

AIDS 845

UNIT 5 ENVIRONMENTAL EXCHANGE

23 The Respiratory System 856

▌ **An Introduction to the Respiratory System** 857

23-1 ▌ **The respiratory system, organized into an upper respiratory system and a lower respiratory system, has several basic functions** 857

Functions of the Respiratory System 857

Organization of the Respiratory System 858

23-2 ▌ **Located outside the thoracic cavity, the upper respiratory system consists of the nose, nasal cavity, paranasal sinuses, and pharynx** 861

The Nose, Nasal Cavity, and Paranasal Sinuses 861

The Pharynx 863

23-3 ▌ **Composed of cartilages, ligaments, and muscles, the larynx produces sound** 863

Cartilages and Ligaments of the Larynx 863

Sound Production 864

The Laryngeal Musculature 865

23-4 ▌ **The trachea and primary bronchi convey air to and from the lungs** 865

The Trachea 865

The Primary Bronchi 866

23-5 ▌ **Enclosed by pleural cavities, the lungs are paired organs containing alveoli, which permit gaseous exchange** 867

Lobes and Surfaces of the Lungs 867

The Bronchi 867

The Bronchioles 867

Alveolar Ducts and Alveoli 869

The Blood Supply to the Lungs 872

The Pleural Cavities and Pleural Membranes 872

23-6 ▌ **External respiration and internal respiration allow gaseous exchange within the body** 873

23-7 ▌ **Pulmonary ventilation—the exchange of air between the atmosphere and the lungs—involves pressure changes, muscle movement, and respiratory rates and volumes** 874

The Movement of Air 874

Pressure Changes during Inhalation and Exhalation 875

The Mechanics of Breathing 878

Respiratory Rates and Volumes 880

23-8 ▌ **Gas exchange depends on the partial pressures of gases and the diffusion of molecules** 882

The Gas Laws 882

Diffusion and Respiratory Function 884

23-9 ▍ **Most oxygen is transported bound to hemoglobin; and carbon dioxide is transported in three ways: as carbonic acid, bound to hemoglobin, or dissolved in plasma** 886

Oxygen Transport 886

Carbon Dioxide Transport 889

Summary: Gas Transport 890

23-10 ▍ **Neurons in the medulla oblongata and pons, along with respiratory reflexes, control respiration** 890

Local Regulation of Gas Transport and Alveolar Function 891

The Respiratory Centers of the Brain 892

Respiratory Reflexes 896

Voluntary Control of Respiration 898

Changes in the Respiratory System at Birth 898

23-11 ▍ **Respiratory performance declines with age** 899

23-12 ▍ **The respiratory system provides oxygen to, and eliminates carbon dioxide from, other organ systems** 899

Chapter Review 902

Spotlights

Respiratory Muscles and Pulmonary Ventilation 879

Control of Respiration 894

Clinical Case

How Long Should a Cough Last? 857

Clinical Notes

Breakdown of the Respiratory Defense System 860

Pneumothorax 878

Decompression Sickness 884

Blood Gas Analysis 886

Carbon Monoxide Poisoning 889

Emphysema and Lung Cancer 901

24 The Digestive System 906

▍ **An Introduction to the Digestive System** 907

24-1 ▍ **The digestive system, consisting of the digestive tract and accessory organs, has overlapping food utilization functions** 907

Functions of the Digestive System 908

The Digestive Organs and the Peritoneum 909

Histology of the Digestive Tract 911

The Movement of Digestive Materials 912

Control of Digestive Functions 913

24-2 ▍ **The oral cavity contains the tongue, salivary glands, and teeth, each with specific functions** 915

The Tongue 916

Salivary Glands 916

The Teeth 918

24-3 ▍ **The pharynx is a passageway between the oral cavity and esophagus** 920

24-4 ▍ **The esophagus is a muscular tube that transports solids and liquids from the pharynx to the stomach** 921

Histology of the Esophagus 921

Swallowing 922

24-5 ▍ **The stomach is a J-shaped organ that receives the bolus from the esophagus and aids in chemical and mechanical digestion** 923

Anatomy of the Stomach 923

Regulation of Gastric Activity 927

Digestion and Absorption in the Stomach 927

24-6 ▍ **The small intestine digests and absorbs nutrients, and associated glandular organs assist with digestive processes** 927

The Small Intestine 927

Histology of the Small Intestine 930

Intestinal Secretions 932

Intestinal Movements 933

The Pancreas 933

The Liver 935

The Gallbladder 940

The Coordination of Secretion and Absorption 940

24-7 ▍ **The large intestine is divided into three parts with regional specialization** 942

The Cecum 943

The Colon 943

The Rectum 945

Histology of the Large Intestine 945

Physiology of the Large Intestine 946

24-8 ▍ **Digestion is the chemical alteration of food that allows the absorption and use of nutrients** 948

The Processing and Absorbing of Nutrients 948

Carbohydrate Digestion and Absorption 948

Lipid Digestion and Absorption 951

Protein Digestion and Absorption 951

Water Absorption 952

Ion Absorption 952

24-9 ▍ **Many age-related changes affect digestion and absorption** 954

24-10 ▍ **The digestive system is extensively integrated with other body systems** 954

Chapter Review 956

Spotlights

Regulation of Gastric Activity 928

Chemical Events of Digestion 949

Clinical Case

An Unusual Transplant 907

Clinical Notes

Peritonitis 909

Epithelial Renewal and Repair 912

Mumps 916

Gastritis and Peptic Ulcers 923

Pancreatitis   933
Cirrhosis   935
Colorectal Cancer   943
Inflammatory and Infectious Disorders of the Digestive System   946

25 Metabolism and Energetics   961

❚ An Introduction to Metabolism and Energetics   962

25-1   ❚ Metabolism refers to all the chemical reactions in the body, and energetics refers to the flow and transformation of energy   962

25-2   ❚ Carbohydrate metabolism involves glycolysis, ATP production, and gluconeogenesis   965
Glycolysis   965
Mitochondrial ATP Production   965
Energy Yield of Glycolysis and Cellular Respiration   971
Gluconeogenesis   972

25-3   ❚ Lipid metabolism involves lipolysis, beta-oxidation, and the transport and distribution of lipids as free fatty acids and lipoproteins   973
Lipid Catabolism   973
Lipid Synthesis   975
Lipid Transport and Distribution   975

25-4   ❚ Protein catabolism involves transamination and deamination, whereas protein synthesis involves amination and transamination   978
Amino Acid Catabolism   978
Protein Synthesis   978

25-5   ❚ The body experiences two patterns of metabolic activity: the absorptive and postabsorptive states   982

25-6   ❚ Adequate nutrition is necessary to prevent deficiency disorders and ensure physiological functioning   983
Food Groups and a Balanced Diet   983
Nitrogen Balance   983
Minerals   985
Vitamins   986
Diet and Disease   987

25-7   ❚ Metabolic rate is the average caloric expenditure, and thermoregulation involves balancing heat-producing and heat-losing mechanisms   988
Energy Gains and Losses   988
Thermoregulation   990

Chapter Review   995
Spotlight
Absorptive and Postabsorptive States   980
Clinical Case
The Miracle Supplement   962
Clinical Notes
Carbohydrate Loading   973

Dietary Fats and Cholesterol   977
Vitamins   986
Alcohol   988
Induced Hypothermia   991
Thermoregulatory Disorders   993

26 The Urinary System   998

❚ An Introduction to the Urinary System   999

26-1   ❚ The kidneys, ureters, urinary bladder, and urethra make up the urinary system, which has three primary functions   999

26-2   ❚ Kidneys are highly vascular organs containing functional units called nephrons, which filter, reabsorb, and secrete   1000
Sectional Anatomy of the Kidneys   1001
Blood Supply and Innervation of the Kidneys   1002
The Nephron   1004

26-3   ❚ Different segments of the nephron form urine by filtration, reabsorption, and secretion   1009
Basic Processes of Urine Formation   1009

26-4   ❚ Hydrostatic and colloid osmotic pressures influence glomerular filtration pressure, which in turn affects the glomerular filtration rate   1013
Filtration Pressures   1014
The Glomerular Filtration Rate   1015
Control of the GFR   1015

26-5   ❚ Countercurrent multiplication, antidiuretic hormone, and aldosterone affect reabsorption and secretion   1018
Reabsorption and Secretion at the PCT   1018
The Nephron Loop and Countercurrent Multiplication   1019
Reabsorption and Secretion at the DCT   1022
Reabsorption and Secretion along the Collecting System   1023
The Control of Urine Volume and Osmotic Concentration   1023
The Function of the Vasa Recta   1027
The Composition of Normal Urine   1027

26-6   ❚ Urine is transported by the ureters, stored in the bladder, and eliminated through the urethra, aided by the micturition reflex   1030
The Ureters   1031
The Urethra   1032
The Micturition Reflex and Urination   1033

26-7   ❚ Age-related changes affect kidney function and the micturition reflex   1035

26-8   ❚ The urinary system is one of several body systems involved in waste excretion   1037

Chapter Review   1037
Spotlight
  Summary of Renal Function   1028
Clinical Case
  A Case of "Hidden" Bleeding   999
Clinical Notes
  Analysis of Renal Blood Flow   1003
  Glomerulonephritis   1008
  Diuretics   1023
  Urinary Obstruction   1034
  Renal Failure and Kidney Transplant   1035

27 Fluid, Electrolyte, and Acid–Base Balance   1041

❚ **An Introduction to Fluid, Electrolyte, and Acid–Base Balance**   1042

27-1 ❚ **Fluid balance, electrolyte balance, and acid–base balance are interrelated and essential to homeostasis**   1042

27-2 ❚ **The ECF and ICF make up the fluid compartments, which also contain cations and anions**   1043
  The ECF and the ICF   1043
  Basic Concepts in the Regulation of Fluids and Electrolytes   1045
  An Overview of the Primary Regulatory Hormones   1046
  The Interplay between Fluid Balance and Electrolyte Balance   1047

27-3 ❚ **Hydrostatic and osmotic pressures regulate the movement of water and electrolytes to maintain fluid balance**   1047
  Fluid Movement within the ECF   1047
  Fluid Gains and Losses   1048
  Fluid Shifts   1048

27-4 ❚ **Sodium, potassium, calcium, magnesium, phosphate, and chloride balance is essential for maintaining homeostasis**   1050
  Sodium Balance   1051
  Potassium Balance   1053
  Balance of Other Electrolytes   1054

27-5 ❚ **In acid–base balance, regulation of hydrogen ions in body fluids involves buffer systems and respiratory and renal compensatory mechanisms**   1056
  The Importance of pH Control   1056
  Types of Acids in the Body   1056
  Mechanisms of pH Control   1058
  Maintenance of Acid–Base Balance   1061

27-6 ❚ **Respiratory acidosis/alkalosis and metabolic acidosis/alkalosis are classes of acid–base balance disturbances**   1064
  Respiratory Acidosis   1064
  Respiratory Alkalosis   1067

  Metabolic Acidosis   1067
  Metabolic Alkalosis   1069
  The Detection of Acidosis and Alkalosis   1069

27-7 ❚ **Aging affects several aspects of fluid, electrolyte, and acid–base balance**   1071

Chapter Review   1072
Spotlight
  The Diagnosis of Acid–Base Disorders   1070
Clinical Case
  When Treatment Makes You Worse   1042
Clinical Notes
  Water and Weight Loss   1049
  Athletes and Salt Loss   1053

UNIT 6   CONTINUITY OF LIFE

28 The Reproductive System   1076

❚ **An Introduction to the Reproductive System**   1077

28-1 ❚ **Basic reproductive system structures are gonads, ducts, accessory glands and organs, and external genitalia**   1077

28-2 ❚ **Spermatogenesis occurs in the testes, and hormones from the hypothalamus, anterior lobe of the pituitary gland, and testes control male reproductive functions**   1078
  The Testes   1078
  Spermatogenesis   1082
  Mitosis and Meiosis   1082
  The Anatomy of a Spermatozoon   1086
  The Male Reproductive Tract   1087
  The Accessory Glands   1088
  Semen   1090
  The External Genitalia   1090
  Hormones and Male Reproductive Function   1092

28-3 ❚ **Oogenesis occurs in the ovaries, and hormones from the pituitary gland and gonads control female reproductive functions**   1094
  The Ovaries   1095
  The Uterine Tubes   1100
  The Uterus   1101
  The Vagina   1105
  The External Genitalia   1106
  The Mammary Glands   1107
  Hormones and Female Reproductive Function   1107

28-4 ❚ **The autonomic nervous system influences male and female sexual function**   1113
  Male Sexual Function   1113
  Female Sexual Function   1114

28-5 ▐ **With age, decreasing levels of reproductive hormones cause functional changes** 1114

    Menopause 1114

    The Male Climacteric 1115

28-6 ▐ **The reproductive system secretes hormones affecting growth and metabolism of all body systems** 1115

Chapter Review 1118

Spotlights

    Regulation of Male Reproduction 1093

    Regulation of Female Reproduction 1110

Clinical Case

    A Post-Game Mystery 1077

Clinical Notes

    Dehydroepiandrosterone (DHEA) 1092

    Prostatic Hypertrophy and Prostate Cancer 1094

    Ovarian Cancer 1100

    Cervical Cancer 1101

    Breast Cancer 1109

29 Development and Inheritance 1121

▐ **An Introduction to Development and Inheritance** 1122

29-1 ▐ **Development, marked by various stages, is a continuous process that occurs from fertilization to maturity** 1122

29-2 ▐ **Fertilization—the fusion of a secondary oocyte and a spermatozoon—forms a zygote** 1123

    The Oocyte at Ovulation 1123

    Oocyte Activation 1125

29-3 ▐ **Gestation consists of three stages of prenatal development: the first, second, and third trimesters** 1125

29-4 ▐ **Cleavage, implantation, placentation, and embryogenesis are critical events of the first trimester** 1126

    Cleavage and Blastocyst Formation 1127

    Implantation 1127

    Placentation 1132

    Embryogenesis 1133

29-5 ▐ **During the second and third trimesters, maternal organ systems support the developing fetus, and the uterus undergoes structural and functional changes** 1138

    Pregnancy and Maternal Systems 1138

    Structural and Functional Changes in the Uterus 1140

29-6 ▐ **Labor consists of the dilation, expulsion, and placental stages** 1141

    Stages of Labor 1141

    Premature Labor 1143

    Difficult Deliveries 1143

    Multiple Births 1143

29-7 ▐ **Postnatal stages are the neonatal period, infancy, childhood, adolescence, and maturity, followed by senescence** 1144

    The Neonatal Period, Infancy, and Childhood 1144

    Adolescence and Maturity 1146

    Senescence 1148

29-8 ▐ **Genes and chromosomes determine patterns of inheritance** 1148

    Patterns of Inheritance 1149

    Sources of Individual Variation 1152

    Sex-Linked Inheritance 1154

    The Human Genome Project and Beyond 1155

Chapter Review 1157

Spotlight

    Extraembryonic Membranes and Placenta Formation 1130

Clinical Case

    The Twins That Looked Nothing Alike 1122

Clinical Notes

    Abortion 1140

    Chromosomal Abnormalities and Genetic Analysis 1156

Appendix

    The Composition of Minor Body Fluids 1161

    The Chemistry of Blood, Cerebrospinal Fluid, and Urine 1162

    The Periodic Table 1163

    Codon Chart 1164

Answers to Checkpoints, End-of-Chapter Questions, and Clinical Case Wrap-Ups 1165

Glossary 1205

Credits 1225

Index 1227

1

An Introduction to Anatomy and Physiology

Learning Outcomes

These Learning Outcomes correspond by number to this chapter's sections and indicate what you should be able to do after completing the chapter.

1-1 ▎ Explain the importance of studying anatomy and physiology.

1-2 ▎ Define anatomy and physiology, describe the origins of anatomical and physiological terms, and explain the significance of *Terminologia Anatomica (International Anatomical Terminology)*.

1-3 ▎ Explain the relationship between anatomy and physiology, and describe various specialties of each discipline.

1-4 ▎ Identify the major levels of organization in organisms, from the simplest to the most complex, and identify major components of each organ system.

1-5 ▎ Explain the concept of homeostasis.

1-6 ▎ Describe how negative feedback and positive feedback are involved in homeostatic regulation, and explain the significance of homeostasis.

1-7 ▎ Use anatomical terms to describe body regions, body sections, and relative positions.

1-8 ▎ Identify the major body cavities and their subdivisions, and describe the functions of each.

SPOTLIGHTS

- Levels of Organization pp. 34–35
- Diagnostic Imaging Techniques pp. 46–47

CLINICAL CASE

- Using A&P to Save a Life p. 28

Clinical Notes

- Autopsies and Cadaver Dissection p. 31
- Auscultation p. 40

Using A&P to Save a Life

The emergency medical technician (EMT) is on the way to the emergency room with a young victim of street violence. A knife with a 6-inch blade had been found next to the bleeding, unconscious man.

"We have a young male with multiple stab wounds. He has lost a lot of blood and we can barely get a blood pressure," the EMT radioes to the triage nurse in the emergency room as the ambulance squeals through traffic. "We started an IV and we are pouring in fluid as fast as we can."

"Where are the wounds?" asks the receiving nurse. "He has a deep wound in his right upper quadrant, just inferior to the diaphragm. I can see bruising from the hub of the knife around the wound, and there is another wound in his anterior right thigh. His pulse is 120 and thready [weak]. His blood pressure is 60 over 30."

"How long has he been down?" questions the nurse. "Less than a half hour. We intubated him [inserted a breathing tube] and started a large bore IV as soon as we got there. We are 10 minutes out now." "Keep the fluids going wide open, keep pressure on the thigh, and take him directly to Trauma Room 1," come the instructions. Meanwhile the nurse orders the trauma team to Trauma Room 1, orders x-ray to be on stand-by in the room, and requests 4 units of type O negative whole blood—the universal donor blood—from the blood bank. **Will the team be ready to save this young man? To find out, turn to the Clinical Case Wrap-Up on p. 51.**

An Introduction to Studying the Human Body

In this textbook we will introduce you to the inner workings of your body, giving information about both its structure (anatomy) and its function (physiology). Many students who use this book are preparing for jobs in health-related fields—but regardless of your career choice, you will find the information within these pages relevant to your future. You do, after all, live in a human body! As a human, you are most likely very curious, and few subjects arouse so much curiosity as our own bodies. The study of anatomy and physiology will provide answers to many questions about the way your body works in both health and disease.

We will focus on the human body, but the principles you will learn apply to other living things as well. Our world contains an enormous diversity of living organisms. They vary widely in appearance and lifestyle. One aim of biology—the science of life—is to discover the unity and the patterns that underlie this diversity, and in this way shed light on what we have in common with other living things.

We can classify animals according to the characteristics that they share. Birds, fish, and humans are placed in a group called the *vertebrates*, characterized by a segmented vertebral column. Their shared characteristics and organizational patterns are clues about how these animals have evolved over time. Many of the complex structures and functions of the human body that we discuss in this book have distant evolutionary origins. When we compare human beings with other creatures, we find two important principles:

- There are obvious structural and functional similarities among vertebrates.
- Form determines function.

In this chapter we explore the structural and functional characteristics of living things. We look at the levels of organization that anatomical structures and physiological processes display. We also introduce *homeostasis*, the state of balance within the body, which is the goal of physiological regulation and the key to survival in a changing environment.

1-1 ▌ Anatomy and physiology directly affect your life

Learning Outcome Explain the importance of studying anatomy and physiology.

Welcome to the field of anatomy and physiology! Anatomy is the study of body structures and physiology is the study of body functions. You will discover how your body works under normal and abnormal conditions. This knowledge will serve as the foundation for understanding all other basic life sciences, and for making common sense decisions about your own life. Basic knowledge of normal physiological function, for example, will prove useful whenever you or a friend or relative becomes ill. We will explain how the body responds to normal and abnormal conditions and maintains homeostasis. As we proceed, you will see how your body copes with injury, disease, or anything that threatens homeostasis.

Anatomy is the oldest medical science. Egyptian drawings from 1600 BCE illustrated basic knowledge of blood vessels.

These drawings show that people have always been fascinated with the human body. Since that time, imaging techniques for studying the human body have evolved, enabling us to describe the locations and functions of body parts. Over the last two decades, the most rapid progress has taken place in the field of molecular biology, which studies processes at the level of individual genes. It uses principles of biology, chemistry, genetics, and biochemistry. Molecular biology helps us understand how the body works at the most fundamental level and reveals the underlying basis for many disorders and diseases.

Medical science is always expanding and affects our everyday lives. We are flooded with health information from the popular press, news media, and advertisements. Medical terms have become part of our common language, and we owe it to ourselves to understand them. This course will significantly expand your vocabulary and enhance your understanding of the origins and meanings of many medical terms.

CHECKPOINT

1. Identify the oldest medical science.

2. Why is studying human anatomy and physiology important?

See the blue Answers tab at the back of the book.

1-2 ┃ Anatomy is structure, and physiology is function

Learning Outcome Define anatomy and physiology, describe the origins of anatomical and physiological terms, and explain the significance of *Terminologia Anatomica (International Anatomical Terminology).*

People have always been interested in the inner workings of the human body. Many medical terms have Greek roots, as do many other anatomical terms and phrases that originated more than 1500 years ago. For example, the term *anatomy* is derived from Greek roots that mean "a cutting open." **Anatomy** is the study of internal and external body structures and their physical relationships among other body parts. In contrast, **physiology**, another Greek term, is the study of how living organisms perform their functions. Someone studying anatomy might, for example, examine how a particular muscle attaches to the skeleton. Someone studying physiology might consider how a muscle contracts or what forces a contracting muscle exerts on the skeleton. You will be studying both anatomy and physiology in this book, so let's look at the relationships between these sciences.

Early anatomists faced serious problems in communication. Saying that a bump is "on the back," for example, does not give very precise information about its location. So anatomists created maps of the human body. Prominent anatomical structures serve as landmarks, distances are measured in

centimeters or inches, and specialized directional terms are used. In effect, anatomy uses a special language that you must learn almost at the start of your study.

That special language, called **medical terminology**, involves using word roots, prefixes, suffixes, and combining forms to build terms related to the body in health and disease. Many of the anatomical and physiological terms you will encounter in this textbook are derived from Greek or Latin. Learning the word parts used in medical terminology will greatly assist in your study of anatomy and physiology, and in your preparation for any health-related career.

There are four basic building blocks—or word parts—of medical terms. *Word roots* are the basic, meaningful parts of a term that cannot be broken down into another term with another definition. *Prefixes* are word elements that are attached to the beginning of words to modify their meaning but cannot stand alone. *Suffixes* are word elements or letters added to the end of a word or word part to form another term. *Combining forms* are independent words or word roots that are used in combination with words, prefixes, suffixes, or other combining forms to build a new term. The table inside the back cover of your textbook lists many commonly used word roots, prefixes, suffixes, and combining forms.

To illustrate the building of medical terms, consider the word *pathology* (puh-THOL-ō-jē). Breaking this word into its basic elements reveals its meaning. The prefix *path-* refers to disease (the Greek term for disease is *pathos*). The suffix *-ology* means "study of." So pathology is the study of disease.

A familiarity with Latin and Greek word roots and patterns makes anatomical terms easier to understand. As we introduce new terms, we will provide notes on pronunciation and relevant word parts.

Latin and Greek terms are not the only ones that have been imported into the anatomical vocabulary over the centuries, and this vocabulary continues to expand. Many anatomical structures and clinical conditions were first named after either the discoverer or, in the case of diseases, the most famous victim. Over the last 100 years, most of these commemorative names, or **eponyms**, have been replaced by more precise terms. Where appropriate, we will give both the eponym and the more precise term, because in clinical medicine, both terms may be used.

It is important for scientists throughout the world to use the same name for each body structure. In 1998, two scientific organizations—the Federative Committee on Anatomical Terminology and the 56 member associations of the International Associations of Anatomists—published *International Anatomical Terminology (Terminologia Anatomica, or TA). Terminologia Anatomica* serves as a worldwide official standard of anatomical vocabulary. Latin continues to be the language of anatomy, but this reference provides an English equivalent term for each anatomical structure. For example, the tendo calcaneus (Latin) is also called the calcaneal tendon (English). You may know the structure better by its eponym, the Achilles tendon. Eponyms are not found in *TA*. We have used *TA* as our standard in preparing this text.

CHECKPOINT

3. Define anatomy.

4. Define physiology.

5. Describe medical terminology.

6. Define eponym.

7. Name the book that serves as the international standard for anatomical vocabulary.

See the blue Answers tab at the back of the book.

1-3 ¦ Anatomy and physiology are closely integrated

Learning Outcome Explain the relationship between anatomy and physiology, and describe various specialties of each discipline.

Anatomy and physiology are closely integrated, both theoretically and practically. Anatomical information provides clues about functions, and physiological processes can be explained only in terms of the underlying anatomy. This is a very important concept in living systems:

- All specific functions are performed by specific structures, and the form of a structure relates to its function. This is known as the *principle of complementarity of structure and function*.

The link between structure and function is always present, but not always understood. For example, the anatomy of the heart was clearly described in the 15th century, but almost 200 years passed before the heart's pumping action was demonstrated.

Anatomists and physiologists approach the relationship between structure and function from different perspectives. To understand the difference, suppose you asked an anatomist and a physiologist to examine a pickup truck and report their findings. The anatomist might begin by measuring and photographing the various parts of the truck and, if possible, taking it apart and putting it back together. The anatomist could then explain its key *structural* relationships—for example, how the pistons are seated in the engine cylinders, how the crankshaft is connected to the pistons, how the transmission links the drive shaft to the axles, and thus to the wheels. The physiologist also would note the relationships among the truck's parts, but he or she would focus mainly on its *functional* characteristics, such as how the combustion of gasoline in the cylinders moves the pistons up and down and makes the drive shaft rotate, and

how the transmission conveys this motion to the axles and wheels so that the car moves. Additionally, he or she might also study the amount of power that the engine could generate, the amount of force transmitted to the wheels in different gears, and so on.

Our basic approach in this text will be to start with the descriptive anatomy of body structures (appearance, size, shape, location, weight, and color) before considering the related functions. Sometimes the group of organs that make up an organ system perform very diverse functions, and in those cases we will consider the functions of each individual organ separately. A good example is our discussion of the digestive system. You will learn about the functions of the salivary glands in one section, and the functions of the tongue in another. In other systems, the organs work together so extensively that we present an overall discussion of their physiology, after we describe the system's anatomy. The lymphatic system and the cardiovascular system are examples of this approach.

Anatomy

How you look at things often determines what you see. You get a very different view of your neighborhood from a satellite photo than from your front yard. Similarly, your method of observation has a dramatic effect on your understanding of the structure of the human body. Based on the degree of structural detail being considered, we divide anatomy into *gross (macroscopic) anatomy* and *microscopic anatomy*.

Gross Anatomy

Gross anatomy, or *macroscopic anatomy*, involves examining relatively large structures. Gross anatomy can be seen without using a microscope and can involve the study of anatomy by dissecting a cadaver. There are many different forms of gross anatomy:

- *Surface anatomy* is the study of general form and superficial (toward the body surface) markings.

- *Regional anatomy* focuses on the anatomical organization of specific areas of the body, such as the head, neck, or trunk. Many advanced courses in anatomy stress a regional approach, because it emphasizes the spatial relationships among structures already familiar to students.

- *Systemic anatomy* is the study of the structure of **organ systems**, which are groups of organs that function together in a coordinated manner. Examples include the *skeletal system*, composed primarily of bones; the *muscular system*, made up of skeletal muscles; and the *cardiovascular system*, consisting of the heart, blood, and vessels. We take a systemic anatomy approach in this book because this format will better clarify the functional relationships among the

Autopsies and Cadaver Dissection There is much to be learned from death. An *autopsy* is a thorough internal and external examination of the body after death to determine the cause of death. Autopsies and cadaver dissections reveal much about life. Historically, humans have been fascinated with the human body. Scientists, physicians, and artists shared this fascination. For example, as early as 304 BCE, Greek anatomist and royal physician Erasistratus and his physician colleague, Herophilus, dissected human cadavers to learn about internal anatomy. In the second century, physician Claudius Galen, often recognized as the "father of anatomy," was the first to show a connection between patient symptoms in life and the autopsy observations in death. In the late 1200s, the University of Bologna was the first institution to perform autopsies. In 1632, the Dutch painter, Rembrandt created *The Anatomy Lesson of Dr. Nicolaes Tulp*. An Italian anatomist, Giovanni Morgagni, advanced the field of anatomical pathology with his 1761 work, *On the Seats and Causes of Diseases*. In 1910, Dr. Richard Cabot, an American physician, studied autopsies and learned that doctors had misdiagnosed their patients 40 percent of the time. Today, using cadavers to learn anatomy is commonplace. Many students consider it to be a profound experience.

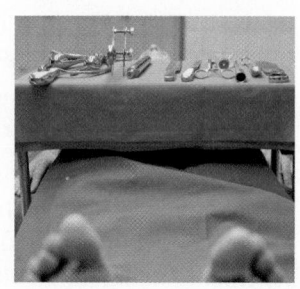

component organs. We will introduce the 11 organ systems in the human body later in the chapter.

- *Clinical anatomy* includes a number of subspecialties important in clinical practice. Examples include *pathological anatomy* (anatomical features that change during illness), *radiographic anatomy* (anatomical structures seen using specialized imaging techniques), and *surgical anatomy* (anatomical landmarks important in surgery).

- *Developmental anatomy* describes the changes in form that take place between conception and adulthood. The techniques of developmental anatomists are similar to those used in gross anatomy and in microscopic anatomy (discussed next) because developmental anatomy considers anatomical structures over a broad range of sizes—from a single cell to an adult human. The most extensive structural changes take place during the first two months of development. The study of these early developmental processes is called **embryology** (em-brē-OL-ō-jē).

Microscopic Anatomy

Microscopic anatomy deals with structures that we cannot see without magnification. The boundaries of microscopic anatomy are set by the limits of the equipment we use. With a dissecting microscope you can see tissue structure. With a light microscope, you can see basic details of cell structure. And with an electron microscope, you can see individual molecules that are only a few nanometers (billionths of a meter) across.

Microscopic anatomy includes two major subdivisions: cytology and histology. **Cytology** (sī-TOL-ō-jē) is the study of the internal structure of individual **cells**, the simplest units of life. Cells are made up of chemical substances in various combinations, and our lives depend on the chemical processes taking place in the trillions of cells in the body. For this reason, we consider basic chemistry (Chapter 2) before we examine cell structure (Chapter 3). **Histology** (his-TOL-ō-jē) is the examination of **tissues**—groups of specialized cells and cell products that work together to perform specific functions (Chapter 4). Tissues combine to form **organs**, such as the heart, kidney, liver, or brain, each with specific functions. Many organs are easy to examine without a microscope, so at the organ level we cross the boundary from microscopic anatomy to gross anatomy. As we proceed through the text, we will consider details at all levels, from microscopic to macroscopic (**Spotlight Figure 1-1**).

Physiology

As noted earlier, physiology is the study of the function of anatomical structures. **Human physiology** is the study of the functions, or workings, of the human body. These functions are complex processes and much more difficult to examine than most anatomical structures. As a result, there are even more specialties in physiology than in anatomy. Examples include:

- *Cell physiology*, the study of the functions of cells, is the cornerstone of human physiology. Cell physiology looks at events involving the atoms and molecules important to life. It includes both chemical processes within cells and chemical interactions among cells.

- *Organ physiology* is the study of the function of specific organs. An example is *cardiac physiology*, the study of heart function—how the heart works.

- *Systemic physiology* includes all aspects of the functioning of specific organ systems. Cardiovascular physiology, respiratory physiology, and reproductive physiology are examples.

- *Pathological physiology* is the study of the effects of diseases on organ functions or system functions. Modern medicine depends on an understanding of both normal physiology and pathological physiology.

Physicians normally use a combination of anatomical, physiological, chemical, and psychological information when they evaluate patients. When a patient presents **signs** (an objective disease indication like a fever) and **symptoms** (a subjective disease indication, such as tiredness) to a physician, the physician will look at the structures affected (gross anatomy), perhaps collect a fluid or tissue sample (microscopic anatomy) for analysis, and ask questions to find out what changes from normal functioning the patient is experiencing. Think back to your last trip to a doctor's office. Not only did the attending physician examine your body, noting any anatomical abnormalities, but he or she also evaluated your physiological processes by asking questions, observing your movements, listening to your body sounds, taking your temperature, and perhaps requesting chemical analyses of fluids such as blood or urine.

In evaluating all these observations to reach a diagnosis, physicians rely on a logical framework based on the scientific method. The **scientific method** is a system of advancing knowledge that begins by proposing a hypothesis to answer a question, and then testing that hypothesis with data collected through observation and experimentation. This method is at the core of all scientific thought, including medical diagnosis.

CHECKPOINT

8. Describe how anatomy and physiology are closely related.

9. What is the difference between gross anatomy and microscopic anatomy?

10. Identify several specialties of physiology.

11. Why is it difficult to separate anatomy from physiology?

See the blue Answers tab at the back of the book.

1-4 ┃ Levels of organization progress from molecules to a complete organism

Learning Outcome Identify the major levels of organization in organisms, from the simplest to the most complex, and identify major components of each organ system.

Over the next three chapters, we will consider the three most basic (chemical, cellular, and tissue) levels of organization of the human body. They interact with more complex structures and vital processes, as we describe in Chapters 5–29. The levels of organization of the human body include:

- *The Chemical Level.* **Atoms** are the smallest stable units of matter. They can combine to form molecules with complex shapes. The atomic components and unique three-dimensional shape of a particular molecule determine its function. For example, complex protein molecules form filaments that produce the contractions of muscle cells in the heart. We explore this level of organization in Chapter 2.

- *The Cellular Level.* **Cells** are the smallest living units in the body. Complex molecules can form various types of larger structures called organelles. Each organelle has a specific function in a cell. Energy-producing organelles provide the energy needed for heart muscle cell contractions. We examine the cellular level of organization in Chapter 3.

- *The Tissue Level.* A **tissue** is a group of cells working together to perform one or more specific functions. Heart muscle cells, also called cardiac muscle cells (*cardium,* heart), interact with other types of cells and with materials outside the cell to form cardiac muscle tissue. We consider the tissue level of organization in Chapter 4.

- *The Organ Level.* **Organs** are made of two or more tissues working together to perform specific functions. Layers of cardiac muscle tissue, in combination with another type of tissue called connective tissue, form the bulk of the wall of the heart, which is a hollow, three-dimensional organ.

- *The Organ System Level.* A group of organs interacting to perform a particular function forms an **organ system.** Each time it contracts, the heart pushes blood into a network of blood vessels. Together, the heart, blood, and blood vessels make up the cardiovascular system, one of 11 organ systems in the body. This system functions to distribute oxygen and nutrients throughout the body.

- *The Organism Level.* An **organism**—in this case, an individual human—is the highest level of organization that we consider. All of the body's organ systems must work together to maintain the life and health of the organism.

The organization at each level determines not only the structural characteristics, but also the functions, of higher levels. For example, the arrangement of atoms and molecules at the chemical level creates the protein filaments and organelles at the cellular level that give individual cardiac muscle cells the ability to contract. At the tissue level, these cells are linked, forming cardiac muscle tissue. The structure of the tissue ensures that the contractions are coordinated, producing a powerful heartbeat. When that beat occurs, the internal anatomy of the heart, an organ, enables it to function as a pump. The heart is filled with blood and connected to the blood vessels, and its pumping action circulates blood through the vessels of the cardiovascular system. Through interactions with the respiratory, digestive, urinary, and other systems, the cardiovascular system performs a variety of functions essential to the survival of the organism.

Something that affects a system will ultimately affect each of the system's parts. For example, after massive blood loss, the heart cannot pump blood effectively. When the heart cannot pump and blood cannot flow, oxygen and nutrients cannot be distributed to the heart or around the body. Very soon, the cardiac muscle tissue begins to break down as individual muscle cells die from oxygen and nutrient starvation. These changes will not be restricted to the cardiovascular system. All cells, tissues, and organs in the body will be damaged. Spotlight Figure 1–1 illustrates the levels of organization and introduces the 11 interdependent, interconnected organ systems in the human body.

The cells, tissues, organs, and organ systems of the body coexist in a relatively small, shared environment, much like the residents of a large city. Just as city dwellers breathe the same air and drink the water supplied by the local water company, cells in the human body absorb oxygen and nutrients from the fluids that surround them. If a city is blanketed in smog or its water supply is contaminated, the people will become ill. Similarly, if the body fluid composition becomes abnormal, cells will be injured or destroyed. For example, suppose the temperature or salt content of the blood changes. The effect on the heart could range from the need for a minor adjustment (heart muscle tissue contracts more often, raising the heart rate) to a total disaster (the heart stops beating, so the individual dies).

CHECKPOINT

12. Identify the major levels of organization of the human body from the simplest to the most complex.

13. Identify the organ systems of the body and cite some major structures of each.

14. At which level of biological organization does a histologist investigate structures?

See the blue Answers tab at the back of the book.

1-5 ▍ Homeostasis is the state of internal balance

Learning Outcome Explain the concept of homeostasis.

Various physiological processes act to prevent harmful changes in the composition of body fluids and the environment inside our cells. **Homeostasis** (hō-mē-o-STĀ-sis; *homeo*, unchanging + *stasis*, standing) refers to the existence of a stable internal environment. Maintaining homeostasis is absolutely vital to an organism's survival. Failure to maintain homeostasis soon leads to illness or even death.

The principle of homeostasis is the central theme of this text and the foundation of all modern physiology. **Homeostatic regulation** is the adjustment of physiological systems to preserve homeostasis. Physiological systems have evolved to maintain homeostasis in an environment that is often inconsistent, unpredictable, and potentially dangerous. An understanding of homeostatic regulation is crucial to making accurate predictions about the body's responses to both normal and abnormal conditions.

Homeostatic regulation involves two general mechanisms: autoregulation and extrinsic regulation.

1. **Autoregulation** is a process that occurs when a cell, a tissue, an organ, or an organ system adjusts in response to some environmental change. For example, when oxygen levels decline in a tissue, the cells release chemicals that widen, or dilate, blood vessels. This dilation increases the rate of blood flow and provides more oxygen to the region.

2. **Extrinsic regulation** is a process that results from the activities of the nervous system or endocrine system. These organ systems detect an environmental change and send an electrical signal (nervous system) or chemical messenger (endocrine system) to control or adjust the activities of another or many other systems simultaneously. For example, when you exercise, your nervous system issues commands that increase your heart rate so that blood will circulate faster. Your nervous system also reduces blood flow to less active organs, such as the digestive tract. The oxygen in circulating blood is then available to the active muscles, which need it most.

In general, the nervous system directs rapid, short-term, and very specific responses. For example, when you accidentally set your hand on a hot stove, the heat produces a painful, localized disturbance of homeostasis. Your nervous system responds by ordering specific muscles to contract and pull your hand away from the stove. These contractions last only as long as the neural activity continues, usually a matter of seconds.

In contrast, the endocrine system releases chemical messengers called *hormones* into the bloodstream. These molecular messengers can affect tissues and organs throughout the body. The responses may not be immediately apparent, but they may persist for days or weeks. Examples of homeostatic regulation dependent on endocrine function include the long-term regulation of blood volume and composition, and the adjustment of organ system function during starvation. Regardless of the system involved, homeostatic regulation always works to keep the internal environment within certain limits. A homeostatic regulatory mechanism consists of three parts: (1) a **receptor**, a sensor that is sensitive to a particular stimulus or environmental change; (2) a **control center**, which receives and processes the information supplied by the receptor and sends out commands; and (3) an **effector**, a cell or organ that responds to the commands of the control center and whose activity either opposes or enhances the stimulus. You are probably already familiar with similar regulatory mechanisms, such as the one involving the thermostat in your house or apartment (Figure 1–2a).

The thermostat is the control center. It receives information about room temperature from an internal or remote thermometer (a receptor). The setting on the thermostat establishes the **set point**, or desired value, which in this case is the temperature you select. (In our example, the set point is 22°C, or about 72°F.) The function of the thermostat is to keep room temperature within acceptable limits, usually within a degree or so of the set point. In summer, the thermostat performs this function by controlling an air conditioner (an effector). When

Interacting atoms form molecules that combine to form the protein filaments of a heart muscle cell. Such cells interlock, creating heart muscle tissue, which makes up most of the walls of the heart, a three-dimensional organ. The heart is only one component of the cardiovascular system, which also includes the blood and blood vessels. The various organ systems must work together to maintain life at the organism level.

Cellular Level

Chemical Level

Atoms in combination

Complex protein molecule

Protein filaments

Heart muscle cell

THE ORGAN SYSTEMS

Integumentary	Skeletal	Muscular	Nervous	Endocrine	Cardiovascular

Major Organs
- Skin
- Hair
- Sweat glands
- Nails

Functions
- Protects against environmental hazards
- Helps regulate body temperature
- Provides sensory information

Major Organs
- Bones
- Cartilages
- Associated ligaments
- Bone marrow

Functions
- Provides support and protection for other tissues
- Stores calcium and other minerals
- Forms blood cells

Major Organs
- Skeletal muscles and associated tendons

Functions
- Provides movement
- Provides protection and support for other tissues
- Generates heat that maintains body temperature

Major Organs
- Brain
- Spinal cord
- Peripheral nerves
- Sense organs

Functions
- Directs immediate responses to stimuli
- Coordinates or moderates activities of other organ systems
- Provides and interprets sensory information about external conditions

Major Organs
- Pituitary gland
- Thyroid gland
- Pancreas
- Adrenal glands
- Gonads
- Endocrine tissues in other systems

Functions
- Directs long-term changes in the activities of other organ systems
- Adjusts metabolic activity and energy use by the body
- Controls many structural and functional changes during development

Major Organs
- Heart
- Blood
- Blood vessels

Functions
- Distributes blood cells, water and dissolved materials including nutrients, waste products, oxygen, and carbon dioxide
- Distributes heat and assists in control of body temperature

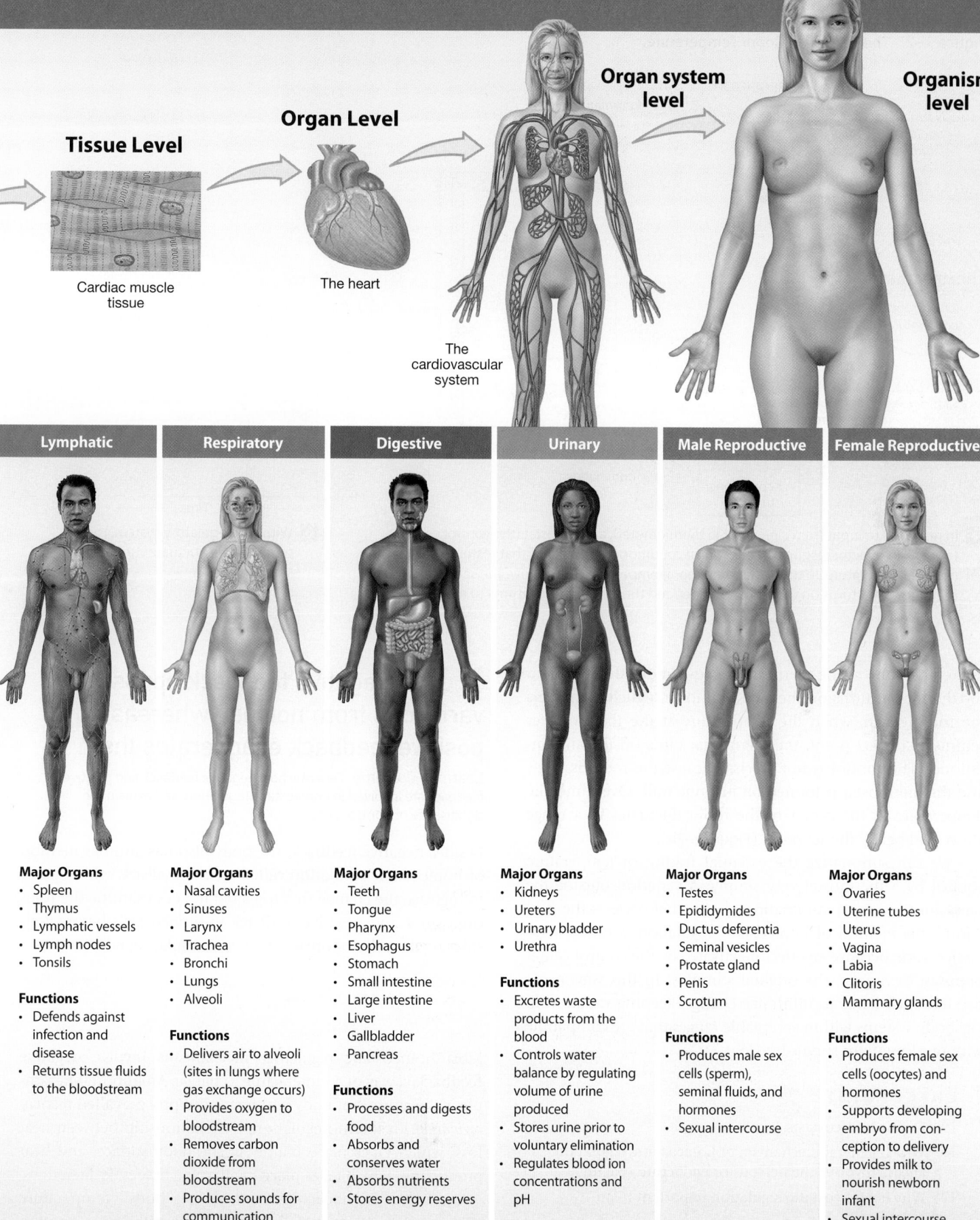

Tissue Level

Cardiac muscle tissue

Organ Level

The heart

Organ system level

The cardiovascular system

Organism level

Lymphatic

Major Organs
- Spleen
- Thymus
- Lymphatic vessels
- Lymph nodes
- Tonsils

Functions
- Defends against infection and disease
- Returns tissue fluids to the bloodstream

Respiratory

Major Organs
- Nasal cavities
- Sinuses
- Larynx
- Trachea
- Bronchi
- Lungs
- Alveoli

Functions
- Delivers air to alveoli (sites in lungs where gas exchange occurs)
- Provides oxygen to bloodstream
- Removes carbon dioxide from bloodstream
- Produces sounds for communication

Digestive

Major Organs
- Teeth
- Tongue
- Pharynx
- Esophagus
- Stomach
- Small intestine
- Large intestine
- Liver
- Gallbladder
- Pancreas

Functions
- Processes and digests food
- Absorbs and conserves water
- Absorbs nutrients
- Stores energy reserves

Urinary

Major Organs
- Kidneys
- Ureters
- Urinary bladder
- Urethra

Functions
- Excretes waste products from the blood
- Controls water balance by regulating volume of urine produced
- Stores urine prior to voluntary elimination
- Regulates blood ion concentrations and pH

Male Reproductive

Major Organs
- Testes
- Epididymides
- Ductus deferentia
- Seminal vesicles
- Prostate gland
- Penis
- Scrotum

Functions
- Produces male sex cells (sperm), seminal fluids, and hormones
- Sexual intercourse

Female Reproductive

Major Organs
- Ovaries
- Uterine tubes
- Uterus
- Vagina
- Labia
- Clitoris
- Mammary glands

Functions
- Produces female sex cells (oocytes) and hormones
- Supports developing embryo from conception to delivery
- Provides milk to nourish newborn infant
- Sexual intercourse

Figure 1–2 **The Control of Room Temperature.**

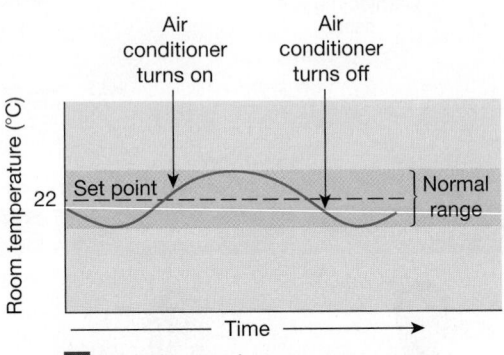

a In response to input from a receptor (a thermometer), a thermostat (the control center) triggers an effector response (either an air conditioner or a heater) that restores normal temperature. In this case, when room temperature rises above the set point, the thermostat turns on the air conditioner, and the temperature returns to normal.

b With this regulatory system, room temperature fluctuates around the set point, 22°C.

the temperature at the thermometer rises above the set point, the thermostat turns on the air conditioner, which then cools the room. Then, when the temperature at the thermometer returns to the set point, the thermostat turns off the air conditioner. The control is not precise because the room is large, and the thermostat is located on just one wall. Over time, the temperature in the center of the room fluctuates in a range above and below the set point (**Figure 1–2b**).

We can summarize the essential feature of temperature control by a thermostat very simply: A variation outside the set point triggers an automatic response that corrects the situation. This method of homeostatic regulation is called *negative feedback*, because an effector activated by the control center opposes, or *negates*, the original stimulus. In this way, negative feedback tends to minimize change, keeping variation in key body systems within acceptable ranges that are compatible with our long-term survival.

CHECKPOINT

15. **Define homeostasis.**

16. **Which general mechanism of homeostatic regulation always involves the nervous or endocrine system?**

17. **Why is homeostatic regulation important to an organism?**

See the blue Answers tab at the back of the book.

1-6 | Negative feedback opposes variations from normal, whereas positive feedback exaggerates them

Learning Outcome Describe how negative feedback and positive feedback are involved in homeostatic regulation, and explain the significance of homeostasis.

Besides negative feedback, the body also has another method of homeostatic regulation called *positive feedback*, which tends to increase the change that triggered it. Let's examine the roles of positive and negative feedback in homeostasis before considering the roles of organ systems in regulating homeostasis.

The Role of Negative Feedback in Homeostasis

Most homeostatic regulatory mechanisms involve **negative feedback**, a way of counteracting a change. An important example is the control of body temperature, a process called *thermoregulation*. In thermoregulation, the relationship between heat loss, which takes place mainly at the body surface, and heat production, which takes place in all active tissues, is altered.

In the homeostatic control of body temperature (**Figure 1–3a**), the control center is in the *hypothalamus*, a region of the brain. This control center receives information from

Figure 1–3 **Negative Feedback: Control of Body Temperature.** In negative feedback, a stimulus produces a response that opposes or negates the original stimulus.

a Events in the regulation of body temperature, which are comparable to those shown in *Figure 1–2*. A control center in the brain (the hypothalamus) functions as a thermostat with a set point of 37°C. If body temperature exceeds 37.2°C, heat loss is increased through enhanced blood flow to the skin and increased sweating.

b The thermoregulatory center keeps body temperature fluctuating within an acceptable range, usually between 36.7°C and 37.2°C.

two sets of temperature receptors, one in the skin and the other within the hypothalamus. At the normal set point, body temperature (as measured with an oral thermometer) is approximately 37°C (98.6°F).

If body temperature rises above 37.2°C, activity in the control center targets two effectors: (1) muscle tissue lining the walls of blood vessels supplying blood to the skin and (2) sweat glands. The muscle tissue relaxes so the blood vessels dilate (widen), increasing blood flow through vessels near the body surface, and the sweat glands speed up their secretion of sweat. The skin then acts like a radiator by losing heat to the environment, and the evaporation of sweat speeds the process.

As body temperature returns to normal, temperature at the hypothalamus declines, and the thermoregulatory control center becomes less active. Blood flow to the skin and sweat gland activity then decrease to previous levels. Body temperature drops below the set point as the secreted sweat evaporates.

Negative feedback is the primary mechanism of homeostatic regulation, and it provides long-term control over the body's internal conditions and systems. Homeostatic mechanisms using negative feedback normally ignore minor variations. They maintain a normal *range* rather than a fixed value.

In our example, body temperature fluctuated around the set-point temperature (**Figure 1–3b**). The regulatory process itself is dynamic. That is, it is constantly changing because the set point may vary with changing environments or differing activity levels. For example, when you are asleep, your thermoregulatory set point is lower. When you work outside on a hot day (or when you have a fever), it is set higher. Body temperature can vary from moment to moment or from day to day for any individual, due to either (1) small fluctuations around the set point or (2) changes in the set point. Comparable variations take place in all other aspects of physiology.

The variability among individuals is even greater than that within an individual. Each of us has homeostatic set points determined by genetic factors, age, gender, general health, and environmental conditions. For this reason, it is impractical to define "normal" homeostatic conditions very precisely. By convention, physiological values are reported either as average values obtained by sampling a large number of individuals, or as a range that includes 95 percent or more of the sample population.

For example, for 95 percent of healthy adults, body temperature ranges between 36.7°C and 37.2°C (98.1°F and 98.9°F).

The other 5 percent of healthy adults have resting body temperatures that are below 36.7°C or above 37.2°C. These temperatures are perfectly normal for them, and the variations have no clinical significance. Physicians must keep this variability in mind when they review lab reports, because unusual values—even those outside the "normal" range—may represent individual variation rather than disease.

The Role of Positive Feedback in Homeostasis

In **positive feedback**, an initial stimulus produces a response that exaggerates or enhances the original change in conditions, rather than opposing it. You seldom encounter positive feedback in your daily life, simply because it tends to produce extreme responses. For example, suppose that the thermostat in **Figure 1–2a** was accidentally connected to a heater rather than to an air conditioner. Now, when room temperature rises above the set point, the thermostat turns on

the heater, causing a further rise in room temperature. Room temperature will continue to increase until someone switches off the thermostat, turns off the heater, or intervenes in some other way. This kind of escalating cycle is often called a **positive feedback loop**.

In the body, positive feedback loops are typically found when a potentially dangerous or stressful process must be completed quickly to restore homeostasis. For example, the immediate danger from a severe cut is the loss of blood, which can lower blood pressure and reduce the efficiency of the heart. The body's response to this blood loss is blood clotting, diagrammed in **Figure 1–4**. We will examine blood clotting more closely in Chapter 19. Labor and delivery are another example of positive feedback in action, as we will discuss in Chapter 29.

The human body is amazingly effective in maintaining homeostasis. Nevertheless, an infection, an injury, or a genetic abnormality can sometimes have effects so severe that homeostatic mechanisms cannot fully compensate for them. One or more variables within the internal environment may then be

Figure 1–4   Positive Feedback: Blood Clotting. In positive feedback, a stimulus produces a response that exaggerates or enhances the original change in conditions, rather than opposing it.

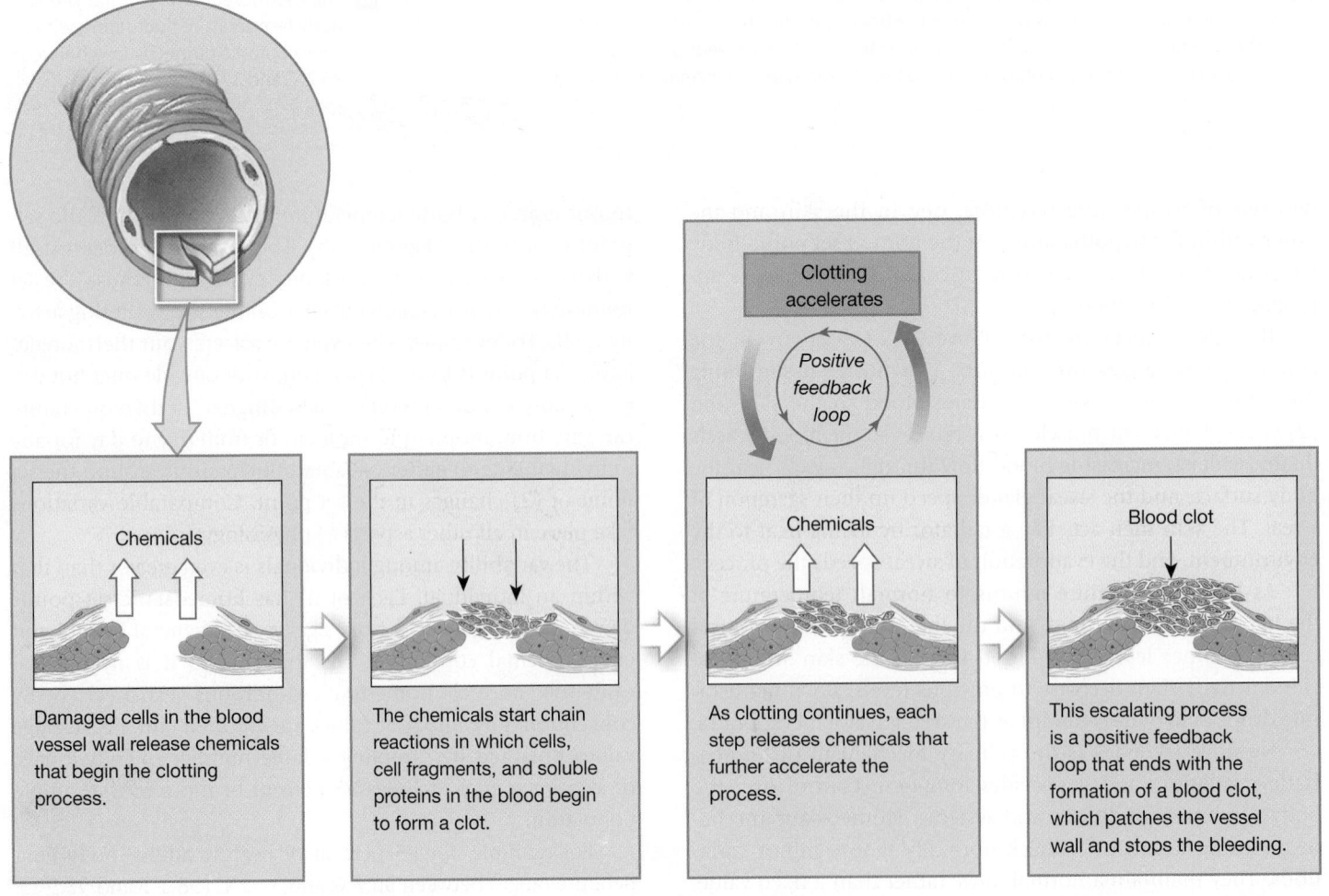

Damaged cells in the blood vessel wall release chemicals that begin the clotting process.

The chemicals start chain reactions in which cells, cell fragments, and soluble proteins in the blood begin to form a clot.

As clotting continues, each step releases chemicals that further accelerate the process.

This escalating process is a positive feedback loop that ends with the formation of a blood clot, which patches the vessel wall and stops the bleeding.

Clotting accelerates

Positive feedback loop

Chemicals

Blood clot

pushed outside their normal range of values. When this happens, organ systems begin to malfunction, producing a state known as illness, or **disease**. In Chapters 5–29, we devote considerable attention to the mechanisms that bring about a variety of human diseases.

Systems Integration, Equilibrium, and Homeostasis

Homeostatic regulation controls aspects of the internal environment that affect every cell in the body. No single organ system has total control over any of these aspects. Instead, such control requires the coordinated efforts of multiple organ systems. In later chapters we will explore the functions of each organ system and see how the systems interact to preserve homeostasis. Table 1–1 lists the roles of various organ systems in regulating several important functions that are subject to homeostatic control. Note that in each case such regulation involves several organ systems.

A **state of equilibrium** exists when opposing processes or forces are in balance. In the case of body temperature, a state of equilibrium exists when the rate of heat loss equals the rate of heat production. Each physiological system functions

to maintain a state of equilibrium that keeps vital conditions within a normal range of values. This is often called a state of **dynamic equilibrium** because physiological systems are continually adapting and adjusting to changing conditions. For example, when muscles become more active, more heat is produced. More heat must then be lost at the skin surface to reestablish a state of equilibrium before body temperature rises outside normal ranges. Yet the adjustments made to control body temperature have other consequences. The sweating that increases heat loss at the skin surface increases losses of both water and salts. Other systems must then compensate for these losses and reestablish an equilibrium state for water and salts.

Note this general pattern: Any adjustments made by one physiological system have direct and indirect effects on a variety of other systems. Maintaining homeostasis is like a juggling act that keeps lots of different objects in the air.

Each organ system interacts with and depends on other organ systems, but introductory students find it easier to learn the basics of anatomy and physiology one system at a time. Chapters 5–29 are organized around individual systems, but remember that these systems all work together. The 11 *System Integrators* in later chapters will help reinforce this message. Each provides an overview of one system's functions

Table 1–1	The Roles of Organ Systems in Homeostatic Regulation	
Internal Stimulus	**Primary Organ Systems Involved**	**Functions of the Organ Systems**
Body temperature	Integumentary system Muscular system Cardiovascular system Nervous system	Heat loss Heat production Heat distribution Coordination of blood flow, heat production, and heat loss
Body fluid composition		
Nutrient concentration	Digestive system Cardiovascular system Urinary system Skeletal system	Nutrient absorption, storage, and release Nutrient distribution Control of nutrient loss in the urine Mineral storage and release
Oxygen, carbon dioxide levels	Respiratory system Cardiovascular system	Absorption of oxygen, elimination of carbon dioxide Internal transport of oxygen and carbon dioxide
Levels of toxins and pathogens	Lymphatic system	Removal, destruction, or inactivation of toxins and pathogens
Body fluid volume	Urinary system Digestive system Integumentary system Cardiovascular system and lymphatic system	Elimination or conservation of water from the blood Absorption of water; loss of water in feces Loss of water through perspiration Distribution of water throughout body tissues
Waste product concentration	Urinary system Digestive system Cardiovascular system	Elimination of waste products from the blood Elimination of waste products from the liver in feces Transport of waste products to sites of excretion
Blood pressure	Cardiovascular system Nervous system and endocrine system	Pressure generated by the heart moves blood through blood vessels Adjustments in heart rate and blood vessel diameter can raise or lower blood pressure

and summarizes its functional relationships with systems covered in previous chapters.

CHECKPOINT

18. Explain the function of negative feedback systems.

19. What happens to the body when homeostasis breaks down?

20. Explain how a positive feedback system works.

21. Why is positive feedback helpful in blood clotting but unsuitable for the regulation of body temperature?

22. Define equilibrium.

23. When the body continuously adapts by utilizing homeostatic systems, it is said to be in a state of _____ equilibrium.

See the blue Answers tab at the back of the book.

1-7 ┃ Anatomical terms describe body regions, anatomical positions and directions, and body sections

Learning Outcome Use anatomical terms to describe body regions, body sections, and relative positions.

Anatomists use anatomical terms to describe body regions, relative positions and directions, and body sections, as well as major body cavities and their subdivisions. In the following sections we will introduce the terms used in superficial anatomy and sectional anatomy.

Superficial Anatomy

Superficial anatomy involves locating structures on or near the body surface. A familiarity with anatomical landmarks (structures that can be felt or palpated), anatomical regions (specific areas used for reference purposes), and terms for anatomical directions will make the material in subsequent chapters easier to understand. As you encounter new terms, create your own mental maps from the information provided in the accompanying anatomical illustrations.

Anatomical Landmarks

Figure 1–5 presents important anatomical landmarks. Understanding the terms and their origins will help you remember both the location of a particular structure and its name. For example, *brachium* refers to the arm, and later we will consider the *brachialis muscle* and the *brachial artery*, which are in the arm, as their names suggest.

The standard anatomical reference for the human form is the **anatomical position.** When the body is in this position, the hands are at the sides with the palms facing forward, and

Clinical Note

Auscultation *Auscultation* (aws-kul-tā-shun) is the action of listening to sounds from the heart, lungs, or other organs with a stethoscope. It is a valuable tool and serves as an important part of the medical diagnosis. These images show the general locations for listening to heart and lung sounds on both the anterior and posterior aspects of the body. *Aspect* refers to the side of the body that faces a particular direction.

the feet are together. **Figure 1–5a** shows an individual in the anatomical position as seen from the front, called an *anterior view.* **Figure 1–5b** shows the body as seen from the back, called a *posterior view.* Unless otherwise noted, all descriptions in this text refer to the body in the anatomical position. A person

Figure 1–5 **Anatomical Landmarks.** Anatomical terms are shown in boldface type and common names are in plain type.

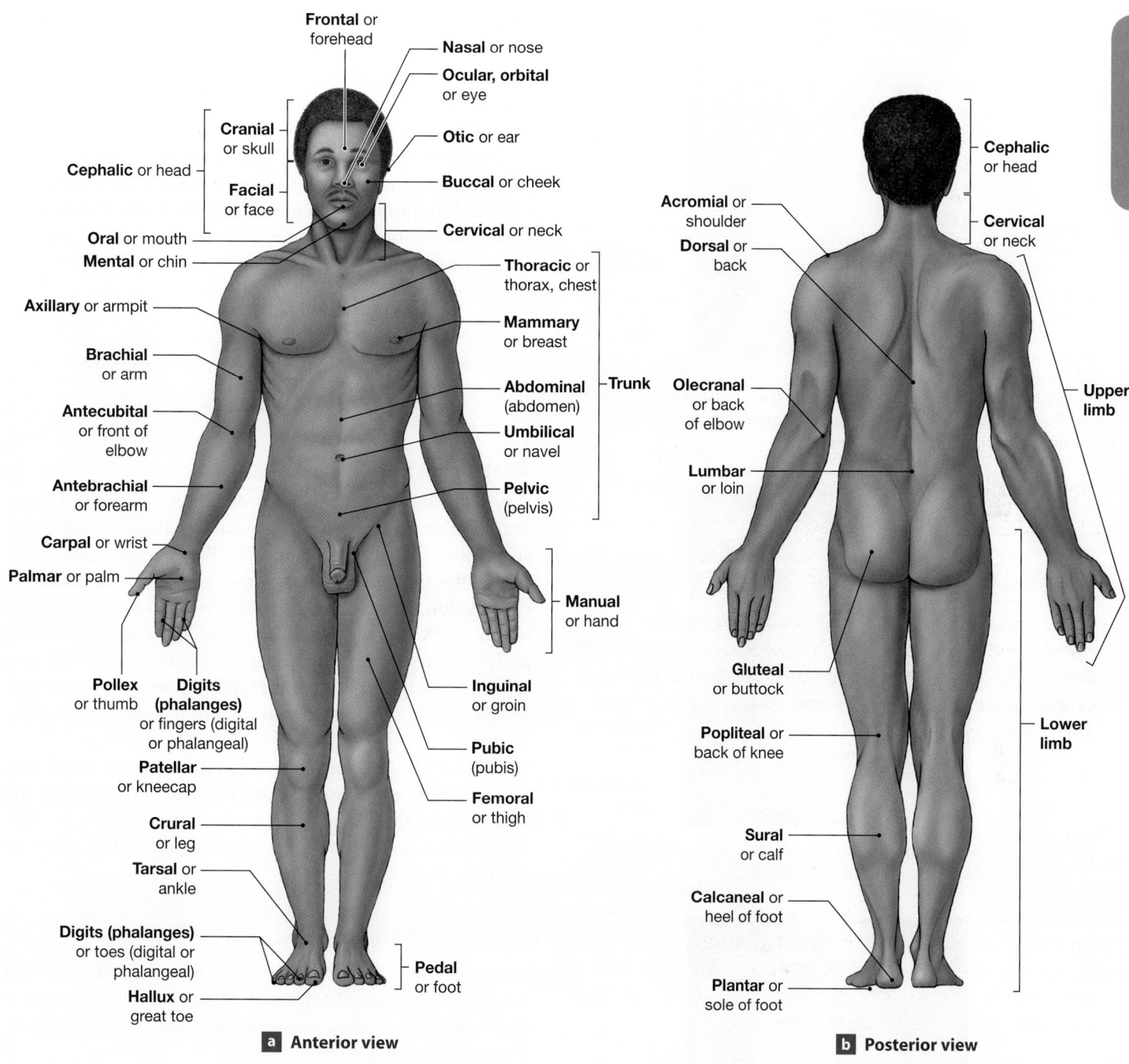

Frontal or forehead

Nasal or nose

Ocular, orbital or eye

Cranial or skull

Otic or ear

Cephalic or head

Buccal or cheek

Facial or face

Cervical or neck

Oral or mouth

Mental or chin

Thoracic or thorax, chest

Axillary or armpit

Mammary or breast

Brachial or arm

Abdominal (abdomen)

Trunk

Antecubital or front of elbow

Umbilical or navel

Antebrachial or forearm

Pelvic (pelvis)

Carpal or wrist

Palmar or palm

Manual or hand

Pollex or thumb **Digits (phalanges)** or fingers (digital or phalangeal)

Inguinal or groin

Patellar or kneecap

Pubic (pubis)

Crural or leg

Femoral or thigh

Tarsal or ankle

Digits (phalanges) or toes (digital or phalangeal)

Pedal or foot

Hallux or great toe

a **Anterior view**

Acromial or shoulder

Cephalic or head

Dorsal or back

Cervical or neck

Olecranal or back of elbow

Upper limb

Lumbar or loin

Gluteal or buttock

Popliteal or back of knee

Lower limb

Sural or calf

Calcaneal or heel of foot

Plantar or sole of foot

b **Posterior view**

lying down in the anatomical position is said to be **supine** (soo-PĪN) when face up, and **prone** when face down.

TIPS & **TRICKS**

Supine means up. In order to carry a bowl of *soup*, your hand must be in the *supine* position.

Anatomical Regions

To describe a general area of interest or injury, clinicians and anatomists often need broader terms in addition to specific landmarks. They use two methods to map the surface of the abdomen and pelvis.

Clinicians refer to four **abdominopelvic quadrants** (Figure 1–6a) formed by a pair of imaginary perpendicular

1

Figure 1–6 **Abdominopelvic Quadrants and Regions.**

Right Upper Quadrant (RUQ)
Left Upper Quadrant (LUQ)
Right Lower Quadrant (RLQ)
Left Lower Quadrant (LLQ)

a **Abdominopelvic quadrants.** The four abdominopelvic quadrants are formed by two perpendicular lines that intersect at the navel. The terms for these quadrants, or their abbreviations, are most often used in clinical discussions.

Right hypochondriac region
Epigastric region
Left hypochondriac region
Right lumbar region
Umbilical region
Left lumbar region
Right inguinal region
Hypogastric (pubic) region
Left inguinal region

b **Abdominopelvic regions.** The nine abdominopelvic regions provide more precise regional descriptions.

Liver
Stomach
Gallbladder
Spleen
Large intestine
Small intestine
Appendix
Urinary bladder

c **Anatomical relationships.** The relationship between the abdominopelvic quadrants and regions and the locations of the internal organs are shown here.

lines that intersect at the umbilicus (navel). This simple method provides useful references for the description of aches, pains, and injuries. The location can help the clinician determine the possible cause. For example, tenderness in the right lower quadrant (RLQ) is a symptom of appendicitis. Tenderness in the right upper quadrant (RUQ), however, may indicate gallbladder or liver problems.

Anatomists prefer more precise terms to describe the location and orientation of internal organs. They recognize nine **abdominopelvic regions** (Figure 1–6b). Figure 1–6c shows the relationships among quadrants, regions, and internal organs.

TIPS&TRICKS

The imaginary lines dividing the abdominopelvic regions resemble a tic-tac-toe game.

Anatomical Directions

Figure 1–7 introduces the main directional terms and some examples of their use. There are many different terms, and some can be used interchangeably. For example, *anterior* refers to the front of the body when viewed in the anatomical position. In humans, this term is equivalent to *ventral*, which refers to the belly. *Posterior* refers to the back of the body; this term is equivalent to *dorsal*. When reading anatomical descriptions, remember that the terms *left* and *right* always refer to the *left* and *right* sides of the *subject*, not of the observer.

Before you read further, analyze the image in detail, and practice using the terms. We start using these terms in the rest of this chapter. If you are familiar with the basic vocabulary, the descriptions throughout this book will be easier to follow.

Sectional Anatomy

Sometimes the only way to understand the relationships among the parts of a three-dimensional object is to slice through it and look at the internal organization. A slice through a three-dimensional object is called a *section*.

An understanding of sectional views is particularly important now that imaging techniques enable us to see inside the living body. These views are sometimes difficult to interpret, but it is worth spending the time required to understand what they show. Once you are able to interpret sectional views, you will have a good mental model for studying the anatomy and physiology of a particular region or system. Radiologists and other medical professionals responsible for interpreting medical scans spend much of their time analyzing sectional views of the body.

Figure 1–7   **Directional References.**

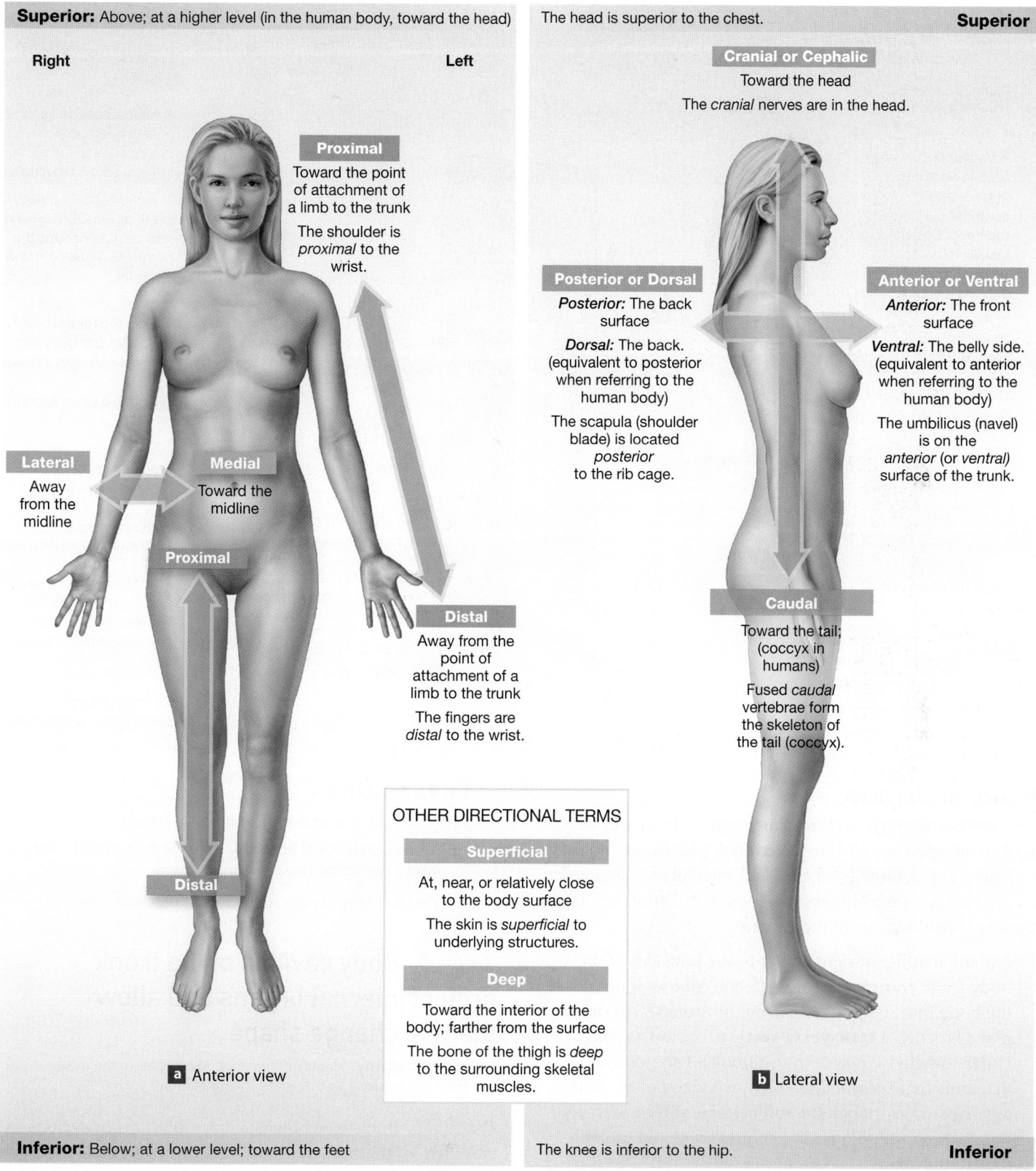

Superior: Above; at a higher level (in the human body, toward the head)

Right

Left

Proximal
Toward the point of attachment of a limb to the trunk

The shoulder is *proximal* to the wrist.

Lateral
Away from the midline

Medial
Toward the midline

Proximal

Distal
Away from the point of attachment of a limb to the trunk

The fingers are *distal* to the wrist.

Distal

OTHER DIRECTIONAL TERMS

Superficial
At, near, or relatively close to the body surface

The skin is *superficial* to underlying structures.

Deep
Toward the interior of the body; farther from the surface

The bone of the thigh is *deep* to the surrounding skeletal muscles.

a Anterior view

The head is superior to the chest.   **Superior**

Cranial or Cephalic
Toward the head
The *cranial* nerves are in the head.

Posterior or Dorsal
Posterior: The back surface
Dorsal: The back. (equivalent to posterior when referring to the human body)
The scapula (shoulder blade) is located *posterior* to the rib cage.

Anterior or Ventral
Anterior: The front surface
Ventral: The belly side. (equivalent to anterior when referring to the human body)
The umbilicus (navel) is on the *anterior* (or *ventral*) surface of the trunk.

Caudal
Toward the tail; (coccyx in humans)
Fused *caudal* vertebrae form the skeleton of the tail (coccyx).

b Lateral view

Inferior: Below; at a lower level; toward the feet

The knee is inferior to the hip.   **Inferior**

Figure 1–8 **Sectional Planes.**

Frontal or coronal plane

Plane is oriented parallel to long axis

A *frontal*, or *coronal*, *section* separates anterior and posterior portions of the body. Coronal usually refers to sections passing through the skull.

Directional term: frontally or coronally

Frontal plane

Sagittal plane

Plane is oriented parallel to long axis

A *sagittal section* separates right and left portions. You examine a sagittal section, but you section sagittally.

In a *midsagittal section*, the plane passes through the midline. It separates the body into equal right and left sides.

A *parasagittal section* misses the midline. It separates the body into unequal right and left sides.

Midsagittal plane

Directional term: sagittally

Transverse, or horizontal, plane

Plane is oriented perpendicular to long axis

A *transverse*, or *cross*, *section* separates superior and inferior portions of the body.

Transverse plane (inferior view)

Directional term: transversely or horizontally

Planes and Sections

Any section through a three-dimensional object can be described in reference to three **sectional planes**, as indicated in Figure 1–8. A *plane* is an axis, and three planes are needed to describe any three-dimensional object. A section is a single view or slice along one of these planes:

- A plane that lies at right angles to the long axis of the body is the **transverse** (or *horizontal*) **plane**. It divides the body into *superior* and *inferior* portions. A cut in this plane is called a **transverse section**, or *cross section*. Unless otherwise noted, in this book all anatomical diagrams that present cross-sectional views of the body are oriented as though the subject were supine with you, the observer, standing at the subject's feet and looking toward the head.

- Two planes are parallel to the long axis of the body. These are the **frontal** (*coronal*) **plane** and the **sagittal plane**.

The atlas that accompanies this text contains images of sections taken through the body in various planes.

1-8 ▎ Body cavities of the trunk protect internal organs and allow them to change shape

Learning Outcome Identify the major body cavities of the trunk and their subdivisions, and describe the functions of each.

The body's trunk is subdivided into three major regions established by the body wall: the thoracic, abdominal, and pelvic regions (p. 41). Most of our vital organs are located within these regions of the trunk. The true **body cavities** are closed, fluid-filled, and lined by a thin tissue layer called a *serous membrane,* or *serosa.* The vital organs of the trunk are suspended within these body cavities; they do not simply lie there. Early

Figure 1–9 **Relationships among the Subdivisions of the Body Cavities of the Trunk.**

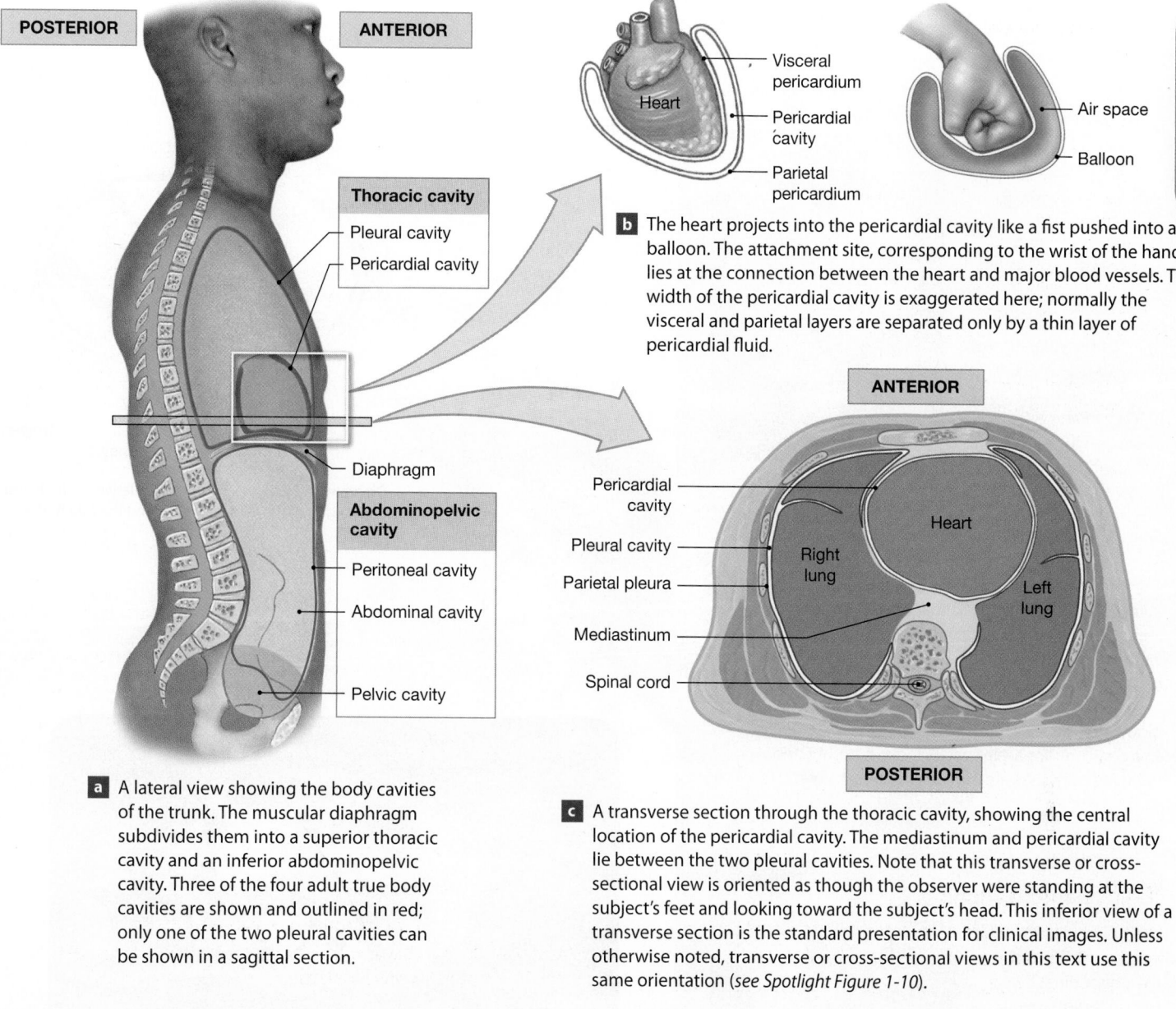

b The heart projects into the pericardial cavity like a fist pushed into a balloon. The attachment site, corresponding to the wrist of the hand, lies at the connection between the heart and major blood vessels. The width of the pericardial cavity is exaggerated here; normally the visceral and parietal layers are separated only by a thin layer of pericardial fluid.

a A lateral view showing the body cavities of the trunk. The muscular diaphragm subdivides them into a superior thoracic cavity and an inferior abdominopelvic cavity. Three of the four adult true body cavities are shown and outlined in red; only one of the two pleural cavities can be shown in a sagittal section.

c A transverse section through the thoracic cavity, showing the central location of the pericardial cavity. The mediastinum and pericardial cavity lie between the two pleural cavities. Note that this transverse or cross-sectional view is oriented as though the observer were standing at the subject's feet and looking toward the subject's head. This inferior view of a transverse section is the standard presentation for clinical images. Unless otherwise noted, transverse or cross-sectional views in this text use this same orientation (*see Spotlight Figure 1-10*).

anatomists used the term "cavity" when referring to internal regions. For example, everything deep to the chest wall of the thoracic region is considered to be within the *thoracic cavity*, and all of the structures deep to the abdominal and pelvic walls are said to lie within the *abdominopelvic cavity*. Internally, the **diaphragm** (DĪ-uh-fram), a flat muscular sheet, separates these anatomical regions. The boundaries of the true body cavities and the regional "cavities" are not identical. For example, the thoracic cavity contains two pleural cavities, each surrounding a lung, the pericardial cavity of the heart, and a large tissue mass, the *mediastinum*, and the peritoneal cavity extends only partway into the pelvic cavity. **Figure 1–9** shows the boundaries between the subdivisions of the thoracic cavity and the abdominopelvic cavity.

The body cavities of the trunk have two essential functions: (1) They protect delicate organs from shocks and impacts; and (2) they permit significant changes in the size and shape of internal organs. For example, the lungs, heart, stomach, intestines, urinary bladder, and many other organs can expand and contract without distorting surrounding tissues or disrupting the activities of nearby organs because they project into body cavities.

The internal organs that are enclosed by these cavities are known as **viscera** (VIS-e-ruh). A delicate serous membrane lines the walls of these internal cavities and covers the surfaces of the enclosed viscera. A watery fluid, called serous fluid, moistens serous membranes, coats opposing surfaces, and reduces friction. The portion of a serous membrane that covers a visceral organ is called the *visceral serosa*. The opposing layer that lines the inner

Over the past several decades, rapid progress has been made in discovering more accurate and more detailed ways to image the human body, both in health and disease.

X-rays

X-rays are the oldest and still the most common method of imaging. X-rays are a form of high-energy radiation that can penetrate living tissues. An x-ray beam travels through the body before striking a photographic plate. Not all of the projected x-rays arrive at the film. The body absorbs or deflects some of those x-rays. The ability to stop the passage of x-rays is referred to as **radiopacity**. When taking an x-ray, these areas that are impenetrable by x-rays appear light or white on the exposed film and are said to be **radiopaque**. In the body, air has the lowest radiopacity. Fat, liver, blood, muscle, and bone are increasingly radiopaque. As a result, radiopaque tissues look white, and less radiopaque tissues are in shades of gray to black.

An x-ray of the skull, taken from the left side

To use x-rays to visualize soft tissues, a very radiopaque substance must be introduced. To study the upper digestive tract, a radiopaque barium solution is ingested by the patient. The resulting x-ray shows the contours of the stomach and intestines.

A **barium-contrast x-ray** of the upper digestive tract

Standard Scanning Techniques

More recently, a variety of **scanning techniques** dependent on computers have been developed to show the less radiopaque, soft tissues of the body in much greater detail.

Diagrammatic views showing the relative position and orientation of the CT scan below and the MRI to the right.

CT scan of the abdomen

CT (computed tomography) **scans** use computers to reconstruct sectional views. A single x-ray source rotates around the body, and the x-ray beam strikes a sensor monitored by the computer. The x-ray source completes one revolution around the body every few seconds. It then moves a short distance and repeats the process. The result is usually displayed as a sectional view in black and white, but it can be colorized for visual effect. CT scans show three-dimensional relationships and soft tissue structures more clearly than do standard x-rays.

- Note that when anatomical diagrams or scans present cross-sectional views, the sections are presented from an inferior perspective, as though the observer were standing at the feet of a person in the supine position and looking toward the head of the subject.

MRI scan of the abdomen

An **MRI** of the same region (in this case, the abdomen) can show soft tissue structure in even greater detail than a CT scan. Magnetic resonance imaging surrounds part or all of the body with a magnetic field 3000 times as strong as that of Earth. This field causes particles within atoms throughout the body to line up in a uniform direction. Energy from pulses of radio waves are absorbed and released by the different atoms. The released energy is used to create a detailed image of the soft tissue structure.

PET scan of the brain

Positron emission tomography (**PET**) is an imaging technique that assesses metabolic and physiological activity of a structure. A PET scan is an important tool in evaluating healthy and diseased brain function.

Ultrasound of the uterus

In **ultrasound** procedures, a small transmitter contacting the skin broadcasts a brief, narrow burst of high-frequency sound and then detects the echoes. The sound waves are reflected by internal structures, and a picture, or **echogram**, is assembled from the pattern of echoes. These images lack the clarity of other procedures, but no adverse effects have been reported, and fetal development can be monitored without a significant risk of birth defects. Special methods of transmission and processing permit analysis of the beating heart without the complications that can accompany dye injections.

Ultrasound transmitter

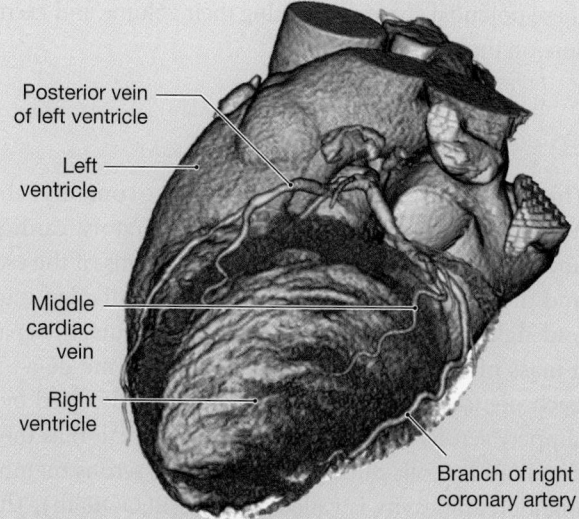

Posterior vein of left ventricle
Left ventricle
Middle cardiac vein
Right ventricle
Branch of right coronary artery

Spiral scan of the heart
[Courtesy of TeraRecon, Inc.]

A **spiral CT scan** is a form of three-dimensional imaging technology that is becoming increasingly important in clinical settings. During a spiral CT scan, the patient is on a platform that advances at a steady pace through the scanner while the imaging source, usually x-rays, rotates continuously around the patient. Because the x-ray detector gathers data quickly and continuously, a higher quality image is generated, and the patient is exposed to less radiation as compared to a standard CT scanner, which collects data more slowly and only one slice of the body at a time.

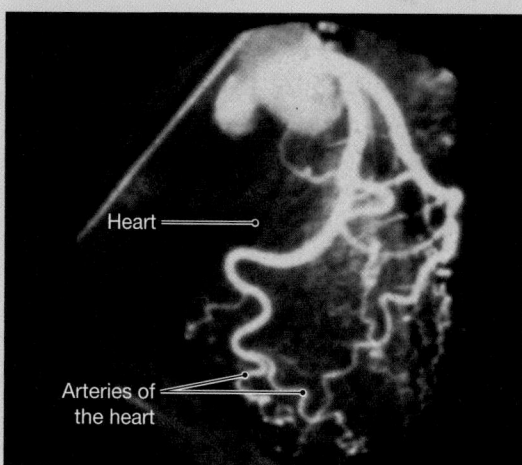

Heart
Arteries of the heart

Digital subtraction angiography of coronary arteries

Digital subtraction angiography (**DSA**) is used to monitor blood flow through specific organs, such as the brain, heart, lungs, and kidneys. X-rays are taken before and after radiopaque dye is administered, and a computer "subtracts" details common to both images. The result is a high-contrast image showing the distribution of the dye.

surface of the body wall or chamber is called the *parietal serosa*. Because the moist parietal and visceral serosae are usually in close contact, the body cavities are called *potential spaces*. In some clinical conditions, however, excess fluid can accumulate within these potential spaces, increasing their volume and exerting pressure on the enclosed viscera.

The Thoracic Cavity

The thoracic cavity (**Figure 1–9a,c**) contains the lungs and heart; associated organs of the respiratory, cardiovascular, and lymphatic systems; the inferior portions of the esophagus; and the thymus. The thoracic cavity is subdivided into the left and right **pleural cavities** (holding the lungs), separated by a mass of tissue called the **mediastinum** (mē-dē-a-STĪ-num). Each pleural cavity surrounds a lung and is lined by a shiny, slippery serous membrane that reduces friction as the lung expands and recoils during breathing. The serous membrane lining a pleural cavity is called a *pleura* (PLOOR-ah). The *visceral pleura* covers the outer surfaces of a lung, and the *parietal pleura* covers the mediastinal surface and the inner body wall.

The mediastinum consists of a mass of connective tissue that surrounds, stabilizes, and supports the esophagus, trachea, and thymus, as well as the major blood vessels that originate or end at the heart. The mediastinum also contains the **pericardial cavity**, a small chamber that surrounds the heart. The relationship between the heart and the pericardial cavity resembles that of a fist pushing into a balloon (**Figure 1–9b**). The wrist corresponds to the *base* (attached portion) of the heart, and the balloon corresponds to the serous membrane that lines the pericardial cavity. The serous membrane associated with the heart is called the *pericardium* (*peri-*, around + *cardium*, heart). The layer covering the heart is the *visceral pericardium*, and the opposing surface is the *parietal pericardium*. During each beat, the heart changes in size and shape. The pericardial cavity permits these changes, and the slippery pericardial serous membrane lining prevents friction between the heart and nearby structures in the thoracic cavity.

The Abdominopelvic Cavity

The abdominopelvic cavity extends from the diaphragm to the pelvis. It is subdivided into a superior **abdominal cavity** and an inferior **pelvic cavity** (**Figure 1–9**). The abdominopelvic cavity contains the *peritoneal* (per-i-tō-NĒ-al) *cavity*, a potential space lined by a serous membrane known as the *peritoneum* (per-i-tō-NĒ-um). The *parietal peritoneum* lines the inner surface of the body wall. A narrow space containing a small amount of fluid separates the parietal peritoneum from the *visceral peritoneum*, which covers the enclosed organs. You are probably already aware of the movements of the organs in this cavity. Most of us have had at least one embarrassing moment when a digestive organ contracted, producing a movement of liquid or

gas and a gurgling or rumbling sound. The peritoneum allows the organs of the digestive system to slide across one another without damage to themselves or the walls of the cavity.

The abdominal cavity extends from the inferior (toward the feet) surface of the diaphragm to the level of the superior (toward the head) margins of the pelvis. This cavity contains the liver, stomach, spleen, small intestine, and most of the large intestine. (**Figure 1–6c** shows the positions of most of these organs.) The organs are partially or completely enclosed by the peritoneal cavity, much as the heart and lungs are enclosed by the pericardial and pleural cavities, respectively. A few organs, such as the kidneys and pancreas, lie between the peritoneal lining and the muscular wall of the abdominal cavity. Those organs are said to be *retroperitoneal* (*retro*, behind).

The pelvic cavity is inferior to the abdominal cavity. The bones of the pelvis form the walls of the pelvic cavity, and a layer of muscle forms its floor. The pelvic cavity contains the urinary bladder, various reproductive organs, and the distal (farthest) portion of the large intestine. In females, the pelvic cavity contains the ovaries, uterine tubes, and uterus. In males, it contains the prostate gland and seminal glands (seminal vesicles). The pelvic cavity also contains the inferior portion of the peritoneal cavity. The peritoneum covers the ovaries and the uterus in females, as well as the superior portion of the urinary bladder in both sexes. Organs such as the urinary bladder and the distal portions of the ureters and large intestine, which extend inferior to the peritoneal cavity, are said to be *infraperitoneal*.

The true body cavities of the trunk in the adult share a common embryological origin. The term "dorsal body cavity" is sometimes used to refer to the internal chamber of the skull (cranial cavity) and the space enclosed by the vertebral bodies and vertebral arches (vertebral cavity). These chambers, which are defined by bony structures, are anatomically and embryologically distinct from true body cavities, and the term "dorsal body cavity" is not encountered in either clinical anatomy or comparative anatomy. For these reasons, we have avoided using that term in our discussion of body cavities.

A partial list of chambers, or spaces, within the body that are not true body cavities would include the cranial cavity, vertebral cavity, oral cavity, digestive cavity, orbits (eye sockets), tympanic cavity of each middle ear, nasal cavities, and paranasal sinuses (air-filled chambers within some cranial bones that are connected to the nasal cavities). These structures will be discussed in later chapters.

Spotlight Figure 1–10 **Diagnostic Imaging Techniques** highlights some clinical tests commonly used for viewing the interior of the body.

CHECKPOINT

26. **Name two essential functions of the body cavities of the trunk.**

27. **Describe the various body cavities of the trunk.**

See the blue Answers tab at the back of the book.

Related Clinical Terms

acute: A disease of short duration but typically severe.

chemotherapy: The treatment of disease or mental disorder by the use of chemical substances, especially the treatment of cancer by cytotoxic and other drugs.

chronic: Illness persisting for a long time or constantly recurring. Often contrasted with acute.

epidemiology: The branch of science that deals with the incidence, distribution, and possible control of diseases and other factors relating to health.

etiology: The science and study of the cause of diseases.

idiopathic: Denoting any disease or condition of unknown cause.

morbidity: The state of being diseased or unhealthy, or the incidence of disease in a population.

pathophysiology: The functional changes that accompany a particular syndrome or disease.

syndrome: a condition characterized by a group of associated symptoms.

Chapter Review

Study Outline

MasteringA&P®

Access more chapter study tools online in the MasteringA&P Study Area:

- Chapter Quizzes, Chapter Practice Test, Art-labeling Activities, Animations, MP3 Tutor Sessions, and Clinical Case Studies
- Practice Anatomy Lab   PAL™
- Interactive Physiology   iP®
- A&P Flix   A&PFlix
- PhysioEx   PhysioEx

An Introduction to Studying the Human Body p. 28

1. Biology is the study of life. One of its goals is to discover the unity and the patterns that underlie the diversity of organisms.

1-1 Anatomy and physiology directly affect your life p. 28

2. This course will help you discover how your body works under normal and abnormal conditions, by serving as a base for understanding other life sciences and expanding your vocabulary.

1-2 Anatomy is structure, and physiology is function p. 29

3. **Anatomy** is the study of internal and external structures of the body and the physical relationships among body parts. **Physiology** is the study of how living organisms perform their vital functions. All physiological functions are performed by specific structures.

4. **Medical terminology** is the use of prefixes, suffixes, word roots, and combining forms to construct anatomical, physiological, or medical terms.

5. *Terminologia Anatomica (International Anatomical Terminology)* was used as the standard in preparing your textbook.

1-3 Anatomy and physiology are closely integrated p. 30

6. All specific functions are performed by specific structures.

7. In **gross** (*macroscopic*) **anatomy**, we consider features that are visible without a microscope. This field includes *surface anatomy* (general form and superficial markings); *regional anatomy* (anatomical organization of specific areas of the body); and *systemic anatomy* (structure of organ systems). In *developmental anatomy*, we examine the changes in form that occur between conception and physical maturity. In *embryology*, we study developmental processes that occur during the first two months of development. *Clinical anatomy* includes anatomical subspecialties important to the practice of medicine.

8. The equipment used determines the limits of *microscopic anatomy*. In **cytology**, we analyze the internal structure of individual cells. In **histology**, we examine **tissues**, groups of cells that perform specific functions. Tissues combine to form **organs**, anatomical structures with multiple functions. *(Spotlight Figure 1–1)*

9. Human physiology is the study of the functions of the human body. It is based on *cell physiology*, the study of the functions of cells. In *organ physiology*, we study the physiology of specific organs. In *systemic physiology*, we consider all aspects of the functioning of specific organ systems. In *pathological physiology*, we study the effects of diseases on organ or system functions.

1-4 Levels of organization progress from molecules to a complete organism p. 32

10. Anatomical structures and physiological mechanisms occur in a series of interacting levels of organization. *(Spotlight Figure 1–1)*

11. The 11 **organ systems** of the body are the integumentary, skeletal, muscular, nervous, endocrine, cardiovascular, lymphatic, respiratory, digestive, urinary, and reproductive systems. *(SpotlightFigure 1–1)*

1-5 Homeostasis is the state of internal balance p. 33

12. **Homeostasis** is the existence of a stable environment within the body.

13. Physiological systems preserve homeostasis through *homeostatic regulation*.

14. **Autoregulation** occurs when a cell, tissue, organ, or organ system adjusts its activities automatically in response to some environmental change. **Extrinsic regulation** results from the activities of the nervous system or endocrine system.

15. Homeostatic regulation mechanisms usually involve a **receptor** that is sensitive to a particular stimulus; a **control center**, which receives and processes the information supplied by the receptor and then sends out commands; and an **effector** that responds to the commands of the control center and whose activity either opposes or enhances the stimulus. *(Figure 1–2)*

1

1-6 ▌ Negative feedback opposes variations from normal, whereas positive feedback exaggerates them p. 36

16. **Negative feedback** is a corrective mechanism involving an action that directly opposes a variation from normal limits. *(Figure 1–3)*

17. In **positive feedback**, an initial stimulus produces a response that exaggerates or enhances the change in the original conditions, creating a *positive feedback loop*. *(Figure 1–4)*

18. No single organ system has total control over the body's internal environment; all organ systems work together. *(Table 1–1)*

1-7 ▌ Anatomical terms describe body regions, anatomical positions and directions, and body sections p. 40

19. The standard arrangement for anatomical reference is called the **anatomical position**. If the person is shown lying down, it can be either **supine** (face up) or **prone** (face down). *(Figure 1–5* and *Spotlight Figure 1-10)*

20. **Abdominopelvic quadrants** and **abdominopelvic regions** represent two approaches to describing anatomical regions of that portion of the body. *(Figure 1–6)*

21. The use of special directional terms provides clarity for the description of anatomical structures. *(Figure 1–7)*

22. The three **sectional planes** (**transverse**, or *horizontal*, **plane**; **frontal**, or *coronal*, **plane**; and **sagittal plane**) describe relationships among the parts of the three-dimensional human body. *(Figure 1–8)*

1-8 ▌ Body cavities in the trunk protect internal organs and allow them to change shape p. 44

23. **Body cavities** protect delicate organs and permit significant changes in the size and shape of internal organs. The body cavities of the trunk surround organs of the respiratory, cardiovascular, digestive, urinary, and reproductive systems. *(Figure 1–9)*

24. The **diaphragm** divides the (superior) **thoracic** and (inferior) **abdominopelvic cavities**. The thoracic cavity consists of two **pleural cavities** (each surrounding a lung) with a central tissue mass known as the **mediastinum**. Within the mediastinum is the **pericardial cavity**, which surrounds the heart. The abdominopelvic cavity consists of the **abdominal cavity** and the **pelvic cavity** and contains the *peritoneal cavity*, a chamber lined by the *peritoneum*, a *serous membrane*. *(Figure 1–9* and *Spotlight Figure 1-10)*

25. **Diagnostic imaging techniques** are used in clinical medicine to view the body's interior. *(Spotlight Figure 1–10)*

Review Questions

See the blue Answers tab at the back of the book.

LEVEL 1 Reviewing Facts and Terms

1. Label the directional terms in the figures below.

(a) _____   (b) _____
(c) _____   (d) _____
(e) _____   (f) _____
(g) _____   (h) _____
(i) _____   (j) _____

Match each numbered item with the most closely related lettered item. Use letters for answers in the spaces provided.

_____ 2. cytology	(a)	study of tissues
_____ 3. physiology	(b)	constant internal environment
_____ 4. histology	(c)	face up position
_____ 5. anatomy	(d)	study of functions
_____ 6. homeostasis	(e)	positive feedback
_____ 7. muscle	(f)	organ system
_____ 8. heart	(g)	study of cells
_____ 9. endocrine	(h)	negative feedback
_____ 10. temperature regulation	(i)	serous membrane
_____ 11. labor and delivery	(j)	study of internal and external
_____ 12. supine		body structures
_____ 13. prone	(k)	diaphragm
_____ 14. divides thoracic and ab-	(l)	tissue
dominopelvic body cavities	(m)	peritoneal cavity
_____ 15. abdominopelvic cavity	(n)	organ
_____ 16. pericardium	(o)	face down position

17. Which of the following is the correct order of organisation in the human body? **(a)** atom, organ system, cell, tissue, organism, **(b)** atom, tissue, cell, organ system, organism, **(c)** atom, cell, tissue, organ system, organism, **(d)** atom, tissue, organ system, cell, organism.

The correct order, from the simplest to the most complex level, is
(a) 2, 4, 1, 3, 6, 5, **(b)** 4, 2, 1, 3, 6, 5, **(c)** 4, 2, 1, 6, 3, 5,
(d) 4, 2, 3, 1, 6, 5, **(e)** 2, 1, 4, 3, 5, 6.

18. Which one of the following is not a form of gross anatomy? **(a)** surface, **(b)** regional, **(c)** systemic, **(d)** microscopic.
19. Pathophysiology is the study of **(a)** organ systems, **(b)** disease origin and process, **(c)** structural dysfunction, **(d)** effects of disease on organ system function.
20. Which is an example of a positive feedback system? **(a)** blood clotting, **(b)** thermoregulation, **(c)** insulin release, **(d)** lipid metabolism.
21. Which of the following sequences best describes a homeostatic process? **(a)** control center, sensor, effector, **(b)** sensor, control center, effector, **(c)** effector, sensor, control center, **(d)** sensor, effector, control center.
22. Which body cavity would enclose each of the following organs? **(a)** heart, **(b)** small intestine, large intestine, **(c)** lung, **(d)** kidneys.
23. Diseases manifest with signs and symptoms. What is the difference between a sign and a symptom?

LEVEL 2 Reviewing Concepts

24. **(a)** Define *anatomy*.
    (b) Define *physiology*.
25. What is the principle of complementarity of structure and function?
26. What distinguishes autoregulation from extrinsic regulation?
27. Why is it important to compare physiological values to normal ranges?

28. Give an example of an organ system involved in the homeostatic control of blood pressure and describe its role.
29. Which of the following is *not* an example of negative feedback? **(a)** Increased pressure in the aorta triggers mechanisms to lower blood pressure, **(b)** A rise in blood calcium levels triggers the release of a hormone that lowers blood calcium levels, **(c)** A rise in estrogen during the menstrual cycle increases the number of progesterone receptors in the uterus, **(d)** Increased blood sugar stimulates the release of a hormone from the pancreas that stimulates the liver to store blood sugar.

LEVEL 3 Critical Thinking and Clinical Applications

30. The hormone *calcitonin* is released from the thyroid gland in response to increased levels of calcium ions in the blood. If this hormone is controlled by negative feedback, what effect would calcitonin have on blood calcium levels?
31. A stroke occurs when there is a disruption in blood flow to the brain, causing brain cells to die. Why might a stroke result in a rise or fall of normal body temperature?
32. John undergoes heart surgery to replace a defective valve. Some days later his blood pressure becomes very low and he feels unwell. An ultrasound of his heart shows fluid in the pericardium. Why has John's blood pressure fallen?

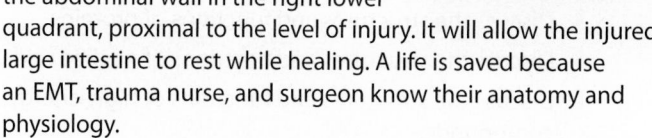

CLINICAL CASE Wrap-Up

Using A&P to Save a Life

The patient is wheeled through the door and into Trauma Room 1. He is barely alive. Because of the location of the abdominal wound, in the right upper quadrant, and the likely depth of the wound, more than 6 inches, the chances are good that the liver has been lacerated. Depending on the angle of the knife, sections of large intestine, containing bacteria that are deadly when released into the abdominal cavity, have probably also been lacerated. Any blood spilled onto the street or into the patient's abdominal cavity is lost from the cardiovascular system and cannot carry oxygen to body tissues. For this reason, the trauma nurse is ready with blood for transfusion.

An x-ray shows free air in the abdominal cavity, a sure sign the large intestine is cut. It also shows a fluid level indicating free blood in the abdominal cavity. Both conditions need immediate surgical attention. The patient is taken directly to the operating room where a surgeon performs open abdominal surgery. The lacerations of the liver and large intestine

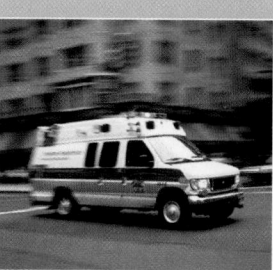

are repaired and the bleeding is stopped. A diverting colostomy is performed, creating an opening from the large intestine through the abdominal wall in the right lower quadrant, proximal to the level of injury. It will allow the injured large intestine to rest while healing. A life is saved because an EMT, trauma nurse, and surgeon know their anatomy and physiology.

1. Besides the liver and most of the large intestine, what other major organs are in the abdominal cavity?

2. If the deep knife wound had been superior to the diaphragm, what body cavity would the knife have entered, and what organs does it contain?

See the blue Answers tab at the back of the book.

2

The Chemical Level of Organization

Learning Outcomes

These Learning Outcomes correspond by number to this chapter's sections and indicate what you should be able to do after completing the chapter.

2-1 ▌ Describe an atom and how atomic structure affects interactions between atoms.

2-2 ▌ Compare the ways in which atoms combine to form molecules and compounds.

2-3 ▌ Distinguish among the major types of chemical reactions that are important for studying physiology.

2-4 ▌ Describe the crucial role of enzymes in metabolism.

2-5 ▌ Distinguish between organic compounds and inorganic compounds.

2-6 ▌ Explain how the chemical properties of water make life possible.

2-7 ▌ Discuss the importance of pH and the role of buffers in body fluids.

2-8 ▌ Describe the physiological roles of inorganic compounds.

2-9 ▌ Discuss the structures and functions of carbohydrates.

2-10 ▌ Discuss the structures and functions of lipids.

2-11 ▌ Discuss the structures and functions of proteins.

2-12 ▌ Discuss the structures and functions of nucleic acids.

2-13 ▌ Discuss the structures and functions of high-energy compounds.

2-14 ▌ Explain the relationship between chemicals and cells.

SPOTLIGHT
• Chemical Notation p. 58

CLINICAL CASE
• What Is Wrong with My Baby? p. 53

Clinical Notes
• Radiation Sickness p. 55
• Fatty Acids and Health p. 74

What Is Wrong with My Baby?

Sean is Maureen's first baby. Maureen and her husband, Conner, had enjoyed an uncomplicated pregnancy and delivery. Maureen had felt healthy throughout the pregnancy, but something was wrong with her baby.

Sean is 1 month old, and not thriving. He seems to have a good appetite and breastfeeds as if he is starving; yet he has dropped 20 percent of his normal birth weight of 7 pounds, 8 ounces. He is down to only 6 pounds at his 1-month checkup, and his skin looks "wrinkly." He is having several greasy, almost foamy-looking stools each day. They remind Maureen of cake

icing. Maureen also notices the funniest thing. Sean's skin tastes salty.

Most alarmingly, he seems to be having some difficulty breathing recently. He sounds wheezy. Maureen and Conner are both from big families, and none of the babies have ever been sickly like this.

What is wrong with Baby Sean? To find out, turn to the Clinical Case Wrap-Up on p. 89.

An Introduction to the Chemical Level of Organization

In this chapter we consider the structure of atoms, the basic chemical building blocks. You will also learn how atoms can combine to form increasingly complex structures.

2-1 Atoms are the basic particles of matter

Learning Outcome Describe an atom and how atomic structure affects interactions between atoms.

Our study of the human body begins at the chemical level of organization. *Chemistry* is the science that deals with the structure of **matter**, defined as anything that takes up space and has mass. **Mass**, the amount of material in matter, is a physical property that determines the weight of an object in Earth's gravitational field. For our purposes, the mass of an object is the same as its weight. However, the two are not always equivalent: In orbit you would be weightless, but your mass would remain unchanged.

The smallest stable units of matter are called **atoms**. Air, elephants, oranges, oceans, rocks, and people are all composed of atoms in varying combinations. The unique characteristics of each object, living or nonliving, result from the types of atoms involved and the ways those atoms combine and interact.

Atoms are composed of **subatomic particles**. Many different subatomic particles exist, but only three, *protons, neutrons, and electrons,* are important for understanding the chemical properties of matter. Protons and neutrons are similar in size and mass, but **protons** (p^+) have a positive electrical charge.

Neutrons (n or n^0) are electrically *neutral*, or uncharged. **Electrons** (e^-) are much lighter than protons—only 1/1836 as massive—and have a negative electrical charge. For this reason, the mass of an atom is determined primarily by the number of protons and neutrons in the **nucleus**, the central region of an atom. The mass of a large object, such as your body, is the sum of the masses of all the component atoms.

Atomic Structure

Atoms normally contain equal numbers of protons and electrons. The number of protons in an atom is known as its **atomic number**. Hydrogen (H) is the simplest atom, with an atomic number of 1. An atom of hydrogen contains one proton, and one electron as well. Hydrogen's proton is located in the center of the atom and forms the nucleus. Hydrogen atoms seldom contain neutrons, but when neutrons are present, they are also located in the nucleus. All atoms other than hydrogen have both neutrons and protons in their nuclei.

Electrons travel around the nucleus at high speed, within a spherical area called the **electron cloud**. We often illustrate atomic structure in the simplified form shown in **Figure 2–1a**. In this two-dimensional representation, the electrons occupy a circular **electron shell**. One reason an electron tends to remain in its electron shell is that the negatively charged electron is attracted to the positively charged proton. The attraction between opposite electrical charges is an example of an *electrical force*. As you will see in later chapters, electrical forces are involved in many physiological processes.

The dimensions of the electron cloud determine the overall size of the atom. To get an idea of the scale involved, consider that if the nucleus were the size of a tennis ball, the

Figure 2–1   **The Structure of Hydrogen Atoms.** Three forms of hydrogen atoms are shown using the two-dimensional electron-shell model, which indicates the electron cloud surrounding the nucleus.

Hydrogen-1	**Hydrogen-2, deuterium**	**Hydrogen-3, tritium**
mass number: 1	mass number: 2	mass number: 3
a A typical hydrogen nucleus contains a proton and no neutrons.	**b** A deuterium (^{2}H) nucleus contains a proton and a neutron.	**c** A tritium (^{3}H) nucleus contains a proton and two neutrons.

electron cloud of a hydrogen atom would have a radius of 10 km (about 6 miles!). In reality, atoms are so small that atomic measurements are most conveniently reported in nanometers (NAN-ō-mē-ter) (nm). One nanometer is 10^{-9} meter (0.000000001 m), or one billionth of a meter. The very largest atoms approach 0.5 nm in diameter (0.0000000005 m, or 0.00000002 in.).

Elements and Isotopes

An **element** is a pure substance composed of atoms of only one kind. Atoms are the smallest particles of an element that still retain the characteristics of that element. As a result, each element has uniform composition and properties. Each element includes all the atoms with the same number of protons, and thus the same atomic number. Only 92 elements exist in nature. Researchers have created about two dozen additional elements through nuclear reactions in laboratories.

Every element has a chemical symbol, an abbreviation recognized by scientists everywhere. Most of the symbols are easy to connect with the English names of the elements (O for oxygen, N for nitrogen, C for carbon, and so on), but a few are abbreviations of their Latin names. For example, the symbol for sodium, Na, comes from the Latin word *natrium*.

Atomic nuclei are unaltered by ordinary chemical processes. So, elements cannot be changed or broken down into simpler substances in chemical reactions. For example, an atom of carbon always remains an atom of carbon, regardless of the chemical events in which it may take part.

Our bodies consist of many elements, and the 13 most abundant elements are listed in Table 2–1. Our bodies also contain atoms of another 14 elements—called *trace elements*—that are present in very small amounts.

Table 2–1	Principal Elements in the Human Body
Element (% of total body weight)	**Significance**
Oxygen, O (65)	A component of water and other compounds; gaseous form is essential for respiration
Carbon, C (18.6)	Found in all organic molecules
Hydrogen, H (9.7)	A component of water and most other compounds in the body
Nitrogen, N (3.2)	Found in proteins, nucleic acids, and other organic compounds
Calcium, Ca (1.8)	Found in bones and teeth; important for membrane function, nerve impulses, muscle contraction, and blood clotting
Phosphorus, P (1.0)	Found in bones and teeth, nucleic acids, and high-energy compounds
Potassium, K (0.4)	Important for proper membrane function, nerve impulses, and muscle contraction
Sodium, Na (0.2)	Important for blood volume, membrane function, nerve impulses, and muscle contraction
Chlorine, Cl (0.2)	Important for blood volume, membrane function, and water absorption
Magnesium, Mg (0.06)	A cofactor for many enzymes
Sulfur, S (0.04)	Found in many proteins
Iron, Fe (0.007)	Essential for oxygen transport and energy capture
Iodine, I (0.0002)	A component of hormones of the thyroid gland
Trace elements: silicon (Si), fluorine (F), copper (Cu), manganese (Mn), zinc (Zn), selenium (Se), cobalt (Co), molybdenum (Mo), cadmium (Cd), chromium (Cr), tin (Sn), aluminum (Al), boron (B), and vanadium (V)	Some function as cofactors; the functions of many trace elements are poorly understood

The atoms of a single element all have the same number of protons, but they can differ in the number of neutrons in the nucleus. For example, most hydrogen nuclei consist solely of a single proton, but 0.015 percent also contains one neutron, and a very small percentage contains two neutrons (**Figure 2–1**). Atoms of the same element whose nuclei contain different numbers of neutrons are called **isotopes**.

Different isotopes of an element have essentially identical chemical properties, and are alike except in mass. The **mass number**—the total number of protons plus neutrons in the nucleus—is used to designate isotopes. For example, hydrogen has three isotopes. Hydrogen-1, or ^{1}H, has one proton and one electron (**Figure 2–1a**). Hydrogen-2, or ^{2}H, also known as *deuterium*, has one proton, one electron, and one neutron (**Figure 2–1b**). Hydrogen-3, or ^{3}H, also known as *tritium*, has one proton, one electron, and two neutrons (**Figure 2–1c**).

The nuclei of some isotopes are unstable. That is, they spontaneously break down and give off subatomic particles or radiation in measurable amounts. Such isotopes are called **radioisotopes**. The breakdown process is called *radioactive decay*. Strongly radioactive isotopes are dangerous, because the particles or radiation they give off can ionize atoms, break apart molecules, and destroy cells and tissues.

Clinical Note

Radiation Sickness *Radiation sickness*, or *acute radiation syndrome*, results from excessive exposure to radiation. It is commonly marked by fatigue, nausea, vomiting, loss of teeth and hair, and in more severe cases by damage to blood-forming tissue. Radiation sickness is also sometimes called radiation poisoning. In a clinical setting, it can occur when large doses of radiation are given to an individual over a short period of time to treat disease, such as cancer. (X-rays and CT scans use low-dose radiation and do not cause radiation sickness.) The amount of radiation received will determine how ill a person can become. Effects range from anorexia (loss of appetite), nausea, and vomiting, to anemia (low red blood cell count), central nervous system damage, and death. Widespread radiation sickness occurred after the 2011 earthquake and tsunami-generated nuclear reactor accident in Fukushima, Japan; the 1986 nuclear reactor accident in Chernobyl, Ukraine; and the atomic bombing of Japan during World War II.

Weakly radioactive isotopes are sometimes used in diagnostic procedures to monitor the structural or functional characteristics of internal organs.

Radioisotopes differ in how rapidly they decay (lose their radioactivity). The decay rate of a radioisotope is commonly expressed as its **half-life:** the time required for half of a given amount of the isotope to decay. The half-lives of radioisotopes range from fractions of a second to billions of years.

Atomic Weights

A typical *oxygen* atom has an atomic number of 8 and contains eight protons and eight neutrons. The mass number of this isotope is therefore 16. The mass numbers of other isotopes of oxygen depend on the number of neutrons present. Mass numbers are useful because they tell us the number of subatomic particles in the nuclei of different atoms. However, they do not tell us the *actual* mass of the atoms. For example, they do not take into account the masses of the electrons or the slight difference between the mass of a proton and that of a neutron. The actual mass of an atom of a specific isotope is known as its *atomic mass*.

The unit used to express atomic mass is the *atomic mass unit* (amu), or *dalton*. By international agreement, one amu is equal to one-twelfth the mass of a carbon-12 atom. One atomic mass unit is very close to the mass of a single proton or neutron. Thus, the atomic mass of an atom of the most common isotope of hydrogen is very close to 1, and that of the most common isotope of oxygen is very close to 16.

The **atomic weight** of an element is an average of the different atomic masses and proportions of its different isotopes. This results in the atomic weight of an element being very close to the mass number of the most common isotope of that element. For example, the mass number of the most common isotope of hydrogen is 1, but the atomic weight of hydrogen is 1.0079, primarily because some hydrogen atoms (0.015 percent) have a mass number of 2, and even fewer have a mass number of 3. (The periodic table of the elements in the Appendix at the back of this book shows the atomic weight of each element rounded to its nearest one-hundredth decimal place. For example, the atomic weight of hydrogen is shown as 1.01, rather than 1.0079.)

Atoms take part in chemical reactions in fixed numerical ratios. To form water, for example, exactly two atoms of hydrogen combine with one atom of oxygen. But individual atoms are far too small and too numerous to be counted, so chemists use a unit called the *mole*. For any element, a **mole** (abbreviated *mol*) is a specific quantity with a weight in grams equal to that element's atomic weight.

The mole is useful because one mole of a given element always contains the same number of atoms as one mole of any other element (just as we use *dozen* to stand for 12 items). That number (called *Avogadro's number*) is 6.023×10^{23}, or about 600 billion trillion. Expressing relationships in moles rather than in grams makes it much easier to keep track of the relative numbers of atoms in chemical samples and processes. For example, if a report stated that a sample contains 0.5 mol of hydrogen atoms and 0.5 mol of oxygen atoms, you would know immediately that the two atoms were present in equal numbers. That would not be so evident if the report stated that there were 0.505 g of hydrogen atoms and 8.00 g of oxygen atoms. Most chemical analyses and clinical laboratory tests report data in moles (mol), millimoles (mmol—1/1000 mol, or 10^{-3} mol), or micromoles (μmol—1/1,000,000 mol, or 10^{-6} mol).

Electrons and Energy Levels

Atoms are electrically neutral. In other words, every positively charged proton is balanced by a negatively charged electron. Thus, each increase in the atomic number has a comparable increase in the number of electrons traveling around the nucleus.

Within the electron cloud, electrons occupy an orderly series of energy levels. The electrons in an energy level may travel in complex patterns around the nucleus, but for our purposes the patterns can be diagrammed as a series of concentric electron shells. The first electron shell is the one closest to the nucleus, and it corresponds to the lowest energy level.

Each energy level is limited in the number of electrons it can hold. The first energy level can hold at most 2 electrons. The next two levels can each hold up to 8 electrons. Note that the maximum number of electrons that may occupy shells 1 through 3 corresponds to the number of elements in rows 1 through 3 of the periodic table of the elements (see Appendix). The electrons in an atom occupy successive shells in an orderly manner: The first energy level fills before any electrons enter the second, and the second energy level fills before any electrons enter the third.

The outermost energy level forms the "surface" of the atom and is called the **valence shell**. The number of electrons in this level determines the chemical properties of the element. Atoms with unfilled energy levels are unstable—that is, they will react with other atoms, usually in ways that result in full outer electron shells. In contrast, atoms with a filled outermost energy level are stable and therefore do not readily react with other atoms.

As indicated in **Figure 2–2a**, a hydrogen atom has one electron in the first energy level, so that level is unfilled. A hydrogen atom readily reacts with other atoms. A helium atom has 2 electrons in its first energy level (**Figure 2–2b**). A helium atom is very stable because its outer energy level is filled. It will not ordinarily react with other atoms. A lithium atom has 3 electrons (**Figure 2–2c**). Its first energy level can hold only

Figure 2–2   The Arrangement of Electrons into Energy Levels. The electron-shell model is also used to indicate the relative energy of electrons in an atom.

The first energy level can hold a maximum of two electrons.

Hydrogen, H
Atomic number: 1
Mass number: 1
1 electron

a **Hydrogen (H).** A typical hydrogen atom has one proton and one electron. The electron orbiting the nucleus occupies the first, or lowest, energy level, diagrammed as an electron shell.

Helium, He
Atomic number: 2
Mass number: 4
(2 protons + 2 neutrons)
2 electrons

b **Helium (He).** An atom of helium has two protons, two neutrons, and two electrons. The two electrons orbit in the same energy level.

The second and third energy levels can each contain up to 8 electrons.

Lithium, Li
Atomic number: 3
Mass number: 6
(3 protons + 3 neutrons)
3 electrons

c **Lithium (Li).** A lithium atom has three protons, three neutrons, and three electrons. The first energy level can hold only two electrons, so the third electron occupies a second energy level.

Neon, Ne
Atomic number: 10
Mass number: 20
(10 protons + 10 neutrons)
10 electrons

d **Neon (Ne).** A neon atom has 10 protons, 10 neutrons, and 10 electrons. The second level can hold up to eight electrons; thus, both the first and second energy levels are filled.

2 of them, so lithium has a single electron in a second, unfilled energy level. Like hydrogen, lithium is also unstable and reactive. The second energy level is filled in a neon atom, which has an atomic number of 10 (**Figure 2–2d**). Neon atoms, like helium atoms and other elements in the last column of the periodic table, are very stable. The atoms of elements that are most

important to biological systems are unstable. (Table 2–1, p. 54). Their instability promotes atom interactions to form larger structures.

CHECKPOINT

1. Define atom.

2. Atoms of the same element that have different numbers of neutrons are called _____.

3. How is it possible for two samples of molecular hydrogen to contain the same number of atoms, yet have different molecular weights?

See the blue Answers tab at the back of the book.

2-2 | Chemical bonds are forces formed by atom interactions

Learning Outcome Compare the ways in which atoms combine to form molecules and compounds.

Elements that do not readily participate in chemical processes are said to be *inert*. Helium, neon, and argon, which are known as noble gases, have filled outermost energy levels. These elements are also called *inert gases*, because their atoms do not react with one another and do not combine with atoms of other elements.

Elements with unfilled outermost energy levels, such as hydrogen and lithium, are called *reactive*, because they readily interact or combine with other atoms. Reactive atoms become stable by gaining, losing, or sharing electrons to fill their outermost energy level. The interactions often involve the formation of **chemical bonds**, which hold the participating atoms together once the reaction has ended. In the sections that follow, we will consider three basic types of chemical bonds: *ionic bonds*, *covalent bonds*, and *hydrogen bonds*.

When chemical bonding takes place, new chemical entities called *molecules* and *compounds* are created. The term **molecule** refers to any chemical structure consisting of atoms held together by shared electrons. A **compound** is a pure chemical substance made up of atoms of two or more different elements in a fixed proportion, regardless of the type of chemical bond joining them. The two categories overlap, but they are not the same. Not all molecules are compounds, because some molecules consist of atoms of only one element. (Two oxygen atoms, for example, can be joined by sharing electrons to form a molecule of oxygen.) And not all compounds consist of molecules, because some compounds, such as ordinary table salt (sodium chloride) are held together by bonds that do not involve shared electrons.

Many substances, however, fit both categories. Take water, for example. Water is a compound because it contains two different elements—hydrogen and oxygen—in a fixed proportion of two hydrogen atoms to one oxygen atom. It also consists of molecules, because the two hydrogen atoms and one oxygen

atom are held together by shared electrons. As we will see in later sections, most biologically important compounds, from carbohydrates to DNA, are molecules.

Regardless of the type of bonding involved, a chemical compound has properties that can be quite different from those of its components. For example, a mixture of hydrogen gas and oxygen gas can explode, but the explosion is a chemical reaction that produces liquid water, a compound used to put out fires.

The **molecular weight** of a molecule or compound is the sum of the atomic weights of its component atoms. It follows from the definition of the mole given previously (p. 55) that the molecular weight of a molecule in grams is equal to the weight of one mole of molecules. Molecular weights are important because we cannot handle individual molecules nor can we easily count the billions of molecules involved in chemical reactions in the body.

As an example of how to calculate a molecular weight, let's use water, or H_2O. The atomic weight of hydrogen is 1.0079. To simplify our calculations, we will round that value up to 1.01. Then, one hydrogen molecule (H_2) with its two atoms has a molecular weight of 2.02 ($1.01 \times 2 = 2.02$). One oxygen atom has an atomic weight of 16.00. Summing up, the molecular weight of 1 molecule of H_2O is 18.02. One mole of water molecules would then have a mass of 18.02 grams.

The human body consists of countless molecules and compounds, so it is a challenge to describe these substances and their varied interactions. Chemists simplify such descriptions through a standardized system of *chemical notation*. The basic rules of this system are described in **Spotlight Figure 2–3**.

Ionic Bonds

An **ion** is an atom or group of atoms that has an electric charge, either positive or negative. Ions with a positive charge (+) are called **cations** (KAT-ī-onz). Ions with a negative charge (−) are called **anions** (AN-ī-onz). **Ionic bonds** are chemical bonds created by the electrical attraction between anions and cations.

TIPS & TRICKS

Think of the *t* in ca**t**ion as a plus sign (+) to remember that a ca**t**ion has a positive charge, and think of the *n* in a**n**ion as standing for *n*egative (–) to remember that a**n**ions have a *n*egative charge.

Ions have an unequal number of protons and electrons. Atoms become ions by losing or gaining electrons. We assign a value of +1 to the charge on a proton. The charge on an electron is −1. When the number of protons is equal to the number of electrons, an atom is electrically neutral. An atom that loses an electron becomes a cation with a charge of +1, because it then has one proton that lacks a corresponding

Before we can consider the specific compounds that occur in the human body, we must be able to describe chemical compounds and reactions effectively. The use of sentences to describe chemical structures and events often leads to confusion. A simple form of "chemical shorthand" makes communication much more efficient. The chemical shorthand we will use is known as **chemical notation**. Chemical notation enables us to describe complex events briefly and precisely; its rules are summarized below.

VISUAL REPRESENTATION | CHEMICAL NOTATION

Atoms

The symbol of an element indicates one atom of that element. A number preceding the symbol of an element indicates more than one atom of that element.

VISUAL REPRESENTATION		CHEMICAL NOTATION	
H one atom of hydrogen	O one atom of oxygen	H one atom of hydrogen	O one atom of oxygen
H H two atoms of hydrogen	O O two atoms of oxygen	2H two atoms of hydrogen	2O two atoms of oxygen

Molecules

A numerical subscript following the symbol of an element indicates the number of atoms of that element in a molecule.

HH
hydrogen molecule
composed of two hydrogen atoms

OO
oxygen molecule
composed of two oxygen atoms

H O H
water molecule
composed of two hydrogen atoms and one oxygen atom

H_2
hydrogen molecule

O_2
oxygen molecule

H_2O
water molecule

Reactions

In a description of a chemical reaction, the participants at the start of the reaction are called reactants, and the reaction generates one or more products. Chemical reactions are represented by chemical equations. An arrow indicates the direction of the reaction, from reactants (usually on the left) to products (usually on the right). In the following reaction, two atoms of hydrogen combine with one atom of oxygen to produce a single molecule of water.

H H + O ⟶ H O H

Chemical reactions neither create nor destroy atoms; they merely rearrange atoms into new combinations. Therefore, the numbers of atoms of each element must always be the same on both sides of the equation for a chemical reaction. When this is the case, the equation is balanced.

$$2H + O \longrightarrow H_2O$$
Balanced equation

$$2H + 2O \longrightarrow H_2O$$
Unbalanced equation

Ions

A superscript plus or minus sign following the symbol of an element indicates an ion. A single plus sign indicates a cation with a charge of +1. (The original atom has lost one electron.) A single minus sign indicates an anion with a charge of −1. (The original atom has gained one electron.) If more than one electron has been lost or gained, the charge on the ion is indicated by a number preceding the plus or minus sign.

Na^+
sodium ion
the sodium atom has lost one electron

Cl^-
chloride ion
the chlorine atom has gained one electron

Ca^{2+}
calcium ion
the calcium atom has lost two electrons

Na^+
sodium ion

Cl^-
chloride ion

Ca^{2+}
calcium ion

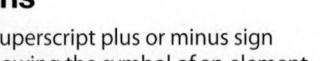

A sodium atom becomes a sodium ion

Electron lost

Sodium atom (Na)

Sodium ion (Na^+)

electron. Losing a second electron would give the cation a charge of +2. Adding an extra electron to a neutral atom produces an anion with a charge of –1. Adding a second electron gives the anion a charge of –2.

In the formation of an ionic bond,

- one atom—the *electron donor*—loses one or more electrons and becomes a cation, with a positive (+) charge.

- another atom—the *electron acceptor*—gains those same electrons and becomes an anion, with a negative (–) charge.

- attraction between the opposite charges then draws the two ions together.

Figure 2–4a illustrates the formation of an ionic bond. The sodium atom diagrammed in ❶ has an atomic number of 11, so this atom normally contains 11 protons and 11 electrons. (Because neutrons are electrically neutral, they do not affect the formation of ions or ionic bonds.) Electrons fill the first and second energy levels, and a single electron occupies the outermost level. Losing that 1 electron would give the sodium atom a full outermost energy level—the second level—and would produce a **sodium ion**, with a charge of +1. (The chemical shorthand for a sodium ion is Na^+.) But a sodium

atom cannot simply throw away the electron: The electron must be donated to an electron acceptor. A chlorine atom has seven electrons in its outermost energy level, so it needs only one electron to achieve stability. A sodium atom can provide the extra electron. In the process (❶), the chlorine atom becomes a **chloride ion** (Cl^-) with a charge of –1.

Both atoms have now become stable ions with filled outermost energy levels. But the two ions do not move apart after the electron transfer, because the positively charged sodium ion is attracted to the negatively charged chloride ion (❷). The combination of oppositely charged ions forms an *ionic compound*—in this case, **sodium chloride**, otherwise known as table salt (❸). Large numbers of sodium and chloride ions interact to form highly structured crystals, held together by the strong electrical attraction of oppositely charged ions (**Figure 2–4b**).

Note that ionic compounds are not called molecules. That is because they consist of an aggregation of ions rather than atoms bonded by shared electrons. Sodium chloride and other ionic compounds are common in body fluids, but they are not present as intact crystals. When placed in water, many ionic compounds dissolve, and some, or all, of the component anions and cations separate.

Figure 2–4   The Formation of Ionic Bonds.

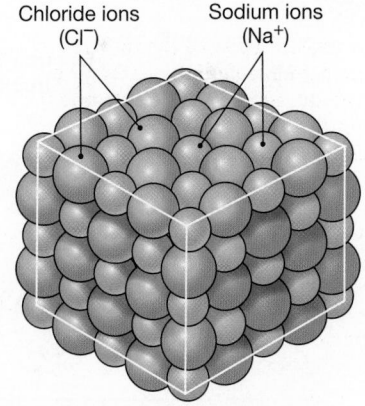

Chloride ions (Cl^-)   Sodium ions (Na^+)

b **Sodium chloride crystal.** Large numbers of sodium and chloride ions form a crystal of sodium chloride (table salt).

c **Photo of sodium chloride crystals** (magnified 0.5X)

a **Formation of an ionic bond.** ❶ A sodium (Na) atom loses an electron, which is accepted by a chlorine (Cl) atom. ❷ Because the sodium ion (Na^+) and chloride ion (Cl^-) have opposite charges, they are attracted to one another. ❸ The association of sodium and chloride ions forms the ionic compound sodium chloride.

2

Covalent Bonds

Some atoms can complete their outer electron shells not by gaining or losing electrons, but by sharing electrons with other atoms. Such sharing creates **covalent** (kō-VĀ-lent) **bonds** between the atoms involved.

Individual hydrogen atoms, as diagrammed in **Figure 2–2a**, do not exist in nature. Instead, we find hydrogen molecules. Molecular hydrogen consists of a pair of hydrogen atoms (**Figure 2–5**). In chemical shorthand, molecular hydrogen is H_2, where H is the chemical symbol for hydrogen, and the subscript 2 indicates the number of atoms. Molecular hydrogen is a gas that is present in the atmosphere in very small quantities. When the two hydrogen atoms share their electrons, each electron whirls around both nuclei. The sharing of one pair of electrons creates a **single covalent bond**, which can be represented by a single line ($-$) in the *structural formula* of a molecule.

Oxygen, with an atomic number of 8, has two electrons in its first energy level and six in its second. The oxygen atoms diagrammed in **Figure 2–5** become stable by sharing two pairs of electrons, forming a **double covalent bond**, represented by two lines ($=$) in its structural formula. Molecular oxygen (O_2) is an atmospheric gas that most organisms need to survive. Our cells would die without a relatively constant supply of oxygen.

Figure 2–5 **Covalent Bonds in Five Common Molecules.**

Molecule	Electron Shell Model and Structural Formula	
Hydrogen (H_2)		H–H
Oxygen (O_2)		O=O
Carbon dioxide (CO_2)		O=C=O
Nitrogen (N_2)		N≡O
Nitric oxide (NO)		N=O

In our bodies, chemical processes that consume oxygen generally also produce **carbon dioxide** (CO_2) as a waste product. Each of the oxygen atoms in a carbon dioxide molecule forms double covalent bonds with the carbon atom, as **Figure 2–5** shows.

A triple covalent bond is the sharing of three pairs of electrons, and is indicated by three lines ($\equiv$) in a structural formula. A triple covalent bond joins two nitrogen atoms to form molecular nitrogen (N_2) (**Figure 2–5**). Molecular nitrogen accounts for about 79 percent of our planet's atmosphere, but our cells ignore it completely. In fact, deep-sea divers live for long periods while breathing artificial air that does not contain nitrogen. (We discuss the reasons for eliminating nitrogen under these conditions in the Decompression Sickness Clinical Note in Chapter 23.)

Covalent bonds usually form molecules in which the outer energy levels of the atoms involved are filled. An atom, ion, or molecule that contains unpaired electrons in its outermost energy level is called a *free radical*. Free radicals are highly reactive. Almost as fast as it forms, a free radical enters additional reactions that are typically destructive. For example, free radicals can damage or destroy vital compounds, such as proteins. Evidence suggests that the cumulative damage from free radicals inside and outside our cells is a major factor in the aging process. Free radicals sometimes form in the course of normal metabolism, but cells have several methods of removing or inactivating them.

Nitric oxide (NO), however, is a free radical with several important functions in the body (**Figure 2–5**). It reacts readily with other atoms or molecules, but it is involved in chemical communication in the nervous system, in the control of blood vessel diameter, in blood clotting, and in the defense against bacteria and other pathogens (disease-causing organisms).

TIPS & TRICKS

Remember this mnemonic for the covalent bonding of hydrogen, oxygen, nitrogen, and carbon atoms: HONC 1234. **H**ydrogen shares *1* pair of electrons (H–), **o**xygen shares *2* pairs ($-$O$-$), **n**itrogen shares *3* pairs $\left(-\overset{|}{N}-\right)$, and **c**arbon shares *4* pairs $\left(-\overset{|}{\underset{|}{C}}-\right)$.

Nonpolar Covalent Bonds

Covalent bonds are very strong, because the shared electrons hold the atoms together. In typical covalent bonds the atoms remain electrically neutral, because each shared electron spends just as much time "at home" as away. (If you and a friend were tossing a pair of baseballs back and forth as fast as you could, on average, each of you would have just one baseball.)

Many covalent bonds involve an equal sharing of electrons. Such bonds are called **nonpolar covalent bonds**. They occur, for instance, between two atoms of the same type. Non-polar covalent bonds are very common. In fact, those involving carbon atoms form most of the structural components of the human body.

Polar Covalent Bonds

Covalent bonds involving different types of atoms may involve an unequal sharing of electrons, because the elements differ in how strongly they attract electrons. An unequal sharing of electrons creates a **polar covalent bond**. For example, in a molecule of water, an oxygen atom forms covalent bonds with two hydrogen atoms (**Figure 2–6**). The oxygen nucleus (with its eight protons) has a much stronger attraction for the shared electrons than the hydrogen atoms (each with a single proton). As a result, the electrons spend more time orbiting the oxygen nucleus than orbiting the hydrogen nuclei.

Because the oxygen atom has two extra electrons most of the time, it develops a slight (partial) negative charge, indicated by δ^-, as shown in **Figure 2–6b**. At the same time, each hydrogen atom develops a slight (partial) positive charge, δ^+, because its electron is away much of the time. (Suppose you and a friend were tossing a pair of baseballs back and forth, but one of you returned them as fast as possible while the

other held onto them for a while before throwing them back. One of you would now, on average, have more than one baseball, and the other would have less than one.)

The unequal sharing of electrons makes polar covalent bonds somewhat weaker than nonpolar covalent bonds. Polar covalent bonds often create *polar molecules*—molecules that have positive and negative ends. Water is the most important polar molecule in the body. We will consider its very interesting characteristics in a later section.

Hydrogen Bonds

Covalent and ionic bonds tie atoms together to form molecules and/or compounds. Other, comparatively weak forces also act between adjacent molecules, and even between atoms within a large molecule. The most important of these weak attractive forces is the hydrogen bond. A **hydrogen bond** is the attraction between a slight positive charge (δ^+) on the hydrogen atom of a polar covalent bond and a slight negative charge (δ^-) on an oxygen, nitrogen, or fluorine atom of another polar covalent bond. The polar covalent bond containing the oxygen, nitrogen, or fluorine atom can be in a different molecule from, or in the same molecule as, the hydrogen atom.

Hydrogen bonds are too weak to create molecules, but they can change the shapes of molecules or pull molecules closer together. For example, hydrogen bonding occurs between water molecules (**Figure 2–7**). At a water surface, this attraction between water molecules slows the rate of evaporation and creates what is known as *surface tension*. Surface tension acts as a barrier that keeps small objects from entering the water. For example, it allows insects to walk across the surface of a pond or puddle. Similarly, the surface tension in a layer of tears on the eye prevents dust particles from touching the surface of the eye. At the cellular level, hydrogen bonds affect the shapes and properties of complex molecules, such as proteins and nucleic acids (including DNA). Hydrogen bonds may also determine the three-dimensional relationships between molecules.

States of Matter

Most matter in our environment exists in one of three states: solid, liquid, or gas. *Solids* keep their volume and their shape at ordinary temperatures and pressures. A lump of granite, a brick, and a textbook are solid objects. *Liquids* have a constant volume, but no fixed shape. The shape of a liquid is determined by the shape of its container. Water, coffee, and soda are liquids. A *gas* has no constant volume and no fixed shape. Gases can be compressed or expanded, and unlike liquids they will fill a container of any size. The most familiar example is the air of our atmosphere.

Figure 2–6   Water Molecules Contain Polar Covalent Bonds.

Hydrogen atom

Hydrogen atom

Oxygen atom

a **Formation of a water molecule.** In forming a water molecule, an oxygen atom completes its outermost energy level by sharing electrons with a pair of hydrogen atoms. The sharing is unequal, because the oxygen atom holds the electrons more tightly than do the hydrogen atoms.

Hydrogen atom

Oxygen atom

δ^+

δ^+

$2\delta^-$

b **Charges on a water molecule.** Because the oxygen atom has two extra electrons much of the time, it develops a slight negative charge, and the hydrogen atoms become weakly positive. The bonds in a water molecule are polar covalent bonds.

2

Figure 2–7   **Hydrogen Bonds Form between Water Molecules.**
The hydrogen atoms of a water molecule have a slight positive charge,
and the oxygen atom has a slight negative charge (*see Figure 2–6b*).
The distances between these molecules have been exaggerated
for clarity.

KEY

●　Hydrogen

⬤　Oxygen

- - - -　Hydrogen bond

What determines whether a substance is a solid, a liquid, or a gas? A matter's state depends on the degree of interaction among its atoms or molecules. The particles of a solid are held tightly together, while those of a gas are very far apart. Water is the only substance that occurs as a solid (ice), a liquid (water), and a gas (water vapor) at temperatures compatible with life. Water exists as a liquid over a broad range of temperatures primarily because of hydrogen bonding among the water molecules. We will talk more about water's unusual properties in a later section.

CHECKPOINT

4.　Define chemical bond and identify several types of chemical bonds.

5.　Which kind of bond holds atoms in a water molecule together? What attracts water molecules to one another?

6.　Both oxygen and neon are gases at room temperature. Oxygen combines readily with other elements, but neon does not. Why?

See the blue Answers tab at the back of the book.

2-3 ┃ Decomposition, synthesis, and exchange reactions are important chemical reactions in physiology

Learning Outcome　Distinguish among the major types of chemical reactions that are important for studying physiology.

Cells stay alive and functional by controlling chemical reactions. In a **chemical reaction**, new chemical bonds form between atoms, or existing bonds between atoms are broken. These changes take place as atoms in the reacting substances, called **reactants**, are rearranged to form different substances, or **products** (**Spotlight Figure 2–3**).

In effect, each cell is a chemical factory. Growth, maintenance and repair, secretion, and contraction all involve complex chemical reactions. Cells use chemical reactions to provide the energy they need to maintain homeostasis and to perform essential functions. All of the reactions under way in the cells and tissues of the body at any given moment make up its **metabolism** (me-TAB-ō-lizm).

Basic Energy Concepts

An understanding of some basic relationships between matter and energy is helpful for any discussion of chemical reactions. **Work** is the movement of an object or a change in the physical structure of matter. In your body, work includes movements like walking or running, and also the synthesis of organic (carbon-containing) molecules and the conversion of liquid water to water vapor (evaporation). **Energy** is the capacity to do work, and movement or physical change cannot take place without energy. The two major types of energy are kinetic energy and potential energy.

1. **Kinetic energy** is the energy of motion—energy that can be transferred to another object and do work. When you fall off a ladder, it is kinetic energy that does the damage.

2. **Potential energy** is stored energy—energy that has the potential to do work. It may derive from an object's position (you standing on a ladder) or from its physical or chemical structure (a stretched spring or a charged battery).

Kinetic energy must be used in climbing the ladder, in stretching the spring, or in charging the battery. The resulting potential energy is converted back into kinetic energy when you descend, the spring recoils, or the battery discharges. The kinetic energy can then be used to perform work. For example, in an MP3 player, the chemical potential energy stored in small batteries is converted to kinetic energy that vibrates the sound-producing membranes in headphones or external speakers.

Energy cannot be destroyed. It can only be converted from one form to another. A conversion between potential energy and kinetic energy is never 100 percent efficient. Each time an energy exchange occurs, some of the energy is released in the form of heat. *Heat* is an increase in random molecular motion, and the temperature of an object is proportional to the average kinetic energy of its molecules. Heat can never be completely converted to work or any other form of energy, and cells cannot capture it or use it to do work.

Cells do work as they synthesize complex molecules and move materials into, out of, and within the cell. The cells of a skeletal muscle at rest, for example, contain potential energy in the form of the positions of protein filaments and the covalent bonds between molecules inside the cells. When a muscle contracts, it performs work. Potential energy is converted into kinetic energy, and heat is released. The amount of heat is proportional to the amount of work done. As a result, when you exercise, your body temperature rises.

Types of Chemical Reactions

Three types of chemical reactions are important to the study of physiology: decomposition reactions, synthesis reactions, and exchange reactions.

Decomposition Reactions

A **decomposition** reaction breaks a molecule into smaller fragments. Here is a diagram of a simple *decomposition reaction*:

$$AB \longrightarrow A + B$$

Decomposition reactions take place outside cells as well as inside them. For example, a typical meal contains molecules of fats, sugars, and proteins that are too large and too complex to be absorbed and used by your body. Decomposition reactions in the digestive tract break these molecules down into smaller fragments that can be absorbed.

Decomposition reactions involving water are important in the breakdown of complex molecules in the body. In **hydrolysis** (hī-DROL-i-sis; *hydro-*, water + *lysis*, a loosening), one of the bonds in a complex molecule is broken, and the components of a water molecule (H and OH) are added to the resulting fragments:

$$A - B + H_2O \longrightarrow A - H + HO - B$$

Collectively, the decomposition reactions of complex molecules within the body's cells and tissues are referred to as **catabolism** (ka-TAB-ō-lizm; *katabole*, a throwing down). When a covalent bond—a form of potential energy—is broken, it releases kinetic energy that can do work. By harnessing the energy released in this way, cells carry out vital functions such as growth, movement, and reproduction.

Synthesis Reactions

Synthesis (SIN-the-sis) is the opposite of decomposition. A *synthesis reaction* assembles smaller molecules into larger molecules. A simple synthetic reaction is diagrammed here:

$$A + B \longrightarrow AB$$

Synthesis reactions may involve individual atoms or the combination of molecules to form even larger products. The formation of water from hydrogen and oxygen molecules is a synthesis reaction. Synthesis always involves the formation of new chemical bonds, whether the reactants are atoms or molecules.

A **dehydration synthesis**, or *condensation reaction*, is the formation of a complex molecule by the removal of a water molecule:

$$A - H + HO - B \longrightarrow A - B + H_2O$$

Dehydration synthesis is the opposite of hydrolysis. We will see examples of both reactions in later sections.

Collectively, the synthesis of new molecules within the body's cells and tissues is known as **anabolism** (a-NAB-ō-lizm; *anabole*, a throwing upward). Anabolism is usually considered an "uphill" process because it takes energy to create a chemical bond (just as it takes energy to push something uphill). Cells must balance their energy budgets, with catabolism providing the energy to support anabolism and other vital functions.

TIPS&TRICKS

To remember the difference between anabolism (synthesis) and catabolism (breakdown), relate the terms to words you already know: *Anabolic* steroids are used to build up muscle tissue, while both *catastrophe* and *catabolism* involve destruction (breakdown).

Exchange Reactions

In an **exchange reaction**, parts of the reacting molecules are shuffled around to produce new products:

$$AB + CD \longrightarrow AD + CB$$

The reactants and products contain the same components (A, B, C, and D), but those components are present in different combinations. In an exchange reaction, the reactant molecules AB and CD must break apart (a decomposition) before they can interact with each other to form AD and CB (a synthesis).

Reversible Reactions

At least in theory, chemical reactions are reversible, so if $A + B \longrightarrow AB$, then $AB \longrightarrow A + B$. Many important biological reactions are freely reversible. Such reactions can be represented as an equation:

$$A + B \rightleftharpoons AB$$

This equation indicates that, in a sense, two reactions are taking place at the same time. One is a synthesis (A + B ⟶ AB) and the other is a decomposition (AB ⟶ A + B).

Recall from Chapter 1 that a state of *equilibrium* exists when opposing processes or forces are in balance. At equilibrium, the rates of the two reactions are in balance. As fast as one molecule of AB forms, another degrades into A + B.

What happens when equilibrium is disturbed—say, if you add more AB? In our example, the rate of the synthesis reaction is directly proportional to the frequency of encounters between A and B. In turn, the frequency of encounters depends on the degree of crowding. (You are much more likely to bump into another person in a crowded room than in a room that is almost empty.) So, adding more AB molecules will increase the rate of conversion of AB to A and B. The amounts of A and B will then increase, leading to an increase in the rate of the reverse reaction—the formation of AB from A and B. Eventually, a balance, or equilibrium, is again established.

TIPS&TRICKS

Jell-O provides an example of a physical reversible reaction. Once Jell-O has been refrigerated, the gelatin sets up and forms a solid, but if it sits without refrigeration for too long, it goes back to a liquid again.

CHECKPOINT

7. When using the rules for chemical notation, how is an ion's electrical charge represented?

8. Using the rules for chemical notation, write the molecular formula for glucose, a compound composed of 6 carbon atoms, 12 hydrogen atoms, and 6 oxygen atoms.

9. Identify and describe three types of chemical reactions important to human physiology.

10. In cells, glucose, a six-carbon molecule, is converted into two three-carbon molecules by a reaction that releases energy. How would you classify this reaction?

See the blue Answers tab at the back of the book.

2-4 ┃ Enzymes catalyze specific biochemical reactions by lowering the energy needed to start them

Learning Outcome Describe the crucial role of enzymes in metabolism.

Most chemical reactions do not take place spontaneously, or if they do, they occur so slowly that they would be of little value to living cells. Before a reaction can proceed, enough energy must be provided to activate the reactants. The amount of energy required to start a reaction is called the **activation energy**. Many reactions can be activated by changes in temperature or

acidity, but such changes are deadly to cells. For example, every day your cells break down complex sugars as part of your normal metabolism. Yet to break down a complex sugar in a laboratory, you must boil it in an acidic solution. Your cells don't have that option! Temperatures that high and solutions that corrosive would immediately destroy living tissues. Instead, your cells use special proteins called *enzymes* to perform most of the complex synthesis and decomposition reactions in your body.

Enzymes promote chemical reactions by lowering the activation energy required (**Figure 2–8**). In doing so, they make it possible for chemical reactions, such as the breakdown of sugars, to proceed under conditions compatible with life. Cells make enzyme molecules, each of which promotes a specific reaction.

Enzymes belong to a class of substances called **catalysts** (KAT-uh-lists; *katalysis*, dissolution), compounds that speed up chemical reactions without themselves being permanently changed or consumed. Enzymatic reactions, which are reversible, can be written as

$$A + B \underset{}{\overset{enzyme}{\rightleftharpoons}} AB$$

An appropriate enzyme can accelerate, or speed up, a reaction, but an enzyme affects only the *rate* of the reaction, not its direction or the products that are formed. An enzyme cannot bring about a reaction that would otherwise be impossible. Enzymatic reactions are generally reversible, and they proceed until equilibrium is reached.

The complex reactions that support life take place in a series of interlocking steps, each controlled by a specific enzyme. Such a reaction sequence is called a *metabolic pathway*. A synthetic pathway can be diagrammed as

$$A \overset{enzyme\ 1}{\underset{Step\ 1}{\longrightarrow}} B \overset{enzyme\ 2}{\underset{Step\ 2}{\longrightarrow}} C \overset{enzyme\ 3}{\underset{Step\ 3}{\longrightarrow}} \text{ and so on}.$$

Figure 2–8 Enzymes Lower Activation Energy. Enzymes lower the activation energy required for a chemical reaction to proceed readily (in order, from 1–4) under conditions in the body.

In many cases, the steps in the synthetic pathway differ from those in the decomposition pathway, and separate enzymes are often involved.

It takes activation energy to start a chemical reaction, but once it has begun, the reaction as a whole may absorb or release energy as it proceeds to completion. If the amount of energy released is greater than the activation energy needed to start the reaction, there will be a net release of energy. Reactions that release energy are said to be **exergonic** (*exo-*, outside + *ergon*, work). Exergonic reactions are relatively common in the body. They generate the heat that maintains your body temperature.

If more energy is required to begin the reaction than is released as it proceeds, the reaction as a whole will absorb energy. Such reactions are called **endergonic** (*endo-*, inside). The synthesis of molecules such as fats and proteins result from endergonic reactions.

CHECKPOINT

11. What is an enzyme?

12. Why are enzymes needed in our cells?

See the blue Answers tab at the back of the book.

2-5 | Inorganic compounds lack carbon, and organic compounds contain carbon

Learning Outcome Distinguish between organic compounds and inorganic compounds.

The human body is very complex, but it contains relatively few elements (Table 2–1, p. 54). Just knowing the identity and quantity of each element in the body will not help you understand the body any more than only memorizing the alphabet will help you understand this textbook. Just as 26 letters can be combined to form thousands of different words in this book, only about 26 elements combine to form thousands of different chemical compounds in our bodies. As we saw in Chapter 1, these compounds make up the living cells that form the framework of the body and carry on all its life processes. Learning about the major classes of chemical compounds will help you to understand the structure and function of the human body.

We now turn our attention to nutrients and metabolites. **Nutrients** are the substances from food that are necessary for normal physiological functions. Nutrients include carbohydrates, proteins, fats, vitamins, minerals, and water. **Metabolites** (me-TAB-ō-līts; *metabole*, change) are substances that are involved in, or a byproduct of, metabolism. We can broadly categorize nutrients and metabolites as either

inorganic or organic. **Inorganic compounds** generally do not contain carbon and hydrogen atoms as their primary structural ingredients. In contrast, carbon and hydrogen always form the basis for **organic compounds**.

The most important inorganic compounds in the body are (1) carbon dioxide, a byproduct of cell metabolism; (2) oxygen, an atmospheric gas required in important metabolic reactions; (3) water, which accounts for most of our body weight; and (4) inorganic acids, bases, and salts—compounds held together partially or completely by ionic bonds. In the next section, we focus on water, its properties, and how those properties establish the conditions necessary for life. Most of the other inorganic molecules and compounds in the body exist in association with water, the primary component of our body fluids. Both carbon dioxide and oxygen, for example, are gas molecules that are transported in body fluids. Also, all the inorganic acids, bases, and salts we will discuss are dissolved in body fluids.

CHECKPOINT

13. Compare organic compounds to inorganic compounds.

See the blue Answers tab at the back of the book.

2-6 | Physiological systems depend on water

Learning Outcome Explain how the chemical properties of water make life possible.

Water (H_2O) is the most important substance in the body. It makes up to two-thirds of total body weight. A change in the body's water content can be fatal because virtually all physiological systems will be affected.

Although water is familiar to everyone, it has some highly unusual properties. These properties are due to the hydrogen bonding between nearby water molecules.

1. *Solubility.* A remarkable number of inorganic and organic molecules are *soluble*, meaning they will dissolve or break up in water. The individual particles become distributed within the water, and the result is a **solution**—a uniform mixture of two or more substances. The liquid in which other atoms, ions, or molecules are distributed is called the **solvent**. The dissolved substances are the **solutes**. In *aqueous solutions*, water is the solvent. The solvent properties of water are so important that we will consider them further in the next section.

2. *Reactivity.* In our bodies, chemical reactions take place in water, but water molecules also take part in some reactions. Hydrolysis and dehydration synthesis are two examples noted earlier in the chapter.

3. *High Heat Capacity.* **Heat capacity** is the quantity of heat required to raise the temperature of a unit mass of a substance 1°C. Water has an unusually high heat capacity. Why? The reason is that water molecules in the liquid state are attracted to one another through hydrogen bonding. Important consequences of this attraction include the following:

■ The temperature of water must be quite high before all the hydrogen bonds are broken between individual water molecules and they have enough energy to break free and become water vapor, a gas. Therefore, water remains a liquid over a broad range of environmental temperatures, and the freezing and boiling points of water are far apart.

■ Water carries a great deal of heat away with it when it changes from a liquid to a gas. This feature explains the cooling effect of perspiration on the skin.

■ An unusually large amount of heat energy is required to change the temperature of 1 g of water by 1°C. As a result, a large mass of water changes temperature slowly. This property is called *thermal inertia.* Thermal inertia helps stabilize body temperature because water accounts for up to two-thirds of the weight of the human body.

4. *Lubrication.* Water is an effective lubricant because there is little friction between water molecules. So, even a thin layer of water between two opposing surfaces will greatly reduce friction between them. (That is why driving on wet roads can be tricky. Your tires may start sliding on a layer of water rather than maintaining contact with the road.) Within joints such as the knee, an aqueous solution prevents friction between the opposing surfaces. Similarly, a small amount of fluid in the body cavities prevents friction between internal organs, such as the heart or lungs, and the body wall. ⤺ p. 45

The Properties of Aqueous Solutions

Water's chemical structure makes it an unusually effective solvent. The covalent bonds in a water molecule are oriented so that the hydrogen atoms are fairly close together. As a result, the water molecule has positive and negative ends, or poles (Figure 2–9a). For this reason, a water molecule is called a **polar molecule**.

Many inorganic compounds are held together partly or completely by ionic bonds. In water, these compounds undergo **dissociation** (di-sō-sē-Ā-shun), or **ionization** (i-on-i-ZĀ-shun). In this process, ionic bonds are broken as the individual ions interact with the positive or negative ends of polar water molecules (Figure 2–9b). The result is a mixture of cations and anions surrounded by water molecules. The water molecules around each ion form a *hydration sphere*.

Figure 2–9 Water Molecules Surround Solutes in Aqueous Solutions.

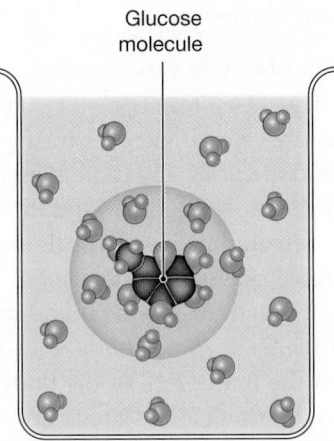

a Water molecule. In a water molecule, oxygen forms polar covalent bonds with two hydrogen atoms. Because both hydrogen atoms are at one end of the molecule, it has an uneven distribution of charges, creating positive and negative poles.

b Sodium chloride in solution. Ionic compounds, such as sodium chloride, dissociate in water as the polar water molecules break the ionic bonds in the large crystal structure. Each ion in solution is surrounded by water molecules, creating hydration spheres.

c Glucose in solution. Hydration spheres also form around an organic molecule containing polar covalent bonds. If the molecule binds water strongly, as does glucose, it will be carried into solution—in other words, it will dissolve. Note that the molecule does not dissociate, as occurs for ionic compounds.

Table 2–2	Important Electrolytes That Dissociate in Body Fluids	
Electrolyte		**Ions Released**
NaCl (sodium chloride)		$\rightarrow Na^+ + Cl^-$
KCl (potassium chloride)		$\rightarrow K^+ + Cl^-$
CaPO$_4$ (calcium phosphate)		$\rightarrow Ca^{2+} + PO_4^{2-}$
NaHCO$_3$ (sodium bicarbonate)		$\rightarrow Na^+ + HCO_3^-$
MgCl$_2$ (magnesium chloride)		$\rightarrow Mg^{2+} + 2Cl^-$
Na$_2$HPO$_4$ (sodium hydrogen phosphate)		$\rightarrow 2Na^+ + HPO_4^{2-}$
Na$_2$SO$_4$ (sodium sulfate)		$\rightarrow 2Na^+ + SO_4^{2-}$

An aqueous solution containing anions and cations will conduct an electrical current. When this happens, cations (+) move toward the negative side, and anions (–) move toward the positive side. Electrical forces across plasma membranes affect the functioning of all cells, and small electrical currents carried by ions are essential to muscle contraction and nerve function. In Chapters 10 and 12 we will discuss these processes in more detail.

Electrolytes and Body Fluids

Soluble inorganic substances whose ions will conduct an electrical current in solution are called **electrolytes** (e-LEK-trō-līts). Sodium chloride in solution is an electrolyte. The dissociation of electrolytes in blood and other body fluids releases a variety of ions. Table 2–2 lists important electrolytes and the ions released when they dissociate.

Changes in the concentrations of electrolytes in body fluids will disturb almost every vital function. For example, declining potassium levels will lead to a general muscular paralysis, and rising concentrations will cause weak and irregular heartbeats. The concentrations of ions in body fluids are carefully regulated, mostly by the coordination of activities at the kidneys (ion excretion), the digestive tract (ion absorption), and the skeletal system (ion storage or release).

Hydrophilic and Hydrophobic Compounds

Some organic molecules contain polar covalent bonds, which also attract water molecules. The hydration spheres that form may then carry these molecules into solution (Figure 2–9c). Molecules that interact readily with water molecules in this way are called **hydrophilic** (hī-drō-FIL-ik; *hydro-*, water + *philos*, loving). Glucose, an important soluble sugar, is one example.

Many other organic molecules have very few or no polar covalent bonds. Such molecules do not have positive and negative ends, and are said to be *nonpolar*. When nonpolar molecules are exposed to water, hydration spheres do not form

and the molecules do not dissolve. Molecules that do not readily interact with water are called **hydrophobic** (hī-drō-FŌB-ik; *hydro-*, water + *phobos*, fear). Fats and oils of all kinds are some of the most familiar hydrophobic molecules. For example, body fat deposits consist of large, hydrophobic droplets trapped in the watery interior of cells. Gasoline and heating oil are hydrophobic molecules not found in the body. When accidentally spilled into lakes or oceans, they form long-lasting oil slicks instead of dissolving.

TIPS & TRICKS

To distinguish between hydrophobic and hydrophilic, remember that a phobia is a fear of something, and that -philic ends with "lic," which resembles "like."

Colloids and Suspensions

Body fluids may contain large and complex organic molecules, such as proteins and protein complexes, that are held in solution by their association with water molecules (Figure 2–9c). A solution containing dispersed proteins or other large molecules is called a **colloid**. Liquid Jell-O is a familiar viscous (thick) colloid. The particles or molecules in a colloid will remain in solution indefinitely.

In contrast, a **suspension** contains large particles in solution, but if undisturbed, its particles will settle out of solution due to the force of gravity. For example, stirring beach sand into a bucket of water creates a temporary suspension that will last only until the sand settles to the bottom. Whole blood is another temporary suspension, because the blood cells are suspended in the blood plasma. If clotting is prevented, the cells in a blood sample will gradually settle to the bottom of the container. Measuring that settling rate, or "sedimentation rate," is a common laboratory test.

CHECKPOINT

14. **Explain how the chemical properties of water make life possible.**

See the blue Answers tab at the back of the book.

2-7 ❙ Body fluid pH is vital for homeostasis

Learning Outcome Discuss the importance of pH and the role of buffers in body fluids.

A hydrogen atom involved in a chemical bond or participating in a chemical reaction can easily lose its electron, to become a hydrogen ion, H^+. Hydrogen ions are extremely reactive in solution. In excessive numbers, they will break chemical bonds, change the shapes of complex molecules, and generally disrupt

2

cell and tissue functions. As a result, the concentration of hydrogen ions in body fluids must be regulated precisely.

A few hydrogen ions are normally present even in a sample of pure water, because some of the water molecules dissociate spontaneously, releasing cations and anions. The dissociation of water is a reversible reaction. We can represent it as:

$$H_2O \rightleftharpoons H^+ + OH^-$$

Notice that the dissociation of one water molecule yields a hydrogen ion (H^+) and a *hydroxide* (hī-DROK-sīd) *ion*, OH^-.

Very few water molecules ionize in pure water, so the number of hydrogen and hydroxide ions is small. The quantities are usually reported in moles, making it easy to keep track of the numbers of hydrogen and hydroxide ions. One liter of pure water contains about 0.0000001 mol of hydrogen ions and an equal number of hydroxide ions. In other words, the concentration of hydrogen ions in a solution of pure water is 0.0000001 mol per liter. This can be written as

$$[H^+] = 1 \times 10^{-7} \text{mol/L}$$

The brackets around the H^+ signify "the concentration of," another example of chemical notation.

The hydrogen ion concentration in body fluids is so important to physiological processes that we use a special shorthand to express it. The **pH** of a solution is defined as the negative logarithm of the hydrogen ion concentration in moles per liter. So instead of using the equation $[H^+] = 1 \times 10^{-7}$ mol/L, we say that the pH of pure water is –(–7), or 7.

Using pH values saves space, but always remember that the pH number is an *exponent* and that the pH scale is logarithmic. For instance, a pH of 6 ($[H^+] = 1 \times 10^{-6}$, or 0.000001 mol/L) means that the concentration of hydrogen ions is *10 times greater than* it is at a pH of 7 ($[H^+] = 1 \times 10^{-7}$, or 0.0000001 mol/L). The pH scale ranges from 0 to 14 (**Figure 2–10**).

Pure water has a pH of 7, but solutions display a wide range of pH values, depending on the nature of the solutes involved.

- A solution with a pH of 7 is said to be **neutral**, because it contains equal numbers of hydrogen and hydroxide ions.
- A solution with a pH below 7 is **acidic** (a-SI-dik), meaning that it contains more hydrogen ions than hydroxide ions.
- A pH above 7 is **basic**, or *alkaline* (AL-kuh-lin), meaning that it has more hydroxide ions than hydrogen ions.

The normal pH of blood ranges from 7.35 to 7.45. Abnormal fluctuations in pH can damage cells and tissues by breaking chemical bonds, changing the shapes of proteins, and altering cellular functions. *Acidosis* is an abnormal physiological state caused by low blood pH (below 7.35). A pH below 7 can produce coma. Likewise, *alkalosis* results from an abnormally high pH (above 7.45). A blood pH above 7.8 generally causes uncontrollable and sustained skeletal muscle contractions.

CHECKPOINT

15. **Define pH, and explain how the pH scale relates to acidity and alkalinity.**

16. **What is the significance of pH in physiological systems?**

See the blue Answers tab at the back of the book.

2-8 | Acids, bases, and salts are inorganic compounds with important physiological roles

Learning Outcome Describe the physiological roles of inorganic compounds.

The body contains both inorganic and organic *acids* and *bases* that may cause acidosis or alkalosis, respectively. An **acid** is any solute that dissociates in solution and releases hydrogen

Figure 2–10   The pH Scale Indicates Hydrogen Ion Concentration. The pH scale is logarithmic; an increase or decrease of one unit corresponds to a tenfold change in H^+ concentration.

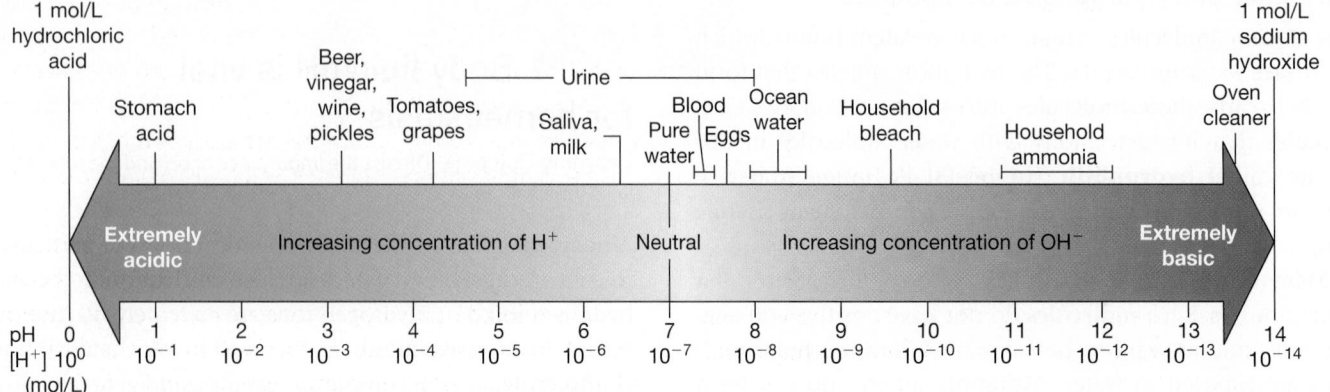

ions, lowering the pH. A hydrogen atom that loses its electron consists solely of a proton, so we often refer to hydrogen ions simply as protons, and to acids as *proton donors*.

A *strong acid* dissociates completely in solution, and the reaction occurs essentially in one direction only. *Hydrochloric acid* (HCl) is a representative strong acid. In water, it ionizes as follows:

$$HCl \longrightarrow H^+ + Cl^-$$

The stomach produces this powerful acid to help break down food. Hardware stores sell HCl under the name muriatic acid, for cleaning concrete and swimming pools.

A **base** is a solute that removes hydrogen ions from a solution, raising the pH. It acts as a *proton acceptor*. In solution, many bases release a hydroxide ion (OH^-). Hydroxide ions have an attraction for hydrogen ions and react quickly with them to form water molecules. A *strong base* dissociates completely in solution. *Sodium hydroxide*, NaOH, is a strong base. In solution, it releases sodium ions and hydroxide ions:

$$NaOH \longrightarrow Na^+ + OH^-$$

Strong bases have a variety of industrial and household uses. Drain openers (Drano) and lye are two familiar examples.

Weak acids and *weak bases* do not dissociate completely. At equilibrium, a significant number of molecules remains intact in the solution. For the same number of molecules in solution, weak acids and weak bases have less impact on pH than do strong acids and strong bases. *Carbonic acid* (H_2CO_3) is a weak acid found in body fluids. In solution, carbonic acid reversibly dissociates into a hydrogen ion and a *bicarbonate ion*, HCO_3^-:

$$H_2CO_3 \rightleftharpoons H^+ + HCO_3^-$$

Salts

A **salt** is an ionic compound containing any cation except a hydrogen ion, and any anion except a hydroxide ion. Because they are held together by ionic bonds, many salts dissociate completely in water, releasing cations and anions. For example, sodium chloride (table salt) dissociates immediately in water, releasing Na^+ and Cl^-. Sodium and chloride are the most abundant ions in body fluids. However, many other ions are present in lesser amounts as a result of the dissociation of other inorganic compounds. Ionic concentrations in the body are regulated in ways we will describe in Chapters 26 and 27.

The ionization of sodium chloride does not affect the local concentrations of hydrogen ions or hydroxide ions. For this reason, NaCl, like many salts, is a "neutral" solute. It does not make a solution more acidic or more basic. Through their interactions with water molecules, however, other salts may indirectly affect the concentrations of H^+ and OH^- ions. Thus, the dissociation of some salts makes a solution slightly acidic or slightly basic.

Buffers and pH Control

Buffers are compounds that stabilize the pH of a solution by removing or replacing hydrogen ions. *Buffer systems* usually involve a weak acid and its related salt, which functions as a weak base. For example, the carbonic acid–bicarbonate buffer system (detailed in Chapter 27) consists of carbonic acid and sodium bicarbonate, $NaHCO_3$, otherwise known as baking soda. Buffers and buffer systems in body fluids help maintain the pH within normal limits. The pH of several body fluids is included in **Figure 2–10**.

The use of antacids such as Alka-Seltzer is one example of the type of reaction that takes place in buffer systems. Alka-Seltzer uses sodium bicarbonate to neutralize excess hydrochloric acid in the stomach.

Note that the effects of neutralization are most evident when you add a strong acid to a strong base. For example, by adding hydrochloric acid to sodium hydroxide, you neutralize both the strong acid and the strong base.

$$HCl + NaOH \longrightarrow H_2O + NaCl$$

This neutralization reaction produces water and a salt—in this case, the neutral salt sodium chloride.

> ### CHECKPOINT
>
> 17. **Define the following terms: acid, base, and salt.**
>
> 18. **How does an antacid help decrease stomach discomfort?**
>
> See the blue Answers tab at the back of the book.

2-9 | Carbohydrates contain carbon, hydrogen, and oxygen in a 1:2:1 ratio

Learning Outcome Discuss the structures and functions of carbohydrates.

Carbohydrates are one type of organic compound. Organic compounds always contain the elements carbon and hydrogen, and generally oxygen as well. Many organic molecules are made up of long chains of carbon atoms linked by covalent bonds. The carbon atoms typically form additional covalent bonds with hydrogen or oxygen atoms and, less commonly, with nitrogen, phosphorus, sulfur, iron, or other elements.

Many organic molecules are soluble in water. Our previous discussion focused on inorganic acids and bases, but there are also important organic acids and bases. For example, active muscle tissues generate *lactic acid*, an organic acid. It must be neutralized by the carbonic acid–bicarbonate buffer system to prevent a potentially dangerous pH decline in body fluids.

Organic compounds are diverse, but certain groupings of atoms occur again and again, even in very different types of molecules. These *functional groups* greatly influence the

Table 2–3 Important Functional Groups of Organic Compounds

Functional Group	Structural Formula*	Importance	Examples
Amino group —NH_2	H $\mid$ $R—N$ $\mid$ H	Acts as a base, accepting H^+, depending on pH; can form bonds with other molecules	Amino acids
Carboxyl group —COOH	OH $\mid$ $R—C=O$	Acts as an acid, releasing H^+ to become $R—COO^-$	Fatty acids, amino acids
Hydroxyl group —OH	$R—O—H$	May link molecules through dehydration synthesis (condensation); hydrogen bonding between hydroxyl groups and water molecules affects solubility	Carbohydrates, fatty acids, amino acids
Phosphate group —PO_4	O $\parallel$ $R—O—P—O^-$ $\mid$ O^-	May link other molecules to form larger structures; may store energy in high-energy bonds	Phospholipids, nucleic acids, high-energy compounds

*A structural formula shows the covalent bonds within a molecule or functional group. The letter R represents the term *R group* and is used to denote the rest of the molecule to which a functional group is attached.

properties of any molecule they are in. Table 2–3 details the functional groups you will study in this chapter.

A **carbohydrate** is an organic molecule that contains carbon, hydrogen, and oxygen in a ratio near 1:2:1. Familiar carbohydrates include the *sugars* and *starches* that make up about half of the typical U.S. diet. Carbohydrates typically account for less than 1 percent of total body weight. Carbohydrates are most important as energy sources that are catabolized. We will focus on *monosaccharides*, *disaccharides*, and *polysaccharides*.

Monosaccharides

A **monosaccharide** (mon-ō-SAK-uh-rīd; *mono-*, single + *sakcharon*, sugar), or *simple sugar*, is a carbohydrate with three to seven carbon atoms. A monosaccharide can be called a *triose* (three-carbon), *tetrose* (four-carbon), *pentose* (five-carbon), *hexose* (six-carbon), or *heptose* (seven-carbon). The hexose **glucose** (GLOO-kōs), $C_6H_{12}O_6$, is the most important metabolic "fuel" in the body. As shown by diagrams known as *structural formulas*, the atoms in a glucose molecule may form either a straight chain (Figure 2–11a) or a ring (Figure 2–11b). In the body, the ring form is more common. A three-dimensional model shows the arrangement of atoms in the ring form most clearly (Figure 2–11c).

The three-dimensional structure of an organic molecule is an important characteristic, because it usually determines the molecule's fate or function. Some molecules have the same molecular formula—in other words, the same types and numbers of atoms—but different structures. Such molecules are called **isomers**. The body usually treats different

Figure 2–11 **The Structures of Glucose.**

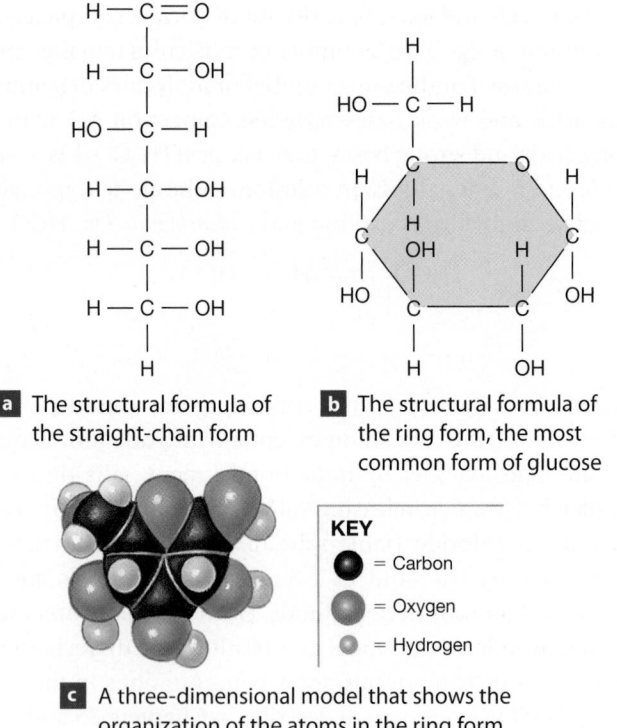

a The structural formula of the straight-chain form

b The structural formula of the ring form, the most common form of glucose

KEY
● = Carbon
● = Oxygen
● = Hydrogen

c A three-dimensional model that shows the organization of the atoms in the ring form

isomers as distinct molecules. For example, the monosaccharides glucose and fructose are isomers. *Fructose* is a hexose found in many fruits and in secretions of the male reproductive tract. It has the same chemical formula as glucose, $C_6H_{12}O_6$, but the arrangement of its atoms differs from that of glucose. As a result, separate enzymes and reaction sequences control its breakdown and synthesis. Monosaccharides such as glucose and fructose dissolve readily in water and are rapidly distributed throughout the body by blood and other body fluids.

Disaccharides and Polysaccharides

Carbohydrates other than simple sugars are complex molecules composed of monosaccharide building blocks. Two monosaccharides joined together form a **disaccharide** (dī-SAK-uh-rīd; *di-*, two). Disaccharides such as *sucrose* (table sugar) have a sweet taste and, like monosaccharides, are quite soluble in water.

The formation of sucrose involves a dehydration synthesis, or condensation reaction (**Figure 2–12a**). Dehydration synthesis reactions link molecules together by the removal of a water molecule. The breakdown of sucrose into simple sugars is an example of hydrolysis, or breakdown by the addition of a water molecule (**Figure 2–12b**). Hydrolysis is the functional opposite of dehydration synthesis.

Many foods contain disaccharides, but all carbohydrates except monosaccharides must be broken apart through hydrolysis before they can provide useful energy. Most popular junk foods (high in calories but otherwise lacking in nutritional content), such as candies and sodas, abound in monosaccharides (commonly fructose) and disaccharides (generally sucrose). Some people cannot tolerate sugar for medical reasons. Others avoid it in an effort to control their weight (because excess sugars are converted to fat for long-term storage). Many of these people use *artificial sweeteners* in their foods and beverages. These compounds have a very sweet taste, but they either cannot be broken down in the body or are used in insignificant amounts.

More complex carbohydrates result when repeated dehydration synthesis reactions add additional monosaccharides or disaccharides. These large molecules are called **polysaccharides** (pol-ē-SAK-uh-rīdz; *poly-*, many). Polysaccharide chains can be straight or highly branched. *Cellulose*, a structural component of many plants, is a polysaccharide that our bodies cannot digest because the particular linkages between the glucose molecules cannot be cleaved by enzymes in the body. Foods such as celery, which contains cellulose, water, and little else, contribute bulk to digestive wastes but are useless as a source of energy.

Starches are large polysaccharides formed from glucose molecules. Most starches are manufactured by plants. Your digestive tract can break these molecules into monosaccharides. Starches such as those in potatoes and grains are a major dietary energy source.

The polysaccharide **glycogen** (GLĪ-kō-jen), or *animal starch*, has many side branches consisting of chains of glucose molecules (**Figure 2–13**). Like most other starches, glycogen does not dissolve in water or other body fluids. Muscle cells make and store glycogen. When muscle cells have a high demand for glucose, glycogen molecules are broken down. When the need is low, these cells absorb glucose from the bloodstream and rebuild glycogen reserves. Table 2–4 (on the next page) summarizes information about carbohydrates.

Figure 2–12 **The Formation and Breakdown of Complex Sugars.** Enzymes perform both these reactions.

a **Formation of the disaccharide sucrose through dehydration synthesis.** During dehydration synthesis, two molecules are joined by the removal of a water molecule.

b **Breakdown of sucrose into simple sugars by hydrolysis.** Hydrolysis reverses the steps of dehydration synthesis; a complex molecule is broken down by the addition of a water molecule.

Table 2–4	Carbohydrates in the Body		
Structural Class	**Examples**	**Primary Function**	**Remarks**
Monosaccharides (simple sugars)	Glucose, fructose	Energy source	Manufactured in the body and obtained from food; distributed in body fluids
Disaccharides	Sucrose, lactose, maltose	Energy source	Sucrose is table sugar, lactose is in milk, and maltose is malt sugar found in germinating grain; all must be broken down to monosaccharides before absorption
Polysaccharides	Glycogen	Storage of glucose	Glycogen is in animal cells; other starches and cellulose are within or around plant cells

Figure 2–13   The Structure of the Polysaccharide Glycogen.
Liver and muscle cells store glucose as the polysaccharide glycogen, a long, branching chain of glucose molecules. Note that this figure uses a kind of shorthand for representing a carbon ring structure: At five corners of each hexagon is a carbon atom. An oxygen atom occupies the remaining corner in each glucose ring.

Glucose molecules

CHECKPOINT

19. **A food contains organic molecules with the elements C, H, and O in a ratio of 1:2:1. What class of compounds do these molecules belong to, and what are their major functions in the body?**

See the blue Answers tab at the back of the book.

2-10 ∎ Lipids often contain a carbon-to-hydrogen ratio of 1:2

Learning Outcome   Discuss the structures and functions of lipids.

Like carbohydrates, **lipids** (*lipos*, fat) contain carbon, hydrogen, and oxygen, and the carbon-to-hydrogen ratio is near 1:2. However, lipids contain much less oxygen than do carbohydrates with the same number of carbon atoms. The hydrogen-to-oxygen ratio is therefore very large. For example, a representative lipid, such as lauric acid (found in coconut, laurel, and palm kernel oils), has a formula of $C_{12}H_{24}O_2$. Lipids may also contain small quantities of phosphorus, nitrogen, or sulfur. Familiar lipids include *fats*, *oils*, and *waxes*. Most lipids are insoluble in water, but special transport mechanisms carry them into the bloodstream.

Lipids form essential structural components of all cells. In addition, lipid deposits are important as energy reserves. On average, lipids provide twice as much energy as carbohydrates do, gram for gram, when broken down in the body. When the supply of lipids exceeds the demand for energy, the excess is stored in fat deposits. For this reason, there has been great interest in developing *fat substitutes* that provide less energy, but have the same desirable taste and texture as the fats found in many foods.

Lipids normally make up 12–18 percent of the total body weight of adult men, and 18–24 percent for adult women. Many kinds of lipids exist in the body. We will consider five classes of lipids: *fatty acids, eicosanoids, glycerides, steroids*, and *phospholipids and glycolipids*.

Fatty Acids

Fatty acids are long carbon chains with hydrogen atoms attached. One end of the carbon chain is always attached to a *carboxyl* (kar-BOK-sil) *group*, —COOH (Table 2–3). The name *carboxyl* should help you remember that a carbon and a hydroxyl (—OH) group are the important structural features of fatty acids. The carbon chain attached to the carboxyl group is known as the hydrocarbon *tail* of the fatty acid. **Figure 2–14a** shows a representative fatty acid, *lauric acid*.

Fatty acids have a very limited solubility in water. When a fatty acid is in solution, only the carboxyl end associates with water molecules, because that is the only hydrophilic portion of the molecule. The hydrocarbon tail is hydrophobic. In general, the longer the hydrocarbon tail, the lower the solubility of the molecule.

Fatty acids may be either saturated or unsaturated (Figure 2–14b). These terms refer to the number of hydrogen atoms bound to the carbon atoms in the hydrocarbon tail. In a *saturated* fatty acid, each carbon atom in the tail has four single covalent bonds (Figure 2–14a). Within the tail, two

of those bonds bind adjacent carbon atoms, and the other two bind hydrogen atoms. The carbon atom at the end of the tail binds three hydrogen atoms. In an *unsaturated* fatty acid, one or more of the single covalent bonds between the carbon atoms have been replaced by a double covalent bond. As a result, the carbon atoms involved will each bind only one hydrogen atom rather than two. This changes the shape of the hydrocarbon tail, giving it a sharp bend, as you can see in Figure 2–14b. The change also affects the way the fatty acid is metabolized. A *monounsaturated* fatty acid has a single double bond in the hydrocarbon tail. A *polyunsaturated* fatty acid contains two or more double bonds.

Eicosanoids

Eicosanoids (ī-KŌ-sa-noydz) are lipids derived from *arachidonic* (ah-rak-i-DON-ik) *acid*, a fatty acid that must be absorbed in the diet because the body cannot synthesize it. The two major classes of eicosanoids are *leukotrienes* and *prostaglandins*. Leukotrienes (lū-kō-TRĪ-ēnz) are produced mostly by cells involved with coordinating the responses to injury or disease. We will consider leukotrienes in Chapters 18 and 22.

Prostaglandins (pros-tuh-GLAN-dinz) are short-chain fatty acids in which five of the carbon atoms are joined in a ring (Figure 2–15). These compounds are released by cells to coordinate or direct local cellular activities, and they are extremely powerful even in small quantities. Virtually all tissues synthesize and respond to them. The effects of prostaglandins vary with their structure and their release site. Prostaglandins released by damaged tissues, for example, stimulate nerve endings and produce the sensation of pain (Chapter 15). Those released in the uterus help trigger the start of labor contractions (Chapter 29).

The body uses several types of chemical messengers. Those that are produced in one part of the body and have effects on distant parts are called *hormones*. Hormones are distributed throughout the body in the bloodstream, but most prostaglandins affect only the area in which they are produced. As a result, prostaglandins are often called *local hormones*. The distinction is not a rigid one, however, as some prostaglandins also enter the bloodstream and affect other areas. We will discuss hormones and prostaglandins in Chapter 18.

Figure 2–14 Fatty Acids.

Lauric acid (C$_{12}$H$_{24}$O$_2$)

a Lauric acid demonstrates two structural characteristics common to all fatty acids: a long chain of carbon atoms and a carboxyl group (—COOH) at one end.

Saturated

Unsaturated

b A fatty acid is either saturated (has single covalent bonds only) or unsaturated (has one or more double covalent bonds). The presence of a double bond causes a sharp bend in the molecule.

Figure 2–15 Prostaglandins. Prostaglandins are unusual short-chain fatty acids.

2

Clinical Note

Fatty Acids and Health Humans love fatty foods. Unfortunately, a diet containing large amounts of saturated fats has been shown to increase the risk of heart disease and other cardiovascular problems. Saturated fats contain only saturated fatty acids. These fats are found in such popular foods as fatty meat and dairy products (including such favorites as butter, cheese, and ice cream).

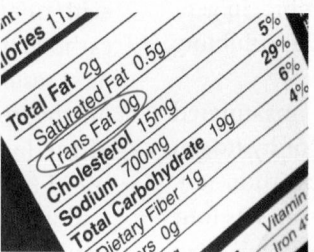

Vegetable oils contain a mixture of monounsaturated and polyunsaturated fatty acids. Recent studies indicate that monounsaturated fats may be more effective than polyunsaturated fats in lowering the risk of heart disease. According to current research, perhaps the healthiest choices are olive and canola oils. These oils contain particularly abundant quantities of oleic acid, an 18-carbon monounsaturated fatty acid. Surprisingly, compounds called *trans* fatty acids, produced from polyunsaturated oils during the manufacturing of some margarines and vegetable shortenings, appear to increase the risk of heart disease. U.S. Food and Drug Administration (FDA) guidelines now require that *trans* fatty acids be declared in the nutrition label of foods and dietary supplements.

The Inuit people of the Arctic regions have lower rates of heart disease than do other populations, even though the typical Inuit diet is high in fats and cholesterol. Interestingly, the main fatty acids in the Inuit diet are omega-3s. These fatty acids have an unsaturated bond three carbons before the last (or omega) carbon, a position known as "omega minus 3." Fish flesh and fish oils, a substantial portion of the Inuit diet, contain an abundance of omega-3 fatty acids. Why does the presence of omega-3 fatty acids in the diet reduce the risks of heart disease, rheumatoid arthritis, and other inflammatory diseases? The search for an answer provides research topics of great interest.

Glycerides

Unlike monosaccharides, individual fatty acids cannot be strung together in a chain by dehydration synthesis. But they can be attached to a modified simple sugar, **glycerol** (GLIS-er-ol), through a similar reaction. The result is a lipid known as a **glyceride** (GLIS-er-īd). Dehydration synthesis can produce a **monoglyceride** (mon-ō-GLI-ser-īd), consisting of glycerol plus one fatty acid. Subsequent reactions can yield a **diglyceride** (glycerol + two fatty acids) and then a **triglyceride** (glycerol + three fatty acids), as in **Figure 2–16**. Hydrolysis breaks the glycerides into fatty acids and glycerol. Comparing **Figure 2–16** with **Figure 2–12** shows that dehydration synthesis and hydrolysis operate the same way, whether the molecules involved are carbohydrates or lipids.

Triglycerides, also known as *triacylglycerols* or *neutral fats*, have three important functions.

1. *Energy Source.* Fat deposits in specialized sites of the body represent a significant energy reserve. In times of need, the triglycerides are taken apart by hydrolysis, yielding fatty acids that can be broken down to provide energy.

2. *Insulation.* Fat deposits under the skin serve as insulation, slowing heat loss to the environment. Heat loss across a layer of lipids is only about one-third of the heat loss through other tissues.

3. *Protection.* A fat deposit around a delicate organ such as a kidney provides a cushion that protects against bumps or jolts.

Triglycerides are stored in the body as lipid droplets within cells. The droplets absorb and accumulate lipid-soluble vitamins, drugs, or toxins that appear in body fluids. This accumulation has both positive and negative effects. For example, the body's lipid reserves retain both valuable lipid-soluble vitamins (A, D, E, K) and potentially dangerous lipid-soluble pesticides, such as the now-banned DDT.

Steroids

Steroids are large lipid molecules that share a distinctive four-ring carbon framework (**Figure 2–17**). They differ in the functional groups that are attached to this basic structure. The steroid **cholesterol** (koh-LES-ter-ol; *chole-*, bile + *stereos*, solid) and related steroids are important for several reasons:

- The outer boundary of all animal cells, called a plasma membrane, contains cholesterol (**Figure 2–17a**). Cells need cholesterol to maintain their plasma membranes, as well as for cell growth and division.

- Steroid hormones are involved in the regulation of sexual function. Examples include the sex hormones, *estrogen* and *testosterone* (**Figure 2–17b,c**).

Figure 2–16 **Triglyceride Formation.** The formation of a triglyceride involves the attachment of fatty acids to a glycerol molecule through dehydration synthesis. In this example, a triglyceride is formed by the attachment of one unsaturated and two saturated fatty acids to a glycerol molecule.

Figure 2–17 **Steroids Have a Complex Four-Ring Structure.** Individual steroids differ in the side chains attached to the carbon rings.

a Cholesterol

b Estrogen **c** Testosterone

Current nutritional advice suggests limiting cholesterol intake to less than 300 mg per day. This amount represents a 40 percent reduction for the average adult in the United States. Unfortunately, blood cholesterol levels can be difficult to control by dietary restriction alone because the body can synthesize cholesterol as well. In fact, the body makes more than enough, so strict vegetarians do not need to eat animal products to ensure adequate amounts of cholesterol.

Phospholipids and Glycolipids

Phospholipids (FOS-fō-lip-idz) and **glycolipids** (GLĪ-kō-lip-idz) are structurally related, and our cells can synthesize both types of lipids, primarily from fatty acids. In a *phospholipid*, a *phosphate group* (PO_4^{3-}) links a diglyceride to a non-lipid group (**Figure 2–18a**). In a *glyco*lipid, a carbohydrate is attached to a diglyceride (**Figure 2–18b**). Note that placing *-lipid* last in these names indicates that the molecule consists primarily of lipid.

The long hydrocarbon tails of phospholipids and glycolipids are hydrophobic, but the opposite ends, the nonlipid *heads*, are hydrophilic. In water, large numbers of these molecules tend to form droplets, or *micelles* (mī-SELZ), with the hydrophilic portions on the outside (**Figure 2–18c**). Most meals contain a mixture of lipids and other organic molecules, and micelles form as the food breaks down in your digestive tract. In addition to phospholipids and glycolipids, micelles may

● Steroid hormones are important in the regulation of tissue metabolism and mineral balance. Examples include *corticosteroids* from the adrenal cortex, which play a role in carbohydrate and protein metabolism, and *calcitriol* from the kidneys, a hormone important in the regulation of the body's calcium ion concentrations.

● Steroid derivatives called *bile salts* are required for the normal processing of dietary fats. The liver produces bile salts and secretes them in bile. They interact with lipids in the intestinal tract and assist the digestion and absorption of lipids.

The body obtains cholesterol in two ways: (1) by absorbing it from animal products in the diet and (2) by synthesizing it. Liver, meat, shellfish, and egg yolks are especially rich dietary sources of cholesterol. A diet high in cholesterol can be harmful, because a strong link exists between high blood cholesterol levels (called *hypercholesterolemia*) and heart disease.

Figure 2–18 **Phospholipids and Glycolipids.**

a The phospholipid *lecithin*. In a phospholipid, a phosphate group links a nonlipid molecule to a diglyceride.

b In a glycolipid, a carbohydrate is attached to a diglyceride.

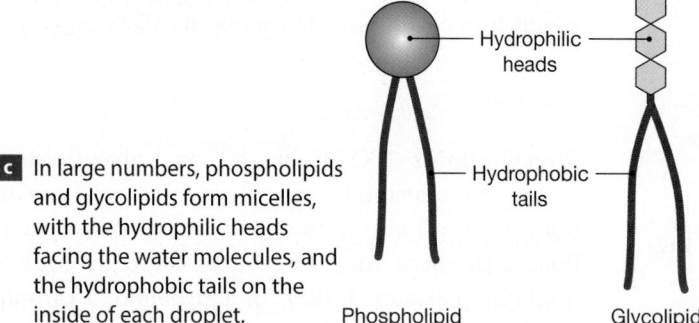

c In large numbers, phospholipids and glycolipids form micelles, with the hydrophilic heads facing the water molecules, and the hydrophobic tails on the inside of each droplet.

contain other insoluble lipids, such as steroids, glycerides, and long-chain fatty acids.

Cholesterol, phospholipids, and glycolipids are called *structural lipids*, because they help form and maintain intracellular structures called membranes. At the cellular level, *membranes* are sheets or layers composed mainly of hydrophobic lipids. For example, the plasma membrane surrounding each cell is composed primarily of phospholipids. It separates the aqueous solution inside the cell from the aqueous solution outside the cell. Also, various internal membranes subdivide the interior of the cell into specialized compartments, each with a distinctive chemical nature and, as a result, a different function.

The five types of lipids and their characteristics are summarized in Table 2–5.

Table 2–5	Representative Lipids and Their Functions in the Body		
Lipid Type	**Example(s)**	**Primary Functions**	**Remarks**
Fatty acids	Lauric acid	Energy source	Absorbed from food or synthesized in cells; transported in the blood
Eicosanoids	Prostaglandins, leukotrienes	Chemical messengers coordinating local cellular activities	Prostaglandins are produced in most body tissues
Glycerides	Monoglycerides, diglycerides, triglycerides	Energy source, energy storage, insulation, and physical protection	Stored in fat deposits; must be broken down to fatty acids and glycerol before they can be used as an energy source
Steroids	Cholesterol	Structural component of plasma membranes, hormones, digestive secretions in bile	All have the same four-carbon ring framework
Phospholipids, glycolipids	Lecithin (a phospholipid)	Structural components of plasma membranes	Derived from fatty acids and nonlipid components

CHECKPOINT

20. Describe lipids.

21. Which lipids would you find in human plasma membranes?

See the blue Answers tab at the back of the book.

2-11 | Proteins contain carbon, hydrogen, oxygen, and nitrogen and are formed from amino acids

Learning Outcome Discuss the structures and functions of proteins.

Proteins are the most abundant organic molecules in the human body and in many ways the most important. The human body contains many different proteins, and they account for about 20 percent of total body weight. All proteins contain carbon, hydrogen, oxygen, and nitrogen. Smaller quantities of sulfur and phosphorus may also be present. Amino acids (discussed shortly) are simple organic compounds that combine to form proteins.

Proteins carry out a variety of essential functions, which we can classify in seven major categories.

1. *Support.* Structural proteins create a three-dimensional framework for the body. They provide strength, organization, and support for cells, tissues, and organs.

2. *Movement.* Contractile proteins bring about muscular contraction. Related proteins are responsible for the movement of individual cells.

3. *Transport.* Special transport proteins bind many substances for transport in the blood, including insoluble lipids, respiratory gases, special minerals such as iron, and several hormones. These substances would not otherwise be transported in the blood. Other specialized proteins move materials from one part of a cell to another.

4. *Buffering.* Proteins provide a *buffering* action and in this way help prevent dangerous changes in the pH of body fluids.

5. *Metabolic Regulation.* Many proteins are *enzymes,* which speed up chemical reactions in cells. The sensitivity of enzymes to environmental factors such as temperature and pH is extremely important in controlling the pace and direction of metabolic reactions.

6. *Coordination and Control.* Protein *hormones* can influence the metabolic activities of every cell in the body or affect the function of specific organs or organ systems.

7. *Defense.* Proteins defend the body in many ways. The tough, waterproof proteins of the skin, hair, and nails protect the body from environmental hazards. Proteins called *antibodies* help protect us from disease by taking part in the *immune response.* Special *clotting proteins* restrict bleeding after an injury.

Protein Structure

Proteins consist of long chains of organic molecules called **amino acids**. Twenty different amino acids occur in significant quantities in the body. All 20 amino acids are small, water-soluble molecules. A typical protein contains 1000 amino acids. The largest protein complexes have 100,000 or more.

Each amino acid consists of five parts (**Figure 2–19**):

- a central carbon atom
- a hydrogen atom
- an *amino group* ($-NH_2$)
- a *carboxyl group* ($-COOH$)
- an *R group* (a variable *side chain* of one or more atoms)

The name *amino acid* refers to the presence of the *amino* group and the carboxyl group, which all amino acids have in common. At physiological pH levels, the carboxyl group can act as

2

Figure 2–19   Amino Acids. Each amino acid consists of a central carbon atom to which four different groups are attached: a hydrogen atom, an amino group ($-NH_2$), a carboxyl group ($-COOH$), and a variable side group designated as R.

Structure of an Amino Acid

- Amino group
- Central carbon
- Carboxyl group
- R group (variable side chain of one or more atoms)

Figure 2–20   The Formation of Peptide Bonds. Peptides form as dehydration synthesis creates a peptide bond between the carboxyl group of one amino acid and the amino group of another. In this example, a peptide bond links the amino acids glycine (for which R = H) and alanine (R = CH_3) to form a dipeptide.

Peptide Bond Formation

Peptide bond

an acid by releasing a hydrogen ion to become a *carboxyl ion* (COO^-). The amino group can act as a base by accepting a hydrogen ion, to become an amino ion ($-NH_3^+$). The result is a molecule, which has both positive and negative charges, but a net charge of zero. Such molecules are called *zwitterions,* derived from the German word that means "hybrid." The different atoms of the R groups distinguish one amino acid from another, giving each its own chemical properties. For example, different R groups are polar, nonpolar, or electrically charged.

A protein begins to form as amino acids are strung together into long chains. **Figure 2–20** shows how dehydration synthesis can link two representative amino acids: *glycine* and *alanine.* This reaction creates a covalent bond between the carboxyl group of one amino acid and the amino group of another. Such a bond is known as a **peptide bond**. Molecules consisting of amino acids held together by peptide bonds are called **peptides**. The molecule created in this example is called a *dipeptide,* because it contains two amino acids.

The chain can be lengthened by the addition of more amino acids. Attaching a third amino acid produces a *tripeptide.* Tripeptides and larger peptide chains are called **polypeptides**. Polypeptides with more than 100 amino acids are usually called proteins. Familiar proteins include *hemoglobin* in red blood cells and *keratin* in fingernails and hair.

There are numerous amino acids, but only 20 form proteins. Amino acids with nonpolar R groups are hydrophobic, while amino acids with polar R groups that form hydrogen bonds with water are hydrophilic. Amino acids with electrically charged R groups are strongly hydrophilic. The properties of the R groups contribute to the overall shape and function of proteins.

Protein Shape

The characteristics of a particular protein are determined in part by the R groups on its amino acids. But the properties of a protein are more than just the sum of the properties of its

parts, for polypeptides can have highly complex shapes that are important to their function. Proteins can have four levels of structural complexity (**Figure 2–21**).

1. **Primary structure** is the sequence of amino acids along the length of a single polypeptide (**Figure 2–21a**). Peptide bonds are responsible for the primary structure of proteins.

2. **Secondary structure** is the shape that results from the presence of hydrogen bonds between atoms at different parts of the polypeptide chain. Hydrogen bonding may create either an *alpha helix* (simple spiral) or *beta sheet* (a flat pleated sheet) (**Figure 2–21b**). Which one forms depends on where hydrogen bonding takes place between the sequence of amino acids in the polypeptide chain. The alpha helix is the more common form, but a given polypeptide chain may have both helical and pleated sections. Ribbon diagrams are used to represent the three-dimensional structure of proteins. An alpha helix is shown as a coiled ribbon, beta sheets as arrows, and less-structured polypeptide chains as round tubes.

3. **Tertiary structure** is the complex coiling and folding that gives a protein its final three-dimensional shape (**Figure 2–21c**). Tertiary structure results primarily from hydrophobic and hydrophilic interactions between the R groups of the polypeptide chain and the surrounding

Figure 2–21 Protein Structure.

a **Primary structure**. The primary structure of a polypeptide is the sequence of amino acids (A1, A2, A3, and so on) along its length.

Linear chain of amino acids

Hydrogen bond

Hydrogen bond

Alpha helix

Beta sheet

OR

b **Secondary structure**. Secondary structure is primarily the result of hydrogen bonding along the length of the polypeptide chain. Such bonding often produces a simple spiral, called an alpha helix (α helix) or a flattened arrangement known as a beta sheet (β sheet).

Alpha helix

Heme units

c **Tertiary structure**. Tertiary structure is the coiling and folding of a polypeptide. Within the cylindrical segments of this globular protein, the polypeptide chain is arranged in an alpha helix.

Hemoglobin (globular protein)

OR

Collagen (fibrous protein)

d **Quaternary structure**. Quaternary structure develops when separate polypeptide subunits interact to form a larger molecule. A single hemoglobin molecule contains four globular subunits. Hemoglobin transports oxygen in the blood; the oxygen binds reversibly to the heme units. In collagen, three helical polypeptide subunits intertwine. Collagen is the principal extracellular protein in most organs.

water molecules, and to a lesser extent from interactions between the R groups of amino acids in different parts of the molecule. Most such interactions are relatively weak. One, however, is very strong: the *disulfide bond*, a covalent bond that may form between the sulfur atoms of two molecules of the amino acid *cysteine* located at different sites along the chain. Disulfide bonds create permanent loops or coils in a polypeptide chain.

4. **Quaternary structure** is the interaction between individual polypeptide chains to form a protein complex (**Figure 2–21d**). Each of the polypeptide subunits has its own secondary and tertiary structures. For example, the protein

2

hemoglobin contains four subunits. Their positions are indicated by differences in the contrast of the colors used for each subunit in the ribbon diagram of hemoglobin. Stronger contrast represents structures at the front, and lower contrast is used for structures at the back. Hemoglobin is found within red blood cells, where it binds and transports oxygen. In *keratin*, two alpha helical polypeptides are wound together like the strands of a rope. Keratin is the tough, water-resistant protein at the surface of the skin and in nails and hair. *Collagen*, composed of three windings of alpha helical polypeptides, is the most abundant structural protein and is found in skin, bones, cartilages, and tendons. Collagen fibers form the framework that supports cells in most tissues.

Fibrous and Globular Proteins

Proteins fall into two general structural classes on the basis of their overall shape and properties:

- **Fibrous proteins** form extended sheets or strands. Fibrous proteins are tough, durable, and generally insoluble in water. They usually play structural roles in the body. Their shapes are usually due to secondary structure (for proteins with the pleated-sheet form) or quaternary structure (as we described for keratin and collagen in the previous section).

- **Globular proteins** are compact, generally rounded, and soluble in water. Many enzymes, hormones, and other molecules that circulate in the bloodstream are globular proteins. These proteins can function only if they remain in solution. The unique shape of each globular protein comes from its tertiary structure. *Myoglobin*, a protein in muscle cells, is a globular protein. Hemoglobin, the oxygen-carrying molecule in red blood cells is also a globular protein. The enzymes that control chemical reactions inside cells are also globular proteins.

Protein Shape and Function

The shape of a protein determines its functional characteristics, and the sequence of amino acids ultimately determines its shape. The 20 common amino acids can be linked in an astonishing number of combinations, creating proteins of enormously varied shape and function. Changing only one of the 10,000 or more amino acids in a protein can significantly alter the way the protein functions. For example, several cancers and *sickle cell anemia*, a blood disorder, result from changing just a single amino acid in the amino acid sequences of complex proteins.

The tertiary and quaternary shapes of complex proteins depend not only on their amino acid sequence, but also on the local environmental conditions. Small changes in the ionic composition, temperature, or pH of their surroundings can affect the function of proteins. Protein shape can also be affected by hydrogen bonding to other molecules in solution.

The significance of these factors is most striking when we consider enzymes, for these proteins are essential to the metabolic operations in every one of our cells.

Enzyme Function

Enzymes are among the most important of all the body's proteins. As noted earlier in this chapter, enzymes speed up (catalyze) the chemical reactions that sustain life. Almost everything that happens inside the human body does so because a specific enzyme makes it possible.

The reactants in enzymatic reactions are called **substrates**. As in other types of chemical reactions, the interactions among substrates yield specific products. Before an enzyme can function as a catalyst—to accelerate a chemical reaction without itself being permanently changed or consumed—the substrates must bind to a special region of the enzyme. This region is called the **active site**. It is typically a groove or pocket into which one or more substrates nestle, like a key fitting into a lock. Weak electrical attractive forces, such as hydrogen bonding, reinforce the physical fit. The tertiary or quaternary structure of the enzyme molecule determines the shape of the active site. Although enzymes are proteins, any organic or inorganic compound that will bind to the active site can be a substrate.

Figure 2–22 presents one example of enzyme structure and function. Substrates bind to the enzyme at its active site (**1**). Substrate binding produces an enzyme-substrate complex (**2**). Substrate binding typically produces a temporary, reversible change in the shape of the enzyme that may place physical stresses on the substrate molecules, leading to product formation (**3**). Product release frees the enzyme, which can then repeat the process (**4**).

Enzymes work quickly, cycling rapidly between substrates and products. For example, an enzyme providing energy during a muscular contraction performs its reaction sequence 100 times per second. Hydrolytic enzymes can work even faster, breaking down almost 20,000 molecules a second!

Figure 2–22 shows an enzyme that catalyzes a synthesis reaction. Other enzymes may catalyze decomposition reactions, reversible reactions, or exchange reactions. Regardless of the reaction they catalyze, all enzymes share three basic characteristics:

1. *Specificity.* Each enzyme catalyzes only one type of reaction, a characteristic called **specificity**. An enzyme's specificity is due to the ability of its active sites to bind only to substrates with particular shapes and charges. For this reason, differences in enzyme structure that do not affect the active site and do not change the response of the enzyme to substrate binding do not affect enzyme function. In fact, different tissues typically contain enzymes that differ slightly in structure, but catalyze the same reaction. Such enzyme variants are called **isozymes**.

Figure 2–22   **A Simplified View of Enzyme Structure and Function.** Each enzyme contains a specific active site somewhere on its exposed surface.

1   Substrates bind to active site of enzyme

2   Once bound to the active site, the substrates are held together and their interaction facilitated

3   Substrate binding alters the shape of the enzyme, and this change promotes product formation

4   Product detaches from enzyme; entire process can now be repeated

2. *Saturation Limits.* The rate of an enzymatic reaction is directly related to the concentrations of substrate molecules and enzymes. An enzyme molecule must encounter appropriate substrates before it can catalyze a reaction. The higher the substrate concentration, the more frequent these encounters. When substrate concentrations are high enough that every enzyme molecule is cycling through its reaction sequence at top speed, further increases in substrate concentration will not affect the rate of reaction unless additional enzyme molecules are provided. The substrate concentration required to have the maximum rate of reaction is called the *saturation limit.* An enzyme that has reached its saturation limit is said to be **saturated**. To increase the reaction rate further, the cell must increase the number of enzyme molecules available. This is one important way that cells promote specific reactions.

3. *Regulation.* Each cell contains an assortment of enzymes, and any particular enzyme may be active under one set of conditions and inactive under another. Virtually anything that changes the tertiary or quaternary shape of an enzyme can turn it "on" or "off" and in this way control reaction rates inside the cell. Because the shape change is immediate, enzyme activation or inactivation is an important method of short-term control over reaction rates and metabolic pathways. Here we will consider only one example of enzyme regulation: the presence or absence of *cofactors.*

Cofactors and Enzyme Function

A **cofactor** is an ion or a molecule that must bind to an enzyme before substrates can also bind. Without a cofactor, the enzyme is intact but nonfunctional. With the cofactor, the enzyme can catalyze a specific reaction. Examples of ionic cofactors include calcium (Ca^{2+}) and magnesium (Mg^{2+}), which bind at the enzyme's active site. Cofactors may also bind at other sites, as long as they produce a change in the shape of the active site that makes substrate binding possible.

Coenzymes are nonprotein organic molecules that function as cofactors. Our bodies convert many vitamins into essential coenzymes. *Vitamins*, detailed in Chapter 25, are structurally related to lipids or carbohydrates, but have unique functional roles. Because the human body cannot synthesize most of the vitamins it needs, you must obtain them from your diet.

Effects of Temperature and pH on Enzyme Function

Each enzyme works best at specific temperatures and pH values. As temperatures rise, protein shape changes and enzyme function deteriorates. Eventually the protein undergoes **denaturation**, a change in tertiary or quaternary structure that makes it nonfunctional. You see permanent denaturation when you fry an egg. As the temperature rises, the proteins in the egg white denature. Eventually, the proteins become completely and irreversibly denatured, forming an insoluble white mass. In the body, death occurs at very high body temperatures (above 43 °C, or 110 °F) because structural proteins and enzymes soon denature, causing irreparable damage to organs and organ systems.

Enzymes are equally sensitive to changes in pH. *Pepsin*, an enzyme that breaks down food proteins in the stomach, works best at a pH of 2.0 (strongly acidic). Your small intestine contains *trypsin*, another enzyme that attacks proteins. Trypsin works only in an alkaline environment, with an optimum pH of 7.7 (weakly basic).

Glycoproteins and Proteoglycans

Glycoproteins (GLĪ-kō-prō-tēnz) and **proteoglycans** (prō-tē-ō-GLĪ-kanz) are combinations of protein and carbohydrate molecules. Glyco*proteins* are large proteins with small carbohydrate groups attached. These molecules may function as enzymes, antibodies, hormones, or protein components of plasma membranes. Glycoproteins in plasma membranes play a major role in identifying normal versus abnormal cells. They are also important in the immune response (Chapter 22). Glycoprotein secretions called *mucins* absorb water to form **mucus**. Mucus coats and lubricates the surfaces of the reproductive and digestive tracts.

Proteo*glycans* are large polysaccharide molecules linked by polypeptide chains. Proteoglycans bind adjacent cells together, and give tissue fluids a syrupy consistency.

> **CHECKPOINT**
>
> 22. Describe a protein.
>
> 23. How does boiling a protein affect its structural and functional properties?
>
> See the blue Answers tab at the back of the book.

2-12 ▍ DNA and RNA are nucleic acids

Learning Outcome Discuss the structures and functions of nucleic acids.

Nucleic (noo-KLĀ-ik) **acids** are large organic molecules composed of carbon, hydrogen, oxygen, nitrogen, and phosphorus. Nucleic acids store and process information at the molecular level inside cells. The two classes of nucleic acid molecules are **deoxyribonucleic** (dē-oks-ē-rī-bō-noo-KLĀ-ik) **acid (DNA)** and **ribonucleic** (rī-bō-noo-KLĀ-ik) **acid (RNA)**. As we will see, these two classes of nucleic acids differ in composition, structure, and function.

The DNA in our cells determines our inherited characteristics, including eye color, hair color, and blood type. DNA affects all aspects of body structure and function, because DNA molecules encode the information needed to build proteins. By directing the synthesis of structural proteins, DNA controls the shape and physical characteristics of our bodies. By controlling the manufacture of enzymes, DNA regulates not only protein synthesis, but also all aspects of cellular metabolism, including the creation and destruction of lipids, carbohydrates, and other vital molecules.

Several forms of RNA cooperate to build specific proteins by using the information provided by DNA. In Chapter 3 we will detail the ways that DNA and RNA work together.

Structure of Nucleic Acids

A nucleic acid consists of one or two long chains that are formed by dehydration synthesis. The individual subunits of a chain are called **nucleotides**. Each nucleotide has three parts (**Figure 2–23a**): (1) a *pentose* (five-carbon sugar) attached to both (2) a phosphate group and (3) a **nitrogenous** (nitrogen-containing) **base**. The pentose is either *ribose* (in RNA) or *deoxyribose* (in DNA).

Five nitrogenous bases occur in nucleic acids: **adenine (A)**, **guanine (G)**, **cytosine (C)**, **thymine (T)**, and **uracil (U)**. Adenine and guanine are double-ringed molecules called *purines* (**Figure 2–23b**). The other three bases are single-ringed molecules called *pyrimidines* (**Figure 2–23c**). Both RNA and DNA contain adenine, guanine, and cytosine. Uracil occurs only in RNA and thymine occurs only in DNA.

A nucleotide forms when a phosphate group binds to a pentose already attached to a nitrogenous base. A nucleic acid forms when nucleotides are joined. Dehydration synthesis attaches the phosphate group of one nucleotide to the sugar of another. The "backbone" of a nucleic acid molecule is a linear sugar-to-phosphate-to-sugar sequence, with the nitrogenous bases projecting to one side.

The primary role of nucleic acids is to store and transfer information—specifically, information essential to the synthesis of proteins within our cells. Regardless of whether we are speaking of DNA or RNA, it is the sequence of nitrogenous bases that carries the information.

RNA and DNA

Important structural differences distinguish RNA from DNA. A molecule of RNA consists of a single chain of nucleotides (**Figure 2–24a**). Its shape depends on the order of the nucleotides and the interactions among them. Our cells have various forms of RNA with different shapes and functions. As we will see in Chapter 3, protein synthesis requires three types of RNA: (1) *messenger RNA (mRNA)*, (2) *transfer RNA (tRNA)*, and (3) *ribosomal RNA (rRNA)*. The roles of other types of RNA, such as microRNAs and small interfering RNAs, in suppressing protein synthesis are under active research.

A DNA molecule consists of a *pair* of nucleotide chains (**Figure 2–24b**). Hydrogen bonding between opposing nitrogenous bases holds the two strands together. The shapes of the nitrogenous bases allow adenine to bond only to thymine, and cytosine to bond only to guanine. As a result, the combinations adenine–thymine (A–T) and cytosine–guanine (C–G) are known as **complementary base pairs**, and the two nucleotide chains of the DNA molecule are known as **complementary strands**. The two strands of DNA twist around one another in a double helix that resembles a spiral staircase. Each step of the staircase corresponds to one complementary base pair (**Figure 2–24b**).

Figure 2–23 Nucleotides and Nitrogenous Bases.

a Nucleotide structure

The nitrogenous base may be a purine or a pyrimidine.

b Purines

c Pyrimidines

Figure 2–24 The Structure of Nucleic Acids. Nucleic acids are long chains of nucleotides. Each nucleic acid molecule starts at the phosphate group of the first nucleotide and ends at the sugar of the last member of the chain. This is also the direction in which DNA is synthesized.

a **RNA molecule.** An RNA molecule has a single nucleotide chain. Its shape is determined by the sequence of nucleotides and by the interactions among them.

b **DNA molecule.** A DNA molecule has a pair of nucleotide chains linked by hydrogen bonding between complementary base pairs.

Through a sequence of events described in Chapter 3, the cell uses one of the two complementary DNA strands to provide the information needed to synthesize a specific protein. Table 2–6 (on the next page) compares RNA with DNA.

CHECKPOINT

24. **Describe a nucleic acid.**

25. **A large organic molecule made of the sugar ribose, nitrogenous bases, and phosphate groups is which kind of nucleic acid?**

See the blue Answers tab at the back of the book.

2

Table 2–6	Comparison of RNA with DNA	
Characteristic	RNA	DNA
Sugar	Ribose	Deoxyribose
Nitrogenous bases	Adenine (A)	Adenine
	Guanine (G)	Guanine
	Cytosine (C)	Cytosine
	Uracil (U)	Thymine (T)
Number of nucleotides in typical molecule	Varies from fewer than 100 nucleotides to about 50,000	Always more than 45 million
Shape of molecule	Varies with hydrogen bonding along the length of the strand; three main types (mRNA, rRNA, tRNA)	Paired strands coiled in a double helix
Function	Performs protein synthesis as directed by DNA	Stores genetic information that controls protein synthesis

2-13 ❚ ATP is a high-energy compound used by cells

Learning Outcome Discuss the structures and functions of high-energy compounds.

To perform their vital functions, cells must use energy, which they obtain by breaking down organic substrates (catabolism). To be useful, that energy must be transferred from molecule to molecule or from one part of the cell to another.

The usual method of energy transfer involves the creation and breakdown of high-energy bonds by enzymes within cells. A *high-energy bond* is a covalent bond whose breakdown releases energy the cell can use directly. In our cells, a high-energy bond generally binds a phosphate group (PO_4^{3-}) to an organic molecule. The product with such a bond is called a **high-energy compound**. Most high-energy compounds are derived from nucleotides, the building blocks of nucleic acids.

The process of attaching a phosphate group to another molecule is called **phosphorylation** (fos-for-i-LĀ-shun). This process does not necessarily produce high-energy bonds. The creation of a high-energy compound requires (1) a phosphate group, (2) enzymes capable of catalyzing the reactions involved, and (3) suitable organic substrates to which the phosphate can be added.

The most important such substrate is the nucleotide *adenosine monophosphate* (*AMP*), which already contains one phosphate group. Attaching a second phosphate group produces **adenosine diphosphate (ADP)**. A significant energy input is required to convert AMP to ADP, and the second phosphate is attached by a high-energy bond. Even more energy is required to add a third phosphate and create the high-energy compound **adenosine triphosphate (ATP)** (**Figure 2–25**).

Figure 2–25 **The Structure of ATP.** A molecule of ATP is formed by attaching two phosphate groups to the nucleotide adenosine monophosphate. These two phosphate groups are connected by high-energy bonds incorporating energy released by catabolism. Cells most often obtain quick energy to power cellular operations by removing one phosphate group from ATP, forming ADP (adenosine diphosphate). ADP can later be reconverted to ATP, and the cycle repeated.

The conversion of ADP to ATP is the most important method of storing energy in our cells. The breakdown of ATP to ADP is the most important method of releasing energy. The relationships involved in this energy transfer can be diagrammed as

$$ADP + \text{phosphate group} + \text{energy} \rightleftharpoons ATP + H_2O$$

The hydrolytic breakdown of ATP to ADP requires an enzyme known as **adenosine triphosphatase (ATPase)**. Throughout life, cells continuously generate ATP from ADP and then use the energy provided by the breakdown of ATP to perform vital functions, such as the synthesis of proteins or the contraction of muscles.

ATP is our most abundant high-energy compound, but there are others. They are typically other nucleotides that have undergone phosphorylation. For example, *guanosine triphosphate (GTP)* and *uridine triphosphate (UTP)* are nucleotide-based high-energy compounds that transfer energy in specific enzymatic reactions.

Table 2–7 summarizes the inorganic and organic compounds covered in this chapter.

CHECKPOINT

26. Describe ATP.

27. What molecule is produced by the phosphorylation of ADP?

See the blue Answers tab at the back of the book.

2-14 ▌ Chemicals and their interactions form functional units called cells

Learning Outcome Explain the relationship between chemicals and cells.

The human body is more than a collection of chemicals. The biochemical building blocks discussed in this chapter form functional units called *cells*. ⟲ p. 32 Each cell behaves like a miniature organism, responding to internal and external stimuli. This is possible only because cells are dynamic structures that adapt to changes in their environment. Such adaptations may involve changes in the chemical organization of the cell. These changes are easily made because organic molecules other than DNA are temporary rather than permanent components of the cell. They are continuously removed and replaced as part of the ongoing process of **metabolic turnover**.

Table 2–7	Classes of Inorganic and Organic Compounds		
Class	**Building Blocks**	**Sources**	**Functions**
INORGANIC			
Water	Hydrogen and oxygen atoms	Absorbed from the diet or generated by metabolism	Solvent; transport medium for dissolved materials and heat; cooling through evaporation; medium for chemical reactions; reactant in hydrolysis
Acids, bases, salts	H^+, OH^-, various anions and cations	Obtained from the diet or generated by metabolism	Structural components; buffers; sources of ions
Dissolved gases	O, C, N, and other atoms	Atmosphere, metabolism	O_2: required for cellular metabolism CO_2: generated by cells as a waste product NO: chemical messenger in cardiovascular, nervous, and lymphatic systems
ORGANIC			
Carbohydrates	C, H, O, in some cases N; CHO in a 1:2:1 ratio	Obtained from the diet or manufactured in the body	Energy source; some structural role when attached to lipids or proteins; energy storage
Lipids	C, H, O, in some cases N or P; CHO not in 1:2:1 ratio	Obtained from the diet or manufactured in the body	Energy source; energy storage; insulation; structural components; chemical messengers; protection
Proteins	C, H, O, N, commonly S	20 common amino acids; roughly half can be manufactured in the body, others must be obtained from the diet	Catalysts for metabolic reactions; structural components; movement; transport; buffers; defense; control and coordination of activities
Nucleic acids	C, H, O, N, and P; nucleotides composed of phosphates, sugars, and nitrogenous bases	Obtained from the diet or manufactured in the body	Storage and processing of genetic information
High-energy compounds	Nucleotides joined to phosphates by high-energy bonds	Synthesized by all cells	Storage or transfer of energy

2

Most of the organic molecules in a cell are replaced at intervals ranging from hours to months. The average time between synthesis and breakdown is known as the *turnover time.* Table 2–8 lists the turnover times of the organic components of representative cells. In the next chapter we will learn more about the functions of these organic components as we explore the cellular level of organization.

CHECKPOINT

28. Identify the biochemical building blocks discussed in this chapter that are the components of cells.

29. Define metabolic turnover.

See the blue Answers tab at the back of the book.

Table 2–8	Turnover Times	
Cell Type	**Component**	**Average Recycling Time***
Liver	Total protein	5–6 days
	Enzymes	1 hour to several days, depending on the enzyme
	Glycogen	1–2 days
	Cholesterol	5–7 days
Muscle cell	Total protein	30 days
	Glycogen	12–24 hours
Neuron	Phospholipids	200 days
	Cholesterol	100+ days
Fat cell	Triglycerides	15–20 days

*Most values were obtained from studies on mammals other than humans.

Related Clinical Terms

artificial sweetener: Organic molecules that can stimulate taste buds and provide a sweet taste to foods without adding substantial amounts of calories to the diet.

heavy metal: The term used for a group of elements on the "heavier" end of the periodic table of elements. Some heavy metals—cobalt, copper, iron, manganese, molybdenum, vanadium, strontium, and zinc—are essential to health in trace amounts. Others are nonessential and can be harmful to health in excessive amounts. These include cadmium, antimony, chromium, mercury, lead, and arsenic.

Chapter Review

Study Outline

MasteringA&P®

Access more chapter study tools online in the MasteringA&P Study Area:

- Chapter Quizzes, Chapter Practice Test, Art-labeling Activities, Animations, MP3 Tutor Sessions, and Clinical Case Studies
- Practice Anatomy Lab **PAL™**
- Interactive Physiology **iP°**
- A&P Flix **A&PFlix**
- PhysioEx **PhysioEx**

An Introduction to the Chemical Level of Organization p. 53

1. Chemicals combine to form complex structures.

2-1 ▍ Atoms are the basic particles of matter p. 53

2. Atoms are the smallest units of matter. They consist of **protons**, **neutrons**, and **electrons**. Protons and neutrons reside in the **nucleus** of an atom. *(Figure 2–1)*

3. The number of protons in an atom is its **atomic number**. Each **element** includes all the atoms that have the same number of protons and thus the same atomic number.

4. Within an atom, an **electron cloud** surrounds the nucleus.

5. The **mass number** of an atom is the total number of protons and neutrons in its nucleus. **Isotopes** are atoms of the same element whose nuclei contain different numbers of neutrons.

6. Electrons occupy an orderly series of **energy levels**, commonly illustrated as **electron shells**. The electrons in the outermost energy level determine an element's chemical properties. *(Figure 2–2)*

2-2 ▍ Chemical bonds are forces formed by atom interactions p. 57

7. Atoms can combine through chemical reactions that create **chemical bonds**. A **molecule** is any chemical structure consisting of atoms held together by covalent bonds. A **compound** is a chemical substance made up of atoms of two or more elements in a fixed proportion.

8. The rules of **chemical notation** are used to describe chemical compounds and reactions. *(Spotlight Figure 2–3)*

9. An **ionic bond** results from the attraction between **ions**, atoms that have gained or lost electrons. **Cations** are positively charged; **anions** are negatively charged. *(Figure 2–4)*

10. Atoms that share electrons to form a molecule are held together by **covalent bonds**. A sharing of one pair of electrons is a **single covalent bond**; a sharing of two pairs is a **double covalent bond**; and a sharing of three pairs is a **triple covalent bond**. A bond with equal sharing of electrons is a **nonpolar covalent bond**; a bond with unequal sharing of electrons is a **polar covalent bond**. *(Figures 2–5, 2–6)*

11. A **hydrogen bond** is a weak, but important, electrical attraction that can affect the shapes and properties of molecules. *(Figure 2–7)*

12. Matter can exist as a *solid*, a *liquid*, or a *gas*, depending on the nature of the interactions among the component atoms or molecules.

13. The **molecular weight** of a molecule is the sum of the atomic weights of its component atoms.

2-3 ❙ **Decomposition, synthesis, and exchange reactions are important chemical reactions in physiology** p. 62

14. A chemical reaction occurs when **reactants** are rearranged to form one or more **products**. Collectively, all the **chemical reactions** in the body constitute its **metabolism**. Through metabolism, cells capture, store, and use energy to maintain homeostasis and to perform essential functions.

15. **Work** is the movement of an object or a change in the physical structure of matter. **Energy** is the capacity to perform work.

16. **Kinetic energy** is the energy of motion. **Potential energy** is stored energy that results from the position or structure of an object. Conversions from potential to kinetic energy (or vice versa) are not 100 percent efficient. Every such energy conversion releases *heat*.

17. A chemical reaction is classified as a **decomposition**, a **synthesis**, or an **exchange reaction**.

18. Cells gain energy to power their functions by **catabolism**, the breakdown of complex molecules. Much of this energy supports **anabolism**, the synthesis of new molecules.

19. All chemical reactions are theoretically reversible. At **equilibrium**, the rates of two opposite reactions are in balance.

2-4 ❙ **Enzymes catalyze specific biochemical reactions by lowering the energy needed to start them** p. 64

20. **Activation energy** is the amount of energy required to start a reaction. **Enzymes** are **catalysts**—compounds that accelerate chemical reactions without themselves being permanently changed or consumed. Enzymes promote chemical reactions by lowering the activation energy needed. *(Figure 2–8)*

21. **Exergonic** reactions release energy. **Endergonic** reactions absorb energy.

2-5 ❙ **Inorganic compounds lack carbon, and organic compounds contain carbon** p. 65

22. **Nutrients** are the essential elements and molecules normally obtained from the diet. **Metabolites**, on the other hand, are molecules that can be synthesized or broken down by chemical reactions inside our bodies. Nutrients and metabolites can be broadly categorized as either **inorganic** or **organic compounds**.

2-6 ❙ **Physiological systems depend on water** p. 65

23. Water is the most important constituent of the body.

24. A **solution** is a uniform mixture of two or more substances. It consists of a medium, or **solvent**, in which atoms, ions, or molecules of another substance, or **solute**, are individually dispersed. In *aqueous solutions*, water is the solvent. *(Figure 2–9)*

25. Many inorganic substances, called **electrolytes**, undergo **dissociation**, or **ionization**, in water to form ions. *(Figure 2–9; Table 2–2)* Molecules that interact readily with water molecules are called **hydrophilic**. Those that do not are called **hydrophobic**.

2-7 ❙ **Body fluid pH is vital for homeostasis** p. 67

26. The **pH** of a solution indicates the concentration of hydrogen ions it contains. Solutions are classified as **neutral**, **acidic**, or **basic** *(alkaline)* on the basis of pH. *(Figure 2–10)*

2-8 ❙ **Acids, bases, and salts are inorganic compounds with important physiological roles** p. 68

27. An **acid** releases hydrogen ions. A **base** removes hydrogen ions from a solution. *Strong acids* and *strong bases* ionize completely. In the case of *weak acids* and *weak bases*, only some of the molecules ionize.

28. A **salt** is an electrolyte whose cation is not a hydrogen ion (H^+) and whose anion is not a hydroxide ion (OH^-).

29. **Buffers** remove or replace hydrogen ions in solution. Buffers and *buffer systems* in body fluids maintain the pH within normal limits.

2-9 ❙ **Carbohydrates contain carbon, hydrogen, and oxygen in a 1:2:1 ratio** p. 69

30. Carbon and hydrogen are the main constituents of **organic compounds**, which generally contain oxygen as well. The properties of the different classes of organic compounds are due to the presence of *functional groups* of atoms. *(Table 2–3)*

31. **Carbohydrates** are most important as an energy source for metabolic processes. The three major types of carbohydrates are **monosaccharides** *(simple sugars)*, **disaccharides**, and **polysaccharides**. Disaccharides and polysaccharides form from monosaccharides by **dehydration synthesis**. *(Figures 2–11 to 2–13; Table 2–4)*

2-10 ❙ **Lipids often contain a carbon-to-hydrogen ratio of 1:2** p. 72

32. **Lipids** include *fats*, *oils*, and *waxes*. Most are insoluble in water. The five important classes of lipids are **fatty acids**, **eicosanoids**, **glycerides**, **steroids**, and **phospholipids** and **glycolipids**. *(Figures 2–14 to 2–18; Table 2–5)*

33. **Triglycerides** *(neutral fats)* consist of three fatty acid molecules attached by dehydration synthesis to a molecule of **glycerol**. **Diglycerides** consist of two fatty acids and glycerol. **Monoglycerides** consist of one fatty acid plus glycerol. *(Figure 2–16)*

34. Steroids (1) are components of plasma membranes, (2) include sex hormones and hormones regulating metabolic activities, and (3) are important in lipid digestion. *(Figure 2–17)*

35. **Phospholipids** and **glycolipids** are structural lipids that are components of *micelles* and plasma membranes.

2-11 ❙ **Proteins contain carbon, hydrogen, oxygen, and nitrogen and are formed from amino acids** p. 77

36. **Proteins** perform a variety of essential functions in the body. Seven important types of proteins are *structural proteins*, *contractile proteins*, *transport proteins*, *buffering proteins*, *enzymes*, *hormones*, and *antibodies*.

37. Proteins are chains of **amino acids**. Each amino acid consists of an *amino group*, a *carboxyl group*, a *hydrogen atom*, and an *R group (side chain)* attached to a central carbon atom. A **polypeptide** is a linear sequence of amino acids held together by **peptide bonds**; **proteins** are polypeptides containing over 100 amino acids. *(Figures 2–19, 2–20)*

38. The four levels of protein structure are **primary structure** (amino acid sequence), **secondary structure** (amino acid interactions, such as hydrogen bonds), **tertiary structure** (complex folding, disulfide bonds, and interaction with water molecules), and **quaternary structure** (formation of protein complexes from individual subunits). **Fibrous proteins**, such as *keratin* and *collagen*, are elongated, tough, durable, and generally insoluble. **Globular proteins**, such as *myoglobin* and *hemoglobin*, are generally rounded and water-soluble. *(Figure 2–21)*

39. The reactants in an enzymatic reaction, called **substrates**, interact to yield a product by binding to the enzyme's **active site**. **Cofactors** are ions or molecules that must bind to the enzyme before the substrates can bind. **Coenzymes** are organic cofactors commonly derived from *vitamins*. *(Figure 2–22)*

40. The shape of a protein determines its functional characteristics. Each protein works best at an optimal combination of temperature and pH and will undergo temporary or permanent **denaturation**, or change in shape, at temperatures or pH values outside the normal range.

2-12 ▌ DNA and RNA are nucleic acids p. 82

41. **Nucleic acids** store and process information at the molecular level. The two kinds of nucleic acids are **deoxyribonucleic acid (DNA)** and **ribonucleic acid (RNA)**. *(Figures 2–23, 2–24; Table 2–6)*

42. Nucleic acids are chains of **nucleotides**. Each nucleotide contains a sugar, a phosphate group, and a **nitrogenous base**. The sugar is *ribose* in RNA and *deoxyribose* in DNA. DNA is a two-stranded double helix containing the nitrogenous bases **adenine, guanine, cytosine,** and **thymine**. RNA consists of a single strand and contains **uracil** instead of thymine.

2-13 ▌ ATP is a high-energy compound used by cells p. 84

43. Cells store energy in the *high-energy bonds* of **high-energy compounds**. The most important high-energy compound is

ATP (adenosine triphosphate). Cells make ATP by adding a phosphate group to **ADP (adenosine diphosphate)** through **phosphorylation**. When ATP is broken down to ADP and phosphate, energy is released. Cells can use this energy to power essential activities. *(Figure 2–25; Table 2–7)*

2-14 ▌ Chemicals and their interactions form functional units called cells p. 85

44. Biochemical building blocks form functional units called *cells*.

45. The continuous removal and replacement of cellular organic molecules (other than DNA), a process called **metabolic turnover**, allows cells to change and to adapt to changes in their environment. *(Table 2–8)*

Review Questions

See the blue Answers tab at the back of the book.

LEVEL 1 Reviewing Facts and Terms

1. An oxygen atom has eight protons. (a) Sketch in the arrangement of electrons around the nucleus of the oxygen atom in the following diagram. (b) How many more electrons will it take to fill the outermost energy level?

Oxygen atom

2. Molecules that interact with water are known as (a) solvent, (b) solute, (c) hydrophobic, (d) hydrophilic.

3. An element's chemical properties are determined by the (a) number of protons and electrons, (b) total number of electrons, (c) number of electrons in the outer shell, (d) number of electron shells.

4. Isotopes of an element differ from each other in the number of (a) protons in the nucleus, (b) neutrons in the nucleus, (c) electrons in the outer shells, (d) a, b, and c are all correct.

5. Atoms that share electrons to form a molecule are held by (a) covalent bonds, (b) hydrogen bonds, (c) ionic bonds, (d) double bonds.

6. All organic compounds in the human body contain all of the following elements *except* (a) hydrogen, (b) oxygen, (c) carbon, (d) calcium, (e) both a and d.

7. The mass of an atom is the total number of (a) protons and electrons, (b) protons and neutrons, (c) neutrons and electrons, (d) protons, neutrons, and electrons.

8. All the chemical reactions that occur in the human body are collectively referred to as (a) anabolism, (b) catabolism, (c) metabolism, (d) homeostasis.

9. Which of the following chemical equations illustrates a typical decomposition reaction?
   (a) A + B $\longrightarrow$ AB
   (b) AB + CD $\longrightarrow$ AD + CB
   (c) $2A_2 + B_2 \longrightarrow 2A_2B$
   (d) AB $\longrightarrow$ A + B

10. The speed, or rate, of a chemical reaction is influenced by (a) the presence of catalysts, (b) the temperature, (c) the concentration of the reactants, (d) a, b, and c are all correct.

11. A pH of 7.8 in the human body typifies a condition referred to as (a) acidosis, (b) alkalosis, (c) dehydration, (d) homeostasis.

12. A(n) _____ is a solute that dissociates to release hydrogen ions, and a(n) _____ is a solute that removes hydrogen ions from solution. (a) base, acid, (b) salt, base, (c) acid, salt, (d) acid, base.

13. Special catalytic molecules called _____ control chemical reactions in the human body. (a) enzymes, (b) cytozymes, (c) cofactors, (d) activators, (e) cytochromes.

14. Which of the following is *not* a function of a protein? (a) support, (b) transport, (c) metabolic regulation, (d) storage of genetic information, (e) movement.

15. Complementary base pairing in DNA includes the pairs (a) adenine–uracil and cytosine–guanine, (b) adenine–thymine and cytosine–guanine, (c) adenine–guanine and cytosine–thymine, (d) guanine–uracil and cytosine–thymine.

16. List four properties of water.

17. What four major classes of organic compounds are found in the body?

18. List three important functions of triglycerides (neutral fats) in the body.

19. List four levels of protein structure.

20. (a) What three basic components make up a nucleotide of DNA?
    (b) What three basic components make up a nucleotide of RNA?

21. What three components are required to create the high-energy compound ATP?

LEVEL 2 Reviewing Concepts

22. If a polypeptide contains 10 peptide bonds, how many amino acids does it contain? (a) 9, (b) 10, (c) 11, (d) 12.

23. A dehydration synthesis reaction between glycerol and a single fatty acid would yield a(n) (a) micelle, (b) omega-3 fatty acid, (c) triglyceride, (d) monoglyceride, (e) diglyceride.

24. Explain how the rate of enzyme reactions can become saturated. How can this be overcome?

25. What is a high-energy bond? How it is used in the formation of ATP?

26. Explain how metabolic turnover allows cells to respond to internal and external stimuli.

27. In an exergonic reaction, (a) large molecules are broken down into smaller ones, (b) small molecules are assembled into larger ones, (c) molecules are rearranged to form new molecules, (d) molecules move from reactants to products and back, (e) energy is released during the reaction.

28. The hydrogen bonding that occurs in water is responsible for all of the following, *except* (a) the high boiling point of water, (b) the low freezing point of water, (c) the ability of water to dissolve nonpolar substances, (d) the ability of water to dissolve inorganic salts, (e) the surface tension of water.

29. A sample that contains an organic molecule has the following constituents: carbon, hydrogen, oxygen, nitrogen, and phosphorus. Is the molecule more likely to be a carbohydrate, a lipid, a protein, or a nucleic acid?

LEVEL 3 Critical Thinking and Clinical Applications

30. An atom of the element calcium has 20 protons and 20 neutrons. Determine the following information about calcium: (a) number of electrons, (b) atomic number, (c) atomic weight, (d) number of electrons in each energy level.

31. Michael is a keen cyclist who often feels hot following exercise. Why would exercise make him hot?

32. An important buffer system in the human body involves carbon dioxide (CO_2) and bicarbonate ion (HCO_3^-) in the reversible reaction

$$CO_2 + H_2O \rightleftharpoons H_2CO_3 \rightleftharpoons H^+ + HCO_3^-$$

If a person becomes excited and exhales large amounts of CO_2, how will the pH of the person's body be affected?

CLINICAL CASE / Wrap-Up

What Is Wrong with My Baby?

Baby Sean has cystic fibrosis (CF), a life-threatening genetic disease. He inherited a defective gene from each of his parents, both mother and father. This faulty gene affects the movement of salt—sodium chloride (NaCl)—in and out of cells. As a result, the body produces thick, sticky secretions, including digestive juices that cannot get out of the pancreas into the small intestine.

Digestive juices contain enzymes that are the catalysts of digestion. These digestive enzymes aid in the decomposition reactions that break down lipids and carbohydrates in food, making them available for absorption and use by the body's cells. Without these enzymes, Sean cannot absorb food, so it passes right through him without being digested.

The defective gene also results in thick, heavy secretions that clog the lungs, making Sean short of breath, wheezy, and susceptible to repeated lung infections. It also makes his skin taste very salty.

The diagnosis of cystic fibrosis is made with a sweat test. A sweat-producing chemical is applied to an area of Sean's skin, and the sweat is collected and tested for salt.

Sean's treatment starts immediately. He is able to absorb his food nearly normally by taking digestive enzymes before each feeding. Maureen and Conner learn to perform special exercises and treatments with Sean several times a day to clear the thick secretions from his lungs. Sean is also carefully monitored for signs of lung infections such as pneumonia. Soon Sean gains weight and begins to thrive.

1. What undigested substances would have made Baby Sean's stools greasy, almost foamy looking?

2. Why are digestive enzymes necessary for life?

See the blue Answers tab at the back of the book.

3

The Cellular Level of Organization

Learning Outcomes

These Learning Outcomes correspond by number to this chapter's sections and indicate what you should be able to do after completing the chapter.

3-1 ▌ List the functions of the plasma membrane and the structural features that enable it to perform those functions.

3-2 ▌ Describe the organelles of a typical cell, and indicate the specific functions of each.

3-3 ▌ Explain the functions of the cell nucleus and discuss the nature and importance of the genetic code.

3-4 ▌ Summarize the role of DNA in protein synthesis, cell structure, and cell function.

3-5 ▌ Describe the processes of cellular diffusion and osmosis, and explain their role in physiological systems.

3-6 ▌ Describe carrier-mediated transport and vesicular transport mechanisms used by cells to facilitate the absorption or removal of specific substances.

3-7 ▌ Explain the origin and significance of the membrane potential.

3-8 ▌ Describe the stages of the cell life cycle, including mitosis, interphase, and cytokinesis, and explain their significance.

3-9 ▌ Discuss the regulation of the cell life cycle.

3-10 ▌ Discuss the relationship between cell division and cancer.

3-11 ▌ Define differentiation, and explain its importance.

SPOTLIGHTS

- Anatomy of a Model Cell pp. 92–93
- Protein Synthesis, Processing, and Packaging pp. 104–105
- Overview of Membrane Transport pp. 124–125
- DNA Replication p. 128
- Stages of a Cell's Life Cycle pp. 130–131

CLINICAL CASE

- When Your Heart Is in the Wrong Place p. 91

Clinical Notes

- Inheritable Mitochondrial Disorders p. 106
- DNA Fingerprinting p. 110
- Mutations p. 112
- Drugs and the Plasma Membrane p. 116
- Telomerase, Aging, and Cancer p. 133
- Parkinson's Disease p. 134

An Introduction to Cells

In this chapter we will see how combinations of chemicals form *cells*, the smallest living units in the human body. We will also look at the chemical events that sustain life, which occur mostly inside cells.

Cells are very small. A typical cell is only about 0.1 mm in diameter. As a result, no one could actually examine the structure of a cell until effective microscopes were invented in the 17th century. In 1665, Robert Hooke inspected thin slices of cork and found that they were made of millions of small, irregular units. In describing what he saw, Hooke used the term *cell* because the many small, bare spaces reminded him of the rooms, or cells, in a prison or monastery. The Latin term, *cella*, means a storeroom or chamber. Hooke saw only the outlines of the cells, and not the cells themselves, but he stimulated broad interest in the microscopic world and in the nature of cellular life. The research that he began about 350 years ago has, over time, produced the *cell theory* in its current form. We can summarize its basic concepts as follows:

- Cells are the building blocks of all plants and animals.
- All cells come from the division of preexisting cells.
- Cells are the smallest units that carry out life's essential physiological functions.
- Each cell maintains homeostasis at the cellular level. Homeostasis at the level of the tissue, organ, organ system, and organism reflects the combined and coordinated actions of many cells.

The human body contains trillions of cells. All our activities—from running to thinking—result from these combined and coordinated actions of millions or even billions of cells. Many insights into human physiology arose from studies of the functioning of individual cells. What we have learned over the years has given us a new understanding of cellular physiology and the mechanisms of homeostatic control. Today, the study of cellular structure and function, or **cytology**, is part of the broader discipline of **cell biology**, which integrates aspects of biology, chemistry, and physics.

The human body contains two general classes of cells: sex cells and somatic cells. **Sex cells** (also called *germ cells* or *reproductive cells*) are either the *sperm* (spermatozoa) of males or the *oocytes* (ŌŌ-sīts) or "eggs" of females. The fusion of a sperm and an oocyte at fertilization is the first step in the creation of a new individual. **Somatic cells** (*soma*, body), or body cells, include all the other cells in the human body. In this chapter, we focus on somatic cells. We will discuss sex cells in Chapters 28 and 29 on the reproductive system and development, respectively.

In the rest of this chapter, we describe the structure of a typical somatic cell, consider some of the ways in which cells interact with their environment, and discuss how somatic cells reproduce. Keep in mind that the "typical" somatic cell is like the "average" person: Any description masks enormous individual variations. **Spotlight Figure 3–1** summarizes the anatomy of a model (representative) cell.

3-1 | The plasma membrane separates the cell from its surrounding environment and performs various functions

Learning Outcome List the functions of the plasma membrane and the structural features that enable it to perform those functions.

When you view a cell through a microscope, the first structure you encounter is its outer boundary, called the **plasma membrane**, or

In our model cell, a *plasma membrane* separates the cell contents, called the *cytoplasm*, from its surroundings. The cytoplasm can be subdivided into the *cytosol*, a fluid, and intracellular structures collectively known as *organelles* (or-ga-NELZ). Organelles are structures suspended within the cytosol that perform specific functions within the cell. They can be subdivided into membranous and nonmembranous organelles. Cells are surrounded by a watery medium known as the **extracellular fluid**. The extracellular fluid in most tissues is called **interstitial** (in-ter-STISH-ul) **fluid**.

 = Plasma membrane

 = Nonmembranous organelles

 = Membranous organelles

Microvilli

Microvilli are extensions of the plasma membrane containing microfilaments.

Function
Increase surface area to facilitate absorption of extracellular materials

Secretory vesicles

CYTOSOL

NUCLEUS

Free ribosomes

Centrosome and Centrioles

Cytoplasm contains two centrioles at right angles; each centriole is composed of 9 microtubule triplets in a 9 + 0 array

Functions
Essential for movement of chromosomes during cell division; organization of microtubules in cytoskeleton

Centrosome

Centrioles

Cytoskeleton

Proteins organized in fine filaments or slender tubes

Functions
Strength and support; movement of cellular structures and materials

Microfilament

Microtubule

Plasma Membrane

Lipid bilayer containing phospholipids, steroids, proteins, and carbohydrates

Functions
Isolation; protection; sensitivity; support; controls entry and exit of materials

Cytosol (distributes materials by diffusion)

Cilia (*see Figure 3-4b*)

Cilia are long extensions of the plasma membrane containing microtubules. There are two types: primary and motile.

Functions
A primary cilium acts as a sensor. Motile cilia move materials over cell surfaces

Golgi apparatus

Stacks of flattened membranes (cisternae) containing chambers

Functions
Storage, alteration, and packaging of secretory products and lysosomal enzymes

Proteasomes

Hollow cylinders of proteolytic enzymes with regulatory proteins at their ends

Functions
Breakdown and recycling of damaged or abnormal intracellular proteins

Mitochondria

Double membrane, with inner membrane folds (cristae) enclosing important metabolic enzymes

Function
Produce 95% of the ATP required by the cell

Ribosomes

RNA + proteins; fixed ribosomes bound to rough endoplasmic reticulum; free ribosomes scattered in cytoplasm

Function
Protein synthesis

Endoplasmic reticulum (ER)

Network of membranous channels extending throughout the cytoplasm

Functions
Synthesis of secretory products; intracellular storage and transport; detoxification of drugs or toxins

Rough ER modifies and packages newly synthesized proteins

Smooth ER synthesizes lipids and carbohydrates

Peroxisomes

Vesicles containing degradative enzymes

Functions
Catabolism of fats and other organic compounds; neutralization of toxic compounds generated in the process

Chromatin

Nuclear envelope

Nucleolus (site of rRNA synthesis and assembly of ribosomal subunits)

NUCLEOPLASM

NUCLEUS

Nucleoplasm containing nucleotides, enzymes, nucleoproteins, and chromatin; surrounded by a double membrane, the nuclear envelope

Functions
Control of metabolism; storage and processing of genetic information; control of protein synthesis

Lysosomes

Vesicles containing digestive enzymes

Function
Intracellular removal of damaged organelles or pathogens

Nuclear pore

3

cell membrane. The plasma membrane is extremely thin, ranging from 6 to 10 nm in thickness. Its general functions include:

- *Physical Isolation.* The plasma membrane is a physical barrier that separates the inside of the cell, or *cytoplasm,* from the surrounding extracellular fluid. For example, the plasma membrane keeps enzymes and structural proteins inside the cell. Conditions inside and outside the cell are very different, and those differences must be maintained to preserve homeostasis.

- *Regulation of Exchange with the Environment.* The plasma membrane acts as a kind of gate-keeper. It controls the entry of ions and nutrients, the elimination of wastes, and the release of secretions.

- *Sensitivity to the Environment.* The plasma membrane is the first part of the cell affected by changes in the composition, concentration, or pH of the extracellular fluid. It also contains a variety of special structural molecules known as *receptors* that allow the cell to recognize and respond to specific molecules in its environment. For instance, the plasma membrane may receive chemical signals from other cells. The binding of just one molecule to a receptor may trigger the activation or deactivation of enzymes that affect many cellular activities.

- *Structural Support.* Specialized connections between plasma membranes, or between membranes and extracellular materials, give tissues stability. For example, the cells at the surface of the skin are tightly bound together, while those in the deepest layers are attached to extracellular protein fibers in underlying tissues.

The plasma membrane contains lipids, proteins, and carbohydrates.

Membrane Lipids

Lipids form most of the surface area of the plasma membrane, but they make up only about 42 percent of its weight. The plasma membrane is called a **phospholipid bilayer**, because the phospholipid molecules in it form two layers (**Figure 3–2**).

Recall from Chapter 2 that a phospholipid has both a hydrophilic end (the phosphate portion) and a hydrophobic end (the lipid portion). ↪ p. 75 In each half of the bilayer, the phospholipids lie with their hydrophilic heads at the membrane surface and their hydrophobic tails on the inside. In this arrangement, the hydrophilic heads of the two layers are in contact with the aqueous environments on both sides of the membrane—the extracellular fluid outside the cell and

Figure 3–2 The Plasma Membrane.

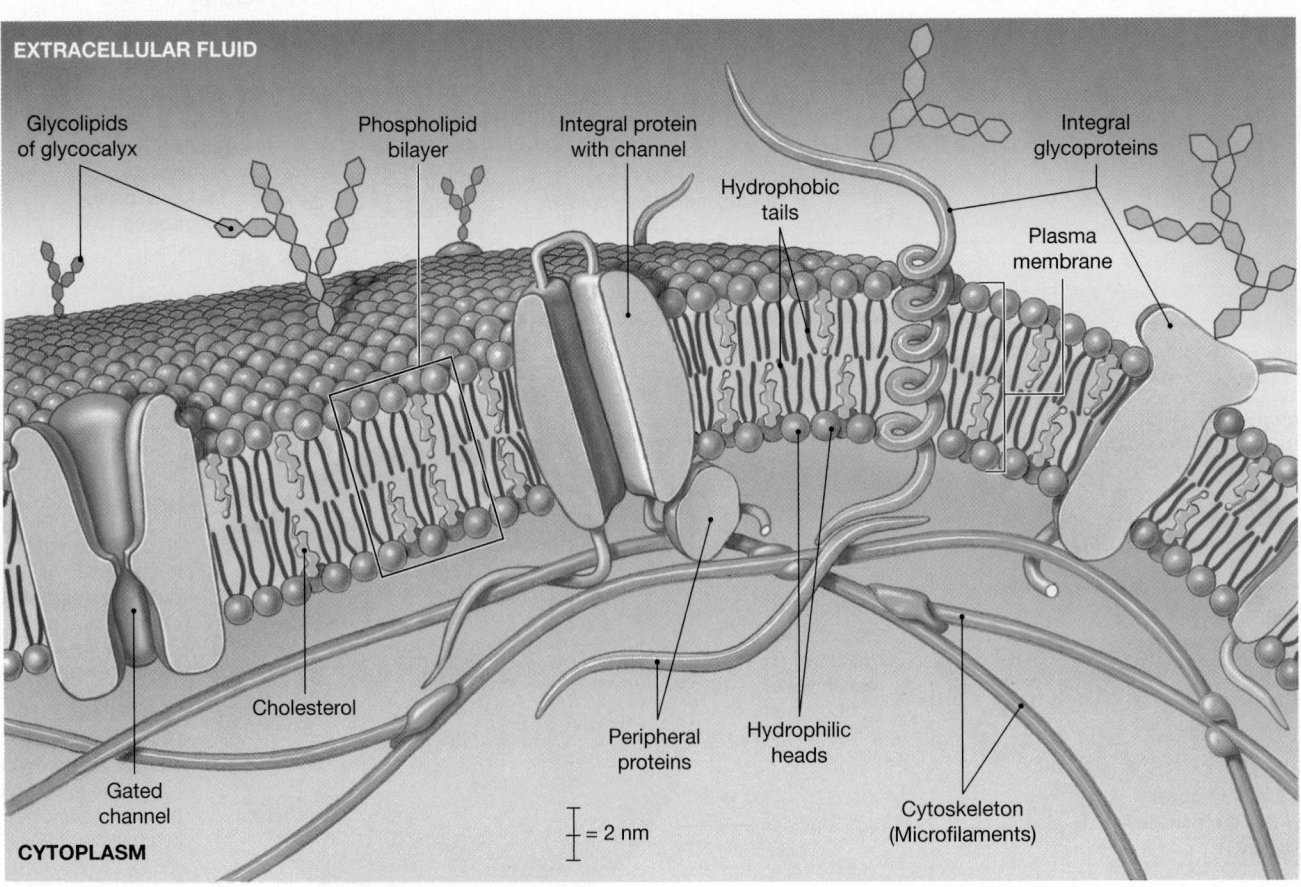

EXTRACELLULAR FLUID

Glycolipids of glycocalyx

Phospholipid bilayer

Integral protein with channel

Hydrophobic tails

Integral glycoproteins

Plasma membrane

Cholesterol

Peripheral proteins

Hydrophilic heads

Cytoskeleton (Microfilaments)

Gated channel

= 2 nm

CYTOPLASM

the intracellular fluid, or *cytosol*, inside the cell. (The cytosol is the fluid component of the cytoplasm.) The hydrophobic tails form the interior of the membrane.

Note the similarities in lipid organization between the plasma membrane and a micelle (**Figure 2–18c**, p. 76). Ions and water-soluble compounds cannot enter a micelle, because the lipid tails of the phospholipid molecules are hydrophobic ("water-fearing") and will not associate with water molecules. For the same reason, such substances cannot cross the lipid portion of the plasma membrane. In this way, the hydrophobic center of the membrane isolates the cytoplasm from the surrounding fluid environment. Such isolation is important because the composition of cytoplasm is very different from that of extracellular fluid, and the cell must keep it that way to survive.

The lipid bilayer also contains cholesterol and other steroids, small quantities of other lipids, proteins, and glycolipids. Cholesterol is an important component of plasma membranes, with almost one cholesterol molecule for each phospholipid molecule. ⊃ p. 74 Cholesterol "stiffens" the plasma membrane, making it less fluid and less permeable.

Membrane Proteins

Proteins, which are much denser than lipids, account for about 55 percent of the weight of a plasma membrane. There are two general structural classes of membrane proteins (**Figure 3–2**). **Integral proteins** are part of the plasma membrane structure and cannot be easily separated from it without damaging or destroying the membrane. Most integral proteins span the width of the membrane one or more times, and are known as *transmembrane proteins*. These proteins contain hydrophobic portions embedded within the hydrophobic lipid bilayer, with their hydrophilic portions extended into the extracellular environment and cytosol. **Peripheral proteins** are bound to the inner or outer surface of the membrane and (like Post-it notes) are easily separated from it. Integral proteins greatly outnumber peripheral proteins.

Membrane proteins carry out a variety of specialized functions. Here are some examples of important types of membrane proteins:

1. *Anchoring Proteins.* **Anchoring proteins** attach the plasma membrane to other structures and stabilize its position. Inside the cell, membrane proteins are bound to the *cytoskeleton*, a network of supporting filaments in the cytoplasm. Outer membrane proteins may attach the cell to extracellular protein fibers or to another cell.

2. *Recognition Proteins (Identifiers).* The cells of the immune system recognize other cells as normal or abnormal based on the presence or absence of characteristic **recognition proteins.** Many important recognition proteins

are glycoproteins. ⊃ p. 82 (We will discuss one group, the *MHC proteins* involved in the immune response, in Chapter 22.)

3. *Enzymes.* Enzymes in plasma membranes may be integral or peripheral proteins. They catalyze reactions in the extracellular fluid or in the cytosol, depending on the location of the protein and its active site. For example, dipeptides are broken down into amino acids by enzymes on the exposed membranes of cells that line the intestinal tract.

4. *Receptor Proteins.* **Receptor proteins** in the plasma membrane are sensitive to the presence of specific extracellular molecules called **ligands** (LĪ-gandz). A ligand can be anything from a small ion, like calcium, to a relatively large and complex hormone. When an extracellular ligand binds to the appropriate receptor, that binding may trigger changes in the activity of the cell. For example, the binding of the hormone *insulin* to a specific membrane receptor protein is the key step that leads to an increase in the cell's rate of glucose absorption. Plasma membranes differ in the type and number of receptor proteins they contain, and these differences account for a cell's sensitivity to specific hormones and other ligands.

5. *Carrier Proteins.* **Carrier proteins** bind solutes and transport them across the plasma membrane. Carrier proteins may require ATP as an energy source. ⊃ p. 84 For example, virtually all cells have carrier proteins that bring glucose into the cytoplasm without expending ATP. Yet these cells must use ATP to transport ions such as sodium and calcium across the plasma membrane and out of the cytoplasm.

6. *Channels.* Some integral proteins contain a **channel**, or central pore, that forms a passageway completely through the plasma membrane. The channel permits water and small solutes to move across the plasma membrane. Ions do not dissolve in lipids, so they cannot cross the phospholipid bilayer. For this reason, ions and other small water-soluble materials can cross the membrane only by passing through channels. Many channels are highly specific and permit only one particular ion to pass. The movement of ions through channels is involved in a variety of physiological mechanisms. Channels account for about 0.2 percent of the total surface area of the plasma membrane, but they are extremely important in physiological processes like muscle contraction and nerve impulse transmission, as we will see in Chapters 10 and 12.

Membranes are neither rigid nor uniform. At each location, the inner and outer surfaces of the plasma membrane may differ in important ways. For example, some enzymes are found only on the inner surface of the membrane, and some receptors are found only on the outer surface.

Some integral proteins are always confined to specific areas of the plasma membrane. These areas, called *rafts*, mark the location of anchoring proteins and some kinds of receptor proteins. Yet because membrane phospholipids are fluid at body temperature, many other integral proteins drift across the surface of the membrane like ice cubes in a bowl of punch. In addition, the composition of the entire plasma membrane can change over time as large areas of the membrane surface are removed and recycled in the ongoing process of metabolic turnover. ⊃ p. 85

Membrane Carbohydrates

Carbohydrates account for about 3 percent of the weight of a plasma membrane. The carbohydrates in the plasma membrane are parts of complex molecules such as *proteoglycans, glycoproteins,* and *glycolipids.* ⊃ pp. 75, 82 The carbohydrate portions of these large molecules extend beyond the outer surface of the membrane, forming a layer known as the **glycocalyx** (gl-ī-kō-KĀ-liks; *calyx,* cup). The glycocalyx has a variety of important functions, including:

- *Lubrication and Protection.* The glycoproteins and glycolipids form a viscous (thick and sticky) layer that lubricates and protects the plasma membrane.

- *Anchoring and Locomotion.* Because its components are sticky, the glycocalyx can help anchor the cell in place. It also takes part in the movement of specialized cells.

- *Specificity in Binding.* Glycoproteins and glycolipids can function as receptors, binding specific extracellular compounds. Such binding can change the properties of the cell surface and indirectly affect the cell's behavior.

- *Recognition.* Cells involved with the immune response recognize glycoproteins and glycolipids as normal or abnormal. The characteristics of the glycocalyx are genetically determined. The body's immune system recognizes its own membrane glycoproteins and glycolipids as "self" rather than as "foreign." This recognition system keeps your immune system from attacking your cells, and allows it to recognize and destroy invading pathogens (disease-causing agents).

The plasma membrane forms a barrier between the cytosol and the extracellular fluid. If the cell is to survive, dissolved substances and larger compounds must be permitted to cross this barrier. Nutrients must be able to enter the cell, and metabolic wastes must be able to leave. The structure of the plasma membrane is ideally suited to this need for selective transport. We will discuss selective transport and other membrane functions further, after we have completed our overview of cellular anatomy.

CHECKPOINT

1. List the general functions of the plasma membrane.
2. Identify the components of the plasma membrane that allow it to carry out its functions.
3. Which component of the plasma membrane is primarily responsible for the membrane's ability to form a physical barrier between the cell's internal and external environments?
4. Which type of integral protein allows water, ions, and small water-soluble solutes to pass through the plasma membrane?

See the blue Answers tab at the back of the book.

3-2 ￼ Organelles within the cytoplasm perform particular functions

Learning Outcome Describe the organelles of a typical cell, and indicate the specific functions of each.

Cytoplasm is a general term for the material between the plasma membrane and the membrane that surrounds the nucleus. It is a colloid with a consistency that varies between that of thin maple syrup and almost-set gelatin, and it contains many more proteins than does extracellular fluid, the watery medium that surrounds cells. ⊃ p. 67 Proteins are so important to the cell that they make up about 30 percent of a typical cell's weight.

The cytoplasm contains cytosol and organelles. **Cytosol,** or *intracellular fluid,* contains dissolved nutrients, ions, soluble and insoluble proteins, and waste products. **Organelles** are structures suspended within the cytosol that perform specific functions for the cell.

The Cytosol

The most important differences between cytosol and extracellular fluid are as follows:

1. *The concentration of potassium ions is much higher in the cytosol than in the extracellular fluid. Conversely, the concentration of sodium ions is much lower in the cytosol than in the extracellular fluid.*

2. *The cytosol contains a much higher concentration of suspended proteins than does extracellular fluid.* Many of the proteins are enzymes that regulate metabolic operations. Others are associated with the various organelles. The consistency of the cytosol is determined in large part by the enzymes and cytoskeletal proteins.

3. *The cytosol usually contains small quantities of carbohydrates, and small reserves of amino acids and lipids.* The extracellular fluid is a transport medium only, and no amino acid or lipids are stored there. The carbohydrates in the cytosol are broken down to provide energy, and the amino acids

are used to manufacture proteins. Lipids, in particular triglycerides, are used primarily as a source of energy when carbohydrates are unavailable.

Masses of insoluble materials may be found in both the cytosol and the extracellular fluid. In the cytosol, these masses are known as **inclusions**. Common cytosol inclusions are stored nutrients, such as glycogen granules in liver or in skeletal muscle cells, lipid droplets in fat cells, and pigment granules such as the brown skin pigment, *melanin*. Melanin is synthesized intracellularly, thus it is a common cytosol inclusion, and then it is secreted into the extracellular fluid as an insoluble material. Another example of an insoluble material in the extracellular fluid is the mineral deposits in bone.

The Organelles

Organelles are the internal structures of cells that perform most of the tasks that keep a cell alive and functioning normally. Each organelle has specific functions related to cell structure, growth, maintenance, and metabolism. We can divide cellular organelles into two broad categories, nonmembranous and membranous. **Nonmembranous organelles** are not completely enclosed by membranes, and all of their components are in direct contact with the cytosol. **Membranous organelles** are isolated from the cytosol by phospholipid membranes, just as the plasma membrane isolates the cytosol from the extracellular fluid.

The cell's nonmembranous organelles include the *cytoskeleton, microvilli, centrioles, cilia, ribosomes,* and *proteasomes.* Membranous organelles include the *endoplasmic reticulum,* the *Golgi apparatus, lysosomes, peroxisomes,* and *mitochondria.* The *nucleus* is also surrounded by a membranous envelope— and is, strictly speaking, a membranous organelle. It has so many vital functions that we will consider it in a separate section.

The Cytoskeleton

The **cytoskeleton** serves as the cell's skeleton. It is an internal protein framework that gives the cytoplasm strength and flexibility. The cytoskeleton of all cells is made of *microfilaments, intermediate filaments,* and *microtubules* (**Figure 3–3**). Muscle cells contain these cytoskeletal parts plus *thick filaments.* The filaments and microtubules of the cytoskeleton form a dynamic network whose continual reorganization affects cell shape and function.

We will consider only a few of the many functions of the cytoskeleton in this section. In addition to the functions we describe here, the cytoskeleton plays a role in the metabolic organization of the cell. It determines where in the cytoplasm key enzymatic reactions take place and where specific proteins are synthesized. For example, many enzymes (especially those involved with metabolism and energy production), ribosomes, and RNA molecules responsible for the synthesis of proteins are attached to the microfilaments and microtubules of the cytoskeleton. The varied metabolic functions of the cytoskeleton are now a subject of intensive research.

Microfilaments

The smallest of the cytoskeletal elements are the **microfilaments**. These protein strands are generally about 5 nm in diameter. Typical microfilaments are made of the protein **actin.** In most cells, actin filaments are common in the periphery of the cell, but rare in the region immediately surrounding the nucleus. In cells that form a layer or lining, such as the lining of the intestinal tract, actin filaments also form a layer, called the *terminal web,* just inside the plasma membrane at the exposed surface of the cell (**Figure 3–3a**).

Microfilaments have three major functions:

1. *Microfilaments anchor the cytoskeleton* to integral proteins of the plasma membrane. They give the cell additional mechanical strength and attach the plasma membrane to the enclosed cytoplasm.

2. *Microfilaments, interacting with other proteins, determine the consistency of the cytoplasm.* Where microfilaments form a dense, flexible network, the cytoplasm has a gelatinous consistency. Where they are widely dispersed, the cytoplasm is more fluid.

3. *Actin can produce movement* of a portion of a cell or change the shape of the entire cell by interacting with the protein *myosin.*

Intermediate Filaments

The protein composition of **intermediate filaments** varies among cell types. These filaments range from 9 to 11 nm in diameter. They are so named because they are intermediate in size between microfilaments and thick filaments. Intermediate filaments, which are insoluble, are the most durable of the cytoskeletal elements.

Intermediate filaments: (1) *strengthen the cell* and help maintain its shape, (2) *stabilize the positions of organelles,* and (3) *stabilize the position of the cell with respect to surrounding cells* through specialized attachment to the plasma membrane. Many cells contain specialized intermediate filaments with unique functions. For example, the keratin fibers in superficial layers of the skin are intermediate filaments that make these layers strong and able to resist stretching.

Microtubules

Most cells contain **microtubules**, hollow tubes built from the globular protein **tubulin**. Microtubules are the largest components of the cytoskeleton, with diameters of about

Figure 3–3   **The Cytoskeleton.**

Microvillus

Microfilaments

Terminal web

Plasma membrane

Terminal web

Mitochondrion

Intermediate filaments

Endoplasmic reticulum

Microtubule

Secretory vesicle

a The cytoskeleton provides strength and structural support for the cell and its organelles. Interactions between cytoskeletal components are also important in moving organelles and in changing the shape of the cell.

Microfilaments and microvilli   SEM × 30,000

b The microfilaments and microvilli of an intestinal cell. Such an image, produced by a scanning electron microscope, is called a scanning electron micrograph.

Microfilaments and microvilli   LM × 3200

c Microtubules (yellow) in living cells, as seen after special fluorescent labeling.

25 nm. Microtubules extend outward into the periphery of the cell from a region near the nucleus called the *centrosome* (**Spotlight Figure 3–1**).

Microtubules play important roles in maintaining the shape of the cell, and they increase in number during mitosis (somatic cell division) and meiosis (germ cell division). Each microtubule forms by the aggregation of tubulin molecules, growing out from its origin at the centrosome. The entire structure persists for a time and then disassembles into individual tubulin molecules again.

Microtubules have the following functions:

1. *Microtubules form the main portions of the cytoskeleton,* giving the cell strength and rigidity and anchoring the position of major organelles.

2. *Microtubules change the shape of the cell, and may assist in cell movement,* through the assembly and disassembly of microtubules

3. *Microtubules can serve as a kind of monorail system to move vesicles or other organelles within the cell.* Proteins called

molecular motors create the movement. These motor proteins bind to the structure being moved and bind to a microtubule and move along its length. The direction of movement depends on the particular motor protein involved. For example, the molecular motors *kinesin* and *dynein* carry materials in opposite directions on a microtubule: Kinesin moves toward one end, dynein toward the other. Regardless of the direction of transport or the nature of the motor, the process requires ATP and is essential to normal cellular function.

4. *During cell division, microtubules distribute duplicated chromosomes* to opposite ends of the dividing cell by forming a network of microtubules called the *spindle apparatus*. We will look at this process in more detail in a later section.

5. *Microtubules form structural components of organelles*, such as *centrioles* and *cilia*.

Thick Filaments

Thick filaments are relatively massive bundles of subunits made of the protein **myosin**. Thick filaments are about 15 nm in diameter and 1.5 μm (micrometers) long. They appear only in all types of muscle cells, where they produce powerful contractions by interacting with actin filaments.

Microvilli

Many cells have small, finger-shaped projections of the plasma membrane on their exposed surfaces (**Figure 3–3b**). These projections, called **microvilli** (singular, *microvillus*), greatly increase the surface area of the cell exposed to the extracellular environment. Accordingly, they cover the surfaces of cells that are actively absorbing materials from the extracellular fluid, such as the cells lining the digestive tract. Microvilli have extensive connections with the cytoskeleton. A core of microfilaments stiffens each microvillus and anchors it to the cytoskeleton at the terminal web.

Centrosome and Centrioles

The **centrosome** is a region of cytoplasm located next to the nucleus in a cell. It is the microtubule organizing center of animal cells, and the heart of the cytoskeletal system. Microtubules of the cytoskeleton generally begin at the centrosome and radiate through the cytoplasm. In all animal cells capable of undergoing cell division, the centrosome surrounds a pair of cylindrical structures called **centrioles**. The centrioles lie perpendicular to each other and are composed of short microtubules (**Figure 3–4a**). The microtubules form nine groups, three in each group. Each of these nine "triplets" is connected to its nearest neighbors on either side. Because there are no central microtubules, this organization is called a *9 + 0 array*.

During cell division, the centrioles aid the formation of the spindle apparatus needed for the movement of DNA strands. Mature red blood cells, skeletal muscle cells, cardiac muscle cells, and typical neurons have no centrioles, and as a result, these cells cannot divide. The centrioles also form the *basal bodies* found at the base of cilia and flagella.

Cilia and Flagella

Cilia (singular, *cilium*) are fairly long, slender extensions of the plasma membrane. Multiple *motile cilia* are found on cells lining both the respiratory and reproductive tracts, and at various other locations in the body. Motile cilia have an internal arrangement similar to that of centrioles. However, in cilia, nine *pairs* of microtubules (rather than triplets) surround a central pair (**Figure 3–4b**)—an organization known as a *9 + 2 array*. The microtubules are anchored to a compact **basal body** situated just beneath the cell surface. The organization of microtubules in the basal body resembles the array of a centriole: nine triplets with no central pair.

A single, nonmotile **primary cilium** is found on the cells of a wide variety of tissues in the body. Its structure is similar to the 9 + 0 microtubule organization of a centriole and basal body. A primary cilium acts as a signal sensor, detecting environmental stimuli and coordinating activities such as embryonic development and homeostasis at the tissue level.

Motile cilia are important because they "beat" rhythmically to move fluids or secretions across the cell surface (**Figure 3–4c**). A motile cilium is relatively stiff during the *power stroke* and flexible during the *return stroke*. The ciliated cells along your trachea beat their cilia in synchronized waves to move sticky mucus and trapped dust particles toward the throat and away from delicate respiratory surfaces. If the cilia are damaged or immobilized by heavy smoking or a metabolic problem, this cleansing action is lost. When irritants are no longer removed, a chronic cough and respiratory infections develop. Ciliated cells also move oocytes along the uterine tubes, and sperm from the testes into the male reproductive tract. Defective primary cilia are known to be responsible for a wide range of human disorders, collectively known as *ciliopathies*.

A **flagellum** (plural, *flagella*) is also an extension of the plasma membrane. Flagella have the same 9 + 2 microtubule organization as motile cilia, but are much longer and beat in a wavelike fashion. The only human cell with a flagellum is a sperm cell.

Ribosomes

Proteins are produced within cells, using information provided by the DNA of the nucleus. **Ribosomes** are the organelles responsible for protein synthesis. The number of ribosomes in a particular cell varies with the type of cell and its demand for new proteins. For example, liver cells, which manufacture blood proteins, contain far more ribosomes than do fat cells, which primarily synthesize lipids.

3

Figure 3–4 **Centrioles and Cilia.**

Microtubules

Plasma membrane

Microtubules

Basal body

b **Motile cilium**. A motile cilium contains nine pairs of microtubules surrounding a central pair (9 + 2 array). The basal body to which it is anchored has a microtubule array similar to that of a centriole.

Centriole   TEM × 185,000

a **Centriole**. A centriole consists of nine microtubule triplets (known as a 9 + 0 array). A pair of centrioles oriented at right angles to one another occupies the centrosome. This micrograph, produced by a transmission electron microscope, is called a TEM.

Power stroke   Return stroke

c **Ciliary movement**. Action of a single motile cilium. During the power stroke, the cilium is relatively stiff. During the return stroke, it bends and returns to its original position.

Individual ribosomes are not visible with the light microscope. In an electron micrograph, they appear as dense granules approximately 25 nm in diameter. Each ribosome is about 60 percent RNA and 40 percent protein.

A functional ribosome consists of two subunits that are normally separate and distinct. One is called a **small ribosomal subunit** and the other a **large ribosomal subunit**. These subunits contain special proteins and **ribosomal RNA (rRNA)**, one of the RNA types introduced in Chapter 2. Before protein synthesis can begin, a small and a large ribosomal subunit in the cytoplasm must join together with a strand of *messenger RNA* (*mRNA*), another type of RNA. Protein synthesis then begins in the cytoplasm.

The two major types of functional ribosomes in cells are free ribosomes and fixed ribosomes. **Free ribosomes** are scattered throughout the cytoplasm. The proteins they manufacture directly enter the cytosol. Ribosomes synthesizing proteins with destinations other than the cytosol become temporarily bound, or fixed, to the *endoplasmic reticulum (ER)*, a membranous organelle. Proteins manufactured by such **fixed ribosomes** enter the ER, where they are modified and packaged for use within the cell or they are secreted from the cell. We will look at ribosomal structure and functions when we discuss the endoplasmic reticulum and protein synthesis in later sections.

Proteasomes

Proteasomes are organelles that contain an assortment of protein-digesting (proteolytic) enzymes, or *proteases*. They are smaller than free ribosomes and their job is to remove proteins from the cytoplasm. Cytoplasmic enzymes attach chains of *ubiquitin*, a molecular "tag," to proteins destined for recycling. Tagged proteins are quickly transported into a proteasome. Once inside, they are rapidly disassembled into amino acids and small peptides, which are released into the cytoplasm.

Proteasomes remove and recycle damaged or denatured proteins. They also break down abnormal proteins, such as those produced within cells infected by viruses. Proteasomes also play a key role in the immune response, as we will see in Chapter 22.

Check Spotlight Figure 3–1 for a review of the characteristics of nonmembranous organelles before reading about membranous organelles in the next sections.

The Endoplasmic Reticulum

The **endoplasmic reticulum** (en-dō-PLAZ-mik re-TIK-ū-lum), or **ER,** is a network of intracellular membranes continuous with the *nuclear envelope*, which surrounds the nucleus. The name *endoplasmic reticulum* is very descriptive. *Endo-* means "within," *plasm* refers to the cytoplasm, and a *reticulum* is a network. The ER has four major functions:

1. *Synthesis.* Specialized regions of the ER synthesize proteins, carbohydrates, and lipids.

2. *Storage.* The ER can store synthesized molecules or materials absorbed from the cytosol without affecting other cellular operations.

3. *Transport.* Materials can travel from place to place within the ER.

4. *Detoxification.* The ER can absorb drugs or toxins and neutralize them with enzymes.

The ER forms hollow tubes, flattened sheets, and chambers called **cisternae** (sis-TUR-nē; singular, *cisterna,* a reservoir for water). Two types of ER exist: *smooth endoplasmic reticulum* and *rough endoplasmic reticulum* (**Figure 3–5**).

Smooth Endoplasmic Reticulum

The term "smooth" refers to the fact that no ribosomes are associated with the smooth endoplasmic reticulum (SER). The SER is involved with the synthesis of lipids, fatty acids, and carbohydrates; the sequestering of calcium ions; and the detoxification of drugs. Important functions of the SER include the following:

- *Synthesis of the phospholipids and cholesterol* needed for maintenance and growth of the plasma membrane, ER, nuclear membrane, and Golgi apparatus in all cells

- *Synthesis of steroid hormones,* such as *androgens* and *estrogens* (the dominant sex hormones in males and in females, respectively) in the reproductive organs

- *Synthesis and storage of glycerides, especially triacylglycerides,* in liver cells and fat cells

- *Synthesis and storage of glycogen* in skeletal muscle and liver cells

In muscle cells, neurons, and many other types of cells, the SER also adjusts the composition of the cytosol by absorbing and storing ions, such as Ca^{2+}, or large molecules. In addition, the SER in liver and kidney cells detoxifies or inactivates drugs.

Figure 3–5   **The Endoplasmic Reticulum.**

a The three-dimensional relationships between the rough and smooth endoplasmic reticula are shown here.

Rough endoplasmic reticulum with fixed (attached) ribosomes

Ribosomes

Cisternae

Nucleus

Free ribosomes

Smooth endoplasmic reticulum

Endoplasmic Reticulum   TEM × 111,000

b Rough endoplasmic reticulum and free ribosomes in the cytoplasm of a cell.

Rough Endoplasmic Reticulum

The rough endoplasmic reticulum (RER) functions as a combination workshop and shipping warehouse. It is where many newly synthesized proteins are chemically modified and packaged for export to their next destination, the *Golgi apparatus.*

The fixed ribosomes on the outer surface of the rough endoplasmic reticulum give the RER a beaded, grainy, or rough appearance (**Figure 3–5**). Both free and fixed ribosomes synthesize proteins using instructions provided by messenger RNA.

The new polypeptide chains produced at fixed ribosomes are released into the cisternae of the RER. Inside the RER, each protein assumes its secondary and tertiary structures. ⟲ p. 78 Some of the proteins are enzymes that will function inside the endoplasmic reticulum. Other proteins are chemically modified by the attachment of carbohydrates, creating glycoproteins.

Most of the proteins and glycoproteins produced by the RER are packaged into small membranous sacs that pinch off from the tips of the cisternae. These **transport vesicles** then deliver their contents to the Golgi apparatus (**Spotlight Figure 3–7**).

The amount of endoplasmic reticulum and the proportion of RER to SER vary with the type of cell and its ongoing activities. For example, pancreatic cells that manufacture digestive enzymes contain an extensive RER, but the SER is relatively small. The situation is just the reverse in the cells of reproductive organs that synthesize steroid hormones.

The Golgi Apparatus

When a transport vesicle carries a newly synthesized protein or glycoprotein that is destined for export from the cell, it travels from the ER to the **Golgi** (GŌL-jē) **apparatus**, or **Golgi complex**, an organelle that looks a bit like a stack of dinner plates (**Figure 3–6**). This organelle typically consists of five or six flattened membranous discs called *cisternae.* A single cell may contain several of these organelles, most often near the nucleus.

The Golgi apparatus has three major functions (**Spotlight Figure 3–7**). It

1. *modifies and packages secretions,* such as hormones or enzymes, for release from the cell;

2. *renews or modifies the plasma membrane;* and

3. *packages special enzymes within vesicles (lysosomes)* for use in the cytoplasm.

Lysosomes

Cells often need to break down and recycle large organic molecules and even complex structures like organelles. The breakdown process requires powerful enzymes, and it often generates toxic chemicals that could damage or kill the cell. **Lysosomes** (LĪ-sō-sōmz; *lyso-,* a loosening + *soma,* body) are special vesicles that provide an isolated environment for potentially dangerous chemical reactions. These vesicles, produced by the Golgi apparatus, contain digestive enzymes. Lysosomes are small, often spherical bodies with contents that look dense and dark in electron micrographs.

Lysosomes have several functions (see **Figure 3–8** on p. 106). One is to remove damaged organelles. *Primary lysosomes* contain inactive enzymes. When these lysosomes fuse with the membranes of damaged organelles (such as mitochondria or fragments of the ER), the enzymes are activated and *secondary lysosomes* are formed. The enzymes then break down the contents. The

Figure 3–6 The Golgi Apparatus.

Secretory vesicles

Secretory product

Transport vesicles

a Here is a three-dimensional view of the Golgi apparatus with a cut edge.

Golgi apparatus TEM × 42,000

b This is a sectional view of the Golgi apparatus of an active secretory cell.

cytosol reabsorbs released nutrients, and the remaining material is expelled from the cell.

Lysosomes also destroy bacteria (as well as liquids and organic debris) that enter the cell from the extracellular fluid. The cell encloses these substances in a small portion of the plasma membrane, which is then pinched off to form a transport vesicle, or *endosome*, in the cytoplasm. (We will discuss this method of transporting substances into the cell, called *endocytosis*, later in this chapter.) Then a primary lysosome fuses with the vesicle, forming a secondary lysosome. Activated enzymes inside break down the contents and release usable substances, such as sugars or amino acids. In this way, the cell both protects itself against harmful substances and obtains valuable nutrients.

Lysosomes also do essential cleanup and recycling inside the cell. For example, when muscle cells are inactive, lysosomes gradually break down their contractile proteins. (This mechanism accounts for the reduction in muscle mass that accompanies aging.) The process is usually precisely controlled, but in a damaged or dead cell, the regulatory mechanism fails. Lysosomes then disintegrate, releasing enzymes that become activated within the cytosol. These enzymes rapidly destroy the cell's proteins and organelles in a process called **autolysis** (aw-TOL-i-sis; *auto-*, self). We do not know how to control lysosomal activities or why the enclosed enzymes do not digest the lysosomal membranes unless the cell is damaged.

Problems in producing lysosomal enzymes cause more than 30 serious diseases affecting children. In these conditions, called *lysosomal storage diseases*, the lack of a specific lysosomal enzyme results in the buildup of waste products and debris that lysosomes normally remove and recycle. Affected individuals may die when vital cells, such as those of the heart, can no longer function.

Peroxisomes

Peroxisomes are smaller than lysosomes and carry a different group of enzymes. In contrast to lysosomes, which are produced at the Golgi apparatus, new peroxisomes are produced by the growth and subdivision of existing peroxisomes. Their enzymes are produced at free ribosomes and transported from the cytosol into the peroxisomes by carrier proteins.

Peroxisomes absorb and break down fatty acids and other organic compounds. As they do so, peroxisomes generate hydrogen peroxide (H_2O_2), a potentially dangerous free radical. ⟲ p. 60 Catalase, the most abundant enzyme within the peroxisome, then breaks down the hydrogen peroxide to oxygen and water. In this way, peroxisomes protect the cell from the potentially damaging effects of free radicals produced during catabolism. Peroxisomes are present in all cells, but their numbers are highest in metabolically active cells, such as liver cells.

Membrane Flow

When the temperature changes markedly, you change your clothes. Similarly, when a cell's environment changes, it changes the structure and properties of its plasma membrane. With the exception of mitochondria, all membranous organelles in the cell are either interconnected or in communication through the movement of vesicles. The membranes of the RER and SER are continuous and are connected to the nuclear envelope. Transport vesicles connect the ER with the Golgi apparatus, and secretory vesicles link the Golgi apparatus with the plasma membrane. Finally, vesicles forming at the exposed surface of the cell remove and recycle segments of the plasma membrane. This continuous movement and exchange of membrane segments is called **membrane flow**. In an actively secreting cell, an area equal to the entire membrane surface may be replaced each hour.

Membrane flow is an example of the dynamic nature of cells. It provides a way for cells to change the characteristics of their plasma membranes by altering their lipids and the proteins serving as receptors, channels, anchors, and enzymes—as they grow, mature, or respond to a specific environmental stimulus.

Mitochondria

The cells of all living things require energy to carry out the functions of life. The organelles that produce energy are the **mitochondria** (mī-tō-KON-drē-ŭh; singular, *mitochondrion*; *mitos*, thread + *chondrion*, granule). These small structures vary widely in shape, from long and slender to short and fat. The number of mitochondria in a particular cell varies with the cell's energy demands. These organelles may account for 30 percent of the volume of a heart muscle cell, yet are absent in red blood cells.

Mitochondria have an unusual double membrane (**Figure 3–9a**). The outer membrane surrounds the organelle. The inner membrane contains numerous folds called **cristae**, which surround the fluid contents, or **matrix**, of the mitochondrion. Cristae increase the membrane surface area in contact with the matrix and the number of attached protein complexes and enzymes involved in making ATP from ADP and P. Metabolic enzymes in the matrix catalyze reactions that release carbon dioxide and provide some additional energy for cellular functions.

Mitochondria contain their own DNA (mtDNA) and ribosomes. The mtDNA codes for small numbers of RNA and polypeptide molecules. The polypeptides are used in enzymes required for energy production. Although mitochondria contain their own genetic system, their functions depend on imported proteins coded by nuclear DNA.

Most of the chemical reactions that release energy take place in the mitochondria, yet most of the cellular activities

Figure 3–7
Protein Synthesis, Processing, and Packaging

The Golgi apparatus plays a major role in modifying and packaging newly synthesized proteins. Some proteins and glycoproteins synthesized in the rough endoplasmic reticulum (RER) are delivered to the Golgi apparatus by transport vesicles. Here's a summary of the process, beginning with DNA.

1 Protein synthesis begins when a gene on DNA produces messenger RNA (mRNA), the template for protein synthesis.

2 The mRNA leaves the nucleus and attaches to a free ribosome in the cytoplasm, or a fixed ribosome on the RER.

3 Proteins constructed on free ribosomes are released into the cytosol for use within the cell.

4 Protein synthesis on fixed ribosomes occurs at the RER. The newly synthesized protein folds into its 3-dimensional shape.

5 The proteins are then modified within the ER. Regions of the ER then bud off forming transport vesicles containing modified proteins and glycoproteins.

3 Protein released into cytoplasm

Ribosome

DNA

1

mRNA

2

Nucleus

Nuclear pore

Rough ER

Cytosol

4

5

Transport vesicle

6 The transport vesicles carry the proteins and glycoproteins generated in the ER toward the Golgi apparatus. The transport vesicles then fuse to create the forming cis face ("receiving side") of the Golgi apparatus.

7 Multiple transport vesicles combine to form cisternae on the cis face. Further protein and glycoprotein modification and packaging occur as the cisternae move toward the maturing (trans) face. Small transport vesicles return resident Golgi proteins to the forming cis face for reuse.

8 The maturing trans face ("shipping side") generates vesicles that carry modified proteins away from the Golgi apparatus. One type of vesicle becomes a lysosome, which contains digestive enzymes.

9 Two other types of vesicles proceed to the plasma membrane: secretory and membrane renewal. **Secretory vesicles** fuse with the plasma membrane and empty their products outside the cell by exocytosis. **Membrane renewal vesicles** add new lipids and proteins to the plasma membrane.

Cisternae

Lysosome

8

6

7

Forming (cis) face

Secretory vesicle

Maturing (trans) face

9

Exocytosis at cell surface

Membrane renewal vesicle

Membrane renewal

105

3

Figure 3–8 **Lysosome Functions.** Primary lysosomes, formed at the Golgi apparatus, contain inactive enzymes. They may be activated under any of the three basic conditions indicated here.

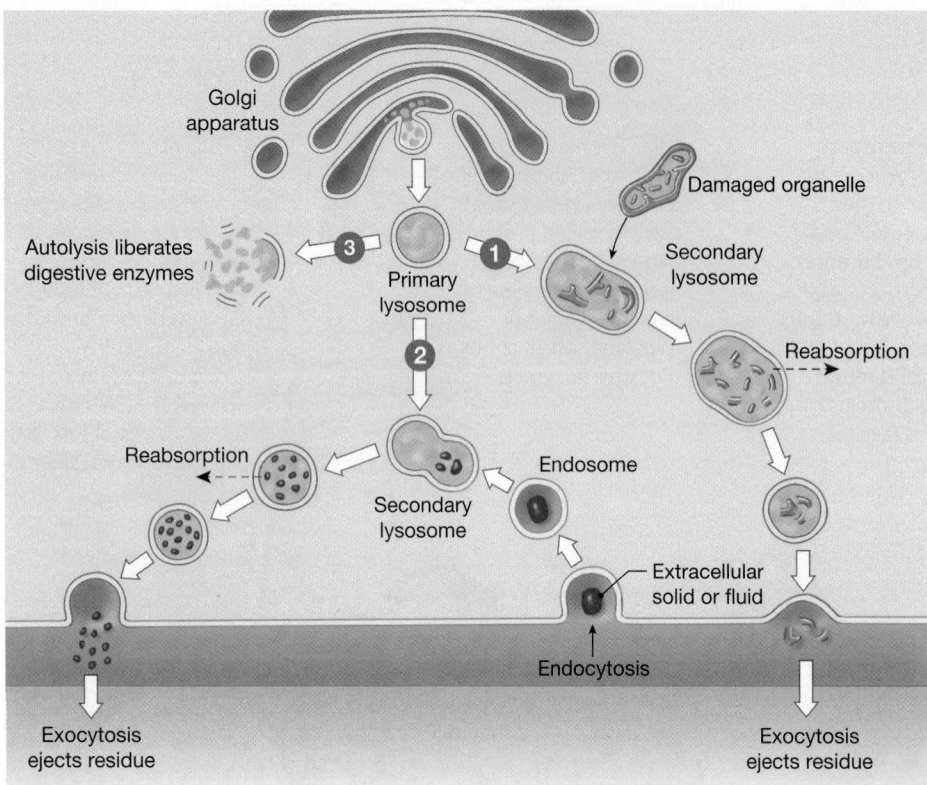

Lysosome activation occurs when:

1 A primary lysosome fuses with the membrane of another organelle, such as a mitochondrion

2 A primary lysosome fuses with an endosome containing fluid or solid materials from outside the cell

3 The lysosomal membrane breaks down during autolysis following injury to, or death of, the cell

that require energy occur in the surrounding cytoplasm. For this reason, cells must store energy in a form that can be moved from place to place. Recall from Chapter 2 that cellular energy is stored and transferred in the form of *high-energy bonds*. The best example is the high-energy bond that attaches a phosphate group (PO_4^{3-}) to adenosine diphosphate (ADP), creating the high-energy compound *adenosine triphosphate* (ATP). Cells can then break this high-energy bond under controlled conditions, reconverting ATP to ADP and phosphate and releasing energy for the cell's use when and where it is needed.

Mitochondrial Energy Production. Most cells generate ATP and other high-energy compounds as they break down carbohydrates, especially glucose. We will examine the entire process in Chapter 25, but a few basic concepts now will help you follow discussions of muscle contraction, neuron function, and endocrine function in Chapters 10–18.

Most ATP is produced inside mitochondria, but the first steps take place in the cytoplasm (**Figure 3–9b**). A reaction sequence called **glycolysis** (*glycos*, sugar + *-lysis*, a loosening) breaks down a glucose molecule into two molecules of *pyruvate*. Mitochondria then absorb the pyruvate molecules.

In the mitochondrial matrix, a carbon dioxide (CO_2) molecule is removed from each absorbed pyruvate molecule. The

Clinical Note

Inheritable Mitochondrial Disorders Several inheritable disorders result from abnormal mitochondrial activity. Mitochondria are not totally self-sufficient, but they do contain DNA and ribosomes to manufacture many of their own specific proteins. The mitochondria involved in congenital diseases contain abnormal DNA. As a result, the enzymes they produce reduce the efficiency of ATP production. Cells throughout the body may be affected. Symptoms involving muscle cells, neurons, and the receptor cells in the eye are most common because these cells have especially high energy demands.

remainder then enters the **citric acid cycle** (also known as the *Krebs cycle* and the *tricarboxylic acid cycle* or *TCA cycle*). The citric acid cycle is an enzymatic pathway that breaks down the absorbed pyruvate.

The remnants of pyruvate molecules contain carbon, oxygen, and hydrogen atoms. The carbon and oxygen atoms are

Figure 3–9 **Mitochondria.**

a Shown here is the three-dimensional organization and a color-enhanced TEM of a typical mitochondrion in longitudinal section.

b This is an overview of the role of mitochondria in energy production. Mitochondria absorb oxygen and short carbon chains, such as pyruvate, and they generate carbon dioxide, ATP, and water.

released as carbon dioxide, which diffuses out of the cell. The hydrogen atoms are delivered to carrier protein complexes in the cristae. There the electrons are removed from the hydrogen atoms and passed along a chain of coenzymes and ultimately transferred to oxygen atoms. The energy released during these steps indirectly supports the enzymatic conversion of ADP to ATP. ⟲ p. 84

Because mitochondrial activity requires oxygen, this method of ATP production is known as **aerobic metabolism** (*aer*, air + *bios*, life), or *cellular respiration*. Aerobic metabolism in mitochondria produces about 95 percent of the ATP needed to keep a cell alive. (Enzymatic reactions, including glycolysis, in the cytoplasm produce the rest.)

CHECKPOINT

5. Differentiate between the cytoplasm and the cytosol.

6. What are the major differences between cytosol and extracellular fluid?

7. Identify the nonmembranous organelles, and cite a function of each.

8. Identify the membranous organelles, and cite their functions.

9. Explain why certain cells in the ovaries and testes contain large amounts of smooth endoplasmic reticulum (SER).

10. What does the presence of many mitochondria imply about a cell's energy requirements?

See the blue Answers tab at the back of the book.

3-3 ❚ The nucleus contains DNA and enzymes essential for controlling cellular activities

Learning Outcome Explain the functions of the cell nucleus and discuss the nature and importance of the genetic code.

The **nucleus** is usually the largest and most conspicuous structure in a cell. If you look at a human cell under a light microscope, it may be the only organelle you see. The nucleus is the control center for cellular operations. A single nucleus stores all the information needed to direct the synthesis of more than 100,000 different proteins in the human body. The nucleus determines the structure of the cell and what functions it can perform by controlling which proteins are synthesized, under what circumstances, and in what amounts. A cell without

a nucleus cannot repair itself, so it will disintegrate within three or four months.

Most cells contain a single nucleus, but exceptions exist. For example, skeletal muscle cells have many nuclei, but mature red blood cells have none. **Figure 3–10** details the structure of a typical nucleus. Surrounding the nucleus and separating it from the cytosol is a **nuclear envelope,** a double membrane with its two layers separated by a narrow **perinuclear space** (*peri-*, around). At several locations, the nuclear envelope is connected to the rough endoplasmic reticulum (**Spotlight Figure 3–1**).

To direct processes that take place in the cytoplasm, the nucleus must receive information about conditions and activities in other parts of the cell. Chemical communication between the nucleus and the cytoplasm takes place through **nuclear pores**. These pores cover about 10 percent of the

Figure 3–10 **The Nucleus.**

Nucleoplasm

Chromatin

Nucleolus

Nuclear envelope

Nuclear pore

Nucleus TEM × 4800

a Important nuclear structures are shown here.

Nuclear pore

Perinuclear space

Nuclear envelope

b A nuclear pore is a large protein complex that spans the nuclear envelope.

Nuclear pores

Inner membrane of nuclear envelope

Broken edge of outer membrane

Outer membrane of nuclear envelope

Nucleus Freeze fracture SEM × 9240

c This cell was frozen and then broken apart to make its internal structures visible. The technique, called *freeze fracture* or *freeze-etching*, provides a unique perspective on the internal organization of cells. The nuclear envelope and nuclear pores are visible. The fracturing process broke away part of the outer membrane of the nuclear envelope, and the cut edge of the nucleus can be seen.

surface of the nucleus. They are large enough to allow ions and small molecules to enter or leave, but are too small for proteins or DNA to pass freely. Each nuclear pore contains regulatory proteins that govern the transport of specific proteins and RNA into or out of the nucleus.

Contents of the Nucleus

The fluid contents of the nucleus are called the *nucleoplasm*. The nucleoplasm contains the **nuclear matrix**, a network of fine filaments that provides structural support and may be involved in the regulation of genetic activity. The nucleoplasm also contains ions, enzymes, RNA and DNA nucleotides, small amounts of RNA, and DNA.

Most nuclei contain several dark-staining areas called **nucleoli** (noo-KLĒ-ō-lī; singular, *nucleolus*). Nucleoli are transient nuclear organelles that synthesize ribosomal RNA. They also assemble the ribosomal subunits, which then enter the cytoplasm through nuclear pores. Nucleoli are composed of RNA, enzymes, and proteins called **histones**. The nucleoli form around portions of DNA that contain the instructions for producing ribosomal proteins and RNA when those instructions are being carried out. Nucleoli are most prominent in cells that manufacture large amounts of proteins, such as liver, nerve, and muscle cells, because those cells need large numbers of ribosomes.

The DNA in the nucleus stores the instructions for protein synthesis. Interactions between the DNA and the histones help determine which information is available to the cell at any moment. The organization of DNA within the nucleus is shown in **Figure 3–11**. At intervals, the DNA strands wind around the histones, forming a complex known as a **nucleosome**. Such winding allows a great deal of DNA to be packaged in a small space.

The entire chain of nucleosomes may coil around other proteins. The degree of coiling varies, depending on whether cell division is underway. In cells that are not dividing, the nucleosomes are loosely coiled within the nucleus, forming a tangle of fine filaments known as **chromatin**. Chromatin gives the nucleus a clumped, grainy appearance. Just before cell division begins, the coiling becomes tighter, forming distinct structures called **chromosomes** (*chroma*, color). In humans, the nuclei of somatic cells contain 23 pairs of chromosomes. One member of each pair is derived from the mother, and one from the father.

Information Storage in the Nucleus

As we saw in Chapter 2, each protein molecule consists of a unique sequence of amino acids. ⟲ p. 77 Any "recipe" for a protein must specify the order of amino acids in the polypeptide chain. This information is stored in the chemical structure of the DNA strands in the nucleus. The chemical

Figure 3–11 **The Organization of DNA within the Nucleus.** DNA strands are coiled around histones to form nucleosomes. Nucleosomes form coils that may be very tight or rather loose. In cells that are not dividing, the DNA is loosely coiled, forming a tangled network known as chromatin. When the coiling becomes tighter, as it does in preparation for cell division, the DNA becomes visible as distinct structures called chromosomes.

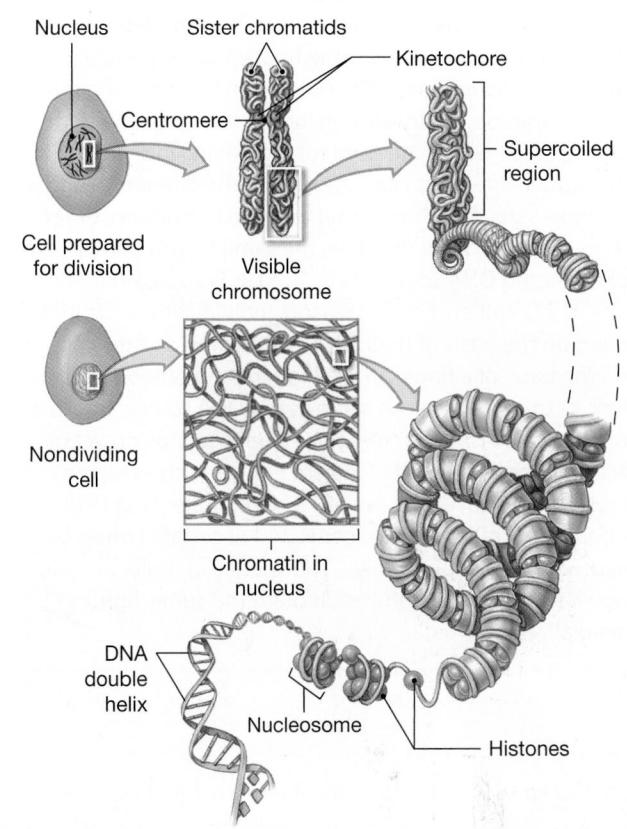

"language" the cell uses is known as the **genetic code**. An understanding of the genetic code has enabled researchers to figure out how cells build proteins and how various structural and functional characteristics are inherited from generation to generation.

How does the genetic code work? Recall the basic structure of nucleic acids described in Chapter 2. ⟲ p. 82 A single DNA molecule consists of a pair of DNA strands held together by hydrogen bonding between complementary nitrogenous bases. Information is stored in the sequence of nitrogenous bases along the length of the DNA strands. Those nitrogenous bases are adenine (A), thymine (T), cytosine (C), and guanine (G).

The genetic code is called a *triplet code*, because a sequence of three nitrogenous bases represents a single amino acid. Thus, the information encoded in the sequence of nitrogenous bases must be read in groups of three. For example, the triplet thymine–guanine–thymine (TGT) on one DNA strand, called

3

DNA Fingerprinting Every nucleated somatic cell in the body carries a set of 46 chromosomes that are copies of the set formed at fertilization. Not all the DNA of these chromosomes codes for proteins. Within the non-protein coding DNA are regions called Short Tandem Repeats (STRs), which contain the same nucleotide sequence repeated over and over. The number of STRs and the number of repetitions vary among individuals. In the United States, 13 core STRs are used for identifying individuals. The individual differences are reflected in the different lengths of these STRs. The chance that any two individuals, other than identical twins, will have the same length and pattern of repeating DNA segments in 13 STRs is approximately one in 575 trillion. For this reason, individuals can be identified on the basis of their DNA pattern, just as they can on the basis of a fingerprint. Skin scrapings, blood, semen, hair, or other tissues can serve as the DNA source. Information from *DNA fingerprinting* has been used to convict or acquit people accused of violent crimes, such as rape or murder. The U.S. Crime Act of 1994 mandates that DNA fingerprints of individuals convicted of certain crimes be kept in a retrievable archive. The science of molecular biology has become a useful addition to the crime-fighting arsenal.

the *coding strand,* is the code for the amino acid cysteine. More than one triplet may stand for the same amino acid, however. For example, the DNA triplet thymine–guanine–cytosine (TGC) also codes for cysteine.

A **gene** is the functional unit of heredity. It is the sequence of nucleotides of a DNA strand that specifies the amino acids needed to produce a specific protein. The needed number of triplet nucleotides, each coding for an amino acid, depends on the size of the polypeptide represented. A relatively short polypeptide chain might need fewer than 100 triplets, but the instructions for building a large protein might involve 1000 or more triplets.

Not all the nucleotide sequences of a DNA strand carry instructions for assembling proteins. Some sequences contain instructions for synthesizing ribosomal RNA or another type that we will discuss shortly, called *transfer RNA.* Others have a regulatory function, and still others have as yet no known function.

CHECKPOINT

11. Describe the contents and structure of the nucleus.

12. What is a gene?

See the blue Answers tab at the back of the book.

3-4 | DNA controls protein synthesis, cell structure, and cell function

Learning Outcome Summarize the role of DNA in protein synthesis, cell structure, and cell function.

The DNA of the nucleus controls the cell by directing the synthesis of specific proteins. Through protein synthesis, the DNA of the nucleus controls cell structure and function. The major events of protein synthesis are gene activation, transcription, and translation.

The Role of Gene Activation in Protein Synthesis

Each DNA molecule contains thousands of genes—the information needed to synthesize thousands of proteins. Normally, the genes are tightly coiled, and histones bound to the DNA keep the genes inactive. Before a gene can affect a cell, the portion of the DNA molecule containing that gene must be uncoiled and the histones temporarily removed in a process called **gene activation**.

The factors controlling this process are only partially understood. We know, however, that every gene contains segments responsible for regulating its own activity. In effect, these are triplets of nucleotides that say "read this message" or "do not read this message," "message starts here," or "message ends here." These "read me," "don't read me," and "start" signals form a special region of DNA called the *promoter*, or control segment, at the start of each gene. Each gene ends with a "stop" signal. Gene activation begins with the temporary disruption of the weak hydrogen bonds between the nitrogenous bases of the two DNA strands and the removal of the histone that guards the promoter.

The Transcription of mRNA

After the complementary DNA strands have separated and the histones have been removed, the enzyme **RNA polymerase** binds to the promoter of the gene. This binding is the first step in the process of **transcription**, the synthesis of RNA from a DNA template. The use of the term *transcription* makes sense, as it means "to copy" or "rewrite."

All three types of RNA are formed through the transcription of DNA, but we will focus here on the transcription of mRNA, which carries the information needed to synthesize proteins. The synthesis of **messenger RNA (mRNA)** is essential, because the DNA is too large to leave the nucleus. Instead, its information is copied to messenger RNA, which is smaller and *can* leave the nucleus and carry the information to the cytoplasm, where protein synthesis takes place.

Recall that the two DNA strands in a gene are complementary. The strand containing the triplets that specify the sequence of amino acids in the polypeptide is the **coding strand**. The other strand, called the **template strand**, contains complementary triplets that will be used as a template for mRNA production. The resulting mRNA will have a nucleotide sequence identical to that of the coding strand, but with uracil substituted for thymine.

Figure 3–12 illustrates the steps in transcription:

1 *RNA polymerase binding.* Once the DNA strands have separated and the promoter has been exposed, transcription can begin. The key event is the binding of RNA polymerase to the template strand.

2 *RNA polymerase forms and joins mRNA codons.* RNA polymerase promotes hydrogen bonding between the nitrogenous bases of the template strand nucleotides and complementary nucleotides in the nucleoplasm. This enzyme begins at a "start" signal in the promoter region. It then strings nucleotides together by covalent bonding. The RNA polymerase travels along the DNA strand, interacting with

only a small portion of the template strand at any one time. The complementary DNA strands separate in front of the enzyme as it moves one nucleotide at a time, and they reassociate behind it.

The enzyme collects additional nucleotides and attaches them one by one to the growing chain. The nucleotides involved are those characteristic of RNA, not of DNA. In other words, RNA polymerase can attach adenine, guanine, cytosine, or uracil, but never thymine. For this reason, wherever an A occurs in the DNA strand, the enzyme attaches a U rather than a T to the growing mRNA strand. In this way, RNA polymerase assembles a complete strand of mRNA.

Notice that the nucleotide sequence of the DNA template strand determines the nucleotide sequence of the mRNA strand. Each DNA triplet corresponds to a sequence of three nucleotides in the mRNA strand. Such a three-base mRNA sequence is called a **codon** (KŌ-don). Codons contain nitrogenous bases that are complementary to those of the triplets in the DNA template strand. For example, if the DNA triplet

Figure 3–12 mRNA Transcription. In this figure, a small portion of a single DNA molecule, containing a single gene, is undergoing transcription. **1** The two DNA strands separate, and RNA polymerase binds to the promoter of the gene. **2** The RNA polymerase moves from one nucleotide to another along the length of the template strand. At each site, complementary RNA nucleotides form hydrogen bonds with the DNA nucleotides of the template strand. The RNA polymerase then strings the arriving nucleotides together into a strand of mRNA. **3** On reaching the stop signal at the end of the gene, the RNA polymerase and the mRNA strand detach, and the two DNA strands reassociate.

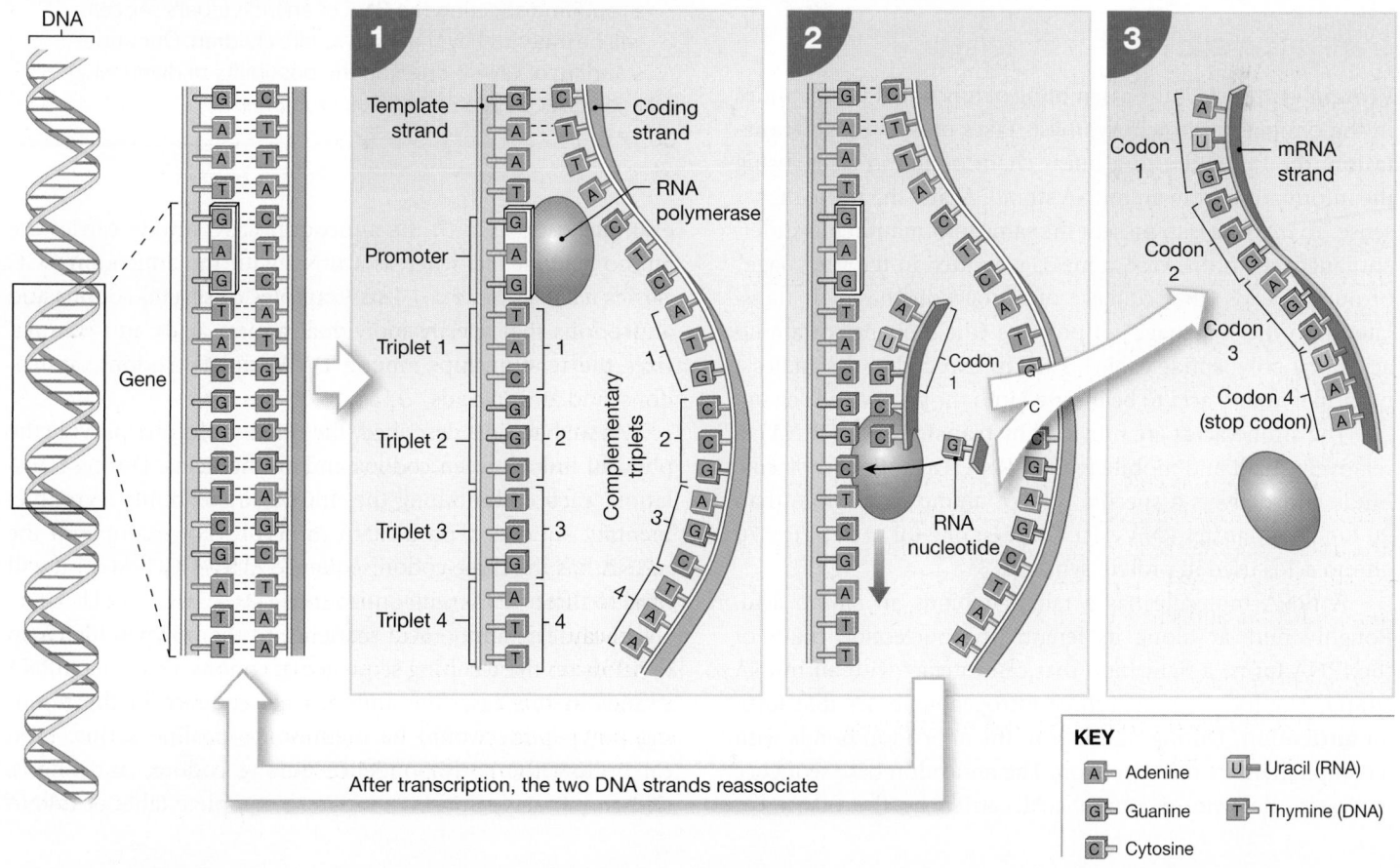

3

is TCG, the corresponding mRNA codon will be AGC. This method of copying ensures that the mRNA exactly matches the coding strand of the gene (except for the substitution of uracil for thymine).

3 *Detachment of mRNA.* At the "stop" signal, the enzyme and the mRNA strand detach from the DNA strand, and transcription ends. The complementary DNA strands now complete their reassociation as hydrogen bonding takes place again between complementary base pairs.

The nucleotide sequence of each gene includes a number of triplets that are not needed to build a functional protein. As a result, the mRNA strand assembled during transcription, sometimes called immature mRNA or *pre-mRNA*, must be "edited" before it leaves the nucleus to direct protein synthesis. In this **RNA processing**, non-coding intervening sequences, called **introns**, are snipped out, and the remaining coding segments, or **exons**, are spliced together. The process creates a much shorter, functional strand of mRNA that then enters the cytoplasm through a nuclear pore.

Intron removal is extremely important and tightly regulated. An error in the editing will produce an abnormal protein with potentially disastrous results. Moreover, we now know that by changing the editing instructions and removing different introns, a single gene can produce mRNAs that code for several different proteins. Some introns, however, act as enzymes to catalyze their own removal. How this variable editing is regulated is unknown.

Translation and Protein Synthesis

Protein synthesis is the assembling of functional polypeptides in the cytoplasm. Protein synthesis takes place through **translation**, the formation of a linear chain of amino acids, using the information from an mRNA strand. Again, the name makes sense: To *translate* is to present the same information in a different language. In this case, a message written in the "language" of nucleic acids (the sequence of nitrogenous bases) is translated into the "language" of proteins (the sequence of amino acids in a polypeptide chain). Each mRNA codon designates a particular amino acid to be inserted into the polypeptide chain.

The amino acids are provided by **transfer RNA (tRNA)**, a relatively small and mobile type of RNA. Each tRNA molecule binds and delivers a specific type of amino acid. More than 20 kinds of transfer RNA exist—at least one for each of the 20 amino acids used in protein synthesis.

A tRNA molecule has a tail that binds an amino acid. Roughly midway along its length, the nucleotide chain of the tRNA forms a tight loop that can interact with an mRNA strand. The loop contains three nitrogenous bases that form an **anticodon**. During translation, the anticodon bonds with a complementary mRNA codon. The anticodon base sequence indicates the type of amino acid carried by the tRNA. For

Clinical Note

Mutations **Mutations** are permanent changes in a cell's DNA that affect the nucleotide sequence of one or more genes.

The simplest is a *point mutation,* a change in a single nucleotide that affects one codon. The triplet code has some flexibility, because several different codons can specify the same amino acid. But a point mutation that produces a codon that specifies a different amino acid will usually change the structure of the completed protein. A single change in the amino acid sequence of a structural protein or enzyme can prove fatal. Certain cancers and two potentially lethal blood disorders discussed in Chapter 19, *thalassemia* and *sickle cell anemia,* result from changes in a single nucleotide.

Several hundred inherited disorders have been traced to abnormalities in enzyme or protein structure that reflect single changes in nucleotide sequence. More extensive mutations, such as additions or deletions of nucleotides, can affect multiple codons in one gene or in several adjacent genes. They can also affect the structure of one or more chromosomes.

Most mutations occur during DNA replication, when cells are duplicating their DNA in preparation for cell division. A single cell, a group of cells, or an entire individual may be affected. This last prospect occurs when the changes are made early in development. For example, a mutation affecting the DNA of an individual's sex cells will be inherited by that individual's children. Our understanding of DNA is opening the possibility of diagnosing and correcting some of these problems.

example, a tRNA with the anticodon GGC always carries the amino acid *proline*, whereas a tRNA with the anticodon CGG carries *alanine*. Table 3–1 lists examples of several codons and anticodons that specify individual amino acids and summarizes the relationships among DNA triplets, codons, anticodons, and amino acids.

As we have just described, the tRNA molecules provide the physical link between codons and amino acids. During translation, each codon along the mRNA strand binds a complementary anticodon on a tRNA molecule. For example, if the mRNA has the three-codon sequence AUG–CCG–AGC, it will bind to three tRNAs with anticodons UAC, GGC, and UCG, respectively. The amino acid sequence of the polypeptide chain depends on the resulting sequence of codons along the mRNA strand. In this case, the amino acid sequence in the resulting polypeptide would be methionine–proline–serine. (You can follow the correspondence among codons, anticodons, and amino acids in Table 3–1. For a complete table of mRNA

Table 3–1	Examples of the Genetic Code			
DNA Triplet				
Template Strand	**Coding Strand**	**mRNA Codon**	**tRNA Anticodon**	**Amino Acid**
AAA	TTT	UUU	AAA	Phenylalanine
AAT	TTA	UUA	AAU	Leucine
ACA	TGT	UGU	ACA	Cysteine
CAA	GTT	GUU	CAA	Valine
TAC	ATG	AUG	UAC	Methionine
TCG	AGC	AGC	UCG	Serine
GGC	CCG	CCG	GGC	Proline
CGG	GCC	GCC	CGG	Alanine

codons, the amino acids, and the start or stop signals they represent, see the Appendix at the back of the book.)

Recall that ribosomes are organelles that play a key role in protein synthesis. A functional ribosome consists of two subunits, a *small ribosomal subunit* and a *large ribosomal subunit*. At the beginning of the translation process in protein synthesis, a small and a large ribosomal subunit must join together with an mRNA strand. **Figure 3–13** illustrates the three phases of the translation process: *initiation, elongation,* and *termination.*

How the Nucleus Controls Cell Structure and Function

As we noted previously, the DNA of the nucleus controls the cell by directing the synthesis of specific proteins. Through the control of protein synthesis, virtually every aspect of cell structure and function can be regulated. Two levels of control are involved:

1. *Direct control.* The DNA of the nucleus has *direct* control over the synthesis of structural proteins, such as cytoskeletal components, membrane proteins (including receptors), and secretory proteins with various functional roles. By issuing appropriate instructions, in the form of mRNA strands, the nucleus can alter the internal structure of the cell, its sensitivity to substances in its environment, or its secretory functions to meet changing needs.

2. *Indirect control.* The DNA of the nucleus has *indirect* control over all other aspects of cellular metabolism, because it regulates the synthesis of enzymes. By ordering or stopping the production of appropriate enzymes, the nucleus can regulate all metabolic activities and functions of the cell. For example, the nucleus can accelerate the rate of glycolysis (p. 106) by increasing the number of needed enzymes in the cytoplasm.

How does the nucleus "know" what genes to activate? We don't have all the answers, but we know that in many cases gene activation or deactivation is triggered by changes in the surrounding cytoplasm. Such changes in the intracellular environment can affect the nucleoplasm enough to turn specific genes on or off. Alternatively, messengers or hormones may enter the nucleus through nuclear pores and bind to specific receptors or promoters along the DNA strands. In other words, continual chemical communication takes place between the cytoplasm and the nucleus. That communication is relatively selective, thanks to the barrier posed by the nuclear envelope and the restrictive characteristics of the nuclear pores.

Of course, continual communication also takes place between the cytoplasm and the extracellular fluid across the plasma membrane, and what crosses the plasma membrane today may alter gene activity tomorrow. In the next section, we will examine how the plasma membrane selectively regulates the passage of materials in and out of the cell.

CHECKPOINT

13. **Define gene activation.**

14. **Describe transcription and translation.**

15. **What process would be affected by the lack of the enzyme RNA polymerase?**

See the blue Answers tab at the back of the book.

3-5 ▌ Diffusion is a passive transport mechanism that assists membrane passage

Learning Outcome Describe the processes of cellular diffusion and osmosis, and explain their role in physiological systems.

The plasma membrane is a barrier that isolates the cytoplasm from the extracellular fluid. Because the plasma membrane is effective as a barrier, conditions inside the cell can be much different from conditions outside the cell. However, the barrier cannot be perfect, because cells are not self-sufficient. Each day they require nutrients to provide the energy they need to stay alive and function normally. They also generate waste products that must be removed. Your body has passageways and openings for nutrients, gases, and wastes, but a continuous, relatively uniform membrane surrounds the cell. So how do materials—whether nutrients or waste products—get across the plasma membrane without damaging it or reducing its effectiveness as a barrier?

The key to answering this question is **permeability**, the property of the plasma membrane that determines precisely which substances can enter or leave the cytoplasm. A membrane through which nothing can pass is **impermeable**. A membrane through which any substance can pass without

Figure 3–13   **The Process of Translation.**

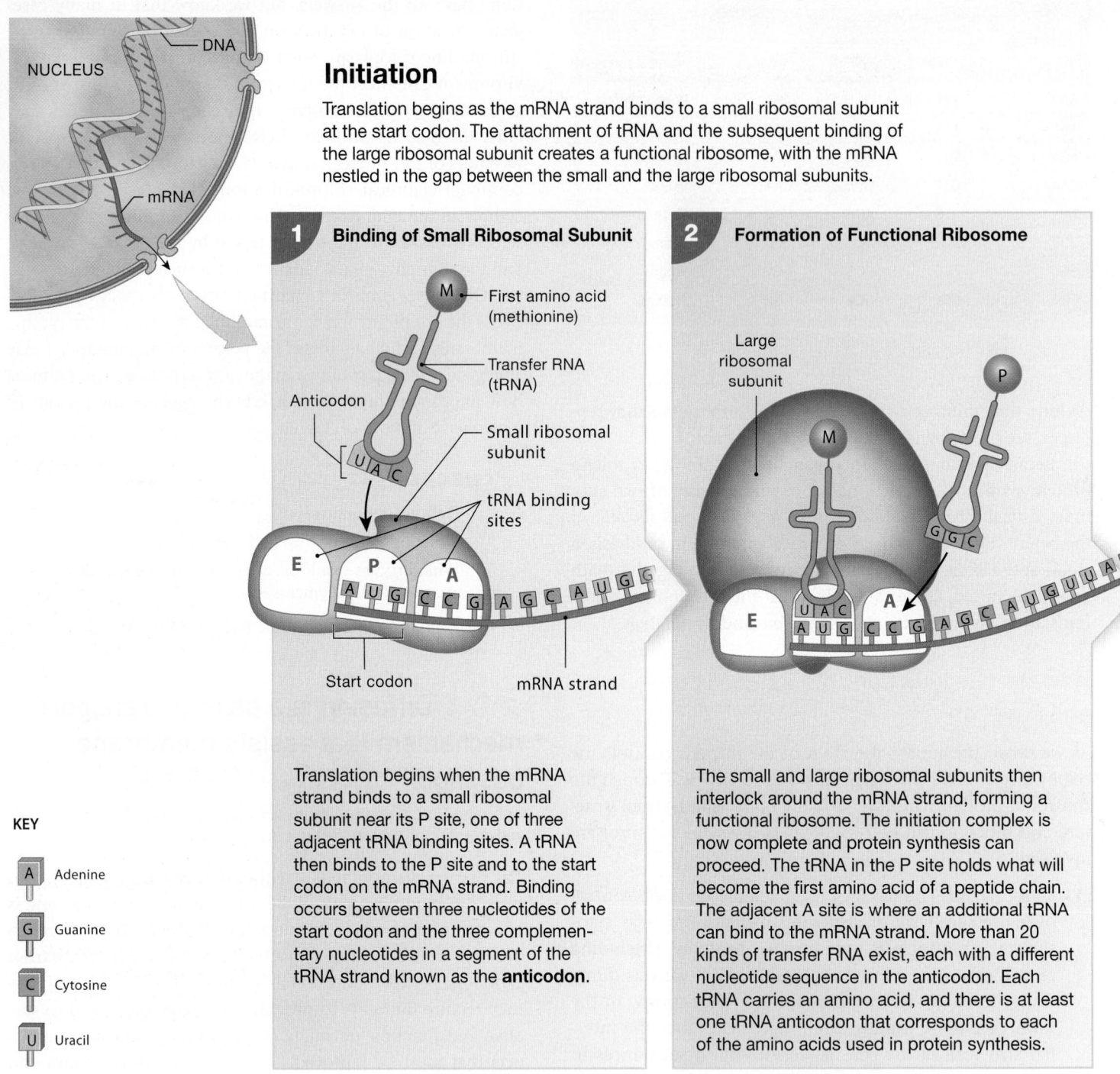

Initiation

Translation begins as the mRNA strand binds to a small ribosomal subunit at the start codon. The attachment of tRNA and the subsequent binding of the large ribosomal subunit creates a functional ribosome, with the mRNA nestled in the gap between the small and the large ribosomal subunits.

1   Binding of Small Ribosomal Subunit

First amino acid (methionine)

Transfer RNA (tRNA)

Anticodon

Small ribosomal subunit

tRNA binding sites

Start codon

mRNA strand

Translation begins when the mRNA strand binds to a small ribosomal subunit near its P site, one of three adjacent tRNA binding sites. A tRNA then binds to the P site and to the start codon on the mRNA strand. Binding occurs between three nucleotides of the start codon and the three complementary nucleotides in a segment of the tRNA strand known as the **anticodon**.

2   Formation of Functional Ribosome

Large ribosomal subunit

The small and large ribosomal subunits then interlock around the mRNA strand, forming a functional ribosome. The initiation complex is now complete and protein synthesis can proceed. The tRNA in the P site holds what will become the first amino acid of a peptide chain. The adjacent A site is where an additional tRNA can bind to the mRNA strand. More than 20 kinds of transfer RNA exist, each with a different nucleotide sequence in the anticodon. Each tRNA carries an amino acid, and there is at least one tRNA anticodon that corresponds to each of the amino acids used in protein synthesis.

KEY

A   Adenine

G   Guanine

C   Cytosine

U   Uracil

difficulty is **freely permeable**. Plasma membranes are called **selectively permeable**, however, because their permeability lies somewhere between those extremes.

A selectively permeable membrane permits the free passage of some materials and restricts the passage of others. The distinction may be based on size, electrical charge, molecular shape, lipid solubility, or other factors. Cells differ in their permeabilities, depending on the lipids and proteins in the plasma membrane and how these components are arranged.

Passage across the membrane is either passive or active. *Passive processes* move ions or molecules across the plasma membrane with no expenditure of energy by the cell. *Active processes* require that the cell expend energy, generally in the form of ATP.

Elongation

In elongation, amino acids are added one by one to the growing polypeptide chain, as the ribosome moves along the mRNA strand. Translation proceeds swiftly, producing a typical protein in about 20 seconds. Although only two mRNA codons are "read" by a ribosome at any one time, many ribosomes can bind to a single mRNA strand, and a multitude of identical proteins may be quickly and efficiently produced.

Termination

Elongation ends when the ribosome reaches the stop codon at the end of the mRNA strand. The mRNA strand remains intact, and it can interact with other ribosomes to create additional copies of the same polypeptide chain.

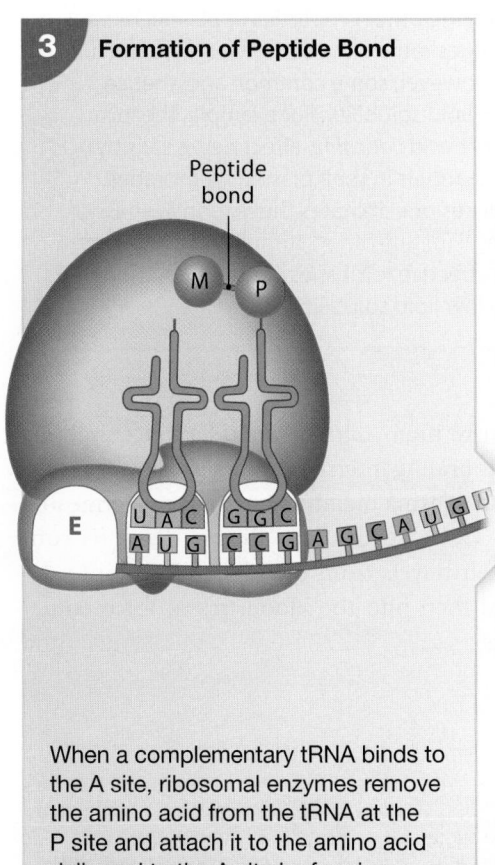

3 **Formation of Peptide Bond**

When a complementary tRNA binds to the A site, ribosomal enzymes remove the amino acid from the tRNA at the P site and attach it to the amino acid delivered to the A site by forming a peptide bond.

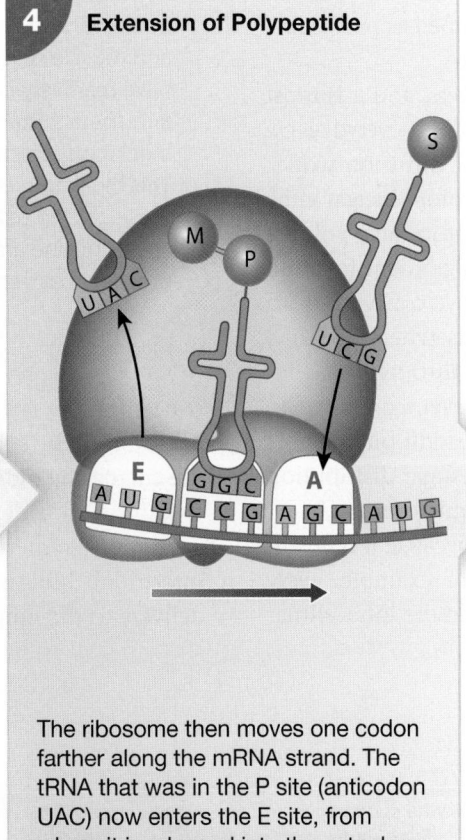

4 **Extension of Polypeptide**

The ribosome then moves one codon farther along the mRNA strand. The tRNA that was in the P site (anticodon UAC) now enters the E site, from where it is released into the cytoplasm. The released tRNA can now bind another amino acid of the same type and repeat the cycle.

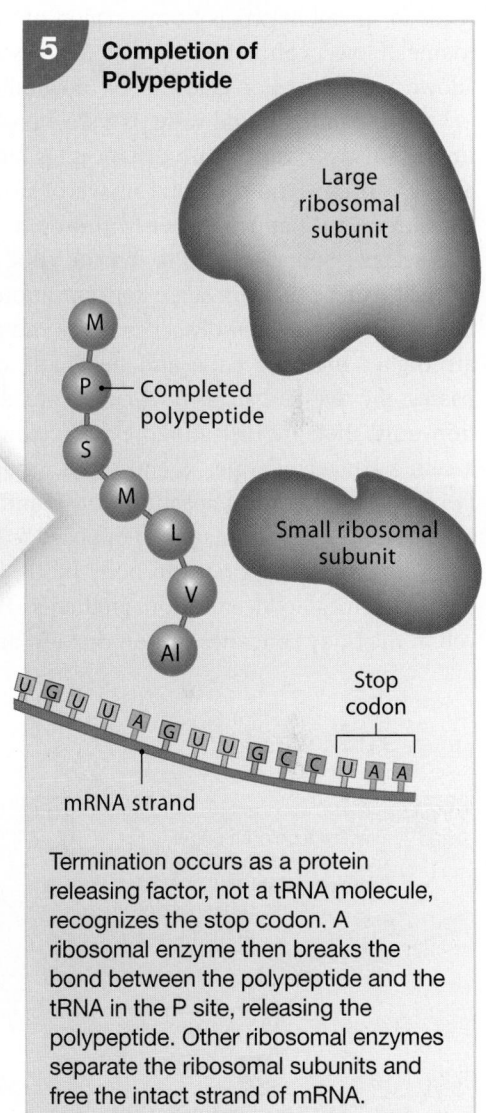

5 **Completion of Polypeptide**

Termination occurs as a protein releasing factor, not a tRNA molecule, recognizes the stop codon. A ribosomal enzyme then breaks the bond between the polypeptide and the tRNA in the P site, releasing the polypeptide. Other ribosomal enzymes separate the ribosomal subunits and free the intact strand of mRNA.

We categorize transport processes according to the mechanism involved: diffusion, carrier-mediated transport, and vesicular transport.

Diffusion

Diffusion is a passive process that results from the random motion and collisions of ions and molecules. Driven by thermal energy, ions and molecules are constantly in motion, colliding and bouncing off one another and off obstacles in their paths. The movement is both passive and random: A molecule can bounce in any direction. One result of this continuous random motion is that, over time, the molecules in any given space will tend to become evenly distributed. This distribution process, the net movement of a substance from an area of higher

3

concentration to an area of lower concentration, is called **diffusion**. The difference between the high and low concentrations is a **concentration gradient** (and thus a potential energy gradient). Diffusion tends to eliminate that gradient.

After the gradient has been eliminated, the molecular motion continues, but net movement no longer occurs in any particular direction. (For simplicity, we restrict use of the term *diffusion* to the directional movement that eliminates concentration gradients—a process sometimes called *net diffusion.*) Because diffusion tends to spread materials from a region of higher concentration to one of lower concentration, it is often described as proceeding "down a concentration gradient" or "downhill."

Diffusion in air and water is a slow process, and it is most important over very short distances. A simple, everyday example can give you a mental image of how diffusion works. Consider what happens when you drop a colored sugar cube into water (**Figure 3–14**). Placing the cube in a large volume of clear water sets up a steep concentration gradient for both ingredients as they dissolve: The sugar and dye concentrations are higher near the cube and negligible elsewhere. As time passes, the dye and sugar molecules spread through the solution until they are distributed evenly. However, compared to a cell, a beaker of water is enormous, and additional factors (which we will ignore) account for dye and sugar distribution over distances of centimeters as opposed to micrometers.

Diffusion is important in body fluids, because it tends to eliminate local concentration gradients. For example, every cell in the body generates carbon dioxide, and the intracellular

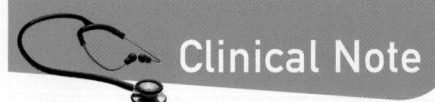

Clinical Note

Drugs and the Plasma Membrane Many clinically important drugs affect the plasma membrane. For some anesthetics, such as chloroform, ether, halothane, and nitrous oxide, potency is directly correlated with their lipid solubility. Presumably, high lipid solubility speeds the drug's entry into cells and enhances its ability to block ion channels or change other properties of plasma membranes. These changes reduce the sensitivity of neurons and muscle cells. However, some common anesthetics have relatively low lipid solubility. For example, the local anesthetics, *procaine* and *lidocaine,* affect nerve cells by blocking sodium channels in their plasma membranes. This blockage reduces or eliminates the responsiveness of these cells to painful (or any other) stimuli. Both procaine and lidocaine are effective local anesthetics, even though procaine has very low lipid solubility.

concentration is fairly high. Carbon dioxide concentrations are lower in the surrounding interstitial fluid, and lower still in the circulating blood. Plasma membranes are freely permeable to carbon dioxide (CO_2), so CO_2 can diffuse down its concentration gradient. It travels from the cell's interior into the interstitial fluid and then into the bloodstream, for eventual delivery to the lungs.

Figure 3–14 Diffusion.

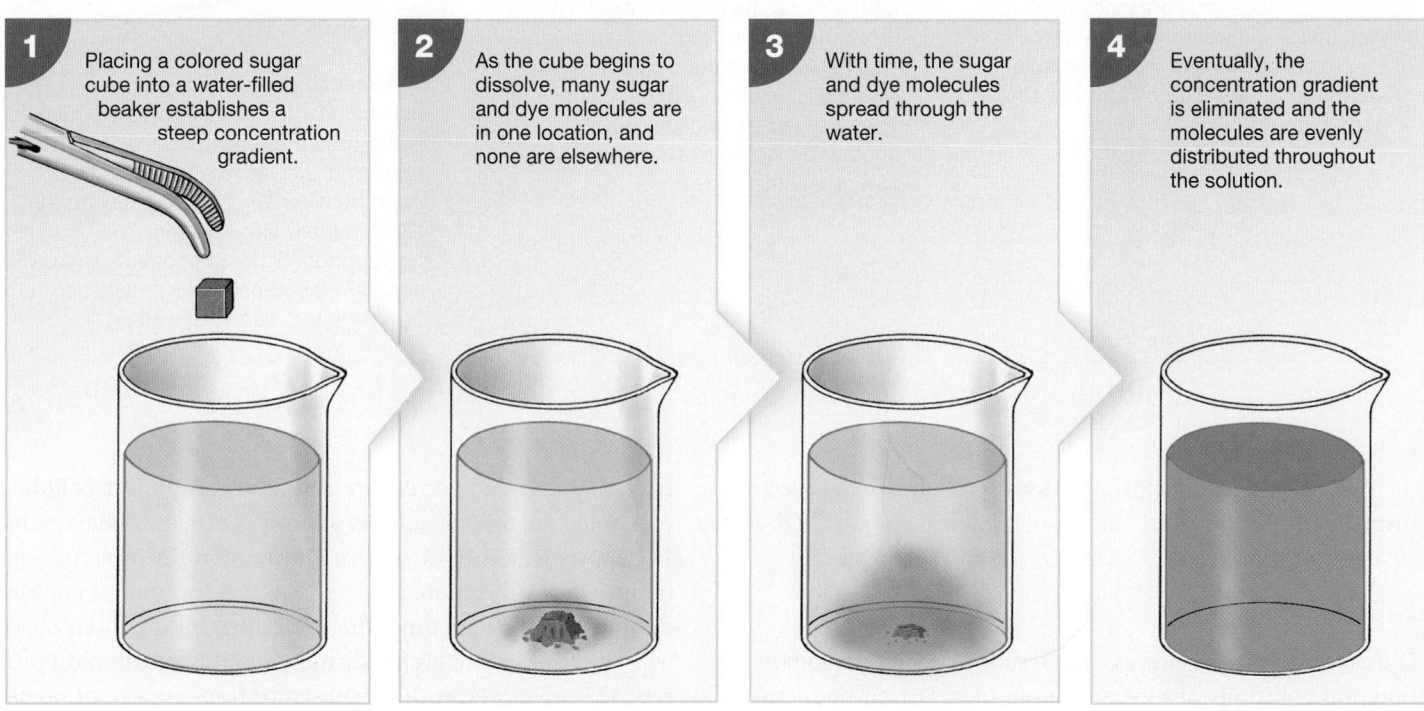

1 Placing a colored sugar cube into a water-filled beaker establishes a steep concentration gradient.

2 As the cube begins to dissolve, many sugar and dye molecules are in one location, and none are elsewhere.

3 With time, the sugar and dye molecules spread through the water.

4 Eventually, the concentration gradient is eliminated and the molecules are evenly distributed throughout the solution.

To be effective, the diffusion of nutrients, waste products, and dissolved gases must keep pace with the demands of active cells. Several important factors influence diffusion rates:

- *Distance.* The shorter the distance, the more quickly concentration gradients are eliminated. In the human body, few cells are farther than 25 μm from a blood vessel.

- *Molecule Size.* The smaller the molecule size, the faster the rate of diffusion. Ions and small organic molecules, such as glucose, diffuse more rapidly than do large proteins.

- *Temperature.* The higher the temperature, the faster the diffusion rate. Diffusion proceeds somewhat more quickly at human body temperature (about 37°C, or 98.6°F) than at cooler environmental temperatures.

- *Concentration Gradient.* The steeper the concentration gradient, the faster diffusion proceeds. When cells become more active, they use more oxygen, so intracellular concentration of oxygen declines. This change increases the concentration gradient for oxygen between the inside of the cell (somewhat low) and the interstitial fluid outside (somewhat high). The rate of oxygen diffusion into the cell then increases.

- *Electrical Forces.* Opposite electrical charges (+ and –) attract each other, and like charges (+ and + *or* – and –) repel each other. The inner cellular cytoplasmic surface of the plasma membrane has a net negative charge relative to the exterior surface, due in part to the high concentration of proteins in the cell. This negative charge tends to attract positive ions from the extracellular fluid into the cell, while repelling the entry of negative ions. For example, interstitial fluid contains higher concentrations of sodium ions (Na^+) and chloride ions (Cl^-) than does cytosol. Both the concentration gradient, or *chemical gradient,* and the electrical gradient favor diffusion of the positively charged sodium ions into the cell. In contrast, diffusion of the negatively charged chloride ions into the cell is favored by the chemical gradient, but opposed by the electrical gradient. For any ion, the net result of the chemical and electrical forces acting on it is called the *electrochemical gradient.*

Diffusion across Plasma Membranes

In extracellular fluids, water and dissolved solutes diffuse freely. A plasma membrane, however, acts as a barrier that selectively restricts diffusion: Some substances pass through easily, but others cannot penetrate the membrane. An ion or a molecule can diffuse across a plasma membrane only by (1) crossing the lipid portion of the membrane by simple diffusion or (2) passing through a membrane channel (**Figure 3–15**).

- *Simple Diffusion.* Alcohol, fatty acids, and steroids can enter cells easily, because they can diffuse through the lipid portions of the membrane. Lipid-soluble drugs, dissolved gases, such as oxygen and carbon dioxide, and

Figure 3–15 **Diffusion across the Plasma Membrane.** The path a substance takes in crossing a plasma membrane depends on the substance's size and lipid solubility.

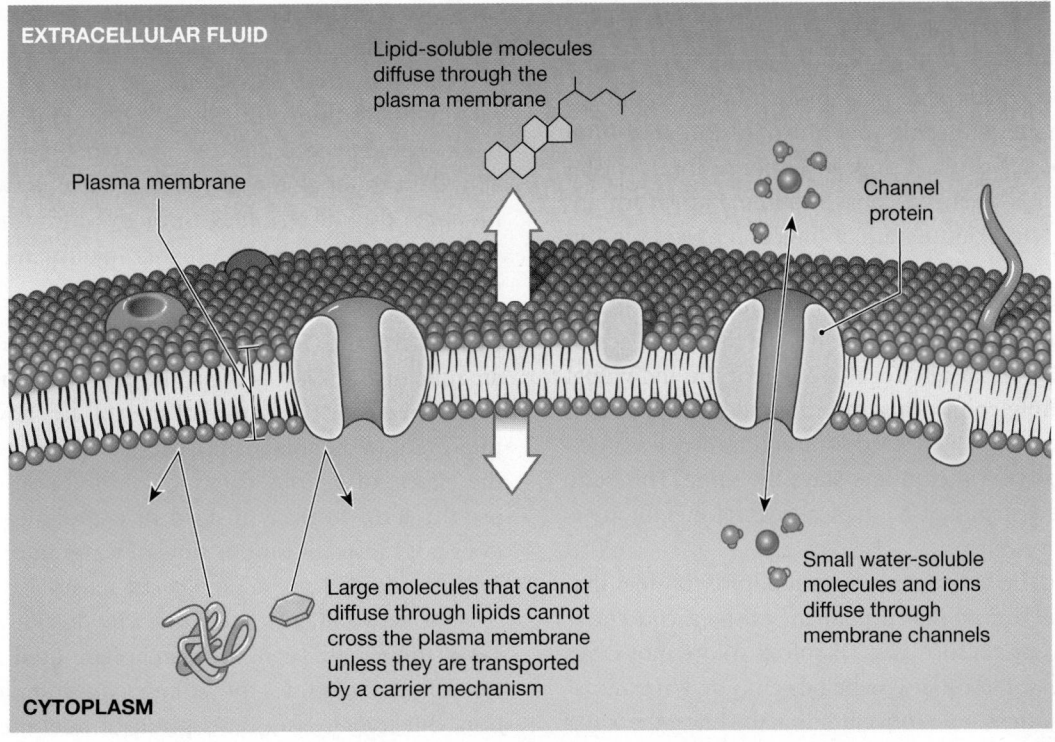

EXTRACELLULAR FLUID

Lipid-soluble molecules diffuse through the plasma membrane

Plasma membrane

Channel protein

Large molecules that cannot diffuse through lipids cannot cross the plasma membrane unless they are transported by a carrier mechanism

Small water-soluble molecules and ions diffuse through membrane channels

CYTOPLASM

3

water molecules also enter and leave our cells by diffusing through the phospholipid bilayer.

The situation is more complicated for ions and water-soluble compounds, which are not lipid-soluble. To enter or leave the cytoplasm, these substances must pass through a membrane channel.

- *Channel-Mediated Diffusion.* Membrane channels are very small passageways created by transmembrane proteins. On average, the channel is about 0.8 nm in diameter. Water molecules can enter or exit freely, but even a small organic molecule, such as glucose, is too big to fit through the channels. Whether an ion can cross a particular membrane channel depends on many factors, including the size and charge of the ion, the size of the hydration sphere, and interactions between the ion and the channel walls. ⟳ p. 66 **Leak channels**, also called passive channels, remain open and allow ions to pass across the plasma membrane.

The mechanics of diffusion through membrane channels is more complex than simple diffusion. For example, the rate at which a particular ion diffuses across the membrane can be limited by the availability of suitable channels. However, many ions, including sodium, potassium, and chloride, move across the plasma membrane at rates comparable to those for simple diffusion.

Osmosis: A Special Case of Diffusion

The net diffusion of water across a membrane is so important that it has a special name: **osmosis** (oz-MŌ-sis; *osmos*, a push). We will always use the term *osmosis* for the movement of water, and the term *diffusion* for the movement of solutes.

Intracellular and extracellular fluids are solutions with a variety of dissolved materials. Each solute diffuses as though it were the only material in solution. The diffusion of sodium ions, for example, takes place only in response to a concentration gradient for sodium. A concentration gradient for another ion has no effect on the rate or direction of sodium ion diffusion.

Some solutes diffuse into the cytoplasm, others diffuse out, and a few (such as proteins) are unable to diffuse across the plasma membrane at all. Yet if we ignore the individual identities and simply count ions and molecules, we find that the *total* concentration of dissolved ions and molecules on either side of the plasma membrane stays the same. This state of equilibrium persists because a typical plasma membrane is freely permeable to water.

To understand the basis for such equilibrium, consider that whenever a solute concentration gradient exists, a concentration gradient for *water* also exists. Dissolved solute molecules occupy space that would otherwise be taken up by water molecules, so the higher the solute concentration, the lower the water

concentration. As a result, *water molecules tend to flow across a membrane toward the solution with the higher solute concentration,* because this movement is *down* the concentration gradient for water. Water will move until water concentrations—and thus solute concentrations—are the same on either side of the membrane. Remember these basic characteristics of osmosis:

- Osmosis is the diffusion of water molecules across a selectively permeable membrane.

- Osmosis takes place across a selectively permeable membrane that is freely permeable to water, but not freely permeable to solutes.

- In osmosis, water flows across a selectively permeable membrane toward the solution that has the higher concentration of solutes, because that is where the concentration of water is lower.

Osmosis and Osmotic Pressure. Figure 3–16 diagrams the process of osmosis. ① shows two solutions (A and B), with different solute concentrations, separated by a selectively permeable membrane. As osmosis takes place, water molecules cross the membrane until the solute concentrations in the two solutions are identical (②). For this reason, the volume of solution B increases while that of solution A decreases. The greater the initial difference in solute concentrations, the stronger is the osmotic flow.

The **osmotic pressure** of a solution is an indication of the force with which pure water moves into that solution as a result of its solute concentration. We can measure a solution's osmotic pressure in several ways. For example, an opposing pressure can prevent the osmotic flow of water into the solution. Pushing against a fluid generates **hydrostatic pressure**. In ③ hydrostatic pressure opposes the osmotic pressure of solution B, so no net osmotic flow occurs.

Osmosis eliminates solute concentration differences more rapidly than solute diffusion. Why? In large part this happens because water molecules can also cross a membrane through abundant water channels called *aquaporins*, which exceed the number of solute channels, through which water can also pass. This difference results in a higher membrane permeability for water than for solutes.

Osmolarity and Tonicity. The total solute concentration in an aqueous solution is the solution's **osmolarity**, or **osmotic concentration**. The nature of the solutes, however, is often as important as the total osmolarity. Therefore, when we describe the effects of various osmotic solutions on cells, we usually use the term **tonicity** instead of osmolarity. A solution that does not cause an osmotic flow of water into or out of a cell is called **isotonic** (*iso-*, same + *tonos*, tension).

The terms *osmolarity* and *tonicity* do not always mean the same thing, although they are often used interchangeably. Osmolarity refers to the solute concentration of the solution, but tonicity is a description of how the solution affects

Figure 3–16 Osmosis. The osmotic pressure of solution B is equal to the amount of hydrostatic pressure required to stop the osmotic flow.

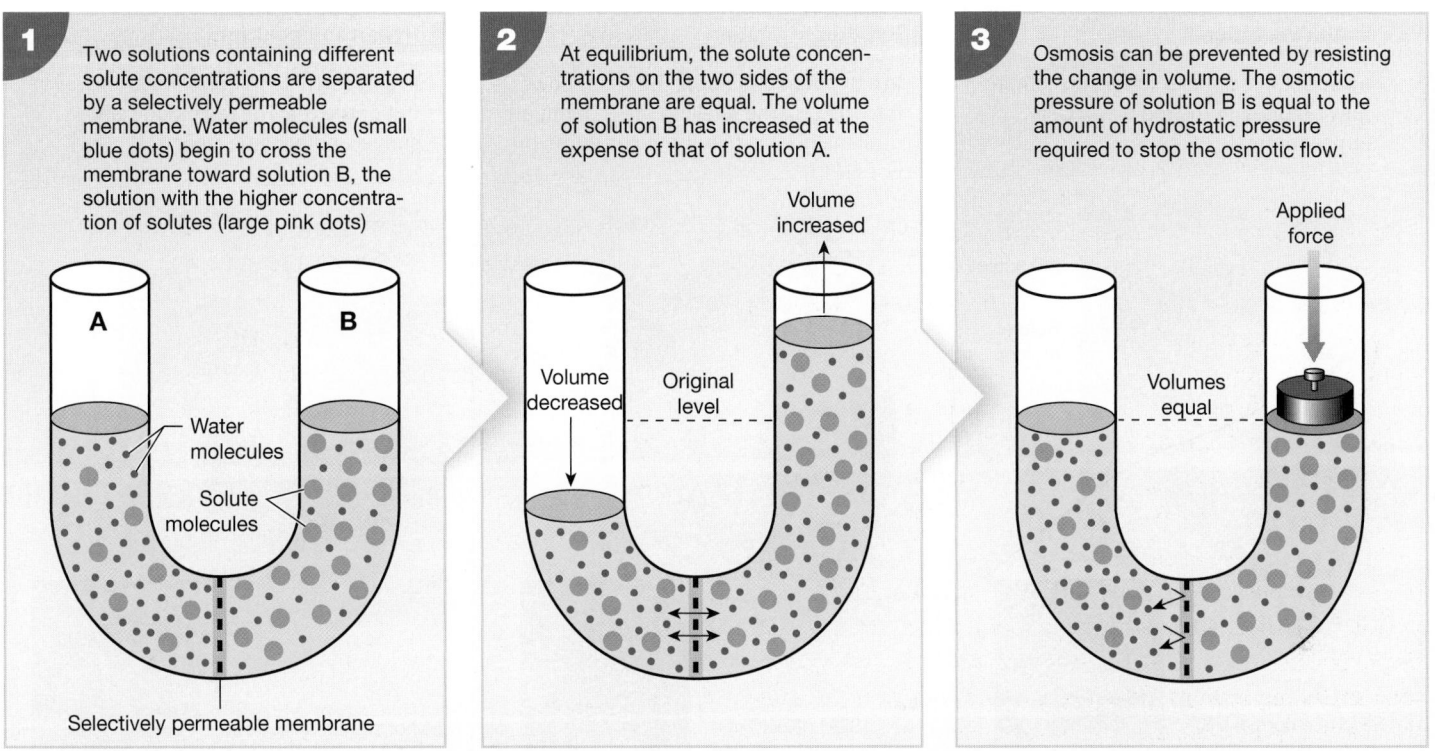

1 Two solutions containing different solute concentrations are separated by a selectively permeable membrane. Water molecules (small blue dots) begin to cross the membrane toward solution B, the solution with the higher concentration of solutes (large pink dots)

2 At equilibrium, the solute concentrations on the two sides of the membrane are equal. The volume of solution B has increased at the expense of that of solution A.

3 Osmosis can be prevented by resisting the change in volume. The osmotic pressure of solution B is equal to the amount of hydrostatic pressure required to stop the osmotic flow.

the shape of a cell. For example, consider a solution with the same osmolarity as the intracellular fluid, but a higher concentration of one or more individual ions. What happens if you put a cell into this solution? If any of those ions can cross the plasma membrane and diffuse into the cell, the osmolarity of the intracellular fluid will increase, and that of the extracellular solution will decrease. Osmosis will then also occur, moving water into the cell. If the process continues, the cell will gradually inflate like a water balloon. In this case, the extracellular solution and the intracellular fluid were initially equal in osmolarity, but they were not isotonic.

Figure 3–17a shows a red blood cell (RBC) in an isotonic solution, where no osmotic flow takes place and the shape of the cell looks normal. But if you put a red blood cell into a **hypotonic** solution, water will flow into the cell, causing it to swell up like a balloon (Figure 3–17b). The cell may eventually burst, releasing its contents. This event is **hemolysis** (*hemo-*, blood + *lysis*, a loosening). In contrast, a cell in a **hypertonic** solution will lose water by osmosis. As it does, the cell shrivels and dehydrates. The shrinking of red blood cells is called **crenation** (Figure 3–17c).

It is often necessary to give patients large volumes of fluid to combat severe blood loss or dehydration. The procedure used in many cases is called an intravenous, or IV, drip and the fluids enter a patient's vein. One fluid frequently administered is a 0.9 percent (0.9 g/dL) solution of sodium chloride (NaCl). This solution, called *normal saline,* approximates the normal

osmotic concentration of extracellular fluids. It is used because sodium and chloride are the most abundant ions in the extracellular fluid. Little net movement of either ion across plasma membranes takes place, so normal saline is essentially isotonic to body cells. An alternative IV treatment involves the use of an isotonic saline solution containing *dextran,* a carbohydrate that cannot cross plasma membranes. The dextran molecules elevate the osmolarity and osmotic pressure of the blood, and as water enters the blood vessels from the surrounding tissue fluid by osmosis, blood volume increases.

CHECKPOINT

16. What is meant by "selectively permeable" when referring to a plasma membrane?

17. Define diffusion.

18. List five factors that influence the diffusion of substances in the body.

19. How would a decrease in the concentration of oxygen in the lungs affect the diffusion of oxygen into the blood?

20. Define osmosis.

21. Some pediatricians recommend using a 10 percent salt solution as a nasal spray to relieve congestion in infants with stuffy noses. What effect would such a solution have on the cells lining the nasal cavity, and why?

See the blue Answers tab at the back of the book.

3

Figure 3–17 **Osmotic Flow across a Plasma Membrane.** The smaller paired arrows indicate an equilibrium with no net water movement. The larger arrows indicate the direction of osmotic water movement.

a Isotonic solution

In an isotonic saline solution, no osmotic flow occurs, and the red blood cells appear normal.

Water molecules

Solute molecules

SEM of a normal RBC in an isotonic solution

b Hypotonic solution

In a hypotonic solution, the water flows into the cell. The swelling may continue until the plasma membrane ruptures, or lyses.

SEM of RBC in a hypotonic solution

c Hypertonic solution

In a hypertonic solution, water moves out of the cell. The red blood cells shrivel and become crenated.

SEM of crenated RBCs in a hypertonic solution

3-6 ∣ Carrier-mediated and vesicular transport assist membrane passage

Learning Outcome Describe carrier-mediated transport and vesicular transport mechanisms used by cells to facilitate the absorption or removal of specific substances.

Besides diffusion, substances are taken into or removed from cells in two additional ways: carrier-mediated transport and vesicular transport. *Carrier-mediated transport* requires specialized integral membrane proteins. It can be passive or active, depending on the substance transported and the nature of the transport mechanism. *Vesicular transport* involves moving materials within small membranous sacs, or *vesicles*. Vesicular transport is always an active process.

Carrier-Mediated Transport

In **carrier-mediated transport**, integral proteins bind specific ions or organic substrates and carry them across the plasma membrane. All forms of carrier-mediated transport have the following characteristics, which they share with enzymes:

- *Specificity.* Each carrier protein in the plasma membrane binds and transports only certain substances. For example,

the carrier protein that transports glucose will not transport other simple sugars.

- *Saturation Limits.* The availability of substrate molecules and carrier proteins limits the rate of transport into or out of the cell, just as the availability of substrates and enzymes limits enzymatic reaction rates. When all the available carrier proteins are operating at maximum speed, the carriers are said to be *saturated*. The rate of transport cannot increase further, regardless of the size of the concentration gradient.

- *Regulation.* Just as enzyme activity often depends on the presence of cofactors, the binding of other molecules, such as hormones, can affect the activity of carrier proteins. For this reason, hormones provide an important means of coordinating carrier protein activity throughout the body. We will examine the interplay between hormones and plasma membranes when we study the endocrine system (Chapter 18) and metabolism (Chapter 25).

Many examples of carrier-mediated transport involve the movement of a single substrate molecule across the plasma membrane. However, a few carrier proteins may move more than one substrate at a time. In **cotransport**, or *symport*, the carrier transports two substances in the same direction simultaneously, either into or out of the cell. In **countertransport**, or *antiport*, one substance moves into the cell and the other moves out.

Let's look at two examples of carrier-mediated transport: *facilitated diffusion* and *active transport*.

Facilitated Diffusion

Many essential nutrients, such as glucose and amino acids, are insoluble in lipids and too large to fit through membrane channels. These substances can be passively transported across the membrane by carrier proteins in a process called **facilitated diffusion** (Figure 3–18). The molecule to be transported must first bind to a **receptor site** on the carrier protein. The shape of the protein then changes, moving the molecule across the plasma membrane and releasing it into the cytoplasm. This takes place without ever creating a continuous open channel between the cell's exterior and interior.

TIPS&TRICKS

To understand the mechanism of cellular channels, think about entering a store that has a double set of automated sliding doors. When you near the first set of automated doors, it opens. As you step onto the mat between the sets of doors, the doors behind you close, trapping you in the vestibule. When you take another step forward, the second set of doors opens, enabling you to enter the store.

As in the case of simple or channel-mediated diffusion, no ATP is expended in facilitated diffusion: The molecules simply move from an area of higher concentration to one of lower concentration. However, once the carrier proteins are saturated, the rate of transport cannot increase, regardless of further increases in the concentration gradient. Facilitated diffusion may occur in either direction across a plasma membrane.

All cells move glucose across their membranes through facilitated diffusion. However, several different carrier proteins are involved. In muscle cells, fat cells, and many other types of cells, the glucose transporter functions only when stimulated by the hormone *insulin*. Inadequate production of this hormone is one cause of *diabetes mellitus*, a metabolic disorder characterized by high blood glucose levels, which we will discuss in Chapter 18.

Active Transport

In **active transport**, a high-energy bond (in ATP or another high-energy compound) provides the energy needed to move ions or molecules across the membrane. Despite the energy cost, active transport offers one great advantage: It does not depend on a concentration gradient. As a result, the cell can import or export specific substrates, *regardless of their intracellular or extracellular concentrations.*

All cells contain carrier proteins called **ion pumps**, which actively transport the cations sodium (Na^+), potassium (K^+), calcium (Ca^{2+}), and magnesium (Mg^{2+}) across their plasma membranes. Specialized cells can transport additional ions, such as iodide (I^-), chloride (Cl^-), and iron (Fe^{2+}).

Many of these carrier proteins move a specific cation or anion in one direction only, either into or out of the cell. Sometimes, one carrier protein will move more than one kind of ion at the same time. If countertransport occurs, the carrier protein is called an **exchange pump**.

Figure 3–18   Facilitated Diffusion. In facilitated diffusion, a molecule, such as glucose, binds to a specific receptor site on a carrier protein. This binding permits the molecule to diffuse across the plasma membrane. If the direction of the glucose concentration gradient is reversed, then facilitated diffusion would occur in the opposite direction.

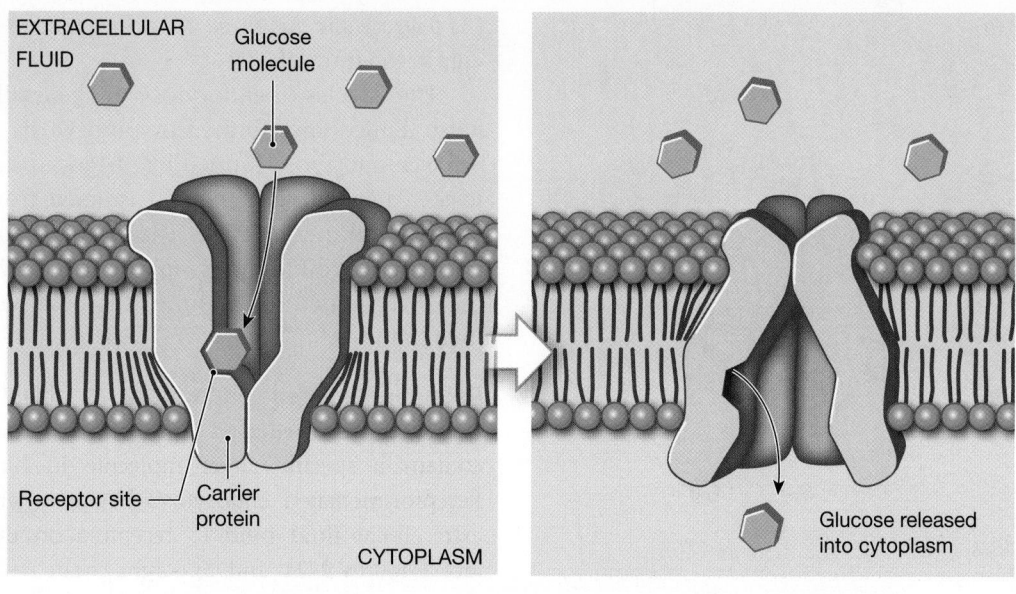

The Sodium–Potassium Exchange Pump. Sodium and potassium ions are the principal cations in body fluids. Sodium ion concentrations are high in the extracellular fluids, but low in the cytoplasm. The distribution of potassium in the body is just the opposite: low in the extracellular fluids and high in the cytoplasm. Because of leak channels in plasma membranes, sodium ions slowly diffuse into the cell, and potassium ions diffuse out. Homeostasis within the cell depends on removing sodium ions and recapturing lost potassium ions. This exchange takes place by a **sodium–potassium exchange pump**. The carrier protein involved in the process is called *sodium–potassium ATPase*.

The sodium–potassium exchange pump exchanges intracellular sodium for extracellular potassium (**Figure 3–19**). On average, for each ATP molecule consumed, three sodium ions are ejected and two potassium ions are reclaimed by the cell. If ATP is readily available, the rate of transport depends on the concentration of sodium ions in the cytoplasm. When the concentration rises, the pump becomes more active. The energy demands are impressive: Sodium–potassium ATPase may use up to 40 percent of the ATP produced by a resting cell!

Secondary Active Transport. In **secondary active transport**, the transport mechanism itself does not require energy from ATP, but the cell often needs to expend ATP at a later time to preserve homeostasis. As with facilitated diffusion, a secondary active transport mechanism moves a specific substrate down its concentration gradient. Unlike the proteins in facilitated diffusion, however, these carrier proteins can also move another substrate at the same time, without regard to its concentration gradient. In effect, the concentration gradient for one substance provides the driving force needed by the carrier protein, and the second substance gets a "free ride."

The concentration gradient for sodium ions most often provides the driving force for cotransport mechanisms that move materials into the cell. For example, sodium-linked cotransport is important in the absorption of glucose and amino acids along the intestinal tract. The initial transport activity proceeds without direct energy expenditure, but the cell must then expend ATP to pump the arriving sodium ions out of the cell by using the sodium–potassium exchange pump (**Figure 3–20**).

Sodium ions are also involved with many countertransport mechanisms. For example, sodium–calcium countertransport keeps intracellular calcium ion concentrations very low.

Vesicular Transport

In **vesicular transport**, materials move into or out of the cell in **vesicles**, small membranous sacs that form at, or fuse with, the plasma membrane. Because these vesicles move tiny droplets of fluid and solutes rather than single molecules, this process is also known as *bulk transport*. The two major types of vesicular transport are *endocytosis* and *exocytosis*.

Endocytosis

We saw earlier in this chapter that extracellular materials can be packaged in vesicles at the cell surface and imported into the cell. This process, called **endocytosis**, involves relatively large volumes of extracellular material. The three major types of endocytosis are (1) receptor-mediated endocytosis, (2) *pinocytosis*, and (3) *phagocytosis*. All three are active processes that require energy in the form of ATP.

The vesicles of endocytosis are generally known as *endosomes*. Endosomes formed by pinocytosis are also called *pinosomes*, and those formed by phagocytosis are called *phagosomes*. Their contents remain isolated from the cytoplasm, trapped within the vesicle. The enclosed materials may move into the surrounding cytoplasm by active transport, simple or facilitated diffusion, or the destruction of the vesicle membrane.

Receptor-Mediated Endocytosis. A highly selective process, **receptor-mediated endocytosis** produces vesicles that contain a specific target molecule in high concentrations. Receptor-mediated endocytosis begins when materials in the extracellular fluid bind to receptors on the membrane surface (**Figures 3–21** and **Spotlight Figure 3–22**). Most receptor

Figure 3–19 **The Sodium–Potassium Exchange Pump.** The operation of the sodium–potassium exchange pump is an example of active transport. For each ATP converted to ADP, this carrier protein pump, also called sodium–potassium ATPase, carries three Na$^+$ out of the cell and two K$^+$ into the cell.

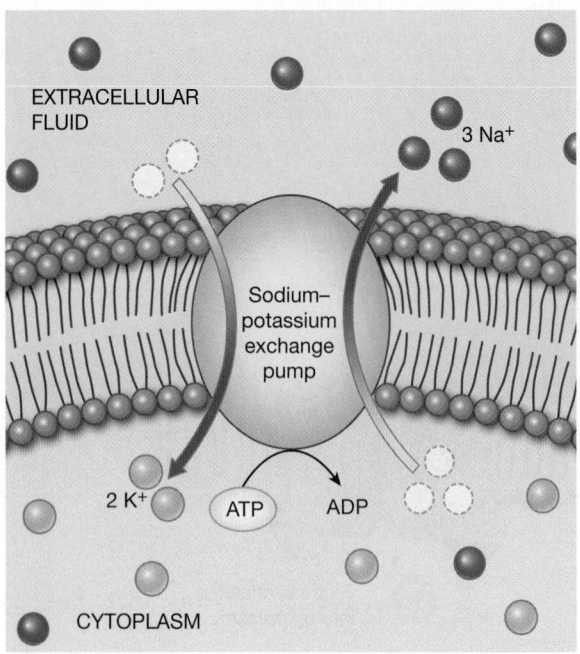

EXTRACELLULAR FLUID

3 Na$^+$

Sodium–potassium exchange pump

2 K$^+$ ATP   ADP

CYTOPLASM

Figure 3–20 **Secondary Active Transport.** In secondary active transport, glucose transport by a carrier protein will take place only after the carrier has bound two sodium ions. In three cycles, three glucose molecules and six sodium ions are transported into the cytoplasm. The cell then pumps the sodium ions across the plasma membrane by the sodium–potassium exchange pump, at a cost of two ATP molecules.

Figure 3–21 **Receptor-Mediated Endocytosis.**

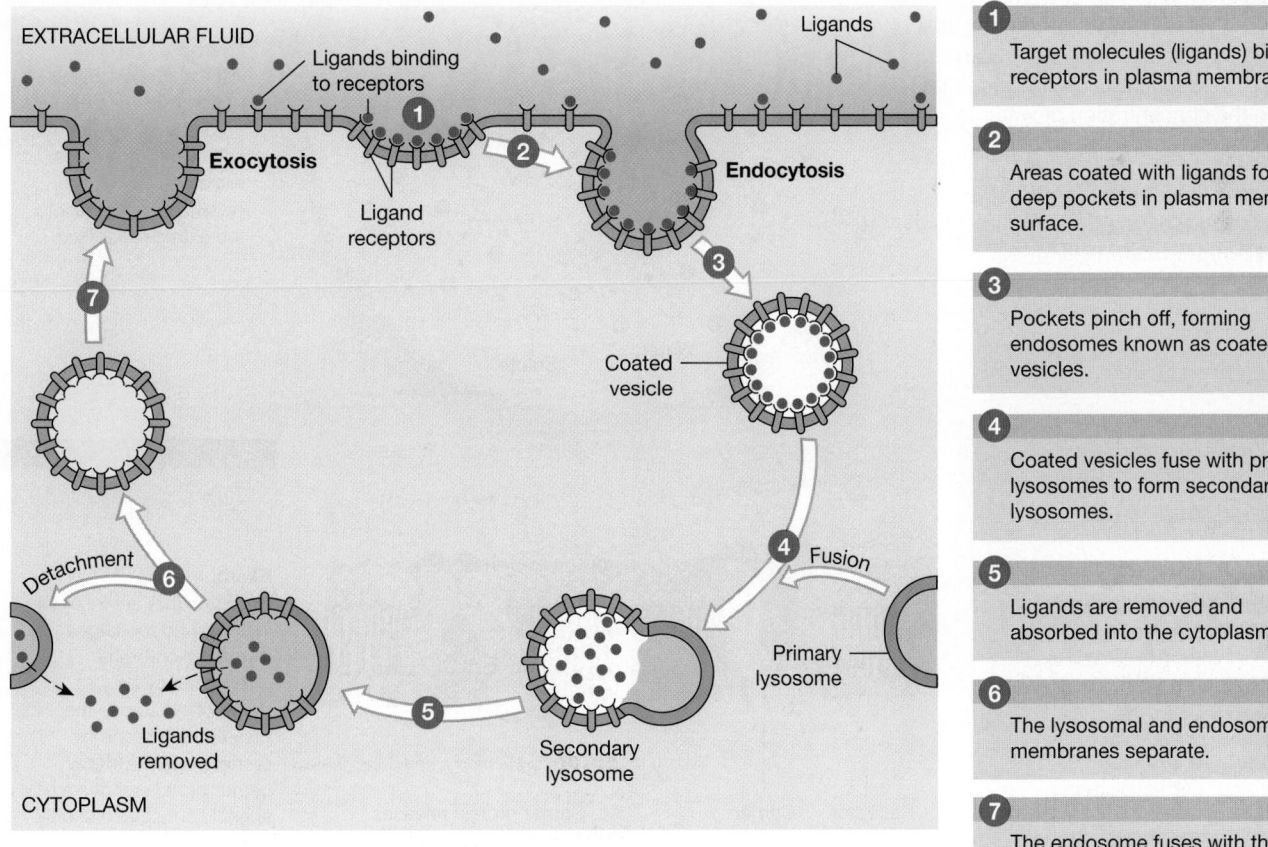

1	Target molecules (ligands) bind to receptors in plasma membrane.
2	Areas coated with ligands form deep pockets in plasma membrane surface.
3	Pockets pinch off, forming endosomes known as coated vesicles.
4	Coated vesicles fuse with primary lysosomes to form secondary lysosomes.
5	Ligands are removed and absorbed into the cytoplasm.
6	The lysosomal and endosomal membranes separate.
7	The endosome fuses with the plasma membrane, and the receptors are again available for ligand binding.

Passive Processes
No cellular energy expended

Diffusion

Diffusion is the movement of molecules from an area of higher concentration to an area of lower concentration. That is, the movement occurs down a concentration gradient (high to low).

Factors Affecting Rate: Size of gradient, molecular size, electric charge, lipid solubility, temperature, and the presence of membrane channel proteins

Substances Involved: Gases, small inorganic ions and molecules, lipid-soluble materials

Plasma membrane

Extracellular fluid

CO_2

Example:
When the concentration of CO_2 inside a cell is greater than outside the cell, the CO_2 diffuses out of the cell and into the extracellular fluid.

Osmosis

Osmosis is the diffusion of water molecules across a selectively permeable membrane. Movement occurs toward higher solute concentration because that is where the concentration of water is lower. Osmosis continues until the concentration gradient is eliminated.

Factors Affecting Rate: Concentration gradient; opposing pressure

Substances Involved: Water only

Water

Solute

Example:
If the solute concentration outside a cell is greater than inside the cell, water molecules will move across the plasma membrane into the extracellular fluid.

Facilitated diffusion

Facilitated diffusion is the movement of materials across a membrane by a carrier protein. Movement follows the concentration gradient.

Factors Affecting Rate: Concentration gradient, availability of carrier proteins

Substances Involved: Glucose and amino acids

Glucose

Receptor site

Carrier protein

Carrier protein releases glucose into cytoplasm

Example:
Nutrients that are insoluble in lipids and too large to fit through membrane channels may be transported across the plasma membrane by carrier proteins. Many carrier proteins move a specific substance in one direction only, either into or out of the cell.

Active Processes
Require ATP

Active transport

Active transport requires carrier proteins that move specific substances across a membrane against their concentration gradient. If the carrier moves one solute in one direction and another solute in the opposite direction, it is called an **exchange pump**.

Factors Affecting Rate: Availability of carrier protein, substrate, and ATP

Substances Involved: Na^+, K^+, Ca^{2+}, Mg^{2+} ; other solutes in special cases

Example:
The **sodium–potassium exchange pump**. One ATP molecule is hydrolyzed for each 3 sodium ions that are ejected from the cell, while 2 potassium ions are reclaimed.

Endocytosis
Endocytosis is the packaging of extracellular materials into a vesicle for transport into the cell.

Receptor-Mediated Endocytosis

In **receptor-mediated endocytosis**, target molecules bind to receptor proteins on the membrane surface, triggering vesicle formation.

Factors Affecting Rate: Number of receptors on the plasma membrane and the concentration of target molecules

Substances Involved: Target molecules called ligands

Example:
Cholesterol and iron ions are transported this way.

Pinocytosis

In **pinocytosis**, vesicles form at the plasma membrane and bring fluids and small molecules into the cell. This process is often called "cell drinking."

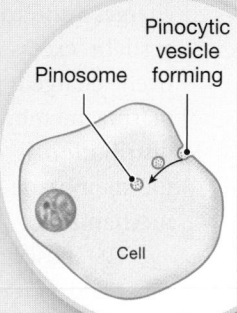

Example:
Once the vesicle is inside the cytoplasm, water and small molecules enter the cell across the vesicle membrane.

Factors Affecting Rate: Stimulus and mechanism not understood

Substances Involved: Extracellular fluid, with dissolved molecules such as nutrients

Phagocytosis

In **phagocytosis**, vesicles form at the plasma membrane to bring solid particles into the cell. This process is often called "cell eating."

Factors Affecting Rate: Presence of pathogens and cellular debris

Substances Involved: Bacteria, viruses, cellular debris, and other foreign material

Example:
Large particles are brought into the cell by cytoplasmic extensions (called **pseudopodia**) that engulf the particle and pull it into the cell.

Exocytosis

In **exocytosis**, intracellular vesicles fuse with the plasma membrane to release fluids and/or solids from the cells.

Factors Affecting Rate: Stimulus and mechanism incompletely understood

Substances Involved: Fluid and cellular wastes; secretory products from some cells

Example:
Cellular wastes in vesicles are ejected from the cell.

3

molecules are glycoproteins, and each binds a specific ligand, or target molecule, such as a transport protein or a hormone. Some receptors are distributed widely over the surface of the plasma membrane, and others are restricted to specific regions or in depressions on the plasma membrane.

Many important substances, including cholesterol and iron ions (Fe²⁺), are distributed through the body attached to special transport proteins. These proteins are too large to pass through membrane pores, but they are able to enter cells by receptor-mediated endocytosis.

Pinocytosis. "Cell drinking," or **pinocytosis** (pi-nō-sī-TŌ-sis), is the formation of endosomes filled with extracellular fluid. Most cells carry out pinocytosis. This process is not as selective as receptor-mediated endocytosis, because no receptor proteins are involved. The target appears to be the fluid contents in general, rather than specific bound ligands. In pinocytosis, a deep groove or pocket forms in the plasma membrane and then pinches off (**Spotlight Figure 3–22**). The steps involved are similar to the steps in receptor-mediated endocytosis, except that ligand binding is not involved.

Phagocytosis. "Cell eating," or **phagocytosis** (fag-ō-sī-TŌ-sis), produces phagosomes containing solid objects that may be as large as the cell itself. In this process, cytoplasmic extensions called **pseudopodia** (soo-dō-PŌ-dē-ah; *pseudo-*, false + *podon*, foot; singular *pseudopodium*) surround the object, and their membranes fuse to form a phagosome (**Spotlight Figure 3–22**). This vesicle then fuses with many lysosomes, and lysosomal enzymes digest its contents. Phagocytosis is performed only by specialized cells, such as the *macrophages*, that protect tissues by engulfing bacteria, cell debris, and other abnormal materials.

Exocytosis

Exocytosis (ek-sō-sī-TŌ-sis), introduced in our discussion of the Golgi apparatus, is the functional reverse of endocytosis. In exocytosis, a vesicle formed inside the cell fuses with, and becomes part of, the plasma membrane. When this takes place, the vesicle contents are released into the extracellular environment (**Spotlight Figure 3–22**). The ejected material may be secretory products, such as mucins or hormones, or waste products, such as those accumulating in endocytic vesicles.

In a few specialized cells, endocytosis produces vesicles on one side of the cell that are discharged through exocytosis on the opposite side. This method of bulk transport is common in cells lining capillaries, which use a combination of pinocytosis and exocytosis to transfer fluid and solutes from the bloodstream into the surrounding tissues. This overall transport process is called *vesicular transport*.

3-7 ▪ The membrane potential results from the unequal distribution of positive and negative charges across the plasma membrane

Learning Outcome Explain the origin and significance of the membrane potential.

Recall that the cell's interior has a slight negative charge with respect to the outside. The cause is a slight excess of positive charges (due to cations) outside the plasma membrane, and a slight excess of negative charges (due primarily to negatively charged proteins) along the cytoplasmic side of the plasma membrane and inside the cell. This unequal charge distribution is created by differences in the permeability of the membrane to various ions, as well as by active transport mechanisms.

The positive and negative charges are attracted to each other and would normally rush together, but the plasma membrane keeps them apart. When positive and negative charges are held apart, a **potential difference** exists between them. We refer to the potential difference across a plasma membrane as the **membrane potential**, also called the **transmembrane potential**.

The unit of measurement of potential difference is the *volt* (V). Most cars, for example, have 12-V batteries. The membrane potentials of cells are much smaller, typically in the vicinity of 0.07 V. Such a value is usually expressed as 70 mV, or 70 *millivolts* (thousandths of a volt).

The membrane potential in an undisturbed cell is called the **resting membrane potential**. Each type of cell has a characteristic resting membrane potential between –10 mV (–0.01 V) and –100 mV (–0.1 V), with the minus sign signifying that the cytoplasm and inner surface of the plasma membrane (cell) contains an excess of negative charges compared with the outside. Examples include fat cells (–40 mV), thyroid cells (–50 mV), neurons (–70 mV), skeletal muscle cells (–85 mV), and cardiac muscle cells (–90 mV).

If the lipid barrier were removed, the positive and negative charges would rush together and the potential difference would be eliminated. Instead, the plasma membrane acts like a dam across a stream. Just as a dam resists the water pressure that builds up on the upstream side, a plasma membrane resists electrochemical forces that would otherwise drive ions into or out of the cell. The water retained behind a dam and the ions held on either side of the plasma membrane have *potential energy*—stored energy that can be released to do work. People have designed many ways to use the potential energy stored behind a dam—for example, turning a mill wheel or a turbine. Similarly, cells have ways of using the potential energy stored in the membrane potential. For example, changes in the membrane potential make possible the transmission of information in the nervous system, and thus our perceptions and thoughts. As we will see in later chapters, changes in the membrane potential also trigger the contractions of muscles and the secretions of glands.

CHECKPOINT

27. **What is the membrane potential, and in what units is it expressed?**

28. **If the plasma membrane were freely permeable to sodium ions (Na^+), how would the membrane potential be affected?**

See the blue Answers tab at the back of the book.

3-8 ▌ Stages of a cell's life cycle include interphase, mitosis, and cytokinesis

Learning Outcome Describe the stages of the cell life cycle, including mitosis, interphase, and cytokinesis, and explain their significance.

The period between fertilization and physical maturity involves tremendous changes in organization and complexity. At fertilization, a single cell is all there is, but at maturity, your body has about 75 trillion cells. This amazing transformation involves a form of cellular reproduction called **cell division**. A single cell divides to produce a pair of **daughter cells**, each half the size of the original. Before dividing, each daughter cell will grow to the size of the original cell.

Even when development is complete, cell division continues to be essential to survival. Cells are highly adaptable, but physical wear and tear, toxic chemicals, temperature changes, and other environmental stresses can damage them. And, like individuals, cells age. The life span of a cell varies from hours to decades, depending on the type of cell and the stresses involved. Many cells apparently self-destruct after a certain period of time because specific "suicide genes" in

the nucleus are activated. This genetically controlled death of cells is called **apoptosis** (ap-op-TŌ-sis; *apo-*, separated from + *ptosis*, a falling). Researchers have identified several genes involved in the regulation of this process. For example, a gene called *bcl-2* produces a regulator protein that appears to prevent apoptosis and to keep a cell alive and functional. If something interferes with the function of this gene, the cell self-destructs.

Because a typical cell does not live nearly as long as a typical person, cell populations must be maintained over time by cell division. For cell division to be successful, the genetic material in the nucleus must be duplicated accurately, and one copy must be distributed to each daughter cell. The duplication of the cell's genetic material is called **DNA replication**, and nuclear division is called **mitosis** (mī-TŌ-sis). Mitosis occurs during the division of somatic cells. The production of sex cells involves a different process, **meiosis** (mī-Ō-sis), described in Chapter 28.

DNA Replication

Recall that each DNA molecule consists of a pair of DNA strands joined by hydrogen bonds between complementary nitrogenous bases. ⊃ p. 82 DNA replication begins when enzymes called *helicases* unwind the strands and disrupt the weak hydrogen bonds between the bases. As the strands unwind, molecules of **DNA polymerase** bind to the exposed nitrogenous bases. This enzyme (1) promotes bonding between the nitrogenous bases of the DNA strand and complementary DNA nucleotides in the nucleoplasm and (2) links the nucleotides by covalent bonds. **Spotlight Figure 3–23** diagrams DNA replication.

Interphase, Mitosis, and Cytokinesis

Spotlight Figure 3–24 (pp. 130–131) depicts the life cycle of a typical cell. That life cycle includes a fairly brief period of mitosis alternating with an *interphase* of variable duration. In a cell preparing to divide, interphase can be divided into the G_1, S, and G_2 phases. DNA replication takes place during the S phase.

Mitosis is the duplication of the chromosomes in the nucleus and their separation into two identical sets in the process of somatic cell division. Although we describe mitosis in stages, *prophase, metaphase, anaphase,* and *telophase,* it is really one continuous process.

TIPS & TRICKS

In order to remember the correct sequence of events during mitosis, imagine the U-shaped contour rug surrounding your toilet as the P-MAT, for **p**rophase, **m**etaphase, **a**naphase, and **t**elophase.

Cytokinesis is the division of the cytoplasm during mitosis. This process usually begins in late anaphase and continues

For cell division to be successful, the genetic material in the nucleus must be duplicated accurately, and one copy must be distributed to each daughter cell. This duplication of the cell's genetic material is called DNA replication.

1 **DNA replication** begins when **helicase** enzymes unwind the strands and disrupt the hydrogen bonds between the bases. As the strands unwind, molecules of **DNA polymerase** bind to the exposed nitrogenous bases. This enzyme (1) promotes bonding between the nitrogenous bases of each separated strand with complementary DNA nucleotides in the nucleoplasm and (2) links the nucleotides by covalent bonds.

KEY

- Adenine
- Guanine
- Cytosine
- Thymine

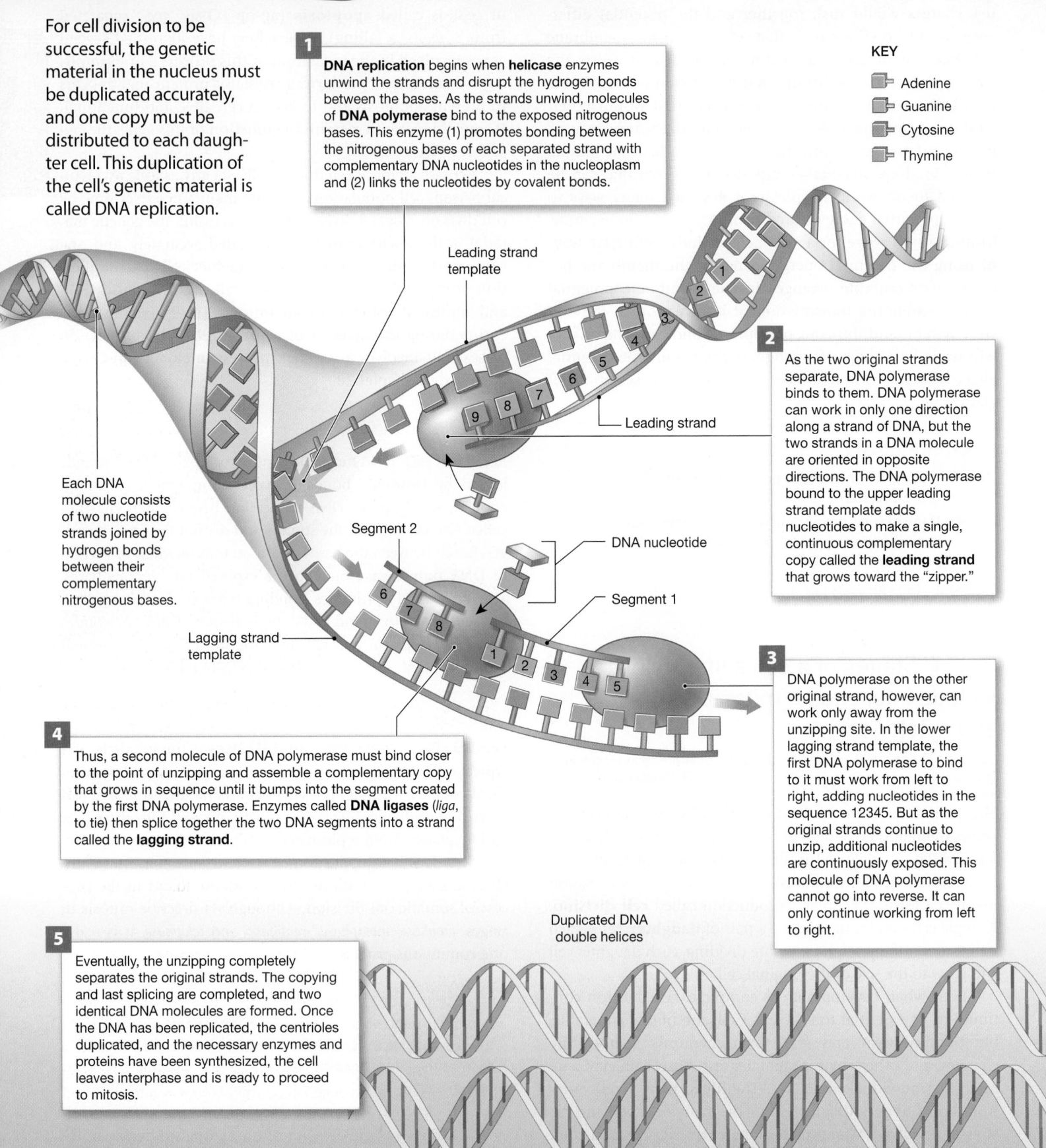

Leading strand template

Leading strand

Each DNA molecule consists of two nucleotide strands joined by hydrogen bonds between their complementary nitrogenous bases.

Segment 2

DNA nucleotide

Segment 1

Lagging strand template

2 As the two original strands separate, DNA polymerase binds to them. DNA polymerase can work in only one direction along a strand of DNA, but the two strands in a DNA molecule are oriented in opposite directions. The DNA polymerase bound to the upper leading strand template adds nucleotides to make a single, continuous complementary copy called the **leading strand** that grows toward the "zipper."

4 Thus, a second molecule of DNA polymerase must bind closer to the point of unzipping and assemble a complementary copy that grows in sequence until it bumps into the segment created by the first DNA polymerase. Enzymes called **DNA ligases** (*liga*, to tie) then splice together the two DNA segments into a strand called the **lagging strand**.

3 DNA polymerase on the other original strand, however, can work only away from the unzipping site. In the lower lagging strand template, the first DNA polymerase to bind to it must work from left to right, adding nucleotides in the sequence 12345. But as the original strands continue to unzip, additional nucleotides are continuously exposed. This molecule of DNA polymerase cannot go into reverse. It can only continue working from left to right.

Duplicated DNA double helices

5 Eventually, the unzipping completely separates the original strands. The copying and last splicing are completed, and two identical DNA molecules are formed. Once the DNA has been replicated, the centrioles duplicated, and the necessary enzymes and proteins have been synthesized, the cell leaves interphase and is ready to proceed to mitosis.

throughout telophase, producing two daughter cells.. The completion of cytokinesis marks the end of cell division, creating two separate and complete cells, each surrounded by its own plasma membrane.

The Mitotic Rate and Energy Use

The preparations for cell division that take place between G_1 and mitosis are difficult to recognize in a light micrograph. However, the start of mitosis is easy to recognize, because the chromosomes become condensed and highly visible. The frequency of cell division in a tissue can be estimated by the number of cells in mitosis at any time. As a result, we often use the term **mitotic rate** when we discuss rates of cell division. In general, the longer the life expectancy of a cell type, the slower the mitotic rate. Long-lived cells, such as muscle cells and neurons, either never divide or do so only under special circumstances. Other cells, such as those covering the surface of the skin or the lining of the digestive tract, are subject to attack by chemicals, pathogens, and abrasion. They survive for only days or even hours. Special cells called **stem cells** maintain these cell populations through repeated cycles of cell division.

Stem cells are relatively unspecialized. Their only function is to produce daughter cells. Each time a stem cell divides, one of its daughter cells develops functional specializations while the other prepares for further stem cell divisions. The rate of stem cell division can vary with the type of tissue and the demand for new cells. In heavily abraded skin, stem cells may divide more than once a day, but stem cells in adult connective tissues may remain inactive for years.

Dividing cells use an unusually large amount of energy. For example, they must synthesize new organic materials and move organelles and chromosomes within the cell. All these processes require ATP in substantial amounts. Cells that do not have adequate energy sources cannot divide. In a person who is starving, normal cell growth and maintenance grind to a halt. For this reason, prolonged starvation stunts childhood growth, slows wound healing, lowers resistance to disease, thins the skin, and changes the lining of the digestive tract.

CHECKPOINT

29. Give the biological terms for (a) cellular reproduction and (b) cellular death.

30. What enzymes must be present for DNA replication to proceed normally?

31. Describe interphase, and identify its stages.

32. A cell is actively manufacturing enough organelles to serve two functional cells. This cell is probably in what phase of its life cycle?

33. Define mitosis, and list its four stages.

34. What would happen if spindle fibers failed to form in a cell during mitosis?

See the blue Answers tab at the back of the book.

3-9 ▮ Several growth factors affect the cell life cycle

Learning Outcome Discuss the regulation of the cell life cycle.

In normal adult tissues, the rate of cell division balances the rate of cell loss or destruction. Mitotic rates are genetically controlled, and many different stimuli may be responsible for activating genes that promote cell division. Some of the stimuli are internal, and many cells set their own pace of mitosis and cell division.

An important internal trigger is the level of **M-phase promoting factor (MPF)**, also known as *maturation-promoting factor*. MPF is assembled from two parts: a cell division cycle protein called *Cdc2* and a second protein called *cyclin*. Cyclin levels climb as the cell life cycle proceeds. When levels are high enough, MPF appears in the cytoplasm and mitosis gets underway.

Various natural body substances, such as hormones, peptides, or nutrients, derived from food can stimulate the division of specific types of cells. These substances are known as **growth factors**. Table 3–2 on p. 132 lists some of these chemical factors and their target tissues. We will discuss them in later chapters.

Researchers have identified genes that inhibit cell division. Such genes are known as *repressor genes*. One gene, called *p53*, synthesizes a protein that binds to DNA in the nucleus. The binding, in turn, activates another gene that directs the production of growth-inhibiting factors inside the cell. Roughly half of all cancers are associated with abnormal forms of the *p53* gene.

There are indications that in humans, the *number* of cell divisions that a cell and its descendants undergoes is regulated at the chromosome level by structures called **telomeres**. Telomeres are terminal segments of DNA with associated proteins. These DNA-protein complexes bend and fold repeatedly to form caps at the ends of chromosomes, much like the plastic sheaths on the tips (aglets) of shoestrings. Telomeres have several functions, notably to attach chromosomes to the nuclear matrix and to protect the ends of the chromosomes from damage during mitosis. The telomeres themselves, however, are subject to wear and tear over the years. Each time a cell divides during adult life, some of the repeating segments break off, and the telomeres get shorter. When they get too short, repressor gene activity signals the cell to stop dividing.

INTERPHASE

Most cells spend only a small part of their time actively engaged in cell division. Somatic cells spend the majority of their functional lives in a state known as **interphase**. During interphase, a cell performs all its normal functions and, if necessary, prepares for cell division.

A cell that is ready to divide first enters the **G₁** phase. In this phase, the cell makes enough mitochondria, cytoskeletal elements, endoplasmic reticula, ribosomes, Golgi membranes, and cytosol for two functional cells. Centriole replication begins in G_1 and commonly continues until G_2. In cells dividing at top speed, G_1 may last just 8–12 hours. Such cells use all their energy on mitosis, and all other activities cease. When G_1 lasts for days, weeks, or months, preparation for mitosis occurs as the cells perform their normal functions.

When the activities of **G₁** have been completed, the cell enters the **S phase**. Over the next 6–8 hours, the cell duplicates its chromosomes. This involves DNA replication and the synthesis of histones and other proteins in the nucleus.

Once DNA replication has ended, there is a brief (2–5-hour) **G₂ phase** devoted to last-minute protein synthesis and to the completion of centriole replication.

6 to 8 hours

S
DNA replication, synthesis of histones

8 or more hours

G₁
Normal cell functions plus cell growth, duplication of organelles, protein synthesis

THE CELL CYCLE

2 to 5 hours

G₂
Protein synthesis

MITOSIS

Prophase

Metaphase

Anaphase

Telophase

1 to 3 hours

CYTOKINESIS

M Phase–MITOSIS AND CYTOKINESIS

G₀

Centrioles in centrosome

Nucleus

Interphase
During interphase, the DNA strands are loosely coiled and chromosomes cannot be seen.

An interphase cell in the **G₀ phase** is not preparing for division, but is performing all of the other functions appropriate for that particular cell type. Some mature cells, such as skeletal muscle cells and most neurons, remain in G_0 indefinitely and never divide. In contrast, stem cells, which divide repeatedly with very brief interphase periods, never enter G_0.

A dividing cell held in place by a sucker pipe to the left and being injected with a needle from the right.

MITOSIS AND CYTOKINESIS

The M phase of the cell cycle includes mitosis and cytokinesis. Mitosis separates the duplicated chromosomes of a cell into two identical nuclei. The term mitosis specifically refers to the division and duplication of the cell's nucleus; division of the cytoplasm to form two distinct new cells involves a separate but related process known as **cytokinesis**.

The two daughter chromosomes are pulled apart and drawn to opposite ends of the cell along the **spindle apparatus** (the complex of spindle fibers).

Centrioles (two pairs) Astral rays and spindle fibers Chromosome with two sister chromatids Chromosomal microtubules Metaphase plate Cleavage furrow Daughter cells

Early prophase

Prophase (PRŌ-fāz; *pro*, before) begins when the chromosomes coil so tightly they become visible as single structures under a light microscope. An array of microtubules called **spindle fibers** extends between the centriole pairs. Smaller microtubules called **astral rays** radiate into the cytoplasm.

Late prophase

As a result of DNA replication during the S phase, two copies of each chromosome now exist. Each copy, called a **chromatid** (KRŌ-ma-tid), is connected to its duplicate copy at a single point, the **centromere** (SEN-trō-mēr). **Kinetochores** (ki-NĒ-tō-korz) are the protein-bound areas of the centromere; they attach to spindle fibers forming **chromosomal microtubules**.

Metaphase

Metaphase (MET-a-fāz; *meta*, after) begins as the chromatids move to a narrow central zone called the metaphase plate. Metaphase ends when all the chromatids are aligned in the plane of the metaphase plate.

Anaphase

Anaphase (AN-a-fāz; *ana-*, apart) begins when the centromere of each chromatid pair splits and the chromatids separate. The two **daughter chromosomes** are now pulled toward opposite ends of the cell along the chromosomal microtubules.

Telophase

During telophase (TĒL-ō-fāz; *telo-*, end), each new cell prepares to return to the interphase state. The nuclear membranes re-form, the nuclei enlarge, and the chromosomes gradually uncoil. This stage marks the end of mitosis.

Cytokinesis

Cytokinesis is the division of the cytoplasm into two daughter cells. Cytokinesis usually begins with the formation of a cleavage furrow and continues throughout telophase. The completion of cytokinesis marks the end of cell division.

3

Table 3–2   Chemical Factors Affecting Cell Division

Factor	Sources	Effects	Targets
M-phase promoting factor (maturation-promoting factor) (MPF)	Forms within cytoplasm from Cdc2 and cyclin	Initiates mitosis	Regulatory mechanism active in all dividing cells
Growth hormone	Anterior lobe of the pituitary gland	Stimulation of growth, cell division, differentiation	All cells, especially in epithelial and connective tissues
Prolactin	Anterior lobe of the pituitary gland	Stimulation of cell growth, division, development	Gland and duct cells of mammary glands
Nerve growth factor (NGF)	Salivary glands; other sources suspected	Stimulation of nerve cell repair and development	Neurons and neuroglia
Epidermal growth factor (EGF)	Duodenal glands; other sources suspected	Stimulation of stem cell divisions and epithelial repairs	Epidermis
Fibroblast growth factor (FGF)	Wide range of cells	Division and differentiation of fibroblasts and related cells	Connective tissues
Erythropoietin (EPO)	Kidneys (primary source)	Stimulation of stem cell divisions and maturation of red blood cells	Red bone marrow
Thymosins and related compounds	Thymus	Stimulation of division and differentiation of lymphocytes (especially T cells)	Thymus and other lymphoid tissues and organs
Chalones	Many tissues	Inhibition of cell division	Cells in the immediate area

CHECKPOINT

35. Define growth factor, and identify several growth factors that affect cell division.

See the blue Answers tab at the back of the book.

3-10 ▮ Tumors and cancers are characterized by abnormal cell growth and division

Learning Outcome   Discuss the relationship between cell division and cancer.

When the rates of cell division and growth exceed the rate of cell death, a tissue begins to enlarge. A **tumor**, or *neoplasm*, is a mass or swelling produced by abnormal cell growth and division. In a *benign tumor*, the cells usually remain within the epithelium (one of the four primary tissue types) or a connective tissue capsule. Such a tumor seldom threatens an individual's life and can usually be surgically removed if its size or position disturbs tissue function.

Cells in a *malignant tumor* no longer respond to normal controls. These cells do not remain confined within the epithelium or a connective tissue capsule, but spread into surrounding tissues. The tumor of origin is called the *primary tumor* (or *primary neoplasm*), and the spreading process is called **invasion**. Malignant cells may also travel to distant tissues and organs and establish *secondary tumors*. This dispersion, called **metastasis** (me-TAS-ta-sis; *meta-*, after + *stasis*, standing still), is very difficult to control.

Cancer is an illness that results from the abnormal proliferation of any of the cells in the body. It is characterized by mutations that disrupt normal cell regulatory controls and produce potentially malignant cells. Normal cells begin the path to becoming malignant when a mutation occurs in a gene involved with cell growth, differentiation, or division. The modified genes are called **oncogenes** (ON-kō-gēnz; *oncōs*, tumor).

Agents that cause a mutation (a change in DNA) are called **mutagens**. Cancer-causing agents, which include many mutagens, are known as **carcinogens**. Examples of carcinogens include radiation, chemicals, bacteria, and viruses. Some chemical carcinogens, such as alcohol and estrogen, are not directly mutagenic. Instead, they promote cancer by promoting rapid cell divisions, which increase the chances that a mutation will occur.

Normal cells become malignant in a multistep process. For example, a mutation may occur in one cell, which is passed on to its daughter cells. One of those cells may undergo another mutation that speeds its rate of cell division. Further growth promoting mutations in one of those cells and its proliferating descendants may produce a primary tumor. The tumor now contains descendants of the original abnormal cell that have been selected for rapid growth. While these tumor cells are ultimately all descended from one abnormal cell, they are not identical to it due to their accumulated mutations.

Cancer cells do not resemble normal cells. They have a different shape and typically become abnormally larger or smaller. Malignant tumor cells are also less subject to controls limiting adhesion and movement between adjacent cells

Figure 3–25   **The Development of Cancer.**

and extracellular tissues. At first, the growth of the primary tumor distorts the tissue, but the basic tissue organization remains intact. Metastasis begins with invasion as malignant cells "break out" of the primary tumor and invade the surrounding tissue. They may then enter the lymphatic system and accumulate in nearby lymph nodes. When metastasis involves the penetration of blood vessels, the cancer cells circulate throughout the body. Cancer progresses in the steps diagrammed in **Figure 3–25**.

Responding to cues that are as yet unknown, cancer cells in the bloodstream ultimately escape out of blood vessels to establish secondary tumors at other sites. The cancer cells in these tumors are extremely active metabolically, and their presence stimulates the growth of new blood vessels (*angiogenesis*) into the area. The increased blood supply provides additional nutrients to the cancer cells and further accelerates tumor growth and metastasis.

As malignant tumors grow, organ function begins to deteriorate. The malignant cells may no longer perform their original functions, or they may perform normal functions in an abnormal way. For example, endocrine cancer cells may produce normal hormones, but in excessively large amounts.

Cancer cells do not use energy very efficiently. They grow and multiply at the expense of healthy tissues, competing for space and nutrients with normal cells. This competition contributes to the starved appearance of many patients in the late stages of cancer. Death may result from the compression of vital organs when nonfunctional cancer cells have killed or replaced the healthy cells in those organs, or when the cancer cells have starved normal tissues of essential nutrients. We will discuss cancer further in later chapters that deal with specific systems.

CHECKPOINT

36. An illness characterized by mutations that disrupt normal control mechanisms and produce potentially malignant cells is termed _____.

37. Define metastasis.

See the blue Answers tab at the back of the book.

Clinical Note

Telomerase, Aging, and Cancer Each telomere contains a sequence of about 8000 nitrogenous bases, but they are multiple copies of the same six-base sequence, TTAGGG, repeated over and over again. Telomeres are not formed by DNA polymerase. Instead, they are created by an enzyme called *telomerase*. Telomerase is functional early in life, but by adulthood it has become inactive. As a result, the telomere segments lost during each mitotic division are not replaced. Eventually, shortening of the telomere reaches a point at which the cell no longer divides.

This mechanism is a factor in the aging process, since many of the signs of age result from the gradual loss of functional stem cell populations. Experiments are in progress to determine whether activating telomerase (or a suspected alternative repair enzyme) can forestall or reverse the effects of aging. This would seem to be a very promising area of research. Activate telomerase, and halt aging—sounds good, doesn't it? Unfortunately, there's a catch: In adults, telomerase activation is a key step in the development of cancer.

If for some reason a cell with short telomeres does *not* respond normally to repressor genes, it will continue to divide. The result is mechanical damage to the DNA strands, chromosomal abnormalities, and mutations. Interestingly, one of the first consequences of such damage is the abnormal activation of telomerase. Once this occurs, the abnormal cells can continue dividing indefinitely. Telomerase is active in at least 90 percent of all cancer cells. Research is underway to find out how to turn off telomerase that has been improperly activated.

Clinical Note

Parkinson's Disease In most cases, differentiation is irreversible. Once genes are turned off, they won't be turned back on. However, some cells, such as stem cells, are relatively undifferentiated. These cells can differentiate into any of several different types of cell, depending on local conditions. For example, if nutrients are abundant, stem cells in many parts of the body can differentiate into fat cells. Researchers are gradually discovering what chemical cues and genes control the differentiation of specific cell types. Recently, this research has resulted in the ability to "turn back the clock" in some types of adult somatic cells and reprogram them into a form of stem cells called *induced pluripotent stem (iPS) cells*. The ability to take a person's stem cells or somatic cells and create new cells or neurons to treat disease may one day revolutionize the practice of medicine.

Parkinson's disease is a neurodegenerative disease characterized by progressive degeneration and loss of dopamine-producing neurons. (Dopamine is one kind of neurotransmitter, a substance that one neuron releases to communicate with other neurons.) Parkinson's disease may be the first disorder suited to treatment by implanting stem cells. Several laboratories have induced either iPS cells or *embryonic stem cells* to differentiate into cells that function as dopamine-producing neurons. In studies on animal models, both iPS and embryonic stem cell–derived dopamine neurons reinnervated the brains of rats with Parkinson's disease. These cells also released dopamine and improved motor function.

3-11 ▮ Differentiation is cellular specialization as a result of gene activation or repression

Learning Outcome Define differentiation, and explain its importance.

An individual's liver cells, fat cells, and neurons all contain the same chromosomes and genes, but in each case a different set of genes has been turned *off*. In other words, your liver cells and fat cells differ because liver cells have one set of genes accessible for transcription, and fat cells another.

When a gene is functionally eliminated, the cell loses the ability to produce a particular protein—and to perform any functions involving that protein. Each time another gene switches off, the cell's functional abilities become more restricted. This

development of specific cellular characteristics and functions that are different from the original cell is called **differentiation**.

Fertilization produces a single cell with all its genetic potential intact. Repeated cell divisions follow, and differentiation begins as the number of cells increases. Differentiation produces specialized cells with limited capabilities. These cells form organized collections known as *tissues*, each with discrete functional roles. In Chapter 4, we will examine the structure and function of tissues and consider the role of tissue interactions in maintaining homeostasis.

> **CHECKPOINT**
>
> 38. **Define differentiation.**
>
> See the blue Answers tab at the back of the book.

Related Clinical Terms

anaplasia: An irreversible change in the size and shape of tissue cells.

dysplasia: A reversible change in the normal shape, size, and organization of tissue cells.

genetic engineering: A general term that encompasses attempts to change the genetic makeup of cells or organisms, including humans.

hyperplasia: An increase in the number of normal cells (not tumor formation) in a tissue or organ, thus enlarging that tissue or organ.

hypertrophy: The enlargement of an organ or tissue due to an increase in the size of its cells.

liposome: A minute spherical sac of lipid molecules enclosing a water droplet. Often formed artificially to carry drugs into the tissues.

necrosis: Death of one or more cells in an organ or tissue due to disease, injury, or inadequate blood supply.

oncologist: Physician who specializes in the identification and treatment of cancers.

prion: A protein particle that is not visible microscopically, contains no nucleic acid, is resistant to destruction, and is thought to be the cause of some brain diseases such as bovine spongiform encephalopathy (BSE), scrapie, and Creutzfeldt-Jakob disease.

scanning electron micrograph (SEM): An image produced by an electron microscope in which a beam of focused electrons moves across an object with that object producing secondary electrons that are scattered and formatted into a three-dimensional image on a cathode-ray tube—also called *scanning microscope*.

transmission electron micrograph (TEM): A cross-sectional image produced by an electron microscope that passes a beam of electrons through an extremely small specimen. After passing through the sample the electrons are focused to form a magnified sectional view.

Chapter Review

Study Outline

MasteringA&P®

Access more chapter study tools online in the MasteringA&P Study Area:
- Chapter Quizzes, Chapter Practice Test, Art-labeling Activities, Animations, MP3 Tutor Sessions, and Clinical Case Studies
- Practice Anatomy Lab   PAL™
- Interactive Physiology   iP®
- A&P Flix   A&PFlix
- PhysioEx   PhysioEx

An Introduction to Cells p. 91

1. Contemporary *cell theory* incorporates several basic concepts: (1) Cells are the building blocks of all plants and animals; (2) cells are produced by the division of preexisting cells; (3) cells are the smallest units that perform all vital physiological functions; and (4) each cell maintains homeostasis at the cellular level. (*Spotlight Figure 3–1*)
2. **Cytology,** the study of cellular structure and function, is part of **cell biology.**
3. The human body contains two types of cells: **sex cells** (*sperm* and *oocytes*) and **somatic cells** (all other cells). (*Spotlight Figure 3–1*)

3-1 ▌ The plasma membrane separates the cell from its surrounding environment and performs various functions p. 91

4. A typical cell is surrounded by **extracellular fluid**—specifically, the **interstitial fluid** of the tissue. The cell's outer boundary is the **plasma membrane (cell membrane).**
5. The plasma membrane's functions include physical isolation, regulation of exchange with the environment, sensitivity to the environment, and structural support. (*Figure 3–2*)
6. The plasma membrane is a **phospholipid bilayer,** which contains other lipids, proteins, and carbohydrates.
7. **Integral proteins** are part of the membrane itself; **peripheral proteins** are attached to, but can separate from, the membrane.
8. Membrane proteins can act as anchors (**anchoring proteins**), identifiers (**recognition proteins**), enzymes, receptors (**receptor proteins**), carriers (**carrier proteins**), or **channels.**
9. The **glycocalyx** on the outer cell surface is formed by the carbohydrate portions of *proteoglycans, glycoproteins,* and *glycolipids.* Functions include lubrication and protection, anchoring and locomotion, specificity in binding, and recognition.

3-2 ▌ Organelles within the cytoplasm perform particular functions p. 96

10. The **cytoplasm** contains the fluid **cytosol** and the **organelles** suspended in the cytosol.
11. Cytosol (intracellular fluid) differs from extracellular fluid in composition and in the presence of **inclusions.**
12. **Nonmembranous organelles** are not completely enclosed by membranes, and all of their components are in direct contact with the cytosol. They include the *cytoskeleton, microvilli, centrioles, cilia, ribosomes,* and *proteasomes.* (*Spotlight Figure 3–1*)
13. **Membranous organelles** are surrounded by phospholipid membranes that isolate them from the cytosol. They include the *endoplasmic reticulum,* the *Golgi apparatus, lysosomes, peroxisomes, mitochondria,* and *nucleus.* (*Spotlight Figure 3–1*)
14. The **cytoskeleton** gives the cytoplasm strength and flexibility. It has four components: **microfilaments** (typically made of **actin**), **intermediate filaments, microtubules** (made of **tubulin**), and **thick filaments** (made of **myosin**). (*Figure 3–3*)
15. **Microvilli** are small projections of the plasma membrane that increase the surface area exposed to the extracellular environment. (*Figure 3–3*)
16. The **centrosome** is a region of the cytoplasm near the nucleus. It is the microtubule organizing center surrounding the paired centrioles. **Centrioles** direct the movement of chromosomes during cell division. (*Figure 3–4*)
17. Multiple motile **cilia,** each anchored by a **basal body,** beat rhythmically to move fluids or secretions across the cell surface. A single **primary cilium** is sensitive to environmental stimuli. (*Figure 3–4*)
18. **Ribosomes,** responsible for manufacturing proteins, are composed of a **small** and a **large ribosomal subunit,** both of which contain **ribosomal RNA (rRNA).** Free ribosomes are in the cytoplasm, and **fixed ribosomes** are attached to the endoplasmic reticulum. (*Spotlight Figure 3–1*)
19. **Proteasomes** remove and break down damaged or abnormal proteins that have been tagged with *ubiquitin.*
20. The **endoplasmic reticulum (ER)** is a network of intracellular membranes that function in synthesis, storage, transport, and detoxification. The ER forms hollow tubes, flattened sheets, and chambers called **cisternae.** Smooth endoplasmic reticulum **(SER)** is involved in lipid synthesis; **rough endoplasmic reticulum (RER)** contains ribosomes on its outer surface and forms **transport vesicles.** (*Figure 3–5; Spotlight Figure 3–7*)
21. The **Golgi apparatus,** or **Golgi complex,** forms **secretory vesicles** and new plasma membrane components, and packages *lysosomes.* Secretions are discharged from the cell by exocytosis. (*Figure 3–6; Spotlight Figure 3–7*)

3

22. **Lysosomes**, vesicles filled with digestive enzymes, are responsible for the **autolysis** of injured cells. (*Figures 3–6, 3–8; Spotlight Figure 3–7*)

23. **Peroxisomes** carry enzymes that neutralize potentially dangerous free radicals.

24. **Membrane flow** refers to the continuous movement and recycling of the membrane among the ER, vesicles, the Golgi apparatus, and the plasma membrane.

25. **Mitochondria** are responsible for ATP production through aerobic metabolism. The **matrix**, or fluid contents of a mitochondrion, lies inside the **cristae**, or folds of an inner membrane. (*Figure 3–9*)

3-3 ▌ The nucleus contains DNA and enzymes essential for controlling cellular activities p. 108

26. The **nucleus** is the control center of cellular operations. It is surrounded by a **nuclear envelope** (a double membrane with a **perinuclear space**), through which it communicates with the cytosol by way of **nuclear pores**. (*Spotlight Figure 3–1; Figure 3–10*)

27. The nucleus contains a supportive **nuclear matrix**; one or more **nucleoli** typically are present.

28. The nucleus controls the cell by directing the synthesis of specific proteins, using information stored in **chromosomes**, which consist of DNA bound to **histones**. In nondividing cells, DNA and associated proteins form a tangle of filaments called **chromatin**. (*Figure 3–11*)

29. The cell's information storage system, the **genetic code**, is called a *triplet code* because a sequence of three nitrogenous bases specifies the identity of a single amino acid. Each **gene** contains all the DNA triplets needed to produce a specific polypeptide chain.

3-4 ▌ DNA controls protein synthesis, cell structure, and cell function p. 110

30. As **gene activation** begins, **RNA polymerase** must bind to the gene.

31. **Transcription** is the production of RNA from a DNA template. After transcription, a strand of **messenger RNA (mRNA)** carries instructions from the nucleus to the cytoplasm. (*Figure 3–12*)

32. During **translation**, a functional polypeptide is constructed using the information contained in the sequence of **codons** along an mRNA strand. The mRNA codons correspond to DNA triplets. The sequence of codons determines the sequence of amino acids in the polypeptide.

33. During translation, complementary base pairing of **anticodons** to mRNA codons occurs, and **transfer RNA (tRNA)** molecules bring amino acids to the ribosomal complex. Translation includes three phases: **initiation**, **elongation**, and **termination**. (*Spotlight Figure 3–13; Table 3–1*)

34. By controlling protein synthesis, the DNA of the nucleus has both direct control (through the synthesis of structural proteins) and indirect control (through the synthesis of enzymes) over the structure and functions of the cell.

3-5 ▌ Diffusion is a passive transport mechanism that assists membrane passage p. 113

35. The **permeability** of a barrier such as the plasma membrane is an indication of the barrier's effectiveness. Nothing can pass through an **impermeable** barrier; anything can pass through a **freely permeable** barrier. Plasma membranes are **selectively permeable**.

36. **Diffusion** is the net movement of a substance from an area of higher concentration to an area of lower concentration. Diffusion occurs until the **concentration gradient** is eliminated. (*Figures 3–14, 3–15*)

37. Most lipid-soluble materials, water, and gases freely diffuse across the phospholipid bilayer of the plasma membrane. Small water-soluble molecules and ions rely on channel-mediated diffusion through a passageway within a transmembrane protein. **Leak channels** are passive channels that allow ions across the plasma membrane.

38. **Osmosis** is the net flow of water across a selectively permeable membrane in response to differences in solute concentration. **Osmotic pressure** of a solution is the force of water movement into that solution resulting from its solute concentration. **Hydrostatic pressure** can oppose osmotic pressure. (*Figure 3–16*)

39. **Tonicity** describes the effects of osmotic solutions on cells. A solution that does not cause an osmotic flow is **isotonic**. A solution that causes water to flow into a cell is **hypotonic** and can lead to **hemolysis** of red blood cells. A solution that causes water to flow out of a cell is **hypertonic** and can lead to **crenation**. (*Figure 3–17*)

3-6 ▌ Carrier-mediated and vesicular transport assist membrane passage p. 120

40. Carrier-mediated transport involves the binding and transporting of specific ions by integral proteins. **Cotransport** moves two substances in the same direction; **countertransport** moves them in opposite directions.

41. In **facilitated diffusion**, compounds are transported across a membrane after binding to a **receptor site** within the channel of a carrier protein. (*Figure 3–18*)

42. **Active transport** mechanisms consume ATP and are not dependent on concentration gradients. Some **ion pumps** are **exchange pumps**. Secondary active transport may involve cotransport or countertransport. (*Figures 3–19, 3–20*)

43. In **vesicular transport**, materials move into or out of the cell in membranous **vesicles**. Movement into the cell is accomplished through **endocytosis**, an active process that can take three forms: **receptor-mediated endocytosis** (by means of **coated vesicles**), **pinocytosis**, or **phagocytosis** (using **pseudopodia**). The ejection of materials from the cytoplasm is accomplished by **exocytosis**. (*Figure 3–21; Spotlight Figure 3–22*)

3-7 ▌ The membrane potential results from the unequal distribution of ions across the plasma membrane p. 126

44. The **potential difference**, measured in volts, between the two sides of a plasma membrane is a **membrane (transmembrane) potential**. The membrane potential in an undisturbed cell is the cell's **resting potential**.

3-8 ▌ Stages of a cell's life cycle include interphase, mitosis, and cytokinesis p. 127

45. **Cell division** is the reproduction of cells. **Apoptosis** is the genetically controlled death of cells. **Mitosis** is the nuclear division of somatic cells. Sex cells are produced by **meiosis**. (*Spotlight Figures 3–23, 3–24*)

46. Most somatic cells spend the majority of their time in **interphase**, which includes the G_1, **S (DNA replication)**, and G_2 phases. (*Spotlight Figures 3–23, 3–24*)

47. Mitosis proceeds in four stages: **prophase, metaphase, anaphase,** and **telophase**. (*Spotlight Figure 3–24*)

48. During **cytokinesis,** the cytoplasm is divided to form two daughter cells and cell division ends. *(Spotlight Figure 3–24)*

49. In general, the longer the life expectancy of a cell type, the slower is the **mitotic rate. Stem cells** undergo frequent mitosis to replace other, more specialized cells.

3-9 ▌ Several growth factors affect the cell life cycle p. 129

50. A variety of hormones and **growth factors** can stimulate cell division and growth. *(Table 3–2)*

3-10 ▌ Tumors and cancers are characterized by abnormal cell growth and division p. 132

51. Produced by abnormal cell growth and division, a **tumor,** or *neoplasm,* can be *benign* or *malignant.* Malignant cells may spread locally (by **invasion**) or to distant tissues and organs (through **metastasis**). The resultant illness is called **cancer.** Modified genes called **oncogenes** often cause malignancy. *(Figure 3–25)*

3-11 ▌ Differentiation is cellular specialization as a result of gene activation or repression p. 134

52. **Differentiation,** a process of cellular specialization, results from the inactivation of particular genes in different cells, producing populations of cells with limited capabilities. Specialized cells form organized collections called *tissues,* each of which has certain functional roles.

Review Questions

See the blue Answers tab at the back of the book.

LEVEL 1 Reviewing Facts and Terms

1. In the following diagram, identify the type of solution (hypertonic, hypotonic, or isotonic) in which the red blood cells are immersed.

Water molecules   Solute molecules

(a) _____   (b) _____   (c) _____

2. What best describes the function of cholesterol within the plasma membrane? **(a)** reduces plasma membrane fluidity and permeability, **(b)** facilitates transport across the plasma membrane, **(c)** anchors and stabilizes plasma membrane proteins, **(d)** lubricates and protects the plasma membrane.

3. Receptors in cell membranes are made of **(a)** proteins, **(b)** carbohydrates, **(c)** phospholipids, **(d)** lipids.

4. _____ ion concentrations are high in extracellular fluids, and _____ ion concentrations are high in the cytoplasm. **(a)** Calcium; magnesium, **(b)** Chloride; sodium, **(c)** Potassium; sodium, **(d)** Sodium; potassium.

5. At resting membrane potential, the inside of the cell is _____, and the cell exterior is _____ **(a)** slightly negative; slightly positive, **(b)** slightly positive; slightly negative, **(c)** slightly positive; neutral; **(d)** slightly negative; neutral.

6. The organelle responsible for a variety of functions centering on the synthesis of lipids and carbohydrates is **(a)** the Golgi apparatus, **(b)** the rough endoplasmic reticulum, **(c)** the smooth endoplasmic reticulum, **(d)** mitochondria.

7. The synthesis of a functional polypeptide by using the information in an mRNA strand is **(a)** translation, **(b)** transcription, **(c)** replication, **(d)** gene activation.

8. Our somatic cell nuclei contain _____ pairs of chromosomes **(a)** 8, **(b)** 16, **(c)** 23, **(d)** 46.

9. Define what is meant by the term "selectively permeable" membrane.

10. The interphase of the cell life cycle is divided into **(a)** prophase, metaphase, anaphase, and telophase, **(b)** G_0, G_1, S, and G_2, **(c)** mitosis and cytokinesis, **(d)** all of these.

11. Discuss aquaporins as an example of transport across the plasma membrane.

12. What are four general functions of the plasma membrane?

13. List seven membranous organelles.

14. By what three major transport mechanisms do substances get into and out of cells?

15. Thick filaments are made of _____ protein subunits. **(a)** actin, **(b)** myosin, **(c)** tubulin, **(d)** kinesin.

16. List three functions of the lysosome and describe its relationship to the Golgi apparatus.

LEVEL 2 Reviewing Concepts

17. Diffusion is important in body fluids, because it tends to **(a)** increase local concentration gradients, **(b)** eliminate local concentration gradients, **(c)** move substances against concentration gradients, **(d)** create concentration gradients.

18. Microvilli are found **(a)** mostly in muscle cells, **(b)** on the inside of plasma membranes, **(c)** in large numbers on cells that secrete hormones, **(d)** in cells that are actively engaged in absorption, **(e)** only on cells lining the reproductive tract.

19. When a cell is placed in a(n) _____ solution, the cell will lose water through osmosis. This process results in the _____ of red blood cells **(a)** hypotonic; crenation, **(b)** hypertonic; crenation, **(c)** isotonic; hemolysis, **(d)** hypotonic; hemolysis.

3

20. Suppose that a DNA segment has the following nucleotide sequence: CTC–ATA–CGA–TTC–AAG–TTA. Which nucleotide sequences would a complementary mRNA strand have? **(a)** GAG–UAU–GAU–AAC–UUG–AAU, **(b)** GAG–TAT–GCT–AAG–TTC–AAT, **(c)** GAG–UAU–GCU–AAG–UUC–AAU, **(d)** GUG–UAU–GGA–UUG–AAC–GGU.

21. How many amino acids are coded in the DNA segment in Review Question 20? **(a)** 18, **(b)** 9, **(c)** 6, **(d)** 3.

22. The sodium–potassium exchange pump **(a)** is an example of facilitated diffusion, **(b)** does not require the input of cellular energy in the form of ATP, **(c)** moves the sodium and potassium ions along their concentration gradients, **(d)** is composed of a carrier protein located in the plasma membrane, **(e)** is not necessary for the maintenance of homeostasis.

23. Describe the process of secondary active transport and provide an example of how it is used in the body.

24. List, in sequence, the phases of the interphase stage of the cell life cycle, and briefly describe what happens in each.

25. Describe how primary tumors spread to other areas of the body during secondary tumor formation.

26. What is angiogenesis? Why is it important for tumors?

LEVEL 3 Critical Thinking and Clinical Applications

27. The transport of a certain molecule exhibits the following characteristics (1) The molecule moves down its concentration gradient; (2) at concentrations above a given level, the rate of transport does not increase; and (3) cellular energy is not required for transport to occur. Which transport process is at work?

28. Solutions A and B are separated by a selectively permeable barrier. Over time, the level of fluid on side A increases. Which solution initially had the higher concentration of solute?

29. A contrast medium can be injected into the blood to help radiographers improve a visual image of a structure. However, some contrast media have very high osmolarity. Using your understanding of osmosis, explain why this might lead to high blood volume.

30. What is the benefit of having some of the cellular organelles enclosed by a membrane similar to the plasma membrane?

CLINICAL CASE Wrap-Up

When Your Heart Is in the Wrong Place

Jackson is diagnosed with primary ciliary dyskinesia (*dys,* something wrong with; *kinesia,* movement), an inherited disorder of cilia in every cell where cilia are found in the body. For this reason, his ears, sinuses, airways, and reproductive tract are all affected. Without the sweeping, wavelike motion of cilia to keep the lining of these areas clear of debris and infecting organisms, frequent infections of the ears, nose, throat, sinuses, and lungs occur.

It is thought that the motion of cilia also has something to do with the arrangement of the organs during fetal development. The tech initially thought she had reversed the right/left markers on Jackson's chest x-ray, but it turns out that his heart really is on the wrong side. In fact, all of the organs in Jackson's

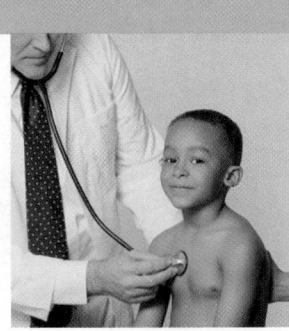

chest are arranged in a mirror-image of normal, a condition known as *situs inversus.*

Because there are only about 25,000 people with primary ciliary dyskinesia in the United States, it is classified as an "orphan" disease (any disease that affects fewer than 200,000 people). Many orphan diseases are genetic and therefore chronic (long-lasting).

1. What other cell, with a "tail" that has the same microtubule structure as motile cilia, might also be affected as a result of Jackson's primary ciliary dyskinesia?

2. What other consequences could result from Jackson's disorder once he reaches adulthood?

See the blue Answers tab at the back of the book.

4 The Tissue Level of Organization

Learning Outcomes

These Learning Outcomes correspond by number to this chapter's sections and indicate what you should be able to do after completing the chapter.

4-1 ▌ Identify the four major types of tissues in the body and describe their roles.

4-2 ▌ Discuss the types and functions of epithelial tissue.

4-3 ▌ Describe the relationship between structure and function for each type of epithelium.

4-4 ▌ Compare the structures and functions of the various types of connective tissue.

4-5 ▌ Describe how cartilage and bone function as a supporting connective tissue.

4-6 ▌ Explain how epithelial and connective tissues combine to form four types of tissue membranes, and specify the functions of each.

4-7 ▌ Describe how connective tissue establishes the framework of the body.

4-8 ▌ Describe the three types of muscle tissue and the special structural features of each type.

4-9 ▌ Discuss the basic structure and role of neural tissue.

4-10 ▌ Describe how injuries affect the tissues of the body.

4-11 ▌ Describe how aging affects the tissues of the body.

SPOTLIGHT
• Inflammation and Regeneration p. 171

CLINICAL CASE
• The Rubber Girl p. 140

Clinical Notes
• Exfoliative Cytology p. 146
• Marfan's Syndrome p. 155

CLINICAL CASE

The Rubber Girl

All her life Anne Marie has felt "delicate." She is a horrible athlete. She once dislocated both shoulders attempting to play high school volleyball. And she has an unstable feeling, as if her kneecaps will dislocate, when she tries dancing or any other twisting activity. She sprains one ankle or the other at least once a year. She has always been "double jointed," with fingers that she can bend back until they nearly touch her dorsal forearm. Her elbows and knees all hyperextend, moving beyond their normal anatomical position.

Anne Marie's skin has always been delicate, pale, and very stretchy. She seems to get abrasions and cuts easily. These heal slowly, leaving huge scars that are also fragile. Once, she had stitches in her face and they cut right through the skin.

Now she has a wide scar over her eyebrow. She bruises easily, too.

Anne Marie thinks she is a lot like her dad. He has the same easy-bruising skin that seems to tear with minimal trauma and heal poorly. He was never very athletic and is also loose jointed. He has twisted both ankles many times. Luckily he has a desk job. **What is the cause of Anne Marie's unusual physical traits? To find out, turn to the Clinical Case Wrap-Up on p. 175.**

An Introduction to the Tissue Level of Organization

In this chapter we discuss how a variety of cell types arranged in various combinations form *tissues*, which are structures with discrete structural and functional properties. Tissues in combination form *organs*, such as the heart or liver, and in turn organs can be grouped into 11 *organ systems.* ⟲ p. 32

4-1 ▌ The four tissue types are epithelial, connective, muscle, and neural

Learning Outcome Identify the four major types of tissues in the body and describe their roles.

The human body contains trillions of cells, but only about 200 *types* of cells. To work efficiently, several different types of cells must coordinate their efforts. Cells working together form **tissues**—collections of specialized cells and cell products that carry out a limited number of functions. The study of tissues is called **histology**. Histologists recognize four basic types of tissue:

1. *Epithelial tissue* covers exposed surfaces, lines internal passageways and chambers, and forms glands.

2. *Connective tissue* fills internal spaces, provides structural support for other tissues, transports materials within the body, and stores energy.

3. *Muscle tissue* is specialized for contraction and includes the skeletal muscles of the body, the muscle of the heart, and the muscular walls of hollow organs.

4. *Neural tissue* carries information from one part of the body to another in the form of electrical impulses.

In this chapter we introduce the basic characteristics of these tissues. This information will help you understand the descriptions of organs and organ systems in later chapters. It will also help you make the connections between anatomical structures and their physiological functions. ATLAS: Embryology Summary 1: The Formation of Tissues

CHECKPOINT

1. **Define histology.**
2. **Identify the four major types of tissues in the body.**

See the blue Answers tab at the back of the book.

4-2 ▌ Epithelial tissue covers body surfaces, lines cavities and tubular structures, and serves essential functions

Learning Outcome Discuss the types and functions of epithelial tissue.

Let's begin our discussion with **epithelial tissue**, because it includes a very familiar feature—the surface of your skin. Epithelial tissue includes *epithelia* and *glands*. **Epithelia** (ep-i-THĒ-lē-a; singular, *epithelium*) are layers of cells that cover internal or external surfaces. **Glands** are structures that produce fluid secretions. They are either attached to or derived from epithelia.

Epithelia cover every exposed surface of your body. Epithelia form the surface of the skin and line the digestive, respiratory, reproductive, and urinary tracts. In fact, they line all passageways that communicate with the outside world. The more delicate epithelia line internal cavities and passageways, such as the chest cavity, fluid-filled spaces in the brain, the inner surfaces of blood vessels, and the chambers of the heart.

Epithelia have several important characteristics:

- *Cellularity.* Epithelia are made almost entirely of cells bound closely together by interconnections known as *cell junctions.* In other tissue types, the cells are often widely separated by extracellular materials.

- *Polarity.* An epithelium has an exposed surface, which faces the exterior of the body or an internal space, and a base, which is attached to underlying tissues. The term **polarity** refers to the presence of structural and functional differences between the exposed and attached surfaces. An epithelium consisting of a single layer of cells has an exposed **apical surface** and an attached **basal surface**. The two surfaces differ in plasma membrane structure and function.

- *Attachment.* The base of an epithelium is bound to a thin, noncellular **basement membrane**, which is also called the **basal lamina**. The basement membrane is a complex structure produced by the basal surface of the epithelium and the underlying connective tissue.

- *Avascularity.* Epithelia are **avascular** (ā-VAS-kū-lar; *a-*, without + *vas*, vessel), which means that they lack blood vessels. Epithelial cells get nutrients by diffusion or absorption across either the exposed or the attached epithelial surface.

- *Regeneration.* Epithelial cells that are damaged or lost at the exposed surface are continuously replaced through stem cell divisions in the epithelium. Regeneration is a characteristic of other tissues as well, but the rates of cell division and replacement are typically much higher in epithelia than in other tissues.

Functions of Epithelial Tissue

Epithelia perform four essential functions:

1. *Provide Physical Protection.* Epithelia protect exposed and internal surfaces from abrasion, dehydration, and destruction by chemical or biological agents.

2. *Control Permeability.* Any substance that enters or leaves your body must cross an epithelium. Some epithelia are relatively impermeable. Others are easily crossed by compounds as large as proteins. Many epithelia contain the molecular "machinery" needed for selective absorption or secretion. The epithelial barrier can also be regulated and modified in response to stimuli. For example, hormones can affect the transport of ions and nutrients through epithelial cells. Even physical stress can alter the structure and properties of epithelia. For example, calluses form on your hands when you do manual labor for some time.

3. *Provide Sensation.* Most epithelia are extremely sensitive to stimulation, because they have a large sensory nerve supply. These sensory nerves continually provide information about the external and internal environments. For example, the lightest touch of a mosquito will stimulate sensory neurons that tell you where to swat. A *neuroepithelium* is an epithelium that is specialized to perform a particular sensory function. Our neuroepithelia contain sensory cells that provide the sensations of smell, taste, sight, equilibrium, and hearing.

4. *Produce Specialized Secretions.* Epithelial cells that produce secretions are called *gland cells.* Individual gland cells are usually scattered among other cell types in an epithelium. In a **glandular epithelium**, most or all of the epithelial cells produce secretions. These cells either discharge their secretions onto the surface of the epithelium (to provide physical protection or temperature regulation) or release them into the surrounding interstitial fluid and blood (to act as chemical messengers).

Specializations of Epithelial Cells

How are epithelial cells different from other body cells? They have several structural specializations. For the epithelium as a whole to perform the functions just listed, individual epithelial cells may be specialized for (1) the movement of fluids over the epithelial surface, providing protection and lubrication; (2) the movement of fluids through the epithelium, to control permeability; or (3) the production of secretions that provide physical protection or act as chemical messengers. Specialized epithelial cells generally possess a strong polarity. Figure 4–1 shows one common type of epithelial polarity.

The epithelial cell is often divided into two functional regions. One is the *apical surface,* where the cell is exposed to an internal or external environment. The other consists of the *basolateral surfaces,* which include both the base, where the cell attaches to underlying epithelial cells or deeper tissues, and the sides, where the cell contacts its neighbors.

Many epithelial cells that line internal passageways have *microvilli* on their exposed surfaces. ⟲ p. 99 Just a few may be present, or microvilli may carpet the entire surface. Microvilli are especially abundant on epithelial surfaces where absorption and secretion take place, such as along portions of the digestive system and kidneys. The epithelial cells in these locations are transport specialists. Each cell has at least 20 times more surface area to transport substances than it would have without microvilli.

Cilia are characteristic of surfaces covered by a **ciliated epithelium**. A typical ciliated cell contains about 250 cilia that beat in a coordinated manner. The synchronized beating of the cilia moves substances over the epithelial surface. For example, the ciliated epithelium that lines the respiratory tract moves mucus up from the lungs and toward the throat, as though on an escalator. The sticky mucus traps inhaled particles, including dust, pollen, and pathogens. The ciliated epithelium then carries the mucus and the trapped debris to the throat. There they are swallowed or expelled by coughing. If abrasion

4

Figure 4–1 **The Polarity of Epithelial Cells.** Many epithelial cells have functional specializations associated with an apical surface that is exposed to either the external environment or internal spaces, and basolateral surfaces (base and sides) that interact with deeper tissues and adjacent cells. Often, the apical surface has microvilli; sometimes it has cilia. This functional polarity is also evident in the uneven distribution of organelles between the free surface (here, the top) and the basement membrane. In some epithelia, such as the lining of the kidney tubules, mitochondria are concentrated near the base of the cell, where energy is in high demand for the cell's transport activities.

- Cilia
- Microvilli
- Apical surface
- Golgi apparatus
- Nucleus
- Mitochondria
- Basement membrane
- Basolateral surfaces

or toxic compounds such as nicotine and carbon monoxide in cigarette smoke injure the cilia or the epithelial cells, ciliary movement can stop. Without the protective upward flow of mucus, infection or disease become more likely.

Maintaining the Integrity of Epithelia

To be effective as a barrier, an epithelium must form a complete cover or lining. Three factors help maintain the physical integrity of an epithelium: (1) intercellular connections, (2) attachment to the basement membrane, and (3) epithelial maintenance and repair.

Intercellular Connections

Cells in an epithelium are firmly attached to one another, and the epithelium as a unit is attached to extracellular fibers of the superficial clear layer of the basement membrane. Many cells in your body form permanent or temporary bonds with other cells or extracellular material. Epithelial cells, however, are specialists in intercellular connection.

Intercellular connections involve either extensive areas of opposing plasma membranes or specialized attachment sites, discussed shortly. Large areas of opposing plasma membranes are interconnected by transmembrane proteins called **cell adhesion molecules (CAMs)**, which bind to each other and to extracellular materials. For example, CAMs on the basolateral surface of an epithelium help bind the cell to the underlying basement membrane. The membranes of adjacent cells may also be bonded by a thin layer of proteoglycans that contain polysaccharide derivatives known as *glycosaminoglycans (GAGs)*, most notably **hyaluronan** *(hyaluronic acid)*.

Cell junctions are specialized areas of the plasma membrane that attach a cell to another cell or to extracellular materials. The three most common types of cell junctions are (1) tight junctions, (2) gap junctions, and (3) desmosomes (**Figure 4–2a**).

Some epithelial functions require rapid intercellular communication. At a **gap junction** (**Figure 4–2b**), two cells are held together by two interlocking transmembrane proteins called *connexons*. Each connexon is composed of six *connexin* proteins that form a cylinder with a central pore. Two aligned connexons form a narrow passageway that lets small molecules and ions pass from cell to cell. Gap junctions are common among epithelial cells, where the movement of ions helps coordinate functions such as the beating of cilia. Gap junctions are also common in other tissues. For example, gap junctions in cardiac muscle tissue and smooth muscle tissue are essential in coordinating muscle cell contractions. A number of human diseases are caused by mutations in the genes that code for different connexin proteins.

At a **tight junction** (also known as an occluding junction), the lipid portions of the two plasma membranes are tightly bound together by interlocking membrane proteins (**Figure 4–2c**). Inferior to the tight junctions, a continuous *adhesion belt* forms a band that encircles cells and binds them to their neighbors. The bands are attached to the microfilaments of the terminal web. ⟲ p. 97 This kind of attachment is so tight that tight junctions largely prevent water and solutes from passing between the cells.

When the epithelium lines a tube, such as the intestinal tract, the apical surfaces of the epithelial cells are exposed to the space inside the tube, a passageway called the **lumen** (LOO-men). Tight junctions effectively isolate the contents of the lumen from the basolateral surfaces of the cell. For example, tight junctions near the apical surfaces of cells that line the digestive tract help keep enzymes, acids, and wastes in the lumen from reaching the basolateral surfaces and digesting or otherwise damaging the underlying tissues and organs.

Most epithelial cells are subject to mechanical stresses—stretching, bending, twisting, or compression—so they must have durable interconnections. At a **desmosome** (DEZ-mō-sōm; *desmos*, ligament + *soma*, body), CAMs and proteoglycans link the opposing plasma membranes. Desmosomes are very strong and can resist stretching and twisting.

Figure 4–2 **Cell Junctions.**

b Gap junctions permit the free diffusion of ions and small molecules between two cells.

Embedded proteins (connexons)

Interlocking junctional proteins

Tight junction

Adhesion belt

Terminal web

Gap junctions

Spot desmosome

Hemidesmosome

a This is a diagrammatic view of an epithelial cell, showing the major types of intercellular connections.

Tight junction

Adhesion belt

c A tight junction is formed by the fusion of the outer layers of two plasma membranes. Tight junctions prevent the diffusion of fluids and solutes between the cells. A continuous adhesion belt lies deep to the tight junction. This belt is tied to the microfilaments of the terminal web.

Clear layer

Dense layer

Basement membrane

e Hemidesmosomes attach a cell to extracellular structures, such as the protein fibers in the basement membrane.

Intermediate filaments

Dense area

Cell adhesion molecules (CAMs)

Proteoglycans

d A spot desmosome ties adjacent cells together.

A typical desmosome is formed by two cells. Within each cell, a complex known as a *dense area* is connected to the cytoskeleton (**Figure 4–2d**). This connection to the cytoskeleton gives the desmosome—and the epithelium—its strength. For example, desmosomes are abundant between cells in the superficial layers of the skin. As a result, damaged skin cells are usually lost in sheets rather than as individual cells. (That is why your skin peels rather than sheds off as a powder after a sunburn.)

There are two types of desmosomes:

- *Spot desmosomes* are small discs connected to bands of intermediate filaments (**Figure 4–2d**). The intermediate filaments stabilize the shape of the cell.

- *Hemidesmosomes* resemble half of a spot desmosome. Rather than attaching one cell to another, a hemidesmosome attaches a cell to extracellular filaments in the basement membrane (**Figure 4–2e**). This attachment helps stabilize the position of the epithelial cell and anchors it to underlying tissues.

Attachment to the Basement Membrane

Not only do epithelial cells hold onto one another, but they also remain firmly connected to the rest of the body. The inner surface of each epithelium is attached to a two-part basement membrane. The layer closer to the epithelium, the *clear layer*, (also called the lamina lucida) contains glycoproteins and a network

4

of fine protein filaments (**Figure 4–2e**). It is secreted by the adjacent layer of epithelial cells. The clear layer acts as a barrier that restricts the movement of proteins and other large molecules from the underlying connective tissue into the epithelium.

The deeper portion of the basement membrane, the *dense layer*, contains bundles of coarse protein fibers produced by connective tissue cells. The dense layer gives the basement membrane its strength. Attachments between the fibers of the clear layer and those of the dense layer hold the two layers together, and hemidesmosomes attach the epithelial cells to the composite basement membrane. The dense layer also acts as a filter that determines what substances can diffuse between the adjacent tissues and the epithelium.

Epithelial Maintenance and Repair

Epithelial cells lead hard lives. They are exposed to disruptive enzymes, toxic chemicals, pathogenic bacteria, and mechanical abrasion. Consider the lining of your small intestine, where epithelial cells are exposed to a variety of enzymes and abraded by partially digested food. In this extreme environment, an epithelial cell may last just a day or two before it is shed or destroyed. The only way the epithelium can maintain its structure over time is by the continual division of *stem cells*. ↺ p. 129 Most epithelial stem cells, also called **germinative cells**, are located near the basement membrane, in a relatively

protected location. ATLAS: Embryology Summary 2: The Development of Epithelia

CHECKPOINT

3. List five important characteristics of epithelial tissue.
4. Identify four essential functions of epithelial tissue.
5. What is the probable function of an epithelial surface whose cells bear many microvilli?
6. Identify the various types of epithelial cell junctions.
7. What is the functional significance of gap junctions?

See the blue Answers tab at the back of the book.

4-3 ▎ Cell shape and number of layers determine the classification of epithelia

Learning Outcome Describe the relationship between structure and function for each type of epithelium.

There are many different specialized types of epithelia. You can easily sort these into categories based on (1) the cell shape, and (2) the number of cell layers between the basement membrane and the exposed surface of the epithelium. Using these two criteria, we can describe almost every epithelium in the body (**Table 4–1**).

Table 4–1	Classifying Epithelia		
	SQUAMOUS	**CUBOIDAL**	**COLUMNAR**
Simple	Simple squamous epithelium	Simple cuboidal epithelium	Simple columnar epithelium
Stratified	Stratified squamous epithelium	Stratified cuboidal epithelium	Stratified columnar epithelium

Classification of Epithelia

Epithelial cells have three basic shapes: *squamous, cuboidal,* and *columnar.* For classification purposes, we look at the superficial cells in a section perpendicular to both the apical surface and the basement membrane. In sectional view, squamous cells appear thin and flat, cuboidal cells look like little boxes, and columnar cells are tall and relatively slender rectangles.

Once you have determined whether the superficial cells are squamous, cuboidal, or columnar, then look at the number of cell layers. There are only two options: *simple* or *stratified.*

If only one layer of cells covers the basement membrane, that layer is a **simple epithelium**. Simple epithelia are necessarily thin. All the cells have the same polarity, so the distance from the nucleus to the basement membrane does not change from one cell to the next.

Because they are so thin, simple epithelia are fragile. A single layer of cells cannot provide much mechanical protection, so simple epithelia are located only in protected areas inside the body. They line internal compartments and passageways, such as the pleural, pericardial, and peritoneal cavities, the heart chambers, and blood vessels.

Simple epithelia are also characteristic of regions in which secretion or absorption occurs. Examples are the lining of the intestines and the gas-exchange surfaces of the lungs. In these places, thinness is an advantage. It reduces the time required for materials to cross the epithelial barrier.

In a **stratified epithelium**, several layers of cells cover the basement membrane. Stratified epithelia are generally located in areas that are exposed to mechanical or chemical stresses. The surface of the skin and the lining of the mouth are examples.

TIPS & TRICKS

To help you remember the meanings of the terms *squamous* and *stratified*, associate the word "**squa**mous" with "**sca**ly," and the word "**strat**ified" with "**strat**osphere," an upper *layer* of Earth's atmosphere.

Squamous Epithelia

The cells in a **squamous epithelium** (SKWĀ-mus; *squama,* plate or scale) are thin, flat, and somewhat irregular in shape, like pieces of a jigsaw puzzle (**Figure 4–3**). From the surface, the

Figure 4–3 Squamous Epithelia.

Simple Squamous Epithelium

LOCATIONS: Mesothelia lining pleural, pericardial, and peritoneal cavities; endothelia lining heart and blood vessels; portions of kidney tubules (thin sections of nephron loops); inner lining of cornea; alveoli of lungs

FUNCTIONS: Reduces friction; controls vessel permeability; performs absorption and secretion

Cytoplasm · Nucleus · Connective tissue · LM × 238

a Lining of peritoneal cavity

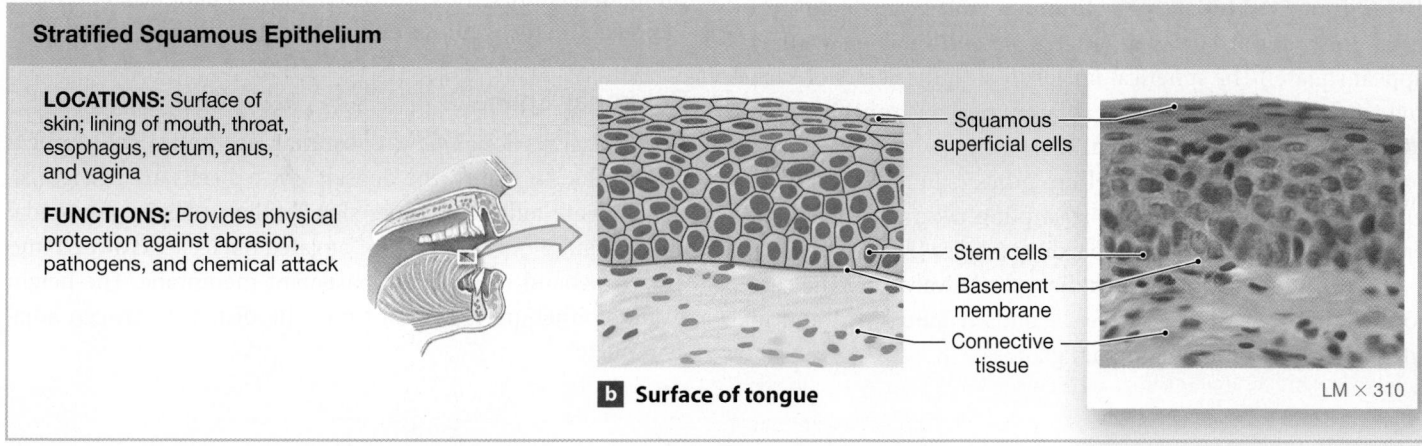

Stratified Squamous Epithelium

LOCATIONS: Surface of skin; lining of mouth, throat, esophagus, rectum, anus, and vagina

FUNCTIONS: Provides physical protection against abrasion, pathogens, and chemical attack

Squamous superficial cells · Stem cells · Basement membrane · Connective tissue · LM × 310

b Surface of tongue

cells resemble fried eggs laid side by side. In sectional view, the disc-shaped nucleus occupies the thickest portion of each cell.

A **simple squamous epithelium** is the body's most delicate type of epithelium. This type of epithelium is located in protected regions where absorption or diffusion takes place, or where a slick, slippery surface reduces friction. Examples are the respiratory exchange surfaces (*alveoli*) of the lungs, the lining of the thoracic and abdominopelvic body cavities of the trunk (**Figure 4–3a**), and the lining of the heart and blood vessels. Smooth linings are extremely important. For example, any irregularity in the lining of a blood vessel may result in the formation of a potentially dangerous blood clot.

Special names have been given to the simple squamous epithelia that line chambers and passageways that do not communicate with the outside world. The simple squamous epithelium that lines the body cavities enclosing the lungs, heart, and abdominal organs is called a **mesothelium** (mez-ō-THĒ-lē-um; *mesos,* middle). The pleura, pericardium, and peritoneum each contain a superficial layer of mesothelium. The simple squamous epithelium lining the inner surface of the heart and all blood vessels is called an **endothelium** (en-dō-THĒ-lē-um; *endo-,* inside).

A **stratified squamous epithelium** (**Figure 4–3b**) is generally located where mechanical stresses are severe. The cells form a series of layers, like the layers in a sheet of plywood. The surface of the skin and the lining of the mouth, esophagus, and anus are areas where this type of epithelium protects against physical and chemical attacks.

On exposed body surfaces, where mechanical stress and dehydration are potential problems, apical layers of epithelial cells are packed with filaments of the protein *keratin.* As a result, superficial layers are both tough and water resistant. Such an epithelium is said to be *keratinized.* A *nonkeratinized* stratified squamous epithelium resists abrasion, but will dry out and deteriorate unless kept moist. Nonkeratinized stratified squamous epithelia are found in the oral cavity, pharynx, esophagus, anus, and vagina.

Cuboidal Epithelia

The cells of a **cuboidal epithelium** resemble hexagonal boxes from their apical surfaces. (In typical sectional views they appear square.) The spherical nuclei are near the center of each cell, and the distance between adjacent nuclei is roughly equal to the height of the epithelium.

A **simple cuboidal epithelium** provides limited protection and occurs where secretion or absorption takes place. Such an epithelium lines portions of the kidney tubules (**Figure 4–4a**).

Stratified cuboidal epithelia are relatively rare. They are located along the ducts of sweat glands (**Figure 4–4b**) and in the larger ducts of the mammary glands.

Clinical Note

Exfoliative Cytology *Exfoliative cytology* (eks-FŌ-lē-a-tiv; *ex-,* from + *folium,* leaf) is the study of cells shed or removed from epithelial surfaces. The cells are examined for a variety of reasons—for example, to check for cellular changes that indicate cancer, or for genetic screening of a fetus. Cells are collected by sampling the fluids that cover the epithelia lining the respiratory, digestive, urinary, or reproductive tract; by removing fluid from the pleural, pericardial, or peritoneal body cavities; or by removing cells from an epithelial surface.

One common sampling procedure is a *Pap test,* named after Dr. George Papanicolaou, who pioneered its use. The most familiar Pap test is that for cervical cancer. This test involves scraping cells from the tip of the *cervix,* the portion of the uterus that projects into the vagina.

Amniocentesis is also based on exfoliative cytology. In this procedure, shed epithelial cells are collected from a sample of *amniotic fluid,* which surrounds and protects a developing fetus. Examination of these cells can determine if the fetus has a chromosomal abnormality, such as occurs in *Down's syndrome.*

Transitional Epithelia

A **transitional epithelium** (**Figure 4–4c**) is an unusual stratified epithelium. Unlike most epithelia, it tolerates repeated cycles of stretching and recoiling (returning to its previous shape) without damage. It is called transitional because the appearance of the epithelium changes as stretching occurs. A transitional epithelium is found in regions of the urinary system, such as the urinary bladder, where large changes in volume occur. In an empty urinary bladder, the superficial cells of the epithelium are typically plump and cuboidal. In the full urinary bladder, when the volume of urine has stretched the lining to its limits, the epithelium appears flattened, and more like a stratified squamous epithelium.

Columnar Epithelia

In a typical sectional view, **columnar epithelial cells** appear rectangular. In reality, the densely packed cells are hexagonal, but they are taller and more slender than cells in a cuboidal epithelium (**Figure 4–5**). The elongated nuclei are crowded into a narrow band close to the basement membrane. The height of the epithelium is several times the distance between adjacent nuclei.

Figure 4–4   **Cuboidal and Transitional Epithelia.**

Simple Cuboidal Epithelium

LOCATIONS: Glands; ducts; portions of kidney tubules; thyroid gland

FUNCTIONS: Limited protection, secretion, absorption

Connective tissue

Nucleus

Cuboidal cells

Basement membrane

LM × 650

a **Kidney tubule**

Stratified Cuboidal Epithelium

LOCATIONS: Lining of some ducts (rare)

FUNCTIONS: Protection, secretion, absorption

Lumen of duct

Stratified cuboidal cells

Basement membrane

Nucleus

Connective tissue

LM × 500

b **Sweat gland duct**

Transitional Epithelium

LOCATIONS: Urinary bladder; renal pelvis; ureters

FUNCTIONS: Permits expansion and recoil after stretching

Epithelium (relaxed)

Basement membrane

Connective tissue and smooth muscle layers

Empty bladder

LM × 400

Epithelium (stretched)

Basement membrane

Connective tissue and smooth muscle layers

Full bladder

LM × 400

c **Urinary bladder**

4

Figure 4–5   **Columnar Epithelia.**

Simple Columnar Epithelium

LOCATIONS: Lining of stomach, intestine, gallbladder, uterine tubes, and collecting ducts of kidneys

FUNCTIONS: Protection, secretion, absorption

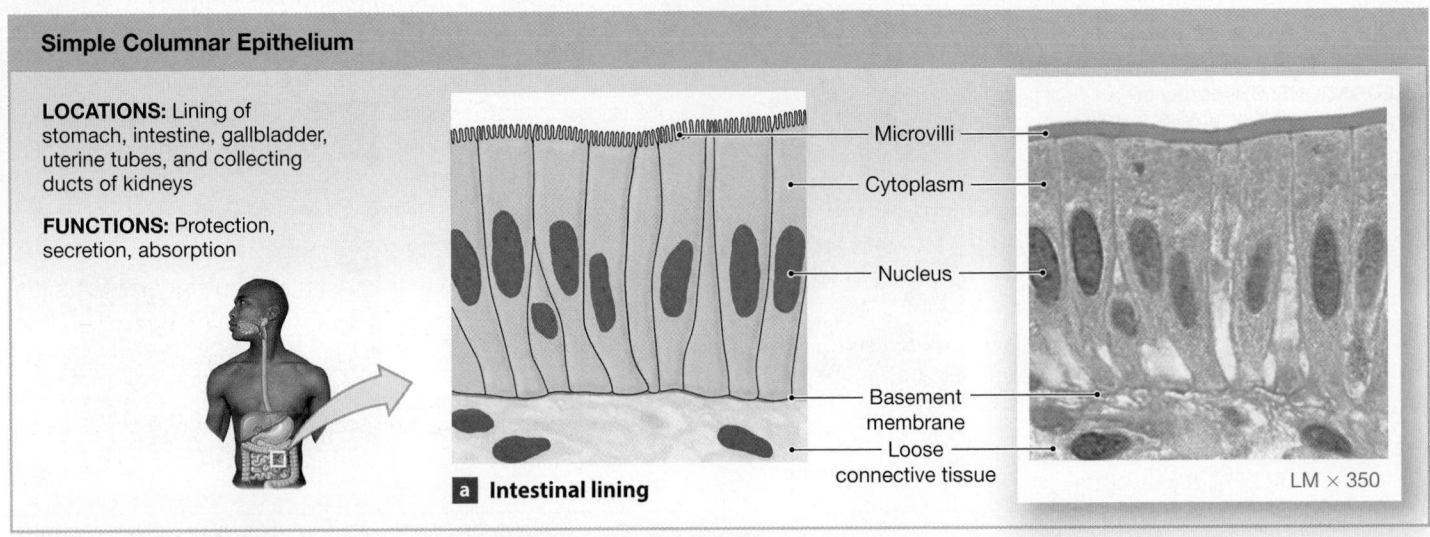

Microvilli

Cytoplasm

Nucleus

Basement membrane

Loose connective tissue

a **Intestinal lining**

LM × 350

Pseudostratified Ciliated Columnar Epithelium

LOCATIONS: Lining of nasal cavity, trachea, and bronchi; portions of male reproductive tract

FUNCTIONS: Protection, secretion, move mucus with cilia

Cilia

Cytoplasm

Nuclei

Basement membrane

Loose connective tissue

b **Trachea**

LM × 350

Stratified Columnar Epithelium

LOCATIONS: Small areas of the pharynx, epiglottis, anus, mammary glands, salivary gland ducts, and urethra

FUNCTION: Protection

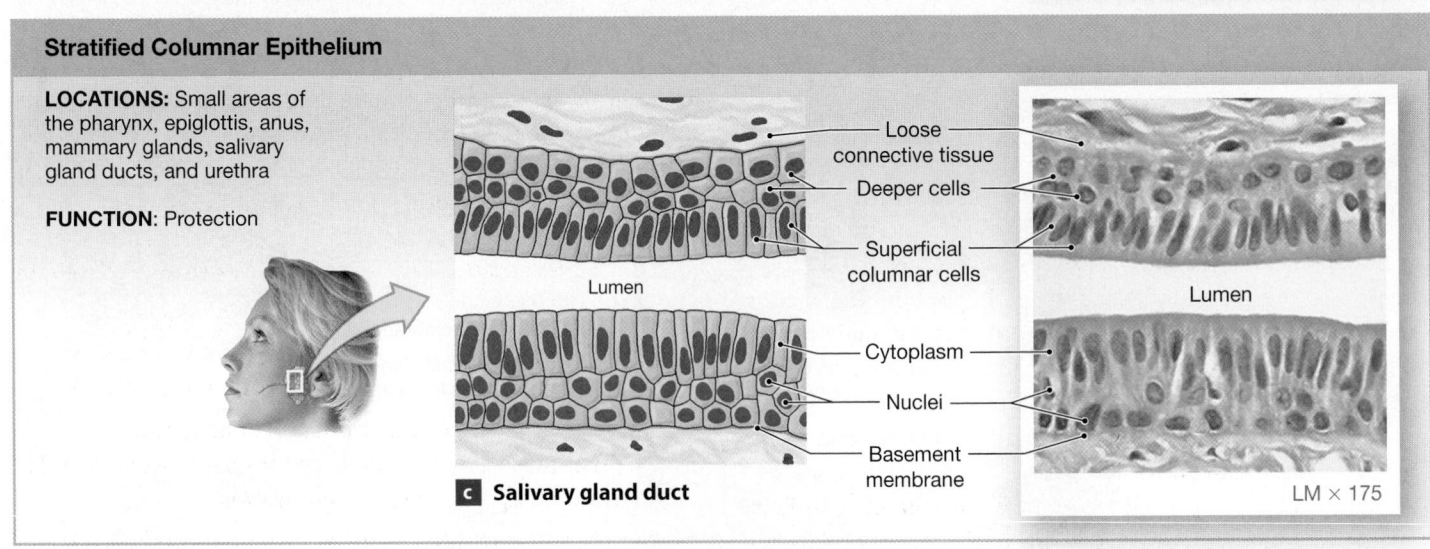

Loose connective tissue

Deeper cells

Superficial columnar cells

Lumen

Cytoplasm

Nuclei

Basement membrane

c **Salivary gland duct**

Lumen

LM × 175

A **simple columnar epithelium** is typically found where absorption or secretion takes place, such as in the small intestine (Figure 4–5a). In the stomach and large intestine, the secretions of simple columnar epithelia protect against chemical stresses.

Portions of the respiratory tract contain a **pseudostratified columnar epithelium**, a columnar epithelium that includes several types of cells with varying shapes and functions. The distances between the cell nuclei and the exposed surface vary, so the epithelium appears to be layered, or stratified (Figure 4–5b). It is not truly stratified, though, because every epithelial cell contacts the basement membrane. Pseudostratified columnar epithelial cells typically have cilia. Epithelia of this type line most of the nasal cavity, the trachea (windpipe), the bronchi (branches of the trachea leading to the lungs), and portions of the male reproductive tract.

Stratified columnar epithelia are relatively rare, providing protection along portions of the pharynx, epiglottis, anus, and urethra, as well as along a few large excretory ducts. The epithelium has either two layers or multiple layers (Figure 4–5c). In the latter case, only the superficial cells are columnar.

Now that we have considered the classes of epithelia, we turn to a specialized type of epithelium: glandular epithelia.

Glandular Epithelia

Many epithelia contain gland cells that are specialized for secretion. Collections of epithelial cells (or structures derived from epithelial cells) that produce secretions are called *glands.* They range from scattered cells to complex glandular organs. Some of these glands, called **endocrine glands**, release their secretions into the interstitial fluid. Others, known as **exocrine glands**, release their secretions into passageways called **ducts** that open onto an epithelial surface.

Endocrine Glands

An endocrine gland produces *endocrine* (*endo-*, inside + *krino*, to separate) *secretions,* which are released directly into the surrounding interstitial fluid. These secretions, also called *hormones,* then enter the bloodstream for distribution throughout the body. Hormones regulate or coordinate the activities of various tissues, organs, and organ systems. Examples of endocrine glands include the thyroid gland and the pituitary gland. Because their secretions are not released into ducts, endocrine glands are often called *ductless glands.*

Endocrine cells may also be part of an epithelial surface, such as the lining of the digestive tract, or they may be found in separate organs, such as the pancreas, thyroid gland, thymus, and pituitary gland. We will consider endocrine cells, organs, and hormones further in Chapter 18.

Exocrine Glands

Exocrine glands produce *exocrine* (*exo-*, outside) *secretions,* which are discharged onto an epithelial surface. Most exocrine secretions reach the surface through tubular ducts, which empty onto the skin surface or onto an epithelium lining an internal passageway that communicates with the exterior. Examples of exocrine secretions are enzymes entering the digestive tract, perspiration on the skin, tears in the eyes, and milk produced by mammary glands.

Exocrine glands have several different methods of secretion. For this reason, they are classified by their mode and type of secretion, and by the structure of the gland cells and associated ducts.

Modes of Secretion. A glandular epithelial cell releases its secretions by (1) merocrine secretion, (2) apocrine secretion, or (3) holocrine secretion.

In **merocrine secretion** (MER-ō-krin; *meros,* part), the product is released from secretory vesicles by exocytosis (Figure 4–6a). ↺ p. 126 This is the most common mode of exocrine secretions. *Mucin* is a merocrine secretion that mixes with water to form **mucus**. Mucus is a lubricant, a protective barrier, and a sticky trap for foreign particles and microorganisms. The mucous secretions of the salivary glands coat food and reduce friction during swallowing. In the skin, merocrine sweat glands produce the watery perspiration that helps cool you on a hot day.

Apocrine secretion (AP-ō-krin; *apo-*, off) involves the loss of cytoplasm as well as the secretory product (Figure 4–6b). The apical portion of the cytoplasm becomes packed with secretory vesicles and is then shed. Milk production in the mammary glands involves a combination of merocrine and apocrine secretions.

Merocrine and apocrine secretions leave a cell relatively intact and able to continue secreting. **Holocrine secretion** (HOL-ō-krin; *holos,* entire), by contrast, destroys the gland cell. During holocrine secretion, the entire cell becomes packed with secretory vesicles and then bursts, releasing the secretion, but killing the cell (Figure 4–6c). Further secretion depends on replacing destroyed gland cells by the division of stem cells. Sebaceous glands, associated with hair follicles, produce an oily hair coating by means of holocrine secretion.

Types of Secretions. We also categorize exocrine glands by the types of secretion produced:

- *Serous glands* secrete a watery solution that contains enzymes. The parotid salivary glands are serous glands.

- *Mucous glands* secrete mucins that hydrate to form mucus. The sublingual salivary glands and the submucosal glands of the small intestine are mucous glands.

- *Mixed exocrine glands* contain more than one type of gland cell. These glands may produce two different exocrine secretions, one serous and the other mucous. The submandibular salivary glands are mixed exocrine glands.

Figure 4–6   **Modes of Glandular Secretion.**

(a) Merocrine secretion

In merocrine secretion, the product is released from secretory vesicles at the apical surface of the gland cell by exocytosis.

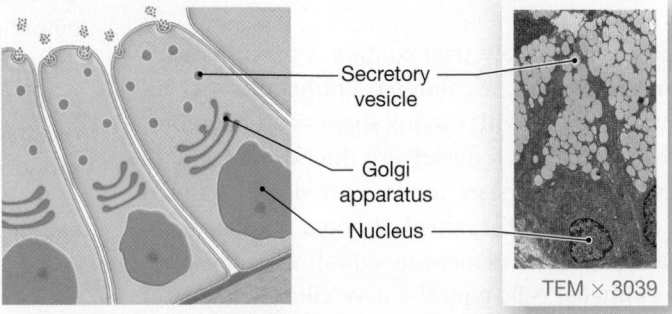

Secretory vesicle

Golgi apparatus

Nucleus

TEM × 3039

(b) Apocrine secretion

Apocrine secretion involves the loss of apical cytoplasm. Inclusions, secretory vesicles, and other cytoplasmic components are shed in the process. The gland cell then grows and repairs itself before it releases additional secretions.

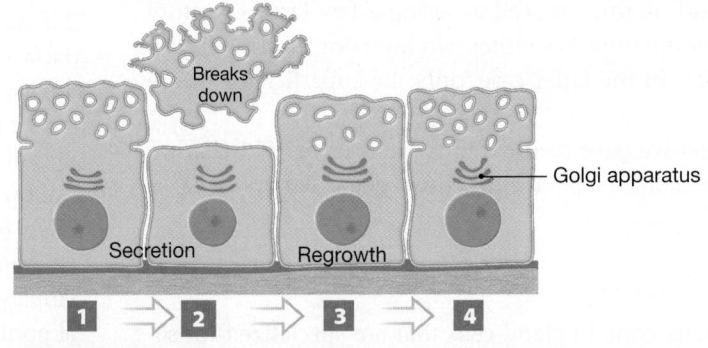

Breaks down

Golgi apparatus

Secretion   Regrowth

1 ⇒ 2 ⇒ 3 ⇒ 4

(c) Holocrine secretion

Holocrine secretion occurs as superficial gland cells burst. Continued secretion involves the replacement of these cells through the mitotic divisions of underlying stem cells.

3 Cells burst, releasing cytoplasmic contents

2 Cells form secretory products and increase in size

1 Cell division replaces lost cells

Stem cell

Salivary gland

Mammary gland

Hair

Sebaceous gland

Hair follicle

Gland Structure. The final method of classifying exocrine glands is by structure. In epithelia that have independent, scattered gland cells, the individual secretory cells are called **unicellular glands**. **Multicellular glands** include glandular epithelia and aggregations of gland cells that produce exocrine or endocrine secretions.

The only **unicellular exocrine glands** in the body are **mucous (goblet) cells**, which secrete mucins. Mucous cells are scattered among other epithelial cells. Both the pseudostratified ciliated columnar epithelium that lines the trachea and the columnar epithelium of the small and large intestines have an abundance of mucous cells.

The simplest **multicellular exocrine gland** is a *secretory sheet*, in which gland cells form an epithelium that releases secretions into an inner compartment. Mucin-secreting cells that line the stomach, for instance, protect that organ from its own acids and enzymes by continuously secreting mucin.

Most other multicellular exocrine glands are in pockets set back from the epithelial surface. Their secretions travel through one or more ducts to the surface. Examples include the salivary glands, which produce mucins and digestive enzymes.

Three characteristics are useful in describing the structure of multicellular exocrine glands (**Figure 4–7**):

1. *The Structure of the Duct.* A gland is *simple* if it has a single duct that does not divide on its way to the gland cells (top row of **Figure 4–7**). The gland is *compound* if the duct divides one or more times on its way to the gland cells (bottom row of **Figure 4–7**).

2. *The Shape of the Secretory Portion of the Gland.* Glands whose glandular cells form tubes are *tubular*; the tubes may be straight or coiled. Those that form blind pockets are *alveolar* (al-VĒ-ō-lar; *alveolus,* sac) or *acinar* (AS-i-nar; *acinus,* chamber). Glands whose secretory cells form both tubes and pockets are called *tubuloalveolar* and *tubuloacinar.*

3. *The Relationship between the Ducts and the Glandular Areas.* A gland is *branched* if several secretory areas (tubular or acinar) share a duct. ("Branched" refers to the glandular areas and not to the duct.)

The vast majority of glands in the body produce either exocrine or endocrine secretions. However, a few complex organs, including the digestive tract and the pancreas, produce both kinds of secretions. We will look at the organization of these glands in Chapters 18 and 24.

Figure 4–7   A Structural Classification of Exocrine Glands.

SIMPLE GLANDS

SIMPLE TUBULAR	SIMPLE COILED TUBULAR	SIMPLE BRANCHED TUBULAR	SIMPLE ALVEOLAR (ACINAR)	SIMPLE BRANCHED ALVEOLAR
Examples: •Intestinal glands	*Examples:* •Merocrine sweat glands	*Examples:* •Gastric glands •Mucous glands of esophagus, tongue, duodenum	*Examples:* •Not found in adult; a stage in development of simple branched glands	*Examples:* •Sebaceous (oil) glands

Duct — Gland cells

COMPOUND GLANDS

COMPOUND TUBULAR	COMPOUND ALVEOLAR (ACINAR)	COMPOUND TUBULOALVEOLAR
Examples: •Mucous glands (in mouth) •Bulbourethral glands (in male reproductive system) •Testes (seminiferous tubules)	*Examples:* •Mammary glands	*Examples:* •Salivary glands •Glands of respiratory passages •Pancreas

4

8. Identify the three cell shapes characteristic of epithelial cells.

9. When classifying epithelial tissue, the number of cell layers determines whether it is simple or stratified. A single layer of cells is termed _____, whereas multiple layers of cells are known as _____.

10. Using a light microscope, a tissue appears as a simple squamous epithelium. Can this be a sample of the skin surface? Why or why not?

11. Why do the pharynx, esophagus, anus, and vagina have a similar epithelial organization?

12. Name the two primary types of glandular epithelia.

13. The secretory cells of sebaceous glands fill with vesicles and then rupture, releasing their contents. Which mode of secretion is this?

14. Which type of gland releases its secretions directly into the interstitial fluid?

See the blue Answers tab at the back of the book.

4-4 | Connective tissue provides a protective structural framework for other tissue types

Learning Outcome Compare the structures and functions of the various types of connective tissue.

It is impossible to discuss epithelial tissue without mentioning an associated type of tissue: **connective tissue**. Recall that the dense layer of the basement membrane of all epithelial tissues is created by connective tissue. In essence, connective tissue connects the epithelium to the rest of the body. Other types of connective tissue include bone, fat, and blood. They provide structure, store energy, and transport materials throughout the body, respectively.

Connective tissues vary widely in appearance and function, but they all share three basic components: (1) specialized cells, (2) extracellular protein fibers, and (3) a fluid known as **ground substance**. Together, the extracellular fibers and ground substance make up the **matrix**, which surrounds the cells. Cells make up the bulk of epithelial tissue, but the matrix typically accounts for most of the volume of connective tissues. ATLAS: Embryology Summary 3: The Origins of Connective Tissues

Connective tissues occur throughout the body, but they are never exposed to the outside environment. Many connective tissues are highly vascular (that is, they have many blood vessels) and contain sensory receptors that detect pain, pressure, temperature, and other stimuli.

Among the specific functions of connective tissues are the following:

▪ Establishing a structural framework for the body.

▪ Transporting fluids and dissolved materials.

▪ Protecting delicate organs.

▪ Supporting, surrounding, and interconnecting other types of tissue.

▪ Storing energy, especially in the form of triglycerides.

▪ Defending the body from invading microorganisms.

Classification of Connective Tissues

We classify connective tissues by their physical properties. The three general categories of connective tissue are connective tissue proper, fluid connective tissues, and supporting connective tissues.

1. **Connective tissue proper** includes those connective tissues with many types of cells and extracellular fibers in a syrupy ground substance. This broad category contains a variety of connective tissues that are grouped into (a) *loose connective tissues* and (b) *dense connective tissues.* These groupings are based on the number of cell types present, and on the relative properties and proportions of fibers and ground substance. Both *adipose tissue* or fat (a loose connective tissue) and *tendons* (a dense connective tissue) are connective tissue proper, but they have very different structural and functional characteristics.

2. **Fluid connective tissues** have distinctive populations of cells suspended in a watery matrix that contains dissolved proteins. The only two types are *blood* and *lymph*.

3. **Supporting connective tissues** differ from connective tissue proper in having a less diverse cell population and a matrix containing much more densely packed fibers. Supporting connective tissues protect soft tissues and support the weight of part or all of the body. The two types of supporting connective tissues are *cartilage* and *bone*. The matrix of cartilage is a gel with characteristics that vary with the predominant type of fiber. The matrix of bone is **calcified**, because it contains mineral deposits (primarily calcium salts) that provide rigidity.

Connective Tissue Proper

Connective tissue proper contains a varied cell population, extracellular fibers, and a viscous (syrupy) ground substance (**Figure 4–8**).The number of cells and cell types in a tissue at any moment varies with local conditions.

Some cells function in local maintenance, repair, and energy storage. They include *fibroblasts, fibrocytes, adipocytes,* and *mesenchymal cells.* These cells are permanent residents of the connective tissue. Other cells defend and repair damaged tissues. These cells are more mobile and include *macrophages, mast cells, lymphocytes, plasma cells,* and *microphages.* They are not permanent residents. Instead they migrate through healthy connective tissues and collect at sites of tissue injury.

Figure 4–8 **The Cells and Fibers of Connective Tissue Proper.** A diagrammatic view of the cell types and fibers of connective tissue proper. (Microphages, not shown, are common only in damaged or abnormal tissues.)

Reticular fibers
Melanocyte
Fixed macrophage
Plasma cell
Blood in vessel
Adipocytes (fat cells)
Ground substance
Mast cell
Elastic fibers
Free macrophage
Collagen fibers
Fibroblast
Mesenchymal cell
Lymphocyte

Connective Tissue Proper Cell Populations

- **Fibroblasts** (FĪ-brō-blasts) are the only cells that are *always* present in connective tissue proper. They are also the most abundant permanent, or fixed, residents of connective tissue proper. Fibroblasts secrete hyaluronan (a polysaccharide derivative) and proteins. (Recall that hyaluronan is one of the ingredients that helps lock epithelial cells together.) In connective tissue proper, extracellular fluid, hyaluronan, and proteins interact to form the proteoglycans that make ground substance viscous. Each fibroblast also secretes protein subunits that assemble to form large extracellular fibers. ⟲ p. 80

- **Fibrocytes** (FĪ-brō-sīts) are the second most abundant fixed cell in connective tissue proper. They differentiate from fibroblasts. Fibrocytes are spindle-shaped cells that maintain the connective tissue fibers of connective tissue proper.

- **Adipocytes** (AD-i-pō-sīts) are also known as fat cells. A typical adipocyte contains a single, enormous lipid droplet. The nucleus, other organelles, and cytoplasm are squeezed to one side, making a sectional view of the cell resemble a class ring. The number of adipocytes varies from one type of connective tissue to another, from one region of the body to another, and among individuals.

- **Mesenchymal cells** are stem cells that are present in many connective tissues. These cells respond to local injury or

infection by dividing to produce daughter cells that differentiate into fibroblasts, macrophages, or other connective tissue cells.

- **Macrophages** (MAK-rō-fā-jez) are large phagocytic cells scattered throughout the matrix. These scavengers engulf damaged cells or pathogens that enter the tissue. (The name literally means "big eater.") Although not abundant, macrophages are important in mobilizing the body's defenses. When stimulated, they release chemicals that activate the immune system and attract large numbers of additional macrophages and other cells involved in tissue defense.

  Macrophages are either *fixed macrophages,* which spend long periods in a tissue, or *free macrophages,* which migrate rapidly through tissues. In effect, fixed macrophages provide a "frontline" defense that can be reinforced by the arrival of free macrophages and other specialized cells.

- **Mast cells** are small, mobile connective tissue cells that are common near blood vessels. The cytoplasm of a mast cell is filled with granules containing **histamine** (HIS-tuh-mēn) and **heparin** (HEP-uh-rin). Histamine, released after injury or infection, stimulates local inflammation. (You are probably familiar with the inflammatory effects of histamine if you've had a cold. People often take antihistamines to reduce cold symptoms.) *Basophils,* blood cells that enter damaged tissues and enhance the inflammation process, also contain granules of histamine and heparin. Heparin levels in the blood are normally low. It is an anticoagulant that enhances local blood flow during inflammation and reduces the development of blood clots in areas of slow-moving blood.

- **Lymphocytes** (LIM-fō-sīts) migrate throughout the body, traveling through connective tissues and other tissues. Their numbers increase markedly wherever tissue damage occurs. Some lymphocytes may develop into **plasma cells**, which produce *antibodies*—proteins involved in defending the body against disease.

- **Microphages** (*neutrophils* and *eosinophils*) are phagocytic blood cells that normally move through connective tissues in small numbers. They are attracted to the site of an infection or injury by chemicals released by macrophages and mast cells.

4

- **Melanocytes** (me-LAN-ō-sīts) synthesize and store the brown pigment **melanin** (MEL-a-nin), which gives tissues a dark color. Melanocytes are common in the epithelium of the skin, where they play a major role in determining skin color. Melanocytes are also abundant in connective tissues of the eye and the dermis of the skin, although the number present differs by body region and among individuals.

Connective Tissue Fibers. Three types of fibers occur in connective tissue: *collagen*, *reticular*, and *elastic* fibers. Fibroblasts form all three by secreting protein subunits that interact in the matrix. Fibrocytes then maintain these fibers.

1. **Collagen fibers** are the most common fibers in connective tissue proper. They are long, straight, and unbranched. Each collagen fiber consists of a bundle of fibrous protein subunits wound together like the strands of a rope. A collagen fiber is flexible like a rope, but it is stronger than steel when pulled from either end. *Tendons*, which connect skeletal muscles to bones, consist almost entirely of collagen fibers. Typical *ligaments* are similar to tendons, but they connect one bone to another bone. Tendons and ligaments can withstand tremendous forces. Uncontrolled muscle contractions or skeletal movements are more likely to break a bone than to snap a tendon or a ligament.

2. **Reticular fibers** (*reticulum*, network) contain the same protein subunits as do collagen fibers, but they are arranged differently. Thinner than collagen fibers, reticular fibers form a branching, interwoven framework that is tough, yet flexible. Reticular fibers resist forces applied from many directions because they form a network rather than share a common alignment. This interwoven network, called a *stroma*, stabilizes the relative positions of the functional cells, or **parenchyma** (pa-RENG-ki-ma), of organs such as the liver. Reticular fibers also stabilize the positions of an organ's blood vessels, nerves, and other structures, despite changing positions and the pull of gravity.

3. **Elastic fibers** contain the protein *elastin*. Elastic fibers are branched and wavy. After stretching, they return to their original length. **Elastic ligaments**, which are dominated by elastic fibers, are rare but have important functions, such as interconnecting vertebrae.

Ground Substance. Ground substance fills the spaces between cells and surrounds connective tissue fibers (**Figure 4–8**). In connective tissue proper, ground substance is clear, colorless, and viscous (due to the presence of proteoglycans and glycoproteins). ⤴ p. 82 Ground substance is viscous enough that

bacteria have trouble moving through it—imagine swimming in molasses. This viscosity slows the spread of pathogens and makes them easier for phagocytes to catch.

Embryonic Connective Tissues

Mesenchyme, or *embryonic connective tissue*, is the first connective tissue to appear in a developing embryo. Mesenchyme contains an abundance of star-shaped stem cells (mesenchymal cells) separated by a matrix with very fine protein filaments (**Figure 4–9a**). Mesenchyme gives rise to all other

Figure 4–9   Embryonic Connective Tissues.

Blood vessel

Mesenchymal cells

Mesenchyme   LM × 136

a This is the first connective tissue to appear in an embryo.

Blood vessel

Mesenchymal cells

Mucous connective tissue   LM × 136
(*Wharton's jelly*)

b This sample was taken from the umbilical cord of a fetus.

connective tissues. **Mucous connective tissue** (Figure 4–9b), or *Wharton's jelly,* is a loose connective tissue found in many parts of the embryo, including the umbilical cord.

Adults have neither form of embryonic connective tissue. Instead, many adult connective tissues contain scattered mesenchymal stem cells that can assist in tissue repair after an injury.

Loose Connective Tissues

Loose connective tissues are the "packing materials" of the body. They fill spaces between organs, cushion and stabilize specialized cells in many organs, and support epithelia. These tissues surround and support blood vessels and nerves, store lipids, and provide a route for the diffusion of materials. Loose connective tissues include mucous connective tissue in embryos and *areolar tissue, adipose tissue,* and *reticular tissue* in adults.

Areolar Tissue. Areolar tissue (*areola,* little space) is the least specialized connective tissue in adults. It may contain all the cells and fibers of any connective tissue proper in a very loosely organized array (Figure 4–10a). Areolar tissue has an open framework. A viscous ground substance provides most of its volume and absorbs shocks. Areolar tissue can distort without damage because its fibers are loosely organized. Elastic fibers make it resilient, so areolar tissue returns to its original shape after external pressure is relieved.

Areolar tissue forms a layer that separates the skin from deeper structures. In addition to providing padding, the elastic properties of this layer allow a considerable amount of independent movement. For this reason, if you pinch the skin of your arm, you will not affect the underlying muscle. Conversely, contractions of the underlying muscle do not pull against your skin: As the muscle bulges, the areolar tissue stretches. The areolar tissue layer under the skin is a common injection site for drugs because it has an extensive blood supply.

The capillaries (thin-walled blood vessels) in areolar tissue deliver oxygen and nutrients and remove carbon dioxide and waste products. They also carry wandering cells to and from the tissue. Epithelia commonly cover areolar tissue, and fibrocytes maintain the dense layer of the basement membrane that separates the two kinds of tissue. The epithelial cells rely on the oxygen and nutrients that diffuse across the basement membrane from capillaries in the underlying connective tissue.

Adipose Tissue. The distinction between areolar tissue and fat, or **adipose tissue**, is somewhat arbitrary. Adipocytes (fat cells) account for most of the volume of adipose tissue (Figure 4–10b), but only a fraction of the volume of areolar tissue. Adipose tissue provides padding, absorbs shocks, acts

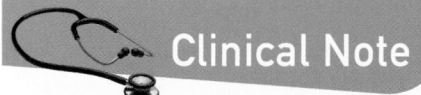

Clinical Note

Marfan's Syndrome *Marfan's syndrome* is an inherited condition caused by the production of an abnormally weak form of *fibrillin,* a glycoprotein that gives strength and elasticity to connective tissues. The effects of this defect are widespread because most organs contain connective tissues. The most visible sign of Marfan's syndrome involves the skeleton. Most individuals with the condition are tall and have abnormally long limbs and fingers. The most serious consequences involve the cardiovascular system. About 90 percent of people with Marfan's syndrome have structural abnormalities in their cardiovascular system. The most dangerous possibility is that the weakened elastic connective tissues in the walls of major arteries, such as the aorta, may burst, causing a sudden, fatal loss of blood.

as an insulator to slow heat loss through the skin, and serves as packing or filler around structures. Adipose tissue is common under the skin of the flanks (between the last rib and the hips), buttocks, and breasts. It fills the bony sockets behind the eyes and surrounds the kidneys. It is also common beneath the mesothelial lining of the pericardial and peritoneal cavities.

Most of the adipose tissue in the body is called **white fat**, because it has a pale, yellow-white color. In infants and young children, however, the adipose tissue between the shoulder blades, around the neck, and possibly elsewhere in the upper body is highly vascularized, and the individual adipocytes contain numerous mitochondria. Together, these characteristics give the tissue a deep, rich color that gives it the name **brown fat**. When these cells are stimulated by the nervous system, lipid breakdown speeds up. The cells do not capture the energy that is released. Instead, the surrounding tissues absorb it as heat. The heat warms the circulating blood, which distributes the heat throughout the body. In this way, an infant can double its metabolic heat generation very quickly. It had been thought that adults had little, if any, active brown fat tissue. Recent evidence has indicated its presence in adults, and that its absence may be connected to obesity.

Adipocytes are not simply storage bins. Instead, they are metabolically active cells. Their lipids are constantly being broken down and replaced. When nutrients are scarce, adipocytes deflate like collapsing balloons as their lipids are broken down and the fatty acids released to support metabolism.

What happens if nutrients become plentiful again? The lost weight can easily be regained in the same areas of the body because the cells are not killed but merely reduced in

Figure 4–10 **Loose Connective Tissues.**

4

Areolar Tissue

LOCATIONS: Within and deep to the dermis of skin, and covered by the epithelial lining of the digestive, respiratory, and urinary tracts; between muscles; around joints, blood vessels, and nerves

FUNCTIONS: Cushions organs; provides support but permits independent movement; phagocytic cells provide defense against pathogens

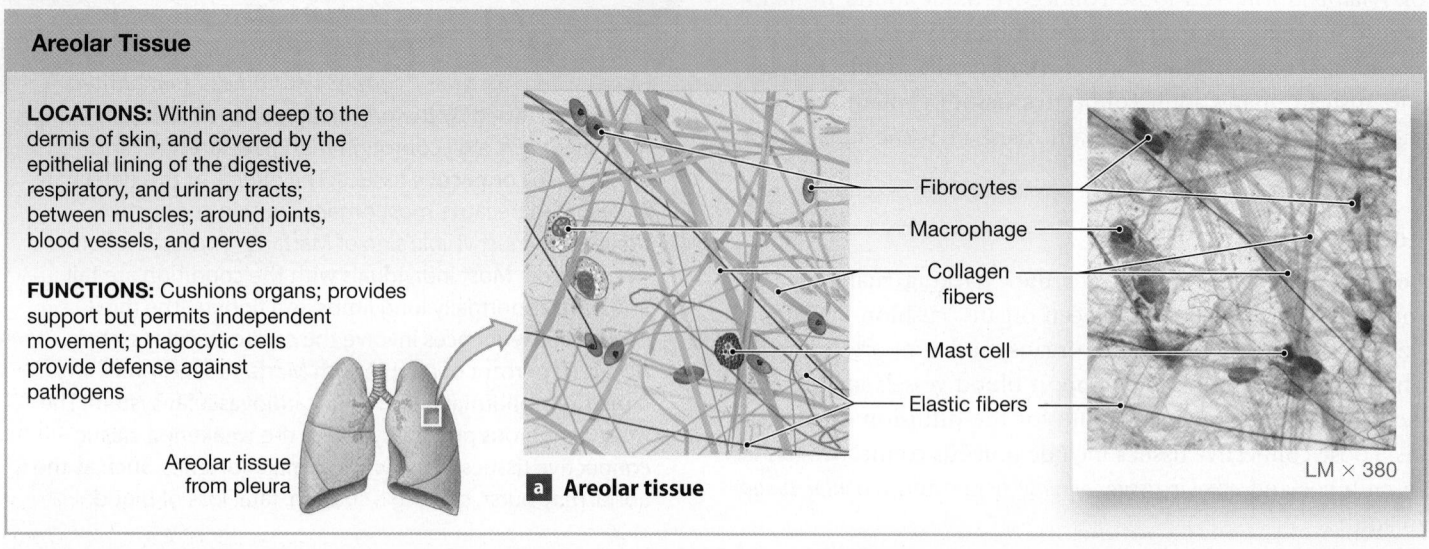

Areolar tissue from pleura

Fibrocytes
Macrophage
Collagen fibers
Mast cell
Elastic fibers

LM × 380

a **Areolar tissue**

Adipose Tissue

LOCATIONS: Deep to the skin, especially at sides, buttocks, and breasts; padding around eyes and kidneys

FUNCTIONS: Provides padding and cushions shocks; insulates (reduces heat loss); stores energy

Adipocytes (white adipose cells)

LM × 300

b **Adipose tissue**

Reticular Tissue

LOCATIONS: Liver, kidney, spleen, lymph nodes, and bone marrow

FUNCTIONS: Provides supporting framework

Reticular tissue from liver

Reticular fibers

LM × 375

c **Reticular tissue**

size. In adults, adipocytes cannot divide. The number of fat cells in peripheral tissues is established in the first few weeks of a newborn's life, perhaps in response to the amount of fats in the diet. However, that is not the end of the story, because loose connective tissues also contain mesenchymal cells. If circulating lipid levels are chronically elevated, the mesenchymal cells will divide, giving rise to cells that differentiate into fat cells. As a result, areas of areolar tissue can become adipose tissue in times of nutritional plenty, even in adults.

Liposuction is a cosmetic surgical procedure that removes excess or unwanted adipose tissue. It turns out that liposuction provides only a temporary and potentially risky solution to the problem of excess weight. Once again, the reason is that adipose tissue can regenerate through the differentiation of mesenchymal cells.

Reticular Tissue. As mentioned earlier, organs such as the spleen and liver contain **reticular tissue**, in which reticular fibers form a complex three-dimensional stroma (**Figure 4–10c**). The stroma supports the functional cells, or parenchyma, of these organs. This fibrous framework is also found in the lymph nodes and bone marrow. Fixed macrophages, fibroblasts, and fibrocytes are associated with the reticular fibers, but these cells are seldom visible, because specialized cells with other functions dominate the organs.

Dense Connective Tissues

Fibers create most of the volume of **dense connective tissues**. These tissues are often called **collagenous** (ko-LAJ-e-nus) **tissues**, because collagen fibers are the dominant type of fiber in them. The body has two types of dense connective tissues: dense regular connective tissue and dense irregular connective tissue.

In **dense regular connective tissue**, the collagen fibers are parallel to each other, packed tightly, and aligned with the forces applied to the tissue. **Tendons** are cords of dense regular connective tissue that attach skeletal muscles to bones (**Figure 4–11a**). The collagen fibers run longitudinally along the tendon and transfer the pull of the contracting muscle to the bone. **Ligaments** resemble tendons, but connect one bone to another or stabilize the positions of internal organs.

An **aponeurosis** (AP-ō-noo-RŌ-sis; plural, *aponeuroses*) is a tendinous sheet that attaches a broad, flat muscle to another muscle or to several bones of the skeleton. It can also stabilize the positions of tendons and ligaments. Aponeuroses are associated with large muscles of the skull, lower back, and abdomen, and with the tendons and ligaments of the palms of the hands and the soles of the feet. Large numbers of fibroblasts are scattered among the collagen fibers of tendons, ligaments, and aponeuroses.

In contrast, the fibers in **dense irregular connective tissue** form an interwoven meshwork in no consistent pattern (**Figure 4–11b**). These tissues strengthen and support areas subjected to stresses from many directions. A layer of dense irregular connective tissue gives skin its strength. Cured leather (animal skin) is an excellent example of the interwoven nature of this tissue. Except at joints, dense irregular connective tissue forms a sheath around cartilages (the *perichondrium*) and bones (the *periosteum*). Dense irregular connective tissue also forms a thick fibrous layer called a **capsule**, which surrounds internal organs such as the liver, kidneys, and spleen and encloses the cavities of joints.

Dense regular and dense irregular connective tissues contain variable amounts of elastic fibers. When elastic fibers outnumber collagen fibers, the tissue has a springy, resilient nature that allows it to tolerate cycles of extension and recoil. Abundant elastic fibers are present in the connective tissue that supports transitional epithelia, in the walls of large blood vessels such as the aorta, and around the respiratory passageways.

Elastic tissue is a dense regular connective tissue made up mainly of elastic fibers. Elastic ligaments are almost completely dominated by elastic fibers. They help stabilize the positions of the vertebrae of the spinal column (**Figure 4–11c**).

Fluid Connective Tissues

Blood and *lymph* are connective tissues with distinctive collections of cells. The fluid matrix that surrounds the cells also includes many types of suspended proteins that do not form insoluble fibers under normal conditions.

In **blood**, the watery matrix is called **plasma**. Plasma contains blood cells and fragments of cells, collectively known as *formed elements*. There are three types of formed elements: red blood cells, white blood cells, and platelets (**Figure 4–12**).

Recall from Chapter 3 that the human body contains a large volume of extracellular fluid. This fluid includes three major subdivisions: *plasma, interstitial fluid,* and *lymph*. Plasma is normally confined to the vessels of the cardiovascular system. Contractions of the heart keep it in motion. **Arteries** carry blood away from the heart and into the tissues of the body. In those tissues, blood pressure forces water and small solutes out of the bloodstream across the walls of **capillaries**, the smallest blood vessels. This process is the origin of the interstitial fluid that bathes the body's cells. The remaining blood flows from the capillaries into **veins** that return it to the heart.

Lymph forms as interstitial fluid enters **lymphatic vessels**. As fluid passes along the lymphatic vessels, cells of the immune system monitor its composition and respond to signs of injury or infection. The lymphatic vessels ultimately return the lymph to large veins near the heart. This recirculation of fluid—from the cardiovascular system, through the interstitial fluid, to the lymph, and then back to the cardiovascular system—is a continuous process that is essential to homeostasis. It helps eliminate local differences in the levels of nutrients, wastes, or toxins; maintains blood volume; and alerts the immune system to infections that may be under way in peripheral tissues.

Figure 4–11   **Dense Connective Tissues.**

Dense Regular Connective Tissue

LOCATIONS: Between skeletal muscles and skeleton (tendons and aponeuroses); between bones or stabilizing positions of internal organs (ligaments); covering skeletal muscles; deep fasciae

FUNCTIONS: Provides firm attachment; conducts pull of muscles; reduces friction between muscles; stabilizes positions of bones

a **Tendon**

Collagen fibers

Fibroblast nuclei

LM × 440

Dense Irregular Connective Tissue

LOCATIONS: Capsules of visceral organs; periostea and perichondria; nerve and muscle sheaths; dermis

FUNCTIONS: Provides strength to resist forces from many directions; helps prevent overexpansion of organs, such as the urinary bladder

b **Deep dermis**

Collagen fiber bundles

LM × 111

Elastic Tissue

LOCATIONS: Between vertebrae of the spinal column (ligamentum flavum and ligamentum nuchae); ligaments supporting penis; ligaments supporting transitional epithelia; in blood vessel walls

FUNCTIONS: Stabilizes positions of vertebrae and penis; cushions shocks; permits expansion and contraction of organs

c **Elastic ligament**

Elastic fibers

Fibroblast nuclei

LM × 887

Figure 4–12 **Formed Elements in the Blood.**

Red blood cells

Red blood cells, or erythrocytes (e-RITH-rō-sīts), are responsible for the transport of oxygen (and, to a lesser degree, of carbon dioxide) in the blood.

Red blood cells account for about half the volume of whole blood and give blood its color.

White blood cells

White blood cells, or leukocytes (LOO-kō-sīts; *leuko-*, white), help defend the body from infection and disease.

Neutrophil

Eosinophil

Basophil

Monocytes are phagocytes similar to the free macrophages in other tissues.

Lymphocytes are uncommon in the blood but they are the dominant cell type in lymph, the second type of fluid connective tissue.

Eosinophils and **neutrophils** are phagocytes. **Basophils** promote inflammation much like mast cells in other connective tissues.

Platelets

Platelets are membrane-enclosed packets of cytoplasm that function in blood clotting.

These cell fragments are involved in the clotting response that seals leaks in damaged or broken blood vessels.

CHECKPOINT

15. Identify several functions of connective tissues.

16. List the three categories of connective tissues.

17. Identify the cells found in connective tissue proper.

18. Lack of vitamin C in the diet interferes with the ability of fibroblasts to produce collagen. What effect might this interference have on connective tissue?

19. Many allergy sufferers take antihistamines to relieve their allergy symptoms. Which cells produce the molecule that this medication blocks?

20. Which type of connective tissue contains primarily triglycerides?

21. Which two types of connective tissue have a fluid matrix?

See the blue Answers tab at the back of the book.

4-5 | Cartilage and bone provide a strong supporting framework

Learning Outcome Describe how cartilage and bone function as a supporting connective tissue.

Cartilage and *bone* are called supporting connective tissues because they provide a strong framework that supports the rest of the body. In these connective tissues, the matrix contains numerous fibers. In bone, it also contains deposits of insoluble calcium salts.

Cartilage

The matrix of **cartilage** is a firm gel that contains polysaccharide derivatives called **chondroitin sulfates** (kon-DROY-tin;

chondros, cartilage). Chondroitin sulfates form complexes with proteins in the ground substance, producing proteoglycans. Cartilage cells, or **chondrocytes** (KON-drō-sīts), are the only cells in the cartilage matrix. They occupy small chambers known as **lacunae** (la-KOO-nē; *lacus,* lake). The physical properties of cartilage depend on the proteoglycans of the matrix, and on the type and abundance of extracellular fibers.

Unlike other connective tissues, cartilage is avascular, so all exchange of nutrients and waste products must take place by diffusion through the matrix. Blood vessels do not grow into cartilage because chondrocytes produce a chemical that discourages their formation. This chemical inhibitor of new blood vessel formation is called **antiangiogenesis factor** (*anti-,* against + *angeion,* vessel + *genesis,* production). Other angiogenesis inhibitors have also been identified in the body or developed as drugs. Various inhibitors are now used in cancer treatments to discourage the formation of new blood vessels to tumors, thereby slowing their growth. Cartilage is generally set apart from surrounding tissues by a fibrous **perichondrium** (per-i-KON-drē-um; *peri-,* around). The perichondrium contains two distinct layers: an outer, fibrous region of dense irregular connective tissue, and an inner, cellular layer. The fibrous layer gives mechanical support and protection and attaches the cartilage to other structures. The cellular layer is important to the growth and maintenance of the cartilage. Blood vessels in the perichondrium provide oxygen and nutrients to the underlying chondrocytes.

Cartilage Growth

Cartilage grows by two mechanisms: *interstitial growth* and *appositional growth* (**Figure 4–13**).

Figure 4–13 **The Growth of Cartilage.**

Matrix
Chondrocyte
Lacuna

Chondrocyte undergoes division within a lacuna surrounded by cartilage matrix.

a **Interstitial growth**

As daughter cells secrete additional matrix, they move apart, expanding the cartilage from within.

New matrix

Dividing stem cell
Fibroblast
Perichondrium
Chondroblasts

Cells in the cellular layer of the perichondrium differentiate into chondroblasts.

b **Appositional growth**

These immature chondroblasts secrete new matrix.

New matrix
Immature chondrocyte
Older matrix
Mature chondrocyte

As the matrix enlarges, more chondroblasts are incorporated; they are replaced by divisions of stem cells in the perichondrium.

Interstitial growth enlarges the cartilage from within. Chondrocytes in the cartilage matrix divide, and the daughter cells produce additional matrix (**Figure 4–13a**). Interstitial growth is most important during development. It begins early in embryonic development and continues through adolescence.

Appositional growth adds new layers of cartilage to the surface (**Figure 4–13b**). In this process, cells of the inner layer of the perichondrium divide repeatedly and become chondroblasts (immature chondrocytes). These cells begin producing cartilage matrix. As they become surrounded by and embedded in new matrix, they differentiate into mature chondrocytes. Appositional growth gradually increases the size of the cartilage by adding to its outer surface.

Both interstitial and appositional growth take place during development, but interstitial growth contributes more to the mass of the adult cartilage. Neither interstitial nor appositional growth takes place in the cartilages of normal adults. However, appositional growth may occur in unusual circumstances, such as after cartilage has been damaged or excessively stimulated by *growth hormone* from the pituitary gland. Minor damage to cartilage can be repaired by appositional growth at the damaged surface. After more severe damage, a dense fibrous patch will replace the injured portion of the cartilage.

Types of Cartilage

The body contains three major types of cartilage: hyaline cartilage, elastic cartilage, and fibrocartilage.

1. **Hyaline cartilage** (HĪ-uh-lin; *hyalos*, glass) is the most common type of cartilage. Except inside joint cavities, a dense perichondrium surrounds hyaline cartilages. Hyaline cartilage is tough but somewhat flexible because its matrix contains closely packed collagen fibers. Because these fibers are not in large bundles and do not stain darkly, they are not always apparent in light microscopy (**Figure 4–14a**). Examples in adults include the connections between the ribs and the sternum; the nasal cartilages and the supporting cartilages along the conducting passageways of the respiratory tract; and the *articular cartilages*, which cover opposing bone surfaces within many joints, such as the elbow and knee.

2. **Elastic cartilage** (**Figure 4–14b**) is extremely resilient and flexible because it contains numerous elastic fibers. These cartilages usually have a yellowish color on gross dissection. Elastic cartilage forms the external flap (the *auricle*, or *pinna*) of the outer ear; the epiglottis, which prevents food and liquids from entering the windpipe when swallowing; a passageway to the middle ear cavity (the *auditory tube*); and small cartilages (the *cuneiform cartilages*) in the larynx, or voice box.

Figure 4–14 **Types of Cartilage.**

Hyaline Cartilage

LOCATIONS: Between tips of ribs and bones of sternum; covering bone surfaces at synovial joints; supporting larynx (voice box), trachea, and bronchi; forming part of nasal septum

FUNCTIONS: Provides stiff but somewhat flexible support; reduces friction between bony surfaces

Chondrocytes in lacunae

Matrix

LM × 500

a Hyaline cartilage

Elastic Cartilage

LOCATIONS: Auricle of external ear; epiglottis; auditory canal; cuneiform cartilages of larynx

FUNCTIONS: Provides support, but tolerates distortion without damage and returns to original shape

Chondrocytes in lacunae

Elastic fibers in matrix

LM × 358

b Elastic cartilage

Fibrocartilage

LOCATIONS: Pads within knee joint; between pubic bones of pelvis; intervertebral discs

FUNCTIONS: Resists compression; prevents bone-to-bone contact; limits movement

Chondrocytes in lacunae

Fibrous matrix

LM × 400

c Fibrocartilage

4

3. **Fibrocartilage** is extremely durable and tough because it has little ground substance and its matrix is dominated by densely interwoven collagen fibers (**Figure 4–14c**). Pads of fibrocartilage called intervertebral discs lie between the spinal vertebrae, between the pubic bones of the pelvis, and around tendons and within or around joints. In these positions, fibrocartilage resists compression, absorbs shocks, and prevents damaging bone-to-bone contact. Cartilage heals poorly. Damaged fibrocartilage in joints such as the knee can interfere with normal movements.

Several complex joints, including the knee, contain both hyaline cartilage and fibrocartilage. The hyaline cartilage covers bony surfaces, and fibrocartilage pads in the joint prevent contact between bones during movement. Injuries to these pads do not heal. After repeated or severe damage, joint mobility is severely reduced. Surgery generally produces only a temporary or incomplete repair.

Bone

Here we focus only on significant differences between cartilage and bone. We will examine the detailed histology of **bone**, or **osseous** (OS-ē-us; *os*, bone) **tissue**, in Chapter 6. The volume of ground substance in bone is very small. About two-thirds of the matrix of bone consists of a mixture of calcium salts—primarily calcium phosphate, with lesser amounts of calcium carbonate. The rest of the matrix is dominated by collagen fibers. This combination gives bone truly remarkable properties. By themselves, calcium salts are hard but rather brittle,

whereas collagen fibers are strong but relatively flexible. In bone, the presence of the minerals surrounding the collagen fibers produces a strong, somewhat flexible combination that is highly resistant to shattering. In its overall properties, bone can compete with the best steel-reinforced concrete. In essence, the collagen fibers in bone act like the steel reinforcing rods, and the mineralized matrix acts like the concrete.

Figure 4–15 shows the general organization of osseous tissue. Lacunae in the matrix contain **osteocytes** (OS-tē-ō-sīts), or bone cells. The lacunae are typically organized around blood vessels that branch through the bony matrix. Diffusion cannot take place through the hard matrix, but osteocytes communicate with the blood vessels and with one another by means of slender cytoplasmic extensions. These extensions run through long, slender passageways in the matrix called **canaliculi** (kan-a-LIK-ū-lē; little canals). These passageways form a branching network for the exchange of materials between blood vessels, interstitial fluid, and osteocytes.

Except in joint cavities, where a layer of hyaline cartilage covers bone, the surfaces are sheathed by a **periosteum** (per-ē-OS-tē-um), a covering composed of fibrous (outer) and cellular (inner) layers. The periosteum helps to attach a bone to surrounding tissues and to associated tendons and ligaments. The cellular layer takes part in appositional bone growth and helps in repairs after an injury.

Unlike cartilage, bone undergoes extensive remodeling throughout life. Complete repairs can be made even after severe damage. Bones also respond to the stresses placed on them. They grow thicker and stronger with exercise and become thin and brittle with inactivity.

Figure 4–15 **Bone.** The osteocytes in bone are generally organized in groups around a central space that contains blood vessels. Bone dust produced during preparation of the section fills the lacunae and the central canal, making them appear dark in the micrograph.

Table 4–2	A Comparison of Cartilage and Bone	
Characteristic	**Cartilage**	**Bone**
STRUCTURAL FEATURES		
Cells	Chondrocytes in lacunae	Osteocytes in lacunae
Ground substance	Chondroitin sulfate (in proteoglycans) and water	A small volume of liquid surrounding insoluble crystals of calcium salts (calcium phosphate and calcium carbonate)
Fibers	Collagen, elastic, and reticular fibers (proportions vary)	Collagen fibers predominate
Vascularity	None	Extensive
Covering	Perichondrium (two layers)	Periosteum (two layers)
Strength	Limited: bends easily, but hard to break	Strong: resists distortion until breaking point
METABOLIC FEATURES		
Oxygen demands	Low	High
Nutrient delivery	By diffusion through matrix	By diffusion through cytoplasm and interstitial fluid in canaliculi
Growth	Interstitial and appositional	Appositional only
Repair capabilities	Limited	Extensive

Table 4–2 summarizes the similarities and differences between cartilage and bone.

CHECKPOINT

22. Identify the two types of supporting connective tissue.

23. Why does bone heal faster than cartilage?

24. If a person has a herniated intervertebral disc, characterized by displacement of the pad of cartilage between the vertebrae, which type of cartilage has been damaged?

See the blue Answers tab at the back of the book.

4-6 | Tissue membranes are physical barriers of four types: mucous, serous, cutaneous, and synovial

Learning Outcome Explain how epithelial and connective tissues combine to form four types of tissue membranes, and specify the functions of each.

A tissue membrane forms a physical barrier. There are many different types of anatomical membranes. You encountered plasma membranes that enclose cells in Chapter 3, and you will find many other kinds of membranes in later chapters. Here we are concerned with the membranes that line or cover body surfaces. Each consists of an epithelium supported by connective tissue. Four such membrane types are found in the body: (1) *mucous membranes*, (2) *serous membranes*, (3) the *cutaneous membrane*, and (4) *synovial membranes* (**Figure 4–16**).

Mucous Membranes

Mucous membranes, or **mucosae** (mū-KŌ-sē), line, passageways and chambers that communicate with the exterior, including those in the digestive, respiratory, reproductive, and urinary tracts (**Figure 4–16a**). The epithelial surfaces of these passageways must be kept moist to reduce friction and, in many cases, to facilitate absorption or secretion. The epithelial surfaces are lubricated either by mucus produced by mucous cells or multicellular glands, or by fluids, such as urine or semen. The areolar tissue component of a mucous membrane is called the **lamina propria** (PRŌ-prē-uh). We will look at the organization of specific mucous membranes in greater detail in later chapters.

Many mucous membranes contain simple epithelia that perform absorptive or secretory functions, such as the simple columnar epithelium of the digestive tract. However, other types of epithelia may be involved. For example, a stratified squamous epithelium is part of the mucous membrane of the mouth, and the mucous membrane along most of the urinary tract contains a transitional epithelium.

Serous Membranes

Serous membranes line the sealed, internal cavities of the trunk—cavities that are not open to the exterior. These membranes consist of a mesothelium supported by areolar tissue (**Figure 4–16b**). As you may recall from Chapter 1, the three types of serous membranes are (1) the *pleura*, which lines the pleural cavities and covers the lungs; (2) the *pericardium*,

4

Figure 4–16   **Types of Membranes.**

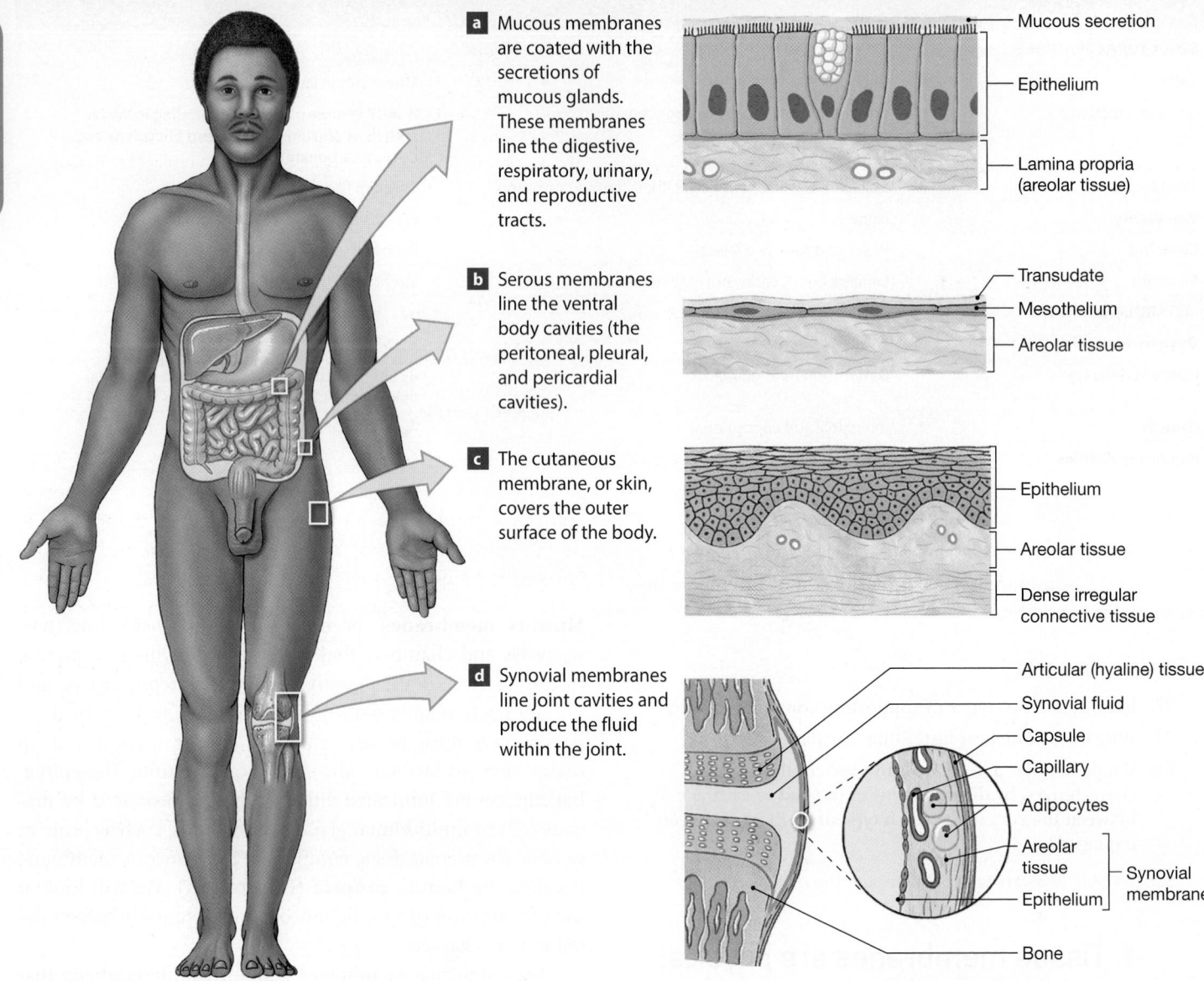

a Mucous membranes are coated with the secretions of mucous glands. These membranes line the digestive, respiratory, urinary, and reproductive tracts.

- Mucous secretion
- Epithelium
- Lamina propria (areolar tissue)

b Serous membranes line the ventral body cavities (the peritoneal, pleural, and pericardial cavities).

- Transudate
- Mesothelium
- Areolar tissue

c The cutaneous membrane, or skin, covers the outer surface of the body.

- Epithelium
- Areolar tissue
- Dense irregular connective tissue

d Synovial membranes line joint cavities and produce the fluid within the joint.

- Articular (hyaline) tissue
- Synovial fluid
- Capsule
- Capillary
- Adipocytes
- Areolar tissue
- Epithelium
- Synovial membrane
- Bone

which lines the pericardial cavity and covers the heart; and (3) the *peritoneum*, which lines the peritoneal cavity and covers the surfaces of the enclosed organs. ⤺ p. 48 Serous membranes are very thin, but they are firmly attached to the body wall and to the organs they cover. When looking at an organ such as the heart or stomach, you are really seeing the tissues of the organ through a transparent serous membrane.

The primary function of any serous membrane is to minimize friction between the surfaces that it covers. Each serous membrane can be divided into a *parietal portion*, which lines the inner surface of the cavity, and an opposing *visceral portion*,

or **serosa**, which covers the outer surfaces of visceral organs. These organs often move or change shape as they carry out their various functions, and the parietal and visceral surfaces of a serous membrane are in close contact at all times. Friction between these surfaces is kept to a minimum because mesothelia are very thin and permeable, and tissue fluids continuously diffuse onto the exposed surfaces, keeping them moist and slippery.

The fluid secreted by the mesothelium of a serous membrane is called *serous fluid*. Serous fluid prevents friction between the cavity walls and the surfaces of internal organs. However, after an injury or in certain disease states where there is an imbalance of hydrostatic and osmotic forces, the volume of serous fluid, now called *transudate* may increase dramatically. This change may complicate existing medical problems or produce new ones.

The Cutaneous Membrane

The **cutaneous membrane** is the skin that covers the surface of your body. It consists of a stratified squamous epithelium and a layer of areolar tissue reinforced by underlying dense irregular connective tissue (**Figure 4–16c**). In contrast to serous and mucous membranes, the cutaneous membrane is thick, relatively waterproof, and usually dry. We will take a closer look at the cutaneous membrane in Chapter 5.

Synovial Membranes

Adjacent bones often interact at joints, or *articulations*. There the two articulating bones are very close together or in contact. Joints that permit significant amounts of movement are complex structures. Such a joint is surrounded by a fibrous capsule, and the ends of the articulating bones lie within a *joint cavity* filled with **synovial** (si-NŌ-vē-ul) **fluid** (**Figure 4–16d**). The synovial fluid is produced by a **synovial membrane**, which lines the joint cavity. A synovial membrane consists of an extensive area of areolar tissue containing a matrix of interwoven collagen fibers, proteoglycans, and glycoproteins. An incomplete layer, or sheet, of macrophages and specialized fibroblasts overlies and separates the areolar tissue from the joint cavity. These cells regulate the composition of the synovial fluid.

This incomplete layer of cells is often called an epithelium, but it differs from true epithelia in four respects: (1) It develops within a connective tissue; (2) no basement membrane is present; (3) gaps of up to 1 mm may separate adjacent cells; and (4) fluid and solutes are continuously exchanged between the synovial fluid and capillaries in the underlying connective tissue.

Even though a smooth layer of articular cartilage covers the ends of the bones, the surfaces must be lubricated to keep friction from damaging the opposing surfaces. Synovial fluid provides the much-needed lubrication. This fluid is similar in composition to the ground substance in loose connective tissues. Synovial fluid circulates from the areolar tissue into the joint cavity and percolates through the articular cartilages, supplying oxygen and nutrients to the chondrocytes. Joint movement is important in stimulating the formation and circulation of synovial fluid: If a synovial joint is immobilized for long periods, the articular cartilages and the synovial membrane undergo degenerative changes.

CHECKPOINT

25. Identify the four types of tissue membranes found in the body.

26. Which cavities in the body are lined by serous membranes?

27. The lining of the nasal cavity is normally moist, contains numerous mucous cells, and rests on a layer of connective tissue called the lamina propria. Which type of membrane is this?

See the blue Answers tab at the back of the book.

4-7 | Connective tissue creates the internal framework of the body

Learning Outcome Describe how connective tissue establishes the framework of the body.

Connective tissue provides the internal structure of the body. Layers of connective tissue surround and support the organs within the trunk cavities and connect them to the rest of the body. These layers (1) provide strength and stability, (2) maintain the relative positions of internal organs, and (3) supply a route for the distribution of blood vessels, lymphatic vessels, and nerves.

Fasciae (FASH-ē-ē; singular, *fascia*) are connective tissue layers and wrappings that support and surround organs. We can divide the fasciae into three types of layers: the superficial fascia, the deep fascia, and the subserous fascia (**Figure 4–17**).

1. The **superficial fascia** (FASH-ē-uh) is also called the **hypodermis** (*hypo*, below + *dermis*, skin) or subcutaneous tissue. This layer of areolar tissue and fat separates the skin from underlying tissues and organs. It also provides insulation and padding, and lets the skin and underlying structures move independently.

2. The **deep fascia** consists of dense irregular connective tissue. The organization of the fibers is like that of plywood: In each layer all the fibers run in the same direction, but the orientation of the fibers changes from layer to layer. This arrangement helps the tissue resist forces coming from many directions. The tough capsules that surround most organs, including the kidneys and the organs enclosed by the thoracic and peritoneal cavities, are bound to the deep fascia. The perichondrium around cartilages, the periosteum around bones and the ligaments that interconnect them, and the connective tissues of muscle (including tendons) are also connected to the deep fascia.

The dense connective tissue components are interwoven. For example, the deep fascia around a muscle blends into the tendon, whose fibers intermingle with those of the

Figure 4–17   **The Fasciae.** The relationships among the connective tissue layers in the body.

periosteum. This arrangement creates a strong, fibrous network and binds structural elements together.

3. The **subserous fascia** is a layer of areolar tissue that lies between the deep fascia and the serous membranes that line body cavities. Because this layer separates the serous membranes from the deep fascia, movements of muscles or muscular organs do not severely distort the delicate body cavity linings.

CHECKPOINT

28. **Name the three layers of fascia and their types of connective tissue.**

See the blue Answers tab at the back of the book.

4-8 | The three types of muscle tissue are skeletal, cardiac, and smooth

Learning Outcome   Describe the three types of muscle tissue and the special structural features of each type.

As we have seen, epithelia cover surfaces and line passageways, and connective tissues support and interconnect parts of the body. Together, these tissues provide a strong, interwoven framework within which the organs of the body can

function. Several vital functions involve movement of one kind or another—materials move along the digestive tract, blood flows within the vessels of the cardiovascular system, and the body itself moves from place to place. **Muscle tissue,** which is specialized for contraction, produces movement. Muscle cells have organelles and properties distinct from those of other cells.

There are three types of muscle tissue: (1) *skeletal muscle,* which forms the large muscles that produce gross body movements; (2) *cardiac muscle,* found in the heart, is responsible for circulating the blood; and (3) *smooth muscle,* found in the walls of visceral organs and a variety of other locations, where it provides elasticity, contractility, and support. The contraction mechanism is similar in all three types of muscle tissue, but the muscle cells differ in internal organization. We will examine only general characteristics at this point and then discuss each type of muscle more fully in Chapter 10.

Skeletal Muscle Tissue

Skeletal muscle tissue contains very large muscle cells—up to 0.3 m (1 ft) or more in length. The individual muscle cells are usually called **muscle fibers** because they are relatively long and slender. Each muscle fiber has several hundred nuclei just inside the plasma membrane (**Figure 4–18a**). For this reason, the fibers are described as *multinucleate.*

Figure 4–18   **Types of Muscle Tissue.**

Skeletal Muscle Tissue

Cells are long, cylindrical, striated, and multinucleate.

LOCATIONS: Combined with connective tissues and neural tissue in skeletal muscles

FUNCTIONS: Moves or stabilizes the position of the skeleton; guards entrances and exits to the digestive, respiratory, and urinary tracts; generates heat; protects internal organs

Striations

Nuclei

Muscle fiber

LM × 180

a **Skeletal muscle**

Cardiac Muscle Tissue

Cells are short, branched, and striated, usually with a single nucleus; cells are interconnected by intercalated discs.

LOCATION: Heart

FUNCTIONS: Circulates blood; maintains blood pressure

Nuclei

Cardiac muscle cells

Intercalated discs

Striations

LM × 450

b **Cardiac muscle**

Smooth Muscle Tissue

Cells are short, spindle-shaped, and nonstriated, with a single, central nucleus.

LOCATIONS: Found in the walls of blood vessels and in digestive, respiratory, urinary, and reproductive organs

FUNCTIONS: Moves food, urine, and reproductive tract secretions; controls diameter of respiratory passageways; regulates diameter of blood vessels

Nuclei

Smooth muscle cells

LM × 235

c **Smooth muscle**

4

Skeletal muscle fibers are incapable of dividing. New muscle fibers are produced instead through the divisions of **myosatellite cells** *(satellite cells)*, stem cells that persist in adult skeletal muscle tissue. As a result, skeletal muscle tissue can at least partially repair itself after an injury.

As noted in Chapter 3, the cytoskeleton contains actin and myosin filaments. ⟲ pp. 97, 99 In skeletal muscle fibers, however, these filaments are organized into repeating patterns that give the cells a *striated*, or banded, appearance. The *striations*, or bands, are easy to see in light micrographs. Skeletal muscle fibers do not usually contract unless stimulated by nerves, and the nervous system provides voluntary control over their activities. Thus, skeletal muscle is called **striated voluntary muscle**.

TIPS & TRICKS

Associate the sound of the word **stri**ated with the sound of the word **stri**ped.

A *skeletal muscle* is an organ of the muscular system. Muscle tissue predominates, but the muscle contains all four types of body tissue. Within a skeletal muscle, adjacent skeletal muscle fibers are tied together by collagen and elastic fibers that blend into the attached tendon or aponeurosis. The tendon or aponeurosis conducts the force of contraction, often to a bone of the skeleton. When the muscles contract, they pull on the attached bone, producing movement.

Cardiac Muscle Tissue

Cardiac muscle tissue is found in the heart. A typical cardiac muscle cell, also known as a **cardiocyte**, is smaller than a skeletal muscle cell (**Figure 4–18b**). A typical cardiac muscle cell has one centrally positioned nucleus, but some cardiocytes have as many as five. Prominent striations resemble those of skeletal muscle because the actin and myosin filaments are arranged the same way in the cells of both tissue types.

Cardiac muscle tissue consists of a branching network of interconnected cardiac muscle cells. The cells are connected at specialized regions known as **intercalated discs**. There the membranes are locked together by desmosomes, proteoglycans, and gap junctions. Ion movement through gap junctions helps synchronize the contractions of the cardiac muscle cells, and the desmosomes and proteoglycans lock the cells together during a contraction.

Cardiac muscle tissue has a very limited ability to repair itself. Some cardiac muscle cells do divide after an injury to the heart, but the repairs are incomplete. Some heart function is usually lost.

Cardiac muscle cells do not rely on nerve activity to start a contraction. Instead, specialized cardiac muscle cells called *pacemaker cells* set a regular rate of contraction. The nervous system can alter the rate of pacemaker cell activity, but it does not provide voluntary control over individual cardiac muscle cells. For this reason, cardiac muscle is called **striated involuntary muscle**.

Smooth Muscle Tissue

Smooth muscle tissue is found in the walls of blood vessels, around hollow organs such as the urinary bladder, and in layers around the respiratory, cardiovascular, digestive, and reproductive tracts. A smooth muscle cell is small and spindle-shaped, with tapering ends and a single, oval nucleus (**Figure 4–18c**). These cells can divide, so smooth muscle tissue can regenerate after an injury.

The actin and myosin filaments in smooth muscle cells are organized differently from those of skeletal and cardiac muscles. One result is that smooth muscle tissue has no striations.

Smooth muscle cells may contract on their own, with gap junctions between adjacent cells coordinating the contractions of individual cells. The contraction of some smooth muscle tissue can be controlled by the nervous system, but contractile activity is not under voluntary control. (Imagine the effort you would need to exert conscious control over the smooth muscles along the 8 m [26 ft.] of digestive tract, not to mention the miles of blood vessels!) Smooth muscle is known as **nonstriated involuntary muscle** because of its appearance and the fact that the nervous system usually does not voluntarily control smooth muscle contractions.

CHECKPOINT

29. Identify the three types of muscle tissue in the body.

30. Which type of muscle tissue has small, tapering cells with single nuclei and no obvious striations?

31. If skeletal muscle cells in adults are incapable of dividing, how is skeletal muscle repaired?

See the blue Answers tab at the back of the book.

4-9 | Neural tissue responds to stimuli and propagates electrical impulses throughout the body

Learning Outcome Discuss the basic structure and role of neural tissue.

Neural tissue, also known as *nervous tissue* or *nerve tissue*, is specialized for the propagation (movement) of electrical impulses from one region of the body to another. Ninety-eight percent of the neural tissue in the body is in the brain and spinal cord, which are the control centers of the nervous system.

Figure 4–19   **Neural Tissue.**

| **NEURONS** | **NEUROGLIA** (supporting cells) |

Nuclei of neuroglia

Cell body

Axon

Dendrites

Nucleolus

Nucleus of neuron

LM × 600

- Maintain physical structure of tissues
- Repair tissue framework after injury
- Perform phagocytosis
- Provide nutrients to neurons
- Regulate the composition of the interstitial fluid surrounding neurons

Dendrites (contacted by other neurons)

Cell body (contains nucleus and major organelles)

Microfibrils and microtubules

Axon (conducts information to other cells)

Contact with other cells

Nucleus

Nucleolus

Mitochondrion

**A representative neuron
(sizes and shapes vary widely)**

Neural tissue contains two basic types of cells: (1) **neurons** (NOOR-onz; *neuro*, nerve) and (2) several kinds of supporting cells, collectively called **neuroglia** (nū-ROG-lē-uh), or *glial cells* (*glia*, glue). Our conscious and unconscious thought processes reflect the communication among neurons in the brain. Such communication involves the propagation of electrical impulses, in the form of reversible changes in the membrane potential. ⟲ p. 126 The frequency and the pattern of the impulses convey information. Neuroglia support and repair neural tissue and supply nutrients to neurons.

Neurons are the longest cells in your body. Many are as long as a meter (39 in.)! Most neurons cannot divide under normal circumstances, so they have a very limited ability to repair themselves after injury. A typical neuron has a large **cell body** with a large nucleus and a prominent nucleolus (**Figure 4–19**). Extending from the cell body are many branching processes (projections or outgrowths) termed **dendrites**

(DEN-drīts; *dendron*, a tree), and one **axon**. The dendrites receive information, typically from other neurons. The axon conducts that information to other cells. Because axons tend to be very long and slender, they are also called **nerve fibers**. In Chapter 12 we will examine the properties of neural tissue more closely.

TIPS & TRICKS

To remember the direction of information flow in a neuron, associate the "t" in dendri**t**e with "**t**o" and the "a" in **a**xon with "**a**way."

CHECKPOINT

32. **A tissue contains irregularly shaped cells with many fibrous projections, some several centimeters long. These are probably which type of cell?**

See the blue Answers tab at the back of the book.

4-10 ❙ The response to tissue injury involves inflammation and regeneration

Learning Outcome Describe how injuries affect the tissues of the body.

Now let's consider what happens after an injury, focusing on the interaction among different tissues. Our example includes connective tissue (blood), an epithelium (the endothelia of blood vessels), a muscle tissue (smooth muscle in the vessel walls), and neural tissue (sensory nerve endings). **Inflammation**, or the **inflammatory response**, isolates the injured area while damaged cells, tissue components, and any dangerous microorganisms, which could cause infection, are cleaned up. **Regeneration** is the repair process that restores normal function after inflammation has subsided. In later chapters, especially Chapters 5 and 22, we will consider inflammation and regeneration in more detail.

Inflammation

Impact, abrasion, distortion, chemical irritation, infection by pathogenic organisms (such as bacteria or viruses), and extreme temperatures (hot or cold) are all stimuli that can produce inflammation. Each of these stimuli kills cells, damages fibers, or injures the tissue in some way. Such changes alter the chemical composition of the interstitial fluid: Damaged cells release prostaglandins, proteins, and potassium ions, and the injury itself may have introduced foreign proteins or pathogens into the body. Damaged connective tissue activates mast cells, which release chemicals that stimulate the inflammatory response.

Tissue conditions soon become even more abnormal. **Necrosis** (ne-KRŌ-sis) is the tissue destruction that takes place after cells have been damaged or killed. It begins several hours after the original injury. Lysosomal enzymes cause the damage. Through widespread autolysis, lysosomes release enzymes that first destroy the injured cells and then attack surrounding tissues. ⊃ p. 102 The result may be **pus**, a collection of debris, fluid, dead and dying cells, and necrotic tissue components. An **abscess** is an accumulation of pus in an enclosed tissue space.

Regeneration

Each organ has a different ability to regenerate after injury. This ability is directly linked to the pattern of tissue organization in the injured organ. Epithelia, connective tissues (except cartilage), and smooth muscle tissue usually regenerate well, but skeletal and cardiac muscle tissues and neural tissue regenerate relatively poorly, if at all. The skin, which is dominated by epithelia and connective tissues, regenerates rapidly and completely after injury. (We will consider the process in Chapter 5.) In contrast, damage to the heart is much more serious. The connective tissues of the heart can be repaired, but the majority of damaged cardiac muscle cells are replaced only by fibrous tissue. The permanent replacement of normal tissue by fibrous tissue is called *fibrosis* (fī-BRŌ-sis). Fibrosis is in muscle and other tissues may occur in response to injury, disease, or aging. **Spotlight Figure 4–20** shows the tissue response to injury and the process of tissue regeneration.

CHECKPOINT

33. Identify the two phases in the response to tissue injury.

34. List the four common signs of inflammation.

See the blue Answers tab at the back of the book.

4-11 ❙ With advancing age, tissue repair declines and cancer rates increase

Learning Outcome Describe how aging affects the tissues of the body.

Aging has two important effects on tissues. The body's ability to repair damage to tissues decreases, and cancer is more likely to occur.

Aging and Tissue Structure

Tissues change with age, and the speed and effectiveness of tissue repairs decrease. Repair and maintenance activities throughout the body slow down, and the rate of energy consumption in general declines. All these changes reflect various hormonal alterations that take place with age, often coupled with reduced physical activity and a more sedentary lifestyle. These factors combine to alter the structure and chemical composition of many tissues.

Epithelia get thinner and connective tissues more fragile. Individuals bruise easily and bones become brittle. Joint pain and broken bones are common in the elderly. Because cardiac muscle cells and neurons are not normally replaced, cumulative damage can eventually cause major health problems, such as cardiovascular disease or deterioration in mental functioning.

In later chapters, we will consider the effects of aging on specific organs and systems. Some of these effects are genetically programmed. For example, the chondrocytes of older individuals produce a slightly different form of proteoglycan than do the chondrocytes of younger people. This difference probably accounts for the thinner and less resilient cartilage of older people.

In some cases, the tissue degeneration can be temporarily slowed or even reversed. Age-related reduction in bone strength, a condition called *osteoporosis*, typically results from a combination of inactivity, low dietary calcium levels, and a reduction in circulating sex hormones. A program of exercise that includes weight-bearing activity, calcium supplements, and in some cases medication can generally maintain healthy bone structure for many years.

Tissues are not isolated, and they combine to form organs with diverse functions. Therefore, any injury affects several types of tissue simultaneously. To preserve homeostasis, the tissues must respond in a coordinated way. The restoration of homeostasis involves two related processes: inflammation and regeneration.

Mast Cell Activation

When an injury damages connective tissue, mast cells release a variety of chemicals. This process, called **mast cell activation**, stimulates inflammation.

Mast cell

Histamine
Heparin
Prostaglandins

stimulates

INFLAMMATION

Inflammation produces several familiar indications of injury, including swelling, redness, heat (warmth), pain, and sometimes loss of function. Inflammation may also result from the presence of pathogens, such as harmful bacteria, within the tissues. The presence of these pathogens constitutes an **infection**.

Exposure to Pathogens and Toxins

Injured tissue contains an abnormal concentration of pathogens, toxins, wastes, and the chemicals from injured cells.

When a tissue is injured, a general defense mechanism is activated.

Increased Blood Flow	**Increased Vessel Permeability**	**Pain**
In response to the released chemicals, blood vessels dilate, increasing blood flow through the damaged tissue.	Vessel dilation is accompanied by an increase in the permeability of the capillary walls. Plasma now diffuses into the injured tissue, so the area becomes swollen.	The abnormal conditions within the tissue and the chemicals released by mast cells stimulate nerve endings that produce the sensation of pain.
		PAIN

Increased Local Temperature	**Increased Oxygen and Nutrients**	**Increased Phagocytosis**	**Removal of Toxins and Wastes**
The increased blood flow and permeability causes the tissue to become warm and red.	Vessel dilation, increased blood flow, and increased vessel permeability result in enhanced delivery of oxygen and nutrients.	Phagocytes in the tissue are activated, and they begin engulfing tissue debris and pathogens.	Enhanced circulation carries away toxins and wastes, distributing them to the kidneys for excretion, or to the liver for inactivation.
	O₂		Toxins and wastes

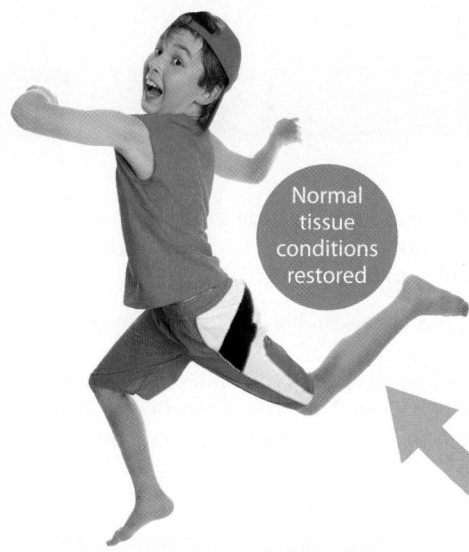

Normal tissue conditions restored

Regeneration

Regeneration is the repair that occurs after the damaged tissue has been stabilized and the inflammation has subsided. Fibroblasts move into the area, laying down a collagenous framework known as **scar tissue**. Over time, scar tissue is usually "remodeled" and gradually assumes a more normal appearance.

Inhibits mast cell activation

Inflammation Subsides

Over a period of hours to days, the cleanup process generally succeeds in eliminating the inflammatory stimuli.

4

Aging and Cancer Incidence

Cancer rates increase with age, and about 25 percent of all people in the United States develop cancer at some point in their lives. It has been estimated that 70–80 percent of cancer cases result from chemical exposure, environmental factors, or some combination of the two, and 40 percent of those cancers are caused by cigarette smoke. Each year in the United States, more than 500,000 individuals die of cancer, making it second only to heart disease as a cause of death. We discussed the development and growth of cancer in Chapter 3. ⤴ p. 132

With this chapter, you have concluded the introductory portion of this text. In combination, the four basic tissue types described here form all of the organs and systems we will discuss in subsequent chapters.

CHECKPOINT

35. Identify some age-related factors that affect tissue repair and structure.

36. What would account for the observed increase in cancer rates with age?

See the blue Answers tab at the back of the book.

Related Clinical Terms

autopsy: An examination of a dead body to discover the cause of death or the extent of disease.

biopsy: An examination of tissue removed from a living body to discover the presence, cause, or extent of disease.

cachexia: Weakness and wasting of the body due to severe chronic illness.

carcinoma: A cancer arising in the epithelial tissue of the skin or of the lining of the internal organs.

immunotherapy: The prevention or treatment of disease with substances that stimulate the immune response.

lesion: A region in an organ or tissue that has suffered damage from injury or disease; a wound, ulcer, abscess, or tumor, for example.

metaplasia: A reversible structural change that alters the character of a tissue.

pathologist: A physician who specializes in the study of disease processes in tissues and body fluids.

remission: Abatement, ending, or lessening in severity of the signs and symptoms of a disease.

sarcoma: A malignant tumor of connective or other nonepithelial tissue.

tissue engineering: Tissue is either grown in or outside of a body to be transplanted into a patient or used for testing.

tissue rejection: Occurs when a transplant recipient's immune system attacks a transplanted organ or tissue.

tissue transplantation: Moving tissues (or organs) from one body and placing them into another body by medical procedures for the purpose of replacing the recipient's damaged or failing tissue (or organ).

tumor grading: A system used to classify cancer cells in terms of how abnormal they look under a microscope and how quickly the tumor is likely to grow and spread.

tumor staging: Defining at what point the patient is in the development of the malignant disease when the diagnosis is made.

xenotransplant: The process of grafting or transplanting organs or tissues between members of different species.

Chapter Review

Study Outline

MasteringA&P®

Access more chapter study tools online in the MasteringA&P Study Area:

- Chapter Quizzes, Chapter Practice Test, Art-labeling Activities, Animations, MP3 Tutor Sessions, and Clinical Case Studies
- Practice Anatomy Lab PAL™
- Interactive Physiology iP®
- A&P Flix A&PFlix™
- PhysioEx PhysioEx™

▌ **An Introduction to the Tissue Level of Organization** p. 140

1. Tissues are structures with discrete structural and functional properties that combine to form organs.

4-1 ▌ **The four tissue types are epithelial, connective, muscle, and neural** p. 140

2. **Tissues** are collections of specialized cells and cell products that perform a relatively limited number of functions. The four *tissue types* are *epithelial tissue, connective tissue, muscle tissue,* and *neural tissue.*

3. **Histology** is the study of tissues.

4-2 ▌ **Epithelial tissue covers body surfaces, lines cavities and tubular structures, and serves essential functions** p. 140

4. **Epithelial tissue** includes epithelia and glands. An **epithelium** is an **avascular** layer of cells that forms a barrier that provides protection and regulates permeability. **Glands** are

secretory structures derived from epithelia. Epithelial cells may show **polarity,** the presence of structural and functional differences between their exposed (apical) and attached (basal) surfaces.

5. A **basement membrane (basal lamina)** attaches epithelia to underlying connective tissues.

6. Epithelia provide physical protection, control permeability, provide sensation, and produce specialized secretions. Gland cells are epithelial cells that produce secretions. In **glandular epithelia,** most cells produce secretions.

7. Epithelial cells are specialized to perform secretory or transport functions and to maintain the physical integrity of the epithelium. *(Figure 4–1)*

8. Many epithelial cells have microvilli.

9. The coordinated beating of the cilia on a **ciliated epithelium** moves materials across the epithelial surface.

10. Cells can attach to other cells or to extracellular protein fibers by means of **cell adhesion molecules (CAMs)** or at specialized attachment sites called **cell junctions.** The three major types of cell junctions are **tight junctions, gap junctions,** and **desmosomes.** *(Figure 4–2)*

11. The inner surface of each epithelium is connected to a two-part basement membrane consisting of a *clear layer* and a *dense layer.* Divisions by **germinative cells** (a kind of stem cell) continually replace the short-lived epithelial cells.

4-3 ▌ Cell shape and number of layers determine the classification of epithelia p. 144

12. Epithelia are classified on the basis of the number of cell layers and the shape of the cells at the apical surface.

13. A **simple epithelium** has a single layer of cells covering the basement membrane; a **stratified epithelium** has several layers. The cells in a **squamous epithelium** are thin and flat. Cells in a **cuboidal epithelium** resemble hexagonal boxes; those in a **columnar epithelium** are taller and more slender. *(Table 4–1; Figures 4–3 to 4–5)*

14. Epithelial cells (or structures derived from epithelial cells) that produce secretions are called *glands.* **Exocrine glands** discharge secretions onto the body surface or into **ducts,** which communicate with the exterior. *Hormones,* the secretions of **endocrine glands,** are released by gland cells into the surrounding interstitial fluid.

15. A glandular epithelial cell may release its secretions by merocrine, apocrine, or holocrine secretory modes. In **merocrine secretion,** the most common mode, the product is released through exocytosis. **Apocrine secretion** involves the loss of both the secretory product and cytoplasm. Unlike the other two methods, **holocrine secretion** destroys the gland cell, which becomes packed with secretions and then bursts. *(Figure 4–6)*

16. In epithelia that contain scattered gland cells, individual secretory cells are called **unicellular glands. Multicellular glands** are organs that contain glandular epithelia that produce exocrine or endocrine secretions.

17. Exocrine glands can be classified on the basis of structure as **unicellular exocrine glands (mucous or goblet cells)** or as **multicellular exocrine glands.** Multicellular exocrine glands can be further classified according to structure. *(Figure 4–7)*

4-4 ▌ Connective tissue provides a protective structural framework for other tissue types p. 152

18. **Connective tissues** are internal tissues with many important functions: establishing a structural framework; transporting fluids and dissolved materials; protecting delicate organs; supporting, surrounding, and interconnecting tissues; storing energy reserves; and defending the body from microorganisms.

19. All connective tissues contain specialized cells and a **matrix,** composed of extracellular protein fibers and a **ground substance.**

20. **Connective tissue proper** is connective tissue that contains varied cell populations and fiber types surrounded by a syrupy ground substance. *(Figure 4–8)*

21. **Fluid connective tissues** have distinctive populations of cells suspended in a watery matrix that contains dissolved proteins. The two types of fluid connective tissues are *blood* and *lymph.*

22. **Supporting connective tissues** have a less diverse cell population than connective tissue proper and a dense matrix with closely packed fibers. The two types of supporting connective tissues are *cartilage* and *bone.*

23. Connective tissue proper contains fibers, a viscous ground substance, and a varied population of cells, including **fibroblasts, fibrocytes, macrophages, adipocytes, mesenchymal cells, melanocytes, mast cells, lymphocytes,** and **microphages.**

24. The three types of fibers in connective tissue are **collagen fibers, reticular fibers,** and **elastic fibers.**

25. The first connective tissue to appear in an embryo is **mesenchyme,** or *embryonic connective tissue.*

26. Connective tissue proper is classified as either **loose connective tissue** or **dense connective tissue.** Loose connective tissues are mesenchyme and **mucous connective tissues** in the embryo; **areolar tissue; adipose tissue,** including **white fat** and **brown fat;** and **reticular tissue.** Most of the volume in dense connective tissue consists of fibers. The two types of dense connective tissue are **dense regular connective tissue** and **dense irregular connective tissue. Elastic tissue** is a dense regular connective tissue. *(Figures 4–9 to 4–11)*

27. **Blood** and **lymph** are connective tissues that contain distinctive collections of cells in a fluid matrix.

28. Blood contains *formed elements:* **red blood cells** (*erythrocytes*), **white blood cells** (*leukocytes*), and **platelets.** The watery matrix of blood is called **plasma.** *(Figure 4–12)*

29. **Arteries** carry blood away from the heart and toward **capillaries,** where water and small solutes move into the interstitial fluid of surrounding tissues. **Veins** return blood to the heart.

30. Lymph forms as interstitial fluid enters the **lymphatic vessels,** which return lymph to the cardiovascular system.

4-5 ▌ Cartilage and bone provide a strong supporting framework p. 159

31. Cartilage and bone are called supporting connective tissues because they support the rest of the body.

32. **Chondrocytes,** or cartilage cells, rely on diffusion through the avascular matrix to obtain nutrients.

33. Cartilage grows by two mechanisms: **interstitial growth** and **appositional growth.** *(Figure 4–13)*

34. The matrix of cartilage is a firm gel that contains **chondroitin sulfates** (used to form proteoglycans) and chondrocytes. Chondrocytes occupy chambers called **lacunae.** A fibrous **perichondrium** separates cartilage from surrounding tissues. The three types of cartilage are **hyaline cartilage, elastic cartilage,** and **fibrocartilage.** *(Figure 4–14)*

35. **Bone,** or **osseous tissue,** consists of **osteocytes,** little ground substance, and a dense, mineralized matrix. Osteocytes are situated in lacunae. The matrix consists of calcium salts and collagen fibers, giving it unique properties. *(Figure 4–15; Table 4–2)*

4

36. Osteocytes depend on diffusion within **canaliculi** for nutrient intake.
37. Each bone is surrounded by a **periosteum** with fibrous and cellular layers.

4-6 ▮ **Tissue membranes are physical barriers of four types: mucous, serous, cutaneous, and synovial** p. 163

38. Membranes form a barrier or interface. Epithelia and connective tissues combine to form membranes that cover and protect other structures and tissues. *(Figure 4–16)*
39. **Mucous membranes** line cavities that communicate with the exterior. They contain areolar tissue called the **lamina propria.**
40. **Serous membranes** line the body's sealed internal cavities of the trunk. They secrete a fluid called a *transudate.*
41. The **cutaneous membrane,** or skin, covers the body surface.
42. **Synovial membranes** form an incomplete lining within the cavities of synovial joints.

4-7 ▮ **Connective tissues create the internal framework of the body** p. 165

43. A network of connective tissue proper ties internal organs and systems together. This network consists of the **superficial fascia** (the **hypodermis,** or subcutaneous tissue, separating the skin from underlying tissues and organs), the **deep fascia** (dense connective tissue), and the **subserous fascia** (the layer between the deep fascia and the serous membranes that line true body cavities). *(Figure 4–17)*

4-8 ▮ **The three types of muscle tissue are skeletal, cardiac, and smooth** p. 166

44. **Muscle tissue** is specialized for contraction. *(Figure 4–18)*
45. The cells of **skeletal muscle tissue** are *multinucleate.* Skeletal muscle, or **striated voluntary muscle,** produces new fibers by the division of **myosatellite cells,** a kind of stem cell.
46. **Cardiocytes,** the cells of **cardiac muscle tissue,** occur only in the heart. Cardiac muscle, or **striated involuntary muscle,** relies on *pacemaker cells* to stimulate regular contraction.

47. **Smooth muscle tissue,** or **nonstriated involuntary muscle,** is not striated. Smooth muscle cells can divide and therefore regenerate after injury.

4-9 ▮ **Neural tissue responds to stimuli and conducts electrical impulses throughout the body** p. 168

48. **Neural tissue** conducts electrical impulses, which convey information from one area of the body to another.
49. Cells in neural tissue are either neurons or neuroglia. **Neurons** transmit information as electrical impulses. Several kinds of **neuroglia** exist, and their basic functions include supporting neural tissue and helping supply nutrients to neurons. *(Figure 4–19)*
50. A typical neuron has a **cell body, dendrites,** and an **axon,** or **nerve fiber.** The axon carries information to other cells.

4-10 ▮ **The response to tissue injury involves inflammation and regeneration** p. 170

51. Any injury affects several types of tissue simultaneously, and they respond in a coordinated manner. Homeostasis is restored by two processes: inflammation and regeneration.
52. **Inflammation,** or the **inflammatory response,** isolates the injured area while damaged cells, tissue components, and any dangerous microorganisms (which could cause **infection**) are cleaned up. **Regeneration** is the repair process that restores normal function. *(Spotlight Figure 4–20)*

4-11 ▮ **With advancing age, tissue repair declines and cancer rates increase** p. 170

53. Tissues change with age. Repair and maintenance become less efficient, and the structure and chemical composition of many tissues are altered.
54. The incidence of cancer increases with age, with roughly three-quarters of all cases caused by exposure to chemicals or by other environmental factors, such as cigarette smoke.

Review Questions

See the blue Answers tab at the back of the book.

LEVEL 1 Reviewing Facts and Terms

1. Identify the six categories of epithelial tissue shown in the drawing below.

(a) _____   (b) _____
(c) _____   (d) _____
(e) _____   (f) _____

2. Which cell type is involved in secreting collagen to repair damaged connective tissues? (a) fibroblasts, (b) osteoclasts, (c) chondroblasts, (d) macrophages.
3. Tissue that is specialized for contraction is (a) epithelial tissue, (b) muscle tissue, (c) connective tissue, (d) neural tissue.
4. A type of cell junction common in cardiac and smooth muscle tissues is the (a) hemidesmosome, (b) basal junction, (c) tight junction, (d) gap junction.
5. The most abundant connections between cells in the superficial layers of the skin are (a) connexons, (b) gap junctions, (c) desmosomes, (d) tight junctions.
6. A _____ membrane has an epithelium that is stratified and supported by dense connective tissue. (a) synovial, (b) serous, (c) cutaneous, (d) mucous.
7. Mucous secretions that coat the passageways of the digestive and respiratory tracts result from _____ secretion. (a) apocrine, (b) merocrine, (c) holocrine, (d) endocrine.
8. Matrix is a characteristic of which type of tissue? (a) epithelial, (b) neural, (c) muscle, (d) connective.
9. Osteocytes and chondrocytes occupy chambers known as (a) lamellae, (b) canaliculi, (c) lacunae, (d) central canal.

10. Which of the following epithelia most easily permits diffusion? **(a)** stratified squamous, **(b)** simple squamous, **(c)** transitional, **(d)** simple columnar.
11. The three major types of cartilage in the body are **(a)** collagen, reticular, and elastic, **(b)** areolar, adipose, and reticular, **(c)** hyaline, elastic, and fibrous, **(d)** tendons, reticular, and elastic.
12. The primary function of serous membranes in the body is to **(a)** minimize friction between opposing surfaces, **(b)** line cavities that communicate with the exterior, **(c)** perform absorptive and secretory functions, **(d)** cover the surface of the body.
13. Which of these is a feature of connective tissue? **(a)** Cells have many ion channels. **(b)** Cells are found in a gel-like matrix. **(c)** Cells have contractile capabilities. **(d)** Cells are bound by desmosomes.
14. The matrix is made up of **(a)** protein fibers and cells, **(b)** ground substance and cells, **(c)** ground substance and protein fibers, **(d)** ground substance, protein fibers, and cells.
15. Axons, dendrites, and a cell body are characteristic of cells located in **(a)** neural tissue, **(b)** muscle tissue, **(c)** connective tissue, **(d)** epithelial tissue.
16. The repair process necessary to restore normal function after inflammation subsides in damaged tissues is **(a)** isolation, **(b)** regeneration, **(c)** reconstruction, **(d)** all of these.
17. What are the four essential functions of epithelial tissue?
18. Differentiate between endocrine and exocrine glands.
19. Explain why skeletal and cardiac muscles are striated but smooth muscles are not.
20. List three basic components of connective tissues.
21. The tendon is classified as dense regular connective tissue. Identify the "cell" and "fiber" type found in dense regular connective tissue.
22. What two cell populations make up neural tissue? What is the function of each?

LEVEL 2 Reviewing Concepts

23. What is the functional significance of tight junctions in epithelial integrity?

24. Explain why tendons and ligaments have a regular arrangement of collagen fibers, whereas in muscle sheaths and the dermis, the arrangement is irregular.
25. Describe the fluid connective tissues in the human body. What are the main differences between fluid connective tissues and supporting connective tissues?
26. Why are infections always a serious threat after a severe burn or abrasion?
27. A layer of glycoproteins and a network of fine protein filaments that prevents the movement of proteins and other large molecules from the connective tissue to the epithelium describes **(a)** interfacial canals, **(b)** the basement membrane, **(c)** the reticular lamina, **(d)** areolar tissue, **(e)** squamous epithelium.
28. Why does damaged cartilage heal slowly? **(a)** Chondrocytes cannot be replaced if killed, and other cell types must take their place, **(b)** Cartilage is avascular, so nutrients and other molecules must diffuse to the site of injury, **(c)** Damaged cartilage becomes calcified, thus blocking the movement of materials required for healing, **(d)** Chondrocytes divide more slowly than other cell types, delaying the healing process, **(e)** Damaged collagen cannot be quickly replaced, thereby slowing the healing process.
29. List the similarities and differences among the three types of muscle tissue.

LEVEL 3 Critical Thinking and Clinical Applications

30. Assuming that you had the necessary materials to perform a detailed chemical analysis of body secretions, how could you determine whether a secretion was merocrine or apocrine?
31. During a lab practical, a student examines a tissue that is composed of densely packed protein fibers that are running parallel and form a cord. There are no striations, but small nuclei are visible. The student identifies the tissue as skeletal muscle. Why is the student's choice wrong, and what tissue is he probably observing?
32. Julie has a sore throat and is finding it difficult to swallow. In the mirror, she observes that her throat is red compared to the rest of her mouth. Explain why Julie's throat is sore and appears red, and why she finds it difficult to swallow.

CLINICAL CASE Wrap-Up

The Rubber Girl

Anne Marie suffers from a variant of Ehlers-Danlos syndrome—a rare inherited disorder of connective tissue. Specifically, the collagen fibers in the matrix of connective tissue are defective. Because connective tissue and collagen occur throughout the body, many tissues are involved, particularly the joints and skin.

Anne Marie's loose joints and dislocations are due to deficient dense connective tissue. Because of abnormal collagen fibers, both her ligaments (which connect bone to bone) and her tendons (which connect muscle to bone) are abnormally loose and heal poorly when overstretched. When she twists her knee and feels as if her kneecap is dislocating, it is, indeed, sliding out of its groove. The ligaments that protect her ankles have been repeatedly sprained.

Thin, translucent skin that is soft, velvety, and stretchy is also typical of Ehlers-Danlos syndrome. Scars appear wide, fragile, and abnormal. Bruises and skin tears occur easily.

1. Connective tissues are made up of specialized cells and a matrix of extracellular protein fibers and a fluid known as ground substance. What part of connective tissue proper is defective in Ehlers-Danlos syndrome?

2. Do you think this genetic syndrome affects just Anne Marie's joints and skin or would you suspect other tisses are affected as well?

See the blue Answers tab at the back of the book.

5

The Integumentary System

Learning Outcomes

These Learning Outcomes correspond by number to this chapter's sections and indicate what you should be able to do after completing the chapter.

5-1   ▌ Describe the main structural features of the epidermis, and explain the functional significance of each.

5-2   ▌ Explain what accounts for individual differences in skin color, and discuss the response of melanocytes to sunlight exposure.

5-3   ▌ Describe the interaction between sunlight and vitamin D_3 production.

5-4   ▌ Describe the roles of epidermal growth factor.

5-5   ▌ Describe the structure and functions of the dermis.

5-6   ▌ Describe the structure and functions of the hypodermis.

5-7   ▌ Describe the mechanisms that produce hair, and explain the structural basis for hair texture and color.

5-8   ▌ Discuss the various kinds of glands in the skin, and list the secretions of those glands.

5-9   ▌ Describe the anatomical structure of nails, and explain how they are formed.

5-10 ▌ Explain how the skin responds to injury and repairs itself.

5-11 ▌ Summarize the effects of aging on the skin.

SPOTLIGHT
- The Epidermis p. 180

CLINICAL CASE
- Skin Cells in Overdrive p. 177

Clinical Notes
- Skin Cancer p. 184
- Decubitus Ulcers p. 187
- Liposuction p. 189
- Burns and Grafts p. 197
- Skin Abnormalities p. 198

An Introduction to the Integumentary System

You are probably more familiar with the skin than with any other organ system. No other organ system is as accessible, large, and underappreciated as the *integumentary system*. Often referred to simply as the **integument** (in-TEG-ū-ment), this system makes up about 16 percent of your total body weight. Its surface, 1.5–2 m² (16.1–21.5 sq ft) in area, is continually abraded, attacked by microorganisms, irradiated by sunlight, and exposed to environmental chemicals. The integumentary system is your body's first line of defense against an often hostile environment. It's the place where you and the outside world meet.

The integumentary system has two major parts: the **cutaneous membrane** or skin, and the **accessory structures** (**Figure 5–1**).

1. The cutaneous membrane has two components: the **epidermis** (*epi-*, above) or superficial epithelium, and the **dermis**, an underlying area of connective tissues.

Figure 5–1   **The Components of the Integumentary System.**

Cutaneous Membrane		Accessory Structures
Epidermis		Hair shaft
Dermis — Papillary layer		Pore of sweat gland duct
Dermis — Reticular layer		Tactile (Meissner's) corpuscle
		Sebaceous gland
		Arrector pili muscle
		Sweat gland duct
Hypodermis		Hair follicle
		Lamellated (pacinian) corpuscle
		Nerve fibers
		Sweat gland
		Artery / Vein — Cutaneous plexus
	Fat	

2. The accessory structures include hair, exocrine glands, and nails. They are located primarily in the dermis and protrude through the epidermis to the skin surface.

The integument does not function in isolation. An extensive network of blood vessels branches through the dermis. Sensory receptors monitor touch, pressure, temperature, and pain, providing valuable information to the central nervous system about the state of the body.

Deep to the dermis is the loose connective tissue of the **hypodermis**, also known as the superficial fascia or *subcutaneous layer*. The hypodermis separates the integument from the deep fascia around other organs, such as muscles and bones. ↻ p. 165 Although the hypodermis is often considered separate from the integument, we will consider it in this chapter because its connective tissue fibers are interwoven with those of the dermis.

The general functions of the integumentary system are summarized in **Figure 5–2** and include:

- *Protection* of underlying tissues and organs against impact, abrasion, fluid loss, and chemical attack.

- *Excretion* of salts, water, and organic wastes by integumentary glands.

- *Maintenance* of normal body temperature through either insulation or evaporative cooling, as needed.

- *Production* of melanin, which protects underlying tissue from ultraviolet radiation.

- *Production* of keratin, which protects against abrasion and serves as a water repellent.

- *Synthesis of vitamin D_3*, a steroid that is converted to calcitriol, a hormone important to normal calcium metabolism.

- *Storage* of lipids in adipocytes in the dermis and in adipose tissue in the hypodermis (subcutaneous layer).

- *Detection* of touch, pressure, pain, vibration, and temperature stimuli and the relaying of that information to the nervous system. (We will consider these *general senses*, which provide information about the external environment, in Chapter 15.)

- *Coordination* of immune response to pathogens and cancers in the skin.

Figure 5–2   **The Cutaneous Membrane and Accessory Structures.**

5-1 ┃ The epidermis is composed of layers with various functions

Learning Outcome Describe the main structural features of the epidermis, and explain the functional significance of each.

The epidermis is a stratified squamous epithelium. Recall from Chapter 4 that such an epithelium provides physical protection and helps keep microorganisms outside the body. ↺ p. 146

Like all other epithelia, the epidermis is avascular. Because there are no local blood vessels, epidermal cells rely on the diffusion of nutrients and oxygen from capillaries within the dermis. As a result, the epidermal cells with the highest metabolic demands are found close to the basement membrane, where the diffusion distance is short. The superficial cells, far removed from the source of nutrients, are dead. **Spotlight Figure 5–3** describes the epidermal layers and functions.

Keratinocytes (ke-RAT-i-nō-sīts), the body's most abundant epithelial cells, dominate the epidermis. These cells form several layers, or *strata*, and contain large amounts of the protein *keratin* (discussed shortly). **Thin skin** covers most of the body surface. **Thick skin** is found on the palms of the hands and the soles of the feet. Note that the terms *thick* and *thin* refer to the relative thickness of the epidermis, not to the cutaneous membrane as a whole.

Spotlight Figure 5–3 shows the five layers of keratinocytes in a section of the epidermis in an area of thick skin. The boundaries between the layers are often difficult to see in a standard light micrograph. Notice that the various layers have Latin names. The word *stratum* (plural, *strata*) means "layer," and the rest of the name refers to the function or appearance of the layer. The strata, in order from the basement membrane toward the free surface, are the *stratum basale*, the *stratum spinosum*, the *stratum granulosum*, the *stratum lucidum*, and the *stratum corneum*.

Stratum Basale

The deepest layer of the epidermis is the **stratum basale** (STRA-tum buh-SAHL-āy) or *stratum germinativum* (STRA-tum jer-mi-na-TĒ-vum). Hemidesmosomes attach the cells of this layer to the basement membrane that separates the epidermis from the areolar tissue of the dermis. ↺ p. 143 The stratum basale and the underlying dermis interlock, increasing the strength of the bond between the epidermis and dermis. The stratum basale forms **epidermal ridges**, which extend into the dermis and are adjacent to dermal projections called **dermal papillae** (singular, *papilla*; a nipple-shaped mound) that project into the epidermis (**Spotlight Figure 5–3**). These ridges and papillae are important because the strength of the attachment is proportional to the surface area of the basement membrane. In other words, the more numerous and deeper the folds, the larger the surface area becomes.

The contours of the skin surface follow the ridge patterns, which vary from small conical pegs (in thin skin) to the complex whorls seen on the thick skin of the palms and soles. Ridges on the palms and soles increase the surface area of the skin and increase friction, ensuring a secure grip. The ridge patterns on the tips of the fingers are the basis of fingerprints. These ridge shapes are determined partially by the interaction of genes and partially by the intrauterine environment. That is, during fetal development our genes determine general characteristics, but the contact with amniotic fluid, another fetus (in the case of twins), or the uterine wall affect the ultimate pattern of a fetus's fingerprints. So, identical twins do not have identical fingerprints. However, the pattern of your epidermal ridges is unique and does not change after birth and during your lifetime. **Basal cells**, or *germinative cells*, dominate the stratum basale. Basal cells are stem cells that divide to replace the more superficial keratinocytes that are shed at the epithelial surface.

Skin surfaces that lack hair also contain specialized epithelial tactile cells known as *Merkel cells* scattered among the cells of the stratum basale. Each Merkel cell together with a sensory nerve terminal is called a **tactile disc**. The tactile cells are sensitive to touch and when compressed release chemicals that stimulate their associated sensory nerve endings. (The skin contains many other kinds of sensory receptors, as we will see in later sections.)

The brown tones of skin result from the synthesis of pigment by cells called *melanocytes*. ↺ p. 154 These pigment cells are distributed throughout the stratum basale, with cell processes extending into more superficial layers.

Stratum Spinosum

Each time a stem cell divides, one of the daughter cells is pushed superficial to the stratum basale into the **stratum spinosum**. This stratum consists of 8 to 10 layers of keratinocytes bound together by desmosomes. ↺ p. 142 The name *stratum spinosum*, which means "spiny layer," refers to the fact that the cells look like miniature pincushions in standard histological sections. They look that way because the keratinocytes were processed with chemicals that shrank the cytoplasm but left the cytoskeletal elements and desmosomes intact. Some of the cells entering this layer from the stratum basale continue to divide, further increasing the thickness of the epithelium.

The stratum spinosum also contains cells that participate in the immune response. Called *dendritic cells* (*dendron*, tree) because of their branching projections, they are also known as *Langerhans cells*. They stimulate defense against (1) microorganisms that manage to penetrate the superficial layers of the epidermis and (2) superficial skin cancers. We will consider dendritic cells and other cells of the immune response in Chapter 22.

BASIC ORGANIZATION OF THE EPIDERMIS

The epidermis consists of stratified squamous epithelium. It is separated from the dermis by a basement membrane. The stratum basale and the underlying dermis interlock, strengthening the bond between the two. The epidermis forms epidermal ridges, which extend into the dermis and are adjacent to dermal papillae that project into the epidermis.

Epidermis
Basement membrane
Epidermal ridge
Dermal papilla
Dermis

Stratum corneum
Basement membrane
Dermis

Thin skin   LM × 200

Thin skin contains four layers of keratinocytes, and is about as thick as the wall of a plastic sandwich bag (about 0.08 mm).

LAYERS OF THE EPIDERMIS

The layers of the epidermis are best shown in a sectional view of thick skin. **Thick skin** contains a fifth layer, the *stratum lucidum*. Because thick skin also has a much thicker superficial layer (the *stratum corneum*), it is about as thick as a standard paper towel (about 0.5 mm).

Surface

Stratum corneum

Stratum lucidum

Dermal papilla

Epidermal ridge

Thick skin   LM × 200

Basement membrane

Epidermis	Characteristics
Stratum corneum	• Multiple layers of flattened, dead, interlocking keratinocytes • Water resistant but not waterproof • Permits slow water loss by insensible perspiration
Stratum lucidum	• Appears as a glassy layer in thick skin only
Stratum granulosum	• Keratinocytes produce keratohyalin and keratin • Keratin fibers develop as cells become thinner and flatter • Gradually the plasma membranes thicken, the organelles disintegrate, and the cells die
Stratum spinosum	• Keratinocytes are bound together by desmosomes
Stratum basale	• Deepest, basal layer • Attachment to basement membrane • Contains basal cells (stem cells), melanocytes, and tactile cells (Merkel cells)

Dermis

EPIDERMAL RIDGES OF THICK SKIN

Fingerprints reveal the pattern of epidermal ridges. This scanning electron micrograph shows the ridges on a fingertip.

Pores of sweat gland ducts

Epidermal ridge

Thick skin   SEM × 20

Stratum Granulosum

Superficial to the stratum spinosum is the **stratum granulosum**, or "grainy layer." It consists of three to five layers of keratinocytes derived from the stratum spinosum. By the time cells are pushed into this layer, most have stopped dividing and have started making large amounts of the proteins **keratin** (KER-a-tin; *keros*, horn) and **keratohyalin** (ker-a-tō-HĪ-a-lin). Keratin is a tough, fibrous protein. It is the basic structural component of hair and nails in humans. ⤴ p. 146 As keratin fibers accumulate, the cells grow thinner and flatter, and their plasma membranes thicken and become less permeable. Keratohyalin forms dense cytoplasmic granules that promote dehydration of the cell as well as aggregation and cross-linking of the keratin fibers. The nuclei and other organelles then disintegrate, and the cells die. Further dehydration creates a tightly interlocked layer of cells that consists of keratin fibers surrounded by keratohyalin.

Stratum Lucidum

In the thick skin of the palms and soles, a glassy **stratum lucidum** ("clear layer") covers the stratum granulosum. The cells in the stratum lucidum are flattened, densely packed, largely without organelles, and filled with keratin.

Stratum Corneum

At the exposed surface of both thick skin and thin skin is the **stratum corneum** (KOR-nē-um; *corneus*, horny). It normally contains 15 to 30 layers of keratinized cells. **Keratinization**, or *cornification*, is the formation of protective, superficial layers of cells filled with keratin. This process takes place on all exposed skin surfaces except the anterior surfaces of the eyes. The dead cells in each layer of the stratum corneum remain tightly interconnected by desmosomes. The connections are so secure that keratinized cells are generally shed in large groups or sheets rather than individually.

It takes 7 to 10 days for a cell to move from the stratum basale to the stratum corneum. The dead cells generally remain in the exposed stratum corneum for an additional two weeks before they are shed or washed away. This arrangement places the deeper portions of the epithelium and underlying tissues beneath a protective barrier of dead, durable, and expendable cells. Normally, the surface of the stratum corneum is relatively dry, so it is unsuitable for the growth of many microorganisms. Lipid secretions from sebaceous glands coat the surface, helping to maintain this barrier against water loss or entry.

The stratum corneum is water resistant, but not waterproof. Water from interstitial fluids slowly penetrates to the surface and evaporates into the surrounding air. You lose about 500 mL (about 1 pt) of water in this way each day. The process is called **insensible perspiration**, because you are unable to see or feel (sense) the water loss. In contrast, you are usually very aware of the **sensible perspiration** produced by active sweat glands.

Damage to the epidermis can increase the rate of insensible perspiration. If the damage breaks connections between superficial and deeper layers of the epidermis, fluid accumulates in pockets, or *blisters*, within the epidermis. (Blisters also form between the epidermis and dermis if the basement membrane is damaged.) If damage to the stratum corneum makes it less effective as a water barrier, the rate of insensible perspiration skyrockets, and a potentially dangerous fluid loss occurs. This loss is a serious consequence of severe burns and a complication in the condition known as *xerosis* (excessively dry skin).

Why do wrinkles form in the skin on the fingers and toes after a long soak in the bath? Previously, it was common to explain this effect as the osmotic movement of water across the epithelium. That is, the fingertips absorb water and swell, making the skin wrinkle. However, such a process failed to explain why wrinkling did not occur if there was nerve damage in the fingers. It turns out that the fingertips do not swell in water, but they actually shrink when they wrinkle, an effect caused by the constriction of underlying blood vessels. The effect is due to the autonomic nervous system, which functions mostly outside our awareness and controls blood vessel diameter. Is there any benefit to having wrinkled fingertips? Recent studies have shown that the wrinkles improve the ability to grip wet or submerged objects, much like the treads of a tire improve traction on a wet road.

CHECKPOINT

1. Identify the layers of the epidermis.

2. Dandruff is caused by excessive shedding of cells from the outer layer of skin in the scalp. So dandruff is composed of cells from which epidermal layer?

3. A splinter that penetrates to the third layer of the epidermis of the palm is lodged in which layer?

4. Why does taking a bath cause wrinkly fingertips and toes?

5. Some criminals sand the tips of their fingers so as not to leave recognizable fingerprints. Would this practice permanently remove fingerprints? Why or why not?

See the blue Answers tab at the back of the book.

5-2 ‖ Factors influencing skin color are epidermal pigmentation and dermal circulation

Learning Outcome Explain what accounts for individual differences in skin color, and discuss the response of melanocytes to sunlight exposure.

One of the most obvious distinguishing characteristics among humans is skin color. What gives skin its various colors? In this section we examine how pigments in the epidermis and blood flow in the dermis influence skin color. We also discuss clinical implications of changes in skin color.

The Role of Epidermal Pigmentation

The epidermis contains variable quantities of two pigments: carotene and melanin. **Carotene** (KAR-uh-tēn) is an orange-yellow pigment that normally accumulates in epidermal cells. It is most apparent in cells of the stratum corneum of light-skinned individuals, but it also accumulates in fatty tissues in the deep dermis and subcutaneous layer. Carotene is found in a variety of orange vegetables, such as carrots and squashes. The skin of someone who eats lots of carrots can actually turn orange from an overabundance of carotene. The color change is very striking in pale-skinned individuals, but less obvious in people with darker skin pigmentation. Carotene can be converted to vitamin A, which is required for both the normal maintenance of epithelia and the synthesis of photoreceptor pigments in the eye.

Melanin is a pigment produced by melanocytes, pigment-producing cells, introduced in Chapter 4. There are two types of melanin, a red-yellow form (pheomelanin) and a brown-black form (eumelanin). The **melanocytes** involved are located in the stratum basale, squeezed between or deep to the epithelial cells (**Figure 5–4**). Melanocytes manufacture both types of melanin from the amino acid *tyrosine* and package it into intracellular vesicles called *melanosomes*. These vesicles, which contain either pheomelanin or eumelanin, travel within the processes of melanocytes and are transferred intact to about 40 basal keratinocytes. The transfer of pigmentation colors the keratinocyte temporarily, until the melanosomes are destroyed by fusion with lysosomes. In individuals with pale skin, this transfer takes place in the stratum basale and stratum spinosum, and the cells of more superficial layers lose their pigmentation. In dark-skinned people, the melanosomes are larger and more numerous, and the transfer may occur in the stratum granulosum as well, making skin pigmentation darker and more persistent.

The ratio of melanocytes to basal cells ranges between 1:4 and 1:20, depending on the region of the body. The skin covering most areas of the body has about 1000 melanocytes per square millimeter. The cheeks and forehead, the nipples, and the genital region (the scrotum of males and the labia majora of females) have higher concentrations (about 2000 per square millimeter). The differences in skin pigmentation among individuals do not reflect different numbers of melanocytes, but rather, different levels of melanin synthesis. A deficiency or absence of melanin production leads to a disorder known as *albinism*. Individuals with this condition have a normal distribution of melanocytes, but the cells are incapable of producing melanin.

There can also be localized differences in the rates of melanin production by your melanocytes. *Freckles* are small, pigmented areas on relatively pale skin. These spots typically have an irregular border. They represent the areas serviced by melanocytes that are producing larger-than-average amounts of melanin. Freckles tend to be most abundant on

Figure 5–4 Melanocytes. The micrograph and accompanying drawing indicate the location and orientation of melanocytes in the stratum basale of a dark-skinned person.

Melanocytes in stratum basale

Melanin pigment

Basement membrane

Melanocytes LM × 600

a This micrograph shows the location and orientation of melanocytes in the stratum basale of a dark-skinned person.

Melanosome

Keratinocyte

Melanin pigment

Melanocyte

Basement membrane

Dermis

b Melanocytes produce and store melanin.

surfaces such as the face, probably due to its greater exposure to the sun. *Lentigos* are similar to freckles but have regular borders and contain abnormal melanocytes. *Senile lentigos*, or *liver spots*, are variably pigmented areas that develop on sun-exposed skin in older individuals with pale skin.

The melanin in keratinocytes protects your epidermis and dermis from the harmful effects of sunlight, which contains significant amounts of **ultraviolet (UV) radiation**. A small amount of UV radiation is beneficial, because it stimulates the epidermis to produce a compound required for calcium ion homeostasis (a process discussed in a later section). However, UV radiation can also damage DNA, causing mutations and promoting the development of cancer. Within keratinocytes, melanosomes become concentrated in the region around the nucleus, where the melanin pigments act like a sunshade to provide some UV protection for the DNA in those cells.

UV radiation can also produce some immediate effects, such as burns. If severe, sunburns can damage both the epidermis and the dermis. Thus, the pigment layers in the epidermis help protect both epidermal and dermal tissues. However, even though melanocytes respond to UV exposure by increasing their activity, the response is not rapid enough to prevent sunburn the first day you spend at the beach. Melanin synthesis accelerates slowly, peaking about 10 days after the initial exposure. Individuals of any skin color can suffer sun damage to the skin, but dark-skinned individuals have greater initial protection.

Over time, cumulative UV damage to the skin can harm fibroblasts, impairing maintenance of the dermis. The result is premature wrinkling. In addition, skin cancers can develop from chromosomal damage in basal cells or melanocytes. One of the major consequences of the global depletion of the ozone layer in Earth's upper atmosphere is likely to be a sharp increase in the rates of skin cancers (such as *malignant melanoma*). In recent years such increased cancer rates have been reported in Australia, which has already experienced a significant loss of ozone, as well as in the United States, Canada, and parts of Europe, which have had a more moderate ozone loss. For this reason, it makes sense to limit your UV exposure through wearing protective clothing and sunscreens during outdoor activities.

The Role of Dermal Circulation

Blood contains red blood cells filled with the pigment *hemoglobin*, which binds and transports oxygen in the bloodstream. When bound to oxygen, hemoglobin is bright red, giving capillaries in the dermis a reddish tint that is most apparent in lightly pigmented individuals. If those vessels are dilated, the red tones become much more pronounced. For example, your skin becomes flushed and red when your body temperature rises because the superficial blood vessels dilate so that the skin can act like a radiator and lose heat. ↺ p. 37

The skin becomes relatively pale when its blood supply is temporarily reduced. A light-skinned person who is frightened may "turn white" due to a sudden drop in blood supply to the skin. During a sustained reduction in circulatory supply, the oxygen levels in the tissues decline. Under these conditions, hemoglobin releases oxygen and turns a much darker red. Seen from the surface, the skin then takes on a bluish coloration called **cyanosis** (sī-uh-NŌ-sis; *kyanos,* blue). In individuals of any skin color, cyanosis is easiest to see in areas of very thin skin, such as the lips or beneath the nails. It can occur in response to extreme cold or as a result of cardiovascular or respiratory disorders, such as heart failure or severe asthma.

Because the skin is easy to observe, changes in the skin's appearance can be useful in diagnosing diseases that primarily affect other body systems. Several diseases can produce secondary effects on skin color and pigmentation:

- In *jaundice* (JAWN-dis), the liver is unable to excrete bile, so a yellowish pigment accumulates in body fluids. In advanced stages, the skin and whites of the eyes turn yellow.

- Some tumors affecting the pituitary gland result in the secretion of large amounts of *melanocyte-stimulating hormone (MSH)*. This hormone causes a darkening of the skin, as if the individual has an extremely deep bronze tan.

- In *Addison's disease,* the pituitary gland secretes large quantities of *adrenocorticotropic hormone (ACTH),* which is structurally similar to MSH. The effect of ACTH on skin color is similar to that of MSH.

- In **vitiligo** (vit-i-LĪ-gō), individuals lose their melanocytes, causing white patches on otherwise normal skin (**Figure 5–5**). The condition develops in about 1 percent of the population. Its incidence increases among individuals with thyroid gland disorders, Addison's disease, or several other disorders. It is suspected that vitiligo develops when the immune defenses malfunction and antibodies attack normal melanocytes. The primary problem with vitiligo is cosmetic, especially for individuals with darkly pigmented skin.

Figure 5–5 **Vitiligo.**

CHECKPOINT

6. **Name the two major pigments in the epidermis.**

7. **Why does exposure to sunlight darken skin?**

8. **Why does the skin of a fair-skinned person appear red during exercise in hot weather?**

See the blue Answers tab at the back of the book.

Clinical Note

Skin Cancer Almost everyone has several benign tumors of the skin. Moles and warts are common examples. However, **skin cancers**, which are more dangerous, are the most common form of cancer.

An *actinic keratosis* is a premalignant scaly area on the skin. It is an indication that sun damage has occurred. In a small proportion of untreated individuals, it can lead to squamous cell carcinoma. In contrast, *basal cell carcinoma*, a cancer that originates in the stratum basale, is the most common skin cancer (**Photo a**). About two-thirds of these cancers appear in body areas subjected to chronic UV exposure. Researchers have identified genetic factors that predispose people to this condition. *Squamous cell carcinomas* are less common, but almost totally restricted to areas of sun-exposed skin. Metastasis (spread to distant sites) seldom occurs in treated squamous cell carcinomas and basal cell carcinomas, and most people survive these cancers. The usual treatment involves the surgical removal of the tumor, and 95 percent of patients survive for five years or longer after treatment. (This statistic, the *5-year survival rate*, is a common method of reporting long-term outcomes.)

Unlike these common and seldom life-threatening cancers, *malignant melanomas* (mel-a-NŌ-muz) are extremely dangerous (**Photo b**). In this condition, cancerous melanocytes grow rapidly and metastasize through the lymphatic system. The outlook for long-term survival is in many cases determined by how early the condition is diagnosed. If the cancer is detected early, while it is still localized, the 5-year survival rate is 99 percent. If it is not detected until extensive metastasis has occurred, the survival rate drops to 14 percent.

To detect melanoma at an early stage, you must examine your skin, and you must know what to look for. The mnemonic ABCD makes it easy to remember this cancer's key characteristics:

- **A** is for *asymmetry:* Melanomas tend to be irregular in shape. Typically, they are raised; they may also ooze or bleed.
- **B** is for *border:* The border of a melanoma is generally irregular, and in some cases notched.
- **C** is for *color:* A melanoma is generally mottled, with any combination of tan, brown, black, red, pink, white, and blue tones.
- **D** is for *diameter:* Any skin growth more than about 5 mm (0.2 in.) in diameter, or approximately the area covered by the eraser on a pencil, is dangerous.

Fair-skinned individuals who live in the tropics are most susceptible to all forms of skin cancer, because their melanocytes are unable to shield them from UV radiation.

Prevent sun damage by avoiding exposure to the sun during the middle of the day and by using a sunscreen (not a tanning oil) before any sun exposure. This practice also delays the cosmetic problems of aging and wrinkling. *Everyone* who spends any time out in the sun should choose a broad-spectrum sunscreen with a sun protection factor (SPF) of at least 15. Blondes, redheads, and people with very pale skin are better off with an SPF of 20 to 30. (The risks are the same for those who spend time in a tanning salon or tanning bed.) The protection offered by these "sunscreens" comes both from organic molecules that absorb UV radiation and from inorganic pigments that absorb, scatter, and reflect UV rays. The higher the SPF, the more of these chemicals the product contains, and the fewer UV rays are able to penetrate to the skin's surface. Wearing a hat with a brim and panels to shield the neck and face, long pants, and long-sleeved shirts provides added protection.

The use of sunscreens will be even more important as the ozone gas in the upper atmosphere is further destroyed by our industrial emissions. Ozone absorbs UV radiation before it reaches Earth's surface. In doing so, ozone assists the melanocytes in preventing skin cancer. Australia, the continent that is most affected by the depletion of ozone above the South Pole (the "ozone hole"), is already reporting an increased incidence of skin cancers.

a **Basal cell carcinoma**

b **Melanoma**

5-3 ▌ Sunlight causes epidermal cells to convert a steroid into vitamin D₃

Learning Outcome Describe the interaction between sunlight and vitamin D₃ production.

Too much sunlight can damage epithelial cells and deeper tissues, but limited exposure to sunlight is beneficial. When exposed to ultraviolet radiation, epidermal cells in the stratum spinosum and stratum basale convert a cholesterol-related steroid into **cholecalciferol** (kō-le-kal-SIF-er-ol), or **vitamin D₃**. The liver then converts cholecalciferol into an intermediary product used by the kidneys to synthesize the hormone **calcitriol** (**Figure 5-6**) (kal-si-TRĪ-ol). Calcitriol is essential for the normal absorption of calcium and phosphorus in the

Figure 5–6   **Sources of Vitamin D₃.**

Figure 5–7   **Rickets.** Rickets, a disease caused by vitamin D₃ deficiency, results in the bending of abnormally weak and flexible bones under the weight of the body, plus other structural changes.

CHECKPOINT

9. Explain the relationship between sunlight exposure and vitamin D₃ synthesis.

10. In some cultures, women must be covered completely, except for their eyes, when they go out in public. Explain why these women may develop bone problems later in life.

See the blue Answers tab at the back of the book.

small intestine. An inadequate supply leads to impaired bone maintenance and growth.

The term *vitamin* is usually reserved for essential organic nutrients that we must get from the diet because the body either cannot make them or makes them in insufficient amounts. Cholecalciferol can be absorbed by the digestive tract, and if the skin cannot make enough cholecalciferol, a dietary supply (from foods such as egg yolks, fish oil, and fortified milk) will maintain normal bone development. Under these circumstances, dietary cholecalciferol acts like a vitamin. For this reason, cholecalciferol is also called *vitamin D₃*.

If cholecalciferol cannot be produced by the skin and is not included in the diet, bone development is abnormal and bone maintenance is inadequate. For example, children who live in areas with overcast skies and whose diet lacks cholecalciferol can have abnormal bone development. This condition, called **rickets** (**Figure 5–7**), has largely been eliminated in the United States because dairy companies are required to add cholecalciferol, usually identified as "vitamin D," to the milk sold in grocery stores. In Chapter 6, we will consider the hormonal control of bone growth in greater detail.

5-4 | Epidermal growth factor has several effects on the epidermis and epithelia

Learning Outcome   Describe the roles of epidermal growth factor.

Epidermal growth factor (EGF) is one of the peptide growth factors introduced in Chapter 3. ⊃ p. 129 It was named for its effects on the epidermis, but we now know that EGF has widespread effects on epithelia throughout the body. EGF is produced by the salivary glands and glands of the duodenum (initial segment of the small intestine). Here are some roles of EGF:

- Promoting the divisions of basal cells in the stratum basale and stratum spinosum

- Accelerating the production of keratin in differentiating keratinocytes

- Stimulating epidermal development and epidermal repair after injury

- Stimulating synthetic activity and secretion by epithelial glands

In the procedure known as tissue culture, cells are grown under laboratory conditions for experimental or therapeutic use. Epidermal growth factor has such a pronounced effect that it can be used in tissue culture to stimulate the growth and division of epidermal cells (or other epithelial cells). Using human epidermal growth factor (hEGF) produced through recombinant DNA technology, it is now possible to grow sheets of epidermal cells for use in treating severe or extensive burns. The burned areas can be covered by epidermal sheets "grown" from a small sample of intact skin from another part of the burn victim's body. (We will consider this treatment later in the chapter when we discuss burns.)

> **CHECKPOINT**
>
> 11. **Name the sources of epidermal growth factor in the body.**
>
> 12. **Identify some roles of epidermal growth factor pertaining to the epidermis.**
>
> See the blue Answers tab at the back of the book.

5-5 ▎ The dermis is the tissue layer that supports the epidermis

Learning Outcome Describe the structure and functions of the dermis.

The dermis lies between the epidermis and the hypodermis (**Figure 5–1**). The dermis has two major components: (1) a superficial *papillary layer* and (2) a deeper *reticular layer.*

The **papillary layer** consists of areolar tissue. It contains the capillaries, lymphatic vessels, and sensory neurons that supply the surface of the skin. The papillary layer gets its name from the dermal papillae that project between the epidermal ridges (**Spotlight Figure 5–3**).

The **reticular layer** lies deep to the papillary layer. It consists of an interwoven meshwork of dense irregular connective tissue containing both collagen and elastic fibers (**Figure 5–8**, p. 157). Bundles of collagen fibers extend superficially beyond the reticular layer to blend into those of the papillary layer, so the boundary between the two layers is indistinct. Collagen fibers of the reticular layer also extend into the deeper hypodermis. In addition to extracellular protein fibers, the dermis contains all the cells of connective tissue proper. ⊃ p. 153

Accessory structures of epidermal origin, such as hair follicles and sweat glands, extend into the dermis. In addition, the reticular and papillary layers of the dermis contain networks of blood vessels and nerve fibers (**Figure 5–1**).

Because of the abundance of sensory receptors in the skin, regional infection or inflammation can be very painful. **Dermatitis** (der-muh-TĪ-tis) is an inflammation of the skin that primarily involves the papillary layer. The inflammation typically begins in a part of the skin exposed to infection or irritated by chemicals, radiation, or mechanical stimuli. Dermatitis may cause no discomfort, or it may produce an annoying itch, as in poison ivy. Other forms of the condition can be quite painful, and the inflammation can spread rapidly across the entire integument.

Dermal Strength and Elasticity

Collagen and elastic fibers give the dermis strength and elasticity. *Collagen fibers* are very strong and resist stretching, but they are easily bent or twisted. *Elastic fibers* permit stretching and then recoil to their original length. The elastic fibers provide flexibility, and the collagen fibers limit that flexibility to prevent damage to the tissue.

The water content of the skin also helps maintain its flexibility and resilience. These properties are collectively known as *skin turgor*. One of the signs of dehydration is the loss of skin turgor, revealed by pinching the skin on the back of the hand. A dehydrated dermis will remain peaked when pinched, but hydrated skin will flatten out.

Aging, hormones, and the destructive effects of ultraviolet radiation permanently reduce the amount of elastin in the dermis. Wrinkles and sagging skin result. The extensive distortion of the dermis that occurs over the abdomen during pregnancy or after substantial weight gain can exceed the elastic limits of the skin. The resulting damage to the dermis prevents it from recoiling to its original size after delivery or weight loss. The skin then wrinkles and creases, creating a network of **stretch marks**.

Tretinoin (Retin-A) is a derivative of vitamin A that can be applied to the skin as a cream or gel. This drug was originally developed to treat acne, but it also increases blood flow to the dermis and stimulates dermal repair. As a result, the rate of wrinkle formation decreases, and existing wrinkles become smaller. The degree of improvement varies among individuals.

Figure 5–8   Reticular Layer of Dermis.

Reticular layer of dermis   SEM × 1500

Cleavage Lines

Most of the collagen and elastic fibers at any location are arranged in parallel bundles oriented to resist the forces applied to the skin during normal movement. The resulting pattern of fiber bundles in the skin establishes **cleavage lines**. Cleavage lines are also called *tension lines* and *Langer lines.* Cleavage lines are clinically significant. A cut parallel to a cleavage line will usually remain closed and heal with little scarring. A cut at right angles to a cleavage line will be pulled open as severed elastic fibers recoil, resulting in greater scarring. For these reasons, surgeons choose to make incisions parallel to cleavage lines (**Figure 5–9**).

The Dermal Blood Supply

Arteries supplying the skin lie deep within the hypodermis. Branches of these arteries form two networks, or *plexuses,* in

Figure 5–9   Cleavage Lines of the Skin. Cleavage lines follow the pattern of collagen fiber bundles in the dermis of the skin.

| ANTERIOR | POSTERIOR |

Clinical Note

Decubitis Ulcers Problems with dermal circulation affect both the epidermis and the dermis. An *ulcer* is a localized shedding of an epithelium. *Decubitis ulcers*, or *bedsores*, affect patients whose circulation is restricted, especially when a splint, a cast, or lying in bed continuously compresses superficial blood vessels. Such sores most commonly affect the skin covering joints or bony prominences, where dermal blood vessels are pressed against deeper structures. The chronic lack of circulation kills epidermal cells, removing a barrier to bacterial infection. Eventually, dermal tissues deteriorate as well. (Cell death and tissue destruction, or *necrosis*, can occur in any tissue deprived of adequate blood flow.) Bedsores can be prevented or treated by frequently changing the position of the body or by placing patients in specially designed beds containing deflating and inflating air coils. Both approaches vary the pressures applied to local blood vessels.

the dermis. The deeper network lies along the border of the hypodermis with the reticular layer of the dermis. This network is called the *cutaneous plexus* (**Figure 5–10**). Tributaries of these arteries supply both the adipose tissues of the subcutaneous layer and the tissues of the integument. As small arteries travel toward the epidermis, branches supply the hair follicles, sweat glands, and other structures in the dermis.

On reaching the papillary layer, the small arteries form another branching network, the *subpapillary plexus* (**Figure 5–10**). It provides arterial blood to capillary loops that follow the contours of the epidermis–dermis boundary. These capillaries empty into small veins of the subpapillary plexus, which drain into veins accompanying the arteries of the cutaneous plexus. This network in turn connects to larger veins in the hypodermis.

Trauma to the skin often results in a *contusion,* or bruise. When dermal blood vessels rupture, blood leaks into the dermis, and the area develops the familiar "black and blue" color.

Innervation of the Skin

The integument is filled with sensory receptors. Anything that comes in contact with the skin—from the lightest touch of a

Figure 5–10 **Dermal Circulation.** Shown are the cutaneous and subpapillary plexuses.

mosquito to the weight of a loaded backpack—initiates a nerve impulse that can reach our conscious awareness. Nerve fibers in the skin control blood flow, adjust gland secretion rates, and monitor sensory receptors in the dermis and the deeper layers of the epidermis.

We have already noted that the deeper layers of the epidermis contain tactile discs. They are fine touch and pressure receptors. The epidermis also contains the extensions of sensory neurons that provide sensations of pain and temperature. The dermis contains similar receptors, as well as other, more specialized receptors. Examples shown in **Figure 5–1** include receptors sensitive to light touch—*tactile (Meissner's) corpuscles,* located in dermal papillae—and receptors sensitive to deep pressure and vibration—*lamellated (pacinian) corpuscles,* in the reticular layer.

Even this partial list of the receptors in the skin is enough to highlight the importance of the integument as a sensory structure. We will return to this topic in Chapter 15, where we consider not only what receptors are present, but how they function.

CHECKPOINT

13. Describe the location of the dermis.

14. Where are the capillaries and sensory neurons that supply the epidermis located?

15. What accounts for the ability of the dermis to undergo repeated stretching?

See the blue Answers tab at the back of the book.

5-6 ∎ The hypodermis connects the dermis to underlying tissues

Learning Outcome Describe the structure and functions of the hypodermis.

The **hypodermis** or *subcutaneous layer* lies deep to the dermis (**Figure 5–1**). The boundary between the two is generally indistinct because the connective tissue fibers of the reticular layer are extensively interwoven with those of the hypodermis. The hypodermis is not a part of the integument, but it is important in stabilizing the position of the skin in relation to underlying tissues, such as skeletal muscles or other organs, while permitting independent movement.

The hypodermis is quite elastic. It consists of areolar tissue and adipose tissue (also known as subcutaneous fat) (**Figure 5–11**). Subcutaneous fat makes up about 80 percent of all body fat. The remaining amount is visceral fat that is associated with visceral organs. Only its superficial region contains large arteries and veins. The venous circulation of this region contains a substantial amount of blood, and much of this volume will shift to the general circulation if these veins constrict. For that reason, we often describe the skin as a *blood reservoir.* The rest of the hypodermis contains a limited number of capillaries and no vital organs. This last characteristic makes **subcutaneous injection**—by means of a **hypodermic needle**—a useful method of administering drugs.

Most infants and small children have extensive "baby fat," which provides insulation and helps reduce heat loss. This subcutaneous fat also serves as a substantial energy reserve and as a shock absorber for the rough-and-tumble activities of our early years.

As we grow, the distribution of subcutaneous fat changes. The greatest changes take place in response to circulating

Figure 5–11 **Hypodermis.**

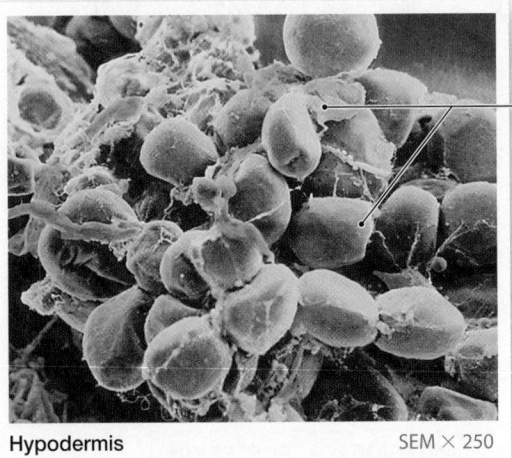

Hypodermis   SEM × 250

Clinical Note

Liposuction The accumulation of excessive amounts of adipose tissue increases a person's risks of diabetes, stroke, and other serious conditions. Dietary restrictions and increased activity levels often promote weight loss and reduce these risks. However, a surgical procedure called **liposuction** or **lipoplasty** often promises a "quick fix." In this procedure, subcutaneous adipose tissue is removed through a tube inserted deep to the skin. Adipose tissue tears relatively easily, and suction applied to the tube rips chunks of adipose tissue from the body. After liposuction, the skin is loose fitting. Until it recoils, a tight-fitting garment is usually worn. Liposuction is relatively common and is increasing in popularity. According to the American Society of Plastic Surgeons, 325,332 liposuction procedures were performed in 2011 (the most recent year for which data are available).

Liposuction might sound like an easy way to remove unwanted fat, but in practice, it can be dangerous. There are risks from anesthesia, bleeding (adipose tissue is quite vascular), sensory loss, infection, and fluid loss. The death rate from liposuction procedures is 1 in 5000. That rate is very high for what is basically cosmetic surgery that provides only a temporary solution to a chronic problem. As noted in Chapter 4, unless there are changes in diet and lifestyle, the damaged adipose tissue will repair itself, and areas of areolar tissue will convert to adipose tissue. Over time, the surgery will have to be repeated.

sex hormones. Beginning at puberty, men accumulate subcutaneous fat at the neck, on the arms, along the lower back, and over the buttocks. In contrast, women accumulate subcutaneous fat at the breasts, buttocks, hips, and thighs. In adults of either gender, the subcutaneous layer of the backs of the hands and the upper surfaces of the feet contains few fat cells, but distressing amounts of adipose tissue can accumulate in the abdominal region, producing a "pot-belly." An excessive amount of such abdominal fat, also called *central adiposity,* is strongly correlated with cardiovascular disease.

CHECKPOINT

16. List the two terms for the tissue that connects the dermis to underlying tissues.

17. Describe the hypodermis.

18. Identify several functions of subcutaneous fat.

See the blue Answers tab at the back of the book.

5-7 ❚ Hair is composed of keratinized dead cells that have been pushed to the surface

Learning Outcome Describe the mechanisms that produce hair, and explain the structural basis for hair texture and color.

Hair and several other structures—hair follicles, sebaceous and sweat glands, and nails—are considered accessory structures of the integument. During embryological development, these structures originate from the epidermis, so they are also known as *epidermal derivatives.* Although located in the dermis, they project to the surface of the skin. ATLAS: Embryology Summary 5: The Development of the Integumentary System

Hairs project above the surface of the skin almost everywhere, except over the sides and soles of the feet, the palms of the hands, the sides of the fingers and toes, the lips, and portions of the external genitalia. The human body has about 2.5 million hairs, and 75 percent of them are on the general body surface, not on the head. Hairs are nonliving structures produced in organs called **hair follicles**.

The hairs and hair follicles on your body have important functions. The 500,000 or so hairs on your head protect your scalp from ultraviolet radiation, help cushion light impacts to the head, and insulate the skull. The hairs guarding the entrances to your nostrils and external ear canals help keep out foreign particles and insects. Your eyelashes do the same for the surface of the eye. Eyebrows are important because they help keep sweat out of your eyes. Hairs are also extremely important as sensory receptors.

Figure 5–12 illustrates important details about the structure of hairs and hair follicles. Each hair follicle opens onto the surface of the epidermis but extends deep into the dermis and usually into the hypodermis. Deep to the epidermis, each follicle is wrapped in a dense connective tissue sheath. A **root hair plexus** of sensory nerves surrounds the base of each hair follicle (**Figure 5–12a**). As a result, you can feel the movement of the shaft of even a single hair. This sensitivity provides an early-warning system that may help prevent injury. For example, you may be able to swat a mosquito before it reaches your skin.

A bundle of smooth muscle cells forms the **arrector pili** (a-REK-tor PI-lē) muscle. It extends from the papillary layer of the dermis to the connective tissue sheath surrounding the hair follicle. When stimulated, the arrector pili muscle contracts, pulling on the follicle and forcing the hair to stand erect. Contraction may be the result of emotional states, such as fear or rage, or a response to cold, producing "goose bumps." In a furry mammal, this action increases the thickness of its insulating coat. Humans do not receive any comparable insulating benefits, but the response persists.

Figure 5–12 **Hair Follicles and Hairs.**

Hair

Sebaceous gland

Arrector pili muscle

Connective tissue sheath

Root hair plexus

a Hair follicles, showing the associated accessory structures

Hair Structure

The medulla, or core, of the hair contains a flexible **soft keratin**.

The cortex contains thick layers of **hard keratin**, which give the hair its stiffness.

The cuticle, although thin, is very tough, and it contains hard keratin.

Follicle Structure

The **internal root sheath** surrounds the hair root and the deeper portion of the shaft. The cells of this sheath disintegrate quickly, and this layer does not extend the entire length of the hair follicle.

The **external root sheath** extends from the skin surface to the hair matrix.

The **glassy membrane** is a thickened, clear layer wrapped in the dense connective tissue sheath of the follicle as a whole.

Connective tissue sheath

b Cross section through a hair follicle and a hair, near the junction between the hair root and hair shaft

Hair shaft

External root sheath

Connective tissue sheath of hair follicle

Internal root sheath

Glassy membrane

Cuticle of hair

Cortex of hair

Medulla of hair

Matrix

Hair papilla

Subcutaneous adipose tissue

Hair follicle

LM × 60

c Histological section along the longitudinal axis of hair follicles

Connective tissue sheath of hair follicle

Glassy membrane

External root sheath

Internal root sheath

Cuticle of hair

Cortex of hair

Medulla of hair

Matrix

Hair papilla

Hair bulb

Subcutaneous adipose tissue

d Diagrammatic view of the base of a hair follicle

Each hair is a long, cylindrical structure that extends outward, past the epidermal surface (Figure 5–12a,c,d). The **hair root** is the portion that anchors the hair into the skin. The root begins at the base of the hair, at the *hair bulb*, and extends distally to the point at which the internal organization of the hair is complete, about halfway to the skin surface. The **hair shaft** extends from this halfway point to the exposed tip of the hair. The hair we see is the protruding part of the hair shaft.

Hair Production

Hair production begins at the bulging base of a hair follicle, a region called the **hair bulb** (Figure 5–12c,d). The hair bulb surrounds a small **hair papilla**, a peg of connective tissue containing capillaries and nerves. A layer of epithelial cells at the base of the hair bulb in contact with the hair papilla produces the hair. This layer is called the **hair matrix**. Basal cells of the hair matrix divide, producing daughter cells that are gradually pushed toward the surface. Daughter cells closest to the center of the matrix form the **medulla**, or core, of the hair (Figure 5–12b,c,d). Daughter cells farther from the center of the hair matrix form the **cortex**, an intermediate layer. Those at the edges of the hair matrix form the **cuticle**, which will be the surface of the hair.

As cell divisions continue at the hair matrix, the daughter cells are pushed toward the surface of the skin, and the hair gets longer. Keratinization is completed by the time these cells approach the surface. At the level that corresponds to the start of the hair shaft, the cells of the medulla, cortex, and cuticle are dead, and the keratinization process is at an end.

The epithelial cells of the follicle walls are organized into several concentric layers. Moving outward from the hair cuticle, these layers include the internal root sheath, the external root sheath, and the glassy membrane (Figure 5–12b,c,d).

The Hair Growth Cycle

Hairs grow and are shed according to a **hair growth cycle**. A hair in the scalp grows for two to five years, at a rate of about 0.33 mm per day. Variations in the growth rate and in the duration of the hair growth cycle account for individual differences in the length of uncut hair.

While hair is growing, the cells of the hair root absorb nutrients and incorporate them into the hair structure. As a result, clipping or collecting hair for analysis can be helpful in diagnosing several disorders. For example, hairs of individuals with lead poisoning or other heavy-metal poisoning contain high levels of those metal ions. Hair samples containing nucleated cells can also be used for identification purposes through DNA fingerprinting. ⤺ p. 110

As a hair grows, the root is firmly attached to the matrix of the follicle. At the end of the growth cycle, the follicle becomes inactive. The hair is now termed a **club hair**. The follicle gets smaller, and over time the connections between the hair matrix and the club hair root break down. When another cycle begins, the follicle produces a new hair. The old club hair is pushed to the surface and is shed.

Healthy adults with a full head of hair typically lose about 100 head hairs each day. Sustained losses of more than 100 hairs per day generally indicate that a net loss of hairs is underway, and hair loss will eventually be noticeable. Temporary increases in hair loss can result from drugs, dietary factors, radiation, an excess of vitamin A, high fever, stress, or hormonal factors related to pregnancy.

In males, changes in the level of the sex hormones circulating in the blood can affect the scalp, causing a shift in the type of hair produced (discussed shortly), beginning at the temples and the crown of the head. This alteration is called *male pattern baldness*. Some cases of male pattern baldness respond to drug therapies, such as the topical application of *minoxidil (Rogaine)*.

Types of Hairs

Hairs first appear after about three months of embryonic development. These hairs, collectively known as *lanugo* (la-NŪ-gō), are extremely fine and unpigmented. Most lanugo hairs are shed before birth. They are replaced by one of two types of hairs in the adult integument: vellus hairs or terminal hairs. **Vellus hairs** are the fine "peach fuzz" hairs located over much of the body surface. **Terminal hairs** are heavy, more deeply pigmented, and sometimes curly. The hairs on your head, including your eyebrows and eyelashes, are terminal hairs that are present throughout life. Hair follicles may alter the structure of the hairs in response to circulating hormones. For example, vellus hairs are present at the armpits, pubic area, and limbs until puberty. After puberty, the follicles produce terminal hairs, in response to circulating sex hormones.

What makes hair curly or straight? The cross-sectional shape of the hair shaft and its hair follicle determine whether hairs are curly or straight. A curly hair and its follicle are oval; a straight hair and its follicle are round.

TIPS & TRICKS

Associate the word *vellus* ("peach fuzz") with "velvet."

Hair Color

What gives your hair its color? Variations in hair color reflect differences in structure and variations in the pigment produced by melanocytes at the hair matrix. Different forms and amounts of melanin give a dark brown, yellow-brown, or red color to the hair. Genes determine these structural and biochemical characteristics, but hormonal and environmental factors also influence the condition of your hair.

5

As pigment production decreases with age, hair color lightens. White hair results from the combination of a lack of pigment and the presence of air bubbles in the medulla of the hair shaft. As the proportion of white hairs increases, the person's overall hair color is described as gray. Because hair itself is dead and inert, any changes in its coloration are gradual.

What happens when we dye our hair? We are able to change our hair color by using chemicals that disrupt the cuticle and permit dyes to enter and stain the cortex and medulla. These color treatments damage the hair's protective cuticle layer and dehydrate and weaken the hair shaft. As a result, the hair becomes thin and brittle. Conditioners and oil treatments attempt to reduce the effects of this structural damage by rehydrating and recoating the shaft.

CHECKPOINT

19. **Describe a typical strand of hair.**

20. **What happens when the arrector pili muscle contracts?**

21. **Once a burn on the forearm that destroys the epidermis and extensive areas of the deep dermis heals, will hair grow again in the affected area?**

See the blue Answers tab at the back of the book.

5-8 | Sebaceous glands and sweat glands are exocrine glands found in the skin

Learning Outcome Discuss the various kinds of glands in the skin, and list the secretions of those glands.

As we have seen, the integumentary system contains various accessory structures that produce exocrine secretions. In this section we will focus on the structure and function of sebaceous glands and sweat glands.

Sebaceous Glands

Sebaceous (se-BĀ-shus) **glands**, or *oil glands*, are holocrine glands that discharge an oily lipid secretion into hair follicles (**Figure 5–13**). The gland cells produce large quantities of lipids as they mature. They release the lipid product through holocrine secretion, a process that involves the rupture of the secretory cells. ⮌ p. 149 Sebaceous glands that communicate with a single hair follicle share a duct and thus are classified as simple branched alveolar glands. ⮌ p. 151

The lipids released from gland cells enter the lumen (open passageway) of the gland. The arrector pili muscles that erect the hair then contract, squeezing the sebaceous gland and forcing the lipids into the hair follicle and onto the surface of the skin. The secreted lipid product, called **sebum** (SĒ-bum), is a mixture of triglycerides, cholesterol, proteins, and electrolytes. Sebum inhibits the growth of bacteria, lubricates and protects the keratin of the hair shaft, and conditions the surrounding skin. Keratin is a tough protein, but dead, keratinized cells become dry and brittle once exposed to the environment. Interestingly, when we shampoo and wash, we remove the oily secretions of sebaceous glands, only to add other lipids to our hair in the form of conditioners, and to our skin in the form of creams and lotions.

Sebaceous follicles are large sebaceous glands that are not associated with hair follicles. Instead, their ducts discharge sebum directly onto the epidermis (**Figure 5–13**). We have sebaceous follicles on the face, back, chest, nipples, and external genitalia.

Surprisingly, sebaceous glands are very active during the last few months of fetal development. Their secretions, mixed with shed epidermal cells, form a protective superficial layer—the *vernix caseosa*—that coats the skin surface. Sebaceous gland activity all but stops after birth, but it increases again at puberty in response to rising levels of sex hormones.

Figure 5–13 **The Structure of Sebaceous Glands and Sebaceous Follicles.**

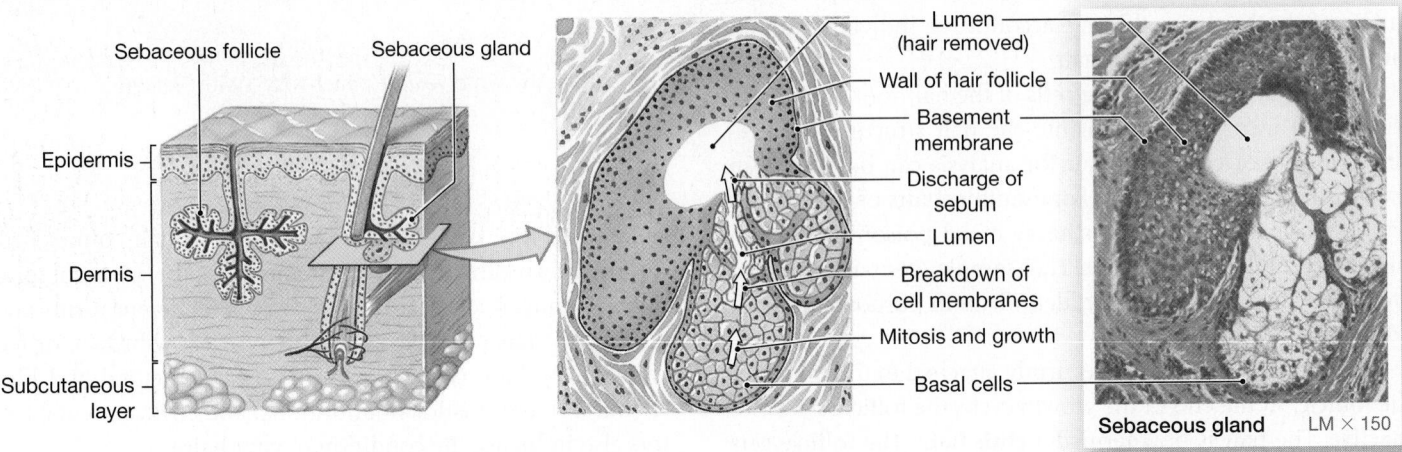

Seborrheic dermatitis is an inflammation around abnormally active sebaceous glands, most often in the scalp. The affected area becomes red and oily, with increased epidermal scaling. In infants, mild cases are called *cradle cap*. Seborrheic dermatitis is a common cause of dandruff in adults. Anxiety, stress, and fungal or bacterial infections can aggravate the problem.

Sweat Glands

The skin contains two types of sweat glands, or **sudoriferous glands** (*sudor*, sweat): *apocrine sweat glands* and *merocrine sweat glands* (Figure 5–14).

Apocrine Sweat Glands

In the armpits (axillae), around the nipples, and in the pubic region, **apocrine sweat glands** secrete their products into hair follicles (Figure 5–14a). These coiled, tubular glands produce a sticky, cloudy, and potentially odorous secretion. The name *apocrine* was originally chosen because it was thought the gland cells used an apocrine method of secretion. ⟲ p. 149 We now know that they rely on merocrine secretion, but the name has not changed.

Apocrine sweat glands begin secreting at puberty. This sweat is a nutrient source for bacteria, whose growth and breakdown intensify its odor. Surrounding the secretory cells in these glands are special **myoepithelial cells** that contract and squeeze the gland, causing the accumulated sweat to discharge into the hair follicles. The secretory activities of the gland cells and the contractions of myoepithelial cells are controlled by the nervous system and by circulating hormones.

Merocrine Sweat Glands

Merocrine sweat glands are also known as **eccrine** (EK-rin) **sweat glands.** These coiled, tubular glands discharge their secretions directly onto the surface of the skin (Figure 5–14b). Merocrine sweat glands are far more numerous and widely distributed than apocrine sweat glands. The adult integument contains 2–5 million merocrine sweat glands, which are smaller than apocrine sweat glands and do not extend as deeply into the dermis. The palms and soles have the highest numbers. The palms have an estimated 500 merocrine sweat glands per square centimeter (3000 per square inch).

As noted earlier, the sweat from merocrine sweat glands is called sensible perspiration. Sweat is 99 percent water, but it also contains some electrolytes (chiefly sodium chloride), a number of organic nutrients, a peptide with antibiotic properties, and various waste products. It has a pH range of 4.0–6.8. Sodium chloride gives sweat a salty taste. (See the

Figure 5–14 Sweat Glands.

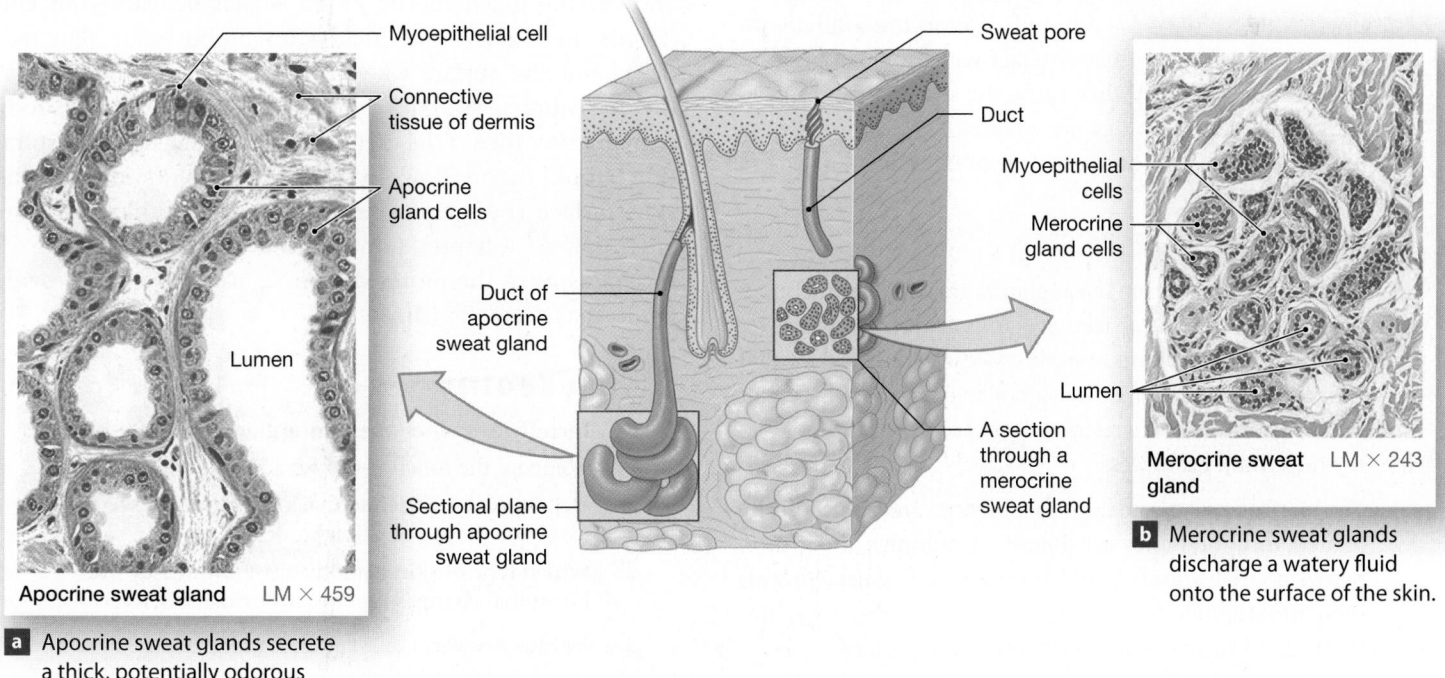

Myoepithelial cell
Connective tissue of dermis
Apocrine gland cells
Lumen
Duct of apocrine sweat gland
Sectional plane through apocrine sweat gland

Apocrine sweat gland LM × 459

a Apocrine sweat glands secrete a thick, potentially odorous fluid into hair follicles.

Sweat pore
Duct
A section through a merocrine sweat gland

Myoepithelial cells
Merocrine gland cells
Lumen

Merocrine sweat gland LM × 243

b Merocrine sweat glands discharge a watery fluid onto the surface of the skin.

Appendix for a complete analysis of the composition of normal sweat.)

The functions of merocrine sweat gland activity include:

- *Cooling the Surface of the Skin to Reduce Body Temperature.* This is the primary function of sensible perspiration (p. 181). The degree of secretory activity is regulated by neural and hormonal mechanisms. When all the merocrine sweat glands are working at their maximum, we can perspire more than a gallon per hour, and dangerous fluid and electrolyte losses can occur. For this reason, athletes in endurance sports must drink fluids at regular intervals.

- *Excreting Water and Electrolytes.* Water, salts (mostly sodium chloride), and a number of metabolized drugs are excreted.

- *Providing Protection from Environmental Hazards.* Sweat dilutes harmful chemicals in contact with the skin. It also discourages the growth of microorganisms in two ways: (1) by either flushing them from the surface or making it difficult for them to adhere to the epidermal surface and (2) through the action of *dermicidin*, a small peptide with powerful antibiotic properties.

The underarms, or armpits, provide consistent heat and moisture, ideal conditions for bacterial growth on the body surface. Deodorants and antiperspirants, singly or in combination, are commonly used to combat underarm bacteria and odor. Deodorants work by temporarily reducing bacteria numbers and by masking the body odor they produce. Antiperspirants target the sweat glands. Antiperspirants contain aluminum compounds that reduce the amount of moisture available to bacteria in two ways: (1) they interact with sweat to plug up sweat gland ducts, and (2) they cause the sweat gland pores to constrict. However, the plugs are eventually lost through the normal shedding of cells at the surface of the skin.

Other Integumentary Glands

As we have seen, merocrine sweat glands are widely distributed across the body surface, sebaceous glands are located wherever there are hair follicles, and apocrine sweat glands are located in relatively restricted areas. The skin also contains a variety of specialized glands that are restricted to specific locations. Here are two important examples:

1. The **mammary glands** of the breasts are anatomically related to apocrine sweat glands. A complex interaction between sex hormones and pituitary hormones controls their development and secretion. We will discuss mammary gland structure and function in Chapter 28.

2. **Ceruminous** (se-RŪ-mi-nus) **glands** are modified sweat glands in the passageway of the external ear. Their secretions combine with those of nearby sebaceous glands, forming a mixture called **cerumen**, or earwax. Together with tiny hairs along the ear canal, earwax helps trap foreign particles, preventing them from reaching the eardrum.

Control of Glandular Secretions and the Homeostatic Role of the Integument

The autonomic nervous system (ANS) is the portion of the nervous system that adjusts our basic life support systems without our conscious control (Chapter 16). It controls the activation and deactivation of sebaceous glands and apocrine sweat glands at the subconscious level. Regional control is not possible. The commands from the ANS affect all the glands of that type, everywhere on the body surface.

Merocrine sweat glands are much more precisely controlled, and the amount of secretion and the area of the body involved can vary independently. For example, when you are nervously awaiting an anatomy and physiology exam, only your palms may begin to sweat.

As we noted earlier, the primary function of sensible perspiration is to cool the surface of the skin and to reduce body temperature. When the environmental temperature is high, this is a key component of *thermoregulation*, the process of maintaining temperature homeostasis. When you sweat in the hot sun, all your merocrine glands are working together. The blood vessels beneath your epidermis are dilated and filled with blood, your skin reddens, and the surface of your skin is warm and wet. As the moisture evaporates, your skin cools. If your body temperature then falls below normal, sensible perspiration stops, blood flow to the skin is reduced, and the skin surface cools and dries, releasing little heat into the environment. Chapter 1 introduced the negative feedback mechanisms of thermoregulation. ⤴ p. 36 We will provide additional details in Chapter 25.

CHECKPOINT

22. Identify two types of exocrine glands found in the skin.

23. What are the functions of sebaceous secretions?

24. Deodorants are used to mask the effects of secretions from which type of skin gland?

25. Which type of skin gland is most affected by the hormonal changes that occur during puberty?

See the blue Answers tab at the back of the book.

5-9 | Nails are keratinized epidermal cells that protect the tips of fingers and toes

Learning Outcome Describe the anatomical structure of nails, and explain how they are formed.

Nails protect the exposed dorsal surfaces of the tips of the fingers and toes (**Figure 5–15a**). They also help limit distortion of the digits from mechanical stress—for example, when you run or grasp objects. The **nail body**, the visible portion of the nail, covers an area of epidermis called the **nail bed** (**Figure 5–15b**). The nail body is recessed deep to the level of the surrounding epithelium and is bordered on either side by **lateral nail grooves** (depressions) and **lateral nail folds**. The **free edge** of the nail—the distal portion that continues past the nail bed—extends over the **hyponychium** (hī-pō-NIK-ē-um), an area of thickened stratum corneum (**Figure 5–15c**).

Nail production takes place at the **nail root**, an epidermal fold not visible from the surface. The deepest portion of the nail root lies very close to the bone of the fingertip. A portion of the stratum corneum of the nail root extends over the exposed nail, forming the **eponychium** (ep-ō-NIK-ē-um; *epi-*, over + *onyx*, nail), or **cuticle**. Underlying blood vessels give the nail its characteristic pink color. Near the root, these vessels may be obscured, leaving a pale crescent known as the **lunula** (LŪ-nū-la; *luna*, moon) (**Figure 5–15a**).

The body of the nail consists of dead, tightly compressed cells packed with keratin. The cells producing the nails can be affected by conditions that alter body metabolism, so changes in the shape, structure, or appearance of the nails can provide useful diagnostic information. For example, the nails may turn yellow in individuals who have chronic respiratory disorders, thyroid gland disorders, or AIDS. Nails may become pitted and distorted as a result of *psoriasis* (a condition marked by rapid stem cell division in the stratum basale), and concave as a result of some blood disorders.

CHECKPOINT

26. What substance makes fingernails hard?

27. What term is used to describe the thickened stratum corneum underlying the free edge of a nail?

28. Where does nail growth occur?

See the blue Answers tab at the back of the book.

Figure 5–15 **The Structure of a Nail.**

a A superficial view

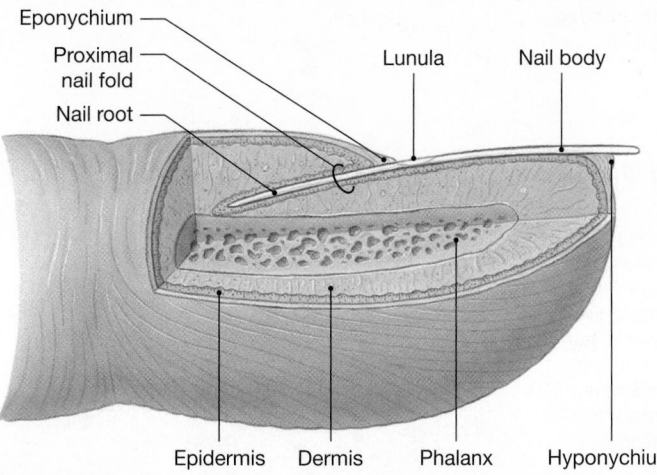

c A longitudinal section

5-10 | Several phases are involved in repairing the integument following an injury

Learning Outcome Explain how the skin responds to injury and repairs itself.

The integumentary system can function independently. It often responds directly and automatically to local influences without the involvement of the nervous or endocrine systems. For example, when the skin is continually subjected to mechanical stresses, stem cells in the stratum basale divide more rapidly, and the depth of the epithelium increases. That is why calluses form on your palms when you perform work with your hands.

A more dramatic display of local regulation can be seen after an injury to the skin. The skin can regenerate effectively, even after considerable damage, because stem cells persist in both the epithelial and connective tissue. Basal cell divisions replace lost epidermal cells, and mesenchymal cell divisions replace lost dermal cells.

The process can be slow. When large surface areas are involved, infection and fluid loss complicate the situation. The speed and effectiveness of skin repair vary with the type of wound. A slender, straight cut, or *incision*, may heal fairly quickly compared with a deep scrape, or *abrasion*, which involves a much greater surface area to be repaired.

Figure 5–16 illustrates the four phases in the regeneration of the skin after an injury:

When damage extends through the epidermis and into the dermis, bleeding, along with swelling, redness, and pain occur. This is the *inflammatory phase* ❶.

A blood clot, or **scab**, forms at the surface as part of the *migratory phase* ❷. The scab temporarily restores the integrity of the epidermis and restricts the entry of additional microorganisms into the area.

The bulk of the clot consists of an insoluble network of *fibrin*, a fibrous protein that forms from blood proteins during the clotting response. The clot's color is due to trapped red blood cells.

If the wound occupies an extensive area or involves a region covered by thin skin, dermal repairs must be underway before epithelial cells can cover the surface. Divisions by fibroblasts and mesenchymal cells produce mobile cells that invade the deeper areas of injury. Endothelial cells of damaged blood vessels also begin to divide, and capillaries follow the fibroblasts, enhancing circulation. The combination of blood clot, fibroblasts, and an extensive capillary network at the base of a wound is called **granulation tissue**.

Over time, deeper portions of the clot dissolve, and the number of capillaries declines in the *proliferation phase* ❸. The repairs do not restore the integument to its original condition, however, because the dermis will contain an abnormally large number of collagen fibers and a few blood vessels. Severely damaged hair follicles, sebaceous or sweat glands, muscle cells, and nerves are seldom repaired. They, too, are replaced by fibrous tissue.

In the *scarring phase*, the formation of rather inflexible, fibrous, noncellular **scar tissue** completes the repair process, but without restoring the tissue to its original condition ❹.

We do not know what regulates the extent of scar tissue formation, and the process is highly variable. For example, surgical procedures performed on a fetus do not leave scars, perhaps because damaged fetal tissues do not produce the same types of growth factors that adult tissues do. In some adults, most often those with dark skin, scar tissue formation may continue beyond the requirements of tissue repair. The result is a thickened mass of scar tissue that begins at the site of injury and grows

Figure 5–16   Repair of Injury to the Integument.

1   Inflammatory Phase

Bleeding occurs at the site of injury immediately after the injury, and mast cells in the region trigger an inflammatory response.

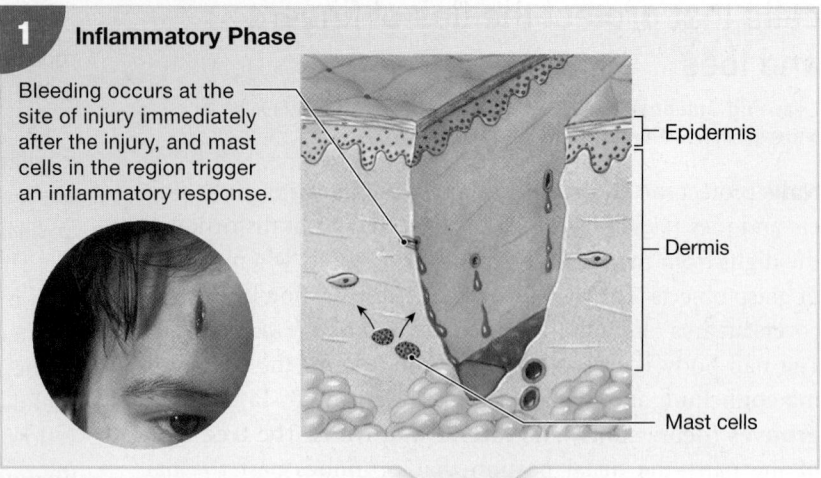

Epidermis

Dermis

Mast cells

2   Migratory Phase

After several hours, a scab has formed and cells of the stratum basale are migrating along the edges of the wound. Phagocytic cells are removing debris, and more of these cells are arriving with the enhanced circulation in the area. Clotting around the edges of the affected area partially isolates the region.

Migrating epithelial cells

Macrophages and fibroblasts

Granulation tissue

3   Proliferation Phase

About a week after the injury, the scab has been undermined by epidermal cells migrating over the collagen fiber meshwork produced by fibroblast proliferation and activity. Phagocytic activity around the site has almost ended, and the fibrin clot is dissolving.

Fibroblasts

4   Scarring Phase

After several weeks, the scab has been shed, and the epidermis is complete. A shallow depression marks the injury site, but fibroblasts in the dermis continue to create scar tissue that will gradually elevate the overlying epidermis.

Scar tissue

Clinical Note

Burns and Grafts Burns are significant injuries. They can damage the integrity of large areas of the skin and compromise many essential functions. Burns result from the exposure of skin to heat, friction, radiation, electrical shock, or strong chemical agents. The severity of the burn depends on the depth of penetration and the total area affected.

First-degree and second-degree burns are also called *partial-thickness burns*, because damage is restricted to the superficial layers of the skin. Only the surface of the epidermis is affected by a *first-degree burn*. In this type of burn, which includes most sunburns, the skin reddens and can be painful. The redness, a sign called **erythema** (er-i-THĒ-muh), results from inflammation of the sun-damaged tissues. In a *second-degree burn*, the entire epidermis and perhaps some of the dermis are damaged. Accessory structures such as hair follicles and glands are generally not affected, but blistering, pain, and swelling occur. If the blisters rupture at the surface, infection can easily develop. Healing typically takes one to two weeks. Some scar tissue may form.

Full-thickness burns, or *third-degree burns*, destroy the epidermis and dermis, extending into the hypodermis. Despite swelling, these burns are less painful than second-degree burns, because sensory nerves are destroyed. Extensive third-degree burns cannot repair themselves, because granulation tissue cannot form and epithelial cells are unable to cover the injury. As a result, the affected area remains open to infection. Extensive third-degree burns often require skin grafts, discussed shortly.

Each year in the United States, roughly 4000 people die from fires and burns. There is a standard reference for calculating the percentage of total surface area damaged. Burns that cover more than 20 percent of the skin surface threaten life, because they affect the following functions:

- *Fluid and Electrolyte Balance.* Even areas with partial-thickness burns lose their effectiveness as barriers to fluid and electrolyte losses. In full-thickness burns, the rate of fluid loss through the skin may reach five times the normal level.
- *Thermoregulation.* Increased fluid loss means increased evaporative cooling. As a result, more energy must be expended to keep body temperature within acceptable limits.
- *Protection from Infection.* The dampness of the epidermal surface, resulting from uncontrolled fluid loss, encourages bacterial growth. If the skin is broken, infection is likely. Widespread bacterial infection, or **sepsis** (*sepsis*, rotting), is the leading cause of death in burn victims.

Effective treatment of full-thickness burns focuses on four procedures:

1. Replacing lost fluids and electrolytes.
2. Providing sufficient nutrients to meet increased metabolic demands for thermoregulation and healing.

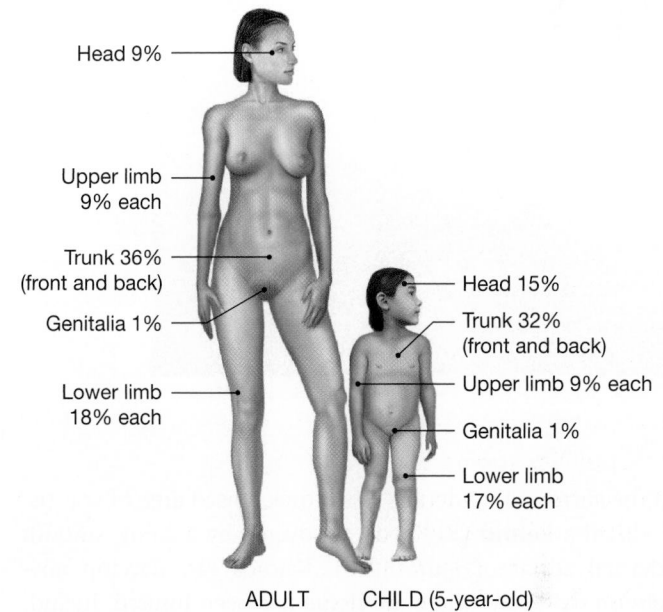

Head 9%
Upper limb 9% each
Trunk 36% (front and back)
Genitalia 1%
Lower limb 18% each

Head 15%
Trunk 32% (front and back)
Upper limb 9% each
Genitalia 1%
Lower limb 17% each

ADULT   CHILD (5-year-old)

3. Preventing infection by cleaning and covering the burn while administering antibiotic drugs.
4. Assisting tissue repair.

This illustration shows a simple method of estimating the surface area affected by burns. This method is called the **rule of nines,** because the surface area in adults is divided into multiples of 9. The rule must be modified for children, because their proportions are quite different.

Because large full-thickness burns cannot heal unaided, surgical procedures are necessary to encourage healing. In a **skin graft,** areas of intact skin are transplanted to cover the site of the burn. A *split-thickness graft* involves a transfer of the epidermis and superficial portions of the dermis. A *full-thickness graft* involves the epidermis and both layers of the dermis.

With fluid-replacement therapies, infection control methods, and grafting techniques, young patients with burns over 80 percent of the body have about a 50 percent chance of recovery. Recent advances in cell culturing may improve survival rates. After a postage-stamp-sized section of undamaged epidermis is removed and grown in the laboratory, basal cell divisions produce large sheets of epidermal cells—up to several square meters in area. These sheets can be transplanted to cover the burn area. Although questions remain about the strength and flexibility of the repairs, skin cultivation is a breakthrough in the treatment of serious burns.

Figure 5–17 **A Keloid.** Keloids are areas of raised fibrous scar tissue.

5

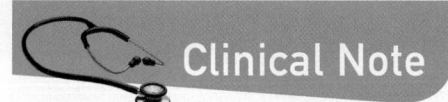

Clinical Note

Skin Abnormalities Because the skin is the most visible organ of the body, abnormalities are easy to recognize. Changes in skin color, tone, and overall condition commonly accompany disease. They can assist in the diagnosis of conditions involving other systems. A bruise, for example, is a swollen, discolored area where blood has leaked through vessel walls. Extensive bruising without any obvious cause may indicate a blood-clotting disorder. Yellowish skin and mucous membranes may signify *jaundice*, which generally indicates some type of liver disorder. The general condition of the skin can also be significant. In addition to color changes, changes in the flexibility, elasticity, dryness, or sensitivity of the skin commonly follow malfunctions in other organ systems.

into the surrounding dermis. This thick, raised area of scar tissue, called a **keloid** (KĒ-loyd), is covered by a shiny, smooth epidermal surface (**Figure 5–17**). Keloids can develop anywhere on the body where the tissue has been injured, including as a result of surgeries. They are harmless and considered by some to be a cosmetic problem. However, some keloids are intentional and some cultures produce keloids as a form of body decoration.

People in societies around the world adorn the skin with culturally significant markings of one kind or another. Tattoos, piercings, keloids and other scar patterns, and even high-fashion makeup are all used to "enhance" the appearance of the integument. Several African cultures perform scarification, creating a series of complex, raised scars on the skin. Polynesian cultures have long used tattoos as a sign of status and beauty. A dark pigment is inserted deep within the dermis of the skin by tapping on a needle, shark tooth, or bit of bone. Because the pigment is inert, the markings remain for life, clearly visible through the overlying epidermis. American popular culture has recently rediscovered tattoos as a fashionable way of adorning the body. Some of the risks associated with tattooing include allergic reactions, blood-borne infections from the use of contaminated equipment, and skin infections.

Tattoos can now be partially or completely removed by laser surgery. The removal process takes time (10 or more sessions may be required to remove a large tattoo), and scars often remain. To remove the tattoo, an intense, narrow beam of light from a laser breaks down the ink molecules in the dermis. Each blast of the laser that destroys the ink also burns the surrounding dermal tissue. Although the burns are minor, they accumulate and result in the formation of localized scar tissue.

CHECKPOINT

29. **What term describes the combination of fibrin clots, fibroblasts, and the extensive network of capillaries in healing tissue?**

30. **Why can skin regenerate effectively even after considerable damage?**

See the blue Answers tab at the back of the book.

5-11 ❙ Effects of aging include skin thinning, wrinkling, and reduced melanocyte activity

Learning Outcome Summarize the effects of aging on the skin.

Aging affects all the components of the integumentary system:

- The epidermis thins as basal cell activity declines. The connections between the epidermis and dermis weaken, making older people more prone to injury, skin tears, and skin infections.

- The number of dendritic cells (Langerhans cells) decreases to about 50 percent of levels seen at maturity (about age 21). This decrease may reduce the sensitivity of the immune system and further encourage skin damage and infection.

- Vitamin D_3 production declines by about 75 percent. The result can be reduced calcium and phosphate absorption, eventually leading to muscle weakness and a reduction in bone strength and density.

- Melanocyte activity declines. In light-skinned individuals the skin becomes very pale. With less melanin in the skin, people become more sensitive to sunlight and more likely to experience sunburn.

- Glandular activity declines. The skin becomes dry and often scaly, because sebum production is reduced. Merocrine sweat glands are also less active. With impaired perspiration, older people cannot lose heat as fast as younger people can. So, the elderly are at greater risk of overheating in warm environments.

- The blood supply to the dermis is reduced. Reduction in blood flow makes the skin become cool, which in turn can stimulate thermoreceptors, making a person feel cold even in a warm room. However, reduced circulation and sweat gland function in the elderly lessens their ability to lose body heat. For this reason, overexertion or exposure to high temperatures (such as those in a sauna or hot tub) can cause body temperatures to soar dangerously high.

- Hair follicles stop functioning or produce thinner, finer hairs. With decreased melanocyte activity, these hairs are gray or white.

- The dermis thins, and the elastic fiber network decreases in size. As a result, the integument becomes weaker and less resilient, and sagging and wrinkling occur. These effects are most noticeable in areas of the body that have been exposed to the sun.

- With changes in levels of sex hormones, secondary sexual characteristics in hair and body fat distribution begin to fade.

- Skin repairs proceed more slowly. Repairs to an uninfected blister might take three to four weeks in a young adult, but the same repairs could take six to eight weeks at age 65–75. Because healing takes place more slowly, recurring infections may result.

The **System Integrator** (Figure 5–18) reviews the integumentary system. System Integrators will appear after each body system has been covered to help build your understanding of the interconnections among all other body systems. Since we have covered only one body system thus far, Figure 5–18 does not include references to body systems to be discussed in upcoming chapters.

CHECKPOINT

31. **Older individuals do not tolerate the summer heat as well as they did when they were young, and they are more prone to heat-related illness. What accounts for these changes?**

32. **Why does hair turn white or gray with age?**

See the blue Answers tab at the back of the book.

SYSTEM INTEGRATOR

The INTEGUMENTARY System

The integumentary system provides physical protection against environmental hazards. It forms the external surface of the body and protects it from dehydration, environmental chemicals, and external forces. The integument (skin) is separated and insulated from the rest of the body by the subcutaneous layer, but it is interconnected with the rest of the body by an extensive circulatory network of blood vessels and lymphatic vessels. As a result, although the protective mechanical functions of the skin can be discussed independently, its physiological activities are always closely integrated with those of other systems.

ABOUT THE SYSTEM INTEGRATORS

Since each body system interacts with every other body system, no one system can be completely understood in isolation. The integration of the various systems allows the human body to function seamlessly, and when disease or injury strikes, multiple systems must respond to heal it.

These charts will introduce the body systems one by one. In so doing, you will see how previously studied systems interact with each other and contribute to overall body homeostasis. As we progress through the organ systems, the complementary nature of these interactions will become clear. Homeostasis depends on the thorough integration of all the body systems working as one.

Skeletal
Page 311

Muscular
Page 406

Nervous
Page 584

Endocrine
Page 673

Cardiovascular
Page 802

Lymphatic
Page 850

Respiratory
Page 900

Digestive
Page 955

Urinary
Page 1036

Reproductive
Page 1116

Figure 5–18   System Integrator.

Related Clinical Terms

carbuncle: A skin infection that often involves a group of hair follicles. The infected material forms a lump, which occurs deep in the skin; the medical term for multiple boils.

cold sore: A lesion that typically occurs in or around the mouth and is caused by a dormant herpes simplex virus that may be reactivated by factors such as stress, fever, or sunburn. Also called *fever blister*.

comedo: The primary sign of acne consisting of an enlarged pore filled with skin debris, bacteria, and sebum (oil); the medical term for a blackhead.

dermatology: The branch of medicine concerned with the diagnosis, treatment, and prevention of diseases of the skin, hair, and nails.

eczema: Rash characterized by inflamed, itchy, dry, scaly, or irritated skin.

frostbite: Injury to body tissues caused by exposure to below-freezing temperatures, typically affecting the nose, fingers, or toes and sometimes resulting in gangrene.

furuncle: A skin infection involving an entire hair follicle and nearby skin tissue; the medical term for a boil.

gangrene: A term that describes dead or dying body tissue that occurs because the local blood supply to the tissue is either lost or is inadequate to keep the tissue alive.

impetigo: An infection of the surface of the skin, caused by staphylococcus ("staph") and streptococcus ("strep") bacteria.

nevus: A benign pigmented spot on the skin such as a mole.

onycholysis: A nail disorder characterized by a spontaneous separation of the nail bed starting at the distal free margin and progressing proximally.

pallor: An unhealthy pale appearance.

porphyria: A rare hereditary disease in which the blood pigment hemoglobin is abnormally metabolized. Porphyrins are excreted in the urine, which becomes dark; other symptoms include mental disturbances and extreme sensitivity of the skin to light.

rosacea: A condition in which certain facial blood vessels enlarge, giving the cheeks and nose a flushed appearance.

scleroderma: An idiopathic chronic autoimmune disease characterized by hardening and contraction of the skin and connective tissue, either locally or throughout the body.

tinea: A skin infection caused by a fungus; also called ringworm.

urticaria: Skin condition characterized by red, itchy, raised areas that appear in varying shapes and sizes; commonly called hives.

Chapter Review

Study Outline

MasteringA&P®

Access more chapter study tools online in the MasteringA&P Study Area:

- Chapter Quizzes, Chapter Practice Test, Art-labeling Activities, Animations, MP3 Tutor Sessions, and Clinical Case Studies
- Practice Anatomy Lab   PAL™
- Interactive Physiology   iP®
- A&P Flix   A&PFlix
- PhysioEx   PhysioEx

An Introduction to the Integumentary System p. 177

1. The **integument**, or **integumentary system**, consists of the **cutaneous membrane**, or *skin* (which includes the **epidermis** and the **dermis**) and the **accessory structures**. Beneath the dermis lies the **hypodermis**, or subcutaneous layer. *(Figures 5–1, 5–2)*

2. Functions of the integument include *protection, excretion, temperature maintenance, vitamin D$_3$ synthesis, nutrient storage, sensory detection,* and *coordination of the immune response. (Figure 5–2)*

5-1 **The epidermis is composed of layers with various functions** p. 179

3. **Thin skin,** consisting of four layers of **keratinocytes,** covers most of the body. Heavily abraded body surfaces may be covered by **thick skin** containing five layers of keratinocytes. *(Spotlight Figure 5–3)*

4. The epidermis provides physical protection, prevents fluid loss, and helps keep microorganisms out of the body.

5. Cell divisions in the **stratum basale,** the deepest epidermal layer, replace more superficial cells. *(Spotlight Figure 5–3)*

6. As epidermal cells age, they pass through the stratum basale, the **stratum spinosum,** the **stratum granulosum,** the **stratum lucidum** (in thick skin), and the **stratum corneum.** In the process, they accumulate large amounts of **keratin.** Ultimately, the cells are shed. *(Spotlight Figure 5–3)*

7. The epidermis and dermis are bonded together by **epidermal ridges** interlocked with **dermal papillae.** *(Spotlight Figure 5–3)*

8. *Dendritic (Langerhans) cells* in the stratum spinosum are part of the immune system. *Tactile (Merkel) cells* in the stratum basale provide sensory information about objects that touch the skin.

5-2 **Factors influencing skin color are epidermal pigmentation and dermal circulation** p. 181

9. The color of the epidermis depends on two factors: epidermal pigmentation and dermal blood supply.

10. The epidermis contains the pigments **carotene** and **melanin. Melanocytes,** which produce melanin, protect us from **ultraviolet (UV) radiation. Vitiligo** is the appearance of nonpigmented white patches of various sizes on otherwise normally-pigmented skin. *(Figures 5–4, 5–5)*

11. Interruptions of the dermal blood supply or poor oxygenation in the lungs can lead to **cyanosis.**

5-3 **Sunlight causes epidermal cells to convert a steroid into vitamin D$_3$** p. 184

12. Epidermal cells synthesize **vitamin D$_3$,** or **cholecalciferol,** when exposed to the UV radiation in sunlight. **Rickets** is a disease caused by vitamin D$_3$ deficiency. *(Figures 5–6, 5–7)*

5-4 ▌ Epidermal growth factor has several effects on the epidermis and epithelia p. 185

13. **Epidermal growth factor (EGF)** promotes growth, division, and repair of the epidermis and epithelial gland synthetic activity and secretion.

5-5 ▌ The dermis is the tissue layer that supports the epidermis p. 186

14. The dermis consists of the superficial **papillary layer** and the deeper **reticular layer**. *(Figures 5–1, 5–2, and 5–8)*
15. The papillary layer of the dermis contains blood vessels, lymphatics, and sensory nerves that supply the epidermis. The reticular layer consists of a meshwork of collagen and elastic fibers oriented to resist tension in the skin.
16. Extensive distension of the dermis can cause **stretch marks.**
17. The pattern of collagen and elastic fiber bundles forms **cleavage lines**. *(Figure 5–9)*
18. Arteries to the skin form the **cutaneous plexus** in the hypodermis. The **subpapillary plexus** is at the base of the papillary layer of the dermis. *(Figure 5–10)*
19. Integumentary sensory receptors detect both light touch and pressure.

5-6 ▌ The hypodermis connects the dermis to underlying tissues p. 188

20. The **hypodermis**, or subcutaneous layer, stabilizes the skin's position against underlying organs and tissues. *(Figure 5–1, 5–11)*

5-7 ▌ Hair is composed of keratinized dead cells that have been pushed to the surface p. 189

21. **Hairs** originate in complex organs called **hair follicles**. Each hair has a **root** and a **shaft**. At the base of the root are a **hair papilla**, surrounded by a **hair bulb**, and a **root hair plexus** of sensory nerves. Hairs have a **medulla**, or core of soft keratin, surrounded by a **cortex** of hard keratin. The **cuticle** is a superficial layer of dead cells that protects the hair. *(Figure 5–12)*
22. Our bodies have both **vellus hairs** ("peach fuzz") and heavy **terminal hairs**. A hair that has stopped growing is called a **club hair**. A curly hair and its follicle are oval; a straight hair and its follicle are round.
23. Each **arrector pili** muscle can erect a single hair. *(Figure 5–12)*

24. Our hairs grow and are shed according to the **hair growth cycle**. A typical hair on the head grows for two to five years and is then shed.

5-8 ▌ Sebaceous glands and sweat glands are exocrine glands found in the skin p. 192

25. A typical **sebaceous gland** discharges waxy **sebum** into a lumen and, ultimately, into a hair follicle. **Sebaceous follicles** are large sebaceous glands that discharge sebum directly onto the epidermis. *(Figure 5–13)*
26. The two types of sweat glands, or **sudoriferous glands**, are apocrine and merocrine sweat glands. **Apocrine sweat glands** produce an odorous secretion. The more numerous **merocrine sweat glands** produce a watery secretion known as sensible perspiration. *(Figure 5–14)*
27. **Mammary glands** of the breasts are structurally similar to apocrine sweat glands. **Ceruminous glands** in the ear produce a waxy substance called **cerumen.**

5-9 ▌ Nails are keratinized epidermal cells that protect the tips of fingers and toes p. 195

28. The **nail body** of a nail covers the **nail bed.** Nail production occurs at the **nail root**, which is covered by the **cuticle**, or **eponychium**. The **free edge** of the nail extends over the **hyponychium**. *(Figure 5–15)*

5-10 ▌ Several phases are involved in repairing the integument following an injury p. 195

29. Based on the division of stem cells, the skin can regenerate effectively even after considerable damage. The process begins with bleeding and inflammation *(inflammatory phase)*, the formation of a **scab** and **granulation tissue** *(migratory phase)*, loss of granulation tissue and undermining of the scab *(proliferation phase)*, and ends with the creation of **scar tissue** *(scarring phase)*. *(Figure 5–16)*
30. **Keloids** are raised areas of fibrous scar tissue. *(Figure 5–17)*

5-11 ▌ Effects of aging include skin thinning, wrinkling, and reduced melanocyte activity p. 198

31. With aging, the integument thins, blood flow decreases, cellular activity decreases, and repairs take place more slowly.
32. The integumentary system provides physical protection for all other body systems. *(Figure 5–18)*

Review Questions

See the blue Answers tab at the back of the book.

LEVEL 1 Reviewing Facts and Terms

1. Identify the different portions **(a–d)** of the cutaneous membrane and the underlying layer of loose connective tissue **(e)** in the diagram to the right.
    (a) _____
    (b) _____
    (c) _____
    (d) _____
    (e) _____
2. The two major components of the integumentary system are **(a)** the cutaneous membrane and the accessory structures, **(b)** the epidermis and the hypodermis, **(c)** the hair and the nails, **(d)** the dermis and the hypodermis.

3. What is the epidermis of the skin mainly made of? **(a)** keratinocytes, **(b)** melanocytes, **(c)** areolar connective tissue, **(d)** dense irregular connective tissue.
4. Sebaceous glands secrete sebum. What is the function of sebum? **(a)** eliminating salts from the body, **(b)** lubricating the skin, **(c)** regulating body temperature, **(d)** promoting wound healing.
5. Exposure of the skin to ultraviolet (UV) radiation **(a)** can result in increased numbers of melanocytes forming in the skin, **(b)** can result in decreased melanin production in melanocytes, **(c)** can cause destruction of vitamin D_3, **(d)** can result in damage to the DNA of cells in the stratum basale, **(e)** has no effect on the skin cells.
6. The small intestine requires _____ for the normal absorption of calcium and phosphorus. **(a)** calcitriol, **(b)** vitamin D_3, **(c)** cholesterol-related steroid, **(d)** sunlight.
7. The cutaneous plexus and subpapillary plexus consist of **(a)** blood vessels providing the dermal blood supply, **(b)** a network of nerves

providing dermal sensations, (c) specialized cells for cutaneous sensations, (d) gland cells that release cutaneous secretions.

8. The accessory structures of the integument include the (a) blood vessels, glands, muscles, and nerves, (b) tactile discs, lamellated corpuscles, and tactile corpuscles, (c) hair, skin, and nails, (d) hair follicles, nails, sebaceous glands, and sweat glands.

9. The portion of the hair follicle where cell divisions occur is the (a) shaft, (b) matrix, (c) root hair plexus, (d) cuticle.

10. The papillary layer of the dermis is made up of (a) areolar tissue, (b) reticular tissue, (c) elastic tissue, (d) smooth muscle.

11. Apocrine sweat glands can be controlled by (a) the autonomic nervous system, (b) regional control mechanisms, (c) the endocrine system, (d) both a and c.

12. The primary function of sensible perspiration is to: (a) get rid of wastes, (b) protect the skin from dryness, (c) maintain electrolyte balance, (d) reduce body temperature.

13. The pale crescent at the base of the nail is known as the (a) eponychium (cuticle), (b) hyponychium, (c) cerumen, (d) lunula.

14. Muscle weakness and a reduction in bone strength in the elderly result from decreased (a) vitamin D_3 production, (b) melanin production, (c) sebum production, (d) dermal blood supply.

15. In which layer(s) of the epidermis does cell division occur?

16. What is the function of the arrector pili muscles?

17. Which of the following is a function of skin? (a) mineral storage, (b) excretion, (c) blood pressure regulation, (d) vitamin C synthesis.

18. Touch is detected in the stratum basale by (a) basal cells, (b) keratinoctyes, (c) Merkel cells, (d) melanocytes.

19. What is meant by the term cyanosis?

LEVEL 2 Reviewing Concepts

20. Contrast insensible perspiration and sensible perspiration.

21. In clinical practice, drugs can be delivered by diffusion across the skin. This delivery method is called transdermal administration. Why are fat-soluble drugs more suitable for transdermal administration than drugs that are water soluble?

22. In our society, a tanned body is associated with good health. However, medical research constantly warns about the dangers of excessive exposure to the sun. Describe how a tan develops and cite a benefit of minimal sun exposure.

23. Describe the composition and development of granulation tissue during integument repair.

24. The fibrous protein that is responsible for the strength and water resistance of the skin surface is (a) collagen, (b) eleidin, (c) keratin, (d) elastin, (e) keratohyalin.

25. The darker an individual's skin color (a) the more melanocytes she has in her skin, (b) the more layers she has in her epidermis, (c) the more melanin her melanocytes produce, (d) the more superficial her blood vessels.

26. In order for bacteria on the skin to cause an infection in the skin, they must accomplish all of the following, *except* (a) survive the bactericidal components of sebum, (b) avoid being flushed from the surface of the skin by sweat, (c) penetrate the stratum corneum, (d) penetrate to the level of the capillaries, (e) escape the dendritic cells.

LEVEL 3 Critical Thinking and Clinical Applications

27. In the elderly, blood supply to the dermis is reduced and sweat glands are less active. This combination of factors would most affect (a) the ability to thermoregulate, (b) the ability to heal injured skin, (c) the ease with which the skin is injured, (d) the physical characteristics of the skin, (e) the ability to grow hair.

28. Two patients are brought to the emergency room. One has cut his finger with a knife; the other has stepped on a nail. Which wound has a greater chance of becoming infected? Why?

29. Exposure to optimum amounts of sunlight is necessary for proper bone maintenance and growth in children. (a) What does sunlight do to promote bone maintenance and growth? (b) If a child lives in an area where exposure to sunlight is rare because of pollution or overcast skies, what can be done to minimize impaired maintenance and growth of bone?

30. A patient, who is otherwise fit and healthy, complains to his doctor that his skin has developed an orange appearance after starting a vegetarian diet. Provide an explanation for the change in skin color.

31. Describe the effect that third-degree burns would have on the integumentary system. Explain why this might be life threatening.

CLINICAL CASE Wrap-Up

Skin Cells in Overdrive

Harold is suffering from psoriasis, a rather common skin disorder that speeds up the life cycle of skin cells. Instead of the keratinocytes maturing and migrating from the stratum basale to the epidermal surface in a normal amount of time, the process takes place too rapidly. Dead skin cells build up on the epidermal surface in scaly patches.

Nearly 10 million Americans suffer from psoriasis. Like nearly half of them, Harold has a family history of the disease. Other risk factors for developing psoriasis include obesity, smoking, and some medications. And, like one-third of psoriasis patients, Harold has a chance of developing psoriatic arthritis, an inflammatory joint disorder that can affect the large and small joints of

the head, neck, and trunk as well as the arms and legs.

Psoriasis is a chronic, or long-term, condition. Harold's dermatologist offered him a variety of treatments. One is phototherapy, which uses ultraviolet light to treat psoriasis. Others include lotions, creams, and shampoos. Still other treatments involve pills and injections aimed at controlling inflammation.

1. What layer of the cutaneous membrane is affected by psoriasis?

2. What is the underlying problem, as seen in the microscope, in psoriasis?

See the blue Answers tab at the back of the book.

6 Osseous Tissue and Bone Structure

Learning Outcomes

These Learning Outcomes correspond by number to this chapter's sections and indicate what you should be able to do after completing the chapter.

6-1 ▐ Describe the primary functions of the skeletal system.

6-2 ▐ Classify bones according to shape and internal organization, giving examples of each type, and explain the functional significance of each of the major types of bone markings.

6-3 ▐ Identify the cell types in bone, and list their major functions.

6-4 ▐ Compare the structures and functions of compact bone and spongy bone.

6-5 ▐ Compare the mechanisms of endochondral ossification and intramembranous ossification.

6-6 ▐ Describe the remodeling and homeostatic mechanisms of the skeletal system.

6-7 ▐ Discuss the effects of exercise, hormones, and nutrition on bone development and on the skeletal system.

6-8 ▐ Explain the role of calcium as it relates to the skeletal system.

6-9 ▐ Describe the types of fractures, and explain how fractures heal.

6-10 ▐ Summarize the effects of the aging process on the skeletal system.

SPOTLIGHTS
- Endochondral Ossification pp. 216–217
- Types of Fractures and Steps in Repair pp. 226–227

CLINICAL CASE
- A Case of Child Abuse? p. 205

Clinical Notes
- Heterotopic Bone Formation p. 214
- Abnormal Bone Development p. 222

A Case of Child Abuse?

Alex's mom has been waiting in the emergency room for over two hours since her son was wheeled off to x-ray. Poor Alex had been carried home from third-grade soccer practice. The coach said she had not witnessed the injury, but it seemed like Alex had twisted and fallen without even having contact with another player. He was unable to bear weight on his left lower leg.

A handsome young man, *not* wearing a white coat, approaches her. "Mrs. Otero, I am the hospital social worker. I am here to speak with you about your son, Alex. Alex has an acute fracture of his tibia and fibula—the bones in his lower leg—but he also has two healed rib fractures and a healing fracture of his upper arm as well. We fear he has been the victim of child abuse, so he is being admitted until we can investigate further."

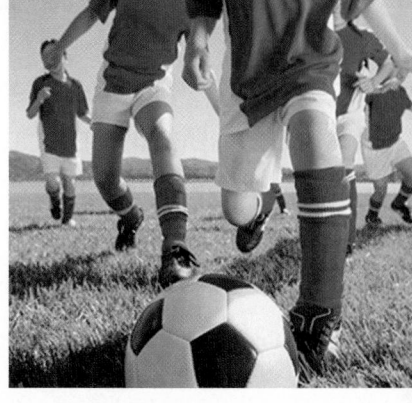

Mrs. Otero is dumbfounded by the suggestion of child abuse. She knows that no one at home has hurt him. Alex had complained about shoulder pain after playing on the monkey bars a few weeks ago, but she had not suspected his upper arm was broken. She considered Alex to be a lot like her. Both were somewhat fragile and prone to breaking bones, although she had not had any more fractures since she was married. They both had soft, discolored teeth. And, oddly enough, the whites of their eyes had a blue tint. **If Alex is not a victim of child abuse, what else could explain his multiple fractures? To find out, turn to the Clinical Case Wrap-Up on p. 231.**

An Introduction to the Skeletal System

In this chapter we expand on the introduction to bone, presented in Chapter 4, by examining the mechanisms involved with the growth, remodeling, and repair of the skeleton. The bones of the skeleton are more than just racks from which muscles hang. They have a variety of vital functions. In addition to supporting the weight of the body, bones work with muscles to maintain body position and to produce controlled, precise movements. Without the skeleton to pull against, contracting muscle fibers could not make us sit, stand, walk, or run.

6-1 ▌ The skeletal system has five primary functions

Learning Outcome Describe the primary functions of the skeletal system.

Your skeletal system includes the bones of the skeleton and the cartilages, ligaments, and other connective tissues that stabilize or interconnect the bones. This system has five primary functions:

1. *Support.* The skeletal system provides structural support for the entire body. Individual bones or groups of bones provide a framework for the attachment of soft tissues and organs.

2. *Storage of Minerals and Lipids.* Minerals are inorganic ions that contribute to the osmotic concentration of body fluids, as you will learn in Chapter 25. Minerals also take part in various physiological processes, and several are important as enzyme cofactors. Calcium is the most abundant mineral in the human body. The calcium salts of bone are a valuable mineral reserve that maintains normal concentrations of calcium and phosphate ions in body fluids. In addition, the bones of the skeleton store energy as lipids in areas filled with *yellow bone marrow.*

3. *Blood Cell Production.* Red blood cells, white blood cells, and other blood elements are produced in *red bone marrow,* which fills the internal cavities of many bones. We will describe blood cell formation when we examine the cardiovascular and lymphatic systems (Chapters 19 and 22).

4. *Protection.* Skeletal structures surround many soft tissues and organs. The ribs protect the heart and lungs, the skull encloses the brain, the vertebrae shield the spinal cord, and the pelvis cradles digestive and reproductive organs.

5. *Leverage.* Many bones function as levers that can change the magnitude and direction of the forces generated by skeletal muscles. The movements produced range from the precise motion of a fingertip to changes in the position of the entire body.

Chapters 6–9 describe the structure and function of the skeletal system. We begin by describing bone, or osseous tissue, a supporting connective tissue introduced in Chapter 4. ⤺ p. 162 All of the features and properties of the skeletal system ultimately depend on the unique and dynamic properties of bone. The bone specimens that you study in lab or that you may have seen in the skeletons of dead animals are only the dry remains of this living tissue. They have the same relationship to the bone in a living organism as a kiln-dried 2-by-4 board has to a living oak tree.

CHECKPOINT

1. **Name the five primary functions of the skeletal system.**

See the blue Answers tab at the back of the book.

6-2 ❚ Bones are classified according to shape and structure, and they have a variety of surface markings

Learning Outcome Classify bones according to shape and internal organization, giving examples of each type, and explain the functional significance of each of the major types of bone markings.

We can classify a bone by its general shape or by its internal tissue organization. Before we consider specific bones of the skeleton, let's look at the classification schemes.

Bone Shapes

The adult skeleton typically contains 206 major bones. We can divide these into six broad categories according to their individual shapes (**Figure 6–1**):

1. **Sutural bones**, or *Wormian bones*, are small, flat, irregularly shaped bones between the flat bones of the skull (**Figure 6–1a**). There are individual variations in the number, shape, and position of the sutural bones.

2. **Irregular bones** have complex shapes (**Figure 6–1b**).

3. **Short bones** are small and boxy, with approximately equal dimensions. (**Figure 6–1c**).

4. **Flat bones** have a thin, flattened shape.

5. **Long bones** are elongated and consist of a shaft (diaphysis) with two ends (epiphyses) that are wider than the shaft (**Figure 6–1e**).

6. **Sesamoid bones** are small, flat, and shaped somewhat like a sesame seed (**Figure 6–1f**). Except for the patellae (kneecaps), there are individual variations in the location and number of sesamoid bones. These variations, along

Figure 6–1 A Classification of Bones by Shape.

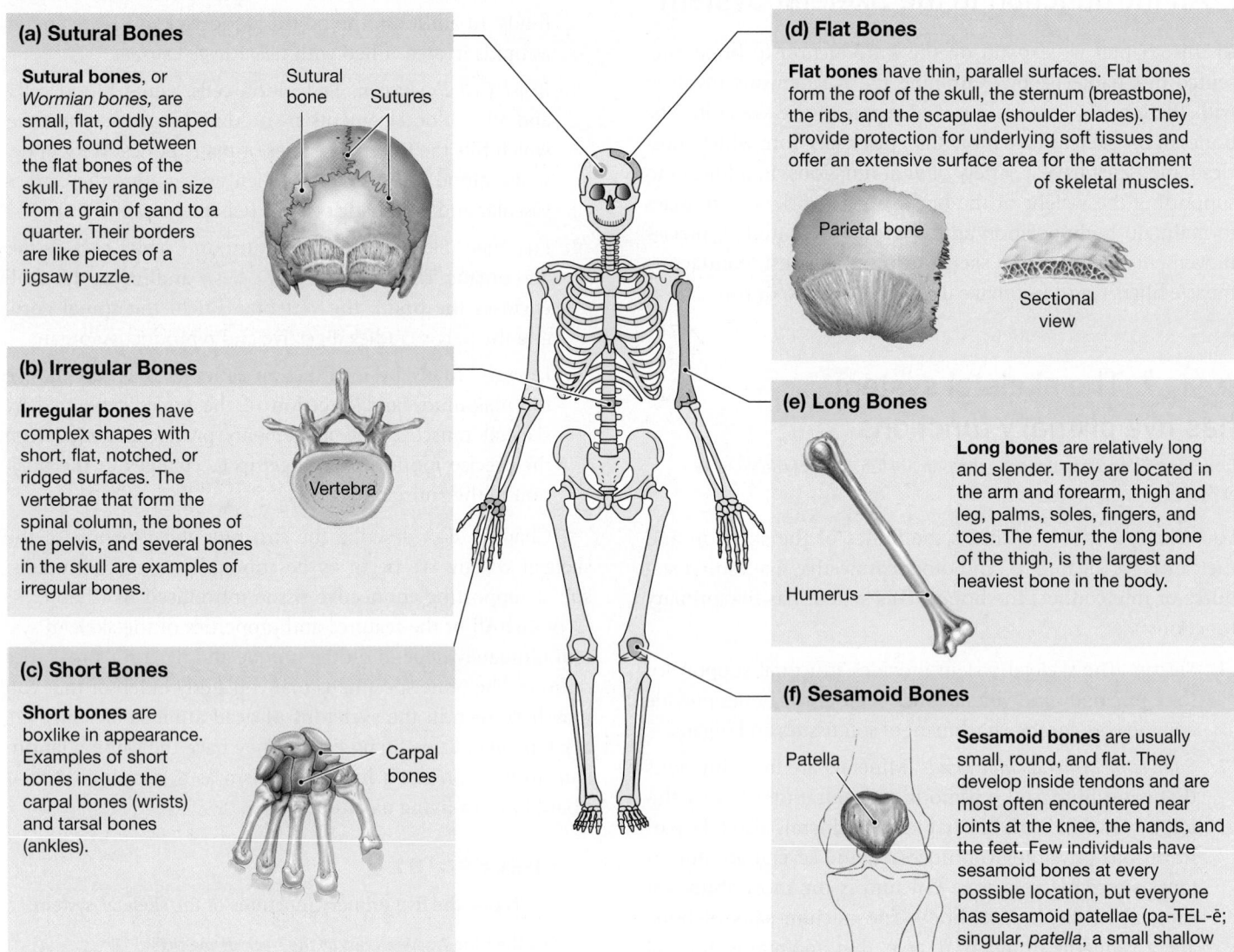

(a) Sutural Bones

Sutural bones, or *Wormian bones,* are small, flat, oddly shaped bones found between the flat bones of the skull. They range in size from a grain of sand to a quarter. Their borders are like pieces of a jigsaw puzzle.

Sutural bone   Sutures

(b) Irregular Bones

Irregular bones have complex shapes with short, flat, notched, or ridged surfaces. The vertebrae that form the spinal column, the bones of the pelvis, and several bones in the skull are examples of irregular bones.

Vertebra

(c) Short Bones

Short bones are boxlike in appearance. Examples of short bones include the carpal bones (wrists) and tarsal bones (ankles).

Carpal bones

(d) Flat Bones

Flat bones have thin, parallel surfaces. Flat bones form the roof of the skull, the sternum (breastbone), the ribs, and the scapulae (shoulder blades). They provide protection for underlying soft tissues and offer an extensive surface area for the attachment of skeletal muscles.

Parietal bone   Sectional view

(e) Long Bones

Long bones are relatively long and slender. They are located in the arm and forearm, thigh and leg, palms, soles, fingers, and toes. The femur, the long bone of the thigh, is the largest and heaviest bone in the body.

Humerus

(f) Sesamoid Bones

Sesamoid bones are usually small, round, and flat. They develop inside tendons and are most often encountered near joints at the knee, the hands, and the feet. Few individuals have sesamoid bones at every possible location, but everyone has sesamoid patellae (pa-TEL-ē; singular, *patella*, a small shallow dish), or kneecaps.

Patella

with varying numbers of sutural bones, account for individual differences in the total number of bones in the skeleton. (Sesamoid bones may form in at least 26 locations.)

Bone Markings

The surfaces of each bone in your body have characteristic features. Elevations or projections form where tendons and ligaments attach, and where adjacent bones articulate (that is, at joints). Depressions, grooves, and tunnels in bone are sites where blood vessels or nerves lie alongside or penetrate the bone. Detailed examination of these **bone markings**, or *surface features*, can yield an abundance of anatomical information. For example, anthropologists, criminologists, and pathologists can often determine the size, age, sex, and general appearance of an individual on the basis of incomplete skeletal remains.

Figure 6–2 introduces the prominent bone markings. It uses specific anatomical terms for the various projections,

Figure 6–2 An Introduction to Bone Markings.

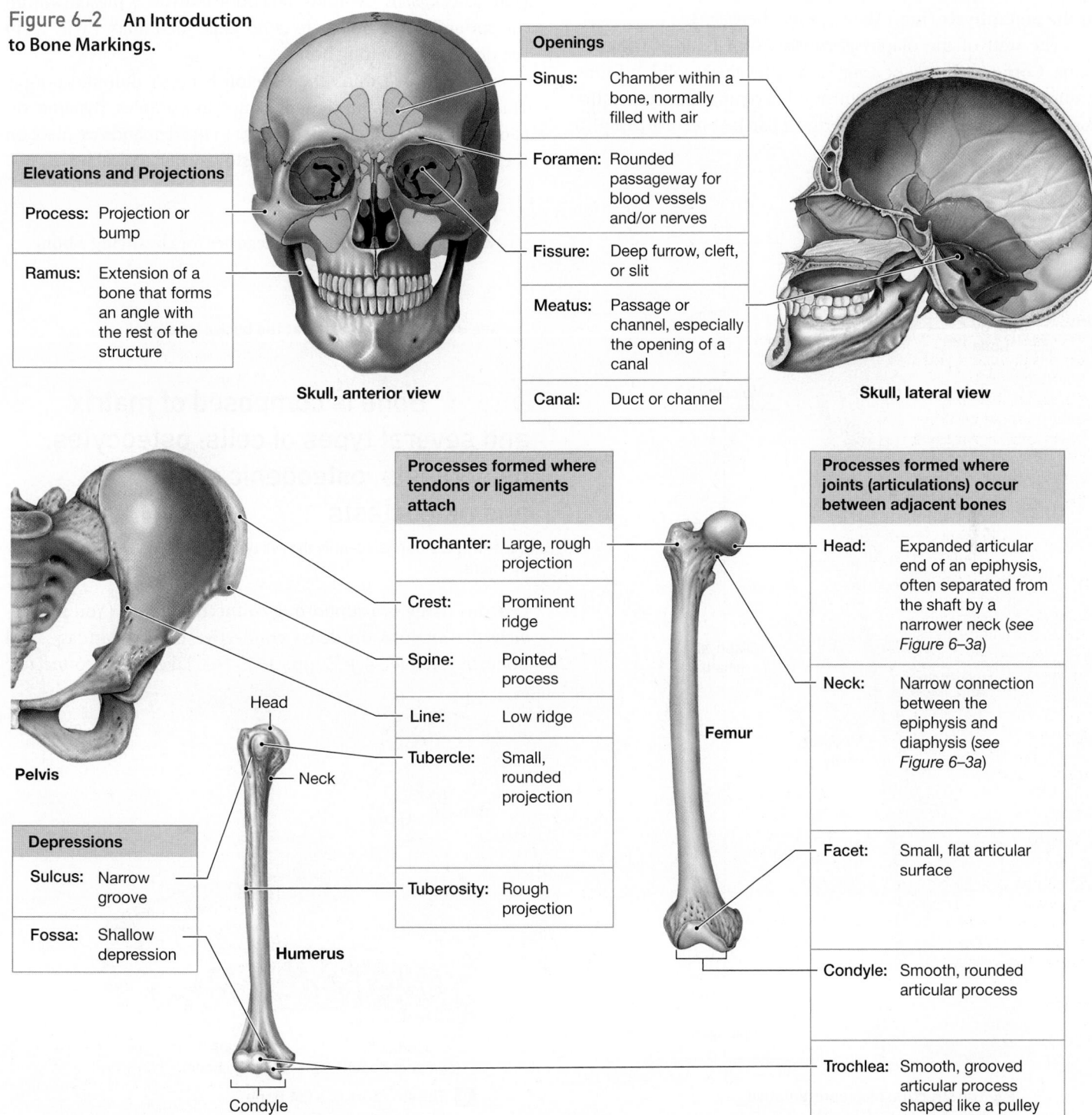

Elevations and Projections

Process:	Projection or bump
Ramus:	Extension of a bone that forms an angle with the rest of the structure

Openings

Sinus:	Chamber within a bone, normally filled with air
Foramen:	Rounded passageway for blood vessels and/or nerves
Fissure:	Deep furrow, cleft, or slit
Meatus:	Passage or channel, especially the opening of a canal
Canal:	Duct or channel

Skull, anterior view

Skull, lateral view

Pelvis

Processes formed where tendons or ligaments attach

Trochanter:	Large, rough projection
Crest:	Prominent ridge
Spine:	Pointed process
Line:	Low ridge
Tubercle:	Small, rounded projection
Tuberosity:	Rough projection

Head

Neck

Depressions

Sulcus:	Narrow groove
Fossa:	Shallow depression

Humerus

Condyle

Femur

Processes formed where joints (articulations) occur between adjacent bones

Head:	Expanded articular end of an epiphysis, often separated from the shaft by a narrower neck (*see Figure 6–3a*)
Neck:	Narrow connection between the epiphysis and diaphysis (*see Figure 6–3a*)
Facet:	Small, flat articular surface
Condyle:	Smooth, rounded articular process
Trochlea:	Smooth, grooved articular process shaped like a pulley

depressions, and openings. These markings provide fixed landmarks that can help us determine the position of the soft-tissue components of other organ systems.

Bone Structure

Figure 6–3a introduces the anatomy of the femur, the long bone of the thigh. This representative long bone has an extended tubular shaft, or **diaphysis** (dī-AF-i-sis). At each end is an expanded area known as the **epiphysis** (ē-PIF-i-sis). The diaphysis is connected to each epiphysis at a narrow zone known as the **metaphysis** (me-TAF-i-sis; *meta*, between).

The wall of the diaphysis consists of a layer of compact bone. **Compact bone**, or *dense bone*, is relatively solid. It forms a sturdy protective layer that surrounds a central space called the **medullary cavity** (*medulla*, innermost part), or *marrow cavity*.

The epiphyses consist largely of spongy bone. **Spongy bone**, also called *cancellous* (KAN-se-lus) or *trabecular bone*, consists of an open network of struts and plates that resembles latticework. The spongy bone of each epiphysis is covered by a thin layer of compact bone, also called *cortical bone*.

Figure 6–3b shows the structure of a flat bone from the skull, such as one of the *parietal bones*. A flat bone resembles a spongy bone sandwich, with layers of compact bone covering a core of spongy bone. Within the cranium, the layer of spongy bone between the layers of compact bone is called the *diploë* (DIP-lō-ē; *diplous*, twofold). Red bone marrow is present within the spongy bone, but there is no large medullary cavity as in the diaphysis of a long bone.

Many people think the skeleton is rather dull and boring. This is far from the truth. Our bones are complex, dynamic organs that constantly change to adapt to the demands we place on them. We will now consider the histology of a typical bone.

CHECKPOINT

2. Identify the six broad categories for classifying a bone according to shape.

3. Define bone marking.

See the blue Answers tab at the back of the book.

6-3 | Bone is composed of matrix and several types of cells: osteocytes, osteoblasts, osteogenic cells, and osteoclasts

Learning Outcome Identify the cell types in bone, and list their major functions.

Osseous tissue is a supporting connective tissue. (You may review the sections on dense connective tissues, cartilage, and bone in Chapter 4.) ⤶ pp. 157–163 Like other connective

Figure 6–3 **Bone Structure.**

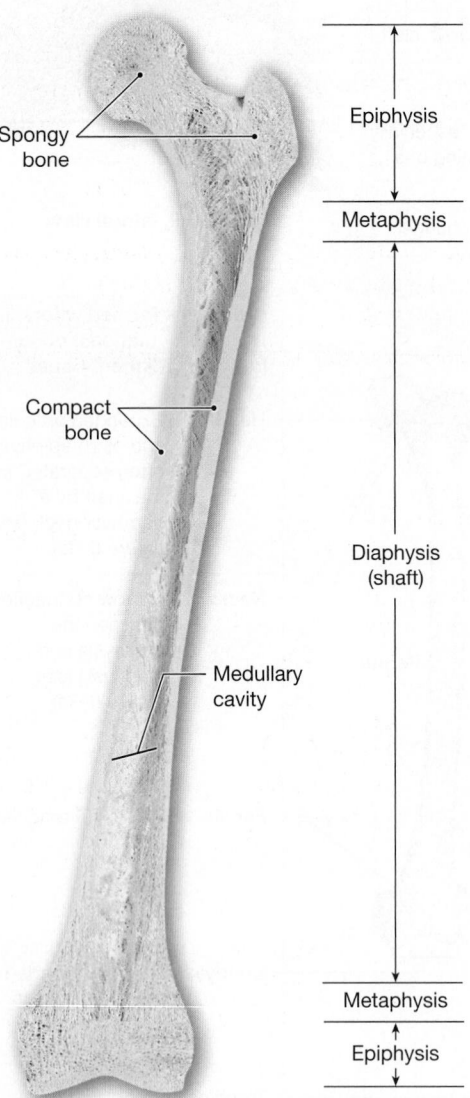

- **a** The structure of a representative long bone (the femur) in longitudinal section

Spongy bone
Epiphysis
Metaphysis
Compact bone
Diaphysis (shaft)
Medullary cavity
Metaphysis
Epiphysis

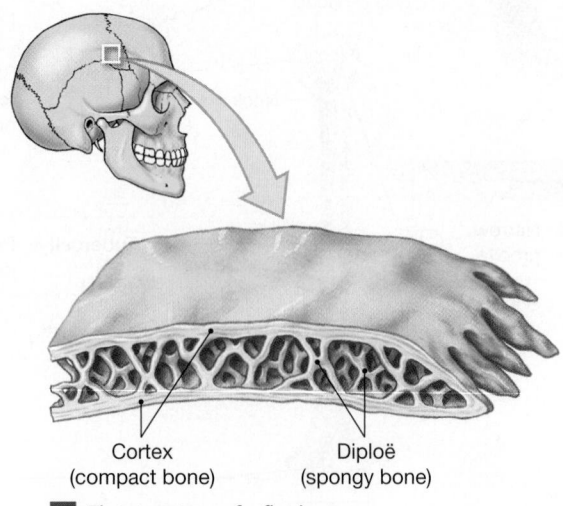

- **b** The structure of a flat bone (the parietal bone)

Cortex (compact bone)　Diploë (spongy bone)

tissues, it contains specialized cells and a matrix consisting of extracellular protein fibers and a ground substance. The matrix of bone tissue is solid and sturdy, due to deposits of calcium salts around the protein fibers.

Recall these characteristics of bone from Chapter 4:

- The matrix of bone is very dense and contains deposits of calcium salts.
- The matrix contains bone cells, or *osteocytes*, within pockets called *lacunae*. (The same term is used for the spaces that chondrocytes occupy in cartilage. ⊃ p. 159) The lacunae of bone are typically organized around blood vessels that branch through the bony matrix.
- *Canaliculi*, narrow passageways through the matrix, extend between the lacunae and nearby blood vessels, forming a branching network for the exchange of nutrients, wastes, and gases.
- Except at joints, a periosteum covers the outer surfaces of bones. It consists of an outer fibrous layer and an inner cellular layer.

We now take a closer look at the organization of the matrix and cells of bone.

Bone Matrix

Calcium phosphate, $Ca_3(PO_4)_2$, makes up almost two-thirds of the weight of bone. Calcium phosphate interacts with calcium hydroxide, $Ca(OH)_2$, to form crystals of **hydroxyapatite** (hī-drok-sē-AP-a-tīt), $Ca_{10}(PO_4)_6(OH)_2$. As these crystals form, they incorporate other calcium salts, such as calcium carbonate ($CaCO_3$), and ions such as sodium, magnesium, and fluoride.

Approximately one-third of the weight of bone is collagen fibers. Cells make up only 2 percent of the mass of a typical bone.

Calcium phosphate crystals are very hard, but relatively inflexible and quite brittle. They can withstand compression but are likely to shatter when exposed to bending, twisting, or sudden impacts. Collagen fibers, by contrast, are remarkably strong. When subjected to tension (pull), they are stronger than steel. They are flexible as well as tough and can easily tolerate twisting and bending, but offer little resistance to compression. When compressed, they simply bend out of the way.

The composition of the matrix in compact bone is the same as that in spongy bone. The collagen fibers provide an organic framework on which hydroxyapatite crystals can form. These crystals form small plates and rods that are locked into the collagen fibers at regular angles. The result is a protein–crystal combination that has the flexibility of collagen and the compressive strength of hydroxyapatite crystals. The protein–crystal interactions allow bone to be strong, somewhat flexible, and highly resistant to shattering.

In its overall properties, bone is on a par with the best steel-reinforced concrete. In fact, bone is far superior to concrete, because it can undergo remodeling (cycles of bone formation and resorption) as needed and can repair itself after injury.

Bone Cells

Bone contains four types of cells: osteocytes, osteoblasts, osteogenic cells, and osteoclasts (**Figure 6–4**).

Osteocytes (OS-tē-ō-sīts) (*osteo-*, bone + *-cyte*, cell) are mature bone cells that make up most of the cell population. Each

Figure 6–4 **Types of Bone Cells.**

Osteocyte: Mature bone cell that maintains the bone matrix

Osteoblast: Immature bone cell that secretes osteoid, the organic component of bone matrix

Osteogenic cell: Stem cell whose divisions produce osteoblasts

Osteoclast: Multinucleate cell that secretes acids and enzymes to dissolve bone matrix osteocyte occupies a lacuna, a pocket sandwiched between layers of matrix. The layers are called **lamellae** (lah-MEL-lē; singular, *lamella*, a thin plate). Osteocytes cannot divide, and a lacuna never contains more than one osteocyte. Narrow passageways called **canaliculi** penetrate the lamellae, radiating through the matrix and connecting lacunae with one another and with sources of nutrients, supplied by blood vessels in the central canal.

Canaliculi contain cytoplasmic extensions of osteocytes. Neighboring osteocytes are linked by gap junctions, which permit the cells to exchange ions and small molecules, including nutrients and hormones. The interstitial fluid that surrounds the osteocytes and their extensions is an additional route for the diffusion of nutrients and waste products.

Osteocytes have two major functions:

1. *Osteocytes maintain the protein and mineral content of the surrounding matrix.* There is continual turnover of matrix components. Osteocytes secrete chemicals that dissolve the adjacent matrix, and the minerals released enter the circulation. Osteocytes then rebuild the matrix, stimulating the deposition of new hydroxyapatite crystals. The turnover rate varies from bone to bone. We will consider this process further in a later section.

2. *Osteocytes take part in the repair of damaged bone.* If released from their lacunae, osteocytes can convert to a less specialized type of cell, such as an osteoblast or an osteogenic cell.

Osteoblasts (OS-tē-ō-blasts; *-blast*, immature precursor cell) produce new bone matrix in a process called **ossification**, or **osteogenesis** (os-tē-ō-JEN-e-sis; *genesis*, production). Osteoblasts make and release the proteins and other organic components of the matrix. Before calcium salts are deposited, this organic matrix is called **osteoid** (OS-tē-oyd). Osteoblasts also help increase local concentrations of calcium phosphate above its solubility limit, triggering the deposition of calcium salts in the organic matrix. This process converts osteoid to bone. Osteocytes develop from osteoblasts that have become completely surrounded by bone matrix.

Bone contains small numbers of mesenchymal cells called **osteogenic** (os-tē-ō-JEN-ik) **cells** or *osteoprogenitor* (os-tē-ō-prō-JEN-i-tor) *cells*. These squamous stem cells divide to produce daughter cells that differentiate into osteoblasts. Osteogenic cells maintain populations of osteoblasts and are important in the repair of a *fracture* (a break or a crack in a bone). Osteogenic cells are found in the inner, cellular layer of the periosteum. They are also found in an inner layer, or *endosteum*, that lines medullary cavities and passageways for blood vessels penetrating the matrix of compact bone.

Osteoclasts (OS-tē-ō-clasts; *klastos*, broken) are cells that absorb and remove bone matrix. They are large cells with 50 or more nuclei. Osteoclasts are not related to osteogenic cells or their descendants. Instead, they are derived from the same stem cells that produce monocytes and macrophages, which are cells involved in the body's defense mechanisms. Acids and protein-digesting enzymes secreted by osteoclasts dissolve the matrix and release the stored minerals. This erosion process is called **osteolysis** (os-tē-OL-i-sis; *osteo-*, bone + *lysis*, a loosening) or *resorption*. It is important in the regulation of calcium and phosphate concentrations in body fluids.

In living bone, osteoclasts are constantly removing matrix, and osteoblasts are always adding to it. The homeostatic balance between the opposing activities of osteoblasts and osteoclasts is very important. When osteoclasts remove calcium salts faster than osteoblasts deposit them, bones weaken. When osteoblast activity predominates, bones become stronger and more massive. This opposition causes some interesting differences in bone structure among individuals. People who subject their bones to muscular stress through weight training or strenuous exercise develop not only stronger muscles, but also stronger bones. Alternatively, declining muscular activity due to immobility leads to reduced bone mass at sites of muscle attachment. We will look at this phenomenon further in a later section of the chapter.

CHECKPOINT

4. Mature bone cells are known as _____, bone-building cells are called _____, and _____ are bone-resorbing cells.

5. How would the compressive strength of a bone be affected if the ratio of collagen to hydroxyapatite increased?

6. If the activity of osteoclasts exceeds the activity of osteoblasts in a bone, how will the mass of the bone be affected?

See the blue Answers tab at the back of the book.

6-4 | Compact bone contains parallel osteons, and spongy bone contains trabeculae

Learning Outcome Compare the structures and functions of compact bone and spongy bone.

Compact bone functions to protect, support, and resist stress. Spongy bone provides some support and stores marrow. In this section, we examine in detail the structures of compact and spongy bone, which allow these types of bone to perform their functions. We also look at the outer and inner coverings of bone, called the periosteum and endosteum, respectively.

Compact Bone Structure

The basic functional unit of mature compact bone is the **osteon** (OS-tē-on), or *Haversian system*. In an osteon, the osteocytes are arranged in concentric layers around a vascular **central canal**, or *Haversian canal* (Figure 6–5). This canal contains one or more blood vessels (normally a capillary and a *venule*, a very small vein) that carry blood to and from the osteon (Figure 6–6a). Central canals generally run parallel to the surface of the bone. Other passageways, known as **perforating canals** or *Volkmann's canals,* extend perpendicular to the surface. Blood vessels in these canals supply blood to osteons deeper in the bone and to tissues of the medullary cavity.

The lamellae of each osteon form a series of nested cylinders around the central canal. In transverse section, these *concentric lamellae* create a targetlike pattern, with the central canal as the bull's-eye. Collagen fibers within each lamella form a spiral pattern that adds strength and resiliency (Figure 6–6b).

Canaliculi radiating through the lamellae interconnect the lacunae of the osteons with one another and with the central canal. *Interstitial lamellae* fill in the spaces between the osteons in compact bone. These lamellae are remnants of osteons whose matrix components have been almost completely recycled by osteoclasts. *Circumferential lamellae* (*circum-*, around + *ferre*, to bear) are found at the outer and inner surfaces of the bone, where they are covered by the periosteum and endosteum, respectively (Figure 6–6a,b). These lamellae are produced during the growth of the bone, as we will describe in a later section.

Compact bone is thickest where stresses arrive from a limited range of directions. All osteons in compact bone are aligned the same, making such bones very strong when stressed along the axis of alignment. You might think of a single osteon as a drinking straw with very thick walls. When you try to push the ends of the straw together or to pull them apart, the straw is quite strong. But if you hold the ends and push from the side, the straw will easily bend at a sharp angle.

The osteons in the diaphysis of a long bone are parallel to the long axis of the shaft. For this reason, the shaft does not bend, even when extreme forces are applied to either end. (The femur can withstand 10–15 times the body's weight without breaking.) Yet a much smaller force applied to the side of the shaft can break the femur. A sudden sideways force, as in a fall or auto accident, causes the majority of breaks in this bone.

Figure 6–5   **The Histology of Compact Bone.**

Canaliculi
Concentric lamellae
Central canal
Osteon
Lacunae

Osteon   LM × 343

a A thin section through compact bone. By this procedure the intact matrix making up the lamellae appear white, and the central canal, lacunae, and canaliculi appear black due to the presence of bone dust.

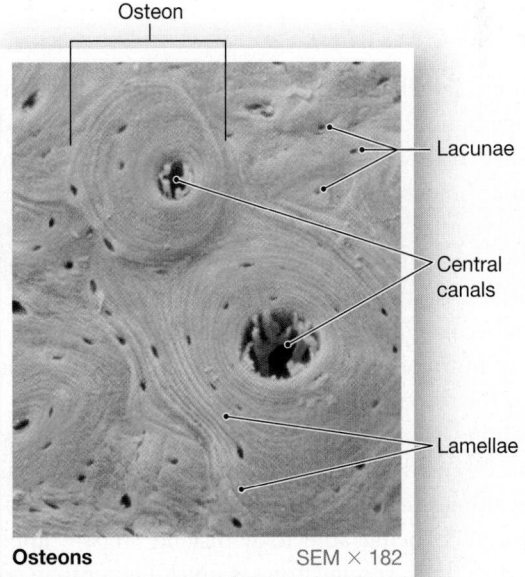

Osteon
Lacunae
Central canals
Lamellae

Osteons   SEM × 182

b Several osteons in compact bone.

Figure 6–6 **The Structure of Compact Bone.**

Circumferential lamellae

Osteons

Perforating fibers

Venule

Capillary

Periosteum

Interstitial lamellae

Concentric lamellae

Trabeculae of spongy bone
(see Figure 6–7)

Vein

Artery   Arteriole   Central canal   Perforating canal

a The organization of osteons and lamellae in compact bone

Central canal

Concentric lamellae

Endosteum

Collagen fiber orientation

b The orientation of collagen fibers in adjacent lamellae

Spongy Bone Structure

In spongy bone, lamellae are not arranged in osteons. The matrix in spongy bone forms a meshwork of supporting bundles of fibers called **trabeculae** (tra-BEK-ū-lē) (**Figure 6–7**). These thin trabeculae branch, creating an open network. There are no capillaries or venules in the matrix of spongy bone. Nutrients reach the osteocytes by diffusion along canaliculi that open onto the surfaces of trabeculae. Red bone marrow is found between the trabeculae of spongy bone. Blood vessels within this tissue deliver nutrients to the trabeculae and remove wastes generated by the osteocytes.

Spongy bone is found where bones are not heavily stressed or where stresses arrive from many directions. The trabeculae are oriented along stress lines and extensively cross-braced.

In addition to being able to withstand stresses applied from many directions, spongy bone is much lighter than compact bone. Spongy bone reduces the weight of the skeleton, making it easier for muscles to move the bones.

Finally, the framework of trabeculae supports and protects the cells of the bone marrow. Spongy bone within the epiphyses of long bones, such as the femur, and the interior of other large bones such as the sternum and ilium (a large bone of the pelvis), contains **red bone marrow**, which forms blood

Figure 6–7 **The Structure of Spongy Bone.**

Trabeculae of spongy bone

Canaliculi opening on surface

Endosteum   Lamellae

cells. At other sites, spongy bone may contain **yellow bone marrow**—adipose tissue important as an energy reserve.

How do compact bone and spongy bone work together to support the body's weight? **Figure 6–8** shows the distribution of forces applied to the femur at the hip joint, and illustrates the functional relationship between compact bone and spongy bone. The head of the femur articulates with a socket on the lateral surface of the pelvis. At the proximal epiphysis of the femur, trabeculae transfer forces from the pelvis to the compact bone of the femoral shaft, across the hip joint. At the distal epiphysis, trabeculae transfer weight from the shaft to the leg, across the knee joint. The femoral

Figure 6–8   The Distribution of Forces on a Long Bone. The femur, or thigh bone, has a diaphysis (shaft) with walls of compact bone and epiphyses filled with spongy bone. The body weight is transferred to the femur at the hip joint. Because the hip joint is off center relative to the axis of the shaft, the body weight is distributed along the bone in a way that compresses the medial (inner) portion of the shaft and stretches the lateral (outer) portion.

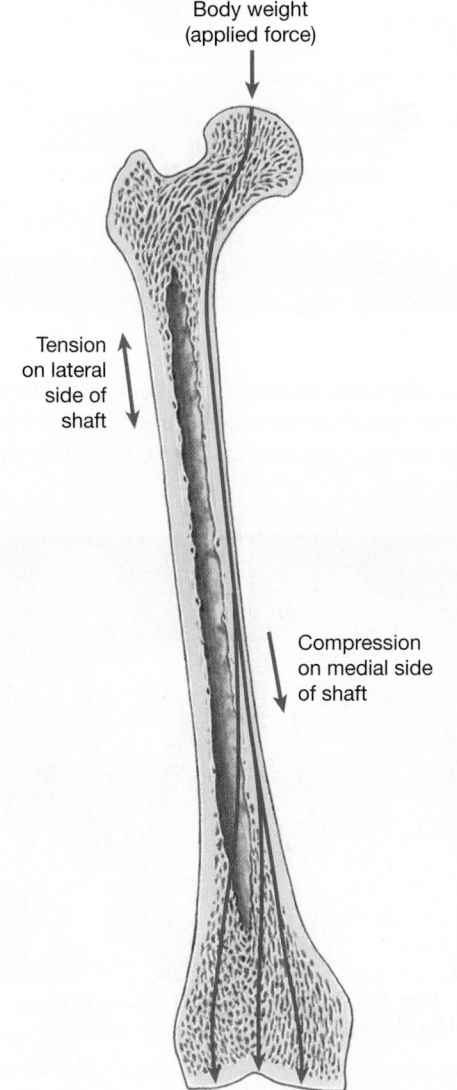

Body weight
(applied force)

Tension
on lateral
side of
shaft

Compression
on medial side
of shaft

head projects medially, and the body weight compresses the medial side of the shaft. However, because the force is applied off center, the bone must also resist the tendency to bend into a lateral bow. So while the medial portion of the shaft is under compression, the lateral portion of the shaft, which resists this bending, is placed under a stretching load, or *tension.* The medullary cavity does not reduce the bone's strength because the center of the bone is not subjected to compression or tension.

The Periosteum and Endosteum

Except within joint cavities, the superficial layer of compact bone that covers all bones is wrapped by a **periosteum**, a membrane with a fibrous outer layer and a cellular inner layer (**Figure 6–9a**). The periosteum (1) isolates the bone from surrounding tissues, (2) provides a route for the blood vessels and nerves, and (3) takes part in bone growth and repair.

Near joints, the periosteum becomes continuous with the connective tissues that hold the bones in place. At a joint that permits significant amounts of movement, called a *synovial joint,* the periosteum is continuous with the joint capsule. The fibers of the periosteum are also interwoven with those of the tendons attached to the bone. As the bone grows, these tendon fibers are cemented into the circumferential lamellae by osteoblasts from the cellular layer of the periosteum. Collagen fibers incorporated into bone tissue from tendons and ligaments, as well as from the superficial periosteum, are called *perforating (Sharpey's) fibers.*

This method of attachment bonds the tendons and ligaments into the general structure of the bone, providing a much stronger attachment than would otherwise be possible. An extremely powerful pull on a tendon or ligament will usually break a bone rather than snap the collagen fibers at the bone surface.

The **endosteum**, an incomplete cellular layer, lines the medullary cavity (**Figure 6–9b**). This layer is active during bone growth, repair, and remodeling. It covers the trabeculae of spongy bone and lines the inner surfaces of the central canals of compact bone. The endosteum consists of a simple flattened layer of osteogenic cells that covers the bone matrix, generally without any intervening connective tissue fibers. Where the cellular layer is not complete, the matrix is exposed. At these exposed sites, osteoclasts and osteoblasts can remove or deposit matrix components. The osteoclasts generally occur in shallow depressions called *osteoclastic crypts (Howship's lacunae)* that they have eroded into the matrix.

CHECKPOINT

7. **Compare the structures and functions of compact bone and spongy bone.**

8. **A sample of bone has lamellae, which are not arranged in osteons. Is the sample most likely taken from the epiphysis or diaphysis?**

See the blue Answers tab at the back of the book.

Figure 6–9   The Periosteum and Endosteum.

Circumferential lamellae

Periosteum

Fibrous layer

Cellular layer

Canaliculi

Osteocyte in lacuna

Perforating fibers

a The periosteum contains outer (fibrous) and inner (cellular) layers. Collagen fibers of the periosteum are continuous with those of the bone, adjacent joint capsules, and attached tendons and ligaments.

Endosteum

Osteoclast

Bone matrix

Osteocyte

Osteogenic cell

Osteoid

Osteoblast

b The endosteum is an incomplete cellular layer containing osteoblasts, osteogenic cells, and osteoclasts.

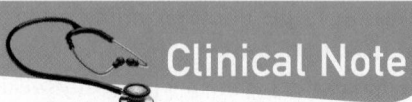

Clinical Note

Heterotopic Bone Formation Bones sometimes form where they do not belong. In response to abnormal stresses, bone may form anywhere in the dermis or within tendons, around joints, in the kidneys, or in skeletal muscles. Dermal bones forming in abnormal locations are called *heterotopic bones* (*hetero-*, different + *topos,* place). These bones can form in very odd places, such as the testes or the whites of the eyes. Physical or chemical events can stimulate the abnormal development of osteoblasts in normal connective tissues, such as sesamoid bones developing within tendons or near points of friction and pressure. Bone can also form within a large blood clot, at an injury site, or within portions of the dermis subjected to overuse.

Persons with the rare genetic disease *fibrodysplasia ossificans progressiva* (FOP) form bone around skeletal muscles and provide the most dramatic demonstrations of heterotopic bone formation. Bone gradually replaces the muscles of the back, neck, and upper limbs. The extent of the conversion can be seen in the photos on the right. **Photo a** shows an adult male with FOP. **Photo b** shows the skeleton of an adult male with advanced FOP. Several of the vertebrae have fused into

a solid mass, and major muscles of the back, shoulders, and hips have undergone extensive ossification. Treatment can be problematic, because any surgical excision may trigger more ossification.

a An adult male with FOP, posterior view

b The skeleton of a man with advanced FOP

6-5 ❙ Bones form through ossification and enlarge through appositional growth and remodeling

Learning Outcome Compare the mechanisms of endochondral ossification and intramembranous ossification.

The growth of the skeleton determines the size and proportions of your body. The bony skeleton begins to form about six weeks after fertilization, when the embryo is approximately 12 mm (0.5 in.) long. (At this stage, the existing skeletal elements are made of cartilage.) During subsequent development, the bones undergo a tremendous increase in size. Bone growth continues through adolescence, and portions of the skeleton generally do not stop growing until about age 25.

In this section, we consider the physical process of bone formation, or *ossification*, and bone growth. Ossification or osteogenesis refers specifically to the formation of bone. The process of **calcification**—the deposition of calcium salts— takes place during ossification, but it can also occur in other tissues. When calcification occurs in tissues other than bone, the result is a calcified tissue (such as calcified cartilage) that does not resemble bone.

Two major forms of ossification exist: endochondral and intramembranous. In *endochondral ossification*, bone replaces existing cartilage. In *intramembranous ossification*, bone develops directly from mesenchyme (loosely organized embryonic connective tissue) or fibrous connective tissue.

Endochondral Ossification

During development, most bones originate as hyaline cartilages that are miniature models of the corresponding bones of the adult skeleton. These cartilage models are gradually replaced by bone through the process of **endochondral** (en-dō-KON-drul) **ossification** (*endo-*, inside + *chondros*, cartilage). The steps in limb bone development provide a good example of this process (**Spotlight Figure 6–11**).

The timing of endochondral ossification can be monitored by comparing the width of the epiphyseal cartilages in successive x-rays. In adults, the former location of this cartilage can often be seen in x-rays as a narrow *epiphyseal line*, which remains after epiphyseal growth has ended (**Figure 6–10**). The completion of epiphyseal growth is called *epiphyseal closure*.

Appositional Growth

A superficial layer of bone, or bone collar, forms early in endochondral ossification (**Spotlight Figure 6–11**). After that, the developing bone increases in diameter through **appositional growth** at the outer surface. In this process, cells of the inner layer of the periosteum differentiate into osteoblasts and deposit superficial layers of bone matrix. Eventually, these osteoblasts become surrounded by matrix and differentiate into osteocytes.

Over much of the surface, appositional growth adds a series of layers that form circumferential lamellae. In time, the deepest circumferential lamellae are recycled and replaced by osteons typical of compact bone. However, blood vessels and collagen fibers of the periosteum can sometimes become enclosed within the matrix produced by osteoblasts. Osteons may then form around the smaller vessels. While bone matrix is being added to the outer surface of the growing bone, osteoclasts are removing bone matrix at the inner surface, but at a slower rate. As a result, the medullary cavity gradually enlarges as the bone gets larger in diameter.

Figure 6–10   **Bone Growth at an Epiphyseal Cartilage.**

a An x-ray of growing epiphyseal cartilages (arrows)

b Epiphyseal lines in an adult (arrows)

SPOTLIGHT

Figure 6–11
Endochondral Ossification

Endochondral ossification begins with the formation of a hyaline cartilage model. This model serves as the pattern for almost all bone development. Limb bone development is a good example of this process. By the time an embryo is six weeks old, the proximal bones of the limbs, the humerus (upper limb) and femur (lower limb), have formed, but they are composed entirely of cartilage. These cartilage models continue to grow by expansion of the cartilage matrix (interstitial growth) and the production of more cartilage at the outer surface (appositional growth).

Initiation of Ossification in the Developing Bone (Steps 1–4)

1 As the cartilage enlarges, chondrocytes near the center of the shaft increase greatly in size. The matrix is reduced to a series of small struts that soon begin to calcify. The enlarged chondrocytes then die and disintegrate, leaving cavities within the cartilage.

2 Blood vessels grow around the edges of the cartilage, and the cells of the perichondrium convert to osteoblasts. The shaft of the cartilage then becomes ensheathed in a superficial layer of bone.

3 Blood vessels penetrate the cartilage and invade the central region. Fibroblasts migrating with the blood vessels differentiate into osteoblasts and begin producing spongy bone at a primary ossification center. Bone formation then spreads along the shaft toward both ends of the former cartilage model.

4 Remodeling occurs as growth continues, creating a medullary cavity. The osseous tissue of the shaft becomes thicker, and the cartilage near each epiphysis is replaced by shafts of bone. Further growth involves increases in length and diameter.

Enlarging chondrocytes within calcifying matrix

Hyaline cartilage

Disintegrating chondrocytes of the cartilage model

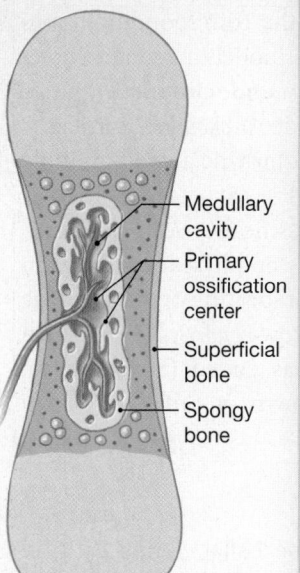

Perichondrium

Epiphysis

Bone collar

Blood vessel

Diaphysis

Periosteum formed from perichondrium

Medullary cavity

Primary ossification center

Superficial bone

Spongy bone

Medullary cavity

Metaphysis

Increasing the Length of a Developing Long Bone (Steps 5–7)

During the initial stages of ossification, osteoblasts move away from the primary ossification center toward the epiphyses. But they do not complete the ossification of the model immediately, because the cartilages of the epiphyses continue to grow. The region where the cartilage is being replaced by bone lies at the metaphysis, the junction between the diaphysis and an epiphysis. On the diaphyseal (shaft) side of the metaphysis, osteoblasts continually invade the cartilage and replace it with bone, while on the epiphyseal side, new cartilage is produced at the same rate. The situation is like a pair of joggers, one in front of the other. As long as they are running at the same speed, they can run for miles without colliding. In this case, the osteoblasts and the epiphysis are both "running away" from the primary ossification center. As a result, the osteoblasts never catch up with the epiphysis, and the bone continues to grow longer and longer.

5

Capillaries and osteoblasts migrate into the epiphyses, creating secondary ossification centers.

Hyaline cartilage

Epiphysis

Metaphysis

Periosteum

Compact bone

Secondary ossification center

6

The epiphyses eventually become filled with spongy bone. The metaphysis, a relatively narrow cartilaginous region called the **epiphyseal cartilage**, or *epiphyseal plate*, now separates the epiphysis from the diaphysis. On the shaft side of the metaphysis, osteoblasts continuously invade the cartilage and replace it with bone. New cartilage is produced at the same rate on the epiphyseal side.

Articular cartilage

Spongy bone

Epiphyseal cartilage

Diaphysis

Within the epiphyseal cartilage, the chondrocytes are organized into zones.

Chondrocytes at the epiphyseal side of the cartilage continue to divide and enlarge.

Chondrocytes degenerate at the diaphyseal side.

Osteoblasts migrate upward from the diaphysis and cartilage is gradually replaced by bone.

7

At puberty, the rate of epiphyseal cartilage production slows and the rate of osteoblast activity accelerates. As a result, the epiphyseal cartilage gets narrower and narrower, until it ultimately disappears. This event is called **epiphyseal closure**. The former location of the epiphyseal cartilage becomes a distinct **epiphyseal line** that remains after epiphyseal growth has ended.

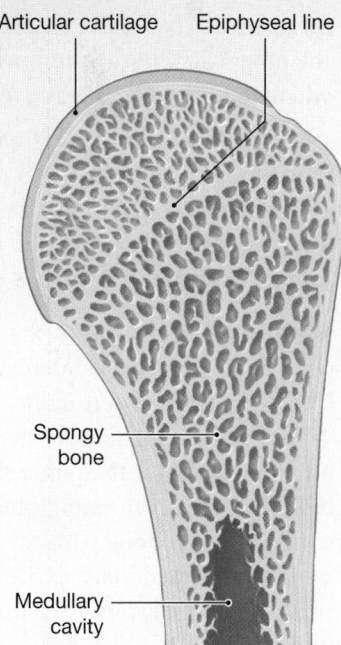

Articular cartilage   Epiphyseal line

Spongy bone

Medullary cavity

A thin cap of the original cartilage model remains exposed to the joint cavity as the **articular cartilage**. This cartilage prevents damaging the joint from bone-to-bone contact.

Intramembranous Ossification

Intramembranous (in-tra-MEM-bra-nus) **ossification** begins when osteoblasts differentiate within a mesenchymal or fibrous connective tissue. This type of ossification is also called *dermal ossification* because it normally takes place in the deeper layers of the dermis. The bones that result are called dermal bones. Examples of dermal bones are the flat bones of the skull, the mandible (lower jaw), and the clavicles (collarbones). The steps of intramembranous ossification are summarized in **Figure 6–12**.

The Blood and Nerve Supplies to Bone

In order for bones to grow and be maintained, they require an extensive blood supply. For this reason, osseous tissue is highly vascular. In a typical bone such as the humerus, three major sets of blood vessels develop (**Figure 6–13**):

1. *The Nutrient Artery and Vein.* The blood vessels that supply the diaphysis form by invading the cartilage model as endochondral ossification begins. Most bones have only one *nutrient artery* and one *nutrient vein*, but a few bones, including the femur, have more than one of each. The vessels enter the bone through one or more round passageways called *nutrient foramina* in the diaphysis. Branches of these large vessels form smaller perforating canals and extend along the length of the shaft into the osteons of the surrounding compact bone.

2. *Metaphyseal Vessels.* The *metaphyseal vessels* supply blood to the inner (diaphyseal) surface of each epiphyseal cartilage, where that cartilage is being replaced by bone.

3. *Periosteal Vessels.* Blood vessels from the periosteum provide blood to the superficial osteons of the shaft. During endochondral bone formation, branches of periosteal vessels also enter the epiphyses, providing blood to the secondary ossification centers.

Following the closure of the epiphyses, all three sets of vessels become extensively interconnected.

The periosteum also contains a network of lymphatic vessels (lymphatics) and sensory nerves. The lymphatics collect lymph from branches that enter the bone and reach individual osteons by the perforating canals. The sensory nerves penetrate the compact bone with the nutrient artery to innervate the endosteum, medullary cavity, and epiphyses. Because of the rich sensory innervation, injuries to bones are usually very painful.

In the next section, we examine the maintenance and replacement of mineral reserves in the adult skeleton.

CHECKPOINT

9. In endochondral ossification, what is the original source of osteoblasts?

10. During intramembranous ossification, which type of tissue is replaced by bone?

11. How could x-rays of the femur be used to determine whether a person has reached full height?

See the blue Answers tab at the back of the book.

6-6 | Bone growth and development depend on a balance between bone formation and bone resorption

Learning Outcome Describe the remodeling and homeostatic mechanisms of the skeletal system.

The process of **remodeling** continuously recycles and renews the organic and mineral components of the bone matrix. Bone remodeling goes on throughout life, as part of normal bone maintenance. Remodeling can replace the matrix but leave the bone as a whole unchanged, or it may change the shape, internal architecture, or mineral content of the bone. Through remodeling, older mineral deposits are removed from bone and released into the circulation at the same time that circulating minerals are being absorbed and deposited.

Bone remodeling involves interplay among the activities of osteocytes, osteoblasts, and osteoclasts. In adults, osteocytes are continuously removing and replacing the surrounding calcium salts. Osteoclasts and osteoblasts also remain active, even after the epiphyseal cartilages have closed. Normally, their activities are balanced: As quickly as osteoblasts form one osteon, osteoclasts remove another by osteolysis.

The turnover rate of bone is quite high. In young adults, almost one-fifth of the skeleton is recycled and replaced each year. Not every part of every bone is affected equally. The rate of turnover differs regionally and even locally. For example, the spongy bone in the head of the femur may be replaced two or three times each year, but the compact bone along the shaft remains largely unchanged.

Because of their biochemical similarity to calcium, heavy-metal ions such as lead, strontium, cobalt, or radioactive uranium or plutonium can be incorporated into the matrix of bone. Osteoblasts do not differentiate between these heavy-metal ions and calcium. This means that any heavy-metal ions present in the bloodstream will be deposited into the bone matrix. Some of these ions are potentially dangerous, and the turnover of bone matrix can have detrimental health effects as ions that are absorbed and accumulated are released into the circulation over a period of years. This was one of the major

Figure 6–12 **Intramembranous Ossification.**

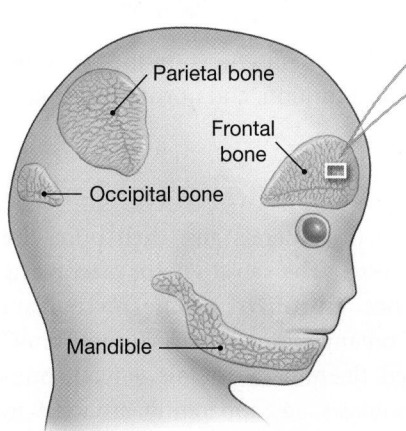

Parietal bone

Frontal bone

Occipital bone

Mandible

Intramembranous ossification starts about the eighth week of embryonic development. This type of ossification occurs in the deeper layers of the dermis, forming **dermal bones**.

Areas of spongy bone are remodeled forming the diploë and a thin covering of compact (cortical) bone.

1 Mesenchymal cells cluster together, differentiate into osteoblasts, and start to secrete the organic components of the matrix. The resulting osteoid then becomes mineralized with calcium salts forming bone matrix.

Bone matrix
Osteoid
Mesenchymal cell
Ossification center
Blood vessel
Osteoblast

2 As ossification proceeds, some osteoblasts are trapped inside bony pockets where they differentiate into osteo-cytes. The developing bone grows outward from the ossification center in small struts called **spicules**.

Spicules

Osteocyte

3 Blood vessels begin to branch within the region and grow between the spicules. The rate of bone growth accelerates with oxygen and a reliable supply of nutrients. As spicules interconnect, they trap blood vessels within the bone.

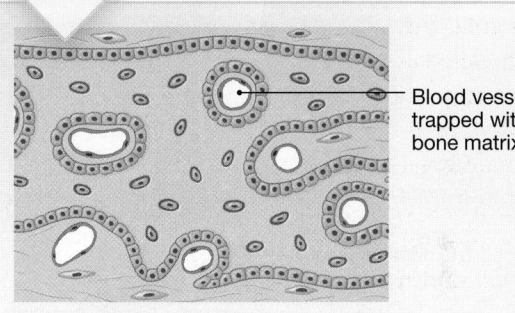

Blood vessel trapped within bone matrix

4 Continued deposition of bone by osteoblasts located close to blood vessels results in a plate of spongy bone with blood vessels weaving throughout.

5 Subsequent remodeling around blood vessels produces osteons typical of compact bone. Osteoblasts on the bone surface along with connective tissue around the bone become the periosteum.

Fibrous periosteum

Blood vessels trapped within bone matrix

Cellular periosteum

6

Figure 6–13 **The Blood Supply to a Mature Bone.**

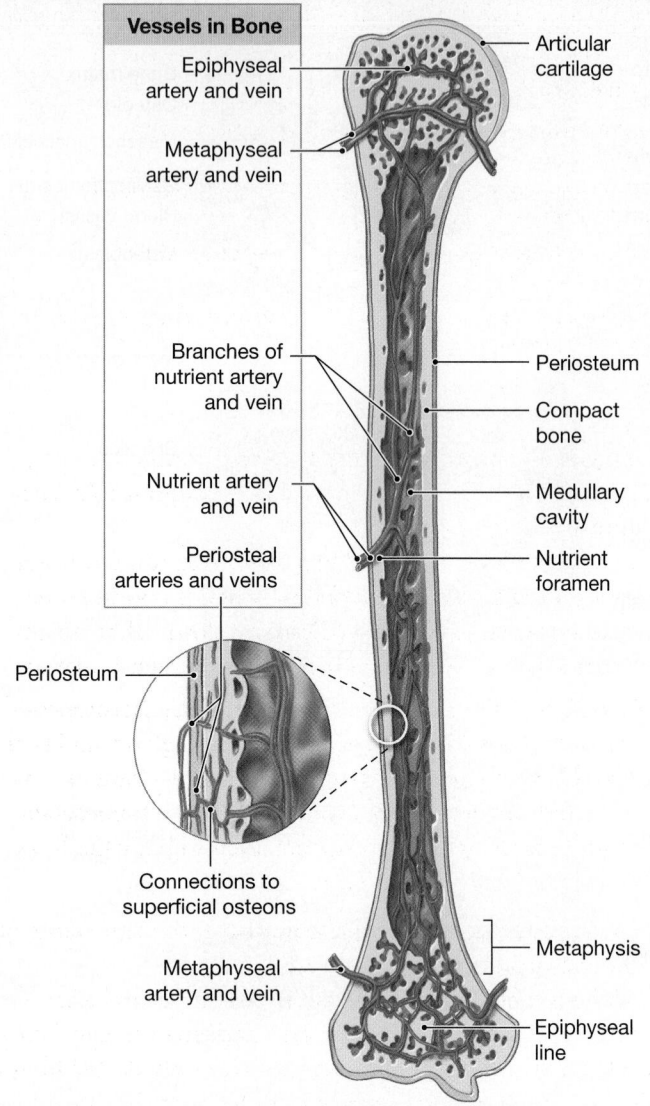

Vessels in Bone

- Epiphyseal artery and vein
- Metaphyseal artery and vein
- Branches of nutrient artery and vein
- Nutrient artery and vein
- Periosteal arteries and veins
- Periosteum
- Connections to superficial osteons
- Metaphyseal artery and vein

- Articular cartilage
- Periosteum
- Compact bone
- Medullary cavity
- Nutrient foramen
- Metaphysis
- Epiphyseal line

complications in the aftermath of the Chernobyl nuclear reactor incident in 1986. Radioactive compounds released in the meltdown of the reactor were deposited into the bones of exposed individuals. Over time, the radiation released by their own bones has caused thyroid cancers, leukemia (cancer of the blood cells, which starts in the red bone marrow), and other potentially fatal cancers.

CHECKPOINT

12. **Describe bone remodeling.**
13. **Explain how heavy-metal ions could be incorporated into bone matrix.**

See the blue Answers tab at the back of the book.

6-7 | Exercise, hormones, and nutrition affect bone development and the skeletal system

Learning Outcome Discuss the effects of exercise, hormones, and nutrition on bone development and on the skeletal system.

What factors have the most important effects on the processes of bone remodeling? Exercise, nutrition, and hormones are key.

The Effects of Exercise on Bone

The turnover and recycling of minerals give each bone the ability to adapt to new stresses. The sensitivity of osteoblasts to electrical events has been theorized as the mechanism that controls the internal organization and structure of bone. Whenever a bone is stressed, the mineral crystals generate minute electrical fields. Osteoblasts are apparently attracted to these electrical fields and, once in the area, begin to produce bone. This finding has led to the successful use of small electric fields in stimulating bone healing.

Because of their adaptability, bone shapes reflect the forces applied to them. For example, bumps and ridges on the surface of a bone mark the sites where tendons are attached. If muscles become more powerful, the corresponding bumps and ridges enlarge to withstand the increased forces. Heavily stressed bones become thicker and stronger, and bones that are not subjected to ordinary stresses become thin and brittle. For this reason, regular exercise is an important stimulus for maintaining normal bone structure. Champion weight lifters typically have massive bones with thick, prominent ridges. In nonathletes (especially "couch potatoes"), moderate amounts of physical activity and weight-bearing activities are essential for stimulating normal bone maintenance and maintaining adequate bone strength.

Degenerative changes in the skeleton take place after relatively brief periods of inactivity. For example, you may use a crutch to take weight off an injured leg while you wear a cast. After a few weeks, your unstressed bones can lose up to a third of their mass. The bones rebuild just as quickly when you resume normal weight loading. However, the removal of calcium salts can be a serious health hazard both for astronauts remaining in a weightless environment and for bedridden or paralyzed patients who spend months or years without stressing their skeleton.

Nutritional and Hormonal Effects on Bone

Normal bone growth and maintenance depend on a combination of nutritional and hormonal factors, such as:

Table 6–1	Hormones Involved in Bone Growth and Maintenance	
Hormone	**Primary Source**	**Effect on Skeletal System**
Calcitriol	Kidneys	Promotes calcium and phosphate ion absorption along the digestive tract
Growth hormone	Pituitary gland	Stimulates osteoblast activity and the synthesis of bone matrix
Thyroxine	Thyroid gland (follicle cells)	With growth hormone, stimulates osteoblast activity and the synthesis of bone matrix
Sex hormones	Ovaries (estrogens) Testes (androgens)	Stimulate osteoblast activity and the synthesis of bone matrix; estrogens stimulate epiphyseal closure earlier than androgens
Parathyroid hormone	Parathyroid glands	Stimulates osteoclast (and osteoblast) activity; increases blood calcium ion concentrations
Calcitonin	Thyroid gland (C cells)	Inhibits osteoclast activity; promotes calcium loss by kidneys; decreases blood calcium ion concentrations

■ *Minerals.* Normal bone growth and maintenance cannot take place without a constant dietary source of calcium and phosphate salts. Lesser amounts of other minerals, such as magnesium, fluoride, iron, and manganese, are also required.

■ *Calcitriol and vitamin D₃.* The hormone *calcitriol*, synthesized in the kidneys, is essential for normal calcium and phosphate ion absorption in the digestive tract. Calcitriol is synthesized from a related steroid, *cholecalciferol* (vitamin D₃), which may be produced in the skin or absorbed from the diet. ⊃ p. 184

■ *Vitamin C.* Adequate levels of vitamin C must be present in the diet. This vitamin, which is required for certain key enzymatic reactions in collagen synthesis, also stimulates osteoblast differentiation. One of the signs of vitamin C deficiency—a condition called *scurvy*—is a loss of bone mass and strength.

■ *Vitamins A, K, and B₁₂.* Three other vitamins have significant effects on bone structure. Vitamin A, which stimulates osteoblast activity, is particularly important for normal bone growth in children. Vitamins K and B₁₂ are required for the synthesis of proteins in normal bone.

■ *Growth hormone and thyroxine.* Growth hormone, produced by the pituitary gland, and *thyroxine,* from the thyroid gland, stimulate bone growth. Growth hormone stimulates protein synthesis and the rates of cell division and cell growth throughout the body. Thyroxine stimulates cell metabolism and increases the rate of osteoblast activity. In proper balance, these hormones maintain normal activity at the epiphyseal cartilages until puberty.

■ *Sex hormones.* At puberty, rising levels of sex hormones (*estrogens* in females and *androgens* in males) stimulate osteoblasts to produce bone faster than the rate at which epiphyseal cartilage expands. Over time, the epiphyseal cartilages narrow and eventually close. The timing of epiphyseal closure differs from bone to bone and from individual to

individual. The toes may complete ossification by age 11, but parts of the pelvis or the wrist may continue to enlarge until about age 25.

Differences in male and female sex hormones account for significant variations in body size and proportions. Because estrogens cause faster epiphyseal closure than do androgens, women are generally shorter than men at maturity.

■ *Calcitonin and parathyroid hormone.* Two other hormones—calcitonin (kal-si-TŌ-nin), from the thyroid gland, and *parathyroid hormone*, from the parathyroid gland—have important effects on bone. They are involved in the homeostatic control of calcium and phosphate levels in body fluids, as we will see in the next section.

The major hormones affecting the growth and maintenance of the skeletal system are summarized in Table 6–1.

The skeletal system is unique in that it persists after life, providing hints to the gender, lifestyle, and environmental conditions experienced by the individual. Not only do the bones reflect the physical stresses placed on the body, but they also provide clues about a person's health and diet. By using the appearance, strength, and composition of bone, forensic scientists and physical anthropologists can detect features characteristic of hormonal deficiencies. Combining the physical clues from the skeleton with modern molecular techniques, such as DNA fingerprinting, a wealth of information can be learned.

CHECKPOINT

14. Why would you expect the arm bones of a weight lifter to be thicker and heavier than those of a jogger?

15. A child who enters puberty several years later than the average age is generally taller than average as an adult. Why?

16. A 7-year-old child has a pituitary gland tumor involving the cells that secrete growth hormone (GH), resulting in increased levels of GH. How will this condition affect the child's growth?

See the blue Answers tab at the back of the book.

Clinical Note

Abnormal Bone Development A variety of endocrine or metabolic problems can result in characteristic skeletal changes. In pituitary dwarfism, inadequate production of growth hormone leads to reduced epiphyseal cartilage activity and abnormally short bones. This condition is becoming increasingly rare in the United States, because children can be treated with synthetic human growth hormone.

Gigantism (also known as *giantism*) results from an overproduction of growth hormone before puberty (**Photo a**). (The world record for height is 272 cm, or 8 ft, 11 in. It was reached by Robert Wadlow, of Alton, Illinois, who died at age 22 in 1940. Wadlow weighed 216 kg, or 475 lb.) If growth hormone levels rise abnormally after epiphyseal cartilages close, the skeleton does not grow longer. Instead, bones get thicker, especially in the face, jaw, and hands. Cartilage growth and alterations in soft-tissue structure lead to changes in physical features, such as the contours of the face. These physical changes take place in the disorder called **acromegaly**.

Several inherited metabolic conditions that affect many systems influence the growth and development of the skeletal system. These conditions produce characteristic variations in body proportions. For example, many individuals with **Marfan's syndrome** are very tall and have long, slender

limbs (**Photo b**). The cause is excessive cartilage formation at the epiphyseal cartilages. These body proportions are not in themselves dangerous. However, the underlying mutation affects the structure of connective tissue throughout the body and commonly causes life-threatening cardiovascular problems.

a Gigantism

b Marfan's syndrome

6-8 | Calcium plays a critical role in bone physiology

Learning Outcome Explain the role of calcium as it relates to the skeletal system.

A dynamic relationship exists between calcium and the skeletal system, and hormones influence calcium balance in the body.

The Skeleton as a Calcium Reserve

A chemical analysis of bone reveals its importance as a mineral reserve (**Figure 6–14**). For the moment, we will focus on the homeostatic regulation of calcium ion concentration in body fluids. (We will consider other minerals in later chapters.) Calcium is the most abundant mineral in the human body. The typical human body contains 1–2 kg (2.2–4.4 lb) of calcium, with nearly 99 percent of it deposited in the skeleton.

Calcium ions play a role in a variety of physiological processes, so the body must tightly control calcium ion concentrations in order to prevent damage to essential physiological systems. Even small variations from the normal concentration affect cellular operations. Larger changes can cause a clinical

crisis. Calcium ions are particularly important to both the plasma membranes and the intracellular activities of neurons and muscle cells, especially cardiac muscle cells. If the calcium concentration of body fluids increases by 30 percent, neurons and muscle cells become unresponsive. If calcium levels decrease by 35 percent, neurons become so excitable that convulsions can occur. A 50 percent reduction in calcium concentration generally causes death. Calcium ion concentration is so closely regulated, however, that daily fluctuations of more than 10 percent are highly unusual.

Hormones and Calcium Balance

Two hormones with opposing effects maintain calcium ion homeostasis. These hormones, parathyroid hormone and calcitonin, coordinate the storage, absorption, and excretion of calcium ions. Three target sites and functions are involved: (1) the bones (storage), (2) the digestive tract (absorption), and (3) the kidneys (excretion).

Figure 6–15a indicates factors that raise calcium levels in the blood. When calcium ion concentrations in the blood fall below normal, cells of the **parathyroid glands**, embedded in the thyroid gland in the neck, release **parathyroid hormone**

Figure 6–14   **A Chemical Analysis of Bone.**

Composition of Bone		Bone Contains
	Calcium 39%	99% of the body's Calcium
	Potassium 0.2%	4% of the body's Potassium
	Sodium 0.7%	35% of the body's Sodium
Organic compounds (mostly collagen) 33%	Magnesium 0.5%	50% of the body's Magnesium
	Carbonate 9.8%	80% of the body's Carbonate
	Phosphate 17%	99% of the body's Phosphate
	Total inorganic 67% components	

(PTH) into the bloodstream. Parathyroid hormone has three major effects, all of which *increase* blood calcium levels:

1. *Stimulating osteoclast activity (indirectly).* Osteoclasts do not have PTH receptors, so PTH binds to adjacent osteoblasts. This binding causes the osteoblasts to release an osteoclast differentiation factor called RANKL (receptor activator of nuclear factor-kB). RANKL activates receptors on pre-osteoclast cells, which in turn causes them to differentiate into mature osteoclasts. These mature osteoclasts release enzymes that promote bone resorption, thereby releasing calcium ions into the bloodstream.

2. *Increasing the rate of intestinal absorption of calcium ions* by enhancing the action of calcitriol. Under normal circumstances, calcitriol is always present, and parathyroid hormone controls its effect on the intestinal epithelium.

3. *Decreasing the rate of excretion of calcium ions by the kidneys.*

Under these conditions, *more* calcium ions *enter* body fluids, and *fewer* are lost. The calcium ion concentration increases to normal levels, and homeostasis is restored.

Figure 6–15b indicates factors that lower blood calcium levels. If the calcium ion concentration of the blood rises above normal, *parafollicular cells (C cells)* in the thyroid gland secrete **calcitonin**. This hormone has two major functions, which together act to *decrease* calcium ion concentrations in the blood:

1. *Inhibiting osteoclast activity.*

2. *Increasing the rate of excretion of calcium ions by the kidneys.*

Under these conditions, *less* calcium *enters* the blood because osteoclasts leave the mineral matrix of bone alone. *More* calcium *leaves* the blood because osteoblasts continue to produce new bone matrix while calcium ion excretion by the kidneys increases. Lower levels of PTH (and calcitriol) also reduce the intestinal absorption of calcium. The net result is a decline in the calcium ion concentration in the blood, restoring homeostasis.

By providing a calcium reserve, the skeleton plays the primary role in the homeostatic maintenance of normal blood calcium ion concentrations. This function can have a direct effect on the shape and strength of the bones in the skeleton. When large numbers of calcium ions are released from the bones, the bones become weaker. When calcium salts are deposited, the bones become denser and stronger.

Changes in mineral content do not necessarily affect the shape of the bone because the bone matrix contains protein fibers as well as mineral deposits. In *osteomalacia* (os-tē-ō-ma-LĀ-shē-uh; *malakia*, softness), the bones appear normal, although they are weak and flexible due to poor mineralization. *Rickets*, a form of osteomalacia affecting children, generally results from a vitamin D_3 deficiency caused by inadequate skin exposure to sunlight and an inadequate dietary supply of the vitamin. ↺ p. 185 The bones of children with rickets are so poorly mineralized that they become very flexible. The bones bend laterally and affected individuals develop a bowlegged appearance because the walls of each femur can no longer resist the tension and compression forces applied by the body weight (**Figure 6–8**). In the United States, homogenized milk is fortified with vitamin D specifically to prevent rickets.

CHECKPOINT

17. **Identify the hormones involved in stimulating and inhibiting the release of calcium ions from bone matrix.**

18. **Why does a child who has rickets have difficulty walking?**

19. **What effect would increased PTH secretion have on blood calcium levels?**

20. **How does calcitonin help lower the calcium ion concentration of blood?**

See the blue Answers tab at the back of the book.

Figure 6–15   **Factors That Alter the Concentration of Calcium Ions in Blood.**

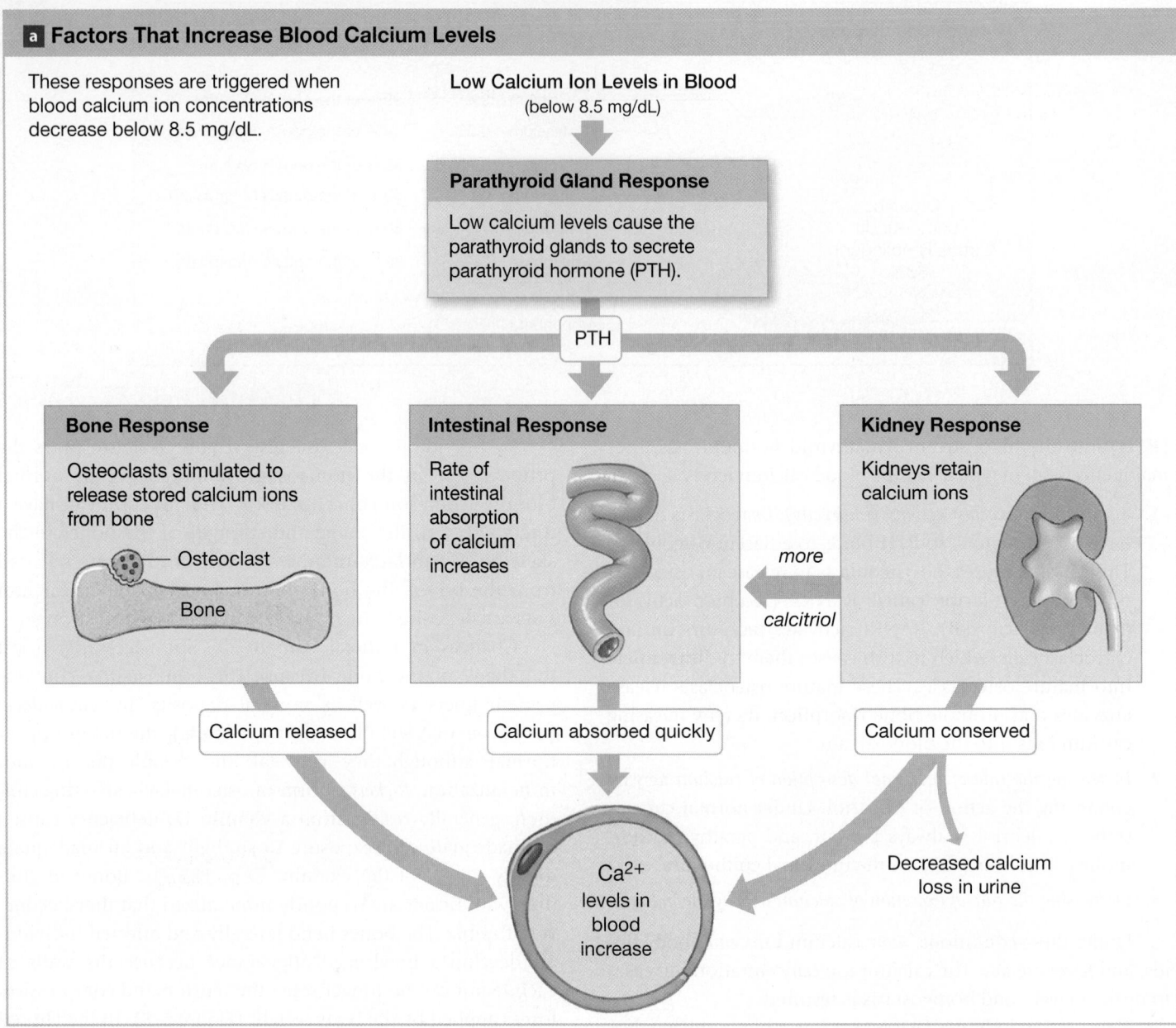

a **Factors That Increase Blood Calcium Levels**

These responses are triggered when blood calcium ion concentrations decrease below 8.5 mg/dL.

Low Calcium Ion Levels in Blood
(below 8.5 mg/dL)

Parathyroid Gland Response

Low calcium levels cause the parathyroid glands to secrete parathyroid hormone (PTH).

PTH

Bone Response

Osteoclasts stimulated to release stored calcium ions from bone

Osteoclast

Bone

Intestinal Response

Rate of intestinal absorption of calcium increases

more
calcitriol

Kidney Response

Kidneys retain calcium ions

Calcium released

Calcium absorbed quickly

Calcium conserved

Ca^{2+} levels in blood increase

Decreased calcium loss in urine

6-9 ┃ A fracture is a crack or break in a bone

Learning Outcome Describe the types of fractures, and explain how fractures heal.

Despite its mineral strength, bone can crack or even break if subjected to extreme loads, sudden impacts, or stresses from unusual directions. The damage is called a **fracture**. Most fractures heal even after severe damage, if the blood supply and the cellular components of the endosteum and periosteum survive. **Spotlight Figure 6–16** on pp. 226–227 illustrates the different types of fractures and the repair process.

CHECKPOINT

21. List the four steps involved in fracture repair, beginning at the onset of the bone break.

22. At which point in fracture repair would you find an external callus?

See the blue Answers tab at the back of the book.

b Factors That Decrease Blood Calcium Levels

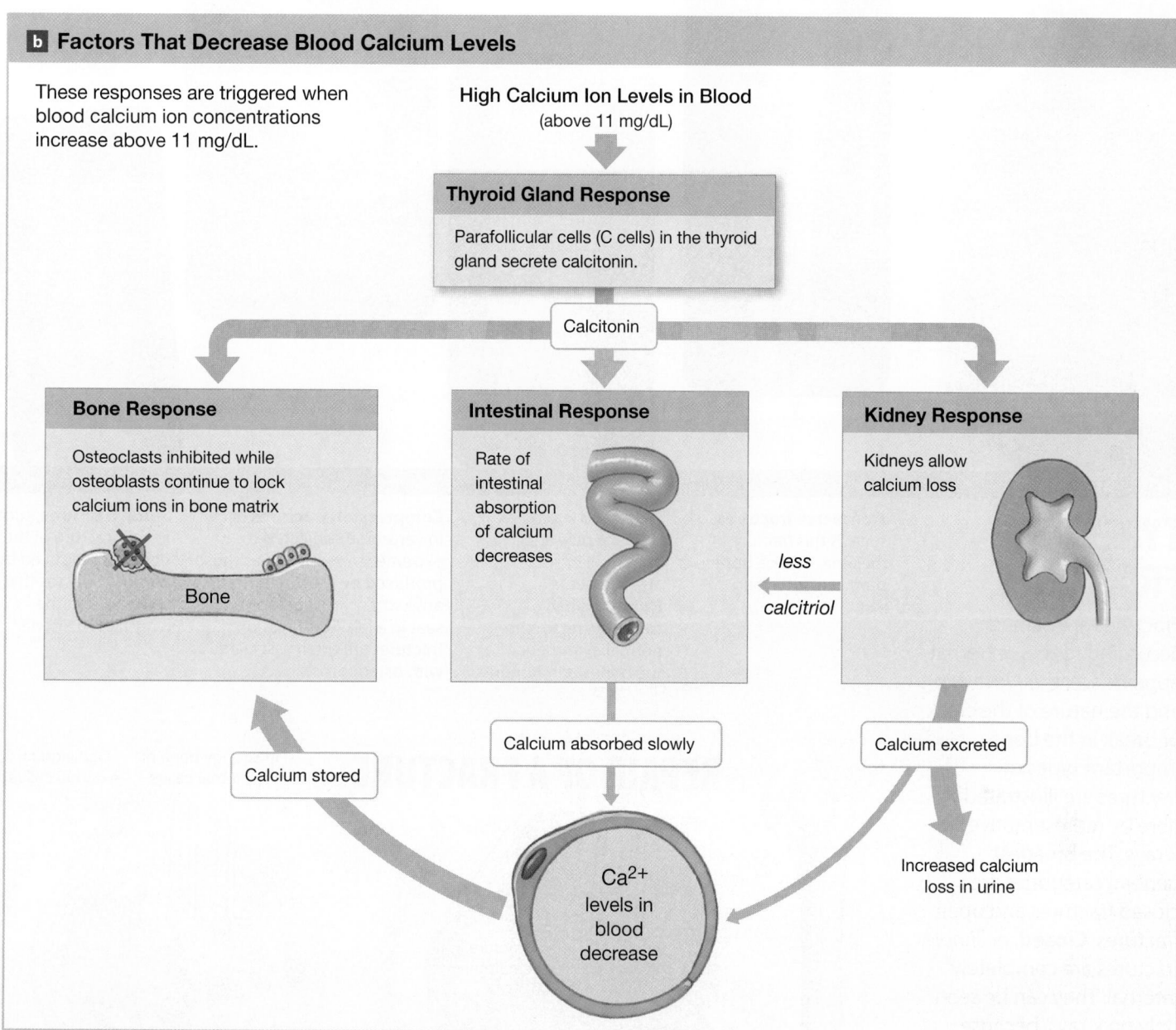

These responses are triggered when blood calcium ion concentrations increase above 11 mg/dL.

High Calcium Ion Levels in Blood
(above 11 mg/dL)

Thyroid Gland Response

Parafollicular cells (C cells) in the thyroid gland secrete calcitonin.

Calcitonin

Bone Response

Osteoclasts inhibited while osteoblasts continue to lock calcium ions in bone matrix

Bone

Intestinal Response

Rate of intestinal absorption of calcium decreases

Kidney Response

Kidneys allow calcium loss

less

calcitriol

Calcium stored

Calcium absorbed slowly

Calcium excreted

Ca^{2+} levels in blood decrease

Increased calcium loss in urine

6-10 ▍ Osteopenia has a widespread effect on aging skeletal tissue

Learning Outcome Summarize the effects of the aging process on the skeletal system.

The bones of the skeleton become thinner and weaker as a normal part of the aging process. Inadequate ossification is called **osteopenia** (os-tē-ō-PĒ-nē-uh; *penia,* lacking). All of us become slightly osteopenic as we age. This reduction in bone mass begins between ages 30 and 40. During that time, osteoblast activity begins to decrease, while osteoclast activity continues at previous levels. Once

the reduction begins, women lose about 8 percent of their skeletal mass every decade. Men lose less—about 3 percent per decade. Not all parts of the skeleton are equally affected. Epiphyses, vertebrae, and the jaws lose more mass than other sites, resulting in fragile limbs, reduction in height, and loss of teeth.

When the reduction in bone mass is sufficient to compromise normal function, the condition is known as **osteoporosis** (os-tē-ō-po-RŌ-sis; *porosus,* porous). The fragile bones that result are likely to break when exposed to stresses that younger individuals could easily tolerate. For example, a hip fracture can occur when a 90-year-old person simply tries to stand. Any

Transverse fracture

Displaced fracture

Compression fracture

Spiral fracture

TYPES OF FRACTURES

Fractures are named according to their external appearance, their location, and the nature of the crack or break in the bone. Important types of fractures are illustrated here by representative x-rays. The broadest general categories are closed fractures and open fractures. **Closed**, or *simple*, fractures are completely internal. They can be seen only on x-rays, because they do not involve a break in the skin. **Open**, or *compound*, fractures project through the skin. These fractures, which are obvious on inspection, are more dangerous than closed fractures, due to the possibility of infection or uncontrolled bleeding. Many fractures fall into more than one category, because the terms overlap.

Transverse fractures, such as this fracture of the ulna, break a bone shaft across its long axis.

Displaced fractures produce new and abnormal bone arrangements. **Nondisplaced fractures** retain the normal alignment of the bones or fragments.

Compression fractures occur in vertebrae subjected to extreme stresses, such as those produced by the forces that arise when you land on your seat in a fall. Compression fractures are often associated with osteoporosis.

Spiral fractures, such as this fracture of the tibia, are produced by twisting stresses that spread along the length of the bone.

REPAIR OF A FRACTURE

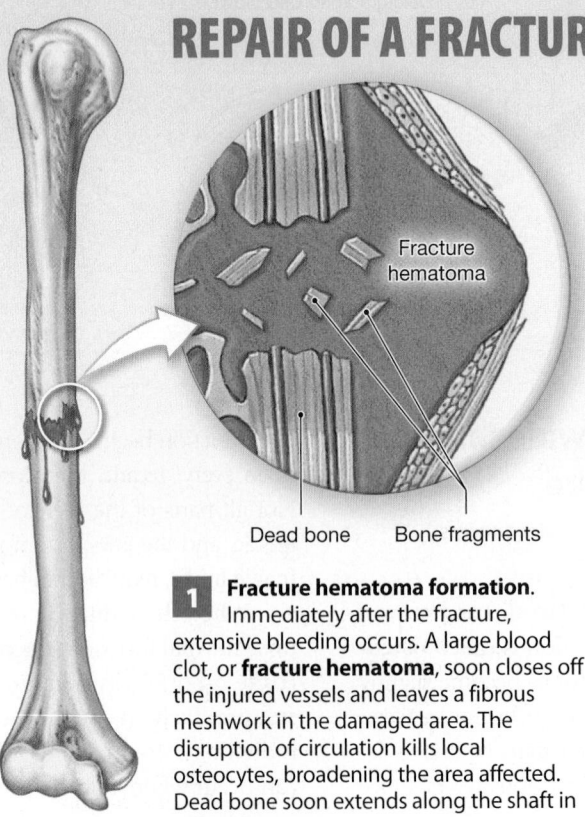

Fracture hematoma

Dead bone   Bone fragments

Spongy bone of internal callus   Cartilage of external callus

Spongy bone of external callus   Periosteum

1 **Fracture hematoma formation.** Immediately after the fracture, extensive bleeding occurs. A large blood clot, or **fracture hematoma**, soon closes off the injured vessels and leaves a fibrous meshwork in the damaged area. The disruption of circulation kills local osteocytes, broadening the area affected. Dead bone soon extends along the shaft in either direction.

2 **Callus formation.** The cells of the intact endosteum and periosteum undergo rapid cycles of cell division, and the daughter cells migrate into the fracture zone. An **internal callus** forms as a network of spongy bone unites the inner edges of the fracture. An **external callus** of cartilage and bone encircles and stabilizes the outer edges of the fracture.

Epiphyseal fracture

Comminuted fracture

Greenstick fracture

Colles fracture

Pott's fracture

Epiphyseal fractures, such as this fracture of the femur, tend to occur where the bone matrix is undergoing calcification and chondrocytes are dying. A clean transverse fracture along this line generally heals well. Unless carefully treated, fractures between the epiphysis and the epiphyseal cartilage can permanently stop growth at this site.

Comminuted fractures, such as this fracture of the femur, shatter the affected area into a multitude of bony fragments.

In a **greenstick fracture**, such as this fracture of the radius, only one side of the shaft is broken, and the other is bent. This type of fracture generally occurs in children, whose long bones have yet to ossify fully.

A **Colles fracture**, a break in the distal portion of the radius, is typically the result of reaching out to cushion a fall.

A **Pott's fracture**, also called a bimalleolar fracture, occurs at the ankle and affects both the medial malleolus of the distal tibia and the lateral malleolus of the distal fibula.

Internal callus   External callus

External callus

3 **Spongy bone formation.** As the repair continues, osteoblasts replace the central cartilage of the external callus with spongy bone, which then unites the broken ends. Fragments of dead bone and the areas of bone closest to the break are resorbed and replaced. The ends of the fracture are now held firmly in place and can withstand normal stresses from muscle contractions.

4 **Compact bone formation.** A swelling initially marks the location of the fracture. Over time, this region will be remodeled by osteoblasts and osteoclasts, and little evidence of the fracture will remain. The repair may be "good as new" or the bone may be slightly thicker and stronger than normal at the fracture site. Under comparable stresses, a second fracture will generally occur at a different site.

Figure 6–17 **The Effects of Osteoporosis on Spongy Bone.**

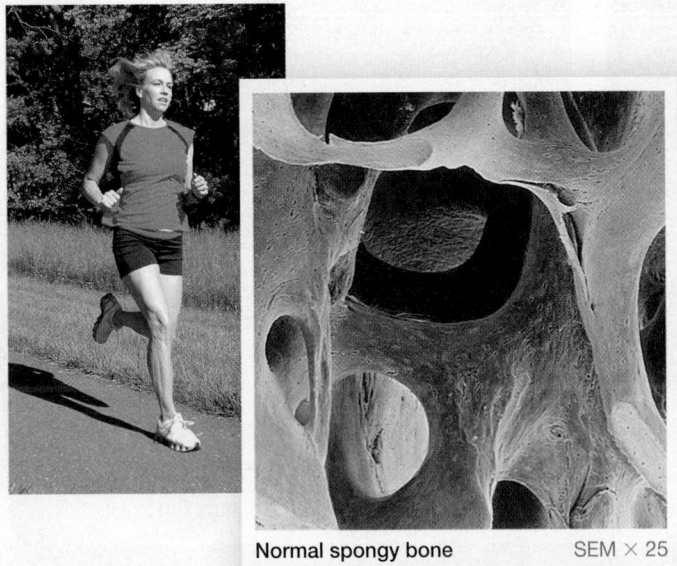

Normal spongy bone   SEM × 25

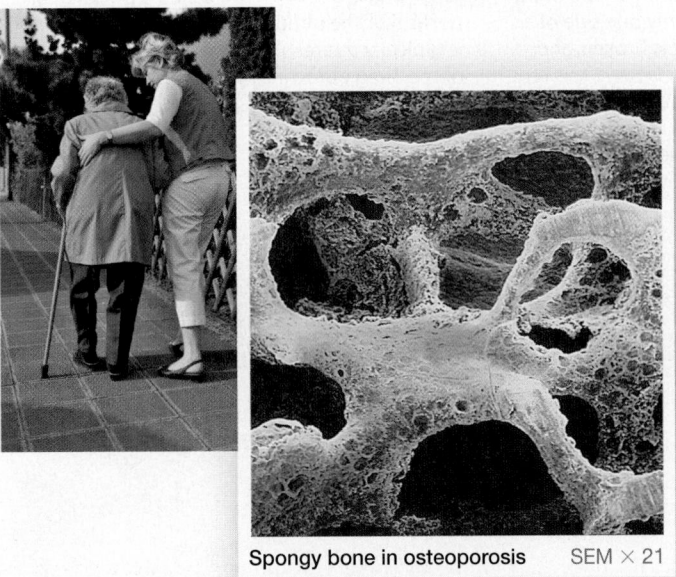

Spongy bone in osteoporosis   SEM × 21

fractures in aged individuals lead to a loss of independence and an immobility that further weakens the skeleton. The extent of the loss of spongy bone mass due to osteoporosis is shown in **Figure 6–17.** The reduction in compact bone mass is equally severe.

Sex hormones are important in maintaining normal rates of bone deposition. Over age 45, an estimated 29 percent of women and 18 percent of men have osteoporosis. In women, the condition accelerates after menopause, due to a decline in circulating estrogens. Severe osteoporosis is less common in men under age 60 than in women of that same age group because men continue to produce androgens until late in life.

Osteoporosis can also develop as a secondary effect of many cancers. Cancers of the bone marrow, breast, or other tissues release a chemical known as **osteoclast-activating factor.** It increases both the number and activity of osteoclasts and produces severe osteoporosis.

CHECKPOINT

23. Define osteopenia.

24. Why is osteoporosis more common in older women than in older men?

See the blue Answers tab at the back of the book.

Related Clinical Terms

achondroplasia: A disorder of bone growth that causes the most common type of dwarfism.

bone marrow transplant: Transferring healthy bone marrow stem cells from one person into another, replacing bone marrow that is either dysfunctional or has been destroyed by chemotherapy or radiation.

bone mineral density test (BMD): A test to predict the risk of bone fractures by measuring how much calcium and other types of minerals are present in the patient's bones.

bone scan: A nuclear scanning test that identifies new areas of bone growth or breakdown. Used to evaluate damage, find

cancer in the bones, and/or monitor the bone's conditions (including infection and trauma).

closed reduction: The correction of a bone fracture by manipulation without incision into the skin.

dual energy x-ray absorptiometry (DEXA): Procedure that uses very small amounts of radiation to measure changes in bone density as small as 1 percent; the test monitors bone density in osteoporosis and osteopenia.

open reduction: The correction of a bone fracture by making an incision into the skin and rejoining the fractured bone parts, often by mechanical means such as a rod, plate, or screw.

orthopedics: The branch of medicine dealing with the correction of deformities of bones or muscles.

osteogenesis imperfecta (OI): An inherited (genetic) disorder characterized by extreme fragility of the bones; also called brittle bone disease.

osteomyelitis: An acute or chronic bone infection.

osteopetrosis: A rare hereditary bone disorder in which the bones become overly dense; it presents in one of three forms: osteopetrosis tarda, osteopetrosis congenita, and "marble bone" disease.

osteosarcoma: A type of cancer that starts in the bones; also called osteogenic sarcoma.

Paget's disease: A chronic disorder that can result in enlarged and misshapen bones due to abnormal bone destruction and regrowth.

traction: The application of a sustained pull on a limb or muscle in order to maintain the position of a fractured bone until healing occurs or to correct a deformity.

Chapter Review

Study Outline

MasteringA&P®

Access more chapter study tools online in the MasteringA&P Study Area:

- Chapter Quizzes, Chapter Practice Test, Art-labeling Activities, Animations, MP3 Tutor Sessions, and Clinical Case Studies
- Practice Anatomy Lab PAL
- Interactive Physiology iP
- A&P Flix A&PFlix
- PhysioEx PhysioEx

An Introduction to the Skeletal System p. 205

1. The skeletal system has a variety of purposes, such as providing a framework for body posture and allowing for precise movements.

6-1 The skeletal system has five primary functions p. 205

2. The skeletal system includes the bones of the skeleton and the cartilages, ligaments, and other connective tissues that stabilize or connect the bones. The functions of the skeletal system include support, storage of minerals and lipids, blood cell production, protection, and leverage.

6-2 Bones are classified according to shape and structure, and they have a variety of surface markings p. 206

3. Bones may be categorized as **sutural bones,** *(Wormian bones)* **irregular bones, short bones, flat bones, long bones,** and **sesamoid bones.** *(Figure 6–1)*

4. Each bone has characteristic **bone markings,** including elevations, projections, depressions, grooves, and tunnels. *(Figure 6–2)*

5. The two types of bone tissue are compact *(dense)* bone and spongy *(cancellous)* bone.

6. A representative long bone has a **diaphysis, epiphyses, metaphyses,** and a **medullary cavity.** *(Figure 6–3)*

7. The medullary cavity and spaces within spongy bone contain either **red bone marrow** (for blood cell formation) or **yellow bone marrow** (for lipid storage).

6-3 Bone is composed of matrix and several types of cells: osteocytes, osteoblasts, osteogenic cells, and osteoclasts p. 208

8. Osseous tissue is a supporting connective tissue with a solid matrix and is ensheathed by a *periosteum.*

9. Bone matrix consists largely of crystals of **hydroxyapatite;** the minerals are deposited in **lamellae.**

10. **Osteocytes,** located in *lacunae,* are mature bone cells. Adjacent osteocytes are interconnected by **canaliculi. Osteoblasts** synthesize the bony matrix by **ossification,** or **osteogenesis. Osteoclasts** dissolve the bony matrix through **osteolysis. Osteogenic cells** differentiate into osteoblasts. *(Figure 6–4)*

6-4 Compact bone contains parallel osteons, and spongy bone contains trabeculae p. 210

11. The basic functional unit of compact bone is the **osteon,** containing osteocytes arranged around a **central canal. Perforating canals** extend perpendicularly to the bone surface. *(Figures 6–5, 6–6)*

12. Compact bone is located where stresses come from a limited range of directions, such as along the diaphysis of long bones.

13. Spongy bone contains **trabeculae,** typically in an open network. *(Figure 6–7)*

14. Spongy bone is located where stresses are few or come from many directions, such as at the epiphyses of long bones. *(Figure 6–8)*

15. A bone is covered by a **periosteum** and lined with an **endosteum.** *(Figure 6–9)*

6-5 Bones form through ossification and enlarge through appositional growth and remodeling p. 215

16. **Ossification** (or **osteogenesis**) is the process of bone formation. **Calcification** is the process of depositing calcium salts within a tissue.

17. **Endochondral ossification** begins with a cartilage model that is gradually replaced by bone at the metaphysis. In this way, bone length increases. *(Spotlight Figure 6–11)*

18. The ends of the cartilage model remain as **articular cartilages.** *(Spotlight Figure 6–11)*

19. The timing of **epiphyseal closure** of the **epiphyseal cartilage** and the formation of an **epiphyseal line** differs among bones and among individuals. *(Figure 6-10, Spotlight Figure 6-11)*
20. Bone diameter increases through **appositional growth**.
21. **Intramembranous ossification** begins when osteoblasts differentiate within connective tissue. The process produces dermal bones. Such ossification begins at an **ossification center**. *(Figure 6–12)*
22. Three major sets of blood vessels provide an extensive supply of blood to bone. *(Figure 6–13)*

6-6 ▌ **Bone growth and development depend on a balance between bone formation and bone resorption** p. 218

23. The organic and mineral components of bone are continuously recycled and renewed through **remodeling**, which involves interplay among the activities of osteocytes, osteoblasts, and osteoclasts.

6-7 ▌ **Exercise, hormones, and nutrition affect bone development and the skeletal system** p. 220

24. The shapes and thicknesses of bones reflect the stresses applied to them. For this reason, regular exercise is important in maintaining normal bone structure.
25. Normal ossification requires a reliable source of minerals, vitamins, and hormones.

26. *Growth hormone* and *thyroxine* stimulate bone growth. Calcitonin and parathyroid hormone control blood calcium levels. *(Table 6–1)*

6-8 ▌ **Calcium plays a critical role in bone physiology** p. 222

27. Calcium is the most abundant mineral in the human body; about 99 percent of it is located in the skeleton. *(Figure 6–14)*
28. Interactions among the bones, digestive tract, and kidneys affect the calcium ion concentration. *(Figure 6–15)*
29. Two hormones, **calcitonin** and **parathyroid hormone (PTH)**, regulate calcium ion homeostasis. Calcitonin leads to a decrease in the calcium ion concentration in blood, whereas parathyroid hormone increases the calcium ion concentration in blood. *(Figure 6–15)*

6-9 ▌ **A fracture is a crack or break in a bone** p. 224

30. A break or crack in a bone is a **fracture**. The repair of a fracture involves the formation of a **fracture hematoma**, an **external callus**, and an **internal callus**. *(Spotlight Figure 6–16)*

6-10 ▌ **Osteopenia has a widespread effect on aging skeletal tissue** p. 225

31. The effects of aging on the skeleton include **osteopenia** and **osteoporosis**. *(Figure 6–17)*

Review Questions

See the blue Answers tab at the back of the book.

LEVEL 1 Reviewing Facts and Terms

1. Which of the following are involved in osteogenesis? (a) fibroblasts, (b) osteoblasts, (c) osteocytes, (d) osteoclasts.
2. The tensile strength of bone is due to (a) hydroxyapatite, (b) calcium phosphate, (c) calcium carbonate, (d) collagen fibers.
3. Osteoprogenitor cells are stem cells that (a) produce new osteoblasts, (b) produce new osteoclasts, (c) differentiate into osteocytes, (d) differentiate into fibroblasts.
4. Which statement best describes perforating (Volkmann's) canals in bone? (a) They are found only in spongy bone. (b) They supply blood to tissues in the medullary cavity. (c) They connect Haversian canals to the periosteum. (d) They produce red blood cells.
5. The vitamins essential for normal adult bone maintenance and repair are (a) A and E, (b) C and D_3, (c) B and E, (d) B complex and K.
6. Select the first event that occurs in the repair of a simple fracture. (a) reabsorption of the internal callus, (b) formation of internal callus, (c) formation of a hematoma, (d) invasion by fibroblasts.
7. Classify the bones in the following diagram according to their shape.

    (a) _____   (b) _____
    (c) _____   (d) _____
    (e) _____   (f) _____

8. The presence of an epiphyseal line indicates (a) epiphyseal growth has ended, (b) epiphyseal growth is just beginning, (c) growth of bone diameter is just beginning, (d) the bone is fractured at the location, (e) no particular event.

9. Why is it important to make collagen fibers within the callus of a fractured bone? (a) seals the fractured bone, preventing blood loss, (b) stimulates macrophages to phagocytose dead tissue, (c) provides a framework for repair, (d) collagen is essential for calcification by osteocytes.

10. The nonpathologic loss of bone that occurs with aging is called (a) osteomyelitis, (b) osteoporosis, (c) osteopenia, (d) osteitis, (e) osteomalacia.

11. What are the five primary functions of the skeletal system?

12. List the four distinctive cell populations of osseous tissue.

13. What are the primary parts of a typical long bone?

14. Which fracture type is typically seen in children? (a) displaced, (b) spiral, (c) comminuted, (d) green stick.

15. List the organic and inorganic components of bone matrix.

16. (a) What nutritional factors are essential for normal bone growth and maintenance? (b) What hormonal factors are necessary for normal bone growth and maintenance?

17. Which three organs or tissues interact to assist in the regulation of calcium ion concentration in body fluids?

18. What major effects of parathyroid hormone oppose those of calcitonin?

LEVEL 2 Reviewing Concepts

19. If spongy bone has no osteons, how do nutrients reach the osteocytes?

20. Why are stresses or impacts to the side of the shaft in a long bone more dangerous than stress applied to the long axis of the shaft?

21. Why do extended periods of inactivity cause degenerative changes in the skeleton?

22. How does the structure of spongy bone relate to its function at the epiphyses in long bones?

23. Why would a physician concerned about the growth patterns of a young child request an x-ray of the hand?

24. Describe what happens during bone remodeling and identify the type of bone which is most at risk of poor remodeling as we age.

25. The process of bone growth at the epiphyseal cartilage is similar to (a) intramembranous ossification, (b) endochondral ossification, (c) the process of osteopenia, (d) the process of healing a fracture, (e) the process of calcification.

26. Describe the stages of fracture repair when osteoblasts and osteoclasts are most active and describe their role during these phases.

LEVEL 3 Critical Thinking and Clinical Applications

27. While playing on her swing set, 10-year-old Sally falls and breaks her right leg. At the emergency room, the doctor tells her parents that the proximal end of the tibia where the epiphysis meets the diaphysis is fractured. The fracture is properly set and eventually heals. During a routine physical when she is 18, Sally learns that her right leg is 2 cm shorter than her left, probably because of her accident. What might account for this difference?

28. Five weeks ago Michael fractured his left tibia. Now the fracture is healing well. Amongst other exercises, partial weight bearing (putting weight back onto the fractured tibia whilst walking) is recommended. Using your knowledge of fracture repair, explain why this has been recommended.

29. Frank does not begin puberty until he is 16. What effect would you predict this will have on his stature? (a) Frank will probably be taller than if he had started puberty earlier, (b) Frank will probably be shorter than if he had started puberty earlier, (c) Frank will probably be a dwarf, (d) Frank will have bones that are heavier than normal, (e) The late onset of puberty will have no effect on Frank's stature

30. In physical anthropology, cultural conclusions can be drawn from a thorough examination of the skeletons of ancient peoples. What sorts of clues might bones provide as to the lifestyles of those individuals?

CLINICAL CASE Wrap-Up

A Case of Child Abuse?

Alex and his mother both suffer from *osteogenesis imperfecta* (OI), which means the imperfect formation of bone. Osteogenesis imperfecta is the most commonly inherited connective tissue disorder in the United States. The inherited defective gene codes for collagen that is made by osteoblasts and released into the bone matrix. This collagen is critical for bone strength and also important in forming teeth, ligaments, and the whites of the eyes. There are several types of OI with varying severities, but the Otero family suffers from the most common, relatively mild type.

Without enough normal collagen in the bone matrix, the bones are brittle and fracture easily, often with minimal trauma. In more severe cases, the skeleton is not strong enough to support the body's weight. Patients with OI have a characteristic heart-shaped face, and the whites of their eyes have a blue tint. They often suffer hearing loss.

Osteogenesis imperfecta seems to have affected people throughout history. An Egyptian mummy dating from 1000 BC has been found with evidence of osteogenesis imperfecta.

Because of the way that this bone disorder is inherited, each of Alex's children will have a 50 percent chance of inheriting osteogenesis imperfecta. At least Alex will be aware of the problem and hopefully will never be accused of child abuse.

1. Why do you think the emergency department personnel suspected child abuse?

2. Do you think Alex should continue to play soccer?

See the blue Answers tab at the back of the book.

7

The Axial Skeleton

Learning Outcomes

These Learning Outcomes correspond by number to this chapter's sections and indicate what you should be able to do after completing the chapter.

7-1 ▌ Identify the bones of the axial skeleton, and specify their functions.

7-2 ▌ Identify the bones, foramina, and fissures of the cranium and face, and explain the significance of the markings on the individual bones.

7-3 ▌ Describe the structure and functions of the orbital complex, nasal complex, and paranasal sinuses.

7-4 ▌ Describe the key structural differences among the skulls of infants, children, and adults.

7-5 ▌ Identify and describe the curvatures of the spinal column, and indicate the function of each.

7-6 ▌ Identify the five vertebral regions, and describe the distinctive structural and functional characteristics of vertebrae in each region.

7-7 ▌ Explain the significance of the articulations between the thoracic vertebrae and the ribs, and between the ribs and sternum.

SPOTLIGHT
- Sectional Anatomy of the Skull pp. 238–239

CLINICAL CASE
- Knocked Out p. 233

Clinical Notes
- Temporomandibular Joint Syndrome p. 248
- Sinusitis p. 251
- Craniostenosis p. 252
- Kyphosis, Lordosis, and Scoliosis p. 255

Knocked Out

Dr. Deb Grady, a member of the American Board of Ringside Medicine, is covering a Golden Gloves boxing event as the ringside physician. She is closely watching a match between two older teenage boys. Up to the third round the fighters seem to be evenly matched. Suddenly, Dr. Grady witnesses a crushing blow delivered to the face of one fighter, knocking him out cold.

Dr. Grady jumps into the ring and carefully rolls the fighter over. She sees immediate and alarming swelling and bruising of the soft tissues around his left eye and profuse bleeding from a cut over the eye. As the young boxer regains consciousness, he sputters and coughs. Dr. Grady feels crackling subcutaneous air (like popping bubble wrap) in the expanding soft tissues around his eye. "How many fingers am I holding up?" she asks the fighter, holding up her index finger. "Two," the confused fighter answers. He is seeing double. "Follow my finger," she instructs, moving her finger up to the level of his forehead. The fighter can track her finger with his right eye, but his injured left eye remains looking straight ahead.

Dr. Grady calls the fight and accompanies the young boxer to the closest emergency room. **What has happened to the boxer's left eye? To find out, turn to the Clinical Case Wrap-Up on p. 266.**

▌ An Introduction to the Axial Skeleton

In this chapter, we turn our attention to the functional anatomy of the bones that form the longitudinal axis of the body.

7-1 ▌ The 80 bones of the head and trunk make up the axial skeleton

Learning Outcome Identify the bones of the axial skeleton, and specify their functions.

The **axial skeleton** forms the longitudinal axis of the body (Figure 7–1). The axial skeleton has 80 bones, about 40 percent of the bones in the human body:

- The *skull* (8 *cranial bones* and 14 *facial bones*).
- Bones associated with the skull (6 *auditory ossicles* and the *hyoid bone*).
- The *vertebral column* (24 *vertebrae*, the *sacrum*, and the *coccyx*).
- The *thoracic cage* (the *sternum* and 24 *ribs*).

The axial skeleton provides a framework that supports and protects the brain, the spinal cord, and the thoracic and abdominal organs. It also provides an extensive surface area for the attachment of muscles that (1) adjust the positions of the head, neck, and trunk; (2) perform respiratory movements; and (3) stabilize or position parts of the **appendicular skeleton**, which supports the limbs. The joints of the axial skeleton allow limited movement, but they are very strong and heavily reinforced with ligaments.

CHECKPOINT

1. Identify the bones of the axial skeleton.
2. List the primary functions of the axial skeleton.

See the blue Answers tab at the back of the book.

7-2 ▌ The skull is composed of 8 cranial bones and 14 facial bones

Learning Outcome Identify the bones, foramina, and fissures of the cranium and face, and explain the significance of the markings on the individual bones.

The **skull** contains 22 bones: 8 form the **cranium**, or *braincase*, and 14 are associated with the face (Figure 7–2).

Seven additional bones are associated with the skull; six auditory ossicles, involved with hearing, and the hyoid bone that supports the voicebox.

The cranium consists of 8 **cranial bones** that enclose the **cranial cavity**, a chamber that supports the brain. Blood vessels, nerves, and membranes that stabilize the position of the brain are attached to the inner surface of the cranium. Its outer surface provides an extensive area for the attachment of muscles that move the eyes, jaws, and head.

Joints, or *articulations*, form where two bones connect. A joint between the occipital bone of the skull and the first vertebra of the neck stabilizes the positions of the brain and spinal cord. The joints between the vertebrae of the neck permit a wide range of head movements.

If you think of the cranium as the house where the brain resides, then the *facial complex* is the front porch (Figure 7–2).

The nine superficial facial bones serve as attachment sites for muscles that control facial expressions and assist in manipulating food.

The five deeper facial bones help separate the oral and nasal cavities, increase the surface area of the nasal cavities, or help form the **nasal septum** (*septum*, wall), which divides the nasal cavity into left and right halves.

Several bones of the skull contain air-filled chambers called **sinuses** (Figures 7–3, 7–4). Sinuses have three major functions: (1) They lessen the weight of the bone; (2) the

Figure 7–1 **The Axial Skeleton.**
ATLAS: Plates 1a,b

SKELETAL SYSTEM 206

APPENDICULAR SKELETON

(see Figure 8–1)

AXIAL SKELETON 80

| 8 | Cranium |
| 14 | Face |

Skull

| 6 | Auditory ossicles |
| 1 | Hyoid |

Associated bones

29 Skull and associated bones

| 1 | Sternum |
| 24 | Ribs |

25 Thoracic cage

24	Vertebrae
1	Sacrum
1	Coccyx

26 Vertebral column

a An anterior view of the entire skeleton, with the axial components highlighted. The numbers in the boxes indicate the number of bones in the adult skeleton.

Skull

Sternum

Costal cartilages

Ribs

Lumbar vertebrae

Sacrum

Coccyx

Cervical vertebrae

Thoracic vertebrae

b Anterior (left) and posterior (right) views of the axial skeleton. The individual bones associated with the skull are not visible.

Figure 7–2   **Cranial and Facial Subdivisions of the Skull.**

SKULL

FACE	14
Maxillae	2
Palatine bones	2
Nasal bones	2
Inferior nasal conchae	2
Zygomatic bones	2
Lacrimal bones	2
Vomer	1
Mandible	1

Nasal bone
Lacrimal bone
Vomer
Zygomatic bone
Maxilla
Mandible

Facial bones protect and support the entrances to the digestive and respiratory tracts.

CRANIUM	8
Occipital bone	1
Parietal bones	2
Frontal bone	1
Temporal bones	2
Sphenoid	1
Ethmoid	1

Frontal bone
Parietal bone
Temporal bone
Ethmoid
Sphenoid
Occipital bone

Cranial bones surround and protect the brain.

ASSOCIATED BONES	7

Hyoid bone	1	Auditory ossicles enclosed in temporal bones (detailed in Chapter 17)	6

Hyoid bone

Auditory ossicles

7

mucous membrane lining them produces mucus that moistens and cleans the air in and adjacent to the sinus; and (3) they serve as resonating chambers in speech production. We will consider the sinuses as we discuss specific bones (Figures 7-5 to 7-12).

The connections between the skull bones of adults are immovable joints called **sutures** (except where the mandible contacts the cranium). At a suture, bones are tied firmly together with dense fibrous connective tissue. Each suture of the skull has a name. These are the four major sutures:

1. *Lambdoid Suture.* The **lambdoid** (LAM-doyd) **suture** (Greek *lambda + eidos*, shape) arches across the posterior surface of the skull (Figure 7–3a). This suture connects the occipital bone with the two parietal bones. One or more **sutural bones** (*Wormian bones*) may be present along the lambdoid suture. ⊃ p. 206

2. *Coronal Suture.* The **coronal suture** attaches the frontal bone to the parietal bones of either side (Figure 7–3b). The occipital, parietal, and frontal bones form the **calvaria** (kal-VA-rē-uh), or skullcap. A cut through the body that parallels the coronal suture produces a *frontal section,* or *coronal section* (Figure 1–8, ⊃ p. 44).

3. *Sagittal Suture.* The **sagittal suture** extends from the lambdoid suture to the coronal suture, between the parietal bones (Figure 7–3b). A cut along the midline of this suture produces a *midsagittal section.* A slice that parallels the sagittal suture produces a *parasagittal section.* ⊃ p. 44

4. *Squamous Sutures.* A **squamous** (SKWA̅-mus) **suture** on each side of the skull joins the temporal bone and the parietal bone of that side. Figure 7–3a shows the intersection between the squamous sutures and the lambdoid suture. Figure 7–3c shows the path of the squamous suture on the right side of the skull.

Figure 7–3 **The Adult Skull.** ATLAS: Plates 4a,b; 5a–e

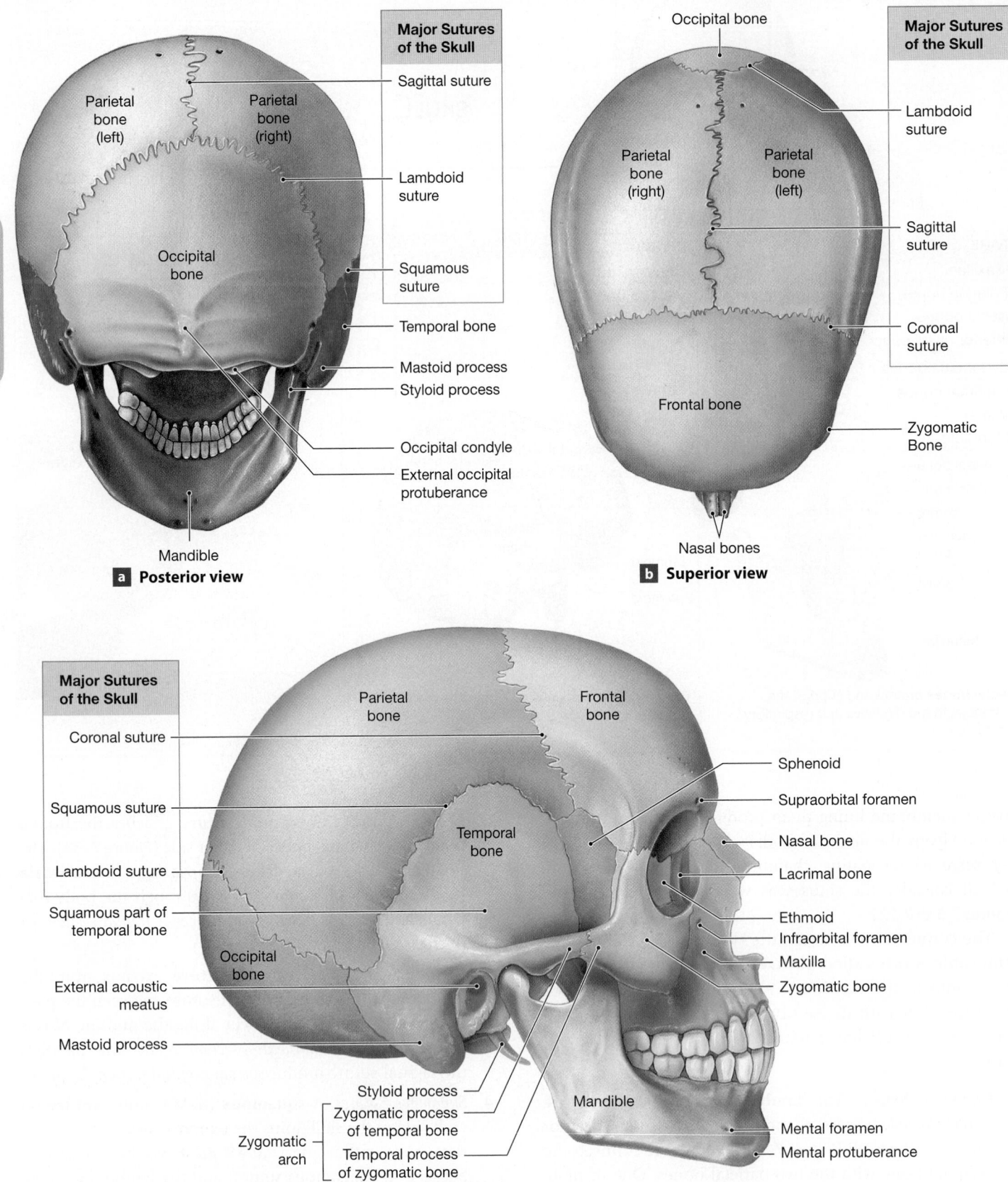

Major Sutures of the Skull
- Sagittal suture
- Lambdoid suture
- Squamous suture

Parietal bone (left)
Parietal bone (right)
Occipital bone
Temporal bone
Mastoid process
Styloid process
Occipital condyle
External occipital protuberance
Mandible

a **Posterior view**

Occipital bone

Major Sutures of the Skull
- Lambdoid suture
- Sagittal suture
- Coronal suture

Parietal bone (right)
Parietal bone (left)
Frontal bone
Zygomatic Bone
Nasal bones

b **Superior view**

Major Sutures of the Skull
- Coronal suture
- Squamous suture
- Lambdoid suture

Squamous part of temporal bone
Occipital bone
External acoustic meatus
Mastoid process
Styloid process
Zygomatic arch
  Zygomatic process of temporal bone
  Temporal process of zygomatic bone

Parietal bone
Frontal bone
Temporal bone
Sphenoid
Supraorbital foramen
Nasal bone
Lacrimal bone
Ethmoid
Infraorbital foramen
Maxilla
Zygomatic bone
Mandible
Mental foramen
Mental protuberance

c **Lateral view**

Figure 7–3 **The Adult Skull** *(continued).*

Foramen or Fissure	Major Structures Using Passageway
Frontal Bone	
Supraorbital foramen	• Supraorbital nerve (branch of N V) • Supraorbital artery
Sphenoid	
Optic canal	• Optic nerve (N II) • Ophthalmic artery
Superior orbital fissure	• Oculomotor nerve (N III), trochlear nerve (N IV), opthalmic branch of trigeminal nerve (N V), and abducens nerve (N VI)
Inferior orbital fissure	• Maxillary branch of trigeminal nerve (N V)
Maxilla	
Infraorbital foramen	• Infraorbital nerve, branch of maxillary nerve (N V) • Infraorbital artery
Mandible	
Mental foramen	• Mental nerve, branch of mandibular nerve (N V) • Mental vessels

Labels (anterior view):
- Sagittal suture
- Parietal bone
- Coronal suture
- Nasal bone
- Ethmoid
- Temporal bone
- Palatine bone
- Lacrimal bone
- Zygomatic bone
- Mastoid process of temporal bone
- Middle nasal concha (part of ethmoid)
- Inferior nasal concha
- **Bony nasal septum**
  - Perpendicular plate of ethmoid
  - Vomer
- Mental protuberance

d **Anterior view**

Foramen or Fissure	Major Structures Using Passageway
Sphenoid	
Foramen lacerum (with temporal and occipital bones)	• Internal carotid artery after leaving carotid canal • Auditory tube
Foramen ovale	• Mandibular branch of trigeminal nerve (N V)
Temporal Bone	
External acoustic meatus	• Air in meatus conducts sound to eardrum
Carotid canal	• Internal carotid artery
Stylomastoid foramen	• Facial nerve (N V)
Occipital Bone	
Foramen magnum	• Medulla oblongata (most caudal portion of brain) • Accessory nerve (XI) • Vertebral arteries
Jugular foramen (with temporal bone)	• Glossopharyngeal, vagus, and accessory nerves (N IX, X, XI)

Labels (inferior view):
- Frontal bone
- Maxilla
- Palatine bone
- Zygomatic bone
- Zygomatic arch
- Medial and lateral pterygoid processes
- Vomer
- Styloid process
- Mandibular fossa
- Occipital condyle
- Mastoid process
- Lambdoid suture
- Occipital bone
- External occipital protuberance

e **Inferior view**

7

THE CRANIAL CAVITY

This sagittal section of the skull was taken just lateral to the nasal septum. It provides the best view of the extent of the cranial cavity. The colored boxes highlight the bones that surround the cranial cavity, and some of their key markings are highlighted within their respective boxes.

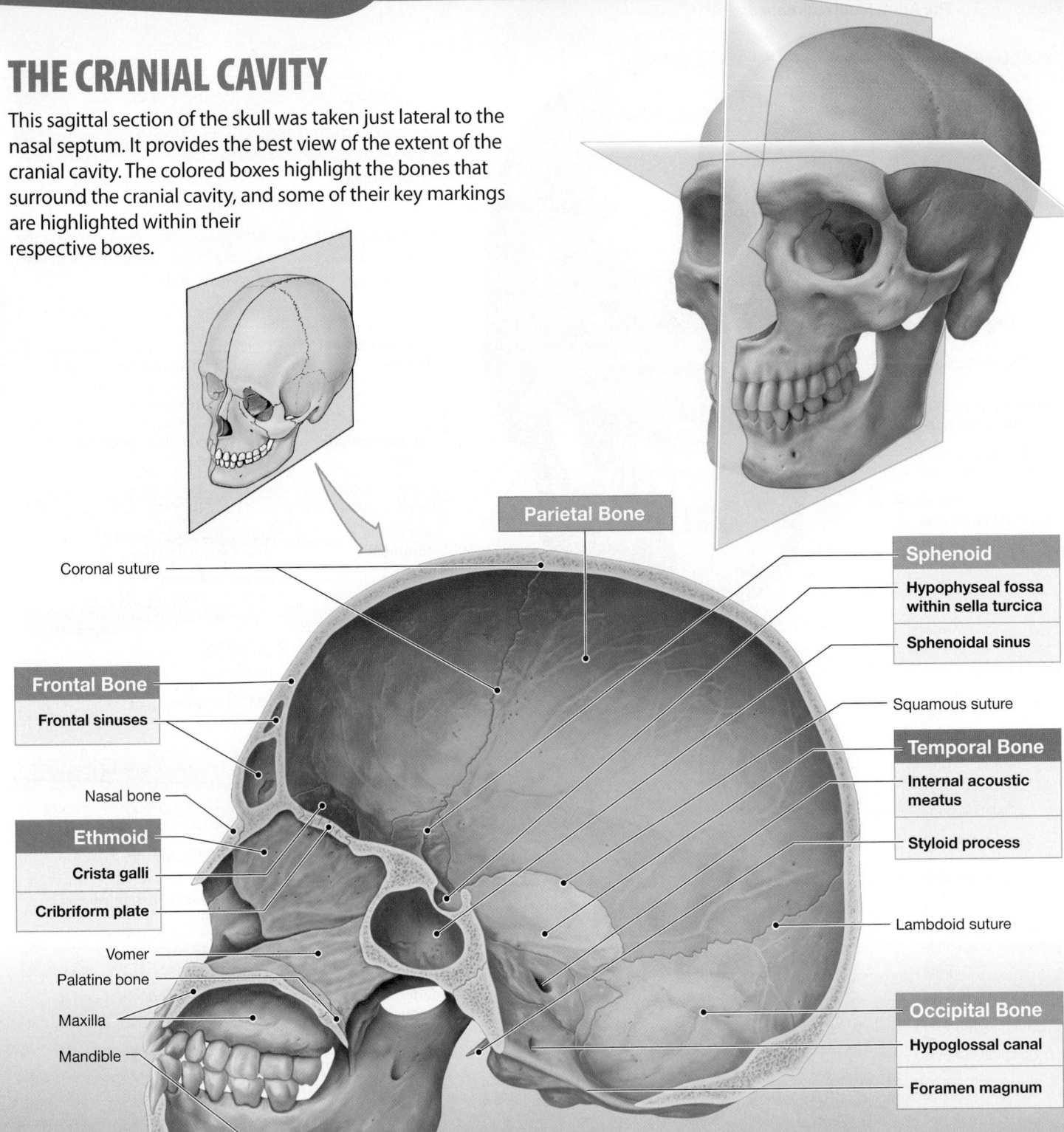

Parietal Bone

Coronal suture

Sphenoid

Hypophyseal fossa within sella turcica

Sphenoidal sinus

Frontal Bone

Frontal sinuses

Squamous suture

Nasal bone

Temporal Bone

Ethmoid

Internal acoustic meatus

Crista galli

Styloid process

Cribriform plate

Lambdoid suture

Vomer

Palatine bone

Maxilla

Occipital Bone

Mandible

Hypoglossal canal

Foramen magnum

a Sagittal section

THE FLOOR OF THE CRANIUM

This superior view of a horizontal section through the skull shows the floor of the cranial cavity. This is where most of the nerves and blood vessels enter or leave the cranial cavity. Compare this figure and chart with *Figure 7–3e* showing the inferior view of the skull and *Figure 14–19* (p. 518) showing the origin of the cranial nerves.

Frontal bone

Crista galli

Cribriform plate

Sella turcica

Parietal bone

Internal occipital crest

b Horizontal section

Understanding skull structure will be especially important to you in later chapters when you study the nervous and cardiovascular systems.

Foramen or Fissure	Major Structures Using Passageway
Ethmoid	
Olfactory foramina	• Olfactory nerve (N I)
Sphenoid	
Optic canal	• Optic nerve (N II) • Ophthalmic artery
Foramen rotundum	• Maxillary branch of trigeminal nerve (N V)
Foramen lacerum	• Internal carotid artery after leaving carotid canal • Auditory tube
Foramen ovale	• Mandibular branch of trigeminal nerve (N V)
Foramen spinosum	• Blood vessels to membranes around central nervous system
Temporal Bone	
Carotid canal	• Internal carotid artery
Internal acoustic meatus	• Vestibulocochlear nerve (N VIII) • Internal acoustic artery • Facial nerve (N VII)
Occipital Bone	
Foramen magnum	• Medulla oblongata (most caudal portion of brain) • Accessory nerve (N XI) • Vertebral arteries
Hypoglossal canal	• Hypoglossal nerve (N XII)
Jugular foramen (with temporal bone)	• Glossopharyngeal, vagus, and accessory nerves (N IX, X, XI) • Internal jugular vein

Focus   The Individual Bones of the Skull

We can explore the surface features of these bones further, using the related images in the *Atlas*. Foramina and fissures create passages for vessels and nerves. The vessels are detailed in Chapter 21. The nerves are shown in the Focus on cranial nerves in Chapter 14.

Cranial Bones

The Occipital Bone (Figure 7–5a)

General Functions: The **occipital bone** forms much of the posterior and inferior surfaces of the cranium.

Articulations: The occipital bone articulates with the parietal bones, the temporal bones, the sphenoid, and the first cervical vertebra (the atlas) (Figure 7–3a–c,e and Spotlight Figure 7–4).

Regions/Markings (Figure 7–5a): The **external occipital protuberance** is a small bump at the midline on the inferior surface.

The **external occipital crest** begins at the external occipital protuberance. This crest marks the attachment of a ligament that helps stabilize the vertebrae of the neck.

The **occipital condyles** are the site of articulation between the skull and the first vertebra of the neck.

The *inferior* and *superior nuchal* (NŪ-kul) *lines* are ridges that intersect the occipital crest. They mark the attachment sites of muscles and ligaments that stabilize the articulation at the occipital condyles and balance the weight of the head over the vertebrae of the neck.

The concave internal surface of the occipital bone (Spotlight Figure 7–4a) closely follows the contours of the brain.

The grooves follow the paths of major blood vessels, and the ridges mark the attachment sites of membranes that stabilize the position of the brain.

Foramina: The **foramen magnum** (Spotlight Figure 7–4b) connects the cranial cavity with the vertebral canal, which is enclosed by the vertebral column. This foramen surrounds the connection between the brain and spinal cord.

The **jugular foramen** lies between the occipital bone and the temporal bone (Figure 7–3e). The *internal jugular vein* passes through this foramen, carrying venous blood from the brain.

The **hypoglossal canals** (Spotlight Figure 7–4b) begin at the lateral base of each occipital condyle and end on the inner surface of the occipital bone near the foramen magnum. The *hypoglossal nerves,* cranial nerves that control the tongue muscles, pass through these canals.

The Parietal Bones (Figure 7–5b)

General Functions: The **parietal bones** form part of the superior and lateral surfaces of the cranium.

Articulations: The parietal bones articulate with one another and with the occipital, temporal, frontal, and sphenoid bones (Figure 7–3a–d and Spotlight Figure 7–4).

Regions/Markings: The *superior* and *inferior temporal lines* are low ridges that mark the attachment sites of the *temporalis muscle,* a large muscle that closes the mouth (Figure 7–5b).

Grooves on the inner surface of the parietal bones mark the paths of cranial blood vessels (Spotlight Figure 7–4a).

Figure 7–5   **The Occipital and Parietal Bones.**

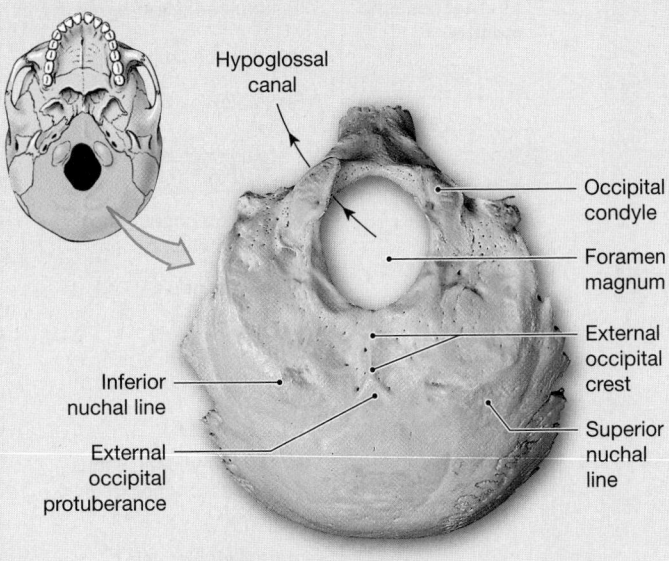

Hypoglossal canal

Inferior nuchal line

External occipital protuberance

Occipital condyle

Foramen magnum

External occipital crest

Superior nuchal line

a Occipital bone, inferior view

Superior temporal line

Inferior temporal line

b Right parietal bone, lateral view

The Frontal Bone (Figure 7–6a,b)

General Functions: The **frontal bone** forms the anterior portion of the cranium and the roof of the *orbits* (eye sockets). Mucus from the *frontal sinuses* within this bone helps flush the surfaces of the nasal cavities.

Articulations: The frontal bone articulates with the parietal, sphenoid, ethmoid, nasal, lacrimal, maxillary, and zygomatic bones (Figure 7–3b–e and Spotlight Figure 7–4).

Regions/Markings: The **frons,** or forehead, forms the anterior, superior portion of the cranium. It provides surface area for the attachment of facial muscles. The *superior temporal line* is continuous with the superior temporal line of the parietal bone.

The **supraorbital margin** is a thickening of the frontal bone that helps protect the eye. The **glabella** (*glabellus,* hairless) is the area between the supraorbital margins and superior to the nasal bones. It marks the most forward projection of the frontal bone.

The **lacrimal fossa** on the superior and lateral surface of the orbit is a shallow depression that marks the location of the *lacrimal* (tear) *gland.* Tears from this gland lubricate the surface of the eye.

The **frontal sinuses** are extremely variable in size and time of appearance. They generally appear after age 6, but some people never develop them. We will describe the frontal sinuses and other sinuses of the cranium and face in a later section.

Foramina: The **supraorbital foramen** provides passage for blood vessels that supply the eyebrow, eyelids, and frontal sinuses.

In some cases, this foramen is incomplete. In that case, the vessels then cross the orbital rim within a **supraorbital notch.**

Remarks: During development, the bones of the cranium form by the fusion of separate centers of ossification. At birth, the fusions are not yet complete: Two frontal bones articulate along the *frontal (metopic) suture.* The suture generally disappears by age 8 as the bones fuse, but the adult skull commonly retains traces of the suture line. This suture, or what remains of it, runs down the center of the frons.

The Temporal Bones (Figure 7–7a,b)

General Functions: The **temporal bones** (1) form part of both the lateral walls of the cranium and the *zygomatic arches,* (2) form the only articulations with the mandible, (3) surround and protect the sense organs of the internal ear, and (4) are attachment sites for muscles that close the jaws and move the head.

Articulations: The temporal bones articulate with the zygomatic, sphenoid, parietal, and occipital bones of the cranium and with the mandible (Figure 7–3 and Spotlight Figure 7–4).

Regions/Markings: The **squamous part** of the temporal bone is the lateral portion that borders the squamous suture. The concave, internal surface parallels the surface of the brain and is called the *cerebral surface.*

The **zygomatic process,** inferior to the temporal squamous part, articulates with the *temporal process* of the zygomatic bone.

Figure 7–6   **The Frontal Bone.**

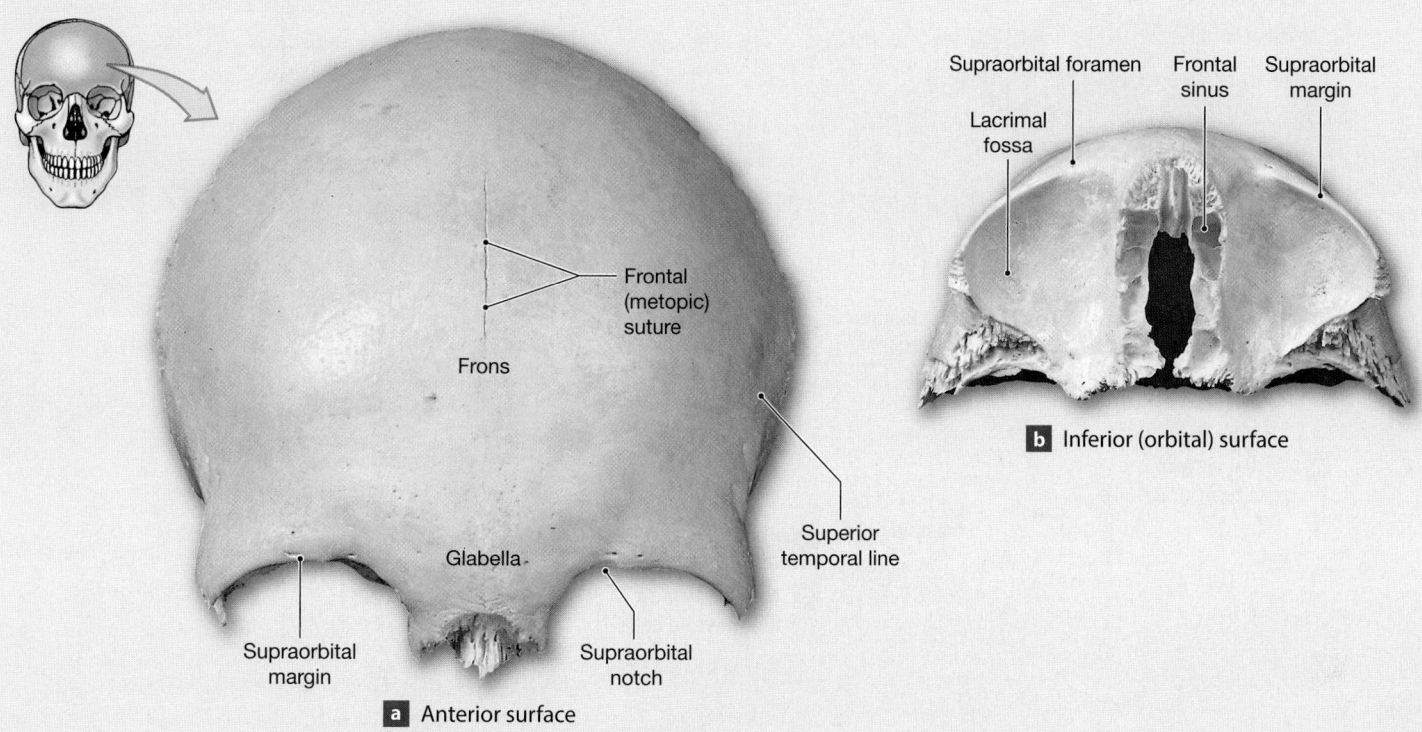

Supraorbital foramen   Frontal sinus   Supraorbital margin

Lacrimal fossa

b Inferior (orbital) surface

Frontal (metopic) suture

Frons

Superior temporal line

Glabella

Supraorbital margin

Supraorbital notch

a Anterior surface

Together, these processes form the **zygomatic arch**, or cheekbone (Figure 7–3c,e).

The **mandibular fossa** on the inferior surface marks the site of articulation with the mandible.

The **mastoid process** (Figure 7–7b,c) is an attachment site for muscles that rotate or extend the head. It contains *mastoid cells* (also called mastoid air cells), small interconnected cavities that connect to the middle ear cavity. If pathogens invade the mastoid cells, *mastoiditis* develops. Signs and symptoms include severe earaches, fever, and swelling behind the ear.

The **styloid** (STĪ-loyd; *stylos,* pillar) **process**, near the base of the mastoid process, is attached to ligaments that support the hyoid bone and to the tendons of several muscles associated with the hyoid bone, the tongue, and the pharynx.

The **petrous part** of the temporal bone, located on its internal surface, encloses the structures of the *internal ear*— sense organs that provide information about hearing and balance.

The **auditory ossicles** are located in the *tympanic cavity,* or *middle ear,* a cavity within the petrous part. These tiny bones— three on each side—transfer sound vibrations from the delicate tympanic membrane, or eardrum, to the internal ear. (We will discuss these bones and their functions in Chapter 17.)

Foramina (Figure 7–3e): The *jugular foramen,* between the temporal and occipital bones, provides passage for the internal jugular vein.

The **carotid canal** provides passage for the internal carotid artery, a major artery to the brain. As it leaves the carotid canal, the internal carotid artery passes through the anterior portion of the foramen lacerum.

The **foramen lacerum** (LA-se-rum; *lacerare,* to tear) is a jagged slit extending between the sphenoid and the petrous part of the temporal bone. This slit contains hyaline cartilage and small arteries that supply the inner surface of the cranium.

The *auditory tube* is an air-filled passageway that connects the pharynx (throat) to the tympanic cavity. This tube passes through the posterior portion of the foramen lacerum.

The **external acoustic meatus**, or *auditory canal,*[1] on the lateral surface ends at the tympanic membrane (which disintegrates during the preparation of a dried skull).

The **stylomastoid foramen** lies posterior to the base of the styloid process. The *facial nerve* passes through this foramen to control the facial muscles.

Figure 7–7 The Temporal Bones.

a Medial view of the right temporal bone

b Lateral view of the right temporal bone

c A cutaway view of the mastoid cells

The **internal acoustic meatus,** or *internal auditory meatus,*[1] begins on the medial surface of the petrous part of the temporal bone. It carries blood vessels and nerves to the internal ear and conveys the facial nerve to the stylomastoid foramen.

[1]The names for these passageways vary widely; the terms *acoustic* and *auditory* are used interchangeably, as are *meatus* and *canal.*

The Sphenoid (Figure 7–8a,b)

General Functions: The **sphenoid** forms part of the floor of the cranium, unites the cranial and facial bones, and acts as a cross-brace that strengthens the sides of the skull. Mucus from the *sphenoidal sinuses* within this bone helps clean the surfaces of the nasal cavities.

Articulations: The sphenoid articulates with the ethmoid and the frontal, occipital, parietal, and temporal bones of the cranium and the palatine bones, zygomatic bones, maxillae, and vomer of the face (Figure 7–3c–e and Spotlight Figure 7–4).

Regions/Markings (Spotlight Figure 7–4 and Figure 7–8):
The shape of the sphenoid has been compared to a bat with its wings extended. This bone is relatively large, but more superficial bones hide much of it.

The **body** forms the central axis of the sphenoid.

The **sella turcica** (TUR-si-kuh), or Turkish saddle, is a bony, saddle-shaped enclosure on the superior surface of the body. Within the sella turcica is a depression known as the **hypophyseal** (hī-pō-FIZ-ē-ul) **fossa** (Spotlight Figure 7–4a). This fossa contains the *pituitary gland.*

Figure 7–8   The Sphenoid.

a Superior surface

b Anterior surface

The **sphenoidal sinuses** (Spotlight Figure 7–4a and Figure 7–8b) are on either side of the body, inferior to the sella turcica.

The **lesser wings** extend horizontally anterior to the sella turcica.

The **greater wings** extend laterally from the body and form part of the cranial floor. A sharp *sphenoidal spine* lies at the posterior, lateral corner of each greater wing. Anteriorly, each greater wing contributes to the posterior wall of the orbit.

The **pterygoid** (TER-i-goyd; *pterygion*, wing) **processes** are vertical projections that originate on either side of the body. Each pterygoid process forms a pair of *pterygoid plates*, which are attachment sites for muscles that move the mandible and soft palate.

Foramina (Spotlight Figure 7–4b and Figure 7–8): The **optic canals** permit passage of the optic nerves from the eyes to the brain.

Penetrating each greater wing are a **superior orbital fissure, foramen rotundum, foramen ovale** (ō-VAH-lē), and **foramen spinosum.** These passages carry blood vessels and nerves to the orbit, face, jaws, and membranes of the cranial cavity, respectively.

The Ethmoid (Figure 7–9a,b)

General Functions: The **ethmoid,** or *ethmoid bone,* forms the anteromedial floor of the cranium, the roof of the nasal cavity, and part of the nasal septum (which subdivides the nasal cavity) and medial orbital wall. Mucus from a network of sinuses, or *ethmoid air cells,* within this bone flushes the surfaces of the nasal cavities.

Articulations: The ethmoid articulates with the frontal bone and sphenoid of the cranium and with the nasal, lacrimal, palatine, and maxillary bones and the inferior nasal conchae and vomer of the face (Figure 7–3c,d and Spotlight Figure 7–4).

Regions/Markings: The ethmoid has three parts: (1) the cribriform plate, (2) the paired lateral masses, and (3) the perpendicular plate.

The **cribriform plate** (*cribrum,* sieve) forms the anteromedial floor of the cranium and the roof of the nasal cavity. The **crista galli** (*crista,* crest + *gallus,* rooster, cock; cock's comb) is a bony ridge that projects superior to the cribriform plate. A membrane that stabilizes the position of the brain, the *falx cerebri,* attaches to this ridge.

The **lateral masses** contain the **ethmoidal labyrinth,** which consists of the interconnected **ethmoid air cells** that open into the nasal cavity on each side. The **superior nasal conchae** (KONG-kē; singular, *concha,* a snail shell) and the **middle nasal conchae** are delicate projections of the lateral masses.

Figure 7–9   **The Ethmoid.**

a Superior surface

b Posterior surface

The **perpendicular plate** forms part of the nasal septum, along with the vomer and a piece of hyaline cartilage.

Foramina: The **olfactory foramina** in the cribriform plate permit passage of the olfactory nerves, which provide the sense of smell.

Remarks: *Olfactory* (smell) *receptors* are located in the epithelium that covers the inferior surfaces of the cribriform plate, the medial surfaces of the superior nasal conchae, and the superior portion of the perpendicular plate.

The nasal conchae break up the airflow in the nasal cavity. They create swirls, turbulence, and eddies that have three major functions: (1) Particles in the air are thrown against the sticky mucus that covers the walls of the nasal cavity; (2) air movement is slowed, providing time for warming, humidification, and dust removal before the air reaches more delicate portions of the respiratory tract; and (3) air is directed toward the superior portion of the nasal cavity, adjacent to the cribriform plate, where the olfactory receptors are located.

Facial Bones

The Maxillae (Figure 7–10a,b)

General Functions: The **maxillae,** or *upper jawbones,* support the upper teeth and form the inferior orbital rim, the lateral margins of the external nares, the upper jaw, and most of the hard palate. The *maxillary sinuses* in these bones produce mucus that flushes the inferior surfaces of the nasal cavities. The maxillae are the largest facial bones, and the maxillary sinuses are the largest sinuses.

Articulations: The maxillae articulate with the frontal bones and ethmoid, with one another, and with all the other facial bones except the mandible (Figure 7–3c–e and Spotlight Figure 7–4a).

Regions/Markings: The **orbital rim** protects the eye and other structures in the orbit. The *anterior nasal spine* is found at the anterior portion of the maxilla, at its articulation with the maxilla of the other side. It is an attachment point for the cartilaginous anterior portion of the nasal septum.

Figure 7–10   **The Maxillae and Palatine Bones.** ATLAS: Plates 8a–d; 12d

Orbital rim

Infraorbital foramen

Anterior nasal spine

Alveolar process

Zygomatic process

a An anterolateral view of the right maxilla.

Palatine bone (horizontal plate)

Palatine process

Alveolar process

Maxillary sinus

b Superior view of a horizontal section through right maxilla and palatine bone; note the size and orientation of the maxillary sinus.

Orbital process

Perpendicular plate

Horizontal plate

c An anterior view of the two palatine bones.

The **alveolar process** of the maxilla supports the upper teeth.

The **palatine processes** form most of the **hard palate,** or bony roof of the mouth. One type of *cleft palate,* a developmental disorder, results when the maxillae fail to meet along the midline of the hard palate. ATLAS: Embryology Summary 6: The Development of the Skull

The **maxillary sinuses** lighten the portion of the maxillae superior to the teeth.

The **nasolacrimal canal** is formed by a maxilla and lacrimal bone. It protects the *lacrimal sac* and the *nasolacrimal duct,* which carries tears from the orbit to the nasal cavity.

Foramina: The **infraorbital foramen** marks the path of a major sensory nerve that reaches the brain by the foramen rotundum of the sphenoid.

The **inferior orbital fissure** (Figure 7–3d) lies between the maxilla and the sphenoid. It permits passage of cranial nerves and blood vessels.

The Palatine Bones (Figure 7–10b,c)

General Functions: The **palatine bones** form the posterior portion of the hard palate. They also contribute to the floor of each orbit.

Articulations: The palatine bones articulate with one another, with the maxillae, with the sphenoid and ethmoid, with the

inferior nasal conchae, and with the vomer (Figure 7–3e and Spotlight Figure 7–4a).

Regions/Markings: The palatine bones are L-shaped. The **horizontal plate** forms the posterior part of the hard palate. The **perpendicular plate** extends from the horizontal plate to the **orbital process,** which forms part of the floor of the orbit. This process contains a small sinus that normally opens into the sphenoidal sinus.

Foramina: Small blood vessels and nerves supplying the roof of the mouth penetrate the lateral portion of the horizontal plate.

The Nasal Bones (Figure 7–11)

General Functions: The paired **nasal bones** support the superior portion of the bridge of the nose. They are connected to cartilages that support the distal portions of the nose. These flexible cartilages, and associated soft tissues, extend to the superior border of the **external nares** (NA-rēz; singular, *naris*), the entrances to the nasal cavity.

Articulations: The paired nasal bones articulate with one another, with the ethmoid, and with the frontal bone and maxillae (Figure 7–3b–d and Spotlight Figure 7–4a).

The Vomer (Figure 7–11)

General Functions: The **vomer** forms the inferior portion of the bony nasal septum.

Figure 7–11   **The Smaller Bones of the Face.**

Supraorbital foramen

Nasal bone

Sphenoid

Temporal bone

Zygomaticofacial foramen

Zygomatic bone

Infraorbital foramen

Maxilla

Lacrimal sulcus

Optic canal

Superior orbital fissure

Lacrimal bone

Middle nasal concha

Inferior nasal concha

Temporal process of zygomatic bone

Mastoid process

Perpendicular plate of ethmoid   Vomer

Bony nasal septum

Articulations: The vomer articulates with the maxillae, sphenoid, ethmoid, and palatine bones, and with the cartilaginous part of the nasal septum, which extends into the fleshy part of the nose (Figure 7–3d,e and Spotlight Figure 7–4a).

The Inferior Nasal Conchae (Figure 7–11)

General Functions: The **inferior nasal conchae** create turbulence in air passing through the nasal cavity. They also increase the epithelial surface area to promote warming and humidification of inhaled air.

Articulations: The inferior nasal conchae articulate with the maxillae, ethmoid, palatine, and lacrimal bones (Figure 7–3d).

The Zygomatic Bones (Figure 7–11)

General Functions: The **zygomatic bones** contribute to the rim and lateral wall of the orbit and form part of the zygomatic arch.

Articulations: The zygomatic bones articulate with the maxillae, and the sphenoid, frontal, and temporal bones (Figure 7–3b–e).

Regions/Markings: The **temporal process** curves posteriorly to meet the zygomatic process of the temporal bone.

Foramina: The **zygomaticofacial foramen** on the anterior surface of each zygomatic bone carries a sensory nerve that innervates the cheek.

The Lacrimal Bones (Figure 7–11)

General Functions: The **lacrimal bones** form part of the medial wall of the orbit.

Articulations: The lacrimal bones are the smallest facial bones. They articulate with the frontal bone and maxillae, and with the ethmoid (Figure 7–3c,d).

Regions/Markings: The **lacrimal sulcus,** a groove along the anterior, lateral surface of the lacrimal bone, marks the location of the lacrimal sac. The lacrimal sulcus leads to the nasolacrimal canal, which begins at the orbit and opens into the nasal cavity. As noted earlier, the lacrimal bone and the maxilla form this canal.

The Mandible (Figure 7–12a,b)

General Functions: The **mandible** forms the lower jaw.

Articulations: The mandible articulates with the mandibular fossae of the temporal bones (Figures 7–3c,e and 7–7b).

Figure 7–12   The Mandible.

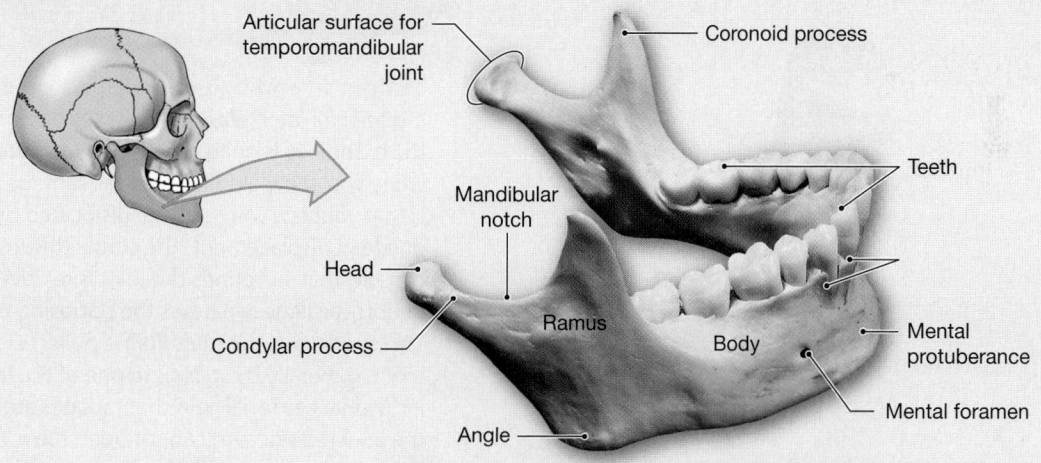

a A lateral and slightly superior view of the mandible

b A medial view of the right mandible

Regions/Markings: The **body** of the mandible is the horizontal portion of that bone.

The **alveolar process** of the mandible supports the lower teeth.

The *mental protuberance* (*mentalis,* chin) is the attachment site for several facial muscles.

A prominent depression on the medial surface marks the position of the *submandibular salivary gland.*

The *mylohyoid line* marks the insertion of the *mylohyoid muscle,* which supports the floor of the mouth.

The **ramus** of the mandible is the ascending part that begins at the *mandibular angle* on either side. On each ramus:

1. The **condylar process** articulates with the temporal bone at the *temporomandibular joint.*

2. The **coronoid** (KOR-ō-noyd) **process** is the insertion point for the *temporalis muscle,* a powerful muscle that closes the jaws.

3. The **mandibular notch** is the depression that separates the condylar and coronoid processes.

Foramina: The **mental foramina** are openings for nerves that carry sensory information from the lips and chin to the brain.

The **mandibular foramen** is the entrance to the *mandibular canal,* a passageway for blood vessels and nerves that service the lower teeth. Dentists typically anesthetize the sensory nerve that enters this canal before they work on the lower teeth.

The Hyoid Bone (Figure 7–13)

General Functions: The **hyoid bone** supports the larynx and is the attachment site for muscles of the larynx, pharynx, and tongue.

Articulations: *Stylohyoid ligaments* connect the *lesser horns* to the styloid processes of the temporal bones.

Regions/Processes: The **body** of the hyoid is an attachment site for muscles of the larynx, tongue, and pharynx.

The **greater horns** help support the larynx and are attached to muscles that move the tongue.

The **lesser horns** are attached to the stylohyoid ligaments. From these ligaments, the hyoid and larynx hang beneath the skull like a child's swing from the limb of a tree.

Figure 7–13 **An anterior view of the hyoid bone**

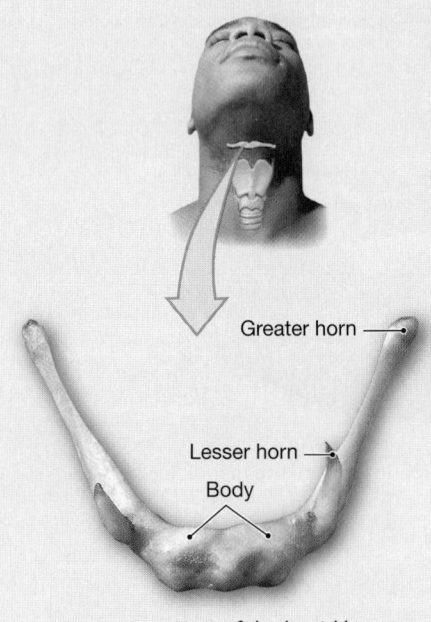

Greater horn

Lesser horn

Body

An anterior view of the hyoid bone

Clinical Note

Temporomandibular Joint Syndrome The *temporomandibular joint (TMJ),* between each temporal bone and the mandible, is quite mobile. It allows your jaw to move while you chew or talk. The disadvantage of such mobility is that your jaw can easily be dislocated by forceful forward or lateral displacement. The connective tissue sheath, or *capsule,* that surrounds the joint is relatively loose. A pad of fibrocartilage separates the opposing bone surfaces. In **TMJ syndrome,** the mandible is pulled slightly out of alignment, generally by spasms in one of the jaw muscles. The individual has facial pain that radiates around the ear on the affected side and cannot open the mouth fully. TMJ syndrome is a repeating cycle of muscle spasm → misalignment → pain → muscle spasm. It has been linked to involuntary behaviors, such as grinding the teeth during sleep *(bruxism),* and to emotional stress. Treatment focuses on breaking the cycle of muscle spasm and pain. When necessary, treatment includes emotional support. Applying heat to the affected joint, coupled with the use of anti-inflammatory drugs, local anesthetics, or both, may help. If teeth grinding is suspected, special mouth guards may be worn during sleep.

See the blue Answers tab at the back of the book.

CHECKPOINT

3. In which bone is the foramen magnum located?

4. Tomás suffers a blow to the skull that fractures the right superior lateral surface of his cranium. Which bone is fractured?

5. Which bone contains the depression called the sella turcica? What is located in this depression?

6. Identify the facial bones.

7. Identify the bone containing the mental foramen, and list the structures using this passageway.

8. Identify the bone containing the optic canal, and cite the structures using this passageway.

9. Name the foramina found in the ethmoid bone.

7-3 ▌ Each orbital complex contains an eye, and the nasal complex encloses the nasal cavities

Learning Outcome Describe the structure and functions of the orbital complex, nasal complex, and paranasal sinuses.

The facial bones not only protect and support the openings of the digestive and respiratory systems, but also protect the sense organs responsible for vision and smell. Together, certain cranial bones and facial bones form an *orbital complex*, which surrounds each eye, and the *nasal complex*, which surrounds the nasal cavities.

The Orbital Complexes

The **orbits** are the bony recesses that contain the eyes. Seven bones of the **orbital complex** form each orbit (Figure 7–14). The frontal bone forms the roof, and the maxilla provides most of the orbital floor. The orbital rim and the first portion of the medial wall are formed by the maxilla, the lacrimal bone, and the lateral mass of the ethmoid. The lateral mass articulates with the sphenoid and a small process of the palatine bone. Several prominent foramina and fissures penetrate the sphenoid or lie between it and the maxilla. Laterally, the sphenoid and maxilla articulate with the zygomatic bone, which forms the lateral wall and rim of the orbit.

The Nasal Complex

The **nasal complex** (Figure 7–15) includes the bones that enclose the nasal cavities and the *paranasal sinuses*, air-filled chambers connected to the nasal cavities. The frontal bone, sphenoid, and ethmoid form the superior wall of the nasal cavities. The lateral walls are formed by the maxillae and the lacrimal bones (not shown), the ethmoid (the superior and middle nasal conchae), and the inferior nasal conchae. The soft tissues of the nose form much of the anterior margin of the nasal cavity, but the maxillae and nasal bones support the bridge of the nose.

The sphenoid, ethmoid, frontal bone, and paired palatine bones and maxillae contain the **paranasal sinuses**. Figure 7–15a shows the locations of the frontal and sphenoid sinuses, ethmoid air cells, and maxillary sinuses. (The tiny palatine sinuses,

Figure 7–14   The Orbital Complex. The right orbital region. ATLAS: Plate 5f

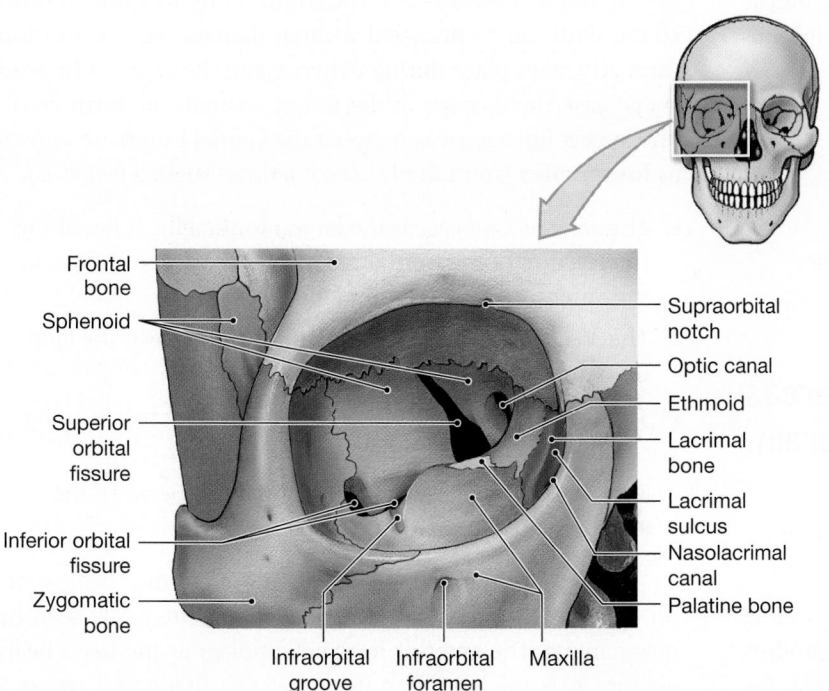

Frontal bone
Sphenoid
Superior orbital fissure
Inferior orbital fissure
Zygomatic bone
Infraorbital groove
Infraorbital foramen
Maxilla
Supraorbital notch
Optic canal
Ethmoid
Lacrimal bone
Lacrimal sulcus
Nasolacrimal canal
Palatine bone

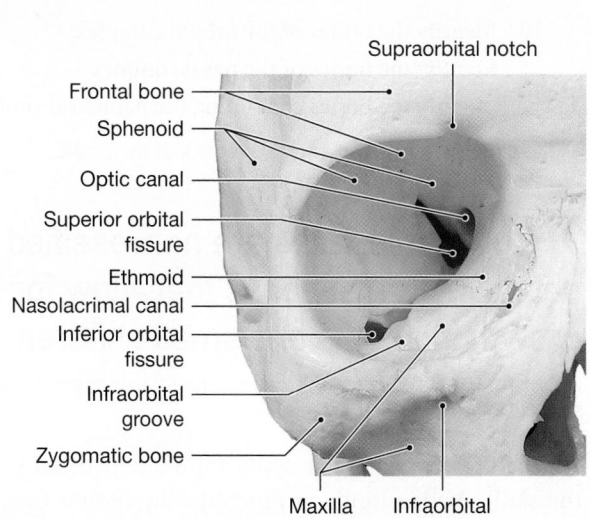

Supraorbital notch
Frontal bone
Sphenoid
Optic canal
Superior orbital fissure
Ethmoid
Nasolacrimal canal
Inferior orbital fissure
Infraorbital groove
Zygomatic bone
Maxilla
Infraorbital foramen

Figure 7–15 **The Nasal Complex.** ATLAS: Plates 11b; 12d; 13b,g

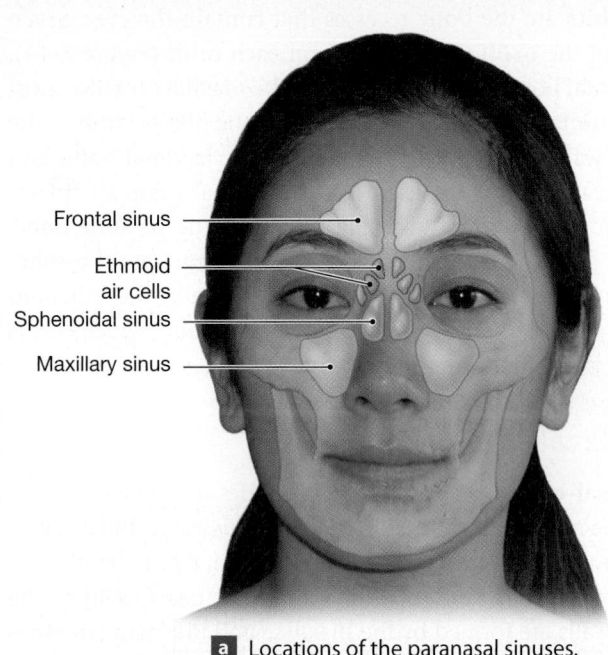

a Locations of the paranasal sinuses.

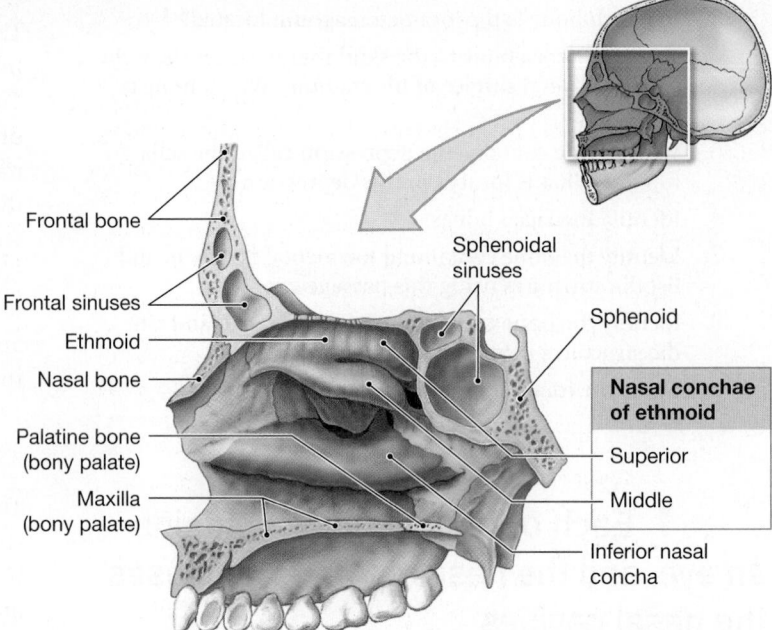

b A sagittal section through the skull, with the nasal septum removed to show major features of the wall of the right nasal cavity. The frontal and sphenoidal sinuses are visible.

not shown, generally open into the sphenoidal sinuses.) The paranasal sinuses lighten the skull bones and provide an extensive area of mucous epithelium, which releases mucus into the nasal cavities. The ciliated epithelium moves the mucus back toward the throat, where it is eventually swallowed or expelled by coughing. Incoming air is humidified and warmed as it flows across this thick carpet of mucus. Foreign particulates, such as dust or microorganisms, become trapped in the sticky mucus and are then swallowed or expelled. This mechanism helps protect the more delicate portions of the respiratory tract.

CHECKPOINT

10. Identify the bones of the orbital complex.

11. Identify the bones of the nasal complex.

12. Identify the bones containing the paranasal sinuses.

See the blue Answers tab at the back of the book.

7-4 ❙ Fontanelles are non-ossified areas between cranial bones that allow for brain growth in infants and small children

Learning Outcome Describe the key structural differences among the skulls of infants, children, and adults.

Many different centers of ossification are involved in forming the skull. As development proceeds, the centers fuse, producing a smaller number of composite bones. For example, the

sphenoid begins as 14 separate ossification centers. At birth, fusion is not yet complete: There are two frontal bones, four occipital bones, and several sphenoid and temporal elements.

The skull organizes around the developing brain. As the time of birth approaches, the brain enlarges rapidly. The bones of the skull are also growing, but they fail to keep pace. At birth, the cranial bones are connected by areas of fibrous connective tissue (**Figure 7–16**). The connections are quite flexible, so the skull can be distorted without damage. Such distortion normally takes place during delivery, and the changes in head shape ease the passage of the infant through the birth canal. The largest fibrous areas between the cranial bones are known as **fontanelles** (fon-tuh-NELZ; sometimes spelled *fontanels*):

- The *anterior fontanelle* is the largest fontanelle. It lies at the intersection of the frontal, sagittal, and coronal sutures in the anterior portion of the skull.

- The *occipital fontanelle* is at the junction between the lambdoid and sagittal sutures.

- The *sphenoidal fontanelles* are at the junctions between the squamous sutures and the coronal suture.

- The *mastoid fontanelles* are at the junctions between the squamous sutures and the lambdoid suture.

The anterior fontanelle is often called the "soft spot" on newborns. It is usually the only fontanelle easily seen by new parents. The anterior fontanelle pulses as the heart beats because it is made of fibrous connective tissue and covers a

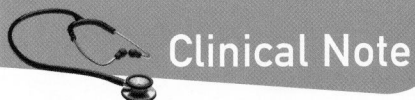

Clinical Note

Sinusitis When an irritant is introduced into the nasal passages, our bodies work to remove it. Large particles or strong chemical agents can cause us to sneeze, expelling a sizable amount of air quickly and forcefully. The sneeze often sweeps the offending particles or chemicals out with the air. Smaller particles or milder irritants trigger the production of mucus by the epithelium of the paranasal sinuses. The mucus stream flushes the nasal surfaces clean, often removing irritants such as pepper, pollen grains, or dust. In *allergic rhinitis,* an overactive immune response to an allergen, such as pollen or house dust mites, causes excessive mucosal swelling, mucus production, and sneezing.

A sinus infection, however, is another matter entirely. A viral, bacterial, or fungal infection produces an inflammation of the mucous membrane of the nasal cavity. As the membrane swells, the small passageways called *ostia* within the paranasal sinuses narrow. Mucus drains more slowly, the sinuses fill, and the individual has headaches and a feeling of pressure within the facial bones. This condition of sinus inflammation and congestion is called **sinusitis**. The maxillary sinuses are commonly involved. Because gravity does little to assist mucus drainage from these sinuses, the flushing action is less effective, and pressure on the sinus walls typically increases.

The relief of pain associated with sinusitis is the basis of a large over-the-counter (OTC) drug market in the United States. Every major pharmaceutical company has at least one product designed specifically to relieve sinus pressure. The active ingredients are compounds that dry the epithelial linings, reduce pain, and restrict further swelling. Antihistamines and vasoconstrictors, which reduce mucus production, combined with pain relievers such as acetaminophen or ibuprofen are often included.

Temporary sinus problems may follow when the mucous epithelium is exposed to chemical irritants or invading microorganisms. Chronic sinusitis may accompany chronic allergies. It may also occur as the result of a **deviated** (nasal) **septum**. In this condition, the nasal septum has a bend in it, generally at the junction between the bony and cartilaginous regions. Septal deviation often blocks the drainage of one or more sinuses, producing chronic cycles of infection and inflammation. A deviated septum can result from injuries to the nose. Many boxers and hockey players suffer from a deviated septum, because their sports subject them to numerous blows to the soft tissues of the nose. It can usually be corrected or improved by surgery.

Frontal sinus

Orbit

Left maxillary sinusitis

Normal right maxillary sinus

---

major blood vessel. This fontanelle is sometimes used to determine whether an infant is dehydrated, as the surface becomes indented when blood volume is low.

The occipital, sphenoidal, and mastoid fontanelles disappear within a month or two after birth. The anterior fontanelle generally persists until the child is nearly 2 years old. Even after the fontanelles disappear, the bones of the skull remain separated by fibrous connections.

The skulls of infants and adults differ in terms of the shape and structure of cranial elements. This difference accounts for variations in proportions as well as in size. The cranium of a young child, compared with the skull as a whole, is relatively larger than that of an adult. The most significant growth in the skull takes place before age 5, because at that age the brain stops growing and the cranial sutures develop. The growth of the cranium is generally coordinated with the expansion of the brain. If one or more sutures form before the brain stops growing, the skull will be abnormal in shape, size, or both.

CHECKPOINT

13. **Define fontanelle, and identify the major fontanelles.**

14. **What purpose does a fontanelle serve?**

See the blue Answers tab at the back of the book.

Figure 7–16 **The Skull of an Infant.**

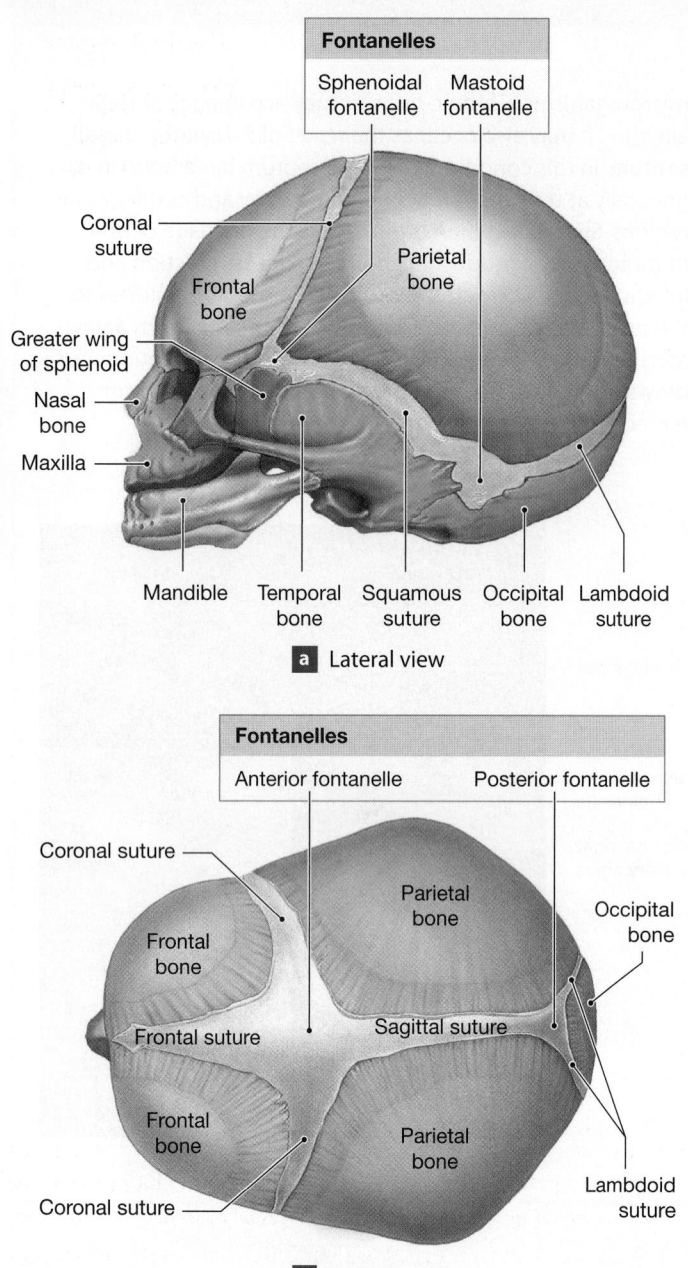

a Lateral view

b Superior view

Clinical Note

Craniostenosis Unusual distortions of the skull result from *craniostenosis* (krā-nē-ō-sten-Ō-sis; *stenosis*, narrowing), the premature closure of one or more fontanelles. As the brain continues to enlarge, the rest of the skull distorts to accommodate it. A long and narrow head is produced by early closure of the sagittal suture. A very broad skull results if the coronal suture forms prematurely. Early closure of all cranial sutures restricts the development of the brain, and surgery must be performed to prevent brain damage. If brain enlargement stops due to genetic or developmental abnormalities, however, skull growth ceases as well. This condition, which results in an undersized head, is called *microcephaly* (mī-krō-SEF-uh-lē; *micro-*, small + *cephalon*, head).

column of support, bearing the weight of the head, neck, and trunk and ultimately transferring the weight to the appendicular skeleton of the lower limbs. The vertebrae also protect the spinal cord and help maintain an upright body position, as when we sit or stand. The total length of the vertebral column of an adult averages 71 cm (28 in.).

Spinal Curvature

The vertebral column is not straight and rigid. A lateral view shows four **spinal curves**: the (1) **cervical curve**, (2) **thoracic curve**, (3) **lumbar curve**, and (4) **sacral curve** (Figure 7–17).

You may have noticed that an infant's body axis forms a C-shape, with the back curving posteriorly. The C-shape results from the thoracic and sacral curves. These are called **primary curves**, because they appear late in fetal development. They are also called **accommodation curves**, because they accommodate the thoracic and abdominopelvic viscera (internal organs). The primary curves are present at birth.

The lumbar and cervical curves, known as **secondary curves**, appear several months after birth. They are also called **compensation curves**, because they help shift the weight to permit an upright posture. The cervical and lumbar secondary curves become accentuated as the toddler learns to walk and run. All four curves are fully developed by age 10.

When you stand, the weight of your body must be transmitted through the vertebral column to the hips and ultimately to the lower limbs. Yet most of your body weight lies anterior to the vertebral column. The various curves bring that weight in line with the body axis. Think about what you do automatically when standing with a heavy object hugged to your chest. You avoid toppling forward by exaggerating the lumbar curve and by keeping the weight back toward the body axis. This

7-5 ┃ The vertebral column has four spinal curves

Learning Outcome Identify and describe the curvatures of the spinal column, and indicate the function of each.

The rest of the axial skeleton consists of the vertebral column, ribs, and sternum. The adult **vertebral column**, or *spine*, consists of 26 bones: the **vertebrae** (24), the **sacrum** (1), and the **coccyx** (1) (KOK-siks), or tailbone. The vertebrae provide a posture can lead to discomfort at the base of the spinal column. For example, many women in the last three months of pregnancy develop chronic back pain from the changes in lumbar curvature that adjust for the increasing weight of the fetus. In many parts of the world, people often carry heavy objects balanced on their head. This practice increases the load on the vertebral column but does not affect the spinal curves because the weight is aligned with the axis of the spine.

Vertebral Anatomy

Each vertebra consists of three basic parts: (1) a *vertebral body*, (2) a *vertebral arch*, and (3) *articular processes* (Figure 7–18a).

The **vertebral body** is the part of a vertebra that transfers weight along the axis of the vertebral column (Figure 7–18a,b,e). The bodies of adjacent vertebrae are interconnected by ligaments, but are separated by pads of fibrocartilage, the **intervertebral discs**.

The **vertebral arch** forms the posterior margin of each **vertebral foramen** (Figure 7–18a,c). The vertebral arch has walls, called **pedicles** (PED-i-kulz), and a roof, formed by flat layers called **laminae** (LAM-i-nē; singular, *lamina*, a thin plate). The pedicles arise along the posterior and lateral margins of the body. The laminae on either side extend dorsally and medially to complete the roof. Together, the vertebral foramina of successive vertebrae form the **vertebral canal**, which encloses the spinal cord (Figure 7–18e).

A **spinous process** projects posteriorly from the point where the vertebral laminae fuse to complete the vertebral arch. You can see—and feel—the spinous processes through the skin of the back when the spine is flexed. **Transverse processes** project laterally or dorsolaterally on both sides from the point where the laminae join the pedicles. These processes are sites of muscle attachment, and they may also articulate with the ribs.

Like the transverse processes, the **articular processes** arise at the junction between the pedicles and the laminae. A **superior** and an **inferior articular process** lie on each side of the vertebra. The superior articular processes articulate with the inferior articular processes of a more superior vertebra (or the occipital condyles, in the case of the first cervical vertebra). The inferior articular processes articulate with the superior articular processes of a more inferior vertebra (or the sacrum, in

the case of the last lumbar vertebra). Each articular process has a smooth concave surface called an **articular facet**. The superior processes have articular facets on their dorsal surfaces, whereas the inferior processes articulate along their ventral surfaces.

Intervertebral discs separate adjacent vertebral bodies, and gaps separate the pedicles of successive vertebrae. These gaps, called **intervertebral foramina**, permit the passage of nerves running to or from the enclosed spinal cord (Figure 7–18e).

Figure 7–17 **The Vertebral Column.** ATLAS: Plate 2b

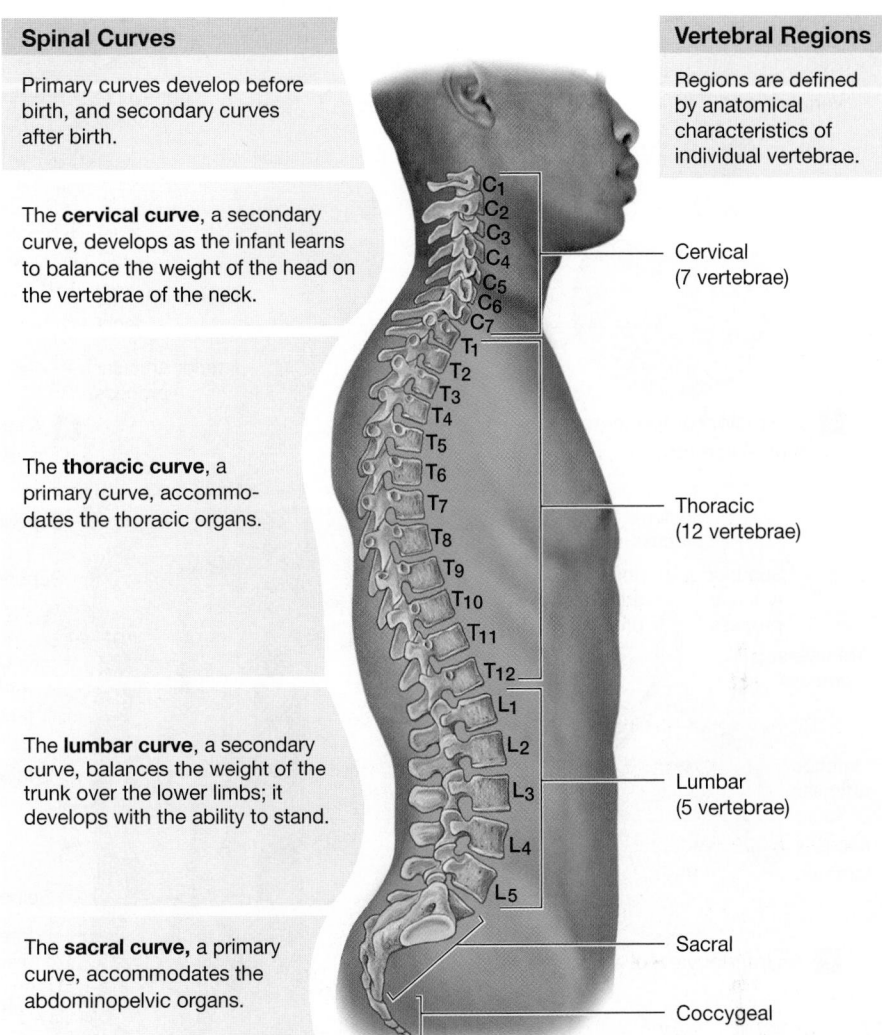

Spinal Curves

Primary curves develop before birth, and secondary curves after birth.

The **cervical curve**, a secondary curve, develops as the infant learns to balance the weight of the head on the vertebrae of the neck.

The **thoracic curve**, a primary curve, accommodates the thoracic organs.

The **lumbar curve**, a secondary curve, balances the weight of the trunk over the lower limbs; it develops with the ability to stand.

The **sacral curve**, a primary curve, accommodates the abdominopelvic organs.

Vertebral Regions

Regions are defined by anatomical characteristics of individual vertebrae.

C1
C2
C3
C4
C5
C6
C7

Cervical (7 vertebrae)

T1
T2
T3
T4
T5
T6
T7
T8
T9
T10
T11
T12

Thoracic (12 vertebrae)

L1
L2
L3
L4
L5

Lumbar (5 vertebrae)

Sacral

Coccygeal

CHECKPOINT

15. What is the importance of the secondary curves of the spine?

16. When you run your finger along a person's spine, what part of the vertebrae are you feeling just beneath the skin?

See the blue Answers tab at the back of the book.

Figure 7–18 **Vertebral Anatomy.**

a The major components of a typical vertebra

b A lateral and slightly inferior view of a vertebra

c An inferior view of a vertebra

d A posterior view of three articulated vertebrae

e A lateral and sectional view of three articulated vertebrae

7-6 ▌ The five vertebral regions are the cervical, thoracic, lumbar, sacral, and coccygeal regions

Learning Outcome Identify the vertebral regions, and describe the distinctive structural and functional characteristics of vertebrae in each region.

As Figure 7–17 shows, the vertebral column is divided into cervical, thoracic, lumbar, sacral, and coccygeal regions. The cervical, thoracic, and lumbar regions consist of individual vertebrae. Seven **cervical vertebrae** (C_1–C_7) constitute the neck and extend inferiorly to the trunk. Twelve **thoracic vertebrae** (T_1–T_{12}) form the superior portion of the back. Each articulates with one or more pairs of ribs. Five **lumbar vertebrae** (L_1–L_5) form the inferior portion of the back. The fifth articulates with the sacrum, which in turn articulates with the coccyx.

During development, the sacrum originates as a group of five vertebrae, and the coccyx begins as three to five very small vertebrae. In general, the vertebrae of the sacrum are completely fused by age 25–30. Ossification of the distal coccygeal vertebrae is not complete before puberty. Thereafter fusion takes place at a variable pace. ATLAS: Embryology Summary 7: The Development of the Vertebral Column

Clinical Note

Kyphosis, Lordosis, and Scoliosis The vertebral column must move, balance, and support the trunk and head. Conditions or events that damage the bones, muscles, and/or nerves can result in distorted shapes and impaired function. In **kyphosis** (kī-FŌ-sis; *kyphos*, humpbacked, bent), the normal thoracic curvature becomes exaggerated posteriorly, producing a "round-back" appearance (**Photo a**). This condition can be caused by (1) osteoporosis with compression fractures affecting the anterior portions of vertebral bodies, (2) chronic contractions in muscles that insert on the vertebrae, or (3) abnormal vertebral growth. In **lordosis** (lor-DŌ-sis; *lordosis*, a bending backward), or "swayback," both the abdomen and buttocks protrude abnormally (**Photo b**). The cause is an anterior exaggeration of the lumbar curvature. This may occur during pregnancy or result from abdominal obesity or weakness

in the muscles of the abdominal wall. **Scoliosis** (skō-lē-Ō-sis; *scoliosis*, crookedness) is an abnormal lateral curvature of the spine in one or more of the movable vertebrae (**Photo c**). Scoliosis is the most common distortion of the spinal curvature. This condition may result from developmental problems from damage to vertebral bodies, or from muscular paralysis affecting one side of the back (as in some cases of polio). In 4 out of 5 cases, the structural or functional cause of the abnormal spinal curvature is impossible to determine. This idiopathic (of no known cause) scoliosis generally appears in girls during adolescence, when periods of growth are most rapid. Small curves may later stabilize once growth is complete. For larger curves, bracing may prevent progression. Severe cases can be treated through surgical straightening with implanted metal rods or cables.

a Kyphosis

b Lordosis

c Scoliosis

Note that when we refer to a specific vertebra, we use a capital letter for the vertebral region: C, T, L, S, and Co represent the cervical, thoracic, lumbar, sacral, and coccygeal regions, respectively. In addition, we use a numerical subscript to indicate the relative position of the vertebra within that region, with 1 indicating the vertebra closest to the skull. For example, C_1 is in contact with the skull and C_3 is the third cervical vertebra. Similarly, T_{12} is in contact with L_1 and L_4 is the fourth lumbar vertebra (**Figure 7–17**). We will use this shorthand throughout the text.

Each vertebra has characteristic markings and articulations, but we will focus on the general characteristics of each

region. We will see how regional variations determine each vertebral group's function.

Cervical Vertebrae

Most mammals—whether giraffes, whales, mice, or humans—have seven cervical vertebrae (**Figure 7–19a**). The cervical vertebrae are the smallest in the vertebral column. They extend from the occipital bone of the skull to the thorax. The body of a cervical vertebra is small compared with the size of the vertebral foramen (**Figure 7–19b**). At this level, the spinal cord still contains most of the axons that connect the brain to the rest of

Figure 7–19   **The Cervical Vertebrae.** ATLAS: Plates 20b; 21a–e

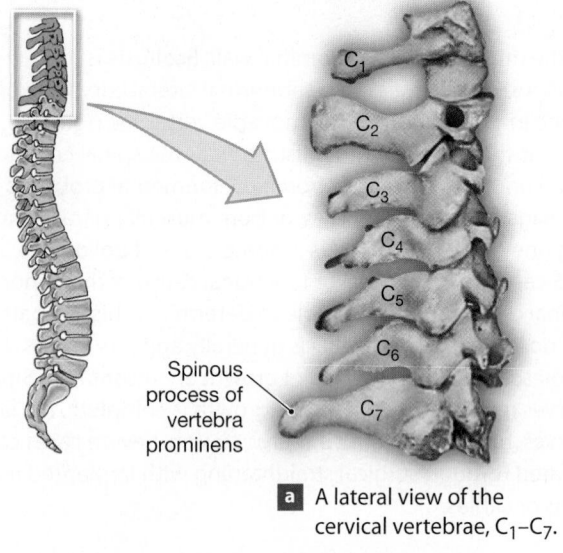

a A lateral view of the cervical vertebrae, C_1–C_7.

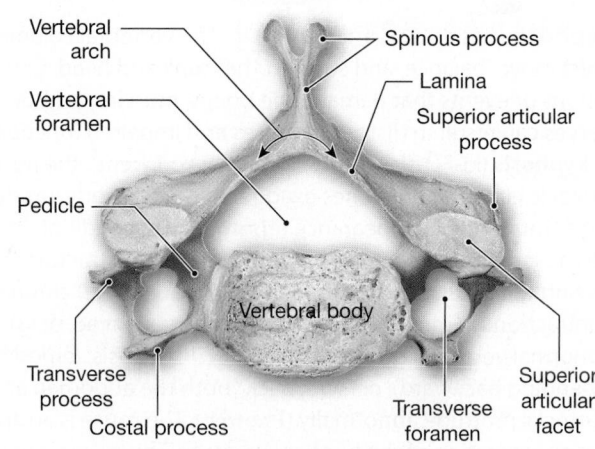

b A superior view of a representative cervical vertebra showing characteristics of C_3–C_6. Notice the typical features listed in *Table 7–1*.

c A lateral view of the same vertebra as in b.

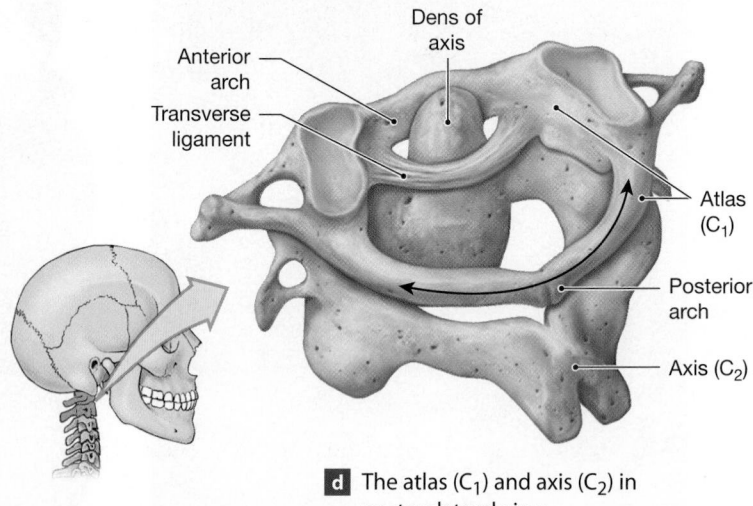

d The atlas (C_1) and axis (C_2) in posterolateral view.

the body. The diameter of the spinal cord decreases as you proceed caudally along the vertebral canal, and so does the diameter of the vertebral arch. However, cervical vertebrae support only the weight of the head, so the vertebral body can be relatively small and light. As you continue toward the sacrum, the loading increases and the vertebral bodies gradually enlarge.

In a typical cervical vertebra, the superior surface of the body is concave from side to side, and it slopes, with the anterior edge inferior to the posterior edge (**Figure 7–19c**). Vertebra C_1 has no spinous process. The spinous processes of the other cervical vertebrae are relatively stumpy, generally shorter than the diameter of the vertebral foramen. In the case of vertebrae C_2–C_6, the tip of each spinous process has a prominent

notch (**Figure 7–19b**). A notched spinous process is said to be **bifid** (BĪ-fid).

Laterally, the transverse processes are fused to the **costal processes**, which originate near the ventrolateral portion of the vertebral body. The costal and transverse processes encircle prominent, round **transverse foramina**. These passageways protect the *vertebral arteries* and *vertebral veins*. These important blood vessels service the brain.

The preceding description is adequate for identifying the cervical vertebrae C_3–C_6. The first two cervical vertebrae are unique, and the seventh is modified. We will describe these vertebrae shortly. The interlocking bodies of articulated C_3–C_7 permit more flexibility than do those of other regions.

Table 7–1	Regional Differences in Vertebral Structure and Function		
	Type (Number)		
Feature	**Cervical Vertebrae (7)**	**Thoracic Vertebrae (12)**	**Lumbar Vertebrae (5)**
Location	Neck	Chest	Inferior portion of back
Body	Small, oval, curved faces	Medium, heart-shaped, flat faces; facets for rib articulations	Massive, oval, flat faces
Vertebral foramen	Large	Medium	Small
Spinous process	Long; split tip; points inferiorly	Long, slender; not split; points inferiorly	Blunt, broad; points posteriorly
Transverse processes	Have transverse foramina	All but two (T_{11}, T_{12}) have facets for rib articulations	Short; no articular facets or transverse foramina
Functions	Support skull, stabilize relative positions of brain and spinal cord, and allow controlled head movement	Support weight of head, neck, upper limbs, and chest; articulate with ribs to allow changes in volume of thoracic cage	Support weight of head, neck, upper limbs, and trunk

Typical appearance (superior view)

Table 7–1 includes a summary of the features of these cervical vertebrae.

Compared with the cervical vertebrae, your head is relatively massive. It sits atop the cervical vertebrae like a soup bowl on the tip of a finger. With this arrangement, small muscles can produce significant effects by tipping the balance one way or another. But if you change position suddenly, as in a fall or during rapid acceleration (a jet takeoff) or deceleration (a car crash), the balancing muscles are not strong enough to stabilize the head. A dangerous partial or complete dislocation of the cervical vertebrae can result, with injury to muscles and ligaments and potential injury to the spinal cord. We use the term **whiplash** to describe such an injury, because the movement of the head resembles the cracking of a whip.

The Atlas (C₁)

The **atlas**, cervical vertebra C_1 (Figure 7–19d), holds up the head. It articulates with the occipital condyles of the skull. This vertebra is named after Atlas, who, according to Greek myth, holds the world on his shoulders. The articulation between the occipital condyles and the atlas is a joint that permits you to nod (such as when you indicate "yes"). The atlas can easily be distinguished from other vertebrae by (1) the lack of a body and spinous process and (2) the presence of a large, round vertebral foramen bounded by **anterior** and **posterior arches**.

The atlas articulates with the second cervical vertebra, the *axis*. This articulation permits rotation (as when you shake your head to indicate "no").

The Axis (C₂)

During development, the body of the atlas fuses to the body of the second cervical vertebra, called the **axis** (C_2) (Figure 7–19d). This fusion creates the prominent **dens** (DENZ; *dens*, tooth), or *odontoid* (ō-DON-toyd; *odontos*, tooth) *process*, of the axis. A transverse ligament binds the dens to the inner surface of the atlas, forming a pivot for rotation of the atlas and skull.

Important muscles controlling the position of the head and neck attach to the especially robust spinous process of the axis.

In children, the fusion between the dens and axis is incomplete. Impacts or even severe shaking can cause dislocation of the dens and severe damage to the spinal cord. In adults, a hit to the base of the skull can be equally dangerous, because a dislocation of the atlas–axis joint can force the dens into the base of the brain, with fatal results.

TIPS & TRICKS

To remember the difference between the atlas and the axis and their respective movements, consider Greek mythology and the Earth. In this case, the head (Earth) is held up by Atlas and is capable of nodding "yes" movements. The Earth rotates on its axis, and the axis allows for us to shake our heads in a "no" movement.

The Vertebra Prominens (C_7)

The transition from one vertebral region to another is not abrupt. The last vertebra of one region generally resembles the first vertebra of the next. The **vertebra prominens**, or seventh cervical vertebra (C_7), has a long, slender spinous process (**Figure 7–19a**) that ends in a broad tubercle that you can feel through the skin at the base of the neck. This vertebra is the interface between the cervical curve, which arches anteriorly, and the thoracic curve, which arches posteriorly (**Figure 7–17**). The transverse processes of C_7 are large, providing additional surface area for muscle attachment.

The **ligamentum nuchae** (lig-uh-MEN-tum-Nū-kē; *nucha*, nape), a stout elastic ligament, begins at the vertebra prominens and extends to an insertion (attachment site) along the occipital crest of the skull. Along the way, it attaches to the spinous processes of the other cervical vertebrae. When your head is upright, this ligament acts like the string on a bow, maintaining the cervical curvature without muscular effort. If you have bent your neck forward, the elasticity in the ligamentum nuchae helps return your head to an upright position.

Thoracic Vertebrae

There are 12 thoracic vertebrae (**Figure 7–20a**). A typical thoracic vertebra has a distinctive heart-shaped body that is more massive than that of a cervical vertebra (**Figure 7–20b**). The vertebral foramen is relatively smaller. The long, slender spinous process projects posteriorly and inferiorly. The spinous processes of T_{10}, T_{11}, and T_{12} increasingly resemble those of the lumbar region as the transition between the thoracic and lumbar curves approaches. Because the inferior thoracic and lumbar vertebrae carry so much weight, the transition between the thoracic and lumbar curves is difficult to stabilize. As a result, compression fractures or compression–dislocation fractures

from a hard fall tend to involve the last thoracic and first two lumbar vertebrae.

Each thoracic vertebra articulates with ribs along the dorsolateral surfaces of the body. The **costal facets** on the vertebral bodies articulate with the heads of the ribs. The location and structure of the articulations vary somewhat among thoracic vertebrae (**Figure 7–20a**). Vertebrae T_1–T_8 each articulate with two pairs of ribs, so their vertebral bodies have two costal facets (*superior* and *inferior*) on each side. Vertebrae T_9–T_{11} have a single costal facet on each side, and each vertebra articulates with a single pair of ribs.

The transverse processes of vertebrae T_1–T_{10} are relatively thick and contain **transverse costal facets** for rib articulation (**Figure 7–20b,c**). Thus, rib pairs 1 through 10 contact their vertebrae at two points: a costal facet and a transverse costal facet. Table 7–1 summarizes the features of thoracic vertebrae.

Lumbar Vertebrae

The five lumbar vertebrae are the largest vertebrae (**Figure 7–21a**). The body of a typical lumbar vertebra (**Figure 7–21b,c**) is thicker than that of a thoracic vertebra, and the superior and inferior surfaces are oval rather than heart shaped. Also note that (1) lumbar vertebrae do not have costal facets; (2) the slender transverse processes, which lack transverse costal facets, project dorsolaterally; (3) the vertebral foramen is triangular; (4) the stumpy spinous processes project dorsally; (5) the superior articular processes face medially ("up and in"); and (6) the inferior articular processes face laterally ("down and out").

The lumbar vertebrae withstand the most weight. Their massive spinous processes provide surface area for the attachment of lower back muscles that reinforce or adjust the lumbar curve. Table 7–1 summarizes the features of lumbar vertebrae.

TIPS & TRICKS

To remember the number of bones in the first three spinal curves, think about mealtimes. You eat breakfast at 7 a.m. (7 cervical vertebrae), lunch at 12 p.m. (12 thoracic vertebrae), and dinner at 5 p.m. (5 lumbar vertebrae).

The Sacrum

The sacrum consists of the fused components of five sacral vertebrae. These vertebrae begin fusing shortly after puberty. In general, they are completely fused at age 25–30. The sacrum protects the reproductive, digestive, and urinary organs. Through paired articulations, it attaches the axial skeleton to the paired hip bones, or pelvic girdle, of the appendicular skeleton (**Figure 7–1a**).

Figure 7–20 **The Thoracic Vertebrae.** Notice the characteristic features listed in *Table 7–1* on p. 257. ATLAS: Plates 22a–c

b Thoracic vertebra, superior view.

a A lateral view of the thoracic region of the vertebral column. The vertebra prominens (C₇) resembles T₁, but lacks facets for rib articulation. Vertebra T₁₂ resembles the first lumbar vertebra (L₁) but has a facet for rib articulation.

c Thoracic vertebra, lateral view.

The broad posterior surface of the sacrum (**Figure 7–22a**) provides an extensive area for the attachment of muscles, especially those that move the thigh. The superior articular processes of the first sacral vertebra articulate with the last lumbar vertebra. The **sacral canal** is a passageway that begins between these articular processes and extends the length of the sacrum. Nerves and membranes that line the vertebral canal in the spinal cord continue into the sacral canal.

Three sacral crests provide surface areas for the attachment of muscles. The **median sacral crest** is a ridge formed by the fused spinous processes of the sacral vertebrae. The laminae of the fifth sacral vertebra fail to contact one another at the midline. They form the **sacral cornua** (KOR-nū-uh; singular,

cornu; cornua, horns). These ridges form the margins of the **sacral hiatus** (hī-Ā-tus), the opening at the inferior end of the sacral canal. This opening is covered by connective tissues. Four pairs of **sacral foramina** open on either side of the median sacral crest. The intervertebral foramina of the fused sacral vertebrae open into these passageways. Each **lateral sacral crest** is a ridge that represents the fused transverse processes of the sacral vertebrae. The sacrum is curved, with a convex posterior surface (**Figure 7–22b**). The degree of curvature is more pronounced in males than in females. The **auricular surface** is a thickened, flattened area lateral and anterior to the superior portion of the lateral sacral crest. The auricular surface is the site of articulation with the pelvic girdle (the *sacroiliac joint*).

Figure 7–21 **The Lumbar Vertebrae.** ATLAS: Plates 23a–c

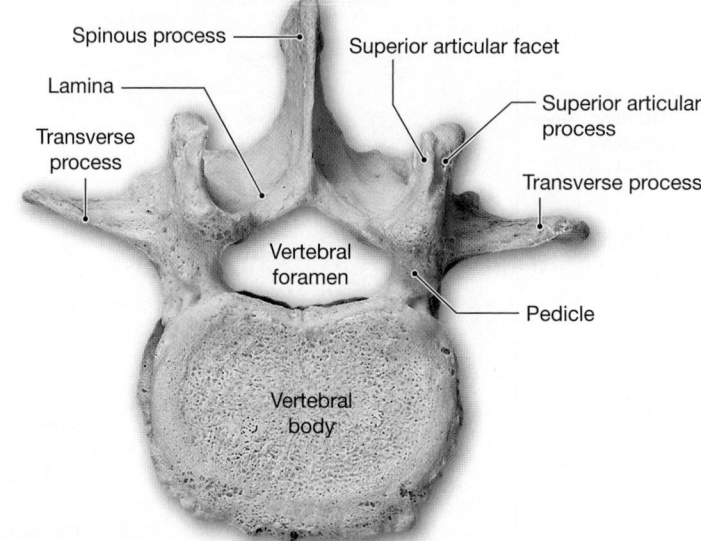

a A lateral view of the lumbar vertebrae, sacrum, and coccyx

b A lateral view of a typical lumbar vertebra

c A superior view of the same vertebra shown in part b

Figure 7–22 **The Sacrum and Coccyx.**

a A posterior view

b A lateral view from the right side

c An anterior view

The **sacral tuberosity** is a roughened area between the lateral sacral crest and the auricular surface. It marks the attachment site of ligaments that stabilize the sacroiliac joint.

The subdivisions of the sacrum are most clearly seen in anterior view (**Figure 7–22c**). The narrow, inferior portion is the sacral **apex.** The broad superior surface is the **base**. The **sacral promontory**, a prominent bulge at the anterior tip of the base, is an important landmark in females during pelvic examinations and during labor and delivery. Prominent *transverse lines* mark the former boundaries of individual vertebrae that fuse during the formation of the sacrum. At the base of the sacrum, a broad sacral **ala**, or *wing*, extends on either side. The anterior and superior surfaces of each ala provide an extensive area for muscle attachment. At the apex, a flattened area marks the site of articulation with the coccyx.

The Coccyx

The small coccyx consists of three to five (typically, four) coccygeal vertebrae that have generally begun fusing by age 26 (**Figure 7–22**). The coccyx provides an attachment site for a number of ligaments and for a muscle that constricts the anal opening. The first two coccygeal vertebrae have transverse processes and unfused vertebral arches. The prominent laminae of the first coccygeal vertebrae are known as the **coccygeal cornua**. These laminae curve to meet the sacral cornua. The coccygeal vertebrae do not fuse completely until late in adulthood. In very old persons, the coccyx may fuse with the sacrum.

CHECKPOINT

17. Why does the vertebral column of an adult have fewer vertebrae than that of a newborn?

18. Joe suffered a hairline fracture at the base of the dens. Which bone is fractured, and where is it located?

19. Examining a human vertebra, you notice that, in addition to the large foramen for the spinal cord, two smaller foramina are on either side of the bone in the region of the transverse processes. From which region of the vertebral column is this vertebra?

20. Why are the bodies of the lumbar vertebrae so large?

See the blue Answers tab at the back of the book.

7-7 | The thoracic cage protects organs in the chest and provides sites for muscle attachment

Learning Outcome Explain the significance of the articulations between the thoracic vertebrae and the ribs, and between the ribs and sternum.

The skeleton of the chest, or **thoracic cage** (**Figure 7–23**), provides bony support to the walls of the thoracic cavity. It consists of the thoracic vertebrae, the ribs, and the sternum (breastbone). The ribs and the sternum form the *rib cage,* whose movements are important in respiration. The thoracic cage as a whole serves two functions:

1. It protects the heart, lungs, thymus, and other structures in the thoracic cavity.

2. It serves as an attachment point for muscles involved in (1) respiration, (2) maintenance of the position of the vertebral column, and (3) movements of the pectoral girdle (both clavicles and scapulae) and upper limbs.

The Ribs

Ribs, or *costae*, are long, curved, flattened bones that originate on or between the thoracic vertebrae and end in the wall of the thoracic cavity. Each of us, regardless of sex, has 12 pairs of ribs (**Figure 7–23**). The first seven pairs are called **true ribs**, or *vertebrosternal ribs.* They reach the anterior body wall and are connected to the sternum by separate cartilaginous extensions, the **costal cartilages**. Beginning with the first rib, the vertebrosternal ribs gradually increase in length and in radius of curvature.

Ribs 8–12 are called **false ribs**, because they do not attach directly to the sternum. The costal cartilages of ribs 8–10, the *vertebrochondral ribs,* fuse together and merge with the cartilages of rib pair 7 before they reach the sternum (**Figure 7–23a**). The last two pairs of ribs (11 and 12) are called *floating ribs* because they have no connection with the sternum. They are also called *vertebral ribs* because they are attached only to the vertebrae (**Figure 7–23b**) and muscles of the body wall.

Figure 7–24a shows the superior surface of a typical rib. The *vertebral end* of the rib articulates with the vertebral column at the **head**, or *capitulum* (ka-PIT-ū-lum). A ridge divides the articular surface of the head into superior and inferior articular facets (**Figure 7–24b**). From the head, a short **neck** leads to the **tubercle**, a small elevation that projects dorsally. The inferior portion of the tubercle contains an articular facet that contacts the transverse process of the thoracic vertebra.

Ribs 1 and 10 originate at costal facets on vertebrae T_1 and T_{10}, respectively, and their tubercular facets articulate with the transverse costal facets on those vertebrae. The heads of ribs 2–9 articulate with costal facets on two adjacent vertebrae. Their tubercular facets articulate with the transverse costal facets of the inferior member of the vertebral pair. Ribs 11 and 12, which originate at T_{11} and T_{12}, do not have tubercular facets and do not contact the transverse processes of T_{11} or T_{12}.

The bend, or *angle,* of the rib is the site where the tubular **body**, or *shaft,* begins curving toward the sternum (**Figure 7–24b**). The internal rib surface is concave, and a prominent *costal groove* along its inferior border marks the path of nerves and blood vessels. The superficial surface is convex and provides an attachment site for muscles of the pectoral girdle and trunk. The *intercostal muscles*, which move the ribs, are attached to the superior and inferior surfaces.

Figure 7–23   **The Thoracic Cage.** ATLAS: Plate 22b

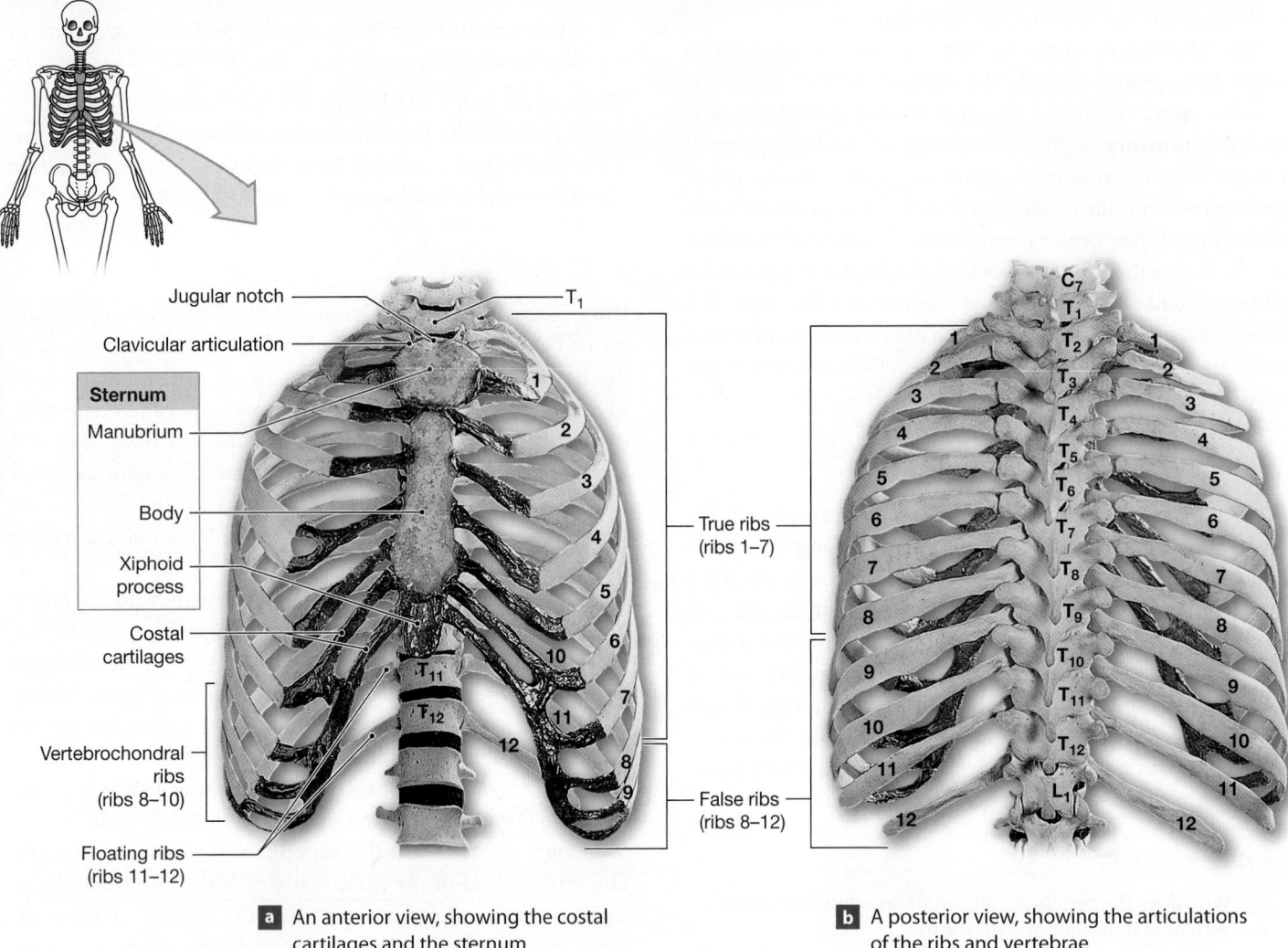

Jugular notch
Clavicular articulation

Sternum
Manubrium
Body
Xiphoid process
Costal cartilages

Vertebrochondral ribs (ribs 8–10)
Floating ribs (ribs 11–12)

True ribs (ribs 1–7)
False ribs (ribs 8–12)

a An anterior view, showing the costal cartilages and the sternum

b A posterior view, showing the articulations of the ribs and vertebrae

The ribs are quite mobile, due to their complex musculature, dual articulations at the vertebrae, and flexible connection to the sternum. Note how the ribs curve away from the vertebral column to angle inferiorly (**Figure 7–23**). Because of the curvature of the ribs, their movements during breathing changes the position of the sternum. Lowering the ribs pulls the sternum inward, and raising the ribs moves it outward. As a result, movements of the ribs change both the width and the depth of the thoracic cage, decreasing or increasing its volume accordingly as we breathe out or in. These movements are important in breathing, as we will see in detail in Chapter 23.

The ribs can bend and move to cushion shocks and absorb hits, but severe or sudden impacts can cause painful rib fractures. A cracked rib can heal without a cast or splint because the ribs are tightly bound in connective tissues. But compound fractures of the ribs can send bone splinters or fragments into the thoracic cavity, with potential damage to internal organs.

Surgery on the heart, lungs, or other organs in the thorax typically involves entering the thoracic cavity. The mobility of the

ribs and the cartilaginous connections with the sternum allow the ribs to be temporarily moved out of the way. "Rib spreaders" are used to push the ribs apart in much the same way that a jack lifts a car off the ground for a tire change. If more extensive access is required, the cartilages of the sternum can be cut and the entire sternum folded out of the way. Once the sternum is replaced, scar tissue reunites the cartilages, and the ribs heal fairly rapidly.

The Sternum

The adult **sternum**, or breastbone, is a flat bone that forms in the anterior midline of the thoracic wall (**Figure 7–23a**). The sternum has three parts:

1. The broad, triangular **manubrium** (ma-NŪ-brē-um) articulates with the *clavicles* (collarbones) and the cartilages of the first pair of ribs. The manubrium is the widest and most superior portion of the sternum. Only the first pair of ribs is attached by cartilage to this portion of the sternum. The **jugular notch**, located between the clavicular

Figure 7–24 **The Ribs.** ATLAS: Plates 22a,b

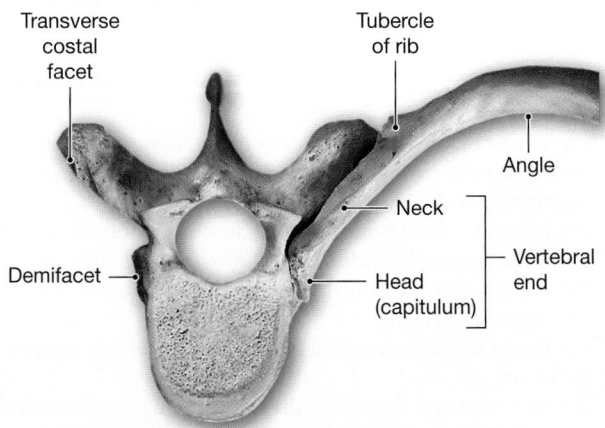

Transverse costal facet

Tubercle of rib

Angle

Neck

Vertebral end

Demifacet

Head (capitulum)

a A superior view of the details of rib structure and the articulations between the ribs and thoracic vertebrae

Head

Neck

Attachment to costal cartilage (sternal end)

Articular facets

Tubercle

Body

Angle

Costal groove

b A posterior view of the head of a representative rib from the right side (ribs 2–9)

articulations, is a shallow indentation on the superior surface of the manubrium.

2. The tongue-shaped **body** attaches to the inferior surface of the manubrium and extends inferiorly along the midline. Individual costal cartilages from rib pairs 2–7 are attached to this portion of the sternum.

3. The **xiphoid** (ZI-foyd) **process** is the smallest part of the sternum. It is attached to the inferior surface of the body. The muscular *diaphragm* and *rectus abdominis muscles* attach to the xiphoid process.

Ossification of the sternum begins at 6 to 10 ossification centers, and fusion is not complete until at least age 25. Before that age, the sternal body consists of four separate bones. In adults, their boundaries appear as a series of transverse lines crossing the sternum. The xiphoid process is generally the last part to ossify and fuse. Its connection to the sternal body can be broken by impact or by strong pressure, creating a spear of bone that can severely damage the liver. The xiphoid process is used as a palpable landmark during the administration of cardiopulmonary resuscitation (CPR), and CPR training strongly emphasizes proper hand positioning to reduce the chances of breaking ribs or the xiphoid process.

CHECKPOINT

21. How are true ribs distinguished from false ribs?
22. Improper administration of cardiopulmonary resuscitation (CPR) can result in a fracture of which bones?
23. What are the main differences between vertebrosternal ribs and vertebrochondral ribs?

See the blue Answers tab at the back of the book.

Related Clinical Terms

craniotomy: The surgical removal of a section of bone (bone flap) from the skull for the purpose of operating on the underlying tissues.

deviated nasal septum: A bent nasal septum (cartilaginous structure dividing the left and right nasal cavities) that slows or prevents sinus drainage.

herniated disc: A disc (fibrocartilage pad) between the vertebrae that slips out of place or ruptures; if it presses on a nerve, it can cause back pain or sciatica.

laminectomy: A surgical operation to remove the posterior vertebral arch on a vertebra, usually to give access to the spinal cord or to relieve pressure on nerves.

spina bifida: A condition resulting from the failure of the vertebral laminae to unite during development; commonly associated with developmental abnormalities of the brain and spinal cord.

spinal fusion: A surgical procedure that stabilizes the spine by joining together (fusing) two or more vertebrae using bone grafts, metal rods, or screws.

Chapter Review

Study Outline

MasteringA&P®

Access more chapter study tools online in the MasteringA&P Study Area:
- Chapter Quizzes, Chapter Practice Test, Art-labeling Activities, Animations, MP3 Tutor Sessions, and Clinical Case Studies
- Practice Anatomy Lab   PAL™
- Interactive Physiology   iP°
- A&P Flix   A&PFlix
- PhysioEx   PhysioEx

7-1 ▌ The 80 bones of the head and trunk make up the axial skeleton p. 233

1. The skeletal system consists of the **axial skeleton** and the appendicular skeleton. The axial skeleton can be divided into the **skull**, the **auditory ossicles** and **hyoid bone**, the **vertebral column**, and the **thoracic cage**. *(Figure 7–1)*

2. The **appendicular skeleton** includes the pectoral and pelvic girdles, which support the upper and lower limbs.

7-2 ▌ The skull is composed of 8 cranial bones and 14 facial bones p. 233

3. The **skull** consists of the **cranium** and the bones of the face. The cranium, composed of **cranial bones,** encloses the **cranial cavity.** The **facial bones** protect and support the entrances to the digestive and respiratory tracts. *(Figure 7–2)*

4. Prominent superficial landmarks on the skull include the **lambdoid, coronal, sagittal,** and **squamous sutures.** *(Figure 7–3)*

5. The bones of the cranium are the **occipital bone,** the two **parietal bones,** the **frontal bone,** the two **temporal bones,** the **sphenoid,** and the **ethmoid.** *(Figures 7–2 to 7–9)*

6. The occipital bone surrounds the **foramen magnum.** *(Figures 7–3 to 7–5)*

7. The frontal bone contains the **frontal sinuses**. *(Spotlight Figure 7–4, Figure 7–6)*

8. The **auditory ossicles** are located in a cavity within the temporal bone. *(Figure 7–7)*

9. The bones of the face are the **maxillae,** the **palatine bones,** the **nasal bones,** the **vomer,** the **inferior nasal conchae,** the **zygomatic bones,** the **lacrimal bones,** and the **mandible.** *(Figures 7–2 to Spotlight Figure 7–4, 7–10 to 7–12)*

10. The left and right maxillae, or *maxillary bones,* are the largest facial bones; they form the upper jaw and most of the **hard palate.** *(Figures 7–3, Spotlight Figure 7–4, 7–10)*

11. The palatine bones are small L-shaped bones that form the posterior portions of the hard palate and contribute to the floor of the orbital cavities. *(Figures 7–3, Spotlight Figure 7–4, 7–10)*

12. The paired nasal bones extend to the superior border of the **external nares**. *(Figures 7–3, Spotlight Figure 7–4, 7–11)*

13. The vomer forms the inferior portion of the **nasal septum.** *(Figures 7–3, Spotlight Figure 7–4, 7–11)*

14. The **temporal process** of the zygomatic bone articulates with the **zygomatic process** of the temporal bone to form the **zygomatic arch.** *(Figures 7–3, 7–7, 7–11)*

15. The paired lacrimal bones, the smallest bones of the face, are situated medially in each **orbit.** *(Figures 7–3, 7–11)*

16. The mandible is the bone of the lower jaw. *(Figures 7–3, Spotlight Figure 7–4, 7–12)*

17. The **hyoid bone,** suspended by *stylohyoid ligaments,* supports the larynx. *(Figure 7–13)*

18. The foramina and fissures of the adult skull are summarized in *Spotlight Figure 7–4.*

7-3 ▌ Each orbital complex contains an eye, and the nasal complex encloses the nasal cavities p. 249

19. Seven bones form each **orbital complex.** *(Figure 7–14)*

20. The **nasal complex** includes the bones that enclose the nasal cavities and the **paranasal sinuses,** hollow airways that connect with the nasal passages. *(Figure 7–15)*

7-4 ▌ Fontanelles are non-ossified fibrous areas between cranial bones that allow for brain growth in infants and small children p. 250

21. Fibrous connective tissue **fontanelles** permit the skulls of infants and children to continue growing after birth. *(Figure 7–16)*

7-5 ▌ The vertebral column has four spinal curves p. 252

22. The **vertebral column** consists of the vertebrae, sacrum, and coccyx. We have 7 **cervical vertebrae** (the first articulates with the skull), 12 **thoracic vertebrae** (which articulate with the ribs), and 5 **lumbar vertebrae** (the last articulates with the sacrum). The **sacrum** and **coccyx** consist of fused vertebrae. *(Figure 7–17)*

23. The spinal column has four **spinal curves.** The **thoracic** and **sacral curves** are called **primary** or **accommodation curves;** the **lumbar** and **cervical curves** are known as **secondary** or **compensation curves.** *(Figure 7–17)*

24. A typical vertebra has a **vertebral body** and a **vertebral arch,** and articulates with adjacent vertebrae at the **superior** and **inferior articular processes.** *(Figure 7–18)*

25. **Intervertebral discs** separate adjacent vertebrae. Spaces between successive **pedicles** form the **intervertebral foramina.** *(Figure 7–18)*

7-6 ▌ The five vertebral regions are the cervical, thoracic, lumbar, sacral, and coccygeal regions p. 254

26. Cervical vertebrae are distinguished by the shape of the body, the relative size of the vertebral foramen, the presence of **costal processes** with **transverse foramina,** and notched **spinous processes.** These vertebrae include the **atlas, axis,** and **vertebra prominens.** *(Figure 7–19; Table 7–1)*

27. Thoracic vertebrae have a distinctive heart-shaped body; long, slender spinous processes; and articulations for the ribs. *(Figures 7–20, 7–23; Table 7–1)*

28. The lumbar vertebrae are the most massive and least mobile of the vertebrae; they are subjected to the greatest strains. *(Figure 7–21; Table 7–1)*

29. The sacrum protects reproductive, digestive, and urinary organs and articulates with the pelvic girdle and with the fused elements of the coccyx. *(Figure 7–22)*

7-7 ❚ The thoracic cage protects organs in the chest and provides sites for muscle attachment p. 261

30. The skeleton of the **thoracic cage** consists of the thoracic vertebrae, the ribs, and the sternum. The **ribs** and **sternum** form the *rib cage. (Figure 7–23)*

31. Ribs 1–7 are **true ribs,** or *vertebrosternal ribs.* Ribs 8–12 are called **false ribs;** they include the *vertebrochondral ribs* (ribs 8–10) and two pairs of *floating (vertebral) ribs* (ribs 11–12). A typical rib has a **head,** or *capitulum;* a **neck;** a **tubercle;** an *angle;* and a **body,** or *shaft.* A *costal groove* marks the path of nerves and blood vessels. *(Figures 7–23, 7–24)*

32. The sternum consists of the **manubrium, body,** and **xiphoid process.** *(Figure 7–23)*

Review Questions

See the blue Answers tab at the back of the book.

LEVEL 1 Reviewing Facts and Terms

1. Identify the cranial and facial bones in the diagram below.

Cranial bones

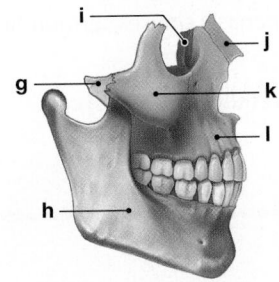

Facial bones

(a) _____	(b) _____
(c) _____	(d) _____
(e) _____	(f) _____
(g) _____	(h) _____
(i) _____	(j) _____
(k) _____	(l) _____

2. Which of the following is the correct list of fontanelles present in an infant's skull? **(a)** sphenoidal, occipital, anterior, mastoid, **(b)** sphenoidal, occipital, anterior, coronal, **(c)** occipital, anterior, sagittal, coronal, **(d)** temporal, occipital, frontal, coronal.

3. Which of the following is the correct list of the four major sutures present in an adult's skull? **(a)** squamous, coronal, lambdoid, occipital, **(b)** squamous, coronal, sagittal, parietal, **(c)** squamous, coronal, sagittal, lambdoid, **(d)** coronal, occipital, sphenoid, frontal.

4. The boundaries between skull bones are immovable joints called **(a)** foramina, **(b)** fontanelles, **(c)** lacunae, **(d)** sutures.

5. The joint between the frontal and parietal bones is correctly called the _____ suture. **(a)** parietal, **(b)** lambdoid, **(c)** squamous, **(d)** coronal.

6. Blood vessels that drain blood from the head pass through the **(a)** jugular foramina, **(b)** hypoglossal canals, **(c)** stylomastoid foramina, **(d)** mental foramina, **(e)** lateral canals.

7. For each of the following vertebrae, indicate its vertebral region.

(a)　　　　　(b)　　　　　(c)

a. _____　　b. _____　　c. _____

8. Cervical vertebrae can usually be distinguished from other vertebrae by the presence of **(a)** transverse processes, **(b)** transverse foramina, **(c)** demifacets on the body, **(d)** the vertebra prominens, **(e)** large spinous processes.

9. The lateral walls of the vertebral foramen are formed by the **(a)** body of the vertebra, **(b)** spinous process, **(c)** pedicles, **(d)** laminae, **(e)** transverse processes.

10. The part of the vertebra that transfers weight along the axis of the vertebral column is the **(a)** vertebral arch, **(b)** lamina, **(c)** pedicles, **(d)** body.

11. Where do articular processes arise in vertebrae?

12. List the three parts of the sternum.

13. Name two functions of the sinuses present in the skull.

14. In addition to the spinal curves, what skeletal element contributes to the flexibility of the vertebral column?

LEVEL 2 Reviewing Concepts

15. What is the relationship between the temporal bone and the ear?

16. What is the relationship between the ethmoid bone and the nasal cavity?

17. List and describe the functions of the thoracic cage.

18. Why is it important to keep your back straight when you lift a heavy object?

19. The transverse processes of cervical vertebrae contain foramen. Trudy thinks the foramina contain vertebral blood vessels. Joseph says they contain cranial nerve X (vagus nerve). Who is correct? What is the function of the transverse foramen?

7

20. Describe the structural relationship of a rib to the vertebral column (T_1–T_8).

21. The secondary spinal curves (a) help position the body weight over the legs, (b) accommodate the thoracic and abdominopelvic viscera, (c) include the thoracic curvature, (d) do all of these, (e) do only a and c.

22. Bell's palsy is the result of inflammation of the nerve that exits through the stylomastoid foramen. Identify the nerve. What is its function? Also state the location of the stylomastoid foramen.

23. Improper administration of CPR (cardiopulmonary resuscitation) can force the _____ into the liver. (a) floating ribs, (b) lumbar vertebrae, (c) manubrium of the sternum, (d) costal cartilage, (e) xiphoid process.

LEVEL 3 Critical Thinking and Clinical Applications

24. Jane has an upper respiratory infection and begins to feel pain in her teeth. This is a good indication that the infection is located in the (a) frontal sinuses, (b) sphenoid bone, (c) temporal bone, (d) maxillary sinuses, (e) zygomatic bones.

25. Jessica is 13. She has had lower back pain for over a year. It is often brought on after prolonged periods of sitting or standing. Suggest a likely cause and give reasons for your answer.

26. After experiencing a minor car accident, Hayley thinks she has sustained whiplash. How does whiplash occur?

CLINICAL CASE Wrap-Up

Knocked Out

The young boxer has sustained a "blowout" fracture of his left orbital complex. The orbit that contains the eye is formed by seven facial bones and is surrounded by four air-filled sinuses. The orbital rim is strong, composed of thick, protective bone. But the very thin medial wall and floor of the orbit were literally blown out by the intact eyeball itself when he was struck with such a forceful blow.

A computerized tomography (CT) scan reveals fractures of the ethmoid, lacrimal, palatine, and maxillary bones that make up the delicate medial wall and floor of the left orbit. Many tiny fracture fragments can be seen within the maxillary sinus.

The opening to the air-filled sinus allowed the air to escape into the soft tissues surrounding the eye when the fighter coughed, causing the subcutaneous air expansion around the

eye. The muscles that move the eye had become entrapped within the fracture fragments, preventing the fighter from gazing upward and giving rise to his double vision.

The young boxer is taken to the operating room. A surgical repair of the orbital floor is performed, using an implant to support the eyeball until fracture healing is complete.

1. What are the seven bones that make up the orbital complex?

2. What are the four paranasal sinuses surrounding the orbital complex?

See the blue Answers tab at the back of the book.

8

The Appendicular Skeleton

Learning Outcomes

These Learning Outcomes correspond by number to this chapter's sections and indicate what you should be able to do after completing the chapter.

8-1 ❙ Identify the bones that form the pectoral girdles, their functions, and their superficial features.

8-2 ❙ Identify the bones of the upper limbs, their functions, and their superficial features.

8-3 ❙ Identify the bones that form the pelvic girdle, their functions, and their superficial features.

8-4 ❙ Identify the bones of the lower limbs, their functions, and their superficial features.

8-5 ❙ Summarize sex differences and age-related changes in the human skeleton.

SPOTLIGHT

• Sex Differences in the Human Skeleton p. 279

CLINICAL CASE

• The Orthopedic Surgeon's Nightmare p. 268

Clinical Notes

• Carpal Tunnel Syndrome p. 275

• Hip Fracture p. 280
• Runner's Knee p. 280
• Stress Fractures p. 284
• Congenital Talipes Equinovarus p. 284

CLINICAL CASE

The Orthopedic Surgeon's Nightmare

It is Caitlyn's first day of "big kids" school. She cries a little when she kisses her Mom good-bye and gets on the bus. She feels lost when she gets to school, but a kind kindergarten teacher meets her and leads her to her new classroom. The best part of her day is getting to go out to the playground.

Caitlyn has seen the climbing structure from outside the playground fence many times, but this is the first time she has the chance to climb all the way to the top. She feels very grown up, perched up so high. Suddenly her foot slips and she feels herself falling to the ground. She stretches out her right hand to break her fall just before she hits the ground. As she lands,

she feels and hears a crack in her elbow. "Help!" Caitlyn cries.

The school nurse examines the child's deformed elbow, notices there is no pulse in her wrist, and springs into action. She splints Caitlyn's arm and calls the orthopedic surgeon. "Doc, we have a five-year-old who has an elbow injury following a fall on her outstretched hand, with no distal pulses."

"Oh no," groans the surgeon. Of all the playground fractures, this is likely to be the most difficult to treat. **What has happened to Caitlyn's elbow? To find out, turn to the Clinical Case Wrap-Up on p. 288.**

An Introduction to the Appendicular Skeleton

In this chapter we focus on the functional anatomy of the bones of the limbs and the bones that connect them to the body. In the previous chapter, we discussed the 80 bones of the axial skeleton. Remaining are 60 percent of the bones that make up the skeletal system. These bones are appended to the axial skeleton. The **appendicular skeleton** includes the bones of the limbs and the supporting bone (pectoral and pelvic) girdles that connect them to the trunk (**Figure 8–1**).

To appreciate the role of the appendicular skeleton in your life, make a mental list of all the things you have done with your arms or legs today. Standing, walking, writing, turning pages, eating, dressing, shaking hands, and texting—the list quickly becomes unwieldy. Your axial skeleton protects and supports internal organs and takes part in vital functions, such as breathing. But your appendicular skeleton lets you manipulate objects and move from place to place.

The appendicular skeleton is dominated by the long bones that support the limbs. ⊃ p. 206 Each long bone shares common features with other long bones. For example, one epiphysis is usually called the *head*, the diaphysis is called the *shaft*, and a *neck* normally separates the head and shaft. For simplicity, an illustration of a single bone will have labels that do not include the name of the bone. For example, a photo of the humerus will have the label *head* rather than *head of the humerus* or *humeral head*. When more than one bone is shown, the label will use the complete name to avoid confusion. In the descriptions in this chapter, we emphasize surface markings that either have functional importance (such as the attachment sites for skeletal muscles and the paths of major nerves and

blood vessels) or provide markings that define areas and locate structures of the body.

8-1 | The pectoral girdles—the clavicles and scapulae—attach the upper limbs to the axial skeleton

Learning Outcome Identify the bones that form the pectoral girdles, their functions, and their superficial features.

Each arm articulates (that is, forms a joint) with the trunk at a **pectoral girdle**, or *shoulder girdle* (**Figure 8–1**). The pectoral girdles consist of two S-shaped *clavicles* (KLAV-i-kulz; collarbones) and two broad, flat *scapulae* (SKAP-ū-lē; singular, *scapula*, SKAP-ū-luh; shoulder blades). The medial, anterior end of each clavicle articulates with the manubrium of the sternum. ⊃ p. 262 These articulations are the *only* direct connections between the pectoral girdles and the axial skeleton. Skeletal muscles support and position the scapulae, which have no direct bony or ligamentous connections to the thoracic cage. As a result, the shoulders are extremely mobile, but not very strong.

Movements of the clavicles and scapulae position the shoulder joints and provide a base for arm movement. The shoulder joints are positioned and stabilized by skeletal muscles that extend between the axial skeleton and the pectoral girdles. Once the joints are in position, other skeletal muscles, including several that originate on the pectoral girdle, move the upper limbs.

The surfaces of the scapulae and clavicles are extremely important as sites for muscle attachment. Bony ridges and projections mark the attachment sites of major muscles. Other bone markings, such as sulci or foramina, indicate the positions of nerves that control the muscles or the passage of blood vessels that nourish the muscles and bones.

Figure 8–1 **An Anterior View of the Appendicular Skeleton.** The numbers in the boxes indicate the total number of bones of each type or within each category. ATLAS: Plates 1a,b

8

The Clavicles

The **clavicles** are S-shaped bones that originate at the superior, lateral border of the manubrium of the sternum, lateral to the jugular notch (**Figure 8–2a**). From the pyramid-shaped **sternal end**, each clavicle curves laterally and posteriorly for about half its length. It then forms a smooth posterior curve to articulate with a process of the scapula, the *acromion* (a-KRŌ-mē-on). The flat, **acromial end** of the clavicle is broader than the sternal end (**Figure 8–2b,c**).

The smooth, superior surface of the clavicle lies just beneath the skin. The acromial end has a rough inferior surface that has prominent lines and tubercles (**Figure 8–2c**). These surface markings are attachment sites for muscles and ligaments of the shoulder. The combination of the direction of curvature and the differences between superior and inferior surfaces makes it easy to distinguish a left clavicle from a right clavicle.

You can explore the connection between your own clavicles and sternum. With your fingers in your jugular notch (the depression just superior to the manubrium), locate the clavicle on either side and find the *sternoclavicular joints* where the sternum articulates with the clavicles. These are the only articulations between the pectoral girdles and the axial skeleton. When you move your shoulders, you can feel the sternal ends of the clavicles change their positions.

The clavicles are small and fragile. It's no surprise that fractures of the clavicle are fairly common. For example, a simple fall can fracture a clavicle if you land on your hand with your arm outstretched. Fortunately, most clavicular fractures heal rapidly with a simple clavicle strap.

The Scapulae

The anterior surface of the **body** of each **scapula** forms a broad triangle (**Figure 8–3a**). The three sides of the triangle are the **superior border**; the **medial border**, or *vertebral border*; and the **lateral border**, or *axillary border* (*axilla*, armpit). Muscles that position the scapula attach along these edges. The corners of the triangle are called the *superior angle*, the *inferior angle*, and the *lateral angle*.

The lateral angle, or *head* of the scapula, forms a broad process that supports the cup-shaped **glenoid cavity** (**Figure 8–3b**). At the glenoid cavity, the scapula articulates with the *humerus*, the proximal bone of the upper limb. This articulation is the shoulder joint, also known as the *glenohumeral joint*. The anterior surface of the body of the scapula is smooth and concave. The depression in the anterior surface is called the **subscapular fossa**.

Two large scapular processes extend beyond the margin of the glenoid cavity (**Figure 8–3b**) superior to the head of the humerus. The smaller, anterior projection is the **coracoid** (KOR-uh-koyd) **process**. The **acromion** is the larger, posterior process. If you run your fingers along the superior surface of the shoulder joint, you will feel this process. The acromion articulates with the clavicle at the *acromioclavicular joint*. Both the coracoid process and the acromion are attached to ligaments and tendons associated with the shoulder joint.

Figure 8–2 **The Right Clavicle.** ATLAS: Plates 26a,b

a The position of the clavicle within the pectoral girdle, anterior view.

b Superior view of the right clavicle.

c Inferior view of the right clavicle. Stabilizing ligaments attach to the conoid tubercle and the costal tuberosity.

Figure 8–3 **The Right Scapula.** ATLAS: Plates 26a,b

a Anterior view

b Lateral view

c Posterior view

The acromion is continuous with the **scapular spine**, a ridge that crosses the posterior surface of the scapular body before ending at the medial border (**Figure 8–3c**). The scapular spine divides the convex posterior surface of the body into two regions. The area superior to this spine is the **supraspinous fossa** (*supra*, above). The region inferior to the spine is the **infraspinous fossa** (*infra*, beneath). Small ridges and lines mark the entire posterior surface where smaller muscles attach to the scapula.

CHECKPOINT

1. **Name the bones of the pectoral girdles.**

2. **How would a broken clavicle affect the mobility of the scapula?**

3. **Which bone articulates with the scapula at the glenoid cavity?**

See the blue Answers tab at the back of the book.

8-2 ▎ The upper limbs are adapted for free movement

Learning Outcome Identify the bones of the upper limbs, their functions, and their superficial features.

The skeleton of the upper limbs consists of the bones of the arms, forearms, wrists, and hands. Note that in anatomical descriptions, the term *arm* refers only to the proximal portion of the upper limb (from shoulder to elbow), not to the entire limb. The distal portion of the upper limb (from elbow to wrist) is the *forearm*. Let's examine the bones of the right upper limb.

The Humerus

The arm, or *brachium*, contains only one bone, the **humerus**, which extends from the scapula to the elbow. At the proximal end of the humerus, the round **head** articulates with the scapula (**Figure 8–4a,b**). The prominent **greater tubercle** is a rounded projection on the lateral surface of the epiphysis, near the margin of the humeral head. The greater tubercle establishes the lateral contour of the shoulder. You can verify its position by feeling for a bump located a few centimeters from the tip of the acromion. The **lesser tubercle** is a smaller projection that lies on the anterior, medial surface of the epiphysis, separated from the greater tubercle by the **intertubercular groove**, or *intertubercular sulcus*. A large tendon runs along the groove. Both tubercles are important sites for muscle attachment.

The **anatomical neck** lies between the tubercles and the articular surface of the head. This neck marks the extent of the joint capsule. The narrower distal **surgical neck** corresponds to the metaphysis of the growing bone. The name reflects the fact that fractures typically occur at this site.

The proximal shaft of the humerus is round in section. The **deltoid tuberosity** is a large, rough elevation on the lateral

Figure 8–4   **The Right Humerus and Elbow Joint.** ATLAS: Plates 31; 34a–d

8

Greater tubercle

Lesser tubercle

Intertubercular groove

Head

Greater tubercle

Anatomical neck

Surgical neck

Deltoid tuberosity

Radial groove

Shaft

Radial fossa

Lateral epicondyle

Coronoid fossa

Capitulum   Trochlea

Condyle

Olecranon fossa

Medial epicondyle

Trochlea

Lateral epicondyle

a **Anterior surface**

b **Posterior surface**

Humerus

Head of radius

Medial epicondyle

Trochlea

Capitulum

Coronoid process of ulna

Radial notch of ulna

c **Elbow joint, anterior view**

Humerus

Medial epicondyle

Olecranon fossa

Olecranon

Trochlea of humerus

Head of radius

Ulna

d **Elbow joint, posterior view**

surface of the shaft, approximately halfway along its length. It is named after the *deltoid muscle,* which attaches to it.

On the posterior surface, the deltoid tuberosity ends at the **radial groove** (Figure 8–4b). This depression marks the path of the *radial nerve,* a large nerve that provides both sensory information from the posterior surface of the limb and motor control over the large muscles that straighten the elbow. Distal to the radial groove, the posterior surface of the humerus is fairly flat.

Near the distal articulation with the bones of the forearm, the shaft expands to either side at the **medial** and **lateral epicondyles.** *Epicondyles* are processes that develop proximal to an articulation. They provide additional surface area for muscle attachment.

The *ulnar nerve* crosses the posterior surface of the medial epicondyle. A hit at the posteromedial surface of the elbow joint can strike this nerve and produce a temporary numbness and paralysis of muscles on the anterior surface of the forearm. Because of the odd sensation, this area is sometimes called the *funny bone.*

At the **condyle,** the humerus articulates with the *radius* and the *ulna,* the bones of the forearm. The condyle is divided into two articular regions: the trochlea and the capitulum (Figure 8–4a). The **trochlea** is the spool-shaped medial portion of the condyle. It extends from the base of the **coronoid fossa** on the anterior surface to the **olecranon** (ō-LEK-ruh-non) **fossa** on the posterior surface. The rounded **capitulum** forms the lateral surface of the condyle (Figure 8–4a,c).

A shallow **radial fossa** superior to the capitulum accommodates a portion of the radial head as the forearm approaches the humerus (Figure 8–4a,c). Projections from the ulnar surface fit into the coronoid fossa and the olecranon fossa as the elbow approaches the limits of its range of motion (Figure 8–4d). The prominent medial head and the differences between the lateral and medial condyles make it easy to tell a left humerus from a right humerus.

The Ulna

The *ulna* and *radius* are parallel bones that support the forearm, or *antebrachium.* In the anatomical position, the **ulna** lies medial to the radius. The **olecranon,** the proximal end of the ulna, is the point of the elbow (Figures 8–4d and 8–5a). On the anterior surface of the proximal epiphysis, the **trochlear notch** of the ulna (Figure 8–5b) articulates with the trochlea of the humerus at the elbow joint. Figure 8–5c shows the trochlear notch of the ulna in a lateral view.

TIPS & TRICKS

The memory tool for olecranon involves the letter "l": The knob you feel at the e*l*bow is the o*l*ecranon of the u*l*na. You can remember that the trochlear notch is a feature of the **u**lna because this notch forms a U in lateral view.

The olecranon forms the superior lip of the trochlear notch. The **coronoid process** forms its inferior lip. At the limit of *extension,* movement that straightens the forearm and arm, the olecranon swings into the olecranon fossa on the posterior surface of the humerus (Figure 8–4d). At the limit of *flexion,* a movement that decreases the angle between the articulating bones, the arm and forearm form a tight V, and the coronoid process projects into the coronoid fossa on the anterior humeral surface. Lateral to the coronoid process, a smooth **radial notch** accommodates the head of the radius at the *proximal radioulnar joint* (Figure 8–5a).

Viewed in cross section, the shaft of the ulna is triangular. The **interosseous membrane,** a fibrous sheet, connects the lateral margin of the ulna to the radius (Figure 8–5a,b). Near the wrist, the shaft of the ulna narrows before ending at a disc-shaped **ulnar head,** or *head of the ulna.* The posterior, lateral surface of the ulnar head has a short **styloid process** (*stylos,* pillar). A triangular cartilage, the *articular disc,* attaches to the styloid process. This cartilage separates the ulnar head from the bones of the wrist. The lateral surface of the ulnar head articulates with the distal end of the radius to form the *distal radioulnar joint* (Figure 8–5b).

The Radius

The **radius** is the lateral bone of the forearm (Figure 8–5). The disc-shaped **radial head,** or *head of the radius,* articulates with the capitulum of the humerus (Figure 8–4c). During flexion, the radial head swings into the radial fossa of the humerus. A narrow neck extends from the radial head to the **radial tuberosity** (Figure 8–5b). This tuberosity marks the attachment site of the *biceps brachii muscle,* a large muscle on the anterior surface of the arm.

The shaft of the radius curves along its length. It also enlarges, and the distal portion of the radius is considerably larger than the distal portion of the ulna. The **ulnar notch** on the medial surface of the distal end of the radius marks the site of articulation with the head of the ulna (Figure 8–5a).

The distal end of the radius articulates with the bones of the wrist. The **styloid process** on the lateral surface of the radius helps stabilize this joint (Figure 8–5a,b). If you are looking at an isolated radius or ulna, you can quickly identify whether it is left or right by finding the radial notch (ulna) or ulnar notch (radius) and remembering that the radius lies lateral to the ulna.

The Carpal Bones

The *carpus,* or wrist, contains eight **carpal bones**. These bones form two rows, one with four **proximal carpal bones** and the other with four **distal carpal bones**.

The proximal carpal bones are the scaphoid, lunate, triquetrum, and pisiform (Figure 8–6a,b).

- The **scaphoid** (*skaphe,* boat) is the proximal carpal bone on the lateral border of the wrist. It is the carpal bone closest to the styloid process of the radius.

Figure 8–5 **The Right Radius and Ulna.** ATLAS: Plates 31; 35f; 36a,b

Olecranon
Proximal radioulnar joint
Radial head
Neck of radius
Radial tuberosity
Ulna
Radius
Interosseous membrane
Ulnar head
Ulnar notch of radius
Styloid process of ulna
Styloid process of radius

a Posterior view

Olecranon
Trochlear notch
Coronoid process
Radial notch
Ulnar tuberosity
Ulna
Distal radioulnar joint
Ulnar head

b Anterior view

Olecranon
Ulna

c Lateral view of ulna, showing trochlear notch

- The comma-shaped **lunate** (*luna*, moon) lies medial to the scaphoid. Like the scaphoid, it articulates with the radius.

- The **triquetrum** (trī-KWĒ-trum; *triquetrus*, three-cornered) is a small, pyramid-shaped bone medial to the lunate. The triquetrum articulates with the articular disc that separates the ulnar head from the wrist.

- The small, pea-shaped **pisiform** (PIS-i-form; *pisum*, pea) sits anterior to the triquetrum.

The distal carpal bones are the trapezium, trapezoid, capitate, and hamate (**Figure 8–6a,b**).

- The **trapezium** (*trapezion*, four-sided with no parallel sides) is the lateral bone of the distal row. Its proximal surface articulates with the scaphoid.

- The wedge-shaped **trapezoid** lies medial to the trapezium. Like the trapezium, it has a proximal articulation with the scaphoid.

- The **capitate** (*caput*, head) is the largest carpal bone. It sits between the trapezoid and the hamate.

- The **hamate** (*hamatum*, hooked) is the medial distal carpal bone.

TIPS & TRICKS

It may help you to identify the eight carpal bones if you remember the sentence "Sam Likes To Push The Toy Car Hard." In lateral-to-medial order, the first four words stand for the proximal carpal bones (scaphoid, lunate, triquetrum, pisiform). The last four stand for the distal carpal bones (trapezium, trapezoid, capitate, hamate).

The Metacarpal Bones and Phalanges

Five **metacarpal** (met-uh-KAR-pul; *metacarpus*, hand) **bones** articulate with the distal carpal bones and support the

Figure 8–6 **Bones of the Right Wrist and Hand.** ATLAS: Plates 38a,b

a Anterior view

b Posterior view

hand (**Figure 8–6**). Roman numerals I–V identify the metacarpal bones, beginning with the lateral metacarpal bone, which articulates with the trapezium. For example, metacarpal I articulates with the proximal bone of the thumb.

Distally, the metacarpal bones articulate with the proximal finger bones. Each hand has 14 finger bones, or **phalanges** (fa-LAN-jēz; singular, *phalanx*). The first finger, known as the **pollex** (POL-eks), or thumb, has two phalanges (*proximal* and *distal*). Each of the other four fingers has three phalanges (*proximal, middle,* and *distal*).

Clinical Note

Carpal Tunnel Syndrome The carpal bones articulate with one another at joints that permit limited sliding and twisting. Ligaments interconnect the carpal bones and help stabilize the wrist joint. The tendons of muscles that flex the fingers pass across the anterior surface of the wrist. These tendons are sandwiched between the intercarpal ligaments and a broad, superficial transverse ligament called the *flexor retinaculum*. Inflammation of the connective tissues between the flexor retinaculum and the carpal bones can compress the tendons and the adjacent *median nerve*. The result is pain, weakness, and reduced wrist mobility. This condition is called *carpal tunnel syndrome*.

Cross section of
right wrist (palm up)

CHECKPOINT

4. List all the bones of an upper limb.

5. The rounded projections on either side of the elbow are parts of which bone?

6. Using the terms *lateral* and *medial*, describe the relationship between the forearm bones when a person is in the anatomical position.

7. Bill accidentally fractures his first distal phalanx with a hammer. Which finger is broken?

See the blue Answers tab at the back of the book.

8-3 ▌ The pelvic girdle—two hip bones—attaches the lower limbs to the axial skeleton

Learning Outcome Identify the bones that form the pelvic girdle, their functions, and their superficial features.

The bones of the **pelvic girdle** attach to the lower limbs and must withstand the stresses involved in weight bearing and mobility. For this reason, they are more massive than the bones of the pectoral girdle. For similar reasons, the bones of the lower limbs are more massive than those of the upper limbs.

In this section we first consider the pelvic girdle, which consists of the two hip bones. Then we examine the *pelvis*, a composite structure that includes the hip bones of the appendicular skeleton plus the sacrum and coccyx of the axial skeleton. ⊃ pp. 258–261

The Pelvic Girdle

The pelvic girdle consists of the paired **hip bones**, also called the **coxal bones** or **pelvic bones**. Each hip bone forms by the fusion of three bones: an **ilium** (IL-ē-um; plural, *ilia*), an **ischium** (IS-kē-um; plural, *ischia*), and a **pubis** (PŪ-bis) (**Figure 8–7**). The ilia have a sturdy articulation with the auricular surfaces of the sacrum, attaching the pelvic girdle to the axial skeleton. ⊃ p. 259 Anteriorly, the medial surfaces of the hip bones are connected by a pad of fibrocartilage at a joint called the *pubic symphysis* (**Figure 8–7b**).

On the lateral surface of each hip bone is an **acetabulum** (as-e-TAB-ū-lum; *acetabulum*, shallow vinegar cup), a concave socket that articulates with the head of the femur (**Figure 8–7a**). A ridge of bone forms the lateral and superior margins of the acetabulum, which has a diameter of about 5 cm (2 in.). The anterior and inferior portion of the ridge is incomplete, leaving a gap called the **acetabular notch**. The smooth, cup-shaped articular surface of the acetabulum is the **lunate surface**.

The ilium, ischium, and pubis meet inside the acetabulum, as though it were a pie sliced into three pieces. Superior to the acetabulum, the ilium forms a broad, curved surface that provides an extensive area for the attachment of muscles,

tendons, and ligaments (**Figure 8–7a**). Bone markings along the margin of the ilium include the *iliac spines*, which mark the attachment sites of important muscles and ligaments; the *gluteal lines*, which mark the attachment of large hip muscles; and the **greater sciatic** (sī-AT-ik) **notch**, through which a major nerve (the *sciatic nerve*) reaches the lower limb.

The ischium forms the posterior, inferior portion of the acetabulum. Posterior to the acetabulum, the prominent **ischial spine** projects superior to the *lesser sciatic notch*. Blood vessels, nerves, and a small muscle pass through this notch. The **ischial tuberosity**, a roughened projection, is located at the posterior and lateral edge of the ischium. When you are sitting, the ischial tuberosities bear your body's weight.

The narrow **ischial ramus** continues until it meets the **inferior ramus** of the pubis. The inferior pubic ramus extends between the ischial ramus and the *pubic tubercle*, a small, elevated area anterior and lateral to the pubic symphysis. There the inferior pubic ramus meets the **superior ramus** of the pubis, which originates near the acetabulum. The *pectineal line*, a ridge that ends at the pubic tubercle, is found on the anterior, superior surface of the superior ramus (**Figure 8–7b**). The pubic ramus and ischial ramus encircle the **obturator** (OB-tū-rā-tor) **foramen** (**Figure 8–7a**). This space is closed by a sheet of collagen fibers whose inner and outer surfaces provide a firm base for the attachment of muscles of the hip.

The broadest part of the ilium extends between the **arcuate line**, which is continuous with the pectineal line, and the **iliac crest** (**Figure 8–7b**). These prominent ridges mark the attachments of ligaments and muscles. The area between the arcuate line and the iliac crest forms a shallow depression known as the **iliac fossa**. The concave surface of the iliac fossa helps support the abdominal organs. It also provides additional area for muscle attachment.

In a medial view of a hip bone, the anterior and medial surface of the pubis has a roughened area that marks the site of articulation with the pubis of the opposite side (**Figure 8–7b**). At this articulation—the **pubic symphysis**—the two pubic bones are attached to a median pad of fibrocartilage.

Posteriorly, the **auricular surface** of the ilium articulates with the auricular surface of the sacrum at the *sacroiliac joint*. ⊃ p. 259 Ligaments arising at the **iliac tuberosity**, a roughened area superior to the auricular surface, stabilize this joint.

The Pelvis

Figure 8–8 shows anterior and posterior views of the **pelvis**, which consists of the two hip bones, the sacrum, and the coccyx. An extensive network of ligaments connects the lateral borders of the sacrum with the iliac crest, the ischial tuberosity, the ischial spine, and the arcuate line. Other ligaments tie the ilia to the posterior lumbar vertebrae. These interconnections increase the stability of the pelvis.

Figure 8–7 **The Right Hip Bone.** The left and right hip bones make up the pelvic girdle.

a Right hip bone, lateral view

b Right hip bone, medial view

The pelvis may be divided into the **true** *(lesser)* **pelvis** and the **false** *(greater)* **pelvis** (Figure 8–9a,b). The true pelvis encloses the *pelvic cavity*, a subdivision of the abdominopelvic cavity. ⤵ p. 48 The superior limit of the true pelvis is a line that extends from either side of the base of the sacrum, along the arcuate line and pectineal line to the pubic symphysis. The bony edge of the true pelvis is called the **pelvic brim**, or *linea terminalis*. The enclosed space is the **pelvic inlet**. The false pelvis consists of the expanded, bladelike portions of each ilium superior to the pelvic brim.

The **pelvic outlet** is the opening bounded by the coccyx, the ischial tuberosities, and the inferior border of the pubic symphysis (Figure 8–9b,c). The surface region bordered by the inferior edges of the pelvis is called the *perineum* (per-i-NĒ-um). Perineal muscles form the floor of the pelvic cavity and support the organs in the true pelvis.

The shape of the pelvis of a female is somewhat different from that of a male. Some of the differences are the result of variations in body size and muscle mass. For example, in females, the pelvis is generally smoother and lighter and has less-prominent markings. Other female pelvic variations appear to be adaptations for childbearing. These adaptations help support the weight of the developing fetus within the uterus, and the passage of the newborn through the pelvic outlet during delivery. In addition, the hormone *relaxin*, produced during pregnancy, loosens the pubic symphysis and sacroiliac ligaments. This loosening allows movement between the hip bones that can further increase the size of the pelvic inlet and outlet. **Spotlight Figure 8–10** compares some major overall differences between adult male and female skeletons.

Figure 8–8 **The Pelvis of an Adult Male.** (See *Figure 7–22*, p. 260, for a detailed view of the sacrum and coccyx.)

8

- Sacrum
- Coccyx
- Ilium
- Pubis
- Ischium
- Hip bone (see Figure 8–7)

a **Anterior view**

- Iliac crest
- L₅
- Iliac fossa
- Sacroiliac joint
- Sacrum
- Ilium
- Arcuate line
- Acetabulum
- Pubis
- Pubic tubercle
- Pubic symphysis
- Ischium
- Obturator foramen

- Iliac crest
- L₅
- Sacral foramina
- Posterior superior iliac spine
- Posterior inferior iliac spine
- Sacrum
- Greater sciatic notch
- Ischial spine
- Ischial tuberosity
- Coccyx

b **Posterior view**

Figure 8–9 **Divisions of the Pelvis.**

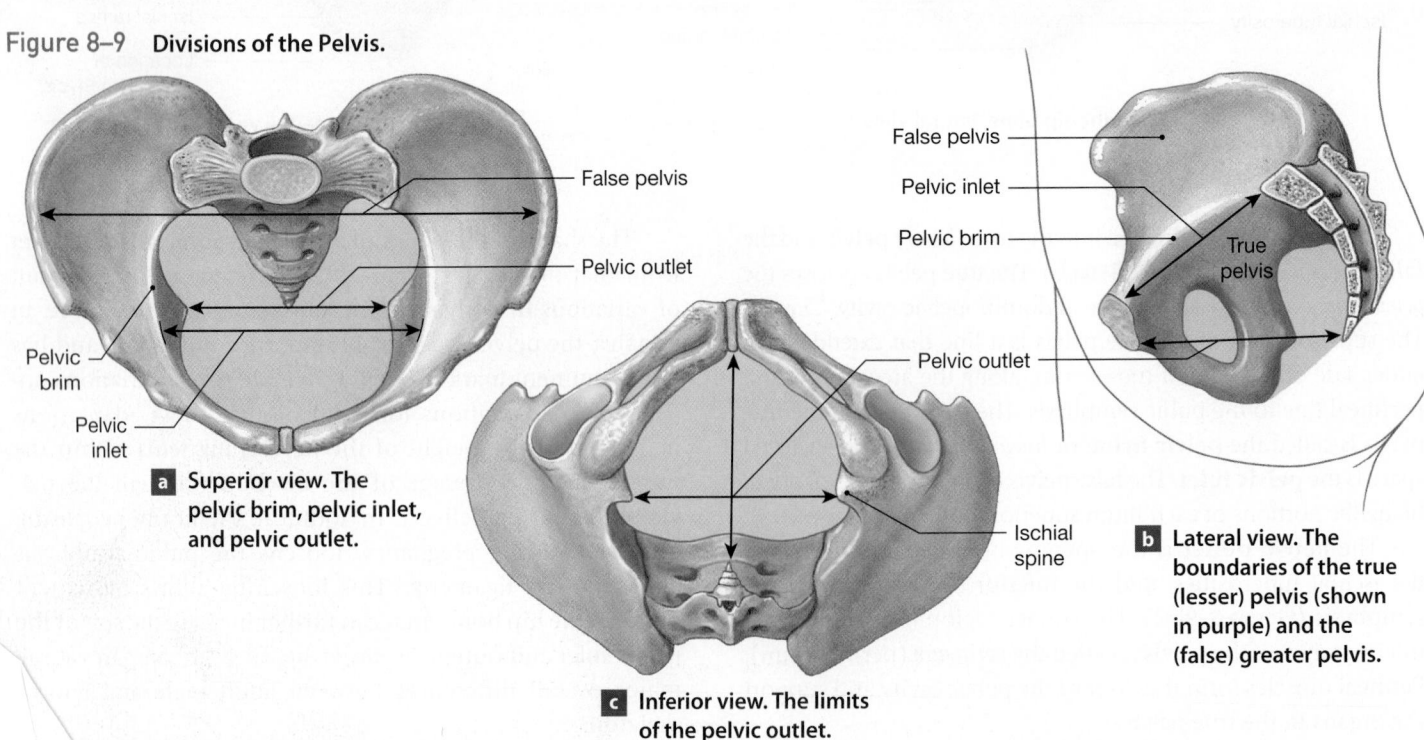

- False pelvis
- Pelvic outlet
- Pelvic brim
- Pelvic inlet

a **Superior view. The pelvic brim, pelvic inlet, and pelvic outlet.**

- Pelvic outlet
- Ischial spine

c **Inferior view. The limits of the pelvic outlet.**

- False pelvis
- Pelvic inlet
- Pelvic brim
- True pelvis
- Pelvic outlet

b **Lateral view. The boundaries of the true (lesser) pelvis (shown in purple) and the (false) greater pelvis.**

The sex of an adult human skeleton can be determined by many of the details seen in the bones. The skull and pelvis are particularly helpful. Not every skeleton shows every feature in classic detail, but this chart gives the basic differences between the sexes.

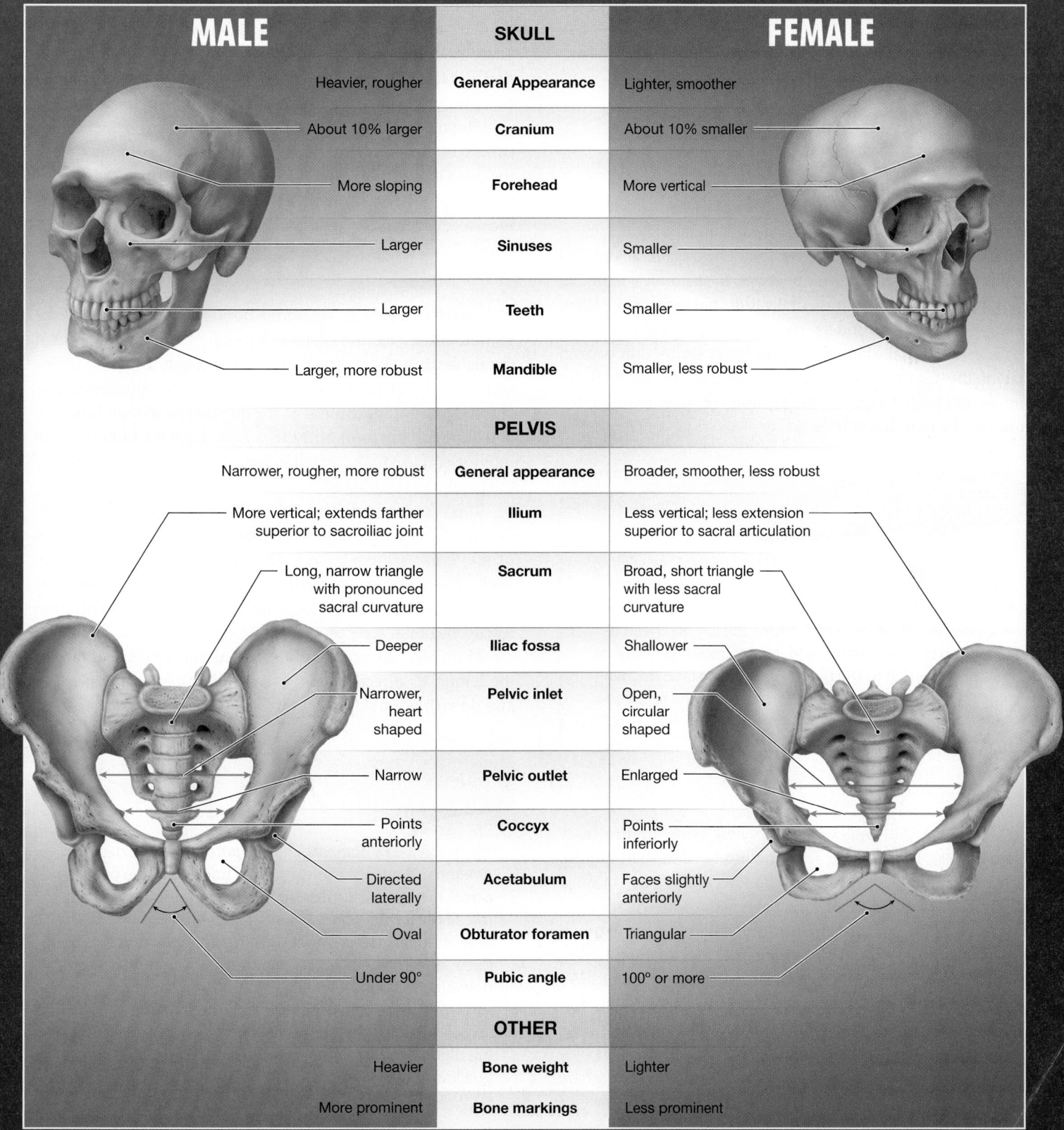

MALE	SKULL	FEMALE
Heavier, rougher	General Appearance	Lighter, smoother
About 10% larger	Cranium	About 10% smaller
More sloping	Forehead	More vertical
Larger	Sinuses	Smaller
Larger	Teeth	Smaller
Larger, more robust	Mandible	Smaller, less robust

MALE	PELVIS	FEMALE
Narrower, rougher, more robust	General appearance	Broader, smoother, less robust
More vertical; extends farther superior to sacroiliac joint	Ilium	Less vertical; less extension superior to sacral articulation
Long, narrow triangle with pronounced sacral curvature	Sacrum	Broad, short triangle with less sacral curvature
Deeper	Iliac fossa	Shallower
Narrower, heart shaped	Pelvic inlet	Open, circular shaped
Narrow	Pelvic outlet	Enlarged
Points anteriorly	Coccyx	Points inferiorly
Directed laterally	Acetabulum	Faces slightly anteriorly
Oval	Obturator foramen	Triangular
Under 90°	Pubic angle	100° or more

MALE	OTHER	FEMALE
Heavier	Bone weight	Lighter
More prominent	Bone markings	Less prominent

8. Name the bones of the pelvic girdle.

9. Which three bones make up a hip bone?

10. How is the pelvis of females adapted for childbearing?

11. When you are seated, which part of the pelvis supports your body's weight?

See the blue Answers tab at the back of the book.

8-4 ▎ The lower limbs are adapted for movement and support

Learning Outcome Identify the bones of the lower limbs, their functions, and their superficial features.

The skeleton of each lower limb consists of a *femur* (thigh), a *patella* (kneecap), a *tibia* and a *fibula* (leg), and the tarsal bones, metatarsal bones, and phalanges of the foot. Once again, anatomical terminology differs from common usage. In anatomical terms, *leg* refers only to the distal portion of the limb, not to the entire lower limb. We will use *thigh* and *leg*, rather than *upper leg* and *lower leg*.

The functional anatomy of the lower limbs differs from that of the upper limbs, primarily because the lower limbs transfer the body weight to the ground. We now examine the bones of the right lower limb.

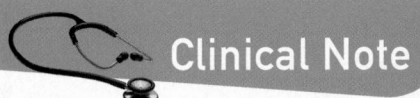

Clinical Note

Hip Fracture A *hip fracture*, or a "broken hip," actually involves the femur, not a hip bone. The two types of hip fractures are femoral neck fractures and intertrochanteric fractures. In a femoral neck fracture, the femoral head is separated from the femur (and its blood supply). There is no disconnection to the blood supply in an intertrochanteric fracture. As a result, such fractures may be repaired with a metal plate and screws.

The Femur

The **femur** is the longest and heaviest bone in the body (Figure 8–11). It articulates with the hip bone at the hip joint and with the tibia of the leg at the knee joint. The rounded epiphysis, or **femoral head**, articulates with the pelvis at the acetabulum. A ligament attaches the acetabulum to the femur at the **fovea capitis**, a small pit in the center of the femoral head. The **neck** of the femur joins the **shaft** at an angle of about 125°.

The **greater** and **lesser trochanters** are large, rough projections that originate at the junction of the neck and shaft. The greater trochanter projects laterally. The lesser trochanter projects posteriorly and medially. These trochanters develop where large tendons attach to the femur. On the anterior surface of the femur, the raised **intertrochanteric** (in-ter-trō-kan-TER-ik) **line** marks the edge of the articular capsule. This line continues around to the posterior surface as the **intertrochanteric crest**.

The **linea aspera** is a rough ridge that runs along the center of the posterior surface of the femur, marking the attachment site of powerful hip muscles (Figure 8–11b). As it approaches the knee joint, the linea aspera divides into a pair of ridges that continue to the **medial** and **lateral epicondyles** (Figure 8–11a,b). These smoothly rounded projections form superior to the **medial** and **lateral condyles**, which are part of the knee joint. The two condyles are separated by a deep **intercondylar fossa**.

The medial and lateral condyles extend across the inferior surface of the femur, but the intercondylar fossa does not reach the anterior surface (Figure 8–11a). The anterior and inferior surfaces of the two condyles are separated by the **patellar surface**, a smooth articular surface over which the patella glides.

The Patella

The **patella** is a large sesamoid bone that forms within the tendon of the *quadriceps femoris*, a group of muscles that extend (straighten) the knee. The patella has a rough, convex anterior surface and a broad **base** (Figure 8–12a). The roughened surface reflects the attachment of the quadriceps tendon (anterior and superior surfaces) and the *patellar ligament* (anterior and inferior surfaces). The patellar ligament connects the **apex** of the patella to the tibia.

The posterior patellar surface (Figure 8–12b) has two concave facets for articulation with the medial and lateral condyles of the femur (Figure 8–12c). The patellae are cartilaginous at birth. They start to ossify after the person begins walking, as thigh and leg movements become more powerful. Ossification usually begins at age 2 or 3 and ends around the time of puberty.

Normally, the patella glides across the patellar surface of the femur. Its direction of movement is superior–inferior (up and down), not medial–lateral (side to side).

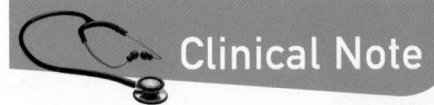

Clinical Note

Runner's Knee *Runner's knee*, or *patellofemoral stress syndrome*, develops from improper tracking of the patella across the patellar surface. In this syndrome, the patella is forced outside its normal track, so that it shifts laterally. The movement is often associated with increased compression forces or with lateral muscles in the quadriceps group overpowering the medial muscles. Running on hard or slanted surfaces (such as the beach shore or the shoulder of a road) and inadequate arch support are often responsible. The misalignment puts lateral pressure on the knee, resulting in swelling and tenderness after exercise.

Figure 8–11   **Bone Markings on the Right Femur.** ATLAS: Plates 32; 75a–d; 77

Neck

Greater trochanter

Femoral head

Intertrochanteric line

Lesser trochanter

Shaft

Patellar surface

Lateral epicondyle

Lateral condyle

Adductor tubercle

Medial epicondyle

Medial condyle

Medial supracondylar ridge

a Anterior surface

Fovea capitis

Femoral head

Neck

Greater trochanter

Intertrochanteric crest

Gluteal tuberosity

Pectineal line

Linea aspera

Lateral supracondylar ridge

Popliteal surface

Intercondylar fossa

Lateral epicondyle

Lateral condyle

b Posterior surface

8

Figure 8–12   **The Right Patella (a,b) and Patella with Femur (c).**

Base of patella

Attachment area for quadriceps tendon

Attachment area for patellar ligament

Apex of patella

a Anterior view

Lateral facet, for lateral condyle of femur

Medial facet, for medial condyle of femur

Articular surface of patella

b Posterior view

Patella

Lateral condyle of femur

Medial condyle of femur

c Inferior view of right femur and patella

8

The Tibia

The **tibia** (TIB-ē-uh), or shinbone, is the large medial bone of the leg (**Figure 8–13a**). The medial and lateral condyles of the femur articulate with the **medial and lateral tibial condyles** at the proximal end of the tibia. The **intercondylar eminence** is a ridge that separates the condyles (**Figure 8–13b**). The anterior surface of the tibia near the condyles has a prominent, rough **tibial tuberosity**, which you can feel through the skin. This tuberosity marks the attachment of the patellar ligament.

The **anterior margin** is a ridge that begins at the tibial tuberosity and extends distally along the anterior tibial surface. You can also easily feel the anterior margin of the tibia through the skin. As it nears the ankle joint, the tibia broadens, and the medial border ends in the **medial malleolus** (ma-LĒ-ō-lus; *malleolus*, hammer), a large process familiar to you as the medial bump at the ankle. The inferior surface of the tibia articulates with the proximal bone of the ankle. The medial malleolus supports this joint medially.

The Fibula

The slender **fibula** (FIB-ū-luh) parallels the lateral border of the tibia (**Figure 8–13a,b**). The head of the fibula articulates with the tibia. The articular facet is located on the anterior, inferior surface of the lateral tibial condyle. The medial border of the thin shaft is bound to the tibia by the **interosseous membrane**, which extends to the lateral margin of the tibia. This membrane helps stabilize the positions of the two bones. It also provides additional surface area for muscle attachment. A cross section of the tibia and fibula is shown in **Figure 8–13c**.

As its small diameter suggests, the fibula does not help transfer weight to the ankle and foot. In fact, it does not even articulate with the femur. However, the fibula is an important site for the attachment of muscles that move the foot and toes. In addition, the distal tip of the fibula extends lateral to the ankle joint. This fibular process, the **lateral malleolus**, gives lateral stability to the ankle. However, forceful movement of the foot

Figure 8–13 **The Right Tibia and Fibula.** ATLAS: Plates 32; 80a,b; 83a,b

Lateral tibial condyle
Head of fibula
Superior tibiofibular joint
Articular surface of medial tibial condyle
Medial tibial condyle
Tibial tuberosity
Interosseous membrane
Anterior margin
Tibia
Fibula
Lateral malleolus (fibula)
Medial malleolus (tibia)
Inferior articular surface

Intercondylar eminence
Articular surface of lateral tibial condyle
Lateral tibial condyle
Head of fibula
Interosseous membrane
Tibia
Fibula
Inferior tibiofibular joint
Lateral malleolus (fibula)

Anterior margin
Tibia
Fibula
Interosseous membrane

c **Transverse section of tibia and fibula**

a **Anterior view**

b **Posterior view**

outward and backward can dislocate the ankle, breaking both the lateral malleolus of the fibula and the medial malleolus of the tibia. This injury is called a *Pott's fracture.* ⟲ p. 227

TIPS & TRICKS

To remember how to distinguish the fibula from the tibia, consider the "fib" and "l" parts of the word *fibula*. To tell a small **li**e is to **fib**. The **fib**ula is sma**ll**er than the tibia, and is **l**ateral to it.

The Tarsal Bones

The ankle, or *tarsus*, consists of seven **tarsal bones** (**Figure 8–14**). The large **talus** transmits the weight of the body from the tibia toward the toes. The articulation between the talus and the tibia occurs across the superior and medial surfaces of the **trochlea**, a spool- or pulley-shaped articular process. The lateral surface of the trochlea articulates with the lateral malleolus of the fibula.

TIPS & TRICKS

To remember where the talus is located, think that the **t**alus is on **t**op of the foot and articulates with the **t**ibia.

The **calcaneus** (kal-KĀ-nē-us), or heel bone, is the largest of the tarsal bones. When you stand normally, most of your weight is transmitted from the tibia, to the talus, to the calcaneus, and then to the ground. The posterior portion of the calcaneus is a rough, knob-shaped projection. This is the attachment site for the *calcaneal tendon (Achilles tendon)*, which arises at the calf muscles. If you are standing, these strong muscles can lift the heel off the ground so that you stand on tiptoes. The superior and anterior surfaces of the calcaneus have smooth facets for articulation with other tarsal bones.

The **cuboid** articulates with the anterior surface of the calcaneus. The **navicular** is anterior to the talus, on the medial side of the ankle. It articulates with the talus and with the three *cuneiform* (kū-NĒ-i-form) *bones*. These are wedge-shaped bones arranged in a row, with articulations between them. They are named according to their position: **medial cuneiform**, **intermediate cuneiform**, and **lateral cuneiform**. Proximally, the cuneiform bones articulate with the anterior surface of the navicular. The lateral cuneiform also articulates with the medial surface of the cuboid. The distal surfaces of the cuboid and the cuneiform bones articulate with the metatarsal bones of the foot.

Figure 8–14   Bones of the Ankle and Foot. ATLAS: Plates 32; 85a; 86a,c; 87a–c; 88

Tarsal bones	
Calcaneus	
Trochlea of talus	
Navicular	
Cuboid	
Cuneiform bones	
Lateral	
Intermediate	
Medial	

| Metatarsal bones | |

Phalanges		Hallux	
Proximal		Proximal phalanx	
Middle			
Distal		Distal phalanx	

V IV III II I

a Superior view, right foot

Talus Cuboid Navicular Cuneiform bones Metatarsal bones Phalanges

Calcaneus

b Lateral view, right foot

Medial cuneiform bone Navicular Talus

Phalanges Metatarsal bones Calcaneus

Transverse arch   Longitudinal arch

c Medial view, right foot

The Metatarsal Bones and Phalanges

The **metatarsal bones** are five long bones that form the distal portion of the foot, or *metatarsus* (Figure 8–14). The metatarsal bones are identified by Roman numerals I–V, proceeding from medial to lateral across the sole. Proximally, metatarsal bones I–III articulate with the three cuneiform bones, and metatarsal bones IV and V articulate with the cuboid. Distally, each metatarsal bone articulates with a different proximal phalanx.

The **phalanges**, or toe bones (Figure 8–14), have the same anatomical organization as the fingers. The toes contain 14 phalanges. The **hallux**, or great toe, has two phalanges (*proximal* and *distal*). The other four toes have three phalanges apiece (*proximal, middle,* and *distal*).

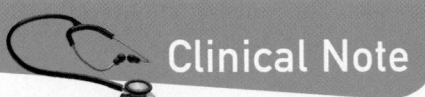

Clinical Note

Stress Fractures Running is beneficial to overall health but places the foot bones under more stress than does walking. *Stress fractures* are hairline fractures that develop in bones subjected to repeated shocks or impacts. Stress fractures of the foot usually involve one of the metatarsal bones. These fractures are caused either by improper placement of the foot while running or by poor arch support. In a fitness regime that includes street running, proper support for the bones of the foot is essential. An entire running-shoe market has arisen around the amateur and professional runner's need for good arch support.

Arches of the Foot

Weight transfer occurs along the **longitudinal arch** of the foot (Figure 8–14c). Ligaments and tendons maintain this arch by tying the calcaneus to the distal portions of the metatarsal bones. However, the lateral, or *calcaneal,* portion of the longitudinal arch has much less curvature than the medial, *talar* portion. Part of the reason is that the talar portion has more elasticity. As a result, the medial plantar surface of the foot remains elevated, so that the muscles, nerves, and blood vessels that supply the inferior surface are not squeezed between the metatarsal bones and the ground. In the condition known as *flatfeet,* normal arches are lost ("fall") or never form.

The elasticity of the talar portion of the longitudinal arch absorbs the shocks from sudden changes in weight loading. For example, the stresses that running or ballet dancing places on the toes are cushioned by the elasticity of this portion of the

Clinical Note

Congenital Talipes Equinovarus The arches of the foot are usually present at birth. Sometimes, however, they fail to develop properly. In **congenital talipes equinovarus** *(clubfoot),* abnormal muscle development distorts growing bones and joints. One or both feet may be involved. The condition can be mild, moderate, or severe. In most cases, the tibia, ankle, and foot are affected. The longitudinal arch is exaggerated, and the feet are turned medially and inverted. If both feet are involved, the soles face one another. This condition affects 1 in 1000 births and is twice as common in boys as girls. Prompt treatment with casts or other supports in infancy helps alleviate the problem. Fewer than half the cases require surgery.

arch. The **transverse arch** describes the degree of curvature change from the medial to the lateral borders of the foot.

When you stand normally, your body weight is distributed evenly between the calcaneus and the distal ends of the metatarsal bones. The amount of weight transferred forward depends on the position of the foot and the placement of your body weight. During *flexion* at the ankle, a movement also called *dorsiflexion,* all your body weight rests on the calcaneus—as when you "dig in your heels." During *extension* at the ankle, also known as *plantar flexion,* when you stand on tiptoe, the talus and calcaneus transfer your weight to the metatarsal bones and phalanges through the more anterior tarsal bones.

CHECKPOINT

12. Identify the bones of the lower limb.

13. The fibula neither participates in the knee joint nor bears weight. When it is fractured, however, walking becomes difficult. Why?

14. While jumping off the back steps at his house, 10-year-old Joey lands on his right heel and breaks his foot. Which foot bone is most likely broken?

15. Which foot bone transmits the weight of the body from the tibia toward the toes?

See the blue Answers tab at the back of the book.

8-5 | Sex differences and age account for individual skeletal variation

Learning Outcome Summarize sex differences and age-related changes in the human skeleton.

A comprehensive study of a human skeleton can reveal important information about an individual. We can estimate

Table 8–1	Age-Related Changes in the Skeleton	
Region and Feature	**Events**	**Age in Years**
GENERAL SKELETON		
Bony matrix	Reduction in mineral content; increased risk of osteoporosis	Begins at age 30–45; values differ for males versus females between ages 45 and 65; similar reductions occur in both sexes after age 65
Bone markings	Reduction in size, roughness	Gradual reduction with increasing age and decreasing muscular strength and mass
SKULL		
Fontanelles	Closure	Completed by age 2
Frontal suture	Fusion	2–8
Occipital bone	Fusion of ossification centers	1–4
Styloid process	Fusion with temporal bone	12–16
Hyoid bone	Complete ossification and fusion	25–30
Teeth	Loss of "baby teeth"; appearance of secondary dentition; eruption of permanent molars	Detailed in Chapter 24 (digestive system)
Mandible	Loss of teeth; reduction in bone mass; change in angle at mandibular notch	Accelerates in later years (60+)
VERTEBRAE		
Curvature	Development of major curves	3 months–10 years
Intervertebral discs	Reduction in size, percentage contribution to height	Accelerates in later years (60+)
LONG BONES		
Epiphyseal cartilages	Fusion	Begins about age 3; ranges vary, but general analysis permits determination of approximate age
PECTORAL AND PELVIC GIRDLES		
Epiphyses	Fusion	Relatively narrow ranges of ages (e.g., 14–16, 16–18, 22–25) increase accuracy of age estimates

a person's muscular development and muscle mass from the appearance of various ridges and from the general bone mass. Details such as the condition of the teeth or the presence of healed fractures give an indication of the individual's medical history. Two important facts, sex and age, can be determined or closely estimated on the basis of measurements indicated in **Spotlight Figure 8–10** and **Table 8–1**. In some cases, the skeleton may provide clues about the individual's nutritional state, handedness, and even occupation.

ATLAS: Embryology Summary 8: The Development of the Appendicular Skeleton

Spotlight Figure 8–10 (p. 279) identifies characteristic differences between the skeletons of males and females. However, not every skeleton shows every feature in classic detail. Many differences, including markings on the skull, cranial capacity, and general skeletal features, reflect differences in average body size, muscle mass, and muscular strength.

The general changes in the skeletal system that take place with age are summarized in **Table 8–1**. Note that these changes

begin at age 3 months and continue throughout life. The epiphyseal cartilages, for example, begin to fuse at about age 3. Degenerative changes in the normal skeletal system, such as a reduction in mineral content in the bony matrix, typically do not begin until age 30–45. The timing of epiphyseal closure is a key factor determining adult body size. Young people whose long bones are still growing should avoid heavy weight training, because they risk crushing the epiphyseal cartilages and thus shortening their stature (**Spotlight Figure 6–11**, p. 217).

CHECKPOINT

16. **Compare and contrast the bones of males and females with respect to weight and bone markings.**

17. **An anthropologist discovered several bones in a deep grave. After close visual inspection and careful measurements, what sort of information could the bones reveal?**

See the blue Answers tab at the back of the book.

Related Clinical Terms

bone graft: A surgical procedure that transplants bone tissue to repair and rebuild diseased or damaged bone.

genu valgum: Deformity in which the knees angle medially and touch one another while standing; commonly called knock-knee.

pelvimetry: Measurement of the dimensions of the female pelvis.

Chapter Review

Study Outline

MasteringA&P®

Access more chapter study tools online in the MasteringA&P Study Area:

- Chapter Quizzes, Chapter Practice Test, Art-labeling Activities, Animations, MP3 Tutor Sessions, and Clinical Case Studies
- Practice Anatomy Lab PAL™
- Interactive Physiology iP®
- A&P Flix A&PFlix
- PhysioEx PhysioEx

An Introduction to the Appendicular Skeleton p. 268

1. The **appendicular skeleton** includes the bones of the upper and lower limbs and the pectoral and pelvic girdles, which connect the limbs to the trunk. *(Figure 8–1)*

8-1 ▌ The pectoral girdles—the clavicles and scapulae—attach the upper limbs to the axial skeleton p. 268

2. Each upper limb articulates with the trunk through a **pectoral girdle,** or *shoulder girdle,* which consists of a **scapula** and a **clavicle.**

3. On each side, a clavicle and scapula position the shoulder joint, help move the upper limb, and provide a base for muscle attachment. *(Figures 8–2, 8–3)*

4. Both the **coracoid process** and the **acromion** of the scapula have attached ligaments and tendons associated with the shoulder joint. *(Figure 8–3)*

8-2 ▌ The upper limbs are adapted for free movement p. 271

5. The scapula articulates with the **humerus** at the shoulder (*glenohumeral*) joint. The **greater** and **lesser tubercles** of the humerus are important sites of muscle attachment. *(Figure 8–4)*

6. The humerus articulates with the **radius** and **ulna,** the bones of the forearm, at the elbow joint. *(Figure 8–5)*

7. The **carpal bones** of the wrist, or *carpus,* form two rows. The distal row articulates with the five **metacarpal bones.** Four of the fingers contain three **phalanges;** the **pollex** (thumb) has only two phalanges. *(Figure 8–6)*

8-3 ▌ The pelvic girdle—two hip bones—attaches the lower limbs to the axial skeleton p. 276

8. The bones of the **pelvic girdle** are more massive than those of the pectoral girdle because of their role in weight bearing and locomotion.

9. The pelvic girdle consists of two **hip bones,** or **coxal bones.** Each hip bone forms through the fusion of an **ilium,** an **ischium,** and a **pubis.** *(Figure 8–7)*

10. The ilium is the largest component of the hip bone. Inside the **acetabulum,** the ilium is fused to the ischium (posteriorly) and the pubis (anteriorly). The *pubic symphysis* limits movement between the pubic bones of the left and right hip bones. *(Figures 8–7, 8–8)*

11. The **pelvis** consists of the hip bones, the sacrum, and the coccyx. It is subdivided into the **false** (*greater*) **pelvis** and the **true** (*lesser*) **pelvis.** *(Figures 8–8, 8–9, Spotlight Figure 8–10)*

8-4 ▌ The lower limbs are adapted for movement and support p. 280

12. The **femur** is the longest and heaviest bone in the body. It articulates with the **tibia** at the knee joint. *(Figures 8–11, 8–13)*

13. The **patella** is a large sesamoid bone. *(Figure 8–12)*

14. The **fibula** parallels the tibia laterally. *(Figure 8–13)*

15. The *tarsus,* or ankle, has seven **tarsal bones.** *(Figure 8–14)*

16. The basic organizational pattern of the **metatarsal bones** and **phalanges** of the foot resembles that of the hand. All the toes have three phalanges, except for the **hallux** (great toe), which has two. *(Figure 8–14)*

17. When a person stands normally, most of the body weight is transferred to the **calcaneus,** and the rest is passed on to the five metatarsal bones. Weight transfer occurs along the **longitudinal arch.** There is also a **transverse arch.** *(Figure 8–14)*

8-5 ▌ Sex differences and age account for individual skeletal variation p. 284

18. Studying a human skeleton can reveal important information, such as the person's weight, sex, body size, muscle mass, and age. *(Spotlight Figure 8–10, Table 8–1)*

19. Age-related changes take place in the skeletal system. These changes begin at about age 1 and continue throughout life. *(Table 8–1)*

Review Questions

See the blue Answers tab at the back of the book.

LEVEL 1 Reviewing Facts and Terms

1. In the following photographs of the scapula, identify the three views (a–c) and the indicated bone markings (d–g).

(a)   (b)   (c)

(a) _____   (b) _____
(c) _____   (d) _____
(e) _____   (f) _____
(g) _____

2. Which of the following is primarily responsible for stabilizing, positioning, and bracing the pectoral girdles? (a) tendons, (b) ligaments, (c) the joint shape, (d) muscles, (e) the shape of the bones within the joint.

3. In the following drawing of the pelvis, label the structures indicated.

(a) _____   (b) _____
(c) _____   (d) _____
(e) _____   (f) _____

4. In anatomical position, the ulna lies (a) medial to the radius, (b) lateral to the radius, (c) inferior to the radius, (d) superior to the radius.

5. The point of the elbow is actually the _____ of the ulna. (a) styloid process, (b) olecranon, (c) coronoid process, (d) trochlear notch.

6. The bones of the hand articulate distally with the (a) carpal bones , (b) ulna and radius, (c) metacarpal bones, (d) phalanges.

7. The proximal carpal bones are the _____, _____, _____, and _____.

8. What is the name of the flexible sheet that interconnects the radius and ulna (and the tibia and fibula)?

9. How is the pelvic girdle attached to the axial skeleton?

10. What makes up the boundaries of the pelvic outlet?

LEVEL 2 Reviewing Concepts

11. The presence of tubercles on bones indicates the positions of (a) tendons and ligaments, (b) muscle attachment, (c) ridges and flanges, (d) a and b.

12. At the glenoid cavity, the scapula articulates with the proximal end of the (a) humerus, (b) radius, (c) ulna, (d) femur.

13. All of the following are structural characteristics of the female pelvic girdle compared to the male pelvic girdle with either one or two exceptions. Identify the exception(s). (a) The female pelvis is adapted for childbearing, (b) The female pelvic girdle is generally lighter than the male pelvic girdle, (c) Relaxin produced by the male pelvic girdle loosens the pubic symphysis and sacroiliac joints, (d) The shape of the female pelvic girdle is the same as that of the male, (e) Answers c and d are exceptions.

14. The large foramen between the pubic ramus and ischial ramus is the (a) foramen magnum, (b) suborbital foramen, (c) acetabulum, (d) obturator foramen.

15. Which of the following skeletal characteristics is an adaption for childbearing? (a) inferior angle of 100° or more between the pubic bones, (b) a relatively broad, low pelvis, (c) less curvature of the sacrum and coccyx, (d) All of these are correct.

16. The fibula (a) forms an important part of the knee joint, (b) articulates with the femur, (c) helps to bear the weight of the body, (d) provides lateral stability to the ankle, (e) does (a) and (b).

17. The tarsal bone that accepts weight and distributes it to the heel or toes is the (a) cuneiform, (b) calcaneus, (c) talus, (d) navicular.

18. What is the difference in skeletal structure between the pelvic girdle and the pelvis?

19. Jack injures himself playing hockey, and the physician who examines him informs him that he has dislocated his pollex. What part of Jack's body did he injure? (a) his arm, (b) his leg, (c) his hip, (d) his thumb, (e) his shoulder.

20. Why would a self-defense instructor advise a student to strike an assailant's clavicle?

21. The pelvis (a) protects the upper abdominal organs, (b) contains bones from both the axial and appendicular skeletons, (c) is composed of the coxal bones, sacrum, and coccyx, (d) does all of these, (e) does (b) and (c).

22. Why is the tibia, but not the fibula, involved in the transfer of weight to the ankle and foot?

23. In determining the age of a skeleton, all of the following pieces of information would be helpful *except* (a) the number of cranial sutures, (b) the size and roughness of the markings of the bones, (c) the presence or absence of fontanelles, (d) the presence or absence of epiphyseal cartilages, (e) the types of minerals deposited in the bones.

LEVEL 3 Critical Thinking and Clinical Applications

24. Explain how ballerinas balance entirely on their toes when executing the pointe technique.

25. While Fred, a fireman, is fighting a fire in a building, part of the ceiling collapses, and a beam strikes him on his left shoulder. He is rescued, but has a great deal of pain in his shoulder. He cannot move his arm properly, especially in the anterior direction. His clavicle is not broken, and his humerus is intact. What is the probable nature of Fred's injury?

26. Michael is experiencing swelling and tenderness in his right knee after running. He follows a circular route and stays on the same side of the road. What condition might Michael have developed? Give reasons for your answer.

8

CLINICAL CASE Wrap-Up

The Orthopedic Surgeon's Nightmare

When Caitlyn arrives at the emergency room, she has no pulse distal to the injury in her arm. Many of the critical nerves and blood vessels to the forearm and hand run anterior to the humerus through the elbow. These structures have been impaled by the sharp fractured edge of the humeral shaft. This is a true surgical emergency.

Of all the orthopedic injuries a child can sustain in a "**f**all **o**n the **o**ut **s**tretched **h**and" (known as a FOOSH injury), a supracondylar distal humerus fracture is the most difficult to reduce (restore to anatomical position) and maintain in a reduced position during healing.

Just proximal to the humeral condyle (the trochlea and capitulum), at the site of the olecranon fossa posteriorly and the radial fossa anteriorly, the humerus is extremely thin. As the elbow extends, the olecranon is wedged into the olecranon fossa. With the force of the falling child's weight, the distal humerus can fracture through this thin bone.

Trying to reduce this fracture in an extremely swollen injured elbow is like trying to perch one razor blade on top of another inside a watermelon. And the reduction is about as stable.

Caitlyn is taken to an operating room where the surgeon performs a closed reduction with the help of x-ray. Her pulse immediately returns. The fracture is held with pins and a long arm posterior splint is applied.

1. What skeletal structures are still attached to the distal humeral fracture fragment?

2. Why would the surgeon want to avoid a circular cast in the immediate post-fracture period and apply a posterior splint instead?

See the blue Answers tab at the back of the book.

9 Joints

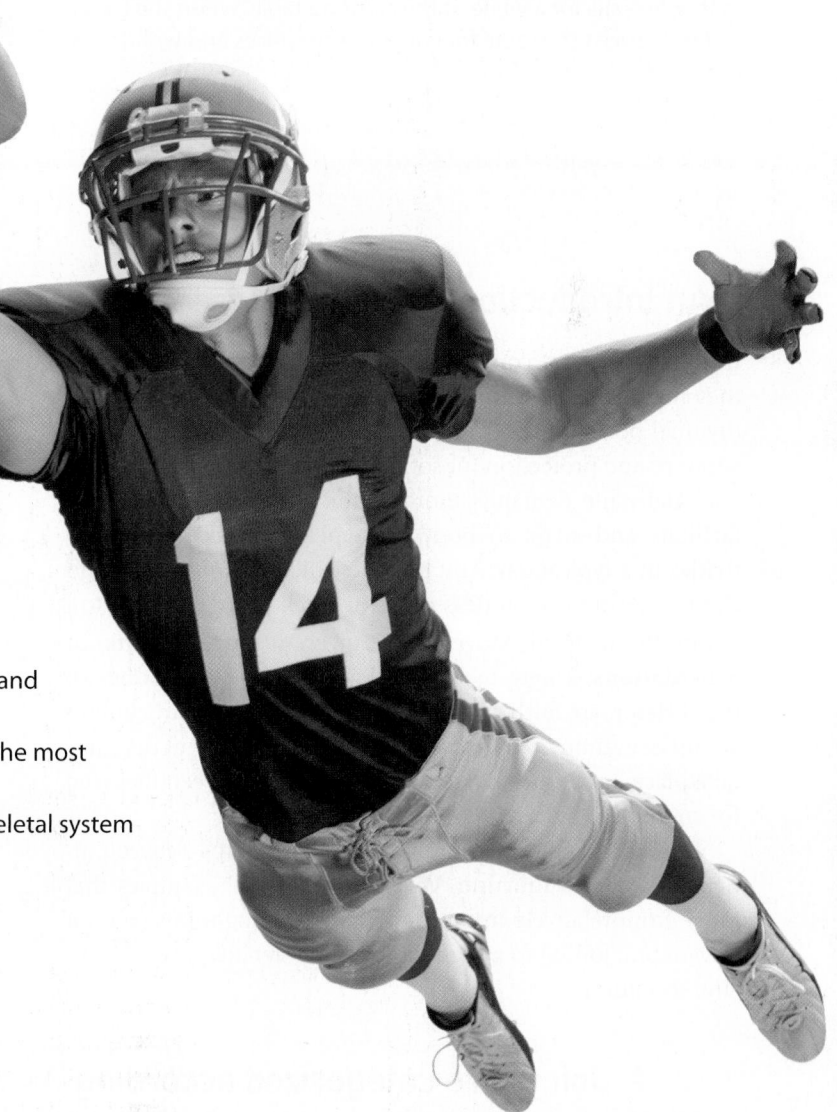

Learning Outcomes

These Learning Outcomes correspond by number to this chapter's sections and indicate what you should be able to do after completing the chapter.

9-1 ❚ Contrast the major categories of joints, and explain the relationship between structure and function for each category.

9-2 ❚ Describe the basic structure of a synovial joint, and describe common synovial joint accessory structures and their functions.

9-3 ❚ Describe how the anatomical and functional properties of synovial joints permit movements of the skeleton.

9-4 ❚ Describe the joints between the vertebrae of the vertebral column.

9-5 ❚ Describe the structure and function of the shoulder joint and the elbow joint.

9-6 ❚ Describe the structure and function of the hip joint and the knee joint.

9-7 ❚ Describe the effects of aging on joints, and discuss the most common age-related clinical problems for joints.

9-8 ❚ Explain the functional relationships between the skeletal system and other body systems.

SPOTLIGHT
• Joint Movement pp. 296–297

CLINICAL CASE
• What's Ailing the Birthday Girl? p. 290

Clinical Notes
• Bursitis and Bunions p. 294
• Dislocation and Subluxation p. 294
• Damage to Intervertebral Discs p. 301
• Knee Injuries p. 309

What's Ailing the Birthday Girl?

Jessica just woke up from her nap at daycare. Earlier this morning the class celebrated her fourth birthday with cupcakes and singing. Now she feels like crying when she tries to roll over. She can't seem to get moving out of her little cot.

"Come on, Jessie, let's play tag," says her best friend. When Jessica finally gets to a sitting position on the side of the cot, she doesn't feel like taking a single step. When she stands, she feels slow and clumsy. She certainly cannot run after her classmates. She sits for a while at the coloring table. When she tries to reach across the table for crayons, she winces and withdraws her hand.

Her teacher notices Jessica's lethargy and offers her a glass of juice. Even her favorite juice does not taste good to her.

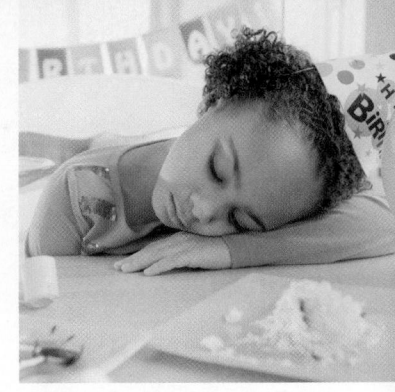

When Jessica's mother comes to pick her up from daycare, the teacher pulls her aside for a quick conference. "Jessica has not been participating in any play activities this whole week," reports the teacher. "Even today, on her birthday, she refused to play tag. She takes a really long time getting up from the floor after story circle. And she has been limping after naptime. I wonder what's going on?" **What's ailing Jessica? To find out, turn to the Clinical Case Wrap-Up on p. 314.**

An Introduction to Joints

In this chapter we consider the ways bones interact wherever they interconnect. In the last two chapters, you learned the individual bones of the skeleton. These bones provide strength, support, and protection for softer tissues of the body. However, your daily life demands more of the skeleton—it must also facilitate and adapt to body movements. Think of your activities in a typical day: You breathe, talk, walk, sit, stand, and change positions countless times. In each case, your skeleton is directly involved. Movements can occur only at **joints**, or **articulations**, where two bones meet, because the bones of the skeleton are fairly inflexible. The characteristic structure of a joint determines the type and amount of movement that may take place. Each joint reflects a compromise between the need for strength and the need for mobility.

In this chapter we compare the relationships between articular form and function. We consider several examples that range from relatively immobile but very strong joints (the intervertebral joints) to a highly mobile but relatively weak joint (the shoulder).

9-1 ▌ Joints are categorized according to their range of motion or structure

Learning Outcome Contrast the major categories of joints, and explain the relationship between structure and function for each category.

We use two classification methods to categorize joints. The first is the one we will use in this chapter. It is a functional scheme because it is based on the amount of movement possible, a property known as the *range of motion* (ROM). Each functional group is further subdivided primarily on the basis of the anatomical structure of the joint (Table 9–1):

1. An *immovable joint* is a **synarthrosis** (sin-ar-THRŌ-sis; *syn*, together + *arthros*, joint). A synarthrosis can be *fibrous* or *cartilaginous*, depending on the nature of the connection. Over time, the two bones may fuse.

2. A *slightly movable joint* is an **amphiarthrosis** (am-fē-ar-THRŌ-sis; *amphi*, on both sides). An amphiarthrosis is either *fibrous* or *cartilaginous*, depending on the nature of the connection between the opposing bones.

3. A *freely movable joint* is a **diarthrosis** (dī-ar-THRŌ-sis; *dia*, through), or *synovial joint*. Diarthroses are subdivided according to the movement permitted.

The second classification scheme relies solely on the anatomy of the joint, without regard to the degree of movement permitted. Using this framework, we classify joints as *fibrous, cartilaginous, bony,* or *synovial*. Bony joints form when fibrous or cartilaginous joints ossify. The ossification may be normal or abnormal, and may occur at various times in life.

The two classification schemes are loosely correlated. We see many anatomical patterns among immovable or slightly movable joints, but there is only one type of freely movable joint—synovial joints. All synovial joints are diarthroses. We will use the functional classification rather than the anatomical one because our primary interest is how joints work.

Table 9–1 Functional and Structural Classifications of Joints

Functional Category	Structural Category and Type		Description
SYNARTHROSIS (no movement)			
At a synarthrosis, the bony edges are quite close together and may even interlock. These extremely strong joints are located where movement between the bones must be prevented.	**FIBROUS**	Suture	A **suture** (*sutura*, a sewing together) is a synarthrotic joint located only between the bones of the skull. The edges of the bones are interlocked and bound together at the suture by dense fibrous connective tissue.
		Gomphosis	A **gomphosis** (gom-FŌ-sis; *gomphos*, bolt) is a synarthrosis that binds the teeth to bony sockets in the maxillae and mandible. The fibrous connection between a tooth and its socket is a *periodontal* (per-ē-ō-DON-tal) *ligament* (*peri*, around + *odontos*, tooth).
	CARTILAGINOUS	Synchondrosis	A **synchondrosis** (sin-kon-DRŌ-sis; *syn*, together+ *chondros*, cartilage) is a rigid, cartilaginous bridge between two articulating bones. The cartilaginous connection between the ends of the first pair of vertebrosternal ribs and manubrium of the sternum is a synchondrosis. Another example is the epiphyseal cartilage, which connects the diaphysis to the epiphysis in a growing long bone.
	BONY	Synostosis	A **synostosis** (sin-os-TŌ-sis) is a totally rigid, immovable joint created when two bones fuse and the boundary between them disappears. The frontal (metopic) suture of the frontal bone, the fusion of an infant's left and right mandibular bones, and the epiphyseal lines of mature long bones are synostoses.
AMPHIARTHROSIS (little movement)			
An amphiarthrosis permits more movement than a synarthrosis, but is much stronger than a freely movable joint. The articulating bones are connected by collagen fibers or cartilage.	**FIBROUS**	Syndesmosis	At a **syndesmosis** (sin-dez-MŌ-sis; *syndesmos*, ligament), bones are connected by a ligament. One example is the distal joint between the tibia and fibula.
	CARTILAGINOUS	Symphysis	At a **symphysis**, the articulating bones are connected by a wedge or pad of fibrocartilage. The joint between the two pubic bones (the *pubic symphysis*) is an example of a symphysis.
DIARTHROSIS (free movement)			
Planes of Movement • **Monoaxial** - movement in one plane; elbow, ankle • **Biaxial** - movement in two planes; ribs and wrist • **Triaxial** - movement in three planes; shoulder, hip	**SYNOVIAL**		**Synovial** (si-NŌ-ve-ul) **joints** permit a wider range of motion than do other types of joints. They are typically located at the ends of long bones, such as those of the upper and lower limbs.

CHECKPOINT

1. Name and describe the three types of joints as classified by their range of motion.

2. What characteristics do typical synarthrotic and amphiarthrotic joints share?

3. In a newborn, the large bones of the skull are joined by fibrous connective tissue. The bones later grow, interlock, and form immovable joints. Structurally, which type of joints are these?

See the blue Answers tab at the back of the book.

9-2 ▮ Synovial joints are freely movable joints containing synovial fluid

Learning Outcome Describe the basic structure of a synovial joint, and describe common synovial joint accessory structures and their functions.

Synovial joints are freely movable and classified as diarthroses. A two-layered **joint capsule**, also called an **articular capsule**, surrounds the synovial joint (**Figure 9–1**). The joint capsule contains an inner *synovial membrane* and an outer *fibrous capsule*. This membrane does not cover the articulating surfaces within the joint. Recall that a synovial membrane consists of areolar tissue covered by an incomplete epithelial layer. The synovial fluid that fills the joint cavity originates in the areolar tissue of the synovial membrane. ⮌ p. 165 Now let's consider the major features of synovial joints.

Articular Cartilage

Under normal conditions, the bony surfaces at a synovial joint cannot contact one another, because special **articular cartilage** covers the articulating surfaces. Articular cartilage resembles hyaline cartilage elsewhere in the body. ⮌ p. 160 However, articular cartilage has no perichondrium (the fibrous sheath described in Chapter 4), and the matrix contains more water than that of other cartilage.

The surface of articular cartilage is slick and smooth. This feature alone can reduce friction during movement at the joint. However, even when pressure is applied across a joint, the smooth articular cartilages do not touch one another because a thin film of synovial fluid within the joint cavity separates them (**Figure 9–1a**). This fluid acts as a lubricant, minimizing friction.

Synovial joints cannot function normally if the articular cartilages are damaged. When such damage occurs, the matrix may begin to break down. The exposed surface will then change from a slick, smooth-gliding surface to a rough abrasive surface of bristly collagen fibers. This abrasive surface drastically increases friction at the joint.

Figure 9–1 **The Structure of a Synovial Joint.**

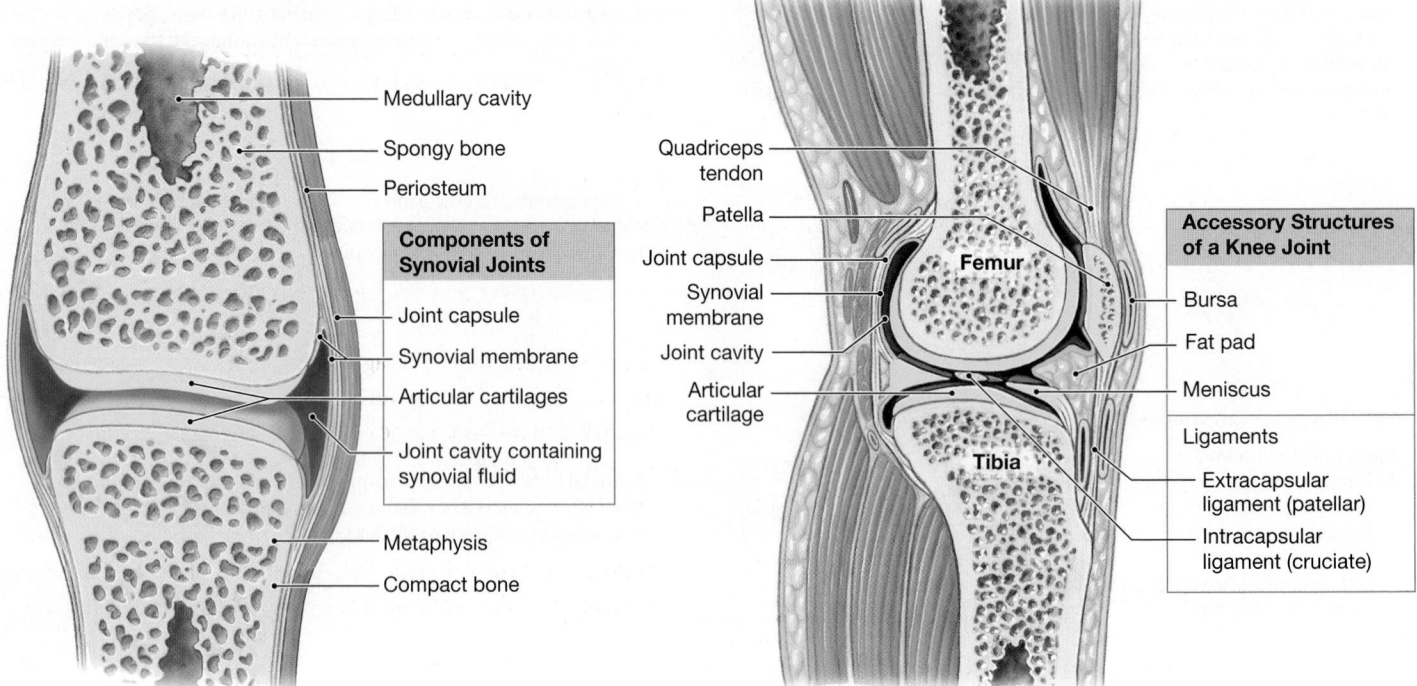

Components of Synovial Joints
Joint capsule
Synovial membrane
Articular cartilages
Joint cavity containing synovial fluid

Labels (left): Medullary cavity, Spongy bone, Periosteum, Metaphysis, Compact bone

Accessory Structures of a Knee Joint
Bursa
Fat pad
Meniscus
Ligaments
Extracapsular ligament (patellar)
Intracapsular ligament (cruciate)

Labels (knee): Quadriceps tendon, Patella, Joint capsule, Synovial membrane, Joint cavity, Articular cartilage, Femur, Tibia

a Synovial joint, sagittal section

b Knee joint, sagittal section

Synovial Fluid

What is synovial fluid? It is a clear, viscous solution with the consistency of egg yolk or heavy molasses. **Synovial fluid** resembles interstitial fluid, but contains proteoglycans with a high concentration of hyaluronan (hyaluronic acid) secreted by fibroblasts of the synovial membrane. ↺ p. 165 Even in a large joint such as the knee, the total quantity of synovial fluid in a joint is small—normally less than 3 mL. The synovial fluid within a joint has three primary functions:

1. *Lubrication.* Articular cartilages act like sponges filled with synovial fluid. When part of an articular cartilage is compressed, some of the synovial fluid is squeezed out of the cartilage and into the space between the opposing surfaces. This thin layer of fluid greatly reduces friction between moving surfaces, just as a thin film of water makes a highway slick by reducing friction between a car's tires and the road. When the compression stops, synovial fluid is pulled back into the articular cartilage.

2. *Nutrient Distribution.* The synovial fluid in a joint must circulate continuously to provide nutrients and a waste disposal route for the chondrocytes of the avascular articular cartilage. It circulates whenever the joint moves. The compression and re-expansion of the articular cartilage pumps synovial fluid into and out of the cartilage matrix. As the synovial fluid flows through the areolar tissue of the synovial membrane, waste products are absorbed and additional nutrients are delivered by diffusion across capillary walls.

3. *Shock Absorption.* Synovial fluid cushions joints that are subjected to compression from shocks and sudden impacts. This occurs because the viscosity of synovial fluid increases with increased pressure. For example, your hip, knee, and ankle joints are compressed as you walk. They are even more severely compressed when you jog or run. When the pressure across a joint suddenly increases, the resulting shock is lessened as the viscosity of synovial fluid increases. As the pressure lessens, viscosity decreases and it regains its lubricating function across the articular surfaces.

Accessory Structures

Synovial joints may have a variety of accessory structures, including pads of cartilage or fat, ligaments, tendons, and bursae (**Figure 9–1b**).

Cartilages and Fat Pads

In several joints, including the knee (**Figure 9–1b**), menisci and fat pads may lie between the opposing articular surfaces. A **meniscus** (me-NIS-kus; a crescent; plural, *menisci*), also called an *articular disc*, is a pad of fibrocartilage located between opposing bones within a synovial joint. Menisci may subdivide a synovial cavity, channel the flow of synovial fluid, or allow for variations in the shapes of the articular surfaces.

Fat pads are localized masses of adipose tissue covered by a layer of synovial membrane. They are commonly superficial to the joint capsule (**Figure 9–1b**). Fat pads protect the articular cartilages and act as packing material for the joint. When the bones move, the fat pads fill in the spaces created as the joint cavity changes shape.

Ligaments

The capsule that surrounds the entire joint is continuous with the periostea of the articulating bones. **Accessory ligaments** support, strengthen, and reinforce synovial joints. *Capsular ligaments,* or *intrinsic ligaments,* are localized thickenings of the joint capsule. *Extrinsic ligaments* are separate from the joint capsule. These ligaments may be located either inside or outside the joint capsule, and are called *intracapsular* or *extracapsular* ligaments, respectively.

Ligaments are very strong. In a **sprain**, a ligament is stretched so much that some of the collagen fibers are torn, but the ligament as a whole survives and the joint is not damaged. With excessive force, one of the attached bones often breaks before the ligament tears. In general, a broken bone heals much more quickly and effectively than does a torn ligament. The reason is that ligaments have no direct blood supply and thus must obtain essential substances by diffusion.

Tendons

Tendons are not part of the joint itself, but tendons passing across or around a joint may limit the joint's range of motion and provide mechanical support for it. For example, tendons associated with the muscles of the arm help brace the shoulder joint.

Bursae

Bursae (BUR-sē; singular, *bursa,* a pouch) are small, thin, fluid-filled pockets in connective tissue. They contain synovial fluid and are lined by a synovial membrane. Bursae may be connected to the joint cavity or separate from it. They form where a tendon or ligament rubs against other tissues. Located around most synovial joints, including the shoulder joint, bursae reduce friction and act as shock absorbers. *Synovial tendon sheaths* are tubular bursae that surround tendons where they cross bony surfaces. Bursae may also appear deep to the skin, covering a bone or lying within other connective tissues exposed to friction or pressure. Bursae that develop in abnormal locations, or because of abnormal stresses, are called *adventitious bursae.*

Factors That Stabilize Synovial Joints

A joint cannot be both highly mobile and very strong. The greater the range of motion at a joint, the weaker it becomes.

A synarthrosis is the strongest type of joint, but it permits no movement. In contrast, a diarthrosis, such as the shoulder joint, is far weaker but has a broad range of motion.

Movement beyond its normal range of motion will damage any mobile diarthrosis. Several factors are responsible for limiting the range of motion, stabilizing the joint, and reducing the chance of injury:

- The collagen fibers of the joint capsule and any accessory, extracapsular, or intracapsular ligaments.
- The shapes of the articulating surfaces and menisci, which may prevent movement in specific directions.
- The presence of other bones, skeletal muscles, or fat pads around the joint.
- Tension in tendons attached to the articulating bones.

When a skeletal muscle contracts and pulls on a tendon, movement in a specific direction may be either encouraged or opposed. The pattern of stabilizing structures varies among joints. For example, the hip joint is stabilized by the shapes of the bones (the head of the femur projects into the acetabulum), a heavy capsule, intracapsular and extracapsular ligaments, tendons, and massive muscles. It is therefore very strong and stable. In contrast, the elbow is another stable

Clinical Note

Bursitis and Bunions Bursae are normally flat and contain a small amount of synovial fluid. When injured, however, they become inflamed and swell with synovial fluid. This causes pain in the affected area whenever the tendon or ligament moves, resulting in a condition called **bursitis**. Inflammation can result from the friction due to repetitive motion, pressure over the joint, irritation by chemical stimuli, infection, or trauma. Bursitis associated with repetitive motion typically occurs at the shoulder. Musicians, golfers, baseball pitchers, and tennis players may develop bursitis there. The most common pressure-related bursitis is a **bunion**. Bunions may form over the base of the great toe as a result of friction and distortion of the first metatarsophalangeal joint from wearing tight shoes, especially narrow shoes with pointed toes.

We have special, older names for bursitis at other locations, indicating the occupations most often associated with them. In "housemaid's knee," which accompanies prolonged kneeling, the affected bursa lies between the patella and the skin. "Student's elbow" is a form of bursitis that can result from propping your head up with your arm on a desk while you read your anatomy and physiology textbook.

Clinical Note

Dislocation and Subluxation In general, the more stable the joint, the more restricted is its range of motion. The shoulder joint is the most mobile synovial joint, but it relies primarily on the surrounding ligaments, muscles, and tendons for stability. For this reason, it is fairly weak.

When reinforcing structures cannot protect a joint from extreme stresses, a **dislocation**, or **luxation** (luk-SĀ-shun), results. In a dislocation, the articulating surfaces are forced out of position. The displacement can damage the articular cartilages, tear ligaments, or distort the joint capsule. The *inside* of a joint has no pain receptors, but nerves that monitor the capsule, ligaments, and tendons are quite sensitive, so dislocations are very painful.

The damage accompanying a partial dislocation, or **subluxation** (sub-luk-SĀ-shun), is less severe. People who are "double jointed" have joints that are weakly stabilized. Their joints permit a greater range of motion than do those of other individuals, but they are more likely to suffer partial or complete dislocations.

joint, but it gains its stability from the interlocking shapes of the articulating bones. The capsule and associated ligaments provide additional support.

CHECKPOINT

4. Describe the components of a synovial joint, and identify the functions of each.
5. Why would improper circulation of synovial fluid lead to the degeneration of articular cartilages in the affected joint?
6. Define subluxation.

See the blue Answers tab at the back of the book.

9-3 | The structure and function of synovial joints enable various skeletal movements

Learning Outcome Describe how the anatomical and functional properties of synovial joints permit movements of the skeleton.

To *describe* human movement, we need a frame of reference that enables accurate and precise communication. In descriptions of movement at synovial joints, phrases such as "bend the leg" or "raise the arm" are not sufficiently precise. Anatomists use descriptive terms that have specific meanings. **Spotlight Figure 9–2** (pp. 296–297) considers these movements with reference to six basic categories of synovial joints.

Types of Movements at Synovial Joints

Gliding Movement

In gliding, two opposing surfaces slide past one another (**Spotlight Figure 9–2**). Gliding occurs between the surfaces of articulating carpal bones, between tarsal bones, and between the clavicles and the sternum. The movement can occur in almost any direction, but the amount of movement is slight, and rotation is generally prevented by the capsule and associated ligaments.

Angular Movement

Examples of angular movement include *flexion, extension, abduction, adduction,* and *circumduction* (**Figure 9–3**). Descriptions of these movements are based on reference to an individual in the anatomical position. ⤴ p. 40

Flexion and Extension. **Flexion** (FLEK-shun) is movement in the anterior–posterior plane that decreases the angle between articulating bones. **Extension** occurs in the same plane, but it increases the angle between articulating bones (**Figure 9–3a**).

Figure 9–3   **Angular Movements.** The red dots indicate the locations of the joints involved in the movements illustrated.

a **Flexion/extension**

b **Abduction/adduction**

c **Adduction/abduction**

d **Circumduction**

SIMPLE MODEL OF ARTICULAR MOTION

Take a pencil as your model and stand it upright on the surface of a desk. The pencil represents a bone, and the desk is an articular surface. A lot of twisting, pushing, and pulling will demonstrate that there are only three ways to move the pencil.

Moving the Point

Linear motion

Possible movement 1 shows the pencil can move. If you hold the pencil upright, without securing the point, you can push the pencil across the surface. This kind of motion is called **gliding**. You could slide the point forward or backward, from side to side, or diagonally.

Changing the Shaft Angle

Angular motion

Possible movement 2 shows the pencil shaft can change its angle with the surface. With the tip held in position, you can move the eraser end of the pencil forward and backward, from side to side, or at some intermediate angle.

Circumduction

Possible movement 3 shows a complex angular movement. Grasp the pencil eraser and move the pencil in any direction until it is no longer vertical. Now, swing the eraser through a complete circle in a movement called **circumduction**.

Rotating the Shaft

Rotation

Possible movement 4 shows that the pencil shaft can rotate. If you keep the shaft vertical and the point at one location, you can still spin the pencil around its longitudinal axis in a movement called **rotation**. No joint can freely rotate because this would tangle blood vessels, nerves, and muscles as they crossed the joint.

AXES OF MOTION

Movement of joints can also be described by the number of axes that they can rotate around. A joint that permits movement around one axis is called **monaxial**, a joint that permits movement around two axes is called **biaxial**, and one that permits movement around three axes is called **triaxial**.

Superior–inferior axis

Lateral–medial axis

Anterior–posterior axis

CLASSIFICATION OF SYNOVIAL JOINTS

Synovial joints are freely movable diarthrotic joints, and they are classified by the type and range of motion permitted. Synovial joints are described as **gliding**, **hinge**, **condylar**, **saddle**, **pivot**, or **ball-and-socket** on the basis of the shapes of the articulating surfaces, which in turn determine the joint movement.

Gliding joint

Gliding joints, or plane joints, have flattened or slightly curved surfaces that slide across one another, but the amount of movement is very slight.

Movement:
Gliding.
Slight nonaxial or multiaxial

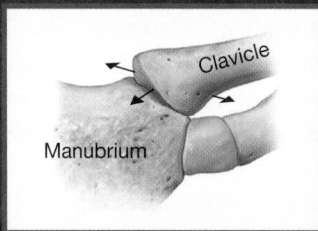

Examples:
• Acromioclavicular and claviculosternal joints
• Intercarpal joint
• Vertebrocostal joint
• Sacroiliac joint

Hinge joint

Hinge joints permit angular motion in a single plane, like the opening and closing of a door.

Movement:
Angular.
Monaxial

Examples:
• Elbow joint
• Knee joint
• Ankle joint
• Interphalangeal joint

Condylar joint

Condylar joints, or ellipsoidal joints, have an oval articular face nestled within a depression on the opposing surface.

Movement:
Angular.
Biaxial

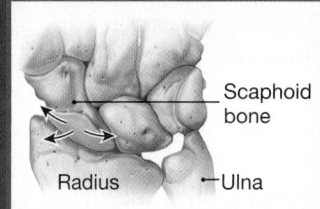

Examples:
• Radiocarpal joint
• Metacarpophalangeal joints 2–5
• Metatarsophalangeal joint

Saddle joint

Saddle joints have complex articular faces and fit together like a rider in a saddle. Each face is concave along one axis and convex along the other.

Movement:
Angular.
Biaxial

Example:
• First carpometacarpal joint

Pivot joint

Pivot joints only permit rotation.

Movement:
Rotation.
Monaxial

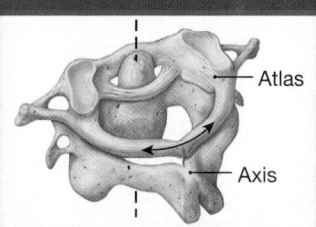

Examples:
• Atlantoaxial joint
• Proximal radioulnar joint

Ball-and-socket joint

In a ball-and-socket joint, the round head of one bone rests within a cup-shaped depression in another.

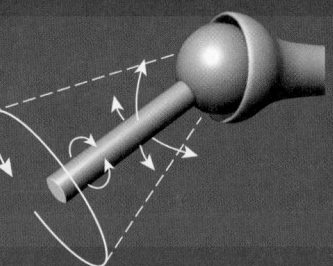

Movement:
Angular, circumduction, and rotation. Triaxial

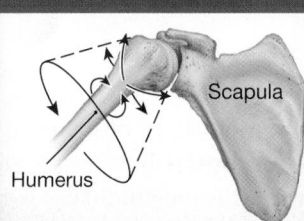

Examples:
• Shoulder joint
• Hip joint

We usually apply these terms to the movements of the long bones of the limbs, but we can also use them to describe movements of the axial skeleton. For example, when you bring your head toward your chest, you flex the intervertebral joints of the neck. When you bend down to touch your toes, you flex the intervertebral joints of the spine. Extension reverses these movements, returning you to the anatomical position.

When a person is in the anatomical position, all of the major joints of the axial and appendicular skeletons (except the ankle) are at full extension. (Shortly we will introduce special terms to describe movements of the ankle joint.) Flexion of the shoulder joint or hip joint moves the limbs anteriorly, whereas extension moves them posteriorly. Flexion of the wrist joint moves the hand anteriorly, and extension moves it posteriorly.

In each of these examples, extension can be continued past the anatomical position. Extension past the anatomical position is called **hyperextension** (Figure 9–3a). When you hyperextend your neck, you can gaze at the ceiling. Ligaments, bony processes, or soft tissues prevent hyperextension of many joints, such as the elbow or the knee.

Abduction and Adduction. Abduction (*ab*, from) is movement *away* from the longitudinal axis of the body in the frontal plane (Figure 9–3b). For example, swinging the upper limb to the side is abduction of the limb. Moving it back to the anatomical position is **adduction** (*ad*, to). Adduction of the wrist moves the heel of the hand and fingers *toward* the body, whereas abduction moves them farther away. Spreading the fingers or toes apart abducts them, because they move *away from* a central digit (Figure 9–3c). Bringing them together is called adduction. (Fingers move toward or away from the middle finger; toes move toward or away from the second toe.) Abduction and adduction always refer to movements of the appendicular skeleton, not to those of the axial skeleton.

TIPS & TRICKS

When someone is **abduct**ed, they are taken away, just as **abduct**ion takes the limb away from the body. During **add**uction, the limb is **add**ed to the body.

Circumduction. Recall the special type of angular movement, circumduction, from our model. Moving your arm in a loop is circumduction, as when you draw a large circle on a whiteboard (Figure 9–3d). Your hand moves in a circle, but your arm does not rotate.

Rotation

We also describe rotational movements with reference to a person in the anatomical position. Rotation of the head may involve **left rotation** or **right rotation** (Figure 9–4a). We describe limb rotation by reference to the longitudinal axis of the

trunk. During medial rotation, also known as *internal rotation* or *inward rotation,* the anterior surface of a limb turns toward the long axis of the trunk (Figure 9–4a). The reverse movement is called **lateral rotation**, *external rotation,* or *outward rotation.*

The proximal joint between the radius and the ulna permits rotation of the radial head. As the shaft of the radius rotates, the distal epiphysis of the radius rolls across the anterior surface of the ulna. This movement, called **pronation**

Figure 9–4 **Rotational Movements.**

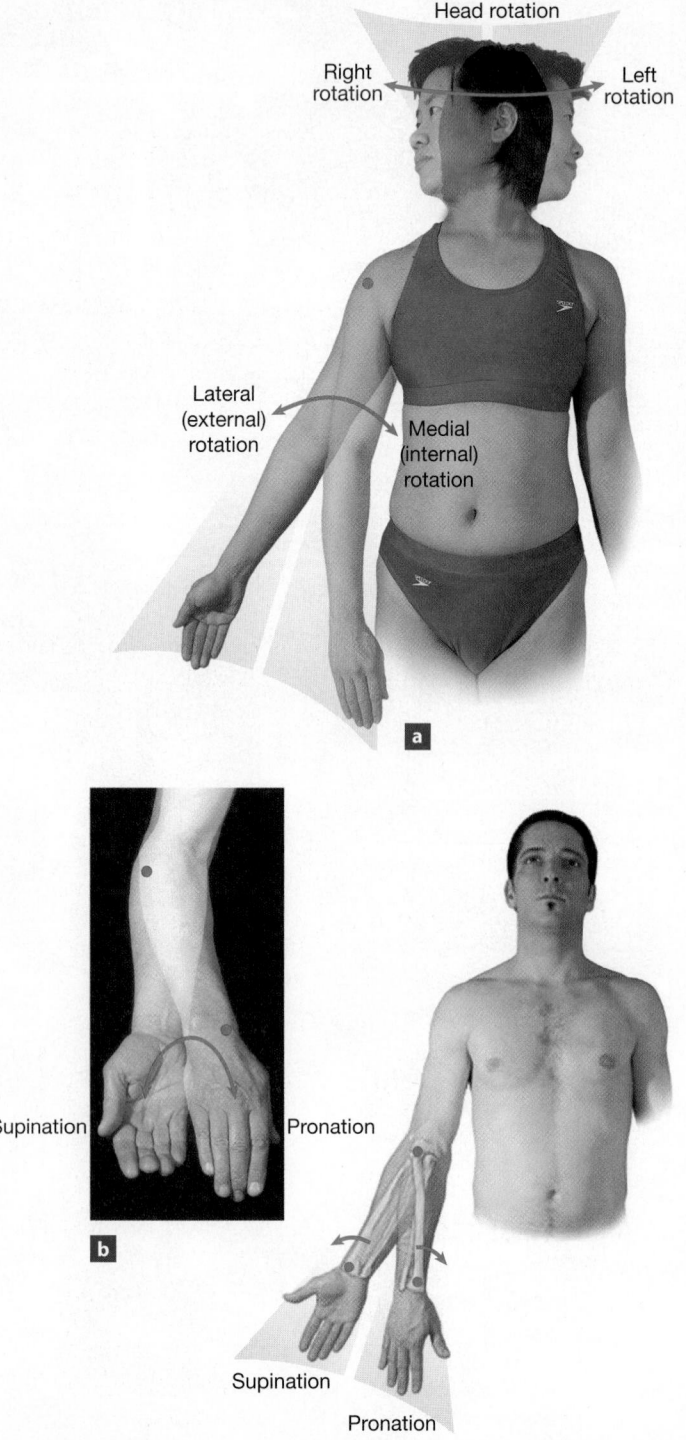

Head rotation

Right rotation

Left rotation

Lateral (external) rotation

Medial (internal) rotation

a

Supination

Pronation

b

Supination

Pronation

(prō-NĀ-shun), turns the wrist and hand from palm facing front to palm facing back (Figure 9–4b). The opposing movement, in which the palm is turned anteriorly, is **supination** (soo-pi-NĀ-shun). The forearm is supinated in the anatomical position. This view makes it easier to follow the path of the blood vessels, nerves, and tendons, which rotate with the radius during pronation.

TIPS & TRICKS

In order to carry a bowl of *soup*, the hand must be *sup*inated.

Special Movements

Several special terms apply to specific joints or unusual types of movement (Figure 9–5):

- **Inversion** (*in*, into + *vertere*, to turn) is a twisting movement of the foot that turns the sole inward, elevating the medial edge of the sole. The opposite movement is called **eversion** (ē-VER-zhun).

- **Dorsiflexion** is flexion at the ankle joint and elevation of the sole, as when you dig in your heel. **Plantar flexion** (*plantaris*, sole), the opposite movement, extends the ankle joint and elevates the heel, as when you stand on tiptoe. However, it is also acceptable (and simpler) to use "flexion and extension at the ankle," rather than "dorsiflexion and plantar flexion."

- **Opposition** is movement of the thumb toward the surface of the palm or the pads of other fingers. Opposition enables you to grasp and hold objects between your thumb and palm. It involves movement at the first carpometacarpal and metacarpophalangeal joints. Flexion at the fifth metacarpophalangeal joint can assist this movement. **Reposition** is the movement that returns the thumb and fingers from opposition.

- **Protraction** is moving a body part anteriorly in the horizontal plane. **Retraction** is the reverse movement. You protract your jaw when you jut your chin forward, and you retract your jaw when you return it to its normal position.

- **Elevation** and **depression** take place when a structure moves in a superior or an inferior direction, respectively. You depress your mandible when you open your mouth, and you elevate your mandible as you close your mouth. Another familiar elevation takes place when you shrug your shoulders.

- **Lateral flexion** occurs when your vertebral column bends to the side. This movement is most pronounced in the cervical and thoracic regions.

Figure 9–5 **Special Movements.**

Eversion   Inversion

Dorsiflexion (ankle flexion)

Plantar flexion (ankle extension)

Opposition

Retraction   Protraction

Depression

Elevation

Lateral flexion

7. Identify the types of synovial joints based on the shapes of the articulating surfaces.

8. When you do jumping jacks, which lower limb movements are necessary?

9. Which movements are associated with hinge joints?

See the blue Answers tab at the back of the book.

9-4 | Intervertebral discs and ligaments are structural components of intervertebral joints

Learning Outcome Describe the joints between the vertebrae of the vertebral column.

The articulations between the superior and inferior articular processes of adjacent vertebrae—called *intervertebral joints*—are gliding joints that permit small movements associated with flexion and rotation of the vertebral column (**Figure 9–6**). Little gliding occurs between adjacent vertebral bodies. From axis to sacrum,

the vertebrae are separated and cushioned by pads of fibrocartilage called **intervertebral discs**. Thus, the bodies of vertebrae form symphyseal joints.

Intervertebral discs and symphyseal joints are not found in the sacrum or coccyx, where vertebrae have fused, or between the first and second cervical vertebrae. Recall that the first cervical vertebra has no vertebral body and no intervertebral disc. The only joint between the first two cervical vertebrae is a pivot joint that allows much more rotation than the symphyseal joints between other cervical vertebrae would allow.

Intervertebral Discs

Each intervertebral disc has a tough outer layer of fibrocartilage, the **anulus fibrosus** (AN-ū-lus fī-BRŌ-sus). The collagen fibers of this layer attach the disc to the bodies of adjacent vertebrae. The anulus fibrosus surrounds the **nucleus pulposus** (pul-PŌ-sus), a soft, elastic, gelatinous core. The nucleus pulposus gives the disc resiliency and enables it to absorb shocks. The superior and inferior surfaces of the disc are almost completely covered by thin **vertebral end plates**. These end plates are composed of hyaline cartilage and fibrocartilage (**Figure 9–6**).

Figure 9–6 **Intervertebral joints.** ATLAS: Plates 20b; 23c

a Anterior view

b Lateral and sectional view

Movement of the vertebral column compresses the nucleus pulposus and displaces it in the opposite direction. This displacement permits smooth gliding movements between vertebrae while maintaining their alignment.

The discs make a significant contribution to an individual's height: They account for about one-quarter the length of the vertebral column superior to the sacrum. As we grow older, the water content of the nucleus pulposus in each disc decreases. The discs gradually become less effective as cushions, and the chances of vertebral injury increase. Water loss from the discs also causes the vertebral column to shorten, accounting for the characteristic decrease in height with advancing age.

Intervertebral Ligaments

Numerous ligaments are attached to the bodies and processes of all vertebrae, binding them together and stabilizing the vertebral column (**Figure 9–6**). Ligaments interconnecting adjacent vertebrae include the following:

- The *anterior longitudinal ligament*, which connects the anterior surfaces of adjacent vertebral bodies.

- The *posterior longitudinal ligament*, which parallels the anterior longitudinal ligament and connects the posterior surfaces of adjacent vertebral bodies.

- The *ligamentum flavum*, which connects the laminae of adjacent vertebrae.

- The *interspinous ligament*, which connects the spinous processes of adjacent vertebrae.

- The *supraspinous ligament*, which interconnects the tips of the spinous processes from C_7 to the sacrum. The *ligamentum nuchae*, which extends from vertebra C_7 to the base of the skull, is continuous with the supraspinous ligament. ⟳ p. 258

TIPS & TRICKS

The inter**spinous** ligament and supra**spinous** ligament get their names because they are attached to the **spinous** processes of the vertebrae.

Vertebral Movements

The following movements can take place across the intervertebral joints of the vertebral column: (1) flexion, or bending

---

Clinical Note

Damage to Intervertebral Discs If the posterior longitudinal ligaments are weakened, as often occurs as we age, intervertebral discs become more vulnerable to damage. The compressed nucleus pulposus of a disc may distort the anulus fibrosus, forcing it partway into the vertebral canal. This condition, seen here in lateral view, is called a **bulging disc** (**Figure a**). If the nucleus pulposus breaks through the anulus fibrosus, it too may protrude into the vertebral canal. This condition, shown here in superior view, is called a **herniated disc** (**Figure b**). When a disc herniates, it compresses spinal nerves and may cause pain.

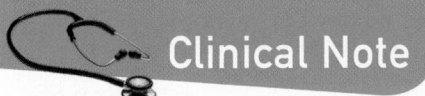

Normal intervertebral disc

Bulging disc

T₁₂

L₁

L₂

Compressed area of spinal nerve

Nucleus pulposus of herniated disc

Spinal nerve

Spinal cord

Anulus fibrosus

a A lateral view of the lumbar region of the spinal column, showing a bulging intervertebral disc

b A sectional view through a herniated disc, showing the release of the nucleus pulposus and its effect on the spinal cord and adjacent spinal nerves anteriorly; (2) extension, or bending posteriorly; (3) lateral flexion, or bending laterally; and (4) rotation. Table 9–2 summarizes information about intervertebral and other joints of the axial skeleton.

CHECKPOINT

10. What types of movements can occur across intervertebral joints?

11. Which regions of the vertebral column lack intervertebral discs? Explain why the absence of discs is significant.

12. Which vertebral movements are involved in (a) bending forward, (b) bending to the side, and (c) moving the head to signify "no"?

See the blue Answers tab at the back of the book.

9-5 ▮ The shoulder is a ball-and-socket joint, and the elbow is a hinge joint

Learning Outcome Describe the structure and function of the shoulder joint and the elbow joint.

In this section we consider the structure and function of two major joints in the upper limb: the shoulder joint and elbow joint.

The Shoulder Joint

The **shoulder joint**, or *glenohumeral joint*, permits the greatest range of motion of any joint. It is also the most frequently

---

Table 9–2	Joints of the Axial Skeleton			
Element	**Joint**	**Type of Joint**	**Movement**	
SKULL				
Cranial and facial bones of skull	Various	Synarthroses (suture or synostosis)	None	
Maxilla/teeth and mandible/teeth	Alveolar	Synarthrosis (gomphosis)	None	
Temporal bone/mandible	Temporomandibular	Combined gliding joint and hinge diarthrosis	Elevation, depression, and lateral gliding	
VERTEBRAL COLUMN				
Occipital bone/atlas	Atlantooccipital	Condylar diarthrosis	Flexion/extension	
Atlas/axis	Atlantoaxial	Pivot diarthrosis	Rotation	
Other vertebral elements	Intervertebral (between vertebral bodies)	Amphiarthrosis (symphysis)	Slight movement	
	Intervertebral (between articular processes)	Gliding diarthrosis	Slight rotation and flexion/extension	
L$_5$/sacrum	Between L$_5$ body and sacral body	Amphiarthrosis (symphysis)	Slight movement	
	Between inferior articular processes of L$_5$ and articular processes of sacrum	Gliding diarthrosis	Slight flexion/extension	
Sacrum/ hip bone	Sacroiliac	Gliding diarthrosis	Slight movement	
Sacrum/coccyx	Sacrococcygeal	Gliding diarthrosis *(may become fused)*	Slight movement	
Coccygeal bones		Synarthrosis (synostosis)	No movement	
THORACIC CAGE				
Bodies of T$_1$–T$_{12}$ and heads of ribs	Costovertebral	Gliding diarthrosis	Slight movement	
Transverse processes of T$_1$–T$_{10}$	Costovertebral	Gliding diarthrosis	Slight movement	
Ribs and costal cartilages		Synarthrosis (synchondrosis)	No movement	
Sternum and first costal cartilage	Sternocostal (1st)	Synarthrosis (synchondrosis)	No movement	
Sternum and costal cartilages 2–7	Sternocostal (2nd–7th)	Gliding diarthrosis*	Slight movement	

*Commonly converts to synchondrosis in elderly individuals.

Figure 9–7 **The Shoulder Joint.** ATLAS: Plate 27d

a Anterior view, frontal section

b Lateral view of pectoral girdle

dislocated joint, demonstrating the principle that stability must be sacrificed to obtain mobility. The shoulder is a ball-and-socket diarthrosis formed by the articulation of the head of the humerus with the glenoid cavity of the scapula (**Figure 9–7a**).

The area of the glenoid cavity is increased by a fibrocartilaginous **glenoid labrum** (*labrum*, lip or edge), which continues beyond the bony rim and deepens the socket (**Figure 9–7b**). The relatively loose articular capsule extends from the scapula, proximal to the glenoid labrum, to the anatomical neck of the humerus. Somewhat oversized, the articular capsule permits an extensive range of motion. The bones of the pectoral girdle provide some stability to the superior surface, because the acromion and coracoid process project laterally superior to the head of the humerus. However, the surrounding skeletal muscles provide most of the stability at this joint, with help from their associated tendons and various ligaments. Bursae reduce friction between the tendons and other tissues at the joint.

The major ligaments that help stabilize the shoulder joint are the *glenohumeral, coracohumeral, coracoacromial, coracoclavicular,* and *acromioclavicular ligaments*. The acromioclavicular ligament reinforces the capsule of the acromioclavicular joint and supports the superior surface of the shoulder. A **shoulder separation** is a relatively common injury involving partial or complete dislocation of the acromioclavicular joint. This injury can result from a hit to the superior surface of the shoulder. The acromion is forcibly depressed while the clavicle is held back by strong muscles. A dislocation of the shoulder joint is different, and involves the glenohumeral joint.

The muscles that move the humerus do more to stabilize the shoulder joint than do all the ligaments and capsular fibers combined. Muscles originating on the trunk, pectoral girdle, and humerus cover the anterior, superior, and posterior surfaces of the capsule. The tendons of the *supraspinatus, infraspinatus, teres minor,* and *subscapularis muscles* reinforce the joint capsule and limit range of movement. These muscles, known as the muscles of the *rotator cuff,* are the primary mechanism for supporting the shoulder joint and limiting its range of movement. Damage to the rotator cuff typically occurs in sports that place severe strains on the shoulder. White-water kayakers, baseball pitchers, and quarterbacks are all at high risk for rotator cuff injuries.

TIPS&TRICKS

The rotator cuff muscles can be remembered by using the acronym **SITS**, for **s**upraspinatus, **i**nfraspinatus, **t**eres minor, and **s**ubscapularis.

The anterior, superior, and posterior surfaces of the shoulder joint are reinforced by ligaments, muscles, and tendons, but the inferior capsule is poorly reinforced. As a result, a dislocation caused by an impact or a violent muscle contraction is most likely to occur at this site. Such a dislocation can tear the inferior capsular wall and the glenoid labrum. The healing process typically leaves a weakness that increases the chances for future dislocations.

As at other joints, bursae at the shoulder reduce friction where large muscles and tendons pass across the joint capsule. The shoulder has several important bursae, such as the *subacromial bursa*, the *subdeltoid bursa*, the *subcoracoid bursa*, and the *subscapular bursa* (Figure 9–7a,b). A tendon of the biceps brachii muscle runs through the shoulder joint. As it passes through the articular capsule, it is surrounded by a tubular bursa that is continuous with the joint cavity. Inflammation of any of these extracapsular bursae can restrict motion and produce the painful symptoms of bursitis. ⤺ p. 293

The Elbow Joint

The elbow joint is a complex hinge joint that involves the humerus, radius, and ulna (Figure 9–8). The largest and strongest articulation at the elbow is the *humeroulnar joint*, where the trochlea of the humerus articulates with the trochlear notch of the ulna. This joint works like a door hinge, with physical limitations imposed on the range of motion. In the case of the elbow, the shape of the trochlear notch of the ulna determines the plane of movement, and the combination of the notch and the olecranon limits the degree of extension permitted. At the smaller *humeroradial joint*, the capitulum of the humerus articulates with the head of the radius.

Muscles that extend the elbow attach to the rough surface of the olecranon. These muscles are mainly under the control of the *radial nerve*, which passes along the radial groove of the humerus. ⤺ p. 273

The large *biceps brachii muscle* covers the anterior surface of the arm. Its distal tendon is attached to the radius at the radial tuberosity. Contraction of this muscle produces supination of the forearm and flexion at the elbow.

The elbow joint is extremely stable because (1) the bony surfaces of the humerus and ulna interlock, (2) a single, thick articular capsule surrounds both the humeroulnar and proximal radioulnar joints, and (3) strong ligaments reinforce the articular capsule.

The *radial collateral ligament* stabilizes the lateral surface of the elbow joint (Figure 9–8a). It extends between the lateral epicondyle of the humerus and the *annular ligament*, which binds the head of the radius to the ulna. The *ulnar collateral ligament* stabilizes the medial surface of the elbow joint. This ligament extends from the medial epicondyle of the humerus anteriorly to the coronoid process of the ulna and posteriorly to the olecranon (Figure 9–8b).

Despite the strength of the capsule and ligaments, the elbow can be damaged by severe impacts or unusual stresses.

Figure 9–8 **The Right Elbow Joint Showing Stabilizing Ligaments.** ATLAS: Plates 35a–g

Humerus • Radial collateral ligament • Radial tuberosity • Antebrachial interosseous membrane • Radius • Ulna • Capitulum • Annular ligament (covering head and neck of radius)

a **Lateral view**

Tendon of biceps brachii muscle • Annular ligament • Humerus • Antebrachial interosseous membrane • Articular capsule • Medial epicondyle • Radius • Ulnar collateral ligament • Ulna • Olecranon of ulna

b **Medial view**

For example, if you fall on your hand with a partially flexed elbow, contractions of muscles that extend the elbow may break the ulna at the center of the trochlear notch. Less violent stresses can produce dislocations or other injuries to the elbow, especially if epiphyseal growth has not been completed. For example, parents in a rush may drag a toddler along behind them, exerting an upward, twisting pull on the elbow joint that can result in a partial dislocation of the radial head from the annular ligament known as *nursemaid's elbow*.

CHECKPOINT

13. Which tissues or structures provide most of the stability for the shoulder joint?

14. Would a tennis player or a jogger be more likely to develop inflammation of the subscapular bursa? Why?

15. A football player received a blow to the upper surface of his shoulder, causing a shoulder separation. What does this mean?

16. Terry suffers an injury to his forearm and elbow. After the injury, he notices an unusually large degree of motion between the radius and the ulna at the elbow. Which ligament did Terry most likely damage?

See the blue Answers tab at the back of the book.

9-6 | The hip is a ball-and-socket joint, and the knee is a hinge joint

Learning Outcome Describe the structure and function of the hip joint and the knee joint.

Next we examine two representative articulations in the lower limb: the hip joint and the knee joint.

The Hip Joint

The hip joint is a sturdy ball-and-socket diarthrosis that permits flexion, extension, adduction, abduction, circumduction, and rotation. **Figure 9–9** introduces the structure of the hip joint.

Figure 9–9   The Right Hip Joint. ATLAS: Plates 71a,b; 72a

Iliofemoral ligament

Articular cartilage

Acetabular labrum

Ligament of the femoral head

Transverse acetabular ligament (spanning acetabular notch)

Acetabulum

Fat pad in acetabular fossa

a A lateral view with the femur removed

Pubofemoral ligament

Greater trochanter

Iliofemoral ligament

Lesser trochanter

b An anterior view

Iliofemoral ligament

Ischiofemoral ligament

Greater trochanter

Lesser trochanter

Ischial tuberosity

c A posterior view, showing additional ligaments that add strength to the capsule

9

The acetabulum, a deep fossa, accommodates the head of the femur. ⤴ p. 276 Within the acetabulum, articular cartilage extends like a horseshoe to either side of the acetabular notch (**Figure 9–9a**). The *acetabular labrum*, a projecting rim of rubbery fibrocartilage, increases the depth of the joint cavity and helps to seal in synovial fluid.

The articular capsule of the hip joint is extremely dense and strong. It extends from the lateral and inferior surfaces of the pelvic girdle to the intertrochanteric line and intertrochanteric crest of the femur, enclosing both the head and neck of the femur. ⤴ p. 280 This arrangement helps to keep the femoral head from moving too far from the acetabulum.

Four broad ligaments reinforce the articular capsule (**Figure 9–9**). Three of them—the *iliofemoral*, *pubofemoral*, and *ischiofemoral ligaments*—are regional thickenings of the capsule. The *transverse acetabular ligament* crosses the acetabular notch, filling in the gap in the inferior border of the acetabulum. A fifth ligament, the *ligament of the femoral head*, or *ligamentum teres* (*teres*, long and round), originates along the transverse acetabular ligament (**Figure 9–9a**) and attaches to the fovea capitis, a small pit at the center of the femoral head. ⤴ p. 280 This ligament tenses only when the hip is flexed and the thigh is undergoing lateral rotation.

Much more important stabilization is provided by the bulk of the surrounding muscles, aided by ligaments and capsular fibers. The combination of an almost complete bony socket, a strong articular capsule, supporting ligaments, and muscular padding makes the hip joint extremely stable. The head of the femur is well supported, but the ball-and-socket joint is not directly aligned with the weight distribution along the shaft. Stress must be transferred at an angle from the joint, along the thin femoral neck to the length of the femur. ⤴ p. 213 Hip fractures (fractures of the femoral neck or between the greater and lesser trochanters of the femur) are more common than hip dislocations. As we noted in Chapter 6, femoral fractures at the hip are common in elderly individuals with severe osteoporosis. ⤴ p. 225

The Knee Joint

The hip joint passes weight to the femur, and then the knee joint transfers the weight from the femur to the tibia. The shoulder is mobile; the hip, stable; and the knee . . . ? If you had to choose one word, it would probably be "complicated." The knee joint functions as a hinge, but the joint is far more complex than the elbow or even the ankle. The rounded condyles of the femur roll across the superior surface of the tibia, so the points of contact are constantly changing. The joint permits flexion, extension, and very limited rotation.

The knee joint contains three separate articulations. Two are between the femur and tibia (medial condyle to medial condyle, and lateral condyle to lateral condyle). One is between the patella and the patellar surface of the femur (**Figure 9–10**).

The Articular Capsule and Joint Cavity

The articular capsule at the knee joint is thin and in some areas incomplete, but various ligaments and tendons of associated muscles strengthen it. A pair of fibrocartilage pads, the medial and lateral menisci, lies between the femoral and tibial surfaces (**Figures 9–1b and 9–10b,d**). The menisci (1) act as cushions, (2) conform to the shape of the articulating surfaces as the femur changes position, and (3) provide lateral stability to the joint. Prominent fat pads cushion the margins of the joint. These fat pads also assist the many bursae in reducing friction between the patella and other tissues.

Supporting Ligaments

A complete dislocation of the knee is very rare, mainly because seven major ligaments stabilize the knee joint:

1. The tendon from the *quadriceps femoris* muscles responsible for extending the knee passes over the anterior surface of the joint. The patella is embedded in this tendon, and the *patellar ligament* continues to its attachment on the anterior surface of the tibia (**Figure 9–10a**). The patellar ligament and two ligamentous bands known as the *patellar retinaculae* support the anterior surface of the knee joint.

2,3. Two *popliteal ligaments* extend between the femur and the heads of the tibia and fibula (**Figure 9–10c**). These ligaments reinforce the knee joint's posterior surface.

4,5. Inside the joint capsule, the *anterior cruciate ligament (ACL)* and *posterior cruciate ligament (PCL)* attach the intercondylar area of the tibia to the condyles of the femur (**Figure 9–10b,d**). *Anterior* and *posterior* refer to the sites of origin of these ligaments on the tibia. They cross one another as they proceed to their destinations on the femur. (The term *cruciate* is derived from the Latin word *crucialis*, meaning a cross.) The ACL and the PCL limit the anterior and posterior movement of the tibia and maintain the alignment of the femoral and tibial condyles.

6,7. The *tibial collateral ligament*, also called the *medial collateral ligament (MCL)*, reinforces the medial surface of the knee joint, and the *fibular collateral ligament*, also called the *lateral collateral ligament (LCL)*, reinforces the lateral surface (**Figure 9–10**). These ligaments tighten only at full extension, the position in which they stabilize the joint.

At full extension, a slight lateral rotation of the tibia tightens the anterior cruciate ligament and jams the lateral meniscus between the tibia and femur. The knee joint is essentially locked in the extended position. With the joint locked, a person can stand for prolonged periods without using (and tiring) the muscles that extend the knee. Unlocking the knee joint requires muscular contractions that medially rotate the tibia or laterally rotate the femur. If the locked knee is struck from the side, the

Figure 9–10 **The Right Knee Joint.** ATLAS: Plates 78a–i; 79a,b; 80a,b

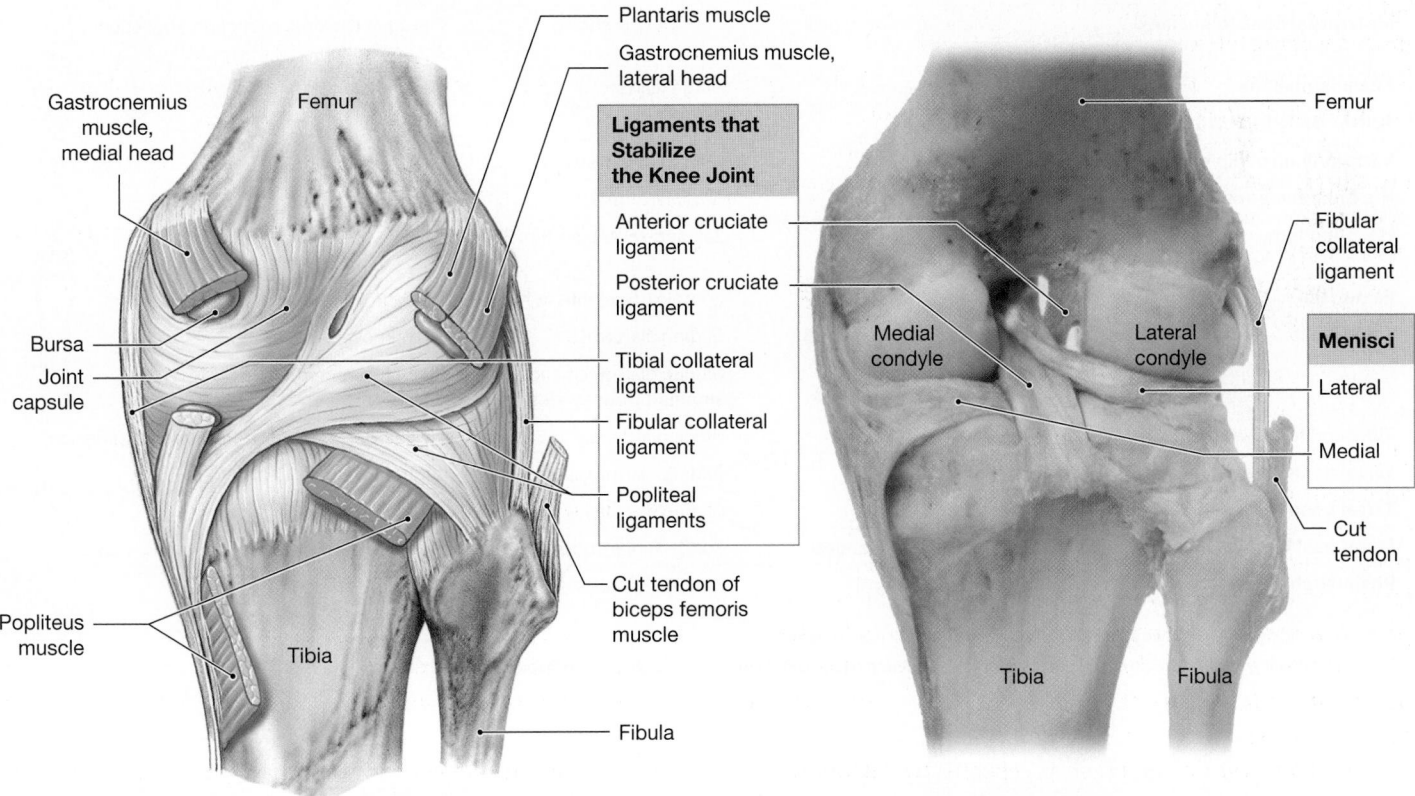

a **Anterior view, superficial layer**

Quadriceps tendon

Joint capsule

Patellar retinaculae

Patella

Patellar ligament

Fibula

Tibia

b **Deep anterior view, flexed**

Patellar surface

Ligaments that Stabilize the Knee Joint

Posterior cruciate ligament

Anterior cruciate ligament

Tibial collateral ligament

Fibular collateral ligament

Cut tendon of biceps femoris muscle

Lateral condyle

Medial condyle

Tibia

Fibula

Menisci

Medial

Lateral

c **Posterior view, superficial layer**

Gastrocnemius muscle, medial head

Femur

Plantaris muscle

Gastrocnemius muscle, lateral head

Ligaments that Stabilize the Knee Joint

Bursa

Joint capsule

Popliteus muscle

Tibia

Anterior cruciate ligament

Posterior cruciate ligament

Tibial collateral ligament

Fibular collateral ligament

Popliteal ligaments

Cut tendon of biceps femoris muscle

Fibula

d **Deep posterior view, extended**

Femur

Fibular collateral ligament

Medial condyle

Lateral condyle

Menisci

Lateral

Medial

Cut tendon

Tibia

Fibula

9

lateral meniscus can tear and the supporting ligaments can be seriously damaged.

The knee joint is structurally complex and is subjected to severe stresses in the course of normal activities. Painful knee injuries are all too familiar to both amateur and professional athletes. Treatment is often costly and prolonged, and repairs seldom make the joint "good as new."

Table 9–3 summarizes information about the joints of the appendicular skeleton.

Table 9–3	Joints of the Appendicular Skeleton			
Articulating Bones	**Joint**	**Type of Joint**	**Movement**	
JOINTS OF EACH PECTORAL GIRDLE AND UPPER LIMB				
Sternum/clavicle	Sternoclavicular	Gliding diarthrosis*	Protraction/retraction, elevation/depression, slight rotation	
Scapula/clavicle	Acromioclavicular	Gliding diarthrosis	Slight movement	
Scapula/humerus	Shoulder, or glenohumeral	Ball-and-socket diarthrosis	Flexion/extension, adduction/abduction, circumduction, rotation	
Humerus/ulna and humerus/radius	Elbow (humeroulnar and humeroradial)	Hinge diarthrosis	Flexion/extension	
Radius/ulna	Proximal radioulnar	Pivot diarthrosis	Rotation	
	Distal radioulnar	Pivot diarthrosis	Pronation/supination	
Radius/carpal bones	Radiocarpal	Condylar diarthrosis	Flexion/extension, adduction/abduction, circumduction	
Carpal bone to carpal bone	Intercarpal	Gliding diarthrosis	Slight movement	
Carpal bone to metacarpal bone (I)	Carpometacarpal of thumb	Saddle diarthrosis	Flexion/extension, adduction/abduction, circumduction, opposition	
Carpal bone to metacarpal bone (II–V)	Carpometacarpal	Gliding diarthrosis	Slight flexion/extension, adduction/abduction	
Metacarpal bone to phalanx	Metacarpophalangeal	Condylar diarthrosis	Flexion/extension, adduction/abduction, circumduction	
Phalanx/phalanx	Interphalangeal	Hinge diarthrosis	Flexion/extension	
JOINTS OF THE PELVIC GIRDLE AND LOWER LIMBS				
Sacrum/ilium of hip bone	Sacroiliac	Gliding diarthrosis	Slight movement	
Hip bone/hip bone	Pubic symphysis	Amphiarthrosis	None†	
Hip bone/femur	Hip	Ball-and-socket diarthrosis	Flexion/extension, adduction/abduction, circumduction, rotation	
Femur/tibia	Knee	Complex, functions as hinge	Flexion/extension, limited rotation	
Tibia/fibula	Tibiofibular (proximal)	Gliding diarthrosis	Slight movement	
	Tibiofibular (distal)	Gliding diarthrosis and amphiarthrotic syndesmosis	Slight movement	
Tibia and fibula with talus	Ankle, or talocrural	Hinge diarthrosis	Flexion/extension (dorsiflexion/plantar flexion)	
Tarsal bone to tarsal bone	Intertarsal	Gliding diarthrosis	Slight movement	
Tarsal bone to metatarsal bone	Tarsometatarsal	Gliding diarthrosis	Slight movement	
Metatarsal bone to phalanx	Metatarsophalangeal	Condylar diarthrosis	Flexion/extension, adduction/abduction	
Phalanx/phalanx	Interphalangeal	Hinge diarthrosis	Flexion/extension	

*A "double-gliding joint," with two joint cavities separated by an articular cartilage.

†During pregnancy, hormones weaken the pubic symphysis and permit movement important to childbirth; see Chapter 29.

Clinical Note

Knee Injuries Athletes place tremendous stresses on their knees. Ordinarily, the medial and lateral menisci move as the position of the femur changes. The knee is most stable when it is not bent. Placing a lot of weight on the knee while it is partially flexed can trap a meniscus between the tibia and femur, resulting in a break or tear in the cartilage. In the most common injury, the lateral surface of the leg is driven medially, tearing the medial meniscus. In addition to being quite painful, the torn cartilage may restrict movement at the joint. It can also lead to chronic problems and the development of a "trick knee"—a knee that feels unstable. Sometimes the meniscus can be heard and felt popping in and out of position when the knee is extended.

Other knee injuries involve tearing one or more stabilizing ligaments or damaging the patella. Torn ligaments can be difficult to correct surgically, and healing is slow. Rupture of the anterior cruciate ligament (ACL) is a common sports injury that affects women two to five times more often than men. ACL injuries are frequently caused by twisting an extended weight-bearing knee. This movement is common in sports such as football, basketball, and skiing. Nonsurgical treatment with exercise and braces is possible, but requires a change in activity patterns. Reconstructive surgery using part of the patellar tendon or a graft from a cadaver tendon may allow a return to active sports. A strong hit or force to the lateral surface of the knee, as in a football tackle or flying hockey puck, may damage both the ACL and the medial collateral ligament (MCL).

The patella can be injured in a number of ways. If the leg is immobilized (as it might be in a football pileup) while you try to extend the knee, the muscles are powerful enough to pull the patella apart. Impacts to the anterior surface of the knee can also shatter the patella. Knee injuries may lead to chronic painful arthritis that impairs walking. Total knee replacement surgery is rarely performed on young people, but it is becoming increasingly common among elderly patients with severe arthritis.

CHECKPOINT

17. Name the bones making up the shoulder joint and knee joint, respectively.

18. At what site are the iliofemoral ligament, pubofemoral ligament, and ischiofemoral ligament located?

19. What signs and symptoms would you expect in a person who has damaged the menisci of the knee joint?

20. Why is "clergyman's knee" (a type of bursitis) common among carpet layers and roofers?

See the blue Answers tab at the back of the book.

9-7 | With advancing age, arthritis and other degenerative changes impair joint mobility

Learning Outcome Describe the effects of aging on joints, and discuss the most common age-related clinical problems for joints.

Joints are subjected to heavy wear and tear throughout our lifetimes. Problems with joint function are common, especially in older individuals. **Rheumatism** (ROO-muh-tiz-um) is a general term for pain and stiffness affecting the *musculoskeletal system* (bones, muscles, and joints). **Arthritis** (ar-THRĪ-tis) means *joint inflammation*, and encompasses all the rheumatic diseases that affect synovial joints. Many types of arthritis are recognized.

Arthritis always involves damage to the articular cartilages, but the specific cause can vary. For example, arthritis can result from bacterial or viral infection, injury to the joint, decreased viscosity or lack of synovial fluid, metabolic problems, or severe physical stresses. The most common type of arthritis is *osteoarthritis* (os-tē-ō-ar-THRĪ-tis), also known as *degenerative arthritis* or *degenerative joint disease (DJD)*. It generally affects people age 60 or older. Osteoarthritis can result from cumulative wear and tear at the joint surfaces or from genetic factors affecting collagen formation. In the U.S. population, 25 percent of women and 15 percent of men over age 60 show signs of this disease.

Rheumatoid arthritis is an inflammatory condition that affects about 0.5–1.0 percent of the adult population, generally those between 40 and 60, and women more than men. Rheumatoid arthritis occurs when the immune response mistakenly attacks the joint tissues. Such a condition, in which the body attacks its own tissues, is called an *autoimmune disease*. Allergies, bacteria, viruses, and genetic factors have all been proposed as contributing to or triggering the destructive inflammation.

In *gouty arthritis*, or *crystal arthritis*, crystals of uric acid form within the synovial fluid of joints. The accumulation of

9

crystals of uric acid over time eventually interferes with normal movement. This form of arthritis is named after the metabolic disorder known as *gout*, discussed further in Chapter 25. In gout, the crystals are derived from uric acid (a metabolic waste product), and the joint most often affected is the metatarsal–phalangeal joint of the great toe. Gout is relatively rare, but other forms of gouty arthritis are much more common. Some degree of calcium salt deposition occurs in the joints in 30–60 percent of those over age 85. The cause is unknown, but the condition appears to be linked to age-related changes in the articular cartilages.

Regular exercise, physical therapy, and drugs that reduce inflammation (such as aspirin) can often slow the progress of osteoarthritis. Surgical procedures can realign or redesign the affected joint. In extreme cases involving the hip, knee, elbow, or shoulder, the defective joint can be replaced by an artificial one.

Degenerative changes comparable to those seen in arthritis may result from joint immobilization. When motion ceases, so does the circulation of synovial fluid, and the cartilages begin to degenerate. **Continuous passive motion (CPM)** of any injured joint appears to encourage the repair process by improving the circulation of synovial fluid. A physical therapist or a machine often performs the movement during the recovery process.

With age, bone mass decreases and bones become weaker, so the risk of fractures increases. ⟲ p. 225 If osteoporosis develops, the bones may weaken to the point at which fractures occur in response to stresses that could easily be tolerated by normal bones. Hip fractures are among the most dangerous fractures seen in elderly people, with or without osteoporosis. These fractures, most often involving persons over age 60, may be accompanied by hip dislocation or by pelvic fractures. Although severe hip fractures are most common among those over age 60, in recent years the frequency of hip fractures has increased dramatically among young, healthy professional athletes.

CHECKPOINT

21. Define rheumatism.
22. Define arthritis.

See the blue Answers tab at the back of the book.

9-8 ┃ The skeletal system supports and stores energy and minerals for other body systems

Learning Outcome Explain the functional relationships between the skeletal system and other body systems.

The bones you study in the lab may seem to be rigid and unchanging structures, but the living skeleton is dynamic and undergoes continuous remodeling. The balance between osteoblast and osteoclast activity is subject to change at a moment's notice. When osteoblast activity predominates, bones thicken and strengthen. When osteoclast activity predominates, bones get thinner and weaker.

The balance between bone formation and bone recycling varies with (1) the age of the individual, (2) the physical stresses applied to the bone, (3) circulating hormone levels, (4) rates of calcium and phosphorus absorption and excretion, and (5) genetic or environmental factors. Most of these variables involve some interaction between the skeletal system and other body systems.

In fact, the skeletal system is intimately associated with other systems. For instance, the bones of the skeleton are attached to the muscular system, extensively connected with the cardiovascular and lymphatic systems, and largely under the physiological control of the endocrine system. The digestive and urinary systems also play important roles in providing the calcium and phosphate minerals needed for bone growth. In return, the skeleton represents a reserve of calcium, phosphate, and other minerals that can compensate for reductions in their supply from the diet. Figure 9–11 reviews the components and functions of the skeletal system, and diagrams the major functional relationships between that system and other systems studied so far.

CHECKPOINT

23. Explain why there must be a balance between osteoclast activity and osteoblast activity.
24. Describe the functional relationship between the skeletal system and the integumentary system.

See the blue Answers tab at the back of the book.

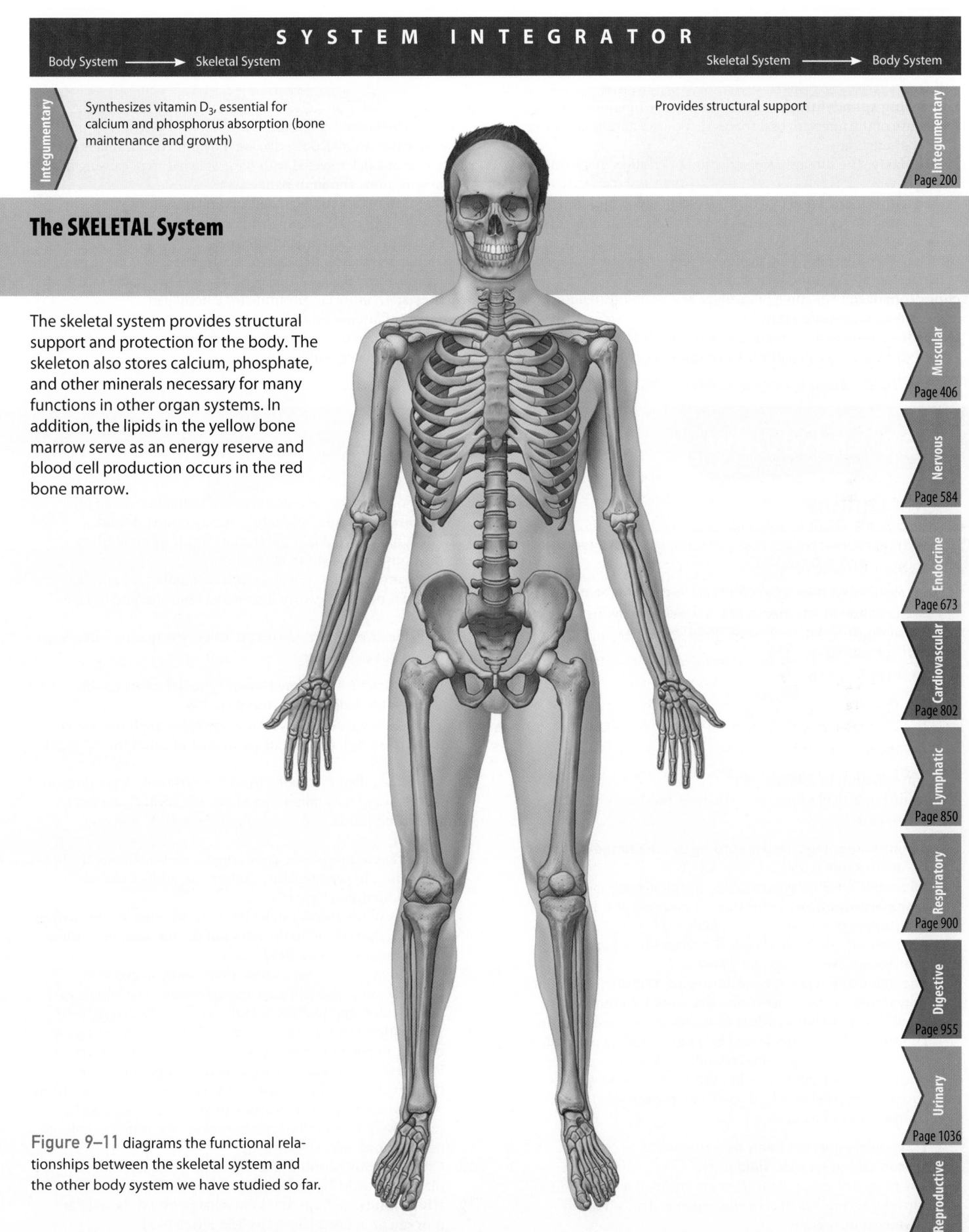

S Y S T E M   I N T E G R A T O R

Body System ⟶ Skeletal System   Skeletal System ⟶ Body System

Integumentary — Synthesizes vitamin D$_3$, essential for calcium and phosphorus absorption (bone maintenance and growth)

Provides structural support — **Integumentary**
Page 200

The SKELETAL System

The skeletal system provides structural support and protection for the body. The skeleton also stores calcium, phosphate, and other minerals necessary for many functions in other organ systems. In addition, the lipids in the yellow bone marrow serve as an energy reserve and blood cell production occurs in the red bone marrow.

Muscular
Page 406

Nervous
Page 584

Endocrine
Page 673

Cardiovascular
Page 802

Lymphatic
Page 850

Respiratory
Page 900

Digestive
Page 955

Urinary
Page 1036

Reproductive
Page 1116

Figure 9–11 diagrams the functional relationships between the skeletal system and the other body system we have studied so far.

Related Clinical Terms

ankylosing spondylitis: A chronic, progressive inflammatory disease of the intervertebral spaces that causes abnormal fusion of the vertebrae.

arthroplasty: The surgical reconstruction or creation of an artificial joint.

arthroscopy: Insertion of a narrow tube containing optical fibers and a tiny camera (arthroscope) directly into the joint for visual examination.

Bouchard's nodes: Bony enlargements on the proximal interphalangeal joints due to osteoarthritis.

chondromalacia: Softening of cartilage as a result of strenuous activity or an overuse injury.

Heberden's nodes: Bony overgrowths on the distal interphalangeal joints due to osteoarthritis that cause the patient to have knobby fingers.

joint mice: Small fibrous, cartilaginous, or bony loose bodies in the synovial cavity of a joint.

Lyme disease: An infectious disease transmitted to humans from the bite of a tick infected with *Borrelia burgdorferi,* causing flu-like symptoms and joint pain.

pannus: Granulation tissue (combination of fibrous connective tissue and capillaries) forming within a synovial membrane, which releases cartilage-destroying enzymes.

prepatellar bursitis: Inflammation of the bursa over the front of the knee just above the kneecap; also known as housemaid's knee.

prosthesis: An artificial substitute for a body part.

synovitis: Inflammation of the synovial membrane.

tophi: Deposits of uric acid crystals often found around joints and usually associated with gout.

Chapter Review

Study Outline

MasteringA&P®

Access more chapter study tools online in the MasteringA&P Study Area:

- Chapter Quizzes, Chapter Practice Test, Art-labeling Activities, Animations, MP3 Tutor Sessions, and Clinical Case Studies
- Practice Anatomy Lab   **PAL**
- Interactive Physiology   **iP**
- A&P Flix   **A&PFlix**
- PhysioEx   **PhysioEx**

▌ An Introduction to Joints p. 290

1. **Joints,** or **articulations,** exist wherever two bones interconnect.

9-1 ▌ Joints are categorized according to their range of motion or structure p. 290

2. *Immovable joints* are **synarthroses;** *slightly movable joints* are **amphiarthroses;** and joints that are *freely movable* are called **diarthroses** or synovial joints. *(Table 9–1)*

3. Alternatively, joints are classified structurally as *fibrous, cartilaginous, bony,* or *synovial. (Table 9–1)*

4. The four major types of synarthroses are a **suture** (skull bones bound together by dense connective tissue), a **gomphosis** (teeth bound to bony sockets by *periodontal ligaments*), a **synchondrosis** (two bones joined by a rigid cartilaginous bridge), and a **synostosis** (two bones completely fused).

5. The two major types of amphiarthroses are a **syndesmosis** (bones connected by a ligament) and a **symphysis** (bones connected by fibrocartilage).

9-2 ▌ Synovial joints are freely movable joints containing synovial fluid p. 292

6. The bony surfaces at diarthroses are enclosed within a **joint capsule,** also called an **articular capsule,** that is lined by a synovial membrane.

7. The bony surfaces within a synovial joint are covered by **articular cartilages,** and lubricated by **synovial fluid.** Functions of synovial fluid include lubrication, nutrient distribution, and shock absorption.

8. Accessory synovial structures include **menisci,** or *articular discs;* **fat pads; accessory ligaments; tendons;** and **bursae.** *(Figure 9–1)*

9. A **dislocation (luxation)** occurs when articulating surfaces are forced out of position.

9-3 ▌ The structure and function of synovial joints enable various skeletal movements p. 294

10. The possible types of articular movements are **linear movement (gliding), angular movement,** and **rotation.** *(Spotlight Figure 9–2)*

11. Joints are called **monaxial, biaxial,** or **triaxial,** depending on the planes of movement they allow. *(Spotlight Figure 9–2)*

12. In **gliding joints,** two opposing surfaces slide past one another.

13. Important terms that describe angular movement are **flexion, extension, hyperextension, abduction, adduction,** and **circumduction.** *(Figure 9–3)*

14. Rotational movement can be **left** or **right, medial** *(internal)* or **lateral** *(external),* or, in the bones of the forearm, **pronation** or **supination.** *(Figure 9–4)*

15. Movements of the foot include **inversion** and **eversion.** The ankle undergoes flexion and extension, also known as **dorsiflexion** and **plantar flexion,** respectively. *(Figure 9–5)*

16. **Opposition** is the thumb movement that enables us to grasp objects. **Reposition** is the opposite of opposition. *(Figure 9–5)*

17. **Protraction** involves moving something anteriorly; **retraction** involves moving it posteriorly. **Depression** and **elevation** occur when we move a structure inferiorly and superiorly, respectively. Lateral flexion occurs when the vertebral column bends to one side. *(Figure 9–5)*

18. **Gliding joints** permit limited movement, generally in a single plane. *(Spotlight Figure 9–2)*

19. **Hinge joints** are monaxial joints that permit only angular movement in one plane. *(Spotlight Figure 9–2)*

20. **Pivot joints** are monaxial joints that permit only rotation. *(Spotlight Figure 9–2)*

21. **Condylar joints** (ellipsoid joints) are biaxial joints with an oval articular face that nestles within a depression in the opposing articular surface. *(Spotlight Figure 9–2)*

22. **Saddle joints** are biaxial joints with articular faces that are concave on one axis and convex on the other. *(Spotlight Figure 9–2)*

23. **Ball-and-socket joints** are triaxial joints that permit rotation as well as other movements. *(Spotlight Figure 9–2)*

9-4 ▌ Intervertebral discs and ligaments are structural components of intervertebral joints p. 300

24. The articular processes of vertebrae form gliding joints with those of adjacent vertebrae. The bodies form symphyseal joints that are separated and cushioned by **intervertebral discs,** which contain an outer **anulus fibrosus** and an inner **nucleus pulposus. Vertebral end plates** cover the superior and inferior surfaces of the disc. Several ligaments stabilize the vertebral column. *(Figure 9–6; Table 9–2)*

9-5 ▌ The shoulder is a ball-and-socket joint, and the elbow is a hinge joint p. 302

25. The **shoulder joint,** or *glenohumeral joint,* is formed by the glenoid cavity and the head of the humerus. This articulation permits the greatest range of motion of any joint. It is a ball-and-socket diarthrosis with various stabilizing ligaments. Strength and stability are sacrificed in favor of mobility. *(Figure 9–7; Table 9–3)*

26. The **elbow joint** permits only flexion–extension. It is a hinge diarthrosis whose capsule is reinforced by strong ligaments. *(Figure 9–8; Table 9–3)*

9-6 ▌ The hip is a ball-and-socket joint, and the knee is a hinge joint p. 305

27. The **hip joint** is a ball-and-socket diarthrosis formed between the acetabulum of the hip bone and the head of the femur. The joint permits flexion–extension, adduction–abduction, circumduction, and rotation. Numerous ligaments stabilize the joint. *(Figure 9–9; Table 9–3)*

28. The **knee joint** is a hinge joint made up of three articulations: two formed between the femur and tibia and one between the patella and femur. The joint permits flexion–extension and limited rotation, and it has various supporting ligaments. *(Figure 9–10; Table 9–3)*

9-7 ▌ With advancing age, arthritis and other degenerative changes impair joint mobility p. 309

29. Problems with joint function are relatively common, especially in older individuals. **Rheumatism** is a general term for pain and stiffness affecting the musculoskeletal system; several major forms exist. **Arthritis** encompasses all the rheumatic diseases that affect synovial joints. Arthritis becomes increasingly common with age.

9-8 ▌ The skeletal system supports and stores energy and minerals for other body systems p. 310

30. Growth and maintenance of the skeletal system are supported by the integumentary system. The skeletal system also interacts with the muscular, cardiovascular, lymphatic, digestive, urinary, and endocrine systems. *(Figure 9–11)*

Review Questions

See the blue Answers tab at the back of the book.

LEVEL 1 Reviewing Facts and Terms

1. Label the structures in the following illustration of a synovial joint.

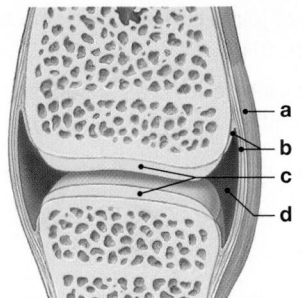

   (a) _____   (b) _____
   (c) _____   (d) _____

2. A synarthrosis located between the bones of the skull is a
   (a) symphysis, **(b)** syndesmosis, **(c)** synchondrosis, **(d)** suture.

3. The joint between adjacent vertebral bodies is a **(a)** syndesmosis, **(b)** symphysis, **(c)** synchondrosis, **(d)** synostosis.

4. The anterior joint between the two pubic bones is a
   (a) synchondrosis, **(b)** synostosis, **(c)** symphysis, **(d)** synarthrosis.

5. Joints typically located between the ends of adjacent long bones are
   (a) synarthroses, **(b)** amphiarthroses, **(c)** diarthroses, **(d)** symphyses.

6. The outer membrane of a bone containing the nerve supply and blood supply to this tissue is called the _____. **(a)** endosteum, **(b)** periosteum, **(c)** perimysium, **(d)** diaphysis.

7. Which of the following is *not* a function of synovial fluid? **(a)** shock absorption, **(b)** nutrient distribution, **(c)** maintenance of ionic balance, **(d)** lubrication of the articular surfaces, **(e)** waste disposal.

8. The structures that limit the range of motion of a joint and provide mechanical support across or around the joint are **(a)** bursae, **(b)** tendons, **(c)** menisci, **(d)** a, b, and c.

9. A partial dislocation of an articulating surface is a **(a)** circumduction, **(b)** hyperextension, **(c)** subluxation, **(d)** supination.

10. _____ leads to the formation of uric acid crystals in the synovial joint. **(a)** Osteoarthritis, **(b)** Rheumatoid arthritis, **(c)** Gouty arthritis, **(d)** Osteoporosis.

11. Rotation of the forearm that makes the palm face posteriorly is **(a)** supination, **(b)** pronation, **(c)** proliferation, **(d)** projection.

12. A saddle joint permits _____ movement but prevents _____ movement. **(a)** rotational; gliding, **(b)** angular; gliding, **(c)** gliding; rotational, **(d)** angular; rotational.

13. Standing on tiptoe is an example of _____ at the ankle. **(a)** elevation, **(b)** flexion, **(c)** extension, **(d)** retraction.

14. Examples of monaxial joints, which permit angular movement in a single plane, are **(a)** the intercarpal and intertarsal joints, **(b)** the shoulder and hip joints, **(c)** the elbow and knee joints, **(d)** all of these.

15. Decreasing the angle between bones is termed **(a)** flexion, **(b)** extension, **(c)** abduction, **(d)** adduction, **(e)** hyperextension.

16. List four ways in which the skeletal system supports other body systems.

17. What is the function of the nucleus pulposus in the intervertebral disc? **(a)** allows attachment to adjacent vertebrae, **(b)** limits rotation, **(c)** shock absorption, **(d)** limits flexion.

18. List three functions of synovial fluid within a synovial joint.

19. List the accessory structures of the synovial joint and describe their functions.

LEVEL 2 Reviewing Concepts

20. Dislocations involving synovial joints are usually prevented by all of the following *except* **(a)** structures such as ligaments that stabilize and support the joint, **(b)** the position of bursae that limits the degree of movement, **(c)** the presence of other bones that prevent certain movements, **(d)** the position of muscles and fat pads that limits the degree of movement, **(e)** the shape of the articular surface.

21. Flexion, extension, abduction, and adduction are all examples of **(a)** linear motion, **(b)** angular motion, **(c)** rotation, **(d)** circumduction.

22. How does a meniscus (articular disc) function in a joint?

23. The elbow joint is an example of which of the following types of articulation? **(a)** monoaxial, **(b)** biaxial, **(c)** triaxial, **(d)** multiaxial.

24. How do articular cartilages differ from other cartilages in the body?

25. Differentiate between osteoarthritis and rheumatoid arthritis.

26. How would you explain to your grandmother the characteristic decrease in height with advancing age?

27. Describe the skeleton's functional role in **(a)** the respiratory system, and **(b)** the integumentary system.

LEVEL 3 Critical Thinking and Clinical Applications

28. While playing tennis, Dave "turns his ankle." He experiences swelling and pain. After being examined, he is told that he has no ruptured ligaments and that the structure of the ankle is not affected. On the basis of the signs and symptoms and the examination results, what happened to Dave's ankle?

29. Articular cartilage is avascular (no blood supply). Explain how articular cartilage is normally supplied with nutrients? Why may joint movements be encouraged during the rehabilitation of patients with osteoarthritis.

30. When playing a contact sport, which injury would you expect to occur more frequently, a dislocated shoulder or a dislocated hip? Why?

CLINICAL CASE   Wrap-Up

What's Ailing the Birthday Girl?

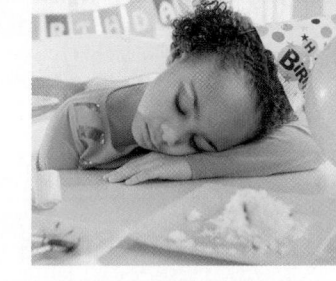

Jessica's mother takes her to see her pediatrician. After hearing about Jessica's symptoms, the pediatrician refers her to a rheumatologist, a doctor who specializes in rheumatic diseases.

This doctor withdraws some synovial fluid using a small needle (a procedure known as *arthrocentesis*) from a warm, swollen, tender knee that has limited motion. Instead of a small amount of clear, viscous fluid with the consistency of heavy molasses, there is a large amount of thin, cloudy, "sick" looking synovial fluid. This fluid is not capable of doing its important jobs of joint lubrication, nutrient distribution, and shock absorption.

Jessica is suffering from juvenile rheumatoid arthritis, the most common type of childhood arthritis.

Arthritis has many forms. The most common type is osteoarthritis, which is thought of as wear and tear of the articular

cartilage. Rheumatoid arthritis, however, is an autoimmune, inflammatory disease of the synovial membrane. For unknown reasons, the immune response mistakenly attacks the synovial membrane, causing it to grow abnormally and produce large amounts of unhealthy synovial fluid. This synovitis eventually destroys the joint cartilage.

1. What category of joints is likely to be affected by juvenile rheumatoid arthritis?

2. If you could look inside Jessica's knee joint, what do you think the synovial membrane would look like?

See the blue Answers tab at the back of the book.

10 Muscle Tissue

Learning Outcomes

These Learning Outcomes correspond by number to this chapter's sections and indicate what you should be able to do after completing the chapter.

10-1 ▌ Specify the functions of skeletal muscle tissue.

10-2 ▌ Describe the organization of muscle at the tissue level.

10-3 ▌ Describe the characteristics of skeletal muscle fibers, and identify the structural components of a sarcomere.

10-4 ▌ Identify the components of the neuromuscular junction, and summarize the events involved in the neural control of skeletal muscle contraction and relaxation.

10-5 ▌ Describe the mechanism responsible for tension production in a muscle fiber, and compare the different types of muscle contraction.

10-6 ▌ Describe the mechanisms by which muscle fibers obtain the energy to power contractions.

10-7 ▌ Relate the types of muscle fibers to muscle performance, and distinguish between aerobic and anaerobic endurance.

10-8 ▌ Identify the structural and functional differences between skeletal muscle fibers and cardiac muscle cells.

10-9 ▌ Identify the structural and functional differences between skeletal muscle fibers and smooth muscle cells, and discuss the roles of smooth muscle tissue in systems throughout the body.

SPOTLIGHTS

- Events at the Neuromuscular Junction pp. 326–327
- Excitation–Contraction Coupling p. 329
- The Contraction Cycle and Cross-Bridge Formation pp. 330–331

CLINICAL CASE

- A Real Eye Opener p. 316

Clinical Notes

- Tetanus p. 332
- Rigor Mortis p. 334
- Delayed-Onset Muscle Soreness p. 349

CLINICAL CASE

A Real Eye Opener

Tamara, a 33-year-old violinist, has waited for this neurology appointment for three weeks. That's how long she has been unable to play gigs with her band. She can still play her violin in the morning, but by afternoon her arms are so tired and weak, they feel like wood. During her last performance, she noticed she had drooled on her instrument, much to her embarrassment. This morning she feared she might choke on her toast. She has been unable to drive because of double vision. Her eyelids have been increasingly droopy, and when she woke up this morning, she couldn't open her left eyelid at all.

Tamara's neurologist takes a complete medical history and does a thorough physical exam. Tamara hadn't realized how weak her legs have gotten until the doctor asks her to stand up from her

chair without using her arms. Walking on her toes is impossible. When the doctor lifts her weak left eyelid, she is not able to track his finger with her gaze and is still seeing double.

"I would like to perform a test," says the neurologist, starting an IV. He injects a small amount of a drug. Suddenly Tamara's strength returns. She opens both eyes wide and no longer sees double. She can toe walk, heel walk, and get up unassisted many times in a row. She feels like she could play her instrument forever. But after 30 minutes, her weakness returns. **What is going on with Tamara?**

To find out, turn to the Clinical Case Wrap-Up on p. 357.

An Introduction to Muscle Tissue

In this chapter we discuss muscle tissue, one of the four primary tissue types, with particular attention to skeletal muscle tissue. We examine the histological and physiological characteristics of skeletal muscle cells, and relate those features to the functions of the entire tissue. We also give an overview of the differences among skeletal, cardiac, and smooth muscle tissues. Reading this chapter will prepare you for our discussion of the muscular system in Chapter 11.

10-1 ▮ Skeletal muscle performs six major functions

Learning Outcome Specify the functions of skeletal muscle tissue.

Muscle tissue consists chiefly of muscle cells that are highly specialized for contraction. Our bodies contain three types of muscle tissue: (1) *skeletal muscle,* (2) *cardiac muscle,* and (3) *smooth muscle.* ⟳ p. 166 Without these muscle tissues, nothing in the body would move, and the body itself could not move. Skeletal muscle tissue moves the body by pulling on our bones, making it possible for us to walk, dance, bite an apple, or play the guitar. Cardiac muscle tissue pumps blood through the cardiovascular system. Smooth muscle tissue pushes fluids and solids along the digestive tract and other internal organs and regulates the diameters of small arteries.

Skeletal muscles are organs composed mainly of skeletal muscle tissue, but they also contain connective tissues, nerves, and blood vessels. Each cell in skeletal muscle tissue is a single muscle *fiber.* Skeletal muscles attach directly or indirectly to bones. Our skeletal muscles perform the following six functions:

1. *Produce Skeletal Movement.* Skeletal muscle contractions pull on tendons to move our bones. Their effects range from simple motions, such as extending the arm or breathing, to the highly coordinated movements of swimming, skiing, or playing a piano.

2. *Maintain Posture and Body Position.* Tension in our skeletal muscles maintains body posture—for example, holding your head still when you read a book or balancing your body weight above your feet when you walk. Without constant muscular activity, we could neither sit upright nor stand.

3. *Support Soft Tissues.* Layers of skeletal muscle make up the abdominal wall and the floor of the pelvic cavity. These muscles support the weight of our visceral organs and shield our internal tissues from injury.

4. *Guard Body Entrances and Exits.* Skeletal muscles encircle the openings of the digestive and urinary tracts. These muscles give us voluntary control over swallowing, defecating, and urinating.

5. *Maintain Body Temperature.* Muscle contractions use energy, and whenever energy is used in the body, some of it is converted to heat. The heat released by working muscles keeps body temperature in the range needed for normal functioning.

6. *Store Nutrients.* When our diet contains too few proteins or calories, the contractile proteins in skeletal muscles are broken down, and their amino acids released into the circulation. The liver can use some of these amino acids to synthesize glucose, and others can be broken down to provide energy.

Let's begin our discussion with the functional anatomy of a typical skeletal muscle, with particular emphasis on the microscopic structural features that make contractions possible.

CHECKPOINT

1. Identify the three types of muscle tissue.
2. Identify the six major functions of skeletal muscle.

See the blue Answers tab at the back of the book.

10-2 ❙ A skeletal muscle contains muscle tissue, connective tissues, blood vessels, and nerves

Learning Outcome Describe the organization of muscle at the tissue level.

Figure 10–1 illustrates the organization of a representative skeletal muscle. Here we consider how connective tissues are organized in skeletal muscle, and how skeletal muscles are supplied

Figure 10–1 **The Organization of Skeletal Muscles.** A skeletal muscle consists of fascicles (bundles of muscle fibers) enclosed by the epimysium. The bundles are separated by connective tissue fibers of the perimysium, and within each bundle, each of the muscle fibers is surrounded by an endomysium. Each muscle fiber has many superficial nuclei, as well as mitochondria and other organelles (*see Figure 10–3*).

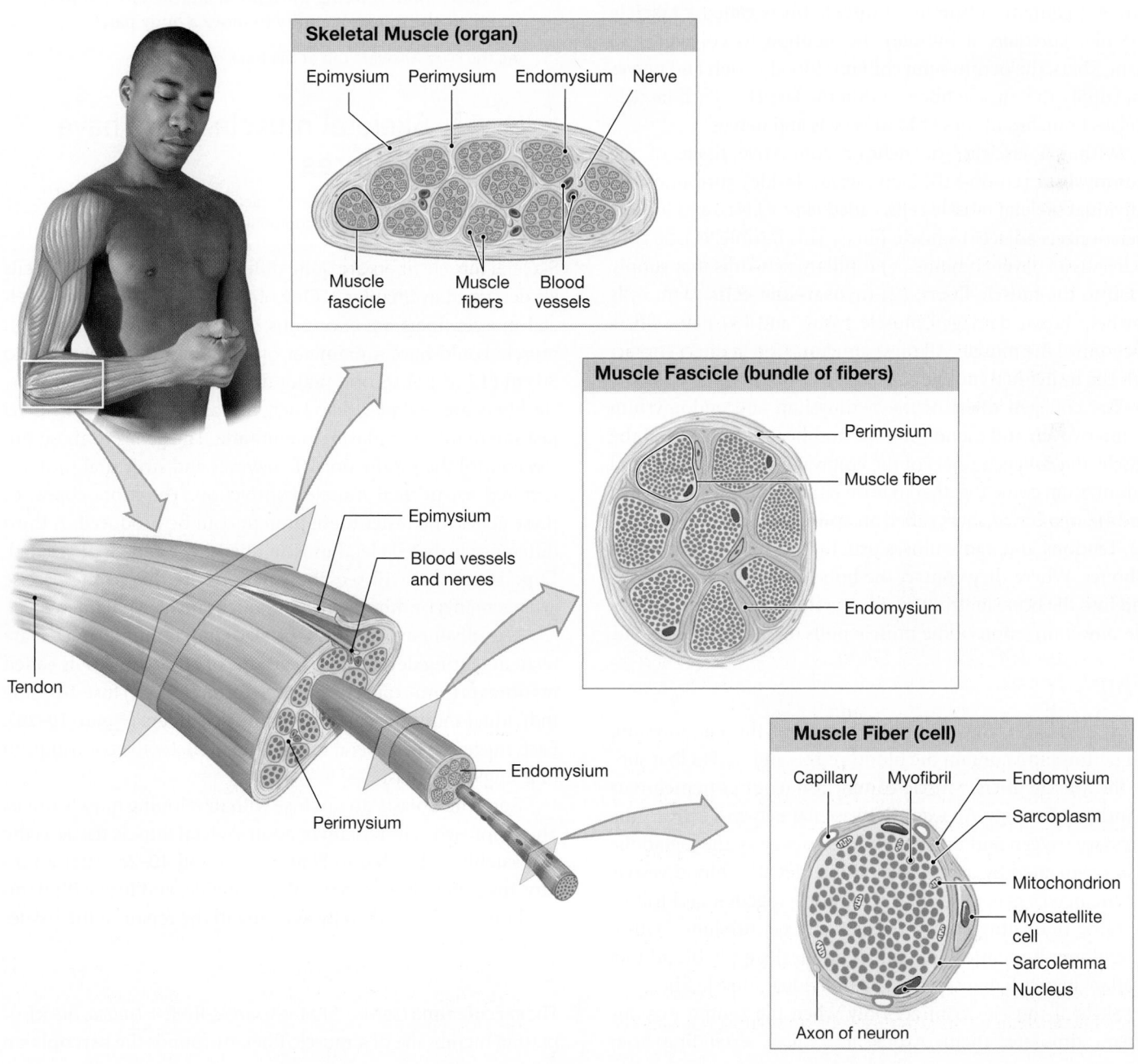

10

with blood vessels and nerves. In the next section we examine skeletal muscle tissue in detail.

Organization of Connective Tissues

As you can see in **Figure 10–1**, each muscle has three layers of connective tissue: (1) an epimysium, (2) a perimysium, and (3) an endomysium.

The **epimysium** (ep-i-MIZ-ē-um; *epi-*, on + *mys*, muscle) is a dense layer of collagen fibers that surrounds the entire muscle. It separates the muscle from nearby tissues and organs. The epimysium is connected to the deep fascia, a dense connective tissue layer. ⊃ p. 165

The **perimysium** (per-i-MIZ-ē-um; *peri-*, around) divides the skeletal muscle into a series of compartments. Each compartment contains a bundle of muscle fibers called a **fascicle** (FAS-i-kl; *fasciculus*, a bundle). In addition to collagen and elastic fibers, the perimysium contains blood vessels and nerves that supply the muscle fibers within the fascicles. Each fascicle receives branches of these blood vessels and nerves.

Within a fascicle, the delicate connective tissue of the **endomysium** (en-dō-MIZ-ē-um; *endo-*, inside) surrounds the individual skeletal muscle cells, called *muscle fibers,* and loosely interconnects adjacent muscle fibers. This flexible, elastic connective tissue layer contains (1) capillary networks that supply blood to the muscle fibers; (2) **myosatellite cells**, stem cells that help repair damaged muscle tissue; and (3) nerve fibers that control the muscle. All these structures are in direct contact with the individual muscle fibers. ⊃ p. 166

The collagen fibers of the perimysium and endomysium are interwoven and blend into one another. At each end of the muscle, the collagen fibers of the epimysium, perimysium, and endomysium come together to form either a bundle known as a **tendon**, or a broad sheet called an **aponeurosis** (ap-ō-nū-RŌ-sis). Tendons and aponeuroses usually attach skeletal muscles to bones. Where they contact the bone, the collagen fibers extend into the bone matrix, providing a firm attachment. As a result, any contraction of the muscle pulls on the attached bone.

Blood Vessels and Nerves

As we have seen, the connective tissues of the endomysium and perimysium contain the blood vessels and nerves that supply the muscle fibers. Muscle contraction requires tremendous quantities of energy. An extensive vascular network delivers the necessary oxygen and nutrients and carries away the metabolic wastes generated by active skeletal muscles. The blood vessels and the nerves generally enter the muscle together and follow the same branching course through the perimysium. Within the endomysium, arterioles (small arteries) supply blood to a capillary network that services the individual muscle fiber.

Skeletal muscles contract only when the central nervous system stimulates them. Axons, *nerve fibers* extending from

neurons, penetrate the epimysium, branch through the perimysium, and enter the endomysium to innervate individual muscle fibers. Skeletal muscles are often called voluntary muscles, because we have voluntary control over their contractions. We may also control many skeletal muscles at a subconscious level. For example, skeletal muscles involved with breathing, such as the *diaphragm,* usually work outside our conscious awareness.

Next, let's examine the microscopic structure of a typical skeletal muscle fiber and relate that microstructure to the physiology of the contraction process.

> **CHECKPOINT**
>
> 3. Describe the three connective tissue layers associated with skeletal muscle tissue.
>
> 4. How would severing the tendon attached to a muscle affect the muscle's ability to move a body part?
>
> See the blue Answers tab at the back of the book.

10-3 ▎ Skeletal muscle fibers have distinctive features

Learning Outcome Describe the characteristics of skeletal muscle fibers, and identify the structural components of a sarcomere.

Skeletal muscle fibers are quite different from the "typical" cells we described in Chapter 3. One obvious difference is size: Skeletal muscle fibers are enormous. A muscle fiber from a thigh muscle could have a diameter of 100 μm and a length up to 30 cm (12 in.). A second major difference is that skeletal muscle fibers are *multinucleate:* Each contains hundreds of nuclei just internal to the plasma membrane. The genes in these nuclei control the production of enzymes and structural proteins required for normal muscle contraction. The more copies of these genes, the faster these proteins can be produced. A third difference is that skeletal muscle fibers are banded, or striated. ⊃ p. 168 As we will see, the striations are due to the precise arrangements of thin (actin) and thick (myosin) filaments.

The distinctive features of size and multiple nuclei are related. During development, groups of embryonic cells called **myoblasts** (*myo-*, muscle + *blastos,* formative cell) fuse, forming individual multinucleate skeletal muscle fibers (**Figure 10–2a**). Each nucleus in a skeletal muscle fiber reflects the contribution of a single myoblast.

Some myoblasts do not fuse with developing muscle fibers. These unfused cells remain in adult skeletal muscle tissue as the myosatellite cells seen in **Figures 10–1** and **10–2a**. After an injury, myosatellite cells may enlarge, divide, and fuse with damaged muscle fibers, thereby assisting in the repair of the tissue.

The Sarcolemma and Transverse Tubules

The **sarcolemma** (sar-kō-LEM-uh; *sarkos,* flesh + *lemma,* husk), or plasma membrane of a muscle fiber, surrounds the **sarcoplasm**

Figure 10–2 **The Formation of a Multinucleate Skeletal Muscle Fiber.**

Muscle fibers develop through the fusion of embryonic cells called myoblasts.

Myoblasts

a A muscle fiber forms by the fusion of myoblasts.

Myosatellite cell

Nuclei

Immature muscle fiber

Muscle fiber LM × 612

Sarcolemma   Striations   Nuclei

Myofibrils

Mitochondria

b A diagrammatic view and a micrograph of one muscle fiber.

Myosatellite cell

Up to 30 cm in length

Mature muscle fiber

(SAR-kō-plazm), or cytoplasm of the muscle fiber (Figure 10–3). Like the plasma membranes of other body cells, the sarcolemma has a characteristic membrane potential, also called the transmembrane potential, due to more negative charges on its cytoplasmic side and more positive charges on its extracellular side. ⊃ p. 126 In a skeletal muscle fiber, a sudden change in the membrane potential is the first step that leads to a contraction.

Even though a skeletal muscle fiber is very large, all regions of the cell must contract at the same time. For this reason, the signal to contract must be distributed quickly throughout the interior of the cell. This signal is conducted through the transverse tubules. **Transverse tubules**, or **T tubules**, are narrow tubes whose surfaces are continuous with the sarcolemma and extend deep into the sarcoplasm (Figure 10–3). They are filled with extracellular fluid and form passageways through the muscle fiber, like a network of tunnels through a mountain. As extensions of the sarcolemma, T tubules share the same special property of being able to conduct an electrical impulse. As a result, electrical impulses conducted by the sarcolemma also travel along the T tubules into the cell interior.

These impulses, called *action potentials*, trigger muscle fiber contraction.

Myofibrils

Each muscle fiber contains hundreds to thousands of cylindrical structures called **myofibrils** (Figure 10–3). A myofibril is 1–2 μm in diameter and as long as the entire cell. Branches of the T tubules encircle each myofibril. The active shortening of myofibrils is responsible for skeletal muscle fiber contraction.

Myofibrils consist of bundles of protein filaments called **myofilaments**. Myofibrils contain two types of myofilaments that we introduced in Chapter 3: **thin filaments** composed primarily of actin, and **thick filaments** composed primarily of myosin. ⊃ pp. 97, 99 In addition, myofibrils contain *titin*, elastic myofilaments associated with the thick filaments. (We will consider the role of titin later in the chapter.)

Myofibrils are anchored to the inner surface of the sarcolemma at each end of a skeletal muscle fiber. In turn, the outer surface of the sarcolemma is attached to collagen fibers of the

Figure 10–3   The Structure and Internal Organization of a Skeletal Muscle Fiber.

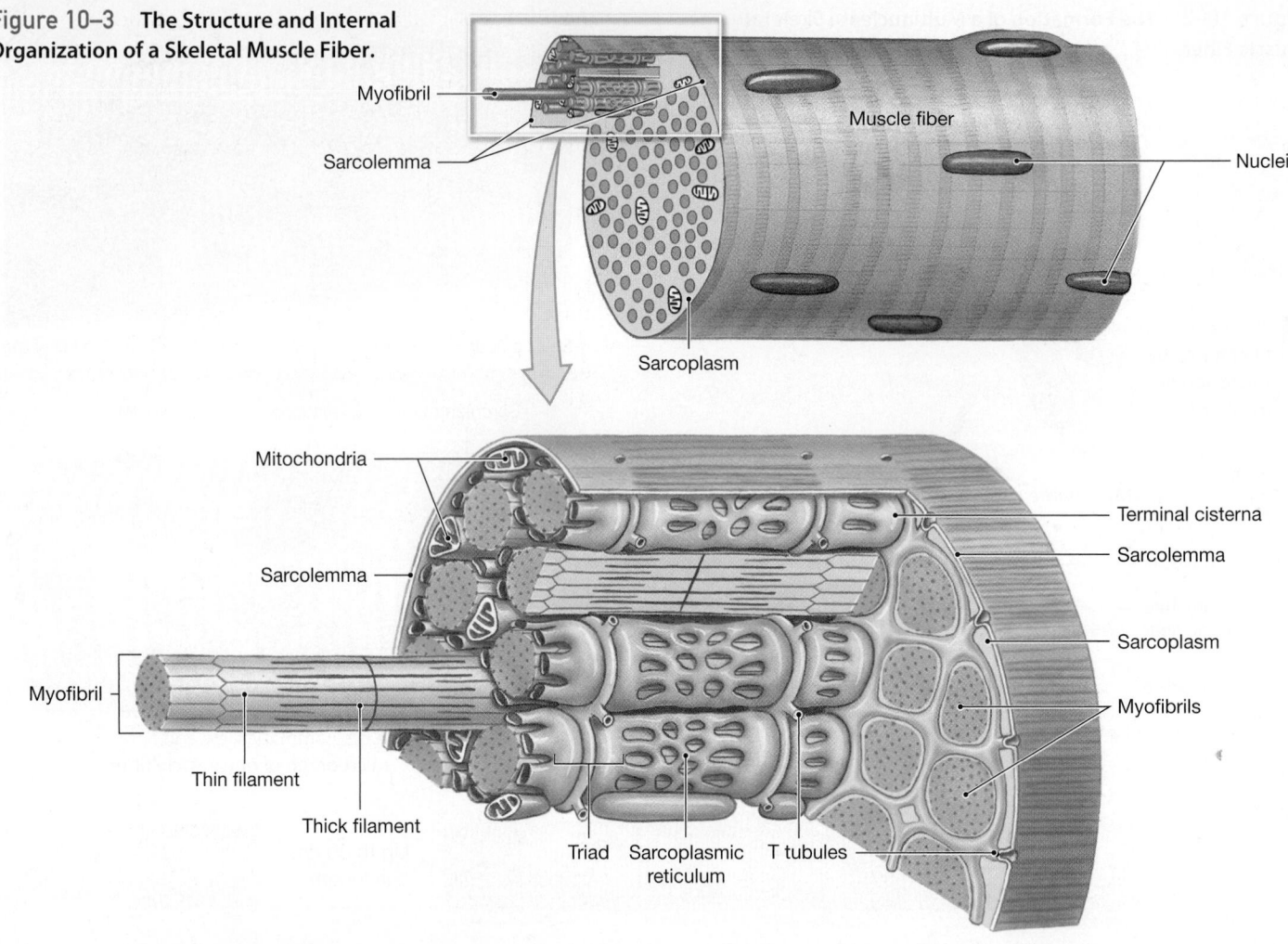

tendon or aponeurosis of the skeletal muscle. As a result, when the myofibrils shorten or contract, the entire cell shortens and pulls on the tendon.

Scattered among the myofibrils are mitochondria and granules of glycogen, the storage form of glucose. Mitochondrial activity and glucose breakdown by glycolysis provide energy in the form of ATP for short-duration, maximum-intensity muscular contractions. ⤺ p. 103

The Sarcoplasmic Reticulum

In skeletal muscle fibers, a membrane complex called the **sarcoplasmic reticulum (SR)** forms a tubular network around each individual myofibril, fitting over it like lacy shirtsleeves (**Figure 10–3**). The SR is similar to the smooth endoplasmic reticulum of other cells.

Wherever a T tubule encircles a myofibril, the tubule is tightly bound to the membranes of the SR. On either side of a T tubule, the tubules of the SR enlarge, fuse, and form expanded chambers called **terminal cisternae** (sis-TUR-nē). This combination of a

pair of terminal cisternae plus a T tubule is known as a **triad**. Although the membranes of the triad are tightly bound together, their fluid contents are separate and distinct.

In Chapter 3, we noted that special ion pumps keep the intracellular concentration of calcium ions (Ca^{2+}) very low. ⤺ p. 122 Most cells pump calcium ions out across their plasma membranes and into the extracellular fluid. Although skeletal muscle fibers do export Ca^{2+} in this way, they also remove calcium ions from the cytosol by actively transporting them into the terminal cisternae of the SR. The cytosol of a resting skeletal muscle fiber contains very low concentrations of Ca^{2+}, around 10^{-7} mmol/L. The free Ca^{2+} concentration levels inside the terminal cisternae may be as much as 1000 times higher. In addition, terminal cisternae contain the protein *calsequestrin*, which reversibly binds Ca^{2+}. Including both the free calcium and the bound calcium, the total concentration of Ca^{2+} inside terminal cisternae can be 40,000 times that of the surrounding cytosol.

A muscle contraction begins when stored calcium ions are released into the cytosol. These ions then diffuse into individual contractile units called *sarcomeres*.

Sarcomeres

As we have seen, myofibrils are bundles of thin and thick myofilaments. These myofilaments are organized into repeating functional units called **sarcomeres** (SAR-kō-mērz; *sarkos,* flesh + *meros,* part). Sarcomeres are the smallest functional units of the muscle fiber. Interactions between the thick and thin filaments within sarcomeres are responsible for muscle contraction.

A myofibril consists of approximately 10,000 sarcomeres, end to end. Each sarcomere has a resting length of about 2 μm. A sarcomere contains (1) thick filaments, (2) thin filaments, (3) proteins that stabilize the positions of the thick and thin filaments, and (4) proteins that regulate the interactions between thick and thin filaments.

Differences in the size, density, and distribution of thick filaments and thin filaments account for the banded appearance of each myofibril (**Figure 10–4**). Each sarcomere has dark bands called **A bands** and light bands called **I bands**. The names of these bands are derived from *anisotropic* and *isotropic,* which refer to their appearance when viewed using polarized light microscopy.

TIPS&TRICKS

You can remember that actin occurs in thin filaments by associating the "tin" in ac*tin* with the word *thin*. Then remember that th*In* filaments look *l*ight and form the *I* band. Similarly, the **A** bands are d**A**rk.

Figure 10–4   Sarcomere Structure, Longitudinal Views.

a A longitudinal view of a sarcomere, showing bands of thick and thin filaments

b A corresponding longitudinal section of a sarcomere in a myofibril from a muscle fiber in the gastrocnemius (calf) muscle of the leg

The A Band

The A bands contain regions of overlapping thick (myosin) and thin (actin) filaments. The thick filaments are at the center of each sarcomere, in the A band. The A band is about as long as a typical thick filament. The A band also includes portions of thin filaments and contains these three subdivisions (**Figure 10–4**):

1. *The M Line.* The **M line** is in the center of the A band. Proteins of the M line connect the central portion of each thick filament to neighboring thick filaments. (The M is for *middle.*) These dark-staining proteins help stabilize the positions of the thick filaments.

2. *The H Band.* In a resting sarcomere, the **H band** is a lighter region on either side of the M line. The H band contains thick filaments, but no thin filaments.

3. *The Zone of Overlap.* The **zone of overlap** is a dark region where thin filaments are located between the thick filaments. Here three thick filaments surround each thin filament, and six thin filaments surround each thick filament.

The longitudinal views in **Figure 10–4** and the cross-sectional views in **Figure 10–5b** should help you to visualize a three-dimensional image of these regions of a sarcomere.

Two T tubules encircle each sarcomere, and the triads containing them are located in the zones of overlap, at the edges of the A band (**Figure 10–3**). As a result, calcium ions released by the SR enter the regions where thick and thin filaments can interact.

The I Band

The I band is a region of the sarcomere that contains thin filaments but no thick filaments. It extends from the A band of one sarcomere to the A band of the next sarcomere (**Figure 10–4**). **Z lines** bisect the I bands and mark the boundary between adjacent sarcomeres. The Z lines consist of proteins called *actinins,* which interconnect thin filaments of adjacent sarcomeres. At both ends of the sarcomere, thin filaments extend from the Z lines toward the M line and into the zone of overlap.

Strands of the elastic protein **titin** extend from the tips of the thick filaments to attachment sites at the Z line (**Figures 10–4a** and **10–5**). Titin helps keep the thick and thin filaments in proper alignment and aids in restoring resting sarcomere length after contraction. It also helps the muscle fiber resist extreme stretching that would otherwise disrupt the contraction mechanism.

Figure 10–5   Sarcomere Structure, Superficial and Cross-Sectional Views.

Sarcomere

Myofibril

a A superficial view of a sarcomere

Thin filament

Thick filament

Actinin filaments

Thin filaments

Titin filament

Thick filaments

Attachment of titin

Z line   I band   M line   H band   Zone of overlap

b Cross-sectional views of different regions of a sarcomere

Each Z line is surrounded by a meshwork of intermediate filaments that interconnect adjacent myofibrils. The myofibrils closest to the sarcolemma, in turn, are bound to attachment sites on the interior surface of the membrane. Because the Z lines of all the myofibrils are aligned in this way, the muscle fiber as a whole has a banded appearance due to the alternating A bands and I bands (**Figure 10–2b**). These bands, or *striations*, are visible with the light microscope, so skeletal muscle tissue is also known as striated muscle. ⟲ p. 168

Figure 10–6 reviews the levels of skeletal muscle organization we have discussed so far. Now let's consider the molecular structure of the thin and thick myofilaments responsible for muscle contraction.

Thin Filaments

A typical thin filament is 5–6 nm in diameter and 1 μm in length (**Figure 10–7a**). A single thin filament contains four proteins: F-actin, nebulin, tropomyosin, and troponin (**Figure 10–7b**).

Filamentous actin, or **F-actin**, is a twisted strand composed of two rows of 300–400 individual globular molecules of **G-actin** (**Figure 10–7b**). A long strand of **nebulin** extends along the F-actin strand in the cleft between the rows of G-actin molecules. Nebulin holds the F-actin strand together. Each G-actin molecule contains an **active site**, where myosin (in the thick filaments) can bind. This is similar to enzymatic reactions in which a substrate molecule binds to the active site of an enzyme. (Recall that a substrate is a substance that is acted upon by an enzyme.) ⟲ p. 80 Under resting conditions, however, the *troponin–tropomyosin complex* prevents myosin binding.

Strands of **tropomyosin** (trō-pō-MĪ-ō-sin; *trope*, turning) cover the active sites on G-actin and prevent actin–myosin interaction. A tropomyosin molecule is a double-stranded protein that covers seven active sites. It is bound to one molecule of troponin midway along its length. A **troponin** (TRŌ-pō-nin) molecule consists of three globular subunits. One subunit binds to tropomyosin, locking them together as a troponin–tropomyosin complex, and a second subunit binds to one G-actin, holding the troponin–tropomyosin complex in position. The third subunit has a receptor that binds two calcium ions. In a resting muscle, intracellular Ca^{2+} concentrations are very low, and that binding site is empty.

A contraction can occur only when the troponin–tropomyosin complex changes position, exposing the active sites on actin. The necessary change in position takes place when calcium ions bind to receptors on the troponin molecules.

The thin filaments are attached to the Z lines at either end of the sarcomere (**Figure 10–7a**). Although it is called a "line" because it looks like a dark line on the surface of the myofibril, the Z line in sectional view is more like a disc with an open meshwork (**Figure 10–5b**). For this reason, the Z line is often called the *Z disc*.

Figure 10–6 Levels of Functional Organization in a Skeletal Muscle.

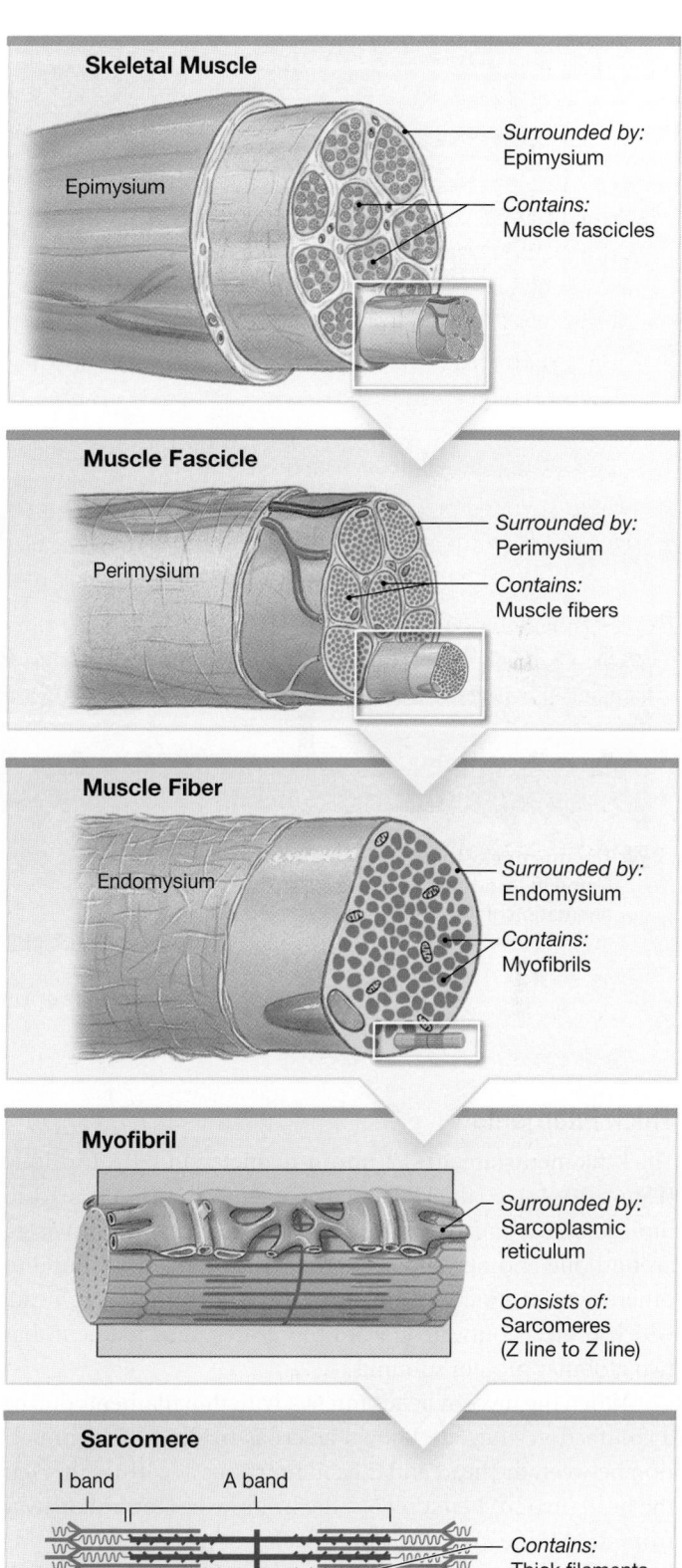

Skeletal Muscle

Epimysium

Surrounded by: Epimysium

Contains: Muscle fascicles

Muscle Fascicle

Perimysium

Surrounded by: Perimysium

Contains: Muscle fibers

Muscle Fiber

Endomysium

Surrounded by: Endomysium

Contains: Myofibrils

Myofibril

Surrounded by: Sarcoplasmic reticulum

Consists of: Sarcomeres (Z line to Z line)

Sarcomere

I band   A band

Contains: Thick filaments

Thin filaments

Z line   M line   Titin   Z line

H band

Figure 10–7 Thin and Thick Filaments.

a The gross structure of a thin filament, showing the attachment at the Z line

b The organization of G-actin subunits in an F-actin strand, and the position of the troponin–tropomyosin complex

c The structure of thick filaments, showing the orientation of the myosin molecules

d The structure of a myosin molecule

Thick Filaments

Thick filaments are 10–12 nm in diameter and 1.6 μm long (**Figure 10–7c**). A thick filament contains about 300 myosin molecules, each made up of a pair of myosin subunits twisted around one another (**Figure 10–7d**). The long tail is bound to other myosin molecules in the thick filament. The free head, which projects outward toward the nearest thin filament, has two globular protein subunits.

When the myosin heads interact with thin filaments during a contraction, they are known as **cross-bridges**. The connection between the head and the tail functions as a hinge that lets the head pivot. When it pivots, the head swings toward or away from the M line. As we will see in a later section, this pivoting is the key step in muscle contraction.

All the myosin molecules are arranged with their tails pointing toward the M line (**Figure 10–7c**). The H band includes a central region where there are no myosin heads. Elsewhere on the thick filaments, the myosin heads are arranged in a spiral, each facing one of the surrounding thin filaments.

Each thick filament has a core of titin. On either side of the M line, a strand of titin extends the length of the thick filament and then continues across the I band to the Z line on that side. The portion of the titin strand exposed within the I band is *elastic,* which means that it will recoil after stretching. In the normal resting sarcomere, the titin strands are completely relaxed. They become tense only when some external force stretches the sarcomere.

Sliding Filaments and Muscle Contraction

What happens to a sarcomere when a skeletal muscle fiber contracts? As you can see in **Figure 10–8**, (1) the H bands and I bands of the sarcomeres narrow, (2) the zones of overlap widen, (3) the Z lines move closer together, and (4) the width of the A band remains constant.

These observations make sense only if the thin filaments are sliding toward the center of each sarcomere, alongside the thick filaments. This explanation is known as the **sliding filament theory**. The contraction weakens with the disappearance of the I bands, at which point the Z lines are in contact with the ends of the thick filaments.

Figure 10–8   **Changes in the Appearance of a Sarcomere during the Contraction of a Skeletal Muscle Fiber.**

Myofibril at rest

a A relaxed sarcomere showing location of the A band, Z lines, and I band.

Contracted myofibril

b During a contraction, the A band stays the same width, but the Z lines move closer together and the I band gets smaller. When the ends of a myofibril are free to move, the sarcomeres shorten simultaneously and the ends of the myofibril are pulled toward its center.

During a contraction, sliding occurs in every sarcomere along the myofibril. As a result, the myofibril gets shorter. Because myofibrils are attached to the sarcolemma at each Z line and at either end of the muscle fiber, when myofibrils get shorter, so does the muscle fiber.

TIPS & TRICKS

To better understand the actions of sliding filaments during a muscle contraction, hold your hands in front of you, palms toward your body and thumbs sticking straight up. Now move your hands together, so that the fingers of one hand move in between the fingers of the other hand. Your fingers represent thin and thick filaments, and your thumbs the Z lines. Notice that finger length stays the same, but your thumbs move closer together.

When muscle cells contract, they pull on the attached tendon fibers the way a line of people might pull on a rope. Let's consider what is happening in light of some basic physical principles that apply to muscle cells. The pull, called *tension*, is an active force: Energy must be expended to produce it. Tension is applied *to* some object, whether a rope, a rubber band, or a book on a tabletop. Tension applied to an object tends to pull the object toward the source of the tension. However, before movement can occur, the applied tension must overcome the object's *load* (or *resistance*), a passive force that opposes movement. The amount of load can depend on the weight of the object, its shape, friction, and other factors. When the applied tension exceeds the load, the object moves.

In contrast, *compression,* or a push applied to an object, tends to force the object away from the source of the compression. Again, no movement can occur until the applied compression exceeds the load of the object. Muscle cells can use energy to

shorten and generate tension through interactions between thick and thin filaments, but not to lengthen and generate compression. In other words, muscle cells can pull, but they cannot push.

CHECKPOINT

5. Describe the structural components of a sarcomere.

6. Why do skeletal muscle fibers appear striated when viewed through a light microscope?

7. Where would you expect to find the greatest concentration of Ca^{2+} in resting skeletal muscle?

See the blue Answers tab at the back of the book.

10-4 | The nervous system communicates with skeletal muscles at the neuromuscular junction

Learning Outcome Identify the components of the neuromuscular junction, and summarize the events involved in the neural control of skeletal muscle contraction and relaxation.

You have seen that the myofilaments in a sarcomere change position during a contraction, but not *how* these changes occur. In this section we first discuss how the plasma membranes of neurons and skeletal muscle fibers propagate electrical impulses. (This topic is explored more fully in Chapter 12.) We then look at how skeletal muscle activity is under neural control (as you will see in **Spotlight Figure 10-9**), how neural stimulation of a muscle fiber is coupled to the contraction of the fiber (presented in **Spotlight Figure 10-10**), and how actin and myosin interact to cause sarcomere contraction (shown in **Spotlight Figure 10-11**). Read the text on p. 328 before studying these three Spotlights.

A single axon may branch to control more than one skeletal muscle fiber, but each muscle fiber has only one neuromuscular junction (NMJ). At the NMJ, the axon terminal of the neuron lies near the motor end plate of the muscle fiber.

Motor neuron

Path of electrical impulse (action potential)

Axon

Neuromuscular junction

Axon terminal

SEE BELOW

Sarcoplasmic reticulum

Motor end plate

Myofibril

Motor end plate

1 The cytoplasm of the axon terminal contains vesicles filled with molecules of acetylcholine, or ACh. Acetylcholine is a **neurotransmitter**, a chemical released by a neuron to change the permeability or other properties of another cell's plasma membrane. The synaptic cleft and the motor end plate contain molecules of the enzyme acetylcholinesterase (AChE), which breaks down ACh.

2 The stimulus for ACh release is the arrival of an electrical impulse, or action potential, at the axon terminal. An action potential is a sudden change in the membrane potential that travels along the length of the axon.

Arriving action potential

Vesicles   ACh

The **synaptic cleft** is a narrow space that separates the axon terminal of the neuron from the opposing motor end plate.

Junctional fold of motor end plate   AChE

Muscle Fiber

The action potential generated at the motor end plate now sweeps across the entire membrane surface. The effects are almost immediate because an action potential is an electrical event that flashes like a spark across the sarcolemmal surface. The effects are brief because the ACh has been removed, and no further stimulus acts upon the motor end plate until another action potential arrives at the axon terminal.

3 When the action potential reaches the neuron's axon terminal, permeability changes in its membrane trigger the exocytosis of ACh into the synaptic cleft. Exocytosis occurs as vesicles fuse with the neuron's plasma membrane.

4 ACh molecules diffuse across the synaptic cleft and bind to ACh receptors on the surface of the motor end plate. ACh binding alters the membrane's permeability to sodium ions. Because the extracellular fluid contains a high concentration of sodium ions, and sodium ion concentration inside the cell is very low, sodium ions rush into the cytosol.

5 The sudden inrush of sodium ions results in the generation of an action potential in the sarcolemma. ACh is removed from the synaptic cleft in two ways. ACh either diffuses away from the synapse, or it is broken down by AChE into acetic acid and choline. This removal inactivates the ACh receptor sites. The muscle fiber pictured above indicates the propagation of the action potential along the sarcolemma.

Motor end plate

ACh receptor site

Na⁺

Na⁺

Na⁺

Action potential

Break down of ACh

AChE

10

Electrical Impulses and Excitable Membranes

As we learned in Chapter 3, all the cells of the body maintain a membrane potential due to an unequal distribution of positive and negative charges across their plasma membrane. ⊃ p. 126 The unequal charge distribution means cells are *polarized*, much like miniature batteries. In cells, the inner surface of the plasma membrane is slightly negative compared to the outer surface. The membrane potential is a measure of cellular polarization (in millivolt units) that compares the cytoplasmic membrane surface charge to the extracellular membrane surface charge. In unstimulated (resting) neurons and skeletal muscle fibers, typical resting membrane potentials are -70 mV and -85 mV, respectively.

Recall that the main contributors of negative charges within a cell are proteins that cannot cross the plasma membrane. There is also an excess of sodium ions outside a cell and an excess of potassium ions inside a cell. These ions can cross a membrane through membrane channels that are specific for each ion.

Various stimuli can lead to a temporary change in the distribution of electrical charges across the plasma membranes of all body cells. Differences in sodium ion and potassium ion membrane permeability underlie such changes. An influx of sodium ions leads to *depolarization* as the membrane potential becomes less negative. The movement of potassium ions out of a cell leads to *hyperpolarization* as the membrane potential becomes more negative. A return to the resting potential is called *repolarization*. In most cells, the depolarization or hyperpolarization of a plasma membrane is a localized change limited by the presence or absence of stimulation. Called a *graded potential*, it does not continue to spread over the plasma membrane.

Neurons and skeletal muscle fibers, however, have electrically **excitable membranes**. Excitable membranes permit rapid communication between different parts of a cell. In neurons and skeletal muscle fibers, the depolarization and repolarization events produce an electrical impulse, or *action potential*, that is propagated along their plasma membranes. Unlike the plasma membranes of most cells, excitable membranes contain *voltage-gated channels* that are activated and inactivated by changes in the membrane potential. These electrical channels become activated when the membranes of neurons and skeletal muscle fibers first depolarize from the resting potential to a *threshold potential* (from -70 to -60 mV for neurons and from -85 to -55 mV for skeletal muscle fibers). Upon reaching the threshold potential, *voltage-gated sodium channels* open and there is a rapid influx of positively charged sodium ions into the cell. The inside of the membrane reverses from a negative to positive charge, and the depolarization peaks at a membrane potential of $+30$ mV with the closure of the voltage-gated sodium channels. *Repolarization* of the membrane then begins as voltage-gated potassium channels open and positively charged potassium ions leave the cell. The loss of more positive charges than entered the cell causes the membrane potential to become negative again. A "resting" *refractory period* follows and the former concentrations of sodium and potassium ions across the plasma membrane are restored through the action of sodium–potassium ion pumps. ⊃ p. 122 A further depolarization cannot occur until the refractory period is over. As a result, the action potential travels in one direction because the refractory period prevents it from propagating back in the direction from where it was initiated.

The Control of Skeletal Muscle Activity

Skeletal muscle fibers begin contraction with the release of their internal stores of calcium ions. That release is under the control of the nervous system. Communication between a neuron and another cell occurs at a *synapse*. When the other cell is a skeletal muscle fiber, the synapse is known as a **neuromuscular junction (NMJ)**, or *myoneural junction*. The NMJ is made up of an *axon terminal* (synaptic terminal) of a neuron, a specialized region of the sarcolemma called the *motor end plate*, and, in between, a narrow space called the *synaptic cleft*.

Motor neurons of the central nervous system (brain and spinal cord) carry instructions in the form of action potentials to skeletal muscle fibers. A motor neuron stimulates a muscle fiber through a series of steps as shown in **Spotlight Figure 10–9** (pp. 326–327).

Excitation–Contraction Coupling

The link between the generation of an action potential in the sarcolemma and the start of a muscle contraction is called **excitation–contraction coupling**. This coupling occurs at the triads. On reaching a triad, an action potential triggers the release of Ca^{2+} from the cisternae of the sarcoplasmic reticulum. The change in the permeability of the SR to Ca^{2+} is temporary, lasting only about 0.03 second. Yet within a millisecond, the Ca^{2+} concentration in and around the sarcomere reaches 100 times resting levels. Because the terminal cisternae are located at the zones of overlap, where the thick and thin filaments interact, the effect of calcium ion release on the sarcomere is almost instantaneous. Troponin is the lock that keeps the active sites inaccessible. Calcium is the key to that lock. Recall from **Figure 10–7b** that troponin binds to both actin and tropomyosin, and that the tropomyosin molecules cover the active sites and prevent interactions between thick filaments and thin filaments. Each troponin molecule also has a binding site for calcium, and this site is empty when the muscle fiber is at rest. Calcium binding changes the shape of the troponin molecule and weakens the bond between troponin and actin. The troponin molecule then changes position, rolling the attached tropomyosin strand away from the active sites (**Spotlight Figure 10–10**). With this change, the **contraction cycle** begins.

The Contraction Cycle

The contraction cycle is a series of molecular events that enable muscle contraction. **Spotlight Figure 10–11**, pp. 330–331, shows the interlocking steps of the contraction cycle.

SPOTLIGHT

Figure 10–10
Excitation-Contraction Coupling

Watch the MasteringA&P® > A&PFlix™ on Excitation-Contraction Coupling.

1 Neural Control

A skeletal muscle fiber contracts when stimulated by a motor neuron at a neuromuscular junction. The stimulus arrives in the form of an action potential at the axon terminal.

2 Excitation

The action potential causes the release of ACh into the synaptic cleft, which leads to excitation—the production of an action potential in the sarcolemma.

3 Release of Calcium Ions

This action potential travels along the sarcolemma and down T tubules to the triads. This triggers the release of calcium ions (Ca^{2+}) from the terminal cisternae of the sarcoplasmic reticulum.

4 Contraction Cycle Begins

The contraction cycle begins when the calcium ions (Ca^{2+}) bind to troponin, resulting in the exposure of the active sites on the thin filaments. This allows cross-bridge formation and will continue as long as ATP is available. (See *Spotlight Figure 10-11* for the details of the contraction cycle.)

5 Sarcomere Shortening

As the thick and thin filaments interact, the sarcomeres shorten, pulling the ends of the muscle fiber closer together.

6 Generation of Muscle Tension

During the contraction, the entire skeletal muscle shortens and produces a pull, or tension, on the tendons at either end.

In a resting sarcomere, the tropomyosin strands cover the active sites on the thin filaments, preventing cross-bridge formation.

When calcium ions enter the sarcomere, they bind to troponin, which rotates and swings the tropomyosin away from the active sites.

Cross-bridge formation then occurs, and the contraction cycle begins.

Muscle fiber contraction

leads to

Tension production

1 Contraction Cycle Begins

The contraction cycle involves a series of interrelated steps. It begins with the arrival of calcium ions (Ca^{2+}) within the zone of overlap in a sarcomere.

2 Active-Site Exposure

Calcium ions bind to troponin, weakening the bond between actin and the troponin–tropomyosin complex. The troponin molecule then changes position, rolling the tropomyosin molecule away from the active sites on actin and allowing interaction with the energized myosin heads.

3 Cross-Bridge Formation

Once the active sites are exposed, the energized myosin heads bind to them, forming cross-bridges.

RESTING SARCOMERE

In the resting sarcomere, each myosin head is already "energized"—charged with the energy that will be used to power a contraction. Each myosin head points away from the M line. In this position, the myosin head is "cocked" like the spring in a mousetrap. Cocking the myosin head requires energy, which is obtained by breaking down ATP; in doing so, the myosin head functions as ATPase, an enzyme that breaks down ATP. At the start of the contraction cycle, the breakdown products, ADP and phosphate (represented as P), remain bound to the myosin head.

M line

Zone of Overlap
(shown in
sequence above)

4 Myosin Head Pivoting

After cross-bridge formation, the energy that was stored in the resting state is released as the myosin head pivots toward the M line. This action is called the power stroke; when it occurs, the bound ADP and phosphate group are released.

5 Cross-Bridge Detachment

When another ATP binds to the myosin head, the link between the myosin head and the active site on the actin molecule is broken. The active site is now exposed and able to form another cross-bridge.

6 Myosin Reactivation

Myosin reactivation occurs when the free myosin head splits ATP into ADP and P. The energy released is used to recock the myosin head.

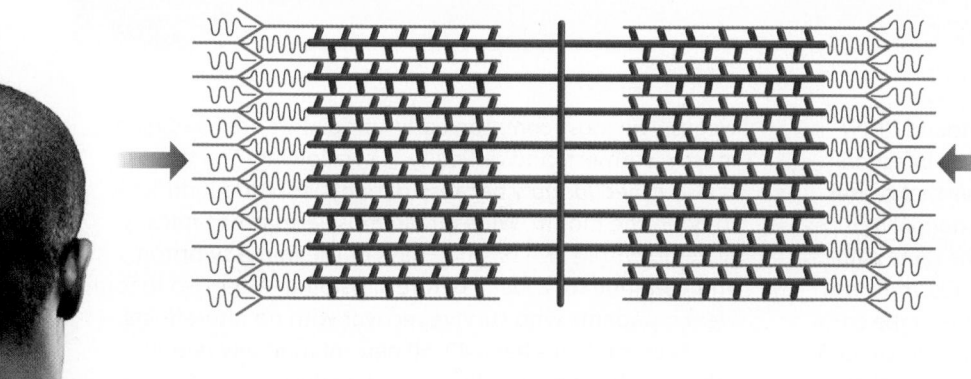

CONTRACTED SARCOMERE

The entire cycle is repeated several times each second, as long as Ca^{2+} concentrations remain elevated and ATP reserves are sufficient. Calcium ion levels will remain elevated only as long as action potentials continue to pass along the T tubules and stimulate the terminal cisternae. Once that stimulus is removed, the calcium channels in the SR close and calcium ion pumps pull Ca^{2+} from the cytosol and store it within the terminal cisternae. Troponin molecules then shift position, swinging the tropomyosin strands over the active sites and preventing further cross-bridge formation.

10

As we have noted, the power stroke refers to the molecular interactions of muscle contraction. Each power stroke shortens the sarcomere by about 0.5 percent. The entire muscle shortens at the same rate because all the sarcomeres contract together. The speed of the shortening depends on the cycling rate (the number of power strokes per second). The greater the load, the slower the cycling rate.

To understand how the contraction cycle produces tension in a muscle fiber, imagine that you are on a tug-of-war team. You reach forward, grab the rope with both hands, and pull it in. This grab-and-pull corresponds to cross-bridge attachment and pivoting. You then let go of the rope, reach forward and grab it, and pull once again. Your actions are not synchronized with the rest of your team. At any given time, some people are reaching and grabbing, some are pulling, and others are letting go. (If everyone let go at the same time, your opponents would pull the rope away.) The amount of tension produced depends on how many people are pulling at the same time.

The situation is similar in a muscle fiber. The myosin heads along a thick filament work together in a similar way to pull a thin filament toward the center of the sarcomere. Each myofibril consists of a string of sarcomeres, and in a contraction all of the thin filaments are pulled toward the centers of the sarcomeres.

If neither end of the myofibril is held in position, both ends move toward the middle, as you can see in **Figure 10–12a**.

Such a contraction seldom occurs in an intact skeletal muscle, because one end of the muscle (the *origin*) is usually fixed in position during a contraction, while the other end (the *insertion*) moves. In that case, the free end moves toward the fixed end (**Figure 10–12b**).

If neither end of the myofibril can move, thick and thin filament interactions consume energy and generate tension, but sliding cannot occur. We will discuss this kind of contraction, called an *isometric contraction,* in a later section.

Figure 10–13 reviews the contraction process, beginning with ACh release and ending with relaxation.

Figure 10–12 **Shortening during a Contraction.**

a When both ends are free to move, the ends of a contracting muscle fiber move toward the center of the muscle fiber.

b When only one end of a myofibril is fixed in position, the free end is pulled toward the fixed end.

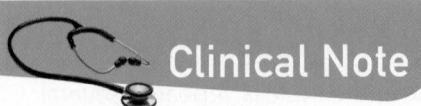

Clinical Note

Tetanus Children are often told to be careful around rusty nails. But it's not the rust or the nail, but instead an infection by the very common bacterium *Clostridium tetani* that causes the disease called **tetanus**. This disease and the normal muscle response to rapid neural stimulation share the same name, but the mechanisms involved are very different. *Clostridium* bacteria occur in soil and virtually everywhere else in the environment, but they can thrive only in tissues with low oxygen levels. For this reason, a deep puncture wound, such as that from a nail, is much more likely to result in tetanus than a shallow, open cut that bleeds freely. When active in body tissues, these bacteria release a powerful toxin called tetanospasmin that affects the central nervous system. Motor neurons, which control skeletal muscles throughout the body, are particularly sensitive to it. The toxin suppresses the mechanism that inhibits motor neuron activity. The result is a sustained, powerful contraction of skeletal muscles throughout the body.

The incubation period (the time between exposure and the development of symptoms) is generally less than two weeks. The most common early complaints are headache, muscle stiffness, and difficulty in swallowing. The disease is also called *lockjaw* because patients also find it difficult to open the mouth. Widespread muscle spasms typically develop within two or three days of the initial symptoms and continue for a week before subsiding. After two to four weeks, patients who survive recover with no aftereffects.

Severe tetanus has a 40–60 percent mortality rate. In other words, for every 100 people who develop severe tetanus, 40 to 60 die. Approximately 500,000 cases of tetanus occur worldwide each year, but only about 100 of them occur in the United States, thanks to an effective immunization program. (Recommended immunization involves a "tetanus shot" followed by a booster shot every 10 years.) In unimmunized patients, early administration of an antitoxin—in most cases, *human tetanus immunoglobulin*—can prevent severe symptoms. However, this treatment does not reduce signs and symptoms that have already appeared.

Figure 10–13   **Steps Involved in Skeletal Muscle Contraction and Relaxation.**

Steps That Initiate a Muscle Contraction

1 ACh released

ACh is released at the neuromuscular junction and binds to ACh receptors on the sarcolemma.

2 Action potential reaches T tubule

An action potential is generated and spreads across the membrane surface of the muscle fiber and along the T tubules.

3 Sarcoplasmic reticulum releases Ca²⁺

The sarcoplasmic reticulum releases stored calcium ions.

4 Active site exposure and cross-bridge formation

Calcium ions bind to troponin, exposing the active sites on the thin filaments. Cross-bridges form when myosin heads bind to those active sites.

5 Contraction cycle begins

The contraction cycle begins as repeated cycles of cross-bridge binding, pivoting, and detachment occur—all powered by ATP.

Steps That End a Muscle Contraction

6 ACh is broken down

ACh is broken down by acetylcholinesterase (AChE), ending action potential generation

7 Sarcoplasmic reticulum reabsorbs Ca²⁺

As the calcium ions are reabsorbed, their concentration in the cytosol decreases.

8 Active sites covered, and cross-bridge formation ends

Without calcium ions, the tropomyosin returns to its normal position and the active sites are covered again.

9 Contraction ends

Without cross-bridge formation, contraction ends.

10 Muscle relaxation occurs

The muscle returns passively to its resting length.

Relaxation

How long does a contraction last? Its duration depends on (1) the period vof stimulation at the neuromuscular junction, (2) the presence of free calcium ions in the cytosol, and (3) the availability of ATP. A single stimulus has only a brief effect on a muscle fiber, because the ACh released as a result of a single action potential is rapidly broken down by AChE. Inside the muscle fiber, the permeability changes in the SR are also very brief. For these reasons, a contraction will continue only if additional action potentials arrive at the axon terminal in rapid succession. When they do, the continual release of ACh into the synaptic cleft produces a series of action potentials in the sarcolemma that keeps Ca^{2+} levels elevated in the cytosol. Under these conditions, the contraction cycle will be repeated over and over.

If just one action potential arrives at the neuromuscular junction, Ca^{2+} concentrations in the cytosol will quickly return to normal resting levels. Two mechanisms are involved in this process: (1) active Ca^{2+} transport across the sarcolemma into the extracellular fluid, and (2) active Ca^{2+} transport into the SR. Of the two, transport into the SR is far more important. Virtually as soon as the calcium ions have been released, the SR returns to its normal permeability and begins to actively take up Ca^{2+} from the surrounding cytosol. As Ca^{2+} concentrations in the cytosol fall, earlier events reverse themselves: (1) Calcium ions detach from troponin, (2) troponin returns to its original position, and (3) the active sites on actin are once again covered by tropomyosin. The contraction has ended.

The sarcomeres do not automatically return to their original length. Sarcomeres shorten actively, but there is no active mechanism for reversing the process. External forces must act on the contracted muscle fiber to stretch the myofibrils and sarcomeres to their original dimensions. We describe those forces in a later section.

10-5 | Sarcomere shortening and muscle fiber stimulation produce tension

Learning Outcome Describe the mechanism responsible for tension production in a muscle fiber, and compare the different types of muscle contraction.

When sarcomeres shorten in a contraction, they shorten the muscle fiber. This shortening exerts tension on the connective tissue fibers attached to the muscle fiber. The tension produced by an individual muscle fiber can vary, and in the next section we consider the specific factors involved. In a subsequent section, we see that the tension produced by an entire skeletal *muscle* can vary even more widely. Why? Not only can tension production vary among the individual muscle fibers, but the number of stimulated muscle fibers can change from moment to moment.

Tension Production by Muscle Fibers

The amount of tension produced by an individual muscle fiber ultimately depends on the number of pivoting cross-bridges. There is no way to regulate the amount of tension produced in that contraction by changing the number of contracting sarcomeres within a muscle fiber. When calcium ions are released, they are released from all triads in the muscle fiber. For this

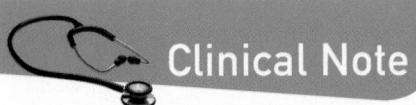 **Clinical Note**

Rigor Mortis When death occurs, circulation ceases and the skeletal muscles are deprived of nutrients and oxygen. Within a few hours, the skeletal muscle fibers run out of ATP and the sarcoplasmic reticulum becomes unable to pump Ca^{2+} out of the cytosol. Calcium ions diffusing into the cytosol from the extracellular fluid or leaking out of the SR then trigger a sustained contraction. Without ATP, the cross-bridges cannot detach from the active sites, so skeletal muscles throughout the body become locked in the contracted position. Because all the skeletal muscles are involved, the body becomes "stiff as a board." This physical state—**rigor mortis**—lasts until the lysosomal enzymes released by autolysis break down the Z lines and titin filaments. Rigor mortis begins two to seven hours after death and ends after one to six days or when decomposition begins. The timing depends on environmental factors, such as temperature. Forensic pathologists can estimate the time of death on the basis of the degree of rigor mortis and environmental conditions.

reason, a muscle fiber is either "on" (producing tension) or "off" (relaxed).

The tension produced by an individual muscle fiber *does* vary, however, for other reasons. It depends on (1) the fiber's resting length at the time of stimulation, which determines the degree of overlap between thick and thin filaments; and (2) the frequency of stimulation, which affects the internal concentration of calcium ions and thus the amount bound to troponin.

Length–Tension Relationships

Figure 10–14 shows the effect of sarcomere length on active tension. When many people pull on a rope, the amount of tension produced is proportional to the number of people pulling. Similarly, in a skeletal muscle fiber, the amount of tension generated during a contraction depends on the number of pivoting cross-bridges in each of the myofibrils. The number of cross-bridges that can form, in turn, depends on the degree of overlap between thick filaments and thin filaments within these sarcomeres. When the muscle fiber is stimulated to contract, only myosin heads in the zones of overlap can bind to active sites and produce tension. For these reasons, we can relate the tension produced by the entire muscle fiber to the structure of individual sarcomeres.

A sarcomere works most efficiently within an optimal range of lengths. When the resting sarcomere length is within this range, the maximum number of cross-bridges can form, and the tension produced is highest. If the resting sarcomere length falls outside the range—if the sarcomere is compressed and shortened, or stretched and lengthened—it cannot produce as much tension when stimulated. The reason is that the number of cross-bridges that form largely determines the amount of tension produced. An increase in sarcomere length reduces the tension produced by reducing the size of the zone of overlap and the number of potential cross-bridges.

Figure 10–14   The Effect of Sarcomere Length on Active Tension.

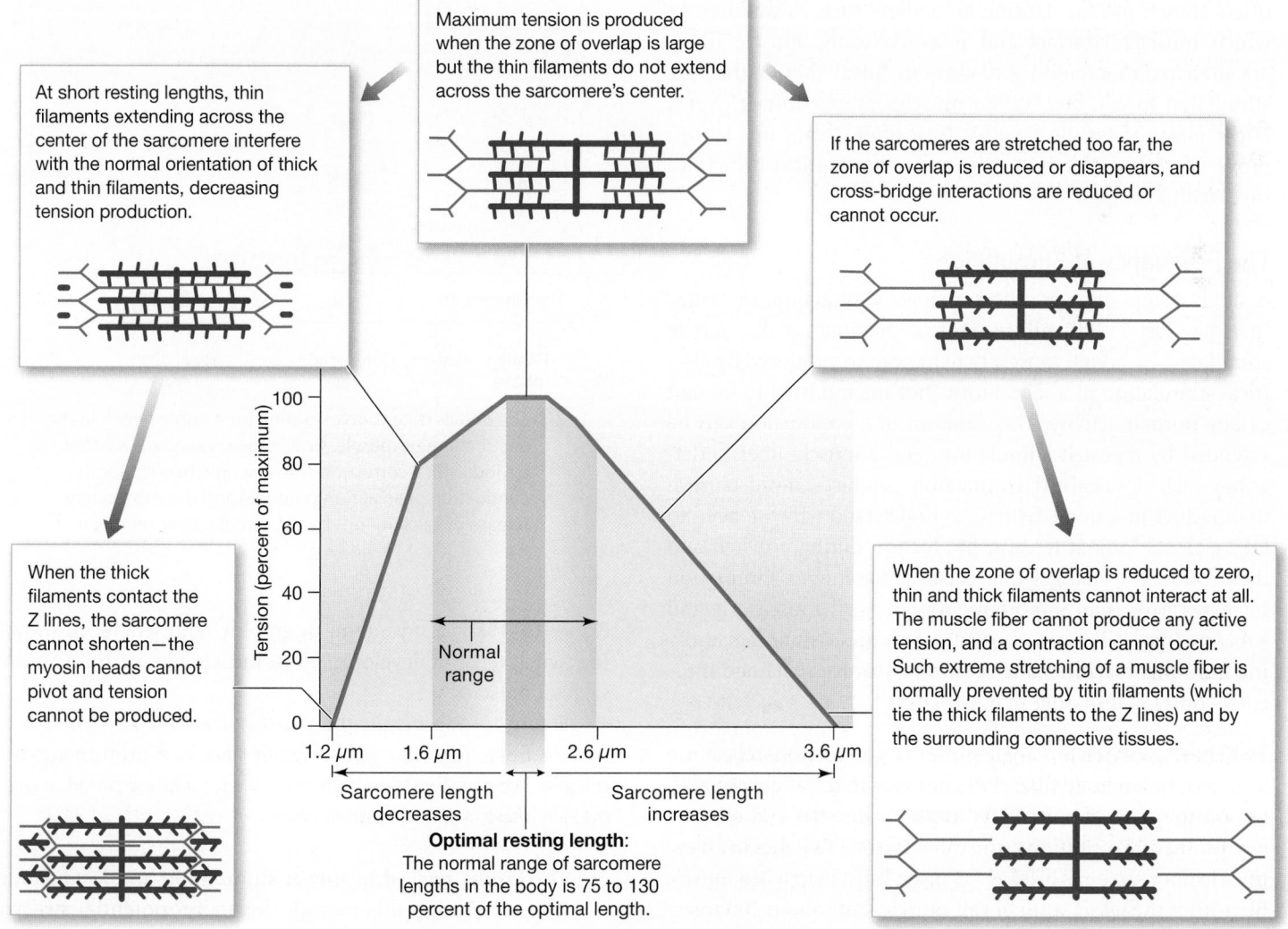

Maximum tension is produced when the zone of overlap is large but the thin filaments do not extend across the sarcomere's center.

At short resting lengths, thin filaments extending across the center of the sarcomere interfere with the normal orientation of thick and thin filaments, decreasing tension production.

If the sarcomeres are stretched too far, the zone of overlap is reduced or disappears, and cross-bridge interactions are reduced or cannot occur.

When the thick filaments contact the Z lines, the sarcomere cannot shorten—the myosin heads cannot pivot and tension cannot be produced.

When the zone of overlap is reduced to zero, thin and thick filaments cannot interact at all. The muscle fiber cannot produce any active tension, and a contraction cannot occur. Such extreme stretching of a muscle fiber is normally prevented by titin filaments (which tie the thick filaments to the Z lines) and by the surrounding connective tissues.

Tension (percent of maximum)

100
80
60
40
20
0

1.2 μm   1.6 μm   2.6 μm   3.6 μm

Normal range

Sarcomere length decreases

Sarcomere length increases

Optimal resting length:
The normal range of sarcomere lengths in the body is 75 to 130 percent of the optimal length.

A decrease in the resting sarcomere length reduces efficiency because the stimulated sarcomere cannot shorten very much before the thin filaments extend across the center of the sarcomere and collide with or overlap the thin filaments of the opposite side. This disrupts the precise three-dimensional relationship between thick and thin filaments and interferes with the binding of myosin heads to active sites and the propagation of the action potential along the T tubules. Because the number of cross-bridges is reduced, tension declines in the stimulated muscle fiber.

Tension production falls to zero when the resting sarcomere is at its shortest length. At this point, the thick filaments are jammed against the Z lines and no further shortening is possible. Although cross-bridge binding can still occur, the myosin heads cannot pivot and generate tension, because the thin filaments cannot move.

In summary, skeletal muscle fibers contract most forcefully when stimulated over a narrow range of lengths (**Figure 10–14**). The arrangement of skeletal muscles, connective tissues, and bones normally prevents extreme compression or excessive stretching. For example, straightening your elbow stretches your *biceps brachii muscle,* but the bones and ligaments of the elbow stop this movement before the muscle fibers stretch too far. During an activity such as walking, in which muscles contract and relax cyclically, muscle fibers are stretched to a length very close to "ideal" before they are stimulated to contract. When muscles must contract over a larger range of resting lengths, they often "team up" to improve efficiency. (We will discuss the mechanical principles involved in Chapter 11.)

The Frequency of Stimulation

A single stimulation produces a single contraction, or *twitch,* that may last 7–100 milliseconds, depending on the muscle stimulated. Although muscle twitches can be produced by electrical stimulation in a laboratory, they are too brief to be part of any normal activity. The duration of a contraction can be extended by repeated stimulation, and a muscle fiber undergoing such a sustained contraction produces more tension than it does in a single twitch. To understand why, we need to take a closer look at tension production during a twitch and then follow the changes that occur as the rate of stimulation increases. This is an important subject, as all consciously and subconsciously directed muscular activities—standing, walking, running, reaching, and so forth—involve sustained muscular contractions rather than twitches.

Twitches. A **twitch** is a single stimulus–contraction–relaxation sequence in a muscle fiber. Twitches vary in duration, depending on the type of muscle, its location, internal and external environmental conditions, and other factors. Twitches in an eye muscle fiber can be as brief as 7.5 msec, but a twitch in a muscle fiber from the *soleus,* a small calf muscle, lasts about 100 msec.

Figure 10–15   **The Development of Tension in a Twitch.**

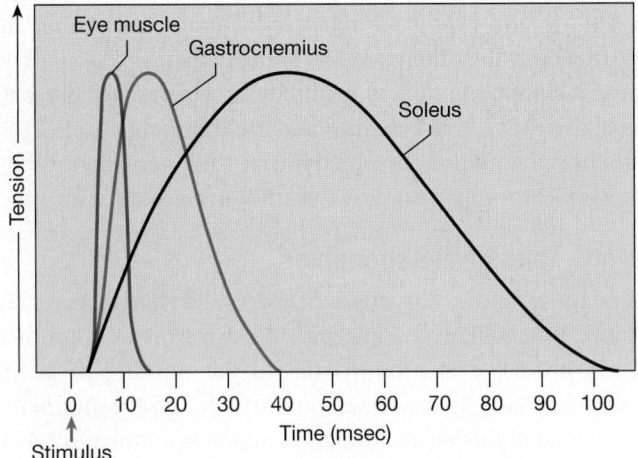

a A myogram showing differences in tension over time for a twitch in different skeletal muscles.

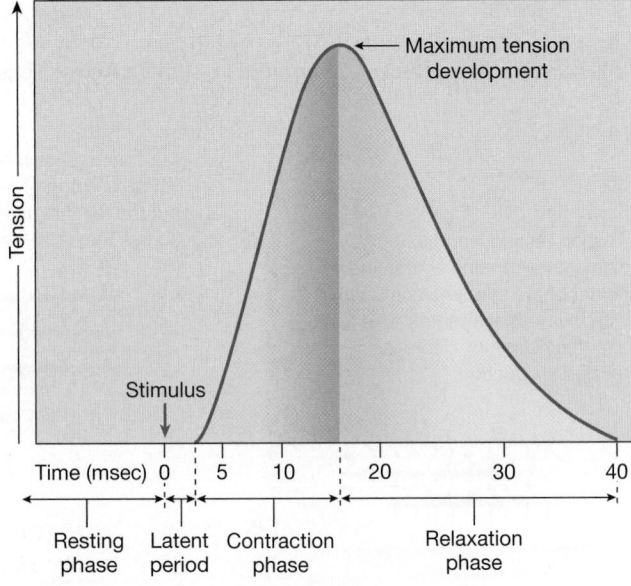

b The details of tension over time for a single twitch in the gastrocnemius muscle. Notice the presence of a latent period, which corresponds to the time needed for the conduction of an action potential and the subsequent release of calcium ions by the sarcoplasmic reticulum.

Figure 10–15a is a myogram, or graphic representation, showing twitch tension development in muscle fibers from various skeletal muscles.

Figure 10–15b details the phases of a 40-msec twitch in a muscle fiber from the *gastrocnemius muscle,* a prominent calf muscle. We can divide a single twitch into a *latent period,* a *contraction phase,* and a *relaxation phase:*

1. The **latent period** begins at stimulation and lasts about 2 msec. During this period, the action potential sweeps

across the sarcolemma, and the SR releases calcium ions. The muscle fiber does not produce tension during the latent period, because the contraction cycle has yet to begin.

2. In the **contraction phase**, tension increases to a peak. As the tension increases, calcium ions are binding to troponin, active sites on thin filaments are being exposed, and cross-bridge interactions are occurring. For this muscle fiber, the contraction phase ends roughly 15 msec after stimulation.

3. The **relaxation phase** lasts about 25 msec. During this period, Ca^{2+} levels are decreasing, active sites are being covered by tropomyosin, and the number of active cross-bridges is declining as they detach. As a result, tension decreases to resting levels.

Treppe. If a skeletal muscle is stimulated a second time immediately after the relaxation phase has ended, the resulting contraction will develop a slightly higher maximum tension than did the first contraction. The increase in peak tension shown in **Figure 10–16a** will continue over the first 30–50 stimulations. After that, the amount of tension produced will remain constant. This pattern is called **treppe** (TREP-eh, German for staircase) because the tension rises in stages, like the steps in a staircase. The rise is thought to result from a gradual increase in the concentration of Ca^{2+} in the cytosol, in part because the calcium ion pumps in the SR have too little time to recapture the ions between stimulations. Most skeletal muscles don't undergo treppe. (However, treppe is a phenomenon in cardiac muscle. It occurs if stimuli of the same intensity are sent to the muscle fiber after a latent period.)

10

Figure 10–16   **Effects of Repeated Stimulations.**

a **Treppe.** Treppe is an increase in peak tension with each successive stimulus delivered shortly after the completion of the relaxation phase of the preceding twitch.

b **Wave summation.** Wave summation occurs when successive stimuli arrive before the relaxation phase has been completed.

c **Incomplete tetanus.** Incomplete tetanus occurs if the stimulus frequency increases further. Tension production rises to a peak, and the periods of relaxation are very brief.

d **Complete tetanus.** During complete tetanus, the stimulus frequency is so high that the relaxation phase is eliminated. Tension plateaus at maximum levels.

10

Wave Summation. If a second stimulus arrives before the relaxation phase has ended, a second, more powerful contraction occurs. The addition of one twitch to another in this way is the summation of twitches, or **wave summation** (Figure 10–16b). The duration of a single twitch determines the maximum time available for wave summation. For example, if a twitch lasts 20 msec (1/50 sec), subsequent stimuli must be separated by less than 20 msec—a stimulation rate of more than 50 stimuli per second. This rate is usually expressed in terms of *stimulus frequency,* which is the number of stimuli per unit time. In this instance, a stimulus frequency of greater than 50 per second produces wave summation, whereas a stimulus frequency of less than 50 per second produces individual twitches and treppe.

Incomplete Tetanus. If the stimulation continues and the muscle is never allowed to relax completely, tension will increase until it reaches a peak value roughly four times the maximum produced by treppe (Figure 10–16c). A muscle producing almost peak tension during rapid cycles of contraction and relaxation is in **incomplete tetanus** (*tetanos,* convulsive tension).

Complete Tetanus. When a higher stimulation frequency eliminates the relaxation phase, **complete tetanus** occurs (Figure 10–16d). Action potentials arrive so rapidly that the SR does not have time to reclaim the Ca^{2+}. The high Ca^{2+} concentration in the cytosol prolongs the contraction, making it continuous.

Tension Production by Skeletal Muscles

Now that you are familiar with the basic mechanisms of muscle contraction at the level of the individual muscle fiber, we can begin to examine the performance of skeletal muscles—the organs of the muscular system. In this section, we will consider the coordinated contractions of an entire population of skeletal muscle fibers. The amount of tension produced in the skeletal muscle *as a whole* is determined by (1) the tension produced by the stimulated muscle fibers and (2) the total number of muscle fibers stimulated.

As muscle fibers actively shorten, they pull on the attached tendons, which become stretched. The tension is transferred in turn to bones, which are moved against an external load. (We will look at the interactions between the muscular and skeletal systems in Chapter 11.)

Muscle studies performed in the laboratory generally measure the tension in a tendon of a muscle. A single twitch is so brief that there isn't enough time to activate a significant percentage of the available cross-bridges. For this reason, twitches are ineffective in performing useful work. However, if a second twitch occurs before the tension returns to zero, tension will peak at a higher level, because additional cross-bridges will

form. Think of pushing a child on a swing: You push gently to start the swing moving, but if you push harder the second time, the child swings higher because the energy of the second push is added to the energy remaining from the first. Likewise, each successive contraction begins before the tension has fallen to resting levels, so the tension continues to increase until it peaks. During a tetanic contraction, there is enough time for essentially all of the potential cross-bridges to form, and tension peaks. Muscles are rarely used this way in the body, but, in a laboratory setting, muscles can be stimulated into tetanic contractions.

Motor Units and Tension Production

The amount of tension produced by the muscle as a whole is the sum of the tensions generated by the individual muscle fibers, since they are all pulling together. For this reason, you can control the amount of tension produced by a skeletal muscle by controlling the number of muscle fibers stimulated.

A typical skeletal muscle contains thousands of muscle fibers. Some motor neurons control a few muscle fibers, but most control hundreds of them. All the muscle fibers controlled by a single motor neuron constitute a **motor unit**. The size of a motor unit is an indication of how fine the control of movement can be. In the muscles of the eye, where precise control is extremely important, a motor neuron may control 4–6 muscle fibers. We have much less precise control over our leg muscles, where a single motor neuron may control 1000–2000 muscle fibers. The muscle fibers of each motor unit are intermingled with those of other motor units (Figure 10–17a). Because of this intermingling, the direction of pull exerted on the tendon does not change when the number of activated motor units changes.

We have all probably seen or felt an involuntary "muscle twitch" under the skin. Such a muscle twitch is called a **fasciculation.** Unlike the twitches discussed earlier, a fasciculation involves more than one muscle fiber. It is the synchronous contraction of a motor unit.

When you decide to move your arm, specific groups of motor neurons in the spinal cord are stimulated. The contraction begins with the activation of the smallest motor units in the stimulated muscle. These motor units generally contain muscle fibers that contract relatively slowly. As the movement continues, larger motor units containing faster and more powerful muscle fibers are activated, and tension rises steeply. The smooth, but steady, increase in muscular tension produced by increasing the number of active motor units is called **recruitment**.

Peak tension occurs when all motor units in the muscle contract in a state of complete tetanus. Such powerful contractions do not last long, however, because the individual muscle

Figure 10–17   **The Arrangement and Activity of Motor Units in a Skeletal Muscle.**

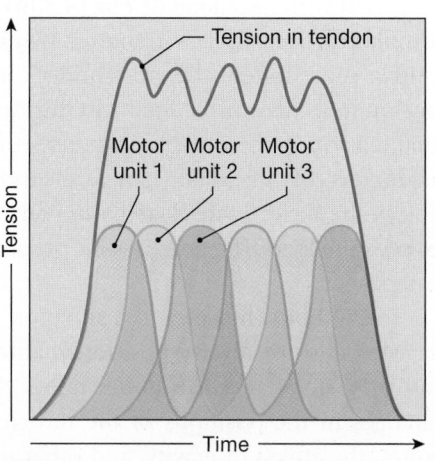

KEY

☐ Motor unit 1

☐ Motor unit 2

☐ Motor unit 3

a Muscle fibers of different motor units are intermingled, so the forces applied to the tendon remain roughly balanced regardless of which motor units are stimulated.

b The tension applied to the tendon remains relatively constant, even though individual motor units cycle between contraction and relaxation.

10

fibers soon use up their available energy reserves. During a sustained contraction, motor units are activated on a rotating basis, so some of them are resting and recovering while others are actively contracting. In this "relay team" approach, called *asynchronous motor unit summation,* each motor unit can recover somewhat before it is stimulated again (**Figure 10–17b**). As a result, when your muscles contract for sustained periods, they produce slightly less than maximum tension.

Muscle Tone

In any skeletal muscle, some motor units are always active, even when the entire muscle is not contracting. Their contractions do not produce enough tension to cause movement, but they do tense and firm the muscle. This resting tension in a skeletal muscle is called **muscle tone**. A muscle with little muscle tone appears limp and flaccid, whereas one with moderate muscle tone is firm and solid. Different motor units are stimulated at different times so some individual muscle fibers can relax while others maintain a constant tension in the attached tendon.

Resting muscle tone stabilizes the positions of bones and joints. For example, in muscles involved with balance and posture, enough motor units are stimulated to produce the tension needed to maintain body position. Muscle tone also helps prevent sudden, uncontrolled changes in the positions of

bones and joints. In addition to bracing the skeleton, the elastic nature of muscles and tendons lets skeletal muscles act as shock absorbers that cushion the impact of a sudden bump or shock. Heightened muscle tone accelerates the recruitment process during a voluntary contraction, because some of the motor units are already stimulated. Strong muscle tone also makes skeletal muscles appear firm and well defined, even at rest.

Activated muscle fibers use energy, so the greater the muscle tone, the higher the "resting" rate of metabolism. Increasing this rate is one of the significant effects of exercise in a weight-loss program. You lose weight when your daily energy use exceeds your daily energy intake in food. Although exercise consumes energy very quickly, the period of activity is usually quite brief. In contrast, elevated muscle tone increases resting energy consumption by a small amount, but the effects are cumulative, and they continue 24 hours per day.

Isotonic and Isometric Contractions

We can classify muscle contractions as *isotonic* or *isometric* on the basis of their pattern of tension production.

Isotonic Contractions. In an **isotonic contraction** (*iso-,* equal + *tonos,* tension), tension increases and the skeletal muscle's length changes. Lifting an object off a desk, walking, and running involve isotonic contractions.

There are two types of isotonic contractions: concentric and eccentric. In a **concentric contraction**, the muscle tension *exceeds* the load and the muscle *shortens*. Consider an experiment with a skeletal muscle that is 1 cm^2 in cross-sectional area and can produce roughly 4 kg (8.8 lb) of tension in complete tetanus. If we hang a load of 2 kg (4.4 lb) from that muscle and stimulate it, the muscle will shorten (**Figure 10–18a**). Before the muscle can shorten, the cross-bridges must produce enough tension to overcome the load—in this case, the 2-kg weight. At first, tension in the muscle fibers rises until the tension in the tendon exceeds the load. Then, as the muscle shortens, the tension in the skeletal muscle remains constant at a value that just exceeds the load. The term *isotonic* originated from this type of experiment.

In the body, however, the situation is more complicated. Muscles are not always positioned directly above the load, and they are attached to bones rather than to static weights. Changes in the positions of the muscle and the articulating bones, the effects of gravity, and other mechanical and physical factors interact to increase or decrease the load the muscle must overcome as a movement proceeds. Nevertheless, during a concentric contraction, the peak tension produced exceeds that load.

The speed of shortening varies with the difference between the amount of tension produced and the size of the load. If all the motor units are stimulated and the load is fairly small, the muscle will shorten very quickly. In contrast, if the muscle barely produces enough tension to overcome the load, it will shorten very slowly.

In an **eccentric contraction**, the peak tension developed is *less than* the load, and the muscle *elongates* due to the contraction of another muscle or the pull of gravity (**Figure 10–18b**). Think of a tug-of-war team trying to stop a moving car. Although everyone pulls as hard as they can, the rope slips through their fingers. The speed of elongation depends on the difference between the amount of tension developed by the active muscle fibers and the size of the load. In our analogy, the team might slow down a small car, but would have little effect on a large truck.

Eccentric contractions are very common, and they are important in a variety of movements. In these movements, you exert precise control over the amount of tension produced. By varying the tension in an eccentric contraction, you can control the rate of elongation, just as you can vary the tension in a concentric contraction.

During physical training, people commonly perform cycles of concentric and eccentric contractions, as when you do biceps curls by holding a weight in your hand and slowly flex and extend your elbow. The flexion involves concentric contractions that exceed the load posed by the weight. During extension, the same muscles are active, but the contractions are eccentric. The tension produced isn't sufficient to overcome the force of gravity, but it is enough to control the speed of movement.

Isometric Contractions. In an **isometric contraction** (*metric*, measure), the muscle as a whole does not change length, and the tension produced never exceeds the load. **Figure 10–18c** shows what happens in our experiment if we attach a weight heavier than 4 kg to the muscle and then stimulate the muscle. Although cross-bridges form and tension rises to peak values, the muscle cannot overcome the load of the weight and so cannot shorten. Examples of isometric contractions include carrying a bag of groceries and holding our heads up. Many of the reflexive muscle contractions that keep your body upright when you stand or sit involve isometric contractions of muscles that oppose the force of gravity.

Notice that when you perform an isometric contraction, the contracting muscle bulges, but not as much as it does during an isotonic contraction. In an isometric contraction, the muscle *as a whole* does not shorten, but the individual muscle fibers shorten as connective tissues stretch. The muscle fibers cannot shorten further, because the tension does not exceed the load. Normal daily activities involve a combination of isotonic and isometric muscular contractions. As you sit and read this text, isometric contractions of postural muscles stabilize your vertebrae and maintain your upright position. When you turn a page, a combination of concentric and eccentric isotonic contractions moves your arm, forearm, hand, and fingers.

TIPS & TRICKS

During an *isotonic* contraction such as lifting a baby, muscle tension remains constant while muscle length changes. During an *isometric* contraction such as holding a baby at arm's length, muscle length remains constant but muscle tension changes.

Load and Speed of Contraction

You can lift a light object more rapidly than you can lift a heavy one because load and the speed of contraction are inversely related. If the load is less than the tension produced, a concentric isotonic contraction will occur, and the muscle will shorten. The heavier the load, the longer it takes for the movement to begin, because muscle tension (which increases gradually) must exceed the load before shortening can occur (**Figure 10–19**). The contraction itself proceeds more slowly. At the molecular level, the speed of cross-bridge pivoting is reduced as the load increases.

For each muscle, an optimal combination of tension and speed exists for any given load. If you have ever ridden a 21-speed bicycle, you are probably already aware of this fact. When you are cruising along comfortably, your thigh and leg muscles are

Figure 10–18 **Concentric, Eccentric, and Isometric Contractions.**

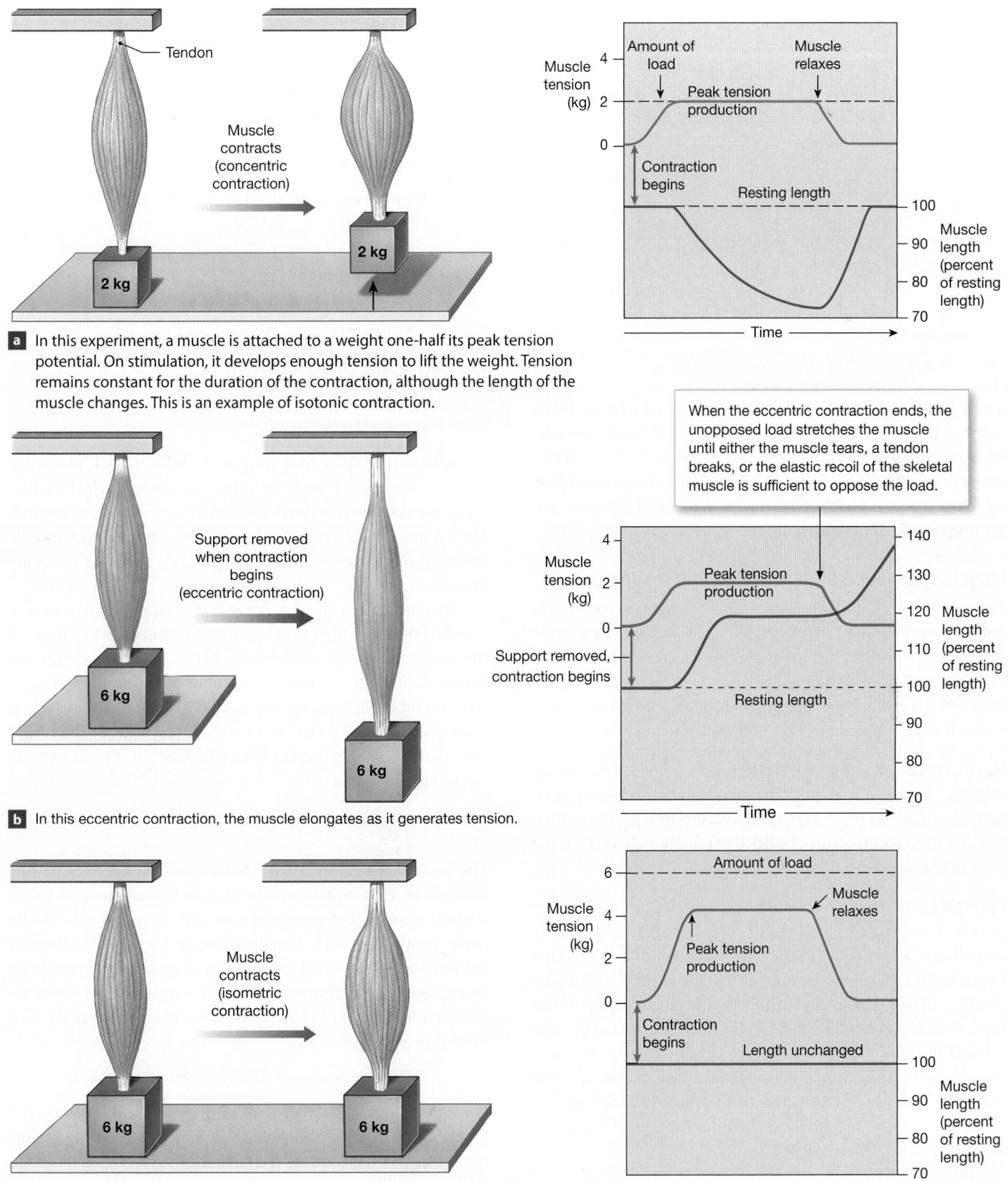

a In this experiment, a muscle is attached to a weight one-half its peak tension potential. On stimulation, it develops enough tension to lift the weight. Tension remains constant for the duration of the contraction, although the length of the muscle changes. This is an example of isotonic contraction.

b In this eccentric contraction, the muscle elongates as it generates tension.

When the eccentric contraction ends, the unopposed load stretches the muscle until either the muscle tears, a tendon breaks, or the elastic recoil of the skeletal muscle is sufficient to oppose the load.

c The same muscle is attached to a weight that exceeds its peak tension capabilities. On stimulation, tension will rise to a peak, but the muscle as a whole cannot shorten. This is an isometric contraction.

10

Figure 10–19 **Load and Speed of Contraction.** The heavier the load on a muscle, the longer it will take for the muscle to begin to shorten and the less the muscle will shorten.

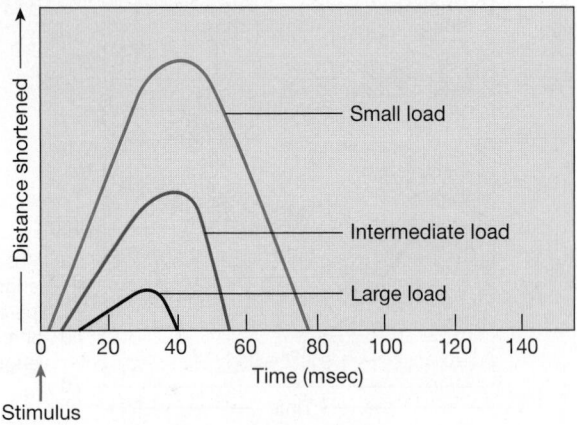

working at an optimal combination of speed and tension. When you start up a hill, the load increases. Your muscles must now develop more tension, and they move more slowly. They are no longer working at optimal efficiency. If you then shift to a lower gear, the load on your muscles decreases and their speed increases, and the muscles are once again working efficiently.

Muscle Relaxation and the Return to Resting Length

As we noted earlier, there is no active mechanism for muscle fiber elongation. The sarcomeres in a muscle fiber can shorten and develop tension, but the power stroke cannot be reversed to push the Z lines farther apart. After a contraction, a muscle fiber returns to its original length through a combination of elastic forces, opposing muscle contractions, and gravity.

Elastic Forces. When a contraction ends, some of the energy initially "spent" in stretching the tendons and distorting intracellular organelles is recovered as they recoil or rebound to their original dimensions. This elasticity gradually helps return the muscle fiber to its original resting length.

Opposing Muscle Contractions. The contraction of opposing muscles can return a muscle to its resting length more quickly than elastic forces can. Consider the muscles of the arm that flex or extend the elbow. Contraction of the *biceps brachii muscle* on the anterior part of the arm flexes the elbow, and contraction of the *triceps brachii muscle* on the posterior part of the arm extends the elbow. When the biceps brachii muscle contracts, the triceps brachii muscle is stretched. When the biceps brachii muscle relaxes, contraction of the triceps brachii muscle extends the elbow and stretches the muscle fibers of the biceps brachii muscle to their original length.

Gravity. Gravity may assist opposing muscle groups in quickly returning a muscle to its resting length after a contraction. For

example, imagine the biceps brachii muscle fully contracted with the elbow pointed at the ground. When the muscle relaxes, gravity will pull the forearm down and stretch the muscle. Although gravity can help in stretching muscles, some active muscle tension is needed to control the rate of movement and to prevent damage to the joint. In our example, eccentric contraction of the biceps brachii muscle can control the movement.

CHECKPOINT

12. Why does a muscle that has been overstretched produce less tension?

13. Can a skeletal muscle contract without shortening?

See the blue Answers tab at the back of the book.

10-6 ▌ ATP provides energy for muscle contraction

Learning Outcome Describe the mechanisms by which muscle fibers obtain the energy to power contractions.

A single muscle fiber may contain 15 billion thick filaments. When that muscle fiber is actively contracting, each thick filament breaks down around 2500 ATP molecules per second. Even a small skeletal muscle contains thousands of muscle fibers, so the ATP demands of a contracting skeletal muscle are enormous.

In practical terms, the demand for ATP in a contracting muscle fiber is so high that it would be impossible to have all the necessary energy available as ATP before the contraction begins. Instead, a resting muscle fiber contains only enough ATP and other high-energy compounds to sustain a contraction until additional ATP can be generated. Throughout the rest of the contraction, the muscle fiber will generate ATP at roughly the same rate as it is used.

ATP and CP Reserves

The primary function of ATP is to transfer energy from one location to another rather than to store it long-term. At rest, a skeletal muscle fiber produces more ATP than it needs. Under these conditions, ATP transfers energy to creatine. *Creatine* (KRĒ-uh-tēn) is a small molecule that muscle cells assemble from fragments of amino acids. The energy transfer creates another high-energy compound, **creatine phosphate (CP)**, also known as phosphocreatine:

$$ATP + creatine \longrightarrow ADP + creatine\ phosphate$$

During a contraction, each myosin head breaks down ATP, producing ADP and phosphate. The energy stored in creatine phosphate is then used to "recharge" ADP, converting it back to ATP through the reverse reaction:

$$ADP + creatine\ phosphate \longrightarrow ATP + creatine$$

The enzyme that catalyzes this reaction is **creatine kinase (CK)**. When muscle cells are damaged, CK leaks across the plasma membranes and into the bloodstream. For this reason, a high blood concentration of CK usually indicates serious muscle damage.

Table 10–1 compares the energy sources of a representative muscle fiber. A resting skeletal muscle fiber contains about six times as much creatine phosphate as readily available ATP, but when a muscle fiber is undergoing a sustained contraction, these energy reserves are exhausted in only about 15 seconds. The muscle fiber must then rely on other mechanisms to generate ATP from ADP.

ATP Generation

As we saw in Chapter 3, most cells in the body generate ATP through (1) aerobic metabolism in mitochondria and (2) glycolysis in the cytoplasm. ⤴ p. 106

Aerobic Metabolism

Aerobic metabolism normally provides 95 percent of the ATP demands of a resting cell. In this process, mitochondria absorb oxygen, ADP, phosphate ions, and organic substrates (such as pyruvate) from the surrounding cytoplasm. The molecules then enter the *citric acid cycle* (also known as the *tricarboxylic acid cycle* or the *Krebs cycle*), an enzymatic pathway that breaks down organic molecules. The carbon atoms are released from the organic substrate as carbon dioxide. The hydrogen atoms are shuttled to respiratory enzymes in the inner mitochondrial membrane, where their electrons are removed. After a series of intermediate steps involving the *electron transport chain*, the protons and electrons are combined with oxygen to form water. Along the way, large amounts of energy are released and used to make ATP. The entire process is very efficient: For each molecule of pyruvate "fed" into the citric acid cycle, the cell gains 17 ATP molecules.

Resting skeletal muscle fibers rely almost exclusively on the aerobic metabolism of fatty acids to generate ATP. These fatty acids are absorbed from the circulation. When the muscle starts contracting, the mitochondria begin breaking down molecules of pyruvate instead of fatty acids. The pyruvate is provided by the enzymatic pathway of glycolysis, which breaks down glucose in the cytoplasm. The glucose can come either from the surrounding interstitial fluid or through the breakdown of glycogen reserves within the sarcoplasm. Because a typical skeletal muscle fiber contains large amounts of glycogen, the shift from fatty acid metabolism to glucose metabolism makes it possible for the cell to continue contracting for an extended period, even without an external source of nutrients.

Glycolysis

Glycolysis is the anaerobic breakdown of glucose to pyruvate in the cytoplasm of a cell. It is an **anaerobic process** because it does not require oxygen. Glycolysis provides a net gain of 2 ATP molecules and generates 2 pyruvate molecules from each glucose molecule. The ATP produced by glycolysis in the cytoplasm is only a small fraction of that produced by aerobic metabolism, in which the breakdown of the 2 pyruvate molecules within mitochondria generates 34 ATP molecules. Thus, when energy demands are relatively low and oxygen is readily available, glycolysis is important only because it provides the substrates for aerobic metabolism in the mitochondria. Yet, because it can proceed in the absence of oxygen, glycolysis becomes an important source of energy when energy demands are at a maximum and the available supply of oxygen limits how quickly the mitochondria can produce ATP.

The glucose broken down under these conditions comes primarily from the reserves of glycogen in the sarcoplasm. Glycogen is a chain of glucose molecules. ⤴ p. 71 It is stored in granules close to the I bands of sarcomeres. Typical skeletal muscle fibers contain large glycogen reserves, which may account for 1.5 percent of the total muscle weight. When the muscle fiber begins to run short of ATP and CP, enzymes split the glycogen molecules, releasing glucose, which can be used to generate more ATP.

When energy demands are low and oxygen is abundant, glycolysis provides substrates for aerobic metabolism, and then aerobic metabolism provides the ATP needed for contraction. However, during peak periods of muscular activity, energy demands are extremely high and oxygen supplies are very limited. During these periods, glycolysis provides most of the ATP needed to sustain muscular contraction.

Table 10–1	Sources of Energy in a Typical Muscle Fiber			
Energy Source	Utilization Process	Initial Available Quantity	Number of Twitches Supported by Each Energy Source Alone	Duration of Isometric Tetanic Contraction Supported by Each Energy Source Alone
ATP	ATP ⟶ ADP + P	3 mmol	10	2 sec
CP	ADP + CP ⟶ ATP + C	20 mmol	70	15 sec
Glycogen	Glycolysis (anaerobic metabolism)	100 mmol	670	130 sec
Glycogen	Aerobic metabolism	100 mmol	12,000	2400 sec (40 min)

10

Energy Use and the Level of Muscular Activity

As physical exertion and muscular activity increases, the pattern of energy production and use changes. **Figure 10–20** illustrates muscle metabolism in a muscle fiber at rest, and at moderate and peak levels of activity.

The anaerobic process of glycolysis enables the cell to generate additional ATP when the mitochondria are unable to meet the energy demands at peak exertion (**Figure 10–20**). Unfortunately, the anaerobic production of energy from glycolysis has two important drawbacks. First, the underutilized pyruvate molecules produced through glycolysis are converted to lactic acid. Lactic acid is an organic acid that dissociates in body fluids. Because it releases hydrogen ions, production of lactic acid can lower the intracellular pH. Buffers in the sarcoplasm resist pH shifts, but these mechanisms are limited. Eventually, changes in pH affect the workings of key enzymes so that the muscle fiber can no longer contract.

Second, glycolysis is a relatively inefficient way to generate ATP. Under anaerobic conditions, each glucose molecule generates 2 pyruvate molecules, which are converted to lactic acid. In return, the cell gains 2 ATP molecules through glycolysis. Had those 2 pyruvate molecules been catabolized aerobically in a mitochondrion, the cell would have produced 34 additional ATP.

Muscle Fatigue

We say an active skeletal muscle is **fatigued** when it can no longer perform at the required level of activity. Many factors are involved in muscle fatigue. For example, muscle fatigue has been correlated with (1) depletion of metabolic reserves within the muscle fibers; (2) damage to the sarcolemma and sarcoplasmic reticulum; (3) a decline in pH within the muscle fibers and the muscle as a whole, which decreases calcium ion binding to troponin and alters enzyme activities; and (4) a sense of weariness and a reduction in the desire to continue the activity, due to the effects of low blood pH and sensations of pain. Muscle fatigue is cumulative—the effects become more pronounced as more neurons and muscle fibers are affected. The result is a gradual reduction in the capabilities and performance of the entire skeletal muscle.

If a muscle fiber is contracting at moderate levels and ATP demands can be met through aerobic metabolism, fatigue will not occur until glycogen, lipid, and amino acid reserves are depleted. This type of fatigue affects the muscles of endurance athletes, such as marathon runners, after hours of exertion. In contrast, sprinters get a different type of muscle fatigue. When a muscle produces a sudden, intense burst of activity at peak levels, glycolysis provides most of the ATP. After just seconds to minutes, tissue pH decreases due to lactic acid buildup and its dissociation, and the muscle can no longer function normally. We will return to the topics of fatigue, athletic training, and metabolic activity later in the chapter.

Normal muscle function requires (1) substantial intracellular energy reserves, (2) a normal circulatory supply, (3) normal blood oxygen levels, and (4) blood pH within normal limits. Anything that interferes with any of these factors will promote premature muscle fatigue. For example, reduced blood flow from tight clothing, heart problems, or blood loss slows the delivery of oxygen and nutrients, accelerates the buildup of lactic acid, and leads to muscle fatigue.

The Recovery Period

When a muscle fiber contracts, conditions in the sarcoplasm change. Energy reserves are consumed, heat is released, and, if the contraction was at peak levels, lactic acid is generated. In the **recovery period**, the conditions in muscle fibers are returned to normal, pre-exertion levels. After a period of moderate activity, muscle fibers may need several hours to recover. After sustained activity at higher levels, complete recovery can take a week.

Lactic Acid Removal and Recycling

Glycolysis enables a skeletal muscle to continue contracting even when mitochondrial activity is limited by the availability of oxygen. As we have seen, however, the anaerobic process of glycolysis and its production of lactic acid is not an ideal way to generate ATP. It wastes the glucose reserves of the muscle fibers, and it is potentially dangerous because lactic acid dissociation can lower the pH of the blood and tissues. During the recovery period, when oxygen is available in abundance, lactate can be converted back to pyruvate. The pyruvate can then be used either by mitochondria to generate ATP or as a substrate for enzyme pathways that synthesize glucose and rebuild glycogen reserves. During exertion, lactate diffuses out of muscle fibers and into the bloodstream. The process continues after the exertion has ended, because intracellular lactate concentrations are still relatively high. The liver absorbs the lactate and converts it to pyruvate. About 20–30 percent of these new pyruvate molecules enter liver mitochondria where they are catabolized, providing the ATP needed to convert the other 70–80 percent of the pyruvate molecules to glucose. (We will cover these processes more fully in Chapter 25.) The glucose molecules are then released into the circulation, where they are absorbed by skeletal muscle fibers and used to rebuild their glycogen reserves. This shuffling of lactate to the liver and glucose back to muscle cells is called the **Cori cycle**.

The Oxygen Debt

During the recovery period, the body's oxygen demand remains elevated above normal resting levels. The more ATP required, the more oxygen will be needed. The amount of oxygen required to restore normal, pre-exertion conditions is called the **oxygen debt**, or **excess postexercise oxygen consumption (EPOC)**.

Figure 10–20 **Muscle Metabolism.**

Muscle Metabolism in a Resting Muscle Fiber

- In a resting skeletal muscle, the demand for ATP is low, and there is more than enough oxygen available for mitochondria to meet that demand.

- Resting muscle fibers absorb fatty acids, which are broken down in the mitochondria creating a surplus of ATP.

- Some mitochondrial ATP is used to convert absorbed glucose to glycogen.

- Mitochondrial ATP is also used to convert creatine to creatine phosphate (CP).

- This results in the buildup of energy reserves (glycogen and CP) in the muscle.

Muscle Metabolism during Moderate Activity

- During moderate levels of activity, the demand for ATP increases.

- There is still enough oxygen for the mitochondria to meet that demand, but no excess ATP is produced.

- The muscle fiber now relies primarily on the aerobic metabolism of glucose from stored glycogen to generate ATP.

- If the glycogen reserves are low, the muscle fiber can also break down other substrates, such as fatty acids.

- All of the ATP now produced is used to power muscle contraction.

Muscle Metabolism during Peak Activity

- During peak levels of activity, the demand for ATP is enormous. Oxygen cannot diffuse into the fiber fast enough for the mitochondria to meet that demand. Only a third of the cell's ATP needs can be met by the mitochondria (not shown).

- The rest of the ATP comes from glycolysis, and when this produces pyruvate faster than the mitochondria can utilize it, the pyruvate builds up in the cytosol. This process is called anaerobic metabolism because no oxygen is used.

- Under these conditions, pyruvate is converted to lactic acid, which dissociates into a lactate ion and a hydrogen ion.

- The buildup of hydrogen ions increases fiber acidity, which inhibits muscle contraction, leading to rapid fatigue.

10

Most of the additional oxygen consumption occurs in skeletal muscle fibers, which must restore ATP, CP, and glycogen concentrations to their former levels, and in liver cells, which generate the ATP needed to convert excess lactate to glucose. However, several other tissues also increase their rate of oxygen consumption and ATP generation during the recovery period. For example, sweat glands increase their secretory activity until normal body temperature is restored. While the oxygen debt is being repaid, breathing rate and depth are increased. As a result, you continue to breathe heavily long after you stop exercising.

Heat Production and Loss

Muscular activity generates considerable heat. During a catabolic process, such as the breakdown of glycogen or the reactions of glycolysis, a muscle fiber captures only a portion of the released energy. ⤺ p. 62 The rest escapes as heat. A resting muscle fiber relying on aerobic metabolism captures about 42 percent of the energy released in catabolism. The other 58 percent warms the sarcoplasm, interstitial fluid, and circulating blood. Active skeletal muscles release roughly 85 percent of the heat needed to maintain normal body temperature.

When muscles become active, their energy consumption increases dramatically. As anaerobic energy production becomes the primary method of ATP generation, muscle fibers become less efficient at capturing energy. At peak levels of exertion, only about 30 percent of the released energy is captured as ATP, and the remaining 70 percent warms the muscle and surrounding tissues. Body temperature soon climbs, and heat loss at the skin accelerates through mechanisms introduced in Chapters 1 and 5. ⤺ pp. 36, 194

Hormones and Muscle Metabolism

Hormones of the endocrine system adjust metabolic activities in skeletal muscle fibers. *Growth hormone* from the pituitary gland and *testosterone* (the primary sex hormone in males) stimulate the synthesis of contractile proteins and the enlargement of skeletal muscles. *Thyroid hormones* elevate the rate of energy consumption in resting and active skeletal muscles. During a sudden crisis, hormones of the adrenal gland, notably *epinephrine* (adrenaline), stimulate muscle metabolism and increase both the duration of stimulation and the force of contraction. We will further examine the effects of hormones on muscle and other tissues in Chapter 18.

CHECKPOINT

14. **How do muscle cells continuously synthesize ATP?**

15. **What is muscle fatigue?**

16. **Define oxygen debt.**

See the blue Answers tab at the back of the book.

10-7 | Muscle performance capabilities depend on muscle fiber type and physical conditioning

Learning Outcome Relate the types of muscle fibers to muscle performance, and distinguish between aerobic and anaerobic endurance.

We can consider muscle performance in terms of **force**, the maximum amount of tension produced by a particular muscle or muscle group, and **endurance**, the amount of time during which the individual can perform a particular activity. Here we consider the factors that determine the performance capabilities of any skeletal muscle: the type, distribution, and size of muscle fibers in the muscle, and physical conditioning or training.

Types of Skeletal Muscle Fibers

The human body has three major types of skeletal muscle fibers: *fast fibers, slow fibers,* and *intermediate fibers* (Table 10–2).

Fast Fibers

Most of the skeletal muscle fibers in the body are called **fast fibers**, because they can reach peak twitch tension in 0.01 sec or less after stimulation. Fast fibers are large in diameter and contain densely packed myofibrils, large glycogen reserves, and relatively few mitochondria. Muscles dominated by fast fibers produce powerful contractions because the tension produced by a muscle fiber is directly proportional to the number of myofibrils. However, fast fibers fatigue rapidly because their contractions use ATP in massive amounts, and they have relatively few mitochondria to generate ATP. As a result, prolonged activity is supported primarily by anaerobic metabolism.

Slow Fibers

Slow fibers have only about half the diameter of fast fibers and take three times as long to reach peak tension after stimulation. These fibers are specialized in ways that enable them to continue contracting long after a fast fiber would have become fatigued. The most important specializations support aerobic metabolism in the numerous mitochondria.

One of the main characteristics of slow muscle fibers is that they are surrounded by a more extensive network of capillaries than is typical of fast muscle tissue. For this reason, they have a dramatically higher oxygen supply to support mitochondrial activity. Slow fibers also contain the red pigment **myoglobin** (MĪ-ō-glō-bin). This globular protein is similar to hemoglobin, the red oxygen-carrying pigment in blood. Both myoglobin and hemoglobin reversibly bind oxygen molecules. Other muscle fiber types contain small amounts of myoglobin, but it is most abundant in slow fibers. As a result, resting slow fibers hold substantial oxygen reserves that can be used during a

Table 10–2	Properties of Skeletal Muscle Fiber Types		
Property	**Fast Fibers**	**Slow Fibers**	**Intermediate Fibers**
Cross-sectional diameter	Large	Small	Intermediate
Time to peak tension	Rapid	Prolonged	Medium
Contraction speed	Fast	Slow	Fast
Fatigue resistance	Low	High	Intermediate
Color	White	Red	Pink
Myoglobin content	Low	High	Low
Capillary supply	Scarce	Dense	Intermediate
Mitochondria	Few	Many	Intermediate
Glycolytic enzyme concentration in sarcoplasm	High	Low	High
Substrates used for ATP generation during contraction (metabolism)	Carbohydrates (anaerobic)	Lipids, carbohydrates, amino acids (aerobic)	Primarily carbohydrates (anaerobic)
Alternative names	Type II-B, FF (fast fatigue), white, fast-twitch glycolytic	Type I, S (slow), red, SO (slow oxidative), slow-twitch oxidative	Type II-A, FR (fast resistant), fast-twitch oxidative

contraction. Skeletal muscles dominated by slow fibers are dark red because slow fibers have both an extensive capillary supply and a high concentration of myoglobin.

With oxygen reserves and a more efficient blood supply, the mitochondria of slow fibers can contribute more ATP during contractions. In addition, demand for ATP is reduced because the cross-bridges in slow fibers cycle more slowly than those of fast fibers. Thus, slow fibers are less dependent on anaerobic metabolism than are fast fibers. Glycogen reserves of slow fibers are also smaller than those of fast fibers, because some of the mitochondrial energy production involves the breakdown of stored lipids rather than glycogen. Figure 10–21 compares the appearance of fast and slow fibers.

Intermediate Fibers

Most properties of **intermediate fibers** are intermediate between those of fast fibers and slow fibers. In appearance, intermediate fibers most closely resemble fast fibers, for they contain little myoglobin and are relatively pale. They have an intermediate capillary network and mitochondrial supply around them and are more resistant to fatigue than are fast fibers.

Muscle Performance and the Distribution of Muscle Fibers

The percentages of fast, intermediate, and slow fibers in a skeletal muscle can be quite variable. In muscles that contain a mixture of fast and intermediate fibers, the proportion can change with physical conditioning. For example, if a muscle is used repeatedly for endurance events, some of the fast fibers will develop the appearance and functional capabilities of intermediate fibers. The muscle as a whole will become more resistant to fatigue.

Muscles dominated by fast fibers appear pale and are often called **white muscles**. Chicken breasts contain "white meat" because chickens use their wings only for brief intervals, as when fleeing a predator, and the power for flight comes from the anaerobic process of glycolysis in the fast fibers of their breast muscles. As we noted earlier, extensive blood vessels and myoglobin give slow fibers a reddish color, so muscles dominated by slow fibers are known as **red muscles**. Chickens walk around all day, and these movements are powered by aerobic metabolism in the slow fibers of the "dark meat" of their legs.

What about human muscles? Most contain a mixture of fiber types and so appear pink. However, there are no slow fibers in muscles of the eye or hand, where swift, but brief, contractions are required. Many back and calf muscles are dominated by slow fibers, and these muscles contract almost continuously to maintain an upright posture. Our genes determine the percentage of fast versus slow fibers in each muscle. As we noted earlier, athletic training can increase the ratio of intermediate fibers to fast fibers.

Muscle Hypertrophy and Atrophy

As a result of repeated, exhaustive stimulation, muscle fibers develop more mitochondria, a higher concentration of glycolytic enzymes, and larger glycogen reserves. Such muscle fibers have more myofibrils than do less-stimulated fibers, and each myofibril contains more thick and thin filaments. The net effect is **hypertrophy**, an enlargement of the stimulated muscle. The number of muscle fibers does not change significantly, but the muscle as a whole enlarges because each muscle fiber increases in diameter.

Figure 10–21 **Fast versus Slow Fibers.** The LM on the right is a longitudinal section of skeletal muscle, showing more mitochondria (M) and a more extensive capillary supply in a slow fiber (R, for red) than in a fast fiber (W, for white).

Slow fibers Smaller diameter, darker color due to myoglobin; fatigue resistant

Fast fibers Larger diameter, paler color; easily fatigued

LM × 170

LM × 170

Capillary

LM × 783

Hypertrophy occurs in muscles that have been repeatedly stimulated to produce near-maximal tension. The intracellular changes that occur increase the amount of tension produced when these muscles contract. The muscles of a bodybuilder are excellent examples of muscular hypertrophy.

A skeletal muscle that is not regularly stimulated by a motor neuron loses muscle tone and mass. The muscle becomes flaccid, and the muscle fibers become smaller and weaker. This reduction in muscle size, tone, and power is called **atrophy**. Individuals paralyzed by spinal cord injuries or other damage to the nervous system gradually lose muscle tone and size in the areas affected. Even a temporary reduction in muscle use can lead to muscular atrophy, as you can easily observe by comparing "before and after" limb muscles in someone who has worn a cast. Muscle atrophy is reversible at first, but dying muscle fibers are not replaced. In extreme atrophy, the functional losses are permanent. That is why physical therapy is crucial for people who are temporarily unable to move normally. Direct electrical stimulation by an external device can substitute for nerve stimulation and prevent or reduce muscle atrophy.

Because skeletal muscles depend on motor neurons for stimulation, disorders that affect the nervous system can indirectly affect the muscular system. In *polio*, a virus attacks motor neurons in the spinal cord and brain, causing muscular paralysis and atrophy.

Physical Conditioning

Physical conditioning and training schedules enable athletes and ordinary people of all ages to improve both power and endurance. In practice, the training schedule varies, depending on whether the activity is supported primarily by anaerobic or aerobic energy production.

Anaerobic endurance is the length of time muscular contraction can continue to be supported by glycolysis and by the existing energy reserves of ATP and CP. Conditioning for anaerobic endurance improves an individual's power. Anaerobic endurance is limited by (1) the amount of ATP and CP available, (2) the amount of glycogen available for breakdown, and (3) the ability of the muscle to tolerate the lactic acid generated during the period of anaerobic metabolism. Typically, the onset of muscle fatigue occurs within 2 minutes of the start of maximal activity.

Activities that require above-average levels of anaerobic endurance include a 50-meter race or swim, pole vaulting, and competitive weight lifting. These activities involve the contractions of fast fibers. The energy for the first 10–20 seconds of activity comes from the ATP and CP reserves of the cytoplasm. As these energy sources decrease, an increase in glycogen breakdown and glycolysis provide additional energy. Athletes training to improve anaerobic endurance perform frequent, brief, intensive workouts that stimulate muscle hypertrophy.

Aerobic endurance is the length of time a muscle can continue to contract while supported by mitochondrial activities. Aerobic activities do not promote muscle hypertrophy. Conditioning for aerobic endurance improves an individual's ability to continue an activity for longer periods of time. Aerobic endurance is determined primarily by the availability of substrates for aerobic respiration, which muscle fibers can obtain by breaking down carbohydrates, lipids, or amino acids. Initially, many of the nutrients broken down by muscle fibers come from reserves in the sarcoplasm. Prolonged aerobic activity, however, must be supported by nutrients provided by the circulating blood.

During exercise, blood vessels in the skeletal muscles dilate, which increases blood flow and the supply of oxygen and nutrients to the active muscle tissue. Warm-up periods are important because they both stimulate circulation in the muscles before the serious workout begins and activate the enzyme that breaks down glycogen to release glucose. Warm-ups also increase muscle temperature and thus accelerate the contraction process. Because glucose is a preferred energy source, endurance athletes such as marathon runners typically "load" or "bulk up" on carbohydrates for the three days before an event. They may also consume glucose-rich "sports drinks" during a competition. (We will consider the risks and benefits of these practices in Chapter 25.)

Training to improve aerobic endurance generally involves sustained low levels of muscular activity. Examples include jogging, distance swimming, and other exercises that do not require peak tension production. Improvements in aerobic endurance result from two factors:

1. *Alterations in the Characteristics of Muscle Fibers.* The proportion of fast and slow fibers in each muscle is genetically determined, and individual differences are significant. These variations affect aerobic endurance, because a person with more slow fibers in a particular muscle will be better able to perform under aerobic conditions than will a person with fewer. However, skeletal muscle cells respond to changes in the pattern of neural stimulation. Fast fibers trained for aerobic competition develop the characteristics of intermediate fibers, and this change improves aerobic endurance.

2. *Improvements in Cardiovascular Performance.* Cardiovascular activity affects muscular performance by delivering oxygen and nutrients to active muscles. Physical training alters cardiovascular function by accelerating blood flow, thus improving oxygen and nutrient availability. Another important benefit of endurance training is an increase in capillaries that serve exercising muscles, providing better blood flow at the cellular level. (We will examine factors involved in improving cardiovascular performance in Chapter 21.)

Physical activity, using a combination of aerobic and anaerobic exercises, should be a lifelong pursuit. It is recommended to alternate an aerobic activity, such as swimming or brisk walking, with an anaerobic activity such as weight lifting or resistance training. This combination, known as *cross-training*, enhances health by increasing muscle mass and improving aerobic endurance.

Clinical Note

Delayed-Onset Muscle Soreness Have you experienced muscle soreness the next day or several days after a period of physical exertion? Considerable controversy exists over the source and significance of this pain, which is known as *delayed-onset muscle soreness* (DOMS). You may have noticed a few of its characteristics:

- DOMS is distinct from the soreness you experience immediately after you stop exercising. The initial short-term soreness is probably related to the biochemical events associated with muscle fatigue.
- DOMS generally begins several hours after exercise and may last three or four days.
- The amount of DOMS is highest when the activity involves eccentric contractions (in which a muscle elongates despite producing tension). Activities dominated by concentric or isometric contractions produce less soreness.
- Levels of CK and myoglobin are elevated in the blood, indicating damage to muscle plasma membranes. The nature of the activity (eccentric, concentric, or isometric) has no effect on these levels. Nor can the levels be used to predict the degree of soreness experienced. Three mechanisms have been proposed to explain DOMS:

1. Small tears may occur in the muscle tissue, leaving muscle fibers with damaged membranes. The sarcolemma of each damaged muscle fiber permits the loss of enzymes, myoglobin, and other chemicals that may stimulate nearby pain receptors.
2. The pain may result from muscle spasms in the affected skeletal muscles. In some studies, stretching the muscle involved after exercise reduces the degree of soreness.
3. The pain may result from tears in the connective tissue framework and tendons of the skeletal muscle.

Some evidence supports each of these mechanisms, but it is unlikely that any one tells the whole story. For example, muscle fiber damage is certainly supported by biochemical findings. But if this damage were the only factor, the type of activity and level of circulating enzymes would be correlated with the level of pain experienced, and such is not the case. DOMS remains a puzzle to be solved.

17. Identify the three types of skeletal muscle fibers.

18. Why would a sprinter experience muscle fatigue before a marathon runner would?

19. Which activity would be more likely to create an oxygen debt: swimming laps or lifting weights?

20. Which type of muscle fibers would you expect to predominate in the leg muscles of someone who excels at endurance activities, such as cycling or long-distance running?

See the blue Answers tab at the back of the book.

10-8 ▮ Cardiac muscle tissue differs structurally and functionally from skeletal muscle tissue

Learning Outcome Identify the structural and functional differences between skeletal muscle fibers and cardiac muscle cells.

We introduced **cardiac muscle tissue** in Chapter 4 and briefly compared its properties with those of other types of muscle. **Cardiac muscle cells** are found only in the heart. We begin a more detailed examination of cardiac muscle tissue by considering its structural characteristics.

Structural Characteristics of Cardiac Muscle Tissue

Like neurons and skeletal muscle fibers, cardiac muscle cells have excitable membranes. ⊃ p. 328 Cardiac muscle cells also contain organized myofibrils, and the presence of many aligned sarcomeres gives the cells a striated appearance. However, there are important structural differences between skeletal muscle fibers and cardiac muscle cells, including the following:

- Cardiac muscle cells are relatively small, averaging 10–20 μm in diameter and 50–100 μm in length.

- A typical cardiac muscle cell has a single, centrally placed nucleus (**Figure 10–22a,b**), although a few may have two or more nuclei. Unlike skeletal muscle, cardiac muscle cells are typically branched.

- The T tubules in a cardiac muscle cell are short and broad, and there are no triads (**Figure 10–22c**). The T tubules encircle the sarcomeres at the Z lines rather than at the zones of overlap.

- The SR of a cardiac muscle cell lacks terminal cisternae, and its tubules contact the plasma membrane as well as the T tubules (**Figure 10–22c**). As in skeletal muscle fibers, an action potential triggers the release of calcium ions from the SR and the contraction of sarcomeres. However, an action potential in a cardiac muscle cell also makes the sarcolemma more permeable to extracellular calcium ions. Because their contractions require both intracellular and

Figure 10–22 **Cardiac Muscle Tissue.**

Cardiac muscle cell
Intercalated discs
Nucleus

Cardiac muscle tissue   LM × 575

a A light micrograph of cardiac muscle tissue.

Cardiac muscle cell (intact)
Intercalated disc (sectioned)

b A diagrammatic view of cardiac muscle. Note the striations and intercalated discs.

Mitochondria
Nucleus
Cardiac muscle cell (sectioned)
Intercalated discs
Myofibrils

Entrance to T tubule
Sarcolemma
Mitochondrion
Contact of sarcoplasmic reticulum with T tubule
Sarcoplasmic reticulum
Myofibrils

c Cardiac muscle tissue showing short, broad T tubules and SR that lacks terminal cisternae.

extracellular calcium ions, cardiac muscle cells are more sensitive to changes in the extracellular calcium ion concentration than are skeletal muscle fibers.

■ Cardiac muscle cells are almost totally dependent on aerobic metabolism for the energy they need to continue contracting. They have energy reserves in the form of glycogen and lipid inclusions. The sarcoplasm of a cardiac muscle cell contains large numbers of mitochondria as well as abundant reserves of myoglobin, which store the oxygen needed to break down energy reserves during times of peak activity.

■ Each cardiac muscle cell contacts several others at specialized sites known as **intercalated** (in-TER-ka-lā-ted) **discs**. ↻ p. 168 Intercalated discs play a vital role in the function of cardiac muscle, as we will see next.

Intercalated Discs

At an intercalated disc (**Figure 10–22a,b**), the sarcolemmas of two adjacent cardiac muscle cells are extensively intertwined and bound together by gap junctions and desmosomes. ↻ p. 142 These connections help stabilize the positions of adjacent cells and maintain the three-dimensional structure of the tissue.

The gap junctions allow ions and small molecules to move from one cell to another. These junctions create a direct electrical connection between two muscle cells. An action potential can travel across an intercalated disc, moving quickly from one cardiac muscle cell to another.

Two muscle cells can "pull together" with maximum efficiency because their myofibrils are essentially locked together at the intercalated disc. There the myofibrils of two interlocking muscle cells are firmly anchored to their plasma membranes.

Because cardiac muscle cells are physically, chemically, and electrically connected to one another, the entire tissue resembles a single, enormous muscle cell. For this reason, cardiac muscle has been called a *functional syncytium* (sin-SISH-ē-um; a fused mass of cells).

Functional Characteristics of Cardiac Muscle Tissue

In Chapter 20, we will examine cardiac muscle physiology in detail, but here we briefly summarize the four major functional specialties of cardiac muscle:

1. Cardiac muscle tissue contracts without neural stimulation. This property is called **automaticity**. Specialized cardiac muscle cells called **pacemaker cells** normally determine the timing of contractions.

2. The nervous system can alter the pace or rate set by the pacemaker cells and adjust the amount of tension produced during a contraction.

3. Cardiac muscle cell contractions last about 10 times as long as do those of skeletal muscle fibers. They also have longer refractory periods and do not readily fatigue.

4. The properties of cardiac muscle sarcolemmas differ from those of skeletal muscle fibers. As a result, individual twitches cannot undergo wave summation, and cardiac muscle tissue cannot produce tetanic contractions. This difference is important, because a heart in a sustained tetanic contraction could not pump blood.

CHECKPOINT

21. **Compare and contrast skeletal muscle tissue and cardiac muscle tissue.**

22. **What feature of cardiac muscle tissue allows the heart to act as a functional syncytium?**

See the blue Answers tab at the back of the book.

10-9 ▌ Smooth muscle tissue differs structurally and functionally from skeletal and cardiac muscle tissue

Learning Outcome Identify the structural and functional differences between skeletal muscle fibers and smooth muscle cells, and discuss the roles of smooth muscle tissue in systems throughout the body.

Smooth muscle tissue forms sheets, bundles, or sheaths around other tissues in almost every organ. Smooth muscles around blood vessels regulate blood flow through vital organs. In the digestive and urinary systems, rings of smooth muscle, called *sphincters,* regulate the movement of materials along internal passageways. Smooth muscles play a variety of other roles in various body systems.

■ *Integumentary System:* Smooth muscles around blood vessels regulate the flow of blood to the superficial dermis; smooth muscles of the arrector pili elevate hairs. ↻ p. 189

■ *Cardiovascular System:* Smooth muscles encircling blood vessels control the distribution of blood and help regulate blood pressure.

■ *Respiratory System:* Smooth muscles contract or relax to alter the diameters of the respiratory passageways and change the resistance to airflow.

■ *Digestive System:* Extensive layers of smooth muscle in the walls of the digestive tract play an essential role in moving materials along the tract. Smooth muscle in the walls of the gallbladder contracts to eject bile into the digestive tract.

■ *Urinary System:* Smooth muscle tissue in the walls of small blood vessels alters the rate of filtration in the kidneys. Layers of smooth muscle in the walls of the ureters transport urine to the urinary bladder, and the contraction of the smooth muscle in the wall of the urinary bladder forces urine out of the body.

- *Reproductive System:* Layers of smooth muscle help move sperm along the reproductive tract in males and cause the ejection of glandular secretions from the accessory glands into the reproductive tract. In females, layers of smooth muscle help move oocytes (and perhaps sperm) along the reproductive tract, and contractions of the smooth muscle in the walls of the uterus expel the fetus at delivery.

Figure 10–23a shows typical smooth muscle tissue as seen under a light microscope. Smooth muscle tissue differs from both skeletal and cardiac muscle tissues in structure and function (Table 10–3).

Structural Characteristics of Smooth Muscle Tissue

All three types of muscle tissue contain actin and myosin. In skeletal and cardiac muscle cells, these proteins are organized in sarcomeres, with thin and thick filaments. The internal organization of a smooth muscle cell is very different:

- Smooth muscle cells are relatively long and slender, ranging from 5 to 10 μm in diameter and from 30 to 200 μm in length.

- Each cell is spindle shaped and has a single, centrally located nucleus.

- A smooth muscle fiber has no T tubules, and the sarcoplasmic reticulum forms a loose network throughout the sarcoplasm.

- Smooth muscle cells lack myofibrils and sarcomeres. As a result, this tissue also has no striations and is called **nonstriated muscle.**

- Thick filaments are scattered throughout the sarcoplasm of a smooth muscle cell. The myosin proteins are organized differently than in skeletal or cardiac muscle cells, and smooth muscle cells have more myosin heads per thick filament.

- The thin filaments in a smooth muscle cell are attached to **dense bodies**, structures distributed throughout the sarcoplasm in a network of intermediate filaments composed of the protein *desmin* (Figure 10–23b). Some of the dense bodies are firmly attached to the sarcolemma. The dense bodies and intermediate filaments anchor the thin filaments such that when sliding occurs between thin and thick

Figure 10–23 **Smooth Muscle Tissue.**

Circular muscle layer

Longitudinal muscle layer

Smooth muscle tissue   LM × 100

a Many visceral organs contain several layers of smooth muscle tissue oriented in different directions. Here, a single sectional view shows smooth muscle cells in both longitudinal (L) and transverse (T) sections.

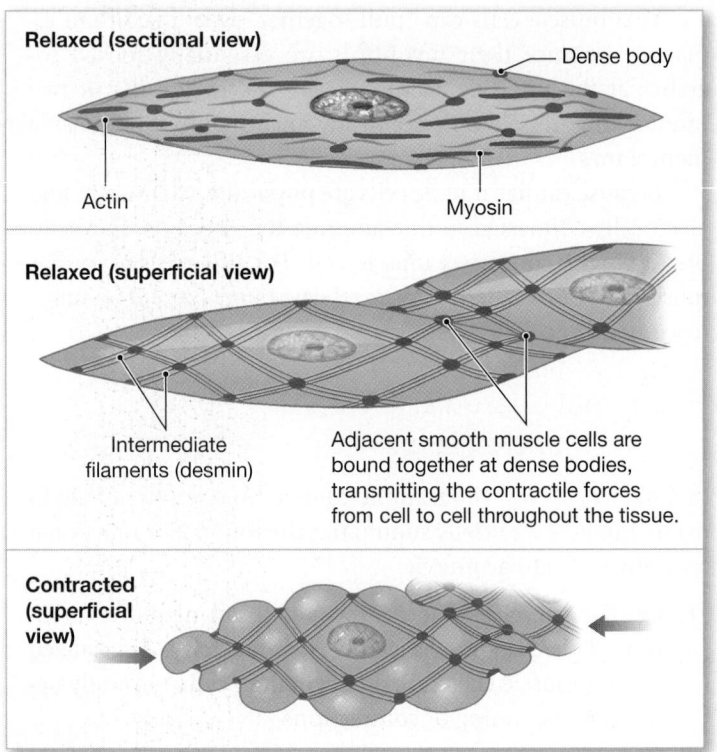

Relaxed (sectional view)

Dense body

Actin

Myosin

Relaxed (superficial view)

Intermediate filaments (desmin)

Adjacent smooth muscle cells are bound together at dense bodies, transmitting the contractile forces from cell to cell throughout the tissue.

Contracted (superficial view)

b A single relaxed smooth muscle cell is spindle shaped and has no striations. Note the changes in cell shape as contraction occurs.

Table 10–3	A Comparison of Skeletal, Cardiac, and Smooth Muscle Tissues		
Property	**Skeletal Muscle**	**Cardiac Muscle**	**Smooth Muscle**
Fiber dimensions (diameter × length)	100 μm × up to 30 cm	10–20 μm × 50–100 μm	5–10 μm × 30–200 μm
Nuclei	Multiple, near sarcolemma	Generally single, centrally located	Single, centrally located
Filament organization	In sarcomeres along myofibrils	In sarcomeres along myofibrils	Scattered throughout sarcoplasm
Sarcoplasmic reticulum (SR)	Terminal cisternae in triads at zones of overlap	SR tubules contact T tubules at Z lines	Dispersed throughout sarcoplasm, no T tubules
Control mechanism	Neural, at single neuromuscular junction on each muscle fiber	Automaticity (pacemaker cells), neural control limited to moderating pace and tension	Automaticity (pacesetter cells), neural or hormonal control (neural control limited to moderating pace and tension)
Ca^{2+} source	Release from SR	Extracellular fluid and release from SR	Extracellular fluid and release from SR
Ca^{2+} regulation	Troponin on thin filaments	Troponin on thin filaments	Calmodulin on myosin heads
Contraction	Rapid onset; may be tetanized; rapid fatigue	Slower onset; cannot be tetanized; resistant to fatigue	Slow onset; may be tetanized; resistant to fatigue
Energy source	Aerobic metabolism at moderate levels of activity; glycolysis (anaerobic during peak activity)	Aerobic metabolism, usually lipid or carbohydrate substrates	Primarily aerobic metabolism

10

filaments, the cell shortens. Dense bodies are not arranged in straight lines, so when a contraction occurs, the muscle cell twists like a corkscrew.

- Adjacent smooth muscle cells are bound together at dense bodies, transmitting the contractile forces from cell to cell throughout the tissue.

- Although smooth muscle cells are surrounded by connective tissue, the collagen fibers never unite to form tendons or aponeuroses, as they do in skeletal muscles.

Functional Characteristics of Smooth Muscle Tissue

Smooth muscle tissue differs from other muscle tissue in (1) excitation–contraction coupling, (2) length–tension relationships, (3) control of contractions, and (4) smooth muscle tone.

Excitation–Contraction Coupling

The trigger for smooth muscle contraction is the appearance of free calcium ions in the cytoplasm. On stimulation, a surge of calcium ions enters the cell from the extracellular fluid, and the sarcoplasmic reticulum releases additional calcium ions. The net result is a rise in calcium ion concentrations throughout the cell.

Once in the sarcoplasm, the calcium ions interact with **calmodulin**, a calcium-binding protein. Calmodulin then activates the enzyme **myosin light chain kinase**, which in turn enables myosin heads to attach to actin. This mechanism is quite different from that in skeletal and cardiac muscles, in which the trigger for contraction is the binding of calcium ions to troponin.

Length–Tension Relationships

Because the thick and thin filaments are scattered and are not organized into sarcomeres in smooth muscle, tension development and resting length are not directly related. A stretched smooth muscle soon adapts to its new length and retains the ability to contract on demand. This ability to function over a wide range of lengths is called **plasticity**. Smooth muscle can contract over a range of lengths four times greater than that of skeletal muscle. Plasticity is especially important in digestive organs, such as the stomach, that change greatly in volume.

Despite the lack of sarcomeres, smooth muscle contractions can be just as powerful as those of skeletal muscles. Like skeletal muscle fibers, smooth muscle cells can undergo sustained contractions.

Control of Contractions

Many smooth muscle cells are not innervated by motor neurons, and the neurons that do innervate smooth muscles are not under voluntary control. We categorize smooth muscle cells as either multiunit or visceral. **Multiunit smooth muscle cells** are innervated in motor units comparable to those of skeletal muscles, but each smooth muscle cell may be connected to more than one motor neuron. In contrast, many **visceral smooth muscle cells** lack a direct contact with any motor neuron.

Multiunit smooth muscle cells have excitable plasma membranes. They resemble skeletal muscle fibers and cardiac muscle cells in that neural activity produces an action potential that travels over the sarcolemma. However, these smooth muscle cells contract more slowly than do skeletal or cardiac muscle cells. We find multiunit smooth muscle cells in the iris of the eye, where they regulate the diameter of the pupil; along portions of the male reproductive tract; within the walls of large arteries; and in the arrector pili muscles of the skin. Multiunit smooth muscle cells do not typically occur in the digestive tract.

Visceral smooth muscle cells are arranged in sheets or layers. Within each layer, adjacent muscle cells are connected by gap junctions. As a result, whenever one muscle cell contracts, the electrical impulse that triggered the contraction can travel to adjacent smooth muscle cells. For this reason, the contraction spreads in a wave that soon involves every smooth muscle cell in the layer. The initial stimulus may be the activation of a motor neuron that contacts one of the muscle cells in the region. But smooth muscle cells also contract or relax in response to chemicals, hormones, local concentrations of oxygen or carbon dioxide, or physical factors such as extreme stretching or irritation.

Many visceral smooth muscle layers show rhythmic cycles of activity in the absence of neural stimulation. For example, these cycles are characteristic of the smooth muscle cells in the wall of the digestive tract, where **pacesetter cells** spontaneously trigger the contraction of entire muscular sheets. Visceral smooth muscle cells are located in the walls of the digestive tract, the gallbladder, the urinary bladder, and many other internal organs.

Smooth Muscle Tone

Both multiunit and visceral smooth muscle tissues have a normal background level of activity, or smooth muscle tone. Neural, hormonal, or chemical factors can also stimulate smooth muscle relaxation, producing a decrease in muscle tone. For example, smooth muscle cells at the entrances to capillaries regulate the amount of blood flow into each vessel. If the tissue becomes starved for oxygen, the smooth muscle cells relax, whereupon blood flow increases, delivering additional oxygen. As conditions return to normal, the smooth muscle regains its normal muscle tone.

CHECKPOINT

23. Identify the structural characteristics of smooth muscle tissue.

24. Which type of muscle tissue is least affected by changes in extracellular Ca^{2+} concentration during contraction?

25. Why can smooth muscle contract over a wider range of resting lengths than skeletal muscle can?

See the blue Answers tab at the back of the book.

Related Clinical Terms

botulism: A severe, potentially fatal paralysis of skeletal muscles, resulting from the consumption of a bacterial toxin.

Duchenne muscular dystrophy (DMD): One of the most common and best understood of the muscular dystrophies.

fibromyalgia: A chronic disorder characterized by widespread musculoskeletal pain, fatigue, and localized tenderness.

muscular dystrophies: A varied collection of inherited diseases that produce progressive muscle weakness and deterioration.

myasthenia gravis: A general muscular weakness resulting from a reduction in the number of ACh receptors on the motor end plate.

myopathy: Disease of muscle tissue.

RICE (rest, ice, compression, and elevation): Acronym for the standard treatment for muscle injuries, bruises, strains, and sprains.

Chapter Review

Study Outline

MasteringA&P®

Access more chapter study tools online in the MasteringA&P Study Area:

- Chapter Quizzes, Chapter Practice Test, Art-labeling Activities, Animations, MP3 Tutor Sessions, and Clinical Case Studies
- Practice Anatomy Lab PAL
- A&P Flix *A&PFlix*
- Interactive Physiology iP®
- PhysioEx PhysioEx

10-1 ▌Skeletal muscle performs six major functions p. 316

1. The three types of muscle tissue are *skeletal muscle, cardiac muscle,* and *smooth muscle.*

2. **Skeletal muscles** attach to bones directly or indirectly. Their functions are to (1) produce skeletal movement, (2) maintain posture and body position, (3) support soft tissues, (4) guard body entrances and exits, (5) maintain body temperature, and (6) store nutrient reserves.

10-2 | A skeletal muscle contains muscle tissue, connective tissues, blood vessels, and nerves p. 317

3. The entire muscle is covered by an **epimysium**. Bundles of muscle fibers (cells) are sheathed by a **perimysium**, and each muscle fiber is surrounded by an **endomysium**. At the ends of the muscle are tendons or aponeuroses that attach the muscle to bones. *(Figure 10–1)*

4. The perimysium and endomysium contain the blood vessels and nerves that supply the muscle fibers.

10-3 | Skeletal muscle fibers have distinctive features p. 318

5. A skeletal muscle fiber has a **sarcolemma** (plasma membrane); **sarcoplasm** (cytoplasm); and **sarcoplasmic reticulum (SR)**, similar to the smooth endoplasmic reticulum of other cells. **Transverse (T) tubules** and **myofibrils** aid in contraction. Filaments in a myofibril are organized into repeating functional units called **sarcomeres**. *(Figures 10–2 to 10–6)*

6. **Myofibrils** contain **myofilaments** called **thin filaments** and **thick filaments.** *(Figures 10–2 to 10–6)*

7. Thin filaments consist of **F-actin, nebulin, tropomyosin,** and **troponin.** Tropomyosin molecules cover **active sites** on the **G-actin** subunits that form the F-actin strand. Troponin binds to G-actin and tropomyosin and holds the tropomyosin in position. *(Figure 10–7)*

8. Thick filaments consist of a bundle of myosin molecules around a **titin** core. Each **myosin** molecule has a long **tail** and a globular **head,** which forms **cross-bridges** with a thin filament during contraction. In a resting muscle cell, tropomyosin prevents the myosin heads from attaching to active sites on G-actin. *(Figure 10–7)*

9. The relationship between thick and thin filaments changes as a muscle fiber contracts. *(Figure 10–8)*

10. When muscle cells contract, they create *tension* and pull on the attached tendons.

10-4 | The nervous system communicates with skeletal muscles at the neuromuscular junction p. 325

11. Neurons and skeletal muscle fibers have **excitable membranes**, plasma membranes that can propagate electrical impulses, or action potentials.

12. A neuron controls the activity of a muscle fiber at a **neuromuscular junction (NMJ).** *(Spotlight Figure 10–9)*

13. At the neuromuscular junction, when an **action potential** arrives at the neuron's **axon** (synaptic) **terminal, acetylcholine (ACh)** is released into the **synaptic cleft.** The binding of ACh to receptors on the motor end plate (and its junctional folds) of the muscle fiber leads to the generation of an action potential in the sarcolemma. *(Spotlight Figure 10–9)*

14. **Excitation–contraction coupling** occurs as the passage of an action potential along a T tubule triggers the release of Ca^{2+} from the cisternae of the SR at triads. *(Spotlight Figure 10–10)*

15. Release of Ca^{2+} initiates a **contraction cycle** of myosin head active-site exposure, cross-bridge formation, pivoting of the myosin head, cross-bridge detachment, and myosin reactivation. The calcium ions bind to troponin, which changes position and moves tropomyosin away from the active sites of actin. Cross-bridges of myosin heads then bind to actin. Next, each myosin head pivots, pulling the actin filament toward the center of the sarcomere, and then detaches. *(Spotlight Figure 10–11)*

16. During a contraction, the free end of a myofibril moves toward the end that is attached or fixed. This occurs in an intact skeletal muscle, when one end of the muscle (the *origin*) is fixed in position during a contraction and the other end (the *insertion*) moves. *(Figure 10–12)*

17. Acetylcholinesterase (AChE) breaks down ACh and limits the duration of muscle stimulation. *(Figure 10–13)*

10-5 | Sarcomere shortening and muscle fiber stimulation produce tension p. 334

18. The amount of tension produced by a muscle fiber depends on the number of cross-bridges formed.

19. Skeletal muscle fibers can contract most forcefully when stimulated over a narrow range of resting lengths. *(Figure 10–14)*

20. A **twitch** is a cycle of contraction and relaxation produced by a single stimulus. *(Figure 10–15)*

21. Repeated stimulation at a slow rate produces **treppe**, a progressive increase in twitch tension. *(Figure 10–16)*

22. Repeated stimulation before the relaxation phase ends may produce **wave summation**, in which one twitch is added to another; **incomplete tetanus**, in which tension peaks and falls at intermediate stimulus rates; or **complete tetanus**, in which the relaxation phase is eliminated by very rapid stimuli. *(Figure 10–16)*

23. The number and size of a muscle's motor units determine how precisely controlled its movements are. *(Figure 10–17)*

24. Resting **muscle tone** stabilizes bones and joints.

25. Normal activities generally include both **isotonic contractions** (in which the tension in a muscle rises and the length of the muscle changes) and **isometric contractions** (in which tension rises, but the length of the muscle stays the same). There are two types of isotonic contractions: **concentric** and **eccentric**. *(Figure 10–18)*

26. Load and speed of contraction are inversely related. *(Figure 10–19)*

27. The return to resting length after a contraction may involve elastic forces, the contraction of opposing muscle groups, and gravity.

10-6 | ATP provides energy for muscle contraction p. 342

28. Muscle contractions require large amounts of energy. *(Table 10–1)*

29. **Creatine phosphate (CP)** can release stored energy to convert ADP to ATP. *(Table 10–1)*

30. At rest or at moderate levels of activity, **aerobic metabolism** (the citric acid cycle and the electron transport chain in the mitochondria) can provide most of the ATP needed to support muscle contractions.

31. At peak levels of activity, the muscle cell relies heavily on **anaerobic metabolism (glycolysis)** to generate ATP, because the mitochondria cannot obtain enough oxygen to meet the existing ATP demands.

32. As muscular activity changes, the pattern of energy production and use changes. *(Figure 10–20)*

33. A fatigued muscle can no longer contract, because of a drop in pH due to the buildup and dissociation of **lactic acid,** the exhaustion of energy resources, or other factors.

34. The **recovery period** begins immediately after a period of muscle activity and continues until conditions inside the muscle have returned to pre-exertion levels. The **oxygen debt, or excess postexercise oxygen consumption (EPOC),** created during exercise is the amount of oxygen required during the recovery period to restore the muscle to its normal condition.

35. Circulating hormones may alter metabolic activities in skeletal muscle fibers.

10-7 ▌ Muscle performance capabilities depend on muscle fiber type and physical conditioning p. 346

36. The three types of skeletal muscle fibers are **fast fibers, slow fibers,** and **intermediate fibers.** (*Table 10–2; Figure 10–21*)
37. Fast fibers, which are large in diameter, contain densely packed myofibrils, large glycogen reserves, and relatively few mitochondria. They produce rapid and powerful contractions of relatively brief duration. (*Figure 10–21*)
38. Slow fibers are about half the diameter of fast fibers and take three times as long to contract after stimulation. Specializations such as abundant mitochondria, an extensive capillary supply, and high concentrations of **myoglobin** enable slow fibers to continue contracting for extended periods. (*Figure 10–21*)
39. Intermediate fibers are very similar to fast fibers, but have a greater resistance to fatigue.
40. Muscles dominated by fast fibers appear pale and are known as **white muscles.**
41. Muscles dominated by slow fibers are rich in myoglobin, appear red, and are known as **red muscles.**
42. Training to develop anaerobic endurance can lead to **hypertrophy** (enlargement) of the stimulated muscles.
43. **Anaerobic endurance** is the time over which muscular contractions can be sustained by glycolysis and reserves of ATP and CP.
44. **Aerobic endurance** is the time over which a muscle can continue to contract while supported by mitochondrial activities.

10-8 ▌ Cardiac muscle tissue differs structurally and functionally from skeletal muscle tissue p. 350

45. **Cardiac muscle tissue** is located only in the heart. **Cardiac muscle cells** are small; have one centrally located nucleus; have short, broad T tubules; and are dependent on aerobic metabolism. **Intercalated discs** connect neighboring cardiac muscle cells where their sarcolemmas connect. (*Figure 10–22; Table 10–3*)
46. Cardiac muscle cells contract without neural stimulation (**automaticity**), and their contractions last longer than those of skeletal muscle.
47. Because cardiac muscle twitches do not exhibit wave summation, cardiac muscle tissue cannot produce tetanic contractions.

10-9 ▌ Smooth muscle tissue differs structurally and functionally from skeletal and cardiac muscle tissue p. 351

48. **Smooth muscle tissue** is nonstriated, involuntary muscle tissue.
49. Smooth muscle cells lack sarcomeres and the resulting striations. The thin filaments are anchored to **dense bodies.** (*Figure 10–23; Table 10–3*)
50. Smooth muscle contracts when calcium ions interact with **calmodulin**, which activates **myosin light chain kinase.**
51. Smooth muscle functions over a wide range of lengths (**plasticity**).
52. In **multiunit smooth muscle cells,** each smooth muscle cell acts relatively independently of other smooth muscle cells in the organ. **Visceral smooth muscle cells** are not always innervated by motor neurons. Neurons that innervate smooth muscle cells are not under voluntary control.

Review Questions

See the blue Answers tab at the back of the book.

◆ LEVEL 1 Reviewing Facts and Terms

1. Identify the structures in the following figure.

(a) _____ (b) _____
(c) _____ (d) _____
(e) _____ (f) _____
(g) _____ (h) _____

2. The connective tissue coverings of a skeletal muscle, listed from superficial to deep, are **(a)** endomysium, perimysium, and epimysium, **(b)** endomysium, epimysium, and perimysium, **(c)** epimysium, endomysium, and perimysium, **(d)** epimysium, perimysium, and endomysium.

3. What substance limits the range of muscle movement? **(a)** actinin, **(b)** myosin, **(c)** actin, **(d)** titin.
4. The detachment of the myosin cross-bridges is directly triggered by **(a)** the repolarization of T tubules, **(b)** the attachment of ATP to myosin heads, **(c)** the hydrolysis of ATP, **(d)** calcium ions.
5. A muscle producing near-peak tension during rapid cycles of contraction and relaxation is said to be in **(a)** incomplete tetanus, **(b)** treppe, **(c)** complete tetanus, **(d)** a twitch.
6. The type of contraction in which the tension rises, but the load does not move, is **(a)** a wave summation, **(b)** a twitch, **(c)** an isotonic contraction, **(d)** an isometric contraction.
7. All motor units are not required to perform a light lifting task, but more units are needed to lift heavier weights. What is this phenomenon called? **(a)** involvement, **(b)** additional support, **(c)** recruitment, **(d)** treppe.
8. An action potential can travel quickly from one cardiac muscle cell to another because of the presence of **(a)** gap junctions, **(b)** tight junctions, **(c)** intercalated discs, **(d)** both a and b.
9. A waiter carries a tray of drinks to a table. The type of muscle contraction that is required for this action is **(a)** isotonic concentric, **(b)** isotonic eccentric, **(c)** isokinetic, **(d)** isometric.

10. What three layers of connective tissue are part of each muscle? What functional role does each layer play?
11. The _____ contains vesicles filled with acetylcholine. (a) axon terminal, (b) motor end plate, (c) neuromuscular junction, (d) synaptic cleft, (e) transverse tubule.
12. What structural feature of a skeletal muscle fiber propagates action potentials into the interior of the cell?
13. What five interlocking steps are involved in the contraction process?
14. What two factors affect the amount of tension produced when a skeletal muscle contracts?
15. List the three energy reserves stored in a typical muscle fiber.
16. What two mechanisms are used to generate ATP from glucose in muscle cells?
17. List three situations that would promote fatigue in muscle fiber.

LEVEL 2 Reviewing Concepts

18. Use ideas of the length-tension relationship to explain why a muscle sarcomere cannot generate maximal tension in inner and outer range.
19. Identify three types of muscle contraction and explain what happens to the muscle that is contracting.
20. During relaxation, muscles return to their original length because of all of the following *except* (a) actin and myosin actively pushing away from one another, (b) the contraction of opposing muscles, (c) the pull of gravity, (d) the elastic nature of the sarcolemma, (e) elastic forces.
21. According to the length–tension relationship, (a) longer muscles can generate more tension than shorter muscles, (b) the greater the zone of overlap in the sarcomere, the greater the tension the muscle can develop, (c) the greatest tension is achieved in sarcomeres where actin and myosin initially do not overlap, (d) there is an optimum range of actin and myosin overlap that will produce the greatest amount of tension, (e) both b and d are correct.

22. What characteristics of slow-twitch fibers make them ideally suited to standing for long periods of time?
23. Describe the role of calcium in muscle contraction.
24. How does cardiac muscle tissue contract without neural stimulation?
25. Atracurium is a drug that blocks the binding of ACh to ACh receptors. Give an example of a site where such binding normally occurs, and predict the physiological effect of this drug.
26. Explain why a murder victim's time of death can be estimated according to the flexibility or rigidity of the body.
27. How does an action potential at the neuromuscular junction ensure that excitation-contraction coupling occurs in the muscle fiber?

LEVEL 3 Critical Thinking and Clinical Applications

28. John is keen to improve his running performance so that he can complete his first marathon. He visits his local health store where you work. He asks you about a creatine supplementation, which is advertising 5–15 percent gains in performance. What advice would you give?
29. Linda's father suffers an apparent heart attack and is rushed to the emergency room of the local hospital. The doctor on call tells her that he has ordered some blood work and that he will be able to tell if her father actually had an attack by looking at the blood levels of CK and cardiac troponin. Why would knowing the level of CK and cardiac troponin help to indicate if a person suffered a heart attack?
30. Bill broke his leg in a football game, and after six weeks the cast is finally removed. As he steps down from the examination table, he loses his balance and falls. Why?

CLINICAL CASE Wrap-Up

A Real Eye Opener

Tamara is suffering from myasthenia gravis, an autoimmune disease of the neuromuscular junction (NMJ). In people with this autoimmune disease, the immune system mistakenly makes antibodies that damage the acetylcholine receptor sites on the surface of the motor end plates of skeletal muscle fibers. The acetylcholine (ACh) that is released by the axon terminal of the motor neuron cannot find enough intact ACh receptors on which to bind. For this reason, the muscle fiber membrane permeability to sodium ions does not change enough, and the cascade of events leading to muscular contraction is feeble. The result is weakness and easy fatiguability. This weakness improves with rest as the ACh store is built up again. Often the muscles of the eyes, including the eyelids, and other facial muscles, are affected first. All voluntary, skeletal muscles can be affected.

The neurologist injected Tamara with Tensilon (edrophonium chloride), a short-acting acetylcholinesterase inhibitor.

The drug blocks acetylcholinesterase (AChE), the enzyme that breaks down ACh. This prolongs muscle stimulation. Sadly, the dramatic effects, such as suddenly being able to open and move both eyes, are short-lived. There are, however, many treatments available for myasthenia gravis, including long-acting acetylcholinesterase inhibitors. Tamara responded well to treatment and was able to resume playing with her band.

1. How would a drug that inhibits AChE make a myasthenia gravis patient stronger?

2. Do you think myasthenia gravis would affect Tamara's heart muscle?

See the blue Answers tab at the back of the book.

11 The Muscular System

Learning Outcomes

These Learning Outcomes correspond by number to this chapter's sections and indicate what you should be able to do after completing the chapter.

11-1 ▮ Describe the arrangement of fascicles in the various types of muscles, and explain the resulting functional differences.

11-2 ▮ Describe the classes of levers, and explain how they make muscles more efficient.

11-3 ▮ Predict the actions of a muscle on the basis of its origin and insertion, and explain how muscles interact to produce or oppose movements.

11-4 ▮ Explain how the name of a muscle can help identify its location, appearance, or function.

11-5 ▮ Identify the principal axial muscles of the body, plus their origins, insertions, actions, and innervation.

11-6 ▮ Identify the principal appendicular muscles of the body, plus their origins, insertions, actions, and innervation, and compare the major functional differences between the upper and lower limbs.

11-7 ▮ Identify age-related changes of the muscular system.

11-8 ▮ Explain the functional relationship between the muscular system and other body systems, and explain the role of exercise in producing various responses in other body systems.

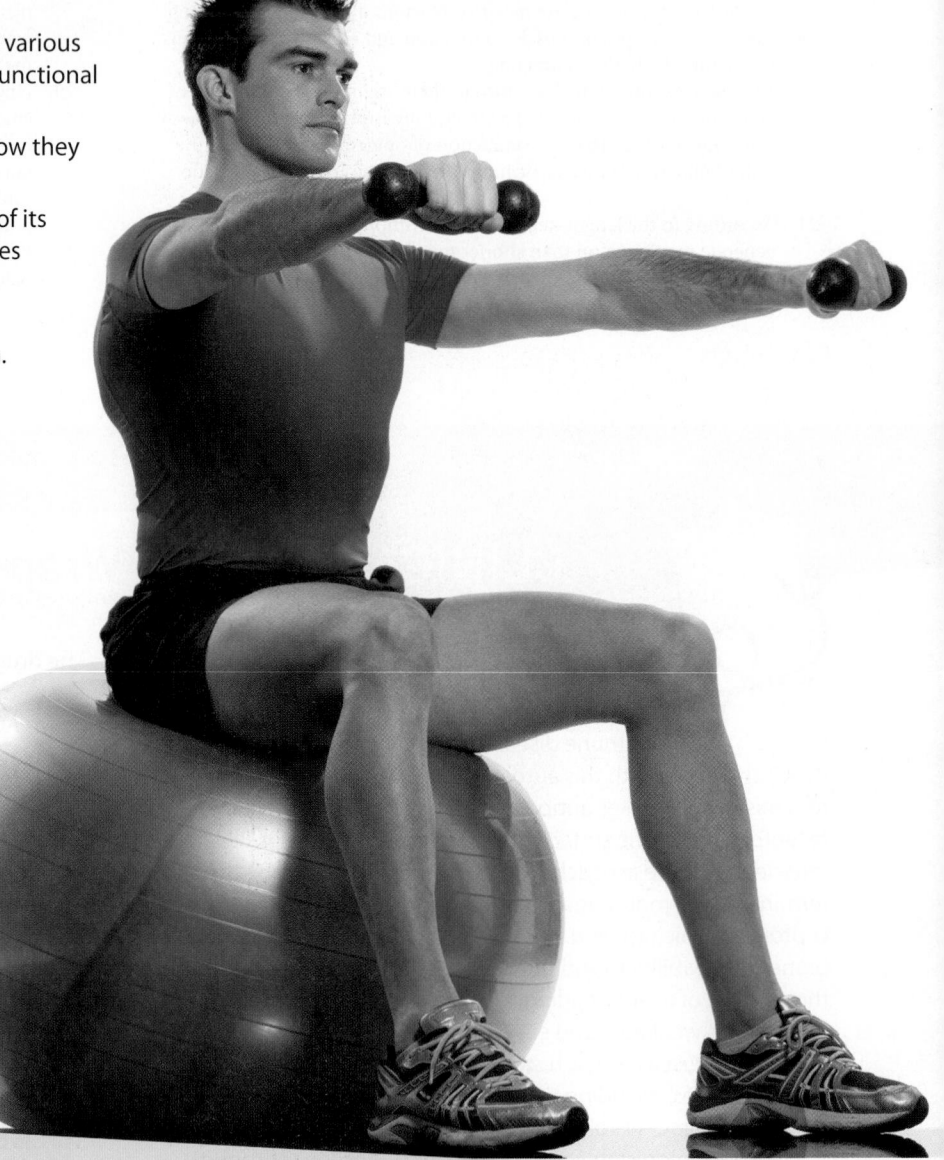

SPOTLIGHT
- Muscle Action p. 364

CLINICAL CASE
- The Weekend Warrior p. 359

Clinical Notes
- Intramuscular Injections p. 370
- Hernia p. 404

The Weekend Warrior

David is a 25-year-old insurance salesman and avid "weekend warrior" water-skier. He is skiing behind a powerful speedboat with both feet strapped into a wake board when the boat driver hits the throttle. David starts to fall awkwardly to his right. He resists the fall and digs into the board with his right foot. Suddenly he feels an extremely painful, incapacitating pop below his right gluteal region and falls into the water.

David is unable to swim; it feels as if his right lower extremity is paralyzed. Fortunately he is wearing a life vest, and within minutes he is pulled into the boat and rescued. Back at the dock, he finds he cannot flex his knee at all, but is able to get around by locking his quads, "hiking" his hip (knee in extension), and doing an "abduction

swing" to place his right foot forward while leaning on a friend for support. He cannot sit on his right side. The next evening, massive bruising appears on the entire posterior surface of his thigh and knee. Three weeks later, the bruising and swelling finally begin to resolve, but David can still feel a ball of muscle behind his right knee.

One month post injury, David is still unable to flex his knee. He visits an orthopedist, who immediately schedules him for surgery. **What kind of injury has David sustained? To find out, turn to the Clinical Case Wrap-Up on p. 410.**

An Introduction to the Muscular System

In this chapter we describe the gross anatomy of the muscular system and consider functional relationships between muscles and bones of the body. Most skeletal muscle fibers contract at similar rates and shorten to the same degree, but variations in microscopic and macroscopic organization can dramatically affect the power, range, and speed of movement produced when muscles contract.

11-1 ▮ Fascicle arrangement is correlated with muscle power and range of motion

Learning Outcome Describe the arrangement of fascicles in the various types of muscles, and explain the resulting functional differences.

Muscle fibers in a skeletal muscle form bundles called *fascicles*. ⊃ p. 318 The muscle fibers in a single fascicle are parallel, but the organization of fascicles in skeletal muscles can vary, as can the relationship between the fascicles and the associated tendon. Based on the patterns of fascicle organization, we can classify skeletal muscles as *parallel muscles, convergent muscles, pennate muscles,* and *circular muscles* (**Figure 11–1**).

Parallel Muscles

In a **parallel muscle**, the fascicles are parallel to the long axis of the muscle. Most of the skeletal muscles in the body are parallel muscles. Some are flat bands with broad attachments *(aponeuroses)* at each end. Others are plump and cylindrical, with tendons at one or both ends. Such a muscle is spindle shaped, with a central **body**, also known as the *belly*. For

example, the *biceps brachii muscle* of the arm is a parallel muscle with a central body (**Figure 11–1a**). When a parallel muscle contracts, it shortens and gets larger in diameter. You can see the bulge of the contracting biceps brachii muscle on the anterior surface of your arm when you flex your elbow. Other parallel muscles include the rectus abdominis and supinator (**Figures 11–1b** and **11–1c**).

A skeletal muscle fiber can contract until it has shortened about 30 percent. The entire parallel muscle shortens by this amount when its fibers contract together, because these fibers are parallel to the long axis of the muscle. Thus, if the muscle is 10 cm (3.94 in.) long and one end is held in place, the other end will move 3 cm when the muscle contracts. The tension developed during this contraction depends on the total number of myofibrils the muscle contains. ⊃ p. 338

We can use the cross-sectional area of the resting muscle to estimate the tension, because the myofibrils are distributed evenly through the sarcoplasm of each cell. For each 6.45 cm² (1 in.²) in cross-sectional area, a parallel muscle can develop approximately 23 kg (50 lb) of isometric tension.

Convergent Muscles

In a **convergent muscle**, muscle fascicles extending over a broad area come together, or converge, on a common attachment site (**Figure 11–1d**). The muscle may pull on a tendon, an aponeurosis, or a slender band of collagen fibers known as a **raphe** (RĀ-fē; seam). The muscle fibers typically spread out, like a fan or a broad triangle, with a tendon at the apex. Examples include the prominent *pectoralis muscles* of the chest.

A convergent muscle is versatile, because the stimulation of different portions of the muscle can change the direction of pull. However, when the entire muscle contracts, the muscle

Figure 11–1   **Muscle Types Based on Pattern of Fascicle Organization.**

Parallel Muscles

a **Parallel muscle**
(Biceps brachii muscle)

Fascicle

Body
(belly)

Cross section

b **Parallel muscle with tendinous bands**
(Rectus abdominis muscle)

c **Wrapping muscle**
(Supinator)

Convergent Muscles

d **Convergent muscle**
(Pectoralis muscles)

Base of muscle

Tendon

Cross section

Pennate Muscles

e **Unipennate muscle**
(Extensor digitorum muscle)

Extended tendon

f **Bipennate muscle**
(Rectus femoris muscle)

g **Multipennate muscle**
(Deltoid muscle)

Tendons

Cross section

Circular Muscles

h **Circular muscle**
(Orbicularis oris muscle)

Contracted

Relaxed

(h)
(d)
(g)
(a)
(b)
(e)
(c)

(f)

11

fibers do not pull as hard on the attachment site as would a parallel muscle of the same size. Why not? The reason is that convergent muscle fibers pull in different directions, rather than all pulling in the same direction as in parallel muscles.

Pennate Muscles

In a **pennate muscle** (*penna*, feather), the fascicles form a common angle with the tendon. Because the muscle fibers pull at an angle, contracting pennate muscles do not move their tendons as far as parallel muscles do. But a pennate muscle contains

more muscle fibers—and thus more myofibrils—than does a parallel muscle of the same size. For this reason, the pennate muscle produces more tension.

If all the muscle fibers are on the same side of the tendon, the pennate muscle is *unipennate*. The *extensor digitorum muscle*, a forearm muscle that extends the finger joints, is unipennate (Figure 11–1e).

More commonly, a pennate muscle has fibers on both sides of the tendon. Such a muscle is called *bipennate*. The *rectus femoris muscle*, a prominent muscle that extends the knee, is bipennate (Figure 11–1f).

If the tendon branches within a pennate muscle, the muscle is said to be *multipennate*. The triangular *deltoid muscle* of the shoulder is multipennate (**Figure 11–1g**).

Circular Muscles

In a **circular muscle**, or **sphincter** (SFINK-ter), the fascicles are concentrically arranged around an opening. When the muscle contracts, the diameter of the opening becomes smaller. Circular muscles guard body entrances and exits of internal passageways, or act as valves along the digestive and urinary tracts. An example is the *orbicularis oris muscle* of the mouth (**Figure 11–1h**).

> **CHECKPOINT**
>
> 1. Based on patterns of fascicle organization, name the four classifications of skeletal muscle tissue.
> 2. Why does a pennate muscle generate more tension than does a parallel muscle of the same size?
> 3. Which type of fascicle arrangement would you expect in a muscle guarding the opening between the stomach and the small intestine?
>
> See the blue Answers tab at the back of the book.

11-2 ▍ The three classes of levers increase muscle efficiency

Learning Outcome Describe the classes of levers, and explain how they make muscles more efficient.

Skeletal muscles do not work in isolation. For muscles attached to the skeleton, the nature and site of the connection determine the force, speed, and range of the movement. These characteristics are interdependent, and the relationships can explain a great deal about the general organization of the muscular and skeletal systems.

Attaching the muscle to a lever can modify the force, speed, or direction of movement produced by muscle contraction. A *lever* is a rigid structure—such as a board, a crowbar, or a bone—that moves on a fixed point called the *fulcrum*. A lever moves when pressure called an *applied force* is sufficient to overcome any load that would otherwise oppose or prevent such movement.

In the body, each bone is a lever and each joint is a fulcrum. Muscles provide the applied force. The load can vary from the weight of an object held in the hand to the weight of a limb or the weight of the entire body, depending on the situation. The important thing about levers is that they can change (1) the direction of an applied force, (2) the distance and speed of movement produced by an applied force, and (3) the effective strength of an applied force.

There are three classes of levers. We find examples of each in the human body (**Figure 11–2**). A pry bar or crowbar is an example of a **first-class lever**. In such a lever, the fulcrum (F) lies between the applied force (AF) and the load (L). The body has few first-class levers. One, involved with extension of the neck, is shown in **Figure 11–2a**.

In a **second-class lever** (**Figure 11–2b**), the load lies between the applied force and the fulcrum. A familiar example is a loaded wheelbarrow. The weight is the load, and the upward lift on the handle is the applied force. In this arrangement, a small force can move a larger weight because the force is always farther from the fulcrum than the load is. That is, the effective force is increased. Notice, however, that when a force moves the handle, the load moves more slowly and covers a shorter distance. In other words, the effective force is increased at the expense of speed and distance. The body has few second-class levers. Ankle extension (plantar flexion) by the calf muscles involves a second-class lever (**Figure 11–2b**).

In a **third-class lever**, such as a catapult, the applied force is between the load and the fulcrum (**Figure 11–2c**). Third-class levers are the most common levers in the body. The effect is the reverse of that for a second-class lever: Speed and distance traveled are increased at the expense of effective force. In the example shown (the biceps brachii muscle, which flexes the elbow), the load is six times farther from the fulcrum than is the applied force. The effective force is reduced to the same degree. The muscle must generate 180 kg (396 lb) of tension at its attachment to the forearm to support 30 kg (66 lb) held in the hand. However, the distance traveled and the speed of movement are increased by that same 6:1 ratio: The load will travel 45 cm (almost 18 in.) when the point of attachment moves 7.5 cm (nearly 3 in.).

Not every muscle is part of a lever system, but the presence of levers provides speed and versatility far in excess of what we would predict for the body on the basis of muscle physiology alone. Skeletal muscle fibers resemble one another closely, and their abilities to contract and generate tension are quite similar. Consider a skeletal muscle that can shorten 1 cm (0.39 in.) while it exerts a 10-kg (22-lb) pull. Without using a lever, this muscle would be performing efficiently only when moving a 10-kg (22-lb) weight a distance of 1 cm (0.39 in.). By using a lever, however, the same muscle operating at the same efficiency could move 20 kg (44 lb) a distance of 0.5 cm (0.2 in.), 5 kg (11 lb) a distance of 2 cm (0.79 in.), or 1 kg (2.2 lb) a distance of 10 cm (3.9 in.).

> **CHECKPOINT**
>
> 4. Define a lever, and describe the three classes of levers.
> 5. The joint between the occipital bone of the skull and the first cervical vertebra (atlas) is an example of which class of lever?
>
> See the blue Answers tab at the back of the book.

Figure 11–2 **The Three Classes of Levers.**

a **First-class lever.**
The fulcrum (F) lies between the applied force (AF) and the load (L).

Example: Pry bar

b **Second-class lever.**
The load (L) lies between the applied force (AF) and the fulcrum (F).

Example: Wheelbarrow

c **Third-class lever.**
The applied force (AF) is between the load (L) and the fulcrum (F).

Example: Catapult

Biceps brachii muscle

11-3 ❙ Muscle origins are at the fixed end of muscles, and insertions are at the movable end of muscles

Learning Outcome Predict the actions of a muscle on the basis of its origin and insertion, and explain how muscles interact to produce or oppose movements.

In this chapter we focus on the functional anatomy of skeletal muscles and muscle groups. Once you are familiar with the basic terminology that we introduce in this section, you will find the names and actions of skeletal muscles easy to understand.

Origins and Insertions

In Chapter 10 we noted that when both ends of a myofibril are free to move, the ends move toward the center during a contraction. ⟳ p. 332 In the body, the ends of a skeletal muscle are always attached to other structures that limit their movement. In most cases one end is fixed in position, and during a contraction the other end moves toward the fixed end. The place where the fixed end attaches to a bone, cartilage, or connective tissue is called the **origin** of the muscle. The site where the movable end attaches to another structure is called the **insertion** of the muscle. The origin is typically proximal to the insertion. When a muscle contracts, it produces a specific **action**, or movement. We describe actions using the terms introduced in Chapter 9 (flexion, extension, adduction, and so forth).

As an example, consider the *gastrocnemius muscle*, a calf muscle that extends from the distal portion of the femur to the calcaneus. As **Figure 11–2b** shows, when the gastrocnemius muscle contracts, it pulls the calcaneus toward the knee. As a result, we say that the gastrocnemius muscle has its origin at the femur and its insertion at the calcaneus. We describe its action as "extension at the ankle" or "plantar flexion."

The decision as to which end is the origin and which is the insertion is usually based on movement from the anatomical position. Part of the fun of studying the muscular system is that you can actually do the movements and think about the muscles involved. As a result, laboratory activities focusing on muscle actions are often like disorganized aerobics classes.

When we cannot easily determine the origins and insertions on the basis of movement from the anatomical position, we have other rules to use. If a muscle extends between a broad aponeurosis and a narrow tendon, the aponeurosis is the origin and the tendon is the insertion. If several tendons are at one end and just one is at the other, the muscle has multiple origins and a single insertion. These simple rules cannot cover every situation. Knowing which end is the origin and which is the insertion is ultimately less important than knowing where the two ends attach and what the muscle accomplishes when it contracts.

Most muscles originate at a bone, but some originate at a connective tissue sheath or band. Examples of these sheaths or bands include *intermuscular septa* (components of the deep fascia that may separate adjacent skeletal muscles), *tendinous inscriptions* that join muscle fibers to form long muscles such as the *rectus abdominis*, the interosseous membranes of the forearm or leg, and the fibrous sheet that spans the obturator foramen of the pelvis.

Actions

Almost all skeletal muscles either originate or insert on the skeleton. When a muscle moves a portion of the skeleton, that movement may involve flexion, extension, adduction, abduction, protraction, retraction, elevation, depression, rotation, circumduction, pronation, supination, inversion, eversion, lateral flexion, opposition, or reposition. (Before proceeding, you may want to review the discussions of planes of motion and **Spotlight Figure 9-2** and **Figures 9–3** to **9–5**.) ⟲ pp. 295–299

We can describe actions in two ways. One way describes actions in terms of the bone or region affected. For example,

we say a muscle such as the biceps brachii performs "flexion of the forearm."

Specialists such as kinesiologists and physical therapists increasingly use the second way. It identifies the joint involved. In this approach, we say the action of the biceps brachii muscle is "flexion at (or of) the elbow." Examples of muscle action are presented in **Spotlight Figure 11–3**.

In complex movements, muscles commonly work in groups rather than individually. Their cooperation improves the efficiency of a particular movement. For example, large muscles of the limbs produce flexion or extension over an extended range of motion. These muscles cannot produce powerful movements at full extension due to the positions of the articulating bones, but they are usually paired with one or more smaller muscles that provide assistance until the larger muscle can perform at maximum efficiency. At the start of the movement, the smaller muscle produces maximum tension, while the larger muscle produces minimum tension. The importance of the smaller "assistant" decreases as the movement proceeds and the effectiveness of the primary muscle increases.

We describe muscles as follows, based on their functions:

- An **agonist**, or **prime mover**, is a muscle whose contraction is mostly responsible for producing a particular movement. The biceps brachii muscle is an agonist that produces flexion at the elbow.

- An **antagonist** is a muscle whose action opposes that of a particular agonist. The *triceps brachii muscle* is an agonist that extends the elbow. For this reason, it is an antagonist of the biceps brachii muscle. Likewise, the biceps brachii is an antagonist of the triceps brachii.

  Agonists and antagonists are functional opposites. If one produces flexion, the other produces extension. When an agonist contracts to produce a particular movement, the corresponding antagonist is stretched, but it usually does not relax completely. Instead, it contracts eccentrically, with just enough tension to control the speed of the movement and ensure its smoothness. ⟲ p. 340

  You may find it easiest to learn about muscles in agonist–antagonist pairs (flexors–extensors, abductors–adductors) that act at a specific joint. This method highlights the functions of the muscles involved, and it can help organize the information into a logical framework. The tables in this chapter are arranged to support such an approach.

- When a **synergist** (*syn-*, together + *ergon*, work) contracts, it helps a larger agonist work efficiently. Synergists may provide additional pull near the insertion or may stabilize the point of origin. Their importance in assisting a particular movement may change as the movement progresses. In many cases, they are most useful at the start, when the agonist is stretched and unable to develop maximum tension.

SPOTLIGHT

Figure 11–3
Muscle Action

Watch over 50 MasteringA&P® > *A&PFlix™* on Group Muscle Actions & Joints.

The action produced by a muscle at any one joint is largely dependent upon the structure of the joint and the location of the insertion of the muscle relative to the axis of movement at the joint. The direction, or geometric paths, of the action produced by a muscle—called *lines of action*—is often represented by an arrow (or more than one arrow in fan-shaped muscles).

Flexion and Extension

At joints that permit flexion and extension, muscles whose lines of action cross the anterior side of a joint are flexors of that joint, and muscles whose lines of action cross the posterior side of a joint are extensors of that joint.

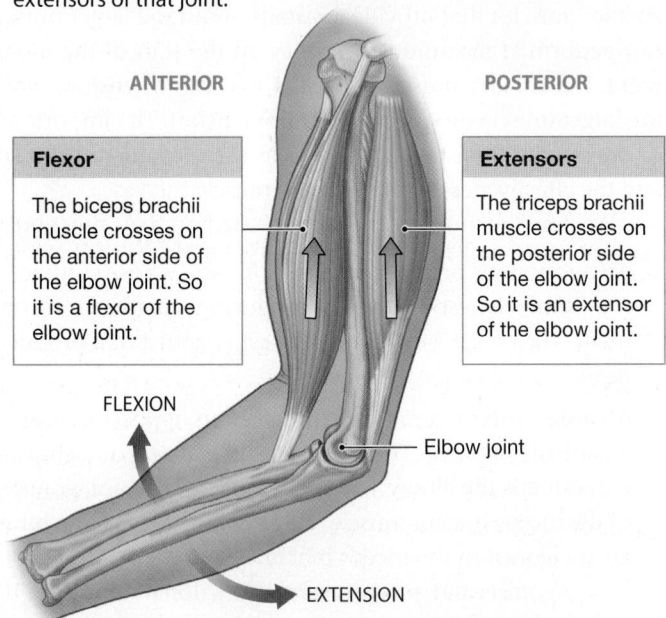

ANTERIOR

POSTERIOR

Flexor

The biceps brachii muscle crosses on the anterior side of the elbow joint. So it is a flexor of the elbow joint.

Extensors

The triceps brachii muscle crosses on the posterior side of the elbow joint. So it is an extensor of the elbow joint.

FLEXION

Elbow joint

EXTENSION

Abduction and Adduction

At joints that permit adduction and abduction, muscles whose lines of action cross the medial side of a joint are adductors of that joint, and muscles whose lines of action cross the lateral side of a joint are abductors of that joint.

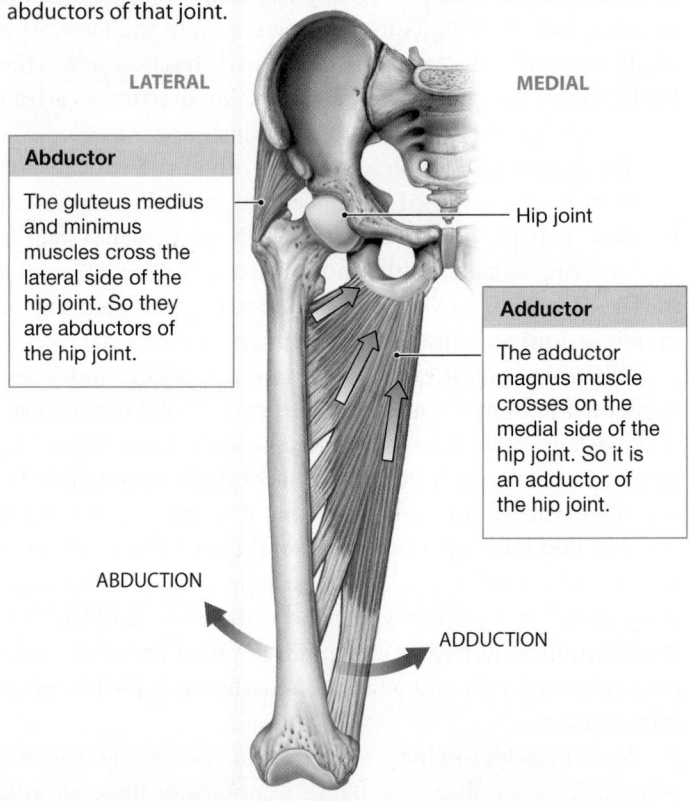

LATERAL

MEDIAL

Abductor

The gluteus medius and minimus muscles cross the lateral side of the hip joint. So they are abductors of the hip joint.

Hip joint

Adductor

The adductor magnus muscle crosses on the medial side of the hip joint. So it is an adductor of the hip joint.

ABDUCTION

ADDUCTION

Medial and Lateral Rotation

At joints that permit rotation, movement or turning of the body part occurs around its axis. The shoulder joint is a ball-and-socket joint that permits rotation. The subscapularis muscle has lines of action that cross the anterior aspect of the shoulder joint. When the subscapularis contracts it produces medial rotation at the joint. The teres minor muscle has lines of action that cross the posterior aspect of the shoulder joint. When the teres minor contracts, it produces lateral rotation at the shoulder.

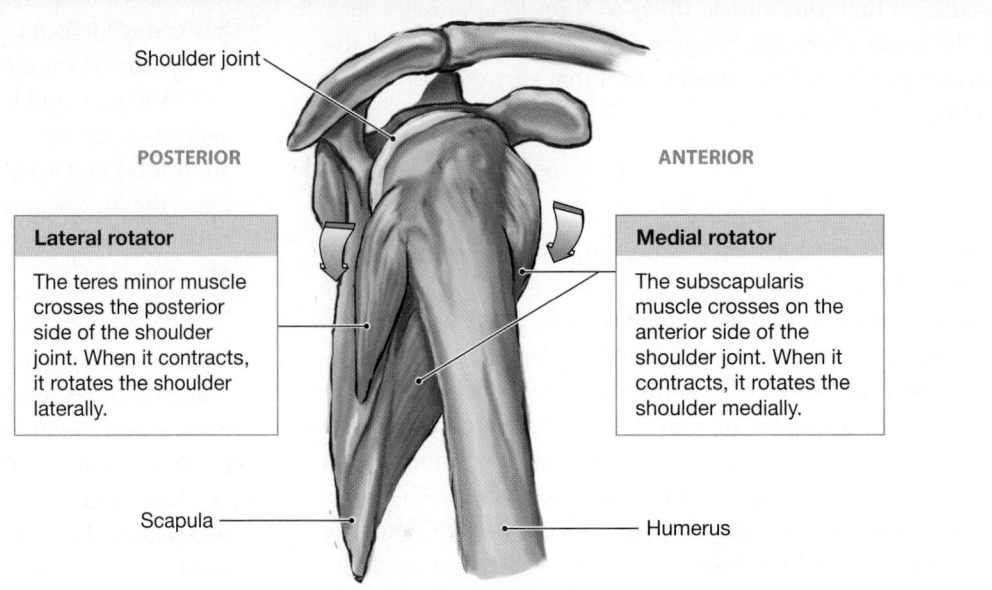

Shoulder joint

POSTERIOR

ANTERIOR

Lateral rotator

The teres minor muscle crosses the posterior side of the shoulder joint. When it contracts, it rotates the shoulder laterally.

Medial rotator

The subscapularis muscle crosses on the anterior side of the shoulder joint. When it contracts, it rotates the shoulder medially.

Scapula

Humerus

For example, the *latissimus dorsi muscle* is a large trunk muscle that extends, adducts, and medially rotates the arm at the shoulder joint. A much smaller muscle, the *teres* (TER-ēz) *major muscle,* assists in starting such movements when the shoulder joint is at full flexion.

■ A **fixator** is a synergist that assists an agonist by preventing movement at another joint, thereby stabilizing the origin of the agonist. Recall that the biceps brachii muscle is an agonist that produces flexion at the elbow. It has two tendons that originate on the scapula and one that inserts on the radius. During flexion, the trapezius and rhomboid muscles act as fixators by stabilizing and preventing the movement of the scapula.

CHECKPOINT

6. The gracilis muscle is attached to the anterior surface of the tibia at one end, and to the pubis and ischium of the pelvis at the other. When the muscle contracts, flexion occurs at the hip. Which attachment point is considered the muscle's origin?

7. Muscle A abducts the humerus, and muscle B adducts the humerus. What is the relationship between these two muscles?

8. Define the term *synergist* as it relates to muscle action.

See the blue Answers tab at the back of the book.

11-4 | Descriptive terms are used to name skeletal muscles

Learning Outcome Explain how the name of a muscle can help identify its location, appearance, or function.

Except for the *platysma* and the *diaphragm,* the complete names of all skeletal muscles include the term *muscle.* The full name, such as the biceps brachii muscle, will usually appear in the text, but for simplicity, only the descriptive name (biceps brachii) will appear in figures and tables.

You need not learn every one of the approximately 700 muscles in the human body, but you will have to become familiar with the most important ones. Fortunately, the names anatomists assigned to the muscles include descriptive terms that can help you remember the names and identify the muscles. When faced with a new muscle name, it is helpful to first identify the descriptive portions of the name. The name of a muscle may include descriptive information about its location in the body, origin and insertion, fascicle organization, position, structural characteristics, and action.

Location in the Body

Table 11–1 includes a useful summary of muscle terminology, including terms that designate specific regions of the body.

Regional terms are most common as modifiers that help identify individual muscles. In a few cases, a muscle is such a prominent feature of a body region that a name referring to the region alone will identify it. Examples include the *temporalis muscle* of the head and the *brachialis* (brā-kē-A-lis) *muscle* of the arm.

Origin and Insertion

Many muscle names include terms for body places that tell you the specific origin and insertion of each muscle. In such cases, the first part of the name indicates the origin, the second part the insertion. The *genioglossus muscle,* for example, originates at the chin *(geneion)* and inserts in the tongue *(glossus).* The names may be long and difficult to pronounce, but Table 11–1 and the anatomical terms introduced in Chapter 1 can help you identify and remember them. ⤺ pp. 40–44

Fascicle Organization

A muscle name may refer to the orientation of the muscle fascicles within a particular skeletal muscle. **Rectus** means "straight," and rectus muscles are parallel muscles whose fibers run along the long axis of the body. Because we have several rectus muscles, the name typically includes a second term that refers to a precise region of the body. For example, the *rectus abdominis muscle* is located on the abdomen, and the *rectus femoris muscle* on the thigh. Other common directional indicators include **transversus** and **oblique**, for muscles whose fibers run across (transversus) or at a slanting (oblique) angle to the longitudinal axis of the body.

Position

Muscles visible at the body surface are often called **externus** or **superficialis**. Deeper muscles are termed **internus** or **profundus**. Superficial muscles that position or stabilize an organ are called **extrinsic**. Muscles located entirely within an organ are **intrinsic**.

Structural Characteristics

Some muscles are named after distinctive structural features. The biceps brachii muscle, for example, is named after its origin. It has two tendons of origin (*bi-,* two + *caput,* head). Similarly, the triceps brachii muscle has three, and the *quadriceps* group has four.

Shape is sometimes an important clue to the name of a muscle. For example, the *trapezius* (tra-PĒ-zē-us), *deltoid, rhomboid* (ROM-boyd), and *orbicularis* (or-bik-ū-LĀ-ris) muscles look like a trapezoid, a triangle, a rhomboid, and a circle, respectively.

11

Table 11–1 Muscle Terminology			
Terms Indicating Specific Regions of the Body	**Terms Indicating Position, Direction, or Fascicle Organization**	**Terms Indicating Structural Characteristics of the Muscle**	**Terms Indicating Actions**
Abdominal (abdomen)	Anterior (front)	**Nature of Origin**	**General**
Ancon (elbow)	External (on the outside)	Biceps (two heads)	Abductor (movement away)
Auricular (ear)	Extrinsic (outside the structure)	Triceps (three heads)	Adductor (movement toward)
Brachial (arm)	Inferior (below)	Quadriceps (four heads)	Depressor (lowering movement)
Capitis (head)	Internal (away from the surface)		Extensor (straightening movement)
Carpi (wrist)	Intrinsic (within the structure)	**Shape**	Flexor (bending movement)
Cervicis (neck)	Lateral (on the side)	Deltoid (triangle)	Levator (raising movement)
Coccygeal (coccyx)	Medial (middle)	Orbicularis (circle)	Pronator (turning into prone position)
Costal (rib)	Oblique (slanting)	Pectinate (comblike)	Supinator (turning into supine position)
Cutaneous (skin)	Posterior (back)	Piriformis (pear-shaped)	Tensor (tensing movement)
Femoris (thigh)	Profundus (deep)	Platysma (flat plate)	
Glossal (tongue)	Rectus (straight)	Pyramidal (pyramid)	**Specific**
Hallux (great toe)	Superficial (toward the surface)	Rhomboid (parallelogram)	Buccinator (trumpeter)
Ilium (groin)	Superior (toward the head)	Serratus (serrated)	Risorius (laugher)
Inguinal (groin)	Transverse (crosswise)	Splenius (bandage)	Sartorius (like a tailor)
Lumbar (lumbar region)		Teres (round and long)	
Nasalis (nose)		Trapezius (trapezoid)	
Nuchal (back of neck)			
Ocular (eye)		**Other Striking Features**	
Oris (mouth)		Alba (white)	
Palpebra (eyelid)		Brevis (short)	
Pollex (thumb)		Gracilis (slender)	
Popliteal (posterior to knee)		Latae (wide)	
Psoas (loin)		Latissimus (widest)	
Radial (forearm)		Longissimus (longest)	
Scapular (scapula)		Longus (long)	
Temporal (temple)		Magnus (large)	
Thoracic (thorax)		Major (larger)	
Tibial (tibia; shin)		Maximus (largest)	
Ulnar (ulna)		Minimus (smallest)	
		Minor (smaller)	
		Vastus (great)	

Many terms refer to muscle size. Long muscles are called **longus** (long) or **longissimus** (longest). **Teres** muscles are both long and round. Short muscles are called **brevis**. Large ones are called **magnus** (big), **major** (bigger), or **maximus** (biggest). Small ones are called **minor** (smaller) or **minimus** (smallest).

Action

Many muscles are named *flexor, extensor, pronator, abductor,* and so on. These are such common actions that the names almost always include other clues as to the appearance or location of the muscle. For example, the *extensor carpi radialis longus muscle* is a long muscle along the radial (lateral) border of the forearm. When it contracts, its primary function is extension at the carpus (wrist).

A few muscles are named after the specific movements associated with special occupations or habits. The *sartorius* (sar-TOR-ē-us) *muscle,* the longest in the body, is active when you cross your legs. Before sewing machines were invented, a tailor would sit on the floor cross-legged. The name of this muscle was derived from *sartor,* the Latin word for "tailor."

The *buccinator* (BUK-si-nā-tor) *muscle* on the face compresses the cheeks—when, for example, you purse your lips and blow forcefully. *Buccinator* translates as "trumpet player."

Another facial muscle, the *risorius* (ri-SOR-ē-us) *muscle,* was supposedly named after the mood expressed. However, the Latin word *risor* means "one who laughs." A more appropriate description for the effect would be "a grimace."

Axial and Appendicular Muscles

The separation of the skeletal system into axial and appendicular divisions serves as a useful guideline for subdividing the muscular system as well:

1. The **axial muscles** arise on the axial skeleton. This category includes approximately 60 percent of the skeletal muscles in the body. They position the head and spinal column and also move the rib cage, assisting in the movements that make breathing possible. They do not play a role in movement or support of either the pectoral or pelvic girdle or the limbs.

2. The **appendicular muscles** stabilize or move components of the appendicular skeleton. These muscles include the remaining 40 percent of all skeletal muscles.

Figure 11–4 provides an overview of the major axial and appendicular muscles of the human body. These are superficial muscles, which tend to be rather large. The superficial muscles cover deeper, smaller muscles that we cannot see unless the overlying muscles are removed or *reflected*—that is, cut and pulled out of the way. Later figures that show deep muscles in specific regions will indicate whether superficial muscles have been removed or reflected.

Next we will study examples of both muscular divisions. Pay attention to patterns of origin, insertion, and action. In the figures in this chapter, you will find that some bony and cartilaginous landmarks are labeled to provide orientation. Should you need further review of skeletal anatomy, figure captions in this chapter indicate the relevant figures in *Martini's Atlas of the Human Body.*

The tables that follow also contain information about the innervation of the individual muscles. **Innervation** is the distribution of nerves to a region or organ. The tables indicate the nerves that control each muscle. Many of the muscles of the head and neck are innervated by *cranial nerves,* which originate at the brain and pass through the foramina of the skull. In addition, *spinal nerves* are connected to the spinal cord and pass through the intervertebral foramina. For example, spinal nerve L_1 passes between vertebrae L_1 and L_2. Spinal nerves may form a complex network called a *plexus* after exiting the spinal cord. One branch of this network may contain axons from several spinal nerves. Many tables identify the spinal nerves involved as well as the names of their specific branches.

CHECKPOINT

9. Identify the kinds of descriptive information used to name skeletal muscles.

10. What does the name *flexor carpi radialis longus* tell you about this muscle?

See the blue Answers tab at the back of the book.

11-5 ▮ Axial muscles are muscles of the head and neck, vertebral column, trunk, and pelvic floor

Learning Outcome Identify the principal axial muscles of the body, plus their origins, insertions, actions, and innervation.

The axial muscles fall into logical groups on the basis of location, function, or both. The groups do not always have distinct anatomical boundaries. For example, a function such as extension of the vertebral column involves muscles along its entire length and movement at each of the intervertebral joints. We will discuss the axial muscles in four groups:

1. *The Muscles of the Head and Neck.* This group includes muscles that move the face, tongue, and larynx. They are responsible for verbal and nonverbal communication—laughing, talking, frowning, smiling, whistling, and so on. You also use these muscles while eating—especially in sucking and chewing—and even while looking for food, as some of them control your eye movements. The group does not include muscles of the neck that are involved with movements of the vertebral column.

2. *The Muscles of the Vertebral Column.* This group includes numerous flexors, extensors, and rotators of the vertebral column.

3. *The Oblique and Rectus Muscles.* This group forms the muscular walls of the trunk between the first thoracic vertebra and the pelvis. In the thoracic area the ribs separate these muscles, but over the abdominal surface the muscles form broad muscular sheets. The neck also has oblique and rectus muscles. They do not form a complete muscular wall, but they share a common developmental origin with the oblique and rectus muscles of the trunk.

4. *The Muscles of the Pelvic Floor.* These muscles extend between the sacrum and pelvic girdle. The group forms the *perineum,* a muscular sheet that closes the pelvic outlet.

Muscles of the Head and Neck

We can divide the muscles of the head and neck into several functional groups. The *muscles of facial expression,* the *muscles of mastication* (chewing), the *muscles of the tongue,* and the *muscles of the pharynx* originate on the skull or hyoid bone.

Figure 11–4   **An Overview of the Major Skeletal Muscles.**

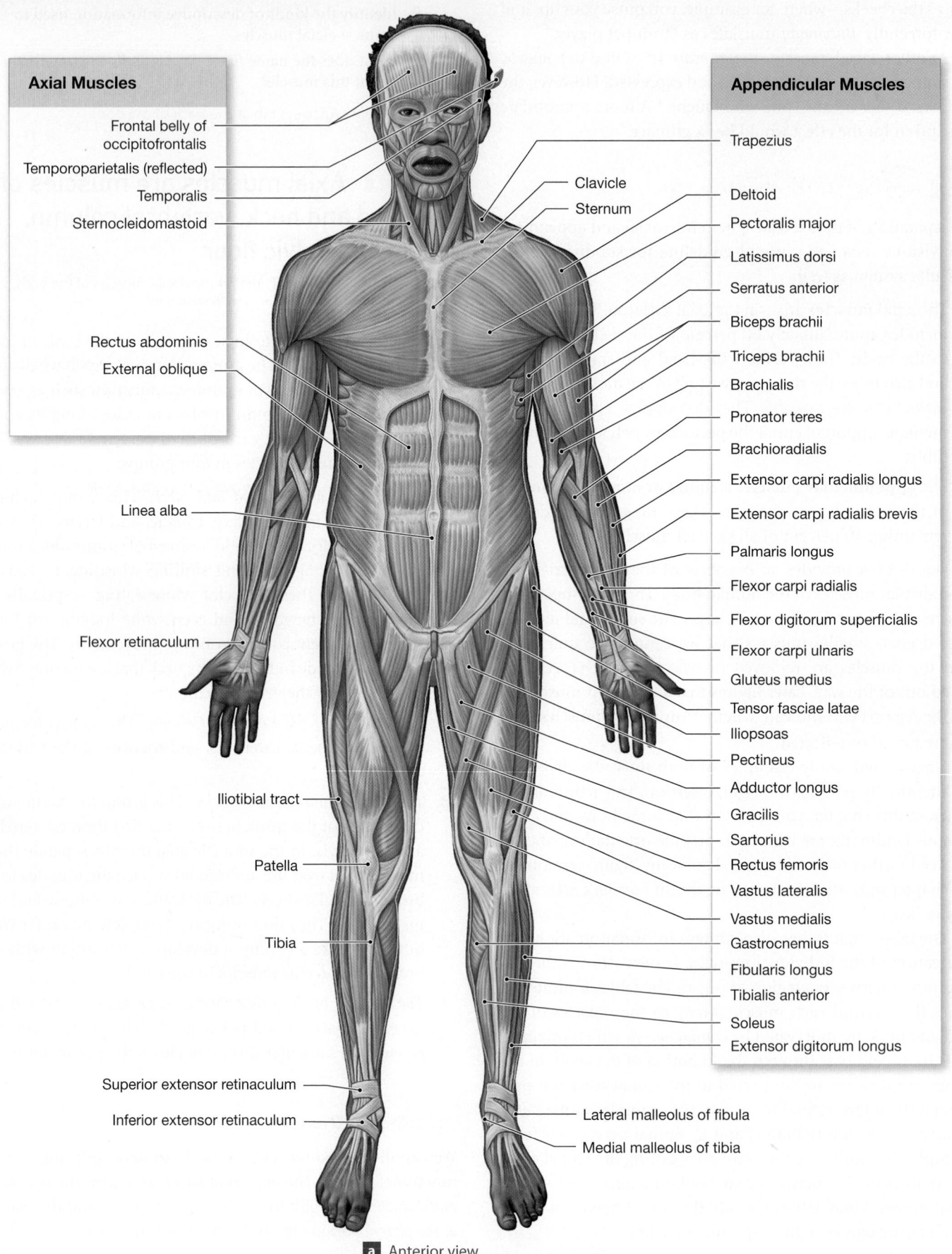

Axial Muscles

Frontal belly of occipitofrontalis
Temporoparietalis (reflected)
Temporalis
Sternocleidomastoid
Rectus abdominis
External oblique
Linea alba
Flexor retinaculum
Iliotibial tract
Patella
Tibia
Superior extensor retinaculum
Inferior extensor retinaculum

Clavicle
Sternum

Appendicular Muscles

Trapezius
Deltoid
Pectoralis major
Latissimus dorsi
Serratus anterior
Biceps brachii
Triceps brachii
Brachialis
Pronator teres
Brachioradialis
Extensor carpi radialis longus
Extensor carpi radialis brevis
Palmaris longus
Flexor carpi radialis
Flexor digitorum superficialis
Flexor carpi ulnaris
Gluteus medius
Tensor fasciae latae
Iliopsoas
Pectineus
Adductor longus
Gracilis
Sartorius
Rectus femoris
Vastus lateralis
Vastus medialis
Gastrocnemius
Fibularis longus
Tibialis anterior
Soleus
Extensor digitorum longus
Lateral malleolus of fibula
Medial malleolus of tibia

a Anterior view
ATLAS: Plates 1a; 39a–d

Figure 11–4   **An Overview of the Major Skeletal Muscles** *(continued)*.

Axial Muscles

Occipital belly of
occipitofrontalis

Sternocleidomastoid

External oblique

Appendicular Muscles

Trapezius

Deltoid

Infraspinatus

Teres minor

Teres major

Rhomboid major

Triceps brachii (long head)

Triceps brachii (lateral head)

Latissimus dorsi

Brachioradialis

Extensor carpi radialis longus

Anconeus

Flexor carpi ulnaris

Extensor digitorum

Extensor carpi ulnaris

Gluteus medius

Tensor fasciae latae

Gluteus maximus

Adductor magnus

Semitendinosus

Semimembranosus

Gracilis

Biceps femoris

Sartorius

Plantaris

Gastrocnemius

Soleus

Iliotibial tract

Calcaneal
tendon

Calcaneus

11

b Posterior view
ATLAS: Plates 1b; 40a,b

11

Muscles involved with sight and hearing also are based on the skull. Here, we will consider the *extrinsic eye muscles*—those associated with movements of the eye. In Chapter 17 we will discuss the intrinsic eye muscles, which control the diameter of the pupil and the shape of the lens, and the tiny skeletal muscles associated with the auditory ossicles.

In the neck, the *extrinsic muscles of the larynx* adjust the position of the hyoid bone and larynx. We will examine the intrinsic laryngeal muscles, including those of the vocal cords, in Chapter 23.

Muscles of Facial Expression

The muscles of facial expression originate on the surface of the skull (**Figure 11–5** and Table 11–2). At their insertions, the fibers of the epimysium are woven into those of the superficial fascia and the dermis of the skin. For this reason, when these muscles contract, the skin moves.

The largest group of facial muscles is associated with the mouth. The **orbicularis oris** muscle constricts the opening, and other muscles move the lips or the corners of the mouth. The **buccinator** muscle has two functions related to eating (in addition to its importance to musicians). During chewing, it cooperates with the masticatory muscles by moving food back across the teeth from the *vestibule*, the space inside the cheeks. In infants, the buccinator provides suction for suckling at the breast.

Smaller groups of muscles control movements of the eyebrows and eyelids, the scalp, the nose, and the external ear. The **epicranium** (ep-i-KRĀ-nē-um; *epi-*, on + *kranion*, skull), or scalp, contains the **temporoparietalis** muscle and

the **occipitofrontalis** muscle, which has a *frontal belly* and an *occipital belly*. The two bellies are separated by the **epicranial aponeurosis**, a thick, collagenous sheet.

The **platysma** (pla-TIZ-muh; *platys,* flat) covers the anterior surface of the neck. This muscle extends from the base of the neck to the periosteum of the mandible and the fascia at the corner of the mouth. One of the effects of aging is the loss of muscle tone in the platysma, resulting in a looseness of the skin of the anterior throat.

Extrinsic Eye Muscles

Six **extrinsic eye muscles**, also known as the *oculomotor muscles*, originate on the surface of the orbit and control the position of each eye. These muscles, shown in **Figure 11–6** and detailed in Table 11–3, are the **inferior rectus**, **medial rectus**, **superior rectus**, **lateral rectus**, **inferior oblique**, and **superior oblique** muscles.

Muscles of Mastication

The muscles of mastication move the mandible at the temporomandibular joint (TMJ) (**Figure 11–7** and Table 11–4). The large **masseter** muscle is the strongest jaw muscle. The **temporalis** muscle assists in elevation of the mandible. You can feel these muscles in action by clenching your teeth while resting your hand on the side of your face below and then above the zygomatic arch. The **pterygoid** muscles, used in various combinations, can elevate, depress, or protract the mandible or slide it from side to side, a movement called *lateral excursion*. These movements are important in making efficient use of your teeth while you chew foods of various consistencies.

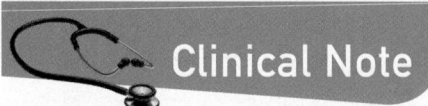

Clinical Note

Intramuscular Injections Drugs are commonly injected into muscle or adipose tissues rather than directly into the bloodstream. (Accessing blood vessels may be technically more complicated.) An **intramuscular (IM) injection** introduces a fairly large volume of a drug, which then enters the circulation gradually. The drug is introduced into the mass of a large skeletal muscle. Uptake is generally faster and accompanied by less tissue irritation than when drugs are administered *intradermally* (injected into the dermis) or *hypodermally* (injected into the hypodermis). Depending on the size of the muscle, up to 5 mL of fluid may be injected at one time. Multiple injections are possible. A decision on the injection technique and the injection site is based on the type of drug and its concentration.

For IM injections, the most common complications involve accidental injection into a blood vessel or piercing of a

nerve. The sudden entry of massive quantities of drug into the bloodstream can have fatal consequences. Damage to a nerve can cause motor paralysis or sensory loss. For these reasons, the site of the injection must be selected with care. Bulky muscles that contain few large vessels or nerves are ideal sites. The gluteus medius muscle or the posterior, lateral, superior part of the gluteus maximus muscle is commonly selected. The deltoid muscle of the arm, about 2.5 cm (1 in.) distal to the acromion, is another effective site. Probably most satisfactory from a technical point of view is the vastus lateralis muscle of the thigh. An injection into this thick muscle will not encounter vessels or nerves, but may cause pain later when the muscle is used in walking. This is the preferred injection site in infants before they start walking, as their gluteal and deltoid muscles are relatively small. The site is also used in elderly patients or others with atrophied gluteal and deltoid muscles.

Figure 11–5   **Muscles of Facial Expression.** ATLAS: Plate 3a–d

Epicranial aponeurosis

Frontal belly of occipitofrontalis

Procerus

Orbicularis oculi

Nasalis

Levator labii superioris

Zygomaticus minor

Levator anguli oris

Zygomaticus major

Mentalis (cut)

Orbicularis oris

Depressor labii inferioris

Depressor anguli oris

Omohyoid

Platysma (cut and reflected)

Temporoparietalis (cut and reflected)

Temporalis

Occipital belly of occipitofrontalis

Masseter

Buccinator

Sternocleidomastoid

Trapezius

a **Lateral view**

Frontal belly of occipitofrontalis

Corrugator supercilii

Temporalis (temporoparietalis removed)

Orbicularis oculi

Nasalis

Zygomaticus minor

Zygomaticus major

Orbicularis oris

Risorius

Platysma

Mentalis (cut)

Thyroid cartilage of the larynx

Epicranial aponeurosis

Temporoparietalis (cut and reflected)

Temporalis

Procerus

Levator labii superioris

Levator anguli oris

Masseter

Buccinator

Depressor anguli oris

Depressor labii inferioris

Sternal head of sternocleidomastoid

Clavicular head of sternocleidomastoid

Trapezius

Clavicle

Platysma (cut and reflected)

b **Anterior view**

Table 11–2 Muscles of Facial Expression (Figure 11–5)

Region and Muscle	Origin	Insertion	Action	Innervation
MOUTH				
Buccinator	Alveolar processes of maxilla and mandible	Blends into fibers of orbicularis oris	Compresses cheeks	Facial nerve (N VII)*
Depressor labii inferioris	Mandible between the anterior midline and the mental foramen	Skin of lower lip	Depresses lower lip	Facial nerve (N VII)
Levator labii superioris	Inferior margin of orbit, superior to the infra-orbital foramen	Orbicularis oris	Elevates upper lip	Facial nerve (N VII)
Levator anguli oris	Maxilla below the infra-orbital foramen	Corner of mouth	Elevates corner of mouth	Facial nerve (N VII)
Mentalis	Incisive fossa of mandible	Skin of chin	Elevates and protrudes lower lip	Facial nerve (N VII)
Orbicularis oris	Maxilla and mandible	Lips	Compresses, purses lips	Facial nerve (N VII)
Risorius	Fascia surrounding parotid salivary gland	Angle of mouth	Draws corner of mouth to the side	Facial nerve (N VII)
Depressor anguli oris	Anterolateral surface of mandibular body	Skin at angle of mouth	Depresses corner of mouth	Facial nerve (N VII)
Zygomaticus major	Zygomatic bone near zygomaticomaxillary suture	Angle of mouth	Retracts and elevates corner of mouth	Facial nerve (N VII)
Zygomaticus minor	Zygomatic bone posterior to zygomaticotemporal suture	Upper lip	Retracts and elevates upper lip	Facial nerve (N VII)
EYE				
Corrugator supercilii	Orbital rim of frontal bone near nasal suture	Eyebrow	Pulls skin inferiorly and anteriorly; wrinkles brow	Facial nerve (N VII)
Levator palpebrae superioris (*Figure 11–6a,b*)	Tendinous band around optic foramen	Upper eyelid	Elevates upper eyelid	Oculomotor nerve (N III)**
Orbicularis oculi	Medial margin of orbit	Skin around eyelids	Closes eye	Facial nerve (N VII)
NOSE				
Procerus	Nasal bones and lateral nasal cartilages	Aponeurosis at bridge of nose and skin of forehead	Moves nose, changes position and shape of nostrils	Facial nerve (N VII)
Nasalis	Maxilla and alar cartilage of nose	Bridge of nose	Compresses bridge, depresses tip of nose; elevates corners of nostrils	Facial nerve (N VII)
EAR				
Temporoparietalis	Fascia around external ear	Epicranial aponeurosis	Tenses scalp, moves auricle of ear	Facial nerve (N VII)
SCALP (EPICRANIUM)				
Occipitofrontalis				
Frontal belly	Epicranial aponeurosis	Skin of eyebrow and bridge of nose	Raises eyebrows, wrinkles forehead	Facial nerve (N VII)
Occipital belly	Occipital bone and mastoid region of temporal bones	Epicranial aponeurosis	Tenses and retracts scalp	Facial nerve (N VII)
NECK				
Platysma	Superior thorax between cartilage of 2nd rib and acromion of scapula	Mandible and skin of cheek	Tenses skin of neck; depresses mandible; pulls lower lip down	Facial nerve (N VII)

*An uppercase N followed by a Roman numeral refers to a cranial nerve.

**This muscle originates in association with the extrinsic eye muscles, so its innervation is unusual.

Figure 11–6   **Extrinsic Eye Muscles.** ATLAS: Plates 12a; 16a,b

	Extrinsic Eye Muscles
	Superior oblique
	Superior rectus
	Lateral rectus
	Medial rectus
	Inferior oblique
	Inferior rectus

a Lateral surface, right eye

b Medial surface, right eye

c Anterior view, right eye

d Anterior view, right orbit

Table 11–3	Extrinsic Eye Muscles (Figure 11–6)			
Muscle	**Origin**	**Insertion**	**Action**	**Innervation**
Inferior rectus	Sphenoid around optic canal	Inferior, medial surface of eyeball	Eye looks inferiorly	Oculomotor nerve (N III)
Medial rectus	Sphenoid around optic canal	Medial surface of eyeball	Eye looks medially	Oculomotor nerve (N III)
Superior rectus	Sphenoid around optic canal	Superior surface of eyeball	Eye looks superiorly	Oculomotor nerve (N III)
Lateral rectus	Sphenoid around optic canal	Lateral surface of eyeball	Eye looks laterally	Abducens nerve (N VI)
Inferior oblique	Maxilla at anterior portion of orbit	Inferior, lateral surface of eyeball	Eye rolls, looks superiorly and laterally	Oculomotor nerve (N III)
Superior oblique	Sphenoid around optic canal	Superior, lateral surface of eyeball	Eye rolls, looks inferiorly and laterally	Trochlear nerve (N IV)

Figure 11–7 **Muscles of Mastication.** ATLAS: Plate 3c,d

Muscles of Mastication	
Temporalis	Lateral pterygoid
Masseter	Medial pterygoid

a **Lateral view.** The temporalis muscle passes medial to the zygomatic arch to insert on the coronoid process of the mandible. The masseter inserts on the angle and lateral surface of the mandible.

b **Lateral view, pterygoid muscles exposed.** The location and orientation of the pterygoid muscles are seen after the overlying muscles and a portion of the mandible are removed.

Table 11–4	Muscles of Mastication (Figure 11–7)			
Muscle	**Origin**	**Insertion**	**Action**	**Innervation**
Masseter	Zygomatic arch	Lateral surface of mandibular ramus	Elevates mandible and closes the jaws	Trigeminal nerve (N V), mandibular branch
Temporalis	Along temporal lines of skull	Coronoid process of mandible	Elevates mandible	Trigeminal nerve (N V), mandibular branch
Pterygoids (medial and lateral)	Lateral pterygoid plate	Medial surface of mandibular ramus	*Medial:* Elevates the mandible and closes the jaws, or slides the mandible from side to side (lateral excursion)	Trigeminal nerve (N V), mandibular branch
			Lateral: Opens jaws, protrudes mandible, or performs lateral excursion	Trigeminal nerve (N V), mandibular branch

TIPS & TRICKS

The medial and lateral *ptery*goid muscles are named for their origin on the pterygoid ("winged") portion of the sphenoid bone. The *ptero*dactyl was a prehistoric winged reptile.

Muscles of the Tongue

The four muscles of the tongue have names ending in *glossus,* the Greek word for "tongue." The **palatoglossus** muscle originates at the palate, the **styloglossus** muscle at the styloid process of the temporal bone, the **genioglossus** muscle at the chin, and the **hyoglossus** muscle at the hyoid bone (**Figure 11–8**). These muscles, used in various combinations, move the tongue in the delicate and complex patterns necessary for speech. They also maneuver food within the mouth in preparation for swallowing (Table 11–5).

Muscles of the Pharynx

The muscles of the pharynx are responsible for initiating the swallowing process (**Figure 11–9** and Table 11–6). The **pharyngeal constrictor** muscles (*superior, middle,* and *inferior*) move food (in the form of a compacted mass called a *bolus*) into the esophagus by constricting the pharyngeal walls. The two **palatal muscles**—the *tensor veli palatini* and the *levator veli palatini*—elevate the soft palate and adjacent portions of the pharyngeal wall and also pull open the entrance to the auditory tube. The **laryngeal elevator** muscles elevate the larynx. As a

Figure 11–8 **Muscles of the Tongue.**

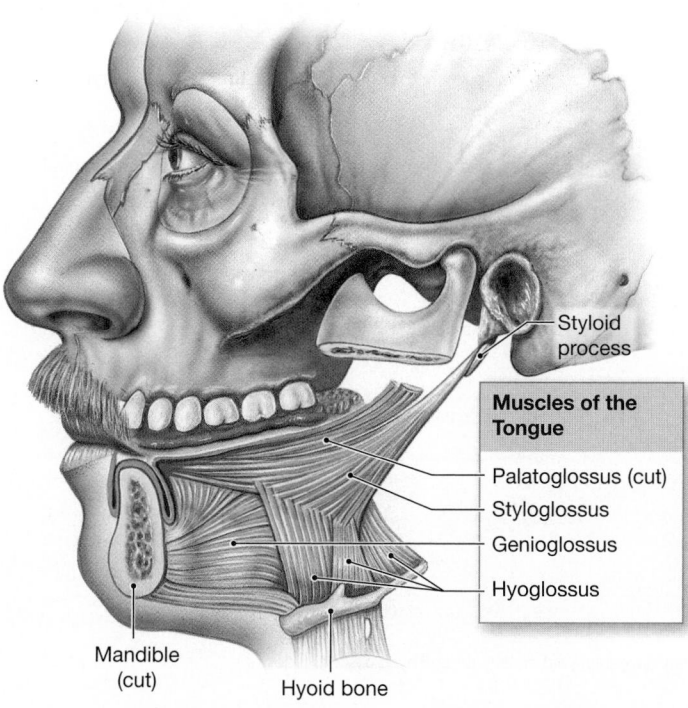

Figure 11–9 **Muscles of the Pharynx.**

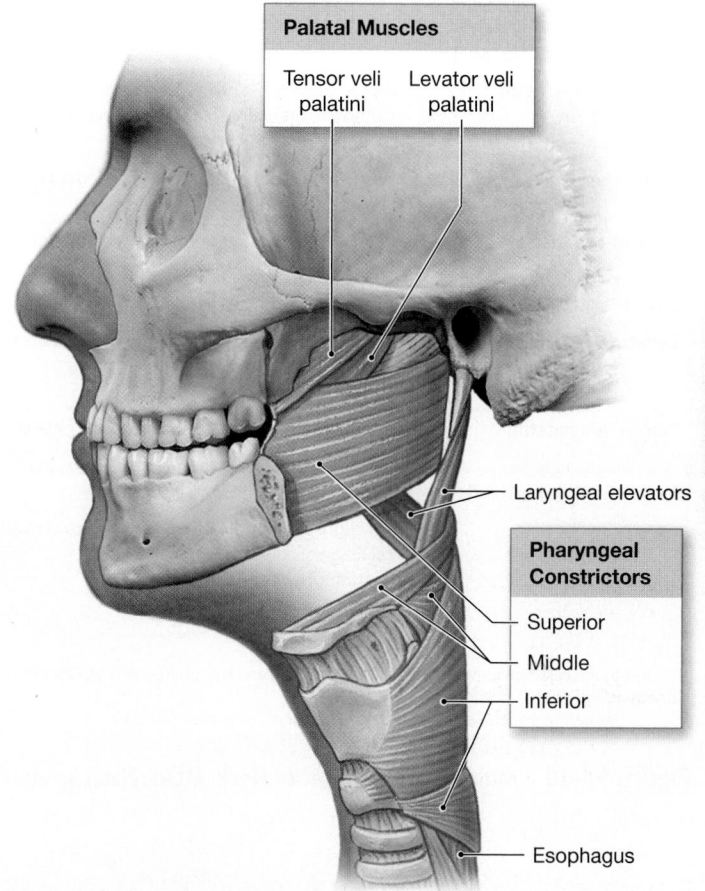

result, swallowing repeatedly can help you adjust to pressure changes when you fly or dive by opening the entrance to the auditory tube.

Anterior Muscles of the Neck

The anterior muscles of the neck include (1) five muscles that control the position of the larynx, (2) muscles that depress the mandible and tense the floor of the mouth, and (3) muscles that provide a stable foundation for muscles of the tongue and pharynx (**Figure 11–10** and Table 11–7).

The **digastric** (dī-GAS-trik) muscle has two bellies, as the name implies (*di-*, two + *gaster*, stomach). One belly extends from the chin to the hyoid bone. The other continues from the hyoid bone to the mastoid portion of the temporal bone. Depending on which belly contracts and whether fixator muscles

are stabilizing the position of the hyoid bone, the digastric muscle can open the mouth by depressing the mandible, or it can elevate the larynx by raising the hyoid bone.

The digastric muscle covers the broad, flat **mylohyoid** muscle, which provides a muscular floor to the mouth, aided by the deeper **geniohyoid** muscles that extend between the hyoid bone and the chin. The **stylohyoid** muscle forms a muscular connection between the hyoid bone and the styloid process of the skull.

The **sternocleidomastoid** (ster-nō-klī-dō-MAS-toyd) muscle extends from the clavicle and the sternum to the mastoid

Table 11–5	Muscles of the Tongue (Figure 11–8)			
Muscle	**Origin**	**Insertion**	**Action**	**Innervation**
Genioglossus	Medial surface of mandible around chin	Body of tongue, hyoid bone	Depresses and protracts tongue	Hypoglossal nerve (N XII)
Hyoglossus	Body and greater horn of hyoid bone	Side of tongue	Depresses and retracts tongue	Hypoglossal nerve (N XII)
Palatoglossus	Anterior surface of soft palate	Side of tongue	Elevates tongue, depresses soft palate	Internal branch of accessory nerve (N XI)
Styloglossus	Styloid process of temporal bone	Along the side to tip and base of tongue	Retracts tongue, elevates side of tongue	Hypoglossal nerve (N XII)

Table 11–6	Muscles of the Pharynx (Figure 11–9)			
Muscle	**Origin**	**Insertion**	**Action**	**Innervation**
PHARYNGEAL CONSTRICTORS				
Superior constrictor	Pterygoid process of sphenoid, medial surfaces of mandible	Median raphe attached to occipital bone	Constricts pharynx to propel bolus into esophagus	Branches of pharyngeal plexus (N X)
Middle constrictor	Horns of hyoid bone	Median raphe	Constricts pharynx to propel bolus into esophagus	Branches of pharyngeal plexus (N X)
Inferior constrictor	Cricoid and thyroid cartilages of larynx	Median raphe	Constricts pharynx to propel bolus into esophagus	Branches of pharyngeal plexus (N X)
PALATAL MUSCLES				
Levator veli palatini	Petrous part of temporal bone; tissues around the auditory tube	Soft palate	Elevates soft palate	Branches of pharyngeal plexus (N X)
Tensor veli palatini	Sphenoidal spine; tissues around the auditory tube	Soft palate	Elevates soft palate	Trigeminal nerve (N V)
LARYNGEAL ELEVATORS*				
	Ranges from soft palate, to cartilage around inferior portion of auditory tube, to styloid process of temporal bone	Thyroid cartilage	Elevate larynx	Branches of pharyngeal plexus (N IX and N X)

*Refers to the palatopharyngeus, salpingopharyngeus, and stylopharyngeus, assisted by the thyrohyoid, geniohyoid, stylohyoid, and hyoglossus muscles, discussed in Tables 11–5 and 11–7.

Figure 11–10 **Muscles of the Anterior Neck.** ATLAS: Plates 3a–d; 17; 18a–c; 25

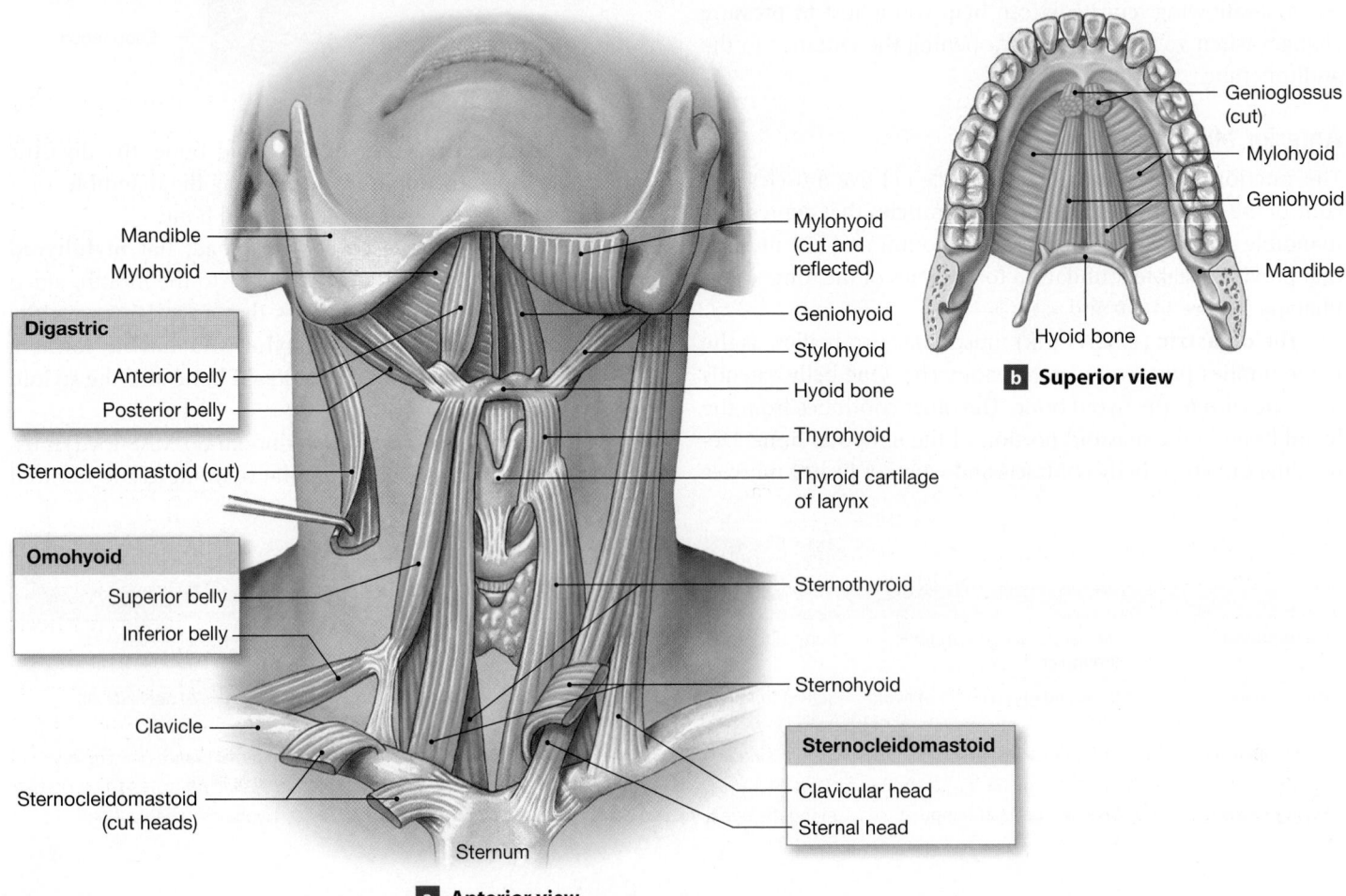

a **Anterior view**

b **Superior view**

Table 11–7	Anterior Muscles of the Neck (Figure 11–10)			
Muscle	**Origin**	**Insertion**	**Action**	**Innervation**
Digastric	Two bellies: *anterior* from inferior surface of mandible at chin; *posterior* from mastoid region of temporal bone	Hyoid bone	Depresses mandible or elevates larynx	*Anterior belly:* Trigeminal nerve (N V), mandibular branch *Posterior belly:* Facial nerve (N VII)
Geniohyoid	Medial surface of mandible at chin	Hyoid bone	As above and pulls hyoid bone anteriorly	Cervical nerve C_1 by hypoglossal nerve (N XII)
Mylohyoid	Mylohyoid line of mandible	Median connective tissue band (raphe) that runs to hyoid bone	Elevates floor of mouth and hyoid bone or depresses mandible	Trigeminal nerve (N V), mandibular branch
Omohyoid (superior and inferior bellies united at central tendon anchored to clavicle and first rib)	Superior border of scapula near scapular notch	Hyoid bone	Depresses hyoid bone and larynx	Cervical spinal nerves C_2–C_3
Sternohyoid	Clavicle and manubrium	Hyoid bone	Depresses hyoid bone and larynx	Cervical spinal nerves C_1–C_3
Sternothyroid	Dorsal surface of manubrium and first costal cartilage	Thyroid cartilage of larynx	Depresses hyoid bone and larynx	Cervical spinal nerves C_1–C_3
Stylohyoid	Styloid process of temporal bone	Hyoid bone	Elevates larynx	Facial nerve (N VII)
Thyrohyoid	Thyroid cartilage of larynx	Hyoid bone	Elevates thyroid, depresses hyoid bone	Cervical spinal nerves C_1–C_2 by hypoglossal nerve (N XII)
Sternocleidomastoid	Two bellies: *clavicular head* attaches to sternal end of clavicle; *sternal head* attaches to manubrium	Mastoid region of skull and lateral portion of superior nuchal line	Together, they flex the neck; alone, one side flexes head toward shoulder and rotates face to opposite side	Accessory nerve (N XI) and cervical spinal nerves (C_2–C_3) of cervical plexus

region of the skull (**Figures 11–5** and **11–10**). The **omohyoid** (ō-mō-HĪ-oyd) muscle attaches to the scapula, the clavicle and first rib, and the hyoid bone. The other members of this group are strap-like muscles that extend between the sternum and larynx (*sternothyroid*) or hyoid bone (*sternohyoid*), and between the larynx and hyoid bone (*thyrohyoid*).

Muscles of the Vertebral Column

The muscles of the vertebral column are covered by more superficial back muscles, such as the trapezius and latissimus dorsi muscles (**Figure 11–4b**). The **erector spinae** muscles include superficial and deep layers. The superficial layer can be divided into **spinalis**, **longissimus**, and **iliocostalis** groups (**Figure 11–11** and **Table 11–8**). In the inferior lumbar and sacral regions, the boundary between the longissimus and iliocostalis muscles is indistinct. When contracting together, the erector spinae extend the vertebral column. When the muscles on only one side contract, the result is lateral flexion of the vertebral column.

Deep to the spinalis muscles, smaller muscles interconnect and stabilize the vertebrae. These muscles include the **semispinalis** group; the **multifidus** muscle; and the **interspinales**, **intertransversarii**, and **rotatores** muscles (**Figure 11–11**). In various combinations, they produce slight extension or rotation of the vertebral column. They are also important in making delicate adjustments in the positions of individual vertebrae, and they stabilize adjacent vertebrae. If injured, these muscles can start a cycle of pain → muscle stimulation → contraction → pain. Resultant pressure on adjacent spinal nerves can lead to sensory losses and mobility limitations. Many of the warm-up and stretching exercises recommended before athletic activity are intended to prepare these small but very important muscles for their supporting role.

The muscles of the vertebral column include many posterior extensors, but few anterior flexors. Why doesn't the vertebral column have a massive series of flexor muscles? One reason is that many of the large trunk muscles flex the vertebral column when they contract. A second reason is that most of the body weight lies anterior to the vertebral column, so gravity tends to flex the spine. However, a few spinal flexors are associated with the anterior surface of the vertebral column. In the neck, the **longus capitis** and the **longus colli** muscles rotate or flex the neck, depending on whether the muscles of one or both sides are contracting (**Figure 11–11**). In the lumbar region, the large **quadratus lumborum** muscles flex the vertebral column and depress the ribs (**Figure 11–11**).

Figure 11–11   **Muscles of the Vertebral Column.**

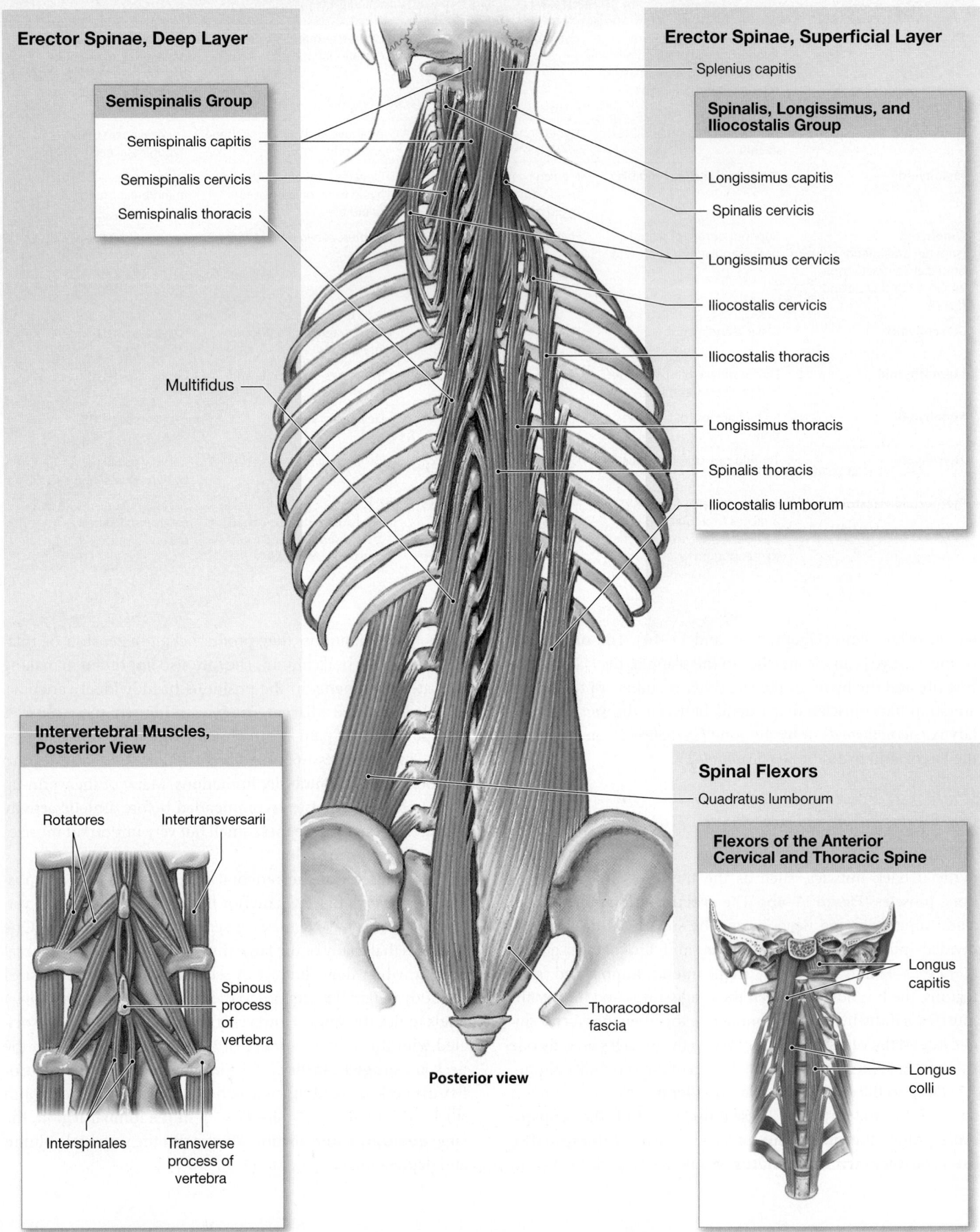

Erector Spinae, Deep Layer

Semispinalis Group

Semispinalis capitis

Semispinalis cervicis

Semispinalis thoracis

Multifidus

Intervertebral Muscles, Posterior View

Rotatores

Intertransversarii

Spinous process of vertebra

Interspinales

Transverse process of vertebra

Erector Spinae, Superficial Layer

Splenius capitis

Spinalis, Longissimus, and Iliocostalis Group

Longissimus capitis

Spinalis cervicis

Longissimus cervicis

Iliocostalis cervicis

Iliocostalis thoracis

Longissimus thoracis

Spinalis thoracis

Iliocostalis lumborum

Spinal Flexors

Quadratus lumborum

Thoracodorsal fascia

Posterior view

Flexors of the Anterior Cervical and Thoracic Spine

Longus capitis

Longus colli

Table 11–8 Muscles of the Vertebral Column (Figure 11–11)

Group and Muscles	Origin	Insertion	Action	Innervation
SUPERFICIAL LAYER				
Splenius (Splenius capitis, splenius cervicis)	Spinous processes and ligaments connecting inferior cervical and superior thoracic vertebrae	Mastoid process, occipital bone of skull, and superior cervical vertebrae	Together, the two sides extend neck; alone, each rotates and laterally flexes neck to that side	Cervical spinal nerves
Erector spinae				
Spinalis group **Spinalis cervicis**	Inferior portion of ligamentum nuchae and spinous process of C$_7$	Spinous process of axis	Extends neck	Cervical spinal nerves
Spinalis group **Spinalis thoracis**	Spinous processes of inferior thoracic and superior lumbar vertebrae	Spinous processes of superior thoracic vertebrae	Extends vertebral column	Thoracic and lumbar spinal nerves
Longissimus group **Longissimus capitis**	Transverse processes of inferior cervical and superior thoracic vertebrae	Mastoid process of temporal bone	Together, the two sides extend head; alone, each rotates and laterally flexes neck to that side	Cervical and thoracic spinal nerves
Longissimus group **Longissimus cervicis**	Transverse processes of superior thoracic vertebrae	Transverse processes of middle and superior cervical vertebrae	Together, the two sides extend head; alone, each rotates and laterally flexes neck to that side	Cervical and thoracic spinal nerves
Longissimus group **Longissimus thoracis**	Broad aponeurosis and transverse processes of inferior thoracic and superior lumbar vertebrae; joins iliocostalis	Transverse processes of superior vertebrae and inferior surfaces of ribs	Extends vertebral column; alone, each produces lateral flexion to that side	Thoracic and lumbar spinal nerves
Iliocostalis group **Iliocostalis cervicis**	Superior borders of vertebrosternal ribs near the angles	Transverse processes of middle and inferior cervical vertebrae	Extends or laterally flexes neck, elevates ribs	Cervical and superior thoracic spinal nerves
Iliocostalis group **Iliocostalis thoracis**	Superior borders of inferior seven ribs medial to the angles	Upper ribs and transverse process of last cervical vertebra	Stabilizes thoracic vertebrae in extension	Thoracic spinal nerves
Iliocostalis group **Iliocostalis lumborum**	Iliac crest, sacral crests, and spinous processes	Inferior surfaces of inferior seven ribs near their angles	Extends vertebral column, depresses ribs	Inferior thoracic and lumbar spinal nerves
DEEP LAYER				
Semispinalis group **Semispinalis capitis**	Articular processes of inferior cervical and transverse processes of superior thoracic vertebrae	Occipital bone, between nuchal lines	Together, the two sides extend head; alone, each extends and laterally flexes neck	Cervical spinal nerves
Semispinalis group **Semispinalis cervicis**	Transverse processes of T$_1$–T$_5$ or T$_6$	Spinous processes of C$_2$–C$_5$	Extends vertebral column and rotates toward opposite side	Cervical spinal nerves
Semispinalis group **Semispinalis thoracis**	Transverse processes of T$_6$–T$_{10}$	Spinous processes of C$_5$–T$_4$	Extends vertebral column and rotates toward opposite side	Thoracic spinal nerves
Semispinalis group **Multifidus**	Sacrum and transverse processes of each vertebra	Spinous processes of the third or fourth more superior vertebrae	Extends vertebral column and rotates toward opposite side	Cervical, thoracic, and lumbar spinal nerves
Semispinalis group **Rotatores**	Transverse processes of each vertebra	Spinous processes of adjacent, more superior vertebra	Extends vertebral column and rotates toward opposite side	Cervical, thoracic, and lumbar spinal nerves
Semispinalis group **Interspinales**	Spinous processes of each vertebra	Spinous processes of more superior vertebra	Extends vertebral column	Cervical, thoracic, and lumbar spinal nerves
Semispinalis group **Intertransversarii**	Transverse processes of each vertebra	Transverse process of more superior vertebra	Laterally flexes the vertebral column	Cervical, thoracic, and lumbar spinal nerves
SPINAL FLEXORS				
Longus capitis	Transverse processes of cervical vertebrae	Base of the occipital bone	Together, the two sides flex the neck; alone, each rotates head to that side	Cervical spinal nerves
Longus colli	Anterior surfaces of cervical and superior thoracic vertebrae	Transverse processes of superior cervical vertebrae	Flexes or rotates neck; limits hyperextension	Cervical spinal nerves
Quadratus lumborum	Iliac crest and iliolumbar ligament	Last rib and transverse processes of lumbar vertebrae	Together, they depress ribs; alone, each side laterally flexes vertebral column	Thoracic and lumbar spinal nerves

11

Oblique and Rectus Muscles

The oblique and rectus muscles lie within the body wall, between the spinous processes of vertebrae and the ventral midline (**Figures 11–4, 11–12,** and Table 11–9). The oblique muscles compress underlying structures or rotate the vertebral column, depending on whether one or both sides contract. The rectus muscles are important flexors of the vertebral column, acting in opposition to the erector spinae. The oblique and rectus muscle groups share embryological origins. We can divide these groups into cervical, thoracic, and abdominal regions.

The oblique group includes the **scalene** muscles of the neck (**Figure 11–12a**) and the **intercostal** and **transversus** muscles of the thorax (**Figure 11–12b,c**). The scalene muscles (*anterior, middle,* and *posterior*) elevate the first two ribs and assist in flexion of the neck. In the thorax, the oblique muscles extend between the ribs, with the **external intercostal** muscles covering the **internal intercostal** muscles. Both

Figure 11–12   Oblique and Rectus Muscles and the Diaphragm. ATLAS: Plates 39b–d; 41a,b; 46

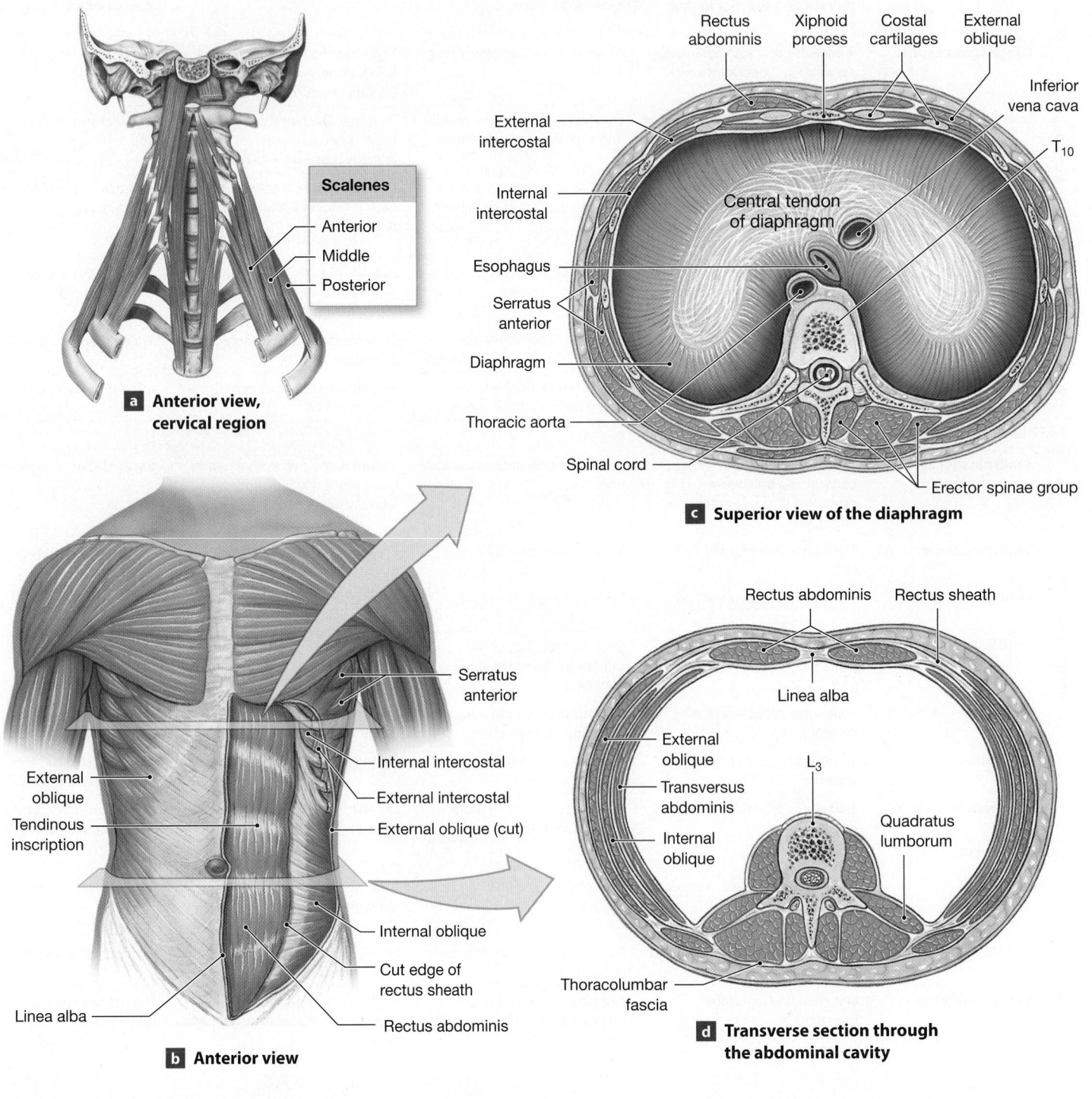

a Anterior view, cervical region

Scalenes
- Anterior
- Middle
- Posterior

Rectus abdominis · Xiphoid process · Costal cartilages · External oblique
Inferior vena cava
External intercostal
Internal intercostal
Central tendon of diaphragm
T10
Esophagus
Serratus anterior
Diaphragm
Thoracic aorta
Spinal cord
Erector spinae group

c Superior view of the diaphragm

Serratus anterior
Internal intercostal
External intercostal
External oblique (cut)
External oblique
Tendinous inscription
Internal oblique
Cut edge of rectus sheath
Linea alba
Rectus abdominis

b Anterior view

Rectus abdominis · Rectus sheath
Linea alba
External oblique
L3
Transversus abdominis
Internal oblique
Quadratus lumborum
Thoracolumbar fascia

d Transverse section through the abdominal cavity

groups of intercostal muscles aid in respiratory movements of the ribs. A small **transversus thoracis** muscle crosses the inner surface of the rib cage and is separated from the pleural cavity by the parietal pleura, a *serous membrane.* ⤴ p. 163 The sternum occupies the place where we might otherwise expect thoracic rectus muscles to be.

The same basic pattern of musculature extends unbroken across the abdominopelvic surface (**Figure 11–12b,d**). Here, the muscles are called the **external oblique**, **internal oblique**, **transversus abdominis**, and **rectus abdominis** muscles (the "abs"). The rectus abdominis muscle inserts at the xiphoid process and originates near the pubic symphysis. This muscle is longitudinally divided by the **linea alba** (white line), a median collagenous partition (**Figure 11–4a**). The rectus abdominis muscle is separated into segments by transverse bands of collagen fibers called **tendinous inscriptions**. Each segment contains muscle fibers that extend longitudinally, originating and inserting on the tendinous inscriptions. Due to the bulging of enlarged muscle fibers between the tendinous inscriptions, bodybuilders often refer to the rectus abdominis as the "six-pack."

The Diaphragm

The term *diaphragm* refers to any muscular sheet that forms a wall. When used without a modifier, however, **diaphragm**, or *diaphragmatic muscle*, specifies the muscular partition that separates the abdominopelvic and thoracic cavities (**Figure 11–12c**). We include this muscle here because it develops in association with the other muscles of the chest wall. The diaphragm is a major muscle used in breathing.

Table 11–9	Oblique and Rectus Muscle Groups (Figure 11–12)			
Group and Muscles	**Origin**	**Insertion**	**Action**	**Innervation***
OBLIQUE GROUP				
Cervical region				
Scalenes (anterior, middle, and posterior)	Transverse and costal processes of cervical vertebrae	Superior surfaces of first two ribs	Elevate ribs or flex neck	Cervical spinal nerves
Thoracic region				
External intercostals	Inferior border of each rib	Superior border of more inferior rib	Elevate ribs	Intercostal nerves (branches of thoracic spinal nerves)
Internal intercostals	Superior border of each rib	Inferior border of the preceding rib	Depress ribs	Intercostal nerves (branches of thoracic spinal nerves)
Transversus thoracis	Posterior surface of sternum	Cartilages of ribs	Depress ribs	Intercostal nerves (branches of thoracic spinal nerves)
Serratus posterior superior (Figure 11–14b)	Spinous processes of C_7–T_3 and ligamentum nuchae	Superior borders of ribs 2–5 near angles	Elevates ribs, enlarges thoracic cavity	Thoracic nerves (T_1–T_4)
Serratus posterior inferior (Figure 11–14b)	Aponeurosis from spinous processes of T_{10}–L_3	Inferior borders of ribs 8–12	Pulls ribs inferiorly; also pulls outward, opposing diaphragm	Thoracic nerves (T_9–T_{12})
Abdominal region				
External oblique	External and inferior borders of ribs 5–12	Linea alba and iliac crest	Compresses abdomen, depresses ribs, flexes or bends spine	Intercostal, iliohypogastric, and ilioinguinal nerves
Internal oblique	Thoracolumbar fascia and iliac crest	Inferior ribs, xiphoid process, and linea alba	Compresses abdomen, depresses ribs, flexes or bends spine	Intercostal, iliohypogastric, and ilioinguinal nerves
Transversus abdominis	Cartilages of ribs 6–12, iliac crest, and thoracolumbar fascia	Linea alba and pubis	Compresses abdomen	Intercostal, iliohypogastric, and ilioinguinal nerves
RECTUS GROUP				
Cervical region	*See muscles in Table 11–6*			
Thoracic region				
Diaphragm	Xiphoid process, cartilages of ribs 4–10, and anterior surfaces of lumbar vertebrae	Central tendinous sheet	Contraction expands thoracic cavity and compresses abdominopelvic cavity	Phrenic nerves (C_3–C_5)
Abdominal region				
Rectus abdominis	Superior surface of pubis around symphysis	Inferior surfaces of costal cartilages (ribs 5–7) and xiphoid process	Depresses ribs, flexes vertebral column, compresses abdomen	Intercostal nerves (T_7–T_{12})

*Where appropriate, spinal nerves involved are given in parentheses.

Muscles of the Pelvic Floor

The muscles of the pelvic floor extend from the sacrum and coccyx to the ischium and pubis (Figure 11–13 and Table 11–10). These muscles (1) support the organs of the pelvic cavity, (2) flex the sacrum and coccyx, and (3) control the movement of materials through the urethra and anus.

The boundaries of the **perineum,** the muscular sheet that forms the pelvic floor, are formed by the inferior margins of the pelvis. A line drawn between the ischial tuberosities divides the perineum into two triangles: an anterior **urogenital triangle** and a posterior **anal triangle** (Figure 11–13). The superficial muscles of the urogenital triangle are the muscles of the external genitalia. They cover deeper muscles that strengthen the

pelvic floor and encircle the urethra. These muscles constitute the **urogenital diaphragm,** a deep muscular layer that extends between the pubic bones.

An even more extensive muscular sheet, the **pelvic diaphragm,** forms the muscular foundation of the anal triangle. This layer, covered by the urogenital diaphragm, extends as far as the pubic symphysis.

The urogenital and pelvic diaphragms do not completely close the pelvic outlet. The urethra, vagina, and anus pass through them to open on the exterior. Muscular sphincters surround the passageways, and the external sphincters permit voluntary control of urination and defecation. Muscles, nerves, and blood vessels also pass through the pelvic outlet as they travel to or from the lower limbs.

Table 11–10	Muscles of the Pelvic Floor (Figure 11–13)			
Group and Muscle	**Origin**	**Insertion**	**Action**	**Innervation***
UROGENITAL TRIANGLE				
Superficial muscles **Bulbospongiosus** Males	Collagen sheath at base of penis; fibers cross over urethra	Median raphe and central tendon of perineum	Compresses base and stiffens penis; ejects urine or semen	Pudendal nerve, perineal branch (S_2–S_4)
Females	Collagen sheath at base of clitoris; fibers run on either side of urethral and vaginal opening	Central tendon of perineum	Compresses and stiffens clitoris; narrows vaginal opening	Pudendal nerve, perineal branch (S_2–S_4)
Ischiocavernosus	Ischial ramus and tuberosity	Pubic symphysis anterior to base of penis or clitoris	Compresses and stiffens penis or clitoris	Pudendal nerve, perineal branch (S_2–S_4)
Superficial transverse perineal	Ischial ramus	Central tendon of perineum	Stabilizes central tendon of perineum	Pudendal nerve, perineal branch (S_2–S_4)
Urogenital diaphragm				
Deep muscles **Deep transverse perineal**	Ischial ramus	Median raphe of urogenital diaphragm	Stabilizes central tendon of perineum	Pudendal nerve, perineal branch (S_2–S_4)
External urethral sphincter Males	Ischial and pubic rami	To median raphe at base of penis; inner fibers encircle urethra	Closes urethra; compresses prostate and bulbo-urethral glands	Pudendal nerve, perineal branch (S_2–S_4)
Females	Ischial and pubic rami	To median raphe; inner fibers encircle urethra	Closes urethra; compresses vagina and greater vestibular glands	Pudendal nerve, perineal branch (S_2–S_4)
ANAL TRIANGLE				
Pelvic diaphragm **Coccygeus**	Ischial spine	Lateral, inferior borders of sacrum and coccyx	Flexes coccygeal joints; tenses and supports pelvic floor	Inferior sacral nerves (S_4–S_5)
Levator ani **Iliococcygeus**	Ischial spine, pubis	Coccyx and median raphe	Tenses floor of pelvis; flexes coccygeal joints; elevates and retracts anus	Pudendal nerve (S_2–S_4)
Pubococcygeus	Inner margins of pubis	Coccyx and median raphe	Tenses floor of pelvis; flexes coccygeal joints; elevates and retracts anus	Pudendal nerve (S_2–S_4)
External anal sphincter	By tendon from coccyx	Encircles anal opening	Closes anal opening	Pudendal nerve, hemorrhoidal branch (S_2–S_4)

*Where appropriate, spinal nerves involved are given in parentheses.

Figure 11–13   **Muscles of the Pelvic Floor.**

Superficial Dissections

Deep Dissections

UROGENITAL TRIANGLE
OF PERINEUM

Vagina

Urethra

Urogenital Diaphragm

External urethral sphincter

Deep transverse perineal

Urogenital Triangle

Ischiocavernosus

Bulbospongiosus

Central tendon of perineum

Superficial
transverse perineal

Pelvic Diaphragm

Pubococcygeus ⎤
⎥ Levator
Iliococcygeus ⎦ ani

Anus

External anal sphincter

Coccygeus

Gluteus maximus

Sacrotuberous ligament

a **Female**

ANAL TRIANGLE

No differences between
deep musculature in
female and male

UROGENITAL TRIANGLE
OF PERINEUM

Testis

Urethra (connecting
segment removed)

Urogenital Diaphragm

External urethral sphincter

Urogenital Triangle

Deep transverse perineal

Ischiocavernosus

Bulbospongiosus

Central tendon of perineum

Superficial
transverse perineal

Pelvic Diaphragm

Anus

Pubococcygeus ⎤
⎥ Levator
Iliococcygeus ⎦ ani

External anal sphincter

Gluteus maximus

Coccygeus

Sacrotuberous ligament

b **Male**

ANAL TRIANGLE

11

CHECKPOINT

11. Describe the location of axial muscles.

12. If you were contracting and relaxing your masseter muscle, what would you probably be doing?

13. Which facial muscle would you expect to be well developed in a trumpet player?

14. Why can swallowing help alleviate the pressure sensations at the eardrum when you are in an airplane that is changing altitude?

15. Damage to the external intercostal muscles would interfere with what important process?

16. If someone hit you in your rectus abdominis muscle, how would your body position change?

17. After spending an afternoon carrying heavy boxes from his basement to his attic, Joe complains that the muscles in his back hurt. Which muscles are most likely sore?

See the blue Answers tab at the back of the book.

11-6 ❙ Appendicular muscles are muscles of the shoulders, upper limbs, pelvis, and lower limbs

Learning Outcome Identify the principal appendicular muscles of the body, plus their origins, insertions, actions, and innervation, and compare the major functional differences between the upper and lower limbs.

The appendicular musculature positions and stabilizes the pectoral and pelvic girdles and moves the upper and lower limbs. There are two major groups of appendicular muscles: (1) *the muscles of the shoulders (pectoral girdles) and upper limbs* and (2) *the muscles of the pelvis (pelvic girdle) and lower limbs*. The

functions and required ranges of motion are very different between these groups.

The muscular connections between the pectoral girdles and the axial skeleton increase the mobility of the arms and also act as shock absorbers. For example, while you jog, you can still perform delicate hand movements, because the muscular connections between the axial and appendicular components of the skeleton smooth out the bounces in your stride.

In contrast, the pelvic girdle has evolved to transfer weight from the axial to the appendicular skeleton. Rigid, bony articulations are essential, because the emphasis is on strength rather than versatility, and a muscular connection would reduce the efficiency of the transfer.

Figure 11–14 provides an introduction to the organization of the appendicular muscles of the trunk. The larger appendicular muscles are often used as injection sites for medications.

Muscles of the Shoulders and Upper Limbs

We can divide the muscles associated with the shoulders and upper limbs into four groups: (1) *muscles that position the pectoral girdle*, (2) *muscles that move the arm*, (3) *muscles that move the forearm and hand*, and (4) *muscles that move the hand and fingers*.

Muscles That Position the Pectoral Girdle

The large, superficial **trapezius** muscles, commonly called the "traps," cover the back and portions of the neck, reaching to the base of the skull. These muscles originate along the midline of the neck and back and insert on the clavicles and the scapular spines (Figures 11–15 and 11–16). The trapezius muscles are innervated by more than one nerve (Table 11–11). For this reason,

Table 11–11	Muscles That Position the Pectoral Girdle (Figures 11–14, 11–15)			
Muscle	**Origin**	**Insertion**	**Action**	**Innervation***
Levator scapulae	Transverse processes of first four cervical vertebrae	Vertebral border of scapula near superior angle	Elevates scapula	Cervical nerves C_3–C_4 and dorsal scapular nerve (C_5)
Pectoralis minor	Anterior-superior surfaces of ribs 2–4, 2–5, or 3–5 depending on anatomical variation	Coracoid process of scapula	Depresses and protracts shoulder; rotates scapula so glenoid cavity moves inferiorly (downward rotation); elevates ribs if scapula is stationary	Medial pectoral nerve (C_8, T_1)
Rhomboid major	Spinous processes of superior thoracic vertebrae	Vertebral border of scapula from spine to inferior angle	Adducts scapula and performs downward rotation	Dorsal scapular nerve (C_5)
Rhomboid minor	Spinous processes of vertebrae C_7–T_1	Vertebral border of scapula near spine	Adducts scapula and performs downward rotation	Dorsal scapular nerve (C_5)
Serratus anterior	Anterior and superior margins of ribs 1–8 or 1–9	Anterior surface of vertebral border of scapula	Protracts shoulder; rotates scapula so glenoid cavity moves superiorly (upward rotation)	Long thoracic nerve (C_5–C_7)
Subclavius	First rib	Clavicle (inferior border)	Depresses and protracts shoulder	Nerve to subclavius (C_5–C_6)
Trapezius	Occipital bone, ligamentum nuchae, and spinous processes of thoracic vertebrae	Clavicle and scapula (acromion and scapular spine)	Depends on active region and state of other muscles; may (1) elevate, retract, depress, or rotate scapula upward, (2) elevate clavicle, or (3) extend neck	Accessory nerve (N XI) and cervical spinal nerves (C_3–C_4)

*Where appropriate, spinal nerves involved are given in parentheses.

Figure 11–14 **An Overview of the Appendicular Muscles of the Trunk.**

Superficial Dissection

Axial Muscles
- Platysma

Appendicular Muscles
- Deltoid
- Pectoralis major
- Latissimus dorsi
- Serratus anterior

Axial Muscles
- External oblique
- Rectus sheath
- Superficial inguinal ring

Deep Dissection

Axial Muscles
- Sternocleidomastoid

Appendicular Muscles
- Trapezius
- Subclavius
- Deltoid (cut and reflected)
- Pectoralis minor
- Subscapularis
- Pectoralis major (cut and reflected)
- Coracobrachialis
- Biceps brachii
- Teres major
- Serratus anterior

Axial Muscles
- External intercostal
- Internal intercostal
- Internal oblique (cut)
- External oblique (cut and reflected)
- Rectus abdominis
- Transversus abdominis

a **Anterior view**
ATLAS: Plates 25; 39b

specific regions can be made to contract independently, and their actions are quite varied.

On the chest, the **serratus anterior** muscle originates along the anterior surfaces of several ribs (**Figures 11–4** and **11–15a,b**). This fan-shaped muscle inserts along the anterior margin of the vertebral border of the scapula. When the serratus anterior muscle contracts, it abducts (protracts) the scapula and swings the shoulder anteriorly.

Two other deep chest muscles arise along the anterior surfaces of the ribs on either side. The **subclavius** (sub-KLĀ-vē-us; *sub-*, below + *clavius*, clavicle) muscle inserts on the inferior border of the clavicle (**Figure 11–15a**). When it contracts, it depresses and protracts the scapular end of the clavicle. Because

ligaments connect this end to the shoulder joint and scapula, those structures move as well. The **pectoralis** (pek-tō-RA-lis) **minor** muscle attaches to the coracoid process of the scapula. The contraction of this muscle generally complements that of the subclavius muscle.

Removing the trapezius muscle reveals the **rhomboid major**, **rhomboid minor**, and **levator scapulae** muscles (**Figure 11–15b**). These muscles are attached to the posterior surfaces of the cervical and thoracic vertebrae. They insert along the vertebral border of each scapula, between the superior and inferior angles. Contraction of a rhomboid muscle adducts (retracts) the scapula on that side. The levator scapulae muscle, as its name implies, elevates the scapula.

Figure 11–14   **An Overview of the Appendicular Muscles of the Trunk** *(Continued)*.

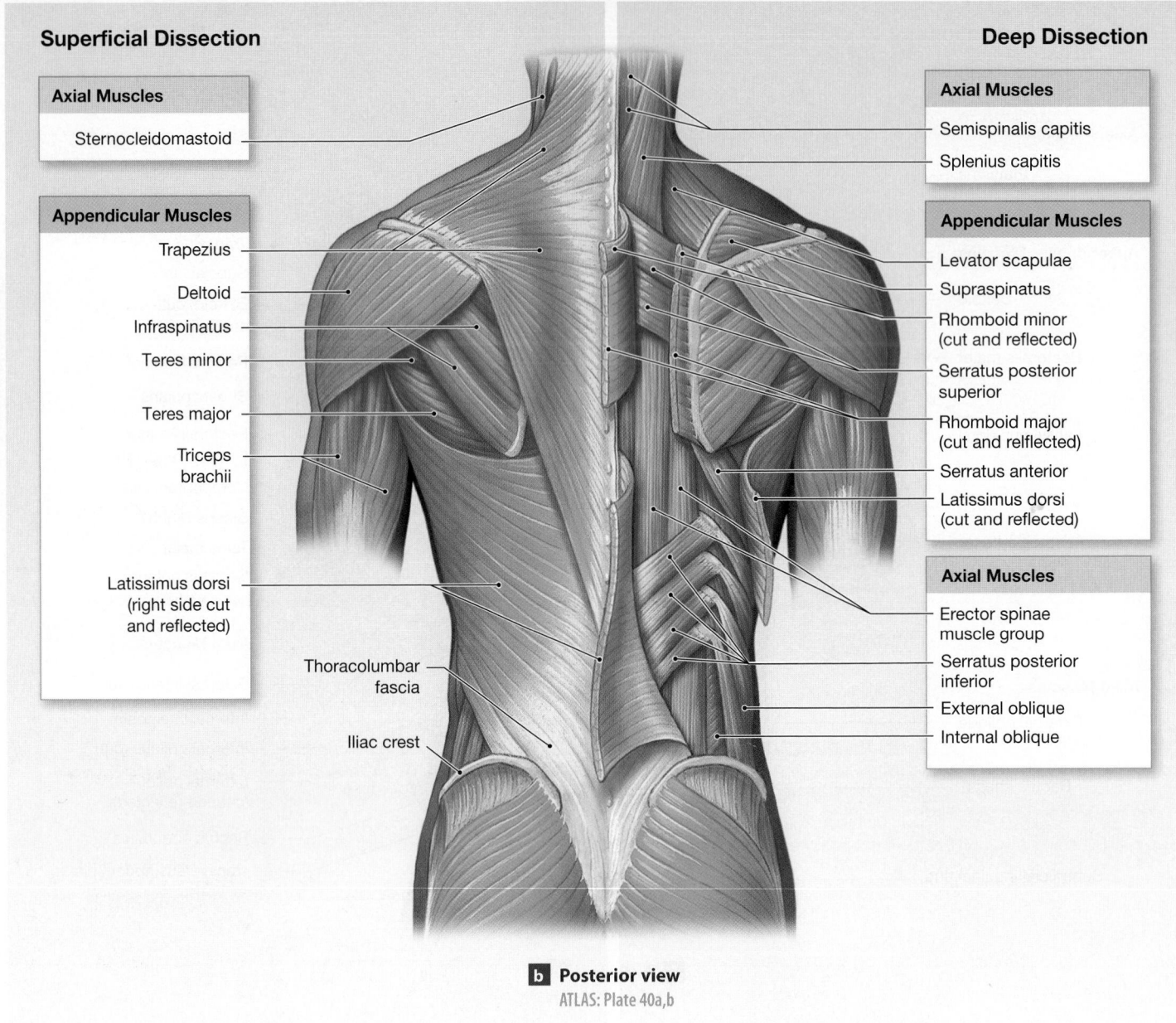

Superficial Dissection

Axial Muscles

Sternocleidomastoid

Appendicular Muscles

Trapezius

Deltoid

Infraspinatus

Teres minor

Teres major

Triceps brachii

Latissimus dorsi (right side cut and reflected)

Thoracolumbar fascia

Iliac crest

Deep Dissection

Axial Muscles

Semispinalis capitis

Splenius capitis

Appendicular Muscles

Levator scapulae

Supraspinatus

Rhomboid minor (cut and reflected)

Serratus posterior superior

Rhomboid major (cut and relflected)

Serratus anterior

Latissimus dorsi (cut and reflected)

Axial Muscles

Erector spinae muscle group

Serratus posterior inferior

External oblique

Internal oblique

b **Posterior view**
ATLAS: Plate 40a,b

Muscles That Move the Arm

The muscles that move the arm (**Figures 11–14, 11–15,** and **11–16**) are easiest to remember when they are grouped by their actions at the shoulder joint (**Table 11–12**). The **deltoid** muscle is the major abductor, but the **supraspinatus** (sū-pra-spī-NĀ-tus) muscle assists at the start of this movement. The **subscapularis** and **teres major** muscles produce medial rotation at the shoulder, whereas the **infraspinatus** and the **teres minor** muscles produce lateral rotation. All these muscles originate on the scapula. The small **coracobrachialis** (KOR-uh-kō-brā-kē-A-lis) muscle is the only muscle attached to the scapula that produces flexion and adduction at the shoulder (**Figure 11–16a**).

TIPS & TRICKS

The supraspinatus and infraspinatus muscles are named for their origins above and below the spine of the scapula, respectively, not because they are located on the spinal column.

The **pectoralis major** muscle extends between the anterior portion of the chest and the crest of the greater tubercle of the humerus (**Figure 11–16a**). The **latissimus dorsi** (la-TIS-i-mus DOR-sē) muscle extends between the thoracic vertebrae at the posterior midline and the intertubercular groove of the humerus (**Figure 11–16b**). The pectoralis major muscle produces flexion at the shoulder joint, and the latissimus dorsi muscle

Figure 11–15   **Muscles That Position the Pectoral Girdle.**

Muscles That Position the Pectoral Girdle

Trapezius
Levator scapulae
Subclavius
Pectoralis minor

Pectoralis major
(cut and reflected)

Internal intercostals

External intercostals

Muscles That Position the Pectoral Girdle

Pectoralis minor
(cut)

Serratus anterior

Biceps brachii, short head

Biceps brachii, long head

T₁₂

a Anterior view
ATLAS: Plates 39a-d; 40a–b

Superficial Dissection

Deep Dissection

Muscles That Position the Pectoral Girdle

Trapezius

Muscles That Position the Pectoral Girdle

Levator scapulae

Rhomboid minor

Rhomboid major

Scapula

Serratus anterior

Triceps brachii

T₁₂ vertebra

b Posterior view
ATLAS: Plates 27b; 40a–b

Figure 11–16 **Muscles That Move the Arm.** ATLAS: Plates 39a–d; 40a–b

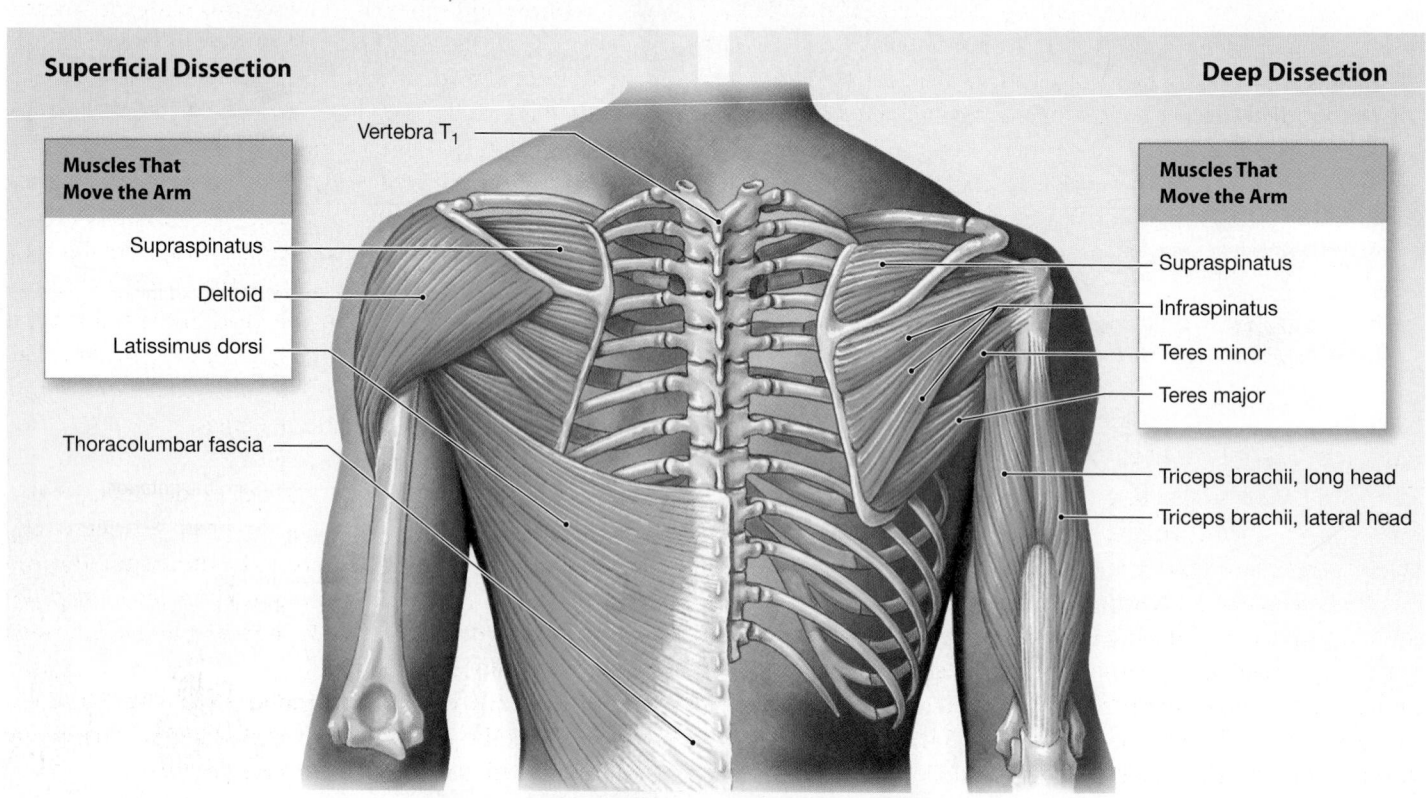

Superficial Dissection

Deep Dissection

Sternum

Clavicle

Ribs (cut)

Muscles That Move the Arm

Deltoid

Pectoralis major

Muscles That Move the Arm

Subscapularis

Coracobrachialis

Teres major

Biceps brachii, short head

Biceps brachii, long head

Vertebra T₁₂

a Anterior view

Superficial Dissection

Deep Dissection

Vertebra T₁

Muscles That Move the Arm

Supraspinatus

Deltoid

Latissimus dorsi

Thoracolumbar fascia

Muscles That Move the Arm

Supraspinatus

Infraspinatus

Teres minor

Teres major

Triceps brachii, long head

Triceps brachii, lateral head

b Posterior view

Table 11–12	Muscles That Move the Arm (Figures 11–14 to 11–16)			
Muscle	**Origin**	**Insertion**	**Action**	**Innervation***
Deltoid	Clavicle and scapula (acromion and adjacent scapular spine)	Deltoid tuberosity of humerus	*Whole muscle:* abduction at shoulder; *anterior part:* flexion and medial rotation; *posterior part:* extension and lateral rotation	Axillary nerve (C_5–C_6)
Supraspinatus	Supraspinous fossa of scapula	Greater tubercle of humerus	Abduction at the shoulder	Suprascapular nerve (C_5)
Subscapularis	Subscapular fossa of scapula	Lesser tubercle of humerus	Medial rotation at shoulder	Subscapular nerves (C_5–C_6)
Teres major	Inferior angle of scapula	Passes medially to reach the medial lip of intertubercular groove of humerus	Extension, adduction, and medial rotation at shoulder	Lower subscapular nerve (C_5–C_6)
Infraspinatus	Infraspinous fossa of scapula	Greater tubercle of humerus	Lateral rotation at shoulder	Suprascapular nerve (C_5–C_6)
Teres minor	Lateral border of scapula	Passes laterally to reach the greater tubercle of humerus	Lateral rotation at shoulder	Axillary nerve (C_5)
Coracobrachialis	Coracoid process	Medial margin of shaft of humerus	Adduction and flexion at shoulder	Musculocutaneous nerve (C_5–C_7)
Pectoralis major	Cartilages of ribs 2–6, body of sternum, and inferior, medial portion of clavicle	Crest of greater tubercle and lateral lip of intertubercular groove of humerus	Flexion, adduction, and medial rotation at shoulder	Pectoral nerves (C_5–T_1)
Latissimus dorsi	Spinous processes of inferior thoracic and all lumbar vertebrae, ribs 8–12, and thoracolumbar fascia	Floor of intertubercular groove of the humerus	Extension, adduction, and medial rotation at shoulder	Thoracodorsal nerve (C_6–C_8)
Triceps brachii (long head)	*See Table 11–13*			

*Where appropriate, spinal nerves involved are given in parentheses.

produces extension. These muscles, commonly known as the "pecs" and the "lats," can also work together to produce adduction and medial rotation of the humerus at the shoulder.

Collectively, the supraspinatus, infraspinatus, teres minor, and subscapularis muscles and their associated tendons form the **rotator cuff**. Sports that involve throwing a ball, such as baseball or football, place considerable strain on the rotator cuff, and rotator cuff injuries are common.

TIPS & TRICKS

The acronym SITS is useful in remembering the four muscles of the rotator cuff.

Muscles That Move the Forearm and Hand

Most of the muscles that insert on the forearm and hand originate on the humerus, but the biceps brachii and triceps brachii muscles are noteworthy exceptions. The **biceps brachii** muscle and the *long head* of the **triceps brachii** muscle originate on the scapula and insert on the bones of the forearm (**Figure 11–17**). The triceps brachii muscle inserts on the olecranon of the ulna. Contraction of the triceps brachii muscle extends the elbow, as when you do push-ups. The biceps brachii muscle inserts on the radial tuberosity, a roughened bump on the anterior surface

of the radius. ↪ p. 273 Contraction of the biceps brachii muscle flexes the elbow and supinates the forearm. With the forearm pronated (palm facing back), the biceps brachii muscle cannot function effectively. As a result, you are strongest when you flex your elbow with a supinated forearm. The biceps brachii muscle then makes a prominent bulge.

The biceps brachii muscle is important in stabilizing the shoulder joint. The short head originates on the coracoid process and supports the posterior surface of the capsule. The long head originates at the supraglenoid tubercle, inside the shoulder joint. ↪ p. 271 After crossing the head of the humerus, it passes along the intertubercular groove. In this position, the tendon helps to hold the head of the humerus within the glenoid cavity while arm movements are underway.

More muscles are shown in **Figure 11–17** and listed in **Table 11–13**. As you study these muscles, notice that, in general, the extensor muscles lie along the posterior and lateral surfaces of the arm, whereas the flexors are on the anterior and medial surfaces. Connective tissue partitions separate major muscle groups, dividing the muscles into *compartments* formed by dense collagenous sheets.

The **brachialis** and **brachioradialis** (BRĀ-kē-ō-rā-dē-A-lis) muscles flex the elbow and are opposed by the **anconeus** muscle and the triceps brachii muscle, respectively.

Figure 11–17   **Muscles That Move the Forearm and Hand.**   ATLAS: Plates 27a–c; 29a; 30; 33a–d; 37a,b

POSTERIOR

Lateral head
Long head } Triceps brachii
Medial head

Humerus
Vein
Artery
Nerve
Brachialis
Biceps brachii

LATERAL

ANTERIOR

Coracoid process of scapula
Humerus
Coracobrachialis

Muscles That Move the Forearm

ACTION AT THE ELBOW

Biceps brachii, short head
Biceps brachii, long head
Triceps brachii, long head
Triceps brachii, medial head
Brachialis
Brachioradialis
Medial epicondyle of humerus
Pronator teres

Muscles That Move the Hand

ACTION AT THE HAND

Flexor carpi radialis
Palmaris longus
Flexor carpi ulnaris

Flexor digitorum superficialis
Pronator quadratus
Flexor retinaculum

b **Anterior view, superficial layer**

Muscles That Move the Forearm

ACTION AT THE ELBOW

Triceps brachii, long head
Triceps brachii, lateral head
Brachioradialis
Anconeus

Olecranon of ulna

Muscles That Move the Hand

ACTION AT THE HAND

Flexor carpi ulnaris
Extensor carpi radialis longus
Extensor carpi ulnaris
Extensor carpi radialis brevis

Extensor retinaculum

Extensor digitorum
Abductor pollicis longus
Extensor pollicis brevis
Ulna

a **Posterior view, superficial layer**

ANTERIOR

Palmaris longus
Flexor digitorum superficialis
Flexor carpi ulnaris
Flexor digitorum profundus
Ulna
Extensor carpi ulnaris

Brachioradialis
Flexor pollicis longus
Radius
Extensor carpi radialis longus
Extensor carpi radialis brevis
Abductor pollicis longus
Extensor digitorum
Extensor pollicis longus
Flexor carpi radialis

Extensor digiti minimi

POSTERIOR

Table 11–13	Muscles That Move the Forearm and Hand (Figure 11–17)			
Muscle	**Origin**	**Insertion**	**Action**	**Innervation**
ACTION AT THE ELBOW				
Flexors — Biceps brachii	*Short head* from the coracoid process; *long head* from the supraglenoid tubercle (both on the scapula)	Tuberosity of radius	Flexion at elbow and shoulder; supination	Musculocutaneous nerve (C_5–C_6)
Brachialis	Anterior, distal surface of humerus	Tuberosity of ulna	Flexion at elbow	Musculocutaneous nerve (C_5–C_6) and radial nerve (C_7–C_8)
Brachioradialis	Ridge superior to the lateral epicondyle of humerus	Lateral aspect of styloid process of radius	Flexion at elbow	Radial nerve (C_5–C_6)
Extensors — Anconeus	Posterior, inferior surface of lateral epicondyle of humerus	Lateral margin of olecranon on ulna	Extension at elbow	Radial nerve (C_7–C_8)
Triceps brachii				
lateral head	Superior, lateral margin of humerus	Olecranon of ulna	Extension at elbow	Radial nerve (C_6–C_8)
long head	Infraglenoid tubercle of scapula	Olecranon of ulna	As above, plus extension and adduction at the shoulder	Radial nerve (C_6–C_8)
medial head	Posterior surface of humerus inferior to radial groove	Olecranon of ulna	Extension at elbow	Radial nerve (C_6–C_8)
PRONATORS/SUPINATORS				
Pronator quadratus	Anterior and medial surfaces of distal portion of ulna	Anterolateral surface of distal portion of radius	Pronation	Median nerve (C_8–T_1)
Pronator teres	Medial epicondyle of humerus and coronoid process of ulna	Midlateral surface of radius	Pronation	Median nerve (C_6–C_7)
Supinator	Lateral epicondyle of humerus, annular ligament, and ridge near radial notch of ulna	Anterolateral surface of radius distal to the radial tuberosity	Supination	Deep radial nerve (C_6–C_8)
ACTION AT THE HAND				
Flexors — Flexor carpi radialis	Medial epicondyle of humerus	Bases of second and third metacarpal bones	Flexion and abduction at wrist	Median nerve (C_6–C_7)
Flexor carpi ulnaris	Medial epicondyle of humerus; adjacent medial surface of olecranon and anteromedial portion of ulna	Pisiform, hamate, and base of fifth metacarpal bone	Flexion and adduction at wrist	Ulnar nerve (C_8–T_1)
Palmaris longus	Medial epicondyle of humerus	Palmar aponeurosis and flexor retinaculum	Flexion at wrist	Median nerve (C_6–C_7)
Extensors — Extensor carpi radialis longus	Lateral supracondylar ridge of humerus	Base of second metacarpal bone	Extension and abduction at wrist	Radial nerve (C_6–C_7)
Extensor carpi radialis brevis	Lateral epicondyle of humerus	Base of third metacarpal bone	Extension and abduction at wrist	Radial nerve (C_6–C_7)
Extensor carpi ulnaris	Lateral epicondyle of humerus; adjacent dorsal surface of ulna	Base of fifth metacarpal bone	Extension and adduction at wrist	Deep radial nerve (C_6–C_8)

The **flexor carpi ulnaris**, **flexor carpi radialis**, and **palmaris longus** muscles are superficial muscles that work together to produce flexion of the wrist. The flexor carpi radialis muscle flexes and *abducts*, and the flexor carpi ulnaris muscle flexes and *adducts*. *Pitcher's arm* is an inflammation at the origins of the flexor carpi muscles at the medial epicondyle of the humerus. This condition results from forcibly flexing the wrist just before releasing a baseball.

The **extensor carpi radialis** muscles and the **extensor carpi ulnaris** muscle have a similar relationship to that between the flexor carpi muscles. That is, the extensor carpi radialis muscles produce extension and *abduction*, whereas the extensor carpi ulnaris muscle produces extension and *adduction*.

The muscles involved in pronation and supination are shown in **Figure 11–18**. The **pronator teres** and **supinator** muscles originate on both the humerus and ulna. These muscles rotate the radius without either flexing or extending the elbow. The **pronator quadratus** muscle originates on the ulna and assists the pronator teres muscle in opposing the actions of

Table 11–14	Muscles That Move the Hand and Fingers (Figure 11–18)			
Muscle	**Origin**	**Insertion**	**Action**	**Innervation**
Abductor pollicis longus	Proximal dorsal surfaces of ulna and radius	Lateral margin of first metacarpal bone	Abduction at joints of thumb and wrist	Deep radial nerve (C_6–C_7)
Extensor digitorum	Lateral epicondyle of humerus	Posterior surfaces of the phalanges, fingers 2–5	Extension at finger joints and wrist	Deep radial nerve (C_6–C_8)
Extensor pollicis brevis	Shaft of radius distal to origin of abductor pollicis longus	Base of proximal phalanx of thumb	Extension at joints of thumb; abduction at wrist	Deep radial nerve (C_6–C_7)
Extensor pollicis longus	Posterior and lateral surfaces of ulna and interosseous membrane	Base of distal phalanx of thumb	Extension at joints of thumb; abduction at wrist	Deep radial nerve (C_6–C_8)
Extensor indicis	Posterior surface of ulna and interosseous membrane	Posterior surface of phalanges of index finger (2), with tendon of extensor digitorum	Extension and adduction at joints of index finger	Deep radial nerve (C_6–C_8)
Extensor digiti minimi	By extensor tendon to lateral epicondyle of humerus and from intermuscular septa	Posterior surface of proximal phalanx of little finger (5)	Extension at joints of little finger	Deep radial nerve (C_6–C_8)
Flexor digitorum superficialis	Medial epicondyle of humerus; adjacent anterior surfaces of ulna and radius	Midlateral surfaces of middle phalanges of fingers 2–5	Flexion at proximal interphalangeal, metacarpophalangeal, and wrist joints	Median nerve (C_7–T_1)
Flexor digitorum profundus	Medial and posterior surfaces of ulna, medial surface of coronoid process, and interosseus membrane	Bases of distal phalanges of fingers 2–5	Flexion at distal interphalangeal joints and, to a lesser degree, proximal interphalangeal joints and wrist	Palmar interosseous nerve, from median nerve, and ulnar nerve (C_8–T_1)
Flexor pollicis longus	Anterior shaft of radius, interosseous membrane	Base of distal phalanx of thumb	Flexion at joints of thumb	Median nerve (C_8–T_1)

the supinator or biceps brachii muscles. During pronation, the tendon of the biceps brachii muscle rotates with the radius. As a result, this muscle cannot assist in flexion of the elbow when the forearm is pronated.

Muscles That Move the Hand and Fingers

Several superficial and deep muscles of the forearm flex and extend the finger joints (**Figure 11–18** and Table 11–14). These large muscles end before reaching the wrist, and only their tendons cross the articulation. This arrangement ensures maximum mobility at both the wrist and hand. The tendons that cross the posterior and anterior surfaces of the wrist pass through **synovial tendon sheaths**, elongated bursae that reduce friction. ↻ p. 293

The muscles of the forearm provide strength and crude control of the hand and fingers. These muscles are known as the *extrinsic muscles of the hand*. Fine control of the hand involves small *intrinsic muscles*, which originate on the carpal and metacarpal

bones. No muscles originate on the phalanges, and only tendons extend across the distal joints of the fingers. The intrinsic muscles of the hand are detailed in **Figure 11–19** and Table 11–15.

The fascia of the forearm thickens on the posterior surface of the wrist, forming the **extensor retinaculum** (ret-i-NAK-ū-lum; plural, *retinacula*), a wide band of connective tissue. The extensor retinaculum holds the tendons of the extensor muscles in place. On the anterior surface, the fascia also thickens to form another wide band of connective tissue, the **flexor retinaculum**, which stabilizes the tendons of the flexor muscles.

Inflammation of the retinacula and synovial tendon sheaths can restrict movement and put pressure on the distal portions of the *median nerve*, a mixed (sensory and motor) nerve that innervates the hand. This condition, known as *carpal tunnel syndrome*, causes tingling, numbness, weakness, and chronic pain. A common cause is repetitive hand or wrist movements. ↻ p. 275

Figure 11–18 **Muscles That Move the Hand and Fingers.**

Muscles That Flex the Fingers and Thumb
- Flexor digitorum superficialis
- Flexor pollicis longus
- Flexor digitorum profundus

Tendon of biceps brachii
Radius
Brachioradialis (retracted)
Median nerve
Pronator teres (cut)
Brachial artery
Flexor carpi ulnaris (retracted)

LATERAL MEDIAL

a **Anterior view, middle layer**

Supinator
Brachialis
Cut tendons of flexor digitorum superficialis
Pronator quadratus

b **Anterior view, deepest layer**

Muscles That Move the Forearm

PRONATORS AND SUPINATORS

Supinator
Pronator teres
Pronator quadratus
Ulna
Radius

Supination

11

Muscles That Extend the Fingers
- Extensor digitorum
- Extensor digiti minimi

Anconeus
Abductor pollicis longus
Extensor pollicis brevis
Tendon of extensor pollicis longus

MEDIAL LATERAL

c **Posterior view, middle layer**

Anconeus
Supinator

Muscles That Move the Thumb
- Extensor pollicis longus
- Abductor pollicis longus
- Extensor pollicis brevis

Extensor indicis
Ulna
Radius
Tendon of extensor digiti minimi (cut)
Tendon of extensor digitorum (cut)

d **Posterior view, deepest layer**

Pronation

e **Pronation and supination**

Figure 11–19 **Intrinsic Muscles of the Hand.** ATLAS: Plates 37b; 38c–f

11

Synovial sheaths

Tendon of flexor digitorum profundus

Tendon of flexor digitorum superficialis

Tendons of flexor digitorum

Tendon of flexor pollicis longus

Intrinsic Muscles of the Hand

Lumbricals
Palmar interosseus
First dorsal interosseus
Abductor digiti minimi
Flexor digiti minimi brevis
Opponens digiti minimi
Palmaris brevis (cut)

Intrinsic Muscles of the Thumb

Adductor pollicis
Flexor pollicis brevis
Opponens pollicis
Abductor pollicis brevis

Flexor retinaculum

Tendon of flexor carpi ulnaris

Tendon of palmaris longus

Tendon of flexor carpi radialis

a **Right hand, anterior (palmar) view**

Tendon of extensor indicis

Intrinsic Muscles of the Hand

First dorsal interosseus
Abductor digiti minimi

Tendons of extensor digitorum

Tendon of extensor digiti minimi

Tendon of extensor pollicis longus
Tendon of extensor pollicis brevis

Tendon of extensor carpi ulnaris

Extensor retinaculum

Tendon of extensor carpi radialis longus

Tendon of extensor carpi radialis brevis

b **Right hand, posterior view**

Table 11–15	Intrinsic Muscles of the Hand (Figure 11–19)			
Muscle	**Origin**	**Insertion**	**Action**	**Innervation**
Palmaris brevis	Palmar aponeurosis	Skin of medial border of hand	Moves skin on medial border toward midline of palm	Ulnar nerve, superficial branch (C_8)
ADDUCTION/ABDUCTION				
Adductor pollicis	Metacarpal and carpal bones	Proximal phalanx of thumb	Adduction of thumb	Ulnar nerve, deep branch (C_8–T_1)
Palmar interosseus* (3–4)	Sides of metacarpal bones II, IV, and V	Bases of proximal phalanges of fingers 2, 4, and 5	Adduction at metacarpophalangeal joints of fingers 2, 4, and 5; flexion at metacarpophalangeal joints; extension at interphalangeal joints	Ulnar nerve, deep branch (C_8–T_1)
Abductor pollicis brevis	Transverse carpal ligament, scaphoid, and trapezium	Radial side of base of proximal phalanx of thumb	Abduction of thumb	Median nerve (C_6–C_7)
Dorsal interosseus (4)	Each originates from opposing faces of two metacarpal bones (I and II, II and III, III and IV, IV and V)	Bases of proximal phalanges of fingers 2–4	Abduction at metacarpophalangeal joints of fingers 2 and 4; flexion at metacarpophalangeal joints; extension at interphalangeal joints	Ulnar nerve, deep branch (C_8–T_1)
Abductor digiti minimi	Pisiform	Proximal phalanx of little finger	Abduction of little finger and flexion at its metacarpophalangeal joint	Ulnar nerve, deep branch (C_8–T_1)
FLEXION				
Flexor pollicis brevis	Flexor retinaculum, trapezium, capitate, and ulnar side of first metacarpal bone	Radial and ulnar sides of proximal phalanx of thumb	Flexion and adduction of thumb	Branches of median and ulnar nerves
Lumbrical (4)	Tendons of flexor digitorum profundus	Tendons of extensor digitorum to digits 2–5	Flexion at metacarpophalangeal joints 2–5; extension at proximal and distal interphalangeal joints, digits 2–5	No. 1 and no. 2 by median nerve; no. 3 and no. 4 by ulnar nerve, deep branch
Flexor digiti minimi brevis	Hamate	Proximal phalanx of little finger	Flexion at joints of little finger	Ulnar nerve, deep branch (C_8–T_1)
OPPOSITION				
Opponens pollicis	Trapezium and flexor retinaculum	First metacarpal bone	Opposition of thumb	Median nerve (C_6–C_7)
Opponens digiti minimi	Trapezium and flexor retinaculum	Fifth metacarpal bone	Opposition of fifth metacarpal bone	Ulnar nerve, deep branch (C_8–T_1)

*The deep, medial portion of the flexor pollicis brevis originating on the first metacarpal bone is sometimes called the *first palmar interosseus muscle;* it inserts on the ulnar side of the phalanx and is innervated by the ulnar nerve.

Muscles of the Pelvis and Lower Limbs

The pelvic girdle is tightly bound to the axial skeleton, permitting little movement. Few muscles of the axial musculature can influence the position of the pelvis. The muscles that position the lower limbs can be divided into three functional groups: (1) *muscles that move the thigh,* (2) *muscles that move the leg,* and (3) *muscles that move the foot and toes.*

Muscles That Move the Thigh

Table 11–16 lists the muscles that move the thigh. **Gluteal muscles** cover the lateral surfaces of the ilia (Figure 11–20a,b,c). The **gluteus maximus** muscle is the largest and most posterior of the gluteal muscles. Its origin includes parts of the ilium; the sacrum, coccyx, and associated ligaments; and the thoracolumbar fascia. Acting alone, this massive muscle produces extension and lateral rotation at the hip joint.

The gluteus maximus shares an insertion with the **tensor fasciae latae** (FASH-ē-ē LAH-tā) muscle, which originates on the iliac crest and the anterior superior iliac spine. Together, these muscles pull on the **iliotibial** (il-ē-ō-TIB-ē-ul) **tract,** a band of collagen fibers that extends along the lateral surface of the thigh and inserts on the tibia. This tract provides a lateral brace for the knee that becomes particularly important when you balance on one foot.

The **gluteus medius** and **gluteus minimus** muscles (Figure 11–20a,b,c) originate anterior to the origin of the gluteus maximus muscle and insert on the greater trochanter of the femur. The anterior gluteal line on the lateral surface of the ilium marks the boundary between these muscles.

Table 11–16	Muscles That Move the Thigh (Figure 11–20)			
Group and Muscle	**Origin**	**Insertion**	**Action**	**Innervation***
GLUTEAL GROUP				
Gluteus maximus	Iliac crest, posterior gluteal line, and lateral surface of ilium; sacrum, coccyx, and thoracolumbar fascia	Iliotibial tract and gluteal tuberosity of femur	Extension and lateral rotation at hip	Inferior gluteal nerve (L_5–S_2)
Gluteus medius	Anterior iliac crest of ilium, lateral surface between posterior and anterior gluteal lines	Greater trochanter of femur	Abduction and medial rotation at hip	Superior gluteal nerve (L_4–S_1)
Gluteus minimus	Lateral surface of ilium between inferior and anterior gluteal lines	Greater trochanter of femur	Abduction and medial rotation at hip	Superior gluteal nerve (L_4–S_1)
Tensor fasciae latae	Iliac crest and lateral surface of anterior superior iliac spine	Iliotibial tract	Flexion and medial rotation at hip; tenses fasciae latae, which laterally supports the knee	Superior gluteal nerve (L_4–S_1)
LATERAL ROTATOR GROUP				
Obturators (externus and internus)	Lateral and medial margins of obturator foramen	Trochanteric fossa of femur (externus); medial surface of greater trochanter (internus)	Lateral rotation at hip	Obturator nerve (externus: L_3–L_4) and special nerve from sacral plexus (internus: L_5–S_2)
Piriformis	Anterolateral surface of sacrum	Greater trochanter of femur	Lateral rotation and abduction at hip	Branches of sacral nerves (S_1–S_2)
Gemelli (superior and inferior)	Ischial spine and tuberosity	Medial surface of greater trochanter with tendon of obturator internus	Lateral rotation at hip	Nerves to obturator internus and quadratus femoris
Quadratus femoris	Lateral border of ischial tuberosity	Intertrochanteric crest of femur	Lateral rotation at hip	Special nerve from sacral plexus (L_4–S_1)
ADDUCTOR GROUP				
Adductor brevis	Inferior ramus of pubis	Linea aspera of femur	Adduction, flexion, and medial rotation at hip	Obturator nerve (L_3–L_4)
Adductor longus	Inferior ramus of pubis anterior to adductor brevis	Linea aspera of femur	Adduction, flexion, and medial rotation at hip	Obturator nerve (L_3–L_4)
Adductor magnus	Inferior ramus of pubis posterior to adductor brevis and ischial tuberosity	Linea aspera and adductor tubercle of femur	Adduction at hip; superior part produces flexion and medial rotation; inferior part produces extension and lateral rotation	Obturator and sciatic nerves
Pectineus	Superior ramus of pubis	Pectineal line inferior to lesser trochanter of femur	Flexion, medial rotation, and adduction at hip	Femoral nerve (L_2–L_4)
Gracilis	Inferior ramus of pubis	Medial surface of tibia inferior to medial condyle	Flexion at knee; adduction and medial rotation at hip	Obturator nerve (L_3–L_4)
ILIOPSOAS GROUP**				
Iliacus	Iliac fossa of ilium	Femur distal to lesser trochanter; tendon fused with that of psoas major	Flexion at hip	Femoral nerve (L_2–L_3)
Psoas major	Anterior surfaces and transverse processes of vertebrae (T_{12}–L_5)	Lesser trochanter in company with iliacus	Flexion at hip or lumbar intervertebral joints	Branches of the lumbar plexus (L_2–L_3)

*Where appropriate, spinal nerves involved are given in parentheses.
**The psoas major and iliacus are often considered collectively as the iliopsoas.

The **lateral rotators** originate at or inferior to the horizontal axis of the acetabulum. There are six lateral rotator muscles in all. The **piriformis** (pir-i-FOR-mis) muscle and the **obturator** muscles are dominant (Figure 11–20c,d).

The **adductors** (Figure 11–20c,d) originate inferior to the horizontal axis of the acetabulum. This muscle group includes the **adductor magnus**, **adductor brevis**, **adductor longus**, **pectineus** (pek-ti-NĒ-us), and **gracilis** (GRAS-i-lis) muscles. All but the adductor magnus originate both anterior and inferior to the joint, so they perform hip flexion as well as adduction. The adductor magnus muscle can produce either adduction and flexion or adduction and extension, depending on the region stimulated. The adductor magnus muscle can also produce medial or lateral rotation at the hip. The other

Figure 11–20 **Muscles That Move the Thigh.** ATLAS: Plates 68a–c; 72a,b; 73a,b

Iliac crest
Sacrum
Gluteus medius (cut)
Gluteus maximus (cut)
Obturator internus

Gluteal Group
Gluteus medius
Gluteus maximus
Gluteus minimus
Tensor fasciae latae

Sartorius
Rectus femoris
Iliotibial tract
Vastus lateralis
Biceps femoris, long head
Biceps femoris, short head
Semimembranosus
Plantaris
Head of fibula
Patella
Patellar ligament

a **Gluteal region, posterior view**

b **Lateral view**

Gluteal Group
| Gluteus maximus (cut) | Gluteus medius (cut) | Gluteus minimus | Tensor fasciae latae |

Iliopsoas Group
Psoas major
Iliacus

Lateral Rotator Group
Piriformis
Superior gemellus
Obturator internus
Obturator externus
Inferior gemellus
Quadratus femoris
Ischial tuberosity
Iliotibial tract

L5
Inguinal ligament

Adductor Group
Pectineus
Adductor brevis
Adductor longus
Adductor magnus
Gracilis

c **Posterior view, deep muscles**

d **Anterior view of the iliopsoas and adductor groups**

muscles, which insert on low ridges along the posterior surface of the femur, produce medial rotation. When an athlete suffers a *pulled groin,* the problem is a *strain*—a muscle tear—in one of these adductor muscles.

A pair of muscles controls the internal surface of the pelvis. The large **psoas** (SŌ-us) **major** muscle originates alongside the inferior thoracic and lumbar vertebrae. Its insertion lies on the lesser trochanter of the femur. Before reaching this insertion, its tendon merges with that of the **iliacus** (il-Ī-ah-kus) muscle, which nestles within the iliac fossa. These two powerful hip flexors are often grouped together and collectively referred to as the **iliopsoas** (il-ē-ō-SŌ-us) muscle (**Figure 11–20d**).

Muscles That Move the Leg

As in the upper limb, muscle distribution in the lower limb has a pattern: Extensor muscles are located along the anterior and lateral surfaces of the leg, and flexors lie along the posterior and medial surfaces (**Figure 11–21** and **Table 11–17**). As in the upper limb, sturdy connective tissue partitions divide the lower limb into separate muscular compartments. The flexors and adductors originate on the pelvic girdle, but most extensors originate on the femoral surface.

The *flexors of the knee* include the **biceps femoris, semi-membranosus** (sem-ē-mem-bra-NŌ-sus), **semitendinosus** (sem-ē-ten-di-NŌ-sus), and **sartorius** muscles (**Figure 11–21**). These muscles originate along the edges of the pelvis and insert on the tibia and fibula. The sartorius muscle is the only knee flexor that originates superior to the acetabulum. Its insertion lies along the medial surface of the tibia. When the sartorius muscle contracts, it produces flexion at the knee and lateral rotation at the hip. This happens when you cross your legs.

Three of these muscles—the biceps femoris, semimembranosus, and semitendinosus muscles—are often called the **hamstrings.** Because they originate on the pelvic surface inferior and posterior to the acetabulum, their contractions produce not only flexion at the knee, but also extension at the hip. A *pulled hamstring* is a common sports injury caused by a strain affecting one of the hamstring muscles.

TIPS & TRICKS

To remember that three muscles make up the ***ham***strings, think "the three little pigs." These three muscles are portions of the cut of meat sold as ham.

The knee joint can be locked at full extension by a slight lateral rotation of the tibia. ⟲ p. 306 The small **popliteus** (pop-LI-tē-us) muscle originates on the femur near the lateral condyle

Table 11–17	Muscles That Move the Leg (Figure 11–21)			
Muscle	**Origin**	**Insertion**	**Action**	**Innervation***
FLEXORS OF THE KNEE				
Biceps femoris	Ischial tuberosity and linea aspera of femur	Head of fibula, lateral condyle of tibia	Flexion at knee; extension and lateral rotation at hip	Sciatic nerve; tibial portion (S_1–S_3; to long head) and common fibular branch (L_5–S_2; to short head)
Semimembranosus	Ischial tuberosity	Posterior surface of medial condyle of tibia	Flexion at knee; extension and medial rotation at hip	Sciatic nerve (tibial portion; L_5–S_2)
Semitendinosus	Ischial tuberosity	Proximal, medial surface of tibia near insertion of gracilis	Flexion at knee; extension and medial rotation at hip	Sciatic nerve (tibial portion; L_5–S_2)
Sartorius	Anterior superior iliac spine	Medial surface of tibia near tibial tuberosity	Flexion at knee; flexion and lateral rotation at hip	Femoral nerve (L_2–L_3)
Popliteus	Lateral condyle of femur	Posterior surface of proximal tibial shaft	Medial rotation of tibia (or lateral rotation of femur); flexion at knee	Tibial nerve (L_4–S_1)
EXTENSORS OF THE KNEE				
Rectus femoris	Anterior inferior iliac spine and superior acetabular rim of ilium	Tibial tuberosity by patellar ligament	Extension at knee; flexion at hip	Femoral nerve (L_2–L_4)
Vastus intermedius	Anterolateral surface of femur and linea aspera (distal half)	Tibial tuberosity by patellar ligament	Extension at knee	Femoral nerve (L_2–L_4)
Vastus lateralis	Anterior and inferior to greater trochanter of femur and along linea aspera (proximal half)	Tibial tuberosity by patellar ligament	Extension at knee	Femoral nerve (L_2–L_4)
Vastus medialis	Entire length of linea aspera of femur	Tibial tuberosity by patellar ligament	Extension at knee	Femoral nerve (L_2–L_4)

*Where appropriate, spinal nerves involved are given in parentheses.

Figure 11–21 **Muscles That Move the Leg.** ATLAS: Plates 69a,b; 70b; 72a,b; 74; 76a,b; 78a–g

Iliac crest

Gluteus medius

Tensor fasciae latae

Gluteus maximus

Adductor magnus

Gracilis

Iliotibial tract

Flexors of the Knee

Biceps femoris, long head

Biceps femoris, short head

Semitendinosus

Semimembranosus

Sartorius

Popliteus

a **Hip and thigh, posterior view**

Anterior superior iliac spine

Inguinal ligament

Iliacus ⎫
Psoas major ⎬ Iliopsoas

Pubic tubercle

Tensor fasciae latae

Pectineus

Adductor longus

Gracilis

Sartorius

Extensors of the Knee (Quadriceps muscles)

Rectus femoris

Vastus lateralis

Vastus medialis

Quadriceps tendon

Patella

Patellar ligament

b **Quadriceps and thigh muscles, anterior view**

POSTERIOR

Sciatic nerve

Femur

Extensors of the Knee (Quadriceps muscles)

Vastus lateralis

Vastus intermedius

Vastus medialis

Rectus femoris

Flexors of the Knee

Semitendinosus

Semimembranosus

Biceps femoris, long head

Biceps femoris, short head

Gracilis

Adductor magnus

Adductor longus

Sartorius

ANTERIOR

c **Sectional view**

11

11

and inserts on the posterior tibial shaft (**Figure 11–22a**). When flexion is started, this muscle contracts to produce a slight medial rotation of the tibia that unlocks the knee joint.

Collectively, four *knee extensors*—the three **vastus muscles,** which originate along the shaft of the femur, and the **rectus femoris muscle**—make up the **quadriceps femoris** (the "quads"). Together, the vastus muscles cradle the rectus femoris muscle the way a bun surrounds a hot dog (**Figure 11–21c**). All four muscles insert on the patella by way of the quadriceps tendon. The force of their contraction is relayed to the tibial tuberosity by way of the patellar ligament. The rectus femoris muscle originates on the anterior inferior iliac spine and the superior acetabular rim—so in addition to extending the knee, it assists in flexion of the hip.

TIPS & TRICKS

Think of the quadriceps muscles as "the four at the fore."

Muscles That Move the Foot and Toes

The extrinsic muscles that move the foot and toes are shown in **Figure 11–22** and listed in Table 11–18. Most of the muscles that move the ankle produce the plantar flexion involved with walking and running movements. The **gastrocnemius** (gas-trok-NĒ-mē-us; *gaster*, stomach + *kneme*, knee) muscle of the calf is an important plantar flexor, but the slow muscle fibers of the underlying **soleus** (SŌ-lē-us) muscle are better suited for making continuous postural adjustments against the pull of gravity.

Table 11–18	Extrinsic Muscles That Move the Foot and Toes (Figure 11–22)			
Muscle	**Origin**	**Insertion**	**Action**	**Innervation**
ACTION AT THE ANKLE				
Flexors (Dorsiflexors)				
Tibialis anterior	Lateral condyle and proximal shaft of tibia	Base of first metatarsal bone and medial cuneiform bone	Flexion (dorsiflexion) at ankle; inversion of foot	Deep fibular nerve (L_4–S_1)
Fibularis tertius	Distal anterior surface of fibula and interosseous membrane	Dorsal surface of fifth metatarsal bone	Flexion (dorsiflexion) at ankle; eversion of foot	Deep fibular nerve (L_5–S_1)
Extensors (Plantar flexors)				
Gastrocnemius	Femoral condyles	Calcaneus by calcaneal tendon	Extension (plantar flexion) at ankle; inversion of foot; flexion at knee	Tibial nerve (S_1–S_2)
Fibularis brevis	Midlateral margin of fibula	Base of fifth metatarsal bone	Eversion of foot and extension (plantar flexion) at ankle	Superficial fibular nerve (L_4–S_1)
Fibularis longus	Lateral condyle of tibia, head and proximal shaft of fibula	Base of first metatarsal bone and medial cuneiform bone	Eversion of foot and extension (plantar flexion) at ankle; supports longitudinal arch	Superficial fibular nerve (L_4–S_1)
Plantaris	Lateral supracondylar ridge	Posterior portion of calcaneus	Extension (plantar flexion) at ankle; flexion at knee	Tibial nerve (L_4–S_1)
Soleus	Head and proximal shaft of fibula and adjacent posteromedial shaft of tibia	Calcaneus by calcaneal tendon (with gastrocnemius)	Extension (plantar flexion) at ankle	Sciatic nerve, tibial branch (S_1–S_2)
Tibialis posterior	Interosseous membrane and adjacent shafts of tibia and fibula	Tarsal and metatarsal bones	Adduction and inversion of foot; extension (plantar flexion) at ankle	Sciatic nerve, tibial branch (S_1–S_2)
ACTION AT THE TOES				
Digital flexors				
Flexor digitorum longus	Posteromedial surface of tibia	Inferior surfaces of distal phalanges, toes 2–5	Flexion at joints of toes 2–5	Sciatic nerve, tibial branch (L_5–S_1)
Flexor hallucis longus	Posterior surface of fibula	Inferior surface, distal phalanx of great toe	Flexion at joints of great toe	Sciatic nerve, tibial branch (L_5–S_1)
Digital extensors				
Extensor digitorum longus	Lateral condyle of tibia, anterior surface of fibula	Superior surfaces of phalanges, toes 2–5	Extension at joints of toes 2–5	Deep fibular nerve (L_4–S_1)
Extensor hallucis longus	Anterior surface of fibula	Superior surface, distal phalanx of great toe	Extension at joints of great toe	Deep fibular nerve (L_4–S_1)

Figure 11–22 **Extrinsic Muscles That Move the Foot and Toes.** ATLAS: Plates 81a,b; 82a,b; 84a,b

Superficial Dissection ⟶ **Deep Dissection**

Ankle Extensors
- Plantaris
- Gastrocnemius, medial head
- Gastrocnemius, lateral head
- Soleus

- Popliteus
- Head of fibula

Ankle Extensors (Deep)
- Tibialis posterior
- Fibularis longus
- Fibularis brevis

Digital Flexors
- Flexor digitorum longus
- Flexor hallucis longus

- Gastrocnemius (cut and removed)
- Calcaneal tendon
- Calcaneus
- Tendon of flexor hallucis longus
- Tendon of flexor digitorum longus
- Tendon of fibularis brevis
- Tendon of fibularis longus

a **Posterior views**

Ankle Extensors
- Gastrocnemius, lateral head
- Fibularis longus
- Soleus
- Fibularis brevis
- Superior extensor retinaculum
- Calcaneal tendon
- Inferior extensor retinaculum
- Tendon of fibularis tertius

- Iliotibial tract
- Head of fibula

Ankle Flexors
- Tibialis anterior

Digital Extensors
- Extensor digitorum longus
- Tendon of extensor hallucis longus

b **Lateral view**

- Patella
- Patellar ligament
- Medial surface of tibial shaft

Ankle Extensors
- Gastrocnemius, medial head
- Soleus
- Tibialis posterior
- Superior extensor retinaculum
- Calcaneal tendon
- Flexor retinaculum
- Inferior extensor retinaculum
- Tendon of tibialis anterior

c **Medial view**

- Patellar ligament
- Fibularis longus
- Tibialis anterior
- Tibia
- Extensor digitorum longus
- Extensor hallucis longus
- Tendon of extensor digitorum longus

d **Anterior view**

11

These muscles are best seen in posterior and lateral views (Figure 11–22a,b). The gastrocnemius muscle arises from two heads located on the medial and lateral epicondyles of the femur just proximal to the knee. The *fabella*, a sesamoid bone, is occasionally present within the lateral head of the gastrocnemius muscle. The gastrocnemius and soleus muscles share a common tendon, the **calcaneal tendon**, commonly known as the *Achilles tendon*.

TIPS&TRICKS

The soleus is so named because it resembles the flat-bodied fish we call sole.

The term "Achilles tendon" comes from Greek mythology. Achilles was a warrior who was invincible but for one vulnerable spot: the calcaneal tendon. Outside mythology, damage to the calcaneal tendon isn't a fatal problem. Although it is among the largest, strongest tendons in the body, its rupture is common. The applied forces increase markedly during rapid acceleration or deceleration. Sprinters can rupture the calcaneal tendon pushing off from the starting blocks, and the elderly often snap this tendon during a stumble or fall. Surgery may be necessary to reposition and reconnect the torn ends of the tendon to promote healing.

Deep to the gastrocnemius and soleus muscles are a pair of **fibularis** muscles, or *peroneus* muscles (Figure 11–22a,b,d). The fibularis muscles produce eversion and extension (plantar flexion) at the ankle. Inversion is caused by the contraction of the **tibialis** (tib-ē-A-lis) muscles. The large **tibialis anterior** muscle flexes the ankle and opposes the gastrocnemius muscle (Figure 11–22b,c,d).

Important digital muscles originate on the surface of the tibia, the fibula, or both (Figure 11–22a,b,c,d). Large synovial tendon sheaths surround the tendons of the tibialis anterior, **extensor digitorum longus**, and **extensor hallucis longus** muscles, where they cross the ankle joint. The positions of these sheaths are stabilized by superior and inferior **extensor retinacula** (Figure 11–22b,c,d).

Intrinsic muscles of the foot originate on the tarsal and metatarsal bones (Figure 11–23 and Table 11–19). Their contractions move the toes and maintain the longitudinal arch of the foot. ⤴ p. 284

Table 11–19	Intrinsic Muscles of the Foot (Figure 11–23)			
Muscle	**Origin**	**Insertion**	**Action**	**Innervation**
FLEXION/EXTENSION				
Flexor hallucis brevis	Cuboid and lateral cuneiform bones	Proximal phalanx of great toe	Flexion at metatarsophalangeal joint of great toe	Medial plantar nerve (L_4–L_5)
Flexor digitorum brevis	Calcaneus (tuberosity on inferior surface)	Sides of middle phalanges, toes 2–5	Flexion at proximal interphalangeal joints of toes 2–5	Medial plantar nerve (L_4–L_5)
Quadratus plantae	Calcaneus (medial, inferior surfaces)	Tendon of flexor digitorum longus	Flexion at joints of toes 2–5	Lateral plantar nerve (L_4–L_5)
Lumbrical (4)	Tendons of flexor digitorum longus	Insertions of extensor digitorum longus	Flexion at metatarsophalangeal joints; extension at proximal interphalangeal joints of toes 2–5	Medial plantar nerve (1), lateral plantar nerve (2–4)
Flexor digiti minimi brevis	Base of metatarsal bone V	Lateral side of proximal phalanx of toe 5	Flexion at metatarsophalangeal joint of toe 5	Lateral plantar nerve (S_1–S_2)
Extensor digitorum brevis	Calcaneus (superior and lateral surfaces)	Dorsal surfaces of toes 1–4	Extension at metatarsophalangeal joints of toes 1–4	Deep fibular nerve (L_5–S_1)
Extensor hallucis brevis	Superior surface of anterior calcaneus	Dorsal surface of the base of proximal phalanx of great toe	Extension of great toe	Deep fibular nerve (L_5–S_1)
ADDUCTION/ABDUCTION				
Adductor hallucis	Bases of metatarsal bones II–IV and plantar ligaments	Proximal phalanx of great toe	Adduction at metatarsophalangeal joint of great toe	Lateral plantar nerve (S_1–S_2)
Abductor hallucis	Calcaneus (tuberosity on inferior surface)	Medial side of proximal phalanx of great toe	Abduction at metatarsophalangeal joint of great toe	Medial plantar nerve (L_4–L_5)
Plantar interosseus (3)	Bases and medial sides of metatarsal bones	Medial sides of toes 3–5	Adduction at metatarsophalangeal joints of toes 3–5	Lateral plantar nerve (S_1–S_2)
Dorsal interosseus (4)	Sides of metatarsal bones	Medial and lateral sides of toe 2; lateral sides of toes 3 and 4	Abduction at metatarsophalangeal joints of toes 3 and 4	Lateral plantar nerve (S_1–S_2)
Abductor digiti minimi	As above	Lateral side of proximal phalanx, toe 5	Abduction at metatarsophalangeal joint of toe 5	Lateral plantar nerve (L_4–L_5)

Figure 11–23 **Intrinsic Muscles of the Foot.** ATLAS: Plates 84a; 85a,b; 86c; 87a–c; 89

Tendon of fibularis brevis

Superior extensor retinaculum

Medial malleolus of tibia

Lateral malleolus of fibula

Tendon of tibialis anterior

Inferior extensor retinaculum

Tendon of fibularis tertius

Tendons of extensor digitorum longus

Intrinsic Muscles of the Great Toe

Extensor hallucis brevis

Abductor hallucis

Intrinsic Muscles of the Foot

Dorsal interossei

Tendons of extensor digitorum brevis

Tendon of extensor hallucis longus

a **Dorsal view**

Superficial Muscles of the Sole of the Foot

Deep Muscles of the Sole of the Foot

Tendons of flexor digitorum brevis

Tendons of flexor digitorum longus

Fibrous tendon sheaths

Intrinsic Muscles of the Foot

Tendon of flexor hallucis longus

Lumbricals

Flexor hallucis brevis

Tendons of flexor digitorum brevis overlying tendons of flexor digitorum longus

Flexor digiti minimi brevis

Abductor hallucis

Tendon of flexor digitorum longus

Quadratus plantae

Flexor digitorum brevis

Tendon of tibialis posterior

Abductor digiti minimi

Tendon of fibularis longus

Plantar aponeurosis (cut)

Calcaneus

b **Plantar view, superficial layer**

c **Plantar view, deep layer**

Clinical Note

Hernia When the abdominal muscles contract forcefully, pressure in the abdominopelvic cavity can increase dramatically. That pressure is applied to internal organs. If the individual exhales at the same time, the pressure is relieved because the diaphragm can move upward as the lungs collapse. But during vigorous isometric exercises or when lifting a weight while holding one's breath, pressure in the abdominopelvic cavity can rise to 106 kg/cm^2, roughly 100 times the normal pressure. A pressure that high can cause a variety of problems, including hernias. A **hernia** develops when a visceral organ or part of an organ protrudes abnormally through an opening in a surrounding muscular wall or partition. There are many types of hernia. Here we will consider only *inguinal (*groin) *hernias* and *diaphragmatic hernias.*

Late in the development of male fetuses, the testes descend into the scrotum. They pass through the abdominal wall at the **inguinal canals.** In adult males, the sperm ducts and associated blood vessels penetrate the abdominal musculature at the inguinal canals as the *spermatic cords,* on their way to the abdominal reproductive organs. In an inguinal hernia, the inguinal canal enlarges. The abdominal contents, such as a portion of the greater omentum, small intestine, or (more rarely) urinary bladder, enter the inguinal canal. If these structures become trapped or twisted, surgery may be required to prevent serious complications. Inguinal hernias are not always caused by unusually high abdominal pressures. Injuries to the abdomen or inherited weakness or distensibility of the canal can have the same effect.

The esophagus and major blood vessels pass through openings in the diaphragm, the muscle that separates the thoracic and abdominopelvic cavities. In a **diaphragmatic hernia** abdominal organs slide into the thoracic cavity. If entry is through the *esophageal hiatus,* the passageway used by the esophagus, a *hiatal hernia* (hī-Ā-tal; *hiatus,* a gap or opening) exists. The severity of the condition depends on the location and size of the herniated organ or organs. Hiatal hernias are very common. Most go unnoticed, although they may increase the severity of gastric acid entry into the esophagus. (This condition is gastroesophageal reflux disease, or GERD, commonly known as heartburn.) Radiologists see hiatal hernias in about 30 percent of individuals whose upper gastrointestinal tracts are examined with barium-contrast techniques.

Sometimes clinical complications other than GERD develop. They generally do so because abdominal organs that have pushed into the thoracic cavity are exerting pressure on structures or organs there. Like inguinal hernias, a diaphragmatic hernia can result from congenital factors or from an injury that weakens or tears the diaphragm. If abdominal organs occupy the thoracic cavity during fetal development, the lungs may be poorly developed at birth.

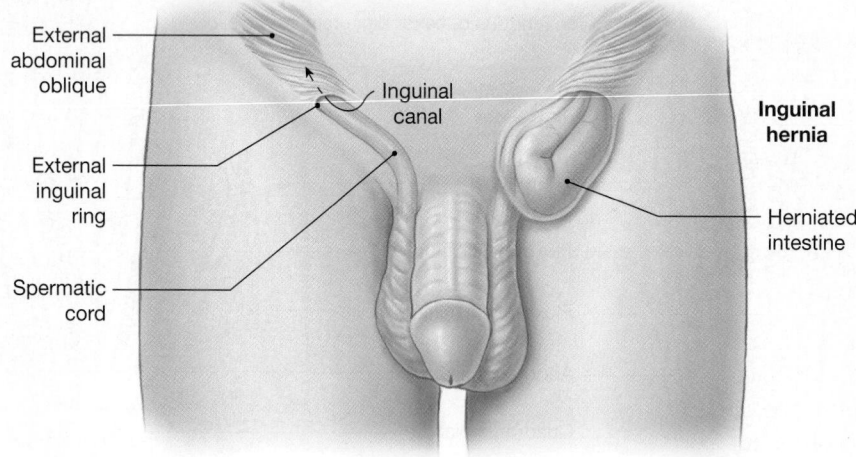

External abdominal oblique

Inguinal canal

External inguinal ring

Spermatic cord

Inguinal hernia

Herniated intestine

CHECKPOINT

18. Shrugging your shoulders uses which muscles?

19. Baseball pitchers sometimes suffer from rotator cuff injuries. Which muscles are involved in this type of injury?

20. An injury to the flexor carpi ulnaris muscle would impair which two movements?

21. Which leg movement would be impaired by injury to the obturator muscle?

22. To what does a "pulled hamstring" refer?

23. How would a torn calcaneal tendon affect movement of the foot?

See the blue Answers tab at the back of the book.

11-7 ┃ With advancing age, the size and power of muscle tissue decrease

Learning Outcome Identify age-related changes of the muscular system.

We can summarize the effects of aging on the muscular system as follows:

■ *Skeletal Muscle Fibers Become Smaller in Diameter.* This reduction in size reflects a decrease in the number of myofibrils. In addition, the muscle fibers contain smaller ATP, CP, and glycogen reserves and less myoglobin. The overall effect is a reduction in skeletal muscle size, strength, and endurance, combined with a tendency to fatigue quickly. Because cardiovascular performance also decreases with age, blood flow to active muscles does not increase with exercise as rapidly as it does in younger people. These factors interact to produce decreases of 30–50 percent in anaerobic and aerobic performance by age 65.

■ *Skeletal Muscles Become Less Elastic.* Aging skeletal muscles develop increasing amounts of fibrous connective tissue, a process called **fibrosis.** Fibrosis makes the muscle less flexible, and the collagen fibers can restrict movement and circulation.

■ *Tolerance for Exercise Decreases.* A lower tolerance for exercise comes in part from tiring quickly and in part from reduced thermoregulation, described in Chapter 5. ⟲ p. 194 Individuals over age 65 cannot eliminate the heat their muscles generate during contraction as effectively as younger people can. For this reason, they are subject to overheating.

■ *The Ability to Recover from Muscular Injuries Decreases.* The number of satellite cells steadily decreases with age, and the amount of fibrous tissue increases. As a result, when an injury occurs, repair capabilities are limited. Scar tissue formation is the usual result.

Regular exercise helps control body weight, strengthens bones, and generally improves the quality of life at all ages. Extremely demanding exercise is not as important as regular exercise. In fact, extreme exercise in the elderly can damage tendons, bones, and joints.

CHECKPOINT

24. Describe general age-related effects on skeletal muscle tissue.

25. Define fibrosis.

See the blue Answers tab at the back of the book.

11-8 ┃ Exercise produces responses in multiple body systems

Learning Outcome Explain the functional relationship between the muscular system and other body systems, and explain the role of exercise in producing various responses in other body systems.

To operate at maximum efficiency, the muscular system must be supported by many other systems. The changes that take place during exercise provide a good example of such interaction. As noted earlier, active muscles consume oxygen and generate carbon dioxide and heat. The immediate effects of exercise on various body systems include the following:

■ *Cardiovascular System:* Blood vessels in active muscles and the skin dilate, and heart rate increases. These adjustments speed up oxygen and nutrient delivery to and carbon dioxide removal from the muscle. They also bring heat to the skin for radiation into the environment.

■ *Respiratory System:* Respiratory rate and depth of respiration increase. Air moves into and out of the lungs more quickly, keeping pace with the increased rate of blood flow through the lungs.

■ *Integumentary System:* Blood vessels dilate, and sweat gland secretion increases. This combination increases evaporation at the skin surface and removes the excess heat generated by muscular activity.

SYSTEM INTEGRATOR

Body System ⟶ Muscular System

Muscular System ⟶ Body System

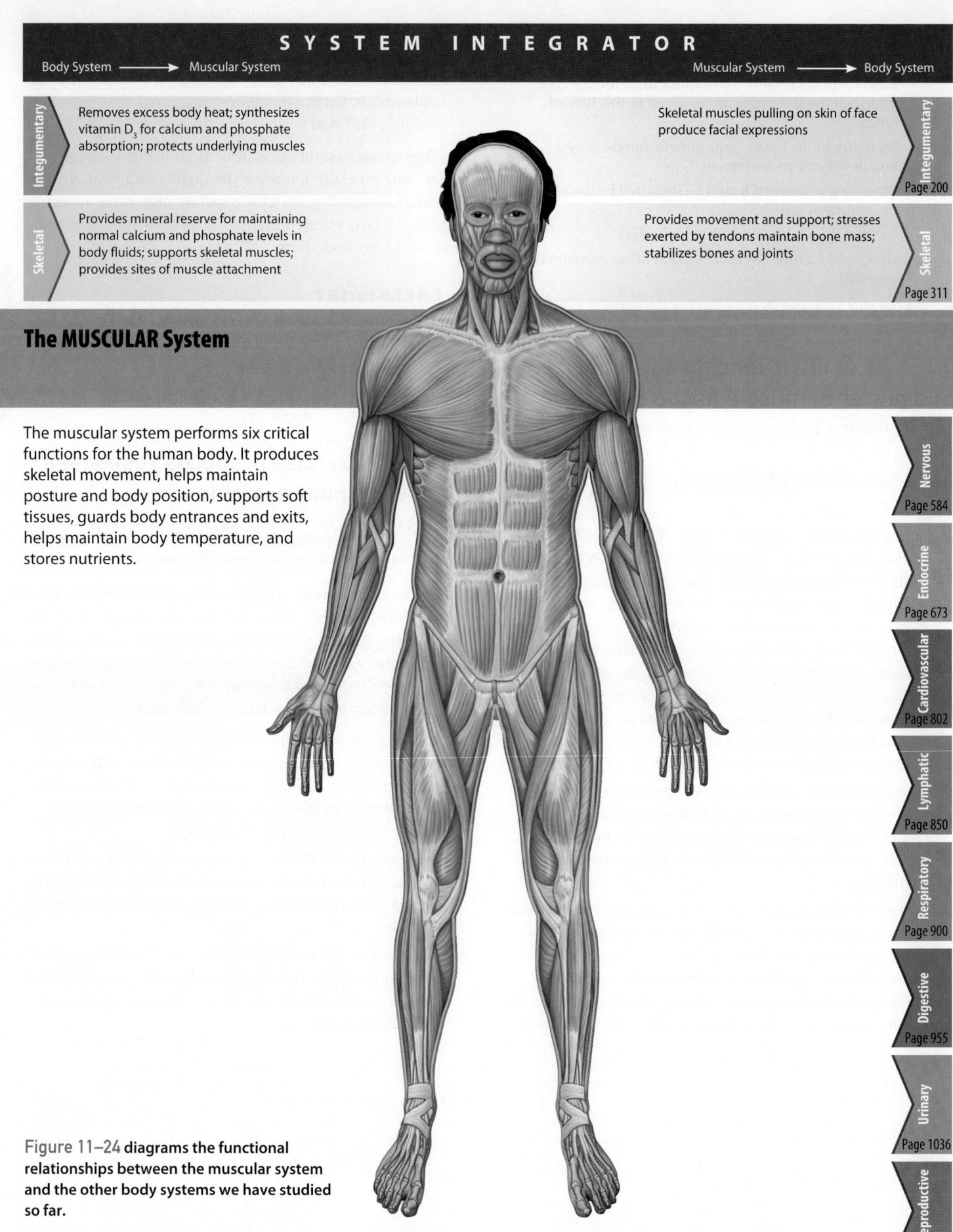

Integumentary
Removes excess body heat; synthesizes vitamin D$_3$ for calcium and phosphate absorption; protects underlying muscles

Integumentary
Skeletal muscles pulling on skin of face produce facial expressions

Page 200

Skeletal
Provides mineral reserve for maintaining normal calcium and phosphate levels in body fluids; supports skeletal muscles; provides sites of muscle attachment

Skeletal
Provides movement and support; stresses exerted by tendons maintain bone mass; stabilizes bones and joints

Page 311

The MUSCULAR System

The muscular system performs six critical functions for the human body. It produces skeletal movement, helps maintain posture and body position, supports soft tissues, guards body entrances and exits, helps maintain body temperature, and stores nutrients.

Nervous Page 584

Endocrine Page 673

Cardiovascular Page 802

Lymphatic Page 850

Respiratory Page 900

Digestive Page 955

Urinary Page 1036

Reproductive Page 1116

Figure 11–24 diagrams the functional relationships between the muscular system and the other body systems we have studied so far.

■ *Nervous and Endocrine Systems:* The responses of other systems are directed and coordinated through neural and endocrine (hormonal) adjustments in heart rate, respiratory rate, sweat gland activity, and mobilization of stored nutrient reserves.

Even when the body is at rest, the muscular system has extensive interactions with other systems. Figure 11–24 summarizes the range of interactions between the muscular system and other body systems studied so far.

CHECKPOINT

26. What major function does the muscular system perform for the body as a whole?

27. Identify the physiological effects of exercise on the cardiovascular, respiratory, and integumentary systems, and indicate the relationship between those physiological effects and the nervous and endocrine systems.

See the blue Answers tab at the back of the book.

Related Clinical Terms

charley horse: Common name for a muscle spasm, especially in the leg.

compartment syndrome: A condition in which increased pressure within the muscle compartment of a limb produces ischemia or "blood starvation."

fibromyositis: Chronic illness characterized by widespread musculoskeletal aches, pains, and stiffness, and soft tissue tenderness.

groin pull: An injury that is due to a strain of the muscles of the inner thigh.

impingement syndrome: Pain on elevation of the shoulder due to an injured or inflamed tendon or bursa coming into contact with the overlying acromial process.

physical therapist: Healthcare professional who uses specially designed exercises and equipment to help patients regain or improve their physical abilities.

plantar fasciitis: Inflammation of the plantar fascia causing foot or heel pain.

shin-splints: Pain along the shinbone (tibia) caused by an overload on the tibia and connective tissues that connect muscle to the bone.

tenosynovitis: Inflammation of a tendon and the sheath that covers it.

torticollis: A shortening or contraction of the muscles of the neck resulting in the head being tipped to one side with the chin turned to the other side.

Chapter Review

Study Outline

MasteringA&P®

Access more chapter study tools online in the MasteringA&P Study Area:

• Chapter Quizzes, Chapter Practice Test, Art-labeling Activities, Animations, MP3 Tutor Sessions, and Clinical Case Studies

• Practice Anatomy Lab **PAL**   • A&P Flix **A&PFlix**

• Interactive Physiology **iP**   • PhysioEx **PhysioEx**

11-1 | Fascicle arrangement is correlated with muscle power and range of motion p. 359

1. Structural variations among skeletal muscles affect their power, range, and speed of movement.

2. A muscle can be classified as a **parallel muscle, convergent muscle, pennate muscle,** or **circular muscle (sphincter)** according to the arrangement of fibers and fascicles in it. A pennate muscle may be *unipennate, bipennate,* or *multipennate. (Figure 11–1)*

11-2 | The three classes of levers increase muscle efficiency p. 361

3. A **lever** is a rigid structure that moves around a fixed point called the **fulcrum.** Levers can change the direction and effective strength of an applied force, and the distance and speed of the movement such a force produces.

4. Levers are classified as **first-class, second-class,** or **third-class levers.** Third-class levers are the most common levers in the body. *(Figure 11–2)*

11-3 | Muscle origins are at the fixed end of muscles, and insertions are at the movable end of muscles p. 362

5. Each muscle can be identified by its *origin, insertion,* and *action.*

6. The site of attachment of the fixed end of a muscle is called the **origin;** the site where the movable end of the muscle attaches to another structure is called the **insertion.**

7. The movement produced when a muscle contracts is its **action.** *(Spotlight Figure 11–3)*

8. According to the function of its action, a muscle can be classified as an **agonist,** or **prime mover;** an **antagonist;** a **synergist;** or a **fixator.**

11

11-4 ▌Descriptive terms are used to name skeletal muscles
p. 365

9. The names of muscles commonly provide clues to their body region, origin and insertion, fascicle organization, position, structural characteristics, and action. *(Table 11–1)*

10. The **axial musculature** arises on the axial skeleton; it positions the head and spinal column and moves the rib cage. The **appendicular musculature** stabilizes or moves components of the appendicular skeleton. *(Figure 11–4)*

11. **Innervation** refers to the distribution of nerves that control a region or organ, including a muscle.

11-5 ▌Axial muscles are muscles of the head and neck, vertebral column, trunk, and pelvic floor p. 367

12. The axial muscles fall into logical groups on the basis of location, function, or both.

13. Important muscles of facial expression include the **orbicularis oris, buccinator, occipitofrontalis** muscles and the **platysma**. *(Figure 11–5; Table 11–2)*

14. Six extrinsic eye muscles (*oculomotor muscles*) control eye movements: the **inferior** and **superior rectus** muscles, the **lateral** and **medial rectus** muscles, and the **inferior** and **superior oblique** muscles. *(Figure 11–6; Table 11–3)*

15. The muscles of mastication (chewing) are the **masseter, temporalis,** and **pterygoid** muscles. *(Figure 11–7; Table 11–4)*

16. The four muscles of the tongue are necessary for speech and swallowing and assist in mastication. They are the **palatoglossus, styloglossus, genioglossus,** and **hyoglossus** muscles. *(Figure 11–8; Table 11–5)*

17. The muscles of the pharynx constrict the pharyngeal walls **(pharyngeal constrictors),** elevate the larynx **(laryngeal elevators),** or raise the soft palate **(palatal muscles).** *(Figure 11–9; Table 11–6)*

18. The anterior muscles of the neck control the position of the larynx, depress the mandible, and provide a foundation for the muscles of the tongue and pharynx. The neck muscles include the **digastric** and **sternocleidomastoid** muscles and seven muscles that originate or insert on the hyoid bone. *(Figure 11–10; Table 11–7)*

19. The superficial muscles of the spine can be classified into the **spinalis, longissimus,** and **iliocostalis** groups. *(Figure 11–11; Table 11–8)*

20. Other muscles of the spine include the **longus capitis** and **longus colli** muscles of the neck, the small intervertebral muscles of the deep layer, and the **quadratus lumborum** muscle of the lumbar region. *(Figure 11–11; Table 11–8)*

21. The oblique muscles include the **scalene** muscles and the **intercostal** and **transversus** muscles. The **external** and **internal intercostal** muscles are important in respiratory (breathing) movements of the ribs. Also important to respiration is the **diaphragm.** *(Figures 11–11, 11–12; Table 11–9)*

22. The **perineum** can be divided into an anterior **urogenital triangle** and a posterior **anal triangle.** The pelvic floor consists of the **urogenital diaphragm** and the **pelvic diaphragm.** *(Figure 11–13; Table 11–10)*

11-6 ▌Appendicular muscles are muscles of the shoulders, upper limbs, pelvis, and lower limbs p. 384

23. The **trapezius** muscle affects the positions of the shoulder girdle, head, and neck. Other muscles inserting on the scapula include the **rhomboid, levator scapulae, serratus anterior, subclavius,** and **pectoralis minor** muscles. *(Figures 11–14, 11–16; Table 11–11)*

24. Among muscles that move the arm, the **deltoid** and the **supraspinatus** muscles are important abductors. The **subscapularis** and **teres major** muscles produce medial rotation at the shoulder; the **infraspinatus** and **teres minor** muscles produce lateral rotation; and the **coracobrachialis** muscle produces flexion and adduction at the shoulder. *(Figures 11–14, 11–15, 11–16; Table 11–12)*

25. The **pectoralis major** muscle flexes the shoulder joint, and the **latissimus dorsi** muscle extends it. *(Figures 11–14, 11–15, 11–16; Table 11–12)*

26. Among muscles that move the forearm and hand, the actions of the **biceps brachii** muscle and the **triceps brachii** muscle (long head) affect the elbow joint. The **brachialis** and **brachioradialis** muscles flex the elbow, opposed by the **anconeus** muscle. The **flexor carpi ulnaris, flexor carpi radialis,** and **palmaris longus** muscles cooperate to flex the wrist. The **extensor carpi radialis** muscles and the **extensor carpi ulnaris** muscle oppose them. The **pronator teres** and **pronator quadratus** muscles pronate the forearm and are opposed by the **supinator** muscle. *(Figures 11–16 to 11–19; Tables 11–13 to 11–15)*

27. Among muscles that move the thigh, **gluteal muscles** cover the lateral surfaces of the ilia. The largest is the **gluteus maximus** muscle, which shares an insertion with the **tensor fasciae latae.** Together, these muscles pull on the **iliotibial tract.** *(Figure 11–20; Table 11–16)*

28. The **piriformis** muscle and the **obturator** muscles are the most important **lateral rotators.** The **adductors** can produce a variety of movements. *(Figure 11–20; Table 11–17)*

29. The **psoas major** and **iliacus** muscles merge to form the **iliopsoas** muscle, a powerful flexor of the hip. *(Figures 11–20, 11–21; Table 11–16)*

30. Among muscles that move the leg, the flexors of the knee include the **biceps femoris, semimembranosus,** and **semitendinosus** muscles (the three **hamstrings**) and the **sartorius** muscle. The **popliteus** muscle unlocks the knee joint. *(Figures 11–21, 11–22; Table 11–17)*

31. Collectively, the knee extensors are known as the **quadriceps femoris.** This group consists of the three **vastus** muscles (intermedius, lateralis, medialis) and the **rectus femoris** muscle. *(Figure 11–21; Table 11–17)*

32. Among muscles that move the foot and toes, the **gastrocnemius** and **soleus** muscles produce plantar flexion (ankle extension). The **fibularis** brevis and longus muscles produce eversion as well as extension (plantar flexion) at the ankle. The fibularis tertius produces eversion and flexion (dorsiflexion) at the ankle. *(Figure 11–22; Table 11–18)*

33. Smaller muscles of the calf and shin position the foot and move the toes. Muscles originating at the tarsal and metatarsal bones provide precise control of the phalanges. *(Figure 11–23; Table 11–19)*

11-7 ▌With advancing age, the size and power of muscle tissue decrease p. 405

34. With advanced age, skeletal muscles undergo **fibrosis,** the tolerance for exercise decreases, and repair of injuries slows.

11-8 ▌Exercise produces responses in multiple body systems p. 405

35. Exercise illustrates the integration of the muscular system with the cardiovascular, respiratory, integumentary, nervous, and endocrine systems. *(Figure 11–24)*

Review Questions

See the blue Answers tab at the back of the book.

LEVEL 1 Reviewing Facts and Terms

1. Name the three pennate muscles in the following figure, and for each muscle indicate the type of pennate muscle based on the relationship of muscle fibers to the tendon.

    (a) _____
        _____

    (b) _____
        _____

    (c) _____
        _____

2. Label the three visible muscles of the rotator cuff in the following posterior view of the deep muscles that move the arm.

    (a) _____
    (b) _____
    (c) _____

3. What is the origin and insertion of a muscle?
4. Levers make muscle action more versatile by all of the following, *except* (a) changing the location of the muscle's insertion, (b) changing the speed of movement produced by an applied force, (c) changing the distance of movement produced by an applied force, (d) changing the strength of an applied force, (e) changing the direction of an applied force.
5. The more movable end of a muscle is the (a) insertion, (b) belly, (c) origin, (d) proximal end, (e) distal end.
6. The muscles of facial expression are innervated by cranial nerve (a) VII, (b) V, (c) IV, (d) VI.
7. Which of these muscles close the mouth and apply biting pressure? (a) pterygoid muscles, (b) masseter, (c) temporalis, (d) buccinators, (e) b and c, (f) a and d.
8. The muscle that rotates the eye medially is the _____ muscle. (a) superior oblique, (b) inferior rectus, (c) medial rectus, (d) lateral rectus.
9. Important flexors of the vertebral column that act in opposition to the erector spinae are the _____ muscles. (a) rectus abdominis, (b) longus capitis, (c) longus colli, (d) scalene.

10. The major flexor of the elbow is the _____ muscle. (a) triceps brachii, (b) biceps brachii, (c) deltoid, (d) subscapularis.
11. The muscles that rotate the radius without producing either flexion or extension of the elbow are the _____ muscles. (a) brachialis and brachioradialis, (b) pronator teres and supinator, (c) biceps brachii and triceps brachii, (d) a, b, and c.
12. Which muscle is active continually when you are awake and supports the weight of the head?
13. Knee extensors known as the quadriceps consist of the (a) three vastus muscles and the rectus femoris muscle, (b) biceps femoris, gracilis, and sartorius muscles, (c) popliteus, iliopsoas, and gracilis muscles, (d) gastrocnemius, tibialis, and peroneus muscles.
14. The sternocleidomastoid connects which structures? What is moved when the sternocleidomastoid contracts?
15. The diaphragm separates which two body compartments?
16. Which four muscle groups make up the axial musculature?
17. What three functions are accomplished by the muscles of the pelvic floor?
18. Where in the muscular system will you find the synovial tendon sheaths?
19. What three functional groups make up the muscles of the lower limbs?

LEVEL 2 Reviewing Concepts

20. What is the function of the *gluteus maximus* muscle?
21. What structures pass through the inguinal canals and produce a weakness in the abdominal muscle wall at that point? If something protrudes through this weakness, what is the condition called?
22. If a muscle becomes traumatised, the sheaths in which the muscle tissue is contained can suffer compartment syndrome. What happens in this condition?
23. What is the alternative name for the calcaneal tendon?
24. Which muscles connect to the hamstring tendons?
25. Why does a convergent muscle exhibit more versatility when contracting than does a parallel muscle?
26. Why can a pennate muscle generate more tension than can a parallel muscle of the same size?
27. Why is it difficult to lift a heavy object when the elbow is at full extension?
28. Which types of movements are affected when the hamstrings are injured?

LEVEL 3 Critical Thinking and Clinical Applications

29. Mary sees Jill coming toward her and immediately contracts her frontalis and procerus muscles. She also contracts her right levator labii muscles. Is Mary glad to see Jill? How can you tell?
30. Mary's newborn is having trouble suckling. The doctor suggests that it may be a problem with a particular muscle. What muscle is the doctor probably referring to? (a) orbicularis oris, (b) buccinator, (c) masseter, (d) risorius, (e) zygomaticus.
31. While unloading her car trunk, Amy strains a muscle and as a result has difficulty moving her arm. The doctor in the emergency room tells her that she strained her pectoralis major. Amy tells you that she thought the pectoralis major was a chest muscle and doesn't understand what that has to do with her arm. What would you tell her?

11

CLINICAL CASE Wrap-Up

The Weekend Warrior

David sustained a complete avulsion, or tearing away, of his hamstrings tendons from their origin on the ischial tuberosity. The long head of the biceps femoris (together with the short head) forms the lateral hamstrings. The semitendinosus and semimembranosus together form the medial hamstrings. When the tendons were forcibly detached from the pelvis, the muscle bellies contracted to about 70 percent of their resting length. With no proximal attachment, they remained contracted.

Under anesthesia, David is placed in a prone position on the operating table. On his right side, the surgeon makes a "7"-shaped incision, traversing the gluteal fold and extending down the posterior thigh. It is very difficult to get tendons to grow back to bone after they have been torn away from their natural attachment. The ischial tuberosity is prepared by scraping away the relatively avascular cortical bone, down to the

bleeding, cancellous bone beneath. This roughening allows for better tendon healing to bone. The tendon mass of the biceps femoris, semitendinosus, and semimembranosus is pulled proximally to the ischial tuberosity, helped by flexing the knee. The tendons are attached with "bone anchoring" sutures until healing can take place. Postoperatively, David is placed in a knee flexion sling (waist band, ankle band, connecting strap that keeps the knee flexed behind him) for 10 weeks.

1. What other muscle, in addition to the hamstrings, assists in flexing the knee?

2. Why does flexing the knee help to bring the avulsed hamstring tendons back up to the ischial tuberosity?

See the blue Answers tab at the back of the book.

12 Neural Tissue

Learning Outcomes

These Learning Outcomes correspond by number to this chapter's sections and indicate what you should be able to do after completing the chapter.

12-1 | Describe the anatomical and functional divisions of the nervous system.

12-2 | Sketch and label the structure of a typical neuron, describe the functions of each component, and classify neurons on the basis of their structure and function.

12-3 | Describe the locations and functions of the various types of neuroglia.

12-4 | Explain how the resting membrane potential is established and maintained.

12-5 | Describe the events involved in the generation and propagation of an action potential.

12-6 | Discuss the factors that affect the speed with which action potentials are propagated.

12-7 | Describe the structure of a synapse, and explain the mechanism involved in synaptic activity.

12-8 | Describe the major types of neurotransmitters and neuromodulators, and discuss their effects on postsynaptic membranes.

12-9 | Discuss the interactions that enable information processing to occur in neural tissue.

SPOTLIGHTS

- Resting Membrane Potential p. 426
- Generation of an Action Potential pp. 434–435
- Propagation of an Action Potential pp. 436–437

CLINICAL CASE

- Did President Franklin D. Roosevelt Really Have Polio? p. 412

Clinical Notes

- Rabies p. 416
- Tumors p. 418
- Demyelination p. 421

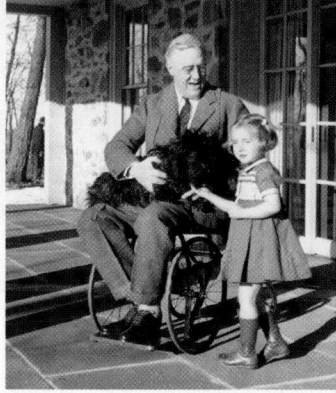
Did Franklin D. Roosevelt Really Have Polio?

On August 10, 1921, at age 39, Franklin D. Roosevelt (FDR) went to bed early feeling tired and complaining of back pain, fever, and chills. The next day, he experienced progressive weakness: first in one leg, then the other, in an ascending pattern. By August 12, both legs were paralyzed and numb, though he still felt pain when lightly touched. Over the following month, the symptoms progressed to involve weakness and abnormal sensations of his upper extremities and his face, and he developed an inability to voluntarily urinate or defecate. His diagnosis of poliomyelitis, or polio, included him in one of the most feared epidemics of the early 20th century.

In the following months, Mr. Roosevelt gradually recovered from the severe weakness and pain. He also regained control of his bodily functions. His legs and buttocks, however, were permanently weakened.

Franklin Roosevelt went on to become the 32nd and longest serving president of the United States. He assisted in the founding of the March of Dimes, an organization that contributed to the elimination of polio from most of the world. **But looking back on his clinical course, did President Franklin D. Roosevelt *really* suffer from poliomyelitis, a viral disease of the motor neurons of the spinal cord? To find out, turn to the Clinical Case Wrap-Up on p. 454.**

12

▌ An Introduction to Neural Tissue

The nervous system includes all the neural tissue in the body. ↺ p. 168 The basic functional units of the nervous system are individual cells called **neurons**. Neurons perform all the communication, information processing, and control functions of the nervous system. Supporting cells, called **neuroglia** (nū-ROG-lē-uh; *glia*, glue) or *glial cells*, have functions essential to the survival and functionality of neurons and to preserving the physical and biochemical structure of neural tissue. They separate and protect the neurons, provide a supportive framework for neural tissue, act as phagocytes, and help regulate the composition of the interstitial fluid. Neuroglia far outnumber neurons.

Neural tissue, with supporting blood vessels and connective tissues, forms the organs of the nervous system. These include the brain; the spinal cord; the receptors in complex sense organs, such as the eye and ear; and the *nerves* that link the nervous system with other systems.

12-1 ▌ The nervous system has anatomical and functional divisions

Learning Outcome Describe the anatomical and functional divisions of the nervous system.

We can look at the nervous system from anatomical and functional perspectives (**Figure 12–1**).

The Anatomical Divisions of the Nervous System

Viewed anatomically, the nervous system has two divisions: the central nervous system and the peripheral nervous system. The **central nervous system (CNS)** consists of the brain and spinal cord. These complex organs include not only neural tissue, but also blood vessels and the various connective tissues that physically protect and support.

What are the general functions of the CNS? It is responsible for integrating, processing, and coordinating sensory data and motor commands. Sensory data convey information about conditions inside or outside the body. Motor commands control or adjust peripheral organs, such as skeletal muscles. When you stumble, for example, the CNS integrates information about your balance and the position of your limbs and then coordinates your recovery by sending motor commands to appropriate skeletal muscles—all in a split second and without your conscious effort. The CNS—specifically, the brain—is also the seat of higher functions, such as intelligence, memory, learning, and emotion.

The **peripheral nervous system (PNS)** includes all the neural tissue outside the CNS. The PNS delivers sensory information to the CNS and carries motor commands to peripheral tissues and systems. Bundles of axons, or *nerve fibers*, carry sensory information and motor commands in the PNS. Such bundles, with associated blood vessels and connective tissues, are called *peripheral nerves*, or simply **nerves**. Nerves connected to the brain are called **cranial nerves**, and those attached to the spinal cord are called **spinal nerves**.

The Functional Divisions of the Nervous System

We can divide the PNS into afferent and efferent divisions, each with different functions. The **afferent division** (*afferens,* to bring to) of the PNS brings sensory information *to* the CNS from receptors in peripheral tissues and organs. **Receptors** are sensory structures that either detect changes in the environment (internal or external) or respond to specific stimuli (Figure 12–1). Our receptors range from the slender cytoplasmic extensions of single cells to complex receptor organs, such as the eye and ear. Receptors may be neurons or specialized cells of other tissues. ⟳ p. 187

The **efferent division** (*effero,* to bring out) of the PNS carries motor commands *from* the CNS to muscles, glands, and adipose tissue. These target organs, which respond by *doing* something, are called **effectors**. The efferent division has both somatic and autonomic components:

- The **somatic nervous system (SNS)** controls skeletal muscle contractions. *Voluntary* contractions are under conscious control. For example, you exert conscious control over your arm as you raise a full glass of water to your lips. *Involuntary* contractions may be simple, automatic responses or complex movements, but they are controlled at the subconscious level, outside your awareness. For instance, if you accidentally place your hand on a hot stove, you will withdraw it immediately, usually before you even notice any pain. This type of automatic response is called a **reflex**.

- The **autonomic nervous system (ANS)**, or *visceral motor system,* automatically regulates smooth muscle, cardiac muscle, glandular secretions, and adipose tissue at the subconscious level. The ANS includes a *sympathetic division* and a *parasympathetic division,* which commonly have antagonistic (opposite) effects. For example, activity of the sympathetic division accelerates the heart rate, whereas parasympathetic activity slows the heart rate.

Figure 12–1   A Functional Overview of the Nervous System.

Now that you have an overview of the nervous system, let's look at the structure of neural tissue and the functional principles that govern neural activities. We begin with neurons, the basic functional units of the nervous system.

12-2 ▌ Neurons are nerve cells specialized for intercellular communication

Learning Outcome Sketch and label the structure of a typical neuron, describe the functions of each component, and classify neurons on the basis of their structure and function.

Let's examine the structure of a representative neuron and see how it is specialized for intercellular communication before

we consider the structural and functional classifications of neurons.

The Structure of Neurons

Neurons have a variety of shapes. **Figure 12–2** shows a *multipolar neuron*, the most common type of neuron in the central nervous system. Each multipolar neuron has a large *cell body*, several short, branched *dendrites*, and a single, long *axon*, ending in terminal branches called *telodendria*.

The Cell Body

The **cell body**, or *soma*, contains a large, round nucleus with a prominent nucleolus (**Figure 12–2**). The cytoplasm surrounding the nucleus is the **perikaryon** (per-i-KAR-ē-on; *peri*, around + *karyon*, nucleus). The cytoskeleton of the perikaryon contains **neurofilaments** and **neurotubules**, which are similar to the intermediate filaments and microtubules of other types of cells. Bundles of neurofilaments, called **neurofibrils**, extend into the dendrites and axon, providing internal support for them.

The perikaryon contains organelles that provide energy and synthesize organic materials, especially the chemical neurotransmitters that are important in cell-to-cell communication. ⤺ p. 326 The numerous mitochondria, free and fixed ribosomes, and membranes of rough endoplasmic reticulum (RER) give the perikaryon a coarse, grainy appearance. Mitochondria generate ATP to meet the high energy

Figure 12–2 **The Anatomy of a Multipolar Neuron.**

a This color-coded figure shows the four general regions of a neuron.

b An understanding of neuron function requires knowing its structural components.

Presynaptic cell

Postsynaptic cell

Direction of action potential

See Figure 12–3

demands of an active neuron. The ribosomes and RER synthesize proteins.

Some areas of the perikaryon contain clusters of RER and free ribosomes. These regions, which stain darkly, are called *Nissl bodies,* after the German microscopist Franz Nissl, who first described them. Nissl bodies give a gray color to areas containing neuron cell bodies—the *gray matter* seen in gross dissection of the brain and spinal cord.

Most neurons lack centrioles, important organelles that help to organize the cytoskeleton and the microtubules that move chromosomes during mitosis. ⊃ p. 131 As a result, typical CNS neurons cannot divide. For this reason, they cannot be replaced if lost to injury or disease. Neural stem cells persist in the adult nervous system, but they are typically inactive except in the nose, where the regeneration of olfactory (smell) receptors maintains our sense of smell, and in the *hippocampus,* a part of the brain involved in storing memories. Researchers are investigating the control mechanisms that trigger neural stem cell activity, with the goal of preventing or reversing neuron loss due to trauma, disease, or aging.

Dendrites and Axons

A variable number of slender, sensitive processes (extensions) known as **dendrites** extend out from the cell body (**Figure 12–2**). Dendrites play key roles in intercellular communication. Typical dendrites are highly branched, and each branch has fine 0.5- to 1-µm-long studded processes called *dendritic spines.* In the CNS, a neuron receives information from other neurons primarily at the dendritic spines, which represent 80–90 percent of the neuron's total surface area.

An **axon** is a long cytoplasmic process capable of propagating an electrical impulse known as an *action potential.* ⊃ p. 326 The **axoplasm** (AK-sō-plazm), or cytoplasm of the axon, contains neurofibrils, neurotubules, small vesicles, lysosomes, mitochondria, and various enzymes. The **axolemma** (*lemma,* husk), a specialized portion of the plasma membrane, surrounds the axoplasm. In the CNS, the axolemma may be exposed to the interstitial fluid or, as we'll see, it may be covered by the cellular processes of neuroglia. The base, or **initial segment**, of the axon in a multipolar neuron joins the cell body at a thickened region known as the **axon hillock** (**Figure 12–2**).

An axon may branch along its length, producing side branches known as **collaterals**. Collaterals enable a single neuron to communicate with several other cells. The main axon trunk and any collaterals end in a series of fine extensions called **telodendria** (tel-ō-DEN-drē-uh; *telo-,* end + *dendron,* tree) or *terminal branches.* The telodendria, in turn, end at **axon terminals** (also called synaptic terminals, synaptic knobs, and synaptic boutons), which play a role in communication with another cell (**Figure 12–2**).

The Synapse

Each axon terminal is part of a **synapse**, a specialized site where the neuron communicates with another cell (**Figure 12–3**). Every synapse involves two cells: (1) the *presynaptic cell,* which sends a message and includes the axon terminal; and (2) the *postsynaptic cell,* which receives the message. Typically, a narrow space called the *synaptic cleft* separates the two cells. ⊃ p. 326

In communication between two cells, the axon terminal of the presynaptic cell most commonly releases chemicals called **neurotransmitters** into the synaptic cleft. Inside the axon terminal, neurotransmitters are contained in *synaptic vesicles.* As we saw in Chapter 10, neurotransmitter release is triggered by electrical events, such as the arrival of an action potential. The neurotransmitters then flood the synaptic cleft and affect the activity of the postsynaptic cell.

The presynaptic cell is usually a neuron. (Specialized receptor cells may form synaptic connections with the dendrites of neurons, a topic we describe in Chapter 15.) The postsynaptic cell can be either a neuron or another type of cell. One neuron may communicate with another at a synapse on a dendrite, on the cell body, or along the length of the axon of the receiving cell. A synapse between a neuron and a muscle cell is called a **neuromuscular junction**. ⊃ p. 325 At a **neuroglandular junction**, a neuron controls or regulates the activity of a secretory (gland) cell. Neurons also *innervate* (are distributed to) a variety of other cell types, such as adipocytes (fat cells). We consider the nature of that innervation in later chapters.

The structure of the axon terminal varies with the type of postsynaptic cell. A relatively simple, round axon terminal occurs where the postsynaptic cell is another neuron. At a synapse, the narrow synaptic cleft separates the **presynaptic membrane**,

Figure 12–3 **The Structure of a Typical Synapse.**

Telodendrion

Axon terminal

Mitochondrion

Synaptic vesicles

Presynaptic membrane

Postsynaptic membrane   Synaptic cleft

12

where neurotransmitters are released, from the **postsynaptic membrane**, which bears receptors for neurotransmitters (**Figure 12–3**). The axon terminal at a neuromuscular junction is much more structurally complex.

Each axon terminal contains mitochondria and thousands of vesicles filled with neurotransmitter molecules. The axon terminal reabsorbs breakdown products of neurotransmitters formed at the synapse and reassembles them. It also receives a continuous supply of neurotransmitters synthesized in the cell body, along with enzymes and lysosomes.

These materials travel the length of the axon on neurotubules, pulled along by protein "molecular motors," called *kinesin* and *dynein,* that run on ATP. The movement of materials between the cell body and axon terminals is called **axoplasmic transport**. Some materials travel slowly, at rates of a few millimeters per day. This transport mechanism is known as the "slow stream." Vesicles containing neurotransmitters move much more rapidly, traveling in the "fast stream" at 5–10 mm per hour.

Axoplasmic transport occurs in both directions. The flow of materials from the cell body to the axon terminal is *anterograde* (AN-ter-ō-grād; *antero-,* forward) *flow,* carried by kinesin. At the same time, other substances are transported from the axon terminal toward the cell body in *retrograde* (RET-rō-grād) *flow* (*retro,* backward), carried by dynein. If debris or unusual chemicals appear in the axon terminal, retrograde flow soon delivers them to the cell body. There they may then alter the activity of the cell by turning certain genes on or off.

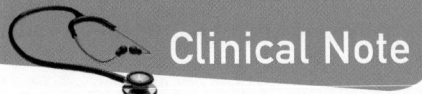

Clinical Note

Rabies *Rabies* is a dramatic example of a clinical condition involving retrograde flow. A bite from a rabid animal injects the rabies virus into peripheral tissues. There the virus particles quickly enter axon terminals and peripheral axons. Retrograde flow then carries the virus into the CNS, with fatal results. Many toxins (including heavy metals), some pathogenic bacteria, and other viruses also bypass CNS defenses through axoplasmic transport.

The Classification of Neurons

We can group neurons by structure or by function.

Structural Classification of Neurons

Neurons are classified as anaxonic, bipolar, unipolar, or multipolar on the basis of the relationship of the dendrites to the cell body and the axon (**Figure 12–4**):

- **Anaxonic** (an-aks-ON-ik) **neurons** are small and have numerous dendrites, but no axon. Anaxonic neurons are located in the brain and in special sense organs. Their functions are poorly understood.

- **Bipolar neurons** have two distinct processes—one dendrite that branches extensively into dendritic branches at its distal tip, and one axon—with the cell body between the two. Bipolar neurons are rare. They occur in special sense organs, where they relay information about sight, smell, or hearing from receptor cells to other neurons. Bipolar neurons are small. The largest measure less than 30 μm from end to end.

- In a **unipolar neuron**, or *pseudounipolar neuron*, the dendrites and axon are continuous—basically, fused—and the cell body lies off to one side. In such a neuron, the initial segment lies where the dendrites converge. The rest of the process, which carries action potentials, is usually considered an axon. Most sensory neurons of the peripheral nervous system are unipolar. Their axons may extend a meter or more, ending at synapses in the central nervous system. The longest carry sensations from the tips of the toes to the spinal cord.

- **Multipolar neurons** have two or more dendrites and a single axon. They are the most common neurons in the CNS. All the motor neurons that control skeletal muscles, for example, are multipolar neurons. The axons of multipolar neurons can be as long as those of unipolar neurons, and the longest carry motor commands from the spinal cord to small muscles that move the toes.

Functional Classification of Neurons

Alternatively, we can categorize neurons by function as (1) sensory neurons, (2) motor neurons, or (3) interneurons.

Sensory Neurons. **Sensory neurons**, or *afferent neurons,* form the afferent division of the PNS. They deliver information from sensory receptors to the CNS. The cell bodies of sensory neurons are located in peripheral *sensory ganglia.* (A **ganglion** is a collection of neuron cell bodies in the PNS.) Sensory neurons are unipolar neurons whose processes, known as **afferent fibers**, extend between a sensory receptor and the central nervous system (spinal cord or brain). The human body's 10 million or so sensory neurons collect information about the external or internal environment. **Somatic sensory neurons** monitor the outside world and our position within it. **Visceral sensory neurons** monitor internal conditions and the status of other organ systems.

Sensory receptors are either the processes of specialized sensory neurons or cells monitored by sensory neurons. We can broadly categorize receptors in three groups:

Figure 12–4 **Structural Classifications of Neurons.**

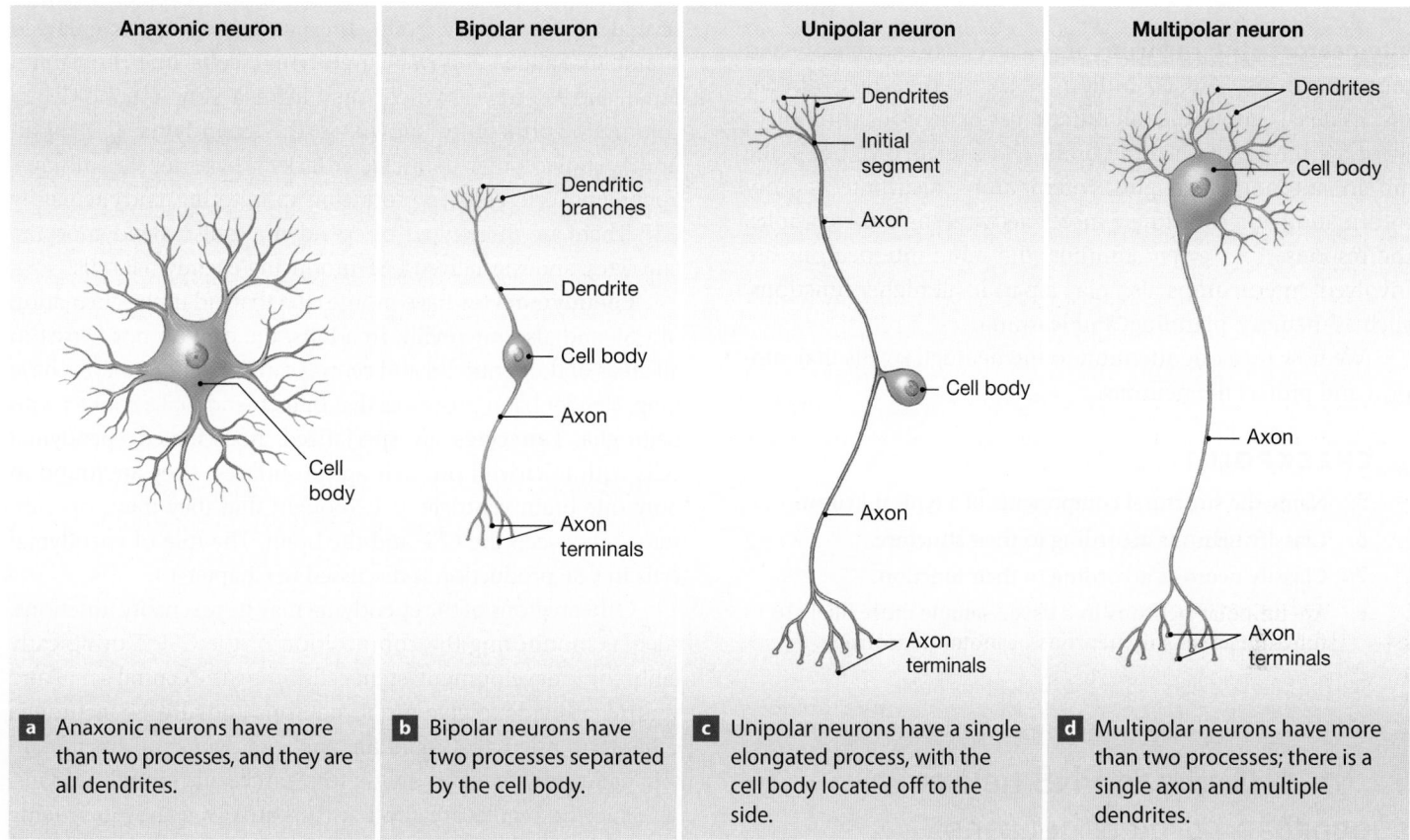

Anaxonic neuron	Bipolar neuron	Unipolar neuron	Multipolar neuron
a Anaxonic neurons have more than two processes, and they are all dendrites.	**b** Bipolar neurons have two processes separated by the cell body.	**c** Unipolar neurons have a single elongated process, with the cell body located off to the side.	**d** Multipolar neurons have more than two processes; there is a single axon and multiple dendrites.

- **Interoceptors** (*intero-*, inside) monitor the digestive, respiratory, cardiovascular, urinary, and reproductive systems and provide sensations of distension (stretch), deep pressure, and pain.

- **Exteroceptors** (*extero-*, outside) provide information about the external environment in the form of touch, temperature, or pressure sensations and the more complex senses of taste, smell, sight, equilibrium (balance), and hearing.

- **Proprioceptors** (prō-prē-ō-SEP-torz) monitor the position and movement of skeletal muscles and joints.

Motor Neurons. **Motor neurons,** or *efferent neurons*, form the efferent division of the PNS. These neurons carry instructions from the CNS to peripheral effectors in a peripheral tissue, organ, or organ system. The human body has about half a million motor neurons. Axons traveling away from the CNS are called **efferent fibers**. As noted earlier, the two major efferent systems are the somatic nervous system (SNS) and the autonomic (visceral) nervous system (ANS).

The somatic nervous system includes all the **somatic motor neurons** that innervate skeletal muscles. You have conscious control over the activity of the SNS. The cell body of a somatic motor neuron lies in the CNS, and its axon extends into the periphery within a peripheral nerve to innervate skeletal muscle fibers at neuromuscular junctions.

You do not have conscious control over the activities of the ANS. **Visceral motor neurons** innervate all peripheral effectors other than skeletal muscles—that is, smooth muscle, cardiac muscle, glands, and adipose tissue throughout the body. The axons of visceral motor neurons in the CNS innervate a second set of visceral motor neurons in peripheral *autonomic ganglia*. The neurons whose cell bodies are located in those ganglia innervate and control peripheral effectors.

TIPS&TRICKS

To distinguish between *afferent* and *efferent*, think of the **SAME** principle: **S** is for **s**ensory, **A** is for **a**fferent, **M** is for **m**otor, and **E** is for **e**fferent. This way, you associate the **S** and **A** together and the **M** and **E** together.

To get from the CNS to a visceral effector such as a smooth muscle cell, a signal must travel along one axon, be relayed across a synapse, and then travel along a second axon to its final destination. The axons extending from the CNS to an autonomic ganglion are called *preganglionic fibers*, and axons connecting the ganglion cells with the peripheral effectors are known as *postganglionic fibers.*

Interneurons. **Interneurons** are located between sensory and motor neurons. The 20 billion or so interneurons (*association neurons*) outnumber all other types of neurons combined. Most are located within the brain and spinal cord, but some are in autonomic ganglia. Interneurons distribute sensory information and coordinate motor activity. The more complex the response to a given stimulus, the more interneurons are involved. Interneurons also play a part in all higher functions, such as memory, planning, and learning.

We now turn our attention to the neuroglia, cells that support and protect the neurons.

CHECKPOINT

5. Name the structural components of a typical neuron.
6. Classify neurons according to their structure.
7. Classify neurons according to their function.
8. Are unipolar neurons in a tissue sample more likely to function as sensory neurons or motor neurons?

See the blue Answers tab at the back of the book.

12-3 ▮ CNS and PNS neuroglia support and protect neurons

Learning Outcome Describe the locations and functions of the various types of neuroglia.

Neuroglia are abundant and diverse, and they make up about half the volume of the nervous system. The organization of neural tissue in the CNS differs from that in the PNS, primarily because the CNS has a greater variety of neuroglial cell types. Histological descriptions have been available for the past century, but the technical problems involved in isolating and manipulating individual glial cells have limited our understanding of their functions. **Figure 12–5** summarizes what we know about the major neuroglial populations in the CNS and PNS.

Neuroglia of the Central Nervous System

The central nervous system has four types of neuroglia: (1) *ependymal cells,* (2) *astrocytes,* (3) *oligodendrocytes,* and (4) *microglia* (**Figure 12–5**).

Ependymal Cells

A fluid-filled central passageway extends along the longitudinal axis of the spinal cord and brain. **Cerebrospinal fluid (CSF)** fills this passageway and also surrounds the brain and spinal cord. This fluid, which circulates continuously, provides a protective cushion and transports dissolved gases, nutrients, wastes, and other materials. The diameter of the internal passageway

varies from one region to another. In the spinal cord, this narrow passageway is called the *central canal* (**Figure 12–6a,b**). In several regions of the brain, the passageway forms enlarged chambers called *ventricles.* **Ependymal cells** line the central canal and ventricles, where they form a simple cuboidal to columnar epithelium known as the **ependyma** (e-PEN-di-muh). Unlike other epithelia, it lacks a basement membrane. Ependymal cells assist in producing, monitoring, and circulating CSF. There are three types of ependymal cells: ependymocytes, tanycytes, and specialized CSF-producing ependymal cells.

Ependymocytes have motile cilia that aid in the circulation of CSF and also microvilli. In adults, the cilia do not persist in all areas of the ventricles and central canal. Ependymocytes have long, slender basal processes that branch and make contact with neuroglia. **Tanycytes** are specialized, non-ciliated ependymal cells with microvilli on their apical surfaces. They are found in only one brain ventricle. It is thought that they transport substances between the CSF and the brain. The role of ependymal cells in CSF production is discussed in Chapter 14.

Other regions of the ependyma may have sensory functions, such as monitoring the composition of the CSF. During early embryonic development, stem cells lining the central canal and ventricles divide to give rise to neurons and all CNS neuroglia other than microglia. In adults, the ependyma appears to contain stem cells that can divide to produce additional neurons. The specific regulatory mechanisms involved are now being investigated.

Clinical Note

Tumors Tumors of the brain, spinal cord, and associated membranes result in approximately 90,000 deaths in the United States each year. *Primary CNS tumors* are tumors that originate in the central nervous system. *Secondary CNS tumors* arise from the metastasis (spread) of cancer cells that originate elsewhere. Roughly 75 percent of CNS tumors are primary tumors. In adults, primary CNS tumors result from the divisions of abnormal neuroglia rather than from the divisions of abnormal neurons, because typical neurons in adults cannot divide. However, through the divisions of stem cells, neurons increase in number until children reach age 4. For this reason, primary CNS tumors involving abnormal neurons can occur in young children. Symptoms of CNS tumors vary with the location affected. Treatment may involve surgery, radiation, chemotherapy, or a combination of these procedures.

Astrocytes

Astrocytes (AS-trō-sīts; *astro-,* star + *-cyte,* cell) are the largest and most numerous neuroglia in the CNS (**Figure 12–6b**).

Figure 12–5 **An Introduction to Neuroglia.**

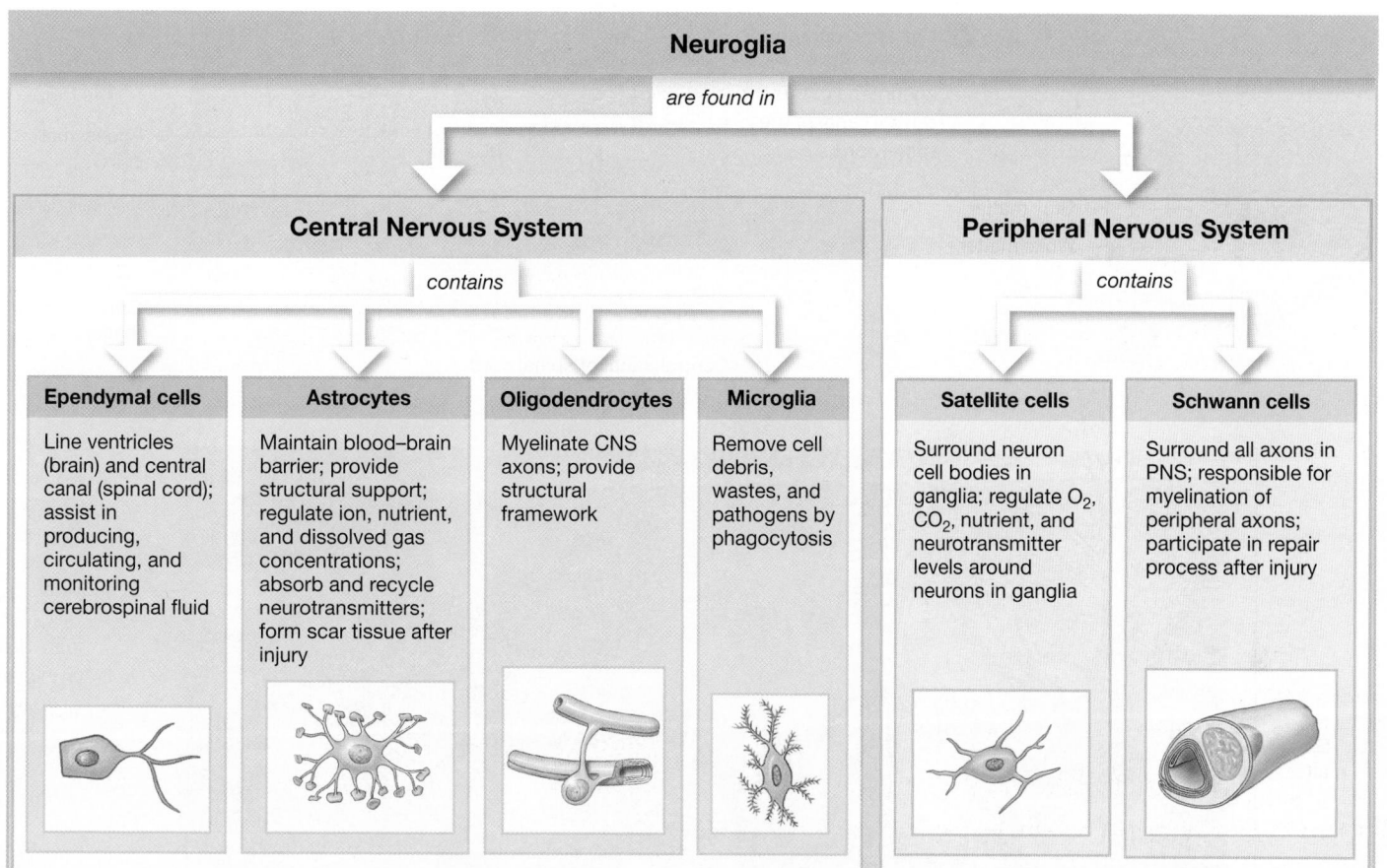

Neuroglia

are found in

Central Nervous System

contains

Ependymal cells	Astrocytes	Oligodendrocytes	Microglia
Line ventricles (brain) and central canal (spinal cord); assist in producing, circulating, and monitoring cerebrospinal fluid	Maintain blood–brain barrier; provide structural support; regulate ion, nutrient, and dissolved gas concentrations; absorb and recycle neurotransmitters; form scar tissue after injury	Myelinate CNS axons; provide structural framework	Remove cell debris, wastes, and pathogens by phagocytosis

Peripheral Nervous System

contains

Satellite cells	Schwann cells
Surround neuron cell bodies in ganglia; regulate O_2, CO_2, nutrient, and neurotransmitter levels around neurons in ganglia	Surround all axons in PNS; responsible for myelination of peripheral axons; participate in repair process after injury

12

These cells have a variety of functions, many of them poorly understood:

- *Maintaining the Blood–Brain Barrier.* Compounds dissolved in circulating blood do not have free access to the interstitial fluid of the CNS. Neural tissue must be physically and biochemically isolated from the general circulation, because hormones, amino acids, or other chemicals in the blood can alter neuron function. The endothelial cells lining CNS capillaries control the chemical exchange between the blood and interstitial fluid. These cells create a **blood–brain barrier (BBB)** that isolates the CNS from the general circulation.

  The slender cytoplasmic extensions of astrocytes end in expanded "feet," processes that wrap around capillaries. These processes form a complete blanket around the capillaries, interrupted only where other neuroglia come in contact with the capillary walls. Astrocytes secrete chemicals that are somehow responsible for maintaining the special permeability characteristics of the capillary endothelial cells. (We discuss the blood–brain barrier further in Chapter 14.)

- *Creating a Three-Dimensional Framework for the CNS.* Astrocytes are packed with microfilaments that extend across the breadth of the cell and its processes. This extensive cytoskeleton helps astrocytes to provide a structural framework for the neurons of the brain and spinal cord.

- *Repairing Damaged Neural Tissue.* In the CNS, damaged neural tissue seldom regains normal function. However, astrocytes that move into an injury site can make structural repairs that stabilize the tissue and prevent further injury. We discuss neural damage and repair in a later section.

- *Guiding Neuron Development.* Astrocytes in the embryonic brain appear to be involved in directing both the growth and interconnection of developing neurons.

- *Controlling the Interstitial Environment.* Astrocytes appear to adjust the composition of interstitial fluid by several means: (1) regulating the concentration of sodium ions, potassium ions, and carbon dioxide; (2) providing a "rapid-transit system" for transporting nutrients, ions, and dissolved gases between capillaries and neurons; (3) controlling the volume of blood flow through the capillaries; (4) absorbing and recycling some neurotransmitters; and (5) releasing chemicals that enhance or suppress communication across axon terminals.

Figure 12–6 **Neuroglia in the CNS.**

a Light micrograph showing the ependyma of the central canal of the spinal cord

Gray matter

White matter

Ependymal cells

Central canal

Gray matter

Central canal of spinal cord   LM × 450

12

Gray matter

White matter

CENTRAL CANAL

Ependymal cells

Neurons

Microglial cell

Myelinated axons

Internode

Myelin (cut)

Oligodendrocyte

Axon

Axolemma

Node

Astrocyte

Unmyelinated axon

Basement membrane

Capillary

b A diagrammatic view of neural tissue in the CNS, showing relationships between neuroglia and neurons

Oligodendrocytes

Like astrocytes, **oligodendrocytes** (ol-i-gō-DEN-drō-sīts; *oligo-*, few) have slender cytoplasmic extensions, but the cell bodies of oligodendrocytes are smaller, with fewer processes, than astrocytes (**Figure 12–6b**). The processes of oligodendrocytes generally are in contact with the exposed surfaces of neurons. The functions of processes ending at the neuron cell body have yet to be determined. Much more is known about the processes that end on the surfaces of axons. Many axons in the CNS are completely sheathed in these processes, which insulate them from contact with the extracellular fluid.

During its maturation, the plasma membrane near the tip of each process of an oligodendrocyte expands to form an enormous pad, and the cytoplasm there becomes very thin. This flattened "pancake" somehow gets wrapped around the axolemma, forming concentric layers of plasma membrane (**Figure 12–6b**). This membranous wrapping, called **myelin** (M Ī-e-lin), serves as electrical insulation and increases the speed at which an action potential travels along the axon. (We describe this mechanism in a later section.)

Many oligodendrocytes cooperate in forming a **myelin sheath** along the length of an axon. Such an axon is said to be **myelinated**. Each oligodendrocyte myelinates segments of several axons. The fairly large areas of the axon that are wrapped in myelin are called **internodes** (*inter*, between). Internodes are typically 1–2 mm in length. The small gaps of a few micrometers that separate adjacent internodes are called **nodes**, or *nodes of Ranvier* (rahn-vē-Ā). An axon's collateral branches originate at nodes.

In dissection, myelinated axons appear glossy white, primarily because of the lipids in the myelin. As a result, regions dominated by myelinated axons are known as the **white matter** of the CNS. Not all axons in the CNS are myelinated, however. **Unmyelinated** axons are not completely covered by oligodendrocytes. Such axons are common where short axons and collaterals form synapses with densely packed neuron cell bodies. Areas containing neuron cell bodies, dendrites, and unmyelinated axons have a dusky gray color, and make up the **gray matter** of the CNS.

In summary, oligodendrocytes play an important role in structural organization by tying clusters of axons together. They also improve the functioning of neurons by wrapping axons within a myelin sheath.

TIPS & TRICKS

The overall color of CNS tissue is related to its structure and function. **Gr**ay matter has a **gr**eat concentration of neuron cell bodies and is a region of inte**gr**ation. **Wh**ite matter has a **wh**ole lot of myelinated axons and **wh**isks nerve impulses.

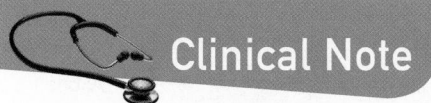

Clinical Note

Demyelination **Demyelination** is the progressive destruction of myelin sheaths, both in the CNS and in the PNS. The result is a loss of sensation and motor control that leaves affected regions numb and paralyzed. Many unrelated conditions that result in the destruction of myelin can cause symptoms of demyelination.

Chronic exposure to heavy-metal ions, such as arsenic, lead, or mercury, can cause heavy-metal poisoning, leading to damage of neuroglia and to demyelination. As demyelination occurs, the affected axons deteriorate, and the condition becomes irreversible.

Diphtheria (dif-THER-ē-uh; *diphthera*, leather + *-ia*, disease) is a disease that results from a bacterial infection. In the nervous system, diphtheria toxin damages Schwann cells and destroys myelin sheaths in the PNS. The resulting demyelination leads to sensory and motor problems that can ultimately produce a fatal paralysis. Due to an effective vaccine, cases are relatively rare in countries with adequate health care.

Multiple sclerosis (skler-Ō-sis; *sklerosis*, hardness), or **MS,** is a disease characterized by recurrent incidents of demyelination that affects axons in the optic nerve, brain, and spinal cord. Common signs and symptoms include partial loss of vision and problems with speech, balance, and general motor coordination, including bowel and urinary bladder control. The time between incidents and degree of recovery vary from case to case. In about one-third of all cases, the disorder is progressive, and each incident leaves a greater degree of functional impairment. The first attack typically occurs in individuals 30–40 years old. The incidence among women is 1.5 times that among men. In some patients, corticosteroid or interferon injections have slowed the progression of the disease.

Guillain-Barré (ghee-yan bah-ray) **syndrome** is an autoimmune disorder characterized by demyelination of peripheral nerves. The signs and symptoms include weakness or tingling of the legs that spreads to the arms. They typically increase in severity, leading to paralysis. When breathing is affected, the patient is placed on a ventilator. The disorder affects both sexes equally. Each year it afflicts 1 of every 100,000 Americans. A virus appears to trigger the syndrome, because the onset is usually within a few days or weeks of a respiratory or gastrointestinal infection, or occasionally after surgery or immunization. Most patients fully recover, but some continue to have residual weakness.

Figure 12–7 **Schwann Cells, Peripheral Axons, and Formation of the Myelin Sheath.**

a A myelinated axon, showing the organization of Schwann cells along the length of the axon.

b The enclosing of a group of unmyelinated axons by a single Schwann cell. A series of Schwann cells is required to cover the axons along their entire length.

1 A Schwann cell first surrounds a portion of the axon within a groove of its cytoplasm.

2 The Schwann cell then begins to rotate around the axon.

3 As the Schwann cell rotates, myelin is wound around the axon in multiple layers, forming a tightly packed membrane.

c Stages in the formation of a myelin sheath by a single Schwann cell along a portion of a single axon.

Microglia

The least numerous and smallest neuroglia in the CNS are phagocytic cells called **microglia** (mī-KRŌG-lē-uh). ⟳ pp. 125–126 Their slender processes have many fine branches (**Figure 12–6b**). These cells can migrate through neural tissue. Microglia appear early in embryonic development, originating from mesodermal stem cells related to stem cells that produce monocytes and macrophages. ⟳ pp. 153, 159 Microglia migrate into the CNS as the nervous system forms. There they remain, acting as a wandering janitorial service and police force by engulfing cellular debris, waste products, and pathogens.

Neuroglia of the Peripheral Nervous System

Recall that the cell bodies of neurons in the PNS are clustered in masses called **ganglia** (singular, *ganglion*). The processes of neuroglia completely insulate neuronal cell bodies and most axons in the PNS from their surroundings. The two types of neuroglia in the PNS are satellite cells and Schwann cells.

Satellite cells surround neuron cell bodies in ganglia. They regulate the environment around the neurons, much as astrocytes do in the CNS.

Schwann cells, or *neurilemma cells*, either form a thick, myelin sheath or indented folds of plasma membrane around peripheral axons (**Figures 12–7, 12–8**). Wherever a Schwann cell covers an axon, the outer surface of the Schwann cell is called the **neurilemma** (nū-ri-LEM-uh). Most axons in the PNS, whether myelinated or unmyelinated, are shielded from contact with interstitial fluids by Schwann cells.

A myelinating Schwann cell can myelinate only one segment of a single axon (**Figure 12–7a**), whereas an oligodendrocyte in the CNS may myelinate portions of several adjacent axons (**Figure 12–6b**). Nonmyelinating Schwann cells can *enclose* segments of several unmyelinated axons. A series of Schwann cells is required to enclose an axon along its entire length (**Figure 12–7b**). Stages in the formation of a myelin sheath are shown in **Figure 12–7c**.

Figure 12–8   Peripheral Nerve Regeneration after Injury.

Site of injury

1 Fragmentation of axon and myelin occurs in distal stump.

Axon   Myelin   Proximal stump   Distal stump

2 Schwann cells form cord, grow into cut, and unite stumps. Macrophages engulf degenerating axon and myelin.

Macrophage

Cord of proliferating Schwann cells

3 Axon sends buds into network of Schwann cells and then starts growing along cord of Schwann cells.

4 Axon continues to grow into distal stump and is enclosed by Schwann cells.

Neural Responses to Injuries

What happens when a neuron is injured? It responds to injury in a very limited way. In the cell body, the Nissl bodies disperse and the nucleus moves away from its centralized location as the cell increases its rate of protein synthesis. If the neuron recovers and regains function, it will regain its normal appearance.

The key to recovery appears to be events in the axon. If, for example, the pressure applied during a crushing injury produces a local decrease in blood flow and oxygen, the affected axonal membrane becomes unexcitable. If the pressure is alleviated after an hour or two, the neuron will recover within a few weeks. More severe or prolonged pressure produces effects similar to those caused by cutting the axon.

In the PNS, Schwann cells play a part in repairing damaged nerves. In the process known as **Wallerian degeneration**, the axon distal to the injury site degenerates, and macrophages migrate into the area to clean up the debris (**Figure 12–8**). The Schwann cells do not degenerate. Instead, they proliferate and form a solid cellular cord that follows the path of the original axon. As the neuron recovers, its axon grows into the site of injury, and the Schwann cells wrap around the axon.

If the axon grows alongside the appropriate cord of Schwann cells, it may eventually reestablish its normal synaptic contacts. However, if it stops growing or wanders off in some new direction, normal function will not return. The growing axon is most likely to arrive at its appropriate destination if the cut edges of the original nerve bundle remain in contact.

Limited regeneration can occur in the CNS, but the situation is more complicated because (1) many more axons are likely to be involved, (2) astrocytes produce scar tissue that can prevent axon growth across the damaged area, and (3) astrocytes release chemicals that block the regrowth of axons.

CHECKPOINT

9. Identify the neuroglia of the central nervous system.

10. Identify the neuroglia of the peripheral nervous system.

11. Which type of neuroglia would increase in number in the brain tissue of a person with a CNS infection?

See the blue Answers tab at the back of the book.

12-4 ▮ The membrane potential is the electrical potential of the cell's interior relative to its surroundings

Learning Outcome Explain how the resting membrane potential is established and maintained.

In Chapter 3, we introduced the concepts of the *membrane potential*, or *transmembrane potential*, and the *resting membrane potential*, two characteristic physiological features of all cells. ↩ p. 126 In this discussion, we focus on the membranes of neurons, but many of the principles discussed apply to other types of cells as well.

The important membrane processes we will be examining are the resting membrane potential, graded potential, action potential, synaptic activity, and information processing.

- *Resting membrane potential* All living cells have a membrane potential that varies from moment to moment depending on the activities of the cell. The *resting membrane potential* is the membrane potential of an unstimulated, resting cell. All neural activities begin with a change in the resting membrane potential of a neuron.

- *Graded potential* A typical stimulus produces a temporary, localized change in the resting membrane potential. The effect, which decreases with distance from the stimulus, is called a *graded potential*.

- *Action potential* If the graded potential is large enough, it triggers an *action potential* in the membrane of the axon. An action potential is an electrical impulse that is propagated (spread) along the surface of an axon and does not diminish as it moves away from its source. This impulse travels along the axon to one or more synapses.

- *Synaptic activity* then produces graded potentials in the plasma membrane of the postsynaptic cell. The presynaptic cell typically releases neurotransmitters, such as ACh. These chemicals bind to receptors on the postsynaptic plasma membrane, changing its permeability and producing graded potentials. The mechanism is comparable to that of the neuromuscular junction, described in Chapter 10. ↩ p. 326

- *Information processing* The response of the postsynaptic cell ultimately depends on what the stimulated receptors do and what other stimuli are influencing the cell at the same time. The integration of stimuli at the level of the individual cell is the simplest form of *information processing* in the nervous system.

When you understand each of these processes, you will know how neurons deal with information and communicate with one another and with peripheral effectors.

The Membrane Potential

Chapter 3 introduced three important concepts regarding the membrane potential:

1. *The extracellular fluid (ECF) and intracellular fluid (cytosol) differ greatly in ionic composition.* The extracellular fluid

contains high concentrations of sodium ions (Na^+) and chloride ions (Cl^-), whereas the cytosol contains high concentrations of potassium ions (K^+) and negatively charged proteins.

2. *Cells have selectively permeable membranes.* If the plasma membrane were freely permeable, diffusion would continue until all the ions were evenly distributed across the membrane and a state of equilibrium existed. But an even distribution does not occur, because cells have selectively permeable membranes. ⊃ p. 113 Ions cannot freely cross the lipid portions of the plasma membrane. They can enter or leave the cell only through membrane channels. Many kinds of membrane channels exist, each with its own properties. At the resting membrane potential, or membrane potential of an undisturbed cell, ions move through *leak channels*—membrane channels that are always open. ⊃ p. 118 Active transport mechanisms, such as the sodium–potassium exchange pump, also move specific ions into or out of the cell. ⊃ p. 122

3. *Membrane permeability varies by ion.* The cell's passive and active transport mechanisms do not ensure an equal distribution of charges across its plasma membrane, because membrane permeability varies by ion. For example, it is easier for K^+ to diffuse out of the cell through a potassium leak channel than it is for Na^+ to enter the cell through a sodium leak channel. Additionally, negatively charged proteins inside the cell are too large to cross the membrane. As a result, the membrane's inner surface has an excess of negative charges with respect to the outer surface.

Both passive and active forces act across the plasma membrane to determine the membrane potential at any moment. Spotlight Figure 12–9 shows the membrane at the normal resting membrane potential.

Passive Forces Acting across the Plasma Membrane

The passive forces acting across the plasma membrane involve both chemical and electrical gradients (Spotlight Figure 12–9). Positive and negative charges attract one another. If nothing separates them, oppositely charged ions will move together and eliminate the potential difference between them. A movement of charges to eliminate a potential difference is called a **current**. If a barrier (such as a plasma membrane) separates the oppositely charged ions, the amount of current depends on how easily the ions can cross the membrane. The **resistance** of the membrane is a measure of how much the membrane restricts ion movement. If the resistance is high, the current is very small, because few ions can cross the membrane. If the resistance is low, the current is very large, because ions flood

across the membrane. The resistance of a plasma membrane can change as ion channels open or close. The changes result in currents carrying ions into or out of the cytoplasm.

The Electrochemical Gradient. Electrical gradients can either reinforce or oppose the chemical gradient for each ion. The **electrochemical gradient** for a specific ion is the sum of the chemical and electrical forces acting on that ion across the plasma membrane. The electrochemical gradients for K^+ and Na^+ are the primary factors affecting the resting membrane potential of most cells, including neurons. Let's consider the forces acting on each ion independently.

The intracellular concentration of potassium ions is relatively high, whereas the extracellular concentration is very low. Therefore, the chemical gradient for potassium ions tends to drive them out of the cell, as indicated by the orange-colored arrow in Figure 12–10a. However, the electrical gradient opposes this movement, because K^+ inside and outside the cell are attracted to the negative charges on the inside of the plasma membrane, and repelled by the positive charges on the outside of the plasma membrane. The white arrow in Figure 12–10a indicates the size and direction of this electrical gradient. The chemical gradient is strong enough to overpower the electrical gradient, but the electrical gradient weakens the force driving K^+ out of the cell. The lighter orange-colored arrow represents the net driving force of the potassium electrochemical gradient.

If the plasma membrane were freely permeable to K^+ but impermeable to other positively charged ions, potassium ions would continue to leave the cell until the electrical gradient (opposing the exit of K^+ from the cell) was as strong as the chemical gradient (driving K^+ out of the cell). The membrane potential at which there is no net movement of a particular ion across the plasma membrane is called the **equilibrium potential** for that ion. For potassium ions, this equilibrium occurs at a membrane potential of about –90 mV, as illustrated in Figure 12–10b. The resting membrane potential is typically –70 mV, a value close to the equilibrium potential for K^+. The difference is due primarily to Na^+ leaking continuously into the cell. The equilibrium potential indicates an ion's contribution to the resting membrane potential.

TIPS & TRICKS

To remember the relative distribution of ions across the resting cell membrane, associate **N**egative with i**N**side and p**O**sitive with the **O**utside.

The sodium ion concentration is relatively high in the extracellular fluid, but extremely low inside the cell. As a result, there is a strong chemical gradient driving Na^+ into the cell (the purple arrow in Figure 12–10c). In addition, the extracellular sodium ions are attracted by the excess of negative charges on

SPOTLIGHT

Figure 12–9
Resting Membrane Potential

Watch the MasteringA&P® > *A&PFlix* on Resting Membrane Potential.

All living cells have a membrane potential that varies from moment to moment depending on the activities of the cell. In an unstimulated, resting cell, the *potential difference* between positive and negative ions on either side of the membrane is called the **resting membrane potential**. Passive and active forces determine the resting membrane potential.

Passive Chemical Gradients

The intracellular concentration of potassium ions (K$^+$) is relatively high, so these ions tend to move out of the cell through potassium leak channels. Similarly, the extracellular concentration of sodium ions (Na$^+$) is relatively high, so sodium ions move into the cell through sodium leak channels. Both of these movements are driven by a concentration gradient, or **chemical gradient**.

Active Na$^+$/K$^+$ Pumps

Sodium–potassium (Na$^+$/K$^+$) exchange pumps maintain the concentration of sodium and potassium ions across the plasma membrane.

Passive Electrical Gradients

Potassium ions leave the cytosol more rapidly than sodium ions enter because the plasma membrane is much more permeable to potassium than to sodium. As a result, there are more positive charges outside the plasma membrane. Negatively charged protein molecules within the cytosol cannot cross the plasma membrane, so there are more negative charges on the cytosol side of the plasma membrane. This results in an **electrical gradient** across the plasma membrane.

Resting Membrane Potential

Whenever positive and negative ions are held apart, a potential difference arises. We measure the size of that potential difference in millivolts (mV). The resting membrane potential for most neurons is about –70 mV. The minus sign shows that the inner surface of the plasma membrane is negatively charged with respect to the exterior.

EXTRACELLULAR FLUID

Cl$^-$

3 Na$^+$

–70 –30 0 +30

mV

K$^+$ leak channel

Sodium–potassium exchange pump

Na$^+$ leak channel

Plasma membrane

ATP

ADP

2 K$^+$

CYTOSOL

Protein

Protein

Protein

Protein

KEY

+ Sodium ion (Na$^+$)

+ Potassium ion (K$^+$)

− Chloride ion (Cl$^-$)

Figure 12–10   **Electrochemical Gradients for Potassium and Sodium Ions.**

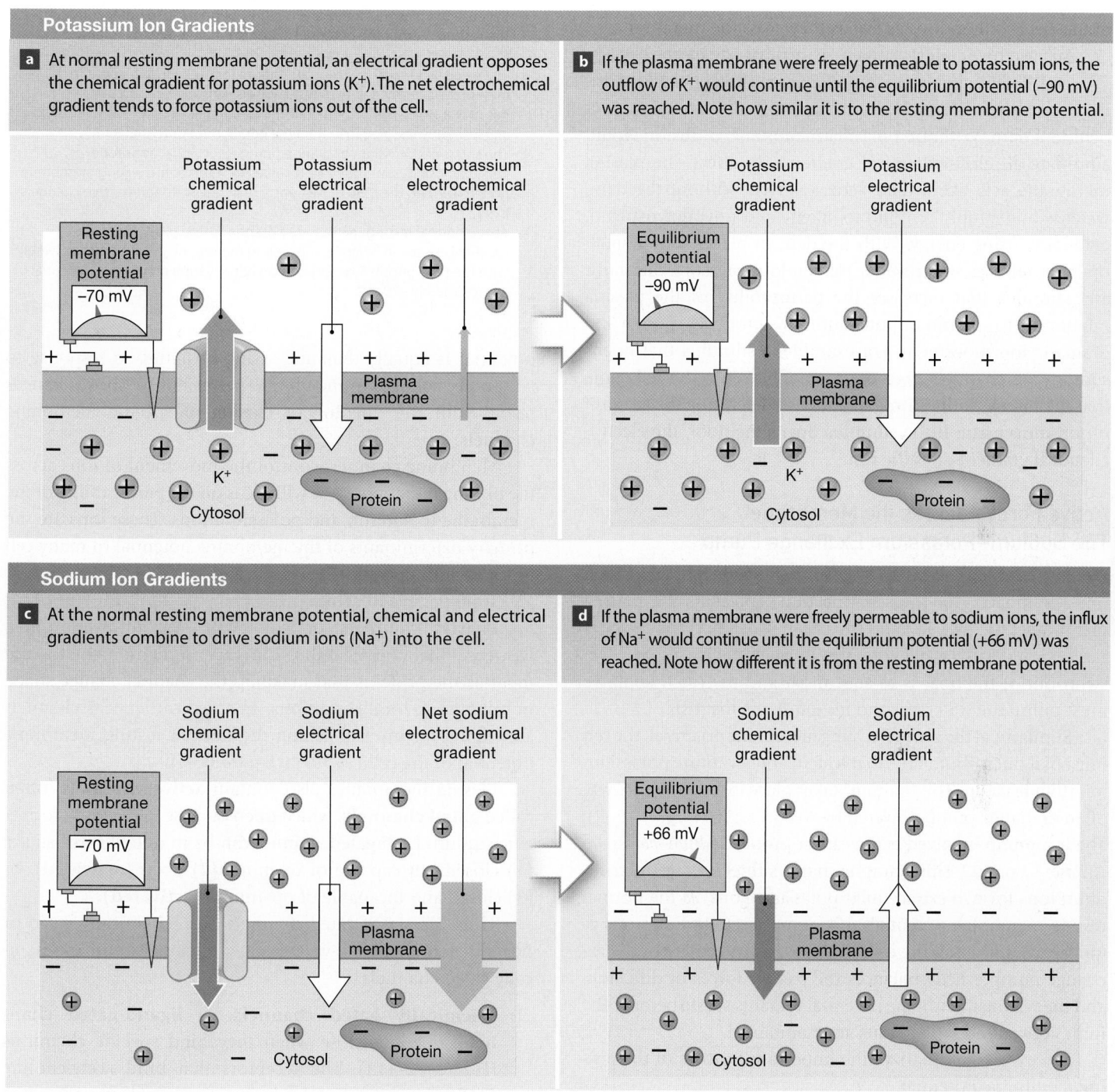

Potassium Ion Gradients

a At normal resting membrane potential, an electrical gradient opposes the chemical gradient for potassium ions (K⁺). The net electrochemical gradient tends to force potassium ions out of the cell.

b If the plasma membrane were freely permeable to potassium ions, the outflow of K⁺ would continue until the equilibrium potential (–90 mV) was reached. Note how similar it is to the resting membrane potential.

Sodium Ion Gradients

c At the normal resting membrane potential, chemical and electrical gradients combine to drive sodium ions (Na⁺) into the cell.

d If the plasma membrane were freely permeable to sodium ions, the influx of Na⁺ would continue until the equilibrium potential (+66 mV) was reached. Note how different it is from the resting membrane potential.

the inner surface of the plasma membrane. The white arrow in Figure 12–10c shows the relative size and direction of this electrical gradient. Both electrical forces and chemical forces drive Na⁺ into the cell, and the lavender arrow represents the net driving force.

If the plasma membrane were freely permeable to Na⁺, these ions would continue to cross it until the interior of the cell contained enough excess positive charges to reverse the electrical gradient. In other words, ion movement would continue until the interior developed such a strongly positive charge that repulsion between the positive charges would prevent any further net movement of Na⁺ into the cell. The equilibrium potential for Na⁺ is approximately +66 mV, as shown in Figure 12–10d. The resting membrane potential is nowhere near that value, because the resting membrane permeability to Na^+ is very low, and because ion pumps in the plasma membrane eject sodium ions as fast as they cross the membrane.

An electrochemical gradient is a form of *potential energy*. ⤴ p. 62 Potential energy is stored energy—the energy of position, as exists in a stretched spring, a charged battery, or water behind a dam. Without a plasma membrane, diffusion would eliminate all electrochemical gradients. In effect, the plasma membrane acts like a dam across a river. Without the dam, water would simply respond to gravity and flow downstream, gradually losing energy. With the dam in place, even a small opening releases water under tremendous pressure. Similarly, any stimulus that increases the permeability of the plasma membrane to sodium or potassium ions produces sudden and dramatic ion movement. For example, a stimulus that opens sodium ion channels triggers a rush of Na^+ into the cell. Note that the nature of the stimulus does not determine the amount of ion movement: If the stimulus opens the door, the electrochemical gradient does the rest.

Active Forces across the Membrane: The Sodium–Potassium Exchange Pump

We can compare a cell to a leaky fishing boat loaded with tiny fish floating in the sea. The hull represents the plasma membrane; the fish, K^+; and the ocean water, Na^+. As the boat rumbles and rolls, water comes in through the cracks, and fish swim out. If the boat is to stay afloat and the catch kept, we must pump the water out and recapture the lost fish.

Similarly, at the normal resting membrane potential, the cell must bail out sodium ions that leak in and recapture potassium ions that leak out. The "bailing" takes place through the activity of an exchange pump powered by ATP (**Spotlight Figure 12–9**). The ion pump involved is the carrier protein *sodium–potassium ATPase*. ⤴ p. 122 This pump exchanges three intracellular sodium ions for two extracellular potassium ions. At the normal resting membrane potential, this pump ejects sodium ions as quickly as they enter the cell. In this way, the activity of the exchange pump exactly balances the passive forces of diffusion, and the resting membrane potential remains stable because the ionic concentration gradients are maintained.

Table 12–1 summarizes the important features of the resting membrane potential.

Changes in the Membrane Potential

As noted previously, the resting membrane potential is the membrane potential of an "undisturbed" cell. Recall that it exists because (1) the cytosol differs from extracellular fluid in chemical and ionic composition, and (2) the plasma membrane is selectively permeable. Yet cells are dynamic structures that continually modify their activities, either in response to external stimuli or to perform specific functions. The membrane

Table 12–1	The Resting Membrane Potential

- Because the plasma membrane is highly permeable to potassium ions, the resting membrane potential of approximately –70 mV is fairly close to –90 mV, the equilibrium potential for K^+.
- The electrochemical gradient for sodium ions is very large, but the membrane's permeability to these ions is very low. Consequently, Na^+ has only a small effect on the normal resting membrane potential, making it just slightly less negative than the equilibrium potential for K^+.
- The sodium–potassium exchange pump ejects 3 Na^+ ions for every 2 K^+ ions that it brings into the cell. It serves to stabilize the resting membrane potential when the ratio of Na^+ entry to K^+ loss through passive channels is 3:2.
- At the normal resting membrane potential, these passive and active mechanisms are in balance. The resting membrane potential varies widely with the type of cell. A typical neuron has a resting membrane potential of approximately –70 mV.

potential is equally dynamic, rising or falling in response to temporary changes in membrane permeability. Those changes result from the opening or closing of specific membrane channels.

Membrane channels control the movement of ions across the plasma membrane. We will focus on the permeability of the membrane to sodium and potassium ions. These ions are the primary determinants of the membrane potential of many cell types, including neurons. Sodium and potassium ion channels are either passive or active.

Passive channels, or **leak channels**, are always open. However, their permeability can vary from moment to moment as the proteins that make up the channel change shape in response to local conditions. As noted earlier, leak channels are important in establishing the normal resting membrane potential of the cell (**Spotlight Figure 12–9**).

Plasma membranes also contain **active channels**, often called **gated channels**, which open or close in response to specific stimuli. Each gated channel can be in one of three states: (1) closed but capable of opening, (2) open (**activated**), or (3) closed and incapable of opening (**inactivated**).

Three classes of gated channels exist: chemically gated or ligand-gated channels, voltage-gated channels, and mechanically gated channels.

1. **Chemically gated channels**, or **ligand-gated channels**, open or close when they bind specific chemicals (**Figure 12–11a**). The receptors that bind acetylcholine (ACh) at the neuromuscular junction are chemically gated channels. ⤴ p. 327 Chemically gated channels are most abundant on the dendrites and cell body of a neuron, the areas where most synaptic communication occurs.

2. **Voltage-gated channels** open or close in response to changes in the membrane potential. They are characteristic of areas of **excitable membrane**, a membrane capable of generating and propagating an action potential. Examples of excitable membranes are the axons of unipolar and multipolar neurons, and the sarcolemma (including

T tubules) of skeletal muscle fibers and cardiac muscle cells. ⊃ pp. 328, 350 The most important voltage-gated channels, for our purposes, are voltage-gated sodium channels, potassium channels, and calcium channels. These sodium channels have two gates that function independently: an *activation gate* that opens on stimulation, letting sodium ions into the cell, and an *inactivation gate* that closes to stop the entry of sodium ions (**Figure 12–11b**).

3. **Mechanically gated channels** open or close in response to physical distortion of the membrane surface (**Figure 12–11c**). Such channels are important in sensory receptors that respond to touch, pressure, or vibration. We discuss these receptors in more detail in Chapter 15.

At the resting membrane potential, most gated channels are closed. When gated channels open, the rate of ion movement

Figure 12–11 **Gated Channels.**

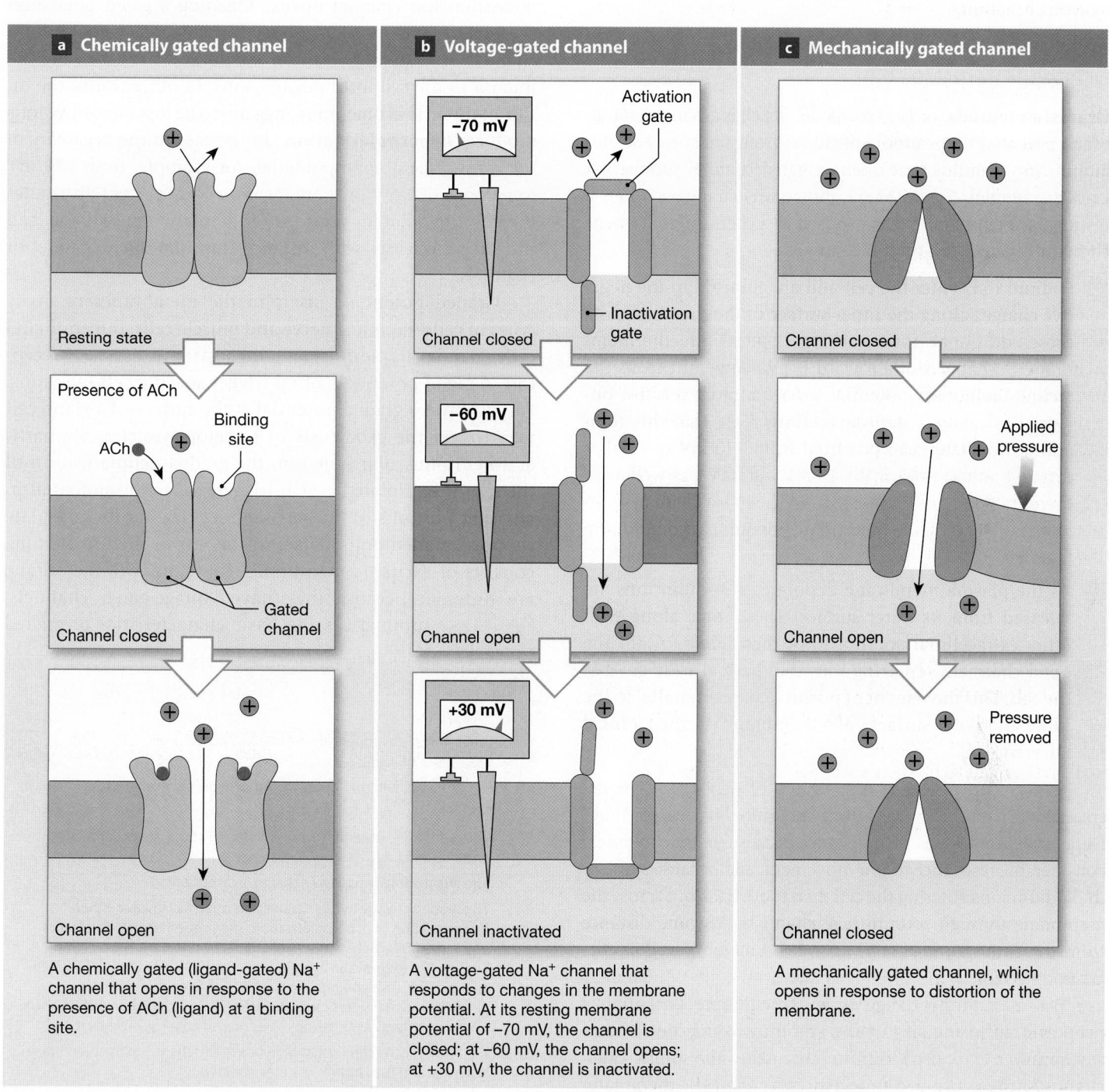

a Chemically gated channel	b Voltage-gated channel	c Mechanically gated channel
Resting state	Channel closed	Channel closed
Presence of ACh / Channel closed	Channel open	Channel open / Applied pressure
Channel open	Channel inactivated	Channel closed / Pressure removed
A chemically gated (ligand-gated) Na⁺ channel that opens in response to the presence of ACh (ligand) at a binding site.	A voltage-gated Na⁺ channel that responds to changes in the membrane potential. At its resting membrane potential of –70 mV, the channel is closed; at –60 mV, the channel opens; at +30 mV, the channel is inactivated.	A mechanically gated channel, which opens in response to distortion of the membrane.

across the plasma membrane increases, changing the membrane potential.

The distribution of membrane channels varies from one region of the plasma membrane to another, affecting how and where a cell responds to specific stimuli. For example, chemically gated sodium channels are widespread on the surfaces of a neuron, but voltage-gated sodium channels are abundant on the axon, its branches, and the axon terminals. Mechanically gated channels are typically located only on the dendrites of sensory neurons. In later sections you will see how these differences in distribution affect the way the neurons function.

Graded Potentials

Graded potentials, or *local potentials*, are changes in the membrane potential that cannot spread far from the site of stimulation. Any stimulus that opens a gated channel produces a graded potential. Figure 12–12 shows what happens when a resting plasma membrane is exposed to a chemical that opens chemically gated sodium ion channels.

1 Sodium ions enter the cell and are attracted to the negative charges along the inner surface of the membrane. As these additional positive charges spread out, the membrane potential shifts toward 0 mV. Any shift from the resting membrane potential toward a more positive potential is called a **depolarization**. Note that this term applies to changes in potential from –70 mV to smaller negative values (–65 mV, –45 mV, –10 mV), as well as to membrane potentials above 0 mV (+10 mV, +30 mV). In all these changes, the membrane potential becomes more positive.

2 As the plasma membrane depolarizes, sodium ions are released from its outer surface. These ions, along with other extracellular sodium ions, then move toward the open channels, replacing ions that have already entered the cell. This movement of positive charges parallel to the inner and outer surfaces of a membrane is called a **local current**.

In a graded potential, the degree of depolarization decreases with distance away from the stimulation site. Why? Depolarization lessens with distance because the cytosol offers considerable resistance to ion movement, and because some of the sodium ions entering the cell then move back out across the membrane through sodium leak channels. At some distance from the entry point, the effects on the membrane potential are undetectable.

The maximum change in the membrane potential is proportional to the size of the stimulus, which determines the number of open sodium channels. The more open channels, the more sodium ions enter the cell, the greater the membrane area affected, and the greater the degree of depolarization.

When a chemical stimulus is removed and normal membrane permeability is restored, the membrane potential soon returns to resting levels. The process of restoring the normal resting membrane potential after depolarization is called **repolarization** (Figure 12–13). Repolarization typically involves a combination of ion movement through membrane channels and the activities of ion pumps, especially the sodium–potassium exchange pump.

What happens to the membrane potential when a gated potassium ion channel opens? Opening a gated potassium channel has the opposite effect from opening a gated sodium channel. The rate of potassium ion outflow increases, and the interior of the cell loses positive ions. In other words, the inside of the cell becomes more negative. The loss of positive ions produces **hyperpolarization**, an increase in the negativity of the resting membrane potential, for example, from –70 mV to perhaps –80 mV or more. Again, a local current distributes the effect to adjacent portions of the plasma membrane, and the effect decreases with distance from the open channel or channels.

Graded potentials occur in the membranes of many types of cells—not just nerve and muscle cells, but epithelial cells, gland cells, adipocytes, and a variety of sensory receptors. Graded potentials often trigger specific cell functions. For example, a graded potential at the surface of a gland cell may trigger the exocytosis of secretory vesicles. Similarly, at a neuromuscular junction, the graded depolarization of the motor end plate by ACh triggers an action potential in adjacent portions of the sarcolemma. The motor end plate supports graded potentials, but the rest of the sarcolemma consists of excitable membrane. These areas of membrane are different because they have voltage-gated channels. Table 12–2 summarizes the basic characteristics of graded potentials.

Table 12–2 Graded Potentials
Graded potentials, whether depolarizing or hyperpolarizing, share four basic characteristics:
1. The membrane potential is most changed at the site of stimulation, and the effect decreases with distance.
2. The effect spreads passively, due to local currents.
3. The graded change in membrane potential may involve either depolarization or hyperpolarization. The properties and distribution of the membrane channels involved determine the nature of the change. For example, in a resting membrane, the opening of sodium channels causes depolarization, whereas the opening of potassium channels causes hyperpolarization. That is, the change in membrane potential reflects whether positive charges enter or leave the cell.
4. The stronger the stimulus, the greater is the change in the membrane potential and the larger is the area affected.

Figure 12–12 **Graded Potentials.** The depolarization radiates in all directions away from the source of stimulation. For clarity, only gated channels are shown. Leak channels are present, but are not responsible for the production of graded potentials. Color changes in the plasma membrane indicate that the resting membrane potential has been disturbed and that the membrane potential is no longer –70 mV.

Resting State

Resting membrane with closed chemically gated sodium ion channels

Extracellular Fluid

–70 mV

Cytosol

Initial segment

1 **Stimulation**

Membrane exposed to chemical that opens the sodium ion channels

–65 mV

Stimulus applied here

2 **Graded Potential**

Spread of sodium ions inside plasma membrane produces a local current that depolarizes adjacent portions of the plasma membrane

Local current

–60 mV   –65 mV   –70 mV

Local current

12

Figure 12–13   **Depolarization, Repolarization, and Hyperpolarization.**

CHECKPOINT

12. Define the resting membrane potential.

13. What effect would a chemical that blocks the voltage-gated sodium channels in the plasma membrane of a neuron have on its ability to depolarize?

14. What effect would decreasing the concentration of extracellular potassium ions have on the membrane potential of a neuron?

See the blue Answers tab at the back of the book.

12-5 ┃ An action potential is an electrical event

Learning Outcome   Describe the events involved in the generation and propagation of an action potential.

Action potentials are propagated changes in the membrane potential that, once initiated, affect an entire excitable membrane. These electrical events are also known as **nerve impulses**. Recall that voltage-gated sodium channels are abundant on the axon, its branches, and its axon terminals. The first step in generating an action potential is the opening of voltage-gated sodium ion channels at one site, usually the initial segment of the axon. The movement of sodium ions into the axon depolarizes adjacent sites, triggering the opening of additional voltage-gated sodium channels. The result is a chain reaction that spreads across the surface of the membrane like a line of falling dominoes. In this way, the action potential is propagated along the length of the axon, ultimately reaching the axon terminals.

The All-or-None Principle

The membrane potential at which an action potential begins is called the **threshold**. Threshold for an axon is typically between –60 mV and –55 mV, corresponding to a depolarization of 10 to 15 mV. A stimulus that shifts the resting membrane potential from –70 mV to –62 mV will not produce an action potential, only a graded depolarization. When such a stimulus is removed, the membrane potential returns to the resting level. The depolarization of the initial segment of the axon is caused by local currents resulting from the graded depolarization of the axon hillock.

The initial depolarization acts like pressure on the trigger of a gun. If you apply a slight pressure, the gun does not fire. It fires only when you apply a certain minimum pressure to the trigger. Once the pressure on the trigger reaches this threshold, the firing pin drops and the gun discharges. At that point, it no longer matters whether you applied the pressure gradually or suddenly or whether you moved just one finger or clenched your entire hand. The speed and range of the bullet that leaves the gun do not change, regardless of the forces that you applied to the trigger.

In the case of an axon or another area of excitable membrane, a graded depolarization is similar to the pressure on the trigger, and the action potential is like the firing of the gun. All stimuli that bring the membrane to threshold generate identical action potentials. In other words, the properties of the action potential are independent of the relative strength of the depolarizing stimulus, as long as that stimulus exceeds the threshold. This concept is the **all-or-none principle**, because a given stimulus either triggers a typical action potential, or none at all. The all-or-none principle applies to all excitable membranes.

Now let's take a closer look at how action potentials are generated and propagated. Generation and propagation are closely related concepts, in terms of both time and space: An action potential must be generated at one site before it can be propagated away from that site.

Generation of Action Potentials

Spotlight Figure 12–14 diagrams the steps involved in generating an action potential from the resting state. At the normal resting membrane potential, the activation gates of the voltage-gated sodium channels are closed. The steps are as follows: (1) depolarization to threshold, (2) activation of voltage-gated sodium channels and rapid depolarization, (3) inactivation of voltage-gated sodium channels and activation of voltage-gated potassium channels, and (4) closing of voltage-gated potassium channels.

The Refractory Period

The plasma membrane does not respond normally to additional depolarizing stimuli from the time an action potential begins until the normal resting membrane potential has stabilized. This period is known as the **refractory period** of the membrane. The membrane cannot respond to further stimulation from the moment the voltage-gated sodium channels open at threshold until sodium channel inactivation ends, because all the voltage-gated sodium channels either are already open or are inactivated. This first part of the refractory period, called the **absolute refractory period**, lasts 0.4–1.0 msec. (The smaller the axon diameter, the longer the duration.) The **relative refractory period** begins when the sodium channels regain their normal resting condition, and continues until the membrane potential stabilizes at resting levels. Another action potential can occur in this period if the membrane is sufficiently depolarized. That depolarization, however, requires a larger-than-normal stimulus, because (1) the local current must deliver enough Na^+ to counteract the exit of positively charged K^+ through open voltage-gated K^+ channels, and (2) the membrane is hyperpolarized to some degree through most of the relative refractory period.

TIPS & TRICKS

Flushing a toilet provides a useful analogy for an action potential. Nothing happens while you press the handle, until the water starts to flow (threshold is reached). After that, the amount of water that is released is independent of how hard or quickly you pressed the handle (all-or-none principle). Finally, you cannot flush the toilet again until the tank refills (refractory period).

The Role of the Sodium–Potassium Exchange Pump

In an action potential, depolarization results from the influx of Na^+, and repolarization involves the loss of K^+. Over time, the sodium–potassium exchange pump returns intracellular and extracellular ion concentrations to prestimulation levels. Compared with the total number of ions inside and outside the cell, however, the number involved in a single action potential is insignificant. Tens of thousands of action potentials can occur before intracellular ion concentrations change enough to disrupt the entire mechanism. For this reason, the exchange pump is not essential to any single action potential.

However, a maximally stimulated neuron can generate action potentials at a rate of 1000 per second. Under these circumstances, the exchange pump is needed to keep ion concentrations within acceptable limits over a prolonged period. The sodium–potassium exchange pump requires energy in the form of ATP. Each time the pump exchanges two extracellular potassium ions for three intracellular sodium ions, one molecule of ATP is broken down to ADP. Recall that the membrane protein of the exchange pump is *sodium–potassium ATPase*, which gets the energy to pump ions by splitting a phosphate group from a molecule of ATP, forming ADP. If the cell runs out of ATP, or if a metabolic poison inactivates sodium–potassium ATPase, the neuron will soon stop functioning.

Propagation of Action Potentials

The events that generate an action potential take place in a small portion of the total plasma membrane surface. But unlike graded potentials, which diminish rapidly with distance, action potentials spread along the entire excitable membrane. To understand how this happens, imagine that you are standing by the doors of a movie theater at the start of a long line. Everyone is waiting for the doors to open. The manager steps outside and says to you, "Let everyone know that we're opening in 15 minutes." How would you spread the news?

If you treated the line as an inexcitable membrane, you would shout, "The doors open in 15 minutes!" as loudly as you could. The closest people in the line would hear the news very clearly, but those farther away might not hear the entire message, and those at the end of the line might not hear you at all.

If, on the other hand, you treated the crowd as an excitable membrane, you would tell the message to the next person in line, with instructions to pass it on. In that way, the message would travel along the line undiminished, until everyone had heard the news. Such a message "moves" as each person repeats it to someone else. Distance is not a factor, and the line can contain 50 people or 5000.

Having each person repeat the message is comparable to the way an action potential spreads along an excitable membrane. An action potential (message) is relayed from one location to another in a series of steps. At each step, the message is repeated. Because the same events take place over and over, the term **propagation** is preferable to the term *conduction*, which suggests a flow of charge similar to that in a conductor such as a copper wire. (In fact, compared to wires, axons are poor conductors of electricity.) Action potentials may travel along an axon by continuous propagation (unmyelinated axons) or by saltatory propagation (myelinated axons).

Continuous Propagation

In an unmyelinated axon, an action potential moves along by **continuous propagation** (Spotlight Figure 12–15).

12

Each neuron receives information in the form of graded potentials on its dendrites and cell body, and graded potentials at the synaptic terminals trigger the release of neurotransmitters. However, the two ends of the neuron may be a meter apart, and even the largest graded potentials affect only a tiny area. Such relatively long-range communication requires a different mechanism—the action potential. **Action potentials** are propagated changes in the membrane potential that, once initiated, affect an entire excitable membrane. Whereas the resting potential depends on leak channels and the graded potential we considered depends on chemically gated channels, action potentials depend on voltage-gated channels.

Axon hillock

Initial segment

Steps in the formation of an action potential at the initial segment of an axon. The first step is a graded depolarizaton caused by the opening of chemically gated sodium ion channels, usually at the axon hillock. Note that when illustrating action potentials, we can ignore both the leak channels and the chemically gated channels, because their properties do not change. The membrane colors in steps 1-4 match the colors of the line graph showing membrane potential changes.

RESTING POTENTIAL

The axolemma contains both voltage-gated sodium channels and voltage-gated potassium channels that are closed when the membrane is at the resting potential.

KEY

⊕ = Sodium ion

⊕ = Potassium ion

1 Depolarization to Threshold

The stimulus that initiates an action potential is a graded depolarization large enough to open voltage-gated sodium channels. The opening of the channels occurs at a membrane potential known as the threshold.

2 Activation of Sodium Channels and Rapid Depolarization

When the sodium channel activation gates open, the plasma membrane becomes much more permeable to Na^+. Driven by the large electrochemical gradient, sodium ions rush into the cytoplasm, and rapid depolarization occurs. The inner membrane surface now contains more positive ions than negative ones, and the membrane potential has changed from –60 mV to a positive value.

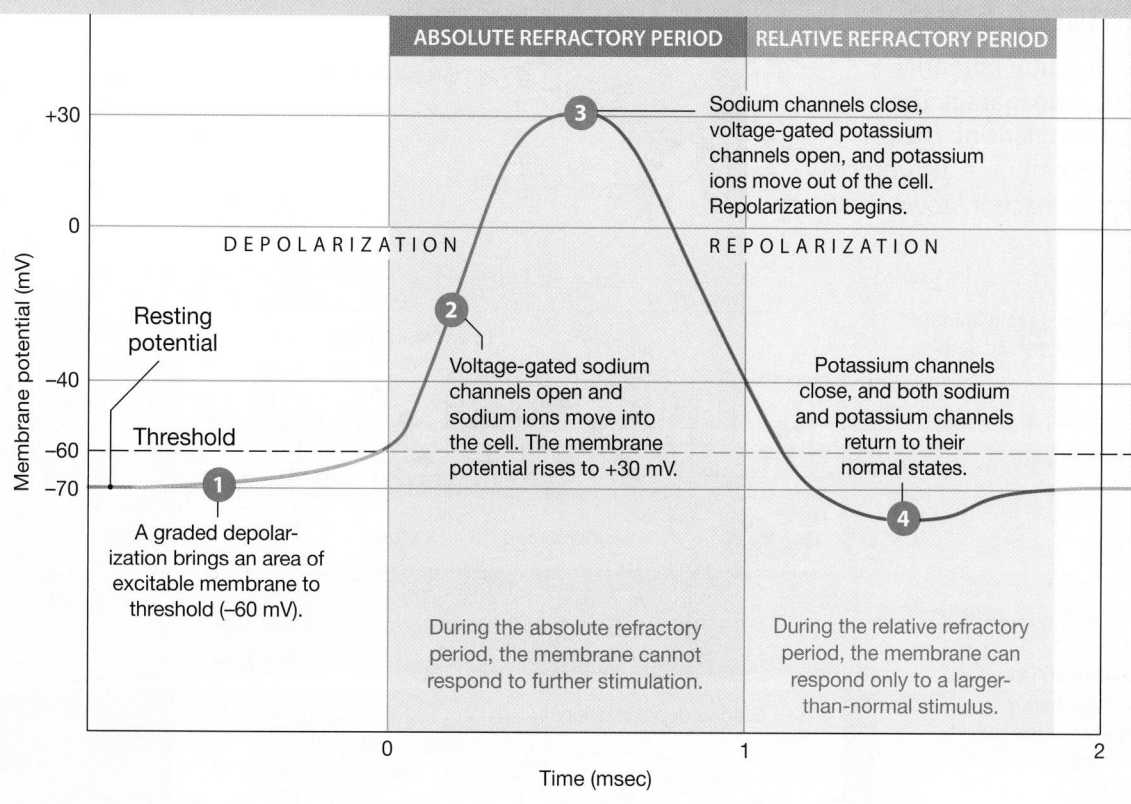

ABSOLUTE REFRACTORY PERIOD | RELATIVE REFRACTORY PERIOD

DEPOLARIZATION

REPOLARIZATION

3 Sodium channels close, voltage-gated potassium channels open, and potassium ions move out of the cell. Repolarization begins.

Resting potential

2 Voltage-gated sodium channels open and sodium ions move into the cell. The membrane potential rises to +30 mV.

Potassium channels close, and both sodium and potassium channels return to their normal states.

Threshold

1 A graded depolarization brings an area of excitable membrane to threshold (–60 mV).

During the absolute refractory period, the membrane cannot respond to further stimulation.

During the relative refractory period, the membrane can respond only to a larger-than-normal stimulus.

Time (msec)

Changes in the membrane potential at one location during the generation of an action potential. The circled numbers in the graph correspond to the steps illustrated below.

---

3 **Inactivation of Sodium Channels and Activation of Potassium Channels**

As the membrane potential approaches +30 mV, the inactivation gates of the voltage-gated sodium channels close. This step is known as **sodium channel inactivation**, and it coincides with the opening of voltage-gated potassium channels. Positively charged potassium ions move out of the cytosol, shifting the membrane potential back toward resting levels. Repolarization now begins.

4 **Closing of Potassium Channels**

The voltage-gated sodium channels remain inactivated until the membrane has repolarized to near threshold levels. At this time, they regain their normal status: closed but capable of opening. The voltage-gated potassium channels begin closing as the membrane reaches the normal resting potential (about –70 mV). Until all of these potassium channels have closed, potassium ions continue to leave the cell. This produces a brief hyperpolarization.

RESTING POTENTIAL

As the voltage-gated potassium channels close, the membrane potential returns to normal resting levels. The action potential is now over, and the membrane is once again at the resting potential.

435

Continuous Propagation along an Unmyelinated Axon

In an unmyelinated axon, an action potential moves along by continuous propagation. The action potential spreads by depolarizing the adjacent region of the axon membrane. This process continues to spread as a chain reaction down the axon.

1 As an action potential develops at the initial segment ①, the membrane potential at this site depolarizes to +30 mV.

2 As the sodium ions entering at ① spread away from the open voltage-gated channels, a graded depolarization quickly brings the membrane in segment ② to threshold.

3 An action potential now occurs in segment ② while segment ① begins repolarization.

4 As the sodium ions entering at segment ② spread laterally, a graded depolarization quickly brings the membrane in segment ③ to threshold, and the cycle is repeated.

Saltatory Propagation along a Myelinated Axon

Because myelin limits the movement of ions across the axon membrane, the action potential must "jump" from node to node during propagation. This results in much faster propagation along the axon.

1 An action potential has occurred at the initial segment ①.

2 A local current produces a graded depolarization that brings the axolemma at the next node to threshold.

3 An action potential develops at node ②.

4 A local current produces a graded depolarization that brings the axolemma at node ③ to threshold.

For convenience, think of the plasma membrane as a series of adjacent segments.

The action potential begins at the axon's initial segment. For a brief moment at the peak of the action potential, the membrane potential becomes positive rather than negative (**1**). A local current then develops as sodium ions begin moving in the cytosol and the extracellular fluid (**2**). The local current spreads in all directions, depolarizing adjacent portions of the membrane. (The axon hillock cannot respond with an action potential because, like the rest of the cell body, it lacks voltage-gated sodium channels.) The process then continues in a chain reaction (**3** and **4**).

Each time a local current develops, the action potential moves forward, but not backward, because the previous segment of the axon is still in the absolute refractory period. As a result, an action potential always proceeds away from the site of generation and cannot reverse direction. Eventually, the most distant portions of the plasma membrane are affected.

As in our "movie line" model, the message is relayed from one location to another. At each step along the way, the message is retold, so distance has no effect on the process. The action potential reaching the axon terminal is identical to the one generated at the initial segment. The net effect is the same as if a single action potential had traveled across the surface of the membrane.

In continuous propagation, an action potential appears to move across the surface of the membrane in a series of tiny steps. Even though the events at any one location take only about a millisecond, they must be repeated at each step along the way. Continuous propagation along unmyelinated axons occurs at a speed of about 1 meter per second (approximately 2 mph). For a second action potential to occur at the same site, a second stimulus must be applied.

TIPS & TRICKS

The "wave" performed by fans in a football stadium illustrates the continuous propagation of an action potential. The "wave" moves, but the people remain in place.

Saltatory Propagation

Saltatory propagation (*saltare,* leaping) in the CNS and PNS carries action potentials along an axon much more rapidly than does continuous propagation. To get the general idea, let's return to the line in front of the movie theater, and assume that it takes 1 second to relay the message to another person. In a model of continuous propagation, the people are jammed together. In 4 seconds, four people would hear the news, and the message would move perhaps 2 meters along the line. In a model of saltatory propagation, in contrast, the people in the line are spaced 5 meters apart. So after 4 seconds the same message would move 20 meters.

Table 12–3	A Comparison of Graded Potentials and Action Potentials
Graded Potentials	**Action Potentials**
Depolarizing or hyperpolarizing	Always depolarizing
No threshold value	Depolarization to threshold must occur before action potential begins
Amount of depolarization or hyperpolarization depends on intensity of stimulus	All-or-none; all stimuli that exceed threshold produce identical action potentials
Passive spread from site of stimulation	Action potential at one site depolarizes adjacent sites to threshold
Effect on membrane potential decreases with distance from stimulation site	Propagated along entire membrane surface without decrease in strength
No refractory period	Refractory period occurs
Occur in most plasma membranes	Occur only in excitable membranes of specialized cells such as neurons and muscle cells

In a myelinated axon, the "people" are the nodes, and the spaces between them are the internodes wrapped in myelin (**Figures 12–6b, 12–7a**). Continuous propagation cannot occur along a myelinated axon, because myelin increases resistance to the flow of ions across the membrane. Ions can readily cross the plasma membrane only at the nodes. As a result, only the nodes with their voltage-gated channels can respond to a depolarizing stimulus.

When an action potential appears at the initial segment of a myelinated axon, the local current skips the internodes and depolarizes the closest node to threshold (**Spotlight Figure 12–15**). Because the nodes may be 1–2 mm apart in a large myelinated axon, the action potential "jumps" from node to node rather than moving along the axon in a series of tiny steps. In addition to being faster, saltatory propagation uses proportionately less energy, because less surface area is involved and fewer sodium ions must be pumped out of the cytoplasm.

Table 12–3 reviews the key differences between graded potentials and action potentials.

CHECKPOINT

15. Define action potential.

16. List the steps involved in the generation and propagation of an action potential.

See the blue Answers tab at the back of the book.

12-6 Axon diameter, in addition to myelin, affects propagation speed

Learning Outcome Discuss the factors that affect the speed with which action potentials are propagated.

As we have seen, myelin greatly increases the propagation speed of action potentials. The diameter of the axon also affects the propagation speed, although less dramatically. Axon diameter is important because ions must move through the cytosol in order to depolarize adjacent portions of the plasma membrane. Cytosol offers much less resistance to ion movement than does the plasma membrane. In this instance, an axon behaves like an electrical cable: The larger the diameter, the lower the resistance. (That is why motors with large current demands, such as the starter on a car, an electric stove, or a big air conditioner, use such thick wires.)

We can classify axons into three groups according to the relationships among the diameter, myelination, and propagation speed:

1. **Type A fibers** are the largest myelinated axons, with diameters ranging from 4 to 20 μm. These fibers carry action potentials at speeds of up to 120 meters per second, or 268 mph.

2. **Type B fibers** are smaller myelinated axons, with diameters of 2–4 μm. Their propagation speeds average around 18 meters per second, or roughly 40 mph.

3. **Type C fibers** are unmyelinated and less than 2 μm in diameter. These axons propagate action potentials at the leisurely pace of 1 meter per second, or a mere 2 mph.

The advantage of myelin becomes clear when you compare Type C to Type A fibers and note that the diameter increases tenfold but the propagation speed increases 120 times.

Type A fibers carry sensory information about position, balance, and delicate touch and pressure sensations from the skin surface to the CNS. The motor neurons that control skeletal muscles also send their commands over large, myelinated Type A axons.

Type B fibers and Type C fibers carry information to and from the CNS. They deliver temperature, pain, and general touch and pressure sensations. They also carry instructions to smooth muscle, cardiac muscle, glands, and other peripheral effectors.

Why isn't every axon in the nervous system large and myelinated? The most likely reason is that it would be physically impossible. If all sensory information were carried by large Type A fibers, your peripheral nerves would be the size of garden hoses, and your spinal cord would be the diameter of a garbage can. Instead, only about one-third of all axons carrying sensory information are myelinated, and most sensory information arrives over slender Type C fibers.

In essence, information transfer in the nervous system represents a compromise between propagation time and available space. Messages are routed according to priority: Urgent news—sensory information about things that threaten survival and motor commands that prevent injury—travels over Type A fibers

(the equivalent of instant messaging, or texting). Less urgent sensory information and motor commands are relayed by Type B fibers (e-mail) or Type C fibers (regular "snail mail").

> ## CHECKPOINT
>
> 17. What is the relationship between myelin and the propagation speed of action potentials?
>
> 18. Which of the following axons is myelinated: one that propagates action potentials at 50 meters per second, or one that carries them at 1 meter per second?
>
> See the blue Answers tab at the back of the book.

12-7 ❘ At synapses, communication occurs among neurons or between neurons and other cells

Learning Outcome Describe the structure of a synapse, and explain the mechanism involved in synaptic activity.

In the nervous system, messages move from one location to another in the form of action potentials (nerve impulses) along axons. To be effective, a message must be not only propagated along an axon but also transferred in some way to another cell. This transfer takes place at synapses.

Synaptic Activity

At a synapse between two neurons, the nerve impulse passes from the **presynaptic neuron** to the **postsynaptic neuron**. Synapses may also involve other types of postsynaptic cells. For example, the neuromuscular junction is a synapse where the postsynaptic cell is a skeletal muscle fiber. Now let's take a closer look at the ways that synapses work.

General Properties of Synapses

A synapse may be *electrical*, with direct physical contact between the cells, or *chemical*, involving a neurotransmitter.

Electrical Synapses

At **electrical synapses**, the presynaptic and postsynaptic membranes are locked together at gap junctions (**Figure 4–2**, p. 143). The lipid portions of opposing membranes, separated by only 2 nm, are held in position by binding between integral membrane proteins called *connexons*. These proteins form pores that permit ions to pass between the cells. Because the two cells are linked in this way, changes in the membrane potential of one cell produce local currents that affect the other cell as if the two shared a common membrane. As a result, an electrical synapse propagates action potentials quickly and efficiently from one cell to the next.

Electrical synapses are extremely rare in both the CNS and PNS. They occur in some areas of the brain, including the vestibular nuclei, the eye, and in at least one pair of PNS ganglia (the ciliary ganglia).

Chemical Synapses

The situation at a **chemical synapse** is far more variable than that at an electrical synapse, because the cells are not directly coupled. For example, an action potential that reaches an electrical synapse is *always* propagated to the next cell. But at a chemical synapse, an arriving action potential *may or may not* release enough neurotransmitter to bring the postsynaptic neuron to threshold. In addition, other factors may intervene and make the postsynaptic cell more or less sensitive to arriving stimuli. In essence, the postsynaptic cell at a chemical synapse is not a slave to the presynaptic neuron, and its activity can be adjusted, or "tuned," by a variety of factors.

Chemical synapses are by far the most abundant type of synapse. Most synapses between neurons, and all communications between neurons and other types of cells, involve chemical synapses. Normally, communication across a chemical synapse takes place in only one direction: from the presynaptic membrane to the postsynaptic membrane.

Acetylcholine (ACh) is the neurotransmitter that has received most of our attention so far, but there are other important chemical transmitters. Based on their effects on postsynaptic membranes, neurotransmitters are often classified as excitatory or inhibitory. **Excitatory neurotransmitters** cause depolarization and promote the generation of action potentials. **Inhibitory neurotransmitters** cause hyperpolarization and suppress the generation of action potentials.

This classification is useful, but not always precise. For example, acetylcholine typically produces a depolarization in the postsynaptic membrane, but acetylcholine released at neuromuscular junctions in the heart has an inhibitory effect, producing a transient hyperpolarization of the postsynaptic membrane. This situation highlights an important aspect of neurotransmitter function: *The effect of a neurotransmitter on the postsynaptic membrane depends on the properties of the receptor, not on the nature of the neurotransmitter.*

Let's continue our discussion of chemical synapses with a look at a synapse that releases the neurotransmitter **acetylcholine (ACh)**. Then we will introduce other important neurotransmitters that you will encounter in later chapters.

Cholinergic Synapses

Synapses that release ACh are known as **cholinergic synapses**. The neuromuscular junction is an example of a cholinergic synapse. ⟳ p. 326 ACh is the most widespread (and best-studied) neurotransmitter. It is released (1) at all neuromuscular junctions involving skeletal muscle fibers, (2) at many synapses in the CNS, (3) at all neuron-to-neuron synapses in the PNS, and (4) at all neuromuscular and neuroglandular junctions in the parasympathetic division of the ANS.

At a cholinergic synapse between two neurons, the presynaptic and postsynaptic membranes are separated by a synaptic cleft that averages 20 nm (0.02 μm) in width. Most of the ACh in the axon terminal is packaged in synaptic vesicles, each containing several thousand molecules of the neurotransmitter. A single axon terminal may contain a million such vesicles.

TIPS & TRICKS

*Cholin*ergic synapses are so named because the neurotransmitter involved is acetyl*cholin*e.

Events at a Cholinergic Synapse

Figure 12–16 diagrams the events that take place at a cholinergic synapse between neurons after an action potential arrives at an axon terminal. For convenience, we will assume that this synapse is adjacent to the initial segment of the axon, an arrangement that is easy to illustrate.

1 *An Action Potential Arrives and Depolarizes the Axon Terminal.* The normal stimulus for neurotransmitter release is the depolarization of the axon terminal by the arrival of an action potential.

2 *Extracellular Calcium Ions Enter the Axon Terminal, Triggering the Exocytosis of ACh.* The depolarization of the axon terminal briefly opens its voltage-gated calcium channels, allowing calcium ions to rush in. Their arrival triggers exocytosis of ACh into the synaptic cleft. The ACh is released in packets of roughly 3000 molecules, the average number of ACh molecules in a single vesicle. ACh release stops very soon, because active transport mechanisms rapidly remove the calcium ions from the terminal cytosol. These ions are either pumped out of the cell or transferred into mitochondria.

3 *ACh Binds to Receptors and Depolarizes the Postsynaptic Membrane.* ACh diffuses across the synaptic cleft toward receptors on the postsynaptic membrane. These ACh receptors are chemically gated ion channels. The primary response is an increased permeability to Na^+, producing a depolarization in the postsynaptic membrane that lasts about 20 msec. (These cation channels also let potassium ions out of the cell, but because sodium ions are driven by a much stronger electrochemical gradient, the net effect is a slight depolarization of the postsynaptic membrane.)

This depolarization is a graded potential. The greater the amount of ACh released at the presynaptic membrane,

Figure 12–16   **Events in the Functioning of a Cholinergic Synapse.**

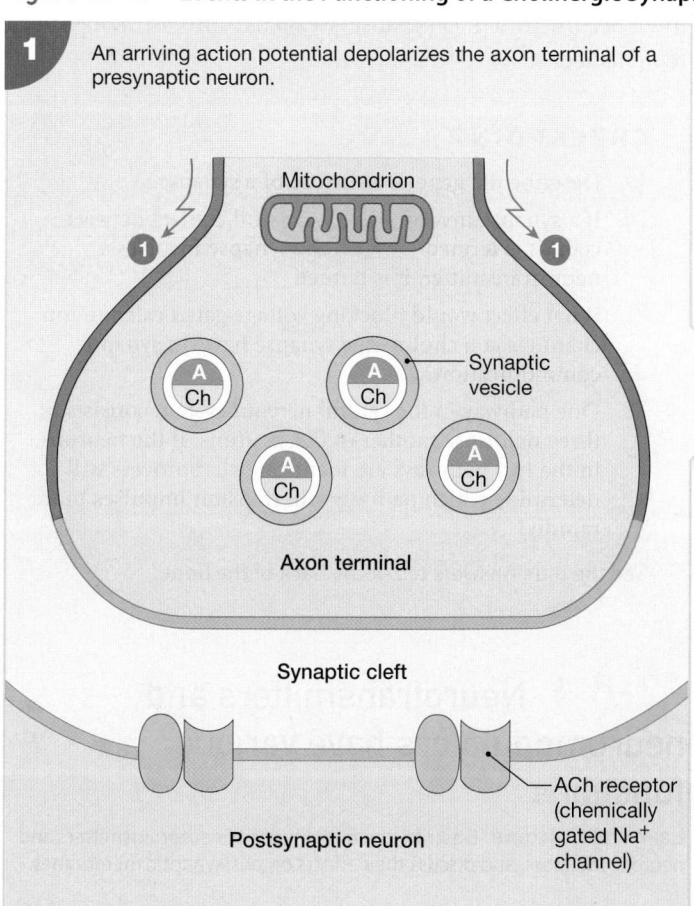

1 An arriving action potential depolarizes the axon terminal of a presynaptic neuron.

2 Calcium ions (Ca^{2+}) enter the cytosol of the axon terminal. This results in ACh release from the synaptic vesicles by exocytosis.

3 ACh diffuses across the synaptic cleft and binds to receptors on the postsynaptic membrane. Sodium channels open, producing a graded depolarization.

4 Depolarization ends as ACh is broken down into acetate and choline by AChE. The axon terminal reabsorbs choline from the synaptic cleft and uses it to resynthesize ACh.

12

the greater the number of open cation channels in the postsynaptic membrane, and so the larger the depolarization. If the depolarization brings an adjacent area of excitable membrane (such as the initial segment of an axon) to threshold, an action potential appears in the postsynaptic neuron.

4 *ACh Is Removed by AChE.* The neurotransmitter's effects on the postsynaptic membrane are temporary, because the synaptic cleft and the postsynaptic membrane contain the enzyme *acetylcholinesterase* (*AChE,* or *cholinesterase*). Roughly half of the ACh released at the presynaptic membrane is broken down before it reaches receptors on the postsynaptic membrane. ACh molecules that bind to receptor sites are generally broken down within 20 msec of their arrival.

AChE breaks down molecules of ACh (by hydrolysis) into **acetate** and **choline**. The choline is actively absorbed by the axon terminal and is used to synthesize more ACh, using acetate provided by *coenzyme A (CoA).* (Recall from Chapter 2 that coenzymes derived from vitamins are required in many enzymatic reactions. ⊃ p. 81) Acetate diffusing away from the synapse can be absorbed and metabolized by the postsynaptic cell or by other cells and tissues.

Synaptic Delay

A **synaptic delay** of 0.2–0.5 msec occurs between the arrival of the action potential at the axon terminal and the effect on the postsynaptic membrane. Most of that delay reflects the time involved in calcium influx and neurotransmitter release, not in the neurotransmitter's diffusion—the synaptic cleft is narrow, and neurotransmitters can diffuse across it in very little time.

Although a delay of 0.5 msec is not very long, in that time an action potential may travel more than 7 cm (about 3 in.) along a myelinated axon. When information is being passed along a chain of interneurons in the CNS, the cumulative synaptic delay may exceed the propagation time along the axons. This is why reflexes are important for survival—they involve only a few synapses and thus provide rapid and automatic responses to stimuli. The fewer synapses involved, the shorter the total synaptic delay and the faster the response. The fastest reflexes have just one synapse, with a sensory neuron directly controlling a motor neuron. The muscle spindle reflexes, discussed in Chapter 13, are arranged in this way.

Synaptic Fatigue

Because ACh molecules are recycled, the axon terminal is not totally dependent on the ACh synthesized in the cell body and delivered by axoplasmic transport. But under intensive stimulation, resynthesis and transport mechanisms may not keep up with the demand for neurotransmitter. **Synaptic fatigue** then occurs, and the synapse weakens until ACh has been replenished.

CHECKPOINT

19. Describe the general structure of a synapse.
20. If a synapse involves direct physical contact between cells, it is termed _____; if the synapse involves a neurotransmitter, it is termed _____.
21. What effect would blocking voltage-gated calcium ion channels at a cholinergic synapse have on synaptic communication?
22. One pathway in the central nervous system consists of three neurons, another of five neurons. If the neurons in the two pathways are identical, what process will determine which pathway will transmit impulses more rapidly?

See the blue Answers tab at the back of the book.

12-8 ❙ Neurotransmitters and neuromodulators have various functions

Learning Outcome Describe the major types of neurotransmitters and neuromodulators, and discuss their effects on postsynaptic membranes.

Now that we have examined the actions of acetylcholine at cholinergic synapses, let's consider the actions of other neurotransmitters, and of neuromodulators, which change the cell's response to neurotransmitters.

The Activities of Other Neurotransmitters

The nervous system relies on a complex form of chemical communication. Each neuron is continuously exposed to a variety of neurotransmitters. Some usually have excitatory effects, while others usually have inhibitory effects. Yet in all cases, the effects depend on the nature of the receptor rather than the structure of the neurotransmitter. (Many drugs affect the nervous system by stimulating receptors that otherwise respond only to neurotransmitters. These drugs can have complex effects on perception, motor control, and emotional states.)

Major categories of neurotransmitters include *biogenic amines, amino acids, neuropeptides, dissolved gases,* and a variety of other compounds. Here we introduce only a few of the most important neurotransmitters. You will encounter additional examples in later chapters.

- **Norepinephrine** (nor-ep-i-NEF-rin), or **NE**, is a neurotransmitter that is widely distributed in the brain and in portions of the ANS. Norepinephrine is also called *noradrenaline,* and synapses that release NE are known as **adrenergic synapses**. Norepinephrine typically has an excitatory, depolarizing effect on the postsynaptic membrane, but the mechanism is quite distinct from that of ACh, as we will see in Chapter 16.

- **Dopamine** (DŌ-puh-mēn) is a CNS neurotransmitter released in many areas of the brain. It may have either inhibitory or excitatory effects. Inhibitory effects play an important role in our precise control of movements. For example, dopamine release in one portion of the brain prevents the overstimulation of neurons that control skeletal muscle tone. If the neurons that produce dopamine are damaged or destroyed, the result can be the characteristic rigidity and stiffness of *Parkinson's disease,* a condition we describe in Chapter 14. At other sites, dopamine release has excitatory effects. Cocaine inhibits the removal of dopamine from synapses in specific areas of the brain. The resulting rise in dopamine concentrations at these synapses is responsible for the "high" experienced by cocaine users.

- **Serotonin** (ser-ō-TŌ-nin) is another important CNS neurotransmitter. Inadequate serotonin production can have widespread effects on a person's attention and emotional states and may be responsible for many cases of severe chronic depression. *Fluoxetine (Prozac), paroxetine (Paxil), sertraline (Zoloft),* and related antidepressant drugs inhibit the reabsorption of serotonin by axon terminals (hence their classification as *selective serotonin reuptake inhibitors,* or *SSRIs*). This inhibition leads to increased serotonin concentrations at synapses, and over time, the increase may relieve the symptoms of depression. Interactions among serotonin, norepinephrine, and other neurotransmitters are thought to be involved in the regulation of sleep and wake cycles.

- **Gamma-aminobutyric** (a-MĒ-nō-bū-TĔR-ik) **acid**, or **GABA**, generally has an inhibitory effect. Roughly 20 percent of the synapses in the brain release GABA, but its functions remain incompletely understood. In the CNS, GABA release appears to reduce anxiety, and some antianxiety drugs work by enhancing this effect.

The functions of many neurotransmitters are not well understood. In a clear demonstration of the principle "the more you look, the more you see," over 100 neurotransmitters have been identified, including certain amino acids, peptides, polypeptides, prostaglandins, and ATP.

In addition, two gases, nitric oxide and carbon monoxide, are known to be important neurotransmitters. **Nitric oxide** (**NO**) is generated by axon terminals that innervate smooth muscle in the walls of blood vessels in the PNS, and at synapses in several regions of the brain. **Carbon monoxide** (**CO**), best known as a component of automobile exhaust, is also generated by specialized axon terminals in the brain, where it functions as a neurotransmitter.

Neuromodulators

It is convenient to discuss each synapse as if it were releasing only one chemical, but axon terminals may release a mixture of active compounds, either through diffusion across the membrane or by exocytosis, along with neurotransmitter molecules. These compounds may have a variety of functions. Those that alter the rate of neurotransmitter release by the presynaptic neuron or change the postsynaptic cell's response to neurotransmitters are called **neuromodulators** (noo-rō-MOD-ū-lā-torz). These substances are typically **neuropeptides**, small peptide chains synthesized and released by the axon terminal. Most neuromodulators act by binding to receptors in the presynaptic or postsynaptic membranes and activating cytoplasmic enzymes.

Neuromodulators called **opioids** (Ō-pē-oydz) have effects similar to those of the drugs *opium* and *morphine,* because they bind to the same group of postsynaptic receptors. Four classes of opioids in the CNS are (1) **endorphins** (en-DOR-finz), (2) **enkephalins** (en-KEF-a-linz), (3) **endomorphins**, and (4) **dynorphins** (DĪ-nor-finz). The primary function of opioids is probably to relieve pain. They inhibit the release of the neurotransmitter *substance P* at synapses that relay pain sensations. Dynorphins have far more powerful pain-relieving effects than morphine or the other opioids.

TIPS & TRICKS

*Endorphin*s are so named because they act like *endo*genous (coming from within the body) m*orphin*e.

In general, neuromodulators (1) have long-term effects that are relatively slow to appear; (2) trigger responses that involve a number of steps and intermediary compounds; (3) may affect the presynaptic membrane, the postsynaptic membrane, or both; and (4) can be released alone or along with a neurotransmitter. Table 12–4 lists major neurotransmitters and neuromodulators of the brain and spinal cord, and their primary effects (if known). In practice, it can be very difficult to distinguish neurotransmitters from neuromodulators on either biochemical or functional grounds: A neuropeptide may function in one site as a neuromodulator and in another as a neurotransmitter. For this reason, Table 12–4 does not distinguish between neurotransmitters and neuromodulators.

Table 12–4   Representative Neurotransmitters and Neuromodulators

Class and Neurotransmitter	Chemical Structure	Mechanism of Action	Location(s)	Comments
Acetylcholine		Primarily direct, through binding to chemically gated channels	CNS: Synapses throughout brain and spinal cord PNS: Neuromuscular junctions; preganglionic synapses of ANS; neuroglandular junctions of parasympathetic division and (rarely) sympathetic division of ANS; amacrine cells of retina	Widespread in CNS and PNS; best known and most studied of the neurotransmitters
BIOGENIC AMINES				
Norepinephrine		Indirect: G proteins and second messengers	CNS: Cerebral cortex, hypothalamus, brain stem, cerebellum, spinal cord PNS: Most neuromuscular and neuroglandular junctions of sympathetic division of ANS	Involved in attention and consciousness, control of body temperature, and regulation of pituitary gland secretion
Epinephrine		Indirect: G proteins and second messengers	CNS: Thalamus, hypothalamus, midbrain, spinal cord	Uncertain functions
Dopamine		Indirect: G proteins and second messengers	CNS: Hypothalamus, midbrain, limbic system, cerebral cortex, retina	Regulation of subconscious motor function; receptor abnormalities have been linked to development of schizophrenia
Serotonin		Primarily indirect: G proteins and second messengers	CNS: Hypothalamus, limbic system, cerebellum, spinal cord, retina	Important in emotional states, moods, and body temperature; several illegal hallucinogenic drugs, such as Ecstasy, target serotonin receptors
Histamine		Indirect: G proteins and second messengers	CNS: Neurons in hypothalamus, with axons projecting throughout the brain	Receptors are primarily on presynaptic membranes; functions in sexual arousal, pain threshold, pituitary hormone secretion, thirst, and blood pressure control
AMINO ACIDS				
Excitatory: **Glutamate**		Indirect: G proteins and second messengers Direct: opens calcium/sodium channels on pre- and postsynaptic membranes	CNS: Cerebral cortex and brain stem	Important in memory and learning; most important excitatory neurotransmitter in the brain
Aspartate		Direct or indirect (G proteins), depending on type of receptor	CNS: Cerebral cortex, retina, and spinal cord	Used by pyramidal cells in the brain that provide voluntary motor control over skeletal muscles
Inhibitory: **Gamma-aminobutyric acid (GABA)**		Direct or indirect (G proteins), depending on type of receptor	CNS: Cerebral cortex, cerebellum, interneurons throughout brain and spinal cord	Direct effects: open Cl^- channels; indirect effects: open K^+ channels and block entry of Ca^{2+}
Glycine		Direct: Opens Cl^- channels	CNS: Interneurons in brain stem, spinal cord, and retina	Produces postsynaptic inhibition; the poison *strychnine* produces fatal convulsions by blocking glycine receptors

(Continued)

Table 12–4 (continued)

Class and Neurotransmitter	Chemical Structure	Mechanism of Action	Location(s)	Comments
NEUROPEPTIDES				
Substance P	Arg-Pro-Lys-Pro-Glu-Glu-Phe-Phe-Gly-Leu-Met	Indirect: G proteins and second messengers	CNS: Synapses of pain receptors within spinal cord, hypothalamus, and other areas of the brain PNS: Enteric nervous system (network of neurons along the digestive tract)	Important in pain pathway, regulation of pituitary gland function, control of digestive tract reflexes
Neuropeptide Y	*36-amino-acid peptide*	Indirect: G proteins and second messengers	CNS: hypothalamus PNS: sympathetic neurons	Stimulates appetite and food intake
Opioids **Endorphins**	*31-amino-acid peptide*	Indirect: G proteins and second messengers	CNS: Thalamus, hypothalamus, brain stem, retina	Pain control; emotional and behavioral effects poorly understood
Enkephalins	Tyr-Gly-Gly-Phe-Met	Indirect: G proteins and second messengers	CNS: Basal nuclei, hypothalamus, midbrain, pons, medulla oblongata, spinal cord	Pain control; emotional and behavioral effects poorly understood
Endomorphin	*9- or 10-amino-acid peptide*	Indirect: G proteins and second messengers	CNS: Thalamus, hypothalamus, basal nuclei	Pain control; emotional and behavioral effects poorly understood
Dynorphin	Tyr-Gly-Gly-Phe-Leu-Arg-Arg-Ile-Arg-Pro-Lys-Leu-Lys-Trp-Asp-Asn-Gln	Indirect: G proteins and second messengers	CNS: Hypothalamus, midbrain, medulla oblongata	Pain control; emotional and behavioral effects poorly understood
PURINES				
ATP, GTP	*(see Figure 2–25)*	Direct or indirect (G proteins), depending on type of receptor	CNS: Spinal cord PNS: Autonomic ganglia	
Adenosine	*(see Figure 2–25)*	Indirect: G proteins and second messengers	CNS: Cerebral cortex, hippocampus, cerebellum	Produces drowsiness; stimulatory effect of caffeine is due to inhibition of adenosine activity
HORMONES				
ADH, oxytocin, insulin, glucagon, secretin, CCK, GIP, VIP, inhibins, ANP, BNP, and many others	Peptide containing fewer than 200 amino acids	Typically indirect: G proteins and second messengers	CNS: Brain (widespread)	Numerous, complex, and incompletely understood
GASES				
Carbon monoxide (CO)	$C{\equiv}O$	Indirect: By diffusion to enzymes activating second messengers	CNS: Brain PNS: Some neuromuscular and neuroglandular junctions	Localization and function poorly understood
Nitric oxide (NO)	$N{\equiv}O$	Indirect: By diffusion to enzymes activating second messengers	CNS: Brain, especially at blood vessels PNS: Some sympathetic neuromuscular and neuroglandular junctions	
LIPIDS				
Anandamide	(structure)	Indirect: G proteins and second messengers	CNS: cerebral cortex, hippocampus, cerebellum	Euphoria, drowsiness, appetite; receptors are targeted by the active ingredient in marijuana

12

How Neurotransmitters and Neuromodulators Work

Functionally, neurotransmitters and neuromodulators fall into one of three groups: (1) *compounds that have a direct effect on membrane potential,* (2) *compounds that have an indirect effect on membrane potential,* or (3) *lipid-soluble gases that exert their effects inside the cell.*

Compounds that have direct effects on membrane potential open or close gated ion channels (**Figure 12–17a**). Examples include ACh and the amino acids *glycine* and *aspartate.* Because these neurotransmitters alter ion movement across the membrane, they are said to have *ionotropic effects.* A few neurotransmitters, notably glutamate, GABA, NE, and serotonin, have both direct and indirect effects, because these compounds target two different classes of receptors. The direct effects are ionotropic. The indirect effects, which involve changes in the metabolic activity of the postsynaptic cell, are called *metabotropic.*

Compounds that have an indirect effect on membrane potential work through intermediaries known as *second messengers.* The neurotransmitter represents a *first messenger,* because it delivers the message to receptors on the plasma membrane or within the cell. Second messengers are ions or molecules that are produced or released inside the cell when a first messenger binds to one of these receptors.

Many neurotransmitters—including epinephrine, norepinephrine, dopamine, serotonin, histamine, and GABA—and many neuromodulators bind to receptors in the plasma membrane. In these instances, the link between the first messenger and the second messenger involves a **G protein**, an enzyme complex coupled to a membrane receptor. The name *G protein* refers to the fact that these proteins bind GTP, a high-energy compound introduced in Chapter 2. ⟳ **p. 85** Several types of G protein exist, but each type includes an enzyme that is "turned on" when an extracellular compound binds to its associated receptor at the cell surface.

Figure 12–17b shows one possible result of this binding: the activation of the enzyme **adenylate cyclase**. This enzyme converts ATP, the energy currency of the cell, to *cyclic-AMP,* a ring-shaped form of the compound AMP that was introduced in Chapter 2. ⟳ **p. 84** The conversion takes place at the inner surface of the plasma membrane. Cyclic-AMP (cAMP) is a second messenger that may open membrane channels, activate intracellular enzymes, or both, depending on the nature of the postsynaptic cell. This is only an overview of the function of one type of G protein. We examine several types of G proteins more closely in later chapters.

Two lipid-soluble gases, nitric oxide (NO) and carbon monoxide (CO), are known to be important neurotransmitters in specific regions of the brain. Because they can diffuse

Figure 12–17 Mechanisms of Neurotransmitter Function.

through lipid membranes, these gases can enter the cell and bind to enzymes on the inner surface of the plasma membrane or elsewhere in the cytoplasm (**Figure 12–17c**). These enzymes then promote the appearance of second messengers that can affect cellular activity.

CHECKPOINT

23. Differentiate between a neurotransmitter and a neuromodulator.

24. Identify the three functional groups of neurotransmitters and neuromodulators.

See the blue Answers tab at the back of the book.

12-9 ∎ Individual neurons process information by integrating excitatory and inhibitory stimuli

Learning Outcome Discuss the interactions that enable information processing to occur in neural tissue.

A single neuron may receive information across thousands of synapses. As we have seen, some of the neurotransmitters arriving at the postsynaptic cell at any moment may be excitatory, and others may be inhibitory. So how does the neuron respond? The net effect on the membrane potential of the cell body—specifically, in the area of the axon hillock—determines how the neuron responds from moment to moment. If the net effect is a depolarization at the axon hillock, that depolarization affects the membrane potential at the initial segment. If threshold is reached at the initial segment, an action potential is generated and propagated along the axon.

Thus it is really the axon hillock that integrates the excitatory and inhibitory stimuli affecting the cell body and dendrites at any given moment. This integration process, which determines the rate of action potential generation at the initial segment, is the simplest level of **information processing** in the nervous system. The excitatory and inhibitory stimuli are integrated through interactions between *postsynaptic potentials*, which we discuss next. Higher levels of information processing involve interactions among neurons and among groups of neurons. We address these topics in later chapters.

Postsynaptic Potentials

Postsynaptic potentials are graded potentials that develop in the postsynaptic membrane in response to a neurotransmitter. (**Figure 12–13** illustrated graded depolarizations and hyperpolarizations.) Two major types of postsynaptic potentials develop at neuron-to-neuron synapses: excitatory postsynaptic potentials and inhibitory postsynaptic potentials.

An **excitatory postsynaptic potential**, or **EPSP**, is a graded depolarization caused by the arrival of a neurotransmitter at the postsynaptic membrane. An EPSP results from the opening of chemically gated membrane channels that lead to depolarization of the plasma membrane. For example, the graded depolarization produced by the binding of ACh is an EPSP. Because it is a graded potential, an EPSP affects only the area immediately surrounding the synapse, as shown in **Figure 12–12**.

We have already noted that not all neurotransmitters have an excitatory (depolarizing) effect. An **inhibitory postsynaptic potential**, or **IPSP**, is a graded hyperpolarization of the postsynaptic membrane. For example, an IPSP may result from the opening of chemically gated potassium channels. While the hyperpolarization continues, the neuron is said to be **inhibited**, because a larger-than-usual depolarizing stimulus is needed to bring the membrane potential to threshold. A stimulus that shifts the membrane potential by 10 mV (from –70 mV to –60 mV) would normally produce an action potential, but if the membrane potential were reset at –85 mV by an IPSP, the same stimulus would depolarize it to only –75 mV, which is below threshold.

Summation

An individual EPSP has a small effect on the membrane potential, typically producing a depolarization of about 0.5 mV at the postsynaptic membrane. Before an action potential will arise in the initial segment, local currents must depolarize that region by at least 10 mV. Therefore, a single EPSP will not result in an action potential, even if the synapse is on the axon hillock. But individual EPSPs combine through the process of **summation**, which integrates the effects of all the graded potentials that affect one portion of the plasma membrane. The graded potentials may be EPSPs, IPSPs, or both. We will consider EPSPs in our discussion. Two forms of summation exist: temporal summation and spatial summation (**Figure 12–18**).

Temporal summation (*tempus,* time) is the addition of stimuli occurring in rapid succession at a *single synapse* that is active *repeatedly.* This form of summation can be likened to using a bucket to fill up a bathtub: You can't fill the tub with a single bucket of water, but you will fill it eventually if you keep repeating the process. In the case of temporal summation, the water in a bucket corresponds to the sodium ions that enter the cytosol during an EPSP. A typical EPSP lasts about 20 msec, but under maximum stimulation an action potential can reach the axon terminal each millisecond. **Figure 12–18a** shows what happens when a second EPSP arrives before the effects of the first EPSP have disappeared: The effects of the two are combined. Every time an action potential arrives, a group of vesicles discharges ACh into the synaptic cleft, and every time more ACh molecules arrive at the postsynaptic membrane, more chemically gated channels open, and the degree of depolarization increases. In this way, a series of small steps can eventually bring the initial segment to threshold.

Spatial summation occurs when simultaneous stimuli applied at different locations have a cumulative effect on the membrane potential. In other words, spatial summation involves *multiple synapses* that are active *simultaneously.* In terms of our bucket analogy, you could fill the bathtub immediately if 50 friends emptied their buckets into it all at the same time.

12

Figure 12–18 **Temporal and Spatial Summation.**

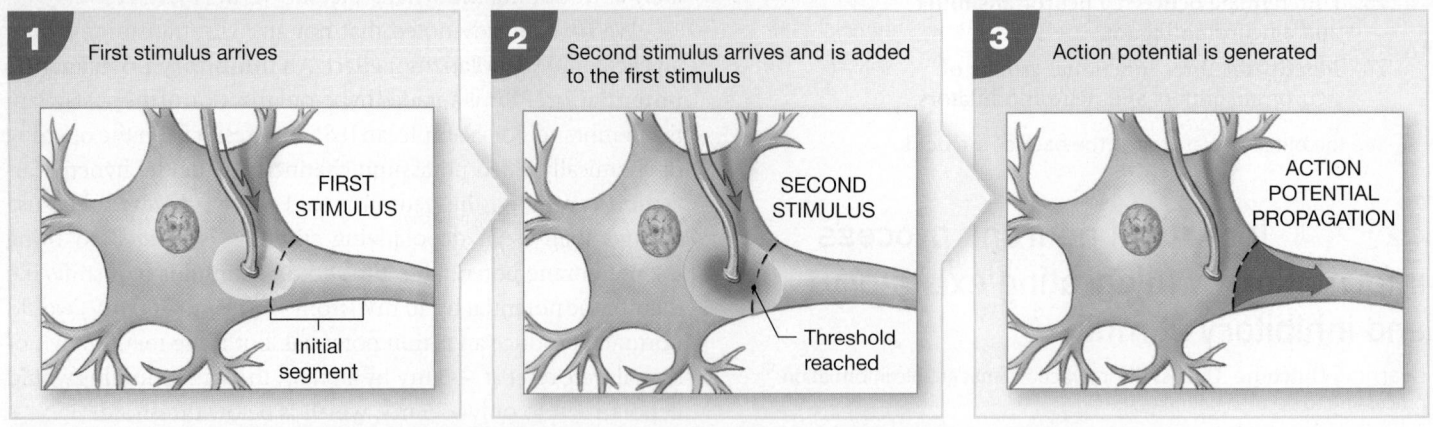

1 First stimulus arrives

FIRST STIMULUS

Initial segment

2 Second stimulus arrives and is added to the first stimulus

SECOND STIMULUS

Threshold reached

3 Action potential is generated

ACTION POTENTIAL PROPAGATION

a **Temporal Summation.** Temporal summation occurs on a membrane that receives two depolarizing stimuli from the same source in rapid succession. The effects of the second stimulus are added to those of the first.

1 Two stimuli arrive simultaneously

TWO SIMULTANEOUS STIMULI

Threshold reached

2 Action potential is generated

ACTION POTENTIAL PROPAGATION

b **Spatial Summation.** Spatial summation occurs when sources of stimulation arrive simultaneously, but at different locations. Local currents spread the depolarizing effects, and areas of overlap experience the combined effects.

In spatial summation, more than one synapse is active at the same time (**Figure 12–18b**), and each "pours" sodium ions across the postsynaptic membrane, producing a graded potential with localized effects. At each active synapse, the sodium ions that produce the EPSP spread out along the inner surface of the membrane and mingle with those entering at other synapses. As a result, the effects on the initial segment are cumulative. The degree of depolarization depends on how many synapses are active at any moment, and on their distance from the initial segment. As in temporal summation, an action potential results when the membrane potential at the initial segment reaches threshold.

Facilitation

Now consider a situation in which summation of EPSPs is under way, but the initial segment has not been depolarized to threshold. The closer the initial segment gets to threshold, the easier it will be for the *next* depolarizing stimulus to trigger an action potential. A neuron whose membrane potential shifts closer to threshold is said to be **facilitated**. The larger the degree of facilitation, the smaller is the additional stimulus needed to trigger an action potential. In a highly facilitated neuron, even a small depolarizing stimulus produces an action potential.

Facilitation can result from the summation of EPSPs or from the exposure of a neuron to certain drugs in the extracellular fluid. For example, the nicotine in cigarettes stimulates postsynaptic ACh receptors, producing prolonged EPSPs that facilitate CNS neurons. Nicotine also increases the release of another neurotransmitter, dopamine, producing feelings of pleasure and reward, thereby leading to addiction.

Summation of EPSPs and IPSPs

Like EPSPs, IPSPs combine spatially and temporally. EPSPs and IPSPs reflect the activation of different types of chemically gated channels, producing opposing effects on the membrane potential. The antagonism between IPSPs and EPSPs is important in cellular information processing. In terms of our bucket

Figure 12–19 **Interactions between EPSPs and IPSPs.** At time 1, a small depolarizing stimulus produces an EPSP. At time 2, a small hyperpolarizing stimulus produces an IPSP of comparable magnitude. If the two stimuli are applied simultaneously, as they are at time 3, summation occurs. Because the two are equal in size but have opposite effects, the membrane potential remains at the resting level. If the EPSP were larger, a net depolarization would result; if the IPSP were larger, a net hyperpolarization would result instead.

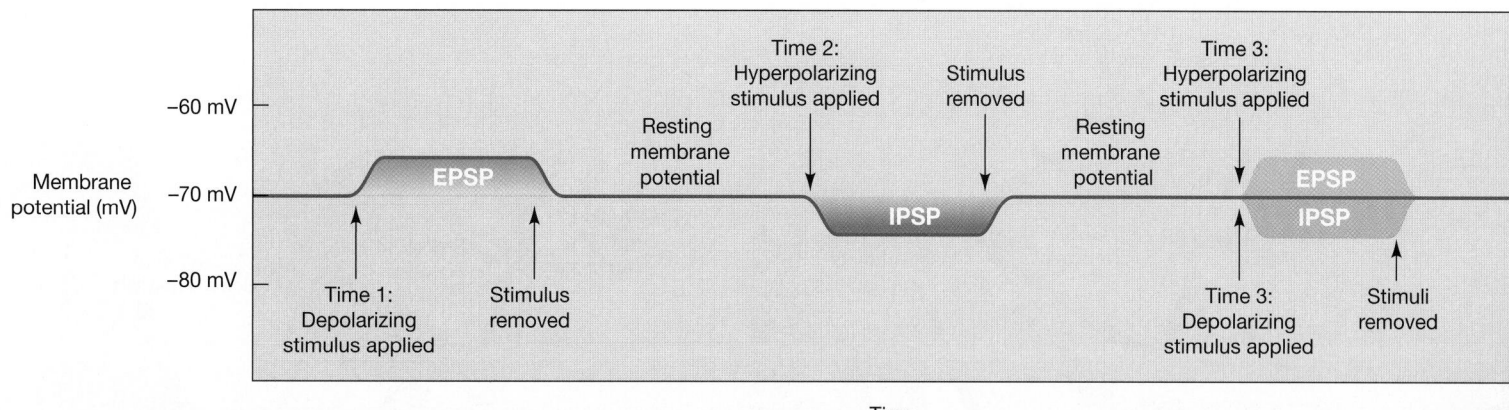

analogy, EPSPs put water into the bathtub, and IPSPs take water out. If more buckets add water than remove water, the water level in the tub rises. If more buckets remove water, the level falls. If a bucket of water is removed every time another bucket is dumped in, the level remains stable. Comparable interactions between EPSPs and IPSPs determine the membrane potential at the boundary between the axon hillock and the initial segment (Figure 12–19).

Neuromodulators, hormones, or both can change the postsynaptic membrane's sensitivity to excitatory or inhibitory neurotransmitters. By shifting the balance between EPSPs and IPSPs, these compounds promote facilitation or inhibition of CNS and PNS neurons.

Presynaptic Inhibition and Presynaptic Facilitation

Inhibitory or excitatory responses may occur not only at synapses involving the cell body and dendrites, but also at synapses found along an axon or its collaterals. At an *axoaxonic* (axon to axon) *synapse,* a synapse occurs between the axons of two neurons. An axoaxonic synapse at the axon terminal can either decrease (inhibit) or increase (facilitate) the rate of neurotransmitter release at the presynaptic membrane. In one form of **presynaptic inhibition**, the release of GABA inhibits the opening of voltage-gated calcium channels in the axon terminal (Figure 12–20a). This inhibition reduces the amount of neurotransmitter released when an action potential arrives there, and thus reduces the effects of synaptic activity on the postsynaptic membrane.

In **presynaptic facilitation**, activity at an axoaxonic synapse increases the amount of neurotransmitter released when

an action potential arrives at the axon terminal (Figure 12–20b). This increase enhances and prolongs the neurotransmitter's effects on the postsynaptic membrane. The neurotransmitter *serotonin* is involved in presynaptic facilitation. In the presence of serotonin released at an axoaxonic synapse, voltage-gated calcium channels remain open longer.

The Rate of Generation of Action Potentials

In the nervous system, complex information is translated into action potentials that are propagated along axons. On arrival, the message is often interpreted solely on the basis of the frequency of action potentials. For example, action potentials arriving at a neuromuscular junction at the rate of 1 per second may produce a series of isolated twitches in the associated skeletal muscle fiber, but at the rate of 100 per second they cause a sustained tetanic contraction. Similarly, you may perceive a few action potentials per second along a sensory fiber as a feather-light touch, but you would perceive hundreds of action potentials per second along that same axon as unbearable pressure. In general, the degree of sensory stimulation or the strength of the motor response is proportional to the frequency of action potentials. In this section, we examine factors that vary the rate of generation of action potentials.

If a graded potential briefly depolarizes the axon hillock such that the initial segment reaches its threshold, an action potential is propagated along the axon. But what happens when the axon hillock *remains* depolarized past threshold for an extended period? The longer the initial segment remains above threshold, the more action potentials it produces. The *frequency* of action potentials depends on the degree of depolarization

Figure 12–20 **Presynaptic Inhibition and Presynaptic Facilitation.**

a **Presynaptic inhibition**

b **Presynaptic facilitation**

above threshold: The greater the degree of depolarization, the higher the frequency of action potentials. The membrane can respond to a second stimulus as soon as the absolute refractory period ends. Holding the membrane above threshold has the same effect as applying a second, larger-than-normal stimulus.

Action potentials can be generated at a maximum rate when the relative refractory period has been completely eliminated. For this reason, the maximum theoretical frequency of action potentials is established by the duration of the absolute refractory period. The absolute refractory period is shortest in large-diameter axons, in which the *theoretical* maximum frequency of action potentials is 2500 per second. However, the highest frequencies recorded from axons in the body range between 500 and 1000 per second.

Table 12–5 summarizes the basic principles of information processing.

Table 12–5	Information Processing

- Information is relayed in the form of action potentials. In general, the degree of sensory stimulation or the strength of the motor response is proportional to the frequency of action potentials.
- The neurotransmitters released at a synapse may have either excitatory or inhibitory effects. The effect on the axon's initial segment reflects a summation of the stimuli that arrive at any moment. The frequency of generation of action potentials is an indication of the degree of sustained depolarization at the axon hillock.
- Neuromodulators can alter either the rate of neurotransmitter release or the response of a postsynaptic neuron to specific neurotransmitters.
- Neurons may be facilitated or inhibited by extracellular chemicals other than neurotransmitters or neuromodulators.
- The response of a postsynaptic neuron to the activation of a presynaptic neuron can be altered by (1) the presence of neuromodulators or other chemicals that cause facilitation or inhibition at the synapse, (2) activity under way at other synapses affecting the postsynaptic cell, and (3) modification of the rate of neurotransmitter release through presynaptic facilitation or presynaptic inhibition.

You are now familiar with the basic components of neural tissue, and the origin and significance of action potentials. In later chapters we consider higher levels of anatomical and functional organization within the nervous system, examine information processing at these levels, and see how a single process—the generation of action potentials—can be responsible for the incredible diversity of sensations and movements that we experience each day.

CHECKPOINT

25. One EPSP depolarizes the initial segment from a resting membrane potential of –70 mV to –65 mV, and threshold is at –60 mV. Will an action potential be generated?

26. Given the situation in Checkpoint 25, if a second, identical EPSP occurs immediately after the first, will an action potential be generated?

27. If the two EPSPs in Checkpoint 26 occurred simultaneously, what form of summation would occur?

See the blue Answers tab at the back of the book.

Related Clinical Terms

anesthetic: An agent that produces a local or general loss of sensation including feelings of pain.

anticholinesterase drug: A drug that blocks the breakdown of ACh by AChE.

atropine: A drug that prevents ACh from binding to the postsynaptic membrane of cardiac muscle cells and smooth muscle cells.

d-tubocurarine: A drug, derived from curare, a South American rainforest vine. It produces paralysis by preventing ACh from binding to the postsynaptic membrane (motor end plate) of skeletal muscle fibers.

dysthymia: A form of clinical depression with a depressed mood for most of the time for at least two years along with at least two of the following signs and symptoms: poor appetite or overeating; insomnia or excessive sleep; low energy or fatigue; low self-esteem; poor concentration or indecisiveness; and hopelessness.

excitotoxicity: Continuous and exaggerated stimulation by a neurotransmitter, especially for the excitatory neurotransmitter

glutamate. The nerve cells can become damaged and killed by the overactivation of receptors.

neuroblastoma: A malignant tumor composed of neuroblasts, most commonly in the adrenal gland; it is the most common cancer in infancy.

neuropathy: Condition that causes tingling, numbness, and/or pain in parts of the body, notably the hands and feet.

neurotoxin: A compound that disrupts normal nervous system function by interfering with the generation or propagation of action potentials. Examples include *tetrodotoxin (TTX), saxitoxin (STX),* and *ciguatoxin (CTX).*

Tay–Sachs disease: A genetic abnormality involving the metabolism of gangliosides, important lipid components of neuron plasma membranes. The result is a gradual deterioration of neurons due to the buildup of metabolic by-products and the release of lysosomal enzymes.

12

Chapter Review

Study Outline

MasteringA&P®

Access more chapter study tools online in the MasteringA&P Study Area:

- Chapter Quizzes, Chapter Practice Test, Art-labeling Activities, Animations, MP3 Tutor Sessions, and Clinical Case Studies
- Practice Anatomy Lab   PAL™
- Interactive Physiology   iP®
- A&P Flix   A&PFlix
- PhysioEx   PhysioEx

An Introduction to Neural Tissue p. 412

1. The nervous system includes all the neural tissue in the body. The basic functional unit is the **neuron**.

12-1 ▌ The nervous system has anatomical and functional divisions p. 412

2. The anatomical divisions of the nervous system are the **central nervous system (CNS)** (the brain and spinal cord) and the **peripheral nervous system (PNS)** (all the neural tissue outside the CNS). Bundles of **axons** *(nerve fibers)* in the PNS are called **nerves**. *(Figure 12–1)*

3. Functionally, the PNS can be divided into an **afferent division**, which brings sensory information from **receptors** to the CNS, and an **efferent division**, which carries motor commands to muscles and glands called **effectors**. *(Figure 12–1)*

4. The efferent division of the PNS includes the **somatic nervous system (SNS)**, which controls skeletal muscle contractions, and the **autonomic nervous system (ANS)**, which controls smooth muscle, cardiac muscle, glandular activity, and adipose tissue. *(Figure 12–1)*

12-2 ▌ Neurons are nerve cells specialized for intercellular communication p. 414

5. The **perikaryon** of a multipolar neuron surrounds the nucleus and contains organelles, including **neurofilaments**, **neurotubules**, and **neurofibrils**. The **axon hillock** connects the **initial segment** of the **axon** to the **cell body**, or *soma*. The **axoplasm** contains numerous organelles. *(Figure 12–2)*

6. **Collaterals** may branch from an axon, with **telodendria** branching from the axon's tip.

7. A **synapse** is a site of intercellular communication. Telodendria end in **axon terminals** (also known as synaptic terminals, synaptic knobs, and synaptic boutons). **Neurotransmitters** released from the axon terminals of the presynaptic cell affect the postsynaptic cell, which may be a neuron or another type of cell. *(Figures 12–2, 12–3)*

8. Neurons are structurally classified as **anaxonic**, **bipolar**, **unipolar**, or **multipolar**. *(Figure 12–4)*

9. The three functional categories of neurons are sensory neurons, motor neurons, and interneurons.

10. **Sensory neurons**, which form the afferent division of the PNS, deliver information received from **interoceptors**, **exteroceptors**, and **proprioceptors** to the CNS.

11. **Motor neurons**, which form the efferent division of the PNS, stimulate or modify the activity of a peripheral tissue, organ, or organ system.

12. **Interneurons** *(association neurons)* are always located in the CNS and may be situated between sensory and motor neurons. They distribute sensory inputs and coordinate motor outputs.

12-3 ▌ CNS and PNS neuroglia support and protect neurons p. 418

13. The four types of **neuroglia**, or *glial cells,* in the CNS are (1) **ependymal cells**, with functions related to the **cerebrospinal fluid (CSF);** (2) **astrocytes**, the largest and most numerous neuroglia; (3) **oligodendrocytes**, which are responsible for the **myelination** of CNS axons; and (4) **microglia**, or phagocytic cells. *(Figures 12–5, 12–6b)*

14. Neuron cell bodies in the PNS are clustered into **ganglia**.

15. **Satellite cells** surround neuron cell bodies within ganglia. **Schwann cells** ensheath axons in the PNS. A single Schwann cell may myelinate one segment of an axon or enclose segments of several unmyelinated axons. *(Figure 12–7)*

16. In the PNS, functional repair of axons may follow **Wallerian degeneration**. In the CNS, many factors complicate the repair process and reduce the chances of functional recovery. *(Figure 12–8)*

12-4 ▌ The membrane potential is the electrical potential of the cell's interior relative to its surroundings p. 424

17. All normal neural signaling depends on events that occur at the plasma membrane. *(Spotlight Figure 12–9)*

18. The **electrochemical gradient** is the sum of all chemical and electrical forces acting across the plasma membrane. *(Figures 12–10, 12–11)*

19. The sodium–potassium exchange pump stabilizes the resting membrane potential at approximately –70 mV. *(Table 12–1)*

20. The plasma membrane contains **passive (leak) channels**, which are always open, and **active (gated) channels**, which open or close in response to specific stimuli. *(Figure 12–10)*

21. The three types of gated channels are **chemically gated channels (ligand-gated channels)**, **voltage-gated channels**, and **mechanically gated channels**. *(Figure 12–11)*

22. A localized **depolarization** or **hyperpolarization** is a **graded potential** (a change in potential that decreases with distance). *(Figures 12–12, 12–13; Table 12–2)*

12-5 ▌ An action potential is an electrical event p. 432

23. An **action potential** arises when a region of excitable membrane depolarizes to its **threshold**. The steps involved, in order, are membrane depolarization to threshold, activation of voltage-gated sodium channels and rapid depolarization, inactivation of voltage-gated sodium channels and activation of voltage-gated potassium channels, and the return to normal permeability. *(Spotlight Figure 12–14; Table 12–3)*

24. The generation of an action potential follows the **all-or-none principle**. The **refractory period** lasts from the time an action potential begins until the normal resting membrane potential has returned. *(Spotlight Figure 12–14; Table 12–3)*

25. In **continuous propagation**, an action potential spreads across the entire excitable membrane surface in a series of small steps. *(Spotlight Figure 12–15)*

26. In **saltatory propagation**, an action potential appears to leap from node to node, skipping the intervening myelinated membrane surface. Saltatory propagation carries nerve impulses many times more rapidly than does continuous propagation. *(Spotlight Figure 12–15)*

12-6 ▌ Axon diameter, in addition to myelin, affects propagation speed p. 438

27. Axons are classified as **Type A fibers**, **Type B fibers**, or **Type C fibers** on the basis of their diameter, myelination, and propagation speed.

12-7 ▌ At synapses, communication occurs among neurons or between neurons and other cells p. 439

28. An action potential traveling along an axon is a **nerve impulse**. At a synapse between two neurons, information passes from the **presynaptic neuron** to the **postsynaptic neuron**.

29. A synapse is either *electrical* (with direct physical contact between cells) or *chemical* (involving a neurotransmitter).

30. **Electrical synapses** occur in the CNS and PNS, but they are rare. At an electrical synapse, the presynaptic and postsynaptic plasma membranes are bound by interlocking membrane proteins at a gap junction. Pores formed by these proteins permit the passage of local currents, and the two neurons act as if they share a common plasma membrane.

31. **Chemical synapses** are far more common than electrical synapses. **Excitatory neurotransmitters** cause depolarization and promote the generation of action potentials, whereas **inhibitory neurotransmitters** cause hyperpolarization and suppress the generation of action potentials.

32. The effect of a neurotransmitter on the postsynaptic membrane depends on the properties of the receptor, not on the nature of the neurotransmitter.

33. **Cholinergic synapses** release the neurotransmitter **acetylcholine (ACh)**. Communication moves from the presynaptic neuron to the postsynaptic neuron across a synaptic cleft. A **synaptic delay** occurs because calcium influx and the release of the neurotransmitter takes an appreciable length of time. *(Figure 12–16)*

34. Choline released during the breakdown of ACh in the synaptic cleft is reabsorbed and recycled by the axon terminal. If stores of ACh are exhausted, **synaptic fatigue** can occur.

12-8 ▌ Neurotransmitters and neuromodulators have various functions p. 442

35. Adrenergic synapses release **norepinephrine (NE)**, also called *noradrenaline*. Other important neurotransmitters include **dopamine**, **serotonin**, and **gamma aminobutyric acid (GABA)**.

36. Neuromodulators influence the postsynaptic cell's response to neurotransmitters. *(Table 12–4)*

37. Neurotransmitters can have a direct or indirect effect on membrane potential, and others are lipid-soluble gases that diffuse across the plasma membrane to exert their effects inside the cell. *(Figure 12–17)*

12-9 ▌ Individual neurons process information by integrating excitatory and inhibitory stimuli p. 447

38. Excitatory and inhibitory stimuli are integrated through interactions between **postsynaptic potentials**. This interaction is the simplest level of **information processing** in the nervous system.

39. A depolarization caused by a neurotransmitter is an **excitatory postsynaptic potential (EPSP)**. Individual EPSPs can combine through **summation**, which can be either **temporal** (occurring at a single synapse when a second EPSP arrives before the effects of the first have disappeared) or **spatial** (resulting from the cumulative effects of multiple synapses at various locations). *(Figure 12–18)*

40. Hyperpolarization of the postsynaptic membrane is an **inhibitory postsynaptic potential (IPSP)**.

41. The most important determinants of neural activity are EPSP–IPSP interactions. *(Figure 12–19)*

42. In **presynaptic inhibition**, GABA release at an *axoaxonic synapse* inhibits the opening of voltage-gated calcium channels in the axon terminal. This inhibition reduces the amount of neurotransmitter released when an action potential arrives at the axon terminal. *(Figure 12–20a)*

43. In **presynaptic facilitation**, activity at an axoaxonic synapse increases the amount of neurotransmitter released when an action potential arrives at the axon terminal. This increase enhances and prolongs the effects of the neurotransmitter on the postsynaptic membrane. *(Figure 12–20b)*

44. The neurotransmitters released at a synapse have excitatory or inhibitory effects. The effect on the initial segment reflects an integration of the stimuli arriving at any moment. The frequency of generation of action potentials depends on the degree of depolarization above threshold at the axon hillock. *(Table 12–5)*

45. Neuromodulators can alter either the rate of neurotransmitter release or the response of a postsynaptic neuron to specific neurotransmitters. Neurons may be facilitated or inhibited by extracellular chemicals other than neurotransmitters or neuromodulators. *(Table 12–5)*

46. The effect of a presynaptic neuron's activation on a postsynaptic neuron may be altered by other neurons. *(Table 12–5)*

47. The greater the degree of sustained depolarization at the axon hillock, the higher the frequency of generation of action potentials. At a frequency of about 1000 per second, the relative refractory period has been eliminated, and further depolarization will have no effect. *(Table 12–5)*

12

Review Questions

See the blue Answers tab at the back of the book.

LEVEL 1 Reviewing Facts and Terms

1. Label the structures in the following diagram of a neuron.

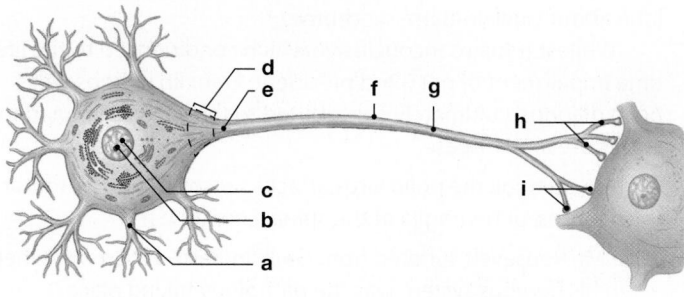

(a) _____	(b) _____
(c) _____	(d) _____
(e) _____	(f) _____
(g) _____	(h) _____
(i) _____	

2. What is the stimulus for the exocytosis of a neurotransmitter into the synaptic cleft? **(a)** an influx of K⁺ ions into the synaptic nob, **(b)** an efflux of K⁺ ions from the postsynaptic neuron, **(c)** an influx of Ca^{2+} ions into the synaptic nob, **(d)** an influx of Ca^{2+} ions into the postsynaptic neuron.

3. A sensory neuron is also known as a/an **(a)** afferent neuron, **(b)** efferent neuron, **(c)** motor neuron, **(d)** visceral neuron.

4. _____ neurons have two or more dendrites and a single axon. **(a)** Bipolar, **(b)** Anaxonic, **(c)** Multipolar, **(d)** Unipolar.

5. The neural cells responsible for the analysis of sensory inputs and coordination of motor outputs are **(a)** neuroglia, **(b)** interneurons, **(c)** sensory neurons, **(d)** motor neurons.

6. Define what is meant by the term depolarization.

7. Myelin increases the speed of electrical conduction during saltatory propagation by **(a)** increasing resistance to flow of ions at the internodes, **(b)** decreasing resistance to flow of ions at the internodes, **(c)** resetting of voltage-gated ion channels, **(d)** resetting of chemically gated ion channels.

8. List the three types of gated channels found in neurons.

9. What are the major components of **(a)** the central nervous system? **(b)** the peripheral nervous system?

10. Which two types of neuroglia insulate neuron cell bodies and axons in the PNS from their surroundings?

11. What three *functional* groups of neurons are found in the nervous system? What is the function of each type of neuron?

LEVEL 2 Reviewing Concepts

12. The typical neuron has a threshold membrane potential of approximately −55mV. How is it exceeded?

13. What happens in the repolarizing phase of an action potential? **(a)** The Na⁺/K⁺ pump returns intracellular and extracellular ion concentrations to prestimulation levels. **(b)** K^+ enters the cell through voltage-gated channels. **(c)** Voltage-gated Na⁺ channels open. **(d)** There is an influx of Ca^{2+} ions in the sarcoplasm.

14. Describe the difference between an action potential and a graded potential.

15. What is the *functional* difference among chemically gated (ligand-gated), voltage-gated, and mechanically gated channels?

16. State the all-or-none principle of action potentials.

17. Describe the steps involved in the generation of an action potential.

18. What is meant by saltatory propagation? How does it differ from continuous propagation?

19. What are the structural and functional differences among type A, B, and C fibers?

20. Describe the events that occur during nerve impulse transmission at a typical cholinergic synapse.

21. What is the difference between temporal summation and spatial summation?

LEVEL 3 Critical Thinking and Clinical Applications

22. Harry has a kidney condition that causes changes in his body's electrolyte levels (concentration of ions in the extracellular fluid). As a result, he is exhibiting tachycardia, an abnormally fast heart rate. Which ion is involved, and how does a change in its concentration cause Harry's symptoms?

23. Twenty neurons synapse with a single receptor neuron. Fifteen of the 20 neurons release neurotransmitters that produce EPSPs at the post-synaptic membrane, and the other five release neurotransmitters that produce IPSPs. Each time one of the neurons is stimulated, it releases enough neurotransmitter to produce a 2-mV change in potential at the postsynaptic membrane. If the threshold of the postsynaptic neuron is 10 mV, how many of the excitatory neurons must be stimulated to produce an action potential in the receptor neuron if all five inhibitory neurons are stimulated? (Assume that spatial summation occurs.)

24. During a regular check-up with the dentist, Beth discovers that she has a cavity which needs to be filled. The dentist injects the anaesthetic lidocaine into her gums. This blocks voltage-gated sodium channels. Why would this prevent pain?

25. What factor determines the maximum frequency of action potentials that could be conducted by an axon?

CLINICAL CASE Wrap-Up

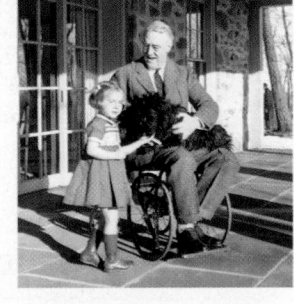

Did Franklin D. Roosevelt Really Have Polio?

Poliomyelitis, also known as infantile paralysis, is caused by a virus that spreads easily in human populations. Most people with polio infections show no symptoms of disease. However, in a small percentage of victims, the virus enters the central nervous system, attacks the motor neurons of the spinal cord, and produces a flaccid paralysis (characterized by weakness and loss of muscle tone).

Certain details about Mr. Roosevelt's case call into question whether he really suffered from polio. Polio is a disease primarily of children. By age 30, most adults have natural immunity, a resistance to infection developed after a previous exposure. In addition, paralysis caused by polio is usually asymmetric, does not appear in an ascending pattern, and progresses for only three to five days—not for nearly a month, as in Mr. Roosevelt's case. Polio almost never involves facial muscles or disturbs urinary bladder and bowel functions. Polio also does not cause pain upon light touch, other pain, or numbness—only motor weakness. Mr. Roosevelt recovered from his paralysis in a descending pattern, opposite to its onset.

The pathological features Mr. Roosevelt experienced are much more characteristic of Guillain-Barré syndrome, an autoimmune disease that involves the demyelination (destruction of the axonal myelin sheath) of peripheral nerves. It causes symmetrical weakness, numbness, and abnormal sensations. In the first half of the 20th century, doctors had extensive experience with polio, but knew little about Guillain-Barré syndrome.

While it remains inconclusive which condition led to the lifetime impairment of our 32nd president, Franklin D. Roosevelt's polio diagnosis ultimately contributed to the virtual elimination of this dreaded disease.

1. Do you think the polio virus attacks sensory neurons, motor neurons, or neuroglia of the spinal cord (CNS)?

2. If Mr. Roosevelt suffered from Guillain-Barré syndrome, where in his nervous system was the pathology taking place?

See the blue Answers tab at the back of the book.

13

The Spinal Cord, Spinal Nerves, and Spinal Reflexes

Learning Outcomes

These Learning Outcomes correspond by number to this chapter's sections and indicate what you should be able to do after completing the chapter.

13-1 ▌ Describe the basic structural and organizational characteristics of the nervous system.

13-2 ▌ Discuss the structure and functions of the spinal cord, and describe the three meningeal layers that surround the central nervous system.

13-3 ▌ Explain the roles of white matter and gray matter in processing and relaying sensory information and motor commands.

13-4 ▌ Describe the major components of a spinal nerve, and relate the distribution pattern of spinal nerves to the regions they innervate.

13-5 ▌ Discuss the significance of neuronal pools, and describe the major patterns of interaction among neurons within and among these pools.

13-6 ▌ Describe the steps in a neural reflex, and classify the types of reflexes.

13-7 ▌ Distinguish among the types of motor responses produced by various reflexes, and explain how reflexes interact to produce complex behaviors.

13-8 ▌ Explain how higher centers control and modify reflex responses.

SPOTLIGHTS

- Peripheral Distribution of Spinal Nerves pp. 466–467
- Spinal Reflexes pp. 476–477

CLINICAL CASE

- Prom Night p. 456

Clinical Notes

- Anesthesia p. 461
- Shingles p. 465
- Sensory Innervation in the Hand p. 470
- Sensory Innervation in the Ankle and Foot p. 473

Prom Night

Joe is still buzzing from champagne as he kisses his girlfriend goodnight and gets into his car to head home. It's been a fun prom. The food was great, the band was awesome, and the afterparty at his friend's house had gone until 3 a.m. Joe turns the car stereo to full blast, leans back, and notices the moon through the trees. That is the last thing he remembers.

The emergency medical technicians (EMTs) at the motor vehicle accident scene note that Joe's car had left the road at a high rate of speed and rolled over at least three times. Joe is still unconscious when he is extricated from the car. He is breathing on his own but has obvious abrasions and bruising on his back. The EMTs splint his entire axial skeleton, using a cervical collar and sandbags on a rigid backboard. They start an IV in the field

and transport Joe to the closest Level I trauma center.

In the emergency department Joe can hear shouting in his ear as he is moved onto a cart. "Wiggle your toes," someone commands, but Joe can move nothing below his waist. "Can you feel this?" a voice asks. Joe can feel someone poking him with a pin from his chest down to his abdomen, just inferior to his navel, but nothing below that. "Can you feel me put this catheter in?" a nurse asks as she inserts a small tube into Joe's bladder. Joe cannot feel the catheter. **What has happened to Joe? To find out, turn to the Clinical Case Wrap-Up on p. 486.**

An Introduction to the Spinal Cord, Spinal Nerves, and Spinal Reflexes

In this chapter we discuss the functional anatomy and organization of the spinal cord and spinal nerves. We also describe simple spinal reflexes.

Organization is usually the key to success in any complex environment. A large corporation, for example, has both a system to distribute messages on specific topics and executive assistants who decide whether an issue can be ignored or easily responded to. Only the most complex and important problems reach the desk of the president.

The nervous system works in much the same way. It has input pathways that route sensations, and processing centers that prioritize and distribute information. There are also several levels that issue motor responses. Your conscious mind (the president) gets involved only in a fraction of the day-to-day activities. The other decisions are handled at lower levels that operate outside your awareness. This very efficient system works only because it is so highly organized.

13-1 ▌ The brain and spinal cord make up the central nervous system (CNS), and the cranial nerves and spinal nerves make up the peripheral nervous system (PNS)

Learning Outcome Describe the basic structural and organizational characteristics of the nervous system.

The nervous system has so many components and does so much that we will cover it in four chapters. If our primary interest were the anatomy of this system, we would probably start with an examination of the central nervous system (brain and spinal cord) and then

move on to the peripheral nervous system (cranial nerves and spinal nerves). However, our primary interest is how the nervous system *functions*, so we will consider the system from a functional perspective. Our basic approach is diagrammed in **Figure 13–1**.

Note that Chapter 12 provided the foundation by considering the function of individual neurons. In Chapter 13, we consider the spinal cord and spinal nerves, and the basic wiring of simple *spinal reflexes*. Reflexes are quick, automatic responses triggered by specific stimuli. Spinal reflexes are controlled in the spinal cord. They can function without any input from the brain. For example, a reflex controlled in the spinal cord makes you drop a frying pan you didn't realize was sizzling hot. Before the information reaches your brain and you become aware of the pain, you've already released the pan. There are much more complex spinal reflexes, but this functional pattern still applies.

Your spinal cord is structurally and functionally integrated with your brain. In Chapter 14 we give an overview of the major components and functions of the brain and cranial nerves. We also discuss the *cranial reflexes*—localized reflex responses comparable in organization and complexity to those of the spinal cord.

In Chapters 15 and 16 we consider the nervous system as an integrated functional unit. In Chapter 15 we deal with the interplay between centers in the brain and spinal cord that takes place in the processing of sensory information. We then examine the conscious and subconscious control of skeletal muscles by the *somatic nervous system* (SNS).

In Chapter 16 we continue with a discussion of the control of visceral functions by the *autonomic nervous system* (ANS). The ANS has processing centers in the brain, spinal cord, and peripheral nervous system. It is responsible for the control of visceral effectors, such as smooth muscles, cardiac muscle, glands, and adipose tissue. We then conclude this section of the book by examining what are often called *higher-order functions*: memory, learning, consciousness, and personality. These fascinating topics are difficult to investigate,

Figure 13–1   **An Overview of Chapters 13 and 14.**

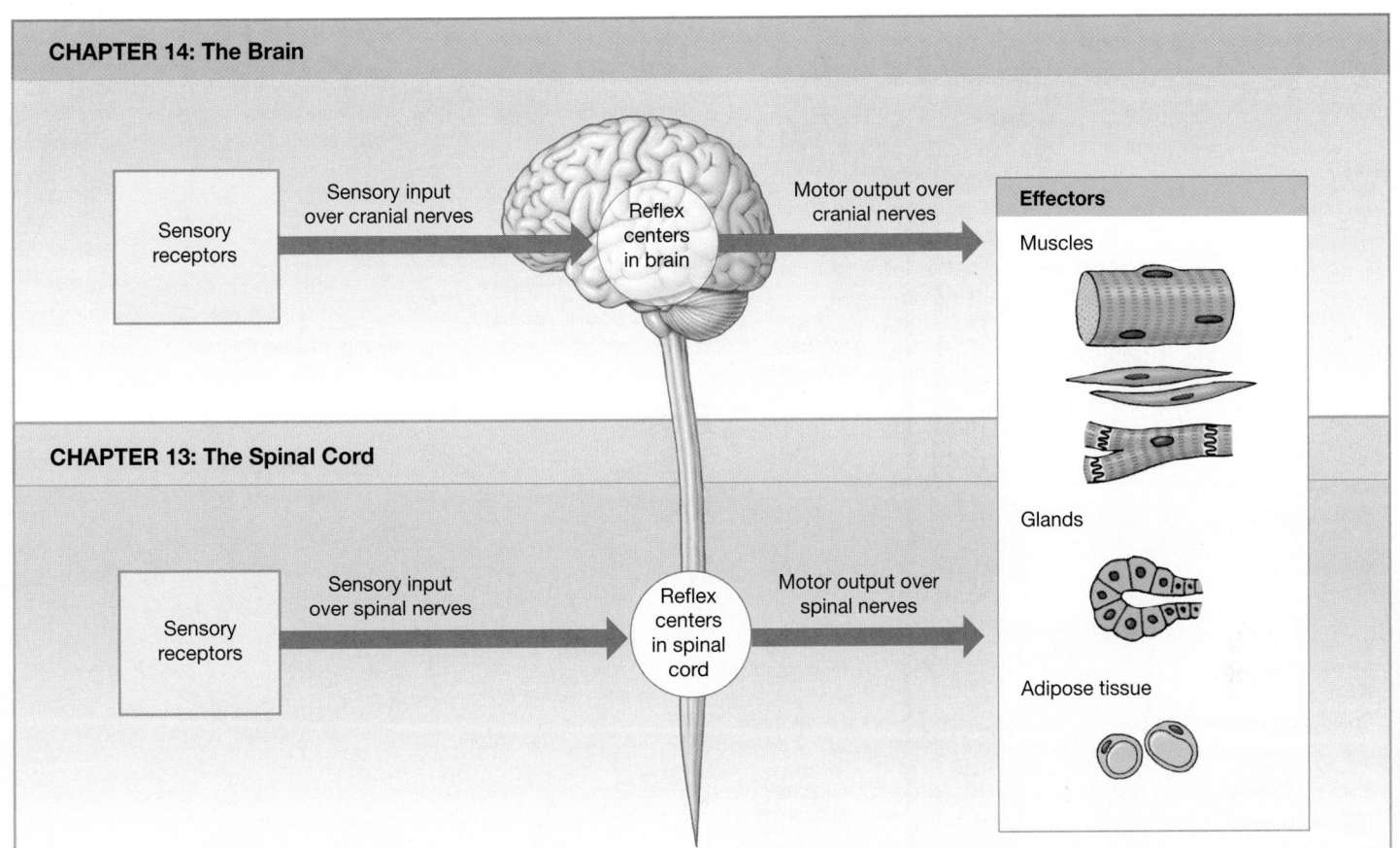

but they can affect activity along the sensory and motor pathways and alter our perception of those activities.

With these basic principles, definitions, and strategies in mind, let's begin our examination of the levels of functional organization in the nervous system.

CHECKPOINT

1. **Name the components of the central nervous system (CNS) and of the peripheral nervous system (PNS).**

2. **Define spinal reflex.**

See the blue Answers tab at the back of the book.

13-2 ▌ The spinal cord is surrounded by three meninges and carries sensory and motor information

Learning Outcome Discuss the structure and functions of the spinal cord, and describe the three meningeal layers that surround the central nervous system.

We begin this section with the gross anatomy of the spinal cord. Then we examine the three membranous layers that surround the spinal cord: the spinal meninges.

Gross Anatomy of the Spinal Cord

The adult spinal cord (Figure 13–2a) is approximately 45 cm (18 in.) long, with a maximum width of roughly 14 mm (0.55 in.). Note that the cord itself is not as long as the vertebral column. Instead, the adult spinal cord ends between vertebrae L_1 and L_2. The posterior (dorsal) surface of the spinal cord has a shallow longitudinal groove, the **posterior median sulcus** (Figure 13–2b). The **anterior median fissure** is a deeper groove along the anterior (ventral) surface.

The amount of gray matter is greatest in segments of the spinal cord dedicated to the sensory and motor control of the limbs. These segments are expanded, forming the **enlargements** of the spinal cord. The **cervical enlargement** supplies nerves to the shoulder and upper limbs. The **lumbar enlargement** provides innervation to structures of the pelvis and lower limbs.

Inferior to the lumbar enlargement, the spinal cord becomes tapered and conical. This region is the **conus medullaris**. The **filum terminale** ("terminal thread"), a slender strand of fibrous tissue, extends from the inferior tip of the conus medullaris. It continues along the length of the vertebral canal as far as the second sacral vertebra, where it provides longitudinal support to the spinal cord as a component of the *coccygeal ligament*.

Figure 13–2   **Gross Anatomy of the Adult Spinal Cord.**   ATLAS: Plates 2a; 20a,b; 24a–c

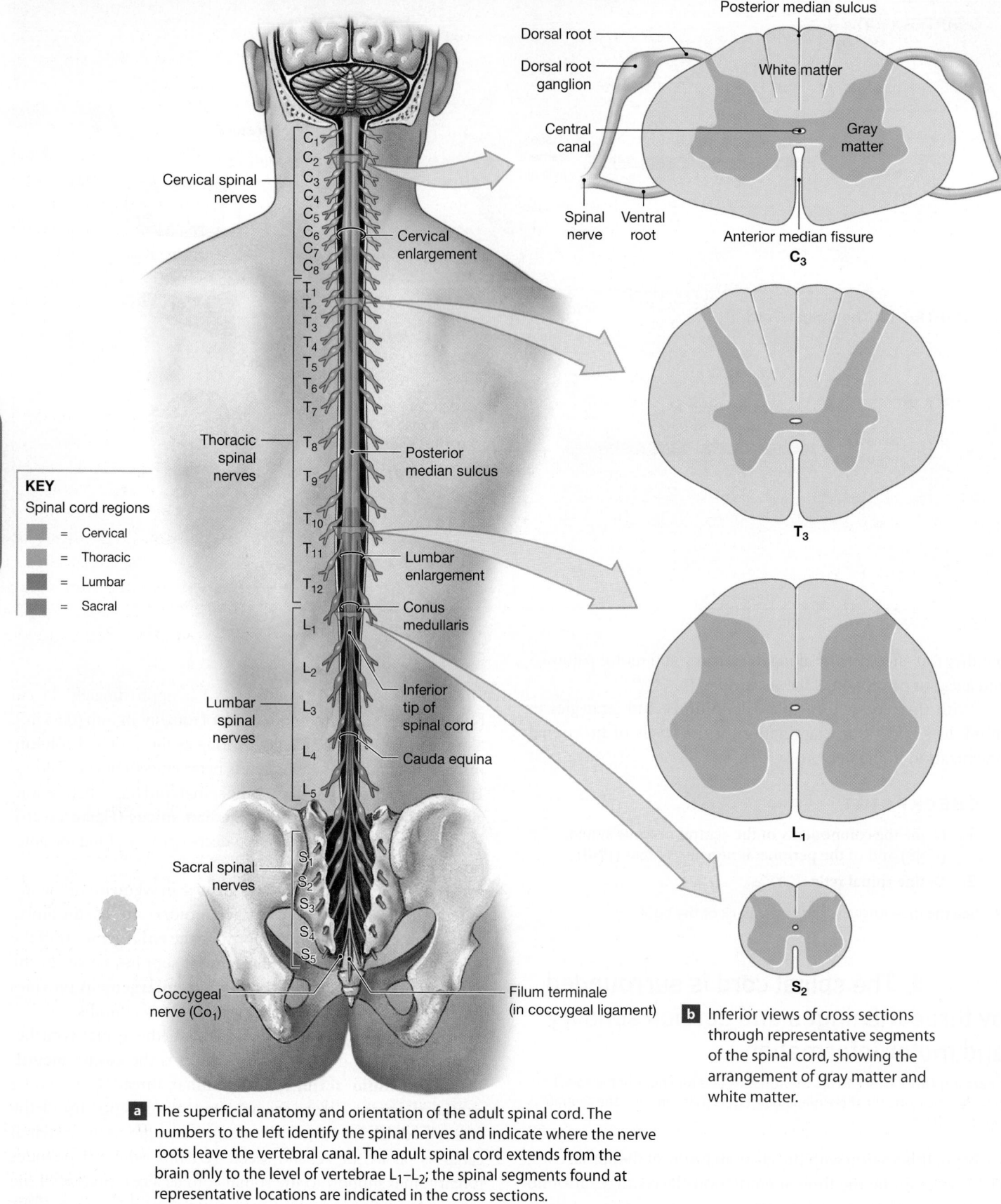

KEY

Spinal cord regions

■ =	Cervical
■ =	Thoracic
■ =	Lumbar
■ =	Sacral

a The superficial anatomy and orientation of the adult spinal cord. The numbers to the left identify the spinal nerves and indicate where the nerve roots leave the vertebral canal. The adult spinal cord extends from the brain only to the level of vertebrae L_1–L_2; the spinal segments found at representative locations are indicated in the cross sections.

b Inferior views of cross sections through representative segments of the spinal cord, showing the arrangement of gray matter and white matter.

The series of sectional views in **Figure 13–2b** illustrate the variations in the proportions of gray matter and white matter in the cervical, thoracic, lumbar, and sacral segments of the spinal cord. The entire spinal cord can be divided into 31 segments on the basis of the origins of the spinal nerves. A letter and number designation, the same method used to identify vertebrae, identify each segment. For example, C_3, the segment in the uppermost section of **Figure 13–2b**, is the third cervical segment.

Every spinal segment is associated with a pair of **dorsal root ganglia**, located near the spinal cord (**Figure 13–2b**). These ganglia contain the cell bodies of sensory neurons. The axons of the neurons form the **dorsal roots**, which bring sensory information into the spinal cord. A pair of **ventral roots** contains the axons of motor neurons that extend into the periphery to control somatic and visceral effectors.

On both sides, the dorsal and ventral roots of each segment pass between the vertebral canal and the periphery at the *intervertebral foramen* between successive vertebrae. The dorsal root ganglion lies between the pedicles of the adjacent vertebrae. (You can review vertebral anatomy in Chapter 7. ⟳ pp. 253–254)

Distal to each dorsal root ganglion, the sensory and motor roots are bound together into a single **spinal nerve**. Spinal nerves are classified as **mixed nerves**—that is, they contain both afferent (sensory) and efferent (motor) fibers. There are 31 pairs of spinal nerves, each identified by its association with adjacent vertebrae. For example, we may speak of "cervical spinal nerves" or even "cervical nerves" when we make a general reference to spinal nerves of the neck. However, when we indicate specific spinal nerves, it is customary to give them a regional number, as indicated in **Figure 13–2**. Each spinal nerve inferior to the first thoracic vertebra takes its name from the vertebra immediately superior to it. For example, spinal nerve T_1 emerges immediately inferior to vertebra T_1, spinal nerve T_2 follows vertebra T_2, and so forth.

The arrangement differs in the cervical region. There the first pair of spinal nerves, C_1, passes between the skull and the first cervical vertebra. For this reason, each cervical nerve takes its name from the vertebra immediately inferior to it. In other words, cervical nerve C_2 *precedes* vertebra C_2, and the same system is used for the rest of the cervical series. The transition from one numbering system to another occurs between the last cervical vertebra and first thoracic vertebra. The spinal nerve found at this location has been designated C_8. Therefore, although there are only seven cervical vertebrae, there are *eight* cervical nerves.

The spinal cord continues to enlarge and elongate until a person is approximately 4 years old. Up to that time, enlargement of the spinal cord keeps pace with the growth of the vertebral column. Throughout this period, the ventral and dorsal roots are very short, and they enter the intervertebral foramina immediately adjacent to their spinal segment. After age 4, the vertebral column continues to elongate, but the spinal cord does not. This vertebral growth moves the intervertebral foramina, and thus the spinal nerves, farther and farther from their original positions relative to the spinal cord. As a result, the dorsal and ventral roots gradually elongate, and the correspondence between the spinal segment and the vertebral segment is lost. For example, in adults, the sacral segments of the spinal cord are at the level of vertebrae L_1–L_2 (**Figure 13–2**).

The adult spinal cord extends only to the level of the first or second lumbar vertebra, but the dorsal and ventral roots of spinal segments L_2 to S_5 extend inferiorly, past the inferior tip of the conus medullaris. When seen in gross dissection, the filum terminale and the long ventral and dorsal roots resemble a horse's tail. For this reason, early anatomists called this complex the **cauda equina** (KAW-duh ek-WĪ-nuh; *cauda*, tail + *equus*, horse).

Spinal Meninges

The vertebral column and its surrounding ligaments, tendons, and muscles isolate the spinal cord from the rest of the body. These structures also protect against bumps and shocks to the skin of the back. The delicate neural tissues must also be protected from damaging contacts with the surrounding bony walls of the vertebral canal. The **spinal meninges** (me-NIN-jēz; singular, *meninx*, membrane), a series of specialized membranes surrounding the spinal cord, provide the necessary physical stability and shock absorption. Blood vessels branching within these layers deliver oxygen and nutrients to the spinal cord.

The relationships among the spinal meninges are shown in **Figure 13–3a**. The spinal meninges consist of three layers: (1) the *dura mater*, (2) the *arachnoid mater*, and (3) the *pia mater*. At the foramen magnum of the skull, the spinal meninges are continuous with the **cranial meninges**, which surround the brain. (The cranial meninges have the same three layers. We discuss them in Chapter 14.)

Bacterial or viral infection can cause **meningitis**, or inflammation of the meningeal membranes. Meningitis is dangerous because it can disrupt the normal circulation of cerebrospinal fluid (CSF), damaging or killing neurons and neuroglia in the affected areas. An initial diagnosis may specify the meninges of the spinal cord *(spinal meningitis)* or brain *(cerebral meningitis)*, but in later stages the entire meningeal system is usually affected.

The Dura Mater

The tough, fibrous **dura mater** (DŪ-ruh MĀ-ter; *durus*, hard + *mater*, mother) is the layer that forms the outermost covering of the spinal cord (**Figure 13–3a**). This layer contains dense collagen fibers that are oriented along the longitudinal axis of the cord. Between the dura mater and the walls of the

Figure 13–3 **The Spinal Cord and Spinal Meninges.**

White matter
Ventral rootlets
of spinal nerve
Ventral root
Dorsal root
Dorsal rootlets
of spinal nerve

Gray matter
Dorsal root
ganglion
Spinal nerve

Meninges
Pia mater
Arachnoid mater
Dura mater

a A posterior view of the spinal cord, showing the meningeal layers, superficial landmarks, and distribution of gray matter and white matter

ANTERIOR

Subarachnoid space
Rami communicantes
Spinal cord
Adipose tissue in epidural space

Vertebral body

Meninges
Dura mater
Arachnoid mater
Pia mater

Autonomic (sympathetic) ganglion
Ventral root of spinal nerve
Ventral ramus
Dorsal ramus

Denticulate ligament
Dorsal root ganglion

POSTERIOR

b A sectional view through the spinal cord and meninges, showing the relationship of the meninges, spinal cord, and spinal nerves

vertebral canal lies the **epidural space**, a region that contains areolar tissue, blood vessels, and a protective padding of adipose tissue (**Figure 13–3b**).

The spinal dura mater does not have extensive, firm connections to the surrounding vertebrae. Attachment sites at either end of the vertebral canal provide longitudinal stability. Cranially, the outer layer of the spinal dura mater fuses with the periosteum of the occipital bone around the margins of the foramen magnum. There, the spinal dura mater becomes continuous with the cranial dura mater. Within the sacral canal, the spinal dura mater tapers from a sheath to a dense cord of collagen fibers that blends with components of the filum terminale to form the **coccygeal ligament** (**Figure 13–2a**). The coccygeal ligament continues along the sacral canal, ultimately blending into the periosteum of the coccyx. Areolar connective tissue and adipose tissue within the epidural space support the spinal dura mater. In addition, this dura mater extends between adjacent vertebrae at each intervertebral foramen, fusing with the connective tissues that surround the spinal nerves.

The Arachnoid Mater

In most anatomical and tissue specimens, a narrow **subdural space** separates the dura mater from deeper meningeal layers. In a healthy living person no such space exists. Instead, the inner surface of the dura mater is in contact with the outer surface of the **arachnoid** (a-RAK-noyd; *arachne*, spider) **mater**, the middle meningeal layer (**Figure 13–3b**). The inner surface of the dura mater and the outer surface of the arachnoid mater are covered by simple squamous epithelia. The arachnoid mater includes this epithelium, called the *arachnoid membrane*, and the *arachnoid trabeculae*, a delicate network of collagen and elastic fibers that extends between the arachnoid membrane and the outer surface of the pia mater. The region between is called the **subarachnoid space**. It is filled with **cerebrospinal fluid (CSF)**, which acts as a shock absorber and a diffusion medium for dissolved gases, nutrients, chemical messengers, and wastes.

The spinal arachnoid mater extends inferiorly as far as the filum terminale, and the dorsal and ventral roots of the cauda equina lie within the fluid-filled subarachnoid space. In adults, the withdrawal of cerebrospinal fluid involves the insertion of a needle into the subarachnoid space in the inferior lumbar region. This procedure is known as a **lumbar puncture** or **spinal tap**.

The Pia Mater

The subarachnoid space extends between the arachnoid epithelium and the innermost meningeal layer, the **pia mater** (*pia*, delicate + *mater*, mother). The pia mater consists of a meshwork of elastic and collagen fibers that is firmly bound to the underlying neural tissue (**Figure 13–3**). These connective-tissue fibers are extensively interwoven with those that span the subarachnoid space, firmly binding the arachnoid to the pia mater. The blood vessels servicing the spinal cord run along the surface of the spinal pia mater, within the subarachnoid space (**Figure 13–4**).

Figure 13–4   The Spinal Cord and Associated Structures. An anterior view of the cervical spinal cord and spinal nerve roots in the vertebral canal. The dura mater and arachnoid mater have been cut and reflected. Notice the blood vessels that run in the subarachnoid space, bound to the outer surface of the delicate pia mater.

Spinal cord

— Anterior median fissure

— Pia mater

— Denticulate ligaments

— Dorsal root

— Ventral root, formed by several "rootlets" from one cervical segment

— Arachnoid mater (reflected)

— Dura mater (reflected)

— Spinal blood vessel

Along the length of the spinal cord, paired **denticulate ligaments** extend from the pia mater through the arachnoid mater to the dura mater (**Figures 13–3b** and **13–4**). Denticulate ligaments, which originate along either side of the spinal cord, prevent lateral (side-to-side) movement. The dural connections at the foramen magnum and the coccygeal ligament prevent longitudinal (superior–inferior) movement.

The spinal meninges accompany the dorsal and ventral roots as these roots pass through the intervertebral foramina. As the sectional view in **Figure 13–3b** shows, the meningeal membranes are continuous with the connective tissues that surround the spinal nerves and their peripheral branches.

CHECKPOINT

3. Identify the three spinal meninges.

4. Damage to which root of a spinal nerve would interfere with motor function?

5. Where is the cerebrospinal fluid that surrounds the spinal cord located?

See the blue Answers tab at the back of the book.

13

13-3 ▮ Gray matter integrates information and initiates commands, and white matter carries information from place to place

Learning Outcome Explain the roles of white matter and gray matter in processing and relaying sensory information and motor commands.

We can relate the functional organization of the spinal cord to its sectional organization (**Figure 13–5**). Together, the anterior median fissure and the posterior median sulcus divide the

---

🩺 Clinical Note

Anesthesia Anesthetics are often injected into the epidural space. Introduced in this way, a drug should affect only the spinal nerves in the immediate area of the injection. The result is a regional pain reliever (analgesic) called an *epidural anesthesia*, which blocks pain sensations. Some sources refer to this as an *epidural block*.

This procedure can be done either peripherally, as when skin lacerations are sewn up, or at sites around the spinal cord for more widespread anesthetic effects, as during labor and delivery. An epidural anesthetic has at least two advantages: (1) It affects only the spinal nerves in the immediate area of the injection; and (2) it provides mainly sensory anesthesia. If a catheter is left in place, continued injection brings about sustained anesthesia.

Local anesthetics can also be introduced as a single dose into the subarachnoid space of the spinal cord. This procedure is commonly called **spinal anesthesia**. The effects include both temporary muscle paralysis and sensory loss. These effects tend to spread as the movement of cerebrospinal fluid distributes the anesthetic along the spinal cord. Problems with overdosing are seldom serious, because controlling the patient's position during administration can limit the distribution of the drug to some degree. Respiration continues even if all thoracic and abdominal segments have been paralyzed. This is possible because the upper cervical spinal nerves control the diaphragm.

Figure 13–5 **The Sectional Organization of the Spinal Cord.**

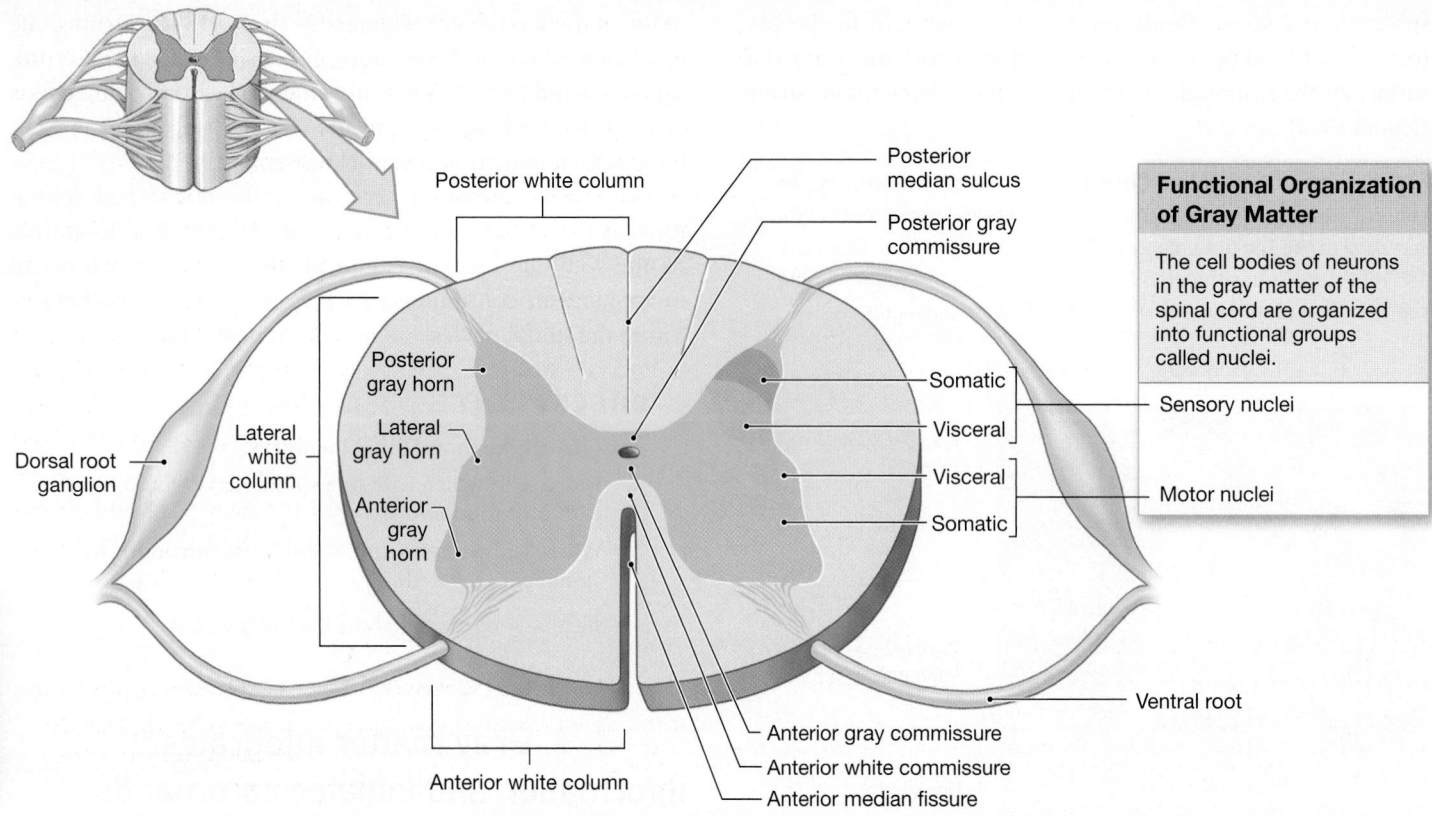

Functional Organization of Gray Matter

The cell bodies of neurons in the gray matter of the spinal cord are organized into functional groups called nuclei.

Sensory nuclei

Motor nuclei

a The left half of this sectional view shows important anatomical landmarks, including the three columns of white matter. The right half indicates the functional organization of the nuclei in the anterior, lateral, and posterior gray horns.

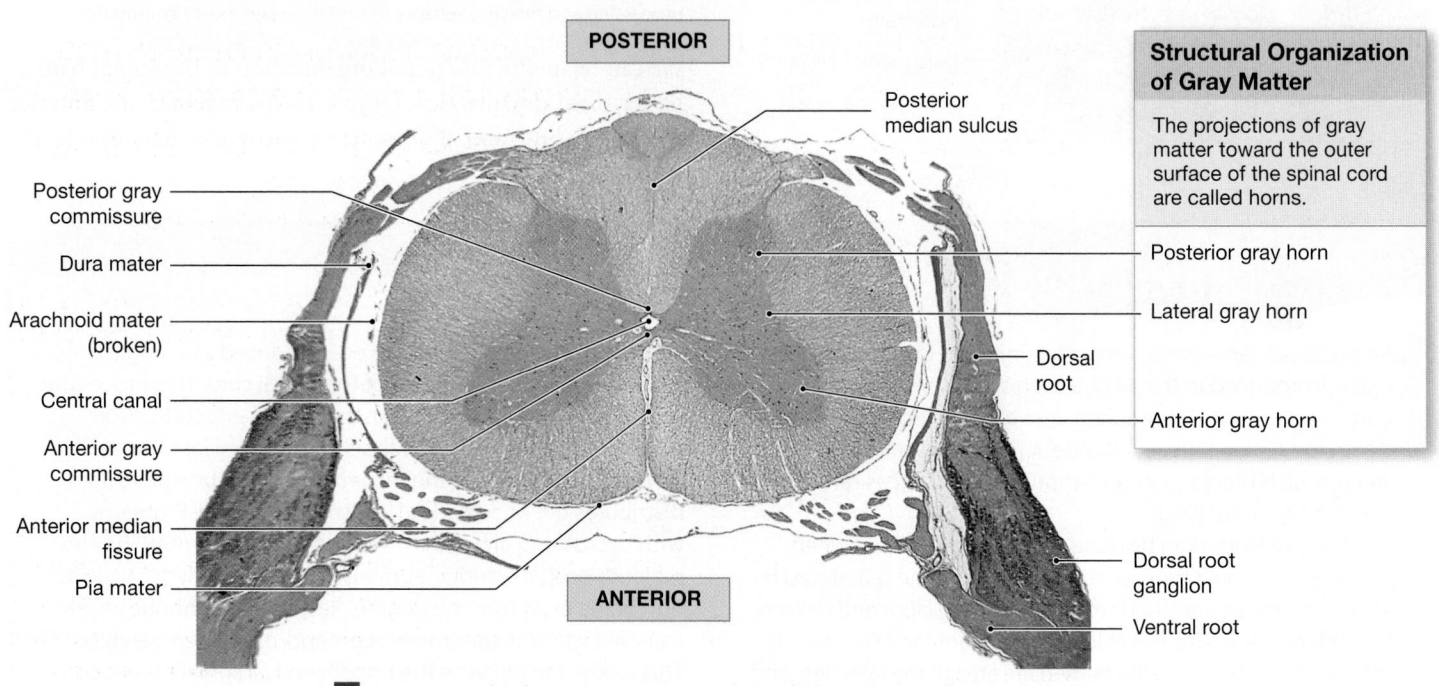

Structural Organization of Gray Matter

The projections of gray matter toward the outer surface of the spinal cord are called horns.

b A micrograph of a transverse section through the spinal cord, showing major landmarks in and surrounding the cord.

spinal cord into left and right sides. The gray matter forms an H or butterfly shape and surrounds the narrow **central canal**. **Horns** are the areas of gray matter on each side of the spinal cord. The gray matter is dominated by the cell bodies of neurons, neuroglia, and unmyelinated axons. The superficial white matter contains large numbers of myelinated and unmyelinated axons.

Organization of Gray Matter

How is the gray matter functionally organized within the nervous system as a whole? **Nuclei** are masses of gray matter within the central nervous system. **Sensory nuclei** receive and relay sensory information from peripheral receptors. **Motor nuclei** issue motor commands to peripheral effectors. Sensory and motor nuclei appear small in transverse section, but they may extend for a considerable distance along the length of the spinal cord.

A frontal section along the length of the central canal of the spinal cord separates the sensory (posterior, or dorsal) nuclei from the motor (anterior, or ventral) nuclei. The **posterior gray horns** contain somatic and visceral sensory nuclei, and the **anterior gray horns** contain somatic motor nuclei. The **lateral gray horns**, located only in the thoracic and lumbar segments, contain visceral motor nuclei. The **gray commissures** (*commissura*, a joining together) posterior to and anterior to the central canal contain axons that cross from one side of the cord to the other before they reach an area in the gray matter.

Figure 13–5a shows the relationship between the function of a particular nucleus (sensory or motor) and its position in the gray matter of the spinal cord. The nuclei within each gray horn are also spatially organized. In the cervical enlargement, for example, the anterior gray horns contain nuclei whose motor neurons control the muscles of the upper limbs. On each side of the spinal cord, in medial to lateral sequence, are somatic motor nuclei that control (1) muscles that position the pectoral girdle, (2) muscles that move the arm, (3) muscles that move the forearm and hand, and (4) muscles that move the hand and fingers. Within each of these regions, the motor neurons that control flexor muscles are grouped separately from those that control extensor muscles. Because the spinal cord is so highly organized, we can predict which muscles will be affected by damage to a specific area of gray matter.

Organization of White Matter

How is the white matter functionally organized? The white matter on each side of the spinal cord can be divided into three regions called **columns**, or *funiculi* (**Figure 13–5a**). The **posterior white columns** lie between the posterior gray horns and the posterior median sulcus. The **anterior white columns** lie between the anterior gray horns and the anterior median fissure. The anterior white columns are interconnected by the **anterior white commissure**, a region where axons cross from one side of the spinal cord to the other. The white matter between the anterior and posterior columns on each side makes up the **lateral white column**.

Each column contains *tracts* whose axons share functional and structural characteristics. A **tract**, or *fasciculus* (fa-SIK-ū-lus; bundle), is a bundle of axons in the CNS that is somewhat uniform in diameter, myelination, and propagation speed. All the axons within a tract relay the same type of information (sensory or motor) in the same direction. Short tracts carry sensory or motor signals between segments of the spinal cord, and longer tracts connect the spinal cord with the brain. **Ascending tracts** carry *sensory* information toward the brain, and **descending tracts** convey *motor* commands to the spinal cord. We describe the major tracts and their functions in Chapters 15 and 16. Because spinal tracts have very specific functions, damage to one produces a characteristic loss of sensation or motor control.

> **CHECKPOINT**
>
> 6. Differentiate between sensory nuclei and motor nuclei.
>
> 7. A person with polio has lost the use of his leg muscles. In which area of his spinal cord would you expect the virus-infected motor neurons to be?
>
> 8. A disease that damages myelin sheaths would affect which portion of the spinal cord?
>
> See the blue Answers tab at the back of the book.

13-4 | Spinal nerves form plexuses that are named according to their level of emergence from the vertebral canal

Learning Outcome Describe the major components of a spinal nerve, and relate the distribution pattern of spinal nerves to the regions they innervate.

In this section we consider the structure and function of spinal nerves and the interwoven networks of spinal nerves called nerve plexuses.

Anatomy of Spinal Nerves

Every segment of the spinal cord is connected to a pair of spinal nerves. Surrounding each spinal nerve is a series of connective tissue layers continuous with those of the associated peripheral nerves (**Figure 13–6**). These layers, best seen in sectional view, are comparable to those associated with skeletal muscles. ⊃ p. 317 The **epineurium**, or outermost layer, consists of a dense network of collagen fibers. The fibers of the **perineurium**, the middle layer, extend inward from the epineurium. These connective tissue partitions divide the nerve into a series

Figure 13–6 **A Peripheral Nerve.** A diagrammatic view and an electron micrograph of a typical spinal nerve. Note the connective tissue layers that are continuous with the associated spinal nerve.

Blood vessels

Connective Tissue Layers

Epineurium covering peripheral nerve

Perineurium (around one fascicle)

Endoneurium

Schwann cell

Myelinated axon

Fascicle

a A typical peripheral nerve and its connective tissue wrappings

b A scanning electron micrograph showing the various layers in great detail (SEM × 340)

13

of compartments that contain bundles of axons, or *fascicles*. Delicate connective tissue fibers of the **endoneurium**, the innermost layer, extend from the perineurium and surround individual axons.

Arteries and veins penetrate the epineurium and branch within the perineurium. Capillaries leaving the perineurium branch in the endoneurium. They supply the axons and Schwann cells of the nerve and the fibroblasts of the connective tissues.

As the spinal nerves extend into the periphery, they branch and interconnect, forming the peripheral nerves that innervate body tissues and organs. The connective tissue sheaths of peripheral nerves are the same as, and continuous with, those of spinal nerves.

If a peripheral *axon* is severed but not displaced, normal function may eventually return as the cut stump grows across the site of injury, away from the cell body and along its former path. ⟳ p. 424 Repairs made after an entire peripheral *nerve* has been damaged are generally incomplete, primarily due to problems with axon alignment and regrowth. Various technologically sophisticated procedures designed to improve nerve regeneration and repair are currently under evaluation.

Peripheral Distribution of Spinal Nerves

A **dermatome** is the specific bilateral region of the skin surface monitored by a single pair of spinal nerves. Each pair of spinal nerves supplies its own dermatome (**Figure 13–7**). Note

that the boundaries of adjacent dermatomes overlap to some degree. Dermatomes are clinically important because damage or infection of a spinal nerve or dorsal root ganglion produces a loss of sensation in the corresponding region of the skin. Additionally, characteristic signs may appear on the skin supplied by that specific nerve.

Peripheral *nerve palsies*, or **peripheral neuropathies**, are regional losses of sensory and motor function most often resulting from nerve trauma or compression. (If your arm or leg has ever "fallen asleep" after you leaned or sat in an uncomfortable position, you have experienced a mild, temporary palsy.) The location of the affected dermatomes provides clues to the location of injuries along the spinal cord, but the information is not precise. More exact conclusions can be drawn if there is loss of motor control, based on the origin and distribution of the peripheral nerves originating at nerve plexuses.

Spotlight Figure 13–8 shows the distribution, or pathway, of a typical spinal nerve that originates from the thoracic or superior lumbar segments (T_1–L_2) of the spinal cord. The spinal nerve forms just lateral to the intervertebral foramen, where the dorsal and ventral roots unite. We consider the pathways of both the sensory information and the motor commands.

Nerve Plexuses

The simple distribution pattern of dorsal and ventral *rami*, or branches, in Spotlight Figure 13–8 applies to spinal nerves T_1–L_2. But in segments controlling the skeletal musculature of

Figure 13–7 **Dermatomes.** Anterior and posterior distributions of dermatomes on the surface of the skin. NV = cranial nerve five (trigeminal nerve).

KEY
Spinal cord regions

	=	Cervical
	=	Thoracic
	=	Lumbar
	=	Sacral

ANTERIOR   **POSTERIOR**

Clinical Note

Shingles The same virus that causes chickenpox, the varicella-zoster virus (VZV), also causes **shingles** (derived from the Latin *cingulum*, girdle). This herpes virus attacks neurons within the dorsal roots of spinal nerves and sensory ganglia of cranial nerves. Shingles produces a painful rash and blisters whose distribution corresponds to that of the affected sensory nerves and follows its dermatome. Anyone who has had chickenpox is at risk of developing shingles. After the initial encounter, the virus remains dormant within neurons of the anterior gray horns of the spinal cord. It is not known what triggers the reactivation of this pathogen. Fortunately for those affected, attacks of shingles usually heal and leave behind only unpleasant memories.

Many people who contract shingles suffer just a single episode in their adult lives. Some experience a painful condition known as *postherpetic neuralgia* after a bout of shingles. Moreover, the problem can recur in people with weakened immune systems, including the elderly and individuals with AIDS or some forms of cancer. Treatment for shingles typically involves large doses of antiviral drugs. In 2006, the U.S. Food and Drug Administration approved a VZV vaccine (Zostavax) for use in people ages 60 and above who have had chickenpox.

the neck, upper limbs, or lower limbs, the situation is more complicated. During development, small skeletal muscles innervated by different ventral rami typically fuse to form larger muscles with compound origins. The anatomical distinctions between the component muscles may disappear, but separate ventral rami continue to provide sensory innervation and motor control to each part of the compound muscle. As they converge, the ventral rami of adjacent spinal nerves blend their fibers, producing a series of compound nerve trunks. Such a complex interwoven network of nerves is called a **nerve plexus** (PLEK-sus; *plexus*, braid).

Only the ventral rami form plexuses. These four major plexuses are the (1) *cervical plexus*, (2) *brachial plexus*, (3) *lumbar plexus*, and (4) *sacral plexus* (**Figure 13–9**). The nerves arising at these plexuses contain sensory as well as motor fibers because these nerves form from the fusion of ventral rami (**Spotlight Figure 13–8**).

SENSORY INFORMATION

A spinal nerve collects sensory information from peripheral structures and delivers it to sensory nuclei in the thoracic or superior lumbar segments of the spinal cord. The dorsal and ventral rami, and the white rami of the rami communicantes, also contain sensory fibers.

From interoceptors of back

From exteroceptors, proprioceptors of back

4 The dorsal root of each spinal nerve carries sensory information to the spinal cord.

Somatic sensory nuclei

3 The dorsal ramus carries sensory information from the skin and skeletal muscles of the back.

2 The ventral ramus carries sensory information from the ventrolateral body surface, structures in the body wall, and the limbs.

From exteroceptors, proprioceptors of body wall, limbs

From interoceptors of body wall, limbs

Dorsal root ganglion

Rami communicantes

Visceral sensory nuclei

Ventral root

KEY

= Somatic sensations

= Visceral sensations

1 The sympathetic nerve carries sensory information from the visceral organs.

From interoceptors of visceral organs

MOTOR COMMANDS

A spinal nerve distributes motor commands that originate in motor nuclei of the thoracic or superior lumbar segments of the spinal cord.

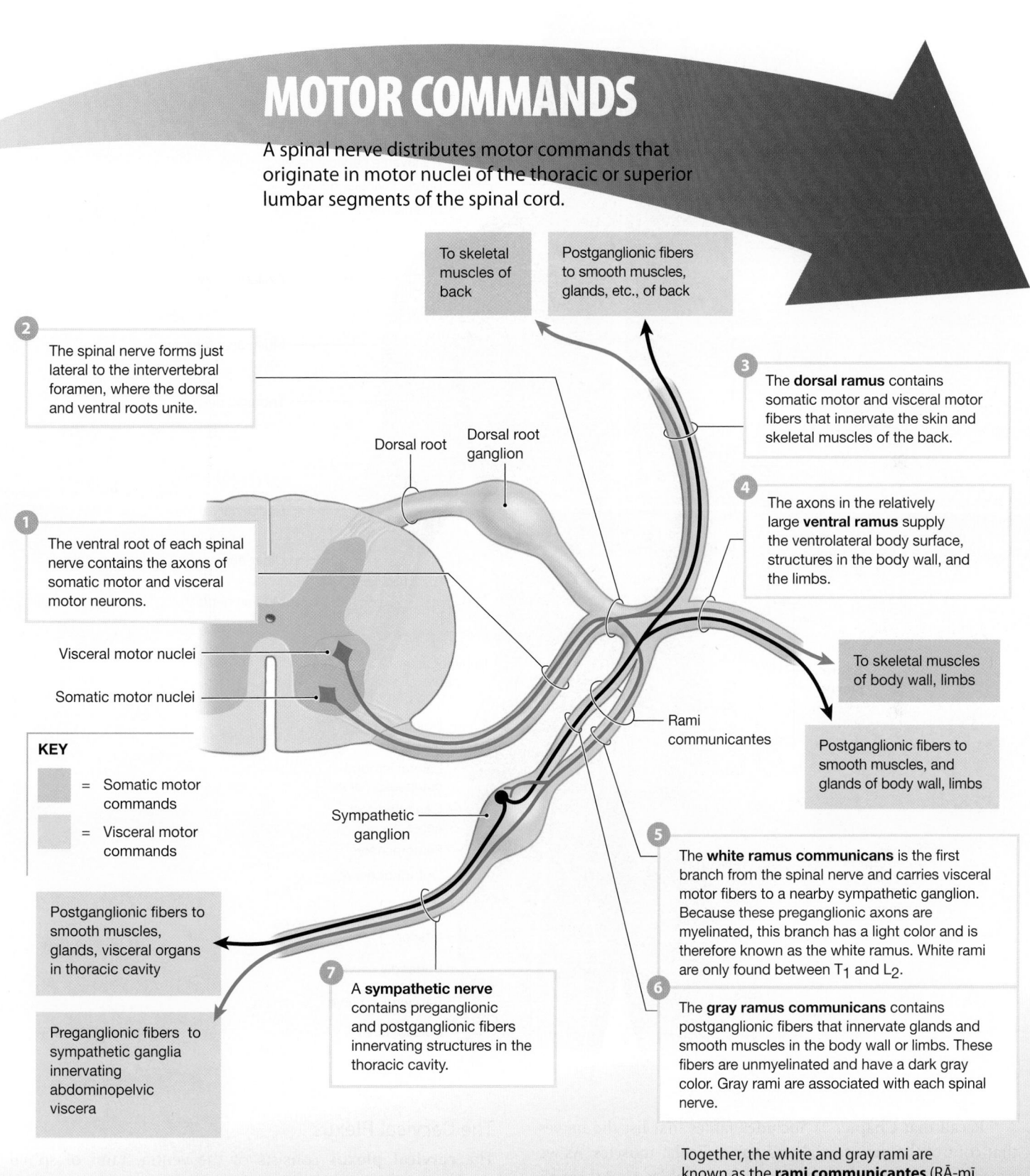

To skeletal muscles of back

Postganglionic fibers to smooth muscles, glands, etc., of back

2 The spinal nerve forms just lateral to the intervertebral foramen, where the dorsal and ventral roots unite.

3 The **dorsal ramus** contains somatic motor and visceral motor fibers that innervate the skin and skeletal muscles of the back.

Dorsal root

Dorsal root ganglion

1 The ventral root of each spinal nerve contains the axons of somatic motor and visceral motor neurons.

4 The axons in the relatively large **ventral ramus** supply the ventrolateral body surface, structures in the body wall, and the limbs.

Visceral motor nuclei

Somatic motor nuclei

To skeletal muscles of body wall, limbs

Rami communicantes

Postganglionic fibers to smooth muscles, and glands of body wall, limbs

KEY

= Somatic motor commands

= Visceral motor commands

Sympathetic ganglion

5 The **white ramus communicans** is the first branch from the spinal nerve and carries visceral motor fibers to a nearby sympathetic ganglion. Because these preganglionic axons are myelinated, this branch has a light color and is therefore known as the white ramus. White rami are only found between T_1 and L_2.

Postganglionic fibers to smooth muscles, glands, visceral organs in thoracic cavity

7 A **sympathetic nerve** contains preganglionic and postganglionic fibers innervating structures in the thoracic cavity.

6 The **gray ramus communicans** contains postganglionic fibers that innervate glands and smooth muscles in the body wall or limbs. These fibers are unmyelinated and have a dark gray color. Gray rami are associated with each spinal nerve.

Preganglionic fibers to sympathetic ganglia innervating abdominopelvic viscera

Together, the white and gray rami are known as the **rami communicantes** (RĀ-mī ko-mū-ni-KAN-tēz), or communicating branches (singular, *ramus communicans*).

Figure 13–9 **Peripheral Nerves and Nerve Plexuses.** ATLAS: Plate 2a

13

Cervical plexus

- C₁
- C₂
- C₃
- C₄
- C₅

Brachial plexus

- C₅
- C₆
- C₇
- C₈
- T₁

- T₂
- T₃
- T₄
- T₅
- T₆
- T₇
- T₈
- T₉
- T₁₀
- T₁₁
- T₁₂

Lumbar plexus

- L₁
- L₂
- L₃
- L₄
- L₅

Sacral plexus

- S₁
- S₂
- S₃
- S₄
- S₅
- Co₁

Lesser occipital nerve
Great auricular nerve
Transverse cervical nerve
Supraclavicular nerve
Phrenic nerve

Axillary nerve

Musculocutaneous nerve

Thoracic nerves

Radial nerve

Ulnar nerve

Median nerve

Iliohypogastric nerve

Ilioinguinal nerve

Lateral femoral cutaneous nerve

Genitofemoral nerve

Femoral nerve

Obturator nerve

Superior ⎫
        ⎬ Gluteal nerves
Inferior ⎭

Pudendal nerve

Saphenous nerve

Sciatic nerve

Recall that Chapter 11 includes tables that list the nerves that control the major axial and appendicular muscles. As we proceed, you may find it helpful to refer to Tables 11–2 through 11–19 in that chapter. ↪ pp. 372–405

The Cervical Plexus

The **cervical plexus** consists of the ventral rami of spinal nerves C_1–C_5 (**Figures 13–9, 13–10**). The branches of the cervical plexus innervate the muscles of the neck and extend into the

Figure 13–10 **The Cervical Plexus.** ATLAS: Plates 3c,d; 18a–c

Cranial Nerves

Accessory nerve (N XI)

Hypoglossal nerve (N XII)

Nerve Roots of Cervical Plexus

C_1
C_2
C_3
C_4
C_5

Clavicle

Nerves of the Cervical Plexus	
Spinal Segments	**Nerve and Distribution**
Great Auricular	
C_2–C_3	Skin over the posterior aspect of the ear and the neck
Lesser Occipital	
C_2	Skin of the neck and the scalp posterior and superior to the ear
Transverse Cervical	
C_3–C_4	Skin of the anterior triangle of the neck
Ansa Cervicalis	
C_1–C_4	Five of the extrinsic laryngeal muscles by way of cranial nerve XII
	Geniohyoid muscle
	Thyrohyoid muscle
	Omohyoid muscle
	Sternothyroid muscle
	Sternohyoid muscle
Phrenic	
C_3–C_5	Diaphragm
Supraclaviculars	
C_3–C_4	Skin of the neck and shoulder

13

thoracic cavity, where they control the diaphragmatic muscles. The **phrenic nerve** is the major nerve of the cervical plexus. The left and right phrenic nerves supply the diaphragm, a key respiratory muscle. Other branches of this nerve plexus are distributed to the skin of the neck and the superior part of the chest.

TIPS & TRICKS

A useful mnemonic for remembering a function of three important cervical nerves is "C_3, C_4, C_5 keep the diaphragm alive."

The Brachial Plexus

The **brachial plexus** innervates the pectoral girdle and upper limb, with contributions from the ventral rami of spinal nerves C_5–T_1 (**Figures 13–9** and **13–11**). The brachial plexus can also have fibers from C_4, T_2, or both.

The nerves that form this plexus originate from trunks and cords. **Trunks** are large bundles of axons from several spinal nerves. **Cords** are smaller branches that originate at trunks. We name both trunks and cords according to their location relative to the *axillary artery*, a large artery supplying the upper limb. Hence we have *superior*, *middle*, and *inferior trunks*, and *lateral*, *medial*, and *posterior cords*. The lateral cord forms the **musculocutaneous nerve** exclusively and, together with the medial cord, contributes to the **median nerve**. The **ulnar nerve** is the other major nerve of the medial cord. The posterior cord gives rise to the **axillary nerve** and the **radial nerve**.

The Lumbar and Sacral Plexuses

The **lumbar plexus** and the **sacral plexus** arise from the lumbar and sacral segments of the spinal cord, respectively. The

Figure 13–11   **The Brachial Plexus.** ATLAS: Plates 27a–c; 29b,c; 30

Brachial Plexus: Major Nerves of the Arm and Forearm

Spinal Segments	Nerve and Distribution
	Musculocutaneous
C_5–T_1	Flexor muscles on the arm (biceps brachii, brachialis, and coracobrachialis); sensory from skin over the lateral surface of the forearm through the lateral antebrachial cutaneous nerve
	Radial
C_5–T_1	Many extensor muscles on the arm and forearm (triceps brachii, anconeus, extensor carpi radialis, extensor carpi ulnaris, and brachioradialis); and abductor pollicis muscle by the deep branch; sensory from skin over the posterolateral surface of the limb through the posterior brachial cutaneus nerve, posterior antebrachial cutaneous nerve, and the superficial branch (radial half of the hand)
	Median
C_6–T_1	Flexor muscles on the forearm (flexor carpi radialis and palmaris longus muscles); pronator quadratus and pronator teres muscles; digital flexors (through the anterior interosseous nerve); sensory from skin over the anterolateral surface of the hand
	Ulnar
C_8–T_1	Flexor carpi ulnaris muscle, flexor digitorum profundus muscle, adductor pollicis muscle, and small digital muscles by the deep branch; sensory from skin over medial surface of the hand through the superficial branch

a Major nerves of the arm, forearm, and hand originating at the right brachial plexus, anterior view

Median nerve

Ulnar nerve

Lateral antebrachial cutaneous nerve

Superficial branch of radial nerve

Deep radial nerve

Radial nerve

Palmar digital nerves

Clinical Note

Sensory Innervation in the Hand Sensory function in the hand is serviced by three nerves of the brachial plexus. In carpal tunnel syndrome compression of one or more nerves at the wrist cause changes or losses of sensation in the areas innervated by the affected nerve(s). ⤵ p. 275

Radial nerve

Ulnar nerve

Median nerve

Anterior   **Posterior**

nerves arising at these plexuses innervate the pelvic girdle and lower limbs (**Figures 13–9** and **13–12**).

The lumbar plexus contains axons from the ventral rami of spinal nerves T_{12}–L_4. The major nerves of this plexus are the **genitofemoral nerve**, the **lateral femoral cutaneous nerve**, and the **femoral nerve**.

The sacral plexus contains axons from the ventral rami of spinal nerves L_4–S_4. Two major nerves arise at this plexus: the **sciatic nerve** and the **pudendal nerve**. The sciatic nerve passes posterior to the femur, deep to the long head of the biceps femoris muscle. As it approaches the knee, the sciatic nerve divides into

Nerves of the Brachial Plexus		Cords	Divisions	Trunks	Ventral Rami Roots
The nerves of the brachial plexus arise from one or more trunks or cords whose names indicate their positions relative to the axillary artery, a large artery supplying the upper limb.				The ventral rami converge to form the superior, middle, and inferior trunks.	The brachial plexus is formed by the ventral rami of spinal nerves C_5–T_1.
Spinal Segments	**Nerve and Distribution**				
	Dorsal Scapular				
C_5	Rhomboid and levator scapulae muscles				
	Nerve to Subclavius				
C_5–C_6	Subclavius muscle				
	Suprascapular				
C_5–C_6	Supraspinatus and infraspinatus muscles; sensory from shoulder joint and scapula				
	Pectoral (medial and lateral)				
C_5–T_1	Pectoralis muscles				
	Subscapular				
C_5–C_6	Subscapularis and teres major muscles				
	Thoracodorsal				
C_6–C_8	Latissimus dorsi muscle				
	Axillary				
C_5–C_6	Deltoid and teres minor muscles				
	Medial Antibrachial Cutaneous				
C_8–T_1	Sensory from skin over anterior and medial surface of the arm and forearm				
	Musculocutaneous (see part a)				
	Median (see part a)				
	Ulnar (see part a)				
	Radial (see part a)				
	Long Thoracic				
C_5–C_7	Serratus anterior muscle				

KEY

- Ventral rami roots
- Trunks
- Divisions
- Cords
- Peripheral nerves

b Right brachial plexus, anterior view

two branches: the **fibular nerve** (or *peroneal nerve*) and the **tibial nerve**. The *sural nerve*, formed by branches of the fibular nerve, is a sensory nerve innervating the lateral portion of the foot. A section of this nerve is often removed for use in nerve grafts.

Discussions of motor performance usually make a distinction between the conscious ability to control motor function—something that requires communication and feedback between the brain and spinal cord—and automatic motor responses coordinated entirely within the spinal cord. These automatic responses, called *reflexes*, are motor responses to specific stimuli. In the rest of this chapter, we look at how sensory neurons, interneurons, and motor neurons interconnect, and

Figure 13–12   The Lumbar and Sacral Plexuses. ATLAS: Plates 70b; 76b; 82b

Lumbar Plexus

Spinal Segments	Nerve and Distribution
	Iliohypogastric
T_{12}–L_1	External and internal oblique and transverse abdominis muscles; skin over the inferior abdomen and buttocks
	Ilioinguinal
L_1	Abdominal muscles (with iliohypogastric nerve); skin over superior, medial thigh and portions of external genitalia
	Genitofemoral
L_1–L_2	Skin over anteromedial thigh and portions of external genitalia
	Lateral Femoral Cutaneous
L_2–L_3	Skin over anterior, lateral, and posterior thigh
	Femoral
L_2–L_4	Quadriceps, sartorius, pectineus, and iliopsoas muscles; skin of the anteromedial thigh, and medial surface of the leg and foot
	Obturator
L_2–L_4	Gracilis, and adductor magnus, brevis and longus muscles; skin from the medial surface of the thigh

Nerve Roots of Lumbar Plexus

The lumbar plexus is formed by the ventral rami of T_{12}–L_4.

T_{12} subcostal nerve

T_{12} · L_1 · L_2 · L_3 · L_4 · L_5

T_{12} · L_1 · L_2 · L_3 · L_4

Lumbosacral trunk

a Lumbar plexus, anterior view

Sacral Plexus

Spinal Segments	Nerve and Distribution
	Superior Gluteal
L_4–S_2	Gluteus minimus, gluteus medius, and tensor fasciae latae muscles
	Inferior Gluteal
L_4–S_2	Gluteus maximus muscle
	Posterior Femoral Cutaneous
S_1–S_3	Skin over perineum and posterior thigh and leg
	Sciatic
L_4–S_3	Semimembranosus, semitendinosus, and adductor magnus muscles; branches into tibial and fibular nerves
	Pudendal
S_2–S_4	Muscles of the perineum; skin of external genitalia, bulbospongiosus and ischiocavernosus muscles

Nerve Roots of Sacral Plexus

The sacral plexus is formed by a branch from L_4 and the ventral rami of L_5–S_4.

Lumbosacral trunk

L_4 · L_5 · S_1 · S_2 · Sacrum · S_3 · S_4 · S_5 · Co_1

b Sacral plexus, anterior view

Iliohypogastric nerve
Ilioinguinal nerve
Genitofemoral nerve
Lateral femoral cutaneous nerve
Femoral nerve
Obturator nerve
Superior gluteal nerve
Inferior gluteal nerve
Pudendal nerve
Posterior femoral cutaneous nerve (cut)
Sciatic nerve
Saphenous nerve
Common fibular nerve
Superficial fibular nerve
Deep fibular nerve

c Nerves of the lumbar and sacral plexuses, anterior view

Clinical Note

Sensory Innervation in the Ankle and Foot This is the normal distribution of sensory nerves innervating the ankle and foot. Mapping of touch and pain perception can be combined with observations of muscle function to detect nerve damage affecting specific peripheral nerves.

Saphenous nerve
Sural nerve
Fibular nerve
Tibial nerve
Saphenous nerve
Sural nerve
Saphenous nerve
Sural nerve
Tibial nerve
Fibular nerve

Superior gluteal nerve
Inferior gluteal nerve
Pudendal nerve
Posterior femoral cutaneous nerve
Sciatic nerve
Tibial nerve
Common fibular nerve
Sural nerve

d Nerves of the sacral plexus, posterior view

13

how these interconnections produce both simple and complex reflexes. ATLAS: Embryology Summary 11: The Development of the Spinal Cord and Spinal Nerves

CHECKPOINT

9. Identify the major networks of nerves known as plexuses.

10. An anesthetic blocks the function of the dorsal rami of the cervical spinal nerves. Which areas of the body will be affected?

11. Injury to which of the nerve plexuses would interfere with the ability to breathe?

12. Compression of which nerve produces the sensation that your leg has "fallen asleep"?

See the blue Answers tab at the back of the book.

13-5 ❙ Interneurons are organized into functional groups called neuronal pools

Learning Outcome Discuss the significance of neuronal pools, and describe the major patterns of interaction among neurons within and among these pools.

Your body has about 10 million sensory neurons, one-half million motor neurons, and 20 *billion* interneurons. The sensory neurons deliver information to the CNS. The motor neurons distribute commands to peripheral effectors, such as skeletal muscles. Interactions among interneurons provide the interpretation, planning, and coordination of incoming and outgoing signals.

The billions of interneurons of the CNS are organized into a much smaller number of **neuronal pools**—functional groups of interconnected neurons. A neuronal pool may be scattered, involving neurons in several regions of the brain, or localized, with neurons restricted to one specific location in the brain or spinal cord. Estimates of the actual number of neuronal pools range between a few hundred and a few thousand.

Each pool has a limited number of input sources and output destinations. Each pool may contain both excitatory and inhibitory neurons. The output of the entire neuronal pool may stimulate or depress activity in other parts of the brain or spinal cord, affecting the interpretation of sensory information or the coordination of motor commands.

The pattern of interaction among neurons provides clues to the functional characteristics of a neuronal pool. We refer to the "wiring diagrams" in Figure 13–13 as *neural circuits*, just as we refer to electrical circuits in the wiring of a house. We can distinguish five circuit patterns:

1. **Divergence** is the spread of information from one neuron to several neurons (Figure 13–13a), or from one pool to multiple pools. Divergence permits the broad distribution of a specific input. Considerable divergence occurs when sensory neurons bring information into the CNS, because the information is distributed to neuronal pools throughout the spinal cord and brain. Visual information arriving from the eyes, for example, reaches your consciousness at the same time it is distributed to areas of the brain that control posture and balance at the subconscious level.

2. In **convergence**, several neurons synapse on a single postsynaptic neuron (Figure 13–13b). Several patterns of activity in the presynaptic neurons can therefore have the same

effect on the postsynaptic neuron. Through convergence, the same motor neurons can be subject to both conscious and subconscious control. For example, the movements of your diaphragm and ribs are now being controlled by your brain at the subconscious level. But you can also consciously control the same motor neurons, as when you take a deep breath and hold it. Two neuronal pools are involved, both synapsing on the same motor neurons.

3. In **serial processing**, information is relayed in a stepwise fashion, from one neuron to another or from one neuronal pool to the next (Figure 13–13c). This pattern occurs as sensory information is relayed from one part of the brain to another. For example, pain sensations on their way to your consciousness make stops at two neuronal pools along the pain pathway.

4. **Parallel processing** occurs when several neurons or neuronal pools process the same information simultaneously (Figure 13–13d). Divergence must take place before parallel processing can occur. As a result, many responses can occur simultaneously. For example, stepping on a sharp object stimulates sensory neurons that distribute the information to several neuronal pools. As a result of parallel processing, you might withdraw your foot, shift your weight, move your arms, feel the pain, and shout "Ouch!" at about the same time.

5. In **reverberation**, collateral branches of axons somewhere along the circuit extend back toward the source of an impulse and further stimulate the presynaptic neurons (Figure 13–13e). Reverberation is like a positive feedback

Figure 13–13   **Neural Circuits: The Organization of Neuronal Pools.**

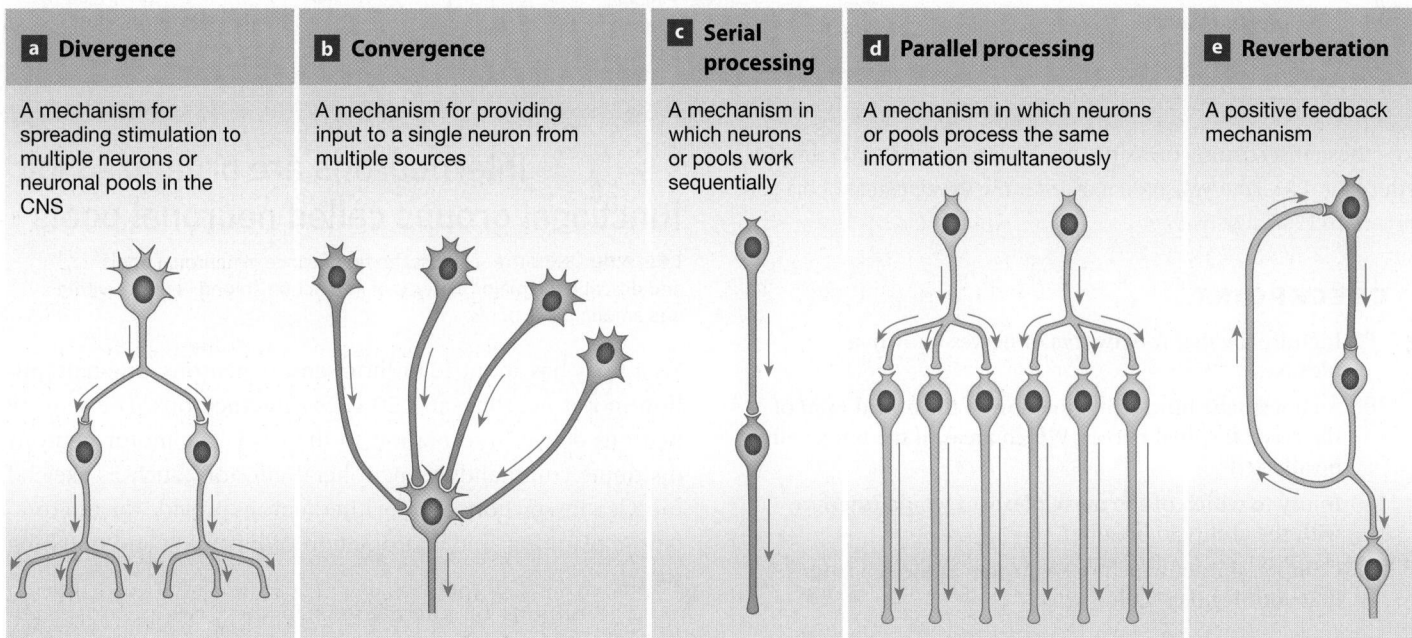

a Divergence	**b** Convergence	**c** Serial processing	**d** Parallel processing	**e** Reverberation
A mechanism for spreading stimulation to multiple neurons or neuronal pools in the CNS	A mechanism for providing input to a single neuron from multiple sources	A mechanism in which neurons or pools work sequentially	A mechanism in which neurons or pools process the same information simultaneously	A positive feedback mechanism

loop involving neurons: Once a reverberating circuit has been activated, it will continue to function until synaptic fatigue or inhibitory stimuli break the cycle. Reverberation can take place within a single neuronal pool, or it may involve a series of interconnected pools. Highly complicated examples of reverberation among neuronal pools in the brain may help maintain consciousness, muscular coordination, and normal breathing.

The functions of the nervous system depend on the interactions among neurons organized in neuronal pools. The most complex neural processing steps take place in the spinal cord and brain. The simplest steps occur within the PNS and the spinal cord. These steps control reflexes that are a bit like Legos: Individually, the reflexes are quite simple, but they can be combined in a great variety of ways to create very complex responses. For this reason, reflexes are the basic building blocks of neural function, as you will see in the next section.

CHECKPOINT

13. Define neuronal pool.

14. List the five circuit patterns found in neuronal pools.

See the blue Answers tab at the back of the book.

13-6 ▍ Reflexes are rapid, automatic responses to stimuli

Learning Outcome Describe the steps in a neural reflex, and classify the types of reflexes.

Conditions inside or outside the body can change rapidly and unexpectedly. **Reflexes** are rapid, automatic responses to specific stimuli. Reflexes preserve homeostasis by making rapid adjustments in the function of organs or organ systems. The response shows little variability: Each time a particular reflex is activated, it normally produces the same motor response.

In Chapter 1, we introduced the basic functional components involved in all types of homeostatic regulation: a *receptor*, a *control center* (or *integration center*), and an *effector*. ⟳ p. 33 Here we consider *neural reflexes*, in which sensory fibers deliver information from peripheral receptors to an integration center in the CNS, and motor fibers carry motor commands to peripheral effectors. In Chapter 18, we examine *endocrine reflexes*, in which hormones in the bloodstream carry the commands to peripheral tissues and organs.

The Reflex Arc

The route followed by nerve impulses to produce a reflex is called a reflex arc. **Spotlight Figure 13–14** diagrams the five steps in a simple neural reflex.

1 *The Arrival of a Stimulus and Activation of a Receptor.* Reflex arcs begin with activation of a *receptor*, which may be a specialized cell or the dendrites of a sensory neuron. Receptors are sensitive to physical or chemical changes in the body or to changes in the external environment. We introduced the general categories of sensory receptors in Chapter 12. ⟳ p. 416 If you lean on a tack, for example, pain receptors in the palm of your hand are activated. These receptors, the dendrites of sensory neurons, respond to stimuli that cause or accompany tissue damage. (We discuss the link between receptor stimulation and sensory neuron activation further in Chapter 15.)

2 *The Activation of a Sensory Neuron.* When the dendrites are stretched, there is a graded depolarization that leads to the formation and propagation of action potentials along the axons of the sensory neurons. This information reaches the spinal cord by way of a dorsal root. In our example, **1** and **2** involve the same cell. However, the two steps may involve different cells. For example, reflexes triggered by loud sounds begin when receptor cells in the internal ear release neurotransmitters that stimulate sensory neurons.

3 *Information Processing in the CNS.* In our example, information processing begins when excitatory neurotransmitter molecules, released by the axon terminal of a sensory neuron, arrive at the postsynaptic membrane of an interneuron. The neurotransmitter produces an excitatory postsynaptic potential (EPSP), which is integrated with other stimuli arriving at the postsynaptic cell at that moment. ⟳ p. 447 The information processing is thus performed by the interneuron. In the simplest reflexes, such as the *stretch reflex*, the sensory neuron innervates a motor neuron directly. In that case, the motor neuron performs the information processing. By contrast, complex reflexes involve several interneurons. Some release excitatory neurotransmitters (*excitatory interneurons*), and others release inhibitory neurotransmitters (*inhibitory interneurons*).

4 *The Activation of a Motor Neuron.* The axons of the stimulated motor neurons carry action potentials into the periphery—in this example, through the ventral root of a spinal nerve.

5 *The Response of a Peripheral Effector.* The release of neurotransmitters by the motor neurons at axon terminals then leads to a response by a peripheral effector—in this case, a skeletal muscle whose contraction pulls your hand away from the tack.

A reflex response generally removes or opposes the original stimulus, and this is an example of *negative feedback*. ⟳ p. 36 However, reflexes may be integrated to produce complex movements that support or enhance the primary response. Two examples are given in **Spotlight Figure 13-14**.

By opposing potentially harmful changes in the internal or external environment, reflexes play an important role in homeostasis. The immediate reflex response is typically not the only response to a stimulus. The other responses, which

Reflexes are rapid, automatic responses to specific stimuli. They preserve homeostasis by making rapid adjustments in the function of organs or organ systems.

Events in a Reflex Arc

The "wiring" of a single reflex is called a **reflex arc**. It begins at a receptor and ends at a peripheral effector such as a muscle fiber or a gland cell. A simple reflex arc, such as this withdrawal reflex, consists of five steps.

1. Arrival of stimulus and activation of receptor

Stimulus

Receptor

2. Activation of a sensory neuron

Dorsal root

Spinal cord

REFLEX ARC

Sensation relayed to the brain by axon collaterals

3. Information processing in the CNS

5. Response by a peripheral effector

Effector

Ventral root

4. Activation of a motor neuron

KEY
— Sensory neuron (stimulated)
— Excitatory interneuron
— Motor neuron (stimulated)

MONOSYNAPTIC REFLEXES

In the simplest reflex arc, a sensory neuron synapses directly on a motor neuron. Beause there is only one synapse, it is called a **monosynaptic reflex**. Transmission across a chemical synapse always involves a synaptic delay, but with only one synapse, the delay between the stimulus and the response is minimized.

Stretch Reflex

The **stretch reflex** is an example of a monosynaptic reflex. Because there is only one synapse, there is little delay between sensory input and motor output. These reflexes control the most rapid motor responses of the nervous system. A stretch reflex provides automatic regulation of skeletal muscle length.

Stretch

Stimulus

Receptor (muscle spindle)

Spinal cord

REFLEX ARC

Contraction

Effector

Response

KEY
— Sensory neuron (stimulated)
— Motor neuron (stimulated)

The **patellar reflex** is an example of a stretch reflex. The stimulus is a tap on the patellar tendon that stretches receptors within the quadriceps muscles. The response is a brief contraction of those muscles, which produces a noticeable kick.

POLYSYNAPTIC REFLEXES

Polysynaptic reflexes can produce far more complicated responses than monosynaptic reflexes, because the interneurons can control motor neurons that activate several muscle groups simultaneously.

Withdrawal Reflex

A **withdrawal reflex** moves affected parts of the body away from a stimulus. A **flexor reflex** is an example of a withdrawal reflex that affects the muscles of a limb. In this example, the stimulus of a hot frying pan causes the contraction of the flexor muscles of the arm, yanking the forearm and hand away from the pan. This response occurs while pain sensations are simultaneously ascending to the brain.

Distribution within gray horns to other segments of the spinal cord

Painful stimulus

Flexors stimulated

Extensors inhibited

Crossed Extensor Reflex

A **crossed extensor reflex** involves a contralateral reflex arc (*contra*, opposite). In other words, a motor response to the stimulus also occurs on a side opposite the stimulus. The crossed extensor reflex complements the flexor reflex, and the two occur simultaneously. When you step on a tack, the flexor reflex pulls the affected foot away from the ground, while the crossed extensor reflex straightens the other leg to support your body weight. In the crossed extensor reflex, interneurons responding to the pain have axons that cross to the other side of the spinal cord. There they stimulate motor neurons that control the extensor muscles of the uninjured leg. As a result, your opposite leg straightens to support the shifting weight. Reverberating circuits use positive feedback to ensure that the movement lasts long enough to be effective. All of this happens without motor commands from higher centers of the brain.

To motor neurons in other segments of the spinal cord

Extensors inhibited

Flexors stimulated

Extensors stimulated

Flexors inhibited

Painful stimulus

KEY

———	Sensory neuron (stimulated)	- - - -	Motor neuron (inhibited)
═══	Excitatory interneuron	▬ ▬ ▬	Inhibitory interneuron
▬▬▬	Motor neuron (stimulated)		

are directed by your brain, involve multiple synapses and take longer to organize and coordinate.

Classification of Reflexes

Reflexes are classified on the basis of (1) their development, (2) the nature of the resulting motor response, (3) the complexity of the neural circuit involved, or (4) the site of information processing. These categories are not mutually exclusive. They represent different ways of describing a single reflex (**Figure 13–15**).

Development of Reflexes

Innate reflexes result from the connections that form between neurons during development. Such reflexes generally appear in a predictable sequence, from the simplest reflex responses (withdrawal from pain) to more complex motor patterns (chewing, suckling, or tracking objects with the eyes). The neural connections responsible for the basic motor patterns of an innate reflex are genetically programmed. Examples include the reflexive removal of your hand from a hot stovetop and blinking when your eyelashes are touched.

More complex, learned motor patterns are called **acquired reflexes**. For example, an experienced driver steps on the brake when trouble appears ahead, or a professional skier makes equally quick adjustments in body position while racing. These motor responses are rapid and automatic, but they were learned rather than preestablished. Repetition enhances such reflexes.

The distinction between innate and acquired reflexes is not absolute. Some people can learn motor patterns more quickly than others. The differences probably have a genetic basis.

Most reflexes, whether innate or acquired, can be modified over time or suppressed through conscious effort. For example, while walking a tightrope over the Grand Canyon, you might ignore a bee sting on your hand. Under other circumstances you would probably withdraw your hand immediately, while shouting and thrashing about as well.

Nature of the Response

Somatic reflexes provide a mechanism for the involuntary control of the muscular system. *Superficial reflexes* are triggered by stimuli at the skin or mucous membranes. *Stretch reflexes* are triggered by the sudden elongation of a tendon, and thus of the muscle to which it attaches. A familiar example is the *patellar*, or *"knee-jerk," reflex* that is usually tested during physical exams. These reflexes are also known as *deep tendon reflexes*, or *myotatic reflexes*.

Visceral reflexes, or *autonomic reflexes*, control the activities of other systems. We consider somatic reflexes in detail in this chapter and visceral reflexes in Chapter 16.

The movements directed by somatic reflexes are neither delicate nor precise. You might wonder why they exist at all, because we have voluntary control over the same muscles. In fact, somatic reflexes are absolutely vital, primarily because they are *immediate*. Making decisions and coordinating voluntary responses take time. In an emergency—when you slip while walking down a flight of stairs, or accidentally press your hand against a knife edge—any delay increases the likelihood of severe injury. Thus, somatic reflexes provide a rapid response that can be modified later, if necessary, by voluntary motor commands.

Complexity of the Circuit

Monosynaptic reflexes involve the simplest reflex arc. Most reflexes are more complicated and have at least one interneuron

Figure 13–15 **The Classification of Reflexes.**

between the sensory neuron and the motor neuron. These types are called *polysynaptic reflexes*. Polysynaptic reflexes have a longer delay between stimulus and response. The length of the delay is proportional to the number of synapses involved. **Spotlight Figure 13–14** highlights these reflexes.

Processing Sites

In **spinal reflexes**, the important interconnections and processing events occur in the spinal cord. We discuss these reflexes further in the next section. We consider reflexes processed in the brain, called **cranial reflexes**, in Chapters 14, 16, and 17.

> ### CHECKPOINT
> 15. Define reflex.
> 16. What is the minimum number of neurons in a reflex arc?
> 17. One of the first somatic reflexes to develop is the suckling reflex. Which type of reflex is this?
>
> See the blue Answers tab at the back of the book.

13-7 ▌ Spinal reflexes vary in complexity

Learning Outcome Distinguish among the types of motor responses produced by various reflexes, and explain how reflexes interact to produce complex behaviors.

Spinal reflexes range in complexity from simple monosynaptic reflexes involving a single segment of the spinal cord to polysynaptic reflexes that involve many segments. In the most complicated spinal reflexes, called **intersegmental reflex arcs**, many segments interact to produce a coordinated, highly variable motor response.

Monosynaptic Reflexes

In monosynaptic reflexes, there is little delay between sensory input and motor output. These reflexes control the most rapid, *stereotyped* (preexisting, mechanically repetitive) *motor responses* of the nervous system to specific stimuli.

The Stretch Reflex

The best-known monosynaptic reflex is the **stretch reflex**, which automatically regulates skeletal muscle length. The knee-jerk or **patellar reflex** is an example. When a physician taps your patellar tendon with a reflex hammer, receptors in the quadriceps muscle are stretched (**Spotlight Figure 13–14**). The distortion of the receptors in turn stimulates sensory neurons that extend into the spinal cord, where they synapse on motor neurons that control the motor units in the stretched muscle. This leads to a reflexive contraction of the stretched muscle that extends the knee in a brief kick.

To summarize: The stimulus (increasing muscle length) activates a sensory neuron, which triggers an immediate motor response (contraction of the stretched muscle) that counteracts the stimulus. The entire reflex is completed within 20–40 msec because the action potentials traveling toward and away from the spinal cord are conducted along large, myelinated Type A fibers.

The receptors in stretch reflexes are called *muscle spindles*. (We describe the sensory mechanism in the next section.) The stretching of muscle spindles produces a sudden burst of activity in the sensory neurons that monitor them. This in turn leads to stimulation of motor neurons that control the motor units in the stretched muscle. The result is rapid muscle shortening, which returns the muscle spindles to their resting length. The rate of action potential generation in the sensory neurons then decreases, causing a drop in muscle tone to resting levels.

Muscle Spindles

The sensory receptors involved in the stretch reflex are **muscle spindles**. Each consists of a bundle of small, specialized skeletal muscle fibers called **intrafusal muscle fibers** (**Figure 13–16**). The muscle spindle is surrounded by larger skeletal muscle fibers, called **extrafusal muscle fibers**. These fibers are responsible for the resting muscle tone and, at greater levels of stimulation, for the contraction of the entire muscle.

Figure 13–16   A Muscle Spindle. The location, structure, and innervation of a muscle spindle.

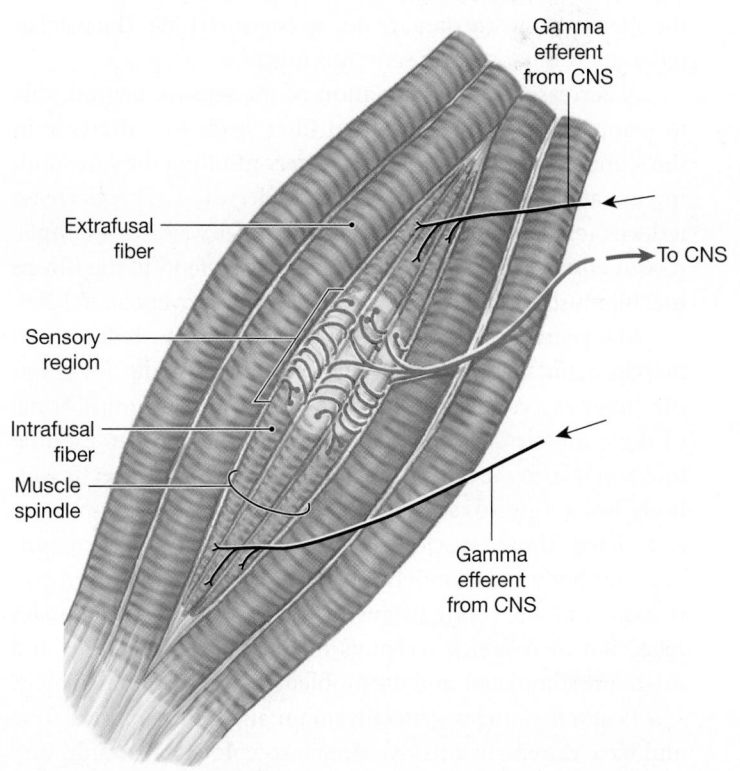

Both sensory and motor neurons innervate each intrafusal fiber. The dendrites of the sensory neuron spiral around the intrafusal fiber in a central sensory region. Axons from spinal motor neurons form neuromuscular junctions on either end of this fiber. Motor neurons innervating intrafusal fibers are called **gamma motor neurons**. Their axons are called **gamma efferents**.

An intrafusal fiber has one set of myofibrils at each end. These myofibrils run from the end of the intrafusal fiber only to the sarcolemma in the central region that is closely monitored by the sensory neuron. The gamma efferents enable the CNS to adjust the sensitivity of the muscle spindle. Before seeing how this is accomplished, let's see how the muscle spindle normally functions as a sensory receptor and look at its effects on the surrounding extrafusal fibers.

The sensory neuron is always active, conducting impulses to the CNS. The axon enters the CNS in a dorsal root and synapses on motor neurons in the anterior gray horn of the spinal cord. Axon collaterals distribute the information to the brain, providing information about the muscle spindle. Stretching the central portion of the intrafusal fiber distorts the dendrites and stimulates the sensory neuron, increasing the frequency of action potential generation. Compressing the central portion inhibits the sensory neuron, decreasing the frequency of action potential generation.

The axon of the sensory neuron synapses on CNS motor neurons that control the extrafusal muscle fibers of the same muscle. An increase in sensory neuron stimulation, caused by stretching of the intrafusal fiber, increases stimulation to the motor neuron controlling the surrounding extrafusal fibers, so muscle tone increases. This increase provides automatic resistance that reduces the chance of muscle damage due to overstretching. The patellar reflex and similar reflexes serve this function.

A decrease in the stimulation of the sensory neuron, due to compression of the intrafusal fiber, leads to a decrease in the stimulation of the motor neuron controlling the surrounding extrafusal fibers, so muscle tone decreases. This decrease reduces resistance to the movement under way. For example, if your elbow is flexed and you let gravity extend it, the triceps brachii muscle, which is compressed by this movement, relaxes.

Many stretch reflexes are **postural reflexes**—reflexes that help us maintain a normal upright posture. Standing, for example, involves a cooperative effort by many muscle groups. Some of these muscles work in opposition to one another, exerting forces that keep the body's weight balanced over the feet. If your body leans forward, stretch receptors in your calf muscles are stimulated. Those muscles then respond by contracting, returning your body to an upright position. If the muscles overcompensate and your body begins to lean back, your calf muscles relax. But then stretch receptors in muscles of your shins and thighs are stimulated, and the problem is corrected immediately.

Postural muscles generally maintain a firm muscle tone and have extremely sensitive stretch receptors. As a result, very fine adjustments are continually being made, and you are not aware of the cycles of contraction and relaxation that occur. Stretch reflexes are only one type of postural reflex. There are many complex polysynaptic postural reflexes.

Now let's return to the role of the gamma efferents. They let the CNS adjust the sensitivity of muscle spindles. Gamma efferents play a vital role whenever voluntary contractions change the length of a muscle. Impulses arriving over gamma efferents cause the contraction of myofibrils in the intrafusal fibers as the biceps brachii muscle shortens. The myofibrils pull on the sarcolemma in the central portion of the intrafusal fiber—the region monitored by the sensory neuron—until that membrane is stretched to its normal resting length. As a result, the muscle spindles remain sensitive to any externally imposed changes in muscle length. For this reason, if someone drops a ball into your palm when your elbow is partially flexed, the muscle spindles automatically adjust the muscle tone to compensate for the increased load.

Polysynaptic Reflexes

Polysynaptic reflexes can produce far more complicated responses than can monosynaptic reflexes. One reason is that the interneurons involved can control several muscle groups. Moreover, these interneurons may produce either excitatory or inhibitory postsynaptic potentials (EPSPs or IPSPs) at CNS motor nuclei. As a result, the response can involve the stimulation of some muscles and the inhibition of others.

The Tendon Reflex

The stretch reflex regulates the length of a skeletal muscle. The **tendon reflex** monitors the external tension produced during a muscular contraction and prevents tearing or breaking of the tendons. The sensory receptors for this reflex have not been identified, but they are distinct from both muscle spindles and proprioceptors in tendons. The receptors are stimulated when the collagen fibers are stretched to a dangerous degree. These receptors activate sensory neurons that stimulate inhibitory interneurons in the spinal cord. These interneurons in turn innervate the motor neurons controlling the skeletal muscle. The greater the tension in the tendon, the greater is the inhibitory effect on the motor neurons. As a result, a skeletal muscle generally cannot develop enough tension to break its tendons.

Withdrawal Reflexes

Withdrawal reflexes move affected parts of the body away from a stimulus. Painful stimuli trigger the strongest withdrawal reflexes, but these reflexes are sometimes initiated by the stimulation of touch receptors or pressure receptors.

The **flexor reflex**, a representative withdrawal reflex, affects the muscles of a limb (**Spotlight Figure 13–14**). Recall from Chapters 9 and 11 that flexion reduces the angle between two articulating bones, and that the contractions of flexor muscles perform this movement. ⮌ pp. 295, 363 If you grab an unexpectedly hot pan on the stove, a dramatic flexor reflex will occur. When the pain receptors in your hand are stimulated, the sensory neurons activate interneurons in the spinal cord that stimulate motor neurons in the anterior gray horns. The result is a contraction of flexor muscles that yanks your hand away from the stove.

When a specific muscle contracts, opposing muscles must relax to permit the movement. For example, flexor muscles that bend the elbow (such as the biceps brachii muscle) are opposed by extensor muscles (such as the triceps brachii muscle) that straighten it out. A potential conflict exists: In theory, the contraction of a flexor muscle should trigger a stretch reflex in the extensors that would cause them to contract, opposing the movement. Interneurons in the spinal cord prevent such competition through **reciprocal inhibition**. When one set of motor neurons is stimulated, those neurons that control antagonistic muscles are inhibited. The term *reciprocal* refers to the fact that the system works both ways. When the flexors contract, the extensors relax, and when the extensors contract, the flexors relax.

Withdrawal reflexes are much more complex than any monosynaptic reflex. They also show tremendous versatility, because the sensory neurons activate many pools of interneurons. If the stimuli are strong, interneurons will carry excitatory and inhibitory impulses up and down the spinal cord, affecting motor neurons in many segments. The end result is always the same: a coordinated movement away from the stimulus. But the distribution of the effects and the strength and character of the motor responses depend on the intensity and location of the stimulus.

Mild discomfort might provoke a brief contraction in muscles of your hand and wrist. More powerful stimuli would produce coordinated muscular contractions affecting the positions of your hand, wrist, forearm, and arm. Severe pain would also stimulate contractions of your shoulder, trunk, and arm muscles. These contractions could last for several seconds, due to the activation of reverberating circuits. In contrast, monosynaptic reflexes are invariable and brief. For example, the patellar reflex is completed in about 20–40 msec.

Crossed Extensor Reflexes

The stretch, tendon, and withdrawal reflexes involve *ipsilateral* (*ipsi*, same + *lateral*, side) *reflex arcs*: The sensory stimulus and the motor response occur on the same side of the body. The crossed extensor reflex involves a *contralateral reflex arc* (*contra*, opposite), because the motor response occurs on the side opposite the stimulus (**Spotlight Figure 13–14**).

General Characteristics of Polysynaptic Reflexes

Polysynaptic reflexes range in complexity from a simple tendon reflex to the complex and variable reflexes associated with standing, walking, and running. Yet all polysynaptic reflexes share the following basic characteristics:

- *They Involve Pools of Interneurons.* Processing takes place in pools of interneurons before motor neurons are activated. The result may be excitation or inhibition. The tendon reflex inhibits motor neurons, and the flexor and crossed extensor reflexes direct specific muscle contractions.

- *They Are Intersegmental in Distribution.* The interneuron pools extend across spinal segments and may activate muscle groups in many parts of the body.

- *They Involve Reciprocal Inhibition.* Reciprocal inhibition coordinates muscular contractions and reduces resistance to movement. In the flexor and crossed extensor reflexes, the contraction of one muscle group is associated with the inhibition of opposing muscles.

- *They Have Reverberating Circuits, Which Prolong the Reflexive Motor Response.* Positive feedback between interneurons that innervate motor neurons and the processing pool maintains the stimulation even after the initial stimulus has faded.

- *Several Reflexes May Cooperate to Produce a Coordinated, Controlled Response.* As a reflex movement gets under way, antagonistic reflexes are inhibited. For example, during the stretch reflex, antagonistic muscles are inhibited. In contrast, in the tendon reflex, antagonistic muscles are stimulated. In complex polysynaptic reflexes, commands may be distributed along the length of the spinal cord, producing a well-coordinated response.

CHECKPOINT

18. Identify the basic characteristics of polysynaptic reflexes.

19. For the patellar (knee-jerk) reflex, how would the stimulation of the muscle spindle by gamma motor neurons affect the speed of the reflex?

20. A weight lifter is straining to lift a 200-kg barbell above his head. Shortly after he lifts it to chest height, his muscles appear to relax and he drops the barbell. Which reflex has occurred?

21. During a withdrawal reflex of the foot, what happens to the limb on the side opposite the stimulus? What is this response called?

See the blue Answers tab at the back of the book.

13-8 ┃ The brain can affect spinal cord–based reflexes

Learning Outcome Explain how higher centers control and modify reflex responses.

Reflex motor behaviors happen automatically, without instructions from higher centers. However, higher centers can have a profound effect on the performance of a reflex. Processing centers in the brain can facilitate or inhibit reflex motor patterns based in the spinal cord. Descending tracts originating in the brain synapse on interneurons and motor neurons throughout the spinal cord. These synapses are continuously active, producing EPSPs or IPSPs at the postsynaptic membrane.

Voluntary Movements and Reflex Motor Patterns

Spinal reflexes produce consistent, stereotyped motor patterns that are triggered by specific external stimuli. However, the same motor patterns can also be activated as needed by centers in the brain. By making use of these preexisting patterns, relatively few descending fibers can control complex motor functions.

For example, neuronal pools in the spinal cord direct the motor patterns for walking, running, and jumping. The descending pathways from the brain provide appropriate facilitation, inhibition, or "fine-tuning" of the established patterns. This is a very efficient system that is similar to an order given in a military drill: A single command triggers a complex, predetermined sequence of events.

Motor control involves a series of interacting levels. At the lowest level are monosynaptic reflexes that are rapid, but stereotyped and relatively inflexible. At the highest level, centers in the brain can modulate or build on reflexive motor patterns.

Reinforcement and Inhibition

A single EPSP may not depolarize the postsynaptic neuron sufficiently to generate an action potential, but it does make that neuron more sensitive to other excitatory stimuli. We introduced this process of *facilitation* in Chapter 12. Alternatively, an IPSP makes the neuron less responsive to excitatory stimulation, through the process of *inhibition*. ⟳ p. 447 By stimulating excitatory or inhibitory interneurons within the brain stem or spinal cord, higher centers can adjust the sensitivity of reflexes by creating EPSPs or IPSPs at the motor neurons involved in reflex responses.

When many of the excitatory synapses are chronically active, the postsynaptic neuron can enter a state of generalized facilitation. This facilitation of reflexes can result in **reinforcement**, an enhancement of spinal reflexes. For example, a method used to overemphasize the patellar reflex is called the Jendrassik maneuver. To do this, the subject hooks the hands together by interlocking the fingers and then tries to pull the hands apart while a light tap is applied to the patella. This reinforcement produces a big kick rather than a twitch. This distractive technique still produces a larger reflex response even if the individual realizes it is just a distraction.

If a stimulus fails to elicit a particular reflex response during a clinical exam, there can be many reasons for the failure.

Figure 13–17   The Babinski Reflexes.

a   The plantar reflex (negative Babinski reflex), a curling of the toes, is seen in healthy adults.

b   The Babinski sign (positive Babinski reflex) occurs in the absence of descending inhibition. It is normal in infants, but pathological in adults.

The person may be consciously suppressing the response, the nerves involved may be damaged, or there may be underlying problems inside the CNS. The clinician may then ask the patient to perform an action designed to provide reinforcement, such as the Jendrassik maneuver. Reinforced reflexes are usually too strong to suppress consciously. If the reflex still fails to appear, the likelihood of nerve or CNS damage is increased. The clinician may then order more sophisticated tests, such as nerve conduction studies or scans.

TIPS & TRICKS

Facilitation and inhibition are similar to what happens when a symphony conductor raises or lowers one hand to control the music's volume while keeping the rhythm going with the baton hand: The basic pattern of beats doesn't change, but the loudness does.

Other descending fibers have an inhibitory effect on spinal reflexes. In adults, stroking the lateral sole of the foot produces a curling of the toes, called a **plantar reflex**, or *negative Babinski reflex*, after about a 1-second delay (**Figure 13–17a**). In contrast, stroking an infant's foot on the lateral sole produces a fanning of the toes known as the **Babinski sign**, or *positive Babinski reflex* (**Figure 13–17b**). This response disappears as descending motor pathways develop. If either the higher centers or the descending tracts are damaged, the Babinski sign reappears in an adult. As a result, clinicians often test this reflex when CNS injury is suspected.

CHECKPOINT

22. **Define reinforcement as it pertains to spinal reflexes.**

23. **After injuring her back, Tina exhibits a positive Babinski reflex. What does this imply about Tina's injury?**

See the blue Answers tab at the back of the book.

Related Clinical Terms

areflexia: Absence of reflexes.

Brown-Sequard syndrome: Loss of sensation and motor function that results from unilateral spinal cord lesions. Proprioception loss and weakness occur ipsilateral to the lesion while pain and temperature loss occur contralateral.

equinovarus: The foot is plantar flexed, inverted, and adducted; also called talipes equinovalgus.

Erb's palsy (Erb-Duchenne palsy): Obstetric condition characterized by paralysis or weakness of a newborn's upper arm muscles caused by a stretch injury to the brachial plexus.

hemiparesis: Slight paralysis or weakness affecting one side of the body.

Kernig's sign: Symptom of meningitis where patient cannot extend the leg at the knee due to stiffness in the hamstring muscles.

myelography: A diagnostic procedure in which a radiopaque dye is introduced into the cerebrospinal fluid to obtain an x-ray image of the spinal cord and cauda equina.

nerve conduction study: Test often performed along with electromyography (EMG); the test stimulates certain nerves and records their ability to send an impulse to the muscle; it can indicate where any blockage of the nerve pathway exists.

nerve growth factor: A peptide that promotes the growth and maintenance of neurons. Other factors that are important to neuron growth and repair include BDNF, NT-3, NT-4, and GAP-43.

paraplegia: Paralysis involving a loss of motor control of the lower, but not the upper, limbs.

quadriplegia: Paralysis involving the loss of sensation and motor control of the upper and lower limbs.

spinal shock: Term applied to all phenomena surrounding physiological or anatomical transection of the spinal cord that results in temporary loss or depression of all or most spinal reflex activity inferior to the level of the injury.

tabes dorsalis: Slow progressive degeneration of the myelin layer of the sensory neurons of the spinal cord that occurs in the tertiary (third) phase of syphilis. Common signs and symptoms are pain, weakness, diminished reflexes, unsteady gait, and loss of coordination.

13

Chapter Review

Study Outline

MasteringA&P®

Access more chapter study tools online in the MasteringA&P Study Area:

- Chapter Quizzes, Chapter Practice Test, Art-labeling Activities, Animations, MP3 Tutor Sessions, and Clinical Case Studies
- Practice Anatomy Lab   PAL™
- Interactive Physiology   iP®
- A&P Flix   A&PFlix
- PhysioEx   PhysioEx

13-1 ▍ The brain and spinal cord make up the central nervous system, and the cranial nerves and spinal nerves make up the peripheral nervous system p. 456

1. The CNS consists of the brain and spinal cord; the remainder of the nervous tissue forms the PNS. (*Figure 13–1*)

13-2 ▍ The spinal cord is surrounded by three meninges and carries sensory and motor information p. 457

2. The adult spinal cord includes two localized **enlargements**, which provide innervation to the limbs. The spinal cord has 31 segments, each associated with a pair of **dorsal roots** and a pair of **ventral roots**. (*Figure 13–2*)

3. The **filum terminale** (a strand of fibrous tissue) originates at the **conus medullaris** and ultimately becomes part of the **coccygeal ligament**. (*Figure 13–2*)

4. **Spinal nerves** are **mixed nerves**: They contain both afferent (sensory) and efferent (motor) fibers.

5. The three **spinal meninges** provide physical stability and shock absorption for neural tissues of the spinal cord; the **cranial meninges** surround the brain. (*Figure 13–3*)

6. The **dura mater**, the outer meningeal layer, covers the spinal cord; inferiorly, it tapers into the **coccygeal ligament**. The **epidural space** separates the dura mater from the walls of the vertebral canal. (*Figures 13–3, 13–4*)

7. Interior to the inner surface of the dura mater are the **subdural space**, the **arachnoid mater** (the second meningeal layer), and the **subarachnoid space**. The subarachnoid space contains **cerebrospinal fluid (CSF)**, which acts as a shock absorber and a diffusion medium for dissolved gases, nutrients, chemical messengers, and waste products. (*Figures 13–3, 13–4*)

8. The **pia mater**, a meshwork of elastic and collagen fibers, is the innermost meningeal layer. **Denticulate ligaments** extend from the pia mater to the dura mater. (*Figures 13–3, 13–4*)

13-3 ▍ Gray matter integrates information and initiates commands, and white matter carries information from place to place p. 461

9. The white matter of the spinal cord contains myelinated and unmyelinated axons, and the gray matter contains cell bodies of neurons and neuroglia and unmyelinated axons. The projections of gray matter toward the outer surface of the spinal cord are called **horns**. (*Figure 13–5*)

10. The **posterior gray horns** contain somatic and visceral sensory nuclei; nuclei in the **anterior gray horns** function in somatic motor control. The **lateral gray horns** contain visceral motor neurons. The **gray commissures** contain axons that cross from one side of the spinal cord to the other. (*Figure 13–5*)

11. The white matter can be divided into six **columns** *(funiculi)*, each of which contains **tracts** *(fasciculi)*. **Ascending tracts** relay information from the spinal cord to the brain, and **descending tracts** carry information from the brain to the spinal cord. *(Figure 13–5)*

13-4 ▍ Spinal nerves form plexuses that are named according to their level of emergence from the vertebral canal p. 463

12. There are 31 pairs of spinal nerves. Each has an **epineurium** (outermost layer), a **perineurium**, and an **endoneurium** (innermost layer). *(Figure 13–6)*

13. Each pair of spinal nerves monitors a region of the body surface called a **dermatome**. *(Figure 13–7)*

14. A typical spinal nerve has a **white ramus** (containing myelinated axons), a **gray ramus** (containing unmyelinated fibers that innervate glands and smooth muscles in the body wall or limbs), a **dorsal ramus** (providing sensory and motor innervation to the skin and muscles of the back), and a **ventral ramus** (supplying the ventrolateral body surface, structures in the body wall, and the limbs). *(Spotlight Figure 13-8)*

15. A complex, interwoven network of nerves is a **nerve plexus**. The four large plexuses are the **cervical plexus**, the **brachial plexus**, the **lumbar plexus**, and the **sacral plexus**. *(Figures 13–9 to 13–12)*

13-5 ▍ Interneurons are organized into functional groups called neuronal pools p. 473

16. The body has sensory neurons, which deliver information to the CNS; motor neurons, which distribute commands to peripheral effectors; and interneurons, which interpret information and coordinate responses.

17. An organized, functional group of interconnected interneurons is a **neuronal pool**.

18. The neural circuit patterns are **divergence**, **convergence**, **serial processing**, **parallel processing**, and **reverberation**. *(Figure 13–13)*

13-6 ▍ Reflexes are rapid, automatic responses to stimuli p. 475

19. A *neural reflex* involves sensory fibers delivering information to the CNS, and motor fibers carrying commands to the effectors by the PNS.

20. A **reflex arc** is the neural "wiring" of a single reflex. *(Spotlight Figure 13–14)*

21. The five steps involved in a neural reflex are (1) the arrival of a stimulus and activation of a receptor, (2) the activation of a sensory neuron, (3) information processing in the CNS, (4) the activation of a motor neuron, and (5) a response by an effector. *(Spotlight Figure 13–14)*

22. Reflexes are classified according to (1) their development, (2) the nature of the resulting motor response, (3) the complexity of the neural circuit involved, and (4) the site of information processing. *(Figure 13–15)*

23. **Innate reflexes** result from the genetically determined connections that form between neurons during development. **Acquired reflexes** are learned and typically are more complex.

24. **Somatic reflexes** control skeletal muscles; **visceral reflexes** *(autonomic reflexes)* control the activities of other systems.

25. In a **monosynaptic reflex**—the simplest reflex arc—a sensory neuron synapses directly on a motor neuron, which acts as the processing center. A **polysynaptic reflex** has at least one interneuron between the sensory afferent and the motor efferent, and there is a longer delay between stimulus and response. *(Spotlight Figure 13–14)*

26. Reflexes processed in the brain are **cranial reflexes**. In a **spinal reflex**, the important interconnections and processing events occur in the spinal cord.

13-7 ▍ Spinal reflexes vary in complexity p. 479

27. Spinal reflexes range from simple monosynaptic reflexes to more complex polysynaptic and **intersegmental reflexes**, in which many spinal segments interact to produce a coordinated motor response.

28. The **stretch reflex** (such as the **patellar**, or **knee-jerk**, **reflex**) is a monosynaptic reflex that automatically regulates skeletal muscle length and muscle tone. The sensory receptors involved are **muscle spindles**. *(Spotlight Figure 13–14; Figure 13–16)*

29. A **postural reflex** maintains a person's normal upright posture.

30. Polysynaptic reflexes can produce more complicated responses than can monosynaptic reflexes. Examples include the **tendon reflex** (which monitors the tension produced during muscular contractions and prevents damage to tendons) and **withdrawal reflexes** (which move affected portions of the body away from a source of stimulation). The **flexor reflex** is a withdrawal reflex affecting the muscles of a limb. The **crossed extensor reflex** complements withdrawal reflexes. *(Spotlight Figure 13–14)*

31. All polysynaptic reflexes (1) involve pools of interneurons, (2) are intersegmental in distribution, (3) involve reciprocal inhibition, (4) have reverberating circuits, which prolong the reflexive motor response, and (5) can cooperate to produce a coordinated response.

13-8 ▍ The brain can affect spinal cord–based reflexes p. 481

32. The brain can facilitate or inhibit reflex motor patterns based in the spinal cord.

33. Motor control involves a series of interacting levels. Monosynaptic reflexes form the lowest level; at the highest level are the centers in the brain that can modulate or build on reflexive motor patterns.

34. Facilitation can produce an enhancement of spinal reflexes known as **reinforcement**. Spinal reflexes may also be inhibited, as when the **plantar reflex** in adults replaces the **Babinski sign** in infants. *(Figure 13–17)*

Review Questions

See the blue Answers tab at the back of the book.

LEVEL 1 Reviewing Facts and Terms

1. Label the anatomical structures of the spinal cord in the following figure.

(a) ____		(b) ____	
(c) ____		(d) ____	
(e) ____		(f) ____	
(g) ____		(h) ____	
(i) ____			

2. The ventral roots of each spinal segment (a) bring sensory information into the spinal cord, (b) control peripheral effectors, (c) contain the axons of somatic motor and visceral motor neurons, (d) do both b and c.

3. Spinal nerves are called mixed nerves because they (a) contain sensory and motor fibers, (b) exit at intervertebral foramina, (c) are associated with a pair of dorsal root ganglia, (d) are associated with dorsal and ventral roots.

4. The adult spinal cord extends only to (a) the coccyx, (b) the sacrum, (c) the third or fourth lumbar vertebra, (d) the first or second lumbar vertebra, (e) the last thoracic vertebra.

5. Which of the following statements is *false* concerning the gray matter of the spinal cord? (a) It is located in the interior of the spinal cord around the central canal, (b) It functions in processing neural information, (c) It is primarily involved in relaying information to the brain, (d) It contains motor neurons, (e) It is divided into regions called horns.

6. The following are the steps involved in a neural reflex.
   (1) activation of a sensory neuron
   (2) activation of a motor neuron
   (3) response by an effector
   (4) arrival of a stimulus and activation of a receptor
   (5) information processing

   The proper sequence of these steps is

   (a) 1, 3, 4, 5, 2.   (b) 4, 5, 3, 1, 2.
   (c) 4, 1, 5, 2, 3.   (d) 4, 3, 1, 5, 2.
   (e) 3, 1, 4, 5, 2.

7. A sensory region monitored by the dorsal rami of a single spinal segment is (a) a ganglion, (b) a fascicle, (c) a dermatome, (d) a ramus.

8. List the three meningeal layers working outwards from the spinal cord.

9. The genitofemoral, femoral, and lateral femoral cutaneous nerves are major nerves of the (a) lumbar plexus, (b) sacral plexus, (c) brachial plexus, (d) cervical plexus.

10. The synapsing of several neurons on the same postsynaptic neuron is called (a) serial processing, (b) reverberation, (c) divergence, (d) convergence.

11. The reflexes that control the most rapid, stereotyped motor responses to stimuli are (a) monosynaptic reflexes, (b) polysynaptic reflexes, (c) tendon reflexes, (d) extensor reflexes.

12. The key structures involved in a stretch reflex, such as the patella reflex, are the patella tendon, muscle spindle stretch receptors, afferent pathway nerve, _____, interneuron, motor neuron, and muscle fibers.

13. The contraction of flexor muscles and the relaxation of extensor muscles illustrate the principle of (a) reverberating circuitry, (b) generalized facilitation, (c) reciprocal inhibition, (d) reinforcement.

14. Reflex arcs in which the sensory stimulus and the motor response occur on the same side of the body are (a) contralateral, (b) paraesthetic, (c) ipsilateral, (d) monosynaptic.

15. Proceeding deep from the most superficial structure, number the following in the correct sequence:

   (a) _____ walls of vertebral canal   (b) _____ pia mater
   (c) _____ dura mater   (d) _____ arachnoid membrane
   (e) _____ subdural space   (f) _____ subarachnoid space
   (g) _____ epidural space   (h) _____ spinal cord

LEVEL 2 Reviewing Concepts

16. Explain the anatomical significance of the fact that spinal cord growth ceases at age 4.

17. List, in sequence, the five steps involved in a neural reflex.

18. Polysynaptic reflexes can produce far more complicated responses than can monosynaptic reflexes because (a) the response time is quicker, (b) the response is initiated by highly sensitive receptors, (c) motor neurons carry impulses at a faster rate than do sensory neurons, (d) the interneurons involved can control several muscle groups.

19. Why do cervical nerves outnumber cervical vertebrae?

20. Why is gray matter gray and white matter white?

21. List all of the CNS sites where cerebrospinal fluid (CSF) is located. What are the functions of CSF?

22. What five characteristics are common to all polysynaptic reflexes?

23. Predict the effects on the body of a spinal cord transection at segment C_7. How would these effects differ from those of a spinal cord transection at segment T_{10}?

24. In relation to neural pathways, what is convergence and divergence?

25. Side-to-side movements of the spinal cord are prevented by the (a) filum terminale, (b) denticulate ligaments, (c) dura mater, (d) pia mater, (e) arachnoid mater.

26. Ascending tracts (a) carry sensory information to the brain, (b) carry motor information to the brain, (c) carry sensory information from the brain, (d) carry motor information from the brain, (e) connect perceptive areas with the brain.

27. In the spinal nerves, what are the epineurium, perineurium, and endoneurium?

13

LEVEL 3 Critical Thinking and Clinical Applications

28. Mary complains that when she wakes up in the morning, her thumb and forefinger are always "asleep." She mentions this condition to her physician, who asks Mary whether she sleeps with her wrists flexed. She replies that she does. The physician tells Mary that sleeping in that position may compress a portion of one of her peripheral nerves, producing her symptoms. Which nerve is involved?

29. The improper use of crutches can produce a condition known as "crutch paralysis," characterized by a lack of response by the extensor muscles of the arm, and a condition known as "wrist drop," consisting of an inability to extend the fingers and wrist. Which nerve is involved?

30. Bowel and urinary bladder control involve spinal reflex arcs that are located in the sacral region of the spinal cord. In both instances, two sphincter muscles—an inner sphincter of smooth muscle and an outer sphincter of skeletal muscle—control the passage of wastes (feces and urine) out of the body. How would a transection of the spinal cord at the L_1 segment level affect an individual's bowel and bladder control?

31. Karen falls down a flight of stairs and suffers lumbar and sacral spinal cord damage due to hyperextension of her back. The injury resulted in edema around the central canal that compressed the anterior horn of the lumbar region. What signs would you expect to observe as a result of this injury?

CLINICAL CASE Wrap-Up

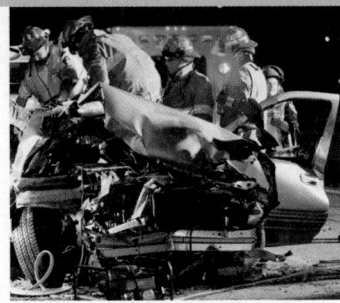

Prom Night

13

The emergency room team "log rolls" Joe over onto his side while maintaining his spinal alignment. The abrasions and bruising of his back are obvious. The doctor can feel that Joe's spinous processes, from the 7th cervical to the 11th thoracic vertebrae are in good alignment. There is an obvious gap, however, between the 11th and 12th thoracic spinous processes. The 12th thoracic spinous process and lumbar spine are displaced lateral to the midline. Palpating in this area causes Joe a great deal of discomfort. Joe has no sensation inferior to his 11th thoracic dermatome, just below the level of the navel. Joe also has no voluntary muscle control of any muscles of his hips or inferior.

X-rays taken in the trauma room show a fracture dislocation of the thoracic spine between the 11th thoracic vertebra (T_{11}) and the 12th thoracic vertebra (T_{12}). An MRI confirms the clinical impression that Joe's spinal cord is completely disrupted at this level.

Joe's spine is surgically realigned and stabilized (fixed in place) with internal fixation. This enables him to begin sitting within the first week after his accident. Within two weeks he is transferred to a spinal rehabilitation facility to begin working on independence as a paraplegic—a person with loss of motor control and sensation to the lower limbs (but not the upper limbs).

1. Do you think Joe will regain motor or sensory activity in his lower extremities? Why or why not?

2. Do you think Joe's Babinski reflexes are normal (toes plantar flexed) or abnormal (toes going up, dorsiflexed)? Why?

See the blue Answers tab at the back of the book.

14

The Brain and Cranial Nerves

Learning Outcomes

These Learning Outcomes correspond by number to this chapter's sections and indicate what you should be able to do after completing the chapter.

14-1 ∥ Name the major brain regions, vesicles, and ventricles, and describe the locations and functions of each.

14-2 ∥ Explain how the brain is protected and supported, and discuss the formation, circulation, and function of cerebrospinal fluid.

14-3 ∥ Describe the anatomical differences between the medulla oblongata and the spinal cord, and identify the main components and functions of the medulla oblongata.

14-4 ∥ List the main components of the pons, and specify the functions of each.

14-5 ∥ List the main components of the cerebellum, and specify the functions of each.

14-6 ∥ List the main components of the midbrain, and specify the functions of each.

14-7 ∥ List the main components of the diencephalon, and specify the functions of each.

14-8 ∥ Identify the main components of the limbic system, and specify the locations and functions of each.

14-9 ∥ Identify the major anatomical subdivisions and functions of the cerebrum, and discuss the origin and significance of the major types of brain waves seen in an electroencephalogram.

14-10 ∥ Describe representative examples of cranial reflexes that produce somatic responses or visceral responses to specific stimuli.

SPOTLIGHT

- Formation and Circulation of Cerebrospinal Fluid p. 494

CLINICAL CASE

- The Neuroanatomist's Stroke p. 488

Clinical Notes

- Epidural and Subdural Hemorrhages p. 493
- Disconnection Syndrome p. 514
- Aphasia and Dyslexia p. 515

The Neuroanatomist's Stroke

Dr. Jill Bolte Taylor, a well-published neuroanatomist, was at the top of her field at age 37 when she awoke to an evolving cerebrovascular accident (CVA), or stroke. It was caused by bleeding into her left cerebral hemisphere near her language center. The irony of her situation was not lost on her. There she was, a brain scientist witnessing a brain-centered medical emergency: her own.

Dr. Taylor's first sign that something was wrong was a throbbing, pulsating headache behind her left eye. She then noticed a shift in her perceptions and consciousness, and experienced a slowing of her thoughts and movements. By the time she was aware that her right arm was paralyzed, she was barely able to call for help. When she arrived at the hospital, she could not

walk, talk, read, write, or recall anything about her life. She felt her spirit surrender and braced for death.

Dr. Taylor awoke later that day, shocked to be alive. She was unable to speak or understand speech. Sounds irritated her. She could not recognize or use numbers. Less than three weeks later, doctors performed open brain surgery to remove a large blood clot that was pressing on the left side of her brain. **Would Dr. Taylor ever fully recover? To find out, turn to the Clinical Case Wrap-Up on p. 533.**

14 ▮ An Introduction to the Brain and Cranial Nerves

In this chapter we introduce the functional organization of the brain and cranial nerves, and describe simple cranial reflexes. The adult human brain contains almost 97 percent of the body's neural tissue. A "typical" brain weighs 1.4 kg (3 lb) and has a volume of 1200 mL (71 in.^{3}).

Brain size varies considerably among individuals. The brains of males are, on average, about 10 percent larger than those of females, due to differences in average body size. No correlation exists between brain size and intelligence. Individuals with the smallest brains (750 mL) and the largest brains (2100 mL) are functionally normal.

14-1 ▮ The brain has several principal structures, each with specific functions

Learning Outcome Name the major brain regions, vesicles, and ventricles, and describe the locations and functions of each.

In this section we introduce the anatomical organization of the brain. We begin with an overview of the brain's major regions and landmarks. Then we discuss the brain's embryonic origins and some prominent internal cavities: the ventricles of the brain.

Major Brain Regions and Landmarks

There are six major brain regions: the cerebrum, cerebellum, diencephalon, midbrain, pons, and medulla oblongata (Figure 14–1). The largest portion of the adult brain is the **cerebrum** (se-RĒ-brum or SER-e-brum). It is made up of large,

paired left and right **cerebral hemispheres**. The brain has an extensive area of **cortex**, a superficial layer of gray matter covering most of its surface. The surfaces of the cerebral hemispheres are highly folded and covered by a neural cortex called the **cerebral cortex** (*cortex*, rind or bark). This cerebral cortex forms a series of elevated ridges, or **gyri** (JĪ-rī; singular, *gyrus*). Gyri increase its surface area. The gyri are separated by shallow depressions called **sulci** (SUL-sī) or by deeper grooves called **fissures**. The cerebrum is the seat of most higher mental functions. Conscious thoughts, sensations, intellect, memory, and complex movements all originate in the cerebrum.

The **cerebellum** (ser-e-BEL-um) is the second-largest part of the brain. It is partially hidden by the cerebral hemispheres. Like the cerebrum, the cerebellum has hemispheres covered by a neural cortex called the *cerebellar cortex*. The cerebellum adjusts ongoing movements by comparing arriving sensations with previously experienced sensations, allowing you to perform the same movements over and over.

The other major anatomical regions of the brain can best be examined after the cerebral and cerebellar hemispheres have been removed (Figure 14–1). The walls of the **diencephalon** (dī-en-SEF-a-lon; *dia*, through + *encephalos*, brain) are composed of the **left thalamus** and **right thalamus** (THAL-a-mus; plural, *thalami*). Each thalamus contains relay and processing centers for sensory information. The **hypothalamus** (*hypo-*, below) is the floor of the diencephalon. It contains centers involved with emotions, autonomic function, and hormone production. The *infundibulum*, a narrow stalk, connects the hypothalamus to the **pituitary gland**, a part of the endocrine system. The hypothalamus and the pituitary gland integrate the nervous and endocrine systems.

The diencephalon is a structural and functional link between the cerebral hemispheres and the brain stem. The

Figure 14–1   **An Introduction to Brain Structures and Functions.**

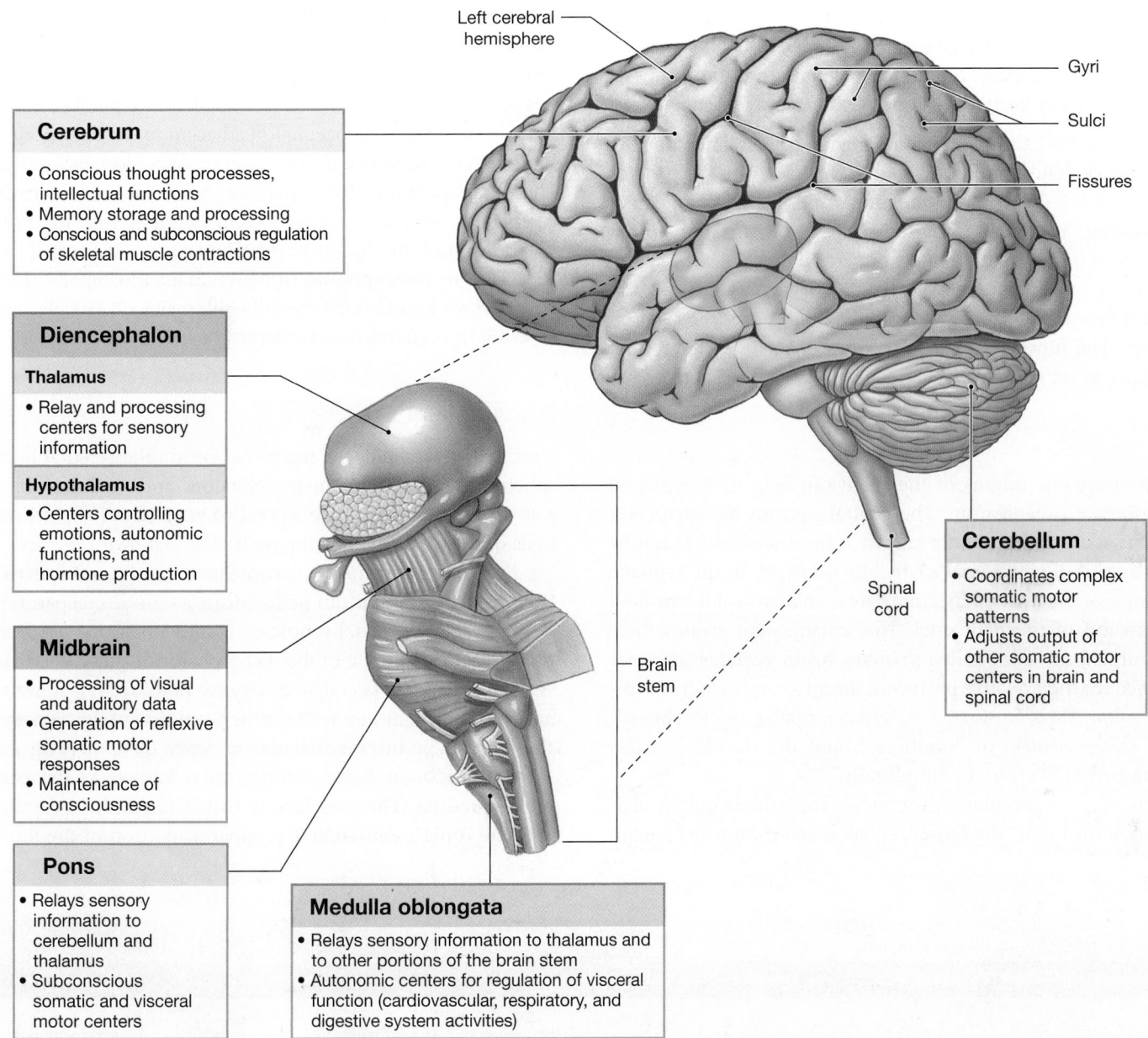

Cerebrum

- Conscious thought processes, intellectual functions
- Memory storage and processing
- Conscious and subconscious regulation of skeletal muscle contractions

Diencephalon

Thalamus

- Relay and processing centers for sensory information

Hypothalamus

- Centers controlling emotions, autonomic functions, and hormone production

Midbrain

- Processing of visual and auditory data
- Generation of reflexive somatic motor responses
- Maintenance of consciousness

Pons

- Relays sensory information to cerebellum and thalamus
- Subconscious somatic and visceral motor centers

Medulla oblongata

- Relays sensory information to thalamus and to other portions of the brain stem
- Autonomic centers for regulation of visceral function (cardiovascular, respiratory, and digestive system activities)

Cerebellum

- Coordinates complex somatic motor patterns
- Adjusts output of other somatic motor centers in brain and spinal cord

Left cerebral hemisphere

Gyri

Sulci

Fissures

Spinal cord

Brain stem

14

---

brain stem contains a variety of important processing centers and nuclei that relay information headed to or from the cerebrum or cerebellum. As **Figure 14–1** shows, the brain stem includes the *midbrain, pons,* and *medulla oblongata.*[1]

- The **midbrain,** or *mesencephalon,* contains nuclei that process visual and auditory information and control reflexes triggered by these stimuli. For example, your immediate,

reflexive responses to a loud, unexpected noise (eye movements and head turning) are directed by nuclei in the midbrain. This region also contains centers that help maintain consciousness.

- The **pons** of the brain connects the cerebellum to the brain stem (*pons* is Latin for "bridge"). In addition to tracts and relay centers, the pons contains nuclei involved with somatic and visceral motor control.

- The **medulla oblongata** connects the brain to the spinal cord. Near the pons, the posterior wall of the medulla oblongata is thin and membranous. The inferior portion of

---

[1]Some sources consider the brain stem to include the diencephalon. We will use the more restrictive definition.

the medulla oblongata resembles the spinal cord in that it has a narrow central canal. The medulla oblongata relays sensory information to the thalamus and to centers in other portions of the brain stem. The medulla oblongata also contains major centers that regulate autonomic function, such as heart rate, blood pressure, and digestion.

Figure 14–1 shows the boundaries and general functions of the diencephalon and brain stem. In considering the parts of the brain, we will begin at the inferior portion of the medulla oblongata. This region has the simplest organization found anywhere in the brain. In many respects, it resembles the spinal cord. We will then ascend to regions of increasing structural and functional complexity. Finally, we will reach the cerebral cortex. The functions and capabilities of this very interesting area are as yet poorly understood.

Embryology of the Brain

The embryonic origins of the brain can help us understand its internal organization. The central nervous system (CNS) begins as a hollow cylinder known as the *neural tube*. This tube has a fluid-filled internal cavity, the *neurocoel*. In the cephalic portion of the neural tube, three areas enlarge rapidly through expansion of the neurocoel. This enlargement creates three prominent divisions called **primary brain vesicles**. They are named for their relative positions: the *prosencephalon* (prōz-en-SEF-a-lon; *proso*, forward + *enkephalos*, brain), or "forebrain"; the *mesencephalon*, or "midbrain"; and the *rhombencephalon* (rom-ben-SEF-a-lon), or "hindbrain."

Table 14–1 summarizes the fates of these three primary divisions of the brain. The prosencephalon and rhombencephalon are subdivided further, forming **secondary brain vesicles**. The prosencephalon forms the **telencephalon** (tel-en-SEF-a-lon; *telos*, end) and the diencephalon. The telencephalon ultimately forms the cerebrum of the brain. The walls of the mesencephalon thicken, and the neurocoel becomes a relatively narrow passageway, much like the central canal of the spinal cord. The portion of the rhombencephalon adjacent to the mesencephalon forms the **metencephalon** (met-en-SEF-a-lon; *meta*, after). The dorsal portion of the metencephalon later becomes the cerebellum, and the ventral portion develops into the pons. The portion of the rhombencephalon closer to the spinal cord forms the **myelencephalon** (mī-el-en-SEF-a-lon; *myelon*, spinal cord), which becomes the medulla oblongata. ATLAS: Embryology Summary 12: The Development of the Brain and Cranial Nerves

Ventricles of the Brain

During development, the neurocoel within the cerebral hemispheres, diencephalon, metencephalon, and medulla oblongata expands to form chambers called **ventricles** (VEN-tri-kls). Cells of the *ependyma* line the ventricles. ᔐ p. 418

Each cerebral hemisphere contains a large **lateral ventricle** (Figure 14–2). The **septum pellucidum**, a thin medial partition, separates the two lateral ventricles. Because there are *two* lateral ventricles, the ventricle in the diencephalon is called the **third ventricle**. The two lateral ventricles are not directly connected, but each communicates with the third ventricle of the diencephalon through an **interventricular foramen** (*foramen of Monro*).

The midbrain has a slender canal known as the **cerebral aqueduct**. This passageway connects the third ventricle with the **fourth ventricle**. The superior portion of the fourth

Table 14–1	Development of the Brain			
Primary Brain Vesicles (3 weeks)		**Secondary Brain Vesicles (6 weeks)**	**Brain Regions at Birth**	**Ventricles**
Prosencephalon		Telencephalon	Cerebrum	Lateral ventricles
		Diencephalon	Diencephalon	Third ventricle
Mesencephalon		Mesencephalon	Midbrain	Cerebral aqueduct
Rhombencephalon		Metencephalon	Cerebellum and Pons	Fourth ventricle
		Myelencephalon	Medulla oblongata	Fourth ventricle

Figure 14–2 **Ventricles of the Brain.** The orientation and extent of the ventricles as they would appear if the brain were transparent. ATLAS: Plates 10; 12a–c; 13a–e

a **Ventricles, lateral view**

b **Ventricles, anterior view**

ventricle lies between the posterior surface of the pons and the anterior surface of the cerebellum. The fourth ventricle extends into the superior portion of the medulla oblongata. There this ventricle narrows and becomes continuous with the central canal of the spinal cord.

The ventricles are filled with cerebrospinal fluid (CSF). The CSF continuously circulates from the ventricles and central canal into the *subarachnoid space* of the surrounding cranial meninges. The CSF passes between the interior and exterior of the CNS through three foramina in the roof of the fourth ventricle. We will describe these foramina in a later section.

> **CHECKPOINT**
>
> 1. **Name the six major regions of the brain.**
> 2. **What brain regions make up the brain stem?**
> 3. **Which primary brain vesicle is destined to form the cerebellum, pons, and medulla oblongata?**
>
> See the blue Answers tab at the back of the book.

14-2 ❙ The brain is protected and supported by the cranial meninges, cerebrospinal fluid, and the blood–brain barrier

Learning Outcome Explain how the brain is protected and supported, and discuss the formation, circulation, and function of cerebrospinal fluid.

The delicate tissues of the brain are protected from physical forces by the bones of the cranium, the *cranial meninges*, and cerebrospinal fluid. In addition, the neural tissue of the brain is biochemically isolated from the general circulation by the *blood–brain barrier*.

Refer to Figures 7–3 and 7–4 (pp. 236–239) for a review of the bones of the cranium. We will discuss the other protective factors here.

The Cranial Meninges

The layers that make up the cranial meninges—the cranial dura mater, arachnoid mater, and pia mater—are continuous with the spinal meninges. ⤺ p. 459 However, the cranial meninges have distinctive anatomical and functional characteristics (**Figure 14–3a**):

- The *dura mater* consists of outer and inner fibrous layers. The outer layer is fused to the periosteum of the cranial bones. As a result, there is no epidural space superficial to the dura mater, as occurs along the spinal cord. The outer, *periosteal*, and inner, *meningeal*, layers of the cranial dura mater are typically separated by a slender gap that contains tissue fluids and blood vessels, including several large venous sinuses. The veins of the brain open into these sinuses, which deliver the venous blood to the *internal jugular veins* of the neck.

- The *arachnoid mater* consists of the arachnoid membrane (an epithelial layer) and the cells and fibers of the arachnoid trabeculae that cross the subarachnoid space to the pia mater. The arachnoid membrane covers the brain, providing a smooth surface that does not follow the brain's underlying folds. This membrane is in contact with the inner epithelial layer of the dura mater. The subarachnoid space extends between the arachnoid membrane and the pia mater.

Figure 14–3 **The Relationship among the Brain, Cranium, and Cranial Meninges.** ATLAS: Plates 7a–d

a A lateral view of the brain, showing its position in the cranium and the organization of the meninges

b A diagrammatic view, showing the orientation of the three largest dural folds: the falx cerebri, tentorium cerebelli, and falx cerebelli

■ The *pia mater* sticks to the surface of the brain, anchored by the processes of astrocytes. It extends into every fold and fissure. It also accompanies the branches of cerebral blood vessels as they penetrate the surface of the brain to reach internal structures.

Dural Folds

In several locations, the inner layer of the dura mater extends into the cranial cavity, forming a sheet that dips inward and then returns. These **dural folds** provide additional stabilization and support to the brain. **Dural sinuses** are large collecting veins located within the dural folds. The three largest dural folds are called the falx cerebri, the tentorium cerebelli, and the falx cerebelli (**Figure 14–3b**):

1. The **falx cerebri** (FALKS SER-e-brī; *falx*, curving or sickle-shaped) is a fold of dura mater that projects between the cerebral hemispheres in the longitudinal fissure. Its inferior

portions attach anteriorly to the crista galli of the ethmoid bone and posteriorly to the *internal occipital crest*, a ridge along the inner surface of the occipital bone. The **superior sagittal sinus** and the **inferior sagittal sinus**, two large venous sinuses, lie within this dural fold. The posterior margin of the falx cerebri intersects the tentorium cerebelli.

2. The **tentorium cerebelli** (ten-TŌ-rē-um ser-e-BEL-ē; *tentorium*, tent) protects the cerebellum and separates the cerebral hemispheres from the cerebellum. It extends across the cranium at right angles to the falx cerebri. The **transverse sinus** lies within the tentorium cerebelli.

3. The **falx cerebelli** divides the two cerebellar hemispheres along the midsagittal line inferior to the tentorium cerebelli.

The Protective Function of the Cranial Meninges

The overall shape of the brain corresponds to that of the cranial cavity (**Figure 14–3a**). The cranial bones provide physical protection by cradling the brain, but they also pose a threat to safety that is countered by the cranial meninges and the CSF. The brain is like a person driving a car: If the car hits a tree, the car protects the driver from contact with the tree, but serious injury will occur unless a seat belt or airbag protects the driver from contact with the car. The tough, fibrous dural folds act like seat belts that hold the brain in position. The cerebrospinal fluid in the subarachnoid space acts like a bumper by cushioning the brain against sudden jolts and shocks.

Cranial trauma is a head injury resulting from impact with another object. Each year in the United States, about 8 million cases of cranial trauma occur, but only 1 case in 8 results in serious brain damage. The percentage is relatively low because the cranial meninges and CSF are so effective in protecting the brain.

Cerebrospinal Fluid

Cerebrospinal fluid (CSF) completely surrounds and bathes the exposed surfaces of the CNS. The CSF has several important functions, including the following:

- *Cushioning Delicate Neural Structures.*

- *Supporting the Brain.* In essence, the brain is suspended inside the cranium and floats in the CSF. A human brain weighs about 1400 g (3.09 lb.) in air, but only about 50 g (1.8 oz.) when supported by CSF.

- *Transporting Nutrients, Chemical Messengers, and Waste Products.* Except at the choroid plexus, where CSF is produced, the ependymal lining is freely permeable. The CSF is in

constant chemical communication with the interstitial fluid that surrounds the neurons and neuroglia of the CNS.

Because free exchange occurs between the interstitial fluid of the brain and the CSF, changes in CNS function can produce changes in the composition of the CSF (**Spotlight Figure 14–4**). As noted in Chapter 13, a *spinal tap* can provide useful clinical information about CNS injury, infection, or disease. ⟳ p. 460

The Formation and Circulation of CSF

A **choroid plexus** (*choroeides*, like a membrane; *plexus*, network) is an area within each ventricle that produces CSF.

Specialized ependymal cells bearing microvilli and interconnected by tight junctions surround the capillaries of a choroid plexus. The ependymal cells secrete CSF into the ventricles. They also remove waste products from the CSF and adjust its composition over time. The differences in composition between CSF and blood plasma (blood with the cellular elements removed) are quite noticeable. For example, blood plasma contains high concentrations of soluble proteins, but the CSF does not. The concentrations of individual ions and the levels of amino acids, lipids, and waste products are also different (**Spotlight Figure 14–4**).

If the normal circulation or resorption of CSF is interrupted, a variety of clinical problems may appear. For example, a problem with the resorption of CSF in infancy causes *hydrocephalus*, or "water on the brain." Infants with this condition have enormously expanded skulls due to the presence of an abnormally large volume of CSF. In adults, a failure of resorption or a blockage of CSF circulation can distort and damage the brain.

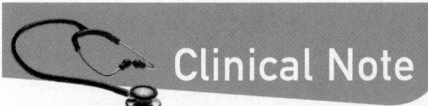

Clinical Note

Epidural and Subdural Hemorrhages A severe head injury may damage meningeal blood vessels and cause bleeding into the cranial cavity. The most serious cases involve an arterial break, because arterial blood pressure is relatively high. If blood is forced between the dura mater and the skull, the condition is known as an **epidural hemorrhage**. The elevated fluid pressure then distorts the underlying tissues of the brain. The individual loses consciousness for a period lasting from minutes to hours after the injury, and death follows in untreated cases. An epidural hemorrhage involving a damaged vein does not produce massive symptoms immediately. The individual may not have neurological problems for hours, days, or even weeks after the original injury. As a result, the problem may not be noticed until the nervous tissue has been

severely damaged. Epidural hemorrhages are rare, occurring in less than 1 percent of head injuries. However, the mortality rate is 100 percent in untreated cases and over 50 percent even after the blood pool has been removed and the damaged vessels have been closed.

A **subdural hemorrhage** is bleeding between the dura mater and the arachnoid mater. Subdural hemorrhages are twice as common as epidural hemorrhages. Small veins that cross the subdural space or one of the dural sinuses causes the bleeding. Because the venous blood pressure in a subdural hemorrhage is lower than that in an arterial epidural hemorrhage, the distortion produced is gradual. The effects on brain function can be quite variable and difficult to diagnose.

Cerebrospinal fluid (CSF) completely surrounds and bathes the exposed surfaces of the CNS. The CSF has three important functions:

- Supporting the brain.
- Cushioning the brain and spinal cord against physical trauma.
- Transporting nutrients, chemical messengers, and wastes.

1 The choroid plexus produces and maintains CSF. Two extensive folds of the choroid plexus originate in the roof of the third ventricle and extend through the interventricular foramina. These folds cover the floors of the lateral ventricles. In the inferior brain stem, a region of the choroid plexus in the roof of the fourth ventricle projects between the cerebellum and the pons.

Interstitial Fluid in thalamus

Nutrients, O_2

Capillaries

Waste products, CO_2

Neuron

Astrocyte

Choroid plexus ependymal cells

Ependymal cells

Removal of waste

Production of CSF

Cerebrospinal Fluid in third ventricle

Tight junction

Choroid plexus

Choroid plexus of third ventricle

2 The CSF circulates from the choroid plexuses through the ventricles and fills the central canal of the spinal cord. As it circulates, materials diffuse between the CSF and the interstitial fluid of the CNS across the ependymal cells.

Choroid plexus of fourth ventricle

Dura mater (periosteal layer)

Arachnoid granulation

Arachnoid trabecula

Cranium

Superior sagittal sinus

CSF fluid movement

Dura mater (meningeal layer)

Subdural space

Arachnoid mater

Cerebral cortex

Subarachnoid space

Pia mater

5 Fingerlike extensions of the arachnoid membrane, called the arachnoid villi, penetrate the meningeal layer of the dura mater and extend into the superior sagittal sinus. In adults, these extensions form large arachnoid granulations. CSF is absorbed into the venous circulation at the arachnoid granulations.

3 The CSF reaches the sub-arachnoid space through two lateral apertures and a single median aperture in the roof of the fourth ventricle.

Spinal cord

Central canal

Dura mater

Cauda equina

Arachnoid mater

Filum terminale

4 Cerebrospinal fluid then flows through the subarachnoid space surrounding the brain, spinal cord, and cauda equina.

The choroid plexuses produce CSF at a rate of about 500 mL, or 2.1 cups, per day. The total volume of CSF at any moment is approximately 150 mL. The entire volume of CSF is replaced about every eight hours. Despite this rapid turnover, the composition of CSF is closely regulated, and the rate of removal normally keeps pace with the rate of production.

The Blood Supply to the Brain

As noted in Chapter 12, neurons have a high demand for energy, but they have no energy reserves in the form of carbohydrates or lipids. Neurons also have no oxygen reserves in the form of myoglobin. Your brain, with billions of neurons, is an extremely active organ with a continuous demand for nutrients and oxygen. These demands are met by an extensive circulatory supply.

Arterial blood reaches the brain through the *internal carotid arteries* and the *vertebral arteries*. Most of the venous blood from the brain leaves the cranium in the *internal jugular veins*, which drain the dural sinuses. A head injury that damages cerebral blood vessels may cause bleeding into the dura mater, either near the dural epithelium or between the outer layer of the dura mater and the skull. These are serious conditions, because the blood entering these spaces compresses and distorts the soft tissues of the brain.

Cerebrovascular diseases are cardiovascular disorders that interfere with the normal blood supply to the brain. The particular distribution of the vessel involved determines the signs and symptoms. The degree of oxygen or nutrient starvation determines their severity. A **cerebrovascular accident (CVA)**, or *stroke*, occurs when the blood supply to a portion of the brain is shut off. Affected neurons begin to die in a matter of minutes.

The Blood–Brain Barrier

Neural tissue in the CNS is isolated from the general circulation by the **blood–brain barrier (BBB)**. This barrier is formed by capillary endothelial cells that are extensively interconnected by tight junctions. These junctions prevent materials from diffusing between endothelial cells. In general, only lipid-soluble compounds (including carbon dioxide; oxygen; ammonia; lipids, such as steroids or prostaglandins; and small alcohols) can diffuse across the membranes of endothelial cells into the interstitial fluid of the brain and spinal cord. Water and ions must pass through channels in the apical and basement plasma membranes. Larger, water-soluble compounds can cross the capillary walls only by active or passive transport.

The restricted permeability of the endothelial lining of brain capillaries is in some way dependent on chemicals secreted by astrocytes. �﹁ p. 419 These glial cells play a key supporting role in the blood–brain barrier because they release chemicals that control the permeabilities of the endothelium to various substances. Astrocytes are in close contact with CNS capillaries. The processes of astrocytes cover the outer surfaces of the endothelial cells. If the astrocytes are damaged or stop stimulating the endothelial cells, the blood–brain barrier disappears.

The choroid plexus is not part of the neural tissue of the brain, so no astrocytes are in contact with the endothelial cells there. Substances do not have free access to the CNS, however, because specialized ependymal cells create a **blood–CSF barrier**. These cells, also interconnected by tight junctions, surround the capillaries of the choroid plexus.

Transport across the blood–brain and blood–CSF barriers is selective and directional. Even the passage of small ions, such as sodium, hydrogen, potassium, or chloride, is controlled. As a result, the pH and concentrations of sodium, potassium, calcium, and magnesium ions in the blood and CSF are different. Some organic compounds are readily transported, and others cross only in very small amounts.

Neurons have a constant need for glucose that must be met regardless of the concentrations in the blood and interstitial fluid. Even when circulating glucose levels are low, endothelial cells continue to transport glucose from the blood to the interstitial fluid of the brain. In contrast, only trace amounts of circulating norepinephrine, epinephrine, dopamine, and serotonin pass into the interstitial fluid or CSF of the brain. This limitation is important, because these compounds are neurotransmitters. Their entry from the bloodstream (where concentrations can be relatively high) could result in the uncontrolled stimulation of neurons throughout the brain.

The blood–brain barrier remains intact throughout the CNS except in small areas known as *circumventricular organs (CVOs)*, which have fenestrated capillaries. Because they are outside the BBB, they provide a direct link between the CNS and the peripheral blood and they play a role in neuroendocrine function. In summary, these noteworthy exceptions include:

1. Portions of the hypothalamus, where the capillary endothelium is extremely permeable. This permeability exposes hypothalamic nuclei to circulating hormones and permits hypothalamic hormones to diffuse into the circulation.

2. Capillaries in the posterior lobe of the pituitary gland, which is continuous with the floor of the hypothalamus. At this site, the hormones antidiuretic hormone and oxytocin, produced by hypothalamic neurons, are released into the circulation.

3. Capillaries in the *pineal gland*. This endocrine gland is located on the posterior, superior surface of the diencephalon. The capillary permeability allows pineal secretions into the general circulation.

4. Capillaries at a choroid plexus. The capillary characteristics of the blood–brain barrier are lost there, but the transport activities of specialized ependymal cells in the choroid plexuses maintain the blood–CSF barrier.

Physicians must sometimes get specific compounds into the interstitial fluid of the brain to fight CNS infections or to treat other neural disorders. To do this, they must understand the limitations of the blood–brain barrier and blood–CSF barrier. For example, when considering possible treatments, the antibiotic *tetracycline* isn't used to treat meningitis or other CNS infections because this drug is excluded from the brain. In contrast, *sulfisoxazole* and *sulfadiazine* enter the CNS very rapidly. Sometimes, chemotherapeutic drugs or imaging dyes may be injected directly into the CSF by lumbar puncture.

CHECKPOINT

4. From superficial to deep, name the layers that make up the cranial meninges.

5. What would happen if the normal circulation or resorption of CSF were blocked?

6. How would decreased diffusion across the arachnoid granulations affect the volume of cerebrospinal fluid in the ventricles?

7. Many water-soluble molecules that are abundant in the blood occur in small amounts or not at all in the extracellular fluid (ECF) of the brain. Why?

See the blue Answers tab at the back of the book.

14-3 | The medulla oblongata is continuous with the spinal cord and contains vital centers

Learning Outcome Describe the anatomical differences between the medulla oblongata and the spinal cord, and identify the main components and functions of the medulla oblongata.

The medulla oblongata is the most inferior of the brain regions. Figure 14–5 shows the position of the medulla oblongata in relation to the other components of the brain stem and the diencephalon. It also illustrates the attachment sites for 11 of the 12 pairs of cranial nerves. The individual cranial nerves are identified by a capital N followed by a Roman numeral. (We will introduce the full names and functions of these nerves in a later section.)

In sectional view, the inferior portion of the medulla oblongata resembles the spinal cord, with a small central canal. However, the gray matter and white matter organization is more complex. Ascending the medulla oblongata, the central canal opens into the fourth ventricle, and the similarity to the spinal cord all but disappears.

The medulla oblongata is a very busy place. All communication between the brain and spinal cord involves tracts that ascend or descend through the medulla oblongata. In addition, the medulla oblongata is a center for the coordination of complex autonomic reflexes and the control of visceral functions.

The medulla oblongata includes three groups of nuclei that we will discuss in later chapters (Figure 14–6a,b):

1. *Autonomic Nuclei Controlling Visceral Activities.* The **reticular formation** is a loosely organized mass of gray matter that contains embedded nuclei. It extends from the medulla oblongata to the midbrain. The portion of the reticular formation in the medulla oblongata contains nuclei and specific reflex, or processing, centers that regulate vital autonomic functions. These **reflex centers** receive inputs from cranial nerves, the cerebral cortex, and the brain stem. Their output controls or adjusts the activities of one or more peripheral systems.

   There are two major groups of reflex centers in the medulla oblongata. The **cardiovascular centers** adjust the heart rate, the strength of cardiac contractions, and the flow of blood through peripheral tissues. (In terms of function, the cardiovascular centers are subdivided into **cardiac** and **vasomotor centers**, but their anatomical boundaries are difficult to determine.)

   The **respiratory rhythmicity centers** set the basic pace for respiratory movements. Their activity is regulated by inputs from the apneustic and pneumotaxic respiratory centers of the pons.

2. *Sensory and Motor Nuclei of Cranial Nerves.* The medulla oblongata contains sensory and motor nuclei associated with five of the cranial nerves (VIII, IX, X, XI, and XII). These cranial nerves provide motor commands to muscles of the pharynx, neck, and back as well as to the visceral organs of the thoracic and peritoneal cavities. Cranial nerve VIII carries sensory information from receptors in the internal ear to the vestibular and cochlear nuclei, which extend from the pons into the medulla oblongata.

3. *Relay Stations along Sensory and Motor Pathways.* The **nucleus gracilis** and the **nucleus cuneatus** pass somatic sensory information to the thalamus. Tracts leaving these brain stem nuclei cross to the opposite side of the brain before reaching their destinations. This crossing over is called a *decussation* (dē-kuh-SĀ-shun; *decussatio*, crossing over), and the site is the **decussation of pyramids**.

The **solitary nucleus** on either side receives visceral sensory information that reaches the CNS from the spinal nerves and cranial nerves. This information is integrated and forwarded to other autonomic centers in the medulla oblongata and elsewhere.

The **olivary nuclei** relay information to the cerebellar cortex about somatic motor commands as they are issued by motor centers at higher levels. The bulk of the olivary nuclei creates the **olives**, prominent olive-shaped bulges along the ventrolateral surface of the medulla oblongata.

Figure 14–6 also contains a summary of the major components of the medulla oblongata and their functions.

Figure 14–5   **The Diencephalon and Brain Stem.**

Diencephalon
- Lateral geniculate nucleus
- Medial geniculate nucleus

Midbrain
- Superior colliculus
- Inferior colliculus

Cerebellar peduncles
- Superior cerebellar peduncle
- Middle cerebellar peduncle
- Inferior cerebellar peduncle

Cerebral peduncle
Optic tract
Thalamus

Cranial nerves
N II
N III
N IV
N V
Pons
N VI
N VII
N VIII
N IX
N X
N XI
N XII

Spinal nerve C₁
Spinal nerve C₂

Medulla oblongata
Spinal cord

a Lateral view

Optic chiasm
Infundibulum
Thalamus

Cranial nerves
N II
N III
N IV
N V
N VI
N VII
N VIII
N IX
N X
N XI
N XII

Optic tract
Cerebral peduncle
Mammillary body
Pons

Medulla oblongata
Ventral roots of spinal nerves C₁ and C₂
Spinal cord

b Anterior view

Choroid plexus in third and fourth ventricles
Thalamus
Third ventricle
Pineal gland

Corpora quadrigemina
- Superior colliculi
- Inferior colliculi

N IV
Cerebral peduncle

Cerebellar peduncles
- Superior
- Middle
- Inferior

Choroid plexus in roof of fourth ventricle

Dorsal roots of spinal nerves C₁ and C₂

c Posterior view

Trochlear nerve (N IV)
Fourth ventricle

d Posterior view, cadaver dissection

14

Figure 14–6   **The Medulla Oblongata.** ATLAS: Plates 9a–c; 11c

a **Anterior view**

The Medulla Oblongata	
Region/Nucleus	**Function**
GRAY MATTER	
Olivary nucleus	Relay information from the red nuclei, other midbrain centers, and the cerebral cortex to the cerebellum

Reflex centers
  Cardiovascular center
  Respiratory rhythmicity center
	Regulate heart rate and force of contraction
	Set the basic pace of respiratory movements

Nucleus gracilis **Nucleus cuneatus**	Relay somatic information to the thalamus
Other nuclei/centers Cranial nerves VIII (in part), IX, X, XI (in part), and XII	Sensory and motor nuclei of five cranial nerves; nuclei relay ascending information from the spinal cord to higher centers
Reticular formation	Contains nuclei and centers that regulate vital autonomic functions; extends into the pons and midbrain
WHITE MATTER	
Ascending and descending tracts within columns	Link the brain with the spinal cord

b **Posterolateral view**

CHECKPOINT

8. Identify the components of the medulla oblongata that are responsible for relaying somatic sensory information to the thalamus.

9. The medulla oblongata is one of the smallest sections of the brain, yet damage there can cause death, whereas similar damage in the cerebrum might go unnoticed. Why?

See the blue Answers tab at the back of the book.

14-4 ❙ The pons contains nuclei and tracts that carry or relay sensory and motor information

Learning Outcome List the main components of the pons, and specify the functions of each.

The pons links the cerebellum with the midbrain, diencephalon, cerebrum, and spinal cord. Important features and regions of the pons are indicated in **Figures 14–5** and **14–7.**

Figure 14–7   **The Pons.** ATLAS: Plates 9a–c; 11c

The Pons	
Region/Nucleus	**Function**
WHITE MATTER	
Descending tracts	Carry motor commands from higher centers to nuclei of cranial or spinal nerves
Ascending tracts	Carry sensory information from brain stem nuclei to the thalamus
Transverse fibers	Interconnect cerebellar hemispheres
GRAY MATTER	
Apneustic and pneumotaxic centers	Adjust activities of the respiratory rhythmicity centers in the medulla oblongata
Reticular formation	Automatic processing of incoming sensations and outgoing motor commands
Cranial nerve nuclei N V, VI, VII, and VIII (in part)	Relay sensory information and issue somatic motor commands
Other nuclei/relay centers	Relay sensory and motor information to the cerebellum

The pons contains four groups of components:

1. *Sensory and Motor Nuclei of Cranial Nerves.* These cranial nerves (V, VI, VII, and VIII) innervate the jaw muscles, the anterior surface of the face, one of the extrinsic eye muscles (the lateral rectus), and the sense organs of the internal ear (the *vestibular* and *cochlear nuclei*).

2. *Nuclei Involved with the Control of Respiration.* On each side of the pons, the reticular formation in this region contains two respiratory centers: the *apneustic center* and the *pneumotaxic center.* These centers modify the activity of the *respiratory rhythmicity centers* in the medulla oblongata.

3. *Nuclei and Tracts That Process and Relay Information Sent to or from the Cerebellum.* The pons links the cerebellum with the brain stem, cerebrum, and spinal cord.

4. *Ascending, Descending, and Transverse Tracts.* Longitudinal tracts interconnect other portions of the CNS. The middle cerebellar peduncles are connected to the **transverse fibers**, which cross the anterior surface of the pons. These fibers are axons that link nuclei of the pons (*pontine nuclei*) with the cerebellar hemisphere of the opposite side.

Figure 14–7 also contains a summary of the major components of the pons and their functions.

CHECKPOINT

10. **Name the four groups of components found in the pons.**

11. **If the respiratory centers of the pons were damaged, what respiratory controls might be lost?**

See the blue Answers tab at the back of the book.

14-5 ┃ The cerebellum coordinates learned and reflexive patterns of muscular activity at the subconscious level

Learning Outcome List the main components of the cerebellum, and specify the functions of each.

The cerebellum is an automatic processing center (**Figure 14–8**). It has two primary functions:

1. *Adjusting the Postural Muscles of the Body.* The cerebellum coordinates rapid, automatic adjustments that maintain balance and equilibrium. It makes these alterations in muscle tone and position by modifying the activities of motor centers in the brain stem.

2. *Programming and Fine-Tuning Movements Controlled at the Conscious and Subconscious Levels.* The cerebellum refines learned movement patterns, such as riding a bicycle or

Figure 14–8 **The Cerebellum.**

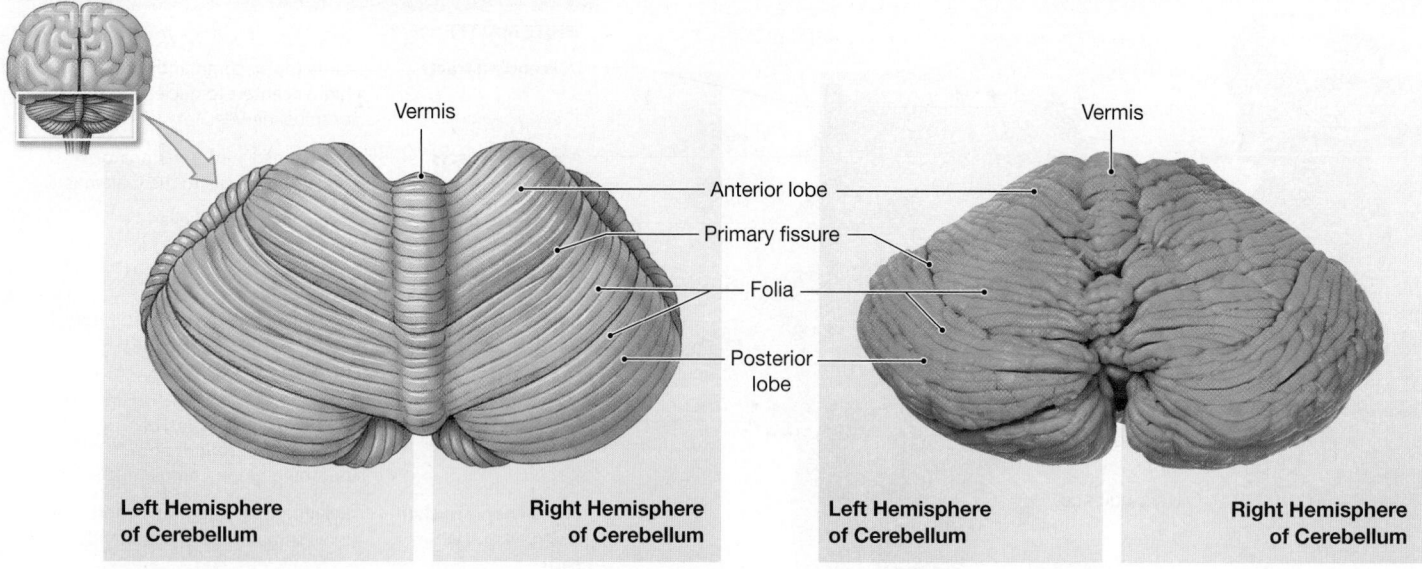

a The posterior, superior surface of the cerebellum, showing major anatomical landmarks and regions

Dendrites

Cell body of Purkinje cell

Axons

Purkinje cells   LM × 320

Midbrain

Choroid plexus of the fourth ventricle

Pons

Flocculonodular lobe

Medulla oblongata

Posterior lobe

Anterior lobe

The Cerebellum	
Region/Nuclei	**Function**
GRAY MATTER	
Cerebellar cortex	Involuntary coordination and control of ongoing body movements
Cerebellar nuclei	Involuntary coordination and control of ongoing body movements
WHITE MATTER	
Arbor vitae	Connects cerebellar cortex and nuclei with cerebellar peduncles
Cerebellar peduncles	
Superior	Link cerebellum with midbrain, diencephalon, and cerebrum
Middle	Carry communications between the cerebellum and pons
Inferior	Link the cerebellum with the medulla oblongata and spinal cord
Transverse fibers	Interconnect pontine nuclei with the opposite cerebellar hemisphere

b A sectional view of the cerebellum, showing the arrangement of gray matter and white matter playing the piano. It performs this function indirectly by regulating activity along motor pathways at the cerebral cortex, basal nuclei, and motor centers in the brain stem. The cerebellum compares the motor commands with proprioceptive information (position sense) and stimulates any adjustments needed to make the movement smooth.

The cerebellum has a complex, highly convoluted surface composed of neural cortex. The **folia** (FŌ-lē-uh; leaves), or folds of the cerebellum surface, are less prominent than the folds in the surfaces of the cerebral hemispheres (Figure 14–8a). The **primary fissure** separates the **anterior** and **posterior lobes**. Along the midline, a narrow band of cortex known as the **vermis** (VER-mis) separates the **cerebellar hemispheres**. The slender **flocculonodular** (flok-ū-lō-NOD-ū-lar) **lobe** lies between the roof of the fourth ventricle and the cerebellar hemispheres and vermis (Figure 14–8b).

Like the cerebrum, the cerebellum has a superficial layer of neural cortex. The cerebellar cortex contains relatively huge, highly branched **Purkinje** (pur-KIN-jē) **cells**. The extensive dendrites of each Purkinje cell receive input from up to 200,000 synapses—more than any other type of brain cell. The internal white matter of the cerebellum forms a branching array that in sectional view resembles a tree. Anatomists call it the **arbor vitae**, or "tree of life" (Figure 14–8b). It connects the cerebellar cortex and nuclei with the cerebellar peduncles, which we discuss below.

The cerebellum receives proprioceptive (position) information from the spinal cord and monitors all proprioceptive, visual, tactile, balance, and auditory sensations received by the brain. Most axons that carry sensory information do not synapse in the cerebellar nuclei but pass through the deeper layers of the cerebellum on their way to the Purkinje cells of the cerebellar cortex. Information about the motor commands issued at the conscious and subconscious levels reaches the Purkinje cells indirectly, after being relayed by nuclei in the pons or by the **cerebellar nuclei** embedded within the arbor vitae.

Tracts that link the cerebellum with the brain stem, cerebrum, and spinal cord leave the cerebellar hemispheres as the superior, middle, and inferior cerebellar peduncles:

- The **superior cerebellar peduncles** link the cerebellum with nuclei in the midbrain, diencephalon, and cerebrum.
- The **middle cerebellar peduncles** are connected to a broad band of fibers that cross the ventral surface of the pons at right angles to the axis of the brain stem. The middle cerebellar peduncles also connect the cerebellar hemispheres with sensory and motor nuclei in the pons.
- The **inferior cerebellar peduncles** communicate between the cerebellum and nuclei in the medulla oblongata and carry ascending and descending cerebellar tracts from the spinal cord.

The cerebellum can be permanently damaged by trauma or stroke. Drugs such as alcohol can also temporarily affect it. The result is **ataxia** (a-TAK-sē-uh; *ataxia*, lack of order), a disturbance in muscular coordination. In severe ataxia, the individual cannot sit or stand without assistance.

CHECKPOINT

12. Identify the components of the cerebellar gray matter.
13. What part of the brain has the arbor vitae? What is its function?

See the blue Answers tab at the back of the book.

14-6 ▌ The midbrain regulates auditory and visual reflexes and controls alertness

Learning Outcome List the main components of the midbrain, and specify the functions of each.

The external anatomy of the midbrain is shown in Figures 14–5 and 14–9, and its major nuclei are listed in Figure 14–9. The **tectum**, or roof of the midbrain, is the region posterior to the cerebral aqueduct. It contains two pairs of sensory nuclei known collectively as the **corpora quadrigemina** (KOR-pōr-uh qua-dri-JEM-i-nuh). These nuclei, the superior and inferior colliculi, process visual and auditory sensations. Each **superior colliculus** (ko-LIK-ū-lus; *colliculus*, hill) receives visual inputs from the lateral geniculate nucleus of the thalamus on that side. Each **inferior colliculus** receives auditory input from nuclei in the medulla oblongata and pons. Some of this information may be forwarded to the medial geniculate nucleus of the thalamus on the same side. The superior colliculi control the reflex movements of the eyes, head, and neck in response to visual stimuli, such as a bright light. The inferior colliculi control reflex movements of the head, neck, and trunk in response to auditory stimuli, such as a loud noise.

The area anterior to the cerebral aqueduct is called the **tegmentum**. On each side, the tegmentum contains a red nucleus and the substantia nigra (Figure 14–9a,c). The **red nucleus** contains numerous blood vessels, which give it a rich red color. This nucleus receives information from the cerebrum and cerebellum and issues subconscious motor commands that affect upper limb position and background muscle tone. The **substantia nigra** (NĪ-gruh; *nigra*, black) is the largest midbrain nucleus. It lies lateral to the red nucleus. The gray matter in this region contains darkly pigmented cells, giving it a black color. The pigment is melanin and here it is a byproduct of dopamine synthesis.

The substantia nigra inhibits activity of the basal nuclei in the cerebrum. The basal nuclei are involved in the subconscious control of muscle tone and learned movements. As discussed in a later section, *Parkinson's disease* is characterized by the loss of neuronal activity in the substantia nigra.

Figure 14–9   **The Midbrain.** ATLAS: Plates 7b; 9a–c; 11c; 12a,c

The Midbrain	
Region/Nuclei	**Function**
GRAY MATTER	
Tectum (roof)	
Superior colliculus	Integrate visual information with other sensory input; initiate reflex responses to visual stimuli
Inferior colliculus	Relay auditory information to medial geniculate nuclei; initiate reflex responses to auditory stimuli
Walls and floor	
Substantia nigra	Regulates activity in the basal nuclei
Red nucleus	Subconscious control of upper limb position and background muscle tone
Reticular formation	Automatic processing of incoming sensations and outgoing motor commands; can initiate involuntary motor responses to stimuli; helps maintain consciousness
Other nuclei/centers	Nuclei associated with cranial nerves III and IV
WHITE MATTER	
Cerebral peduncles	Connect primary motor cortex with motor neurons in brain and spinal cord; carry ascending sensory information to thalamus

a A posterior view. The underlying nuclei are colored only on the right.

b Posterior view of a cadaver dissection of the diencephalon and brain stem.

c A superior view of a transverse section at the level of the midbrain.

The nerve fiber bundles on the ventrolateral surfaces of the midbrain (**Figures 14–5** and **14–9c**) are the **cerebral peduncles** (*peduncles*, little feet). They contain (1) descending fibers that go to the cerebellum by way of the pons and (2) descending fibers that carry voluntary motor commands issued by the cerebral hemispheres.

The midbrain also contains the *reticular activating system (RAS)*, a specialized component of the reticular formation. Stimulation of the RAS makes you more alert and attentive; damage to the RAS produces unconsciousness. We will consider the role of the RAS in maintaining consciousness in Chapter 16.

14. Identify the sensory nuclei within the corpora quadrigemina.

15. Which area of the midbrain controls reflexive movements of the eyes, head, and neck in response to visual stimuli?

See the blue Answers tab at the back of the book.

14-7 ❚ The diencephalon integrates sensory information with motor output at the subconscious level

Learning Outcome List the main components of the diencephalon, and specify the functions of each.

The diencephalon is a division of the brain that consists of the epithalamus, thalamus, and hypothalamus. **Figure 14–5** shows its position and its relationship to landmarks on the brain stem.

The *epithalamus* is the roof of the diencephalon superior to the third ventricle. The anterior portion of the epithalamus contains an extensive area of choroid plexus that extends through the interventricular foramina into the lateral ventricles. The posterior portion of the epithalamus contains the **pineal gland** (**Figure 14–5c**), an endocrine structure that secretes the hormone **melatonin**. Melatonin is important in regulating day–night cycles and reproductive functions. (We will describe the role of melatonin in Chapter 18.)

Most of the neural tissue in the diencephalon is concentrated in the *left thalamus* and *right thalamus*, which form the lateral walls, and the *hypothalamus*, which forms the floor. Ascending sensory information from the spinal cord and cranial nerves (other than the olfactory tract) synapses in a nucleus in the left or right thalamus before reaching the cerebral cortex and our conscious awareness. The hypothalamus contains centers involved with emotions and visceral processes that affect the cerebrum as well as other components of the brain stem. It also controls a variety of autonomic functions and forms the link between the nervous and endocrine systems.

The Thalamus

On each side of the diencephalon, the thalamus is the final relay point for sensory information ascending to the primary sensory cortex. It acts as a filter, passing on only a small portion of the arriving sensory information. The thalamus also coordinates the activities of the basal nuclei (in the cerebrum) and the cerebral cortex by relaying information between them.

The third ventricle separates the left thalamus and right thalamus. Each thalamus consists of a rounded mass of *thalamic nuclei* (**Figure 14–10**). Viewed in a midsagittal section through the brain (**Figure 14–11a**), each thalamus extends from the anterior commissure to the inferior base of the pineal gland. A projection of gray matter called an **interthalamic adhesion**

Figure 14–10 **The Thalamus.**

Frontal lobe · Parietal lobe · Occipital lobe

a A lateral view of the brain, color coded to indicate the regions that receive input from the thalamic nuclei shown in part (b)

Limbic system · Frontal lobes · Parietal lobe and cingulate gyrus · Association areas of cerebral cortex

Anterior group · Medial group · Lateral group · Posterior group · Pulvinar nucleus · Auditory input · Medial geniculate nucleus · Lateral geniculate nucleus · Ventral group · Basal nuclei · Cerebellum · General sensory input · Visual input

b An enlarged view of the thalamic nuclei of the left side

extends into the ventricle from the thalamus on either side (**Figures 14–11b** and **14–12**), but no fibers cross the midline.

Functions of Thalamic Nuclei

The thalamic nuclei deal primarily with the relay of sensory information to the basal nuclei (parts of the cerebrum which we will discuss shortly) and cerebral cortex. The five major groups of thalamic nuclei are shown in **Figure 14–10b** and listed in **Table 14–2**. They are the anterior, medial, ventral, posterior, and lateral groups:

1. The *anterior group* includes the **anterior nuclei**, which are part of the *limbic system*. This system is involved with emotion and motivation. We will discuss it in a later section.

Table 14–2	The Thalamus
Group/Nuclei	**Function**
Anterior Group	Part of the limbic system
Medial Group	Integrates sensory information for projection to the frontal lobes
Ventral Group	Projects sensory information to the primary sensory cortex; relays information from cerebellum and basal nuclei to motor area of cerebral cortex
Posterior Group	
Pulvinar nuclei	Integrates sensory information for projection to association areas of cerebral cortex
Lateral geniculate nuclei	Project visual information to the visual cortex
Medial geniculate nuclei	Project auditory information to the auditory cortex
Lateral Group	Integrates sensory information and influences emotional states

2. The nuclei of the *medial group* provide an awareness of emotional states by connecting emotional centers in the hypothalamus with the *frontal lobes* of the cerebral hemispheres. The medial group also receives and relays sensory information from other portions of the thalamus.

3. The nuclei of the *ventral group* relay information from the *basal nuclei* of the cerebrum and the cerebellum to somatic motor areas of the cerebral cortex. Ventral group nuclei also relay sensory information about touch, pressure, pain, temperature, and proprioception (position) to the sensory areas of the cerebral cortex.

4. The *posterior group* includes the pulvinar and geniculate nuclei. The **pulvinar nuclei** integrate sensory information for projection to the cerebral cortex. The **lateral geniculate** (je-NIK-ū-lāt; *genicula,* little knee) **nucleus** of each thalamus receives visual information over the *optic tract,* which originates at the eyes. The output of the lateral geniculate nucleus goes to the *occipital lobes* of the cerebral hemispheres and to the midbrain. The **medial geniculate nucleus** relays auditory information to the appropriate area of the cerebral cortex from specialized receptors of the internal ear.

5. The nuclei of the *lateral group* form feedback loops with the limbic system and the *parietal lobes* of the cerebral hemispheres. The lateral group affects emotional states and the integration of sensory information.

The Hypothalamus

The hypothalamus extends from the area superior to the *optic chiasm,* a crossover where the optic tracts from the eyes arrive at the brain, to the posterior margins of the **mammillary** (*mammilla,* nipple) **bodies** (**Figure 14–11a**). The mammillary bodies process sensory information, including olfactory sensations, and contain motor nuclei that control reflex movements associated with eating, such as chewing, licking, and swallowing. They are also involved in memory.

Immediately posterior to the optic chiasm, a narrow stalk called the **infundibulum** (in-fun-DIB-ū-lum; *infundibulum,* funnel) extends inferiorly, connecting the floor of the hypothalamus to the pituitary gland (**Figure 14–11b**).

The floor of the hypothalamus between the infundibulum and the mammillary bodies is the **tuberal area** (*tuber,* swelling). The tuberal area contains nuclei that are involved with the control of pituitary gland function.

Functions of the Hypothalamus

The hypothalamus contains important control and integrative centers, in addition to those associated with the limbic system. These centers are shown in **Figure 14–11b**, along with a summary of their functions.

Hypothalamic centers may be stimulated by (1) sensory information from the cerebrum, brain stem, and spinal cord; (2) changes in the compositions of the CSF and interstitial fluid; or (3) chemical stimuli in the circulating blood that move rapidly across highly permeable capillaries to enter the hypothalamus (where there is no blood–brain barrier).

The hypothalamus performs the following functions:

1. *The Subconscious Control of Skeletal Muscle Contractions.* The hypothalamus directs somatic motor patterns associated with rage, pleasure, pain, and sexual arousal by stimulating centers in other portions of the brain. For example, hypothalamic centers control the changes in facial expression that accompany rage and the basic movements associated with sexual activity.

2. *The Control of Autonomic Function.* The hypothalamus adjusts and coordinates the activities of autonomic centers in the pons and medulla oblongata that regulate heart rate, blood pressure, respiration, and digestive functions.

3. *The Coordination of Activities of the Nervous and Endocrine Systems.* The hypothalamus coordinates neural and endocrine activities by inhibiting or stimulating endocrine cells in the pituitary gland through the production of *regulatory hormones.* These hormones are produced at the tuberal area. They are released into local capillaries for transport to the anterior lobe of the pituitary gland.

4. *The Secretion of Two Hormones.* The hypothalamus secretes *antidiuretic hormone* (ADH, also called vasopressin) and *oxytocin* (OXT). The **supraoptic nucleus** produces ADH, which restricts water loss by the kidneys. The **paraventricular nucleus** produces oxytocin, which stimulates smooth muscle contractions in the uterus and mammary glands of females and the prostate gland of males. These hormones are transported along axons that pass through the infundibulum to the posterior lobe of the pituitary gland. There

Figure 14–11 **The Hypothalamus in Sagittal Section.**

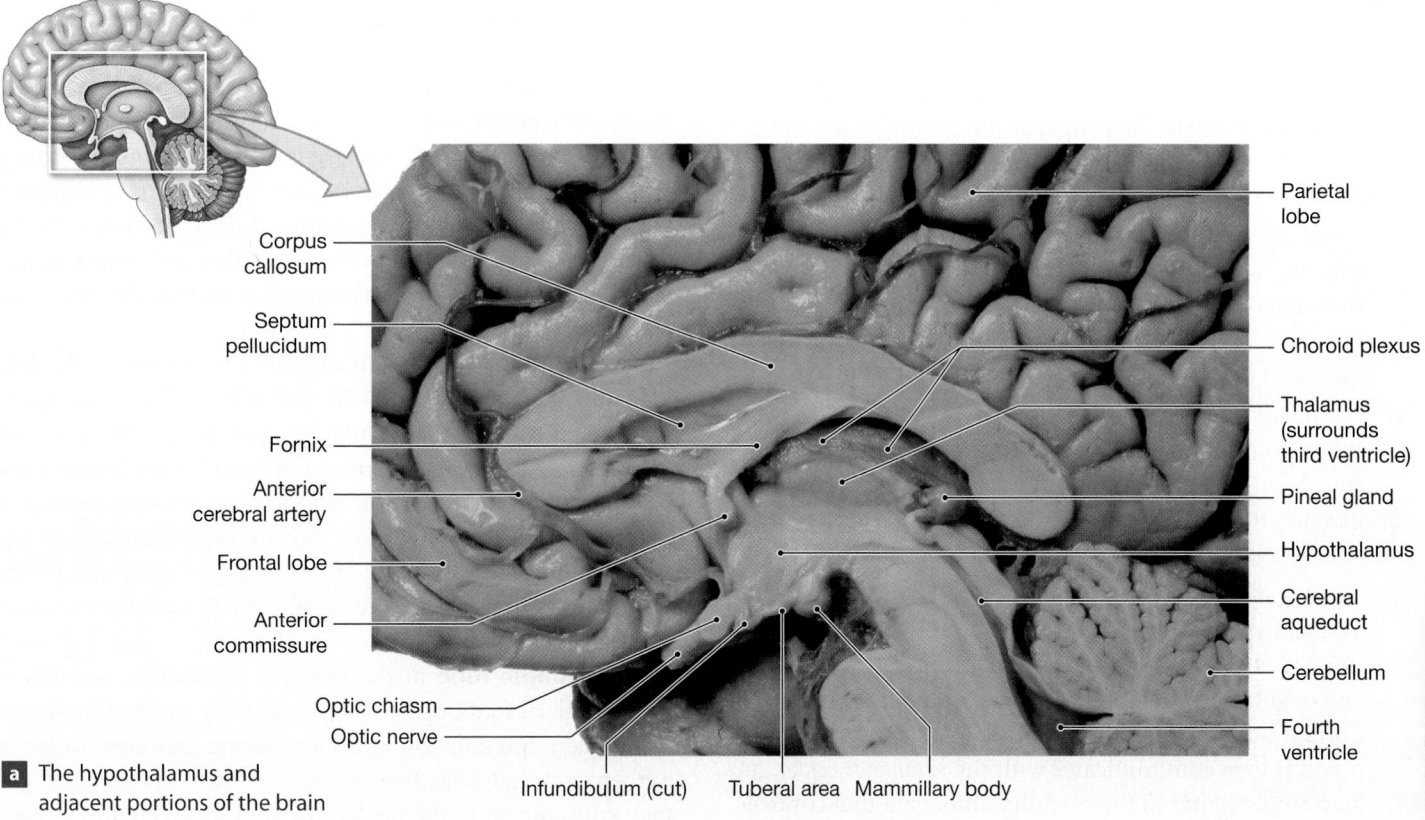

a The hypothalamus and adjacent portions of the brain

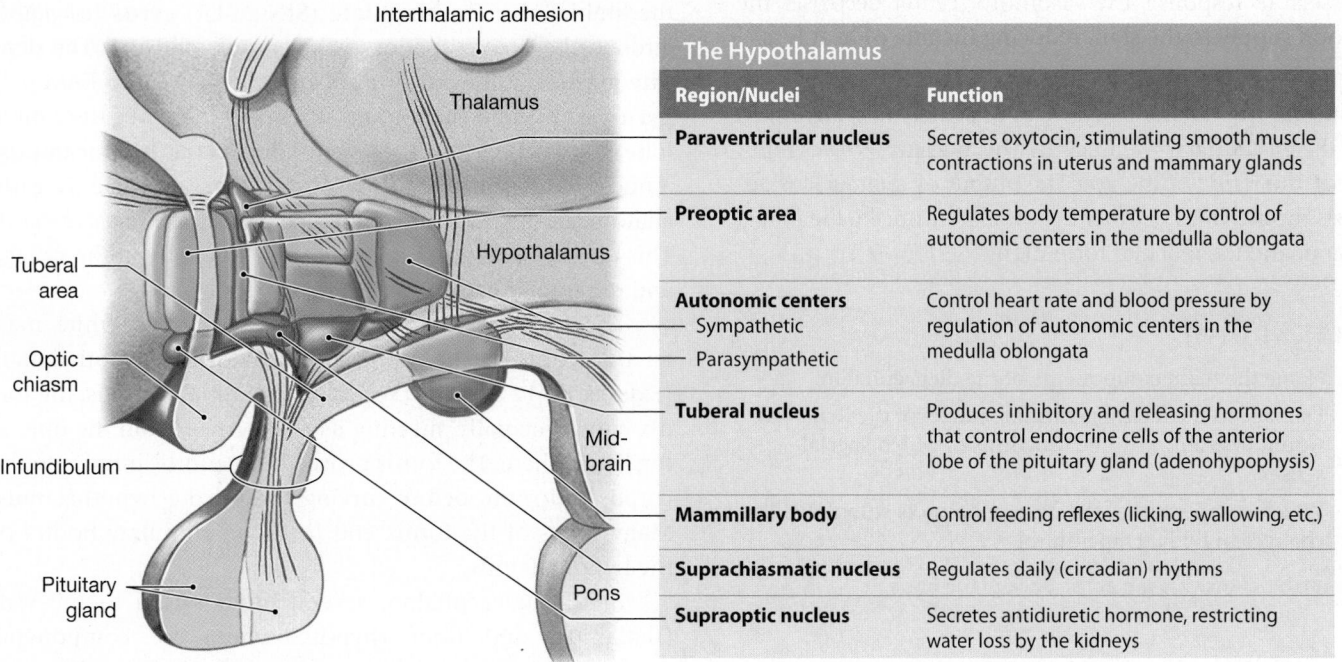

b A diagrammatic view of the hypothalamus, showing the locations and functions of major nuclei and centers

The Hypothalamus	
Region/Nuclei	**Function**
Paraventricular nucleus	Secretes oxytocin, stimulating smooth muscle contractions in uterus and mammary glands
Preoptic area	Regulates body temperature by control of autonomic centers in the medulla oblongata
Autonomic centers Sympathetic Parasympathetic	Control heart rate and blood pressure by regulation of autonomic centers in the medulla oblongata
Tuberal nucleus	Produces inhibitory and releasing hormones that control endocrine cells of the anterior lobe of the pituitary gland (adenohypophysis)
Mammillary body	Control feeding reflexes (licking, swallowing, etc.)
Suprachiasmatic nucleus	Regulates daily (circadian) rhythms
Supraoptic nucleus	Secretes antidiuretic hormone, restricting water loss by the kidneys

14

the hormones are released into the blood for distribution throughout the body.

5. *The Production of Emotions and Behavioral Drives.* Specific hypothalamic centers produce sensations that lead to conscious or subconscious changes in behavior. For example, stimulation of the **feeding center** produces the sensation of hunger. Stimulation of the **thirst center** produces the sensation of thirst. The **satiety center** is the region that regulates food intake. These unfocused "impressions" originating in the hypothalamus are called **drives**. The conscious sensations are only part of the hypothalamic response. For instance, the thirst center also orders the release of ADH by neurons in the supraoptic nucleus.

6. *Coordination between Voluntary and Autonomic Functions.* When you think about a dangerous or stressful situation, your heart rate and respiratory rate go up and your body prepares for an emergency. The hypothalamus makes these autonomic adjustments.

7. *The Regulation of Body Temperature.* The **preoptic area** of the hypothalamus is responsible for thermoregulation. It coordinates the activities of other CNS centers and regulates other physiological systems to maintain normal body temperature. If body temperature falls, the preoptic area communicates with the *vasomotor center*, an autonomic center in the medulla oblongata that controls blood flow by regulating the diameter of peripheral blood vessels. In response, the vasomotor center decreases the blood supply to the skin, reducing the rate of heat loss.

8. *The Control of Circadian Rhythms.* The **suprachiasmatic nucleus** coordinates daily cycles of activity that are linked to the 24-hour day–night cycle. This nucleus receives input from the retina of the eye. The output of this nucleus adjusts the activities of other hypothalamic nuclei, the pineal gland, and the reticular formation.

CHECKPOINT

16. Name the main components of the diencephalon.

17. Damage to the lateral geniculate nuclei of the thalamus would interfere with the functions of which special sense?

18. Which component of the diencephalon is stimulated by changes in body temperature?

See the blue Answers tab at the back of the book.

14-8 ❘ The limbic system is a group of tracts and nuclei that function in emotion, motivation, and memory

Learning Outcome Identify the main components of the limbic system, and specify the locations and functions of each.

The **limbic system** (*limbus*, border) includes nuclei and tracts along the border between the cerebrum and diencephalon. This system is a functional grouping rather than an anatomical one. Functions of the limbic system include (1) establishing emotional states; (2) linking the conscious, intellectual functions of the cerebral cortex with the unconscious and autonomic functions of the brain stem; and (3) facilitating memory storage and retrieval. The sensory cortex, motor cortex, and association areas of the cerebral cortex enable you to perform complex tasks, but it is largely the limbic system that makes you *want* to do them. For this reason, the limbic system is also known as the *motivational system*.

Figure 14–12 focuses on major components of the limbic system. The **amygdaloid** (ah-MIG-da-loyd; *amygdale*, almond) **body** is commonly referred to as the *amygdala* (**Figure 14–12b**). It appears to act as an interface between the limbic system, the cerebrum, and various sensory systems. It plays a role in regulating heart rate, in responding to fear and anxiety and controlling the "fight or flight" response by the sympathetic division of the ANS, and in linking emotions with specific memories.

The **limbic lobe** of the cerebral hemisphere consists of the superficial folds, or *gyri*, and underlying structures adjacent to the diencephalon. The gyri curve along the *corpus callosum*, a fiber tract that links the two cerebral hemispheres The gyri then continue on to the medial surface of the cerebrum lateral to the diencephalon (**Figure 14–12a**). There are three gyri in the limbic lobe. The **cingulate** (SIN-gū-lāt) **gyrus** (*cingulum*, girdle or belt) sits superior to the corpus callosum. The **dentate gyrus** and the **parahippocampal** (pa-ra-hip-ō-KAM-pal) **gyrus** form the posterior and inferior portions of the limbic lobe (**Figure 14–12b**). These gyri conceal the **hippocampus**, a nucleus inferior to the floor of the lateral ventricle. To early anatomists, this structure resembled a sea horse (*hippocampus*). This nucleus is important in learning, especially in the storage and retrieval of new long-term memories.

The **fornix** (FOR-niks, arch) is a tract of white matter that connects the hippocampus with the hypothalamus (**Figures 14–12 and 14–15b**). From the hippocampus, the fornix curves medially, meeting its counterpart from the opposing hemisphere. The fornix proceeds anteriorly, inferior to the corpus callosum, before curving toward the hypothalamus. Many fibers of the fornix end in the mammillary bodies of the hypothalamus.

In the diencephalon, several other nuclei in the wall (thalamus) and floor (hypothalamus) are components of the limbic 'system. The *anterior nucleus* of the thalamus (**Figure 14–12b**) relays information from the mammillary body (of the hypothalamus) to the cingulate gyrus on that side. The boundaries between the hypothalamic nuclei of the limbic system are often poorly defined, but experimental stimulation has outlined a number of important hypothalamic

Figure 14–12 **The Limbic System.**

Corpus callosum

Fornix

Pineal gland

Cingulate gyrus (superior portion of limbic lobe)

Anterior group of thalamic nuclei

Hypothalamus

Mammillary body

Temporal lobe of cerebrum

Parahippocampal gyrus (inferior portion of limbic lobe)

Hippocampus (within dentate gyrus)

a A diagrammatic sagittal section through the cerebrum, showing the cortical areas associated with the limbic system. The parahippocampal gyrus is shown as though transparent to make deeper limbic components visible.

The Limbic System

Diencephalic Components

Thalamus
  Anterior nuclei group

Hypothalamus
  Hypothalamic nuclei
  Mammillary body

Other Components

Reticular formation (not shown)

Corpus callosum

Olfactory tract

The Limbic System

Cerebral Components

Limbic lobe cortical areas
  Cingulate gyrus (superior portion of limbic lobe)
  Parahippocampal gyrus (inferior portion of limbic lobe)
  Dentate gyrus

Tracts
  Fornix

Nuclei
  Amygdaloid body

  Hippocampus (within dentate gyrus)

b A three-dimensional reconstruction of the limbic system, showing the relationships among the major components.

centers responsible for the emotions of rage, fear, pain, sexual arousal, and pleasure. The stimulation of specific regions of the hypothalamus can also produce heightened alertness and a generalized excitement or generalized lethargy and sleep. These responses are caused by the stimulation or inhibition of the reticular formation. Although the reticular formation extends the length of the brain stem, its headquarters reside in the midbrain.

CHECKPOINT

19. **What are the primary functions of the limbic system?**

20. **Damage to the amygdaloid body would interfere with regulation of what division of the autonomic nervous system (ANS)?**

See the blue Answers tab at the back of the book.

14-9 ▌ The cerebrum, the largest region of the brain, contains motor, sensory, and association areas

Learning Outcome Identify the major anatomical subdivisions and functions of the cerebrum, and discuss the origin and significance of the major types of brain waves seen in an electroencephalogram.

The cerebrum is the largest region of the brain. Conscious thoughts and all intellectual functions originate in the cerebral hemispheres. Much of the cerebrum is involved in processing somatic sensory and motor information. Gray matter in the cerebrum is located in the *cerebral cortex* and in deeper *basal nuclei*. The white matter of the cerebrum lies deep to the cerebral cortex and around the basal nuclei.

The Cerebral Cortex

A layer of cerebral cortex ranging from 1 to 4.5 mm thick covers the paired cerebral hemispheres, which dominate the superior and lateral surfaces of the cerebrum. The gyri increase the surface area of the cerebral hemispheres, and thus the number of cortical neurons they contain. The total surface area of the cerebral hemispheres is roughly equivalent to 2200 cm^2 (2.5 ft^2) of flat surface.

The entire brain has enlarged over the course of human evolution, but the cerebral hemispheres have enlarged at a much faster rate than has the rest of the brain. This enlargement reflects the large numbers of neurons needed for complex analytical and integrative functions. Since the neurons involved are in the superficial layer of cortex, it is there that the expansion has been most pronounced. The only solution available, other than an enlargement of the entire skull, was for the cortical layer to fold like a crumpled piece of paper.

Landmarks and features on the surface of one cerebral hemisphere are shown in **Figure 14–13a,b**. (The two cerebral hemispheres are almost completely separated by a deep **longitudinal fissure**, seen in **Figure 14–14b**.) Each cerebral hemisphere can be divided into *lobes*, or regions, named after the overlying bones of the skull.

Your brain has a unique pattern of sulci and gyri, as individual as a fingerprint, but the boundaries between lobes are reliable landmarks. On each hemisphere, the **central sulcus**, a deep groove, divides the anterior **frontal lobe** from the more posterior **parietal lobe**. The horizontal **lateral sulcus** separates the frontal lobe from the **temporal lobe**. The **insula** (IN-sū-luh; *insula*, island), an "island" of cortex, lies medial to the lateral sulcus (**Figure 14–13c**). The more posterior **parieto-occipital sulcus** separates the parietal lobe from the **occipital lobe** (**Figure 14–13d**).

Each lobe contains functional regions whose boundaries are less clearly defined. Some of these regions deal with sensory information and others with motor commands. Keep in mind three points about the cerebral lobes:

1. *Each cerebral hemisphere receives sensory information from, and sends motor commands to, the opposite side of the body.* For example, the motor areas of the left cerebral hemisphere control muscles on the right side, and the right cerebral hemisphere controls muscles on the left side. This crossing over has no known functional significance.

2. *The two hemispheres have different functions, even though they look almost identical.* We discuss these differences in a later section.

3. *The correspondence between a specific function and a specific region of the cerebral cortex is imprecise.* Because the boundaries are indistinct and have considerable overlap, one region may have several functions. Some aspects of cortical function, such as consciousness, cannot easily be assigned to any single region. However, we know that normal individuals use all portions of the brain.

The White Matter of the Cerebrum

The interior of the cerebrum consists mostly of white matter. We can classify the axons of white matter as association fibers, commissural fibers, and projection fibers (**Figure 14–14**).

- **Association fibers** interconnect areas of cerebral cortex within a single cerebral hemisphere. Shorter association fibers are called **arcuate** (AR-kū-āt) **fibers**, because they curve in an arc to pass from one gyrus to another. Longer association fibers are organized into discrete bundles, or *fasciculi*. The **longitudinal fasciculi** connect the frontal lobe to the other lobes of the same hemisphere.

- **Commissural** (kom-i-SŪR-ul; *commissura*, crossing over) **fibers** interconnect and permit communication between the cerebral hemispheres. Bands of commissural fibers linking the hemispheres include the **corpus callosum** and the **anterior commissure**. The corpus callosum alone contains more than 200 million axons, carrying some 4 billion impulses per second!

- **Projection fibers** link the cerebral cortex to the diencephalon, brain stem, cerebellum, and spinal cord. All projection fibers pass through the diencephalon. There, axons heading to sensory areas of the cerebral cortex pass among the axons descending from motor areas of the cortex. In gross dissection, the ascending fibers and descending fibers look alike. The entire collection of projection fibers is known as the **internal capsule**.

The Basal Nuclei

While your cerebral cortex is consciously directing a complex movement or solving some intellectual puzzle, other centers of your cerebrum, diencephalon, and brain stem are processing

Figure 14–13 **The Brain in Lateral View.** ATLAS: Plates 11a,b; 13b–e; 14a–d

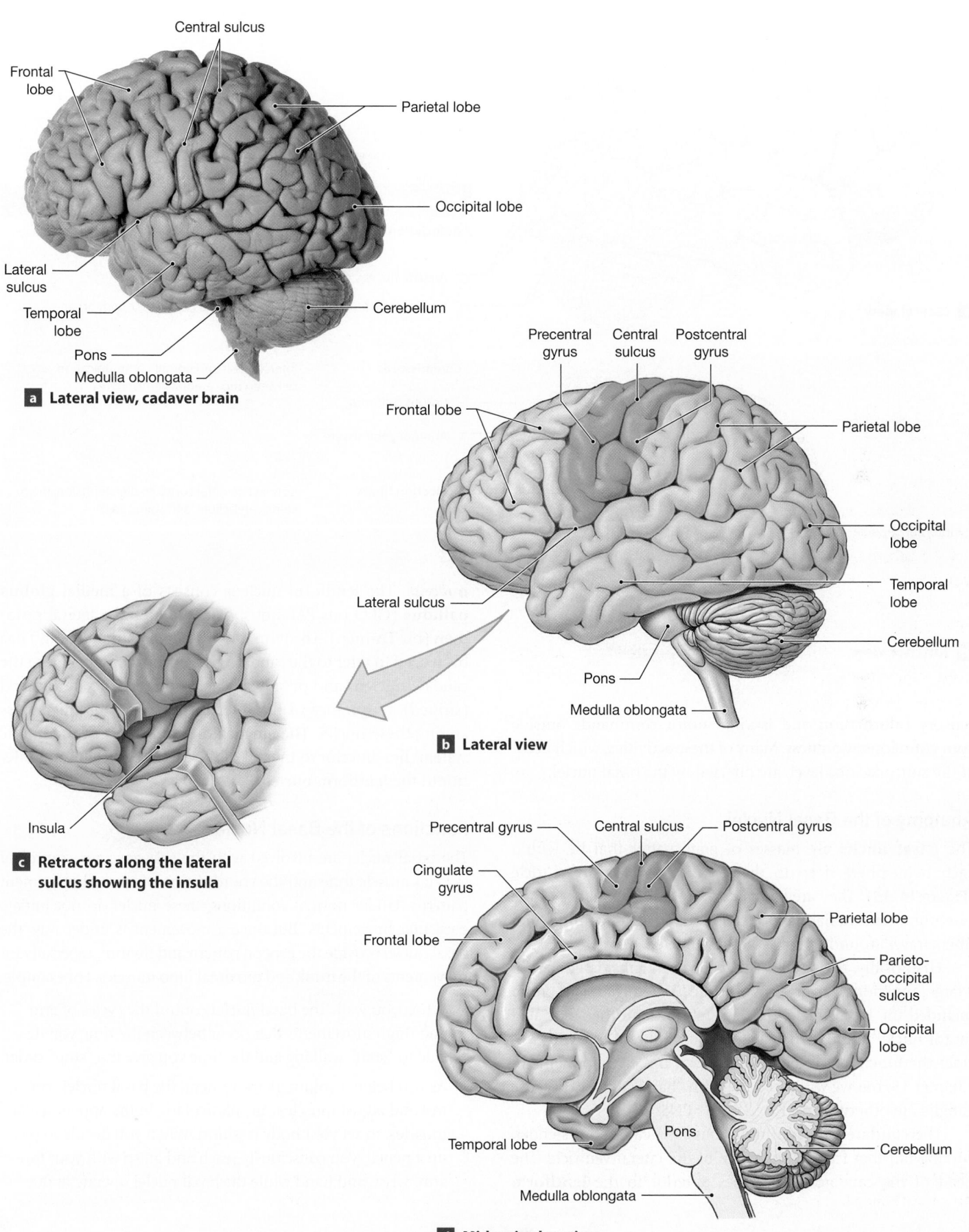

a **Lateral view, cadaver brain**

c **Retractors along the lateral sulcus showing the insula**

b **Lateral view**

d **Midsagittal section**

Figure 14–14   **Fibers of the White Matter of the Cerebrum.**

a **Lateral view**

Longitudinal fissure

Internal capsule

b **Anterior view**

Fibers of the White Matter of the Cerebrum	
Fibers/Tracts	**Function**
Association fibers	Interconnect cortical areas within the same hemisphere
Arcuate fibers	Interconnect gyri within a lobe
Longitudinal fasciculi	Interconnect the frontal lobe with other cerebral lobes
Commissures	Interconnect and permit communication between the cerebral hemispheres
Corpus callosum	
Anterior commissure	
Projection fibers	Connect cerebral cortex to diencephalon, brain stem, cerebellum, and spinal cord

sensory information and issuing motor commands outside your conscious awareness. Many of these activities, which occur at the subconscious level, are directed by the basal nuclei.

Anatomy of the Basal Nuclei

The **basal nuclei** are masses of gray matter that lie within each hemisphere deep to the floor of the lateral ventricle (**Figure 14–15**). They are embedded in the white matter of the cerebrum. The radiating projection fibers and commissural fibers travel around or between these nuclei.

Historically, the basal nuclei have been considered part of a larger functional group known as the *basal ganglia*. This group included the basal nuclei of the cerebrum and the associated motor nuclei in the diencephalon and midbrain. We will consider the functional interactions among these components in Chapter 15, but we will avoid the term "basal ganglia" because ganglia are otherwise restricted to the PNS.

The **caudate nucleus** has a massive head and a slender, curving tail that follows the curve of the lateral ventricle. The head of the caudate nucleus lies anterior to the **lentiform**

nucleus. The lentiform nucleus consists of a medial **globus pallidus** (GLŌ-bus PAL-i-dus; pale globe) and a lateral **putamen** (pū-TĀ-men). The term *corpus striatum* (striated body) has been used to refer to the caudate and lentiform nuclei, or to the caudate nucleus and putamen. The name refers to the striated (striped) appearance of the internal capsule as its fibers pass among these nuclei. The amygdaloid body, part of the limbic system, lies anterior to the tail of the caudate nucleus and inferior to the lentiform nucleus.

Functions of the Basal Nuclei

The basal nuclei are involved with the subconscious control of skeletal muscle tone and the coordination of learned movement patterns. Under normal conditions, these nuclei do not initiate particular movements. But once a movement is under way, the basal nuclei provide the general pattern and rhythm, especially for movements of the trunk and proximal limb muscles. For example:

- When you walk, the basal nuclei control the cycles of arm and thigh movements that occur between the time you decide to "start" walking and the time you give the "stop" order.

- As you begin a voluntary movement, the basal nuclei control and adjust muscle tone, particularly in the appendicular muscles, to set your body position. When you decide to pick up a pencil, you consciously reach and grasp with your forearm, wrist, and hand while the basal nuclei operate at the

Figure 14–15 **The Basal Nuclei.** ATLAS: Plates 11c; 13a,f

Head of caudate nucleus

Internal capsule

Putamen

Thalamus

Choroid plexus

Pineal gland

Corpus callosum

Lateral ventricle (anterior horn)

Septum pellucidum

Fornix (cut edge)

Third ventricle

Fornix

Lateral ventricle (posterior horn)

Head of caudate nucleus

Lentiform nucleus

Tail of caudate nucleus

Thalamus

Amygdaloid body

a The relative positions of the basal nuclei in the intact brain, lateral view

b A superior view of two transverse sections at different levels

Corpus callosum

Lateral ventricle

Septum pellucidum

Internal capsule

Lateral sulcus

Insula

Anterior commissure

Tip of inferior horn of lateral ventricle

Putamen

Globus pallidus

c Frontal section

The Basal Nuclei	
Nuclei	**Function**
Caudate nucleus	Subconscious adjustment and modification of voluntary motor commands

Lentiform nucleus
  Putamen
  Globus pallidus
	Subconscious adjustment and modification of voluntary motor commands

Claustrum	Plays a role in the subconscious processing of visual information
Amygdaloid body	Component of limbic system

14

subconscious level to position your shoulder and stabilize your arm.

How do the basal nuclei carry out their functions? They alter the motor commands issued by the cerebral cortex through a feedback loop. Information arrives at the caudate nucleus and putamen from sensory, motor, and integrative areas of the cerebral cortex. Lateral processing then takes place in these nuclei and in the adjacent globus pallidus. Most of the output of the basal nuclei leaves the globus pallidus and synapses in the thalamus. Nuclei in the thalamus then project the information to appropriate areas of the cerebral cortex.

Activity of the basal nuclei is inhibited by neurons in the substantia nigra of the midbrain, which release the neurotransmitter *dopamine*. ⤶ p. 443 If the substantia nigra is damaged or the neurons secrete less dopamine, basal nuclei become more active. The result is a gradual, generalized increase in muscle tone and the appearance of symptoms characteristic of *Parkinson's disease*. ⤶ p. 134 People with Parkinson's disease have difficulty starting voluntary movements, because opposing muscle groups do not relax. Instead, they must be overpowered. Once a movement is under way, every aspect must be voluntarily controlled through intense effort and concentration.

Motor and Sensory Areas of the Cortex

The major motor and sensory regions of the cerebral cortex are shown in **Figure 14–16** and listed in **Table 14–3**. The central

Figure 14–16 **Motor and Sensory Regions of the Cerebral Cortex.**

a Major anatomical landmarks on the surface of the left cerebral hemisphere. The lateral sulcus has been pulled apart to expose the insula.

b The left hemisphere generally contains the general interpretive area and the speech center. The prefrontal cortex of each hemisphere is involved with conscious intellectual functions.

c Regions of the cerebral cortex as determined by histological analysis. Several of the 50 Brodmann areas are shown for comparison with the results of functional mapping.

Table 14–3	The Cerebral Cortex
Lobe/Region	**Function**
FRONTAL LOBE	
Primary motor cortex	Voluntary control of skeletal muscles
PARIETAL LOBE	
Primary sensory cortex	Conscious perception of touch, pressure, pain, vibration, taste, and temperature
OCCIPITAL LOBE	
Visual cortex	Conscious perception of visual stimuli
TEMPORAL LOBE	
Auditory cortex and olfactory cortex	Conscious perception of auditory (hearing) and olfactory (smell) stimuli
ALL LOBES	
Association areas	Integration and processing of sensory data; processing and initiation of motor activities

sulcus separates the motor and sensory areas of the cortex. The **precentral gyrus** of the frontal lobe forms the anterior border of the central sulcus. The surface of this gyrus is the **primary motor cortex**. Neurons of the primary motor cortex direct voluntary movements by controlling somatic motor neurons in the brain stem and spinal cord. These cortical neurons are called **pyramidal cells**, because their cell bodies resemble little pyramids.

The primary motor cortex is like the keyboard of a piano. If you strike a specific piano key, you produce a specific sound. Similarly, if you stimulate a specific motor neuron in the primary motor cortex, you generate a contraction in a specific skeletal muscle.

Like the monitoring gauges in the dashboard of a car, the sensory areas of the cerebral cortex report key information. At each location, sensory information is reported in the pattern of neuron activity in the cortex. The **postcentral gyrus** of the parietal lobe forms the posterior border of the central sulcus, and its surface contains the **primary sensory cortex**. Neurons in this region receive general somatic sensory information from receptors for touch, pressure, pain, vibration, or temperature. You are aware of these sensations only when nuclei in the thalamus relay the information to the primary sensory cortex.

Sensations of sight, sound, smell, and taste arrive at other portions of the cerebral cortex (**Figure 14–16a**). The **visual cortex** of the occipital lobe receives visual information. The **auditory cortex** and **olfactory cortex** of the temporal lobe receive information about hearing and smell, respectively. The **gustatory cortex**, which receives information from taste receptors of the tongue and pharynx, lies in the anterior portion of the insula and adjacent portions of the frontal lobe.

Association Areas

The sensory and motor regions of the cortex are connected to nearby **association areas**, regions of the cortex that interpret incoming data or coordinate a motor response (**Figure 14–16a**). Like the information provided by the gauges in a car, the arriving information must be noticed and interpreted before the driver can take appropriate action. *Sensory association areas* are cortical regions that monitor and interpret the information that arrives at the sensory areas of the cortex. Examples include the somatic sensory association area, visual association area, and auditory association area.

The **somatic sensory association area** monitors activity in the primary sensory cortex. The somatic sensory association area allows you to recognize a touch as light as a mosquito landing on your arm (and gives you a chance to swat the mosquito before it bites).

The special senses of smell, sight, and hearing involve separate areas of the sensory cortex, and each has its own association area. These areas monitor and interpret arriving sensations. For example, the **visual association area** monitors the patterns of activity in the visual cortex and interprets the results. You see the symbols *c*, *a*, and *r* when the stimulation of receptors in your eyes leads to the stimulation of neurons in your visual cortex. Your visual association area recognizes that these are letters and that *c* + *a* + *r* = *car*. An individual with a damaged visual association area could scan the lines of a printed page and see rows of symbols that are clear, but would perceive no meaning from the symbols. Similarly, the **auditory association area** monitors sensory activity in the auditory cortex. Word recognition takes place in this association area.

The **premotor cortex**, or **somatic motor association area**, coordinates learned movements. The primary motor cortex does nothing on its own, any more than a piano keyboard can play itself. The neurons in the primary motor cortex must be stimulated by neurons in other parts of the cerebrum. When you perform a voluntary movement, the premotor cortex relays the instructions to the primary motor cortex. With repetition, the proper pattern of stimulation becomes stored in your premotor cortex. You can then perform the movement smoothly and easily by triggering the *pattern* rather than by controlling the individual neurons. This principle applies to any learned movement, from something as simple as picking up a glass to something as complex as playing the piano. One area of the premotor cortex, the *frontal eye field*, controls learned eye movements, such as when you scan these lines of type (**Figure 14–16b**). Individuals with damage to the frontal eye field can understand written letters and words but cannot read, because their eyes cannot follow the lines on a printed page.

Integrative Centers

Integrative centers are areas that receive information from many association areas and direct extremely complex motor activities.

14

These centers also perform complicated analytical functions. For example, consider the *prefrontal cortex* of the frontal lobe (**Figure 14–16b**). It integrates information from sensory association areas and performs abstract intellectual functions, such as predicting the consequences of possible responses.

Integrative centers are located in the lobes and cortical areas of both cerebral hemispheres. Integrative centers concerned with complex processes, such as speech, writing, mathematical computation, and understanding spatial relationships, are largely restricted to either the left or the right hemisphere. These centers include the *general interpretive area* and the *speech center*. The corresponding regions on the opposite hemisphere are also active, but their functions are less well defined.

The General Interpretive Area. The **general interpretive area** also called the *Wernicke's area*, is present in only one hemisphere, typically the left (**Figure 14–16b**). This analytical center receives information from all the sensory association areas. It plays an essential role in your personality by integrating

Clinical Note

Disconnection Syndrome The functional differences between the hemispheres become apparent if the corpus callosum is cut. Such a procedure is sometimes performed to treat epileptic seizures that cannot be controlled by other methods. This surgery produces symptoms of **disconnection syndrome**. In this condition, the two hemispheres function independently, each "unaware" of stimuli or motor commands that involve the other.

Individuals with this syndrome exhibit some rather interesting changes in their mental abilities. For example, they can recognize but cannot verbally identify objects that they touch with the left hand, because the sensory information arrives at the right hemisphere but the speech center is on the left. They can verbally identify the object if they feel it with the right hand, but they cannot say whether it is the same object they previously touched with the left hand. Sensory information from the left side of the body arrives at the right hemisphere and cannot reach the general interpretive area. For this reason, conscious decisions are made without regard to sensations from the left side.

Two years after a surgical sectioning of the corpus callosum, the most striking behavioral abnormalities have disappeared. The person may test normally. In addition, individuals born without a functional corpus callosum do not have sensory, motor, or intellectual problems. In some way, the CNS adapts to these situations, probably by increasing the amount of information transferred across the anterior commissure.

sensory information and coordinating access to complex visual and auditory memories.

Damage to the general interpretive area affects your ability to interpret what you see or hear, even though you understand the words as individual entities. For example, if your general interpretive area were damaged, you might still understand the meaning of the spoken words *sit* and *here*, because word recognition takes place in the auditory association areas. But you would be totally bewildered by the request *sit here*. Damage to another portion of the general interpretive area might leave you able to see a chair clearly, and to know that you recognize it, but you would be unable to name it because the connection to your visual association area has been disrupted.

The Speech Center. Some of the neurons in the general interpretive area innervate the **speech center**, also called the *Broca's area* or the *motor speech area* (**Figure 14–16b**). This center lies along the edge of the premotor cortex in the same hemisphere as the general interpretive area (usually the left). The speech center regulates the patterns of breathing and vocalization needed for normal speech. This regulation involves coordinating the activities of the respiratory muscles, the laryngeal and pharyngeal muscles, and the muscles of the tongue, cheeks, lips, and jaws. A person with a damaged speech center can make sounds but not words.

The motor commands issued by the speech center are adjusted by feedback from the auditory association area, also called the *receptive speech area*. Damage to the related sensory areas can cause a variety of speech-related problems. (See the discussion of *aphasia* on p. 515.) Some affected individuals have difficulty speaking although they know exactly which words to use. Others talk constantly but use all the wrong words.

The Prefrontal Cortex. The **prefrontal cortex** of the frontal lobe coordinates information relayed from the association areas of the entire cortex (**Figure 14–16b**). In doing so, it performs such abstract intellectual functions as predicting the consequences of events or actions. The prefrontal cortex doesn't fully develop until the early 20s. Damage to the prefrontal cortex leads to difficulties in estimating temporal relationships between events. Questions such as "How long ago did this happen?" or "What happened first?" become difficult to answer.

The prefrontal cortex has extensive connections with other cortical areas and with other portions of the brain. Feelings of frustration, tension, and anxiety are generated as the prefrontal cortex interprets ongoing events and makes predictions about future situations or consequences. If the connections between the prefrontal cortex and other brain regions are physically severed, the frustrations, tensions, and anxieties are removed. During the middle of the 20th century, this rather drastic procedure, called **prefrontal lobotomy**, was used to "cure" a variety of mental illnesses, especially those associated with violent or antisocial behavior. After a lobotomy, the patient would no longer be concerned about what had previously been a major problem, whether psychological (hallucinations) or physical (severe pain). However, the individual was often equally unconcerned about tact, decorum, and toilet training. Since then, drugs that target specific pathways and regions of the CNS have been developed, so lobotomies are no longer used to change behavior.

Brodmann Areas. Early in the 20th century, numerous researchers attempted to describe and classify regional differences in the histological organization of the cerebral cortex. They hoped to correlate the patterns of cellular organization with specific functions. By 1919, at least 200 patterns had been described, but most of the classification schemes have since been abandoned. However, the cortical map prepared by Korbinian Brodmann in 1909 has proved useful to neuroanatomists. Brodmann, a German neurologist, described about 50 patterns of cellular organization in the cerebral cortex.

Several of these *Brodmann areas* are shown in **Figure 14–16c**. Some correspond to known functional areas. For example, Brodmann area 44 corresponds to the speech center, and area 41 to the auditory cortex. Area 4 follows the contours of the primary motor cortex. In other cases, the correspondence is less precise. For instance, Brodmann area 42 forms only a small portion of the auditory association area.

Hemispheric Lateralization

Each of the two cerebral hemispheres is responsible for specific functions that are not ordinarily performed by the opposite hemisphere. This regional specialization is called *hemispheric lateralization*. **Figure 14–17** indicates the major functional differences between the hemispheres. In most people, the left hemisphere contains the general interpretive and speech centers and is responsible for language-based skills. For example, reading, writing, and speaking depend on processing done in the left cerebral hemisphere. In addition, the premotor cortex involved with the control of hand movements is larger on the left side for right-handed individuals than for left-handed ones. The left hemisphere is also important in performing analytical tasks, such as mathematical calculations and logical decision-making. For these reasons, the left cerebral hemisphere has been called the *dominant hemisphere*.

The right cerebral hemisphere analyzes sensory information and relates the body to the sensory environment. Interpretive centers in this hemisphere permit you to identify familiar objects by touch, smell, sight, taste, or feel. For example, the right hemisphere plays a dominant role in recognizing faces and in understanding three-dimensional relationships. It is also important in analyzing the emotional context of a conversation—for instance, distinguishing between the threat "Get lost!" and the question "Get lost?" Individuals with a damaged

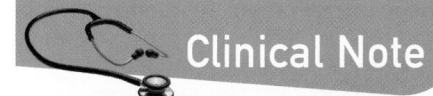

Aphasia and Dyslexia **Aphasia** (*a-*, without + *phasia*, speech) is a disorder affecting the ability to speak or read. *Global aphasia* results from extensive damage to the general interpretive area or to the associated sensory tracts. Affected individuals are unable to speak, read, or understand speech. Global aphasia often accompanies a severe stroke or tumor that affects a large area of cortex, including the speech and language areas. Recovery is possible when the condition results from edema or hemorrhage, but the process often takes months or even years.

Dyslexia (*dys-*, difficult, faulty + *lexis*, diction) is a disorder affecting the comprehension and use of written words. *Developmental dyslexia* affects children. Estimates are that up to 15 percent of children in the United States have some degree of dyslexia. Children with dyslexia have difficulty reading and writing, although their other intellectual functions may be normal or above normal. Their writing looks uneven and disorganized. They typically write letters in the wrong order (*dig* becomes *gid*) or reverse them (*E* becomes ∃). Recent evidence suggests that at least some forms of dyslexia result from problems in processing, sorting, and integrating visual or auditory information.

right hemisphere may be unable to add emotional inflections to their own words.

Left-handed people represent 9 percent of the human population. In most cases, the primary motor cortex of the right hemisphere controls motor function for the dominant (left) hand. However, the centers involved with speech and analytical function are in the left hemisphere, just as they are in right-handed people. Interestingly, an unusually high percentage of musicians and artists are left-handed. The complex motor activities performed by these individuals are directed by the primary motor cortex and association areas of the right cerebral hemisphere, near the association areas involved with spatial visualization and emotions.

Monitoring Brain Activity: The Electroencephalogram

Various methods have led to maps of brain activity and brain function. The primary sensory cortex and the primary motor cortex have been mapped by direct stimulation in patients undergoing brain surgery. The functions of other regions of the cerebrum can be revealed by the behavioral changes that follow localized injuries or strokes. The activities of specific regions can be visualized by a PET scan or through functional magnetic resonance imaging (fMRI). ⊃ p. 47

Clinicians commonly monitor the electrical activity of the brain to assess brain activity. Neural function depends on electrical events of the plasma membrane of neurons.

Figure 14–17 **Hemispheric Lateralization.** Functional differences between the left and right cerebral hemispheres.

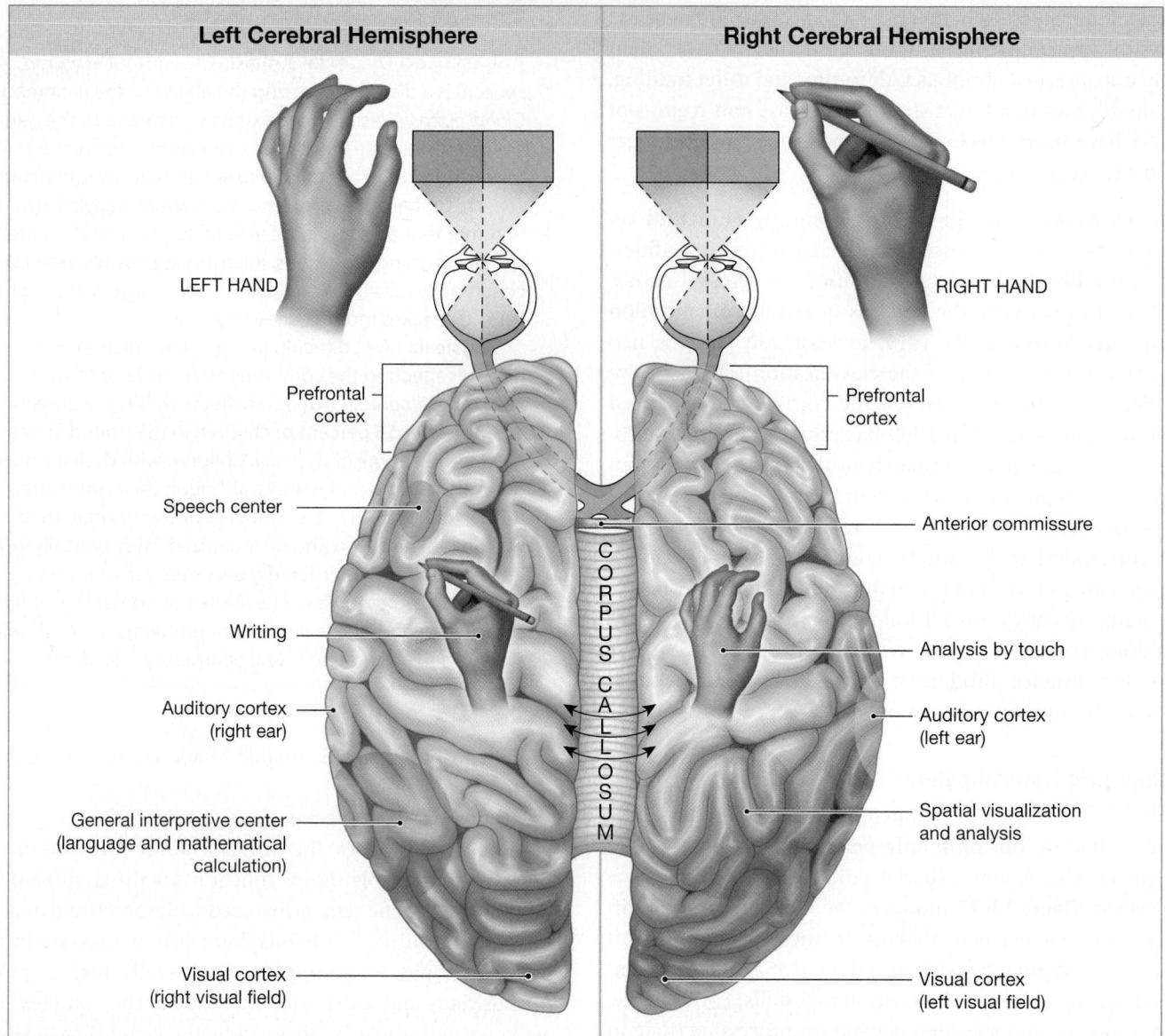

The brain contains billions of neurons, and their activity generates an electrical field that can be measured by electrodes placed on the brain or on the outer surface of the skull. The electrical activity changes constantly, as nuclei and cortical areas are stimulated or quiet down. A printed recording of the electrical activity of the brain is called an **electroencephalogram (EEG)**. The electrical patterns observed are called *brain waves*.

Typical brain waves are shown in **Figure 14–18a–d**. **Alpha waves** occur in the brains of healthy, awake adults who are resting with their eyes closed. Alpha waves disappear during sleep, but they also vanish when the individual begins to concentrate on some specific task.

During attention to stimuli or tasks, higher-frequency **beta waves** replace alpha waves. Beta waves are typical of individuals

who are either concentrating on a task, under stress, or in a state of psychological tension.

Theta waves may appear transiently during sleep in normal adults but are most often observed in children and in intensely frustrated adults. The presence of theta waves under other circumstances may indicate the presence of a brain disorder, such as a tumor.

Delta waves are very-large-amplitude, low-frequency waves. They are normally seen during deep sleep in individuals of all ages. Delta waves are also seen in the brains of infants (whose brains are still developing) and in awake adults when a tumor, vascular blockage, or inflammation has damaged portions of the brain.

Electrical activity in the two hemispheres is generally synchronized by a "pacemaker" mechanism that appears to

Figure 14–18 **Brain Waves.** The four electrical patterns revealed by electroencephalograms (EEGs). The heights (amplitudes) of the four waves are not drawn to the same scale.

Patient being wired
for EEG monitoring

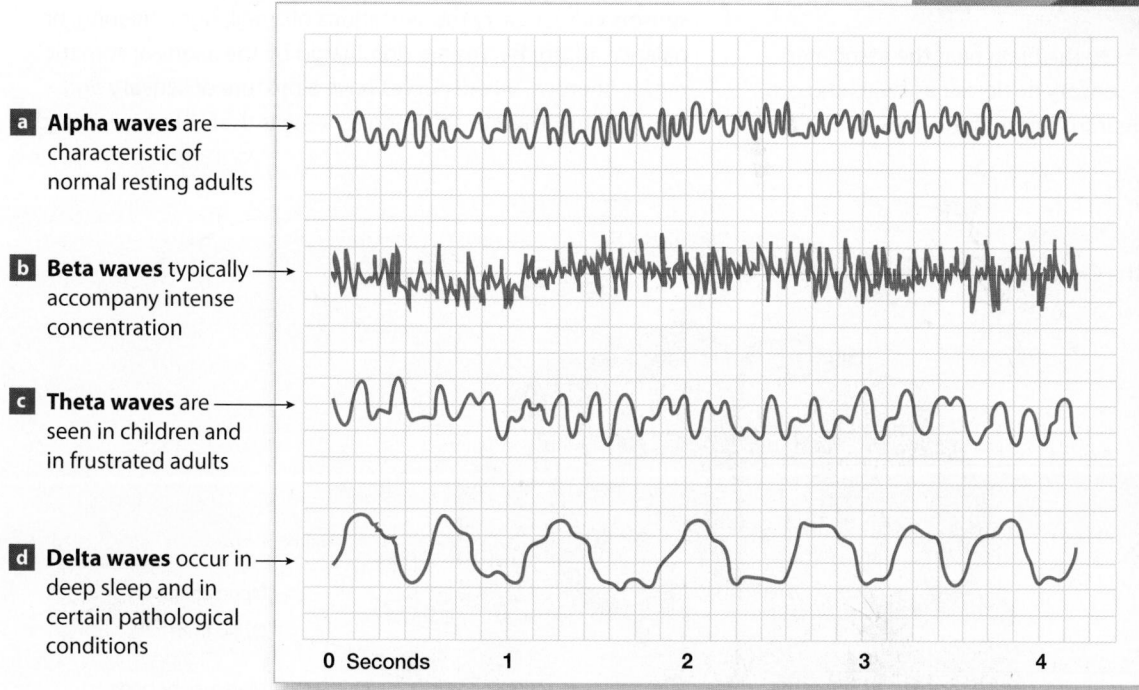

a Alpha waves are characteristic of normal resting adults

b Beta waves typically accompany intense concentration

c Theta waves are seen in children and in frustrated adults

d Delta waves occur in deep sleep and in certain pathological conditions

0 Seconds 1 2 3 4

involve the thalamus. Asynchrony between the hemispheres can indicate localized damage or other cerebral abnormalities. For example, a tumor or injury affecting one hemisphere typically changes the pattern in that hemisphere, and the patterns of the two hemispheres are no longer aligned.

A **seizure** is a temporary cerebral disorder accompanied by abnormal movements, unusual sensations, inappropriate behavior, or some combination of these symptoms. Clinical conditions characterized by seizures are known as seizure disorders, or *epilepsies*. Seizures of all kinds are accompanied by a marked change in the pattern of electrical activity recorded in an electroencephalogram. The change begins in one portion of the cerebral cortex but may then spread across the entire cortical surface, like a wave on the surface of a pond.

The nature of the signs and symptoms produced depends on the region of the cortex involved. If a seizure affects the primary motor cortex, movements will occur. If it affects the auditory cortex, the individual will hear strange sounds.

CHECKPOINT

21. What name is given to fibers carrying information between the brain and spinal cord, and through which brain regions do they pass?

22. What symptoms would you expect to observe in an individual who has damage to the basal nuclei?

23. A patient suffers a head injury that damages her primary motor cortex. Where is this area located?

24. Which senses would be affected by damage to the temporal lobes of the cerebrum?

25. After suffering a stroke, a patient is unable to speak. He can understand what is said to him, and he can understand written messages, but he cannot express himself verbally. Which part of his brain has been affected by the stroke?

26. A patient is having a difficult time remembering facts and recalling long-term memories. Which part of his cerebrum is probably involved?

See the blue Answers tab at the back of the book.

Focus | Cranial Nerves

Cranial nerves are PNS components that connect directly to the brain. The 12 pairs of cranial nerves are visible on the ventral surface of the brain (Figure 14–19). Each has a name related to its distribution or its function.

The number assigned to a cranial nerve corresponds to the nerve's position along the longitudinal axis of the brain, beginning at the cerebrum. Roman numerals preceded by the letter N are usually used. (You may sometimes encounter these numerals preceded by the letters CN.)

Each cranial nerve attaches to the brain near the associated sensory or motor nuclei. The sensory nuclei act as switching centers, with the postsynaptic neurons relaying the information to

other nuclei or to processing centers in the cerebral or cerebellar cortex. In a similar way, the motor nuclei receive convergent inputs from higher centers or from other nuclei along the brain stem.

In this section, we classify cranial nerves as primarily sensory, special sensory, motor, or mixed (sensory and motor). In this classification, sensory nerves carry somatic sensory information, including touch, pressure, vibration, temperature, or pain. Special sensory nerves carry the sensations of smell, sight, hearing, or balance. Motor nerves are dominated by the axons of somatic motor neurons. Mixed nerves have a mixture of sensory and motor fibers. This classification scheme is useful, but it is based

Figure 14–19 **Origins of the Cranial Nerves.** An inferior view of the brain.

Olfactory bulb: termination of olfactory nerve (N I)

Olfactory tract

Optic nerve (N II)

Infundibulum

Oculomotor nerve (N III)

Basilar artery

Pons

Vertebral artery

Medulla oblongata

Cerebellum

Spinal cord

Optic chiasm

Optic tract

Mammillary body

Trochlear nerve (N IV)

Trigeminal nerve (N V)

Abducens nerve (N VI)

Facial nerve (N VII)

Vestibulocochlear nerve (N VIII)

Glossopharyngeal nerve (N IX)

Vagus nerve (N X)

Hypoglossal nerve (N XII)

Accessory nerve (N XI)

on the primary function, and a cranial nerve can have important secondary functions. Three examples are worth noting:

1. The olfactory receptors, the visual receptors, and the receptors of the internal ear are innervated by cranial nerves that are dedicated almost entirely to carrying special sensory information. In contrast, the sensation of taste, considered to be one of the special senses, is carried by axons that form only a small part of large cranial nerves that have other primary functions.

2. As elsewhere in the PNS, a nerve containing tens of thousands of motor fibers that lead to a skeletal muscle also contains sensory fibers from muscle spindles and Golgi tendon organs (proprioceptive sensory nerve endings) in that muscle. We assume that these sensory fibers are present but ignore them in classifying the nerve.

3. Regardless of their other functions, several cranial nerves (III, VII, IX, and X) distribute autonomic fibers to peripheral ganglia, just as spinal nerves deliver them to ganglia along the spinal cord. We will note the presence of small numbers of autonomic fibers (and will discuss them further in Chapter 16) but ignore them in the classification of the nerve.

Cranial nerves are clinically important, in part because they can provide clues to underlying CNS problems. Clinicians assess cranial nerve function with a number of standardized tests.

TIPS&TRICKS

Two useful mnemonics for remembering the names of the cranial nerves in order are "**Oh Oh Oh, To Touch And Feel Very Green Vegetables, Ah Heaven!**" and "**Oh, Once One Takes The Anatomy Final, Very Good Vacations Are Heavenly!**"

The Olfactory Nerves (I)

Primary function: Special sensory (smell)

Origin: Receptors of olfactory epithelium

Pass through: Olfactory foramina in cribriform plate of ethmoid ⟲pp. 239, 244

Destination: Olfactory bulbs

The first pair of cranial nerves carries special sensory information responsible for the sense of smell (**Figure 14–20**). The olfactory receptors are specialized neurons in the epithelium covering the roof of the nasal cavity, the superior nasal conchae, and the

Figure 14–20 **The Olfactory Nerve.**

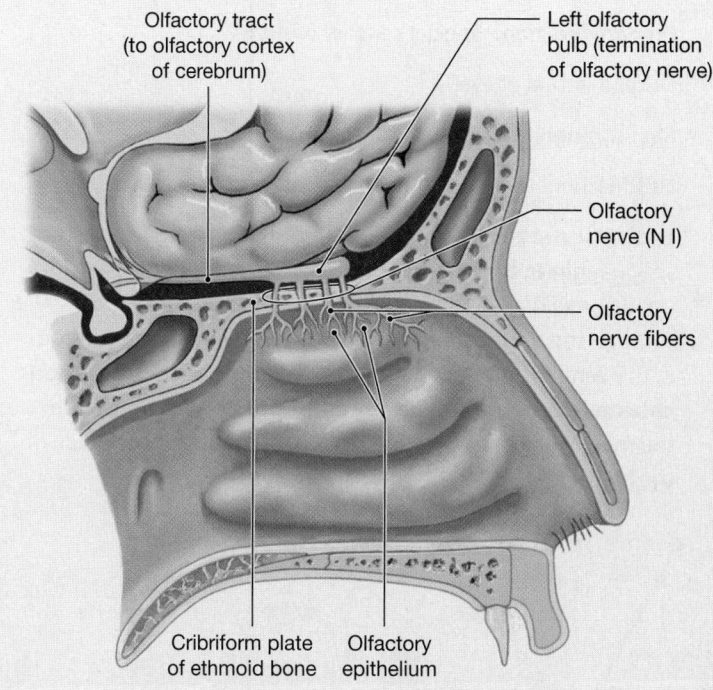

Olfactory tract (to olfactory cortex of cerebrum)

Left olfactory bulb (termination of olfactory nerve)

Olfactory nerve (N I)

Olfactory nerve fibers

Cribriform plate of ethmoid bone

Olfactory epithelium

superior parts of the nasal septum. Axons from these sensory neurons collect to form 20 or more bundles that penetrate the cribriform plate of the ethmoid bone. These bundles are components of the **olfactory nerves** (I). Almost at once these bundles enter the **olfactory bulbs**, neural masses on either side of the crista galli. The olfactory afferents synapse within the olfactory bulbs. The axons of the postsynaptic neurons proceed to the cerebrum along the slender **olfactory tracts** (**Figures 14–19** and **14–20**).

Because the olfactory tracts look like typical peripheral nerves, anatomists about a century ago misidentified these tracts as the first cranial nerve. Later studies demonstrated that the olfactory tracts and bulbs are part of the cerebrum, but by then the numbering system was already firmly established. Anatomists were left with a forest of tiny olfactory nerve bundles lumped together as cranial nerve I.

The olfactory nerves are the only cranial nerves attached directly to the cerebrum. The rest originate or terminate within nuclei of the diencephalon or brain stem, and the ascending sensory information synapses in the thalamus before reaching the cerebrum.

The Optic Nerves (II)

Primary function: Special sensory (vision)

Origin: Retina of eye

Pass through: Optic canals of sphenoid ⊃ p. 243

Destination: Diencephalon by way of the optic chiasm

The **optic nerves** (II) carry visual information from special sensory ganglia in the eyes (Figure 14–21). These nerves contain about 1 million sensory nerve fibers. The optic nerves pass through the optic canals of the sphenoid. Then they converge at the ventral, anterior margin of the diencephalon, at the **optic chiasm** (*chiasma*, a crossing). At the optic chiasm, fibers from the medial half of each retina cross over to the opposite side of the brain.

Figure 14–21 The Optic Nerve.

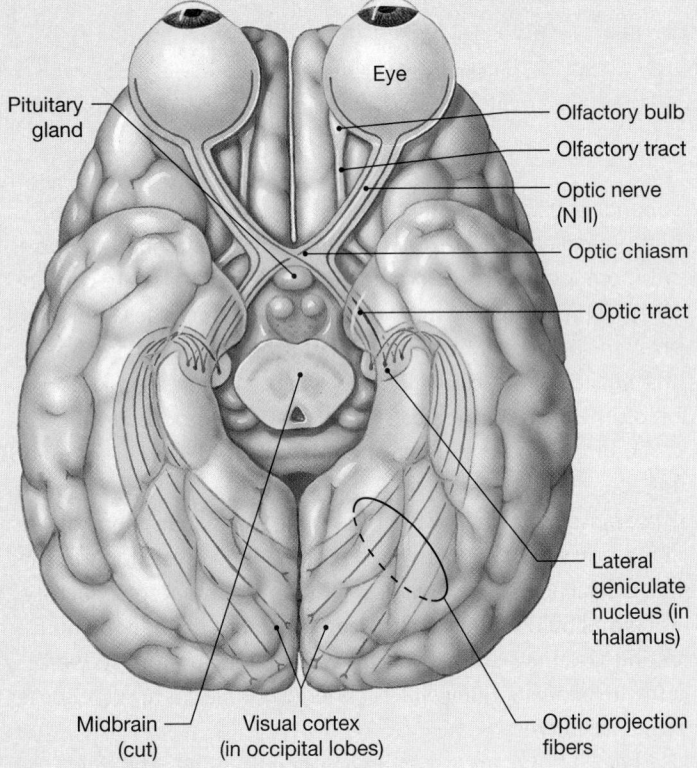

Pituitary gland

Eye

Olfactory bulb

Olfactory tract

Optic nerve (N II)

Optic chiasm

Optic tract

Lateral geniculate nucleus (in thalamus)

Midbrain (cut)

Visual cortex (in occipital lobes)

Optic projection fibers

The reorganized axons continue toward the lateral geniculate nuclei of the thalamus as the **optic tracts** (Figures 14–19 and 14–21). After synapsing in the lateral geniculates, projection fibers deliver the information to the visual cortex of the occipital lobes. With this arrangement, each cerebral hemisphere receives visual information from the lateral half of the retina of the eye on that side and from the medial half of the retina of the eye of the opposite side (Figure 14–17). A few axons in the optic tracts bypass the lateral geniculate nuclei and synapse in the superior colliculi of the midbrain. We will consider that pathway in Chapter 17.

The Oculomotor Nerves (III)

Primary function: Motor (eye movements)

Origin: Midbrain

Pass through: Superior orbital fissures of sphenoid ⊃ pp. 237, 244, 246, 249

Destination: *Somatic motor:* superior, inferior, and medial rectus muscles; inferior oblique muscle; levator palpebrae superioris muscle. *Visceral motor:* intrinsic eye muscles

The midbrain contains the motor nuclei controlling the third and fourth cranial nerves. Each **oculomotor nerve** (III) innervates four of the six extrinsic muscles that move the eye, and the levator palpebrae superioris muscle, which raises the upper eyelid (Figure 14–22). On each side of the brain, nerve III emerges from the ventral surface of the midbrain and penetrates the posterior wall of the orbit at the superior orbital fissure. Individuals with damage to this nerve often complain of pain over the eye, droopy eyelids, and double vision, because the movements of the left and right eyes cannot be coordinated properly.

The oculomotor nerve also delivers preganglionic autonomic fibers to neurons of the **ciliary ganglion**. These neurons control intrinsic eye muscles. These muscles change the diameter of the pupil, adjusting the amount of light entering the eye. They also change the shape of the lens to focus images on the retina.

Figure 14–22   Cranial Nerves Controlling the Extrinsic Eye Muscles. ATLAS: Plates 16a,b

The Trochlear Nerves (IV)

Primary function: Motor (eye movements)

Origin: Midbrain

Pass through: Superior orbital fissures of sphenoid � pp. 237, 243, 246, 249

Destination: Superior oblique muscle

A **trochlear** (TRŌK-lē-ar; *trochlea*, a pulley) **nerve** (IV), the smallest cranial nerve, innervates the superior oblique muscle of each eye (Figure 14–22). The trochlea is a pulley-shaped, ligamentous sling. Each superior oblique muscle passes through a trochlea on its way to its insertion on the surface of the eye. An individual with damage to cranial nerve IV or to its nucleus has difficulty looking down and to the side.

The Abducens Nerves (VI)

Primary function: Motor (eye movements)

Origin: Pons

Pass through: Superior orbital fissures of sphenoid ↺ pp. 237, 243, 246, 249

Destination: Lateral rectus muscle

The **abducens** (ab-DŪ-senz) **nerves** (VI) innervate the lateral rectus muscles, the sixth pair of extrinsic eye muscles. Contraction of the lateral rectus muscle makes the eye look to the side. In essence, the *abducens* causes *abduction* of the eye. Each abducens nerve emerges from the inferior surface of the brain stem at the border between the pons and the medulla oblongata (Figure 14–22). Along with the oculomotor and trochlear nerves from that side, it reaches the orbit through the superior orbital fissure.

The Trigeminal Nerves (V)

Primary function: Mixed (sensory and motor) to face

Origin: *Ophthalmic branch* (sensory): orbital structures, cornea, nasal cavity, skin of forehead, upper eyelid, eyebrow, nose (part). *Maxillary branch* (sensory): lower eyelid, upper lip, gums, and teeth; cheek; nose, palate, and pharynx (part). *Mandibular branch* (mixed): sensory from lower gums, teeth, and lips; palate and tongue (part); motor from motor nuclei of pons

Pass through (on each side): Ophthalmic branch through superior orbital fissure, maxillary branch through foramen rotundum, mandibular branch through foramen ovale ⊃ pp. 237, 239, 243, 249

Destination: Ophthalmic, maxillary, and mandibular branches to sensory nuclei in pons; mandibular branch also innervates muscles of mastication ⊃ p. 374

The pons contains the nuclei associated with three cranial nerves (V, VI, and VII) and contributes to a fourth (VIII). The **trigeminal** (trī-JEM-i-nal) **nerves** (V), the largest cranial nerves, are mixed nerves. Each provides both somatic sensory information from the head and face, and motor control over the muscles of mastication. Sensory (dorsal) and motor (ventral) roots originate on the lateral surface of the pons (Figure 14–23). The sensory branch is larger, and the enormous **semilunar ganglion** contains the cell bodies of the sensory neurons.

As the name implies, the trigeminal has three major branches. The small motor root contributes to only one of the three. **Tic douloureux** (doo-luh-ROO; *douloureux*, painful), or *trigeminal neuralgia*, is a painful condition affecting the area innervated by the maxillary and mandibular branches of the trigeminal nerve. Sufferers complain of debilitating pain triggered by contact with the lip, tongue, or gums. The cause of the condition is unknown.

The trigeminal nerve branches are associated with the *ciliary, sphenopalatine, submandibular,* and *otic ganglia*. These are autonomic (parasympathetic) ganglia whose neurons innervate structures of the face. Although its nerve fibers may pass around or through these ganglia, the trigeminal nerve does not contain visceral motor fibers. We discussed the ciliary ganglion on page 494 and will describe the other ganglia next, with the branches of the *facial nerves* (VII) and the *glossopharyngeal nerves* (IX).

Figure 14–23   **The Trigeminal Nerve.**

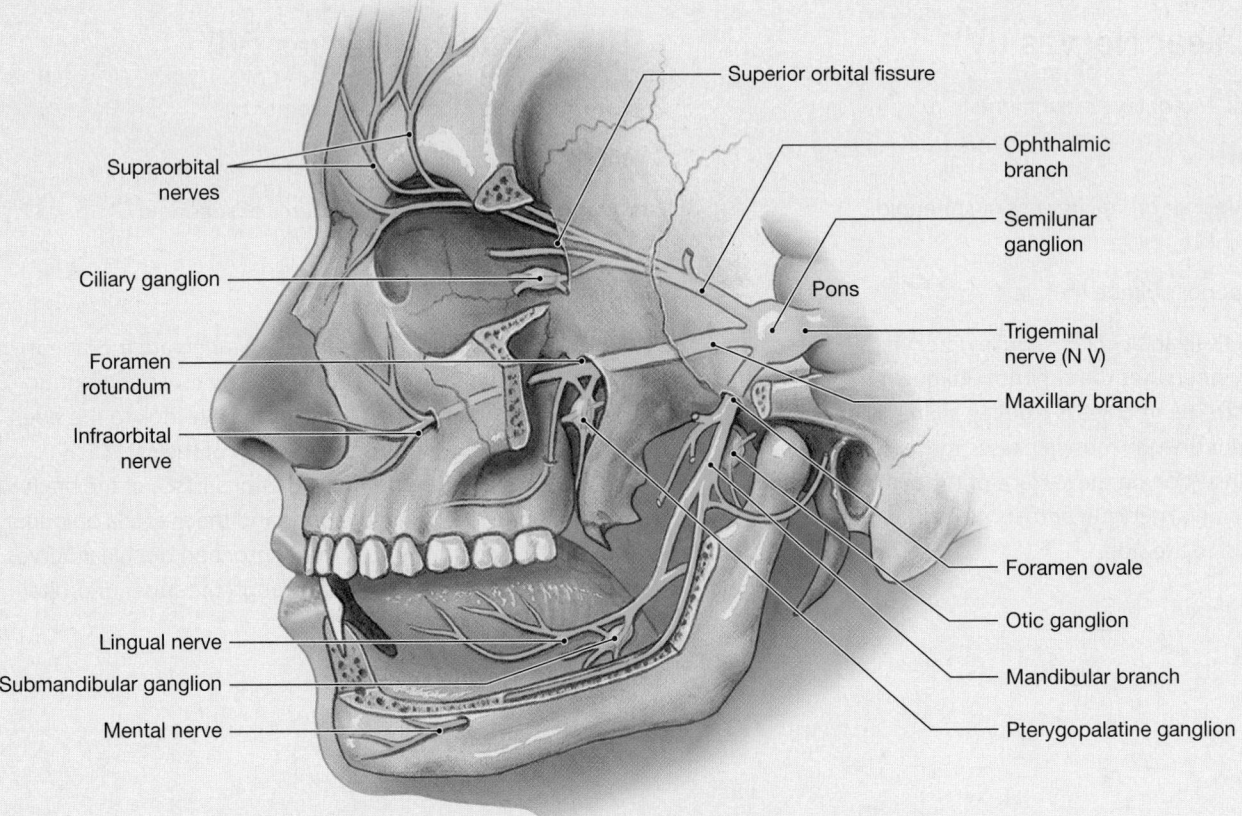

Superior orbital fissure

Ophthalmic branch

Semilunar ganglion

Pons

Trigeminal nerve (N V)

Maxillary branch

Supraorbital nerves

Ciliary ganglion

Foramen rotundum

Infraorbital nerve

Foramen ovale

Otic ganglion

Mandibular branch

Pterygopalatine ganglion

Lingual nerve

Submandibular ganglion

Mental nerve

The Facial Nerves (VII)

Primary function: Mixed (sensory and motor) to face

Origin: *Sensory*: taste receptors on anterior two-thirds of tongue. *Motor*: motor nuclei of pons

Pass through: Internal acoustic meatus to canals leading to the stylomastoid foramina ⤵ pp. 237, 239, 243

Destination: *Sensory*: sensory nuclei of pons. *Somatic motor*: muscles of facial expression. ⤵ p. 370 *Visceral motor*: lacrimal (tear) gland and nasal mucous glands by way of the pterygopalatine ganglion; submandibular and sublingual salivary glands by way of the submandibular ganglion

The **facial nerves** (VII) are mixed nerves. The cell bodies of the sensory neurons are located in the **geniculate ganglia**, and the motor nuclei are in the pons. On each side, the sensory and motor roots emerge from the pons and enter the internal acoustic meatus of the temporal bone. Each facial nerve then passes through the facial canal to reach the face by way of the stylomastoid foramen. The nerve then splits to form the temporal, zygomatic, buccal, mandibular, and cervical branches (Figure 14–24).

The sensory neurons monitor proprioceptors in the facial muscles and provide deep pressure sensations over the face. They also receive taste information from receptors along the anterior two-thirds of the tongue. Somatic motor fibers control the superficial muscles of the scalp and face and deep muscles near the ear.

The facial nerves carry preganglionic autonomic fibers to the pterygopalatine and submandibular ganglia. Postganglionic fibers from the **pterygopalatine ganglia** innervate the lacrimal glands and small glands of the nasal cavity and pharynx. The **submandibular ganglia** innervate the *submandibular* and *sublingual* (*sub-*, under + *lingual*, pertaining to the tongue) *salivary glands*.

Bell's palsy is a cranial nerve disorder that results from an inflammation of a facial nerve. The condition is probably due to a viral infection. Signs and symptoms include paralysis of facial muscles on the affected side and loss of taste sensations from the anterior two-thirds of the tongue. The condition is usually painless. In most cases the symptoms fade after a few weeks or months.

Figure 14–24   **The Facial Nerve.**

a The origin and branches of the facial nerve

b The superficial distribution of the five major branches of the facial nerve

The Vestibulocochlear Nerves (VIII)

Primary function: Special sensory: balance and equilibrium (vestibular branch) and hearing (cochlear branch)

Origin: Monitor receptors of the internal ear (vestibule and cochlea)

Pass through: Internal acoustic meatus of temporal bones �. 239, 243

Destination: Vestibular and cochlear nuclei of pons and medulla oblongata

The **vestibulocochlear nerves** (VIII) are also known as the *acoustic nerves*, the *auditory nerves*, and the *stato-acoustic nerves*. We will use *vestibulocochlear*, because this term indicates the names of the two major branches: the vestibular branch and the cochlear branch. Each vestibulocochlear nerve lies posterior to the origin of the facial nerve, straddling the boundary between the pons and the medulla oblongata (Figure 14–25). This nerve

reaches the sensory receptors of the internal ear by entering the internal acoustic meatus in company with the facial nerve.

Each vestibulocochlear nerve has two distinct bundles of sensory fibers. The **vestibular** (*vestibulum*, **cavity**) **branch** originates at the receptors of the *vestibule*, the portion of the internal ear concerned with balance sensations. The sensory neurons are located in an adjacent sensory ganglion, and their axons target the **vestibular nuclei** of the pons and medulla oblongata. These afferents convey information about the orientation and movement of the head. The **cochlear** (*cochlea*, snail shell) **branch** monitors the receptors in the *cochlea*, the portion of the internal ear that provides the sense of hearing. The cell bodies of the sensory neurons are located within a peripheral ganglion (the *spiral ganglion*), and their axons synapse within the **cochlear nuclei** of the pons and medulla oblongata. Axons leaving the vestibular and cochlear nuclei relay the sensory information to other centers or initiate reflexive motor responses. We discuss balance and the sense of hearing in Chapter 17.

Figure 14–25 **The Vestibulocochlear Nerve.**

The Glossopharyngeal Nerves (IX)

Primary function: Mixed (sensory and motor) to head and neck

Origin: *Sensory*: posterior one-third of the tongue, part of the pharynx and palate, carotid arteries of the neck. *Motor*: motor nuclei of medulla oblongata

Pass through: Jugular foramina between the occipital bone and the temporal bones ↷ pp. 237, 239

Destination: *Sensory*: sensory nuclei of medulla oblongata. *Somatic motor*: pharyngeal muscles involved in swallowing. *Visceral motor*: parotid salivary gland by way of the otic ganglion

The medulla oblongata contains the sensory and motor nuclei of cranial nerves IX, X, XI, and XII, in addition to the vestibular nucleus of nerve VIII. The **glossopharyngeal** (glos-ō-fah-RIN-jē-al; *glossum*, tongue) **nerves** (IX) innervate the tongue and pharynx. Each glossopharyngeal nerve penetrates the cranium within the jugular foramen, with cranial nerves X and XI.

The glossopharyngeal nerves are mixed nerves, but sensory fibers are most abundant. The sensory neurons on each side are in the **superior** (*jugular*) **ganglion** and **inferior** (*petrosal*) **ganglion** (Figure 14–26). The sensory fibers carry general sensory information from the lining of the pharynx and the soft palate to a nucleus in the medulla oblongata. These nerves also provide taste sensations from the posterior third of the tongue. Additionally, they have special receptors that monitor the blood pressure and dissolved gas concentrations in the carotid arteries, major blood vessels in the neck.

The somatic motor fibers control the pharyngeal muscles involved in swallowing. Visceral motor fibers synapse in the **otic ganglion**. Postganglionic fibers innervate the parotid salivary gland of the cheek.

The Vagus Nerves (X)

Primary function: Mixed (sensory and motor), widely distributed in the thorax and abdomen

Origin: *Sensory*: pharynx (part), auricle and external acoustic meatus (a portion of the external ear), diaphragm, and visceral organs in thoracic and abdominopelvic cavities. *Motor*: motor nuclei in medulla oblongata

Pass through: Jugular foramina between the occipital bone and the temporal bones ↷ pp. 237, 239

Figure 14–26   **The Glossopharyngeal Nerve.**

- Otic ganglion
- Inferior (petrosal) ganglion
- Lingual branch

- Pons
- N VII
- N VI
- N V
- N VIII
- Glossopharyngeal nerve (N IX)
- Medulla oblongata
- Superior (jugular) ganglion
- Pharyngeal branches
- Parotid salivary gland
- Carotid sinus branch

- Carotid body
- Carotid sinus
- Common carotid artery

Figure 14–27 **The Vagus Nerve.**

Labels on figure:
Pharyngeal branch
Vagus nerve (N X)
Pons
Medulla oblongata
Auricular branch to external ear
Superior ganglion of vagus nerve
Inferior ganglion of vagus nerve
Pharyngeal branch
Superior laryngeal nerve
Superior laryngeal nerve
Internal branch
External branch
Recurrent laryngeal nerve
Cardiac branches
Cardiac plexus
Right lung
Left lung
Liver
Anterior vagal trunk
Spleen
Stomach
Celiac plexus
Pancreas
Colon
Small intestine
Hypogastric plexus

Destination: *Sensory*: sensory nuclei and autonomic centers of medulla oblongata. *Visceral motor*: muscles of the palate, pharynx, digestive, respiratory, and cardiovascular systems in the thoracic and abdominal cavities

The **vagus** (VĀ-gus) **nerves** (X) arise immediately posterior to the attachment of the glossopharyngeal nerves. Many small rootlets contribute to their formation. Developmental studies indicate that these nerves probably represent the fusion of several smaller cranial nerves during our evolutionary history. *Vagus* means wandering, and the vagus nerves branch and radiate extensively. Figure 14–27 shows only the general pattern of distribution.

Sensory neurons are located in the **superior** (jugular) **ganglion** and the **inferior** (nodose; NŌ-dōs) **ganglion**. Each vagus nerve provides somatic sensory information about the external acoustic meatus and the diaphragm, and special sensory information from pharyngeal taste receptors. But most of the vagal afferents carry visceral sensory information from receptors along the esophagus, respiratory tract, and abdominal viscera as distant as the last portions of the large intestine. This visceral sensory information is vital to the autonomic control of visceral function.

The motor components of the vagus are equally diverse. Each vagus nerve carries preganglionic autonomic (parasympathetic) fibers that affect the heart and control smooth muscles and glands within the areas monitored by its sensory fibers, including the stomach, intestines, and gallbladder. Difficulty in swallowing is one of the most common signs of damage to either nerve IX or X. Damage to either one prevents the coordination of the swallowing reflex.

The Accessory Nerves (XI)

Primary function: Motor to muscles of the neck and upper back

Origin: Motor nuclei of spinal cord and medulla oblongata

Pass through: Jugular foramina between the occipital bone and the temporal bones ⮌ pp. 237, 239

Destination: Internal branch innervates voluntary muscles of palate, pharynx, and larynx; external branch controls sternocleidomastoid and trapezius muscles

The **accessory nerves** (XI) are also known as the *spinal accessory nerves* or the *spinoaccessory nerves*. Unlike other cranial nerves, each accessory nerve has some motor fibers that originate in the lateral part of the anterior gray horns of the first five cervical segments of the spinal cord (**Figure 14–28**). These somatic motor

fibers form the **spinal root** of nerve XI. They enter the cranium through the foramen magnum. They then join the motor fibers of the **cranial root**, which originates at a nucleus in the medulla oblongata. The composite nerve leaves the cranium through the jugular foramen and divides into two branches.

The **internal branch** of nerve XI joins the vagus nerve and innervates the voluntary swallowing muscles of the soft palate and pharynx and the intrinsic muscles that control the vocal cords. The **external branch** of nerve XI controls the sternocleidomastoid and trapezius muscles of the neck and back. ⮌ pp. 377, 384 The motor fibers of this branch originate in the lateral gray part of the anterior horns of cervical spinal nerves C_1 to C_5.

The Hypoglossal Nerves (XII)

Primary function: Motor (tongue movements)

Origin: Motor nuclei of medulla oblongata

Pass through: Hypoglossal canals of occipital bone ⮌ pp. 238, 239, 240

Destination: Muscles of the tongue ⮌ p. 375

Figure 14–28 The Accessory and Hypoglossal Nerves. ATLAS: Plates 18a–c; 25

527

Each **hypoglossal** (hī-pō-GLOS-al) **nerve** (XII) leaves the cranium through the hypoglossal canal. The nerve then curves to reach the skeletal muscles of the tongue (**Figure 14–28**). This cranial nerve provides voluntary motor control over movements of the tongue. Its condition is checked by having you stick out your tongue. Damage to one hypoglossal nerve or to its associated nuclei causes the tongue to veer toward the affected side.

Table 14–4 summarizes the basic distribution and function of each cranial nerve.

Table 14–4	Cranial Nerve Branches and Functions				
Cranial Nerve (Number)	**Sensory Ganglion**	**Branch**	**Primary Function**	**Foramen**	**Innervation**
Olfactory (N I)			Special sensory	Olfactory foramina of ethmoid	Olfactory epithelium
Optic (N II)			Special sensory	Optic canal	Retina of eye
Oculomotor (N III)			Motor	Superior orbital fissure	Inferior, medial, superior rectus, inferior oblique and levator palpebrae superioris muscles; intrinsic eye muscles
Trochlear (N IV)			Motor	Superior orbital fissure	Superior oblique muscle
Trigeminal (N V)	Semilunar		Mixed	Superior orbital fissure	Areas associated with the jaws
		Ophthalmic	Sensory	Superior orbital fissure	Orbital structures, nasal cavity, skin of forehead, upper eyelid, eyebrows, nose (part)
		Maxillary	Sensory	Foramen rotundum	Lower eyelid; superior lip, gums, and teeth; cheek, nose (part), palate, and pharynx (part)
		Mandibular	Mixed	Foramen ovale	*Sensory*: inferior gums, teeth, lips, palate (part), and tongue (part) *Motor*: muscles of mastication
Abducens (N VI)			Motor	Superior orbital fissure	Lateral rectus muscle
Facial (N VII)	Geniculate		Mixed	Internal acoustic meatus to facial canal; exits at stylomastoid foramen	*Sensory*: taste receptors on anterior 2/3 of tongue *Motor*: muscles of facial expression, lacrimal gland, submandibular gland, sublingual salivary glands
Vestibulocochlear (N VIII)		Cochlear Vestibular	Special sensory	Internal acoustic meatus	Cochlea (receptors for hearing) Vestibule (receptors for motion and balance)
Glossopharyngeal (N IX)	Superior (jugular) and inferior (petrosal)		Mixed	Jugular foramen	*Sensory*: posterior 1/3 of tongue; pharynx and palate (part); receptors for blood pressure, pH, oxygen, and carbon dioxide concentrations *Motor*: pharyngeal muscles and parotid salivary gland
Vagus (N X)	Superior (jugular) and inferior (nodose)		Mixed	Jugular foramen	*Sensory*: pharynx; auricle and external acoustic meatus; diaphragm; visceral organs in thoracic and abdominopelvic cavities *Motor*: palatal and pharyngeal muscles and visceral organs in thoracic and abdominopelvic cavities
Accessory (N XI)		Internal	Motor	Jugular foramen	Skeletal muscles of palate, pharynx, and larynx (with vagus nerve)
		External	Motor	Jugular foramen	Sternocleidomastoid and trapezius muscles
Hypoglossal (N XII)			Motor	Hypoglossal canal	Tongue musculature

14-10 ❙ Cranial reflexes involve sensory and motor fibers of cranial nerves

Learning Outcome Describe representative examples of cranial reflexes that produce somatic responses or visceral responses to specific stimuli.

Cranial reflexes are automatic responses to stimuli that involve the sensory and motor fibers of cranial nerves. They include both monosynaptic and polysynaptic reflex arcs. Later chapters contain numerous examples of cranial reflexes. In this section we will simply provide an overview and general introduction.

Table 14–5 lists representative examples of cranial reflexes and their functions. These reflexes are clinically important because they provide a quick and easy method for observing the condition of cranial nerves and specific nuclei and tracts in the brain. The somatic reflexes mediated by the cranial nerves are seldom more complex than the somatic reflexes of the spinal cord that we discussed in Chapter 13. ↺ p. 478 These cranial reflexes are often used to check for damage to the cranial nerves or the associated processing centers in the brain.

The brain stem contains many reflex centers that control visceral motor activity. The motor output of these reflexes is distributed by the autonomic nervous system. As you will see in Chapter 16, the cranial nerves carry most of the commands issued by the parasympathetic division of the ANS. Spinal nerves T_1–L_2 carry the sympathetic commands. Many of the centers that coordinate autonomic reflexes are located in the medulla oblongata. These centers can direct very complex visceral motor responses that are essential to the control of respiratory, digestive, and cardiovascular functions.

CHECKPOINT

27. What are cranial reflexes?

28. Which reflex and type are you testing if you shine a light in a person's eye?

See the blue Answers tab at the back of the book.

Table 14–5	Cranial Reflexes				
Reflex	**Stimulus**	**Afferents**	**Central Synapse**	**Efferents**	**Response**
SOMATIC REFLEXES					
Corneal reflex	Contact with corneal surface	N V (trigeminal)	Motor nucleus for N VII (facial)	N VII (facial)	Blinking of eyelids
Tympanic reflex	Loud noise	N VIII (vestibulocochlear)	Inferior colliculus	N VII (facial)	Reduced movement of auditory ossicles
Auditory reflexes	Loud noise	N VIII (vestibulocochlear)	Motor nuclei of brain stem and spinal cord	N III, IV, VI, VII, X, and cervical nerves	Eye and/or head movements triggered by sudden sounds
Vestibulo-ocular reflexes	Rotation of head	N VIII (vestibulocochlear)	Motor nuclei controlling eye muscles	N III, IV, VI	Opposite movement of eyes to stabilize field of vision
VISCERAL REFLEXES					
Direct light reflex	Light striking photoreceptors	N II (optic)	Superior colliculus	N III (oculomotor)	Constriction of ipsilateral pupil
Consensual light reflex	Light striking photoreceptors	N II (optic)	Superior colliculus	N III (oculomotor)	Constriction of contralateral pupil

Related Clinical Terms

attention deficit hyperactivity disorder (ADHD): Disorder occurring mainly in children characterized by hyperactivity, inability to concentrate, and impulsive or inappropriate behavior.

autism: Any of a range of behavioral disorders occurring primarily in children, including such symptoms as poor concentration, hyperactivity, and impulsivity.

concussion: A brain injury often caused by a hit to the head that may result in a bad headache, altered levels of alertness, loss of memory, or unconsciousness.

Creutzfeldt-Jakob disease (CJD): A rare, degenerative, invariably fatal brain disorder that is marked by rapid mental deterioration. The disease, which is caused by a prion (an infectious protein particle),

typically starts by causing mental and emotional problems, then progresses to affect motor skills, such as walking and talking.

delirium: An acutely disturbed state of mind that occurs in fever, intoxication, and other disorders and is characterized by restlessness, hallucinations, and incoherence of thought and speech.

dementia: A chronic or persistent disorder of the mental processes caused by brain disease or injury and marked by memory disorders, personality changes, and impaired reasoning.

Glasgow coma scale: The most widely used scoring system to quantify the level of consciousness of a victim of a traumatic brain injury. It rates three functions: eye opening, verbal response, and motor response.

migraine: A type of headache marked by severe debilitating head pain lasting several hours or longer.

myoclonus: A quick, involuntary muscle jerk or contraction; persistent myoclonus usually indicates a nervous system disorder.

pallidectomy: The destruction of all or part of the globus pallidus by chemicals or freezing; used in the treatment of Parkinson's disease.

prosopagnosia: The inability to recognize other humans by their faces.

psychosis: A severe mental disorder in which thought and emotions are so impaired that contact with reality is lost.

stupor: A state of near-unconsciousness or insensibility.

transient ischemic attack (TIA): An episode in which a person has stroke-like symptoms that last less than 24 hours and result in no permanent injury to the brain, but may be a warning sign of the potential for a major stroke.

Chapter Review

Study Outline

MasteringA&P®

Access more chapter study tools online in the MasteringA&P Study Area:

- Chapter Quizzes, Chapter Practice Test, Art-labeling Activities, Animations, MP3 Tutor Sessions, and Clinical Case Studies
- Practice Anatomy Lab   PAL™
- Interactive Physiology   iP®
- A&P Flix   A&PFlix
- PhysioEx   PhysioEx

An Introduction to the Brain and Cranial Nerves p. 488

1. The adult human brain contains almost 97 percent of the body's neural tissue and averages 1.4 kg (3 lb) in weight and 1200 mL (71 in.^{3}) in volume.

14-1 ▌ The brain has several principal structures, each with specific functions p. 488

2. The six regions in the adult brain are the **cerebrum, cerebellum, diencephalon, midbrain** (mesencephalon), **pons**, and **medulla oblongata**. *(Figure 14–1)*

3. The brain contains extensive areas of **neural cortex**, a layer of gray matter on the surfaces of the cerebrum (**cerebral cortex**) and cerebellum.

4. The brain forms from three swellings at the superior tip of the developing *neural tube*: the *prosencephalon*, the *mesencephalon*, and the *rhombencephalon*. The prosencephalon ("forebrain") forms the **telencephalon** (which becomes the cerebrum) and diencephalon; the rhombencephalon ("hindbrain") forms the **metencephalon** (cerebellum and pons) and **myelencephalon** (medulla oblongata). *(Table 14–1)*

5. During development, the central passageway of the brain expands to form chambers called **ventricles**, which contain cerebrospinal fluid (CSF). *(Figure 14–2)*

14-2 ▌ The brain is protected and supported by the cranial meninges, cerebrospinal fluid, and the blood–brain barrier p. 491

6. The cranial meninges (the *dura mater, arachnoid mater*, and *pia mater*) are continuous with the meninges of the spinal cord.

7. Folds of dura mater, including the **falx cerebri, tentorium cerebelli**, and **falx cerebelli**, stabilize the position of the brain. *(Figure 14–3)*

8. Cerebrospinal fluid (CSF) (1) protects delicate neural structures, (2) supports the brain, and (3) transports nutrients, chemical messengers, and waste products.

9. Cerebrospinal fluid is produced at the **choroid plexuses**, reaches the subarachnoid space through the **lateral** and **median apertures**, and diffuses across the **arachnoid granulations** into the **superior sagittal sinus**. *(Spotlight Figure 14–4)*

10. The **blood–brain barrier (BBB)** isolates neural tissue from the general circulation.

11. The blood–brain barrier is incomplete in parts of the **hypothalamus**, the **pituitary gland**, the **pineal gland**, and the **choroid plexus**.

14-3 ▌ The medulla oblongata is continuous with the spinal cord and contains vital centers p. 496

12. The medulla oblongata connects the brain and spinal cord. It contains relay stations such as the **olivary nuclei**, and **reflex centers**, including the **cardiovascular** and **respiratory rhythmicity centers**. The **reticular formation** begins in the medulla oblongata and extends into more superior portions of the brain stem. *(Figures 14–5, 14–6)*

14-4 ▌ The pons contains nuclei and tracts that carry or relay sensory and motor information p. 498

13. The pons contains (1) sensory and motor nuclei for four cranial nerves; (2) nuclei that help control respiration; (3) nuclei and tracts linking the cerebellum with the brain stem, cerebrum, and spinal cord; and (4) ascending, descending, and transverse tracts. *(Figure 14–7)*

14-5 ▌ The cerebellum coordinates learned and reflexive patterns of muscular activity at the subconscious level p. 499

14. The cerebellum adjusts postural muscles and programs and tunes ongoing movements. The two **cerebellar hemispheres** consist of the **anterior** and **posterior lobes**, the **vermis**, and the **flocculonodular lobe**. *(Figure 14–8)*

15. The **superior, middle**, and **inferior cerebellar peduncles** link the cerebellum with the brain stem, diencephalon, cerebrum, and spinal cord and interconnect the two cerebellar hemispheres.

14-6 ❚ **The midbrain regulates auditory and visual reflexes and controls alertness** p. 501

16. The **tectum** (roof of the midbrain) contains the **corpora quadrigemina** (**superior colliculi** and **inferior colliculi**). The tegmentum contains the **red nucleus**, the **substantia nigra**, the **cerebral peduncles**, and the headquarters of the reticular activating system (RAS). (*Figure 14–9*)

14-7 ❚ **The diencephalon integrates sensory information with motor output at the subconscious level** p. 503

17. The diencephalon is composed of the epithalamus, the hypothalamus, and the thalamus. (*Figures 14–10, 14–11*)

18. The thalamus is the final relay point for ascending sensory information and coordinates the activities of the basal nuclei and cerebral cortex. (*Figures 14–10, 14–11; Table 14–2*)

19. The hypothalamus can (1) control somatic motor activities at the subconscious level, (2) control autonomic function, (3) coordinate activities of the nervous and endocrine systems, (4) secrete certain hormones, (5) produce emotions and behavioral **drives**, (6) coordinate voluntary and autonomic functions, (7) regulate body temperature, and (8) coordinate circadian cycles of activity. (*Figure 14–11*)

14-8 ❚ **The limbic system is a group of tracts and nuclei that function in emotion, motivation, and memory** p. 506

20. The **limbic system**, or *motivational system*, includes the **amygdaloid body, cingulate gyrus, dentate gyrus, parahippocampal gyrus, hippocampus,** and **fornix**. The functions of the limbic system involve emotional states, behavioral drives, and memory. (*Figure 14–12*)

14-9 ❚ **The cerebrum, the largest region of the brain, contains motor, sensory, and association areas** p. 508

21. The cortical surface contains **gyri** (elevated ridges) separated by **sulci** (shallow depressions) or **fissures** (deeper grooves). The **longitudinal fissure** separates the two **cerebral hemispheres**. The **central sulcus** separates the **frontal** and **parietal lobes**. Other sulci form the boundaries of the **temporal** and **occipital lobes**. (*Figure 14–13*)

22. The white matter of the cerebrum contains **association fibers, commissural fibers,** and **projection fibers**. (*Figure 14–14*)

23. The **basal nuclei** include the **caudate nucleus, globus pallidus,** and **putamen**; they control muscle tone and coordinate learned movement patterns and other somatic motor activities. (*Figure 14–15*)

24. The **primary motor cortex** of the **precentral gyrus** directs voluntary movements. The **primary sensory cortex** of the **postcentral gyrus** receives somatic sensory information from touch, pressure, pain, vibration, taste, and temperature receptors. (*Figure 14–16; Table 14–3*)

25. **Association areas**, such as the **somatic sensory association area, visual association area,** and **premotor cortex (somatic motor association area)**, control our ability to understand sensory information and coordinate a motor response. (*Figure 14–16*)

26. The **general interpretive area** receives information from all the sensory association areas. It is present in only one hemisphere—generally the left. (*Figures 14–16b, 14–17*)

27. The **speech center** regulates the patterns of breathing and vocalization needed for normal speech. (*Figures 14–16b, 14–17*)

28. The **prefrontal cortex** coordinates information from the secondary and special association areas of the entire cortex and performs abstract intellectual functions. (*Figure 14–16*)

29. The left hemisphere typically contains the general interpretive and speech centers and is responsible for language-based skills. The right hemisphere is typically responsible for spatial relationships and analyses. (*Figure 14–17*)

30. Brain activity is measured using an **electroencephalogram**. **Alpha waves** appear in healthy resting adults; **beta waves** occur when adults are concentrating; **theta waves** appear in children; and **delta waves** are normal during sleep. (*Figure 14–18*)

Focus: Cranial Nerves p. 518

31. We have 12 pairs of cranial nerves. Except for N I and N II, each nerve attaches to the ventrolateral surface of the brain stem near the associated sensory or motor nuclei. (*Figure 14–19*)

32. The **olfactory nerves** (N I) carry sensory information responsible for the sense of smell. The olfactory afferents synapse within the **olfactory bulbs**. (*Figures 14–19, 14–20*)

33. The **optic nerves** (N II) carry visual information from special sensory receptors in the eyes. (*Figures 14–19, 14–21*)

34. The **oculomotor nerves** (N III) are the primary source of innervation for four of the extrinsic eye muscles. (*Figure 14–22*)

35. The **trochlear nerves** (N IV), the smallest cranial nerves, innervate the superior oblique muscles of the eyes. (*Figure 14–22*)

36. The **trigeminal nerves** (N V), the largest cranial nerves, are mixed nerves with *ophthalmic, maxillary,* and *mandibular branches*. (*Figure 14–23*)

37. The **abducens nerves** (N VI) innervate the lateral rectus muscles of the eyes. (*Figure 14–22*)

38. The **facial nerves** (N VII) are mixed nerves that control muscles of the scalp and face. They provide pressure sensations over the face and receive taste information from the tongue. (*Figure 14–24*)

39. The **vestibulocochlear nerves** (N VIII) contain the **vestibular branch**, which monitors sensations of balance, position, and movement, and the **cochlear branch**, which monitors hearing receptors. (*Figure 14–25*)

40. The **glossopharyngeal nerves** (N IX) are mixed nerves that innervate the tongue and pharynx and control the action of swallowing. (*Figure 14–26*)

41. The **vagus nerves** (N X) are mixed nerves that are vital to the autonomic control of visceral function in the thorax and abdomen. (*Figure 14–27*)

42. The **accessory nerves** (N XI) have **internal branches**, which innervate voluntary swallowing muscles of the soft palate and pharynx, and **external branches**, which control muscles associated with the pectoral girdle. (*Figure 14–28*)

43. The **hypoglossal nerves** (N XII) provide voluntary motor control over tongue movements. (*Figure 14–28*)

44. The branches and functions of the cranial nerves are summarized in *Table 14–4*.

14-10 ❚ **Cranial reflexes involve sensory and motor fibers of cranial nerves** p. 529

45. Cranial reflexes are monosynaptic and polysynaptic reflex arcs that involve sensory and motor fibers of cranial nerves. (*Table 14–5*)

Review Questions

See the blue Answers tab at the back of the book.

LEVEL 1 Reviewing Facts and Terms

1. Identify the six major regions of the brain in the following diagram.

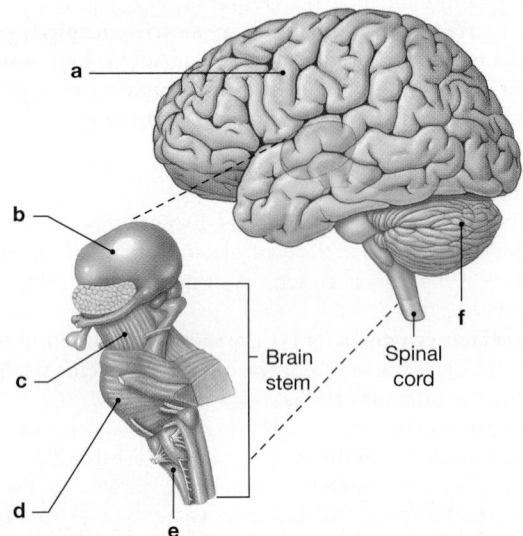

(a) _____ (b) _____
(c) _____ (d) _____
(e) _____ (f) _____

2. Identify the layers of the cranial meninges in the following diagram.

(a)_____ (b)_____
(c)_____ (d)_____
(e)_____ (f)_____
(g)_____

3. The term *higher brain centers* refers to those areas of the brain involved in higher-order functions. These centers would probably include nuclei, centers, and cortical areas of (a) the cerebrum, (b) the cerebellum, (c) the diencephalon, (d) all of these, (e) a and c only.
4. Which of the following is the site of cerebrospinal fluid production? (a) dural sinus, (b) choroid plexus, (c) falx cerebri, (d) tentorium cerebelli, (e) insula.
5. The area of the pons which controls blood pressure is also known as the _____ center.
6. The dural fold that divides the two cerebellar hemispheres is the (a) transverse sinus, (b) falx cerebri, (c) tentorium cerebelli, (d) falx cerebelli.

7. Cerebrospinal fluid is produced and secreted by (a) neurons, (b) ependymal cells, (c) Purkinje cells, (d) basal nuclei.
8. The two major groups of reflex centers in the medulla oblongata are the _____ _____ and the respiratory rhythmicity center.
9. The centers in the pons that modify the activity of the respiratory rhythmicity centers in the medulla oblongata are the (a) apneustic and pneumotaxic centers, (b) inferior and superior peduncles, (c) cardiac and vasomotor centers, (d) nucleus gracilis and nucleus cuneatus.
10. The final relay point for ascending sensory information that will be projected to the primary sensory cortex is the (a) hypothalamus, (b) thalamus, (c) spinal cord, (d) medulla oblongata.
11. Which major emergency autonomic response is controlled by the limbic system?
12. Coordination of learned movement patterns at the subconscious level is performed by (a) the cerebellum, (b) the substantia nigra, (c) association fibers, (d) the hypothalamus.
13. The two cerebral hemispheres are functionally different, even though anatomically they appear the same. (a) true, (b) false.
14. What is the corpus callosum?
15. Which three areas in the brain are not isolated from the general circulation by the blood–brain barrier?
16. Using the mnemonic device "Oh, Once One Takes The Anatomy Final, Very Good Vacations Are Heavenly," list the names of the 12 pairs of cranial nerves.

LEVEL 2 Reviewing Concepts

17. Which reflex will you be testing if you shine a light in a person's eye?
18. What is the role of the cerebellum?
19. The only cranial nerves that are attached to the cerebrum are the _____ nerves. (a) optic, (b) oculomotor, (c) trochlear, (d) olfactory, (e) abducens.
20. If symptoms characteristic of Parkinson's disease appear, which part of the midbrain is inhibited from secreting a neurotransmitter? Which neurotransmitter is it?
21. What is the common term used for the condition cerebrovascular accident?
22. Which type of brain waves occur in adults who are awake and resting with their eyes closed?
23. Which structure in the brain would your A&P instructor be referring to when talking about a nucleus that resembles a sea horse and that is important in the storage and retrieval of long-term memories? In which functional system of the brain is it located?
24. What are the principal functional differences between the right and left cerebral hemispheres?
25. Damage to the vestibular nucleus would lead to (a) loss of sight, (b) loss of hearing, (c) inability to sense pain, (d) difficulty in maintaining balance, (e) inability to swallow.
26. A cerebrovascular accident occurs when (a) the reticular activating system fails to function, (b) the prefrontal lobe is damaged, (c) the blood supply to a portion of the brain is cut off, (d) a descending tract in the spinal cord is severed, (e) brain stem nuclei hypersecrete dopamine.
27. What kinds of problems are associated with the presence of lesions in the Wernicke's area and the Broca's area?

LEVEL 3 Critical Thinking and Clinical Applications

28. Smelling salts can sometimes help restore consciousness after a person has fainted. The active ingredient of smelling salts is ammonia, and it acts by irritating the lining of the nasal cavity. Propose a mechanism

by which smelling salts would raise a person from the unconscious state to the conscious state.

29. A police officer has just stopped Bill on suspicion of driving while intoxicated. The officer asks Bill to walk the yellow line on the road and then to place the tip of his index finger on the tip of his nose. How would these activities indicate Bill's level of sobriety? Which part of the brain is being tested by these activities?

30. Colleen falls down a flight of stairs and bumps her head several times. Soon after, she develops a headache and blurred vision. Diagnostic tests at the hospital reveal an epidural hematoma in the temporoparietal area. The hematoma is pressing against the brain stem. What other signs and symptoms might she experience as a result of the injury?

31. Cerebral meningitis is a condition in which the meninges of the brain become inflamed as the result of viral or bacterial infection. This condition can be life threatening. Why?

32. Infants have little to no control of the movements of their head. One of the consequences of this is that they are susceptible to shaken baby syndrome, caused by vigorous shaking of an infant or young child by the arms, legs, chest, or shoulders. Forceful shaking can cause brain damage leading to mental retardation, speech and learning disabilities, paralysis, seizures, hearing loss, and even death. Damage to which areas of the brain would account for the clinical signs observed in this syndrome?

CLINICAL CASE ▸ Wrap-Up

The Neuroanatomist's Stroke

Dr. Taylor's full recovery took about eight years. While the left side of her brain was injured and shut down, the right side continued functioning. Because language and thoughts are controlled in the left hemisphere, Dr. Taylor "sat in an absolutely silent mind" for the first month after her stroke. Similarly, since the general interpretive center for mathematical calculation is in the left hemisphere, she had to learn to use numbers all over again. And because the primary motor cortex governing the right side of the body resides in the precentral gyrus of the left hemisphere, she had to learn to use her right arm again.

The stroke had injured some cells in her brain beyond repair. But some cells were dormant, capable of recovery and of making new neuronal connections. This ability of nerve cells to make these new connections is neuroplasticity, which permits healing as the brain reorganizes itself after an injury.

Dr. Taylor wants anatomy and physiology students to know two things. First, "if you study the brain, you will never be bored."

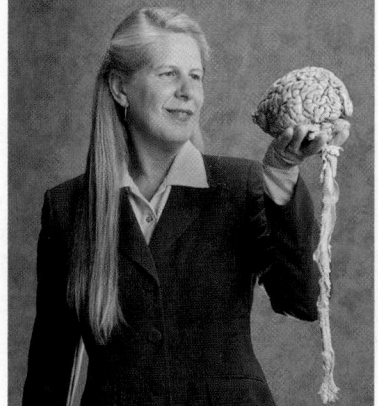

Second, "if you treat stroke patients like they will recover, they are more likely to recover."

She has written a best-selling memoir about her personal experience with stroke, *My Stroke of Insight: A Brain Scientist's Personal Journey*. She offers suggestions for the stroke survivor, caregiver, and everyone who deals with stroke.

1. How would you know, just based on signs and symptoms, which side of Dr. Taylor's brain has been injured by a stroke?

2. What is neuroplasticity and why was it important in Dr. Taylor's recovery?

See the blue Answers tab at the back of the book.

15

Sensory Pathways and the Somatic Nervous System

Learning Outcomes

These Learning Outcomes correspond by number to this chapter's sections and indicate what you should be able to do after completing the chapter.

15-1 ▎ Specify the components of the afferent and efferent divisions of the nervous system, and explain what is meant by the somatic nervous system.

15-2 ▎ Explain why receptors respond to specific stimuli, and how the organization of a receptor affects its sensitivity.

15-3 ▎ Identify the receptors for the general senses, and describe how they function.

15-4 ▎ Identify the major sensory pathways, and explain how it is possible to distinguish among sensations that originate in different areas of the body.

15-5 ▎ Describe the components, processes, and functions of the somatic motor pathways, and the levels of information processing involved in motor control.

SPOTLIGHT
• Somatic Sensory Pathways pp. 546–547

CLINICAL CASE
• Living with Cerebral Palsy p. 535

Clinical Notes
• Assessment of Tactile Sensitivities p. 543
• Amyotrophic Lateral Sclerosis p. 552
• Cerebral Palsy p. 553

An Introduction to Sensory Pathways and the Somatic Nervous System

In this chapter we examine how the nervous system works as an integrated unit. We consider sensory receptors, sensory processing centers in the brain, and conscious and subconscious motor functions. The basic events that occur along sensory and motor pathways are shown in **Figure 15–1**. In our discussion, we will focus on the "general senses" that provide information about the body and its environment. The "special senses"—smell, taste, sight, equilibrium (balance), and hearing—are considered in Chapter 17.

15-1 ▮ Sensory information from all parts of the body is routed to the somatosensory cortex

Learning Outcome Specify the components of the afferent and efferent divisions of the nervous system, and explain what is meant by the somatic nervous system.

Sensory receptors are specialized cells or cell processes that monitor specific conditions in the body or the external environment. When stimulated, a receptor passes information to the CNS in the form of action potentials along the axon of a sensory neuron. Such axons are parts of *sensory pathways*—the nerves, nuclei, and tracts that deliver somatic and visceral sensory information to their final destinations inside the CNS. Taken together, the receptors, sensory neurons, and sensory pathways make up the afferent division of the nervous system. ⊃ p. 413

Somatic and visceral sensory information often travels along the same pathway. Somatic sensory information is distributed to *sensory processing centers in the brain*—either the primary sensory cortex of the cerebral hemispheres or appropriate areas of the cerebellar hemispheres. Visceral sensory information is distributed primarily to reflex centers in the brain stem and diencephalon.

In this chapter we also consider the somatic motor portion of the efferent division—the nuclei, motor tracts, and motor neurons that control peripheral effectors. Whether somatic motor commands arise at the conscious or subconscious levels, they travel from motor centers in the brain along *somatic motor pathways*, which consist of motor nuclei, tracts, and nerves. The motor neurons and pathways that control skeletal muscles form the somatic nervous system (SNS).

In Chapter 16, we will discuss the visceral motor portion of the efferent division. All visceral motor commands are carried into the PNS by the autonomic nervous system (ANS). Both somatic and visceral motor commands may be issued in response to arriving sensory information. These commands may be modified on the basis of planning, memories, and learning—the so-called *higher-order functions* of the brain. We will also consider these higher-order functions in Chapter 16.

CHECKPOINT

1. **What do we call the body's specialized cells that monitor specific internal or external conditions?**

2. **Is it possible for somatic motor commands to occur at the subconscious level?**

See the blue Answers tab at the back of the book.

Figure 15–1   **An Overview of Events Occurring Along the Sensory and Motor Pathways.**

Immediate Involuntary Response

Processing centers in the spinal cord or brain stem may direct an immediate reflex response even before sensations reach the cerebral cortex.

Motor Pathway (involuntary)

Sensory Pathway

Arriving stimulus

Depolarization of Receptor

A stimulus produces a graded change in the membrane potential of a receptor cell.

Action Potential Generation

If the stimulus depolarizes the receptor cell to threshold, action potentials develop in the initial segment.

Propagation

Axons of sensory neurons carry information about the type of stimulus (touch, pressure, temperature) as action potentials to the CNS.

CNS Processing

Information processing occurs at every relay synapse. Sensory information may be distributed to multiple nuclei and centers in the spinal cord and brain.

Voluntary Response

The voluntary response, which is not immediate, can moderate, enhance, or supplement the relatively simple involuntary reflexive response.

Motor Pathway (voluntary)

Perception

Only about 1 percent of arriving sensations are relayed to the primary sensory cortex.

15

15-2 ▮ Sensory receptors connect our internal and external environments with the nervous system

Learning Outcome Explain why receptors respond to specific stimuli, and how the organization of a receptor affects its sensitivity.

As we noted, sensory receptors are specialized cells or cell processes that provide your central nervous system with information about conditions inside or outside the body. We use the term **general senses** to describe our sensitivity to temperature, pain, touch, pressure, vibration, and proprioception (body position). You have general sensory receptors throughout your body. They are fairly simple in structure. Some of the information they send to the CNS reaches the primary sensory cortex and your awareness.

As noted in Chapter 12, the CNS interprets sensory information on the basis of the frequency of arriving action potentials. ⟲ p. 449 For example, when pressure sensations are arriving, the harder the pressure, the higher the frequency of action potentials. The arriving information is called a **sensation**. The conscious awareness of a sensation is called a **perception**.

The **special senses** are **olfaction** (smell), **vision** (sight), **gustation** (taste), **equilibrium** (balance), and **hearing**. These

sensations are provided by receptors that are structurally more complex than those of the general senses. Special sensory receptors are located in **sense organs** such as the eye or ear, where surrounding tissues protect the receptors. The information these receptors provide is distributed to specific areas of the cerebral cortex (the auditory cortex, the visual cortex, and so forth) and to centers throughout the brain stem. We consider the special senses in Chapter 17.

Sensory receptors represent the interface between your nervous system and your internal and external environments. A sensory receptor detects an arriving stimulus and translates it into an action potential that can be conducted to the CNS. This translation, or conversion, process is called *transduction*. If transduction does not take place, then as far as you are concerned, the stimulus doesn't exist. For example, bees can see ultraviolet light that you can't see. Dogs can respond to sounds you can't hear. In each case the stimuli are there—but your receptors cannot detect them. Now let's examine the basic concepts of receptor function and sensory processing.

The Detection of Stimuli

How do receptors detect stimuli? Each receptor has a characteristic sensitivity. For example, a touch receptor is very sensitive to pressure but relatively insensitive to chemical stimuli. A taste receptor is sensitive to dissolved chemicals but insensitive to pressure. This feature is called *receptor specificity.*

Specificity may result from the structure of the receptor cell, or from the presence of accessory cells or structures that shield the receptor cell from other stimuli. The simplest receptors are the dendrites of sensory neurons. The branching tips of these dendrites, called **free nerve endings**, are not protected by accessory structures. Free nerve endings extend through a tissue the way grass roots extend into the soil. Many different stimuli can stimulate them. For this reason, free nerve endings show little receptor specificity. For example, chemicals, pressure, temperature changes, or trauma may stimulate free nerve endings that respond to tissue damage by providing pain sensations.

Complex receptors, such as the eye's visual receptors, are protected by accessory cells and connective tissue layers. These cells are seldom exposed to any stimulus other than light and so provide very specific information.

The area monitored by a single receptor cell is its *receptive field* (**Figure 15–2**). Whenever a sufficiently strong stimulus arrives in the receptive field, the CNS receives the information "stimulus arriving at receptor X." The larger the receptive field, the poorer your ability to localize a stimulus. A touch receptor on the general body surface, for example, may have a receptive field 7 cm (2.5 in.) in diameter. As a result, you can describe a light touch there as affecting only a general area, not an exact spot. On your tongue or fingertips, the receptive fields are less than a millimeter in diameter. In these areas, you can be very precise about the location of a stimulus.

An arriving stimulus can take many forms. It may be a physical force (such as pressure), a dissolved chemical, a sound, or light. Regardless of the nature of the stimulus, however,

sensory information is sent to the CNS only in the form of action potentials, which are electrical events.

As noted earlier, transduction is the translation of an arriving stimulus into an action potential by a sensory receptor. Transduction begins when a stimulus changes the membrane potential of the receptor cell. This change, called a *receptor potential*, is either a graded depolarization or a graded hyperpolarization. The stronger the stimulus, the larger the receptor potential.

The typical receptors for the general senses are the dendrites of sensory neurons. In these cases, the sensory neuron is the receptor cell. Any receptor potential that depolarizes the plasma membrane will bring the membrane closer to threshold. A depolarizing receptor potential in a neural receptor is called a *generator potential.*

Specialized receptor cells that communicate with sensory neurons across chemical synapses give us sensations of taste, hearing, equilibrium, and vision. In these cases, the receptor potential and the generator potential occur in different cells. The receptor potential develops in the receptor cell, and the generator potential appears later, in the sensory neuron. The receptor cells develop graded receptor potentials in response to stimulation, and the change in membrane potential alters the rate of neurotransmitter release at the synapse. The result is a depolarization or hyperpolarization of the sensory neuron. If sufficient depolarization occurs, an action potential appears in the sensory neuron.

Whenever a sufficiently large generator potential appears, action potentials develop in the axon of a sensory neuron. For reasons discussed in Chapter 12, the greater the degree of sustained depolarization at the axon hillock, the higher the frequency of action potentials in the afferent fiber. ⊃ p. 436 The CNS then processes and interprets this arriving information at the conscious and subconscious levels.

The Interpretation of Sensory Information

Sensory information that arrives at the CNS is routed according to the location and nature of the stimulus. In previous chapters we emphasized the fact that axons in the CNS are organized in bundles with specific origins and destinations. Along sensory pathways, a series of neurons relays information from one point (the receptor) to another (a neuron at a specific site in the cerebral cortex). For example, sensations of touch, pressure, pain, and temperature arrive at the primary sensory cortex. Visual, auditory, gustatory (taste), and olfactory (smell) sensations reach their respective visual, auditory, gustatory, and olfactory regions of the cortex.

The link between peripheral receptor and cortical neuron is called a **labeled line**. Each labeled line consists of axons carrying information about one *modality*, or type of stimulus (touch, pressure, light, and sound). The CNS interprets the modality

Figure 15–2   Receptors and Receptive Fields. Each receptor cell monitors a specific area known as the receptive field.

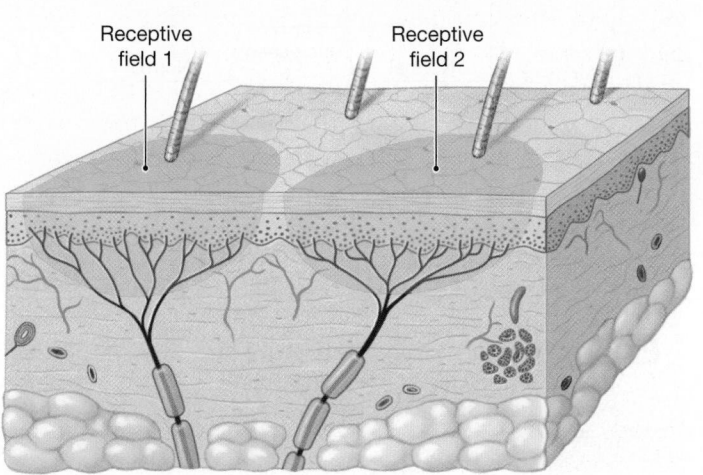

Receptive field 1

Receptive field 2

entirely on the basis of the labeled line over which it arrives. For this reason, you cannot tell the difference between a true sensation and a false one generated somewhere along the line. For example, when you rub your eyes, you commonly see flashes of light. The stimulus is mechanical (pressure) rather than visual, but *any* activity along the optic nerve is projected to the visual cortex. You experience it as a visual perception.

The identity of the active labeled line indicates the type of stimulus. Where it arrives within the sensory cortex determines its perceived location. For example, if activity in a labeled line that carries touch sensations stimulates the facial region of your primary sensory cortex, you perceive a touch on the face.

All other characteristics of the stimulus—its strength, duration, and variation—are conveyed by the frequency and pattern of action potentials. The translation of complex sensory information into meaningful patterns of action potentials is called *sensory coding*.

Receptors can be categorized by the nature of their response to stimulation (**Figure 15–3**). Some sensory neurons, called **tonic receptors**, are always active (**Figure 15–3a**). Other receptors, called **phasic receptors**, are usually not active. Phasic receptors provide information about the intensity and rate of change of a stimulus (**Figure 15–3b**). Receptors that combine phasic and tonic coding can convey extremely complicated sensory information.

Adaptation

Adaptation is a reduction in sensitivity in the presence of a constant stimulus. You seldom notice the rumble of the tires when you ride in a car, or the background noise of the air conditioner. The reason is that your nervous system quickly adapts to stimuli that are painless and constant. *Peripheral adaptation*

occurs when the level of receptor activity changes. The receptor responds strongly at first, but then its activity gradually declines, in part because the size of the generator potential gradually decreases. This response is characteristic of phasic receptors, which are also called **fast-adapting receptors**. For example, temperature receptors (*thermoreceptors*) are phasic receptors. You seldom notice room temperature unless it changes suddenly.

Tonic receptors show little peripheral adaptation and so are called **slow-adapting receptors**. Pain receptors (*nociceptors*) are slow-adapting receptors. This is one reason why pain sensations remind you of an injury long after the initial damage has taken place.

Adaptation also occurs along sensory pathways inside the CNS. For example, a few seconds after you have been exposed to a new smell, awareness of the stimulus virtually disappears, although the sensory neurons are still quite active. This process is known as *central adaptation*. Central adaptation generally involves the inhibition of nuclei along a sensory pathway.

Peripheral adaptation reduces the amount of information that reaches the CNS. Central adaptation at the subconscious level further restricts the amount of detail that arrives at the cerebral cortex. Most of the incoming sensory information is processed in centers along the spinal cord or brain stem at the subconscious level. This processing can produce reflexive motor responses, but you are seldom consciously aware of either the stimuli or the responses.

The output from higher centers can increase receptor sensitivity or facilitate transmission along a sensory pathway. The reticular activating system in the midbrain helps focus our attention and heightens or reduces our awareness of arriving sensations. ↺ p. 502 This adjustment of sensitivity can take place under conscious or subconscious direction. When you "listen

Figure 15–3 Tonic and Phasic Sensory Receptors.

a **Tonic receptors** are always active and generate action potentials at a frequency that reflects the background level of stimulation. When the stimulus increases or decreases, the rate of action potential generation changes accordingly.

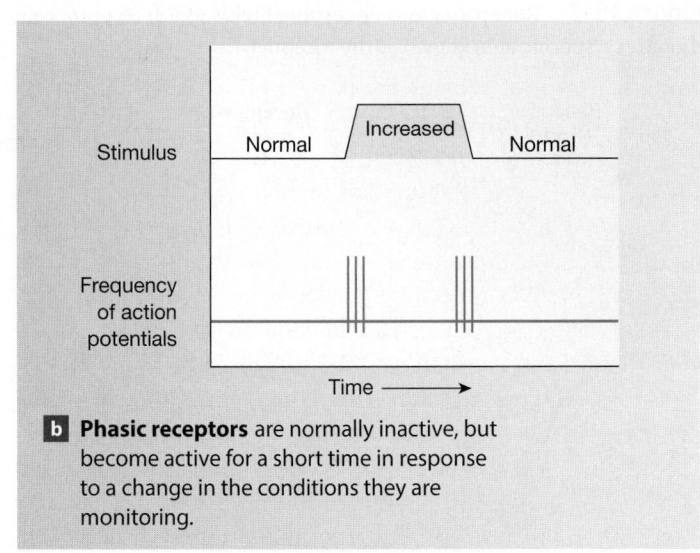

b **Phasic receptors** are normally inactive, but become active for a short time in response to a change in the conditions they are monitoring.

carefully," your sensitivity and awareness of auditory stimuli increase. Output from higher centers can also inhibit transmission along a sensory pathway. Such inhibition occurs when you enter a noisy factory or walk along a crowded city street. You automatically tune out the high level of background noise. Next let's consider how these basic concepts of receptor function and sensory processing apply to the general senses.

> **CHECKPOINT**
>
> 3. **Receptor A has a circular receptive field on the skin surface with a diameter of 2.5 cm. Receptor B has a circular receptive field 7.0 cm in diameter. Which receptor provides more precise sensory information?**
>
> 4. **Define adaptation.**
>
> See the blue Answers tab at the back of the book.

15-3 | General sensory receptors are classified by the type of stimulus that excites them

Learning Outcome Identify the receptors for the general senses, and describe how they function.

Receptors for the general senses are scattered throughout the body. They are simple in structure. Using the simple classification scheme introduced in Chapter 12, we divide them into three types. ⟳ p. 417 *Exteroceptors* provide information about the external environment. *Proprioceptors* report the positions and movements of skeletal muscles and joints. *Interoceptors* monitor visceral organs and functions.

Using a more detailed classification system, we can divide the general sensory receptors into four types by the nature of the stimulus that excites them: *nociceptors* (pain), *thermoreceptors* (temperature), *mechanoreceptors* (physical distortion), and *chemoreceptors* (chemical concentration). Each class of receptors has distinct structural and functional characteristics. The difference between a somatic receptor and a visceral receptor is its location, not its structure. A pain receptor in the gut looks and acts like a pain receptor in the skin, but the two sensations are delivered to separate locations in the CNS.

Proprioception, however, is a purely somatic sensation. There are no proprioceptors in the visceral organs of the thoracic and abdominopelvic cavities. Your mental map of your body doesn't include these organs. You cannot tell, for example, where your spleen, appendix, or pancreas is at the moment. The visceral organs also have fewer pain, temperature, and touch receptors than we find elsewhere in the body. The sensory information you receive is poorly localized because the receptive fields are very large and may be widely separated.

General sensations are widely distributed in the CNS, but most of the processing takes place in centers along the sensory pathways in the spinal cord or brain stem. Only about 1 percent of the information provided by afferent fibers reaches the cerebral cortex and our awareness. For example, you usually do not feel the clothes you wear or hear the hum of the engine when riding in a car.

Nociceptors

Pain receptors, or **nociceptors** (*noxa*, harm), are especially common in the superficial portions of the skin, in joint capsules, within the periostea of bones, and around the walls of blood vessels. Other deep tissues and most visceral organs have few nociceptors. Pain receptors are free nerve endings with large receptive fields (**Figure 15–2**). As a result, it is often difficult to determine the exact source of a painful sensation.

Different nociceptors may be sensitive to (1) extremes of temperature, (2) mechanical damage, and (3) dissolved chemicals, such as chemicals released by injured cells. Very strong stimuli, however, will excite all three receptor types. For that reason, people describing very painful sensations—whether caused by acids, heat, or a deep cut—use similar descriptive terms, such as "burning."

Stimulation of the dendrites of a nociceptor causes depolarization. When the initial segment of the axon reaches threshold, an action potential heads toward the CNS.

Two types of axons—Type A and Type C fibers—carry painful sensations. ⟳ p. 439 Myelinated Type A fibers carry sensations of **fast pain**, or *prickling pain*. An injection or a deep cut produces this type of pain. These sensations reach the CNS quickly, where they often trigger somatic reflexes. They are also relayed to the primary sensory cortex and so receive conscious attention. In most cases, the arriving information allows you to localize the stimulus to an area several inches in diameter.

Slower, Type C fibers carry sensations of **slow pain**, or *burning and aching pain*. These sensations cause a generalized activation of the reticular formation and thalamus. You become aware of the pain but only have a general idea of the area affected.

Pain receptors are tonic receptors (**Figure 15–3a**). Significant peripheral adaptation does not occur, and the receptors continue to respond as long as the painful stimulus remains. Painful sensations stop only after tissue damage has ended. However, central adaptation may reduce the *perception* of the pain while pain receptors remain stimulated. This effect involves the inhibition of centers in the thalamus, reticular formation, lower brain stem, and spinal cord.

An understanding of the origins of pain sensations and an ability to control or reduce pain levels have always been among the most important aspects of medical treatment. After all, it is usually pain that causes us to seek treatment. We usually tolerate or ignore conditions that are not painful. Although we often use the term *pain pathways*, it is clear that pain distribution and perception are extremely complex.

The sensory neurons that bring pain sensations into the CNS release *glutamate* and/or *substance P* as neurotransmitters.

15

These neurotransmitters facilitate neurons along the pain pathways. As a result, the level of pain experienced (especially chronic pain) can be out of proportion to the amount of painful stimuli or the apparent tissue damage. This effect may be one reason why people differ so widely in their perception of the pain associated with childbirth, headaches, or back pain. This facilitation is also presumed to play a role in phantom limb pain (discussed shortly): the sensory neurons may be inactive, but the hyperexcitable interneurons may continue to generate pain sensations.

The release of endorphins and enkephalins within the CNS can reduce the level of pain that a person feels. As noted in Chapter 12, endorphins and enkephalins are neuromodulators that inhibit activity along pain pathways in the brain. ⊃ p. 443 These compounds are structurally similar to morphine. They are found in the limbic system, hypothalamus, and reticular formation. The pain centers in these areas also use substance P as a neurotransmitter. Endorphins bind to the presynaptic membrane and prevent the release of substance P, reducing the conscious perception of pain, although the painful stimulus remains.

TIPS & TRICKS

The **P** in substance **P** stands for **p**eptide and is involved with **p**ain, which it transmits **p**eripherally.

Thermoreceptors

Temperature receptors, or **thermoreceptors**, are free nerve endings located in the dermis, in skeletal muscles, in the liver, and in the hypothalamus. Cold receptors are three or four times more numerous than warm receptors. There are no structural differences between warm and cold thermoreceptors. Temperature sensations are conducted along the same pathways that carry pain sensations. They are sent to the reticular formation, the thalamus, and (to a lesser extent) the primary sensory cortex.

Thermoreceptors are phasic receptors (**Figure 15–3b**). They are very active when the temperature is changing, but they quickly adapt to a stable temperature. When you enter an air-conditioned classroom on a hot summer day or a warm lecture hall on a brisk fall evening, the temperature change seems extreme at first, but you quickly become comfortable as adaptation occurs.

Mechanoreceptors

Mechanoreceptors are sensitive to stimuli that distort their plasma membranes. These membranes contain *mechanically gated ion channels*. The gates open or close in response to stretching, compression, twisting, or other distortions of the membrane. There are three classes of mechanoreceptors:

1. **Tactile receptors** provide the closely related sensations of touch, pressure, and vibration. Touch sensations provide information about shape or texture, whereas pressure sensations indicate the degree of mechanical distortion. Vibration sensations indicate a pulsing pressure. The receptors involved may be specialized in some way. For example, rapidly adapting tactile receptors are best suited for detecting vibration. But your interpretation of a sensation as touch rather than pressure is typically a matter of the degree of stimulation, and not of differences in the type of receptor stimulated.

2. **Baroreceptors** (bar-ō-rē-SEP-torz; *baro-*, pressure) detect pressure changes in the walls of blood vessels and in portions of the digestive, respiratory, and urinary tracts.

3. **Proprioceptors** monitor the positions of joints and skeletal muscles. They are the most structurally and functionally complex of the general sensory receptors.

Tactile Receptors

Fine touch and pressure receptors provide detailed information about a source of stimulation, including its exact location, shape, size, texture, and movement. These receptors are extremely sensitive and have narrow receptive fields. **Crude touch and pressure receptors** provide poor localization. They give little additional information about the stimulus because they have large receptive fields.

Tactile receptors range in complexity from free nerve endings to specialized sensory complexes with accessory cells and supporting structures. **Figure 15–4** shows six types of tactile receptors in the skin:

1. *Free nerve endings* sensitive to touch and pressure are located between epidermal cells (**Figure 15–4a**). There appear to be no structural differences between these receptors and the free nerve endings that provide temperature or pain sensations. These are the only sensory receptors on the corneal surface of the eye, but in other portions of the body surface, more specialized tactile receptors are probably more important. Free nerve endings that provide touch sensations are tonic receptors with small receptive fields.

2. Wherever hairs are located, the nerve endings of the **root hair plexus** monitor distortions and movements across the body surface (**Figure 15–4b**). When a hair is displaced, the movement of its follicle distorts the sensory dendrites and produces action potentials. These receptors adapt rapidly, so they are best at detecting initial contact and subsequent movements. For example, you generally feel your clothing only when you move or when you consciously focus on tactile sensations from the skin.

3. **Tactile discs**, or *Merkel* (MER-kel) *discs*, are fine touch and pressure receptors associated with sensitivity to shape and texture (**Figure 15–4c**). They are extremely sensitive tonic receptors, with very small receptive fields. The flattened dendritic processes of a single myelinated afferent fiber

Figure 15–4 **Tactile Receptors in the Skin.**

a **Free nerve endings**

b **Root hair plexus**

Hair

Tactile disc

Tactile corpuscle

Free nerve ending

Root hair plexus

Lamellated corpuscle

Sensory nerve

Ruffini corpuscle

c **Tactile discs**

Merkel cell

Nerve terminal (dendrite)

Tactile disc

Afferent nerve fiber

d **Tactile corpuscle**

Tactile corpuscle

Epidermis

Capsule

Dendrites

Dermis

Sensory nerve fiber

Tactile corpuscle   LM × 330

e **Lamellated corpuscle**

Dermis

Dendritic process

Accessory cells (specialized fibroblasts)

Concentric layers (lamellae) of collagen fibers separated by fluid

Lamellated corpuscle (cross section)   LM × 125

f **Ruffini corpuscle**

Collagen fibers

Sensory nerve fiber

Capsule

Dendrites

15

make close contact with unusually large epithelial cells in the stratum basale of the skin. We described these *Merkel cells* in Chapter 5. ⟳ p. 179

4. **Tactile corpuscles**, or *Meissner's* (MĪS-nerz) *corpuscles*, give us sensations of fine touch and pressure and low-frequency vibration. They adapt to stimulation within a second after contact. Tactile corpuscles are fairly large structures, measuring about 100 µm in length and 50 µm in width. These receptors are most abundant in the eyelids, lips, fingertips, nipples, and external genitalia. The dendrites are highly coiled and interwoven, and modified Schwann cells surround them. A fibrous capsule surrounds the entire complex and anchors it within the dermis (**Figure 15–4d**).

TIPS & TRICKS

To remember that Meissner's corpuscles perceive pressure sensations, associate the *m* and *ss* in "***M**ei**ss**ner*" with ***mass**age.*

5. **Lamellated** (LAM-e-lāt-ed; lamella, thin plate) **corpuscles**, or *pacinian* (pa-SIN-ē-an) *corpuscles*, are sensitive to deep pressure. Because they are fast-adapting receptors, they are most sensitive to pulsing or high-frequency vibrating stimuli. A single dendrite lies within a series of concentric layers of collagen fibers and supporting cells (specialized fibroblasts) (**Figure 15–4e**). The entire corpuscle may reach 4 mm in length and 1 mm in diameter. The concentric layers, separated by interstitial fluid, shield the dendrite from virtually every source of stimulation other than direct pressure. Lamellated corpuscles adapt quickly because distortion of the capsule soon relieves pressure on the sensory process. Somatic sensory information is provided by lamellated corpuscles located throughout the dermis, notably in the fingers, mammary glands, and external genitalia; in the superficial and deep fasciae; and in joint capsules. Visceral sensory information is provided by lamellated corpuscles in mesenteries, in the pancreas, and in the walls of the urethra and urinary bladder.

6. **Ruffini** (rū-FĒ-nē) **corpuscles** are also sensitive to pressure and distortion of the skin, but they are located in the reticular (deep) dermis. These receptors are tonic and show little if any adaptation. The capsule surrounds a core of collagen fibers that are continuous with those of the surrounding dermis (**Figure 15–4f**). In the capsule, a network of dendrites is intertwined with the collagen fibers. Any tension or distortion of the dermis tugs or twists the capsular fibers, stretching or compressing the attached dendrites and altering the activity in the myelinated afferent fiber.

Your sensitivity to tactile sensations may be altered by infection, disease, or damage to sensory neurons or pathways. For this reason, mapping tactile responses can sometimes aid clinical assessment. Sensory losses with clear regional boundaries

indicate trauma to spinal nerves. For example, sensory loss within the boundaries of a dermatome can help identify the affected spinal nerve. (**Figure 13–7**, p. 465)

Tickle and itch sensations are closely related to the sensations of touch and pain. The receptors involved are free nerve endings. Unmyelinated Type C fibers carry the information. People usually (but not always) describe tickle sensations as pleasurable. A light touch that moves across the skin typically produces a tickle sensation. Psychological factors influence the interpretation of tickle sensations, and tickle sensitivity differs greatly from person to person.

Itching is probably produced by the stimulation of the same receptors. Specific "itch spots" can be mapped in the skin, the inner surfaces of the eyelids, and the mucous membrane of the nose. Itch sensations are absent from other mucous membranes and from deep tissues and viscera. Itching is extremely unpleasant, even more unpleasant than pain. Individuals with extreme itching will scratch even when pain is the result. Itch receptors can be stimulated by the injection of histamine or proteolytic enzymes into the epidermis and superficial dermis. The precise receptor mechanism is unknown.

Baroreceptors

Baroreceptors monitor changes in pressure in an organ. A baroreceptor consists of free nerve endings that branch within the elastic tissues in the wall of a distensible organ, such as a blood vessel or a portion of the respiratory, digestive, or urinary tract. When the pressure changes, the elastic walls of the tract recoil or expand. This movement distorts the dendritic branches and alters the rate of action potential generation. Baroreceptors respond immediately to a change in pressure, but they adapt rapidly, and the output along the afferent fibers gradually returns to normal.

Baroreceptors monitor blood pressure in the walls of major vessels. These vessels include the carotid arteries (at the *carotid sinuses*) and the aorta (at the *aortic sinuses*). The information plays a major role in regulating cardiac function and adjusting blood flow to vital tissues.

Baroreceptors in the lungs monitor the degree of lung expansion. This information is relayed to the respiratory rhythmicity centers, which set the pace of respiration. Comparable stretch receptors at various sites in the digestive and urinary tracts trigger a variety of visceral reflexes, including urination and defecation reflexes. We describe those baroreceptor reflexes in the digestive and urinary system chapters.

Proprioceptors

Proprioceptors monitor the position of joints, the tension in tendons and ligaments, and the state of muscular contraction. There are three major groups of proprioceptors:

1. *Muscle Spindles.* Muscle spindles monitor skeletal muscle length and trigger stretch reflexes. ⟳ p. 479

2. *Golgi Tendon Organs.* **Golgi tendon organs** are similar in function to Ruffini corpuscles but are located at the junction between a skeletal muscle and its tendon. In a Golgi tendon organ, dendrites branch repeatedly and wind around the densely packed collagen fibers of the tendon. These receptors are stimulated by tension in the tendon. They monitor the external tension developed during muscle contraction.

3. *Receptors in Joint Capsules.* Joint capsules are richly innervated by free nerve endings that detect pressure, tension, and movement at the joint. Your sense of body position results from the integration of information from these receptors with information provided by muscle spindles, Golgi tendon organs, and the receptors of the internal ear.

Proprioceptors do not adapt to constant stimulation, and each receptor continuously sends information to the CNS. Most proprioceptive information is processed at subconscious levels. Only a small proportion of the arriving proprioceptive information reaches your awareness.

Chemoreceptors

Specialized nerve cells, called **chemoreceptors**, can detect small changes in the concentration of specific chemicals or compounds. In general, chemoreceptors respond only to water-soluble and lipid-soluble substances that are dissolved in body fluids (interstitial fluid, blood plasma, and CSF). These receptors exhibit peripheral adaptation over a period of seconds. Central adaptation may also take place.

The chemoreceptors included in the general senses do not send information to the primary sensory cortex, so you are not consciously aware of the sensations they provide. The arriving sensory information is routed to brain stem centers that deal with the autonomic control of respiratory and cardiovascular functions. Neurons in the respiratory centers of the brain respond to the concentration of hydrogen ions (pH) and levels of carbon dioxide molecules in the cerebrospinal fluid.

Chemoreceptors are also located in the **carotid bodies**, near the origin of the internal carotid arteries on each side of the neck, and in the **aortic bodies**, between the major branches of the aortic arch. These receptors monitor the pH and the carbon dioxide and oxygen levels in arterial blood. The afferent fibers leaving the carotid or aortic bodies reach the respiratory centers by traveling within cranial nerves IX (glossopharyngeal) and X (vagus).

CHECKPOINT

5. List the four types of general sensory receptors, and identify the nature of the stimulus that excites each type.

6. Identify the three classes of mechanoreceptors.

7. What would happen to you if the information from proprioceptors in your legs were blocked from reaching the CNS?

See the blue Answers tab at the back of the book.

Clinical Note

Assessment of Tactile Sensitivities Regional sensitivity to light touch can be checked by gentle contact with a fingertip or a slender wisp of cotton. The **two-point discrimination test** provides a more detailed sensory map of tactile receptors. Two fine points of a bent paper clip or another object are applied to the skin surface simultaneously. The person then describes the contact. When the points fall within a single receptive field, the person will report only one point of contact. A normal person loses two-point discrimination at 1 mm (0.04 in.) on the surface of the tongue, at 2–3 mm (0.08–0.12 in.) on the lips, at 3–5 mm (0.12–0.20 in.) on the backs of the hands and feet, and at 4–7 cm (1.6–2.75 in.) over the general body surface.

Applying the base of a tuning fork to the skin tests vibration receptors. Damage to an individual spinal nerve produces insensitivity to vibration along the paths of the related sensory nerves. If the sensory loss results from spinal cord damage, the injury site can typically be located by walking the tuning fork down the spinal column, resting its base on the vertebral spinous processes.

Various descriptive terms indicate the degree of sensitivity in an area. *Anesthesia* implies a total loss of sensation. The person cannot perceive touch, pressure, pain, or temperature sensations in that area. *Hypesthesia* (hypoesthesia) is a reduction in sensitivity. *Paresthesia* is the presence of abnormal sensations such as the pricking sensation when an arm or leg "falls asleep" as a result of pressure on a peripheral nerve.

15-4 ❙ Separate pathways carry somatic sensory and visceral sensory information

Learning Outcome Identify the major sensory pathways, and explain how it is possible to distinguish among sensations that originate in different areas of the body.

A sensory neuron that delivers sensations to the CNS is called a **first-order neuron**. The cell body of a first-order general sensory neuron is located in a dorsal root ganglion or cranial nerve ganglion. In the CNS, the axon of that sensory neuron synapses on an interneuron known as a **second-order neuron**, which may be located in the spinal cord or brain stem. If the sensation is to reach our awareness, the second-order neuron synapses on a **third-order neuron** in the thalamus.

Somewhere along its length, the axon of the second-order neuron crosses over to the opposite side of the CNS. As a result, the right side of the thalamus receives sensory information from the left side of the body, and the left side of the thalamus receives sensory information from the right side of the body. The reason for this crossover is unknown. Although it has no apparent functional benefit, crossover occurs along sensory and motor pathways in all vertebrates.

The axon of the third-order neuron then ascends (without crossing back over) and synapses on neurons of the primary sensory cortex of the cerebral hemisphere. As a result, the right cerebral hemisphere receives sensory information from the left side of the body, and the left cerebral hemisphere receives sensations from the right side.

Somatic Sensory Pathways

Somatic sensory pathways carry sensory information from the skin and muscles of the body wall, head, neck, and limbs. We will consider three major somatic sensory pathways: (1) the *spinothalamic pathway*, (2) the *posterior column pathway*, and (3) the *spinocerebellar pathway*. These pathways are made up of pairs of spinal tracts, symmetrically arranged on opposite sides of the spinal cord. All the axons within a tract share a common origin and destination.

Figure 15–5 shows the relative positions of the spinal tracts in each somatic sensory pathway. Note that tract names often give clues to their function. For example, if the name of a tract begins with *spino-*, the tract *starts* in the spinal cord and *ends* in the brain. It must therefore be an ascending tract that carries sensory information. The rest of the name indicates the tract's destination. Thus, a *spinothalamic tract* begins in the spinal cord and carries sensory information to the thalamus.

If, on the other hand, the name of a tract ends in *-spinal*, the tract *ends* in the spinal cord and *starts* in a higher center of the brain. It must therefore be a descending tract that carries motor commands. The first part of the name indicates the nucleus or cortical area of the brain where the tract originates. For example, a *corticospinal tract* carries motor commands from the cerebral cortex to the spinal cord. We consider such tracts later in the chapter. Table 15–1 summarizes the somatic sensory pathways discussed in this section.

The Spinothalamic Pathway

The *spinothalamic pathway* carries sensations of poorly localized touch, pressure, pain, and temperature. This pathway includes

Figure 15–5 Sensory Pathways and Ascending Tracts in the Spinal Cord. A cross-sectional view of the spinal cord showing the locations of the major ascending (sensory) tracts. For information about these tracts, see Table 15–1.

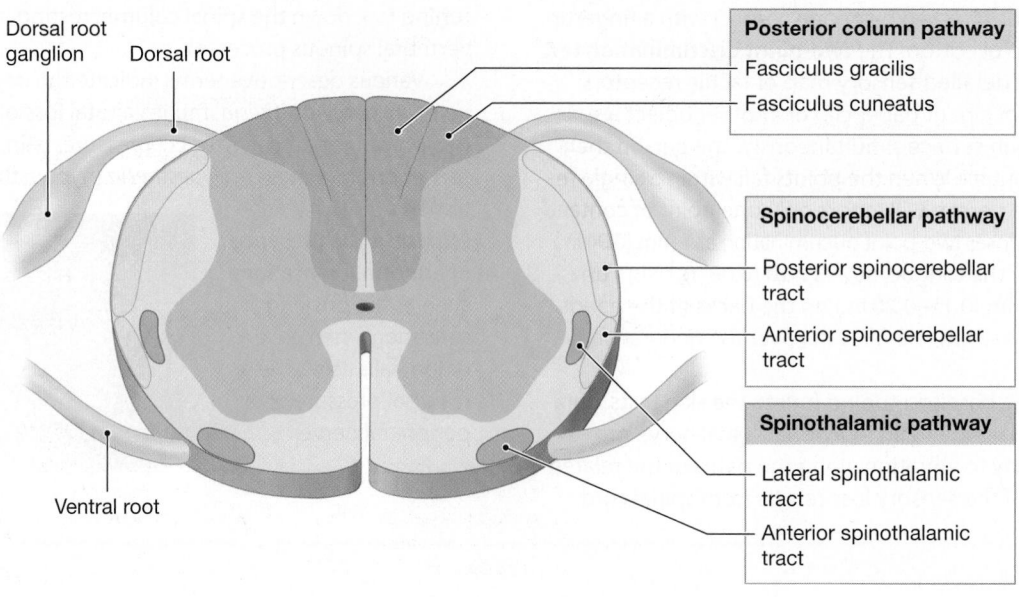

Posterior column pathway
- Fasciculus gracilis
- Fasciculus cuneatus

Spinocerebellar pathway
- Posterior spinocerebellar tract
- Anterior spinocerebellar tract

Spinothalamic pathway
- Lateral spinothalamic tract
- Anterior spinothalamic tract

Dorsal root ganglion Dorsal root

Ventral root

small tracts that deliver sensations to reflex centers in the brain stem as well as larger tracts that carry sensations destined for the cerebral cortex. We will ignore the smaller tracts in this discussion.

Sensations bound for the cerebral cortex ascend within the anterior spinothalamic tract or lateral spinothalamic tract. These tracts end at third-order neurons in the ventral nucleus group of the thalamus. After the sensations have been sorted and processed, they are relayed to the primary sensory cortex (**Spotlight Figure 15–6**).

The perception that an arriving stimulus is painful rather than cold, hot, or vibrating depends on which second-order and third-order neurons are stimulated. The ability to localize that stimulus to a specific location in the body depends on the stimulation of an appropriate area of the primary sensory cortex. Any abnormality along the pathway can result

in inappropriate sensations or inaccurate localization of the source. Consider these examples:

- A person can experience painful sensations that are not real. For example, a person may continue to experience pain in an amputated limb. This **phantom limb pain** is caused by activity in the sensory neurons or interneurons along the spinothalamic pathway. The neurons involved were once part of the labeled line that monitored conditions in the intact limb. These labeled lines and pathways are developmentally programmed. For this reason, even people born without limbs can have phantom limb pain.

- A person can feel pain in an uninjured part of the body when the pain actually originates at another location. For example, strong visceral pain sensations arriving at a segment of the spinal cord can stimulate interneurons that

Table 15–1	Principal Ascending (Sensory) Pathways					
		Location of Neuron Cell Bodies			**Location of Neuron Cell Bodies**	
Pathway/Tract	**Sensation(s)**	**First-Order**	**Second-Order**	**Third-Order**	**Final Destination**	**Site of Crossover**
SPINOTHALAMIC PATHWAY						
Lateral spinothalamic tracts	Pain and temperature	Dorsal root ganglia; axons enter CNS in dorsal roots	Interneurons in posterior gray horn; axons enter lateral spinothalamic tract on opposite side	Ventral nuclei of thalamus	Primary sensory cortex on side opposite stimulus	Axons of second-order neurons at level of entry
Anterior spinothalamic tracts	Crude touch and pressure	Dorsal root ganglia; axons enter CNS in dorsal roots	Interneurons in posterior gray horn; axons enter anterior spinothalamic tract on opposite side	Ventral nuclei of thalamus	Primary sensory cortex on side opposite stimulus	Axons of second-order neurons at level of entry
POSTERIOR COLUMN PATHWAY						
Fasciculus gracilis	Proprioception and fine touch, ventral pressure, and vibration from inferior half of body	Dorsal root ganglia of inferior half of body; axons enter CNS in dorsal roots and join fasciculus gracilis	Nucleus gracilis of medulla oblongata; axons cross over before entering medial lemniscus	Ventral nuclei of thalamus	Primary sensory cortex on side opposite stimulus	Axons of second-order neurons before entering the medial lemniscus
Fasciculus cuneatus	Proprioception and fine touch, ventral pressure, and vibration from superior half of body	Dorsal root ganglia of superior half of body; axons enter CNS in dorsal roots and join fasciculus cuneatus	Nucleus cuneatus of medulla oblongata; axons cross over before entering medial lemniscus	Ventral nuclei of thalamus	Primary sensory cortex on side opposite stimulus	Axons of second-order neurons before entering the medial lemniscus
SPINOCEREBELLAR PATHWAY						
Posterior spinocerebellar tracts	Proprioception	Dorsal root ganglia; axons enter CNS in dorsal roots	Interneurons in posterior gray horn; axons enter posterior spinothalamic tract on same side	Not present	Cerebellar cortex on side of stimulus	None
Anterior spinocerebellar tracts	Proprioception	Dorsal root ganglia; axons enter CNS in dorsal roots	Interneurons in same spinal section; axons enter anterior spinocerebellar tract on the same or opposite side	Not present	Cerebellar cortex on side opposite (and side of) stimulus	Axons of most second-order neurons cross over before entering tract; many re-cross at cerebellum

SPINOTHALAMIC PATHWAY

The **spinothalamic pathway** provides conscious sensations of poorly localized ("crude") touch, pressure, pain, and temperature. In this pathway, the axons of first-order neurons enter the spinal cord and synapse on second-order neurons within the posterior gray horns. The axons of these interneurons cross to the opposite side of the spinal cord before ascending to the thalamus. The third-order neuron synapses in the primary sensory cortex.

KEY
- Axon of first-order neuron
- Second-order neuron
- Third-order neuron

A Sensory Homunculus

A **sensory homunculus** ("little human") is a functional map of the primary sensory cortex. The proportions are very different from those of any individual because the area of sensory cortex devoted to a particular body region is proportional to the number of sensory neurons it contains.

Primary Sensory Cortex

The anterior spino-thalamic tracts of the spinothalamic pathway carry crude touch and pressure sensations.

Midbrain

Medulla oblongata

Anterior spinothalamic tract

Spinal cord

Crude touch and pressure sensations from right side of body

Primary Sensory Cortex

The lateral spinothalamic tracts of the spinothalamic pathway carry pain and temperature sensations.

Midbrain

Medulla oblongata

Lateral spinothalamic tract

Spinal cord

Pain and temperature sensations from right side of body

POSTERIOR COLUMN PATHWAY

The **posterior column pathway** carries sensations of highly localized ("fine") touch, pressure, vibration, and proprioception. This pathway is also known as the dorsal column-medial lemniscus pathway. It begins at a peripheral receptor and ends at the primary sensory cortex of the cerebral hemispheres.

Primary Sensory Cortex

Ventral nuclei in thalamus

Midbrain

Nucleus gracilis and nucleus cuneatus

Medial lemniscus

Medulla oblongata

Fasciculus gracilis and fasciculus cuneatus

Dorsal root ganglion

Spinal cord

Fine-touch, vibration, pressure, and proprioception sensations from right side of body

SPINOCEREBELLAR PATHWAY

The cerebellum receives proprioceptive information about the position of skeletal muscles, tendons, and joints along the **spinocerebellar pathway**. The **posterior spinocerebellar tracts** contain axons that do not cross over to the opposite side of the spinal cord. These axons reach the cerebellar cortex by the inferior cerebellar peduncle of that side. The **anterior spinocerebellar tracts** are dominated by axons that have crossed over to the opposite side of the spinal cord.

PONS

Cerebellum

Medulla oblongata

Spinocerebellar pathway

Posterior spinocerebellar tract

Anterior spinocerebellar tract

Spinal cord

Proprioceptive input from Golgi tendon organs, muscle spindles, and joint capsule receptors

15

are part of the spinothalamic pathway. Activity in these interneurons leads to the stimulation of the primary sensory cortex, so the individual feels pain in a specific part of the body surface. This phenomenon is called **referred pain**. One familiar example is the pain of a heart attack, which is frequently felt in the left arm. Women may experience pain in the jaw, neck, shoulder, stomach, or back. Another is the pain of appendicitis, which is generally felt first in the area around the navel and then in the right lower quadrant. These and additional examples for both anterior and posterior aspects are shown in **Figure 15–7**.

The Posterior Column Pathway

The *posterior column pathway* carries sensations of precise touch and vibrations, and proprioception. The spinal tracts involved are the left and right **fasciculus gracilis** (*gracilis*, slender) and the left and right **fasciculus cuneatus** (*cuneus*, wedge-shaped). On each side of the posterior median sulcus, the fasciculus gracilis is medial to the fasciculus cuneatus (**Figure 15–5**).

The axons of the first-order neurons reach the CNS within the dorsal roots of spinal nerves and the sensory roots of cranial nerves. The axons ascending within the posterior column are organized according to the region innervated. Axons carrying sensations from the inferior half of the body ascend within the fasciculus gracilis and synapse in the nucleus gracilis of the medulla oblongata. Axons carrying sensations from the superior half of the trunk, upper limbs, and neck ascend in the fasciculus cuneatus and synapse in the nucleus cuneatus. ⊃ p. 496

Axons of the second-order neurons of the nucleus gracilis and nucleus cuneatus ascend to the thalamus. As they ascend, these axons cross over to the opposite side of the brain stem. The crossing of an axon from the left side to the right side, or from the right side to the left side, is called **decussation**. Once on the opposite side of the brain, the axons enter a tract called the **medial lemniscus** (*lemniskos*, ribbon). As it ascends, the medial lemniscus runs alongside a smaller tract that carries sensory information from the face, relayed from the sensory nuclei of the trigeminal nerve (V) (**Spotlight Figure 15–6**).

The axons in these tracts synapse on third-order neurons in one of the ventral nuclei of the thalamus. ⊃ p. 503 These nuclei sort the arriving information according to (1) the nature of the stimulus and (2) the region of the body involved. Processing in the thalamus determines whether you perceive a given sensation as fine touch, pressure, or vibration.

Our ability to localize the sensation—to determine precisely where on the body a specific stimulus originated—depends on the projection of information from the thalamus to the primary sensory cortex. Sensory information from the toes arrives at the medial end of the primary sensory cortex, and information from the head arrives at the lateral end. When

Figure 15–7 Referred Pain. Pain sensations from visceral organs are often perceived as arising from areas distant from the actual origin, because the body surface is innervated by the same spinal segments. Each region of perceived pain is labeled according to the organ at which the pain originates.

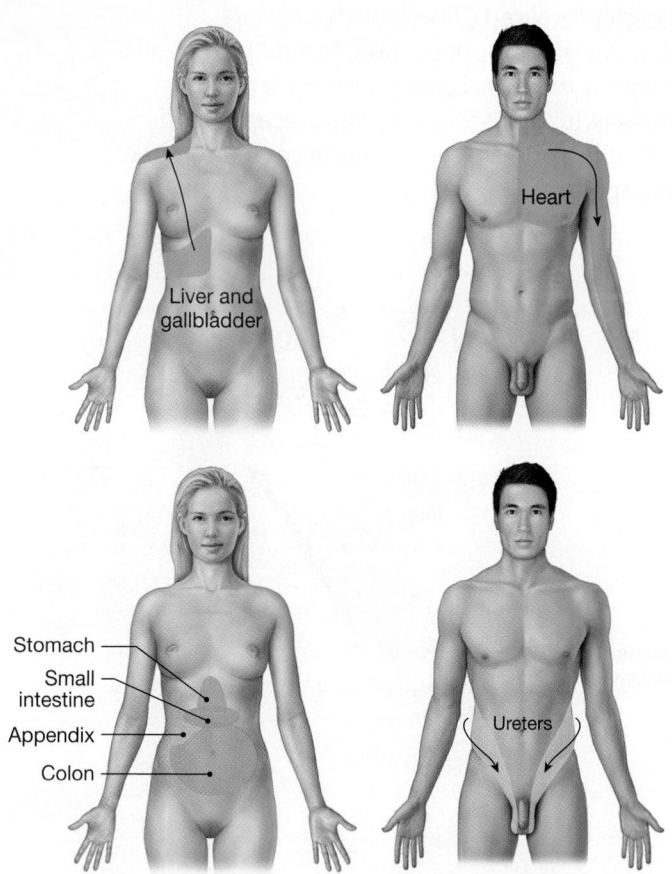

neurons in one portion of your primary sensory cortex are stimulated, you become aware of sensations originating at a specific location. If your primary sensory cortex were damaged or the projection fibers were cut, you could detect a light touch but would be unable to determine its source.

The same sensations are reported whether the cortical neurons are activated by axons ascending from the thalamus or by direct electrical stimulation. Researchers have electrically stimulated the primary sensory cortex in awake individuals during brain surgery and asked the subjects where they thought the stimulus originated. The results were used to create a functional map of the primary sensory cortex known as a *sensory homunculus*.

In a sensory homunculus, the body features are distorted. For example, the face is huge, with enormous lips and tongue, but the back is relatively tiny. These distortions take place because the area of sensory cortex devoted to a particular body region is proportional to the *number of sensory neurons* in the region, not to the region's absolute size. In other words, many more cortical neurons are required to process sensory information from the tongue, which has tens of thousands of taste and touch receptors, than to process sensations from the back, where touch receptors are few and far between.

The Spinocerebellar Pathway

The *spinocerebellar pathway* conveys information about muscle, tendon, and joint positions from the spine to the cerebellum (**Spotlight Figure 15–6**). This information does not reach our awareness. The axons of first-order sensory neurons synapse on interneurons in the dorsal gray horns of the spinal cord. The axons of these second-order neurons ascend in one of the spinocerebellar tracts: the posterior spinocerebellar tracts or the anterior spinocerebellar tracts.

The sensations carried by the anterior spinocerebellar tracts reach the cerebellar cortex by the superior cerebellar peduncle. Interestingly, many of the axons that cross over and ascend to the cerebellum then cross over again within the cerebellum, synapsing on the same side as the original stimulus. The functional significance of this "double cross" is unknown.

The information carried by the spinocerebellar pathway ultimately arrives at the *Purkinje cells* of the cerebellar cortex. ↺ p. 501 Proprioceptive information from each part of the body is relayed to a specific portion of the cerebellar cortex. We will consider the integration of proprioceptive information and the role of the cerebellum in somatic motor control in a later section.

Visceral Sensory Pathways

Visceral sensory information is collected by interoceptors monitoring visceral tissues and organs, primarily within the thoracic and abdominopelvic cavities. These interoceptors include nociceptors, thermoreceptors, tactile receptors, baroreceptors, and chemoreceptors, although none of them is as numerous as they are in somatic tissues. The axons of the first-order neurons usually travel in company with autonomic motor fibers innervating the same visceral structures.

Cranial nerves V, VII, IX, and X carry visceral sensory information from the mouth, palate, pharynx, larynx, trachea, esophagus, and associated vessels and glands. ↺ pp. 552–527 This information is delivered to the **solitary nucleus**, a large nucleus on each side of the medulla oblongata. The solitary nucleus is a major processing and sorting center for visceral sensory information. It has extensive connections with the various cardiovascular and respiratory centers as well as with the reticular formation.

The dorsal roots of spinal nerves T_1–L_2 carry visceral sensory information provided by receptors in organs located between the diaphragm and the pelvic cavity. The dorsal roots of spinal nerves S_2–S_4 carry visceral sensory information from organs in the inferior portion of the pelvic cavity. These organs include

the last segment of the large intestine, the urethra and base of the urinary bladder, and the prostate gland (males) or the cervix of the uterus and adjacent portions of the vagina (females).

The first-order neurons deliver the visceral sensory information to interneurons whose axons ascend within the spinothalamic pathway. Most of the sensory information is delivered to the solitary nucleus. Because it never reaches the primary sensory cortex, you remain unaware of these sensations.

CHECKPOINT

8. As a result of pressure on her spinal cord, Jill cannot feel fine touch or pressure on her lower limbs. Which spinal tract is being compressed?

9. Which spinal tract carries action potentials generated by nociceptors?

10. Which cerebral hemisphere receives impulses carried by the right fasciculus gracilis?

See the blue Answers tab at the back of the book.

15-5 ▌ The somatic nervous system is an efferent division that controls skeletal muscles

Learning Outcome Describe the components, processes, and functions of the somatic motor pathways, and the levels of information processing involved in motor control.

Motor commands issued by the CNS are distributed by the somatic nervous system (SNS) and the autonomic nervous system (ANS). The somatic nervous system, also called the *somatic motor system*, controls the contractions of skeletal muscles. The output of the SNS is under voluntary control.

The autonomic nervous system, or *visceral motor system*, controls visceral effectors, such as smooth muscle, cardiac muscle, glands, and adipocytes. We examine the organization of the ANS in Chapter 16. Our interest here is the structure of the SNS. Throughout this discussion we will use the terms *motor neuron* and *motor control* to refer specifically to somatic motor neurons and pathways that control skeletal muscles.

Somatic motor pathways always involve at least two motor neurons: an **upper motor neuron**, whose cell body lies in a CNS processing center, and a **lower motor neuron**, whose cell body lies in a nucleus of the brain stem or spinal cord. The upper motor neuron synapses on the lower motor neuron, which in turn innervates a single motor unit in a skeletal muscle. ↺ p. 338

Activity in the upper motor neuron may facilitate or inhibit the lower motor neuron. Activation of the lower motor neuron triggers a contraction in the innervated muscle. Only the axon of the lower motor neuron extends outside the CNS. Destruction of or damage to a lower motor neuron eliminates voluntary and reflex control over the innervated motor unit.

Figure 15–8 indicates the positions of the associated motor (descending) tracts in the spinal cord. Conscious and subconscious motor commands control skeletal muscles by traveling over three integrated motor pathways: the *corticospinal pathway,* the *medial pathway,* and the *lateral pathway.* The basal nuclei and cerebellum monitor and adjust activity within these motor pathways. The output of these centers stimulates or inhibits the activity of either (1) motor nuclei or (2) the primary motor cortex. Table 15–2 summarizes the descending (motor) tracts discussed in this section.

The Corticospinal Pathway

The **corticospinal pathway** (Figure 15–9), sometimes called the *pyramidal system,* provides voluntary control over skeletal muscles. This system begins at the *pyramidal cells* of the primary motor cortex. ⊃ p. 513 The axons of these upper motor neurons descend into the brain stem and spinal cord to synapse on lower motor neurons that control skeletal muscles. In general, the corticospinal pathway is direct: The upper motor neurons synapse directly on the lower motor neurons. However, the corticospinal pathway also works indirectly, as it innervates centers of the medial and lateral pathways.

The corticospinal pathway contains three pairs of descending tracts: (1) the *corticobulbar tracts,* (2) the *lateral corticospinal tracts,* and (3) the *anterior corticospinal tracts.* These tracts enter the white matter of the internal capsule, descend into the brain stem, and emerge on either side of the midbrain as the *cerebral peduncles* ⊃ p. 502.

Figure 15–8 Descending (Motor) Tracts in the Spinal Cord. A cross-sectional view showing the locations of the major descending (motor) tracts that contain the axons of upper motor neurons. The origins and destinations of these tracts are listed in Table 15–2. Ascending sensory tracts (shown in *Figure 15-5*) appear in dashed outline.

The Corticobulbar Tracts

Axons in the **corticobulbar** (kor-ti-kō-BUL-bar; *bulbar,* brain stem) **tracts** synapse on lower motor neurons in the motor nuclei of cranial nerves III, IV, V, VI, VII, IX, XI, and XII. The corticobulbar tracts provide conscious control over skeletal muscles that move the eye, jaw, and face, and some muscles of the neck and pharynx. The corticobulbar tracts also innervate the motor centers of the medial and lateral pathways.

Table 15–2	Principal Descending (Motor) Pathways				
Tract	**Location of Upper Motor Neurons**	**Destination**	**Site of Crossover**	**Action**	
CORTICOSPINAL PATHWAY					
Corticobulbar tracts	Primary motor cortex (cerebral hemisphere)	Lower motor neurons of cranial nerve nuclei in brain stem	Brain stem	Conscious motor control of skeletal muscles	
Lateral corticospinal tracts	Primary motor cortex (cerebral hemisphere)	Lower motor neurons of anterior gray horns of spinal cord	Pyramids of medulla oblongata	Conscious motor control of skeletal muscles	
Anterior corticospinal tracts	Primary motor cortex (cerebral hemisphere)	Lower motor neurons of anterior gray horns of spinal cord	Level of lower motor neuron	Conscious motor control of skeletal muscles	
MEDIAL PATHWAY					
Vestibulospinal tracts	Vestibular nuclei (at border of pons and medulla oblongata)	Lower motor neurons of anterior gray horns of spinal cord	None (uncrossed)	Subconscious regulation of balance and muscle tone	
Tectospinal tracts	Tectum (midbrain: superior and inferior colliculi)	Lower motor neurons of anterior gray horns (cervical spinal cord only)	Brain stem (midbrain)	Subconscious regulation of eye, head, neck, and upper limb position in response to visual and auditory stimuli	
Reticulospinal tracts	Reticular formation (network of nuclei in brain stem)	Lower motor neurons of anterior gray horns of spinal cord	None (uncrossed)	Subconscious regulation of reflex activity	
LATERAL PATHWAY					
Rubrospinal tracts	Red nuclei of midbrain	Lower motor neurons of anterior gray horns of spinal cord	Brain stem (midbrain)	Subconscious regulation of upper limb muscle tone and movement	

15

Figure 15–9 **The Corticospinal Pathway.** The corticospinal pathway originates at the primary motor cortex. The corticobulbar tracts end at the motor nuclei of cranial nerves on the opposite side of the brain. Most fibers in this pathway cross over in the medulla and enter the lateral corticospinal tracts; the rest descend in the anterior corticospinal tracts and cross over after reaching target segments in the spinal cord.

KEY

→ Axon of upper-motor neuron

→ Lower-motor neuron

Motor homunculus on primary motor cortex of left cerebral hemisphere

Primary Motor Cortex

Corticobulbar tract

To skeletal muscles

Motor nuclei of cranial nerves

To skeletal muscles

Decussation of pyramids

Lateral corticospinal tract

To skeletal muscles

Midbrain

Cerebral peduncle

Medulla oblongata

Pyramids

Anterior corticospinal tract

Spinal cord

The Corticospinal Tracts

Axons in the **corticospinal tracts** synapse on lower motor neurons in the anterior gray horns of the spinal cord. As they descend, the corticospinal tracts are visible along the ventral surface of the medulla oblongata as a pair of thick bands, the **pyramids**. Along the length of the pyramids, roughly 85 percent of the axons cross the midline (decussate) to enter the descending **lateral corticospinal tracts** on the opposite side of the spinal cord.

The other 15 percent continue uncrossed along the spinal cord as the **anterior corticospinal tracts**. Then, at the spinal segment it targets, an axon in the anterior corticospinal tract crosses over to the opposite side of the spinal cord in the anterior white commissure. The axon synapses on lower motor neurons in the anterior gray horns.

The Motor Homunculus

The activity of pyramidal cells in a specific portion of the primary motor cortex results in the contraction of specific peripheral muscles. Which muscles are stimulated depends on the region of motor cortex that is active. As in the primary sensory cortex, the primary motor cortex corresponds point by point with specific regions of the body. The cortical areas have been mapped out in diagrammatic form, creating a **motor homunculus**. Figure 15–9 shows the motor homunculus of the left cerebral hemisphere and the corticospinal pathway controlling skeletal muscles on the right side of the body.

The proportions of the motor homunculus are quite different from those of the actual body, because the area of primary motor cortex devoted to a specific body region is proportional to the number of motor units involved in the region's control, not to the region's actual size. For this reason, the homunculus indicates the degree of fine motor control available. For example, the hands, face, and tongue are all capable of varied and complex movements, so they appear very large. In contrast, the trunk is relatively small. These proportions are similar to those of the sensory homunculus (Spotlight Figure 15–6).

The sensory and motor homunculi differ, however, with respect to some other areas of the body. For example, some highly sensitive regions, such as the sole of the foot, contain few motor units. Also, some areas with an abundance of motor units, such as the eye muscles, are not particularly sensitive.

The Medial and Lateral Pathways

Several centers in the cerebrum, diencephalon, and brain stem may issue somatic motor commands as a result of processing performed at a subconscious level. These centers and their associated tracts were long known as the *extrapyramidal system (EPS)*. It was thought that they operated independently of, and in parallel with, the *pyramidal system* (corticospinal pathway). This classification scheme is both inaccurate and misleading, because motor control is integrated at all levels through extensive feedback loops and interconnections.

Instead, it is more appropriate to group these nuclei and tracts in terms of their primary functions: The components of the **medial pathway** help control gross movements of the

15

Clinical Note

Amyotrophic Lateral Sclerosis **Amyotrophic lateral sclerosis (ALS)** is a progressive, degenerative disorder that affects motor neurons in the spinal cord, brain stem, and cerebral hemispheres. The degeneration affects both upper and lower motor neurons. A defect in axonal transport is thought to underlie the disease. Because a motor neuron and its dependent muscle fibers are so intimately related, the destruction of CNS neurons causes the associated skeletal muscles to atrophy. ALS is commonly known as Lou Gehrig's disease, named after the famous New York Yankees player who died of the disorder. Noted physicist Stephen Hawking is also afflicted with this condition.

trunk and proximal limb muscles, and those of the **lateral pathway** help control the distal limb muscles that perform more precise movements.

The medial and lateral pathways can modify or direct skeletal muscle contractions by stimulating, facilitating, or inhibiting lower motor neurons. Note that the axons of upper motor neurons in the medial and lateral pathways synapse on the same lower motor neurons innervated by the corticospinal pathway. This means that the various motor pathways interact not only within the brain, through connections between the primary motor cortex and motor centers in the brain stem, but also at the level of the lower motor neuron, through excitatory or inhibitory interactions.

The Medial Pathway

The medial pathway is primarily concerned with the control of muscle tone and gross movements of the neck, trunk, and proximal limb muscles. The upper motor neurons of the medial pathway are located in the *vestibular nuclei*, the *superior* and *inferior colliculi*, and the *reticular formation*.

The vestibular nuclei (located at the border of the pons and medulla) receive information over the vestibulocochlear nerve (N VIII). This nerve has receptors in the internal ear that monitor the position and movement of the head. These nuclei respond to changes in the orientation of the head, sending motor commands that alter the muscle tone, extension, and position of the neck, eyes, head, and limbs. The primary goal is to maintain posture and balance. The descending fibers in the spinal cord constitute the **vestibulospinal tracts** (**Figure 15–8**).

The superior and inferior colliculi are located in the *tectum*, or roof of the midbrain (**Figure 14–9**, p. 502). The superior colliculi receive visual sensations and the inferior colliculi receive auditory sensations. Axons of upper motor neurons in the colliculi descend in the **tectospinal tracts**. These axons cross to the opposite side immediately, before descending to synapse

on lower motor neurons in the brain stem or spinal cord. Axons in the tectospinal tracts direct reflexive changes in the position of the head, neck, and upper limbs in response to bright lights, sudden movements, or loud noises.

The reticular formation is a loosely organized network of neurons that extends throughout the brain stem. ⟲ p. 496 The reticular formation receives input from almost every ascending and descending pathway. It also has extensive interconnections with the cerebrum, the cerebellum, and brain stem nuclei. Axons of upper motor neurons in the reticular formation descend into the **reticulospinal tracts** without crossing to the opposite side. The region stimulated determines the effects of reticular formation stimulation. For example, the stimulation of upper motor neurons in one portion of the reticular formation produces eye movements. The stimulation of another portion activates respiratory muscles.

The Lateral Pathway

The lateral pathway is primarily concerned with the control of muscle tone and the more precise movements of the distal parts of the limbs. The upper motor neurons of the lateral pathway lie within the red nuclei of the midbrain. ⟲ p. 498 Axons of upper motor neurons in the red nuclei cross to the opposite side of the brain and descend into the spinal cord in the **rubrospinal tracts** (*ruber*, red). In humans, the rubrospinal tracts are small and extend only to the cervical spinal cord. There they provide motor control over distal muscles of the upper limbs. Normally, their role is insignificant as compared with that of the lateral corticospinal tracts. However, the rubrospinal tracts can be important in maintaining motor control and muscle tone in the upper limbs if the lateral corticospinal tracts are damaged.

The Basal Nuclei and Cerebellum

The basal nuclei and cerebellum are responsible for coordination and feedback control over muscle contractions. Those contractions can be consciously or subconsciously directed.

The Basal Nuclei

The basal nuclei provide the background patterns of movement involved in voluntary motor activities. For example, they may control muscles that determine the background position of the trunk or limbs, or they may direct rhythmic cycles of movement, as in walking or running. These nuclei do not exert direct control over lower motor neurons. Instead, they adjust the activities of upper motor neurons in the various motor pathways based on input from all portions of the cerebral cortex, as well as from the substantia nigra.

The basal nuclei adjust or establish patterns of movement by two major pathways:

1. One group of axons synapses on thalamic neurons, whose axons extend to the premotor cortex, the motor association

Cerebral Palsy The term **cerebral palsy (CP)** refers to a number of disorders that affect voluntary motor performance. The motor dysfunction is not progressive. It appears during infancy or childhood and persists throughout life. The cause may be trauma due to premature or unusually stressful birth, maternal exposure to drugs (including alcohol), or a genetic defect that causes the motor pathways to develop improperly. Problems during labor and delivery may compress or interrupt placental circulation or oxygen supplies. If the oxygen concentration in fetal blood declines significantly for as little as 5–10 minutes, CNS function can be permanently impaired. The cerebral cortex, cerebellum, basal nuclei, hippocampus, and thalamus are likely to be affected. The result is difficulty with motor skills, posture and balance, memory, speech, and learning.

area that directs activities of the primary motor cortex. This arrangement creates a feedback loop that changes the sensitivity of the pyramidal cells and alters the pattern of instructions carried by the corticospinal tracts.

2. A second group of axons synapses in the reticular formation, altering the excitatory or inhibitory output of the reticulospinal tracts.

Two distinct populations of interneurons exist in the basal nuclei: one that stimulates neurons by releasing acetylcholine (ACh), and another that inhibits neurons by releasing gamma aminobutyric acid (GABA). Under normal conditions, the excitatory interneurons are kept inactive, and the axons leaving the basal nuclei have an inhibitory effect on upper motor neurons. In *Parkinson's disease*, the excitatory neurons become more active, leading to problems with the voluntary control of movement. ⟲ p. 134

If the primary motor cortex is damaged, the person loses fine motor control over skeletal muscles. However, the basal nuclei can still control some voluntary movements. In effect, the medial and lateral pathways function as they usually do, but the corticospinal pathway cannot fine-tune the movements. For example, after damage to the primary motor cortex, the basal nuclei can still receive information about planned movements from the prefrontal cortex and can perform preparatory movements of the trunk and limbs. But because the corticospinal pathway is inoperative, precise movements of the forearms, wrists, and hands cannot occur. An individual in this condition can stand, maintain balance, and even walk, but all movements are hesitant, awkward, and poorly controlled.

The Cerebellum

The cerebellum monitors proprioceptive (position) sensations, visual information from the eyes, and vestibular (balance) sensations from the internal ear as movements are under way. Axons within the spinocerebellar tracts deliver proprioceptive information to the cerebellar cortex. Visual information is relayed from the superior colliculi, and balance information is relayed from the vestibular nuclei. The output of the cerebellum affects upper motor neuron activity in the corticospinal, medial, and lateral pathways.

All motor pathways send information to the cerebellum when motor commands are issued. As the movement proceeds, the cerebellum monitors proprioceptive and vestibular information, comparing the arriving sensations with those experienced during previous movements. It then adjusts the activities of the upper motor neurons involved. In general, any voluntary movement begins with the activation of far more motor units than are required. The cerebellum acts like a brake, inhibiting the number of motor commands used for that movement. The pattern and degree of inhibition changes from moment to moment, making the movement efficient, smooth, and precisely controlled.

The patterns of cerebellar activity are learned by trial and error, over many repetitions. Many of the basic patterns are established early in life. Examples include the fine balancing adjustments you make while standing and walking. The ability to fine-tune a complex pattern of movement improves with practice, until the movements become fluid and automatic. Consider the relaxed, smooth movements of acrobats, golfers, and sushi chefs. These people move without thinking about the details of their movements. Think about what happens when you concentrate on voluntary control of various movements. The rhythm and pattern of the movement usually fall apart as your primary motor cortex starts overriding the commands of the basal nuclei and cerebellum.

Levels of Processing and Motor Control

All sensory and motor pathways involve a series of synapses, one after the other. Along the way, the information is distributed to processing centers operating at the subconscious level. Consider what happens when you stumble. You often recover your balance even as you become aware that a problem exists. Long before your cerebral cortex could assess the situation, evaluate possible responses (shift weight *here*, move leg *there*, and so on), and issue appropriate motor commands, monosynaptic and polysynaptic reflexes, perhaps adjusted by the brain stem and cerebellum, successfully prevent a fall. This is a general pattern: Spinal and cranial reflexes provide rapid, involuntary, preprogrammed responses that preserve homeostasis over the short term. Voluntary responses are more complex and require more time to prepare and execute.

Cranial and spinal reflexes control the most basic motor activities. Integrative centers in the brain perform more elaborate processing. As we move from the medulla oblongata to the cerebral cortex, the motor patterns become increasingly complex and variable. The primary motor cortex of the cerebral hemispheres directs the most complex and variable motor activities.

During development, the spinal reflexes and cranial reflexes appear first. More complex reflexes and motor patterns develop as CNS neurons multiply, enlarge, and interconnect. The process proceeds slowly, as billions of neurons establish trillions of synaptic connections. At birth, neither the cerebral nor the cerebellar cortex is fully functional. Centers in the diencephalon and brain stem primarily direct the behavior of newborn infants.

CHECKPOINT

11. What is the anatomical basis for the fact that the left side of the brain controls motor function on the right side of the body?

12. An injury involving the superior portion of the motor cortex affects which region of the body?

13. What effect would increased stimulation of the motor neurons of the red nucleus have on muscle tone?

See the blue Answers tab at the back of the book.

Related Clinical Terms

analgesia: The inability to feel pain while still conscious.

flaccid paralysis: Weakness or loss of muscle tone as a result of disease or injury to the nerves innervating the muscles.

hyperalgesia: Increased sensitivity to pain that may be caused by damage to nociceptors or peripheral nerves.

pain threshold: The lowest intensity of stimulation at which pain is experienced. It can vary among individuals.

pain tolerance: The maximum pain level an individual can withstand. It can vary among individuals.

spastic paralysis: The chronic pathological condition in which muscles are affected by persistent spasms and exaggerated tendon reflexes due to damage to motor nerves of the central nervous system.

syphilis: A sexually transmitted disease (STD) caused by the bacterium *Treponema pallidum*. Many of the signs and symptoms are indistinguishable from those of other diseases. Late-stage untreated syphilis can cause progressive degeneration of the posterior column pathway, spinocerebellar pathway, and dorsal roots of the spinal cord.

Chapter Review

Study Outline

MasteringA&P®

Access more chapter study tools online in the MasteringA&P Study Area:

- Chapter Quizzes, Chapter Practice Test, Art-labeling Activities, Animations, MP3 Tutor Sessions, and Clinical Case Studies
- Practice Anatomy Lab PAL
- Interactive Physiology iP®
- A&P Flix *A&PFlix*
- PhysioEx PhysioEx

An Introduction to Sensory Pathways and the Somatic Nervous System p. 535

1. The nervous system works as an integrated unit. This chapter considers sensory pathways (sensory receptors and sensory processing centers in the brain) and motor pathways (including both conscious and subconscious motor functions). *(Figure 15–1)*

15-1 | Sensory information from all parts of the body is routed to the somatosensory cortex p. 535

2. The brain, spinal cord, and peripheral nerves continuously communicate with each other and with the internal and external environments. Information arrives by sensory receptors and ascends within the afferent division, while motor commands descend and are distributed by the efferent division.

15-2 | Sensory receptors connect our internal and external environments with the nervous system p. 536

3. A sensory receptor is a specialized cell or cell process that monitors specific conditions within the body or in the external environment. Arriving information is a **sensation**; awareness of a sensation is a **perception**.

4. The **general senses** are our sensitivity to pain, temperature, touch, pressure, vibration, and proprioception. Receptors for these senses are distributed throughout the body. **Special senses**, located in specific **sense organs**, are structurally more complex.

5. Each receptor cell monitors a specific *receptive field*. *Transduction* begins when a large enough stimulus depolarizes the *receptor potential* or *generator potential* to the point where action potentials are produced. *(Figure 15–2)*

6. A **labeled line** is a link between a peripheral receptor and a cortical neuron. **Tonic receptors** are always active. **Phasic receptors** provide information about the intensity and rate of change of a stimulus. **Adaptation** is a reduction in sensitivity in the presence of a constant stimulus. Tonic receptors are

slow-adapting receptors, while phasic receptors are **fast-adapting receptors**. *(Figure 15–3)*

15-3 ▎ General sensory receptors are classified by the type of stimulus that excites them p. 539

7. Three types of **nociceptors** in the body provide information on pain as related to extremes of temperature, mechanical damage, and dissolved chemicals. Myelinated Type A fibers carry **fast pain**. Slower Type C fibers carry **slow pain**.

8. **Thermoreceptors** are found in the dermis. **Mechanoreceptors** are sensitive to distortion of their membranes, and include **tactile receptors**, **baroreceptors**, and **proprioceptors**. There are six types of tactile receptors in the skin, and three kinds of proprioceptors. Chemoreceptors include **carotid bodies** and **aortic bodies**. *(Figure 15–4)*

15-4 ▎ Separate pathways carry somatic sensory and visceral sensory information p. 544

9. Sensory neurons that deliver sensation to the CNS are referred to as **first-order neurons**. These synapse on **second-order neurons** in the brain stem or spinal cord. The next neuron in this chain is a **third-order neuron**, found in the thalamus.

10. Three major somatic sensory pathways carry sensory information from the skin and musculature of the body wall, head, neck, and limbs: the *spinothalamic pathway*, the *posterior column pathway*, and the *spinocerebellar pathway*. *(Figure 15–5)*

11. The **spinothalamic pathway** carries poorly localized ("crude") sensations of touch, pressure, pain, and temperature. The axons involved decussate in the spinal cord and ascend within the **anterior** and **lateral spinothalamic tracts** to the ventral nuclei of the thalamus. Abnormalities along the spinothalamic pathway can lead to **phantom limb pain**, painful sensations that are perceived as real, and **referred pain**, inaccurate localizations of the source of pain. *(Spotlight Figure 15–6, Figure 15–7; Table 15–1)*

12. The **posterior column pathway** carries fine touch, pressure, and proprioceptive sensations. The axons ascend within the **fasciculus gracilis** and **fasciculus cuneatus** and relay information to the thalamus by the **medial lemniscus**. Before the axons enter the medial lemniscus, they cross over to the opposite side of the brain stem. This crossing over is called **decussation**. *(Spotlight Figure 15–6; Table 15–1)*

13. The **spinocerebellar pathway**, including the **posterior** and **anterior spinocerebellar tracts**, carries sensations to the cerebellum concerning the position of muscles, tendons, and joints. *(Spotlight Figure 15–6; Table 15–1)*

14. Visceral sensory pathways carry information collected by interoceptors. Sensory information from cranial nerves V, VII, IX, and X is delivered to the **solitary nucleus** in the medulla oblongata. Dorsal roots of spinal nerves T_1–L_2 carry visceral sensory information from organs between the diaphragm and the pelvic cavity. Dorsal roots of spinal nerves S_2–S_4 carry sensory information from more inferior structures.

15-5 ▎ The somatic nervous system is an efferent division that controls skeletal muscles p. 549

15. Somatic motor (descending) pathways always involve an **upper motor neuron** (whose cell body lies in a CNS processing center) and a **lower motor neuron** (whose cell body is located in a nucleus of the brain stem or spinal cord). *(Figure 15–8)*

16. The neurons of the primary motor cortex are *pyramidal cells*. The **corticospinal pathway** provides voluntary skeletal muscle control. The **corticobulbar tracts** end at the cranial nerve nuclei. The **corticospinal tracts** synapse on lower motor neurons in the anterior gray horns of the spinal cord. The corticospinal tracts are visible along the medulla oblongata as a pair of thick bands, the **pyramids**, where most of the axons decussate to enter the descending **lateral corticospinal tracts**. Those that do not cross over enter the **anterior corticospinal tracts**. The corticospinal pathway provides a rapid, direct mechanism for controlling skeletal muscles. *(Figure 15–9; Table 15–2)*

17. The **medial** and **lateral pathways** include several other centers that issue motor commands as a result of processing performed at a subconscious level. *(Table 15–2)*

18. The medial pathway primarily controls gross movements of the neck, trunk, and proximal limbs. It includes the vestibulospinal, tectospinal, and reticulospinal tracts. The **vestibulospinal tracts** carry information related to maintaining balance and posture. Commands carried by the **tectospinal tracts** change the position of the head, neck, and upper limbs in response to bright lights, sudden movements, or loud noises. Motor commands carried by the **reticulospinal tracts** vary according to the region stimulated. *(Table 15–2)*

19. The lateral pathway consists of the **rubrospinal tracts**, which primarily control muscle tone and movements of the distal muscles of the upper limbs. *(Table 15–2)*

20. The basal nuclei adjust the motor commands issued in other processing centers and provide background patterns of movement involved in voluntary motor activities.

21. The cerebellum monitors proprioceptive sensations, visual information, and vestibular (balance) sensations. The integrative activities performed by neurons in the cortex and nuclei of the cerebellum are essential for the precise control of movements.

22. Spinal and cranial reflexes provide rapid, involuntary, preprogrammed responses that preserve homeostasis. Voluntary responses are more complex and require more time to prepare and execute.

23. During development, the spinal and cranial reflexes are first to appear. Complex reflexes develop over years, as the CNS matures and the brain grows in size and complexity.

Review Questions

See the blue Answers tab at the back of the book.

LEVEL 1 Reviewing Facts and Terms

1. The larger the receptive field, the **(a)** larger the stimulus needed to stimulate a sensory receptor, **(b)** fewer sensory receptors there are, **(c)** harder it is to locate the exact point of stimulation, **(d)** larger the area of the somatosensory cortex in the brain that deals with the area, **(e)** closer together the receptor cells.

2. _____ receptors are normally inactive, but become active for a short time whenever there is a change in the modality that they monitor.

3. In sensory organ function what is adaptation?

4. The area of primary sensory cortex devoted to a body region is relative to the **(a)** size of the body area, **(b)** distance of the body area from the brain, **(c)** number of motor units in the area of the body, **(d)** number

of sensory receptors in the area of the body, (e) size of the nerves that serve the area of the body.

5. Identify and shade in the locations of all the ascending sensory tracts in the following diagram of the spinal cord.

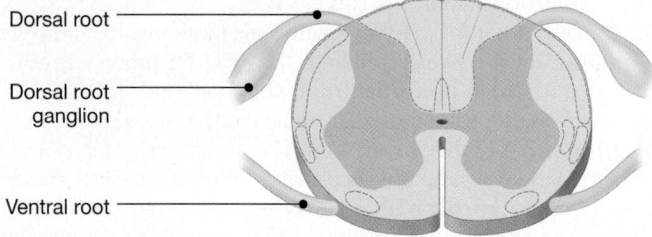

Dorsal root

Dorsal root ganglion

Ventral root

15

6. Identify six types of tactile receptors located in the skin, and describe their sensitivities.
7. What three types of mechanoreceptors respond to stretching, compression, twisting, or other distortions of their plasma membrane?
8. What are the three major somatic sensory pathways and their functions?
9. Which three pairs of descending tracts make up the corticospinal pathway?
10. Which three motor tracts make up the medial pathway?
11. In pain perception, describe the roles of the (a) Type A, and (b) Type C fibers.
12. The corticospinal tract (a) carries motor commands from the cerebral cortex to the spinal cord, (b) carries sensory information from the spinal cord to the brain, (c) starts in the spinal cord and ends in the brain, (d) does all of these.
13. List the general senses and the special senses.

LEVEL 2 Reviewing Concepts

14. Efferent output from the brain stimulates which tissues and organs?
15. What is a motor homunculus? How does it differ from a sensory homunculus?
16. Describe the relationship among first-, second-, and third-order neurons in a sensory pathway.
17. Damage to the posterior spinocerebellar tract on the left side of the spinal cord at the L_1 level would interfere with the coordinated movement of which limb(s)?
18. What effect does injury to the primary motor cortex have on peripheral muscles?
19. What is muscle tone? When does muscle tone fail?
20. Explain the phenomenon of *referred pain* in terms of labeled lines and organization of sensory tracts and pathways.

LEVEL 3 Critical Thinking and Clinical Applications

21. Kayla is having difficulty controlling her eye movements and has lost some control of her facial muscles. After an examination and testing, Kayla's physician tells her that her cranial nerves are perfectly normal but that a small tumor is putting pressure on certain fiber tracts in her brain. This pressure is the cause of Kayla's symptoms. Where is the tumor most likely located?
22. Harry, a construction worker, suffers a fractured skull when a beam falls on his head. Diagnostic tests indicate severe damage to the motor cortex. His wife is anxious to know if he will ever be able to move or walk again. What would you tell her?
23. Denzel had to have his arm amputated at the elbow after an accident. He tells you that he can sometimes still feel pain in his fingers even though the hand is gone. He says this is especially true when he bumps the stub. How can this be?

CLINICAL CASE   Wrap-Up

Living with Cerebral Palsy

Communication between the periphery and the central nervous system follows neural pathways. Somatic sensory tracts are ascending, ending in the brain. Somatic motor tracts are descending, beginning in the brain. Between the input and the output, nerve impulses are analyzed, influenced, and modified in various parts of the brain, including the cerebrum, cerebellum, basal nuclei, hippocampus, and thalamus. These were the very areas of Michael's brain that were injured during birth.

The damage affected the sensory and motor processing centers in the brain. As a result, Michael's voluntary movements, muscle tone, postural muscle control, and balance were all affected. These deficits were stable and did not get worse. There was no limit to the development of the rest of Michael's brain, or how brilliant he could become.

In spite of his disabilities, Michael chose to define himself in terms of his abilities. He earned his Ph.D. in special education.

He was a pioneer and became a catalyst in introducing the concept of self-determination to the field of education. He has been an educator, writer, and motivator his entire life. He is not just living, but thriving despite his cerebral palsy.

1. Name the principal descending motor pathways that carry information that would be affected by the brain damage causing Michael's cerebral palsy.

2. Besides aspiration pneumonia, what other difficulties could result from poor motor control of the muscles used in swallowing?

See the blue Answers tab at the back of the book.

16

The Autonomic Nervous System and Higher-Order Functions

Learning Outcomes

These Learning Outcomes correspond by number to this chapter's sections and indicate what you should be able to do after completing the chapter.

16-1 ❚ Compare the organization of the autonomic nervous system with that of the somatic nervous system.

16-2 ❚ Describe the structures and functions of the sympathetic division of the autonomic nervous system.

16-3 ❚ Describe the mechanisms of sympathetic neurotransmitter release and their effects on target organs and tissues.

16-4 ❚ Describe the structures and functions of the parasympathetic division of the autonomic nervous system.

16-5 ❚ Describe the mechanisms of parasympathetic neurotransmitter release and their effects on target organs and tissues.

16-6 ❚ Discuss the functional significance of dual innervation and autonomic tone.

16-7 ❚ Describe the hierarchy of interacting levels of control in the autonomic nervous system, including the significance of visceral reflexes.

16-8 ❚ Explain how memories are created, stored, and recalled, and distinguish among the levels of consciousness and unconsciousness.

16-9 ❚ Describe some of the ways in which the interactions of neurotransmitters influence brain function.

16-10 ❚ Summarize the effects of aging on the nervous system and give examples of interactions between the nervous system and other organ systems.

SPOTLIGHT
- Overview of the Autonomic Nervous System pp. 560–561

CLINICAL CASE
- The First Day in Anatomy Lab p. 558

Clinical Notes
- Amnesia p. 579
- Categorizing Nervous System Disorders p. 581
- Alzheimer's Disease p. 583

The First Day in Anatomy Lab

It is Tim's first day in anatomy lab and he is very excited. He has been dreaming of this day for years. He has a new, white lab coat and a great anatomy and physiology textbook, and he is ready to go.

The first thing Tim notices upon entering the lab is the body lying on the table. A human cadaver certainly resembles a human being, but there is no mistaking the two. There are bright lights and a smothering smell of formaldehyde from the embalming fluid used in the cadaver.

Tim suddenly feels hot and sweaty. He has a vague discomfort in his chest and feels sick to his stomach. His instructor says something to him but he cannot understand the words. He tries to respond, but is unable to speak. When his instructor looks at him, Tim appears pale and confused. Then he becomes light-headed. His vision blurs and he hears ringing in his ears. The last thing Tim remembers is a sensation of profound weakness.

Tim hits the floor with a thud, landing in a crumpled, prone position. His instructor rolls him over, loosens his tie and belt, and props his lower legs up on a chair. Within ten seconds Tim regains consciousness but feels exhausted. It is as if he is climbing out of a very deep sleep. He is very embarrassed. **What happened to Tim? To find out, turn to the Clinical Case Wrap-Up on p. 588.**

---

16 | An Introduction to the Autonomic Nervous System and Higher-Order Functions

In this chapter we focus on the autonomic nervous system (ANS), which adjusts our basic life support systems without our conscious control. We also consider aspects of higher-order functions such as consciousness, learning, and intelligence. Finally, we look at the effects of aging on the nervous system and conclude with an overview of interactions between the nervous system and other body systems.

16-1 | The autonomic nervous system is involved in the unconscious regulation of visceral functions and has sympathetic and parasympathetic divisions

Learning Outcome Compare the organization of the autonomic nervous system with that of the somatic nervous system.

Your conscious thoughts, plans, and actions are a tiny fraction of what the nervous system does. If all consciousness were eliminated, your vital physiological processes would continue virtually unchanged. After all, a night's sleep is not a life-threatening event. Longer, deeper states of unconsciousness are not necessarily more dangerous, as long as you get nourishment and other basic care. People with severe brain injuries can survive in a coma for decades.

How do people survive under these conditions? Their survival is possible because the **autonomic nervous system** (ANS) makes routine homeostatic adjustments in physiological systems. The ANS coordinates cardiovascular, respiratory, digestive, urinary, and reproductive functions. It adjusts internal water, electrolyte, nutrient, and dissolved gas concentrations in body fluids—without instructions or interference from the conscious mind.

The practice of medicine has benefitted greatly from our understanding of the ANS. For example, in 1960, the five-year survival rate for patients after their first heart attack was very low. The problem was that it was difficult and sometimes impossible to control high blood pressure. Today, many heart attack survivors lead normal lives. What changed? We learned to manipulate the ANS with medicine and clinical procedures. Let's begin our examination of the ANS by comparing its organization with that of the somatic nervous system (SNS).

Organization of the ANS

Figure 16–1 compares the organization of the somatic and autonomic nervous systems. Both are efferent divisions that carry motor commands. The SNS controls skeletal muscles, and the ANS controls visceral effectors such as smooth muscle, glands, cardiac muscle, and adipocytes. The primary structural difference between the two is that in the SNS, motor neurons of the central nervous system exert direct control over skeletal muscles (**Figure 16–1a**). In the ANS, by contrast, motor neurons of the central nervous system synapse on visceral motor neurons in autonomic ganglia, and these ganglionic neurons control visceral effectors (**Figure 16–1b**).

The hypothalamus contains the integrative centers for autonomic activity. The neurons in these centers are comparable to the upper motor neurons in the SNS. Visceral motor neurons

Figure 16–1   **The Organization of the Somatic and Autonomic Nervous Systems.**

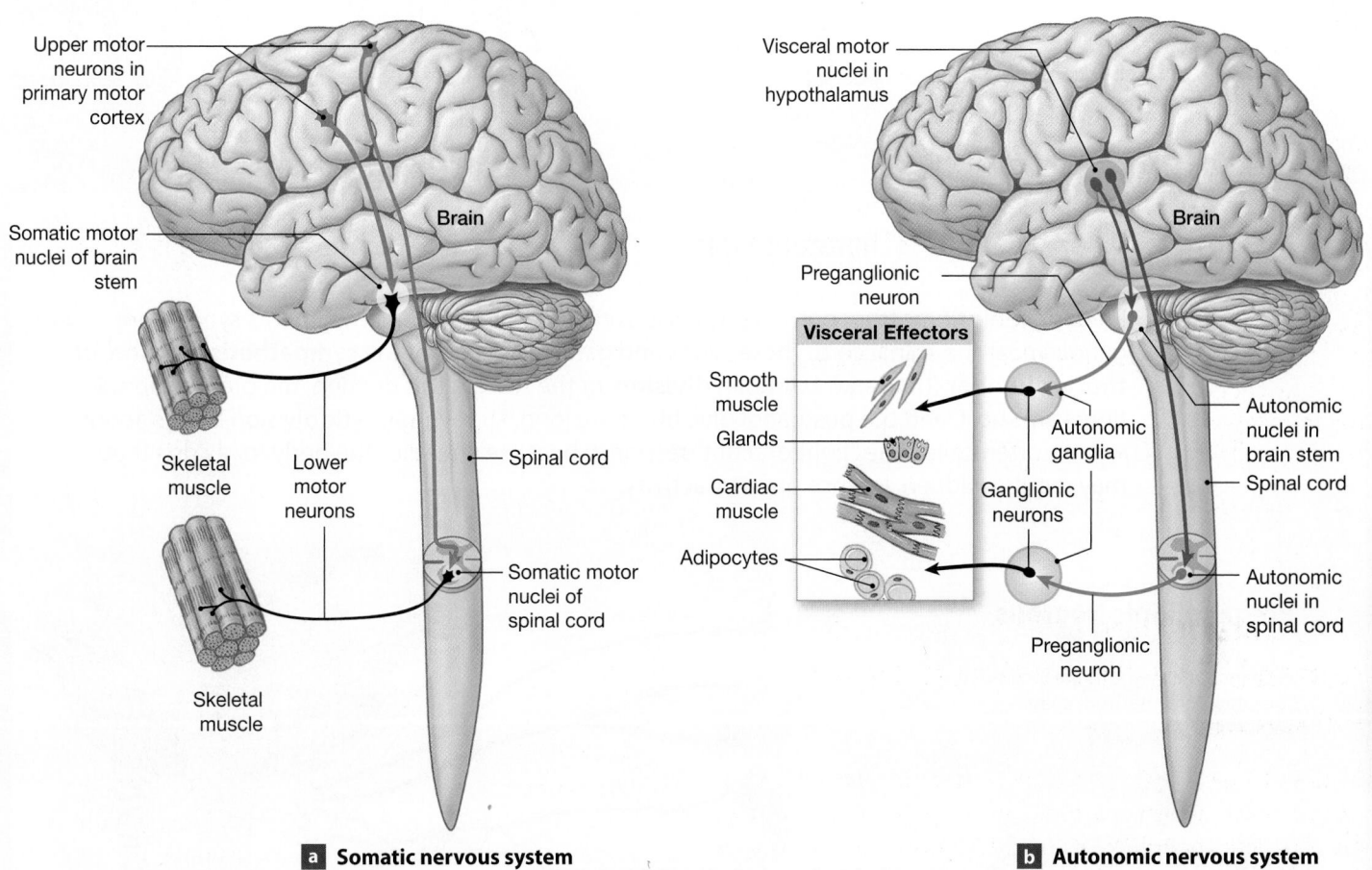

a **Somatic nervous system**

b **Autonomic nervous system**

in the brain stem and spinal cord are known as **preganglionic neurons** because they extend to ganglia. These neurons are part of visceral reflex arcs. Most of their activities represent direct reflex responses, rather than responses to commands from the hypothalamus. The axons of preganglionic neurons are called **preganglionic fibers**.

Preganglionic fibers leave the CNS and synapse on **ganglionic neurons**—visceral motor neurons in peripheral ganglia. These ganglia, which contain hundreds to thousands of ganglionic neurons, are called **autonomic ganglia**. Ganglionic neurons innervate visceral effectors such as smooth muscle, glands, cardiac muscle, and adipocytes. The axons of ganglionic neurons are called **postganglionic fibers**, because they begin at the autonomic ganglia and extend to the peripheral target organs.

Somatic or visceral sensory information can trigger visceral reflexes, and the ANS distributes the motor commands of those reflexes. Sometimes those motor commands control the activities of target organs. For example, in cold weather, the ANS stimulates the arrector pili muscles and gives you "goosebumps." ⤶ p. 189 In other cases, the motor commands may alter some ongoing activity. A sudden, loud noise can startle

you and make you jump, but thanks to the ANS, that sound can also increase your heart rate dramatically and temporarily stop all digestive gland secretion. These changes in visceral activity take place in response to neurotransmitters released by postganglionic fibers. As noted in Chapter 12, a specific neurotransmitter may stimulate or inhibit activity, depending on the response of particular plasma membrane receptors. We consider the major types of receptors later in the chapter. Now let's turn to the anatomy and physiology of the ANS.

TIPS&**TRICKS**

Each autonomic ganglion functions somewhat like a baton handoff in a relay race. Within the ganglion, one runner (the preganglionic fiber) hands off the baton (a neurotransmitter) to the next runner (the postganglionic fiber), who then continues on toward the finish line (the target effector).

Divisions of the ANS

You are probably already familiar with the names of the two main subdivisions of the ANS: the *sympathetic division* and the *parasympathetic division* (**Spotlight Figure 16-2**).

AUTONOMIC NERVOUS SYSTEM

Consists of two divisions

Sympathetic Division (Thoracolumbar)

Preganglionic fibers from both the thoracic and upper lumbar spinal segments synapse in ganglia near the spinal cord. These axons and ganglia are part of the **sympathetic division**, or **thoracolumbar** (thor-a-kō-LUM-bar) **division**, of the ANS. In this division, the preganglionic fibers are short, and the postganglionic fibers are long. The sympathetic division brings about what is often called the "fight or flight" response because it readies the body for a crisis that may require sudden, intense physical activity.

Preganglionic Neurons

Preganglionic neurons are located in the lateral gray horns of spinal segments T_1–L_2.

Ganglia

Ganglia are located near the spinal cord. Preganglionic fibers release acetylcholine (Ach), stimulating ganglionic neurons.

Target Organs

Most postganglionic fibers release norepinephrine (NE) at neuroeffector junctions.

Sympathetic activation

"Fight or flight" response

KEY

— Preganglionic fibers
— Postganglionic fibers

Parasympathetic Division (Craniosacral)

Preganglionic fibers originating in either the brain stem (cranial nerves III, VII, IX, and X) or the sacral segments of the spinal cord are part of the **parasympathetic division**, or **craniosacral division**, of the ANS. The preganglionic fibers synapse in terminal ganglia, located close to the target organs, or intramural (*murus*, wall) ganglia, within the tissues of the target organs. In this division, the preganglionic fibers are long, and the postganglionic fibers are short. The parasympathetic division brings about what is often called the "rest and digest" response because it conserves energy and promotes sedentary activities, such as digestion.

Preganglionic Neurons

Preganglionic neurons in brain stem and in lateral portion of anterior gray horns of S_2–S_4.

Ganglia

Ganglia are in or near the target organ. Preganglionic fibers release acetylcholine (Ach), stimulating ganglionic neurons.

Target Organs

All postganglionic fibers release ACh at neuroeffector junctions.

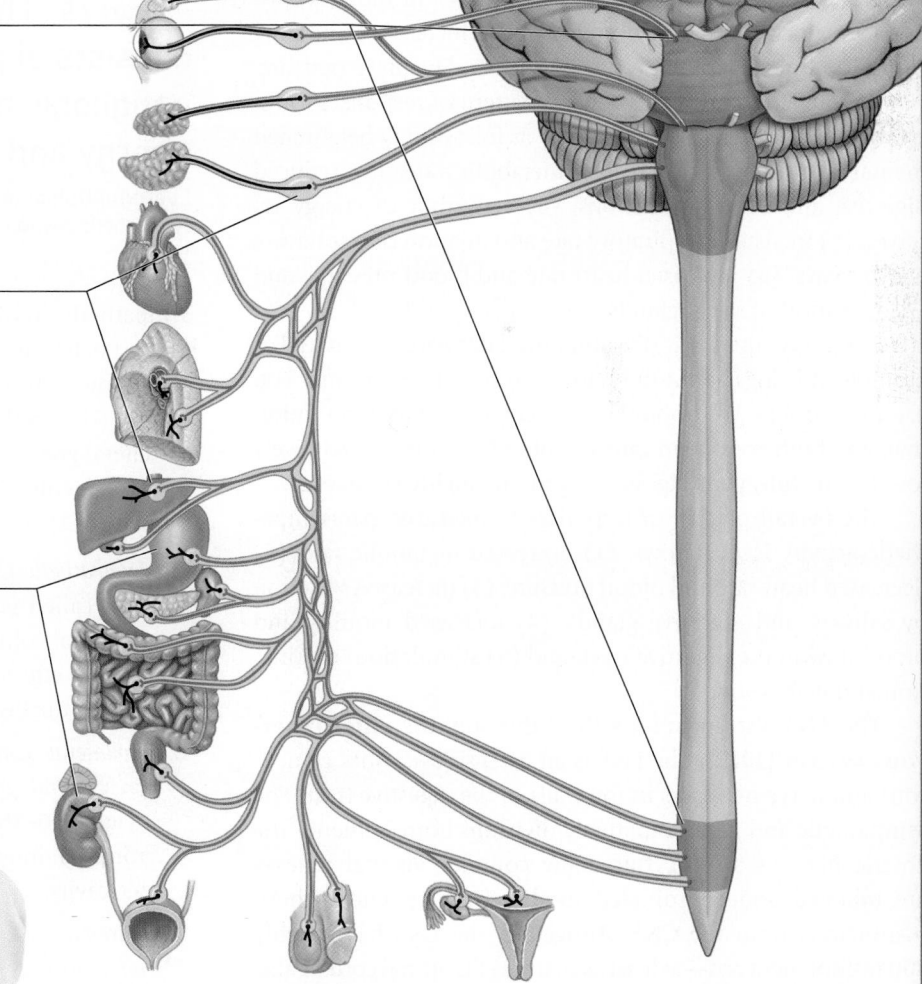

Parasympathetic stimulation

"Rest and digest" response

KEY

— Preganglionic fibers
— Postganglionic fibers

The two divisions work in several ways:

1. **Most often, these two divisions have opposing effects.** If the sympathetic division causes excitation, the parasympathetic causes inhibition.

2. **The two divisions may also work independently.** Only one division innervates some structures.

3. **The two divisions may work together, with each controlling one stage of a complex process.**

In general, the sympathetic division "kicks in" only during exertion, stress, or emergency, and the parasympathetic division predominates under resting conditions (**Spotlight Figure 16–2**).

An increase in sympathetic activity prepares the body to deal with emergencies. Imagine walking down a long, dark alley and hearing strange noises in the darkness ahead. Your body responds right away, and you become more alert and aware of your surroundings. Your metabolic rate rises quickly, up to twice its resting level. Your digestive and urinary processes stop temporarily, and more blood flows to your skeletal muscles. You begin breathing more quickly and more deeply. Both your heart rate and blood pressure increase, circulating your blood more rapidly. You feel warm and begin to perspire.

We can summarize this general pattern of responses to increased levels of sympathetic activity as follows: (1) heightened mental alertness, (2) increased metabolic rate, (3) reduced digestive and urinary functions, (4) activation of energy reserves, (5) increased respiratory rate and dilation of respiratory passageways, (6) increased heart rate and blood pressure, and (7) activation of sweat glands.

The parasympathetic division stimulates visceral activity. For example, it brings about the state of "rest and digest" after you eat a big dinner. Your body relaxes, energy demands are minimal, and both your heart rate and blood pressure are relatively low. Meanwhile, your digestive organs are highly stimulated.

The overall pattern of responses to increased parasympathetic activity is as follows: (1) decreased metabolic rate, (2) decreased heart rate and blood pressure, (3) increased secretion by salivary and digestive glands, (4) increased motility and blood flow in the digestive tract, and (5) stimulation of urination and defecation.

The ANS also includes a third division: the **enteric nervous system (ENS)**. The ENS is an extensive network of neurons and nerve networks in the walls of the digestive tract. The sympathetic and parasympathetic divisions both influence the enteric nervous system, but many complex visceral reflexes are initiated and coordinated locally. They operate without instructions from the CNS. Altogether, the ENS has roughly 100 million neurons—at least as many as the spinal cord. It also uses the same neurotransmitters found in the brain.

In this chapter, we focus on the sympathetic and parasympathetic divisions, which integrate and coordinate visceral functions throughout the body. We consider the enteric nervous system later in this chapter, when we discuss visceral reflexes, and again when we examine the control of digestion in Chapter 24.

CHECKPOINT

1. Identify the two major divisions of the autonomic nervous system.

2. How many motor neurons are needed to carry an action potential from the spinal cord to smooth muscles in the wall of the intestine?

3. While out for a walk, Julie suddenly meets an angry dog. Which division of the autonomic nervous system is responsible for the physiological changes that occur in Julie as she turns and runs?

4. On the basis of anatomy, how could you distinguish the sympathetic division from the parasympathetic division of the ANS?

See the blue Answers tab at the back of the book.

16-2 ┃ The sympathetic division consists of preganglionic neurons and ganglionic neurons involved in using energy and increasing metabolic rate

Learning Outcome Describe the structures and functions of the sympathetic division of the autonomic nervous system.

Spotlight Figure 16–2 shows the overall organization of the sympathetic division of the ANS. Its preganglionic neurons are located between segments T_1 and L_2 of the spinal cord, and its ganglionic neurons are located in ganglia near the vertebral column. The cell bodies of the preganglionic neurons are in the lateral gray horns, and their axons enter the ventral roots of these segments. The ganglionic neurons are in three locations (**Figure 16–3**):

1. *Sympathetic Chain Ganglia.* **Sympathetic chain ganglia**, also called *paravertebral ganglia*, lie on both sides of the vertebral column (**Figure 16–3a**). Neurons in these ganglia control effectors in the body wall, inside the thoracic cavity, and in the head and limbs.

2. *Collateral Ganglia.* **Collateral ganglia**, also known as *prevertebral ganglia*, are anterior to the vertebral bodies (**Figure 16–3b**). Collateral ganglia contain ganglionic neurons that innervate tissues and organs in the abdominopelvic cavity.

3. *Adrenal Medullae.* The center of each adrenal (*ad-*, near + *renal*, kidney) gland is known as the **adrenal medulla**, or *suprarenal medulla*. It is a modified sympathetic ganglion (**Figure 16–3c**). The ganglionic neurons of the adrenal medullae have very short axons. When stimulated, they release their neurotransmitters into the bloodstream, not at a

Figure 16–3 **Sites of Ganglia in Sympathetic Pathways.** Superior views of sections through the thoracic spinal cord, showing the three major patterns of distribution for preganglionic and postganglionic fibers.

a SYMPATHETIC CHAIN GANGLIA

Spinal nerve

Preganglionic neuron

Autonomic ganglion of right sympathetic chain

Autonomic ganglion of left sympathetic chain

Innervates visceral effectors by spinal nerves

Sympathetic nerve (postganglionic fibers)

White ramus

Ganglionic neuron

Gray ramus

Innervates visceral organs in thoracic cavity by sympathetic nerves

Note: Both innervation patterns occur on each side of the body.

KEY
— Preganglionic neurons
— Ganglionic neurons

synapse. This mode of release allows the neurotransmitters to function as hormones, affecting target cells throughout the body.

In the sympathetic division, the preganglionic fibers are relatively short because the ganglia are located near the spinal cord. In contrast, the postganglionic fibers are relatively long, except at the adrenal medullae.

Organization and Anatomy of the Sympathetic Division

The ventral roots of spinal segments T_1 to L_2 contain sympathetic preganglionic fibers that originate from the lateral gray horns. We described the basic pattern of sympathetic innervation in these regions in **Spotlight Figure 13–8**, pp. 466–467. After passing through the intervertebral foramen, each ventral root gives rise to a myelinated *white ramus*, which carries myelinated preganglionic fibers into a nearby sympathetic chain ganglion. These fibers may synapse within the sympathetic chain ganglia, and/ or at one of the collateral ganglia, or in the adrenal medullae (**Figure 16–3**). They diverge extensively, with one preganglionic fiber synapsing on two dozen or more ganglionic neurons. Preganglionic fibers running between the sympathetic chain ganglia interconnect them, making the chain look like a string of pearls. Each ganglion in the sympathetic chain innervates a particular body segment or group of segments.

Sympathetic Chain Ganglia

If a preganglionic fiber carries motor commands targeting structures in the body wall, thoracic cavity, head, neck, or limbs, it synapses in one or more sympathetic chain ganglia. The unmyelinated postganglionic

b COLLATERAL GANGLIA

Lateral gray horn

Splanchnic nerve (preganglionic fibers)

White ramus

Postganglionic fibers

Collateral ganglion

Innervates visceral organs in abdominopelvic cavity

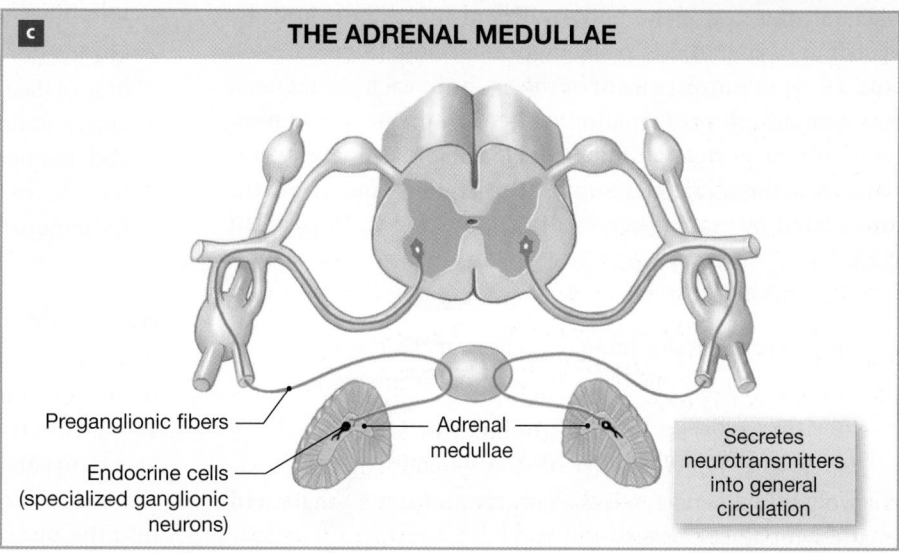

c THE ADRENAL MEDULLAE

Preganglionic fibers

Adrenal medullae

Endocrine cells (specialized ganglionic neurons)

Secretes neurotransmitters into general circulation

16

fibers then follow differing paths, depending on where their targets lie:

- Postganglionic fibers that control visceral effectors in the body wall, head, neck, or limbs enter the *gray ramus* and return to the spinal nerve for subsequent distribution (**Figure 16–3a**, right). These postganglionic fibers innervate sweat glands, smooth muscles in superficial blood vessels, and arrector pili muscles.

- Postganglionic fibers innervating structures in the thoracic cavity, such as the heart and lungs, form bundles known as **sympathetic nerves** (**Figure 16–3a**).

For the sake of clarity, **Figure 16–3a** shows sympathetic nerves on the left side and spinal nerve distribution on the right but in reality *both* innervation patterns occur on *each* side of the body.

Figure 16–4 provides a more detailed view of the structure of the ganglion chain and the sympathetic division as a whole. The left side of the figure shows the distribution to the skin (and to skeletal muscles and other tissues of the body wall). The right side shows the innervation of visceral organs.

Each sympathetic chain contains 3 cervical, 10–12 thoracic, 4–5 lumbar, and 4–5 sacral ganglia, plus 1 coccygeal ganglion. (The numbers vary because adjacent ganglia sometimes fuse.) Preganglionic neurons are limited to spinal cord segments T_1–L_2, and these spinal nerves have both white rami (myelinated preganglionic fibers) and gray rami (unmyelinated postganglionic fibers). The neurons in the cervical, inferior lumbar, and sacral sympathetic chain ganglia are innervated by preganglionic fibers that run along the axis of the chain. In turn, these chain ganglia provide postganglionic fibers, through gray rami, to the cervical, lumbar, and sacral spinal nerves. As a result, only spinal nerves T_1–L_2 have white rami, and every spinal nerve has a gray ramus that carries sympathetic postganglionic fibers for distribution in the body wall.

The spinal nerves provide somatic motor innervation to skeletal muscles of the body wall and limbs, but they also distribute sympathetic postganglionic fibers (**Figures 16–3a** and **16–4**). About 8 percent of the axons in each spinal nerve are sympathetic postganglionic fibers. In the head and neck, sympathetic postganglionic fibers leaving the superior cervical sympathetic ganglia supply the regions and structures innervated by cranial nerves III, VII, IX, and X. ⊃ pp. 520, 523, 525

In summary:

- The cervical, inferior lumbar, and sacral chain ganglia receive preganglionic fibers from spinal segments T_1–L_2.

- Only the thoracic and superior lumbar ganglia (T_1–L_2) receive preganglionic fibers from white rami.

- Every spinal nerve receives a gray ramus from a ganglion of the sympathetic chain.

Collateral Ganglia

The abdominopelvic viscera receive sympathetic innervation by sympathetic preganglionic fibers that synapse in separate collateral ganglia (**Figures 16–3b**). These fibers pass through the sympathetic chain without synapsing. They form the **splanchnic** (SPLANK-nik) **nerves**, which lie in the posterior wall of the abdominal cavity. In adults the collateral ganglia are typically single rather than paired. They originate as paired ganglia (left and right), but the two usually fuse.

Postganglionic fibers leaving the collateral ganglia extend throughout the abdominopelvic cavity, innervating a variety of visceral tissues and organs. The general function is to (1) reduce blood flow and energy use by organs that are not important to immediate survival (such as the digestive tract) and (2) release stored energy.

The splanchnic nerves innervate three collateral ganglia (**Figure 16–3**). Preganglionic fibers from the seven inferior thoracic spinal segments end at either the **celiac** (SĒ-lē-ak) **ganglion** or the **superior mesenteric ganglion**. These ganglia are embedded in an extensive network of autonomic nerves. Preganglionic fibers from the lumbar segments form splanchnic nerves that end at the **inferior mesenteric ganglion**. All three ganglia are named after nearby arteries:

- The celiac ganglion is named after the *celiac trunk*, a major artery and its branches supplying the stomach, spleen, and liver. The celiac ganglion commonly consists of a pair of interconnected masses of gray matter located at the base of that artery. The celiac ganglion may also form a single mass or many small, interwoven masses. Postganglionic fibers from this ganglion innervate the stomach, liver, gallbladder, pancreas, and spleen.

- The superior mesenteric ganglion is found near the base of the *superior mesenteric artery*, which provides blood to the stomach, small intestine, and pancreas. Postganglionic fibers leaving the superior mesenteric ganglion innervate the small intestine and the proximal two-thirds of the large intestine.

- The inferior mesenteric ganglion is located near the base of the *inferior mesenteric artery*, which supplies the large intestine and other organs in the inferior portion of the abdominopelvic cavity. Postganglionic fibers from this ganglion provide sympathetic innervation to the kidney, urinary bladder, the terminal portions of the large intestine, and the sex organs.

Adrenal Medullae

Preganglionic fibers entering an adrenal gland proceed to its center, a region called the adrenal medulla (**Figures 16–3c** and **16–4**). The adrenal medulla is a modified sympathetic ganglion where preganglionic fibers synapse on *neuroendocrine cells*, specialized neurons that secrete hormones (chemical messengers) into the bloodstream. These neuroendocrine cells secrete the

Figure 16–4 **The Distribution of Sympathetic Innervation.** The distribution of sympathetic fibers is the same on both sides of the body. For clarity, the innervation of somatic structures is shown on the left, and the innervation of visceral structures on the right.

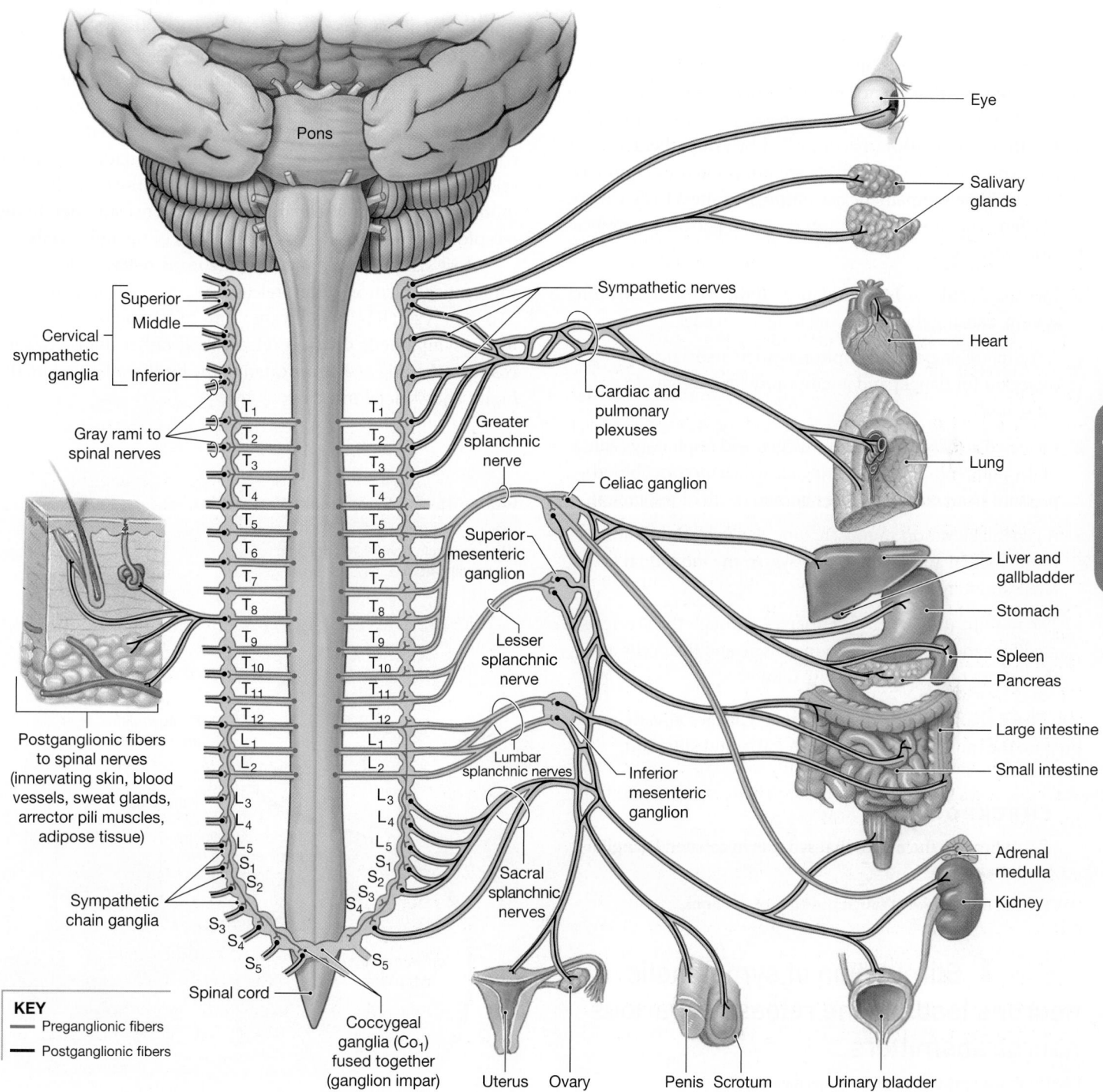

neurotransmitters *epinephrine (E)* and *norepinephrine (NE)*. Epinephrine, or *adrenaline*, makes up 75–80 percent of the secretory output. The rest is NE, or *noradrenaline*.

The bloodstream carries the neurotransmitters throughout the body, where they cause changes in the metabolic activities of many different cells. These effects resemble those produced by the stimulation of sympathetic postganglionic fibers. They differ, however, in two respects: (1) Cells not innervated by sympathetic postganglionic fibers are affected; and (2) the effects last much longer than those produced by direct sympathetic innervation, because the hormones continue to diffuse out of the bloodstream for an extended period.

Sympathetic Activation

The sympathetic division can change the activities of tissues and organs by releasing NE at peripheral synapses, and by distributing E and NE throughout the body in the bloodstream. The visceral motor fibers that target specific effectors, such as smooth muscle fibers in blood vessels of the skin, can be activated in reflexes that do not involve other visceral effectors. In a crisis, however, the entire division responds. This event, called **sympathetic activation**, is controlled by sympathetic centers in the hypothalamus. The effects are not limited to peripheral tissues, because sympathetic activation also alters CNS activity.

When sympathetic activation occurs, a person experiences the following changes:

- Increased alertness by stimulation of the reticular activating system, causing the individual to feel "on edge."

- A feeling of energy and euphoria, often associated with a disregard for danger and a temporary insensitivity to painful stimuli.

- Increased activity in the cardiovascular and respiratory centers of the pons and medulla oblongata, which increases blood pressure, heart rate, breathing rate, and depth of respiration.

- A general elevation in muscle tone through stimulation of the medial and lateral pathways, so the individual *looks* tense and may begin to shiver.

- The mobilization of energy reserves, through the accelerated breakdown of glycogen in muscle and liver cells and the release of lipids by adipose tissues.

These changes, plus the peripheral changes already noted, prepare the individual to cope with a stressful situation.

CHECKPOINT

5. Where do the nerves that synapse in collateral ganglia originate?

See the blue Answers tab at the back of the book.

16-3 | Stimulation of sympathetic neurons leads to the release of various neurotransmitters

Learning Outcome Describe the mechanisms of sympathetic neurotransmitter release and their effects on target organs and tissues.

We have examined the structure of the sympathetic division of the ANS and the general effects of sympathetic activation. Now let's consider the cellular basis of these effects on peripheral organs.

When stimulated, sympathetic preganglionic neurons release ACh at synapses with ganglionic neurons. Synapses that use ACh as a transmitter are called *cholinergic*. ⊃ p. 440 The effect on the ganglionic neurons is always excitatory.

These ganglionic neurons then release neurotransmitters at specific target organs. The axon terminals differ from neuromuscular junctions of the somatic nervous system. In the sympathetic nervous system, the telodendria form a branching network, with each branch resembling a string of pearls. Each "pearl" is a swollen segment called a **varicosity** and is packed with neurotransmitter vesicles (**Figure 16–5**). Chains of varicosities pass along or near the surfaces of effector cells. These cells have no specialized postsynaptic membranes, such as the motor end plates in skeletal muscle cells. Instead, membrane receptors are scattered across the surfaces of the target cells.

Most sympathetic ganglionic neurons release NE at their varicosities. Neurons that release NE are called *adrenergic*. ⊃ p. 443 A small but significant number of ganglionic neurons in the sympathetic division release ACh rather than NE. Varicosities releasing ACh are located in the body wall, the skin, the brain, and skeletal muscles.

Figure 16–5 Sympathetic Varicosities.

Preganglionic fiber (myelinated)

Ganglionic neuron

Ganglion

Postganglionic fiber (unmyelinated)

Varicosities

Vesicles containing norepinephrine (NE)

Mitochondrion

Schwann cell cytoplasm

5 μm

Smooth muscle cells

Varicosities

The NE released by varicosities affects its targets until it is reabsorbed or inactivated by enzymes. Fifty to 80 percent of the NE is reabsorbed by varicosities and is either reused or broken down by the enzyme *monoamine oxidase (MAO)*. The rest of the NE diffuses out of the area or is broken down by the enzyme *catechol-O-methyltransferase (COMT)* in surrounding tissues.

In general, the effects of NE on the postsynaptic membrane last for a few seconds, significantly longer than the 20-msec duration of ACh effects. (As usual, the responses of the target cells vary with the nature of their receptors.) When the adrenal medullae release NE or E into the bloodstream, the effects last even longer because (1) the bloodstream does not contain MAO or COMT, and (2) most tissues contain relatively low concentrations of those enzymes. After the adrenal medullae are stimulated, tissue concentrations of NE and E throughout the body may remain elevated for as long as 30 seconds, and the effects may persist for several minutes.

Sympathetic Stimulation and the Release of NE and E

The effects of sympathetic stimulation result primarily from the interactions of NE and E with adrenergic membrane receptors. There are two classes of these receptors: *alpha receptors* and *beta receptors*. In general, norepinephrine (NE) stimulates alpha receptors to a greater degree than it does beta receptors, and epinephrine (E) stimulates both classes of receptors. For this reason, localized sympathetic activity, involving the release of NE at varicosities, primarily affects nearby alpha receptors. By contrast, generalized sympathetic activation and the release of E by the adrenal medullae affect alpha and beta receptors throughout the body.

Alpha receptors and beta receptors are *G protein-coupled receptors*. As we saw in Chapter 12, the effects of stimulating such a receptor depend on the production of *second messengers*, intracellular intermediaries with varied functions. ⟲ p. 446

The stimulation of **alpha (α) receptors** activates their associated G proteins on the cytoplasmic surface of the plasma membrane. There are two types of alpha receptors: alpha-1 (α_1) and alpha-2 (α_2).

- α_1 is the more common type of alpha receptor. Its stimulation results in G protein activation that brings about the release of intracellular calcium ions into the cytosol from reserves in the endoplasmic reticulum. This action generally has an excitatory effect on the target cell. For example, the stimulation of α_1 receptors on smooth muscle cells causes peripheral blood vessels to constrict and sphincters along the digestive tract and urinary bladder to close.

- Stimulation of α_2 receptors and G protein activation results in a lowering of cyclic-AMP (cAMP) levels in the cytoplasm. Cyclic-AMP is an important second messenger that can activate or inactivate key enzymes. ⟲ p. 446 This reduction generally has an inhibitory effect on the cell. The presence

of α_2 receptors in the parasympathetic division helps coordinate sympathetic and parasympathetic activities. When the sympathetic division is active, the NE released binds to α_2 receptors at parasympathetic neuromuscular and neuroglandular junctions and inhibits their activity.

Beta (β) receptors are located on the plasma membranes of cells in many organs, including skeletal muscles, the lungs, the heart, and the liver. The stimulation of beta receptors and G protein activation triggers changes in the metabolic activity of the target cell. These changes occur indirectly, as G protein activation results in an increase in intracellular cAMP levels. There are three major types of beta receptors: beta-1 (β_1), beta-2 (β_2), and beta-3 (β_3).

- The stimulation of β_1 receptors leads to an increase in metabolic activity. For example, the stimulation of β_1 receptors in skeletal muscles accelerates the metabolic activities of the muscles. The stimulation of β_1 receptors in the heart increases heart rate and force of contraction.

- The stimulation of β_2 receptors causes inhibition, triggering a relaxation of smooth muscles along the respiratory tract. As a result, respiratory passageways dilate, making breathing easier. The inhalers used to treat asthma trigger this response.

- A third type of beta receptor, beta-3 (β_3), is found in adipose tissue. Stimulation of β_3 receptors leads to *lipolysis*, the breakdown of triglycerides stored within adipocytes. The fatty acids generated through lipolysis are released into the bloodstream for use by other tissues.

Sympathetic Stimulation and the Release of ACh and NO

The vast majority of sympathetic postganglionic fibers are adrenergic (release NE), but as we noted, a few are cholinergic (release ACh). These postganglionic fibers innervate skin sweat glands and the blood vessels to skeletal muscles and the brain. The activation of these sympathetic fibers stimulates sweat gland secretion and dilates the blood vessels.

In other regions of the body, the parasympathetic division (rather than the sympathetic division) releases ACh. However, neither the body wall nor skeletal muscles are innervated by the parasympathetic division, and in these areas both ACh and NE are needed to regulate visceral functions with precision. For example, ACh causes most small peripheral arteries to dilate (*vasodilation*), and NE causes them to constrict (*vasoconstriction*). This means that the sympathetic division can increase blood flow to skeletal muscles, by releasing ACh to activate cholinergic terminals. At the same time, it can reduce the blood flow to other tissues in the body wall, by releasing NE to stimulate adrenergic terminals.

The sympathetic division also includes *nitroxidergic* synapses, which release *nitric oxide (NO)* as a neurotransmitter. Such synapses occur where neurons innervate smooth muscles in the walls of blood vessels in many regions, notably in skeletal

16

muscles and the brain. The activity of these synapses produces vasodilation and increased blood flow through these regions.

Summary: The Sympathetic Division

To summarize our discussion of the sympathetic division:

1. The sympathetic division of the ANS includes two sets of sympathetic chain ganglia, one on each side of the vertebral column; three collateral ganglia anterior to the vertebral column; and two adrenal medullae.

2. The preganglionic fibers are short, because the ganglia are close to the spinal cord. The postganglionic fibers are longer and extend a considerable distance to their target organs. (In the case of the adrenal medullae, very short axons end at capillaries that carry their secretions to the bloodstream.)

3. The sympathetic division shows extensive divergence. That is, a single preganglionic fiber may innervate two dozen or more ganglionic neurons in different ganglia. As a result, a single sympathetic motor neuron in the CNS can control a variety of visceral effectors and can produce a complex and coordinated response.

4. All preganglionic neurons release ACh at their synapses with ganglionic neurons. Most postganglionic fibers release NE, but a few release ACh or NO.

5. The effector response depends on the second messengers triggered by G protein activation when NE or E binds to alpha receptors or beta receptors.

CHECKPOINT

6. How would a drug that stimulates acetylcholine receptors affect the sympathetic nervous system?

7. An individual with high blood pressure is given a medication that blocks beta receptors. How could this medication help correct that person's condition?

See the blue Answers tab at the back of the book.

16-4 ❙ The parasympathetic division consists of preganglionic neurons and ganglionic neurons involved in conserving energy and lowering metabolic rate

Learning Outcome Describe the structures and functions of the parasympathetic division of the autonomic nervous system.

The parasympathetic division of the ANS (**Figure 16–6**) consists of:

1. *Preganglionic Neurons in the Brain Stem and in Sacral Segments of the Spinal Cord.* The midbrain, pons, and medulla oblongata contain autonomic nuclei associated with cranial nerves III, VII, IX, and X. In sacral segments of the spinal cord, the parasympathetic nuclei lie in the lateral gray horns of spinal segments S_2–S_4.

2. *Ganglionic Neurons in Peripheral Ganglia within or Adjacent to the Target Organs.* Preganglionic fibers of the parasympathetic division do not diverge as extensively as do those of the sympathetic division. A typical preganglionic fiber synapses on six to eight ganglionic neurons. All are in the same ganglion. The postganglionic fibers then extend to the same target organ. The ganglion may be a **terminal ganglion,** located near the target organ, or an **intramural** (*murus,* wall) **ganglion,** embedded in the tissues of the target organ. Note that this pattern makes the effects of parasympathetic stimulation very specific and localized. Terminal ganglia are usually paired. Examples include the parasympathetic ganglia associated with the cranial nerves. Intramural ganglia typically consist of interconnected masses and clusters of ganglion cells.

Organization and Anatomy of the Parasympathetic Division

Parasympathetic preganglionic fibers leave the brain in cranial nerves III (oculomotor), VII (facial), IX (glossopharyngeal), and X (vagus) (**Figure 16–6**). These fibers carry the cranial parasympathetic output to visceral structures in the head. These fibers synapse in the *ciliary, pterygopalatine, submandibular,* and *otic ganglia.* ↺ pp. 520, 522, 523 Short postganglionic fibers continue to their peripheral targets.

The vagus nerve alone provides roughly 75 percent of all parasympathetic outflow. It supplies preganglionic parasympathetic innervation to structures in the neck and in the thoracic and abdominopelvic cavities as distant as the distal portion of the large intestine. The many branches of the vagus nerve mingle with preganglionic and postganglionic fibers of the sympathetic division, forming plexuses comparable to those formed by spinal nerves innervating the limbs. We consider these plexuses in a later section.

The preganglionic fibers in the sacral segments of the spinal cord carry the sacral parasympathetic output. These fibers do not join the ventral roots of the spinal nerves. Instead, they form distinct **pelvic nerves,** which innervate intramural ganglia in the walls of the kidneys, urinary bladder, terminal portions of the large intestine, and sex organs.

Parasympathetic Activation

The functions of the parasympathetic division center on relaxation, food processing, and nutrient absorption. This division has been called the *anabolic system* (*anabole,* a raising up), because its stimulation leads to a general increase in the nutrient content of the blood. In response to this increase, cells throughout the body absorb nutrients and use them to support growth

Figure 16–6   **The Distribution of Parasympathetic Innervation.**

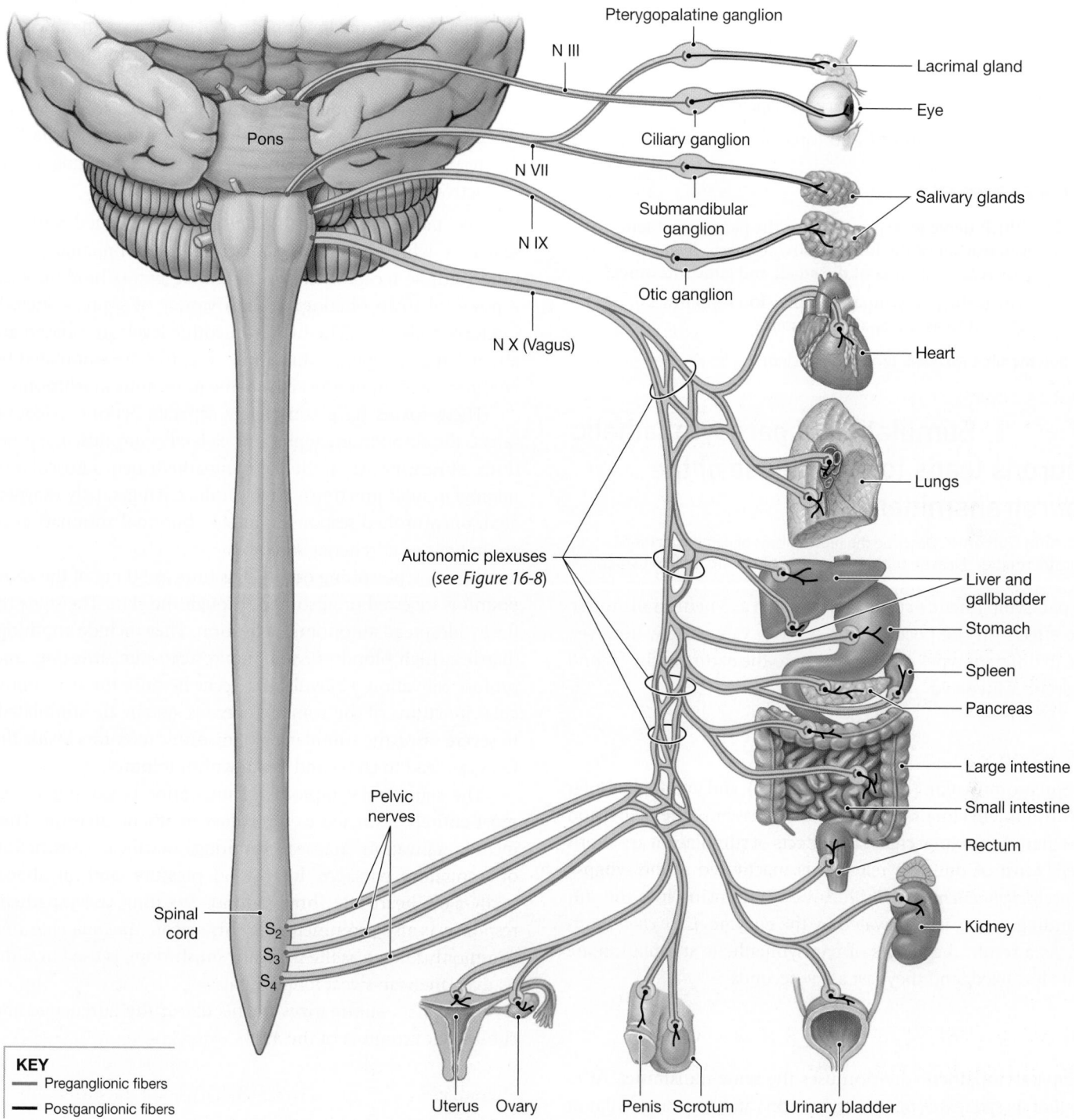

KEY
— Preganglionic fibers
— Postganglionic fibers

and cell division and to create energy reserves in the form of lipids or glycogen.

The major effects of the parasympathetic division include the following:

- Constriction of the pupils (to restrict the amount of light that enters the eyes) and focusing of the lenses of the eyes on nearby objects.

- Secretion by digestive glands, including salivary glands, gastric glands, duodenal glands, intestinal glands, the pancreas (exocrine and endocrine), and the liver.

- The secretion of hormones that promote the absorption and use of nutrients by peripheral cells.

- Changes in blood flow and glandular activity associated with sexual arousal.

16

- An increase in smooth muscle activity along the digestive tract.
- The stimulation and coordination of defecation.
- Contraction of the urinary bladder during urination.
- Constriction of the respiratory passageways.
- A reduction in heart rate and force of contraction.

> ### CHECKPOINT
>
> 8. Which nerve is responsible for the parasympathetic innervation of the lungs, heart, stomach, liver, pancreas, and parts of the small and large intestines?
>
> 9. Why is the parasympathetic division sometimes referred to as the anabolic system?
>
> See the blue Answers tab at the back of the book.

16-5 ▌ Stimulation of parasympathetic neurons leads to the release of the neurotransmitter ACh

Learning Outcome Describe the mechanisms of parasympathetic neurotransmitter release and their effects on target organs and tissues.

All parasympathetic neurons release ACh as a neurotransmitter. The effects on the postsynaptic cell can vary widely, however, due to different types of receptors or to the nature of the second messenger involved.

Neurotransmitter Release

The neuromuscular (neuron to muscle) and neuroglandular (neuron to effector) junctions of the parasympathetic division have narrow synaptic clefts. The effects of stimulation are short-lived. Most of the ACh released is inactivated at the synapse by *acetylcholinesterase (AChE)*. Any ACh diffusing into the surrounding tissues is inactivated by the enzyme *tissue cholinesterase*. As a result, the effects of parasympathetic stimulation are quite localized, and they last a few seconds.

Membrane Receptors and Responses

The parasympathetic division uses the same transmitter, ACh, at all of its synapses (neuron to neuron) and neuromuscular or neuroglandular junctions. Two types of ACh receptors occur on the postsynaptic membranes:

1. **Nicotinic** (nik-ō-TIN-ik) **receptors** occur on ganglion cells of both the parasympathetic and sympathetic divisions. They also occur at neuromuscular junctions of the somatic nervous system. ACh always causes excitation of the ganglionic neuron or muscle fiber. It works by opening chemically gated Na^+ channels in the postsynaptic membrane.

2. **Muscarinic** (mus-ka-RIN-ik) **receptors** occur at cholinergic neuromuscular or neuroglandular junctions in the parasympathetic division. They also occur at the few cholinergic junctions in the sympathetic division. Muscarinic receptors are G protein–coupled receptors. Their stimulation and G protein activation produces longer-lasting effects than does the stimulation of nicotinic receptors. The response can be excitatory or inhibitory, depending on the activation or inactivation of specific enzymes.

The names *nicotinic* and *muscarinic* originated with researchers who found that dangerous environmental toxins bind to these receptor sites. Nicotinic receptors bind *nicotine*, a powerful toxin obtained from a variety of sources, including tobacco leaves. (The highest nicotine levels in tobacco are about 3 mg per gram.) Muscarinic receptors are stimulated by *muscarine*, a toxin produced by some poisonous mushrooms.

These toxins have completely separate actions. Nicotine targets the autonomic ganglia and skeletal neuromuscular junctions. Muscarine acts at the parasympathetic neuromuscular or neuroglandular junctions. They produce dangerously exaggerated, uncontrolled responses due to abnormal stimulation of cholinergic or adrenergic receptors.

Nicotine poisoning occurs if as little as 50 mg of the compound is ingested or absorbed through the skin. The signs reflect widespread autonomic activation. They include vomiting, diarrhea, high blood pressure, rapid heart rate, sweating, and profuse salivation. Convulsions occur because the neuromuscular junctions of the somatic nervous system are stimulated. In severe cases, the stimulation of nicotinic receptors inside the CNS can lead to coma and death within minutes.

The signs and symptoms of muscarine poisoning are almost entirely restricted to the parasympathetic division. They include salivation, nausea, vomiting, diarrhea, constriction of respiratory passages, low blood pressure, and an abnormally slow heart rate (bradycardia). Sweating, a sympathetic response, is also prominent. For this reason, *Amanita muscaria*, commonly known as the fly-agaric mushroom, is used in some Native American sweat lodge rituals.

Table 16–1 summarizes details about the adrenergic and cholinergic receptors of the ANS.

Summary: The Parasympathetic Division

In summary:

- The parasympathetic division includes visceral motor nuclei associated with cranial nerves III, VII, IX, and X, and with sacral segments S_2–S_4.
- Ganglionic neurons are located in ganglia within or next to their target organs.
- The parasympathetic division innervates areas serviced by the cranial nerves and organs in the thoracic and abdominopelvic cavities.

Table 16–1	Adrenergic and Cholinergic Receptors of the ANS		
Receptor	**Location**	**Response**	**Mechanism**
ADRENERGIC			
α_1	Widespread, found in most tissues	Excitation, stimulation of metabolism	Enzyme activation; intracellular release of Ca^{2+}
α_2	Sympathetic neuromuscular or neuroglandular junctions	Inhibition of effector cell	Reduction of cAMP concentrations
α_2	Parasympathetic neuromuscular or neuroglandular junctions	Inhibition of neurotransmitter release	Reduction of cAMP concentrations
β_1	Heart, kidneys, liver, adipose tissue*	Stimulation, increased energy consumption	Enzyme activation
β_2	Smooth muscle in vessels of heart and skeletal muscle; smooth muscle layers in intestines, lungs, bronchi	Inhibition, relaxation	Enzyme activation
CHOLINERGIC			
Nicotinic	All autonomic synapses between preganglionic and ganglionic neurons; neuromuscular junctions of SNS	Stimulation, excitation; muscular contraction	Opening of chemically gated Na^+ channels
Muscarinic	All parasympathetic and cholinergic sympathetic neuromuscular or neuroglandular junctions	Variable	Enzyme activation causing changes in membrane permeability to K^+

*Adipocytes also contain an additional receptor type, β_3, not found in other tissues. Stimulation of β_3 receptors causes lipolysis.

16

- All parasympathetic neurons are cholinergic. Ganglionic neurons have nicotinic receptors, which are excited by ACh. Muscarinic receptors at neuromuscular or neuroglandular junctions produce either excitation or inhibition, depending on the action of enzymes triggered by G protein activation when ACh binds to the receptor.

- The effects of parasympathetic stimulation are generally brief and restricted to specific organs and sites.

CHECKPOINT

10. What neurotransmitter is released by all parasympathetic neurons?

11. Name the two types of cholinergic receptors on the postsynaptic membranes of parasympathetic neurons.

12. How would the stimulation of muscarinic receptors in cardiac muscle affect the heart?

See the blue Answers tab at the back of the book.

16-6 ┃ The sympathetic and parasympathetic divisions interact, creating dual innervation

Learning Outcome Discuss the functional significance of dual innervation and autonomic tone.

Figure 16–7 and **Table 16–2** compare key structural features of the sympathetic and parasympathetic divisions of the ANS. The differences in structure lead to differences in functions. The sympathetic division has widespread impact, reaching

organs and tissues throughout the body. The parasympathetic division innervates only visceral structures that are supplied by the cranial nerves or that lie within the abdominopelvic cavity.

Some organs are innervated by just one division, but most vital organs receive **dual innervation**, so they receive instructions from both the sympathetic and parasympathetic divisions. Where dual innervation exists, the two divisions commonly have opposing effects. Dual innervation with opposing effects is most obvious in the digestive tract, heart, and lungs. At other sites, the responses may be separate or complementary. **Table 16–3** provides a functional comparison of the two divisions, noting the effects of sympathetic or parasympathetic activity on specific organs and systems.

Anatomy of Dual Innervation

Parasympathetic postganglionic fibers from the ciliary, pterygopalatine, submandibular, and otic ganglia of the head travel by the cranial nerves to their peripheral destinations. Sympathetic innervation reaches the same structures by traveling directly from the superior cervical ganglia of the sympathetic chain.

In the thoracic and abdominopelvic cavities, the sympathetic postganglionic fibers mingle with parasympathetic preganglionic fibers, forming a series of nerve networks collectively called *autonomic plexuses*. They include the cardiac plexus, the pulmonary plexus, the esophageal plexus, the celiac plexus, the inferior mesenteric plexus, and the hypogastric plexus (**Figure 16–8**). Nerves leaving these networks travel with the blood vessels and lymphatic vessels that supply visceral organs.

Autonomic fibers entering the thoracic cavity intersect at the **cardiac plexus** and the **pulmonary plexus**. These plexuses contain sympathetic and parasympathetic fibers to the heart and lungs, respectively, as well as the parasympathetic ganglia whose output affects those organs. The **esophageal plexus** contains descending branches of the vagus nerves and splanchnic nerves leaving the sympathetic chain on either side.

Parasympathetic preganglionic fibers of the vagus nerves enter the abdominopelvic cavity with the esophagus. There the fibers enter the **celiac plexus**, also known as the *solar plexus.* The celiac plexus and associated smaller plexuses, such as the **inferior mesenteric plexus**, innervate viscera within the abdominal cavity. The **hypogastric plexus** innervates the digestive, urinary, and reproductive organs of the pelvic cavity. This plexus contains the parasympathetic outflow of the pelvic nerves, sympathetic postganglionic fibers from the inferior mesenteric ganglion, and splanchnic nerves from the sacral sympathetic chain.

Autonomic Tone

Even without stimuli, autonomic motor neurons show a resting level of spontaneous activity. This background level of activity determines an individual's **autonomic tone**. Autonomic tone is an important aspect of ANS function, just as muscle tone is a key aspect of SNS function. If a nerve is absolutely inactive under normal conditions, then all it can do is increase its activity on demand. But if the nerve maintains a background level of activity, then it can increase or decrease its activity, providing a greater range of control options.

Autonomic tone is significant where dual innervation occurs and the two ANS divisions have opposing effects. It is even more important where dual innervation does not occur. To

Figure 16–7 **Summary: The Anatomical Differences between the Sympathetic and Parasympathetic Divisions.**

Table 16–2	A Structural Comparison of the Sympathetic and Parasympathetic Divisions of the ANS	
Characteristic	**Sympathetic Division**	**Parasympathetic Division**
Location of CNS visceral motor neurons	Lateral gray horns of spinal segments T_1–L_2	Brain stem and spinal segments S_2–S_4
Location of PNS ganglia	Near vertebral column	Typically intramural
Preganglionic fibers		
Length	Relatively short	Relatively long
Neurotransmitter released	Acetylcholine	Acetylcholine
Postganglionic fibers		
Length	Relatively long	Relatively short
Neurotransmitter released	Normally NE; sometimes NO or ACh	Acetylcholine
Neuromuscular or neuroglandular junction	Varicosities and enlarged axonic terminals that release transmitter near target cells	Junctions that release transmitter to special receptor surface
Degree of divergence from CNS to ganglion cells	Approximately 1:32	Approximately 1:6
General function	Stimulates metabolism; increases alertness; prepares for emergency ("fight or flight")	Promotes relaxation, nutrient uptake, energy storage ("rest and digest")

Table 16–3	A Functional Comparison of the Sympathetic and Parasympathetic Divisions of the ANS	
Structure	**Sympathetic Effects (receptor or synapse type)**	**Parasympathetic Effects (all muscarinic receptors)**
EYE	Dilation of pupil (α_1); accommodation for distance vision (β_2)	Constriction of pupil; accommodation for close vision
Lacrimal glands	None (not innervated)	Secretion
SKIN		
Sweat glands	Increased secretion, palms and soles (α_1); generalized increase in secretion (cholinergic)	None (not innervated)
Arrector pili muscles	Contraction; erection of hairs (α_1)	None (not innervated)
CARDIOVASCULAR SYSTEM		
Blood vessels		None (not innervated)
To skin	Dilation (β_2 and cholinergic); constriction (α_1)	None (not innervated)
To skeletal muscles	Dilation (β_2 and cholinergic; nitroxidergic)	None (not innervated)
To heart	Dilation (β_2); constriction (α_1, α_2)	None (not innervated)
To lungs	Dilation (α_2); constriction (α_1)	None (not innervated)
To digestive viscera	Constriction (β_2); dilation (α_2)	None (not innervated)
To kidneys	Constriction, decreased urine production (α_1, α_2); dilation, increased urine production (β_1, β_2)	None (not innervated)
To brain	Dilation (cholinergic and nitroxidergic)	None (not innervated)
Veins	Constriction (α_1, β_2)	None (not innervated)
Heart	Increased heart rate, force of contraction, and blood pressure (α_1, β_1)	Decreased heart rate, force of contraction, and blood pressure
ENDOCRINE SYSTEM		
Adrenal gland	Secretion of epinephrine, norepinephrine by adrenal medulla	None (not innervated)
Posterior lobe of pituitary gland	Secretion of ADH (β_1, β_2)	None (not innervated)
Pancreas	Decreased insulin secretion (α_2)	Increased insulin secretion
Pineal gland	Increased melatonin secretion (β_1, β_2)	Inhibition of melatonin synthesis
RESPIRATORY SYSTEM		
Airways	Increased airway diameter (β_2)	Decreased airway diameter
Secretory glands	Mucous secretion (α_1)	None (not innervated)
DIGESTIVE SYSTEM		
Salivary glands	Production of viscous secretion (α_1, β_1) containing mucins and enzymes	Production of copious, watery secretion
Sphincters	Constriction (α_1)	Dilation
General level of activity	Decreased (α_2, β_2)	Increased
Secretory glands	Inhibition (α_2)	Stimulation
Liver	Glycogen breakdown, glucose synthesis and release (α_1, β_2)	Glycogen synthesis
Pancreas	Decreased exocrine secretion (α_1)	Increased exocrine secretion
SKELETAL MUSCLES		
	Increased force of contraction, glycogen breakdown (β_2)	None (not innervated)
	Facilitation of ACh release at neuromuscular junction (α_2)	None (not innervated)
ADIPOSE TISSUE	Lipolysis, fatty acid release (α_1, β_1, β_3)	None (not innervated)
URINARY SYSTEM		
Kidneys	Secretion of renin (β_1)	Uncertain effects on urine production
Urinary bladder	Constriction of internal sphincter; relaxation of urinary bladder (α_1, β_2)	Tensing of urinary bladder, relaxation of internal sphincter to eliminate urine
MALE REPRODUCTIVE SYSTEM	Increased glandular secretion and ejaculation (α_1)	Erection
FEMALE REPRODUCTIVE SYSTEM		
	Increased glandular secretion; contraction of pregnant uterus (α_1)	Variable (depending on hormones present)
	Relaxation of nonpregnant uterus (β_2)	Variable (depending on hormones present)

16

Figure 16–8   **The Autonomic Plexuses and Ganglia.**

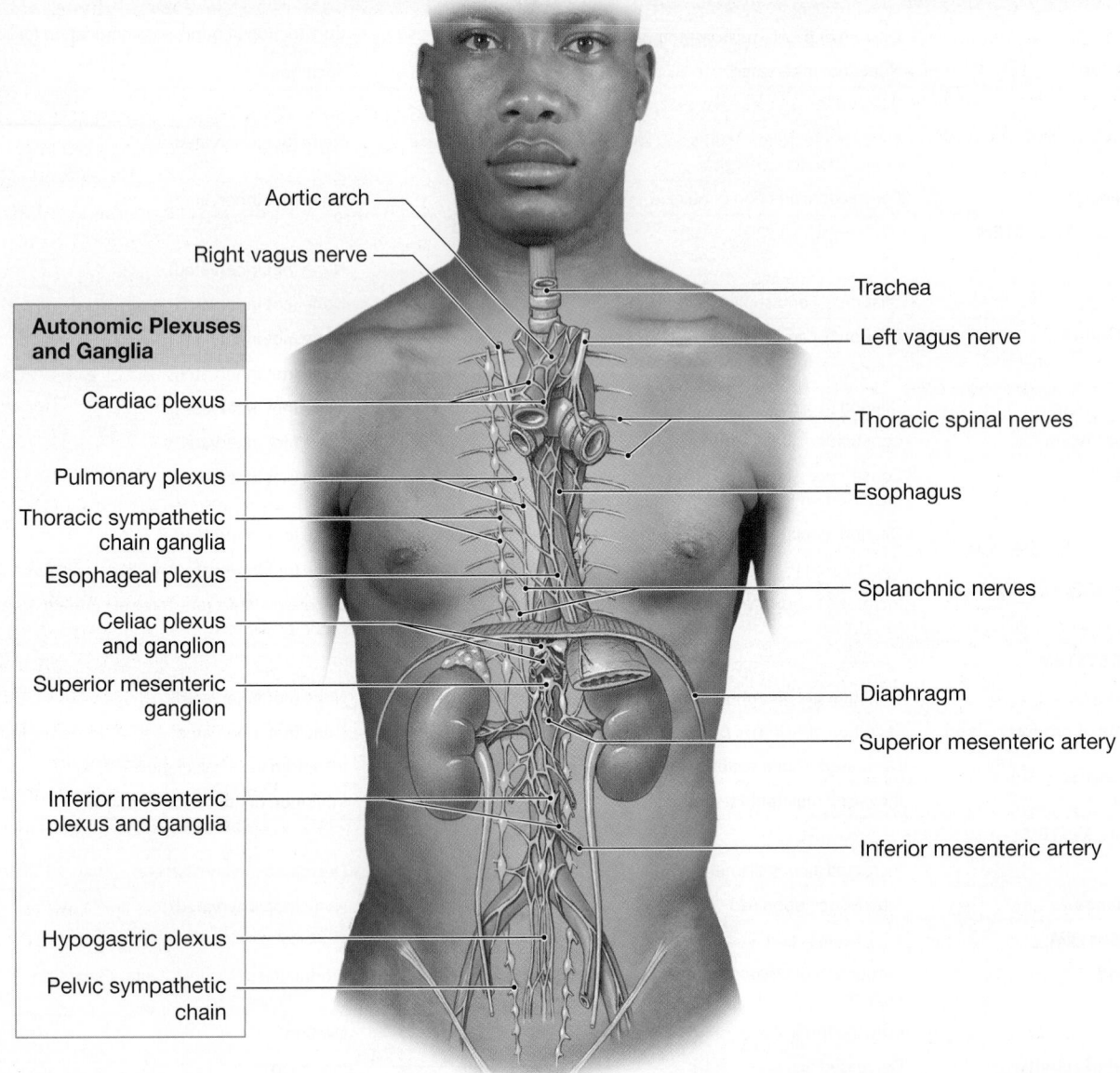

demonstrate how autonomic tone affects ANS function, let's consider one example of each situation.

The heart receives dual innervation. Recall that the heart consists of cardiac muscle tissue, and that specialized pacemaker cells trigger its contractions. ⮌ p. 168 The two autonomic divisions have opposing effects on heart function. Acetylcholine, released by postganglionic fibers of the parasympathetic division, causes a reduction in heart rate. Norepinephrine, released by varicosities of the sympathetic division, accelerates heart rate. Small amounts of both of these neurotransmitters are released continuously, creating autonomic tone. However, the parasympathetic division dominates under resting conditions. Heart rate can be controlled very precisely to meet the demands

of active tissues through small adjustments in the balance between parasympathetic stimulation and sympathetic stimulation. In a crisis, stimulation of the sympathetic innervation and inhibition of the parasympathetic innervation accelerate the heart rate to the maximum extent possible.

The sympathetic control of blood vessel diameter demonstrates how autonomic tone allows fine adjustment of peripheral activities in target organs that are innervated by only one ANS division. Blood flow to specific organs must be controlled to meet the tissue demands for oxygen and nutrients. When a blood vessel dilates, blood flow through it increases, and when it constricts, blood flow is reduced. Sympathetic postganglionic fibers release NE at the smooth muscle cells in the walls of peripheral vessels. This background sympathetic tone keeps these muscles partially contracted, so the blood vessels are ordinarily at about half their maximum diameter. When increased blood flow is needed, the rate of NE release decreases and sympathetic cholinergic fibers are stimulated. As a result, the smooth muscle cells relax, the vessels dilate, and blood flow increases. By adjusting sympathetic tone and the activity of cholinergic fibers, the sympathetic division can exert precise control of vessel diameter over its entire range.

> **CHECKPOINT**
>
> 13. **What effect would the loss of sympathetic tone have on blood flow to a tissue?**
>
> 14. **What physiological changes would you expect in a patient who is about to undergo a dental root canal and is quite anxious about the procedure?**
>
> See the blue Answers tab at the back of the book.

16-7 | Visceral reflexes play a role in the integration and control of autonomic functions

Learning Outcome Describe the hierarchy of interacting levels of control in the autonomic nervous system, including the significance of visceral reflexes.

Recall that centers involved in somatic motor control are found in all portions of the CNS. The lowest level of regulatory control consists of the lower motor neurons involved in cranial and spinal reflex arcs. The highest level consists of the pyramidal motor neurons of the primary motor cortex, operating with feedback from the cerebellum and basal nuclei.

Similarly, the ANS is also organized into a series of interacting levels. At the bottom are visceral motor neurons in the lower brain stem and spinal cord that are involved in cranial and spinal visceral reflexes. **Visceral reflexes** are autonomic reflexes initiated in the viscera. They provide automatic motor responses that can be modified, facilitated, or inhibited by higher centers, especially those of the hypothalamus.

For example, when a light is shone in one of your eyes, a visceral reflex constricts the pupils of *both* eyes (the *consensual light reflex*). The visceral motor commands are distributed by parasympathetic fibers. In darkness, your pupils dilate, but this *pupillary reflex* is directed by sympathetic fibers. However, the motor nuclei directing pupillary constriction or dilation are also controlled by hypothalamic centers concerned with emotional states. For example, when you are queasy or nauseated, your pupils constrict. When you are sexually aroused, your pupils dilate.

Visceral Reflexes

Each **visceral reflex arc** consists of a receptor, a sensory neuron, a processing center (one or more interneurons), and two visceral motor neurons (**Figure 16–9**). All visceral reflexes are polysynaptic. They can be long reflexes or short reflexes.

Long reflexes are the autonomic equivalents of the polysynaptic reflexes introduced in Chapter 13. ⟲ p. 480 Visceral sensory neurons deliver information to the CNS. It travels along the dorsal roots of spinal nerves, within the sensory branches of cranial nerves, and within the autonomic nerves that innervate visceral effectors. The processing steps involve interneurons within the CNS. The ANS then carries the motor commands to the appropriate visceral effectors. Long reflexes typically coordinate the activities of an entire organ.

Short reflexes bypass the CNS entirely. They involve sensory neurons and interneurons whose cell bodies lie in autonomic ganglia. These interneurons synapse on ganglionic neurons, and then postganglionic fibers distribute the motor commands. Short reflexes control very simple motor responses with localized effects. In general, short reflexes may control patterns of activity in one small part of a target organ.

In most organs, long reflexes are most important in regulating visceral activities, but this is not the case with the digestive

Figure 16–9 Visceral Reflexes. Visceral reflexes have the same basic components as somatic reflexes, but all visceral reflexes are polysynaptic. Note that short visceral reflexes bypass the CNS altogether.

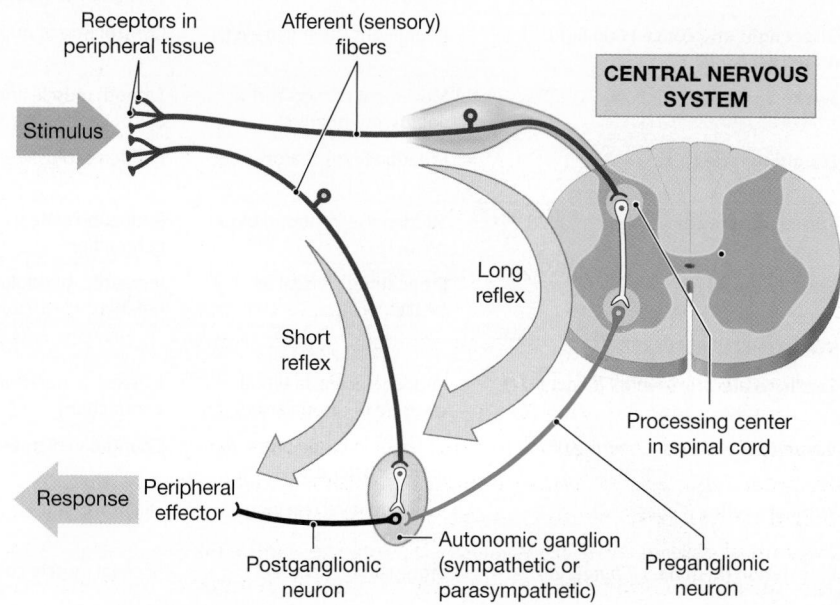

tract and its associated glands. In these areas, short reflexes provide most of the control and coordination for normal functioning. The neurons involved form the *enteric nervous system*, introduced on p. 562. The ganglia in the walls of the digestive tract contain the cell bodies of visceral sensory neurons, interneurons, and visceral motor neurons, and all their axons form extensive nerve nets. Parasympathetic innervation by the visceral motor neurons can stimulate and coordinate various digestive activities, but the enteric nervous system is quite capable of controlling digestive functions independent of the central nervous system. We consider the functions of the enteric nervous system further in Chapter 24.

As we examine other body systems in later chapters, you will come across many examples of autonomic reflexes involved in respiration, cardiovascular function, and other visceral activities. Look at Table 16–4 to preview some of the most important ones. Notice that the parasympathetic division participates in a variety of reflexes that affect individual organs and systems. This specialization reflects its relatively specific and restricted pattern of innervation.

In contrast, fewer sympathetic reflexes exist. The sympathetic division is typically activated as a whole. One reason is that it has such a high degree of divergence. Another reason is that the release of hormones by the adrenal medullae produces widespread peripheral effects.

Higher Levels of Autonomic Control

Centers in the brain stem that regulate specific visceral functions control the levels of activity in the sympathetic and parasympathetic divisions of the ANS. As in the SNS, in the ANS simple reflexes based in the spinal cord provide rapid and automatic responses to stimuli. Processing centers in the medulla oblongata coordinate more complex sympathetic and parasympathetic reflexes. In addition to the cardiovascular and respiratory centers, the medulla oblongata contains centers and nuclei involved with salivation, swallowing, digestive secretions, peristalsis, and urinary function. These centers are in turn subject to regulation by the hypothalamus. ⤺ p. 504

The term *autonomic* was originally applied because the regulatory centers involved with the control of visceral function were thought to operate autonomously—that is, independent of other CNS activities. This view has been drastically revised in light of later research. Because the hypothalamus interacts with all other portions of the brain, activity in the limbic system,

Table 16–4	Representative Visceral Reflexes		
Reflex	**Stimulus**	**Response**	**Comments**
PARASYMPATHETIC REFLEXES			
Gastric and intestinal reflexes (*Chapter 24*)	Pressure and physical contact	Smooth muscle contractions that propel food materials and mix with secretions	By vagus nerves
Defecation (*Chapter 24*)	Distention of rectum	Relaxation of internal anal sphincter	Requires voluntary relaxation of external anal sphincter
Urination (*Chapter 26*)	Distention of urinary bladder	Contraction of walls of urinary bladder; relaxation of internal urethral sphincter	Requires voluntary relaxation of external urethral sphincter
Direct light and consensual light reflexes (*Chapter 14*)	Bright light shining in eye(s)	Constriction of pupils of both eyes	
Swallowing reflex (*Chapter 24*)	Movement of food and liquids into pharynx	Smooth muscle and skeletal muscle contractions	Coordinated by medullary swallowing center
Coughing reflex (*Chapter 23*)	Irritation of respiratory tract	Sudden explosive ejection of air	Coordinated by medullary coughing center
Baroreceptor reflex (*Chapters 17, 20, 21*)	Sudden rise in carotid blood pressure	Reduction in heart rate and force of contraction	Coordinated in cardiac centers of medulla oblongata
Sexual arousal (*Chapter 28*)	Erotic stimuli (visual or tactile)	Increased glandular secretions, sensitivity, erection	
SYMPATHETIC REFLEXES			
Cardioacceleratory reflex (*Chapter 21*)	Sudden decline in blood pressure in carotid artery	Increase in heart rate and force of contraction	Coordinated in cardiac centers of medulla oblongata
Vasomotor reflexes (*Chapter 21*)	Changes in blood pressure in major arteries	Changes in diameter of peripheral vessels	Coordinated in vasomotor center in medulla oblongata
Pupillary reflex (*Chapter 17*)	Low light level reaching visual receptors	Dilation of pupil	
Ejaculation (in males) (*Chapter 28*)	Erotic stimuli (tactile)	Skeletal muscle contractions ejecting semen	

Table 16–5	A Comparison of the ANS and SNS	
Characteristic	**ANS**	**SNS**
Innervation	Visceral effectors, including cardiac muscle, smooth muscle, glands, fat cells	Skeletal muscles
Activation	In response to sensory stimuli or from commands of higher centers	In response to sensory stimuli or from commands of higher centers
Relay and processing centers	Brain stem	Brain stem and thalamus
Headquarters	Hypothalamus	Cerebral cortex
Feedback received from	Limbic system and thalamus	Cerebellum and basal nuclei
Control method	Adjustment of activity in brain stem processing centers that innervate preganglionic neurons	Direct (corticospinal) and indirect (medial and lateral) pathways that innervate lower motor neurons
Reflexes	Polysynaptic (short and long)	Monosynaptic and polysynaptic (always long)

thalamus, or cerebral cortex can have dramatic effects on autonomic function. For example, what happens when you become angry? Your heart rate accelerates, your blood pressure rises, and your respiratory rate increases. What happens when you think about your next meal? Your stomach "growls" and your mouth waters.

The Integration of SNS and ANS Activities

Figure 16–10 and **Table 16–5** show how the activities of the somatic nervous system (discussed in Chapter 15) and the autonomic nervous system are integrated. We have considered somatic and visceral motor pathways separately, but the two have many parallels, in terms of both organization and function. Integration takes place at the level of the brain stem, and both systems are under the influence of higher centers.

CHECKPOINT

15. Define visceral reflex.

16. Luke has a brain tumor that is interfering with the function of his hypothalamus. Explain why this tumor would interfere with autonomic function.

See the blue Answers tab at the back of the book.

Figure 16–10   A Comparison of Somatic and Autonomic Function. The SNS and ANS have parallel organization and are integrated at the level of the brain stem. Blue arrows indicate ascending sensory information; red arrows, descending motor commands; dashed lines indicate pathways of communication and feedback among higher centers.

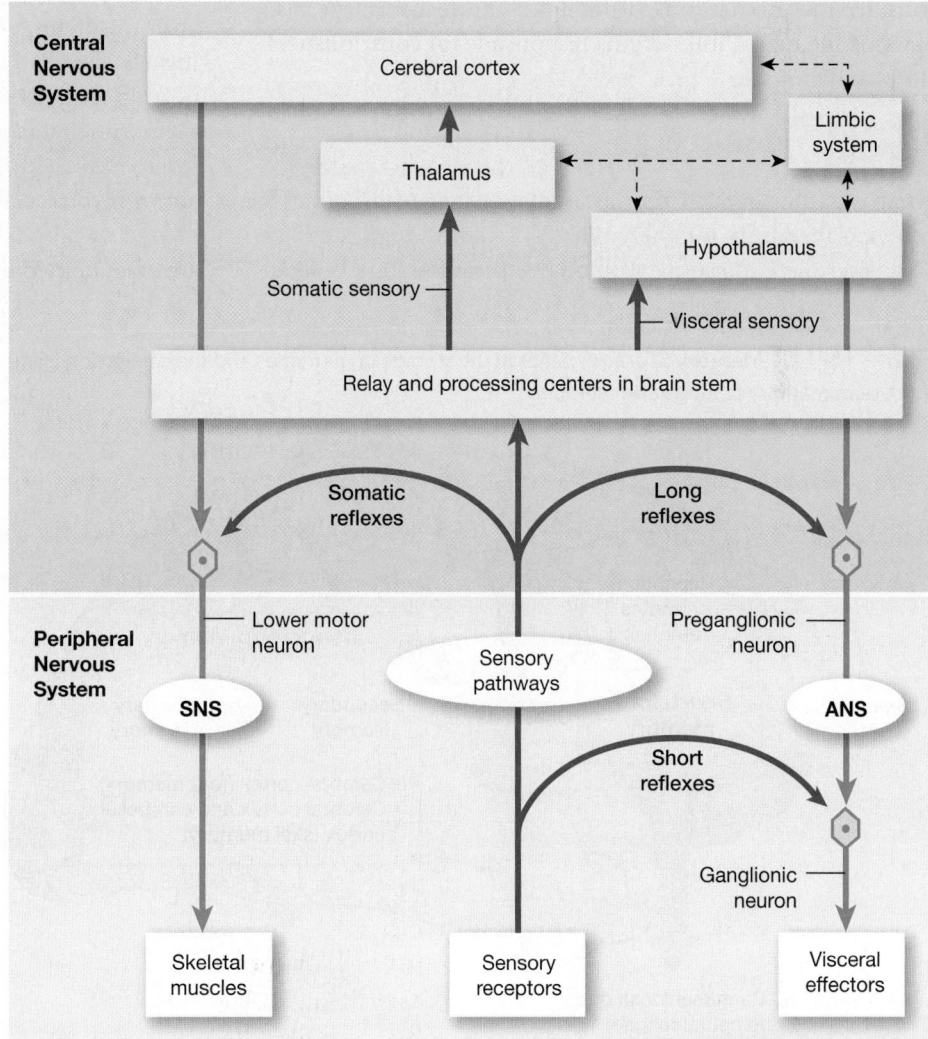

16-8 ❙ Higher-order functions include memory and states of consciousness

Learning Outcome Explain how memories are created, stored, and recalled, and distinguish among the levels of consciousness and unconsciousness.

Higher-order functions share three characteristics:

1. The cerebral cortex is required for their performance. They involve complex interactions among areas of the cortex and between the cerebral cortex and other areas of the brain.

2. They involve both conscious and unconscious information processing.

3. They are not part of the programmed "wiring" of the brain. For this reason, higher-order functions are subject to adjustment over time.

In Chapter 14, we considered functional areas of the cerebral cortex and the regional specializations of the left and right cerebral hemispheres. ⤺ pp. 512–513 In this section, we consider the mechanisms of memory and learning. We also describe the neural interactions responsible for consciousness, sleep, and arousal.

Memory

What was the topic of the last sentence you read? What is your social security number? What does a hot dog taste like? To answer these questions, you access *memories,* stored bits of information gathered through experience. **Fact memories** are specific bits of information, such as the color of a stop sign or the smell of a perfume. **Skill memories** are learned motor behaviors. You can probably remember how to light a match or open a screw-top jar, for example. With repetition, skill memories become incorporated at the unconscious level. Examples include the complex motor patterns involved in skiing, playing the violin, and similar activities. Skill memories related to programmed behaviors, such as eating, are stored in appropriate portions of the brain stem. Complex skill memories involve the integration of motor patterns in the basal nuclei, cerebral cortex, and cerebellum.

Two classes of memories are recognized. **Short-term memories** do not last long, but while they persist the information can be recalled immediately. Short-term memories contain small bits of information, such as a person's name or a telephone number. Repeating a phone number or other bit of information reinforces the original short-term memory and helps to make sure it gets converted to a long-term memory.

Long-term memories last much longer, in some cases for an entire lifetime. The conversion from short-term to long-term memory is called **memory consolidation**. There are two types of long-term memory: (1) *Secondary memories* are long-term memories that fade with time and may require considerable effort to recall. (2) *Tertiary memories* are long-term memories that are with you for a lifetime, such as your name or the contours of your own body. Relationships among these memory classes are diagrammed in **Figure 16–11**.

Figure 16–11 **Memory Storage.** Steps in the storage of memories and the conversion from short-term memory to long-term memory.

Sensory input

Repetition promotes retention

Short-Term Memory

Consolidation

Long-Term Memory

Secondary Memory → Tertiary Memory

• Cerebral cortex (fact memory)
• Cerebral cortex and cerebellar cortex (skill memory)

Permanent loss due to neural fatigue, shock, interference by other stimuli

Temporary loss

Permanent loss

Brain Regions Involved in Memory Consolidation and Access

The amygdaloid body (amygdala) and the hippocampus, two components of the limbic system (Figure 14–12, p. 867), are essential to memory consolidation. Damage to the hippocampus leads to an inability to convert short-term memories to new long-term memories, although existing long-term memories remain intact and accessible. Tracts leading from the amygdaloid body to the hypothalamus may link memories to specific emotions.

The **nucleus basalis**, a cerebral nucleus near the diencephalon, plays an uncertain role in memory storage and retrieval. Tracts connect this nucleus with the hippocampus, amygdaloid body, and all areas of the cerebral cortex. Damage to this nucleus is associated with changes in emotional states, memory, and intellectual function (as we will see in the discussion of Alzheimer's disease later in this chapter).

Most long-term memories are stored in the cerebral cortex. Conscious motor and sensory memories are referred to the appropriate association areas. For example, visual memories are stored in the visual association area, and memories of voluntary motor activity are stored in the premotor cortex. Special portions of the occipital and temporal lobes are crucial to the memories of faces, voices, and words.

In at least some cases, a specific memory probably depends on the activity of a single neuron. For example, in one portion of the temporal lobe an individual neuron responds to the sound of one word and ignores others. A specific neuron may also be activated by the proper combination of sensory stimuli associated with a particular individual, such as your grandmother. As a result, these neurons are called "grandmother cells."

Information on one subject is parceled out to many different regions of the brain. Your memories of cows are stored in the visual association area (what a cow looks like, that the letters *c-o-w* mean "cow"), the auditory association area (the "moo" sound and how the word *cow* sounds), the speech center (how to say the word *cow*), and the frontal lobes (how big cows are, what they eat). Related information, such as how you feel about cows and what milk tastes like, is stored in other locations. If one of those storage areas is damaged, your memory will be incomplete in some way. How these memories are accessed and assembled on demand remains a mystery.

Cellular Mechanisms of Memory Formation and Storage

Memory consolidation at the cellular level involves anatomical and physiological changes in neurons and synapses. For legal, ethical, and practical reasons, scientists do not conduct much research on these mechanisms with human subjects. Research on other animals, commonly those with relatively simple nervous systems, has indicated that the following mechanisms may be involved:

- *Increased Neurotransmitter Release.* A synapse that is frequently active increases the amount of neurotransmitter it stores, and it releases more with each stimulation. The more neurotransmitter released, the greater the effect on the postsynaptic neuron.

- *Facilitation at Synapses.* When a neural circuit is repeatedly activated, the axon terminals begin continuously releasing neurotransmitter in small quantities. The neurotransmitter binds to receptors on the postsynaptic membrane, producing a graded depolarization that brings the membrane closer to threshold. The facilitation that results affects all neurons in the circuit.

- *The Formation of Additional Synaptic Connections.* Evidence indicates that when one neuron repeatedly communicates with another, the axon tip branches and forms additional synapses on the postsynaptic neuron. As a result, the presynaptic neuron has a greater effect on the membrane potential of the postsynaptic neuron.

These processes create anatomical changes that facilitate communication along a specific neural circuit. This facilitated

16

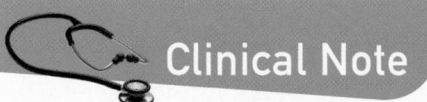

Clinical Note

Amnesia **Amnesia** is the loss of memory as a result of trauma or disease. The type of memory loss depends on the specific regions of the brain affected. For example, damage to the auditory association areas can make it difficult to remember sounds. Recovery may be partial or complete. In *retrograde amnesia* (*retro-*, behind), the person loses memories of events that occurred before the trauma or the onset of the disease. Some degree of retrograde amnesia commonly follows a head injury. After a fall or a car wreck, many victims are unable to remember the moments preceding the accident. In *anterograde amnesia* (*antero-*, ahead), a person is unable to store additional memories, but earlier memories remain intact and accessible. The problem appears to involve an inability to generate long-term memories. Some degree of anterograde amnesia is a common sign of senility, a condition discussed

further on p. 583. At least two drugs—*diazepam* (*Valium*) and *triazolam* (*Halcion*)—have been known to cause brief periods of anterograde amnesia. Brain injuries can cause more prolonged memory problems. A person with permanent anterograde amnesia lives in surroundings that are always new. The person can read magazines, chuckle over them, and reread them a few minutes later with equal pleasure. Clinicians must introduce themselves at every meeting, even if they have been treating the patient for years.

In *post-traumatic amnesia (PTA)*, a head injury produces a combination of retrograde and anterograde amnesias. The individual can neither remember the past nor consolidate memories of the present. The extent and duration of the amnesia varies with the severity of the injury.

communication is thought to be the basis of memory storage. A single circuit that corresponds to a single memory has been called a **memory engram**. This definition is based on function rather than structure. We know too little about the organization and storage of memories to be able to describe the neural circuits involved. Memory engrams form as the result of experience and repetition. Repetition is crucial—that's why you probably need to read these chapters more than once before an exam.

Efficient conversion of a short-term memory into a memory engram takes time, usually at least an hour. Whether that conversion will occur depends on several factors. They include the nature, intensity, and frequency of the original stimulus. Very strong, repeated, or exceedingly pleasant or unpleasant events are most likely to be converted to long-term memories. Drugs that stimulate the CNS, such as caffeine and nicotine, may enhance memory consolidation through facilitation. We discussed the membrane effects of those drugs in Chapter 12. ↺ pp. 445, 448

The hippocampus plays a key role in consolidating memories. The mechanism remains unknown, but it is linked to the presence of *NMDA* (N-methyl D-aspartate) *receptors*, which are chemically gated calcium channels. When activated by the neurotransmitter *glutamate*, the gates open and calcium enters the cell. Blocking NMDA receptors in the hippocampus prevents long-term memory formation.

States of Consciousness

The difference between a conscious individual and an unconscious one might seem obvious. A conscious individual is alert and attentive, and an unconscious individual is not. Yet there are many degrees of both states. *Conscious* implies an awareness of and attention to external events and stimuli, but a healthy conscious person can be nearly asleep, wide awake, or high-strung and jumpy. *Unconscious* can refer to conditions ranging from the deep, unresponsive state induced by anesthesia before major surgery, to deep sleep, to the light, drifting "nod" that occasionally plagues students who are reading anatomy and physiology textbooks.

A person's degree of wakefulness at any moment indicates the level of ongoing CNS activity. When you are asleep, you are unconscious but can still be awakened by normal sensory stimuli. Healthy individuals cycle between the alert, conscious state and sleep each day. When CNS function becomes abnormal or depressed, the state of wakefulness can be affected. An individual in a *coma*, for example, is unconscious and cannot be awakened, even by strong stimuli.

Sleep

We recognize two general levels of sleep, each with characteristic patterns of brain wave activity (**Figure 16–12a**):

Figure 16–12   **States of Sleep.**

a An EEG of the various sleep states. The EEG pattern during REM sleep resembles the alpha waves typical of awake adults.

b Typical pattern of sleep stages in a healthy young adult during a single night's sleep.

1. In **deep sleep**, also called *slow wave* or *non-REM (NREM) sleep*, your entire body relaxes, and activity at the cerebral cortex is at a minimum. Heart rate, blood pressure, respiratory rate, and energy use decline by up to 30 percent.

2. During **rapid eye movement (REM) sleep**, you dream actively and your blood pressure and respiratory rate change. Although the EEG resembles that of the awake state, you become even less receptive to outside stimuli than in deep sleep, and your muscle tone decreases markedly. Intense inhibition of somatic motor neurons probably prevents you from acting out the responses you envision while dreaming. The neurons controlling your eye muscles escape this inhibitory influence, and your eyes move rapidly as dream events unfold.

Periods of REM and deep sleep alternate throughout the night, beginning with deep sleep for about an hour and a half (Figure 16–12b). Rapid eye movement periods initially average about 5 minutes, but over an eight-hour night they gradually increase to about 20 minutes. Each night we probably spend less than two hours dreaming, but variation among individuals is significant. For example, children spend more time in REM sleep than do adults, and extremely tired individuals have very short and infrequent REM periods.

Sleep produces only minor changes in the physiological activities of other organs and systems, and none of these changes appear to be essential to normal function. The significance of sleep must lie in its impact on the CNS, but the physiological or biochemical basis remains to be determined. We do know that protein synthesis in neurons increases during sleep.

Extended periods without sleep lead to a variety of disturbances in mental function. Roughly 25 percent of the U.S. population experiences some form of *sleep disorder.* Examples include abnormal patterns or duration of REM sleep or unusual

behaviors during sleep, such as sleepwalking. In some cases, these problems affect the individual's conscious activities. Slowed reaction times, irritability, and behavioral changes may result. Memory loss has also been linked to sleep disorders.

Arousal and the Reticular Activating System

Arousal, or awakening from sleep, appears to be one of the functions of the reticular formation. The reticular formation is especially well suited for providing "watchdog" services, because it has extensive interconnections with the sensory, motor, and integrative nuclei and pathways all along the brain stem.

Your state of consciousness results from complex interactions between the reticular formation and the cerebral cortex. One of the most important brain stem components is a diffuse network in the reticular formation known as the **reticular activating system (RAS).** ⤸ p. 502 This network extends from the medulla oblongata to the midbrain (Figure 16–13). The output of the RAS projects to thalamic nuclei that influence large areas of the cerebral cortex. When the RAS is inactive, so is the cerebral cortex, and stimulation of the RAS produces a widespread activation of the cerebral cortex.

The midbrain portion of the RAS appears to be the "headquarters" of the system. Stimulating this area produces the most pronounced and long-lasting effects on the cerebral cortex. Stimulating other portions of the RAS seems to have an effect only to the degree that it changes the activity of the midbrain

Clinical Note

Categorizing Nervous System Disorders Neural tissue is extremely delicate. The characteristics of its extracellular environment must be kept within narrow homeostatic limits. Signs and symptoms of neurological disorders appear when homeostatic regulatory mechanisms break down. This may happen under the stress of genetic or environmental factors, infection, or trauma.

Literally hundreds of disorders affect the nervous system. They can be categorized as:

- *Infections,* which include diseases such as rabies, meningitis, and polio
- *Congenital disorders,* such as spina bifida and hydrocephalus
- *Degenerative disorders,* such as Parkinson's disease and Alzheimer's disease
- *Tumors* of neural origin
- *Trauma,* such as spinal cord injuries and concussions
- *Toxins,* such as heavy metals and the neurotoxins found in certain seafoods
- *Secondary disorders,* which are problems resulting from dysfunction in other systems, for example, strokes and several demyelination disorders

A standard physical examination includes a neurological component. The physician checks the general status of the CNS and PNS. In *neurological examinations,* a physician attempts to trace the location of a specific problem by evaluating the sensory, motor, behavioral, and cognitive functions of the nervous system.

Figure 16–13   The Reticular Activating System. The midbrain "headquarters" of the reticular formation receives collateral inputs from a variety of sensory pathways. Stimulation of this region produces arousal and heightened states of attentiveness.

region. The greater the stimulation to the midbrain region of the RAS, the more alert and attentive the individual will be to incoming sensory information. The thalamic nuclei associated with the RAS may also play an important role in focusing attention on specific mental processes.

Sleep may be ended by any stimulus sufficient to activate the reticular formation and RAS. Arousal occurs rapidly, but the effects of a single stimulation of the RAS last less than a minute. After that, consciousness can be maintained by positive feedback, because activity in the cerebral cortex, basal nuclei, and sensory and motor pathways will continue to stimulate the RAS.

After many hours of activity, the reticular formation becomes less responsive to stimulation. You become less alert and more lethargic. The precise mechanism remains unknown, but neural fatigue probably plays a relatively minor role in the decreasing RAS activity. Evidence suggests that the regulation of sleep–wake cycles involves an interplay between brain stem nuclei that use different neurotransmitters. One group of nuclei stimulates the RAS with norepinephrine and maintains the awake, alert state. Another group depresses RAS activity with serotonin, promoting deep sleep. These "dueling" nuclei are located in the brain stem.

CHECKPOINT

17. List three characteristics of higher-order functions.

18. As you recall facts while you take your A&P test, which type of memory are you using?

19. Name the two general levels of sleep that have characteristic patterns of brain wave activity.

20. You are asleep. What would happen if your reticular activating system (RAS) were suddenly stimulated?

See the blue Answers tab at the back of the book.

16-9 | Neurotransmitters influence brain chemistry and behavior

Learning Outcome Describe some of the ways in which the interactions of neurotransmitters influence brain function.

Changes in the normal balance between two or more neurotransmitters can profoundly affect brain function. Recall from the previous section that the interplay between populations of neurons releasing norepinephrine and serotonin appears to be involved in regulating sleep–wake cycles. Another example concerns *Huntington's disease*. The primary problem in this inherited disease is the destruction of ACh-secreting and GABA-secreting neurons in the basal nuclei. The reason for this destruction is unknown. Symptoms appear as the basal nuclei and frontal lobes slowly degenerate. An individual with

Huntington's disease has difficulty controlling movements, and intellectual abilities gradually decline. In many cases, the importance of a specific neurotransmitter was revealed while searching for the mechanism of action for other drugs. Three examples follow.

1. It is now clear that an extensive network of tracts delivers serotonin to nuclei and higher centers throughout the brain, and variations in serotonin levels affect sensory interpretation and emotional states. Compounds that enhance the effects of serotonin produce hallucinations. For instance, *lysergic acid diethylamide (LSD)* is a powerful hallucinogenic drug that activates serotonin receptors in the brain stem, hypothalamus, and limbic system. Compounds that merely enhance the effects of serotonin also produce hallucinations, whereas compounds that inhibit serotonin production or block its action cause severe depression and anxiety. An effective antidepressive drug now in widespread use, *fluoxetine (Prozac)*, slows the removal of serotonin at synapses, causing an increase in serotonin concentrations at the postsynaptic membrane. Such drugs are classified as selective serotonin reuptake inhibitors (SSRIs). Other important SSRIs include *Celexa, Luvox, Paxil,* and *Zoloft.*

2. Norepinephrine is another important neurotransmitter with pathways throughout the brain. Drugs that stimulate NE release cause exhilaration, and those that depress NE release cause depression. One inherited form of depression has been linked to a defective enzyme involved in NE synthesis.

3. Disturbances in dopamine transmission have been linked to several neurological disorders. We have already seen that inadequate dopamine causes the motor problems of Parkinson's disease. ↺ p. 512 Excessive production of dopamine may be associated with *schizophrenia*, a psychological disorder marked by pronounced disturbances of mood, thought patterns, and behavior. Amphetamines, or "speed," stimulate dopamine secretion and, in large doses, can produce symptoms resembling those of schizophrenia. Dopamine is thus important not only in the nuclei involved in the control of intentional movements, but in many other centers of the diencephalon and cerebrum.

CHECKPOINT

21. What would be an effect of a drug that substantially increases the amount of serotonin released in the brain?

22. Identify the neurotransmitters thought to be involved with the regulation of sleep-wake cycles.

23. Amphetamines stimulate the secretion of which neurotransmitter?

See the blue Answers tab at the back of the book.

16-10 ❘ Aging produces various structural and functional changes in the nervous system

Learning Outcome Summarize the effects of aging on the nervous system and give examples of interactions between the nervous system and other organ systems.

The aging process affects all body systems, and the nervous system is no exception. Anatomical and physiological changes probably begin by age 30 and accumulate over time. An estimated 85 percent of people above age 65 lead relatively normal lives, but they exhibit noticeable changes in mental performance and in CNS function. Common age-related anatomical changes in the nervous system include the following:

■ *A Reduction in Brain Size and Weight.* This reduction results primarily from a decrease in the volume of the cerebral cortex. The brains of elderly individuals have narrower gyri and wider sulci than do those of young people, and the subarachnoid space is larger.

■ *A Reduction in the Number of Neurons.* Brain shrinkage has been linked to a loss of cortical neurons, although evidence indicates that neurons are not lost (at least to the same degree) in brain stem nuclei.

■ *A Decrease in Blood Flow to the Brain.* With age, fatty deposits gradually build up in the walls of blood vessels. Just as a clog in a drain reduces water flow, these deposits reduce the rate of blood flow through arteries. (This process, called *arteriosclerosis*, affects arteries throughout the body, as we discuss further in Chapter 21.) Even if the reduction in blood flow is not large enough to damage neurons, arteriosclerosis increases the chances that the affected vessel wall will rupture, damaging the surrounding neural tissue and producing signs and symptoms of a *cerebrovascular accident (CVA)*, also called a stroke.

■ *Changes in the Synaptic Organization of the Brain.* In many areas, the number of dendritic branches, spines, and interconnections appears to decrease. Synaptic connections are lost, and the rate of neurotransmitter production declines.

■ *Intracellular and Extracellular Changes in CNS Neurons.* Many neurons in the brain accumulate abnormal intracellular deposits, including lipofuscin and neurofibrillary tangles. **Lipofuscin** is a granular pigment with no known function. **Neurofibrillary tangles** are masses of neurofibrils that form dense mats inside the cell body and axon. **Plaques** are extracellular accumulations of fibrillar proteins, surrounded by abnormal dendrites and axons. Both plaques and tangles contain deposits of several peptides—primarily two forms of **amyloid** β **(Aβ) protein**, fibrillar and soluble. They appear in brain regions such as the hippocampus, specifically associated with memory

16

Clinical Note

Alzheimer's Disease **Alzheimer's disease (AD)** is a progressive disorder characterized by the loss of higher-order cerebral functions. It is the most common cause of senility. Symptoms may appear at 50–60 years of age or later, although the disease occasionally affects younger people. Alzheimer's disease has widespread impact. An estimated 5.2 million people in the United States—including about 11 percent of those over age 65, and 38 percent over age 85—have some form of the condition. It causes approximately 100,000 deaths each year.

The link is uncertain, but microscopic examination of the brain shows abnormal plaques and neurofibrillary tangles in regions involved with memory, emotions, and intellectual function. It is still unknown whether these deposits cause AD or are secondary signs of ongoing metabolic alterations with an environmental, hereditary, or infectious basis.

Genetic factors certainly play a major role. The late-onset form of AD has been traced to a gene on chromosome 19 that codes for proteins involved in cholesterol transport. Fewer than 5 percent of people with Alzheimer's disease have the early-onset form, which may develop before age 50. This form of AD has been linked to genes on chromosomes 1, 14, 19, and 21. Interestingly, the majority of individuals with *Down's syndrome* develop AD relatively early in life. (Down's syndrome results from an extra copy of chromosome 21. We discuss this condition further in Chapter 29.)

There is no cure for Alzheimer's disease. A few medications and supplements slow its progress in many patients. Diagnosis involves excluding metabolic disorders, chemical/drug interactions, nutritional deficiencies, and anatomical conditions that can mimic dementia; taking a detailed history and doing a thorough physical exam; and evaluating mental function and mood (depression may mimic dementia).

SYSTEM INTEGRATOR

Body System ——→ Nervous System

Nervous System ——→ Body System

Integumentary
Provides sensations of touch, pressure, pain, vibration, and temperature; hair provides some protection and insulation for skull and brain; protects peripheral nerves

Integumentary
Controls contraction of arrector pili muscles and secretion of sweat glands

Page 200

Skeletal
Provides calcium ions for neural function; protects brain and spinal cord

Skeletal
Controls skeletal muscle contractions that results in bone thickening and maintenance and determine bone position

Page 311

Muscular
Facial muscles express emotional state; intrinsic laryngeal muscles permit communication; muscle spindles provide proprioceptive sensations

Muscular
Controls skeletal muscle contractions; coordinates respiratory and cardiovascular activities

Page 406

The Nervous System

The nervous system is closely integrated with other body systems. Every moment of your life, billions of neurons in your nervous system are exchanging information across trillions of synapses and performing the most complex integrative functions in the body. As part of this process, the nervous system monitors all other systems and issues commands that adjust their activities. However, the significance and impact of these commands varies greatly from one system to another. The normal functions of the muscular system, for example, simply cannot be performed without instructions from the nervous system. By contrast, the cardiovascular system is relatively independent—the nervous system merely coordinates and adjusts cardiovascular activities to meet the circulatory demands of other systems. In the final analysis, the nervous system is like the conductor of an orchestra, directing the rhythm and balancing the performances of each section to produce a symphony, instead of simply a very loud noise.

Endocrine Page 673

Cardiovascular Page 802

Lymphatic Page 850

Respiratory Page 900

Digestive Page 955

Urinary Page 1036

Reproductive Page 1116

Figure 16–14 diagrams the functional relationships between the nervous system and other body systems we have studied so far.

processing. Their significance is unknown. Evidence indicates that they appear in all aging brains. In excess, they seem to be associated with clinical abnormalities.

The anatomical changes associated with aging of the nervous system affect all neural functions. For example, memory consolidation typically becomes more difficult, and secondary memories, especially those of the recent past, become harder to access. The sensory systems of elderly people—notably, hearing, balance, vision, smell, and taste—become less sensitive. Lights must be brighter, sounds louder, and smells stronger before they are perceived. Reaction rates are slowed, and reflexes—even some withdrawal reflexes—weaken or disappear. The precision of motor control decreases, and it takes longer for an elderly person to perform a given motor pattern than it did 20 years earlier.

For roughly 85 percent of the elderly population, these changes do not interfere with their abilities to function in society. But for reasons yet unknown, some elderly individuals become incapacitated by progressive CNS changes. These degenerative changes, which can include memory loss, anterograde amnesia, and emotional disturbances, are often lumped together as *senile dementia,* or **senility**. By far the most common and disabling form of senile dementia is Alzheimer's disease. The relationships between the nervous system and the integumentary, skeletal, and muscular systems are shown in Figure 16–14.

CHECKPOINT

24. Memory retrieval deteriorates with aging. What are some possible reasons for these changes?

25. Identify several common age-related anatomical changes in the nervous system.

26. Name the most common form of senile dementia.

27. Identify the relationships between the nervous system and the body systems studied so far.

See the blue Answers tab at the back of the book.

Related Clinical Terms

alpha₁-receptor agonists: Drugs used to treat hypotension (low-blood pressure) by stimulating α_1 receptors to cause vasoconstriction of blood vessels.

alpha₂-receptor agonists: Drugs used to treat hypertension (high blood pressure) by stimulating α_2 adrenergic receptors to inhibit sympathetic vasomotor centers.

beta-adrenergic blockers: Drugs that decrease heart rate and force of contraction, lowering peripheral blood pressure by acting on beta adrenergic receptors to diminish the effects of epinephrine.

parasympathetic blocking agents: Drugs that target the muscarinic receptors at neuromuscular or neuroglandular junctions.

parasympathomimetic drugs: Drugs that mimic parasympathetic stimulation and increase the activity along the digestive tract.

sympathetic blocking agents: Drugs that bind to receptor sites, preventing a normal response to neurotransmitters or sympathomimetic drugs.

sympathomimetic drugs: Drugs that mimic the effects of sympathetic stimulation.

Chapter Review

Study Outline

MasteringA&P®

Access more chapter study tools online in the MasteringA&P Study Area:

- Chapter Quizzes, Chapter Practice Test, Art-labeling Activities, Animations, MP3 Tutor Sessions, and Clinical Case Studies
- Practice Anatomy Lab PAL™
- Interactive Physiology iP®
- A&P Flix A&PFlix
- PhysioEx PhysioEx

▌ **An Introduction to the Autonomic Nervous System and Higher-Order Functions** p. 558

1. The autonomic nervous system (ANS) adjusts our basic life support systems without conscious control.

16-1 ▌ **The autonomic nervous system is involved in the unconscious regulation of visceral functions and has sympathetic and parasympathetic divisions** p. 558

2. The **autonomic nervous system (ANS)** coordinates cardiovascular, respiratory, digestive, urinary, and reproductive functions.

3. **Preganglionic neurons** in the CNS send axons to synapse on **ganglionic neurons** in **autonomic ganglia** outside the CNS. (*Figure 16–1*)

16

4. Preganglionic fibers from the thoracic and lumbar segments form the **sympathetic division**, or *thoracolumbar division* ("fight or flight" system), of the ANS. Preganglionic fibers leaving the brain and sacral segments form the **parasympathetic division**, or *craniosacral division* ("rest and digest" system) (*Spotlight Figure 16–2*).

16-2 ▌ The sympathetic division consists of preganglionic neurons and ganglionic neurons involved in using energy and increasing metabolic rate p. 562

5. The sympathetic division consists of preganglionic neurons between segments T_1 and L_2, ganglionic neurons in ganglia near the vertebral column, and specialized neurons in the adrenal glands. (*Spotlight Figure 16–2*)

6. The two types of sympathetic ganglia are **sympathetic chain ganglia** (*paravertebral ganglia*) and **collateral ganglia** (*prevertebral ganglia*). (*Figure 16–3*)

7. In spinal segments T_1–L_2, ventral roots give rise to the myelinated white ramus, which, in turn, leads to the sympathetic chain ganglia. (*Figures 16–3, 16–4*)

8. **Postganglionic fibers** targeting structures in the body wall and limbs rejoin the spinal nerves and reach their destinations by way of the dorsal and ventral rami. (*Figures 16–3, 16–4*)

9. Postganglionic fibers targeting structures in the thoracic cavity form **sympathetic nerves**, which go directly to their visceral destinations. Preganglionic fibers run between the sympathetic chain ganglia and interconnect them. (*Figures 16–3, 16–4*)

10. The abdominopelvic viscera receive sympathetic innervation by preganglionic fibers that synapse within collateral ganglia. The preganglionic fibers that innervate the collateral ganglia form the **splanchnic nerves**. (*Figures 16–3, 16–4*)

11. The **celiac ganglion** innervates the stomach, liver, gallbladder, pancreas, and spleen; the **superior mesenteric ganglion** innervates the small intestine and initial segments of the large intestine; and the **inferior mesenteric ganglion** innervates the kidneys, urinary bladder, the terminal portions of the large intestine, and the sex organs. (*Figures 16–4, 16–8*)

12. Preganglionic fibers entering an adrenal gland synapse within the **adrenal medulla**. (*Figures 16–3, 16–4*)

13. In a crisis, the entire sympathetic division responds—an event called **sympathetic activation**. Its effects include increased alertness, a feeling of energy and euphoria, increased cardiovascular and respiratory activities, a general elevation in muscle tone, and a mobilization of energy reserves.

16-3 ▌ Stimulation of sympathetic neurons leads to the release of various neurotransmitters p.566

14. The stimulation of the sympathetic division has two distinctive results: the release of either ACh or *norepinephrine (NE)* at specific locations, and the secretion of *epinephrine (E)* and NE into the general circulation.

15. Sympathetic ganglionic neurons end in telodendria studded with **varicosities** containing neurotransmitters. (*Figure 16–5*)

16. The two types of sympathetic receptors are **alpha receptors** and **beta receptors**.

17. Most postganglionic fibers are *adrenergic*; a few are *cholinergic* or *nitroxidergic*. (*Table 16–1*)

18. The sympathetic division includes two sympathetic chain ganglia, three collateral ganglia, and two adrenal medullae. (*Figure 16–7; Tables 16–2, 16–3*)

16-4 ▌ The parasympathetic division consists of preganglionic neurons and ganglionic neurons involved in conserving energy and lowering metabolic rate p. 568

19. The parasympathetic division includes preganglionic neurons in the brain stem and sacral segments of the spinal cord, and ganglionic neurons in peripheral ganglia located within (**intramural**) or next to (**terminal**) target organs. (*Spotlight Figure 16–2*)

20. Preganglionic fibers leave the brain as components of cranial nerves III, VII, IX, and X. Those leaving the sacral segments form **pelvic nerves**. (*Figure 16–6*)

21. The effects produced by the parasympathetic division center on relaxation, food processing, and energy absorption.

16-5 ▌ Stimulation of parasympathetic neurons leads to the release of the neurotransmitter ACh p. 568

22. All parasympathetic preganglionic and postganglionic fibers release ACh. The effects are short-lived, because ACh is inactivated by *acetylcholinesterase (AChE)* and by *tissue cholinesterase*. (*Figure 16–7*)

23. Postsynaptic membranes have two types of ACh receptors. The stimulation of **muscarinic receptors** produces a longer-lasting effect than does the stimulation of **nicotinic receptors**. (*Table 16–1*)

24. The parasympathetic division innervates areas serviced by cranial nerves and organs in the thoracic and abdominopelvic cavities. (*Figure 16–6*)

16-6 ▌ The sympathetic and parasympathetic divisions interact, creating dual innervation p. 570

25. The sympathetic division has widespread influence on visceral and somatic structures.

26. The parasympathetic division innervates only visceral structures that are serviced by cranial nerves or lying within the abdominopelvic cavity. Organs with dual innervation receive input from both divisions. (*Tables 16–2, 16–3*)

27. The parasympathetic and sympathetic nerves intermingle in the thoracic and abdominopelvic cavities to form a series of characteristic *autonomic plexuses* (nerve networks): the **cardiac**, **pulmonary**, **esophageal**, **celiac**, **inferior mesenteric**, and **hypogastric plexuses**. (*Figure 16–8*)

28. Important physiological and functional differences exist between the sympathetic and parasympathetic divisions. (*Figure 16–7; Tables 16–2, 16–3*)

29. Even when stimuli are absent, autonomic motor neurons show a resting level of activation, the **autonomic tone.**

16-7 ▌ Visceral reflexes play a role in the integration and control of autonomic functions p. 571

30. **Visceral reflex arcs** perform the simplest function of the ANS, and can be either **long reflexes** (with interneurons) or **short reflexes** (bypassing the CNS). (*Figure 16–9*)

31. Parasympathetic reflexes govern respiration, cardiovascular functions, and other visceral activities. (*Table 16–4*)

32. Levels of activity in the sympathetic and parasympathetic divisions of the ANS are controlled by centers in the brain stem that regulate specific visceral functions.

33. The SNS and ANS are organized in parallel. Integration occurs at the level of the brain stem and higher centers. (*Figure 16–10; Table 16–5*)

16-8 | Higher-order functions include memory and states of consciousness p. 578

34. Higher-order functions (1) are performed by the cerebral cortex and involve complex interactions among areas of the cerebral cortex and between the cortex and other areas of the brain, (2) involve conscious and unconscious information processing, and (3) are subject to modification and adjustment over time.
35. Memories can be classified as **short term** or **long term**.
36. The conversion from short-term to long-term memory is **memory consolidation**. *(Figure 16–11)*
37. **Amnesia** is the loss of memory as a result of disease or trauma.
38. In **deep sleep** (*slow wave* or *non-REM sleep*), the body relaxes and cerebral cortex activity is low. In **rapid eye movement (REM) sleep**, active dreaming occurs. *(Figure 16–12)*
39. The **reticular activating system (RAS)**, a network in the reticular formation, is most important to arousal and the maintenance of consciousness. *(Figure 16–13)*

16-9 | Neurotransmitters influence brain chemistry and behavior p. 582

40. Changes in the normal balance between two or more neurotransmitters can profoundly affect brain function.

16-10 | Aging produces various structural and functional changes in the nervous system p. 583

41. Age-related changes in the nervous system include a reduction in brain size and weight, a reduction in the number of neurons, a decrease in blood flow to the brain, changes in the synaptic organization of the brain, and intracellular and extracellular changes in CNS neurons.
42. The nervous system monitors all other systems and issues commands that adjust their activities. The efficiency of these activities typically decreases with aging. *(Figure 16–14)*

Review Questions

See the blue Answers tab at the back of the book.

LEVEL 1 Reviewing Facts and Terms

1. The preganglionic and ganglionic neurons are missing from the diagram below showing the distribution of sympathetic innervation. Indicate their distribution using red for the preganglionic neurons and black for the ganglionic neurons.

2. The autonomic division of the nervous system directs (a) voluntary motor activity, (b) conscious control of skeletal muscles, (c) unconscious control of skeletal muscles, (d) processes that maintain homeostasis, (e) sensory input from the skin.

3. Which division of the autonomic nervous system speeds up the heart rate and which division slows it down?
4. Effects produced by the parasympathetic branch of the autonomic nervous system include (a) dilation of the pupils, (b) increased secretion by digestive glands, (c) dilation of respiratory passages, (d) increased heart rate, (e) increased breakdown of glycogen by the liver.
5. A progressive disorder characterized by the loss of higher-order cerebral functions is (a) Parkinson's disease, (b) parasomnia, (c) Huntington's disease, (d) Alzheimer's disease.
6. The _____ and _____ are two limbic system structures that are key in memory consolidation.
7. Which four ganglia serve as origins for postganglionic fibers involved in control of visceral structures in the head?
8. What are the components of a visceral reflex arc?
9. What cellular mechanisms identified in animal studies are thought to be involved in memory formation and storage?
10. What physiological activities distinguish non-REM sleep from REM sleep?
11. When the reticular activating system becomes active, what effect does it have on an individual?
12. All preganglionic autonomic fibers release _____ at their axon terminals, and the effects are always _____. (a) norepinephrine, inhibitory, (b) norepinephrine, excitatory, (c) acetylcholine, excitatory, (d) acetylcholine, inhibitory.
13. The neurotransmitter at all synapses and neuromuscular or neuroglandular junctions in the parasympathetic division of the ANS is (a) epinephrine, (b) norepinephrine, (c) cyclic-AMP, (d) acetylcholine.
14. What type of a signaling chemical is cyclic-AMP (cAMP)? What can it activate or inactivate?
15. What are the alternative names for the neurotransmitters epinephrine and norepinephrine?
16. The sympathetic nervous system, when very active, stimulates the bulk of the tissues and organs it innervates. Why is it not so for the parasympathetic nervous system?
17. Which four pairs of cranial nerves are associated with the cranial segment of the parasympathetic division of the ANS?
18. Which six plexuses in the thoracic and abdominopelvic cavities innervate visceral organs, and what are the effects of sympathetic versus parasympathetic stimulation?
19. Which neurotransmitter is released by postganglionic fibers in the sympathetic nervous system?

LEVEL 2 Reviewing Concepts

20. Dual innervation refers to situations in which (**a**) vital organs receive instructions from both sympathetic and parasympathetic fibers, (**b**) the atria and ventricles of the heart receive autonomic stimulation from the same nerves, (**c**) sympathetic and parasympathetic fibers have similar effects, (**d**) a, b, and c are correct.

21. Damage to the hippocampus, a component of the limbic system, leads to (**a**) a loss of emotion due to forgetfulness, (**b**) a loss of consciousness, (**c**) a loss of long-term memory, (**d**) an immediate loss of short-term memory.

22. Why does sympathetic function remain intact even when the ventral roots of the cervical spinal nerves are damaged?

23. During sympathetic stimulation, a person may begin to feel "on edge"; this is the result of (**a**) increased energy metabolism by muscle tissue, (**b**) increased cardiovascular activity, (**c**) stimulation of the reticular activating system, (**d**) temporary insensitivity to painful stimuli, (**e**) decreased levels of epinephrine in the blood.

24. Under which of the following circumstances would the diameter of peripheral blood vessels be greatest? (**a**) increased sympathetic stimulation, (**b**) decreased sympathetic stimulation, (**c**) increased parasympathetic stimulation, (**d**) decreased parasympathetic stimulation,(**e**) both increased parasympathetic and sympathetic stimulation.

25. A possible side effect of a drug used to open the airways of someone suffering from an asthma attack is (**a**) decreased activity of the digestive system, (**b**) diarrhea, (**c**) profuse urination, (**d**) increased blood pressure, (**e**) decreased heart rate.

26. List some key physiological effects which occur during sympathetic activation (the "fight or flight" reaction).

27. Why is autonomic tone a significant part of ANS function?

28. Nicotine stimulates cholinergic receptors of the ANS. Based on this information, how would cigarette smoking affect the cardiovascular system?

29. The sight of a tragic accident or very bad news can cause people to faint. How does this happen?

LEVEL 3 Critical Thinking and Clinical Applications

30. Phil is stung on his cheek by a wasp. Because Phil is allergic to wasp venom, his throat begins to swell and his respiratory passages constrict. Would acetylcholine or epinephrine be more helpful in relieving his condition? Why?

31. While studying the activity of smooth muscle in blood vessels, Shelly discovers that, when applied to a muscle plasma membrane, a molecule chemically similar to a neurotransmitter triggers an increase in intracellular calcium ions. Which neurotransmitter is the molecule mimicking, and to which receptors is it binding?

CLINICAL CASE Wrap-Up

The First Day in Anatomy Lab

Tim has suffered a vasovagal (*vaso*, vascular; *vagal*, vagus nerve) loss of consciousness, commonly known as fainting. Environmental triggers, including the smell of the anatomy lab and the sight of the cadaver, caused a momentary malfunction of his autonomic nervous system (ANS). The sympathetic division of the ANS, sometimes called the "fight or flight" division, failed him. The parasympathetic division of the ANS, sometimes called the "rest and digest" or "faint and freeze" division, went into overdrive.

Parasympathetic stimulation of the vagus nerve caused it to release ACh at the cardiac plexus. Meanwhile, there was no counteracting stimulation from the sympathetic nervous system. This caused Tim's heart rate to slow, the force of contractions to decrease, and his blood pressure to drop. As a result, blood flow to the brain decreased, causing the fainting episode. If the brain does not get enough blood, it shuts down quickly.

Tim's rapid recovery begins as soon as the sympathetic division of the ANS takes over. Norepinephrine is released at the cardiac plexus and throughout the body. His heart rate and blood pressure increase, as does his level of consciousness.

"Don't worry, Tim," says his instructor. "Everybody is allowed one fainting episode. And everybody can overcome this. Next time just lie down before you fall down."

1. Why would the instructor tell Tim to lie down, or want him to get his head below his heart, the next time he feels faint?

2. In addition to the cardiac plexus, where else in the body would the sympathetic nervous system act during Tim's recovery?

See the blue Answers tab at the back of the book.

17

The Special Senses

Learning Outcomes

These Learning Outcomes correspond by number to this chapter's sections and indicate what you should be able to do after completing the chapter.

17-1 ▍ Describe the sensory organs of smell, trace the olfactory pathways to their destinations in the brain, and explain the physiological basis of olfactory discrimination.

17-2 ▍ Describe the sensory organs of taste, trace the gustatory pathways to their destinations in the brain, and explain the physiological basis of gustatory discrimination.

17-3 ▍ Identify the internal and accessory structures of the eye, and explain the functions of each.

17-4 ▍ Explain color and depth perception, describe how light stimulates the production of nerve impulses, and trace the visual pathways to their destinations in the brain.

17-5 ▍ Describe the structures of the external, middle, and internal ear, explain their roles in equilibrium and hearing, and trace the pathways for equilibrium and hearing to their destinations in the brain.

SPOTLIGHTS
- Olfaction and Gustation pp. 592–593
- Refractive Problems p. 608
- Photoreception pp. 610–611

CLINICAL CASE
- A Chance to See p. 590

Clinical Notes
- Diabetic Retinopathy p. 600
- Detached Retina p. 600
- Glaucoma p. 603
- Motion Sickness p. 619

A Chance to See

Makena is a 12-year-old Kenyan girl living with her family in a small village in sub-Saharan Africa. Their village is miles away from roads, electricity, and health care. Her family is very poor, and lives on less than $1.25 per day. Makena's daily chores include walking two hours to collect water, every morning before school and again after school. Water is so precious in this part of the world that nobody in Makena's village has ever taken a shower.

Makena is a bright girl, but she does not do well in school. She can't see the chalkboard in her classroom. She can't see the soccer ball when she tries to play with her classmates. She even has trouble seeing the path that leads to the dry riverbed where

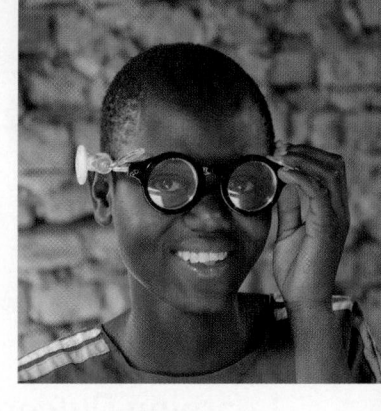

she collects water from a deep hole. She has no idea that other people can see these things better.

No one in Makena's village wears eyeglasses, and most do not even know what they are. Makena's teacher, however, studied at the University of Nairobi and knows that glasses can correct vision problems. The teacher thinks perhaps nearsightedness is a problem for Makena. **Can a desperately poor child, many miles away from health care of any kind, obtain glasses to correct her poor vision? To find out, turn to the Clinical Case Wrap-Up on p. 633.**

An Introduction to the Special Senses

Our knowledge of the world around us is limited to those characteristics that stimulate our sensory receptors. We may not realize it, but our picture of the environment is incomplete. Colors we cannot see guide insects to flowers. Sounds we cannot hear and smells we cannot detect give dolphins, dogs, and cats key information about their surroundings.

What we *do* perceive varies considerably with the state of our nervous system. For example, during sympathetic activation, you experience a heightened awareness of sensory information. You hear sounds that normally you would not notice. Yet, when concentrating on a difficult problem, you may be unaware of fairly loud noises.

Finally, our perception of any stimulus reflects activity in the cerebral cortex, but that activity can be inappropriate. In cases of phantom limb pain, for example, a person feels pain in a missing limb. During an epileptic seizure, a person may experience sights, sounds, or smells that have no physical basis.

In our discussion of the general senses and sensory pathways in Chapter 15, we introduced basic principles of receptor function and sensory processing. We now turn to the five *special senses:* olfaction (smell), gustation (taste), vision, equilibrium (balance), and hearing. The sense organs involved are structurally more complex than those of the general senses, but the same basic principles of receptor function apply. ATLAS: Embryology Summary 13: The Development of Special Sense Organs

17-1 ▌ Olfaction, the sense of smell, involves olfactory receptors responding to chemical stimuli

Learning Outcome Describe the sensory organs of smell, trace the olfactory pathways to their destinations in the brain, and explain the physiological basis of olfactory discrimination.

The sense of smell, called *olfaction,* is made possible by paired **olfactory organs**. These organs are located in the nasal cavity on either side of the nasal septum (**Figure 17–1a**). The olfactory organs are made up of two layers: the olfactory epithelium and the lamina propria.

The **olfactory epithelium** contains the **olfactory receptor cells**, supporting cells, and regenerative **basal cells** (*stem cells*) (**Figure 17–1b**). This epithelium covers the inferior surface of the cribriform plate, the superior portion of the perpendicular plate, and the superior nasal conchae of the ethmoid. ⮌ p. 244

The second layer, the underlying lamina propria, consists of areolar tissue, numerous blood vessels, and nerves. This layer also contains **olfactory glands**. Their secretions absorb water and form thick, pigmented mucus.

What happens when you inhale through your nose? The air swirls within your nasal cavity. This turbulence brings airborne compounds to your olfactory organs. A normal, relaxed inhalation carries a small sample (about 2 percent) of the inhaled air to the olfactory organs. If you sniff repeatedly, you increase the flow of air across the olfactory epithelium, increasing the stimulation of the olfactory receptors. However, only the molecules of water-soluble and lipid-soluble materials that can diffuse into the overlying mucus can stimulate those receptors.

Olfactory Receptors

Olfactory receptor cells are highly modified neurons. The exposed tip of each receptor cell forms a prominent knob that projects beyond the epithelial surface (**Figure 17–1b**). The knob is a base for up to 20 cilia-shaped dendrites that extend into the surrounding mucus. These dendrites lie parallel to the epithelial surface, exposing their considerable surface area to dissolved compounds called *odorants.* Olfactory reception begins

Figure 17–1 **The Olfactory Organs.**

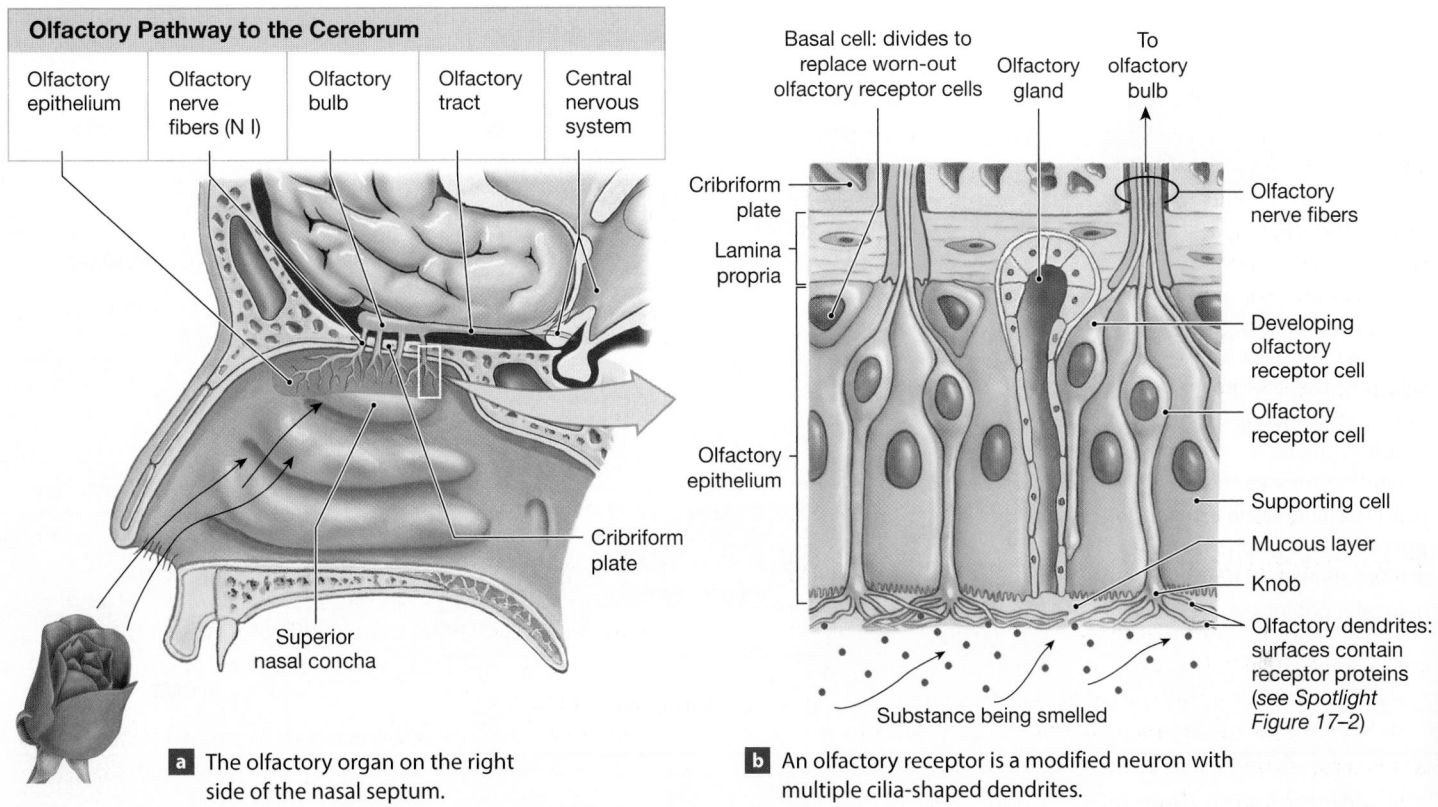

Olfactory Pathway to the Cerebrum

Olfactory epithelium	Olfactory nerve fibers (N I)	Olfactory bulb	Olfactory tract	Central nervous system

Superior nasal concha

Cribriform plate

a The olfactory organ on the right side of the nasal septum.

Basal cell: divides to replace worn-out olfactory receptor cells

Olfactory gland

To olfactory bulb

Cribriform plate

Lamina propria

Olfactory epithelium

Olfactory nerve fibers

Developing olfactory receptor cell

Olfactory receptor cell

Supporting cell

Mucous layer

Knob

Olfactory dendrites: surfaces contain receptor proteins (*see Spotlight Figure 17–2*)

Substance being smelled

b An olfactory receptor is a modified neuron with multiple cilia-shaped dendrites.

with the binding of an odorant to a G protein-coupled receptor in the plasma membrane of an olfactory receptor cell dendrite (**Spotlight Figure 17–2**).

Between 10 and 20 million olfactory receptors fill an area of roughly 5 cm^2 (0.8 in.^{2}). If we take into account the exposed dendritic surfaces, the actual sensory area probably approaches that of the entire body surface. Yet, our olfactory sensitivities cannot compare with those of other vertebrates such as dogs, cats, or fishes. A German shepherd dog sniffing for smuggled drugs or explosives has an olfactory receptor surface 72 times greater than that of the nearby customs inspector!

Olfactory Pathways

The olfactory system is very sensitive. However, the activation of an afferent fiber does not guarantee an awareness of the stimulus. Considerable convergence takes place along the olfactory pathway, and inhibition at the intervening synapses can prevent the sensations from reaching the *olfactory cortex* of the cerebral hemispheres. ⊃ p. 513 The olfactory receptors themselves adapt very little to an ongoing stimulus. Rather, central adaptation ensures that you quickly lose awareness of a new smell but remain sensitive to others.

Axons leaving the olfactory epithelium collect into 20 or more bundles that penetrate the cribriform plate of the ethmoid

bone to reach the *olfactory bulbs* of the cerebrum, where the first synapse occurs (**Figure 17–1**). Efferent fibers from nuclei elsewhere in the brain also innervate neurons of the olfactory bulbs. This arrangement provides a mechanism for central adaptation or facilitation of olfactory sensitivity. Axons leaving the olfactory bulb travel along the olfactory tract to the olfactory cortex, the hypothalamus, and portions of the limbic system.

Olfactory stimulation is the only type of sensory information that reaches the cerebral cortex directly. All other sensations are relayed from processing centers in the thalamus. Certain smells can trigger profound emotional and behavioral responses, as well as the memories, due to the parallel distribution of olfactory information to the limbic system and hypothalamus. The perfume industry understands the practical implications of these connections and works to develop odors that trigger sexual responses.

Olfactory Discrimination

The human olfactory system can make subtle distinctions among 2000–4000 chemical stimuli. No apparent structural differences exist among the olfactory receptor cells, but the epithelium as a whole contains receptor populations with distinct sensitivities. At least 50 "primary smells" are known. It is almost impossible to describe these sensory impressions

Olfaction and gustation are special senses that give us vital information about our environment. Although the sensory information is diverse and complex, each special sense originates at receptor cells that may be neurons or specialized receptor cells that communicate with sensory neurons.

OLFACTION

Olfactory receptors are the dendrites of specialized neurons. When dissolved chemicals contact the dendritic processes, there is a depolarization, called a **generator potential**. This graph shows the action potentials produced by a generator potential.

GUSTATION

The receptors for the senses of taste, vision, equilibrium, and hearing are specialized cells that have unexcitable membranes and form synapses with the processes of sensory neurons. When stimulated, the membrane of the receptor cell undergoes a graded depolarization that triggers the release of chemical transmitters at the synapse. These transmitters then depolarize the sensory neuron, creating a generator potential and action potentials that are propagated to the CNS. Because a synapse is involved, there is a slight synaptic delay. However, this arrangement permits modification of the sensitivity of the receptor cell by presynaptic facilitation or inhibition.

Olfactory reception occurs on the surface membranes of the olfactory dendrites. **Odorants**—dissolved chemicals that stimulate olfactory receptors—interact with receptors called odorant-binding proteins on the membrane surface.

In general, odorants are small organic molecules. The strongest smells are associated with molecules of either high water or high lipid solubilities. As few as four odorant molecules can activate an olfactory receptor.

1 The binding of an **odorant** to its receptor protein leads to the activation of adenylyl cyclase, the enzyme that converts ATP to cyclic AMP (cAMP).

2 The cAMP opens sodium ion channels in the plasma membrane, which then begins to depolarize.

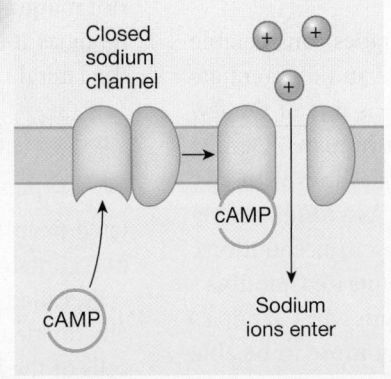

3 If sufficient depolarization occurs, an action potential is triggered in the axon, and the information is relayed to the CNS.

Some 90 percent of the gustatory receptor cells respond to two or more different taste stimuli. The different tastes involve different receptor mechanisms. A salty stimulus involves the diffusion of Na^+ ions through a sodium ion leak channel common in epithelial cells. Stimuli for a sour or acidic taste include H^+ ions that diffuse through the same epithelial Na^+ channel. The intracellular increase in cations leads to depolarization and neurotransmitter release. Sweet, bitter, and umami stimuli bind to specific G protein-coupled receptors. The resulting multiple chemical pathways lead to depolarization and neurotransmitter release.

Salt and Sour Channels

The diffusion of sodium ions from salt solutions or hydrogen ions from acids or sour solutions into the receptor cell leads to depolarization.

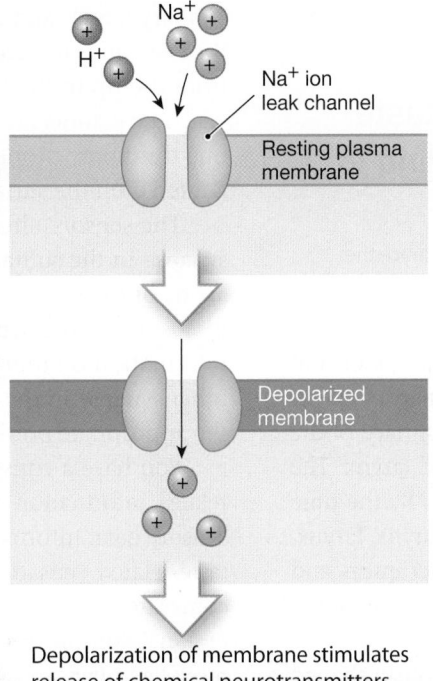

Depolarization of membrane stimulates release of chemical neurotransmitters.

Sweet, Bitter, and Umami Receptors

Receptors responding to stimuli that produce sweet, bitter, and umami sensations are linked to G proteins called **gustducins** (GUST-doos-inz)—protein complexes that use second messengers to produce their effects.

Activation of second messengers stimulates release of chemical neurotransmitters.

effectively. It appears likely that the CNS interprets each smell on the basis of the overall pattern of receptor activity.

The human olfactory organs can discriminate among many smells, but acuity varies widely, depending on the nature of the odorant. We detect many odorants in amazingly small concentrations. One example is mercaptan, an odorant commonly added to natural gas, propane, and butane, which are otherwise odorless. Because we can smell mercaptan in extremely low concentrations (a few parts per billion), its addition enables us to detect a gas leak almost at once and take steps to prevent an explosion.

The olfactory receptor population undergoes considerable turnover. Basal cells in the epithelium divide and differentiate to produce new receptor cells. This turnover is one of the few examples of neuronal replacement in adult humans. Despite this process, the total number of receptors declines with age, and those that remain become less sensitive. As a result, elderly people have difficulty detecting odors in low concentrations. This decline in the number of receptors accounts for Grandma's tendency to use too much perfume and explains why Grandpa's aftershave seems so strong: They must apply more to be able to smell it.

CHECKPOINT

1. Define olfaction.

2. Trace the olfactory pathway, beginning at the olfactory epithelium.

3. When you first enter the A&P lab for dissection, you are very aware of the odor of preservatives. By the end of the lab period, the smell doesn't seem to be nearly as strong. Why?

See the blue Answers tab at the back of the book.

17-2 ▎ Gustation, the sense of taste, involves taste receptors responding to chemical stimuli

Learning Outcome Describe the sensory organs of taste, trace the gustatory pathways to their destinations in the brain, and explain the physiological basis of gustatory discrimination.

Gustation, or taste, provides information about the foods and liquids we eat and drink. **Taste receptors**, or *gustatory* (GUS-ta-tor-ē) *receptors,* are distributed over the superior surface of the tongue and adjacent portions of the pharynx and larynx. The most important taste receptors are on the tongue. By the time we reach adulthood, the taste receptors on the pharynx, larynx, and epiglottis have decreased in number. Taste receptors and specialized epithelial cells form sensory structures called **taste buds**. An adult has about 10,000 taste buds.

The surface of the tongue has numerous variously shaped epithelial projections called *lingual papillae* (pa-PIL-ē; *papilla,* a

nipple-shaped mound). The human tongue has four types of lingual papillae (**Figure 17–3b,c**): (1) **filiform** (*filum*, thread) **papillae**, (2) **fungiform** (*fungus*, mushroom) **papillae**, (3) **vallate** (VAL-āt; *vallum*, wall) **papillae**, and (4) **foliate** (FŌ-lē-āt) **papillae**.

The distribution of these lingual papillae varies by region. Filiform papillae provide friction that helps the tongue move objects around in the mouth, but do not contain taste buds. Each small fungiform papilla contains about five taste buds. There are up to 12 vallate papillae that form a V near the posterior margin of the tongue. Each large vallate papilla contains as many as 100 taste buds. The foliate papillae are folds found on the lateral margins of the posterior tongue.

Taste Receptors

Taste buds are recessed into the surrounding epithelium, isolated from the unprocessed contents of the mouth. Each taste bud contains about 40–100 receptor cells and many small stem cells, called basal cells (**Figure 17–3b,c**). The basal cells continually divide to produce daughter cells that mature in stages. The cells of the last stage are called **gustatory receptor cells**. Cells at all three stages—basal, transitional, and mature—are innervated by sensory neurons. Each gustatory receptor cell extends microvilli, sometimes called *taste hairs*, into the surrounding fluids through the **taste pore**, a narrow opening. Despite this relatively protected position, it's still a hard life: A typical gustatory receptor cell survives for only about 10 days before it is replaced.

Gustatory Pathways

Taste buds are innervated by cranial nerves VII (facial), IX (glossopharyngeal), and X (vagus). The facial nerve innervates all the taste buds located on the anterior two-thirds of the tongue, from the tip to the line of vallate papillae. The glossopharyngeal nerve innervates the vallate papillae and the posterior one-third of the tongue. The vagus nerve innervates taste buds scattered on the surface of the epiglottis.

The sensory afferent fibers carried by these cranial nerves synapse in the solitary nucleus of the medulla oblongata. The axons of the postsynaptic neurons then enter the medial lemniscus. There, the neurons join axons that carry somatic sensory information on touch, pressure, and proprioception. After another synapse in the thalamus, the information is projected to the appropriate portions of the gustatory cortex of the insula.

You have a conscious perception of taste as the brain correlates information received from the taste buds with other sensory data. Information about the texture of food, along with taste-related sensations such as "peppery" or "burning hot," comes from sensory afferents in the trigeminal cranial nerve (V).

In addition, the level of stimulation from the olfactory receptors plays an overwhelming role in taste perception. The combination of taste and smell is what provides the *flavor*, or

Figure 17–3 **Gustatory Receptors.**

Water receptors (pharynx)   Umami

a Location of tongue papillae.

Taste buds

Vallate papilla

Foliate papillae

Fungiform papilla

Filiform papillae

b The structure and representative locations of the four types of lingual papillae. Taste receptors are located in taste buds, which form pockets in the epithelium of fungiform, foliate, and vallate papillae.

Taste buds

Taste buds   LM × 280

Transitional cell
Gustatory cell
Basal cell
Taste hairs (microvilli)
Taste pore

c Taste buds in a vallate papilla. A diagrammatic view of a taste bud, showing gustatory (receptor) cells and supporting cells.

distinctive quality of a particular food or drink. You are several thousand times more sensitive to "tastes" when your olfactory organs are fully functional. By contrast, when you have a cold and your nose is stuffed up, airborne molecules cannot reach your olfactory receptors, so meals taste dull and unappealing. The reduction in flavor takes place even though the taste buds may be responding normally. Without accompanying olfactory sensations, you are now limited to the basic taste sensations provided by your gustatory receptors.

Gustatory Discrimination

You are probably already familiar with the four **primary taste sensations**: sweet, salty, sour, and bitter. There is some evidence for differences in sensitivity to tastes along the axis of the tongue, with greatest sensitivity to salty–sweet anteriorly and sour–bitter posteriorly. However, there are no differences

in the structure of the taste buds. Taste buds in all portions of the tongue provide all four primary taste sensations.

Humans have two additional taste sensations that are less widely known:

- *Umami.* **Umami** (oo-MAH-mē, derived from Japanese meaning "deliciousness") is a pleasant, savory taste corresponding to the flavor of beef broth, chicken broth, and Parmesan cheese. Umami is due to the presence of free glutamates (an amino acid) in food. The distribution of these receptors is not known in detail, but they are present in taste buds of the vallate papillae.

- *Water.* Most people say that water has no flavor. However, research on humans and other vertebrates has demonstrated the presence of **water receptors**, especially in the pharynx. The sensory output of these receptors is processed in the hypothalamus and affects several systems that affect water balance and the regulation of blood volume. For example, minor reductions in the secretion of antidiuretic hormone (a hormone that regulates urination) occur each time you take a long drink.

Gustation reception is described in **Spotlight Figure 17–2**. The threshold for receptor stimulation varies for each of the primary taste sensations. Also, the taste receptors respond more readily to unpleasant than to pleasant stimuli. For example, we are almost a thousand times more sensitive to acids, which taste sour, than to either sweet or salty chemicals. We are a hundred times more sensitive to bitter compounds than to acids. This sensitivity has survival value, because acids can damage the mucous membranes of the mouth and pharynx, and many potent biological toxins have an extremely bitter taste.

Taste sensitivity differs significantly among individuals. Many conditions related to taste sensitivity are inherited. The best-known example involves sensitivity to the compound *phenylthiocarbamide* (PTC). This substance tastes bitter to some people, but is tasteless to others. PTC is not found in foods, but compounds related to PTC are found in Brussels sprouts, broccoli, cabbage, and cauliflower.

Our tasting abilities change with age. Many elderly people find that their food tastes bland and unappetizing, while children tend to find the same foods too spicy. What accounts for these differences?

We begin life with more than 10,000 taste buds, but the number begins declining dramatically by age 50. The sensory loss becomes especially significant because, as we have already noted, older individuals also experience a decline in the number of olfactory receptors.

CHECKPOINT

4. **Define gustation.**

5. **If you completely dry the surface of your tongue and then place salt or sugar crystals on it, you can't taste them. Why not?**

6. **Your grandfather can't understand why foods he used to enjoy just don't taste the same anymore. How would you explain this to him?**

See the blue Answers tab at the back of the book.

17-3 ▮ Internal eye structures contribute to vision, while accessory eye structures provide protection

Learning Outcome Identify the internal and accessory structures of the eye, and explain the functions of each.

We rely more on vision than on any other special sense. Our eyes contain our visual receptors. The eyes are elaborate structures that enable us not only to detect light, but also to create detailed visual images. We begin our discussion of these fascinating organs with the *accessory structures* of the eye, which provide protection, lubrication, and support.

Accessory Structures of the Eye

The **accessory structures** of the eye include the eyelids and the superficial epithelium of the eye, as well as the structures involved with the production, secretion, and removal of tears. **Figure 17–4** shows the superficial anatomy of the eye and its accessory structures.

Eyelids and Superficial Epithelium of the Eye

The eyelids, or **palpebrae** (pal-PĒ-brē), are a continuation of the skin. Their continual blinking keeps the surface of the eye lubricated. They also act like windshield wipers, removing dust and debris. The eyelids can also close firmly to protect the delicate surface of the eye.

The **palpebral fissure** is the gap that separates the free margins of the upper and lower eyelids. The two eyelids are connected, however, at the **medial canthus** (KAN-thus) and the **lateral canthus** (Figure 17–4a). The **eyelashes**, along the margins of the eyelids, are very robust hairs. They help prevent foreign matter (including insects) from reaching the surface of the eye.

The eyelashes are associated with unusually large sebaceous glands. Along the inner margin of the lid are small, modified sebaceous glands called **tarsal glands**, or *Meibomian* (mī-BŌ-mē-an) *glands.* They secrete a lipid-rich product that helps keep the eyelids from sticking together. At the medial canthus, the **lacrimal caruncle** (KAR-ung-kul), a mass of soft tissue, contains glands producing the thick secretions that contribute to the gritty deposits that sometimes appear after a good night's sleep.

Bacteria occasionally invade and infect these various glands. A *chalazion* (kah-LĀ-zē-on; small lump) is a cyst that results from an infection of a tarsal gland. An infection in a sebaceous gland associated with an eyelash produces a painful localized swelling known as a *sty.*

The skin covering the visible surface of the eyelid is very thin. Deep to the skin lie the muscle fibers of the *orbicularis oculi* and *levator palpebrae superioris muscles.* ⟲ p. 373 These skeletal muscles close the eyelids and raise the upper eyelid, respectively.

The epithelium covering the inner surfaces of the eyelids and the outer surface of the eyeball is called the **conjunctiva** (kon-junk-TĪ-vuh). It is a mucous membrane covered by a specialized stratified squamous epithelium. The **palpebral conjunctiva** covers the inner surface of the eyelids. The **ocular conjunctiva**, or *bulbar conjunctiva*, covers the anterior surface of the eye (**Figure 17–4b**).

Figure 17–4   External Features and Accessory Structures of the Eye.
ATLAS: Plates 3c; 12a; 16a,b

a Gross and superficial anatomy of the accessory structures

Labels (part a): Lateral canthus, Sclera, Eyelashes, Pupil, Palpebra, Palpebral fissure, Medial canthus, Lacrimal caruncle, Corneal limbus

Labels (part b): Superior rectus muscle, Tendon of superior oblique muscle, Lacrimal gland ducts, Lacrimal gland, Ocular conjunctiva, Lateral canthus, Lower eyelid, Orbital fat, Inferior rectus muscle, Inferior oblique muscle, Lacrimal punctum, Lacrimal caruncle, Superior lacrimal canaliculus, Medial canthus, Inferior lacrimal canaliculus, Lacrimal sac, Nasolacrimal duct, Inferior nasal concha, Opening of nasolacrimal duct

b The organization of the lacrimal apparatus

The ocular conjunctiva extends to the edges of the **cornea** (KOR-nē-uh), a transparent part of the outer fibrous layer of the eye. The cornea is covered by a very delicate squamous *corneal epithelium*, five to seven cells thick, that is continuous with the ocular conjunctiva. A constant supply of fluid washes over the surface of the eyeball, keeping the ocular conjunctiva and cornea moist and clean. Mucous cells in the epithelium assist the accessory glands in lubricating the surfaces of the conjunctiva to prevent drying out and friction.

Conjunctivitis, or *pinkeye*, is an inflammation of the conjunctiva. The most obvious sign, redness, is due to the dilation of blood vessels deep to the conjunctival epithelium. This condition may be caused by pathogenic infection or by physical, allergic, or chemical irritation of the conjunctival surface.

The Lacrimal Apparatus

A constant flow of tears keeps conjunctival surfaces moist and clean. Tears reduce friction, remove debris, prevent bacterial infection, and provide nutrients and oxygen to portions of the conjunctival epithelium. The **lacrimal apparatus** produces, distributes, and removes tears. The lacrimal apparatus of each eye consists of (1) a *lacrimal gland* with associated ducts, (2) paired *lacrimal canaliculi*, (3) a *lacrimal sac*, and (4) a *nasolacrimal duct* (**Figure 17–4b**).

The **fornix** of the eye is a pocket created where the palpebral conjunctiva becomes continuous with the ocular conjunctiva (**Figure 17–5a**). The lateral portion of the superior fornix receives 10–12 ducts from the **lacrimal gland**, or tear gland (**Figure 17–4b**). This gland is about the size and shape of an almond, measuring roughly 12–20 mm (0.5–0.75 in.). It nestles within a depression in the frontal bone, just inside the orbit and superior and lateral to the eyeball. ⟲ p. 241

The lacrimal gland normally provides the key ingredients and most of the volume of the tears that bathe the conjunctival surfaces. The lacrimal secretions supply nutrients and oxygen to the corneal cells by diffusion. The lacrimal secretions are watery and slightly alkaline. They contain the antibacterial enzyme **lysozyme** and antibodies that attack pathogens before they enter the body.

The lacrimal gland produces about 1 mL of tears each day. Once the lacrimal secretions have reached the ocular surface, they mix with the products of accessory glands and the oily secretions of the tarsal glands. The result is a superficial "oil slick" that aids lubrication and slows evaporation.

17

Figure 17–5   **The Sectional Anatomy of the Eye.** ATLAS: Plates 12a; 16a,b

a Sagittal section of left eye

- Fornix
- Palpebral conjunctiva
- Eyelash
- Ocular conjunctiva
- Ora serrata
- Cornea
- Pupil
- Iris
- Corneal limbus

Optic nerve
Lens
Fovea
Retina
Choroid
Sclera

Fibrous layer
- Cornea
- Sclera

Anterior cavity

Vascular layer (uvea)
- Iris
- Ciliary body
- Choroid

Inner layer (retina)
- Neural part
- Pigmented part

Posterior cavity

b Horizontal section of right eye

17

Visual axis

Anterior cavity

| Posterior chamber | Anterior chamber |

Edge of pupil

Nose

Lacrimal punctum
Lacrimal caruncle
Medial canthus
Ciliary processes
Ciliary body
Ora serrata

Ethmoidal labyrinth
Medial rectus muscle
Optic disc
Optic nerve
Central artery and vein

Lens

Cornea
Iris
Ciliary zonule
Corneal limbus
Conjunctiva
Lower eyelid
Lateral canthus

Sclera
Choroid
Retina
Posterior cavity
Lateral rectus muscle
Fovea

Orbital fat

c Horizontal dissection of right eye

Blinking sweeps the tears across the ocular surface. Tears accumulate at the medial canthus in an area known as the *lacrimal lake,* or "lake of tears." The lacrimal lake covers the lacrimal caruncle, which bulges anteriorly. The **lacrimal puncta** (singular, *punctum*), two small pores, drain the lacrimal lake. They empty into the **lacrimal canaliculi**, small canals that in turn lead to the **lacrimal sac**, which nestles within the lacrimal sulcus of the orbit (**Figure 17–4b**). ↻ p. 247

From the inferior portion of the lacrimal sac, the **nasolacrimal duct** passes through the *nasolacrimal canal,* formed by the lacrimal bone and the maxilla. This duct delivers tears to the nasal cavity on that side. The duct empties into the *inferior meatus,* a narrow passageway inferior and lateral to the inferior nasal concha. When a person cries, tears rushing into the nasal cavity produce a runny nose. If the lacrimal puncta can't provide enough drainage, the lacrimal lake overflows and tears stream across the face.

The Eye

Your eyes are extremely sophisticated visual instruments. They are more versatile and adaptable than the most expensive cameras, yet compact and durable. Each eye is a slightly irregular spheroid, a little smaller than a Ping-Pong ball. An eye has an average diameter of 24 mm (almost 1 inch) and a weight of about 8 g (0.28 oz). Within the orbit, the eyeball shares space with the extrinsic eye muscles, the lacrimal gland, and the cranial nerves and blood vessels that supply the eye and adjacent portions of the orbit and face. **Orbital fat** cushions and insulates the eye (**Figures 17–4b** and **17–5c**).

The wall of the eye has three distinct layers, formerly called tunics (**Figure 17–5b**): (1) an outer *fibrous layer,* (2) an intermediate *vascular layer (uvea),* and (3) a deep *inner layer (retina).* The visual receptors, or *photoreceptors,* are located in the inner layer.

The eyeball itself is hollow and filled with fluid. Its interior can be divided into two cavities, anterior and posterior (**Figure 17–5c**). The **anterior cavity** has two chambers, the **anterior chamber** (between the cornea and the colored portion known as the iris) and the **posterior chamber** (between the iris and the lens). A clear, watery fluid called *aqueous humor* fills the entire anterior cavity. The **posterior cavity**, or *vitreous chamber,* contains a gelatinous substance called the *vitreous body.* The fluid part of this body is *vitreous humor.* The vitreous body and the aqueous humor help to stabilize the shape of the eye.

The Fibrous Layer

The **fibrous layer**, the outermost layer of the eye, consists of the *sclera* (SKLER-uh) and the *cornea.* The fibrous layer (1) supports and protects, (2) serves as an attachment site for the extrinsic eye muscles, and (3) contains structures that assist in focusing.

The **sclera**, or "white of the eye," covers most of the ocular surface (**Figure 17–5b,c**). The sclera consists of a dense fibrous connective tissue containing both collagen and elastic fibers. This layer is thickest over the posterior surface of the eye, near the exit of the optic nerve. The sclera is thinnest over the anterior surface. The six extrinsic eye muscles insert on the sclera, blending their collagen fibers with those of the fibrous layer. ↻ p. 370

The surface of the sclera contains small blood vessels and nerves that penetrate the sclera to reach internal structures. The network of small vessels interior to the ocular conjunctiva generally does not carry enough blood to lend an obvious color to the sclera. On close inspection, however, the vessels are visible as red lines against the white background of collagen fibers.

The transparent cornea is structurally continuous with the sclera. The border between the two is called the **corneal limbus** (**Figures 17–4a** and **17–5a,c**). Deep to the delicate corneal epithelium, the cornea consists primarily of a dense matrix containing multiple layers of collagen fibers, organized so as not to interfere with the passage of light. The cornea has no blood vessels. Its superficial epithelial cells obtain oxygen and nutrients by diffusion from the tears that flow across their free surfaces. The cornea also has numerous free nerve endings, making it the most sensitive portion of the eye.

Damage to the cornea may cause blindness even though the functional components of the eye—including the photoreceptors—are perfectly normal. The cornea has a very restricted ability to repair itself. For this reason, corneal injuries must be treated immediately to prevent serious vision losses.

Restoring vision after corneal scarring generally requires replacing the cornea through a corneal transplant. Corneal replacement is probably the most common form of transplant surgery. Such transplants can be performed between unrelated individuals, because there are no blood vessels to carry white blood cells, which attack foreign tissues, into the area. Corneal grafts are obtained from donor eyes. For best results, the tissues must be removed within five hours after the donor's death.

The Vascular Layer

The **vascular layer**, or **uvea** (Ū-vē-uh), is a pigmented region that includes the iris, ciliary body, and choroid (**Figure 17–5b,c**). It contains numerous blood vessels, lymphatic vessels, and the intrinsic (smooth) muscles of the eye. This middle layer (1) provides a route for blood vessels and lymphatics that supply tissues of the eye; (2) regulates the amount of light that enters the eye; (3) secretes and reabsorbs the aqueous humor that circulates within the chambers of the eye; and (4) controls the shape of the lens, an essential part of the focusing process.

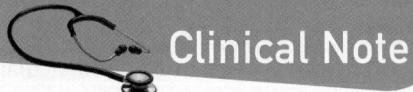 **Clinical Note**

Diabetic Retinopathy A *retinopathy* is a disease of the retina. **Diabetic retinopathy** develops in many people with *diabetes mellitus,* an endocrine disorder that interferes with glucose metabolism. Diabetes affects many systems, but serious cardiovascular problems are particularly common. Diabetic retinopathy develops over years. It results from the blockage of small retinal blood vessels, followed by excessive growth of abnormal blood vessels. These blood vessels invade the retina (inner layer) and extend into the space between its two parts—the outer pigmented part and the inner neural part. Visual acuity is gradually lost through damage to photoreceptors (which are deprived of oxygen and nutrients), leakage of blood into the posterior cavity, and the overgrowth of blood vessels. Laser therapy can seal leaking vessels and block new vessel growth. The posterior cavity can be drained and the cloudy fluid replaced by a suitably clear substitute. This procedure is called a *vitrectomy.* These are only temporary, imperfect fixes. Diabetic retinopathy is a leading cause of blindness in the United States.

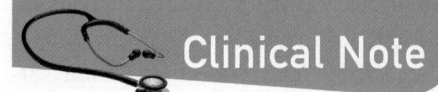 **Clinical Note**

Detached Retina Photoreceptors depend entirely on the diffusion of oxygen and nutrients from blood vessels in the choroid. In a **detached retina,** the neural part of the retina becomes separated from the pigmented part. This condition can result from a variety of factors, including a sudden hard impact to the eye. Unless the two parts of the inner layer (retina) are reattached, the photoreceptors will degenerate and vision will be lost. Reattachment involves "welding" the two parts together using laser beams focused through the cornea. These beams heat the two parts, fusing them together at several points around the retina. However, the procedure destroys the photoreceptors and other cells at the "welds," producing permanent blind spots.

The Iris. The **iris** is visible through the transparent corneal surface. The iris contains blood vessels, pigment cells, and two layers of smooth muscle fibers called *pupillary muscles.* When these muscles contract, they change the diameter of the **pupil**, the central opening of the iris. There are two types of pupillary muscles: dilators and constrictors (**Figure 17–6**).

The autonomic nervous system controls both muscle groups. For example, parasympathetic activation in response to bright light causes the pupils to constrict (the *consensual light reflex*). Sympathetic activation in response to dim light causes the pupils to dilate. ⊃ p. 571

The body of the iris consists of highly vascular, pigmented, loose connective tissue. The anterior surface has no epithelial covering. Instead, it has an incomplete layer of fibroblasts and melanocytes. Melanocytes are also scattered within the body of the iris. The posterior surface is covered by a pigmented epithelium that is part of the inner layer and contains melanin granules.

Figure 17–6 **The Pupillary Muscles.**

Pupillary constrictor (sphincter)

Pupil

Pupillary dilator (radial)

The **pupillary dilator muscles** extend radially away from the edge of the pupil. Contraction of these muscles enlarges the pupil.

The **pupillary constrictor muscles** form a series of concentric circles around the pupil. When these sphincter muscles contract, the diameter of the pupil decreases.

Decreased light intensity
Increased sympathetic stimulation

Increased light intensity
Increased parasympathetic stimulation

How is eye color determined? Our genes influence the density and distribution of melanocytes on the anterior surface and interior of the iris, as well as the density of the pigmented epithelium. When the connective tissue of the iris contains few melanocytes, light passes through it and bounces off the pigmented epithelium. The eye then appears blue. Individuals with green, brown, or black eyes have increasing numbers of melanocytes in the body and on the surface of the iris. The eyes of people with albinism, whose cells do not produce melanin, appear a very pale gray or blue-gray.

The Ciliary Body. At its periphery, the iris attaches to the anterior portion of the **ciliary body**, a thickened region that begins deep to the junction between the cornea and the sclera. The ciliary body extends posteriorly to the level of the **ora serrata** (Ō-ra ser-RA-tuh; serrated mouth), the serrated anterior edge of the thick, neural part of the inner layer, or retina (**Figure 17–5a,c**). The bulk of the ciliary body consists of the **ciliary muscle**, a ring of smooth muscle that projects into the interior of the eye. The epithelium covering this muscle has numerous folds called **ciliary processes**. The **ciliary zonule** (*suspensory ligament*) is the ring of fibers that attaches the lens to the ciliary processes. The connective tissue fibers hold the lens in place posterior to the iris and centered on the pupil. As a result, any light passing through the pupil also passes through the lens.

The Choroid. The **choroid** is a vascular layer that separates the fibrous layer and the inner layer posterior to the ora serrata (**Figure 17–5c**). The choroid is covered by the sclera and attached to the outermost layer of the retina. An extensive capillary network in the choroid delivers oxygen and nutrients to the retina. The choroid also contains melanocytes, which are especially numerous near the sclera.

The Inner Layer

The **inner layer**, or **retina**, is the innermost layer of the eye. It consists of a thin, outer layer called the *pigmented part*, and a thick, inner layer called the *neural part*. The pigmented part of the retina absorbs light that passes through the neural part, preventing light from bouncing back through the neural part and producing visual "echoes." The pigment cells also have important biochemical interactions with the retina's light receptors, which are located in the neural part of the retina. In addition to light receptors, the neural part of the retina contains supporting cells and neurons that do preliminary processing and integration of visual information.

The two layers of the retina are normally very close together, but not tightly interconnected. The pigmented part of the retina continues over the ciliary body and iris, but the neural part extends anteriorly only as far as the ora serrata. The neural part of the retina forms a cup that establishes the posterior and lateral boundaries of the posterior cavity (**Figure 17–5b,c**).

Organization of the Retina. In sectional view, the neural part of the retina contains several layers of cells (**Figure 17–7a**). The outermost layer, closest to the pigmented part of the retina, contains the **photoreceptors**, the cells that detect light.

The eye has two main types of photoreceptors: rods and cones. **Rods** do not discriminate among colors of light. Highly sensitive to light, they enable us to see in dimly lit rooms, at twilight, and in pale moonlight. **Cones** give us color vision. We have three types of cones, and each type contains a different visual pigment: red, green, and blue. A person who sees these three primary colors has normal color vision and thus is referred to as a *trichromat* (*tri-*, three + *chroma*, color). Their stimulation in various combinations enables us to perceive different colors. Cones give us sharper, clearer images than rods do, but cones require more intense light. If you sit outside at sunset with your textbook open to a colorful illustration, you can detect the gradual shift in your visual system from cone-based vision (a clear image in full color) to rod-based vision (a relatively grainy image in black and white). A third type of photoreceptor is the *intrinsically photosensitive retinal ganglion cell* (ipRGC). The photopigment in the ipRGC is melanopsin. These cells are known to respond to different levels of brightness and influence the body's 24-hour circadian rhythm (biological clock).

TIPS & TRICKS

Associate the "r" in *r*od with the "r" in da*r*k, and associate the "c" in *c*ones with the "c" in *c*olor and in a*c*uity.

Rods and cones are not evenly distributed across the retina. Approximately 125 million rods form a broad band around the periphery of the retina. As you move away from the periphery, toward the center of the retina, the density of rods gradually decreases. In contrast, most of the roughly 6 million cones are concentrated in the area where a visual image arrives after it passes through the cornea and lens. This region, known as the **macula** (MAK-ū-luh; spot), has no rods. The very highest concentration of cones occurs at the center of the macula, an area called the **fovea centralis** (FŌ-vē-uh; shallow depression). It is usually simply referred to as the **fovea** (**Figure 17–7c**). The fovea is the site of sharpest vision. When you look directly at an object, its image falls on this portion of the retina. An imaginary line drawn from the center of that object through the center of the lens to the fovea establishes the **visual axis** of the eye (**Figure 17–5c**).

What are some of the visual consequences of this distribution of photoreceptors? When you look directly at an object, you are placing its image on the fovea, the center of color vision. You see a very good image as long as there is enough light to stimulate the cones. But in very dim light, cones cannot function. That is why you can't see a dim star if you stare directly at it,

Figure 17–7 **The Organization of the Retina.**

a The cellular organization of the retina. The photoreceptors are closest to the choroid, rather than near the posterior cavity (vitreous chamber).

b The optic disc in diagrammatic sagittal section.

c A photograph of the retina as seen through the pupil.

but you can see it if you shift your gaze to one side or the other. Shifting your gaze moves the image of the star from the fovea, where it does not provide enough light to stimulate the cones, to the periphery, where it can affect the more sensitive rods.

Rods and cones synapse with about 6 million neurons called **bipolar cells** (**Figure 17–7a**). These cells in turn synapse within the layer of neurons called **ganglion cells**, which lie

adjacent to the posterior cavity. A network of **horizontal cells** extends across the neural part of the retina at the level of the synapses between photoreceptors and bipolar cells. A comparable layer of **amacrine** (AM-a-krin) **cells** occurs where bipolar cells synapse with ganglion cells. Horizontal and amacrine cells can facilitate or inhibit communication between photoreceptors and ganglion cells, altering the sensitivity of the retina.

These cells play an important role in the eye's adjustment to dim or brightly lit environments.

The Optic Disc. Axons from an estimated 1 million ganglion cells converge on the **optic disc**, a circular region just medial to the fovea. The optic disc is the origin of the optic nerve (N II). From this point, the axons turn, penetrate the wall of the eye, and proceed toward the diencephalon (**Figure 17–7b**). The *central retinal artery* and *central retinal vein,* which supply the retina, pass through the center of the optic nerve and emerge on the surface of the optic disc (**Figure 17–7b,c**). The optic disc has no photoreceptors or other structures typical of the rest of the retina. The optic disc is commonly called the **blind spot** because light striking this area goes unnoticed. Why don't you see a blank spot in your field of vision? The reason is that involuntary eye movements keep the visual image moving and allow your brain to fill in the missing information. However, try the simple activity in **Figure 17–8** to prove that a blind spot really exists in your field of vision.

The Chambers of the Eye

Recall that the ciliary body and lens divide the interior of the eye into a large posterior cavity and a smaller anterior cavity made up of anterior and posterior chambers (**Figure 17–5c**). Both chambers are filled with the fluid *aqueous humor.* The posterior cavity also contains aqueous humor, but a gelatinous substance known as the *vitreous body* takes up most of its volume.

Aqueous Humor. Aqueous humor is a fluid that circulates within the anterior cavity. It passes from the posterior chamber to the anterior chamber through the pupil (**Figure 17–9**). It also freely diffuses through the posterior cavity and across the

Figure 17–8   A Demonstration of the Presence of a Blind Spot. Close your left eye and stare at the plus sign with your right eye, keeping the plus sign in the center of your field of vision. Begin with the page a few inches away from your eye, and gradually increase the distance. The dot will disappear when its image falls on the blind spot, at your optic disc. To check the blind spot in your left eye, close your right eye and repeat the sequence while you stare at the dot.

surface of the retina. Its composition resembles that of cerebrospinal fluid.

Aqueous humor forms through active secretion by epithelial cells of the ciliary body's ciliary processes. The epithelial cells regulate its composition. Because aqueous humor circulates, it provides an important route for nutrient and waste transport. It also serves as a fluid cushion.

The eye is filled with fluid, and fluid pressure provided by the aqueous humor helps retain the eye's shape. Fluid pressure also stabilizes the position of the retina, pressing the neural part against the pigmented part. In effect, the aqueous humor acts like the air inside a balloon.

The eye's **intraocular pressure** can be measured in the anterior chamber, where the fluid pushes against the inner surface of the cornea. Intraocular pressure is most often checked by applanation tonometry. This procedure involves placing a small, flat disk on the anesthetized cornea to measure the tension. Normal intraocular pressure ranges from 12 to 21 mm Hg.

Aqueous humor is secreted into the posterior chamber at a rate of 1–2 mL per minute. It leaves the anterior chamber at the same rate. After filtering through a network of connective tissues located near the base of the iris, aqueous humor enters the **scleral venous sinus** (*canal of Schlemm*), a passageway that

17

Clinical Note

Glaucoma If aqueous humor cannot drain into the scleral venous sinus, intraocular pressure rises due to the continued production of aqueous humor, and **glaucoma** results. The sclera is a fibrous coat, so it cannot expand like an inflating balloon, but it does have one weak point—the optic disc, where the optic nerve penetrates the wall of the eye. Gradually the increasing pressure pushes the optic nerve outward, damaging its nerve fibers. When intraocular pressures have risen to roughly twice normal levels, the distortion of the optic nerve fibers begins to interfere with the propagation of action potentials, and peripheral vision begins to deteriorate. If this condition is not corrected, tunnel vision and then complete blindness may result.

Glaucoma affects about 2 percent of people over age 35. In most cases, the primary factors responsible cannot be determined. Glaucoma is a relatively common condition, with more than 2 million cases in the United States alone. For this reason, most eye exams include a test of intraocular pressure. Glaucoma may be treated by the topical application of drugs that constrict the pupil and tense the edge of the iris, making the surface more permeable to aqueous humor. Surgical correction involves perforating the wall of the anterior chamber to encourage drainage. This procedure is now performed by laser surgery on an outpatient basis.

Figure 17–9   **The Circulation of Aqueous Humor.** Aqueous humor, which is secreted at the ciliary body, circulates through the posterior and anterior chambers before it is reabsorbed through the scleral venous sinus.

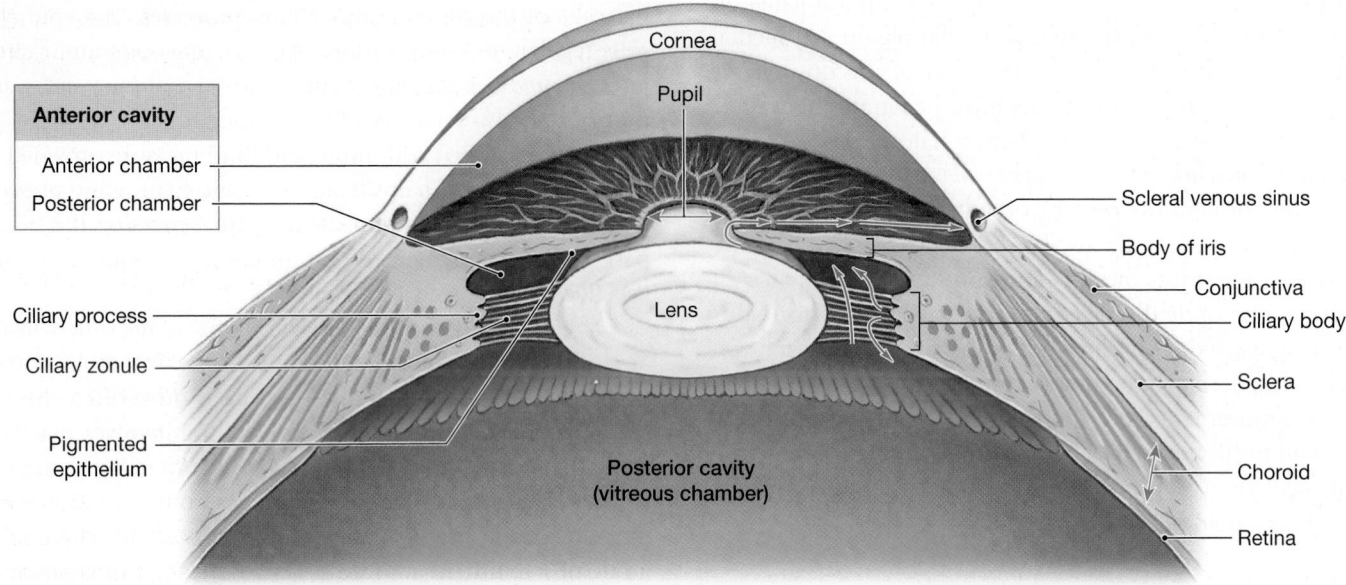

extends completely around the eye at the level of the corneal limbus. Collecting channels deliver the aqueous humor from this channel to veins in the sclera.

Aqueous humor is removed and recycled within a few hours of its formation. The rate of removal normally keeps pace with the rate of generation at the ciliary processes.

The Vitreous Body. The posterior cavity of the eye contains the **vitreous body**, a gelatinous mass. Its fluid portion is called *vitreous humor*. The vitreous body helps stabilize the shape of the eye. Otherwise, the eye might distort as the extrinsic eye muscles change its position within the orbit. Specialized cells embedded in the vitreous body produce the collagen fibers and proteoglycans that account for the gelatinous consistency of this mass. Unlike the aqueous humor, the vitreous body is formed during development of the eye and is not replaced.

The Lens

The **lens** lies posterior to the cornea and is held in place by the ciliary zonule that originates on the ciliary body of the choroid (**Figures 17–5b** and **17–9**). The primary function of the lens is to focus the visual image on the photoreceptors. The lens does so by changing its shape.

The lens consists of concentric layers of cells that are precisely organized. A dense fibrous capsule covers the entire lens. Many of the capsular fibers are elastic. Unless an outside force is applied, they will contract and make the lens spherical. Around the edges of the lens, these capsular fibers intermingle with those of the ciliary zonule.

The cells in the interior of the lens are called **lens fibers**. These highly specialized cells have lost their nuclei and other organelles. They are long and slender and are filled with transparent proteins called **crystallins**. These proteins give the lens both its clarity and its focusing power. Crystallins are extremely stable proteins. They remain intact and functional for a lifetime.

The transparency of the lens depends on the crystallin proteins maintaining a precise combination of structural and biochemical characteristics. Because these proteins are not renewed, any modifications they undergo over time can accumulate and result in the lens losing its transparency. This abnormality is known as a **cataract**. Cataracts can result from injuries, radiation, or reaction to drugs. **Senile cataracts**, however, are a natural consequence of aging and are the most common form.

Over time, the lens turns yellowish and eventually begins to lose its transparency. As the lens becomes "cloudy," the individual needs brighter and brighter light for reading. Visual clarity begins to fade. If the lens becomes completely opaque, the person will be functionally blind, even though the photoreceptors are normal.

Cataract surgery involves removing the lens, either intact or after it has been shattered with high-frequency sound waves. The missing lens is then replaced by an artificial substitute. Vision is then fine-tuned with glasses or contact lenses.

Refraction. The retina has about 130 million photoreceptors. Each one monitors the light striking a specific site on the retina. The processing of information from all the receptors produces a visual image.

The eye is often compared to a camera. To provide useful information, the lens of the eye, like a camera lens, must focus the arriving image. To say that an image is "in focus" means that the rays of light arriving from an object strike the sensitive surface of the retina (or the semiconductor device that records light electronically in digital cameras) in precise order so as to form a miniature image of the object. If the rays are not perfectly focused, the image is blurry. Focusing normally occurs in two steps, as light passes first through the cornea and then the lens.

Light is **refracted**, or bent, when it passes from one medium to another medium with a different density. You can see this effect if you stick a pencil into a glass of water. Notice that the shaft of the pencil appears to bend sharply at the air–water interface. This effect occurs when light is refracted as it passes into the air from the much denser water.

In the human eye, the greatest amount of refraction occurs when light passes from the air into the corneal tissues, which have a density close to that of water. When you open your eyes under water, you cannot see clearly because refraction at the corneal surface has been largely eliminated. Light passes unbent from one watery medium to another.

Additional refraction takes place when the light passes from the aqueous humor into the relatively dense lens. The lens provides the extra refraction needed to focus the light rays from an object toward a **focal point**—a specific point of intersection on the retina.

The distance between the center of the lens and its focal point is the **focal distance** of the lens. Whether in the eye or in a camera, the focal distance is determined by two factors:

1. *The distance of the object from the lens.* The closer an object is to the lens, the greater the focal distance (**Figure 17–10a,b**).

2. *The shape of the lens.* The rounder the lens, the more refraction occurs. So, a very round lens has a shorter focal distance than a flatter one (**Figure 17–10b,c**).

Accommodation. Accommodation is the automatic adjustment of the eye to give us clear vision (**Figure 17–11**). During accommodation, the lens becomes rounder to focus the image

Figure 17–10 **Factors Affecting Focal Distance.** Light rays from a source are refracted when they reach the lens of the eye. The rays are then focused onto a single focal point.

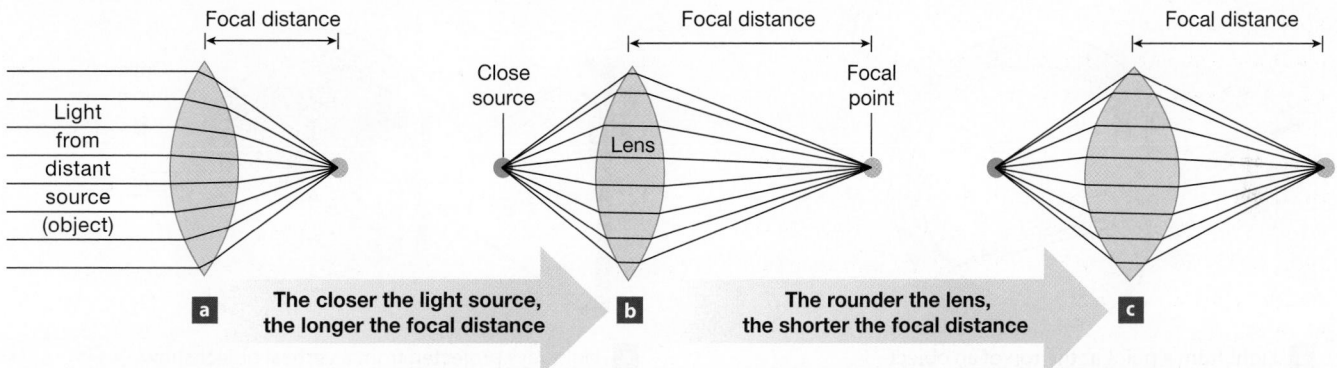

Figure 17–11 **Accommodation.** For the eye to form a sharp image, the focal distance must equal the distance between the center of the lens and the retina.

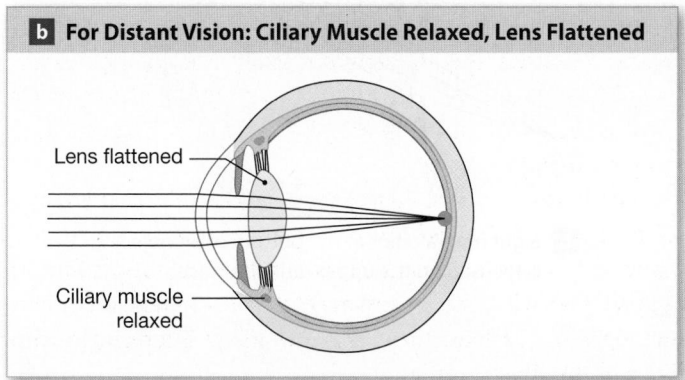

of a nearby object on the retina. The lens becomes flatter to focus the image of a distant object on the retina.

How does the lens change shape? The lens is held in place by the ciliary zonule that originates at the ciliary body. Smooth muscle fibers in the ciliary body act like sphincter muscles. When the ciliary muscle contracts, the ciliary body moves toward the lens, thereby reducing the tension in the ciliary zonule. The elastic capsule then pulls the lens into a rounder shape that increases the refractive (bending) power of the lens. This enables it to bring light from nearby objects into focus on the retina (**Figure 17–11a**). When the ciliary muscle relaxes, the ciliary zonule pulls at the circumference of the lens, making the lens flatter (**Figure 17–11b**).

The greatest amount of refraction is required to view objects that are very close to the lens. The inner limit of clear vision, known as the *near point of vision*, is determined by the degree of elasticity in the lens. Children can usually focus on something 7–9 cm (3–4 in.) from the eye, but over time the lens tends to become stiffer and less responsive. A young adult can usually focus on objects 15–20 cm (6–8 in.) away. As we age, this distance gradually increases. The near point at age 60 is typically about 83 cm (33 in.).

If light passing through the cornea and lens is not refracted properly, the visual image will be distorted. In the condition called **astigmatism**, the degree of curvature in the cornea or lens varies from one axis to another. Minor astigmatism is very common. The image distortion may be so minimal that people are unaware of the condition.

Image Reversal. We have been discussing light that originates at a single point, either near or far from the viewer. An object we see is really a complex light source that must be treated as a number of individual points. Light from each point is focused on the retina as in **Figure 17–12a,b**. The result is a miniature image of the original, but the image arrives upside down and reversed from left to right.

Why does an image arrive upside down? Consider **Figure 17–12c**, a sagittal section through an eye that is looking at a telephone pole. The image of the top of the pole lands at the bottom of the retina, and the image of the bottom hits the top of the retina. Now consider **Figure 17–12d**, a horizontal section through an eye that is looking at a picket fence. The image of the left edge of the fence falls on the right side of the retina, and the image of the right edge falls on the left side of the retina.

Figure 17–12 **Image Formation.**

a Light from a point at the top of an object is focused on the lower retinal surface.

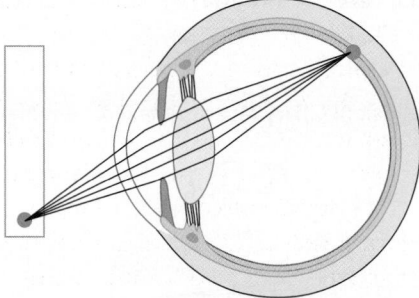

b Light from a point at the bottom of an object is focused on the upper retinal surface.

c Light rays projected from a vertical object show why the image arrives upside down. (Note that the image is also reversed.)

d Light rays projected from a horizontal object show why the image arrives with a left and right reversal. The image also arrives upside down. (As noted in the text, these representations are not drawn to scale.)

The brain compensates for this image reversal, so you are not aware of any difference between the orientation of the image on the retina and that of the object. (Note that these illustrations are not drawn to scale because the fovea occupies a small area of the retina, and the projected images are very tiny. As a result, the cross-over of light rays is shown in the lens, but it actually occurs very close to the fovea.)

Visual Acuity. How well you see, or your **visual acuity**, is rated by comparison to a "normal" standard. The standard vision rating of 20/20 is defined as the level of detail seen at a distance of 20 feet by a person with normal vision. That is, a person with a visual acuity of 20/20 sees clearly at 20 feet what should normally be seen at 20 feet. Vision rated as 20/15 is better than average, because at 20 feet the person is able to see details that would be clear to a normal eye only at a distance of 15 feet. Conversely, a person with 20/30 vision must be 20 feet from an object to discern details that a person with normal vision could make out at a distance of 30 feet.

When visual acuity falls below 20/200, even with the help of glasses or contact lenses, the individual is considered to be legally blind. There are probably fewer than 400,000 legally blind people in the United States, and more than half are over 65 years old. The term *blindness* implies a total absence of vision due to damage to the eyes or to the optic pathways. Common causes of blindness include diabetes mellitus, cataracts, glaucoma, corneal scarring, retinal detachment, accidental injuries, and hereditary factors that are still not understood.

An abnormal blind spot, or **scotoma** (skō-TŌ-muh), may appear in the field of vision at positions other than at the optic disc. A scotoma is permanent and may result from compression of the optic nerve, damage to photoreceptors, or central damage along the visual pathway.

Floaters are small spots that drift across the field of vision. They are generally temporary and result from blood cells or cellular debris in the vitreous body. You may have seen floaters if you have ever stared at a blank wall or a white sheet of paper and noticed these little spots. **Spotlight Figure 17–13** describes common visual abnormalities.

CHECKPOINT

7. Which layer of the eye would be affected first by inadequate tear production?

8. When the lens of your eye is more rounded, are you looking at an object that is close to you or far from you?

9. As Sue enters a dimly lit room, most of the available light becomes focused on the fovea of her eye. Will she be able to see very clearly?

10. How would a blockage of the scleral venous sinus affect your vision?

See the blue Answers tab at the back of the book.

17-4 | Photoreceptors respond to light and change it into electrical signals essential to visual physiology

Learning Outcome Explain color and depth perception, describe how light stimulates the production of nerve impulses, and trace the visual pathways to their destinations in the brain.

How does our special sense of vision work? Let's begin to answer this question by examining visual physiology, the way in which photoreceptors function. Then we will consider the structure and function of the visual pathways.

Visual Physiology

The rods and cones of the retina are called *photoreceptors* because they detect *photons*, basic units of visible light. Light energy is a form of *radiant energy* that travels in waves with characteristic *wavelengths* (distances between wave peaks).

Our eyes are sensitive to wavelengths of 700–400 nm, the spectrum of visible light. Remember this spectrum, as seen in a rainbow, with the acronym ROY G. BIV (*Red, Orange, Yellow, Green, Blue, Indigo, Violet*). Photons of red light carry the least energy and have the longest wavelength. Photons from the violet portion of the spectrum contain the most energy and have the shortest wavelength.

Rods provide the central nervous system with information about the presence or absence of photons, with little regard to their wavelength. Cones provide information about the wavelength of arriving photons, giving us the perception of color.

Anatomy of Rods and Cones

Figure 17–14a compares the structures of rods and cones. The names *rod* and *cone* refer to the shape of each photoreceptor's outer segment.

Visual Pigments. Both rods and cones contain special organic compounds called **visual pigments**. These pigments are located in each photoreceptor's outer segment, in flattened membranous plates called *discs*.

The first key step in the process of *photoreception*—the detection of light—is the absorption of photons by visual pigments. Visual pigments are derivatives of the compound **rhodopsin** (rō-DOP-sin), or *visual purple*, the visual pigment found in rods (**Figure 17–14b**). Rhodopsin consists of a protein, **opsin**, bound to the pigment **retinal** (RET-i-nal), or *retinene*, which is synthesized from **vitamin A**. All rods contain the same form of opsin. Cones contain the same retinal pigment that rods do, but the retinal is attached to different forms of opsin.

We have three types of cones—**blue cones**, **green cones**, and **red cones**. Each type has a different form of opsin and is sensitive to a different range of wavelengths. Their stimulation

A camera focuses an image by moving the lens toward or away from the film. This method cannot work in our eyes, because the distance from the lens to the macula cannot change. We focus images on the retina by changing the shape of the lens to keep the focal distance constant.

The eye has a fixed focal distance and focuses by varying the shape of the lens.

A camera lens has a fixed size and shape and focuses by varying the distance to the film.

Emmetropia (normal vision)

In the healthy eye, when the ciliary muscle is relaxed and the lens is flattened, a distant image will be focused on the retina's surface. This condition is called **emmetropia** (*emmetro-*, proper + *opia*, vision).

Myopia (nearsightedness)

If the eyeball is too deep or the resting curvature of the lens is too great, the image of a distant object is projected in front of the retina. Myopic people see distant objects as blurry and out of focus. Vision at close range will be normal because the lens is able to round as needed to focus the image on the retina.

Myopia corrected with a diverging, concave lens

Diverging lens

Hyperopia (farsightedness)

If the eyeball is too shallow or the lens is too flat, hyperopia results. The ciliary muscle must contract to focus even a distant object on the retina. And at close range the lens cannot provide enough refraction to focus an image on the retina. Older people become farsighted as their lenses lose elasticity, a form of hyperopia called **presbyopia** (*presbys*, old man).

Hyperopia corrected with a converging, convex lens

Converging lens

Surgical Correction

Variable success at correcting myopia and hyperopia has been achieved by surgery that reshapes the cornea. In **photorefractive keratectomy** (PRK) a computer-guided laser shapes the cornea to exact specifications. The entire procedure can be done in less than a minute. A variation on PRK is called *LASIK* (*Laser-Assisted in-Situ Keratomileusis*). In this procedure the interior layers of the cornea are reshaped and then re-covered by the flap of original outer corneal epithelium. Roughly 70 percent of LASIK patients achieve normal vision, and LASIK has become the most common form of refractive surgery.

Even after surgery, many patients still need reading glasses, and both immediate and long-term visual problems can occur.

Figure 17–14   **Structure of Rods, Cones, and Rhodopsin Molecule.**

Pigment Epithelium

The pigment epithelium absorbs photons that are not absorbed by visual pigments. It also phagocytizes old discs shed from the tip of the outer segment.

Melanin granules

In a cone, the discs are infoldings of the plasma membrane, and the outer segment tapers to a blunt point.

In a rod, each disc is an independent entity, and the outer segment forms an elongated cylinder.

Outer Segment

The outer segment of a photoreceptor contains flattened membranous plates, or **discs,** that contain the visual pigments.

Discs

Connecting stalks

Inner Segment

The inner segment contains the photoreceptor's major organelles and is responsible for all cell functions other than photoreception. It also releases neurotransmitters.

Mitochondria

Golgi apparatus

Nuclei

Cone

Rods

Rhodopsin molecule

Retinal

Opsin

b Structure of rhodospin molecule

Each photoreceptor synapses with a bipolar cell.

Bipolar cell

LIGHT

a Structure of rods and cones

in various combinations is the basis for color vision. In an individual with normal vision, the cone population consists of 16 percent blue cones, 10 percent green cones, and 74 percent red cones. Their sensitivities overlap, but each type is most sensitive to a specific portion of the visual spectrum (**Figure 17–15**).

New discs containing visual pigment are continuously assembled at the base of the outer segment of both rods and cones. A completed disc then moves toward the tip of the segment. After about 10 days, the disc is shed in a small droplet of cytoplasm. The pigment cells absorb droplets with shed discs, break down the disc membrane's components, and reconvert the retinal to vitamin A. The pigment cells then store the vitamin A for later transfer to the photoreceptors.

Photoreceptor functioning is described in **Spotlight Figure 17–16.**

Figure 17–15   **Cone Types and Sensitivity to Color.** This graph compares the absorptive characteristics of blue, green, and red cones with those of typical rods.

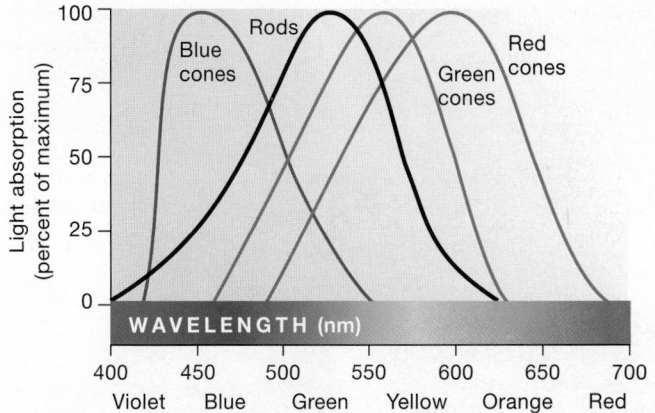

RESTING STATE
DARKNESS

−40 mV

Sodium entry through gated channels produces dark current

The plasma membrane in the outer segment of the photoreceptor contains chemically gated sodium ion channels. In darkness, these gated channels are kept open in the presence of cGMP (cyclic guanosine monophosphate), a derivative of the high-energy compound guanosine triphosphate (GTP). Because the channels are open, the membrane potential is approximately −40 mV, rather than the −70 mV typical of resting neurons. At the −40 mV membrane potential, the photoreceptor is continuously releasing neurotransmitters (in this case, glutamate) across synapses at the inner segment. The inner segment also continuously pumps sodium ions out of the cytoplasm.

Na+

Na+

Rod

The movement of ions into the outer segment, on to the inner segment, and out of the cell is known as the **dark current.**

Neurotransmitter Release

Bipolar cell

Cytosol

Rhodopsin

cGMP

Disc

Na+

Gated Na+ Channel

1 Opsin activation occurs

The bound retinal molecule has two possible configurations: the 11-*cis* form and the 11-*trans* form.

Photon

Rhodopsin

11-*cis* retinal

Opsin

11-*trans* retinal

Normally, the molecule is in the curved 11-*cis* form; on absorbing light it changes to the more linear 11-*trans* form. This change activates the opsin molecule.

ACTIVE STATE
IN LIGHT

2

Opsin activates transducin, which in turn activates phosphodiesterase (PDE)

Transducin is a G protein—a membrane-bound enzyme complex

In this case, opsin activates transducin, and transducin in turn activates **phosphodiesterase** (PDE).

3

Cyclic GMP levels decline and gated sodium channels close

Phosphodiesterase is an enzyme that breaks down cGMP.

The removal of cGMP from the gated sodium channels results in their inactivation. The rate of Na⁺ entry into the cytoplasm then decreases.

4

Dark current is reduced and rate of neurotransmitter release declines

The reduction in the rate of Na⁺ entry reduces the dark current. At the same time, active transport continues to export Na⁺ from the cytoplasm. When the sodium channels close, the membrane potential drops toward –70 mV. As the plasma membrane hyperpolarizes, the rate of neurotransmitter release decreases. This decrease signals the adjacent bipolar cell that the photoreceptor has absorbed a photon. After absorbing a photon, retinal does not spontaneously revert to the 11-*cis* form. Instead, the entire rhodopsin molecule must be broken down into retinal and opsin, in a process called *bleaching*. It is then reassembled.

Figure 17–17   **Bleaching and Regeneration of Visual Pigments.**

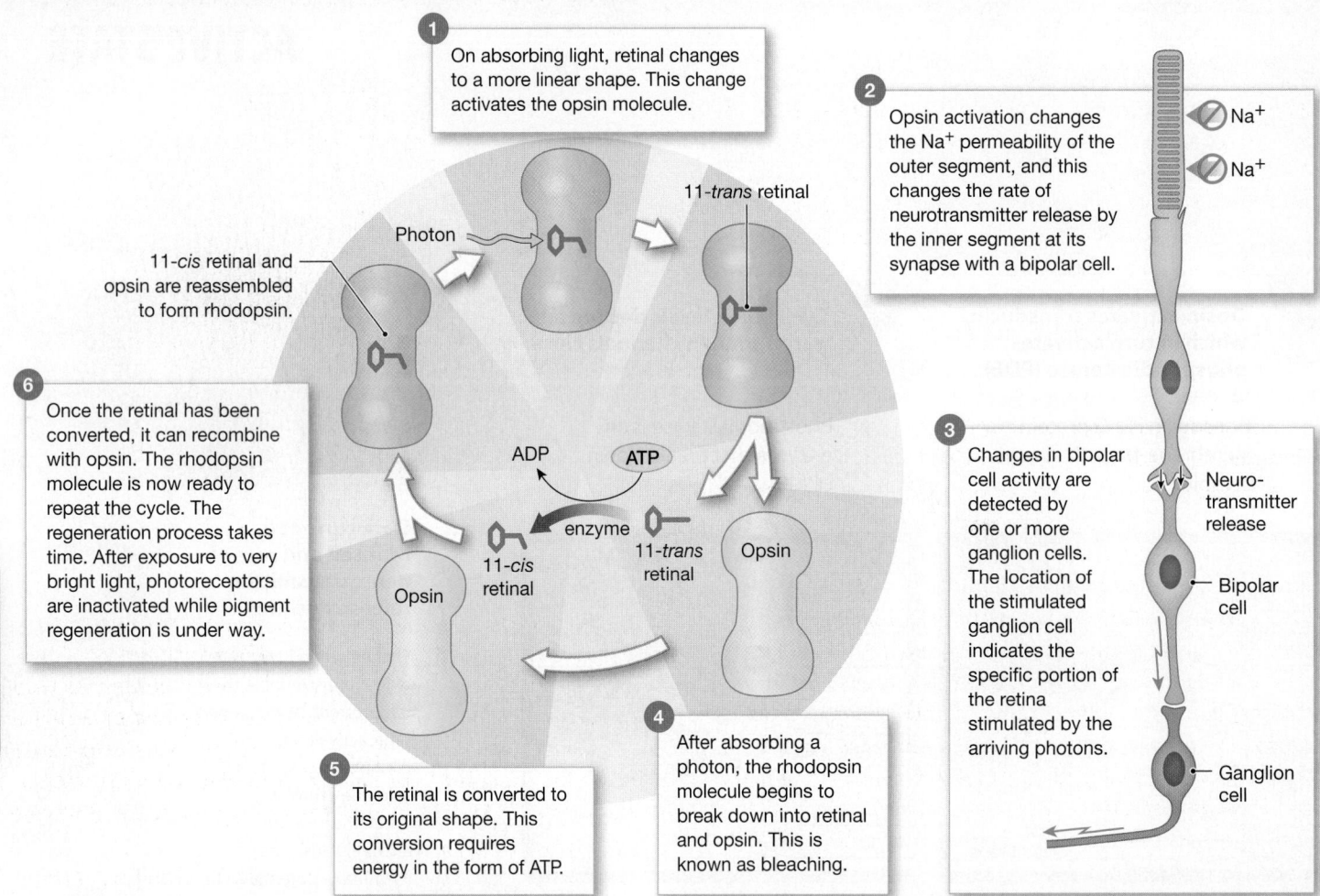

1. On absorbing light, retinal changes to a more linear shape. This change activates the opsin molecule.

2. Opsin activation changes the Na⁺ permeability of the outer segment, and this changes the rate of neurotransmitter release by the inner segment at its synapse with a bipolar cell.

11-*trans* retinal

Photon

11-*cis* retinal and opsin are reassembled to form rhodopsin.

3. Changes in bipolar cell activity are detected by one or more ganglion cells. The location of the stimulated ganglion cell indicates the specific portion of the retina stimulated by the arriving photons.

Neuro-transmitter release

Bipolar cell

Ganglion cell

6. Once the retinal has been converted, it can recombine with opsin. The rhodopsin molecule is now ready to repeat the cycle. The regeneration process takes time. After exposure to very bright light, photoreceptors are inactivated while pigment regeneration is under way.

ADP   ATP

enzyme

11-*cis* retinal

11-*trans* retinal

Opsin

Opsin

4. After absorbing a photon, the rhodopsin molecule begins to break down into retinal and opsin. This is known as bleaching.

5. The retinal is converted to its original shape. This conversion requires energy in the form of ATP.

Bleaching and Regeneration of Visual Pigments. Rhodopsin is broken down through a process called **bleaching** (Figure 17–17). Before it can recombine with opsin, the retinal must be enzymatically converted from the 11-*trans* form to the 11-*cis* form. This conversion requires energy in the form of ATP (adenosine triphosphate), and it takes time.

Bleaching contributes to the lingering visual impression you have after you see a camera's flash. Following intense exposure to light, a photoreceptor cannot respond to further stimulation until its rhodopsin molecules have been regenerated. As a result, a "ghost" image remains on the retina. We seldom notice bleaching under ordinary circumstances, because our eyes are constantly making small, involuntary changes in position that move the image across the retina's surface.

While the rhodopsin molecule is being reassembled, membrane permeability of the outer segment is returning to normal. Opsin is inactivated when bleaching occurs, and the

breakdown of cGMP halts as a result. As other enzymes generate cGMP in the cytoplasm, the chemically gated sodium channels reopen.

As previously noted, the visual pigments of the photoreceptors are synthesized from vitamin A. The body contains vitamin A reserves sufficient for several months. A significant amount is stored in the cells of the pigmented part of the retina. If dietary sources are inadequate, these reserves are gradually used up and the amount of visual pigment in the photoreceptors begins to decline. Daylight vision is affected, but in daytime the light is usually bright enough to stimulate any visual pigments that remain within the densely packed cone population of the fovea. As a result, the problem first becomes apparent at night, when the dim light proves insufficient to activate the rods. This condition, known as **night blindness** or **nyctalopia**, can be treated by eating a diet rich in vitamin A. The body can convert the carotene pigments in many vegetables to vitamin A. Carrots are a particularly good source of carotene, leading to the old adage that carrots are good for your eyes.

Light and Dark Adaptation

The sensitivity of your visual system varies with the intensity of illumination. After 30 minutes or more in the dark, almost all visual pigments will have recovered from photobleaching. They are fully receptive to stimulation. This is the **dark-adapted state**. When dark-adapted, the visual system is extremely sensitive. For example, a single rod will hyperpolarize in response to a single photon of light. Even more remarkable, if as few as seven rods absorb photons at one time, you will see a flash of light.

When the lights come on, at first they seem almost unbearably bright, but over the next few minutes your sensitivity decreases as bleaching occurs. Eventually, the rate of bleaching is balanced by the rate at which the visual pigments reassemble. This condition is the **light-adapted state**. If you moved from the depths of a cave to the full sunlight of midday, your receptor sensitivity would decrease by a factor of 25,000.

A variety of central responses further adjust light sensitivity. The pupils constrict, by the *pupillary constrictor reflex*, reducing the amount of light entering your eye to one-thirtieth the maximum dark-adapted level. Dilating the pupil fully can produce a thirtyfold increase in the amount of light entering the eye, and facilitating some of the synapses along the visual pathway can perhaps triple its sensitivity. In these ways, the sensitivity of the entire system may increase by a factor of more than 1 million.

Retinitis pigmentosa (RP) is an inherited disease characterized by progressive retinal degeneration. As the visual receptors gradually deteriorate, blindness eventually results. The mutations that are responsible change the structure of the photoreceptors—specifically, the visual pigments of the membrane discs. We do not know how the altered pigments lead to the destruction of photoreceptor cells.

Color Vision

An ordinary lightbulb or the sun emits photons of all wavelengths. These photons stimulate both rods and cones. Your eyes also detect photons that reach your retina after they bounce off objects around you. If photons of all colors bounce off an object, the object will appear white to you. If the object absorbs all the photons (so that none reaches the retina), the object will appear black.

An object appears to have a particular color when it reflects (or transmits) photons from one portion of the visible spectrum and absorbs the rest. Color discrimination takes place through the integration of information arriving from all three

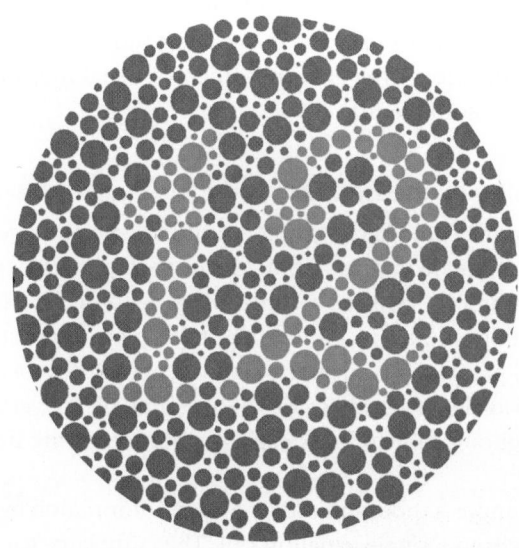

Figure 17–18 **A Standard Test for Color Vision.** People who lack one or more types of cones cannot see the number 12 in this pattern.

types of cones. For example, you perceive yellow from a combination of inputs from highly stimulated green cones, less strongly stimulated red cones, and relatively unaffected blue cones (**Figure 17–15**, p. 609). If all three cone populations are stimulated, we perceive the color as white. We also perceive white if rods, rather than cones, are stimulated. For this reason, everything appears black-and-white when you enter dimly lit surroundings or walk by starlight.

Persons who are unable to distinguish certain colors have a form of **color blindness**. The standard tests for color vision involve picking numbers or letters out of a complex colored picture such as the one in **Figure 17–18**. Color blindness occurs when one or more types of cones are nonfunctional. The cones may be absent, or they may be present but unable to manufacture the necessary visual pigments. In the most common type of color blindness (red–green color blindness), the red cones are missing, so the individual cannot distinguish red light from green light. Inherited color blindness involving one or two cone pigments is not unusual. Ten percent of all males show some color blindness, but only about 0.67 percent of all females have color blindness. Total color blindness is extremely rare. Only 1 person in 300,000 does not manufacture any cone pigments. We consider the inheritance of color blindness in Chapter 29.

The Visual Pathways

The visual pathways begin at the photoreceptors and end at the visual cortex of the cerebral hemispheres. In other sensory pathways we have examined, only one synapse lies between a receptor and a sensory neuron that delivers information to the CNS. In the visual pathways, the message must cross two synapses (photoreceptor to bipolar cell, and bipolar cell to ganglion cell). Only then does it head toward the brain. The extra synapse increases the synaptic delay, but it provides an opportunity for the processing and integration of visual information before it leaves the retina.

Processing by the Retina

Each photoreceptor in the retina monitors a specific receptive field. The retina contains about 130 million photoreceptors, 6 million bipolar cells, and 1 million ganglion cells. Thus, a considerable amount of convergence takes place at the start of the visual pathway. The degree of convergence differs between rods and cones. Regardless of the amount of convergence, each ganglion cell monitors a specific portion of the field of vision.

As many as a thousand rods may pass information by their bipolar cells to a single ganglion cell. The fairly large ganglion cells that monitor rods are called **M cells** (*magnocells; magnus,* great). They provide information about the general form of an object, motion, and shadows in dim lighting. Because so much convergence occurs, the activation of an M cell indicates that light has arrived in a general area rather than at a specific location.

The loss of specificity due to convergence is partially overcome by the fact that the activity of ganglion cells varies according to the pattern of activity in their receptive field, which is usually shaped like a circle. Typically, a ganglion cell responds differently to stimuli that arrive in the center of its receptive field than to stimuli that arrive at the edges (**Figure 17–19**). Some ganglion cells, called **on-center neurons**, are excited by light arriving in the center of their sensory field and are inhibited when light strikes the edges of their receptive field. Others, known as **off-center neurons**, are inhibited by light in the central zone, but are stimulated by illumination at the edges. On-center and off-center neurons provide information about which portion of their receptive field is illuminated. This kind of retinal processing within ganglion receptive fields improves the detection of the edges of objects within the visual field.

Cones typically show very little convergence. In the fovea, the ratio of cones to ganglion cells is 1:1. The ganglion cells that monitor cones, called **P cells** (*parvo cells; parvus,* small), are smaller and more numerous than M cells. P cells are active in bright light, and they provide information about edges, fine detail, and color. Because little convergence occurs, the activation of a P cell means that light has arrived at one specific location. As a result, cones provide more precise information about a visual image than do rods. We could say that images formed

Figure 17–19   Convergence and Ganglion Cell Function. Photoreceptors are organized in groups within a receptive field. Each ganglion cell monitors a well-defined portion of that field. Some ganglion cells (on-center neurons, labeled A) respond strongly to light arriving at the center of their receptive field. Others (off-center neurons, labeled B) respond most strongly to illumination of the edges of their receptive field.

by rods have a coarse, grainy, pixelated appearance that blurs details. By contrast, images produced by cones are sharp, clear, and of high resolution.

Central Processing of Visual Information

Axons from the entire population of ganglion cells converge on the optic disc, penetrate the wall of the eye, and proceed toward the diencephalon as the optic nerve (II). The two optic nerves, one from each eye, reach the diencephalon at the optic chiasm (**Figure 17–20**). From that point, approximately half the fibers proceed toward the lateral geniculate nucleus of the same side of the brain, whereas the other half cross over to reach the lateral geniculate nucleus of the opposite side. ⊃ p. 501 From each lateral geniculate nucleus, visual information travels to the occipital cortex of the cerebral hemisphere on that side.

Figure 17–20 **The Visual Pathways.** The crossover of some nerve fibers occurs at the optic chiasm. As a result, each hemisphere receives visual information from the medial half of the field of vision of the eye on that side, and from the lateral half of the field of vision of the eye on the opposite side. Visual association areas integrate this information to develop a composite picture of the entire field of vision.

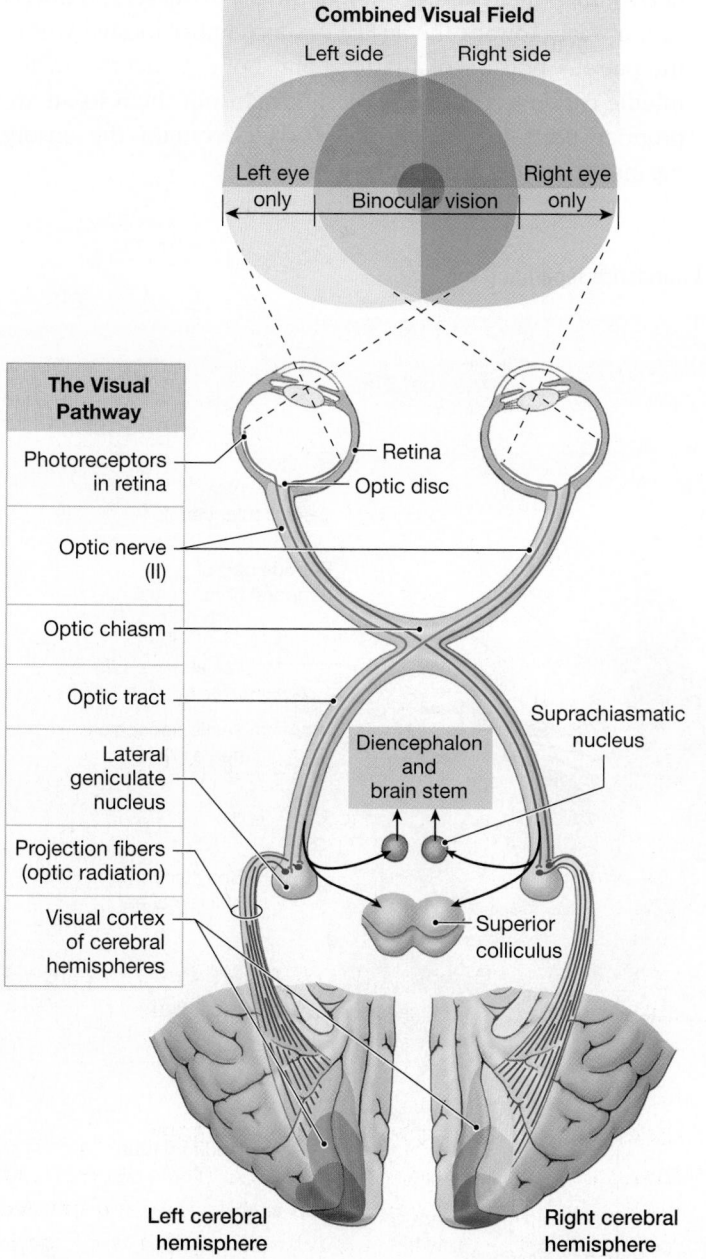

Combined Visual Field

Left side — Right side

Left eye only — Binocular vision — Right eye only

The Visual Pathway

- Photoreceptors in retina
- Retina
- Optic disc
- Optic nerve (II)
- Optic chiasm
- Optic tract
- Lateral geniculate nucleus
- Projection fibers (optic radiation)
- Visual cortex of cerebral hemispheres

Diencephalon and brain stem

Suprachiasmatic nucleus

Superior colliculus

Left cerebral hemisphere — Right cerebral hemisphere

The bundle of projection fibers linking the lateral geniculates with the visual cortex is known as the **optic radiation**. Collaterals from the fibers synapsing in the lateral geniculate continue to subconscious processing centers in the diencephalon and brain stem. For example, the pupillary reflexes and reflexes that control eye movement are triggered by collaterals carrying information to the superior colliculi.

The Field of Vision. You perceive a visual image due to the integration of information that arrives at the visual cortex of the occipital lobes. Each eye receives a slightly different visual image. One reason is that the foveae in your two eyes are 5–7.5 cm (2–3.0 in.) apart. Another reason is that your nose and eye socket block the view of the opposite side. **Depth perception** is the ability to judge depth or distance by interpreting the three-dimensional relationships among objects in view. Your brain obtains it by comparing the relative positions of objects within the images received by your two eyes.

When you look straight ahead, the visual images from your left and right eyes overlap (**Figure 17–20**). The image received by the fovea of each eye lies in the center of the region of overlap. A vertical line drawn through the center of this region marks the division of visual information at the optic chiasm. Visual information from the left half of the combined field of vision reaches the visual cortex of your right occipital lobe. Visual information from the right half of the combined field of vision arrives at the visual cortex of your left occipital lobe.

The cerebral hemispheres thus contain a map of the entire field of vision. As in the case of the primary sensory cortex, the map does not faithfully duplicate the areas within the sensory field. For example, the area assigned to the macula and fovea covers about 35 times the surface it would cover if the map were proportionally accurate. The map is also upside down and reversed, duplicating the orientation of the visual image at the retina.

The Brain Stem and Visual Processing. Many centers in the brain stem receive visual information, either from the lateral geniculate nuclei or through collaterals from the optic tracts. Collaterals that bypass the lateral geniculates synapse in the superior colliculi or in the hypothalamus. The superior colliculi of the midbrain issue motor commands that control unconscious eye, head, or neck movements in response to visual stimuli.

Visual inputs to the suprachiasmatic nucleus of the hypothalamus affect the function of other brain stem nuclei. ⊃ p. 506 The suprachiasmatic nucleus and the *pineal gland* of the epithalamus receive visual information and use it to establish a **circadian rhythm** (*circa*, about + *dies*, day), which is a daily pattern of visceral activity that is tied to the day–night cycle. This circadian rhythm affects your metabolic rate, endocrine function, blood pressure, digestive activities, sleep–wake cycle, and other physiological and behavioral processes.

CHECKPOINT

11. If you had been born without cones in your eyes, would you still be able to see? Explain.

12. How could a diet deficient in vitamin A affect vision?

13. What effect would a decrease in phosphodiesterase activity in photoreceptor cells have on vision?

See the blue Answers tab at the back of the book.

17-5 | Equilibrium sensations originate within the internal ear, while hearing involves the detection and interpretation of sound waves

Learning Outcome Describe the structures of the external, middle, and internal ear, explain their roles in equilibrium and hearing, and trace the pathways for equilibrium and hearing to their destinations in the brain.

The special senses of equilibrium and hearing are provided by the *internal ear*, a receptor complex located in the petrous part of the temporal bone of the skull. *Equilibrium* sensations inform us of the position of the head in space by monitoring gravity, linear acceleration, and rotation. *Hearing* enables us to detect and interpret sound waves.

The basic receptor mechanism for both senses is the same. The receptors, called *hair cells*, are mechanoreceptors. The complex structure of the internal ear and the different arrangement of accessory structures enable hair cells to respond to different stimuli, providing the input for both senses.

Anatomy of the Ear

The ear is divided into three anatomical regions: the external ear, the middle ear, and the internal ear (**Figure 17–21**). The *external ear*—the visible portion of the ear—collects and directs sound waves toward the *middle ear*, a chamber located within the petrous portion of the temporal bone. Structures of the middle ear collect sound waves and transmit them to an appropriate portion of the *internal ear*, which contains the sensory organs for both hearing and equilibrium.

Figure 17–21   The Anatomy of the Ear. The dashed lines indicate the boundaries separating the three anatomical regions of the ear (external, middle, and internal).

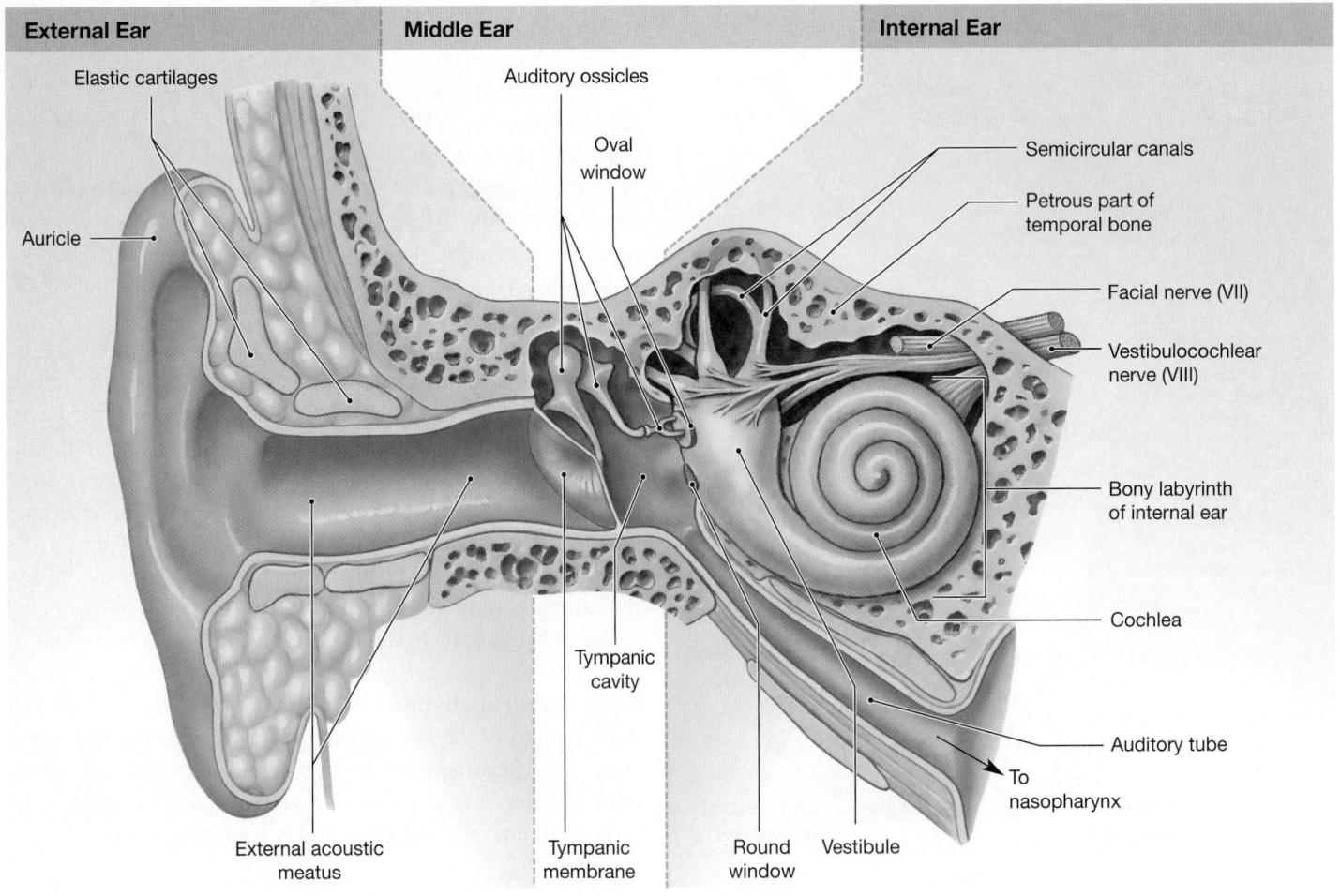

| External Ear | Middle Ear | Internal Ear |

Elastic cartilages

Auditory ossicles

Oval window

Semicircular canals

Petrous part of temporal bone

Auricle

Facial nerve (VII)

Vestibulocochlear nerve (VIII)

Bony labyrinth of internal ear

Cochlea

Auditory tube

To nasopharynx

Tympanic cavity

External acoustic meatus

Tympanic membrane

Round window

Vestibule

The External Ear

The **external ear** includes the outer fleshy and cartilaginous **auricle**, or *pinna*, which surrounds a passageway called the **external acoustic meatus**, or *auditory canal*. The auricle protects the opening of the canal and provides directional sensitivity. Sounds coming from behind the head are blocked by the auricle, but sounds coming from the side or front are collected and channeled into the external acoustic meatus. (When you "cup" your ear with your hand to hear a faint sound more clearly, you are exaggerating this effect.) The external acoustic meatus ends at the **tympanic membrane** (*tympanon*, drum) or *eardrum*. This thin, semitransparent sheet separates the external ear from the middle ear.

The tympanic membrane is very delicate. The auricle and the narrow external acoustic meatus provide some protection from accidental injury. In addition, **ceruminous glands**— integumentary glands along the external acoustic meatus— secrete a waxy material that helps keep out foreign objects or small insects. The canal is also lined with many small, outwardly projecting hairs. These hairs trap debris and also provide increased tactile sensitivity through their root hair plexuses. The slightly waxy secretion of the ceruminous glands, called **cerumen**, also slows the growth of microorganisms and reduces the chances of infection.

The Middle Ear

The **middle ear**, or **tympanic cavity**, is an air-filled chamber separated from the external acoustic meatus by the tympanic membrane. The middle ear communicates both with the *nasopharynx* (the superior portion of the pharynx), through the **auditory tube**, and with the mastoid air cells, through a number of small connections (**Figures 17–21** and **17–22**). The auditory tube is also called the *pharyngotympanic tube* or the *Eustachian tube*. About 4 cm (1.6 in.) long, it consists of two portions. The portion near the connection to the middle ear is narrow and is supported by elastic cartilage. The portion near the opening into the nasopharynx is broad and funnel shaped. The auditory tube equalizes pressure on either side of the tympanic membrane. Unfortunately, the auditory tube can also allow microorganisms to travel from the nasopharynx into the middle ear. Invasion by microorganisms can lead to an unpleasant middle ear infection known as *otitis media*.

The Auditory Ossicles. The middle ear contains three tiny ear bones, collectively called **auditory ossicles**. These ear bones connect the tympanic membrane with one of the receptor complexes of the internal ear (**Figures 17–21** and **17–22**). The three auditory ossicles are the malleus, the incus, and the stapes. The **malleus** (*malleus*, hammer) attaches at three points to the interior surface of the tympanic membrane. The **incus** (*incus*, anvil), the middle ossicle, attaches the malleus to the **stapes** (*stapes*, stirrup), the inner ossicle. The edges of the base of the stapes are bound to the edges of the *oval window*, an opening in the temporal bone that surrounds the internal ear.

The articulations between the auditory ossicles are the smallest synovial joints in the body. Each has a tiny joint capsule and supporting extracapsular ligaments.

Vibration of the tympanic membrane converts arriving sound waves into mechanical movements. The auditory ossicles act as levers that conduct those vibrations to the internal ear. The ossicles are connected in such a way that an in–out movement of the tympanic membrane produces a rocking motion of the stapes. In this way, the ossicles function as a lever system that collects the force applied to the tympanic membrane and focuses it on the oval window. Considerable amplification occurs, so we can hear very faint sounds, because the tympanic membrane is 22 times larger and heavier than the oval window.

That degree of amplification can be a problem when we are exposed to very loud noises. In the middle ear, two small muscles protect the tympanic membrane and ossicles from violent movements under very noisy conditions:

1. The **tensor tympani** (TEN-sor tim-PAN-ē) **muscle** is a short ribbon of muscle originating on the petrous part of the temporal bone and the auditory tube, and inserting on the "handle" of the malleus. When the tensor tympani contracts, it pulls the malleus medially, stiffening the tympanic membrane. This increased stiffness reduces the amount of movement possible. The tensor tympani muscle is innervated by motor fibers of the mandibular branch of the trigeminal nerve (V).

2. The **stapedius** (sta-PĒ-dē-us) **muscle**, innervated by the facial nerve (VII), originates from the posterior wall of the middle ear and inserts on the stapes. Contraction of the stapedius pulls the stapes, reducing movement of the stapes at the oval window.

The Internal Ear

Receptors in the **internal ear** provide us with the senses of equilibrium and hearing (**Figures 17–21** and **17–23**).

The superficial contours of the internal ear are formed by a layer of dense bone known as the **bony labyrinth** (*labyrinthos*, network of canals). The walls of the bony labyrinth are continuous with the surrounding temporal bone. The inner contours of the bony labyrinth closely follow the contours of the **membranous labyrinth**, a delicate, interconnected network of fluid-filled tubes. The receptors of the internal ear are found within those tubes.

Between the bony and membranous labyrinths flows **perilymph** (PER-i-limf), a liquid that closely resembles cerebrospinal fluid. The membranous labyrinth contains **endolymph** (EN-dō-limf), a fluid with electrolyte concentrations that differ from those of typical body fluids. The physical relationships

Figure 17–22 **The Middle Ear.**

Auditory Ossicles

| Malleus | Incus | Stapes |

Temporal bone
(petrous part)

Stabilizing
ligaments

Branch of facial
nerve VII (cut)

External
acoustic meatus

Tympanic cavity
(middle ear)

Tympanic
membrane

Oval window

**Muscles of
the Middle Ear**

Tensor tympani
muscle

Stapedius muscle

Round window

Auditory tube

a The structures of the middle ear

Malleus attached to
tympanic membrane

Malleus

Tendon of tensor
tympani muscle

Incus

Base of stapes
at oval window

Stapes

Stapedius muscle

Inner surface of
tympanic membrane

b The tympanic membrane and auditory ossicles

Incus

Malleus

Points of
attachment
to tympanic
membrane

Stapes

Base of stapes

c The isolated auditory ossicles

are shown in **Figure 17–23a**. (See the Appendix for a chemical analysis of perilymph, endolymph, and other body fluids.)

We can subdivide the bony labyrinth into the *vestibule*, three *semicircular canals*, and the *cochlea* (**Figure 17–23b**). The **vestibule** (VES-ti-būl) consists of a pair of membranous sacs: the **saccule** (SAK-ūl) and the **utricle** (Ū-tri-kul). Receptors in these sacs provide sensations of gravity and linear acceleration.

The **semicircular canals** enclose slender *semicircular ducts*. Receptors in the semicircular ducts are stimulated by rotation of the head. The combination of vestibule and semicircular canals is called the **vestibular complex**. The fluid-filled chambers

within the vestibule are continuous with those of the semicircular canals.

The **cochlea** (KOK-lē-uh; *cochlea*, a snail shell) is a spiral-shaped, bony chamber that contains the **cochlear duct** of the membranous labyrinth. Receptors within the cochlear duct provide the sense of hearing. The duct is sandwiched between a pair of perilymph-filled chambers. The entire complex spirals around a central bony hub, much like a snail shell.

The walls of the bony labyrinth consist of dense bone everywhere except at two small areas near the base of the cochlear spiral (**Figure 17–21**). The **round window** is a thin,

Figure 17–23 **The Internal Ear.**

KEY	
	Membranous labyrinth
	Bony labyrinth

Semicircular ducts
- Anterior
- Lateral
- Posterior

Vestibule

Semicircular canal

Cristae within ampullae

Maculae

Endolymphatic sac

Cochlea

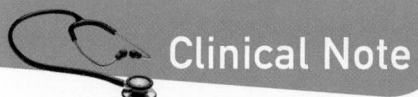

- Perilymph
- Bony labyrinth
- Endolymph
- Membranous labyrinth

Utricle
Saccule
Vestibular duct
Cochlear duct

Tympanic duct   Spiral organ

a A section through one of the semicircular canals, showing the relationship between the bony and membranous labyrinths, and the boundaries of perilymph and endolymph.

b The bony and membranous labyrinths. Areas of the membranous labyrinth containing sensory receptors (cristae, maculae, and spiral organ) are shown in purple.

Clinical Note

Motion Sickness The signs and symptoms of **motion sickness** are exceedingly unpleasant. They include headache, sweating, facial flushing, nausea, vomiting, and various changes in mental perspective. Motion sickness may result when central processing stations, such as the tectum of the midbrain, receive conflicting sensory information. Why and how these conflicting reports result in nausea, vomiting, and other signs and symptoms are not known. Sitting belowdecks on a moving boat or reading in a car or airplane tends to provide the necessary conditions. Your eyes (which are tracking lines on a page) report that your position in space is not changing, but your semicircular ducts report that your body is lurching and turning. To counter this effect, seasick sailors watch the horizon rather than their immediate surroundings, so that their eyes will provide visual confirmation of the movements detected by their internal ears. It is not known why some individuals are almost immune to motion sickness, but others find travel by boat or plane almost impossible.

Drugs commonly used to prevent motion sickness include dimenhydrinate (*Dramamine*), scopolamine (*Transderm Scop*), and promethazine. These drugs appear to depress activity at the vestibular nuclei. Sedatives, such as prochlorperazine (*Compazine*), may also be effective.

membranous partition that separates the perilymph of the cochlear chambers from the air-filled middle ear. Collagen fibers connect the bony margins of the opening known as the **oval window** to the base of the stapes.

Equilibrium

As just noted, receptors of the vestibular complex provide you with equilibrium sensations. The semicircular ducts convey information about rotational movements of the head. For example, when you turn your head to the left, receptors stimulated in the semicircular ducts tell you how rapid the movement is, and in which direction. The saccule and the utricle convey information about your position with respect to gravity. If you stand with your head tilted to one side, these receptors report the angle involved and whether your head tilts forward or backward. The saccule and the utricle are also stimulated by sudden acceleration. When your car accelerates from a stop, the saccular and utricular receptors give you the sensation of increasing speed.

The Semicircular Ducts

Sensory receptors in the semicircular ducts respond to rotational movements of the head. These **hair cells** are active during a movement, but are quiet when the body is motionless. The **anterior**, **posterior**, and **lateral semicircular ducts** are continuous with the utricle (**Figure 17–24a**). Each semicircular

Figure 17–24   **The Semicircular Ducts.**

a An anterior view of the right semicircular ducts, the utricle, and the saccule, showing the locations of sensory receptors.

17

b A cross section through the ampulla of a semicircular duct.

c Endolymph movement along the length of the duct moves the cupula and stimulates the hair cells.

d A representative hair cell (receptor) from the vestibular complex. Bending the sterocilia toward the kinocilium depolarizes the cell and stimulates the sensory neuron. Displacement in the opposite direction inhibits the sensory neuron.

duct contains an **ampulla**, an expanded region that contains the receptors. The region in the wall of the ampulla that contains the receptors is known as a **crista ampullaris** or *crista* (Figure 17–24b). Each crista ampullaris is bound to a **cupula** (KŪ-pū-luh), a gelatinous structure that extends the full width of the ampulla. The receptors in the cristae ampullares are hair cells (Figure 17–24b,d).

Hair cells are the receptors in other portions of the membranous labyrinth as well. Regardless of location, they are always surrounded by supporting cells and monitored by the dendrites of sensory neurons. The free surface of each hair cell supports 80–100 long **stereocilia**, which resemble very long microvilli (Figure 17–24d). Each hair cell in the vestibule also contains a single large cilium called a **kinocilium** (kī-nō-SIL-ē-um). Hair cells do not actively move their kinocilia or stereocilia. However, when an external force pushes against these processes, the distortion of the plasma membrane alters the rate at which the hair cell releases chemical transmitters.

Hair cells provide information about the direction and strength of mechanical stimuli. The stimuli involved, however, depend on the hair cell's location: gravity or acceleration in the vestibule, rotation in the semicircular canals, and sound in the cochlea. The sensitivities of the hair cells differ, because each of these regions has different accessory structures that determine which stimulus will provide the force to deflect the kinocilia and stereocilia.

At a crista, the kinocilia and stereocilia of the hair cells are embedded in the cupula (Figure 17–24b). Because the cupula has a density very close to that of the surrounding endolymph, it essentially floats above the receptor surface. When your head rotates in the plane of the duct, the movement of endolymph along the length of the semicircular duct pushes the cupula to the side and distorts the receptor processes (Figure 17–24c). Movement of fluid in one direction stimulates the hair cells, and movement in the opposite direction inhibits them. When the endolymph stops moving, the elastic nature of the cupula makes it return to its normal position.

Even the most complex movement can be analyzed in terms of motion in three rotational planes. Each semicircular duct responds to one of these rotational movements. A horizontal rotation, as in shaking your head "no," stimulates the hair cells of the lateral semicircular duct. Nodding "yes" excites the anterior duct, and tilting your head from side to side activates receptors in the posterior duct.

The Utricle and Saccule

The utricle and saccule provide equilibrium sensations, whether the body is moving or is stationary. A slender passageway that is continuous with the narrow endolymphatic duct connects the two chambers. The **endolymphatic duct** ends in a closed cavity called the **endolymphatic sac** (Figure 17–24a). This sac projects through the dura mater that lines the temporal bone and into the subarachnoid space, where a capillary network surrounds it. Portions of the cochlear duct secrete endolymph continuously, and excess fluid returns to the general circulation at the endolymphatic sac. There the capillaries absorb endolymph removed by a combination of active transport and vesicular transport.

The hair cells of the utricle and saccule are clustered in oval structures called **maculae** (MAK-ū-lē; *macula*, spot) (Figure 17–25a). The macula in the utricle is sensitive to changes in horizontal movement. The macula in the saccule is sensitive to changes in vertical movement. As in the ampullae, the hair cell processes are embedded in a gelatinous otolithic membrane whose surface contains densely packed calcium carbonate crystals. These crystals are called **otoliths** ("ear stones").

The macula of the saccule is diagrammed in Figure 17–25b. The function of the macula in the utricle is shown in Figure 17–25c. When your head is in the normal, upright position, the otoliths sit atop the otolithic membrane of the utricular macula (❶). Their weight presses on the macular surface, pushing the hair cell processes down rather than to one side or another. When your head is tilted, the pull of gravity on the otoliths shifts them to the side, distorting the hair cell processes and stimulating the macular receptors (❷). The change in receptor activity tells the CNS that your head is no longer level.

A similar mechanism accounts for your perception of linear acceleration when you are in a car that speeds up suddenly. The otoliths lag behind, and the effect on the hair cells is comparable to tilting your head back. Under normal circumstances, your nervous system distinguishes between the sensations of tilting and linear acceleration by integrating vestibular sensations with visual information. Many amusement park rides confuse your sense of equilibrium by combining rapid rotation with changes in position and acceleration while providing restricted or misleading visual information.

Pathways for Equilibrium Sensations

Hair cells of the vestibule and semicircular ducts are monitored by sensory neurons located in adjacent **vestibular ganglia**. Sensory fibers from these ganglia form the **vestibular branch** of the vestibulocochlear nerve (VIII). ⟲ p. 524 These fibers innervate neurons within the pair of **vestibular nuclei** at the boundary between the pons and the medulla oblongata.

The two vestibular nuclei have four functions:

1. Integrating sensory information about balance and equilibrium that arrives from both sides of the head.

2. Relaying information from the vestibular complex to the cerebellum.

Figure 17–25 **The Saccule and Utricle.**

a The location of the maculae

b The structure of an individual macula

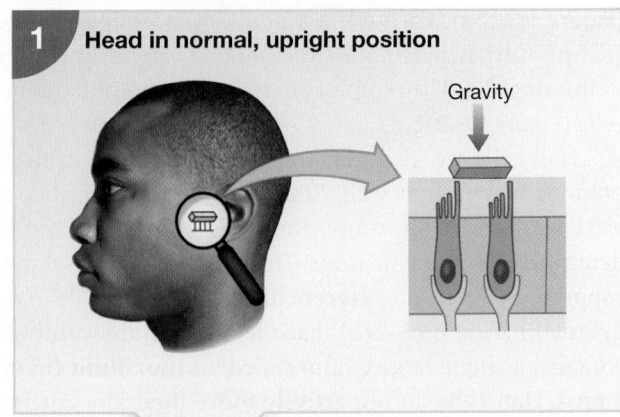

1 **Head in normal, upright position**

Gravity

2 **Head tilted posteriorly**

Gravity

Receptor output increases

Otolith moves "downhill," distorting hair cell processes

c A diagrammatic view of utricular macular function when the head is held normally **1** and then tilted back **2**

3. Relaying information from the vestibular complex to the cerebral cortex, providing a conscious sense of head position and movement.

4. Sending commands to motor nuclei in the brain stem and spinal cord.

The reflexive motor commands issued by the vestibular nuclei are distributed to the motor nuclei for cranial nerves involved with eye, head, and neck movements (N III, N IV, N VI, and N XI). Instructions descending in the *vestibulospinal tracts* of the spinal cord adjust peripheral muscle tone and complement the reflexive movements of the head or neck. ⤺ p. 552 These pathways are indicated in **Figure 17–26**.

The automatic movements of the eyes in response to sensations of motion are directed by the *superior colliculi* of the midbrain. ⤺ p. 501 These movements attempt to keep your gaze focused on a specific point in space, despite changes in body position and orientation. If your body is turning or spinning rapidly, your eyes will fix on one point for a moment and then jump ahead to another in a series of short, jerky movements.

If either the brain stem or the internal ear is damaged, this type of eye movement can occur even when the body is stationary. Individuals with this condition, which is called **nystagmus** (nis-TAG-mus), have trouble controlling their eye movements. Physicians commonly check for nystagmus by asking patients to watch a small penlight as it is moved across the field of vision.

Hearing

The receptors of the cochlear duct provide a sense of hearing that enables us to detect the quietest whisper, yet remain functional in a noisy room. The receptors responsible for auditory sensations are hair cells similar to those of the vestibular complex. However, their placement within the cochlear duct and the organization of the surrounding accessory structures shield them from stimuli other than sound.

Figure 17–26   Pathways for Equilibrium Sensations.

In conveying vibrations from the tympanic membrane to the oval window, the auditory ossicles convert pressure fluctuations in air into much greater pressure fluctuations in the perilymph of the cochlea. These fluctuations stimulate hair cells along the cochlear spiral. The *frequency* of the perceived sound is determined by *which part* of the cochlear duct is stimulated. The *intensity* (volume) of the perceived sound is determined by *how many* of the hair cells at that location are stimulated. Let's take a closer look at the mechanics of this remarkably elegant process.

The Cochlear Duct

In sectional view (**Figures 17–27** and **17–28**), the cochlear duct lies between a pair of perilymphatic chambers or *scalae*: the **scala vestibule** (SKĀ-luh ves-TIB-yū-lē), or *vestibular duct*, and the **scala tympani** (SKĀ-luh TIM-pa-nē), or *tympanic duct*. The outer surfaces of these ducts are encased by the bony labyrinth everywhere except at the oval window (the base of the scala vestibuli) and the round window (the base of the scala tympani). These scalae really form one long and continuous perilymphatic chamber because they are interconnected at the tip of the cochlear spiral. This chamber begins at the oval window; extends through the scala vestibuli, around the top of the cochlea, and along the scala tympani; and ends at the round window.

The cochlear duct is an elongated tubelike structure suspended between the vestibular duct and the tympanic duct. The hair cells of the cochlear duct are located in a structure called the **spiral organ** (*organ of Corti*) (**Figures 17–27b** and **17–28a,b**). This sensory structure sits on the **basilar membrane**, a membrane that separates the cochlear duct from the scala tympani. The hair cells are arranged in a series of longitudinal rows. They lack kinocilia, and their stereocilia are in contact with the overlying **tectorial** (tek-TOR-ē-al; *tectum*, roof) **membrane**. This membrane is firmly attached to the inner wall of the cochlear duct. When a portion of the basilar membrane bounces up and down, the stereocilia of the hair cells are pressed against the tectorial membrane and become distorted. The basilar membrane moves in response to pressure fluctuations within the perilymph. These pressure changes are triggered by sound waves arriving at the tympanic membrane. The basic properties of sound are key to understanding this process.

An Introduction to Sound

Hearing is the perception of sound, but what is sound? It consists of waves of pressure conducted through a medium such as air or water. In air, each *pressure wave* consists of a region where the air molecules are crowded together and an adjacent zone

Figure 17–27 **The Cochlea.**

a The structure of the cochlea

b Diagrammatic and sectional views of the cochlear spiral

where they are farther apart (**Figure 17–29a**). These waves are sine waves—that is, S-shaped curves that repeat in a regular pattern. They travel through the air at about 1235 km/h (768 mph).

The *wavelength* of sound is the distance between two adjacent wave crests (peaks), or the distance between two adjacent wave troughs (**Figure 17–29b**). The **frequency** of a sound is the number of waves that pass a fixed reference point in a given time. Physicists use the term **cycles** rather than *waves*. We measure the frequency of a sound in terms of the number of cycles

per second (cps), a unit called **hertz (Hz)**. Wavelength and frequency are inversely related.

What we perceive as the **pitch** of a sound is our sensory response to its frequency. A *high-frequency* sound (high pitch, short wavelength) might have a frequency of 15,000 Hz or more. A very *low-frequency* sound (low pitch, long wavelength) could have a frequency of 100 Hz or less.

It takes energy to produce sound waves. When you strike a tuning fork, it vibrates, pushing against the surrounding air.

Figure 17–28 **The Spiral Organ.**

Bony cochlear wall
Scala vestibuli
Vestibular membrane
Cochlear duct
Tectorial membrane
Basilar membrane
Scala tympani
Spiral organ

Spiral ganglion

Cochlear branch of N VIII

a A three-dimensional section of the cochlea, showing the compartments, tectorial membrane, and spiral organ

Tectorial membrane
Outer hair cell
Basilar membrane   Inner hair cell   Nerve fibers

Cochlear duct
Vestibular membrane
Tectorial membrane
Scala tympani   Basilar membrane   Hair cells of spiral organ   Spiral ganglion cells of cochlear nerve
Spiral organ   LM × 125

b Diagrammatic and sectional views of the receptor hair cell complex of the spiral organ

It produces sound waves whose frequency depends on the instrument's frequency of vibration.

The harder you strike the tuning fork, the more energy you provide. The energy increases the **amplitude**, or height, of the sound wave (**Figure 17–29b**). The amount of energy in a sound wave, or *intensity*, determines how loud it seems. The greater the energy content, the larger the amplitude, and the louder the sound. We report sound energy in **decibels** (DES-i-belz, dB). Table 17–1 shows the decibel levels of familiar sounds.

When sound waves strike an object, their energy is a physical pressure. You may have seen windows move in a room where

a stereo is blasting. The more flexible the object, the more easily it will respond to the pressure of sound waves. Even soft stereo music will vibrate a sheet of paper held in front of the speaker. Given the right combination of frequencies and amplitudes, an object will begin to vibrate at the same frequency as the sound, a phenomenon called *resonance*. For you to be able to hear any sound, your thin, flexible tympanic membrane must vibrate in resonance with the sound waves.

Probably more than 6 million people in the United States have at least a partial hearing deficit. It is due to problems with either the transfer of vibrations by the auditory ossicles or damage to the receptors or the auditory pathways.

Figure 17–29 **The Nature of Sound.**

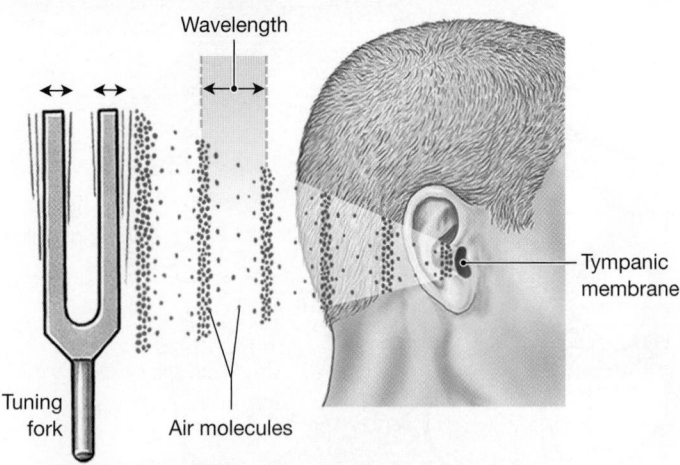

a Sound waves (here, generated by a tuning fork) travel through the air as pressure waves.

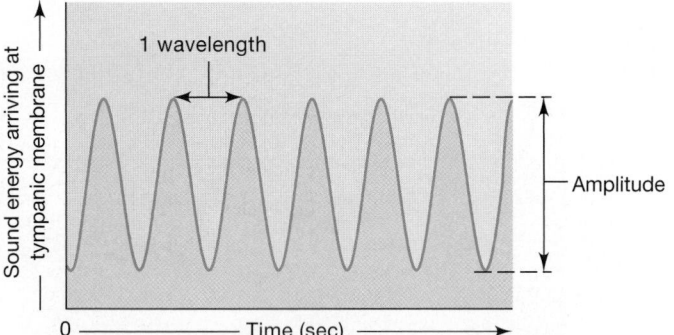

b A graph showing the sound energy arriving at the tympanic membrane. The distance between wave peaks is the wavelength. The number of waves arriving each second is the frequency, which we perceive as pitch. Frequencies are reported in cycles per second (cps), or hertz (Hz). The amount of energy carried by the wave is its amplitude. The greater the amplitude, the louder the sound.

The Hearing Process

We divide the process of hearing into six basic steps (Figure 17–30):

1 *Sound waves arrive at the tympanic membrane.* Sound waves enter the external acoustic meatus and travel toward the tympanic membrane. The orientation of the auditory canal provides some directional sensitivity. Sound waves approaching a particular side of the head have direct access to the tympanic membrane on that side. Sounds arriving from another direction must bend around corners or pass through the auricle or other body tissues.

2 *Movement of the tympanic membrane causes displacement of the auditory ossicles.* The tympanic membrane provides a

Table 17–1	Intensity of Representative Sounds	
Typical Decibel Level	**Example**	**Dangerous Time Exposure**
0	Lowest audible sound	
30	Quiet library; soft whisper	
40	Quiet office; living room; bedroom away from traffic	
50	Light traffic at a distance; refrigerator; gentle breeze	
60	Air conditioner at 20 feet; conversation; sewing machine in operation	
70	Busy traffic; noisy restaurant	Some damage if continuous
80	Subway; heavy city traffic; alarm clock at 2 feet; factory noise	More than 8 hours
90	Truck traffic; noisy home appliances; shop tools; gas lawn mower	Less than 8 hours
100	Chain saw; boiler shop; pneumatic drill	2 hours
120	"Heavy metal" rock concert; sandblasting; thunderclap nearby	Immediate danger
140	Gunshot; jet plane	Immediate danger
160	Rocket launching pad	Hearing loss inevitable

surface for the collection of sound. The membrane vibrates in resonance to sound waves with frequencies between approximately 20 and 20,000 Hz. When the tympanic membrane vibrates, so do the malleus, incus, and stapes. In this way, the sound is amplified.

3 *Movement of the stapes at the oval window establishes pressure waves in the perilymph of the scala vestibuli.* Liquids are not compressible: If you push down on one part of a waterbed, the bed bulges somewhere else. Because the rest of the cochlea is sheathed in bone, pressure applied at the oval window can be relieved only at the round window. The stapes actually has a rocking movement, but the in–out component is easiest to visualize and describe. Basically, when the stapes moves inward, the round window bulges outward, into the middle ear cavity. As the stapes moves in and out, vibrating at the frequency of the sound arriving at the tympanic membrane, it creates pressure waves within the perilymph.

4 *The pressure waves distort the basilar membrane on their way to the round window of the scala tympani.* The stapes create pressure waves that travel through the perilymph of the scala vestibuli and scala tympani to reach the round window. In doing so, the waves distort the basilar membrane. The location of maximum distortion varies with the frequency of the sound, due to regional differences in the width and

Figure 17–30 **Sound and Hearing.** Steps in the reception and transduction of sound energy.

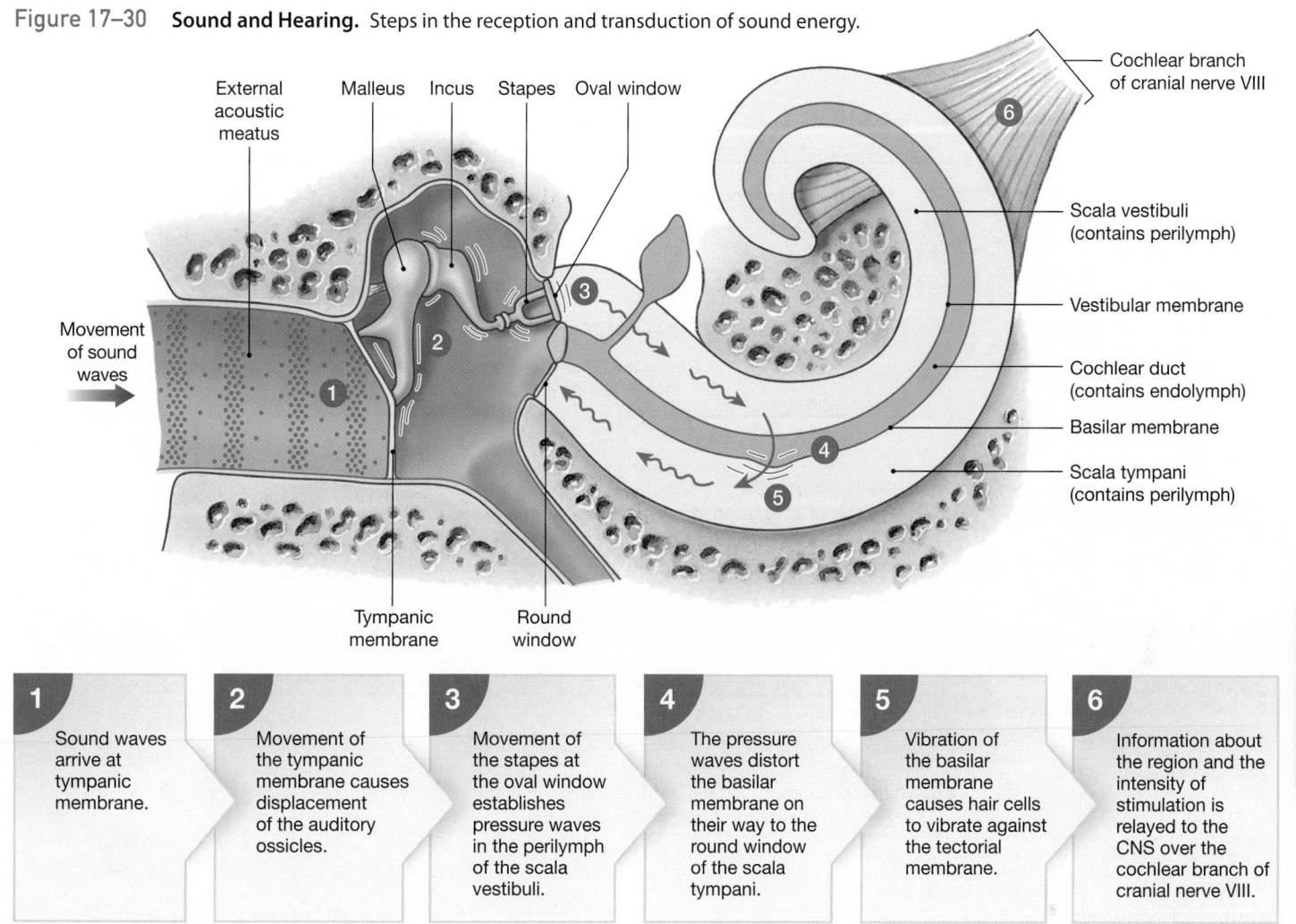

1	2	3	4	5	6
Sound waves arrive at tympanic membrane.	Movement of the tympanic membrane causes displacement of the auditory ossicles.	Movement of the stapes at the oval window establishes pressure waves in the perilymph of the scala vestibuli.	The pressure waves distort the basilar membrane on their way to the round window of the scala tympani.	Vibration of the basilar membrane causes hair cells to vibrate against the tectorial membrane.	Information about the region and the intensity of stimulation is relayed to the CNS over the cochlear branch of cranial nerve VIII.

flexibility of the basilar membrane along its length. High-frequency sounds, which have a very short wavelength, vibrate the basilar membrane near the oval window. The lower the frequency of the sound, the longer the wavelength, and the farther from the oval window is the area of maximum distortion (**Figure 17–31**). In this way, information about frequency is translated into information about *position* along the basilar membrane.

The *amount* of movement at a given location depends on the amount of force applied by the stapes, which in turn depends on the energy content of the sound. The louder the sound, the more the basilar membrane moves.

⑤ ***Vibration of the basilar membrane causes hair cells to vibrate against the tectorial membrane.*** Vibration of the affected region of the basilar membrane moves hair cells against the tectorial membrane. This movement leads to the displacement of the stereocilia, which in turn opens ion channels in the plasma membranes of the hair cells. The resulting inrush of ions depolarizes

the hair cells, leading to the release of neurotransmitters that stimulate sensory neurons.

The hair cells of the spiral organ are arranged in several rows. A very soft sound may stimulate only a few hair cells in a portion of one row. As the intensity of a sound increases, not only do these hair cells become more active, but additional hair cells—first in the same row and then in adjacent rows—are stimulated as well. The number of hair cells responding in a given region of the spiral organ provides information on the intensity of the sound.

⑥ ***Information about the region and the intensity of stimulation is relayed to the CNS over the cochlear branch of cranial nerve VIII.*** The cell bodies of the bipolar sensory neurons that monitor the cochlear hair cells are located at the center of the bony cochlea, in the **spiral ganglion** (**Figure 17–28a**). From there, the information is carried by the cochlear branch of cranial nerve VIII to the cochlear nuclei of the medulla oblongata for distribution to other centers in the brain.

Figure 17–31 **Frequency Discrimination.**

a The flexibility of the basilar membrane varies along its length, so pressure waves of different frequencies affect different parts of the membrane.

b The effects of a vibration of the stapes at a frequency of 6000 Hz. When the stapes moves inward, as shown here, the basilar membrane distorts toward the round window, which bulges into the cavity of the middle ear.

c When the stapes moves outward, as shown here, the basilar membrane rebounds and distorts toward the oval window.

Auditory Pathways

Stimulation of hair cells activates sensory neurons whose cell bodies are in the adjacent spiral ganglion (**Figure 17–32 1**). The afferent fibers of those neurons form the **cochlear branch** of the vestibulocochlear nerve (VIII). These axons enter the medulla oblongata, where they synapse at the **cochlear nucleus** on that side **2**. From there, information ascends to the **superior olivary nucleus** of the pons and both inferior colliculi of the midbrain **3**. This midbrain processing center coordinates a number of responses to acoustic stimuli, including auditory reflexes that involve skeletal muscles of the head, face, and trunk **4**. These reflexes automatically change the position of your head in response to a sudden loud noise. You usually turn your head and your eyes toward the source of the sound.

Before reaching the cerebral cortex and your awareness, ascending auditory sensations synapse in the medial geniculate nucleus of the thalamus **5**. Projection fibers then deliver the information to the auditory cortex of the temporal lobe **6**. Information travels to the cortex over labeled lines: High-frequency sounds activate one portion of the cortex, low-frequency sounds another. In effect, the auditory cortex contains a map of the spiral organ. So, information about *frequency*,

translated into information about *position* on the basilar membrane, is projected in that form onto the auditory cortex. There it is interpreted to produce your subjective sensation of pitch.

Most of the auditory information from one cochlea is projected to the auditory complex of the cerebral hemisphere on the opposite side of the brain. However, each auditory cortex also receives information from the cochlea on that side. These interconnections play a role in localizing left/right sounds. They can also reduce the functional impact of damage to a cochlea or ascending pathway.

An individual whose auditory cortex is damaged will respond to sounds and have normal acoustic reflexes, but will find it difficult or impossible to interpret the sounds and recognize a pattern in them. Damage to the adjacent association area leaves the ability to detect the tones and patterns intact, but produces an inability to comprehend their meaning.

Auditory Sensitivity

Our hearing abilities are remarkable, but it is difficult to assess the absolute sensitivity of the system. The range from the softest audible sound to the loudest tolerable blast represents a trillionfold increase in power. The receptor mechanism is

Figure 17–32 **Pathways for Auditory Sensations.**

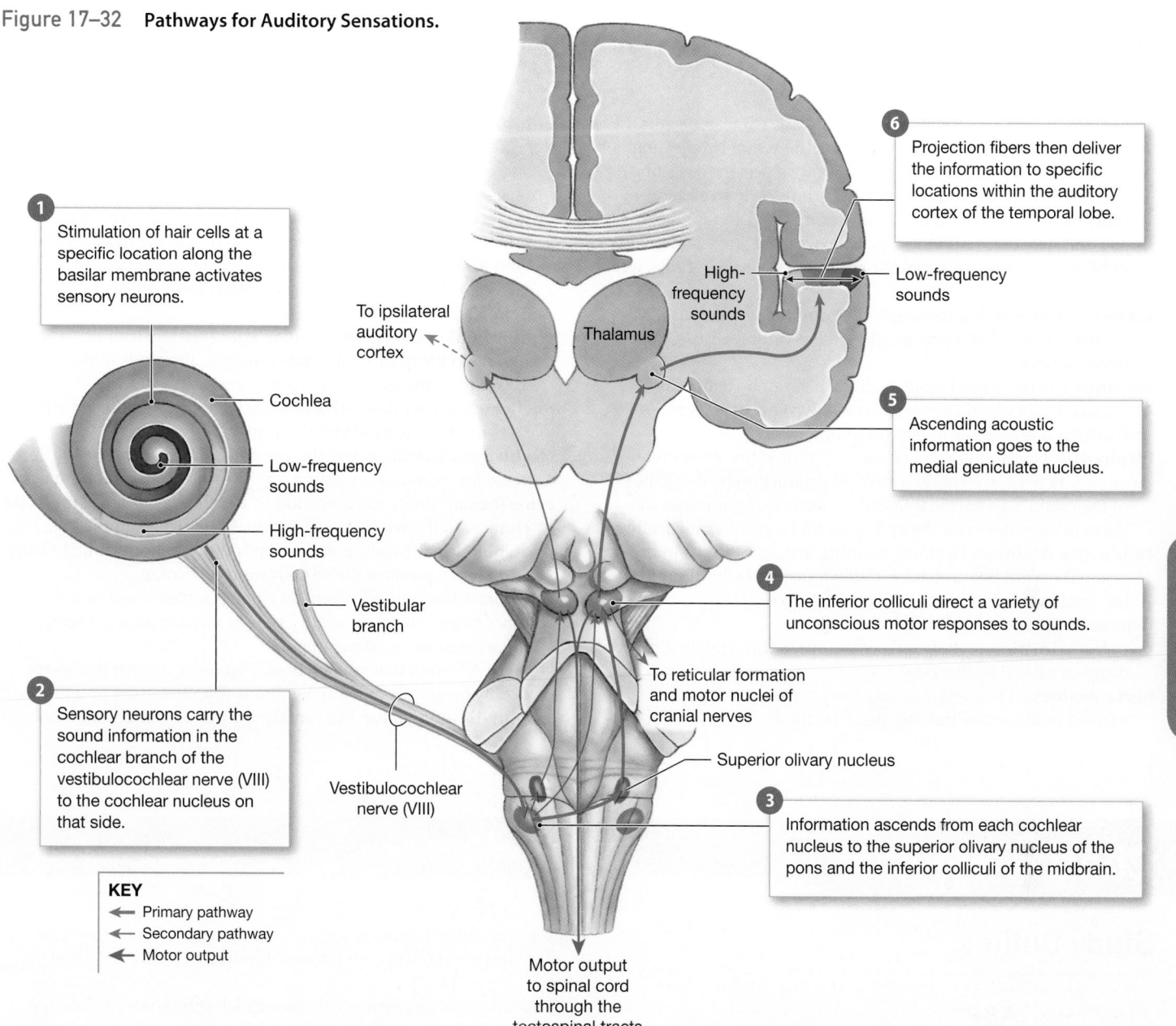

1 Stimulation of hair cells at a specific location along the basilar membrane activates sensory neurons.

2 Sensory neurons carry the sound information in the cochlear branch of the vestibulocochlear nerve (VIII) to the cochlear nucleus on that side.

3 Information ascends from each cochlear nucleus to the superior olivary nucleus of the pons and the inferior colliculi of the midbrain.

4 The inferior colliculi direct a variety of unconscious motor responses to sounds.

5 Ascending acoustic information goes to the medial geniculate nucleus.

6 Projection fibers then deliver the information to specific locations within the auditory cortex of the temporal lobe.

To ipsilateral auditory cortex

Cochlea

Low-frequency sounds

High-frequency sounds

Vestibular branch

Vestibulocochlear nerve (VIII)

Thalamus

High-frequency sounds

Low-frequency sounds

To reticular formation and motor nuclei of cranial nerves

Superior olivary nucleus

Motor output to spinal cord through the tectospinal tracts

KEY
← Primary pathway
← Secondary pathway
← Motor output

so sensitive that, if we were to remove the stapes, we could, in theory, hear air molecules bouncing off the oval window. We never use the full potential of this system, because body movements and our internal organs produce squeaks, groans, thumps, and other sounds that are tuned out by central and peripheral adaptation. When other environmental noises fade away, the level of adaptation drops and the system becomes increasingly sensitive. For example, when you relax in a quiet room, your heartbeat seems to get louder and louder as the auditory system adjusts to the level of background noise.

Young children have the greatest hearing range: They can detect sounds ranging from a 20-Hz buzz to a 20,000-Hz whine. With age, damage due to loud noises or other injuries

accumulates. The tympanic membrane gets less flexible, the articulations between the ossicles stiffen, and the round window may begin to ossify. As a result, older individuals show some degree of hearing loss.

CHECKPOINT

14. If the round window were not able to bulge out with increased pressure in the perilymph, how would the perception of sound be affected?

15. How would the loss of stereocilia from hair cells of the spiral organ affect hearing?

16. Why would blockage of the auditory tube produce an earache?

See the blue Answers tab at the back of the book.

Related Clinical Terms

age-related macular degeneration: A disease associated with aging that gradually destroys sharp, central vision by affecting the macula, the part of the eye that allows a person to see fine detail. It is painless and in some cases can slowly worsen over time, causing little concern to the person, or it can rapidly progress and may cause blindness in both eyes. It is the leading cause of blindness in persons over the age of 60. It has two forms, wet and dry.

ageusia: A rare inability to taste. More common is hypogeusia, a disorder in which the person affected has trouble distinguishing between tastes.

anosmia: The complete loss of smell, which can be temporary (caused by an obstruction such as a polyp) or permanent (perhaps due to aging or a brain tumor).

blepharitis: Common and persistent inflammation of the eyelid caused by poor hygiene, excessive oil production by the glands of the eyelid, or a bacterial infection. Signs and symptoms include itching, flakes on the eyelashes, and a gritty, sandy feeling.

conductive deafness: Deafness resulting from conditions in the external or middle ear that block the transfer of vibrations from the tympanic membrane to the oval window.

hyposmia: A lessened sensitivity to odors.

mydriasis: Dilation of the pupils of the eye induced by medical eye drops or caused by disease.

nerve deafness: Deafness resulting from problems within the cochlea or along the auditory nerve pathway.

ophthalmologist: A physician who specializes in ophthalmology, which is the branch of medicine dealing with the diseases and surgery of the visual pathways, including the eye, brain, and areas surrounding the eye.

optometrist: A primary eye care doctor who diagnoses, manages, and treats disorders of the visual system and eye diseases and who also measures vision for the purpose of correcting one's vision problems with specifically prescribed lenses.

otalgia: Pain in the ear; an earache.

photophobia: An oversensitivity to light possibly leading to tearing, discomfort, or pain. Causes include abrasions to the corneal area, inflammation, disease, and some medications.

Snellen chart: A printed chart of block letters in graduated type sizes used to measure visual acuity.

strabismus: The abnormal alignment of one or both eyes that prevents the person from gazing on the same point with both eyes.

synesthesia: Abnormal condition in which sensory nerve messages connect to the wrong centers of the brain. For example, touching an object may produce the perception of a sound, while hearing a tone may produce the visualization of a color.

tinnitus: A buzzing, whistling, or ringing sound heard in the absence of an external stimulus. Causes include injury, disease, inflammation, or some drugs.

vertigo: A feeling that you are dizzily spinning or that things are dizzily turning about you. Vertigo is usually caused by a problem with the internal ear, but can also be due to vision problems.

Chapter Review

Study Outline

MasteringA&P®

Access more chapter study tools online in the MasteringA&P Study Area:

- Chapter Quizzes, Chapter Practice Test, Art-labeling Activities, Animations, MP3 Tutor Sessions, and Clinical Case Studies
- Practice Anatomy Lab **PAL™**   • A&P Flix **A&PFlix™**
- Interactive Physiology **iP®**   • PhysioEx **PhysioEx™**

17-1 ▌ Olfaction, the sense of smell, involves olfactory receptors responding to chemical stimuli p. 590

1. The **olfactory organs** contain the **olfactory epithelium** with **olfactory receptors**, supporting cells, and **basal (stem) cells**. The surfaces of the olfactory organs are coated with the secretions of the **olfactory glands**. *(Figure 17–1)*
2. The olfactory receptors are highly modified neurons.
3. In olfaction, the arriving information reaches the information centers without first synapsing in the thalamus. *(Figure 17–1)*

4. Olfactory reception involves detecting dissolved chemicals as they interact with odorant-binding proteins. *(Spotlight Figure 17–2)*
5. The olfactory system can distinguish thousands of chemical stimuli. The CNS interprets smells by the pattern of receptor activity.
6. The olfactory receptor population shows considerable turnover. The number of olfactory receptors declines with age.

17-2 ▌ Gustation, the sense of taste, involves taste receptors responding to chemical stimuli p. 594

7. **Taste** (gustatory) **receptors** are clustered in **taste buds**.
8. Taste buds are associated with epithelial projections (*lingual papillae*) on the tongue. *(Figure 17–3)*
9. Each taste bud contains **basal cells** (stem cells) and **gustatory cells**, which extend *taste hairs* through a narrow **taste pore**. *(Figure 17–3)*
10. The taste buds are monitored by cranial nerves that synapse within the solitary nucleus of the medulla oblongata. Postsynaptic neurons carry the nerve impulses on to the thalamus, where third-order neurons project to the primary sensory cortex.

17

11. The **primary taste sensations** are sweet, salty, sour, and bitter. Receptors also exist for **umami** and **water**. (*Spotlight Figure 17–2*)

12. Taste sensitivity exhibits significant individual differences, some of which are inherited.

13. The number of taste buds declines with age.

17-3 ▌ Internal eye structures contribute to vision, while accessory eye structures provide protection p. 596

14. The **accessory structures** of the eye include the **palpebrae** (eyelids), separated by the **palpebral fissure**, the **eyelashes**, and the **tarsal glands**. (*Figures 17–4, 17–5*)

15. An epithelium called the **conjunctiva** covers most of the exposed surface of the eye. The **cornea** is transparent. (*Figures 17–4, 17–5*)

16. The secretions of the **lacrimal gland** contain **lysozyme**. Tears collect in the *lacrimal lake* and reach the inferior meatus of the nose after they pass through the **lacrimal puncta**, the **lacrimal canaliculi**, the **lacrimal sac**, and the **nasolacrimal duct**. (*Figure 17–4*)

17. The eye has three layers, formerly called tunics: an outer **fibrous layer**, a middle **vascular layer** (uvea), and a deeper, **inner layer** (retina). (*Figure 17–5*)

18. The fibrous layer consists of the **sclera**, the cornea, and the **corneal limbus**. (*Figure 17–5*)

19. The vascular layer, or **uvea**, includes the **iris**, the **ciliary body**, and the **choroid**. The iris contains muscle fibers that change the diameter of the **pupil**. The ciliary body contains the **ciliary muscle** and the **ciliary processes**, which attach to the **ciliary zonule** (*suspensory ligament*) of the lens. (*Figures 17–5, 17–6*)

20. The **inner layer**, or **retina**, consists of an outer *pigmented part* and an inner *neural part* which contains visual receptors and associated neurons. (*Figures 17–5, 17–7*)

21. The retina contains two types of **photoreceptors: rods** and **cones**.

22. Cones are densely clustered in the **fovea (fovea centralis)**, at the center of the **macula**. (*Figure 17–7*)

23. The direct line to the CNS proceeds from the photoreceptors to **bipolar cells**, then to **ganglion cells**, and, finally, to the brain by the optic nerve. The axons of ganglion cells converge at the **optic disc**, or **blind spot**. **Horizontal cells** and **amacrine cells** modify the signals passed among other components of the retina. (*Figures 17–7, 17–8*)

24. The ciliary body and lens divide the interior of the eye into a large **posterior cavity**, or *vitreous chamber*, and a smaller **anterior cavity**. The anterior cavity is subdivided into the **anterior chamber**, which extends from the cornea to the iris, and a **posterior chamber**, between the iris and the ciliary body and lens. (*Figure 17–9*)

25. The fluid **aqueous humor** circulates within the eye and re-enters the circulation after diffusing through the walls of the anterior chamber and into the **scleral venous sinus** (*canal of Schlemm*). (*Figure 17–9*)

26. The **lens** lies posterior to the cornea and forms the anterior boundary of the posterior cavity. This cavity contains the **vitreous body**, a clear, gelatinous mass that helps stabilize the shape of the eye and support the retina. (*Figure 17–9*)

27. The lens focuses a visual image on the photoreceptors. A **cataract** is the condition in which a lens has lost its transparency.

28. Light is **refracted** (bent) when it passes through the cornea and lens. During **accommodation**, the shape of the lens changes to focus an image on the retina. "Normal" **visual acuity** is rated 20/20. (*Figures 17–10 to Spotlight Figure 17–13*)

17-4 ▌ Photoreceptors respond to light and change it into electrical signals essential to visual physiology p. 607

29. The two types of photoreceptors are rods, which respond to almost any photon, regardless of its energy content, and cones, which have characteristic ranges of sensitivity. (*Figure 17–14*)

30. Each photoreceptor contains an **outer segment** with membranous **discs**. A narrow stalk connects the outer segment to the **inner segment**. Light absorption occurs in the **visual pigments**, which are derivatives of **rhodopsin** (opsin plus the pigment retinal, which is synthesized from vitamin A). (*Figure 17–14*)

31. Color sensitivity depends on the integration of information from **red**, **green**, and **blue** cones. **Color blindness** is the inability to detect certain colors. (*Figures 17–15, 17–18*)

32. In the absence of photons, a photoreceptor produces a dark current and constantly releases neurotransmitter. Photon absorption reduces the dark current and decreases the release of neurotransmitter to the bipolar cell. (*Spotlight Figure 17–16*)

33. During the process of **bleaching**, rhodopsin molecules are broken down to regenerate retinal back to its photon-absorbing 11-*cis* form. (*Spotlight Figure 17–16, Figure 17–17*)

34. In the **dark-adapted state**, most visual pigments are fully receptive to stimulation. In the **light-adapted state**, the pupil constricts and bleaching of the visual pigments takes place.

35. The ganglion cells that monitor rods, called **M cells** (*magnocells*), are relatively large. The ganglion cells that monitor cones, called **P cells** (*parvo cells*), are smaller and more numerous. (*Figure 17–19*)

36. Visual data from the left half of the combined field of vision arrive at the visual cortex of the right occipital lobe. Data from the right half of the combined field of vision arrive at the visual cortex of the left occipital lobe. (*Figure 17–20*)

37. **Depth perception** is obtained by comparing relative positions of objects between the left- and right-eye images. (*Figure 17–20*)

38. Visual inputs to the suprachiasmatic nucleus of the hypothalamus affect the function of other brain stem nuclei. This nucleus establishes a visceral **circadian rhythm**, which is tied to the day–night cycle and affects other metabolic processes.

17-5 ▌ Equilibrium sensations originate within the internal ear, while hearing involves the detection and interpretation of sound waves p. 616

39. The receptors of the internal ear provide the senses of equilibrium and hearing.

40. The ear is divided into the **external ear**, the **middle ear**, and the **internal ear**. (*Figure 17–21*)

41. The external ear includes the **auricle**, or *pinna*, which surrounds the entrance to the **external acoustic meatus**, which ends at the **tympanic membrane** (*eardrum*). (*Figure 17–21*)

42. The middle ear communicates with the nasopharynx by the **auditory** (*pharyngotympanic*) **tube**. The middle ear encloses and protects the **auditory ossicles**. (*Figures 17–21, 17–22*)

43. The **membranous labyrinth** (the chambers and tubes) of the internal ear contains the fluid **endolymph**. The **bony labyrinth** surrounds and protects the membranous labyrinth and can be subdivided into the **vestibule**, the **semicircular canals**, and the **cochlea**. (*Figures 17–21, 17–23*)

44. The vestibule of the internal ear encloses the **saccule** and **utricle**. The semicircular canals contain the **semicircular ducts**. The cochlea contains the **cochlear duct**, an elongated portion of the membranous labyrinth. (*Figure 17–23*)

17

45. The **round window** separates the **perilymph** from the air spaces of the middle ear. The **oval window** is connected to the base of the stapes. *(Figure 17–21)*

46. The basic receptors of the internal ear are **hair cells**, which provide information about the direction and strength of mechanical stimuli. *(Figure 17–24)*

47. The **anterior**, **posterior**, and **lateral semicircular ducts** are continuous with the utricle. Each duct contains an **ampulla** with a gelatinous **cupula** and associated sensory receptors. *(Figures 17–23, 17–24)*

48. The saccule and utricle are connected by a passageway that is continuous with the **endolymphatic duct**. This duct terminates in the **endolymphatic sac**. In the saccule and utricle, hair cells cluster within **maculae**, where their hair cells contact the gelatinous otolithic membrane. The surface of the membrane contains calcium carbonate crystals called **otoliths**. *(Figures 17–24, 17–25)*

49. The vestibular receptors activate sensory neurons of the **vestibular ganglia**. The axons form the **vestibular branch** of the vestibulocochlear nerve (VIII), synapsing within the **vestibular nuclei**. *(Figure 17–26)*

50. The **cochlear duct** lies between the **scala vestibuli** (*vestibular duct*) and the **scala tympani** (*tympanic duct*). The hair cells of the cochlear duct lie within the **spiral organ** (**organ of Corti**). *(Figures 17–27, 17–28)*

51. The energy content of a sound determines its *intensity*, or loudness, measured in **decibels**. Sound waves travel toward the tympanic membrane, which vibrates, and the auditory ossicles then conduct these vibrations to the internal ear. Movement at the oval window applies pressure to the perilymph of the scala vestibuli. *(Figures 17–29, 17–30; Table 17–1)*

52. Pressure waves distort the **basilar membrane** and push the hair cells of the spiral organ against the **tectorial membrane**. The **tensor tympani** and **stapedius muscles** contract to reduce the amount of motion when very loud sounds arrive. *(Figures 17–30, 17–31)*

53. The sensory neurons are located in the **spiral ganglion** of the cochlea. The afferent fibers of these neurons form the **cochlear branch** of the vestibulocochlear nerve (VIII), synapsing at their respective left or right **cochlear nucleus**. *(Figure 17–32)*

17

Review Questions

See the blue Answers tab at the back of the book.

LEVEL 1 Reviewing Facts and Terms

1. Identify the structures in the following horizontal section of the eye.

Anterior cavity

Posterior cavity

(a) _____ (b) _____
(c) _____ (d) _____
(e) _____ (f) _____
(g) _____ (h) _____
(i) _____ (j) _____

2. A reduction in sensitivity in the presence of a constant stimulus is (a) transduction, (b) sensory coding, (c) line labeling, (d) adaptation.

3. A blind spot occurs in the retina where (a) the fovea is located, (b) ganglion cells synapse with bipolar cells, (c) the optic nerve attaches to the retina, (d) rod cells are clustered to form the macula, (e) amacrine cells are located.

4. Taste receptors are made up of (a) basal cells, (b) filiform papillae, (c) gustatory cells, (d) fungiform papillae.

5. Umami taste receptors detect (a) glutamate, (b) glucose, (c) water, (d) sodium chloride.

6. The retina is also called (a) the vascular layer, (b) the fibrous layer, (c) the inner layer, (d) all of these.

7. At sunset, your visual system adapts to (a) fovea vision, (b) rod-based vision, (c) macular vision, (d) cone-based vision.

8. A better-than-average visual acuity rating is (a) 20/20, (b) 20/30, (c) 15/20, (d) 20/15.

9. The malleus, incus, and stapes are the tiny bones located in the (a) external ear, (b) middle ear, (c) internal ear, (d) membranous labyrinth.

10. Identify the structures of the external, middle, and internal ear in the following figure.

EXTERNAL EAR   MIDDLE EAR   INTERNAL EAR

(a) _____ (b) _____
(c) _____ (d) _____
(e) _____ (f) _____
(g) _____ (h) _____
(i) _____

11. Receptors in the saccule and utricle provide sensations of (a) angular acceleration, (b) hearing, (c) vibration, (d) gravity and linear acceleration.
12. The spiral organ is located in the _____ of the internal ear. (a) utricle, (b) bony labyrinth, (c) vestibule, (d) cochlea.
13. Auditory information about the frequency and intensity of stimulation is relayed to the CNS over the cochlear branch of cranial nerve (a) IV, (b) VI, (c) VIII, (d) X.
14. What are the four types of papillae on the human tongue?
15. (a) What structures make up the fibrous layer of the eye? (b) What are the functions of the fibrous layer?
16. What structures make up the vascular layer of the eye?
17. What are the three auditory ossicles in the middle ear, and what are their functions?

LEVEL 2 Reviewing Concepts

18. Trace the olfactory pathway from the time an odor reaches the olfactory epithelium until nerve impulses reach their final destination in the brain.
19. Explain why we are so much more sensitive to acids than to sweet chemicals.
20. What is the usual result if a sebaceous gland of an eyelash or a tarsal gland becomes infected?
21. Displacement of stereocilia toward the kinocilium of a hair cell (a) produces a depolarization of the membrane, (b) produces a hyperpolarization of the membrane, (c) decreases the membrane permeability to sodium ions, (d) increases the membrane permeability to potassium ions, (e) does not affect the membrane potential of the cell.

22. Damage to the cupula of the lateral semicircular duct would interfere with the perception of (a) the direction of gravitational pull, (b) linear acceleration, (c) horizontal rotation of the head, (d) vertical rotation of the head, (e) angular rotation of the head.
23. Photoreceptor signals in the retina can be modified by (a) bipolar cells, (b) ganglion cells, (c) amacrine cells, (d) horizontal cells, (e) a and b, (f) c and d, (g) a, b, c, and d.

LEVEL 3 Critical Thinking and Clinical Applications

24. You are at a park watching some deer 35 feet away from you. A friend taps you on the shoulder to ask a question. As you turn to look at your friend, who is standing just 2 feet away, what changes would your eyes undergo?
25. Your friend Shelly suffers from myopia (nearsightedness). You remember from your physics class that concave lenses cause light waves to spread or diverge and that convex lenses cause light waves to converge. What type of corrective lenses would you suggest to your friend? (a) concave lenses, (b) convex lenses.
26. Tom has surgery to remove polyps (growths) from his sinuses. After he heals from the surgery, he notices that his sense of smell is not as keen as it was before the surgery. Can you suggest a reason for this?
27. For a few seconds after you ride the express elevator from the 20th floor to the ground floor, you still feel as if you are descending, even though you have come to a stop. Why?
28. Juan tells his physician that he has been feeling dizzy, especially when he closes his eyes. He is asked to stand with his feet together and arms extended forward. As long as he keeps his eyes open, he exhibits very little movement. But when he closes his eyes, his body begins to sway a great deal, and his arms tend to drift together toward the left side of his body. Why does this occur?

CLINICAL CASE Wrap-Up

A Chance to See

Makena suffers from *myopia,* or nearsightedness. Due to the elongated shape of Makena's eyeballs, distant images are projected in front of her retina and appear blurry and out of focus to her. Glasses can correct this easily, but nobody in Makena's village has ever met an eye doctor of any kind. The closest optical shop is far out of this family's reach, both in distance and cost.

Makena's teacher sets up a makeshift eye clinic with the help of volunteers from the Centre for Vision in the Developing World, a group dedicated to correcting vision for poor children. To test her vision, Makena attempts to read from an eye chart taped to the wall of the school building 20 feet away. Makena's visual acuity tests at 20/200.

Next, Makena puts on a simple pair of universal glasses, made with a clear membrane filled with silicone oil, held between two

plastic discs on each side. A removable syringe filled with silicone oil is attached to each side. With her right eye covered, Makena uses the syringe to inject more silicone into the left lens until her left eye can read the eye chart perfectly. Then she repeats the process for her right eye. The final step is to remove the syringes. Now she can see things she never knew she had been missing—and she gets to keep the glasses.

1. What does Makena's initial eye test finding of 20/200 mean?

2. If Makena's vision is corrected to a state of emmetropia, what does this mean?

See the blue Answers tab at the back of the book.

18

The Endocrine System

Learning Outcomes

These Learning Outcomes correspond by number to this chapter's sections and indicate what you should be able to do after completing this chapter.

18-1 ▌ Explain the importance of intercellular communication, describe the mechanisms involved, and compare the modes of intercellular communication that occur in the endocrine and nervous systems.

18-2 ▌ Compare the cellular components of the endocrine system with those of other systems, contrast the major structural classes of hormones, and explain the general mechanisms of hormonal action on target organs.

18-3 ▌ Describe the location, hormones, and functions of the pituitary gland, and discuss the effects of abnormal pituitary hormone production.

18-4 ▌ Describe the location, hormones, and functions of the thyroid gland, and discuss the effects of abnormal thyroid hormone production.

18-5 ▌ Describe the location, hormone, and functions of the parathyroid glands, and discuss the effects of abnormal parathyroid hormone production.

18-6 ▌ Describe the location, structure, hormones, and general functions of the adrenal glands, and discuss the effects of abnormal adrenal hormone production.

18-7 ▌ Describe the location of the pineal gland, and discuss the functions of the hormone it produces.

18-8 ▌ Describe the location, structure, hormones, and functions of the pancreas, and discuss the effects of abnormal pancreatic hormone production.

18-9 ▌ Describe the functions of the hormones produced by the kidneys, heart, thymus, testes, ovaries, and adipose tissue.

18-10 ▌ Explain how hormones interact to produce coordinated physiological responses and influence behavior, describe the role of hormones in the general adaptation syndrome, and discuss how aging affects hormone production and give examples of interactions between the endocrine system and other organ systems.

SPOTLIGHTS

- Structural Classification of Hormones p. 639
- G Proteins and Second Messengers p. 641
- Diabetes Mellitus p. 664
- The General Adaptation Syndrome p. 671

CLINICAL CASE

- Stones, Bones, and Groans p. 635

Clinical Notes

- Diabetes Insipidus p. 649
- Endocrine Disorders p. 668
- Hormones and Athletic Performance p. 672

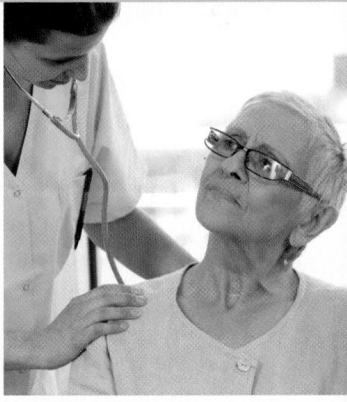

Stones, Bones, and Groans

Monica is back in her doctor's office, following up on her lab results. She feels a little crazy, seeing a new family practitioner with so many vague complaints. She has a history of high blood pressure (hypertension) and heart palpitations (irregular heartbeats). She has had several kidney stones. She has osteopenia and suffers from nagging bone pain. Recently she has experienced vague abdominal pain with loss of appetite and constipation. Furthermore, she has been feeling chronic fatigue, some of which she attributes to having trouble falling asleep and regularly awakening in the middle of the night. Monica has also noticed that her memory is getting worse and that she feels increasingly depressed. Worst of all, her husband of 40 years has complained that she is getting more irritable and cranky.

"Monica," says her doctor, "I think I know what is causing your symptoms. The answer is right here in your lab results." **What could be causing Monica's many complaints? To find out, turn to the Clinical Case Wrap-Up on p. 677.**

An Introduction to the Endocrine System

The human body contains about 30 chemical messengers known as hormones, which regulate activities such as sleep, body temperature, hunger, and stress management. These hormones are products of the endocrine system, which along with the nervous system, controls and coordinates our body processes.

In this chapter we examine the structural and functional organization of the endocrine system and compare it to the nervous system. After an overview of the endocrine system and the characteristics of the hormones it produces, we consider the structure and function of the body's various endocrine glands. Finally, we look at the ways in which hormones modify metabolic operations, and the interactions between the endocrine system and other body systems. Let's begin by considering the role of intercellular communication in maintaining homeostasis.

18-1 ┃ Homeostasis is preserved through intercellular communication

Learning Outcome Explain the importance of intercellular communication, describe the mechanisms involved, and compare the modes of intercellular communication that occur in the endocrine and nervous systems.

To preserve homeostasis, cellular activities must be coordinated throughout the body. Neurons monitor or control specific cells or groups of cells. Only a small fraction of all the cells in the body are innervated, however, and the commands from the nervous system are very specific and relatively short-lived. Yet many life processes are not short-lived. For example, your body takes decades to reach adult height. The body continually controls and maintains its reproductive capabilities for at least 30 years in a typical female, and even longer in males. Long-term processes, such as growth, development, or reproduction, involve or affect metabolic activities in virtually every cell and tissue. There is no way that the nervous system can regulate such processes. Instead, the endocrine system provides this type of regulation. It uses chemical messengers to relay information and instructions between cells. To understand how these messages are generated and interpreted, let's take a closer look at how cells communicate with one another.

In a few specialized cases, adjacent cells coordinate cellular activities by exchanging ions and molecules across gap junctions. This **direct communication** occurs between two cells of the same type, and the cells must be in extensive physical contact. The two cells communicate so closely that they function as a single entity. Gap junctions (1) coordinate ciliary movement among epithelial cells, (2) coordinate the contractions of cardiac muscle cells, and (3) facilitate the propagation of action potentials from one neuron to the next at electrical synapses.

Most communication between cells involves the release and receipt of chemical messages. Each cell continuously "talks" to its neighbors by releasing chemicals into the extracellular fluid. These chemicals tell cells what their neighbors are doing at any moment. The result is the coordination of tissue function at the local level. The use of chemical messengers to transfer information from cell to cell within a single tissue is called **paracrine communication**. The chemicals involved are called *paracrine factors*, also known as *local hormones*. Examples of paracrine factors include the prostaglandins, introduced in Chapter 2, and the various growth factors, discussed in Chapter 3. ⤶ pp. 73, 129

Paracrine factors enter the bloodstream, but their concentrations are usually so low that distant cells and tissues are not affected. However, some paracrine factors, including several of the prostaglandins and related chemicals, have primary effects in their tissues of origin and secondary effects in other tissues and organs. When these secondary effects occur, the paracrine factors are also acting as **hormones**—chemical messengers that are released in one tissue and transported in the bloodstream to alter the activities of specific cells in other tissues. Most cells release paracrine factors, but only specialized cells produce typical hormones.

Nevertheless, the difference between paracrine factors and hormones is mostly a matter of degree. Paracrine factors can diffuse out of their tissue of origin and have widespread effects, and hormones can affect their tissues of origin as well as distant cells. By convention, a substance with effects outside its tissue of origin is called a *hormone* if its chemical structure is known, and a *factor* if that structure remains to be determined.

In intercellular communication, hormones are like messages and the cardiovascular system is e-mail. A hormone released into the bloodstream is distributed throughout the body. Each hormone has **target cells**, specific cells that have the receptors needed to bind and "read" the hormonal message when it arrives. But hormones are really like e-mail spam—cells throughout the body are exposed to them whether or not they have the necessary receptors. At any moment, each individual cell can respond to only a few of the hormones present. The cell ignores other hormones, because it lacks the receptors to read the messages they contain. The activity of hormones in coordinating cellular activities in tissues in distant portions of the body is called **endocrine communication**.

How do hormones work? They alter the operations of target cells by changing the types, quantities, or activities of important enzymes and structural proteins. A hormone may

- stimulate the synthesis of an enzyme or a structural protein not already present in the cytoplasm by activating appropriate genes in the cell nucleus;
- increase or decrease the rate of synthesis of a particular enzyme or other protein by changing the rate of transcription or translation; or
- turn an existing enzyme or membrane channel "on" or "off" by changing its shape or structure.

Through one or more of these mechanisms, a hormone can modify the physical structure or biochemical properties of its target cells. Because the target cells can be anywhere in the body, a single hormone can alter the metabolic activities of multiple tissues and organs at the same time. These effects may be slow to appear, but they typically last for days. Consequently, hormones are effective in coordinating cell, tissue, and organ activities on a sustained, long-term basis. For example, circulating hormones keep body water content and levels of electrolytes and organic nutrients within normal limits 24 hours a day throughout our entire lives.

Cells can respond to several different hormones simultaneously. Gradual changes in the quantities and identities of circulating hormones can produce complex changes in the body's physical structure and physiological capabilities. Examples include the processes of embryonic and fetal development, growth, and puberty. Hormonal regulation is quite suitable for

Table 18–1	Mechanisms of Intercellular Communication		
Mechanism	**Transmission**	**Chemical Mediators**	**Distribution of Effects**
Direct communication	Through gap junctions	Ions, small solutes, lipid-soluble materials	Usually limited to adjacent cells of the same type that are interconnected by connexons
Paracrine communication	Through extracellular fluid	Paracrine factors	Primarily limited to a local area, where paracrine factor concentrations are relatively high. Target cells must have appropriate receptors
Endocrine communication	Through the bloodstream	Hormones	Target cells are primarily in other tissues and organs and must have appropriate receptors
Synaptic communication	Across synapses	Neurotransmitters	Limited to very specific area; target cells must have appropriate receptors

directing gradual, coordinated processes, but it is totally unable to handle situations requiring split-second responses. That kind of crisis management is the job of the nervous system.

The nervous system, too, relies primarily on chemical communication, but it does not send messages through the bloodstream. Instead, as we have seen, neurons release a neurotransmitter at a synapse very close to target cells that have the appropriate receptors. The command to release the neurotransmitter rapidly travels from one location to another in the form of action potentials propagated along axons. The nervous system thus acts like a telecommunications company, with a cable network carrying high-speed "messages" to specific destinations throughout the body. The effects of neural stimulation are generally short-lived, and they tend to be restricted to specific target cells—primarily because the neurotransmitter is rapidly broken down or recycled. This **synaptic communication** is ideal for crisis management: If you are in danger of being hit by a speeding bus, the nervous system can coordinate and direct your leap to safety. Once the crisis is over and the neural circuits quiet down, things soon return to normal.

Table 18-1 summarizes the four ways cells and tissues communicate with one another. Viewed from a general perspective, the differences between the nervous system and endocrine system seem relatively clear. In fact, these broad organizational and functional distinctions are the basis for treating them as two separate systems. Yet when we consider them in detail, we see that the two systems are similarly organized:

- Both systems rely on the release of chemicals that bind to specific receptors on their target cells.

- The two systems share many chemical messengers. For example, norepinephrine and epinephrine are called *hormones* when released into the bloodstream, but *neurotransmitters* when released across synapses.

- Both systems are regulated mainly by negative feedback control mechanisms.

- The two systems share a common goal: to preserve homeostasis by coordinating and regulating the activities of other cells, tissues, organs, and systems.

Next we introduce the components and functions of the endocrine system and further explore the interactions between the nervous and endocrine systems. We consider specific endocrine organs, hormones, and functions in detail in later sections.

CHECKPOINT

1. Define hormone.
2. Describe paracrine communication.
3. Identify four mechanisms of intercellular communication.

See the blue Answers tab at the back of the book.

18-2 ┃ The endocrine system regulates physiological processes through the binding of hormones to receptors

Learning Outcome Compare the cellular components of the endocrine system with those of other systems, contrast the major structural classes of hormones, and explain the general mechanisms of hormonal action on target organs.

The **endocrine system** includes all the endocrine cells and tissues of the body that produce hormones or paracrine factors with effects beyond their tissues of origin. As noted in Chapter 4, *endocrine cells* are glandular secretory cells that release their secretions into the extracellular fluid. This characteristic distinguishes them from *exocrine cells*, which secrete their products onto epithelial surfaces, generally by way of ducts. ↺ p. 149 The chemicals released by endocrine cells may affect only nearby cells, as in the case of most paracrine factors, or they may affect cells throughout the body.

Figure 18-1 introduces the tissues, organs, and hormones of the endocrine system. Some of these organs, such as the pituitary gland, have endocrine secretion as a primary function. Others, such as the pancreas, have many other functions in addition to endocrine secretion. We consider such endocrine organs in more detail in chapters on other systems.

Classes of Hormones

We can divide hormones into three groups on the basis of their chemical structure: (1) *amino acid derivatives*, (2) *peptide hormones*, and (3) *lipid derivatives*.

Amino acid derivatives, sometimes known as *biogenic amines*, are relatively small molecules that are structurally related to amino acids, the building blocks of proteins. ↺ p. 77 These hormones are synthesized from the amino acids *tyrosine* (TĪ-rō-sēn) and *tryptophan* (TRIP-tō-fan). Those made from tyrosine include (1) thyroid hormones, produced by the thyroid gland, and (2) the compounds epinephrine (E), norepinephrine (NE), and dopamine, which are sometimes called *catecholamines* (kat-e-KŌ-la-mēnz). The primary hormone made from tryptophan is melatonin (mel-a-TŌ-nin), produced by the pineal gland.

We can divide peptide hormones into two groups. One group consists of glycoproteins, and the other group is made up of short polypeptides and small proteins.

There are two classes of lipid derivatives: (1) *eicosanoids* and (2) *steroid hormones*. Eicosanoids are signaling molecules and include leukotrienes, prostaglandins, thromboxanes, and

Figure 18–1 **Organs and Tissues of the Endocrine System.**

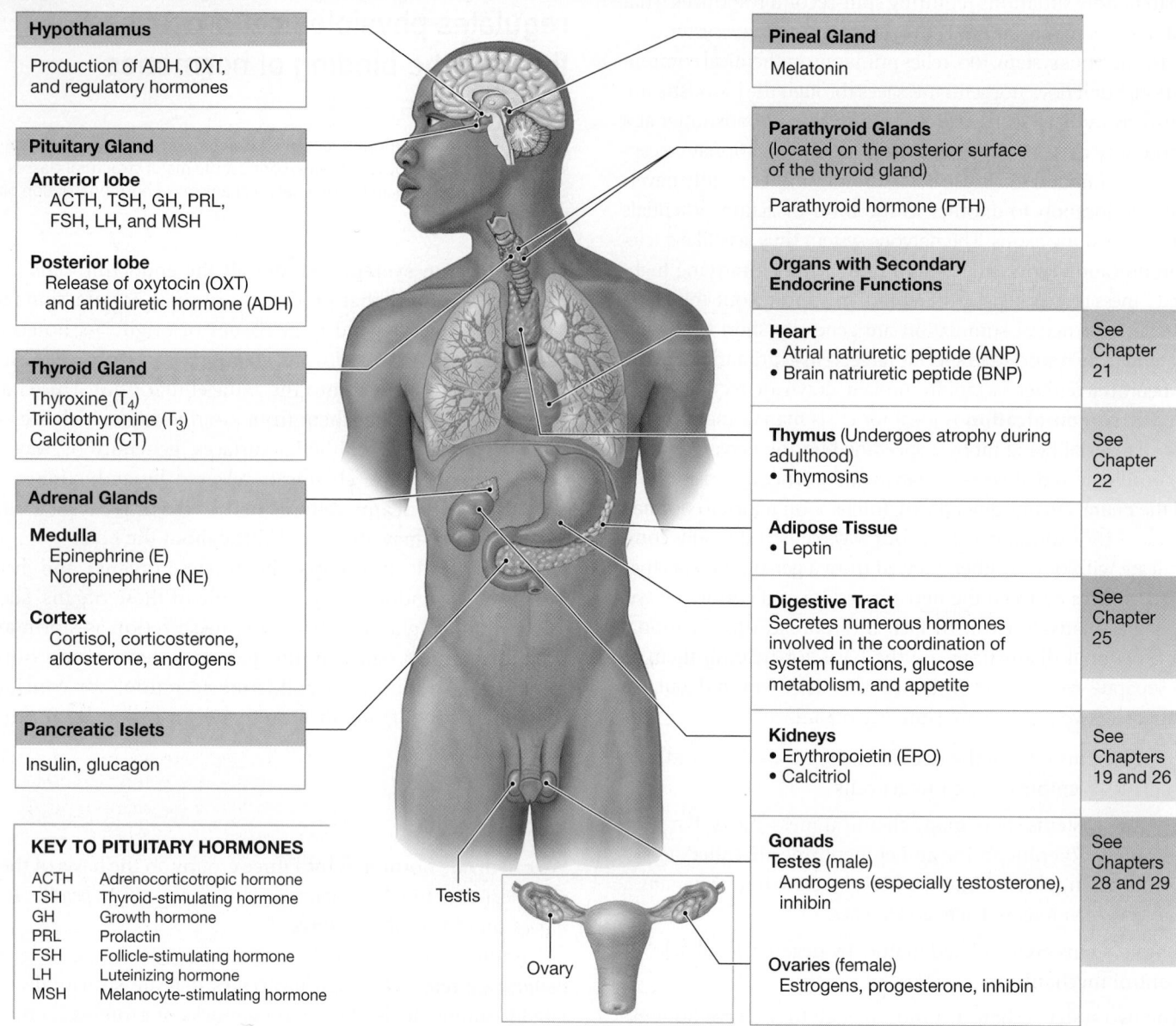

Hypothalamus

Production of ADH, OXT, and regulatory hormones

Pituitary Gland

Anterior lobe
ACTH, TSH, GH, PRL, FSH, LH, and MSH

Posterior lobe
Release of oxytocin (OXT) and antidiuretic hormone (ADH)

Thyroid Gland

Thyroxine (T_4)
Triiodothyronine (T_3)
Calcitonin (CT)

Adrenal Glands

Medulla
Epinephrine (E)
Norepinephrine (NE)

Cortex
Cortisol, corticosterone, aldosterone, androgens

Pancreatic Islets

Insulin, glucagon

KEY TO PITUITARY HORMONES

ACTH	Adrenocorticotropic hormone
TSH	Thyroid-stimulating hormone
GH	Growth hormone
PRL	Prolactin
FSH	Follicle-stimulating hormone
LH	Luteinizing hormone
MSH	Melanocyte-stimulating hormone

Testis

Ovary

Pineal Gland

Melatonin

Parathyroid Glands
(located on the posterior surface of the thyroid gland)

Parathyroid hormone (PTH)

Organs with Secondary Endocrine Functions

Heart
• Atrial natriuretic peptide (ANP)
• Brain natriuretic peptide (BNP)

See Chapter 21

Thymus (Undergoes atrophy during adulthood)
• Thymosins

See Chapter 22

Adipose Tissue
• Leptin

Digestive Tract
Secretes numerous hormones involved in the coordination of system functions, glucose metabolism, and appetite

See Chapter 25

Kidneys
• Erythropoietin (EPO)
• Calcitriol

See Chapters 19 and 26

Gonads
Testes (male)
Androgens (especially testosterone), inhibin

See Chapters 28 and 29

Ovaries (female)
Estrogens, progesterone, inhibin

prostacyclins. Steroid hormones are lipids structurally similar to cholesterol (**Figure 2–17a**, p. 75). The individual hormones differ in the side chains attached to the basic ring structure. Over time, the liver gradually absorbs these steroids and converts them to a soluble form that can be excreted in the bile or urine (**Spotlight Figure 18–2**).

In this chapter we focus on circulating hormones that function primarily to coordinate activities in many tissues and organs. We consider eicosanoids in chapters that discuss individual tissues and organs, including Chapters 19 (the blood), 22 (the lymphatic system), and 28 (the reproductive system).

Secretion and Distribution of Hormones

Hormones are typically released where capillaries are abundant, and the hormones quickly enter the bloodstream for distribution throughout the body. Within the blood, hormones may circulate freely or travel bound to special carrier proteins. A freely circulating hormone remains functional for less than one hour, and sometimes for as little as two minutes. It is inactivated when (1) it diffuses out of the bloodstream and binds to receptors on target cells, (2) it is absorbed and broken down by cells of the liver or kidneys, or (3) it is broken down by enzymes in the blood or interstitial fluids.

HORMONES

The hormones of the body can be divided into three groups on the basis of their chemical structure.

Amino Acid Derivatives

Amino acid derivatives are small molecules that are structurally related to amino acids, the building blocks of proteins.

Derivatives of Tyrosine

Thyroid Hormones

Thyroxine (T₄)

Catecholamines

Epinephrine

Sources of tyrosine include meat, dairy, and fish.

Derivative of Tryptophan

Melatonin

Turkey is a well-known source of tryptophan. Other sources include chocolate, oats, bananas, dried dates, milk, cottage cheese, and peanuts.

Peptide Hormones

Peptide hormones are chains of amino acids. Most peptide hormones are synthesized as **prohormones**—inactive molecules that are converted to active hormones before or after they are secreted.

Glycoproteins

These proteins are more than 200 amino acids long and have carbohydrate side chains. The glycoproteins include thyroid-stimulating hormone (TSH), luteinizing hormone (LH), and follicle-stimulating hormone (FSH) from the anterior lobe of the pituitary gland, as well as several hormones produced in other organs.

Short Polypeptides/ Small Proteins

This group of peptide hormones is large and diverse. It includes hormones that range from **short chain polypeptides**, such as antidiuretic hormone (ADH) and oxytocin (OXT) (each 9 amino acids long), to **small proteins**, such as insulin (51 amino acids), growth hormone (GH; 191 amino acids), and prolactin (PRL; 198 amino acids). This group includes all the hormones secreted by the hypothalamus, heart, thymus, digestive tract, pancreas, and posterior lobe of the pituitary gland, as well as several hormones produced in other organs.

Structure of insulin

— Chain B

— Chain A

Lipid Derivatives

There are two classes of lipid derivatives: **eicosanoids**, derived from arachidonic acid, a 20-carbon fatty acid; and **steroid hormones**, derived from cholesterol.

Eicosanoids

Eicosanoids (Ī-kō-sa-noydz) are important paracrine factors that coordinate cellular activities and affect enzymatic processes (such as blood clotting) in extracellular fluids. Some eicosanoids, such as **leukotrienes** (lū-kō-TRĪ-ēns), have secondary roles as hormones. A second group of eicosanoids—**prostaglandins**—are involved primarily in coordinating local cellular activities. In some tissues, prostaglandins are converted to **thromboxanes** (throm-BOX-ānz) and **prostacyclins** (pros-ta-SĪ-klinz), which also have strong paracrine effects.

Prostaglandin E

Aspirin suppresses the production of prostaglandins.

Steroid Hormones

Steroid hormones are released by the reproductive organs (androgens by the testes in males, estrogens and progesterone by the ovaries in females), by the cortex of the adrenal glands (corticosteroids), and by the kidneys (calcitriol). Because circulating steroid hormones are bound to specific transport proteins in the plasma, they remain in circulation longer than do secreted peptide hormones.

Estrogen

Thyroid hormones and steroid hormones remain in circulation much longer, because when these hormones enter the bloodstream, more than 99 percent of them become attached to special transport proteins. For each hormone an equilibrium state exists between its free and bound forms. As the free hormones are removed and inactivated, bound hormones are released to replace them. At any given time, the bloodstream contains a substantial reserve (several weeks' supply) of bound hormones.

Mechanisms of Hormone Action

Hormones coordinate cell, tissue, and organ activities on a sustained basis. They circulate in the extracellular fluid and bind to specific receptors on plasma membranes or within target cells. Binding modifies cellular activities by altering membrane permeability, activating or inactivating key enzymes, or changing genetic activity.

To affect a target cell, a hormone must first interact with an appropriate receptor. A hormone receptor, like a neurotransmitter receptor, is a protein molecule to which a particular molecule binds strongly. Each cell has receptors for several different hormones, but cells in different tissues have different combinations of receptors. This arrangement is one reason hormones have different effects on different tissues. For every cell, the presence or absence of a specific receptor determines the cell's hormonal sensitivities. If a cell has a receptor that can bind a particular hormone, that cell responds to the hormone. If a cell lacks that receptor, the hormone has no effect on that cell.

Hormone receptors are located either on the plasma membrane or inside the cell. Using a few specific examples, let's consider the basic mechanisms involved.

Hormones and Plasma Membrane Receptors

The receptors for catecholamines (E, NE, and dopamine), peptide hormones, and eicosanoids are in the plasma membranes of their target cells. Catecholamines and peptide hormones cannot penetrate a plasma membrane because they are not lipid soluble. Instead, these hormones bind to receptor proteins at the *outer* surface of the plasma membrane (extracellular receptors). Eicosanoids *are* lipid soluble. They diffuse across the plasma membrane to reach receptor proteins on the *inner* surface of the membrane (intracellular receptors).

First and Second Messengers

Communication between the hormone and the cell uses first and second messengers. A *first messenger* is a hormone that binds to a receptor on the plasma membrane surface. A second messenger is an intermediary molecule that appears due to a hormone-receptor interaction. Important second messengers are (1) *cyclic AMP (cAMP)*, a derivative of ATP; (2) *cyclic GMP (cGMP)*, a derivative of GTP, another high-energy compound;

and (3) calcium ions (Ca^{2+}). **Spotlight Figure 18–3** focuses on cAMP and Ca^{2+}.

When a small number of hormone molecules binds to membrane receptors, thousands of second messengers may appear in a cell. This process, called *amplification*, magnifies the effect of a hormone on the target cell. Moreover, the arrival of a single hormone may promote the release of more than one type of second messenger, or the production of a linked sequence of enzymatic reactions known as a *receptor cascade*. Through such mechanisms, the hormone can alter many aspects of cell function at the same time.

The presence or absence of a hormone can also affect the nature and number of hormone receptor proteins in the plasma membrane. **Down-regulation** is a process in which the presence of a hormone triggers a decrease in the number of hormone receptors. In down-regulation, cells become *less* sensitive to high levels of a particular hormone. In contrast, **up-regulation** is a process in which the absence of a hormone triggers an increase in the number of hormone receptors. In up-regulation, cells become *more* sensitive to low levels of a particular hormone.

The link between the first messenger and the second messenger generally involves a **G protein**, an enzyme complex coupled to a membrane receptor. The name *G protein* refers to the fact that these proteins bind GTP. ⤵ p. 446 A G protein is activated when a hormone binds to its receptor at the membrane surface. What happens next depends on the nature of the G protein and its effects on second messengers in the cytoplasm. **Spotlight Figure 18–3** diagrams three major patterns of response to G protein activation. Roughly 80 percent of prescription drugs target receptors coupled to G proteins.

G Proteins and cAMP. Spotlight Figure 18–3 shows the steps involved in *increasing* cAMP levels:

- The activated G protein activates the enzyme **adenylate cyclase**.
- Adenylate cyclase converts ATP to the ring-shaped molecule *cyclic AMP*.
- Cyclic AMP then functions as a second messenger, typically by activating a *kinase* (KĪ-nās). A kinase is an enzyme that attaches a high-energy phosphate group ($-PO_4^{3-}$) to another molecule in a process called *phosphorylation*.
- Generally, cyclic AMP activates kinases that phosphorylate proteins. The effect on the target cell depends on the nature of these proteins. The phosphorylation of plasma membrane proteins, for example, can open ion channels. In the cytoplasm, many important enzymes can be activated only by phosphorylation. One important example is the enzyme that releases glucose from glycogen reserves in skeletal muscles and the liver.

Many hormones, including calcitonin, parathyroid hormone, ADH, ACTH, epinephrine, FSH, LH, TSH, and glucagon,

MECHANISM OF HORMONE ACTION
First and Second Messengers

A hormone that binds to receptors in the plasma membrane cannot directly affect the activities inside the target cell. For example, it cannot begin building a protein or catalyzing a specific reaction. Instead, the hormone uses an intracellular intermediary to bring about its effects. The hormone, or **first messenger**, does something that leads to the appearance of a **second messenger** in the cytoplasm. The second messenger may act as an enzyme activator, inhibitor, or cofactor. The net result is a change in the rates of various metabolic reactions. The two most important second messengers are cyclic AMP (cAMP), a derivative of ATP, and calcium ions (Ca^{2+}).

Watch the MasteringA&P® > *A&PFlix*™ on
Mechanism of Hormone Action: Second Messenger cAMP

The first messenger (a peptide hormone, catecholamine, or eicosanoid) binds to a membrane receptor and activates a G protein.

A **G protein** is an enzyme complex coupled to a membrane receptor that serves as a link between the first and second messenger.

Effects on cAMP Levels

Many G proteins, once activated, exert their effects by changing the concentration of cyclic AMP, which acts as the second messenger within the cell.

If levels of cAMP increase, enzymes may be activated or ion channels may be opened, accelerating the metabolic activity of the cell.

In some instances, G protein activation results in decreased levels of cAMP in the cytoplasm. This decrease has an inhibitory effect on the cell.

First Messenger Examples
- Epinephrine and norepinephrine (β receptors)
- Calcitonin
- Parathyroid hormone
- ADH, ACTH, FSH, LH, TSH

First Messenger Examples
- Epinephrine and norepinephrine (α_2 receptors)

Effects on Ca²⁺ Levels

Some G proteins use Ca^{2+} as a second messenger.

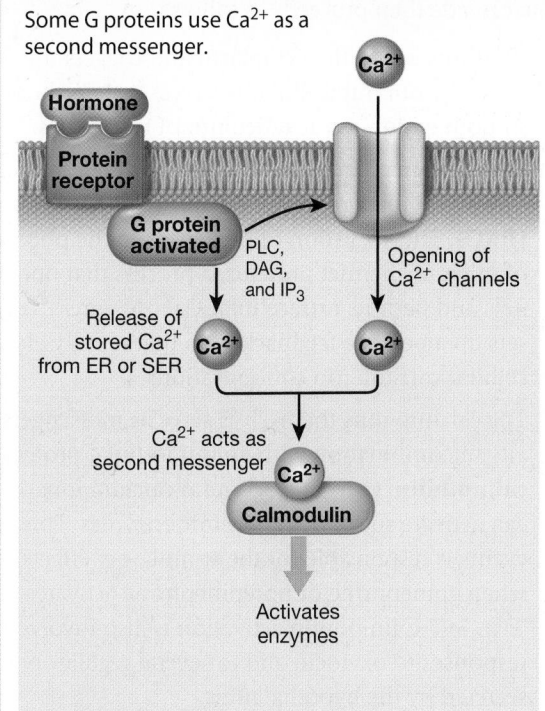

First Messenger Examples
- Epinephrine and norepinephrine (α_1 receptors)
- Oxytocin
- Regulatory hormones of hypothalamus
- Several eicosanoids

produce their effects by this mechanism. The increase in cAMP levels is usually short-lived, because the cytoplasm contains another enzyme, **phosphodiesterase (PDE)**, which inactivates cyclic AMP by converting it to AMP (adenosine monophosphate).

Spotlight Figure 18–3 (center) depicts one way the activation of a G protein can *lower* the concentration of cAMP within the cell. In this case, the activated G protein stimulates PDE activity and inhibits adenylate cyclase activity. Levels of cAMP then decline, because cAMP breakdown accelerates while cAMP synthesis is prevented. The decline has an inhibitory effect on the cell, because without phosphorylation, key enzymes remain inactive. This mechanism is responsible for the inhibitory effects that follow when epinephrine and norepinephrine stimulate α_2 adrenergic receptors, as discussed in Chapter 16. ↪ p. 567

G Proteins and Calcium Ions. An activated G protein can trigger either the opening of calcium ion channels in the plasma membrane or the release of calcium ions from intracellular compartments. Spotlight Figure 18–3 (right panel) diagrams the steps involved. The G protein first activates the enzyme *phospholipase C (PLC)*. This enzyme triggers a receptor cascade that begins with the production of **diacylglycerol (DAG)** and **inositol triphosphate (IP$_3$)** from membrane phospholipids. The cascade then proceeds as follows:

- IP$_3$ diffuses into the cytoplasm and triggers the release of Ca^{2+} from intracellular reserves, such as those in the smooth endoplasmic reticulum of many cells.

- The combination of DAG and intracellular calcium ions activates another membrane protein: **protein kinase C (PKC)**. The activation of PKC leads to the phosphorylation of calcium channel proteins, a process that opens the channels and permits extracellular Ca^{2+} to enter the cell. This sets up a positive feedback loop that rapidly elevates intracellular calcium ion concentrations.

- The calcium ions themselves serve as messengers, generally in combination with an intracellular protein called **calmodulin**. Once it has bound calcium ions, calmodulin can activate specific cytoplasmic enzymes. This chain of events is responsible for the stimulatory effects that follow when epinephrine or norepinephrine activates α_1 receptors. ↪ p. 567 Calmodulin activation is also involved in the responses to oxytocin and to several regulatory hormones secreted by the hypothalamus.

Hormones and Intracellular Receptors

Steroid hormones diffuse across the lipid part of the plasma membrane and bind to receptors in the cytoplasm or nucleus. The hormone–receptor complexes then activate or deactivate specific genes (Figure 18–4a). By this mechanism, steroid hormones can alter the rate of DNA transcription in the nucleus.

In this way, they change the pattern of protein synthesis. Alterations in the synthesis of enzymes or structural proteins directly affect both the metabolic activity and the structure of the target cell. For example, the sex hormone *testosterone* stimulates the production of enzymes and structural proteins in skeletal muscle fibers, causing muscle size and strength to increase.

Thyroid hormones cross the plasma membrane primarily by a transport mechanism. Once in the cytoplasm, these hormones bind to receptors within the nucleus and on mitochondria (Figure 18–4b). The hormone–receptor complexes in the nucleus activate specific genes or change the rate of transcription. The change in transcription rate affects the metabolic activities of the cell by increasing or decreasing the concentration of specific enzymes. Thyroid hormones bound to mitochondria increase the mitochondrial rates of ATP production.

Control of Endocrine Activity by Endocrine Reflexes

As noted earlier, the functional organization of the nervous system parallels that of the endocrine system in many ways. In Chapter 13, we considered the basic operation of neural reflex arcs, the simplest organizational units in the nervous system. ↪ p. 475 The most direct arrangement was a monosynaptic reflex, such as the stretch reflex. Polysynaptic reflexes provide more complex and variable responses to stimuli. Higher centers, which integrate multiple inputs, can facilitate or inhibit these reflexes as needed.

Endocrine reflexes are the functional counterparts of neural reflexes. Endocrine reflexes can be triggered by (1) *humoral stimuli* (changes in the composition of the extracellular fluid), (2) *hormonal stimuli* (the arrival or removal of a specific hormone), or (3) *neural stimuli* (the arrival of neurotransmitters at neuroglandular junctions). In most cases, negative feedback controls endocrine reflexes: A stimulus triggers the production of a hormone, and the direct or indirect effects of the hormone reduce the intensity of the stimulus.

A simple endocrine reflex involves only one hormone. The endocrine cells involved respond directly to changes in the composition of the extracellular fluid. The secreted hormone adjusts the activities of target cells and restores homeostasis. Simple endocrine reflexes control hormone secretion by the heart, pancreas, parathyroid glands, and digestive tract.

More complex endocrine reflexes involve one or more intermediary steps and two or more hormones.

The hypothalamus provides the highest level of endocrine control. It integrates the activities of the nervous and endocrine systems in three ways (Figure 18–5):

1. The hypothalamus itself acts as an endocrine organ. Hypothalamic neurons synthesize hormones and transport them along axons to the posterior lobe of the

Figure 18–4   **Effects of Intracellular Hormone Binding.**

a Steroid hormones diffuse through the plasma membrane and bind to receptors in the cytoplasm or nucleus. The complex then binds to DNA in the nucleus, activating specific genes.

b Thyroid hormones enter the cytoplasm and bind to receptors in the nucleus to activate specific genes. They also bind to receptors on mitochondria and accelerate ATP production.

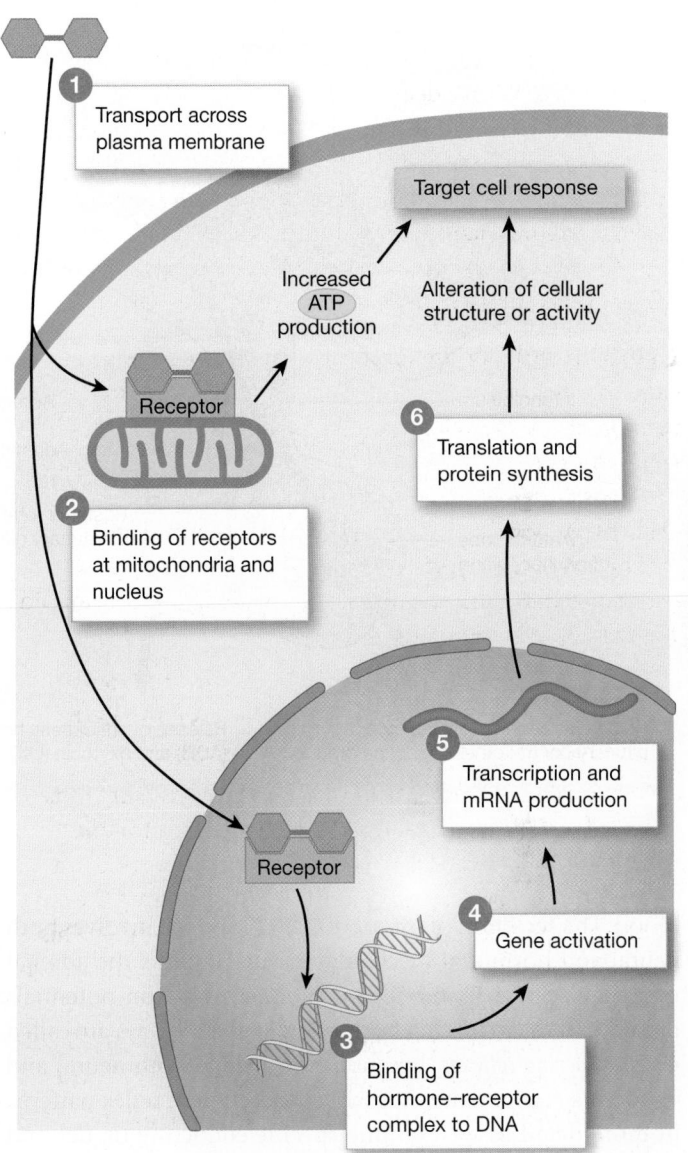

pituitary gland, where they are released into the circulation. We introduced two of these hormones, antidiuretic hormone (ADH) and oxytocin (OXT), in Chapter 14. ⤶ p. 504

2. The hypothalamus secretes **regulatory hormones**, special hormones that control endocrine cells in the pituitary gland. The hypothalamic regulatory hormones control the secretory activities of endocrine cells in the anterior lobe of the pituitary gland. In turn, the hormones from the

anterior lobe control the activities of endocrine cells in the thyroid, adrenal cortex, and reproductive organs.

3. The hypothalamus contains autonomic centers that exert direct neural control over the endocrine cells of the adrenal medullae. When the sympathetic division is activated, the adrenal medullae release the hormones epinephrine (E) and norepinephrine (NE) into the bloodstream.

The hypothalamus secretes regulatory hormones and ADH in response to changes in the composition of the circulating

Figure 18–5 **Three Mechanisms of Hypothalamic Control over Endocrine Function.**

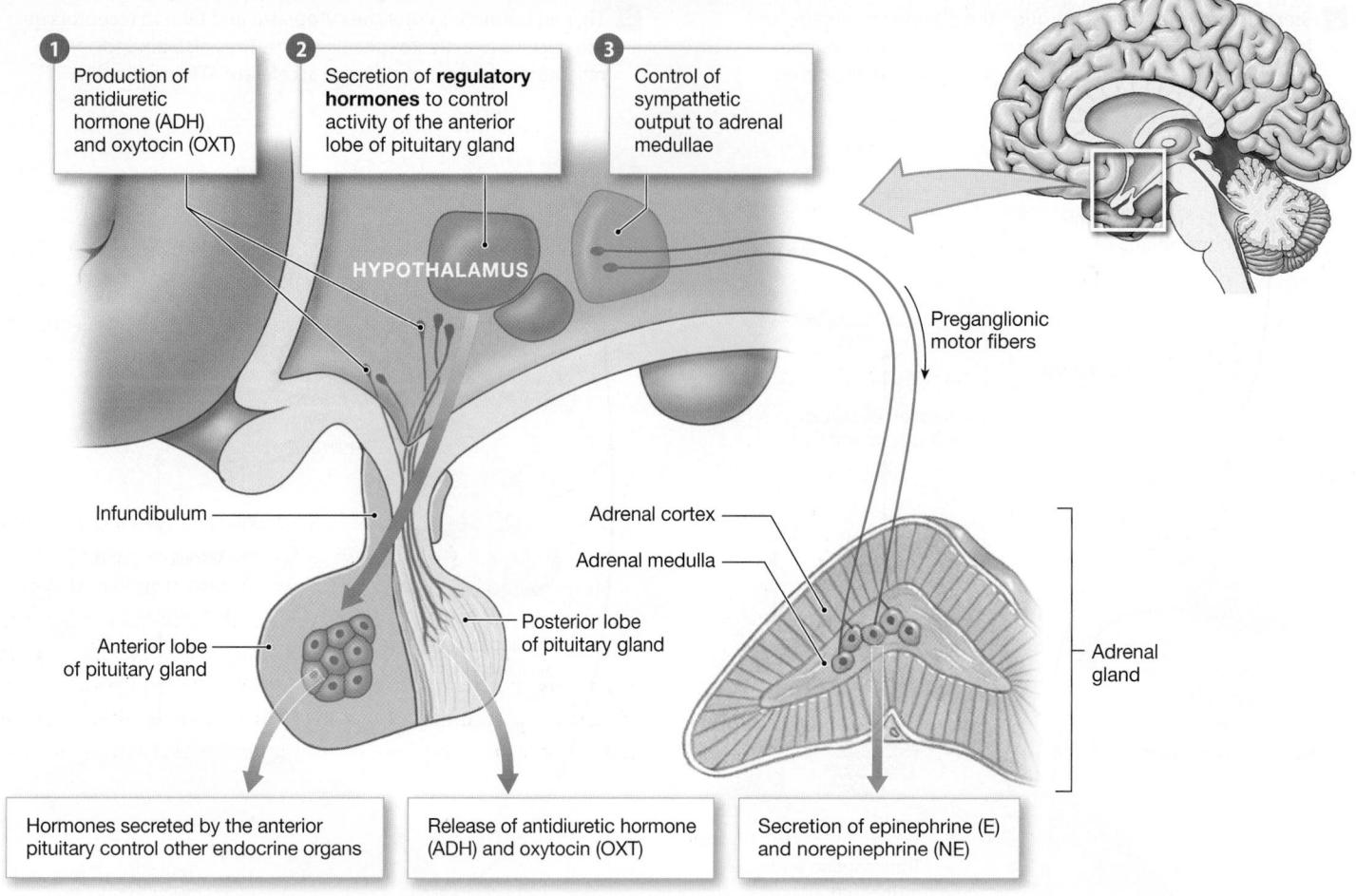

1 Production of antidiuretic hormone (ADH) and oxytocin (OXT)

2 Secretion of **regulatory hormones** to control activity of the anterior lobe of pituitary gland

3 Control of sympathetic output to adrenal medullae

HYPOTHALAMUS

Preganglionic motor fibers

Infundibulum

Adrenal cortex

Adrenal medulla

Posterior lobe of pituitary gland

Anterior lobe of pituitary gland

Adrenal gland

Hormones secreted by the anterior pituitary control other endocrine organs

Release of antidiuretic hormone (ADH) and oxytocin (OXT)

Secretion of epinephrine (E) and norepinephrine (NE)

blood. The secretion of oxytocin OXT, E, and NE involves both neural and hormonal mechanisms. For example, the adrenal medullae secrete E and NE in response to action potentials rather than to circulating hormones. Such pathways are called *neuroendocrine reflexes*, because they include both neural and endocrine components. We will consider these reflex patterns in more detail as we examine specific endocrine tissues and organs.

In Chapter 15, we noted that sensory receptors provide complex information by varying the frequency and pattern of action potentials in a sensory neuron. In a similar way, the endocrine system sends complex commands by changing the amount of hormone secreted and the pattern of hormone release. In a simple endocrine reflex, hormones are released continuously, but the rate of secretion rises and falls in response to humoral stimuli. For example, when blood glucose levels climb, the pancreas increases its secretion of *insulin*, a hormone that stimulates glucose uptake and utilization. As insulin levels rise, glucose levels decline, reducing the stimulation of the insulin-secreting cells. As glucose levels return to normal, the rate of

insulin secretion returns to resting levels. (We introduced this regulatory pattern, called *negative feedback*, in Chapter 1 when we considered the control of body temperature. ⟳ **p. 36**)

In this example, the responses of the target cells change over time, because the effect of insulin is proportional to its concentration. However, the relationship between hormone concentration and target cell response is not always predictable. For instance, a hormone can have one effect at low concentrations and more exaggerated effects—or even different effects—at high concentrations. (We consider specific examples later in the chapter.)

Several hypothalamic and pituitary hormones are released in sudden bursts called *pulses*, rather than continuously. When hormones arrive in pulses, target cells may vary their response with the frequency of the pulses. For example, the target cell response to one pulse every three hours can differ from the response when pulses arrive every 30 minutes. The most complicated hormonal instructions from the hypothalamus involve changes in the frequency of pulses *and* in the amount secreted in each pulse.

CHECKPOINT

4. How could you distinguish between a neural response and an endocrine response on the basis of response time and duration?

5. How would the presence of a substance that inhibits the enzyme adenylate cyclase affect the activity of a hormone that produces its cellular effects by way of the second messenger cAMP?

6. What primary factor determines each cell's hormonal sensitivities?

See the blue Answers tab at the back of the book.

18-3 ▮ The bilobed pituitary gland is an endocrine organ that releases nine peptide hormones

Learning Outcome Describe the location, hormones, and functions of the pituitary gland, and discuss the effects of abnormal pituitary hormone production.

Figure 18–6 shows the anatomy of the **pituitary gland**, or **hypophysis** (hī-POF-i-sis). This small, oval gland lies nestled within the *sella turcica*, a depression in the sphenoid bone (**Figure 7–8**, p. 243). The pituitary gland is inferior to the hypothalamus. It is connected by the slender, funnel-shaped structure called the **infundibulum** (in-fun-DIB-ū-lum; funnel). The base of the infundibulum lies between the optic chiasm and the mammillary bodies. Cradled by the sella turcica, the pituitary gland is held in position by the *sellar diaphragm*, a dural sheet that encircles the infundibulum. The sellar diaphragm locks the pituitary gland in position and isolates it from the cranial cavity.

The pituitary gland has anterior and posterior lobes that differ in function and developmental anatomy. It releases nine important peptide hormones. Seven come from the anterior lobe and two from the posterior lobe. All nine hormones bind to membrane receptors, and all nine use cAMP as a second messenger. ATLAS: Embryology Summary 14: The Development of the Endocrine System

The Anterior Lobe of the Pituitary Gland

The **anterior lobe** of the pituitary gland, also called the **adenohypophysis** (ad-e-no-hī-POF-i-sis), contains a variety of endocrine cells. The anterior lobe has three regions: (1) the **pars distalis** (dis-TAL-is; distal part), the largest and most anterior portion of the pituitary gland; (2) an extension called the **pars tuberalis**, which wraps around the adjacent portion of the infundibulum; and (3) the slender **pars intermedia**, a narrow band bordering the posterior lobe (**Figure 18–6**). An extensive

Figure 18–6   **The Anatomy and Orientation of the Pituitary Gland.**

a Relationship of the pituitary gland to the hypothalamus

b Histology of the pituitary gland showing the anterior and posterior lobes capillary network radiates through these regions, giving every endocrine cell immediate access to the bloodstream.

The Hypophyseal Portal System

The hypothalamus controls the production of hormones in the anterior lobe of the pituitary gland by secreting specific regulatory hormones. At the *median eminence*, a swelling near the attachment of the infundibulum, hypothalamic neurons release regulatory factors into the surrounding interstitial fluids. Their secretions enter the bloodstream quite easily, because the endothelial cells lining the capillaries in this region are unusually permeable. These **fenestrated** (FEN-es-trā-ted; *fenestra,* window) **capillaries** allow relatively large molecules to enter or leave the bloodstream. ⟲ p. 495

The capillary networks in the median eminence are supplied by the *superior hypophyseal artery* (**Figure 18–7**). Before leaving the hypothalamus, the capillary networks unite to form a series of larger vessels that spiral around the infundibulum to reach the anterior lobe. In the anterior lobe, these vessels form a second capillary network that branches among the endocrine cells.

This vascular arrangement is unusual because it carries blood from one capillary network to another. In contrast, a typical artery conducts blood from the heart to a capillary network, and a typical vein carries blood from a capillary network back to the heart. Blood vessels that link two capillary networks are called **portal vessels**. In this case, they have the histological structure of veins, so they are also called portal veins. The entire complex is a **portal system**. Portal systems are named for their destinations. This particular network is the **hypophyseal** (hī-po-FI-sē-al) portal system.

Portal systems are an efficient means of chemical communication. This one ensures that all the hypothalamic hormones entering the portal vessels reach the target cells in the anterior lobe before being diluted through mixing with the general circulation. The communication is strictly one way, however. Any chemicals released by the cells "downstream" must do a complete circuit of the cardiovascular system before they reach the capillaries of the portal system.

Hypothalamic Control of the Anterior Lobe

Two classes of hypothalamic regulatory hormones exist: releasing hormones and inhibiting hormones. A **releasing hormone (RH)** stimulates the synthesis and secretion of one or more hormones at the anterior lobe. In contrast, an **inhibiting**

Figure 18–7 The Hypophyseal Portal System and the Blood Supply to the Pituitary Gland.

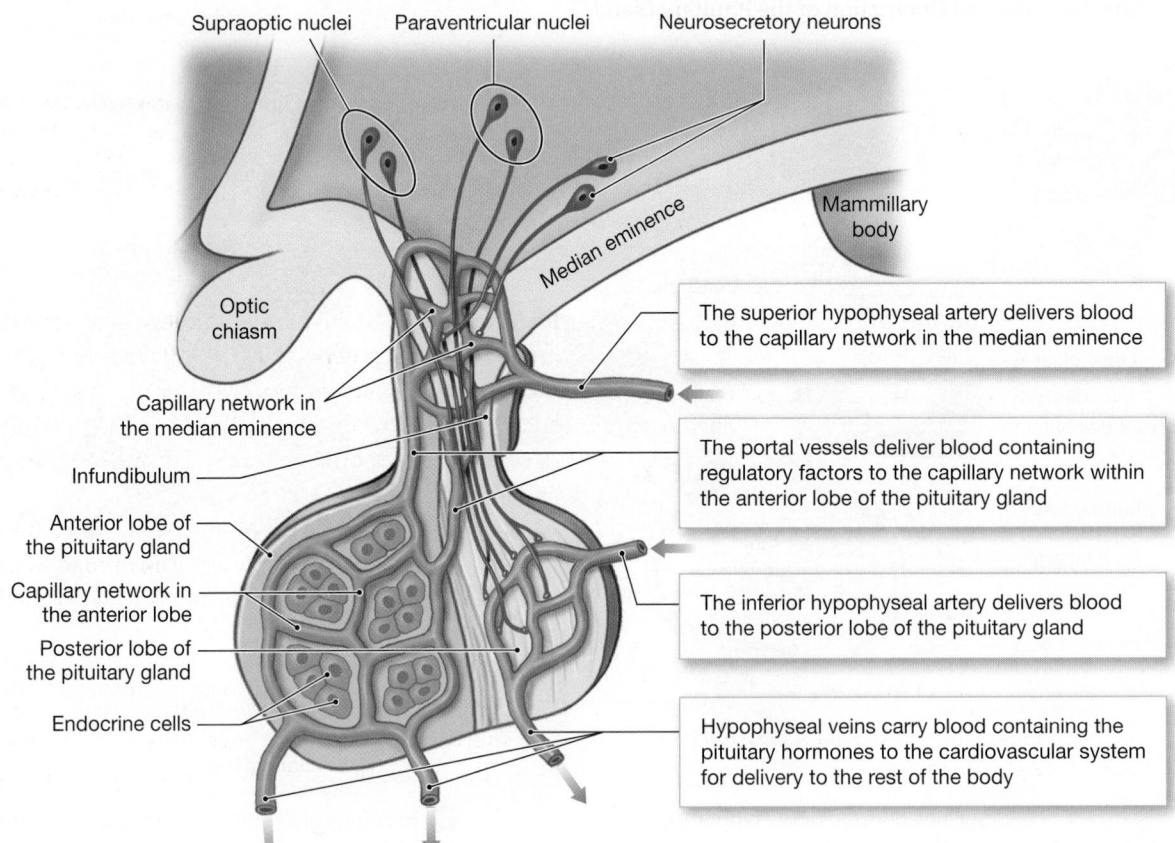

Supraoptic nuclei   Paraventricular nuclei   Neurosecretory neurons

Median eminence

Mammillary body

Optic chiasm

Capillary network in the median eminence

Infundibulum

Anterior lobe of the pituitary gland

Capillary network in the anterior lobe

Posterior lobe of the pituitary gland

Endocrine cells

The superior hypophyseal artery delivers blood to the capillary network in the median eminence

The portal vessels deliver blood containing regulatory factors to the capillary network within the anterior lobe of the pituitary gland

The inferior hypophyseal artery delivers blood to the posterior lobe of the pituitary gland

Hypophyseal veins carry blood containing the pituitary hormones to the cardiovascular system for delivery to the rest of the body

Figure 18–8 **Feedback Control of Endocrine Secretion.**

Feedback Control of Endocrine Secretion at the Hypothalamus

Typical Pattern of Regulation when Multiple Endocrine Organs Are Involved

Hypothalamus

RH

Pituitary gland

Anterior lobe

Hormone 1

Negative feedback

Endocrine organ

Hormone 2

Target cells

KEY

→ Stimulation

⊣ Inhibition

a The hypothalamus produces a releasing hormone (RH) to stimulate hormone production by other glands. Control occurs by negative feedback.

Variations on the Typical Pattern of Regulation of Endocrine Organs by the Hypothalamus and Anterior Pituitary Lobe

PIH

Stimulation

PRF

Inhibition

Anterior lobe

PRL

Stimulates mammary glands

b The regulation of prolactin (PRL) production by the anterior lobe. In this case, the hypothalamus produces both a releasing factor (PRF) and an inhibiting hormone (PIH). When one is stimulated, the other is inhibited.

GH–IH

Stimulation

GH–RH

Inhibition

Anterior lobe

GH

Epithelia, adipose tissue, liver

Liver

Somatomedins

Stimulate growth of skeletal muscle, cartilage, and many other tissues

c The regulation of growth hormone (GH) production by the anterior lobe. When GH–RH release is inhibited, GH–IH release is stimulated.

Releasing hormone (RH)	Hormone 1 (from pituitary)	Endocrine target organ	Hormone 2 (from target organ)
TRH	TSH	Thyroid gland	Thyroid hormones
CRH	ACTH	Adrenal cortex	Glucocorticoids
GnRH	FSH	Testes	Inhibin
		Ovaries	Inhibin Estrogens
	LH	Testes	Androgens
		Ovaries	Progesterone Estrogens

d Table showing the hypothalamic releasing hormones that follow the typical pattern of regulation shown in **a** above.

hormone (IH) prevents the synthesis and secretion of hormones from the anterior lobe. Releasing hormones, inhibiting hormones, or some combination of the two may control an endocrine cell in the anterior lobe. The regulatory hormones released at the hypothalamus travel directly to the anterior lobe by the hypophyseal portal system.

Negative feedback controls the rate at which the hypothalamus secretes regulatory hormones. The primary regulatory patterns are diagrammed in Figure 18–8, and we will refer to them as we examine specific pituitary hormones.

Hormones of the Anterior Lobe

The functions and control mechanisms of seven hormones from the anterior lobe are reasonably well understood:

thyroid-stimulating hormone; adrenocorticotropic hormone; two gonadotropins called *follicle-stimulating hormone* and *luteinizing hormone; prolactin; growth hormone;* and *melanocyte-stimulating hormone.* Of the six hormones produced by the pars distalis, four regulate the production of hormones by other endocrine glands. The names of these hormones indicate their activities, but many of the phrases are so long that we often use abbreviations instead.

The hormones of the anterior lobe are also called *tropic hormones* (*trope,* a turning). They "turn on" endocrine glands or support the functions of other organs. (Some sources call them *trophic hormones* [*trophe,* nourishment] instead.)

Thyroid-Stimulating Hormone. Thyroid-stimulating hormone (TSH), or *thyrotropin,* targets the thyroid gland and triggers the release of thyroid hormones. TSH is released in response to *thyrotropin-releasing hormone (TRH)* from the hypothalamus. Then as circulating concentrations of thyroid hormones rise, the rates of TRH and TSH production decline (**Figure 18–8a**).

Adrenocorticotropic Hormone. Adrenocorticotropic hormone (ACTH), also known as *corticotropin,* stimulates the release of steroid hormones by the *adrenal cortex,* the outer portion of the adrenal gland. ACTH specifically targets cells that produce *glucocorticoids* (glū-kō-KOR-ti-koydz), hormones that affect glucose metabolism. ACTH release occurs under the stimulation of **corticotropin-releasing hormone (CRH)** from the hypothalamus. As glucocorticoid levels increase, the rates of CRH release and ACTH release decline (**Figure 18–8a**).

Gonadotropins. The hormones called **gonadotropins** (gō-nad-ō-TRŌ-pinz) regulate the activities of the *gonads.* (These organs—the testes in males and the ovaries in females—produce reproductive cells as well as hormones.) Gonadotropin-releasing hormone (GnRH) from the hypothalamus stimulates production of gonadotropins. An abnormally low production of gonadotropins results in **hypogonadism.** Children with this condition do not mature sexually, and adults with hypogonadism cannot produce functional sperm (males) or oocytes (females). The two gonadotropins are follicle-stimulating hormone and luteinizing hormone.

■ **Follicle-stimulating hormone (FSH),** or *follitropin,* promotes follicle development in females and, in combination with luteinizing hormone, stimulates the secretion of *estrogens* (ES-trō-jenz) by ovarian cells. *Estradiol* is the most important estrogen. In males, FSH stimulates *nurse cells,* specialized cells in the seminiferous tubules where sperm differentiate. In response, the nurse cells promote the physical maturation of developing sperm. FSH production is inhibited by *inhibin,* a peptide hormone released by cells in the testes and ovaries (**Figure 18–8a**). (The role of inhibin in suppressing the release of GnRH as well as FSH is under debate.)

■ **Luteinizing** (LŪ-tē-in-ī-zing) **hormone (LH),** or *lutropin,* induces *ovulation,* the production of reproductive cells in females. It also promotes ovarian secretion of estrogens and *progesterone,* which prepare the body for possible pregnancy. In males, this gonadotropin is sometimes called *interstitial cell–stimulating hormone (ICSH),* because it stimulates the production of sex hormones by the *interstitial cells* of the testes. These male sex hormones are called **androgens** (AN-drō-jenz; *andros,* man). The most important one is *testosterone.* LH production, like FSH production, is stimulated by GnRH from the hypothalamus. Estrogens, progesterone, and androgens inhibit GnRH production (**Figure 18–8a**).

Prolactin. Prolactin (*pro-,* before + *lac,* milk) **(PRL)** works with other hormones to stimulate mammary gland development. In pregnancy and during the nursing period that follows delivery, PRL also stimulates milk production by the mammary glands. The functions of PRL in males are poorly understood, but evidence indicates that PRL helps regulate androgen production by making interstitial cells more sensitive to LH.

Prolactin production is inhibited by the neurotransmitter dopamine, also known as **prolactin-inhibiting hormone (PIH).** The hypothalamus also secretes several *prolactin-releasing factors (PRF).* Few of these have been identified. Circulating PRL stimulates PIH release and inhibits the secretion of PRF (**Figure 18–8b**).

Although PRL exerts the dominant effect on the glandular cells, normal development of the mammary glands is regulated by the interaction of several hormones. Prolactin, estrogens, progesterone, glucocorticoids, pancreatic hormones, and hormones produced by the placenta cooperate in preparing the mammary glands for secretion. Milk ejection occurs only in response to oxytocin release at the posterior lobe of the pituitary gland. We describe the functional development of the mammary glands in Chapter 28.

Growth Hormone. Growth hormone (GH), or **somatotropin,** stimulates cell growth and replication by accelerating the rate of protein synthesis. Skeletal muscle cells and chondrocytes (cartilage cells) are particularly sensitive to GH, although virtually every tissue responds to some degree.

The stimulation of growth by GH involves two mechanisms. The primary mechanism, which is indirect, is best understood. Liver cells respond to GH by synthesizing and releasing **somatomedins,** known also as **insulin-like growth factors (IGFs).** These peptide hormones stimulate tissue growth by binding to receptors on a variety of plasma membranes (**Figure 18–8b**). In skeletal muscle fibers, cartilage cells, and other target cells, somatomedins increase the uptake of amino acids and their incorporation into new proteins.

These effects develop almost immediately after GH is released. They are particularly important after a meal, when the blood contains high concentrations of glucose and amino acids. In functional terms, cells can now obtain ATP easily through the aerobic metabolism of glucose, and amino acids are readily available for protein synthesis. Under these conditions, GH, acting through the somatomedins, stimulates protein synthesis and cell growth.

The direct actions of GH are more selective. They tend to appear after blood glucose and amino acid concentrations have returned to normal levels:

- In epithelia and connective tissues, GH stimulates stem cell divisions and the differentiation of daughter cells. (Somatomedins then stimulate the growth of these daughter cells.)

- In adipose tissue, GH stimulates the breakdown of stored triglycerides by adipocytes (fat cells), which then release fatty acids into the blood. As circulating fatty acid levels rise, many tissues stop breaking down glucose to generate ATP and instead start breaking down fatty acids. This process is termed a **glucose-sparing effect**.

- In the liver, GH stimulates the breakdown of glycogen reserves by liver cells, which then release glucose into the bloodstream. Because most other tissues are now metabolizing fatty acids rather than glucose, blood glucose concentrations begin to climb to levels significantly higher than normal. The elevation of blood glucose levels by GH has been called a **diabetogenic effect**, because *diabetes mellitus*, an endocrine disorder we consider later in the chapter, is characterized by abnormally high blood glucose concentrations.

The production of GH is regulated by **growth hormone–releasing hormone** (**GH–RH**, or *somatocrinin*) and **growth hormone–inhibiting hormone** (**GH–IH**, or *somatostatin*) from the hypothalamus. Somatomedins inhibit GH–RH and stimulate GH–IH (**Figure 18–8b**).

Melanocyte-Stimulating Hormone. The pars intermedia may secrete two forms of **melanocyte-stimulating hormone** (**MSH**), or *melanotropin*. As the name indicates, MSH stimulates the melanocytes of the skin, increasing their production of melanin, a brown, black, or yellow-brown pigment. ⊃ p. 182 Dopamine inhibits the release of MSH.

Melanocyte-stimulating hormone from the pituitary gland is important in the control of skin pigmentation in fishes, amphibians, reptiles, and many mammals other than primates. In humans, MSH is produced locally, within sun-exposed skin. The pars intermedia in adult humans is virtually nonfunctional, and the circulating blood usually does not contain MSH. However, the human pars intermedia secretes MSH (1) during fetal development, (2) in very young children,

(3) in pregnant women, and (4) in the course of some diseases. The functional significance of MSH secretion under these circumstances is not known. The administration of a synthetic form of MSH causes the skin to darken, so MSH has been suggested as a means of obtaining a "sunless tan."

The Posterior Lobe of the Pituitary Gland

The **posterior lobe** of the pituitary gland, also called the **neurohypophysis** (noo-rō-hī-POF-i-sis), contains the axons of hypothalamic neurons. Neurons of the **supraoptic** and **paraventricular nuclei** manufacture antidiuretic hormone (ADH) and oxytocin (OXT), respectively. These hormones move along axons in the infundibulum to axon terminals, which end on the basement membranes of capillaries in the posterior lobe. They travel by means of axoplasmic transport. ⊃ p. 416

Antidiuretic Hormone

Antidiuretic hormone (ADH), also known as *vasopressin (VP)*, is released in response to a variety of stimuli, most notably a rise in the solute concentration in the blood or a fall in blood volume or blood pressure. A rise in the solute concentration stimulates specialized neurons in the hypothalamus. These neurons are called *osmoreceptors* because they respond to a change in the osmotic concentration of body fluids. The osmoreceptors then stimulate the neurosecretory neurons that release ADH.

Clinical Note

Diabetes Insipidus *Diabetes* occurs in several forms. All are characterized by excessive urine production (polyuria). Most forms result from endocrine abnormalities, but physical damage to the kidneys can also cause diabetes. The two most prevalent forms are diabetes insipidus and diabetes mellitus. (We describe diabetes mellitus in the Spotlight on page 638.) **Diabetes insipidus** generally develops because the posterior lobe of the pituitary gland no longer releases adequate amounts of ADH. Water conservation by the kidneys is impaired, and excessive amounts of water are lost in the urine. As a result, the person is constantly thirsty, but the body does not retain the fluids consumed.

Mild cases of diabetes insipidus may not require treatment if fluid and electrolyte intake keep pace with urinary losses. In severe cases, the fluid losses can reach 10 liters per day, and dehydration and electrolyte imbalances are fatal without treatment. This condition can be effectively treated with desmopressin, a synthetic form of ADH.

The primary function of ADH is to act on the kidneys to retain water and decrease urination. With losses minimized, any water absorbed from the digestive tract will be retained, reducing the concentrations of electrolytes in the extracellular fluid. In high concentrations, ADH also causes *vasoconstriction*, a narrowing of peripheral blood vessels that helps elevate blood pressure. Alcohol inhibits ADH release, which explains why people find themselves making frequent trips to the bathroom after consuming alcoholic beverages.

Oxytocin

In women, **oxytocin** (*oxy-*, quick + *tokos*, childbirth) **(OXT)** stimulates smooth muscle contraction in the wall of the uterus, promoting labor and delivery. After delivery, oxytocin promotes the ejection of milk by stimulating the contraction of myoepithelial cells around the secretory alveoli and the ducts of the mammary glands.

Until the last stages of pregnancy, the uterine smooth muscles are relatively insensitive to oxytocin, but they become more sensitive as the time of delivery approaches. The trigger for normal labor and delivery is probably a sudden rise in oxytocin levels at the uterus. There is good evidence, however, that the uterus and fetus secrete most of the oxytocin involved. Oxytocin from the posterior lobe plays only a supporting role.

Oxytocin secretion and milk ejection are part of a neuroendocrine reflex called the *milk let-down reflex*. The normal stimulus is an infant suckling at the breast, and sensory nerves innervating the nipples relay the information to the hypothalamus. Oxytocin is then released into the circulation at the posterior lobe, and the myoepithelial cells respond by squeezing milk from the secretory alveoli into large collecting ducts. Any factor that affects the hypothalamus can modify this reflex. For example, anxiety, stress, and other factors can

Figure 18–9 Pituitary Hormones and Their Targets.

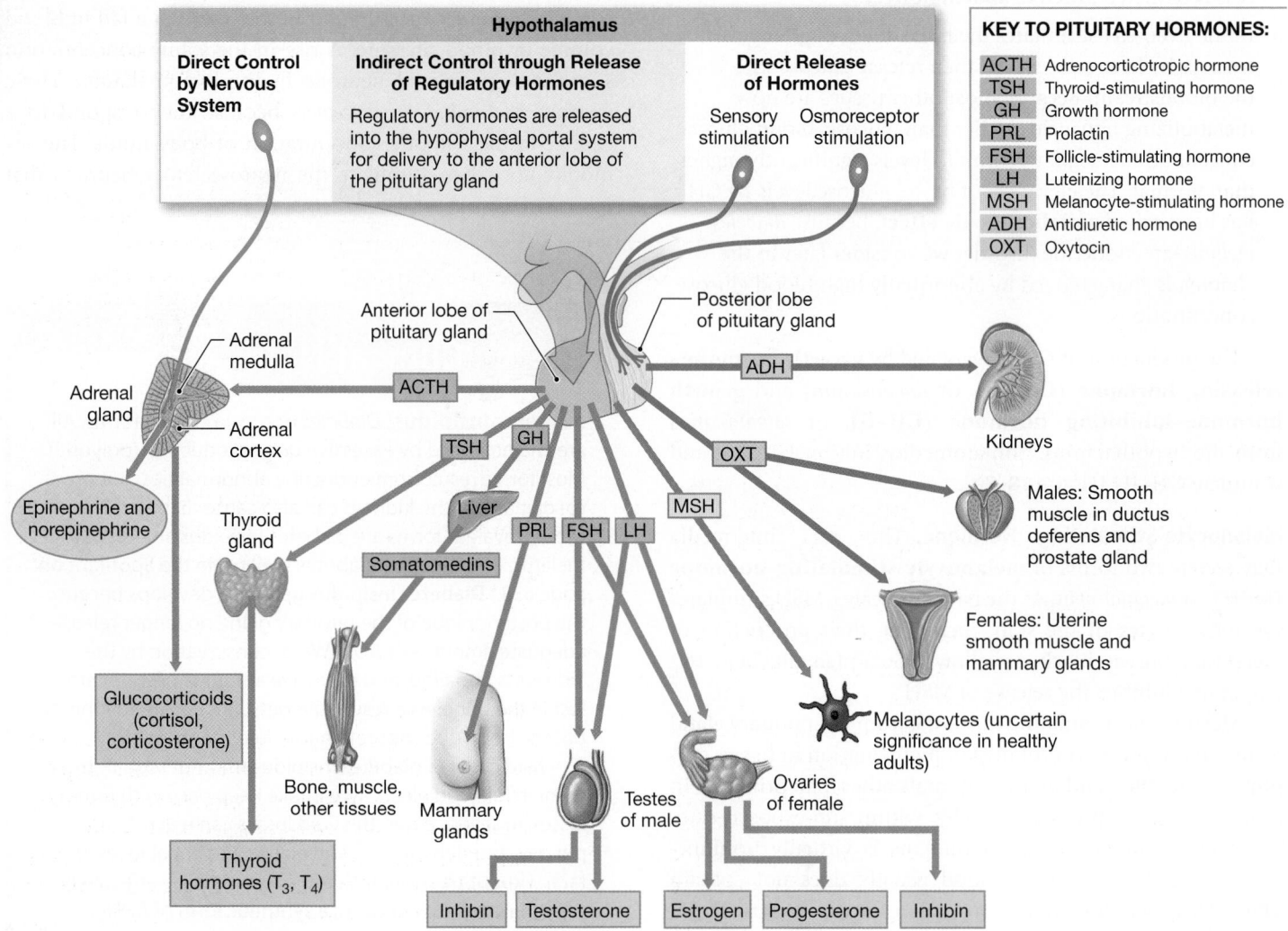

Table 18–2	The Pituitary Hormones			
	Hormone	**Target**	**Hormonal Effect**	**Hypothalamic Regulatory Hormone**
ANTERIOR LOBE				
Pars distalis	**Thyroid-stimulating hormone (TSH)**	Thyroid gland	Secretion of thyroid hormones	Thyrotropin-releasing hormone (TRH)
	Adrenocorticotropic hormone (ACTH)	Adrenal cortex (zona fasciculata)	Secretion of glucocorticoids (cortisol, corticosterone)	Corticotropin-releasing hormone (CRH)
	Gonadotropins:			
	Follicle-stimulating hormone (FSH)	Follicle cells of ovaries	Secretion of estrogen, follicle development	Gonadotropin-releasing hormone (GnRH)
		Nurse cells of testes	Stimulation of sperm maturation	Gonadotropin-releasing hormone (GnRH)
	Luteinizing hormone (LH)	Follicle cells of ovaries	Ovulation, formation of corpus luteum, secretion of progesterone	Gonadotropin-releasing hormone (GnRH)
		Interstitial cells of testes	Secretion of testosterone	Gonadotropin-releasing hormone (GnRH)
	Prolactin (PRL)	Mammary glands	Production of milk	Prolactin-releasing factor (PRF)
				Prolactin-inhibiting hormone (PIH)
	Growth hormone (GH)	All cells	Growth, protein synthesis, lipid mobilization and catabolism	Growth hormone–releasing hormone (GH–RH)
				Growth hormone–inhibiting hormone (GH–IH)
Pars intermedia (not active in normal adults)	**Melanocyte-stimulating hormone (MSH)**	Melanocytes	Increased melanin synthesis in epidermis	Melanocyte-stimulating hormone–inhibiting hormone (MSH–IH)
POSTERIOR LOBE				
	Antidiuretic hormone (ADH)	Kidneys	Reabsorption of water, elevation of blood volume and pressure	None: Transported along axons from supraoptic nucleus to the posterior lobe of the pituitary gland
	Oxytocin (OXT)	Uterus, mammary glands (females)	Labor contractions, milk ejection	None: Transported along axons from paraventricular nucleus to the posterior lobe of the pituitary gland
		Ductus deferens and prostate gland (males)	Contractions of ductus deferens and prostate gland	

18

prevent the flow of milk, even when the mammary glands are fully functional. In contrast, nursing mothers can become conditioned to associate a baby's crying with suckling. In these women, milk let-down may begin as soon as they hear a baby cry.

The functions of oxytocin in sexual activity remain uncertain, but it is known that circulating concentrations of oxytocin rise during sexual arousal and peak at orgasm in both sexes. Evidence indicates that in men, oxytocin stimulates smooth muscle contractions in the walls of the *ductus deferens* (sperm duct) and prostate gland. These actions may be important in *emission*—the ejection of sperm, secretions of the prostate gland, and the secretions of other glands into the male reproductive tract before ejaculation. Studies suggest that the oxytocin released in females during intercourse may stimulate smooth muscle contractions in the uterus and vagina that promote the transport of sperm toward the uterine tubes.

Summary: The Hormones of the Pituitary Gland

Figure 18–9 and Table 18–2 summarize important information about the hormones of the pituitary gland. Review them carefully before considering the structure and function of other endocrine organs.

CHECKPOINT

7. Identify the two lobes of the pituitary gland.

8. If a person were dehydrated, how would the amount of ADH released by the posterior pituitary lobe change?

9. The release of which pituitary hormone would lead to an increased level of somatomedins in the blood?

10. What effect would elevated circulating levels of cortisol, a steroid hormone from the adrenal cortex, have on the pituitary secretion of ACTH?

See the blue Answers tab at the back of the book.

18-4 ▎ The thyroid gland lies inferior to the larynx and requires iodine for hormone synthesis

Learning Outcome Describe the location, hormones, and functions of the thyroid gland, and discuss the effects of abnormal thyroid hormone production.

The **thyroid gland** curves across the anterior surface of the trachea just inferior to the *thyroid* ("shield-shaped") *cartilage*, which forms most of the anterior surface of the larynx (**Figure 18–10a**). The two **lobes** of the thyroid gland are united by a slender connection, the **isthmus** (IS-mus). You can feel the gland with your fingers. When something goes wrong with this gland, it typically becomes visible as it enlarges and distorts the surface of the neck. The size of the gland varies, depending on heredity and environmental and nutritional factors, but its average weight is about 34 g (1.2 oz). An extensive blood supply gives the thyroid gland a deep red color.

Thyroid Follicles and Thyroid Hormones

The thyroid gland contains large numbers of **thyroid follicles**, hollow spheres lined by a simple cuboidal epithelium (**Figure 18–10b,c**). The follicle cells surround a **follicle cavity** that holds a viscous *colloid*, a fluid containing large quantities of dissolved proteins. A network of capillaries surrounds each follicle, delivering nutrients and regulatory hormones to the glandular cells and accepting their secretory products and metabolic wastes.

TIPS & TRICKS

> The structure of a thyroid follicle is similar to that of a gel capsule: The simple cuboidal epithelium is comparable to the capsule itself, and the colloid to the capsule's viscous contents.

Follicle cells synthesize a globular protein called **thyroglobulin** (thī-rō-GLOB-ū-lin) and secrete it into the colloid of the thyroid follicles (**Figure 18–10c**). Thyroglobulin molecules contain the amino acid *tyrosine*, the building block of thyroid hormones. The formation of thyroid hormones involves the following basic steps (**Figure 18–11a**, p. 654):

1 Iodide ions are absorbed from the diet by the digestive tract, and the bloodstream delivers them to the thyroid gland. TSH-sensitive carrier proteins in the basement membrane of the follicle cells actively transport iodide ions (I^-) into the cytoplasm. Normally, the follicle cells maintain intracellular concentrations of iodide that are many times higher than those in the extracellular fluid.

2 The iodide ions diffuse to the apical surface of each follicle cell, where they lose an electron and are converted to an atom of iodine (I^0) by the enzyme *thyroid peroxidase*. This reaction sequence, which occurs at the apical membrane surface, also attaches one or two iodine atoms (I^0) to the tyrosine portions of a thyroglobulin molecule within the follicle cavity.

3 Tyrosine molecules with attached iodine atoms (I^0) become linked by covalent bonds, forming molecules of thyroid hormones that remain incorporated into thyroglobulin. Thyroid peroxidase probably carries out the pairing process. The hormone **thyroxine** (thi-ROKS-ēn), also known as *tetraiodothyronine* or T_4, contains four iodine atoms. A related molecule called **triiodothyronine**, or T_3, contains three iodine atoms. Eventually, each molecule of thyroglobulin contains four to eight molecules of T_3 or T_4 hormones or both.

The major factor controlling the rate of thyroid hormone release is the concentration of TSH in the circulating blood (**Figure 18–11b**). TSH stimulates iodide transport into the follicle cells and stimulates the production of thyroglobulin and thyroid peroxidase. TSH also stimulates the release of thyroid hormones. Under the influence of TSH, the following steps occur (**Figure 18–11a**):

4 Follicle cells remove thyroglobulin from the follicles by endocytosis.

5 Lysosomal enzymes break down the thyroglobulin, and the amino acids and thyroid hormones enter the cytoplasm. The amino acids are then recycled and used to synthesize more thyroglobulin.

6 The released molecules of T_3 and T_4 diffuse across the basement membrane and enter the bloodstream. About 90 percent of all thyroid secretions is T_4, and T_3 is secreted in comparatively small amounts.

7 About 75 percent of the T_4 molecules and 70 percent of the T_3 molecules entering the bloodstream become attached to transport proteins called **thyroid-binding globulins (TBGs)**. Most of the rest of the T_4 and T_3 in the circulation is attached to **transthyretin**, also known as *thyroid-binding prealbumin (TBPA)*, or to *albumin*, one of the plasma proteins. Only the relatively small quantities of thyroid hormones that remain unbound—roughly 0.3 percent of the circulating T_3 and 0.03 percent of the circulating T_4—are free to diffuse into peripheral tissues.

Equilibrium exists between the amount of bound and unbound thyroid hormones. At any moment, free thyroid hormones are being bound to carriers at the same rate at which bound hormones are being released. When unbound thyroid hormones diffuse out of the bloodstream and into other tissues, the equilibrium is disturbed. The carrier proteins then release additional thyroid hormones until a new equilibrium is reached. The bound thyroid hormones make up a substantial reserve: The bloodstream normally has more than a week's supply of thyroid hormones.

Figure 18–10 **The Thyroid Gland.**

Hyoid bone

Superior thyroid artery

Thyroid cartilage of larynx

Superior thyroid vein

Common carotid artery

Right lobe of thyroid gland

Middle thyroid vein

Thyrocervical trunk

Trachea

Outline of clavicle

Outline of sternum

Internal jugular vein

Cricoid cartilage of larynx

Left lobe of thyroid gland

Isthmus of thyroid gland

Inferior thyroid artery

Inferior thyroid veins

a Location and anatomy of the thyroid gland

Thyroid follicles

The thyroid gland   LM × 122

b Histological organization of the thyroid

Capillary

Capsule

Follicle cavities

Thyroid follicle

C cell

C cell

Cuboidal epithelium of follicle

Thyroid follicle

Thyroglobulin stored in colloid of follicle

Follicles of the thyroid gland   LM × 260

c Histological details of the thyroid gland showing thyroid follicles and both cell types in the follicular epithelium ATLAS: Plate 18c

18

Figure 18–11   **The Thyroid Follicles.**

a The synthesis, storage, and secretion of thyroid hormones. The numbered events are explained in the text.

b The regulation of thyroid secretion.

TSH plays a key role in both the synthesis and the release of thyroid hormones. In the absence of TSH, the thyroid follicles become inactive, and neither synthesis nor secretion occurs. TSH binds to plasma membrane receptors and, by stimulating adenylate cyclase, it activates key enzymes involved in thyroid hormone production (**Spotlight Figure 18–3**).

Functions of Thyroid Hormones

Thyroid hormones affect almost every cell in the body. They enter target cells by means of an energy-dependent transport system. Inside a target cell, they bind to receptors (1) in the cytoplasm, (2) on the surfaces of mitochondria, and (3) in the nucleus.

- Thyroid hormones bound to cytoplasmic receptors are essentially held in storage. If intracellular levels of thyroid hormones decline, the bound thyroid hormones are released into the cytoplasm.

- The thyroid hormones binding to mitochondria increase the rates of mitochondrial ATP production.

- The binding to receptors in the nucleus activates genes that control the synthesis of enzymes involved in energy transformation and utilization. One specific effect is the accelerated production of sodium–potassium ATPase. Recall that this membrane protein ejects intracellular sodium and recovers extracellular potassium. As noted in Chapter 3, this exchange pump uses large amounts of ATP. ⟳ p. 122

Thyroid hormones also activate genes that code for enzymes involved in glycolysis and ATP production. Coupled with the direct effect of thyroid hormones on mitochondria, this effect increases the metabolic rate of the cell. The effect is called the **calorigenic effect** (*calor*, heat) of thyroid hormones because the cell consumes more energy and generates more heat. In young children, TSH production increases in cold weather, and the calorigenic effect may help them adapt to cold climates. (This response does not occur in adults.) In growing children, thyroid hormones are also essential to normal development of the skeletal, muscular, and nervous systems.

The thyroid gland produces large amounts of T_4, but T_3 is primarily responsible for the observed effects of thyroid hormones: a strong, immediate, and short-lived increase in the rate of cellular metabolism. At any moment, T_3 released from the thyroid gland accounts for only 10–15 percent of the T_3 in peripheral tissues. However, enzymes in the liver, kidneys, and other tissues convert T_4 to T_3. Roughly 85–90 percent of the T_3 that reaches the target cells comes from the conversion of T_4 within peripheral tissues. Table 18–3 summarizes the effects of thyroid hormones on major organs and systems.

Iodine and Thyroid Hormones

We obtain iodine from the food we eat. Each day the follicle cells in the thyroid gland absorb 120–150 μg of I^-, the minimum dietary amount needed to maintain normal thyroid function. The iodine ions are actively transported into the thyroid follicle cells. As a result, the concentration of I^- inside thyroid follicle cells is generally about 30 times higher than that in the blood. If blood I^- levels rise, so do levels inside the follicle cells.

The thyroid follicles contain most of the iodide reserves in the body. TSH stimulates the active transport of iodide. The resulting increase in the rate of iodide movement into the cytoplasm accelerates the formation of thyroid hormones.

The typical diet in the United States provides approximately 500 μg of iodide per day, about three times the minimum daily requirement. Much of the excess is due to the I^- added to table salt (iodized salt), which is consumed in large quantities in processed foods. For this reason, iodide deficiency is seldom responsible for limiting the rate of thyroid hormone production. (This is not necessarily the case in other countries.) The kidneys remove excess I^- from the blood, and each day the liver excretes a small amount of I^- (about 20 μg) into the *bile*, an exocrine product stored in the gallbladder. Iodide excreted by the kidneys is eliminated in urine. The I^- excreted in bile is eliminated in feces. The losses in the bile continue even if the diet contains less than the minimum iodide requirement and

can gradually deplete the iodide reserves in the thyroid. Thyroid hormone production then declines, regardless of the circulating levels of TSH.

The C Cells of the Thyroid Gland and Calcitonin

The thyroid also contains a second population of endocrine cells. These cells are the **C (clear) cells**, or *parafollicular cells* (Figure 18–10c). They lie sandwiched between the cuboidal follicle cells and their basement membrane. They are larger than the cells of the follicular epithelium and do not stain as clearly.

C cells produce the hormone **calcitonin (CT)**, which helps to regulate Ca^{2+} concentrations in body fluids. We introduced the functions of this hormone in Chapter 6. ⟳ p. 223 The net effect of calcitonin release is a drop in the Ca^{2+} concentration in body fluids. Calcitonin (1) inhibits osteoclasts, which slows the rate of Ca^{2+} release from bone, and (2) stimulates Ca^{2+} excretion by the kidneys.

The control of calcitonin secretion is an example of direct endocrine regulation: Neither the hypothalamus nor the pituitary gland is involved. The C cells respond directly to an elevation in the Ca^{2+} concentration of blood. When the concentration rises, calcitonin secretion increases. The Ca^{2+} concentration then drops, eliminating the stimulus and "turning off" the C cells.

Calcitonin is probably most important during childhood, when it stimulates bone growth and mineral deposition in the skeleton. It also appears to be important in reducing the loss of bone mass (1) during prolonged starvation and (2) in the late stages of pregnancy. At that time the maternal skeleton competes with the developing fetus for calcium ions from the diet. The role of calcitonin in the healthy nonpregnant adult is unclear.

In several chapters, you have seen the importance of Ca^{2+} in controlling muscle cell and neuron activities. Calcium ion concentrations also affect the sodium permeabilities of excitable membranes. At high Ca^{2+} concentrations, sodium permeability decreases and membranes become less responsive. Such problems are relatively rare.

Problems caused by lower-than-normal Ca^{2+} concentrations are equally dangerous and much more common. When calcium ion concentrations decline, sodium permeabilities increase. As a result, cells become extremely excitable. If calcium levels fall too far, convulsions or muscular spasms can result. The *parathyroid glands* and *parathyroid hormone* are largely responsible for maintaining adequate calcium levels.

Table 18–3	Effects of Thyroid Hormones on Peripheral Tissues

1. Elevates rates of oxygen consumption and energy consumption; in children, may cause a rise in body temperature
2. Increases heart rate and force of contraction; generally results in a rise in blood pressure
3. Increases sensitivity to sympathetic stimulation
4. Maintains normal sensitivity of respiratory centers to changes in oxygen and carbon dioxide concentrations
5. Stimulates red blood cell formation and thus enhances oxygen delivery
6. Stimulates activity in other endocrine tissues
7. Accelerates turnover of minerals in bone

CHECKPOINT

11. Identify the hormones of the thyroid gland.

12. What signs and symptoms would you expect to see in an individual whose diet lacks iodine?

13. When a person's thyroid gland is removed, signs of decreased thyroid hormone concentration do not appear until about one week later. Why?

See the blue Answers tab at the back of the book.

18-5 ❚ The four parathyroid glands, embedded in the posterior surface of the thyroid gland, secrete parathyroid hormone to elevate blood Ca²⁺

Learning Outcome Describe the location, hormone, and functions of the parathyroid glands, and discuss the effects of abnormal parathyroid hormone production.

There are normally two pairs of **parathyroid glands** embedded in the posterior surfaces of the thyroid gland (**Figure 18–12a**). A dense capsule surrounding each parathyroid gland separates it from the cells of the thyroid gland. All together, the four parathyroid glands weigh a mere 1.6 g (0.06 oz). The histological appearance of a single parathyroid gland is shown in **Figure 18–12b,c**. The parathyroid glands have at least two cell populations. The **parathyroid (chief) cells** produce parathyroid hormone. The functions of the other cells, called *oxyphils*, are unknown.

Like the C cells of the thyroid gland, the parathyroid cells monitor the circulating concentration of calcium ions. When the Ca^{2+} concentration of the blood falls below normal, the parathyroid cells secrete **parathyroid hormone (PTH)**, or *parathormone*. The net result of PTH secretion is an increase in Ca^{2+} concentration in body fluids. Parathyroid hormone has three major effects:

1. It mobilizes calcium from bone by affecting osteoblast and osteoclast activity. PTH stimulates osteoblasts to secrete a growth factor known as RANKL. Osteoclasts have no PTH receptors, but both precursor and mature osteoclasts have RANKL receptors. This growth factor results in an increase in osteoclasts and osteoclast activity. With more active osteoclasts, the rates of mineral turnover and Ca^{2+} release accelerate. As bone matrix erodes, blood Ca^{2+} rises.

2. It enhances the resorption of Ca^{2+} by the kidneys, reducing urinary losses.

Figure 18–12 **The Parathyroid Glands.**

Connective tissue capsule of parathyroid gland

Blood vessel

Thyroid follicle

Parathyroid gland   LM × 94

b Both parathyroid and thyroid tissues.

Left lobe of thyroid gland

Parathyroid glands

a Thyroid gland, posterior view. The location of the parathyroid glands on the posterior surface of the thyroid lobes. (The thyroid lobes are located anterior to the trachea.)

Parathyroid (chief) cells

Oxyphil cells

Parathyroid cells and oxyphil cells   LM × 600

c Parathyroid gland cells.

Figure 18–13 The Homeostatic Regulation of Calcium Ion Concentrations.

18-6 I The adrenal glands, consisting of a cortex and medulla, cap the kidneys and secrete several hormones

Learning Outcome Describe the location, structure, hormones, and general functions of the adrenal glands, and discuss the effects of abnormal adrenal hormone production.

A yellow, pyramid-shaped **adrenal** (ad-RĒ-nal) **gland** (*ad-*, near + *renes*, kidneys), or *suprarenal gland*, sits on the superior border of each kidney (**Figure 18–14**). Each adrenal gland lies near the level of the 12th rib. A dense fibrous capsule firmly attaches each adrenal gland to the superior portion of each kidney. The adrenal gland on each side nestles among the kidney, the diaphragm, and the major arteries and veins that run along the posterior wall of the abdominopelvic cavity. The adrenal glands project into the peritoneal cavity, and their anterior surfaces are covered by a layer of parietal peritoneum. Like other endocrine glands, the adrenal glands are highly vascularized.

A typical adrenal gland weighs about 5.0 g (0.18 oz), but its size can vary greatly as secretory demands change. The adrenal gland has two parts with separate endocrine functions: a superficial **adrenal cortex** and an inner **adrenal medulla** (**Figure 18–14b**).

The Adrenal Cortex

The yellowish color of the adrenal cortex is due to stored lipids, especially cholesterol and various fatty acids. The adrenal cortex produces more than two dozen steroid hormones. They are collectively called **corticosteroids**. In the bloodstream, these hormones are bound to transport proteins called *transcortins*.

Corticosteroids are vital. If the adrenal glands are destroyed or removed, the person will die unless corticosteroids are administered. Like other steroid hormones, corticosteroids exert their effects by turning on transcription of certain genes in the nuclei of their target cells and determining their transcription rates. The resulting changes in the nature and concentration of enzymes in the cytoplasm affect cellular metabolism.

Deep to the adrenal capsule are three distinct zones in the adrenal cortex (**Figure 18–14c**): (1) an outer *zona glomerulosa*; (2) a middle *zona fasciculata*; and (3) an inner *zona reticularis*. Each zone synthesizes specific steroid hormones.

3. It stimulates the formation and secretion of *calcitriol* by the kidneys. In general, the effects of calcitriol complement or enhance those of PTH, but calcitriol also enhances Ca^{2+} and PO_4^{3-} absorption by the digestive tract. ⟳ p. 221

Figure 18–13 illustrates the roles of calcitonin and PTH in regulating Ca^{2+} concentrations. It is likely that PTH, aided by calcitriol, is the primary regulator of circulating calcium ion concentrations in healthy adults. Information about the hormones of the thyroid gland and parathyroid glands is summarized in **Table 18–4**.

Table 18–4	Hormones of the Thyroid Gland and Parathyroid Glands			
Gland/Cells	**Hormone**	**Target**	**Hormonal Effect**	**Regulatory Control**
THYROID GLAND				
Follicular epithelium	Thyroxine (T_4) Triiodothyronine (T_3)	Most cells	Increases energy utilization, oxygen consumption, growth, and development	Stimulated by TSH from the anterior lobe of the pituitary gland
C cells	Calcitonin (CT)	Bone, kidneys	Decreases Ca^{2+} concentrations in body fluids	Stimulated by elevated blood Ca^{2+} levels; actions opposed by PTH
PARATHYROID GLANDS				
Parathyroid (chief) cells	Parathyroid hormone (PTH)	Bone, kidneys	Increases Ca^{2+} concentrations in body fluids	Stimulated by low blood Ca^{2+} levels; PTH effects enhanced by calcitriol and opposed by calcitonin

The Zona Glomerulosa

The **zona glomerulosa** (glō-mer-ū-LŌ-suh) is the outer region of the adrenal cortex. It produces **mineralocorticoids**, steroid hormones that affect the electrolyte composition of body fluids. **Aldosterone** is the main mineralocorticoid of the zona glomerulosa.

The zona glomerulosa accounts for about 15 percent of the volume of the adrenal cortex (**Figure 18–14c**). A *glomerulus* is a little ball, and as the term *zona glomerulosa* implies, the endocrine cells in this region form small, dense knots or clusters. This zone extends from the capsule to the radiating cords of the deeper zona fasciculata.

Aldosterone. Aldosterone stimulates the conservation of sodium ions and the elimination of potassium ions. This hormone targets cells that regulate the ionic composition of excreted fluids. It causes the retention of sodium ions by the kidneys, sweat glands, salivary glands, and pancreas, preventing Na^+ loss in urine, sweat, saliva, and digestive secretions. A loss of K^+ accompanies this retention of Na^+. As a secondary effect, the retention of Na^+ enhances the osmotic reabsorption of water by the kidneys, sweat glands, salivary glands, and pancreas. The effect at the kidneys is most dramatic when normal levels of ADH are present. In addition, aldosterone increases the sensitivity of salt receptors in the taste buds of the tongue. As a result, a person's interest in (and consumption of) salty food increases.

Aldosterone secretion occurs in response to a drop in blood Na^+ content, blood volume, or blood pressure, or to a rise in blood K^+ concentration. Changes in either Na^+ or K^+ concentration have a direct effect on the zona glomerulosa, but the secretory cells are most sensitive to changes in potassium levels. A rise in potassium levels is very effective in stimulating the release of aldosterone. Aldosterone release also occurs in response to *angiotensin II*. We discuss this hormone, part of the *renin-angiotensin-aldosterone system*, later in this chapter.

The Zona Fasciculata

The **zona fasciculata** (fa-sik-ū-LA-tuh; *fasciculus*, little bundle) produces steroid hormones collectively known as **glucocorticoids**, due to their effects on glucose metabolism. This zone begins at the inner border of the zona glomerulosa and extends toward the adrenal medulla (**Figure 18–14c**). It contributes about 78 percent of the cortical volume. The endocrine cells are larger and contain more lipids than those of the zona glomerulosa, and the lipid droplets give the cytoplasm a pale, foamy appearance. The cells of the zona fasciculata form individual cords composed of stacks of cells. Adjacent cords are separated by flattened blood vessels (sinusoids) with fenestrated walls.

The Glucocorticoids. When stimulated by ACTH from the anterior lobe of the pituitary gland, the zona fasciculata secretes primarily **cortisol** (KOR-ti-sol), also called *hydrocortisone*, along with smaller amounts of the related steroid **corticosterone** (kor-ti-KOS-te-rōn). The liver converts some of the circulating cortisol to **cortisone**, another active glucocorticoid. Negative feedback regulates glucocorticoid secretion: The glucocorticoids released have an inhibitory effect on the production of corticotropin-releasing hormone (CRH) in the hypothalamus, and on ACTH in the anterior pituitary lobe (**Figure 18–8a**).

Effects of Glucocorticoids. Glucocorticoids speed up the rates of glucose synthesis and glycogen formation, especially in the liver. Adipose tissue responds by releasing fatty acids into the blood, and other tissues begin to break down fatty acids and proteins instead of glucose. This process is another example of a glucose-sparing effect (p. 649).

Glucocorticoids also show **anti-inflammatory** effects. That is, they inhibit the activities of white blood cells and other components of the immune system. "Steroid creams" are commonly used to control irritating allergic rashes, such as poison ivy rash, and injections of glucocorticoids may be used

Figure 18–14 **The Adrenal Gland.** ATLAS: Plates 61a, b; 62b

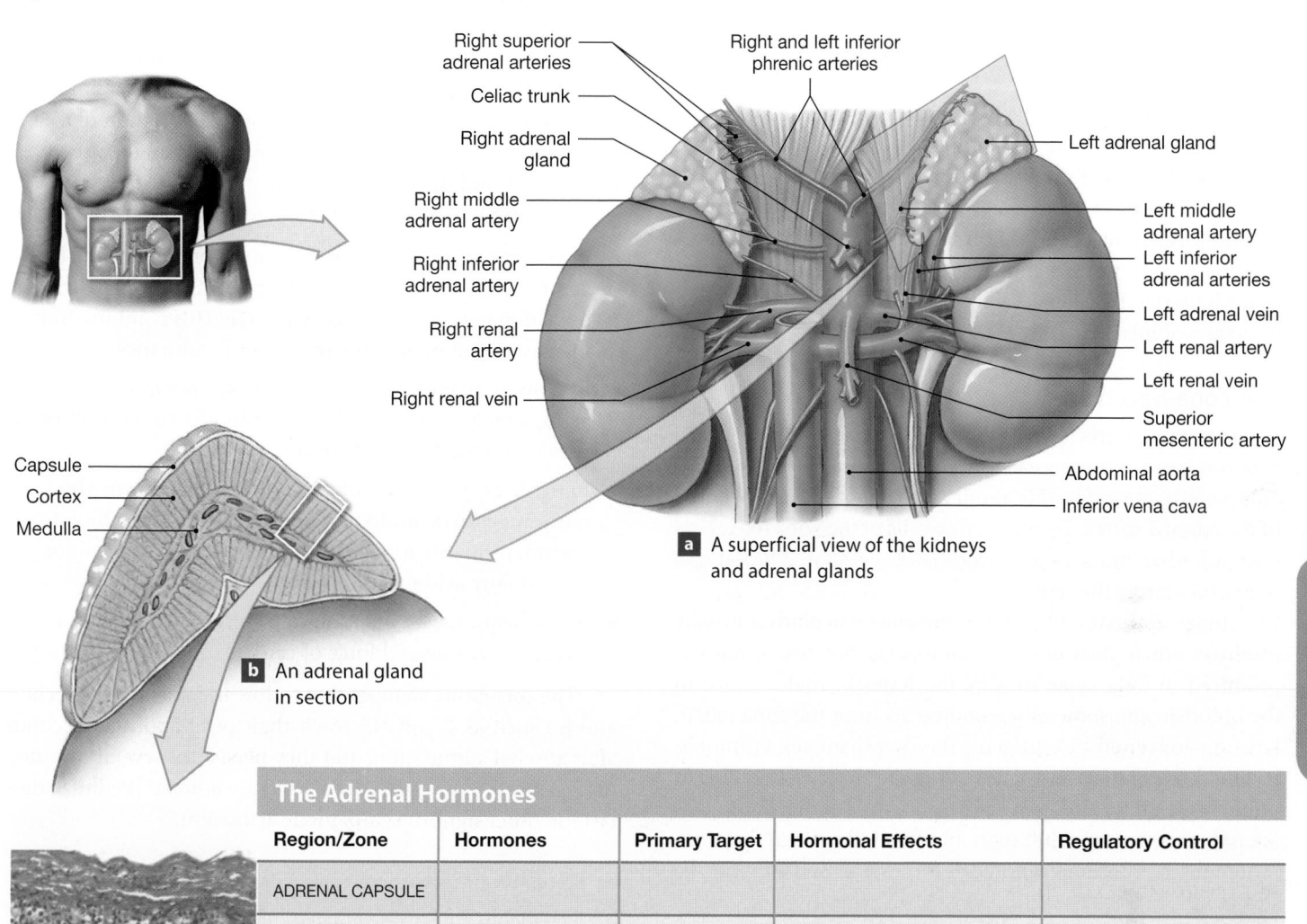

a A superficial view of the kidneys and adrenal glands

Labels (panel a):
Right superior adrenal arteries
Celiac trunk
Right adrenal gland
Right middle adrenal artery
Right inferior adrenal artery
Right renal artery
Right renal vein
Right and left inferior phrenic arteries
Left adrenal gland
Left middle adrenal artery
Left inferior adrenal arteries
Left adrenal vein
Left renal artery
Left renal vein
Superior mesenteric artery
Abdominal aorta
Inferior vena cava

Labels (panel b):
Capsule
Cortex
Medulla

b An adrenal gland in section

18

The Adrenal Hormones

Region/Zone	Hormones	Primary Target	Hormonal Effects	Regulatory Control
ADRENAL CAPSULE				
ADRENAL CORTEX **Zona glomerulosa**	Mineralocorticoids, primarily aldosterone	Kidneys	Increase renal reabsorption of Na⁺ and water (especially in the presence of ADH), and accelerate urinary loss of K⁺	Stimulated by angiotensin II, elevated blood K⁺ or fall in blood Na⁺; inhibited by ANP and BNP
Zona fasciculata	Glucocorticoids (cortisol [hydrocortisone], corticosterone)	Most cells	Increase rates of glucose and glycogen formation by the liver; release of amino acids from skeletal muscles, and lipids from adipose tissues; promote peripheral utilization of lipids; anti-inflammatory effects	Stimulated by ACTH from the anterior lobe of the pituitary gland
Zona reticularis	Androgens	Most cells	Adrenal androgens stimulate the development of pubic hair in boys and girls before puberty.	Androgen secretion is stimulated by ACTH.
ADRENAL MEDULLA	Epinephrine (E), norepinephrine (NE)	Most cells	Increases cardiac activity, blood pressure, glycogen breakdown, blood glucose levels; releases lipids by adipose tissue	Stimulated by sympathetic preganglionic fibers

Adrenal gland   LM × 140

c The major regions and zones of an adrenal gland and the hormones they produce to control more severe allergic reactions. How do these treatments work? Glucocorticoids slow the migration of phagocytic cells into an injury site and cause phagocytic cells already in the area to become less active. In addition, mast cells exposed to these steroids are less likely to release histamine and other chemicals that promote inflammation. ⤴ pp. 170–171 As a result, swelling and further irritation are dramatically reduced. On the negative side, the rate of wound healing decreases, and the weakening of the region's defenses makes it more susceptible to infectious organisms. For that reason, the topical steroids used to treat superficial rashes should never be applied to open wounds.

The Zona Reticularis

The **zona reticularis** (re-tik-ū-LAR-is; *reticulum*, network) forms a narrow band bordering each adrenal medulla (**Figure 18–14c**). This zone accounts for only about 7 percent of the total volume of the adrenal cortex. The endocrine cells of the zona reticularis form a folded, branching network, and fenestrated blood vessels wind among the cells.

Under stimulation by ACTH, the zona reticularis normally produces small quantities of androgens, the sex hormones produced in large quantities by the testes in males. Once in the bloodstream, some of the androgens from the zona reticularis are converted to estrogens, the dominant sex hormones in females. Adrenal androgens stimulate the development of pubic hair in boys and girls before puberty. Adrenal androgens are not important in adult men, but in adult women they promote muscle mass and blood cell formation, and support the sex drive.

The Adrenal Medulla

The boundary between the adrenal cortex and the adrenal medulla is irregular, and the supporting connective tissues and blood vessels are extensively interconnected. The adrenal medulla is pale gray or pink, due in part to the many blood vessels in the area. It contains large, rounded cells—similar to cells in sympathetic ganglia that are innervated by preganglionic sympathetic fibers. The sympathetic division of the autonomic nervous system controls secretory activities of the adrenal medullae. ⤴ p. 562

The adrenal medulla contains two populations of secretory cells: One produces epinephrine (adrenaline), the other norepinephrine (noradrenaline). Evidence suggests that the two types of cells are distributed in different areas and that their secretory activities can be independently controlled. The secretions are packaged in vesicles that form dense clusters just inside plasma membranes. The hormones in these vesicles are continuously released at low levels by exocytosis. Sympathetic stimulation dramatically accelerates the rate of exocytosis and hormone release.

Epinephrine and Norepinephrine

Epinephrine makes up 75–80 percent of the secretions from the adrenal medullae. The rest is norepinephrine. These hormones interact with alpha and beta receptors on plasma membranes, as we described in Chapter 16. ⤴ p. 567 Stimulation of α_1 and β_1 receptors, the most common types, speeds up the use of cellular energy and the mobilization of energy reserves.

Activation of the adrenal medullae has the following effects:

- In skeletal muscles, epinephrine and norepinephrine trigger mobilization of glycogen reserves and accelerate the breakdown of glucose to provide ATP. This combination increases both muscular strength and endurance.

- In adipose tissue, stored fats are broken down into fatty acids, which are released into the bloodstream for other tissues to use for ATP production.

- In the liver, glycogen molecules are broken down. The resulting glucose molecules are released into the bloodstream, primarily for use by neural tissue, which cannot shift to fatty acid metabolism.

- In the heart, the stimulation of β_1 receptors triggers an increase in the rate and force of cardiac muscle contraction.

The metabolic changes that follow the release of catecholamines such as E and NE reach their peak about 30 seconds after adrenal stimulation, and they persist for several minutes. As a result, the effects of stimulating the adrenal medullae outlast the other signs of sympathetic activation.

CHECKPOINT

17. Identify the two regions of the adrenal gland, and cite the hormones secreted by each.

18. List the three zones of the adrenal cortex.

19. What effect would elevated cortisol levels have on blood glucose levels?

See the blue Answers tab at the back of the book.

18-7 ┃ The pineal gland, attached to the roof of the third ventricle, secretes melatonin

Learning Outcome Describe the location of the pineal gland, and discuss the functions of the hormone it produces.

The **pineal gland**, part of the epithalamus, lies in the posterior portion of the roof of the third ventricle. The pineal gland contains neurons, neuroglia, and special secretory cells called **pinealocytes** (pin-Ē-al-ō-sīts). These cells synthesize the hormone **melatonin** from molecules of the neurotransmitter *serotonin* (**Figure 18–15**). Collaterals from the visual pathways enter the pineal gland and affect the rate of melatonin production. This rate is lowest during daylight hours and highest at night.

Figure 18–15 **The Pineal Gland.**

Pinealocytes

Pineal gland

LM × 400

■ *Influencing Circadian Rhythms.* Because its activity is cyclical, the pineal gland may also be involved in maintaining basic *circadian rhythms*—daily changes in physiological processes that follow a regular day–night pattern. ⤴ p. 506 Increased melatonin secretion in darkness has been suggested as a primary cause of *seasonal affective disorder (SAD)*. This condition can develop during the winter in people who live at high latitudes, where sunlight is scarce or lacking. It is characterized by changes in mood, eating habits, and sleeping patterns.

CHECKPOINT

20. Identify the hormone-secreting cells of the pineal gland.

21. Increased amounts of light would inhibit the production of which hormone?

22. List three possible functions of melatonin.

See the blue Answers tab at the back of the book.

Among the functions suggested for melatonin in humans are the following:

■ *Inhibiting Reproductive Functions.* In some mammals, melatonin slows the maturation of sperm, oocytes, and reproductive organs by reducing the rate of GnRH secretion. The significance of this effect in humans remains unclear. Circumstantial evidence suggests that melatonin may play a role in the timing of human sexual maturation. Melatonin levels in the blood decline at puberty, and pineal tumors that eliminate melatonin production cause premature puberty in young children.

■ *Protecting against Damage by Free Radicals.* Melatonin is a very effective *antioxidant*. It may protect CNS neurons from free radicals, such as nitric oxide (NO) or hydrogen peroxide (H_2O_2), that may form in active neural tissue.

18-8 ❚ The pancreas is both an exocrine organ and endocrine gland

Learning Outcome Describe the location, structure, hormones, and functions of the pancreas, and discuss the effects of abnormal pancreatic hormone production.

The **pancreas** is mostly retroperitoneal and lies in the loop between the inferior border of the stomach and the proximal portion of the small intestine (**Figure 18–1**). It is a slender, pale organ with a nodular (lumpy) texture (**Figure 18–16a**). The pancreas is 20–25 cm (8–10 in.) long and weighs about 80 g (2.8 oz) in adults. We will consider its anatomy further in Chapter 24, because it is primarily an exocrine organ that makes digestive enzymes.

18

Figure 18–16 **The Pancreas.** ATLAS: Plate 49e

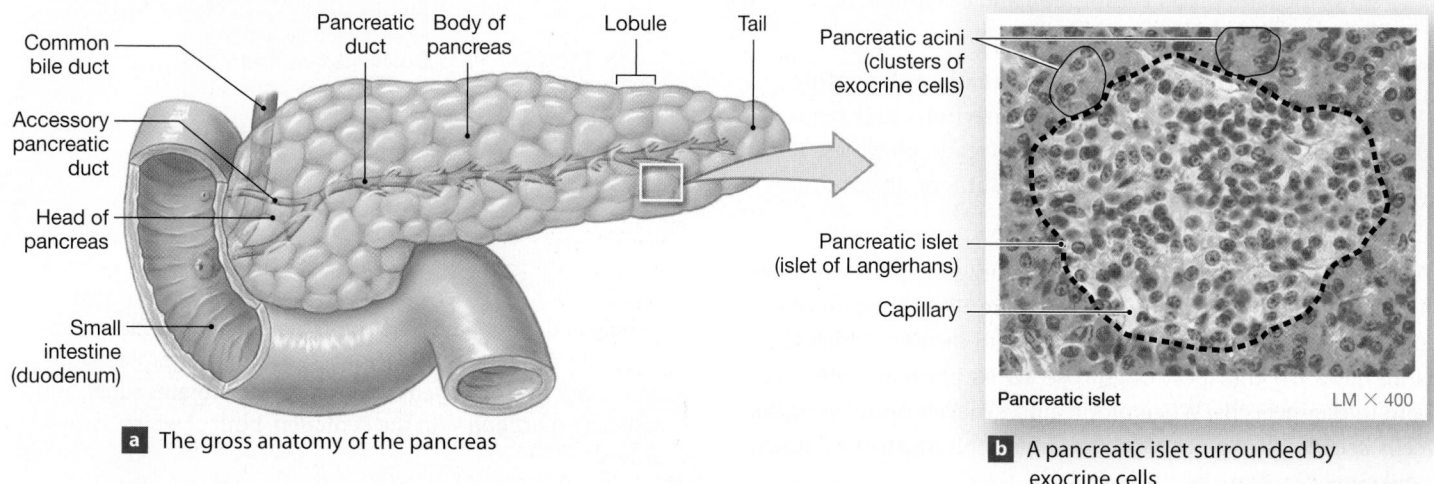

Common bile duct

Pancreatic duct

Body of pancreas

Lobule

Tail

Accessory pancreatic duct

Head of pancreas

Small intestine (duodenum)

Pancreatic acini (clusters of exocrine cells)

Pancreatic islet (islet of Langerhans)

Capillary

Pancreatic islet

LM × 400

a The gross anatomy of the pancreas

b A pancreatic islet surrounded by exocrine cells

The **exocrine pancreas** consists of clusters of gland cells, called *pancreatic acini*, and their attached ducts. The exocrine pancreas takes up roughly 99 percent of the pancreatic volume. Together the gland and duct cells secrete large quantities of an alkaline, enzyme-rich fluid that reaches the lumen of the digestive tract through a network of secretory ducts.

The **endocrine pancreas** consists of small groups of cells scattered among the exocrine cells. The endocrine clusters are known as **pancreatic islets**, or the *islets of Langerhans* (LAN-ger-hanz) (**Figure 18–16b**). Pancreatic islets account for only about 1 percent of all cells in the pancreas. Nevertheless, a typical pancreas contains approximately 2 million pancreatic islets and their secretions are vital to our survival.

The Pancreatic Islets

The pancreatic islets are surrounded by an extensive, fenestrated capillary network that carries pancreatic hormones into the bloodstream. Each islet contains four types of cells:

1. **Alpha cells** produce the hormone glucagon (GLŪ-ka-gon). Glucagon raises blood glucose levels by increasing the rates of glycogen breakdown and glucose release by the liver.

2. **Beta cells** produce the hormone insulin (IN-suh-lin). Insulin lowers blood glucose levels by increasing the rate of glucose uptake and utilization by most body cells, and by increasing glycogen synthesis in skeletal muscles and the liver. Beta cells also secrete *amylin*, a peptide hormone whose level rises after a meal.

3. **Delta cells** produce a peptide hormone identical to growth hormone–inhibiting hormone (GH–IH), a hypothalamic regulatory hormone. GH–IH suppresses the release of glucagon and insulin by other islet cells and slows the rates of food absorption and enzyme secretion along the digestive tract.

4. **F cells** produce the hormone **pancreatic polypeptide (PP)**. PP inhibits gallbladder contractions and regulates the production of some pancreatic enzymes. It may also help control the rate of nutrient absorption by the digestive tract.

We will focus on insulin and glucagon, the hormones that regulate blood glucose levels (**Figure 18–17**). When blood glucose levels increase, beta cells secrete insulin, which then stimulates the transport of glucose across plasma membranes and into target cells. When blood glucose levels decrease, alpha cells secrete glucagon, which stimulates glycogen breakdown and glucose release by the liver.

Figure 18–17   The Regulation of Blood Glucose Concentrations.

TIPS & TRICKS

To help in differentiating between the polysaccharide storage molecule "glycogen" and the pancreatic hormone "glucagon," remember that glyco**gen** literally means **gen**erates sugar; and associate gluca**gon** with the Penta**gon**, both of which "issue orders."

Insulin

Insulin is a peptide hormone released by beta cells when blood glucose concentrations exceed normal levels (70–110 mg/dL). Elevated levels of some amino acids, including arginine and leucine, also stimulate secretion of insulin. This hormone affects cellular metabolism in a series of steps that begins when insulin binds to receptor proteins on the plasma membrane of a target cell. Binding activates the receptor, which functions as a kinase, attaching phosphate groups to intracellular enzymes. These enzymes then produce primary and secondary effects in the cell. The biochemical details of these effects remain unresolved.

One of the most important effects is the enhancement of glucose absorption and utilization. Insulin receptors are present in most plasma membranes, and cells that have them are called *insulin dependent*. However, cells in the brain and kidneys, cells in the lining of the digestive tract, and red blood cells lack insulin receptors. These cells are called *insulin independent*, because they can absorb and utilize glucose without insulin stimulation.

The effects of insulin on its target cells include the following:

- *Accelerating Glucose Uptake (All Target Cells).* This effect results from an increase in the number of glucose transport proteins in the plasma membrane. These proteins move glucose into the cell by facilitated diffusion, which follows the concentration gradient for glucose and does not require ATP.

- *Accelerating Glucose Utilization (All Target Cells) and Enhanced ATP Production.* This effect occurs for two reasons: (1) The rate of glucose use is proportional to its availability, so when more glucose enters the cell, more is used. (2) Second messengers activate a key enzyme involved in the initial steps of glycolysis.

- *Stimulating Glycogen Formation (Skeletal Muscles and Liver Cells).* When excess glucose enters these cells, it is stored as glycogen.

- *Stimulating Amino Acid Absorption and Protein Synthesis.* This helps to maintain glucose levels by preventing the conversion of amino acids to glucose.

- *Stimulating Triglyceride Formation in Adipose Tissue.* Insulin stimulates the absorption of fatty acids and glycerol by adipocytes, which store these components as triglycerides. Adipocytes also increase their absorption of glucose, and excess glucose is used in the synthesis of additional triglycerides.

To summarize, the pancreas secretes insulin when glucose is abundant. The hormone stimulates glucose utilization to support growth and to build carbohydrate (glycogen) and lipid (triglyceride) reserves. The accelerated use of glucose soon brings circulating glucose levels within normal limits.

TIPS & TRICKS
The function of *in*sulin is to get glucose *in*to cells.

Glucagon

When glucose concentrations decrease below normal, alpha cells release glucagon to mobilize energy reserves. When glucagon binds to a receptor in the target cell's plasma membrane, the hormone activates adenylate cyclase. As we have seen, cAMP acts as a second messenger that activates cytoplasmic enzymes (p. 640). The primary effects of glucagon are as follows:

- *Stimulating the Breakdown of Glycogen in Skeletal Muscle and Liver Cells.* The glucose molecules released are either metabolized for energy (in skeletal muscle fibers) or released into the bloodstream (by liver cells).

- *Stimulating the Breakdown of Triglycerides in Adipose Tissue.* The adipocytes then release the fatty acids into the bloodstream for use by other tissues.

- *Stimulating the Production and Release of Glucose by the Liver.* Liver cells absorb amino acids from the bloodstream, convert them to glucose, and release the glucose into the circulation. This process of glucose synthesis in the liver is called *gluconeogenesis* (gloo-kō-nē-ō-JEN-e-sis).

The results are a reduction in glucose use and the release of more glucose into the bloodstream. Blood glucose concentrations soon increase toward normal levels.

Pancreatic alpha cells and beta cells monitor blood glucose concentrations, and they secrete glucagon and insulin without endocrine or nervous instructions. Yet because the alpha cells and beta cells are highly sensitive to changes in blood glucose levels, any hormone that affects blood glucose concentrations indirectly affects the production of both insulin and glucagon. Autonomic activity also influences insulin production: Parasympathetic stimulation enhances insulin release, and sympathetic stimulation inhibits it. Information about insulin, glucagon, and other pancreatic hormones is summarized in Table 18–5.

Diabetes Mellitus

Whether glucose is absorbed by the digestive tract or manufactured and released by the liver, very little glucose leaves the body once it has entered the bloodstream. The kidneys reabsorb virtually all glucose, so glucose does not appear in the urine. However, in diabetes mellitus, glucose accumulates in the blood and urine as a result of faulty glucose metabolism.

Diabetes mellitus can be caused by genetic abnormalities, and some of the genes responsible have been identified. Mutations that result in inadequate insulin production, the synthesis of abnormal insulin molecules, or the production of defective receptor proteins produce comparable symptoms. Under these

18

Untreated diabetes mellitus disrupts metabolic activities throughout the body. Clinical problems arise because the tissues involved are experiencing an energy crisis—in essence, most of the tissues are responding as they would during chronic starvation, breaking down lipids and even proteins because they are unable to absorb glucose from their surroundings. Problems involving abnormal changes in blood vessel structure are particularly dangerous. An estimated 25.8 million people in the United States have some form of diabetes.

Retinal Damage

The proliferation of capillaries and hemorrhaging at the retina may cause partial or complete blindness. This condition is called **diabetic retinopathy**.

Early Heart Attacks

Degenerative blockages in cardiac circulation can lead to early heart attacks. For a given age group, heart attacks are three to five times more likely in people with diabetes than in nondiabetic people.

Kidney Degeneration

Degenerative changes in the kidneys, a condition called **diabetic nephropathy**, can lead to kidney failure.

Peripheral Nerve Problems

Abnormal blood flow to neural tissues is probably responsible for a variety of neural problems with peripheral nerves, including abnormal autonomic function. These disorders are collectively termed **diabetic neuropathies**.

Diabetes Mellitus

Diabetes mellitus (mel-Ī-tus; *mellitum*, honey), is characterized by glucose concentrations that are high enough to overwhelm the reabsorption capabilities of the kidneys. (The presence of abnormally high glucose levels in the blood in general is called **hyperglycemia** [hī-per-glī-SĒ-mē-ah].) Glucose appears in the urine (**glycosuria**; glī-kō-SYŪ-rē-a), and urine volume generally becomes excessive (**polyuria**).

subdivided into

Type 1 Diabetes

Type 1 is characterized by inadequate insulin production by the pancreatic beta cells. Persons with Type 1 diabetes require insulin to live and usually require multiple injections daily, or continuous infusion through an insulin pump or other device. This form of diabetes accounts for approximately 5% of cases. It usually develops in children and young adults.

Type 2 Diabetes

Type 2 is the most common form of diabetes mellitus. Most people with this form of diabetes produce normal amounts of insulin, at least initially, but their tissues do not respond properly, a condition known as insulin resistance. Type 2 diabetes is associated with obesity, and weight loss through diet and exercise can be an effective treatment, especially when coupled with oral medicines.

Peripheral Tissue Damage

Blood flow to the distal portions of the limbs is reduced, and peripheral tissues may suffer as a result. For example, a reduction in blood flow to the feet can lead to tissue death, ulceration, infection, and loss of toes or a major portion of one or both feet.

Table 18–5	Hormones Produced by the Pancreatic Islets			
Structure/Cells	**Hormone**	**Primary Targets**	**Hormonal Effect**	**Regulatory Control**
PANCREATIC ISLETS				
Alpha cells	Glucagon	Liver, adipose tissue	Mobilizes lipid reserves; promotes glucose synthesis and glycogen breakdown in liver; elevates blood glucose concentrations	Stimulated by low blood glucose concentrations; inhibited by GH–IH from delta cells
Beta cells	Insulin	Most cells	Facilitates uptake of glucose by target cells; stimulates formation and storage of lipids and glycogen	Stimulated by high blood glucose concentrations, parasympathetic stimulation, and high levels of some amino acids; inhibited by GH–IH from delta cells and by sympathetic activation
Delta cells	GH–IH (somatostatin)	Other islet cells, digestive epithelium	Inhibits insulin and glucagon secretion; slows rates of nutrient absorption and enzyme secretion along digestive tract	Stimulated by protein-rich meal; mechanism unclear
F cells	Pancreatic polypeptide (PP)	Digestive organs	Inhibits gallbladder contraction; regulates production of pancreatic enzymes; influences rate of nutrient absorption by digestive tract	Stimulated by protein-rich meal and by parasympathetic stimulation

conditions, obesity accelerates the onset and severity of the disease. Diabetes mellitus can also result from other pathological conditions, injuries, immune disorders, or hormonal imbalances.

The two major types of diabetes mellitus are Type 1 diabetes (previously known as juvenile diabetes) and Type 2 diabetes. This disorder is described in Spotlight Figure 18–18.

CHECKPOINT

23. Identify the types of cells in the pancreatic islets and the hormones produced by each.

24. Why does a person with Type 1 or Type 2 diabetes urinate frequently and have increased thirst?

25. What effect would increased levels of glucagon have on the amount of glycogen stored in the liver?

See the blue Answers tab at the back of the book.

18-9 | Many organs have secondary endocrine functions

Learning Outcome Describe the functions of the hormones produced by the kidneys, heart, thymus, testes, ovaries, and adipose tissue.

As we noted earlier, many organs of other body systems have secondary endocrine functions. Examples are the intestines (digestive system), the kidneys (urinary system), the heart (cardiovascular system), the thymus (lymphatic system), and the *gonads*—the testes in males and the ovaries in females (reproductive system).

Several new hormones from these endocrine tissues have been identified. In many cases, their structures and modes of action remain uncertain, and we have not described them in this chapter. However, in one instance, researchers traced a significant new hormone to an unexpected site of origin, leading

to the realization that the body's adipose tissue has important endocrine functions. We include the endocrine functions of adipose tissue in this section, although all the details have yet to be worked out. Table 18–6 provides an overview of some of the hormones those organs of other systems produce.

The Intestines

The intestines process and absorb nutrients. They release a variety of hormones that coordinate the activities of the digestive system. Most digestive processes are hormonally controlled locally, although the autonomic nervous system can affect the pace of digestive activities. We describe these hormones in Chapter 24.

The Kidneys

The kidneys release the steroid hormone *calcitriol*, the peptide hormone *erythropoietin*, and the enzyme *renin*. Calcitriol is important for calcium ion homeostasis. Erythropoietin and renin are involved in the regulation of blood volume and blood pressure.

Calcitriol

Calcitriol is a steroid hormone secreted by the kidneys in response to parathyroid hormone (PTH) (Figure 18–19a). *Cholecalciferol* (vitamin D_3) is a related steroid that is synthesized in the skin or absorbed from the diet. Cholecalciferol is converted to calcitriol, although not directly. The term *vitamin D* applies to the entire group of related steroids, including calcitriol, cholecalciferol, and various intermediate products.

The best-known function of calcitriol is to stimulate calcium and phosphate ion absorption along the digestive tract. The effects of PTH on Ca^{2+} absorption result primarily from stimulation of calcitriol release. Calcitriol's other effects on calcium metabolism include (1) stimulating the formation and

Table 18–6	Representative Hormones Produced by Organs of Other Systems		
Organ	**Hormone**	**Primary Target**	**Hormonal Effect**
Intestines	Many (secretin, gastrin, cholecystokinin, etc.)	Other regions and organs of the digestive system	Coordinate digestive activities
Kidneys	Erythropoietin (EPO) Calcitriol	Red bone marrow Intestinal lining, bone, kidneys	Stimulates red blood cell production Stimulates calcium and phosphate absorption; stimulates Ca^{2+} release from bone; inhibits PTH secretion
Heart	Natriuretic peptides (ANP and BNP)	Kidneys, hypothalamus, adrenal gland	Increase water and salt loss at kidneys; decrease thirst; suppress secretion of ADH and aldosterone
Thymus	Thymosins (many)	Lymphocytes and other cells of the immune response	Coordinate and regulate immune response
Gonads	*See Table 18–7*		
Adipose tissues	Leptin	Hypothalamus	Suppression of appetite; permissive effects on GnRH and gonadotropin synthesis

differentiation of osteogenic cells and osteoclasts, (2) stimulating bone resorption by osteoclasts, (3) stimulating Ca^{2+} reabsorption by the kidneys, and (4) suppressing PTH production. Evidence indicates that calcitriol also affects lymphocytes and keratinocytes in the skin, but these effects have nothing to do with regulating calcium levels.

Erythropoietin

Erythropoietin (e-rith-rō-POY-e-tin; *erythros*, red + *poiesis*, making), or **EPO**, is a peptide hormone released by the kidneys in response to low oxygen levels in kidney tissues. EPO stimulates the red bone marrow to produce red blood cells. The increase in the number of red blood cells elevates blood volume. Because these cells transport oxygen, this increase also improves oxygen delivery to peripheral tissues. We will consider EPO again in Chapter 19.

Renin

Specialized kidney cells called *juxtaglomerular cells* release **renin** (RĒ-nin) in response to (1) sympathetic stimulation or (2) a decline in renal blood flow. Once in the bloodstream, renin functions as an enzyme that starts an enzymatic cascade known as the *renin-angiotensin-aldosterone system (RAAS)* (Figure 18–19b). RAAS is involved with blood pressure regulation and electrolyte metabolism. First, renin converts **angiotensinogen**, a plasma protein produced by the liver, to angiotensin I. In the capillaries of the lungs, angiotensin-converting enzyme (ACE) modifies **angiotensin I** to the hormone **angiotensin II**. This hormone, in turn, stimulates the secretion of aldosterone by the adrenal cortex and ADH by the posterior lobe of the pituitary gland. The combination of aldosterone and ADH retains salt and water by restricting their loss by the kidneys. Angiotensin II also stimulates thirst and elevates blood pressure. For these reasons, drugs known as *ACE inhibitors* are used to treat hypertension.

Because renin plays such a key role in the renin-angiotensin-aldosterone system, many physiological and endocrinological

references consider renin to be a hormone. We will take a closer look at the renin-angiotensin-aldosterone system when we examine the control of blood pressure and blood volume in Chapter 21.

The Heart

The endocrine cells in the heart are cardiac muscle cells in the walls of the *atria* (chambers that receive blood from the veins) and the *ventricles* (chambers that pump blood to the rest of the body). If blood volume becomes too great, these cells are stretched excessively, to the point at which they begin to secrete **natriuretic peptides** (nā-trē-ū-RET-ik; *natrium*, sodium + *ouresis*, urination). In general, the effects of natriuretic peptides oppose those of angiotensin II: Natriuretic peptides promote the loss of Na^+ and water by the kidneys, and inhibit renin release and the secretion of ADH and aldosterone. They also suppress thirst and prevent angiotensin II and norepinephrine from elevating blood pressure. The net result is a reduction in both blood volume and blood pressure, thereby reducing the stretching of the cardiac muscle cells in the heart walls. We discuss two natriuretic peptides—*ANP* (atrial natriuretic peptide) and *BNP* (brain natriuretic peptide)—when we consider the control of blood pressure and volume in Chapters 21 and 26.

The Thymus

The **thymus** is located in the mediastinum, generally just deep to the sternum. This gland produces several hormones that are important in developing and maintaining immune defenses. **Thymosin** (THĪ-mō-sin) is the name originally given to an extract from the thymus that promotes the development and maturation of *lymphocytes*, the white blood cells responsible for immunity. The extract actually contains a blend of several complementary hormones. The term *thymosins* is sometimes used to refer to all thymic hormones. We consider the histological organization of the thymus and the functions of the thymosins in Chapter 22.

Figure 18–19 **Endocrine Functions of the Kidneys.**

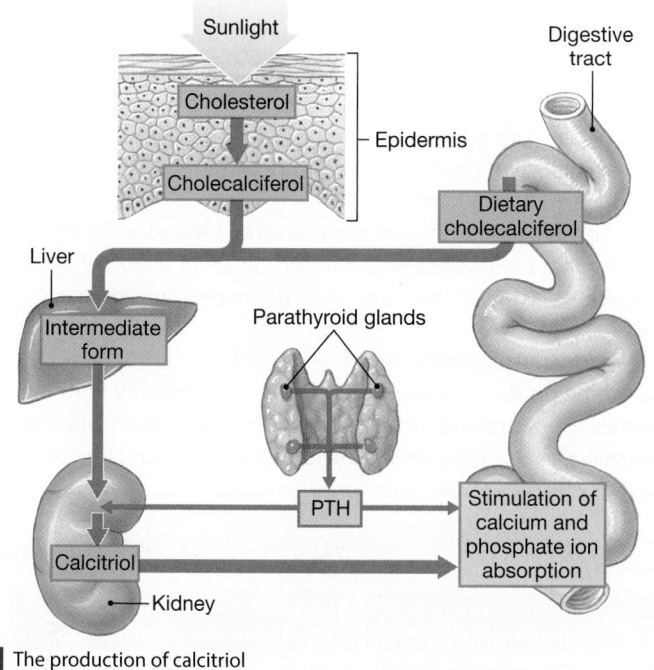

a The production of calcitriol

b The release of renin and erythropoietin, and an overview of the renin-angiotensin-aldosterone system beginning with the activation of angiotensinogen by renin

18

The Gonads

Information about the reproductive hormones of the testes and ovaries is presented in Table 18–7. In males, the **interstitial cells** of the testes produce the male hormones known as androgens. **Testosterone** (tes-TOS-ter-ōn) is the most important androgen. During embryonic development, the production of testosterone affects the development of CNS structures, including hypothalamic nuclei, which will later

Table 18–7	Hormones of the Reproductive System			
Structure/Cells	**Hormone**	**Primary Target**	**Hormonal Effect**	**Regulatory Control**
TESTES				
Interstitial cells	Androgens	Most cells	Support functional maturation of sperm, protein synthesis in skeletal muscles, male secondary sex characteristics, and associated behaviors	Stimulated by LH from the anterior lobe of the pituitary gland
Nurse cells	Inhibin	Pituitary gland	Inhibits secretion of FSH	Stimulated by FSH from the anterior lobe
OVARIES				
Follicular cells	Estrogens	Most cells	Support follicle maturation, female secondary sex characteristics, and associated behaviors	Stimulated by FSH and LH from the anterior lobe of the pituitary gland
	Inhibin	Pituitary gland	Inhibits secretion of FSH	Stimulated by FSH from anterior lobe
Corpus luteum	Progesterone	Uterus, mammary glands	Prepares uterus for implantation; prepares mammary glands for secretory activity	Stimulated by LH from the anterior lobe of the pituitary gland

Clinical Note

Endocrine Disorders Regulation of hormone levels often involves negative feedback control mechanisms. These mechanisms involve the endocrine organ, neural regulatory factors, and the target tissues. Hormone overproduction (hypersecretion) or underproduction (hyposecretion) may cause problems. Abnormal cellular sensitivity to the hormone can also lead to problems.

Primary disorders arise from problems within the endocrine organ. The underlying cause may be a metabolic factor. Hypothyroidism due to a lack of dietary iodine is an example. An endocrine organ may also malfunction due to physical damage that destroys cells or disrupts the normal blood supply. Congenital problems may also affect the regulation, production, or release of hormones by endocrine cells.

Secondary disorders are due to problems in other organs or target tissues. Such disorders often involve the hypothalamus or pituitary gland. For example, secondary hypothyroidism occurs when the hypothalamus or pituitary gland doesn't produce enough TRH and/or TSH.

Abnormalities in target cells can affect their sensitivity or responsiveness to a particular hormone. For example, Type 2 diabetes is due to the target cells' decreased sensitivity to insulin.

Endocrine disorders often reflect either abnormal hormone production or abnormal cellular sensitivity to hormones. The signs and symptoms highlight the significance of normally "silent" hormonal contributions. The characteristics of these disorders are summarized in Table 18–8.

Table 18–8	Clinical Implications of Endocrine Malfunctions			
Hormone	**Underproduction or Tissue Insensitivity**	**Principal Signs and Symptoms**	**Overproduction or Tissue Hypersensitivity**	**Principal Signs and Symptoms**
Growth hormone (GH)	Pituitary growth failure	Retarded growth, abnormal fat distribution, low blood glucose hours after a meal	Gigantism, acromegaly	Excessive growth
Antidiuretic hormone (ADH) or vasopressin (VP)	Diabetes insipidus	Polyuria, dehydration, thirst	SIADH (syndrome of inappropriate ADH secretion)	Increased body weight and water content
Thyroxine (T_4), triiodothyronine (T_3)	Hypothyroidism, infantile hypothyroidism, myxedema	Low metabolic rate; low body temperature; impaired physical and mental development	Hyperthyroidism, Graves disease	High metabolic rate and body temperature
Parathyroid hormone (PTH)	Hypoparathyroidism	Muscular weakness, neurological problems, formation of dense bones, tetany due to low blood $Ca2^+$ concentrations	Hyperparathyroidism	Neurological, mental, muscular problems due to high blood Ca^{2+} concentrations; weak and brittle bones
Insulin	Diabetes mellitus (Type 1)	High blood glucose, impaired glucose utilization, dependence on lipids for energy; glycosuria	Excess insulin production or administration	Low blood glucose levels, possibly causing coma
Mineralocorticoids (MCs)	Hypoaldosteronism	Polyuria, low blood volume, high blood K^+, low blood Na^+ concentrations	Aldosteronism	Increased body weight due to Na^+ and water retention; low blood K^+ concentration
Glucocorticoids (GCs)	Addison's disease	Inability to tolerate stress, mobilize energy reserves, or maintain normal blood glucose concentrations	Cushing's disease	Excessive breakdown of tissue proteins and lipid reserves; impaired glucose metabolism
Epinephrine (E), norepinephrine (NE)	None identified		Pheochromocytoma	High metabolic rate, body temperature, and heart rate; elevated blood glucose levels
Estrogens (females)	Hypogonadism	Sterility, lack of secondary sex characteristics	Adrenogenital syndrome	Overproduction of androgens by zona reticularis of adrenal cortex leads to masculinization
			Precocious puberty	Premature sexual maturation and related behavioral changes
Androgens (males)	Hypogonadism	Sterility, lack of secondary sex characteristics	Adrenogenital syndrome (gynecomastia)	Abnormal production of estrogen, sometimes due to adrenal or interstitial cell tumors; leads to breast enlargement
			Precocious puberty	Premature sexual maturation and related behavioral changes

influence sexual behaviors. **Nurse cells** in the testes support the differentiation and physical maturation of sperm. Under FSH stimulation, these cells secrete the hormone **inhibin**. It inhibits the secretion of FSH at the anterior lobe of the pituitary gland and perhaps suppresses GnRH release at the hypothalamus.

In females, steroid hormones called **estrogens** are produced in the ovaries under FSH and LH stimulation. **Estradiol** is the main estrogen. Circulating FSH stimulates the secretion of inhibin by ovarian cells, and inhibin suppresses FSH release through a feedback mechanism comparable to that in males.

At ovulation, follicles in the ovary release an immature gamete, or oocyte. The remaining follicle cells then reorganize into a *corpus luteum* (LOO-tē-um; "yellow body") that releases a mixture of estrogens and **progesterone** (prō-JES-ter-ōn). During pregnancy, the placenta and uterus produce additional hormones that interact with those produced by the ovaries and the pituitary gland to promote normal fetal development and delivery. We consider the hormonal aspects of pregnancy in Chapter 29.

Adipose Tissue

Recall from Chapter 4 that adipose tissue is a type of loose connective tissue. ⟳ p. 155 Adipose tissue produces a peptide hormone called **leptin**, which has several functions. Its best known function is feedback control of appetite. When we eat, adipose tissue absorbs glucose and lipids and synthesizes triglycerides for storage. At the same time, it releases leptin into the bloodstream. Leptin binds to hypothalamic neurons involved with emotion and appetite control. The result is a sense of fullness (satiation) and the suppression of appetite.

Leptin was first discovered in a strain of obese mice that had a defective leptin gene. When treated with leptin, these overweight mice quickly turned into slim, athletic animals. The initial hope that leptin could be used to treat human obesity was soon dashed, however. Most obese people appear to have defective leptin receptors (or leptin pathways) in the appetite centers of the CNS. Their circulating leptin levels are already several times higher than those in individuals of normal body weight. Additional leptin would have no effect. Researchers are now investigating the structure of the receptor protein and the biochemistry of the pathway triggered by leptin binding.

Leptin must be present for normal levels of GnRH and gonadotropin synthesis to take place. This explains why (1) thin girls commonly enter puberty relatively late, (2) an increase in body fat can improve fertility, and (3) women stop menstruating when their body fat content becomes very low.

CHECKPOINT

26. Identify two hormones secreted by the kidneys.
27. Identify a hormone released by adipose tissue.
28. Describe the action of renin in the bloodstream.

See the blue Answers tab at the back of the book.

18-10 ┃ Hormones interact to produce coordinated physiological responses

Learning Outcome Explain how hormones interact to produce coordinated physiological responses and influence behavior, describe the role of hormones in the general adaptation syndrome, and discuss how aging affects hormone production and give examples of interactions between the endocrine system and other organ systems.

We usually study hormones individually, but the extracellular fluids contain a mixture of hormones whose concentrations change daily or even hourly. As a result, cells never respond to only one hormone. Instead, they respond to multiple hormones. When a cell receives instructions from two hormones at the same time, four outcomes are possible:

1. The two hormones may have opposing or **antagonistic effects**, as in the case of PTH and calcitonin, or insulin and glucagon. The net result depends on the balance between the two hormones. In general, when two antagonistic hormones are present, the observed effects are weaker than those produced by either hormone acting unopposed.

2. The two hormones may have additive effects, so that the net result is greater than the effect that each would produce acting alone. In some cases, the net result is greater than the *sum* of the hormones' individual effects. This interaction is a **synergistic effect** (sin-er-JIS-tik; *synairesis*, a drawing together). An example is the glucose-sparing action of GH and glucocorticoids.

3. One hormone can have a **permissive effect** on another. In such cases, the first hormone is needed for the second to produce its effect. For example, epinephrine does not change energy consumption unless thyroid hormones are also present in normal concentrations.

4. Finally, hormones may produce different, but complementary, results in specific tissues and organs. These **integrative effects** are important in coordinating the activities of diverse physiological systems. The differing effects of calcitriol and parathyroid hormone on tissues involved in calcium metabolism are an example.

When multiple hormones regulate a complex process, it is very difficult to determine whether a hormone has synergistic, permissive, or integrative effects. Next we consider three examples of processes regulated by complex hormonal interactions: growth, the response to stress, and behavior.

Role of Hormones in Growth

Several endocrine organs work together to bring about normal growth. Several hormones—GH, thyroid hormones, insulin, PTH, calcitriol, and reproductive hormones—are especially important. Many others have secondary effects on growth. The circulating concentrations of these hormones are regulated independently. Every time the hormonal mixture changes, metabolic operations are modified to some degree. The modifications vary in duration and intensity, producing unique individual growth patterns.

- *Growth Hormone (GH).* The effects of GH on protein synthesis and cellular growth are most apparent in children. GH supports their muscular and skeletal development. In adults, growth hormone helps to maintain normal blood glucose concentrations and to mobilize lipid reserves in adipose tissue. GH is not the primary hormone involved, however. An adult with a GH deficiency but normal levels of thyroxine (T_4), insulin, and glucocorticoids will have no physiological problems.

- *Thyroid Hormones.* Normal growth also requires appropriate levels of thyroid hormones. If these hormones are absent during fetal development or for the first year after birth, the nervous system fails to develop normally, and developmental delay results. If T_4 concentrations decline later in life but before puberty, normal skeletal development does not continue.

- *Insulin.* Growing cells need adequate supplies of energy and nutrients. Without insulin, the passage of glucose and amino acids across plasma membranes stops or is drastically reduced.

- *Parathyroid Hormone (PTH) and Calcitriol.* Parathyroid hormone and calcitriol promote the absorption of calcium salts from the bloodstream for deposition in bone. Without adequate levels of both hormones, bones can still enlarge, but are poorly mineralized, weak, and flexible. For example, *rickets* is a condition typically caused by inadequate calcitriol production due to vitamin D deficiency in growing children. As a result, the lower limb bones are so weak that they bend under the body's weight. ⊃ p. 185

- *Reproductive Hormones.* The presence or absence of reproductive hormones (androgens in males, estrogens in females) affects the activity of osteoblasts in key locations and the growth of specific cell populations. Androgens and estrogens stimulate cell growth and differentiation in their target tissues, but their targets differ. The differential growth induced by each accounts for gender-related differences in skeletal proportions and secondary sex characteristics.

The Hormonal Responses to Stress

Any condition—physical or emotional—that threatens homeostasis is a form of **stress**. Specific homeostatic adjustments oppose many stresses. For example, a decrease in body temperature leads to shivering or changes in the pattern of blood flow, which can restore normal body temperature.

In addition, the body has a *general* response to stress that can occur while other, more specific responses are under way. A wide variety of stress-causing factors produce the same general pattern of hormonal and physiological adjustments. These responses are part of the **general adaptation syndrome (GAS)**, also known as the **stress response**. Hans Selye first described GAS in 1936. It has three phases: the *alarm phase*, the *resistance phase*, and the *exhaustion phase* (**Spotlight Figure 18–20**).

The Effects of Hormones on Behavior

As we have seen, the hypothalamus regulates many endocrine functions, and hypothalamic neurons monitor the levels of many circulating hormones. Other portions of the CNS are also quite sensitive to hormonal stimulation.

We can see the behavioral effects of specific hormones most clearly in individuals whose endocrine glands are oversecreting or undersecreting. But even normal changes in circulating hormone levels can cause behavioral changes. For example, in *precocious* (premature) *puberty*, sex hormones are produced at an inappropriate time, perhaps as early as age 5 or 6. Not only does an affected child begin to develop adult secondary sex characteristics, but the child's behavior also changes. The "nice little kid" disappears, and the child becomes aggressive and assertive due to the effects of sex hormones on CNS function. In normal teenagers, these behaviors are usually attributed to environmental stimuli, such as peer pressure, but here we can see that they have a physiological basis as well. In adults, changes in the mixture of hormones reaching the CNS can affect intellectual capabilities, memory, learning, and emotional states.

We now briefly turn to the effects of aging on hormone production.

Aging and Hormone Production

The endocrine system undergoes relatively few functional changes with age. The most dramatic exception is the decline in the concentrations of reproductive hormones. We noted the effects of these hormonal changes on the skeletal system in Chapter 6 ⊃ p. 228, and we will continue the discussion in Chapter 29.

Blood and tissue concentrations of many other hormones, including TSH, thyroid hormones, ADH, PTH, prolactin, and glucocorticoids, do not change with advancing age. Circulating hormone levels may remain within normal limits, but some endocrine tissues become less responsive to stimulation. For example, in elderly individuals, smaller amounts of GH and insulin are secreted after a carbohydrate-rich meal. The

ALARM

Alarm Phase ("Fight or Flight")

During the **alarm phase**, an immediate response to the stress occurs. The sympathetic division of the autonomic nervous system directs this response. In the alarm phase, (1) energy reserves are mobilized, mainly in the form of glucose, and (2) the body prepares to deal with the stress-causing factor by "fight or flight" responses. Epinephrine is the dominant hormone of the alarm phase. Its secretion is part of a generalized sympathetic activation.

Brain

General sympathetic activation

Sympathetic stimulation

Adrenal medulla

Epinephrine, norepinephrine

Immediate Short-Term Responses to Crises

- Increased mental alertness
- Increased energy use by all cells
- Mobilization of glycogen and lipid reserves
- Changes in circulation
- Decreased digestive activity and urine production
- Increased sweat gland secretion
- Increased heart rate and respiratory rate

RESISTANCE

Resistance Phase

If a stress lasts longer than a few hours, the person enters the **resistance phase** of GAS. Glucocorticoids are the dominant hormones of the resistance phase. Epinephrine, GH, and thyroid hormones are also involved. Energy demands in the resistance phase remain higher than normal, due to the combined effects of these hormones. Neural tissue has a high demand for energy, and requires a reliable supply of glucose. If blood glucose levels fall too low, neural function deteriorates. Glycogen reserves can meet neural demand during the alarm phase, but become depleted after several hours. Hormones of the resistance phase mobilize lipids and amino acids as energy sources to conserve glucose for use by neural tissue.

Sympathetic stimulation

Pancreas

Growth hormone

Glucagon

ACTH   Adrenal cortex

Glucocorticoids

Kidney

Mineralocorticoids (with ADH)

Renin-angiotensin-aldosterone system

Long-Term Metabolic Adjustments

- Mobilization of remaining energy reserves: Lipids are released by adipose tissue; amino acids are released by skeletal muscle
- Conservation of glucose: Peripheral tissues (except neural) break down lipids to obtain energy
- Elevation of blood glucose concentrations: Liver synthesizes glucose from other carbohydrates, amino acids, and lipids
- Conservation of salts and water, loss of K^+ and H^+

EXHAUSTION

Exhaustion Phase

The body's lipid reserves are sufficient to maintain the resistance phase for weeks or even months. But when the resistance phase ends, homeostatic regulation breaks down and the **exhaustion phase** begins. Unless corrective actions are taken almost immediately, the failure of one or more organ systems will prove fatal. The production of aldosterone throughout the resistance phase results in a conservation of Na^+ at the expense of K^+. As the body's K^+ content decreases, a variety of cells begin to malfunction. The underlying problem of the exhaustion phase is the body's inability to sustain the endocrine and metabolic adjustments of the resistance phase.

Collapse of Vital Systems

- Exhaustion of lipid reserves
- Cumulative structural or functional damage to vital organs
- Inability to produce glucocorticoids
- Failure of electrolyte balance

Clinical Note

Hormones and Athletic Performance The use of hormones to improve athletic performance is widely banned. The International Olympic Committee, the U.S. Olympic Committee, the National Collegiate Athletic Association, Major League Baseball, and the National Football League all ban it. The American Medical Association and the American College of Sports Medicine condemn the practice. Yet a significant number of amateur and professional athletes persist in this dangerous practice. Athletes most often use synthetic forms of testosterone. They might also use any combination of testosterone, GH, EPO, and a variety of synthetic hormones.

Androgen Abuse
The use of *anabolic steroids*, or androgens, has become popular with many amateur and professional athletes. The goal of steroid use is to increase muscle mass, endurance, and "competitive spirit." A steroid, such as *androstenedione*, is converted to testosterone in the body. One supposed justification for this steroid use is the unfounded opinion that compounds made in the body are not only safe, but good for you. In reality, the administration of natural or synthetic androgens in abnormal amounts carries unacceptable health risks. Androgens are known to produce several complications. These problems include (1) premature closure of epiphyseal cartilages, (2) various liver dysfunctions (such as jaundice and liver tumors), (3) prostate gland enlargement and urinary tract obstruction, and (4) testicular atrophy and infertility. Links to heart attacks, impaired cardiac function, and strokes have also been suggested.

The normal regulation of androgen production involves a feedback mechanism comparable to that described for adrenal steroids earlier in this chapter. GnRH stimulates the production of LH, and LH stimulates the secretion of testosterone and other androgens by the interstitial cells of the testes. Circulating androgens, in turn, inhibit the production of both GnRH and LH. For this reason, high doses of synthetic androgens can suppress the normal production of testosterone and depress the manufacture of GnRH by the hypothalamus. *This suppression of GnRH release can be permanent.*

The use of androgens as "bulking agents" by female bodybuilders may add muscle mass, but it can also alter muscular proportions and secondary sex characteristics. For example, women taking steroids can develop irregular menstrual periods and changes in body hair distribution (such as baldness). Finally, androgen abuse can depress the immune system.

Erythropoietin Abuse
Erythropoietin (EPO) is readily available because it is now synthesized by recombinant DNA techniques. Endurance athletes, such as cyclists and marathon runners, sometimes use it to boost the number of oxygen-carrying red blood cells in the bloodstream. This effect increases the oxygen content of blood, but it also makes blood more viscous. For this reason, the heart must work harder to push the "thick" blood through the blood vessels. This effort can result in death due to heart failure or stroke in young and otherwise healthy individuals. The Tour de France bicycle races have been tainted by competitors testing positive for banned substances. These include testosterone, erythropoietin, and an illegal blood transfusion.

Androgens and EPO are hormones with reasonably well-understood effects. Drug testing is now widespread in amateur and professional sports, but athletes interested in "getting an edge" are experimenting with drugs not easily detected by standard tests. The long-term and short-term effects of these drugs are difficult to predict.

Gamma-hydroxybutyrate Use
Another drug used by amateur athletes is **gamma-hydroxybutyrate (GHB)**. It was tested for use as an anesthetic in the 1960s but rejected, in part because it was linked to seizures. In 1990, the drug appeared in health-food stores. It was sold as an anabolic agent and diet aid. It has also been used as a "date rape" drug. According to the FDA, GHB and related compounds—sold or distributed under the names Renewtrient, Revivarant, Blue Nitro, Firewater, and Serenity—have recently been responsible for serious illnesses and even deaths. Signs and symptoms include reduced heart rate, lowered body temperature, confusion, hallucinations, seizures, and coma at doses from 0.25 teaspoon to 4 tablespoons.

---

reduced levels of GH and other tropic hormones affect tissues throughout the body. These hormonal effects involve the reductions in bone density and muscle mass noted in earlier chapters.

Finally, age-related changes in peripheral tissues may make them less responsive to some hormones. This loss of sensitivity has been documented in the case of glucocorticoids and ADH.

Extensive integration occurs between the endocrine system and other body systems. For all systems, the endocrine system adjusts metabolic rates and substrate utilization. It also regulates growth and development. **Figure 18–21** shows the functional relationships between the endocrine system and other systems studied so far.

CHECKPOINT

29. **Insulin decreases blood glucose levels and glucagon increases blood glucose levels. This is an example of which type of hormonal interaction?**

30. **The lack of which hormones would inhibit skeletal formation?**

31. **Why do levels of GH–RH and CRH rise during the resistance phase of the general adaptation syndrome?**

32. **Discuss the general role of the endocrine system in the functioning of other body systems.**

33. **Discuss the functional relationship between the endocrine system and the muscular system.**

See the blue Answers tab at the back of the book.

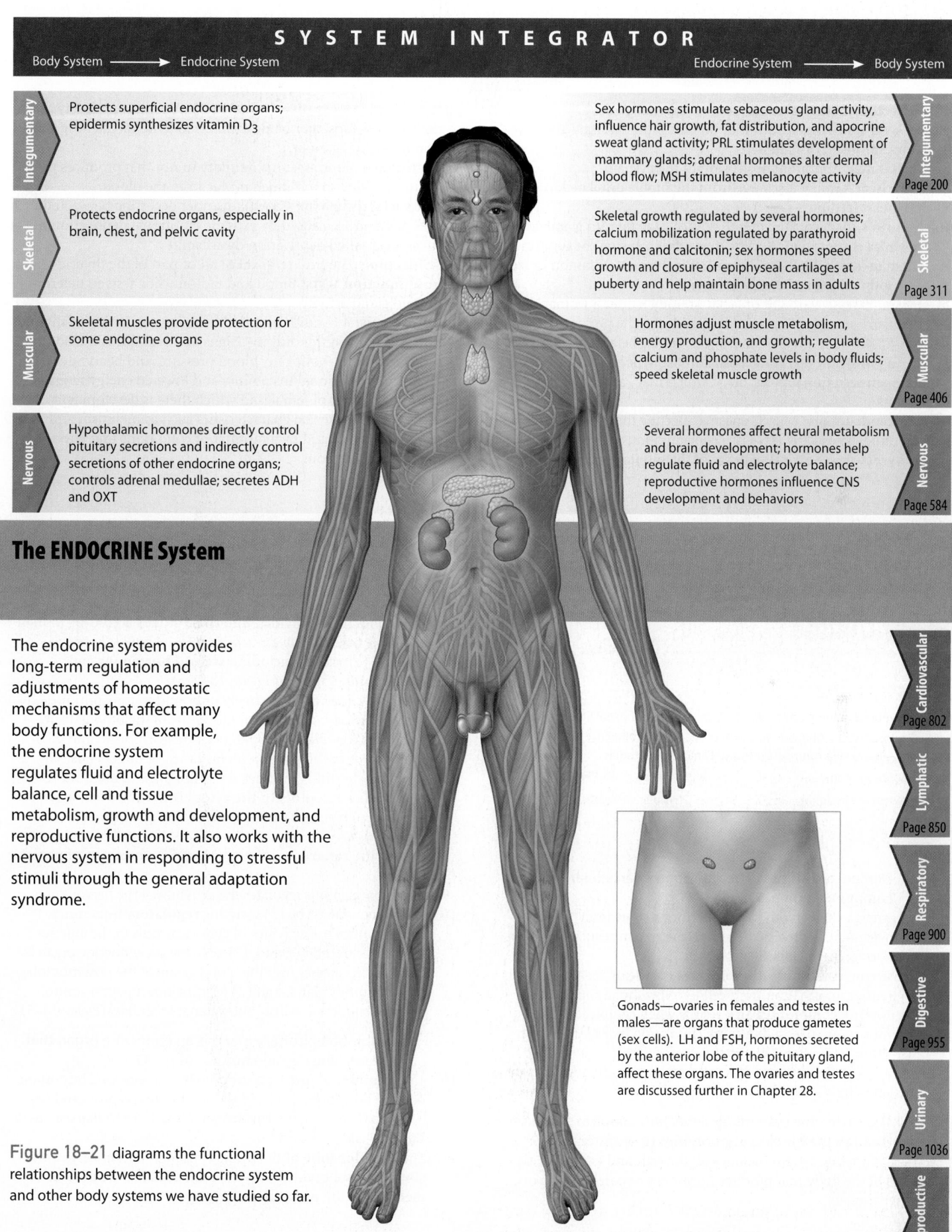

SYSTEM INTEGRATOR

Body System ⟶ Endocrine System   Endocrine System ⟶ Body System

Integumentary — Protects superficial endocrine organs; epidermis synthesizes vitamin D₃

Integumentary — Sex hormones stimulate sebaceous gland activity, influence hair growth, fat distribution, and apocrine sweat gland activity; PRL stimulates development of mammary glands; adrenal hormones alter dermal blood flow; MSH stimulates melanocyte activity
Page 200

Skeletal — Protects endocrine organs, especially in brain, chest, and pelvic cavity

Skeletal — Skeletal growth regulated by several hormones; calcium mobilization regulated by parathyroid hormone and calcitonin; sex hormones speed growth and closure of epiphyseal cartilages at puberty and help maintain bone mass in adults
Page 311

Muscular — Skeletal muscles provide protection for some endocrine organs

Muscular — Hormones adjust muscle metabolism, energy production, and growth; regulate calcium and phosphate levels in body fluids; speed skeletal muscle growth
Page 406

Nervous — Hypothalamic hormones directly control pituitary secretions and indirectly control secretions of other endocrine organs; controls adrenal medullae; secretes ADH and OXT

Nervous — Several hormones affect neural metabolism and brain development; hormones help regulate fluid and electrolyte balance; reproductive hormones influence CNS development and behaviors
Page 584

The ENDOCRINE System

The endocrine system provides long-term regulation and adjustments of homeostatic mechanisms that affect many body functions. For example, the endocrine system regulates fluid and electrolyte balance, cell and tissue metabolism, growth and development, and reproductive functions. It also works with the nervous system in responding to stressful stimuli through the general adaptation syndrome.

Cardiovascular Page 802

Lymphatic Page 850

Respiratory Page 900

Digestive Page 955

Urinary Page 1036

Reproductive Page 1116

Gonads—ovaries in females and testes in males—are organs that produce gametes (sex cells). LH and FSH, hormones secreted by the anterior lobe of the pituitary gland, affect these organs. The ovaries and testes are discussed further in Chapter 28.

Figure 18–21 diagrams the functional relationships between the endocrine system and other body systems we have studied so far.

Related Clinical Terms

adrenalectomy: Surgical removal of an adrenal gland.

empty sella syndrome: Condition in which the pituitary gland becomes shrunken or flattened.

exophthalmos: Abnormal protrusion of the eyeballs.

galactorrhea: A milky discharge from the nipple unrelated to normal breast feeding.

Hashimoto's disease: Disorder that affects the thyroid gland, also known as chronic lymphocytic thyroiditis, causing the immune system to attack the thyroid gland. It is the most common cause of hypothyroidism in the United States.

hirsutism: Excessive growth of facial or body hair in a woman. Hirsutism is a sign of hyperandrogenism, or the presence of abnormally high levels of androgens. It may be a sign of polycystic ovarian syndrome, congenital adrenal hyperplasia (CAH), or androgen-secreting tumors, all of which may cause infertility in women.

hypocalcemic tetany: Muscle spasms affecting the face and upper extremities; caused by low Ca^{2+} concentrations in body fluids.

hypophysectomy: Surgical removal of the pituitary gland.

posttraumatic stress disorder (PTSD): A common anxiety disorder that develops after being exposed to a life-threatening situation or terrifying event.

prolactinoma: Noncancerous pituitary tumor that produces prolactin, resulting in too much prolactin in the blood.

psychosocial dwarfism: Growth disorder occurring between the ages of 2 and 15, caused by extreme emotional deprivation or stress; also called deprivation dwarfism.

thyroidectomy: Surgical removal of all or part of the thyroid gland.

thyroid function tests: Blood and radionuclide tests to determine thyroid gland activity.

thyrotoxicosis: A condition caused by the oversecretion of thyroid hormones (hyperthyroidism). Signs and symptoms include increases in metabolic rate, blood pressure, and heart rate; excitability and emotional instability; and lowered energy reserves.

virilism: A disorder of females in which there is development of secondary male sexual characteristics such as hirsutism and lowered voice caused by a number of conditions that affect hormone regulation.

Chapter Review

Study Outline

MasteringA&P®

Access more chapter study tools online in the MasteringA&P Study Area:

- Chapter Quizzes, Chapter Practice Test, Art-labeling Activities, Animations, MP3 Tutor Sessions, and Clinical Case Studies
- Practice Anatomy Lab   PAL™
- Interactive Physiology   iP®
- A&P Flix   *A&PFlix*
- PhysioEx   PhysioEx™

18-1 ▌ Homeostasis is preserved through intercellular communication p. 635

1. In general, the nervous system performs short-term "crisis management," whereas the endocrine system regulates longer-term, ongoing metabolic processes.

2. **Paracrine communication** involves the use of chemical signals to transfer information from cell to cell within a single tissue.

3. **Endocrine communication** results when chemicals, called hormones, are released into the circulation by *endocrine cells*. The hormones alter the metabolic activities of many tissues and organs simultaneously by modifying the activities of **target cells**. (*Table 18–1*)

18-2 ▌ The endocrine system regulates physiological processes through the binding of hormones to receptors p. 637

4. The endocrine system includes all the cells and endocrine tissues of the body that produce hormones or paracrine factors. (*Figure 18–1*)

5. Hormones can be divided into three groups according to their chemical structure: *amino acid derivatives*; *peptide hormones*; and *lipid derivatives*, including **steroid hormones** and **eicosanoids**. (*Spotlight Figure 18–2*)

6. Hormones may circulate freely or bound to transport proteins. Free hormones are rapidly removed from the bloodstream.

7. Receptors for *catecholamines*, peptide hormones, and eicosanoids are in the plasma membranes of target cells. Thyroid and steroid hormones cross the plasma membrane and bind to receptors in the cytoplasm or nucleus, activating or inactivating specific genes. (*Spotlight Figure 18–3, Figure 18–4*)

8. **Endocrine reflexes** are the functional counterparts of neural reflexes.

9. The hypothalamus regulates the activities of the nervous and endocrine systems by (1) secreting **regulatory hormones**, which control the activities of endocrine cells in the anterior lobe of the pituitary gland; (2) acting as an endocrine organ by releasing hormones into the bloodstream at the posterior lobe of the pituitary gland; and (3) exerting direct neural control over the endocrine cells of the adrenal medullae. (*Figure 18–5*)

18-3 ▌ The bilobed pituitary gland is an endocrine organ that releases nine peptide hormones p. 645

10. The **pituitary gland**, or **hypophysis**, releases nine important peptide hormones. All bind to membrane receptors and use cyclic AMP as a second messenger. (*Figures 18–6 through 18–9; Table 18–2*)

11. The **anterior lobe** of the pituitary gland, or **adenohypophysis**, can be subdivided into the **pars distalis**, the **pars intermedia**, and the **pars tuberalis**. (*Figure 18–6*)

12. At the median eminence of the hypothalamus, neurons release regulatory factors (either **releasing hormones**, **RH**, or **inhibiting hormones**, **IH**) into the surrounding interstitial fluids through **fenestrated capillaries**. (*Figure 18–7*)

13. The **hypophyseal portal system** ensures that these regulatory factors reach the intended target cells in the pituitary before they enter the general circulation. (*Figure 18–7*)

14. **Thyroid-stimulating hormone (TSH)** triggers the release of thyroid hormones. *Thyrotropin-releasing hormone (TRH)* from the hypothalamus promotes the pituitary's secretion of TSH. (*Figure 18–8*)

15. **Adrenocorticotropic hormone (ACTH)** stimulates the release of *glucocorticoids* by the adrenal cortex. Corticotropin-releasing hormone (CRH) from the hypothalamus causes the pituitary to secrete ACTH. (*Figure 18–8*)

16. **Follicle-stimulating hormone (FSH)** stimulates follicle development and estrogen secretion in females and sperm production in males. **Luteinizing hormone (LH)** causes *ovulation* and *progesterone* production in females, and androgen production in males. Gonadotropin-releasing hormone (GnRH) from the hypothalamus promotes the pituitary's secretion of both FSH and LH. (*Figure 18–8*)

17. **Prolactin (PRL)** from the pituitary, together with other hormones, stimulates both the development of the mammary glands and milk production. (*Figure 18–8*)

18. **Growth hormone (GH, or somatotropin)** from the pituitary stimulates cell growth and replication through the release of **somatomedins** or **IGF**s from liver cells. The production of GH is regulated by **growth hormone–releasing hormone (GH–RH)** and **growth hormone–inhibiting hormone (GH–IH)** from the hypothalamus. (*Figure 18–8*)

19. **Melanocyte-stimulating hormone (MSH)** may be secreted by the pars intermedia of the pituitary during fetal development, early childhood, pregnancy, or certain diseases. This hormone stimulates melanocytes to produce melanin.

20. The **posterior lobe** of the pituitary gland, or **neurohypophysis**, contains the unmyelinated axons of hypothalamic neurons. Neurons of the **supraoptic** and **paraventricular nuclei** manufacture **antidiuretic hormone (ADH)** and **oxytocin (OXT)**, respectively. ADH decreases the amount of water lost at the kidneys and, in higher concentrations, elevates blood pressure. In women, oxytocin stimulates contractile cells in the mammary glands and has a stimulatory effect on smooth muscles in the uterus. (*Figure 18–9; Table 18–2*)

18-4 ▌ **The thyroid gland lies inferior to the larynx and requires iodine for hormone synthesis** p. 652

21. The thyroid gland lies anterior to the *thyroid cartilage* of the larynx and consists of two **lobes** connected by a narrow **isthmus**. (*Figure 18–10*)

22. The thyroid gland contains numerous **thyroid follicles**. Thyroid follicles release several hormones, including **thyroxine** and **triiodothyronine**. (*Figures 18–10, 18–11; Table 18–4*)

23. Most of the thyroid hormones entering the bloodstream are attached to special **thyroid-binding globulins (TBGs)**; the rest are attached to either **transthyretin** or albumin. (*Figure 18–11*)

24. In target cells, thyroid hormones are held in storage in the cytoplasm, bound to mitochondria (where they increase ATP production), or bound to receptors activating genes that control energy utilization. They also exert a **calorigenic effect**. (*Table 18–3*)

25. The **C cells** of the thyroid follicles produce **calcitonin (CT)**, which helps regulate Ca^{2+} concentrations in body fluids, especially during childhood and pregnancy. (*Figure 18–10; Table 18–4*)

18-5 ▌ **The four parathyroid glands, embedded in the posterior surface of the thyroid gland, secrete parathyroid hormone to elevate blood Ca^{2+}** p. 656

26. Four **parathyroid glands** are embedded in the posterior surface of the thyroid gland. **Parathyroid chief cells** produce **parathyroid hormone (PTH)** in response to lower-than-normal concentrations of Ca^{2+}. The parathyroid glands, aided by *calcitriol*, are the primary regulators of blood calcium levels in healthy adults. (*Figures 18–12, 18–13; Table 18–4*)

18-6 ▌ **The adrenal glands, consisting of a cortex and medulla, cap the kidneys and secrete several hormones** p. 657

27. One **adrenal** (*suprarenal*) **gland** lies along the superior border of each kidney. The gland is subdivided into the superficial **adrenal cortex** and the inner **adrenal medulla**. (*Figure 18–14*)

28. The adrenal cortex manufactures steroid hormones called **corticosteroids**. The cortex can be subdivided into three areas: (1) the **zona glomerulosa**, which releases **mineralocorticoids**, principally **aldosterone**; (2) the **zona fasciculata**, which produces **glucocorticoids**, notably **cortisol** and **corticosterone**; and (3) the **zona reticularis**, which produces androgens under ACTH stimulation. (*Figure 18–14*)

29. The adrenal medulla produces epinephrine (75–80 percent of medullary secretion) and norepinephrine (20–25 percent). (*Figure 18–14*)

18-7 ▌ **The pineal gland, attached to the roof of the third ventricle, secretes melatonin** p. 660

30. The **pineal gland** contains **pinealocytes**, which synthesize **melatonin**. Suggested functions include inhibiting reproductive functions, protecting against damage by free radicals, and setting circadian rhythms. (*Figure 18–15*)

18-8 ▌ **The pancreas is both an exocrine organ and endocrine gland** p. 661

31. The pancreas contains both exocrine and endocrine cells. Cells of the endocrine pancreas form clusters called **pancreatic islets** (*islets of Langerhans*). These islets contain **alpha cells**, which secrete the hormone glucagon; **beta cells**, which secrete **insulin**; **delta cells**, which secrete **somatostatin (GH–IH)**; and **F cells**, which secrete **pancreatic polypeptide**. (*Figure 18–16; Table 18–5*)

32. Insulin lowers blood glucose by increasing the rate of glucose uptake and utilization by most body cells; glucagon raises blood glucose by increasing the rates of glycogen breakdown and glucose manufacture in the liver. (*Figure 18–17; Table 18–5*)

33. **Diabetes mellitus** is an endocrine disorder characterized by high blood glucose levels. (*Spotlight Figure 18–18*)

18-9 ▌ **Many organs have secondary endocrine functions** p. 665

34. The intestines produce hormones important in coordinating digestive activities. (*Table 18–6*)

35. Endocrine cells in the kidneys produce the hormones *calcitriol* and *erythropoietin* and the enzyme *renin*. (*Table 18–6*)

36. **Calcitriol** stimulates calcium and phosphate ion absorption along the digestive tract. (*Figure 18–19*)

37. **Erythropoietin (EPO)** stimulates red blood cell production by the red bone marrow. (*Figure 18–19*)

18

18

38. In the *renin-angiotensin-aldosterone system (RAAS)*, **renin** converts **angiotensinogen** to **angiotensin I**. In the capillaries of the lungs, angiotensin-converting enzyme (ACE) converts angiotensin I to **angiotensin II**, a hormone that (1) stimulates the adrenal production of aldosterone, (2) stimulates the pituitary release of ADH, (3) promotes thirst, and (4) elevates blood pressure. *(Figure 18–19)*

39. Specialized muscle cells in the heart produce **natriuretic peptides** (*ANP* and *BNP*) when blood volume becomes excessive. In general, their actions oppose those of angiotensin II. *(Table 18–6)*

40. The thymus produces several hormones, collectively known as **thymosins**, which play a role in developing and maintaining normal immune defenses. *(Table 18–6)*

41. The **interstitial cells** of the testes produce androgens. **Testosterone** is the most important sex hormone in males. *(Table 18–7)*

42. In females, *oocytes* develop in follicles, and follicle cells produce **estrogens**, especially **estradiol**. After ovulation, the remaining follicle cells reorganize into a *corpus luteum*. Those cells release a mixture of estrogens and **progesterone**. *(Table 18–7)*

43. Adipose tissue secretes **leptin** (a feedback control for appetite).

18-10 ▌ Hormones interact to produce coordinated physiological responses p. 669

44. Endocrine system hormones often interact, producing (1) **antagonistic** (opposing) **effects**; (2) **synergistic** (additive) **effects**; (3) **permissive effects**, in which one hormone is necessary for another to produce its effect; or (4) **integrative effects**, in which hormones produce different, but complementary, results.

45. Normal growth requires the cooperation of several endocrine organs. Several hormones are especially important: GH, thyroid hormones, insulin, PTH, calcitriol, and reproductive hormones.

46. Any condition that threatens homeostasis is a **stress**. Our bodies respond to a variety of stress-causing factors through the **general adaptation syndrome (GAS)**, or **stress response**.

47. The GAS can be divided into three phases: (1) the **alarm phase** (an immediate, "fight or flight" response, under the direction of the sympathetic division of the ANS); (2) the **resistance phase**, dominated by glucocorticoids; and (3) the **exhaustion phase**, the eventual breakdown of homeostatic regulation and failure of one or more organ systems. *(Spotlight Figure 18–20)*

48. Many hormones affect the CNS. Changes in the normal mixture of hormones can significantly alter intellectual capabilities, memory, learning, and emotional states.

49. The endocrine system undergoes few functional changes with advanced age. The major changes include a decline in the concentration of growth hormone and reproductive hormones.

50. The endocrine system provides long-term regulation and homeostatic adjustments that affect many body systems. *(Figure 18–21)*

Review Questions

See the blue Answers tab at the back of the book.

LEVEL 1 Reviewing Facts and Terms

1. The use of a chemical messenger to transfer information from cell to cell within a single tissue is referred to as _____ communication. **(a)** direct, **(b)** paracrine, **(c)** hormonal, **(d)** endocrine.

2. Cyclic AMP functions as a second messenger to **(a)** build proteins and catalyze specific reactions, **(b)** activate adenylate cyclase, **(c)** open ion channels and activate key enzymes in the cytoplasm, **(d)** bind the hormone–receptor complex to DNA segments.

3. Adrenocorticotropic hormone (ACTH) stimulates the release of **(a)** thyroid hormones by the hypothalamus, **(b)** gonadotropins by the adrenal glands, **(c)** growth hormones by the hypothalamus, **(d)** steroid hormones by the adrenal glands.

4. FSH production in males supports **(a)** the maturation of sperm by stimulating nurse cells, **(b)** the development of muscles and strength, **(c)** the production of male sex hormones, **(d)** an increased desire for sexual activity.

5. The two hormones released by the posterior lobe of the pituitary gland are **(a)** GH and gonadotropin, **(b)** estrogen and progesterone, **(c)** GH and prolactin, **(d)** ADH and oxytocin.

6. All of the following are true of the endocrine system, *except* that it **(a)** releases chemicals into the bloodstream for distribution throughout the body, **(b)** releases hormones that simultaneously alter the metabolic activities of many different tissues and organs, **(c)** produces effects that can last for hours, days, and even longer, **(d)** produces rapid, local, brief-duration responses to specific stimuli, **(e)** functions to control ongoing metabolic processes.

7. A cell's hormonal sensitivities are determined by the **(a)** chemical nature of the hormone, **(b)** quantity of circulating hormone, **(c)** shape of the hormone molecules, **(d)** presence or absence of appropriate receptors, **(e)** thickness of its plasma membrane.

8. Identify the endocrine glands and tissues in the following diagram.

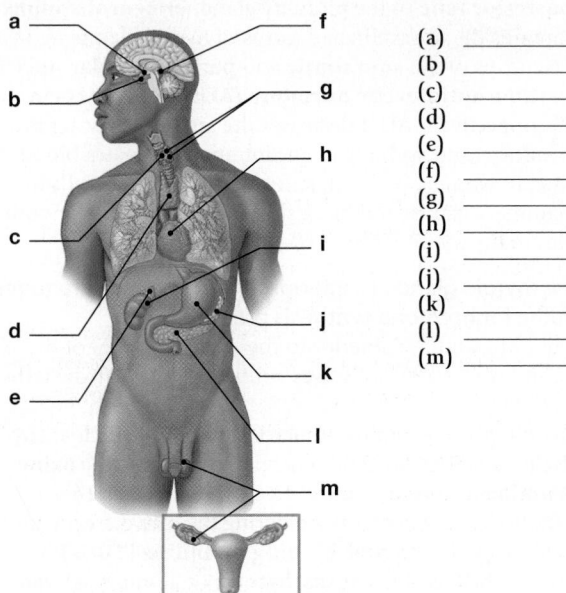

(a) _____
(b) _____
(c) _____
(d) _____
(e) _____
(f) _____
(g) _____
(h) _____
(i) _____
(j) _____
(k) _____
(l) _____
(m) _____

9. Hormones are an example of which type of communication? **(a)** direct, **(b)** paracrine, **(c)** endocrine, **(d)** synaptic.

10. Insulin is secreted by which of the following cells? **(a)** alpha cells, **(b)** f cells, **(c)** beta cells, **(d)** delta cells.

11. List three distinct hormone classes.
12. What five primary effects result from the action of thyroid hormones?
13. What effects do calcitonin and parathyroid hormone have on blood calcium levels?
14. What three zones make up the adrenal cortex, and what kind of hormones does each zone produce?
15. Which two hormones are released by the kidneys, and what is the importance of each hormone?
16. What are the four opposing effects of natriuretic peptides and angiotensin II?
17. What four cell populations make up the endocrine pancreas? Which hormone does each type of cell produce?

LEVEL 2 Reviewing Concepts

18. What is the primary difference in the way the nervous and endocrine systems communicate with their target cells?
19. In what ways can a hormone modify the activities of its target cells?
20. Differentiate between paracrine and endocrine communication.
21. How would blocking the activity of phosphodiesterase (POE) affect a cell that responds to hormonal stimulation by the cAMP second-messenger system?
22. A cell receives instructions from two separate hormones. Describe the possible outcomes.
23. A researcher observes that stimulation by a particular hormone induces a marked increase in the activity of G proteins in the target plasma membrane. The hormone being studied is probably (a) a steroid, (b) a peptide, (c) testosterone, (d) estrogen, (e) aldosterone.
24. Increased blood calcium levels would result in *increased* (a) secretion of calcitonin, (b) secretion of PTH, (c) retention of calcium by the kidneys, (d) osteoclast activity, (e) excitability of neural membranes.

25. In Type 2 diabetes, insulin levels are frequently normal, yet the target cells are less sensitive to the effects of insulin. This suggests that the target cells (a) are impermeable to insulin, (b) may lack enough insulin receptors, (c) cannot convert insulin to an active form, (d) have adequate internal supplies of glucose, (e) both b and c.

LEVEL 3 Critical Thinking and Clinical Applications

26. Roger has been extremely thirsty. He drinks numerous glasses of water every day and urinates a great deal. Name two disorders that could produce these signs and symptoms. What test could a clinician perform to determine which disorder Roger has?
27. The sarcoplasmic reticulum of the heart can be modified by phosphorylation to increase the amount of stored calcium. The system uses the second messenger cAMP. Explain how caffeine would affect cardiac function as a phosphodiesterase inhibitor.
28. Sherry tells her physician that she has been restless and irritable lately. She has a hard time sleeping and complains of diarrhea and weight loss. During the examination, her physician notices a higher-than-normal heart rate and a fine tremor in her outstretched fingers. What tests could the physician perform to make a positive diagnosis of Sherry's condition?
29. What are two benefits of having a portal system connect the median eminence of the hypothalamus with the anterior lobe of the pituitary gland?
30. Pamela and her teammates are considering taking testosterone supplements (anabolic steroids) to enhance their competitive skills. What natural effects of this hormone are they hoping to gain? What additional side effects might these women expect should they begin an anabolic steroid regime?

18

CLINICAL CASE Wrap-Up

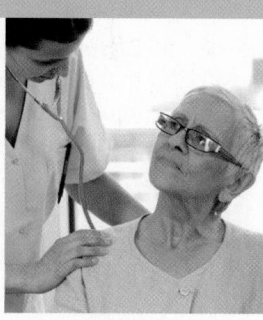

Stones, Bones, and Groans

The doctor jokes with Monica that she has a case of "stones, bones, and groans." He then shows her the lab results, which indicate elevated blood calcium (Ca^{2+}) levels—a pivotal key to diagnosing Monica's condition.

Calcium is critical in the control of many systems. It provides the means for cholinergic synapses to work, including those in the CNS, all neuron-to-neuron synapses in the PNS, and all neuromuscular junctions. Calcium also plays an essential role in the muscular contraction cycles of all skeletal, smooth, and cardiac muscles. It also strengthens the skeleton, which acts like a Ca^{2+} storage system. Calcium is the most closely regulated element in our bodies.

The parathyroid glands function to increase blood calcium. Normally four small parathyroid glands, each about the size of a grain of rice, produce parathyroid hormone (PTH). They are located on the posterior surface of the thyroid gland. Their chief cells monitor the circulating concentration of Ca^{2+} and produce PTH in a negative feedback control system.

It turns out that one of Monica's parathyroid glands has a benign tumor that is producing PTH without feedback control. The excess PTH causes the elevated blood Ca^{2+} levels, osteopenia, and bone pain. The elevated blood Ca^{2+} is responsible for Monica's heart palpitations, kidney stones, muscle fatigue, memory loss, depression, and irritability.

Outpatient surgery is performed to remove the tumor. The remaining three parathyroid glands take over normal Ca^{2+} regulation. Monica soon begins to feel like her old self.

1. In addition to mobilizing calcium from bone, how else does PTH increase blood calcium levels?

2. What hormone counteracts PTH and decreases blood Ca^{2+} levels?

See the blue Answers tab at the back of the book.

19

Blood

Learning Outcomes

These Learning Outcomes correspond by number to this chapter's sections and indicate what you should be able to do after completing the chapter.

19-1 ▌ Describe the components and major functions of blood, identify blood collection sites, and list the physical characteristics of blood.

19-2 ▌ Specify the composition and functions of plasma.

19-3 ▌ List the characteristics and functions of red blood cells, describe the structure and functions of hemoglobin, describe how red blood cell components are recycled, and explain erythropoiesis.

19-4 ▌ Explain the importance of blood typing, and the basis for ABO and Rh incompatibilities.

19-5 ▌ Categorize white blood cell types based on their structures and functions, and discuss the factors that regulate the production of each type.

19-6 ▌ Describe the structure, function, and production of platelets.

19-7 ▌ Discuss the mechanisms that control blood loss after an injury, and describe the reaction sequences responsible for blood clotting.

SPOTLIGHTS
- The Composition of Whole Blood pp. 680–681
- Hemolytic Disease of the Newborn pp. 694–695

CLINICAL CASE
- A Mysterious Blood Disorder p. 679

Clinical Notes
- Collecting Blood for Analysis p. 682
- Plasma Expanders p. 683
- Abnormal Hemoglobin p. 686

A Mysterious Blood Disorder

Today is Jasmine's first day at her new job. She just graduated from college with a degree in biology. Becoming a researcher for a pharmaceutical company is a dream come true. But today, she can't believe how bad she feels.

Jasmine is severely fatigued, even though she slept for 10 hours last night. Her gums bled when she brushed her teeth this morning. She is so short of breath that she can barely climb the flight of stairs leading into the office building. She feels her heart thumping rapidly with the slightest exertion.

Jasmine's menstrual periods have been lasting up to three weeks, and she has suffered several nosebleeds in the last month. She has big bruises on her legs that she cannot explain, and she has had a sore throat for the past six weeks.

Last week Jasmine saw her primary care physician, who drew blood for a complete blood count (CBC). Her physician calls

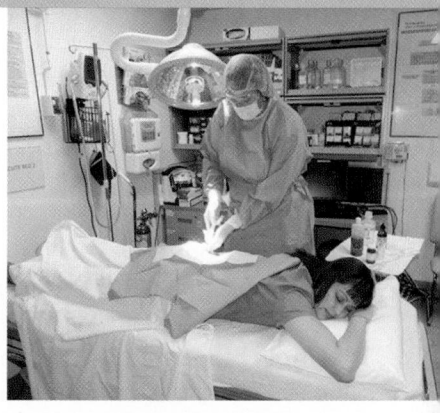

her back with some bad news. Instead of having 4.2–5.5 million red blood cells (RBCs) per μL of blood, Jasmine has fewer than 800,000. Instead of having 5000–10,000 white blood cells (WBCs) per μL of blood, she has fewer than 900. And instead of having 150,000–500,000 platelets per μL of blood, she has only 25,000.

Jasmine's doctor refers her to a hematologist, a physician specializing in diseases of the blood. The hematologist evaluates Jasmine and recommends a bone marrow biopsy. **What will Jasmine's bone marrow biopsy reveal? To find out, turn to the Clinical Case Wrap-Up on p. 709.**

An Introduction to Blood and the Cardiovascular System

In this chapter we discuss the nature of blood, the fluid component of the **cardiovascular system**. This body system also includes a pump (the heart) that circulates the fluid and a series of conducting hoses (the blood vessels) that carry it throughout the body. In Chapter 18, we noted the importance of this system for transporting hormones, but that is only one of its many vital roles.

In adults, circulating blood provides each of the body's roughly 75 trillion cells a source of nutrients and oxygen, and a way of removing wastes. The blood also transports specialized cells that defend tissues from infection and disease. These services are essential—so much so that cells deprived of circulation may die in a matter of minutes. In this chapter we take a close look at the structure and functions of blood, a fluid connective tissue with remarkable properties. We discuss the heart in Chapter 20 and the blood vessels in Chapter 21.

19-1 | Blood has several important functions and unique physical characteristics

Learning Outcome Describe the components and major functions of blood, identify blood collection sites, and list the physical characteristics of blood.

Blood is a specialized fluid connective tissue that contains cells suspended in a fluid matrix. In Chapter 4 we introduced the components and properties of this connective tissue. ⁀ p. 157

The functions of blood include the following:

- *Transporting Dissolved Gases, Nutrients, Hormones, and Metabolic Wastes.* Blood carries oxygen from the lungs to

peripheral tissues, and carbon dioxide from those tissues back to the lungs. Blood distributes nutrients absorbed by the digestive tract or released from storage in adipose tissue or in the liver. It carries hormones from endocrine glands toward their target cells. It also absorbs the wastes produced by tissue cells and carries them to the kidneys for excretion.

- *Regulating the pH and Ion Composition of Interstitial Fluids.* Diffusion between interstitial fluids and blood eliminates local deficiencies or excesses of ions, such as calcium or potassium. Blood also absorbs and neutralizes acids generated by active tissues, such as lactic acid produced by skeletal muscles.

- *Restricting Fluid Losses at Injury Sites.* Blood contains enzymes and other substances that respond to breaks in vessel walls by initiating the process of *clotting*. A blood clot acts as a temporary patch that prevents further blood loss.

- *Defending against Toxins and Pathogens.* Blood transports *white blood cells*, specialized cells that migrate into other tissues to fight infections or remove debris. Blood also delivers *antibodies*, proteins that specifically attack invading organisms or foreign compounds.

- *Stabilizing Body Temperature.* Blood absorbs the heat generated by active skeletal muscles and redistributes it to other tissues. If body temperature is already high, that heat will be lost across the surface of the skin. If body temperature is too low, the warm blood is directed to the brain and to other temperature-sensitive organs.

Spotlight Figure 19–1 (p. 680) describes the composition of whole blood, which is made up of fluid *plasma* and *formed elements* (cells and cell fragments).

19

A Fluid Connective Tissue

Blood is a fluid connective tissue with a unique composition. It consists of a matrix called **plasma** (PLAZ-muh) and formed elements (cells and cell fragments). The term **whole blood** refers to the combination of plasma and the formed elements. The cardiovascular system of an adult male contains 5–6 liters (5.3–6.4 quarts) of whole blood. An adult female contains 4–5 liters (4.2–5.3 quarts). The sex differences in blood volume primarily reflect differences in average body size.

PLASMA

Plasma, the fluid matrix of blood, makes up 46–63% of the volume of whole blood. In many respects, the composition of plasma resembles that of interstitial fluid. This similarity exists because water, ions, and small solutes are continuously exchanged between plasma and interstitial fluids across the walls of capillaries. The primary differences between plasma and interstitial fluid involve (1) the levels of respiratory gases (oxygen and carbon dioxide, due to the respiratory activities of tissue cells), and (2) the concentrations and types of dissolved proteins (because plasma proteins cannot cross capillary walls).

Plasma	Plasma Proteins	7%
55% (Range 46-63%)	Other Solutes	1%
	Water	92%

consists of +

Formed Elements	Platelets	< .1%
45% (Range 37–54%)	White Blood Cells	
	Red Blood Cells	< .1%
		99.9%

The **hematocrit** (he-MAT-ō-krit) is the percentage of formed elements in a sample of blood. The normal hematocrit, or **packed cell volume (PCV)**, in adult males is 46 and in adult females is 42. The sex difference in hematocrit primarily reflects the fact that androgens (male hormones) stimulate red blood cell production, whereas estrogens (female hormones) do not.

Formed elements are blood cells and cell fragments that are suspended in plasma. These elements account for 37–54% of the volume of whole blood. Three types of formed elements exist: platelets, white blood cells, and red blood cells. Formed elements are produced through the process of **hemopoiesis** (hē-mō-poy-Ē-sis), also called **hematopoiesis**. Two populations of stem cells—myeloid stem cells and lymphoid stem cells—are responsible for the production of formed elements.

FORMED ELEMENTS

Plasma Proteins

Plasma proteins are in solution rather than forming insoluble fibers like those in other connective tissues, such as loose connective tissue or cartilage. On average, each 100 mL of plasma contains 7.6 g of protein, almost five times the concentration in interstitial fluid. The large size and globular shapes of most blood proteins prevent them from crossing capillary walls, so they remain trapped within the bloodstream. The liver synthesizes and releases more than 90% of the plasma proteins, including all albumins and fibrinogen, most globulins, and various prohormones.

Albumins
(al-BŪ-minz) constitute about 60% of the plasma proteins. As the most abundant plasma proteins, they are major contributors to the osmotic pressure of plasma.

Globulins
(GLOB-ū-linz) account for approximately 35% of the proteins in plasma. Important plasma globulins include antibodies and transport globulins. **Antibodies**, also called **immunoglobulins** (i-mū-nō-GLOB-ū-linz), attack foreign proteins and pathogens. **Transport globulins** bind small ions, hormones, and other compounds.

Fibrinogen
(fī-BRIN-ō-jen) functions in clotting, and normally accounts for roughly 4% of plasma proteins. Under certain conditions, fibrinogen molecules interact, forming large, insoluble strands of **fibrin** (FĪ-brin) that form the basic framework for a blood clot.

Plasma also contains enzymes and hormones whose concentrations vary widely.

Other Solutes

Other solutes are generally present in concentrations similar to those in the interstitial fluids. However, because blood is a transport medium there may be differences in nutrient and waste product concentrations between arterial blood and venous blood.

Organic Nutrients:
Organic nutrients are used for ATP production, growth, and cell maintenance. This category includes lipids (fatty acids, cholesterol, glycerides), carbohydrates (primarily glucose), and amino acids.

Electrolytes:
Normal extracellular ion composition is essential for vital cellular activities. The major plasma electrolytes are Na^+, K^+, Ca^{2+}, Mg^{2+}, Cl^-, HCO_3^-, HPO_4^-, and SO_4^{2-}.

Organic Wastes:
Wastes are carried to sites for break down or excretion. Examples of organic wastes include urea, uric acid, creatinine, bilirubin, and ammonium ions.

Platelets

Platelets (PLĀT-lets) are small, membrane-bound cell fragments that contain enzymes and other substances important for clotting.

White Blood Cells

White blood cells (**WBCs**), or **leukocytes** (LŪ-kō-sīts; *leukos*, white + *-cyte*, cell), play a role in the body's defense mechanisms. There are five classes of leukocytes, each with slightly different functions that will be explored later in the chapter.

Neutrophils

Eosinophils

Basophils

Lymphocytes

Monocytes

Red Blood Cells

Red blood cells (**RBCs**), or **erythrocytes** (e-RITH-rō-sīts; *erythros*, red + *-cyte*, cell), are the most abundant blood cells. These specialized cells are essential for the transport of oxygen in the blood.

The components of whole blood can be **fractionated**, or separated, for analytical or clinical purposes. We will see examples of uses for fractionated blood later in the chapter.

Whole blood from any source—veins, capillaries, or arteries—has the same basic physical characteristics:

- Blood temperature is about 38°C (100.4°F), slightly above normal body temperature.

- Blood is five times as viscous as water—that is, five times as sticky, five times as cohesive, and five times as resistant to flow as water. The high viscosity results from interactions among dissolved proteins, formed elements, and water molecules in plasma.

- Blood is slightly alkaline, with a pH between 7.35 and 7.45 (average: 7.4).

Adult males typically have more blood than do adult females. We can estimate blood volume in liters for an individual of either sex by calculating 7 percent of the body weight in kilograms. For example, a 75-kg (165-lb) person would have a blood volume of approximately 5.25 liters (5.4 quarts).

Clinical Note

19

Collecting Blood for Analysis Fresh whole blood is generally collected from a superficial vein, such as the *median cubital vein* on the anterior surface of the elbow. The procedure is called **venipuncture** (VĒN-i-punk-chur; *vena*, vein + *punctura*, a piercing). It is a common sampling technique because (1) superficial veins are easy to locate, (2) the walls of veins are thinner than those of comparably sized arteries, and (3) blood pressure in the venous system is relatively low, so the puncture wound seals quickly. The most common clinical procedures examine venous blood.

A small drop of blood can be used to prepare a *blood smear*, a thin film of blood on a microscope slide. The blood smear is then stained with special dyes to show each type of formed element. Blood from peripheral capillaries can be obtained by puncturing the tip of a finger, an earlobe, or (in infants) the great toe or heel. Small amounts of capillary blood can also be used to test (among other items) glucose, cholesterol, and hemoglobin levels. This method is valuable when veins are difficult to access.

An **arterial puncture**, or "arterial stick," can be used for checking the efficiency of gas exchange at the lungs. Samples are generally drawn from the *radial artery* at the wrist or the *brachial artery* at the elbow.

CHECKPOINT

1. List five major functions of blood.
2. Identify the composition of the formed elements in blood.
3. What two components make up whole blood?
4. Why is venipuncture a common technique for obtaining a blood sample?

See the blue Answers tab at the back of the book.

19-2 | Plasma, the fluid portion of blood, contains significant quantities of plasma proteins

Learning Outcome Specify the composition and functions of plasma.

As shown in **Spotlight Figure 19–1**, plasma makes up about 55 percent of the volume of whole blood. The components of plasma include plasma proteins, other solutes, and water.

Plasma Proteins

Plasma contains significant quantities of dissolved proteins, namely albumins, globulins, and fibrinogen. These three types make up more than 99 percent of the plasma proteins. The others are circulating enzymes, hormones, and prohormones.

Albumins

Albumins make up the majority of the plasma proteins. For this reason, they are the major contributors to plasma osmolarity and osmotic pressure. Albumins are also important for transporting fatty acids, thyroid hormones, some steroid hormones, and other substances.

Globulins

Globulins are the second most-abundant proteins in plasma. Examples of plasma globulins include antibodies (immunoglobulins) that aid in body defense, and transport globulins. Transport globulins bind small ions, hormones, and compounds that might otherwise be removed by the kidneys or that have very low solubility in water. Important examples of transport globulins include:

- *Hormone-binding proteins*, which provide a reserve of hormones in the bloodstream. Examples include *thyroid-binding globulin* and *transthyretin*, which transport thyroid hormones, and *transcortin*, which transports ACTH. ⊃ pp. 652, 657

- *Metalloproteins*, which transport metal ions. *Transferrin*, for example, transports iron (Fe^{2+}).

- *Apolipoproteins* (ap-ō-lip-ō-PRŌ-tēnz), which carry triglycerides and other lipids in blood. When bound to lipids, an apolipoprotein becomes a **lipoprotein** (LĪ-pō-prō-tēn).

- *Steroid-binding proteins,* which transport steroid hormones in blood. For example, *testosterone-binding globulin (TeBG)* binds and transports testosterone.

Fibrinogen

The third major type of plasma protein, fibrinogen, functions in clotting. If steps are not taken to prevent clotting in a blood sample, the conversion of *fibrinogen* (a soluble protein) to *fibrin* (an insoluble protein) will take place. This conversion removes the clotting proteins, leaving a fluid known as **serum**. The clotting process also removes calcium ions and other materials from solution, so plasma and serum differ in several significant ways. (See Appendix.) For this reason, the results of a blood test generally indicate whether the sample was plasma or serum.

Other Plasma Proteins

The remaining 1 percent of plasma proteins is composed of specialized proteins whose levels vary widely. Peptide hormones—including insulin, prolactin (PRL), and the glycoproteins thyroid-stimulating hormone (TSH), follicle-stimulating hormone (FSH), and luteinizing hormone (LH)—are normally present in circulating blood. Their plasma concentrations rise and fall from day to day or even hour to hour.

Origins of the Plasma Proteins

The liver synthesizes and releases more than 90 percent of the plasma proteins, including all albumins and fibrinogen, most globulins, and various prohormones. For this reason, liver disorders can alter the composition and functional properties of blood. For example, some forms of liver disease can lead to uncontrolled bleeding due to the inadequate synthesis of fibrinogen and other proteins involved in clotting.

> ### CHECKPOINT
>
> 5. List the three major types of plasma proteins.
> 6. What would be the effects of a decrease in the amount of plasma proteins?
> 7. Which specific plasma protein would you expect to be elevated during a viral infection?
>
> See the blue Answers tab at the back of the book.

Clinical Note

Plasma Expanders *Plasma expanders* can be used to increase blood volume temporarily, over a period of hours. They are often used to buy time for lab work to determine a person's blood type. (Transfusion of the wrong blood type can kill the recipient.) Isotonic electrolyte solutions such as normal (physiological) saline can be used as plasma expanders, but their effects are short-lived due to diffusion into interstitial fluid and cells. This fluid loss is slowed by the addition of solutes that cannot freely diffuse across plasma membranes. One example is lactated *Ringer's solution,* an isotonic saline also containing lactate, potassium chloride, and calcium chloride ions. The effects of Ringer's solution fade gradually as the liver, skeletal muscles, and other tissues absorb and metabolize the lactate ions. Another option is isotonic saline solution containing purified human albumin. However, the plasma expanders in clinical use often contain large carbohydrate molecules, rather than proteins, to maintain proper osmotic concentration. These carbohydrates are not metabolized, but phagocytes gradually remove them from the bloodstream, and blood volume slowly declines. Plasma expanders are easily stored. Their sterile preparation avoids viral or bacterial contamination, which can be a problem with donated plasma. Note that plasma expanders provide a temporary solution to low blood volume, but they do not increase the amount of oxygen carried by the blood. Red blood cells are needed for that function.

19-3 ┃ Red blood cells, formed by erythropoiesis, contain hemoglobin that can be recycled

Learning Outcome List the characteristics and functions of red blood cells, describe the structure and functions of hemoglobin, describe how red blood cell components are recycled, and explain erythropoiesis.

The formed elements (blood cells and cell fragments) make up about 45 percent of whole blood volume (**Spotlight Figure 19-1**). The most abundant blood cells are the red blood cells (RBCs), which account for 99.9 percent of the formed elements. These cells give whole blood its deep red color because they contain the red pigment *hemoglobin* (HĒ-mō-glō-bin), which binds and transports oxygen and carbon dioxide.

Abundance of RBCs

RBCs make up roughly one-third of all cells in the human body. A standard blood test reports the number of RBCs per microliter (μL) of whole blood as the *red blood cell count.* In adult males, 1 microliter, or 1 *cubic millimeter* (mm^3), of whole blood contains 4.5–6.3 million RBCs. In adult females, 1 microliter contains 4.2–5.5 million. A single drop of whole blood contains approximately 260 million RBCs. The blood of an average adult has 25 trillion RBCs.

The percentage of a blood sample that consists of formed elements (most of which are red blood cells) is known as the hematocrit (**Spotlight Figure 19-1**). The hematocrit is determined by centrifuging a blood sample so that all the formed elements come out of suspension. Whole blood contains about

Figure 19–2 **The Anatomy of Red Blood Cells.**

Blood smear   LM × 477

Red blood cells   Colorized SEM × 2100

0.45–1.16 μm   2.31–2.85 μm

7.2–8.4 μm

a When viewed in a standard blood smear, RBCs appear as two-dimensional objects, because they are flattened against the surface of the slide.

b The three-dimensional shape of RBCs.

c A sectional view of a mature RBC, showing the normal ranges for its dimensions.

Red blood cell (RBC)

Rouleau (stacked RBCs)

Nucleus of endothelial cell

Blood vessels (viewed in longitudinal section)

Sectioned capillaries   LM × 1430

d When traveling through relatively narrow capillaries, RBCs may stack like dinner plates.

1000 red blood cells for each white blood cell. After centrifugation, the white blood cells and platelets form a very thin *buffy coat* above a thick layer of RBCs.

Many conditions can affect the hematocrit. For example, the hematocrit increases during dehydration, due to a reduction in plasma volume, or after *erythropoietin* (*EPO*) stimulation. ⟳ p. 666 The hematocrit decreases as a result of internal bleeding or problems with RBC formation. So, the hematocrit alone does not provide specific diagnostic information. Still, an abnormal hematocrit is an indication that other, more specific tests are needed. (We consider some of those tests later in the chapter.)

Structure of RBCs

Red blood cells are among the most specialized cells of the body. A red blood cell is very different from the "typical cell"

we discussed in Chapter 3. Each RBC is a biconcave disc with a thin central region and a thicker outer margin (**Figure 19–2**). An average RBC has a diameter of 7.8 μm and a maximum thickness of 2.85 μm, although the center narrows to about 0.8 μm.

This unusual shape has three important effects on RBC function:

1. *Gives Each RBC a Large Surface-Area-to-Volume Ratio.* Each RBC carries oxygen bound to intracellular proteins. That oxygen must be absorbed or released quickly as the RBC passes through the capillaries of the lungs or peripheral tissues. The greater the surface area per unit volume, the faster the exchange between the RBC's interior and the surrounding plasma. The total surface area of all the RBCs in the blood of a typical adult is about 3800 square meters, (nearly 4600 square yards), some 2000 times the total surface area of the body.

2. *Enables RBCs to Form Stacks, Like Dinner Plates, That Smooth the Flow through Narrow Blood Vessels.* These stacks, known as *rouleaux* (roo-LŌ; *rouleau*, singular), form and dissociate repeatedly without affecting the cells involved. An entire stack can pass along a blood vessel that is only slightly larger than the diameter of a single RBC. In contrast, individual cells would bump the walls, bang together, and form logjams that could restrict or prevent blood flow. Such stacks are shown in **Figure 19–2d**.

3. *Enables RBCs to Bend and Flex When Entering Small Capillaries and Branches.* Red blood cells are very flexible. By changing shape, individual RBCs can squeeze through capillaries as narrow as 4 μm.

When they differentiate, the RBCs of humans and other mammals lose most of their organelles, including their nuclei. The cells retain only the cytoskeleton. (The RBCs of vertebrates other than mammals have nuclei.) Without nuclei and ribosomes, circulating mammalian RBCs cannot divide or synthesize structural proteins or enzymes. As a result, the RBCs cannot perform repairs, so their life span is relatively short—normally less than 120 days. With few organelles and no ability to synthesize proteins, their energy demands are low. Without mitochondria, they obtain the energy they need through the anaerobic metabolism of glucose that is absorbed from the surrounding plasma. The lack of mitochondria ensures that absorbed oxygen will be carried to peripheral tissues, not "stolen" by mitochondria in the RBC.

Hemoglobin

A developing red blood cell loses any organelle not directly associated with the cell's primary function: the transport of respiratory gases. Molecules of **hemoglobin (Hb; Hgb)** make up more than 95 percent of its intracellular proteins. Hemoglobin is responsible for the cell's ability to transport oxygen and carbon dioxide.

The hemoglobin content of whole blood is reported in grams of Hb per deciliter (100 mL) of whole blood (g/dL). Normal ranges are 14–18 g/dL in males and 12–16 g/dL in females.

Hemoglobin Structure

Hb molecules have a complex quaternary structure. ⊃ p. 79 Each Hb molecule has two *alpha* (α) *chains* and two *beta* (β)

chains of polypeptides (Figure 19–3). Each chain is a globular protein subunit that resembles the myoglobin in skeletal and cardiac muscle cells. Like myoglobin, each Hb chain contains a single molecule of **heme**, a nonprotein pigment complex. Each heme unit holds an iron ion in such a way that the iron can interact with an oxygen molecule, forming **oxyhemoglobin**, **HbO_2**. Blood that contains RBCs filled with oxyhemoglobin is bright red.

The binding of an oxygen molecule to the iron in a heme unit is completely reversible. The iron–oxygen interaction is very weak. The iron and oxygen can easily dissociate without damaging the heme unit or the oxygen molecule. A hemoglobin molecule whose iron is not bound to oxygen is called **deoxyhemoglobin**. Blood containing RBCs filled with deoxyhemoglobin is dark red—almost burgundy.

The RBCs of an embryo or a fetus contain a different form of hemoglobin, known as *fetal hemoglobin*. It binds oxygen more readily than does the hemoglobin of adults. For this reason, a developing fetus can "steal" oxygen from the maternal bloodstream at the placenta. The conversion from fetal hemoglobin to the adult form begins shortly before birth and continues over the next year. The production of fetal hemoglobin can be stimulated in adults by the administration of drugs such as *hydroxyurea* or *butyrate*. This is one method of treatment for conditions, such as *sickle cell anemia* or *thalassemia*, that result from the production of abnormal forms of adult hemoglobin.

Hemoglobin Function

Each RBC contains about 280 million Hb molecules. Because one Hb molecule contains four heme units, each RBC can

Figure 19–3 **The Structure of Hemoglobin.** Hemoglobin consists of four globular protein subunits. Each subunit contains a single molecule of heme—a nonprotein ring surrounding a single ion of iron.

Hemoglobin molecule

Abnormal Hemoglobin Abnormal hemoglobin characterizes several inherited disorders. Two of the best known are thalassemia and sickle cell anemia (SCA).

The various forms of **thalassemia** (thal-ah-SĒ-mē-uh) result from an inability to produce adequate amounts of alpha or beta chains of hemoglobin (Hb). As a result, the rate of RBC production is slowed, and mature RBCs are fragile and short-lived. The scarcity of healthy RBCs reduces the oxygen-carrying capacity of the blood. It also leads to problems with the development and growth of systems throughout the body. Individuals with severe thalassemia must periodically undergo *transfusions*—the administration of blood components—to keep adequate numbers of RBCs in the bloodstream.

Sickle cell anemia results from a mutation affecting the amino acid sequence of the beta chains of the Hb molecule. When blood contains abundant oxygen, the Hb molecules and the RBCs that carry them appear normal. However, when the defective hemoglobin gives up enough of its bound oxygen, the adjacent Hb molecules interact and the cells become stiff and curved. This "sickling" makes the RBCs fragile and easily damaged. Moreover, an RBC that has folded to squeeze into a narrow capillary delivers its oxygen to the surrounding tissue, but the cell can become stuck as sickling occurs. A circulatory blockage results, and nearby tissues become starved for oxygen.

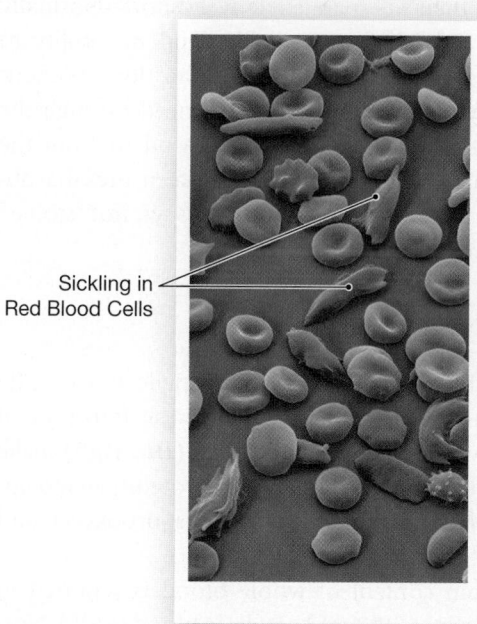

Sickling in Red Blood Cells

potentially carry more than a billion molecules of oxygen at a time. Roughly 98.5 percent of the oxygen carried by the blood is bound to Hb molecules inside RBCs.

The amount of oxygen bound to hemoglobin depends mostly on the oxygen content of the plasma. When plasma oxygen levels are low, hemoglobin releases oxygen. Under these conditions, typical of peripheral capillaries, plasma carbon dioxide levels are elevated. The alpha and beta chains of hemoglobin then bind carbon dioxide, forming **carbaminohemoglobin**. In the capillaries of the lungs, plasma oxygen levels are high and carbon dioxide levels are low. Upon reaching these capillaries, RBCs absorb oxygen (which is then bound to hemoglobin) and release carbon dioxide. We will revisit these processes in Chapter 23.

Normal activity levels can be sustained only when tissue oxygen levels are kept within normal limits. If the hematocrit is low or the Hb content of the RBCs is reduced, a condition known as **anemia** results. Anemia interferes with oxygen delivery to peripheral tissues. Every system is affected as organ function deteriorates due to oxygen starvation. Anemic individuals become weak and lethargic. They may also become confused, because the brain is affected as well.

RBC Formation and Turnover

An RBC is exposed to severe mechanical stresses. A single round trip from the heart, through the peripheral tissues, and back to the heart usually takes less than a minute. In that time, the RBC gets pumped out of the heart and forced along vessels, where it bounces off the walls and collides with other RBCs. It forms stacks, contorts, and squeezes through tiny capillaries. Then it is rushed back to the heart to make another round trip.

With all this wear and tear and no repair mechanisms, a typical RBC has a short life span. After it travels about 700 miles in 120 days, either its plasma membrane ruptures or some other damage is detected by phagocytes of the spleen, liver, and red bone marrow, which engulf the RBC. The continuous elimination of RBCs usually goes unnoticed, because new ones enter the bloodstream at a comparable rate. About 1 percent of the circulating RBCs are replaced each day, and in the process approximately 3 million new RBCs enter the bloodstream *each second*!

Hemoglobin Conservation and Recycling

Macrophages of the spleen, liver, and red bone marrow play a role in recycling red blood cell components. These phagocytes also detect and remove Hb molecules from **hemolyzed**, or ruptured, red blood cells. Hemoglobin remains intact only under the conditions inside RBCs. If the Hb released by hemolysis is not phagocytized, its components will not be recycled. When hemolysis occurs, the Hb breaks down, and the alpha and beta chains are filtered by the kidneys and eliminated in urine. When abnormally large numbers of RBCs break down in the bloodstream, urine may turn red or brown. This condition is called **hemoglobinuria**. The presence of intact RBCs in urine—a sign called **hematuria** (hē-ma-TOO-rē-uh)—occurs only after kidney damage or damage to vessels along the urinary tract.

Once a phagocytic cell has engulfed and broken down an RBC, each part of the Hb molecule has a different fate (**Figure 19–4**). The globular proteins are broken apart into their component amino acids, which are then either metabolized by the cell or released into the bloodstream for use by other cells. Each heme unit is stripped of its iron and converted to **biliverdin** (bil-i-VER-din), an organic compound with a green color. (Bad bruises commonly develop a greenish tint when biliverdin forms in the blood-filled tissues.) Biliverdin is then converted to **bilirubin** (bil-i-ROO-bin), an orange-yellow pigment, and released into the bloodstream. There, the bilirubin binds to albumin and is transported to the liver for excretion in bile.

If the bile ducts are blocked or the liver cannot absorb or excrete bilirubin, circulating levels of the compound climb rapidly. Bilirubin then diffuses into peripheral tissues, giving them a yellow color that is most apparent in the skin and over the sclera of the eyes. This combination of signs (yellow skin and eyes) is called **jaundice** (JAWN-dis).

In the large intestine, bacteria convert bilirubin to related pigments called *urobilinogens* (ūr-ō-bī-LIN-ō-jens) and *stercobilinogens* (ster-kō-bī-LIN-ō-jens). Some of the urobilinogens are absorbed into the bloodstream and are then excreted into urine. When exposed to oxygen, some of the urobilinogens and stercobilinogens are converted to **urobilins** (ūr-ō-BĪ-lins) and **stercobilins** (ster-kō-BĪ-lins).

Iron

Large quantities of free iron are toxic to cells, so in the body iron is generally bound to transport or storage proteins. Iron ions (Fe^{2+}) extracted from heme molecules may be bound and stored in a phagocytic cell or released into the bloodstream, where it binds to **transferrin** (trans-FER-in), a plasma protein. Red blood cells developing in the red bone marrow absorb the amino acids and transferrins from the bloodstream and use them to synthesize new Hb molecules. Excess transferrins are removed in the liver and spleen, and the iron is stored in two special protein–iron complexes: **ferritin** (FER-i-tin) and **hemosiderin** (hē-mō-SID-e-rin).

This recycling system is remarkably efficient. Although roughly 26 mg of iron is incorporated into new Hb molecules

each day, a dietary supply of 1–2 mg can keep pace with the incidental losses that occur at the kidneys and digestive tract.

Any impairment in iron uptake or metabolism can cause serious clinical problems, because RBC formation will be affected. For example, *iron-deficiency anemia* results from a lack of iron in the diet or from problems with iron absorption. Too much iron can also cause problems, due to excessive buildup in secondary storage sites, such as the liver and cardiac muscle tissue. Excessive iron deposition in cardiac muscle cells has been linked to heart disease.

RBC Production

Embryonic blood cells appear in the bloodstream during the third week of development. These cells divide repeatedly, rapidly increasing in number. Blood forms primarily in the vessels of the embryonic *yolk sac* during the first eight weeks of development. As other organ systems appear, some of the embryonic blood cells move out of the bloodstream and into the liver, spleen, thymus, and bone marrow. These embryonic cells differentiate into stem cells that divide to produce blood cells.

The liver and spleen are the primary sites of hemopoiesis from the second to fifth months of development, but as the skeleton enlarges, the bone marrow becomes increasingly important. In adults, red bone marrow is the only site of red blood cell production, as well as the primary site of white blood cell formation.

Red blood cell formation, or **erythropoiesis** (e-rith-rō-poy-Ē-sis), takes place only in *red bone marrow*, or **myeloid** (MĪ-e-loyd; *myelos*, marrow) **tissue**. This tissue is located in portions of the vertebrae, sternum, ribs, skull, scapulae, pelvis, and proximal limb bones. Other marrow areas contain a fatty tissue known as *yellow bone marrow*. ↺ p. 213 Under extreme stimulation, such as severe and sustained blood loss, areas of yellow marrow can convert to red marrow, increasing the rate of RBC formation.

Stages in RBC Maturation

A red blood cell matures in a series of stages. Blood specialists, known as **hematologists** (hē-ma-TOL-o-jists), have named key stages. Divisions of **hemocytoblasts** (*hemo-*, blood + *cyte*, cell + *blastos*, precursor), or **hematopoietic stem cells (HSCs)**, in the red bone marrow produce (1) **myeloid stem cells**, which in turn divide to produce red blood cells and several classes of white blood cells, and (2) **lymphoid stem cells**, which divide to produce the various classes of lymphocytes. Cells destined to become RBCs first differentiate into **proerythroblasts** and then proceed through various **erythroblast** stages (**Figure 19–5**). Erythroblasts, which actively synthesize hemoglobin, are named based on total size, amount of hemoglobin, and size and appearance of the nucleus.

19

Figure 19–4   **Recycling of Red Blood Cell Components.**

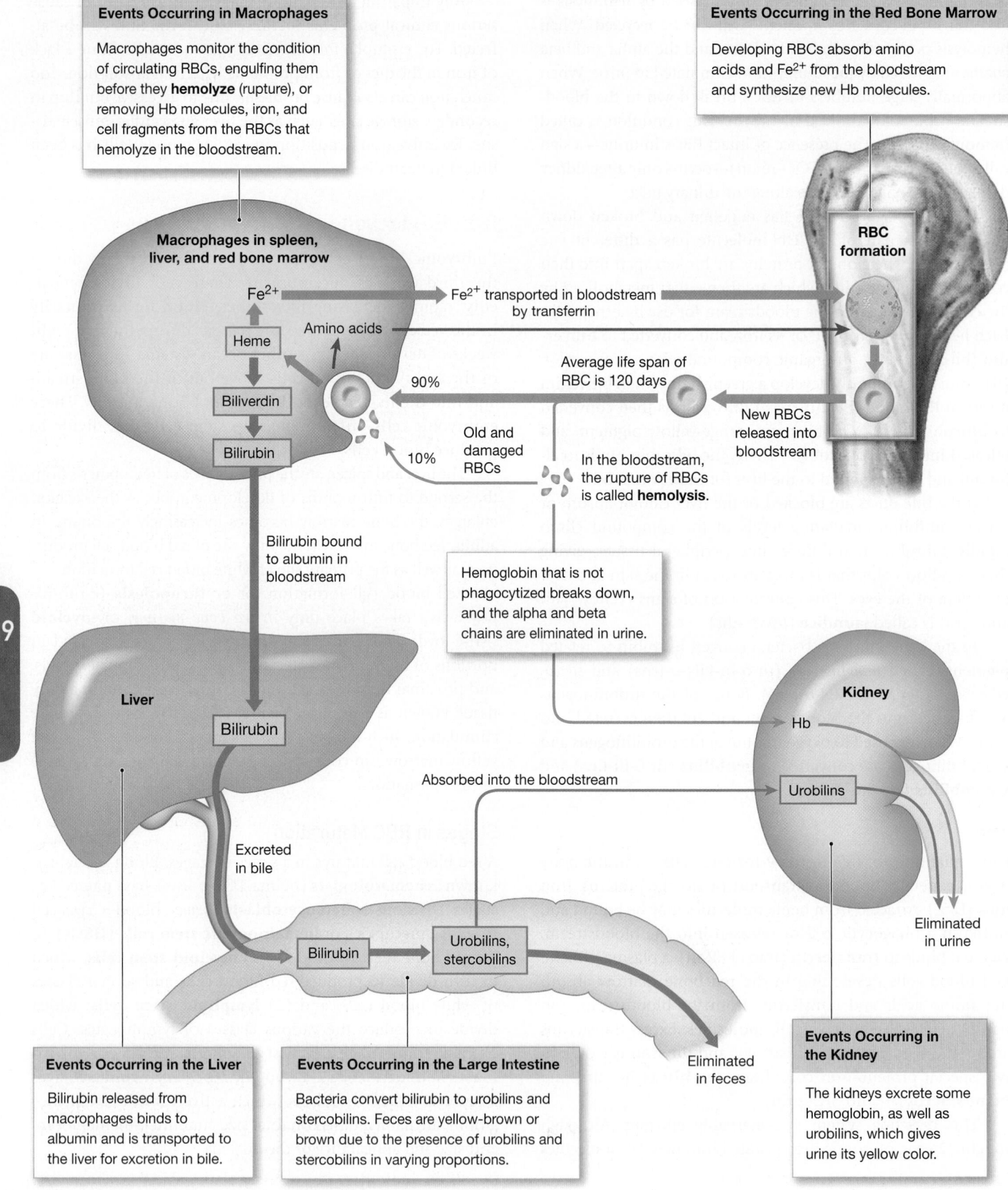

Events Occurring in Macrophages

Macrophages monitor the condition of circulating RBCs, engulfing them before they **hemolyze** (rupture), or removing Hb molecules, iron, and cell fragments from the RBCs that hemolyze in the bloodstream.

Events Occurring in the Red Bone Marrow

Developing RBCs absorb amino acids and Fe^{2+} from the bloodstream and synthesize new Hb molecules.

Macrophages in spleen, liver, and red bone marrow

Fe^{2+}

Heme

Biliverdin

Bilirubin

Amino acids

Fe^{2+} transported in bloodstream by transferrin

RBC formation

Average life span of RBC is 120 days

90%

10%

Old and damaged RBCs

New RBCs released into bloodstream

In the bloodstream, the rupture of RBCs is called **hemolysis**.

Bilirubin bound to albumin in bloodstream

Hemoglobin that is not phagocytized breaks down, and the alpha and beta chains are eliminated in urine.

Liver

Bilirubin

Excreted in bile

Kidney

Hb

Urobilins

Absorbed into the bloodstream

Eliminated in urine

Bilirubin

Urobilins, stercobilins

Eliminated in feces

Events Occurring in the Liver

Bilirubin released from macrophages binds to albumin and is transported to the liver for excretion in bile.

Events Occurring in the Large Intestine

Bacteria convert bilirubin to urobilins and stercobilins. Feces are yellow-brown or brown due to the presence of urobilins and stercobilins in varying proportions.

Events Occurring in the Kidney

The kidneys excrete some hemoglobin, as well as urobilins, which gives urine its yellow color.

19

Figure 19–5 **Stages of RBC Maturation.** Red blood cells are produced in the red bone marrow. The color density in the cytoplasm indicates the abundance of hemoglobin. Note the reductions in the sizes of the cell and nucleus leading up to the formation of a reticulocyte.

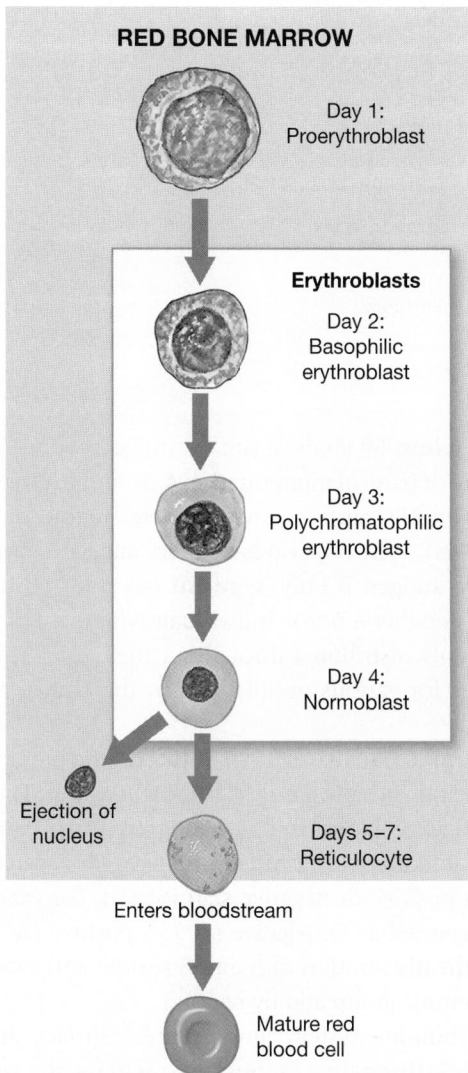

RED BONE MARROW

Day 1:
Proerythroblast

Erythroblasts

Day 2:
Basophilic
erythroblast

Day 3:
Polychromatophilic
erythroblast

Day 4:
Normoblast

Ejection of
nucleus

Days 5–7:
Reticulocyte

Enters bloodstream

Mature red
blood cell

Regulation of Erythropoiesis

For erythropoiesis to proceed normally, the red bone marrow needs adequate supplies of amino acids, iron, and vitamins (including B_{12}, B_6, and folic acid) for protein synthesis. We obtain **vitamin B_{12}** from dairy products and meat. In order to absorb vitamin B_{12}, we need *intrinsic factor*, which is produced in the stomach. Without vitamin B_{12} from the diet, normal stem cell divisions cannot occur and *pernicious anemia* results. Thus, pernicious anemia is caused by a vitamin B_{12} deficiency, a problem with the production of intrinsic factor, or a problem with the absorption of vitamin B_{12} bound to intrinsic factor.

Erythropoiesis is stimulated directly by the peptide hormone erythropoietin (⟲ p. 666) and indirectly by several hormones, including thyroxine, androgens, and growth hormone. As noted earlier in **Spotlight Figure 19–1** on pp. 680–681, estrogens do not stimulate erythropoiesis. This fact accounts for the differences in hematocrit values between males and females.

Erythropoietin (EPO), also called **erythropoiesis-stimulating hormone**, is a glycoprotein. It is formed by the kidneys and liver. EPO appears in the plasma when peripheral tissues, especially the kidneys, are exposed to low oxygen concentrations. A low oxygen level in tissues is called **hypoxia** (hī-POKS-ē-uh; *hypo-*, below + *oxy-*, presence of oxygen). Erythropoietin is released (1) during anemia; (2) when blood flow to the kidneys declines; (3) when the oxygen content of air in the lungs declines, due to disease or high altitude; and (4) when the respiratory surfaces of the lungs are damaged. Once in the bloodstream, EPO travels to the red bone marrow, where it stimulates stem cells and developing RBCs.

Erythropoietin has two major effects: (1) It stimulates cell division rates in erythroblasts and in the stem cells that produce erythroblasts, and (2) it speeds up the maturation of RBCs, mainly by accelerating Hb synthesis. Under maximum EPO stimulation, bone marrow can increase RBC formation tenfold, to about 30 million cells per second.

The ability to increase the rate of blood formation quickly and dramatically is important to a person recovering from a severe blood loss. But if EPO is administered to a healthy individual, as in the case of the cyclists mentioned in Chapter 18, the hematocrit may rise to 65 or more. ⟲ p. 672 Such an increase can place an intolerable strain on the heart. Similar problems can occur after **blood doping**, a practice in which athletes elevate their hematocrits by reinfusing packed RBCs that were removed and stored at an earlier date. The goal is to improve oxygen delivery to muscles, thereby enhancing performance. The strategy can be dangerous, however, because it makes blood more viscous and increases the workload on the heart.

Blood tests provide information about the general health of an individual, usually with a minimum of trouble and expense. Several common blood tests focus on RBCs. These *RBC tests* assess the number, size, shape, and maturity of circulating RBCs, indicating the erythropoietic activities under way.

After roughly four days of differentiation, the erythroblast, now called a *normoblast*, sheds its nucleus and becomes a **reticulocyte** (re-TIK-ū-lō-sīt), which contains 80 percent of the Hb of a mature RBC. Hb synthesis then continues for two to three more days. During this period, while the cell is synthesizing hemoglobin and other proteins, its cytoplasm still contains RNA, which can be seen under the microscope with certain stains. After two days in the red bone marrow, the reticulocyte enters the bloodstream. At this time, reticulocytes normally account for about 0.8 percent of the RBC population in the blood and can still be detected by staining. After 24 hours in circulation, the reticulocytes complete their maturation and become indistinguishable from other mature RBCs.

19

Table 19–1	RBC Tests and Related Terminology		
		Terms Associated with Abnormal Values	
Test	**Determines**	**Elevated**	**Depressed**
Hematocrit (Hct)	Percentage of formed elements in whole blood Normal = 37–54%	Polycythemia (may reflect erythrocytosis or leukocytosis)	Anemia
Reticulocyte count (Retic.)	Percentage of circulating reticulocytes Normal = 0.8%	Reticulocytosis	Diminished erythropoiesis
Hemoglobin concentration (Hb; Hgb)	Concentration of hemoglobin in blood Normal = 12–18 g/dL	Polycythemia	Anemia
RBC count	Number of RBCs per μL of whole blood Normal = 4.2–6.3 million cells/μL	Erythrocytosis/polycythemia	Anemia
Mean corpuscular volume (MCV)	Average volume of one RBC Normal = 82–101 μm^3/cell (normocytic)	Macrocytic	Microcytic
Mean corpuscular hemoglobin concentration (MCHC)	Average weight of Hb in one RBC Normal = 27–34 ug/RBC (normochromic)	Hyperchromic	Hypochromic

The tests can also be useful in detecting problems, such as internal bleeding, that may not produce other obvious signs or symptoms. Table 19–1 lists examples of important blood tests and related terms.

CHECKPOINT

8. Describe hemoglobin.

9. How would the hematocrit change after an individual suffered a significant blood loss?

10. Dave develops a blockage in his renal arteries that restricts blood flow to the kidneys. What effect will this have on his hematocrit?

11. In what way would a disease that causes damage to the liver affect the level of bilirubin in the blood?

See the blue Answers tab at the back of the book.

19-4 ▮ The ABO blood types and Rh system are based on antigen–antibody responses

Learning Outcome Explain the importance of blood typing, and the basis for ABO and Rh incompatibilities.

Antigens are substances that can trigger a protective defense mechanism called an *immune response*. Most antigens are proteins, but some other types of organic molecules are antigens as well. Your plasma membranes contain **surface antigens**, substances that your immune system recognizes as "normal." In other words, your immune system ignores these substances rather than attacking them as "foreign."

Your **blood type** is determined by the presence or absence of specific surface antigens in RBC plasma membranes. The surface antigens involved are integral membrane glycoproteins or glycolipids whose characteristics are genetically determined. Red blood cells have at least 50 kinds of surface antigens, but three surface antigens are of particular importance: **A**, **B**, and **Rh** (or **D**).

Based on RBC surface antigens, there are four blood types (**Figure 19–6a**): **Type A** blood has surface antigen A only, **type B** has surface antigen B only, **type AB** has both A and B, and **type O** has neither A nor B. Individuals with these blood types are not evenly distributed throughout the world. The average percentages for various populations in the United States are given in Table 19–2.

The term **Rh positive (Rh⁺)** indicates the presence of the Rh surface antigen, commonly called the *Rh factor*. The absence of this antigen is indicated as **Rh negative (Rh⁻)**. When the complete blood type is recorded, the term *Rh* is usually omitted, and a positive or negative sign is used. For example, the data are reported as O negative (O⁻), A positive (A⁺), and so on. As in the distribution of A and B surface antigens, Rh type differs by ethnic group and by region (Table 19–2).

Your immune system ignores these surface antigens—called **agglutinogens** (a-glū-TIN-ō-jenz)—on your own RBCs. However, your plasma contains antibodies, sometimes called *agglutinins* (a-GLŪ-ti-ninz), that will attack the antigens on "foreign" RBCs. When these antibodies attack, the foreign cells **agglutinate**, or clump together. This process is called **agglutination**. If you have type A blood, your plasma contains anti-B antibodies, which will attack type B surface antigens. If you have type B blood, your plasma contains anti-A antibodies. The RBCs of an individual with type O blood have neither A nor B surface antigens, and that person's plasma contains both anti-A and anti-B antibodies. A type AB individual has RBCs with both A and B surface antigens, and the plasma does not contain anti-A or anti-B antibodies. The presence of anti-A and/or anti-B antibodies is genetically determined. These antibodies are present throughout life, regardless of whether the individual has ever been exposed to foreign RBCs.

Figure 19–6   **Blood Types and Cross-Reactions.**

Type A	**Type B**	**Type AB**	**Type O**
Type A blood has RBCs with surface antigen A only.	**Type B** blood has RBCs with surface antigen B only.	**Type AB** blood has RBCs with both A and B surface antigens.	**Type O** blood has RBCs lacking both A and B surface antigens.

Surface antigen A

Surface antigen B

| If you have type A blood, your plasma contains anti-B antibodies, which will attack type B surface antigens. | If you have type B blood, your plasma contains anti-A antibodies, which will attack type A surface antigens. | If you have type AB blood, your plasma has neither anti-A nor anti-B antibodies. | If you have type O blood, your plasma contains both anti-A and anti-B antibodies. |

a Blood type depends on the presence of surface antigens (agglutinogens) on RBC surfaces. The plasma contains antibodies (agglutinins) that will react with foreign surface antigens.

RBC

Surface antigens   Opposing antibodies   Agglutination (clumping)   Hemolysis

b In a cross-reaction, antibodies react with their target antigens causing agglutination and hemolysis of the affected RBCs.

Table 19–2	Differences in Blood Group Distribution				
	Percentage with Each Blood Type				
Population	**O**	**A**	**B**	**AB**	**Rh⁺**
U.S. (AVERAGE)	46	40	10	4	85
African American	49	27	20	4	95
Caucasian	45	40	11	4	85
Chinese American	42	27	25	6	100
Filipino American	44	22	29	6	100
Hawaiian	46	46	5	3	100
Japanese American	31	39	21	10	100
Korean American	32	28	30	10	100
NATIVE NORTH AMERICAN	79	16	4	1	100
NATIVE SOUTH AMERICAN	100	0	0	0	100
AUSTRALIAN ABORIGINE	44	56	0	0	100

19

In contrast, the plasma of an Rh-negative individual does not contain anti-Rh antibodies. These antibodies are present only if the individual has been **sensitized** by previous exposure to Rh⁺ RBCs. Such exposure can occur accidentally during a transfusion, but it can also accompany a seemingly normal pregnancy involving an Rh⁻ mother and an Rh⁺ fetus. (See Spotlight Figure 19–8 on pp. 694–695.)

Cross-Reactions in Transfusions

When an antibody meets its specific surface antigen, the RBCs agglutinate and may also hemolyze. This is called a **cross-reaction** (Figure 19–6b). For instance, an anti-A antibody that encounters A surface antigens will cause the RBCs bearing the A surface antigens to clump or even break up. Clumps and fragments of RBCs under attack form drifting masses that can plug small blood vessels in the kidneys, lungs, heart, or brain, damaging or destroying affected tissues. Such cross-reactions, or *transfusion reactions*, can be prevented by making sure that the blood types of the donor and the recipient are **compatible**—that is, that the donor's blood cells and the recipient's plasma will not cross-react.

In practice, the surface antigens on the donor's cells are more important in determining compatibility than are the antibodies in the donor's plasma. Why? Unless large volumes of whole blood or plasma are transferred, cross-reactions between the donor's plasma and the recipient's blood cells will not produce significant agglutination. This is because the donated plasma is diluted quickly through mixing with the larger plasma volume of the recipient. (One unit of whole blood, 500 mL, contains roughly 275 mL of plasma, only about 10 percent of normal plasma volume.) When the goal is to increase the blood's oxygen-carrying capacity rather than its plasma volume, packed RBCs, with a minimal amount of plasma, are often transfused. This practice minimizes the risk of a cross-reaction.

Testing for Transfusion Compatibility

Extra care must be taken to avoid potentially life-threatening cross-reactions between the donor's cells and the recipient's plasma. For this reason, a compatibility test is usually performed in advance. This process normally involves two steps: (1) a determination of blood type and (2) a cross-match test.

The standard test for blood type considers only the three surface antigens most likely to produce dangerous cross-reactions: A, B, and Rh (Figure 19–7). The test involves taking drops of blood and mixing them separately with solutions containing anti-A, anti-B, and anti-Rh (anti-D) antibodies. Any cross-reactions are then recorded. For example, if an individual's RBCs clump together when exposed to anti-A and to anti-B antibodies, the individual has type AB blood. If no reactions occur after exposure, that person must have type O blood.

The presence or absence of the Rh surface antigen is also noted, and the individual is classified as Rh positive or Rh

Figure 19–7   Blood Type Testing. Test results for blood samples from four individuals. Drops are mixed with solutions containing antibodies to the surface antigens A, B, AB, and D (Rh). Clumping occurs when the sample contains the corresponding surface antigen(s). The individuals' blood types are shown at right.

Anti-A	Anti-B	Anti-D	Blood type
			A⁺
			B⁺
			AB⁺
			O⁻

negative on that basis. Type O⁺ is the most common blood type. The RBCs of type O⁺ individuals lack surface antigens A and B but have the Rh antigen.

Spotlight Figure 19–8 describes a serious condition known as hemolytic disease of the newborn (HDN). HDN involves Rh incompatibility between a pregnant mother and her developing fetus.

Standard blood-typing of both donor and recipient can be completed in a matter of minutes. However, in an emergency, there may not be time for preliminary testing. For example, a person with a severe gunshot wound may require 5 *liters* or more of blood before the damage can be repaired. Under these circumstances, type O blood (preferably O⁻) will be administered. Because the donated RBCs lack both A and B surface antigens, the recipient's blood can have anti-A antibodies, anti-B antibodies, or both and still not cross-react with the donor's blood. Because cross-reactions with type O blood are very unlikely, type O individuals are sometimes called *universal donors*. Type AB individuals were once called *universal recipients*, because they lack anti-A and anti-B antibodies that would attack donated RBCs, and so can safely receive blood of any type. However, now that blood supplies are usually adequate and compatibility testing is regularly done, the term has largely been dropped. If the recipient's blood type is known to be AB, type AB blood will be administered.

It is now possible to use enzymes to strip off the A or B surface antigens from RBCs and create type O blood in the laboratory. The procedure is expensive and time-consuming and has limited use in emergency treatment. Still, cross-reactions can occur, even to type O⁻ blood, because at least 48 other surface antigens are present. For this reason, whenever time and facilities permit, further testing is performed to ensure complete compatibility between donor blood and recipient blood. **Cross-match testing** involves exposing the donor's RBCs to a sample of the recipient's plasma under controlled conditions. This procedure reveals significant cross-reactions involving surface antigens other than A, B, or Rh. Another way to avoid compatibility problems is to replace lost blood with synthetic blood substitutes, which do not contain surface antigens that can trigger a cross-reaction.

Because blood groups are inherited, blood tests are also used as paternity tests and in crime detection. The blood collected cannot prove that a particular individual *is* a certain child's father or *is* guilty of a specific crime, but it can prove that the individual is *not* involved. It is impossible, for example, for an adult with type AB blood to be the parent of an infant with type O blood. Testing for additional surface antigens, other than the standard ABO groups, can increase the accuracy of the conclusions. When available, DNA identity testing, which has nearly 100 percent accuracy, has replaced blood type identity testing.

CHECKPOINT

12. What is the function of surface antigens on RBCs?

13. Which blood type can be safely transfused into a person with type O blood?

14. Why can't a person with type A blood safely receive blood from a person with type B blood?

See the blue Answers tab at the back of the book.

19-5 ▌ The various types of white blood cells contribute to the body's defenses

Learning Outcome Categorize white blood cell types based on their structures and functions, and discuss the factors that regulate the production of each type.

Unlike red blood cells, white blood cells (WBCs) have nuclei and other organelles, and they lack hemoglobin. White blood cells, or leukocytes, help defend the body against invasion by pathogens. They also remove toxins, wastes, and abnormal or damaged cells.

Several types of WBCs can be distinguished microscopically in a blood smear by using either of two standard stains: *Wright's stain* or *Giemsa stain*. Traditionally, WBCs have been divided into two groups based on their appearance after staining: (1) *granular leukocytes*, or *granulocytes* (with abundant stained granules)—the *neutrophils*, *eosinophils*, and *basophils*; and (2) *agranular leukocytes*, or *agranulocytes* (with few, if any, stained

granules)—the *monocytes* and *lymphocytes*. This categorization is convenient but somewhat misleading. In reality, the granules in "granular leukocytes" are secretory vesicles and lysosomes, and the "agranular leukocytes" also contain vesicles and lysosomes, but they are just smaller and difficult to see with the light microscope.

A typical microliter of blood contains 5000 to 10,000 WBCs, compared with 4.2 to 6.3 million RBCs. Most of the WBCs in the body at any moment are in connective tissue proper or in organs of the lymphatic system. Circulating WBCs are only a small fraction of the total WBC population.

WBC Circulation and Movement

Unlike RBCs, WBCs circulate for only a short time in their life span. White blood cells migrate through the loose and dense connective tissues of the body. They use the bloodstream to travel from one organ to another and for rapid transportation to areas of infection or injury. As they travel along the miles of capillaries, WBCs can detect the chemical signs of damage to surrounding tissues. When these cells detect problems, they leave the bloodstream and enter the damaged area.

Circulating WBCs have four characteristics:

1. *All Can Migrate Out of the Bloodstream.* When WBCs in the bloodstream are activated, they contact and adhere to the vessel walls in a process called *margination*. After further interaction with endothelial cells, the activated WBCs squeeze between adjacent endothelial cells and enter the surrounding tissue. This process is called *emigration*, or *diapedesis* (*dia*, through + *pedesis*, a leaping).

2. *All Are Capable of Amoeboid Movement. Amoeboid movement* is a gliding motion made possible by the flow of cytoplasm into slender cellular processes extended in the direction of movement. (The movement is so named because it is similar to that of an *amoeba*, a type of protozoan.) The mechanism is not fully understood, but it involves the continuous rearrangement of bonds between actin filaments in the cytoskeleton. It also requires calcium ions and ATP. This mobility allows WBCs to move through the endothelial lining and into peripheral tissues.

3. *All Are Attracted to Specific Chemical Stimuli.* This characteristic, called **positive chemotaxis** (kē-mō-TAK-sis), guides WBCs to invading pathogens, damaged tissues, and other active WBCs.

4. *Neutrophils, Eosinophils, and Monocytes Are Capable of Phagocytosis.* These cells may engulf pathogens, cell debris, or other materials. Neutrophils and eosinophils are sometimes called *microphages*, to distinguish them from the larger macrophages in connective tissues. Macrophages are monocytes that have moved out of the bloodstream and have become actively phagocytic. ⊃ p. 153

Hemolytic disease of the newborn is an RBC-related disorder caused by a cross-reaction between fetal and maternal blood types. Genes controlling the presence or absence of any surface antigen in the plasma membrane of a red blood cell are provided by both parents, so a child can have a blood type different from that of either parent. During pregnancy, when fetal and maternal vascular systems are closely intertwined, the mother's antibodies may cross the placenta, attacking and destroying fetal RBCs. The resulting condition, called **hemolytic disease of the newborn (HDN)**, has many forms, some quite dangerous and others so mild as to remain undetected.

Rh⁻ mother

Rh⁺ fetus

The most common form of hemolytic disease of the newborn develops after an Rh⁻ woman has carried an Rh⁺ fetus.

FIRST PREGNANCY

1 Problems seldom develop during a first pregnancy, because very few fetal cells enter the maternal bloodstream then, and thus the mother's immune system is not stimulated to produce anti-Rh antibodies.

During First Pregnancy

Maternal blood supply and tissue

Rh⁻   Rh⁻
Rh⁻ ← Rh⁻

Rh⁺ ← Rh⁺

Fetal blood supply and tissue

Rh⁺   Rh⁺

Placenta

2 Exposure to fetal red blood cell antigens generally occurs during delivery, when bleeding takes place at the placenta and uterus. Such mixing of fetal and maternal blood can stimulate the mother's immune system to produce anti-Rh antibodies, leading to sensitization.

Hemorrhaging at Delivery

Rh⁺   Maternal blood supply and tissue   Rh⁻

Rh⁺   Rh⁻
Rh⁺ ← Rh⁻

Rh⁺

Fetal blood supply and tissue

Rh⁺

Rh antigen on fetal red blood cells

3 About 20% of Rh⁻ mothers who carried Rh⁺ children become sensitized within six months of delivery. Because the anti-Rh antibodies are not produced in significant amounts until after delivery, a woman's first infant is not affected.

Maternal Antibody Production

Rh⁻   Maternal blood supply and tissue

Rh⁻   Rh⁻   Rh⁻

Maternal antibodies to Rh antigen

SECOND PREGNANCY

Rh⁻ mother

Rh⁺ fetus

4

If a future pregnancy involves an Rh⁺ fetus, maternal anti-Rh antibodies produced after the first delivery cross the placenta and enter the fetal bloodstream. These antibodies destroy fetal RBCs, producing a dangerous anemia. The fetal demand for blood cells increases, and they leave the bone marrow and enter the bloodstream before completing their development. Because these immature RBCs are erythroblasts, HDN is also known as **erythroblastosis fetalis** (e-rith-rō-blas-TŌ-sis fē-TAL-is). Without treatment, the fetus may die before delivery or shortly thereafter. A newborn with severe HDN is anemic, and the high concentration of circulating bilirubin produces jaundice. Because the maternal antibodies remain active in the newborn for one to two months after delivery, the infant's entire blood volume may require replacement to remove the maternal anti-Rh antibodies, as well as the damaged RBCs. Fortunately, the mother's anti-Rh antibody production can be prevented if such antibodies (available under the name RhoGAM) are administered to the mother in weeks 26–28 of pregnancy and during and after delivery. These antibodies destroy any fetal RBCs that cross the placenta before they can stimulate a maternal immune response. Because maternal sensitization does not occur, no anti-Rh antibodies are produced. In the United States, this relatively simple procedure has almost entirely eliminated HDN mortality caused by Rh incompatibilities.

During Second Pregnancy

Maternal blood supply and tissue

Rh⁻

Rh⁻

Rh⁻

Rh⁻

Rh⁺

Rh⁺

Rh⁺

Maternal anti-Rh antibodies

Fetal blood supply and tissue

Hemolysis of fetal RBCs

Types of WBCs

Neutrophils, eosinophils, basophils, and monocytes are part of the body's *nonspecific defenses*. A variety of stimuli activate such defenses, which do not discriminate between one type of threat and another. Lymphocytes, in contrast, are responsible for *specific defenses*, which mount a counterattack against specific types of invading pathogens or foreign proteins. We discuss the interactions among WBCs and the relationships between specific and nonspecific defenses in Chapter 22.

Neutrophils

Fifty to 70 percent of the circulating WBCs are **neutrophils** (NOO-trō-filz). This name reflects the fact that the granules of these WBCs are chemically neutral and thus are difficult to stain with either acidic or basic (alkaline) dyes. A mature neutrophil has a very dense, segmented nucleus with two to five lobes resembling beads on a string (**Figure 19–9a**). This structure has given neutrophils another name: **polymorphonuclear** (pol-ē-mor-fō-NOO-klē-ar) **leukocytes** (*poly*, many + *morphe*, form), or *PMNs*. "Polymorphs," or "polys," as they are often called, are roughly 12 μm in diameter. Their cytoplasm is packed with pale granules containing lysosomal enzymes and bactericidal (bacteria-killing) compounds.

Neutrophils are highly mobile. They are generally the first of the WBCs to arrive at the site of an injury. These very active cells specialize in attacking and digesting bacteria that have been "marked" with antibodies or with *complement proteins*—plasma proteins involved in tissue defenses. (We discuss the complement system in Chapter 22.)

Upon encountering a bacterium, a neutrophil quickly engulfs it, and the metabolic rate of the neutrophil increases dramatically. Along with this *respiratory burst*, the neutrophil produces highly reactive, destructive chemical agents, including *hydrogen peroxide* (H_2O_2) and *superoxide anions* (O_2^-), which can kill bacteria.

Meanwhile, the vesicle containing the engulfed pathogen fuses with lysosomes that contain digestive enzymes and small peptides called **defensins**. This process is called **degranulation** because it reduces the number of granules in the cytoplasm. Defensins kill a variety of pathogens, including bacteria, fungi, and enveloped viruses, by combining to form large channels in their plasma membranes. The digestive enzymes then break down the bacterial remains.

While actively attacking bacteria, a neutrophil releases prostaglandins and leukotrienes. ⟲ p. 639 The prostaglandins increase capillary permeability in the affected region, contributing to local inflammation and restricting the spread of injury and infection. Leukotrienes are hormones that attract other phagocytes and help coordinate the immune response.

Most neutrophils have a short life span. They survive in the bloodstream for only about 10 hours. When actively engulfing debris or pathogens, they may last 30 minutes or less. A neutrophil dies after engulfing one to two dozen bacteria, but its breakdown releases chemicals that attract other neutrophils to the site. A mixture of dead neutrophils, cellular debris, and other waste products form the *pus* associated with infected wounds.

TIPS & TRICKS

Remember that **neu**trophils are the most **nu**merous of the white blood cells.

Eosinophils

Eosinophils (ē-ō-SIN-ō-filz) have granules that stain darkly with *eosin*, a red dye. The granules also stain with other acid dyes, so the name **acidophils** (a-SID-ō-filz) applies as well. Eosinophils generally represent 2–4 percent of the circulating WBCs. They are similar in size to neutrophils. However, the combination of deep red granules and a typically bilobed (two-lobed) nucleus makes eosinophils easy to identify (**Figure 19–9b**).

Figure 19–9 **White Blood Cells.**

| **a** Neutrophil LM × 1500 | **b** Eosinophil LM × 1500 | **c** Basophil LM × 1500 | **d** Monocyte LM × 1500 | **e** Lymphocyte LM × 1500 |

Eosinophils attack objects that are coated with antibodies. They will engulf antibody-marked bacteria, protozoa, or cellular debris, but their primary mode of attack is the exocytosis of toxic compounds, including nitric oxide and cytotoxic enzymes. This is particularly effective against multicellular parasites that are too big to engulf, such as flukes or parasitic roundworms. The number of circulating eosinophils increases dramatically during a parasitic infection.

Eosinophils increase in number during allergic reactions as well, because they are sensitive to circulating *allergens* (materials that trigger allergies). Eosinophils are also attracted to sites of injury, where they release enzymes that reduce inflammation produced by mast cells and neutrophils. This controls the spread of inflammation to adjacent tissues.

Basophils

Basophils (BĀ-sō-filz; *baso-*, base + *phileo*, to love) have numerous granules that stain darkly with basic dyes. In a standard blood smear, the inclusions are deep purple or blue (**Figure 19–9c**). Measuring 8–10 μm in diameter, basophils are smaller than neutrophils or eosinophils. They are also relatively rare, accounting for less than 1 percent of the circulating WBC population.

Basophils migrate to injury sites and cross the capillary endothelium to accumulate in the damaged tissues. There they discharge their granules into the interstitial fluids. The granules contain *histamine*, which dilates blood vessels, and *heparin*, a compound that prevents blood clotting. Stimulated basophils release these chemicals into the interstitial fluids to enhance the local inflammation initiated by mast cells. ↻ p. 153 Although mast cells release the same compounds in damaged connective tissues, mast cells and basophils are distinct populations with separate origins. Stimulated basophils also release other chemicals that attract eosinophils and other basophils to the area.

Monocytes

Monocytes (MON-ō-sīts) are spherical cells that may exceed 15 μm in diameter, nearly twice the diameter of a typical RBC. When flattened in a blood smear, they look even larger, so monocytes are fairly easy to identify. The nucleus is large and tends to be oval or kidney bean–shaped rather than lobed (**Figure 19–9d**). Monocytes normally account for 2–8 percent of circulating WBCs.

An individual monocyte is transported in the bloodstream. It remains in circulation for only about 24 hours before entering peripheral tissues to become a tissue macrophage. Macrophages are aggressive phagocytes. They often attempt to engulf items as large as or larger than themselves. When active, they release chemicals that attract and stimulate neutrophils, monocytes, and other phagocytic cells. Active macrophages also secrete substances that draw fibroblasts into the region. The fibroblasts then begin producing scar tissue to wall off the injured area.

Lymphocytes

Typical **lymphocytes** (LIM-fō-sīts) are slightly larger than RBCs and lack abundant, deeply staining granules. In blood smears, lymphocytes typically have a large, round nucleus surrounded by a thin halo of cytoplasm (**Figure 19–9e**).

Lymphocytes account for 20–40 percent of the circulating WBC population. Lymphocytes continuously migrate from the bloodstream, through peripheral tissues, and back to the bloodstream. Circulating lymphocytes represent only a small fraction of all lymphocytes. At any moment, most of your body's lymphocytes are in other connective tissues and in organs of the lymphatic system.

The circulating blood contains three functional classes of lymphocytes. They cannot be distinguished with a light microscope:

1. **T cells** are responsible for *cell-mediated immunity*, a specific defense mechanism against invading foreign cells, and for the coordination of the immune response. T cells either enter peripheral tissues and attack foreign cells directly, or control the activities of other lymphocytes.

2. **B cells** are responsible for *humoral immunity*, a specific defense mechanism that involves the production of antibodies. These antibodies are distributed by blood, lymph, and interstitial fluid and are capable of attacking foreign antigens throughout the body. Activated B cells differentiate into **plasma cells**, which are specialized to synthesize and secrete antibodies. The T cells responsible for cellular immunity must migrate to their targets, but the antibodies produced by plasma cells in one location can destroy antigens almost anywhere in the body.

3. **Natural killer (NK) cells** carry out *immune surveillance*—the detection and subsequent destruction of abnormal cells. NK cells, sometimes known as *large granular lymphocytes*, are important in preventing cancer.

The Differential Count and Changes in WBC Profiles

A variety of conditions, including infection, inflammation, and allergic reactions, cause characteristic changes in circulating populations of WBCs. By examining a stained blood smear, we can obtain a **differential count** of the WBC population. The values reported indicate the number of each type of cell in a sample of 100 WBCs. The normal range of abundance for each type of WBC is shown in Table 19–3.

The term **leukopenia** (loo-kō-PĒ-nē-uh; *penia*, poverty) indicates inadequate numbers of WBCs. **Leukocytosis** (loo-kō-si-TŌ-sis) refers to excessive numbers of WBCs. A modest leukocytosis is normal during an infection. Extreme leukocytosis (100,000/μL or more) generally indicates the presence of some form of **leukemia** (loo-KĒ-mē-uh), a cancer of the WBCs. The endings -*penia* and -*osis* can also indicate low or high

numbers of specific types of WBCs. For example, *lymphopenia* means too few lymphocytes, and *lymphocytosis* means too many.

WBC Production

Stem cells that produce WBCs originate in the red bone marrow, with the divisions of hemocytoblasts (**Figure 19–10**). As we have noted, hemocytoblast divisions produce myeloid stem cells and lymphoid stem cells. Myeloid stem cells divide to create **progenitor cells**, which give rise to all the formed elements except lymphocytes. One type of progenitor cell produces

Table 19–3	Formed Elements of the Blood			
Cell	**Abundance (average number per μL)**	**Appearance in a Stained Blood Smear**	**Functions**	**Remarks**
RED BLOOD CELLS	5.2 million (range: 4.2–6.3 million)	Flattened, circular cell; no nucleus, mitochondria, or ribosomes; red	Transport oxygen from lungs to tissues and carbon dioxide from tissues to lungs	Remain in bloodstream; 120-day life expectancy; amino acids and iron recycled; produced in red bone marrow
WHITE BLOOD CELLS	7000 (range: 5000–10,000)			
Neutrophils	4150 (range: 1800–7300) Differential count: 50–70%	Round cell; nucleus lobed and may resemble a string of beads; cytoplasm contains large, pale inclusions	Phagocytic: Engulf pathogens or debris in tissues, release cytotoxic enzymes and chemicals	Move into tissues after several hours; may survive minutes to days, depending on tissue activity; produced in red bone marrow
Eosinophils	165 (range: 0–700) Differential count: 2–4%	Round cell; nucleus generally in two lobes; cytoplasm contains large granules that generally stain bright red	Phagocytic: Engulf antibody-labeled materials, release cytotoxic enzymes, reduce inflammation; increase in allergic and parasitic situations	Move into tissues after several hours; survive minutes to days, depending on tissue activity; produced in red bone marrow
Basophils	44 (range: 0–150) Differential count: <1%	Round cell; nucleus generally cannot be seen through dense, blue-stained granules in cytoplasm	Enter damaged tissues and release histamine and other chemicals that promote inflammation	Survival time unknown; assist mast cells of tissues in producing inflammation; produced in red bone marrow
Monocytes	456 (range: 200–950) Differential count: 2–8%	Very large cell; kidney bean–shaped nucleus; abundant pale cytoplasm	Enter tissues to become macrophages; engulf pathogens or debris	Move into tissues after 1–2 days; survive for months or longer; produced primarily in red bone marrow
Lymphocytes	2185 (range: 1500–4000) Differential count: 20–40%	Generally round cell, slightly larger than RBC; round nucleus; very little cytoplasm	Cells of lymphatic system, providing defense against specific pathogens or toxins	Survive for months to decades; circulate from blood to tissues and back; produced in red bone marrow and lymphatic tissues
PLATELETS	350,000 (range: 150,000–500,000)	Round to spindle-shaped cytoplasmic fragment; contain enzymes, proenzymes, actin, and myosin; no nucleus	Hemostasis: Clump together and stick to vessel wall (platelet phase); activate intrinsic pathway of coagulation phase	Remain in bloodstream or in vascular organs; remain intact for 7–12 days; produced by megakaryocytes in red bone marrow

Figure 19–10   **The Origins and Differentiation of Formed Elements.** Hemocytoblast divisions give rise to myeloid stem cells or lymphoid stem cells. Lymphoid stem cells produce the various lymphocytes. Myeloid stem cells produce progenitor cells that divide to produce the other classes of formed elements. The targets of EPO and the four colony-stimulating factors (CSFs) are indicated.

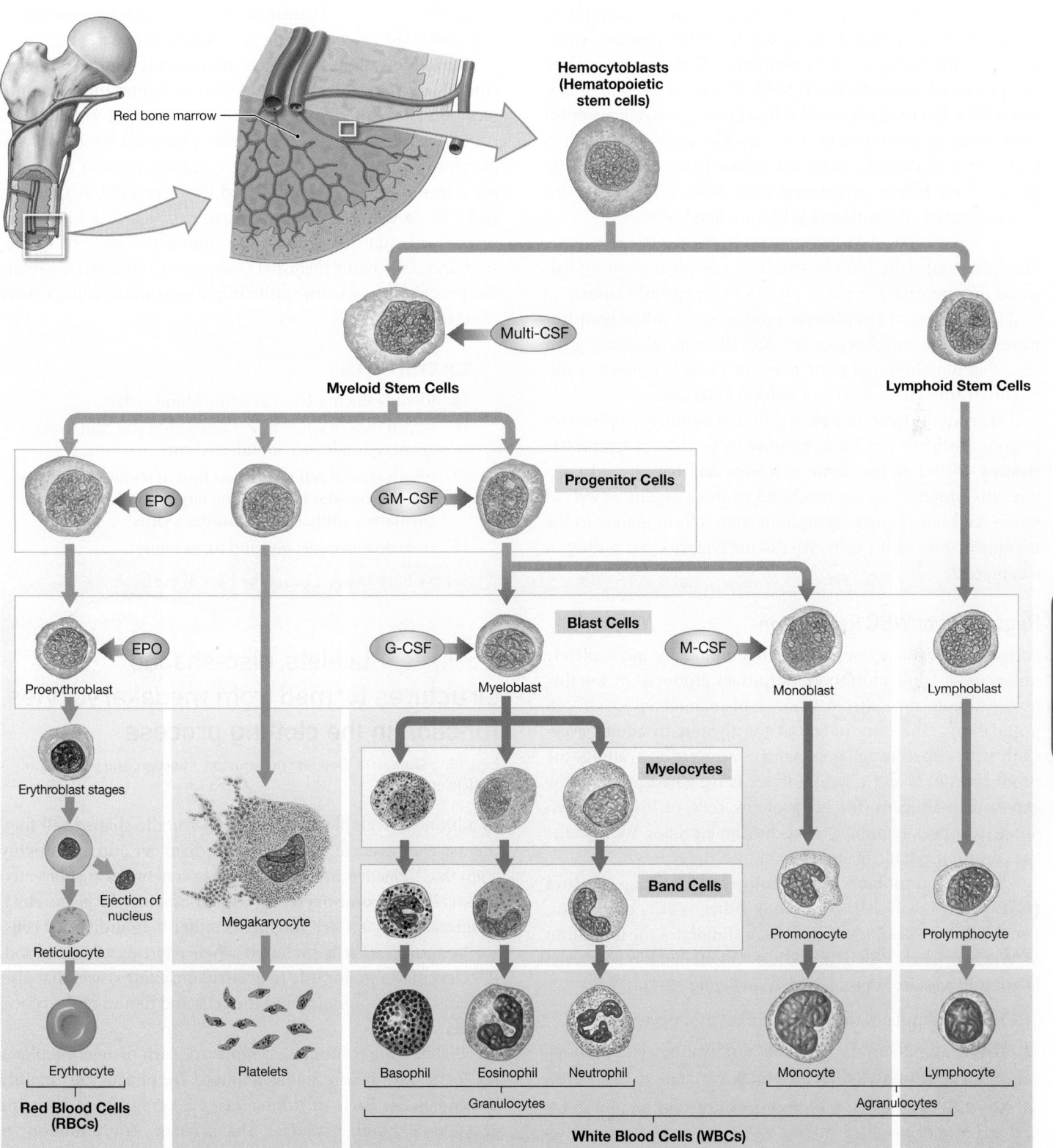

19

daughter cells that mature into RBCs. A second type produces cells that manufacture platelets. A third type produces daughter cells that develop into neutrophils, eosinophils, basophils, and monocytes.

Granulocytes (basophils, eosinophils, and neutrophils) complete their development in the red bone marrow. These WBCs go through a series of maturational stages, proceeding from *blast cells* to *myelocytes* to *band cells* before becoming mature WBCs. For example, a cell differentiating into a neutrophil goes from a myeloblast to a *neutrophilic myelocyte* and then becomes a *neutrophilic band cell*. Some band cells enter the bloodstream before completing their maturation. Normally, 3–5 percent of all circulating WBCs are band cells.

Monocytes begin to differentiate in the red bone marrow. Then they enter the bloodstream and complete development when they become free macrophages in peripheral tissues.

The process of lymphocyte production is called **lympho-poiesis**. Some lymphocytes are derived from lymphoid stem cells that remain in red bone marrow. These lymphocytes differentiate into either B cells or natural killer cells.

Many of the lymphoid stem cells that produce lymphocytes migrate from the red bone marrow to peripheral **lymphatic tissues**, including the thymus, spleen, and lymph nodes. As a result, lymphocytes are produced in these organs as well as in the red bone marrow. Lymphoid stem cells migrating to the thymus mature into T cells. We discuss lymphocytes further in Chapter 22.

Regulation of WBC Production

Factors that regulate lymphocyte maturation are incompletely understood. Until adulthood, hormones produced by the thymus promote the differentiation and maintenance of T cell populations. The importance of the thymus in adults, especially with respect to aging, remains controversial. In adults, the production of B and T lymphocytes is regulated primarily by exposure to antigens (foreign proteins, cells, or toxins). When antigens appear, lymphocyte production escalates. We describe the control mechanisms in Chapter 22.

Several hormones, called **colony-stimulating factors (CSFs)**, are involved in regulating other WBC populations. Four CSFs have been identified. Each stimulates the formation of WBCs or both WBCs and RBCs. The designation for each factor indicates its target, as shown in **Figure 19–10**:

1. **M-CSF** stimulates the production of monocytes.

2. **G-CSF** stimulates the production of granulocytes (neutrophils, eosinophils, and basophils).

3. **GM-CSF** stimulates the production of both granulocytes and monocytes.

4. **Multi-CSF** accelerates the production of granulocytes, monocytes, platelets, and RBCs.

Chemical communication between lymphocytes and other WBCs helps to coordinate the immune response. For example, active macrophages release chemicals that make lymphocytes more sensitive to antigens and also accelerate the development of specific immunity. In turn, active lymphocytes release multi-CSF and GM-CSF, reinforcing nonspecific defenses.

Immune system hormones are currently under intensive study because of their potential clinical importance. The molecular structures of many of the stimulating factors have been identified, and several can be produced by genetic engineering. The U.S. Food and Drug Administration approved the administration of synthesized forms of EPO, G-CSF, and GM-CSF to stimulate the production of specific blood cell lines. For instance, a genetically engineered form of G-CSF, sold under the name *filgrastim* (*Neupogen*), is used to stimulate the production of neutrophils in patients undergoing cancer chemotherapy.

CHECKPOINT

15. Identify the five types of white blood cells.

16. Which type of white blood cell would you find in the greatest numbers in an infected cut?

17. Which type of cell would you find in elevated numbers in a person who is producing large amounts of circulating antibodies to combat a virus?

18. How do basophils respond to an injury?

See the blue Answers tab at the back of the book.

19-6 ▌ Platelets, disc-shaped structures formed from megakaryocytes, function in the clotting process

Learning Outcome Describe the structure, function, and production of platelets.

In a blood smear, platelets appear as spindle-shaped cell fragments. They average about 4 μm in diameter and are roughly 1 μm thick. Platelets in nonmammalian vertebrates are nucleated cells called **thrombocytes** (THROM-bō-sīts; *thrombos*, clot). In humans they are cell fragments rather than individual cells, so the term *platelet* is preferred when referring to our blood. Platelets play a major role in a vascular *clotting system* that also includes plasma proteins and the cells and tissues of the blood vessels.

Platelets are continuously replaced. Each platelet circulates for 9–12 days before being removed by phagocytes, mainly in the spleen. Each microliter of circulating blood contains 150,000–500,000 platelets. The average concentration is 350,000/μL. About one-third of the platelets in the body at any moment are in the spleen and other vascular organs, rather than in the bloodstream. These reserves are mobilized during a circulatory crisis, such as severe bleeding.

An abnormally low platelet count (80,000/μL or less) is known as **thrombocytopenia** (throm-bō-sī-tō-PĒ-nē-uh). Thrombocytopenia generally indicates excessive platelet destruction or inadequate platelet production. Signs include bleeding along the digestive tract, within the skin, and occasionally inside the CNS.

In **thrombocytosis** (throm-bō-sī-TŌ-sis), platelet counts can exceed 1,000,000/μL. Thrombocytosis usually results from accelerated platelet formation in response to infection, inflammation, or cancer.

Platelet Functions

The functions of platelets include:

- *Releasing Chemicals Important to the Clotting Process.* By releasing enzymes and other factors at the appropriate times, platelets help initiate and control the clotting process.

- *Forming a Temporary Patch in the Walls of Damaged Blood Vessels.* Platelets clump together at an injury site, forming a *platelet plug*, which can slow blood loss while clotting takes place.

- *Reducing the Size of a Break in the Vessel Wall.* Platelets contain filaments of actin and myosin. After a blood clot has formed, platelet filaments contract to shrink the clot and reduce the size of the break in the vessel wall.

Platelet Production

Platelet production, or **thrombocytopoiesis**, takes place in the red bone marrow. Normal red bone marrow contains **megakaryocytes** (meg-a-KAR-ē-ō-sīts; *mega-*, big + *karyon*, nucleus + *-cyte*, cell), enormous cells (up to 160 μm in diameter) with large nuclei (**Figure 19–10**). During their development and growth, megakaryocytes manufacture structural proteins, enzymes, and membranes. They then begin shedding cytoplasm in small membrane-enclosed packets. These packets are the platelets that enter the bloodstream. A mature megakaryocyte gradually loses all of its cytoplasm, producing about 4000 platelets before phagocytes engulf its nucleus for breakdown and recycling.

Three substances influence the rate of megakaryocyte activity and platelet formation. They are: (1) *thrombopoietin* (TPO), or *thrombocyte-stimulating factor*, a peptide hormone produced in the kidneys (and perhaps other sites) that accelerates platelet formation and stimulates the production of megakaryocytes; (2) *interleukin-6* (IL-6), a hormone that stimulates platelet formation; and (3) multi-CSF, which stimulates platelet production by promoting the formation and growth of megakaryocytes.

CHECKPOINT

19. Define thrombocytopoiesis.
20. Explain the difference between platelets and thrombocytes.
21. List the three primary functions of platelets.

See the blue Answers tab at the back of the book.

19-7 | Hemostasis involves vascular spasm, platelet plug formation, and blood coagulation

Learning Outcome Discuss the mechanisms that control blood loss after an injury, and describe the reaction sequences responsible for blood clotting.

The process of **hemostasis** (*haima*, blood + *stasis*, halt), the stopping of bleeding, halts the loss of blood through the walls of damaged vessels. At the same time, it establishes a framework for tissue repairs. Hemostasis has three phases: the *vascular phase*, the *platelet phase*, and the *coagulation phase*. The boundaries of these phases are somewhat arbitrary. In reality, hemostasis is a complex *cascade*, or chain reaction, in which many things happen at once, and all of them interact to some degree.

The Vascular Phase

Cutting the wall of a blood vessel triggers a contraction in the smooth muscle fibers of the vessel wall (**Figure 19–11 ❶**). This local contraction of the vessel is a **vascular spasm**, which decreases the diameter of the vessel at the site of injury. Such a constriction can slow or even stop the loss of blood through the wall of a small vessel. The vascular spasm lasts about 30 minutes, a period called the **vascular phase** of hemostasis.

During the vascular phase, changes take place in the endothelium of the vessel at the injury site:

- *The Endothelial Cells Contract and Expose the Underlying Basement Membrane to the Bloodstream.*

- *The Endothelial Cells Begin Releasing Chemical Factors and Local Hormones.* We will discuss several of these factors, including *ADP, tissue factor*, and *prostacyclin*, in later sections. Endothelial cells also release peptide hormones called **endothelins**. These local hormones (1) stimulate smooth muscle contraction and promote vascular spasms, and (2) stimulate the division of endothelial cells, smooth muscle cells, and fibroblasts to accelerate the repair process.

- *The Endothelial Plasma Membranes Become "Sticky."* A tear in the wall of a small artery or vein may be partially sealed off by the attachment of endothelial cells on either side of the break. In small capillaries, endothelial cells on opposite sides of the vessel may stick together and prevent

19

Figure 19–11   **The Vascular, Platelet, and Coagulation Phases of Hemostasis and Clot Retraction.**

1 Vascular Phase

The vascular phase of hemostasis lasts for about 30 minutes after the injury occurs. The endothelial cells contract and release endothelins, which stimulate smooth muscle contraction and endothelial division. The endothelial cells become "sticky" and adhere to platelets and each other.

Knife blade

Blood vessel injury

Vascular spasm

2 Platelet Phase

The platelet phase of hemostasis begins with the attachment of platelets to sticky endothelial surfaces, to the basement membrane, to exposed collagen fibers, and to each other. As they become activated, platelets release a variety of chemicals that promote aggregation, vascular spasm, clotting, and vessel repair.

Release of chemicals (ADP, PDGF, Ca^{2+}, platelet factors)

Plasma in vessel lumen

Platelet aggregation

Platelet adhesion to damaged vessel

Endothelium

Basement membrane

Vessel wall

Platelet plug may form

Contracted smooth muscle cells

Interstitial fluid

Cut edge of vessel wall

19

blood flow along the damaged vessel. The stickiness is also important because it helps platelets to attach as the platelet phase gets under way.

The Platelet Phase

The attachment of platelets to sticky endothelial surfaces, to the basement membrane, and to exposed collagen fibers marks the start of the **platelet phase** of hemostasis (**Figure 19–11 2**). The attachment of platelets to exposed surfaces is called **platelet adhesion**. As more and more platelets arrive, they begin sticking to one another as well. This process, called **platelet aggregation**, forms a **platelet plug** that may close the break in the vessel wall if the damage is not severe or the vessel is relatively small. Platelet aggregation begins within 15 seconds after an injury occurs.

As they arrive at the injury site, the platelets become activated. The first sign of activation is that they become more spherical and develop cytoplasmic processes that extend toward nearby platelets. At this time, the platelets begin releasing a wide variety of compounds. These compounds include (1) *adenosine diphosphate (ADP)*, which stimulates platelet aggregation and secretion; (2) *thromboxane A$_2$* and *serotonin*, which stimulate vascular spasms; (3) *clotting factors*, proteins that play a role in blood clotting; (4) *platelet-derived growth factor (PDGF)*, a peptide that promotes vessel repair; and (5) calcium ions, which are required for platelet aggregation and in several steps in the clotting process.

The platelet phase proceeds quickly, as ADP, thromboxane, and calcium ions released from each arriving platelet stimulate further aggregation. This positive feedback loop ultimately produces a platelet plug that will be reinforced as clotting occurs.

3 **Coagulation Phase**

Coagulation, or blood clotting, involves a complex sequence of steps leading to the conversion of circulating fibrinogen (a soluble protein) into fibrin (an insoluble protein). As the fibrin network grows, blood cells and additional platelets are trapped in the fibrous tangle, forming a **blood clot** that seals off the damaged portion of the vessel.

4 **Clot Retraction**

Once the fibrin meshwork has formed, platelets and red blood cells stick to the fibrin strands. The platelets then contract, and the entire clot begins to undergo clot retraction, a process that continues over 30–60 minutes.

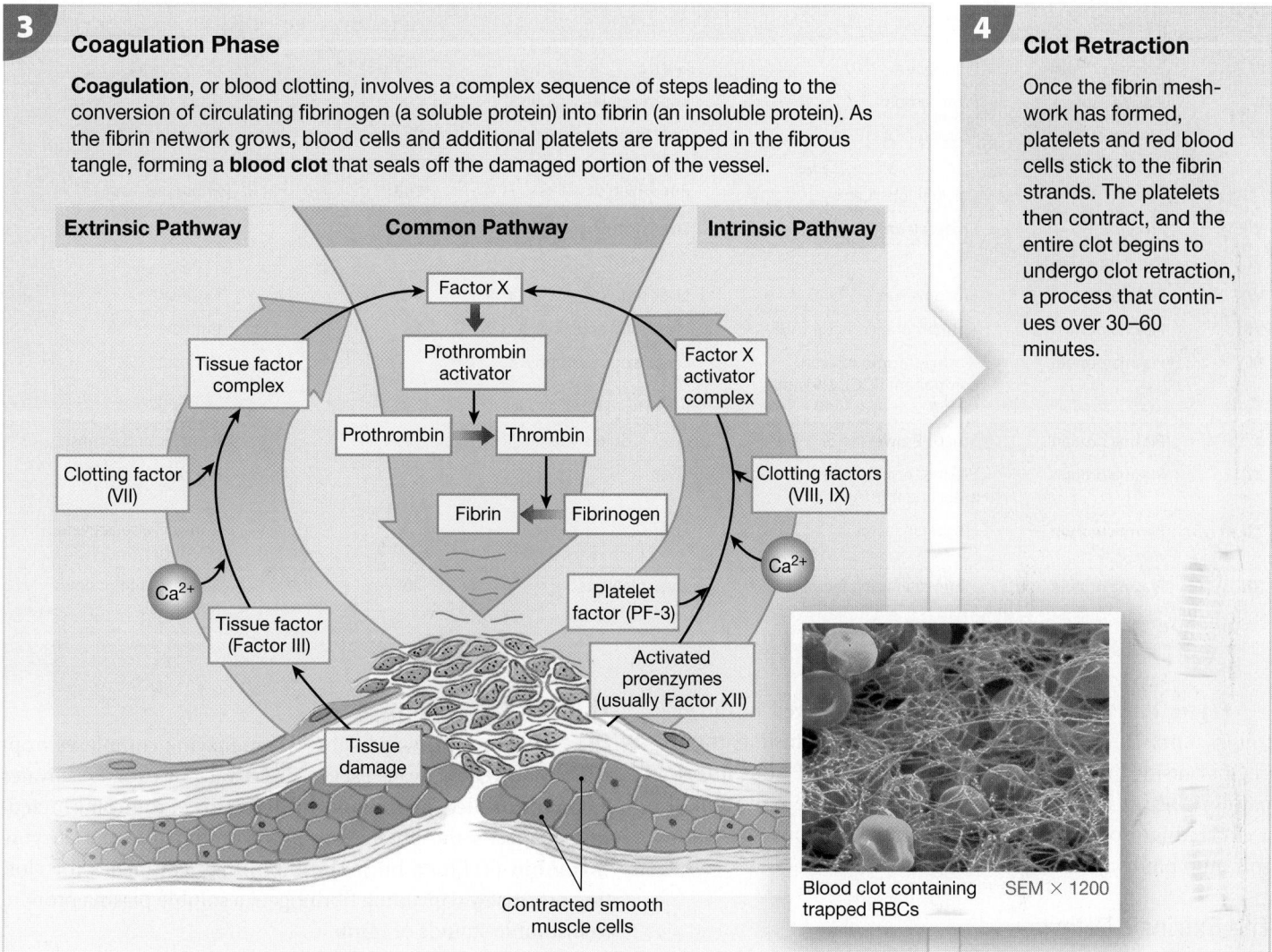

Blood clot containing trapped RBCs   SEM × 1200

However, platelet aggregation must be controlled and restricted to the injury site.

Several key factors limit the growth of the platelet plug: (1) **prostacyclin**, a prostaglandin that inhibits platelet aggregation and is released by endothelial cells; (2) inhibitory compounds released by WBCs entering the area; (3) circulating plasma enzymes that break down ADP near the plug; (4) compounds that, when abundant, inhibit plug formation (for example, serotonin, which at high concentrations blocks the action of ADP); and (5) the development of a blood clot, which reinforces the platelet plug, but isolates it from the general circulation.

The Coagulation Phase

The vascular and platelet phases begin within a few seconds after the injury. The **coagulation** (cō-ag-ū-LĀ-shun) **phase**, or blood clotting, does not start until 30 seconds or more after the vessel has been damaged. (Figure 19–11 **3**) shows the sequence of steps involved in the formation and structure of a blood clot.

Clotting Factors

Normal blood clotting depends on the presence of **clotting factors**, or **procoagulants**, in the plasma. Important clotting factors include Ca^{2+} and 11 different proteins (Table 19–4).

Many of the proteins are **proenzymes**, or inactive enzymes. When converted to active enzymes, these proteins direct essential reactions in the clotting response. During the coagulation phase, enzymes and proenzymes interact successively. The activation of one proenzyme commonly creates an enzyme that activates a second proenzyme, and so on in a cascade.

Table 19–4	Clotting Factors				
Factor	Structure	Name	Source	Concentration in Plasma (μg/mL)	Pathway
I	Plasma protein	Fibrinogen	Liver	2500–3500	Common
II	Plasma protein	Prothrombin	Liver, requires vitamin K	100	Common
III	Glycoprotein	Tissue factor (TF)	Damaged tissue, activated platelets	0	Extrinsic
IV	Ion	Calcium ions (Ca^{2+})	Bone, diet, platelets	100	Entire process
V	Protein	Proaccelerin	Liver, platelets	10	Extrinsic and intrinsic
VI	(No longer used)				
VII	Plasma protein	Proconvertin	Liver, requires vitamin K	0.5	Extrinsic
VIII	Plasma protein	Antihemophilic factor (AHF)	Liver, lung endothelial cells	15	Intrinsic
IX	Plasma protein	Plasma thromboplastin component (PTC), Christmas factor	Liver, requires vitamin K	3	Intrinsic
X	Plasma protein	Stuart–Prower factor	Liver, requires vitamin K	10	Extrinsic and intrinsic
XI	Plasma protein	Plasma thromboplastin antecedent (PTA)	Liver	<5	Intrinsic
XII	Plasma protein	Hageman factor	Liver	<5	Intrinsic; also activates plasmin
XIII	Plasma protein	Fibrin-stabilizing factor (FSF)	Liver, platelets	20	Stabilizes fibrin, slows fibrinolysis

Figure 19–11 ③ shows the cascades involved in the *extrinsic, intrinsic,* and *common pathways.* The extrinsic pathway begins outside the bloodstream, in the vessel wall. The intrinsic pathway begins inside the bloodstream, with the activation of a circulating proenzyme. These two pathways converge at the common pathway.

The Extrinsic Pathway

The **extrinsic pathway** begins when damaged endothelial cells or peripheral tissues release **Factor III**, also known as **tissue factor (TF)**. The greater the damage, the more tissue factor is released and the faster clotting takes place. Tissue factor then combines with Ca^{2+} and another clotting factor (Factor VII) to form an enzyme complex capable of activating Factor X, the first step in the common pathway.

The Intrinsic Pathway

The **intrinsic pathway** begins with the activation of proenzymes (usually Factor XII) exposed to collagen fibers at the injury site (or to a glass surface of a slide or collection tube). This pathway proceeds with the assistance of **PF-3**, a platelet factor released by aggregating platelets. Platelets also release a variety of other factors that accelerate the reactions of the intrinsic pathway. After a series of linked reactions, activated Factors VIII and IX combine to form an enzyme complex capable of activating Factor X.

The Common Pathway

The **common pathway** begins when enzyme complexes from either pathway activate Factor X. Activated Factor X activates a complex called **prothrombin activator**. Prothrombin activator converts the proenzyme prothrombin into the enzyme **thrombin** (THROM-bin). Thrombin then completes the clotting process by converting fibrinogen, a soluble plasma protein, to insoluble strands of fibrin.

Interactions among the Pathways

When a blood vessel is damaged, both the extrinsic and the intrinsic pathways respond. The extrinsic pathway is shorter and faster than the intrinsic pathway, and it is usually the first to initiate clotting. In essence, the extrinsic pathway produces a small amount of thrombin very quickly. This quick patch is reinforced by the intrinsic pathway, which later produces more thrombin.

The time required to complete clot formation varies with the site and the nature of the injury. In tests of the clotting system, blood held in fine glass tubes normally clots in 8–18 minutes (the *coagulation time*). A small puncture wound typically stops bleeding in 1–4 minutes (the *bleeding time*).

Feedback Control of Blood Clotting

The activity of the common pathway stimulates both the intrinsic and extrinsic pathways. This positive feedback loop accelerates the clotting process, and speed can be very important in reducing blood loss after a severe injury. Thrombin generated in the common pathway stimulates blood clotting by (1) stimulating the formation of tissue factor (TF) and (2) stimulating the release of PF-3 by platelets (Figure 19–11 ❸).

How is clotting controlled or stopped? Blood clotting is restricted by substances that either deactivate or remove clotting factors and other stimulatory agents from the blood. Examples include the following:

- Normal plasma contains several **anticoagulants**— enzymes that inhibit clotting. One, **antithrombin-III**, inhibits several clotting factors, including thrombin.

- **Heparin**, a compound released by basophils and mast cells, is a cofactor that accelerates the activation of antithrombin-III. Heparin is also used clinically to impede or prevent clotting.

- **Thrombomodulin** is released by endothelial cells. This protein binds to thrombin and converts it to an enzyme that activates protein C. **Protein C** is a plasma protein that inactivates several clotting factors and stimulates the formation of *plasmin*, an enzyme that gradually breaks down fibrin strands.

- Prostacyclin released during the platelet phase inhibits platelet aggregation and opposes the stimulatory action of thrombin, ADP, and other factors.

- Other plasma proteins with anticoagulant properties include *alpha-2-macroglobulin*, which inhibits thrombin, and C_1 *inactivator*, which inhibits several clotting factors involved in the intrinsic pathway.

The clotting process involves a complex chain of events, and disorders that affect any individual clotting factor can disrupt the entire process. For this reason, managing many clinical conditions involves controlling or manipulating the clotting response. Aspirin, for example, is a common drug that interferes with blood clotting and prolongs bleeding time. It acts by inhibiting platelets from producing prostaglandins and thromboxane A_2, which stimulates new platelet activation and aggregation.

Calcium Ions, Vitamin K, and Blood Clotting

Calcium ions and **vitamin K** affect almost every aspect of the clotting process. All three pathways (intrinsic, extrinsic, and common) require Ca^{2+}, so any disorder that lowers plasma Ca^{2+} concentrations will impair blood clotting.

Adequate amounts of vitamin K must be present for the liver to synthesize four of the clotting factors, including prothrombin. Vitamin K is a fat-soluble vitamin, present in green vegetables, grain, and organ meats, that is absorbed with dietary lipids. We obtain roughly half of our daily requirement from the diet. Bacteria in the large intestine manufacture the other half. A vitamin K deficiency may arise from a diet inadequate in fats or in vitamin K, or a disorder that affects fat digestion and absorption (such as problems with bile production), or prolonged use of antibiotics that kill normal intestinal bacteria. Vitamin K deficiency will cause the eventual breakdown of the common pathway due to a lack of clotting factors. Ultimately, the deficiency can deactivate the entire clotting system.

Clot Retraction

Clot retraction, or *syneresis* (si-NER-e-sis; "a drawing together"), pulls the torn edges of the vessel closer together. This process reduces residual bleeding and stabilizes the injury site. The process also reduces the size of the damaged area, making it easier for fibroblasts, smooth muscle cells, and endothelial cells to complete repairs (Figure 19–11 ❹).

Fibrinolysis

As the repairs proceed, the clot gradually dissolves in a process called **fibrinolysis** (fi-bri-NOL-i-sis). The process begins when two enzymes activate the proenzyme **plasminogen**. These enzymes are thrombin, produced by the common pathway, and **tissue plasminogen activator** (t-PA), released by damaged tissues at the site of injury. The activation of plasminogen produces the enzyme **plasmin** (PLAZ-min), which begins digesting the fibrin strands and eroding the clot.

To perform its vital functions, blood must be kept in motion. On average, an RBC completes two circuits around the cardiovascular system each minute. Blood begins to circulate in the third week of embryonic development and continues throughout life. If the blood supply is cut off, dependent tissues may die in a matter of minutes. In Chapter 20, we examine the structure and function of the heart—the pump that maintains this vital blood flow.

19

CHECKPOINT

22. **A sample of red bone marrow has unusually few megakaryocytes. What body process would you expect to be impaired as a result?**

23. **Vitamin K is fat soluble, and some dietary fat is required for its absorption. How could a diet of fruit juice and water have an effect on blood clotting?**

24. **Unless chemically treated, whole blood will coagulate in a test tube. This clotting process begins when Factor XII becomes activated. Which clotting pathway is involved in this process?**

See the blue Answers tab at the back of the book.

Related Clinical Terms

arterial stick: The taking of a blood sample from an artery rather than a vein. It is usually more painful due to arteries being deeper, having more nerves, and having thicker walls.

blood bank: Place where blood is collected, typed, separated into components, stored, and prepared for transfusion to recipients.

bone marrow biopsy: The removal of a small piece of bone marrow for either laboratory analysis, to diagnose and stage some forms of cancer, to diagnose other blood disorders, to find the source of unexplained fever, or to diagnose fibrosis of bone marrow or myeloma, a tumor composed of cells normally found in the bone marrow.

disseminated intravascular coagulation: A serious disorder in which the proteins that control blood clotting become abnormally active, causing small blood clots to form, which can prevent blood from reaching vital organs.

dyscrasia: An abnormal condition, especially of the blood.

ecchymosis: Skin discoloration caused by the escape of blood into tissues from ruptured blood vessels.

embolism: A condition in which a drifting blood clot (an embolus) becomes stuck in a blood vessel, blocking circulation to the area downstream.

hematology: The science concerned with the medical study of blood and blood-producing organs.

hemochromatosis: A rare metabolic disorder wherein the skin has a bronze coloration; accompanied by cirrhosis and severe diabetes mellitus; caused by the deposit of the protein-iron complex hemosiderin in tissues.

hemophilia: Inherited disorders characterized by the inadequate production of clotting factors.

hemopoietic growth factor: A group of proteins that cause blood cells to grow and mature.

hemosiderosis: An increase in tissue iron stores without any associated damage.

hypervolemic: Having an excessive blood volume.

hypovolemic: Having a low blood volume.

iron overload: Pathology in which iron accumulates in the tissues; characterized by bronzed skin, enlarged liver, diabetes mellitus, and abnormalities of the pancreas.

myeloproliferative disorder: A group of slow-growing blood cancers, including chronic myelogenous leukemia, characterized by large numbers of abnormal RBCs, WBCs, or platelets growing and spreading in the bone marrow and the peripheral blood.

normovolemic: Referring to a normal blood volume.

phlebotomist: Medical technician who extracts blood by venipuncture for treatment or for laboratory analysis.

plaque: An abnormal accumulation of large quantities of lipids within a blood vessel wall.

plasmapheresis: A procedure consisting of the removal of blood from a person, separating the blood cells from plasma, and returning these blood cells to the person's circulation, diluted with fresh plasma or a substitute. Used to treat autoimmune disorders.

Schilling test: The test to determine whether the body absorbs vitamin B_{12} normally.

septicemia: Systemic toxic illness due to bacterial invasion of the bloodstream from a local infection. Signs and symptoms include chills, fever, and exhaustion. The disorder is treated with massive doses of antibiotics. Also known as blood poisoning.

thrombolytic: An agent that causes the breakup of a thrombus (clot).

thrombus: A blood clot attached to the luminal (inner) surface of a blood vessel.

Chapter Review

Study Outline

MasteringA&P®

Access more chapter study tools online in the MasteringA&P Study Area:

- Chapter Quizzes, Chapter Practice Test, Art-labeling Activities, Animations, MP3 Tutor Sessions, and Clinical Case Studies
- Practice Anatomy Lab PAL
- Interactive Physiology iP
- A&P Flix A&PFlix
- PhysioEx PhysioEx

An Introduction to Blood and the Cardiovascular System
p. 679

1. The **cardiovascular system** enables the rapid transport of nutrients, respiratory gases, waste products, and cells within the body.

19-1 ▌ Blood has several important functions and unique physical characteristics p. 679

2. **Blood** is a specialized fluid connective tissue. Its functions include (1) transporting dissolved gases, nutrients, hormones, and metabolic wastes; (2) regulating the pH and ion composition of interstitial fluids; (3) restricting fluid losses at injury sites; (4) defending the body against toxins and pathogens; and (5) regulating body temperature by absorbing and redistributing heat.

3. Blood contains **plasma** and **formed elements—red blood cells (RBCs), white blood cells (WBCs),** and **platelets**. The plasma and formed elements make up **whole blood**, which can be **fractionated** for analytical or clinical purposes. (*Spotlight Figure 19–1*)

4. **Hemopoiesis**, or **hematopoiesis**, is the process of blood cell formation. Myeloid and lymphoid stem cells divide to form all types of blood cells. (*Spotlight Figure 19-1*)

5. Whole blood from any region of the body has roughly the same temperature, viscosity, and pH.

19-2 ▌ Plasma, the fluid portion of blood, contains significant quantities of plasma proteins p. 682

6. Plasma accounts for 55 percent of the volume of blood; roughly 92 percent of plasma is water. *(Spotlight Figure 19–1)*

7. Plasma differs from interstitial fluid in terms of its oxygen and carbon dioxide levels and the concentrations and types of dissolved proteins.

8. The three major types of plasma proteins are *albumins*, *globulins*, and *fibrinogen*.

9. **Albumins** make up about 60 percent of plasma proteins. **Globulins** constitute roughly 35 percent of plasma proteins; they include **antibodies (immunoglobulins)**, which attack foreign proteins and pathogens, and **transport globulins**, which bind ions, hormones, and other compounds. **Fibrinogen** molecules are converted to **fibrin** in the clotting process. **Serum** is plasma without fibrinogen.

10. The liver synthesizes and releases more than 90 percent of the plasma proteins.

19-3 ▌ Red blood cells, formed by erythropoiesis, contain hemoglobin that can be recycled p. 683

11. Red blood cells make up slightly less than half of the blood volume and 99.9 percent of the formed elements. The **hematocrit** value indicates the percentage of formed elements within whole blood. It is commonly reported as the *volume of packed red cells (VPRC)* or the *packed cell volume (PCV)*. *(Spotlight Figure 19–1; Table 19–1)*

12. Each RBC is a biconcave disc, providing a large surface-to-volume ratio. This shape allows RBCs to stack, bend, and flex. *(Figure 19–2)*

13. Red blood cells lack most organelles, including mitochondria and nuclei, retaining only the cytoskeleton. They typically degenerate after about 120 days in the bloodstream.

14. Molecules of **hemoglobin (Hb; Hgb)** make up more than 95 percent of the proteins in RBCs. Hemoglobin is a globular protein formed from two pairs of polypeptide subunits. Each subunit contains a single molecule of **heme**, which also has an iron atom that can reversibly bind an oxygen molecule. Phagocytes recycle damaged or dead RBCs. *(Figures 19–3, 19–4)*

15. Damaged RBCs are continuously replaced at a rate of approximately 3 million new RBCs entering the bloodstream per second. They are replaced before they **hemolyze**.

16. The components of hemoglobin are individually recycled. The heme is stripped of its iron and converted to **biliverdin**, which is converted to **bilirubin**. If bile ducts are blocked, bilirubin builds up in skin and eyes, resulting in **jaundice**. *(Figure 19–4)*

17. Iron is recycled by being stored in phagocytic cells or transported through the bloodstream, bound to **transferrin**.

18. **Erythropoiesis**, the formation of red blood cells, occurs only in *red bone marrow* **(myeloid tissue)**. The process speeds up under stimulation by erythropoietin (EPO or **erythropoiesis-stimulating hormone)**. Stages in RBC development include **erythroblasts** and **reticulocytes**. *(Figure 19–5)*

19-4 ▌ The ABO blood types and Rh system are based on antigen–antibody responses p. 690

19. **Blood type** is determined by the presence or absence of specific **surface antigens** (*agglutinogens*) in the RBC plasma membranes: antigens **A**, **B**, and **Rh (D)**. Antibodies (*agglutinins*) in the plasma will react with RBCs that have different surface antigens. When an antibody meets its specific surface antigen, a **cross-reaction** results. *(Figures 19–6 to Spotlight Figure 19–8; Table 19–2)*

19-5 ▌ The various types of white blood cells contribute to the body's defenses p. 693

20. White blood cells, or **leukocytes**, have nuclei and other organelles. They defend the body against pathogens and remove toxins, wastes, and abnormal or damaged cells.

21. White blood cells are capable of *margination*, amoeboid movement, and **positive chemotaxis**. Some WBCs are also capable of *phagocytosis*.

22. *Granular leukocytes* (*granulocytes*) are subdivided into **neutrophils**, **eosinophils**, and **basophils**. Fifty to 70 percent of circulating WBCs are neutrophils, which are highly mobile phagocytes. The much less common eosinophils are phagocytes attracted to foreign compounds that have reacted with circulating antibodies. The fairly rare basophils migrate to damaged tissues and release *histamine* and *heparin*, aiding the inflammatory response. *(Figure 19–9)*

23. *Agranular leukocytes* (*agranulocytes*) include **monocytes** and **lymphocytes**. Monocytes that migrate into peripheral tissues become tissue macrophages. Lymphocytes, the primary cells of the lymphatic system, include **T cells** (which enter peripheral tissues and attack foreign cells directly, or affect the activities of other lymphocytes); **B cells** (which produce antibodies); and **natural killer (NK) cells** (which destroy abnormal cells). *(Figure 19–9; Table 19–3)*

24. A **differential count** of the WBC population can indicate a variety of disorders. **Leukemia** is indicated by extreme **leukocytosis**—that is, excessive numbers of WBCs. *(Table 19–3)*

25. Granulocytes and monocytes are produced by myeloid stem cells in the red bone marrow that divide to create **progenitor cells**. Lymphoid stem cells also originate in the red bone marrow, but many migrate to peripheral **lymphatic tissues**. *(Figure 19–10)*

26. Factors that regulate lymphocyte maturation are not completely understood. Several **colony-stimulating factors** (CSFs) are involved in regulating other WBC populations and in coordinating RBC and WBC production. *(Figure 19–10)*

19-6 ▌ Platelets, disc-shaped structures formed from megakaryocytes, function in the clotting process p. 700

27. Platelets are spindle-shaped cell fragments. They circulate for 9–12 days before being removed by phagocytes. *(Figure 19–11)*

28. The functions of platelets include (1) transporting and releasing chemicals important to the clotting process, (2) forming a temporary patch in the walls of damaged blood vessels, and (3) reducing the size of a break in the vessel wall.

29. During **thrombocytopoiesis**, **megakaryocytes** in the red bone marrow release packets of cytoplasm (platelets) into the circulating blood. The rate of platelet formation is stimulated by thrombopoietin or thrombocyte-stimulating factor, interleukin-6, and multi-CSF.

19-7 ▌ Hemostasis involves vascular spasm, platelet plug formation, and blood coagulation p. 701

30. **Hemostasis** halts the loss of blood through the walls of damaged vessels. It consists of three phases: the *vascular phase*, the *platelet phase*, and the *coagulation phase*.

31. The **vascular phase** is a period of local blood vessel constriction, or **vascular spasm**, at the injury site. *(Figure 19–11)*

19

32. The **platelet phase** follows as platelets are activated, aggregate at the site, and adhere to the damaged surfaces. *(Figure 19–11)*
33. The **coagulation phase** occurs as factors released by platelets and endothelial cells interact with **clotting factors** (through either the **extrinsic pathway**, the **intrinsic pathway**, or the **common pathway**) to form a **blood clot**. In this reaction sequence, soluble fibrinogen is converted to large, insoluble fibers of fibrin. *(Figure 19–11; Table 19–4)*
34. During **clot retraction**, platelets contract and pull the torn edges of the damaged vessel closer together. *(Figure 19–11)*
35. During **fibrinolysis**, the clot gradually dissolves through the action of **plasmin**, the activated form of circulating **plasminogen**.

Review Questions

See the blue Answers tab at the back of the book.

LEVEL 1 Reviewing Facts and Terms

1. Identify the five types of white blood cells in the following photographs,

(a)   (b)   (c)

(d)   (e)

(a) _____   (b) _____
(c) _____   (d) _____
(e) _____

2. Hemopoiesis is the production of more (a) red blood cells, (b) white blood cells, (c) platelets, (d) a, b, and c.
3. The major difference between plasma and interstitial fluids is the concentration of (a) respiratory gases, (b) proteins, (c) electrolytes, (d) water, (e) organic wastes, (f) a and b, (g) d and e.
4. Erythropoietin is (a) a hormone that stimulates monocyte production, (b) a hormone that stimulates red blood cell production, (c) a plasma protein that carries triglycerides, (d) a plasma protein that carries steroid hormones.
5. Serum is (a) the same as blood plasma, (b) plasma minus the formed elements, (c) plasma minus the proteins, (d) plasma minus fibrinogen, (e) plasma minus the electrolytes.
6. A hemoglobin molecule is composed of (a) two protein chains, (b) three protein chains, (c) four protein chains and nothing else, (d) four protein chains and four heme groups, (e) four heme groups but no protein.
7. The following is a list of the phases involved in the process of hemostasis.
   (1) coagulation
   (2) fibrinolysis
   (3) vascular spasm
   (4) retraction
   (5) platelet phase
   The correct sequence of these phases is
   (a) 5, 1, 4, 2, 3, (b) 3, 5, 1, 4, 2, (c) 2, 3, 5, 1, 4, (d) 3, 5, 4, 1, 2, (e) 4, 3, 5, 2, 1.

8. Stem cells responsible for lymphopoiesis are located in (a) the thymus and spleen, (b) the lymph nodes, (c) the red bone marrow, (d) all of these structures.
9. _____ and _____ affect almost every aspect of the clotting process. (a) Calcium, vitamin K, (b) Calcium, vitamin B_{12}, (c) Sodium, vitamin K, (d) Sodium, vitamin B_{12}.
10. State three reasons why superficial veins are used in the collection of fresh whole blood.
11. Name the three major types of plasma proteins and identify their functions,
12. Which type of antibodies does plasma contain for each of the following blood types? (a) type A, (b) type B, (c) type AB, (d) type O.
13. What four characteristics of WBCs are important to their response to tissue invasion or injury?
14. Describe the structure and functions of platelets.
15. Name the three types of lymphocytes and identify their functions,
16. What is the difference between prothrombin and thrombin?
17. What four conditions cause the release of erythropoietin?
18. What contribution from the intrinsic and the extrinsic pathways is necessary for the common pathway to begin?

LEVEL 2 Reviewing Concepts

19. Dehydration would (a) cause an increase in the hematocrit, (b) cause a decrease in the hematocrit, (c) have no effect on the hematocrit, (d) cause an increase in plasma volume.
20. Erythropoietin directly stimulates RBC formation by (a) increasing rates of mitotic divisions in erythroblasts, (b) speeding up the maturation of red blood cells, (c) accelerating the rate of hemoglobin synthesis, (d) a, b, and c.
21. The waste product bilirubin is formed from (a) transferrin, (b) globin, (c) heme, (d) hemosiderin, (e) ferritin.
22. A difference between the A, B, and O blood types and the Rh factor is (a) Rh agglutinogens are not found on the surface of red blood cells, (b) Rh agglutinogens do not produce a cross-reaction, (c) individuals who are Rh⁻ do not carry agglutinins to Rh factor unless they have been previously sensitized, (d) Rh agglutinogens are found free in the plasma, (e) Rh agglutinogens are found bound to plasma proteins.
23. What is the relation of the biconcave structure of red blood cells to their function?
24. How do elements of blood defend against toxins and pathogens in the body?
25. What is the role of blood in the stabilization and maintenance of body temperature?
26. Explain how red blood cell components are recycled.
27. Using an example, demonstrate the significance of colony stimulating factors (CSF) in the regulation of white blood cell populations.

19

LEVEL 3 Critical Thinking and Clinical Applications

28. A test for prothrombin time is used to identify deficiencies in the extrinsic clotting pathway; prothrombin time is prolonged if any of the factors are deficient. A test for activated partial thromboplastin time is used in a similar fashion to detect deficiencies in the intrinsic clotting pathway. Which factor would be deficient if a person had a prolonged prothrombin time but a normal partial thromboplastin time?

29. In the disease mononucleosis ("mono"), the spleen enlarges because of increased numbers of phagocytes and other cells. Common signs and symptoms of this disease include pale complexion, a tired feeling, and a lack of energy sometimes to the point of not being able to get out of bed. What might cause these signs and symptoms?

30. Almost half of our vitamin K is synthesized by bacteria that inhabit the large intestine. Based on this information, how could taking a broad-spectrum antibiotic for a long time cause frequent nosebleeds?

31. Roger is suffering from chronic obstructive pulmonary disease. In his last review, his blood oxygen saturations had fallen to 92%. Explain how this could change his hematocrit and what problems this could lead to.

CLINICAL CASE   Wrap-Up

A Mysterious Blood Disorder

Jasmine's hematologist takes a thorough medical history. Jasmine has no known family history or close relatives because she was adopted from an orphanage as a young girl. Her health was completely normal until six months ago. On physical examination, she appears pale, tired, and has a rapid heartbeat. Her bruises are clearly visible.

The hematologist performs a bone marrow aspirate and biopsy by inserting a large bore needle into the red bone marrow of Jasmine's posterior iliac crest and extracting cells. The specimen shows that there are not enough blood-making cells in Jasmine's bone marrow. Normally 30 percent of the red bone marrow is cellular and is involved in making all the blood cell precursors. But Jasmine's red bone marrow shows very little hemopoiesis and virtually no hemocytoblasts, myeloid stem cells, or lymphoid stem cells. The hemopoietic marrow was replaced by adipose tissue. No wonder she has so few red cells, white cells, and platelets in her blood. None of these cells are being made.

The hematologist diagnoses Jasmine with aplastic anemia. This serious red bone marrow failure often develops for no apparent reason (idiopathic). Without any known blood relatives, Jasmine appeals to the National Marrow Donor Program to find a compatible, unrelated donor for a bone marrow transplant, which is a type of stem cell transplant.

1. Why does Jasmine bleed and bruise so easily?
2. Why is Jasmine short of breath climbing stairs?

See the blue Answers tab at the back of the book.

19

20

The Heart

Learning Outcomes

These Learning Outcomes correspond by number to this chapter's sections and indicate what you should be able to do after completing the chapter.

20-1 ❚ Describe the anatomy of the heart, including vascular supply and pericardium structure, and trace the flow of blood through the heart, identifying the major blood vessels, chambers, and heart valves.

20-2 ❚ Explain the events of an action potential in cardiac muscle, indicate the importance of calcium ions to the contractile process, describe the conducting system of the heart, and identify the electrical events associated with a normal electrocardiogram.

20-3 ❚ Explain the events of the cardiac cycle, including atrial and ventricular systole and diastole, and relate the heart sounds to specific events in the cycle.

20-4 ❚ Define cardiac output, describe the factors that influence heart rate and stroke volume, and explain how adjustments in stroke volume and cardiac output are coordinated at different levels of physical activity.

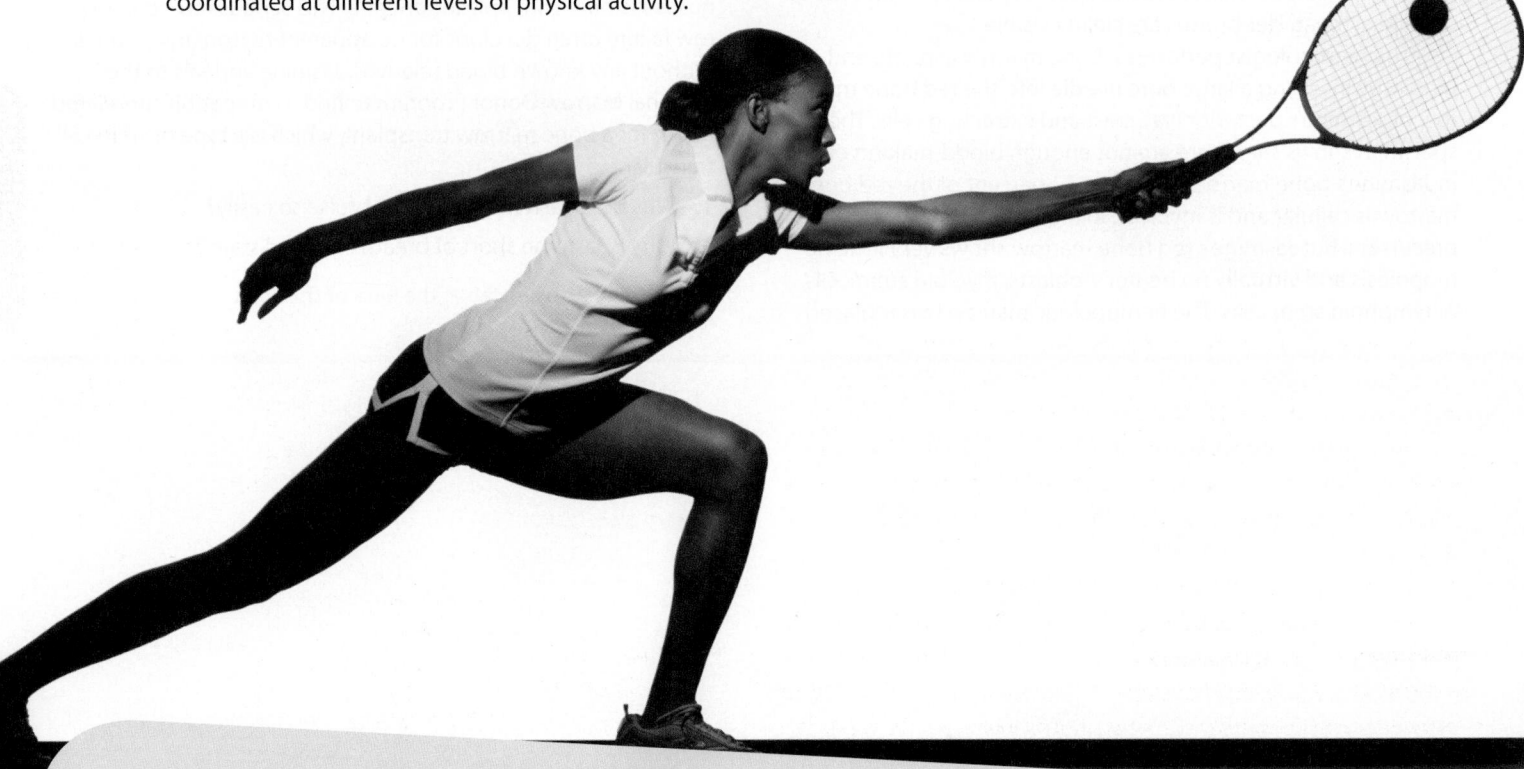

SPOTLIGHTS

- Heart Disease and Heart Attacks pp. 724–725
- Cardiac Arrhythmias p. 730

CLINICAL CASE

- A Needle to the Chest p. 711

Clinical Note

- Abnormal Conditions Affecting Cardiac Output p. 734

A Needle to the Chest

It has been a slow night in the emergency department. Dr. Jim is reading an emergency medicine journal and nodding off. He perks up when he hears the speakerphone crackle.

"We have a male, mid-40s, driver of a vehicle involved in a head-on collision with a truck," reports an EMT. "He was unconscious at the scene, but is now semi-alert, complaining of severe chest pain. His respirations are 30 and shallow. His blood pressure is 80/58. Pulse 130. Carotid pulses are present, but barely palpable. He has massive jugular venous distension (the veins in his neck are backed up with blood). The seat belt was broken and he has a steering wheel contusion on his anterior chest."

Dr. Jim instructs the EMTs to take the victim directly into Trauma Room 1. The patient is loaded supine on the trauma cart and Dr. Jim quickly introduces himself to the barely conscious man. A chest x-ray shows a fractured sternum and several fractured ribs, but the heart shadow (outline on the x-ray) appears normal. The heart sounds are distant and muffled through a stethoscope. After a rapid bedside ultrasound, Dr. Jim puts on sterile gloves, does a skin prep of the patient's anterior chest, and sticks a large bore needle with a 100-mL syringe attached into the man's sub-xiphoid space, directing the needle toward his left shoulder. The syringe fills with dark blood twice. **What is Dr. Jim doing? To find out, turn to the Clinical Case Wrap-Up on p. 748.**

An Introduction to the Cardiovascular System

In this chapter we consider the structure and function of the heart. This extraordinary organ beats approximately 100,000 times each day.

Blood flows through a network of blood vessels that extend between the heart and peripheral tissues. Those blood vessels make up a **pulmonary circuit**, which carries blood to and from the gas exchange surfaces of the lungs, and a **systemic circuit**, which transports blood to and from the rest of the body (**Figure 20–1**). Each circuit begins and ends at the heart, and blood travels through these circuits in sequence. Thus, blood returning to the heart from the systemic circuit must complete the pulmonary circuit before re-entering the systemic circuit.

Arteries, or *efferent vessels*, carry blood away from the heart, and **veins**, or *afferent vessels*, return blood to the heart. Microscopic thin-walled vessels called **capillaries** interconnect the smallest arteries and the smallest veins. Capillaries are called **exchange vessels**, because their thin walls permit the exchange of nutrients, dissolved gases, and wastes between the blood and surrounding tissues.

Each day the heart pumps about 8000 liters of blood—enough to fill forty 55-gallon drums, or 8800 quart-sized milk cartons. Try transferring a gallon of water by using a squeeze pump, and you'll appreciate just how hard the heart has to work to keep you alive. Despite its impressive workload, the heart is a small organ, about the size of a clenched fist.

The heart has four muscular chambers, two associated with each circuit. The **right atrium** (Ā-trē-um; entry chamber; plural, *atria*) receives blood from the systemic circuit and passes it to the **right ventricle** (VEN-tri-kl; little belly), which then pumps blood into the pulmonary circuit. The **left atrium**

Figure 20–1   An Overview of the Cardiovascular System. Driven by the pumping of the heart, blood flows through the pulmonary and systemic circuits in sequence. Each circuit begins and ends at the heart and contains arteries, capillaries, and veins.

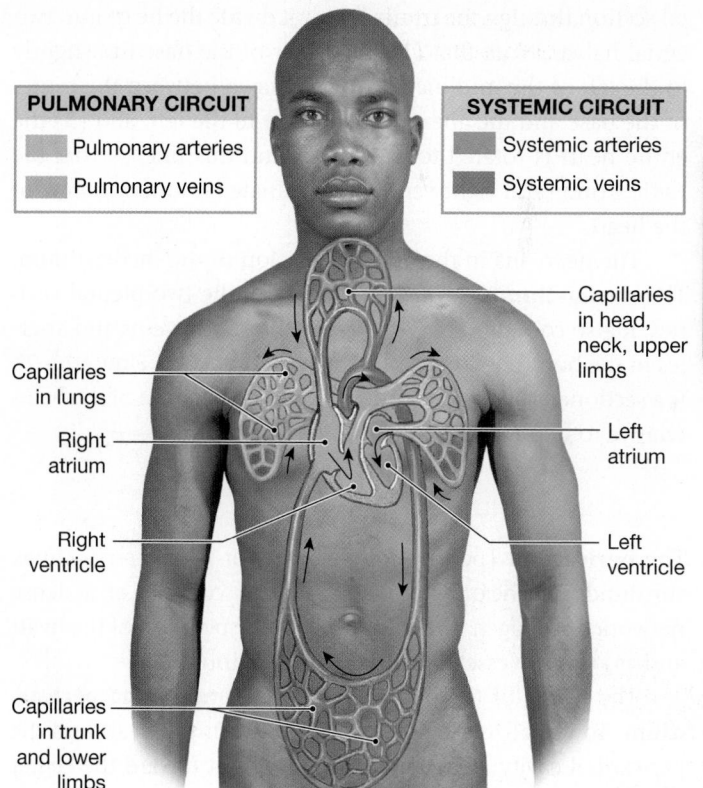

PULMONARY CIRCUIT
- Pulmonary arteries
- Pulmonary veins

SYSTEMIC CIRCUIT
- Systemic arteries
- Systemic veins

Capillaries in lungs

Right atrium

Right ventricle

Capillaries in trunk and lower limbs

Capillaries in head, neck, upper limbs

Left atrium

Left ventricle

collects blood from the pulmonary circuit and empties it into the **left ventricle**, which pumps blood into the systemic circuit. When the heart beats, first the atria contract, and then the ventricles contract. The two ventricles contract at the same time and eject equal volumes of blood into the pulmonary and systemic circuits.

20-1 ❙ The heart is a four-chambered organ, supplied by the coronary circulation, that pumps oxygen-poor blood to the lungs and oxygen-rich blood to the rest of the body

Learning Outcome Describe the anatomy of the heart, including vascular supply and pericardium structure, and trace the flow of blood through the heart, identifying the major blood vessels, chambers, and heart valves.

The heart is located near the anterior chest wall, directly posterior to the sternum (**Figure 20–2a**). The great veins and arteries are connected to the superior end of the heart at its base. The base sits posterior to the sternum at the level of the third costal cartilage, centered about 1.2 cm (0.5 in.) to the left side. The inferior, pointed tip of the heart is the **apex** (Ā-peks). A typical adult heart measures approximately 12.5 cm (5 in.) from the base to the apex, which reaches the fifth intercostal space approximately 7.5 cm (3 in.) to the left of the midline. A midsagittal section through the trunk does not divide the heart into two equal halves. Note that (1) the center of the base lies slightly to the left of the midline, (2) a line drawn between the center of the base and the apex points further to the left, and (3) the entire heart is rotated to the left around this line, so that the right atrium and right ventricle dominate an anterior view of the heart.

The heart sits in the anterior portion of the mediastinum. The **mediastinum** is the region between the two pleural cavities. It also contains the *great vessels* (the largest veins and arteries in the body), thymus, esophagus, and trachea. **Figure 20–2b** is a sectional, superior view that shows the position of the heart relative to other structures in the mediastinum.

The Pericardium

The **pericardial** (per-i-KAR-dē-al) **sac**, or *fibrous pericardium*, surrounds the heart. The pericardial sac consists of a dense network of collagen fibers. It stabilizes the position of the heart and associated vessels within the mediastinum.

The lining of the pericardial cavity is called the **pericardium**. To visualize the relationship between the heart and the pericardial cavity, imagine pushing your fist toward the center of a large, partially inflated balloon (**Figure 20–2c**). The balloon represents the pericardium, and your fist is the heart. Your wrist, where the balloon folds back on itself, corresponds to the **base** of the heart, where the great vessels are attached. The air space inside the balloon corresponds to the pericardial cavity.

The pericardium is lined by a delicate serous membrane that can be subdivided into two portions. The **visceral pericardium**, or *epicardium*, covers and adheres closely to the outer surface of the heart. The **parietal pericardium** lines the inner surface of the tough pericardial sac surrounding the heart (**Figure 20–2c**).

The potential space between the parietal and visceral surfaces is the pericardial cavity. It normally contains 15–50 mL of **pericardial fluid**, secreted by the pericardial membranes. This fluid acts as a lubricant, reducing friction between the opposing surfaces as the heart beats. Pathogens can infect the pericardium, producing inflammation and the condition **pericarditis**. The inflamed pericardial surfaces rub against one another, making a distinctive scratching sound that can be heard through a stethoscope. The pericardial inflammation also commonly results in increased production of pericardial fluid. Fluid then collects in the pericardial cavity, restricting the movement of the heart. This condition, called **cardiac tamponade** (tam-po-NĀD; *tampon*, plug), can also result from traumatic injuries (such as stab wounds) that produce bleeding into the pericardial cavity.

Superficial Anatomy of the Heart

You can easily identify the position of the heart within the rib cage and the four chambers of the heart in a superficial view (**Figure 20–3a,b**). The two atria have relatively thin muscular walls and are highly expandable. When not filled with blood, the outer portion of each atrium deflates and becomes a lumpy, wrinkled flap. This expandable extension of an atrium is called an *atrial appendage*, or an **auricle** (AW-ri-kl; *auris*, ear), because it reminded early anatomists of the external ear (**Figure 20–3b**). The **coronary sulcus**, a deep groove, marks the border between the atria and the ventricles. The **anterior interventricular sulcus** and the **posterior interventricular sulcus** are shallower depressions that mark the boundary between the left and right ventricles (**Figure 20–3b,c**).

Substantial amounts of fat generally lie in the coronary and interventricular sulci. In fresh or preserved hearts, this fat must be stripped away to expose the underlying grooves. These sulci also contain the arteries and veins that carry blood to and from the cardiac muscle.

The Heart Wall

A section through the wall of the heart reveals three distinct layers: an outer epicardium, a middle myocardium, and an inner endocardium. **Figure 20–4a** illustrates these three layers:

1. The **epicardium** is the visceral pericardium that covers the outer surface of the heart. This serous membrane consists of an exposed mesothelium and an underlying layer of loose areolar connective tissue that is attached to the myocardium.

Figure 20–2 **The Location of the Heart in the Thoracic Cavity.** ATLAS: Plates 47a,b

Trachea

Thyroid gland

First rib (cut)

Base of heart

Right lung

Left lung

Parietal pericardium (cut)

Apex of heart

Diaphragm

a An anterior view of the chest, showing the position of the heart and major blood vessels relative to the ribs, lungs, and diaphragm.

Posterior mediastinum

Esophagus

Aorta (arch segment removed)

Right pleural cavity

Left pulmonary artery

Right lung

Left lung

Left pleural cavity

Bronchus of lung

Left pulmonary vein

Right pulmonary artery

Aortic arch

Pulmonary trunk

Right pulmonary vein

Left atrium

Superior vena cava

Left ventricle

Right atrium

Pericardial cavity

Right ventricle

Epicardium

Anterior mediastinum

Pericardial sac

Sternum

b A superior view of the organs in the mediastinum; portions of the lungs have been removed to reveal blood vessels and airways. The heart is located in the anterior part of the mediastinum, immediately posterior to the sternum.

Base of heart

Cut edge of parietal pericardium

Wrist (corresponds to base of heart)

Fibrous tissue of pericardial sac

Inner wall (corresponds to epicardium)

Parietal pericardium

Areolar tissue

Air space (corresponds to pericardial cavity)

Mesothelium

Fibrous attachment to diaphragm

Cut edge of epicardium

Outer wall (corresponds to parietal pericardium)

Apex of heart

Balloon

c The relationship between the heart and the pericardial cavity; compare with the fist-and-balloon example.

20

Figure 20–3 **The Position and Superficial Anatomy of the Heart.**

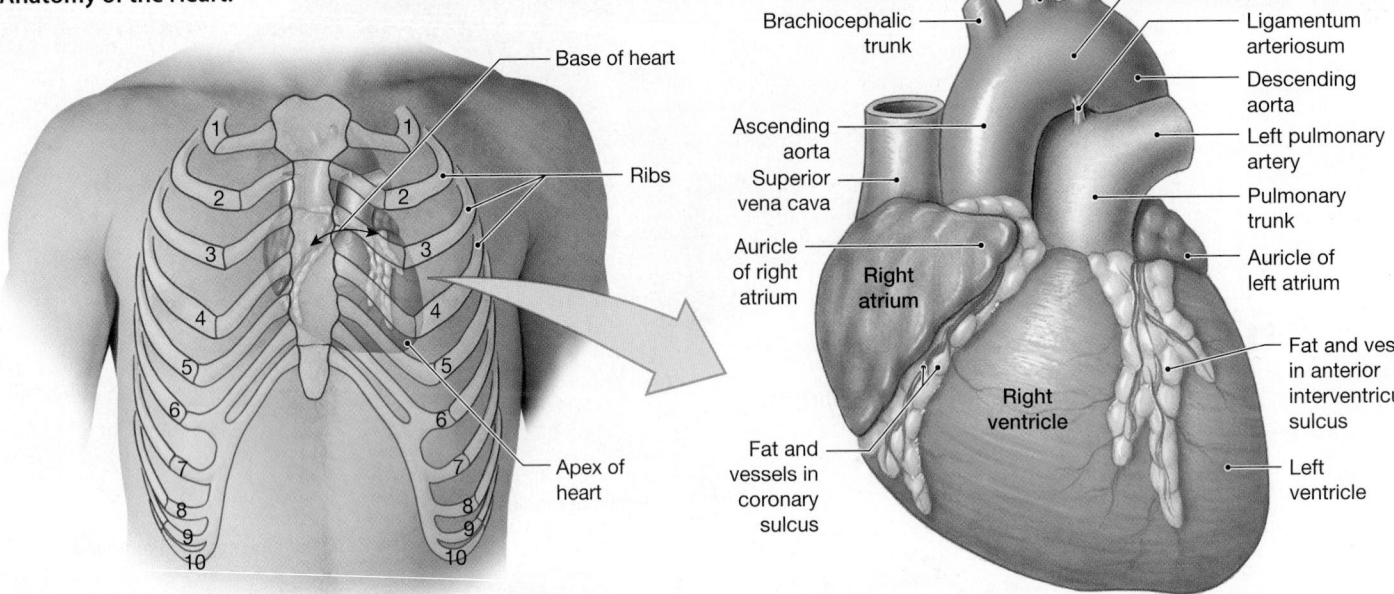

a Heart position relative to the rib cage.

b Major anatomical features on the anterior surface.

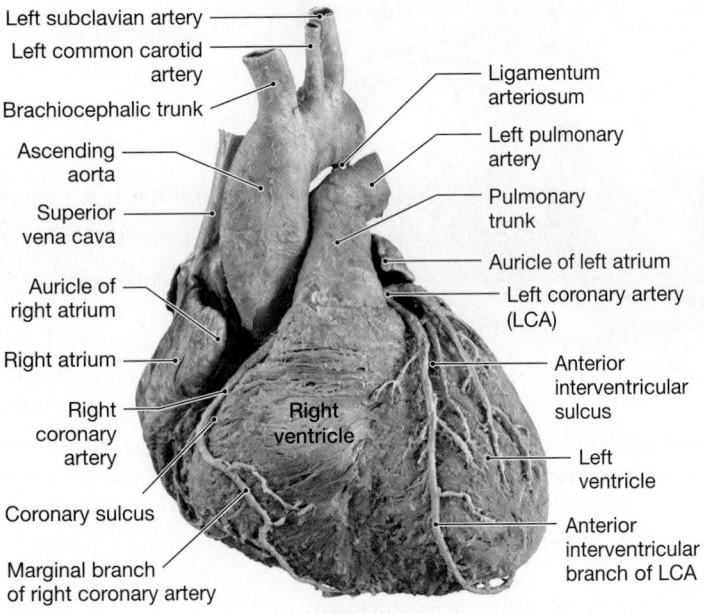

c Anterior surface of the heart, cadaver dissection.

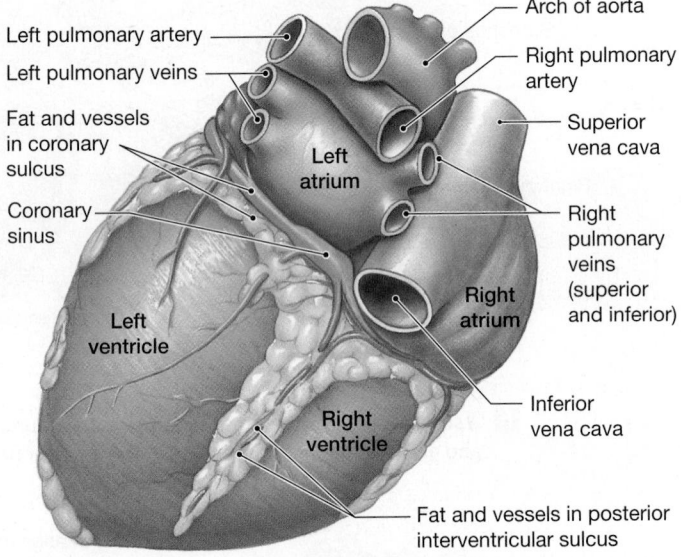

d Major landmarks on the posterior surface. Coronary arteries (which supply the heart itself) are shown in red; coronary veins are shown in blue.

2. The **myocardium**, or muscular wall of the heart, forms the atria and ventricles. This layer contains cardiac muscle tissue, blood vessels, and nerves. The myocardium consists of concentric layers of cardiac muscle tissue. The atrial myocardium contains muscle bundles that wrap around the atria and form figure eights that encircle the great vessels (**Figure 20–4b**). Superficial ventricular muscles wrap around both ventricles, and deeper muscle layers spiral around and between the ventricles toward the apex in a figure-eight pattern.

3. The **endocardium** covers the inner surfaces of the heart, including those of the heart valves. This simple squamous epithelium is continuous with the endothelium of the attached great vessels.

Figure 20–4   **The Heart Wall.**

Myocardium (cardiac muscle tissue)
- Cardiac muscle cells
- Connective tissues

Pericardial cavity

Parietal pericardium
- Dense fibrous layer
- Areolar tissue
- Mesothelium
- Artery
- Vein

Endocardium
- Endothelium
- Areolar tissue

Epicardium (visceral pericardium)
- Mesothelium
- Areolar tissue

Heart wall

Atrial musculature

Ventricular musculature

b Cardiac muscle tissue forms concentric layers that wrap around the atria or spiral within the walls of the ventricles.

a A diagrammatic section through the heart wall, showing the relative positions of the epicardium, myocardium, and endocardium. The proportions are not to scale; the thickness of the myocardial wall has been greatly reduced.

Cardiac Muscle Tissue

As noted in Chapter 10, **cardiac muscle cells** are interconnected by **intercalated discs** (Figure 20–5). At an intercalated disc, the interlocking membranes of adjacent cells are held together by desmosomes and linked by gap junctions (Figure 20–5b). Intercalated discs transfer the force of contraction from cell to cell and propagate action potentials. Table 20–1 provides a quick review of the structural and functional differences between cardiac muscle cells and skeletal muscle fibers. Histological characteristics that distinguish cardiac muscle cells from skeletal muscle fibers include (1) small size; (2) a single, centrally located nucleus; (3) branching interconnections between cells; and (4) the presence of intercalated discs.

TIPS & TRICKS

The term *intercalated* means "inserted between other elements." So, intercalated discs appear to have been inserted between cardiac muscle cells.

Internal Anatomy and Organization

Next let's examine the major landmarks and structures visible on the interior surface of the heart. In a sectional view, you can see that the right atrium communicates with the right ventricle, and the left atrium with the left ventricle (Figure 20–6a,c). The atria are separated by the **interatrial septum** (*septum*, wall), and the ventricles are separated by the much thicker **interventricular septum**. Each septum is a muscular partition. **Atrioventricular (AV) valves** are folds of fibrous tissue that extend into the openings between the atria and ventricles. These valves permit blood to flow only in one direction: from the atria to the ventricles.

The Right Atrium

The right atrium receives blood from the systemic circuit through the two great veins: the **superior vena cava** (VĒ-na KĀ-vuh; *venae cavae*, plural) and the **inferior vena cava**. The superior vena cava opens into the posterior and superior portion of the right atrium. It delivers blood to the right atrium from the head, neck, upper limbs, and chest. The inferior vena cava opens into the posterior and inferior portion of the right atrium. It carries blood to the right atrium from the rest of the trunk, the viscera, and the lower limbs. The *cardiac veins* draining the myocardium return blood to the **coronary sinus**, a

20

Figure 20–5 **Cardiac Muscle Cells.**

a Cardiac muscle cells

Cardiac muscle cell
Mitochondria
Intercalated disc (sectioned)
Nucleus
Cardiac muscle cell (sectioned)
Bundles of myofibrils
Intercalated discs

Intercalated disc
Gap junction
Z-lines bound to opposing plasma membranes
Desmosomes

b Structure of an intercalated disc

Cardiac muscle tissue   LM × 575

c Cardiac muscle tissue

Table 20–1	Structural and Functional Differences between Cardiac Muscle Cells and Skeletal Muscle Fibers	
Feature	**Cardiac Muscle Cells**	**Skeletal Muscle Fibers**
Size	10–20 μm × 50–100 μm	100 μm × up to 40 cm
Nuclei	Typically 1 (rarely 2–5)	Multiple (hundreds)
Contractile proteins	Sarcomeres along myofibrils	Sarcomeres along myofibrils
Internal membranes	Short T tubules; no triads formed with sarcoplasmic reticulum	Long T tubules form triads with cisternae of the sarcoplasmic reticulum
Mitochondria	Abundant (25% of cell volume)	Much less abundant
Inclusions	Myoglobin, lipids, glycogen	Little myoglobin, few lipids, but extensive glycogen reserves
Blood supply	Very extensive	More extensive than in most connective tissues, but sparse compared with supply to cardiac muscle cells
Metabolism (resting)	Not applicable	Aerobic, primarily lipid-based
Metabolism (active)	Aerobic, primarily using lipids and carbohydrates	Anaerobic, through breakdown of glycogen reserves
Contractions	Twitches with brief relaxation periods; long refractory period prevents tetanic contractions	Usually sustained contractions
Stimulus for contraction	Autorhythmicity of pacemaker cells generates action potentials	Activity of somatic motor neuron generates action potentials in sarcolemma
Trigger for contraction	Calcium entry from the ECF and calcium release from the sarcoplasmic reticulum	Calcium release from the sarcoplasmic reticulum
Intercellular connections	Branching network with plasma membranes locked together at intercalated discs; connective tissue fibers tie adjacent layers together	Adjacent fibers tied together by connective tissue fibers

20

Figure 20–6   **The Sectional Anatomy of the Heart.**

b The papillary muscles and chordae tendineae support the right AV (tricuspid) valve. The photograph was taken from inside the right ventricle, looking toward a light shining from the right atrium.

a A diagrammatic frontal section through the heart, showing major landmarks and the path of blood flow (marked by arrows) through the atria, ventricles, and associated vessels.

c Anterior view of a frontally sectioned heart showing internal features and valves.

20

large, thin-walled vein that opens into the right atrium inferior to the connection with the superior vena cava.

The opening of the coronary sinus lies near the posterior edge of the interatrial septum. From the fifth week of embryonic development until birth, an oval opening called the **foramen ovale** penetrates the interatrial septum and connects the two atria of the fetal heart. Before birth, the foramen ovale permits blood to flow from the right atrium to the left atrium while the lungs are developing. At birth, the foramen ovale closes, and the opening is permanently sealed off within three months of delivery. (If the foramen ovale does not close, serious cardiovascular problems may result. We consider these in Chapter 21.) A small, shallow depression called the **fossa ovalis** remains at this site in the adult heart (**Figure 20–6a**). ATLAS: Embryology Summary 15: The Development of the Heart

The posterior walls of the right atrium and the interatrial septum have smooth surfaces. In contrast, the anterior atrial wall and the inner surface of the auricle contain prominent muscular ridges called the **pectinate muscles** (*pectin*, comb), or *musculi pectinati* (**Figure 20–6a**).

The Right Ventricle

Blood travels from the right atrium into the right ventricle through a broad opening bordered by three fibrous flaps. These flaps, called **cusps**, are part of the **right atrioventricular (AV) valve**, also known as the **tricuspid** (trī-KUS-pid; *tri*, three) **valve**. The free edge of each cusp is attached to connective tissue fibers called the **chordae tendineae** (KOR-dē TEN-di-nē-ē; tendinous cords). The fibers originate at the **papillary** (PAP-i-ler-ē) **muscles**, conical muscular projections that arise from the inner surface of the right ventricle (**Figure 20–6a,b,c**). The right AV valve closes when the right ventricle contracts, preventing the backflow of blood into the right atrium. Without the chordae tendineae to anchor their free edges, the cusps would be like swinging doors that permit blood flow in both directions.

The internal surface of the ventricle also contains a series of muscular ridges: the **trabeculae carneae** (tra-BEK-ū-lē KAR-nē-ē; *carneus*, fleshy). The *moderator band* is a muscular ridge that extends horizontally from the inferior portion of the interventricular septum and connects to the anterior papillary muscle. This ridge contains part of the *conducting system*, an internal network that coordinates the contractions of cardiac muscle cells. The moderator band delivers the stimulus for contraction to the papillary muscles. As a result, they begin tensing the chordae tendineae before the rest of the ventricle contracts.

TIPS & TRICKS

The saying "To tug on your heartstrings" may help you remember the functions of the papillary muscles and the chordae tendineae: Contractions of the papillary muscles pull on the chordae tendineae, which "tug" on your heart's valves.

The superior end of the right ventricle tapers to the **conus arteriosus**, a cone-shaped pouch that ends at the **pulmonary valve**, or *pulmonary semilunar valve*. The pulmonary valve consists of three semilunar (half-moon-shaped) cusps of thick connective tissue. Blood flowing from the right ventricle passes through this valve into the **pulmonary trunk**, the start of the pulmonary circuit. The cusps prevent backflow as the right ventricle relaxes. Once in the pulmonary trunk, blood flows into the **left pulmonary arteries** and the **right pulmonary arteries**. These vessels branch repeatedly within the lungs before supplying the capillaries, where gas exchange occurs.

The Left Atrium

From the respiratory capillaries, blood collects into small veins that ultimately unite to form the four pulmonary veins. The posterior wall of the left atrium receives blood from two **left** and two **right pulmonary veins**. Like the right atrium, the left atrium has an auricle. A valve, the **left atrioventricular (AV) valve**, or **bicuspid** (bī-KUS-pid) **valve**, guards the entrance to the left ventricle (**Figure 20–6a,c**). As the name *bicuspid* implies, the left AV valve contains two cusps, not three. Clinicians often call this valve the **mitral** (MĪ-tral; *mitre*, a bishop's hat) **valve**. The left AV valve permits blood to flow from the left atrium into the left ventricle, but it prevents backflow when the left ventricle contracts.

TIPS & TRICKS

To remember the locations of the tricuspid and bicuspid (mitral) valves, think **"try** to be **right"** for the *tri*cuspid, and associate the *l* in mitra*l* with the *l* in *l*eft.

The Left Ventricle

Even though the two ventricles hold and pump equal amounts of blood, the left ventricle is much larger than the right ventricle. What's the reason? It has thicker walls. These thick, muscular walls enable the left ventricle to push blood through the large systemic circuit. In contrast, the right ventricle needs to pump blood, at lower pressure, only about 15 cm (6 in.) to and from the lungs.

The internal organization of the left ventricle resembles that of the right ventricle, but it has no moderator band (**Figure 20–6a**). The trabeculae carneae are prominent. A pair of large papillary muscles tenses the chordae tendineae that anchor the cusps of the AV valve and prevent blood from flowing back into the left atrium.

Blood leaves the left ventricle through the **aortic valve**, or *aortic semilunar valve*, and goes into the **ascending aorta**. The arrangement of cusps in the aortic valve is the same as that in the pulmonary valve. Once the blood has been pumped out of the heart and into the systemic circuit, the aortic valve prevents backflow into the left ventricle. From the ascending aorta, blood flows through the **aortic arch** and into the **descending aorta** (Figure 20–6a). The pulmonary trunk is attached to the aortic arch by the *ligamentum arteriosum*, a fibrous band left over from an important fetal blood vessel that once linked the pulmonary and systemic circuits.

Structural Differences between the Left and Right Ventricles

The function of the atria is to collect blood that is returning to the heart and to convey it to the ventricles. The demands on the right and left atria are similar, and the two chambers look almost identical. The demands on the right and left ventricles, however, are very different, and the two have significant structural differences.

Anatomical differences between the left and right ventricles are easiest to see in a three-dimensional view (Figure 20–7a). The lungs are close to the heart, and the pulmonary blood vessels are relatively short and wide. For these reasons, the right ventricle normally does not need to work very hard to push blood through the pulmonary circuit. Accordingly, the muscular wall of the right ventricle is relatively thin. In sectional view, it resembles a pouch attached to the massive wall of the left ventricle. When the right ventricle contracts, it acts like a bellows, squeezing the blood against the thick wall of the left ventricle. This action moves blood very efficiently with minimal effort, but it develops relatively low pressures.

A comparable pumping arrangement would not work well for the left ventricle. Four to six times as much pressure must be exerted to push through the systemic circuit as through the pulmonary circuit. The left ventricle has an extremely thick muscular wall and is round in cross section (Figure 20–7a). When this ventricle contracts, it shortens and narrows. In other words, (1) the distance between the base and apex decreases, and (2) the diameter of the ventricular chamber decreases. The effect is similar to simultaneously squeezing and rolling up the end of a toothpaste tube. The pressure generated is more than enough to open the aortic valve and eject blood into the ascending aorta.

As the powerful left ventricle contracts, it bulges into the right ventricular cavity (Figure 20–7b). This action makes the right ventricle more efficient. Individuals with severe damage to the right ventricle may survive, because the contraction of the left ventricle helps push blood into the pulmonary circuit. We return to this topic in Chapter 21, where we consider the integrated functioning of the cardiovascular system.

The Heart Valves

As we have seen, the heart has two pairs of one-way valves that prevent the backflow of blood as the chambers contract. Let's look at the structure and function of these heart valves.

The Atrioventricular Valves. The atrioventricular (AV) valves prevent the backflow of blood from the ventricles to the atria when the ventricles are contracting. The chordae tendineae and papillary muscles play important roles in the normal function of the AV valves. When the ventricles are relaxed, the chordae tendineae are loose, and the AV valves offer no resistance as blood flows from the atria into the ventricles (Figure 20–8a). When the ventricles contract, blood moving back toward the atria swings the cusps together, closing the valves (Figure 20–8b). At the same time, the contraction of the papillary muscles tenses the chordae tendineae, stopping the cusps before they swing into the atria. If the chordae tendineae were cut or the papillary muscles were damaged, backflow, called **regurgitation**, of blood into the atria would occur each time the ventricles contracted.

Figure 20–7 **Structural Differences between the Left and Right Ventricles.** ATLAS: Plate 45d

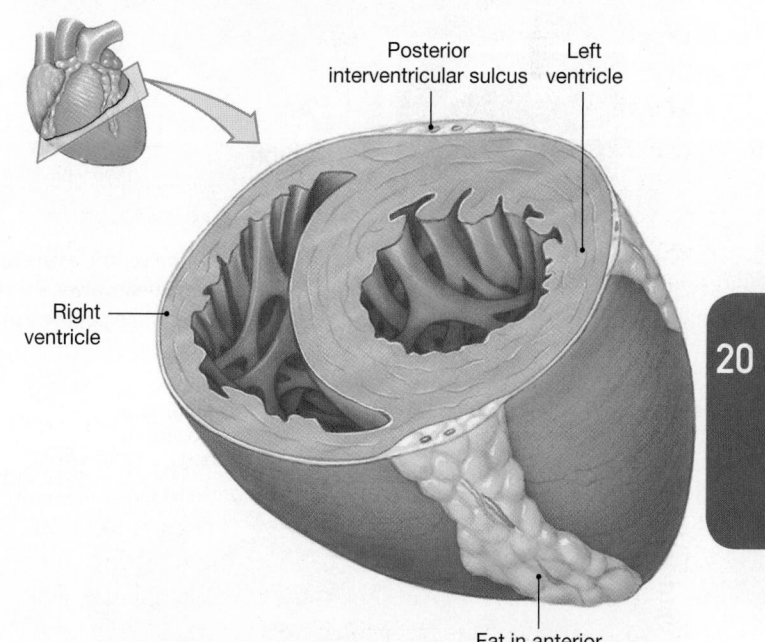

a A diagrammatic sectional view through the heart, showing the relative thicknesses of the two ventricles. Notice the pouchlike shape of the right ventricle and the greater thickness of the left ventricle.

b Diagrammatic views of the ventricles just before a contraction (dilated) and just after a contraction (contracted).

Figure 20–8 **Valves of the Heart.** Red (oxygenated) and blue (deoxygenated) arrows indicate blood flow into or out of a ventricle; black arrows show blood flow into an atrium; and green arrows indicate contraction of the ventricles.

Transverse Sections, Superior View, Atria and Vessels Removed

Frontal Sections through Left Atrium and Ventricle

POSTERIOR

Cardiac skeleton — Left AV (bicuspid) valve (open)

RIGHT VENTRICLE — LEFT VENTRICLE

Right AV (tricuspid) valve (open)

Aortic valve (closed)
Pulmonary valve (closed)

ANTERIOR

Aortic valve closed

Pulmonary veins
LEFT ATRIUM

Aortic valve (closed)

Left AV (bicuspid) valve (open)

Chordae tendineae (loose)

Papillary muscles (relaxed)

LEFT VENTRICLE (relaxed and filling with blood)

a When the ventricles are relaxed, the AV valves are open and the semilunar valves are closed. The chordae tendineae are loose, and the papillary muscles are relaxed.

Relaxed ventricles

20

Contracting ventricles

Right AV (tricuspid) valve (closed)
Cardiac skeleton
Left AV (bicuspid) valve (closed)

RIGHT VENTRICLE — LEFT VENTRICLE

Aortic valve (open)
Pulmonary valve (open)

Aortic valve open

Aorta
Aortic sinus
Aortic valve (open)

LEFT ATRIUM

Left AV (bicuspid) valve (closed)

Chordae tendineae (tense)

Papillary muscles (contracted)

Left ventricle (contracted)

b When the ventricles are contracting, the AV valves are closed and the semilunar valves are open. In the frontal section notice the attachment of the left AV valve to the chordae tendineae and papillary muscles.

The Semilunar Valves. The pulmonary and aortic valves prevent the backflow of blood from the pulmonary trunk and aorta into the right and left ventricles, respectively. Unlike the AV valves, the semilunar valves do not need muscular braces, because the arterial walls do not contract and the relative positions of the cusps are stable. When the semilunar valves close, the three symmetrical cusps support one another like the legs of a tripod (Figure 20–8a).

Adjacent to each cusp of the aortic valve are saclike expansions of the base of the ascending aorta. These sacs, called **aortic sinuses**, prevent the individual cusps from sticking to the wall of the aorta when the valve opens (Figure 20–8b). The *right* and *left coronary arteries*, which deliver blood to the myocardium, originate at the right and left aortic sinuses. (The posterior aortic sinus does not give rise to any blood vessel.)

Serious valve problems can interfere with the working of the heart. If valve function deteriorates to the point at which the heart cannot maintain adequate circulatory flow, symptoms of **valvular heart disease (VHD)** appear. Congenital malformations may be responsible, but in many cases the condition develops after **carditis**, an inflammation of the heart, occurs. One important cause of carditis is **rheumatic** (rū-MAT-ik) **fever**, an inflammatory autoimmune response to an infection by streptococcal bacteria. It most often occurs in children.

Connective Tissues and the Cardiac Skeleton

The connective tissues of the heart include large numbers of collagen and elastic fibers. Each cardiac muscle cell is wrapped in a strong, but elastic, sheath. Adjacent cells are tied together by fibrous cross-links, or "struts." These fibers are, in turn, interwoven into sheets that separate the superficial and deep muscle layers. The connective tissue fibers (1) provide physical support for the cardiac muscle fibers, blood vessels, and nerves of the myocardium; (2) help distribute the forces of contraction; (3) add strength and prevent overexpansion of the heart; and (4) provide elasticity that helps return the heart to its original size and shape after a contraction.

The **cardiac skeleton** (sometimes called the *fibrous skeleton*) of the heart consists of four dense bands of tough elastic tissue that encircle the heart valves and the bases of the pulmonary trunk and aorta (Figure 20–8). These bands stabilize the positions of the heart valves and ventricular muscle cells. They also electrically insulate the ventricular cells from the atrial cells.

The Blood Supply to the Heart

The heart works continuously, so cardiac muscle cells need reliable supplies of oxygen and nutrients. A great volume of blood flows through the chambers of the heart, but the myocardium has its own, separate blood supply. The **coronary circulation** supplies blood to the muscle tissue of the heart. During maximum exertion, the heart's demand for oxygen rises considerably. The blood flow to the myocardium may then increase to nine times that of resting levels. The coronary circulation includes an extensive network of coronary blood vessels (Figure 20–9).

The Coronary Arteries

The left and right **coronary arteries** originate at the base of the ascending aorta, at the aortic sinuses (Figure 20–9a). Blood pressure here is the highest in the systemic circuit. Each time the left ventricle contracts, it forces blood into the aorta. The arrival of this blood at high pressures stretches the elastic walls of the aorta. When the left ventricle relaxes, blood no longer flows into the aorta, pressure declines, and the walls of the aorta recoil. This recoil, called *elastic rebound*, pushes blood both forward, into the systemic circuit, and backward, through the left and right aortic sinuses and then into the respective coronary arteries. In this way, the combination of elevated blood pressure and elastic rebound ensures a continuous flow of blood to meet the demands of active cardiac muscle tissue. Yet myocardial blood flow is not steady. It peaks while the heart muscle is relaxed, and almost ceases while it contracts.

The **right coronary artery** follows the coronary sulcus around the heart. It supplies blood to (1) the right atrium, (2) portions of both ventricles, and (3) portions of the conducting system of the heart, including the *sinoatrial (SA) node* and the *atrioventricular (AV) node*. The cells of these nodes are essential to establishing the normal heart rate. We focus on them in a later section.

Inferior to the right atrium, the right coronary artery generally gives rise to one or more **marginal arteries**, which extend across the surface of the right ventricle (Figure 20–9a,b). The right coronary artery then continues across the posterior surface of the heart. It supplies the **posterior interventricular artery**, or *posterior descending artery*, which runs toward the apex within the posterior interventricular sulcus (Figure 20–9b,c). The posterior interventricular artery supplies blood to the interventricular septum and adjacent portions of the ventricles.

The **left coronary artery** supplies blood to the left ventricle, left atrium, and interventricular septum. As it reaches the anterior surface of the heart, it gives rise to a circumflex branch and an anterior interventricular branch. The **circumflex artery** curves to the left around the coronary sulcus. It eventually meets and fuses with small branches of the right coronary artery (Figure 20–9a–c). The much larger **anterior interventricular artery**, or *left anterior descending artery (LAD)*, swings around the pulmonary trunk and runs along the surface within the anterior interventricular sulcus (Figure 20–9a).

20

Figure 20–9 **The Coronary Circulation.** ATLAS: Plates 45b,c

a Coronary vessels supplying and draining the anterior surface of the heart.

b Coronary vessels supplying and draining the posterior surface of the heart.

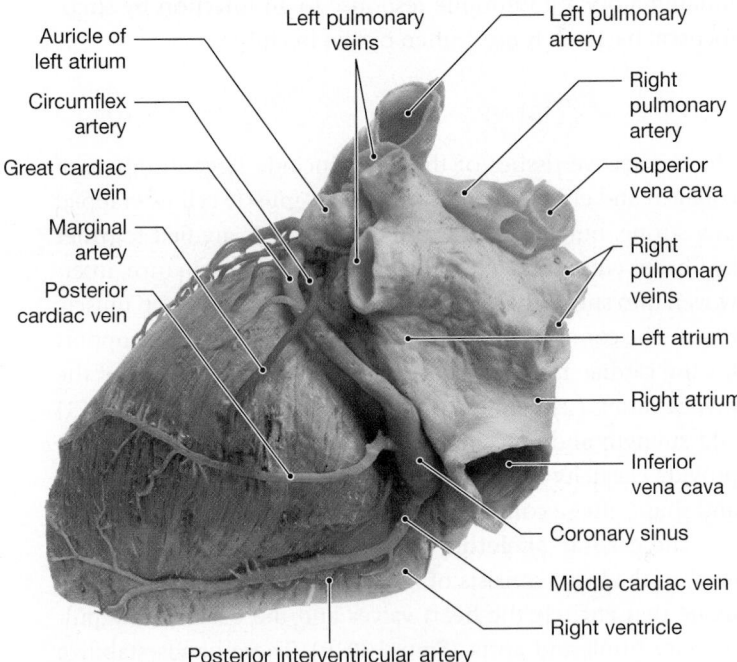

c A posterior view of the heart; the vessels have been injected with colored latex (liquid rubber).

The anterior interventricular artery supplies small tributaries continuous with those of the posterior interventricular artery. Such interconnections between arteries are called **arterial anastomoses** (a-nas-tō-MŌ-sēz; *anastomosis*, outlet). Because the arteries are interconnected in this way, the blood supply to the cardiac muscle remains relatively constant despite pressure fluctuations in the left and right coronary arteries as the heart beats.

The Cardiac Veins

The various cardiac veins are shown in **Figure 20–9**. The **great cardiac vein** begins on the anterior surface of the ventricles, along the interventricular sulcus. This vein drains blood from the region supplied by the anterior interventricular artery, a branch of the left coronary artery. The great cardiac vein reaches the level of the atria and then curves around the left side of the heart within the coronary sulcus. The vein empties into the coronary sinus, which lies in the posterior portion of the coronary sulcus. The coronary sinus opens into the right atrium near the base of the inferior vena cava.

Other cardiac veins empty into the great cardiac vein or the coronary sinus. These veins include (1) the **posterior cardiac vein**, draining the area served by the circumflex artery; (2) the **middle cardiac vein**, draining the area supplied by the posterior interventricular artery; and (3) the **small cardiac vein**, which receives blood from the posterior surfaces of the right atrium and ventricle. The **anterior cardiac veins**, which drain the anterior surface of the right ventricle, empty directly into the right atrium.

Coronary artery disease is characterized by interrupted blood flow to the myocardium. **Spotlight Figure 20–10** describes this condition, along with myocardial infarction.

1. Damage to the semilunar valve of the right ventricle would affect blood flow into which vessel?

2. What prevents the AV valves from swinging into the atria?

3. Why is the left ventricle more muscular than the right ventricle?

See the blue Answers tab at the back of the book.

20-2 ❚ The conducting system distributes electrical impulses through the heart, and an electrocardiogram records the associated electrical events

Learning Outcome Explain the events of an action potential in cardiac muscle, indicate the importance of calcium ions to the contractile process, describe the conducting system of the heart, and identify the electrical events associated with a normal electrocardiogram.

Next we look at several aspects of the contraction of the heart. We begin with an overview of how the heart works—cardiac physiology. Then we examine the structure and function of the conducting system, the electrical events as recorded in an electrocardiogram, and the functioning of contractile cells.

Cardiac Physiology

In a single cardiac contraction, or heartbeat, the entire heart contracts in series—first the atria and then the ventricles. Two types of cardiac muscle cells are involved in a normal heartbeat. (1) Specialized muscle cells of the *conducting system* control and coordinate the heartbeat, and (2) *contractile cells* produce the powerful contractions that propel blood.

Each heartbeat begins with an action potential generated at a pacemaker called the *SA node*, which is part of the conducting system. The conducting system then propagates and distributes this electrical impulse to stimulate contractile cells to push blood in the right direction at the proper time. A procedure known as *electrocardiography* can monitor the electrical events of the conducting system from the surface of the body. The printed record of the result is called an *electrocardiogram* (*ECG* or *EKG*).

The arrival of an electrical impulse at a cardiac muscle cell's plasma membrane produces an action potential that is comparable to an action potential in a skeletal muscle fiber. As in a skeletal muscle fiber, this action potential triggers the contraction of the cardiac muscle cell. Thanks to the coordination provided by the conducting system, the atria contract first, driving blood into the ventricles through the AV valves, and the ventricles contract next, driving blood out of the heart through the semilunar valves.

The SA node generates impulses at regular intervals, and one heartbeat follows another throughout your life. After each heartbeat comes a brief pause—less than half a second—before the next heartbeat begins. The period from the start of one heartbeat to the start of the next is called the *cardiac cycle*.

A heartbeat lasts only about 370 msec. Although brief, it is a very busy period! Let's follow the steps that produce a single heartbeat, from the generation of an action potential at the SA node through the contractions of the atria and ventricles.

The Conducting System

Unlike skeletal muscle, cardiac muscle tissue contracts on its own, without neural or hormonal stimulation. This property is called **automaticity**, or *autorhythmicity*. The cells that initiate and distribute the stimulus to contract are part of the heart's **conducting system**, also known as the *cardiac conduction system* or the *nodal system*. This system is a network of specialized cardiac muscle cells that initiates and distributes electrical impulses. The actual contraction lags behind the beginning of an electrical impulse (the action potential). The delay comes from the time it takes for calcium ions to enter the sarcoplasm and activate the contraction process, as described in Chapter 10. ⮌ p. 335

The conducting system includes the following elements (**Figure 20–11a**):

- The *sinoatrial (SA) node*, located in the wall of the right atrium.

- The *atrioventricular (AV) node*, located at the junction between the atria and ventricles.

- *Conducting cells* interconnect the two nodes and distribute the contractile stimulus throughout the myocardium. In the atria, conducting cells are found in **internodal pathways**, which distribute the contractile stimulus to atrial muscle cells as this electrical impulse travels from the SA node to the AV node. (The importance of these pathways in relaying the signal to the AV node remains in dispute, because an impulse can also spread from contractile cell to contractile cell, reaching the AV node at about the same time as an impulse that travels an internodal pathway.) In the ventricles, conducting cells include those in the *AV bundle* and the *bundle branches*, as well as the *Purkinje* (pur-KIN-jē) *fibers*, which distribute the stimulus to the ventricular myocardium.

Most of the cells of the conducting system are smaller than the contractile cells of the myocardium and contain very few myofibrils. Purkinje cells, however, are much larger in diameter than the contractile cells. As a result, they conduct action potentials more quickly than other conducting cells.

Conducting cells of the SA and AV nodes share a special characteristic. Their excitable membranes do not have a stable resting potential. Each time it repolarizes, the membrane then drifts toward threshold. This gradual depolarization is called a

20

Coronary Artery Disease

HEART DISEASE

The term **coronary artery disease (CAD)** refers to areas of partial or complete blockage of coronary circulation. Cardiac muscle cells need a constant supply of oxygen and nutrients, so any reduction in blood flow to the heart muscle produces a corresponding reduction in cardiac performance. Such reduced circulatory supply, known as **coronary ischemia** (is-KĒ-mē-uh), generally results from partial or complete blockage of the coronary arteries. The usual cause is the formation of a fatty deposit, or atherosclerotic plaque, in the wall of a coronary vessel. The plaque, or an associated thrombus (clot), then narrows the passageway and reduces blood flow. Spasms in the smooth muscles of the vessel wall can further decrease or even stop blood flow. One of the first symptoms of CAD is commonly **angina pectoris** (an-JĪ-nuh PEK-tor-is; *angina*, pain spasm + *pectoris*, of the chest). In its most common form, a temporary ischemia develops when the workload of the heart increases. Although the individual may feel comfortable at rest, exertion or emotional stress can produce a sensation of pressure, chest constriction, and pain that may radiate from the sternal area to the arms, back, and neck.

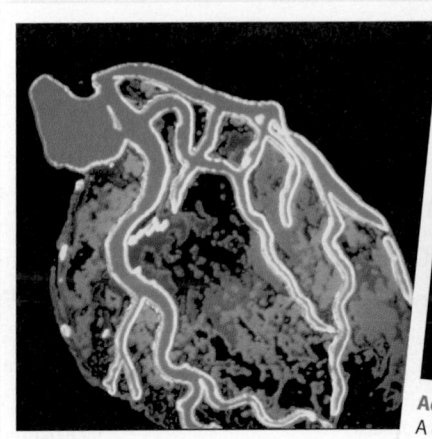

Normal Heart
A color-enhanced **digital subtraction angiography (DSA)** scan of a normal heart.

Advanced Coronary Artery Disease
A color-enhanced DSA scan showing advanced coronary artery disease. Blood flow to the ventricular myocardium is severely restricted.

*Plaques may be visible by **angiography** or high-resolution ultrasound, and the effects on coronary blood flow can be detected in digital subtraction angiography (DSA) scans of the heart as shown above.*

Normal Artery

Tunica externa

Tunica media

Cross section

Narrowing of Artery

Lipid deposit of plaque

Cross section

Left: *Chest pain or discomfort is a common symptom of angina pectoris. Although similar symptoms may accompany a heart attack, during a heart attack, pain often radiates down the left arm.*

Below: *The presence of any two of these coronary risk factors more than doubles the risk of heart attack. It has been estimated that their reduction could prevent 150,000 deaths each year in the United States alone.*

Risk Factors for CAD and Myocardial Infarction

SMOKING • HIGH BLOOD PRESSURE • HIGH CHOLESTEROL •
• MALE OVER AGE 70 • SEVERE EMOTIONAL STRESS •

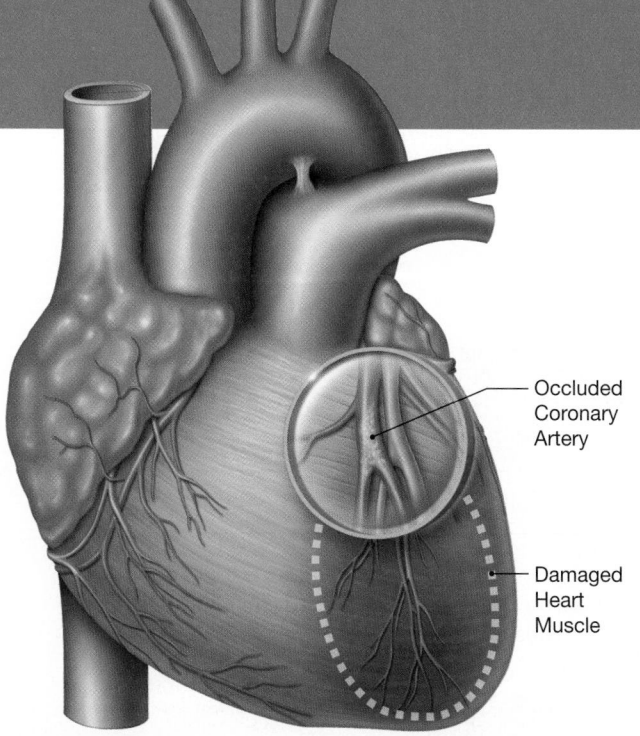

Occluded
Coronary
Artery

Damaged
Heart
Muscle

Myocardial Infarction
HEART ATTACK

In a **myocardial** (mī-ō-KAR-dē-al) **infarction (MI)**, or heart attack, part of the coronary circulation becomes blocked, and cardiac muscle cells die from lack of oxygen. The death of affected tissue creates a nonfunctional area known as an infarct. Heart attacks most commonly result from severe coronary artery disease (CAD). The consequences depend on the site and nature of the circulatory blockage. If it occurs near the start of one of the coronary arteries, the damage will be widespread and the heart may stop beating. If the blockage involves one of the smaller arterial branches, the individual may survive the immediate crisis but may have many complications such as reduced contractility and cardiac arrhythmias.

A crisis often develops as a result of thrombus formation at a plaque (the most common cause of an MI), a condition called **coronary thrombosis**. A vessel already narrowed by plaque formation may also become blocked by a sudden spasm in the smooth muscles of the vascular wall.

Individuals having an MI experience intense pain, similar to that felt in angina, but persisting even at rest. However, pain does not always accompany a heart attack, and silent heart attacks may be even more dangerous than more apparent attacks, because the condition may go undiagnosed and may not be treated before a fatal MI occurs. A myocardial infarction can usually be diagnosed with an ECG and blood studies. Damaged myocardial cells release enzymes into the circulation, and these elevated enzymes can be measured in diagnostic blood tests. The enzymes include **cardiac troponin T**, **cardiac troponin I**, and a special form of creatinine phosphokinase, **CK-MB**.

Treatment of CAD and Myocardial Infarction

About 25% of MI patients die before obtaining medical assistance, and 65% of MI deaths among those under age 50 occur within an hour after the initial infarction.

Risk Factor Modification

Stop smoking, high blood pressure treatment, dietary modification to lower cholesterol and promote weight loss, stress reduction, and increased physical activity (where appropriate)

Drug Treatment

- Drugs that reduce coagulation and therefore the risk of thrombosis, such as aspirin and coumadin
- Drugs that block sympathetic stimulation (propranolol or metoprolol)
- Drugs that cause vasodilation, such as **nitroglycerin** (nī-trō-GLIS-er-in)
- Drugs that block calcium movement into the cardiac and vascular smooth muscle cells (calcium channel blockers)
- In a myocardial infarction, drugs to relieve pain, fibrinolytic agents to help dissolve clots, and oxygen

Noninvasive Surgery

- **Atherectomy**. Blockage by a single, soft plaque may be reduced with the aid of a long, slender **catheter** (KATH-e-ter) inserted into a coronary artery to the plaque. A variety of surgical tools can be slid into the catheter, and the plaque can then be removed.
- **Balloon angioplasty** (AN-jē-ō-plas-tē; *angeion*, vessel). In balloon angioplasty, the tip of the catheter contains an inflatable balloon. Once in position, the balloon is inflated, pressing the plaque against the vessel walls. Because plaques commonly redevelop after angioplasty, a fine tubular wire mesh called a stent may be inserted into the vessel, holding it open.

Coronary Artery Bypass Graft (CABG)

In a coronary artery bypass graft, a small section is removed from either a small artery or a peripheral vein and is used to create a detour around the obstructed portion of a coronary artery. As many as four coronary arteries can be rerouted this way during a single operation. The procedures are named according to the number of vessels repaired, so we speak of single, double, triple, or quadruple coronary bypasses.

Figure 20–11   **The Conducting System of the Heart.**

b Changes in the membrane potential of a pacemaker cell in the SA node that is establishing a heart rate of 72 beats per minute. Note the presence of a prepotential, a gradual spontaneous depolarization.

a Components of the conducting system.

prepotential or *pacemaker potential* (Figure 20–11b). The prepotential results from a slow inflow of Na⁺ without a compensating outflow of K⁺.

The rate of spontaneous depolarization differs in various parts of the conducting system. It is fastest at the SA node. Without neural or hormonal stimulation, the SA node generates action potentials at a rate of 80–100 per minute. Isolated cells of the AV node depolarize more slowly, generating 40–60 action potentials per minute. Because the SA node reaches threshold first, it establishes the heart rate. In other words, the impulse generated by the SA node brings the AV nodal cells to threshold faster than does the prepotential of the AV nodal cells. The normal resting heart rate is somewhat slower than 80–100 beats per minute, however, due to the effects of parasympathetic innervation. (We discuss the influence of autonomic innervation on heart rate in a later section.)

If the SA node or any of the atrial pathways becomes damaged, the heart continues to beat, but at a slower rate, usually 40–60 beats per minute, as dictated by the AV node. Certain cells in the Purkinje fiber network depolarize spontaneously at an even slower rate. If the rest of the conducting system is damaged, these cells can stimulate a heart rate of 20–40 beats per minute. Under normal conditions, cells of the AV bundle, the bundle branches, and most Purkinje fibers do not depolarize spontaneously. If, due to damage or disease, these cells *do* begin depolarizing spontaneously, the heart may no longer pump blood effectively. Death can result if the problem persists.

Now let's trace the path of an impulse from its initiation at the SA node, examining its effects on the surrounding myocardium as we proceed.

The Sinoatrial (SA) Node

The **sinoatrial** (sī-nō-Ā-trē-al) **node (SA node)** is embedded in the posterior wall of the right atrium, near the entrance of the superior vena cava (Figure 20–12 **1**). The SA node contains **pacemaker cells**, which establish the heart rate. As a result, the SA node is also known as the *cardiac pacemaker* or the *natural pacemaker*.

The SA node is connected to the larger AV node by the internodal pathways in the atrial walls. An action potential takes approximately 50 msec to travel from the SA node to the AV node along these pathways. Along the way, the conducting cells pass the stimulus to contractile cells of both atria. The action potential then spreads across the atrial surfaces by cell-to-cell contact (Figure 20–12 **2**). The stimulus affects only the atria, because the cardiac skeleton isolates the atrial myocardium from the ventricular myocardium.

The Atrioventricular (AV) Node

The relatively large **atrioventricular (AV) node** (Figure 20–12 **2**) sits within the floor of the right atrium near the opening of the coronary sinus. The impulse slows as it leaves the internodal pathways and enters the AV node, because the nodal cells are smaller in diameter than the conducting cells. (Chapter 12 discussed the relationship between diameter and propagation speed. ⊃ p. 439) In addition, the connections between nodal cells are less efficient than those between conducting cells at relaying the impulse from one cell to another. As a result, the impulse takes about 100 msec to pass through the AV node (Figure 20–12 **3**). This delay is important because it allows the atria to contract before the ventricles do. Otherwise, contraction

Figure 20–12   Impulse Conduction through the Heart.

1 SA node activity and atrial activation begin.

SA node

Time = 0

2 Stimulus spreads across the atrial surfaces and reaches the AV node.

AV node

Elapsed time = 50 msec

3 There is a 100-msec delay at the AV node. Atrial contraction begins.

AV bundle

Bundle branches

Elapsed time = 150 msec

4 The impulse travels along the interventricular septum within the AV bundle and the bundle branches to the Purkinje fibers and, by the moderator band, to the papillary muscles of the right ventricle.

Moderator band

Elapsed time = 175 msec

5 The impulse is distributed by Purkinje fibers and relayed throughout the ventricular myocardium. Atrial contraction is completed, and ventricular contraction begins.

Elapsed time = 225 msec   Purkinje fibers

of the powerful ventricles would close the AV valves and prevent blood flow from the atria into the ventricles.

After this brief delay, the impulse is conducted along the atrioventricular bundle and the bundle branches to the Purkinje fibers and the papillary muscles (**Figure 20–12 ❹**). The Purkinje fibers then distribute the impulse to the ventricular myocardium, and ventricular contraction begins (**Figure 20–12 ❺**).

The cells of the AV node can conduct impulses at a maximum rate of 230 per minute. Because each impulse results in a ventricular contraction, this value is the maximum normal heart rate. Even if the SA node generates impulses at a faster rate, the ventricles will still contract at 230 beats per minute (bpm). This limitation is important, because mechanical factors (discussed later) begin to decrease the pumping efficiency of the heart at rates above approximately 180 bpm. Rates above 230 bpm occur only when the heart or the conducting system has been damaged or stimulated by drugs. As ventricular rates increase toward their theoretical maximum limit of 300–400 bpm, pumping effectiveness becomes dangerously, if not fatally, reduced.

A number of clinical problems result from abnormal pacemaker function. **Bradycardia** (brād-ē-KAR-dē-uh; *bradys*, slow) is a condition in which the heart rate is slower than normal. **Tachycardia** (tak-ē-KAR-dē-uh; *tachys*, swift) is a faster-thannormal heart rate. These terms are relative, and in clinical practice the definitions vary with the normal resting heart rate of the individual.

The AV Bundle, Bundle Branches, and Purkinje Fibers

The connection between the AV node and the **AV bundle**, also called the *bundle of His* (hiss), is normally the only electrical connection between the atria and the ventricles. Once an impulse enters the AV bundle, it travels to the interventricular septum and enters the **right** and **left bundle branches**. The left bundle branch, which supplies the massive left ventricle, is much larger than the right bundle branch. Both branches extend toward the apex of the heart, turn, and fan out deep to the endocardial surface. As the branches diverge, they conduct the impulse to **Purkinje fibers** and, through the moderator band, to the papillary muscles of the right ventricle.

Purkinje fibers conduct action potentials very rapidly—as fast as small myelinated axons. Within about 75 msec, the signal to begin a contraction has reached all the ventricular cardiac muscle cells. By this time, the atria have completed their contractions and ventricular contraction can safely occur. The entire process, from the generation of an impulse at the SA node to the complete depolarization of the ventricular myocardium, normally takes around 225 msec.

Because the bundle branches deliver the impulse across the moderator band to the papillary muscles directly, rather

20

than by way of Purkinje fibers, the papillary muscles begin contracting before the rest of the ventricular musculature does. Contraction of the papillary muscles applies tension to the chordae tendineae, bracing the AV valves. By limiting the movement of the cusps, tension in the chordae tendineae prevents the backflow of blood into the atria when the ventricles contract.

The Purkinje fibers radiate from the apex toward the base of the heart. As a result, the ventricles contract in a wave that begins at the apex and spreads toward the base. The contraction pushes blood toward the base of the heart, into the aorta and pulmonary trunk.

Damage to the conducting pathways disturbs the normal rhythm of the heart. The resulting problems are called *conduction deficits*. If the SA node or internodal pathways are damaged, the AV node assumes command. The heart continues beating normally, but at a slower rate.

If an abnormal conducting cell or ventricular muscle cell begins generating action potentials at a higher rate, the impulses can override those of the SA or AV node. The origin of these abnormal signals is called an **ectopic** (ek-TOP-ik; out of place) **pacemaker**. The activity of an ectopic pacemaker partially or completely bypasses the conducting system, disrupting the timing of ventricular contraction. The result may be a dangerous reduction in the pumping efficiency of the heart. Such conditions are commonly diagnosed with the aid of an *electrocardiogram*.

The Electrocardiogram

The electrical events in the heart are powerful enough to be detected by electrodes on the surface of the body. A recording of these events is an **electrocardiogram** (ē-lek-trō-KAR-dē-ō-gram), also called an **ECG** or **EKG**. Each time the heart beats, a wave of depolarization spreads through the atria, pauses at the AV node, then travels down the interventricular septum to the apex, turns, and spreads through the ventricular myocardium toward the base (**Figure 20–12**).

An ECG integrates electrical information from electrodes placed at different locations on the body surface. Clinicians can use an ECG to assess the performance of specific nodal, conducting, and contractile components. When a portion of the heart has been damaged by a heart attack, for example, the ECG reveals an abnormal pattern of impulse conduction.

The appearance of the ECG varies with the placement of the monitoring electrodes, or *leads*. **Figure 20–13a** shows the leads in one of the standard configurations. **Figure 20–13b** depicts the important features of an ECG recorded with that configuration. Note the following ECG features:

- The small **P wave**, which accompanies the depolarization of the atria. Depolarization of the atria causes atrial contraction.

- The **QRS complex**, which appears as the ventricles depolarize. This electrical signal is relatively strong, because the ventricular muscle is much more massive than that of the atria. It is also a complex signal, largely because of the complex pathway that the spread of depolarization takes through the ventricles. The ventricles begin contracting shortly after the peak of the **R wave**.

- The smaller **T wave**, which indicates ventricular repolarization. Atrial repolarization is not apparent, because it takes place while the ventricles are depolarizing, and the QRS complex masks the electrical events.

To analyze an ECG, you must measure the size of the voltage changes and determine the durations and temporal (time) relationships of the various components. The amount of depolarization during the P wave and the QRS complex is particularly important in making a diagnosis. For example, an excessively large QRS complex often indicates that the heart has become enlarged. A smaller-than-normal electrical signal may mean that the mass of the heart muscle has decreased (although monitoring problems are more often responsible). The size and shape of the T wave may also be affected by any condition that slows ventricular repolarization. For example, starvation and low cardiac energy reserves, coronary ischemia, or abnormal ion concentrations reduce the size of the T wave.

The times between waves are reported as *segments* and *intervals*. Segments generally extend from the end of one wave to the start of another. Intervals are more variable, but always include at least one entire wave. Commonly used segments and intervals are labeled in **Figure 20–13b**. The names, however, can be somewhat misleading. For example:

- The **P–R interval** extends from the start of atrial depolarization to the start of the QRS complex (ventricular depolarization) rather than to R, because in abnormal ECGs the peak at R can be difficult to determine. Extension of the P–R interval to more than 200 msec can indicate damage to the conducting pathways or AV node.

- The **Q–T interval** indicates the time required for the ventricles to undergo a single cycle of depolarization and repolarization. It is usually measured from the end of the P–R interval rather than from the bottom of the Q wave. The Q–T interval can be lengthened by electrolyte disturbances, some medications, conduction problems, coronary ischemia, or myocardial damage. A congenital heart defect that can cause sudden death without warning may be detectable as a prolonged Q–T interval.

An *arrhythmia* (ā-RITH-mē-uh) is an irregularity in the normal rhythm or force of the heartbeat. Serious arrhythmias may indicate damage to the myocardium, injuries to the pacemakers or conduction pathways, exposure to drugs, or

Figure 20–13 **An Electrocardiogram.**

a Electrode placement for recording a standard ECG.

b An ECG printout is a strip of graph paper containing a record of the electrical events monitored by the electrodes. The placement of electrodes on the body surface affects the size and shape of the waves recorded. The example is a normal ECG; the enlarged section indicates the major components of the ECG and the measurements most often taken during clinical analysis.

abnormalities in the electrolyte composition of extracellular fluids. **Spotlight Figure 20–14** describes cardiac arrhythmias.

Contractile Cells

The Purkinje fibers distribute the stimulus to the **contractile cells**, which form the bulk of the atrial and ventricular walls. These cells account for about 99 percent of the muscle cells in the heart. In both cardiac muscle cells and skeletal muscle fibers, (1) an action potential leads to the appearance of Ca^{2+} among the myofibrils, and (2) the binding of Ca^{2+} to troponin on the thin filaments initiates the contraction. But skeletal and cardiac muscle cells differ in terms of the nature of the action potential, the source of the Ca^{2+}, and the duration of the resulting contraction. ⟲ p. 353

The Action Potential in Cardiac Muscle Cells

The resting potential of a ventricular contractile cell is approximately –90 mV, comparable to that of a resting skeletal muscle fiber (–85 mV). (The resting potential of an atrial

contractile cell is about –80 mV, but the basic principles described here apply to atrial cells as well.) An action potential begins when the membrane of the ventricular muscle cell reaches threshold, usually at about –75 mV. Threshold is normally reached in a portion of the membrane next to an intercalated disc. The typical stimulus is the excitation of an adjacent muscle cell. Once threshold has been reached, the action potential proceeds in three basic steps (**Figure 20–15a**):

1 *Rapid Depolarization.* The stage of *rapid depolarization* in a cardiac muscle cell resembles that in a skeletal muscle fiber. At threshold, voltage-gated sodium channels open, and the membrane suddenly becomes permeable to Na^+. A massive influx of sodium ions rapidly depolarizes the sarcolemma. The channels involved are called **fast sodium channels**, because they open quickly and remain open for only a few milliseconds.

2 *The Plateau.* As the membrane potential approaches +30 mV, the voltage-gated sodium channels close. They

Figure 20–14
Cardiac Arrhythmias

Despite the variety of sophisticated equipment available to assess or visualize cardiac function, in the majority of cases the ECG provides the most important diagnostic information. ECG analysis is especially useful in detecting and diagnosing **cardiac arrhythmias** (ā-RITH-mē-uz)—abnormal patterns of cardiac electrical activity. Momentary arrhythmias are not inherently dangerous, but clinical problems appear when arrhythmias reduce the pumping efficiency of the heart.

Premature Atrial Contractions (PACs)

Premature atrial contractions (PACs) often occur in healthy individuals. In a PAC, the normal atrial rhythm is momentarily interrupted by a "surprise" atrial contraction. Stress, caffeine, and various drugs may increase the incidence of PACs, presumably by increasing the permeabilities of the SA pacemakers. The impulse spreads along the conduction pathway, and a normal ventricular contraction follows the atrial beat.

Paroxysmal Atrial Tachycardia (PAT)

In **paroxysmal** (par-ok-SIZ-mal) **atrial tachycardia**, or **PAT**, a premature atrial contraction triggers a flurry of atrial activity. The ventricles are still able to keep pace, and the heart rate jumps to about 180 beats per minute.

Atrial Fibrillation (AF)

During **atrial fibrillation** (fib-ri-LĀ-shun), the impulses move over the atrial surface at rates of perhaps 500 beats per minute. The atrial wall quivers instead of producing an organized contraction. The ventricular rate cannot follow the atrial rate and may remain within normal limits. Even though the atria are now nonfunctional, their contribution to ventricular end-diastolic volume (the maximum amount of blood the ventricles can hold at the end of atrial contraction) is so small that the condition may go unnoticed in older individuals.

Premature Ventricular Contractions (PVCs)

Premature ventricular contractions (PVCs) occur when a Purkinje cell or ventricular myocardial cell depolarizes to threshold and triggers a premature contraction. Single PVCs are common and not dangerous. The cell responsible is called an **ectopic pacemaker**. The frequency of PVCs can be increased by exposure to epinephrine, to other stimulatory drugs, or to ionic changes that depolarize cardiac muscle plasma membranes.

Ventricular Tachycardia (VT)

Ventricular tachycardia is defined as four or more PVCs without intervening normal beats. It is also known as **VT** or **V-tach**. Multiple PVCs and VT may indicate that serious cardiac problems exist.

Ventricular Fibrillation (VF)

Ventricular fibrillation (VF) is responsible for the condition known as **cardiac arrest**. VF is rapidly fatal, because the ventricles quiver and stop pumping blood.

Figure 20–15   **The Action Potentials in Skeletal and Cardiac Muscle.**

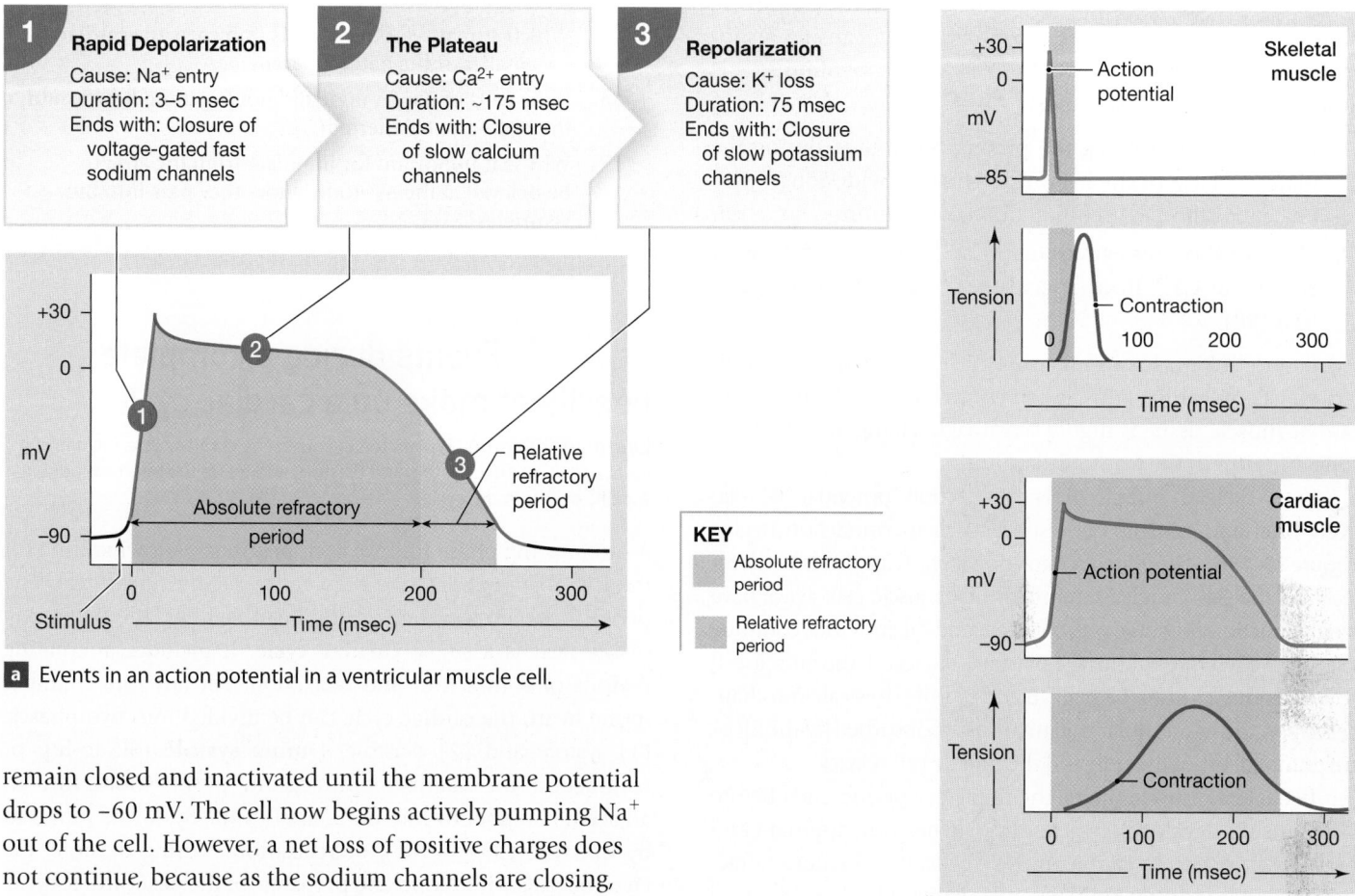

1 Rapid Depolarization
Cause: Na⁺ entry
Duration: 3–5 msec
Ends with: Closure of
voltage-gated fast
sodium channels

2 The Plateau
Cause: Ca²⁺ entry
Duration: ~175 msec
Ends with: Closure
of slow calcium
channels

3 Repolarization
Cause: K⁺ loss
Duration: 75 msec
Ends with: Closure
of slow potassium
channels

KEY
Absolute refractory period
Relative refractory period

a Events in an action potential in a ventricular muscle cell.

b Action potentials and twitch contractions in a skeletal muscle (above) and cardiac muscle (below). The shaded areas indicate the durations of the absolute (blue) and relative (beige) refractory periods.

remain closed and inactivated until the membrane potential drops to –60 mV. The cell now begins actively pumping Na⁺ out of the cell. However, a net loss of positive charges does not continue, because as the sodium channels are closing, voltage-gated calcium channels are opening. These channels are called **slow calcium channels**, because they open slowly and remain open for a relatively long period—roughly 175 msec. While the slow calcium channels are open, calcium ions enter the cytosol. The entry of positive charges through the calcium channels, in the form of Ca²⁺, roughly balances the loss of positive ions through the active transport of Na⁺, and the membrane potential remains near 0 mV for an extended period. This portion of the action potential curve is called the *plateau*. The presence of a plateau is the major difference between action potentials in cardiac muscle cells and in skeletal muscle fibers. In a skeletal muscle fiber, rapid depolarization is immediately followed by rapid repolarization.

3 *Repolarization.* As the plateau continues, slow calcium channels begin closing, and **slow potassium channels** begin opening. As these channels open, potassium ions (K⁺) rush out of the cell, and the net result is a period of rapid repolarization that restores the resting potential.

The Refractory Period. As with skeletal muscle contractions, the membrane will not respond normally to a second stimulus

for some time after an action potential begins. This time is called the refractory period. Initially, in the absolute refractory period, the membrane cannot respond at all, because the sodium channels are either already open or closed and inactivated. In a ventricular muscle cell, the absolute refractory period lasts approximately 200 msec. It includes the plateau and the initial period of rapid repolarization.

The absolute refractory period is followed by a shorter (50 msec) relative refractory period. During this period, the voltage-gated sodium channels are closed, but can open. The membrane will respond to a stronger-than-normal stimulus by initiating another action potential.

In total, an action potential in a ventricular contractile cell lasts 250–300 msec, about 30 times as long as a typical action potential in a skeletal muscle fiber.

The Role of Calcium Ions in Cardiac Contractions

The appearance of an action potential in the cardiac muscle plasma membrane produces a contraction by causing an increase in the concentration of Ca^{2+} around the myofibrils. This process takes place in two steps:

1. Calcium ions crossing the plasma membrane during the plateau phase of the action potential provide roughly 20 percent of the Ca^{2+} required for a contraction.

2. The arrival of this extracellular Ca^{2+} triggers the release of additional Ca^{2+} from reserves in the sarcoplasmic reticulum (SR).

Extracellular calcium ions thus have both direct and indirect effects on cardiac muscle cell contraction. For this reason, cardiac muscle tissue is highly sensitive to changes in the Ca^{2+} concentration of the extracellular fluid.

In a skeletal muscle fiber, the action potential is relatively brief and ends as the resulting twitch contraction begins (**Figure 20–15b**). The twitch contraction is short and ends as the SR reclaims the Ca^{2+} it released. In a cardiac muscle cell, as we have seen, the action potential is prolonged, and calcium ions continue to enter the cell throughout the plateau. As a result, the muscle cell actively contracts until the plateau ends. As the slow calcium channels close, the intracellular calcium ions are absorbed by the SR or are pumped out of the cell, and the muscle cell relaxes.

In skeletal muscle fibers, the refractory period ends before peak tension develops. As a result, twitches can summate and tetanus can occur. In cardiac muscle cells, the absolute refractory period continues until relaxation is under way. Because summation is not possible, tetanic contractions cannot occur in a normal cardiac muscle cell, regardless of the frequency or intensity of stimulation. This feature is vital: A heart in tetany could not pump blood. With a single twitch lasting 250 msec or longer, a normal cardiac muscle cell could reach 300–400 contractions per minute under maximum stimulation. This rate is not reached in a normal heart, due to limitations imposed by the conducting system.

The Energy for Cardiac Contractions

When a normal heart is beating, it gets energy as the mitochondria break down fatty acids (stored as lipid droplets) and glucose (stored as glycogen). These aerobic reactions can occur only when oxygen is readily available. ⊃ p. 343

In addition to obtaining oxygen from the coronary circulation, cardiac muscle cells maintain their own sizable reserves of oxygen. In these cells, oxygen molecules are bound to the heme units of myoglobin molecules. (We discussed this globular protein, which reversibly binds oxygen molecules, and its function in muscle fibers in Chapter 10.) ⊃ p. 346 Normally, the combination of circulatory supplies plus myoglobin reserves is enough to meet the oxygen demands of the heart, even when it is working at maximum capacity.

CHECKPOINT

4. Define automaticity.

5. Which structure of the heart is known as the cardiac pacemaker or the natural pacemaker?

6. If the cells of the SA node did not function, how would the heart rate be affected?

7. Why is it important for impulses from the atria to be delayed at the AV node before they pass into the ventricles?

See the blue Answers tab at the back of the book.

20-3 ❚ Events during a complete heartbeat make up a cardiac cycle

Learning Outcome Explain the events of the cardiac cycle, including atrial and ventricular systole and diastole, and relate the heart sounds to specific events in the cycle.

A brief resting phase follows each heartbeat, allowing time for the chambers to relax and prepare for the next heartbeat. The period between the start of one heartbeat and the beginning of the next is a single **cardiac cycle**. It includes alternating periods of contraction and relaxation. For any one chamber in the heart, the cardiac cycle can be divided into two phases: (1) systole and (2) diastole. During **systole** (SIS-tō-lē), or contraction, the chamber contracts and pushes blood into an adjacent chamber or into an arterial trunk. Systole is followed by **diastole** (dī-AS-tō-lē), or relaxation. During diastole, the chamber fills with blood and prepares for the next cardiac cycle.

TIPS & TRICKS

To remember that systole is contraction, relate "**syst**" to "**system**," as in during systole, a contraction sends blood out of the heart and into the bloodstream. The word part "**di**" can mean two, so during **di**astole, the heart "puts two feet up" and relaxes.

Fluids move from an area of higher pressure to an area of lower pressure. In the cardiac cycle, the pressure within each chamber rises during systole and falls during diastole. Valves between adjacent chambers help ensure that blood flows in the required direction, but blood flows from one chamber to another only if the pressure in the first chamber exceeds that in the second. This basic principle governs the movement of blood between atria and ventricles, between ventricles and arterial trunks, and between major veins and atria.

The correct pressure relationships depend on the careful timing of contractions. For example, blood could not move in the proper direction if an atrium and its attached ventricle contracted at precisely the same moment. The heart's elaborate pacemaking and conducting systems normally provide the required spacing between atrial systole and ventricular systole.

At a representative heart rate of 75 bpm, a sequence of systole and diastole in either the atria or the ventricles lasts 800 msec.

Phases of the Cardiac Cycle

The phases of the cardiac cycle—atrial systole, atrial diastole, ventricular systole, and ventricular diastole—are diagrammed in **Figure 20–16** for a heart rate of 75 bpm. When the cardiac cycle begins, all four chambers are relaxed, and the ventricles are partially filled with blood. Let's start by focusing on the atria. During atrial systole, the atria contract, filling the ventricles completely with blood (**Figure 20–16a,b**). Atrial systole lasts 100 msec. Over this period, blood cannot flow into the atria because atrial pressure exceeds venous pressure. Yet there is very little backflow into the veins, even though the connections with the venous system lack valves. The reason is that blood takes the path of least resistance. Resistance to blood flow through the broad AV connections and into the ventricles is less than that through the smaller, angled openings of the large veins.

Figure 20–16   Phases of the Cardiac Cycle. Thin black arrows indicate blood flow, green arrows indicate contractions, and the red-hued areas indicate which heart chambers are in systole.

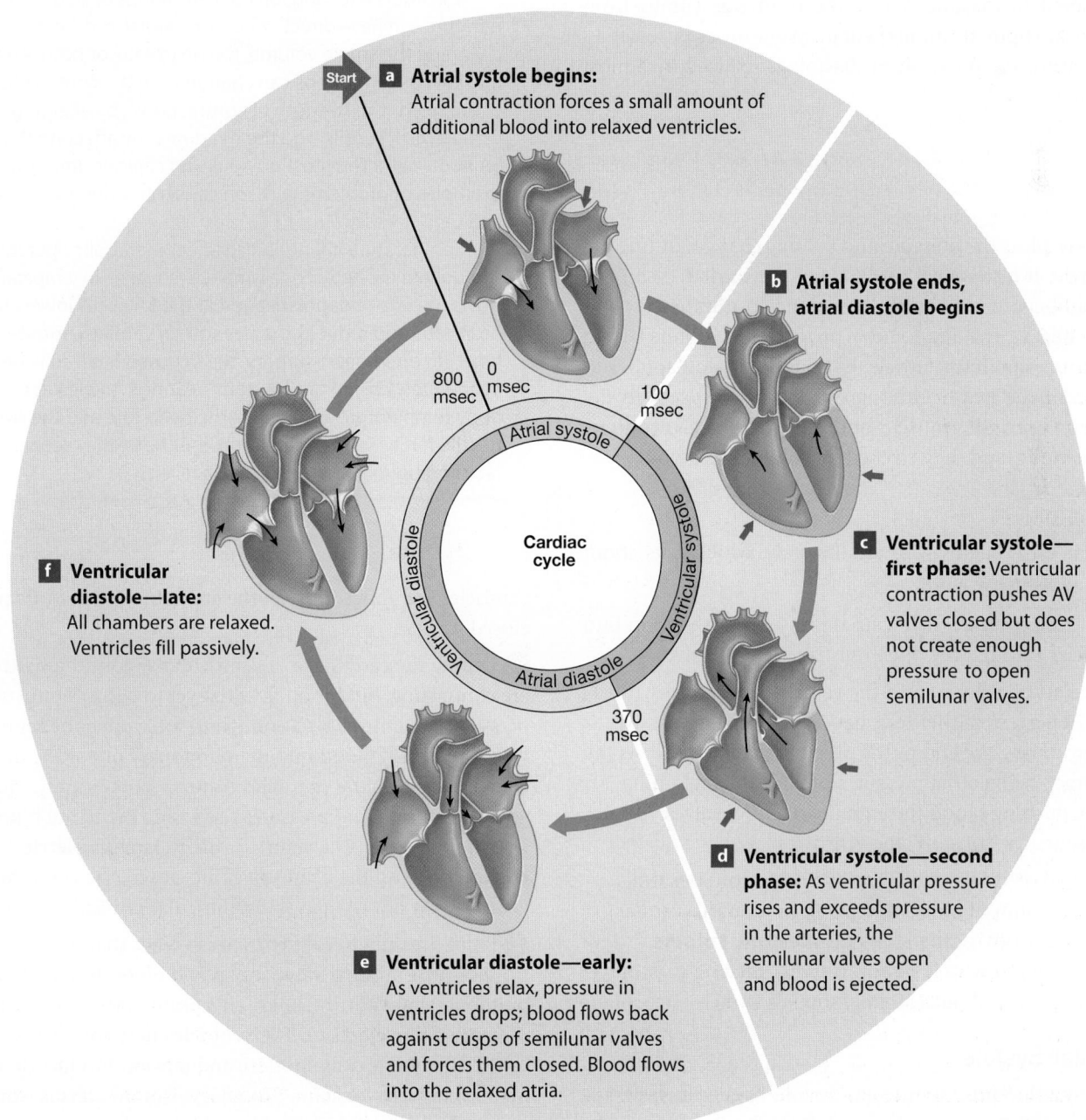

Start **a** **Atrial systole begins:** Atrial contraction forces a small amount of additional blood into relaxed ventricles.

b **Atrial systole ends, atrial diastole begins**

c **Ventricular systole— first phase:** Ventricular contraction pushes AV valves closed but does not create enough pressure to open semilunar valves.

d **Ventricular systole—second phase:** As ventricular pressure rises and exceeds pressure in the arteries, the semilunar valves open and blood is ejected.

e **Ventricular diastole—early:** As ventricles relax, pressure in ventricles drops; blood flows back against cusps of semilunar valves and forces them closed. Blood flows into the relaxed atria.

f **Ventricular diastole—late:** All chambers are relaxed. Ventricles fill passively.

Cardiac cycle

Atrial systole
Ventricular systole
Atrial diastole
Ventricular diastole

800 msec
0 msec
100 msec
370 msec

20

20

The atria next enter atrial diastole, which continues until the start of the next cardiac cycle. Atrial diastole and ventricular systole begin at the same time. Ventricular systole lasts 270 msec. During this period, the ventricles push blood through the systemic and pulmonary circuits and toward the atria (**Figure 20–16c,d**). The heart then enters ventricular diastole (**Figure 20–16e,f**), which lasts 530 msec (the 430 msec remaining in this cardiac cycle, plus the first 100 msec of the next, when the atria are again contracting). For the rest of this cycle, filling occurs passively, and both the atria and the ventricles are relaxed. The next cardiac cycle begins with atrial systole, which completes the filling of the ventricles.

When the heart rate increases, all the phases of the cardiac cycle are shortened. The greatest reduction occurs in the length of time spent in diastole. When the heart rate climbs from 75 bpm to 200 bpm, the time spent in systole drops by less than 40 percent, but the duration of diastole is reduced by almost 75 percent.

Pressure and Volume Changes in the Cardiac Cycle

Figure 20–17 plots the pressure and volume changes during the cardiac cycle. It also shows an ECG for the cardiac cycle. The circled numbers in the figure correspond to numbered paragraphs in the text. The figure shows pressure and volume within the left atrium and left ventricle, but our discussion applies to both sides of the heart. Although pressures are lower in the right atrium and right ventricle, both sides of the heart contract at the same time, and they eject equal volumes of blood.

Atrial Systole

The cardiac cycle begins with atrial systole, which lasts about 100 msec in a resting adult:

1 As the atria contract, rising atrial pressures push blood into the ventricles through the open right and left AV valves.

2 At the start of atrial systole, the ventricles are already filled to about 70 percent of their normal capacity, due to passive blood flow during the end of the previous cardiac cycle. As the atria contract, rising atrial pressures provide the remaining 30 percent by pushing blood through the open AV valves. Atrial systole essentially "tops off" the ventricles.

3 At the end of atrial systole, each ventricle contains the maximum amount of blood that it will hold in this cardiac cycle. That quantity is called the **end-diastolic volume (EDV)**. In an adult who is standing at rest, the end-diastolic volume is typically about 130 mL (about 4.4 oz).

Ventricular Systole

As atrial systole ends, ventricular systole begins. It lasts approximately 270 msec in a resting adult. As the pressures in the

Clinical Note

Abnormal Conditions Affecting Cardiac Output

Various drugs, abnormal variations in extracellular ion concentrations, and changes in body temperature can alter the basic rhythm of contraction established by the SA node. In Chapter 12, we noted that several drugs, including caffeine and nicotine, have a stimulatory effect on the excitable membranes in the nervous system. ⊃ p. 448 These drugs also cause an increase in heart rate. Caffeine acts directly on the conducting system. It increases the rate of depolarization at the SA node. Nicotine directly stimulates the activity of sympathetic neurons that innervate the heart.

Disorders affecting ion concentrations or body temperature can have direct effects on cardiac output by changing the stroke volume, the heart rate, or both. Abnormal ion concentrations can change both the contractility of the heart (the strength of contractions), by affecting the cardiac muscle cells, and the heart rate, by affecting the SA nodal cells. The most obvious and clinically important examples of problems with ion concentrations involve K^+ and Ca^{2+}.

Temperature changes also affect metabolic operations throughout the body. For example, a reduction in temperature slows the rate of depolarization at the SA node, lowers the heart rate, and reduces the strength of cardiac contractions. (In open-heart surgery, the exposed heart may be deliberately chilled until it stops beating.) An elevated body temperature accelerates the heart rate and the contractile force. That is one reason your heart may seem to race and pound when you have a fever.

ventricles rise above those in the atria, the AV valves are pushed closed.

4 During the early stage of ventricular systole, the ventricles are contracting, but blood flow has yet to occur. Ventricular pressures are not yet high enough to force open the semilunar valves and push blood into the pulmonary or aortic trunk. Over this period, the ventricles contract isometrically. They generate tension and pressures rise inside them, but blood does not flow out. The ventricles are in **isovolumetric contraction**: All the heart valves are closed, the volumes of the ventricles do not change, and ventricular pressures are rising.

5 Once pressure in the ventricles exceeds that in the arterial trunks, the semilunar valves open and blood flows into the pulmonary and aortic trunks. This point marks the beginning of **ventricular ejection**. The ventricles now contract isotonically: The muscle cells shorten, and tension production remains relatively constant. (To review isotonic versus isometric contractions, see **Figure 10–18**, p. 341.)

Figure 20–17 **Pressure and Volume Relationships in the Cardiac Cycle.** Major features of the cardiac cycle are shown for a heart rate of 75 bpm. The circled numbers correspond to those in the associated box. For further details, see the numbered list in the text.

After reaching a peak, ventricular pressures gradually decline near the end of ventricular systole. **Figure 20–17** shows values for the left ventricle and aorta. The right ventricle also goes through periods of isovolumetric contraction and ventricular ejection, but pressures in the right ventricle and pulmonary trunk are much lower.

During ventricular ejection, each ventricle ejects 70–80 mL of blood, the **stroke volume (SV)** of the heart. The stroke volume at rest is roughly 60 percent of the end-diastolic volume. This percentage, known as the *ejection fraction*, varies in response to changing demands on the heart. (We discuss the regulatory mechanisms involved in the next section.)

6 As the end of ventricular systole approaches, ventricular pressures fall rapidly. Blood in the aorta and pulmonary trunk now starts to flow back toward the ventricles, and this movement closes the semilunar valves. As the backflow begins, pressure decreases in the aorta. When the semilunar valves close, pressure rises again as the elastic arterial walls recoil. This small, temporary rise produces a valley in the pressure tracing, called a *dicrotic* (dī-KROT-ik; *dikrotos*, double beating) *notch*. The amount of blood remaining in the ventricle when the semilunar valve closes is the **end-systolic volume (ESV)**. At rest, the end-systolic volume is 50 mL, about 40 percent of the end-diastolic volume.

Ventricular Diastole

The period of ventricular diastole lasts for the 430 msec remaining in the current cardiac cycle and continues through atrial systole in the next cycle.

7 All the heart valves are now closed, and the ventricular myocardium is relaxing. Because ventricular pressures are still higher than atrial pressures, blood cannot flow into the ventricles. This is **isovolumetric relaxation**. Ventricular pressures drop rapidly over this period, because the elasticity of the connective tissues of the heart and cardiac skeleton helps re-expand the ventricles toward their resting dimensions.

8 When ventricular pressures fall below those of the atria, the atrial pressures force the AV valves open. Blood now flows from the atria into the ventricles. Both the atria and the ventricles are in diastole, but the ventricular pressures continue to fall as the ventricular chambers expand. Throughout this period, pressures in the ventricles are so far below those in the major veins that blood pours through the relaxed atria and on through the open AV valves into the ventricles. This passive mechanism is the primary method of ventricular filling. The ventricles become nearly three-quarters full before the cardiac cycle ends.

The relatively minor contribution that atrial systole makes to ventricular volume explains why individuals can survive quite normally when their atria have been so severely damaged that they can no longer function. In contrast, damage to one or both ventricles can leave the heart unable to pump enough blood through peripheral tissues and organs. A condition of **heart failure** then exists.

Heart Sounds

Listening to the heart, a technique called *auscultation*, is a simple and effective method of cardiac assessment. Clinicians use an instrument called a **stethoscope** to listen for normal and abnormal heart sounds. Where the stethoscope is placed depends on which valve is under examination (**Figure 20–18a**).

Figure 20–18 **Heart Sounds.**

a Placements of a stethoscope for listening to the different sounds produced by individual valves

b The relationship between heart sounds and key events in the cardiac cycle

Valve sounds must pass through the pericardium, surrounding tissues, and the chest wall, and some tissues muffle sounds more than others. For this reason, the placement of the stethoscope differs somewhat from the position of the valve under review.

There are four heart sounds, named S_1 through S_4 (**Figure 20–18b**). If you listen to your own heart with a stethoscope, you will clearly hear the *first* and *second heart sounds*. These sounds accompany the closing of your heart valves. The first heart sound, known as "lubb" (S_1), lasts a little longer than the second, called "dupp" (S_2). S_1 marks the start of ventricular contraction, when the AV valves close and the semilunar valves open. S_2 occurs at the beginning of ventricular filling, when the semilunar valves close and the AV valves open.

Third and *fourth heart sounds* are usually very faint and are seldom audible in healthy adults. These sounds are associated with blood flowing into the ventricles (S_3) and atrial contraction (S_4), rather than with valve action.

If the valve cusps are malformed or there are problems with the papillary muscles or chordae tendineae, the heart valves may not close properly. AV valve regurgitation then occurs during ventricular systole. The surges, swirls, and eddies that accompany regurgitation create a rushing, gurgling sound known as a **heart murmur**. Minor heart murmurs are common and inconsequential.

CHECKPOINT

8. **Give the technical terms for heart contraction and heart relaxation.**

9. **List the phases of the cardiac cycle.**

10. **Is the heart always pumping blood when pressure in the left ventricle is rising? Explain.**

11. **What could cause an increase in the size of the QRS complex in an electrocardiogram?**

See the blue Answers tab at the back of the book.

20-4 ▮ Cardiodynamics examines the factors that affect cardiac output

Learning Outcome Define cardiac output, describe the factors that influence heart rate and stroke volume, and explain how adjustments in stroke volume and cardiac output are coordinated at different levels of physical activity.

The term **cardiodynamics** refers to the movements and forces generated during cardiac contractions. Each time the heart beats, the two ventricles eject equal amounts of blood. Earlier we introduced these terms:

- *End-Diastolic Volume (EDV):* The amount of blood in each ventricle at the end of ventricular diastole (the start of ventricular systole).

- *End-Systolic Volume (ESV):* The amount of blood remaining in each ventricle at the end of ventricular systole (the start of ventricular diastole).

- *Stroke Volume (SV):* The amount of blood pumped out of each ventricle during a single beat. It can be expressed as $SV = EDV - ESV$.

- *Ejection Fraction:* The percentage of the EDV represented by the SV.

Stroke volume is the most important factor in an examination of a single cardiac cycle. If the heart were a simple hand pump, the stroke volume would be the amount of water pumped in one up–down cycle of the handle (**Figure 20–19**). Where you stop when you lift the handle determines how much water the pump contains—the end-diastolic volume. How far down you push the handle determines how much water remains in the pump at the end of the cycle—the end-systolic volume. You pump the maximum amount of water when you pull the handle all the way to the top and then push it all the way to the bottom. In other words, you get the largest stroke volume when the EDV is as large as it can be and the ESV is as small as it can be.

When considering cardiac function over time, physicians generally are most interested in the **cardiac output (CO)**, the amount of blood pumped by the left ventricle in one minute. In essence, cardiac output is an indication of the blood flow through peripheral tissues—and without adequate blood flow, homeostasis cannot be maintained. The cardiac output provides a useful indication of ventricular efficiency over time. We can calculate it by multiplying the heart rate (HR) by the average stroke volume (SV):

CO	=	HR	×	SV
cardiac		heart		stroke
output		rate		volume
(mL/min)		(beats/min)		(mL/beat)

For example, if the heart rate is 75 bpm and the stroke volume is 80 mL per beat, the cardiac output is

CO = 75 bpm × 80 mL/beat = 6000 mL/min (6 L/min)

The body precisely adjusts cardiac output to supply peripheral tissues with enough blood as conditions change. When necessary, the heart rate can increase by 250 percent, and stroke volume in a normal heart can almost double.

Overview: Factors Affecting Cardiac Output

Figure 20–20 summarizes the factors involved in the normal regulation of cardiac output. Cardiac output can be adjusted by changes in either heart rate or stroke volume. For convenience, we will consider these independently as we discuss the

Figure 20–19 **A Simple Model of Stroke Volume.** The stroke volume of the heart can be compared to the amount of water ejected from a simple pump. The amount of water ejected varies with the amount of movement of the pump handle.

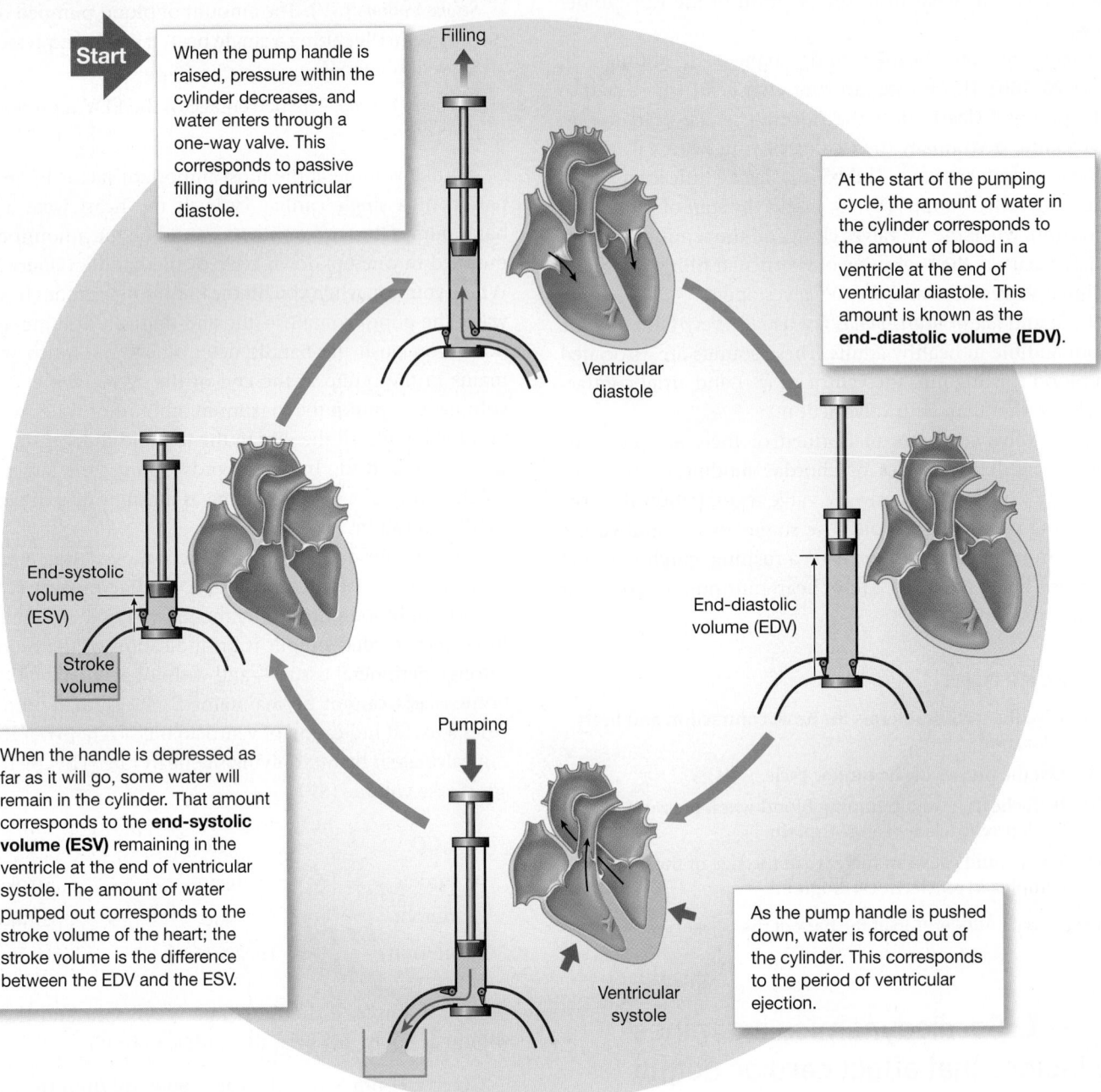

Start

Filling

When the pump handle is raised, pressure within the cylinder decreases, and water enters through a one-way valve. This corresponds to passive filling during ventricular diastole.

At the start of the pumping cycle, the amount of water in the cylinder corresponds to the amount of blood in a ventricle at the end of ventricular diastole. This amount is known as the **end-diastolic volume (EDV)**.

Ventricular diastole

End-diastolic volume (EDV)

End-systolic volume (ESV)

Stroke volume

Pumping

When the handle is depressed as far as it will go, some water will remain in the cylinder. That amount corresponds to the **end-systolic volume (ESV)** remaining in the ventricle at the end of ventricular systole. The amount of water pumped out corresponds to the stroke volume of the heart; the stroke volume is the difference between the EDV and the ESV.

As the pump handle is pushed down, water is forced out of the cylinder. This corresponds to the period of ventricular ejection.

Ventricular systole

individual factors involved. However, changes in cardiac output generally reflect changes in both heart rate and stroke volume.

The heart rate can be adjusted by the activities of the autonomic nervous system or by circulating hormones. The stroke volume can be adjusted by changes in the end-diastolic volume (how full a ventricle is when it starts to contract), the end-systolic volume (how much blood remains in a ventricle after it contracts), or both. As we saw in **Figure 20–19**, stroke volume peaks when EDV is highest and ESV is lowest. A variety

of other factors can influence cardiac output under abnormal circumstances, and we consider several examples in the Clinical Note on p. 734.

Factors Affecting the Heart Rate

Under normal circumstances, autonomic activity and circulating hormones make homeostatic adjustments to the heart rate as cardiovascular demands change. These factors act by modifying the natural rhythm of the heart. Even a heart removed for a

Figure 20–20 **Factors Affecting Cardiac Output.**

Figure 20–21 **Autonomic Innervation of the Heart.**

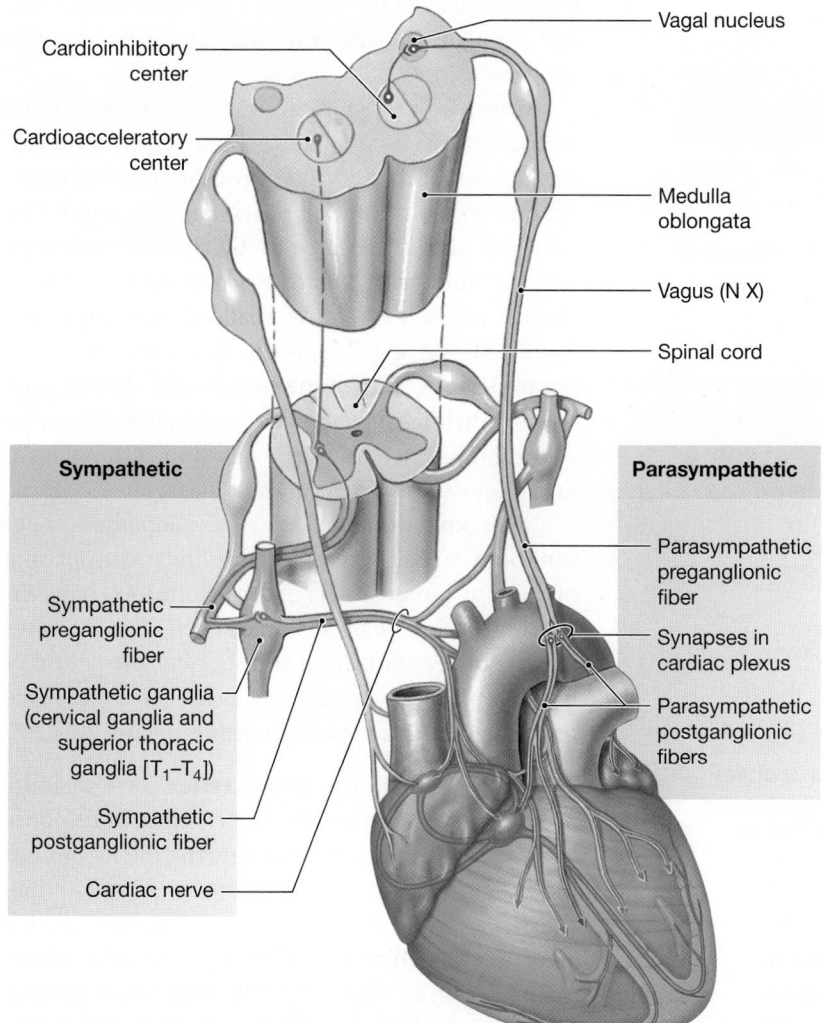

heart transplant continues to beat unless it is kept chilled in a preservation solution.

Autonomic Innervation

The sympathetic and parasympathetic divisions of the autonomic nervous system innervate the heart by means of the nerve network known as the *cardiac plexus* (**Figure 16–8**, p. 574, and **Figure 20–21**). Postganglionic sympathetic neurons are located in the cervical and upper thoracic ganglia. The vagus nerves (N X) carry parasympathetic preganglionic fibers to small ganglia in the cardiac plexus. Both ANS divisions innervate the SA and AV nodes and the atrial muscle cells. Both divisions also innervate ventricular muscle cells, but sympathetic fibers far outnumber parasympathetic fibers there.

The *cardiac centers* of the medulla oblongata contain the autonomic headquarters for cardiac control. ⊃ p. 496 The **cardioacceleratory center** controls sympathetic neurons that increase the heart rate. The adjacent **cardioinhibitory center** controls the parasympathetic neurons that slow the heart rate. Reflex pathways regulate the cardiac centers. They also receive input from higher centers, especially from the parasympathetic and sympathetic headquarters in the hypothalamus.

TIPS&TRICKS

To remember the effect of the sympathetic nervous system on cardiac performance, remember that **s**ympathetic input **s**peeds and **s**trengthens the heartbeat.

Cardiac Reflexes. Information about the status of the cardiovascular system arrives over visceral sensory fibers accompanying the vagus nerve and the sympathetic nerves of the cardiac plexus. The cardiac centers monitor baroreceptors and chemoreceptors innervated by the glossopharyngeal (N IX) and vagus (N X) nerves. ⊃ pp. 525, 526 On the basis of the information received, the cardiac centers adjust the heart's activity to maintain adequate circulation to vital organs, such as the brain.

The cardiac centers respond to changes in blood pressure as reported by baroreceptors, and to changes in arterial concentrations of dissolved oxygen and carbon dioxide as reported by chemoreceptors. For example, a decline in blood pressure or oxygen concentrations or an increase in carbon dioxide levels generally means that the heart must work harder to meet the demands of peripheral tissues. The cardiac centers then call for an increase in cardiac activity. We detail these reflexes and their effects on the heart and peripheral vessels in Chapter 21.

20

Figure 20–22 **Autonomic Regulation of Pacemaker Function.**

a Pacemaker cells have membrane potentials closer to threshold than those of other cardiac muscle cells (–60 mV versus –90 mV). Their plasma membranes undergo spontaneous depolarization to threshold, producing action potentials at a frequency determined by (1) the membrane potential and (2) the rate of depolarization.

b Parasympathetic stimulation releases ACh, which extends repolarization and decreases the rate of spontaneous depolarization. The heart rate slows.

c Sympathetic stimulation releases NE, which shortens repolarization and accelerates the rate of spontaneous depolarization. As a result, the heart rate increases.

Autonomic Tone. Like other organs with dual innervation, the heart has a resting autonomic tone. Both autonomic divisions are normally active at a steady background level, releasing ACh and NE at the nodes and into the myocardium. For this reason,

cutting the vagus nerves increases the heart rate, and sympathetic blocking agents slow the heart rate.

Parasympathetic effects dominate in a healthy, resting individual. Without autonomic innervation, the pacemaker cells of the SA node establish the heart rate. Such a heart beats at a rate of 80–100 bpm. At rest, a typical adult heart with normal innervation beats more slowly, at 70–80 bpm, due to activity in the parasympathetic nerves innervating the SA node. If parasympathetic activity increases, the heart rate declines further. Conversely, the heart rate increases if parasympathetic activity decreases, or if sympathetic activation occurs. Through dual innervation and adjustments in autonomic tone, the ANS can make very delicate adjustments in cardiovascular function to meet the demands of other systems.

Effects on the SA Node. How do the sympathetic and parasympathetic divisions alter the heart rate? They do so by changing the ionic permeabilities of cells in the conducting system. The most dramatic effects take place at the SA node, which affects the heart rate through changes in the rate at which impulses are generated.

Consider the SA node of a resting individual whose heart is beating at 75 bpm (**Figure 20–22a**). Any factor that changes the rate of spontaneous depolarization or the duration of repolarization in nodal cells will alter the heart rate by changing the time required for these cells to reach threshold. Acetylcholine released by parasympathetic neurons opens chemically gated K^+ channels in the plasma membrane. Then K^+ leaves the nodal cells, dramatically slowing their rate of spontaneous depolarization and also slightly extending their duration of repolarization (**Figure 20–22b**). As a result, heart rate declines.

Norepinephrine released by sympathetic neurons binds to beta-1 receptors, leading to the opening of sodium ion channels and calcium ion channels. Then an influx of positively charged ions increases the rate of depolarization and shortens the period of repolarization. The nodal cells reach threshold more quickly, and the heart rate increases (**Figure 20–22c**).

The Atrial Reflex. The **atrial reflex**, or *Bainbridge reflex*, involves adjustments in heart rate in response to an increase in the **venous return** (the amount of blood returning to the heart through veins). When the walls of the right atrium are stretched, stretch receptors there trigger a reflexive increase in heart rate by stimulating sympathetic activity (**Figure 20–22**). Thus, when the rate of venous return to the heart increases, so does the heart rate, and for this reason the cardiac output increases as well.

Hormones

Epinephrine, norepinephrine, and thyroid hormone increase heart rate by their effect on the SA node. The effects of epinephrine on the SA node are similar to those of norepinephrine. Epinephrine also affects the contractile cells. After massive sympathetic stimulation of the adrenal medullae, the myocardium may become so excitable that abnormal contractions occur.

Venous Return

In addition to its indirect effect on heart rate by the atrial reflex, venous return also directly affects nodal cells. When venous return increases, the atria receive more blood and the walls are stretched. Stretching of the cardiac muscle fibers of the SA node leads to more rapid depolarization and an increase in the heart rate.

Factors Affecting the Stroke Volume

The stroke volume is the difference between the end-diastolic volume and the end-systolic volume. Changes in either EDV or ESV can change the stroke volume, and thus cardiac output. The factors involved in the regulation of stroke volume are indicated in **Figure 20–23**.

The EDV

Recall that the EDV is the amount of blood in a ventricle at the end of diastole, just before a contraction begins. Two factors affect this volume: the filling time and the venous return. **Filling time** is the duration of ventricular diastole. It depends entirely

on the heart rate: The faster the heart rate, the shorter is the time available for filling. Venous return is variable over this period. It varies in response to changes in cardiac output, blood volume, patterns of peripheral circulation, skeletal muscle activity, and other factors that affect the rate of blood flow back to the heart. (We explore these factors in Chapter 21.)

Preload. The degree of stretching in ventricular muscle cells during ventricular diastole is called the **preload**. The preload is directly proportional to the EDV: The greater the EDV, the larger the preload. Preload matters because it affects the ability of muscle cells to produce tension. As sarcomere length increases past resting length, the amount of force produced during systole increases.

The amount of preload, and hence the degree of myocardial stretching, varies with the demands on the heart. When you are standing at rest, your EDV is low. The ventricular muscle is stretched very little, and the sarcomeres are relatively short. During ventricular systole, the cardiac muscle cells develop little power, and the ESV (the amount of blood in the ventricle after contraction) is relatively high because the muscle cells contracted only a short distance. If you begin exercising, venous return increases and more blood flows into your heart. Your EDV increases, and the myocardium stretches further. As the sarcomeres approach optimal lengths, the ventricular muscle cells can contract more efficiently and produce more forceful contractions. They also shorten more, and more blood is pumped out of your heart.

Figure 20–23   **Factors Affecting Stroke Volume.** The arrows indicate the nature of the effects:
↑ = increases, ↓ = decreases.

The EDV and Stroke Volume. In general, the greater the EDV, the larger the stroke volume. Stretching the cardiac muscle cells *past* their optimal length would *reduce* the force of contraction, but this degree of stretching does not normally take place. Myocardial connective tissues, the cardiac skeleton, and the pericardial sac all limit the expansion of the ventricles.

The relationship between the amount of ventricular stretching and the contractile force means that, within normal physiological limits, increasing the EDV results in a corresponding increase in the stroke volume. This general rule of "more in = more out" was first proposed by Ernest H. Starling based on his studies and research by Otto Frank. The relationship is known as the **Frank–Starling principle**, or *Starling's law of the heart*.

Autonomic adjustments to cardiac output normally make the effects of the Frank–Starling principle difficult to see. However, we can see the effects more clearly in individuals who have received a heart transplant, because the implanted heart is not innervated by the ANS. The most obvious effect of the Frank–Starling principle in these hearts is that the outputs of the left and right ventricles remain balanced under a variety of conditions.

Consider, for example, an individual at rest, with the two ventricles ejecting equal volumes of blood. Although the ventricles contract together, they work in series: When the heart contracts, blood leaving the right ventricle heads to the lungs. During the next ventricular diastole, that volume of blood passes through the left atrium, to be ejected by the left ventricle at the next contraction. If the venous return decreases, the EDV of the right ventricle will decline. During ventricular systole, the right ventricle will then pump less blood to the lungs. In the next cardiac cycle, the EDV of the left ventricle will be reduced, and that ventricle will eject a smaller volume of blood. The output of the two ventricles will again be in balance, but both will have smaller stroke volumes than they did initially.

The ESV

After the ventricle has contracted and ejected the stroke volume, the amount of blood that remains in the ventricle at the end of ventricular systole is the ESV. Three factors that influence the ESV are the *preload* (discussed earlier), the *contractility* of the ventricle, and the *afterload*.

Contractility. **Contractility** is the amount of force produced during a contraction, at a given preload. Under normal circumstances, autonomic innervation or circulating hormones can alter contractility. Under special circumstances, drugs or abnormal ion concentrations in the extracellular fluid can alter contractility.

Factors that increase contractility are said to have a *positive inotropic action* (*ino-*, fiber). Factors that decrease contractility have a *negative inotropic action*. Positive inotropic agents typically stimulate Ca^{2+} entry into cardiac muscle cells, thus increasing the force and duration of ventricular contractions. Negative inotropic agents may block Ca^{2+} movement or depress cardiac muscle metabolism. Positive and negative inotropic factors include ANS activity, hormones, and changes in extracellular ion concentrations.

Effects of Autonomic Activity on Contractility. Autonomic activity alters the degree of contraction and changes the ESV in the following ways:

- Sympathetic stimulation has a positive inotropic effect. It causes the release of norepinephrine (NE) by postganglionic fibers of the cardiac nerves and the secretion of epinephrine (E) and NE by the adrenal medullae. These hormones affect heart rate, as we will discuss shortly. They also stimulate alpha and beta receptors in cardiac muscle plasma membranes. This stimulation increases cardiac muscle cell metabolism and the force and degree of contraction. The net effect is that the ventricles contract more forcefully, increasing the ejection fraction and decreasing the ESV.

- Parasympathetic stimulation from the vagus nerves has a negative inotropic effect. The primary effect of acetylcholine (ACh) is at the membrane surface, where it produces hyperpolarization and inhibition. As a result, the force of cardiac contractions is reduced. The atria show the greatest changes in contractile force because the ventricles are not extensively innervated by the parasympathetic division. However, under strong parasympathetic stimulation or after the administration of drugs that mimic the actions of ACh, the ventricles contract less forcefully, the ejection fraction decreases, and the ESV becomes larger.

Hormones. Many hormones affect the contractility of the heart. For example, epinephrine, norepinephrine, and thyroid hormones all have positive inotropic effects. Glucagon also has a positive inotropic effect. Before synthetic inotropic agents were available, glucagon was widely used to stimulate cardiac function. It is still used in cardiac emergencies and to treat some forms of heart disease.

The drugs *isoproterenol*, *dopamine*, and *dobutamine* mimic the action of E and NE by stimulating beta-1 receptors on cardiac muscle cells. ⊃ p. 567 Dopamine (at high doses) and dobutamine also stimulate Ca^{2+} entry through alpha-1 receptor stimulation. *Digitalis* and related drugs elevate intracellular Ca^{2+} concentrations, but by a different mechanism. They interfere with the removal of Ca^{2+} from the cytosol of cardiac muscle cells.

Many of the drugs used to treat hypertension (high blood pressure) have a negative inotropic action. Beta-blocking drugs such as *propranolol, timolol, metoprolol, atenolol,* and *labetalol* block beta receptors, alpha receptors, or both, and prevent sympathetic stimulation of the heart. Calcium channel blockers such as *nifedipine* or *verapamil* also have a negative inotropic effect.

Afterload. The **afterload** is the amount of tension that the contracting ventricle must produce to force open the semilunar valve and eject blood. Afterload increases with increased resistance to blood flow out of the ventricle. The greater the afterload, the longer the period of isovolumetric contraction, the shorter the duration of ventricular ejection, and the larger the ESV. In other words, as the afterload increases, the stroke volume decreases.

Any factor that restricts blood flow through the arterial system increases afterload. For example, either the constriction of peripheral blood vessels or a circulatory blockage elevates arterial blood pressure and increases the afterload. If the afterload is too great, the ventricle cannot eject blood. Such a high afterload is rare in a normal heart. However, damage to the heart muscle can weaken the myocardium enough that even a modest rise in arterial blood pressure can reduce stroke volume

to dangerously low levels, producing symptoms of heart failure. Damage to the semilunar valve that restricts blood flow will also increase afterload.

Summary: The Control of Cardiac Output

Figure 20–24 summarizes the factors that regulate heart rate and stroke volume, which interact to determine cardiac output under normal conditions.

The heart rate is influenced by the autonomic nervous system, circulating hormones, and the venous return.

- Sympathetic stimulation increases the heart rate, and parasympathetic stimulation decreases it. Under resting conditions, parasympathetic tone dominates, and the heart rate is slightly slower than the intrinsic heart rate. When activity levels rise, venous return increases and triggers the atrial reflex. The result is an increase in sympathetic tone and an increase in heart rate.

Figure 20–24   A Summary of the Factors Affecting Cardiac Output.

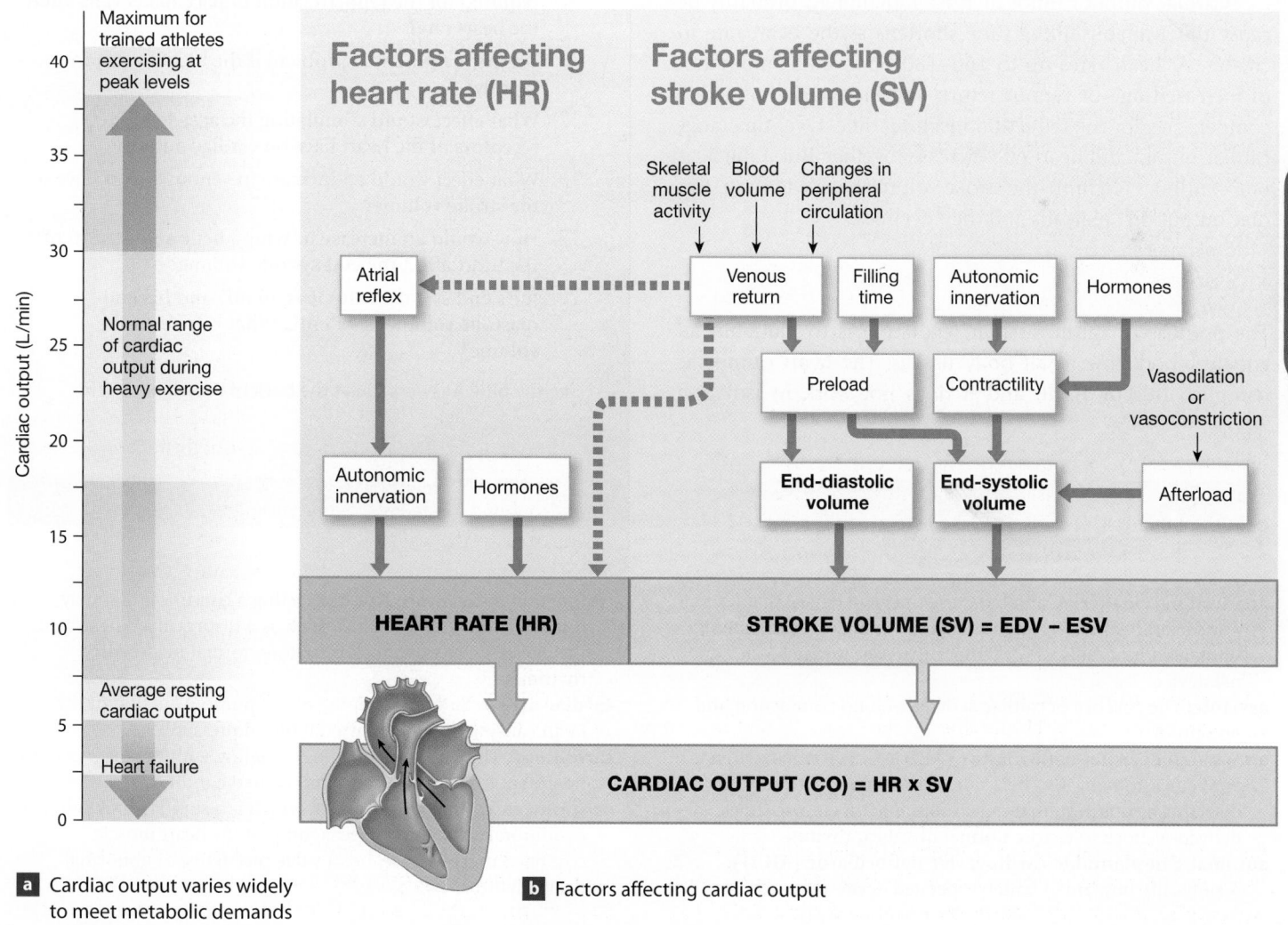

a   Cardiac output varies widely to meet metabolic demands

b   Factors affecting cardiac output

- Circulating hormones, specifically E, NE, and T_3, accelerate heart rate.

- An increase in venous return stretches the nodal cells and increases heart rate.

- The stroke volume is the difference between the end-diastolic volume (EDV) and the end-systolic volume (ESV).

- The EDV is determined by the available filling time and the rate of venous return.

- The ESV is determined by the amount of preload (the degree of myocardial stretching), the degree of contractility (adjusted by hormones and autonomic innervation), and the afterload (the arterial resistance to blood flow out of the heart).

In most healthy people, increasing both the stroke volume and the heart rate, such as occurs during heavy exercise, can raise the cardiac output by 300–500 percent, to 18–30 L/min. The difference between resting and maximal cardiac outputs is the **cardiac reserve**. Trained athletes exercising at maximal levels may increase cardiac output by nearly 700 percent, to 40 L/min.

Cardiac output cannot increase indefinitely, primarily because the available filling time shortens as the heart rate increases. At heart rates up to 160–180 bpm, the combination of increased rate of venous return and increased contractility compensates for the reduction in filling time. Over this range, cardiac output and heart rate increase together. But if the heart rate continues to climb, the stroke volume begins to drop. Cardiac output first plateaus and then declines.

The Heart and the Cardiovascular System

The purpose of cardiovascular regulation is to maintain adequate blood flow to all body tissues. The heart cannot accomplish this by itself, and it does not work in isolation. For example, when blood pressure changes, the cardiovascular centers adjust not only the heart rate but also the diameters of peripheral blood vessels. These adjustments work together to keep the blood pressure within normal limits and to maintain circulation to vital tissues and organs. In Chapter 21 we complete this story by detailing the cardiovascular responses to changing activity patterns and circulatory emergencies. We then conclude our discussion of the cardiovascular system by examining the anatomy of the pulmonary and systemic circuits.

CHECKPOINT

12. Define cardiac output.

13. Caffeine has effects on conducting cells and contractile cells that are similar to those of NE. What effect would drinking large amounts of caffeinated drinks have on the heart?

14. If the cardioinhibitory center of the medulla oblongata were damaged, which part of the autonomic nervous system would be affected, and how would the heart be influenced?

15. How does a drug that increases the length of time required for the repolarization of pacemaker cells affect the heart rate?

16. Why is it a potential problem if the heart beats too rapidly?

17. What effect would stimulating the acetylcholine receptors of the heart have on cardiac output?

18. What effect would an increase in venous return have on the stroke volume?

19. How would an increase in sympathetic stimulation of the heart affect the end-systolic volume?

20. Joe's end-systolic volume is 40 mL, and his end-diastolic volume is 125 mL. What is Joe's stroke volume?

See the blue Answers tab at the back of the book.

Related Clinical Terms

artificial pacemaker: A small, battery-operated device that keeps one's heart beating in a regular rhythm. It may be permanently implanted internally or for temporary usage it may be an external device.

asystole: The absence of cardiac activity with no contraction and no output.

automated external defibrillator (AED): A device that, when applied, automatically checks the function of the heart. Upon detecting a condition that may respond to an electric shock, it delivers a shock to restore normal heartbeat rhythm.

automatic implantable cardioverter defibrillator (AICD): A surgically implanted battery-operated device that monitors the function of the heart. Upon detecting a condition that may respond to an electric shock, such as a disorganized heartbeat, the device delivers a shock to restore normal heartbeat rhythm.

cardiac arrest: Sudden stopping of the pumping action of the heart causing the loss of arterial blood pressure.

cardiology: The branch of medicine dealing with the diagnosis and treatment of heart disorders and related conditions.

cardiomegaly: An enlarged heart, which is a sign of some other condition such as stress, weakening of the heart muscle, coronary artery disease, heart valve problems, or abnormal heart rhythms.

cardiomyoplasty: A surgical procedure that uses stimulated latissimus dorsi muscle to assist with cardiac function. The latissimus dorsi muscle is relocated and wrapped around the left and right ventricles and stimulated to contract during cardiac systole by means of an implanted burst-stimulator.

commotio cordis: Sudden cardiac arrest as the result of a blunt hit or impact to the chest.

congestive heart failure: The heart condition of weakness, edema, and shortness of breath caused by the inability of the heart to maintain adequate blood circulation in the peripheral tissues and the lungs.

cor pulmonale: Weakness of the right ventricle of the heart due to prolonged high blood pressure in the pulmonary artery and right ventricle; or any disease or malfunction that affects the pulmonary circuit in the lungs.

echocardiography: A noninvasive diagnostic test that uses ultrasound to make images of the heart chambers, valves, and surrounding structures. This diagnostic tool can also measure cardiac output, detect inflammation around the heart, identify abnormal anatomy, and detect infections of the heart valves.

endocarditis: Inflammation or infection of the endocardium, the inner lining of the heart muscle.

fibrillation: Fast twitching of the heart muscle fibers with little or no movement of the muscle as a whole. Atrial fibrillation occurs in the atria of the heart and is characterized by chaotic quivers and irregular ventricular beating with both atria and ventricles being out of sync.

heart block: Delay in the normal electrical pulses that cause the heart to beat.

mitral valve prolapse: A condition in which the mitral (bicuspid) valve cusps do not close properly and are pushed back toward the left atrium.

myocarditis: Inflammation of the myocardium, the middle layer of the heart wall tissue.

palpitation: Irregular and rapid beating of the heart.

percutaneous transluminal coronary angioplasty (PTCA): The surgical use of a balloon-tipped catheter to enlarge a narrowed artery.

sick sinus syndrome: A group of heart rhythm disorders or problems in which the sinoatrial node does not work properly to regulate the heart rhythms.

Chapter Review

Study Outline

MasteringA&P®

Access more chapter study tools online in the MasteringA&P Study Area:

- Chapter Quizzes, Chapter Practice Test, Art-labeling Activities, Animations, MP3 Tutor Sessions, and Clinical Case Studies
- Practice Anatomy Lab   PAL™
- Interactive Physiology   iP®
- A&P Flix   A&PFlix
- PhysioEx   PhysioEx

❚ An Introduction to the Cardiovascular System p. 711

1. The blood vessels can be subdivided into the **pulmonary circuit** (which carries blood to and from the lungs) and the **systemic circuit** (which transports blood to and from the rest of the body).

2. **Arteries** carry blood away from the heart; **veins** return blood to the heart. **Capillaries**, or *exchange vessels*, are thin-walled, narrow-diameter vessels that connect the smallest arteries and veins. (*Figure 20–1*)

3. The heart has four chambers: the **right atrium** and **right ventricle**, and the **left atrium** and **left ventricle**.

20-1 ❚ The heart is a four-chambered organ, supplied by the coronary circulation, that pumps oxygen-poor blood to the lungs and oxygen-rich blood to the rest of the body p. 712

4. The heart is surrounded by the **pericardial cavity** and lies within the anterior portion of the **mediastinum**, which separates the two pleural cavities. (*Figure 20–2*)

5. The pericardial cavity is lined by the **pericardium**. The **visceral pericardium (epicardium)** covers the heart's outer surface, and the **parietal pericardium** lines the inner surface of the **pericardial sac**, which surrounds the heart. (*Figure 20–2*)

6. The **coronary sulcus**, a deep groove, marks the boundary between the atria and the ventricles. Other surface markings also provide useful reference points in describing the heart and associated structures. (*Figure 20–3*)

7. The bulk of the heart consists of the muscular **myocardium**. The **endocardium** lines the inner surfaces of the heart, and the **epicardium** covers the outer surface. (*Figure 20–4*)

8. **Cardiac muscle cells** are interconnected by **intercalated discs**, which convey the force of contraction from cell to cell and conduct action potentials. (*Figure 20–5; Table 20–1*)

9. The atria are separated by the **interatrial septum**, and the ventricles are divided by the **interventricular septum**. The right atrium receives blood from the systemic circuit via two large veins, the **superior vena cava** and the **inferior vena cava**. (The atrial walls contain the **pectinate muscles**, prominent muscular ridges.) (*Figure 20–6*)

10. Blood flows from the right atrium into the right ventricle by the **right atrioventricular (AV) valve (tricuspid valve)**. This opening is bounded by three **cusps** of fibrous tissue braced by the **chordae tendineae**, which are connected to **papillary muscles**. (*Figure 20–6*)

11. Blood leaving the right ventricle enters the **pulmonary trunk** after passing through the **pulmonary valve**. The pulmonary trunk divides to form the **left** and **right pulmonary arteries**. The **left** and **right pulmonary veins** return blood from the lungs to the left atrium. Blood leaving the left atrium flows into the left ventricle by the **left atrioventricular (AV) valve (bicuspid**, or **mitral, valve)**. Blood leaving the left ventricle passes through the **aortic valve** and into the systemic circuit via the **ascending aorta**. (*Figure 20–6*)

20

12. Anatomical differences between the ventricles reflect the functional demands placed on them. The wall of the right ventricle is relatively thin, whereas the left ventricle has a massive muscular wall. *(Figure 20–7)*

13. Valves normally permit blood flow in only one direction, preventing the **regurgitation** (backflow) of blood. *(Figure 20–8)*

14. The connective tissues of the heart (mainly collagen and elastic fibers) and the **cardiac skeleton** support the heart's contractile cells and valves. *(Figure 20–8)*

15. The **coronary circulation** meets the high oxygen and nutrient demands of cardiac muscle cells. The **coronary arteries** originate at the base of the ascending aorta. Interconnections between arteries, called **arterial anastomoses**, ensure a constant blood supply. The **great**, **posterior**, **small**, **anterior**, and **middle cardiac veins** are epicardial vessels that carry blood from the coronary capillaries to the **coronary sinus**. *(Figure 20–9)*

16. In **coronary artery disease (CAD)**, portions of the coronary circulation undergo partial or complete blockage. A **myocardial infarction (MI)**, or heart attack, occurs when part of the coronary circulation becomes blocked and muscle tissue dies when it cannot be oxygenated. *(Spotlight Figure 20–10)*

20-2 ▍ The conducting system distributes electrical impulses through the heart, and an electrocardiogram records the associated electrical events p. 723

17. Two general classes of cardiac muscle cells are involved in the normal **heartbeat**: cells of the **conducting system** and **contractile cells**.

18. The **conducting system** is composed of the *sinoatrial node*, the *atrioventricular node*, and *conducting cells*. The conducting system initiates and distributes electrical impulses within the heart. Nodal cells establish the rate of cardiac contraction, and conducting cells distribute the contractile stimulus from the SA node to the atrial myocardium and the AV node (along *internodal pathways*), and from the AV node to the ventricular myocardium. *(Figure 20–11)*

19. Unlike skeletal muscle, cardiac muscle contracts without neural or hormonal stimulation. **Pacemaker cells** in the **sinoatrial (SA) node** (*cardiac pacemaker*) normally establish the rate of contraction. From the SA node, the stimulus travels to the **atrioventricular (AV) node**, and then to the **AV bundle**, which divides into **bundle branches**. From there, **Purkinje fibers** convey the impulses to the ventricular myocardium. *(Figures 20–11, 20–12)*

20. A recording of electrical activities in the heart is an **electrocardiogram (ECG or EKG)**. Important landmarks of an ECG include the **P wave** (atrial depolarization), the **QRS complex** (ventricular depolarization), and the **T wave** (ventricular repolarization). *(Figure 20–13)*

21. **Cardiac arrhythmias** are abnormal patterns of electrical activity in the heart. *(Spotlight Figure 20–14)*

22. **Contractile cells** form the bulk of the atrial and ventricular walls. Cardiac muscle cells have a long refractory period, so rapid stimulation produces twitches rather than tetanic contractions. *(Figure 20–15)*

20-3 ▍ Events during a complete heartbeat make up a cardiac cycle p. 732

23. The **cardiac cycle** includes periods of **atrial** and **ventricular systole** (contraction) and **atrial** and **ventricular diastole** (relaxation). *(Figure 20–16)*

24. When the heart beats, the two ventricles eject equal volumes of blood. *(Figure 20–17)*

25. The closing of valves and rushing of blood through the heart cause characteristic heart sounds, which can be heard during *auscultation*. *(Figure 20–18)*

20-4 ▍ Cardiodynamics examines the factors that affect cardiac output p. 737

26. The amount of blood ejected by a ventricle during a single beat is the **stroke volume (SV)**. The amount of blood pumped by a ventricle each minute is the **cardiac output (CO)**. *(Figure 20–19)*

27. Cardiac output can be adjusted by changes in either stroke volume or heart rate. *(Figure 20–20)*

28. The **cardioacceleratory center** in the medulla oblongata activates sympathetic neurons; the **cardioinhibitory center** controls the parasympathetic neurons that slow the heart rate. These cardiac centers receive inputs from higher centers and from receptors monitoring blood pressure and the concentrations of dissolved gases. *(Figure 20–21)*

29. The basic heart rate is established by the pacemaker cells of the SA node, but it can be modified by the autonomic nervous system. The **atrial reflex** accelerates the heart rate when the walls of the right atrium are stretched. *(Figure 20–22)*

30. Sympathetic activity produces more powerful contractions that reduce the ESV. Parasympathetic stimulation slows the heart rate, reduces the contractile strength, and raises the ESV.

31. Cardiac output is affected by various factors, including autonomic innervation and hormones. *(Figure 20–22)*

32. The stroke volume is the difference between the **end-diastolic volume (EDV)** and the **end-systolic volume (ESV)**. The **filling time** and **venous return** interact to determine the EDV. Normally, the greater the EDV, the more powerful is the succeeding contraction (the **Frank–Starling principle**). *(Figure 20–23)*

33. The difference between resting and maximal cardiac outputs is the **cardiac reserve**. *(Figure 20–24)*

34. The heart does not work in isolation in maintaining adequate blood flow to all tissues.

20

Review Questions

LEVEL 1 Reviewing Facts and Terms

1. Identify the superficial structures in the following diagram of the heart.

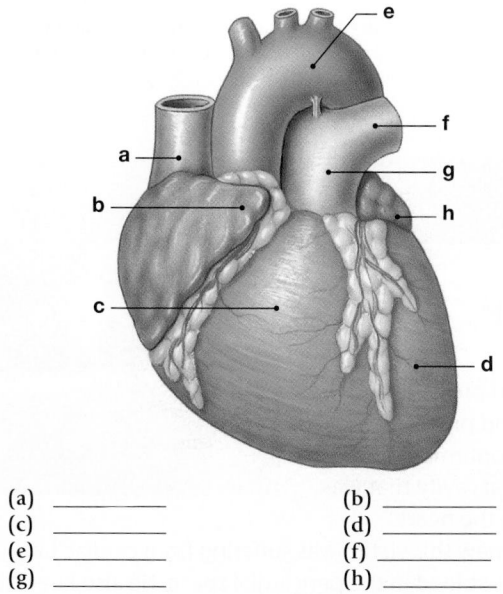

(a) _____   (b) _____
(c) _____   (d) _____
(e) _____   (f) _____
(g) _____   (h) _____

2. Identify the structures in the following diagram of a sectional view of the heart.

(a) _____   (b) _____
(c) _____   (d) _____
(e) _____   (f) _____
(g) _____   (h) _____
(i) _____   (j) _____
(k) _____   (l) _____
(m) _____

3. The great cardiac vein drains blood from the heart muscle to the (a) left ventricle, (b) right ventricle, (c) right atrium, (d) left atrium.

4. The pericardium is an example of which type of membrane? (a) cutaneous, (b) serous, (c) mucous, (d) epithelium.

5. In an ECG, atrial repolarizartion is (a) represented by a P wave, (b) represented by a QRS complex, (c) represented by a T wave, (d) masked by the QRS event.

6. Which valve separates the right atria and ventricle? (a) bicuspid, (b) aortic semilunar, (c) tricuspid, (d) pulmonary semilunar.

7. The heart is surrounded by the (a) pleural cavity, (b) peritoneal cavity, (c) abdominopelvic cavity, (d) mediastinum, (e) abdominal cavity.

8. The cardiac skeleton of the heart has which *two* of the following functions? (a) It physically isolates the muscle fibers of the atria from those of the ventricles, (b) It maintains the normal shape of the heart, (c) It helps distribute the forces of cardiac contraction, (d) It allows more rapid contraction of the ventricles, (e) It strengthens and helps prevent overexpansion of the heart.

9. Cardiac output is equal to the (a) difference between the end-diastolic volume and the end-systolic volume, (b) product of heart rate and stroke volume, (c) difference between the stroke volume at rest and the stroke volume during exercise, (d) stroke volume less the end-systolic volume, (e) product of heart rate and blood pressure.

10. The second heart sound midway through the cardiac cycle is the (a) sudden flow of blood through aortic semilunar valves, (b) sudden flow of blood through AV valves, (c) blood rebounding off closed AV valves, (d) blood rebounding off closed aortic semilunar valves.

11. During the cardiac cycle, the amount of blood ejected from the left ventricle when the semilunar valve opens is the (a) stroke volume (SV), (b) end-diastolic volume (EDV), (c) end-systolic volume (ESV), (d) cardiac output (CO).

12. What role do the chordae tendineae and papillary muscles play in the normal function of the AV valves?

13. Explain what is meant by the terms systole and diastole.

14. What are the valves in the heart, and what is the function of each?

15. Trace the normal pathway of an electrical impulse through the conducting system of the heart.

16. What is the cardiac cycle? What phases and events are necessary to complete a cardiac cycle?

17. What are stroke volume and cardiac output?

LEVEL 2 Reviewing Concepts

18. An increase in afterload (a) increases end systolic volume and stroke volume, (b) increases end systolic volume but decreases stroke volume, (c) decreases end systolic volume and stroke volume, (d) decreases end systolic volume but increases stroke volume.

19. Which of the following is *longer*? (a) the refractory period of cardiac muscle, (b) the refractory period of skeletal muscle.

20. If the papillary muscles fail to contract, (a) the ventricles will not pump blood, (b) the atria will not pump blood, (c) the semilunar valves will not open, (d) the AV valves will not close properly, (e) none of these happen.

21. When electrical impulses reach the atrioventricular node, there is a delay of 100 milliseconds (0.1 second). Provide an explanation for this delay.

22. Describe the function of the SA node in the cardiac cycle. How does this function differ from that of the AV node?

23. What are the sources and significance of the four heart sounds?

24. Explain why the action potential of cardiac muscle cell has a plateau phase, whereas the action potential of skeletal muscle does not.

25. What factors influence cardiac output?

26. What effect does sympathetic stimulation have on the heart? What effect does parasympathetic stimulation have on the heart?

27. Describe the effects of epinephrine, norepinephrine, glucagon, and thyroid hormones on the contractility of the heart.

20

LEVEL 3 Critical Thinking and Clinical Applications

28. Lisa often feels light-headed for a few seconds if she stands up quickly after sitting for too long. An app on her smart phone records that her seated resting heart rate is 65 bpm but it rises to 107 bpm when she stands. Why does this happen?

29. Harvey has a heart murmur in his left ventricle that produces a loud "gurgling" sound at the beginning of systole. Which valve is probably faulty?

30. The following measurements were made on two individuals (the values recorded remained stable for one hour):

    Person 1: heart rate, 75 bpm; stroke volume, 60 mL
    Person 2: heart rate, 90 bpm; stroke volume, 95 mL

    Which person has the greater venous return? Which person has the longer ventricular filling time?

31. Mr. Sinha is 73 and has acute lung injury. He is on a mechanical ventilator, which uses positive pressure to help improve lung ventilation. But, cardiac output begins to fall. Explain how this might occur.

CLINICAL CASE Wrap-Up

A Needle to the Chest

With the force of the head-on collision, the patient's chest first broke the seat belt and then collided with the steering wheel. This caused fractures of the anterior chest wall, including the sternum and ribs.

The heart, located just deep to the sternum, then hit the steering wheel hard enough to cause bleeding into the pericardial sac. As the delicate parietal and visceral pericardial linings bled into the pericardial cavity, the tough, fibrous pericardial sac itself did not expand.

The blood in the pericardial cavity compressed the heart, squeezing the atria and ventricles. During atrial diastole, the atria could not fill. This caused a backup of the venous system that was obvious in the veins of the patient's neck. The compressed ventricles could not fill with blood during ventricular diastole. Because the end-diastolic volume (EDV) was so low, the stroke volume (SV) was also very low. This was evident in the blood pressure of 80/58, with a narrow difference between the

systolic blood pressure and the diastolic blood pressure. The ultrasound confirmed fluid in the pericardial cavity that was compressing the heart.

Dr. Jim knew this victim was suffering from cardiac tamponade due to blood in the pericardial space. He also knew the problem was severe enough that with the decreasing stroke volume, the heart could not pump adequate blood to the patient's brain. So he removed the blood from the pericardial space with a needle, a procedure known as pericardiocentesis, and saved this patient's life.

1. Why do you think the heart sounds were muffled and distant?

2. What other cardiac structures were likely to be injured by this collision?

See the blue Answers tab at the back of the book.

20

21

Blood Vessels and Circulation

Learning Outcomes

These Learning Outcomes correspond by number to this chapter's sections and indicate what you should be able to do after completing the chapter.

21-1 ▍ Distinguish among the types of blood vessels based on their structure and function, and describe how and where fluid and dissolved materials enter and leave the cardiovascular system.

21-2 ▍ Explain the mechanisms that regulate blood flow through vessels, describe the factors that influence blood pressure, and discuss the mechanisms that regulate movement of fluids between capillaries and interstitial spaces.

21-3 ▍ Describe the control mechanisms that regulate blood flow and pressure in tissues, and explain how the activities of the cardiac, vasomotor, and respiratory centers are coordinated to control blood flow through the tissues.

21-4 ▍ Explain the cardiovascular system's homeostatic response to exercise and hemorrhaging, and identify the principal blood vessels and functional characteristics of the special circulation to the brain, heart, and lungs.

21-5 ▍ Describe the three general functional patterns seen in the pulmonary and systemic circuits of the cardiovascular system.

21-6 ▍ Identify the major arteries and veins of the pulmonary circuit.

21-7 ▍ Identify the major arteries and veins of the systemic circuit.

21-8 ▍ Identify the differences between fetal and adult circulation patterns, and describe the changes in the patterns of blood flow that occur at birth.

21-9 ▍ Discuss the effects of aging on the cardiovascular system, and give examples of interactions between the cardiovascular system and other organ systems.

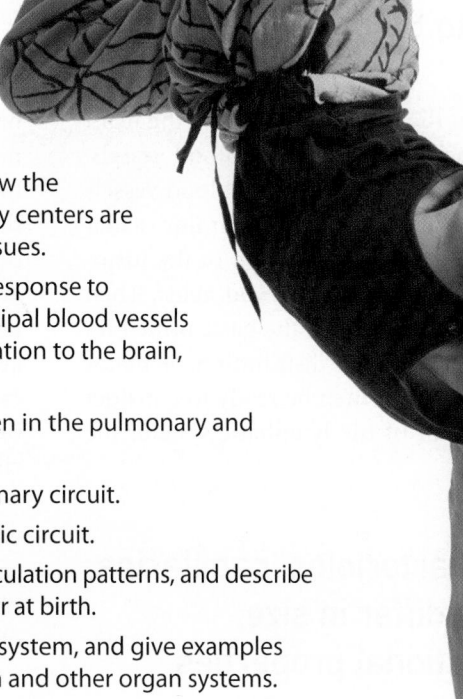

SPOTLIGHT
- Congenital Heart Problems p. 800

CLINICAL CASE
- Did Ancient Mummies Have Atherosclerosis? p. 750

Clinical Notes
- Arteriosclerosis pp. 754–755
- Edema p. 767

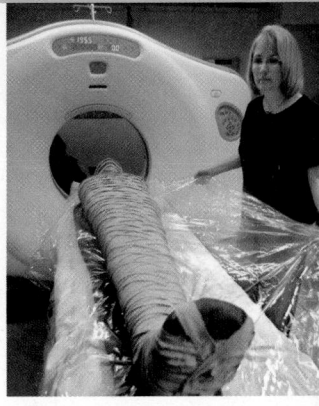
Dr. Iacullo is part of a research team studying ancient mummies from Egypt, Peru, New Mexico, and the Aleutian Islands in Alaska. The study spans 4000 years of human history and cultures from three different continents. Some of the mummies were once members of Egypt's upper class, while others were likely common people from the Americas. Their diets varied greatly and included nuts, berries, farmed corn, domesticated animals, wild game, sea otters, and whales. Their activity patterns varied—the wealthy Egyptians, for instance, were much more sedentary than the hunter-gatherers of the Americas.

Dr. Iacullo is studying the incidence of atherosclerosis (the buildup of arterial plaque) in mummies. Her team uses computerized tomography (CT) scans to look for calcifications in the walls of the mummies' arteries. They discover that one-third of the mummies have CT evidence of plaque buildup. This finding spans all four geographic populations, all diets, and all lifestyle activity levels. **Why do you think these ancient people developed atherosclerosis, which today we usually associate with the unhealthy diet and inactivity of a modern lifestyle? To find out, turn to the Clinical Case Wrap-Up on p. 806.**

An Introduction to Blood Vessels and Circulation

Blood circulates throughout the body, moving from the heart through the tissues and back to the heart, in blood vessels. In this chapter we examine the organization of blood vessels and consider the integrated functions of the cardiovascular system as a whole. We begin with a description of the histological organization of arteries, capillaries, and veins. Then we explore the functions of these vessels, the basic principles of cardiovascular regulation, and the distribution of major blood vessels in the body. We will then be ready to consider the organization and function of the lymphatic system, the focus of Chapter 22.

21-1 ❘ Arteries, arterioles, capillaries, venules, and veins differ in size, structure, and functional properties

Learning Outcome Distinguish among the types of blood vessels based on their structure and function, and describe how and where fluid and dissolved materials enter and leave the cardiovascular system.

The cardiovascular system has five general classes of blood vessels. **Arteries** carry blood away from the heart. As they enter peripheral tissues, arteries branch repeatedly, and the branches decrease in diameter. The smallest arterial branches are called **arterioles** (ar-TER-ē-ōls). From the arterioles, blood moves into **capillaries**, where diffusion takes place between blood and interstitial fluid. From the capillaries, blood enters small **venules** (VEN-ūls), which unite to form larger **veins** that return blood to the heart.

Blood leaves the heart through the pulmonary trunk, which originates at the right ventricle, and the aorta, which originates at the left ventricle. Each of these arterial trunks has an internal diameter of about 2.5 cm (1 in.). The pulmonary arteries that branch from the pulmonary trunk carry blood to the lungs. The systemic arteries that branch from the aorta distribute blood to all other organs. Within these organs, the vessels branch into several hundred million tiny arterioles that provide blood to more than 10 billion capillaries within their own branching networks. These capillaries are barely the diameter of a single red blood cell. If all the capillaries in your body were placed end to end, their combined length would be more than 25,000 miles, enough to circle the planet.

The vital functions of the cardiovascular system depend entirely on events at the capillary level: All chemical and gaseous exchange between blood and interstitial fluid takes place across capillary walls. Cells rely on capillary diffusion to obtain nutrients and oxygen and to remove metabolic wastes, such as carbon dioxide and urea. Diffusion takes place very rapidly, because the distances involved are very short. Few cells lie farther than 125 μm (0.005 in.) from a capillary. As we will see, homeostatic mechanisms operating at the local, regional, and systemic levels adjust blood flow through the capillaries to meet the demands of peripheral tissues.

Blood vessels must be resilient enough to withstand changes in pressure, and flexible enough to move with underlying tissues and organs. The pressures inside vessels vary with distance from the heart, and the structures of different vessels reflect this fact. The arteries, veins, and capillaries also differ in function, and these functional differences are associated with distinctive anatomical features.

The Structure of Vessel Walls

The walls of arteries and veins have three distinct layers—the tunica intima, tunica media, and tunica externa (**Figure 21–1**):

1. The **tunica intima** (IN-ti-muh), or *tunica interna*, is the inner layer of a blood vessel. It includes the endothelial lining and a surrounding layer of connective tissue with a variable number of elastic fibers. In arteries, the outer margin of the tunica intima contains a thick layer of elastic fibers called the **internal elastic membrane**.

2. The **tunica media** is the middle layer of a blood vessel. It contains concentric sheets of smooth muscle tissue in a framework of loose connective tissue. The collagen fibers bind the tunica media to the tunica intima and tunica externa. The tunica media is commonly the thickest layer in a small artery. It is separated from the surrounding tunica externa by a thin band of elastic fibers called the **external elastic membrane**. The smooth muscle cells of the tunica media encircle the endothelium that lines the lumen of the blood vessel. When these smooth muscles contract, the vessel decreases in diameter, and when they relax, the diameter increases. Large arteries also contain layers of longitudinally arranged smooth muscle cells.

3. The **tunica externa** (eks-TER-nuh), or *tunica adventitia* (ad-ven-TISH-uh), is the outer layer of a blood vessel. It is a connective tissue sheath. In arteries, it contains collagen fibers with scattered bands of elastic fibers. In veins, it is generally thicker than the tunica media and contains networks of elastic fibers and bundles of smooth muscle cells. The connective tissue fibers of the tunica externa typically blend into those of adjacent tissues, stabilizing and anchoring the blood vessel.

Figure 21–1   Comparisons of a Typical Artery and a Typical Vein.

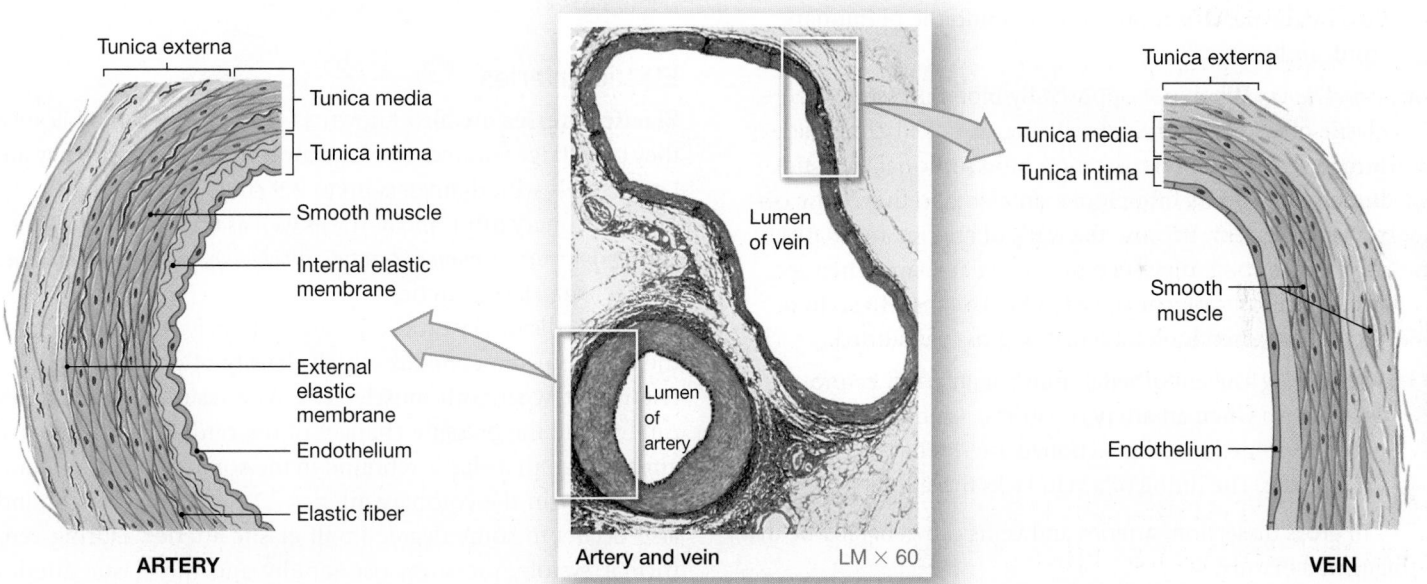

Feature	Typical Artery	Typical Vein
GENERAL APPEARANCE IN SECTIONAL VIEW	Usually round, with relatively thick wall	Usually flattened or collapsed, with relatively thin wall
TUNICA INTIMA		
Endothelium	Usually rippled, due to vessel constriction	Often smooth
Internal elastic membrane	Present	Absent
TUNICA MEDIA	Thick, dominated by smooth muscle cells and elastic fibers	Thin, dominated by smooth muscle cells and collagen fibers
External elastic membrane	Present	Absent
TUNICA EXTERNA	Collagen and elastic fibers	Collagen and elastic fibers and smooth muscle cells

Their layered walls give arteries and veins considerable strength. The muscular and elastic components also permit controlled changes in diameter as blood pressure or blood volume changes. However, the walls of arteries and veins are too thick to allow diffusion between the bloodstream and surrounding tissues, or even between the blood and the tissues of the vessel itself. For this reason, the walls of large vessels contain small arteries and veins that supply the smooth muscle cells and fibroblasts of the tunica media and tunica externa. These blood vessels are called the **vasa vasorum** ("vessels of vessels").

Differences between Arteries and Veins

Arteries and veins supplying the same region lie side by side (**Figure 21–1**). In sectional view, you can distinguish arteries and veins by the following features:

- *Vessel walls* In general, the walls of arteries are thicker than those of veins. The tunica media of an artery contains more smooth muscle and elastic fibers than does that of a vein. These components help resist the arterial pressure generated by the heart as it pumps blood into the pulmonary trunk and aorta.

- *Vessel lumen* When not opposed by blood pressure, the elastic fibers in the arterial walls recoil, constricting the lumen. Thus, seen on dissection or in sectional view, the lumen of an artery often looks smaller than that of the corresponding vein. Because the walls of arteries are relatively thick and strong, they keep their circular shape when sectioned. In contrast, cut veins tend to collapse. In section, these veins often look flattened or grossly distorted.

- *Vessel lining* The endothelial lining of an artery cannot contract, so when an artery constricts, its endothelium becomes folded and the sectioned arteries have a pleated appearance. The lining of a vein lacks these folds.

In gross dissection, arteries and veins can generally be distinguished because:

- The thicker walls of arteries can be felt if the vessels are compressed.

- Arteries usually keep their cylindrical shape, but veins often collapse.

- Arteries are more resilient: When stretched, they keep their shape and lengthen. When released, they snap back. A small vein cannot tolerate as much distortion without collapsing or tearing.

- Veins typically contain *valves*—internal structures that prevent the backflow of blood toward the capillaries. In an intact vein, the location of each valve is marked by a slight distension of the vessel wall. (We consider valve structure in a later section.)

Arteries

Their relatively thick, muscular walls make arteries elastic and contractile. Elasticity permits the vessel diameter to change passively in response to changes in blood pressure. For example, it allows arteries to absorb the surging pressure waves that accompany the contractions of the ventricles.

The contractility of the arterial walls enables them to actively change diameter. This change takes place primarily under the control of the sympathetic division of the autonomic nervous system. When stimulated, arterial smooth muscles contract, constricting the artery—a process called **vasoconstriction**. When these smooth muscles relax, the diameter of the lumen increases—a process called **vasodilation**. Vasoconstriction and vasodilation affect (1) the afterload on the heart, (2) peripheral blood pressure, and (3) capillary blood flow. We explore these effects in a later section. Contractility is also important during the vascular phase of hemostasis, when the contraction of a damaged vessel wall helps reduce bleeding. ⟲ p. 701

In traveling from the heart to peripheral capillaries, blood passes through *elastic arteries*, *muscular arteries*, and *arterioles* (**Figure 21–2**).

Elastic Arteries

Elastic arteries are also known as *conducting arteries* because they carry large volumes of blood away from the heart. They are large vessels with diameters up to 2.5 cm (1 in.) (**Figure 21–2**). The pulmonary trunk and aorta, as well as their major branches (the *pulmonary, common carotid, subclavian*, and *common iliac arteries*), are elastic arteries.

The walls of elastic arteries are extremely resilient because the tunica media contains a high density of elastic fibers and relatively few smooth muscle cells. As a result, elastic arteries can tolerate the pressure changes of the cardiac cycle. We have already seen that elastic rebound in the aorta helps to maintain blood flow in the coronary arteries. ⟲ p. 721 Elastic rebound also occurs to some degree in all elastic arteries. During ventricular systole, pressures rise rapidly and the elastic arteries expand as the stroke volume is ejected. During ventricular diastole, blood pressure within the arterial system falls and the elastic fibers recoil to their original dimensions. Their expansion cushions the sudden rise in pressure during ventricular systole, and their recoil slows the drop in pressure during ventricular diastole. In this way, elastic arteries help to make blood flow continuous.

This function is important because blood pressure is the driving force behind blood flow: The greater the fluctuations in pressure, the greater the changes in blood flow. The elasticity of the arterial system dampens the pressure peaks and valleys that accompany the heartbeat. By the time blood reaches the arterioles, the pressure fluctuations have disappeared, and blood flow is continuous.

Figure 21–2 **Histological Structure of Blood Vessels.** Representative diagrammatic views of arteries, veins, and capillaries.

Veins

Large Vein

Large veins include the superior and inferior venae cavae (also termed the great veins), and their branches within the abdominopelvic and thoracic cavities.

- Tunica externa
- Tunica media
- Endothelium
- Tunica intima

Medium-sized Vein

Medium-sized veins range from 2 to 9 mm in internal diameter and correspond in general size to medium-sized arteries. In these veins, the tunica media is thin, and it contains relatively few smooth muscle fibers.

- Tunica externa
- Tunica media
- Endothelium
- Tunica intima

Venule

Venules collect blood from capillaries. They vary widely in size and character. The smallest venules resemble expanded capillaries, and venules smaller than 50 μm in total diameter lack a tunica media altogether.

- Tunica externa
- Endothelium

Fenestrated Capillary

Fenestrated capillaries are capillaries that contain "windows," or pores in their walls, due to an incomplete or perforated endothelial lining.

- Pores
- Endothelial cells
- Basement membrane

Arteries

Elastic Artery

The walls of elastic arteries are not very thick relative to the vessel diameter, but they are extremely resilient. The tunica media of these vessels contains relatively few smooth muscle fibers and a high density of elastic fibers.

- Internal elastic membrane ⎤
- Endothelium ⎦ Tunica intima
- Tunica media
- Tunica externa

Muscular Artery

A typical muscular artery has a diameter of approximately 0.4 cm (0.16 in.). Muscular arteries have a thicker tunica media with a greater percentage of smooth muscle fibers than elastic arteries.

- Tunica externa
- Tunica media
- Endothelium
- Tunica intima

Arteriole

These vessels are considerably smaller than muscular arteries. Arterioles have an average diameter of about 30 μm. They have a poorly defined tunica externa, and their tunica media consists of scattered smooth muscle fibers that may not form a complete layer.

- Smooth muscle cells (tunica media)
- Endothelium
- Basement membrane

Capillaries

Continuous Capillary

Continuous capillaries are found in most regions of the body. In these capillaries, the endothelium is a complete lining, and the endothelial cells are connected by tight junctions and desmosomes

- Endothelial cells
- Basement membrane

21

Clinical Note

Arteriosclerosis **Arteriosclerosis** (ar-tēr-ē-ō-skler-Ō-sis; *arterio-*, artery + *sklerosis*, hardness) is a thickening and toughening of arterial walls. Complications related to arteriosclerosis account for about half of all deaths in the United States. For example, arteriosclerosis of coronary vessels is responsible for *coronary artery disease (CAD)*, and arteriosclerosis of arteries supplying the brain can lead to strokes. ⟲ p. 698 Arteriosclerosis takes two major forms:

1. **Focal calcification** is the deposition of calcium salts following the gradual degeneration of smooth muscle in the tunica media. Some focal calcification is a part of the aging process. It may also develop in association with atherosclerosis (described next).

2. **Atherosclerosis** (ath-er-ō-skler-Ō-sis; *athero-*, fatty degeneration) is the formation of lipid deposits in the tunica media associated with damage to the endothelial lining. It is the most common form of arteriosclerosis.

Atherosclerosis tends to develop in people whose blood contains elevated levels of plasma lipids—specifically, cholesterol. Circulating cholesterol is transported to peripheral tissues in *lipoproteins*, which are protein–lipid complexes. (We will discuss the various types of lipoproteins in Chapter 25.)

When plasma cholesterol levels are chronically elevated, cholesterol-rich lipoproteins remain in circulation for an extended period. Circulating monocytes then begin removing them from the bloodstream. Eventually, the monocytes become filled with lipid droplets. Now called *foam cells*, they attach themselves to the endothelial lining of blood vessels. There they release cytokines. These growth factors stimulate smooth muscle cells near the tunica intima to divide, thickening the vessel wall.

Other monocytes then invade the area, migrating between the endothelial cells. As these changes take place, the monocytes, smooth muscle cells, and endothelial cells begin phagocytizing lipids as well. The result is an atherosclerotic **plaque**, a fatty mass of tissue that projects into the lumen of the vessel. At this point, the plaque has a relatively simple structure, and evidence suggests that the process can be reversed with appropriate dietary adjustments.

If the conditions persist, the endothelial cells become swollen with lipids, and gaps appear in the endothelial lining. Platelets now begin sticking to the exposed collagen fibers, forming a localized blood clot that restricts arterial blood flow. The structure of the plaque is now relatively complex.

A typical plaque is shown in **Figures a** and **b**. Elderly people—especially elderly men—are most likely to develop atherosclerotic plaques. Estrogen may slow plaque formation. This may account for the lower incidence of CAD, myocardial infarctions (MIs), and strokes in women. After menopause, when estrogen production decreases, the risks of CAD, MIs, and strokes in women increase markedly.

In addition to advanced age and male sex, other important risk factors for atherosclerosis include high blood cholesterol levels, high blood pressure, and cigarette smoking. About 20 percent of middle-aged men have all three of these risk factors. They are four times more likely to experience an MI or a cardiac arrest than other men in their age group. Fewer women develop atherosclerotic plaques, but elderly female smokers with high blood cholesterol and high blood pressure are at much greater risk than other women. Diabetes mellitus, obesity, and stress can promote the development of atherosclerosis in both men and women. Evidence also indicates that at least some forms of atherosclerosis may be linked

Muscular Arteries

Muscular arteries, or *medium-sized arteries*, are also known as *distribution arteries* because they distribute blood to the body's skeletal muscles and internal organs. Most of the vessels of the arterial system are muscular arteries. They are characterized by a thick tunica media. It contains more smooth muscle cells than does the tunica media of elastic arteries (**Figures 21–1** and **21–2**). A typical muscular artery has a lumen diameter of approximately 0.4 cm (0.16 in.), but some have diameters as small as 0.05 cm. The *external carotid arteries* of the neck, the *brachial arteries* of the arms, the *mesenteric arteries* of the abdomen, and the *femoral arteries* of the thighs are examples of muscular arteries. Superficial muscular arteries are important as *pressure points*—places in the body where muscular arteries can be pressed against deeper bones to reduce blood flow and control severe bleeding. Major arterial pressure points are the

common carotid, radial, brachial, femoral, popliteal, posterior tibial, and dorsal pedal.

Arterioles

Arterioles, with an internal diameter of 30 μm or less, are considerably smaller than muscular arteries. Arterioles have a poorly defined tunica externa. In the larger arterioles, the tunica media consists of one or two layers of smooth muscle cells (**Figure 21–2**). In the smallest arterioles, the tunica media contains scattered smooth muscle cells that do not form a complete layer.

The diameters of smaller muscular arteries and arterioles change in response to local conditions or to sympathetic or endocrine stimulation. For example, arterioles in most tissues vasodilate when oxygen levels are low. Also, as we saw in Chapter 16, arterioles vasoconstrict under sympathetic

Tunica externa

Lipid deposits (plaque)

Tunica media

Coronary artery LM × 6

a A cross-sectional view of a large plaque

Plaque deposit in vessel wall

b A section of a coronary artery narrowed by a plaque

to chronic infection with *Chlamydia pneumoniae*, a bacterium responsible for several types of respiratory infections, including some forms of pneumonia.

We discussed potential treatments for atherosclerotic plaques, such as catheterization with balloon angioplasty and stenting, and bypass surgery, in Chapter 20. ↺ p. 725 In the many cases where changes in diet do not lower circulating cholesterol levels sufficiently, drug therapies can bring them under control. Genetic engineering techniques have been used to treat an inherited form of *hypercholesterolemia* (high blood cholesterol) linked to extensive plaque formation. (People with this condition are unable to absorb and recycle

cholesterol in the liver.) In this experimental procedure, circulating cholesterol levels declined after copies of appropriate genes were inserted into some of the individual's liver cells.

Without question, the best approach to atherosclerosis is to avoid it by eliminating or reducing associated risk factors. Suggestions include (1) reducing your intake of dietary cholesterol, saturated fats, and trans fatty acids by restricting consumption of fatty meats (such as beef, lamb, and pork), egg yolks, and cream; (2) not smoking; (3) checking your blood pressure and taking steps to lower it if necessary; (4) having your blood cholesterol levels checked annually; (5) controlling your weight; and (6) exercising regularly.

stimulation. ↺ p. 567 Changes in their diameter affect the amount of force required to push blood around the cardiovascular system: More pressure is required to push blood through a constricted vessel than through a dilated one. The force opposing blood flow is called **resistance (R)**, so arterioles are also called **resistance vessels**.

Vessel characteristics change gradually with distance from the heart. Each type of vessel described here actually represents the midpoint in a portion of a continuum. Thus, the largest muscular arteries contain a considerable amount of elastic tissue, and the smallest resemble heavily muscled arterioles.

Arteries carry blood under great pressure, and their walls are adapted to handle that pressure. Occasionally, local arterial pressure exceeds the capacity of the elastic components of the tunics. The result is an **aneurysm** (AN-yū-rizm), or bulge in the weakened wall of an artery. The bulge is like a bubble in the wall of a tire—and like a bad tire, the artery can suffer a catastrophic

blowout. The most dangerous aneurysms occur in arteries of the brain (where they cause strokes) or in the aorta (where a rupture will cause fatal bleeding in a matter of minutes).

Capillaries

When we think of the cardiovascular system, we think first of the heart or the great blood vessels connected to it. But the microscopic capillaries that permeate most tissues do the real work of the cardiovascular system. These delicate vessels weave throughout active tissues, forming intricate networks that surround muscle fibers. Capillaries radiate through connective tissues, and branch beneath the basement membrane of epithelia.

Capillaries are the *only* blood vessels whose walls permit exchange between the blood and the surrounding interstitial fluids. Exchange can take place quickly because capillary walls are thin and diffusion distances are short. In addition, blood flows through capillaries relatively slowly, allowing sufficient time for the diffusion or active transport of materials across the capillary walls. In this way, the histological structure of capillaries permits a two-way exchange of substances between blood and interstitial fluid.

A typical capillary consists of an endothelial tube inside a thin basement membrane. Neither a tunica media nor a tunica externa is present (**Figure 21–2**). The average diameter of a capillary is a mere 8 μm, very close to that of a single red blood cell. The two major types of capillaries are *continuous capillaries* and *fenestrated capillaries*.

Continuous Capillaries

Continuous capillaries supply most regions of the body. In a **continuous capillary**, the endothelium is a complete lining. A cross section through a large continuous capillary cuts across several endothelial cells (**Figure 21–3a**). In a small continuous capillary, a single endothelial cell may completely encircle the lumen.

Continuous capillaries are located in all tissues except epithelia and cartilage. Continuous capillaries permit the water, small solutes, and lipid-soluble materials to diffuse into the

interstitial fluid. At the same time, they prevent the loss of blood cells and plasma proteins. In addition, some exchange may occur between blood and interstitial fluid by *bulk transport*—the movement of materials by endocytosis (by endosomes) or exocytosis at the inner endothelial surface. ↪ p. 122

In specialized continuous capillaries in most of the central nervous system and in the thymus, the endothelial cells are bound together by tight junctions. These capillaries have very restricted permeability. We discussed one example—the capillaries responsible for the *blood–brain barrier*—in Chapters 12 and 14. ↪ pp. 419, 495

Fenestrated Capillaries

Fenestrated (FEN-es-trā-ted; *fenestra*, window) **capillaries** contain "windows," or pores, that penetrate the endothelial lining (**Figure 21–3b**). The pores allow rapid exchange of water and solutes between blood and interstitial fluid. Examples of fenestrated capillaries include the *choroid plexus* of the brain and the blood vessels in a variety of endocrine organs, such as the hypothalamus and the pituitary, pineal, and thyroid glands. Fenestrated capillaries are also found along absorptive areas of

Figure 21–3 Capillary Structure.

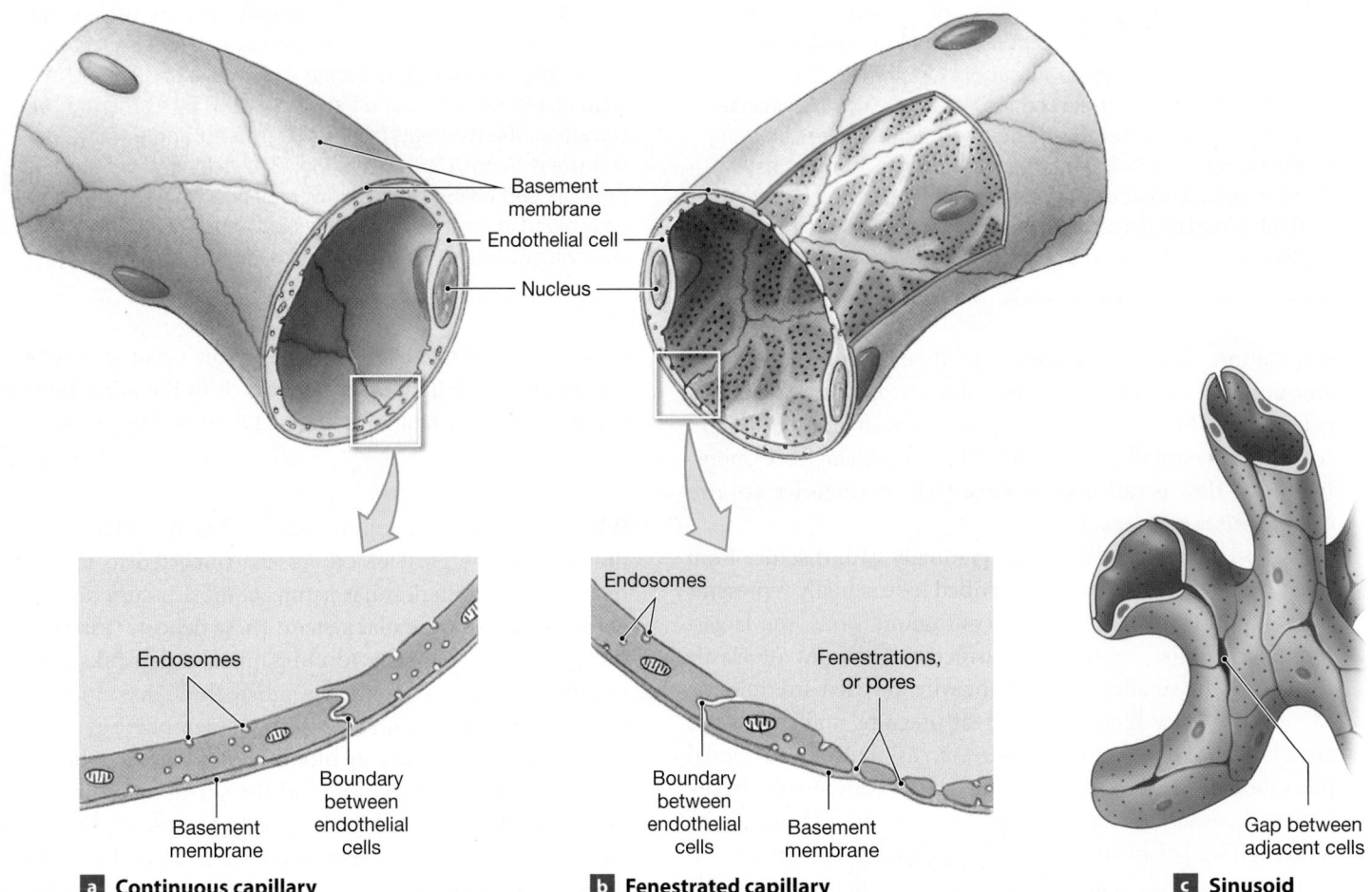

Basement membrane
Endothelial cell
Nucleus
Endosomes
Boundary between endothelial cells
Basement membrane

a Continuous capillary

Endosomes
Fenestrations, or pores
Boundary between endothelial cells
Basement membrane

b Fenestrated capillary

Gap between adjacent cells

c Sinusoid

21

the intestinal tract and at filtration sites in the kidneys. Both the number of pores and their permeability characteristics may vary from one region of the capillary to another.

Sinusoids (SĪ-nuh-soydz), also called **sinusoidal capillaries**, resemble fenestrated capillaries that are flattened and irregularly shaped (**Figure 21–3c**). In addition to being fenestrated, sinusoids commonly have gaps between adjacent endothelial cells, and the basement membrane is either thinner or absent. As a result, sinusoids permit the free exchange of water and solutes as large as plasma proteins between blood and interstitial fluid.

Blood moves through sinusoids relatively slowly, maximizing the time available for exchange across the sinusoidal walls. Sinusoids occur in the liver, bone marrow, spleen, and many endocrine organs, including the pituitary and adrenal glands. At liver sinusoids, plasma proteins secreted by liver cells enter the bloodstream. Along sinusoids of the liver, spleen, and bone marrow, phagocytic cells monitor the passing blood, engulfing damaged red blood cells, pathogens, and cellular debris.

Capillary Beds

Capillaries function not as individual units, but rather, as part of an interconnected network called a **capillary bed**, or

capillary plexus (**Figure 21–4**). A single arteriole generally gives rise to dozens of capillaries. They empty into several *venules*, the smallest vessels of the venous system. A **precapillary sphincter** guards the entrance to each capillary. Contraction of the smooth muscle cells of this sphincter narrows the capillary entrance, reducing or stopping the flow of blood. When one precapillary sphincter constricts, blood is diverted into other branches of the network. When a precapillary sphincter relaxes, the entrance dilates, and blood flows into the capillary.

A capillary bed contains several direct connections between arterioles and venules. The wall in the first part of such a passageway contains smooth muscle that can change its diameter. This segment is called a **metarteriole** (met-ar-TĒR-ē-ōl) or **precapillary arteriole**. The rest of the passageway resembles a typical capillary in structure and is called a **thoroughfare channel**.

More than one artery may supply blood to a capillary bed. The multiple arteries are called **collaterals**. They fuse before giving rise to arterioles. The fusion of two collateral arteries that supply a capillary bed is an example of an **arterial anastomosis**. (An *anastomosis* is the joining of blood vessels.) The interconnections between the *anterior* and *posterior interventricular arteries* of the heart are arterial anastomoses. ⮌ p. 722 An arterial

Figure 21–4 The Organization of a Capillary Bed.

b A micrograph of a number of capillary beds.

a A typical capillary bed. Solid arrows indicate consistant blood flow; dashed arrows indicate variable or pulsating blood flow.

anastomosis acts like an insurance policy: If one artery is compressed or blocked, capillary circulation will continue.

Arteriovenous (ar-tēr-ē-o-VĒ-nus) **anastomoses** are direct connections between arterioles and venules. When an arteriovenous anastomosis is dilated, blood bypasses the capillary bed and flows directly into the venous circulation. The pattern of blood flow through an arteriovenous anastomosis is regulated primarily by sympathetic innervation under the control of the cardiovascular center of the medulla oblongata.

Angiogenesis (an-jē-ō-JEN-e-sis; *angio-*, blood vessel + *genesis*, production) is the formation of new blood vessels from pre-existing vessels and occurs under the direction of **vascular endothelial growth factor (VEGF)**. Angiogenesis occurs in the embryo as tissues and organs develop. It may also occur at other times in any body tissue in response to factors released by cells that are *hypoxic*, or oxygen-starved. Clinically, angiogenesis is probably most important in cardiac muscle, where it takes place in response to a chronically constricted or occluded vessel.

Vasomotion

Although blood normally flows from arterioles to venules at a constant rate, the flow within each capillary is quite variable. Each precapillary sphincter alternately contracts and relaxes, perhaps a dozen times per minute. These rhythmic changes in vessel diameter are called **vasomotion**. As a result, the blood flow within any capillary occurs in pulses rather than as a steady and constant stream. The net effect is that blood may reach the venules by one route now and by a different route later.

Vasomotion is controlled locally by changes in the concentrations of chemicals and dissolved gases in the interstitial fluids. For example, when dissolved oxygen concentrations decrease within a tissue, the capillary sphincters relax, so blood flow to the area increases. This process is an example of capillary *autoregulation*. We focus on it in a later section.

When you are at rest, blood flows through about 25 percent of the vessels within a typical capillary bed in your body. Your cardiovascular system does not contain enough blood to maintain adequate blood flow to all the capillaries in all the capillary beds in your body at the same time. As a result, when many tissues become active, the blood flow through capillary beds must be coordinated. We describe the mechanisms by which the cardiovascular center performs this coordination later in the chapter.

Veins

Veins collect blood from all tissues and organs and return it to the heart. The walls of veins can be thinner than those of corresponding arteries because the blood pressure in veins is lower than that in arteries. We classify veins on the basis of their size. Even though their walls are thinner, in general veins are larger in diameter than their corresponding arteries. (Review **Figure 21–2** to compare typical arteries and veins.)

Venules

Venules are the smallest venous vessels. They collect blood from capillary beds. They vary widely in size and structure. An average venule has an internal diameter of roughly 20 *μ*m. Venules smaller than 50 *μ*m lack a tunica media, and the smallest venules resemble expanded capillaries.

Medium-Sized Veins

Medium-sized veins are comparable in size to muscular arteries. They range from 2 to 9 mm in internal diameter. Their tunica media is thin and contains relatively few smooth muscle cells. The thickest layer of a medium-sized vein is the tunica externa, which contains longitudinal bundles of elastic and collagen fibers.

Large Veins

Large veins include the superior and inferior venae cavae and their branches within the abdominopelvic and thoracic cavities. All large veins have all three layers. The slender tunica media is surrounded by a thick tunica externa composed of a mixture of elastic and collagen fibers.

Venous Valves

The arterial system is a high-pressure system: Almost all the force developed by the heart is required to push blood along the network of arteries and through miles of capillaries. Blood pressure in a peripheral venule is only about 10 percent of that in the ascending aorta, and pressures continue to fall along the venous system.

The blood pressure in venules and medium-sized veins is so low that it cannot overcome the force of gravity. In the limbs, veins of this size contain **valves**, folds of the tunica intima that project from the vessel wall and point in the direction of blood flow. These valves, like those in the heart, permit blood flow in one direction only. Venous valves prevent blood from moving back toward the capillaries (**Figure 21–5**).

As long as the valves function normally, any movement that distorts or compresses a vein pushes blood toward the heart. This effect improves *venous return*, the rate of blood flow to the heart. ⤾ p. 740 The mechanism is particularly important when you are standing, because blood returning from your feet must overcome gravity to ascend to the heart. Valves compartmentalize the blood within the veins, dividing the weight of the blood among the compartments. Any contraction of the surrounding skeletal muscles squeezes the blood toward the heart. Although you are probably not aware of it, when you stand, rapid cycles of contraction and relaxation occur within your leg muscles, helping to push blood toward the trunk. When you lie down, venous valves play a smaller part in venous return, because your heart and major vessels are at the same level.

Figure 21–5   **The Function of Valves in the Venous System.**

Valve closed

Valve opens superior to contracting muscle

Valve closed

Valve closes inferior to contracting muscle

Figure 21–6   **The Distribution of Blood in the Cardiovascular System.**

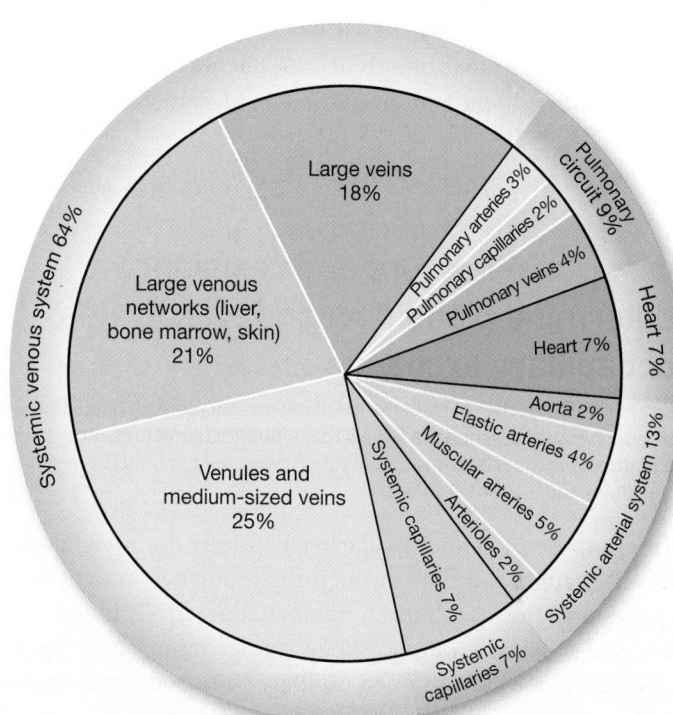

If the walls of the veins near the valves weaken or become stretched and distorted, the valves may not work properly. Blood then pools in the veins, and the vessels become grossly distended. The effects range from mild discomfort and a cosmetic problem, as in superficial **varicose veins** in the thighs and legs, to painful distortion of adjacent tissues, as in **hemorrhoids**.

The Distribution of Blood

Our total blood volume is unevenly distributed among arteries, veins, and capillaries (**Figure 21–6**). The heart, arteries, and capillaries in the pulmonary and systemic circuits normally contain 30–35 percent of the blood volume (roughly 1.5 liters of whole blood). The venous system contains the rest (65–70 percent, or about 3.5 liters). About one-third of the blood in the venous system (about a liter) is circulating within the liver, bone marrow, and skin. These organs have extensive venous networks that at any moment contain large volumes of blood.

Veins are much more *distensible,* or expandable, than arteries because their walls are thinner, with less smooth muscle. For a given rise in blood pressure, a typical vein stretches about eight times as much as a corresponding artery. The *capacitance* of a blood vessel is the relationship between the volume of blood it contains and the blood pressure. If a vessel behaves

like a balloon, expanding easily at low pressures, it has high capacitance. If it behaves more like a truck tire, expanding only at high pressures, it has low capacitance. Veins, which expand easily, are called **capacitance vessels**. Because veins have high capacitance, they can accommodate large changes in blood volume. If the blood volume rises or falls, the elastic walls stretch or recoil, changing the volume of blood in the venous system.

If serious hemorrhaging (blood loss) occurs, the *vasomotor center* of the medulla oblongata stimulates sympathetic nerves that innervate smooth muscle cells in the walls of medium-sized veins. This activity has two major effects:

1. *Systemic veins constrict.* This process, called **venoconstriction** (vē-nō-kon-STRIK-shun), decreases the amount of blood within the venous system, increasing the volume within the arterial system and capillaries. Venoconstriction can keep the blood volume within the arteries and capillaries at near-normal levels despite a significant blood loss.

2. *The constriction of veins in the liver, skin, and lungs redistributes a significant proportion of the total blood volume.* As a result, blood flow to delicate organs (such as the brain) and to active skeletal muscles can be increased or maintained after blood loss. The amount of blood that can be shifted from veins in the liver, skin, and lungs to the general circulation is called the **venous reserve**. It is normally about 20 percent of total blood volume.

21

CHECKPOINT

1. List the five general classes of blood vessels.

2. Which type of vessel is characterized by thin walls with very little smooth muscle tissue in the tunica media?

3. Why are valves located in veins but not in arteries?

4. Where in the body would you find fenestrated capillaries?

See the blue Answers tab at the back of the book.

21-2 ▮ Pressure and resistance determine blood flow and affect rates of capillary exchange

Learning Outcome Explain the mechanisms that regulate blood flow through vessels, describe the factors that influence blood pressure, and discuss the mechanisms that regulate movement of fluids between capillaries and interstitial spaces.

The purpose of cardiovascular regulation is the maintenance of adequate blood flow through the capillaries in peripheral tissues and organs. Under normal circumstances, blood flow is equal to cardiac output. When cardiac output goes up, so does the blood flow through capillary beds, and when cardiac output declines, capillary blood flow is reduced.

Capillary blood flow is determined by the interplay between *pressure (P)* and *resistance (R)* in the cardiovascular network. To keep blood moving, the heart must generate enough pressure to overcome the resistance to blood flow in the pulmonary and systemic circuits. In general terms, flow (F) is directly proportional to the pressure (increased pressure → increased flow), and inversely proportional to resistance (increased resistance → decreased flow). However, the absolute pressure is less important than the pressure *gradient*—the difference in pressure from one end of the vessel to the other. This relationship can be summarized as

$$F \propto \frac{\Delta P}{R}$$

where the symbol ∝ means "is proportional to" and Δ means "the difference in." The largest pressure gradient is found between the base of the aorta and the proximal ends of peripheral capillary beds. The cardiovascular center can alter this pressure gradient, and change the rate of capillary blood flow, by adjusting cardiac output and peripheral resistance.

Blood leaving the peripheral capillaries enters the venous system. The pressure gradient across the venous system is relatively small, but venous resistance is very low. The low venous blood pressure—aided by valves, skeletal muscle contraction, gravity, and other factors—is enough to return the blood to the heart. When necessary, cardiovascular control centers can raise venous pressure (through venoconstriction) to improve venous return and maintain adequate cardiac output.

We will begin this section by examining blood pressure and resistance more closely. We will then consider the mechanisms of *capillary exchange*, the transfer of liquid and solutes between the blood and interstitial fluid. Capillary exchange provides tissues with oxygen and nutrients and removes the carbon dioxide and waste products generated by active cells.

Active tissues need more blood flow than inactive ones. Even something as simple as a change in position—going from sitting to standing, for instance—triggers a number of cardiovascular changes. We will end this section with a discussion of what those changes are and how they are coordinated.

Pressure

When talking about cardiovascular pressures, three values are usually reported:

1. *Blood Pressure.* The term **blood pressure (BP)** refers to arterial pressure, usually reported in millimeters of mercury (mm Hg). Systemic arterial pressures in adults average 120 mm Hg at the entrance to the aorta to roughly 35 mm Hg at the start of a capillary network.

2. *Capillary Hydrostatic Pressure.* Hydrostatic pressure is the force exerted by a fluid pressing against a wall. **Capillary hydrostatic pressure (CHP)**, or *capillary pressure*, is the pressure within capillary walls. Along the length of a typical capillary, pressures decline from roughly 35 mm Hg to about 18 mm Hg.

3. *Venous Pressure.* **Venous pressure** is the pressure within the venous system. Venous pressure is quite low: The pressure gradient from the venules to the right atrium is only about 18 mm Hg.

The difference in pressure (ΔP) across the entire systemic circuit, sometimes called the *circulatory pressure*, averages about 100 mm Hg. For circulation to occur, the circulatory pressure must overcome the **total peripheral resistance**—the resistance of the entire cardiovascular system. The arterial network has by far the largest pressure gradient (85 mm Hg), and this primarily reflects the relatively high resistance of the arterioles.

Total Peripheral Resistance

The total peripheral resistance of the cardiovascular system reflects a combination of factors: *vascular resistance*, *blood viscosity*, and *turbulence* (**Figure 21–7**).

Vascular Resistance

Vascular resistance, the forces that oppose blood flow in the blood vessels, is the largest component. The most important factor in vascular resistance is friction between blood and the vessel walls. The amount of friction depends on two factors: vessel length and vessel diameter.

Figure 21–7   **Factors Affecting Friction and Vascular Resistance.**

Vessel Length. Increasing the length of a blood vessel increases friction: The longer the vessel, the greater the surface area in contact with blood. You can easily blow the water out of a snorkel that is 2.5 cm (1 in.) in diameter and 25 cm (10 in.) long, but you cannot blow the water out of a 15 m (16 yard) long garden hose, because the total friction is too great. The most dramatic changes in blood vessel length occur between birth and maturity, as individuals grow to adult size.

In adults, vessel length can increase or decrease gradually when individuals gain or lose weight, but on a day-to-day basis this component of vascular resistance can be considered constant.

Vessel Diameter. The effects of friction on blood act in a narrow zone closest to the vessel wall. In a small-diameter vessel, friction with the walls slows nearly all the blood. Resistance is therefore relatively high. Blood near the center of a large-diameter vessel does not encounter friction with the walls, so the resistance in large vessels is fairly low.

Differences in diameter have much more significant effects on resistance than do differences in length. If two vessels are equal in radius (one-half the diameter) but one is twice as long as the other, the longer vessel offers twice as much resistance to blood flow. But for two vessels of equal length, one twice the radius of the other, the narrower one offers 16 times as much resistance to blood flow. This relationship, expressed in terms of the vessel radius r and resistance R, can be summarized as $R \propto 1/r^4$.

More significantly, there is no way to control vessel length, but vessel diameter can change quickly through vasoconstriction or vasodilation. Most of the peripheral resistance occurs in arterioles, the smallest vessels of the arterial system. As noted earlier in the chapter, arterioles are extremely muscular: The wall of an arteriole with an inner diameter of 30 μm can have a 20-μm-thick layer of smooth muscle. When these smooth muscles contract or relax, peripheral resistance increases or decreases. Because a small change in diameter produces a large change in resistance, mechanisms that alter the diameters of arterioles provide control over peripheral resistance and blood flow.

Blood Viscosity

Viscosity is the resistance to flow caused by interactions among molecules and suspended materials in a liquid. Liquids of low viscosity, such as water (viscosity 1.0), flow at low pressures. Thick, syrupy fluids, such as molasses (viscosity 300), flow only under higher pressures. Whole blood has a viscosity about five times that of water, due to its plasma proteins and blood cells. Under normal conditions, the viscosity of blood remains stable. Anemia, polycythemia, and other disorders that affect the hematocrit also change blood viscosity, and thus peripheral resistance.

Turbulence

High flow rates, irregular surfaces, and sudden changes in vessel diameter upset the smooth flow of blood, creating eddies and swirls. This phenomenon, called **turbulence**, increases resistance and slows blood flow.

Turbulence normally occurs when blood flows between the atria and the ventricles, and between the ventricles and the aortic and pulmonary trunks. It also develops in large arteries, such as the aorta, when cardiac output and arterial flow rates are very high. However, turbulence seldom occurs in smaller vessels unless their walls are damaged. For example, an atherosclerotic plaque creates abnormal turbulence and restricts blood flow. Because turbulence makes a distinctive sound, or *bruit* (brū-Ē), plaques in large blood vessels can often be detected with a stethoscope.

Table 21–1 provides a review of the terms and relationships discussed in this section.

An Overview of Cardiovascular Pressures

Look at the graphs in **Figure 21–8** for an overview of the vessel diameters, cross-sectional areas, pressures, and velocity of blood flow in the systemic circuit.

- *Vessel Diameters.* As blood proceeds from the aorta toward the capillaries, vessels diverge. The arteries branch repeatedly, and each branch is smaller in diameter than the preceding one (**Figure 21–8a**). As blood proceeds from the capillaries toward the venae cavae, vessels converge. Vessel diameters increase as venules combine to form small and medium-sized veins.

- *Total Cross-Sectional Areas.* Although the arterioles, capillaries, and venules are small in diameter, the body has large numbers of them. All the blood flowing through the aorta also flows through peripheral capillaries. Blood pressure

and the speed of blood flow are proportional to the cross-sectional area of the vessels involved. What is important is not the cross-sectional area of each individual vessel, but the *combined* cross-sectional area of *all* the vessels (**Figure 21–8b**). In effect, your blood moves from one big pipe (the aorta, with a cross-sectional area of 4.5 cm^2) into countless tiny ones (the peripheral capillaries, with a total cross-sectional area of 5000 cm^2), and then blood travels back to the heart through two large pipes (the venae cavae).

- *Pressures.* As arteries branch, their total cross-sectional area increases, and blood pressure falls rapidly (**Figure 21–8c**). Most of the decline takes place in the small arteries and arterioles. Venous pressures are relatively low.

- *Velocity of Blood Flow.* As the total cross-sectional area of the vessels increases from the aorta toward the capillaries, the velocity of blood flow decreases (**Figure 21–8d**). Blood flow velocity then increases as the total cross-sectional area drops from the capillaries toward the venae cavae.

Figure 21–9 graphs the blood pressure throughout the cardiovascular system. Systemic pressures are highest in the aorta, peaking at about 120 mm Hg. Pressures reach a minimum of 2 mm Hg at the entrance to the right atrium. Pressures in the pulmonary circuit are much lower than those in the systemic circuit. The right ventricle does not ordinarily develop high pressures because the pulmonary vessels are much shorter and more distensible than the systemic vessels, thus providing less resistance to blood flow.

Table 21–1	Key Terms and Relationships Pertaining to Blood Circulation
Blood Flow (*F*)	The volume of blood flowing per unit of time through a vessel or a group of vessels; may refer to circulation through a capillary, a tissue, an organ, or the entire vascular network. Total blood flow is equal to cardiac output.
Blood Pressure (*BP*)	The hydrostatic pressure in the arterial system that pushes blood through capillary beds.
Circulatory Pressure	The pressure difference between the base of the ascending aorta and the entrance to the right atrium.
Hydrostatic Pressure	A pressure exerted by a liquid in response to an applied force.
Peripheral Resistance (*PR*)	The resistance of the arterial system; affected by such factors as vascular resistance, viscosity, and turbulence.
Resistance (*R*)	A force that opposes movement (in this case, blood flow).
Total Peripheral Resistance	The resistance of the entire cardiovascular system.
Turbulence	A resistance due to the irregular, swirling movement of blood at high flow rates or exposure to irregular surfaces.
Vascular Resistance	A resistance due to friction within a blood vessel, primarily between the blood and the vessel walls. Increases with increasing length or decreasing diameter; vessel length is constant, but vessel diameter can change.
Venous Pressure	The hydrostatic pressure in the venous system.
Viscosity	A resistance to flow due to interactions among molecules within a liquid.
RELATIONSHIPS AMONG THE PRECEDING TERMS	
$F \propto P$	Flow is proportional to the pressure gradient.
$F \propto 1/R$	Flow is inversely proportional to resistance.
$F \propto P/R$	Flow is directly proportional to the pressure gradient, and inversely proportional to resistance.
$F \propto BP/PR$	Flow is directly proportional to blood pressure, and inversely proportional to peripheral resistance.
$R \propto 1/r^4$	Resistance is inversely proportional to the fourth power of the vessel radius.

21

Figure 21–8   **Relationships among Vessel Diameter, Cross-Sectional Area, Blood Pressure, and Blood Velocity within the Systemic Circuit.**

Figure 21–9   **Pressures within the Systemic Circuit.** Notice the general reduction in circulatory pressure within the systemic circuit and the elimination of the pulse pressure within the arterioles.

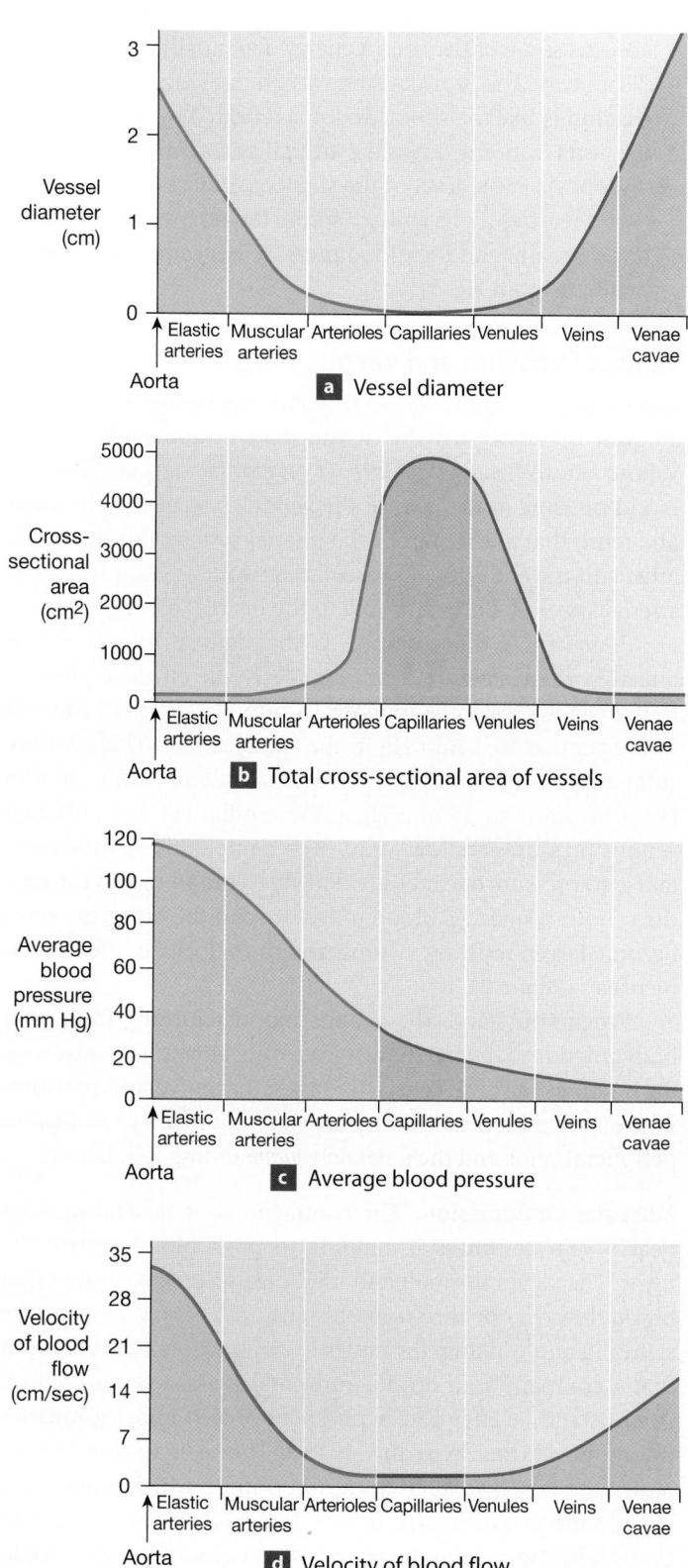

a Vessel diameter

b Total cross-sectional area of vessels

c Average blood pressure

d Velocity of blood flow

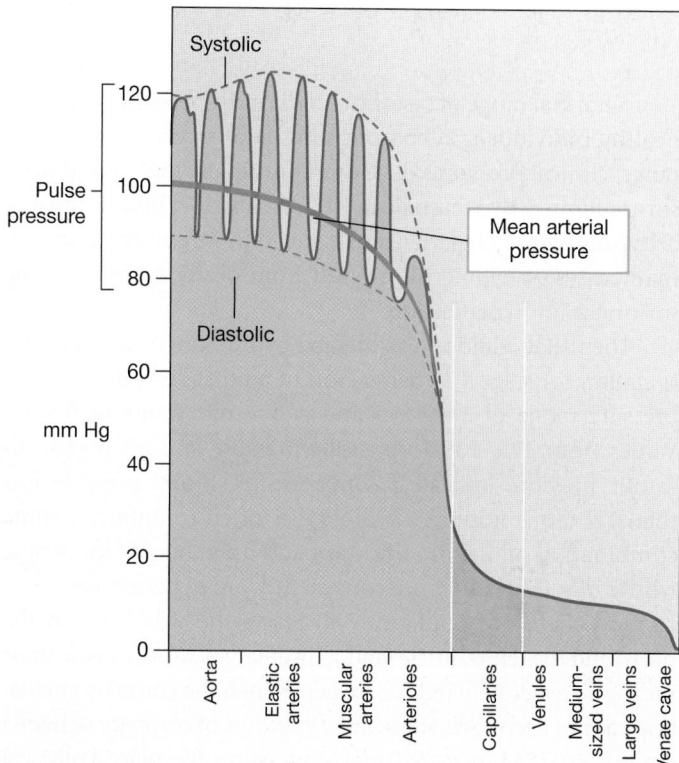

Arterial Blood Pressure

Arterial pressure is important because it maintains blood flow through capillary beds. To do this, it must always be high enough to overcome the peripheral resistance. Arterial pressure is not constant. Rather, it rises during ventricular systole and falls during ventricular diastole. The peak blood pressure measured during ventricular systole is called **systolic pressure**, and the minimum blood pressure at the end of ventricular diastole is called **diastolic pressure**. In recording blood pressure, we separate systolic and diastolic pressures by a slash, as in "120/80" ("one-twenty over eighty") or "110/75."

A *pulse* is a rhythmic fluctuation in pressure that accompanies each heartbeat. The difference between the systolic and diastolic pressures is the **pulse pressure** (Figure 21–9). To report a single blood pressure value, we use the **mean arterial pressure (MAP)**. It is calculated by adding one-third of the pulse pressure to the diastolic pressure:

$$\text{MAP} = \text{diastolic pressure} + \frac{\text{pulse pressure}}{3}$$

21

For a systolic pressure of 120 mm Hg and a diastolic pressure of 90 mm Hg, we calculate MAP as follows:

$$MAP = 90 + \frac{(120 - 90)}{3} = 90 + 10 = 100 \text{ mm Hg}$$

A normal range of systolic and diastolic pressures occurs in healthy individuals. When pressures shift outside the normal range, clinical problems develop. Abnormally high blood pressure is termed **hypertension**. Abnormally low blood pressure is **hypotension**. Hypertension is much more common. In fact, many cases of hypotension result from overly aggressive drug treatment for hypertension.

The usual criterion established by the American Heart Association for stage 1 hypertension in adults is a systolic blood pressure range of 140–159 and a diastolic range of 90–99. Values from 120–139 for systolic pressure and 80–89 for diastolic pressure indicate *pre-hypertension*. Blood pressure less than 120/80 is normal. Cardiologists often recommend some combination of diet modification and drug therapy for people whose blood pressures are consistently pre-hypertensive.

Hypertension significantly increases the workload on the heart, and the left ventricle gradually enlarges. More muscle mass means a greater demand for oxygen. When the coronary circulation cannot keep pace, signs and symptoms of coronary ischemia appear. ⤺ p. 724 Increased arterial pressures also place a physical stress on the walls of blood vessels throughout the body. This stress promotes or accelerates the development of arteriosclerosis. It also increases the risk of aneurysms, heart attacks, and strokes.

Elastic Rebound

As systolic pressure climbs, the arterial walls stretch, just as an extra puff of air expands a partially inflated balloon. This expansion allows the arterial system to accommodate some of the blood provided by ventricular systole. When diastole begins and blood pressures fall, the arteries recoil to their original dimensions. This phenomenon is called **elastic rebound**. Some blood is forced back toward the left ventricle, closing the aortic valve and helping to drive additional blood into the coronary arteries. However, most of the push from elastic rebound forces blood toward the capillaries. This maintains blood flow along the arterial network while the left ventricle is in diastole.

Pressures in Small Arteries and Arterioles

The mean arterial pressure and the pulse pressure become smaller as the distance from the heart increases (**Figure 21–9**):

- The mean arterial pressure declines as the arterial branches become smaller and more numerous. In essence, blood pressure decreases as it overcomes friction and produces blood flow.

- The pulse pressure lessens due to the cumulative effects of elastic rebound along the arterial system. The effect can be likened to a series of ever-softer echoes following a loud shout. Each time an echo is produced, the reflecting surface absorbs some of the sound energy. Eventually, the echo disappears. The pressure surge accompanying ventricular ejection is like the shout, and it is reflected by the wall of the aorta, echoing down the arterial system until it finally disappears at the level of the small arterioles. By the time blood reaches a precapillary sphincter, no pressure fluctuations remain, and the blood pressure is steady at approximately 35 mm Hg.

Venous Pressure and Venous Return

Venous pressure, although low, determines venous return—the amount of blood arriving at the right atrium each minute. Venous return has a direct impact on cardiac output. ⤺ p. 741 Blood pressure at the start of the venous system is only about one-tenth that at the start of the arterial system, but the blood must still travel through a vascular network as complex as the arterial system before returning to the heart.

Pressures at the entrance to the right atrium fluctuate, but they average about 2 mm Hg. Thus, the effective pressure in the venous system is roughly 16 mm Hg (from 18 mm Hg in the venules to 2 mm Hg in the venae cavae). This pressure compares with 65 mm Hg in the arterial system (from 100 mm Hg at the aorta to 35 mm Hg at the capillaries). Yet, although venous pressures are low, veins offer comparatively little resistance, so pressure declines very slowly as blood moves through the venous system. As blood moves toward the heart, the veins become larger, resistance drops, and the velocity of blood flow increases (**Figure 21–8**).

When you stand, the venous blood returning from your body inferior to the heart must overcome gravity as it travels up the inferior vena cava. Two factors assist the low venous pressures in propelling blood toward your heart: *muscular compression* of peripheral veins and the *respiratory pump* during inhalation.

Muscular Compression. The contractions of skeletal muscles near a vein compress it, helping to push blood toward the heart. The valves in small and medium-sized veins ensure that blood flows in one direction only (**Figure 21–5**). When you are standing and walking, the cycles of contraction and relaxation that accompany your normal movements assist venous return. If you stand at attention, with knees locked and leg muscles immobilized, that assistance is lost. The reduction in venous return then leads to a fall in cardiac output, which reduces the blood supply to the brain. This decline is sometimes enough to cause **fainting**, a temporary loss of consciousness. You would then collapse, but while you were in the horizontal position, both venous return and cardiac output would return to normal.

The Respiratory Pump. As you inhale, your thoracic cavity expands, reducing the pressure within the pleural cavities. This drop in pressure pulls air into your lungs. At the same time, it also pulls blood into the inferior vena cava and right atrium from the smaller veins of your abdominal cavity and lower body. The effect on venous return through the superior vena cava is less pronounced, because blood in that vessel is normally assisted by gravity. As you exhale, your thoracic cavity decreases in size. Internal pressure then rises, forcing air out of your lungs and pushing venous blood into the right atrium. This mechanism is called the **respiratory pump**. Such pumping action becomes more important during heavy exercise, when respirations are deep and frequent.

Capillary Pressures and Capillary Exchange

Capillary exchange plays a key role in homeostasis. The most important processes that move materials across typical capillary walls are *diffusion*, *filtration*, and *reabsorption*.

Diffusion

As we saw in Chapter 3, *diffusion* is the net movement of ions or molecules from an area where their concentration is higher to an area where their concentration is lower. ⊃ p. 115 The difference between the high and low concentrations represents a *concentration gradient*. Diffusion tends to eliminate that gradient. Diffusion occurs most rapidly when (1) the distances involved are short, (2) the concentration gradient is large, and (3) the ions or molecules involved are small.

Different substances diffuse across capillary walls by different routes:

1. *Water, ions, and small organic molecules, such as glucose, amino acids, and urea,* can usually enter or leave the bloodstream by diffusion between adjacent endothelial cells or through the pores of fenestrated capillaries.

2. *Many ions, including sodium, potassium, calcium, and chloride,* can diffuse across endothelial cells by passing through channels in plasma membranes.

3. *Large water-soluble compounds* are unable to enter or leave the bloodstream except at fenestrated capillaries, such as those of the hypothalamus, the kidneys, many endocrine organs, and the intestinal tract.

4. *Lipids, such as fatty acids and steroids, and lipid-soluble materials, including soluble gases such as oxygen and carbon dioxide,* can cross capillary walls by diffusion through the endothelial plasma membranes.

5. *Plasma proteins* are normally unable to cross the endothelial lining anywhere except in sinusoids, such as those of the liver, where plasma proteins enter the bloodstream.

Filtration

Filtration is the removal of solutes as a solution flows across a porous membrane. Solutes too large to pass through the pores are filtered out of the solution. The driving force for filtration is hydrostatic pressure. As we saw earlier, it pushes water from an area of higher pressure to an area of lower pressure.

In *capillary filtration*, water and small solutes are forced across a capillary wall, leaving larger solutes and suspended proteins in the bloodstream (**Figure 21–10**). The solute molecules that leave the bloodstream are small enough to pass between adjacent endothelial cells or through the pores in a fenestrated capillary. Filtration takes place primarily at the arterial end of a capillary, where capillary hydrostatic pressure (CHP) is highest.

Reabsorption

Reabsorption occurs as the result of osmosis. *Osmosis* is a special term for the diffusion of water across a selectively permeable membrane that separates two solutions of differing solute concentrations. Water molecules tend to diffuse across a membrane *toward* the solution containing the higher solute concentration (**Figure 3–16**, p. 119).

Figure 21–10   Capillary Filtration. Capillary hydrostatic pressure (CHP) forces water and solutes through the gaps between adjacent endothelial cells in continuous capillaries. The sizes of solutes that move across the capillary wall are determined by the dimensions of the gaps.

The **osmotic pressure (OP)** of a solution is an indication of the force of osmotic water movement. In other words, it represents the pressure that must be applied to prevent osmotic movement across a membrane. The higher the solute concentration of a solution, the greater is the solution's osmotic pressure. The presence of suspended proteins that cannot cross capillary walls creates an osmotic pressure called *blood colloid osmotic pressure (BCOP)*. Clinicians often use the term *oncotic pressure* (*onkos*, a swelling) when referring to the colloid osmotic pressure of body fluids. The two terms are equivalent. Osmotic water movement continues until either the solute concentrations are equalized or an opposing hydrostatic pressure prevents the movement.

Now let's look at the interplay between filtration and reabsorption along the length of a typical capillary. In this discussion, remember that hydrostatic pressure forces water *out of* a solution, and osmotic pressure draws water *into* a solution.

The Interplay between Filtration and Reabsorption

The continuous movement of water out of the capillaries, through peripheral tissues, and then back to the bloodstream by way of the lymphatic system has four important functions:

1. It ensures that plasma and interstitial fluid, two major components of extracellular fluid, are in constant communication and mutual exchange.

2. It accelerates the distribution of nutrients, hormones, and dissolved gases throughout tissues.

3. It assists in the transport of insoluble lipids and tissue proteins that cannot enter the bloodstream by crossing the capillary walls.

4. It has a flushing action that carries bacterial toxins and other chemical stimuli to lymphatic tissues and organs responsible for providing immunity to disease.

Capillary blood pressure declines as blood flows from the arterial end to the venous end of a capillary. As a result, the rates of filtration and reabsorption gradually change as blood passes along the length of a capillary. The factors involved are diagrammed in **Figure 21–11**.

Net hydrostatic pressure is the difference between the pressure inside the capillary wall and the hydrostatic pressure outside the capillary. The *net capillary hydrostatic pressure* tends to push water and solutes out of capillaries and into the interstitial fluid. Factors that contribute to the net hydrostatic pressure include

1. the *capillary hydrostatic pressure (CHP)*, which ranges from 35 mm Hg at the arterial end of a capillary to 18 mm Hg at the venous end, and

2. the *interstitial fluid hydrostatic pressure (IHP)*. Measurements of IHP have yielded very small values that differ from tissue

Figure 21–11 **Forces Acting across Capillary Walls.** At the arterial end of the capillary, capillary hydrostatic pressure (CHP) is greater than blood colloid osmotic pressure (BCOP), so fluid moves out of the capillary (filtration). Near the venule, CHP is lower than BCOP, so fluid moves into the capillary (reabsorption). In this model, interstitial fluid colloid osmotic pressure (ICOP) and interstitial fluid hydrostatic pressure (IHP) are assumed to be 0 mm Hg and so are not shown.

to tissue—from +6 mm Hg in the brain to −6 mm Hg in subcutaneous tissues. A positive IHP opposes CHP, and the tissue hydrostatic pressure must be overcome before fluid can move out of a capillary. A negative IHP assists CHP, and additional fluid will be pulled out of the capillary. However, under normal circumstances the average IHP is 0 mm Hg, and we can assume that the net hydrostatic pressure is equal to CHP. (For this reason, IHP is not included in **Figure 21–11**.)

Plasma proteins in capillary blood create capillary colloid osmotic pressure. The *net capillary colloid osmotic pressure* tends to pull water and solutes into a capillary from the interstitial fluid. The net colloid osmotic pressure is the difference between

1. the *blood colloid osmotic pressure (BCOP)*, which is about 25 mm Hg, and

2. the *interstitial fluid colloid osmotic pressure (ICOP)*. The ICOP is as variable and low as the IHP, because the interstitial fluid in most tissues contains negligible quantities of suspended proteins. Reported values of ICOP are from 0 to 5 mm Hg, within the range of pressures recorded for the IHP. It is thus safe to assume that under normal circumstances the net colloid osmotic pressure is equal to the BCOP. (For this reason, ICOP is not included in **Figure 21–11**.)

The **net filtration pressure (NFP)** is the difference between the net hydrostatic pressure and the net osmotic pressure. In terms of the factors just listed, this means that

net filtration   =   net hydrostatic   −   net colloid
pressure   pressure   osmotic pressure

NFP   =   (CHP − IHP)   −   (BCOP − ICOP)

At the arterial end of a capillary, the net filtration pressure can be calculated as follows:

NFP = (35 − 0) − (25 − 0) = 35 − 25 = 10 mm Hg

Because this value is positive, it indicates that fluid will tend to move *out of* the capillary and into the interstitial fluid. At the venous end of the capillary, the net filtration pressure will be

NFP = (18 − 0) − (25 − 0) = 18 − 25 = −7 mm Hg

The minus sign indicates that fluid tends to move *into* the capillary; that is, reabsorption is occurring.

The transition between filtration and reabsorption occurs where the CHP is 25 mm Hg, because at that point the hydrostatic and osmotic forces are equal—that is, the NFP is 0 mm Hg. If the maximum filtration pressure at the arterial end of the capillary were equal to the maximum reabsorption pressure at the venous end, this transition point would lie midway along the length of the capillary. Under these circumstances, filtration would occur along the first half of the capillary, and an identical

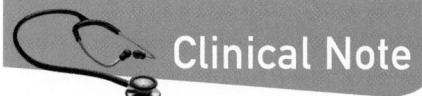

Clinical Note

Edema **Edema** (e-DĒ-muh) is an abnormal accumulation of interstitial fluid. Edema has many causes, and we will encounter specific examples in later chapters. The underlying problem in all types of edema is a disturbance in the normal balance between hydrostatic and osmotic forces at the capillary level. For instance:

- You usually have swelling at a bruise. When a capillary is damaged, plasma proteins can cross the capillary wall and enter the interstitial fluid. The resulting rise in the interstitial fluid colloid osmotic pressure (ICOP) reduces the rate of capillary reabsorption and produces a localized edema.

- In starvation, the liver cannot synthesize enough plasma proteins to maintain normal concentrations in the blood. Blood colloid osmotic pressure (BCOP) declines. Fluids then begin moving from the blood into peripheral tissues. In children, fluid builds up in the abdominopelvic cavity, producing the swollen bellies typical of starvation victims. A reduction in BCOP also takes place after severe burns and in several types of liver and kidney diseases.

- In the U.S. population, most serious cases of edema result from increases in arterial blood pressure, venous pressure, or total circulatory pressure. The increase may be due to heart problems such as heart failure, venous blood clots that elevate venous pressures, or other cardiovascular abnormalities. The net result is an increase in capillary hydrostatic pressure (CHP) that accelerates fluid movement into the tissues.

amount of reabsorption would occur along the second half. However, the maximum filtration pressure is higher than the maximum reabsorption pressure, so the transition point between filtration and reabsorption normally lies closer to the venous end of the capillary than to the arterial end. As a result, more filtration than reabsorption occurs along the capillary. Of the nearly 24 liters of fluid that move out of the plasma and into the interstitial fluid each day, 20.4 liters (85 percent) are reabsorbed. The remainder (3.6 liters) flows through the tissues and into lymphatic vessels, for eventual return to the venous system (**Figure 21–11**).

Any condition that affects hydrostatic or osmotic pressures in the blood or tissues will shift the balance between hydrostatic and osmotic forces. We can then predict the effects on the basis of an understanding of capillary dynamics. For example,

- If hemorrhaging occurs, both blood volume and blood pressure decrease. This reduction in CHP lowers the NFP and increases the amount of reabsorption. The result is a decrease in the volume of interstitial fluid and an increase in the circulating plasma volume. This process is known as a *recall of fluids*.

■ If dehydration occurs, the plasma volume decreases due to water loss, and the concentration of plasma proteins increases. The increase in BCOP accelerates reabsorption and a recall of fluids that delays the onset and severity of clinical signs and symptoms.

■ If the CHP rises or the BCOP declines, fluid moves out of the blood and builds up in peripheral tissues, a condition called *edema*.

CHECKPOINT

5. Identify the factors that contribute to total peripheral resistance.

6. In a healthy person, where is blood pressure greater: at the aorta or at the inferior vena cava? Explain.

7. While standing in the hot sun, Sally begins to feel light-headed and faints. Explain what happened.

8. Mike's blood pressure is 125/70. What is his mean arterial pressure?

See the blue Answers tab at the back of the book.

21-3 ▐ Cardiovascular regulatory mechanisms involve autoregulation, neural mechanisms, and endocrine responses

Learning Outcome Describe the control mechanisms that regulate blood flow and pressure in tissues, and explain how the activities of the cardiac, vasomotor, and respiratory centers are coordinated to control blood flow through the tissues.

Homeostatic mechanisms regulate cardiovascular activity to ensure that blood flow through tissues, called **tissue perfusion**, meets the demand for oxygen and nutrients. The factors that affect tissue perfusion are (1) cardiac output, (2) peripheral resistance, and (3) blood pressure. We discussed cardiac output in Chapter 20 (p. 737). We considered peripheral resistance and blood pressure earlier in this chapter.

Most cells are relatively close to capillaries. When a group of cells becomes active, the circulation to that region must increase to bring the necessary oxygen and nutrients, and to carry away the waste products and carbon dioxide they generate. The purpose of cardiovascular regulation is to ensure that these blood flow changes occur (1) at an appropriate time, (2) in the right area, and (3) without drastically changing blood pressure and blood flow to vital organs.

The regulatory mechanisms focus on controlling cardiac output and blood pressure to restore adequate blood flow after

blood pressure decreases. We can group these mechanisms as follows:

■ *Autoregulation.* Local factors change the pattern of blood flow within capillary beds as precapillary sphincters open and close in response to chemical changes in interstitial fluids. This is an example of autoregulation at the tissue level. Autoregulation causes immediate, localized homeostatic adjustments. If autoregulation fails to normalize conditions at the tissue level, neural mechanisms and endocrine factors are activated.

■ *Neural Mechanisms.* Neural mechanisms respond to changes in arterial pressure or blood gas levels sensed at specific sites. When those changes occur, the cardiovascular center of the autonomic nervous system adjusts cardiac output and peripheral resistance to maintain blood pressure and ensure adequate blood flow.

■ *Endocrine Mechanisms.* The endocrine system releases hormones that enhance short-term adjustments and that direct long-term changes in cardiovascular performance.

Now let's see how each of these regulatory mechanisms responds to inadequate perfusion of skeletal muscles. The regulatory relationships are diagrammed in **Figure 21–12**.

Autoregulation of Blood Flow within Tissues

Under normal resting conditions, cardiac output remains stable, and peripheral resistance within individual tissues is adjusted to control local blood flow.

Factors that promote the dilation of precapillary sphincters are called **vasodilators**. **Local vasodilators** act at the tissue level to accelerate blood flow through their tissue of origin. Examples of local vasodilators include the following:

■ Decreased tissue oxygen levels or increased CO_2 levels.

■ Lactic acid or other acids generated by tissue cells.

■ Nitric oxide (NO) released from endothelial cells.

■ Rising concentrations of potassium ions or hydrogen ions in the interstitial fluid.

■ Chemicals released during local inflammation, including histamine and NO. ⊃ p. 170

■ Elevated local temperature.

These factors work by relaxing the smooth muscle cells of the precapillary sphincters. All of them indicate that tissue conditions are in some way abnormal. An increase in blood flow, which brings oxygen, nutrients, and buffers, may be sufficient to restore homeostasis.

As noted in Chapter 19, aggregating platelets and damaged tissues produce compounds that stimulate precapillary sphincters to constrict. These compounds are **local vasoconstrictors**. Examples include prostaglandins and thromboxanes released

Figure 21–12   Short-Term and Long-Term Cardiovascular Responses. This diagram indicates general mechanisms that compensate for a reduction in blood pressure and blood flow.

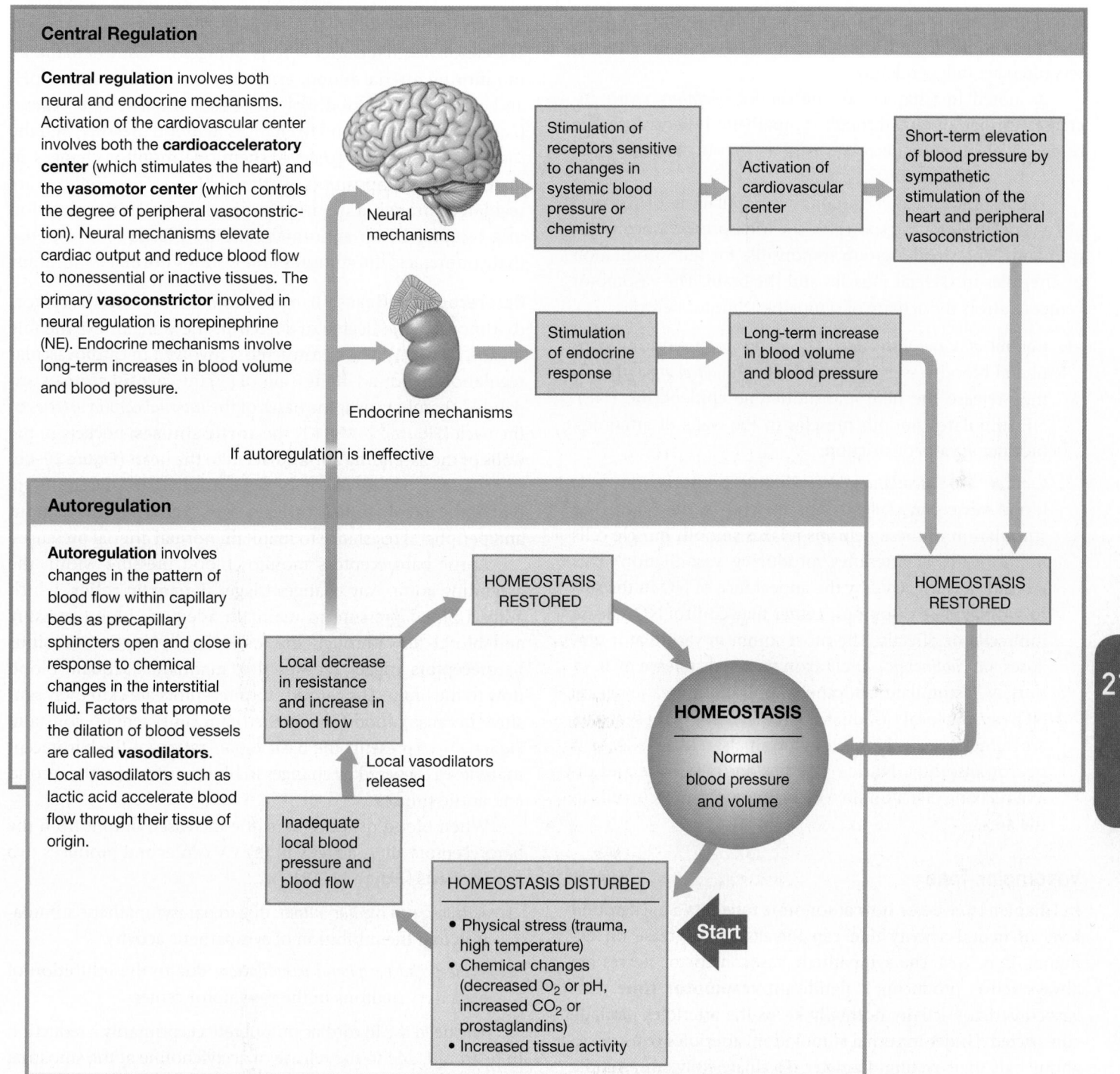

Central Regulation

Central regulation involves both neural and endocrine mechanisms. Activation of the cardiovascular center involves both the **cardioacceleratory center** (which stimulates the heart) and the **vasomotor center** (which controls the degree of peripheral vasoconstriction). Neural mechanisms elevate cardiac output and reduce blood flow to nonessential or inactive tissues. The primary **vasoconstrictor** involved in neural regulation is norepinephrine (NE). Endocrine mechanisms involve long-term increases in blood volume and blood pressure.

Neural mechanisms

Stimulation of receptors sensitive to changes in systemic blood pressure or chemistry

Activation of cardiovascular center

Short-term elevation of blood pressure by sympathetic stimulation of the heart and peripheral vasoconstriction

Endocrine mechanisms

Stimulation of endocrine response

Long-term increase in blood volume and blood pressure

If autoregulation is ineffective

Autoregulation

Autoregulation involves changes in the pattern of blood flow within capillary beds as precapillary sphincters open and close in response to chemical changes in the interstitial fluid. Factors that promote the dilation of blood vessels are called **vasodilators**. Local vasodilators such as lactic acid accelerate blood flow through their tissue of origin.

HOMEOSTASIS RESTORED

HOMEOSTASIS RESTORED

Local decrease in resistance and increase in blood flow

Local vasodilators released

Inadequate local blood pressure and blood flow

HOMEOSTASIS

Normal blood pressure and volume

HOMEOSTASIS DISTURBED

• Physical stress (trauma, high temperature)
• Chemical changes (decreased O_2 or pH, increased CO_2 or prostaglandins)
• Increased tissue activity

Start

21

by activated platelets and white blood cells, and the endothelins released by damaged endothelial cells.

Local vasodilators and vasoconstrictors control blood flow within a single capillary bed (**Figure 21–4**). In high concentrations, these factors also affect arterioles, increasing or decreasing blood flow to all the capillary beds in a given area.

Neural Mechanisms

The nervous system adjusts cardiac output and peripheral resistance to maintain adequate blood flow to vital tissues and organs. Centers responsible for these regulatory activities include the *cardiac centers* and the *vasomotor center* of the medulla oblongata. ⟳ p. 496

It is difficult to distinguish the cardiac and vasomotor centers anatomically. They are often considered to form a complex **cardiovascular (CV) center**, composed of the cardioinhibitory center, cardioacceleratory center, and the vasomotor center. In functional terms, however, the cardiac and vasomotor centers often act independently.

As noted in Chapter 20, the *cardioacceleratory center* increases cardiac output through sympathetic innervation. The *cardioinhibitory center* decreases cardiac output through parasympathetic innervation. ⊃ p. 739

The vasomotor center contains two populations of neurons: (1) a very large group responsible for widespread vasoconstriction and (2) a smaller group responsible for the vasodilation of arterioles in skeletal muscles and the brain. The vasomotor center controls the activity of sympathetic motor neurons:

1. *Control of Vasoconstriction.* The neurons innervating peripheral blood vessels in most tissues are *adrenergic*; that is, they release the neurotransmitter norepinephrine (NE). NE stimulates smooth muscles in the walls of arterioles, producing vasoconstriction.

2. *Control of Vasodilation.* Vasodilator neurons innervate blood vessels in skeletal muscles and in the brain. The stimulation of these neurons relaxes smooth muscle cells in the walls of arterioles, producing vasodilation. This relaxation is triggered by the appearance of NO in the surroundings. The vasomotor center may control NO release indirectly or directly. The most common vasodilator synapses are *cholinergic*—their axon terminals release ACh. In turn, ACh stimulates endothelial cells in the area to release NO, causing local vasodilation. Other vasodilator synapses are *nitroxidergic*—their axon terminals release NO as a neurotransmitter. Nitric oxide has an immediate and direct relaxing effect on the vascular smooth muscle cells in the area.

Vasomotor Tone

In Chapter 16, we saw how autonomic tone sets a background level of neural activity that can increase or decrease on demand. ⊃ p. 574 The sympathetic vasoconstrictor nerves are always active, producing a significant **vasomotor tone**. This vasoconstrictor activity normally keeps the arterioles partially constricted. Under maximal stimulation, arterioles constrict to about half their resting diameter. To dilate fully, an arteriole increases its resting diameter by about 1.5 times.

Constriction has a large effect on resistance, because, as we saw earlier, resistance increases sharply as luminal diameter decreases. The resistance of a maximally constricted arteriole is roughly 80 *times* that of a fully dilated arteriole. Because blood pressure varies directly with peripheral resistance, the vasomotor center can control arterial blood pressure very effectively by making modest adjustments in vessel diameters. Extreme

stimulation of the vasomotor center also produces venoconstriction and mobilizes the venous reserve.

Reflex Control of Cardiovascular Function

The cardiovascular center detects changes in tissue demand by monitoring arterial blood, especially its blood pressure, pH, and concentrations of dissolved gases. The *baroreceptor reflexes* (*baro-*, pressure) respond to changes in blood pressure, and the *chemoreceptor reflexes* (*chemo-*, chemical) monitor changes in the chemical composition of arterial blood. These reflexes are regulated through a negative feedback loop: The stimulation of a receptor by an abnormal condition leads to a response that counteracts the stimulus and restores normal conditions.

Baroreceptor Reflexes. Baroreceptors are specialized receptors that monitor the degree of stretch in the walls of expandable organs. ⊃ p. 545 The baroreceptors involved in cardiovascular regulation are found in the walls of (1) the **carotid sinuses**, expanded chambers near the bases of the *internal carotid arteries* of the neck (**Figure 21–21**); (2) the **aortic sinuses**, pockets in the walls of the ascending aorta adjacent to the heart (**Figure 20–8b**, p. 720); and (3) the wall of the right atrium. These receptors are part of the **baroreceptor reflexes**, which adjust cardiac output and peripheral resistance to maintain normal arterial pressures.

Aortic baroreceptors monitor blood pressure within the ascending aorta. Any changes trigger the **aortic reflex**, which adjusts blood pressure to maintain adequate blood pressure and blood flow through the systemic circuit. Carotid sinus baroreceptors trigger reflexes that maintain adequate blood flow to the brain. The carotid sinus receptors are extremely sensitive because blood flow to the brain must remain constant. **Figure 21–13** presents the basic organization of the baroreceptor reflexes triggered by changes in blood pressure at the carotid and aortic sinuses.

When blood pressure rises, the increased output from the baroreceptors alters activity in the CV center and produces two major effects (**Figure 21–13**):

1. *A decrease in cardiac output*, due to parasympathetic stimulation and the inhibition of sympathetic activity.

2. *Widespread peripheral vasodilation*, due to the inhibition of excitatory neurons in the vasomotor center.

The decrease in cardiac output reflects primarily a reduction in heart rate due to the release of acetylcholine at the sinoatrial (SA) node. ⊃ p. 740 The widespread vasodilation lowers peripheral resistance, and this effect, combined with a reduction in cardiac output, leads to a fall in blood pressure to normal levels.

When blood pressure falls below normal, baroreceptor output is reduced accordingly (**Figure 21–13**). This change has two major effects working together to raise blood pressure:

1. *An increase in cardiac output*, through the stimulation of sympathetic innervation to the heart. This results from

Figure 21–13 **Baroreceptor Reflexes of the Carotid and Aortic Sinuses.**

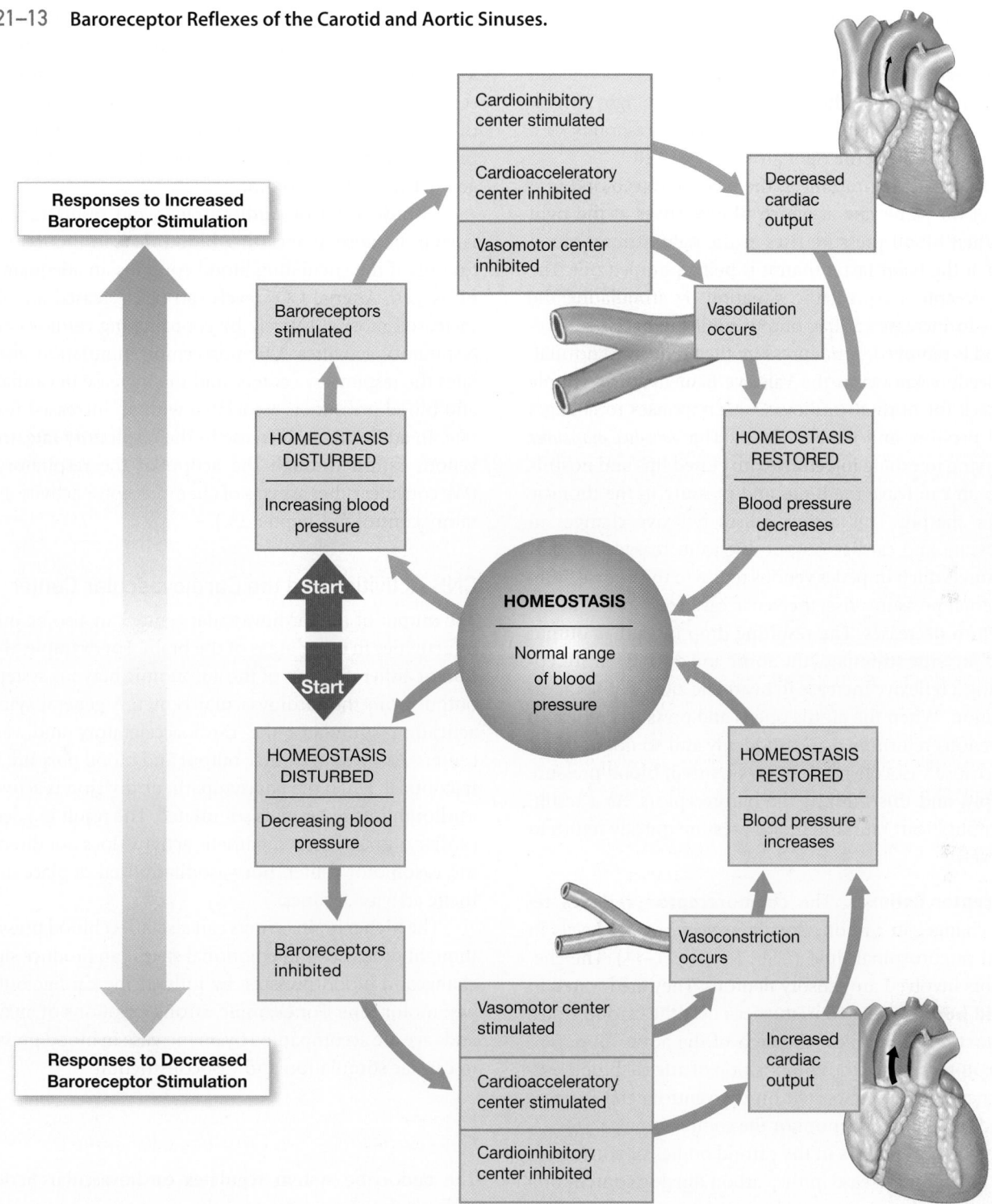

the stimulation of the cardioacceleratory center and is accompanied by an inhibition of the cardioinhibitory center.

2. *Widespread peripheral vasoconstriction*, caused by the stimulation of sympathetic vasoconstrictor neurons by the vasomotor center.

The effects on the heart result from the release of NE by sympathetic neurons innervating the SA node, the atrioventricular (AV) node, and the general myocardium. In a crisis, sympathetic activation occurs, and its effects are enhanced by the release of both NE and epinephrine (E) from the adrenal medullae. The net effect is an immediate increase in heart rate and stroke volume, and a corresponding increase in cardiac output.

The vasoconstriction, which also results from the release of NE by sympathetic neurons, increases peripheral resistance.

These adjustments—increased cardiac output and increased peripheral resistance—work together to raise blood pressure.

Atrial baroreceptors monitor blood pressure at the end of the systemic circuit—at the venae cavae and the right atrium. Recall from Chapter 20 that the **atrial reflex** responds to a stretching of the wall of the right atrium. ⤵ p. 740

Under normal circumstances, the heart pumps blood into the aorta at the same rate at which blood arrives at the right atrium. When blood pressure rises at the right atrium, blood is arriving at the heart faster than it is being pumped out. The atrial baroreceptors correct the situation by stimulating the CV centers to increase cardiac output until the backlog of venous blood is removed. Atrial pressure then returns to normal.

A procedure known as the Valsalva maneuver is a simple way to check for normal cardiovascular responses to changes in arterial pressure and venous return. The *Valsalva maneuver* involves trying to exhale forcefully with closed lips and nostrils so that no air can leave the lungs and pressure in the thoracic cavity rises sharply. This action causes reflexive changes in blood pressure and cardiac output due to increased intrathoracic pressure, which impedes venous return to the right atrium. When internal pressures rise, the venae cavae collapse, and the venous return decreases. The resulting drop in cardiac output and blood pressure stimulates the aortic and carotid baroreceptors, causing a reflexive increase in heart rate and peripheral vasoconstriction. When the glottis opens and pressures return to normal, venous return increases suddenly and so does cardiac output. Because vasoconstriction has occurred, blood pressure rises sharply, and this inhibits the baroreceptors. As a result, cardiac output, heart rate, and blood pressure quickly return to normal levels.

Chemoreceptor Reflexes. The **chemoreceptor reflexes** respond to changes in carbon dioxide, oxygen, or pH levels in blood and cerebrospinal fluid (CSF) (**Figure 21–14**). The chemoreceptors involved are sensory neurons. They are located in the **carotid bodies**, situated in the neck near the carotid sinus, and the **aortic bodies**, near the arch of the aorta. ⤵ p. 543 These receptors monitor the composition of arterial blood. Additional chemoreceptors located on the ventrolateral surfaces of the medulla oblongata monitor the composition of CSF.

When chemoreceptors in the carotid bodies or aortic bodies detect either an increase in the carbon dioxide content or a decrease in the pH of the arterial blood, the cardioacceleratory and vasomotor centers are stimulated. At the same time, the cardioinhibitory center is inhibited. This dual effect causes an increase in cardiac output, peripheral vasoconstriction, and an increase in blood pressure. A decrease in the oxygen level at the aortic bodies has the same effects. Strong stimulation of the carotid or aortic chemoreceptors causes widespread sympathetic activation, with more dramatic increases in heart rate and cardiac output.

The chemoreceptors of the medulla oblongata are involved primarily with the control of respiratory function, and secondarily with regulating blood flow to the brain. For example, a steep rise in CSF carbon dioxide levels triggers the vasodilation of cerebral vessels, but produces vasoconstriction in most other organs. The result is increased blood flow—and increased oxygen delivery—to the brain.

Coordination of cardiovascular and respiratory activities is vital, because accelerating blood flow in the tissues is useful only if the circulating blood contains an adequate amount of oxygen. Arterial CO_2 levels can be decreased and O_2 levels increased most effectively by coordinating cardiovascular and respiratory activities. Chemoreceptor stimulation also stimulates the respiratory centers, and the increase in cardiac output and blood pressure is associated with an increased respiratory rate. In addition, an increase in the respiratory rate accelerates venous return through the action of the respiratory pump. (We consider other aspects of chemoreceptor activity and respiratory control in Chapter 23.)

CNS Activities and the Cardiovascular Center

The output of the cardiovascular center can also be influenced by activities in other areas of the brain. For example, the activation of either division of the autonomic nervous system affects output from the cardiovascular center. A general sympathetic activation stimulates the cardioacceleratory and vasomotor centers. As a result, cardiac output and blood pressure increase. In contrast, when the parasympathetic division is activated, the cardioinhibitory center is stimulated. The result is a decrease in cardiac output. Parasympathetic activity does not directly affect the vasomotor center, but vasodilation takes place as sympathetic activity declines.

The higher brain centers can also affect blood pressure. Our thought processes and emotional states can produce significant changes in blood pressure by influencing cardiac output and vasomotor tone. For example, strong emotions of anxiety, fear, and rage are accompanied by an increase in blood pressure, due to cardiac stimulation and vasoconstriction.

Hormones and Cardiovascular Regulation

The endocrine system regulates cardiovascular performance in both the short term and the long term. As we have seen, E and NE from the adrenal medullae stimulate cardiac output and peripheral vasoconstriction. Other hormones important in regulating cardiovascular function include (1) antidiuretic hormone (ADH), (2) angiotensin II, (3) erythropoietin (EPO), and (4) the natriuretic peptides (ANP and BNP). ⤵ p. 666 All four are concerned primarily with the long-term regulation of blood volume (**Figure 21–15**). ADH and angiotensin II also affect blood pressure.

Figure 21–14   **The Chemoreceptor Reflexes.**

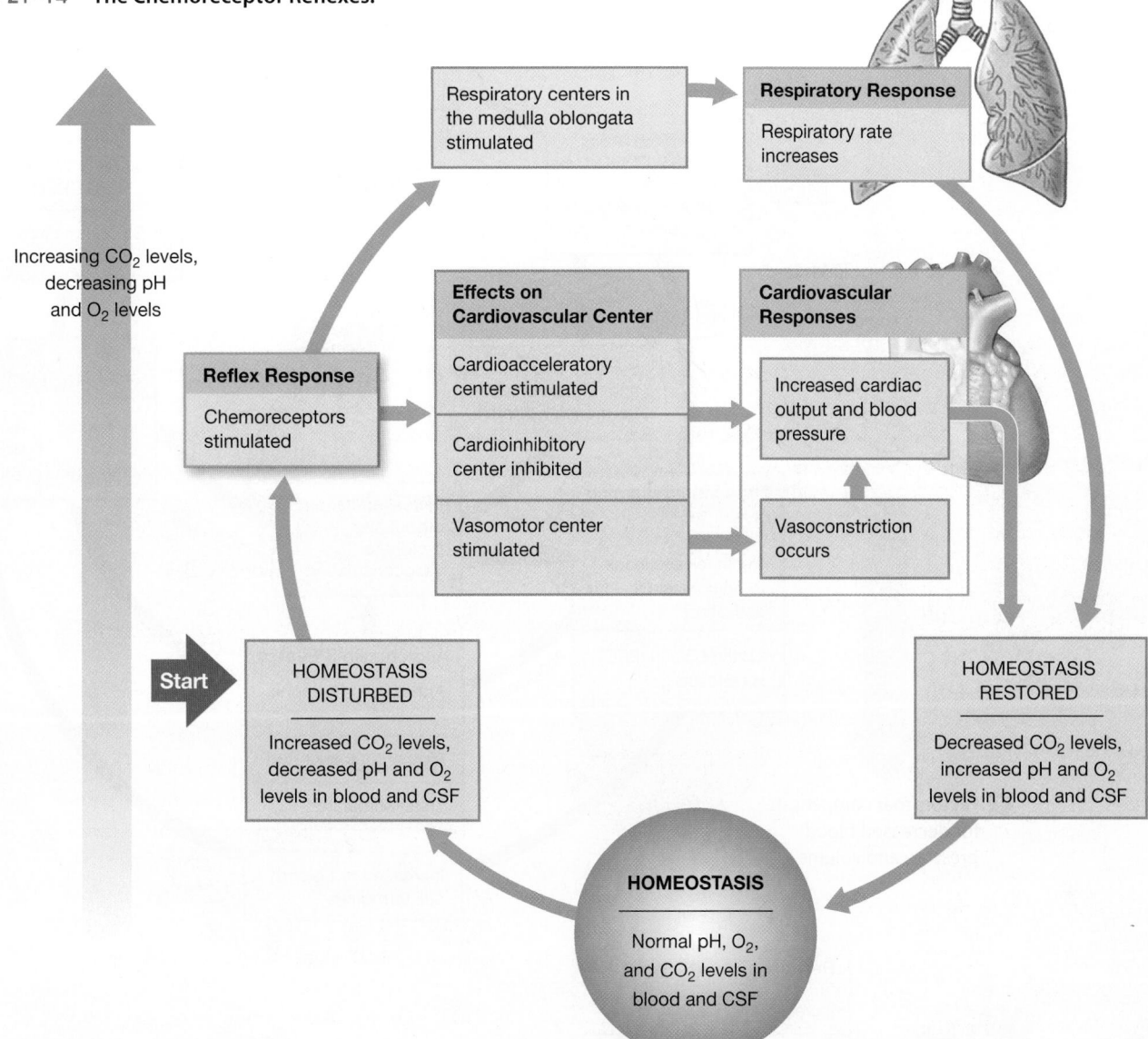

Antidiuretic Hormone

Antidiuretic hormone (ADH) is released at the posterior lobe of the pituitary gland in response to a decrease in blood volume, to an increase in the osmotic concentration of the plasma, or (secondarily) to circulating angiotensin II. It brings about a peripheral vasoconstriction that elevates blood pressure. This hormone also stimulates water conservation by the kidneys, thus preventing a reduction in blood volume that would further reduce blood pressure (**Figure 21–15a**).

Angiotensin II

Angiotensin II appears in the blood when specialized kidney cells, called *juxtaglomerular cells*, release the enzyme *renin* (RĒ-nin) in response to a fall in renal blood pressure (**Figure 21–15a**). Once in the bloodstream, renin starts an enzymatic chain reaction. In the first step, renin converts *angiotensinogen*, a plasma protein produced by the liver, to *angiotensin I*. In the capillaries of the lungs, *angiotensin-converting enzyme (ACE)* then modifies angiotensin I to angiotensin II, an active hormone with diverse effects.

Angiotensin II has four important functions: (1) It stimulates the adrenal production of aldosterone, causing Na^+ retention and K^+ loss by the kidneys; (2) it stimulates the secretion of ADH, in turn stimulating water reabsorption by the kidneys and complementing the effects of aldosterone; (3) it stimulates thirst, resulting in increased fluid consumption (the presence of ADH and aldosterone ensures that the additional water consumed will be retained, increasing blood volume); and (4) it stimulates cardiac output and triggers the constriction of arterioles, in turn increasing the systemic blood pressure. The effect of angiotensin II on blood pressure is four to eight times greater than the effect of norepinephrine.

Figure 21–15 **The Hormonal Regulation of Blood Pressure and Blood Volume.**

HOMEOSTASIS
Normal blood pressure and volume

HOMEOSTASIS DISTURBED
Blood pressure and volume decrease

Start

HOMEOSTASIS RESTORED
Blood pressure and volume increase

Decreasing blood pressure and volume

Short-term

Long-term

Sympathetic activation and release of adrenal hormones E and NE

Combined Short-Term and Long-Term Effects

Increased blood pressure	Increased blood volume

Endocrine Response of Kidneys

Renin release leads to angiotensin II activation

Erythropoietin (EPO) is released

Angiotensin II

Increased cardiac output and peripheral vasoconstriction

Angiotensin II Effects

Antidiuretic hormone released

Aldosterone secreted

Thirst stimulated

Increased red blood cell formation

a Factors that compensate for decreased blood pressure and volume

Responses to ANP and BNP

Increased Na+ loss in urine

Increased water loss in urine

Decreased thirst

Inhibition of ADH, aldosterone, epinephrine, and norepinephrine release

Peripheral vasodilation

Combined Effects

Decreased blood volume

Natriuretic peptides released by the heart

HOMEOSTASIS DISTURBED
Increasing blood pressure and volume

HOMEOSTASIS RESTORED
Decreasing blood pressure and volume

HOMEOSTASIS
Normal blood pressure and volume

Increasing blood pressure and volume

b Factors that compensate for increased blood pressure and volume

Erythropoietin

The kidneys release *erythropoietin (EPO)* if blood pressure falls or if the oxygen content of the blood becomes abnormally low (**Figure 21–15a**). EPO acts directly on blood vessels, causing vasoconstriction, thereby increasing blood pressure. EPO also stimulates the production and maturation of red blood cells. These cells increase the volume and viscosity of the blood and improve its oxygen-carrying capacity.

Natriuretic Peptides

Cardiac muscle cells in the wall of the right atrium of the heart produce *atrial natriuretic peptide* (nā-trē-ū-RET-ik; *natrium*, sodium + *ouresis*, urination), or *ANP*, in response to excessive stretching during diastole. Ventricular muscle cells exposed to comparable stimuli produce a related hormone called *brain natriuretic peptide*, or *BNP*. These peptide hormones reduce blood volume and blood pressure. They do so by (1) increasing sodium ion excretion by the kidneys; (2) promoting water losses by increasing the volume of urine produced; (3) reducing thirst; (4) blocking the release of ADH, aldosterone, epinephrine, and norepinephrine; and (5) stimulating peripheral vasodilation (**Figure 21–15b**). As blood volume and blood pressure decrease, the stresses on the walls of the heart are removed, and natriuretic peptide production ceases.

> **CHECKPOINT**
>
> 9. Describe the actions of vasodilators and local vasodilators.
>
> 10. How would applying slight pressure to the common carotid artery affect your heart rate?
>
> 11. What effect would the vasoconstriction of the renal artery have on blood pressure and blood volume?
>
> See the blue Answers tab at the back of the book.

21-4 ∣ The cardiovascular system adapts to physiological stress and maintains a special vascular supply to the brain, heart, and lungs

Learning Outcome Explain the cardiovascular system's homeostatic response to exercise and hemorrhaging, and identify the principal blood vessels and functional characteristics of the special circulation to the brain, heart, and lungs.

In this chapter and the previous two, we have considered the blood, the heart, and the cardiovascular system as individual entities. Yet in our day-to-day lives, the cardiovascular system operates as an integrated complex. The interactions are fascinating and very important when physical or physiological conditions are changing rapidly.

In this section we begin by looking at the patterns of cardiovascular responses to exercise and blood loss. These two common stresses provide examples of the cardiovascular system's adaptability in maintaining homeostasis. The homeostatic responses involve interplay among the cardiovascular system, the endocrine system, and other systems. These responses are aided by autoregulation at the tissue level. Then we consider the patterns of blood supply to the brain, heart, and lungs, in which blood flow is controlled by separate mechanisms.

The Cardiovascular Response to Exercise

At rest, cardiac output averages about 5.8 liters per minute. That amount changes dramatically during exercise. In addition, the pattern of blood distribution changes markedly, as detailed in Table 21–2.

Light Exercise

Before you begin to exercise, your heart rate increases slightly due to a general rise in sympathetic activity as you think about the workout ahead. As you begin light exercise, three interrelated changes take place:

- *Extensive vasodilation occurs* as skeletal muscles consume oxygen more quickly. Peripheral resistance decreases, blood flow through the capillaries increases, and blood enters the venous system at a faster rate.

- *The venous return increases* as skeletal muscle contractions squeeze blood along the peripheral veins and faster breathing pulls blood into the venae cavae by the respiratory pump.

- *Cardiac output rises*, primarily in response to (1) the rise in venous return (the Frank–Starling principle ⤴ p. 742) and (2) atrial stretching (the atrial reflex). Some sympathetic stimulation occurs, leading to increases in heart rate and contractility, but there is no massive sympathetic activation. The increased cardiac output keeps pace with the elevated demand, and arterial pressures are maintained despite the drop in peripheral resistance.

Table 21–2	Changes in Blood Distribution during Exercise		
	Tissue Blood Flow (mL/min)		
Organ	**Rest**	**Light Exercise**	**Strenuous Exercise**
Skeletal muscles	1200	4500	12,500
Heart	250	350	750
Brain	750	750	750
Skin	500	1500	1900
Kidney	1100	900	600
Abdominal viscera	1400	1100	600
Miscellaneous	600	400	400
Total cardiac output	5800	9500	17,500

This regulation by venous feedback produces a gradual increase in cardiac output to about double resting levels. The increase supports accelerated blood flow to skeletal muscles, cardiac muscle, and the skin. The flow to skeletal and cardiac muscles increases as arterioles and precapillary sphincters dilate in response to local factors. The flow to the skin increases in response to the rise in body temperature.

Heavy Exercise

At higher levels of exertion, other physiological adjustments take place as the cardiac and vasomotor centers activate the sympathetic nervous system. Cardiac output increases toward maximal levels. Major changes in the peripheral distribution of blood improve blood flow to active skeletal muscles.

Under massive sympathetic stimulation, the cardioacceleratory center can increase cardiac output to levels as high as 20–25 liters per minute. But that is still not enough to meet the demands of active skeletal muscles unless the vasomotor center severely restricts the blood flow to "nonessential" organs, such as those of the digestive system. During exercise at maximal levels, your blood essentially races between the skeletal muscles and the lungs and heart. Although blood flow to most tissues is diminished, skin perfusion increases further, because body temperature continues to climb. Only the blood supply to the brain remains unaffected.

Exercise, Cardiovascular Fitness, and Health

Cardiovascular performance improves significantly with training. Table 21–3 compares the cardiac performance of athletes with that of nonathletes. Trained athletes have larger hearts and greater stroke volumes than do nonathletes, and these are important functional differences.

Recall that cardiac output is equal to the stroke volume times the heart rate. For the same cardiac output, the person with a greater stroke volume has a slower heart rate. An athlete at rest can maintain normal blood flow to peripheral tissues at a heart rate as low as 32 bpm (beats per minute). When necessary, the athlete's cardiac output can increase to levels 50 percent higher than those of nonathletes. Thus, a trained athlete can tolerate sustained levels of activity that are well beyond the capabilities of nonathletes.

Exercise and Cardiovascular Disease

Regular exercise has several beneficial effects. Even a moderate exercise routine (jogging 5 miles a week, for example) can lower total blood cholesterol levels. Exercise lowers cholesterol by stimulating enzymes that help move low-density lipoproteins (LDLs, or so-called "bad cholesterol") from the blood to the liver. In the liver, the cholesterol is converted to bile and excreted from the body. Exercise also increases the size of the lipoprotein particles that carry cholesterol, making it harder for small proteins to lodge in the vessel walls. A high cholesterol level is one of the major risk factors for atherosclerosis, which leads to cardiovascular disease and strokes. In addition, a healthy lifestyle that includes regular exercise, a balanced diet, weight control, and not smoking reduces stress, lowers blood pressure, and slows the formation of plaques.

Regular moderate exercise (30 minutes most days of the week) may cut the incidence of heart attacks almost in half. However, only an estimated 8 percent of adults in the United States currently exercise at recommended levels. Exercise also speeds recovery after a heart attack. Regular light-to-moderate exercise (such as walking, jogging, or bicycling), coupled with a low-fat diet and a low-stress lifestyle, reduces symptoms of coronary artery disease (such as angina). Such exercise also improves a person's mood and overall quality of life. However, exercise does not remove underlying medical problems. For example, atherosclerotic plaques, described on p. 754, do not disappear and seldom grow smaller with exercise.

There is no evidence that *intense* athletic training lowers the incidence of cardiovascular disease. On the contrary, the strains placed on all body systems—including the cardiovascular system—during an ultramarathon, Ironman Triathlon, or other extreme athletic event can be severe. People with congenital aneurysms, cardiomyopathy, or cardiovascular disease risk fatal cardiovascular problems, such as an arrhythmia or heart attack, during severe exercise. Even healthy individuals can develop acute physiological disorders, such as kidney failure, after extreme exercise. We discuss the effects of exercise on other systems in later chapters.

Table 21–3	Effects of Training on Cardiovascular Performance				
Subject	Heart Weight (g)	Stroke Volume (mL)	Heart Rate (bpm)	Cardiac Output (L/min)	Blood Pressure (systolic/diastolic)
Nonathlete (rest)	300	60	83	5.0	120/80
Nonathlete (maximum)		104	192	19.9	187/75
Trained athlete (rest)	500	100	53	5.3	120/80
Trained athlete (maximum)		167	182	30.4	200/90*

* Diastolic pressures in athletes during maximal activity have not been accurately measured.

The Cardiovascular Response to Hemorrhaging

In Chapter 19, we considered the local cardiovascular reaction to a break in the wall of a blood vessel. ⟲ p. 701 When hemostasis fails to prevent significant blood loss, the entire cardiovascular system makes adjustments (**Figure 21–16**). The immediate problem is to maintain adequate blood pressure and peripheral blood flow. The long-term problem is to restore normal blood volume.

Short-Term Elevation of Blood Pressure

Almost as soon as the pressures start to decrease, several short-term responses appear:

- In the initial neural response, carotid and aortic reflexes increase cardiac output and cause peripheral vasoconstriction.

With blood volume reduced, an increase in heart rate, typically up to 180–200 bpm, maintains cardiac output.

- The combination of stress and anxiety stimulates the sympathetic nervous system headquarters in the hypothalamus, which in turn triggers a further increase in vasomotor tone, constricting the arterioles and raising blood pressure. At the same time, venoconstriction mobilizes the venous reserve and quickly improves venous return.

- Short-term hormonal effects also occur. For instance, sympathetic activation causes the adrenal medullae to secrete E and NE. These hormones increase cardiac output and extend peripheral vasoconstriction. In addition, the release of ADH by the posterior lobe of the pituitary gland and the production of angiotensin II enhance vasoconstriction as part of the long-term response.

Figure 21–16   Cardiovascular Responses to Blood Loss. These mechanisms can cope with blood losses equivalent to approximately 30 percent of total blood volume.

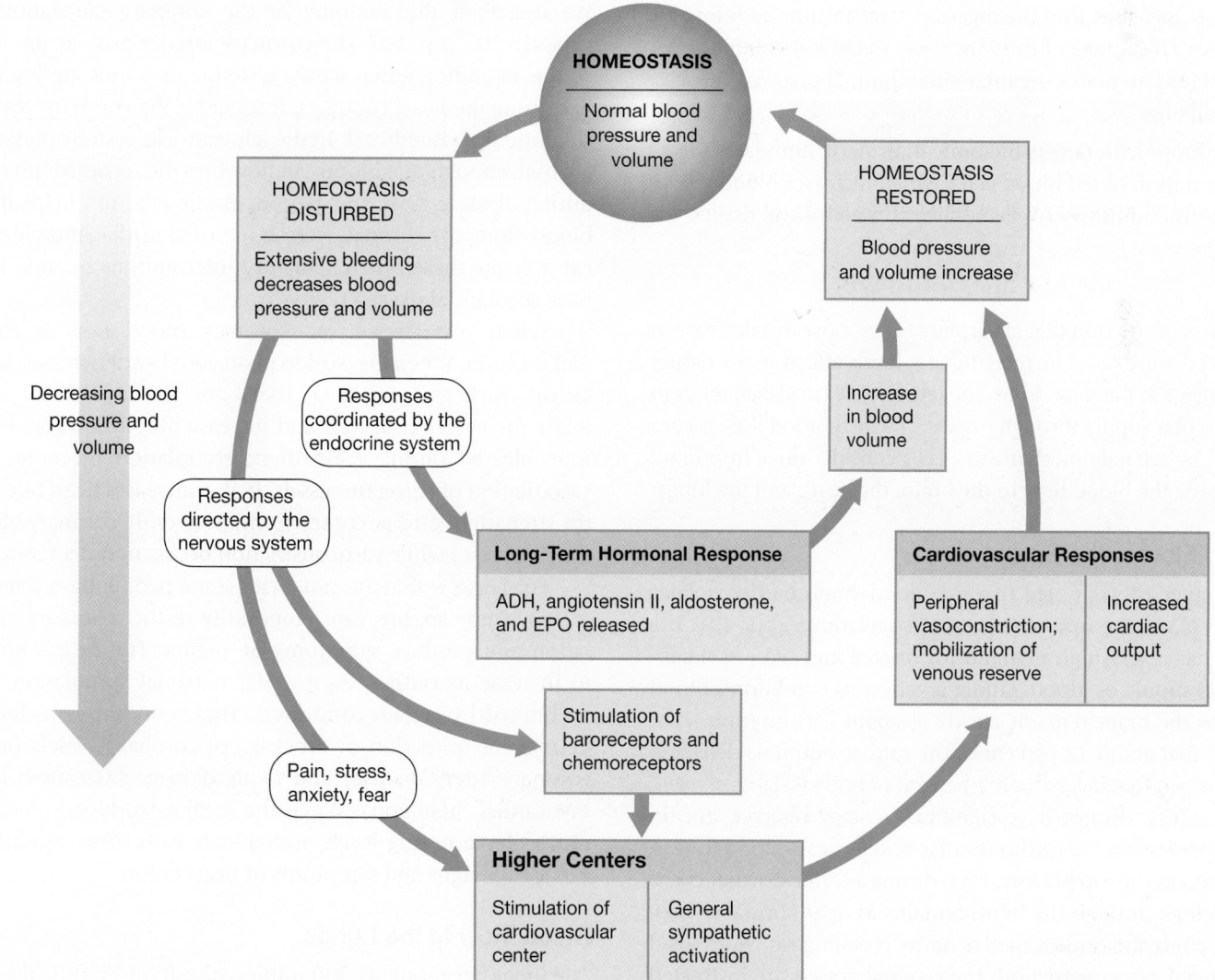

This combination of short-term responses elevates blood pressure and improves peripheral blood flow. It often restores normal arterial pressures and peripheral circulation after blood losses of up to 20 percent of total blood volume. Such adjustments are more than enough to compensate for the blood loss when you donate blood. (Most blood banks collect 500 mL of whole blood, roughly 10 percent of your total blood volume.) If compensatory mechanisms fail, the individual develops signs of *shock*.

Long-Term Restoration of Blood Volume

Short-term responses temporarily compensate for a reduction in blood volume. Long-term responses are geared to restoring normal blood volume. This process can take several days after a serious hemorrhage. The steps include the following:

- The decrease in capillary blood pressure triggers a recall of fluids from the interstitial spaces.

- Aldosterone and ADH promote fluid retention and reabsorption at the kidneys, preventing further reductions in blood volume.

- Thirst increases, and the digestive tract absorbs additional water. This intake of fluid increases the blood volume and ultimately replaces the interstitial fluids "borrowed" at the capillaries.

- Erythropoietin targets the bone marrow. It stimulates the maturation of red blood cells, which increases blood volume and improves oxygen delivery to peripheral tissues.

21 Vascular Supply to Special Regions

The vasoconstriction that takes place in response to a decrease in blood pressure or an increase in CO_2 levels affects many tissues and organs at the same time. The term *special circulation* refers to the vascular supply through organs in which blood flow is controlled by separate mechanisms. Let's consider three important examples: the blood flow to the brain, the heart, and the lungs.

Blood Flow to the Brain

In Chapter 14, we noted that the blood–brain barrier isolates most CNS tissue from the general circulation. ⟲ p. 495 The brain has a very high demand for oxygen and receives a substantial supply of blood. Under a variety of conditions, blood flow to the brain remains steady at about 750 mL/min. That means that about 12 percent of the cardiac output is delivered to an organ that is less than 2 percent of body weight.

Neurons do not have significant energy reserves, and in functional terms the cardiovascular system treats blood flow to the brain as the top priority. Even during a cardiovascular crisis, blood flow through the brain remains as near normal as possible: While the cardiovascular center is calling for widespread peripheral vasoconstriction, the cerebral vessels are instructed to dilate.

Total blood flow to the brain remains relatively constant, but blood flow to specific regions of the brain changes from moment to moment. These changes occur in response to local changes in the composition of interstitial fluid that accompany neural activity. When you read, write, speak, or walk, specific regions of your brain become active. Blood flow to those regions increases almost instantaneously. These changes ensure that the active neurons will receive the oxygen and nutrients they require.

The brain receives arterial blood through four arteries. An interruption of flow in any one of these large vessels does not significantly reduce blood flow to the brain as a whole because these arteries form anastomoses inside the cranium. However, a plaque or a blood clot may still block a small artery, and weakened arteries may rupture. Such incidents temporarily or permanently shut off blood flow to a localized area of the brain, damaging or killing the dependent neurons. Signs and symptoms of a *stroke*, or *cerebrovascular accident (CVA)*, then appear.

Blood Flow to the Heart

We described the anatomy of the coronary circulation in Chapter 20. ⟲ p. 721 The coronary arteries arise at the base of the ascending aorta, where systemic pressures are highest. Each time the heart contracts, it squeezes the coronary vessels, so blood flow is reduced. In the left ventricle, systolic pressures are high enough that blood can flow into the myocardium only during diastole. Over this period, elastic rebound helps drive blood along the coronary vessels. Normal cardiac muscle cells can tolerate these brief circulatory interruptions because they have substantial oxygen reserves.

When you are at rest, coronary blood flow is about 250 mL/min. When the workload on your heart increases, local factors, such as reduced O_2 levels and increased lactic acid, dilate the coronary vessels and increase blood flow. Epinephrine released during sympathetic stimulation promotes the vasodilation of coronary vessels. It also increases heart rate and the strength of cardiac contractions. As a result, coronary blood flow increases while vasoconstriction occurs in other tissues.

For reasons that are not clear, some people have *coronary spasms*. These spasms can temporarily restrict coronary circulation and produce symptoms of angina. The heart's ability to increase its output, even under maximal stimulation, can be limited by certain conditions. These conditions include a permanent restriction or blockage of coronary vessels (as in coronary artery disease) and tissue damage (as caused by a myocardial infarction). When the cardiac workload increases much above resting levels, individuals with these conditions experience signs and symptoms of heart failure.

Blood Flow to the Lungs

The lungs have roughly 300 million *alveoli* (al-VĒ-ō-lī; *alveolus*, sac), delicate epithelial pockets where gas exchange takes place.

An extensive capillary network surrounds each alveolus. Local responses to oxygen levels within individual alveoli regulate blood flow through the lungs. How does this local regulation work? When an alveolus contains plenty of oxygen, its blood vessels dilate. Blood flow then increases, promoting the absorption of oxygen from air inside the alveolus. When the oxygen content of the air is very low, the vessels constrict. They shunt blood to other alveoli that contain more oxygen. This local mechanism makes the respiratory system very efficient. There is no benefit in circulating blood through the capillaries of an alveolus that contains little oxygen.

This mechanism is precisely the opposite of that in other tissues, where a decline in oxygen levels causes local vasodilation rather than vasoconstriction. The difference makes functional sense, but its physiological basis remains a mystery.

Blood pressure in pulmonary capillaries (average: 10 mm Hg) is lower than that in systemic capillaries. The BCOP (25 mm Hg) is the same as elsewhere in the bloodstream. As a result, reabsorption exceeds filtration in pulmonary capillaries. Fluid moves continuously into the pulmonary capillaries across the alveolar surfaces. This flow prevents fluid from building up in the alveoli and interfering with the diffusion of respiratory gases. If the blood pressure in pulmonary capillaries rises above 25 mm Hg, fluid enters the alveoli, causing *pulmonary edema*.

CHECKPOINT

12. Why does blood pressure increase during exercise?

13. Name the immediate and long-term problems related to the cardiovascular response to hemorrhaging.

14. Explain the role of aldosterone and ADH in long-term restoration of blood volume.

See the blue Answers tab at the back of the book.

21-5 ❙ The pulmonary and systemic circuits of the cardiovascular system exhibit three general functional patterns

Learning Outcome Describe the three general functional patterns seen in the pulmonary and systemic circuits of the cardiovascular system.

You already know that the cardiovascular system consists of the *pulmonary circuit* and the *systemic circuit*. The pulmonary circuit consists of arteries and veins that transport blood between the heart and the lungs. This circuit begins at the right ventricle and ends at the left atrium. From the left ventricle, the arteries of the systemic circuit transport oxygenated blood and nutrients to all organs and tissues. Veins of the systemic circuit ultimately return deoxygenated blood to the right atrium. **Figure 21–17** summarizes the main distribution routes within the pulmonary and systemic circuits.

Figure 21–17 **A Schematic Overview of the Pattern of Circulation.** *RA* stands for right atrium, *LA* for left atrium.

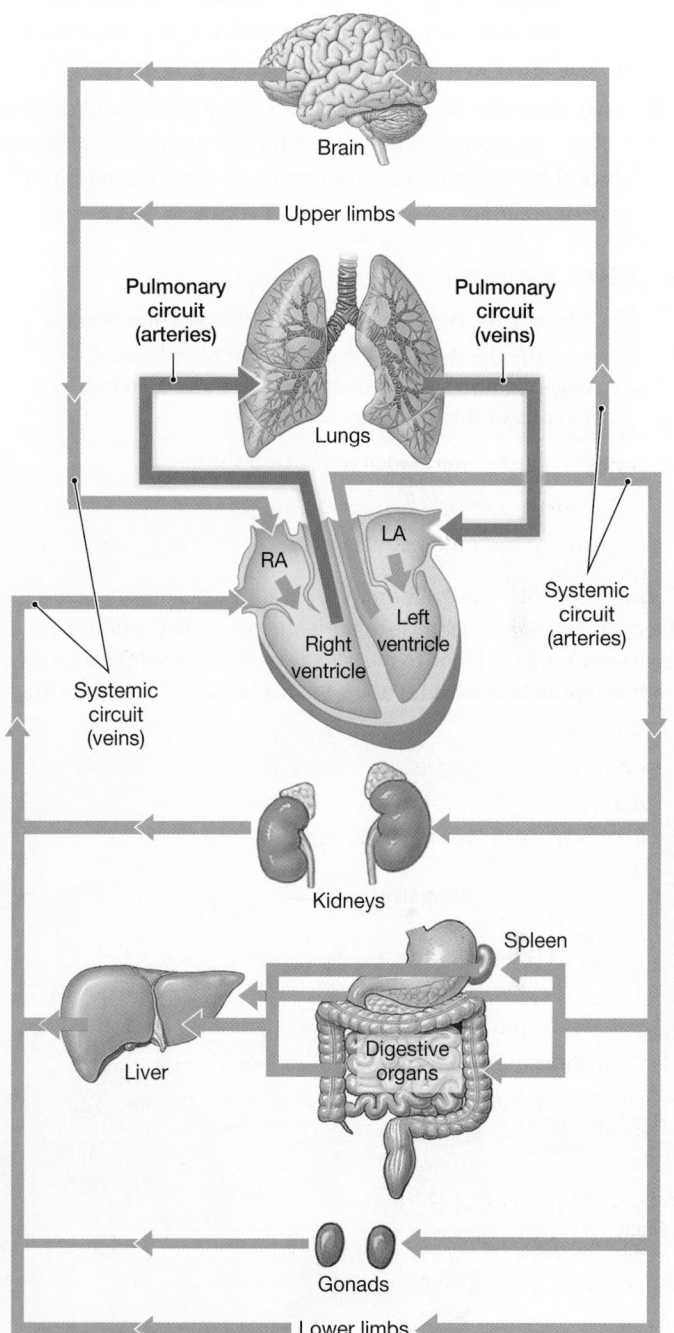

In the pages that follow, we examine the vessels of the pulmonary and systemic circuits further. Three major patterns of blood vessel organization are worth noting:

1. The peripheral distributions of arteries and veins on the body's left and right sides are generally identical, except near the heart, where the largest vessels connect to the atria or ventricles. Corresponding arteries and veins usually follow the same path. For example, the left and right subclavian *arteries* parallel the left and right subclavian *veins*.

2. A single vessel may have several names as it crosses specific anatomical boundaries. These names make accurate anatomical descriptions possible when the vessel extends far into the periphery. For example, the *external iliac artery* becomes the *femoral artery* as it leaves the trunk and enters the lower limb.

3. Several arteries and veins usually service tissues and organs. Often, anastomoses between adjacent arteries or veins reduce the impact of a temporary or even permanent blockage, or *occlusion*, in a single blood vessel.

CHECKPOINT

15. Identify the two circuits of the cardiovascular system.

16. Identify the three major patterns of blood vessel organization seen in the pulmonary and systemic circuits of the cardiovascular system.

See the blue Answers tab at the back of the book.

21-6 | In the pulmonary circuit, deoxygenated blood enters the lungs in arteries, and oxygenated blood leaves the lungs by veins

Learning Outcome Identify the major arteries and veins of the pulmonary circuit.

Blood entering the right atrium has just returned from the peripheral capillary beds, where it released oxygen and absorbed carbon dioxide. After traveling through the right atrium and ventricle, this deoxygenated blood enters the pulmonary trunk, the start of the pulmonary circuit (**Figure 21–18**). At the lungs, oxygen is replenished, and carbon dioxide is released. The oxygenated blood returns to the heart for distribution by the systemic circuit.

Figure 21–18 **The Pulmonary Circuit.** The pulmonary circuit consists of pulmonary arteries, which deliver deoxygenated blood from the right ventricle to the lungs; pulmonary capillaries, where gas exchange occurs; and pulmonary veins, which deliver oxygenated blood to the left atrium. As the enlarged view shows, diffusion across the capillary walls at alveoli removes carbon dioxide and provides oxygen to the blood. ATLAS: Plates 42a; 44c; 47b

Ascending aorta
Superior vena cava
Right lung
Right pulmonary arteries
Right pulmonary veins
Inferior vena cava
Descending aorta
Aortic arch
Pulmonary trunk
Left lung
Left pulmonary arteries
Left pulmonary veins
Alveolus
Capillary
O_2
CO_2

Compared with the systemic circuit, the pulmonary circuit is short: The base of the pulmonary trunk and the lungs are only about 15 cm (6 in.) apart.

The arteries of the pulmonary circuit differ from those of the systemic circuit in that they carry deoxygenated blood. (For this reason, most color-coded diagrams show the pulmonary arteries in blue, the same color as systemic veins.) As the pulmonary trunk curves over the superior border of the heart, it gives rise to the **left** and **right pulmonary arteries**. These large arteries enter the lungs before branching repeatedly, giving rise to smaller and smaller arteries. The smallest branches, the *pulmonary arterioles*, provide blood to capillary networks that surround *alveoli*. The walls of these small air pockets are thin enough for gas to be exchanged between the capillary blood and inspired air. As oxygenated blood leaves the alveolar capillaries, it enters venules that in turn unite to form larger vessels carrying blood toward the **pulmonary veins**. These four veins, two from each lung, empty into the left atrium, to complete the pulmonary circuit.

TIPS & TRICKS

Arteries and veins are defined by the direction of blood flow relative to the heart, not by the oxygen content of the blood they carry. So if you remember that *a*rteries carry blood *a*way from the heart, and veins carry blood toward the heart, you can remember that the pulmonary *a*rteries carry oxygen-poor blood *a*way from the heart to the lungs, and the pulmonary veins deliver oxygen-rich blood to the heart.

CHECKPOINT

17. Name the blood vessels that enter and exit the lungs, and whether they contain primarily oxygenated or deoxygenated blood.

18. Trace the path of a drop of blood through the lungs, beginning at the right ventricle and ending at the left atrium.

See the blue Answers tab at the back of the book.

21-7 ❙ The systemic circuit carries oxygenated blood from the left ventricle to tissues and organs other than the pulmonary exchange surfaces, and returns deoxygenated blood to the right atrium

Learning Outcome Identify the major arteries and veins of the systemic circuit.

The systemic circuit supplies the capillary beds in all parts of the body not serviced by the pulmonary circuit. This circuit begins at the left ventricle and ends at the right atrium. At any moment the systemic circuit contains about 84 percent of total blood volume.

Systemic Arteries

Figure 21-19 provides an overview of the systemic arterial system and shows the relative locations of major systemic arteries. **Figures 21-20** to **21-25** show the detailed distribution of these vessels and their branches. By convention, several large arteries are called *trunks*. Examples are the *pulmonary*, *brachiocephalic*, *thyrocervical*, and *celiac trunks*. Most of the major arteries are paired, with one artery of each pair on either side of the body. For this reason, the terms *right* and *left* appear in figures only when the arteries on both sides are labeled.

The Ascending Aorta

The **ascending aorta** (**Figure 21-20**) begins at the aortic valve of the left ventricle. The left and right coronary arteries originate in the aortic sinus at the base of the ascending aorta, just superior to the aortic valve. The distribution of coronary vessels was described in Chapter 20 and illustrated in **Figure 20-9**, p. 722.

The Aortic Arch

The **aortic arch** curves like the handle of a cane across the superior surface of the heart, connecting the ascending aorta with the *descending aorta* (**Figure 21-19**). Three elastic arteries originate along the aortic arch and deliver blood to the head, neck, shoulders, and upper limbs: (1) the **brachiocephalic** (brā-kē-ō-se-FAL-ik) **trunk**, (2) the **left common carotid artery**, and (3) the **left subclavian artery** (**Figures 21-20** and **21-21**). The brachiocephalic trunk, also called the *innominate artery* (i-NOM-i-nat; unnamed), ascends for a short distance before branching to form the **right subclavian artery** and the **right common carotid artery**.

We have only one brachiocephalic trunk, with the left common carotid and left subclavian arteries arising separately from the aortic arch. However, in terms of their peripheral distribution, the vessels on the left side are mirror images of those on the right side. **Figures 21-20** and **21-22** illustrate the major branches of these arteries.

The Subclavian Arteries. The subclavian arteries supply blood to the arms, chest wall, shoulders, back, and CNS (**Figures 21-19** and **21-20**). Three major branches arise before a subclavian artery leaves the thoracic cavity: (1) the **internal thoracic artery**, supplying the pericardium and anterior wall of the chest; (2) the **vertebral artery**, which provides blood to the brain and spinal cord; and (3) the **thyrocervical trunk**, which provides blood to muscles and other tissues of the neck, shoulder, and upper back.

After leaving the thoracic cavity and passing across the superior border of the first rib, the subclavian is called the **axillary artery**. This artery crosses the axilla to enter the arm, where it gives

Figure 21–19 **An Overview of the Major Systemic Arteries.**

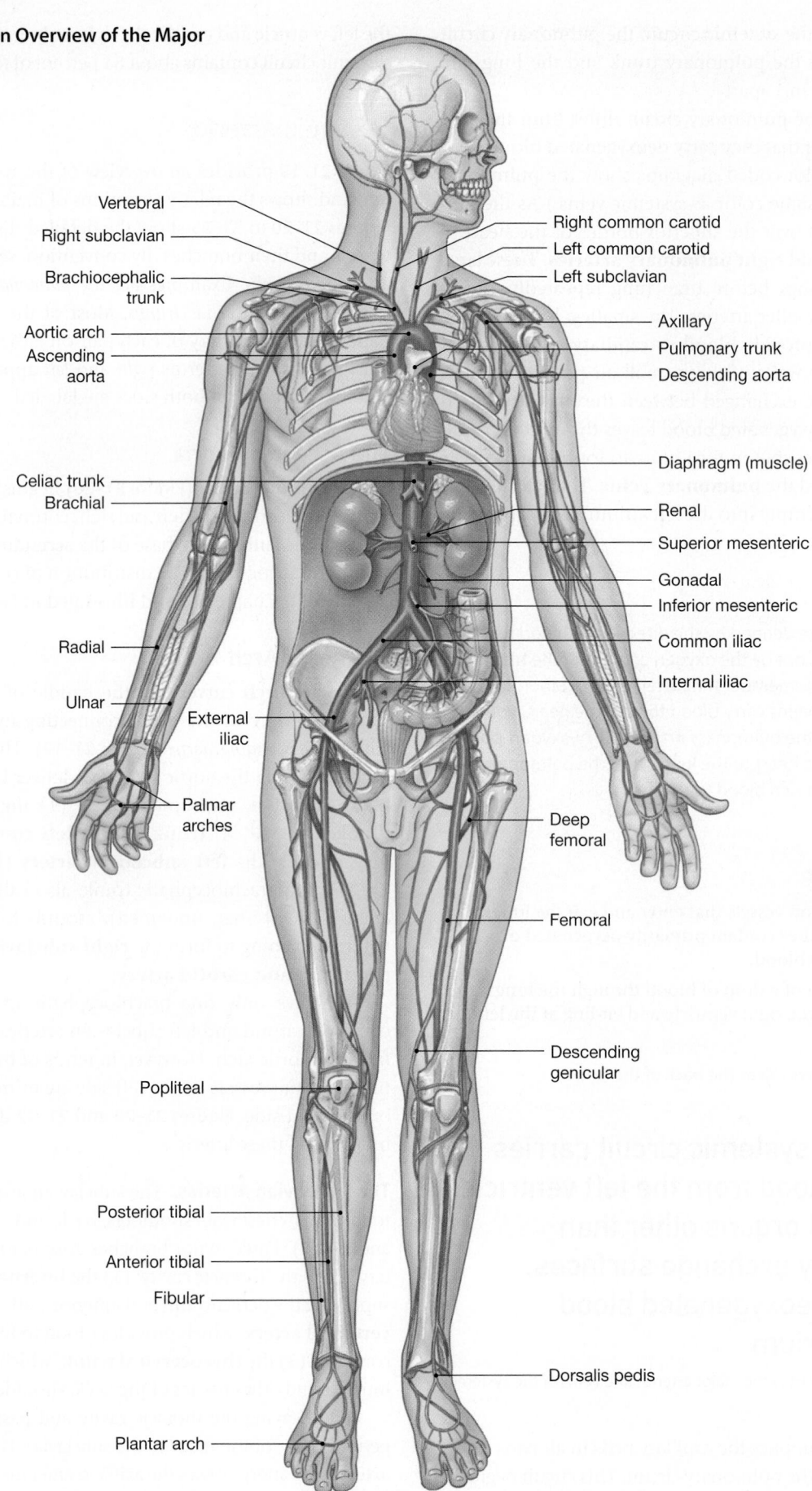

Vertebral

Right subclavian

Brachiocephalic trunk

Aortic arch

Ascending aorta

Celiac trunk

Brachial

Radial

Ulnar

External iliac

Palmar arches

Right common carotid

Left common carotid

Left subclavian

Axillary

Pulmonary trunk

Descending aorta

Diaphragm (muscle)

Renal

Superior mesenteric

Gonadal

Inferior mesenteric

Common iliac

Internal iliac

Deep femoral

Femoral

Descending genicular

Popliteal

Posterior tibial

Anterior tibial

Fibular

Dorsalis pedis

Plantar arch

21

Figure 21–20   **Arteries of the Chest and Upper Limb.** ATLAS: Plates 27a–c; 29c; 30; 45a

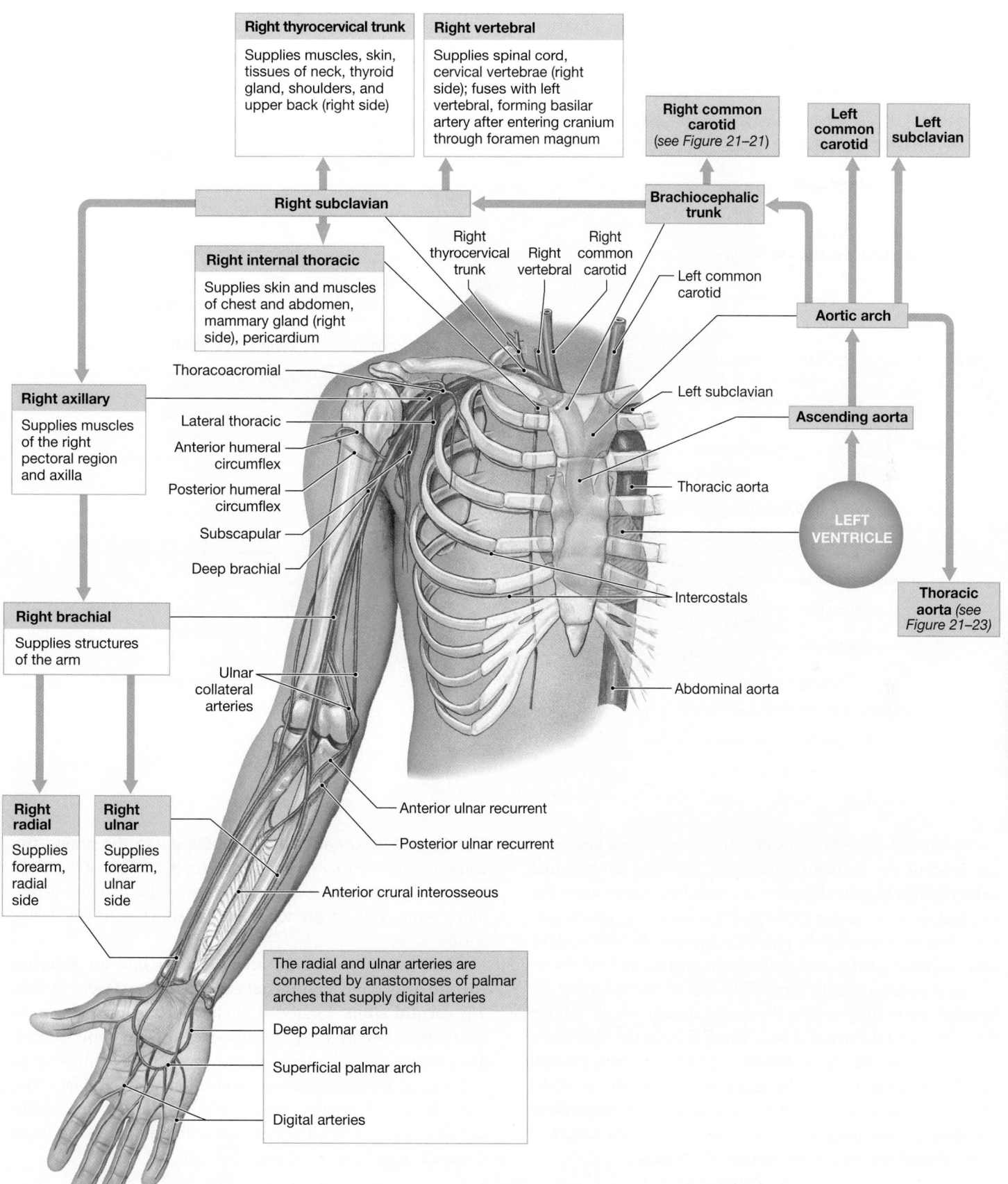

Right thyrocervical trunk
Supplies muscles, skin, tissues of neck, thyroid gland, shoulders, and upper back (right side)

Right vertebral
Supplies spinal cord, cervical vertebrae (right side); fuses with left vertebral, forming basilar artery after entering cranium through foramen magnum

Right common carotid
(see Figure 21–21)

Left common carotid

Left subclavian

Right subclavian

Brachiocephalic trunk

Right thyrocervical trunk

Right vertebral

Right common carotid

Left common carotid

Aortic arch

Right internal thoracic
Supplies skin and muscles of chest and abdomen, mammary gland (right side), pericardium

Left subclavian

Ascending aorta

Thoracoacromial

Right axillary
Supplies muscles of the right pectoral region and axilla

Lateral thoracic

Anterior humeral circumflex

Posterior humeral circumflex

Subscapular

Deep brachial

Thoracic aorta

LEFT VENTRICLE

Right brachial
Supplies structures of the arm

Intercostals

Thoracic aorta (see Figure 21–23)

Ulnar collateral arteries

Abdominal aorta

Right radial
Supplies forearm, radial side

Right ulnar
Supplies forearm, ulnar side

Anterior ulnar recurrent

Posterior ulnar recurrent

Anterior crural interosseous

The radial and ulnar arteries are connected by anastomoses of palmar arches that supply digital arteries

Deep palmar arch

Superficial palmar arch

Digital arteries

21

Figure 21–21 **Arteries of the Neck and Head.** Shown as seen from the right side. ATLAS: Plates 3c,d; 15b; 18a–c; 45a

Anterior cerebral

Middle cerebral

Ophthalmic

Cerebral arterial circle

Carotid canal

Posterior cerebral

Basilar

Internal carotid

Carotid sinus

Vertebral

Inferior thyroid

Thyrocervical trunk

Transverse cervical

Suprascapular

Subclavian

Axillary

Internal thoracic

Second rib

Branches of the External Carotid

Superficial temporal

Maxillary

Occipital

Facial

Lingual

External carotid

Common carotid

Brachiocephalic trunk

Clavicle

First rib

rise to *humeral circumflex arteries*, which supply structures near the head of the humerus. Distally, it becomes the **brachial artery**, which supplies blood to the rest of the upper limb. The brachial artery gives rise to the *deep brachial artery*, which supplies deep structures on the posterior aspect of the arm, and the *ulnar collateral arteries*, which supply the area around the elbow.

As it approaches the coronoid fossa of the humerus, the brachial artery divides into the **radial artery**, which follows the radius, and the **ulnar artery**, which follows the ulna to the wrist. These arteries supply blood to the forearm and, through the *ulnar recurrent arteries*, the region around the elbow. At the wrist, the radial and ulnar arteries fuse to form the **superficial** and **deep palmar arches**, which supply blood to the hand and to the **digital arteries** of the thumb and fingers.

The Carotid Artery and the Blood Supply to the Brain. The common carotid arteries ascend deep in the tissues of the neck. You can usually locate the carotid artery by pressing gently along either side of the windpipe (trachea) until you feel a strong pulse.

Each common carotid artery divides into an **external carotid artery** and an **internal carotid artery** (Figure 21–21). The **carotid sinus**, located at the base of the internal carotid artery, may extend along a portion of the common carotid. The external carotid arteries supply blood to the structures of the neck, esophagus, pharynx, larynx, lower jaw, and face. The internal carotid arteries enter the skull through the carotid canals of the temporal bones, delivering blood to the brain (Figure 7–3 and Spotlight Figure 7–4, pp. 236–239).

The internal carotid arteries ascend to the level of the optic nerves, where each artery divides into three branches: (1) an **ophthalmic artery**, which supplies the eyes; (2) an **anterior cerebral artery**, which supplies the frontal and parietal lobes of the brain; and (3) a **middle cerebral artery**, which supplies the midbrain and lateral surfaces of the cerebral hemispheres (**Figures 21–21** and **21–22**).

The brain is extremely sensitive to changes in blood supply. An interruption of blood flow for several seconds produces unconsciousness. After four minutes some permanent neural damage can occur. Such crises are rare, because blood reaches the brain through the vertebral arteries as well as by way of the internal carotid arteries. The left and right vertebral arteries arise from the subclavian arteries and ascend within the transverse foramina of the cervical vertebrae (**Figure 7–19b,c**, p. 256). The vertebral arteries enter the cranium at the foramen magnum, where they fuse along the ventral surface of the medulla oblongata to form the **basilar artery**. The vertebral arteries and the basilar artery supply blood to the spinal cord, medulla

Figure 21–22   **Arteries of the Brain.** ATLAS: Plates 15a–c

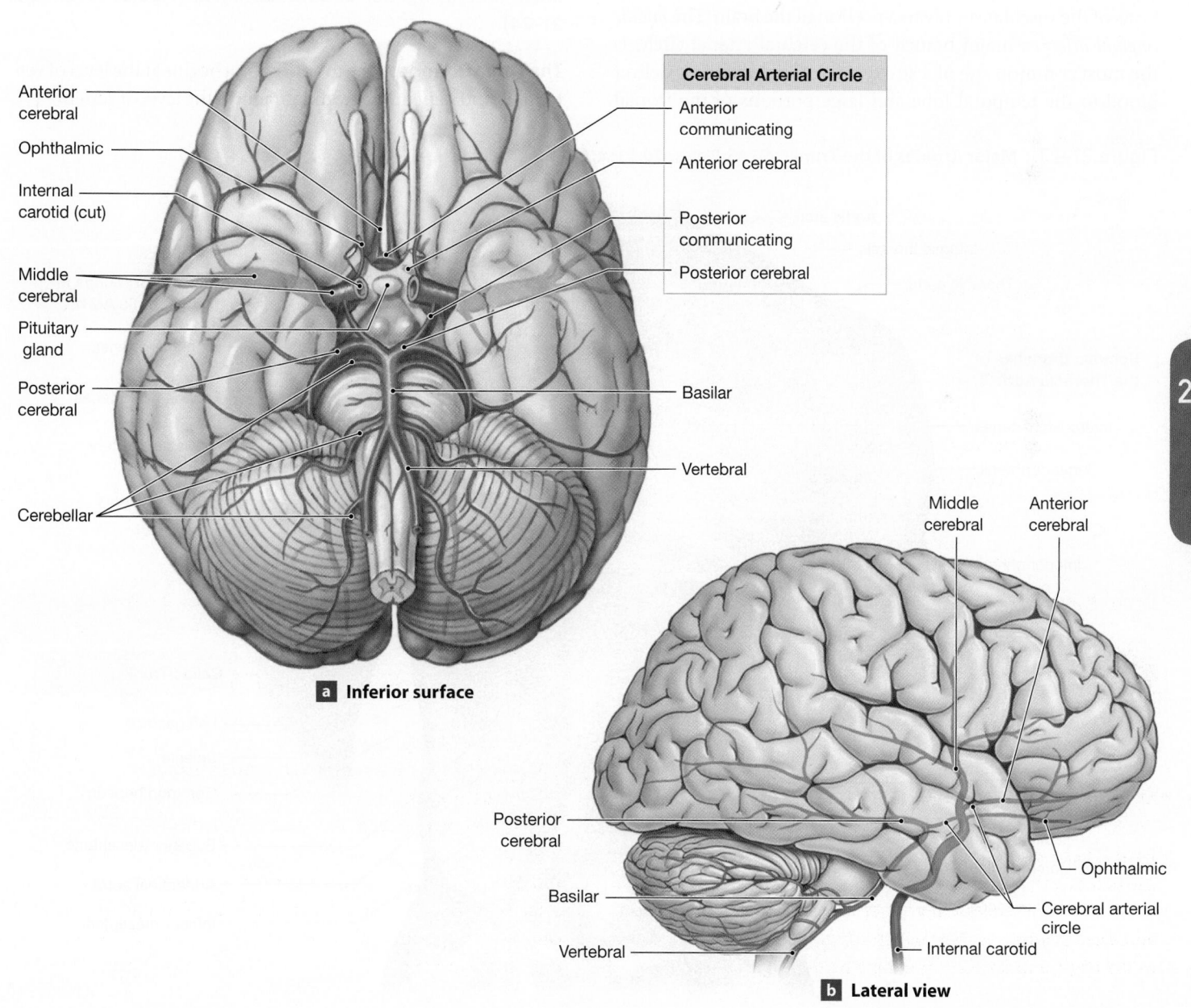

a Inferior surface

b Lateral view

oblongata, pons, and cerebellum. They then divide into the **posterior cerebral arteries**, which in turn branch off into the **posterior communicating arteries** (Figure 21–22).

The internal carotid arteries normally supply the arteries of the anterior half of the cerebrum, and the rest of the brain receives blood from the vertebral arteries. But this pattern of blood flow can easily change, because the internal carotid arteries and the basilar artery are interconnected. They form a ring-shaped anastomosis called the **cerebral arterial circle**, or *circle of Willis*, which encircles the infundibulum of the pituitary gland (Figure 21–22). With this arrangement, the brain can receive blood from either the carotid or the vertebral arteries, reducing the likelihood of a serious interruption of circulation.

Strokes, or *cerebrovascular accidents (CVAs)*, are interruptions of the vascular supply to a portion of the brain. The *middle cerebral artery*, a major branch of the cerebral arterial circle, is the most common site of a stroke. Superficial branches deliver blood to the temporal lobe and large portions of the frontal

and parietal lobes. Deep branches supply the basal nuclei and portions of the thalamus. If a stroke blocks the middle cerebral artery on the left side of the brain, aphasia and a sensory and motor paralysis of the right side of the body result. In a stroke affecting the middle cerebral artery on the right side, the individual experiences a loss of sensation and motor control over the left side of the body and has difficulty drawing or interpreting spatial relationships. Strokes affecting vessels that supply the brain stem also produce distinctive symptoms. Strokes affecting the lower brain stem are commonly fatal.

The Descending Aorta

The **descending aorta** is continuous with the aortic arch. The diaphragm divides the descending aorta into a superior **thoracic aorta** and an inferior **abdominal aorta** (Figures 21–23 and 21–24).

The Thoracic Aorta. The thoracic aorta begins at the level of vertebra T_5 and penetrates the diaphragm at the level of vertebra T_{12}.

Figure 21–23 **Major Arteries of the Trunk.** ATLAS: Plates 47d; 53c,e; 62a,b

Somatic Branches of the Thoracic Aorta
- Intercostal arteries
- Superior phrenic

- Aortic arch
- Internal thoracic
- Thoracic aorta

Visceral Branches of the Thoracic Aorta
- Bronchial arteries
- Esophageal arteries
- Mediastinal artery
- Pericardial artery

- Inferior phrenic
- Diaphragm
- Adrenal
- Renal
- Gonadal
- Lumbar
- Terminal segment of the aorta
- Common iliac
- Median sacral

Celiac Trunk
- Left gastric
- Splenic
- Common hepatic

- Superior mesenteric
- Abdominal aorta
- Inferior mesenteric

a A diagrammatic view, with most of the thoracic and abdominal organs removed

Figure 21–23 **Major Arteries of the Trunk** *(Continued)*

b A flowchart showing major arteries of the trunk

It travels within the mediastinum, on the posterior thoracic wall, slightly to the left of the vertebral column. This vessel supplies blood to branches that service the tissues and organs of the mediastinum, the muscles of the chest and the diaphragm, and the thoracic spinal cord.

We group the branches of the thoracic aorta anatomically as either visceral or somatic (**Figure 21–23**):

- *Visceral branches* supply the organs of the chest. The **bronchial arteries** supply the tissues of the lungs not involved in gas exchange. The **pericardial arteries** supply the pericardium. The **esophageal arteries** supply the esophagus, and the **mediastinal arteries** supply the tissues of the mediastinum.

- *Somatic branches* supply the chest wall. The **intercostal arteries** supply the chest muscles and the vertebral column area. The **superior phrenic** (FREN-ik) **arteries** deliver blood to the superior surface of the diaphragm, which separates the thoracic and abdominopelvic cavities.

The Abdominal Aorta. The abdominal aorta is a continuation of the thoracic aorta (**Figure 21–23**). It begins immediately inferior to the diaphragm and descends slightly to the left of the vertebral column but posterior to the peritoneal cavity. A cushion of adipose tissue commonly surrounds the abdominal aorta. At the level of vertebra L_4, the abdominal aorta splits into two major arteries—the *left* and *right common iliac arteries*—that supply deep pelvic structures and the lower limbs. The region where the abdominal aorta splits is called the *terminal segment of the aorta*.

The abdominal aorta delivers blood to all the abdominopelvic organs and structures. The major branches to visceral organs are unpaired. They arise on the anterior surface of the abdominal aorta and extend into the mesenteries. By contrast, branches to the body wall, the kidneys, the urinary bladder, and other structures outside the peritoneal cavity are paired. They originate along the lateral surfaces of the abdominal aorta. **Figure 21–23** shows the major arteries of the trunk after most thoracic and abdominal organs have been removed. **Figure 21–24** shows the distribution of those arteries to abdominopelvic organs.

TIPS&TRICKS

The aorta resembles a walking cane: the ascending aorta, aortic arch, and the start of the descending aorta form the cane's handle, while the thoracic and abdominal segments of the descending aorta form the cane's shaft.

The abdominal aorta gives rise to three unpaired arteries (**Figures 21–23** and **21–24**).

21

Figure 21–24 **Arteries Supplying the Abdominopelvic Organs.** *(See also Figure 24–24, p. 918.)* ATLAS: Plates 53a–e; 54c; 55

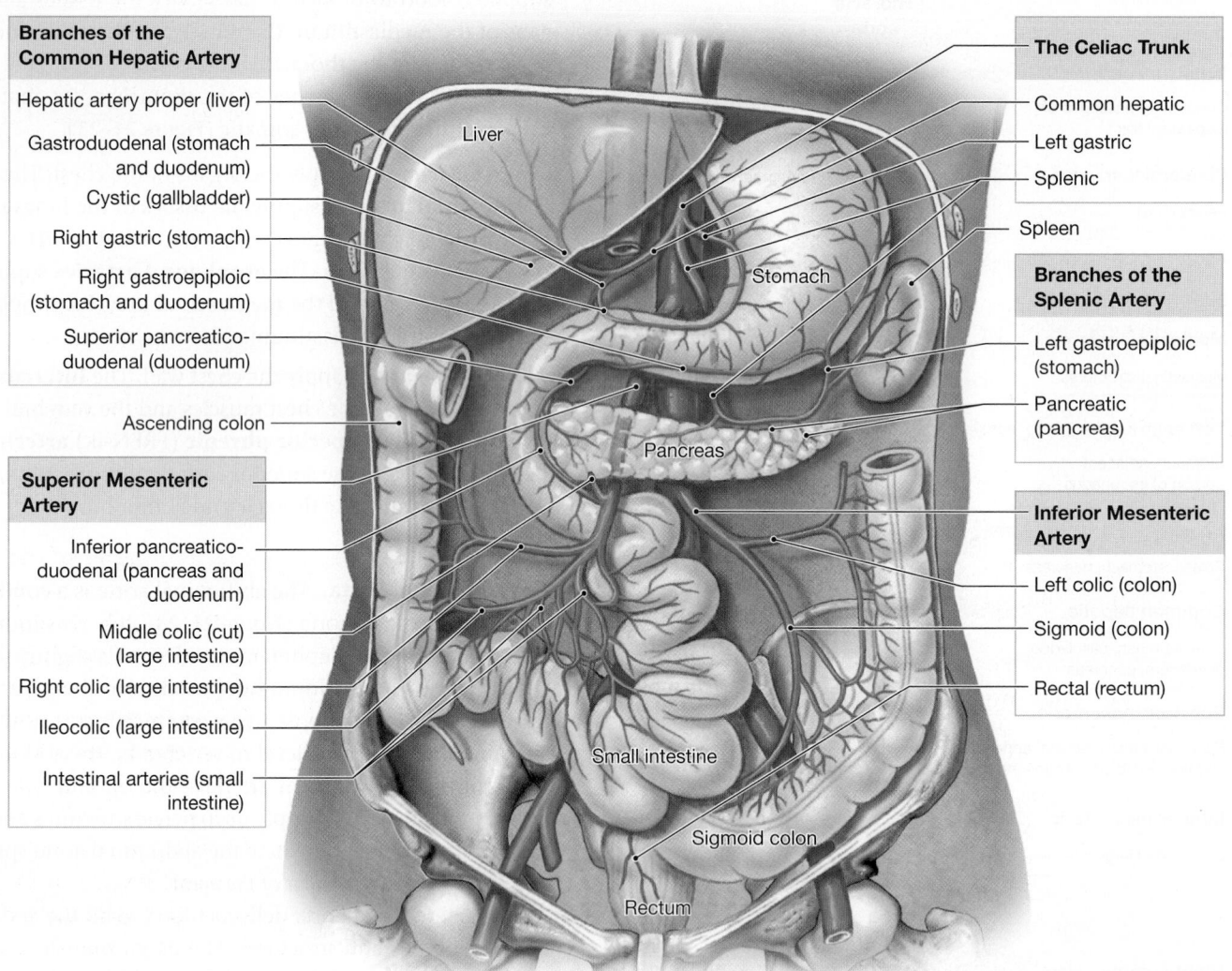

Branches of the Common Hepatic Artery
- Hepatic artery proper (liver)
- Gastroduodenal (stomach and duodenum)
- Cystic (gallbladder)
- Right gastric (stomach)
- Right gastroepiploic (stomach and duodenum)
- Superior pancreatico-duodenal (duodenum)
- Ascending colon

Superior Mesenteric Artery
- Inferior pancreatico-duodenal (pancreas and duodenum)
- Middle colic (cut) (large intestine)
- Right colic (large intestine)
- Ileocolic (large intestine)
- Intestinal arteries (small intestine)

The Celiac Trunk
- Common hepatic
- Left gastric
- Splenic
- Spleen

Branches of the Splenic Artery
- Left gastroepiploic (stomach)
- Pancreatic (pancreas)

Inferior Mesenteric Artery
- Left colic (colon)
- Sigmoid (colon)
- Rectal (rectum)

Liver — Stomach — Pancreas — Small intestine — Sigmoid colon — Rectum

21

1. The **celiac** (SĒ-lē-ak) **trunk** delivers blood to the liver, stomach, and spleen. The celiac trunk divides into three branches: (a) The **left gastric artery** supplies the stomach and the inferior portion of the esophagus. (b) The **splenic artery** supplies the spleen and arteries to the stomach (*left gastroepiploic artery*) and pancreas (*pancreatic arteries*). (c) The **common hepatic artery** supplies arteries to the liver (*hepatic artery proper*), stomach (*right gastric artery*), gallbladder (*cystic artery*), and duodenal area (*gastroduodenal, right gastroepiploic,* and *superior pancreaticoduodenal arteries*).

2. The **superior mesenteric** (mez-en-TER-ik) **artery** arises about 2.5 cm (1 in.) inferior to the celiac trunk. It supplies arteries to the pancreas and duodenum (*inferior pancreaticoduodenal artery*), small intestine (*intestinal arteries*), and most of the large intestine (*right* and *middle colic* and the *ileocolic arteries*).

3. The **inferior mesenteric artery** arises about 5 cm (2 in.) superior to the terminal aorta. It delivers blood to the terminal portions of the colon (*left colic* and *sigmoid arteries*) and the rectum (*rectal arteries*).

The abdominal aorta also gives rise to five paired arteries (**Figure 21–23**):

1. The **inferior phrenic arteries** supply the inferior surface of the diaphragm and the inferior portion of the esophagus.

2. The **adrenal arteries** originate on either side of the aorta near the base of the superior mesenteric artery. Each adrenal artery supplies one adrenal gland, which caps the superior part of a kidney.

3. The short (about 7.5 cm) **renal arteries** arise along the posterolateral surface of the abdominal aorta, about 2.5 cm (1 in.) inferior to the superior mesenteric artery. They

travel posterior to the peritoneal lining to reach the adrenal glands and kidneys. We consider the branches of the renal arteries in Chapter 26.

4. The **gonadal** (gō-NAD-al) **arteries** originate between the superior and inferior mesenteric arteries. In males, they are called *testicular arteries* and are long, thin arteries that supply blood to the testes and scrotum. In females, they are termed *ovarian arteries* and supply blood to the ovaries, uterine tubes, and uterus. The distribution of gonadal vessels (both arteries and veins) differs by gender. We describe the differences in Chapter 28.

5. Small **lumbar arteries** arise on the posterior surface of the aorta. They supply the vertebrae, spinal cord, and abdominal wall.

Arteries of the Pelvis and Lower Limbs

Near the level of vertebra L_4, the terminal segment of the abdominal aorta divides to form a pair of elastic arteries—the **right** and **left common iliac** (IL-ē-ak) **arteries**—plus the small **median sacral artery** (Figure 21–23). The common iliac arteries carry blood to the pelvis and lower limbs. They descend posterior to the cecum and sigmoid colon along the inner surface of the ilium. At the level of the lumbosacral joint, each common iliac divides to form an **internal iliac artery** and an **external iliac artery** (Figure 21–24). The internal iliac arteries enter the pelvic cavity to supply the urinary bladder, the internal and external walls of the pelvis, the external genitalia, the medial side of the thigh, and, in females, the uterus and vagina. The major branches of the internal iliac artery are the *gluteal, internal pudendal, obturator,* and *lateral sacral arteries*. The external iliac arteries supply blood to the lower limbs. They are much larger in diameter than the internal iliac arteries.

Arteries of the Thigh and Leg. Each external iliac artery crosses the surface of an iliopsoas muscle and penetrates the abdominal wall midway between the anterior superior iliac spine and the pubic symphysis on that side. It emerges on the anterior, medial surface of the thigh as the **femoral artery** (Figure 21–25a,b). Roughly 5 cm (2 in.) distal to the emergence of the femoral artery, the **deep femoral artery** branches off its lateral surface. The deep femoral artery supplies blood to the ventral and lateral regions of the skin and deep muscles of the thigh. It gives rise to the *femoral circumflex arteries*.

The femoral artery continues inferiorly and posterior to the femur. As it approaches the knee, it gives rise to the *descending genicular artery*, which supplies the area around the knee. At the popliteal fossa, posterior to the knee joint, the femoral artery becomes the **popliteal** (pop-LIT-ē-al) **artery**, which then branches to form the **posterior** and **anterior tibial arteries**. The posterior tibial artery gives rise to the **fibular artery**, or *peroneal* (perone, fibula) *artery*, and then continues inferiorly along the posterior surface of the tibia. The anterior tibial artery

passes between the tibia and fibula. It emerges on the anterior surface of the tibia. As it descends toward the foot, the anterior tibial artery provides blood to the skin and muscles of the anterior portion of the leg.

Arteries of the Foot. At the ankle, the anterior tibial artery becomes the **dorsalis pedis artery**. It then branches repeatedly, supplying the ankle and dorsal portion of the foot (Figure 21–25a,b). Figure 21–25c charts the flow of blood from the external iliac artery to the lower limbs.

At the ankle, the posterior tibial artery divides to form the **medial** and **lateral plantar arteries**. They supply blood to the plantar surface of the foot. These arteries are connected to the dorsalis pedis artery through a pair of anastomoses. The arrangement produces a **dorsal arch** (*arcuate arch*) and a **plantar arch**. Small arteries branching off these arches supply the distal portions of the foot and the toes.

Systemic Veins

Veins collect blood from the tissues and organs of the body by means of an elaborate venous network that drains into the right atrium of the heart by the superior and inferior venae cavae (Figure 21–26). The branching pattern of peripheral veins is much more variable than is the branching pattern of arteries. We base the discussion that follows on the most common arrangement of veins. Complementary arteries and veins commonly run side by side. In many cases they have comparable names.

One significant difference between the arterial and venous systems involves the distribution of major veins in the neck and limbs. Arteries in these areas are deep beneath the skin, protected by bones and surrounding soft tissues. In contrast, the neck and limbs generally have two sets of peripheral veins, one superficial and the other deep. This dual venous drainage is important for controlling body temperature. In hot weather, venous blood flows through superficial veins, where heat can easily be lost. In cold weather, blood is routed to the deep veins to minimize heat loss.

The Superior Vena Cava

All the body's systemic veins (except the cardiac veins) ultimately drain into either the superior vena cava or the inferior vena cava. The **superior vena cava (SVC)** receives blood from the tissues and organs of the head, neck, chest, shoulders, and upper limbs.

Venous Return from the Cranium. Numerous veins drain the cerebral hemispheres. The *superficial cerebral veins* and small veins of the brain stem empty into a network of dural sinuses (Figure 21–27a,b). These sinuses include the *superior* and *inferior sagittal sinuses*, the *petrosal sinuses*, the *occipital sinus*, the *left* and *right transverse sinuses*, and the *straight sinus* (Figure 21–27c).

21

Figure 21–25 **Arteries of the Lower Limb.** ATLAS: Plates 68c; 70b; 78b–g

Common iliac
External iliac
Superior gluteal
Inguinal igament
Deep femoral
Lateral femoral circumflex

Internal iliac
Lateral sacral
Internal pudendal
Obturator
Medial femoral circumflex

Femoral
Descending genicular
Popliteal

Anterior tibial
Posterior tibial
Fibular

Dorsalis pedis
Medial plantar
Lateral plantar
Dorsal arch
Plantar arch

a **Anterior view**

Superior gluteal

b **Posterior view**

Right external iliac

Femoral
Thigh

Deep femoral
Hip joint, femoral head, deep muscles of the thigh

Lateral femoral circumflex
Quadriceps muscles

Medial femoral circumflex
Adductor and obturator muscles, hip joint

Descending genicular
Skin of leg, knee joint

Popliteal
Leg and foot

Posterior tibial	**Anterior tibial**
Connected by anastomoses of dorsalis pedis, dorsal arch, and plantar arch, which supply distal portions of the foot and the toes	

Fibular

c A flowchart of blood flow to a lower limb

Figure 21–26 **An Overview of the Major Systemic Veins.**

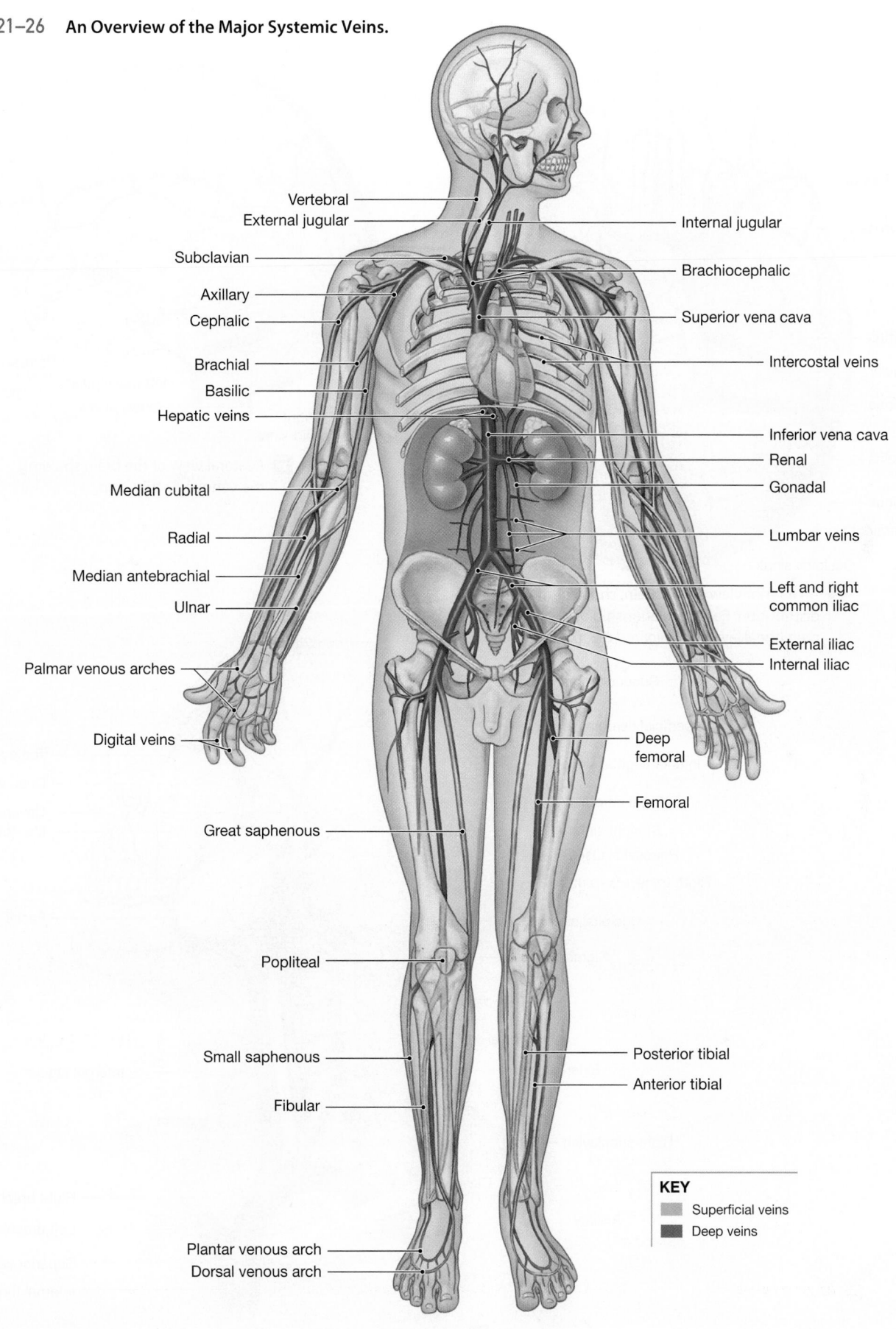

Vertebral

External jugular

Subclavian

Axillary

Cephalic

Brachial

Basilic

Hepatic veins

Median cubital

Radial

Median antebrachial

Ulnar

Palmar venous arches

Digital veins

Great saphenous

Popliteal

Small saphenous

Fibular

Plantar venous arch

Dorsal venous arch

Internal jugular

Brachiocephalic

Superior vena cava

Intercostal veins

Inferior vena cava

Renal

Gonadal

Lumbar veins

Left and right common iliac

External iliac

Internal iliac

Deep femoral

Femoral

Posterior tibial

Anterior tibial

KEY

Superficial veins

Deep veins

21

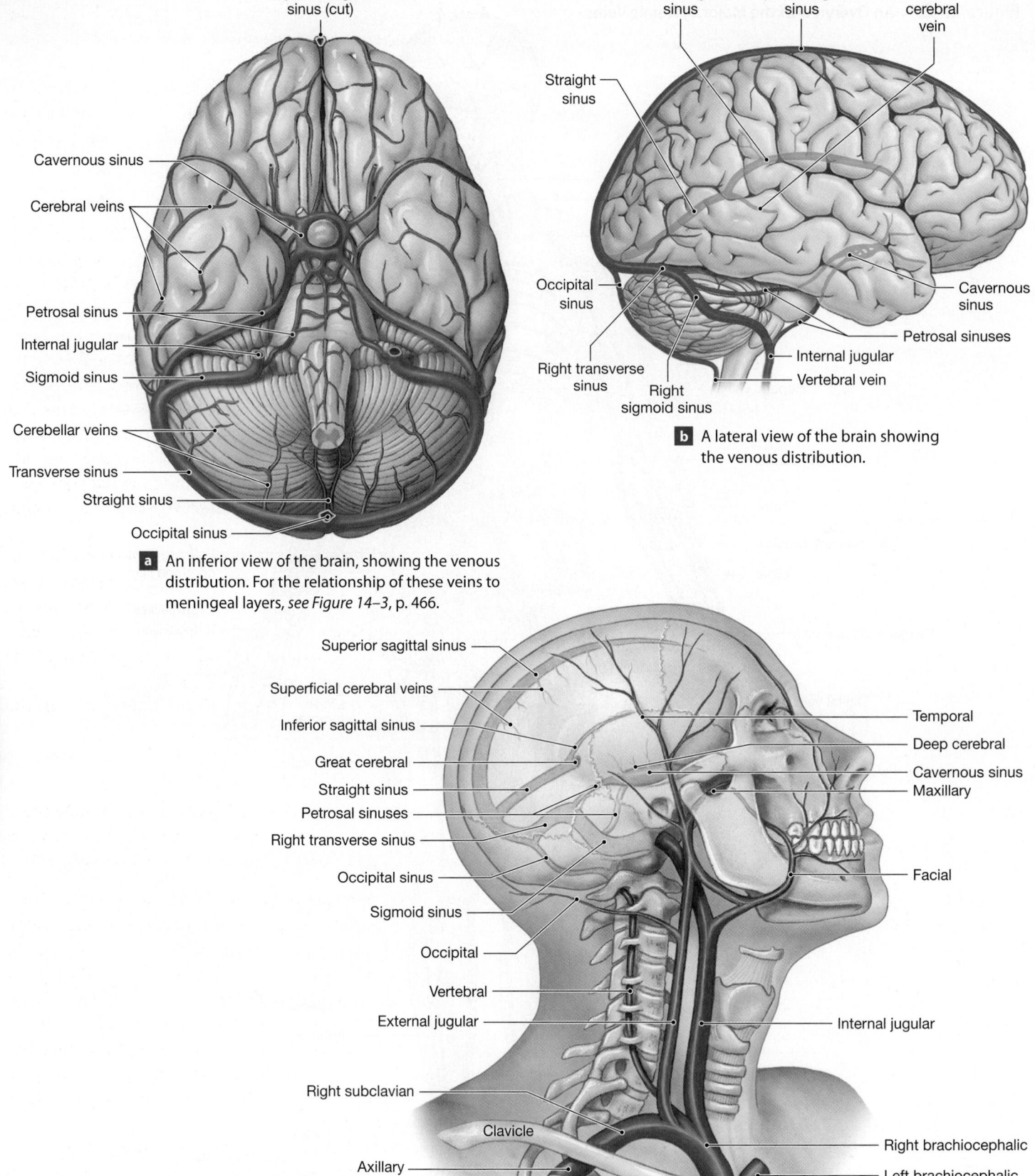

Superior sagittal sinus (cut)

Cavernous sinus

Cerebral veins

Petrosal sinus

Internal jugular

Sigmoid sinus

Cerebellar veins

Transverse sinus

Straight sinus

Occipital sinus

a An inferior view of the brain, showing the venous distribution. For the relationship of these veins to meningeal layers, *see Figure 14–3*, p. 466.

Inferior sagittal sinus

Superior sagittal sinus

Great cerebral vein

Straight sinus

Occipital sinus

Right transverse sinus

Right sigmoid sinus

Cavernous sinus

Petrosal sinuses

Internal jugular

Vertebral vein

b A lateral view of the brain showing the venous distribution.

Superior sagittal sinus

Superficial cerebral veins

Inferior sagittal sinus

Great cerebral

Straight sinus

Petrosal sinuses

Right transverse sinus

Occipital sinus

Sigmoid sinus

Occipital

Vertebral

External jugular

Right subclavian

Clavicle

Axillary

First rib

Temporal

Deep cerebral

Cavernous sinus

Maxillary

Facial

Internal jugular

Right brachiocephalic

Left brachiocephalic

Superior vena cava

Internal thoracic

c Veins draining the brain and the superficial and deep portions of the head and neck.

21

The largest, the **superior sagittal sinus**, is in the falx cerebri (Figure 14–3, p. 492). Most of the *inferior cerebral veins* converge within the brain to form the **great cerebral vein**. It delivers blood from the interior of the cerebral hemispheres and the choroid plexus to the **straight sinus**. Other cerebral veins drain into the **cavernous sinus** with numerous small veins from the orbit. Blood from the cavernous sinus reaches the internal jugular vein through the petrosal sinuses.

The venous sinuses converge within the dura mater in the region of the lambdoid suture. The left and right transverse sinuses begin at the confluence of the occipital, sagittal, and straight sinuses. Each transverse sinus drains into a **sigmoid sinus**, which penetrates the jugular foramen and leaves the skull as the **internal jugular vein**. It descends parallel to the common carotid artery in the neck. ⟳ p. 783

Vertebral veins drain the cervical spinal cord and the posterior surface of the skull. These vessels descend within the transverse foramina of the cervical vertebrae, along with the vertebral arteries. The vertebral veins empty into the *brachiocephalic veins* of the chest (discussed later in the chapter).

Superficial Veins of the Head and Neck. The superficial veins of the head converge to form the **temporal**, **facial**, and **maxillary veins** (Figure 21–27c). The temporal vein and the maxillary vein drain into the **external jugular vein**. The facial vein drains into the internal jugular vein. A broad anastomosis between the external and internal jugular veins at the angle of the mandible provides dual venous drainage of the face, scalp, and cranium. The external jugular vein descends toward the chest just deep to the skin on the anterior surface of the sternocleidomastoid muscle. Posterior to the clavicle, the external jugular vein empties into the *subclavian vein*. In healthy individuals, the external jugular vein is easily palpable. A *jugular venous pulse (JVP)* is sometimes detectable at the base of the neck.

Venous Return from the Upper Limbs. The **digital veins** empty into the **superficial** and **deep palmar veins** of the hand, which interconnect to form the **palmar venous arches** (Figure 21–28). The superficial arch empties into the **cephalic vein**, which ascends along the radial side of the forearm; the **median antebrachial vein**; and the **basilic vein**, which ascends on the ulnar side. Anterior to the elbow is the superficial **median cubital vein**, which passes from the cephalic vein, medially and at an oblique angle, to connect to the basilic vein. (Venous blood samples are typically collected from the median cubital.) From the elbow, the basilic vein passes superiorly along the medial surface of the biceps brachii muscle.

The deep palmar veins drain into the **radial vein** and the **ulnar vein**. These veins fuse to form the **brachial vein**, running parallel to the brachial artery. As the brachial vein continues toward the trunk, it merges with the basilic vein and becomes the **axillary vein**, which enters the axilla.

Formation of the Superior Vena Cava. The cephalic vein joins the axillary vein on the lateral surface of the first rib, forming the **subclavian vein**, which continues into the chest. The subclavian vein passes superior to the first rib and along the superior margin of the clavicle. It merges with the external and internal jugular veins of that side. This fusion creates the **brachiocephalic vein**, or *innominate vein*, which penetrates the body wall and enters the thoracic cavity.

Each brachiocephalic vein receives blood from the vertebral vein of the same side, which drains the back of the skull and spinal cord. Near the heart, at the level of the first and second ribs, the left and right brachiocephalic veins join, creating the superior vena cava. Close to the point of fusion, the **internal thoracic vein** empties into the brachiocephalic vein.

The **azygos** (AZ-ī-gos) **vein** is the major branch of the superior vena cava. This vein ascends from the lumbar region over the right side of the vertebral column to enter the thoracic cavity through the diaphragm. The azygos vein joins the superior vena cava at the level of vertebra T_2. On the left side, the azygos receives blood from the smaller **hemiazygos vein**, which in many people also drains into the left brachiocephalic vein through the *highest intercostal vein*.

The azygos and hemiazygos veins are the chief collecting vessels of the thorax. They receive blood from (1) **intercostal veins**, which in turn receive blood from the chest muscles; (2) **esophageal veins**, which drain blood from the inferior portion of the esophagus; and (3) smaller veins draining other mediastinal structures.

Figure 21–29 diagrams the venous branches of the superior vena cava.

The Inferior Vena Cava

The **inferior vena cava (IVC)** collects most of the venous blood from organs inferior to the diaphragm. (A small amount reaches the superior vena cava by the azygos and hemiazygos veins.)

Veins Draining the Lower Limbs. Blood leaving capillaries in the sole of each foot collects into a network of **plantar veins**, which supply the **plantar venous arch** (Figure 21–30a). The plantar network sends blood to the deep veins of the leg: the **anterior tibial vein**, the **posterior tibial vein**, and the **fibular vein**. The **dorsal venous arch** collects blood from capillaries on the superior surface of the foot and the **digital veins** of the toes. The plantar arch and the dorsal arch are extensively interconnected, and the path of blood flow can easily shift from superficial to deep veins.

The dorsal venous arch is drained by two superficial veins: the **great saphenous** (sa-FĒ-nus; saphenes, prominent) **vein** and the **small saphenous vein**. The great saphenous vein ascends along the medial aspect of the leg and thigh, draining into the *femoral vein* near the hip joint. The small saphenous

21

Figure 21–28 **The Venous Drainage of the Abdomen and Chest.** ATLAS: Plates 27c; 29c; 47b,d; 61a; 62a,b

KEY
- Superficial veins
- Deep veins

21

vein arises from the dorsal venous arch and ascends along the posterior and lateral aspect of the calf. This vein then enters the popliteal fossa, where it meets the **popliteal vein**, formed by the union of the fibular and both tibial veins (**Figure 21–30a,b**). The popliteal vein is easily palpated in the popliteal fossa adjacent to the adductor magnus muscle. At the femur, the popliteal vein becomes the **femoral vein**, which ascends along the thigh, next to the femoral artery. Immediately before penetrating the abdominal wall, the femoral vein receives blood from (1) the great saphenous vein; (2) the **deep femoral vein**, which collects blood from deeper structures in the thigh; and (3) the **femoral circumflex vein**, which drains the region around the neck and head of the femur. The femoral vein penetrates the body wall and emerges in the pelvic cavity as the **external iliac vein**.

Veins Draining the Pelvis. The external iliac veins receive blood from the lower limbs, the pelvis, and the lower abdomen.

As the left and right external iliac veins cross the inner surface of the ilium, they are joined by the **internal iliac veins**, which drain the pelvic organs (**Figure 21–29**). The internal iliac veins are formed by the fusion of the *gluteal, internal pudendal, obturator,* and *lateral sacral veins* (**Figure 21–30a**).

Figure 21–29 **Flowchart of Circulation to the Superior and Inferior Venae Cavae.**

Figure 21–30 **Venous Drainage from the Lower Limb.** ATLAS: Plates 70b; 74; 78a–g

Superior gluteal

Femoral circumflex

Femoral

Great saphenous

Small saphenous

Fibular

Common iliac

Internal iliac

External iliac

Inferior gluteal
Lateral sacral
Internal pudendal
Obturator

Femoral

Deep femoral

Collects blood from the thigh

Small saphenous

Collects blood from superficial veins of the leg and foot

Popliteal

Posterior tibial

Anterior tibial   **Fibular**

The dorsal and plantar venous arches collect blood from the foot and toes

Dorsal venous arch

Plantar venous arch
Digital

Internal pudendal

Superior gluteal
Inferior gluteal
Obturator

Femoral circumflex

Deep femoral

Femoral

Great saphenous

Collects blood from the superficial veins of the lower limb

Small saphenous

Anterior tibial

KEY
☐ Superficial veins
☐ Deep veins

a Anterior view   **b** Posterior view

The union of external and internal iliac veins forms the **common iliac vein**. Its right and left branches ascend at an oblique angle. The left common iliac vein receives blood from the *median sacral vein*, which drains the area supplied by the middle sacral artery (**Figure 21–28**). Anterior to vertebra L_5, the common iliac veins unite to form the inferior vena cava.

Veins Draining the Abdomen. The inferior vena cava ascends posterior to the peritoneal cavity, parallel to the aorta. The abdominal portion of the inferior vena cava collects blood from six major veins (**Figures 21–28** and **21–29**):

1. **Lumbar veins** drain the lumbar portion of the abdomen, including the spinal cord and body wall muscles. Superior branches of these veins are connected to the azygos vein

(right side) and hemiazygos vein (left side), which empty into the superior vena cava.

2. **Gonadal** (*ovarian* or *testicular*) **veins** drain the ovaries or testes. The right gonadal vein empties into the inferior vena cava. The left gonadal vein generally drains into the left renal vein.

3. **Hepatic veins** from the liver empty into the inferior vena cava at the level of vertebra T_{10}.

4. **Renal veins**, the largest branches of the inferior vena cava, collect blood from the kidneys.

5. **Adrenal veins** drain the adrenal glands. In most people, only the right adrenal vein drains into the inferior vena cava. The left adrenal vein drains into the left renal vein.

Figure 21–31   **The Hepatic Portal System.** *(See also Figure 24–24, p. 918.)* ATLAS: Plates 53b; 54a–c; 55; 57a,b

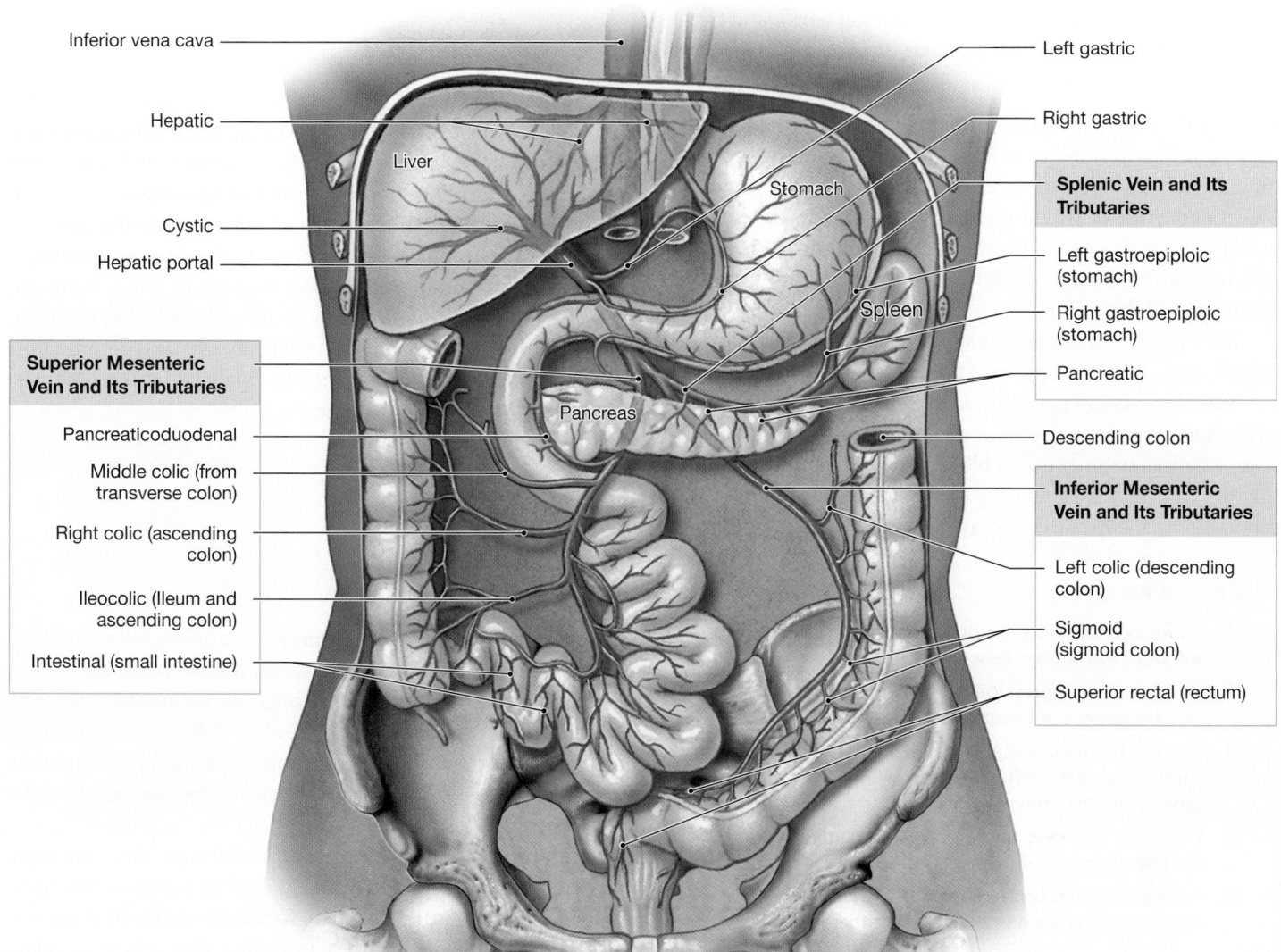

Inferior vena cava

Hepatic

Liver

Cystic

Hepatic portal

Superior Mesenteric Vein and Its Tributaries

Pancreaticoduodenal

Middle colic (from transverse colon)

Right colic (ascending colon)

Ileocolic (Ileum and ascending colon)

Intestinal (small intestine)

Pancreas

Stomach

Spleen

Left gastric

Right gastric

Splenic Vein and Its Tributaries

Left gastroepiploic (stomach)

Right gastroepiploic (stomach)

Pancreatic

Descending colon

Inferior Mesenteric Vein and Its Tributaries

Left colic (descending colon)

Sigmoid (sigmoid colon)

Superior rectal (rectum)

21

6. **Phrenic veins** drain the diaphragm. Only the right phrenic vein drains into the inferior vena cava. The left drains into the left renal vein.

   Figure 21–29 diagrams the branches of the inferior vena cava.

The Hepatic Portal System

The **hepatic portal system** begins in the capillaries of the digestive organs and ends in the liver sinusoids (**Figure 21–31**). (As you may recall from Chapter 18, a blood vessel connecting two capillary beds is called a *portal vessel*. The network is a *portal system*.) Blood flowing in the hepatic portal system is quite different from blood in other systemic veins, because it contains substances absorbed from the stomach and intestines. For example, levels of blood glucose and amino acids in the hepatic portal vein often exceed those found anywhere else in the

cardiovascular system. The hepatic portal system delivers these and other absorbed compounds directly to the liver for storage, metabolic conversion, or excretion.

   The largest vessel of the hepatic portal system is the **hepatic portal vein** (**Figure 21–31**). It delivers venous blood to the liver. It receives blood from three large veins draining organs within the peritoneal cavity:

- The **inferior mesenteric vein** collects blood from capillaries along the inferior portion of the large intestine. Its branches include the *left colic vein* and the *superior rectal veins*. They drain the descending colon, sigmoid colon, and rectum.

- The **splenic vein** is formed by the union of the inferior mesenteric vein and veins from the spleen, the lateral border of the stomach (*left gastroepiploic vein*), and the pancreas (*pancreatic veins*).

- The **superior mesenteric vein** collects blood from veins draining the stomach *(right gastroepiploic vein)*, the small intestine *(intestinal* and *pancreaticoduodenal veins)*, and two-thirds of the large intestine *(ileocolic, right colic,* and *middle colic veins)*.

The hepatic portal vein forms through the fusion of the superior mesenteric, inferior mesenteric, and splenic veins. The superior mesenteric vein normally contributes the greater volume of blood and most of the nutrients. As it proceeds, the hepatic portal vein receives blood from the left and right **gastric veins**, which drain the medial border of the stomach, and from the **cystic vein** of the gallbladder.

After passing through liver sinusoids, blood collects in the hepatic veins, which empty into the inferior vena cava.

The composition of the blood in the systemic circuit is relatively stable despite changes in diet and digestive activity. The reason for this stability is that blood from the intestines goes to the liver first, and the liver regulates the nutrient content of the blood before it enters the inferior vena cava.

CHECKPOINT

19. A blockage of which branch from the aortic arch would interfere with blood flow to the left arm?

20. Why would compression of the common carotid arteries cause a person to lose consciousness?

21. Grace is in an automobile accident, and her celiac trunk is ruptured. Which organs will be affected most directly by this injury?

22. Whenever Tim gets angry, a large vein bulges in the lateral region of his neck. Which vein is this?

23. A thrombus that blocks the popliteal vein would interfere with blood flow in which other veins?

See the blue Answers tab at the back of the book.

21-8 ❙ Modifications of fetal and maternal cardiovascular systems promote the exchange of materials, and independence occurs at birth

Learning Outcome Identify the differences between fetal and adult circulation patterns, and describe the changes in the patterns of blood flow that occur at birth.

The fetal cardiovascular system differs from the adult cardiovascular system because the fetus and the adult have different sources of respiratory and nutritional support. Most strikingly, the fetal lungs are collapsed and nonfunctional, and the digestive tract has nothing to digest. Instead, diffusion across the placenta provides for the respiratory and nutritional needs of the fetus.

Placental Blood Supply

Fetal patterns of blood flow are diagrammed in **Figure 21–32a**. Blood flows to the placenta through a pair of **umbilical arteries**. They arise from the internal iliac arteries and enter the umbilical cord. Blood returns from the placenta in the single **umbilical vein**, bringing oxygen and nutrients to the developing fetus. The umbilical vein drains into the **ductus venosus**, a vascular connection to an intricate network of veins within the developing liver. The ductus venosus collects blood from the veins of the liver and from the umbilical vein, and empties into the inferior vena cava. When the placental connection is broken at birth, blood stops flowing in the umbilical vessels, and they soon degenerate. Remnants of these vessels persist throughout life as fibrous cords.

Fetal Circulation in the Heart and Great Vessels

One of the most interesting aspects of cardiovascular development reflects the differences between the life of an embryo or fetus and that of an infant. Throughout embryonic and fetal life, the lungs are collapsed. Yet after delivery, the newborn infant must extract oxygen from inhaled air rather than across the placenta. ATLAS: Embryology Summary 16: The Development of the Cardiovascular System

The interatrial and interventricular septa develop early in fetal life, but the interatrial partition remains functionally incomplete until birth. The **foramen ovale**, or *interatrial opening*, is associated with a long flap that acts as a valve. Blood can flow freely from the right atrium to the left atrium, but any backflow closes the valve and isolates the two chambers from one another. Thus, blood entering the heart at the right atrium can bypass the pulmonary circuit. A second short-circuit exists between the pulmonary and aortic trunks. This connection, the **ductus arteriosus**, consists of a short, muscular vessel.

With the lungs collapsed, the capillaries are compressed and little blood flows through the lungs. During diastole, blood enters the right atrium and flows into the right ventricle, but it also passes into the left atrium through the foramen ovale. About 25 percent of the blood arriving at the right atrium bypasses the pulmonary circuit in this way. In addition, more than 90 percent of the blood leaving the right ventricle passes through the ductus arteriosus and enters the systemic circuit rather than continuing to the lungs.

Figure 21–32 **Fetal Circulation.**

a Blood flow to and from the placenta in full-term fetus (before birth)

b Blood flow through the neonatal (newborn) heart after delivery

Cardiovascular Changes at Birth

At birth, dramatic changes take place. When an infant takes the first breath, the lungs expand, and so do the pulmonary vessels. The resistance in the pulmonary circuit declines suddenly, and blood rushes into the pulmonary vessels. Within a few seconds, rising O_2 levels stimulate the constriction of the ductus arteriosus, isolating the pulmonary and aortic trunks from one another. As pressures rise in the left atrium, the valvular flap closes the foramen ovale. In adults, the interatrial septum bears the *fossa ovalis*, a shallow depression that marks the site of the foramen ovale (**Figure 20–6a**, p. 717). The remnants of the ductus arteriosus persist throughout life as the *ligamentum arteriosum*, a fibrous cord.

If the proper cardiovascular changes do not take place at birth or shortly after, problems eventually develop. Their severity depends on which connection remains open and on the size of the opening. Treatment may involve surgery to close the foramen ovale, the ductus arteriosus, or both. Other congenital heart defects result from abnormal cardiac development or inappropriate connections between the heart and major arteries and veins. **Spotlight Figure 21–33** focuses on congenital heart problems.

CHECKPOINT

24. **Name the three vessels that carry blood to and from the placenta.**

25. **A blood sample taken from the umbilical cord contains high levels of oxygen and nutrients, and low levels of carbon dioxide and waste products. Is this sample from an umbilical artery or from the umbilical vein? Explain.**

26. **Name the structures in the fetal circulation that stop functioning at birth. What becomes of these structures?**

See the blue Answers tab at the back of the book.

Although minor individual variations in the vascular network are quite common, congenital heart problems serious enough to threaten homeostasis are relatively rare.

Patent Foramen Ovale and Patent Ductus Arteriosus

If the foramen ovale remains open, or patent, blood recirculates through the pulmonary circuit instead of entering the left ventricle. The movement, driven by the relatively high systemic pressure, is called a "left-to-right shunt." Arterial oxygen content is normal, but the left ventricle must work much harder than usual to provide adequate blood flow through the systemic circuit. Hence, pressures rise in the pulmonary circuit. If the pulmonary pressures rise enough, they may force blood into the systemic circuit through the ductus arteriosus. This condition—a patent ductus arteriosus—creates a "right-to-left shunt." Because the circulating blood is not adequately oxygenated, it develops a deep red color. The skin then develops the blue tones typical of cyanosis and the infant is known as a "blue baby."

Tetralogy of Fallot

The tetralogy of Fallot (fa-LŌ) is a complex group of heart and circulatory defects that affect 0.10% of newborn infants. In this condition, (1) the pulmonary trunk is abnormally narrow (pulmonary stenosis), (2) the interventricular septum is incomplete, (3) the aorta originates where the interventricular septum normally ends, and (4) the right ventricle is enlarged and both ventricles thicken in response to the increased workload.

Ventricular Septal Defect

A ventricular septal defect is an abnormal opening in the wall (septum) between the left and right ventricles. It affects 0.12% of newborns. The opening between the two ventricles has an effect similar to a connection between the atria: When the more powerful left ventricle beats, it ejects blood into the right ventricle and pulmonary circuit.

Atrioventricular Septal Defect

In an atrioventricular septal defect, both the atria and ventricles are incompletely separated. The results are quite variable, depending on the extent of the defect and the effects on the atrioventricular valves. This type of defect most commonly affects infants with Down's syndrome, a disorder caused by the presence of an extra copy of chromosome 21.

Transposition of the Great Vessels

In the transposition of the great vessels, the aorta is connected to the right ventricle instead of to the left ventricle, and the pulmonary artery is connected to the left ventricle instead of the right ventricle. This malformation affects 0.05% of newborn infants.

Normal Heart Structure
Most congenital heart problems result from abnormal formation of the heart or problems with the connections between the heart and the great vessels.

21-9 ▎ Aging affects the blood, heart, and blood vessels

Learning Outcome Discuss the effects of aging on the cardiovascular system, and give examples of interactions between the cardiovascular system and other organ systems.

The capabilities of the cardiovascular system gradually decline. As you age, your cardiovascular system undergoes the following major changes:

- **Age-related changes in blood** may include (1) a decreased hematocrit; (2) constriction or blockage of peripheral veins by a stationary blood clot called a *thrombus*, which can become detached, pass through the heart, and become wedged in a small artery (commonly in the lungs), causing *pulmonary embolism*; and (3) pooling of blood in the veins of the legs because valves are not working effectively.

- **Age-related changes in the heart** include (1) a reduction in maximum cardiac output, (2) changes in the activities of nodal and conducting cells, (3) a reduction in the elasticity of the cardiac (fibrous) skeleton, (4) progressive atherosclerosis that can restrict coronary circulation, and (5) replacement of damaged cardiac muscle cells by scar tissue.

- **Age-related changes in blood vessels** may be linked to arteriosclerosis: (1) The inelastic walls of arteries become less tolerant of sudden pressure increases, which can lead to an *aneurysm*, whose rupture may (depending on the vessel) cause a stroke, myocardial infarction, or massive blood loss; (2) calcium salts can be deposited on weakened vascular walls, increasing the risk of a stroke or myocardial infarction; (3) lipid deposits in the tunica media associated with a damaged endothelial lining and calcium salts can form atherosclerotic plaques; and (4) thrombi can form at atherosclerotic plaques.

The cardiovascular system is both anatomically and functionally linked to all other systems. **Figure 21–34** shows the relationships between the cardiovascular system and the other body systems we have studied so far.

CHECKPOINT

27. Identify components of the cardiovascular system that are affected by age.

28. Define thrombus.

29. Define aneurysm.

30. Describe what the cardiovascular system provides for all other body systems.

31. What is the relationship between the skeletal system and the cardiovascular system?

See the blue Answers tab at the back of the book.

Related Clinical Terms

angiogram: An x-ray of a blood vessel that becomes visible due to a prior injection of dye into the subject's bloodstream.

carotid sinus massage: A procedure that involves rubbing the large part of the arterial wall at the point where the common carotid artery divides into its two main branches.

deep vein thrombosis (DVT): A blood clot in a major vein, usually in the legs. They often occur after extended periods of inactivity, such as long airplane flights. The clot can break free and travel as an embolus to the lungs, where it can cause respiratory distress or failure.

intermittent claudication: A limp that results from cramping leg pain that is typically caused by obstruction of the arteries.

normotensive: Having normal blood pressure.

orthostatic hypotension: A form of low blood pressure that occurs when you stand up from sitting or lying down. It can cause dizziness or a light-headed feeling.

phlebitis: Inflammation of a vein.

Raynaud's phenomenon: A condition resulting in the discoloration of the fingers and/or the toes when a person is subjected to changes in temperature or to emotional stress.

sclerotherapy: The treatment of varicose veins in which an irritant is injected to cause inflammation, coagulation of blood, and a narrowing of the blood vessel wall.

sounds of Korotkoff: Distinctive sounds, caused by turbulent arterial blood flow, heard through the stethoscope while measuring blood pressure.

sphygmomanometer: A device that measures blood pressure using an inflatable cuff placed around a limb.

syncope: A temporary loss of consciousness due to a sudden drop in blood pressure.

thrill: A vibration felt in a blood vessel that usually occurs due to abnormal blood flow. It is also often noticed at the fistula of a hemodialysis patient.

thrombophlebitis: An inflammation in a vein associated with the formation of a thrombus (clot).

vascular murmur: Periodic abnormal sounds heard upon auscultation that are produced as a result of turbulent blood flow.

white coat hypertension: A short-term increase in blood pressure triggered by the sight of medical personnel in white coats or other medical attire.

21

SYSTEM INTEGRATOR

Body System ⟶ Cardiovascular System   Cardiovascular System ⟶ Body System

Integumentary
Stimulation of mast cells produces localized changes in blood flow and capillary permeability

Integumentary
Delivers immune system cells to injury sites; clotting response seals breaks in skin surface; carries away toxins from sites of infection; provides heat
Page 200

Skeletal
Provides calcium needed for normal cardiac muscle contraction; protects blood cells developing in red bone marrow

Skeletal
Transports calcium and phosphate for bone deposition; delivers EPO to red bone marrow, parathyroid hormone, and calcitonin to osteoblasts and osteoclasts
Page 311

Muscular
Skeletal muscle contractions assist in moving blood through veins; protects superficial blood vessels, especially in neck and limbs

Muscular
Delivers oxygen and nutrients, removes carbon dioxide, lactic acid, and heat during skeletal muscle activity
Page 406

Nervous
Controls patterns of circulation in peripheral tissues; modifies heart rate and regulates blood pressure; releases ADH

Nervous
Endothelial cells maintain blood–brain barrier; helps generate CSF
Page 584

Endocrine
Erythropoietin regulates production of RBCs; several hormones elevate blood pressure; epinephrine stimulates cardiac muscle, elevating heart rate and contractile force

Endocrine
Distributes hormones throughout the body; heart secretes ANP and BNP
Page 673

The CARDIOVASCULAR System

The section on vessel distribution demonstrated the extent of the anatomical connections between the cardiovascular system and other organ systems. This figure summarizes some of the physiological relationships involved.

The most extensive communication occurs between the cardiovascular and lymphatic systems. Not only are the two systems physically interconnected, but cells of the lymphatic system also move from one part of the body to another within the vessels of the cardiovascular system. We examine the lymphatic system in detail, including its role in the immune response, in the next chapter.

Lymphatic
Page 850

Respiratory
Page 900

Digestive
Page 955

Urinary
Page 1036

Reproductive
Page 1116

Figure 21–34 diagrams the functional relationships between the cardiovascular system and the other body systems we have studied so far.

Chapter Review

Study Outline

MasteringA&P®

Access more chapter study tools online in the MasteringA&P Study Area:

- Chapter Quizzes, Chapter Practice Test, Art-labeling Activities, Animations, MP3 Tutor Sessions, and Clinical Case Studies
- Practice Anatomy Lab **PAL™**
- Interactive Physiology **iP°**
- A&P Flix **A&PFlix**
- PhysioEx **PhysioEx**

21-1 ▊ Arteries, arterioles, capillaries, venules, and veins differ in size, structure, and functional properties p. 750

1. Blood flows through a network of arteries, veins, and capillaries. All chemical and gaseous exchange between blood and interstitial fluid takes place across capillary walls.

2. **Arteries** and **veins** form an internal distribution system through which the heart propels blood. Arteries branch repeatedly, decreasing in size until they become **arterioles**. From the arterioles, blood enters **capillary** networks. Blood flowing from the capillaries enters small **venules** before entering larger veins.

3. The walls of arteries and veins contain three layers: the innermost **tunica intima**, the **tunica media**, and the outermost **tunica externa**. (*Figure 21–1*)

4. In general, the walls of arteries are thicker than those of veins. Arteries constrict when blood pressure does not distend them, but veins constrict very little. The endothelial lining cannot contract, so when constriction occurs, the lining of an artery becomes folded. (*Figure 21–1*)

5. The arterial system includes the large **elastic arteries**, medium-sized **muscular arteries**, and smaller arterioles. As blood proceeds toward the capillaries, the number of vessels increases, but the diameters of the individual vessels decrease and the walls become thinner. (*Figure 21–2*)

6. **Atherosclerosis**, a type of **arteriosclerosis**, is associated with changes in the endothelial lining of arteries. Fatty masses of tissue called **plaques** typically develop during atherosclerosis.

7. Capillaries are the only blood vessels whose walls are thin enough to permit an exchange between blood and interstitial fluid. Capillaries are **continuous** or **fenestrated**. **Sinusoids** have fenestrated walls and form elaborate networks that allow very slow blood flow. Sinusoids are located in the liver and in various endocrine organs. (*Figure 21–3*)

8. Capillaries form interconnected networks called **capillary beds (capillary plexuses)**. A band of smooth muscle, the **precapillary sphincter**, adjusts the blood flow into each capillary. Blood flow in a capillary changes as **vasomotion** occurs. The entire capillary bed may be bypassed by blood flow through **arteriovenous anastomoses**. (*Figure 21–4*)

9. Venules collect blood from the capillaries and merge into **medium-sized veins** and then **large veins**. The arterial system is a high-pressure system; blood pressure in veins is much lower. **Valves** in veins prevent the backflow of blood. (*Figures 21–1, 21–2, 21–5*)

10. Peripheral **venoconstriction** helps maintain adequate blood volume in the arterial system after a hemorrhage. The **venous reserve** normally accounts for about 20 percent of total blood volume. (*Figure 21–6*)

21-2 ▊ Pressure and resistance determine blood flow and affect rates of capillary exchange p. 760

11. Cardiovascular regulation involves the manipulation of blood pressure and resistance to control the rates of blood flow and capillary exchange.

12. Blood flows from an area of higher pressure to one of lower pressure, and blood flow is proportional to the pressure gradient. The *circulatory pressure* is the pressure gradient across the systemic circuit. It is reported as three values: arterial **blood pressure (BP)**, **capillary hydrostatic pressure (CHP)**, and **venous pressure**.

13. The **resistance (R)** determines the rate of blood flow through the systemic circuit. The major determinant of blood flow rate is the **peripheral resistance**—the resistance of the arterial system. Neural and hormonal control mechanisms regulate blood pressure and peripheral resistance.

14. **Vascular resistance** is the resistance of blood vessels. It is the largest component of peripheral resistance and depends on vessel length and vessel diameter. (*Figure 21–7*)

15. **Viscosity** and **turbulence** also contribute to peripheral resistance. (*Table 21–1*)

16. The high arterial pressures overcome peripheral resistance and maintain blood flow through peripheral tissues. Capillary pressures are normally low, and small changes in capillary pressure determine the rate of movement of fluid into or out of the bloodstream. Venous pressure, normally low, determines *venous return* and affects cardiac output and peripheral blood flow. (*Figures 21–8, 21–9; Table 21–1*)

17. Arterial blood pressure rises during ventricular systole and falls during ventricular diastole. The difference between these two blood pressures is the **pulse pressure**. Blood pressure is measured at the brachial artery with the use of a sphygmomanometer. (*Figures 21–8, 21–9*)

18. Valves, muscular compression, and the **respiratory pump** help the relatively low venous pressures propel blood toward the heart. (*Figures 21–5, 21–8*)

19. At the capillaries, blood pressure forces water and solutes out of the plasma, across capillary walls. Water moves out of the capillaries, through the peripheral tissues, and back to the bloodstream by way of the lymphatic system. Water movement across capillary walls is determined by the interplay between osmotic pressures and hydrostatic pressures. (*Figure 21–10*)

20. **Osmotic pressure (OP)** is a measure of the pressure that must be applied to prevent osmotic movement across a membrane. Osmotic water movement continues until either solute concentrations are equalized or the movement is prevented by an opposing hydrostatic pressure.

21

21. The rates of filtration and reabsorption gradually change as blood passes along the length of a capillary, as determined by the **net filtration pressure** (the difference between the net hydrostatic pressure and the net osmotic pressure). *(Figure 21–11)*

21-3 ▎ Cardiovascular regulatory mechanisms involve autoregulation, neural mechanisms, and endocrine responses p. 768

22. Homeostatic mechanisms ensure that **tissue perfusion** (blood flow) delivers adequate oxygen and nutrients.

23. Autoregulation, neural mechanisms, and endocrine mechanisms influence the coordinated regulation of cardiovascular function. Autoregulation involves local factors changing the pattern of blood flow within capillary beds in response to chemical changes in interstitial fluids. Neural mechanisms respond to changes in arterial pressure or blood gas levels. Hormones can assist in short-term adjustments (changes in cardiac output and peripheral resistance) and long-term adjustments (changes in blood volume that affect cardiac output and gas transport). *(Figure 21–12)*

24. Peripheral resistance is adjusted at the tissues by local factors that result in the dilation or constriction of precapillary sphincters. *(Figure 21–4)*

25. **Cardiovascular (CV) centers** of the medulla oblongata are responsible for adjusting cardiac output and peripheral resistance to maintain adequate blood flow. The vasomotor center contains one group of neurons responsible for controlling vasoconstriction, and another group responsible for controlling vasodilation.

26. **Baroreceptor reflexes** monitor the degree of stretch within expandable organs. Baroreceptors are located in the **carotid sinuses**, the **aortic sinuses**, and the right atrium. *(Figure 21–13)*

27. **Chemoreceptor reflexes** respond to changes in the oxygen or CO_2 levels in the blood. They are triggered by sensory neurons located in the **carotid bodies** and the **aortic bodies**. *(Figure 21–14)*

28. The endocrine system provides short-term regulation of cardiac output and peripheral resistance with epinephrine and norepinephrine from the adrenal medullae. Hormones involved in the long-term regulation of blood pressure and volume are *antidiuretic hormone (ADH)*, *angiotensin II, erythropoietin (EPO)*, and *natriuretic peptides (ANP and BNP)*. *(Figure 21–15)*

21-4 ▎ The cardiovascular system adapts to physiological stress and maintains a special vascular supply to the brain, heart, and lungs p. 775

29. During exercise, blood flow to skeletal muscles increases at the expense of blood flow to nonessential organs, and cardiac output rises. Cardiovascular performance improves with training. Athletes have larger stroke volumes, slower resting heart rates, and larger cardiac reserves than do nonathletes. *(Tables 21–2, 21–3)*

30. Blood loss lowers blood volume and venous return and decreases cardiac output. Compensatory mechanisms include an increase in cardiac output, mobilization of venous reserves, peripheral vasoconstriction, and the release of hormones that promote the retention of fluids and the manufacture of erythrocytes. *(Figure 21–16)*

31. The blood–brain barrier, the coronary circulation, and the circulation to alveolar capillaries in the lungs are examples of special circulations, in which cardiovascular dynamics and regulatory mechanisms differ from those in other tissues.

21-5 ▎ The pulmonary and systemic circuits of the cardiovascular system exhibit three general functional patterns p. 779

32. The peripheral distributions of arteries and veins are generally identical on both sides of the body, except near the heart. *(Figure 21–17)*

21-6 ▎ In the pulmonary circuit, deoxygenated blood enters the lungs in arteries, and oxygenated blood leaves the lungs by veins p. 780

33. The pulmonary circuit includes the pulmonary trunk, the **left** and **right pulmonary arteries**, and the **pulmonary veins**, which empty into the left atrium. *(Figure 21–18)*

21-7 ▎ The systemic circuit carries oxygenated blood from the left ventricle to tissues and organs other than the pulmonary exchange surfaces, and returns deoxygenated blood to the right atrium p. 781

34. The **ascending aorta** gives rise to the coronary circulation. The **aortic arch** communicates with the **descending aorta**. *(Figures 21–19 to 21–25)*

35. Three elastic arteries originate along the aortic arch: the **left common carotid artery**, the **left subclavian artery**, and the **brachiocephalic trunk**. *(Figures 21–20 to 21–22)*

36. The remaining major arteries of the body originate from the **descending aorta**. *(Figures 21–23 to 21–25)*

37. Arteries in the neck and limbs are deep beneath the skin; in contrast, there are generally two sets of peripheral veins, one superficial and one deep. This dual venous drainage is important for controlling body temperature. *(Figure 21–26)*

38. The **superior vena cava** receives blood from the head, neck, chest, shoulders, and arms. *(Figures 21–26 to 21–29)*

39. The **inferior vena cava** collects most of the venous blood from organs inferior to the diaphragm. *(Figures 21–28 to 21–30)*

40. The **hepatic portal system** directs blood from the other digestive organs to the liver before the blood returns to the heart. *(Figure 21–31)*

21-8 ▎ Modifications of fetal and maternal cardiovascular systems promote the exchange of materials, and independence occurs at birth p. 798

41. Blood flows to the placenta in a pair of **umbilical arteries** and from the placenta in a single **umbilical vein**. *(Figure 21–32)*

42. The interatrial partition remains functionally incomplete until birth. The **foramen ovale** allows blood to flow freely from the right to the left atrium, and the **ductus arteriosus** short-circuits the pulmonary trunk.

43. The foramen ovale closes, leaving the fossa ovalis. The ductus arteriosus constricts, leaving the ligamentum arteriosum. *(Figure 21–32)*

44. Congenital cardiovascular problems generally reflect abnormalities of the heart or of interconnections between the heart and great vessels. *(Spotlight Figure 21–33)*

21

21-9 | Aging affects the blood, heart, and blood vessels p. 801

45. Age-related changes in the blood include (1) a decreased hematocrit, (2) constriction or blockage of peripheral veins by a *thrombus* (stationary blood clot), and (3) pooling of blood in the veins of the legs because valves are not working effectively.

46. Age-related changes in the heart include (1) a reduction in the maximum cardiac output, (2) changes in the activities of nodal and conducting cells, (3) a reduction in the elasticity of the cardiac skeleton, (4) progressive atherosclerosis that can restrict coronary circulation, and (5) the replacement of damaged cardiac muscle cells by scar tissue.

47. Age-related changes in blood vessels, commonly related to arteriosclerosis, include (1) a weakening in the walls of arteries, potentially leading to the formation of an *aneurysm*, (2) deposition of calcium salts on weakened vascular walls, increasing the risk of a stroke or myocardial infarction, (3) lipid deposits in the tunica media associated with damaged endothelial lining and calcium salts forming atherosclerotic plaques, and (4) the formation of a thrombus at atherosclerotic plaques.

48. The cardiovascular system is anatomically and functionally connected to all other body systems. *(Figure 21–34)*

Review Questions

See the blue Answers tab at the back of the book.

LEVEL 1 Reviewing Facts and Terms

1. Identify the major arteries in the following diagram.

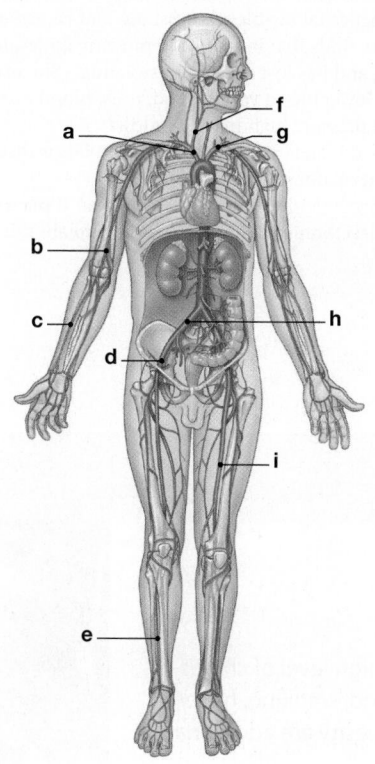

(a) _____   (b) _____
(c) _____   (d) _____
(e) _____   (f) _____
(g) _____   (h) _____
(i) _____

2. The blood vessels that play the most important role in regulating blood pressure and blood flow to a tissue are the **(a)** arteries, **(b)** arterioles, **(c)** veins, **(d)** venules, **(e)** capillaries.

3. Cardiovascular function is regulated by all of the following *except* **(a)** local factors, **(b)** neural factors, **(c)** endocrine factors, **(d)** venous return, **(e)** conscious control.

4. Baroreceptors that function in the regulation of blood pressure are located in the **(a)** left ventricle, **(b)** brain stem, **(c)** carotid sinus, **(d)** common iliac artery, **(e)** pulmonary trunk.

5. The two-way exchange of substances between blood and body cells occurs only through **(a)** arterioles, **(b)** capillaries, **(c)** venules, **(d)** a, b, and c.

6. Large molecules such as peptides and proteins move into and out of the bloodstream by way of **(a)** continuous capillaries, **(b)** fenestrated capillaries, **(c)** thoroughfare channels, **(d)** metarterioles.

7. The local control of blood flow due to the action of precapillary sphincters is **(a)** vasomotion, **(b)** autoregulation, **(c)** selective resistance, **(d)** turbulence.

8. Blood is transported through the venous system by means of **(a)** muscular contractions, **(b)** increasing blood pressure, **(c)** the respiratory pump, **(d)** a and c.

9. Identify the major veins in the following diagram.

(a) _____   (b) _____
(c) _____   (d) _____
(e) _____   (f) _____
(g) _____   (h) _____
(i) _____

10. The most important factor in vascular resistance is **(a)** the viscosity of the blood, **(b)** the diameter of the lumen of blood vessels, **(c)** turbulence due to irregular surfaces of blood vessels, **(d)** the length of the blood vessels.

11. Net hydrostatic pressure forces water _____ a capillary; net osmotic pressure forces water _____ a capillary. **(a)** into, out of, **(b)** out of, into, **(c)** out of, out of, **(d)** into, into.

12. Capillaries with pores that penetrate the endothelial lining are called **(a)** sinusoidal, **(b)** continuous, **(c)** fenestrated, **(d)** thoroughfare.

13. The unpaired arteries supplying blood to the visceral organs include **(a)** the adrenal, renal, and lumbar arteries, **(b)** the iliac, gonadal, and femoral arteries, **(c)** the celiac and superior and inferior mesenteric arteries, **(d)** a, b, and c.

14. Net filtration pressure may be calculated as **(a)** net hydrostatic pressure − net colloid pressure, **(b)** net colloid pressure − net hydrostatic pressure, **(c)** capillary hydrostatic pressure − blood colloid osmotic pressure, **(d)** blood colloid osmotic pressure − capillary hydrostatic pressure.

15. Total peripheral resistance is determined by **(a)** vessel length, **(b)** vessel diameter, **(c)** blood viscosity, **(d)** blood turbulence, **(e)** all of the above.

16. List three factors that aid venous return and describe the mechanism of their action.

17. What are the primary forces that cause fluid to move **(a)** out of a capillary at its arterial end and into the interstitial fluid, **(b)** into a capillary at its venous end from the interstitial fluid?

18. List the three layers of a typical blood vessel and describe the function of each.

LEVEL 2 Reviewing Concepts

19. A major difference between the arterial and venous systems is that **(a)** arteries are usually more superficial than veins, **(b)** in the limbs there is dual venous drainage, **(c)** veins are usually less branched compared to arteries, **(d)** veins exhibit a much more orderly pattern of branching in the limbs, **(e)** veins are not found in the abdominal cavity.

20. An arteriole has capillary hydrostatic pressure of 40 mm Hg and blood colloid osmotic pressure of 30 mm Hg. The net filtration pressure will be **(a)** −10 mm Hg and filtration occurs, **(b)** 10 mm Hg and reabsorption occurs, **(c)** 10 mm Hg and filtration occurs, **(d)** 10 mm Hg and reabsorption occurs.

21. Which of the following is *greater*? **(a)** the osmotic pressure of the interstitial fluid during inflammation, **(b)** the osmotic pressure of the interstitial fluid during normal conditions, **(c)** neither is greater.

22. How can local factors match blood flow to tissue oxygen demand?

23. Why do capillaries permit the diffusion of materials, whereas arteries and veins do not?

24. Why does cardiovascular function decline with age?

25. How do pressure and resistance affect cardiac output and peripheral blood flow?

26. Why is blood flow to the brain relatively continuous and constant?

27. Compare the effects of the cardioacceleratory and cardioinhibitory centers on cardiac output and blood pressure.

LEVEL 3 Critical Thinking and Clinical Applications

28. Bob is sitting outside on a warm day and is sweating profusely. Mary wants to practice taking blood pressures, and he agrees to play the patient. Mary finds that Bob's blood pressure is elevated, even though he is resting and has lost fluid from sweating. (She reasons that fluid loss should lower blood volume and, thus, blood pressure.) Why is Bob's blood pressure high instead of low?

29. Explain how the dietary intake of saturated fats is instrumental in the development of atherosclerotic plaques.

30. Scientists have hypothesized that angiotensin II plays an important role in the development of hypertension. Explain this hypothesis.

21

CLINICAL CASE Wrap-Up

Did Ancient Mummies Have Atherosclerosis?

The majority of American men and women today have atherosclerotic plaques in their arteries by the age of 50. The disease is usually thought of as a modern one, the result of too many cheeseburgers and hours spent in front of flat-screen TVs. But what could account for its incidence in the ancient mummies?

The degree of atherosclerosis discovered in the mummies turns out to be correlated with their age at death. The mean age at death for mummies without visible atherosclerosis is only 32. Among the mummies of people over age 40 that were examined, 50 percent had atherosclerosis.

Atherosclerosis is a vascular disease that progresses over many years. It is influenced by both genetics and environment. The most important risk factor, however, seems to be increasing age. Sex, diet, high level of cholesterol in the blood, smoking, hypertension, and obesity are additional risk factors.

The prevalence of atherosclerosis in varied, preindustrial populations raises the possibility of a basic, human predisposition to the disease. Atherosclerosis doesn't seem to be a purely modern health problem, caused exclusively by poor diet and lifestyle choices. It appears to have been stalking humans for millennia.

1. Why does the finding of calcium, which is easily seen on a CT scan, within systemic arteries indicate the presence of atherosclerosis?

2. Why do you think atherosclerosis is never seen in veins?

See the blue Answers tab at the back of the book.

22 The Lymphatic System and Immunity

Learning Outcomes

These Learning Outcomes correspond by number to this chapter's sections and indicate what you should be able to do after completing the chapter.

22-1 ▮ Distinguish between innate (nonspecific) and adaptive (specific) defenses, and explain the role of lymphocytes in the immune response.

22-2 ▮ Identify the major components of the lymphatic system, describe the structure and functions of each component, and discuss the importance of lymphocytes.

22-3 ▮ List the body's innate (nonspecific) defenses, and describe the components, mechanisms, and functions of each.

22-4 ▮ Define adaptive (specific) defenses, identify the forms and properties of immunity, and distinguish between cell-mediated (cellular) immunity and antibody-mediated (humoral) immunity.

22-5 ▮ Discuss the types of T cells and their roles in the immune response, and describe the mechanisms of T cell activation and differentiation.

22-6 ▮ Discuss the mechanisms of B cell activation and differentiation, describe the structure and function of antibodies, and explain the primary and secondary responses to antigen exposure.

22-7 ▮ Describe the development of immunocompetence, list and explain examples of immune disorders and allergies, and discuss the effects of stress on immune function.

22-8 ▮ Describe the effects of aging on the lymphatic system and the immune response.

22-9 ▮ Give examples of interactions between the lymphatic system and other organ systems we have studied so far and explain how the nervous and endocrine systems influence the immune response.

SPOTLIGHT
- Cytokines of the Immune System pp. 846–847

CLINICAL CASE
- Isn't There a Vaccine for That? p. 808

Clinical Notes
- Cancer and the Lymphatic System p. 814
- Graft Rejection and Immunosuppression p. 832
- AIDS p. 845

Isn't There a Vaccine for That?

Baby Ruthie is standing and holding on to a play table when her mother arrives to pick her up from day care. Ruthie is 11 months old and not yet walking. She hasn't had an appetite since yesterday, and last night she had a fever. When her mother gives Ruthie a hug, she notices a small blister near the corner of Ruthie's mouth. She also notices that Ruthie seems feverish again.

"Ruthie didn't do much today," reports the day care attendant. "Did you notice this little blister on her face?" asks Ruthie's mother. "No. It must have just appeared," the attendant replies. "Tommy, one of our older kids, had a little blister like that on his cheek just last week. He hasn't been back since."

That night, Ruthie's fever worsens. She won't eat and seems to have a sore throat when trying to swallow. She is listless. When her mother undresses her for bed, she notices more small blisters on Ruthie's face, abdomen, and back. She takes her straight to the Urgent Care Center.

"This looks like chickenpox," says the doctor. "Thanks to vaccines, we don't see this disease very often any more." **Ruthie's mother is certain that Ruthie is current on all her vaccinations. How could Ruthie have come down with chickenpox? To find out, turn to the Clinical Case Wrap-Up on p. 855.**

An Introduction to the Lymphatic System and Immunity

Many organs and systems work together to keep us alive and healthy. The lymphatic system plays a central role in the ongoing struggle to maintain health. In this chapter we discuss the components of the lymphatic system and the ways those components interact.

The world is not always kind to the human body. Accidental bumps, cuts, and scrapes; chemical and thermal burns; extreme cold; and ultraviolet radiation are just a few of the hazards in our physical environment. Making matters worse, the world around us contains an assortment of viruses, bacteria, fungi, and parasites capable of not only surviving but thriving inside our bodies—and potentially causing us great harm. These disease-causing organisms, called **pathogens**, are responsible for many diseases in humans. Each pathogen has a different lifestyle and attacks the body in a specific way. For example, viruses spend most of their time hidden within cells, which they often eventually destroy. Some of the largest parasites actually burrow through internal organs. Many bacteria multiply in interstitial fluids, where they release foreign proteins—enzymes or toxins—that can damage cells, tissues, even entire organ systems. And as if that were not enough, we are constantly at risk from renegade body cells that have the potential to produce lethal cancers. ⊃ p. 132

22-1 | Surface barriers and internal defenses make up innate defenses, and lymphocytes provide adaptive defenses

Learning Outcome Distinguish between innate (nonspecific) and adaptive (specific) defenses, and explain the role of lymphocytes in the immune response.

The **lymphatic system** includes the cells, tissues, and organs responsible for defending the body. This system acts both against environmental hazards, such as various pathogens, and against internal threats, such as cancer cells. We introduced *lymphocytes*, the primary cells of the lymphatic system, in Chapters 4 and 19. ⊃ pp. 159, 697 These cells are vital to the body's ability to resist or overcome infection and disease. Lymphocytes respond to invading pathogens (such as bacteria or viruses), abnormal body cells (such as virus-infected cells or cancer cells), and foreign proteins (such as the toxins released by some bacteria). They act to eliminate these threats or render them harmless through a combination of physical and chemical attacks.

The ability to resist infection and disease is called **immunity**. We have two forms of defense systems that work independently or together to provide immunity. These forms are *innate (nonspecific) defenses* and *adaptive (specific) defenses*. Surface barriers prevent or slow the entry of infectious organisms, or attack them if they do enter. These defenses are called *innate* because you are born with them and *nonspecific* because they do not distinguish one potential threat from another. In contrast, lymphocytes respond specifically. If a bacterial pathogen invades peripheral tissues, lymphocytes organize a defense against that particular type of bacterium. For this reason, we say that lymphocytes provide an *adaptive* defense. Defense against specific antigens is known as the **immune response**. All the cells and tissues involved in producing immunity are considered part of the *immune system*. The immune system includes parts of the integumentary, skeletal, lymphatic, cardiovascular, respiratory, and digestive systems.

We begin this chapter by examining the organization of the lymphatic system. We then consider the body's two forms of defenses. Finally, we will see how the lymphatic system interacts with cells and tissues of other systems to defend the body against infection and disease.

CHECKPOINT

1. Define pathogen.
2. Explain the difference between nonspecific defense and specific defense.

See the blue Answers tab at the back of the book.

22-2 ❙ Lymphatic vessels, lymphocytes, lymphoid tissues, and lymphoid organs function in body defenses

Learning Outcome Identify the major components of the lymphatic system, describe the structure and functions of each component, and discuss the importance of lymphocytes.

The lymphatic system consists of (1) **lymph**, a fluid that resembles plasma but contains a much lower concentration of suspended proteins; (2) a network of **lymphatic vessels**, often called **lymphatics**, which begin in peripheral tissues and connect to veins; (3) an array of **lymphoid tissues** and **lymphoid organs** scattered throughout the body; and (4) lymphocytes and smaller numbers of phagocytes and other cells.

Primary lymphoid tissues and organs are sites where lymphocytes are formed and mature. They include red bone marrow and the thymus gland. Recall that red bone marrow is also where other defense cells, the monocytes and macrophages, are also formed. **Secondary lymphoid tissues and organs** are where lymphocytes are activated and cloned (production of identical cellular copies). These structures include the appendix, spleen, lymph nodes, tonsils, and MALT. **Figure 22–1** provides a general overview of the primary vessels, tissues, and organs of this system.

Figure 22–1   **An Overview of the Lymphatic System.**

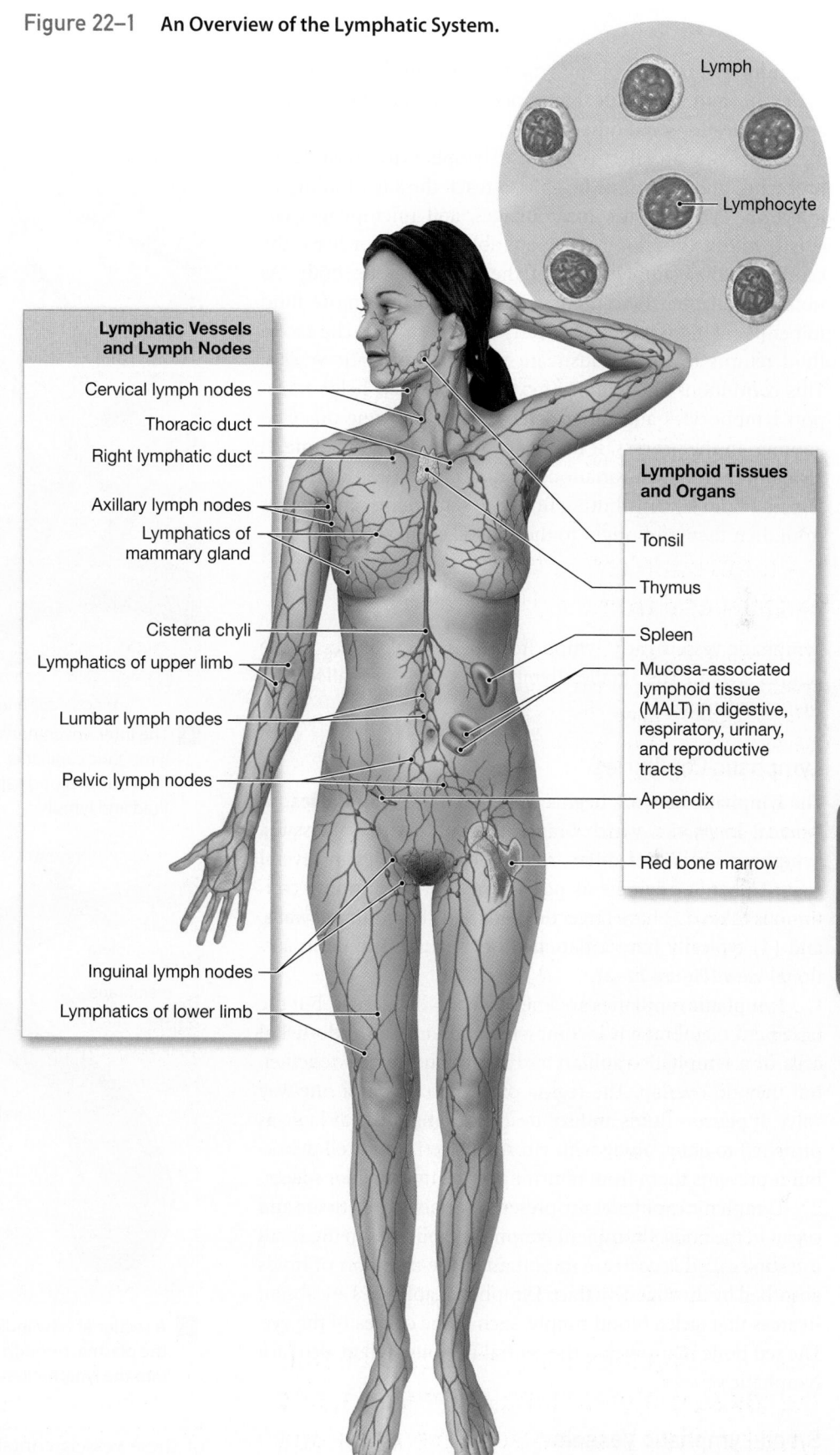

Lymph

Lymphocyte

Lymphatic Vessels and Lymph Nodes

Cervical lymph nodes
Thoracic duct
Right lymphatic duct
Axillary lymph nodes
Lymphatics of mammary gland
Cisterna chyli
Lymphatics of upper limb
Lumbar lymph nodes
Pelvic lymph nodes
Inguinal lymph nodes
Lymphatics of lower limb

Lymphoid Tissues and Organs

Tonsil
Thymus
Spleen
Mucosa-associated lymphoid tissue (MALT) in digestive, respiratory, urinary, and reproductive tracts
Appendix
Red bone marrow

22

Functions of the Lymphatic System

The primary function of the lymphatic system is to produce, maintain, and distribute lymphocytes that provide defense against infections and other environmental hazards.

To provide an effective defense, lymphocytes must detect problems, and they must be able to reach the site of injury or infection. Lymphocytes, macrophages, and microphages circulate within the blood. They are able to enter or leave the capillaries that supply most of the tissues of the body. As noted in Chapter 21, capillaries normally deliver more fluid to peripheral tissues than they carry away. ⟲ p. 766 The excess fluid returns to the bloodstream through lymphatic vessels. This continuous circulation of extracellular fluid helps transport lymphocytes and other defense cells from one organ to another. In the process, it maintains normal blood volume. It also eliminates local variations in the composition of the interstitial fluid by distributing hormones, nutrients, and wastes from their tissues of origin to the general circulation.

Lymphatic Vessels

Lymphatic vessels carry lymph from peripheral tissues to the venous system. The smallest lymphatic vessels are called *lymphatic capillaries*.

Lymphatic Capillaries

The lymphatic network begins with **lymphatic capillaries**, or *terminal lymphatics*, which branch through peripheral tissues. Lymphatic capillaries differ from blood capillaries in several ways. They (1) originate as pockets rather than forming continuous tubes, (2) have larger diameters, (3) have thinner walls, and (4) typically have a flattened or irregular outline in sectional view (**Figure 22–2**).

Lymphatic capillaries are lined by endothelial cells, but the basement membrane is incomplete or absent. The endothelial cells of a lymphatic capillary are not bound tightly together, but they do overlap. The region of overlap acts as a one-way valve. It permits fluids and solutes (including those as large as proteins) to enter, along with viruses, bacteria, and cell debris, but it prevents them from returning to the intercellular spaces.

Lymphatic capillaries are present in almost every tissue and organ in the body. Prominent lymphatic capillaries in the small intestine called *lacteals* are important in the transport of lipids absorbed by the digestive tract. Lymphatic capillaries are absent in areas that lack a blood supply, such as the cornea of the eye. The red bone marrow and the central nervous system also lack lymphatic vessels.

Small Lymphatic Vessels

From the lymphatic capillaries, lymph flows into larger lymphatic vessels that lead toward the body's trunk. The walls

Figure 22–2 Lymphatic Capillaries.

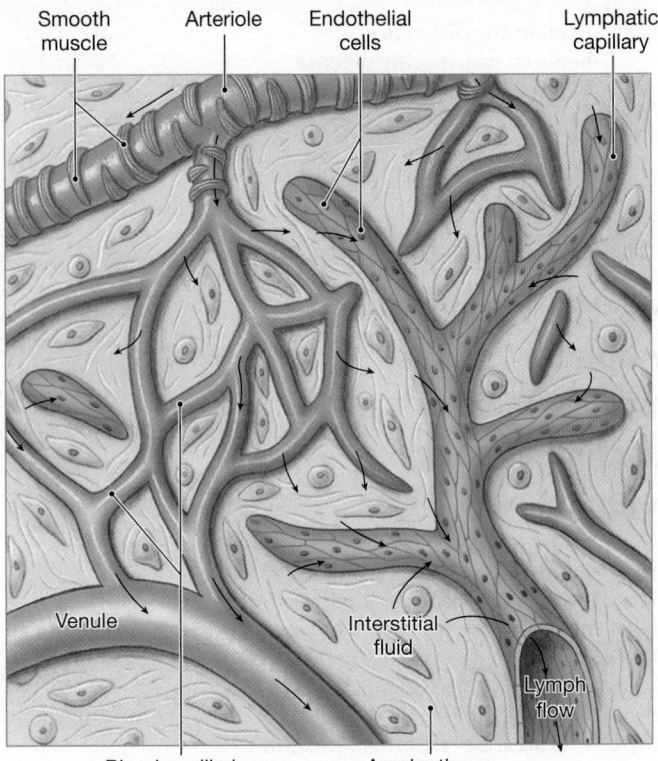

a The interwoven network formed by blood capillaries and lymphatic capillaries. Arrows indicate the movement of fluid out of blood capillaries and the net flow of interstitial fluid and lymph.

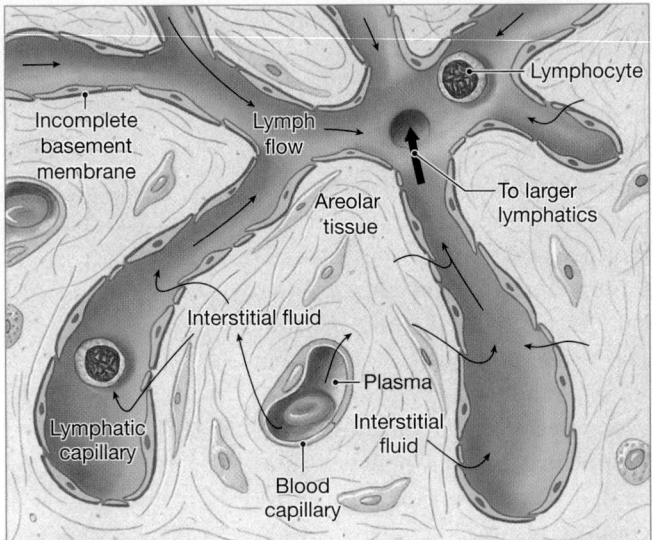

b A sectional view indicating the movement of fluid from the plasma, through the tissues as interstitial fluid, and into the lymphatic system as lymph.

of these vessels contain layers comparable to those of veins. Like veins, the larger lymphatic vessels also contain valves (**Figure 22–3**). The valves are quite close together, and produce

Figure 22–3 **Lymphatic Vessels and Valves.**

Vein

Artery

Lymphatic vessel

Lymphatic valve

From lymphatic capillaries

Toward venous system

Artery

Vein

Lymphatic vessel

a A diagrammatic view of areolar connective tissue containing small blood vessels and a lymphatic vessel. The cross-sectional view emphasizes their structural differences.

Lymphatic valve

Lymphatic vessel

b Like valves in veins, each lymphatic valve consists of a pair of flaps that permit movement of fluid in only one direction.

Lymphatic vessel and valve LM × 63

noticeable bulges. As a result, large lymphatic vessels have a beaded appearance (**Figure 22–3a**). The valves prevent the backflow of lymph within lymphatic vessels, especially in the limbs. Pressures within the lymphatic system are minimal, and the valves are essential to maintaining normal lymph flow toward the thoracic cavity. Contractions of skeletal muscles surrounding the lymphatic vessels aid the flow of lymph.

Lymphatic vessels are often associated with blood vessels (**Figure 22–3a**). Differences in size, general appearance, and branching pattern distinguish lymphatic vessels from arteries and veins. We can see characteristic color differences in living tissues. Most arteries are bright red, veins are dark red (although usually illustrated as blue to distinguish them from arteries), and lymphatic vessels are a pale golden color. In general, a tissue contains many more lymphatic vessels than veins, but the lymphatic vessels are much smaller.

Major Lymph-Collecting Vessels

Two sets of lymphatic vessels collect lymph from the lymphatic capillaries: superficial lymphatics and deep lymphatics.

Superficial lymphatics are located in the subcutaneous layer deep to the skin; in the areolar tissues of the mucous membranes lining the digestive, respiratory, urinary, and reproductive tracts; and in the areolar tissues of the serous membranes lining the pleural, pericardial, and peritoneal cavities. **Deep lymphatics** are larger lymphatic vessels that accompany deep arteries and veins supplying skeletal muscles and other organs of the neck, limbs, and trunk, and the walls of visceral organs.

Superficial and deep lymphatics converge to form even larger vessels called **lymphatic trunks**. The trunks in turn empty into two large collecting vessels: the thoracic duct and the right lymphatic duct. The **thoracic duct** collects lymph from the body inferior to the diaphragm and from the left side of the body superior to the diaphragm. The smaller **right lymphatic duct** collects lymph from the right side of the body superior to the diaphragm (**Figure 22–4a**).

The thoracic duct begins inferior to the diaphragm at the level of vertebra L_2 (**Figure 22–4b**). The base of the thoracic duct is an expanded, saclike chamber called the **cisterna chyli**

Figure 22–4 **The Relationship between the Lymphatic Ducts and the Venous System.** ATLAS: Plates 48a,b

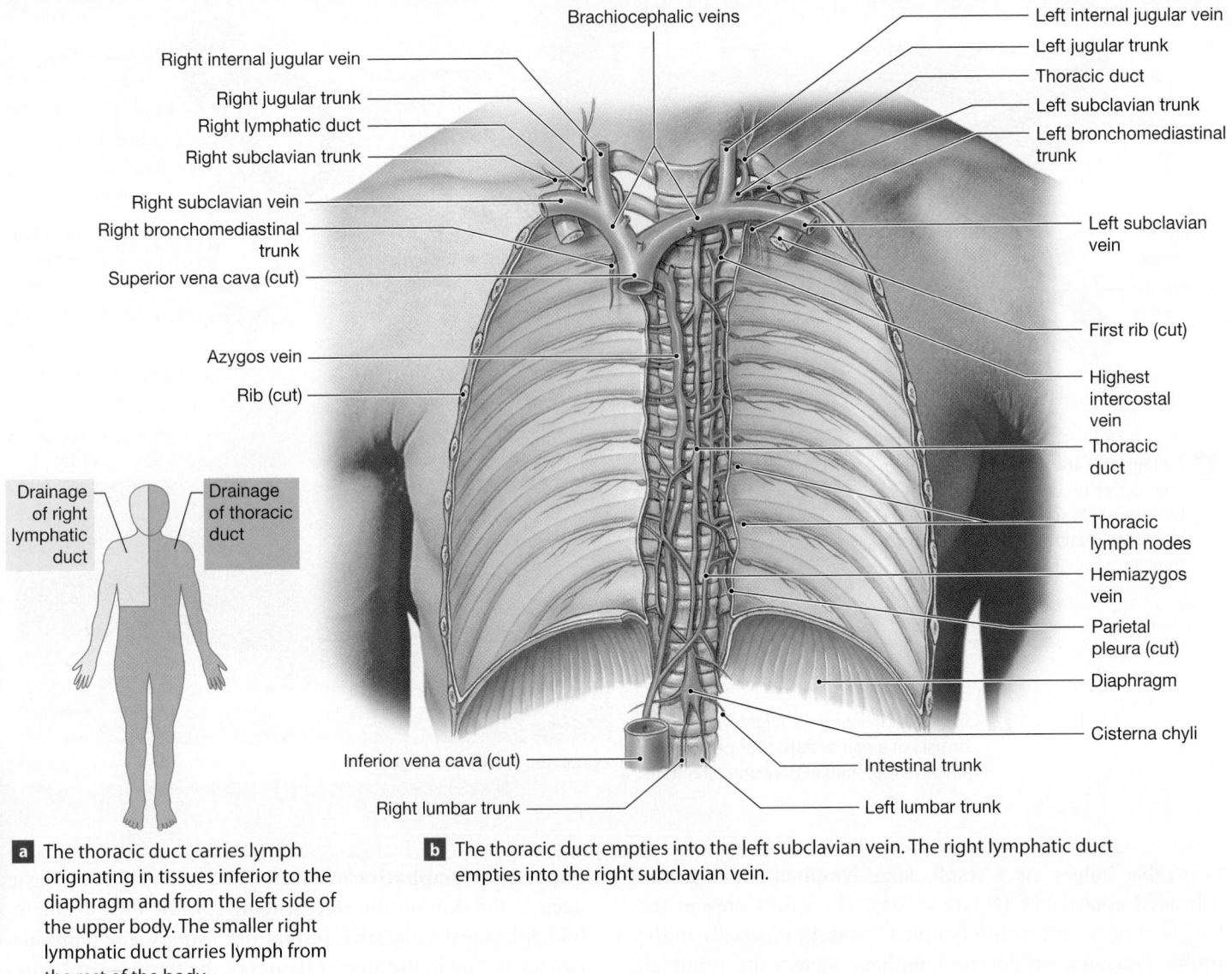

a The thoracic duct carries lymph originating in tissues inferior to the diaphragm and from the left side of the upper body. The smaller right lymphatic duct carries lymph from the rest of the body.

b The thoracic duct empties into the left subclavian vein. The right lymphatic duct empties into the right subclavian vein.

(KĪ-lī; *chylos*, juice). The cisterna chyli receives lymph from the inferior part of the abdomen, the pelvis, and the lower limbs by way of the *right* and *left lumbar trunks* and the *intestinal trunk*.

The inferior segment of the thoracic duct lies anterior to the vertebral column. From the second lumbar vertebra, it passes posterior to the diaphragm alongside the aorta. It then ascends along the left side of the vertebral column to the level of the left clavicle. It collects lymph from the *left bronchomediastinal trunk*, the *left subclavian trunk*, and the *left jugular trunk*, and then empties into the left subclavian vein near the left internal jugular vein (**Figure 22–4b**). In this way, lymph re-enters the venous circulation from the left side of

the head, neck, and thorax, as well as from the entire body inferior to the diaphragm.

The *right lymphatic duct* is formed by the merging of the *right jugular, right subclavian*, and *right bronchomediastinal trunks* in the area near the right clavicle. This duct empties into the right subclavian vein, delivering lymph from the right side of the body superior to the diaphragm.

Blockage of the lymphatic drainage from a limb produces **lymphedema** (limf-e-DĒ-muh). In this condition, interstitial fluids accumulate and the limb gradually becomes swollen and grossly distended. If the condition persists, the connective tissues lose their elasticity and the swelling becomes permanent.

Lymphedema by itself does not pose a major threat to life. The danger comes from the constant risk that an uncontrolled infection will develop in the affected area. Because the interstitial fluids are essentially stagnant, toxins and pathogens can accumulate and overwhelm local defenses without fully activating the immune system.

Lymphocytes

Lymphocytes account for 20–40 percent of circulating leukocytes. However, circulating lymphocytes are only a small fraction of the total lymphocyte population. The body contains some 10^{12} lymphocytes, with a combined weight of more than a kilogram (2.2 lb).

Types of Lymphocytes

Three classes of lymphocytes circulate in blood: (1) **T** (**t**hymus-dependent) **cells**, (2) **B** (**b**one marrow–derived) **cells**, and (3)

NK (**n**atural **k**iller) **cells**. Each type has distinctive biochemical and functional characteristics (**Figure 22–5**).

Most lymphocytes are T cells, and the primary types of T cells include cytotoxic T (T_C) cells, helper T (T_H) cells, suppressor T (T_S) cells, and memory T cells. Cytotoxic T cells are involved in direct cellular attack. These lymphocytes are the primary cells involved in the production of *cell-mediated immunity*, or *cellular immunity*. The interplay between helper T cells and suppressor T cells helps establish and control the sensitivity of the immune response. For this reason, these cells are also known as *regulatory T cells*.

We will examine these T cells throughout this chapter. Other types of T cells also take part in the immune response. For example, *inflammatory T cells* stimulate regional inflammation and local defenses in an injured tissue. *Suppressor/inducer T cells* suppress B cell activity but stimulate other T cells.

Under proper stimulation, B cells differentiate into **plasma cells** that secrete antibodies. These antibodies are soluble

Figure 22–5 **Classes of Lymphocytes.**

proteins, also known as *immunoglobulins*. ↪ p. 683 B cells are responsible for *antibody-mediated immunity*, which is also known as *humoral* ("liquid") *immunity* because antibodies are found in body fluids.

Antibodies bind to specific chemical targets called **antigens**, which stimulate the immune response. Most antigens are pathogens, parts or products of pathogens, or other foreign substances. Antigens are usually proteins, but some lipids, polysaccharides, and nucleic acids are also antigens. The binding of an antibody to its target antigen starts a chain reaction leading to the destruction of the target compound or organism.

Natural killer cells are also known as **large granular lymphocytes**. NK cells provide innate (nonspecific) immunity. They attack foreign cells, cells infected with viruses, and cancer cells that appear in normal tissues. Their continuous "policing" of peripheral tissues has been called *immune surveillance* (Figure 22–5).

Life Span and Circulation of Lymphocytes

The various types of lymphocytes are not evenly distributed in the blood, bone marrow, spleen, thymus, and peripheral lymphoid tissues. The ratio of B cells to T cells varies among tissues and organs. For example, B cells are seldom found in the thymus, but T cells outnumber B cells in blood by a ratio of 8:1.

The lymphocytes in these organs are visitors, not residents. All types of lymphocytes move throughout the body. They wander through tissues and then enter blood vessels or lymphatic vessels for transport.

T cells move quickly. For example, a wandering T cell may spend about 30 minutes in the blood, 5–6 hours in the spleen, and 15–20 hours in a lymph node. B cells, which are responsible for antibody production, move more slowly. A typical B cell spends about 30 hours in a lymph node, a small lymph-filtering structure, before moving on.

Lymphocytes have relatively long life spans. About 80 percent survive 4 years, and some last 20 years or more. Throughout your life, you maintain normal lymphocyte populations by producing new lymphocytes in your red bone marrow and lymphoid tissues.

Lymphocyte Production

In Chapter 19, we discussed *hemopoiesis*—the formation of the cellular elements of blood. ↪ pp. 682, 687, 698 In adults, red blood cell formation, or *erythropoiesis*, is normally confined to red bone marrow. In contrast, lymphocyte production, or **lymphopoiesis** (lim-fō-poy-Ē-sis), involves the red bone marrow, thymus, and peripheral lymphoid tissues (Figure 22–6).

Red bone marrow plays the primary role in maintaining normal lymphocyte populations. Hemocytoblasts divide in the bone marrow of adults to generate the lymphoid stem cells that produce all types of lymphocytes. The red bone marrow produces two distinct populations of lymphoid stem cells.

One group of lymphoid stem cells remains in the red bone marrow and the other group migrates to the thymus (Figure 22–6). Lymphoid stem cells in the red bone marrow divide to produce immature B cells and NK cells. B cell development involves intimate contact with large **stromal** (*stroma*, a bed) **cells** in the red bone marrow. The cytoplasmic extensions of stromal cells contact or even wrap around the developing B cells. Stromal cells produce an immune system hormone, or *cytokine*, called *interleukin-7*. It promotes the differentiation of B cells. (We consider cytokines and their varied effects in a later section.)

As they mature, B cells and NK cells enter the bloodstream and migrate to peripheral tissues (Figure 22–6). Most of the B cells move into lymph nodes, the spleen, or other lymphoid tissues. The NK cells patrol the body, moving through peripheral tissues in search of abnormal cells.

The second group of lymphoid stem cells migrates to the thymus (Figure 22–6). These stem cells and their descendants develop and mature in an environment that is isolated from the general circulation by the **blood–thymus barrier**. Under the influence of thymic hormones, the lymphoid stem cells divide repeatedly, producing the various kinds of T cells. These T cells undergo a selection process to ensure that they will not react to the body's own healthy cells and cellular products. During this process up to 98 percent of developing T cells are deselected and undergo apoptosis. When the development of the surviving T cells nears completion, they re-enter the bloodstream and travel to peripheral tissues, including lymphoid tissues and organs, such as the spleen (Figure 22–6).

Clinical Note

Cancer and the Lymphatic System Metastasizing cancer cells commonly spread within lymphatic vessels. These vessels occur in almost all portions of the body except the central nervous system, and lymphatic capillaries offer little resistance to the passage of cancer cells. As a result, the lymph nodes serve as "way stations" for migrating cancer cells. For this reason, an analysis of lymph nodes can provide information on the spread of cancer cells. Such information helps determine the appropriate therapies. *Lymphomas* are one group of cancers originating in the cells of the lymphatic system.

Figure 22–6   **The Origin and Distribution of Lymphocytes.**

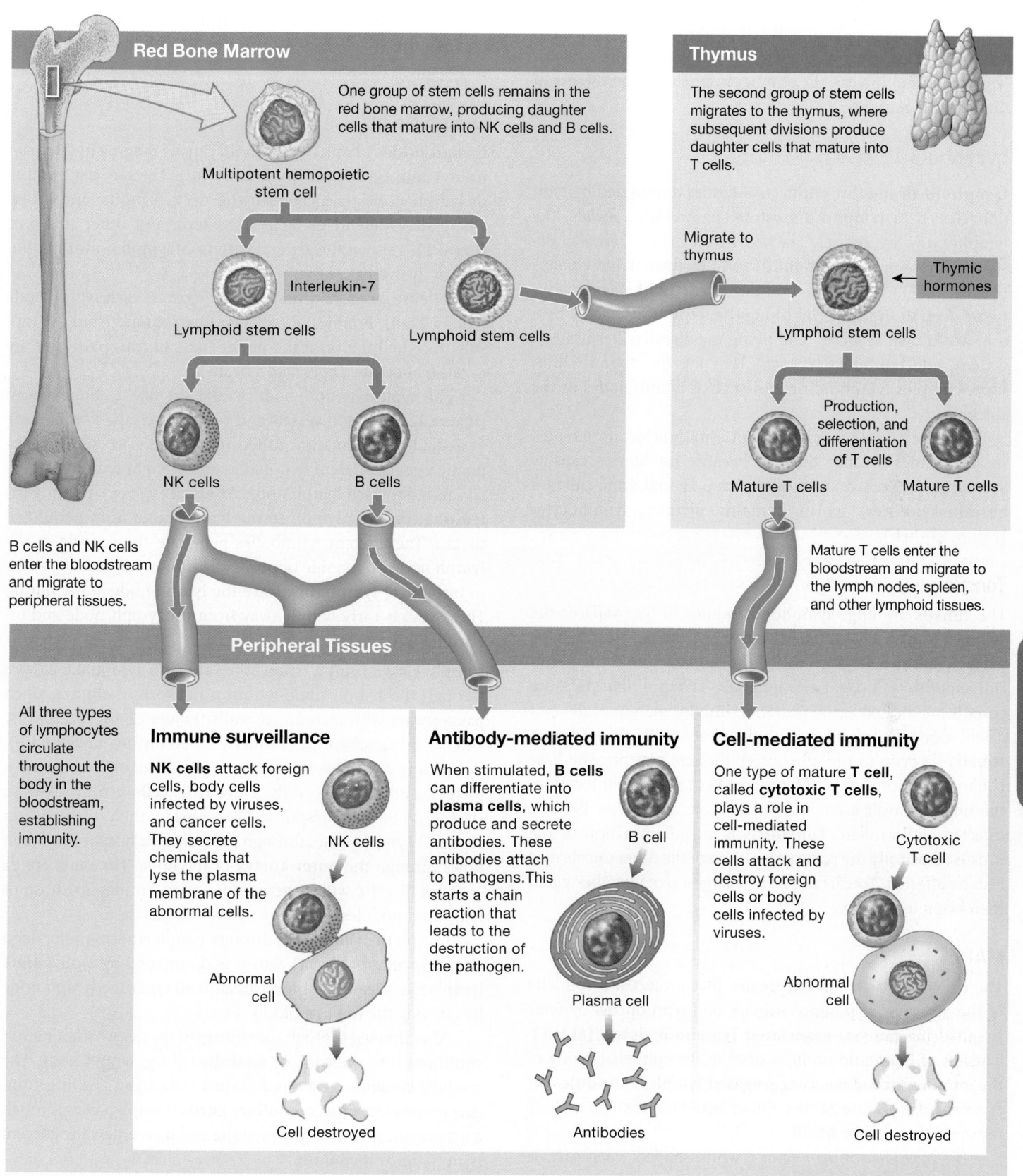

Red Bone Marrow

One group of stem cells remains in the red bone marrow, producing daughter cells that mature into NK cells and B cells.

Multipotent hemopoietic stem cell

Interleukin-7

Lymphoid stem cells

Lymphoid stem cells

NK cells

B cells

B cells and NK cells enter the bloodstream and migrate to peripheral tissues.

Thymus

The second group of stem cells migrates to the thymus, where subsequent divisions produce daughter cells that mature into T cells.

Migrate to thymus

Thymic hormones

Lymphoid stem cells

Production, selection, and differentiation of T cells

Mature T cells

Mature T cells

Mature T cells enter the bloodstream and migrate to the lymph nodes, spleen, and other lymphoid tissues.

Peripheral Tissues

All three types of lymphocytes circulate throughout the body in the bloodstream, establishing immunity.

Immune surveillance

NK cells attack foreign cells, body cells infected by viruses, and cancer cells. They secrete chemicals that lyse the plasma membrane of the abnormal cells.

NK cells

Abnormal cell

Cell destroyed

Antibody-mediated immunity

When stimulated, **B cells** can differentiate into **plasma cells**, which produce and secrete antibodies. These antibodies attach to pathogens. This starts a chain reaction that leads to the destruction of the pathogen.

B cell

Plasma cell

Antibodies

Cell-mediated immunity

One type of mature **T cell**, called **cytotoxic T cells**, plays a role in cell-mediated immunity. These cells attack and destroy foreign cells or body cells infected by viruses.

Cytotoxic T cell

Abnormal cell

Cell destroyed

22

The T cells and B cells that migrate from their sites of origin retain the ability to divide, producing daughter cells of the same type. For example, a dividing B cell produces other B cells, not T cells or NK cells. As we will see, the ability of specific types of lymphocytes to increase in number is crucial to the success of the immune response.

Lymphoid Tissues

Lymphoid tissues are connective tissues dominated by lymphocytes. In a **lymphoid nodule**, or *lymphatic nodule*, the lymphocytes are densely packed in an area of areolar tissue. In many areas, lymphoid nodules form large clusters (**Figure 22–7**). Lymphoid nodules occur in the connective tissue deep to the epithelia lining the respiratory tract, where they are known as *tonsils*, and along the digestive, respiratory, urinary, and reproductive tracts. They are also found within more complex lymphoid organs, such as lymph nodes or the spleen.

A single nodule averages about a millimeter in diameter. Its boundaries are not distinct, because no fibrous capsule surrounds it. Each nodule often has a central zone called a **germinal center**, which contains dividing lymphocytes (**Figure 22–7a,b**).

Tonsils

The **tonsils** are large lymphoid nodules in the walls of the pharynx (**Figure 22–7a**). Most people have five tonsils. A single **pharyngeal tonsil**, often called the *adenoid*, lies in the posterior superior wall of the nasopharynx. Left and right **palatine tonsils** are located at the posterior, inferior margin of the oral cavity, along the boundary with the pharynx. A pair of **lingual tonsils** lie deep to the mucous epithelium covering the base (pharyngeal portion) of the tongue. Because of their location, the lingual tonsils are usually not visible unless they become infected and swollen. **Tonsillitis** is an inflammation of the tonsils (especially the palatines) although the other tonsils may also be affected. Tonsils reach their largest size by puberty and then begin to atrophy.

MALT

The collection of lymphoid tissues that protect the epithelia of the digestive, respiratory, urinary, and reproductive systems is called the **mucosa-associated lymphoid tissue (MALT)**. Clusters of lymphoid nodules deep to the epithelial lining of the intestine are known as **aggregated lymphoid nodules**, or *Peyer's patches* (**Figure 22–7b**). Other examples of MALT include the appendix and the tonsils.

The *appendix*, or *vermiform* ("worm-shaped") *appendix*, is a blind pouch that originates near the junction between the small and large intestines. Its walls contain a mass of fused lymphoid nodules.

Lymphoid Organs

A fibrous connective tissue capsule separates lymphoid organs—the *lymph nodes*, the *thymus*, and the *spleen*—from surrounding tissues.

Lymph Nodes

Lymph nodes are small lymphoid organs ranging in diameter from 1 mm to 25 mm (to about 1 in.). The greatest number of lymph nodes is located in the neck, armpits, and groin, where they defend us against bacteria and other invaders. **Figure 22–1** shows the general pattern of lymph node distribution in the body.

A dense connective tissue capsule covers each lymph node (**Figure 22–8**). Bundles of collagen fibers extend from the capsule into the interior of the node. These fibrous partitions are called **trabeculae** (*trabecula*, a beam).

The typical lymph node is shaped like a kidney bean (**Figure 22–8**). Blood vessels and nerves reach the lymph node at a shallow indentation called the **hilum**. Two sets of lymphatic vessels, afferent lymphatics and efferent lymphatics, are connected to each lymph node. **Afferent** (*afferens*, to bring to) **lymphatics** bring lymph to the lymph node from peripheral tissues. The afferent lymphatics penetrate the capsule of the lymph node on the side opposite the hilum. **Efferent** (*efferens*, to bring out) **lymphatics** leave the lymph node at the hilum. These vessels carry lymph away from the lymph node and toward the venous circulation.

Lymph Flow. Lymph from the afferent lymphatics flows through the lymph node within a network of sinuses, open passageways with incomplete walls (**Figure 22–8**). Lymph first enters a *subcapsular space* (formerly called the *subcapsular sinus*). It contains a meshwork of branching reticular fibers, macrophages, and dendritic cells. **Dendritic cells** are involved in starting an immune response. (We consider their role in a later section.) Lymph passes through the subcapsular space and then flows through the **outer cortex** of the node. The outer cortex contains B cells within germinal centers similar to those of lymphoid nodules.

Lymph then continues through lymph sinuses in the **deep cortex** (*paracortical area*), which is dominated by T cells. Here lymphocytes leave the bloodstream and enter the lymph node by crossing the walls of blood vessels.

After flowing through the sinuses of the deep cortex, lymph continues into the core, or **medulla**, of the lymph node. The medulla contains B cells and plasma cells organized into elongate masses known as **medullary cords**. Lymph passes through a network of sinuses in the medulla and then enters the efferent lymphatics at the hilum.

Lymph Node Function. A lymph node functions like a kitchen water filter. It purifies lymph before it reaches the veins. As lymph flows through a lymph node, at least 99 percent of the antigens in the lymph are removed. Fixed macrophages in the walls of the lymphatic sinuses engulf debris or pathogens in lymph as it flows past. Antigens removed in this way are then processed by the macrophages and "presented" to nearby lymphocytes. Other antigens bind to receptors on the surfaces of dendritic cells, where they can stimulate lymphocyte activity. This process—*antigen presentation*—is generally the first step in the activation of the immune response.

In addition to filtering, lymph nodes function as an early-warning system. Any infection or other abnormality in a peripheral tissue puts antigens into the interstitial fluid, and thus into the lymph leaving the area. These antigens then

Figure 22–7 **Lymphoid Nodules.**

Pharyngeal tonsil

Palate

Palatine tonsil

Lingual tonsil

a The locations of the tonsils

Pharyngeal epithelium

Germinal centers within nodules

Pharyngeal tonsil LM × 40

Mucous membrane of intestinal wall

Intestinal lumen

Germinal center

Aggregated lymphoid nodule in intestinal mucosa

Underlying connective tissue

b Diagrammatic view of aggregated lymphoid nodule

Aggregated lymphoid nodules LM × 20

22

Figure 22–8 **The Structure of a Lymph Node.** ATLAS: Plate 70a

stimulate macrophages and lymphocytes in nearby lymph nodes.

To protect a house against intruders, you might guard all the entrances and exits or place traps by the windows and doors. The distribution of lymphoid tissues and lymph nodes follows such a pattern. The largest lymph nodes are located where peripheral lymphatics connect with the trunk, such as in the groin, the axillae, and the base of the neck. These nodes are often called *lymph glands*. Because lymph is monitored in these nodes, potential problems can be detected and dealt with before they affect the vital organs of the trunk. The mesenteries of the gut also have aggregations of lymph nodes, located near the trachea and passageways leading to the lungs, and in association with the thoracic duct. These lymph nodes protect against pathogens and other antigens within the digestive and respiratory systems.

A minor injury commonly produces a slight enlargement of the nodes along the lymphatic vessels draining the region. This sign is often called "swollen glands." It typically indicates inflammation in peripheral structures. The enlargement generally results from an increase in the number of lymphocytes and phagocytes in the node in response to a minor, localized infection. Chronic or excessive enlargement of lymph nodes is **lymphadenopathy** (lim-fad-e-NOP-a-thē). This condition may occur in response to bacterial or viral infections, endocrine disorders, or cancer.

The Thymus

The **thymus** is a pink, grainy organ located in the mediastinum, generally just posterior to the sternum (**Figure 22–9a,b**). In newborn infants and young children, the thymus is relatively large. It commonly extends from the base of the neck to the superior border of the heart. The thymus reaches its greatest size relative to body size in the first year or two after birth. (The organ continues to increase in mass throughout childhood, but the body as a whole grows even faster, so the size of the thymus relative to that of the other organs in the mediastinum gradually decreases.)

The thymus reaches its maximum absolute size, at a weight of about 40 g (1.4 oz), just before puberty. After puberty, it gradually diminishes in size and becomes increasingly fibrous, a process called *involution*. By the time a person reaches age 50, the thymus may weigh less than 12 g (0.3 oz). The gradual decrease in the size of functional tissue and secretory abilities of the thymus may make elderly people more susceptible to disease.

The capsule that covers the thymus divides it into two **thymic lobes** (**Figure 22–9b**). Fibrous partitions called **septa** (singular, *septum*) originate at the capsule and divide the lobes into **lobules** averaging 2 mm in diameter (**Figure 22–9b,c**). Each lobule consists of a densely packed outer thymic **cortex** and a paler, central **medulla**. The cortex contains actively dividing lymphocytes (T cells). As the T cells mature, they migrate into the medulla. After about three weeks, these T cells leave the thymus by entering one of the medullary blood vessels.

Lymphocytes in the cortex are arranged in clusters that are completely surrounded by **thymic epithelial cells (TECs)**. Thymic epithelial cells regulate T cell development and function.

Figure 22–9 The Thymus. ATLAS: Plate 47a

a The appearance and position of the thymus in relation to other organs in the chest.

b Anatomical landmarks on the thymus.

A thymic corpuscle LM × 550

d Higher magnification reveals the unusual structure of thymic corpuscles. The small cells are lymphocytes in various stages of development.

The thymus gland LM × 50

c Fibrous septa divide the tissue of the thymus into lobules resembling interconnected lymphoid nodules.

22

Epithelial cells also encircle the blood vessels of the cortex. These cells maintain the blood–thymus barrier.

Maturing T cells leave the cortex and enter the medulla of the thymus. The medulla has no blood–thymus barrier. The thymic epithelial cells in the medulla cluster together in concentric layers, forming distinctive structures known as **thymic (Hassall's) corpuscles** (Figure 22–9d). Despite their imposing appearance, the function of thymic corpuscles remains unknown. T cells in the medulla can enter or leave the bloodstream across the walls of blood vessels in this region or within one of the efferent lymphatics that collect lymph from the thymus.

The thymus produces several hormones that are important to the development and maintenance of normal immunological defenses. *Thymosin* (THĪ-mō-sin) is the name originally given to an extract from the thymus that promotes the development and maturation of lymphocytes. This extract actually contains several complementary hormones. They include *thymosin-a*, *thymosin-b*, *thymosin V*, *thymopoietin*, *thymulin*, and others. The plural term, *thymosins*, is now sometimes used to refer to all thymic hormones.

The Spleen

The adult **spleen** contains the largest collection of lymphoid tissue in the body. In essence, the spleen performs the same functions for blood that lymph nodes perform for lymph. Functions of the spleen can be summarized as (1) removing abnormal blood cells and other blood components by phagocytosis, (2) storing iron recycled from red blood cells, and (3) initiating immune responses by B cells and T cells in response to antigens in circulating blood.

Anatomy of the Spleen. The spleen is about 12 cm (5 in.) long and weighs, on average, nearly 160 g (5.6 oz). In gross dissection, the spleen is deep red, due to the blood it contains. The spleen lies along the curving lateral border of the stomach, extending between the 9th and 11th ribs on the left side. It is attached to the lateral border of the stomach by the **gastrosplenic ligament**, a broad band of mesentery (Figure 22–10a).

The spleen has a soft texture, so its shape primarily reflects the shapes of the structures around it. The spleen is in contact with the muscular diaphragm, the stomach, and the left kidney. The *diaphragmatic surface* is smooth and convex, conforming to the shape of the diaphragm and body wall. The *visceral surface* contains indentations that conform to the shape of the stomach (the *gastric area*) and the kidney (the *renal area*) (Figure 22–10b). Splenic blood vessels (the *splenic artery* and *splenic vein*) and lymphatic vessels communicate with the spleen on the visceral surface at the **hilum**, a groove marking the border between the gastric and renal areas.

Histology of the Spleen. The spleen is surrounded by a capsule containing collagen and elastic fibers.[1] The cellular components within make up the **pulp** of the spleen (Figure 22–10c). **Red pulp** contains large quantities of red blood cells, and **white pulp** resembles lymphoid nodules.

The splenic artery enters at the hilum and branches to produce a number of arteries that radiate outward toward the capsule. These **trabecular arteries** in turn branch extensively, and their finer branches are surrounded by areas of white pulp. Capillaries then discharge the blood into the red pulp.

The cell population of the red pulp includes all the normal components of circulating blood, plus fixed and free macrophages. The structural framework of the red pulp consists of a network of reticular fibers. The blood passes through this meshwork and enters large sinusoids, also lined by fixed macrophages. The sinusoids empty into small veins, which ultimately collect into **trabecular veins** that continue toward the hilum.

This circulatory arrangement gives the phagocytes in the spleen an opportunity to identify and engulf damaged or infected cells in circulating blood. Macrophages are scattered throughout the red pulp, and the area surrounding the white pulp has a high concentration of lymphocytes and dendritic cells. For this reason, any microorganism or other antigen in the blood quickly triggers an immune response.

The spleen tears so easily that a seemingly minor hit to the left side of the abdomen can rupture the capsule. The result is serious internal bleeding and eventual circulatory shock. Such an injury is a known risk of contact sports (such as football, hockey, and rugby) and of more individual athletic activities, such as skiing and sledding.

Because the spleen is so fragile, it is very difficult to repair surgically. (Sutures typically tear out before they have been tensed enough to stop the bleeding.) A severely ruptured spleen is removed, a process called a **splenectomy** (splē-NEK-tō-mē). A person can survive without a spleen but lives with an increased risk of bacterial infection (particularly involving pneumococcal bacteria).

The Lymphatic System and Body Defenses

As we have noted, the human body has multiple defense mechanisms. Together they provide *resistance*—the natural or acquired ability to maintain immunity. We can sort body defenses into two general categories:

1. **Innate (nonspecific) immunity** does not distinguish one type of threat from another. Their response is the same, regardless of the type of invader. These innate defenses

---

[1]The spleens of dogs, cats, and other mammals of the order *Carnivora* have extensive layers of smooth muscle that can contract to eject blood into the bloodstream. The human spleen lacks those muscle layers and cannot contract.

Figure 22–10   **The Spleen.**   ATLAS: Plates 49e; 55; 56e,f; 57a,b

a A transverse section through the trunk, showing the typical position of the spleen projecting into the peritoneal cavity. The shape of the spleen roughly conforms to the shapes of adjacent organs.

b A posterior view of the surface of an intact spleen, showing major anatomical landmarks.

c Spleen histology. White pulp is dominated by lymphocytes; it appears purple because the nuclei of lymphocytes stain very darkly. Red pulp contains a large number of red blood cells.

The spleen   LM × 50

---

are present at birth. They include *physical barriers, phagocytic cells, immune surveillance, interferons, complement, inflammation,* and *fever.* They provide a defensive capability known as **nonspecific resistance**.

2. **Adaptive (specific) immunity** protects against particular threats. For example, an adaptive defense may protect against one type of bacterium, but not other bacteria and viruses. Many specific defenses develop after birth as a result of accidental or deliberate exposure to environmental hazards. *Adaptive defenses depend on the activities of specific lymphocytes.* B cells and T cells are part of our adaptive defenses, which provides protection known as **specific resistance**.

Innate and adaptive defenses work together. Both must function normally to provide adequate resistance to infection and disease.

CHECKPOINT

3. List the components of the lymphatic system.

4. How would blockage of the thoracic duct affect lymph circulation?

5. If the thymus failed to produce thymic hormones, which population of lymphocytes would be affected?

6. Why do lymph nodes enlarge during some infections?

See the blue Answers tab at the back of the book.

22-3 | Innate (nonspecific) defenses do not discriminate between potential threats and respond the same regardless of the invader

Learning Outcome List the body's innate (nonspecific) defenses, and describe the components, mechanisms, and functions of each.

Innate (nonspecific) defenses prevent the approach, deny the entry, or limit the spread of microorganisms or other environmental hazards. Seven major categories of innate defenses are summarized in **Figure 22–11.**

Physical Barriers

To cause trouble, an antigenic substance or a pathogen must enter body tissues. In other words, it must cross an epithelium—either at the skin or across a mucous membrane. The epithelial covering of the skin has multiple layers, a keratin coating, and a network of desmosomes that lock adjacent cells together. ↺ pp. 142–143, 179–181 These barriers are very effective in protecting underlying tissues. Even along the more delicate internal passageways of the respiratory, digestive, urinary, and reproductive tracts, epithelial cells are tied together by tight junctions. These cells generally are supported by a dense and fibrous basement membrane.

In addition, specialized accessory structures and secretions protect most epithelia. The hairs on most areas of your body provide some protection against mechanical abrasion, especially on the scalp. They often prevent hazardous materials or insects from contacting your skin. The epidermal surface also receives the secretions of sebaceous and sweat glands. These secretions flush the surface and wash away microorganisms and chemicals. Such secretions may also contain chemicals that kill bacteria, destructive enzymes called *lysozymes*, and antibodies.

The epithelia lining the digestive, respiratory, urinary, and reproductive tracts are more delicate, but they are equally well defended. Mucus bathes most surfaces of your digestive tract. Your stomach contains a powerful acid that can destroy many pathogens. Mucus moves across the respiratory tract lining, urine flushes the urinary passages, and glandular secretions do the same for the reproductive tract. Special enzymes, antibodies, and an acidic pH add to the effectiveness of these secretions.

Phagocytes

Phagocytes serve as janitors and police in peripheral tissues. They remove cellular debris and respond to invasion by foreign substances or pathogens. Phagocytes are the "first line of cellular defense" against pathogenic invasion. Many phagocytes attack and remove microorganisms even before lymphocytes

detect them. The human body has two general classes of phagocytic cells: *microphages* and *macrophages*.

Microphages

Microphages are the neutrophils and eosinophils that normally circulate in the blood. These phagocytic cells leave the bloodstream and enter peripheral tissues that have been subjected to injury or infection. As noted in Chapter 19, neutrophils are abundant, mobile, and quick to phagocytize cellular debris or invading bacteria. ↺ p. 696 Eosinophils are less abundant. They target foreign substances or pathogens that have been coated with antibodies.

Macrophages

Macrophages are large, actively phagocytic cells. Your body contains several types of macrophages. Most are derived from the monocytes of the circulating blood. Typically, macrophages are either fixed in position or freely mobile. As a result they are usually classified as fixed macrophages or free macrophages, but the distinction is not absolute. During an infection, fixed macrophages may lose their attachments and begin roaming around the damaged tissue.

No organs or tissues are dominated by phagocytes, but almost every tissue in the body shelters resident or visiting macrophages. This relatively diffuse collection of phagocytes has been called the **monocyte–macrophage system**, or the *reticuloendothelial system*.

How does an activated macrophage respond to a pathogen? It may:

- engulf a pathogen or other foreign object and destroy it with lysosomal enzymes.

- bind to or remove a pathogen from the interstitial fluid, but be unable to destroy the invader until assisted by other cells.

- destroy its target by releasing toxic chemicals, such as *tumor necrosis factor*, nitric oxide, or hydrogen peroxide, into the interstitial fluid.

We consider these responses further in a later section.

Fixed Macrophages. **Fixed macrophages**, or *histiocytes*, reside in specific tissues and organs. These cells are normally incapable of movement, so their targets must diffuse or otherwise move through the surrounding tissue until they are within range. Fixed macrophages are scattered among connective tissues, usually in close association with collagen or reticular fibers. They are found in the papillary and reticular layers of the dermis, in the subarachnoid space of the meninges, and in bone marrow. In some organs, the fixed macrophages have special names. **Microglia** are macrophages in the central nervous system, and **Kupffer cells** are macrophages located in and around the liver sinusoids.

Figure 22–11   **Innate Defenses.** Innate (nonspecific) defenses deny pathogens access to the body or destroy them without distinguishing among specific types.

Innate Defenses	
Physical barriers keep hazardous organisms and materials outside the body.	Duct of eccrine sweat gland — Hair — Secretions — Epithelium
Phagocytes engulf pathogens and cell debris.	Fixed macrophage Neutrophil Free macrophage Eosinophil Monocyte
Immune surveillance is the destruction of abnormal cells by NK cells in peripheral tissues.	Natural killer cell — Lysed abnormal cell
Interferons are chemical messengers that coordinate the defenses against viral infections.	Interferons released by activated lymphocytes, macrophages, or virus-infected cells
Complement is a system of circulating proteins that assist antibodies in the destruction of pathogens.	Complement — Lysed pathogen
Inflammation is a localized, tissue-level response that tends to limit the spread of an injury or infection.	Mast cell — 1. Blood flow increased / 2. Phagocytes activated / 3. Capillary permeability increased / 4. Complement activated / 5. Clotting reaction walls off region / 6. Regional temperature increased / 7. Adaptive defenses activated
Fever is an elevation of body temperature that accelerates tissue metabolism and body defenses.	Body temperature rises above 37.2 °C in response to pyrogens

Free Macrophages. **Free macrophages**, or *wandering macrophages*, travel throughout the body. They arrive at the site of an injury by migrating through adjacent tissues or by leaving the circulating blood. Some tissues contain free macrophages. For example, **alveolar macrophages**, also known as *phagocytic dust cells*, monitor the exchange surfaces of the lungs for respiratory pathogens.

22

Movement and Phagocytosis

Free macrophages and microphages function in similar ways:

- Both can move through capillary walls by squeezing between adjacent endothelial cells. This process is known as *emigration,* or *diapedesis.* ⤺ p. 693 The endothelial cells in an injured area develop membrane "markers" that signal passing blood cells that something is wrong. The phagocytic cells then attach to the endothelial lining and migrate into the surrounding tissues.

- Both may be attracted to or repelled by chemicals in the surrounding fluids, a phenomenon called **chemotaxis**. They are particularly sensitive to cytokines released by other body cells and to chemicals released by pathogens.

- For both, phagocytosis begins with **adhesion**, the attachment of the phagocyte to its target. In adhesion, receptors on the plasma membrane of the phagocyte bind to the surface of the target. Adhesion is followed by the formation of a vesicle containing the bound target (**Figure 3–21**, p. 123). The contents of the vesicle are digested once the vesicle fuses with lysosomes or peroxisomes.

All phagocytic cells function in much the same way, although their targets may differ. The life span of an actively phagocytic cell can be rather brief. For example, most neutrophils die before they have engulfed more than 25 bacteria, and during an infection a neutrophil may attack that many in an hour.

TIPS&**TRICKS**

Membrane markers and chemotaxis are like putting up the flag on your mailbox: They signal the need for action.

Immune Surveillance

The immune system generally ignores the body's own cells unless they become abnormal in some way. Natural killer (NK) cells are lymphocytes responsible for recognizing and destroying abnormal cells when they appear in peripheral tissues. The constant monitoring of normal tissues by NK cells is called **immune surveillance**.

The plasma membrane of an abnormal cell generally contains antigens that are not found on the membranes of normal cells. NK cells recognize an abnormal cell by detecting those antigens. NK cells are much less selective about their targets than are other lymphocytes: They respond to a *variety* of abnormal antigens that may appear anywhere on a plasma membrane. They also attack *any* membrane containing abnormal antigens. As a result, NK cells are highly versatile. A single NK cell can attack bacteria in the interstitial fluid, body cells infected with viruses, or cancer cells.

NK cells respond much more rapidly than T cells or B cells. The activation of T cells and B cells involves a complex and time-consuming sequence of events. NK cells respond immediately on contact with an abnormal cell.

NK Cell Activation

Activated NK cells react in a predictable way (**Figure 22–12**):

1 *Recognition and adhesion* If a cell has unusual components in its plasma membrane, an NK cell recognizes that cell as abnormal. This recognition activates the NK cell, which then adheres to its target cell.

2 *Realignment of Golgi apparatus* The Golgi apparatus moves around the nucleus until the maturing face points directly toward the abnormal cell. The process might be compared to the rotation of a tank turret to point the cannon toward the enemy. The Golgi apparatus then produces a flood of secretory vesicles that contain proteins called **perforins**. These vesicles move through the cytosol toward the cell surface.

3 *Secretion of perforin* The perforins are released at the NK cell surface by exocytosis and diffuse across the narrow gap between the NK cell and its target.

4 *Lysis of abnormal cell* On reaching the opposing plasma membrane, perforin molecules interact with one another and with the membrane to create a network of pores in it. These pores are large enough to allow the free passage of ions, proteins, and other intracellular materials. As a result, the target cell can no longer maintain its internal environment, and it quickly disintegrates.

TIPS&**TRICKS**

*Perfor*in gets its name because it *perfor*ates the target cell.

Why doesn't perforin affect the plasma membrane of the NK cell? The answer is not clear, but NK cell membranes contain a second protein, called *protectin*. It may bind and inactivate perforin.

NK cells attack cancer cells and body cells infected with viruses. Cancer cells probably appear throughout life, but their plasma membranes generally contain unusual proteins called **tumor-specific antigens**, which NK cells recognize as abnormal. The NK cells then destroy the cancer cells, preserving tissue integrity. Unfortunately, some cancer cells avoid detection, perhaps because they lack tumor-specific antigens or because these antigens are covered in some way. Other cancer cells are able to destroy the NK cells that detect them. This process of avoiding detection or neutralizing body defenses is called **immunological escape**. Once immunological escape has occurred, cancer cells can multiply and spread without interference by NK cells.

In viral infections, the viruses replicate inside cells, beyond the reach of circulating antibodies. However, infected cells incorporate viral antigens into their plasma membranes,

Figure 22–12   **How Natural Killer Cells Kill Cellular Targets.**

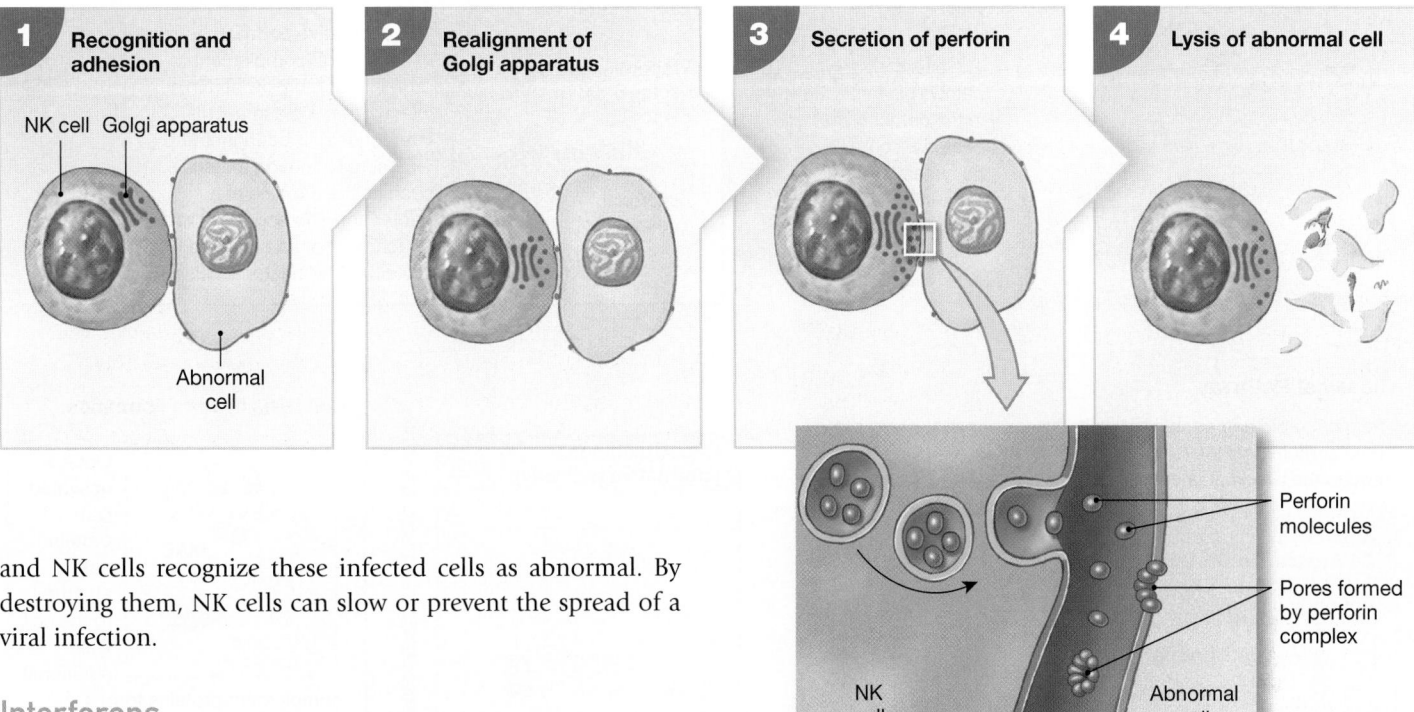

and NK cells recognize these infected cells as abnormal. By destroying them, NK cells can slow or prevent the spread of a viral infection.

Interferons

Interferons (in-ter-FĒR-onz) **(IFNs)** are small proteins released by activated lymphocytes and macrophages, and by tissue cells infected with viruses. An interferon binds to surface receptors on the membrane of a normal cell and, by second messengers, triggers the production of **antiviral proteins** in the cytoplasm. Antiviral proteins do not prevent viruses from entering the cell. Instead, they interfere with viral replication inside the cell. In addition to slowing the spread of viral infections, IFNs stimulate the activities of macrophages and NK cells.

At least three types of interferons exist: **interferon alpha** (α), **interferon beta** (β), and **interferon gamma** (γ) (Figure 22–13). Most cells other than lymphocytes and macrophages respond to viral infection by secreting interferon beta.

Interferons are examples of **cytokines** (SĪ-tō-kīnz), which are chemical messengers that tissue cells release to coordinate local activities. Most cells produce cytokines only for paracrine communication—that is, cell-to-cell communication within one tissue. However, defense cells release cytokines that also act as hormones, affecting cells and tissues throughout the body. We discuss their role in regulating adaptive (specific) defenses in a later section.

Complement System

Plasma contains over 30 special **complement (C) proteins** that form the **complement system**. The term *complement* refers to the fact that this system complements the action of antibodies.

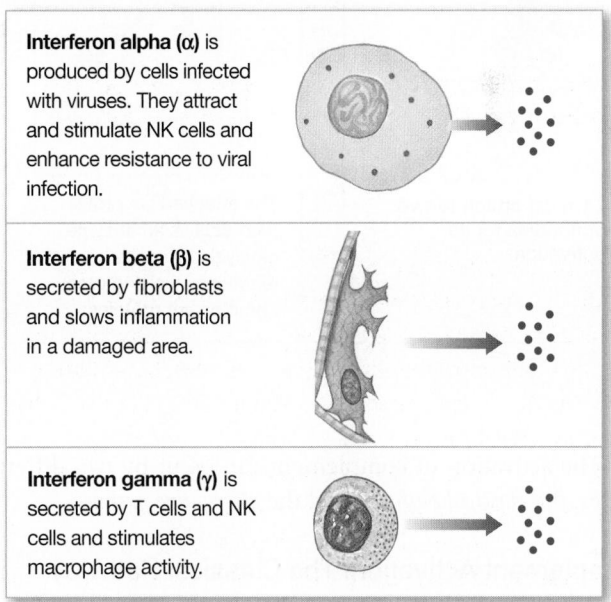

Figure 22–13   **Interferons.**

Interferon alpha (α) is produced by cells infected with viruses. They attract and stimulate NK cells and enhance resistance to viral infection.

Interferon beta (β) is secreted by fibroblasts and slows inflammation in a damaged area.

Interferon gamma (γ) is secreted by T cells and NK cells and stimulates macrophage activity.

The complement proteins interact with one another in chain reactions, or *cascades*, reminiscent of those of the clotting system. **Figure 22–14** provides an overview of the complement system.

22

Figure 22–14 Pathways of Complement Activation.

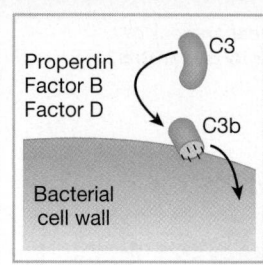

The alternative pathway is important in the defense against bacteria, some parasites, and virus-infected cells.

Alternative Pathway

Properdin
Factor B
Factor D

C3

C3b

Bacterial
cell wall

The alternative pathway begins when several complement proteins, notably properdin, interact in the plasma. This interaction can be triggered by exposure to foreign materials, such as the capsule of a bacterium. The end result is the attachment of an activated C3b protein to the bacterial cell wall.

Classical Pathway

The most rapid and effective activation of the complement system occurs through the classical pathway.

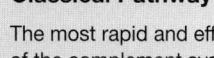

Antibody Binding and C1 Attachment

Antibody binding

Antibodies

Bacterial
cell wall

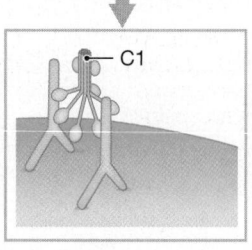

C1

C1 must attach to two antibodies for its activation.

Activation and Cascade

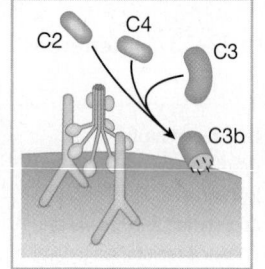

C2   C4

C3

C3b

The attached C1 protein then acts as an enzyme, catalyzing a series of reactions involving other complement proteins.

C3b Attachment (alternative pathway)

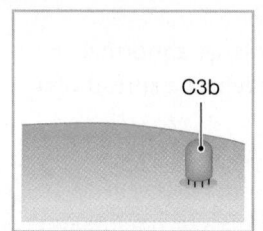

C3b

C3b Attachment (classical pathway)

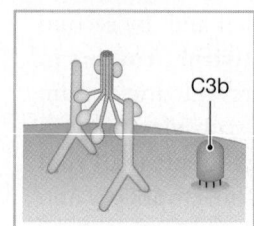

C3b

The classical pathway ends with the conversion of an inactive C3 to an activated C3b that attaches to the cell wall.

Cell Lysis by Pore Formation

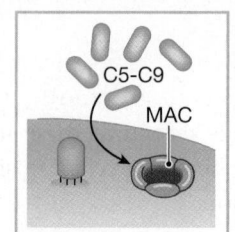

C5-C9

MAC

Once an activated C3b protein has attached to the cell wall, additional complement proteins form a **membrane attack complex (MAC)** in the membrane that destroys the integrity of the target cell.

Multiple pores
in bacterium

Cell lysis

Enhanced Phagocytosis

A coating of complement proteins and antibodies both attracts phagocytes and makes the target cell easier to engulf. This enhancement of phagocytosis, a process called **opsonization**, occurs because macrophage membranes contain receptors that detect and bind to complement proteins and bound antibodies.

Histamine Release

Release of histamine by mast cells and basophils increases the degree of local inflammation and accelerates blood flow to the region.

The activation of complement can occur by two different routes: the *classical pathway* and the *alternative pathway*.

Complement Activation: The Classical Pathway and the Alternative Pathway

The **classical pathway** activates the complement system most rapidly and effectively. It begins with the binding of complement protein C1 to two nearby antibodies already attached to its specific antigen, such as a bacterial cell wall. In the absence of antibody molecules, the **alternative pathway**, or *properdin pathway*, activates the complement system. The alternative pathway is slower and less effective than the classical pathway. It begins when several complement proteins—including **properdin** (*factor P*), *factor B*, and *factor D*—interact in the plasma. Exposure to foreign materials, such as the cell wall of a bacterium, can trigger this interaction. Like the classical pathway, the alternative pathway ends with the conversion of inactive C3 protein to the activated C3b protein. Complement activation brings about the following effects: pore formation, enhanced phagocytosis, and histamine release. Figure 22–14 shows the pathways of complement activation and its effects.

Inflammation

Inflammation, or the *inflammatory response*, is a localized tissue response to injury. ⊃ p. 170 Inflammation produces local redness *(rubor)*, swelling *(tumor)*, heat *(calor)*, and pain *(dolor)*. These are known as the so-called *cardinal signs and symptoms of inflammation*. Sometimes a fifth is included: lost function *(functio laesa)*.

Many stimuli can produce inflammation. They include impact, abrasion, distortion, chemical irritation, infection by pathogens, and extreme temperatures (hot or cold). Each of these stimuli kills cells, damages connective tissue fibers, or injures the tissue in some other way. The changes alter the chemical composition of the interstitial fluid. Damaged cells release prostaglandins, proteins, and potassium ions. The injury itself may have introduced foreign proteins or pathogens. These changes in the interstitial environment trigger the complex process of inflammation.

Inflammation has several effects:

- The injury is temporarily repaired, and additional pathogens are prevented from entering the wound.
- The spread of pathogens away from the injury is slowed.
- Local, regional, and systemic defenses are mobilized to overcome the pathogens and facilitate permanent repairs. This repair process is called *regeneration*.

The Response to Injury

Mast cells play a pivotal role in the inflammatory response. Figure 22–15 summarizes the events of inflammation in the skin. Comparable events take place in almost any tissue subjected to physical damage or infection.

When stimulated by mechanical stress or chemical changes in the local environment, mast cells release histamine, heparin, prostaglandins, and other chemicals into interstitial fluid. The histamine makes capillaries more permeable and speeds up blood flow through the area. The combination of abnormal tissue conditions and chemicals released by mast cells stimulates local sensory neurons, producing sensations of pain. The person then may take steps to limit the damage, such as removing a splinter or cleaning a wound.

The increased blood flow reddens the area and raises the local temperature. These changes increase the rate of enzymatic reactions and accelerate the activity of phagocytes. The rise in

Figure 22–15   **Inflammation and the Steps in Tissue Repair.**

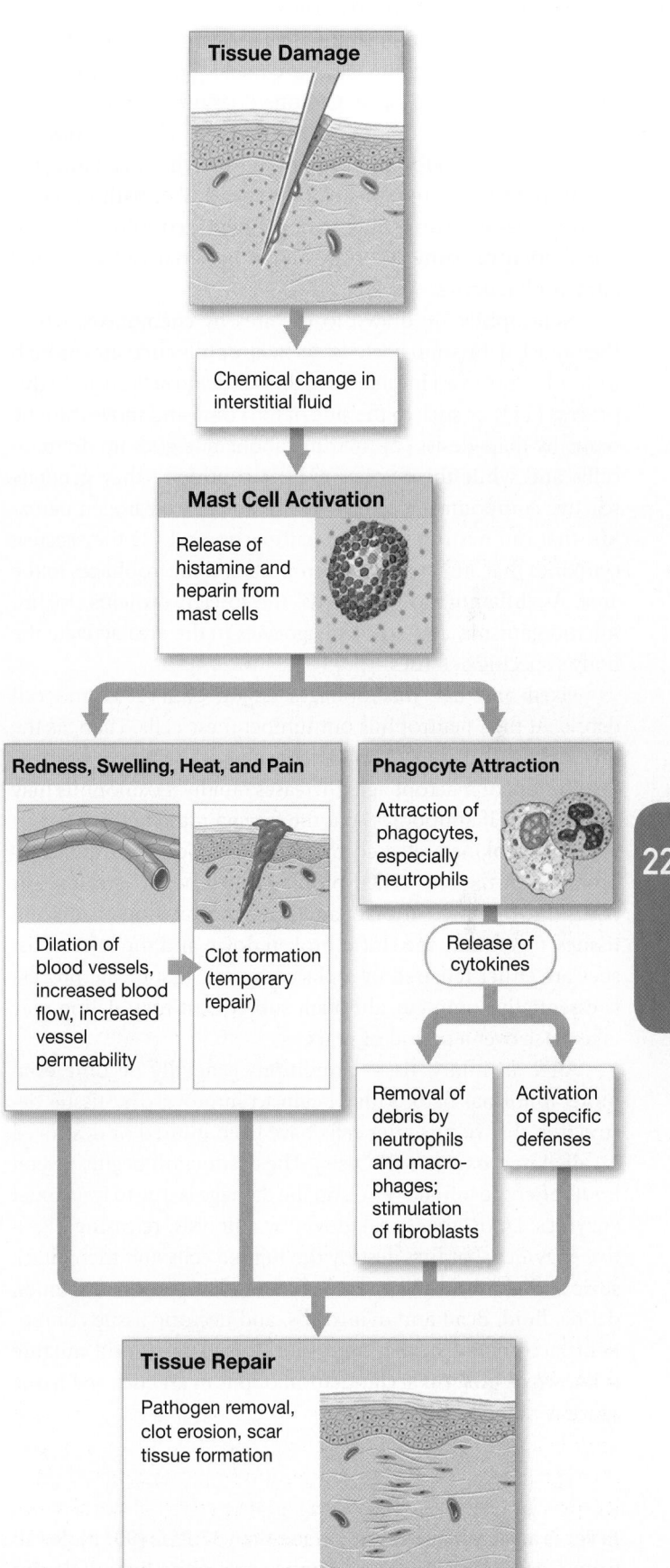

temperature may also denature foreign proteins or vital enzymes of invading microorganisms.

Because vessel permeability has increased, clotting factors and complement proteins can leave the bloodstream and enter the injured or infected area. Clotting does not take place at the actual site of injury, due to the presence of heparin. However, a clot soon forms around the damaged area, both isolating the region and slowing the spread of the chemical or pathogen into healthy tissues. Meanwhile, complement activation through the alternative pathway breaks down bacterial cell walls and attracts phagocytes.

Neutrophils are drawn to the area by chemotaxis, where they attack debris and bacteria. As neutrophils circulate through a blood vessel in an injured area, they undergo *activation*. In this process (1) they stick to the side of the vessel and move into the tissue by diapedesis; (2) their metabolic rate goes up dramatically, and while this *respiratory burst* continues, they generate reactive compounds, such as nitric oxide and hydrogen peroxide, that can destroy engulfed pathogens; and (3) they secrete cytokines that attract other neutrophils and macrophages to the area. As inflammation proceeds, the foreign proteins, toxins, microorganisms, and active phagocytes in the area activate the body's specific defenses.

Fixed and free macrophages engulf pathogens and cell debris. At first, neutrophils outnumber these cells. Then, as the macrophages and neutrophils continue to secrete cytokines, the number of macrophages increases rapidly. Eosinophils may get involved if antibodies coat the foreign materials.

The cytokines released by active phagocytes stimulate fibroblasts in the area. The fibroblasts then begin forming scar tissue that reinforces the clot and slows the invasion of adjacent tissues. Over time, the clot is broken down and the injured tissues are either repaired or replaced by scar tissue. The process is essentially complete, although subsequent remodeling may take place over a period of years.

After an injury, tissue conditions generally become even more abnormal before they begin to improve. The tissue destruction that occurs after cells have been injured or destroyed is called **necrosis** (ne-KRŌ-sis). The destruction begins several hours after the initial event, and the damage is due to lysosomal enzymes. Lysosomes break down by autolysis, releasing digestive enzymes that first destroy the injured cells and then attack surrounding tissues. ⮌ p. 102 As local inflammation continues, debris, fluid, dead and dying cells, and necrotic tissue components accumulate at the injury site. This viscous fluid mixture is known as **pus**. An accumulation of pus in an enclosed tissue space is called an **abscess**.

Fever

Fever is a body temperature greater than 37.2°C (99°F). Recall from Chapter 14 that the preoptic area of the hypothalamus contains the temperature-regulating center. ⮌ p. 506 Circulating proteins called **pyrogens** (PĪ-rō-jenz; *pyro-*, fever or heat + *-gen*, substance) can reset this thermostat and raise body temperature.

A variety of stimuli either act as pyrogens themselves or stimulate macrophages to release pyrogens. These stimuli include pathogens, bacterial toxins, and antigen–antibody complexes. Active macrophages release a cytokine called **tumor necrosis factor (TNF)**, which directly causes fever or stimulates the release of **endogenous pyrogen**, or **interleukin-1** (in-ter-LOO-kin), abbreviated **IL-1**.

Within limits, a fever can be beneficial. High body temperatures may inhibit some viruses and bacteria. The most likely beneficial effect is on body metabolism. For each 1°C rise in body temperature, metabolic rate increases by 10 percent. Cells can move faster, and enzymatic reactions take place more quickly. As a result, tissue defenses can be mobilized more rapidly and the repair process speeds up.

CHECKPOINT

7. List the body's innate (nonspecific) defenses.

8. What types of cells would be affected by a decrease in the number of monocyte-forming cells in red bone marrow?

9. A rise in the level of interferon in the body suggests what kind of infection?

10. What effects do pyrogens have in the body?

See the blue Answers tab at the back of the book.

22-4 | Adaptive (specific) defenses respond to individual threats and are either cell-mediated or antibody-mediated

Learning Outcome Define adaptive (specific) defenses, identify the forms and properties of immunity, and distinguish between cell-mediated (cellular) immunity and antibody-mediated (humoral) immunity.

Adaptive (specific) defenses result from the coordinated activities of T cells and B cells. These cells respond to the presence of specific antigens. In general, T cells bring about **cell-mediated immunity**, or *cellular immunity*, which defends against abnormal cells and pathogens inside cells. B cells provide **antibody-mediated immunity**, or *humoral immunity*, which defends against antigens and pathogens in body fluids.

Both kinds of immunity are important because they come into play under different circumstances. Activated T cells do not respond to antigens in solution, and antibodies (produced by activated B cells) cannot cross plasma membranes. Moreover, helper T cells play a crucial role in antibody-mediated immunity by stimulating the activity of B cells.

Figure 22–16   **Forms of Immunity.**

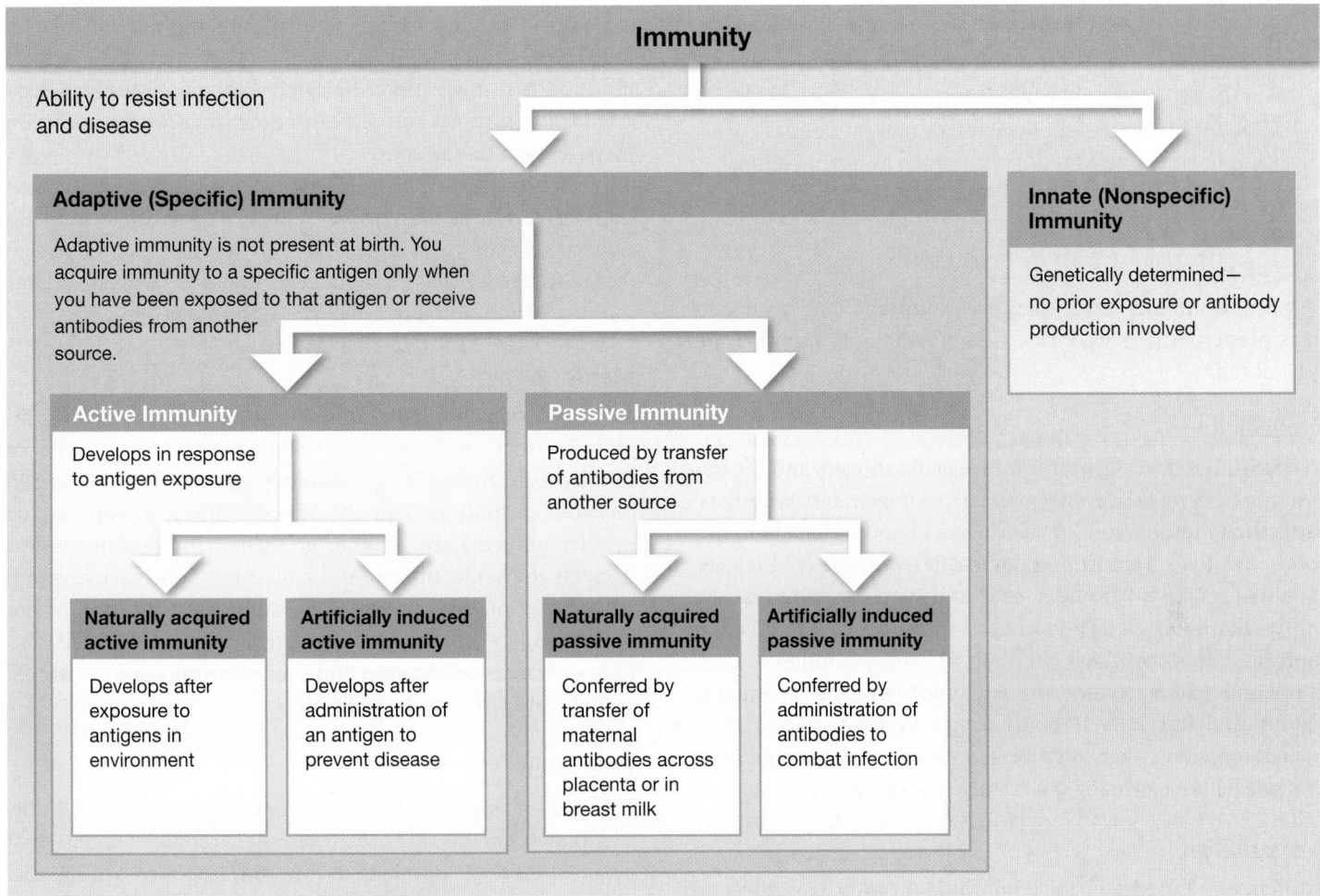

Forms of Immunity

We can classify immunity as either innate or adaptive. (**Figure 22–16**). As we stated earlier, **innate immunity** is genetically determined. It is present at birth and is not related to previous exposures to a particular antigen. For example, people do not get the same diseases that goldfish do. Innate immunity breaks down only in the case of AIDS or other conditions that depress all aspects of specific resistance.

Adaptive immunity is not present at birth. Instead, you develop immunity to a specific antigen only when you have been exposed to that antigen. Adaptive immunity can be *active* or *passive*. These forms of immunity can be either naturally acquired or artificially induced.

Active immunity develops after exposure to an antigen. It is a consequence of the immune response. In active immunity, the body responds to an antigen by making its own antibody. The immune system is *capable* of defending against a huge number of antigens. However, the appropriate defenses are mounted only after you encounter a particular antigen. Active immunity can

result from natural exposure to an antigen in the environment (*naturally acquired active immunity*) or from deliberate exposure to an antigen (*artificially induced active immunity*).

- **Naturally acquired active immunity** normally begins to develop after birth. It continues to build as you encounter "new" pathogens or other antigens. You might compare this process to the development of a child's vocabulary: The child begins with a few basic common words and learns new ones as they are encountered.

- **Artificially induced active immunity** stimulates the body to produce antibodies under controlled conditions so that you will be able to overcome natural exposure to the pathogen in the future. This is the basic principle behind *immunization*, or *vaccination*, to prevent disease. A **vaccine** is a preparation designed to induce an immune response. It contains either a dead or an inactive pathogen, or antigens derived from that pathogen.

Passive immunity is produced by transferring antibodies from another source.

- In **naturally acquired passive immunity**, a baby receives antibodies from the mother, either during gestation (by crossing the placenta) or in early infancy (through breast milk).

- In **artificially induced passive immunity**, a person receives antibodies to fight infection or prevent disease. For example, someone who has been bitten by a rabid animal gets injections containing antibodies against the rabies virus.

Properties of Adaptive Immunity

Regardless of the form, adaptive immunity has four general properties: (1) *specificity*, (2) *versatility*, (3) *memory*, and (4) *tolerance*.

Specificity

A specific defense is activated by a specific antigen, and the immune response targets that particular antigen and no others. **Specificity** results from the activation of appropriate lymphocytes and the production of antibodies with targeted effects. Specificity occurs because T cells and B cells respond to the molecular structure of an antigen. The shape and size of the antigen determine which lymphocytes will respond to it. Each T cell or B cell has receptors that will bind to one specific antigen, ignoring all others. The response of an activated T cell or B cell is equally specific. Either lymphocyte will destroy or inactivate that antigen without affecting other antigens or normal tissues.

Versatility

Millions of antigens in the environment can pose a threat to health. Over a normal lifetime, you encounter only a fraction of that number—perhaps tens of thousands of antigens. Your immune system, however, has no way of anticipating which antigens it will encounter. It must be ready to confront *any* antigen at *any* time. **Versatility** results in part from the large diversity of lymphocytes present in the body, and in part from variability in the structure of synthesized antibodies.

During development, cells in the lymphatic system differentiate to produce a huge number of lymphocytes with varied antigen sensitivities. The trillion or more T cells and B cells in your body include millions of different lymphocyte populations, distributed throughout the body. Each population contains several thousand cells with receptors in their membranes that differ from the receptors of other lymphocyte populations. As a result, each population of lymphocytes responds to a different antigen.

Several thousand lymphocytes are not enough to overcome a pathogenic invasion. However, when activated by an appropriate antigen, a lymphocyte begins to divide, producing more lymphocytes with the same specificity. All the cells produced by these divisions make up a **clone**, and all the members of that clone are sensitive to the same specific antigen.

To understand how this system works, think about running a commercial kitchen with only samples on display. You can display a wide selection because the samples don't take up much space, and you don't have to expend energy now preparing food that might never be eaten. When a customer selects one of your samples and places an order for several dozen, you prepare them on the spot.

The same principle applies to the lymphatic system. Your body contains a small number of many different kinds of lymphocytes. When an antigen arrives, lymphocytes sensitive to it are "selected." These lymphocytes divide to generate a large number of additional lymphocytes of the same type.

Memory

As we just saw, during the initial response to an antigen, lymphocytes that are sensitive to it undergo repeated cycles of cell division. Immunologic **memory** exists because those cell divisions produce two groups of cells. One group attacks the invader immediately. Another group remains inactive unless it meets the same antigen at a later date. This inactive group is made up of *memory cells* that enable your immune system to "remember" an antigen it has previously encountered, and to launch a faster, stronger, and longer-lasting counterattack if such an antigen appears again.

Tolerance

The immune system does not respond to all antigens. All cells and tissues in the body, for example, contain antigens that normally do not stimulate an immune response. We say that the immune system exhibits **tolerance** toward such antigens.

The immune response targets foreign cells and compounds, but it generally ignores normal tissues. During their differentiation in the red bone marrow (B cells) and thymus (T cells), any cells that react to antigens that are normally present in the body are destroyed. As a result, mature B cells and T cells ignore normal antigens, also called *self-antigens*, but attack foreign antigens, or *nonself antigens*. Tolerance can also develop over time in response to chronic exposure to an antigen in the environment. Such tolerance lasts only as long as the exposure continues.

An Introduction to the Immune Response

Figure 22–17 provides an overview of the immune response. When an antigen triggers an immune response, it usually activates both T cells and B cells. The activation of T cells generally occurs first, but only after phagocytes have been exposed to the antigen. Once activated, T cells attack the antigen and stimulate the activation of B cells. Activated B cells mature into cells that produce antibodies. Antibodies in the bloodstream then bind to and attack the antigen. We examine these processes more closely in the sections that follow.

Figure 22–17   **An Overview of the Immune Response.**

Cell-Mediated Immunity

Phagocytes activated → T cells activated

Direct Physical and Chemical Attack

Activated T cells find the pathogens and attack them through phagocytosis or the release of chemical toxins.

Adaptive Immunity

Antigen presentation triggers specific defenses, or an immune response.

Communication and feedback

Antibody-Mediated Immunity

Activated B cells give rise to cells that produce antibodies.

Attack by Circulating Antibodies

Destruction of antigens

CHECKPOINT

11. Explain the difference between cell-mediated (cellular) immunity and antibody-mediated (humoral) immunity.

12. Identify the two forms of active immunity and the two forms of passive immunity.

13. List the four general properties of adaptive immunity.

See the blue Answers tab at the back of the book.

22-5 ∎ T cells play a role in initiating, maintaining, and controlling the immune response

Learning Outcome Discuss the types of T cells and their roles in the immune response, and describe the mechanisms of T cell activation and differentiation.

The role of T cells in the immune response is varied and important in resisting pathogens. We have already noted four major types of T cells:

1. *Cytotoxic T* (T_C) *cells* are responsible for cell-mediated immunity. These cells enter peripheral tissues and directly attack antigens physically and chemically.

2. *Memory T cells* respond to antigens they have already encountered by cloning more lymphocytes to ward off the invader.

3. *Helper T* (T_H) *cells* stimulate the responses of both T cells and B cells. Helper T cells are absolutely vital to the

immune response, because they must activate B cells before the B cells can produce antibodies. The reduction in the helper T cell population that occurs in AIDS is largely responsible for the loss of immunity. (We discuss AIDS on p. 845.)

4. *Suppressor T* (T_S) *cells* inhibit T cell and B cell activities and moderate the immune response.

Before an immune response can begin, T cells must be activated by exposure to an antigen. This activation seldom occurs through direct interaction between a T cell and the antigen. Foreign substances or pathogens entering a tissue commonly fail to stimulate an immediate immune response.

Antigen Presentation

For T cells to recognize an antigen, the antigen must be bound to glycoproteins in the plasma membranes of another cell. Recall that glycoproteins are integral membrane components. ⊃ p. 96 **Antigen presentation** occurs when an antigen–glycoprotein combination capable of activating T cells appears in a plasma membrane.

The structure of these glycoproteins is genetically determined. The genes controlling their synthesis are located along one portion of chromosome 6, in a region called the **major histocompatibility complex (MHC)**. These membrane glycoproteins are called **MHC proteins**, or *human leukocyte antigens (HLAs)*.

The amino acid sequences and the shapes of MHC proteins differ among individuals. Each MHC molecule has a

Clinical Note

Graft Rejection and Immunosuppression Organ transplant may be a treatment option for patients with severe disorders of the bone marrow, kidneys, liver, heart, lungs, or pancreas. Finding a suitable donor is the first major problem. In the United States, each day many people die while waiting for an organ transplant, and dozens are added to the transplant waiting list.

After transplant surgery, the major problem is **graft rejection**. In graft rejection, T cells are activated by contact with MHC proteins on plasma membranes in the donated tissues. The cytotoxic T cells that develop then attack and destroy the foreign cells.

Transplant success can be improved by *immunosuppression*, a reduction in the sensitivity of the immune system. Until recently, the drugs used to produce immunosuppression did not selectively target the immune response. For example, **prednisone** (PRED-ni-sōn), a corticosteroid, was used because it has anti-inflammatory effects that reduce the number of circulating white blood cells and depress the immune response. Unfortunately, corticosteroid use also causes undesirable side effects in systems other than the immune system.

An understanding of the communication among T cells, macrophages, and B cells has now led to the development of drugs with more selective effects. **Cyclosporin A**, a compound derived from a fungus, was the most important *immunosuppressive drug* developed in the 1980s. This compound suppresses all aspects of the immune response. It acts primarily by suppressing helper T cell activity while leaving suppressor T cells relatively unaffected. Even more narrowly focused drugs are available. These include monoclonal antibodies that prevent antigen recognition by T cell receptors.

distinct three-dimensional shape with a relatively narrow central groove. An antigen that fits into this groove can be held in position by hydrogen bonding.

Two major classes of MHC proteins are known: *Class I* and *Class II*. An antigen bound to a Class I MHC protein acts like a red flag that in effect tells the immune system "Hey, I'm an abnormal cell—kill me!" An antigen bound to a Class II MHC protein tells the immune system "Hey, this antigen is dangerous—get rid of it!"

Class I MHC proteins are in the plasma membranes of all nucleated cells. These proteins are continuously synthesized and exported to the plasma membrane in vesicles created at the Golgi apparatus. As they form, Class I proteins pick up small peptides from the surrounding cytoplasm and carry them to the cell surface. If the cell is healthy and the peptides are normal, T cells ignore them. If the cytoplasm contains abnormal

(nonself) peptides or viral proteins (**Figure 22–18a**), they soon appear in the plasma membrane, and T cells will recognize them as foreign and be activated. Ultimately, their activation leads to the destruction of the abnormal cells. This is the primary reason that donated organs are commonly rejected by the recipient. Despite preliminary cross-match testing, the recipient's T cells recognize the transplanted tissue as foreign.

Class II MHC proteins are present only in the plasma membranes of antigen-presenting cells and lymphocytes. **Antigen-presenting cells (APCs)** are specialized cells responsible for activating T cell defenses against foreign cells (including bacteria) and foreign proteins. Antigen-presenting cells include all the phagocytic cells of the monocyte–macrophage group discussed in other chapters, including (1) free and fixed macrophages in connective tissues, (2) the Kupffer cells of the liver, and (3) the microglia in the central nervous system (Chapter 12). ⊃ pp. 153, 423 The dendritic (Langerhans) cells of the skin and the dendritic cells of the lymph nodes and spleen are APCs that are not phagocytic. ⊃ p. 179

Phagocytic APCs engulf and break down pathogens or foreign antigens. This **antigen processing** creates fragments of the antigen, which are then bound to Class II MHC proteins and inserted into the plasma membrane (**Figure 22–18b,c**). *Class II MHC proteins appear in the plasma membrane only when the cell is processing antigens.* Exposure to an APC membrane containing a processed antigen can stimulate appropriate T cells.

The dendritic cells remove antigenic materials from their surroundings by pinocytosis rather than phagocytosis. However, their plasma membranes still present antigens bound to Class II MHC proteins. More important is that, unlike macrophages that do not migrate very far from an infection site, dendritic cells travel with their foreign antigens to lymphocyte-packed lymph nodes.

Antigen Recognition

How do T cells recognize antigens? Inactive T cells have receptors that can bind either Class I or Class II MHC proteins. These receptors also have binding sites for a specific target antigen. If an MHC protein contains any antigen other than the specific target of a particular kind of T cell, the T cell remains inactive. If the MHC protein contains the antigen that the T cell is programmed to detect, binding occurs. This process is called **antigen recognition**, because the T cell recognizes that it has found an appropriate target.

Some T cells can recognize antigens bound to Class I MHC proteins, whereas others can recognize antigens bound to Class II MHC proteins. Whether a T cell responds to antigens held by one class or the other depends on a type of protein in the T cell's own plasma membrane. The membrane proteins involved are members of a larger group of proteins called **CD** (*cluster of differentiation*) **markers**.

Figure 22–18 **Antigens and MHC Proteins.**

1 Antigen presentation by Class I MHC proteins is triggered by viral or bacterial infection of a body cell.

2 The infection results in the appearance of abnormal peptides in the cytoplasm.

3 The abnormal peptides are incorporated into Class I MHC proteins as they are synthesized at the endoplasmic reticulum.

Plasma membrane

Viral or bacterial pathogen

Transport vesicle

Endoplasmic reticulum

Nucleus

5 The abnormal peptides are displayed by Class I MHC proteins on the plasma membrane.

4 After export to the Golgi apparatus, the MHC proteins reach the plasma membrane within transport vesicles.

a Infected cell.

b Cells exposed to fluorescent-tagged antibodies. The green cell is an APC and the red cell is a lymphocyte.

1 Phagocytic APCs engulf the extracellular pathogens.

2 Lysosomal action produces antigenic fragments.

Plasma membrane

Lysosome

Nucleus

Endoplasmic reticulum

5 Antigenic fragments are displayed by Class II MHC proteins on the plasma membrane.

4 Antigenic fragments are bound to Class II MHC proteins.

3 The endoplasmic reticulum produces Class II MHC proteins.

c Phagocytic antigen-presenting cell.

Lymphocytes, macrophages, and other related cells have CD markers. More than 70 types of CD markers exist, and each type is designated by a number. All T cells have a **CD3 receptor complex** in their plasma membranes, and this complex ultimately activates the T cell. Either of two other CD markers may be bound to the CD3 receptor complex, and these two CD markers are especially important in specific groups of T cells:

1. **CD8** markers are found on cytotoxic T cells and suppressor T cells, which together are often called *CD8 T cells* or *CD8 + T cells*. CD8 T cells respond to antigens presented by Class I MHC proteins.

2. **CD4** markers are found on helper T cells, often called *CD4 T cells* or *CD4 + T cells*. CD4 T cells respond to antigens presented by Class II MHC proteins.

Figure 22–19   **Antigen Recognition and Activation of Cytotoxic T Cells.**

T cells are not usually activated upon their first encounter with the antigen. Antigen recognition simply prepares the cell for activation.

Costimulation

In order to proceed from recognition to activation, a T cell must also bind to the stimulating cell at a second site. This vital secondary binding process, called *costimulation*, essentially confirms the initial activation signal.

Costimulation is like the safety on a gun: It helps prevent T cells from mistakenly attacking normal (self) tissues. If a cell displays an unusual antigen but does not display the "I am an active phagocyte" or "I am infected" signal, T cell activation will not occur. Costimulation is important only in determining whether a T cell will become activated. Once activation has occurred, the "safety" is off and the T cell will attack any cells that carry the target antigens.

Activation of CD8 T Cells

Two different types of CD8 T cells are activated by exposure to antigens bound to Class I MHC proteins. One type responds quickly, giving rise to large numbers of *cytotoxic T cells* and

memory T cells (**Figure 22–19**). The other type responds more slowly and produces relatively small numbers of *suppressor T cells*.

Cytotoxic T Cells

Cytotoxic T (T_C) cells seek out and destroy abnormal and infected cells. They are highly mobile cells that roam throughout injured tissues. When a cytotoxic T cell encounters its target antigen bound to Class I MHC proteins, it immediately destroys the target cell (**Figure 22–19**). The T cell may (1) release perforin to destroy the target cell's plasma membrane, (2) secrete a poisonous **lymphotoxin** (lim-fō-TOK-sin) to kill the target cell, or (3) activate genes in the target cell's nucleus that tell that cell to die. (We introduced genetically programmed cell death, called *apoptosis*, in Chapter 3.) ↻ p. 127

The entire sequence of events, from the appearance of the antigen in a tissue to cell destruction by cytotoxic T cells, takes a significant amount of time. After the first exposure to an antigen, two days or more may pass before the concentration of cytotoxic T cells reaches effective levels at the site of injury or infection. Over this period, the damage or infection may spread, making it more difficult to control.

Memory T_C Cells

The same cell divisions that produce cytotoxic T cells also produce thousands of **memory T_C cells**. These cells do not differentiate further the first time the antigen triggers an immune response. However, if the same antigen appears a second time, memory T cells *immediately* differentiate into cytotoxic T cells. They produce a prompt, effective cellular response that can overwhelm the invader before it becomes well established in the tissues.

Suppressor T Cells

Suppressor T (T_S) cells suppress the responses of other T cells and of B cells by secreting inhibitory cytokines called *suppression factors*. Suppression does not take place right away, because activation takes much longer for suppressor T cells than for other types of T cells. In addition, upon activation, most of the CD8 T cells in the bloodstream produce cytotoxic T cells rather than suppressor T cells. As a result, suppressor T cells act *after* the initial immune response. In effect, these cells limit the degree of immune system activation from a single stimulus.

Activation of CD4 T Cells

Upon activation, helper T cells (T_H cells) with CD4 markers undergo a series of divisions that produce both active helper T cells and **memory helper T cells**, also called **memory T_H cells** (Figure 22–20). The memory helper T cells remain in reserve. The active helper T cells secrete a variety of cytokines that coordinate specific and nonspecific defenses and stimulate cell-mediated and antibody-mediated immunities. These cytokines do the following:

1. stimulate the T cell divisions that produce memory helper T cells and speed the maturation of cytotoxic T cells;

2. enhance nonspecific defenses by attracting macrophages to the affected area, preventing their departure, and stimulating their phagocytic activity and effectiveness;

3. attract and stimulate the activity of cytotoxic T cells, providing another means of destroying abnormal cells and pathogens; and

4. promote the activation of B cells, leading ultimately to antibody production.

Figure 22–21 provides a review of the methods of antigen presentation and T cell stimulation. The plasma membranes of infected or otherwise abnormal cells trigger an immune response when CD8 T cells recognize antigens bound to Class I MHC proteins. Extracellular pathogens or foreign proteins trigger an immune response when CD4 T cells recognize antigens displayed by Class II MHC proteins. In the next section, we will see how the helper T cells derived from activated CD4 T cells

Figure 22–20 Antigen Recognition and Activation of Helper T Cells. Inactive CD4 T cells (T_H cells) must be exposed to appropriate antigens bound to Class II MHC proteins. The T_H cells then undergo activation, dividing to produce active T_H cells and memory T_H cells.

in turn activate B cells that are sensitive to the same specific antigen.

CHECKPOINT

14. Identify the four major types of T cells.

15. How can the presence of an abnormal peptide in the cytoplasm of a cell initiate an immune response?

16. A decrease in the number of cytotoxic T cells would affect which type of immunity?

17. How would a lack of helper T cells affect the antibody-mediated immune response?

See the blue Answers tab at the back of the book.

Figure 22–21 **A Summary of the Pathways of T Cell Activation.**

a **Activation by Class I MHC proteins**

Antigen bound to Class I MHC protein

Indicates that the cell is infected or otherwise abnormal

CD8 T Cells

Cytotoxic T Cells	Memory T_C Cells	Suppressor T Cells
Attack and destroy infected and abnormal cells displaying antigen	Await reappearance of the antigen	Control or moderate immune response by T cells and B cells

b **Activation by Class II MHC proteins**

Antigen bound to Class II MHC protein

Indicates presence of pathogens, toxins, or foreign proteins

CD4 T Cells

Helper T Cells	Memory T_H Cells
Stimulate immune response by T cells and B cells	Await reappearance of the antigen

22-6 | B cells respond to antigens by producing specific antibodies

Learning Outcome Discuss the mechanisms of B cell activation and differentiation, describe the structure and function of antibodies, and explain the primary and secondary responses to antigen exposure.

B cells are responsible for launching a chemical attack on antigens. They produce appropriate specific *antibodies*.

B Cell Sensitization and Activation

As noted earlier, the body has millions of B cell populations. Each B cell carries its own particular type of antibody molecules in its plasma membrane. If corresponding antigens appear in the interstitial fluid, they interact with these superficial antibodies. When binding occurs, the B cell prepares to undergo activation. This preparatory process is called **sensitization** (**Figure 22–22**). Because B cells migrate throughout the body, pausing briefly in one lymphoid tissue or another, sensitization typically takes place within the lymph node nearest the site of infection or injury.

Recall that B cell plasma membranes contain Class II MHC proteins. During sensitization, antigens are brought into the cell by endocytosis. The antigens subsequently appear on the surface of the B cell, bound to Class II MHC proteins. (The mechanism is comparable to that shown in **Figure 22–18b**). The sensitized B cell is then on "standby" but generally does not undergo activation unless it receives the "OK" from a helper T cell. The need for activation by a helper T cell helps prevent inappropriate activation, the same way that costimulation acts as a "safety" in cell-mediated immunity.

What happens when a sensitized B cell meets a helper T cell that has already been activated by antigen presentation? The helper T cell binds to the B cell's MHC complex, recognizes the antigen, and begins secreting cytokines that promote B cell activation (**Figure 22–22**). After activation of the B cell, these same cytokines stimulate B cell division, speed plasma cell formation, and enhance antibody production.

The activated B cell typically divides several times, producing daughter cells that differentiate into plasma cells and *memory B cells* (**Figure 22–22**). The plasma cells begin synthesizing and secreting large quantities of antibodies into the interstitial fluid. These antibodies have the same target as the antibodies on the surface of the sensitized B cell. When stimulated by cytokines from helper T cells, a plasma cell can secrete up to 100 million antibody molecules each hour.

Memory B cells perform the same role in antibody-mediated immunity that memory T cells perform in cell-mediated immunity. Memory B cells do not respond to a threat on first exposure. Instead, they remain in reserve to deal with subsequent injuries or infections that involve the same antigens. On subsequent exposure, the memory B cells divide and differentiate into plasma cells that secrete antibodies in massive quantities.

Figure 22–22 **The Sensitization and Activation of B Cells.** A B cell is sensitized by exposure to antigens. Once antigens are bound to antibodies in the B cell plasma membrane, the B cell displays those antigens on Class II MHC proteins in its plasma membrane. Activated helper T cells encountering the antigens release cytokines that costimulate the sensitized B cell and trigger its activation. The activated B cell then divides, producing memory B cells and plasma cells that secrete antibodies.

Antibody Structure

A Y-shaped antibody molecule consists of two pairs of polypeptide chains: one pair of **heavy chains** and one pair of **light chains** (Figure 22–23). Each chain contains both *constant segments* and *variable segments*.

The constant segments of the heavy chains form the base of the antibody molecule (Figure 22–23a,b). B cells produce only five types of constant segments. These are the basis of a classification scheme that identifies antibodies as *IgG, IgE, IgD, IgM*, or *IgA*, as we discuss in the next section. The structure of the constant segments of the heavy chains determines the way the antibody is secreted and how it is distributed within the body. For example, antibodies in one class circulate in body fluids, whereas those of another class bind to the membranes of basophils and mast cells.

The heavy-chain constant segments, which are bound to constant segments of the light chains, also contain binding sites that can activate the complement system. These binding sites are covered when the antibody is secreted but become exposed when the antibody binds to an antigen.

The specificity of an antibody molecule depends on the amino acid sequence of the variable segments of the light and heavy chains. The free tips of the two variable segments form the **antigen binding sites** of the antibody molecule (Figure 22–23a). These sites can interact with an antigen in the same way that the active site of an enzyme interacts with a substrate molecule. ⤴ p. 80

Small differences in the structure of the variable segments affect the precise shape of the antigen binding site. These differences make antibodies specific for different antigens. The distinctions are the result of minor genetic variations that occur during the production, division, and differentiation of B cells. A normal adult body contains roughly 10 trillion B cells, which can produce an estimated 100 million types of antibodies, each with a different specificity.

The Antigen–Antibody Complex

An **antigen–antibody complex** forms when an antibody molecule binds to its corresponding antigen molecule. Once the two molecules are in position, hydrogen bonding and other weak chemical forces lock them together.

Figure 22–23   **Antibody Structure and Function.**

a A diagrammatic view of the structure of an antibody.

b A computer-generated image of a typical antibody.

c Antibodies bind to portions of an antigen called antigenic determinant sites, or epitopes.

d Antibody molecules can bind a hapten (partial antigen) once it has become a complete antigen by combining with a carrier molecule.

Antibodies do not bind to the entire antigen. Instead, they bind to specific portions of its exposed surface—regions called **antigenic determinant sites**, or **epitopes** (Figure 22–23c). The specificity of the binding depends initially on the three-dimensional "fit" between the variable segments of the antibody molecule and the corresponding antigenic determinant sites. A **complete antigen** has at least two antigenic determinant sites, one for each of the antigen binding sites on an antibody molecule. Exposure to a complete antigen can lead to B cell sensitization and a subsequent immune response. Most environmental antigens have multiple antigenic determinant sites, and entire microorganisms may have thousands.

Haptens, or *partial antigens*, do not ordinarily cause B cell activation and antibody production. Haptens include short peptide chains, steroids and other lipids, and several drugs, including antibiotics such as *penicillin*. However, haptens may

become attached to carrier molecules, forming combinations that can function as complete antigens (**Figure 22–23d**). In some cases, the carrier contributes an antigenic determinant site. Antibodies will then attack both the hapten and the carrier molecule. If the carrier molecule is normally present in the tissues, the antibodies may begin attacking and destroying normal cells. This process is the basis for several drug reactions, including allergies to penicillin.

Classes and Actions of Antibodies

Body fluids may contain five classes of antibodies, or **immunoglobulins (Igs)**: *IgG, IgE, IgD, IgM,* and *IgA* (Table 22–1). The classes are determined by differences in the structure of the heavy-chain constant segments. For this reason, the classes have no effect on the antibody's specificity, which is determined by the antigen binding sites.

Table 22–1	Classes of Antibodies

Secretory piece

IgG is the largest and most diverse class of antibodies. They account for 80 percent of all antibodies. IgG antibodies are responsible for resistance against many viruses, bacteria, and bacterial toxins. These antibodies can cross the placenta, and maternal IgG provides passive immunity to the fetus during embryological development. However, the anti-Rh antibodies produced by Rh-negative mothers are also IgG antibodies and produce *hemolytic disease of the newborn*.

IgE attaches as an individual molecule to the exposed surfaces of basophils and mast cells. When a suitable antigen is bound by IgE molecules, the cell is stimulated to release histamine and other chemicals that accelerate inflammation in the immediate area. IgE is also important in the allergic response.

IgD is an individual molecule on the surfaces of B cells, where it can bind antigens in the extracellular fluid. This binding can play a role in the sensitization of the B cell involved.

IgM is the first class of antibody secreted after an antigen is encountered. IgM concentration declines as IgG production accelerates. Although plasma cells secrete individual IgM molecules, IgM circulates as a five-antibody starburst. The anti-A and anti-B antibodies responsible for the agglutination of incompatible blood types are IgM antibodies. IgM antibodies may also attack bacteria that are insensitive to IgG.

IgA is found primarily in glandular secretions such as mucus, tears, saliva, and semen. These antibodies attack pathogens before they gain access to internal tissues. IgA antibodies circulate in blood as individual molecules or in pairs. Epithelial cells absorb them from the blood and attach a *secretory piece*, which confers solubility, before secreting the IgA molecules onto the epithelial surface.

The formation of an antigen–antibody complex may cause the elimination of the antigen in seven ways:

1. *Neutralization.* Both viruses and bacterial toxins have specific sites that must bind to target regions on body cells before they can enter or injure those cells. Antibodies may bind to those sites, making the virus or toxin incapable of attaching itself to a cell. This mechanism is known as **neutralization**.

2. *Precipitation and Agglutination.* Each antibody molecule has two antigen binding sites, and most antigens have many antigenic determinant sites. If individual antigens (such as macromolecules or bacterial cells) are far apart, an antibody molecule will necessarily bind to two antigenic sites on the same antigen. However, if antigens are close together, an antibody can bind to antigenic determinant sites on two separate antigens. In this way, antibodies can link large numbers of antigens together. The three-dimensional structure created by such binding is known as an **immune complex**. When the antigen is a soluble molecule, such as a toxin, this process may create complexes that are too large to remain in solution. The formation of insoluble immune complexes is called **precipitation**. When the target antigen is on the surface of a cell or virus, the formation of large complexes is called **agglutination**. For example, the clumping of erythrocytes that takes place when incompatible blood types are mixed is an agglutination reaction. ⤷ p. 692

3. *Activation of the Complement System.* When an antibody molecule binds to an antigen, portions of the antibody molecule change shape. This change exposes areas that bind complement proteins. The bound complement molecules then activate the complement system, which destroys the antigen (as discussed previously).

4. *Attraction of Phagocytes.* Antigens covered with antibodies attract eosinophils, neutrophils, and macrophages. These cells phagocytize pathogens and destroy foreign or abnormal plasma membranes.

5. *Opsonization.* A coating of antibodies and complement proteins increases the effectiveness of phagocytosis. This effect is called *opsonization*. Some bacteria have slick plasma membranes or capsules, but opsonization makes it easier for phagocytes to hold onto their prey before they engulf it. Phagocytes can bind more easily to antibodies and complement proteins than they can to the bare surface of a pathogen.

6. *Stimulation of Inflammation.* Antibodies may promote inflammation by stimulating basophils and mast cells.

7. *Prevention of Bacterial and Viral Adhesion.* Antibodies dissolved in saliva, mucus, and perspiration coat epithelia, adding an additional layer of defense. A covering of antibodies makes it difficult for pathogens to attach to and penetrate body surfaces.

TIPS & TRICKS

The classes of immunoglobulins—Ig**M**,Ig**A**, Ig**D**, Ig**G**, Ig**E**— spell **MADGE**.

Primary and Secondary Responses to Antigen Exposure

The initial immune response to an antigen is called the **primary response**. When the antigen appears again, it triggers a more extensive and prolonged **secondary response**. This response is due to the presence of large numbers of memory cells that are primed for the arrival of the antigen. Primary and secondary responses occur in both cell-mediated and antibody-mediated immunities. Let's look at the pattern of antibody production over time to see the differences between the primary and secondary responses.

The Primary Response

The primary response takes time to develop because the antigen must activate the appropriate B cells. These cells must then differentiate into plasma cells. As plasma cells differentiate and begin secreting, the concentration of circulating antibodies makes a gradual, sustained rise (**Figure 22–24a**).

During the primary response, the **antibody titer**, or level of antibodies in the plasma, does not peak until one to two weeks after the initial exposure. If the individual is no longer exposed to the antigen, the antibody concentration then declines. The antibody titer declines because (1) plasma cells have very high metabolic rates and survive for only a short time, and (2) suppressor T cells release suppression factors that inhibit further production of plasma cells. However, suppressor T cell activity does not begin immediately after exposure to the antigen. Also, under normal conditions helper T cells outnumber suppressor T cells by more than 3 to 1. As a result, many B cells are activated before suppressor T cell activity has a noticeable effect.

Activated B cells start dividing immediately. At each cycle of division, some of the daughter cells differentiate into plasma cells, while others continue to divide. Molecules of *immunoglobulin M*, or *IgM*, are the first to appear in the bloodstream. The plasma cells that produce IgM differentiate after only a few cycles of B cell division. Levels of *immunoglobulin G*, or *IgG*, rise more slowly, because the plasma cells responsible differentiate only after repeated cell divisions that also generate large numbers of memory B cells. In general, IgM is less effective as a defense than IgG. However, IgM provides an immediate defense that can fight the infection until massive quantities of IgG can be produced.

The Secondary Response

Unless memory B cells are exposed to the same antigen a second time, they do not differentiate into plasma cells. If and when that exposure occurs, the memory B cells respond right away— faster than the B cells stimulated during the initial exposure. This response is immediate in part because memory B cells are activated at relatively low antigen concentrations. In addition, these cells synthesize more effective and destructive antibodies. Activated memory B cells divide and differentiate into plasma cells that secrete these antibodies in massive quantities. This secretion is the secondary response to antigen exposure.

During the secondary response, antibody concentrations and titers increase more rapidly and reach levels many times higher than they did in the primary response (**Figure 22–24b**). The secondary response appears even if the second exposure occurs years after the first. The reason is that memory cells may survive for 20 years or more.

The primary response develops slowly and does not produce antibodies in massive quantities. For these reasons, it

Figure 22–24 **The Primary and Secondary Responses in Antibody-Mediated Immunity.**

a The primary response takes about two weeks to develop peak antibody levels (titers). IgM and IgG antibody levels do not remain elevated.

b The secondary response has a very rapid increase in IgG antibody concentration and rises to levels much higher than those of the primary response. Antibody levels remain elevated for an extended period after the second exposure to the antigen.

may not prevent an infection the first time a pathogen appears in the body. However, a person who survives the first infection will probably be resistant to that pathogen in the future, thanks to a rapid and overwhelming secondary response. The effectiveness of the secondary response is one of the basic principles behind the use of immunization to prevent disease.

TIPS&TRICKS

The antibody response is like ordering a custom suit. The first suit (the primary antibody response) takes time to make because the tailor (an activated B cell) must first make a pattern (a clone of memory cells). Subsequent suits (secondary responses) are made much more quickly because the pattern already exists.

Summary of the Immune Response

We have now examined the basic cellular and chemical interactions that follow the appearance of a foreign antigen in the body. Table 22–2 reviews the cells that participate in tissue defenses and Figure 22–25 gives a timeline for their appearance at the site of a bacterial infection.

In the early stages of infection, neutrophils and NK cells migrate into the threatened area and destroy bacteria. Over time, cytokines draw increasing numbers of phagocytes into the region. Cytotoxic T cells appear as arriving T cells are activated by antigen presentation. Last of all, the population of plasma cells rises as activated B cells differentiate. This rise is followed

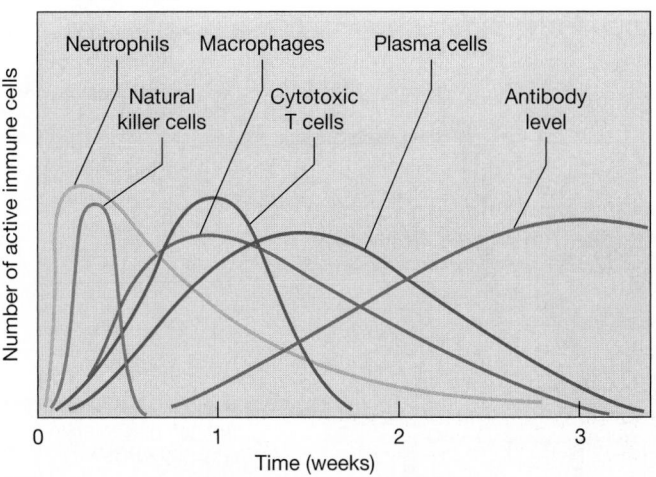

Figure 22–25 **The Course of the Body's Response to a Bacterial Infection.** The basic sequence of events begins with the appearance of bacteria in peripheral tissues at time 0.

by a gradual, sustained increase in the activity (titer) of circulating antibodies.

Figure 22–26 provides an integrated review of the relationship between innate and adaptive immunity. The basic sequence of events is similar when a viral infection occurs. The initial steps are different, however, because cytotoxic T cells provide an adaptive defense and NK cells provide an innate defense. Figure 22–27 contrasts the events involved in defending against bacterial infection with those involved in defending against viral infection.

Table 22–2	Cells That Participate in Tissue Defenses
Cell	**Functions**
Neutrophils	Phagocytosis; stimulation of inflammation
Eosinophils	Phagocytosis of antigen–antibody complexes; suppression of inflammation; participation in allergic response
Mast cells and basophils	Stimulation and coordination of inflammation by release of histamine, heparin, leukotrienes, prostaglandins
ANTIGEN-PRESENTING CELLS	
Macrophages (free and fixed macrophages, Kupffer cells, microglia, etc.)	Phagocytosis; antigen processing; antigen presentation with Class II MHC proteins; secretion of cytokines, especially interleukins and interferons
Dendritic cells	Pinocytosis; antigen processing; antigen presentation bound to Class II MHC proteins
LYMPHOCYTES	
NK cells	Destruction of plasma membranes containing abnormal antigens
Cytotoxic T cells (T_C, CD8 marker)	Lysis of plasma membranes containing antigens bound to Class I MHC proteins; secretion of perforins, defensins, lymphotoxins, and other cytokines
Helper T cells (T_H, CD4 marker)	Secretion of cytokines that stimulate cell-mediated and antibody-mediated immunity; activation of sensitized B cells
B cells	Differentiation into plasma cells, which secrete antibodies and provide antibody-mediated immunity
Suppressor T cells (T_S, CD8 marker)	Secretion of suppression factors that inhibit the immune response
Memory cells (T_S, T_H, B)	Produced during the activation of T cells and B cells; remain in tissues awaiting rearrival of antigens

22

Figure 22–26   **An Integrated Summary of the Immune Response.**

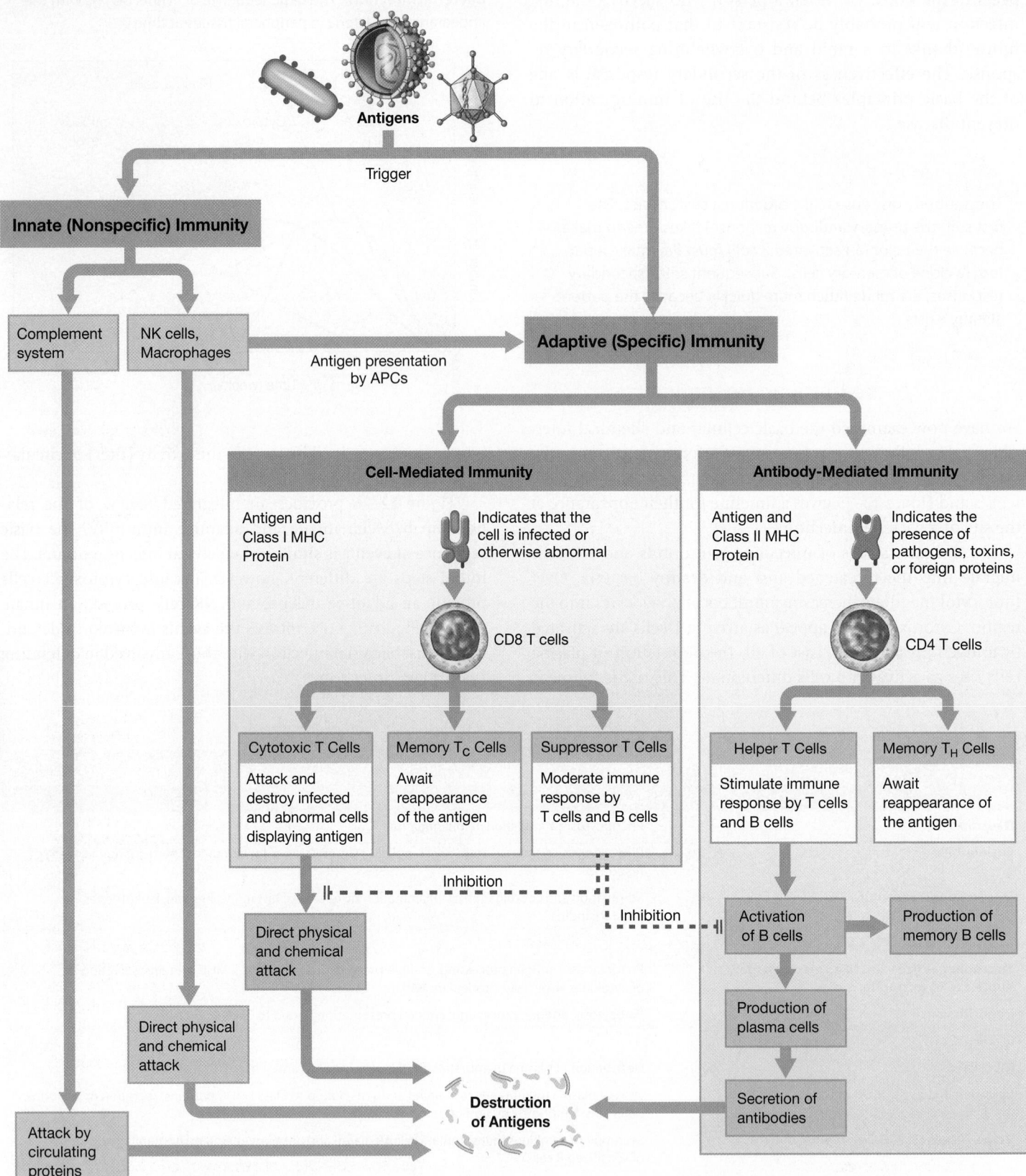

Figure 22–27   **Defenses against Bacterial and Viral Pathogens.**

a Defenses against bacteria involve phagocytosis and antigen presentation by APCs.

b Defenses against viruses involve direct contact with virus-infected cells and antigen presentation by APCs.

CHECKPOINT

18. Define sensitization.

19. A sample of lymph contains an elevated number of plasma cells. Would you expect the number of antibodies in the blood to be increasing or decreasing? Why?

20. Describe the structure of an antibody.

21. Which would be more negatively affected—the primary response or the secondary response—by a lack of memory B cells for a particular antigen?

See the blue Answers tab at the back of the book.

22-7 | Immunocompetence enables a normal immune response; abnormal responses result in immune disorders

Learning Outcome Describe the development of immunocompetence, list and explain examples of immune disorders and allergies, and discuss the effects of stress on immune function.

We begin this section by exploring **immunocompetence**, the ability to produce an immune response after exposure to an antigen. Then we consider some of the disorders that result when immune responses go wrong. Finally, we examine the role of stress in immune responses.

The Development of Immunocompetence

Cell-mediated immunity can be demonstrated as early as the third month of fetal development. Active antibody-mediated immunity follows about one month later.

The first cells that leave the fetal thymus migrate to the skin and into the epithelia lining the mouth, the digestive tract, and the uterus and vagina in females. These cells take up residence in these tissues as antigen-presenting cells, such as the dendritic cells of the skin, whose primary function will be to help activate T cells. T cells that leave the thymus later in development populate lymphoid organs throughout the body.

The plasma membranes of the first B cells produced in the liver and bone marrow carry IgM antibodies. Sometime after the fourth month the fetus may produce IgM antibodies if exposed to specific pathogens. It is uncommon for the fetus to produce antibodies, however, because the developing fetus has naturally acquired passive immunity due to IgG antibodies from the mother's bloodstream. These are the only antibodies that can cross the placenta. They include the antibodies responsible for the clinical problems due to Rh incompatibility, discussed in Chapter 19. ↺ p. 694 Problems with incompatibilities involving the ABO blood groups rarely occur. The reason is that anti-A and anti-B antibodies are IgM antibodies, which cannot cross the placenta.

The natural immunity provided by maternal IgG may not be enough to protect the fetus if the maternal defenses are overwhelmed by a bacterial or viral infection. For example, the microorganisms responsible for syphilis and rubella ("German measles") can cross from the maternal to the fetal bloodstream, producing a congenital infection that leads to the production of fetal antibodies. IgM provides only a partial defense, and these infections can result in severe developmental problems for the fetus.

Delivery eliminates the maternal supply of IgG. Although the mother provides IgA antibodies in breast milk, the infant gradually loses its passive immunity. The amount of maternal IgG in the infant's bloodstream declines rapidly over the first two months after birth. During this period, the infant becomes vulnerable to infection by bacteria or viruses that were previously overcome by maternal antibodies. The infant's immune system begins to respond to infections, environmental antigens, and vaccinations. As a result, the infant begins producing its own IgG. It has been estimated that, from birth to age 12, children encounter a "new" antigen every six weeks. (This fact explains why most parents, who were exposed to the same antigens when they were children, remain healthy while their children develop runny noses and colds.) Over this period, the concentration of circulating antibodies gradually rises toward normal adult levels. Populations of memory B cells and T cells also increase.

Skin tests are sometimes used to determine whether an individual has been exposed to a particular antigen. In these procedures, small quantities of antigen are injected into the skin, generally on the anterior surface of the forearm. If resistance has developed, the region becomes inflamed over the next two to four days. Many states require a tuberculosis test, called a *tuberculin skin test*, before children enter public school. If the test is positive, further tests must then be performed to determine whether an infection is currently under way. Skin tests are also used to check for allergies to environmental antigens.

Cytokines of the Immune System

Cytokines are chemical messengers involved in cellular immunity. They include hormones and paracrine-like glycoproteins important to the immune response. Interferons, interleukins, and tumor necrosis factors are examples of cytokines (**Spotlight Figure 22–28**).

Immune Disorders

Because the immune response is so complex, many opportunities exist for things to go wrong. A variety of clinical conditions result from disorders of the immune function. **Autoimmune disorders** develop when the immune response inappropriately targets normal body cells and tissues. In an **immunodeficiency disease**, either the immune system fails to develop normally or the immune response is blocked in some way. Autoimmune disorders and immunodeficiency diseases are fairly rare—clear evidence of the effectiveness of the immune system's control mechanisms. A far more common (and generally far less dangerous) class of immune disorders is **allergies**. Next we consider examples of each type of immune disorder.

Autoimmune Disorders

Autoimmune disorders affect an estimated 5 percent of adults in North America and Europe. In previous chapters we have cited many examples of the effects of autoimmune disorders on the function of major systems.

The immune system usually recognizes but ignores antigens normally found in the body—self-antigens. When the recognition system does not work correctly, however, activated B cells make antibodies against other body cells and tissues.

Clinical Note

AIDS **Acquired immunodeficiency syndrome (AIDS)**, or *late-stage HIV disease*, is caused by the **human immunodeficiency virus (HIV)**. This virus is a *retrovirus*: It carries its genetic information in RNA rather than in DNA. The virus enters human leukocytes by receptor-mediated endocytosis. ⸠ p. 123 Specifically, the virus binds to CD4 T cells. HIV also infects several types of antigen-presenting cells, including those of the monocyte–macrophage line. It is the infection of helper T cells that leads to clinical problems.

Once the virus is inside a cell, the viral enzyme *reverse transcriptase* synthesizes a complementary strand of DNA. This strand is then incorporated into the cell's genetic material. When these inserted viral genes are activated, the infected cell begins synthesizing viral proteins. In effect, the viral genes take over the cell's synthetic machinery and force the cell to produce additional viruses. These new viruses are then shed at the cell surface.

Cells infected with HIV are ultimately destroyed by (1) formation of pores in the plasma membrane as the viruses are shed, (2) interruption of cell maintenance due to the continuing synthesis of viral components, (3) autolysis, or (4) stimulation of programmed cell death (apoptosis).

The gradual destruction of helper T cells impairs the immune response, because these cells play a central role in coordinating cell-mediated and antibody-mediated responses to antigens. To make matters worse, suppressor T cells are relatively unaffected by the virus. Over time the excess of suppressing factors "turns off" the normal immune response. Circulating antibody levels decline and cell-mediated immunity is reduced. The body is left with impaired defenses against a wide variety of microbial invaders. Microorganisms that would ordinarily be harmless can now initiate lethal *opportunistic infections*. The risk of cancer also increases because immune surveillance is depressed.

Infection with HIV occurs through intimate contact with the body fluids of infected individuals. The major routes of transmission involve contact with blood, semen, or vaginal secretions, although all body fluids may contain the virus. Worldwide, most individuals with AIDS become infected through sexual contact with an HIV-infected person (who may *not* necessarily show clinical signs of AIDS). The next largest group of infected individuals is intravenous drug users who shared contaminated needles. Relatively few individuals have become infected with the virus after receiving a transfusion of contaminated blood or blood products. Finally, an increasing number of infants are born with AIDS acquired from infected mothers.

AIDS continues to be a public health problem of massive proportions. Current statistics from the Centers for Disease Control and Prevention (CDC) show that more than 1.1 million people in the United States are infected with HIV. Moreover, nearly 1 in 5 people are not aware that they are infected.

The most effective ways to prevent HIV infection include abstinence, mutual monogamy of uninfected partners, male circumcision (to reduce the risk of transmission), and the avoidance of needle sharing. Circumcised men are up to 70 percent less likely to become infected than uncircumcised men. They are also less likely to infect their female partners. The use of latex condoms labeled "for the prevention of disease" has been shown to prevent the passage of HIV as well as the hepatitis and herpes viruses. Natural condoms, also called lambskin condoms, are made from the intestinal membrane of a lamb. They are not effective against disease transmission because infectious agents can pass through the naturally occurring pores. (The pores are too small for sperm to pass through, so they are effective as a contraceptive.) Synthetic condoms, usually made of polyurethane, are considered to offer similar protection as latex, but differences in their manufacture cause variable reliability.

Clinical signs and symptoms of AIDS may not appear until 5–10 years or more after infection. When they do appear, they are commonly mild, consisting of swollen lymph nodes and chronic, but nonfatal, infections. So far as is known, however, AIDS is almost always fatal. Most people who carry the virus eventually die from complications of the disease. (A handful of infected individuals have been able to tolerate the virus without apparent illness for many years.)

Despite intensive efforts, a vaccine has yet to be developed that prevents HIV infection in an uninfected person exposed to the virus. The survival rate for AIDS patients has been steadily increasing. That is, more people are living longer before dying of AIDS or its complications. New drugs and drug combinations slow the progression of the disease, and improved antibiotic therapies help combat secondary infections.

HIV (green) budding from an infected T$_H$ cell (blue)　　SEM × 40,000

Innate (nonspecific) immunity and adaptive (specific) immunity are coordinated by both physical interactions and the release of chemical messengers.

The Immune System
is coordinated by

Physical Interactions

Chemical Messengers

CYTOKINES

An example of the release of chemical messengers is the secretion of cytokines by many cell types involved in the immune response. The six groups of cytokines shown here merit special attention.

Interleukins

Interleukins may be the most diverse and important chemical messengers in the immune system. Nearly 20 types of interleukins have been identified. Lymphocytes and macrophages are the primary sources of interleukins, but endothelial cells, fibroblasts, and astrocytes also produce certain interleukins, such as interleukin-1 (IL-1).

The functions of interleukins include the following:

1. **Increasing T Cell Sensitivity to Antigens Exposed on Macrophage Membranes**. Heightened sensitivity speeds up the production of cytotoxic and regulatory T cells.

2. **Stimulating B Cell Activity, Plasma Cell Formation, and Antibody Production**. These events promote the production of antibodies and the development of antibody-mediated immunity.

3. **Enhancing Innate (Nonspecific) Immunity**. Known effects of interleukin production include (a) stimulation of inflammation, (b) formation of scar tissue by fibroblasts, (c) increasing body temperature by the preoptic nucleus of the hypothalamus, (d) stimulation of mast cell formation, and (e) promotion of adrenocorticotropic hormone (ACTH) secretion by the anterior lobe of the pituitary gland.

4. **Moderating the Immune Response**. Some interleukins help suppress immune function and shorten the immune response.

Two interleukins, IL-1 and IL-2, are important in stimulating and maintaining the immune response. When released by activated macrophages and lymphocytes, these cytokines stimulate the activities of other immune cells and of the secreting cell. The result is a positive feedback loop that helps to recruit additional immune cells.

Although mechanisms exist to control the degree of stimulation, the regulatory process sometimes breaks down. Massive production of interleukins can cause problems at least as severe as those of the primary infection. For example, in *Lyme disease*, activated macrophages release IL-1 in response to a localized bacterial infection. This release produces fever, pain, skin rash, and arthritis throughout the entire body.

Some interleukins enhance the immune response, while others suppress it. The quantities secreted at any moment affect the nature and intensity of the response to an antigen. During a typical infection, the pattern of interleukin secretion changes constantly. Whether stimulatory interleukins or suppressive interleukins predominate plays a part in determining whether the individual overcomes the infection. For this reason, interleukins and their interactions are now the focus of intensive research.

Lymphocyte

Interleukin

Fixed macrophage

Interleukin

Interferons

Interferons (IFNs) make the cell that synthesizes them, and that cell's neighbors, resistant to viral infection. In this way, IFNs slow the spread of a virus. They may have other beneficial effects as well. For example, interferon alpha (α) and interferon gamma (γ) attract and stimulate NK cells. Interferon beta (β) slows the progress of inflammation associated with viral infection. Interferon gamma also stimulates macrophages, making them more effective at killing bacterial or fungal pathogens.

Because they stimulate NK cell activity, interferons can be used to fight some cancers. For example, interferon alpha has been used to treat malignant melanoma, bladder cancer, ovarian cancer, and some forms of leukemia. Interferon alpha or gamma may be used to treat Kaposi's sarcoma, a cancer that typically develops in patients with AIDS.

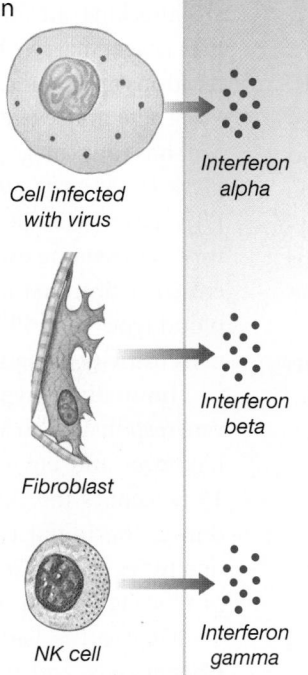

Cell infected with virus → Interferon alpha

Fibroblast → Interferon beta

NK cell → Interferon gamma

Phagocyte-Activating Chemicals

Several cytokines coordinate immune defenses by adjusting the activities of phagocytic cells. These cytokines include factors that attract free macrophages and microphages and prevent their premature departure from the site of an injury.

Colony-Stimulating Factors

Colony-stimulating factors (CSFs) were introduced in Chapter 19. p. 700 These factors are produced by active T cells, cells of the monocyte–macrophage system, endothelial cells, and fibroblasts. CSFs stimulate the production of blood cells in red bone marrow and lymphocytes in lymphoid tissues and organs.

Miscellaneous Cytokines

This general category includes many chemicals discussed in earlier chapters. Examples include leukotrienes, lymphotoxins, perforin, hemopoiesis-stimulating factor, and suppression factors.

NOTE: Cytokines are often classified according to their origins. For example, *lymphokines* are produced by lymphocytes. *Monokines* are secreted by active monocytes, macrophages, and other antigen-presenting cells. These terms are misleading, however, because lymphocytes and macrophages may secrete the same cytokines. In addition, cells involved in adaptive immunity and tissue repair can also secrete cytokines.

Tumor Necrosis Factors

Tumor necrosis factors (TNFs) slow the growth of a tumor and kill sensitive tumor cells. Activated macrophages secrete one type of TNF and carry the molecules in their plasma membranes. Cytotoxic T cells produce a different type of TNF. In addition to their effects on tumor cells, TNFs stimulate granular leukocyte production, promote eosinophil activity, act as pyrogens and cause fever, and increase T cell sensitivity to interleukins.

Macrophage → TNF

Cytotoxic T cell → TNF

These "misguided" antibodies are called **autoantibodies**. The trigger may be a reduction in suppressor T cell activity, the excessive stimulation of helper T cells, tissue damage that releases large quantities of antigenic fragments, haptens bound to compounds normally ignored, viral or bacterial toxins, or a combination of factors.

The resulting condition depends on the antigen that the autoantibodies attack. For example:

- The inflammation of *thyroiditis* is due to autoantibodies against thyroglobulin.
- *Rheumatoid arthritis* occurs when autoantibodies form immune complexes in connective tissues around the joints.
- Type 1 diabetes develops when autoantibodies attack cells in the pancreatic islets.

Many autoimmune disorders appear to be cases of mistaken identity. For example, proteins associated with the measles, Epstein–Barr, influenza, and other viruses contain amino acid sequences that are similar to those of myelin proteins. As a result, antibodies that target these viruses may also attack myelin sheaths. This accounts for the neurological complications that sometimes follow a vaccination or a viral infection. It also may be responsible for *multiple sclerosis*.

For unknown reasons, the risk of autoimmune problems increases if an individual has an unusual type of MHC protein. At least 50 clinical conditions have been linked to specific variations in MHC structure.

Immunodeficiency Diseases

Immunodeficiency diseases result from (1) problems with the embryonic development of lymphoid organs and tissues; (2) an infection with a virus, such as HIV, that depresses immune function; or (3) treatment with, or exposure to, immunosuppressive agents, such as radiation or drugs.

Individuals born with **severe combined immunodeficiency disease (SCID)** fail to develop either cell-mediated or antibody-mediated immunity. Their lymphocyte populations are low, and normal B cells and T cells are absent. Such infants cannot produce an immune response, so even a mild infection can prove fatal. Total isolation offers protection but at great cost—extreme restrictions on lifestyle. Bone marrow transplants from a compatible donor, normally a close relative, have been used to colonize lymphoid tissues with functional lymphocytes. Gene-splicing techniques have led to therapies that can treat at least one form of SCID.

AIDS is an immunodeficiency disease that results from a viral infection that targets helper T cells. As the number of T cells declines, normal immune control breaks down. When a subsequent infection occurs, suppressor factors from suppressor T cells inhibit an immune response before the few surviving helper T cells can stimulate the formation of cytotoxic T cells

or plasma cells in adequate numbers. We discussed AIDS in the Clinical Note (p. 845).

Immunosuppressive drugs have been used for many years to prevent graft rejection after transplant surgery. Unfortunately, immunosuppressive agents can destroy stem cells and lymphocytes and lead to a complete failure of the immune response.

Allergies

Allergies are inappropriate or excessive immune responses to antigens. The sudden increase in cellular activity or antibody titers can have several unpleasant side effects. For example, neutrophils or cytotoxic T cells may destroy normal cells while they are attacking an antigen. Or the antigen–antibody complex may trigger massive inflammation. Antigens that set off allergic reactions are often called **allergens**.

There are several types of allergies. A complete classification has four categories. They include *immediate hypersensitivity (Type I)*, *cytotoxic reactions (Type II)*, *immune complex disorders (Type III)*, and *delayed hypersensitivity (Type IV)*. In Chapter 19 we discussed one example of a cytotoxic (Type II) reaction: the cross-reaction that follows the transfusion of an incompatible blood type. ⊃ p. 692 Here we focus on immediate (Type I) hypersensitivity. It is probably the most common type of allergy.

Immediate hypersensitivity is a rapid and especially severe response to an antigen. One form, *allergic rhinitis*, includes hay fever and environmental allergies. This form may affect 15 percent of the U.S. population. Sensitization to an allergen during the initial exposure leads to the production of large quantities of IgE. Genes may determine a person's tendency to produce IgE in response to particular allergens.

Due to the lag time needed to activate B cells, produce plasma cells, and synthesize antibodies, the first exposure to an allergen does not produce signs and symptoms. Instead, it sets the stage for the next encounter. After sensitization, the IgE molecules become attached to basophils and mast cells throughout the body. When the individual meets the same allergen again, the bound antibodies stimulate these cells to release histamine, heparin, several cytokines, prostaglandins, and other chemicals into the surrounding tissues. This results in a sudden, massive inflammation of the affected tissues.

The cytokines and other mast cell secretions draw basophils, eosinophils, T cells, and macrophages into the area. These cells release their own chemicals, extending and intensifying the responses initiated by mast cells. The severity of the allergic reaction depends on the individual's sensitivity and on the location involved. If allergen exposure occurs at the body surface, the response may be restricted to that area. If the allergen enters the bloodstream, the response could be lethal.

In **anaphylaxis** (an-a-fi-LAK-sis; *ana-*, again + *phylaxis*, protection), a circulating allergen affects mast cells throughout the body (**Figure 22–29**). (In drug reactions, such as allergies to penicillin, IgE antibodies are produced in response to a hapten [partial antigen] bound to a larger molecule that is widely distributed within the body because the combination acts as an allergen.) A wide range of signs and symptoms can develop within minutes. Changes in capillary permeabilities produce swelling and edema in the dermis, and raised welts, or *hives*, appear on the skin. Smooth muscles along the respiratory passageways contract, and the narrowed passages make breathing extremely difficult. In severe cases, an extensive peripheral vasodilation occurs, producing a drop in blood pressure that can lead to a circulatory collapse. This response is **anaphylactic shock**.

The prompt administration of drugs that block the action of histamine, known as **antihistamines** (an-tē-HIS-ta-mēnz), can prevent many of the signs and symptoms of immediate hypersensitivity. *Benadryl (diphenhydramine hydrochloride)* is a popular antihistamine that is available over the counter. Severe anaphylaxis is treated with injections of antihistamines, corticosteroids, and epinephrine.

Stress and the Immune Response

One of the normal effects of interleukin-1 secretion is the stimulation of adrenocorticotropic hormone (ACTH) production by the anterior lobe of the pituitary gland. This hormone in turn prompts the secretion of glucocorticoids by the adrenal cortex. ⤸ p. 658 The anti-inflammatory effects of the glucocorticoids may help control the extent of the immune response.

It is clear, however, that chronic stress depresses the immune system and can be a serious threat to health. The long-term secretion of glucocorticoids, as in the resistance phase of the *stress response*, can inhibit the immune response and lower a person's resistance to disease. ⤸ p. 671 The effects of glucocorticoids that alter the effectiveness of innate and adaptive defenses include the following:

- *Depression of Inflammation.* Glucocorticoids inhibit mast cells and make capillaries less permeable. Inflammation becomes less likely. When it does occur, the reduced permeability of the capillaries slows the entry of fibrinogen, complement proteins, and cellular defenders into tissues.

- *Reduction in the Abundance and Activity of Phagocytes in Peripheral Tissues.* This reduction further impairs innate defense mechanisms. It also interferes with the processing and presentation of antigens to lymphocytes.

- *Inhibition of Interleukin Secretion.* A reduction in interleukin production depresses the response of lymphocytes, even to antigens bound to MHC proteins.

Figure 22–29   **The Mechanism of Anaphylaxis.**

The mechanisms that bring about these changes are still under investigation.

CHECKPOINT

22. **Which kind of immunity protects a developing fetus, and how is that immunity produced?**

23. **Define autoimmune disorder.**

24. **How does increased stress reduce the effectiveness of the immune response?**

See the blue Answers tab at the back of the book.

SYSTEM INTEGRATOR

Body System ⟶ Lymphatic System

Lymphatic System ⟶ Body System

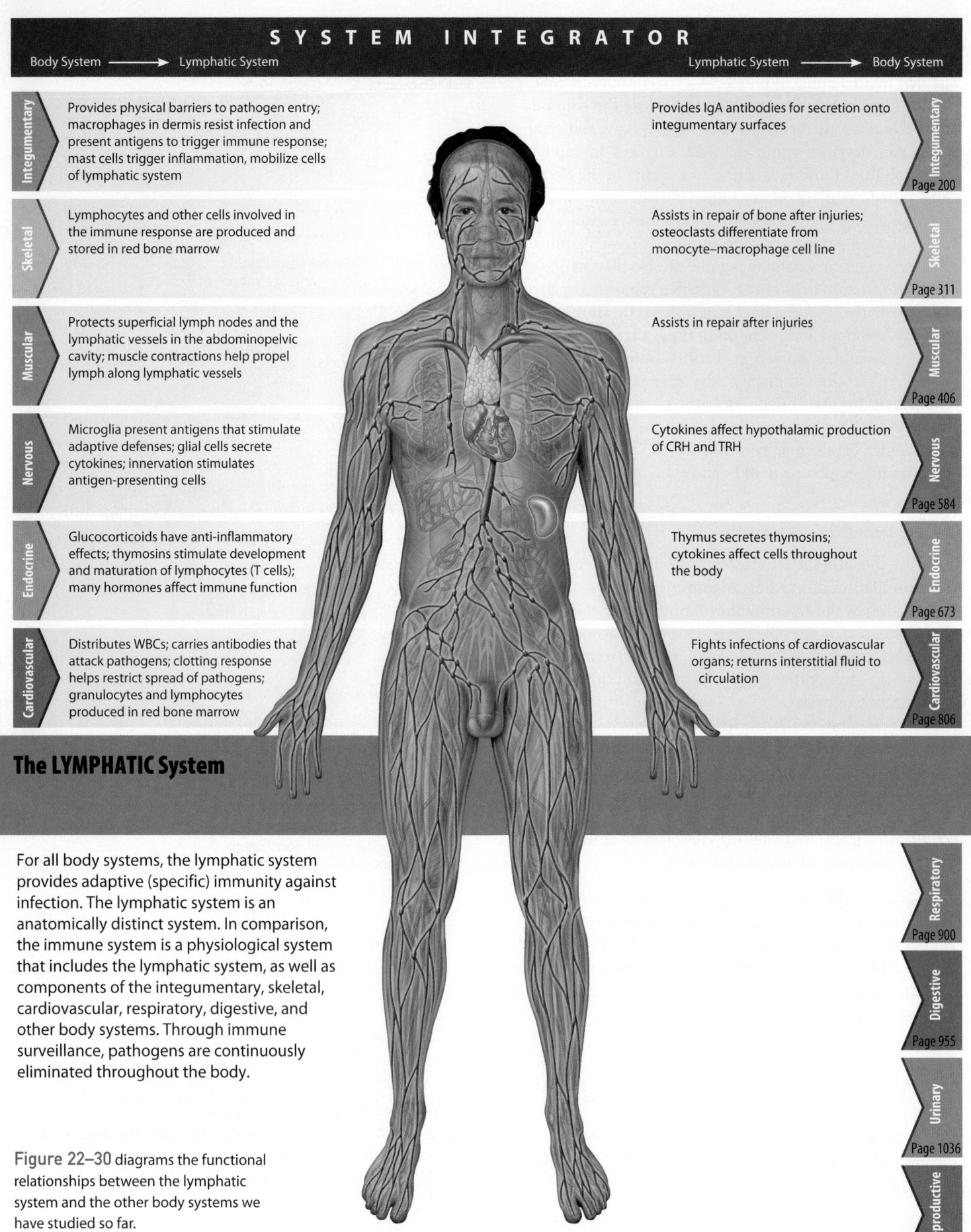

Integumentary
Provides physical barriers to pathogen entry; macrophages in dermis resist infection and present antigens to trigger immune response; mast cells trigger inflammation, mobilize cells of lymphatic system

Skeletal
Lymphocytes and other cells involved in the immune response are produced and stored in red bone marrow

Muscular
Protects superficial lymph nodes and the lymphatic vessels in the abdominopelvic cavity; muscle contractions help propel lymph along lymphatic vessels

Nervous
Microglia present antigens that stimulate adaptive defenses; glial cells secrete cytokines; innervation stimulates antigen-presenting cells

Endocrine
Glucocorticoids have anti-inflammatory effects; thymosins stimulate development and maturation of lymphocytes (T cells); many hormones affect immune function

Cardiovascular
Distributes WBCs; carries antibodies that attack pathogens; clotting response helps restrict spread of pathogens; granulocytes and lymphocytes produced in red bone marrow

Integumentary
Provides IgA antibodies for secretion onto integumentary surfaces
Page 200

Skeletal
Assists in repair of bone after injuries; osteoclasts differentiate from monocyte–macrophage cell line
Page 311

Muscular
Assists in repair after injuries
Page 406

Nervous
Cytokines affect hypothalamic production of CRH and TRH
Page 584

Endocrine
Thymus secretes thymosins; cytokines affect cells throughout the body
Page 673

Cardiovascular
Fights infections of cardiovascular organs; returns interstitial fluid to circulation
Page 806

The LYMPHATIC System

For all body systems, the lymphatic system provides adaptive (specific) immunity against infection. The lymphatic system is an anatomically distinct system. In comparison, the immune system is a physiological system that includes the lymphatic system, as well as components of the integumentary, skeletal, cardiovascular, respiratory, digestive, and other body systems. Through immune surveillance, pathogens are continuously eliminated throughout the body.

Figure 22–30 diagrams the functional relationships between the lymphatic system and the other body systems we have studied so far.

Respiratory
Page 900

Digestive
Page 955

Urinary
Page 1036

Reproductive
Page 1116

22-8 ┃ The immune response diminishes as we age

Learning Outcome Describe the effects of aging on the lymphatic system and the immune response.

As we age, the immune system becomes less effective at combating disease. T cells become less responsive to antigens, so fewer cytotoxic T cells respond to an infection. This effect may be due to the gradual shrinking of the thymus and to lower circulating levels of thymic hormones. Because the number of helper T cells is also reduced, B cells are less responsive, so antibody levels do not rise as quickly after antigen exposure. The net result is an increased susceptibility to viral and bacterial infections. For this reason, vaccinations for acute viral diseases such as the flu (influenza), and for pneumococcal pneumonia, are strongly recommended for elderly individuals. The increased incidence of cancer in elderly people reflects the fact that immune surveillance declines, so tumor cells are not eliminated as effectively.

CHECKPOINT

25. **Why are elderly people more susceptible to viral and bacterial infections?**

26. **What may account for the increased incidence of cancer among elderly people?**

See the blue Answers tab at the back of the book.

22-9 ┃ The nervous and endocrine systems influence the immune response

Learning Outcome Give examples of interactions between the lymphatic system and other organ systems we have studied so far and explain how the nervous and endocrine systems influence the immune response.

Figure 22–30 summarizes the interactions between the lymphatic system and other physiological systems we have studied so far. Interactions among elements of the immune response and the nervous and endocrine systems are now the focus of intense research. For example,

- The thymus secretes oxytocin, ADH, and endorphins as well as thymic hormones. The effects on the CNS are not known, but removal of the thymus lowers brain endorphin levels.

- Both thymic hormones and cytokines help establish the normal levels of CRH and TRH produced by the hypothalamus.

- Other thymic hormones affect the anterior lobe of the pituitary gland directly, stimulating the secretion of prolactin and GH.

Conversely, the nervous system can apparently adjust the sensitivity of the immune response:

- The PNS innervates dendritic cells in the lymph nodes, spleen, skin, and other antigen-presenting cells. The nerve endings release neurotransmitters that heighten local immune responses. For this reason, some skin conditions, such as *psoriasis*, worsen when a person is under stress.

- Neuroglia in the CNS produce cytokines that promote an immune response.

- A sudden decline in immune function can occur after even a brief period of emotional distress.

CHECKPOINT

27. **What is the relationship between the endocrine system and the lymphatic system?**

28. **Identify the role of the lymphatic system for all body systems.**

See the blue Answers tab at the back of the book.

Related Clinical Terms

adenitis: Inflammation of the adenoid (pharyngeal tonsil).

allograft: Transplant between compatible recipient and donor of the same species.

anamnestic response: An immune response that is initiated by memory cells.

autograft: A transplant of tissue that is taken from the same person.

Burkitt's lymphoma: A malignant cancer of B lymphocytes.

chronic fatigue syndrome: A complicated disorder most often characterized by extreme fatigue that does not improve with rest, and which may worsen with physical activity.

congenital thymic aplasia: Congenital (present at birth) absence of the thymus and parathyroid glands and a deficiency of immunity.

Coombs test: A medical test to detect antibodies or complement in the blood.

dermatomyositis: An autoimmune disease characterized by inflammation of the skin and muscles.

eczema: A genetic inflammatory skin disorder, often with crusts, papules, and leaky eruptions.

Hodgkin's lymphoma: A malignant lymphoma affecting lymph nodes and lymph organs.

host versus graft disease: A pathological condition in which cells from the transplanted tissue of a donor initiate an immunological response, attacking the cells and tissue of the recipient.

hybridoma: A tissue culture composed of cancer cells fused to lymphocytes to mass-produce a specific antibody.

immunology: Branch of biomedicine concerned with the structure and function of the immune system.

infectious mononucleosis: An acute disease caused by the Epstein–Barr virus, producing fever, swelling of the lymph nodes, sore throat, and increased lymphocytes in the bloodstream.

latex allergy: Hypersensitivity to products made of the sap of the rubber plant.

polymyositis: An autoimmune disease characterized by inflammation and atrophy of muscles.

sentinel node: The first lymph node to receive drainage from a tumor. It is used to determine if there is lymphatic metastasis in some types of cancer.

splenomegaly: Enlargement of the spleen.

systemic lupus erythematosus (SLE): An autoimmune disease in which a person's immune system attacks and injures its own organs and tissues in virtually every system of the body.

xenograft: A transplant that is made between two different species.

Chapter Review

Study Outline

MasteringA&P®

Access more chapter study tools online in the MasteringA&P Study Area:

- Chapter Quizzes, Chapter Practice Test, Art-labeling Activities, Animations, MP3 Tutor Sessions, and Clinical Case Studies
- Practice Anatomy Lab **PAL**
- Interactive Physiology **iP**
- A&P Flix **A&PFlix**
- PhysioEx **PhysioEx**

22-1 ▌ Surface barriers and internal defenses make up innate defenses, and lymphocytes provide adaptive defenses p. 808

1. The cells, tissues, and organs of the **lymphatic system** play a central role in the body's defenses against a variety of **pathogens**, or disease-causing organisms.

2. *Lymphocytes*, the primary cells of the lymphatic system, are central to an **immune response** against specific threats to the body. **Immunity** is the ability to resist infection and disease.

22-2 ▌ Lymphatic vessels, lymphocytes, lymphoid tissues, and lymphoid organs function in body defenses p. 809

3. The lymphatic system includes a network of **lymphatic vessels**, or **lymphatics**, that carry **lymph** (a fluid similar to plasma, but with a lower concentration of proteins). **Primary lymphoid tissues and organs** are sites where lymphocytes are formed and mature. **Secondary lymphoid tissues and organs** are where lymphocytes are activated and cloned. An array of **lymphoid tissues** and **lymphoid organs** is connected to lymphatics. *(Figure 22–1)*

4. The lymphatic system produces, maintains, and distributes lymphocytes (which attack invading organisms, abnormal cells, and foreign proteins); it also helps maintain blood volume and eliminate local variations in the composition of interstitial fluid.

5. Lymph flows along a network of lymphatic vessels, the smallest of which are the **lymphatic capillaries** (*terminal lymphatics*). The lymphatic vessels empty into the **thoracic duct** and the **right lymphatic duct**. *(Figures 22–2 to 22–4)*

6. The three classes of lymphocytes are **T** (**t**hymus-dependent) **cells**, **B** (**b**one marrow–derived) **cells**, and **NK** (**n**atural **k**iller) **cells**. *(Figure 22–5)*

7. **Cytotoxic T cells** attack foreign cells or body cells infected by viruses and provide **cell-mediated (cellular) immunity**. *Regulatory T cells* (**helper T cells** and **suppressor T cells**) regulate and coordinate the immune response.

8. B cells can differentiate into **plasma cells**, which produce and secrete *antibodies* that react with specific chemical targets called **antigens**. Antibodies in body fluids are called *immunoglobulins*. B cells are responsible for **antibody-mediated (humoral) immunity**.

9. NK cells (also called **large granular lymphocytes**) attack foreign cells, normal cells infected with viruses, and cancer cells. NK cells provide *immune surveillance*.

10. Lymphocytes continuously migrate into and out of the blood through the lymphoid tissues and organs. **Lymphopoiesis** (lymphocyte production) involves the red bone marrow, thymus, and peripheral lymphoid tissues. *(Figure 22–6)*

11. **Lymphoid tissues** are connective tissues dominated by lymphocytes. In a **lymphoid nodule**, the lymphocytes are densely packed in an area of areolar tissue. The lymphoid tissue that protects the epithelia of the digestive, respiratory, urinary, and reproductive tracts is called **mucosa-associated lymphoid tissue (MALT)**. *(Figure 22–7)*

12. Important lymphoid organs include the **lymph nodes**, the **thymus**, and the **spleen**. Lymphoid tissues and organs are distributed in areas that are especially vulnerable to injury or invasion.

13. Lymph nodes are encapsulated masses of lymphoid tissue. The **deep cortex** is dominated by T cells; the **outer cortex** and **medulla** contain B cells. *(Figure 22–8)*

14. The thymus lies behind the sternum, in the anterior mediastinum. **Thymic epithelial cells (TECs)** regulate T cell development and function. *(Figure 22–9)*

15. The adult spleen contains the largest mass of lymphoid tissue in the body. The cellular components form the **pulp** of the spleen. **Red pulp** contains large numbers of red blood cells, and **white pulp** resembles lymphoid nodules. *(Figure 22–10)*

16. The lymphatic system is a major component of the body's defenses, which are classified as either (1) **innate (nonspecific) immunity**, which protects without distinguishing one threat from another, or (2) **adaptive (specific) immunity**, which protects against particular threats only.

22-3 ❚ Innate (nonspecific) defenses do not discriminate between potential threats and respond the same regardless of the invader p. 822

17. Innate (nonspecific) defenses prevent the approach, deny the entry, or limit the spread of living or nonliving hazards. *(Figure 22–11)*

18. Physical barriers include skin, mucous membranes, hair, epithelia, and various secretions of the integumentary and digestive systems.

19. The two types of phagocytic cells are **microphages** and **macrophages** (cells of the **monocyte–macrophage system**). Microphages are neutrophils and eosinophils in circulating blood.

20. **Phagocytes** leave the bloodstream by *emigration*, or *diapedesis* (migration between adjacent endothelial cells), and exhibit **chemotaxis** (sensitivity and orientation to chemical stimuli).

21. **Immune surveillance** involves constant monitoring of normal tissues by NK cells that are sensitive to abnormal antigens on the surfaces of otherwise normal cells. NK cells kill cancer cells that have **tumor-specific antigens** on their surfaces. *(Figure 22–12)*

22. **Interferons (IFNs)**—small proteins released by cells infected with viruses—trigger the production of **antiviral proteins**, which interfere with viral replication inside the cell. IFNs are **cytokines**, which are chemical messengers released by tissue cells to coordinate local activities. *(Figure 22–13)*

23. At least 30 **complement proteins** make up the **complement system**. These proteins interact with each other in cascades to destroy target plasma membranes, stimulate inflammation, attract phagocytes, or enhance phagocytosis. The complement system can be activated by either the **classical pathway** or the **alternative pathway**. *(Figure 22–14)*

24. **Inflammation** is a localized tissue response to injury. *(Figure 22–15)*

25. A **fever** (body temperature greater than 37.2°C [99°F]) can inhibit pathogens and accelerate metabolic processes. **Pyrogens** can reset the body's thermostat and raise the temperature.

22-4 ❚ Adaptive (specific) defenses respond to individual threats and are either cell-mediated or antibody-mediated p. 828

26. T cells are responsible for **cell-mediated (cellular) immunity**. B cells provide **antibody-mediated (humoral) immunity**.

27. Forms of immunity include **innate immunity** (genetically determined and present at birth) or **adaptive immunity** (produced by prior exposure to an antigen or antibody production). The two types of adaptive immunity are **active immunity** (which appears after exposure to an antigen) and **passive immunity** (produced by the transfer of antibodies from another source). *(Figure 22–16)*

28. Immunity exhibits four general properties: **specificity**, **versatility**, **memory**, and **tolerance**. *Memory cells* enable the immune system to "remember" previous target antigens. Tolerance is the ability of the immune system to ignore some antigens, such as those of normal body cells.

29. The immune response is triggered by the presence of an antigen and includes cell-mediated and antibody-mediated defenses. *(Figure 22–17)*

22-5 ❚ T cells play a role in initiating, maintaining, and controlling the immune response p. 831

30. **Antigen presentation** occurs when an antigen–glycoprotein combination appears in a plasma membrane (typically of a macrophage or dendritic cell). T cells sensitive to this combination are activated if they contact the membrane of the antigen-presenting cell.

31. All body cells have plasma membrane glycoproteins. The genes controlling their synthesis make up a chromosomal region called the **major histocompatibility complex (MHC)**. The membrane glycoproteins are called **MHC proteins**. **APCs (antigen-presenting cells)** are involved in antigen stimulation.

32. Lymphocytes are not activated by free antigens, but respond instead to an antigen bound to either a **Class I** or a **Class II** MHC protein in a process called **antigen recognition**. *(Figure 22–18)*

33. Class I MHC proteins are in all nucleated body cells. Class II MHC proteins are only in antigen-presenting cells (APCs) and lymphocytes.

34. Whether a T cell responds to antigens held in Class I or Class II MHC proteins depends on the structure of the T cell plasma membrane. T cell plasma membranes contain proteins called **CD** (*cluster of differentiation*) **markers**. **CD3 markers** are present on all T cells. **CD8 markers** are on cytotoxic and suppressor T cells. **CD4 markers** are on all helper T cells.

35. One type of CD8 cell responds quickly to a Class I MHC-bound antigen, giving rise to large numbers of cytotoxic T cells and memory T cells. The other type of CD8 cell responds more slowly, giving rise to small numbers of suppressor T cells.

36. Cytotoxic T cells seek out and destroy abnormal and infected cells, using three different methods, including the secretion of **lymphotoxin**. *(Figure 22–19)*

37. Cell-mediated immunity (cellular immunity) results from the activation of CD8 T cells by antigens bound to Class I MHCs. When activated, most of these T cells divide to generate cytotoxic T cells and **memory T_C cells**, which remain in reserve to guard against future such attacks. Suppressor T cells depress the responses of other T cells and of B cells. *(Figures 22–19, 22–21)*

38. Helper, or CD4, T cells respond to antigens presented by Class II MHC proteins. When activated, helper T cells secrete cytokines that aid in coordinating adaptive and innate defenses, and regulate cell-mediated and antibody-mediated immunity. *(Figures 22–20, 22–21)*

22-6 ❚ B cells respond to antigens by producing specific antibodies p. 836

39. B cells become **sensitized** when antibody molecules in their membranes bind antigens. The antigens are then displayed on the Class II MHC proteins of the B cells, which become activated by helper T cells activated by the same antigen. *(Figure 22–22)*

40. An active B cell may differentiate into a plasma cell or produce daughter cells that differentiate into plasma cells and **memory B cells**. Antibodies are produced by plasma cells. *(Figure 22–22)*

41. A Y-shaped antibody molecule consists of two parallel pairs of polypeptide chains containing *constant* and *variable* segments. *(Figure 22–23)*

42. When antibody molecules bind to an antigen, they form an **antigen–antibody complex**. Effects that appear after binding include **neutralization** (antibody binding that prevents viruses or bacterial toxins from binding to body cells); **precipitation** (formation of an insoluble **immune complex**) and **agglutination** (formation of large complexes); *opsonization* (coating of pathogens with antibodies and complement

22

proteins to enhance phagocytosis); stimulation of inflammation; and prevention of bacterial or viral adhesion. *(Figure 22–23)*

43. The five classes of antibodies **(immunoglobulins, Ig)** in body fluids are (1) **IgG**, responsible for resistance against many viruses, bacteria, and bacterial toxins; (2) **IgE**, which releases chemicals that accelerate local inflammation; (3) **IgD**, located on the surfaces of B cells; (4) **IgM**, the first type of antibody secreted after an antigen arrives; and (5) **IgA**, found in glandular secretions. *(Table 22–1)*

44. In humoral immunity, the antibodies first produced by plasma cells are the agents of the **primary response**. The maximum antibody level and **antibody titer** appears during the **secondary response** to antigen exposure. *(Figure 22–24)*

45. The initial steps in the immune responses to viral and bacterial infections differ. *(Figures 22–25 to 22–27; Table 22–2)*

46. Cytokines are chemical messengers coordinated by the immune system. **Interleukins** increase T cell sensitivity to antigens exposed on macrophage membranes; stimulate B cell activity, plasma cell formation, and antibody production; enhance innate defenses; and moderate the immune response. *(Spotlight Figure 22–28)*

47. Interferons (IFNs) slow the spread of a virus by making the synthesizing cell and its neighbors resistant to viral infections. *(Spotlight Figure 22–28)*

48. **Tumor necrosis factors (TNFs)** slow tumor growth and kill tumor cells, and act as a pyrogen. *(Spotlight Figure 22–28)*

49. Several cytokines adjust the activities of phagocytic cells to coordinate innate and adaptive defenses. *(Spotlight Figure 22–28)*

50. **Colony-stimulating factors (CSFs)** are factors produced by active T cells, cells of the monocyte–macrophage group, endothelial cells, and fibroblasts. *(Spotlight Figure 22–28)*

22-7 ▌ Immunocompetence enables a normal immune response; abnormal responses result in immune disorders p. 844

51. **Immunocompetence** is the ability to produce an immune response after exposure to an antigen. A developing fetus receives passive immunity from the maternal bloodstream. After delivery, the infant begins developing active immunity following exposure to environmental antigens.

52. **Autoimmune disorders** develop when the immune response inappropriately targets normal body cells and tissues.

53. In an **immunodeficiency disease**, either the immune system does not develop normally or the immune response is blocked.

54. **Allergies** are inappropriate or excessive immune responses to **allergens** (antigens that trigger allergic reactions). The four types of allergies are *immediate hypersensitivity (Type I), cytotoxic reactions (Type II), immune complex disorders (Type III)*, and *delayed hypersensitivity (Type IV)*.

55. In **anaphylaxis**, a circulating allergen affects mast cells throughout the body. *(Figure 22–29)*

56. Interleukin-1 released by active macrophages triggers the release of ACTH by the anterior pituitary gland. Glucocorticoids produced by the adrenal cortex moderate the immune response, but their long-term secretion can lower a person's resistance to disease.

22-8 ▌ The immune response diminishes as we age p. 851

57. With aging, the immune system becomes less effective at combating disease.

22-9 ▌ The nervous and endocrine systems influence the immune response p. 851

58. The lymphatic system has extensive interactions with the nervous and endocrine systems. *(Figure 22–30)*.

Review Questions

See the blue Answers tab at the back of the book.

LEVEL 1 Reviewing Facts and Terms

1. Identify the structures of the lymphatic system in the following diagram.

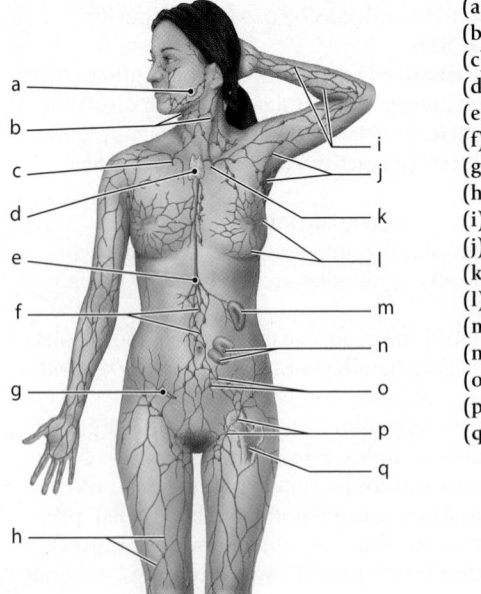

(a)	_____
(b)	_____
(c)	_____
(d)	_____
(e)	_____
(f)	_____
(g)	_____
(h)	_____
(i)	_____
(j)	_____
(k)	_____
(l)	_____
(m)	_____
(n)	_____
(o)	_____
(p)	_____
(q)	_____

2. Lymph from the right arm, the right half of the head, and the right chest is received by the **(a)** cisterna chyli, **(b)** right lymphatic duct, **(c)** right thoracic duct, **(d)** aorta.

3. When interstitial fluids accumulate in a limb, and it gradually becomes swollen and grossly distended, the condition is known as _____.

4. The specificity of an antibody is determined by the **(a)** fixed segment, **(b)** antigenic determinants, **(c)** variable region, **(d)** size of the antibody, **(e)** antibody class.

5. The major histocompatibility complex (MHC) **(a)** is responsible for forming lymphocytes, **(b)** produces antibodies in lymph glands, **(c)** is a group of genes that codes for human leukocyte antigens, **(d)** is a membrane protein that can recognize foreign antigens, **(e)** is the antigen found on bacteria that stimulates an immune response.

6. In the liver, where are the Kupffer cells and what type of cell are they?

7. Phagocytes move through capillary walls by squeezing between adjacent endothelial cells, a process known as **(a)** diapedesis, **(b)** chemotaxis, **(c)** adhesion, **(d)** perforation.

8. Perforins are proteins associated with the activity of **(a)** T cells, **(b)** B cells, **(c)** NK cells, **(d)** plasma cells.

9. Complement activation **(a)** stimulates inflammation, **(b)** attracts phagocytes, **(c)** enhances phagocytosis, **(d)** achieves a, b, and c.

10. The most beneficial effect of fever is that it **(a)** inhibits the spread of some bacteria and viruses, **(b)** increases the metabolic rate by up to 10 percent, **(c)** stimulates the release of pyrogens, **(d)** achieves a and b.

11. CD4 markers are associated with **(a)** cytotoxic T cells, **(b)** suppressor T cells, **(c)** helper T cells, **(d)** a, b, and c.

12. List the specific functions of each of the body's primary and secondary lymphoid tissues and organs.
13. Give a function for each of the following: (a) cytotoxic T cells, (b) helper T cells, (c) suppressor T cells, (d) plasma cells, (e) NK cells, (f) stromal cells, (g) thymic epithelial cells, (h) interferons, (i) pyrogens, (j) T cells, (k) B cells, (l) interleukins, (m) tumor necrosis factor, (n) colony-stimulating factors.
14. What are the three classes of lymphocytes, and where does each class originate?
15. How does the stomach protect the body from infections?

LEVEL 2 Reviewing Concepts

16. Compared with innate defenses, adaptive defenses(a) do not distinguish between one threat and another, (b) are always present at birth,(c) protect against threats on an individual basis, (d) deny the entry of pathogens to the body.
17. Blocking the antigen receptors on the surface of lymphocytes would interfere with(a) phagocytosis of the antigen, (b) that lymphocyte's ability to produce antibodies,(c) antigen recognition, (d) the ability of the lymphocyte to present antigen, (e) opsonization of the antigen.
18. A *decrease* in which population of lymphocytes would impair all aspects of an immune response?(a) cytotoxic T cells, (b) helper T cells, (c) suppressor T cells, (d) B cells, (e) plasma cells.
19. Where is the hormone thymosin secreted from? What is its effect?
20. Explain what the membrane attack complex is in complement fixation.
21. What is apoptosis? Which immune system cell is responsible for initiating this mechanism in the target cells?
22. Natural killer cells secrete perforins. How does this kill the target cell?
23. Name three antigen-presenting cells.

24. An anesthesia technician is advised that she should be vaccinated against hepatitis B, which is caused by a virus. She is given one injection and is told to come back for a second injection in a month and a third injection after six months. Why is this series of injections necessary?

LEVEL 3 Critical Thinking and Clinical Applications

25. An investigator at a crime scene discovers some body fluid on the victim's clothing. The investigator carefully takes a sample and sends it to the crime lab for analysis. On the basis of the analysis of antibodies, could the crime lab determine whether the sample is blood plasma or semen? Explain.
26. Ted finds out that he has been exposed to measles. He is concerned that he might have contracted the disease, so he goes to see his physician. The physician takes a blood sample and sends it to a lab for antibody levels and titers. The results show an elevated level and activity of IgM antibodies to rubella (measles) virus but very few IgG antibodies to the virus. Has Ted contracted the disease?
27. While walking along the street, you and your friend see an elderly woman whose left arm appears to be swollen to several times its normal size. Your friend wonders aloud what might be its cause. You say that it may be likely that the woman had a radical mastectomy (the removal of a breast because of cancer). Explain the rationale behind your answer.
28. Paula's grandfather is diagnosed as having lung cancer. His physician takes biopsies of several lymph nodes from neighboring regions of the body, and Paula wonders why, since his cancer is in the lungs. What would you tell her?
29. Willy is allergic to ragweed pollen and tells you that he read about a medication that can help his condition by blocking certain antibodies. Do you think that this treatment could help Willy? Explain.

CLINICAL CASE Wrap-Up

Isn't There a Vaccine for That?

Immunizations against infectious diseases are one of the most successful medical advancements in the history of public health. A vaccine, containing weakened or dead forms of the infecting microbe (antigen), is administered orally, intranasally, subcutaneously, or intramuscularly. The antigens are not virulent enough to cause the disease, but they provoke the adaptive immune response in the body. This immune response involves B cells that mature into plasma cells and make antibodies. It also involves cytotoxic (killer) T cells and helper (defensive) T cells. Suppressor T cells are the regulators. Memory T cells and memory B cells help the body remember the antigen so when the protected person is exposed to the real disease, a strong secondary immune response occurs.

Immunizations are as varied as the diseases they protect against. Some are given as live, attenuated (greatly weakened) forms of a virus. Others are given as inactivated antigens, made from small pieces of killed bacteria. Some vaccines require repeat doses (booster shots) to boost the secondary immune response and keep immune memory alive.

Baby Ruthie is due for her chickenpox vaccination next month, at her 12-month checkup. She is not yet immune to the varicella-zoster virus (VZV) that causes chickenpox. Disease control in babies not yet immunized depends on the rest of the population being immunized. This is known as "herd" immunity. Ruthie's day care classmate, Tommy, had never been vaccinated and was spreading chickenpox to other unprotected children in the class.

1. Is immunity to chickenpox, developed after an immunization, innate immunity or adaptive immunity? Why?

2. Does baby Ruthie still need her chickenpox vaccination at her 12-month checkup?

See the blue Answers tab at the back of the book.

23

The Respiratory System

Learning Outcomes

These Learning Outcomes correspond by number to this chapter's sections and indicate what you should be able to do after completing the chapter.

23-1 ▌ Describe the primary functions of the respiratory system, and explain how the delicate respiratory exchange surfaces are protected from pathogens, debris, and other hazards.

23-2 ▌ Identify the organs of the upper respiratory system, and describe their functions.

23-3 ▌ Describe the structure of the larynx, and discuss its roles in normal breathing and in sound production.

23-4 ▌ Discuss the structures of the extrapulmonary airways.

23-5 ▌ Describe the superficial anatomy of the lungs, the structure of a pulmonary lobule, and the functional anatomy of alveoli.

23-6 ▌ Define and compare the processes of external respiration and internal respiration.

23-7 ▌ Summarize the physical principles controlling the movement of air into and out of the lungs, and describe the origins and actions of the muscles responsible for respiratory movements.

23-8 ▌ Summarize the physical principles governing the diffusion of gases into and out of the blood and body tissues.

23-9 ▌ Describe the structure and function of hemoglobin, and the transport of oxygen and carbon dioxide in the blood.

23-10 ▌ List the factors that influence respiration rate, and discuss reflex respiratory activity and the brain centers involved in the control of respiration.

23-11 ▌ Describe age-related changes in the respiratory system.

23-12 ▌ Give examples of interactions between the respiratory system and other organ systems studied so far.

SPOTLIGHTS
- Respiratory Muscles and Pulmonary Ventilation p. 879
- Control of Respiration pp. 894–895

CLINICAL CASE
- How Long Should a Cough Last? p. 857

Clinical Notes
- Breakdown of the Respiratory Defense System p. 860

- Pneumothorax p. 878
- Decompression Sickness p. 884
- Blood Gas Analysis p. 886
- Carbon Monoxide Poisoning p. 889
- Emphysema and Lung Cancer p. 901

How Long Should a Cough Last?

Andrea is in her A&P lab dissecting a cadaver when one of her lab partners suddenly sneezes. With Andrea's next breath, she inhales droplets from the sneeze, which contains rhinovirus—one of the viruses that cause the common cold. The virus quickly spreads across her respiratory mucosa, overtakes the innate immunity of her respiratory epithelium, and enters the epithelial cells of her pharynx and trachea. Once inside the respiratory epithelial cells, the virus begins to multiply.

Three days later, the infected respiratory cells burst, releasing rhinovirus to infect other cells throughout Andrea's body.

Andrea feels terrible. She has a fever of 101°F, a runny nose, sore throat, general fatigue, and an unrelenting cough. After a few days, her cough becomes productive with thick, yellow mucus.

After a week, Andrea is still coughing. She makes an appointment to see her family doctor. She insists she needs a prescription for antibiotics because her cough is lasting longer than a week. The last time she had a bad cough, a doctor prescribed antibiotics, and her cough finally subsided after a second week. **Is a cough lasting longer than a week cause for concern? To find out, turn to the Clinical Case Wrap-Up on p. 905.**

An Introduction to the Respiratory System

When we think of the respiratory system, we generally think of breathing—pulling air into and out of our bodies. However, an efficient respiratory system must do more than merely move air. Cells need energy for maintenance, growth, defense, and division. Our cells obtain that energy mainly through aerobic mechanisms that require oxygen and produce carbon dioxide.

Many aquatic organisms can obtain oxygen and excrete carbon dioxide by diffusion across the surface of the skin or specialized structures, such as the gills of a fish. But such arrangements are poorly suited for life on land. The exchange surfaces must be very thin to permit rapid diffusion. In air, these exposed membranes collapse, evaporation and dehydration reduce blood volume, and the delicate surfaces become vulnerable to attack by pathogens. The respiratory exchange surfaces of humans are just as thin and delicate as those of an aquatic organism, but they are confined to the inside of the *lungs*—a warm, moist, protected environment. Under these conditions, diffusion can take place between the air and the blood.

The cardiovascular system is the link between your interstitial fluids and the exchange surfaces of your lungs. Circulating blood carries oxygen from the lungs to peripheral tissues. Your blood also accepts and transports the carbon dioxide generated by those tissues, delivering it to the lungs. In this chapter we describe how air enters the lungs as a result of the actions of respiratory muscles, and how oxygen and carbon dioxide are exchanged across delicate epithelial surfaces within the lungs.

23-1 | The respiratory system, organized into an upper respiratory system and a lower respiratory system, has several basic functions

Learning Outcome Describe the primary functions of the respiratory system, and explain how the delicate respiratory exchange surfaces are protected from pathogens, debris, and other hazards.

The **respiratory system** is composed of structures involved in ventilation (providing air) and gas exchange. In this section we consider this body system's functions and structural organization.

Functions of the Respiratory System

The respiratory system has five basic functions:

1. Providing an extensive surface area for gas exchange between air and circulating blood.

2. Moving air to and from the exchange surfaces of the lungs along the respiratory passageways.

3. Protecting respiratory surfaces from dehydration, temperature changes, or other environmental variations, and defending the respiratory system and other tissues from invasion by pathogens.

4. Producing sounds for speaking, singing, and other forms of communication.

5. Facilitating the detection of odors by olfactory receptors in the superior portions of the nasal cavity.

In addition, the capillaries of the lungs indirectly help to regulate blood volume and blood pressure. They do so through the conversion of angiotensin I to angiotensin II. ⤺ p. 666

23

Organization of the Respiratory System

We can organize the respiratory system from either an anatomical or a functional perspective. Anatomically, we can divide the system into an upper respiratory system and a lower respiratory system (**Figure 23–1**). The **upper respiratory system** consists of the nose, nasal cavity, paranasal sinuses, and pharynx (throat). These passageways filter, warm, and humidify incoming air, protecting the more delicate surfaces of the lower respiratory system. They also reabsorb heat and water from the outgoing air. The **lower respiratory system** includes the larynx (voice box), trachea (windpipe), bronchi, bronchioles, and alveoli of the lungs.

TIPS & TRICKS

To recall the boundary between the upper and lower respiratory systems, remember that the *l*ower respiratory system begins at the *l*arynx.

The term **respiratory tract** refers to the passageways that carry air to and from the exchange surfaces of the lungs. The *conducting portion* of the respiratory tract begins at the entrance to the nasal cavity and extends through the pharynx, larynx, trachea, bronchi, and larger bronchioles. The *respiratory portion* of the tract includes the smallest, thinnest *bronchioles* and the associated **alveoli** (al-VĒ-ō-lī), air-filled pockets within the lungs where all gas exchange between air and blood takes place.

Gas exchange can take place quickly and efficiently because the distance between the blood in an alveolar capillary and the air inside an alveolus is generally less than 1 μm. In some cases this distance is as short as 0.1 μm. To meet the metabolic requirements of peripheral tissues, the surface area for gas exchange in the lungs must be very large. It is about 35 times the surface area of your body. Estimates of the surface area involved in gas exchange range from 70 m^2 to 140 m^2 (753 ft^2 to 1506 ft^2).

Filtering, warming, and humidifying inhaled air begin at the entrance to the respiratory tract and continue as air

Figure 23–1   The Structures of the Respiratory System. Only the conducting portion of the respiratory tract is shown; the smaller bronchioles and alveoli have been omitted. ATLAS: Plates 47a,b

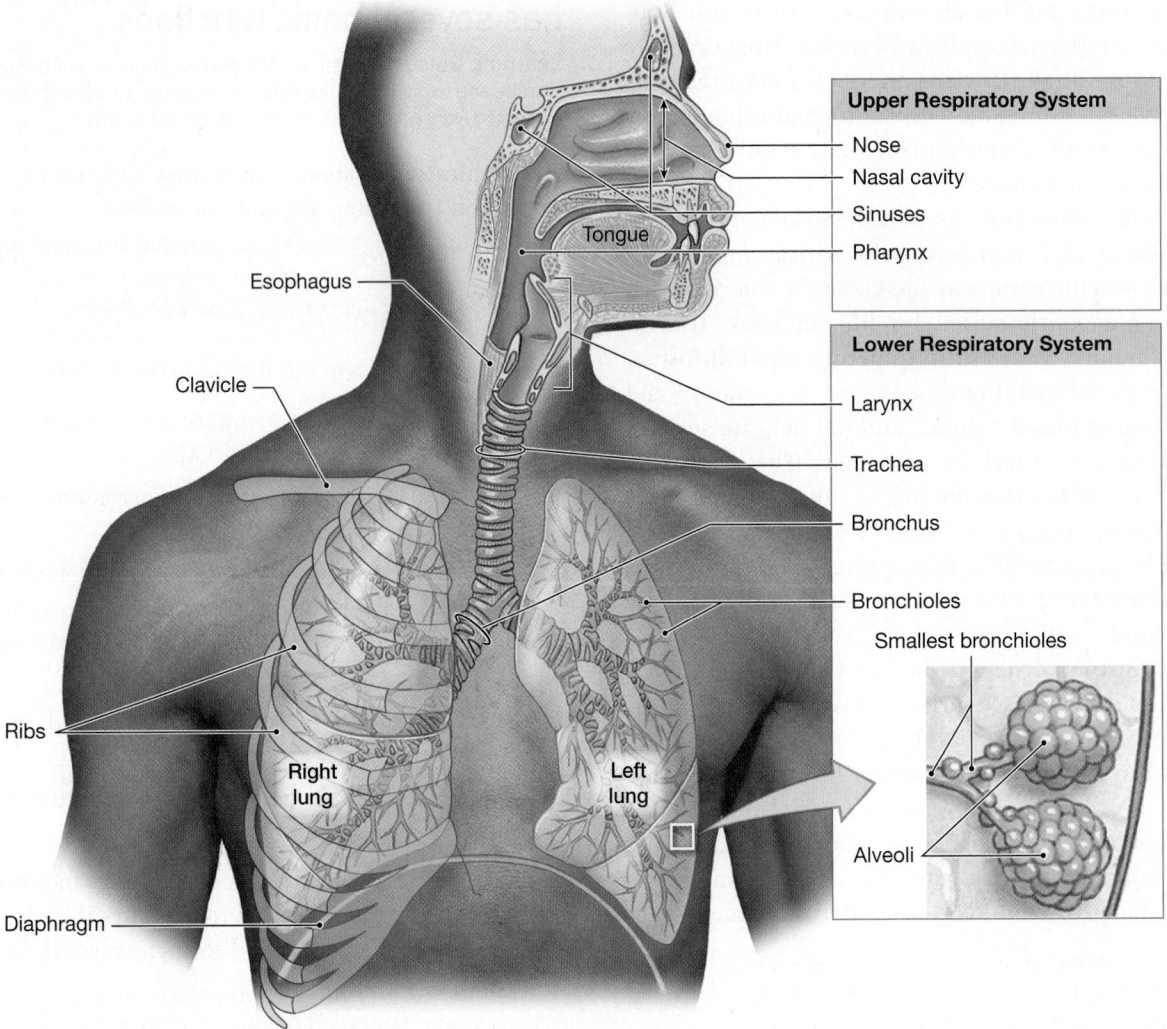

Upper Respiratory System
- Nose
- Nasal cavity
- Sinuses
- Pharynx

Tongue

Esophagus

Clavicle

Lower Respiratory System
- Larynx
- Trachea
- Bronchus
- Bronchioles
- Smallest bronchioles
- Alveoli

Ribs

Right lung

Left lung

Diaphragm

23

passes along the conducting portion. By the time air reaches the alveoli, most foreign particles and pathogens have been removed, and the humidity and temperature are within acceptable limits. The success of this "conditioning process" is due to the respiratory mucosa.

The Respiratory Mucosa

The **respiratory mucosa** (mū-KŌ-suh) lines the conducting portion of the respiratory system. A *mucosa* is a *mucous membrane*, one of the four types of membranes introduced in Chapter 4. It consists of an epithelium and an underlying layer of areolar tissue. ↺ p. 163 The **lamina propria** (LAM-i-nuh PRŌ-prē-uh) is the underlying layer of areolar tissue that supports the respiratory epithelium. In the upper respiratory system, trachea, and bronchi, the lamina propria contains mucous glands that discharge their secretions onto the epithelial surface (**Figure 23–2**). The lamina propria in the conducting portions of the lower respiratory system contains bundles of smooth muscle cells. At the bronchioles, the smooth muscles form thick bands that encircle or spiral around the lumen.

Figure 23–2   **The Respiratory Epithelium of the Nasal Cavity and Conducting System.**

Superficial view   SEM × 1647

a A surface view of the epithelium. The cilia of the epithelial cells form a dense layer that resembles a shag carpet. The movement of these cilia propels mucus across the epithelial surface.

Movement of mucus to pharynx

Ciliated columnar epithelial cell

Mucous cell

Stem cell

Mucus layer

Lamina propria

b A diagrammatic view of the respiratory epithelium of the trachea, showing the direction of mucus transport inferior to the pharynx.

Cilia

Lamina propria

Nucleus of columnar epithelial cell

Mucous cell

Basement membrane

Stem cell

c A sectional view of the respiratory epithelium, a pseudostratified ciliated columnar epithelium.

23

The structure of the respiratory epithelium changes along the respiratory tract. A pseudostratified ciliated columnar epithelium with numerous mucous cells lines the nasal cavity and the superior portion of the pharynx. ⤴ p. 149 The epithelium lining inferior portions of the pharynx is a stratified squamous epithelium similar to that of the oral cavity. These portions of the pharynx conduct air to the larynx and also convey food to the esophagus. The pharyngeal epithelium must therefore protect against abrasion and chemical attack.

A pseudostratified ciliated columnar epithelium, comparable to that of the nasal cavity, lines the superior portion of the lower respiratory system. The smaller bronchioles have a cuboidal epithelium with scattered cilia. The exchange surfaces of the alveoli are lined by a very delicate simple squamous epithelium. Other, more specialized cells are scattered among the squamous cells and together they form the *alveolar epithelium*.

The Respiratory Defense System

Debris or pathogens in inhaled air can severely damage the delicate exchange surfaces of the respiratory system. A series of filtration mechanisms that make up the **respiratory defense system** prevent such contamination.

Along much of the respiratory tract, mucous cells in the epithelium and mucous glands in the lamina propria produce a sticky mucus that bathes exposed surfaces. In the nasal cavity, cilia sweep that mucus and any trapped debris or microorganisms toward the pharynx. There it is swallowed and exposed to the acids and enzymes of the stomach. In the lower respiratory system, the cilia beat toward the pharynx, moving a carpet of mucus in that direction and cleaning the respiratory surfaces. This process is often called a *mucus escalator* (**Figure 23–2b,c**).

TIPS&TRICKS

The mucus layer of the respiratory epithelium functions like sticky flypaper, but instead of flies, it traps particles and debris from the air that moves past it.

Filtration in the nasal cavity removes virtually all particles larger than about 10 μm from the inhaled air. Smaller particles may be trapped by the mucus of the nasopharynx or by secretions of the pharynx. The rate of mucus production in the nasal cavity and paranasal sinuses speeds up upon exposure to unpleasant stimuli, such as noxious vapors, large quantities of dust and debris, allergens, or pathogens. (The familiar signs and symptoms of the "common cold" appear when any of more than 200 types of viruses invades the respiratory epithelium.)

Most particles 1–5 μm in diameter are trapped in the mucus coating the respiratory bronchioles or in the liquid covering the alveolar surfaces. These areas are outside of the mucus escalator, but the foreign particles can be engulfed by alveolar macrophages. Most particles smaller than about 0.5 μm remain suspended in the air.

Clinical Note

Breakdown of the Respiratory Defense System Large quantities of airborne particles may overload the respiratory defenses and produce a variety of illnesses. For example, irritants in the lining of the conducting passageways can provoke the formation of mucus plugs that block airflow and reduce pulmonary function. Damage to the epithelium in the affected area may allow irritants to enter the surrounding tissues of the lung. The irritants then produce local inflammation. Airborne irritants—such as those in cigarette smoke—are known to promote the development of lung cancer (p. 901).

Aggressive pathogens can also overwhelm respiratory defenses. **Tuberculosis** (tū-ber-kū-LŌ-sis), or **TB**, results from an infection of the lungs by the bacterium *Mycobacterium tuberculosis*. Other organs may be invaded as well. Bacteria may colonize the respiratory passageways, the interstitial spaces, the alveoli, or a combination of the three. Signs and symptoms are variable. They generally include coughing and chest pain, plus fever, night sweats, fatigue, and weight loss. In 1900, TB, then known as "consumption," was the leading cause of death. It remains among the most common and serious infectious diseases. An estimated one-third of the world's population is infected with TB. According to the Centers for Disease Control and Prevention (CDC), there are nearly 2 million TB-related deaths worldwide each year.

The respiratory defense system can also fail due to inherited congenital defects affecting mucus production or transport. For example, **cystic fibrosis (CF)** is the most common lethal inherited disease in individuals of Northern European descent. It occurs in 1 birth in 2500. The respiratory mucosa in these individuals produces dense, viscous mucus that cannot be transported by the respiratory defense system. The mucus escalator stops working, leading to frequent infections. Mucus also blocks the smaller respiratory passageways, making breathing difficult. For people who have CF and live in the United States, the median predicted age of survival is the mid-30s. Death generally results from heart failure associated with a massive chronic bacterial infection of the lungs.

CHECKPOINT

1. Identify several functions of the respiratory system.
2. List the two anatomical divisions of the respiratory system.
3. What membrane lines the conducting portion of the respiratory tract?

See the blue Answers tab at the back of the book.

23-2 ┃ Located outside the thoracic cavity, the upper respiratory system consists of the nose, nasal cavity, paranasal sinuses, and pharynx

Learning Outcome Identify the organs of the upper respiratory system, and describe their functions.

As we have noted, the upper respiratory system consists of the nose, nasal cavity, paranasal sinuses, and pharynx (**Figures 23–1** and **23–3**).

The Nose, Nasal Cavity, and Paranasal Sinuses

The nose is the primary passageway for air entering the respiratory system. Air normally enters through the paired **external nares** (NĀ-res), or *nostrils* (**Figure 23–3a**), which open into the *nasal cavity*. The **nasal vestibule** is the space contained within the flexible tissues of the nose (**Figure 23–3c**). The epithelium of the vestibule contains coarse hairs that extend across the external nares. Large airborne particles, such as sand, sawdust, or even insects, are trapped in these hairs and prevented from entering the nasal cavity.

The **nasal septum** divides the nasal cavity into left and right portions (**Figure 23–3b**). The bony portion of the nasal septum is formed by the fusion of the perpendicular plate of the ethmoid bone and the plate of the vomer (**Figure 7–3d**, p. 237). The anterior portion of the nasal septum is formed of hyaline cartilage. This cartilaginous plate supports the *dorsum nasi* (DOR-sum NĀ-zī), or bridge, and *apex* (tip) of the nose.

The maxillary, nasal, frontal, ethmoid, and sphenoid bones form the lateral and superior walls of the nasal cavity. The mucous secretions produced in the *paranasal sinuses* (sinuses of the frontal, sphenoid, ethmoid, and paired maxillary and palatine bones) help keep the surfaces of the nasal cavity moist and clean (**Figure 7–15**, p. 250). The tears draining through the nasolacrimal ducts do so as well.

The *olfactory region* is the superior portion of the nasal cavity. It includes the areas lined by olfactory epithelium: (1) the inferior surface of the cribriform plate, (2) the superior portion of the nasal septum, and (3) the superior nasal conchae. Receptors in the olfactory epithelium provide your sense of smell. ↺ p. 591

The *superior, middle,* and *inferior nasal conchae* project toward the nasal septum from the lateral walls of the nasal cavity. (Some references use the terms *superior, middle,* and *inferior, turbinates.*) ↺ pp. 244, 247 To pass from the vestibule to the internal nares, air tends to flow between adjacent conchae, through the **superior, middle,** and **inferior meatuses** (mē-Ā-tus-ez; *meatus,* a passage) (**Figure 23–3b**). These are narrow grooves rather than open passageways. The incoming air bounces off the conchal surfaces and churns like a stream flowing over rocks. This turbulence serves several purposes. As the air swirls, small airborne particles are likely to come into contact with the mucus that coats the lining of the nasal cavity. In addition, the turbulence provides extra time for warming and humidifying incoming air. It also creates circular air currents that bring olfactory stimuli to the olfactory receptors.

The bony **hard palate** is made up of portions of the maxillary and palatine bones. The hard palate forms the floor of the nasal cavity and separates it from the oral cavity. A fleshy **soft palate** extends posterior to the hard palate, marking the boundary between the superior *nasopharynx* (nā-zō-FAR-ingks) and the rest of the pharynx. The nasal cavity opens into the nasopharynx through a connection known as the **internal nares**.

The Nasal Mucosa

The mucosa of the nasal cavity prepares inhaled air for arrival at the lower respiratory system. Throughout much of the nasal cavity, the lamina propria contains an abundance of arteries, veins, and capillaries that bring nutrients and water to the secretory cells. The lamina propria of the nasal conchae also contains an extensive network of highly expandable veins. This vascularization warms and humidifies the incoming air (and cools and dehumidifies the outgoing air as well). As cool, dry air passes inward over the exposed surfaces of the nasal cavity, the warm epithelium radiates heat, and water in the mucus evaporates. In this way, air moving from your nasal cavity to your lungs is heated almost to body temperature. It is also nearly saturated with water vapor. These changes protect more delicate respiratory surfaces from chilling or drying out—two potentially disastrous events. If you breathe through your mouth, you eliminate much of this preliminary filtration, heating, and humidifying of the inhaled air. To avoid alveolar damage, patients breathing on a respirator (mechanical ventilator), which utilizes a tube to conduct air directly into the trachea, must receive air that has been externally filtered and humidified.

As air moves out of the respiratory tract, it again passes over the epithelium of the nasal cavity. This air is warmer and more humid than the air that enters. It warms the nasal mucosa, and moisture condenses on the epithelial surfaces. In this way, breathing through your nose helps prevent heat loss and water loss.

The extensive vascularization of the nasal cavity and the vulnerable position of the nose make a nosebleed, or *epistaxis* (ep-i-STAK-sis), a fairly common event. (The leading cause of epistaxis in children is nose-picking.) This bleeding generally involves vessels of the mucosa covering the cartilaginous portion of the septum. Possible causes include trauma (such as a punch in the nose), drying, infections, allergies, or clotting disorders. Hypertension can also bring on a nosebleed by rupturing small vessels of the lamina propria.

23

Figure 23–3   **The Structures of the Upper Respiratory System.** ATLAS: Plate 19

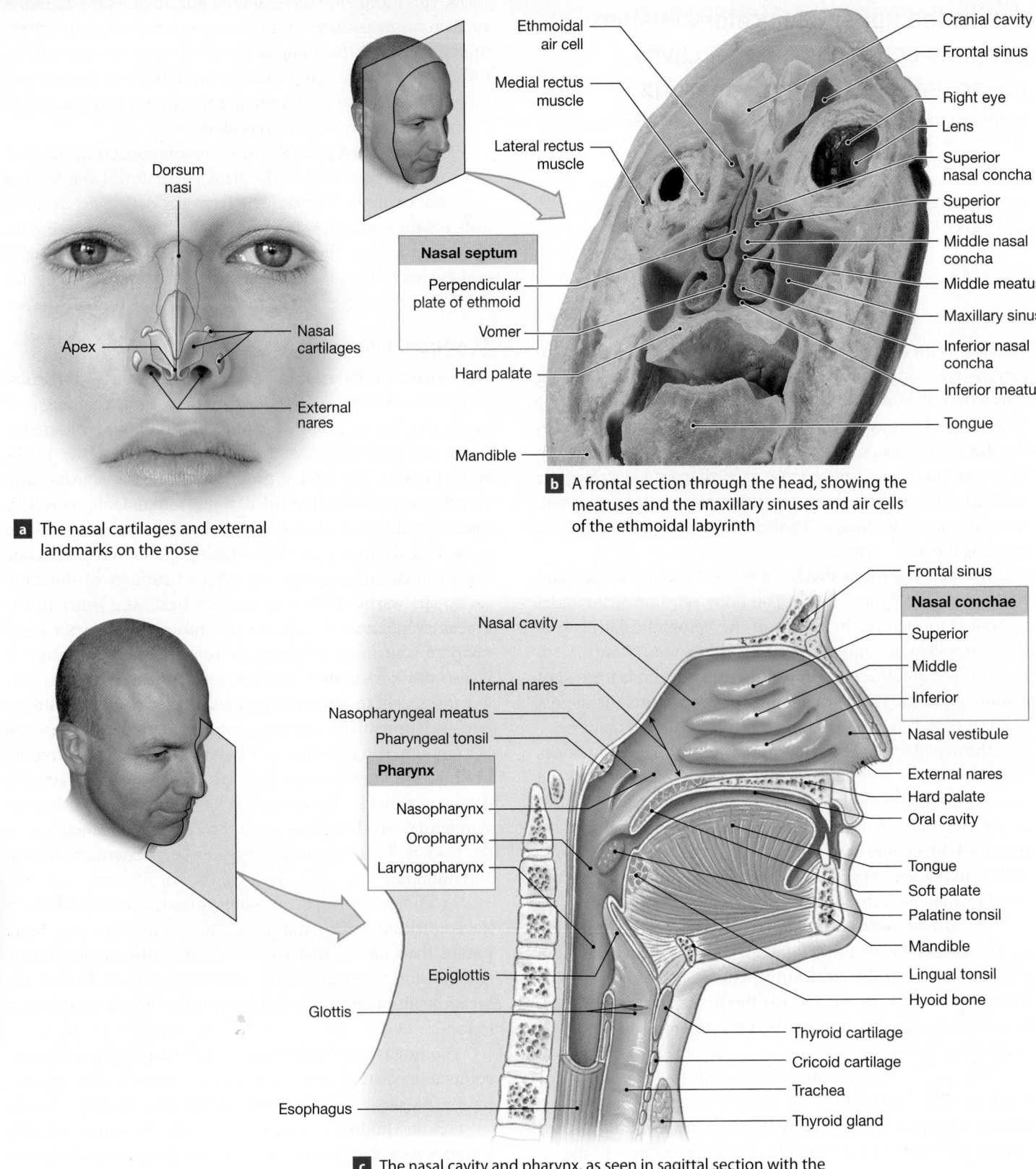

Dorsum nasi

Apex

Nasal cartilages

External nares

a The nasal cartilages and external landmarks on the nose

Ethmoidal air cell

Medial rectus muscle

Lateral rectus muscle

Cranial cavity

Frontal sinus

Right eye

Lens

Superior nasal concha

Superior meatus

Middle nasal concha

Middle meatus

Maxillary sinus

Inferior nasal concha

Inferior meatus

Tongue

Nasal septum	
Perpendicular plate of ethmoid	
Vomer	

Hard palate

Mandible

b A frontal section through the head, showing the meatuses and the maxillary sinuses and air cells of the ethmoidal labyrinth

Nasal cavity

Internal nares

Nasopharyngeal meatus

Pharyngeal tonsil

Pharynx	
Nasopharynx	
Oropharynx	
Laryngopharynx	

Epiglottis

Glottis

Esophagus

Frontal sinus

Nasal conchae	
Superior	
Middle	
Inferior	

Nasal vestibule

External nares

Hard palate

Oral cavity

Tongue

Soft palate

Palatine tonsil

Mandible

Lingual tonsil

Hyoid bone

Thyroid cartilage

Cricoid cartilage

Trachea

Thyroid gland

c The nasal cavity and pharynx, as seen in sagittal section with the nasal septum removed

23

The Pharynx

The **pharynx** (FAR-ingks), or throat, is a chamber shared by the digestive and respiratory systems. It extends between the internal nares and the entrances to the larynx and esophagus. The curving superior and posterior walls of the pharynx are closely bound to the axial skeleton, but the lateral walls are flexible and muscular.

We can divide the pharynx into the nasopharynx, the oropharynx, and the laryngopharynx (Figure 23–3c):

1. The **nasopharynx** is the superior portion of the pharynx. It is connected to the posterior portion of the nasal cavity through the internal nares. The soft palate separates it from the oral cavity. The nasopharynx is lined by the same pseudostratified ciliated columnar epithelium as in the nasal cavity. The *pharyngeal tonsil* is located on the posterior wall of the nasopharynx. Each auditory tube opens into the nasopharynx at the **nasopharyngeal meatus** on either side of this tonsil. ⊃ pp. 617, 816

2. The **oropharynx** (*oris*, mouth) extends between the soft palate and the base of the tongue at the level of the hyoid bone. The posterior portion of the oral cavity communicates directly with the oropharynx, as does the posterior inferior portion of the nasopharynx. At the boundary between the nasopharynx and the oropharynx, the epithelium changes from pseudostratified columnar epithelium to stratified squamous epithelium.

3. The narrow **laryngopharynx** (la-rin-gō-FAR-ingks) is the inferior part of the pharynx. It includes that portion of the pharynx between the hyoid bone and the entrance to the larynx and esophagus. Like the oropharynx, the laryngopharynx is lined with a stratified squamous epithelium that resists abrasion, chemical attack, and invasion by pathogens.

CHECKPOINT

4. Name the structures of the upper respiratory system.

5. Why is the vascularization of the nasal cavity important?

6. Why is the lining of the nasopharynx different from that of the oropharynx and the laryngopharynx?

See the blue Answers tab at the back of the book.

23-3 ▌ Composed of cartilages, ligaments, and muscles, the larynx produces sound

Learning Outcome Describe the structure of the larynx, and discuss its roles in normal breathing and in sound production.

Inhaled air leaves the pharynx and enters the larynx through a narrow opening of the *glottis*. The glottis is the vocal apparatus, or "voice box," of the larynx. The **larynx** (LAR-ingks) is a cartilaginous tube that surrounds and protects the glottis.

The larynx begins at the level of vertebra C_4 or C_5 and ends at the level of vertebra C_6. Essentially a cylinder, the larynx has incomplete cartilaginous walls that are stabilized by ligaments and skeletal muscles (Figure 23–4).

Cartilages and Ligaments of the Larynx

Three large, unpaired cartilages form the larynx: (1) the thyroid cartilage, (2) the cricoid cartilage, and (3) the epiglottis (Figure 23–4). The **thyroid cartilage** (*thyroid*, shield shaped) is the largest laryngeal cartilage. Made of hyaline cartilage, it forms most of the anterior and lateral walls of the larynx. In section, this cartilage is U-shaped, and posteriorly, it is incomplete. You can easily see and feel the prominent anterior surface of the thyroid cartilage, called the *laryngeal prominence* or *Adam's apple*. The inferior surface articulates with the cricoid cartilage. The superior surface has ligamentous attachments to the hyoid bone and to the epiglottis and smaller laryngeal cartilages.

The thyroid cartilage sits superior to the **cricoid** (KRĪ-koyd; ring shaped) **cartilage**, another hyaline cartilage. The posterior portion of the cricoid is greatly expanded, providing support in the absence of the thyroid cartilage. The cricoid and thyroid cartilages protect the glottis and the entrance to the trachea. Their broad surfaces provide sites for the attachment of important laryngeal muscles and ligaments. Ligaments attach the inferior surface of the cricoid cartilage to the first tracheal cartilage. The superior surface of the cricoid cartilage articulates with the small, paired *arytenoid cartilages*.

The shoehorn-shaped **epiglottis** (ep-i-GLOT-is) projects superior to the glottis and forms a lid over it. The epiglottis is composed of elastic cartilage. It has ligamentous attachments to the anterior and superior borders of the thyroid cartilage and the hyoid bone. During swallowing, the larynx is elevated and the epiglottis folds back over the glottis, preventing both liquids and solid food from entering the respiratory tract.

The larynx also contains three pairs of smaller hyaline cartilages: (1) The **arytenoid** (ar-i-TĒ-noyd; ladle shaped) **cartilages** articulate with the superior border of the enlarged portion of the cricoid cartilage. (2) The **corniculate** (kor-NIK-ū-lāt; horn shaped) **cartilages** articulate with the arytenoid cartilages. The corniculate and arytenoid cartilages function in the opening and closing of the glottis and the production of sound. (3) Elongated, curving **cuneiform** (kū-NĒ-i-form; wedge shaped) **cartilages** lie within folds of tissue (the *aryepiglottic folds*) that extend between the lateral surface of each arytenoid cartilage and the epiglottis (Figures 23–4c and 23–5).

Ligaments bind together the various laryngeal cartilages. Additional ligaments attach the thyroid cartilage to the hyoid bone, and the cricoid cartilage to the trachea (cricotracheal ligament) (Figure 23–4a,b). The median cricothyroid ligament attaches the thyroid cartilage to the cricoid cartilage. Because it has easily identifiable landmarks and is clear of the thyroid gland,

Figure 23–4 **The Anatomy of the Larynx.**

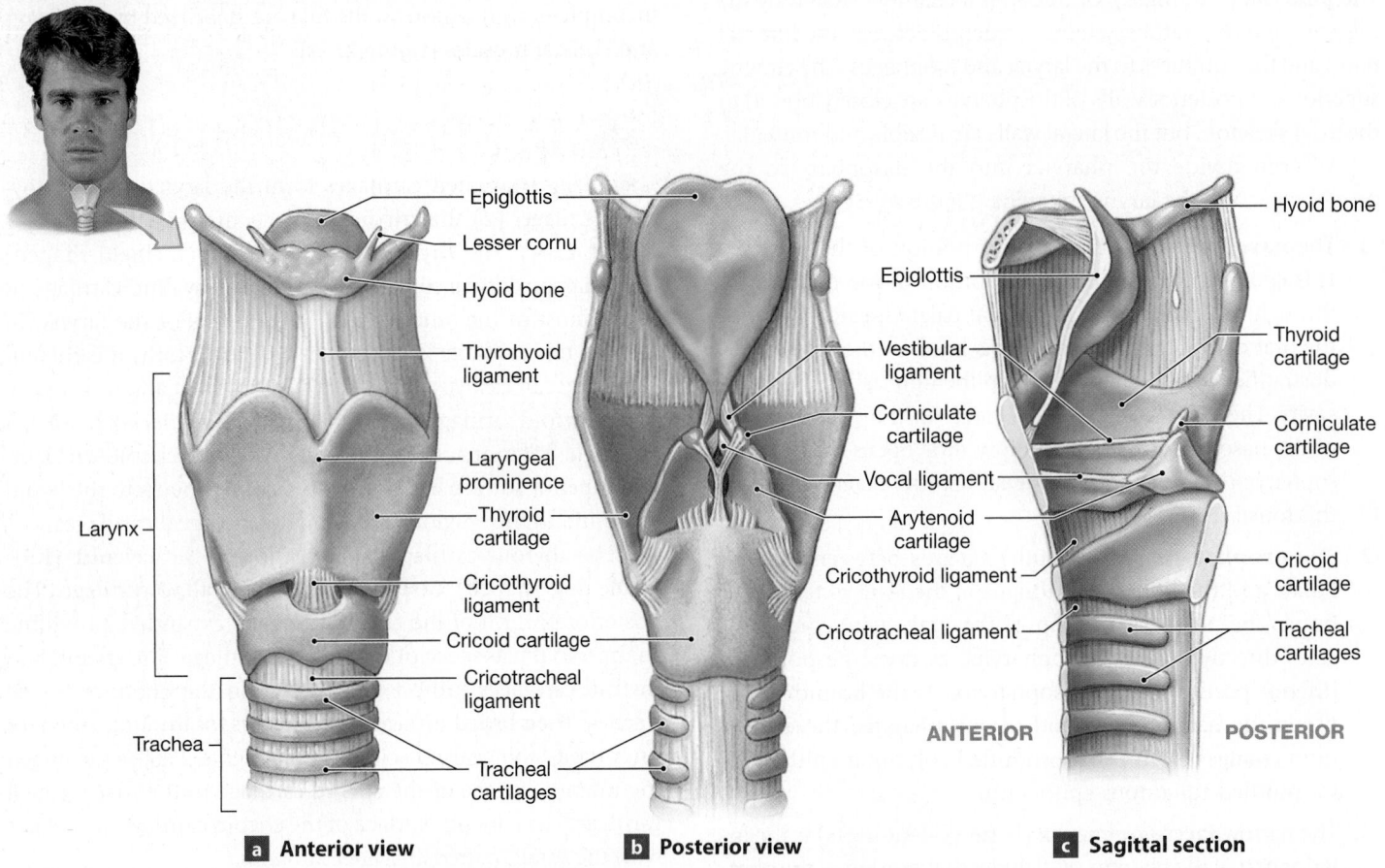

a **Anterior view**

b **Posterior view**

c **Sagittal section**

this ligament is a common placement site for an emergency tracheostomy, a tracheal incision to bypass an airway obstruction. The **vestibular ligaments** and the **vocal ligaments** extend between the thyroid cartilage and the arytenoid cartilages.

The vestibular and vocal ligaments are covered by folds of laryngeal epithelium. The vestibular ligaments lie within the superior pair of folds, known as the **vestibular folds** (**Figure 23–5**). These folds are fairly inelastic and lie laterally to the glottis. The **glottis** (GLOT-is) is made up of the **vocal folds** and the space between them, called the *rima glottidis* (**Figure 23–5**). The vestibular folds help prevent foreign objects from entering the open glottis. They also protect the more inferior, delicate vocal folds of the glottis.

The vocal folds are highly elastic, because the vocal ligaments consist of elastic tissue. The vocal folds are involved with the production of sound. For this reason they are known as the **vocal cords**.

Sound Production

How do you produce sounds? Air passing through your open glottis vibrates its vocal folds and produces sound waves. The pitch of the sound depends on the diameter, length, and tension in your vocal folds. The diameter and length are directly related to the size of your larynx. You control the tension by contracting voluntary muscles that reposition the arytenoid cartilages relative to the thyroid cartilage. When the distance increases, your vocal folds tense and the pitch rises. When the distance decreases, your vocal folds relax and the pitch falls.

Children have slender, short vocal folds, so their voices tend to be high-pitched. At puberty, the larynx of males enlarges much more than does that of females. The vocal cords of an adult male are thicker and longer. They produce lower tones than those of an adult female.

Sound production at the larynx is called *phonation* (fō-NĀ-shun; *phone*, voice). Phonation is one part of speech production. Clear speech also requires *articulation*, the modification of those sounds by other structures, such as the tongue, teeth, and lips to form words. In a stringed instrument, such as a guitar, the quality of the sound produced does not depend solely on the nature of the vibrating string. Rather, the entire instrument becomes involved as the walls vibrate and the composite sound echoes within the hollow body. Similar amplification and resonance take place within your pharynx, oral cavity, nasal cavity, and paranasal sinuses. The combination gives you the particular and distinctive sound of your voice. That sound changes when you have a sinus infection and your nasal cavity and paranasal sinuses are filled with mucus rather than air.

Figure 23–5   **The Glottis and Surrounding Structures.**

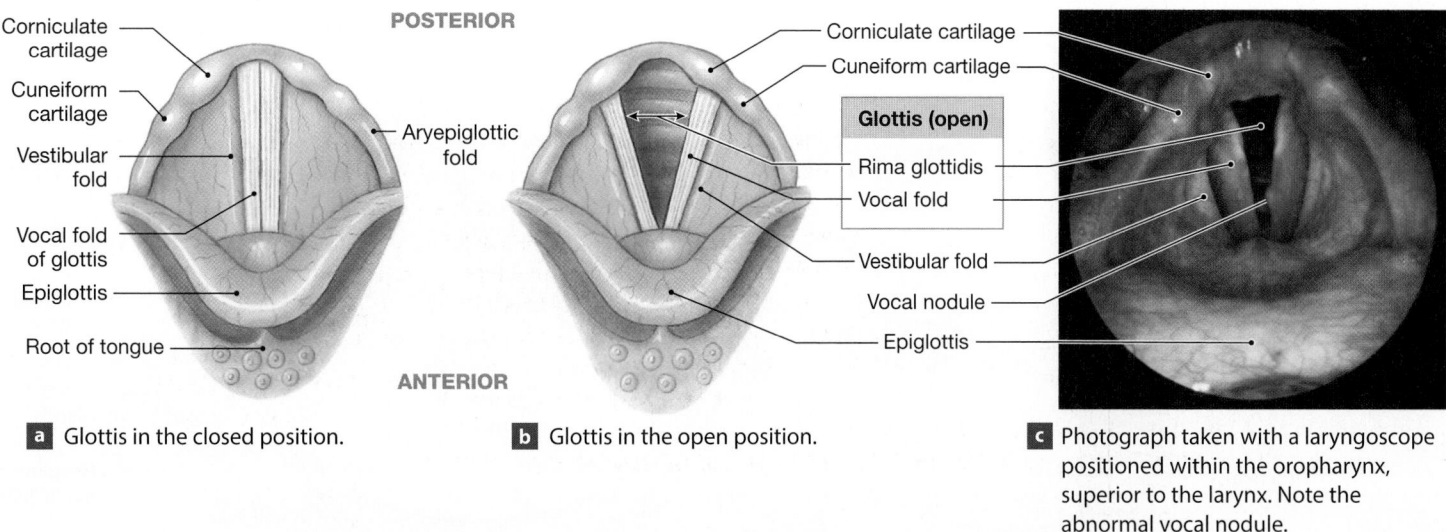

a Glottis in the closed position.

b Glottis in the open position.

c Photograph taken with a laryngoscope positioned within the oropharynx, superior to the larynx. Note the abnormal vocal nodule.

The final production of distinct words depends further on voluntary movements of your tongue, lips, and cheeks. An infection or inflammation of the larynx is known as *laryngitis* (lar-in-JĪ-tis). It commonly affects the vibrational qualities of the vocal folds. Hoarseness is the most familiar result. Mild cases are temporary and seldom serious. However, bacterial or viral infections of the epiglottis can be very dangerous. The resulting swelling may close the glottis and cause suffocation. This condition, *acute epiglottitis* (ep-i-glot-TĪ-tis), can develop rapidly after a bacterial infection of the throat. Young children are most likely to be affected.

The Laryngeal Musculature

The larynx is associated with two sets of muscles. They include (1) muscles of the neck and pharynx, which position and stabilize the larynx (↺ pp. 374–376), and (2) smaller intrinsic muscles that control tension in the glottal vocal folds or open and close the glottis. These smaller muscles insert on the thyroid, arytenoid, and corniculate cartilages. The opening or closing of the glottis involves rotational movements of the arytenoid cartilages.

When you swallow, both sets of muscles work together to prevent food or drink from entering the glottis. Food is crushed and chewed into a pasty mass, known as a *bolus*, before being swallowed. Muscles of the neck and pharynx then elevate the larynx, bending the epiglottis over the glottis, so that the bolus can glide across the epiglottis rather than falling into the larynx. While this movement is under way, the glottis is closed.

Food or liquids that touch the vestibular folds or glottis trigger the *coughing reflex*. In a cough, the glottis is kept closed while the chest and abdominal muscles contract, compressing the lungs. When the glottis is opened suddenly, a blast of air from the trachea ejects material that blocks the entrance to the glottis.

CHECKPOINT

7. **Identify the paired and unpaired cartilages associated with the larynx.**

8. **What are the highly elastic vocal folds of the glottis better known as?**

9. **When the tension in your vocal folds increases, what happens to the pitch of your voice?**

See the blue Answers tab at the back of the book.

23-4 ▎ The trachea and primary bronchi convey air to and from the lungs

Learning Outcome Discuss the structures of the extrapulmonary airways.

Three large, extrapulmonary airways are associated with the lungs: the trachea and the right and left primary bronchi.

The Trachea

The **trachea** (TRĀ-kē-uh), or windpipe, is a tough, flexible tube with a diameter of about 2.5 cm (1 in.) and a length of about 11 cm (4.33 in.) (Figure 23–6). The trachea begins anterior to vertebra C_6 in a ligamentous attachment to the cricoid cartilage. It ends in the mediastinum, at the level of vertebra T_5, where it branches to form the *right* and *left primary bronchi*.

23

Figure 23–6 **The Anatomy of the Trachea.** ATLAS: Plates 42b,c

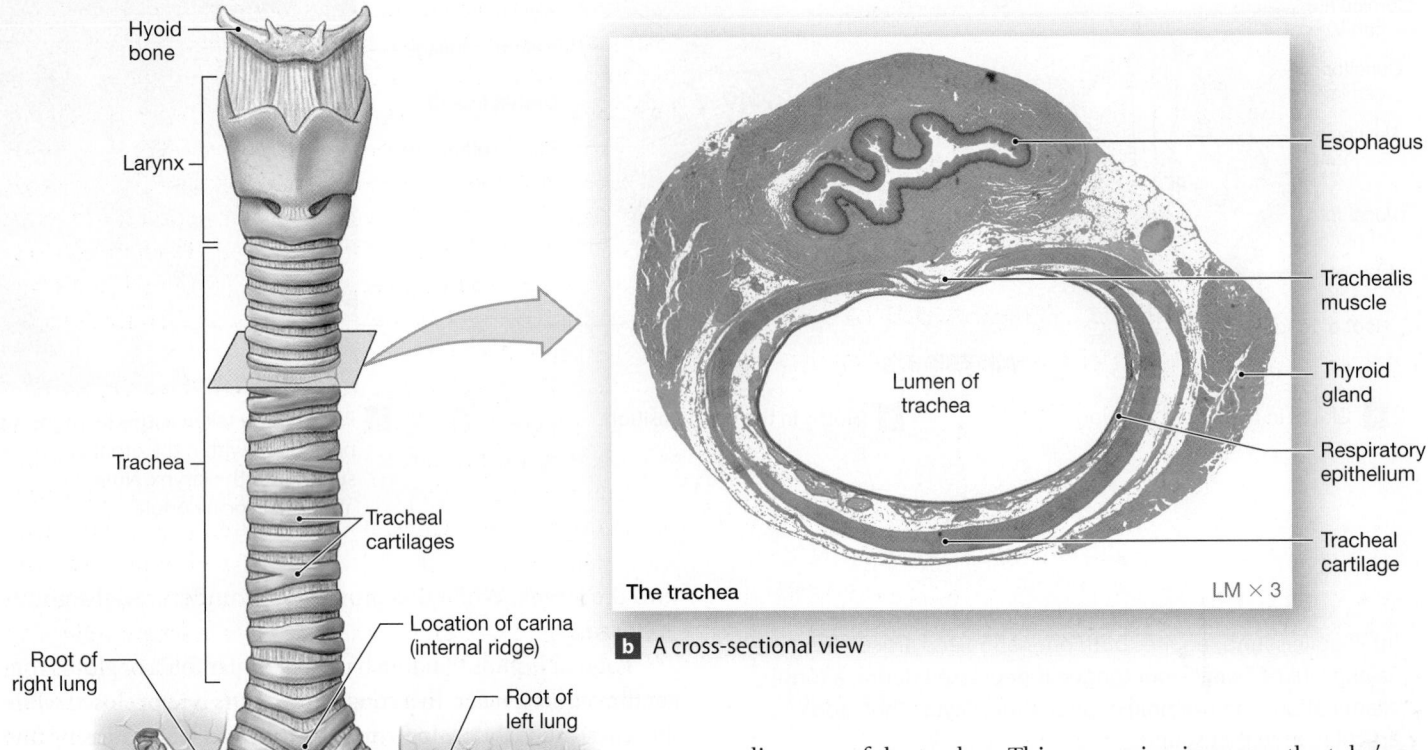

Hyoid bone

Larynx

Trachea

Tracheal cartilages

Location of carina (internal ridge)

Root of right lung

Root of left lung

Lung tissue

Primary bronchi

Secondary bronchi

RIGHT LUNG

LEFT LUNG

Esophagus

Trachealis muscle

Thyroid gland

Respiratory epithelium

Tracheal cartilage

Lumen of trachea

The trachea

LM × 3

b A cross-sectional view

a A diagrammatic anterior view showing the plane of section for part (b)

The epithelium of the trachea is continuous with that of the larynx. The mucosa of the trachea resembles that of the nasal cavity and nasopharynx (**Figure 23–2a**). The **submucosa** (sub-mū-KŌ-suh), a thick layer of connective tissue, surrounds the mucosa. The submucosa contains mucous glands that communicate with the epithelial surface through a number of secretory ducts. The trachea contains 15–20 **tracheal cartilages**, which serve to stiffen the tracheal walls and protect the airway (**Figure 23–6a**). They also prevent it from collapsing or overexpanding as pressures change in the respiratory system.

Each tracheal cartilage is C-shaped. The closed portion of the C protects the anterior and lateral surfaces of the trachea. The open portion of the C faces posteriorly, toward the esophagus (**Figure 23–6b**). Because these cartilages are not continuous, the posterior tracheal wall can easily distort when you swallow, allowing large masses of food to pass through the esophagus.

An elastic ligament and the **trachealis muscle**, a band of smooth muscle, connect the ends of each tracheal cartilage (**Figure 23–6b**). Contraction of the trachealis muscle reduces the

diameter of the trachea. This narrowing increases the tube's resistance to airflow. The normal diameter of the trachea changes from moment to moment, primarily under the control of the sympathetic division of the ANS. Sympathetic stimulation increases the diameter of the trachea and makes it easier to move large volumes of air along the respiratory passageways.

The Primary Bronchi

The trachea branches within the mediastinum into the **right** and **left primary bronchi** (BRONG-kī; singular, *bronchus*). An internal ridge called the **carina** (ka-RĪ-nuh) separates the two bronchi (**Figure 23–6a**). Like the trachea, the primary bronchi have C-shaped rings, but the ends of the C overlap. The right primary bronchus supplies the right lung, and the left supplies the left lung. The right primary bronchus is larger in diameter than the left, and descends toward the lung at a steeper angle. For these reasons, most foreign objects that enter the trachea find their way into the right bronchus rather than the left.

Before branching further, each primary bronchus travels to a groove along the medial surface of its lung. This groove, the **hilum** of the lung, also provides access for entry to pulmonary vessels, nerves, and lymphatics (**Figure 23–7c**). The entire array is firmly anchored in a meshwork of dense connective tissue. This complex is the **root of the lung** (**Figure 23–6a**). The root attaches to the mediastinum and fixes the positions of the major nerves, blood vessels, and lymphatic vessels. The roots of the lungs are anterior to vertebrae T_5 (right) and T_6 (left).

CHECKPOINT

10. List functions of the trachea.

11. Why are the cartilages that reinforce the trachea C-shaped?

12. If food accidentally enters the bronchi, in which bronchus is it more likely to lodge? Why?

See the blue Answers tab at the back of the book.

23-5 ▍ Enclosed by pleural cavities, the lungs are paired organs containing alveoli, which permit gaseous exchange

Learning Outcome Describe the superficial anatomy of the lungs, the structure of a pulmonary lobule, and the functional anatomy of alveoli.

The left and right lungs are surrounded by the left and right pleural cavities, respectively (**Figure 23–7**). Each lung is a blunt cone. Its tip, or apex, points superiorly. The apex on each side extends superior to the first rib. The broad concave inferior portion, or base, of each lung rests on the superior surface of the diaphragm.

Lobes and Surfaces of the Lungs

The lungs have distinct **lobes** that are separated by deep fissures (**Figure 23–7**). The right lung has three lobes—*superior*, *middle*, and *inferior*—separated by the *horizontal* and *oblique* fissures. The left lung has only two lobes—*superior* and *inferior*—separated by the *oblique fissure*. The right lung is broader than the left, because most of the heart and great vessels project into the left thoracic cavity. However, the left lung is longer than the right lung, because the diaphragm rises on the right side to accommodate the mass of the liver. The lateral and medial lung surfaces are shown in **Figure 23–7b,c.**

The heart is located to the left of the midline, so its corresponding impression is larger in the left lung than in the right. In anterior view, the medial edge of the right lung forms a vertical line, but the medial margin of the left lung is indented at the **cardiac notch** (**Figure 23–7**). **Figure 23–8** shows the relationship between the heart and the lungs.

The Bronchi

The primary bronchi and their branches form the **bronchial tree**. Because the left and right primary bronchi are outside the lungs, they are called *extrapulmonary bronchi*. As the primary bronchi enter the lungs, they divide to form smaller passageways (**Figure 23–6a**). The branches within the lungs are collectively called the *intrapulmonary bronchi*.

Each primary bronchus divides to form **secondary bronchi**, also known as *lobar bronchi*. In each lung, one secondary bronchus goes to each lobe, so the right lung has three secondary bronchi, and the left lung has two.

Figure 23–9 depicts the branching pattern of the left primary bronchus as it enters the lung. (The number of branches has been reduced for clarity.) In each lung, the secondary bronchi branch to form **tertiary bronchi**, or *segmental bronchi*. The branching pattern differs between the two lungs, but each tertiary bronchus ultimately supplies air to a single **bronchopulmonary segment**, a specific region of one lung (**Figure 23–9a**). The right lung has 10 bronchopulmonary segments. During development, the left lung also has 10 segments, but adjacent tertiary bronchi fuse, generally reducing the number to 8 or 9.

The walls of the primary, secondary, and tertiary bronchi contain progressively less cartilage. In the secondary and tertiary bronchi, the cartilages form plates arranged around the lumen. These cartilages serve the same structural purpose as the rings of cartilage in the trachea and primary bronchi. As the amount of cartilage decreases, the amount of smooth muscle increases. With less cartilaginous support, the amount of tension in those smooth muscles has a greater effect on bronchial diameter and the resistance to airflow. During a respiratory infection, the bronchi and bronchioles can become inflamed and constricted, increasing resistance. In this condition, called **bronchitis**, the individual has difficulty breathing.

The Bronchioles

Each tertiary bronchus branches several times within a bronchopulmonary segment, forming many **bronchioles**. These bronchioles then branch into the finest conducting branches, called **terminal bronchioles**. Nearly 6500 terminal bronchioles arise from each tertiary bronchus. The lumen of each terminal bronchiole has a diameter of 0.3–0.5 mm.

The walls of bronchioles lack cartilage but are dominated by smooth muscle tissue (**Figure 23–9c**). In functional terms, bronchioles are to the respiratory system what arterioles are to the cardiovascular system. Changes in the diameter of the bronchioles control the resistance to airflow and the distribution of air in the lungs.

The autonomic nervous system controls the diameter of the bronchioles. It does so by regulating the activity in the smooth muscle layer. Sympathetic activation leads to **bronchodilation**, the enlargement of the diameter of the airway. Parasympathetic stimulation leads to **bronchoconstriction**, a reduction in the diameter of the airway. Bronchoconstriction also takes place during allergic reactions such as anaphylaxis, in response to histamine released by activated mast cells and basophils. ⤺ p. 849

Bronchodilation and bronchoconstriction are ways of adjusting the resistance to airflow. These actions direct airflow toward or away from specific portions of the respiratory exchange

23

Figure 23–7 **The Gross Anatomy of the Lungs.** ATLAS: Plates 42–47

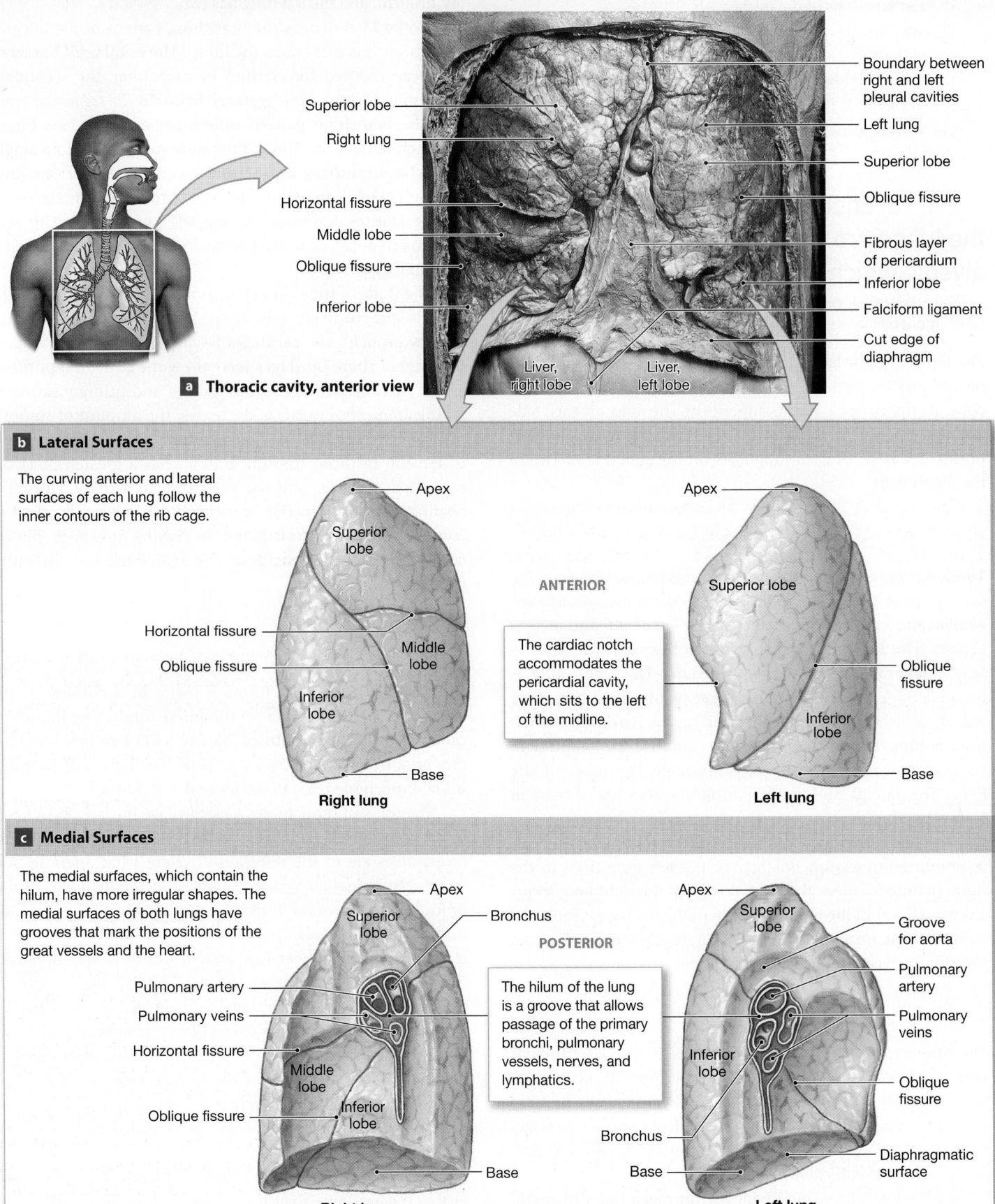

a **Thoracic cavity, anterior view**

Superior lobe

Right lung

Horizontal fissure

Middle lobe

Oblique fissure

Inferior lobe

Boundary between right and left pleural cavities

Left lung

Superior lobe

Oblique fissure

Fibrous layer of pericardium

Inferior lobe

Falciform ligament

Cut edge of diaphragm

Liver, right lobe

Liver, left lobe

b **Lateral Surfaces**

The curving anterior and lateral surfaces of each lung follow the inner contours of the rib cage.

Apex

Superior lobe

ANTERIOR

Apex

Superior lobe

Horizontal fissure

Oblique fissure

Middle lobe

The cardiac notch accommodates the pericardial cavity, which sits to the left of the midline.

Oblique fissure

Inferior lobe

Inferior lobe

Base

Base

Right lung

Left lung

c **Medial Surfaces**

The medial surfaces, which contain the hilum, have more irregular shapes. The medial surfaces of both lungs have grooves that mark the positions of the great vessels and the heart.

Apex

Superior lobe

Bronchus

POSTERIOR

Apex

Superior lobe

Groove for aorta

Pulmonary artery

Pulmonary veins

Horizontal fissure

The hilum of the lung is a groove that allows passage of the primary bronchi, pulmonary vessels, nerves, and lymphatics.

Pulmonary artery

Pulmonary veins

Middle lobe

Inferior lobe

Inferior lobe

Oblique fissure

Oblique fissure

Bronchus

Base

Diaphragmatic surface

Base

Right lung

Left lung

23

Figure 23–8 **The Relationship between the Lungs and Heart.** This transverse section was taken at the level of the cardiac notch.

Labels (left side):
- Pericardial cavity
- Right lung, middle lobe
- Oblique fissure
- Right pleural cavity
- Atria
- Esophagus
- Aorta
- Right lung, inferior lobe
- Spinal cord

Labels (right side):
- Body of sternum
- Ventricles
- Rib
- Left lung, superior lobe
- Visceral pleura
- Left pleural cavity
- Parietal pleura
- Bronchi
- Mediastinum
- Left lung, inferior lobe

surfaces. Tension in the smooth muscles commonly causes the bronchiole mucosa to form a series of folds that limits airflow. Excessive stimulation, as in **asthma** (AZ-muh), can almost completely prevent airflow along the terminal bronchioles.

Pulmonary Lobules

The connective tissues of the root of each lung extend into the lung's *parenchyma*, or functional cells. These fibrous partitions, or *trabeculae*, contain elastic fibers, smooth muscles, and lymphatic vessels. The trabeculae branch repeatedly, dividing the lobes into ever-smaller compartments. The branches of the conducting passageways, pulmonary vessels, and nerves of the lungs follow these trabeculae.

The finest partitions, or **interlobular septa** (*septum*, a wall), divide the lung into **pulmonary lobules** (LOB-ūlz). Branches of the pulmonary arteries, pulmonary veins, and respiratory passageways supply each lobule (**Figure 23–9c**). The connective tissues of the septa are, in turn, continuous with those of the *visceral pleura*, the serous membrane covering the lungs.

Each terminal bronchiole delivers air to a single pulmonary lobule. Within the lobule, the terminal bronchiole branches to form several **respiratory bronchioles**. The respiratory bronchioles are the thinnest and most delicate branches of the bronchial tree. They deliver air to the gas exchange surfaces of the lungs.

Before incoming air moves beyond the terminal bronchioles, it has been filtered and humidified. A cuboidal epithelium lines the terminal bronchioles and respiratory bronchioles. There are only scattered cilia and no mucous cells or underlying mucous glands. If particulates or pathogens reach this part of the respiratory tract, there is little to prevent them from damaging the delicate exchange surfaces of the lungs.

Alveolar Ducts and Alveoli

Respiratory bronchioles are connected to individual alveoli and to multiple alveoli along regions called **alveolar ducts**. Alveolar ducts end at **alveolar sacs**, common chambers connected to multiple individual alveoli. Each lung contains about 150 million alveoli. They give the lungs an open, spongy appearance (**Figures 23–9c,d** and **23–10a,b**).

Each alveolus is associated with an extensive network of capillaries (**Figure 23–10a**). A network of elastic fibers surrounds the capillaries. These fibers help maintain the relative positions of the alveoli and respiratory bronchioles. When these fibers recoil during exhalation, they reduce the size of the alveoli and help push air out of the lungs.

The alveolar epithelium consists mainly of simple squamous epithelium (**Figure 23–10c**). The squamous epithelial cells, called **type I pneumocytes** (or *type I alveolar cells*), are unusually thin and are the sites of gas diffusion. Roaming **alveolar macrophages**, or *dust cells*, patrol the epithelial surface. They phagocytize any particles that have eluded other defenses. Large **type II pneumocytes**, also called *type II alveolar cells*, are scattered among the squamous cells. Type II pneumocytes produce **surfactant** (sur-FAK-tant), an oily secretion containing phospholipids and proteins. They secrete surfactant onto the alveolar surfaces, where it forms a superficial coating over a thin layer of water.

23

Figure 23–9 **The Bronchi, Lobules, and Alveoli of the Lung.** ATLAS: Plates 42b,c; 47b–d

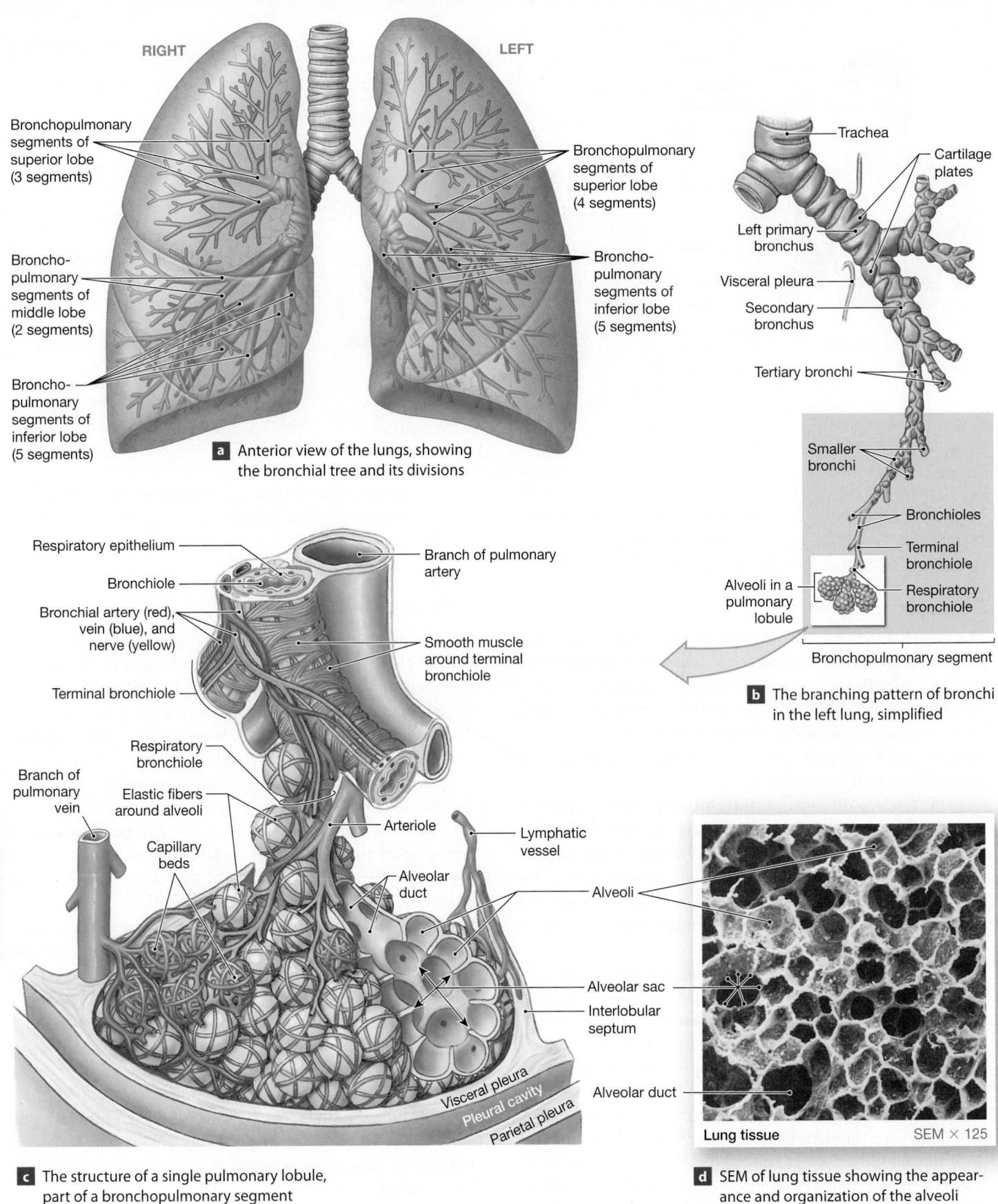

RIGHT

LEFT

Bronchopulmonary segments of superior lobe (3 segments)

Broncho-pulmonary segments of middle lobe (2 segments)

Broncho-pulmonary segments of inferior lobe (5 segments)

Bronchopulmonary segments of superior lobe (4 segments)

Broncho-pulmonary segments of inferior lobe (5 segments)

a Anterior view of the lungs, showing the bronchial tree and its divisions

Trachea

Cartilage plates

Left primary bronchus

Visceral pleura

Secondary bronchus

Tertiary bronchi

Smaller bronchi

Bronchioles

Terminal bronchiole

Alveoli in a pulmonary lobule

Respiratory bronchiole

Bronchopulmonary segment

b The branching pattern of bronchi in the left lung, simplified

Respiratory epithelium

Bronchiole

Bronchial artery (red), vein (blue), and nerve (yellow)

Terminal bronchiole

Branch of pulmonary vein

Respiratory bronchiole

Elastic fibers around alveoli

Capillary beds

Branch of pulmonary artery

Smooth muscle around terminal bronchiole

Arteriole

Lymphatic vessel

Alveolar duct

Alveoli

Alveolar sac

Interlobular septum

Alveolar duct

Visceral pleura

Pleural cavity

Parietal pleura

Lung tissue   SEM × 125

c The structure of a single pulmonary lobule, part of a bronchopulmonary segment

d SEM of lung tissue showing the appearance and organization of the alveoli

23

Figure 23–10 **Alveolar Organization.**

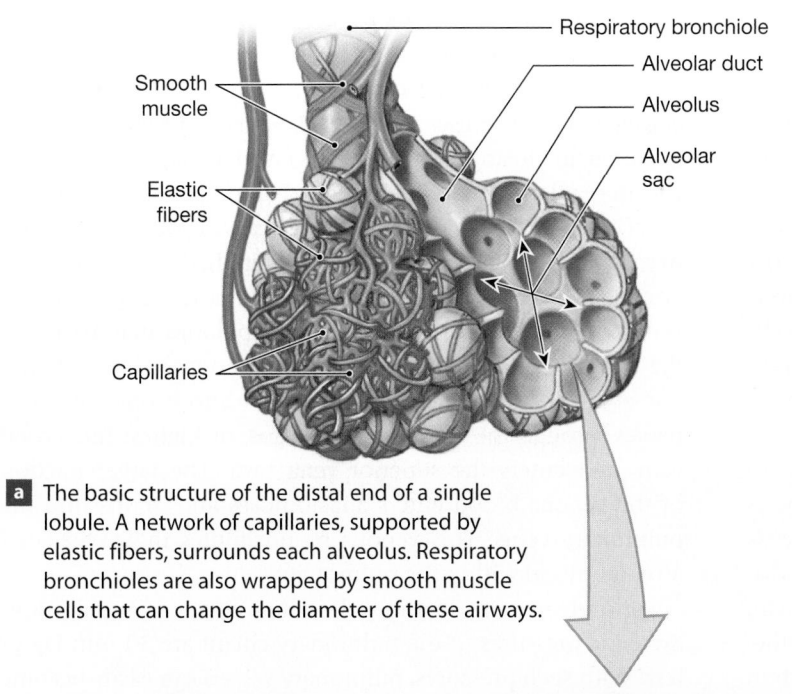

a The basic structure of the distal end of a single lobule. A network of capillaries, supported by elastic fibers, surrounds each alveolus. Respiratory bronchioles are also wrapped by smooth muscle cells that can change the diameter of these airways.

Histology of the lung   LM × 14

b Low-power micrograph of lung tissue.

c A diagrammatic view of alveolar structure. A single capillary may be involved in gas exchange with several alveoli simultaneously.

d The respiratory membrane, which consists of an alveolar epithelial cell, a capillary endothelial cell, and their fused basement membranes.

23

Surfactant plays a key role in keeping the alveoli open. It reduces surface tension in the liquid coating the alveolar surface. Recall from Chapter 2 that *surface tension* results from the attraction between water molecules at an air–water boundary. ↻ p. 61 Surface tension creates a barrier that keeps small objects from entering the water, but it also tends to collapse small air bubbles. Without surfactant, the surface tension would collapse the alveoli in much the same way. Surfactant forms a thin

TIPS&TRICKS

The term ***surfactant*** is derived from its purpose as a ***surf***ace ***act***ive ***ag***e***nt***.

surface layer that interacts with the water molecules, reducing the surface tension and keeping the alveoli open.

If type II pneumocytes produce inadequate amounts of surfactant due to injury or genetic abnormalities, respiration becomes difficult. The alveoli collapse after each exhalation. With each breath, the inhalation must be forceful enough to pop open the alveoli. A person without enough surfactant is soon exhausted by the effort of inflating and deflating the lungs. This condition is called *respiratory distress syndrome.*

Gas exchange occurs across the **respiratory membrane** of the alveoli. The respiratory membrane has three layers (**Figure 23–10d**). It contains (1) the squamous epithelial cells lining the alveolus, (2) the endothelial cells lining an adjacent capillary, and (3) the fused basement membranes that lie between the alveolar and endothelial cells.

At the respiratory membrane, only a very short distance separates alveolar air from blood. The total distance can be as little as 0.1 μm, but averages about 0.5 μm. Diffusion proceeds very rapidly across the respiratory membrane because the distance is short and both oxygen and carbon dioxide are small, lipid-soluble molecules. The plasma membranes of the epithelial and endothelial cells do not prevent oxygen and carbon dioxide from moving between blood and alveolar air.

Certain diseases can compromise the function of the respiratory membrane. **Pneumonia** (nū-MŌ-nē-uh) develops from an infection or any other stimulus that causes inflammation of the lobules of the lung. As inflammation occurs, fluids leak into the alveoli. The respiratory bronchioles swell, narrowing passageways and restricting the passage of air. Respiratory function deteriorates as a result. When bacteria are involved, they are generally types that normally inhabit the mouth and pharynx but have managed to evade the respiratory defenses.

Pneumonia becomes more likely when the respiratory defenses have already been compromised by other factors. Such factors include epithelial damage from smoking and the breakdown of the immune system in AIDS. The respiratory defenses of healthy individuals prevent infection and tissue damage, but the breakdown of those defenses in AIDS can lead to a massive, potentially fatal lung infection. The most common pneumonia that develops in individuals with AIDS results from infection by the fungus *Pneumocystis jiroveci* (formerly known as *Pneumocystis carinii*).

The Blood Supply to the Lungs

Two circuits nourish lung tissue. One supplies the *respiratory* portion of the lungs. The other perfuses the *conducting* portion.

The respiratory exchange surfaces receive blood from arteries of the pulmonary circuit. The pulmonary arteries carry deoxygenated blood. They enter the lungs at the hilum and branch with the bronchi as they approach the lobules. Each lobule receives an arteriole and a venule, and a network of capillaries surrounds each alveolus as part of the respiratory membrane. Oxygen-rich blood from the alveolar capillaries passes through the pulmonary venules and then enters the pulmonary veins, which deliver the blood to the left atrium.

In addition to providing for gas exchange, the endothelial cells of the alveolar capillaries are the primary source of *angiotensin-converting enzyme (ACE)*, which converts circulating angiotensin I to angiotensin II. This enzyme plays an important role in regulating blood volume and blood pressure. ⊃ p. 773

The tissues of conducting passageways of your lungs receive oxygen and nutrients from capillaries supplied by the bronchial arteries, which branch from the thoracic aorta. ⊃ p. 787 The venous blood from these bronchial capillaries may return to the heart by either the systemic or pulmonary circuit. About one-eighth of the venous blood drains into bronchial veins, passes through the azygos, hemiazygos, or highest intercostal vein, and enters the superior vena cava. The larger fraction of the venous blood enters anastomoses and drains into the pulmonary veins. This venous blood dilutes the oxygenated alveolar blood within the pulmonary veins.

Blood pressure in the pulmonary circuit is usually low. Systemic pressures in the pulmonary circuit are 30 mm Hg or less. With such pressures, pulmonary vessels can easily become blocked by small blood clots, fat masses, or air bubbles in the pulmonary arteries. Because the lungs receive the entire cardiac output, any such objects drifting in blood are likely to be trapped in the pulmonary arterial or capillary networks. Very small blood clots occasionally form in the venous system. These are usually trapped in the pulmonary capillary network, where they soon dissolve. Larger emboli are much more dangerous. The blockage of a branch of a pulmonary artery stops blood flow to a group of lobules or alveoli. This condition is called **pulmonary embolism**. If a pulmonary embolism is in place for several hours, the alveoli will permanently collapse. If the blockage occurs in a major pulmonary vessel rather than a minor branch, pulmonary resistance increases. The resistance places extra strain on the right ventricle, which may be unable to maintain cardiac output, and congestive heart failure can result.

The Pleural Cavities and Pleural Membranes

The thoracic cavity has the shape of a broad cone. Its walls are the rib cage, and the muscular diaphragm forms its floor. The two **pleural cavities** are separated by the mediastinum (**Figure 23–8**). Each lung is surrounded by a single pleural cavity, which is lined by a serous membrane called the **pleura** (PLOOR-uh; plural, *pleurae*). The pleura consists of two layers: the parietal pleura and the visceral pleura. The **parietal pleura** covers the inner surface of the thoracic wall and extends over the diaphragm and mediastinum. The **visceral pleura** covers the outer surfaces of the lungs, extending into the fissures between the lobes. Each pleural cavity actually represents a potential space rather than an open chamber, because the parietal and visceral pleurae are usually in close contact.

Both pleurae secrete a small amount of **pleural fluid**. Pleural fluid forms a moist, slippery coating that provides lubrication. It reduces friction between the parietal and visceral surfaces as you breathe. Samples of pleural fluid, obtained through a long needle inserted between the ribs, are sometimes needed for diagnostic purposes. This sampling procedure is called *thoracentesis* (thōr-a-sen-TĒ-sis; *thora-*, thoracic + *centesis*, puncture). The extracted fluid is examined for bacteria, blood cells, or other abnormal components.

In some diseases, the normal coating of pleural fluid does not prevent friction between the pleural surfaces. The result is pain and pleural inflammation, a condition called *pleurisy*. When pleurisy develops, the secretion of pleural fluid may be excessive, or the inflamed pleurae may adhere to one another, limiting movement. In either case, breathing becomes difficult, and prompt medical attention is required.

CHECKPOINT

13. What would happen to the alveoli if surfactant were not produced?

14. Trace the path air takes in flowing from the glottis to the respiratory membrane.

15. Which arteries supply blood to the gas exchange surfaces and conducting portions of the respiratory system?

16. List the functions of the pleura.

See the blue Answers tab at the back of the book.

23-6 ❙ External respiration and internal respiration allow gaseous exchange within the body

Learning Outcome Define and compare the processes of external respiration and internal respiration.

The general term *respiration* includes two integrated processes: *external respiration* and *internal respiration*. The definitions of these terms vary among references. In our discussion, **external respiration** includes all the processes involved in the exchange of oxygen and carbon dioxide between the body's interstitial fluids and the external environment. The purpose of external respiration, and the primary function of the respiratory system, is to meet the respiratory demands of cells. **Internal respiration** is the absorption of oxygen and the release of carbon dioxide by those cells. We consider the biochemical pathways responsible for oxygen consumption and for the generation of carbon dioxide by mitochondria—pathways known collectively as *cellular respiration*—in Chapter 25.

Our discussion here focuses on three integrated steps in external respiration (Figure 23–11):

1. *Pulmonary ventilation*, or breathing, physically moves air into and out of the lungs.

2. *Gas diffusion* across the respiratory membrane between alveolar air spaces and alveolar capillaries, and across capillary walls between blood and other tissues.

3. *Transport of oxygen and carbon dioxide* between alveolar capillaries and capillary beds in other tissues.

Figure 23–11 An Overview of the Key Steps in Respiration.

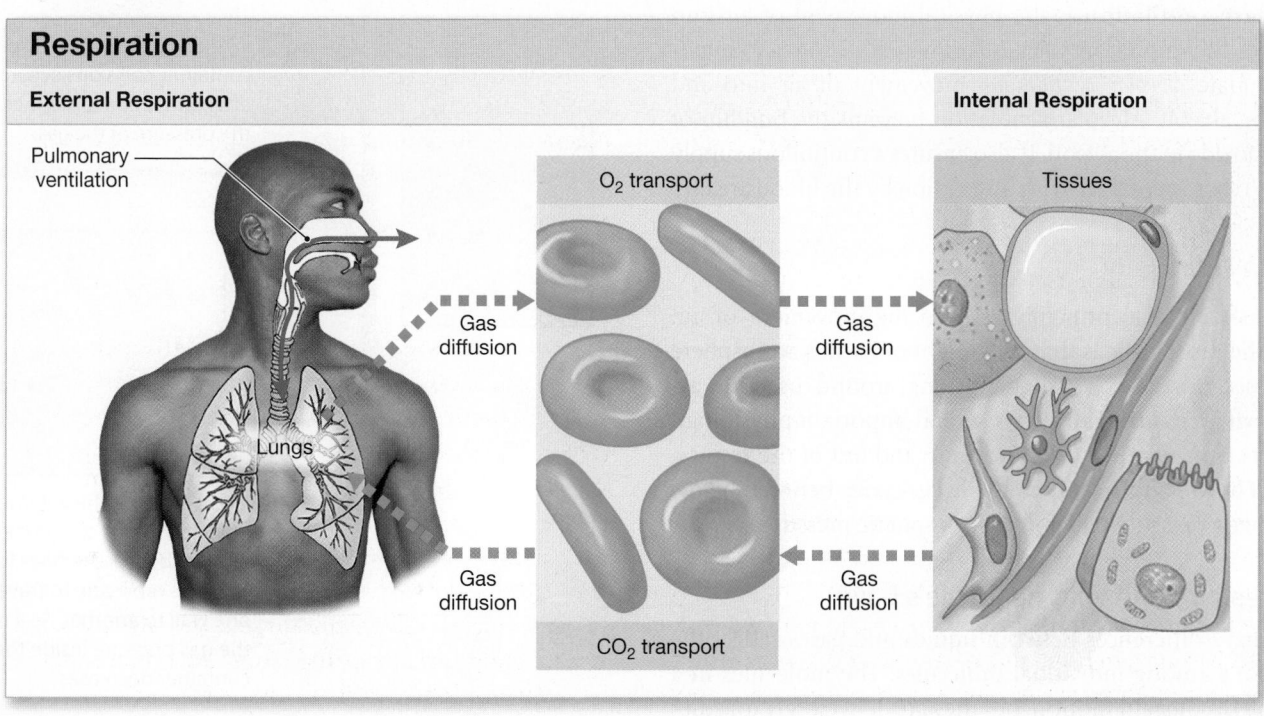

Respiration

External Respiration | Internal Respiration

Pulmonary ventilation

Lungs

O_2 transport

Gas diffusion

Gas diffusion

Tissues

Gas diffusion

Gas diffusion

CO_2 transport

Abnormalities affecting any of the steps involved in external respiration ultimately affect the concentrations of gases in interstitial fluids, and thus cellular activities as well. If the oxygen level declines, the affected tissues will become starved for oxygen. **Hypoxia**, or low tissue oxygen levels, places severe limits on the metabolic activities of the affected area. For example, the effects of coronary ischemia result from chronic hypoxia affecting cardiac muscle cells. ⤶ p. 724 If the oxygen supply is cut off completely, the condition called **anoxia** (an-OK-sē-uh; *an-*, without + *oxy-*, oxygen) results. Anoxia kills cells very quickly. Much of the damage from strokes and heart attacks results from local anoxia.

In the sections that follow, we examine each of the processes involved in external respiration in greater detail.

CHECKPOINT

17. **Define external respiration and internal respiration.**

18. **Name the integrated steps involved in external respiration.**

See the blue Answers tab at the back of the book.

23-7 ▌ Pulmonary ventilation—the exchange of air between the atmosphere and the lungs—involves pressure changes, muscle movement, and respiratory rates and volumes

Learning Outcome Summarize the physical principles controlling the movement of air into and out of the lungs, and describe the origins and actions of the muscles responsible for respiratory movements.

Pulmonary ventilation is the physical movement of air into and out of the respiratory tract. Its primary function is to maintain adequate *alveolar ventilation*—movement of air into and out of the alveoli. Alveolar ventilation prevents the buildup of carbon dioxide in the alveoli. It also ensures a continuous supply of oxygen that keeps pace with absorption by the bloodstream.

The Movement of Air

Some basic physical principles govern the movement of air. One of the most basic is that the weight of Earth's atmosphere compresses our bodies and everything around us. This **atmospheric pressure (atm)** has several important physiological effects. For example, air moves into and out of the respiratory tract as the air pressure in the lungs cycles between below atmospheric pressure and above atmospheric pressure.

Gas Pressure and Volume (Boyle's Law)

The primary differences between liquids and gases reflect the interactions among individual molecules. The molecules in a liquid are in constant motion, but they are held closely together by weak interactions, such as the hydrogen bonding between adjacent water molecules. ⤶ p. 61 Yet because the electrons of adjacent atoms tend to repel one another, liquids tend to resist compression. If you squeeze a balloon filled with water, it will change into a different shape, but the volumes of the two shapes will be the same.

In a gas, such as air, the molecules bounce around as independent objects. At normal atmospheric pressures, gas molecules are much farther apart than the molecules in a liquid. The forces acting between gas molecules are minimal because the molecules are too far apart for weak interactions to take place, so an applied pressure can push them closer together. Consider a sealed container of air at atmospheric pressure. The pressure exerted by the gas inside results from gas molecules bumping into the walls of the container. The greater the number of collisions, the higher the pressure.

You can change the gas pressure within a sealed container by changing the volume of the container, giving the gas molecules more or less room in which to bounce around. If you decrease the volume of the container, pressure increases (**Figure 23–12a**). If you increase the volume of the container, pressure decreases (**Figure 23–12b**).

Figure 23–12 **The Relationship between Gas Pressure and Volume.**

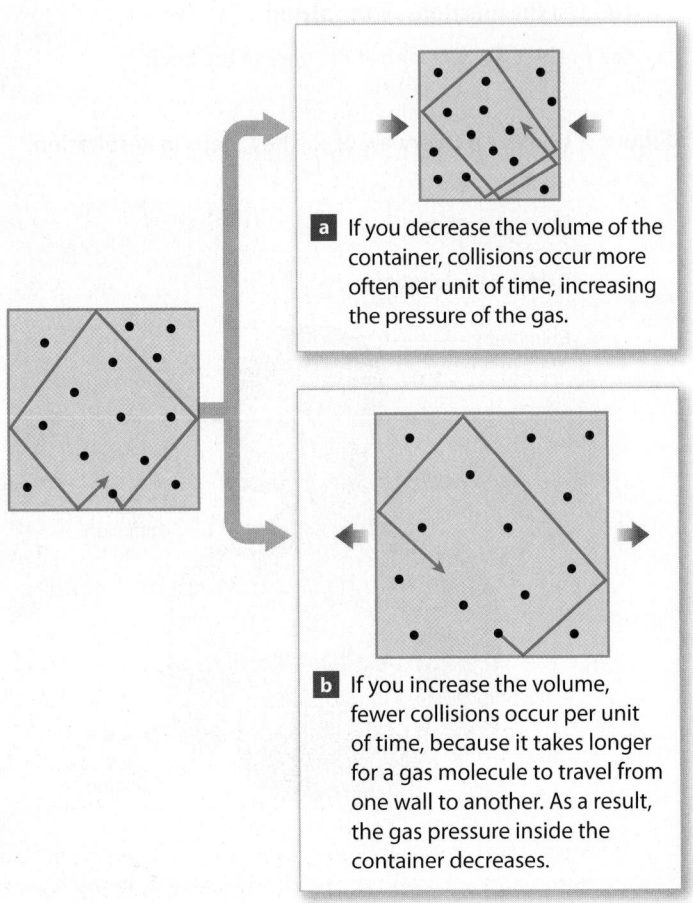

a If you decrease the volume of the container, collisions occur more often per unit of time, increasing the pressure of the gas.

b If you increase the volume, fewer collisions occur per unit of time, because it takes longer for a gas molecule to travel from one wall to another. As a result, the gas pressure inside the container decreases.

23

pressures and peritoneal pressure from increasing so high that an alveolar rupture or hernia could occur.

The Intrapleural Pressure

Intrapleural pressure is the pressure in the pleural cavity, between the parietal and visceral pleurae. Intrapleural pressure averages about −4 mm Hg (**Figure 23–14b**). During a powerful inhalation, it can reach −18 mm Hg. This pressure is below atmospheric pressure, due to the relationship between the lungs and the body wall. Earlier, we noted that the lungs are highly elastic. In fact, they would collapse to about 5 percent of their normal resting volume if their elastic fibers could recoil completely. The elastic fibers cannot recoil so much, however. The reason is that they are not strong enough to overcome the fluid bond between the parietal and visceral pleurae.

The elastic fibers continuously oppose that fluid bond and pull the lungs away from the chest wall and diaphragm, lowering the intrapleural pressure slightly. The elastic fibers remain stretched even after a full exhalation. For this reason, intrapleural pressure remains below atmospheric pressure throughout normal cycles of inhalation and exhalation.

The cyclical changes in the intrapleural pressure create the *respiratory pump* that assists the venous return to the heart. ⟲ p. 765

The Respiratory Cycle

A **respiratory cycle** is a single cycle of inhalation and exhalation. The curves in **Figure 23–14a,b** show the intrapulmonary and intrapleural pressures during a single respiratory cycle of an individual at rest. The graph in **Figure 23–14c** plots the **tidal volume**, the amount of air you move into or out of your lungs during a single respiratory cycle.

TIPS & TRICKS

Tidal volume floods and ebbs like the ocean *tides*.

At the start of the respiratory cycle, the intrapulmonary and atmospheric pressures are equal, and no air is moving. Inhalation begins with the fall of intrapleural pressure that takes place when the thoracic cavity expands. This pressure gradually falls to approximately −6 mm Hg. Over the period, intrapulmonary pressure drops to just under −1 mm Hg. It then begins to rise as air flows into the lungs.

Figure 23–14 **Pressure and Volume Changes during Inhalation and Exhalation.** One sequence of inhalation and exhalation makes up a single respiratory cycle.

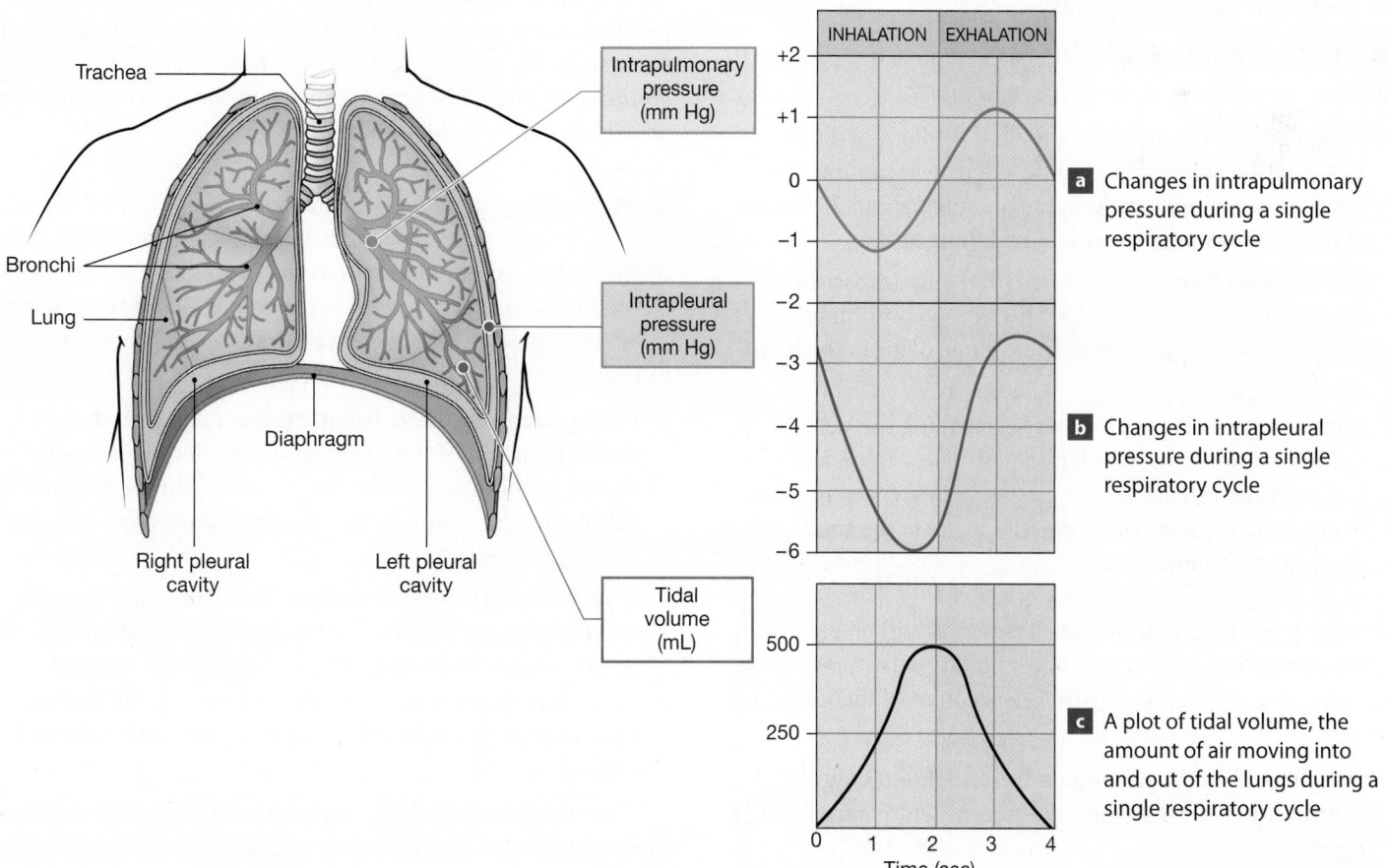

a Changes in intrapulmonary pressure during a single respiratory cycle

b Changes in intrapleural pressure during a single respiratory cycle

c A plot of tidal volume, the amount of air moving into and out of the lungs during a single respiratory cycle

When exhalation begins, intrapleural and intrapulmonary pressures rise rapidly, forcing air out of the lungs. At the end of exhalation, air again stops moving when the difference between intrapulmonary and atmospheric pressures has been eliminated. The amount of air moved into the lungs during inhalation equals the amount moved out of the lungs during exhalation. That amount is the tidal volume.

The Mechanics of Breathing

As we have just seen, you move air into and out of the respiratory system by changing the volume of the lungs. Those changes alter the pressure relationships, producing air movement. The changes of volume in the lungs take place through the contraction of skeletal muscles—specifically, those that insert on the rib cage—and the *diaphragm*, which separates the thoracic and abdominopelvic cavities. Because of the nature of their articulations with the vertebrae, when the ribs are elevated they swing outward, increasing the depth of the thoracic cavity. ↺ p. 262 We can compare this movement to raising the handle of a bucket. Respiratory muscles and pulmonary ventilation are featured in **Spotlight Figure 23–15**.

The Respiratory Muscles

We introduced the skeletal muscles involved in respiratory movements in Chapter 11. Of those muscles, the most important are the *diaphragm* and the *external intercostal muscles*. ↺ pp. 380–381

Muscles Used in Inhalation. Inhalation is an active process. It involves one or more of the following actions:

- Contraction of the diaphragm flattens the floor of the thoracic cavity, increasing its volume and drawing air into the lungs. Contraction of the diaphragm brings about 75 percent of the air movement in normal breathing at rest.

- Contraction of the external intercostal muscles assists in inhalation by raising the ribs. This action contributes approximately 25 percent of the volume of air in the lungs at rest.

- Contraction of accessory muscles, including the sternocleidomastoid, serratus anterior, pectoralis minor, and scalene muscles, can assist the external intercostal muscles in elevating the ribs. These muscles increase the speed and amount of rib movement.

Muscles Used in Exhalation. Exhalation is either passive or active, depending on the level of respiratory activity. When exhalation is active, it may involve one or more of the following actions:

- The internal intercostal and transversus thoracic muscles depress the ribs. This action reduces the width and depth of the thoracic cavity.

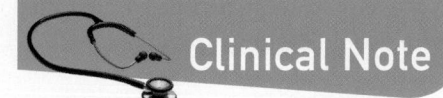

Clinical Note

Pneumothorax Air can enter the pleural cavity due to an injury to the chest wall that penetrates the parietal pleura, or a rupture of the alveoli that breaks through the visceral pleura. This condition, called **pneumothorax** (noo-mō-THOR-aks; *pneumo-*, air), breaks the fluid bond between the pleurae and allows the elastic fibers to recoil, resulting in a "collapsed lung," or **atelectasis** (at-e-LEK-ta-sis; *ateles,* incomplete + *ektasis,* extension). The opposite lung is not affected due to compartmentalization. The treatment for a collapsed lung involves removing as much of the air as possible from the affected pleural cavity and then sealing the opening. This treatment lowers the intrapleural pressure and reinflates the lung.

- The abdominal muscles, including the external and internal oblique, transversus abdominis, and rectus abdominis muscles, can assist the internal intercostal muscles in exhalation by compressing the abdomen. This action forces the diaphragm upward.

Modes of Breathing

We use the respiratory muscles in various combinations, depending on the volume of air that must be moved into or out of the system. We usually classify respiratory movements as *quiet breathing* or *forced breathing* according to the pattern of muscle activity during a single respiratory cycle.

Quiet Breathing. In **quiet breathing**, or **eupnea** (ŪP-nē-uh; *eu-*, normal + *-pnea*, breath), inhalation involves muscular contractions, but exhalation is a passive process. Inhalation usually involves contracting both the diaphragm and the external intercostal muscles. The relative contributions of these muscles can vary:

- During **diaphragmatic breathing**, or **deep breathing**, contraction of the diaphragm provides the necessary change in thoracic volume. Air is drawn into the lungs as the diaphragm contracts. Air is exhaled passively when the diaphragm relaxes.

- In **costal breathing**, or **shallow breathing**, the thoracic volume changes because the rib cage alters its shape. Inhalation takes place when contractions of the external intercostal muscles raise the ribs and enlarge the thoracic cavity. Exhalation takes place passively when these muscles relax.

During quiet breathing, expansion of the lungs stretches their elastic fibers. In addition, elevation of the rib cage stretches

The Respiratory Muscles

The most important skeletal muscles involved in respiratory movements are the diaphragm and the external intercostals. These muscles are the **primary respiratory muscles** and are active during normal breathing at rest. The **accessory respiratory muscles** become active when the depth and frequency of respiration must be increased markedly.

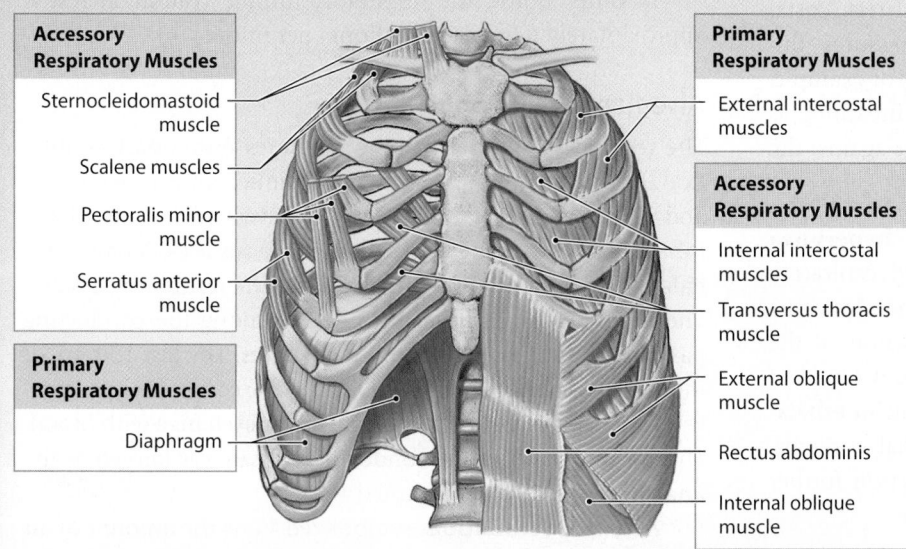

Accessory Respiratory Muscles
- Sternocleidomastoid muscle
- Scalene muscles
- Pectoralis minor muscle
- Serratus anterior muscle

Primary Respiratory Muscles
- Diaphragm

Primary Respiratory Muscles
- External intercostal muscles

Accessory Respiratory Muscles
- Internal intercostal muscles
- Transversus thoracis muscle
- External oblique muscle
- Rectus abdominis
- Internal oblique muscle

The Mechanics of Breathing

Pulmonary ventilation, air movement into and out of the respiratory system, occurs by changing the volume of the lungs. The changes in volume take place through the contraction of sleketal muscles. As the ribs are elevated or the diaphragm is depressed, the volume of the thoracic cavity increases and air moves into the lungs. The outward movement of the ribs as they are elevated resembles the outward swing of a raised bucket handle.

Ribs and sternum elevate

Diaphragm contracts

Respiratory Movements

Respiratory muscles may be used in various combinations, depending on the volume of air that must be moved in or out of the lungs. In **quiet breathing**, inhalation involves muscular contractions, but exhalation is a passive process. **Forced breathing** calls upon the accessory muscles to assist with inhalation, and exhalation involves contraction by the transversus thoracis, internal intercostal, and rectus abdominis muscles.

Inhalation

Inhalation is an active process. It primarily involves the diaphragm and the external intercostal muscles, with assistance from the accessory respiratory muscles as needed.

Accessory Respiratory Muscles (Inhalation)
- Sternocleidomastoid muscle
- Scalene muscles
- Pectoralis minor muscle
- Serratus anterior muscle

Primary Respiratory Muscles (Inhalation)
- External intercostal muscles
- Diaphragm

Exhalation

During forced exhalation, the tranversus thoracis and internal intercostal muscles actively depress the ribs, and the abdominal muscles (external and internal obliques, transversus abdominis, and rectus abdominis) compress the abdomen and push the diaphragm up.

Accessory Respiratory Muscles (Exhalation)
- Transversus thoracis muscle
- Internal intercostal muscles
- Rectus abdominis

KEY
- ⇨ = Movement of rib cage
- ⇨ = Movement of diaphragm
- ⇨ = Muscle contraction

opposing skeletal muscles and elastic fibers in the connective tissues of the body wall. When the muscles of inhalation relax, these elastic components recoil, returning the diaphragm, the rib cage, or both to their original positions. This action is called **elastic rebound**.

We typically use diaphragmatic breathing at minimal levels of activity. As we need increased volumes of air, our inspiratory movements become larger and the contribution of rib movement increases. Even when we are at rest, costal breathing can predominate when abdominal pressures, fluids, or masses restrict diaphragmatic movements. For example, pregnant women rely more and more on costal breathing as the enlarging uterus pushes the abdominal organs against the diaphragm.

Forced Breathing. **Forced breathing**, or **hyperpnea** (hī-PERP-nē-uh), involves active inspiratory and expiratory movements. In forced breathing, our accessory muscles assist with inhalation, and exhalation involves contraction of the internal intercostal muscles. At absolute maximum levels of forced breathing, our abdominal muscles take part in exhalation. Their contraction compresses the abdominal contents, pushing them up against the diaphragm. This action further reduces the volume of the thoracic cavity.

Respiratory Rates and Volumes

The respiratory system is extremely adaptable. You can be breathing slowly and quietly one moment, rapidly and deeply the next. The respiratory system adapts to meet the oxygen demands of the body by varying both the number of breaths per minute and the amount of air moved per breath. When you exercise at peak levels, the amount of air moving into and out of the respiratory tract can be 50 times the amount moved at rest.

Respiratory Rate

Your **respiratory rate** is the number of breaths you take each minute. As you read this, you are probably breathing quietly, with a low respiratory rate. The normal respiratory rate of a resting adult ranges from 12 to 18 breaths each minute, roughly one for every four heartbeats. Children breathe more rapidly, at rates of about 18–20 breaths per minute.

The Respiratory Minute Volume

We can calculate the amount of air moved each minute, symbolized $\dot{V}_E$, by multiplying the respiratory rate f by the tidal volume V_T. This value is called the **respiratory minute volume**. The respiratory rate at rest averages 12 breaths per minute, and the tidal volume at rest averages around 500 mL per breath. On that basis, we calculate the respiratory minute volume as follows:

$$\dot{V}_E = f \times V_T$$

$$\left(\begin{array}{c}\text{Volume of air moved}\\\text{each minute}\end{array}\right) = \left(\begin{array}{c}\text{breaths per}\\\text{minute}\end{array}\right) \times \left(\begin{array}{c}\text{tidal}\\\text{volume}\end{array}\right)$$

$$= 12 \text{ per minute} \times 500 \text{ mL}$$
$$= 6000 \text{ mL per minute}$$
$$= 6.0 \text{ liters per minute}$$

In other words, the respiratory minute volume at rest is approximately 6 liters (1.6 gallons) per minute.

Alveolar Ventilation

The respiratory minute volume measures pulmonary ventilation. It provides an indication of how much air is moving into and out of the respiratory tract. However, only some of the inhaled air reaches the alveolar exchange surfaces. A typical inhalation pulls about 500 mL of air into the respiratory system. The first 350 mL of inhaled air travels along the conducting passageways and enters the alveolar spaces. The last 150 mL of inhaled air never gets that far. It stays in the conducting passageways and thus does not participate in gas exchange with blood. The volume of air in the conducting passages is known as the **anatomic dead space**, denoted V_D.

Alveolar ventilation, symbolized $\dot{V}_A$, is the amount of air reaching the alveoli each minute. The alveolar ventilation is less than the respiratory minute volume, because some of the air never reaches the alveoli, but instead remains in the dead space of the lungs. We can calculate alveolar ventilation by subtracting the dead space from the tidal volume:

$$\dot{V}_A = f \times (V_T - V_D)$$

$$\left(\begin{array}{c}\text{alveolar}\\\text{ventilation}\end{array}\right) = \left(\begin{array}{c}\text{breaths}\\\text{per minute}\end{array}\right) \times \left(\begin{array}{cc}\text{tidal} & \text{anatomic}\\\text{volume} & \text{dead space}\end{array}\right)$$

At rest, alveolar ventilation rates are approximately 4.2 liters per minute (12 per minute × 350 mL). However, the gas arriving in the alveoli is significantly different from the surrounding atmosphere. The reason is that inhaled air always mixes with "used" air in the conducting passageways (the anatomic dead space) on its way in. The air in alveoli thus contains less oxygen and more carbon dioxide than does atmospheric air.

Relationships among V_T, $\dot{V}_E$, and $\dot{V}_A$

The respiratory minute volume can be increased by increasing either the tidal volume or the respiratory rate. In other words, you can breathe more deeply or more quickly or both. Under maximum stimulation, the tidal volume can increase to roughly 4.8 liters. At peak respiratory rates of 40–50 breaths per minute and maximum cycles of inhalation and exhalation, the respiratory minute volume can approach 200 liters (about 55 gal) per minute.

In functional terms, the alveolar ventilation rate is more important than the respiratory minute volume, because it determines the rate of delivery of oxygen to the alveoli. The respiratory rate and the tidal volume together determine the alveolar ventilation rate:

- For a given respiratory rate, increasing the tidal volume (breathing more deeply) increases the alveolar ventilation rate.

- For a given tidal volume, increasing the respiratory rate (breathing more quickly) increases the alveolar ventilation rate.

The alveolar ventilation rate can change independently of the respiratory minute volume. In our previous example, the respiratory minute volume at rest was 6 liters and the alveolar ventilation rate was 4.2 L/min. If the respiratory rate rises to 20 breaths per minute, but the tidal volume drops to 300 mL, the respiratory minute volume remains the same ($20 \times 300 = 6000$). However, the alveolar ventilation rate drops to only 3 L/min ($20 \times [300 - 150] = 3000$). For this reason, whenever the demand for oxygen increases, both the tidal volume *and* the respiratory rate must be regulated closely. (We focus on the mechanisms involved in a later section.)

Respiratory Performance and Volume Relationships

We exchange only a small proportion of the air in our lungs during a single quiet respiratory cycle. We can increase the tidal volume by inhaling more vigorously and exhaling more completely.

We can divide the total volume of the lungs into a series of *volumes* and *capacities* (each the sum of various volumes), as indicated in **Figure 23–16**. These measurements are obtained by an instrument called a **spirometer**. Spirometry values are useful in diagnosing problems with pulmonary ventilation. Adult females, on average, have smaller bodies and thus smaller lung volumes than do males. As a result, there are sex-related differences in respiratory volumes and capacities. The table in the figure shows representative values for both sexes.

Pulmonary volumes include the following:

- The **tidal volume (V_T)** is the amount of air you move into or out of your lungs during a single respiratory cycle. The tidal volume averages about 500 mL in both males and females.

- The **expiratory reserve volume (ERV)** is the amount of air that you can voluntarily expel after you have completed a normal, quiet respiratory cycle. As an example, with maximum use of the accessory muscles, males can expel an additional 1000 mL of air, on average. Female expiratory reserve volume averages 700 mL.

Figure 23–16 **Pulmonary Volumes and Capacities.** The red line indicates the volume of air within the lungs during breathing.

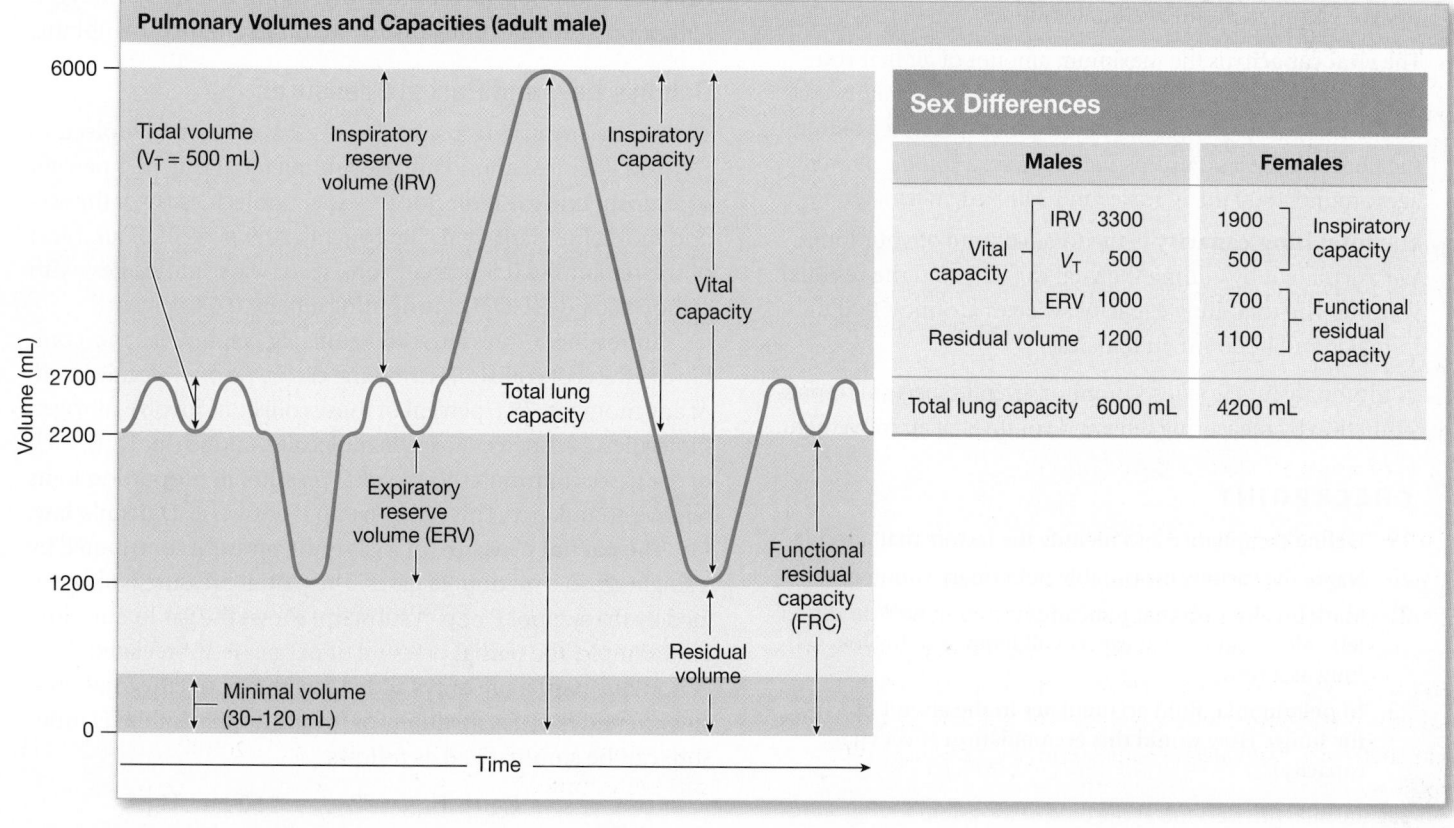

- The **residual volume** is the amount of air that remains in your lungs even after a maximal exhalation—typically about 1200 mL in males and 1100 mL in females. The **minimal volume**, a component of the residual volume, is the amount of air that would remain in your lungs if they were allowed to collapse. The minimal volume ranges from 30 to 120 mL. Unlike other volumes, it cannot be measured in a healthy person. The minimal volume and the residual volume are very different, because the fluid bond between the lungs and the chest wall normally prevents the recoil of the elastic fibers in the lungs. Some air remains in the lungs, even at minimal volume, because the surfactant coating the alveolar surfaces prevents their collapse.

- The **inspiratory reserve volume (IRV)** is the amount of air that you can take in over and above the tidal volume. On average, the lungs of males are larger than those of females. For this reason, the inspiratory reserve volume of males averages 3300 mL, compared with 1900 mL in females.

We can calculate respiratory capacities by adding the values of various volumes. Examples include the following:

- The **inspiratory capacity** is the amount of air that you can draw into your lungs after you have completed a quiet respiratory cycle. The inspiratory capacity is the sum of the tidal volume and the inspiratory reserve volume.

- The **functional residual capacity (FRC)** is the amount of air remaining in your lungs after you have completed a quiet respiratory cycle. The FRC is the sum of the expiratory reserve volume and the residual volume.

- The **vital capacity** is the maximum amount of air that you can move into or out of your lungs in a single respiratory cycle. The vital capacity is the sum of the expiratory reserve volume, the tidal volume, and the inspiratory reserve volume. It averages around 4800 mL in males and 3400 mL in females.

- The **total lung capacity** is the total volume of your lungs. We calculate it by adding the vital capacity and the residual volume. The total lung capacity averages around 6000 mL in males and 4200 mL in females.

Pulmonary function tests monitor several aspects of respiratory function by measuring rates and volumes of air movement.

CHECKPOINT

19. Define compliance and identify the factors that affect it.

20. Name the various measurable pulmonary volumes.

21. Mark breaks a rib that punctures the chest wall on his left side. What do you expect will happen to his left lung as a result?

22. In pneumonia, fluid accumulates in the alveoli of the lungs. How would this accumulation affect vital capacity?

See the blue Answers tab at the back of the book.

23-8 | Gas exchange depends on the partial pressures of gases and the diffusion of molecules

Learning Outcome Summarize the physical principles governing the diffusion of gases into and out of the blood and body tissues.

Pulmonary ventilation ensures both that your alveoli are supplied with oxygen and that the carbon dioxide arriving from your bloodstream is removed. The actual process of gas exchange takes place between blood and alveolar air across the respiratory membrane. To understand these events, let's first consider (1) the *partial pressures* of the gases involved and (2) the diffusion of molecules between a gas and a liquid. Then we can discuss the movement of oxygen and carbon dioxide across the respiratory membrane.

The Gas Laws

Gases are exchanged between the alveolar air and the blood through diffusion, which takes place in response to concentration gradients. As we saw in Chapter 3, the rate of diffusion varies in response to a variety of factors, including the size of the concentration gradient and the temperature. ⊃ p. 117 The principles that govern the movement and diffusion of gas molecules, such as those in the atmosphere, are relatively straightforward. These principles, known as *gas laws*, have been understood for about 250 years. You have already heard about Boyle's law, which determines the direction of air movement in pulmonary ventilation. In this section, you will learn about gas laws and other factors that determine the rate of oxygen and carbon dioxide diffusion across the respiratory membrane.

Dalton's Law and Partial Pressures

The air we breathe is a mixture of gases. Nitrogen molecules (N_2) are the most abundant, accounting for about 78.6 percent of atmospheric gas molecules. Oxygen molecules (O_2), the second most abundant, make up roughly 20.9 percent of air. Most of the remaining 0.5 percent consists of water molecules, with carbon dioxide (CO_2) contributing a mere 0.04 percent.

Atmospheric pressure, 760 mm Hg, represents the combined effects of collisions involving each type of molecule in air. At any moment, 78.6 percent of those collisions involve nitrogen molecules, 20.9 percent oxygen molecules, and so on. Thus, each of the gases contributes to the total pressure in proportion to its relative abundance. This relationship is known as **Dalton's law**.

The **partial pressure** of a gas is the pressure contributed by a single gas in a mixture of gases. The partial pressure is abbreviated by the symbol P or p. A subscript shows the gas in question. For example, the partial pressure of oxygen is abbreviated P_{O_2}.

All the partial pressures added together equal the total pressure exerted by a gas mixture. For the atmosphere, this relationship can be summarized as follows:

$$P_{N_2} + P_{O_2} + P_{H_2O} + P_{CO_2} = 760 \text{ mm Hg}$$

We can easily calculate the partial pressure of each gas because we know the individual percentages of each gas in air. For example, the partial pressure of oxygen, P_{O_2}, is 20.9 percent of 760 mm Hg, or roughly 159 mm Hg. Table 23–2 includes the partial pressures of other atmospheric gases.

Diffusion between Liquids and Gases (Henry's Law)

Differences in pressure, which move gas molecules from one place to another, also affect the movement of gas molecules into and out of solution. At a given temperature, the amount of a particular gas in solution is directly proportional to the partial pressure of that gas. This principle is known as **Henry's law**.

When a gas under pressure contacts a liquid, the pressure tends to force gas molecules into solution. At a given pressure, the number of dissolved gas molecules rises until an equilibrium is established. At equilibrium, gas molecules diffuse out of the liquid as quickly as they enter it, so the total number of gas molecules in solution remains constant. If the partial pressure increases, more gas molecules go into solution. If the partial pressure decreases, gas molecules come out of solution.

You see Henry's law in action whenever you open a can of soda. The soda was put into the can under pressure, and the gas (carbon dioxide) is in solution (Figure 23–17a). When you open the can, the pressure decreases and the gas molecules begin coming out of solution (Figure 23–17b). Theoretically, the process will continue until an equilibrium develops between the surrounding air and the gas in solution. In fact, the volume of the can is so small, and the volume of the atmosphere so great, that within a half hour or so virtually all the carbon dioxide comes out of solution. You are left with "flat" soda.

The actual *amount* of a gas in solution at a given partial pressure and temperature depends on the solubility of the gas in that particular liquid. In body fluids, carbon dioxide is highly soluble, and oxygen is somewhat less soluble. Nitrogen has very limited solubility. The dissolved gas content is usually reported in milliliters of gas per 100 mL (1 dL) of solution. To see the differences in relative solubility of these three gases, we can compare the gas content of blood in the pulmonary veins with the partial pressure of each gas in the alveoli. In

Table 23–2	Partial Pressures (mm Hg) and Normal Gas Concentrations (%) in Air			
Source of Sample	Nitrogen (N₂)	Oxygen (O₂)	Carbon Dioxide (CO₂)	Water Vapor (H₂O)
Inhaled air (dry)	597 (78.6%)	159 (20.9%)	0.3 (0.04%)	3.7 (0.5%)
Alveolar air (saturated)	573 (75.4%)	100 (13.2%)	40 (5.2%)	47 (6.2%)
Exhaled air (saturated)	569 (74.8%)	116 (15.3%)	28 (3.7%)	47 (6.2%)

Figure 23–17 Henry's Law and the Relationship between Solubility and Pressure.

a Increasing the pressure drives gas molecules into solution until an equilibrium is established.

Example Soda is put into the can under pressure, and the gas (carbon dioxide) is in solution at equilibrium.

b When the gas pressure decreases, dissolved gas molecules leave the solution until a new equilibrium is reached.

Example Opening the can of soda relieves the pressure, and bubbles form as the dissolved gas leaves the solution.

Clinical Note

Decompression Sickness *Decompression sickness* is a painful condition that develops when a person is exposed to a sudden drop in atmospheric pressure. Nitrogen is the gas responsible for the problems experienced, due to its high partial pressure in air. When the pressure drops, nitrogen comes out of solution, forming bubbles like those in a shaken can of soda. The bubbles may form in joint cavities, in the bloodstream, and in the cerebrospinal fluid. Individuals with decompression sickness typically curl up from the pain in affected joints. This reaction accounts for the condition's common name: *the bends*.

Decompression sickness most commonly affects scuba divers who return to the surface too quickly after breathing air under greater-than-normal pressure while submerged. It can also develop in airline passengers subject to sudden losses of cabin pressure.

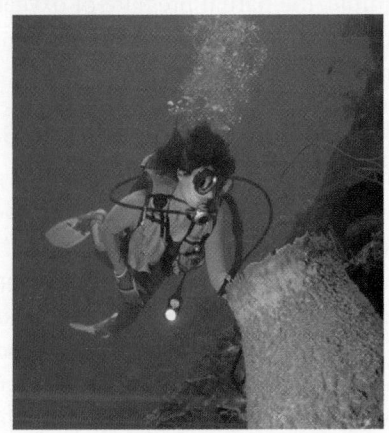

a pulmonary vein, plasma generally contains 2.62 mL/dL of dissolved CO_2 (P_{CO_2} = 40 mm Hg), 0.29 mL/dL of dissolved O_2 (P_{O_2} = 100 mm Hg), and 1.25 mL/dL of dissolved N_2 (P_{N_2} = 573 mm Hg). Thus even though the partial pressure of carbon dioxide is less than one-tenth the partial pressure of nitrogen, the plasma contains more than twice as much carbon dioxide as nitrogen in solution.

Diffusion and Respiratory Function

The gas laws apply to the diffusion of oxygen, carbon dioxide, and nitrogen between a gas and a liquid. We will now consider how differing partial pressures and solubilities determine the direction and rate of diffusion across the respiratory membrane that separates the air within the alveoli from the blood in alveolar capillaries.

The Composition of Alveolar Air

As soon as air enters the respiratory tract, its characteristics begin to change. In passing through the nasal cavity, inhaled air becomes warmer, and the amount of water vapor increases. Humidification and filtration continue as the air travels through the pharynx, trachea, and bronchial passageways. On reaching the alveoli, the incoming air mixes with air remaining in the alveoli from the previous respiratory cycle. For this reason, alveolar air contains more carbon dioxide and less oxygen than does atmospheric air.

The last 150 mL of inhaled air never gets farther than the conducting passageways. It remains in the anatomic dead space of the lungs. During the next exhalation, the departing alveolar air mixes with air in the dead space, producing yet another mixture that differs from both atmospheric and alveolar samples. The differences in composition between atmospheric (inhaled) and alveolar air are given in Table 23–2.

Efficiency of Diffusion at the Respiratory Membrane

Gas exchange at the respiratory membrane is efficient for the following five reasons:

1. *The differences in partial pressure across the respiratory membrane are substantial.* This fact is important, because the greater the difference in partial pressure, the faster the rate of gas diffusion. Conversely, if P_{O_2} in alveoli decreases, the rate of oxygen diffusion into blood drops. This is why many people feel light-headed at altitudes of 3000 m or more. The partial pressure of oxygen in their alveoli has dropped so low that the rate of oxygen absorption is significantly reduced.

2. *The distances involved in gas exchange are short.* The fusion of capillary and alveolar basement membranes reduces the distance for gas exchange to an average of 0.5 μm. Inflammation of the lung tissue or a buildup of fluid in alveoli increases the diffusion distance and impairs alveolar gas exchange.

3. *The gases are lipid soluble.* Both oxygen and carbon dioxide diffuse readily through the surfactant layer and the alveolar and endothelial plasma membranes.

4. *The total surface area is large.* The combined alveolar surface area at peak inhalation may approach 140 m^2 (1506 ft^2). This area is slightly bigger than half of a tennis court. Damage to alveolar surfaces, which occurs in emphysema, reduces the available surface area and the efficiency of gas transfer.

5. *Blood flow and airflow are coordinated.* This close coordination makes both pulmonary ventilation and pulmonary circulation more efficient. For example, blood flow is greatest around alveoli with the highest P_{O_2} values, where oxygen uptake can proceed with maximum efficiency. If

Figure 23–18   **An Overview of Respiratory Processes and Partial Pressures in Respiration.**

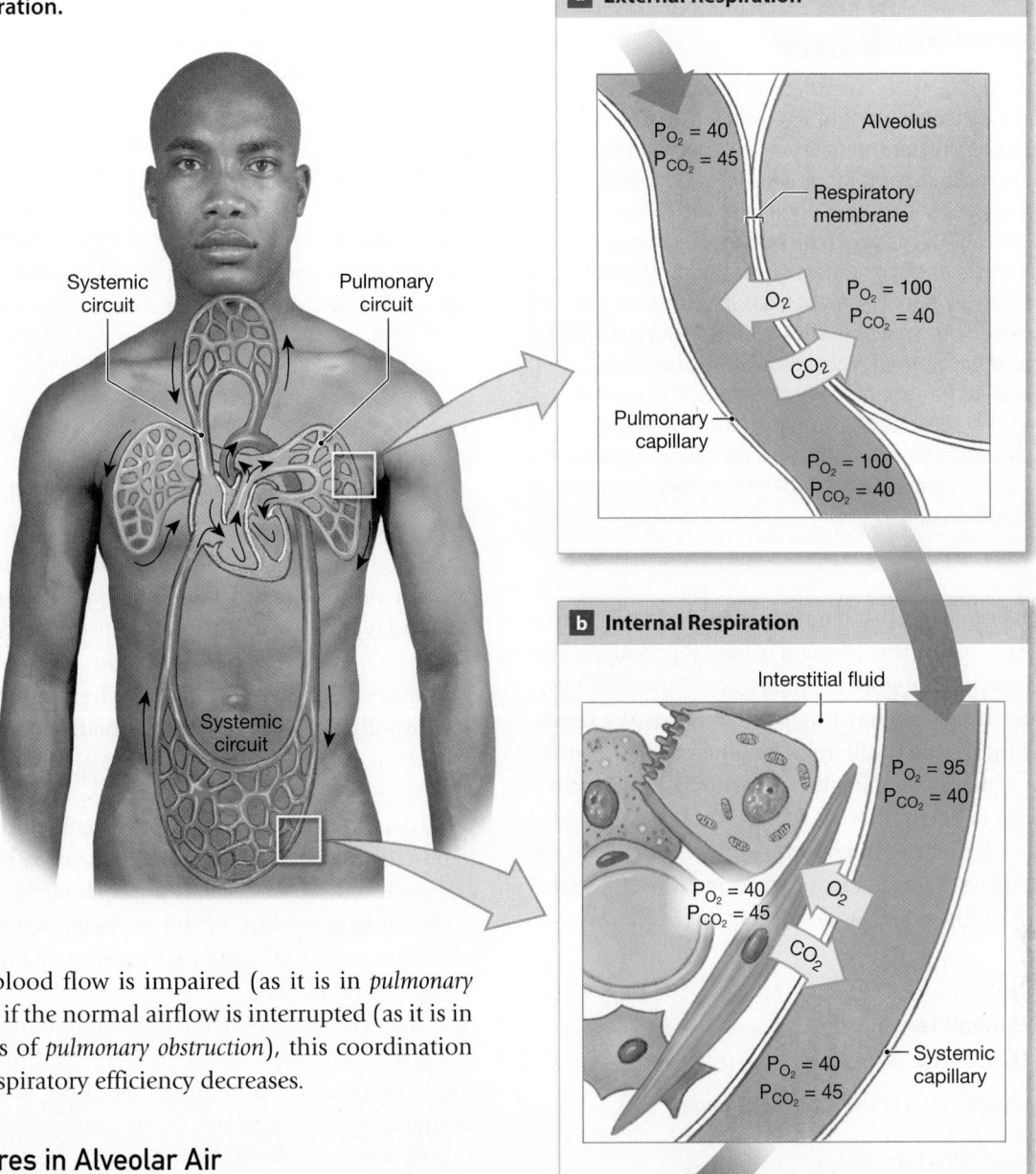

the normal blood flow is impaired (as it is in *pulmonary embolism*), or if the normal airflow is interrupted (as it is in various forms of *pulmonary obstruction*), this coordination is lost and respiratory efficiency decreases.

Partial Pressures in Alveolar Air and Alveolar Capillaries

Figure 23–18 illustrates the partial pressures of oxygen and carbon dioxide in the pulmonary and systemic circuits. Notice that blood arriving from the pulmonary arteries has a lower P_{O_2} and a higher P_{CO_2} than does alveolar air (**Figure 23–18a**). Diffusion between the alveolar gas mixture and the alveolar capillaries then increases the P_{O_2} of blood and decreases its P_{CO_2}. By the time the blood enters the pulmonary venules, it has reached equilibrium with the alveolar air. Blood departs the alveoli with a P_{O_2} of about 100 mm Hg and a P_{CO_2} of roughly 40 mm Hg.

Diffusion between alveolar air and blood in the alveolar capillaries occurs very rapidly. When you are at rest, a red blood cell moves through one of your alveolar capillaries in about 0.75 second. When you exercise, the passage takes less than 0.3

second. This amount of time is usually sufficient to reach an equilibrium between the alveolar air and the blood.

Partial Pressures in the Systemic Circuit

Oxygenated blood leaves the alveolar capillaries to return to the heart, to be pumped into the systemic circuit. As this blood enters the pulmonary veins, it mixes with blood that flowed through capillaries around conducting passageways. Because gas exchange in the lungs occurs only at alveoli, the blood leaving the conducting passageways carries relatively little oxygen. The partial pressure of oxygen in the pulmonary veins therefore

23

Clinical Note

Blood Gas Analysis Blood samples can be analyzed to determine their concentrations of dissolved gases. The usual tests include the determination of pH, P_{CO_2}, and P_{O_2} in an arterial sample. Such samples provide information about the degree of oxygenation in peripheral tissues. For example, if the arterial P_{CO_2} is very high and the P_{O_2} is very low, tissues are not receiving adequate oxygen. This problem may be solved by giving the patient a gas mixture that has a high P_{O_2} (or even pure oxygen, with a P_{O_2} of 760 mm Hg).

Blood gas measurements also provide information on the efficiency of gas exchange at the lungs. If the arterial P_{O_2} remains low despite the administration of oxygen, or if the P_{CO_2} is very high, pulmonary exchange problems may exist. Such problems may include pulmonary edema, asthma, or pneumonia.

drops to about 95 mm Hg. This is the P_{O_2} in the blood that enters the systemic circuit. No further changes in partial pressure occur until the blood reaches the systemic capillaries (**Figure 23–18b**).

Normal interstitial fluid has a P_{O_2} of 40 mm Hg. As a result, oxygen diffuses out of the capillaries until the capillary partial pressure is the same as that in the adjacent tissues. Inactive peripheral tissues normally have a P_{CO_2} of about 45 mm Hg, whereas blood entering peripheral capillaries normally has a P_{CO_2} of 40 mm Hg. As a result, carbon dioxide diffuses into the blood as oxygen diffuses out (**Figure 23–18b**).

CHECKPOINT

23. **Define Dalton's law.**

24. **Define Henry's law.**

See the blue Answers tab at the back of the book.

23-9 ▌ Most oxygen is transported bound to hemoglobin; and carbon dioxide is transported in three ways: as carbonic acid, bound to hemoglobin, or dissolved in plasma

Learning Outcome Describe the structure and function of hemoglobin, and the transport of oxygen and carbon dioxide in the blood.

Oxygen and carbon dioxide have limited solubilities in blood plasma. For example, at the normal P_{O_2} of alveoli, 100 mL of plasma absorbs only about 0.3 mL of oxygen. The limited solubilities of these gases are a problem. The peripheral tissues need more oxygen and generate more carbon dioxide than the plasma alone can absorb and transport.

Red blood cells (RBCs) solve this problem. They remove dissolved oxygen and CO_2 molecules from plasma and bind them (in the case of oxygen) or use them to manufacture soluble compounds (in the case of carbon dioxide). As these reactions remove dissolved gases from blood plasma, these gases continue to diffuse into the blood, but never reach equilibrium.

The important thing about these reactions is that they are both temporary and completely reversible. When oxygen or carbon dioxide concentrations in blood plasma are high, RBCs remove the excess molecules. When plasma concentrations are falling, the RBCs release their stored reserves.

Oxygen Transport

Each 100 mL of blood leaving the alveolar capillaries carries away about 20 mL of oxygen. Only 1.5 percent (about 0.3 mL) of this amount is oxygen molecules in solution. The rest of the oxygen molecules are bound to *hemoglobin (Hb) molecules*. Recall that the hemoglobin molecule consists of four globular protein subunits, each containing a heme unit. The oxygen molecules bind to the iron ions in the center of heme units. ↺ p. 685 Thus, each hemoglobin molecule can bind four molecules of oxygen, forming **oxyhemoglobin (HbO$_2$)**. This reaction is reversible. We can summarize it as

$$Hb + O_2 \rightleftharpoons HbO_2.$$

This reversible reaction allows hemoglobin to pick up oxygen in the lungs and then release it to body tissues elsewhere.

Each red blood cell has approximately 280 million molecules of hemoglobin. With four heme units per hemoglobin molecule, each RBC potentially can carry more than a billion molecules of oxygen.

The percentage of heme units containing bound oxygen at any given moment is called the **hemoglobin saturation**. If all the Hb molecules in the blood are fully loaded with oxygen, saturation is 100 percent. If, on average, each Hb molecule carries two O_2 molecules, saturation is 50 percent.

In Chapter 2, we saw that the shape and functional properties of a protein change in response to changes in its environment. ↺ p. 80 Hemoglobin is no exception. Any changes in shape can affect oxygen binding. Under normal conditions, the most important environmental factors affecting hemoglobin are (1) the P_{O_2} of blood, (2) blood pH, (3) temperature, and (4) ongoing metabolic activity within RBCs.

Hemoglobin and P_{O_2}

Hemoglobin in RBCs carries most of the oxygen in the bloodstream and releases it in response to changes in the partial pressure of oxygen in the surrounding plasma. If the P_{O_2} increases, hemoglobin binds oxygen. If the P_{O_2} decreases, hemoglobin releases oxygen.

An **oxygen–hemoglobin saturation curve**, or *oxygen–hemoglobin dissociation curve*, is a graph that relates the hemoglobin saturation to the partial pressure of oxygen (**Figure 23–19**). The binding and dissociation, or release, of oxygen to hemoglobin is a typical reversible reaction. At equilibrium, oxygen molecules bind to heme at the same rate that other oxygen molecules are being released. If the P_{O_2} increases, then more oxygen molecules bind to hemoglobin, and fewer are released. Referring to the equation given in the previous section, we say that the reaction shifts to the right. If the P_{O_2} decreases, more oxygen molecules are released from hemoglobin, while fewer oxygen molecules bind. In this case, we say that the reaction shifts to the left.

Notice that the graph of this relationship between P_{O_2} and hemoglobin saturation is a curve rather than a straight line. It is a curve because the shape of the hemoglobin molecule changes slightly each time it binds an oxygen molecule. This change in shape enhances the ability of hemoglobin to bind *another* oxygen molecule. In other words, the binding of the first oxygen molecule makes it easier to bind the second; binding the second promotes binding of the third; and binding the third enhances binding of the fourth.

Because each arriving oxygen molecule makes it easier for hemoglobin to bind the *next* oxygen molecule, the saturation curve takes the form shown in **Figure 23–19**. Once the first oxygen molecule binds to the hemoglobin, the slope rises rapidly until it levels off near 100 percent saturation. Where the slope is steep, a very small change in plasma P_{O_2} results in a large change in the amount of oxygen bound to Hb or released from HbO_2. Notice that hemoglobin will be more than 90 percent saturated if exposed to an alveolar P_{O_2} above 60 mm Hg. Thus, near-normal oxygen transport can continue even when the oxygen content of alveolar air decreases below normal, or $P_{O_2} = 100$ mm Hg. Without this ability, you could not survive at high altitudes. Conditions that significantly reduce pulmonary ventilation would be immediately fatal.

At normal alveolar pressures ($P_{O_2} = 100$ mm Hg) the hemoglobin saturation is very high (97.5 percent), but complete saturation does not occur until the P_{O_2} reaches excessively high levels (about 250 mm Hg). In functional terms, the maximum saturation is not as important as the ability of hemoglobin to provide oxygen over the normal P_{O_2} range in body tissues. Over that range—from 100 mm Hg at the alveoli to perhaps 15 mm Hg in active tissues—the saturation drops from 97.5 percent to less than 20 percent, and a small change in P_{O_2} makes a big difference in terms of the amount of oxygen bound to hemoglobin.

Note that the relationship between P_{O_2} and hemoglobin saturation remains valid whether the P_{O_2} is rising or falling. If the P_{O_2} increases, the saturation goes up and hemoglobin stores oxygen. If the P_{O_2} decreases, hemoglobin releases oxygen into its surroundings. When oxygenated blood arrives in the peripheral capillaries, the blood P_{O_2} declines rapidly due to gas exchange with the interstitial fluid. As the P_{O_2} falls, hemoglobin gives up its oxygen.

The relationship between the P_{O_2} and hemoglobin saturation provides a mechanism for automatic regulation of oxygen delivery. Inactive tissues have little demand for oxygen, and the local P_{O_2} is usually about 40 mm Hg. Under these conditions, hemoglobin does not release much oxygen. As it passes through the capillaries, it goes from 97 percent saturation ($P_{O_2} = 95$ mm Hg) to 75 percent saturation ($P_{O_2} = 40$ mm Hg). Because hemoglobin still retains three-quarters of its oxygen, venous blood has a relatively large oxygen reserve. This reserve is important, because it can be mobilized if tissue oxygen demands increase.

Active tissues use oxygen at an accelerated rate, so their P_{O_2} may drop to 15–20 mm Hg. Hemoglobin passing through these capillaries goes from 97 percent saturation to about 20 percent saturation. In practical terms, this means that as blood circulates through peripheral capillaries, active tissues receive 3.5 times as much oxygen as do inactive tissues.

Hemoglobin and pH

At a given P_{O_2}, hemoglobin releases additional oxygen if the pH decreases. The oxygen–hemoglobin saturation curve in **Figure 23–19** was determined in normal blood, with a pH of 7.4 and a temperature of 37 °C. In addition to consuming oxygen, active tissues generate acids that lower the pH of the interstitial fluid. When the pH decreases, the shape of hemoglobin molecules changes. As a result of this change,

Figure 23–19 **An Oxygen–Hemoglobin Saturation Curve.** The saturation characteristics of hemoglobin at various partial pressures of oxygen under normal conditions (body temperature of 37°C and blood pH of 7.4).

P_{O_2} (mm Hg)	% saturation of Hb
10	13.5
20	35
30	57
40	75
50	83.5
60	89
70	92.7
80	94.5
90	96.5
100	97.5

23

the molecules release their oxygen reserves more readily, so the slope of the hemoglobin saturation curve changes (**Figure 23–20a**). In other words, as pH drops, the saturation declines. At a tissue P_{O_2} of 40 mm Hg, for example, you can see in **Figure 23–20a** that a pH decrease from 7.4 to 7.2 changes hemoglobin saturation from 75 percent to 60 percent. This means that hemoglobin molecules release 20 percent more oxygen in peripheral tissues at a pH of 7.2 than they do at a pH of 7.4. This effect of pH on the hemoglobin saturation curve is called the **Bohr effect**.

Carbon dioxide is the primary compound responsible for the Bohr effect. When CO_2 diffuses into the blood, it rapidly diffuses into red blood cells. There, an enzyme called **carbonic anhydrase** catalyzes the reaction of CO_2 with water molecules:

$$CO_2 + H_2O \overset{\text{carbonic anhydrase}}{\rightleftharpoons} H_2CO_3 \rightleftharpoons H^+ + HCO_3^-$$

The product of this enzymatic reaction, H_2CO_3, is called *carbonic acid*, because it dissociates into a hydrogen ion (H^+) and a bicarbonate ion (HCO_3^-). The rate of carbonic acid formation depends on the amount of carbon dioxide in solution, which, as noted earlier, depends on the P_{CO_2}. When the P_{CO_2} rises, the rate of carbonic acid formation accelerates and the reaction proceeds from left to right. The hydrogen ions that are generated diffuse out of the RBCs, and the pH of the plasma drops. When the P_{CO_2} declines, hydrogen ions diffuse out of the plasma and into the RBCs. As a result, the pH of the plasma increases as the reaction proceeds from right to left.

Hemoglobin and Temperature

At a given P_{CO_2}, hemoglobin releases additional oxygen if the temperature increases. Changes in temperature affect the slope of the hemoglobin saturation curve (**Figure 23–20b**). As the temperature increases, hemoglobin releases more oxygen. As the temperature decreases, hemoglobin holds oxygen more tightly. Temperature effects are significant only in active tissues that are generating large amounts of heat. For example, active skeletal muscles generate heat, and the heat warms blood that flows through these organs. As the blood warms, the Hb molecules release more oxygen than can be used by the active muscle fibers.

Hemoglobin and BPG

Red blood cells lack mitochondria. These cells produce ATP only by glycolysis. As a result, lactic acid is formed, as we saw in Chapter 10. ⤶ p. 344 The metabolic pathways involved in glycolysis in an RBC also generate the compound **2,3-bisphosphoglycerate** (biz-fos-fō-GLIS-er-āt), or **BPG**. This compound has a direct effect on oxygen binding and release. For any partial pressure of oxygen, the higher the concentration of BPG, the greater the release of oxygen by Hb molecules. Normal RBCs always contain BPG.

Both BPG synthesis and the Bohr effect improve oxygen delivery when the pH changes: BPG levels rise when the pH increases, and the Bohr effect appears when the pH decreases. The concentration of BPG can be increased by high blood pH, thyroid hormones, growth hormone, epinephrine, and androgens. When plasma P_{O_2}

Figure 23–20 **The Effects of pH and Temperature on Hemoglobin Saturation.**

a **Effect of pH.** When the pH decreases below normal levels, more oxygen is released; the oxygen–hemoglobin saturation curve shifts to the right. When the pH increases, less oxygen is released; the curve shifts to the left.

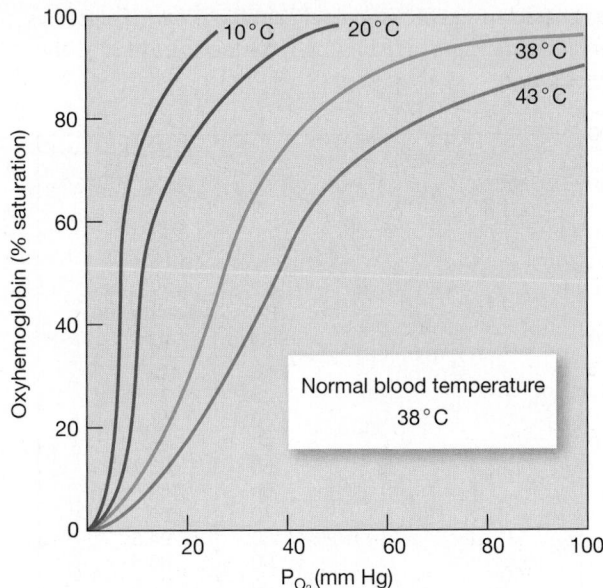

b **Effect of temperature.** When the temperature increases, more oxygen is released; the oxygen–hemoglobin saturation curve shifts to the right.

levels are low for an extended time, red blood cells generate more BPG. These factors improve oxygen delivery to the tissues, because when BPG levels are elevated, hemoglobin releases about 10 percent more oxygen at a given P_{O_2} than it would do otherwise.

The production of BPG decreases as RBCs age. For this reason, the level of BPG can determine how long a blood bank can store fresh whole blood. When BPG levels get too low, hemoglobin becomes firmly bound to the available oxygen. The blood is then useless for transfusions, because the RBCs will no longer release oxygen to peripheral tissues, even at a disastrously low P_{O_2}.

Fetal Hemoglobin

The RBCs of a developing fetus contain **fetal hemoglobin**. The structure of fetal hemoglobin differs from that of adult hemoglobin, giving it a much higher affinity for oxygen. At the same P_{O_2}, fetal hemoglobin binds more oxygen than does adult hemoglobin (**Figure 23–21**). This trait is key to transferring oxygen across the placenta.

A fetus obtains oxygen from the maternal bloodstream. At the placenta, maternal blood has a relatively low P_{O_2}, ranging from 35 to 50 mm Hg. If maternal blood arrives at the placenta with a P_{O_2} of 40 mm Hg, hemoglobin saturation is roughly 75 percent. The fetal blood arriving at the placenta has a P_{O_2} close to 20 mm Hg. However, because fetal hemoglobin has a higher affinity for oxygen, it is still 58 percent saturated.

As diffusion takes place between fetal blood and maternal blood, oxygen enters the fetal bloodstream until the P_{O_2} reaches equilibrium at 30 mm Hg. At this P_{O_2}, the maternal hemoglobin is less than 60 percent saturated, but the fetal hemoglobin is over 80 percent saturated, as you can see on **Figure 23–21**. The steep slope of the saturation curve for fetal hemoglobin means that when fetal RBCs reach peripheral tissues of the fetus, the Hb molecules release a large amount of oxygen in response to a very small change in P_{O_2}.

Carbon Dioxide Transport

Carbon dioxide is generated by aerobic metabolism in peripheral tissues. Carbon dioxide travels in the bloodstream in three different ways. After entering the blood, a CO_2 molecule either (1) is converted to a molecule of carbonic acid, (2) binds to hemoglobin within red blood cells, or (3) dissolves in plasma. All three reactions are completely reversible, allowing carbon dioxide to be picked up from body tissues and then delivered to the alveoli. Let's consider the events that take place as blood enters peripheral tissues in which the P_{CO_2} is 45 mm Hg.

Carbonic Acid Formation

Roughly 70 percent of the carbon dioxide absorbed by blood is transported as molecules of carbonic acid. Carbon dioxide is converted to carbonic acid through the activity of the enzyme carbonic anhydrase in RBCs. The carbonic acid molecules immediately dissociate into a hydrogen ion and a bicarbonate ion, as described earlier (p. 888). We can ignore the intermediate steps in this sequence and summarize the reaction as

$$CO_2 + H_2O \xrightleftharpoons{\text{carbonic anhydrase}} H^+ + HCO_3^-$$

Figure 23–21   A Functional Comparison of Fetal and Adult Hemoglobin.

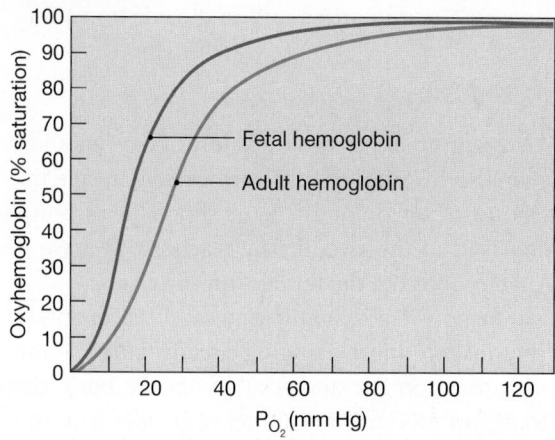

This reaction is completely reversible. In peripheral capillaries, it proceeds vigorously, tying up large numbers of CO_2 molecules. The reaction continues as carbon dioxide diffuses out of the interstitial fluids.

The hydrogen ions and bicarbonate ions have different fates. Most of the hydrogen ions bind to hemoglobin molecules, forming HbH^+. The Hb molecules function as pH buffers, tying up the released hydrogen ions before the ions can leave the RBCs and lower the pH of the plasma. The bicarbonate ions move into the plasma with the aid of a countertransport mechanism that exchanges intracellular bicarbonate ions (HCO_3^-) for extracellular chloride ions (Cl^-). This exchange trades one anion for another and does not require ATP. The result is a mass movement of chloride ions into the RBCs, an event known as the **chloride shift**.

CO_2 Binding to Hemoglobin

About 23 percent of the carbon dioxide carried by blood is bound to the protein portions of Hb molecules inside RBCs. These CO_2 molecules are attached to exposed amino groups ($-NH_2$) of the Hb molecules. The resulting compound is called **carbaminohemoglobin** (kar-BAM-i-nō-hē-mō-glō-bin), $HbCO_2$. We can summarize the reversible reaction as follows:

$$CO_2 + HbNH_2 \rightleftharpoons HbNHCOOH$$

We can abbreviate this reaction without the amino groups as

$$CO_2 + Hb \rightleftharpoons HbCO_2$$

Transport in Plasma

Plasma becomes saturated with carbon dioxide quite rapidly. Only about 7 percent of the carbon dioxide absorbed at peripheral capillaries is transported as dissolved gas molecules. RBCs absorb the rest and convert it using carbonic anhydrase or store it as carbaminohemoglobin.

Figure 23–22 summarizes carbon dioxide transport.

Summary: Gas Transport

Figure 23–23 summarizes the transport of oxygen and carbon dioxide in the respiratory and cardiovascular systems. Note that the bottom portion of the figure shows the carbon dioxide being delivered to the alveoli. The reactions we have just discussed then proceed in the reverse direction.

Gas transport is a dynamic process. It is capable of varying its responses to meet changing circumstances. Some of the responses are automatic and result from the basic chemistry of the transport mechanisms. Other responses require coordinated adjustments in the activities of the cardiovascular and respiratory systems. We consider those levels of control and regulation next.

Figure 23–22 **Carbon Dioxide Transport in Blood.**

respiratory systems. We consider those levels of control and regulation next.

CHECKPOINT

25. Identify the three ways that carbon dioxide is transported in the bloodstream.

26. As you exercise, hemoglobin releases more oxygen to active skeletal muscles than it does when those muscles are at rest. Why?

27. How would blockage of the trachea affect blood pH?

See the blue Answers tab at the back of the book.

23-10 ▮ Neurons in the medulla oblongata and pons, along with respiratory reflexes, control respiration

Learning Outcome List the factors that influence respiration rate, and discuss reflex respiratory activity and the brain centers involved in the control of respiration.

Figure 23–23   **A Summary of the Primary Gas Transport Mechanisms.**

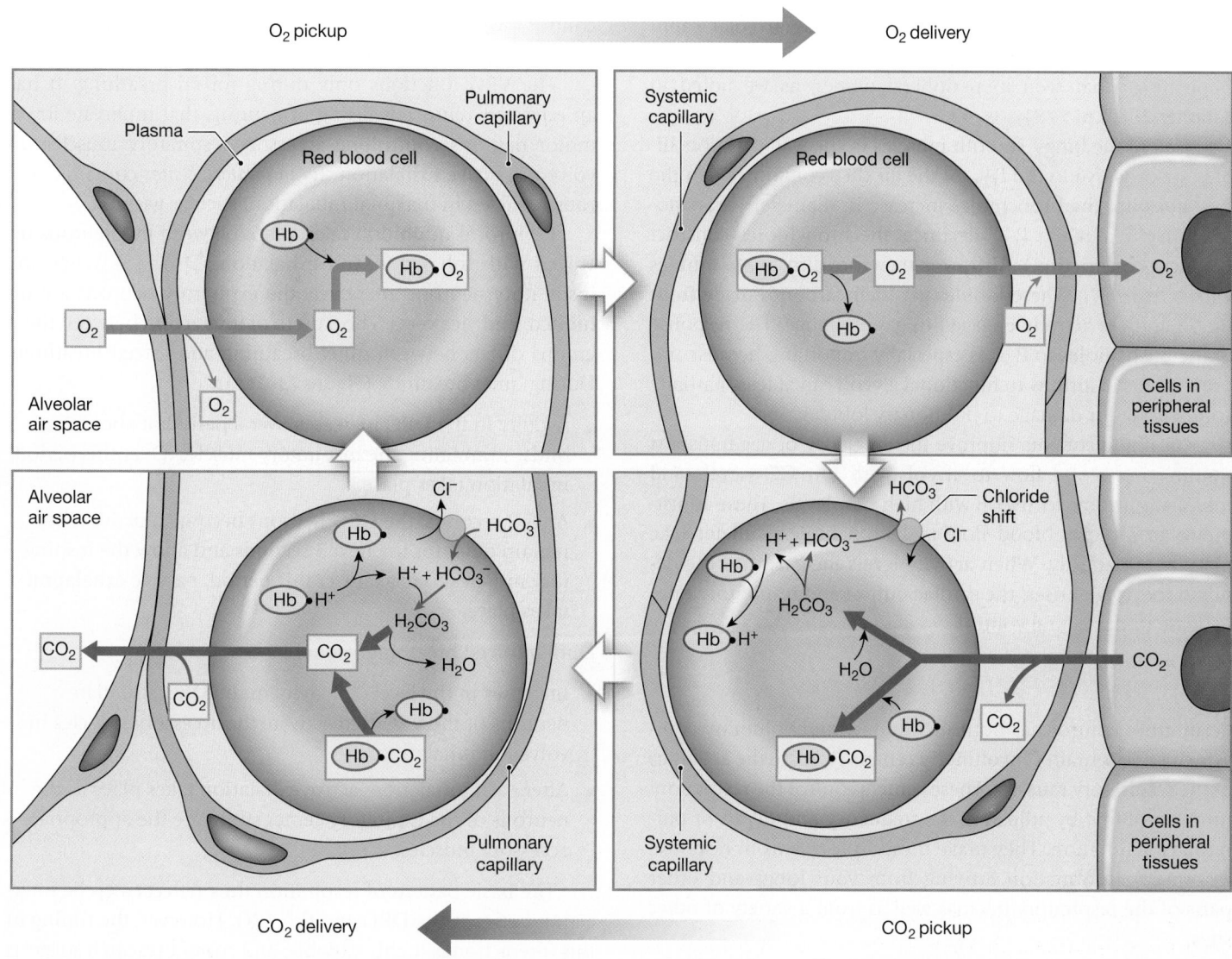

Peripheral cells continuously absorb oxygen and generate carbon dioxide. Under normal conditions, the cellular rates of absorption and generation are matched by the capillary rates of delivery and removal. Both rates are identical to those of oxygen absorption and carbon dioxide excretion at the lungs.

If diffusion rates at the peripheral and alveolar capillaries become unbalanced, homeostatic mechanisms intervene to restore equilibrium. Such mechanisms involve (1) changes in blood flow and oxygen delivery that are regulated at the local level and (2) changes in the depth and rate of respiration under the control of the brain's respiratory centers. The activities of the respiratory centers are coordinated with changes in cardiovascular function, such as fluctuations in blood pressure and cardiac output.

Local Regulation of Gas Transport and Alveolar Function

The rate of oxygen delivery in each tissue and the efficiency of oxygen pickup at the lungs are largely regulated at the local level. For example, when a peripheral tissue becomes more active, the interstitial P_{O_2} falls and the P_{CO_2} rises. This change increases the difference between the partial pressures in the tissues and in the arriving blood. As a result, more oxygen is delivered and more carbon dioxide is carried away. In addition, the rising P_{CO_2} levels cause the relaxation of smooth muscles in the walls of arterioles and capillaries in the area, increasing local blood flow.

At the lungs, local factors coordinate (1) *lung perfusion*, or blood flow to the alveoli, with (2) *alveolar ventilation*, or airflow. This local coordination takes place over a wide range

23

of conditions and activity levels. As blood flows toward the alveolar capillaries, it is directed toward lobules with a relatively high P_{O_2}. This movement takes place because alveolar capillaries constrict when the local P_{O_2} is low. (This response is the opposite of that seen in peripheral tissues, as we noted in Chapter 21. ⮌ p. 778)

Also in the lungs, smooth muscles in the walls of bronchioles are sensitive to the P_{CO_2} of the air they contain. When the P_{CO_2} goes up, the bronchioles increase in diameter (bronchodilation). When the P_{CO_2} declines, the bronchioles constrict (bronchoconstriction). Airflow is therefore directed to lobules with a high P_{CO_2}. These lobules get their carbon dioxide from blood and are actively engaged in gas exchange. The response of the bronchioles to P_{CO_2} is especially important, because improvements in airflow to functional alveoli can at least partially compensate for damage to pulmonary lobules.

Local adjustments improve the efficiency of gas transport by directing blood flow to alveoli with low CO_2 levels and increasing airflow to alveoli with high CO_2 levels. These adjustments in alveolar blood flow and bronchiole diameter take place automatically. When activity levels increase and the demand for oxygen rises, the cardiac output and respiratory rates increase under neural control.

The Respiratory Centers of the Brain

Respiratory control has both involuntary and voluntary components. Your brain's involuntary centers regulate the activities of the respiratory muscles. These centers control the respiratory minute volume by adjusting the frequency and depth of pulmonary ventilation. They make these adjustments in response to sensory information arriving from your lungs and other parts of the respiratory tract, as well as from a variety of other sites.

The voluntary control of respiration reflects activity in the cerebral cortex. This activity affects the output of either the respiratory centers in the medulla oblongata and pons or motor neurons in the spinal cord that control respiratory muscles. The **respiratory centers** are three pairs of nuclei in the reticular formation of the medulla oblongata and pons. The motor neurons in the spinal cord are generally controlled by *respiratory reflexes*, but they can also be controlled voluntarily through commands delivered by the corticospinal pathway. ⮌ p. 550

Respiratory Centers in the Medulla Oblongata

We introduced the paired *respiratory rhythmicity centers* of the medulla oblongata in Chapter 14; these centers play a key role in establishing the pace of respiration. ⮌ p. 496 Each respiratory rhythmicity center can be subdivided into a *dorsal respiratory group (DRG)* and a *ventral respiratory group (VRG)*. The DRG is chiefly concerned with inspiration whereas the VRG is primarily associated with expiration.

The DRG functions in every respiratory cycle, whether quiet or forced. The DRG's *inspiratory center* contains neurons that control lower motor neurons innervating the external intercostal muscles and the diaphragm.

The VRG functions only during forced breathing. It has an *expiratory center* consisting of neurons that innervate lower motor neurons controlling accessory respiratory muscles involved in active exhalation. Its *inspiratory center* contains neurons involved in maximal inhalation, such as gasping.

Reciprocal inhibition takes place between the neurons involved with inhalation and exhalation. ⮌ p. 481 When the inspiratory neurons are active, the expiratory neurons are inhibited, and vice versa. The pattern of interaction between these groups differs between quiet breathing and forced breathing. During quiet breathing (**Figure 23–24a**):

- Activity in the DRG increases over a period of about 2 seconds, stimulating the inspiratory muscles. Over this period, inhalation takes place.

- After 2 seconds, the DRG neurons become inactive. They remain quiet for the next 3 seconds and allow the inspiratory muscles to relax. Over this period, passive exhalation takes place.

During forced breathing (**Figure 23–24b**):

- Increases in the level of activity in the DRG stimulate neurons of the VRG that activate the accessory muscles involved in inhalation.

- After each inhalation, active exhalation takes place as the neurons of the expiratory center stimulate the appropriate accessory muscles.

The basic pattern of respiration thus reflects a cyclic interaction between the DRG and the VRG. However, the timing of this interaction is highly variable, and current research suggests that respiratory rate and rhythm cannot be attributed to a single area. The nuclei involved in respiratory control form a complex network that includes (1) the rhythmicity centers, (2) centers in the pons that are modulated by peripheral and higher central inputs, and (3) an area in the ventrolateral medulla, known as the *central pattern generator* or the *pre-Bötzinger complex*, that is essential to all forms of breathing. There are reciprocal pathways between all these nuclei and deciphering the communication is extremely difficult and the patterns change with different situations. Central nervous system stimulants, such as amphetamines or even caffeine, increase the respiratory rate by facilitating the respiratory centers. These actions are opposed by CNS depressants, such as barbiturates or opiates.

Respiratory control is an area of intense study and our understanding of the interactions and mechanisms remains incomplete. Due to the brain stem locations, it is nearly impossible to study this mechanism in humans, and much of the work has been done in cats and newborn rodents.

Figure 23–24 **Basic Regulatory Patterns of Respiration.**

The Apneustic and Pneumotaxic Centers of the Pons

The apneustic centers and pneumotaxic centers of the pons regulate the depth and rate of respiration in response to sensory stimuli or input from other centers in the brain. Each apneustic center provides continuous stimulation to the DRG on that side of the brain stem. During quiet breathing, stimulation from the apneustic center helps increase the intensity of inhalation over the next 2 seconds. Under normal conditions, after 2 seconds the apneustic center is inhibited by signals from the pneumotaxic center on that side. During forced breathing, the apneustic centers also respond to sensory input from the vagus nerves regarding the amount of lung inflation.

The pneumotaxic centers inhibit the apneustic centers and promote passive or active exhalation. Centers in the hypothalamus and cerebrum can alter the activity of the pneumotaxic centers, as well as the respiratory rate and depth. However, essentially normal respiratory cycles continue even if the brain stem superior to the pons has been severely damaged.

In some cases, the inhibitory output of the pneumotaxic centers is cut off by a stroke or other damage to the brain stem, or sensory innervation from the lungs is eliminated due to damage to the vagus nerves. In these cases, the person inhales to maximum capacity and maintains that state for 10–20 seconds at a time. Intervening exhalations are brief, and little pulmonary ventilation occurs.

The CNS regions involved with respiratory control are diagrammed in **Spotlight Figure 23–25.** Interactions between the DRG and the VRG establish the basic pace and depth of respiration. The pneumotaxic centers modify that pace: An increase in pneumotaxic output quickens the pace of respiration by shortening the duration of each inhalation. A decrease in pneumotaxic output slows the respiratory pace, but increases the depth of respiration, because the apneustic centers are more active.

Sudden infant death syndrome (SIDS), also known as *crib death*, is the leading cause of death for babies 1–12 months old and kills an estimated 2250 infants each year in the United States alone. Most crib deaths occur between midnight and 9:00 a.m., in the late fall or winter, and involve infants two to four months old. Eyewitness accounts indicate that the sleeping infant suddenly stops breathing, turns blue, and relaxes. Genetic factors appear to be involved, but controversy remains as to the relative importance of other factors. The age at the time of death corresponds with a period when the pacemaker complex and respiratory centers are establishing connections with other portions of the brain. It has been suggested that SIDS results from a problem in the interconnection process that disrupts the reflexive respiratory pattern.

Respiratory control involves multiple levels of regulation. Most of the regulatory activities occur outside of our awareness.

LEVEL 3

Higher Centers

Higher centers in the hypothalamus, limbic system, and cerebral cortex can alter the activity of the pneumotaxic centers, but essentially normal respiratory cycles continue even if the brain stem superior to the pons has been severely damaged.

Higher Centers
- Cerebral cortex
- Limbic system
- Hypothalamus

LEVEL 2

Apneustic and Pneumotaxic Centers

The **apneustic** (ap-NŪ-stik) **centers** and the **pneumotaxic** (nū-mō-TAKS-ik) **centers** of the pons are paired nuclei that adjust the output of the respiratory rhythmicity centers.

The pneumotaxic centers inhibit the apneustic centers and promote passive or active exhalation. An increase in pneumotaxic output quickens the pace of respiration by shortening the duration of each inhalation. A decrease in pneumotaxic output slows the respiratory pace but increases the depth of respiration, because the apneustic centers are more active.

The apneustic centers promote inhalation by stimulating the DRG. During forced breathing, the apneustic centers adjust the degree of stimulation in response to sensory information from N X (the vagus nerve) concerning the amount of lung inflation.

Pons

LEVEL 1

Respiratory Rhythmicity Centers

The most basic level of respiratory control involves pacemaker cells in the medulla oblongata. These neurons generate cycles of contraction and relaxation in the diaphragm. The **respiratory rhythmicity centers** set the pace of respiration by adjusting the activities of these pace-makers and coordinating the activities of additional respiratory muscles. Each rhythmicity center can be subdivided into a **dorsal respiratory group (DRG)** and a **ventral respiratory group (VRG)**. The DRG is mainly concerned with inspiration and the VRG is primarily associated with expiration. The DRG modifies its activities in response to input from chemoreceptors and baroreceptors that monitor O_2, CO_2, and pH in the blood and CSF and from stretch receptors that monitor the degree of stretching in the walls of the lungs.

Medulla oblongata

To diaphragm ◄

The inspiratory center of the DRG contains neurons that control lower motor neurons innervating the external intercostal muscles and the diaphragm. This center functions in every respiratory cycle.

To external ◄
intercostal
muscles

To accessory ◄
inspiratory
muscles

The VRG has inspiratory and expiratory centers that function only when breathing demands increase and accessory muscles become involved.

To accessory ◄
expiratory
muscles

In addition to the centers in the pons, the DRG, and the VRG, the pre-Bötzinger complex in the medulla is essential to all forms of breathing. Its mechanisms are poorly understood.

Respiratory Centers and Reflex Controls

This illustration shows the locations and relationships between the major respiratory centers in the pons and medulla oblongata and other factors important to the reflex control of respiration. Pathways for conscious control over respiratory muscles are not shown.

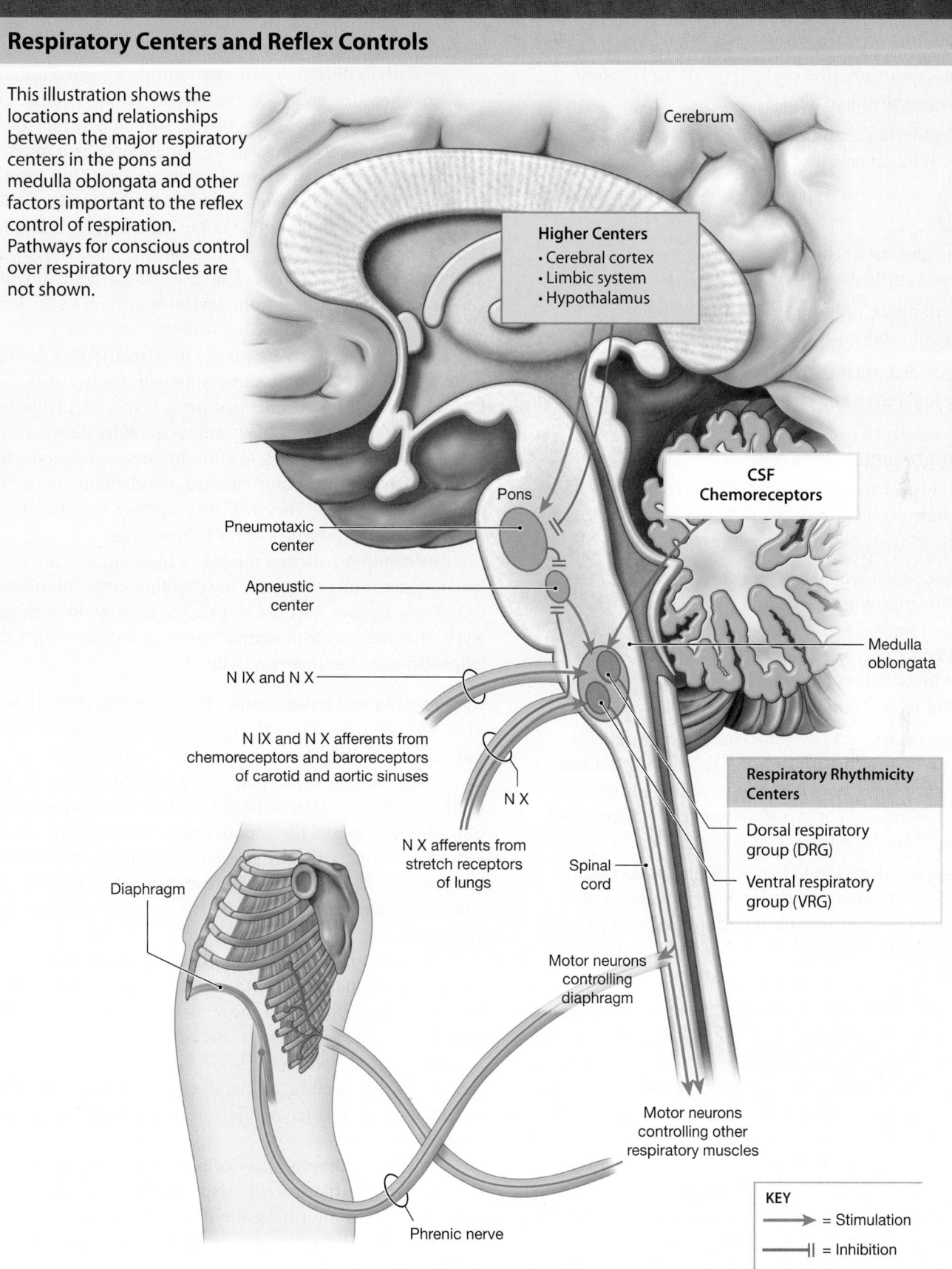

Cerebrum

Higher Centers
- Cerebral cortex
- Limbic system
- Hypothalamus

CSF Chemoreceptors

Pons

Pneumotaxic center

Apneustic center

N IX and N X

N IX and N X afferents from chemoreceptors and baroreceptors of carotid and aortic sinuses

N X

N X afferents from stretch receptors of lungs

Diaphragm

Phrenic nerve

Medulla oblongata

Respiratory Rhythmicity Centers

Dorsal respiratory group (DRG)

Ventral respiratory group (VRG)

Spinal cord

Motor neurons controlling diaphragm

Motor neurons controlling other respiratory muscles

KEY

→ = Stimulation

⊣⊢ = Inhibition

Respiratory Reflexes

The activities of the respiratory centers are modified by sensory information from several sources:

- Chemoreceptors sensitive to the P_{CO_2}, pH, or P_{O_2} of the blood or cerebrospinal fluid.

- Baroreceptors in the aortic or carotid sinuses sensitive to changes in blood pressure.

- Stretch receptors that respond to changes in the volume of the lungs.

- Irritating physical or chemical stimuli in the nasal cavity, larynx, or bronchial tree.

- Other sensations, including pain, changes in body temperature, and abnormal visceral sensations.

Information from these receptors alters the pattern of respiration. The induced changes have been called *respiratory reflexes*.

The Chemoreceptor Reflexes

The respiratory centers are strongly influenced by chemoreceptor inputs from cranial nerves IX and X, and from receptors that monitor the composition of the cerebrospinal fluid (CSF):

- The glossopharyngeal nerves (N IX) carry chemoreceptive information from the carotid bodies, adjacent to the carotid sinus. ⟲ **pp. 543, 781** The carotid bodies are stimulated by a decrease in the pH or P_{O_2} of blood. Because changes in P_{CO_2} affect pH, an increase in the P_{CO_2} indirectly stimulates these receptors.

- The vagus nerves (N X) monitor chemoreceptors in the aortic bodies, near the aortic arch. ⟲ **pp. 543, 781** These receptors are sensitive to the same stimuli as the carotid bodies. Carotid and aortic body receptors are often called *peripheral chemoreceptors*.

- Chemoreceptors are located on the ventrolateral surface of the medulla oblongata in a region known as the *chemosensitive area*. The neurons in that area respond only to P_{CO_2} the and pH of the CSF. They are often called *central chemoreceptors*.

We discussed chemoreceptors and their effects on cardiovascular function in Chapters 15 and 21. ⟲ **pp. 543, 772–773** Stimulation of these chemoreceptors leads to an increase in the depth and rate of respiration. Under normal conditions, a drop in arterial P_{O_2} has little effect on the respiratory centers, until the arterial P_{O_2} drops by about 40 percent, to below 60 mm Hg. If the P_{O_2} of arterial blood drops to 40 mm Hg (the level in peripheral tissues), the respiratory rate increases by only 50–70 percent. In contrast, a rise of just 10 percent in the arterial P_{CO_2} causes the respiratory rate to double, even if the P_{O_2} remains completely normal. Carbon dioxide levels are therefore responsible for regulating respiratory activity under normal conditions.

Although the receptors monitoring CO_2 levels are more sensitive, oxygen and carbon dioxide receptors work together in a crisis. Carbon dioxide is generated during oxygen consumption, so when oxygen concentrations are falling rapidly, CO_2 levels are usually increasing. This cooperation breaks down only under unusual circumstances. For example, you can hold your breath longer than normal by taking deep, full breaths, but the practice is very dangerous. The danger lies in the fact that the increased ability is due not to extra oxygen, but to less carbon dioxide. If the P_{CO_2} is driven down far enough, your ability to hold your breath can increase to the point at which you become unconscious from oxygen starvation in the brain without ever feeling the urge to breathe.

The chemoreceptors are subject to adaptation—a decrease in sensitivity after chronic stimulation—if the P_{O_2} or P_{CO_2} remains abnormal for an extended period. This adaptation can complicate the treatment of chronic respiratory disorders. For example, if the P_{O_2} remains low for an extended period while the P_{CO_2} remains chronically elevated, the chemoreceptors will reset to those values. They will then oppose any attempts to return the partial pressures to the proper range.

Any condition altering the pH of blood or CSF affects respiratory performance because the chemoreceptors monitoring CO_2 levels are also sensitive to pH. The increase in lactic acid levels after exercise, for example, causes a decrease in pH that helps stimulate respiratory activity.

Hypercapnia and Hypocapnia. Hypercapnia is an increase in the P_{CO_2} of arterial blood. **Figure 23–26a** diagrams the central response to hypercapnia. Stimulation of chemoreceptors in the carotid and aortic bodies triggers this response. It is reinforced by the stimulation of CNS chemoreceptors. Carbon dioxide crosses the blood–brain barrier quite rapidly. For this reason, a rise in arterial P_{CO_2} almost immediately increases CO_2 levels in the CSF, decreasing the pH of the CSF and stimulating the chemoreceptive neurons of the medulla oblongata.

These chemoreceptors stimulate the respiratory centers to increase the rate and depth of respiration. Your breathing becomes more rapid, and more air moves into and out of your lungs with each breath. Because more air moves into and out of the alveoli each minute, alveolar concentrations of carbon dioxide decline, accelerating the diffusion of carbon dioxide out of alveolar capillaries. In this way, homeostasis is restored.

The most common cause of hypercapnia is hypoventilation. In **hypoventilation**, the respiratory rate remains abnormally low and cannot meet the demands for normal oxygen delivery and carbon dioxide removal. Carbon dioxide then accumulates in the blood.

Figure 23–26   **The Chemoreceptor Response to Changes in P_{CO_2}.**

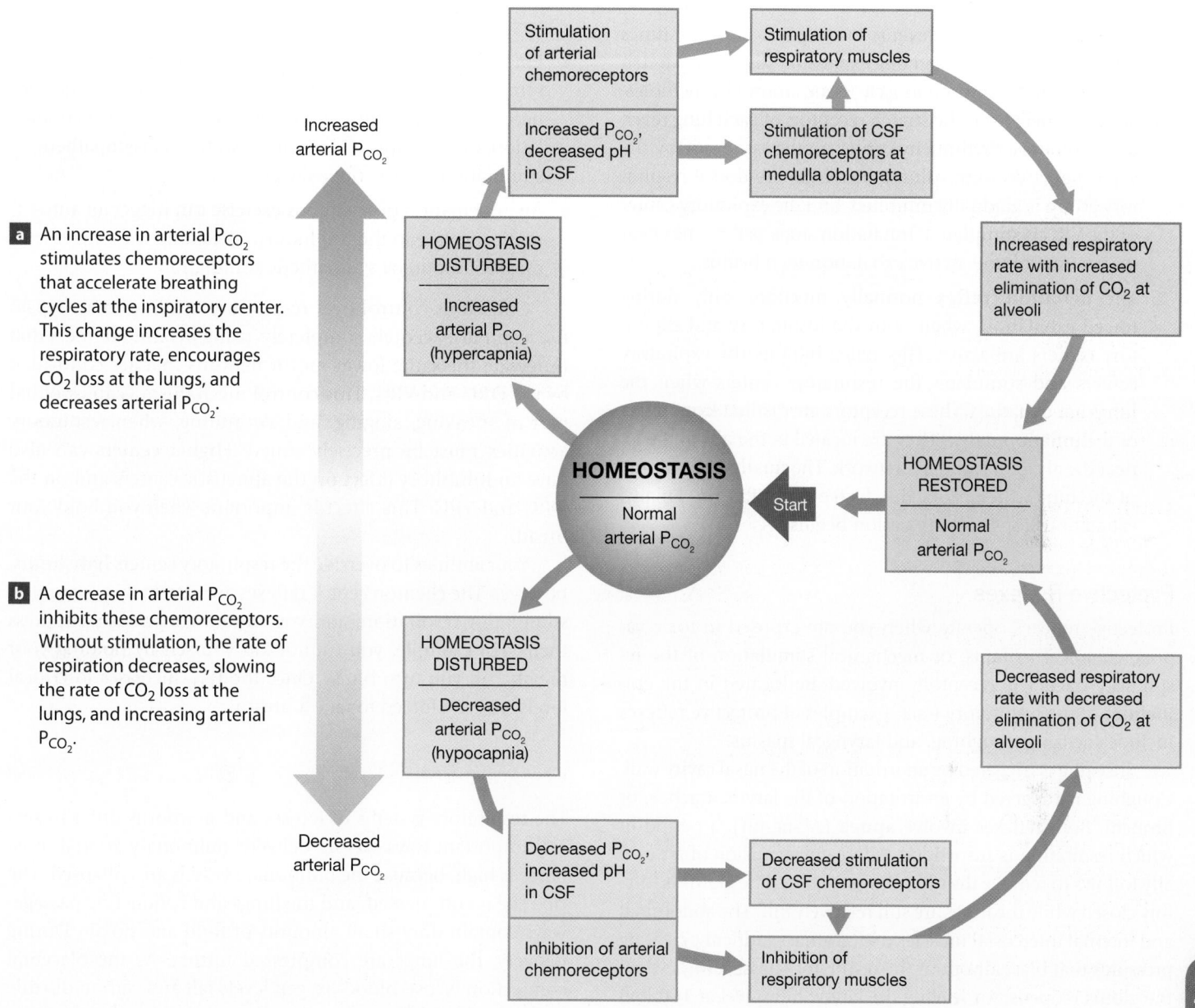

a An increase in arterial P_{CO_2} stimulates chemoreceptors that accelerate breathing cycles at the inspiratory center. This change increases the respiratory rate, encourages CO_2 loss at the lungs, and decreases arterial P_{CO_2}.

b A decrease in arterial P_{CO_2} inhibits these chemoreceptors. Without stimulation, the rate of respiration decreases, slowing the rate of CO_2 loss at the lungs, and increasing arterial P_{CO_2}.

Increased arterial P_{CO_2}

Decreased arterial P_{CO_2}

Stimulation of arterial chemoreceptors

Stimulation of respiratory muscles

Increased P_{CO_2}, decreased pH in CSF

Stimulation of CSF chemoreceptors at medulla oblongata

Increased respiratory rate with increased elimination of CO_2 at alveoli

HOMEOSTASIS DISTURBED
___
Increased arterial P_{CO_2} (hypercapnia)

HOMEOSTASIS
___
Normal arterial P_{CO_2}

Start

HOMEOSTASIS RESTORED
___
Normal arterial P_{CO_2}

HOMEOSTASIS DISTURBED
___
Decreased arterial P_{CO_2} (hypocapnia)

Decreased respiratory rate with decreased elimination of CO_2 at alveoli

Decreased P_{CO_2}, increased pH in CSF

Decreased stimulation of CSF chemoreceptors

Inhibition of arterial chemoreceptors

Inhibition of respiratory muscles

23

If the rate and depth of respiration exceed the demands for oxygen delivery and carbon dioxide removal, the condition called **hyperventilation** exists. Hyperventilation gradually leads to **hypocapnia**, an abnormally low P_{CO_2}. If the arterial P_{CO_2} drops below normal levels, chemoreceptor activity decreases and the respiratory rate falls (**Figure 23–26b**). This situation continues until the P_{CO_2} returns to normal and homeostasis is restored.

The Baroreceptor Reflexes

We described the effects of carotid and aortic baroreceptor stimulation on systemic blood pressure in Chapter 21. ⊃pp. 770–771

The output from these baroreceptors also affects the respiratory centers. When blood pressure falls, the respiratory rate increases. When blood pressure rises, the respiratory rate decreases. These adjustments result from the stimulation or inhibition of the respiratory centers by sensory fibers in the glossopharyngeal (N IX) and vagus (N X) nerves.

The Hering–Breuer Reflexes

The **Hering–Breuer reflexes** are named after the physiologists who described them in 1865. The sensory information from these reflexes goes to the apneustic centers and the ventral respiratory group. The Hering–Breuer reflexes are not involved in normal quiet breathing (eupnea) or in tidal volumes under 1000 mL. There are two such reflexes:

1. The **inflation reflex** prevents overexpansion of the lungs during forced breathing. Lung expansion stimulates stretch receptors in the smooth muscle tissue around bronchioles. Sensory fibers from the stretch receptors of each lung reach the respiratory rhythmicity center on the same side by the vagus nerve. As lung volume increases, the dorsal respiratory group is gradually inhibited, and the expiratory center of the VRG is stimulated. Inhalation stops as the lungs near maximum volume. Active exhalation then begins.

2. The **deflation reflex** normally functions only during forced exhalation, when both the inspiratory and expiratory centers are active. This reflex inhibits the expiratory centers and stimulates the inspiratory centers when the lungs are deflating. These receptors are distinct from those of the inflation reflex. They are located in the alveolar wall near the alveolar capillary network. The smaller the volume of the lungs, the greater the degree of inhibition. Finally, exhalation stops and inhalation begins.

Protective Reflexes

Protective reflexes operate when you are exposed to toxic vapors, chemical irritants, or mechanical stimulation of the respiratory tract. The receptors involved are located in the epithelium of the respiratory tract. Examples of protective reflexes include sneezing, coughing, and laryngeal spasms.

Sneezing is triggered by an irritation of the nasal cavity wall. Coughing is triggered by an irritation of the larynx, trachea, or bronchi. Both reflexes involve **apnea** (AP-nē-uh), a period in which respiration is suspended. A forceful expulsion of air usually follows to remove the offending stimulus. The glottis is forcibly closed while the lungs are still relatively full. The abdominal and internal intercostal muscles then contract suddenly, creating pressures that blast air out of the respiratory passageways when the glottis reopens. Air leaving the larynx can travel at 160 kph (99 mph), carrying mucus, foreign particles, and irritating gases out of the respiratory tract via the nose or mouth.

Laryngeal spasms result when chemical irritants, foreign objects, or fluids enter the area around the glottis. This reflex generally closes the airway temporarily. A very strong stimulus, such as a toxic gas, could close the glottis so powerfully that you could lose consciousness and die without taking another breath. Fine chicken bones or fish bones that pierce the laryngeal walls can also stimulate laryngeal spasms, swelling, or both, restricting the airway.

Voluntary Control of Respiration

Activity of the cerebral cortex has an indirect effect on the respiratory centers, as the following examples show:

- Conscious thought processes tied to strong emotions, such as rage or fear, affect the respiratory rate by stimulating centers in the hypothalamus.

- Emotional states can affect respiration by activating the sympathetic or parasympathetic division of the autonomic nervous system. Sympathetic activation causes bronchodilation and increases the respiratory rate. Parasympathetic stimulation has the opposite effect.

- An anticipation of strenuous exercise can trigger an automatic increase in the respiratory rate, along with increased cardiac output, by sympathetic stimulation.

Conscious control over respiratory activities may bypass the respiratory centers completely, using pyramidal fibers that innervate the same lower motor neurons that are controlled by the DRG and VRG. This control mechanism is an essential part of speaking, singing, and swimming, when respiratory activities must be precisely timed. Higher centers can also have an inhibitory effect on the apneustic centers and on the DRG and VRG. This effect is important when you hold your breath.

Your abilities to override the respiratory centers have limits, however. The chemoreceptor reflexes are extremely powerful in stimulating respiration, and you cannot consciously suppress them. For example, you cannot kill yourself by holding your breath "till you turn blue." Once the P_{CO_2} increases to critical levels, you are forced to take a breath.

Changes in the Respiratory System at Birth

The respiratory systems of fetuses and newborns differ in several important ways. Before delivery, pulmonary arterial resistance is high, because the pulmonary vessels are collapsed. The rib cage is compressed, and the lungs and conducting passageways contain only small amounts of fluid and no air. During delivery, the lungs are compressed further. As the placental connection is lost, blood oxygen levels fall and carbon dioxide levels climb rapidly.

At birth, the newborn infant takes a truly heroic first breath through powerful contractions of the diaphragmatic and external intercostal muscles. The inhaled air must enter the respiratory passageways with enough force to overcome surface tension and inflate the bronchial tree and most of the alveoli. The same drop in pressure that pulls air into the lungs also pulls blood into the pulmonary circulation. The changes in blood flow and rise in oxygen levels lead to the closure of the *foramen ovale*, an interatrial connection, and the *ductus arteriosus*, the fetal connection between the pulmonary trunk and the aorta. ⟳ pp. 798–799 ATLAS: Embryology Summary 18: The Development of the Respiratory System

The exhalation that follows does not empty the lungs completely, because the rib cage does not return to its former, fully compressed state. Cartilages and connective tissues keep the conducting passageways open, and surfactant covering the alveolar surfaces prevents their collapse. The next breaths complete the inflation of the alveoli.

Pathologists sometimes use these physical changes to determine whether a newborn infant died before delivery or shortly thereafter. Before the first breath, the lungs are completely filled with amniotic fluid, and extracted lungs will sink if placed in water. After the infant's first breath, even the collapsed lungs contain enough air to keep them afloat.

CHECKPOINT

28. What effect does exciting the pneumotaxic centers have on respiration?

29. Are peripheral chemoreceptors more or less sensitive to levels of carbon dioxide than they are to levels of oxygen? Why?

30. Little Johnny is angry with his mother, so he tells her that he will hold his breath until he turns blue and dies. Should Johnny's mother worry that this will happen?

See the blue Answers tab at the back of the book.

23-11 | Respiratory performance declines with age

Learning Outcome Describe age-related changes in the respiratory system.

Many factors interact to reduce the efficiency of the respiratory system in elderly individuals. Here are three examples:

1. With age, elastic tissue deteriorates throughout the body. These changes reduce the compliance of the lungs and lower vital capacity.

2. Chest movements are restricted by arthritic changes in the rib articulations and by decreased flexibility at the costal cartilages. Along with elastic tissue deterioration, the stiffening and reduction in chest movement effectively limit the respiratory minute volume. This restriction contributes to the reduction in exercise performance and capabilities with increasing age.

3. Some degree of emphysema is normal in individuals over age 50. However, the extent varies widely with the lifetime exposure to cigarette smoke and other respiratory irritants. **Figure 23–27** compares the respiratory performance of individuals who have never smoked with individuals who have smoked for various periods of time. The message is quite clear: Some decrease in respiratory performance is inevitable, but you can prevent serious respiratory deterioration by stopping smoking or never starting.

Figure 23–27 Decline in Respiratory Performance with Age and Smoking. The relative respiratory performances of people who have never smoked, people who quit smoking at age 45, people who quit smoking at age 65, and lifelong smokers.

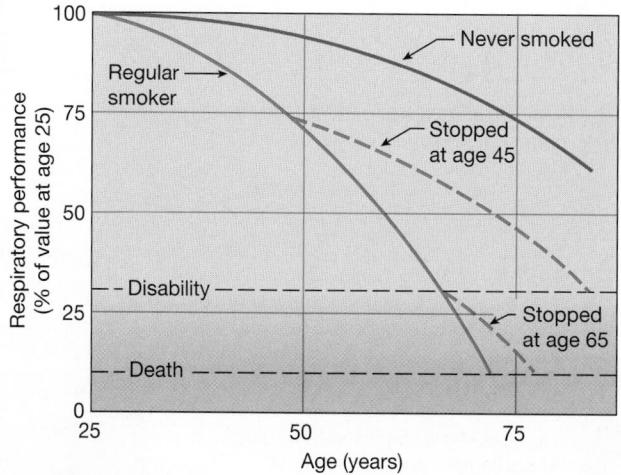

CHECKPOINT

31. Name several age-related factors that affect the respiratory system.

See the blue Answers tab at the back of the book.

23-12 | The respiratory system provides oxygen to, and eliminates carbon dioxide from, other organ systems

Learning Outcome Give examples of interactions between the respiratory system and other organ systems studied so far.

The goal of respiratory activity is to maintain homeostatic oxygen and carbon dioxide levels in peripheral tissues. Changes in respiratory activity alone are seldom enough to accomplish this. Coordinated changes in cardiovascular activity must also take place.

Consider these examples of the integration between the respiratory and cardiovascular systems:

- At the local level, changes in lung perfusion in response to changes in alveolar P_{O_2} improve the efficiency of gas exchange within or among lobules.

- Chemoreceptor stimulation not only increases the respiratory drive, but it also causes blood pressure to rise and cardiac output to increase.

- The stimulation of baroreceptors in the lungs has secondary effects on cardiovascular function. For example, the stimulation of airway stretch receptors not only triggers the inflation reflex, but also increases heart rate. Thus, as

23

SYSTEM INTEGRATOR

Body System ———➤ Respiratory System   Respiratory System ———➤ Body System

Integumentary
Protects portions of upper respiratory tract; hairs guard entry to external nares

Integumentary
Provides oxygen to nourish tissues and removes carbon dioxide
Page 200

Skeletal
Movements of ribs important in breathing; axial skeleton surrounds and protects lungs

Skeletal
Provides oxygen to skeletal structures and disposes of carbon dioxide
Page 311

Muscular
Muscular activity generates carbon dioxide; respiratory muscles fill and empty lungs; other muscles control entrances to respiratory tract; intrinsic laryngeal muscles control airflow through larynx and produce sounds

Muscular
Provides oxygen needed for muscle contractions and disposes of carbon dioxide generated by active muscles
Page 406

Nervous
Monitors respiratory volume and blood gas levels; controls pace and depth of respiration

Nervous
Provides oxygen needed for neural activity and disposes of carbon dioxide
Page 584

Endocrine
Epinephrine and norepinephrine stimulate respiratory activity and dilate respiratory passageways

Endocrine
Angiotensin-converting enzyme (ACE) from capillaries of lungs converts angiotensin I to angiotensin II
Page 673

Cardiovascular
Circulates the red blood cells that transport oxygen and carbon dioxide between lungs and peripheral tissues

Cardiovascular
Bicarbonate ions contribute to buffering capability of blood; activation of angiotensin II by ACE important in regulation of blood pressure and volume
Page 802

Lymphatic
Tonsils protect against infection at entrance to respiratory tract; lymphatic vessels monitor lymph drainage from lungs and mobilize adaptive immunity when infection occurs

Lymphatic
Alveolar phagocytes present antigens to trigger specific defenses; mucous membrane lining the nasal cavity and upper pharynx traps pathogens, protects deeper tissues
Page 850

The RESPIRATORY System

The respiratory system provides oxygen and eliminates carbon dioxide for our cells. Stabilizing the concentrations of these gases involves a continual exchange of materials with the outside world. The respiratory system is therefore crucial to maintaining homeostasis for all body systems.

Figure 23–28 diagrams the functional relationships between the respiratory system and the other body systems we have studied so far.

Digestive
Page 955

Urinary
Page 1036

Reproductive
Page 1116

Clinical Note

Emphysema and Lung Cancer Emphysema and lung cancer are two relatively common disorders. Both are often associated with cigarette smoking. **Emphysema** (em-fi-ZĒ-muh) is a chronic, progressive condition. It is characterized by shortness of breath and an inability to tolerate physical exertion. The underlying problem is the destruction of alveolar surfaces and inadequate surface area for oxygen and carbon dioxide exchange. In essence, respiratory bronchioles and alveoli no longer function. The alveoli gradually expand, and adjacent alveoli merge to form larger air spaces supported by fibrous tissue without alveolar capillary networks. As connective tissues are eliminated, compliance increases, so air moves into and out of the lungs more easily than before. But the loss of respiratory surface area restricts oxygen absorption, so the individual becomes short of breath.

Emphysema has been linked to breathing air that contains fine particles or toxic vapors, such as those in cigarette smoke. Genetic factors also predispose individuals to the condition. Some degree of emphysema is a normal consequence of aging. An estimated 66 percent of adult males and 25 percent of adult females have detectable areas of emphysema in their lungs.

Lung cancer, or *bronchopulmonary carcinoma*, is an aggressive class of malignancies originating in the bronchial passageways or alveoli. These cancers affect the epithelial cells that line conducting passageways, mucous glands, or alveoli. Signs and symptoms generally do not appear until tumors restrict airflow or compress adjacent structures. Chest pain, shortness of breath, a cough or a wheeze, and weight loss are common. Treatment varies with the cellular organization of the tumor and whether metastasis (cancer cell migration) has occurred. Surgery, radiation, or chemotherapy may be involved.

According to the CDC, more people die from lung cancer than any other type of cancer. In 2009, lung cancer affected an estimated 110,000 men and 96,000 women in the United States, and accounted for 12.6 percent of new cancer cases in both men and women.

the lungs fill, cardiac output rises and more blood flows through the alveolar capillaries.

The adaptations that take place at high altitudes are an excellent example of the functional interplay between the respiratory and cardiovascular systems. Atmospheric pressure decreases with increasing altitude, and so do the partial pressures of the component gases, including oxygen. People living in Denver or Mexico City function normally with alveolar oxygen pressures in the 80–90 mm Hg range. At higher elevations, alveolar P_{O_2} is even lower. At 3300 meters (10,826 ft), an altitude many hikers and skiers have experienced, alveolar P_{O_2} is about 60 mm Hg.

Despite the low alveolar P_{O_2}, millions of people live and work at altitudes this high or higher. Important physiological adjustments include increased respiratory rate, increased heart rate, increased BPG levels, and elevated hematocrit. Thus, even though the hemoglobin is not fully saturated, the bloodstream holds more of it, and the round trip between the lungs and the peripheral tissues takes less time. However, most of these adaptations take days to weeks to develop. As a result, athletes planning to compete in events at high altitude must begin training under such conditions well in advance.

The respiratory system is functionally linked to all other body systems as well. **Figure 23–28** illustrates these interrelationships with the systems studied so far.

CHECKPOINT

32. Identify the functional relationship between the respiratory system and all other organ systems.
33. Describe the functional relationship between the respiratory system and the lymphatic system.

See the blue Answers tab at the back of the book.

Related Clinical Terms

asbestosis: Pneumoconiosis, disease of the lungs, caused by the inhalation of asbestos particles over time.

asphyxia: Impaired oxygen–carbon dioxide exchange that results in suffocation.

aspirate: Drawing fluid from the body by suction; foreign material accidentally sucked into the lungs.

bronchography: A procedure in which radiopaque materials are introduced into the airways to improve x-ray imaging of the bronchial tree.

bronchoscope: A fiber-optic bundle small enough to be inserted into the trachea and finer airways; the procedure is called *bronchoscopy*.

cardiopulmonary resuscitation (CPR): The application of cycles of compression to the rib cage and mouth-to-mouth breathing to restore cardiovascular and respiratory function.

Cheyne-Stokes breathing: Hyperpnea (deep, fast breathing) alternating with apnea (absence of breathing).

chronic obstructive pulmonary disease (COPD): A general term describing temporary or permanent lung disease of the bronchial tree.

dyspnea: The condition of labored breathing.

endotracheal tube: Tube that is passed through the mouth or nose to the trachea.

Heimlich maneuver, or *abdominal thrust*: Sudden compression applied to the abdomen just inferior to the diaphragm, to force air out of the lungs to clear a blocked trachea or larynx.

hemoptysis: Coughing up blood or bloody mucus.

hemothorax: The condition of having blood in the pleural cavity.

hyperbaric oxygenation: Therapy to force more oxygen into the blood by use of a pressure chamber.

nasal polyps: Benign growths on the mucous lining of the nasal cavity.

orthopnea: Condition in which one has breathing difficulty except when in an upright position.

otorhinolaryngology: Branch of medicine dealing with disease and treatment of the ear, nose, and throat.

rales: Abnormal hissing or other respiratory sounds.

rhinitis: Inflammation of the nasal cavity.

rhinoplasty: Plastic surgery of the nose.

severe acute respiratory syndrome (SARS): A harsh viral respiratory illness caused by a coronavirus that typically progresses to pneumonia.

sputum: A mixture of saliva and mucus coughed up from the respiratory tract, often as the result of an infection.

stridor: Harsh vibrating breathing sound caused by an obstruction in the windpipe or larynx.

stuttering: To speak with a continued involuntary repetition of sounds.

tachypnea: Rapid rate of breathing.

tussis: A cough.

wheeze: An audible whistling sound when breathing.

Chapter Review

Study Outline

MasteringA&P®

Access more chapter study tools online in the MasteringA&P Study Area:

- Chapter Quizzes, Chapter Practice Test, Art-labeling Activities, Animations, MP3 Tutor Sessions, and Clinical Case Studies
- Practice Anatomy Lab **PAL™**
- A&P Flix **A&PFlix**
- Interactive Physiology **iP®**
- PhysioEx **PhysioEx**

An introduction to the respiratory system p. 857

1. Body cells must obtain oxygen and eliminate carbon dioxide. Gas exchange takes place at respiratory surfaces inside the lungs.

23-1 ‖ The respiratory system, organized into an upper respiratory system and a lower respiratory system, has several basic functions p. 857

2. The functions of the **respiratory system** include (1) providing an area for gas exchange between air and circulating blood; (2) moving air to and from exchange surfaces; (3) protecting respiratory surfaces from environmental variations and defending the respiratory system and other tissues from invasion by pathogens; (4) producing sounds; and (5) facilitating the detection of olfactory stimuli.

3. The respiratory system includes the **upper respiratory system**, composed of the nose, nasal cavity, paranasal sinuses, and pharynx, and the **lower respiratory system**, which includes the larynx, trachea, bronchi, bronchioles, and alveoli of the lungs. *(Figure 23–1)*

4. The **respiratory tract** consists of the conducting airways that carry air to and from the **alveoli**. The passageways of the upper respiratory tract filter and humidify incoming air. The lower respiratory tract includes delicate conduction passages and the exchange surfaces of the alveoli.

5. The **respiratory mucosa** (respiratory epithelium and underlying connective tissue) lines the conducting portion of the respiratory tract.

6. The respiratory epithelium changes in structure along the respiratory tract. It is supported by the **lamina propria**, a layer of areolar tissue. *(Figure 23–2)*

7. Contamination of the respiratory system is prevented by the mucus and cilia of the **respiratory defense system**. *(Figure 23–2)*

23-2 ‖ Located outside the thoracic cavity, the upper respiratory system consists of the nose, nasal cavity, paranasal sinuses, and pharynx p. 861

8. The upper respiratory system consists of the nose, nasal cavity, paranasal sinuses, and pharynx. *(Figures 23–1, 23–3)*

9. Air normally enters the respiratory system through the **external nares**, which open into the *nasal cavity*. The **nasal vestibule** (entryway) is guarded by hairs that screen out large particles. *(Figure 23–3)*

10. Incoming air flows through the **superior**, **middle**, and **inferior meatuses** (narrow grooves) and bounces off the conchal surfaces. *(Figure 23–3)*

11. The **hard palate** separates the oral and nasal cavities. The **soft palate** separates the superior nasopharynx from the rest of the pharynx. The connections between the nasal cavity and nasopharynx are the **internal nares**.

12. The nasal mucosa traps particles, warms and humidifies incoming air, and cools and dehumidifies outgoing air.

13. The **pharynx**, or throat, is a chamber shared by the digestive and respiratory systems. The **nasopharynx** is the superior part of the pharynx. The **oropharynx** is continuous with the oral cavity. The **laryngopharynx** includes the narrow zone between the hyoid bone and the entrance to the esophagus. *(Figure 23–3)*

23-3 ‖ Composed of cartilages, ligaments, and muscles, the larynx produces sound p. 863

14. Inhaled air passes through the glottis en route to the lungs; the **larynx** surrounds and protects the glottis, or vocal apparatus. *(Figure 23–4)*

15. The cylindrical larynx is composed of three large cartilages (the **thyroid cartilage**, **cricoid cartilage**, and **epiglottis**) and three smaller pairs of cartilages (the **arytenoid**, **corniculate**, and **cuneiform cartilages**). The epiglottis projects into the pharynx. *(Figures 23–4, 23–5)*

16. A pair of inelastic **vestibular folds** surround the glottis. The **glottis** includes the elastic **vocal folds** and the space between them, the *rima glottidis*. (*Figure 23–5*)

17. Air passing through the open glottis vibrates its vocal folds, producing sound. The pitch of the sound depends on the diameter, length, and tension of the vocal folds.

18. The muscles of the neck and pharynx position and stabilize the larynx. The smaller intrinsic muscles regulate tension in the vocal folds or open and close the glottis. During swallowing, both sets of muscles help prevent particles from entering the glottis.

23-4 ▌ The trachea and primary bronchi convey air to and from the lungs p. 865

19. The **trachea** extends from the sixth cervical vertebra to the fifth thoracic vertebra. The **submucosa** contains C-shaped **tracheal cartilages**, which stiffen the tracheal walls and protect the airway. The posterior tracheal wall can distort to permit large masses of food to pass through the esophagus. (*Figure 23–6*)

20. The trachea branches within the mediastinum to form the **right** and **left primary bronchi**. Each bronchus enters a lung at the **hilum** (a groove). The **root of the lung** is a connective tissue mass that includes the bronchus, pulmonary vessels, and nerves. (*Figures 23–6, 23–7*)

23-5 ▌ Enclosed by pleural cavities, the lungs are paired organs containing alveoli, which permit gaseous exchange p. 867

21. The **lobes** of the lungs are separated by fissures. The right lung has three lobes, the left lung two. (*Figure 23–7*)

22. The anterior and lateral surfaces of the lungs follow the inner contours of the rib cage. The concavity of the medial surface of the left lung is the **cardiac notch**, which conforms to the shape of the pericardium. (*Figures 23–7, 23–8*)

23. The primary bronchi and their branches form the **bronchial tree**. The **secondary** and **tertiary bronchi** are branches within the lungs. As they branch, the amount of cartilage in their walls decreases and the amount of smooth muscle increases. (*Figure 23–9*)

24. Each tertiary bronchus supplies air to a single **bronchopulmonary segment**. (*Figure 23–9*)

25. **Bronchioles** within the bronchopulmonary segments ultimately branch into **terminal bronchioles**. Each terminal bronchiole delivers air to a single **pulmonary lobule** in which the terminal bronchiole branches into **respiratory bronchioles**. The connective tissues of the root of the lung extend into the parenchyma of the lung as a series of *trabeculae* (partitions) that branch to form **interlobular septa**, which divide the lung into lobules. (*Figure 23–9*)

26. The respiratory bronchioles open into **alveolar ducts**, where many alveoli are interconnected. The respiratory exchange surfaces are extensively connected to the circulatory system by the capillaries of the pulmonary circuit. (*Figure 23–9*)

27. The **respiratory membrane** consists of a simple squamous epithelium, the endothelial cell lining an adjacent capillary, and their fused basement membranes. **Type II pneumocytes** (type II alveolar cells) scattered in the respiratory membrane produce **surfactant** that reduces surface tension and keeps the alveoli from collapsing. **Alveolar macrophages** patrol the epithelium and engulf foreign particles. (*Figure 23–10*)

28. The conducting portions of the respiratory tract receive blood from the external carotid arteries, the thyrocervical trunks, and the bronchial arteries. Venous blood flows into the pulmonary veins, bypassing the rest of the systemic circuit and diluting the oxygenated blood leaving the alveoli.

29. Each lung is surrounded by a single **pleural cavity** lined by a **pleura** (serous membrane). The two types of pleurae are the **parietal pleura**, covering the inner surface of the thoracic wall, and the **visceral pleura**, covering the lungs.

23-6 ▌ External respiration and internal respiration allow gaseous exchange within the body p. 873

30. Respiratory physiology focuses on a series of integrated processes. **External respiration** is the exchange of oxygen and carbon dioxide between interstitial fluid and the external environment and includes **pulmonary ventilation** (breathing). **Internal respiration** is the exchange of oxygen and carbon dioxide between interstitial fluid and cells. If the oxygen content declines, the affected tissues suffer from **hypoxia**; if the oxygen supply is completely shut off, **anoxia** and tissue death result. (*Figure 23–11*)

23-7 ▌ Pulmonary ventilation—the exchange of air between the atmosphere and the lungs—involves pressure changes, muscle movement, and respiratory rates and volumes p. 874

31. **Pulmonary ventilation** is the physical movement of air into and out of the respiratory tract.

32. As pressure on a gas decreases, its volume expands; as pressure increases, gas volume contracts. This inverse relationship is **Boyle's law**. (*Figure 23–12; Table 23–1*)

33. Movement of the diaphragm and ribs changes lung volume.

34. The relationship between **intrapulmonary pressure** (the pressure inside the respiratory tract) and **atmospheric pressure** (**atm**) determines the direction of airflow. **Intrapleural pressure** is the pressure in the potential space between the parietal and visceral pleurae. (*Figures 23–13, 23–14*)

35. A **respiratory cycle** is a single cycle of inhalation and exhalation. The amount of air moved in one respiratory cycle is the **tidal volume**. A **spirometer** is an instrument used to measure the capacity of the lungs. (*Figure 23–14*)

36. The diaphragm and the external and internal intercostal muscles are involved in normal **quiet breathing**, or **eupnea**. Accessory muscles become active during the active inspiratory and expiratory movements of **forced breathing**, or **hyperpnea**. (*Spotlight Figure 23–15*)

37. **Alveolar ventilation** is the amount of air reaching the alveoli each minute. The **vital capacity** includes the **tidal volume** plus the **expiratory** and **inspiratory reserve volumes**. The air left in the lungs at the end of maximum exhalation is the **residual volume**. (*Figure 23–16*)

23-8 ▌ Gas exchange depends on the partial pressures of gases and the diffusion of molecules p. 882

38. In a mixed gas, the individual gases exert a pressure proportional to their abundance in the mixture (**Dalton's law**). The pressure contributed by a single gas is its **partial pressure**. (*Table 23–2*)

39. The amount of a gas in solution is directly proportional to the partial pressure of that gas (**Henry's law**). (*Figure 23–17*)

40. Alveolar air and atmospheric air differ in composition. Gas exchange across the respiratory membrane is efficient due to differences in partial pressures, the short diffusion distance, lipid-soluble gases, the large surface area of all the alveoli combined, and the coordination of blood flow and airflow. (*Figure 23–18*)

23

23-9 ❙ Most oxygen is transported bound to hemoglobin; and carbon dioxide is transported in three ways: as carbonic acid, bound to hemoglobin, or dissolved in plasma p. 886

41. Blood entering peripheral capillaries delivers oxygen and absorbs carbon dioxide. The transport of oxygen and carbon dioxide in blood involves reactions that are completely reversible.

42. Oxygen is carried mainly by RBCs, reversibly bound to hemoglobin. At alveolar P_{O_2}, hemoglobin is almost fully saturated; at the P_{O_2} of peripheral tissues, it releases oxygen but still retains a substantial oxygen reserve. The effect of pH on the hemoglobin saturation curve is called the **Bohr effect**. When low plasma P_{O_2} continues for extended periods, red blood cells generate more **2,3-bisphosphoglycerate (BPG)**, which reduces hemoglobin's affinity for oxygen. *(Figures 23–19, 23–20)*

43. **Fetal hemoglobin** has a stronger affinity for oxygen than does adult hemoglobin, aiding the removal of oxygen from maternal blood. *(Figure 23–21)*

44. Aerobic metabolism in peripheral tissues generates CO_2. About 7 percent of the CO_2 transported in blood is dissolved in the plasma, 23 percent is bound as **carbaminohemoglobin**, and 70 percent is converted to carbonic acid, which dissociates into H^+ and HCO_3^-. *(Figure 23–22)*

45. Driven by differences in partial pressure, oxygen enters the blood at the lungs and leaves it in peripheral tissues; similar forces drive carbon dioxide into the blood at the tissues and into the alveoli at the lungs. *(Figure 23–23)*

23-10 ❙ Neurons in the medulla oblongata and pons, along with respiratory reflexes, control respiration p. 890

46. Normally, the cellular rates of gas absorption and generation are matched by the capillary rates of delivery and removal and are identical to the rates of oxygen absorption and carbon dioxide removal at the lungs. When these rates are unbalanced, homeostatic mechanisms restore equilibrium.

47. Local factors regulate alveolar blood flow *(lung perfusion)* and airflow *(alveolar ventilation)*. Alveolar capillaries constrict under conditions of low oxygen, and bronchioles dilate under conditions of high carbon dioxide.

48. The **respiratory centers** include three pairs of nuclei in the reticular formation of the pons and medulla oblongata. The *respiratory rhythmicity centers* set the pace for respiration; the **apneustic centers** cause strong, sustained inspiratory movements; and the **pneumotaxic centers** inhibit the apneustic centers and promote exhalation. *(Figure 23–24, Spotlight Figure 23–25)*

49. Stimulation of the chemoreceptor reflexes is based on the level of carbon dioxide in the blood and CSF. *(Spotlight Figure 23–25, Figure 23–26)*

50. The **inflation reflex** prevents overexpansion of the lungs during forced breathing. The **deflation reflex** stimulates inhalation when the lungs are collapsing.

51. Conscious and unconscious thought processes can affect respiration by affecting the respiratory centers.

52. Before delivery, the fetal lungs are filled with body fluids and are collapsed. At the first breath, the lungs inflate and do not collapse completely thereafter.

23-11 ❙ Respiratory performance declines with age p. 899

53. The respiratory system is generally less efficient in the elderly because (1) elastic tissue deteriorates, lowering compliance and the vital capacity of the lungs; (2) movements of the chest are restricted by arthritic changes and decreased flexibility of costal cartilages; and (3) some degree of emphysema is generally present. *(Figure 23–27)*

23-12 ❙ The respiratory system provides oxygen to, and eliminates carbon dioxide from, other organ systems p. 899

54. The respiratory system has extensive anatomical and physiological connections to the cardiovascular system. *(Figure 23–28)*

Review Questions

See the blue Answers tab at the back of the book.

23

LEVEL 1 Reviewing Facts and Terms

1. Which alveolar cell type produces surfactant? **(a)** goblet, **(b)** type II pneumocyte, **(c)** type I pneumocytes, **(d)** Kuppfer cells.

2. The hard palate separates the **(a)** nasal cavity from the larynx, **(b)** left and right sides of the nasal cavity, **(c)** nasal cavity and the oral cavity, **(d)** external nares from the internal nares, **(e)** soft palate from the nasal cavity.

3. Air moves into the lungs because **(a)** the gas pressure in the lungs is less than atmospheric pressure, **(b)** the volume of the lungs decreases with inspiration, **(c)** the thorax is muscular, **(d)** contraction of the diaphragm decreases the volume of the thoracic cavity, **(e)** the respiratory control center initiates active expansion of the thorax.

4. The glottis closes partway through an exhalation. The abdominal and internal intercostal muscles then contract suddenly, creating pressure that blasts the air out of the respiratory passages. This describes a **(a)** sneeze, **(b)** hiccough, **(c)** cough, **(d)** laryngeal spasm, **(e)** gag.

5. When the diaphragm and external intercostal muscles contract, **(a)** exhalation occurs, **(b)** intrapulmonary pressure increases, **(c)** intrapleural pressure decreases, **(d)** the volume of the lungs decreases, **(e)** the size of the thoracic cavity increases.

6. Identify the structures of the respiratory system in the figure at left:

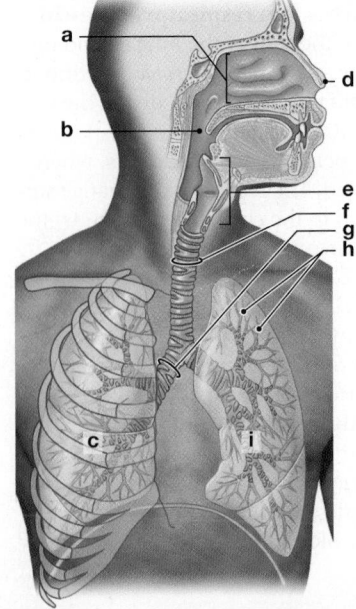

(a) _____
(b) _____
(c) _____
(d) _____
(e) _____
(f) _____
(g) _____
(h) _____
(i) _____

7. During the winter, Brad sleeps in a dorm room that lacks any humidifier for the heated air. In the mornings he notices that his nose is "stuffy" similar to when he has a cold, but after showering and drinking some water, the stuffiness disappears until the next morning. What might be the cause of Brad's nasal condition?

8. During inhalation (**a**) intrapulmonary pressure rises, (**b**) intrapulmonary pressure becomes positive, (**c**) intrapleural pressure falls, (**d**) a, b, and c.

9. Name the three regions of the pharynx. Where is each region located?

10. Define internal and external respiration.

11. What three integrated steps are involved in external respiration?

12. During quiet breathing the inspiratory muscles are stimulated by which respiratory group? (**a**) dorsal, (**b**) ventral, (**c**) pneumotaxic, (**d**) apneustic.

13. By what three ways is carbon dioxide transported in the bloodstream?

LEVEL 2 Reviewing Concepts

14. Which of the following does *not* occur in internal respiration? (**a**) Oxygen diffuses from the blood to the interstitial spaces, (**b**) Carbon dioxide diffuses from the interstitial spaces to the blood, (**c**) Hemoglobin binds more oxygen, (**d**) Bicarbonate ions are formed in red blood cells, (**e**) Chloride ions diffuse into red blood cells as bicarbonate ions diffuse out.

15. Gas exchange at the respiratory membrane is efficient because (**a**) the differences in partial pressure are substantial, (**b**) the gases are lipid soluble, (**c**) the total surface area is large, (**d**) of a, b, and c.

16. For any partial pressure of oxygen, if the concentration of 2,3-bisphosphoglycerate (BPG) increases, (**a**) the amount of oxygen released by hemoglobin will decrease, (**b**) the oxygen levels in hemoglobin will be unaffected, (**c**) the amount of oxygen released by hemoglobin will increase, (**d**) the amount of carbon dioxide carried by hemoglobin will increase.

17. A loss of surfactant would (**a**) increase compliance making the lungs easier to expand, (**b**) decrease compliance making the lungs easier to expand, (**c**) increase compliance making the lungs more difficult to expand, (**d**) decrease compliance making the lungs more difficult to expand.

18. Explain how the structure of the respiratory membrane is related to its efficiency in gas exchange.

19. How does diaphragm contraction lead to pressure changes in the thorax and inward air flow during inspiration?

20. How are type II pneumocytes involved with keeping the alveoli from collapsing?

21. How does pulmonary ventilation differ from alveolar ventilation, and what is the function of each type of ventilation?

22. What is the significance of (**a**) Boyle's law, (**b**) Dalton's law, and (**c**) Henry's law to the process of respiration?

23. How does the efficiency of the respiratory system decline with age?

24. What are the differences between pulmonary volumes and respiratory capacities? How are pulmonary volumes and respiratory capacities determined?

25. What is the functional difference between the dorsal respiratory group (DRG) and the ventral respiratory group (VRG) of the medulla oblongata?

LEVEL 3 Critical Thinking and Clinical Applications

26. While holidaying in Peru, Jonathan visits an Inca city high in the Andean mountains. The atmospheric pressure drops from 760 mm Hg (101 k Pa) to 550 mm Hg (73 k Pa) at approximately 9190 ft (2800 m). His hemoglobin oxygen saturation measures 93%. Calculate the reduction in partial pressure of oxygen from 0 ft to 9190 ft, and explain why hemoglobin saturations don't fall by the same degree.

27. Mr. B. has had chronic advanced emphysema for 15 years. While hospitalized with a respiratory infection, he goes into respiratory distress. Without thinking, his nurse immediately administers pure oxygen, which causes Mr. B. to stop breathing. Why?

28. Cary hyperventilates for several minutes before diving into a swimming pool. After he enters and begins swimming underwater, he blacks out and almost drowns. What caused this to happen?

29. Why do individuals who are anemic generally not exhibit an increase in respiratory rate or tidal volume, even though their blood is not carrying enough oxygen?

30. Lung compliance would be increased by (**a**) scar tissue around the ribcage, (**b**) narrowed airways, (**c**) increased mucus production, (**d**) loss of elastic tissue.

CLINICAL CASE Wrap-Up

23

How Long Should a Cough Last?

Andrea, like many people, thinks a typical cough should only last about a week. In reality, however, coughs caused by colds and flu tend to last about 18 days. Andrea thinks she got over her previous cough due to antibiotics, but antibiotics only work against bacteria. If her previous cough was caused by a virus, her body likely resolved the cough on its own during the normal course of the illness. It's worth noting that viruses cause 90 percent of acute cough illnesses.

Fever is the body's innate immunity at work. Excessive mucus production, leading to a productive cough and runny nose, indicates that the immune system is working. The excessive mucus produced by the respiratory mucosa builds up and irritates the trachea and bronchi. This causes parasympathetic reflexive coughing.

The explosive ejection of air and mucus against a suddenly open glottis helps clear away cold viruses. It also helps spread these viruses to new hosts.

Andrea's doctor tells her to drink plenty of fluids, rest, and use a vaporizer and saline nasal spray. He will not prescribe antibiotics because they are ineffective against cold viruses. He also advises Andrea to wear a mask to avoid spreading this virus to her lab partners for the duration of her cough.

1. Do you think Andrea should take medicine to suppress her acute, productive cough?

2. What anatomical structures of Andrea's respiratory system are infected by the rhinovirus?

See the blue Answers tab at the back of the book.

24 The Digestive System

Learning Outcomes

These Learning Outcomes correspond by number to this chapter's sections and indicate what you should be able to do after completing the chapter.

24-1 ▮ Identify the organs of the digestive system, list their major functions, describe the histology of the digestive tract, and outline the mechanisms that regulate digestion.

24-2 ▮ Discuss the anatomy of the oral cavity, and list the functions of its major structures and regions.

24-3 ▮ Describe the structure and functions of the pharynx.

24-4 ▮ Describe the structure and functions of the esophagus.

24-5 ▮ Describe the anatomy of the stomach, including its histology, and discuss its roles in digestion and absorption.

24-6 ▮ Describe the anatomy and histology of the small intestine, explain the functions and regulation of intestinal secretions, and describe the structure, functions, and regulation of the accessory digestive organs.

24-7 ▮ Describe the gross and histological structure of the large intestine, including its regional specializations and role in nutrient absorption.

24-8 ▮ List the nutrients required by the body, describe the chemical events responsible for the digestion of organic nutrients, and describe the mechanisms involved in the absorption of organic and inorganic nutrients.

24-9 ▮ Summarize the effects of aging on the digestive system.

24-10 ▮ Give examples of interactions between the digestive system and other organ systems studied so far.

SPOTLIGHTS

- Regulation of Gastric Activity pp. 928–929
- Chemical Events of Digestion p. 949

CLINICAL CASE

- An Unusual Transplant p. 907

Clinical Notes

- Peritonitis p. 909
- Epithelial Renewal and Repair p. 912
- Mumps p. 916
- Gastritis and Peptic Ulcers p. 923
- Pancreatitis p. 933
- Cirrhosis p. 935
- Colorectal Cancer p. 943
- Inflammatory and Infectious Disorders of the Digestive System p. 946

An Unusual Transplant

Mirasol has been in the hospital and on antibiotics for a stubborn case of pneumonia for over a week. She desperately misses her grandchildren, and is feeling worse than ever. Her cough has resolved, but she has had watery diarrhea 10 times today. The last few times, there was blood in the stool. She has lost her appetite and feels crampy abdominal pain. She is dehydrated and unable to drink enough liquids to keep up with her fluid loss from diarrhea. She also has a fever and has lost weight.

Her doctor enters the room with results of recent tests performed on her stool. "I'm afraid the normal, helpful bacteria that live in your colon have been killed off by the antibiotics used to cure your pneumonia. Now, harmful bacteria have taken over and are causing your bloody diarrhea," reports the doctor. "Your colon is overgrown with *Clostridium difficile* bacteria. These bacteria are producing toxins that are causing your colitis

(inflammation of the colon). The usual treatment for this condition is more antibiotics."

"Doctor," cries Mirasol, "I don't think I can take any more antibiotics. I feel worse now than when I arrived."

"Well, I have an idea that may work," the doctor replies. "A transplant, from one of your healthy, adult grandchildren. It's a little unusual, however. You need to keep an open mind." **What kind of transplant can possibly help Mirasol? To find out, turn to the Clinical Case Wrap-Up on p. 960.**

An Introduction to the Digestive System

The digestive system may not have the visibility of the integumentary system or the glamour of the reproductive system, but it is certainly just as important. All living organisms must get nutrients from their environment to sustain life. These nutrients are used as raw materials for synthesizing essential compounds (anabolism). They are also broken down to provide the energy that cells need to continue functioning (catabolism). ⤷ pp. 62–64, 343–344 Catabolic reactions require two essential ingredients: oxygen and organic molecules (such as carbohydrates, fats, or proteins) that can be broken down by enzymes inside cells. Obtaining oxygen and organic molecules can be straightforward for a single-celled organism like an amoeba. The process is much more complicated for animals as large and complex as humans. Along with increasing size and complexity come a division of labor within the body and the need to coordinate organ system activities.

In this chapter we discuss the structure and function of the digestive tract and several digestive glands, notably the liver and pancreas. We look at the process of digestion and how it breaks down large and complex organic molecules into smaller fragments that can be absorbed by the digestive epithelium. We also see how a few organic wastes are removed from the body.

24-1 �The digestive system, consisting of the digestive tract and accessory organs, has overlapping food utilization functions

Learning Outcome Identify the organs of the digestive system, list their major functions, describe the histology of the digestive tract, and outline the mechanisms that regulate digestion.

In our bodies, the respiratory system works with the cardiovascular system to supply the oxygen needed for catabolism. The *digestive system*, working with the cardiovascular and lymphatic systems, provides the needed organic molecules. In effect, the digestive system supplies both the fuel that keeps all the body's cells running and the building blocks needed for cell growth and repair.

The **digestive system** consists of a muscular tube, the **digestive tract**, also called the *gastrointestinal (GI) tract* or *alimentary canal*, and various **accessory organs**. The *oral cavity* (mouth), *pharynx* (throat), *esophagus, stomach, small intestine,* and *large intestine* make up the digestive tract. Accessory digestive organs include the teeth, tongue, and various *glandular organs*, such as the salivary glands, liver, gallbladder, and pancreas. The glandular organs secrete their products into ducts that empty into the digestive tract. These secretions contain water, enzymes, buffers, and other substances. Food enters the digestive tract and passes along its length. Along the way, the secretions of the glandular

Figure 24–1   **The Components of the Digestive System.**

Major Organs of the Digestive Tract

Oral Cavity (Mouth)

Ingestion, mechanical processing with accessory organs (teeth and tongue), moistening, mixing with salivary secretions

Pharynx

Muscular propulsion of materials into the esophagus

Esophagus

Transport of materials to the stomach

Stomach

Chemical breakdown of materials by acid and enzymes; mechanical processing through muscular contractions

Small Intestine

Enzymatic digestion and absorption of water, organic substrates, vitamins, and ions

Large Intestine

Dehydration and compaction of indigestible materials in preparation for elimination

Accessory Organs of the Digestive System

Teeth

Mechanical processing by chewing (mastication)

Tongue

Assists mechanical processing with teeth, sensory analysis

Salivary Glands

Secretion of lubricating fluid containing enzymes that break down carbohydrates

Liver

Secretion of bile (important for lipid digestion), storage of nutrients, many other vital functions

Gallbladder

Storage and concentration of bile

Pancreas

Exocrine cells secrete buffers and digestive enzymes; endocrine cells secrete hormones

Anus

24

organs prepare organic and inorganic nutrients for absorption across the epithelium of the digestive tract.

Figure 24–1 shows the major parts of the digestive system. The digestive tract begins at the oral cavity. It continues through the pharynx, esophagus, stomach, small intestine, and large intestine, which opens to the exterior at the anus. These structures have overlapping functions, but each has certain areas of specialization and shows distinctive histological characteristics.

Functions of the Digestive System

We can regard digestive functions as a series of six integrated steps:

1. **Ingestion** takes place when materials enter the oral cavity of the digestive tract. Ingestion is an active process involving conscious choice and decision making.

2. **Mechanical processing** is crushing and shearing that makes materials easier to propel along the digestive tract.

It also increases their surface area, making them more susceptible to attack by enzymes. Mechanical processing may or may not be required before ingestion. You can swallow liquids immediately, but you must chew most solids first. Tearing and mashing with the teeth, followed by squashing and compacting by the tongue, are examples of preliminary mechanical processing. Swirling, mixing, and churning motions of the stomach and intestines provide mechanical processing after ingestion.

3. **Digestion** refers to the chemical breakdown of food into small organic fragments suitable for absorption by the digestive epithelium. Simple molecules in food, such as glucose, can be absorbed intact. However, epithelial cells have no way to absorb molecules with the size and complexity of proteins, polysaccharides, or triglycerides. Digestive enzymes must first disassemble these molecules. For example, the starches in a potato are of no nutritional value until enzymes have broken them down to simple sugars

that the digestive epithelium can absorb for distribution to body cells.

4. **Secretion** is the release of water, acids, enzymes, buffers, and salts by the epithelium of the digestive tract and by glandular organs.

5. **Absorption** is the movement of organic molecules, electrolytes (inorganic ions), vitamins, and water across the digestive epithelium and into the interstitial fluid of the digestive tract.

6. **Excretion** is the removal of wastes from the body. The digestive tract and glandular organs discharge wastes in secretions that enter the lumen of the tract. Most of these wastes mix with the indigestible residue of the digestive process and then leave the body. The process called **defecation** (def-e-KĀ-shun) ejects materials from the digestive tract, eliminating them as matter called **feces**.

The lining of the digestive tract also plays a protective role. It safeguards surrounding tissues against (1) the corrosive effects of digestive acids and enzymes; (2) mechanical stresses, such as abrasion; and (3) bacteria that either are swallowed with food or live in the digestive tract. The digestive epithelium and its secretions provide a nonspecific defense against these bacteria. When bacteria reach the underlying layer of areolar tissue, the *lamina propria*, macrophages and other cells of the immune system attack them.

We explore specific functions in more detail as we consider the individual regions and components of the system. First, however, let's look at several structural and functional characteristics of the system as a whole.

The Digestive Organs and the Peritoneum

The abdominopelvic cavity contains the *peritoneal cavity*, which is lined by a serous membrane. This membrane consists of a superficial mesothelium covering a layer of areolar tissue. ⤸ pp. 48, 163 We divide this serous membrane into two parts. The serosa or *visceral peritoneum* covers organs that project into the peritoneal cavity. The *parietal peritoneum* lines the inner surfaces of the body wall.

The serous membrane lining the peritoneal cavity continuously produces peritoneal fluid, which lubricates the surfaces. Because a thin layer of peritoneal fluid separates them, the parietal and visceral surfaces can slide without friction and resulting irritation. The membrane secretes and reabsorbs about 7 liters (7.4 quarts) of fluid each day, but the volume within the peritoneal cavity at any one time is very small. Liver disease, kidney disease, and heart failure can increase the rate at which fluids move into the peritoneal cavity. The buildup of fluid creates a characteristic abdominal swelling called **ascites** (a-SĪ-tēz). This fluid can distort internal organs and cause symptoms such as heartburn, indigestion, and lower back pain.

Mesenteries

Portions of the digestive tract are suspended within the peritoneal cavity by sheets of serous membrane that connect the parietal peritoneum with the visceral peritoneum. These **mesenteries** (MEZ-en-ter-ēz) are double sheets of peritoneal membrane. The areolar tissue between the mesothelial surfaces provides a route to and from the digestive tract for blood vessels, nerves, and lymphatic vessels. Mesenteries stabilize the positions of the attached organs. The mesenteries also prevent the intestines from becoming entangled during digestive movements or sudden changes in body position.

During embryonic development, the digestive tract and accessory organs are suspended within the peritoneal cavity by *dorsal* and *ventral mesenteries* (**Figure 24–2a**). The ventral mesentery later disappears along most of the digestive tract. It persists in adults in only two places: on the ventral surface of the stomach, between the stomach and the liver (the *lesser omentum*) (**Figure 24–2b,d**); and between the liver and the anterior abdominal wall (the *falciform ligament*) (**Figure 24–2c,d**). The **lesser omentum** (ō-MEN-tum; *omentum*, fat skin) stabilizes the position of the stomach and provides an access route for blood vessels and other structures entering or leaving the liver. The **falciform** (FAL-si-form; *falx*, sickle + *forma*, form) **ligament** helps stabilize the position of the liver relative to the diaphragm and abdominal wall. ATLAS: Embryology Summary 19: The Development of the Digestive System

As the digestive tract elongates, it twists and turns within the crowded peritoneal cavity. The dorsal mesentery of the stomach becomes greatly enlarged and forms an enormous pouch that extends inferiorly between the body wall and the anterior surface of the small intestine. This pouch is the **greater omentum** (**Figure 24–2b,d**). It hangs like an apron from the lateral and inferior borders of the stomach. Adipose tissue in the greater omentum conforms to the shapes of the surrounding

---

Clinical Note

Peritonitis An inflammation of the peritoneal membrane (peritoneum) is called **peritonitis** (per-i-tō-NĪ-tis). This painful condition interferes with the normal functions of the affected organs. Physical damage, chemical irritation, and bacterial invasion of the peritoneum can lead to severe cases of peritonitis. Such cases can even be fatal. In untreated appendicitis, the appendix may rupture and release bacteria into the peritoneal cavity. This event may cause peritonitis. Peritonitis can also be a complication of any surgery that opens the peritoneal cavity. There is also danger of peritonitis in any disease or injury that perforates the stomach or intestines.

24

Figure 24–2 **Mesenteries.**

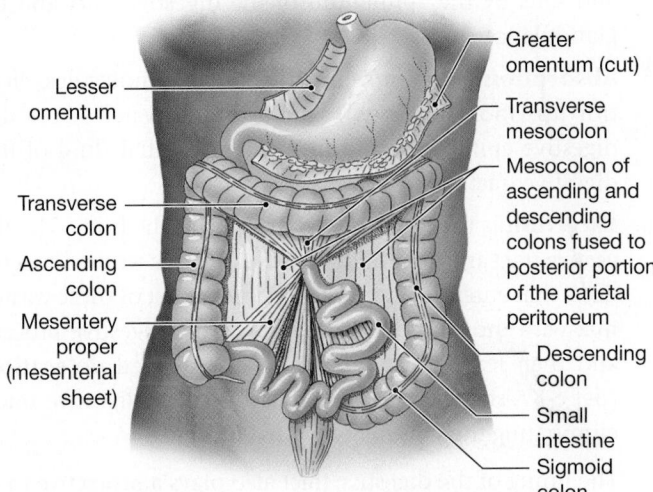

a During embryonic development, the digestive tube is initially suspended by dorsal and ventral mesenteries. In adults, the ventral mesentery is lost, except where it connects the stomach to the liver (at the lesser omentum) and the liver to the anterior body wall and diaphragm (at the falciform ligament).

b A diagrammatic view of the organization of mesenteries in an adult. As the digestive tract enlarges, mesenteries associated with the proximal portion of the small intestine, the pancreas, and the ascending and descending portions of the colon fuse to the body wall.

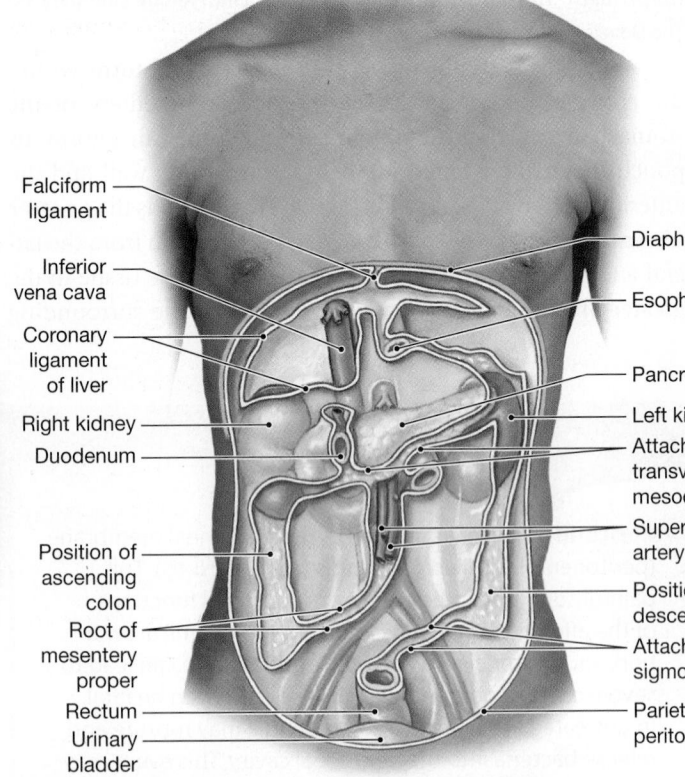

c An anterior view of the empty peritoneal cavity, showing the attachment of mesenteries to the posterior body wall. Some visceral organs that were originally suspended within the peritoneal cavity are now retroperitoneal due to fusion of the serosa with the parietal peritoneum.

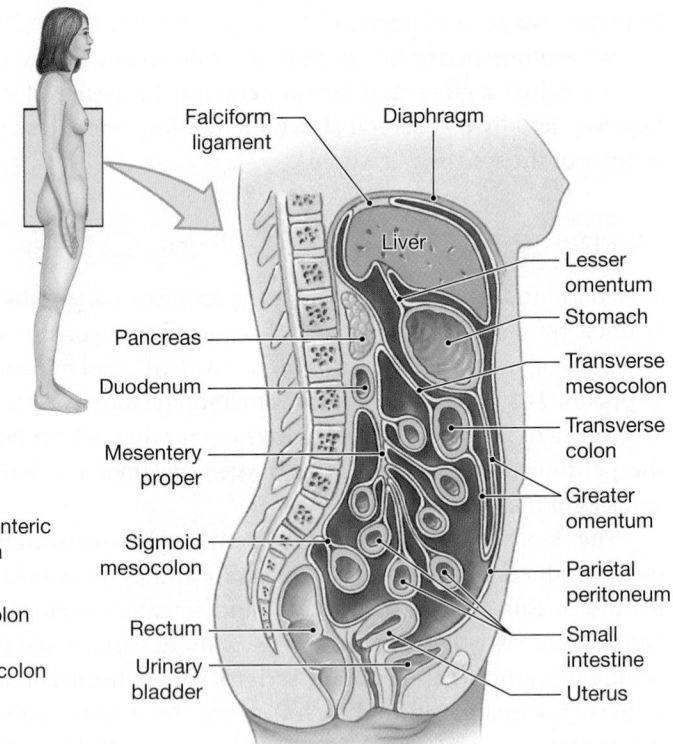

d A sagittal section showing the mesenteries of an adult. Notice that the pancreas, duodenum, and rectum are retroperitoneal.

organs, providing padding and protection across the anterior and lateral surfaces of the abdomen. When an individual gains weight, this adipose tissue contributes to the characteristic "beer belly." The lipids in the adipose tissue are an important energy reserve. The greater omentum also insulates to reduce heat loss across the anterior abdominal wall.

All but the first 25 cm (10 in.) of the small intestine is suspended by the **mesentery proper**, a thick mesenterial sheet. It provides stability, but permits some independent movement. The mesentery associated with the initial portion of the small intestine (the *duodenum*) and the pancreas fuses with the posterior abdominal wall and locks those structures in place. Only their anterior surfaces remain covered by peritoneum. We describe these organs as **retroperitoneal** (*retro*, behind) because their mass lies posterior to, rather than surrounded by, the peritoneal cavity.

A **mesocolon** is a mesentery associated with a portion of the large intestine. During normal development, the mesocolon of the *ascending colon*, the *descending colon*, and the *rectum* of the large intestine fuse to the posterior body wall. These regions become locked in place. Thereafter, these organs are retroperitoneal. The visceral peritoneum covers only their anterior surfaces and portions of their lateral surfaces (**Figure 24–2b,c,d**). The **transverse mesocolon**, which supports the transverse colon, and the **sigmoid mesocolon**, which supports the sigmoid colon, are all that remain of the original embryonic mesocolon.

Histology of the Digestive Tract

The four major layers of the digestive tract are (1) the *mucosa*, (2) the *submucosa*, (3) the *muscularis externa*, and (4) the *serosa*. The structure of these layers varies by region. **Figure 24–3** is a composite view. It most closely resembles the small intestine, the longest segment of the digestive tract.

The Mucosa

The inner lining, or **mucosa**, of the digestive tract is a *mucous membrane*. It consists of an epithelium, moistened by glandular secretions, and a *lamina propria* of areolar tissue.

The Digestive Epithelium. The mucosal epithelium is either simple or stratified, depending on its location and the stresses placed on it. Mechanical stresses are most severe in the oral cavity, pharynx, and esophagus. These structures are lined by a stratified squamous epithelium. In contrast, the stomach, the small intestine, and almost the entire length of the large intestine (where absorption occurs) have a simple columnar epithelium that contains mucous cells. Scattered among the columnar cells are **enteroendocrine cells**. They secrete hormones that coordinate the activities of the digestive tract and the accessory glands.

The lining of the digestive tract appears as longitudinal folds, which disappear as the tract fills. The lining also

Figure 24–3 The Structure of the Digestive Tract. A diagrammatic view of a representative portion of the digestive tract. The features illustrated are typical of those of the small intestine.

Clinical Note

Epithelial Renewal and Repair The life span of a typical epithelial cell varies from two to three days in the esophagus to six days in the large intestine. The divisions of epithelial stem cells continuously renew the lining of the entire digestive tract. These divisions normally keep pace with the rates of cell destruction and loss at epithelial surfaces. This high rate of cell division explains why radiation and anticancer drugs that inhibit mitosis have drastic effects on the digestive tract. Lost epithelial cells are no longer replaced. The cumulative damage to the epithelial lining quickly leads to problems in absorbing nutrients. In addition, the exposure of the lamina propria to digestive enzymes can cause internal bleeding and other serious problems.

has permanent transverse folds called *circular folds*, or *plicae* (PLĪ-sē; folds) *circulares* (**Figure 24–3**). The folding increases the surface area available for absorption. The secretions of gland cells in the mucosa and submucosa—or in accessory glandular organs—are carried to the epithelial surfaces by ducts.

The Lamina Propria. The lamina propria is a layer of areolar tissue that also contains blood vessels, sensory nerve endings, lymphatic vessels, smooth muscle cells, and scattered lymphatic tissue. In the oral cavity, pharynx, esophagus, stomach, and *duodenum* (the proximal portion of the small intestine), the lamina propria also contains the secretory cells of mucous glands.

In most areas of the digestive tract, the lamina propria contains a narrow sheet of smooth muscle and elastic fibers. This sheet is called the **muscularis** (mus-kū-LAIR-is) **mucosae** (mū-KŌ-sē) (**Figure 24–3**). The smooth muscle cells in the muscularis mucosae are arranged in two concentric layers. The inner layer encircles the lumen (the *circular muscle*), and the outer layer contains muscle cells arranged parallel to the long axis of the tract (the *longitudinal layer*). Contractions in these layers alter the shape of the lumen and move the epithelial pleats and folds.

The Submucosa

The **submucosa** is a layer of dense irregular connective tissue that binds the mucosa to the muscularis externa (**Figure 24–3**). The submucosa has numerous blood vessels and lymphatic vessels. In some regions it also contains exocrine glands that secrete buffers and enzymes into the lumen of the digestive tract.

Along its outer margin, the submucosa contains a network of intrinsic nerve fibers and scattered neurons. This network is

the **submucosal plexus**, or *Meissner's plexus*. It contains sensory neurons, parasympathetic ganglionic neurons, and sympathetic postganglionic fibers that innervate the mucosa and submucosa.

The Muscularis Externa

The submucosal plexus lies along the inner border of the **muscularis externa**, also called the **muscularis**. Smooth muscle cells dominate this region. Like the smooth muscle cells in the muscularis mucosae, those in the muscularis externa are arranged in an inner circular layer and an outer longitudinal layer. These layers play an essential role in mechanical processing and in moving materials along the digestive tract.

The movements of the digestive tract are coordinated primarily by the sensory neurons, interneurons, and motor neurons of the enteric nervous system (ENS). ⟳ p. 562 The ENS is primarily innervated by the parasympathetic division of the ANS. Sympathetic postganglionic fibers also synapse here. Many of these fibers continue onward to innervate the mucosa and the **myenteric** (mī-en-TER-ik) **plexus** (*mys*, muscle + *enteron*, intestine), or *Auerbach's plexus*. This plexus is a network of parasympathetic ganglia, sensory neurons, interneurons, and sympathetic postganglionic fibers. It lies sandwiched between the circular and longitudinal muscle layers. In general, parasympathetic stimulation *increases* muscle tone and activity. Sympathetic stimulation decreases muscle tone and activity.

TIPS & TRICKS

Because the parasympathetic nervous system plays a dominant role in the digestive process, it is often referred to as the "rest and digest" division.

The Serosa

A serous membrane known as the **serosa** covers the muscularis externa along most portions of the digestive tract inside the peritoneal cavity (**Figure 24–3**).

There is no serosa covering the muscularis externa of the oral cavity, pharynx, esophagus, and rectum. Instead, a dense network of collagen fibers firmly attaches the digestive tract to adjacent structures. This fibrous sheath is called an *adventitia* (ad-ven-TISH-uh).

The Movement of Digestive Materials

The muscular layers of the digestive tract consist of *visceral smooth muscle tissue*. We introduced this type of smooth muscle in Chapter 10. ⟳ p. 353 The smooth muscle along the digestive tract has rhythmic cycles of activity due to *pacesetter cells*. These

24

smooth muscle cells undergo spontaneous depolarization, triggering a wave of contraction that spreads throughout the entire muscular sheet. Pacesetter cells are located in the muscularis mucosae and muscularis externa, which surround the lumen of the digestive tract. The coordinated contractions of the muscularis externa play a vital role in moving materials along the tract, through *peristalsis*, and in mechanical processing, through *segmentation*.

Peristalsis

The muscularis externa propels materials from one portion of the digestive tract to another by contractions known as **peristalsis** (per-i-STAL-sis). Peristalsis consists of waves of muscular contractions that move a **bolus** (BŌ-lus), or soft rounded ball of digestive contents, along the length of the digestive tract (**Figure 24–4**). During a peristaltic movement, the circular muscles contract behind the bolus while circular muscles ahead of the bolus relax. Longitudinal muscles ahead of the bolus then contract, shortening adjacent segments. A wave of contraction in the circular muscles then forces the bolus forward.

TIPS & TRICKS

Squeezing toothpaste out of a tube is similar to peristalsis: Your squeezing hand (contracting circular muscles) forces toothpaste (the bolus) along and out of the tube (the digestive tract).

Segmentation

Most areas of the small intestine and some portions of the large intestine undergo cycles of contraction that churn and fragment the bolus, mixing the contents with intestinal secretions. This activity, called **segmentation**, does not follow a set pattern. For this reason, segmentation does not push materials along the tract in any one direction.

Control of Digestive Functions

Local factors interact with neural and hormonal mechanisms to regulate the activities of the digestive system (**Figure 24–5**, p. 914).

Local Factors

The initial regulation of digestive function occurs at the local level (❶). Local environmental factors such as the pH, volume, or chemical composition of the intestinal contents can have a direct effect on digestive activity in that segment of the digestive tract. Some of these local factors have a direct effect on local digestive activities; for example, stretching of the intestinal wall can stimulate localized contractions of smooth muscles. In other cases, the local factors stimulate the release of chemicals.

Figure 24–4 **Peristalsis.** Peristalsis propels materials along the length of the digestive tract.

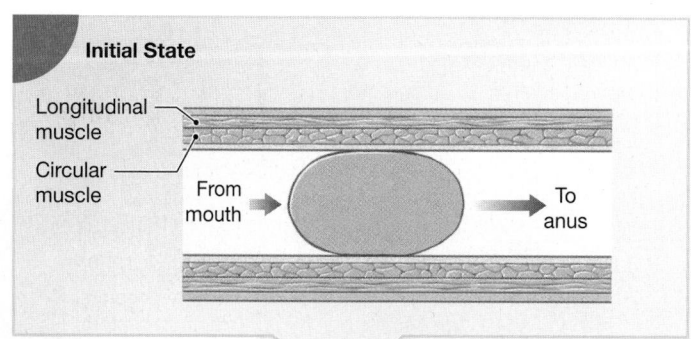

Initial State

Longitudinal muscle
Circular muscle
From mouth
To anus

1 **Contraction of circular muscles behind bolus**

Contraction

2 **Contraction of longitudinal muscles ahead of bolus**

Contraction

Contraction

3 **Contraction in circular muscle layer forces bolus forward**

Prostaglandins, histamine, and other chemicals released into interstitial fluid may affect adjacent cells within a small segment of the tract. For example, the release of histamine in the lamina propria of the stomach stimulates the secretion of acid by cells in the adjacent epithelium.

Figure 24–5   **The Regulation of Digestive Activities.**

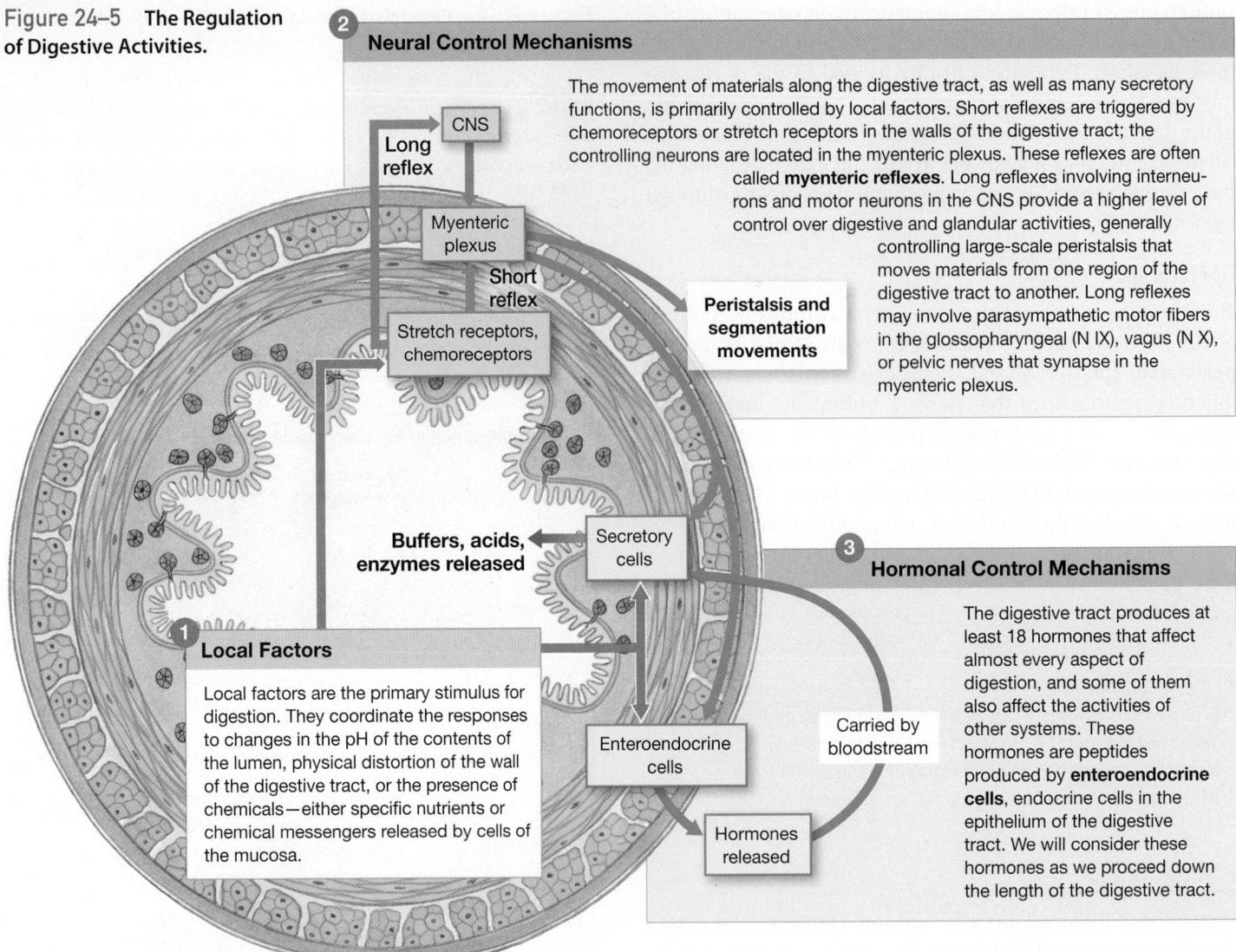

② Neural Control Mechanisms

The movement of materials along the digestive tract, as well as many secretory functions, is primarily controlled by local factors. Short reflexes are triggered by chemoreceptors or stretch receptors in the walls of the digestive tract; the controlling neurons are located in the myenteric plexus. These reflexes are often called **myenteric reflexes**. Long reflexes involving interneurons and motor neurons in the CNS provide a higher level of control over digestive and glandular activities, generally controlling large-scale peristalsis that moves materials from one region of the digestive tract to another. Long reflexes may involve parasympathetic motor fibers in the glossopharyngeal (N IX), vagus (N X), or pelvic nerves that synapse in the myenteric plexus.

CNS

Long reflex

Myenteric plexus

Short reflex

Stretch receptors, chemoreceptors

Peristalsis and segmentation movements

Buffers, acids, enzymes released

Secretory cells

① Local Factors

Local factors are the primary stimulus for digestion. They coordinate the responses to changes in the pH of the contents of the lumen, physical distortion of the wall of the digestive tract, or the presence of chemicals—either specific nutrients or chemical messengers released by cells of the mucosa.

Enteroendocrine cells

Carried by bloodstream

Hormones released

③ Hormonal Control Mechanisms

The digestive tract produces at least 18 hormones that affect almost every aspect of digestion, and some of them also affect the activities of other systems. These hormones are peptides produced by **enteroendocrine cells**, endocrine cells in the epithelium of the digestive tract. We will consider these hormones as we proceed down the length of the digestive tract.

Neural Mechanisms

Neural mechanisms control digestive tract movement (②). For example, sensory receptors in the walls of the digestive tract trigger peristaltic movements that are limited to a few centimeters. The motor neurons that control smooth muscle contraction and glandular secretion are located in the myenteric plexus. These neurons are usually considered parasympathetic, because some of them are innervated by parasympathetic preganglionic fibers. However, the plexus also contains sensory neurons, motor neurons, and interneurons responsible for local reflexes that operate entirely outside the control of the central nervous system. As we noted in Chapter 16, local reflexes are called *short reflexes.* ⤶ p. 575

In general, short reflexes control localized activities that involve small segments of the digestive tract. For example, they may coordinate local peristalsis and trigger secretion by digestive glands in response to the arrival of a bolus. Many neurons are involved. The enteric nervous system has about as many neurons as the spinal cord, and as many neurotransmitters as the brain. The specific functions and interactions of these neurotransmitters in the enteric nervous system are unknown.

Sensory information from receptors in the digestive tract is also distributed to the CNS. There it can trigger *long reflexes,* which involve interneurons and motor neurons in the CNS. ⤶ p. 575

Hormonal Mechanisms

Digestive hormones can enhance or diminish the sensitivity of the smooth muscle cells to neural commands (③). These hormones, produced by enteroendocrine cells in the digestive tract, travel through the bloodstream to reach their target organs.

24

1. Identify the organs of the digestive system.
2. List and define the six primary functions of the digestive system.
3. What is the importance of the mesenteries?
4. Name the layers of the gastrointestinal tract beginning at the surface of the lumen.
5. Which is more efficient in propelling intestinal contents from one place to another: peristalsis or segmentation?
6. What effect would a drug that blocks parasympathetic stimulation of the digestive tract have on peristalsis?

See the blue Answers tab at the back of the book.

24-2 ▍ The oral cavity contains the tongue, salivary glands, and teeth, each with specific functions

Learning Outcome Discuss the anatomy of the oral cavity, and list the functions of its major structures and regions.

Let's continue our exploration of the digestive tract by following the path of ingested food. We begin at the **oral cavity** (mouth), or **buccal** (BUK-ul) **cavity** (**Figure 24–6**). The functions of the oral cavity include (1) *sensory analysis* of food before swallowing; (2) *mechanical processing* through the actions of the teeth, tongue, and palatal surfaces; (3) *lubrication* by mixing with mucus and saliva; and (4) limited *digestion* of carbohydrates and lipids.

The oral cavity is lined by the **oral mucosa**, which has a stratified squamous epithelium. A layer of keratinized cells covers regions exposed to severe abrasion, such as the superior surface of the tongue and the opposing surface of the hard palate (part of the roof of the mouth). The epithelial lining of the cheeks, lips, and inferior surface of the tongue is relatively thin and nonkeratinized. Nutrients are not absorbed in the oral cavity, but the mucosa inferior to the tongue is thin enough and vascular enough to permit the rapid absorption of lipid-soluble drugs. *Nitroglycerin* may be administered by this route to treat acute angina pectoris. ⊃ p. 725

The mucosae of the **cheeks**, or lateral walls of the oral cavity, are supported by pads of fat and the buccinator muscles. Anteriorly, the mucosa of each cheek is continuous with that of the lips, or **labia** (LĀ-bē-uh; singular, *labium*). The **vestibule** is the space between the cheeks (or lips) and the teeth. The **gingivae** (JIN-ji-vē), or *gums*, are ridges of oral mucosa that surround the base of each tooth on the alveolar processes of the maxillary bones and mandible. In most regions, the gingivae are firmly bound to the periostea of the underlying bones. The *frenulum of lower lip* is a fold of mucosa that extends from the gingiva to the lower lip, attaching the lip to the gum.

The hard and soft palates form the roof of the oral cavity. The tongue dominates its floor (**Figure 24–6b**). The floor of the mouth inferior to the tongue gets extra support from the

Figure 24–6 The Oral Cavity. ATLAS: Plates 11a; 19

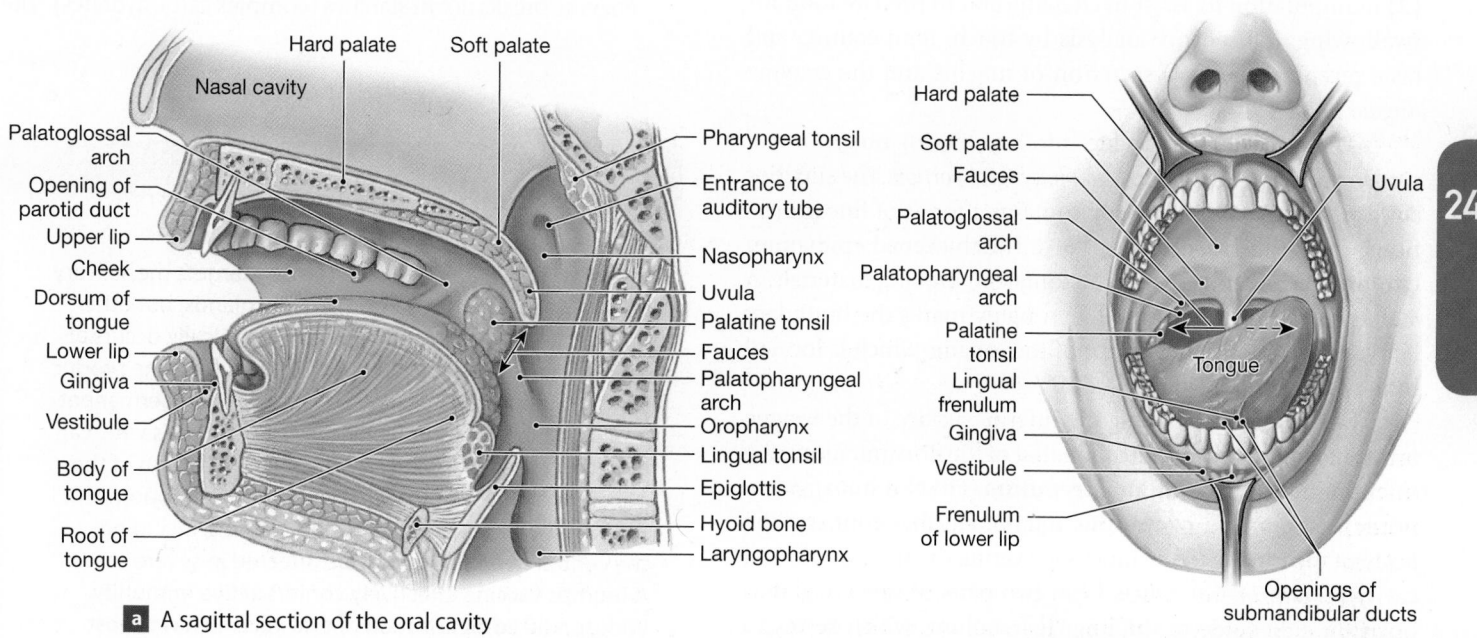

a A sagittal section of the oral cavity

b An anterior view of the oral cavity

geniohyoid and mylohyoid muscles. ↺ p. 375 The palatine processes of the maxillary bones and the horizontal plates of the palatine bones form the *hard palate*. A prominent central ridge, or *raphe* (RĀ-fee), extends along the midline of the hard palate. The mucosa lateral and anterior to the raphe is thick, with complex ridges. When your tongue compresses food against the hard palate, these ridges provide traction. The *soft palate* lies posterior to the hard palate. A thinner and more delicate mucosa covers the posterior margin of the hard palate and extends onto the soft palate.

The posterior margin of the soft palate supports the **uvula** (Ū-vū-luh), a dangling process that helps prevent food from entering the pharynx too soon (**Figure 24–6a**). On either side of the uvula are two pairs of muscular *pharyngeal arches* (**Figure 24–6b**). The more anterior **palatoglossal** (pal-a-tō-GLOS-al) **arch** extends between the soft palate and the base of the tongue. A curving line that connects the palatoglossal arches and uvula forms the boundaries of the **fauces** (FAW-sēz), the arched opening between the soft palate and the base of the tongue. The fauces serve as the passageway between the oral cavity and the oropharynx. The more posterior **palatopharyngeal** (pal-a-tō-fa-RIN-jē-al) **arch** extends from the soft palate to the pharyngeal wall. A palatine tonsil lies between the palatoglossal and palatopharyngeal arches on either side.

The Tongue

The **tongue** (**Figure 24–6**) manipulates food inside the mouth and occasionally brings in foods (such as ice cream on a cone). The primary functions of the tongue are (1) mechanical processing by compression, abrasion, and distortion; (2) manipulation to assist in chewing and to prepare food for swallowing; (3) sensory analysis by touch, temperature, and taste receptors; and (4) secretion of mucins and the enzyme *lingual lipase*.

We can divide the tongue into an anterior **body**, or *oral portion*, and a posterior **root**, or *pharyngeal portion*. The superior surface, or *dorsum*, of the body contains a forest of fine projections, the *lingual papillae*. ↺ p. 594 The thickened epithelium covering each papilla assists the tongue in moving materials. A V-shaped line of vallate papillae roughly marks the boundary between the body and the root of the tongue, which is located in the oropharynx (**Figure 24–6a**).

The epithelium covering the inferior surface of the tongue is thinner and more delicate than that of the dorsum. Along the inferior midline is the **lingual frenulum** (FREN-ū-lum; *frenum*, bridle), a thin fold of mucous membrane that connects the body of the tongue to the mucosa covering the floor of the oral cavity (**Figure 24–6a**). Ducts from two pairs of salivary glands open on either side of the lingual frenulum, which serves to prevent extreme movements of the tongue. However, an overly restrictive lingual frenulum hinders normal eating or speech.

Properly diagnosed, this condition, called *ankyloglossia* (ang-ki-lō-GLOS-ē-uh), can be corrected surgically.

The tongue's epithelium is flushed by the secretions of small glands that extend into the underlying lamina propria. These secretions contain water, mucins, and the enzyme **lingual lipase**. This enzyme works over a broad pH range (3.0–6.0), enabling it to start lipid digestion immediately. Because lingual lipase tolerates an acid environment, it can continue to break down lipids—specifically, triglycerides—for a considerable time after the food reaches the stomach.

The tongue contains two groups of skeletal muscles. The large **extrinsic tongue muscles** perform all gross movements of the tongue. ↺ p. 374 The smaller **intrinsic tongue muscles** change the shape of the tongue and assist the extrinsic muscles during precise movements, as in speech. Both intrinsic and extrinsic tongue muscles are under the control of the hypoglossal cranial nerves (N XII).

Salivary Glands

Three pairs of salivary glands secrete into the oral cavity (**Figure 24–7a**). Each pair has a distinctive cellular organization and produces *saliva*, a mixture of glandular secretions, with slightly different properties:

1. The large **parotid** (pa-ROT-id) **salivary glands** lie inferior to the zygomatic arch deep to the skin covering the lateral and posterior surface of the mandible. Each gland has an irregular shape. It extends from the mastoid process of the temporal bone across the outer surface of the masseter muscle. The parotid salivary glands produce a serous secretion containing large amounts of *salivary amylase*. This enzyme breaks down starches (complex carbohydrates). The

Clinical Note

Mumps The *mumps virus* most often targets the salivary glands, especially the parotid salivary glands, but other organs can also be infected. Infection typically occurs at 5–9 years of age. The first exposure stimulates the production of antibodies and, in most cases, confers permanent immunity. In postadolescent males, the mumps virus can also infect the testes and cause sterility. Infection of the pancreas by the mumps virus can produce temporary or permanent diabetes. Other organ systems, including the central nervous system, are affected in severe cases. A mumps vaccine effectively confers active immunity. Widespread administration of that vaccine has almost eliminated mumps in the United States.

Figure 24–7 **The Salivary Glands.** ATLAS: Plates 3c,d; 18a,b

Parotid duct
Openings of sublingual ducts
Lingual frenulum
Opening of left submandibular duct
Submandibular duct

Salivary Glands
Parotid salivary gland
Sublingual salivary gland
Submandibular salivary gland

a A lateral view, showing the relative positions of the salivary glands and ducts on the left side of the head. For clarity, the left ramus and body of the mandible have been removed. For the positions of the parotid and submandibular ducts in the oral cavity, see *Figure 24–6*.

Duct — Mucous cells — Serous cells

Submandibular salivary gland   LM × 300

b The submandibular gland secretes a mixture of mucins, produced by mucous cells, and enzymes, produced by serous cells.

secretions of each parotid gland are drained by a **parotid duct**, which empties into the vestibule at the second upper molar.

2. The **sublingual** (sub-LING-gwal) **salivary glands** are covered by the mucous membrane of the floor of the mouth. These glands produce a mucous secretion that acts as a buffer and lubricant. Numerous **sublingual ducts** open along either side of the lingual frenulum.

3. The **submandibular salivary glands** lie along the inner surfaces of the mandible within a depression called the *mandibular groove*. Cells of the submandibular glands (**Figure 24–7b**) secrete a mixture of buffers, glycoproteins called *mucins*, and salivary amylase. The **submandibular ducts** open into the mouth on either side of the lingual frenulum immediately posterior to the teeth (**Figure 24–6b**).

Saliva

The salivary glands produce 1.0–1.5 liters of saliva each day. Saliva is 99.4 percent water. The remaining 0.6 percent includes electrolytes (principally Na^+, Cl^-, and HCO_3^-), buffers, glycoproteins, antibodies, enzymes, and waste products. The glycoproteins, called **mucins**, give saliva its lubricating action. About 70 percent of saliva comes from the submandibular salivary glands. Another 25 percent comes from the parotid glands, and 5 percent from the sublingual salivary glands.

Saliva continuously flushes the oral surfaces, helping to keep them clean. Buffers in the saliva keep the pH of your mouth near 7.0. They prevent the buildup of acids produced by bacteria. In addition, saliva contains antibodies (IgA) and *lysozyme*. Both help control populations of oral bacteria. A reduction in or elimination of salivary secretions—caused by radiation, emotional distress, certain drugs, sleep, or other factors—triggers a bacterial population explosion. This proliferation rapidly leads to recurring infections and progressive erosion of the teeth and gums.

The saliva produced when you eat has a variety of functions, including the following:

- Lubricating the mouth.
- Moistening and lubricating food in the mouth.
- Dissolving chemicals that can stimulate the taste buds and provide sensory information about the food.
- Beginning the digestion of complex carbohydrates before the food is swallowed. The enzyme involved is **salivary amylase**, also known as *ptyalin* or *alpha-amylase*. Saliva also contains a small amount of lingual lipase secreted by the glands of the tongue. The digestive process begins in the oral cavity, but it is not completed there. No absorption of nutrients takes place across the lining of the oral cavity.

Control of Salivary Secretions

The autonomic nervous system normally controls salivary secretions. Each salivary gland has parasympathetic and sympathetic innervation. The parasympathetic outflow originates in the **salivatory nuclei** of the medulla oblongata and synapses in the submandibular and otic ganglia. ⊃ pp. 523, 525

24

Any object in your mouth can trigger a salivary reflex. It stimulates receptors monitored by the trigeminal nerve (N V) or taste buds innervated by cranial nerve VII, IX, or X. Parasympathetic stimulation speeds up secretion by all the salivary glands. As a result, you produce large amounts of saliva. The role of sympathetic innervation is unclear. Evidence suggests that it provokes the secretion of small amounts of very thick saliva.

The salivatory nuclei are also influenced by other brain stem nuclei, as well as by the activities of higher centers. For example, chewing with an empty mouth, smelling food, or even thinking about food begins an increase in salivary secretion rates. That is why chewing gum keeps your mouth moist. Irritating stimuli in the esophagus, stomach, or intestines also speed up production of saliva, as does nausea. Increased saliva production in response to unpleasant stimuli helps reduce the stimulus by dilution, by rinsing, or by buffering strong acids or bases.

The Teeth

Tongue movements are important in passing food across the opposing surfaces, or *occlusal surfaces*, of the **teeth**. These surfaces carry out chewing, or **mastication** (mas-ti-KĀ-shun), of food. Mastication breaks down tough connective tissues in meat and the plant fibers in vegetables. It also helps saturate the food with salivary secretions and enzymes.

Figure 24–8   Teeth.

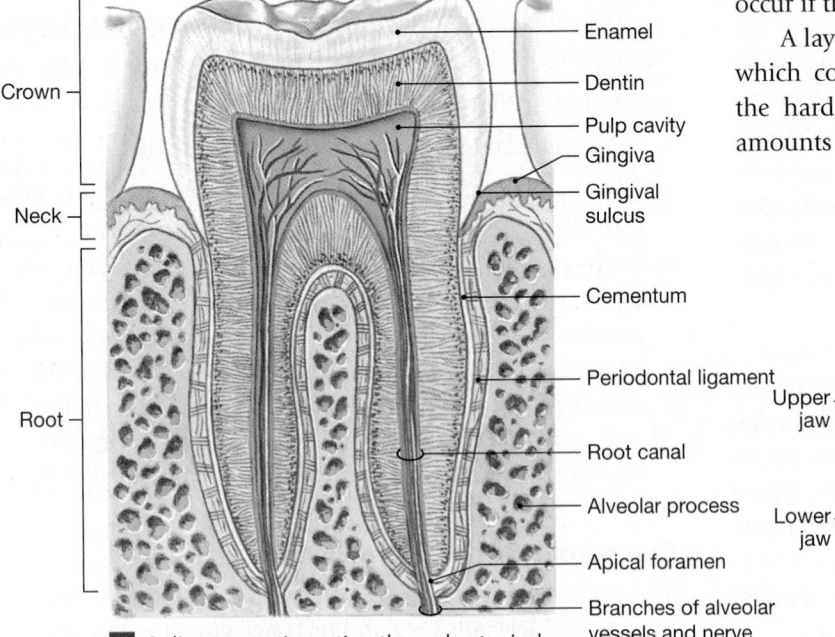

a A diagrammatic section through a typical adult tooth.

Figure 24–8a is a sectional view through an adult tooth. The bulk of each tooth consists of a mineralized matrix similar to that of bone. This material, called **dentin**, differs from bone in that it does not contain cells. Instead, cytoplasmic processes extend into the dentin from cells in the central **pulp cavity**, an interior chamber. The pulp cavity receives blood vessels and nerves through the **root canal**, a narrow tunnel located at the **root**, or base, of the tooth. Blood vessels and nerves enter the root canal through an opening called the **apical foramen** to supply the pulp cavity.

The root of each tooth sits in a bony cavity or socket called the *tooth socket* or an *alveolus*. Collagen fibers of the **periodontal ligament** extend from the dentin of the root to the alveolar bone, creating a strong articulation known as a gomphosis. ⤺ p. 291 A layer of **cementum** (se-MEN-tum) covers the dentin of the root. Cementum protects and firmly anchors the periodontal ligament. Cementum is histologically similar to bone and is less resistant to erosion than is dentin.

The **neck** of the tooth marks the boundary between the root and the **crown**, the exposed portion of the tooth that projects beyond the soft tissue of the gingiva. A shallow groove called the **gingival sulcus** surrounds the neck of each tooth. The mucosa of the gingival sulcus is very thin and is not tightly bound to the periosteum. The epithelium is bound to the tooth at the base of the sulcus. This epithelial attachment prevents bacterial access to the lamina propria of the gingiva and the relatively soft cementum of the root. When you brush and massage your gums, you stimulate the epithelial cells and strengthen the attachment. A condition called *gingivitis*, an inflammation of the gingivae caused by bacterial infection, can occur if the attachment breaks down.

A layer of **enamel** covers the dentin of the crown. Enamel, which contains calcium phosphate in a crystalline form, is the hardest biologically manufactured substance. Adequate amounts of calcium, phosphates, and vitamin D_3 during

b The adult teeth from the right side of the upper and lower jaws. *Figure 24–9a,b* provides a view of the occlusal surfaces.

childhood are essential if the enamel coating is to be complete and resistant to decay.

Tooth decay generally results from the action of bacteria that live in your mouth. Bacteria adhering to the surfaces of the teeth produce a sticky matrix that traps food particles and creates deposits known as *dental plaque*. Over time, this organic material can become calcified, forming a hard layer of *tartar*, or *dental calculus*, which can be difficult to remove. Tartar deposits most commonly develop at or near the gingival sulcus, where brushing cannot remove the soft plaque deposits.

Types of Teeth

The alveolar processes of the maxillae and the mandible form the *maxillary* and *mandibular arcades*, or upper and lower dental arches, respectively. These arcades contain four types of teeth, each with specific functions (**Figure 24–8b**):

1. **Incisors** (in-SĪ-zerz) are blade-shaped teeth located at the front of the mouth. Incisors are useful for clipping or cutting, as when you nip off the tip of a carrot stick. These teeth have a single root.

2. The **cuspids** (KUS-pidz), or *canines*, are conical, with a sharp ridgeline and a pointed tip. Also known as the "eyeteeth," because they lie directly under the eye, the cuspids are used for tearing or slashing. You might weaken a tough piece of celery using the clipping action of the incisors and then take advantage of the shearing action of the cuspids. Cuspids have a single root.

3. **Bicuspids** (bī-KUS-pidz), or **premolars**, have flattened crowns with prominent ridges. They crush, mash, and grind. Bicuspids have one or two roots.

4. **Molars** have very large, flattened crowns with prominent ridges adapted for crushing and grinding. You can usually shift a tough nut to your bicuspids and molars for successful crunching. Molars in the upper jaw typically have three roots while those in the lower jaw usually have two roots.

Dental Succession

Two sets of teeth form during development. The first to appear are the **deciduous teeth** (de-SID-ū-us; *deciduus*, falling off), the temporary teeth of the **primary dentition**. Deciduous teeth are also called *primary teeth, milk teeth,* or *baby teeth*. Most children have 20 deciduous teeth (**Figure 24–9a**). On each side of the upper or lower jaw, the primary dentition consists of two incisors, one cuspid, and a pair of deciduous molars. These teeth are later replaced by the **secondary dentition**, or *permanent dentition* of the larger adult jaws (**Figure 24–9b**). Three additional molars appear on each side of the upper and lower jaws

as the person ages. These molars extend the rows of teeth posteriorly and bring the permanent tooth count to 32.

TIPS&TRICKS

Here's how to remember the timing of dental succession: The deciduous (primary) teeth are present during the primary grades, and the secondary (permanent) teeth are present during and after the secondary grades.

As replacement proceeds, the periodontal ligaments and roots of the primary teeth erode. The deciduous teeth either fall out or are pushed aside by the **eruption**, or emergence, of the secondary teeth (**Figure 24–9c**). The adult premolars take the place of the deciduous molars. The adult molars extend the tooth rows as the jaw enlarges.

The third molars, or *wisdom teeth*, may not erupt before age 21. Wisdom teeth may not erupt because they develop in inappropriate positions or because there is not enough space. Any teeth that develop in locations that do not allow them to erupt are called *impacted teeth*. Impacted wisdom teeth can be surgically removed to prevent the formation of abscesses.

Mastication

The *muscles of mastication* close your jaws and slide or rock your lower jaw from side to side. ↺ p. 370 Chewing is not a simple process. It can involve any combination of mandibular elevation/depression, protraction/retraction, and medial/lateral movement. (Try classifying the movements involved the next time you eat.)

During mastication, you force food from the oral cavity to the vestibule and back, crossing and recrossing the **occlusal** (biting) **surfaces**. This movement results in part from the action of the muscles of mastication. Control would be impossible, however, without help from the muscles of the cheeks, lips, and tongue. Once you have shredded or torn the food to a satisfactory consistency and have moistened it with salivary secretions, your tongue compacts the debris into a moist, cohesive bolus that is fairly easy to swallow.

CHECKPOINT

7. Name the structures associated with the oral cavity.

8. Which type of epithelium lines the oral cavity?

9. The digestion of which nutrient would be affected by damage to the parotid salivary glands?

10. Which type of tooth is most useful for chopping off bits of rigid food?

11. Where are the fauces located?

See the blue Answers tab at the back of the book.

Figure 24–9 **Primary and Secondary Dentitions.**

Central incisors (7.5 mo)
Lateral incisor (9 mo)
Cuspid (18 mo)
Deciduous 1st molar (14 mo)
Deciduous 2nd molar (24 mo)

Deciduous 2nd molar (20 mo)
Deciduous 1st molar (12 mo)
Cuspid (16 mo)
Lateral incisor (7 mo)
Central incisors (6 mo)

a The primary teeth, with the age at eruption given in months

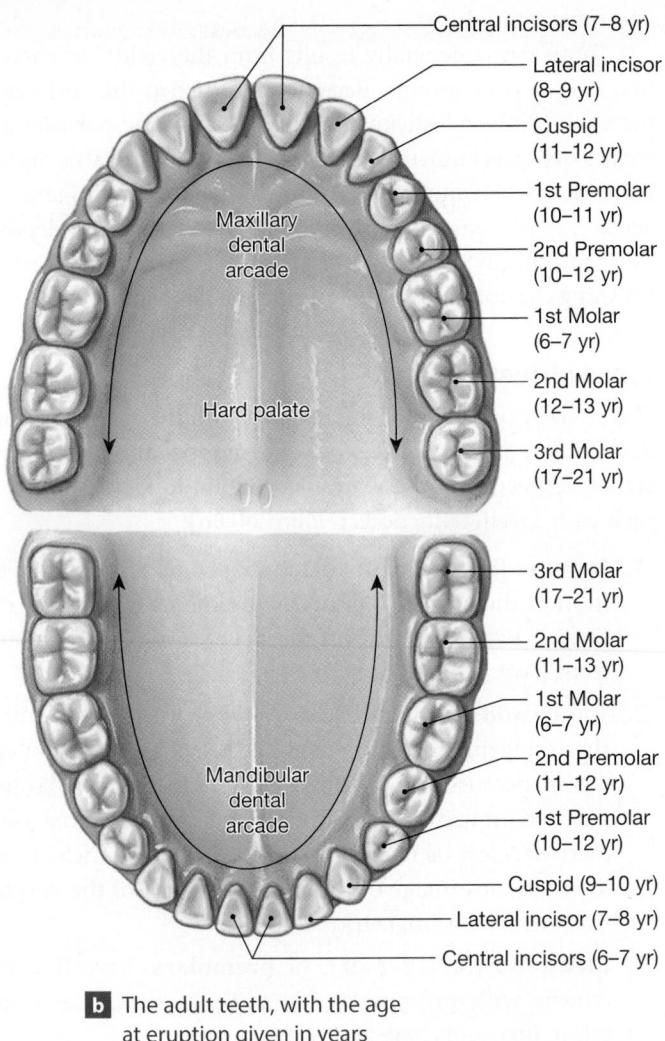

Central incisors (7–8 yr)
Lateral incisor (8–9 yr)
Cuspid (11–12 yr)
1st Premolar (10–11 yr)
2nd Premolar (10–12 yr)
1st Molar (6–7 yr)
2nd Molar (12–13 yr)
3rd Molar (17–21 yr)

Maxillary dental arcade

Hard palate

3rd Molar (17–21 yr)
2nd Molar (11–13 yr)
1st Molar (6–7 yr)
2nd Premolar (11–12 yr)
1st Premolar (10–12 yr)
Cuspid (9–10 yr)
Lateral incisor (7–8 yr)
Central incisors (6–7 yr)

Mandibular dental arcade

b The adult teeth, with the age at eruption given in years

Maxilla exposed to show developing permanent teeth
Erupted deciduous teeth
First and second molars
Mandible exposed to show developing permanent teeth

c Maxilla and mandible with unerupted teeth exposed

24-3 ▮ The pharynx is a passageway between the oral cavity and esophagus

Learning Outcome Describe the structure and functions of the pharynx.

The **pharynx** (FAR-ingks), or throat, is an anatomical space that serves as a common passageway for solid food, liquids, and air. We described the epithelial lining and regions of the pharynx—nasopharynx, oropharynx, and laryngopharynx—in Chapter 23. ⟲ p. 863 Food normally passes through the oropharynx and laryngopharynx on its way to the esophagus. These regions have a stratified squamous epithelium similar to that of the oral cavity. The lamina propria contains scattered mucous glands and the lymphatic tissue of the pharyngeal, palatal, and lingual tonsils. Deep to the lamina propria lies a dense layer of elastic fibers, bound to the underlying skeletal muscles.

We described the specific pharyngeal muscles involved in swallowing in Chapter 11. ⟲ p. 374

▪ The *pharyngeal constrictor muscles* push the bolus toward and into the esophagus.

▪ The *palatopharyngeus* and *stylopharyngeus muscles* elevate the larynx.

▪ The *palatal muscles* elevate the soft palate and adjacent portions of the pharyngeal wall.

These muscles work with muscles of the oral cavity and esophagus to start swallowing, which pushes the bolus along the esophagus and into the stomach.

CHECKPOINT

12. Describe the structure and function of the pharynx.

13. Identify the muscles associated with the pharynx.

See the blue Answers tab at the back of the book.

24-4 ┃ The esophagus is a muscular tube that transports solids and liquids from the pharynx to the stomach

Learning Outcome Describe the structure and functions of the esophagus.

The **esophagus** (**Figure 24–10**) is a hollow muscular tube that conveys solid food and liquids to the stomach. Its length is approximately 25 cm (10 in.) and its diameter is about 2 cm (0.80 in.) at its widest point.

The esophagus begins posterior to the cricoid cartilage, at the level of vertebra C_6. It is narrowest at this point. The esophagus descends toward the thoracic cavity posterior to the trachea, continuing inferiorly along the posterior wall of the mediastinum. It then enters the abdominopelvic cavity through the **esophageal hiatus** (hī-Ā-tus), an opening in the diaphragm. The esophagus empties into the stomach anterior to vertebra T_7.

The esophagus is innervated by parasympathetic and sympathetic fibers from the esophageal plexus. ↺ p. 574 Resting muscle tone in the circular muscle layer in the superior 3 cm (1.2 in.) of the esophagus normally prevents air from entering

the esophagus. A comparable zone at the inferior end of the esophagus normally remains in a state of active contraction. This state prevents the backflow (reflux) of materials from the stomach into the esophagus. Neither region has a well-defined sphincter muscle. Nevertheless, we often use the terms *upper esophageal sphincter* and *lower esophageal sphincter (cardiac sphincter)* to describe these regions, which are similar in function to other sphincters.

TIPS & TRICKS

Just as security gates control the passage of people by opening and closing, sphincters control the passage of material through them by dilating and constricting.

Histology of the Esophagus

The wall of the esophagus contains mucosal, submucosal, and muscularis layers comparable to those shown in **Figure 24–3**. Distinctive features of the esophageal wall include the following (**Figure 24–10**):

- The mucosa of the esophagus contains a nonkeratinized, stratified squamous epithelium similar to that of the pharynx and oral cavity.

- The mucosa and submucosa are packed into large folds that extend the length of the esophagus. These folds allow for expansion during the passage of a large bolus. Muscle tone in the walls keeps the lumen closed, except when you swallow.

Figure 24–10 **The Esophagus.**

Stratified squamous epithelium

Lamina propria

Muscularis mucosae

Esophageal mucosa LM × 275

Muscularis mucosae

Mucosa

Submucosa

Muscularis externa

Adventitia

a A transverse section through an empty esophagus (LM × 5).

b This light micrograph illustrates the extreme thickness of the epithelial portion of the esophageal mucosal layer.

- The muscularis mucosae consists of an irregular layer of smooth muscle.

- The submucosa contains scattered *esophageal glands*. They produce mucus that reduces friction between the bolus and the esophageal lining.

- The muscularis externa has the usual inner circular and outer longitudinal layers. However, in the superior third of the esophagus, these layers contain skeletal muscle fibers. The middle third contains a mixture of skeletal and smooth muscle tissue. Along the inferior third, only smooth muscle occurs.

- There is no serosa, but an adventitia of connective tissue outside the muscularis externa anchors the esophagus to the posterior body wall. Over the 1–2 cm (0.4–0.8 in.) between the diaphragm and stomach, the esophagus is retroperitoneal. Peritoneum covers the anterior and left lateral surfaces.

Swallowing

Swallowing, or **deglutition** (dē-glū-TISH-un), is a complex process that can be initiated voluntarily but proceeds automatically once it begins. You take conscious control over swallowing when you eat or drink, but swallowing is also controlled at the subconscious level. The **swallowing reflex** begins when tactile receptors on the palatal arches and uvula are stimulated by the passage of the bolus. The information is relayed to the **swallowing center** of the medulla oblongata over the trigeminal (N V) and glossopharyngeal (N IX) cranial nerves. Motor commands from this center then signal the pharyngeal musculature, producing a coordinated and stereotyped pattern of muscle contraction. It takes less than a second for the pharyngeal muscles to propel the bolus into the esophagus. During this period, the respiratory centers are inhibited and breathing stops. Swallowing takes place at regular intervals as saliva collects at the back of the mouth. Each day you swallow approximately 2400 times.

We divide swallowing into buccal, pharyngeal, and esophageal phases, detailed in **Figure 24–11**.

Primary peristaltic waves are peristaltic movements coordinated by afferent and efferent fibers in the glossopharyngeal (N IX) and vagus (N X) nerves. For a typical bolus, the entire trip along the esophagus takes about 9 seconds. Liquids may make the journey in a few seconds, flowing ahead of the peristaltic contractions with the assistance of gravity.

A dry or poorly lubricated bolus travels much more slowly. A series of *secondary peristaltic waves* may be required to push it all the way to the stomach. Secondary peristaltic waves are local reflexes triggered by the stimulation of sensory receptors in the esophageal walls.

Figure 24–11 The Swallowing Process. This sequence, based on a series of x-rays, shows the phases of swallowing and the movement of a bolus from the mouth to the stomach.

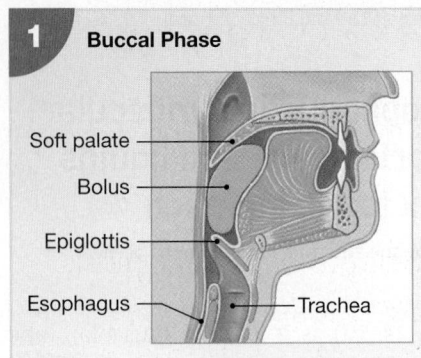

1 Buccal Phase

Soft palate
Bolus
Epiglottis
Esophagus — Trachea

The **buccal phase** begins with the compression of the bolus against the hard palate. Subsequent retraction of the tongue then forces the bolus into the oropharynx and assists in the elevation of the soft palate, thereby sealing off the nasopharynx. Once the bolus enters the oropharynx, reflex responses begin and the bolus is moved toward the stomach.

2 Pharyngeal Phase

Tongue
Bolus

The **pharyngeal phase** begins as the bolus comes into contact with the palatoglossal and palatopharyngeal arches and the posterior pharyngeal wall. Elevation of the larynx and folding of the epiglottis direct the bolus past the closed glottis. At the same time, the uvula and soft palate block passage back to the nasopharynx.

3 Esophageal Phase

Peristalsis in esophagus
Trachea

The **esophageal phase** begins as the contraction of pharyngeal muscles forces the bolus through the entrance to the esophagus. Once in the esophagus, the bolus is pushed toward the stomach by peristalsis.

4 Bolus Enters Stomach

Thoracic cavity
Lower esophageal sphincter
Stomach

The approach of the bolus triggers the opening of the lower esophageal sphincter. The bolus then continues into the stomach.

CHECKPOINT

14. Name the structure connecting the pharynx to the stomach.

15. Compared to other segments of the digestive tract, what is unusual about the muscularis externa of the esophagus?

16. What is occurring when the soft palate and larynx elevate and the glottis closes?

See the blue Answers tab at the back of the book.

24-5 ▌ The stomach is a J-shaped organ that receives the bolus from the esophagus and aids in chemical and mechanical digestion

Learning Outcome Describe the anatomy of the stomach, including its histology, and discuss its roles in digestion and absorption.

The **stomach** has four major functions: (1) storage of ingested food; (2) mechanical breakdown of ingested food; (3) disruption of chemical bonds in food through the action of acid and enzymes; and (4) production of *intrinsic factor*, a glycoprotein needed for the absorption of vitamin B_{12} in the small intestine. Ingested substances combine with the secretions of the glands of the stomach, producing a viscous, highly acidic, soupy mixture of partially digested food called **chyme** (KĪM).

Anatomy of the Stomach

The stomach has the shape of an expanded J (**Figure 24–12**). A short **lesser curvature** forms the medial surface of the organ, and a long **greater curvature** forms the lateral surface. The anterior and posterior surfaces are smoothly rounded. The shape and size of the stomach can vary greatly from individual to individual and even from one meal to the next. In an "average" stomach, the lesser curvature is approximately 10 cm (4 in.) long, and the greater curvature measures about 40 cm (16 in.).

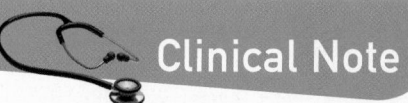

Clinical Note

Gastritis and Peptic Ulcers A superficial inflammation of the gastric mucosa is called *gastritis* (gas-TRĪ-tis). The condition can develop after a person has swallowed drugs, including beverage alcohol and aspirin. Gastritis is also associated with smoking, severe emotional or physical stress, bacterial infection of the gastric wall, or ingestion of strongly acidic or alkaline chemicals. Over time, gastritis can lead to the erosion of the stomach or duodenal lining, and a *peptic ulcer* may form.

The stomach typically extends between the levels of vertebrae T_7 and L_3.

We divide the stomach into four regions (**Figure 24–12**):

1. *The Cardia.* The **cardia** (KAR-dē-uh) is the smallest part of the stomach. It consists of the superior, medial portion of the stomach within 3 cm (1.2 in.) of the junction between the stomach and the esophagus. The cardia contains abundant mucous glands. Their secretions coat the connection with the esophagus and help protect that tube from the acid and enzymes of the stomach.

2. *The Fundus.* The **fundus** (FUN-dus) is the portion of the stomach that is superior to the junction between the stomach and the esophagus. The fundus contacts the inferior, posterior surface of the diaphragm (**Figure 24–12a**).

3. *The Body.* The area of the stomach between the fundus and the curve of the J is the **body**, the largest region of the stomach. The body acts as a mixing tank for ingested food and secretions produced in the stomach. *Gastric* (*gaster*, stomach) *glands* in the fundus and body secrete most of the acid and enzymes involved in gastric digestion.

4. *The Pylorus.* The **pylorus** (pī-LOR-us) forms the sharp curve of the J. The pylorus is divided into a **pyloric antrum** (*antron*, cavity), which is connected to the body, and a **pyloric canal**, which empties into the *duodenum*, the proximal segment of the small intestine. As mixing movements take place during digestion, the pylorus frequently changes shape. A muscular **pyloric sphincter** regulates the release of chyme into the duodenum. Glands in the pylorus secrete mucus and important digestive hormones, including *gastrin*, a hormone that stimulates gastric glands.

TIPS&TRICKS

The stomach squeezes chyme into the small intestine just as you squeeze cake frosting out of a pastry bag.

The stomach's volume increases while you eat and then decreases as chyme enters the small intestine. When the stomach is relaxed (empty), the mucosa has prominent folds called **rugae** (ROO-gē; wrinkles). These temporary features let the gastric lumen expand (**Figure 24–12b**). The stomach can stretch up to 50 times its empty size. As the stomach fills, the rugae gradually flatten out until, at maximum distension, they almost disappear. (The world record, set in 2013, for eating hot dogs with buns in 10 minutes is 69!) When empty, the stomach resembles a muscular tube with a narrow, constricted lumen. When full, it can contain 1–1.5 liters of material.

The muscularis mucosae and muscularis externa of the stomach contain extra layers of smooth muscle cells in addition to the usual circular and longitudinal layers. The muscularis mucosae generally contain an outer, circular layer of muscle cells.

24

Figure 24–12 **The Stomach.** ATLAS: Plates 49a–c; 50a–c

Esophagus

Diaphragm

Left gastric artery

Vagus nerve (N X)

Fundus

Cardia

Spleen

Greater curvature with greater omentum attached

Greater omentum

Liver, right lobe

Liver, left lobe

Body

Lesser curvature

Common hepatic artery

Gallbladder

Bile duct

Pyloric sphincter

Pylorus

a The position and external appearance of the stomach, showing superficial landmarks

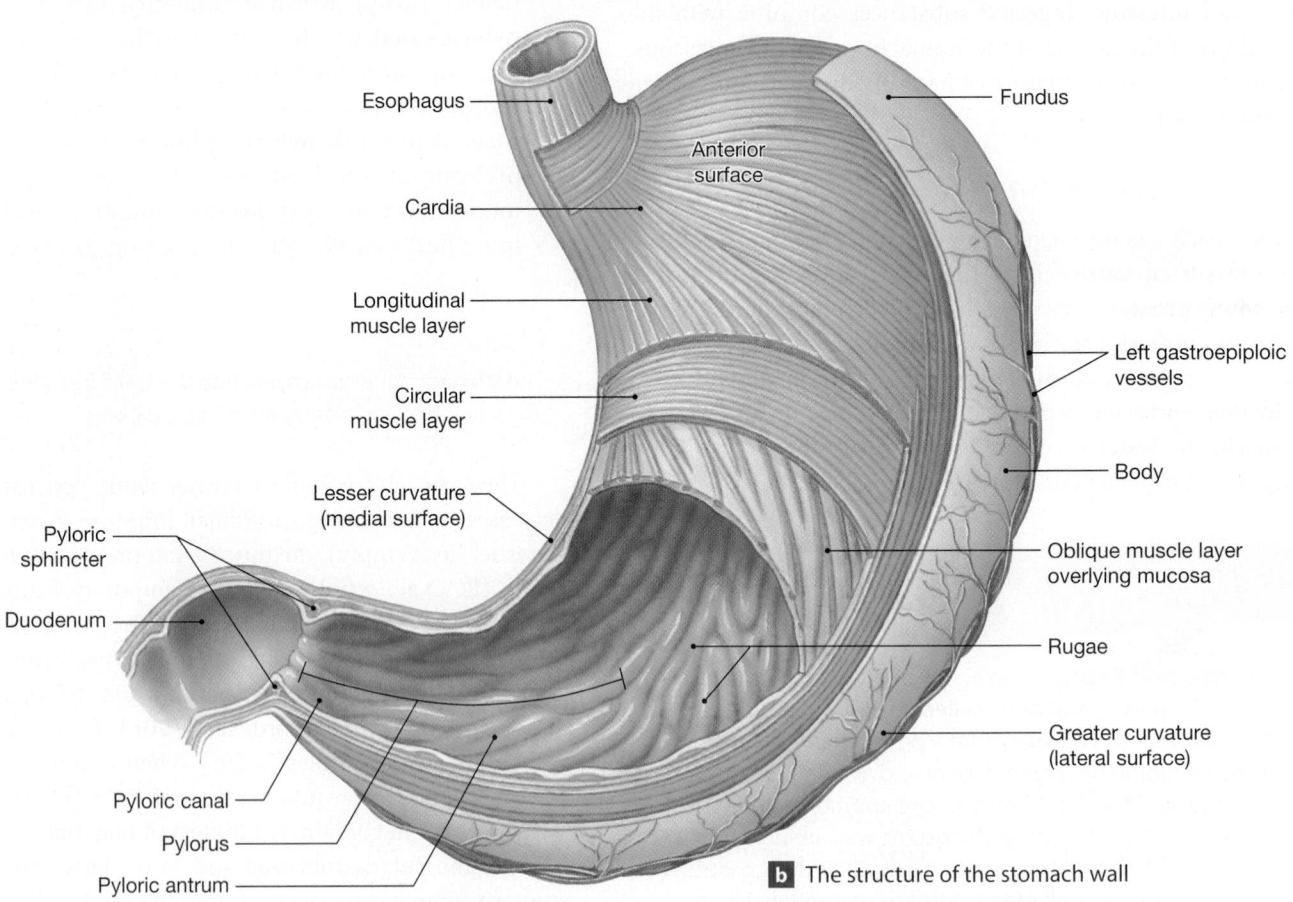

Esophagus

Cardia

Longitudinal muscle layer

Circular muscle layer

Lesser curvature (medial surface)

Pyloric sphincter

Duodenum

Pyloric canal

Pylorus

Pyloric antrum

Fundus

Anterior surface

Left gastroepiploic vessels

Body

Oblique muscle layer overlying mucosa

Rugae

Greater curvature (lateral surface)

b The structure of the stomach wall

The muscularis externa has an inner, **oblique layer** of smooth muscle (**Figure 24–12b**). The extra layers of smooth muscle strengthen the stomach wall and assist in the mixing and churning essential to the formation of chyme.

TIPS & TRICKS

Both circular folds and rugae allow a large surface area to fit within a small volume, just as crumpling a sheet of paper rearranges its surface area so that it occupies a smaller space. When the stomach fills, its rugae allow its volume to expand, so that the stomach's lining can hold a larger volume of chyme.

Histology of the Stomach

Simple columnar epithelium lines all portions of the stomach (**Figure 24–13a**). The epithelium is a *secretory sheet*, which produces a carpet of mucus that covers the interior surface of the stomach. The alkaline mucous layer protects epithelial cells against the acid and enzymes in the gastric lumen.

Shallow depressions called **gastric pits** open onto the gastric surface (**Figure 24–13b**). The mucous cells at the base, or *neck*, of each gastric pit actively divide, replacing superficial cells that are shed into the chyme. A typical gastric epithelial cell has a life span of three to seven days. Exposure to alcohol or other chemicals that damage or kill epithelial cells increases cell turnover.

Gastric Glands

In the fundus and body of the stomach, each gastric pit communicates with several **gastric glands**, which extend deep into the underlying lamina propria (**Figure 24–13b**). Gastric glands are dominated by two types of secretory cells: *parietal cells* and *chief cells*. Together, they secrete about 1500 mL (1.6 qt.) of **gastric juice** each day.

Parietal cells are especially common along the proximal portions of each gastric gland (**Figure 24–13b**). These cells secrete **intrinsic factor**. This glycoprotein helps absorb **vitamin B$_{12}$** across the intestinal lining. (Recall from Chapter 19 that this vitamin is essential for normal erythropoiesis.) ⊃ p. 689

Figure 24–13   The Stomach Lining.

Layers of the Stomach Wall

Mucosa
- Gastric pit (opening to gastric gland)
- Mucous epithelium
- Lamina propria
- Muscularis mucosae

Submucosa

Muscularis externa
- Oblique muscle
- Circular muscle
- Longitudinal muscle

Serosa

- Artery and vein
- Lymphatic vessel
- Myenteric plexus

a Stomach wall

- Gastric pit
- Gastric gland

- Lamina propria
- Mucous cells
- Neck

Cells of Gastric Glands
- Parietal cells
- G cell
- Chief cells
- Smooth muscle cell

b Gastric gland

24

Figure 24–14   **The Secretion of Hydrochloric Acid.**

Parietal cells also secrete *hydrochloric acid* (HCl). They do not produce HCl in the cytoplasm, however. This acid is so strong that it would erode a secretory vesicle and destroy the cell. Instead, H$^+$ and Cl$^-$, the two ions that form HCl, are transported independently by different mechanisms (**Figure 24–14**).

The initial step in HCl production is the formation of carbonic acid within parietal cells. The dissociation of carbonic acid releases bicarbonate and hydrogen ions. The H$^+$ are actively transported into the lumen of the gastric gland. The bicarbonate ions are exchanged for chloride ions from the interstitial fluid. When gastric glands are actively secreting, enough bicarbonate ions diffuse into the bloodstream from the interstitial fluid to increase the pH of the blood significantly. This sudden influx of bicarbonate ions has been called the *alkaline tide*.

The secretions of the parietal cells can keep the stomach contents at pH 1.5–2.0. Although this highly acidic environment does not by itself digest chyme, it has four important functions:

1. The acidity of gastric juice kills most of the microorganisms ingested with food.

2. The acidity denatures proteins and inactivates most of the enzymes in food.

3. The acidity helps break down plant cell walls and the connective tissues in meat.

4. An acidic environment is essential for the activation and function of *pepsin*, a protein-digesting enzyme secreted by chief cells.

Chief cells, also called *zymogenic cells*, are most abundant near the base of a gastric gland (**Figure 24–13b**). These cells secrete **pepsinogen** (pep-SIN-ō-jen), an inactive proenzyme. Acid in the gastric lumen converts pepsinogen to **pepsin**, an active *proteolytic*, or protein-digesting, enzyme. Pepsin functions most effectively at a strongly acidic pH of 1.5–2.0. In addition, the stomachs of newborn infants (but not of adults) produce **rennin**, also known as *chymosin*, and **gastric lipase**. These enzymes are important for the digestion of milk. Rennin coagulates milk proteins. Gastric lipase initiates the digestion of milk fats.

Pyloric Glands

Glands in the pylorus produce primarily a mucous secretion, rather than enzymes or acid. In addition, several types of enteroendocrine cells are scattered among the mucus-secreting cells. These enteroendocrine cells produce at least seven hormones, most notably **gastrin** (GAS-trin). Gastrin is produced by *G cells*, which are most abundant in the gastric pits of the pyloric antrum. Gastrin stimulates secretion by both parietal and chief cells, as well as contractions of the gastric wall that mix and stir the gastric contents.

The pyloric glands also contain *D cells*, which release **somatostatin**, a hormone that inhibits the release of gastrin. D cells continuously release their secretions into the interstitial fluid adjacent to the G cells. Neural and hormonal stimuli can override this inhibition of gastrin production when the stomach is preparing for digestion or is already engaged in digestion.

Several other hormones play a role in hunger and satiety (a feeling of not being hungry). Levels of *ghrelin*, a hormone produced by *P/D1 cells* lining the fundic region of the stomach, increase before meals to initiate hunger. Ghrelin levels decrease shortly after eating to curb appetite. Ghrelin is also antagonistic to *leptin*, a hormone derived from fat tissue that induces satiety. Another hormone from the stomach and small intestine, *obestatin*, is thought to decrease appetite and inhibit thirst. The same gene encodes both ghrelin and obestatin.

Regulation of Gastric Activity

The production of acid and enzymes by the gastric mucosa can be (1) controlled by the CNS; (2) regulated by short reflexes of the enteric nervous system, coordinated in the wall of the stomach; and (3) regulated by hormones of the digestive tract. Gastric control proceeds in three overlapping phases. They are named according to the location of the control center: the *cephalic phase*, the *gastric phase*, and the *intestinal phase* (**Spotlight Figure 24–15**).

Digestion and Absorption in the Stomach

The stomach carries out preliminary digestion of proteins by pepsin. For a variable period, it allows the digestion of carbohydrates and lipids by salivary amylase and lingual lipase. These enzymes continue to work until the pH throughout the contents of the stomach falls below 4.5. They generally remain active one to two hours after a meal.

As the stomach contents become more fluid and the pH approaches 2.0, pepsin activity increases. Protein disassembly begins. Protein digestion is not completed in the stomach, because time is limited and pepsin attacks only specific types of peptide bonds, not all of them. However, pepsin generally has enough time to break down complex proteins into smaller peptide and polypeptide chains before the chyme enters the duodenum.

Although digestion takes place in the stomach, nutrients are not absorbed there. This is because: (1) The epithelial cells are covered by a blanket of alkaline mucus and are not directly exposed to chyme; (2) the epithelial cells lack the specialized transport mechanisms of cells that line the small intestine; (3) the gastric lining is somewhat impermeable to water; and (4) digestion has not been completed by the time chyme leaves the stomach. At this stage, most carbohydrates, lipids, and proteins are only partially broken down.

Some drugs can be absorbed in the stomach. For example, ethyl alcohol can diffuse through the mucous barrier and penetrate the lipid membranes of the epithelial cells. As a result, beverage alcohol is absorbed in your stomach before any nutrients in a meal reach the bloodstream. Meals containing large amounts of fat slow the rate of alcohol absorption. Why? The reason is that alcohol is lipid soluble, and some of it will be dissolved in fat droplets in the chyme. Aspirin

is another lipid-soluble drug that can enter the bloodstream across the gastric mucosa. Such drugs alter the properties of the mucous layer and can promote epithelial damage by stomach acid and enzymes. Prolonged use of aspirin can cause gastric bleeding, so individuals with stomach ulcers usually avoid aspirin.

> **CHECKPOINT**
>
> 17. Name the four major regions of the stomach.
>
> 18. Discuss the significance of the low pH in the stomach.
>
> 19. How does a large meal affect the pH of blood leaving the stomach?
>
> 20. When a person suffers from chronic gastric ulcers, the branches of the vagus nerves that supply the stomach are sometimes cut in an attempt to provide relief. Why might this be an effective treatment?
>
> See the blue Answers tab at the back of the book.

24-6 ┃ The small intestine digests and absorbs nutrients, and associated glandular organs assist with digestive processes

Learning Outcome Describe the anatomy and histology of the small intestine, explain the functions and regulation of intestinal secretions, and describe the structure, functions, and regulation of the accessory digestive organs.

The stomach is a holding tank in which food is saturated with gastric juices and exposed to stomach acid and the digestive effects of pepsin. Most of the important digestive and absorptive steps of digestion take place in the small intestine, where chemical digestion is completed and the products of digestion are absorbed. The mucosa of the small intestine produces only a few of the enzymes involved. The pancreas provides digestive enzymes, as well as buffers that help neutralize chyme. The liver secretes *bile*, a solution stored in the gallbladder for discharge into the small intestine. Bile contains buffers and *bile salts*, compounds that facilitate the digestion and absorption of lipids.

The Small Intestine

The **small intestine** plays the key role in the digestion and absorption of nutrients. Ninety percent of nutrient absorption takes place in the small intestine. Most of the rest occurs in the large intestine. The small intestine averages 6 m (19.7 ft) in length (range: 4.5–7.5 m; 14.8–24.6 ft). Its diameter ranges from 4 cm (1.6 in.) at the stomach to about 2.5 cm (1 in.) at the junction with the large intestine. It occupies all abdominal regions except the right and left hypochondriac and epigastric regions (see **Figure 1–6b**, p. 42). The small intestine has

24

The duodenum plays a key role in controlling digestive function because it monitors the contents of the chyme and adjusts the activities of the stomach and accessory glands to protect the delicate absorptive surfaces of the jejunum. This pivotal role of the duodenum is apparent when you consider the three phases of gastric secretion. The phases are named according to the location of the control center involved.

1 CEPHALIC PHASE

The **cephalic phase** of gastric secretion begins when you see, smell, taste, or think of food. This phase, which is directed by the CNS, prepares the stomach to receive food. The neural output proceeds by way of the parasympathetic division of the autonomic nervous system. The vagus nerves innervate the submucosal plexus of the stomach. Next, postganglionic parasympathetic fibers innervate mucous cells, chief cells, parietal cells, and G cells of the stomach. In response to stimulation, the production of gastric juice speeds up, reaching rates of about 500 mL/h, or about 2 cups per hour. This phase generally lasts only minutes.

Sight, smell, taste, or thoughts of food

KEY
→ Stimulation
→ Stimulation

Emotional states can exaggerate or inhibit the cephalic phase. For example, anger or hostility leads to excessive gastric secretion. On the other hand, anxiety, stress, or fear decreases gastric secretion and gastric contractions, or *motility*.

2 GASTRIC PHASE

The **gastric phase** begins with the arrival of food in the stomach and builds on the stimulation provided during the cephalic phase. This phase may continue for three to four hours while the acid and enzymes process the ingested materials. The stimuli that initiate the gastric phase are (1) distension of the stomach, (2) an increase in the pH of the gastric contents, and (3) the presence of undigested materials in the stomach, especially proteins and peptides. The gastric phase consists of the following mechanisms:

Local Response

Distension of the gastric wall stimulates the release of histamine in the lamina propria, which binds to receptors on the parietal cells and stimulates acid secretion.

Neural Response

The stimulation of stretch receptors and chemoreceptors triggers short reflexes coordinated in the submucosal and myenteric plexuses. This in turn activates the stomach's secretory cells. The stimulation of the myenteric plexus produces powerful contractions called **mixing waves** in the muscularis externa.

ATLAS: Plate 50c

Hormonal Response

Neural stimulation and the presence of peptides and amino acids in chyme stimulate the secretion of the hormone gastrin, primarily by G cells. Gastrin travels in the bloodstream to parietal and chief cells, whose increased secretions reduce the pH of the gastric juice. In addition, gastrin also stimulates gastric motility.

INTESTINAL PHASE

The **intestinal phase** of gastric secretion begins when chyme first enters the small intestine. The function of the intestinal phase is to control the rate of gastric emptying to ensure that the secretory, digestive, and absorptive functions of the small intestine can proceed with reasonable efficiency. Although here we consider the intestinal phase as it affects stomach activity, the arrival of chyme in the small intestine also triggers other neural and hormonal events that coordinate the activities of the intestinal tract, pancreas, liver, and gallbladder.

Neural Responses

Chyme leaving the stomach decreases the distension in the stomach, thereby reducing the stimulation of stretch receptors. Distension of the duodenum by chyme stimulates stretch receptors and chemoreceptors that trigger the **enterogastric reflex**. This reflex inhibits both gastrin production and gastric contractions and stimulates the contraction of the pyloric sphincter, which prevents further discharge of chyme. At the same time, local reflexes at the duodenum stimulate mucus production, which helps protect the duodenal lining from the arriving acid and enzymes.

Hormonal Responses

The arrival of chyme in the duodenum triggers hormonal responses:
- Arrival of lipids and carbohydrates stimulates the secretion of cholecystokinin (CCK) and gastric inhibitory peptide (GIP).
- A decrease in pH below 4.5 stimulates the secretion of secretin.
- Partially digested proteins in the duodenum stimulate G cells that secrete gastrin, which circulates to the stomach and speeds gastric processing.

CENTRAL REFLEXES

Two central reflexes, the gastroenteric reflex and gastroileal reflex, are also triggered by the stimulation of stretch receptors in the stomach wall as it fills. These reflexes accelerate movement along the small intestine while the enterogastric reflex controls the rate of chyme entry into the duodenum.

Central Gastric Reflexes

The **gastroenteric reflex** stimulates motility and secretion along the entire small intestine.

The **gastroileal** (gas-trō-IL-ē-al) **reflex** triggers the opening of the ileocecal valve, allowing materials to pass from the small intestine into the large intestine.

The **ileocecal valve** controls the passage of materials into the large intestine.

In general, the rate of movement of chyme into the small intestine is highest when the stomach is greatly distended and the meal contains little protein. A large meal containing small amounts of protein, large amounts of carbohydrates (such as rice or pasta), wine (alcohol), or after-dinner coffee (caffeine) will leave your stomach very quickly. One reason is that both alcohol and caffeine stimulate gastric secretion and motility.

The *vomiting* reflex occurs in response to irritation of the fauces, pharynx, esophagus, stomach, or proximal segment of the small intestine. These sensations are relayed to the vomiting center of the medulla oblongata, which coordinates motor responses. In preparation for vomiting, the pyloric sphincter relaxes and the contents of the duodenum are discharged back into the stomach by strong peristaltic waves that travel toward the stomach. Vomiting, or **emesis** (EM-e-sis), then occurs as the gastroesophageal sphincter relaxes and the stomach regurgitates its contents through the esophagus and pharynx and out through the mouth.

Figure 24–16   **Segments of the Intestine.**   ATLAS: Plates 49a,b,d; 51a,b

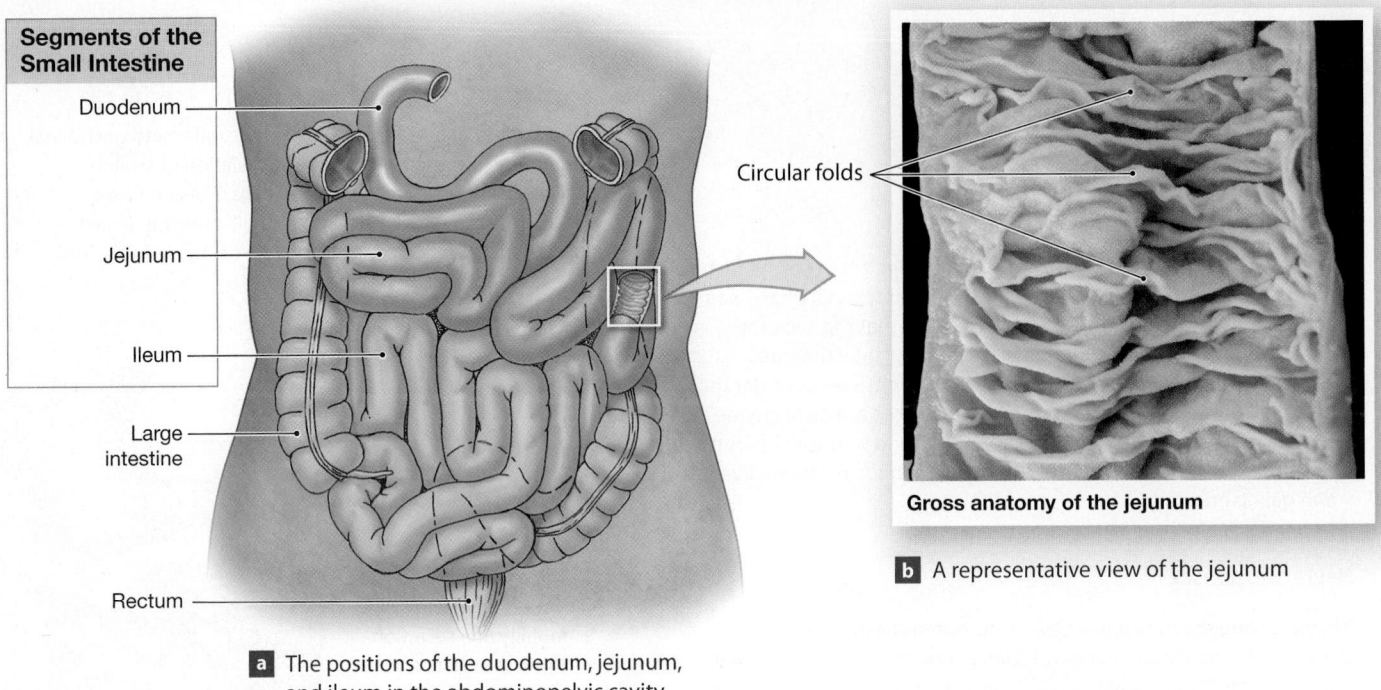

Segments of the Small Intestine

Duodenum

Jejunum

Ileum

Large intestine

Rectum

Circular folds

Gross anatomy of the jejunum

b A representative view of the jejunum

a The positions of the duodenum, jejunum, and ileum in the abdominopelvic cavity

three segments: the duodenum, the jejunum, and the ileum (**Figure 24–16a**).

The **duodenum** (dū-ō-DĒ-num), 25 cm (10 in.) in length, is the segment closest to the stomach. This portion of the small intestine is a "mixing bowl." It receives chyme from the stomach and digestive secretions from the pancreas and liver. From its connection with the stomach, the duodenum curves in a C that encloses the pancreas. Except for the proximal 2.5 cm (1 in.), the duodenum is in a retroperitoneal position between vertebrae L_1 and L_4 (**Figure 24–2d**).

A rather abrupt bend marks the boundary between the duodenum and the **jejunum** (je-JOO-num). At this junction, the small intestine reenters the peritoneal cavity, supported by a sheet of mesentery. The jejunum is about 2.5 meters (8.2 ft) long. The bulk of chemical digestion and nutrient absorption occurs there.

The **ileum** (IL-ē-um), the final segment of the small intestine, is also the longest. It averages 3.5 meters (11.5 ft) in length. The ileum ends at the **ileocecal** (il-ē-ō-SĒ-kal) **valve**. This sphincter controls the flow of material from the ileum into the *cecum* of the large intestine.

TIPS & TRICKS

To remember the order of the small intestine segments, beginning at the stomach, use this mnemonic: **D**on't **j**ump **i**n—**d**uodenum, **j**ejunum, and **i**leum. Also, do not confuse ileum, the last segment of the small intestine, with ilium, which is a bone.

The small intestine fills much of the peritoneal cavity. Its position is stabilized by the mesentery proper, a broad mesentery attached to the posterior body wall (**Figure 24–2c,d**). The stomach, large intestine, abdominal wall, and pelvic girdle restrict movement of the small intestine during digestion. Blood vessels, lymphatic vessels, and nerves reach the segments of the small intestine within the connective tissue of the mesentery. The primary blood vessels involved are branches of the superior mesenteric artery and the superior mesenteric vein. ↪ pp. 788, 798

The segments of the small intestine—the duodenum, jejunum, and ileum—are distinguished by both histological specialization and primary function.

Histology of the Small Intestine

The intestinal lining has a series of transverse folds called **circular folds** or *plicae circulares* (**Figure 24–16b**). Unlike the rugae in the stomach, the circular folds are permanent features. They do not disappear when the small intestine fills. The small intestine contains about 800 circular folds—roughly 2 per centimeter. They greatly increase the surface area available for absorption.

Intestinal Villi

The mucosa of the small intestine has a series of fingerlike projections, the **intestinal villi** (**Figure 24–17a,b**). The villi are covered by simple columnar epithelium that is carpeted with

Figure 24–17 **The Intestinal Wall.** ATLAS: Plates 51a–d

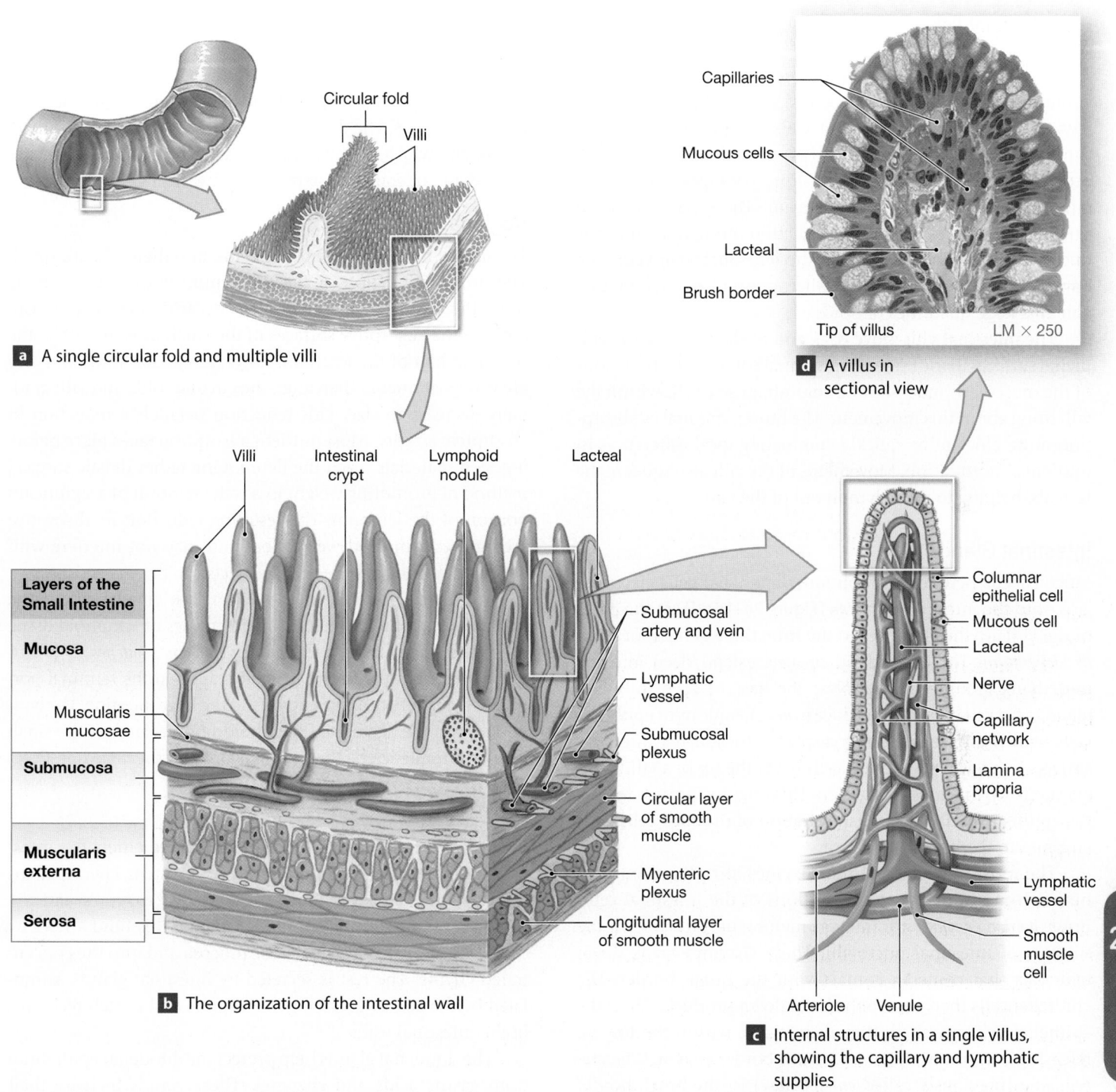

a A single circular fold and multiple villi

d A villus in sectional view

Capillaries

Mucous cells

Lacteal

Brush border

Tip of villus　LM × 250

b The organization of the intestinal wall

Villi　Intestinal crypt　Lymphoid nodule　Lacteal

Layers of the Small Intestine

Mucosa

Muscularis mucosae

Submucosa

Muscularis externa

Serosa

Submucosal artery and vein

Lymphatic vessel

Submucosal plexus

Circular layer of smooth muscle

Myenteric plexus

Longitudinal layer of smooth muscle

c Internal structures in a single villus, showing the capillary and lymphatic supplies

Columnar epithelial cell

Mucous cell

Lacteal

Nerve

Capillary network

Lamina propria

Lymphatic vessel

Smooth muscle cell

Arteriole　Venule

microvilli. The cells are said to have a *brush border* because the microvilli project from the epithelium like the bristles on a brush (**Figure 24–17d**).

If the small intestine were a simple tube with smooth walls, it would have a total absorptive area of about 3300 cm^2 (3.6 ft^2). Instead, the mucosa contains circular folds. Each circular fold supports a forest of villi, and epithelial cells bearing

microvilli cover each villus. This arrangement increases the total area for absorption by a factor of more than 600, to approximately 2 million cm^2 (more than 2200 ft^2). This area is nearly the size of 10 lanes in a bowling alley.

The lamina propria of each villus contains an extensive network of capillaries that originate in a vascular network within the submucosa (**Figure 24–17c**). These capillaries carry absorbed nutrients to the hepatic portal circulation for delivery to the liver. The liver then adjusts the nutrient concentrations of blood before the blood reaches the general systemic circulation.

In addition to capillaries and nerve endings, each villus contains a lymphatic capillary called a **lacteal** (LAK-tē-ul; *lacteus*, milky) (**Figure 24–17b,c**). Lacteals transport materials that cannot enter blood capillaries. For example, absorbed fatty acids are assembled into protein–lipid packages that are too large to diffuse into the bloodstream. These packets, called *chylomicrons*, reach the venous circulation through the thoracic duct, which delivers lymph into the left subclavian vein. The name *lacteal* refers to the pale, milky appearance of lymph that contains large quantities of lipids.

The intestinal villi move back and forth, exposing the epithelial surfaces to the liquefied intestinal contents. Contractions of the muscularis mucosae and smooth muscle cells within the villi bring about this movement. The movement makes absorption more efficient by quickly eliminating local differences in nutrient concentration. Movements of the villi also squeeze the lacteals, helping to move lymph out of the villi.

Intestinal Glands

Mucous cells between the columnar epithelial cells eject mucins onto the intestinal surfaces (**Figure 24–17c,d**). At the bases of the villi are the entrances to the **intestinal glands**, or *crypts of Lieberkühn*. These glandular pockets extend deep into the underlying lamina propria. Near the base of each intestinal gland, stem cell divisions produce new generations of epithelial cells, which are continuously displaced toward the intestinal surface. In a few days the new cells reach the tip of a villus and are shed into the intestinal lumen. This ongoing process renews the epithelial surface. The disintegration of the shed cells adds enzymes to the lumen.

The most important of the enzymes introduced into the lumen come from the apical portions of the intestinal cells. *Brush border enzymes* are integral membrane proteins on the surfaces of intestinal microvilli. These enzymes break down materials that come in contact with the brush border. The epithelial cells then absorb the breakdown products. Once the epithelial cells are shed, they disintegrate within the lumen, releasing both intracellular and brush border enzymes. *Enteropeptidase* (previously called *enterokinase*) is one brush border enzyme that enters the lumen in this way. It does not directly participate in digestion. Instead, it activates a key pancreatic proenzyme, trypsinogen. (We consider the functions of enteropeptidase and other brush border enzymes in a later section.) Intestinal glands also contain enteroendocrine cells that produce several intestinal hormones, including gastrin, cholecystokinin, and secretin.

The duodenum has numerous mucous glands, both in the epithelium and deep to it. In addition to intestinal glands, its submucosa contains **duodenal glands**, also called *submucosal glands* or *Brunner's glands*. Duodenal glands produce copious quantities of mucus when chyme arrives from the stomach. The mucus protects the epithelium from the acidity of chyme and also contains bicarbonate ions that help raise the pH of the chyme. As chyme travels the length of the duodenum, its pH increases from 1–2 to 7–8. The duodenal glands also secrete the hormone *urogastrone*, which inhibits gastric acid production and stimulates the division of epithelial stem cells along the digestive tract.

Regional Specializations

The duodenum has few circular folds, and their villi are small. The primary function of the duodenum is to receive chyme from the stomach and neutralize its acids before they can damage the absorptive surfaces of the small intestine. Over the proximal half of the jejunum, however, circular folds and villi are very prominent. Thereafter, the circular folds and villi gradually decrease in size. This reduction parallels a reduction in absorptive activity: Most nutrient absorption takes place before ingested materials reach the ileum. One rather drastic surgical method of promoting weight loss is the removal of a significant portion of the jejunum. The resulting reduction in absorptive area causes a marked weight loss and may not interfere with adequate nutrition, but the side effects can be very serious.

The distal portions of the ileum lack circular folds. The lamina propria there contains 20–30 masses of lymphoid tissue called aggregated lymphoid nodules, or *Peyer's patches*. p. 816 These lymphoid tissues are most abundant in the terminal portion of the ileum, near the entrance to the large intestine. The lymphocytes in the aggregated lymphoid nodules protect the small intestine from bacteria that normally inhabit the large intestine.

Intestinal Secretions

Roughly 1.8 liters of watery **intestinal juice** enters the intestinal lumen each day. Intestinal juice moistens chyme, helps buffer acids, and keeps both the digestive enzymes and the products of digestion in solution. Much of this fluid arrives by osmosis, as water flows out of the mucosa and into the concentrated chyme. The rest is secreted by intestinal glands, stimulated by the activation of touch receptors and stretch receptors in the intestinal walls.

The duodenal glands help protect the duodenal epithelium from gastric acids and enzymes. These glands increase their secretion in response to (1) local reflexes, (2) the release of the hormone *enterocrinin* by enteroendocrine cells of the duodenum, and (3) parasympathetic stimulation through the vagus nerves. The first two mechanisms operate only after chyme arrives in the duodenum. However, the duodenal glands begin secreting during the cephalic phase of gastric secretion, long before chyme reaches the pyloric sphincter. They do so because vagus nerve activity triggers their secretion. Thus, the duodenal lining has protection in advance.

Sympathetic stimulation inhibits the duodenal glands, leaving the duodenal lining unprepared for the arrival of chyme. This effect probably explains why chronic stress or other factors that promote sympathetic activation can cause duodenal ulcers.

Intestinal Movements

After chyme has arrived in the duodenum, weak peristaltic contractions move it slowly toward the jejunum. The contractions are myenteric reflexes that are not under CNS control. Their effects are limited to within a few centimeters of the site of the original stimulus. Motor neurons in the submucosal and myenteric plexuses control these short reflexes. In addition, some of the smooth muscle cells contract periodically, even without stimulation, establishing a basic contractile rhythm that then spreads from cell to cell.

The stimulation of the parasympathetic system increases the sensitivity of the weak myenteric reflexes and speeds up both local peristalsis and segmentation. More elaborate reflexes coordinate activities along the entire length of the small intestine. The gastroenteric and gastroileal reflexes speed up movement along the small intestine (**Spotlight Figure 24–15**). This effect is the opposite from that of the enterogastric reflex.

Hormones released by the digestive tract can enhance or suppress reflexes. For example, the gastroileal reflex is triggered by stretch receptor stimulation. However, the degree of ileocecal valve relaxation is enhanced by gastrin, which is secreted in large quantities when food enters the stomach.

The Pancreas

The **pancreas** lies posterior to the stomach. It extends laterally from the duodenum toward the spleen. The pancreas is an elongate, pinkish-gray organ about 15 cm (6 in.) long and weighing about 80 g (3 oz) (**Figure 24–18a**). The broad **head** of the pancreas lies within the loop formed by the duodenum as it leaves the pylorus. The slender **body** of the pancreas extends toward the spleen, and the **tail** is short and bluntly rounded. The pancreas is retroperitoneal and is firmly bound to the posterior wall of the abdominal cavity. The surface of the pancreas has a lumpy, lobular texture. A thin, transparent capsule of connective tissue wraps the entire organ. The pancreatic lobules, associated blood vessels, and excretory ducts are visible through the anterior capsule and the overlying layer of peritoneum. Arterial blood reaches the pancreas by way of branches of the splenic, superior mesenteric, and common hepatic arteries. The pancreatic arteries and pancreaticoduodenal arteries are the major branches from these vessels. The splenic vein and its branches drain the pancreas.

The pancreas is primarily an exocrine organ. It produces digestive enzymes and buffers. The large **pancreatic duct** *(duct of Wirsung)* delivers these secretions to the duodenum. (In 3–10 percent of the population, a small **accessory pancreatic duct** [*Santorini duct*] branches from the pancreatic duct.) The pancreatic duct extends within the attached mesentery to reach the duodenum, where it meets the *common bile duct* from the liver and gallbladder (**Figure 24–21b**). The two ducts then empty into the *duodenal ampulla (ampulla of Vater)*, a chamber located roughly halfway along the length of the duodenum. When present, the accessory pancreatic duct usually empties into the duodenum independently, outside the duodenal ampulla.

Histological Organization

Partitions of connective tissue divide the interior of the pancreas into distinct lobules. The blood vessels and tributaries of the pancreatic ducts are situated within these connective tissue septa (**Figure 24–18b**). The pancreas is an example of a *compound tubuloalveolar gland*, a structure described in Chapter 4. � p. 151 In each lobule, the ducts branch repeatedly before ending in blind pockets called **pancreatic acini** (AS-i-nī). Each pancreatic acinus is lined with simple cuboidal epithelium. *Pancreatic islets*, the endocrine tissues of the pancreas, are scattered among the acini (**Figure 24–18b,c**). The islets account for only about 1 percent of the cell population of the pancreas.

The pancreas has two distinct functions, one endocrine and the other exocrine. The endocrine cells of the pancreatic islets secrete insulin and glucagon into the bloodstream to control blood sugar. The exocrine cells include the acinar cells and the epithelial cells that line the duct system. Together, these exocrine cells secrete **pancreatic juice**—an alkaline mixture of digestive enzymes, water, and ions—into the small intestine. Acinar cells secrete pancreatic enzymes, which do most of the digestive work in the small intestine. Pancreatic enzymes break

Clinical Note

Pancreatitis *Pancreatitis* (pan-krē-a-TĪ-tis) is an inflammation of the pancreas. The condition is extremely painful. The factors that may produce it include a blockage of the excretory ducts, bacterial or viral infections, ischemia, and drug reactions—especially those involving alcohol. These stimuli provoke a crisis by injuring exocrine cells in at least a portion of the organ. Then lysosomes in the damaged cells activate the proenzymes, and autolysis begins. The proteolytic enzymes digest the surrounding, undamaged cells, activating their enzymes and starting a chain reaction. In most cases, only a portion of the pancreas is affected. The condition usually subsides in a few days. In 10–15 percent of cases, the process does not subside. The enzymes can then ultimately destroy the pancreas. If the islet cells are damaged, diabetes mellitus may result. ↺ p. 663

Figure 24–18   **The Pancreas.**   ATLAS: Plates 54d; 55; 57a

Pancreatic duct

Connective tissue septum

Exocrine cells in pancreatic acini

Endocrine cells in pancreatic islet

b Diagram of the cellular organization of the pancreas.

Common bile duct

Pancreatic duct

Lobules

Tail of pancreas

Body of pancreas

Accessory pancreatic duct

Head of pancreas

Duodenal papilla

Duodenum

Duct

Pancreatic acini (exocrine)

Pancreatic islet (endocrine)

Pancreas   LM × 75

a The gross anatomy of the pancreas. The head of the pancreas is tucked in to a C-shaped curve of the duodenum that begins at the pylorus of the stomach.

c Light micrograph of the cellular organization of the pancreas.

down ingested materials into small molecules suitable for absorption. The water and ions, secreted primarily by the cells lining the pancreatic ducts, help dilute and neutralize acid in the chyme.

Physiology of the Pancreas

Each day, the pancreas secretes about 1000 mL (1 qt) of pancreatic juice. Hormones from the duodenum control these secretory activities. When chyme arrives in the duodenum, secretin is released. This hormone triggers the pancreatic secretion of a watery buffer solution with a pH of 7.5–8.8. Among its other components, the secretion contains bicarbonate and phosphate buffers that help raise the pH of the chyme.

Another duodenal hormone, cholecystokinin, stimulates the production and secretion of pancreatic enzymes. Stimulation by the vagus nerves also increases the secretion of

pancreatic enzymes. Recall that this stimulation takes place during the cephalic phase of gastric regulation, so the pancreas starts to synthesize enzymes before food even reaches the stomach. This head start is important, because enzyme synthesis takes much longer than the production of buffers. By starting early, the pancreatic cells are ready to meet the demand when chyme arrives in the duodenum.

The pancreatic enzymes include the following:

- **Pancreatic alpha-amylase**, a **carbohydrase** (kar-bō-HĪ-drās)—an enzyme that breaks down certain starches. Pancreatic alpha-amylase is almost identical to salivary amylase.

- **Pancreatic lipase**, which breaks down certain complex lipids, releasing products (such as fatty acids) that can be easily absorbed.

- **Nucleases**, which break down RNA or DNA.

- **Proteolytic enzymes**, which break apart certain proteins. These enzymes include **proteases**, which break apart large protein complexes, and **peptidases**, which break small peptide chains into individual amino acids.

Proteolytic enzymes account for about 70 percent of total pancreatic enzyme production. These enzymes are secreted as inactive proenzymes. They are activated only after they reach the small intestine. Proenzymes discussed earlier include pepsinogen, angiotensinogen, plasminogen, fibrinogen, and many of the clotting factors and enzymes of the complement system. ⊃ pp. 666, 681, 705, 825 As in the stomach, the release of a proenzyme rather than an active enzyme protects the secretory cells from the destructive effects of their own products. Among the proenzymes secreted by the pancreas are **trypsinogen** (trip-SIN-ō-jen), **chymotrypsinogen** (kī-mo-trip-SIN-ō-jen) **procarboxypeptidase** (prō-kar-bok-sē-PEP-ti-dās), and **pro-elastase** (pro-ē-LAS-tās).

Inside the duodenum, enteropeptidase in the brush border and the lumen triggers the conversion of trypsinogen to **trypsin**, an active protease. Trypsin then activates the other proenzymes, producing **chymotrypsin**, **carboxypeptidase**, and **elastase**. Each enzyme attacks peptide bonds linking specific amino acids and ignores others. Together, these enzymes break down proteins into a mixture of dipeptides, tripeptides, and amino acids.

The Liver

The **liver** is the largest visceral organ. It is one of our most versatile organs and the center for metabolic regulation in the body. Most of its mass lies in the right hypochondriac and epigastric regions, but it may extend into the left hypochondriac and umbilical regions as well. The liver weighs about 1.5 kg (3.3 lb). This large, firm, reddish-brown organ performs essential metabolic and synthetic functions.

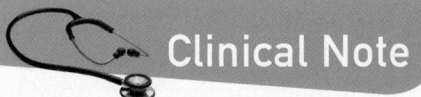

Clinical Note

Cirrhosis Any condition that severely damages the liver is a serious threat to life. The liver has a limited ability to regenerate itself after injury. Even so, liver function does not fully recover unless the normal vascular pattern is restored. Examples of important types of liver disease include **cirrhosis**, which is characterized by the replacement of lobules by fibrous tissue, and various forms of *hepatitis* caused by viral infections. In some cases, liver transplants are used to treat liver failure. Unfortunately, the supply of suitable donor tissue is limited. The success rate is highest in young, otherwise healthy individuals. Clinical trials are now under way to test an artificial liver known as *ELAD* (*e*xtracorporeal *l*iver *a*ssist *d*evice). It may prove beneficial for the long-term support of individuals with chronic liver disease.

Anatomy of the Liver

The liver is wrapped in a tough fibrous capsule and is covered by a layer of visceral peritoneum. On the anterior surface, the **falciform ligament** marks the division between the organ's left and right lobes (Figure 24–19a,b). A thickening in the posterior margin of the falciform ligament is the **round ligament**, or *ligamentum teres*. This fibrous band marks the path of the fetal umbilical vein.

On the posterior surface of the liver, the impression left by the inferior vena cava marks the division between the right lobe and the small **caudate** (KAW-dāt) **lobe** (Figure 24–19a,c). Inferior to the caudate lobe lies the **quadrate lobe**, sandwiched between the left lobe and the gallbladder. Afferent blood vessels and other structures reach the liver by traveling within the connective tissue of the lesser omentum. They converge at a region called the **porta hepatis** ("doorway to the liver").

We discussed the circulation to the liver in Chapter 21 and summarized it in Figures 21–24 and 21–31, pp. 788, 797. Nearly one-third of the blood supply to the liver is arterial blood from the hepatic artery proper. The rest is venous blood from the hepatic portal vein, which begins in the capillaries of the esophagus, stomach, small intestine, and most of the large intestine. Liver cells, called **hepatocytes** (HEP-a-tō-sīts), adjust circulating levels of nutrients through selective absorption and secretion. The blood leaving the liver returns to the systemic circuit through the hepatic veins. These veins open into the inferior vena cava.

Histological Organization of the Liver

Connective tissue divides each lobe of the liver into approximately 100,000 **liver lobules**, the basic functional units of the liver. The histological organization and structure of a typical liver lobule are shown in Figure 24–20.

Each lobule is roughly 1 mm in diameter. Adjacent lobules are separated by an *interlobular septum*. The hepatocytes in a liver lobule form a series of irregular plates arranged like the spokes of a wheel (Figure 24–20a,b). The plates are only one cell thick. Exposed hepatocyte surfaces are covered with short microvilli. Within a lobule, sinusoids between adjacent plates empty into the **central vein**. (We introduced sinusoids in Chapter 21. ⊃ p. 757) The liver sinusoids lack a basement membrane, so large openings between the endothelial cells allow solutes—even those as large as plasma proteins—to pass out of the bloodstream and into the spaces surrounding the hepatocytes.

The lining of the sinusoids contains typical endothelial cells and a large number of **Kupffer** (KOOP-fer) **cells**, also known as *stellate reticuloendothelial cells*. ⊃ p. 822 These phagocytic cells are part of the monocyte–macrophage system. They engulf pathogens, cell debris, and damaged blood cells. Kupffer cells also store iron, some lipids, and heavy metals (such as tin or mercury) that are absorbed by the digestive tract.

24

Figure 24–19   **The Anatomy of the Liver.**   ATLAS: Plates 49a,b,e; 54a–c; 57a,b

Liver

Falciform ligament

Porta hepatis

Right lobe of liver

Caudate lobe of liver

Inferior vena cava

Pleural cavity

Cut edge of diaphragm

Sternum

Left lobe of liver

Stomach

Lesser omentum

Aorta

Spleen

a A transverse section through the superior abdomen (diagrammatic view)

Coronary ligament

Right lobe

Left lobe

Falciform ligament

Round ligament

Gallbladder

b The anterior surface of the liver

Left hepatic vein

Inferior vena cava

Caudate lobe

Left lobe

Coronary ligament

Porta Hepatis

Hepatic portal vein

Hepatic artery proper

Common bile duct

Quadrate lobe

Gallbladder

Right lobe

c The posterior surface of the liver

Blood enters the liver sinusoids from small branches of the hepatic portal vein and hepatic artery proper. A typical liver lobule has a hexagonal shape in cross section (**Figure 24–20a**). There are six **portal areas**, or *portal triads*, one at each corner of the lobule. A portal area contains three structures: (1) a branch of the hepatic portal vein, (2) a branch of the hepatic artery proper, and (3) a small branch of the bile duct (**Figure 24–20a–c**).

Branches from the arteries and veins deliver blood to the sinusoids of adjacent liver lobules (**Figure 24–20a,b**). As blood flows through the sinusoids, hepatocytes absorb solutes from the plasma and secrete materials such as plasma proteins. Blood then leaves the sinusoids and enters the central vein of the lobule. The central veins ultimately merge to form the hepatic veins, which then empty into the inferior vena cava. Liver diseases, such as the various forms of *hepatitis*, and conditions such as alcoholism, can lead to degenerative changes in the liver tissue and constriction of the circulatory supply.

Pressures in the hepatic portal system are usually low, averaging 10 mm Hg or less. This pressure can increase markedly, however, if blood flow through the liver is restricted by a blood clot or damage to the organ. Such a rise in portal pressure is called *portal hypertension*. As pressures rise, small peripheral veins and capillaries in the portal system become distended. If they rupture, extensive bleeding can take place. Portal hypertension can also force fluid into the peritoneal cavity across the serosal surfaces of the liver and viscera, producing ascites.

The Bile Duct System

The liver secretes a fluid called **bile** into a network of narrow channels between the opposing membranes of adjacent liver cells. These passageways, called **bile canaliculi**, extend outward, away from the central vein (**Figure 24–20b**). Eventually, they connect with fine **bile ductules** (DUK-tūlz), which carry

24

Figure 24–20   **Liver Histology.**

a A diagrammatic view of liver structure, showing relationships among lobules

1 mm

Interlobular septum   Bile duct   Branch of hepatic portal vein   Portal area   Bile ductules

Branch of hepatic artery proper

Branch of hepatic portal vein (containing blood)

Hepatocytes

Sinusoids

Bile duct

Portal area   LM × 320

c A sectional view showing the vessels and ducts within a portal area

Sinusoid

Hepatocytes

Central vein

Kupffer cells

Bile canaliculi

Portal Area
Bile duct
Branch of hepatic portal vein
Branch of hepatic artery proper

b A single liver lobule and its cellular components

24

bile to the bile ducts in the nearest portal area (**Figure 24–20a**). The **right** and **left hepatic ducts** (**Figure 24–21a,c**) collect bile from all the bile ducts of the liver lobes. These ducts unite to form the **common hepatic duct**, which leaves the liver. The bile in the common hepatic duct either flows into the *common bile duct*, which empties into the duodenal ampulla, or enters the *cystic duct*, which leads to the gallbladder (**Figure 24–21a,b,c**).

The **common bile duct** is formed by the union of the **cystic duct** and the common hepatic duct. The common bile duct passes within the lesser omentum toward the stomach, turns,

and penetrates the wall of the duodenum to meet the pancreatic duct at the duodenal ampulla (**Figure 24–21b**).

The Physiology of the Liver

The liver carries out more than 200 functions. They fall into three general categories: (1) *metabolic regulation*, (2) *hematological regulation*, and (3) *bile production*. In this discussion we provide a general overview.

Metabolic Regulation. The liver is the primary organ involved in regulating the composition of circulating blood. All blood

Figure 24–21   **The Anatomy and Physiology of the Gallbladder and Bile Ducts.**   ATLAS: Plates 49c,e; 51a; 54b–d

a A view of the inferior surface of the liver, showing the position of the gallbladder and ducts that transport bile from the liver to the gallbladder and duodenum. A portion of the lesser omentum has been cut away.

b A sectional view through a portion of the duodenal wall, showing the duodenal ampulla and related structures.

c A radiograph (cholangiogram, anterior-posterior view) of the gallbladder, biliary ducts, and pancreatic ducts.

1 The liver secretes bile continuously— about 1 liter per day.

2 Bile becomes more concentrated the longer it remains in the gallbladder.

3 The release of CCK by the duodenum triggers dilation of the hepatopancreatic sphincter and contraction of the gallbladder. This ejects bile into the duodenum through the duodenal ampulla.

4 In the lumen of the digestive tract, bile salts break the lipid droplets apart by emulsification.

d Physiology of the gallbladder.

leaving the absorptive surfaces of the digestive tract enters the hepatic portal system and flows into the liver. Liver cells extract nutrients or toxins from the blood before it reaches the systemic circulation through the hepatic veins. The liver removes and stores excess nutrients. It corrects nutrient deficiencies by

mobilizing stored reserves or performing synthetic activities. The liver's regulatory activities affect the following:

- *Carbohydrate Metabolism.* The liver stabilizes blood glucose levels at about 90 mg/dL. If blood glucose levels decrease, hepatocytes break down glycogen reserves and release glucose into the bloodstream. They also synthesize glucose from other carbohydrates or from available amino acids. The synthesis of glucose from other compounds is called *gluconeogenesis.* If blood glucose levels climb, hepatocytes remove glucose from the bloodstream. They either store it as glycogen or use it to synthesize lipids that can be stored in the liver or other tissues. Circulating hormones, such as insulin and glucagon, regulate these metabolic activities. ⟲ pp. 662–663

- *Lipid Metabolism.* The liver regulates circulating levels of triglycerides, fatty acids, and cholesterol. When those levels decline, the liver breaks down its lipid reserves and releases the breakdown products into the bloodstream. When the levels are high, the lipids are removed for storage. However, this regulation takes place only after lipid levels have risen within the general circulation, because most lipids absorbed by the digestive tract bypass the hepatic portal circulation.

- *Amino Acid Metabolism.* The liver removes excess amino acids from the bloodstream. These amino acids can be used to synthesize proteins or can be converted to lipids or glucose for energy storage.

- *Waste Product Removal.* When converting amino acids to lipids or carbohydrates, or when breaking down amino acids to get energy, the liver strips off the amino groups. This process is called *deamination.* Ammonia, a toxic waste product, is formed. The liver neutralizes ammonia by converting it to *urea,* a fairly harmless compound excreted by the kidneys. The liver also removes other waste products, circulating toxins, and drugs from the blood for inactivation, storage, or excretion.

- *Vitamin Storage.* Fat-soluble vitamins (A, D, E, and K) and vitamin B_{12} are absorbed from the blood and stored in the liver. These reserves are used when your diet contains inadequate amounts of those vitamins.

- *Mineral Storage.* The liver converts iron reserves to ferritin and stores this protein–iron complex. ⟲ p. 687

- *Drug Inactivation.* The liver removes and breaks down circulating drugs, limiting the duration of their effects. When physicians prescribe a particular drug, they must take into account the rate at which the liver removes that drug from the bloodstream. For example, a drug that is absorbed relatively quickly must be administered every few hours to keep blood concentrations at therapeutic levels.

Hematological Regulation. The liver receives about 25 percent of cardiac output. It is also the largest blood reservoir in your body. As blood passes through it, the liver performs the following functions:

- *Phagocytosis and Antigen Presentation.* Kupffer cells in the liver sinusoids engulf old or damaged red blood cells, cellular debris, and pathogens, removing them from the bloodstream. Kupffer cells are antigen-presenting cells (APCs) that can stimulate an immune response. ⟲ p. 832

- *Synthesis of Plasma Proteins.* Hepatocytes synthesize and release most of the plasma proteins. These proteins include the albumins (which contribute to the osmotic concentration of the blood), the various types of transport proteins, clotting proteins, and complement proteins.

- *Removal of Circulating Hormones.* The liver is the primary site for the absorption and recycling of epinephrine, norepinephrine, insulin, thyroid hormones, and steroid hormones, such as the sex hormones (estrogens and androgens) and corticosteroids. The liver also absorbs cholecalciferol (vitamin D_3) from the blood. Liver cells then convert the cholecalciferol, which may be synthesized in the skin or absorbed in the diet, into an intermediary product, 25-hydroxy-D_3, that is released back into the bloodstream. The kidneys absorb this intermediary and use it to generate calcitriol, a hormone important to Ca^{2+} metabolism. ⟲ p. 665

- *Removal of Antibodies.* The liver absorbs and breaks down antibodies, releasing amino acids for recycling.

- *Removal or Storage of Toxins.* The liver absorbs lipid-soluble toxins in the diet, such as the insecticide DDT (banned in the United States since 1972, but still found in the environment), and stores them in lipid deposits, where they do not disrupt cellular functions. The liver removes other toxins from the bloodstream and either breaks them down or excretes them in the bile.

- *The Synthesis and Secretion of Bile.* The liver synthesizes bile and excretes it into the lumen of the duodenum. Hormonal and neural mechanisms regulate bile secretion. Bile consists mostly of water, with minor amounts of ions, *bilirubin* (a pigment derived from hemoglobin), cholesterol, and an assortment of lipids collectively known as **bile salts**. (Bile salts play a role in the digestion of lipids, as we discuss in the next section.) The water and ions help dilute and buffer acids in chyme as it enters the small intestine. Bile salts are synthesized from cholesterol in the liver. Several related compounds are involved. The most abundant are derivatives of the steroids *cholate* and *chenodeoxycholate.*

The Functions of Bile. Most dietary lipids are not water soluble. Mechanical processing in the stomach creates large drops containing a variety of lipids. Pancreatic lipase is not lipid soluble, so the enzymes can interact with lipids only at the surface of a lipid droplet. The larger the droplet, the more lipids are inside, isolated and protected from these enzymes. Bile salts break the droplets apart in a process called **emulsification** (ē-mul-si-fi-KĀ-shun), which dramatically increases the surface area accessible to enzymes.

24

Emulsification creates tiny *emulsion droplets* with a superficial coating of bile salts. The formation of tiny droplets increases the surface area available for enzymatic attack. In addition, the layer of bile salts facilitates interaction between the lipids and lipid-digesting enzymes from the pancreas.

After lipid digestion has been completed, bile salts promote the absorption of lipids by the intestinal epithelium. More than 90 percent of the bile salts are themselves reabsorbed, primarily in the ileum, as lipid digestion is completed. The reabsorbed bile salts enter the hepatic portal circulation. The liver then collects and recycles them. The cycling of bile salts from the liver to the small intestine and back is called the **enterohepatic circulation of bile**.

TIPS & TRICKS

Oil and water ordinarily don't mix. The emulsification process helps mix oils and water-soluble enzymes so that fats can be broken down.

The Gallbladder

The **gallbladder** is a hollow, pear-shaped organ that stores and concentrates bile prior to its excretion into the small intestine. This muscular sac is located in a fossa, or recess, in the posterior surface of the liver's right lobe (**Figure 24–21a**). The gallbladder is divided into three regions: (1) the **fundus**, (2) the **body**, and (3) the **neck**. The cystic duct extends from the gallbladder to the point where it unites with the common hepatic duct to form the common bile duct. At the duodenum, the common bile duct meets the pancreatic duct before emptying into a chamber called the **duodenal ampulla** (am-PUL-luh) (**Figure 24–21b**), which receives buffers and enzymes from the pancreas and bile from the liver and gallbladder. The duodenal ampulla opens into the duodenum at the **duodenal papilla**, a small mound.

The muscular **hepatopancreatic sphincter** (*sphincter of Oddi*) encircles the lumen of the common bile duct and, generally, the pancreatic duct and duodenal ampulla as well.

Physiology of the Gallbladder

A major function of the gallbladder is *bile storage*, but it is released into the duodenum only under the stimulation of the intestinal hormone CCK (**Figure 24–21d**). Without CCK, the hepatopancreatic sphincter remains closed, so bile exiting the liver in the common hepatic duct cannot flow through the common bile duct and into the duodenum. Instead, it enters the cystic duct and is stored within the expandable gallbladder. Whenever chyme enters the duodenum, CCK is released, relaxing the hepatopancreatic sphincter and stimulating contractions of the gallbladder that push bile into the small intestine. The amount of CCK secreted increases markedly when the chyme contains large amounts of lipids.

The gallbladder also functions in *bile modification*. When full, the gallbladder contains 40–70 mL of bile. The composition of bile gradually changes as it remains in the gallbladder: Much of the water is absorbed, and the bile salts and other components of bile become increasingly concentrated.

If bile becomes too concentrated, crystals of insoluble minerals and salts begin to form. These deposits are called *gallstones*. Small gallstones are not a problem so long as they can be flushed down the bile duct and excreted. In *cholecystitis* (kō-lē-sis-TĪ-tis; *chole*, bile + *kystis*, bladder + *-itis*, inflammation), the gallstones are so large that they can damage the wall of the gallbladder or block the cystic duct or common bile duct. In that case, the gallbladder may need to be surgically removed. This does not seriously impair digestion, because bile production continues at normal levels. However, the bile is more dilute, and its entry into the small intestine is not as closely tied to the arrival of food in the duodenum.

The Coordination of Secretion and Absorption

A combination of neural and hormonal mechanisms coordinates the activities of the digestive glands. These regulatory mechanisms are centered on the duodenum, where acids must be neutralized and appropriate enzymes added.

Neural mechanisms involving the CNS prepare the digestive tract for activity (through parasympathetic innervation) or inhibit its activity (through sympathetic innervation). Neural mechanisms also coordinate the movement of materials along the length of the digestive tract (through the enterogastric, gastroenteric, and gastroileal reflexes).

In addition, motor neurons synapsing in the digestive tract release a variety of neurotransmitters. Many of these chemicals are also released in the CNS. In general, their functions are poorly understood. Examples of neurotransmitters that may be important include substance P, enkephalins, and endorphins.

We will now summarize the information presented thus far on the regulation of intestinal and glandular function. We will also consider some additional details about the regulatory mechanisms involved.

Intestinal Hormones

The intestinal tract secretes a variety of peptide hormones with similar chemical structures. Many of these hormones have multiple effects in several regions of the digestive tract, and in the accessory glandular organs as well. The origins and primary effects of these important digestive hormones are shown in **Figure 24–22**.

Other intestinal hormones are produced in relatively small quantities. Examples include *motilin*, which stimulates intestinal contractions; *villikinin*, which promotes the movement of villi and the associated lymph flow; and *somatostatin*, which inhibits gastric secretion. Functional interactions among gastrin, GIP, secretin, CCK, and VIP are diagrammed in **Figure 24–23**.

Figure 24–22   **Major Hormones of the Duodenum.**

Major Hormones of the Duodenum
Gastrin
Gastrin is secreted by G cells in the duodenum when they are exposed to large quantities of incompletely digested proteins. The functions of gastrin include promoting increased stomach motility and stimulating the production of gastric acids and enzymes. (Gastrin is also produced by the stomach.)
Secretin
Secretin is released when chyme arrives in the duodenum. Secretin's primary effect is an increase in the secretion of bile (by the liver) and buffers (by the pancreas), which in turn act to increase the pH of the chyme. Among its secondary effects, secretin reduces gastric motility and secretory rates.
Gastric Inhibitory Peptide (GIP)
Gastric inhibitory peptide is secreted when fats and carbohydrates—especially glucose—enter the small intestine. The inhibition of gastric activity is accompanied by the stimulation of insulin release at the pancreatic islets. GIP has several secondary effects, including stimulating duodenal gland activity, stimulating lipid synthesis in adipose tissue, and increasing glucose use by skeletal muscles.
Cholecystokinin (CCK)
Cholecystokinin is secreted when chyme arrives in the duodenum, especially when the chyme contains lipids and partially digested proteins. In the pancreas, CCK accelerates the production and secretion of all types of digestive enzymes. It also causes a relaxation of the hepatopancreatic sphincter and contraction of the gallbladder, resulting in the ejection of bile and pancreatic juice into the duodenum. Thus, the net effects of CCK are to increase the secretion of pancreatic enzymes and to push pancreatic secretions and bile into the duodenum. The presence of CCK in high concentrations has two additional effects: It inhibits gastric activity, and it appears to have CNS effects that reduce the sensation of hunger.
Vasoactive Intestinal Peptide (VIP)
Vasoactive intestinal peptide stimulates the secretion of intestinal glands, dilates regional capillaries, and inhibits acid production in the stomach. By dilating capillaries in active areas of the intestinal tract, VIP provides an efficient mechanism for removing absorbed nutrients.
Enterocrinin
Enterocrinin is released when chyme enters the duodenum. It stimulates mucin production by the submucosal glands.

Intestinal Absorption

On average, it takes about five hours for materials to pass from the duodenum to the end of the ileum. The first of the materials to enter the duodenum after you eat breakfast may leave the small intestine at lunchtime. Along the way, the organ's absorptive effectiveness is enhanced by the fact that so much of the mucosa is movable. The microvilli can be moved by their supporting microfilaments, the individual villi by smooth muscle cells, groups of villi by the muscularis mucosae, and the circular folds by the muscularis mucosae and the muscularis externa. These movements stir and mix the intestinal contents, changing the environment around each epithelial cell from moment to moment.

CHECKPOINT

21. Name the three regions of the small intestine from proximal to distal.

22. How is the small intestine adapted for nutrient absorption?

23. Does a high-fat meal raise or lower the level of cholecystokinin in the blood?

24. How would the pH of the intestinal contents be affected if the small intestine did not produce secretin?

25. The digestion of which nutrient would be most impaired by damage to the exocrine pancreas?

See the blue Answers tab at the back of the book.

24

Figure 24–23 **The Functions of Major Digestive Tract Hormones.** The primary actions of gastrin, GIP, secretin, CCK, and VIP are shown.

24-7 ∣ The large intestine is divided into three parts with regional specialization

Learning Outcome Describe the gross and histological structures of the large intestine, including its regional specializations and role in nutrient absorption.

The horseshoe-shaped **large intestine** begins at the end of the ileum and ends at the anus. The large intestine lies inferior to the stomach and liver and almost completely frames the small intestine (**Figure 24–1**). The large intestine stores digestive wastes and reabsorbs water. Resident bacteria in the large intestine are an important source of vitamins, especially vitamin K, biotin, and vitamin B_5.

The large intestine, also known as the *large bowel*, has an average length of about 1.5 meters (4.9 ft) and a width of 7.5 cm (3 in.). We can divide it into three parts: (1) the pouchlike *cecum*, the first portion of the large intestine;

(2) the *colon*, the largest portion; and (3) the *rectum*, the last 15 cm (6 in.) of the large intestine and the end of the digestive tract (**Figure 24–24a**).

The Cecum

Material arriving from the ileum first enters an expanded pouch called the **cecum** (SĒ-kum). The ileum attaches to the medial surface of the cecum and opens into the cecum at the *ileocecal valve* (**Figure 24–24a,b**). The cecum collects and stores materials from the ileum and begins the process of compaction.

The slender, hollow **appendix**, or *vermiform appendix* (*vermis*, worm), is attached to the posteromedial surface of the cecum (**Figure 24–24a,b**). The appendix is normally about 9 cm (3.6 in.) long, but its size and shape are quite variable. A small mesentery called the **mesoappendix** connects the appendix to the ileum and cecum. Lymphoid nodules dominate the mucosa and submucosa of the appendix. The primary function of the appendix is as an organ of the lymphatic system. Inflammation of the appendix is known as *appendicitis*.

The Colon

The **colon** has a larger diameter and a thinner wall than the small intestine. Distinctive features of the colon include the following (**Figure 24–24a**):

- The wall of the colon forms a series of pouches, or **haustra** (HAWS-truh; singular, *haustrum*). The creases between the haustra affect the mucosal lining as well, producing a series of internal folds. Haustra permit the colon to expand and elongate.

- Three separate longitudinal bands of smooth muscle— called the **teniae coli** (TĒ-nē-ē KŌ-lē)—run along the outer surfaces of the colon just deep to the serosa. These bands correspond to the outer layer of the muscularis externa in other portions of the digestive tract. Muscle tone within the teniae coli is what creates the haustra.

- The serosa of the colon contains numerous teardrop-shaped sacs of fat called **omental appendices**, also called *fatty* or *epiploic* (ep-i-PLŌ-ik; *epiploon*, omentum) *appendages*.

We can subdivide the colon into four regions: the ascending colon, transverse colon, descending colon, and sigmoid colon (**Figure 24–24a**).

1. The **ascending colon** begins at the superior border of the cecum and ascends along the right lateral and posterior wall of the peritoneal cavity to the inferior surface of the liver. There, the colon bends sharply to the left at the **right colic flexure**, or *hepatic flexure*. This bend marks the end of the ascending colon and the beginning of the transverse colon.

2. The **transverse colon** curves anteriorly from the right colic flexure and crosses the abdomen from right to left. The transverse colon is supported by the transverse mesocolon. It is separated from the anterior abdominal wall by the layers of the greater omentum. At the left side of the body, the transverse colon passes inferior to the greater curvature of the stomach. Near the spleen, the colon makes a 90° turn at the **left colic flexure**, or *splenic flexure*, and becomes the descending colon.

3. The **descending colon** proceeds inferiorly along the left side until reaching the iliac fossa formed by the inner surface of the left ilium. The descending colon is retroperitoneal and firmly attached to the abdominal wall. At the iliac fossa, the descending colon curves at the **sigmoid flexure** and becomes the sigmoid colon.

4. The sigmoid flexure is the start of the **sigmoid** (SIG-moyd) **colon** (*sigmeidos*, the Greek letter *S*), an S-shaped segment that is only about 15 cm (6 in.) long. The sigmoid colon lies posterior to the urinary bladder, suspended from the sigmoid mesocolon. The sigmoid colon empties into the *rectum*.

Clinical Note

Colorectal Cancer Colorectal cancer is relatively common in the United States. Aside from skin cancers, colorectal cancer is the third most common cancer in the United States, affecting both men and women. At the writing of this edition, the American Cancer Society's statistics showed that there were 102,480 new cases of colon cancer and 40,340 new cases of rectal cancer. The death rate has declined over the past 20 years for both men and women. The best defense appears to be early detection and prompt treatment. Standard screening involves checking the feces for blood. This simple procedure can be performed easily on a stool (fecal) sample as part of a routine physical. For individuals at increased risk because of family history, associated disease, or older age, visual inspection of the lumen by fiber-optic colonoscopy is prudent. Such examination can find polyps before they develop into cancers. The 5-year survival rate for people whose cancer is found at an early stage and treated immediately is greater than 90 percent.

24

Figure 24–24 **The Large Intestine.** ATLAS: Plates 49a–c; 58a–c; 59; 64; 65

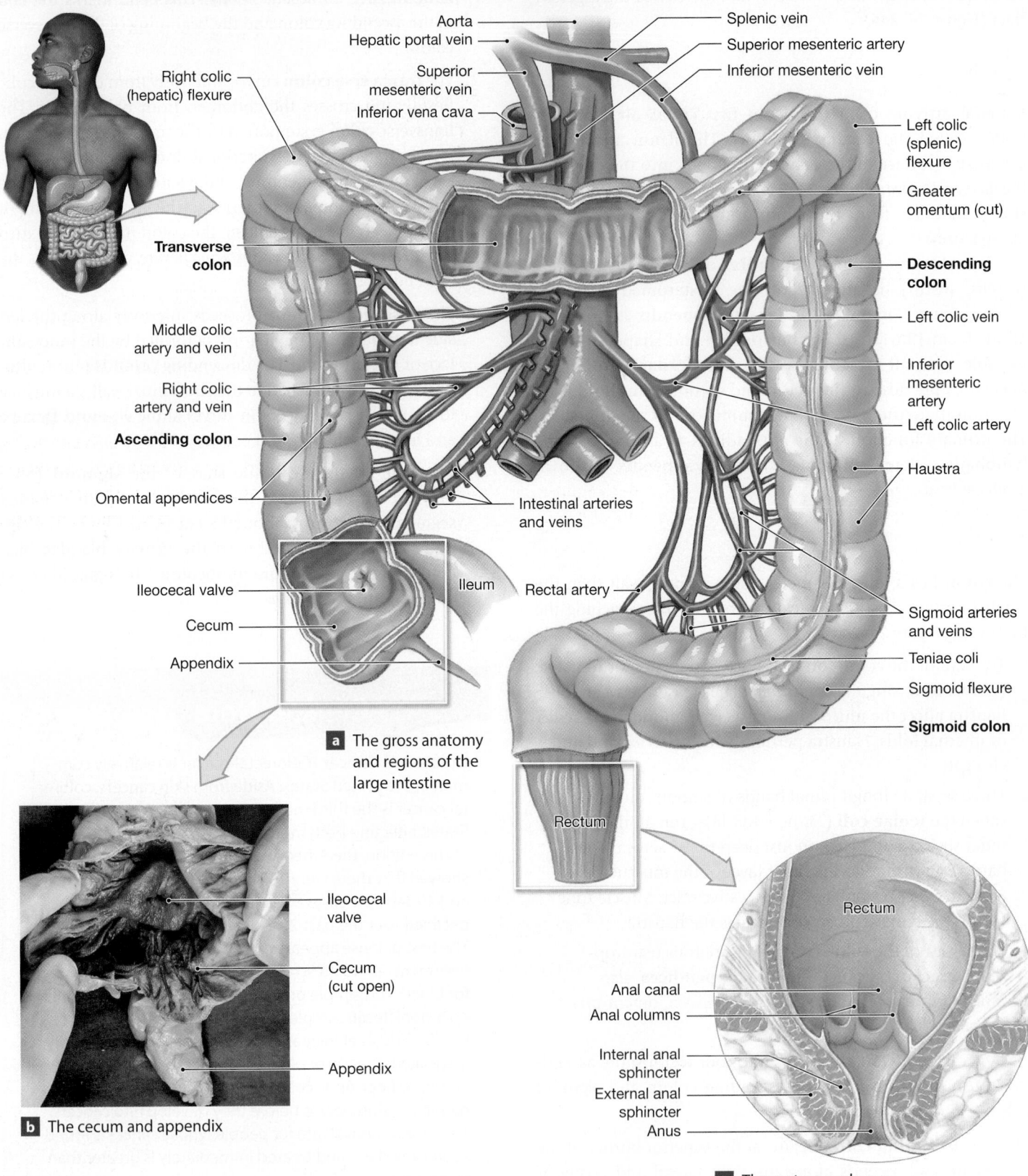

Aorta

Hepatic portal vein

Superior mesenteric vein

Inferior vena cava

Splenic vein

Superior mesenteric artery

Inferior mesenteric vein

Right colic (hepatic) flexure

Transverse colon

Middle colic artery and vein

Right colic artery and vein

Ascending colon

Omental appendices

Ileocecal valve

Cecum

Appendix

Ileum

Rectal artery

Intestinal arteries and veins

Left colic (splenic) flexure

Greater omentum (cut)

Descending colon

Left colic vein

Inferior mesenteric artery

Left colic artery

Haustra

Sigmoid arteries and veins

Teniae coli

Sigmoid flexure

Sigmoid colon

Rectum

a The gross anatomy and regions of the large intestine

Ileocecal valve

Cecum (cut open)

Appendix

b The cecum and appendix

Rectum

Anal canal

Anal columns

Internal anal sphincter

External anal sphincter

Anus

c The rectum and anus

24

The large intestine receives blood from branches of the superior mesenteric and inferior mesenteric arteries. The superior mesenteric and inferior mesenteric veins collect venous blood from the large intestine. ⊃ p. 788, 797

The Rectum

The **rectum** (REK-tum) forms the last 15 cm (6 in.) of the digestive tract (**Figure 24–24a,c**). It is an expandable organ for the temporary storage of feces. The movement of fecal material into the rectum triggers the urge to defecate (expel feces).

The last portion of the rectum, the **anal canal**, contains small longitudinal folds called **anal columns**. The distal margins of these columns are joined by transverse folds that mark the boundary between the columnar epithelium of the proximal rectum and a stratified squamous epithelium like that in the oral cavity. The **anus**, or *anal orifice*, is the exit of the anal canal. There, the epidermis becomes keratinized and identical to the surface of the skin.

The circular muscle layer of the muscularis externa in this region forms the **internal anal sphincter** (**Figure 24–24c**). The smooth muscle cells of this sphincter are not under voluntary control. The **external anal sphincter**, which guards the anus,

consists of a ring of skeletal muscle fibers that encircles the distal portion of the anal canal. This sphincter consists of skeletal muscle and is under voluntary control.

The lamina propria and submucosa of the anal canal contain a network of veins. If venous pressures there rise too high due to straining during defecation or pregnancy, the veins can become distended, producing *hemorrhoids*.

Histology of the Large Intestine

The diameter of the colon is about three times that of the small intestine, but its wall is much thinner. The major characteristics of the colon are the lack of villi, the abundance of mucous cells, and the presence of distinctive intestinal glands (**Figure 24–25**). The glands in the large intestine are deeper than those of the small intestine and are dominated by mucous cells. The mucosa of the large intestine does not produce enzymes. Any digestion that occurs results from enzymes introduced in the small intestine or from bacterial action. The mucus provides lubrication as the fecal material becomes drier and more compact. Mucus is secreted as local stimuli, such as friction or exposure to harsh chemicals, trigger short reflexes involving local nerve plexuses. Large lymphoid nodules are scattered throughout the lamina propria and submucosa.

Figure 24–25 **The Mucosa and Glands of the Colon.**

Teniae coli
Omental appendices
Haustrum

Aggregated lymphoid nodule

| Layers of the Large Intestine |
| **Mucosa** |
| Muscularis mucosae |
| **Submucosa** |
| **Muscularis externa** |
| Circular layer |
| Longitudinal layer (teniae coli) |
| **Serosa** |

Simple columnar epithelium
Mucous cells
Intestinal crypt
Muscularis mucosae
Submucosa

The colon LM × 110

a Diagrammatic view of the colon wall

b Colon histology showing detail of mucosal layer

24

The muscularis externa of the large intestine is unusual, because the longitudinal layer has been reduced to the muscular bands of the teniae coli. However, the mixing and propulsive contractions of the colon resemble those of the small intestine.

Physiology of the Large Intestine

Less than 10 percent of the nutrient absorption under way in the digestive tract occurs in the large intestine. Nevertheless, the absorptive operations in this segment of the digestive tract are important. The large intestine also prepares fecal material for defecation.

Absorption in the Large Intestine

The reabsorption of water is an important function of the large intestine. Roughly 1500 mL of material enters the colon each day, but only about 200 mL of feces is ejected. To appreciate how efficient our digestion is, consider the average composition of feces: 75 percent water, 5 percent bacteria, and the rest a mixture of indigestible materials, small quantities of inorganic matter, and the remains of epithelial cells.

In addition to reabsorbing water, the large intestine absorbs a number of other substances that remain in the feces or were secreted into the digestive tract along its length. Examples include useful compounds such as bile salts and vitamins; organic waste products such as urobilinogen; and various toxins generated by bacterial action. Most of the bile salts entering the large intestine are promptly reabsorbed in the cecum and transported in blood to the liver for secretion into bile.

Vitamins. Vitamins are organic molecules that are important as cofactors or coenzymes in many metabolic pathways. The normal bacterial residents of the colon generate three vitamins that supplement our diets:

1. *Vitamin K*, a fat-soluble vitamin the liver requires for synthesizing four clotting factors, including prothrombin. Intestinal bacteria produce about half of your daily vitamin K requirements.

2. *Biotin*, a water-soluble vitamin important in various reactions, notably those of glucose metabolism.

3. *Vitamin B_5* (pantothenic acid), a water-soluble vitamin required in the manufacture of steroid hormones and some neurotransmitters.

Vitamin K deficiencies lead to impaired blood clotting. They result from either (1) not enough lipids in the diet, which impairs the absorption of all fat-soluble vitamins; or (2) problems affecting lipid processing and absorption, such as inadequate bile production or chronic diarrhea (frequent, watery bowel movements). Disorders due to deficiencies of biotin or vitamin B_5 are extremely rare after infancy. The intestinal bacteria generally produce sufficient amounts to supplement any dietary shortage.

Organic Wastes. We discussed the fate of bilirubin, a breakdown product of heme, in Chapter 19. ⤵ p. 687 In the large intestine, bacteria convert bilirubin to *urobilinogens* and *stercobilinogens*. Some urobilinogens are absorbed into the bloodstream and then excreted in urine. The urobilinogens and stercobilinogens remaining within the colon are converted to **urobilins** and **stercobilins** by exposure to oxygen. These pigments in various proportions give feces a yellow-brown or brown color.

Bacterial action breaks down peptides that remain in the feces. This action generates (1) ammonia, in the form of soluble *ammonium ions* (NH_4^+); (2) *indole* and *skatole*, two nitrogen-containing compounds that are primarily responsible for the odor of feces; and (3) hydrogen sulfide (H_2S), a gas with a "rotten egg" odor. Significant amounts of ammonia and smaller amounts of other toxins cross the colonic epithelium and enter the hepatic portal circulation. The liver removes these toxins and converts them to relatively nontoxic compounds that can be released into the blood and excreted at the kidneys.

Intestinal enzymes do not alter indigestible carbohydrates. These materials arrive in the colon virtually intact. These

Clinical Note

Inflammatory and Infectious Disorders of the Digestive System Digestive system disorders are both very diverse and relatively common because the system has so many parts, and each part has so many functions.

The largest category of digestive disorders includes those resulting from inflammation or infection of the digestive tract. In part, this is because the epithelium that lines most of the digestive tract has two properties that are difficult to reconcile. On one hand, it must be thin enough to absorb nutrients rapidly and efficiently. On the other hand, it must resist damage from ingested materials and enzymes.

The delicacy of the epithelium makes it susceptible to damage from chemicals or abrasion. For example, *peptic ulcers* develop if acids and enzymes contact and erode the gastric or duodenal lining. Pathogens in food, including bacteria, viruses, and multicellular parasites, may also get through the epithelial barriers and cause infections. Small battles are continually being fought. In fact, 80 percent of the body's plasma cells are normally located within the lamina propria of the digestive tract. This fact indicates how often antigens of one kind or another somehow cross the epithelial barriers.

High rates of cell division and exposure to strong chemicals are both correlated with an increased risk of cancer. As a result, cancers of the digestive tract are relatively common. Predictably, most are epithelial cancers that develop in the stem cell populations responsible for epithelial cell renewal.

complex polysaccharides provide a reliable nutrient source for bacteria in the colon. The metabolic activities of these bacteria create small amounts of **flatus**, or intestinal gas. Foods with large amounts of indigestible carbohydrates (such as beans) stimulate bacterial gas production. Distension of the colon, cramps, and the frequent discharge of intestinal gases can result.

Movements of the Large Intestine

The gastroileal and gastroenteric reflexes move materials into the cecum while you eat. Movement from the cecum to the transverse colon is very slow, allowing hours for water absorption to convert the already thick material into a sludgy paste. Peristaltic waves move material along the length of the colon.

Segmentation movements, called *haustral churning*, mix the contents of adjacent haustra. Powerful peristaltic contractions called **mass movements** occur a few times each day. They move material from the transverse colon through the rest of the large intestine. The stimulus is distension of the stomach and duodenum. The commands are relayed over the intestinal nerve plexuses. The contractions force feces into the rectum and produce the conscious urge to defecate.

The rectal chamber is usually empty, except when a powerful peristaltic contraction forces feces out of the sigmoid colon. Distension of the rectal wall then triggers the **defecation reflex**. This reflex involves two positive feedback loops (Figure 24–26). Both loops are triggered by the stimulation of stretch receptors in the rectum.

Figure 24–26 **The Defecation Reflex.**

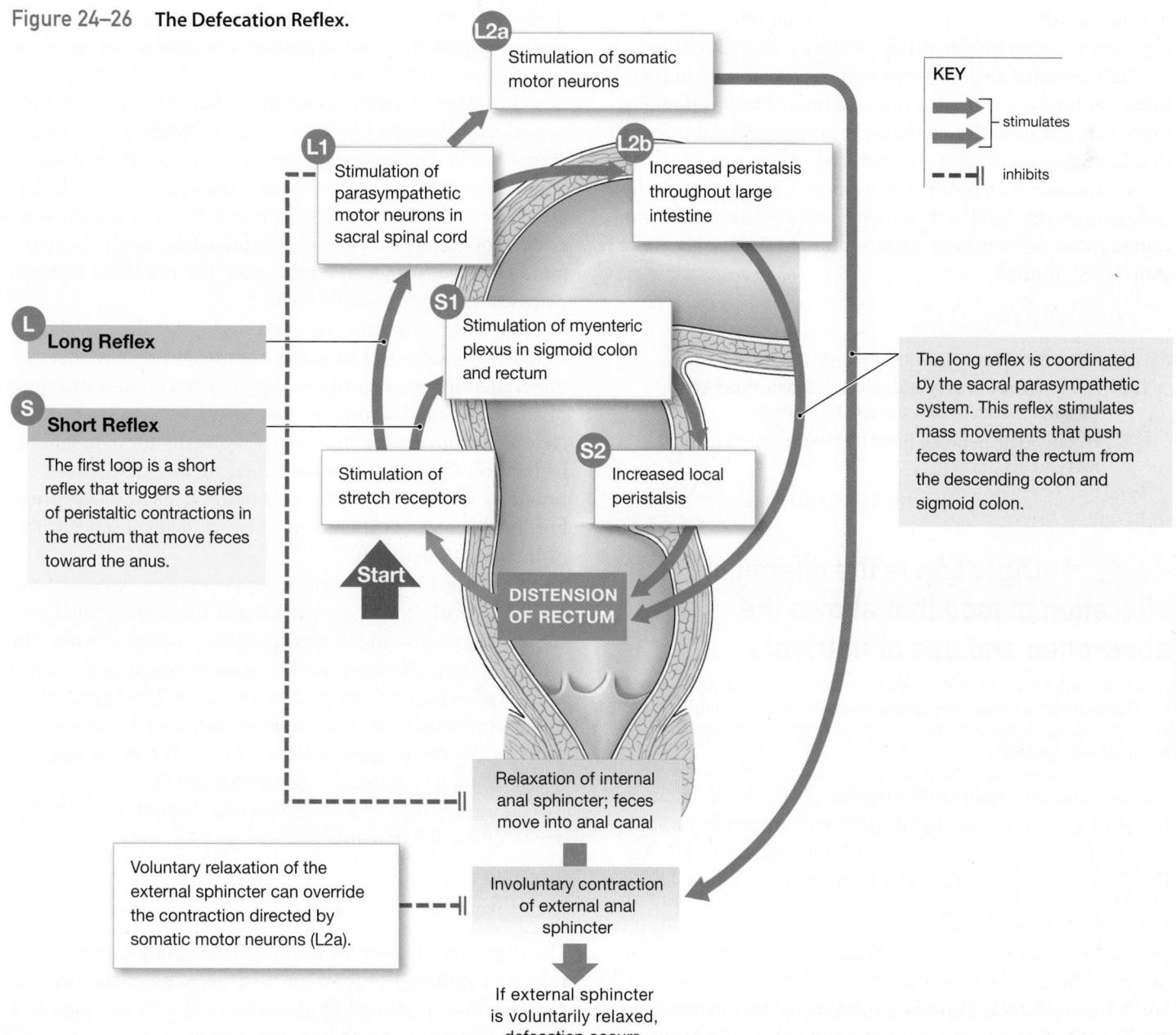

KEY

→→ stimulates

- - -⊣ inhibits

L2a Stimulation of somatic motor neurons

L1 Stimulation of parasympathetic motor neurons in sacral spinal cord

L2b Increased peristalsis throughout large intestine

S1 Stimulation of myenteric plexus in sigmoid colon and rectum

L **Long Reflex**

S **Short Reflex**

The first loop is a short reflex that triggers a series of peristaltic contractions in the rectum that move feces toward the anus.

S2 Increased local peristalsis

Stimulation of stretch receptors

Start

DISTENSION OF RECTUM

The long reflex is coordinated by the sacral parasympathetic system. This reflex stimulates mass movements that push feces toward the rectum from the descending colon and sigmoid colon.

Relaxation of internal anal sphincter; feces move into anal canal

Voluntary relaxation of the external sphincter can override the contraction directed by somatic motor neurons (L2a).

Involuntary contraction of external anal sphincter

If external sphincter is voluntarily relaxed, defecation occurs.

24

Rectal stretch receptors also trigger two reflexes important to the *voluntary* control of defecation. One is a long reflex mediated by parasympathetic innervation within the pelvic nerves. This reflex causes the internal anal sphincter to relax. This smooth muscle sphincter controls the movement of feces into the anal canal. The second is a somatic reflex that stimulates the immediate contraction of the external anal sphincter, a skeletal muscle. ↺ pp. 381–383 The pudendal nerves carry the motor commands.

Both the internal and external anal sphincters must relax for feces to be eliminated. However, the two reflexes just mentioned open the internal sphincter but close the external sphincter. The actual release of feces requires a conscious effort to open the external sphincter. In addition, other consciously directed activities can raise intra-abdominal pressures and help force fecal material out of the rectum. These activities include tensing the abdominal muscles or elevating intra-abdominal pressures by attempting to exhale forcibly with a closed glottis.

If the external anal sphincter remains constricted, the peristaltic contractions cease. However, additional rectal expansion triggers further defecation reflexes. The urge to defecate usually develops when rectal pressure reaches about 15 mm Hg. If this pressure is more than 55 mm Hg, the external anal sphincter involuntarily relaxes and defecation takes place. This mechanism brings about defecation in infants, and in adults with severe spinal cord injuries.

CHECKPOINT

26. **Identify the four regions of the colon.**

27. **Name some major histological differences between the large intestine and the small intestine.**

28. **Differentiate between haustral churning and mass movements.**

See the blue Answers tab at the back of the book.

24-8 ▮ Digestion is the chemical alteration of food that allows the absorption and use of nutrients

Learning Outcome List the nutrients required by the body, describe the chemical events responsible for the digestion of organic nutrients, and describe the mechanisms involved in the absorption of organic and inorganic nutrients.

A balanced diet contains all the ingredients needed to maintain homeostasis. These ingredients include six nutrients: carbohydrates, lipids, proteins, vitamins, minerals, and water. This section describes the chemical events involved in the processing and absorbing of these nutrients.

The Processing and Absorbing of Nutrients

Food contains large organic molecules, many of them insoluble. The digestive system first breaks down the physical structure of the ingested material and then disassembles the component molecules into smaller fragments. This disassembly eliminates any antigenic properties, so that the fragments do not trigger an immune response after absorption. Cells absorb the molecules released into the bloodstream and either (1) break them down to provide energy for the synthesis of ATP or (2) use these molecules to synthesize carbohydrates, proteins, and lipids. In this section we focus on the mechanics of digestion and absorption. The fates of the compounds inside cells are the focus in Chapter 25.

Most ingested organic materials are complex chains of simpler molecules. In a typical dietary carbohydrate, the basic molecules are simple sugars. In a protein, the building blocks are amino acids. In lipids, they are generally fatty acids. And in nucleic acids, they are nucleotides. Digestive enzymes break the bonds between the component molecules of carbohydrates, proteins, lipids, and nucleic acids in a process called *hydrolysis*. ↺ p. 63

The classes of digestive enzymes differ with respect to their targets. *Carbohydrases* break the bonds between simple sugars, *proteases* split the linkages between amino acids, and *lipases* separate fatty acids from glycerides. Some enzymes in each class are even more selective, breaking bonds between specific molecules. For example, a particular carbohydrase might break the bond between two glucose molecules, but not those between glucose and another simple sugar.

Digestive enzymes secreted by the salivary glands, tongue, stomach, and pancreas are mixed into the ingested material as it passes along the digestive tract. These enzymes break down large carbohydrates, proteins, lipids, and nucleic acids into smaller fragments. Typically these fragments in turn must be broken down even further before absorption can occur. The final enzymatic steps involve brush border enzymes, which are attached to the exposed surfaces of intestinal microvilli.

Nucleic acids are broken down into their component nucleotides. Brush border enzymes digest these nucleotides into sugars, phosphates, and nitrogenous bases that are absorbed by active transport. However, nucleic acids represent only a small fraction of all the nutrients absorbed each day. The digestive fates of carbohydrates, lipids, and proteins, the major dietary components, are shown in **Spotlight Figure 24–27**. The major digestive enzymes and their functions are summarized in Table 24–1. Next we take a closer look at the digestion and absorption of carbohydrates, lipids, and proteins.

Carbohydrate Digestion and Absorption

The digestion of complex carbohydrates (simple polysaccharides and starches) proceeds in two steps. One step involves carbohydrases produced by the salivary glands and pancreas. The other step uses brush border enzymes.

A typical meal contains carbohydrates, proteins, lipids, water, minerals (electrolytes), and vitamins. The digestive system handles each component differently. Large organic molecules must be chemically broken down before they can be absorbed. Water, minerals, and vitamins can be absorbed without processing, but they may require special transport mechanisms.

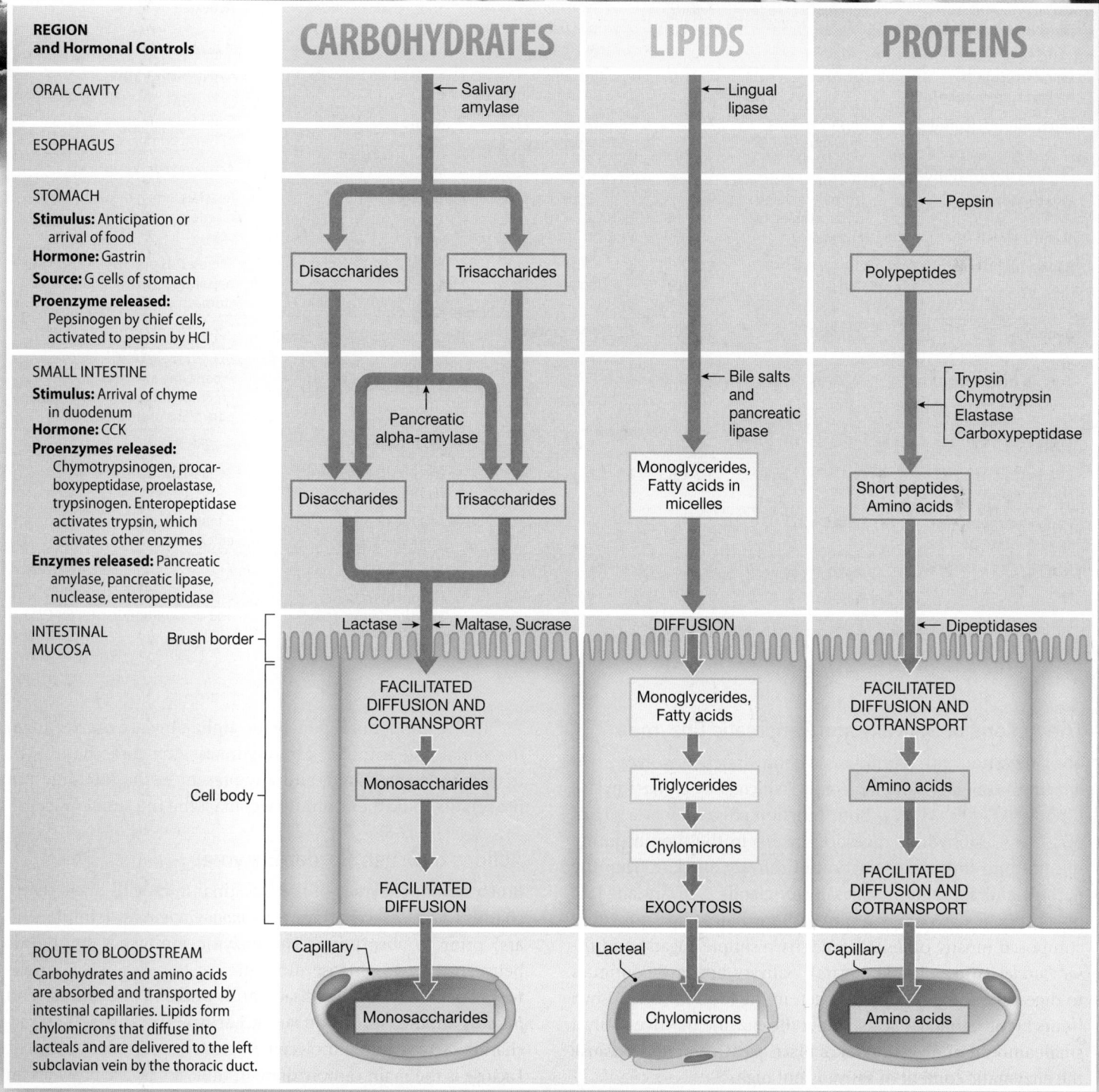

REGION and Hormonal Controls	CARBOHYDRATES	LIPIDS	PROTEINS
ORAL CAVITY	← Salivary amylase	← Lingual lipase	
ESOPHAGUS			
STOMACH **Stimulus:** Anticipation or arrival of food **Hormone:** Gastrin **Source:** G cells of stomach **Proenzyme released:** Pepsinogen by chief cells, activated to pepsin by HCl	Disaccharides Trisaccharides		← Pepsin Polypeptides
SMALL INTESTINE **Stimulus:** Arrival of chyme in duodenum **Hormone:** CCK **Proenzymes released:** Chymotrypsinogen, procarboxypeptidase, proelastase, trypsinogen. Enteropeptidase activates trypsin, which activates other enzymes **Enzymes released:** Pancreatic amylase, pancreatic lipase, nuclease, enteropeptidase	Pancreatic alpha-amylase Disaccharides Trisaccharides	← Bile salts and pancreatic lipase Monoglycerides, Fatty acids in micelles	Trypsin Chymotrypsin Elastase Carboxypeptidase Short peptides, Amino acids
INTESTINAL MUCOSA — Brush border	Lactase → ← Maltase, Sucrase FACILITATED DIFFUSION AND COTRANSPORT	DIFFUSION Monoglycerides, Fatty acids	← Dipeptidases FACILITATED DIFFUSION AND COTRANSPORT
Cell body	Monosaccharides FACILITATED DIFFUSION	Triglycerides Chylomicrons EXOCYTOSIS	Amino acids FACILITATED DIFFUSION AND COTRANSPORT
ROUTE TO BLOODSTREAM Carbohydrates and amino acids are absorbed and transported by intestinal capillaries. Lipids form chylomicrons that diffuse into lacteals and are delivered to the left subclavian vein by the thoracic duct.	Capillary Monosaccharides	Lacteal Chylomicrons	Capillary Amino acids

Table 24–1	Digestive Enzymes and Their Functions				
Enzyme (proenzyme)	**Source**	**Optimal pH**	**Target**	**Products**	**Remarks**
CARBOHYDRASES					
Maltase, sucrase, lactase	Brush border of small intestine	7–8	Maltose, sucrose, lactose	Monosaccharides	Found in membrane surface of microvilli
Pancreatic alpha-amylase	Pancreas	6.7–7.5	Complex carbohydrates	Disaccharides and trisaccharides	Breaks bonds between simple sugars
Salivary amylase	Salivary glands	6.7–7.5	Complex carbohydrates	Disaccharides and trisaccharides	Breaks bonds between simple sugars
PROTEASES					
Carboxypeptidase (procarboxypeptidase)	Pancreas	7–8	Proteins, polypeptides, amino acids	Short-chain peptides	Activated by trypsin
Chymotrypsin (chymotrypsinogen)	Pancreas	7–8	Proteins, polypeptides	Short-chain peptides	Activated by trypsin
Dipeptidases, peptidases	Brush border of small intestine	7–8	Dipeptides, tripeptides	Amino acids	Found in membrane surface of brush border
Elastase (proelastase)	Pancreas	7–8	Elastin	Short-chain peptides	Activated by trypsin
Enteropeptidase	Brush border and lumen of small intestine	7–8	Trypsinogen	Trypsin	Reaches lumen through disintegration of shed epithelial cells
Pepsin (pepsinogen)	Chief cells of stomach	1.5–2.0	Proteins, polypeptides	Short-chain polypeptides	Secreted as proenzyme pepsinogen; activated by H^+ in stomach acid
Rennin	Stomach	3.5–4.0	Milk proteins		Secreted only in infants; causes protein coagulation
Trypsin (trypsinogen)	Pancreas	7–8	Proteins, polypeptides	Short-chain peptides	Proenzyme activated by enteropeptidase; activates other pancreatic proteases
LIPASES					
Lingual lipase	Glands of tongue	3.0–6.0	Triglycerides	Fatty acids and monoglycerides	Begins lipid digestion
Pancreatic lipase	Pancreas	7–8	Triglycerides	Fatty acids and monoglycerides	Bile salts must be present for efficient action
NUCLEASES					
	Pancreas	7–8	Nucleic acids	Nitrogenous bases and simple sugars	Includes ribonuclease for RNA and deoxy-ribonuclease for DNA

24

The Actions of Salivary and Pancreatic Enzymes

The digestion of complex carbohydrates involves two enzymes—salivary amylase and pancreatic alpha-amylase (Spotlight Figure 24–27). Both function effectively at a pH of 6.7–7.5. Carbohydrate digestion begins in the mouth during mastication, through the action of salivary amylase from the parotid and submandibular salivary glands. Salivary amylase breaks down starches (complex carbohydrates) into a mixture composed mostly of *disaccharides* (two simple sugars) and *trisaccharides* (three simple sugars). Salivary amylase continues to digest the starches and glycogen in the food for one to two hours before stomach acids render the enzyme inactive. Only a small amount of digestion takes place over this period because the enzymatic content of saliva is not high.

In the duodenum, pancreatic alpha-amylase breaks down the remaining complex carbohydrates. Any disaccharides or trisaccharides produced, and any present in the food, are not broken down further until they contact the intestinal mucosa.

Actions of Brush Border Enzymes

Brush border enzymes of the intestinal microvilli break disaccharides and trisaccharides into *monosaccharides* (simple sugars) prior to absorption. The enzyme **maltase** splits bonds between the two glucose molecules of the disaccharide **maltose**. **Sucrase** breaks the disaccharide **sucrose** into glucose and *fructose*, another six-carbon sugar. **Lactase** hydrolyzes the disaccharide **lactose** into a molecule of glucose and one of *galactose*. Lactose is the main carbohydrate in milk, so lactase provides an essential function in infancy and early childhood by breaking down lactose. If the intestinal mucosa stops producing lactase, the person becomes **lactose intolerant**. After ingesting milk and other dairy products, lactose-intolerant people can experience a variety of unpleasant digestive problems, including lower abdominal pain, gas, diarrhea, and vomiting.

Absorption of Monosaccharides

The intestinal epithelium then absorbs the monosaccharides by facilitated diffusion and cotransport mechanisms (see Figure 3–18, p. 121). Both methods involve a carrier protein. Facilitated diffusion and cotransport differ in three major ways:

1. *Facilitated diffusion moves only one molecule or ion through the plasma membrane, whereas cotransport moves more than one molecule or ion through the membrane at the same time.* In cotransport, the transported substances move in the same direction: down the concentration gradient for at least one of them.

2. *Facilitated diffusion does not require ATP.* Cotransport by itself does not consume ATP, but the cell must often expend ATP to preserve homeostasis. For example, the process may bring in sodium ions that must later be pumped out of the cell.

3. *Facilitated diffusion does not take place if there is an opposing concentration gradient for the particular molecule or ion.* By contrast, cotransport can take place despite an opposing concentration gradient for one of the transported substances. For example, cells lining the small intestine continue to absorb glucose when glucose concentrations inside the cells are much higher than they are in the intestinal contents.

The cotransport system that takes up glucose also brings sodium ions into the cell. This passive process resembles facilitated diffusion, except that both a sodium ion and a glucose molecule must bind to the carrier protein before they can move into the cell. Glucose cotransport is an example of sodium-linked cotransport. ⟲ p. 123 Comparable cotransport mechanisms exist for other simple sugars and for some amino acids. These mechanisms deliver valuable nutrients to the cytoplasm, but they also bring in sodium ions that must be ejected by the sodium–potassium exchange pump.

The simple sugars that are transported into the cell at its apical surface diffuse through the cytoplasm. They then reach the interstitial fluid by facilitated diffusion across the basolateral surfaces. These monosaccharides diffuse into the capillaries of the villus for eventual transport to the liver in the hepatic portal vein.

Lipid Digestion and Absorption

Lipid digestion involves lingual lipase from glands of the tongue, and pancreatic lipase from the pancreas (Spotlight Figure 24–27).

The most important and abundant dietary lipids are triglycerides. They consist of three fatty acids attached to a single molecule of glycerol (see Figure 2–16, p. 75). The lingual and pancreatic lipases break off two of the fatty acids, leaving monoglycerides.

Lipases are water-soluble enzymes, and lipids tend to form large drops that exclude water molecules. As a result, lipases can attack only the exposed surfaces of the lipid drops. Lingual lipase begins breaking down triglycerides in the mouth and continues for a variable time within the stomach. The lipid drops are so large, however, and the available time so short, that only about 20 percent of the lipids have been digested by the time the chyme enters the duodenum.

Bile salts improve chemical digestion by emulsifying the lipid drops into tiny emulsion droplets, thereby providing better access for pancreatic lipase. The emulsification takes place only after the chyme has been mixed with bile in the duodenum. Pancreatic lipase then breaks apart the triglycerides to form a mixture of fatty acids and monoglycerides. As these molecules are released, they interact with bile salts in the surrounding chyme to form small lipid–bile salt complexes called **micelles** (mī-SELZ). A micelle is only about 2.5 nm (0.0025 μm) in diameter.

When a micelle contacts the intestinal epithelium, the lipids diffuse across the plasma membrane and enter the cytoplasm. The intestinal cells synthesize new triglycerides from the monoglycerides and fatty acids. These triglycerides, in company with absorbed steroids, phospholipids, and fat-soluble vitamins, are then coated with proteins. The resulting complexes are known as **chylomicrons** (kī-lō-MĪ-kronz; *chylos*, juice + *mikros*, small).

The intestinal cells then secrete the chylomicrons into interstitial fluid by exocytosis. The protein coating keeps the chylomicrons suspended in the interstitial fluid, but they are generally too large to diffuse into capillaries. Most of the chylomicrons diffuse into the intestinal lacteals, which lack basement membranes and have large gaps between adjacent endothelial cells. From the lacteals, the chylomicrons proceed along the lymphatic vessels and through the thoracic duct. They finally enter the bloodstream at the left subclavian vein.

Most of the bile salts within micelles are reabsorbed by sodium-linked cotransport. Only about 5 percent of the bile salts secreted by the liver enters the colon. Only about 1 percent is lost in feces.

Protein Digestion and Absorption

Proteins have very complex structures, so protein digestion is both complex and time-consuming. Proteolytic enzymes must access individual proteins, so the three-dimensional organization of food must first be disrupted. This step involves mechanical processing in the oral cavity, through mastication,

24

and chemical processing in the stomach, through the action of hydrochloric acid. The strong acid in the stomach kills pathogens and breaks down plant cell walls and animal connective tissues. Acid also disrupts tertiary and secondary protein structure, exposing peptide bonds to enzymatic attack.

The acidic content of the stomach also provides the proper environment for the activity of pepsin, the proteolytic enzyme secreted in an inactive form by chief cells of the stomach (**Spotlight Figure 24–27**). Pepsin works effectively at a pH of 1.5–2.0. It breaks the peptide bonds within a polypeptide chain.

When chyme enters the duodenum, enteropeptidase from the small intestine triggers the conversion of trypsinogen to trypsin. Buffers increase the pH to 7–8. Pancreatic proteases can now begin working. Trypsin, chymotrypsin, and elastase are like pepsin in that they break specific peptide bonds within a polypeptide. For example, trypsin breaks peptide bonds involving the amino acids *arginine* or *lysine*. Chymotrypsin targets peptide bonds involving *tyrosine* or *phenylalanine*.

Carboxypeptidase also acts in the small intestine. This enzyme chops off the last amino acid of a polypeptide chain, no matter which amino acids are involved. Thus, while the other peptidases generate a variety of short peptides, carboxypeptidase produces free amino acids.

The epithelial surfaces of the small intestine contain several peptidases, notably **dipeptidases**. These enzymes break short peptide chains into individual amino acids. (Dipeptidases break apart *dipeptides*.) These amino acids, as well as those produced by the pancreatic enzymes, are absorbed through both facilitated diffusion and cotransport mechanisms.

The amino acids diffuse through the cell to its basolateral surface. There they are released into interstitial fluid by facilitated diffusion and cotransport. Once in the interstitial fluid, the amino acids diffuse into intestinal capillaries for transport to the liver by means of the hepatic portal vein.

Water Absorption

Cells cannot actively absorb or secrete water. All movement of water across the lining of the digestive tract involves passive water flow down osmotic gradients. The production of glandular secretions also involves passive water flow down osmotic gradients. Recall that when a selectively permeable membrane separates two solutions, water tends to flow into the solution that has the higher concentration of solutes. ⤳ p. 119 Osmotic movements are rapid, so interstitial fluid and the fluids in the intestinal lumen always have the same osmotic concentration of solutes, or osmolarity.

Intestinal epithelial cells continuously absorb nutrients and ions. These activities gradually lower the solute concentration in the lumen. As the solute concentration drops, water

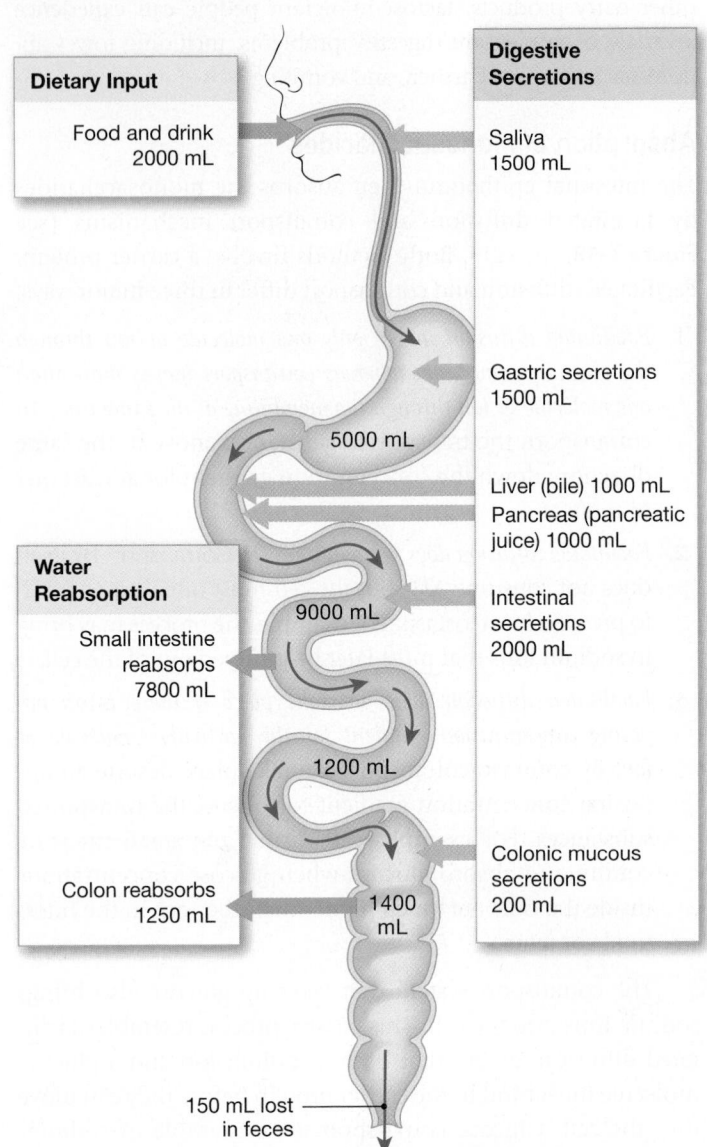

Figure 24–28 **Digestive Secretion and Water Reabsorption.** The purple arrows indicate secretion, and the blue arrows show water ingestion and reabsorption.

moves into the surrounding tissues, maintaining osmotic equilibrium.

Each day, about 2000 mL of water enters the digestive tract in the form of food or drink. The salivary, gastric, intestinal, colonic, pancreatic, and bile secretions provide an additional 7200 mL. Of that total, only about 150 mL is lost in feces. The sites of secretion and absorption of water are shown in **Figure 24–28**.

Ion Absorption

Osmosis does not distinguish among solutes. All that matters is the total concentration of solutes. To maintain homeostasis,

Table 24–2	The Absorption of Ions and Vitamins	
Ion or Vitamin	**Transport Mechanism**	**Regulatory Factors**
Na^+	Channel-mediated diffusion, cotransport, or active transport	Increased when sodium-linked cotransport is under way; stimulated by aldosterone
Ca^{2+}	Active transport	Stimulated by calcitriol and PTH
K^+	Channel-mediated diffusion	Follows concentration gradient
Mg^{2+}	Active transport	
Fe^{2+}	Active transport	
Cl^-	Channel-mediated diffusion or carrier-mediated transport	
I^-	Channel-mediated diffusion or carrier-mediated transport	
HCO_3^-	Channel-mediated diffusion or carrier-mediated transport	
NO_3^-	Channel-mediated diffusion or carrier-mediated transport	
PO_4^{3-}	Active transport	
SO_4^{2-}	Active transport	
Water-soluble vitamins (except B_{12})	Channel-mediated diffusion	Follows concentration gradient
Vitamin B_{12}	Active transport	Must be bound to intrinsic factor prior to absorption
Fat-soluble vitamins	Diffusion	Absorbed from micelles along with dietary lipids

however, the concentrations of specific ions must be closely regulated. Thus, each ion must be handled individually, and the rate of intestinal absorption of each must be tightly controlled (Table 24–2). Many of the regulatory mechanisms controlling the rates of absorption are poorly understood.

Sodium ions (Na^+) are usually the most abundant cations in food. They may enter intestinal cells by diffusion, by cotransport with another nutrient, or by active transport. These ions are then pumped into interstitial fluid across the base of the cell.

The rate of Na^+ uptake from the lumen is generally proportional to the concentration of Na^+ in the intestinal contents. As a result, eating heavily salted foods leads to increased sodium ion absorption and an associated gain of water through osmosis. The rate of sodium ion absorption by the digestive tract is increased by aldosterone, a steroid hormone from the adrenal cortex. ⤶ p. 658

Calcium ion (Ca^{2+}) absorption involves active transport at the epithelial surface. Calcitriol speeds up the rate of transport. ⤶ p. 657

As other solutes move out of the lumen, the concentration of potassium ions (K^+) increases. These ions can diffuse into the epithelial cells, driven by the concentration gradient. The absorption of magnesium (Mg^{2+}), iron (Fe^{2+}), and other cations involves specific carrier proteins. The cell must use ATP to obtain and transport these ions to interstitial fluid. Regulatory factors controlling their absorption are poorly understood.

The anions chloride (Cl^-), iodide (I^-), bicarbonate (HCO_3^-), and nitrate (NO_3^-) are absorbed by diffusion or

carrier-mediated transport. Phosphate (PO_4^{3-}) and sulfate (SO_4^{2-}) ions enter epithelial cells only by active transport.

Vitamin Absorption

Vitamins are organic compounds required in very small quantities. There are two major groups of vitamins: fat-soluble vitamins and water-soluble vitamins. Vitamins A, D, E, and K are **fat-soluble vitamins**. Their structure allows them to dissolve in lipids. The nine **water-soluble vitamins** include the B vitamins, common in milk and meats, and vitamin C, found in citrus fruits. We consider the functions of vitamins and associated nutritional problems in Chapter 25.

All but one of the water-soluble vitamins are easily absorbed by diffusion across the digestive epithelium. By itself vitamin B_{12} cannot be absorbed by the intestinal mucosa in normal amounts. This vitamin must be bound to *intrinsic factor*, a glycoprotein secreted by the parietal cells of the stomach. The combination is then absorbed through active transport.

Fat-soluble vitamins in the diet enter the duodenum in fat droplets, mixed with triglycerides. The vitamins remain in association with these lipids as they form emulsion droplets and, after further digestion, micelles. The fat-soluble vitamins are then absorbed from the micelles along with the fatty acids and monoglycerides. Vitamin K produced in the colon is absorbed with other lipids released through bacterial action. Taking supplements of fat-soluble vitamins while you have an empty stomach, are fasting, or are on a low-fat diet is relatively ineffective. The reason is that proper absorption of these vitamins requires the presence of other lipids.

24

CHECKPOINT

29. Name the nutrients the body requires.

30. What component of food would increase the number of chylomicrons in the lacteals?

31. The absorption of which vitamin would be impaired by the removal of the stomach?

32. Why is it that diarrhea is potentially life threatening, but constipation (infrequent defecation) is not?

See the blue Answers tab at the back of the book.

24-9 | Many age-related changes affect digestion and absorption

Learning Outcome Summarize the effects of aging on the digestive system.

Normal digestion and absorption take place in elderly people. However, many changes in the digestive system parallel age-related changes we have already discussed in connection with other systems:

- *The division rate of epithelial stem cells declines.* The digestive epithelium becomes more susceptible to damage by abrasion, acids, or enzymes. Peptic ulcers therefore become more likely. Stem cells in the epithelium divide less frequently with age, so tissue repair is less efficient. In the mouth, esophagus, and anus, the stratified epithelium becomes thinner and more fragile.

- *Smooth muscle tone decreases.* General motility decreases, and peristaltic contractions are weaker as a result of a decrease in smooth muscle tone. These changes slow the rate of fecal movement and promote constipation. Sagging and inflammation of the haustra in the colon can occur. Straining to eliminate compacted feces can stress the less resilient walls of blood vessels, producing hemorrhoids. Problems are not restricted to the lower digestive tract. For example, weakening of muscular sphincters can lead to esophageal reflux and frequent bouts of "heartburn."

- *The effects of cumulative damage become apparent.* A familiar example is the gradual loss of teeth due to *dental caries* (cavities) or gingivitis. Cumulative damage can involve internal organs as well. Toxins such as alcohol and other injurious chemicals that are absorbed by the digestive tract are transported to the liver for processing. The cells of the liver are not immune to these toxic compounds, and chronic exposure can lead to cirrhosis or other types of liver disease.

- *Cancer rates increase.* Cancers are most common in organs in which stem cells divide to maintain epithelial cell populations. Rates of colon cancer and stomach cancer rise with age. Oral, esophageal, and pharyngeal cancers are particularly common among elderly smokers.

- *Dehydration is common among the elderly.* One reason is that osmoreceptor sensitivity declines with age.

- *Changes in other systems have direct or indirect effects on the digestive system.* For example, reduction in bone mass and calcium content in the skeleton is associated with erosion of the tooth sockets and eventual tooth loss. The decline in olfactory and gustatory sensitivities with age can lead to dietary changes that affect the entire body.

CHECKPOINT

33. Identify general digestive system changes that occur with aging.

See the blue Answers tab at the back of the book.

24-10 | The digestive system is extensively integrated with other body systems

Learning Outcome Give examples of interactions between the digestive system and other organ systems studied so far.

Figure 24–29 summarizes the physiological relationships between the digestive system and other organ systems we have studied so far. The digestive system has particularly extensive anatomical and physiological connections to the nervous, cardiovascular, endocrine, and lymphatic systems. As we have seen, the digestive tract is also an endocrine organ that produces a variety of hormones. Many of these hormones, and some of the neurotransmitters produced by the digestive system, can enter the circulation, cross the blood–brain barrier, and alter CNS activity. In this way, a continual exchange of chemical information takes place among these systems.

CHECKPOINT

34. Identify the functional relationships between the digestive system and other body systems.

35. What body systems may be affected by inadequate calcium absorption?

See the blue Answers tab at the back of the book.

Integumentary

Provides vitamin D₃ needed for the absorption of calcium and phosphorus

Provides lipids for storage by adipocytes in subcutaneous layer

Integumentary — Page 200

Skeletal

Skull, ribs, vertebrae, and pelvic girdle support and protect parts of digestive tract; teeth are important in mechanical processing of food

Absorbs calcium and phosphate ions for incorporation into bone matrix; provides lipids for storage in yellow marrow

Skeletal — Page 311

Muscular

Protects and supports digestive organs in abdominal cavity; controls entrances and exits of digestive tract

Liver regulates blood glucose and fatty acid levels, metabolizes lactate from active muscles

Muscular — Page 406

Nervous

ANS regulates movement and secretion; reflexes coordinate passage of materials along tract; control over skeletal muscles regulates ingestion and defecation; hypothalamic centers control hunger, satiation, and feeding

Provides substrates essential for neurotransmitter synthesis

Nervous — Page 584

Endocrine

Epinephrine and norepinephrine stimulate constriction of sphincters and depress digestive activity; hormones coordinate activity along digestive tract

Provides nutrients and substrates to endocrine cells; endocrine cells of pancreas secrete insulin and glucagon; liver produces angiotensinogen

Endocrine — Page 673

Cardiovascular

Distributes hormones of the digestive tract; carries nutrients, water, and ions from sites of absorption; delivers nutrients and toxins to liver

Absorbs fluid to maintain normal blood volume; absorbs vitamin K; liver excretes heme (as bilirubin), synthesizes coagulation proteins

Cardiovascular — Page 802

Lymphatic

Tonsils and other lymphoid nodules defend against infection and toxins absorbed from the digestive tract; lymphatic vessels carry absorbed lipids to venous system

Secretions of digestive tract (acids and enzymes) provide innate defense against pathogens

Lymphatic — Page 850

Respiratory

Increased thoracic and abdominal pressure through contraction of respiratory muscles can assist in defecation

Pressure of digestive organs against the diaphragm can assist in exhalation and limit inhalation

Respiratory — Page 900

The DIGESTIVE System

For all systems, the digestive system absorbs organic substrates, vitamins, ions, and water required by all cells.

Figure 24–29 diagrams the functional relationships between the digestive system and the other body systems we have studied so far.

Urinary — Page 1036

Reproductive — Page 1116

Related Clinical Terms

borborygmus: A rumbling or gurgling sound made by the movement of fluids and gases in the intestines.

cathartics: Drugs that promote defecation.

cholelithiasis: The presence of gallstones in the gallbladder.

cholera: A bacterial infection of the digestive tract that causes massive fluid losses through diarrhea.

colitis: A general term for a condition characterized by inflammation of the colon.

Crohn's disease: An incurable chronic inflammatory bowel disease that can affect any part of the digestive tract, from the mouth to the anus. The presence of strictures, fistulas, and fissures is common.

diverticulitis: An infection and inflammation of mucosal pockets of the large intestine (diverticula).

diverticulosis: The formation of diverticula, generally along the sigmoid colon.

dysphagia: Difficulty or discomfort in swallowing due to disease.

esophageal varices: Swollen and fragile esophageal veins that result from portal hypertension.

fecal occult blood test: Test to check for hidden blood in feces.

gastrectomy: The surgical removal of the stomach, generally to treat advanced stomach cancer.

gastroesophageal reflux disease (GERD): Chronic condition in which the lower esophageal sphincter allows gastric acids to backflow into the esophagus, causing heartburn, acid indigestion, and possible injury to the esophageal lining.

gastroscope: A fiber-optic instrument inserted into the mouth and directed along the esophagus and into the stomach; used to examine the interior of the stomach and to perform minor surgical procedures.

halitosis: Bad breath that may be due to poor oral hygiene, an infection, diabetes, or other disease.

insoluble fiber: Indigestible plant carbohydrates that do not dissolve in water and pass through the GI tract unchanged. Found in many vegetables and the skins of fruits, insoluble fiber speeds up the passage of material in the GI tract. Individuals consuming diets rich in insoluble fiber decrease their risk for developing diabetes, atherosclerosis, and colorectal cancers, among other diseases.

irritable bowel syndrome (IBS): A common disorder affecting the large intestine, accompanied by cramping, abdominal pain, bloating, gas, diarrhea, and constipation.

pancreatic cancer: Malignancy of the pancreas that does not cause symptoms in its early stages, leading to late detection and a survival rate of only 4 percent.

periodontal disease: A loosening of the teeth within the alveolar sockets caused by erosion of the periodontal ligaments by acids produced through bacterial action.

polyps: Small growths with a stalk protruding from a mucous membrane that is usually benign.

pulpitis: An infection of the pulp of a tooth; treatment may involve a root canal procedure.

pyloric stenosis: Uncommon condition in which the muscle of the lower end of the stomach enlarges and prevents food from entering the small intestine.

pylorospasm: Spasm of the pyloric sphincter, accompanied by pain and vomiting.

root canal: Removal of the alveolar nerve in a severely damaged tooth.

soluble fiber: Indigestible plant carbohydrates found in beans, oats, and citrus fruits that dissolve in water when eaten, forming a gel within the digestive tract to slow the passage of material. Diets rich in soluble fiber lower blood cholesterol levels.

Chapter Review

Study Outline

24

MasteringA&P®

Access more chapter study tools online in the MasteringA&P Study Area:

- Chapter Quizzes, Chapter Practice Test, Art-labeling Activities, Animations, MP3 Tutor Sessions, and Clinical Case Studies
- Practice Anatomy Lab PAL™
- Interactive Physiology iP®
- A&P Flix A&PFlix
- PhysioEx PhysioEx

24-1 ▌ The digestive system, consisting of the digestive tract and accessory organs, has overlapping food utilization functions p. 907

1. The **digestive system** consists of the muscular **digestive tract** and various **accessory organs**. *(Figure 24–1)*

2. Double sheets of peritoneal membrane called **mesenteries** suspend the digestive tract. The **greater omentum** lies anterior to the abdominal viscera. Its adipose tissue provides padding, protection, insulation, and an energy reserve. *(Figure 24–2)*

3. The *lamina propria* and epithelium form the **mucosa** (mucous membrane) of the digestive tract. Proceeding outward is the **submucosa**, the **muscularis externa**, and a layer of areolar tissue called the *adventitia*. For viscera projecting into the peritoneal cavity, the muscularis externa is covered by a serous membrane called the **serosa**. *(Figure 24–3)*

4. The muscularis externa propels materials through the digestive tract by the contractions of **peristalsis**. **Segmentation** movements in the small intestine churn digestive materials. *(Figure 24–4)*

5. Neural reflexes, hormones, and local mechanisms control digestive tract activities. *(Figure 24–5)*

24-2 ▌ The oral cavity contains the tongue, salivary glands, and teeth, each with specific functions p. 915

6. The functions of the **oral cavity**, or **buccal cavity**, are (1) *sensory analysis* of foods; (2) *mechanical processing* by the

teeth, tongue, and palatal surfaces; (3) *lubrication*, by mixing with mucus and salivary gland secretions; and (4) limited *digestion* of carbohydrates and lipids.

7. The oral cavity is lined by the **oral mucosa**. The *hard* and *soft palates* form the roof of the oral cavity, and the *tongue* forms its floor. *(Figure 24–6)*

8. **Intrinsic** and **extrinsic tongue muscles** are controlled by the hypoglossal nerves. *(Figure 24–6)*

9. The **parotid**, **sublingual**, and **submandibular salivary glands** discharge their secretions into the oral cavity. *(Figure 24–7)*

10. **Mastication** (chewing) of the **bolus** occurs through the contact of the **occlusal** (opposing) **surfaces** of the **teeth**. The **periodontal ligament** anchors each tooth in an *alveolus*, or bony socket. **Dentin** forms the basic structure of a tooth. The **crown** is coated with **enamel**, the **root** with **cementum**. *(Figure 24–8)*

11. During childhood and early adulthood, the 20 primary teeth, or **deciduous teeth**, are replaced by the 32 teeth of the **secondary dentition**. *(Figure 24–9)*

24-3 | **The pharynx is a passageway between the oral cavity and esophagus** p. 920

12. Propulsion of the bolus through the **pharynx** results from contractions of the *pharyngeal constrictor muscles* and the *palatal muscles*, and from elevation of the larynx.

24-4 | **The esophagus is a muscular tube that transports solids and liquids from the pharynx to the stomach** p. 921

13. The **esophagus** carries solids and liquids from the pharynx to the stomach through the **esophageal hiatus**, an opening in the diaphragm. *(Figure 24–10)*

14. The esophageal mucosa consists of a stratified epithelium. Mucous secretion by esophageal glands of the submucosa reduces friction during the passage of foods. The proportions of skeletal and smooth muscle of the muscularis externa change from the pharynx to the stomach. *(Figure 24–10)*

15. Swallowing, or **deglutition**, can be divided into **buccal, pharyngeal**, and **esophageal phases**. Swallowing begins with the compaction of a bolus and its movement into the pharynx, followed by the elevation of the larynx, reflection of the *epiglottis*, and closure of the *glottis*. After the *upper esophageal sphincter* is opened, peristalsis moves the bolus down the esophagus to the *lower esophageal sphincter*. *(Figure 24–11)*

24-5 | **The stomach is a J-shaped organ that receives the bolus from the esophagus and aids in chemical and mechanical digestion** p. 923

16. The **stomach** has four major functions: (1) storage of ingested food, (2) mechanical breakdown of food, (3) disruption of chemical bonds by acid and enzymes, and (4) production of *intrinsic factor*.

17. The four regions of the stomach are the **cardia, fundus, body**, and **pylorus**. The **pyloric sphincter** guards the exit from the stomach. In a relaxed state, the stomach lining contains numerous **rugae** (ridges and folds). *(Figure 24–12)*

18. Within the **gastric glands**, *parietal cells* secrete *intrinsic factor* and *hydrochloric acid*. **Chief cells** secrete **pepsinogen**, which is converted by acids in the gastric lumen to the enzyme **pepsin**. **Enteroendocrine** cells of the stomach secrete several compounds, notably the hormone **gastrin**. *(Figures 24–13, 24–14)*

19. Gastric control involves (1) the **cephalic phase**, which prepares the stomach to receive ingested materials; (2) the **gastric phase**, which begins with the arrival of food in the stomach;

and (3) the **intestinal phase**, which controls the rate of gastric emptying. Vomiting, or **emesis**, is reverse peristalsis. *(Spotlight Figure 24–15)*

24-6 | **The small intestine digests and absorbs nutrients, and associated glandular organs assist with digestive processes** p. 927

20. Most of the important digestive and absorptive functions occur in the **small intestine**. The pancreas, liver, and gallbladder provide digestive secretions and buffers.

21. The small intestine consists of the **duodenum**, the **jejunum**, and the **ileum**. A sphincter, the **ileocecal valve**, marks the transition between the small and large intestines. *(Figure 24–16)*

22. The intestinal mucosa has transverse folds called **circular folds** (*plicae circulares*) and small projections called **intestinal villi**. These folds and projections increase the surface area for absorption. Each villus contains a terminal lymphatic called a **lacteal**. Pockets called **intestinal glands** are lined by enteroendocrine, mucous, and stem cells. *(Figures 24–16, 24–17)*

23. **Intestinal juice** moistens chyme, helps buffer acids, and holds digestive enzymes and digestive products in solution.

24. The **duodenal** (*submucosal* or *Brunner's*) **glands** of the duodenum produce mucus, bicarbonate ions, and the hormone *urogastrone*. The ileum contains masses of lymphoid tissue called *aggregated lymphoid nodules*, or *Peyer's patches*, near the entrance to the large intestine.

25. The **gastroenteric reflex**, initiated by stretch receptors in the stomach, stimulates motility and secretion along the entire small intestine. The **gastroileal reflex** triggers the relaxation of the ileocecal valve.

26. The **pancreatic duct** penetrates the wall of the duodenum. Within each lobule of the **pancreas**, ducts branch repeatedly before ending in the **pancreatic acini** (blind pockets). *(Figure 24–18)*

27. The pancreas has two functions: endocrine (secreting insulin and glucagon into the blood) and exocrine (secreting **pancreatic juice** into the small intestine). Pancreatic enzymes include **carbohydrases, lipases, nucleases**, and **proteolytic enzymes**.

28. The **liver** performs metabolic and hematological regulation and produces **bile**. The bile ducts from all the **liver lobules** unite to form the **common hepatic duct**. That duct meets the **cystic duct** to form the **common bile duct**, which empties into the duodenum. *(Figures 24–19 to 24–21)*

29. The liver lobule is the organ's basic functional unit. **Hepatocytes** form irregular plates arranged in the form of spokes of a wheel. **Bile canaliculi** carry bile to the **bile ductules**, which lead to **portal areas**. *(Figure 24–20)*

30. In **emulsification**, **bile salts** break apart large drops of lipids, making the lipids accessible to lipases secreted by the pancreas.

31. The liver is the primary organ involved in regulating the composition of circulating blood. All the blood leaving the absorptive surfaces of the digestive tract flows into the liver before entering the systemic circulation. The liver regulates metabolism as it removes and stores excess nutrients, vitamins, and minerals from the blood; mobilizes stored reserves; synthesizes needed nutrients; and removes waste products.

32. The liver's hematological activities include the monitoring of circulating blood by phagocytes and antigen-presenting cells; the synthesis of plasma proteins; the removal of circulating hormones and antibodies; and the removal or storage of toxins.

24

33. The liver synthesizes bile, which is composed of water, ions, bilirubin, cholesterol, and bile salts (an assortment of lipids).

34. The **gallbladder** stores, modifies, and concentrates bile. *(Figure 24–21)*

35. Neural and hormonal mechanisms coordinate the activities of the digestive glands. Gastrointestinal activity is stimulated by parasympathetic innervation and inhibited by sympathetic innervation. The **enterogastric**, **gastroenteric**, and **gastroileal reflexes** coordinate movement from the stomach to the large intestine.

36. Intestinal hormones include **secretin**, **cholecystokinin (CCK)**, **gastric inhibitory peptide (GIP)**, **vasoactive intestinal peptide (VIP)**, **gastrin**, and **enterocrinin**. *(Figures 24–22, 24–23)*

24-7 ▌ **The large intestine is divided into three parts with regional specialization** p. 942

37. The main functions of the **large intestine** are to (1) reabsorb water and compact materials into feces, (2) absorb vitamins produced by bacteria, and (3) store fecal material prior to defecation. The large intestine consists of the *cecum*, *colon*, and *rectum*. *(Figure 24–24)*

38. The **cecum** collects and stores material from the ileum and begins the process of compaction. The **appendix** is attached to the cecum. *(Figure 24–24)*

39. The **colon** has a larger diameter and a thinner wall than the small intestine. The colon bears **haustra** (pouches), **teniae coli** (longitudinal bands of muscle), and sacs of fat **(omental appendices)**. *(Figure 24–24)*

40. The four regions of the colon are the **ascending colon**, **transverse colon**, **descending colon**, and **sigmoid colon**. *(Figure 24–24)*

41. The **rectum** terminates in the **anal canal**, leading to the **anus**. *(Figure 24–24)*

42. Histological characteristics of the colon include the absence of villi and the presence of mucous cells and deep intestinal glands. *(Figure 24–25)*

43. The large intestine reabsorbs water and other substances such as vitamins, urobilinogen, bile salts, and toxins. Bacteria are responsible for intestinal gas, or **flatus**.

44. The gastroileal reflex moves materials from the ileum into the cecum while you eat. Distension of the stomach and duodenum stimulates **mass movements** of materials from the transverse colon through the rest of the large intestine and into the rectum. Muscular sphincters control the passage of fecal material to the anus. Distension of the rectal wall triggers the **defecation reflex**. *(Figure 24–26)*

24-8 ▌ **Digestion is the chemical alteration of food that allows the absorption and use of nutrients** p. 948

45. The digestive system first breaks down the physical structure of the ingested material and then disassembles the component molecules into smaller fragments by *hydrolysis*. *(Spotlight Figure 24–27; Table 24–1)*

46. Salivary and pancreatic amylases break down complex carbohydrates into *disaccharides* and *trisaccharides*. These in turn are broken down into *monosaccharides* by enzymes at the epithelial surface. The monosaccharides are then absorbed by the intestinal epithelium by facilitated diffusion or cotransport. *(Spotlight Figure 24–27)*

47. *Triglycerides* are emulsified into lipid droplets that interact with bile salts to form **micelles**. The fatty acids and monoglycerides resulting from the action of pancreatic lipase diffuse from the micelles across the intestinal epithelium. Triglycerides are then synthesized and released into the interstitial fluid, for transport to the general circulation by way of the lymphatic system. *(Spotlight Figure 24–27)*

48. Protein digestion involves a low pH (which destroys tertiary and quaternary structure), the gastric enzyme pepsin, and various pancreatic proteases. Peptidases liberate amino acids that are absorbed and exported to interstitial fluid. *(Spotlight Figure 24–27)*

49. About 2000 mL of water is ingested each day, and digestive secretions provide another 7200 mL. Nearly all is reabsorbed by osmosis. *(Figure 24–28)*

50. Various processes, including diffusion, cotransport, and carrier-mediated and active transport, are responsible for the movements of cations (sodium, calcium, potassium, and so on) and anions (chloride, iodide, bicarbonate, and so on) into epithelial cells. *(Table 24–2)*

51. The **water-soluble vitamins** (except B_{12}) diffuse easily across the digestive epithelium. **Fat-soluble vitamins** are enclosed within fat droplets and absorbed with the products of lipid digestion. *(Table 24–2)*

24-9 ▌ **Many age-related changes affect digestion and absorption** p. 954

52. Age-related changes include a thinner and more fragile epithelium due to a reduction in epithelial stem cell divisions, weaker peristaltic contractions as smooth muscle tone decreases, the effects of cumulative damage, increased cancer rates, and increased dehydration.

24-10 ▌ **The digestive system is extensively integrated with other body systems** p. 954

53. The digestive system has extensive anatomical and physiological connections to the nervous, endocrine, cardiovascular, and lymphatic systems. *(Figure 24–29)*

Review Questions

See the blue Answers tab at the back of the book.

LEVEL 1 Reviewing Facts and Terms

1. The enzymatic breakdown of large molecules into their basic building blocks is called **(a)** absorption, **(b)** secretion, **(c)** mechanical processing, **(d)** chemical digestion.

2. The outer layer of the digestive tract is known as the **(a)** serosa, **(b)** mucosa, **(c)** submucosa, **(d)** muscularis.

3. The four regions of the stomach are **(a)** caecum, ileum, taenia coul, haustra, **(b)** greater curvature, lesser curvature, lesser omentum, greater omentum, **(c)** cardia, fundus, body, pylorus.

4. A branch of the hepatic portal vein, hepatic artery proper, and branch of the bile duct form **(a)** a liver lobule, **(b)** the sinusoids, **(c)** a portal area, **(d)** the hepatic duct, **(e)** the pancreatic duct.

5. Label the digestive system structures in the following figure.

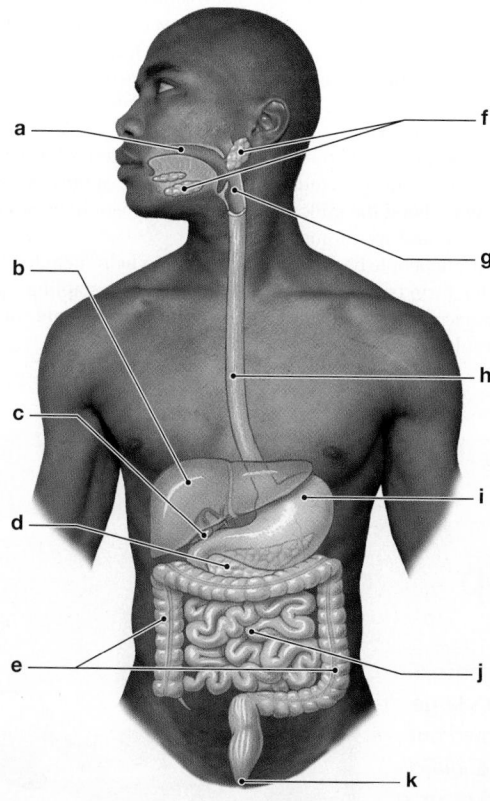

(a) _____	(b) _____
(c) _____	(d) _____
(e) _____	(f) _____
(g) _____	(h) _____
(i) _____	(j) _____
(k) _____	

6. Label the four layers of the digestive tract in the following figure.

Mesenteric
artery and vein

| (a) _____ | (b) _____ |
| (c) _____ | (d) _____ |

7. Which of the following will cause the release of pancreatic fluid into the duodenum during digestion? **(a)** high blood glucose, **(b)** ghrelin and leptin, **(c)** somatostatin and obestatin, **(d)** secretin and cholecystokinin.

8. Regional movements that occur in the small intestine and function to churn and fragment the digestive material are called **(a)** segmentation, **(b)** pendular movements, **(c)** peristalsis, **(d)** mass movements, **(e)** mastication.

9. The primary function of bile is to **(a)** break down proteins, **(b)** emulsify fats, **(c)** stimulate the enterogastric reflex, **(d)** deactivate pancreatic enzymes.

10. Label the three segments of the small intestine in the following figure.

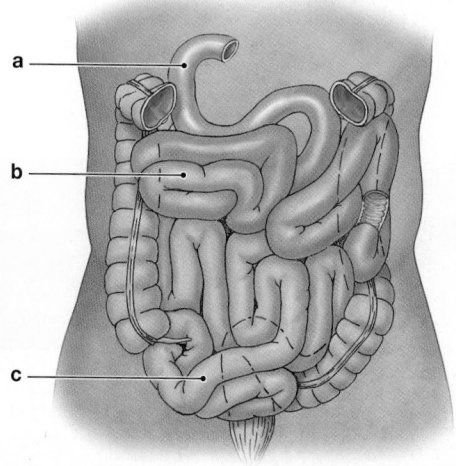

| (a) _____ | (b) _____ |
| (c) _____ | |

11. What is the function of Kupffer cells? **(a)** remove circulating hormones and antibodies, **(b)** metabolize drugs into less harmful metabolites, **(c)** phagocytize pathogens, cell debris, and old red blood cell, **(d)** remove excess amino acids and triglycerides.

12. Vitamins generated by bacteria in the colon are **(a)** vitamins A, D, and E, **(b)** B complex vitamins and vitamin C, **(c)** vitamin K, biotin, and pantothenic acid, **(d)** niacin, thiamine, and riboflavin.

13. The final enzymatic steps in the digestive process are accomplished by **(a)** brush border enzymes of the intestinal microvilli, **(b)** enzymes secreted by the stomach, **(c)** enzymes secreted by the pancreas, **(d)** the action of bile from the gallbladder.

14. List the main changes to digestion and absorption that occur with age.

15. Nutrients move from the small intestine into the body by **(a)** carrier mediated transport, **(b)** active transport, **(c)** co-transport, **(d)** all of the above.

16. What three basic mechanisms regulate the activities of the digestive tract?

17. What are the three phases of swallowing, and how are they controlled?

18. What are the primary digestive functions of the pancreas, liver, and gallbladder?

19. Which hormones produced by duodenal enteroendocrine cells effectively coordinate digestive functions?

20. Describe two ways in which the digestive and respiratory systems interact.

21. What two positive feedback loops are involved in the defecation reflex?

LEVEL 2 Reviewing Concepts

22. During defecation, **(a)** stretch receptors in the rectal wall initiate a series of peristaltic contractions in the colon and rectum, **(b)** stretch receptors in the rectal wall activate parasympathetic centers in the sacral region of the spinal cord, **(c)** the internal anal sphincter relaxes while the external anal sphincter contracts, **(d)** all of these occur, **(e)** only a and b occur.

24

23. Describe the role of bile in the digestion of food and explain how its release is normally regulated.
24. How does stratified squamous epithelium relate to the function of the pharynx?
25. Describe how peristalsis works to propel food through the gastrointestinal tract.
26. Describe the role of the stomach in the chemical breakdown of food.
27. How does each of the three phases of gastric secretion promote and facilitate gastric control?
28. Nutritionists have found that after a heavy meal, the pH of blood increases slightly, especially in the veins that carry blood away from the stomach. What causes this increase in blood pH?

LEVEL 3 Critical Thinking and Clinical Applications

29. Some people with gallstones develop pancreatitis. How could this occur?
30. Rayee is 42 and has been clinically obese most of his life. Despite having cirrhosis of the liver from non-alcoholic fatty liver disease, he still has fast food and sugary drinks. Blood was recently found in his stools. His doctor explains that liver problems are causing his esophagus to bleed. How could this happen?
31. A condition known as lactose intolerance is characterized by painful abdominal cramping, gas, and diarrhea. The cause of the problem is an inability to digest the milk sugar lactose. How would this cause the observed signs and symptoms?
32. Recently, more people have turned to surgery to help them lose weight. One form of weight control surgery involves stapling a portion of the stomach shut, creating a smaller volume. How would such a surgery result in weight loss?

CLINICAL CASE Wrap-Up

An Unusual Transplant

Normally the lining of the colon is slimy with mucus and covered with millions of beneficial bacteria. These friendly, "good" bacteria generate vitamins K, B_5, and biotin. They also convert bilirubin to urobilinogens and stercobilinogens. The bacteria also break down the remaining peptides in feces.

When these normal, beneficial intestinal bacteria have been killed by antibiotics, they are always replaced by other bacteria. Often the replacing bacteria are harmful, producing toxins, irritating the intestinal lining, and causing colitis. This is the case with *Clostridium difficile* infections.

One way of treating *C. difficile* infections is with a fecal transplant, as Mirasol's doctor suggests. Normal feces from a healthy donor contain populations of good bacteria that usually colonize large intestines. With a minimal amount of processing, the donated feces can be infused through a nose tube (nasogastric or NG tube). Then the normal intestinal bacteria can

take up residence in the recipient's large intestine, crowd out the harmful *C. difficile* bacteria, and resume their usual work of protecting the lining of the large intestine, producing vitamins, and recycling.

After overcoming her initial disgust, Mirasol agrees to the transplant. She recovers completely and leaves the hospital two days after her treatment.

1. If Mirasol's *Clostridium difficile* colitis is not treated with a fecal transplant, what kind of health problems might she continue to experience?

2. If you could look inside Mirasol's large intestine with a colonoscope, what do you think you would see?

See the blue Answers tab at the back of the book.

24

25

Metabolism and Energetics

Learning Outcomes

These Learning Outcomes correspond by number to this chapter's sections and indicate what you should be able to do after completing the chapter.

25-1 ▎ Define energetics and metabolism, and explain why cells must synthesize new organic components.

25-2 ▎ Describe the basic steps in glycolysis, the citric acid cycle, and the electron transport system, and summarize the energy yields of glycolysis and cellular respiration.

25-3 ▎ Describe the pathways involved in lipid metabolism, and summarize the mechanisms of lipid transport and distribution.

25-4 ▎ Summarize the main processes of protein metabolism, and discuss the use of protein as an energy source.

25-5 ▎ Differentiate between the absorptive and postabsorptive metabolic states, and summarize the characteristics of each.

25-6 ▎ Explain what makes up a balanced diet and why such a diet is important.

25-7 ▎ Define metabolic rate, discuss the factors involved in determining an individual's BMR, and discuss the homeostatic mechanisms that maintain a constant body temperature.

SPOTLIGHT

• Absorptive and Postabsorptive States pp. 980–981

CLINICAL CASE

• The Miracle Supplement p. 962

Clinical Notes

• Carbohydrate Loading p. 973

• Dietary Fats and Cholesterol p. 977
• Vitamins p. 986
• Alcohol p. 988
• Induced Hypothermia p. 991
• Thermoregulatory Disorders p. 993

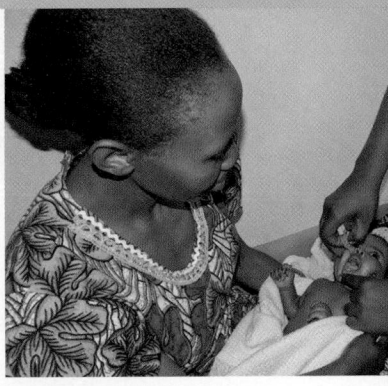
The Miracle Supplement

Justine, a pregnant 30-year-old Kenyan woman, is walking home from the clinic after her final pre-natal visit. As the sun sets, she realizes she is suddenly no longer able to see. Terrified, she manages to walk home blind, in rural Kenya, after dark. The next morning, her vision returns.

Justine knows children who have experienced night blindness. Many of them died young; in fact, every family in Justine's village has experienced childhood death. Several of the village's children also have stunted growth and slow skeletal development. They seem to have diminished ability to fight infections, and commonly develop malarial, respiratory, and diarrheal infections.

Fortunately, Justine goes on to deliver a healthy baby at the clinic. Shortly after giving birth, doctors treat Justine with a single high-dose vitamin supplement. She is told this will help her night blindness and ensure that there is enough of this vitamin in her breast milk to keep her baby healthy too.

When Justine's baby is six months old, Justine takes her for her first dose of vitamin supplement. This will be repeated every six months until the child is five. Costing only two cents per dose and reducing overall childhood mortality (death rate) by 23 percent, this vitamin supplement has been called the best public health investment ever. **What could this critical vitamin be? To find out, turn to Clinical Case Wrap-Up on p. 997.**

An Introduction to Metabolism and Energetics

The amount and type of nutrients you get from meals can vary widely. Your body builds energy reserves when nutrients are abundant. It mobilizes these reserves when nutrients are in short supply. The nervous and endocrine systems adjust and coordinate the metabolic activities of the body's tissues. These systems control the storage and mobilization of these energy reserves. In this chapter we examine (1) how the body obtains energy from the breakdown of organic molecules, stores it in the form of the energy transfer molecule ATP, and uses it to support intracellular operations such as the construction of new organic molecules, and (2) how the body balances heat gains and losses to maintain homeostasis.

25-1 ▎ Metabolism refers to all the chemical reactions in the body, and energetics refers to the flow and transformation of energy

Learning Outcome Define energetics and metabolism, and explain why cells must synthesize new organic components.

Cells are chemical factories that break down organic molecules to obtain energy. Our cells can then use this energy to generate ATP. Reactions within mitochondria provide most of the energy a typical cell needs. ⟲ p. 106 To carry out these reactions, cells must have a reliable supply of oxygen and nutrients, including water, vitamins, mineral ions, and organic substrates (the reactants in enzymatic reactions).

The lungs absorb oxygen. The other substances are absorbed by the digestive tract. The cardiovascular system then carries these substances throughout the body. They diffuse from the bloodstream into the tissues, where our cells can absorb and use them.

Mitochondria break down the organic nutrients to provide energy for cell growth, cell division, contraction, secretion, and other functions. Each tissue contains a unique mixture of various kinds of cells. As a result, the energy and nutrient requirements of any two tissues, such as loose connective tissue and cardiac muscle, can be quite different. Also, activity levels can change rapidly within a tissue, and such changes affect the requirements of the body. For example, when skeletal muscles start contracting, the tissue demand for oxygen increases dramatically. Thus, the energy and nutrient requirements of the body vary from moment to moment (resting versus exercising), hour to hour (asleep versus awake), and year to year (growing child versus adult). Understanding energy requirements is one aspect of **energetics**, the study of the flow of energy and its change(s) from one form to another.

The term **metabolism** (me-TAB-ō-lizm) refers to all the chemical reactions that take place in an organism. Chemical reactions within cells, collectively known as *cellular metabolism*, provide the energy needed to maintain homeostasis and to perform essential functions. Such functions include (1) *metabolic turnover*, the periodic breakdown and replacement of the organic components of a cell; (2) growth and cell division; and (3) special processes, such as secretion, contraction, and the propagation of action potentials.

Figure 25–1 provides a broad overview of the processes involved in cellular metabolism. The cell absorbs organic molecules from the surrounding interstitial fluids. Amino acids,

Figure 25–1   **An Introduction to Cellular Metabolism.** Cells obtain organic molecules from the interstitial fluid and break them down to produce ATP. Only about 40 percent of the energy released by catabolism is captured in the form of ATP; the rest is lost as heat. The ATP generated by catabolism provides energy for all vital cellular activities, including anabolism.

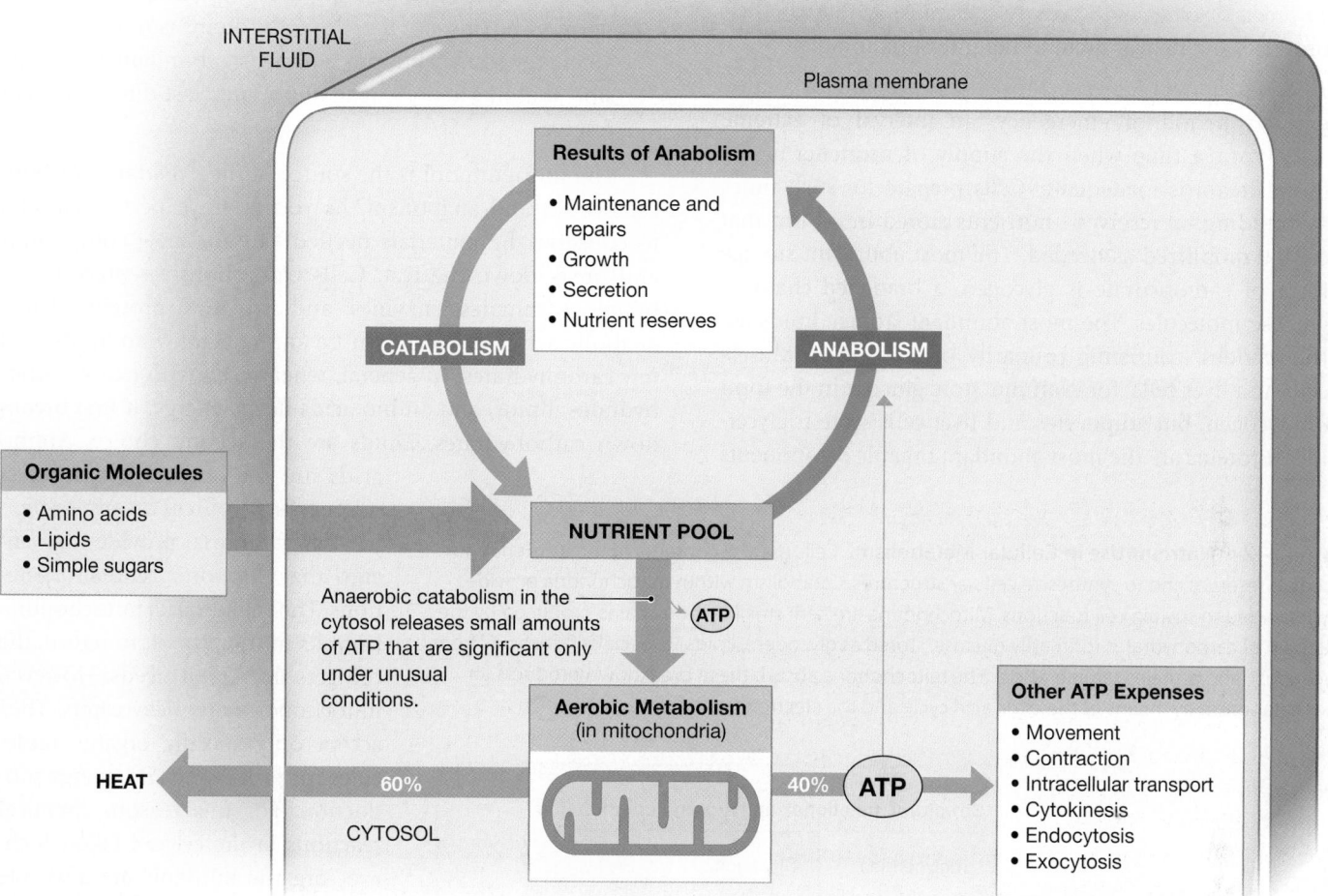

lipids, and simple sugars cross the plasma membrane and join nutrients already in the cytosol. All the cell's organic building blocks collectively form a *nutrient pool* that the cell draws on to provide energy and to create new intracellular components.

The breakdown of organic molecules is called **catabolism**. This process releases energy that can be used to synthesize ATP or other high-energy compounds. Catabolism proceeds in a series of steps. In general, the first steps take place in the cytosol. Enzymes there break down large organic molecules previously assembled by the cell (such as glycogen, triglycerides, or proteins) into smaller fragments that join the nutrient pool. For example, carbohydrates are broken down into simple sugars, triglycerides are split into fatty acids and glycerol, and proteins are broken down to individual amino acids.

These preparatory steps produce relatively little ATP. However, further catabolic activity yields smaller organic molecules that mitochondria can absorb and process. Mitochondrial activity releases significant amounts of energy. As mitochondrial enzymes break the covalent bonds that hold these molecules

together, they capture about 40 percent of the energy released and use it to convert ADP to ATP. The other 60 percent escapes as heat that warms the cell and the surrounding tissues.

The ATP produced by mitochondria provides energy to support both **anabolism**, which is the synthesis of new organic molecules, and other cell functions. Those functions vary among cell types, and they include ciliary or cell movement, contraction, active transport, and cell division. For example, muscle fibers need ATP to provide energy for contraction, but gland cells need ATP to synthesize and transport secretions. We have considered such specialized functions in other chapters, so here we focus on anabolic processes.

In terms of energy, anabolism is an "uphill" process that involves the formation of new chemical bonds. Cells synthesize new organic components for four basic reasons:

1. *To Carry Out Structural Maintenance or Repairs.* All cells must expend energy for ongoing maintenance and repairs, because most structures in the cell are temporary rather

than permanent. Their removal and replacement are part of the process of *metabolic turnover.* ⤳ p. 85

2. *To Support Growth.* Cells preparing to divide increase in size and synthesize extra proteins and organelles.

3. *To Produce Secretions.* Secretory cells must synthesize their products and deliver them to the interstitial fluid.

4. *To Store Nutrient Reserves.* Most cells "prepare for a rainy day"—a period of emergency, an interval of extreme activity, or a time when the supply of nutrients in the bloodstream is inadequate. Cells prepare for such times by building up reserves—nutrients stored in a form that can be mobilized as needed. The most abundant storage form of carbohydrate is glycogen, a branched chain of glucose molecules. The most abundant storage lipids are triglycerides, consisting primarily of fatty acids. Muscle cells and liver cells, for example, store glucose in the form of glycogen, but adipocytes and liver cells store triglycerides. Proteins are the most abundant organic components

in the body. They perform a variety of vital functions for the cell, and when energy is available, cells synthesize additional proteins. However, when glucose or fatty acids are unavailable, cells can break down proteins into their component amino acids, and use the amino acids as an energy source. The primary function of proteins is not to serve as an energy source, but proteins are so abundant and accessible that they represent an important "last-ditch" nutrient reserve.

The nutrient pool is the source of the substrates for both catabolism and anabolism. As you might expect, cells tend to conserve the materials needed to build new compounds and break down the rest. Cells continuously replace membranes, organelles, enzymes, and structural proteins. These anabolic activities require more amino acids than lipids, and few carbohydrates. In general, when a cell with excess carbohydrates, lipids, and amino acids needs energy, it first breaks down carbohydrates. Lipids are the second choice. Amino acids are seldom broken down if other energy sources are available.

Mitochondria provide the energy that supports cellular operations. The cell feeds its mitochondria from its nutrient pool. In return, the cell gets the ATP it needs. However, mitochondria are picky eaters: They accept only specific organic molecules for processing and energy production. For this reason, chemical reactions in the cytosol take whichever organic nutrients are available and break them down into smaller fragments that the mitochondria can process. The mitochondria then break the fragments down further, generating carbon dioxide, water, and ATP (Figure 25–2). This mitochondrial activity involves two pathways: the *citric acid cycle* and the *electron transport system.* We describe these important catabolic and anabolic cellular reactions in the next section.

Figure 25–2 **Nutrient Use in Cellular Metabolism.** Cells use the contents of the nutrient pool to build up reserves and to synthesize cellular structures. Catabolism within mitochondria provides the ATP needed to sustain cell functions. Mitochondria are "fed" small carbon chains produced by the breakdown of carbohydrates (primarily glucose, stored as glycogen), lipids (especially fatty acids from triglycerides), and proteins (amino acids). The mitochondria absorb these breakdown products for further catabolism by means of the citric acid cycle and the electron transport system.

CHECKPOINT

1. **Define metabolism.**

2. **Define catabolism.**

3. **Define anabolism.**

See the blue Answers tab at the back of the book.

25-2 ❙ Carbohydrate metabolism involves glycolysis, ATP production, and gluconeogenesis

Learning Outcome Describe the basic steps in glycolysis, the citric acid cycle, and the electron transport system, and summarize the energy yields of glycolysis and cellular respiration.

Most cells generate ATP and other high-energy compounds by breaking down carbohydrates, especially glucose. We can summarize the complete reaction as follows:

$$\underset{\substack{\text{glucose}}}{C_6H_{12}O_6} + \underset{\substack{\text{oxygen}}}{6O_2} \longrightarrow \underset{\substack{\text{carbon} \\ \text{dioxide}}}{6CO_2} + \underset{\substack{\text{water}}}{6H_2O}$$

The breakdown takes place in a series of small steps. Several of these steps release sufficient energy to convert ADP to ATP. The complete catabolism of one molecule of glucose provides a typical body cell a net gain of 36 molecules of ATP.

Most ATP production takes place inside mitochondria, but the first steps take place in the cytosol. In Chapter 10 we introduced the process of *glycolysis*, which breaks down glucose in the cytosol into smaller molecules that mitochondria can absorb and utilize. ⸠ p. 343 These reactions are said to be *anaerobic* because glycolysis does not require oxygen. The subsequent reactions take place in mitochondria. These reactions use oxygen and are considered *aerobic*. The mitochondrial activity responsible for ATP production is called **aerobic metabolism**, also known as **cellular respiration**.

Glycolysis

Glycolysis (glī-KOL-i-sis; *glykus*, sweet + *lysis*, a loosening) is the breakdown of glucose to **pyruvic acid**. In this process, a series of enzymatic steps breaks the six-carbon glucose molecule ($C_6H_{12}O_6$) into two 3-carbon molecules of pyruvic acid (CH_3—CO—COOH). At the normal pH inside cells, each pyruvic acid molecule loses a hydrogen ion and exists as the negatively charged ion CH_3—CO—COO⁻. This ionized form is called **pyruvate**.

Glycolysis requires (1) glucose molecules; (2) appropriate cytosolic enzymes; (3) ATP and ADP; (4) inorganic phosphates; and (5) **NAD** (*nicotinamide adenine dinucleotide*), a coenzyme that removes hydrogen atoms during one of the enzymatic reactions. Recall from Chapter 2 that coenzymes are organic molecules that are essential to enzyme function. ⸠ p. 81 If any of these participants is missing, glycolysis cannot take place.

Figure 25–3 provides an overview of the steps in glycolysis. Glycolysis begins when an enzyme *phosphorylates*—that is, attaches a phosphate group—to the last (sixth) carbon atom of a glucose molecule, creating **glucose-6-phosphate**. This step "costs" the cell 1 ATP molecule. It has two important

results: (1) It traps the glucose molecule within the cell, because phosphorylated glucose cannot cross the plasma membrane; and (2) it prepares the glucose molecule for further biochemical reactions.

A second phosphorylation takes place in the cytosol before the 6-carbon chain is broken into two 3-carbon fragments. Then energy benefits begin to appear as these fragments are converted to pyruvate. Two of the steps release enough energy to generate ATP from ADP and inorganic phosphate (PO_4^{3-} or P_i). In addition, 2 molecules of NAD are converted to NADH. The net reaction of glycolysis looks like this:

Glucose + 2NAD + 2ADP + 2 P_i ⟶

2 Pyruvate + 2 NADH + 6ATP

This reaction sequence is anaerobic (without oxygen) and provides the cell a net gain of 2 molecules of ATP for each glucose molecule converted to two pyruvate molecules. A few highly specialized cells, such as red blood cells, lack mitochondria and derive all their ATP through glycolysis. Skeletal muscle fibers rely on glycolysis for ATP production during periods of active contraction. Most cells can survive for brief periods using the ATP provided by glycolysis alone. However, when oxygen is readily available, mitochondrial activity provides most of the ATP that cells need.

Mitochondrial ATP Production

For the cell, glycolysis yields an immediate net gain of 2 ATP molecules for each glucose molecule it breaks down. However, a great deal of additional energy is still stored in the chemical bonds of pyruvate. The cell's ability to capture that energy depends on the availability of oxygen. If oxygen supplies are adequate, mitochondria absorb the pyruvate molecules and break them down. As you will see in the next section, the hydrogen atoms of each pyruvate molecule (CH_3—CO—COO⁻) are removed by coenzymes and are ultimately the source of most of the cell's energy gain. The carbon and oxygen atoms are removed and released as carbon dioxide in a process called **decarboxylation** (dē-kar-boks-i-LĀ-shun).

Recall that a double membrane surrounds each mitochondrion. The *outer membrane* contains large pores that are permeable to ions and small organic molecules such as pyruvate. Ions and molecules easily enter the *intermembrane space* separating the outer membrane from the *inner membrane*. The inner membrane contains a carrier protein that moves pyruvate into the mitochondrial matrix, where it is broken down through the citric acid cycle.

The Citric Acid Cycle

In the mitochondrion, a pyruvate molecule participates in a complex reaction involving NAD and another coenzyme called **coenzyme A**, or **CoA**. This reaction yields 1 molecule of carbon dioxide, 1 of NADH, and 1 of **acetyl-CoA** (a-SĒ-til KŌ-ā)—a 2-carbon **acetyl group** (CH_3CO) bound to coenzyme A. This sets the stage for a sequence of enzymatic reactions

25

Figure 25–3   Glycolysis. Glycolysis breaks down a 6-carbon glucose molecule into two 3-carbon molecules of pyruvate through a series of enzymatic steps. Further catabolism of pyruvate begins with its entry into a mitochondrion.

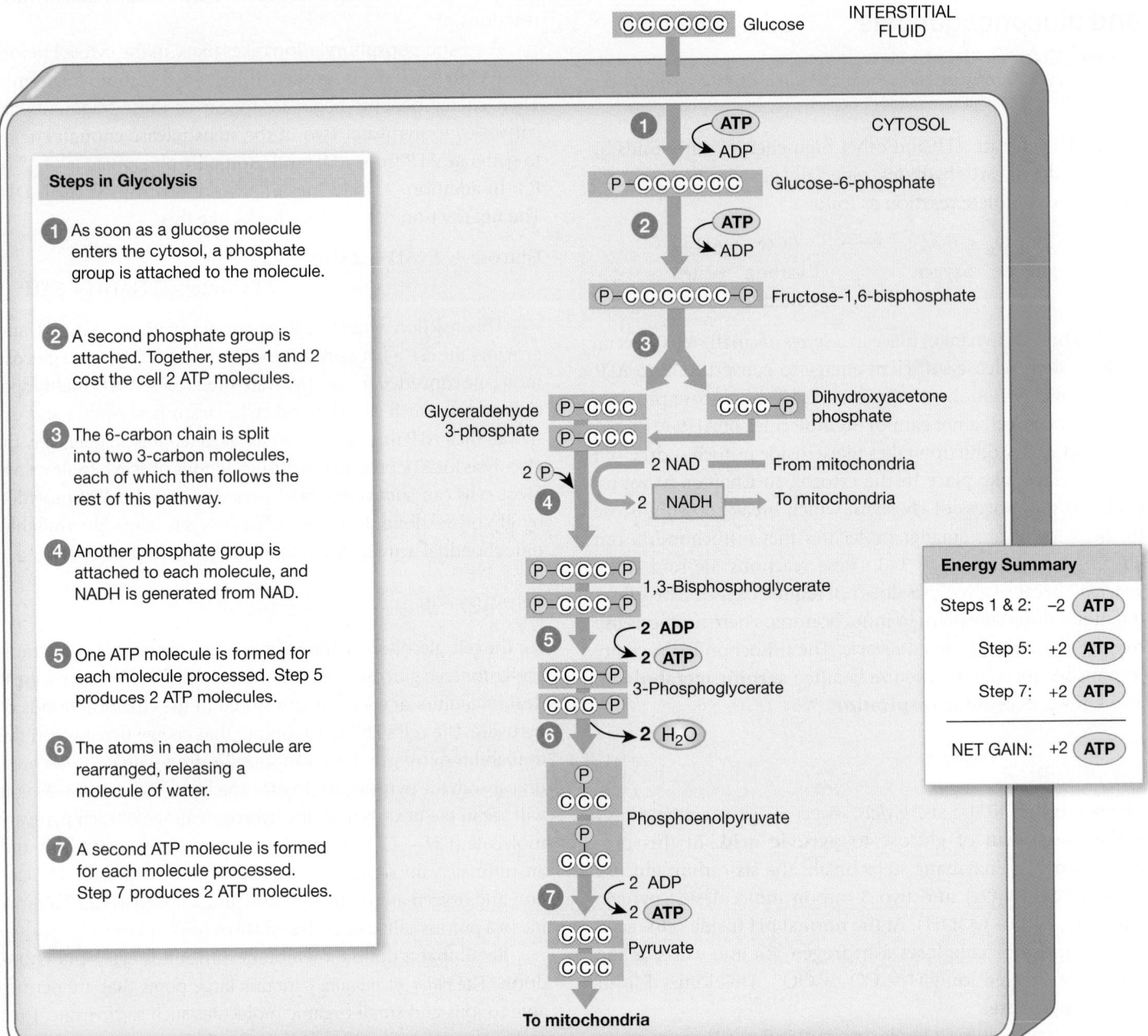

Steps in Glycolysis

1 As soon as a glucose molecule enters the cytosol, a phosphate group is attached to the molecule.

2 A second phosphate group is attached. Together, steps 1 and 2 cost the cell 2 ATP molecules.

3 The 6-carbon chain is split into two 3-carbon molecules, each of which then follows the rest of this pathway.

4 Another phosphate group is attached to each molecule, and NADH is generated from NAD.

5 One ATP molecule is formed for each molecule processed. Step 5 produces 2 ATP molecules.

6 The atoms in each molecule are rearranged, releasing a molecule of water.

7 A second ATP molecule is formed for each molecule processed. Step 7 produces 2 ATP molecules.

Energy Summary

Steps 1 & 2:	–2 **ATP**
Step 5:	+2 **ATP**
Step 7:	+2 **ATP**
NET GAIN:	+2 **ATP**

called the **citric acid cycle**. It is also known as the *tricarboxylic* (trī-kar-bok-SIL-ik) *acid cycle* (TCA cycle) and the *Krebs cycle*. Because citric acid is the first substrate of the cycle, we'll use citric acid cycle as the preferred term.

The function of the citric acid cycle is to remove hydrogen atoms from organic molecules and transfer them to coenzymes. The overall pattern of the citric acid cycle is shown in **Figure 25–4**.

At the start of the citric acid cycle, the 2-carbon acetyl group carried by CoA is transferred to a 4-carbon oxaloacetic acid molecule to make the 6-carbon compound citric acid. Coenzyme A is released intact and can thus bind another acetyl group. A

complete revolution or "turn" of the citric acid cycle removes 2 carbon atoms, regenerating the 4-carbon chain. (This is why the reaction sequence is called a *cycle*.) We can summarize the fate of the atoms in the acetyl group as follows:

- The 2 carbon atoms are removed in enzymatic reactions that incorporate 4 oxygen atoms and form 2 molecules of carbon dioxide, a waste product.

- The hydrogen atoms are removed by the coenzyme NAD or a related coenzyme called **FAD** (*f*lavin *a*denine *d*inucleotide) (**Figure 25–4**).

Figure 25–4 **The Citric Acid Cycle.**

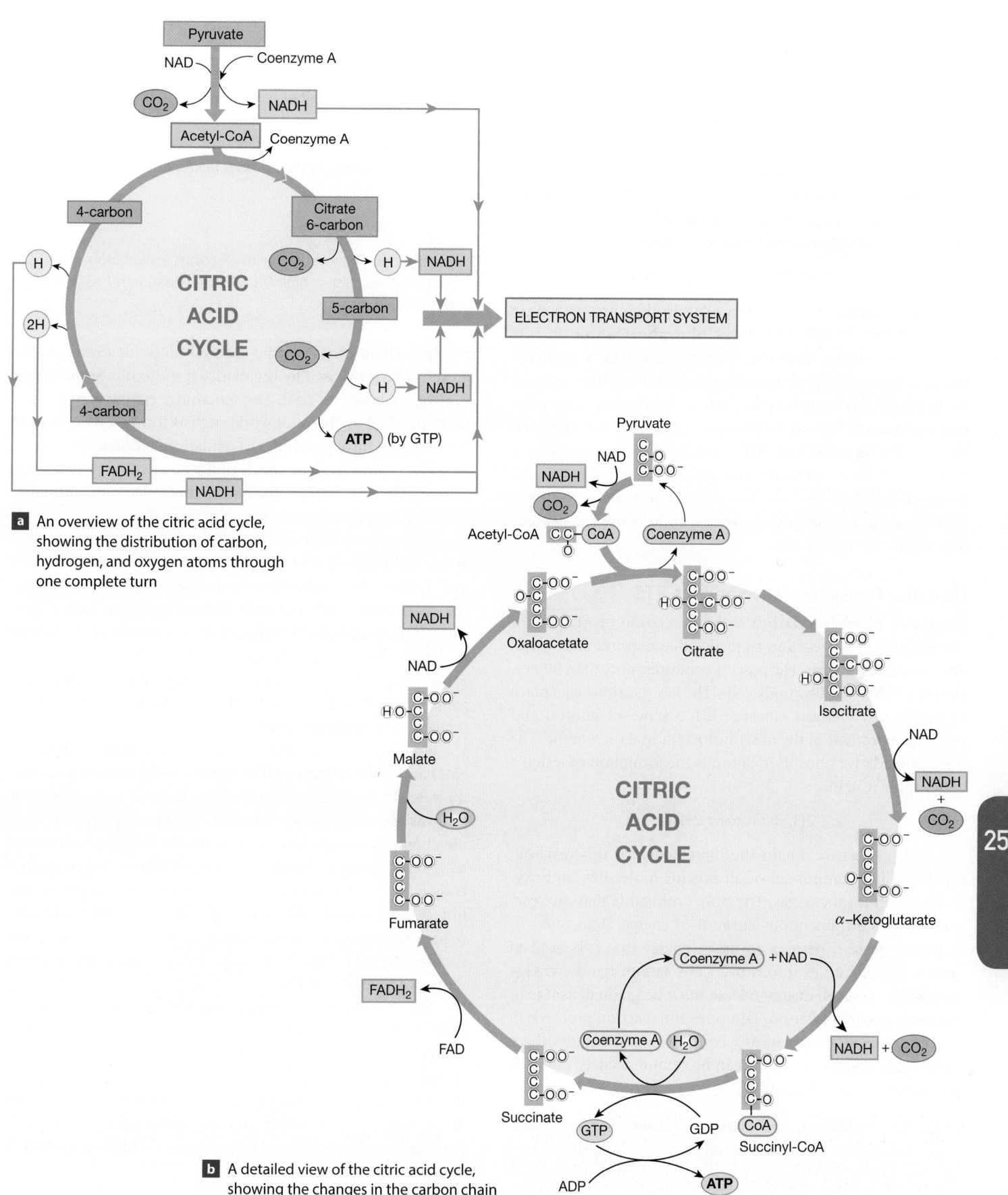

a An overview of the citric acid cycle, showing the distribution of carbon, hydrogen, and oxygen atoms through one complete turn

b A detailed view of the citric acid cycle, showing the changes in the carbon chain

Several of the steps in a turn of the citric acid cycle involve more than one reaction and require more than one enzyme. Water molecules are tied up in two of those steps. We can summarize the entire sequence as follows:

$$CH_3CO - CoA + 3NAD + FAD + GDP + P_i + 2H_2O \longrightarrow$$
$$CoA + 2CO_2 + 3NADH + FADH_2 + 2H^+ + GTP$$

The only immediate energy benefit of one turn of the citric acid cycle is the formation of a single molecule of GTP (*guanosine triphosphate*) from GDP (*guanosine diphosphate*) and P_i. In practical terms, GTP is the equivalent of ATP, because GTP readily transfers a phosphate group to ADP, producing ATP:

$$GTP + ADP \longrightarrow GDP + ATP$$

The formation of GTP from GDP in the citric acid cycle is an example of **substrate-level phosphorylation**. In this process, an enzyme uses the energy released by a chemical reaction to transfer a phosphate group to a suitable acceptor molecule. GTP is formed in the citric acid cycle, but many reaction pathways in the cytosol phosphorylate ADP and form ATP directly. For example, the ATP produced during glycolysis is generated through substrate-level phosphorylation. Normally, however, substrate-level phosphorylation provides a relatively small amount of energy compared with *oxidative phosphorylation*, which we discuss next.

Oxidative Phosphorylation and the ETS

Oxidative phosphorylation is the generation of ATP within mitochondria in a reaction sequence that requires coenzymes and consumes oxygen. The process produces more than 90 percent of the ATP used by body cells. The key reactions take place in the *electron transport system (ETS)*, a series of integral and peripheral proteins in the inner mitochondrial membrane. The basis of oxidative phosphorylation is the formation of water, a very simple reaction:

$$2H_2 + O_2 \longrightarrow 2H_2O$$

Cells can easily obtain the ingredients for this reaction. Hydrogen is a component of all organic molecules, and oxygen is an atmospheric gas. The only problem is that the reaction releases a tremendous amount of energy all at once. In fact, this reaction releases so much energy that it is used to launch space shuttles into orbit. Cells cannot handle energy explosions. Instead, energy release must be gradual, as it is in oxidative phosphorylation. This powerful reaction proceeds in a series of small, enzymatically controlled steps. Under these controlled conditions, energy can be captured safely, and ATP generated.

Oxidation, Reduction, and Energy Transfer. The enzymatic steps of oxidative phosphorylation involve oxidation and reduction. (There are different types of oxidation and reduction reactions, but for our purposes the most important are those involving the transfer of electrons.) The loss of electrons is a form of **oxidation**. The gain of electrons is a form of **reduction**. The two reactions are always paired. When electrons pass from one molecule to another, the electron donor is oxidized and the electron recipient is reduced. Oxidation and reduction are important because electrons carry chemical energy. In a typical oxidation–reduction reaction, *the reduced molecule gains energy at the expense of the oxidized molecule.*

In such an exchange, the reduced molecule does not gain all the energy released by the oxidized molecule. Some energy is always released as heat. The remaining energy may be used to do physical or chemical work, such as forming ATP. By passing electrons through a series of oxidation–reduction reactions, cells can capture and use much of the energy that is released as water is formed. Water is formed as these electrons ultimately combine with hydrogen ions and oxygen atoms.

Coenzymes play a key role in this process. A coenzyme acts as an intermediary that accepts electrons from one molecule and transfers them to another molecule. In the citric acid cycle, the coenzymes NAD and FAD remove hydrogen atoms from organic molecules. Each hydrogen atom consists of an electron (e^-) and a proton (a hydrogen ion, H^+). Thus, when a coenzyme accepts hydrogen atoms, the coenzyme is reduced and gains energy. The donor molecule loses electrons and energy as it gives up its hydrogen atoms.

Note that NADH and $FADH_2$ are the reduced forms of NAD and FAD. As indicated in **Figure 25–4a**, they then transfer their newly acquired hydrogen atoms to other coenzymes. The protons are subsequently released. The electrons, which carry the chemical energy, enter a sequence of oxidation–reduction reactions known as the *electron transport system*. This sequence ends with the transfer of electrons to oxygen and the formation of a water molecule. At several steps along the oxidation–reduction sequence, enough energy is released to support the synthesis of ATP from ADP. Now let's consider that reaction sequence in greater detail.

The coenzyme FAD accepts two hydrogen atoms from the citric acid cycle. In doing so, FAD gains 2 electrons, forming $FADH_2$. The oxidized form of the coenzyme NAD has a positive charge (NAD^+). This coenzyme also gains 2 electrons as 2 hydrogen atoms are removed from the donor molecule, resulting in the formation of NADH and the release of a proton (H^+). For this reason, the reduced form of NAD is often described as "NADH + H^+."

The Electron Transport System. The **electron transport system (ETS)**, or *respiratory chain*, is a sequence of proteins called **cytochromes** (SĪ-tō-krōmz; *cyto-*, cell + *chroma*, color). Each cytochrome has two parts: a protein and a pigment. The protein is embedded in the inner membrane of a mitochondrion. The protein surrounds the pigment complex, which contains a metal ion, either iron (Fe^{3+}) or copper (Cu^{2+}). We will consider four cytochromes: *b*, *c*, *a*, and a_3.

Figure 25–5 summarizes the basic steps in oxidative phosphorylation. Let's first look at the general path taken by the electrons that are captured and delivered by coenzymes (Figure 25–5a):

1 *A coenzyme strips a pair of hydrogen atoms from a substrate molecule.* As we have seen, different coenzymes are used for different substrate molecules. Recall that during glycolysis, which takes place in the cytosol, NAD is reduced to NADH. Within mitochondria, both NAD and FAD are reduced through reactions of the citric acid cycle, producing NADH and $FADH_2$, respectively.

2 *NADH and $FADH_2$ deliver hydrogen atoms to coenzymes embedded in the inner mitochondrial membrane.* The electrons carry the energy, and the protons that accompany them are released before the electrons are transferred to the ETS. As indicated in Figure 25–5a, the electrons travel to the ETS by one of two paths. Which one depends on whether the donor is NADH or $FADH_2$. The path from NADH involves the coenzyme **FMN** (*flavin mononucleotide*). The path from $FADH_2$ proceeds directly to **coenzyme Q** (*ubiquinone*). Both FMN and coenzyme Q are bound to the inner mitochondrial membrane.

3 *Coenzyme Q releases hydrogen ions and passes electrons to cytochrome b.*

4 *Electrons are passed along the electron transport system, losing energy in a series of small steps.* The sequence is cytochrome *b* to *c* to *a* to a_3.

5 *At the end of the ETS, an oxygen atom accepts the electrons and combines with hydrogen ions to form water.*

Notice that this reaction sequence starts with the removal of 2 hydrogen atoms from a substrate molecule and ends with the formation of water from 2 hydrogen ions and 1 oxygen ion. This is the reaction we described earlier as releasing too much energy if performed in a single step. However, because the reaction takes place in a series of small steps, the hydrogen and oxygen combine safely rather than explosively.

ATP Generation and the ETS. As we noted in Chapter 3, concentration gradients across membranes represent a form of potential energy that can be harnessed by the cell. The electron transport system does not produce ATP directly. Instead, it creates the conditions necessary for ATP production. It creates a steep concentration gradient across the inner mitochondrial membrane. The electrons that travel along the ETS release energy as they pass from coenzyme to cytochrome and from cytochrome to cytochrome. The energy released at each of several steps drives hydrogen ion pumps. These pumps move hydrogen ions from the mitochondrial matrix into the space between the inner and outer mitochondrial membranes. The pumps create a large concentration gradient for hydrogen ions across the inner membrane. This concentration gradient provides the energy to convert ADP to ATP.

Despite the concentration gradient, hydrogen ions cannot diffuse into the matrix because they are not lipid soluble. However, hydrogen ion channels in the inner membrane permit H^+ to diffuse into the matrix. These ion channels and their attached *coupling factors* make up a protein complex called *ATP synthase*. The kinetic energy of passing hydrogen ions generates ATP in a process known as *chemiosmosis* (kem-ē-oz-MŌ-sis), or *chemiosmotic phosphorylation*.

Figure 25–5b diagrams the mechanism of ATP generation. Hydrogen ions are pumped at three places, as (1) FMN reduces coenzyme Q; (2) cytochrome *b* reduces cytochrome *c*; and (3) electrons are passed from cytochrome *a* to cytochrome a_3.

For each pair of electrons removed from a substrate in the citric acid cycle by NAD, 6 hydrogen ions are pumped across the inner membrane of the mitochondrion and into the intermembrane space (Figure 25–5b). Their reentry into the matrix provides the energy to generate three molecules of ATP.

Alternatively, for each pair of electrons removed from a substrate in the citric acid cycle by FAD, 4 hydrogen ions are pumped across the inner membrane and into the intermembrane space. Their reentry into the matrix provides the energy to generate 2 molecules of ATP.

The Importance of Oxidative Phosphorylation. Oxidative phosphorylation is the most important cellular mechanism for generating ATP. In most cases, if oxidative phosphorylation slows or stops, the cell dies. If many cells are affected, the individual may die. Oxidative phosphorylation requires both oxygen and electrons. For this reason, the rate of ATP generation is limited by the availability of either oxygen or electrons.

Cells obtain oxygen by diffusion from the extracellular fluid. If the supply of oxygen is cut off, mitochondrial ATP production stops, because reduced cytochrome a_3 will have no acceptor for its electrons. With the last reaction stopped, the entire ETS comes to a halt, like cars at a washed-out bridge. Oxidative phosphorylation can no longer take place, and cells quickly succumb to energy starvation. Because neurons have a high demand for energy, the brain is one of the first organs to be affected.

Hydrogen cyanide gas is sometimes used as a pesticide to kill rats or mice. In some states where capital punishment is legal, it is used to execute criminals. The cyanide ion (CN^-) binds to cytochrome a_3 and prevents the transfer of electrons to oxygen. As a result, cells die from energy starvation.

25

Figure 25–5 **Oxidative Phosphorylation.**

Steps in Oxidative Phosphorylation

1 A coenzyme strips 2 hydrogen atoms from a substrate molecule.

2 NADH and FADH$_2$ deliver hydrogen atoms to coenzymes embedded in the inner membrane of a mitochondrion.

3 Coenzyme Q releases hydrogen ions and passes electrons to cytochrome b.

4 Electrons are passed along the Electron Transport System, losing energy in a series of small steps.

5 Oxygen accepts the low-energy electrons, and with hydrogen ions, forms water.

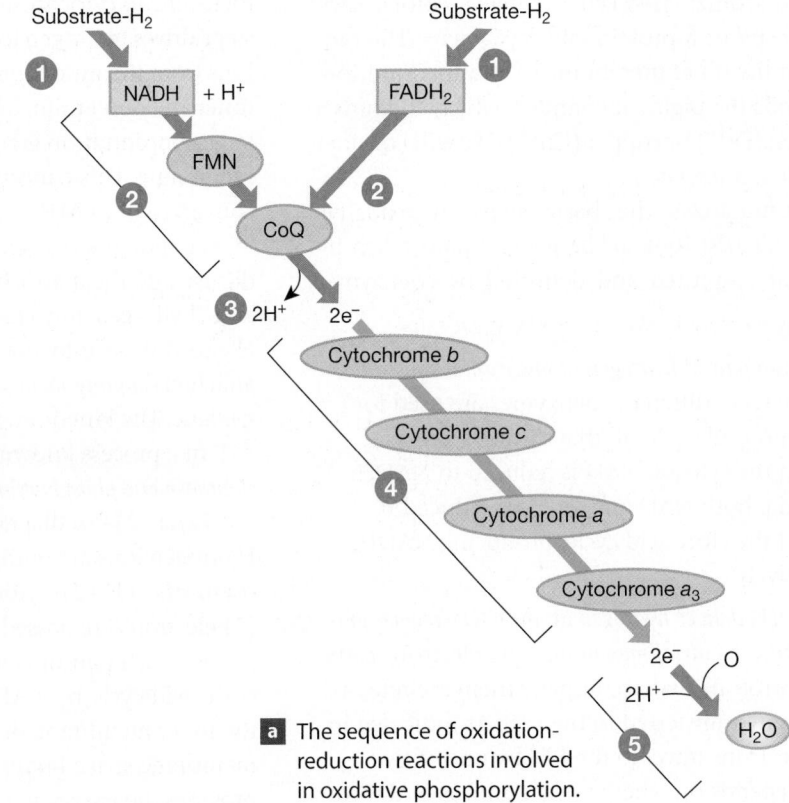

a The sequence of oxidation-reduction reactions involved in oxidative phosphorylation.

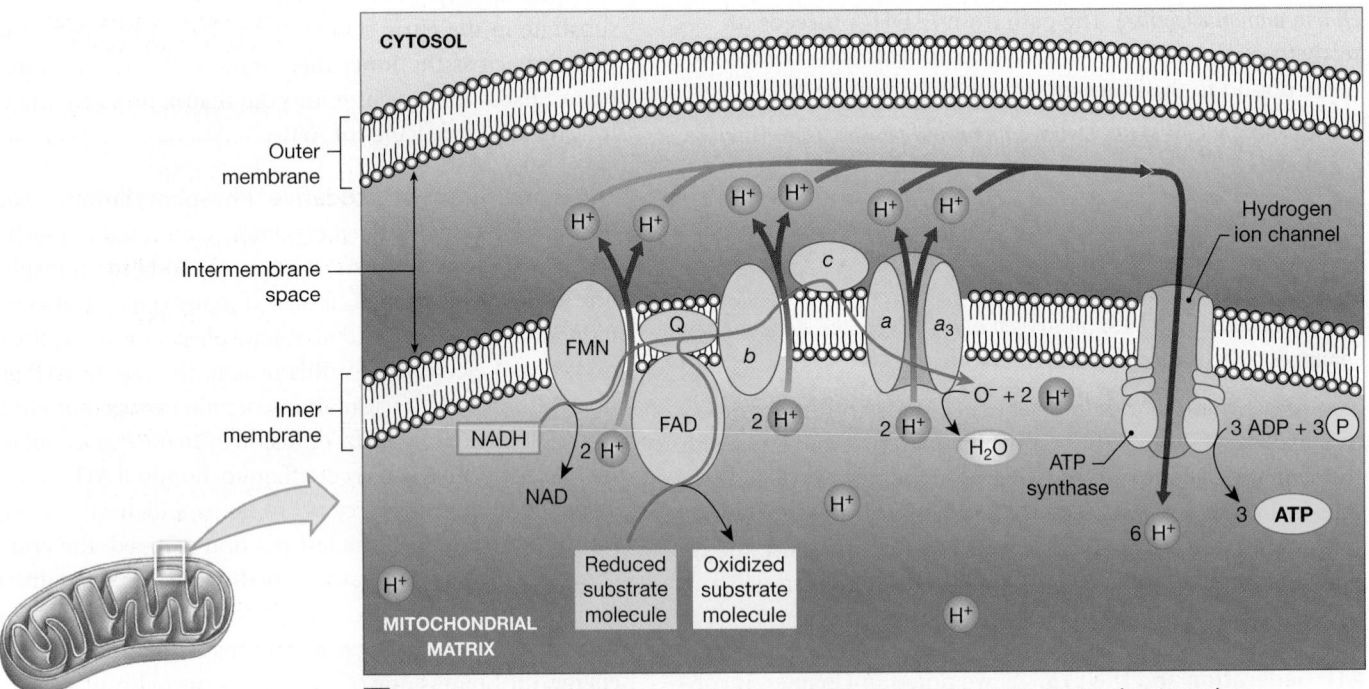

b The locations of the coenzymes and the electron transport system. Notice the sites where hydrogen ions are pumped into the intermembrane space, providing the concentration gradient essential to the generation of ATP. The red line indicates the path taken by the electrons.

Energy Yield of Glycolysis and Cellular Respiration

For most cells, the main method of generating ATP is the complete reaction pathway that begins with glucose and ends with carbon dioxide and water. **Figure 25–6** summarizes the process in terms of energy production:

■ *Glycolysis.* During glycolysis, the cell gains a net 2 molecules of ATP for each glucose molecule broken down anaerobically to 2 molecules of pyruvate. Two molecules of NADH are also produced. In most cells, electrons are passed from NADH to FAD by an intermediate in the intermembrane space, and then to CoQ and the electron transport system. This sequence of events ultimately provides an additional 4 ATP molecules.

■ *The Electron Transport System.* The citric acid cycle breaks down the 2 pyruvate molecules, transferring hydrogen atoms to NADH and $FADH_2$. These coenzymes provide electrons to the ETS. Each of the 8 molecules of NADH yields 3 molecules of ATP and 1 water molecule. Each of the 2 $FADH_2$ molecules yields 2 ATP molecules and 1 water molecule. Thus, the shuffling from the citric acid cycle to the ETS yields 28 molecules of ATP.

■ *The Citric Acid Cycle.* Each of the two revolutions of the citric acid cycle required to break down both pyruvate

Figure 25–6   A Summary of the Energy Yield of Aerobic Metabolism. For each glucose molecule broken down by glycolysis, only 2 molecules of ATP (net) are produced. However, glycolysis, the formation of acetyl-CoA, and the citric acid cycle all yield molecules of reduced coenzymes (NADH or $FADH_2$). Many additional ATP molecules are produced when electrons from these coenzymes pass through the electron transport system. In most cells, each of the 2 NADH molecules produced in glycolysis provides another 2 ATP molecules. Each of the 8 NADH molecules produced in the mitochondria yields 3 ATP molecules, for a total of 24. Another 2 ATP molecules are gained from each of the 2 $FADH_2$ molecules generated in the mitochondria. The citric acid cycle generates an additional 2 ATP molecules in the form of GTP.

Figure 25–7   Carbohydrate Breakdown and Synthesis. The pathways for glycolysis and gluconeogenesis. Many of the reactions are freely reversible, but separate regulatory enzymes control the key steps, which are indicated by colored arrows. Some amino acids, carbohydrates, lactate, and glycerol can be converted to glucose. The enzymatic reaction that converts pyruvate to acetyl-CoA cannot be reversed.

molecules completely yields 1 molecule of ATP by way of GTP. This cycling provides an additional gain of 2 molecules of ATP.

Summing up, for each glucose molecule processed, the cell gains 36 molecules of ATP: 2 from glycolysis, 4 from the NADH generated in glycolysis, 2 from the citric acid cycle (by means of GTP), and 28 from the ETS.

Cardiac muscle cells and liver cells are able to gain an additional 2 ATP molecules for each glucose molecule broken down. They do so by increasing the energy yield from the 2 NADH generated during glycolysis from 4 to 6 ATP molecules. Although the NADH produced during glycolysis does not enter mitochondria in these cells, each NADH molecule passes its electrons to an intermediate molecule that can cross the mitochondrial membrane. Inside the matrix, NAD strips electrons from the intermediate molecule, forming NADH. The subsequent transfer of electrons to FMN, CoQ, and the ETS results in the production of 6 ATP molecules, just as if the 2 NADH molecules formed during glycolysis had been generated in the citric acid cycle.

Gluconeogenesis

Some of the steps in the breakdown of glucose, or glycolysis, are essentially irreversible. For this reason, cells cannot generate glucose simply by using the same enzymes and reversing the steps of glycolysis (**Figure 25–7**). Glycolysis and the production of glucose involve different sets of regulatory enzymes. As a result, the two processes are regulated independently.

Some 3-carbon molecules other than pyruvate can be used to synthesize glucose. For this reason, a cell can create glucose molecules from other carbohydrates, lactate, glycerol, or some amino acids. However, cells cannot use acetyl-CoA to make glucose. The reason is that the decarboxylation step (removal of CO_2) between pyruvate and acetyl-CoA cannot be reversed. **Gluconeogenesis** (glū-kō-nē-ō-JEN-e-sis) is the synthesis of glucose from noncarbohydrate precursors, such as lactate, glycerol, or amino acids. Fatty acids and many amino acids cannot be used for gluconeogenesis, because their catabolic pathways produce acetyl-CoA.

Glucose molecules formed by gluconeogenesis can be used to manufacture other simple sugars, complex carbohydrates, proteoglycans, or nucleic acids. In the liver and in skeletal muscle, glucose molecules are stored as glycogen. Glycogen is an important energy reserve that can be broken down when the cell cannot obtain enough glucose from interstitial fluid. Glycogen molecules are large, but glycogen reserves take up very little space because they form compact, insoluble granules.

The formation of glycogen from glucose, known as **glycogenesis**, is a complex process. It involves several steps and

Clinical Note

Carbohydrate Loading Eating carbohydrates just before exercise does not improve your performance. In fact, it can decrease your endurance by slowing the mobilization of existing energy reserves. Runners or swimmers preparing for lengthy endurance events, such as a marathon or 5-km swim, do not eat immediately before participating. For two hours before the event, they also limit their intake to just drinking water.

Performance in endurance sports improves if muscles have large stores of glycogen. Endurance athletes try to build these stores by eating carbohydrate-rich diets for three days before competing. This practice is called *carbohydrate loading*. Studies in Sweden, Australia, and South Africa have recently shown that attempts to deplete stores by exercising to exhaustion before carbohydrate loading, a practice called *carbohydrate depletion/loading*, are less effective than three days of rest or minimal exercise during carbohydrate loading. The less intense approach improves mood and reduces the risks of muscle and kidney damage.

requires the high-energy compound *uridine triphosphate (UTP)*. The breakdown of glycogen, called **glycogenolysis**, takes place quickly and involves a single enzymatic step.

TIPS & TRICKS

To make several similar-sounding terms easier to tell apart, learn these word parts: *genesis* means "the formation of," *lysis* means "a loosening," *glyco* and *gluco* refer to glucose, and *neo* means "new." So, **gluconeogenesis** is the formation of new glucose, **glycogenesis** is the formation of glycogen (the storage form of glucose), **glycogenolysis** is the breakdown of glycogen to glucose, and **glycolysis** is the breakdown of glucose to pyruvate.

CHECKPOINT

4. What is the primary role of the citric acid cycle in the production of ATP?

5. Each NADH produced by glycolysis in skeletal muscle fibers leads to the production of 2 ATP molecules in the mitochondria, but each NADH produced by glycolysis in cardiac muscle cells leads to the production of 3 ATP molecules. Why?

6. How would a decrease in the level of NAD in the cytosol affect ATP production in mitochondria?

See the blue Answers tab at the back of the book.

25-3 ❙ Lipid metabolism involves lipolysis, beta-oxidation, and the transport and distribution of lipids as free fatty acids and lipoproteins

Learning Outcome Describe the pathways involved in lipid metabolism, and summarize the mechanisms of lipid transport and distribution.

Like carbohydrates, lipid molecules contain carbon, hydrogen, and oxygen, but the atoms are present in different proportions. Triglycerides are the most abundant lipid in the body, so our discussion focuses on pathways for triglyceride breakdown and synthesis.

Lipid Catabolism

During lipid catabolism, or **lipolysis**, lipids are broken down into pieces that can be either converted to pyruvate or channeled directly into the citric acid cycle (**Figure 25–8**). A triglyceride is first split into its component parts by hydrolysis, yielding 1 molecule of glycerol and 3 fatty acid molecules. Enzymes in the cytosol convert glycerol to pyruvate, which then enters the citric acid cycle. The catabolism of fatty acids involves a completely different set of enzymes that generate acetyl-CoA directly.

Beta-Oxidation

Beta-oxidation is a sequence of reactions in which fatty acid molecules are broken down into 2-carbon acetic acid fragments, and FAD and NAD^+ are reduced. Each acetic acid ultimately forms acetyl-CoA. This process takes place inside mitochondria, so the carbon chains can enter the citric acid cycle immediately as acetyl-CoA. **Figure 25–8** diagrams one step in the process of beta-oxidation. Each step generates molecules of acetyl-CoA, NADH, and $FADH_2$ and leaves a shorter carbon chain bound to coenzyme A.

Beta-oxidation has substantial energy benefits. For each 2-carbon fragment removed from the fatty acid, the cell gains 12 ATP molecules from the processing of acetyl-CoA in the citric acid cycle, plus 5 ATP molecules from the NADH and $FADH_2$. The cell can therefore gain 144 ATP molecules from the breakdown of one 18-carbon fatty acid molecule. By comparison, this number of ATP molecules yields almost 1.5 times the energy obtained by the complete breakdown of three 6-carbon glucose molecules. The catabolism of other lipids follows similar patterns, generally ending with the formation of acetyl-CoA.

Lipids and Energy Production

Lipids are important energy reserves because they can provide large amounts of ATP. Lipids can be stored in compact droplets in the cytosol because they are insoluble in water. This

25

Figure 25–8 **Beta-Oxidation.** During beta-oxidation, the carbon chains of fatty acids are broken down to yield molecules of acetyl-CoA, which can be used in the citric acid cycle. The reaction also donates hydrogen atoms to coenzymes, which then deliver them to the electron transport system.

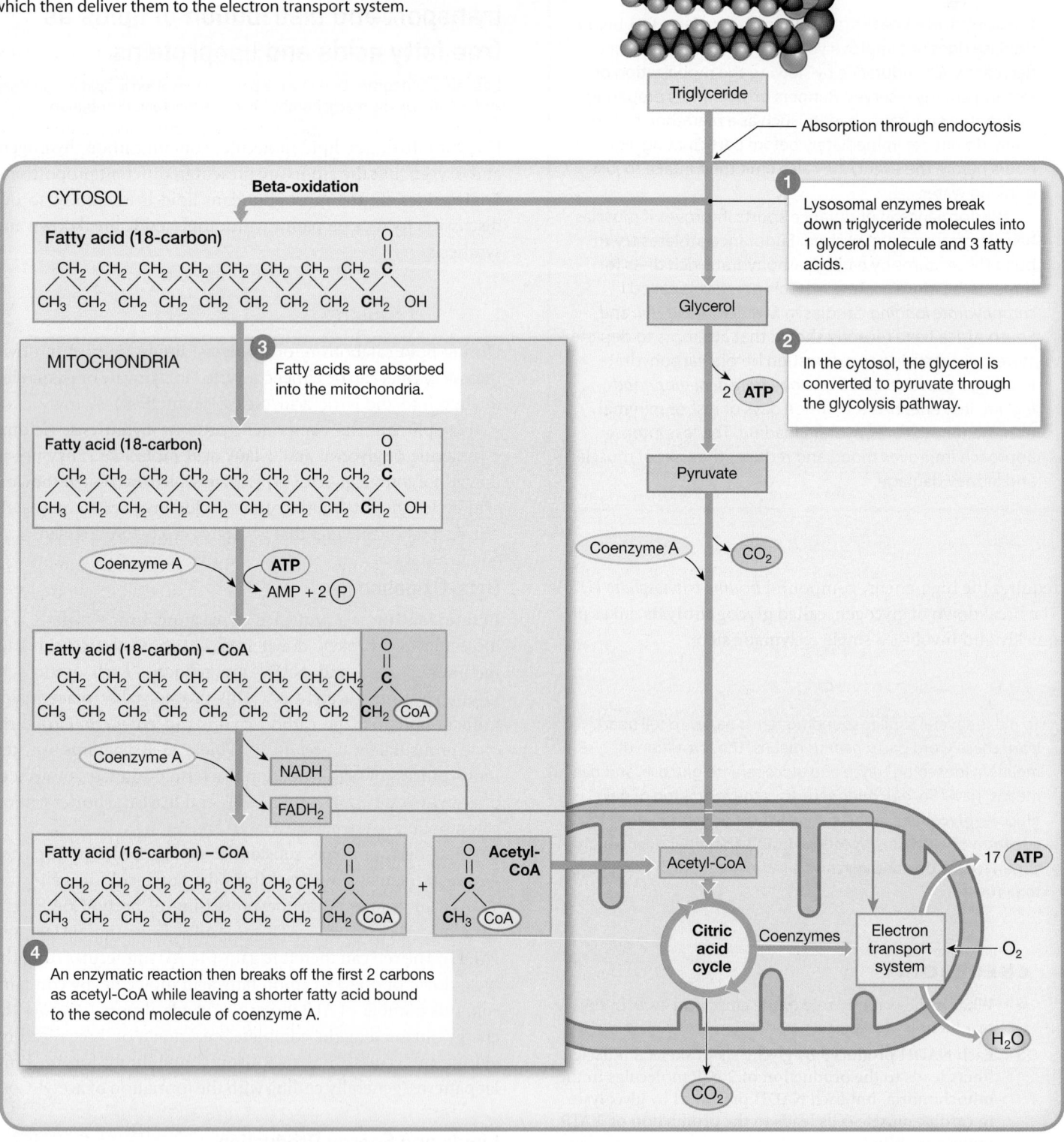

storage method saves space, but when the lipid droplets are large, it is difficult for water-soluble enzymes to get at them. For this reason, lipid reserves are more difficult to access than carbohydrate reserves. Also, most lipids are processed inside

mitochondria, and mitochondrial activity is limited by the availability of oxygen.

The net result is that lipids cannot provide large amounts of ATP quickly. However, cells with modest energy demands can shift to lipid-based energy production when glucose supplies are limited. Skeletal muscle fibers normally cycle between lipid metabolism and carbohydrate metabolism. At rest (when energy demands are low), these cells break down fatty acids. During activity (when energy demands are high and immediate), skeletal muscle fibers shift to glucose metabolism.

Lipid Synthesis

Lipogenesis (lip-ō-JEN-e-sis) is the synthesis of lipids. Glycerol is synthesized from *dihydroxyacetone phosphate*, one of the 3-carbon intermediate products shared by the pathways of glycolysis and gluconeogenesis. The synthesis of most types of lipids, including nonessential fatty acids and steroids, begins with acetyl-CoA. Lipogenesis can use almost any organic substrate, because lipids, amino acids, and carbohydrates can be converted to acetyl-CoA. In other words, the body can turn just about anything we eat into fat.

Fatty acid synthesis involves a reaction sequence quite distinct from that of beta-oxidation. Body cells cannot *build* every fatty acid they can break down. For example, **linoleic acid** and **linolenic acid** are both 18-carbon unsaturated fatty acids synthesized by plants. They cannot be synthesized in the human body. They are considered **essential fatty acids**, because they must be included in your diet. These fatty acids are also needed to synthesize prostaglandins and some of the phospholipids in plasma membranes throughout the body.

Lipid Transport and Distribution

Like glucose, lipids are needed throughout the body. All cells need lipids to maintain their plasma membranes. Important steroid hormones must reach target cells in many different tissues. Free fatty acids make up a small percentage of the total circulating lipids. Because most lipids are not soluble in water, special transport mechanisms carry them from one region of the body to another. Most lipids circulate through the bloodstream as lipoproteins (**Figure 25–9**).

Free Fatty Acids

Free fatty acids (FFAs) are lipids that can diffuse easily across plasma membranes. In the blood, free fatty acids are generally bound to albumin, the most abundant plasma protein. Sources of free fatty acids in the blood include the following:

- Fatty acids that are not used in the synthesis of triglycerides, but diffuse out of the intestinal epithelium and into the blood.
- Fatty acids that diffuse out of lipid reserves (such as those in the liver and adipose tissue) when triglycerides are broken down.

Liver cells, cardiac muscle cells, skeletal muscle fibers, and many other body cells can metabolize free fatty acids. These lipids are an important energy source during periods of starvation, when glucose supplies are limited.

Lipoproteins

Lipoproteins are lipid–protein complexes that contain large insoluble glycerides and cholesterol. A superficial coating of phospholipids and proteins makes the entire complex soluble. Exposed proteins of the complexes bind to specific membrane receptors. For this reason, these membrane proteins determine which cells absorb the associated lipids.

Lipoproteins are usually classified into five major groups according to size and the relative proportions of lipid and protein:

1. *Chylomicrons.* Chylomicrons are the largest lipoproteins, ranging in diameter from 0.03 to 0.5 μm. About 95 percent of the weight of a chylomicron consists of triglycerides. They are produced by intestinal epithelial cells from the fats in food. ⊃ p. 932 Chylomicrons carry absorbed lipids from the intestinal tract into the lymph and then to the bloodstream. The liver is the primary source of all the other groups of lipoproteins, which shuttle lipids among various tissues.

2. *Very Low Density Lipoproteins (VLDLs).* Very low density lipoproteins contain triglycerides manufactured by the liver, plus small amounts of phospholipids and cholesterol. The primary function of VLDLs is to transport these triglycerides to skeletal muscles and adipose tissues. The VLDLs range in diameter from 25 to 75 nm (0.025–0.075 μm).

3. *Intermediate-Density Lipoproteins (IDLs).* Intermediate-density lipoproteins are intermediate in size and lipid composition between VLDLs and low-density lipoproteins (LDLs). IDLs contain smaller amounts of triglycerides than do VLDLs. IDLs also have more phospholipids and cholesterol than LDLs.

4. *Low-Density Lipoproteins (LDLs).* Low-density lipoproteins contain cholesterol, lesser amounts of phospholipids, and very few triglycerides. These lipoproteins are about 25 nm in diameter. They deliver cholesterol to peripheral tissues. LDL cholesterol is often called "bad cholesterol" because the cholesterol may wind up in arterial plaques.

5. *High-Density Lipoproteins (HDLs).* High-density lipoproteins have roughly equal amounts of lipid and protein. The lipids are largely cholesterol and phospholipids. HDLs are about 10 nm in diameter. Their primary function is to transport excess cholesterol from peripheral tissues back to the liver for storage or excretion in the bile. HDL cholesterol is called "good cholesterol" because it is returning from peripheral tissues and does not cause circulatory problems. Actually, applying the terms *good* and *bad* to cholesterol can be misleading, for cholesterol metabolism is complex and variable. (For more details, see the Clinical Note "Dietary

25

Figure 25–9 **Lipid Transport and Utilization.**

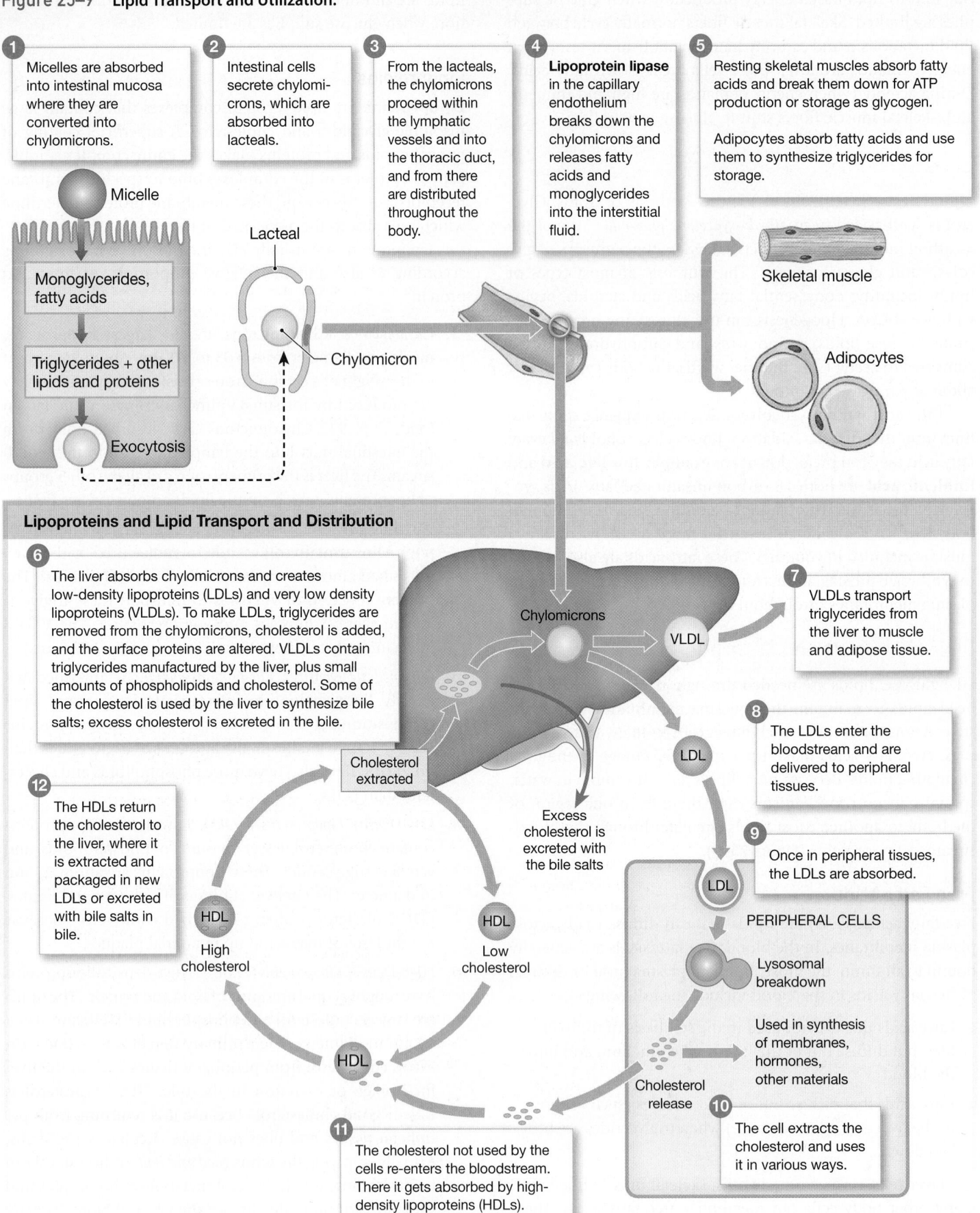

1 Micelles are absorbed into intestinal mucosa where they are converted into chylomicrons.

2 Intestinal cells secrete chylomicrons, which are absorbed into lacteals.

3 From the lacteals, the chylomicrons proceed within the lymphatic vessels and into the thoracic duct, and from there are distributed throughout the body.

4 **Lipoprotein lipase** in the capillary endothelium breaks down the chylomicrons and releases fatty acids and monoglycerides into the interstitial fluid.

5 Resting skeletal muscles absorb fatty acids and break them down for ATP production or storage as glycogen.

Adipocytes absorb fatty acids and use them to synthesize triglycerides for storage.

Micelle

Monoglycerides, fatty acids

Triglycerides + other lipids and proteins

Exocytosis

Lacteal

Chylomicron

Skeletal muscle

Adipocytes

Lipoproteins and Lipid Transport and Distribution

6 The liver absorbs chylomicrons and creates low-density lipoproteins (LDLs) and very low density lipoproteins (VLDLs). To make LDLs, triglycerides are removed from the chylomicrons, cholesterol is added, and the surface proteins are altered. VLDLs contain triglycerides manufactured by the liver, plus small amounts of phospholipids and cholesterol. Some of the cholesterol is used by the liver to synthesize bile salts; excess cholesterol is excreted in the bile.

Chylomicrons

VLDL

7 VLDLs transport triglycerides from the liver to muscle and adipose tissue.

8 The LDLs enter the bloodstream and are delivered to peripheral tissues.

LDL

9 Once in peripheral tissues, the LDLs are absorbed.

LDL

PERIPHERAL CELLS

Cholesterol extracted

Excess cholesterol is excreted with the bile salts

12 The HDLs return the cholesterol to the liver, where it is extracted and packaged in new LDLs or excreted with bile salts in bile.

HDL
High cholesterol

HDL
Low cholesterol

Lysosomal breakdown

Used in synthesis of membranes, hormones, other materials

Cholesterol release

10 The cell extracts the cholesterol and uses it in various ways.

HDL

11 The cholesterol not used by the cells re-enters the bloodstream. There it gets absorbed by high-density lipoproteins (HDLs).

25

Clinical Note

Dietary Fats and Cholesterol Elevated cholesterol levels are associated with the development of coronary artery disease (CAD) ⤴ p. 724 and atherosclerosis ⤴ p. 754. Nutritionists now recommend that you keep your cholesterol intake under 300 mg per day. This amount represents a 40 percent reduction for the average American adult. Due to rising concerns about cholesterol, such phrases as "low cholesterol," "contains no cholesterol," and "cholesterol free" are now widely used in the advertising and packaging of foods. Cholesterol content alone, however, does not tell the entire story. Consider the following basic information about cholesterol and about lipid metabolism in general:

- *Cholesterol Has Many Vital Functions in the Human Body.* Cholesterol serves as a waterproofing for the epidermis, a lipid component of all plasma membranes, a key constituent of bile, and the precursor of several steroid hormones and one vitamin (vitamin D_3). Because cholesterol is so important, the goal of dietary restrictions is not to eliminate it from the diet or from the circulating blood. The goal is to keep cholesterol levels within acceptable limits.

- *The Cholesterol Content of the Diet Is Not the Only Source of Circulating Cholesterol.* The human body can manufacture cholesterol from acetyl-CoA obtained through glycolysis or the beta-oxidation of other lipids. Such cholesterol probably accounts for only about 20 percent of the cholesterol in the bloodstream. The rest comes from metabolism of saturated fats in the diet. If the diet contains an abundance of saturated fats, cholesterol levels in the blood rise because excess lipids are broken down to acetyl-CoA and used to synthesize cholesterol. Consequently, individuals trying to lower serum cholesterol levels by dietary control must also restrict other lipids, especially saturated fats.

- *Genetic Factors Affect Each Individual's Cholesterol Level.* If you reduce the cholesterol in your diet, your body synthesizes more to maintain "acceptable" concentrations in the blood. But what is an "acceptable" level? It depends on your genetic makeup. Because individuals have different genes, their cholesterol levels can vary, even on similar diets. In virtually all instances, however, dietary restrictions can lower blood cholesterol significantly.

- *Cholesterol Levels Vary with Age and Physical Condition.* In general, as we age, our cholesterol levels gradually climb. At age 19, three out of four males have fasting cholesterol levels (levels measured 8–12 hours after a meal) below 170 mg/dL. Cholesterol levels in females of this age are slightly higher, typically at or below 175 mg/dL. As age

increases, the cholesterol levels gradually climb. Over age 70, the levels are 230 mg/dL (males) and 250 mg/dL (females). Cholesterol levels are considered unhealthy if they are higher than those of 90 percent of the population in that age group. For males, this level ranges from 185 mg/dL at age 19 to 250 mg/dL at age 70. For females, the comparable levels are 190 mg/dL and 275 mg/dL respectively.

The levels that are considered optimal to reduce risk of atherosclerosis and coronary artery disease depend on the presence or absence of associated risk factors. These factors include, but are not limited to, hypertension, diabetes mellitus, personal and family history of heart disease, and smoking. The recommendations and guidelines vary depending on the reference consulted. However, everyone should be screened for high blood cholesterol levels as they age. Men should be screened at a younger age than women.

When ordering a blood test for cholesterol, most physicians also request information about circulating triglycerides. In fasting individuals, triglycerides are usually present at levels of 40–150 mg/dL. (After a person has consumed a fatty meal, triglyceride levels may be temporarily elevated.)

When cholesterol levels are high, or when an individual has a family history of atherosclerosis or CAD, further tests and calculations may be performed. The HDL level is measured, and the LDL level is calculated as:

$$LDL = cholesterol - HDL - \frac{triglycerides}{5}$$

A high total cholesterol value linked to a high LDL level spells trouble. In effect, too much cholesterol is being exported to peripheral tissues. Problems can also exist in individuals with high total cholesterol—or even normal total cholesterol—but low HDL levels (below 35 mg/dL). In such cases, excess cholesterol delivered to the tissues cannot easily be returned to the liver for excretion. In either event, the amount of cholesterol in peripheral tissues—and especially in arterial walls—is likely to increase.

For years, LDL:HDL ratios were considered valid predictors of the risk of developing atherosclerosis. Risk-factor analysis and LDL levels are now thought to be more accurate indicators. For males with more than one risk factor, many clinicians recommend dietary changes and drug therapy if LDL levels exceed 130 mg/dL, or even 100 mg/dL, regardless of the total cholesterol or HDL levels.

25

Fats and Cholesterol" on p. 977.) **Figure 25–9** also shows the relationship between LDL and HDL lipoproteins and the transport of cholesterol.

We now know that exercise lowers cholesterol levels. It is recommended that individuals get 30 minutes per day of moderate to vigorous exercise. Recent studies have also shown that vigorous exercise—the equivalent of 20 miles of jogging per week—lowered LDL levels even more than moderate exercise.

CHECKPOINT

7. Define beta-oxidation.

8. Identify the five major groups of lipoproteins.

9. Why are high-density lipoproteins (HDLs) considered beneficial to health?

See the blue Answers tab at the back of the book.

25-4 | Protein catabolism involves transamination and deamination, whereas protein synthesis involves amination and transamination

Learning Outcome Summarize the main processes of protein metabolism, and discuss the use of protein as an energy source.

The body can synthesize 100,000 to 140,000 different proteins. They have various functions and structures. Yet, each protein contains some combination of the same 20 amino acids. Under normal conditions, cellular proteins are continuously recycled in the cytosol. Peptide bonds are broken, and the free amino acids are used in new proteins.

If other energy sources are inadequate, mitochondria can generate ATP by breaking down amino acids in the citric acid cycle. Not all amino acids enter the cycle at the same point, however, so the ATP benefits vary. Nonetheless, the average ATP yield per gram is comparable to that of carbohydrate catabolism.

Amino Acid Catabolism

The first step in amino acid catabolism is the removal of the amino group ($-NH_2$). This process requires a coenzyme derivative of **vitamin B_6** *(pyridoxine)*. The amino group is removed by *transamination* (tranz-am-i-NĀ-shun) or *deamination* (dē-am-i-NĀ-shun). We consider other aspects of amino acid catabolism in a later section.

Transamination

Transamination attaches the amino group of an amino acid to a **keto acid**, which resembles an amino acid except that the second carbon binds an oxygen atom rather than an amino group (**Figure 25–10a**). This transfer converts the keto acid into an amino acid that can leave the mitochondrion and enter the

cytosol. There it can be used for protein synthesis. In the process, the original amino acid becomes a keto acid that can be broken down in the citric acid cycle.

Cells in many different tissues perform transaminations, enabling cells to synthesize many of the amino acids needed for protein synthesis. Cells of the liver, skeletal muscles, heart, lung, kidney, and brain are particularly active in protein synthesis. They carry out many transaminations.

Deamination

Deamination prepares an amino acid for breakdown in the citric acid cycle (**Figure 25–10b**). Deamination is the removal of an amino group and a hydrogen atom in a reaction that generates an ammonium ion (NH_4^+). Ammonium ions are highly toxic, even in low concentrations.

Liver cells are the primary sites of deamination. They also have enzymes that remove ammonium ions by synthesizing **urea**, a fairly harmless water-soluble compound excreted in urine. The **urea cycle** is the reaction sequence that produces urea (**Figure 25–10c**).

When glucose supplies are low and lipid reserves are inadequate, liver cells break down internal proteins and absorb additional amino acids from the blood. The amino acids are deaminated, and their carbon chains are broken down to provide ATP.

Proteins and ATP Production

Three factors make protein catabolism an impractical source of quick energy:

1. Proteins are more difficult to break apart than are complex carbohydrates or lipids.

2. One of the by-products, ammonium ions, is toxic to cells.

3. Proteins form the most important structural and functional components of any cell. Extensive protein catabolism threatens homeostasis at the cellular and system levels.

Several inherited metabolic disorders result from an inability to produce specific enzymes involved in amino acid metabolism. For example, individuals with *phenylketonuria* (fen-il-kē-tō-NOO-rē-uh), or PKU, have a defect in the enzyme *phenylalanine hydroxylase*. They cannot convert phenylalanine to tyrosine. This reaction is an essential step in the synthesis of norepinephrine, epinephrine, dopamine, and melanin. If PKU is not detected in infancy, phenylalanine builds up and central nervous system development is inhibited. Severe brain damage results. The condition is common enough that a warning is printed on the packaging of products that contain phenylalanine, such as diet drinks.

Protein Synthesis

We discussed the basic mechanism for protein synthesis in Chapter 3 (**Figures 3–12** and **3–13**, pp. 111, 114–115). Your body

Figure 25–10 **Amino Acid Catabolism and Synthesis.**

Amino Acid Catabolism

a Transamination

In transamination, the amino group (–NH₂) of one amino acid is transferred to a keto acid.

Glutamic acid + Keto acid 1 →(Transaminase)→ Keto acid 2 + Tyrosine

b Deamination

In deamination, the amino group is removed and an ammonium ion is released.

Glutamic acid →(Deaminase, H₂O → H⁺)→ Organic acid + NH₄⁺ (Ammonium ion)

c Urea cycle. The urea cycle takes two metabolic waste products—ammonium ions and carbon dioxide—and produces urea, a relatively harmless, soluble compound that is excreted in the urine.

2 NH₄⁺ (Ammonium ions) + CO₂ (Carbon dioxide) → Urea cycle → Urea (H₂N–C(=O)–NH₂)

Amino Acid Synthesis

d Amination

In amination, an ammonium ion (NH₄⁺) is used to form an amino group that is attached to a molecule, yielding an amino acid.

α–Ketoglutarate + NH₄⁺ (→ H⁺) → Glutamic acid + H₂O

can synthesize about half of the various amino acids needed to build proteins. There are 10 **essential amino acids**, which must come from the diet. Your body cannot synthesize eight of them (*isoleucine, leucine, lysine, threonine, tryptophan, phenylalanine, valine,* and *methionine*). The other two (*arginine* and *histidine*) can be synthesized, but not in the amounts that growing children need.

Other amino acids are called **nonessential amino acids** because the body can make them on demand. Your body cells can readily synthesize their carbon frameworks. Then an amino group can be added by transamination or by **amination**—using an ammonium ion as a reactant (**Figure 25–10d**).

Protein deficiency diseases develop in people who do not eat adequate amounts of all essential amino acids. All amino acids must be available if protein synthesis is to take place. Every transfer RNA molecule must appear at the active ribosome in the proper sequence, bearing its individual amino acid. If that does not happen, the entire process comes to a halt. A person will be malnourished to some degree if the diet is deficient in essential amino acids, regardless of its energy content. Examples of protein deficiency diseases include *marasmus* and *kwashiorkor*. More than 100 million children worldwide have symptoms of these disorders, although none of these conditions is common in the United States today.

Spotlight Figure 25–11 summarizes the metabolic pathways for lipids, carbohydrates, and proteins. The diagrams present the reactions in a "typical" cell. Note, however, that no one

	8 am	9 am	10 am	11 am	12 noon	1 pm	2 pm	3 pm	4 pm	5 pm	6 pm
Glucose Levels											
Lipid Levels											
Amino Acid Levels											
MEALS		Breakfast				Lunch					Dinner

ABSORPTIVE STATE

The **absorptive state** is the time following a meal, when nutrient absorption is under way. Nutrient absorption was outlined in *Spotlight Figure 24–27*. After a typical meal, the absorptive state lasts for about 4 hours. If you are fortunate enough to eat three meals a day, you spend 12 out of every 24 hours in the absorptive state. Insulin is the primary hormone of the absorptive state. Various other hormones stimulate amino acid uptake (growth hormone) and protein synthesis (growth hormone, androgens, and estrogens).

Blood glucose levels elevated

Insulin (binds to carrier protein)

LIPIDS

CARBOHYDRATES

PROTEINS

Triglycerides

Glycogen

Proteins

Insulin

Glucose

Insulin

Insulin
Androgens
Estrogens
Growth hormone

Fatty acids

Glycerol

Glycolysis

ATP

Amino acids

Blood lipid levels elevated

Blood amino acid levels elevated

Insulin, Growth hormone

Pyruvate

KEY

→ Catabolic pathway

→ Anabolic pathway

→ Stimulation

Insulin

CO_2

Acetyl-CoA

ATP

Citric acid cycle

Coenzymes

Electron transport system

O_2 ← O_2

MITOCHONDRIA

H_2O

CO_2

The **postabsorptive state** is the time when nutrients are not being absorbed and your body must rely on internal energy reserves to meet its energy demands. You spend about 12 hours each day in the postabsorptive state. If you skip meals, however, you can extend that time considerably. Metabolic activity in the postabsorptive state is focused on mobilizing energy reserves and maintaining normal blood glucose levels. Several hormones coordinate these activities. These hormones include glucagon, epinephrine, glucocorticoids, and growth hormone.

POSTABSORPTIVE STATE

Glucose released into the bloodstream by the liver

LIPIDS

Triglycerides

CARBOHYDRATES

Glycogen

Glucagon →
Epinephrine →

Glucose

PROTEINS

Proteins

Glucagon,
Epinephrine,
Glucocorticoids,
Growth hormone

Fatty acids released into the bloodstream by adipocytes

Fatty acids

Glycerol

ATP

Glucocorticoids,
Growth hormone

Amino acids

← Glucagon

Amino acids released into the bloodstream by the liver

Gluconeogenesis

Glucocorticoids stimulate lipid catabolism in all tissues other than neural tissue, which can only metabolize glucose. The shift in metabolic activity elsewhere preserves blood glucose for use by the nervous system.

Pyruvate

Glucocorticoids,
Growth hormone

CO_2

Acetyl-CoA

ATP

Liver cells break down lipids and amino acids, and as the levels of acetyl-CoA increase, substances called **ketone bodies** begin to form. Liver cells do not catabolize ketone bodies, which diffuse into the general circulation. Cells in peripheral tissues absorb them, convert them to acetyl-CoA, and use them as an energy source.

Citric acid cycle

Coenzymes

Electron transport system

O_2 ← O_2

H_2O

MITOCHONDRIA

Ketone bodies

CO_2

981

cell can carry out all the anabolic and catabolic operations and interconversions required by the body as a whole. As cells differentiate, each type develops its own set of enzymes. These enzymes determine the cell's metabolic capabilities. In the face of such cellular diversity, homeostasis can be preserved only when the metabolic activities of tissues, organs, and organ systems are coordinated.

CHECKPOINT

10. Define transamination.

11. Define deamination.

12. How would a diet that is deficient in pyridoxine (vitamin B$_6$) affect protein metabolism?

See the blue Answers tab at the back of the book.

25-5 ▎ The body experiences two patterns of metabolic activity: the absorptive and postabsorptive states

Learning Outcome Differentiate between the absorptive and postabsorptive metabolic states, and summarize the characteristics of each.

The nutrient requirements of each tissue vary with the types and quantities of enzymes present in the cytosol of cells. From a metabolic standpoint, we consider the body to have five distinctive components: the liver, adipose tissue, skeletal muscle, neural tissue, and other peripheral tissues:

1. *The Liver.* The liver is the focal point of metabolic regulation and control. Liver cells contain a great diversity of enzymes, so they can break down or synthesize most of the carbohydrates, lipids, and amino acids needed by other body cells. Liver cells have an extensive blood supply, so they are in an excellent position to monitor and adjust the nutrient composition of circulating blood. The liver also contains significant energy reserves in the form of glycogen.

2. *Adipose Tissue.* Adipose tissue stores lipids, primarily as triglycerides. Adipocytes are located in many areas: in areolar tissue, in mesenteries, within red and yellow bone marrows, in the epicardium, and around the eyes and the kidneys.

3. *Skeletal Muscle.* Skeletal muscle accounts for almost half of a healthy person's body weight. Skeletal muscle fibers maintain substantial glycogen reserves. In addition, if other nutrients are unavailable, their contractile proteins can be broken down and the amino acids used as an energy source.

4. *Neural Tissue.* Neural tissue has a high demand for energy, but the cells do not maintain reserves of carbohydrates, lipids, or proteins. Neurons must have a reliable supply of glucose, because they are generally unable to metabolize

other molecules. If blood glucose levels become too low, neural tissue in the central nervous system cannot continue to function, and the person falls unconscious.

5. *Other Peripheral Tissues.* Other peripheral tissues do not maintain large metabolic reserves, but they are able to metabolize glucose, fatty acids, or other substrates. Their preferred source of energy varies according to instructions from the endocrine system.

To understand the interrelationships among these five components, let's consider events over a typical 24-hour period. During this time, the body experiences two broad patterns of metabolic activity: the *absorptive state* and the *postabsorptive state* (Spotlight Figure 25–11).

In the absorptive state that follows a meal, cells absorb nutrients to be used for growth, maintenance, and energy reserves. Hours later, in the postabsorptive state, metabolic reactions are focused on maintaining blood glucose levels that meet the needs of neural tissue.

During the postabsorptive state, liver cells conserve glucose and break down lipids and amino acids. Both lipid catabolism and amino acid catabolism generate acetyl-CoA. As the concentration of acetyl-CoA rises, compounds called *ketone bodies* begin to form. A **ketone body** is an organic compound produced by fatty acid metabolism that dissociates in solution, releasing a hydrogen ion. There are three such compounds: (1) **acetoacetate** (as-e-tō-AS-e-tāt), (2) **acetone** (AS-e-tōn), and (3) **betahydroxybutyrate** (bā-ta-hī-droks-ē-BŪ-te-rāt). Liver cells do not catabolize any of the ketone bodies. Instead, these compounds diffuse through the cytosol and into the general circulation. Cells in peripheral tissues then absorb the ketone bodies and reconvert them to acetyl-CoA for breakdown in the citric acid cycle.

The normal concentration of ketone bodies in the blood is about 30 mg/dL. Very few of these compounds normally appear in urine. During even a brief period of fasting, the increased production of ketone bodies results in *ketosis* (kē-TŌ-sis), a high concentration of ketone bodies in body fluids. The appearance of ketone bodies in the bloodstream—*ketonemia*—lowers blood pH, which must be controlled by buffers. During prolonged starvation, ketone levels continue to rise. Eventually, buffering capacities are exceeded and a dangerous drop in pH takes place. This acidification of the blood by ketone bodies is called *ketoacidosis* (kē-tō-as-i-DŌ-sis). In severe ketoacidosis, the circulating concentration of ketone bodies can reach 200 mg/dL, and the pH may drop below 7.05. A pH that low can disrupt tissue activities and cause coma, cardiac arrhythmias, and death.

In summary, during the postabsorptive state, the liver acts to stabilize blood glucose concentrations. It does so first by the breakdown of glycogen reserves and later by gluconeogenesis. Over the remainder of the postabsorptive state, the combination of lipid and amino acid catabolism provides the necessary ATP. These processes generate large quantities of ketone bodies that diffuse into the bloodstream. Amino acid catabolism results in the release of toxic ammonium ions, which become tied up in the formation of urea.

Changes in the activity of the liver, adipose tissue, skeletal muscle, and other peripheral tissues ensure a steady supply of glucose to the nervous system. The supply remains steady despite daily or even weekly changes in nutrient availability. Only after a prolonged period of starvation does neural tissue begin to metabolize ketone bodies and lactate molecules, as well as glucose.

TIPS&TRICKS

The absorptive state is like harvest time: Food is being gathered and stored. The postabsorptive state corresponds to the time between harvests, when stored food is used for nourishment.

CHECKPOINT

13. **What process in the liver increases after you have eaten a high-carbohydrate meal?**

14. **Why do blood levels of urea increase during the postabsorptive state?**

15. **If a cell accumulates more acetyl-CoA than it can metabolize in the citric acid cycle, what products are likely to form?**

See the blue Answers tab at the back of the book.

25-6 ▌ Adequate nutrition is necessary to prevent deficiency disorders and ensure physiological functioning

Learning Outcome Explain what makes up a balanced diet and why such a diet is important.

The postabsorptive state can be maintained for a considerable period. For homeostasis to be maintained indefinitely, however, the digestive tract must regularly absorb enough fluids, organic substrates, minerals, and vitamins to keep pace with cellular demands. The absorption of nutrients from food is called **nutrition**.

The body's requirement for each nutrient varies from day to day and from person to person. *Nutritionists* attempt to analyze a diet in terms of its ability to meet the needs of a specific individual. A **balanced diet** contains all the ingredients needed to maintain homeostasis. Such a diet must include adequate essential amino acids and fatty acids, minerals, and vitamins, and substrates for energy generation. In addition, the diet must include enough water to replace losses in urine, feces, and evaporation. A balanced diet prevents **malnutrition**, an unhealthy state resulting from inadequate or excessive absorption of one or more nutrients.

Food Groups and a Balanced Diet

One way of avoiding malnutrition is to consume a diet based on the ChooseMyPlate plan (**Figure 25–12** and Table 25–1). The United States Department of Agriculture created this diagram to offer personalized eating plans and food assessments based on the current Dietary Guidelines for Americans. The color-coded food group slices indicate the proportions of food we should consume from each of the **five basic food groups**: grains (orange), vegetables (green), fruits (red), dairy products (blue), and protein (purple). Oils should be used sparingly in addition to the five basic food groups. For more information, visit www.choosemyplate.gov.

What is important concerning the food you eat is that you take in nutrients in sufficient *quantity* (adequate to meet your energy needs) and *quality* (including essential amino acids, fatty acids, vitamins, and minerals). There is nothing magical about five groups. At various times since 1940, the U.S. government has advocated 11, 7, 4, or 6 food groups. The key is making intelligent choices about what you eat. Poor choices can lead to malnutrition even if all five groups are represented.

For example, consider the essential amino acids. The liver cannot synthesize any of these amino acids, so you must obtain them from your diet. Some foods in the dairy and protein groups—specifically, beef, fish, poultry, eggs, and milk—provide all the essential amino acids in sufficient quantities. They are said to contain **complete proteins**. Many plants also supply adequate *amounts* of protein but contain **incomplete proteins**, which are deficient in one or more of the essential amino acids. People who follow a vegetarian diet, which is largely restricted to the grains, vegetables, and fruits groups (with or without the dairy products group), must become adept at varying their food choices to include combinations of ingredients that meet all their amino acid requirements. Even with a proper balance of amino acids, people who eat a vegan diet, which avoids all animal products, face a significant problem, because vitamin B_{12} is obtained only from animal products or from fortified cereals or tofu. (Some health-food products, such as *Spirulina*, are marketed as sources of this vitamin. Unfortunately, humans cannot utilize the form of B_{12} that they contain.)

Nitrogen Balance

A variety of important compounds in the body contain nitrogen atoms. These **N compounds** include:

- Amino acids, which are part of the framework of all proteins and protein derivatives, such as glycoproteins and lipoproteins.

- Purines and pyrimidines, the nitrogenous bases of RNA and DNA.

25

Figure 25–12 **MyPlate Plan.** The proportion of each food group is indicated by the size of its color-coded area. Foods should be consumed in proportions based on both the food group and the person's level of activity.

Find your balance between food and physical activity

- Be sure to stay within your daily calorie needs.
- Be physically active for at least 30 minutes most days of the week.
- About 60 minutes a day of physical activity may be needed to prevent weight gain.
- For sustaining weight loss, at least 60 to 90 minutes a day of physical activity may be required.
- Children and teenagers should be physically active for 60 minutes most days.

Know the limits on fats, sugars, and salt (sodium)

- Make most of your fat sources from fish, nuts, and vegetable oils.
- Limit solid fats like butter, margarine, shortening, and lard.
- Check the nutrition facts label to keep saturated fats, trans fats, and sodium low.
- Choose food and beverages low in added sugars. Added sugars contribute calories with few, if any, nutrients.

Table 25–1 Basic Food Groups and Their General Effects on Health

Nutrient Group	Provides	Health Effects
Grains (recommended: at least half of the total eaten as whole grains)	Carbohydrates; vitamins E, thiamine, niacin, folate; calcium; phosphorus; iron; sodium; dietary fiber	Whole grains prevent rapid rise in blood glucose levels, and consequent rapid rise in insulin levels
Vegetables (recommended: especially dark-green and orange vegetables)	Carbohydrates; vitamins A, C, E, folate; dietary fiber; potassium	Reduce risk of cardiovascular disease; protect against colon cancer (folate) and prostate cancer (lycopene in tomatoes)
Fruits (recommended: a variety of fruit each day)	Carbohydrates; vitamins A, C, E, folate; dietary fiber; potassium	Reduce risk of cardiovascular disease; protect against colon cancer (folate)
Dairy (recommended: low-fat or fat-free milk, yogurt, and cheese)	Complete proteins; fats; carbohydrates; calcium; potassium; magnesium; sodium; phosphorus; vitamins A, B_{12}, pantothenic acid, thiamine, riboflavin	Whole milk: High in calories, may cause weight gain; saturated fats correlated with heart disease
Protein (recommended: lean meats, fish, poultry, eggs, dry beans, nuts, legumes)	Complete proteins; fats; calcium; potassium; phosphorus; iron; zinc; vitamins E, thiamine, B_6	Fish and poultry lower risk of heart disease and colon cancer (compared to red meat). Consumption of up to one egg per day does not appear to increase incidence of heart disease; nuts and legumes improve blood cholesterol ratios, lower risk of heart disease and diabetes

25

- *Creatine*, important in energy storage in muscle tissue (as creatine phosphate).

- *Porphyrins*, complex ring-shaped molecules that bind metal ions and are essential to the function of hemoglobin, myoglobin, and the cytochromes.

Despite the importance of nitrogen to these compounds, your body neither stores nitrogen nor maintains reserves of N compounds, as it does carbohydrates (glycogen) and lipids (triglycerides). Your body can synthesize the carbon chains of the N compounds, but you must obtain nitrogen atoms either by recycling N compounds already in the body or by absorbing nitrogen from your diet. You are in **nitrogen balance** when the amount of nitrogen you absorb from your diet balances the amount you lose in urine and feces. This is the normal condition, and it means that the rates of synthesis and breakdown of N compounds are equivalent.

Growing children, athletes, people recovering from an illness or injury, and pregnant or lactating women actively synthesize N compounds. These individuals must absorb more nitrogen than they excrete. Such individuals are in a state of **positive nitrogen balance**.

When excretion exceeds ingestion, a **negative nitrogen balance** exists. This is an extremely unsatisfactory situation. The body contains only about a kilogram of nitrogen tied up in N compounds, and a decrease of one-third can be fatal. Even when energy reserves are mobilized (as during starvation), carbohydrates and lipid reserves are broken down first and N compounds are conserved.

Like N compounds, minerals and vitamins are essential parts of the diet. Your body cannot synthesize minerals, and your cells can generate only a small quantity of a very few vitamins. We consider minerals and vitamins next.

Minerals

Minerals are inorganic ions released through the dissociation of electrolytes. Minerals are important for three reasons:

1. *Ions such as sodium and chloride determine the osmotic concentrations of body fluids.* Potassium is important in maintaining the osmotic concentration of the cytosol inside body cells.

2. *Ions in various combinations play major roles in important physiological processes.* As we have seen, these processes include the maintenance of membrane potentials; the construction and maintenance of the skeleton; muscle contraction; the generation of action potentials; the release of neurotransmitters; hormone production; blood clotting; the transport of respiratory gases; buffer systems; fluid absorption; and waste removal.

3. *Ions are essential cofactors in a variety of enzymatic reactions.* For example, calcium-dependent ATPase in skeletal muscle also requires the presence of magnesium ions. Another ATPase required for the conversion of glucose to pyruvate needs both potassium and magnesium ions. Carbonic anhydrase is important in CO_2 transport, buffering systems, and gastric acid secretion. This enzyme requires the presence of zinc ions. Finally, each component of the electron transport system requires an iron atom. The final cytochrome (a_3) of the ETS must bind a copper ion as well.

The major minerals and a summary of their functional roles are listed in Table 25–2. Your body contains significant reserves of several important minerals. These reserves help reduce the effects of variations in the dietary supply. However, the reserves are often small. Chronic dietary reductions can lead to a variety of clinical problems. Alternatively, a dietary excess of mineral ions can be equally dangerous, because storage capabilities are limited.

Table 25–2	Minerals and Mineral Reserves			
Mineral	**Significance**	**Total Body Content**	**Primary Route of Excretion**	**Recommended Daily Allowance (RDA) in mg**
BULK MINERALS				
Sodium	Major cation in body fluids; essential for normal membrane function	110 g, primarily in body fluids	Urine, sweat, feces	1500
Potassium	Major cation in cytoplasm; essential for normal membrane function	140 g, primarily in cytoplasm	Urine	4700
Chloride	Major anion in body fluids; functions in forming HCl	89 g, primarily in body fluids	Urine, sweat	2300
Calcium	Essential for normal muscle and neuron function and normal bone structure	1.36 kg, primarily in skeleton	Urine, feces	1000–1200
Phosphorus	In high-energy compounds, nucleic acids, and bone matrix (as phosphate)	744 g, primarily in skeleton	Urine, feces	700
Magnesium	Cofactor of enzymes, required for normal membrane functions	29 g (skeleton, 17 g; cytoplasm and body fluids, 12 g)	Urine	310–400
TRACE MINERALS				
Iron	Component of hemoglobin, myoglobin, and cytochromes	3.9 g (1.6 g stored as ferritin or hemosiderin)	Urine (traces)	8–18
Zinc	Cofactor of enzyme systems, notably carbonic anhydrase	2 g	Urine, hair (traces)	8–11
Copper	Required as cofactor for hemoglobin synthesis	127 mg	Urine, feces (traces)	0.9
Manganese	Cofactor for some enzymes	11 mg	Feces, urine (traces)	1.8–2.3
Cobalt	Cofactor for transaminations; mineral in vitamin B_{12} (cobalamin)	1.1 g	Feces, urine	0.0001
Selenium	Antioxidant	Variable	Feces, urine	0.055
Chromium	Cofactor for glucose metabolism	0.0006 mg	Feces, urine	0.02–0.035

25

Problems involving iron are particularly common. The body of a healthy man contains about 3.5 g of iron in the ionic form Fe^{2+}. Of that amount, 2.5 g is bound to the hemoglobin in circulating red blood cells. The rest is stored in the liver and red bone marrow. In women, the total body iron content averages 2.4 g, with roughly 1.9 g incorporated into red blood cells. Thus, a woman's iron reserves consist of only 0.5 g, half that of a typical man. For this reason, if the diet contains inadequate amounts of iron, premenopausal women are more likely to develop signs of iron deficiency than are men.

Vitamins

A vitamin is an essential organic nutrient that functions as a coenzyme in vital enzymatic reactions. Vitamins are assigned to either of two groups based on their chemical structure and characteristics: *fat-soluble vitamins* or *water-soluble vitamins*.

Fat-Soluble Vitamins

Vitamins A, D, E, and K are the **fat-soluble vitamins**. Fat-soluble vitamins dissolve in lipids. You absorb these vitamins primarily from the digestive tract, along with the lipid contents of micelles. However, when exposed to sunlight, your skin can synthesize small amounts of vitamin D, and intestinal bacteria produce some vitamin K.

The modes of action of these vitamins require further study. Vitamin A has long been recognized as a structural component of the visual pigment retinal. Its more general metabolic effects are not well understood. Vitamin D is ultimately converted to calcitriol. This hormone binds to cytoplasmic receptors within the intestinal epithelium and helps to increase the rate of intestinal absorption of calcium and phosphorus. Vitamin E is thought to stabilize intracellular membranes. Vitamin K is necessary in a reaction that is essential to the synthesis of several

Clinical Note

Vitamins "If a little is good, a lot must be better" is a common—but dangerously incorrect—attitude about vitamins. When you take too much of a fat-soluble vitamin, tissue lipids absorb the excess. Because these vitamins later diffuse back into the bloodstream, the signs and symptoms of hypervitaminosis, once apparent, are likely to persist. When taken in massive amounts (from ten to thousands of times the recommended daily allowance), fat-soluble vitamins can produce acute symptoms of *vitamin toxicity*. Vitamin A toxicity is the most common condition. It afflicts some children whose parents are overanxious about proper nutrition and vitamins. A single enormous overdose can produce nausea, vomiting, headache, dizziness, lethargy, and even death. Chronic overdose can lead to hair loss, joint pain, hypertension, weight loss, and liver enlargement.

proteins, including at least three of the clotting factors. Current information about the fat-soluble vitamins is summarized in Table 25–3.

Fat-soluble vitamins normally diffuse into plasma membranes and other lipids in the body, including the lipid inclusions in the liver and adipose tissue. As a result, your body contains a significant reserve of these vitamins. Normal metabolic operations can continue for several months after dietary sources have been cut off. For this reason, a dietary insufficiency of fat-soluble vitamins rarely causes signs and symptoms of **hypovitaminosis** (hī-pō-vi-ta-min-Ō-sis) or *vitamin deficiency disease*. However, other factors can cause hypovitaminosis involving either fat-soluble or water-soluble vitamins. Problems

Table 25–3   The Fat-Soluble Vitamins

Vitamin	Significance	Sources	Recommended Daily Allowance (RDA) in mg	Effects of Deficiency	Effects of Excess
A	Maintains epithelia; required for synthesis of visual pigments; supports immune system; promotes growth and bone remodeling	Leafy green and yellow vegetables	0.7–0.9	Retarded growth, night blindness, deterioration of epithelial membranes	Liver damage, skin paling, CNS effects (nausea, anorexia)
D (also known as D₃)	Required for normal bone growth, intestinal calcium and phosphorus absorption, and retention of these ions at the kidneys	Synthesized in skin exposed to sunlight	0.005–0.015*	Rickets, skeletal deterioration	Calcium deposits in many tissues, disrupting functions
E	Prevents breakdown of vitamin A and fatty acids	Meat, milk, vegetables	15	Anemia, other problems suspected	Nausea, stomach cramps, blurred vision, fatigue
K	Essential for liver synthesis of prothrombin and other clotting factors	Vegetables; production by intestinal bacteria	0.09–0.12	Bleeding disorders	Liver dysfunction, jaundice

* Unless exposure to sunlight is inadequate for extended periods and alternative sources are unavailable.

25

	Table 25–4	The Water-Soluble Vitamins				
Vitamin	**Significance**	**Sources**	**Recommended Daily Allowance (RDA) in mg**	**Effects of Deficiency**	**Effects of Excess**	
B₁ (thiamine)	Coenzyme in many pathways	Milk, meat, bread	1.1–1.2	Muscle weakness, CNS and cardiovascular problems, including heart disease; called *beriberi*	Hypotension	
B₂ (riboflavin)	Part of FAD, involved in multiple pathways, including glycolysis and citric acid cycle	Milk, meat, eggs and cheese	1.1–1.3	Epithelial and mucosal deterioration	Itching, tingling	
B₃ (niacin)	Part of NAD, involved in multiple pathways	Meat, bread, potatoes	14–16	CNS, GI, epithelial, and mucosal deterioration; called *pellagra*	Itching, burning; vasodilation; death after large dose	
B₅ (pantothenic acid)	Coenzyme A, in multiple pathways	Milk, meat	10	Retarded growth, CNS disturbances	None reported	
B₆ (pyridoxine)	Coenzyme in amino acid and lipid metabolism	Meat, whole grains, vegetables, orange juice, cheese and milk	1.3–1.7	Retarded growth, anemia, convulsions, epithelial changes	CNS alterations, perhaps fatal	
B₉ (folic acid)	Coenzyme in amino acid and nucleic acid metabolism	Leafy vegetables, some fruits, liver, cereal and bread	0.2–0.4	Retarded growth, anemia, gastrointestinal disorders, developmental abnormalities	Few noted, except at massive doses	
B₁₂ (cobalamin)	Coenzyme in nucleic acid metabolism	Milk, meat	0.0024	Impaired RBC production, causing *pernicious anemia*	Polycythemia	
B₇ (biotin)	Coenzyme in many pathways	Eggs, meat, vegetables	0.03	Fatigue, muscular pain, nausea, dermatitis	None reported	
C (ascorbic acid)	Coenzyme in many pathways	Citrus fruits	75–90; Smokers add 35 mg	Epithelial and mucosal deterioration; called *scurvy*	Kidney stones	

may be due to an inability to absorb a vitamin from the digestive tract, inadequate storage, or excessive demand.

Too much of a vitamin can also have harmful effects. **Hypervitaminosis** (hī-per-vī-tuh-min-Ō-sis) occurs when dietary intake exceeds the body's ability to use, store, or excrete a particular vitamin. This condition usually involves one of the fat-soluble vitamins, because the excess is retained and stored in body lipids.

Water-Soluble Vitamins

Most of the **water-soluble vitamins** (Table 25–4) are components of coenzymes. For example, NAD is derived from niacin, FAD from vitamin B₂ (riboflavin), and coenzyme A from vitamin B₅ (pantothenic acid).

Water-soluble vitamins are rapidly exchanged between the fluid compartments of the digestive tract and the circulating blood. Excessive amounts are readily excreted in urine. For this reason, hypervitaminosis involving water-soluble vitamins is relatively uncommon, except among people who take large doses of vitamin supplements. Only vitamins B₁₂ and C are stored in significant quantities. Insufficient intake of other water-soluble vitamins can lead to initial signs and symptoms of vitamin deficiency within a period of days to weeks.

The bacteria that live in the intestines help prevent deficiency diseases. They produce small amounts of five of the nine water-soluble vitamins, in addition to fat-soluble vitamin K. The intestinal epithelium can easily absorb all the water-soluble vitamins except B₁₂. The B₁₂ molecule is large. It must be bound to *intrinsic factor* from the gastric mucosa before absorption can take place, as we discussed in Chapter 24. ⤴ p. 925

Diet and Disease

Diet has a profound influence on a person's general health. We have already considered the effects of too many or too few nutrients, above-normal or below-normal concentrations of minerals, and hypervitaminosis or avitaminosis. More subtle long-term problems can occur when the diet includes the wrong proportions or combinations of nutrients. The average diet in the United States contains too much sodium and too many calories. Lipids—particularly saturated fats—provide too great a proportion of those calories. This diet increases the incidence of obesity, heart disease, atherosclerosis, hypertension, and diabetes in the U.S. population.

CHECKPOINT

16. Identify the two classes of vitamins.

17. Would an athlete in intensive training try to maintain a positive or a negative nitrogen balance?

18. How would a decrease in the amount of bile salts in the bile affect the amount of vitamin A in the body?

See the blue Answers tab at the back of the book.

Clinical Note

Alcohol Alcohol production and sales are big business throughout the world. We see beer commercials on television, billboards advertising various brands of liquor, and TV or movie characters enjoying a drink. All demonstrate the prominence of alcohol in many societies. Many people are unaware of the medical consequences of this cultural fondness for alcohol. Problems with alcohol are usually divided into those stemming from alcohol abuse and those involving alcoholism. The boundary between these conditions is hazy. *Alcohol abuse* is the general term for overuse and its behavioral and physical effects. *Alcoholism* is chronic alcohol abuse accompanied by the physiological changes associated with addiction to other CNS-active drugs. Alcoholism has received the most attention in recent years, but alcohol abuse—especially when combined with driving a car—is also in the spotlight.

Consider these statistics provided by the CDC and the National Council on Alcoholism and Drug Dependence for people living in the United States:

- One in four children—over 7 million—grows up in a home where somebody abuses alcohol.
- 17.6 million people—1 in every 12 adults—abuse alcohol.
- 79,000 deaths are attributed to alcohol use annually.
- Alcoholism remains the third leading cause of lifestyle-related deaths.
- Some major clinical conditions are caused almost entirely by alcohol consumption. For example, alcohol is responsible for 60–90 percent of all liver disease in the United States.
- Alcohol affects all physiological systems. Major clinical symptoms of alcoholism include (1) disorientation and confusion (nervous system); (2) ulcers, diarrhea, and cirrhosis (digestive system); (3) cardiac arrhythmias, cardiomyopathy, and anemia (cardiovascular system); (4) depressed sexual drive and testosterone levels (reproductive system); and (5) itching and angiomas (benign blood vessel or lymphatic vessel tumors) (integumentary system).
- The toll on newborn infants has risen steadily since the 1960s as the number of female drinkers has increased.

Women who consume 1 ounce of alcohol per day during pregnancy have a higher rate of spontaneous abortion and bear children with lower birth weights than do women who consume no alcohol. Women who drink heavily may have children with *fetal alcohol syndrome (FAS)*. Facial abnormalities, a small head, slow growth, and mental retardation characterize this condition.

- Alcohol abuse is considerably more widespread than alcoholism. The medical effects are less well documented, but they are clearly significant.

Several factors interact to produce alcoholism. The primary risk factors are sex (males are more likely to become alcoholics than are females) and a family history of alcoholism. There does appear to be a genetic component: A gene on chromosome 11 has been implicated in some inherited forms of alcoholism. The relative importance of genes versus social environment has been difficult to assess. It is likely that alcohol abuse and alcoholism result from a variety of factors.

Treatment may consist of counseling and behavior modification. To be successful, treatment must include total avoidance of alcohol. Support groups, such as Alcoholics Anonymous (AA), can be very helpful in providing a social framework for abstinence. The drug *disulfiram (Antabuse)* sensitizes an individual to alcohol such that a drink produces intense nausea. Unfortunately, it has not proved to be as effective a deterrent as originally anticipated. Clinical tests indicated that it could increase the time between drinks but could not prevent drinking altogether.

Another drug, naltrexone (ReVia, Depade, and injectable Vivitrol), has shown promise in reducing the frequency and severity of drinking relapses in alcoholics. Naltrexone is an opioid receptor antagonist. It blocks the part of the brain that elicits pleasurable feelings after the consumption of alcohol. The result is a decreased craving for alcohol, enabling individuals to stop drinking more readily. Unlike Antabuse, it does not cause nausea.

25-7 | Metabolic rate is the average caloric expenditure, and thermoregulation involves balancing heat-producing and heat-losing mechanisms

Learning Outcome Define metabolic rate, discuss the factors involved in determining an individual's BMR, and discuss the homeostatic mechanisms that maintain a constant body temperature.

A person's daily energy expenditures vary widely with activity. For example, a person leading a sedentary life may have minimal energy demands, but a single hour of swimming can significantly increase daily energy use. If your daily energy intake exceeds your total energy demands, you store the excess energy, primarily as triglycerides in adipose tissue. If your daily energy expenditures exceed your intake, the result is a net reduction in your body's energy reserves and a corresponding loss in weight. In this section, first we consider aspects of energy intake and expenditure. Then we turn to the topic of thermoregulation.

Energy Gains and Losses

When chemical bonds are broken, energy is released. Inside cells, a significant amount of energy may be used to synthesize

ATP, but much of it is lost to the environment as heat. The process of *calorimetry* (kal-ō-RIM-e-trē) measures the total amount of energy released when the bonds of organic molecules are broken. The unit of measurement is the **calorie** (KAL-ō-rē) (cal), defined as the amount of energy required to raise the temperature of 1 g of water 1 degree Celsius. One gram of water is not a very practical measure when you are interested in the metabolic processes that keep a 70-kg human alive, so we use the **kilocalorie** (KIL-ō-kal-ō-rē) (kcal), or **Calorie** (with a capital C), also known as the "large calorie," instead. One Calorie is the amount of energy needed to raise the temperature of 1 *kilogram* of water 1 degree Celsius. Calorie-counting guides list the caloric value of foods in Calories, not calories.

The Energy Content of Food

In cells, organic molecules are oxidized to carbon dioxide and water. Oxidation also takes place when something burns, and this process can be experimentally controlled. A known amount of food is placed in a chamber called a **calorimeter** (kal-ō-RIM-e-ter), which is filled with oxygen and surrounded by a known volume of water. Once the food is inside, the chamber is sealed and the contents are electrically ignited. When the material has completely oxidized and only ash remains in the chamber, the number of Calories released can be determined by comparing the water temperatures before and after the test.

The energy potential of food is usually expressed in Calories per gram (Cal/g). The catabolism of lipids releases a considerable amount of energy, roughly 9.46 Cal/g. The catabolism of carbohydrates or proteins is not as productive, because many of the carbon and hydrogen atoms are already bound to oxygen. Their average yields are comparable: 4.18 Cal/g for carbohydrates and 4.32 Cal/g for proteins. Most foods are mixtures of fats, proteins, and carbohydrates, so the calculated values listed in a "calorie counter" vary.

Energy Expenditure: Metabolic Rate

Clinicians can examine your metabolic state and determine how many Calories you are utilizing. The result can be expressed as Calories per hour, Calories per day, or Calories per unit of body weight per day. What is actually measured is the sum of all the varied anabolic and catabolic processes taking place in your body—your **metabolic rate** at that time. The metabolic rate changes according to the activity under way. For instance, measurements taken while a person is sprinting are quite different from those taken while a person is sleeping. **Figure 25–13** shows the caloric expenditures for various common activities. In an attempt to reduce the variations in studies of energetics and metabolism, physiologists and clinicians standardize the testing conditions so as to determine the **basal metabolic rate (BMR)**. Ideally, the BMR is the minimum resting energy expenditure of an awake, alert person.

A direct method of determining the BMR involves monitoring respiratory activity. In resting individuals, energy use is proportional to oxygen consumption. If we assume that average amounts of carbohydrates, lipids, and proteins are being catabolized, 4.825 Calories are expended per liter of oxygen consumed.

An average individual has a BMR of 70 Cal per hour, or about 1680 Cal per day. Although the test conditions are standardized, many uncontrollable factors can influence the BMR. These factors include age, gender, physical condition, body weight, and genetic differences.

Rather than measuring the actual metabolic rate, clinicians usually monitor the concentration of thyroid hormones. Clinicians take this approach because the BMR is technically difficult to measure, and because circulating thyroid hormone levels have a profound effect on the BMR. The results are then compared with normal values to obtain an index of metabolic activity. One such test, the T_4 **assay**, measures the amount of thyroxine in the blood.

Obesity is defined as body weight more than 20 percent above the ideal weight for a given individual. Currently, obesity is considered an epidemic and is taking its toll on the health of adults and children. Diseases associated with obesity include heart disease, cancer, and diabetes. For this reason, Calorie counting and exercise are important in a weight-control program.

The control of appetite is poorly understood. Stretch receptors along the digestive tract, especially in the stomach, play a role, but other factors are probably more significant. Social factors, psychological pressures, and dietary habits are important. Evidence also indicates that complex hormonal stimuli interact to affect appetite. For example, the hormones *cholecystokinin (CCK)* and *adrenocorticotropic hormone (ACTH)* suppress the appetite. The hormone *leptin*, released by adipose tissues, also plays a role. During the absorptive state, adipose tissues release leptin into the bloodstream as they synthesize triglycerides. When leptin binds to CNS neurons that function in emotion

Figure 25–13   Caloric Expenditures for Various Activities.

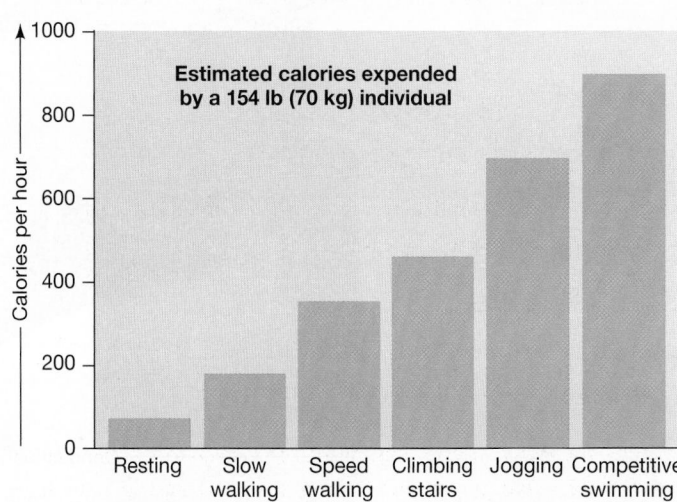

and the control of appetite, the result is a sense of satiation and the suppression of appetite. *Ghrelin* (GREL-in), a hormone secreted from the gastric mucosa, stimulates appetite. Ghrelin levels decline when the stomach is full, and increase during the fasting state. Blood ghrelin levels tend to be low in obese individuals, suggesting that it plays a role in obesity. Other effects of ghrelin are currently being studied. ⊃ p. 926

Thermoregulation

The BMR estimates the rate of energy use by the body. The energy that cells do not capture and harness is released as heat. This heat serves an important homeostatic purpose. Humans are subject to vast changes in environmental temperatures, but our complex biochemical systems have a major limitation. Our enzymes operate over only a relatively narrow temperature range. Accordingly, our bodies have anatomical and physiological mechanisms that keep body temperatures within acceptable limits, regardless of environmental conditions. This homeostatic process is called **thermoregulation**. Failure to control body temperature can result in a series of physiological changes. For example, a body temperature below 36°C (97°F) or above 40°C (104°F) can cause disorientation. A temperature above 42°C (108°F) can cause convulsions and permanent cell damage.

We continuously produce heat as a by-product of metabolism. When energy use increases due to physical activity, or when our cells are more active metabolically (as they are during the absorptive state), additional heat is generated. The heat produced by biochemical reactions is retained by water, which makes up nearly two-thirds of body weight. Water is an excellent conductor of heat, so the heat produced in one region of the body is rapidly distributed by diffusion, as well as through the bloodstream. If body temperature is to remain constant, that heat must be lost to the environment at the same rate it is generated. When environmental conditions rise above or fall below "ideal," the body must control the gains or losses to maintain homeostasis.

Mechanisms of Heat Transfer

Heat exchange with the environment involves four basic processes: (1) *radiation*, (2) *convection*, (3) *evaporation*, and (4) *conduction* (**Figure 25–14**).

Objects warmer than the environment lose heat as **radiation**. When you feel the heat from the sun, you are experiencing radiant heat. Your body loses heat the same way, but in proportionately smaller amounts. More than 50 percent of the heat you lose indoors is lost through radiation. The exact amount varies with both body temperature and skin temperature.

Convection is heat loss to the cooler air that moves across the surface of your body. As your body loses heat to the air next to your skin, that air warms and rises, moving away from the surface of the skin. Cooler air replaces it, and this air in turn becomes warmed. This pattern then repeats. Convection accounts for roughly 15 percent of the body's heat loss indoors but is insignificant as a mechanism of heat gain.

When water evaporates, it changes from a liquid to a vapor. **Evaporation** absorbs energy—about 0.58 Cal per gram of water evaporated—and cools the surface where evaporation occurs. The rate of evaporation at your skin is highly variable. Each hour, 20–25 mL of water crosses epithelia and evaporates from the alveolar surfaces and the surface of the skin. This *insensible water loss* remains relatively constant. At rest, it accounts for roughly 20 percent of your body's average indoor heat loss. The sweat glands responsible for *sensible perspiration* have a tremendous scope of activity, ranging from virtual inactivity to secretory rates of 2–4 liters (2.1–4.2 quarts) per hour. ⊃ p. 181

Conduction is the direct transfer of energy through physical contact. When you come into an air-conditioned classroom and sit on a cold plastic chair, you are immediately aware of this process. Conduction is generally not an effective way of gaining or losing heat. Its impact depends on the temperature of the object and the amount of skin surface area involved. When you are lying on

Figure 25–14 **Mechanisms of Heat Transfer.**

- Radiation
- Evaporation
- Convection
- Conduction

cool sand in the shade, conductive losses can be considerable. When you are standing on the same sand, conductive losses are negligible.

The Regulation of Heat Gain and Heat Loss

Heat loss and heat gain involve the activities of many systems. Those activities are coordinated by the **heat-loss center** and **heat-gain center**, respectively, in the preoptic area of the anterior hypothalamus. ⤺ p. 506 These centers modify the activities of other hypothalamic nuclei. The overall effect is to control temperature by influencing two processes: the rate of heat production and the rate of heat loss to the environment. Changes in behavior may also support these processes.

Mechanisms for Increasing Heat Loss. When the temperature at the preoptic nucleus rises above its set point, the heat-loss center is stimulated, producing three major effects:

1. *The inhibition of the vasomotor center causes peripheral vasodilation, and warm blood flows to the surface of the body.* The skin takes on a reddish color, skin temperatures rise, and radiative and convective losses increase.

2. *As blood flow to the skin increases, sweat glands are stimulated to increase their secretory output.* The perspiration flows across the body surface, and evaporative heat losses accelerate. Maximal secretion, if completely evaporated, would remove 2320 Cal per hour.

3. *The respiratory centers are stimulated, and the depth of respiration increases.* Often, the individual begins breathing through an open mouth rather than through the nasal passageways. This increases evaporative heat losses through the lungs.

Mechanisms for Promoting Heat Gain. The function of the heat-gain center of the brain is to prevent **hypothermia** (hī-pō-THER-mē-uh), or below-normal body temperature. When the temperature at the preoptic nucleus drops below acceptable levels, the heat-loss center is inhibited and the heat-gain center is activated.

Heat Conservation. The sympathetic vasomotor center decreases blood flow to the dermis, thereby reducing losses by radiation, convection, and conduction. The skin cools. With blood flow restricted, the skin may take on a bluish or pale color. The epithelial cells are not damaged, because they can tolerate extended periods at temperatures as low as 25°C (77°F) or as high as 49°C (120°F).

In addition, blood returning from the limbs is shunted into a network of deep veins. ⤺ p. 789 Under warm conditions, blood flows in a superficial venous network (**Figure 25–15a**). In cold conditions, blood is diverted to a network of veins that lie deep to an insulating layer of subcutaneous fat (**Figure 25–15b**). This venous network wraps around the deep arteries, and

heat is conducted from the warm blood flowing outward to the limbs to the cooler blood returning from the periphery (**Figure 25–15c**). This arrangement traps the heat close to the body core and restricts heat loss. Such exchange between fluids that are moving in opposite directions is called *countercurrent exchange*. (We return to this topic in Chapter 26.)

Heat Generation. We can divide the mechanisms for generating heat into two broad categories: shivering thermogenesis and nonshivering thermogenesis. In **shivering thermogenesis** (ther-mō-JEN-e-sis), a gradual increase in muscle tone increases the energy consumption of skeletal muscle tissue throughout your body. The more energy used, the more heat is produced. Both agonist and antagonist muscles are involved. The degree of stimulation varies with the demand.

If the heat-gain center is extremely active, muscle tone increases to the point at which stretch receptor stimulation will produce brief, oscillatory contractions of antagonistic muscles. In other words, you begin to **shiver**. Shivering increases the workload of the muscles and further elevates oxygen and energy consumption. The heat that is produced warms the deep vessels, to which blood has been shunted by the sympathetic vasomotor center. Shivering can elevate body temperature quite effectively, increasing the rate of heat generation by as much as 400 percent.

The intermittent cycle of capillary vasoconstriction and vasodilation in our hands when they are exposed to cold temperature is called the *Lewis Wave* or the *hunter's response*. This is an autonomic response to reduce heat loss at the extremities.

Nonshivering thermogenesis involves the release of hormones that increase the metabolic activity of all tissues:

- The heat-gain center stimulates the adrenal medullae through the sympathetic division of the autonomic nervous system. As a result, epinephrine is released. It increases the rates of glycogenolysis in liver and skeletal muscle, and the metabolic rate of most tissues. The effects are immediate.

- The preoptic nucleus regulates the production of thyrotropin-releasing hormone (TRH) by the hypothalamus. In children, when body temperatures are below normal, additional TRH is released. This hormone stimulates the release of thyroid-stimulating hormone (TSH) by the anterior lobe of the pituitary gland. In response to this release of TSH, the thyroid gland increases the rate of thyroxine release into the blood. Thyroxine increases not only the rate of carbohydrate catabolism, but also the rate of catabolism of all other nutrients. These effects develop gradually, over a period of days to weeks.

Sources of Individual Variation in Thermoregulation

The timing of thermoregulatory responses differs from person to person. A person may undergo **acclimatization** (a-klī-ma-ti-ZĀ-shun)—a physiological adjustment to a particular environment over time. For example, natives of Tierra del Fuego (off the southernmost tip of South America) once lived naked in the snow, but Hawaii residents often unpack their sweaters when the temperature drops below 22°C (72°F).

Another source of variation is body size. Although heat *production* takes place within the mass of the body, heat *loss*

Figure 25–15   Vascular Adaptations for Heat Loss and Conservation.

WARM ENVIRONMENT

Brachial vein (deep)
Basilic vein (superficial)
Brachial artery (deep)
Cephalic vein (superficial)
Median cubital vein (superficial)
Radial artery (deep)
Basilic vein (superficial)
Radial vein (deep)
Ulnar artery (deep)
Ulnar vein (deep)

COLD ENVIRONMENT

a Circulation through the blood vessels of the forearm in a warm environment. Blood enters the limb in a deep artery and returns to the trunk in a network of superficial veins that radiate heat to the environment through the overlying skin.

b Circulation through the blood vessels of the forearm in a cold environment. Blood now returns to the trunk by a network of deep veins that flow around the artery. The amount of heat loss is decreased, as shown in part (c).

Warm blood from trunk
Warm blood returns to trunk
37°C
36.5°–37°C
24°C
23°C
Cooled blood to distal capillaries
Cool blood returns to trunk

c Countercurrent heat exchange occurs as heat radiates from the warm arterial blood into the cooler venous blood flowing in the opposite direction. By the time the arterial blood reaches distal capillaries, where most heat loss to the environment occurs, it is already 13°C cooler than it was when it left the trunk. This mechanism decreases the rate of heat loss by conserving body heat within the trunk.

takes place across a body surface. As an object (or person) gets larger, its surface area increases at a much slower rate than does its total volume. This relationship affects thermoregulation, because heat generated by the "volume" (that is, by internal tissues) is lost at the body surface. For this reason, small individuals lose heat more readily than do large individuals.

Thermoregulatory Problems in Infants. During embryonic development, the mother's body is at normal body temperature. At birth, the infant's temperature-regulating mechanisms are not fully functional. Infants also lose heat quickly as a result of their small size. Consequently, newborns must be dried promptly and kept bundled up. Those born prematurely need a thermally regulated incubator. Infants' body temperatures are also less stable than those of adults.

Infants cannot shiver, but they have a different mechanism for raising body temperature rapidly. In infants, the adipose tissue between the shoulder blades, around the neck, and possibly elsewhere in the upper body, is histologically and functionally different from most of the adipose tissue in adults. The tissue is highly vascularized, and individual adipocytes contain numerous mitochondria. Together, these characteristics give the tissue a deep, rich color that is the source of the name **brown fat**. ⊃ p. 155

Individual adipocytes in brown fat are innervated by sympathetic fibers. When these nerves are stimulated, lipolysis speeds up in the adipocytes. The cells do not capture the energy released through fatty acid catabolism. Instead, the energy radiates into the surrounding tissues as heat. This heat quickly warms the blood passing through the surrounding network of vessels, and it is then distributed throughout the body. In this way, an infant can very quickly speed up metabolic heat generation by 100 percent. In contrast, nonshivering thermogenesis in an adult raises heat production by only 10–15 percent after a period of weeks.

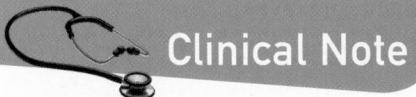

Clinical Note

Thermoregulatory Disorders Heat exhaustion and heat stroke are malfunctions of thermoregulatory mechanisms. With **heat exhaustion**, also known as *heat prostration*, excessive fluids are lost in perspiration, and the individual has difficulty maintaining blood volume. The heat-loss center stimulates sweat glands, whose secretions moisten the surface of the skin to provide evaporative cooling. As fluid losses mount, blood volume decreases. The resulting decline in blood pressure is not countered by peripheral vasoconstriction, because the heat-loss center is actively stimulating peripheral vasodilation. As blood flow to the brain declines, headache, nausea, and eventual collapse follow. Treatment is straightforward: Provide fluids, salts, and a cooler environment.

In **heat stroke**, body temperature rises uncontrollably because the thermoregulatory center stops functioning. The sweat glands are inactive, and the skin becomes hot and dry. Heat stroke is more serious and can follow an untreated case of heat exhaustion. Predisposing factors include any preexisting condition that affects peripheral circulation, such as heart disease or diabetes. Unless the problem is recognized in time, body temperature may climb to 41°C–45°C (106°F–113°F). Temperatures in this range quickly disrupt a variety of vital physiological systems and destroy brain, liver, skeletal muscle, and kidney cells. Effective treatment involves lowering the body temperature as rapidly as possible.

Hypothermia is below-normal body temperature. If body temperature drops significantly below normal levels, thermoregulatory mechanisms become less sensitive and less effective. Cardiac output and respiratory rate decrease. If the core temperature falls below 28°C (82°F), cardiac arrest is likely. The individual then has no heartbeat, no respiratory rate, and no response to external stimuli—even painful ones. The body temperature continues to decline, and the skin turns blue or pale and cold.

At this point, people commonly assume that the person has died. But because metabolic activities have decreased system-wide, the victim may still be saved, even after several hours. Treatment consists of cardiopulmonary support and gradual rewarming, both external and internal. The skin can be warmed up to 45°C (113°F) without damage. Warm baths or blankets can be used. One effective method of raising internal temperatures involves the introduction of warm saline solution into the peritoneal cavity.

Hypothermia is a significant risk for people engaged in water sports. It may complicate the treatment of a drowning victim. Water absorbs heat about 27 times as fast as air does, and the body's heat-gain mechanisms are unable to keep pace over long periods or when faced with a large temperature gradient. But hypothermia in cold water does have a positive side. On several occasions, small children who have "drowned" in cold water have been successfully revived after periods of up to four hours. Children lose body heat quickly, and their systems soon stop functioning as their body temperature declines. This rapid drop in temperature prevents the oxygen starvation and tissue damage that would otherwise take place when breathing stops.

Resuscitation is not attempted if the person has actually frozen. Water expands roughly 7 percent during ordinary freezing. The process destroys plasma membranes throughout the body. Very small organisms can be frozen and subsequently thawed without ill effects, because their surface-to-volume ratio is enormous, and the freezing process occurs so rapidly that ice crystals never form.

With increasing age and size, an infant's body temperature becomes more stable. As a result, brown fat becomes less important. With increased body size, skeletal muscle mass, and insulation, shivering thermogenesis becomes more effective in raising body temperature. Adults have little, if any, brown fat.

Thermoregulatory Variations among Adults. Adults of a given body weight may differ in their thermal responses due to variations in body mass and tissue distribution. Adipose tissue is an excellent insulator. It conducts heat at only about one-third the rate of other tissues. As a result, people with a more substantial layer of subcutaneous fat may not begin to shiver until long after their thinner companions.

Two otherwise similar individuals may also differ in their response to temperature changes because their hypothalamic "thermostats" are at different settings. We all experience daily oscillations in body temperature. Temperatures peak during the day or early evening. They fall 1°C–2°C (1.8°F–3.6°F) at night. The ovulatory cycle also causes temperature fluctuations in women, as you will see in Chapter 28.

Individuals vary in the timing of their maximum temperature setting. Some have a series of peaks, with an afternoon low. The origin of these patterns is unclear. The patterns do not result from daily activity regimens. For example, the temperatures of people who work at night still peak during the same range of times as the rest of the population.

Fevers

Any elevated body temperature is called **pyrexia** (pī-REK-sē-uh). Pyrexia is usually temporary. A **fever** is a body temperature greater than 37.2°C (99°F). We discussed fevers when we examined innate (nonspecific) immunity. ↻ p. 828 Fevers occur for a variety of reasons, not all of them pathological. In young children, transient fevers with no ill effects can result from exercise in warm weather. Similar exercise-related elevations were rarely seen in adults until running marathons became popular. Temperatures from 39°C to 41°C (103°F to 109°F) may result. For this reason, competitions are usually held when the air temperature is below 28°C (82°F).

Fevers can also result from the following factors:

- Abnormalities that affect the entire thermoregulatory mechanism, such as heat exhaustion or heat stroke.

- Clinical problems that restrict blood flow, such as congestive heart failure.

- Conditions that impair sweat gland activity, such as drug reactions and some skin conditions.

- The resetting of the hypothalamic "thermostat" by circulating *pyrogens*—most notably, interleukin-1.

> ### CHECKPOINT
>
> 19. How would the BMR of a pregnant woman compare with her own BMR before she became pregnant?
>
> 20. What effect does peripheral vasoconstriction on a hot day have on an individual's body temperature?
>
> 21. Why do infants have greater problems with thermoregulation than adults?
>
> See the blue Answers tab at the back of the book.

Related Clinical Terms

anorexia: Persistent loss of appetite.

anorexia nervosa: Eating disorder occurring primarily among girls and women characterized by a desire to lose, or not gain, weight through starvation, due to a distorted view of the person's own body. There are typically two types: strict diet and exercise, and binging and purging.

binge-purge syndrome: Eating disorder characterized by excessive eating followed by periods of fasting or self-induced vomiting.

bulimia: An eating disorder usually characterized by episodic binge eating that is followed by feelings of guilt, depression, and self-condemnation. It is often associated with steps taken to lose weight, such as self-induced vomiting, the use of laxatives, dieting, or fasting.

eating disorders: Psychological problems that result in inadequate or excessive food consumption. Examples include anorexia nervosa and bulimia.

familial hypercholesterolemia: The most common inherited type of hyperlipidemia characterized by high levels of lipids in one's blood. It affects one in every 500 children born, who then present with LDL levels in excess of 190.

heat cramps: A condition that usually follows heavy sweating due to strenuous activity that causes a loss of salt in the body and results in painful muscle spasms in the abdomen, arms, or legs.

hyperuricemia: A condition in which the plasma uric acid level exceeds 7.4 mg/dL. It can result in the condition called *gout*.

ketonuria: The presence of ketone bodies in the urine.

kwashiorkor: A form of malnutrition due to a protein deficiency in the diet that typically affects young children in tropical regions.

marasmus: Severe malnourishment that causes a child's weight to fall significantly below the standards set for children of similar age.

orexigenic: Having a stimulating effect on the appetite.

pica: Tendency or craving for substances other than normal food. Possible organic causes are iron deficiency, lead encephalopathy, pregnancy, and zinc deficiency.

protein-calorie malnutrition (PCM): A severe deficiency of protein in the diet in addition to inadequate caloric intake. It results in the condition termed *kwashiorkor*.

skin-fold test: Test to estimate the amount of body fat on a person.

25

Chapter Review

Study Outline

MasteringA&P®

Access more chapter study tools online in the MasteringA&P Study Area:
- Chapter Quizzes, Chapter Practice Test, Art-labeling Activities, Animations, MP3 Tutor Sessions, and Clinical Case Studies
- Practice Anatomy Lab PAL™   • A&P Flix A&PFlix
- Interactive Physiology iP®   • PhysioEx PhysioEx

25-1 ▎ **Metabolism refers to all the chemical reactions in the body, and energetics refers to the flow and transformation of energy** p. 962

1. **Energetics** is the study of the flow of energy and its change(s) from one form to another. Its focus includes understanding a range of energy requirements from cells to the whole body.
2. In general, during *cellular metabolism*, cells break down excess carbohydrates first and then lipids, while conserving amino acids. Only about 40 percent of the energy released through **catabolism** is captured in ATP; the rest is released as heat. *(Figure 25–1)*
3. Cells synthesize new compounds **(anabolism)** to (1) perform structural maintenance or repairs, (2) support growth, (3) produce secretions, and (4) build and store nutrient reserves.
4. Cells "feed" small organic molecules to their mitochondria; in return, the cells get the ATP they need to perform cellular functions. *(Figure 25–2)*

25-2 ▎ **Carbohydrate metabolism involves glycolysis, ATP production, and gluconeogenesis** p. 965

5. Most cells generate ATP and other high-energy compounds through the breakdown of carbohydrates.
6. **Glycolysis** and **aerobic metabolism**, or *cellular respiration*, provide most of the ATP used by typical cells. Glycogen can be broken down to glucose molecules. In glycolysis, each molecule of glucose yields 2 molecules of **pyruvic acid** (as **pyruvate** ions), a net 2 molecules of ATP, and 2 **NADH** molecules. *(Figure 25–3)*
7. In the presence of oxygen, pyruvate molecules enter mitochondria, where they are broken down completely in the **citric acid cycle**. Carbon and oxygen atoms are lost as carbon dioxide **(decarboxylation)**; hydrogen atoms are passed to coenzymes, which initiate the oxygen-consuming and ATP-generating reaction **oxidative phosphorylation**. *(Figure 25–4)*
8. **Cytochromes** of the **electron transport system (ETS)** pass electrons to oxygen, resulting in the formation of water. As this transfer occurs, the ETS generates ATP. *(Figure 25–5)*
9. For each glucose molecule processed through glycolysis, the citric acid cycle, and the ETS, most cells gain 36 molecules of ATP. *(Figure 25–6)*
10. Cells can break down other nutrients to provide substrates for the citric acid cycle if supplies of glucose are limited.
11. **Gluconeogenesis**, the synthesis of glucose from noncarbohydrate precursors such as lactate, glycerol, or amino acids, enables a liver cell to synthesize glucose molecules when carbohydrate reserves are depleted. **Glycogenesis** is the process of glycogen formation. Glycogen is an important energy reserve

when cells cannot obtain enough glucose from interstitial fluid. *(Figure 25–7)*

25-3 ▎ **Lipid metabolism involves lipolysis, beta-oxidation, and the transport and distribution of lipids as free fatty acids and lipoproteins** p. 973

12. During **lipolysis** (lipid catabolism), lipids are broken down into pieces that can be converted into pyruvate or channeled into the citric acid cycle.
13. Triglycerides, the most abundant lipids in the body, are split into glycerol and fatty acids. The glycerol enters the glycolytic pathways (glycolysis and gluconeogenesis), and the fatty acids enter the mitochondria.
14. **Beta-oxidation** is the breakdown of a fatty acid molecule into 2-carbon fragments that can be used in the citric acid cycle. The steps of beta-oxidation cannot be reversed, and the body cannot manufacture all the fatty acids needed for normal metabolic operations. *(Figure 25–8)*
15. Lipids cannot provide large amounts of ATP quickly. However, cells can shift to lipid-based energy production when glucose reserves are limited.
16. In **lipogenesis** (the synthesis of lipids), almost any organic substrate can be used to form glycerol. **Essential fatty acids** are those that cannot be synthesized and must be included in the diet.
17. Most lipids circulate as **lipoproteins** (lipid–protein complexes that contain large glycerides and cholesterol). The largest lipoproteins, chylomicrons, carry absorbed lipids from the intestinal tract to the bloodstream. All other lipoproteins are derived from the liver and carry lipids to and from various tissues of the body. *(Figure 25–9)*
18. Capillary walls of adipose tissue, skeletal muscle, cardiac muscle, and the liver contain **lipoprotein lipase**, an enzyme that breaks down lipids, releasing a mixture of fatty acids and monoglycerides into the interstitial fluid. *(Figure 25–9)*

25-4 ▎ **Protein catabolism involves transamination and deamination, whereas protein synthesis involves amination and transamination** p. 978

19. If other energy sources are inadequate, mitochondria can break down amino acids in the citric acid cycle to generate ATP. In the mitochondria, the amino group can be removed by either **transamination** (an exchange reaction) or **deamination**. *(Figure 25–10)*
20. Protein catabolism is impractical as a source of quick energy.
21. Roughly half the amino acids needed to build proteins can be synthesized. There are 10 **essential amino acids**, which must be acquired through the diet. **Amination**, the attachment of an amino group to a carbon framework, is an important step in the synthesis of **nonessential amino acids**. *(Figure 25–10, Spotlight Figure 25–11)*

25-5 ▎ **The body experiences two patterns of metabolic activity: the absorptive and postabsorptive states** p. 982

22. No one cell of a human can perform all the anabolic and catabolic operations necessary to support life. Homeostasis can be preserved only when metabolic activities of different tissues are coordinated.
23. The body has five metabolic components: the liver, adipose tissue, skeletal muscle, neural tissue, and other peripheral

25

tissues. The liver is the focal point for metabolic regulation and control. Adipose tissue stores lipids, primarily in the form of triglycerides. Skeletal muscle contains substantial glycogen reserves, and the contractile proteins can be degraded and the amino acids used as an energy source. Neural tissue does not contain energy reserves; glucose must be supplied to it for energy. Other peripheral tissues are able to metabolize glucose, fatty acids, or other substrates under the direction of the endocrine system.

24. For about four hours after a meal, nutrients enter the blood as intestinal absorption proceeds. (*Spotlight Figure 25–11*)

25. The liver closely regulates the circulating levels of glucose and amino acids.

26. The **absorptive state** exists when nutrients are being absorbed by the digestive tract. Adipocytes remove fatty acids and glycerol from the bloodstream and synthesize new triglycerides to be stored for later use. (*Spotlight Figure 25–11*)

27. During the absorptive state, glucose molecules are catabolized and amino acids are used to build proteins. Skeletal muscles may also catabolize circulating fatty acids, and the energy obtained is used to increase glycogen reserves.

28. The **postabsorptive state** extends from the end of the absorptive state to the next meal. (*Spotlight Figure 25–11*)

29. When blood glucose levels fall, the liver begins breaking down glycogen reserves and releasing glucose into the bloodstream. As the time between meals increases, liver cells synthesize glucose molecules from smaller carbon fragments and from glycerol molecules. Fatty acids undergo beta-oxidation; the fragments enter the citric acid cycle or combine to form **ketone bodies**. (*Spotlight Figure 25–11*)

30. Some amino acids can be converted to pyruvate and used for gluconeogenesis; others, including most of the essential amino acids, are converted to acetyl-CoA and are either catabolized or converted to ketone bodies.

31. Neural tissue continues to be supplied with glucose as an energy source until blood glucose levels become very low.

25-6 ▌ **Adequate nutrition is necessary to prevent deficiency disorders and ensure physiological functioning** p. 983

32. **Nutrition** is the absorption of nutrients from food. A **balanced diet** contains all the ingredients needed to maintain homeostasis; a balanced diet prevents **malnutrition**.

33. The five **basic food groups** are grains, vegetables, fruits, dairy, and protein. These are arranged on a *plate* to reflect the recommended daily food consumption balanced with daily physical activity. (*Figure 25–12; Table 25–1*)

34. Amino acids, purines, pyrimidines, creatine, and porphyrins are **N compounds**, which contain nitrogen atoms. An adequate dietary supply of nitrogen is essential, because the body does not maintain large nitrogen reserves. **Nitrogen balance**

is a state in which the amount of nitrogen absorbed equals that lost in urine and feces.

35. **Minerals** act as cofactors in a variety of enzymatic reactions. They also contribute to the osmotic concentration of body fluids and play a role in membrane potentials, action potentials, the release of neurotransmitters, muscle contraction, skeletal construction and maintenance, gas transport, buffer systems, fluid absorption, and waste removal. (*Table 25–2*)

36. Vitamins are needed in very small amounts. Vitamins A, D, E, and K are **fat-soluble vitamins**; taken in excess, they can lead to **hypervitaminosis**. **Water-soluble vitamins** are not stored in the body; a lack of adequate dietary supplies may lead to **hypovitaminosis** (*vitamin deficiency disease*). (*Tables 25–3, 25–4*)

37. A balanced diet can improve one's general health.

25-7 ▌ **Metabolic rate is the average caloric expenditure, and thermoregulation involves balancing heat-producing and heat-losing mechanisms** p. 988

38. The energy content of food is usually expressed in **kilocalories (kcal)** or as **Calories** per gram (Cal/g).

39. The catabolism of lipids releases 9.46 Cal/g, about twice the amount released by equivalent weights of carbohydrates or proteins.

40. The sum of all the anabolic and catabolic processes under way is an individual's **metabolic rate**. The **basal metabolic rate (BMR)** is the rate of energy utilization by a person at rest. (*Figure 25–13*)

41. The homeostatic regulation of body temperature is **thermoregulation**. Heat exchange with the environment involves four processes: **radiation**, **convection**, **evaporation**, and **conduction**. (*Figure 25–14*)

42. The *preoptic area* of the hypothalamus acts as the body's thermostat, affecting the **heat-loss center** and the **heat-gain center**.

43. Mechanisms for increasing heat loss include both physiological mechanisms (peripheral vasodilation, increased perspiration, and increased respiration) and behavioral modifications.

44. Responses that conserve heat include a decreased blood flow to the dermis and *countercurrent exchange*. (*Figure 25–15*)

45. Heat is generated by **shivering thermogenesis** and **nonshivering thermogenesis**.

46. Thermoregulatory responses differ among individuals. One important source of variation is **acclimatization** (a physiological adjustment to an environment over time).

47. **Pyrexia** is an elevated body temperature. **Fever**, a body temperature above 37.2°C (99°F), can result from problems with the thermoregulatory mechanism, circulation, or sweat gland activity, or from the resetting of the hypothalamic "thermostat" by circulating pyrogens.

Review Questions

See the blue Answers tab at the back of the book.

LEVEL 1 Reviewing Facts and Terms

1. The complete catabolism of one glucose molecule typically yields a net energy gain of **(a)** 32 ATP, **(b)** 36 ATP, **(c)** 36 ATP and 2 GTP, **(d)** 34 ATP.

2. The breakdown of glucose to pyruvate is an _____ process. **(a)** anaerobic, **(b)** aerobic, **(c)** anabolic, **(d)** oxidative, **(e)** both a and d.

3. The sequence of cytochromes in the electron transfer chain is **(a)** b, c, a, c, a3, **(b)** a, b, c, a3, **(c)** a3, a, c, b, **(d)** c, b, a, a3.

4. Beta oxidation is the **(a)** attachment of an amino group to a keto acid, **(b)** breakdown of fatty acids into 2C fragments, **(c)** removal of an amino group that produces ammonium, **(d)** conversion of ammonium to urea.

5. The four functions that are supported by anabolism in cellular metabolism are structural maintenance, cellular growth, _____, and nutrient storage.

6. State the functional roles of minerals within the body.

7. Use the following word bank to fill in the missing terms in the Nutrient Use in Cellular Metabolism diagram below.

Word Bank: citric acid cycle, amino acids, glucose, electron transport system, fatty acids, ATP, CO_2, two-carbon chains, H_2O.

(a) _____   (b) _____
(c) _____   (d) _____
(e) _____   (f) _____
(g) _____   (h) _____
(i) _____

8. What is meant by the term metabolic rate?
9. List three mechanisms by which heat loss can be increased in the body.

LEVEL 2 Reviewing Concepts

10. What is oxidative phosphorylation? Explain how the electron transport system is involved in this process.
11. How are lipids catabolized in the body? How is beta-oxidation involved in lipid catabolism?
12. How do the absorptive and postabsorptive states maintain normal blood glucose levels?
13. Explain why protein is not necessarily a good source of energy.
14. Explain how cholesterol in very low-density lipids (i) becomes part of the plasma membrane, and (ii) is excreted in bile.
15. Explain why low-density lipids are known as "bad cholesterol."

LEVEL 3 Critical Thinking and Clinical Applications

16. A man is found dead in his apartment by police, who note that some parts of his body look sunburnt. Toxicology reports show that he died from cyanide poisoning. Why did he appear to be sunburnt?
17. Charlie has a blood test that shows a normal level of LDLs but an elevated level of HDLs in his blood. Given that his family has a history of cardiovascular disease, he wonders if he should modify his lifestyle. What would you tell him?
18. Jill suffers from anorexia nervosa. One afternoon she is rushed to the emergency room because of cardiac arrhythmias. Her breath smells fruity, and her blood and urine samples contain high levels of ketone bodies. Why do you think she is having the arrhythmias?

CLINICAL CASE Wrap-Up

The Miracle Supplement

Vitamin A is the miracle vitamin supplement that can reduce overall childhood mortality by 23 percent in high-risk areas like rural Kenya. Preschool-age children and pregnant women in their last trimester are the most vulnerable to vitamin A deficiency.

Like other fat-soluble vitamins, vitamin A is an essential organic nutrient that functions as a coenzyme in vital enzymatic reactions. In the retina of the eye, retinal (made from vitamin A) combines with opsin to form rhodopsin, which is necessary for low-light and color vision. Vitamin A deficiency is the leading cause of preventable childhood blindness, affecting as many as half a million children worldwide. As a coenzyme, vitamin A is critical in many steps in the immune system. Vitamin A deficiency leads to a decreased immune response and increased vulnerability to malaria, respiratory infections, and diarrhea. Vitamin A is also critical for normal bone growth and development, and its deficiency can lead to stunting and skeletal deformities.

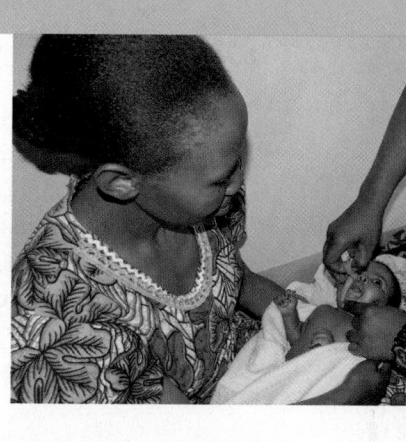

Vitamin A is found naturally in leafy green, yellow, and orange vegetables and fruits. Vitamin A is also available in milk, liver, eggs, and some fatty fish. None of these food items are available to Justine, who is among the very poor living in sub-Saharan Africa. But thanks to inexpensive vitamin A supplements, Justine's daughter will be protected from the health problems caused by vitamin A deficiency.

1. Other than diet and vitamin supplementation, is there any other available source of vitamin A?

2. If Justine's diet—which includes very little milk, liver, eggs, fish, leafy green, yellow, and orange vegetables, and fruits—is deficient in vitamin A, what other vitamin deficiencies might be expected?

See the blue Answers tab at the back of the book.

26

The Urinary System

Learning Outcomes

These Learning Outcomes correspond by number to this chapter's sections and indicate what you should be able to do after completing the chapter.

26-1 ▌ Identify the organs of the urinary system, and describe the functions of the system.

26-2 ▌ Describe the location and structures of the kidneys, identify major blood vessels associated with each kidney, trace the path of blood flow through a kidney, describe the structure of a nephron, and identify the functions of each segment of the nephron and collecting system.

26-3 ▌ Describe the basic processes that form urine.

26-4 ▌ Describe the factors that influence glomerular filtration pressure and the rate of filtrate formation.

26-5 ▌ Identify the types and functions of transport mechanisms found along each segment of the nephron, explain the role of countercurrent multiplication, describe hormonal influence on the volume and concentration of urine, and describe the characteristics of a normal urine sample.

26-6 ▌ Describe the structures and functions of the ureters, urinary bladder, and urethra, discuss the voluntary and involuntary regulation of urination, and describe the micturition reflex.

26-7 ▌ Describe the effects of aging on the urinary system.

26-8 ▌ Give examples of interactions between the urinary system and other organ systems studied so far.

SPOTLIGHT
- Summary of Renal Function pp. 1028–1029

CLINICAL CASE
- A Case of "Hidden" Bleeding p. 999

Clinical Notes
- Analysis of Renal Blood Flow p. 1003
- Glomerulonephritis p. 1008
- Diuretics p. 1023
- Urinary Obstruction p. 1034
- Renal Failure and Kidney Transplant p. 1035

A Case of "Hidden" Bleeding

The motorcycle accident scene is horrific. The driver is dead, and his passenger—a 14-year-old high school freshman named Mike—is unconscious more than 50 feet from the wreckage. When emergency responders deliver Mike to the trauma center, he has no measurable blood pressure, in spite of the 4 liters of fluids they have infused. The trauma team begins administering whole blood transfusions, but Mike is bleeding faster than replacement blood can be supplied.

Noting Mike's bloated abdominal appearance and the abrasions on his left posterior chest and flank, the trauma team suspects that the massive blood loss is due to intra-abdominal injuries. An x-ray of Mike's chest and abdomen shows fractured left 8th through 12th ribs and fractures of the left transverse processes of L_1, L_2, and L_3 vertebrae. A nurse inserts a urinary catheter into

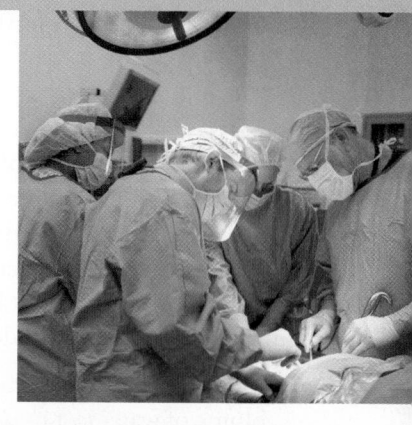

Mike's bladder, which drains clear urine.

Upon opening Mike's abdomen through a midline incision, the surgeon encounters a copious amount of free blood. Clearing this blood quickly, and moving the abdominal contents out of the way, he finds a shattered spleen that is bleeding profusely. The surgeon performs an emergency splenectomy. Rinsing the abdominal contents reveals no further visible bleeding, but the surgical team cannot stabilize Mike's blood pressure, no matter how quickly they transfuse blood. **Something is still bleeding, but from where? To find out, turn to the Clinical Case Wrap-Up on p. 1040.**

An Introduction to the Urinary System

The urinary system removes most physiological wastes. In this chapter, we consider the functional organization of the urinary system and describe how the kidneys remove metabolic wastes from the circulation to produce urine. We also explain the major regulatory mechanisms controlling urine production and concentration, and identify how urine is transported to the urinary bladder and released from the body through the urinary tract passageways.

26-1 | The kidneys, ureters, urinary bladder, and urethra make up the urinary system, which has three primary functions

Learning Outcome Identify the organs of the urinary system, and describe the functions of the system.

The **urinary system** has three major functions: (1) *excretion*, the removal of organic waste products from body fluids; (2) *elimination*, the discharge of these waste products into the environment; and (3) *homeostatic regulation* of the volume and solute concentration of blood.

The organs of the urinary system include the paired kidneys, paired ureters, urinary bladder, and urethra (**Figure 26–1**). The two **kidneys** perform the excretory functions of the urinary system. These organs produce **urine**, which is a fluid that contains water, ions, and small soluble compounds. Urine leaving the kidneys flows along the **urinary tract**, which consists of paired tubes called **ureters** (ū-RĒ-terz), to the **urinary**

Figure 26–1   An Introduction to the Urinary System. An anterior view of the urinary system, showing the positions of its organs.

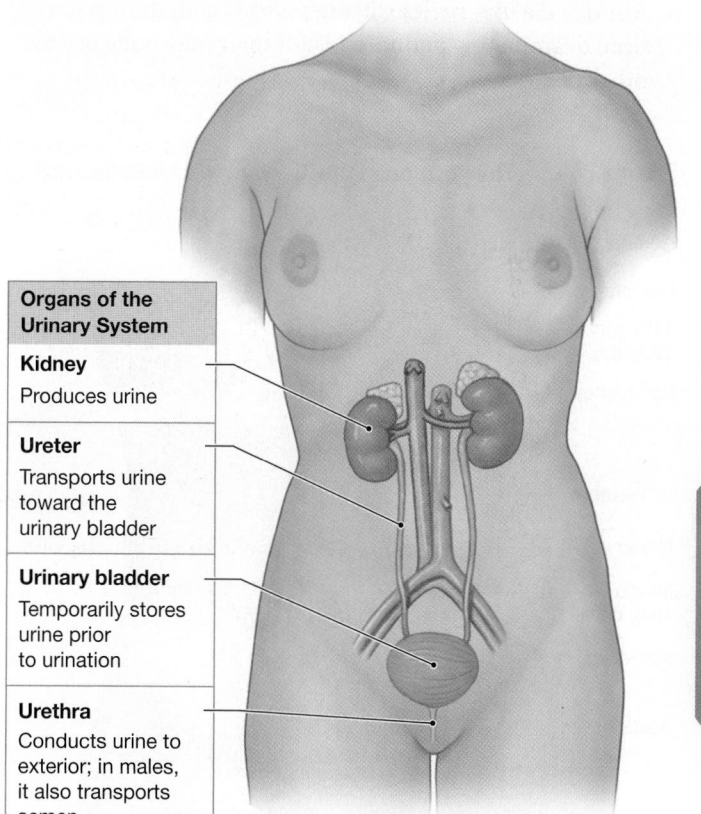

Organs of the Urinary System
Kidney Produces urine
Ureter Transports urine toward the urinary bladder
Urinary bladder Temporarily stores urine prior to urination
Urethra Conducts urine to exterior; in males, it also transports semen

Anterior view

bladder, a muscular sac for temporary storage of urine. After leaving the urinary bladder, urine passes through the **urethra** (ū-RĒ-thra), which conducts the urine to the exterior.

26

The urinary bladder and the urethra eliminate urine. This process is called **urination** or **micturition** (mik-chū-RISH-un). In this process, the muscular urinary bladder contracts and forces urine through the urethra and out of the body. ATLAS: Embryology Summary 20: The Development of the Urinary System

The urinary system removes wastes generated by cells throughout the body, but it has several other essential homeostatic functions. They include the following:

- *Regulating blood volume and blood pressure,* by adjusting the volume of water lost in urine, releasing erythropoietin, and releasing renin.

- *Regulating plasma concentrations of sodium, potassium, chloride, and other ions,* by influencing the quantities lost in urine. The kidneys also control calcium ion levels through the synthesis of calcitriol.

- *Helping to stabilize blood pH,* by controlling the loss of hydrogen ions and bicarbonate ions in urine.

- *Conserving valuable nutrients,* by preventing their loss in urine while removing organic wastes—especially nitrogenous wastes such as *urea* and *uric acid*.

- *Assisting the liver* in detoxifying poisons and, during starvation, deaminating amino acids for their metabolic use by other tissues.

These activities are carefully regulated to keep the composition of blood within acceptable limits. A disruption of any one of them has immediate consequences and can be fatal.

> **CHECKPOINT**
>
> 1. **Name the three major functions of the urinary system.**
> 2. **Identify the organs of the urinary system.**
> 3. **Define micturition.**
>
> See the blue Answers tab at the back of the book.

26-2 ┃ Kidneys are highly vascular organs containing functional units called nephrons, which filter, reabsorb, and secrete

Learning Outcome Describe the location and structures of the kidneys, identify major blood vessels associated with each kidney, trace the path of blood flow through a kidney, describe the structure of a nephron, and identify the functions of each region of the nephron and collecting system.

The kidneys are located on either side of the vertebral column, between vertebrae T_{12} and L_3 (**Figure 26–2a**). The left kidney lies slightly superior to the right kidney. The right kidney is slightly inferior due to the position of the right lobe of the liver.

The superior surface of each kidney is capped by an adrenal gland. The kidneys and adrenal glands lie between the muscles of the posterior body wall and the parietal peritoneum, in a retroperitoneal position (**Figure 26–2b**).

Figure 26–2   The Position of the Kidneys. ATLAS: Plates 57a,b

Adrenal gland
Diaphragm
11th and 12th ribs
Left kidney
L_1 vertebra
Ureter
Inferior vena cava
Iliac crest
Aorta
Urinary bladder
Urethra
Right kidney
Renal artery and vein

a A posterior view of the trunk

External oblique
Parietal peritoneum
Stomach
Renal vein
Renal artery
Aorta
Pancreas
Ureter
Spleen
Left kidney
Vertebra

Connective tissue layers
Fibrous capsule
Perinephric fat
Renal fascia

Quadratus lumborum
Psoas major
Inferior vena cava

b A superior view of a transverse section at the level indicated in part (a)

The kidneys are held in position within the abdominal cavity by (1) the overlying peritoneum, (2) contact with adjacent visceral organs, and (3) supporting connective tissues. Three concentric layers of connective tissue protect and stabilize each kidney (**Figure 26–2b**):

1. The **fibrous capsule**, a layer of collagen fibers that covers the outer surface of the entire organ.

2. The **perinephric fat**, a thick layer of adipose tissue that surrounds the fibrous capsule.

3. The **renal fascia**, a dense, fibrous outer layer that anchors the kidney to surrounding structures. Collagen fibers extend outward from the fibrous capsule through the perinephric fat to this layer. Posteriorly, the renal fascia fuses with the deep fascia surrounding the muscles of the body wall. Anteriorly, the renal fascia forms a thick layer that fuses with the peritoneum.

In effect, each kidney hangs suspended by collagen fibers from the renal fascia and is packed in a soft cushion of adipose tissue. This arrangement prevents the jolts and shocks of day-to-day living from disturbing normal kidney function. If the suspensory fibers break or become detached, a slight bump or hit can displace the kidney and stress the attached vessels and ureter. This condition is called a *floating kidney*. It may cause pain or other problems from the distortion of the ureter or blood vessels during movement.

A typical adult kidney is reddish-brown and about 10 cm (4 in.) long, 5.5 cm (2.2 in.) wide, and 3 cm (1.2 in.) thick (**Figures 26–3** and **26–4**). Each kidney weighs about 150 g (5.25 oz). The **hilum**, a prominent medial indentation, is the point of entry for the *renal artery* and *renal nerves*. The hilum is also the point of exit for the *renal vein* and the ureter.

Sectional Anatomy of the Kidneys

The fibrous capsule covering the outer surface of the kidney also lines the **renal sinus**, an internal cavity within the kidney

Figure 26–3   The Gross Anatomy of the Urinary System. The abdominopelvic cavity (with the digestive organs removed), showing the kidneys, ureters, urinary bladder, and blood supply to the urinary structures. ATLAS: Plates 61a; 62a,b

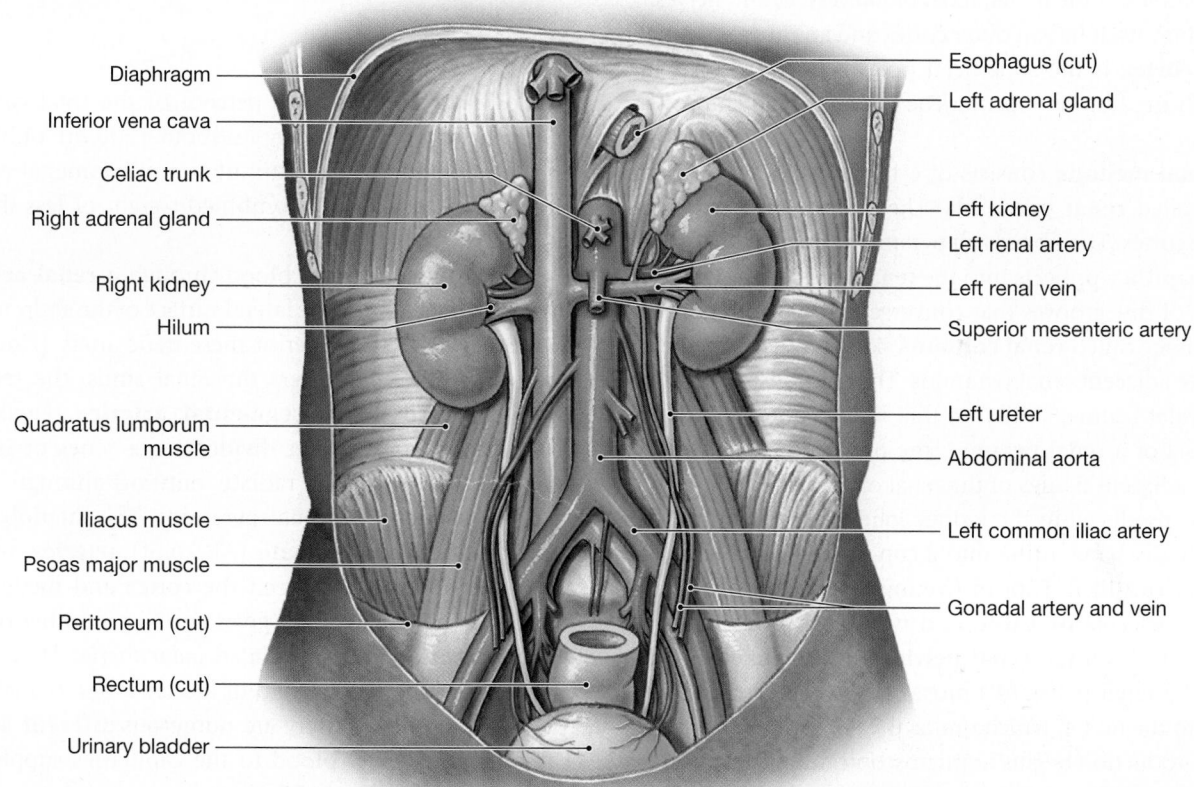

Diaphragm
Inferior vena cava
Celiac trunk
Right adrenal gland
Right kidney
Hilum
Quadratus lumborum muscle
Iliacus muscle
Psoas major muscle
Peritoneum (cut)
Rectum (cut)
Urinary bladder

Esophagus (cut)
Left adrenal gland
Left kidney
Left renal artery
Left renal vein
Superior mesenteric artery
Left ureter
Abdominal aorta
Left common iliac artery
Gonadal artery and vein

Anterior view

26

Figure 26–4   **The Structure of the Kidney.** ATLAS: Plates 57a,b; 61b

a A diagrammatic view of a frontal section through the left kidney

b A frontal section of the left kidney (cadaver)

(**Figure 26–4a,b**). The fibrous capsule is bound to the outer surfaces of the structures within the renal sinus. In this way, it stabilizes the positions of the ureter, renal blood vessels, and nerves.

The kidney itself has an outer cortex and an inner medulla. The **renal cortex** is the superficial portion of the kidney, in contact with the fibrous capsule. The cortex is reddish brown and granular.

The **renal medulla** consists of 6 to 18 distinct triangular structures called **renal pyramids**. The base of each pyramid touches the cortex. The tip of each pyramid—a region known as the **renal papilla**—projects into the renal sinus. Each pyramid has a series of fine grooves that converge at the papilla. Bands of cortical tissue called **renal columns** extend into the medulla and separate adjacent renal pyramids. The columns have a distinctly granular texture, similar to that of the cortex. A **kidney lobe** consists of a renal pyramid, the overlying area of renal cortex, and adjacent tissues of the renal columns.

Urine is produced in the kidney lobes. Ducts within each renal papilla discharge urine into a cup-shaped drain called a **minor calyx** (KĀ-liks). Four or five minor calyces (KAL-i-sēz) merge to form a **major calyx**, and two or three major calyces combine to form the **renal pelvis**, a large, funnel-shaped chamber. The renal pelvis fills most of the renal sinus and is connected to the ureter, which drains the kidney.

Urine production begins in microscopic, tubular structures called *nephrons* (NEF-ronz) in the cortex of each kidney lobe.

Each kidney has about 1.25 million nephrons, with a combined length of about 145 km (85 miles).

Blood Supply and Innervation of the Kidneys

The kidneys receive 20–25 percent of the total cardiac output. In normal, healthy people, about 1200 mL of blood flow through the kidneys each minute—a phenomenal amount of blood for organs with a combined weight of less than 300 g (10.5 oz)!

Each kidney receives blood through a **renal artery**. This vessel originates along the lateral surface of the abdominal aorta near the level of the superior mesenteric artery (**Figure 21–23**, pp. 786–787). As it enters the renal sinus, the renal artery provides blood to the **segmental arteries** (**Figure 26–5a**). Segmental arteries further divide into a series of **interlobar arteries**. These arteries radiate outward through the renal columns between the renal pyramids. The interlobar arteries supply blood to the **arcuate** (AR-kū-āt) **arteries**, which arch along the boundary between the cortex and medulla of the kidney. Each arcuate artery gives rise to a number of **cortical radiate arteries**, also called *interlobular arteries*. They supply the cortical portions of the adjacent kidney lobes. Branching from each cortical radiate artery are numerous **afferent arterioles**. These vessels deliver blood to the capillaries supplying individual nephrons (**Figure 26–5b,c**).

Figure 26–5   **The Blood Supply to the Kidneys.** ATLAS: Plates 53c,d; 61a–c

b Circulation in a single kidney lobe

a A sectional view, showing major arteries and veins

c A flowchart of renal circulation

Clinical Note

Analysis of Renal Blood Flow The rate of blood flow through the kidneys can be estimated by administering the compound *para-aminohippuric acid (PAH)*. This compound is removed by the nephrons and eliminated in urine. Virtually all the PAH contained in the blood that arrives at the kidneys is removed before the blood leaves in the renal veins. For this reason, renal blood flow can be approximated by comparing plasma concentrations of PAH with the amount secreted in urine. In practice, however, it is usually easier to measure the glomerular filtration rate (p. 1015).

After passing through the capillaries of the nephrons, blood enters a network of venules and small veins that converge on the **cortical radiate veins**, also called *interlobular veins* (**Figure 26–5a,c**). The cortical radiate veins deliver blood to **arcuate veins**. These veins in turn empty into **interlobar veins**, which drain directly into the **renal vein**. There are no segmental veins.

26

Renal nerves innervate the kidneys and ureters. Most of the nerve fibers involved are sympathetic postganglionic fibers from the celiac plexus and the inferior splanchnic nerves. ⊃ pp. 564, 574 A renal nerve enters each kidney at the hilum and follows the branches of the renal arteries to reach individual nephrons. The sympathetic innervation (1) adjusts rates of urine formation by changing blood flow and blood pressure at the nephron; and (2) stimulates the release of renin, which ultimately restricts water and salt loss in the urine by stimulating reabsorption by the nephron. When a substance is reabsorbed, it is "reclaimed," eventually reentering the blood.

The Nephron

Recall that the kidneys remove wastes from the blood and help to regulate blood volume and blood pressure, blood ion levels, and blood pH. In the kidneys, the functional units—the smallest structures that can carry out all the functions of a system—are the **nephrons** (NEF-ronz). Each nephron consists of a renal corpuscle and a renal tubule (**Figure 26–6**). The **renal corpuscle** (KOR-pus-ul) is a spherical structure consisting of the *glomerular* (Bowman's) *capsule*, a cup-shaped chamber approximately 200 μm in diameter, and a capillary network known as

Figure 26–6 **The Functional Anatomy of a Representative Nephron and the Collecting System.**
The major functions and characteristic cells of each segment of the nephron (purple) and the collecting system (tan) are noted. In this representative figure, the nephron has been significantly shortened and some components rearranged.

the *glomerulus*. The **renal tubule** begins at the renal corpuscle. It is a long tubular passageway that may be 50 mm (1.97 in.) in length.

Blood arrives at the renal corpuscle by way of an afferent arteriole. This arteriole delivers blood to the **glomerulus** (glo-MER-ū-lus; plural, *glomeruli*), which consists of about 50 intertwined capillaries. The glomerulus projects into the glomerular capsule much as the heart projects into the pericardial cavity. Blood leaves the glomerulus in an **efferent arteriole**. It flows into a network of capillaries called the *peritubular capillaries*, which surround the renal tubule. These capillaries in turn drain into small venules that return the blood to the venous system (**Figure 26–7b,c**).

The process of *filtration* takes place in the renal corpuscle. In this process, blood pressure forces water and dissolved solutes out of the glomerular capillaries and into a chamber—the *capsular space*—that is continuous with the lumen of the renal tubule (**Figure 26–6**). Filtration produces an essentially protein-free solution, known as a **filtrate**, that is otherwise similar to blood plasma. From the renal corpuscle, filtrate enters the renal tubule. The renal tubule has three crucial functions: (1) reabsorbing all the useful organic nutrients in the filtrate; (2) reabsorbing more than 90 percent of the water in the filtrate; and (3) secreting into the tubule lumen any wastes that did not pass into the filtrate at the glomerulus.

The renal tubule has two convoluted (coiled or twisted) segments—the *proximal convoluted tubule* (PCT) and the *distal convoluted tubule* (DCT). They are separated by a simple U-shaped tube, the *nephron loop*, also called the *loop of Henle* (HEN-lē). The convoluted segments lie in the cortex of the kidney, and the nephron loop dips at least partially into the medulla. For clarity, the nephron shown in **Figure 26–6** has been shortened and straightened.

The regions of the nephron vary by structure and function. As the filtrate travels along the renal tubule, it is now called **tubular fluid**, and it gradually changes in composition. The changes that take place and the characteristics of the urine that result are due to the activities under way in each segment of the nephron. **Figure 26–6** and **Table 26–1** survey the regional specializations.

Figure 26–7   The Locations and Structures of Cortical and Juxtamedullary Nephrons.

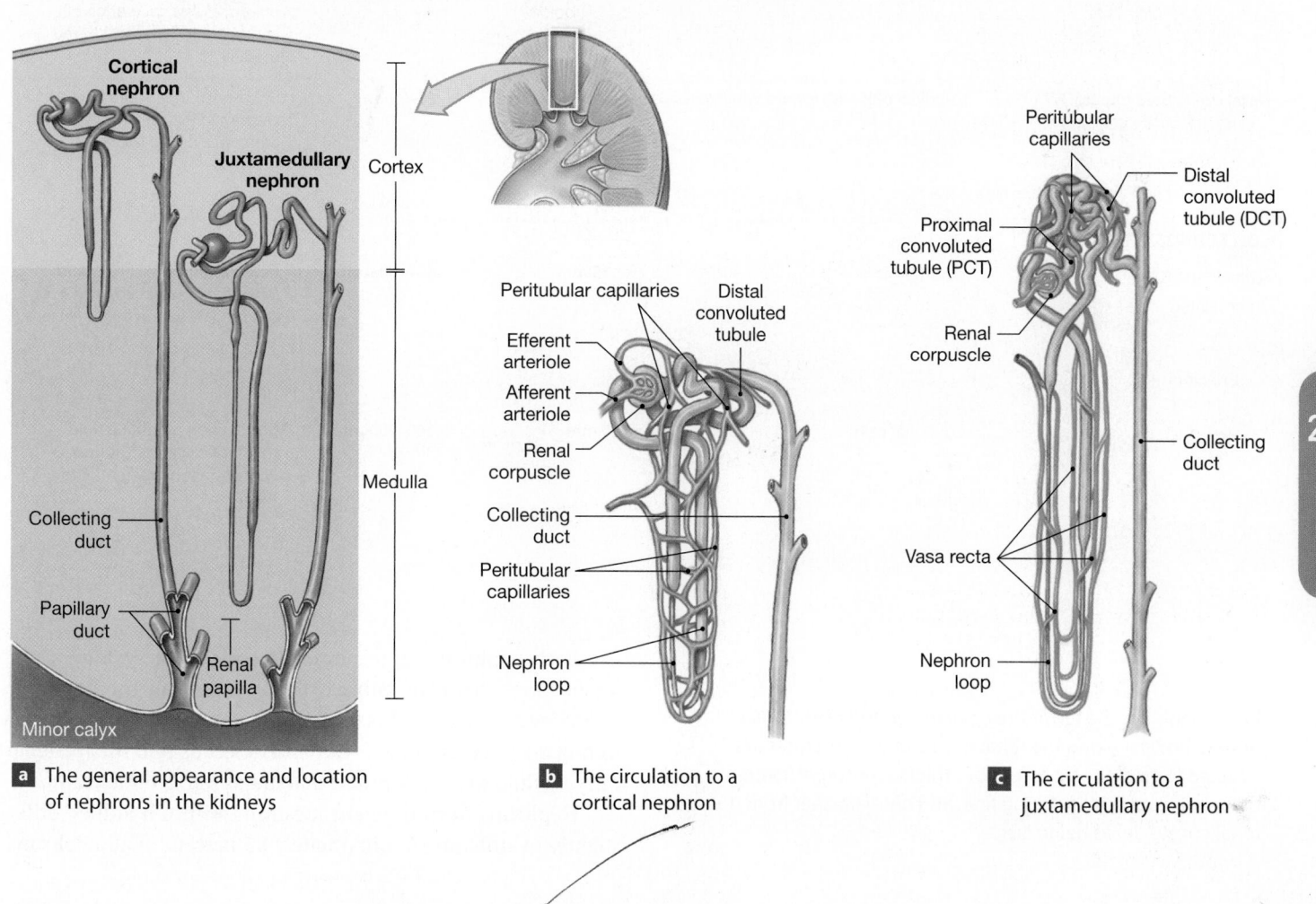

a The general appearance and location of nephrons in the kidneys

b The circulation to a cortical nephron

c The circulation to a juxtamedullary nephron

26

Table 26–1	The Organization of the Nephron and Collecting System				
Region	**Histological Characteristics**	**Length**	**Diameter**	**Primary Function**
NEPHRON				
Renal corpuscle Capsular space Capsular epithelium Capillaries of glomerulus Visceral epithelium	Glomerulus (capillary knot), mesangial cells, and dense layer, enclosed by the glomerular capsule; visceral epithelium (podocytes) and capsular epithelium separated by capsular space	150–250 μm (spherical)	150–250 μm	Filtration of blood plasma
Renal tubule				
Proximal convoluted tubule (PCT) Microvilli	Cuboidal cells with microvilli	14 mm	60 μm	Reabsorption of ions, organic molecules, vitamins, water; secretion of drugs, toxins, acids
Nephron loop	Squamous or low cuboidal cells	30 mm	15 μm	Descending limb: reabsorption of water from tubular fluid
			30 μm	Ascending limb: reabsorption of ions; assists in creation of a concentration gradient in the medulla
Distal convoluted tubule (DCT)	Cuboidal cells with few if any microvilli	5 mm	30–50 μm	Reabsorption of sodium ions and calcium ions; secretion of acids, ammonia, drugs, toxins
COLLECTING SYSTEM				
Collecting duct Intercalated cell Principal cell	Cuboidal cells	15 mm	50–100 μm	Reabsorption of water, sodium ions; secretion or reabsorption of bicarbonate ions or hydrogen ions
Papillary duct	Columnar cells	5 mm	100–200 μm	Conduction of tubular fluid to minor calyx; contributes to concentration gradient of the medulla

TIPS & TRICKS

In the nephron, the terms *proximal tubule* and *distal tubule* refer to how far along the renal tubule these structures are situated from the renal corpuscle. Think of *proximal* (nearer the renal corpuscle) as being first, and *distal* (farther from the renal corpuscle) as being last.

Each nephron empties into the **collecting system**, a series of tubes that carry tubular fluid away from the nephron. *Collecting ducts* receive this fluid from many nephrons. Each collecting duct begins in the cortex and descends into the medulla, carrying fluid to a *papillary duct* that drains into a minor calyx.

Nephrons from different locations within a kidney differ slightly in structure. Approximately 85 percent of all nephrons are **cortical nephrons**, located almost entirely within the superficial cortex of the kidney (Figure 26–7a,b). In a cortical nephron, the nephron loop is relatively short, and the efferent arteriole delivers blood to a network of **peritubular capillaries**, which surround the entire renal tubule. These capillaries drain into small venules that carry blood to the cortical radiate veins (Figure 26–5c).

The remaining 15 percent of nephrons, termed **juxtamedullary** (juks-tuh-MED-u-lār-ē; *juxta*, near) **nephrons**, have long nephron loops that extend deep into the medulla (Figure 26–7a,c). In juxtamedullary nephrons, the peritubular capillaries are connected to the **vasa recta** (*vasa*, vessel + *recta*, straight)—long, straight capillaries that parallel the nephron loop.

Cortical nephrons do most of the reabsorption and secretion in the kidneys because they are more numerous than juxtamedullary nephrons. However, as you will see later in the chapter, it is the juxtamedullary nephrons that enable the kidneys to produce concentrated urine.

Next let's examine the structure of each segment of a representative nephron.

The Renal Corpuscle

Each renal corpuscle is 150–250 μm in diameter (Figure 26–8). It includes both the **glomerular capsule** and the capillary network known as the *glomerulus* (Figure 26–8a). The glomerular capsule is connected to the initial segment of the renal tubule and forms the outer wall of the renal corpuscle. It encapsulates the glomerular capillaries.

The outer wall of the capsule is lined by a simple squamous **capsular epithelium** (Figure 26–8a). This layer is continuous with the **visceral epithelium**, which covers the glomerular capillaries. The **capsular space** separates the capsular and visceral epithelia. The two epithelial layers are continuous where the glomerular capillaries are connected to the afferent arteriole and efferent arteriole.

The visceral epithelium consists of large cells with complex processes, or "feet," that wrap around the specialized *dense layer*

Figure 26–8 **The Renal Corpuscle.**

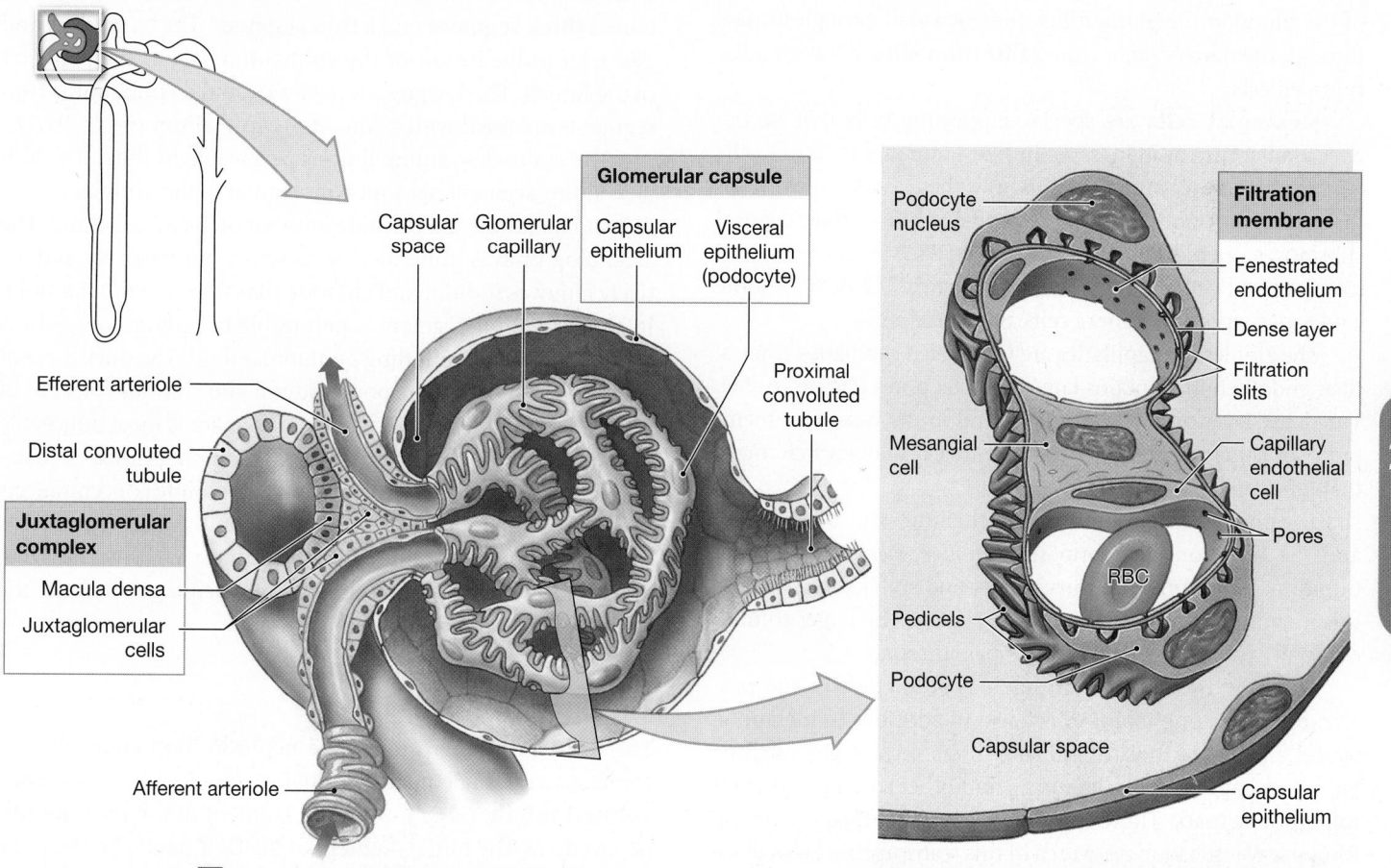

a Important structural features of a renal corpuscle.

b This cross section through a portion of the glomerulus shows the components of the filtration membrane of the nephron.

26

Clinical Note

Glomerulonephritis Glomerulonephritis (glo-mer-ū-lō-nef-RĪ-tis) is an inflammation of the glomeruli that affects filtration in the kidneys. The condition is often an *immune complex disorder*. It may develop after an infection involving *Streptococcus* bacteria. ⤻ p. 848 The kidneys are not the sites of infection, but as the immune system responds, the number of circulating antigen–antibody complexes skyrockets. These complexes are small enough to pass through the dense layer, but too large to fit through the filtration slits of the filtration membrane. The complexes clog up the filtration mechanism. As a result, filtrate production decreases. Any condition that leads to a massive immune response, including viral infections and autoimmune disorders, can cause glomerulonephritis.

of the glomerular capillaries. These unusual cells are called **podocytes** (PŌ-dō-sīts; *podos*, foot + *-cyte*, cell). Their feet are known as **pedicels** (Figure 26–8b). Materials passing out of the blood at the glomerulus must be small enough to pass through the narrow gaps, called **filtration slits**, between adjacent pedicels.

Mesangial cells are special supporting cells that lie between adjacent capillaries. Actin-like filaments in these cells enable them to contract. In this way these cells control capillary diameter and the rate of capillary blood flow. Several substances, including angiotensin II, ADH (vasopressin), and histamine, affect mesangial cell contraction. ⤻ p. 666 Some evidence suggests that these cells also make renin.

The glomerular capillaries are fenestrated capillaries. That is, their endothelium contains large-diameter pores (Figure 26–8b). The dense layer differs from that found in the basement membrane of other capillary networks in that it may encircle more than one capillary.

Together, the fenestrated endothelium, the dense layer, and the filtration slits form the *filtration membrane*. During filtration, blood pressure forces water and small solutes across this membrane and into the capsular space. The larger solutes, especially plasma proteins, do not pass through.

Filtration by the renal corpuscle is both effective and passive, but it has one major drawback: In addition to metabolic wastes and excess ions, useful substances such as glucose, free fatty acids, amino acids, vitamins, and other solutes also enter the capsular space. These substances are reabsorbed before the filtrate leaves the kidneys. Much of this reabsorption takes place in the proximal convoluted tubules.

The Proximal Convoluted Tubule

Recall that the renal tubule consists of three segments: the proximal convoluted tubule, the nephron loop, and the distal convoluted tubule. The **proximal convoluted tubule (PCT)** is the first segment (Figure 26–6). Its entrance lies almost directly opposite the point where the afferent and efferent arterioles connect to the glomerulus. The lining of the PCT is a simple cuboidal epithelium whose apical surfaces have microvilli (Table 26–1). The tubular cells reabsorb organic nutrients, ions, water, and plasma proteins (if present) from the tubular fluid and release them into the **peritubular fluid**, the interstitial fluid surrounding the renal tubule. The reabsorbed substances in the peritubular fluid eventually reenter the blood. *Reabsorption* is the primary function of the PCT, but the epithelial cells can also secrete substances into the lumen of the renal tubule.

The Nephron Loop (Loop of Henle)

The PCT makes an acute bend that turns the renal tubule toward the renal medulla. This turn leads to the **nephron loop**, or *loop of Henle* (Figures 26–6, 26–7). We divide the nephron loop into a **descending limb** and an **ascending limb**. Fluid in the descending limb flows toward the renal pelvis. Fluid in the ascending limb flows toward the renal cortex. Each limb contains a **thick segment** and a **thin segment**. The terms *thick* and *thin* refer to the height of the epithelium, not to the diameter of the lumen: Thick segments have a cuboidal epithelium. Thin segments are lined with a squamous epithelium (Table 26–1).

The entire descending limb is permeable to water. Its thick descending segment has functions similar to those of the PCT: It reabsorbs sodium and chloride ions out of the tubular fluid. The ascending limb is impermeable to water, but passively and actively removes sodium and chloride ions from the tubular fluid. Its thin ascending segment is permeable to sodium ions, which diffuse into the surrounding peritubular fluid. The thick ascending segment actively transports sodium and chloride ions out of the tubular fluid. The effect of this movement is most noticeable in the renal medulla, where the long ascending limbs of juxtamedullary nephrons create unusually high solute concentrations in the peritubular fluid. The descending limb is freely permeable to water, but not to solutes, such as sodium and chloride ions. Water moves out of the descending limb, helping to increase the solute concentration of the tubular fluid.

The Distal Convoluted Tubule

The thick ascending limb of the nephron loop ends where it forms a sharp angle near the renal corpuscle. The **distal convoluted tubule (DCT)**, the third segment of the renal tubule, begins there. The initial segment of the DCT passes between the afferent and efferent arterioles (Figure 26–8a).

In sectional view, the DCT differs from the PCT in that the DCT has a smaller diameter and its epithelial cells lack microvilli (Table 26–1). The DCT is an important site for three vital processes: (1) the active secretion of ions, acids, drugs, and toxins into the tubule; (2) the selective reabsorption of sodium ions and calcium ions from tubular fluid; and (3) the selective reabsorption of water, which assists in concentrating the tubular fluid.

The Juxtaglomerular Complex. The epithelial cells of the DCT near the renal corpuscle are taller than those elsewhere along the DCT, and their nuclei are clustered together. This region is called the **macula densa** (MAK-ū-la DEN-sa) (Figure 26–8a). The cells of the macula densa are closely associated with unusual smooth muscle fibers in the wall of the afferent arteriole. These fibers are known as **juxtaglomerular cells**. Together, the macula densa and juxtaglomerular cells form the **juxtaglomerular complex (JGC)**, an endocrine structure that secretes the hormone *erythropoietin* and the enzyme *renin*. ⟲ p. 666

The Collecting System

The distal convoluted tubule, the last segment of the nephron, opens into the collecting system (Figure 26–6). Individual nephrons drain into a nearby **collecting duct**. Several collecting ducts then converge into a larger **papillary duct**, which in turn empties into a minor calyx. The epithelium lining the papillary duct is typically columnar (Table 26–1).

Two main types of cells are found in the collecting duct: intercalated cells and principal cells. Intercalated cells are cuboidal cells with microvilli. Alpha-intercalated cells and beta-intercalated cells make up the population of intercalated cells. Alpha-intercalated cells secrete hydrogen ions and reabsorb bicarbonate ions, while beta-intercalated cells secrete bicarbonate ions and reabsorb hydrogen ions. Principal cells are cuboidal cells that reabsorb water and secrete potassium. Together, these cells regulate the acid–base balance in the blood.

The collecting system does more than simply transport tubular fluid from the nephrons to the renal pelvis. It also adjusts the fluid's composition and determines the final osmotic concentration and volume of urine, the final product. We consider these activities of the collecting system later in the chapter.

CHECKPOINT

4. Which portions of a nephron are in the renal cortex?

5. Why don't plasma proteins pass into the capsular space under normal circumstances?

6. Damage to which part of a nephron would interfere with the hormonal control of blood pressure?

See the blue Answers tab at the back of the book.

26-3 ❙ Different segments of the nephron form urine by filtration, reabsorption, and secretion

Learning Outcome Describe the basic processes that form urine.

Most people don't commonly think of it this way, but the goal of urine production is to maintain homeostasis by regulating the volume and composition of blood. This process involves the excretion of solutes—specifically, metabolic wastes. Our bodies form three important organic wastes:

1. *Urea.* Urea is the most abundant organic waste. You generate approximately 21 g of urea each day, most of it through the breakdown of amino acids.

2. *Creatinine.* Skeletal muscle tissue generates creatinine through the breakdown of *creatine phosphate*, a high-energy compound that plays an important role in muscle contraction. ⟲ p. 342 Your body generates roughly 1.8 g of creatinine each day. Virtually all of it is excreted in urine.

3. *Uric Acid.* **Uric acid** is a waste formed during the recycling of the nitrogenous bases from RNA molecules. You produce approximately 480 mg of uric acid each day.

These wastes are dissolved in the bloodstream. They can be eliminated only when dissolved in urine. For this reason, their removal involves an unavoidable water loss. The kidneys are usually capable of producing concentrated urine with an osmotic concentration of 1200–1400 mOsm/L, more than four times that of plasma. (We discuss methods of reporting solute concentrations in a later section.) If the kidneys were unable to concentrate the filtrate produced by glomerular filtration, fluid losses would lead to fatal dehydration in a matter of hours. The kidneys also ensure that the fluid that *is* lost does not contain potentially useful organic substrates that are present in blood plasma, such as sugars or amino acids. These valuable materials must be reabsorbed and retained for use by other tissues.

Basic Processes of Urine Formation

In forming urine, the kidneys use three distinct processes that we have already mentioned:

1. *Filtration.* In **filtration**, blood pressure forces water and solutes across the wall of the glomerular capillaries and into the capsular space. Solute molecules small enough to pass through the filtration membrane are carried by the surrounding water molecules.

2. *Reabsorption.* **Reabsorption** is the removal of water and solutes from the filtrate, and their movement across the tubular epithelium and into the peritubular fluid. Reabsorption

26

takes place after filtrate has left the renal corpuscle. Most of the reabsorbed materials are nutrients the body can use. Filtration is based solely on size, but reabsorption is a selective process. Reabsorption involves either simple diffusion or the activity of carrier proteins in the tubular epithelium. The reabsorbed substances in the peritubular fluid eventually reenter the blood. Water reabsorption takes place passively, through osmosis.

3. *Secretion.* **Secretion** is the transport of solutes from the peritubular fluid, across the tubular epithelium, and into the tubular fluid. Secretion is necessary because filtration does not force all the dissolved materials out of the plasma. Tubular secretion, which removes substances from the blood, can further lower the plasma concentration of undesirable materials. It provides a backup process for filtration. Secretion is often the primary method of preparing substances, including many drugs, for excretion.

Together, these processes produce a fluid that is very different from other body fluids. Table 26–2 shows the efficiency of the renal system by comparing the concentrations of some substances in urine and plasma.

TIPS&TRICKS

Secretion by the urinary system takes place when cells produce and then discharge substances into the urine, whereas *excretion* is the elimination of wastes from the body in the form of urine, sweat, and feces.

Table 26–2	Normal Laboratory Values for Solutes in Plasma and Urine	
Solute	**Plasma**	**Urine**
IONS (mEq/L)		
Sodium (Na$^+$)	135–145	40–220
Potassium (K$^+$)	3.5–5.0	25–100
Chloride (Cl$^-$)	100–108	110–250
Bicarbonate (HCO$_3^-$)	20–28	1–9
METABOLITES AND NUTRIENTS (mg/dL)		
Glucose	70–110	0.009
Lipids	450–1000	0.002
Amino acids	40	0.188
Proteins	6–8 g/dL	0.000
NITROGENOUS WASTES (mg/dL)		
Urea	8–25	1800
Creatinine	0.6–1.5	150
Uric acid	2–6	40
Ammonia	<0.1	60

An Overview of Renal Function

Figure 26–9 summarizes the general functions of the various segments of the nephron and collecting system in the formation of urine. Most regions carry out a combination of reabsorption and secretion. Note that the balance between the two processes shifts from one region to another.

Normal kidney function can continue only as long as filtration, reabsorption, and secretion function within relatively narrow limits. A disruption in kidney function has immediate effects on the composition of the circulating blood. If both kidneys are affected, death follows within a few days unless medical assistance is provided.

Before we take a closer look at the functions of the individual portions of the nephron, let's briefly examine each of the three major processes involved in urine formation.

Filtration

In filtration, hydrostatic pressure forces water through membrane pores, and solute molecules small enough to pass through those pores are carried along. Filtration takes place as larger solutes and suspended materials are left behind. We can see filtration in action in a drip coffee machine. Gravity forces hot water through the filter, and the water carries a variety of dissolved compounds into the pot. The large coffee grounds never reach the pot, because they cannot fit through the pores of the filter. In other words, they are "filtered out" of the solution; the coffee we drink is the filtrate.

In the body, the heart pushes blood around the cardiovascular system and generates hydrostatic pressure. As you have seen, filtration takes place across the walls of capillaries as water and dissolved materials are pushed into the interstitial fluids of the body (see Figure 21–10, p. 765). In some sites (for example, the liver), the pores are so large that even plasma proteins can enter the interstitial fluids. At the renal corpuscle, however, the specialized filtration membrane restricts the passage of even the smallest circulating proteins.

Reabsorption and Secretion

The processes of reabsorption and secretion by the kidneys involve a combination of diffusion, osmosis, channel-mediated diffusion, and carrier-mediated transport. We considered diffusion and osmosis in several other chapters. Here we briefly review carrier-mediated transport mechanisms.

Types of Carrier-Mediated Transport. In previous chapters, we looked at four major types of *carrier-mediated transport*:

■ In *facilitated diffusion*, a carrier protein transports a molecule across the plasma membrane without expending energy (see Figure 3–18, p. 121). Such transport always follows the concentration gradient for the ion or molecule transported.

Figure 26–9 An Overview of Urine Formation.

Proximal convoluted tubule

Distal convoluted tubule

Glomerulus

Glomerular capsule

Collecting duct

Nephron loop

Urine storage
and elimination

KEY

Filtration occurs exclusively in the renal corpuscle, across the filtration membrane.

Water reabsorption occurs primarily along the PCT and the descending limb of the nephron loop, but also to a variable degree in the DCT and collecting system.

Variable water reabsorption occurs in the DCT and collecting system.

Solute reabsorption occurs along the PCT, the ascending limb of the nephron loop, the DCT, and the collecting system.

Variable solute reabsorption or secretion occurs at the PCT, the DCT, and the collecting system.

- *Active transport* is driven by the hydrolysis of ATP to ADP on the inner plasma membrane surface (see **Figure 3–19**, p. 122). Exchange pumps and other carrier proteins are active along the kidney tubules. Active transport can operate despite an opposing concentration gradient.

- In *cotransport*, carrier protein activity is not directly linked to the hydrolysis of ATP (see **Figure 3–20**, p. 123). Instead, two substrates (ions, molecules, or both) cross the membrane while bound to the carrier protein. The movement

of the substrates always follows the concentration gradient of at least one of the transported substances. Cotransport is used for the reabsorption of organic and inorganic compounds from the tubular fluid.

- *Countertransport* resembles cotransport, except that the two transported ions move in *opposite* directions (see **Figures 23–23**, p. 891, and **24–14**, p. 926). Countertransport operates in the PCT, DCT, and collecting system.

Characteristics of Carrier-Mediated Transport. All carrier-mediated processes share five features that are important for an understanding of kidney function:

1. *A specific substrate binds to a carrier protein that facilitates movement across the membrane.* The movement may be active or passive.

2. *A given carrier protein typically works in one direction only.* In facilitated diffusion, the concentration gradient of the substance being transported determines that direction. In active transport, cotransport, and countertransport, the location and orientation of the carrier proteins determine whether a particular substance is reabsorbed or secreted. The carrier protein that transports amino acids from the tubular fluid to the cytosol, for example, will not carry amino acids back into the tubular fluid.

3. *The distribution of carrier proteins can vary among portions of the cell surface.* Transport between tubular fluid and interstitial (peritubular) fluid involves two steps. First, the material enters the cell at its apical surface. Then the material leaves the cell at its basolateral surface and enters the peritubular fluid. Each step involves a different carrier protein. For example, the apical surfaces of cells along the proximal convoluted tubule contain carrier proteins that bring amino acids, glucose, and many other nutrients into these cells by sodium-linked cotransport. In contrast, the basolateral surfaces contain carrier proteins that move those nutrients out of the cell by facilitated diffusion.

4. *The membrane of a single tubular cell contains many types of carrier proteins.* Each cell can have multiple functions. A cell that reabsorbs one compound can secrete another.

5. *Carrier proteins, like enzymes, can be saturated.* Recall that when an enzyme is *saturated*, further increases in substrate concentration have no effect on the rate of reaction. p. 81 Likewise, when a carrier protein is saturated, further increases in substrate concentration have no effect on the rate of transport across the plasma membrane. For any substance, the concentration at saturation is called the **transport maximum (T_m)** or *tubular maximum.* In healthy people, carrier proteins involved in tubular secretion seldom become saturated. However, carriers involved in tubular reabsorption are often at risk of saturation, especially during the absorptive state following a meal.

26

T_m and the Renal Threshold. Normally, any plasma proteins and nutrients, such as amino acids and glucose, are removed from the tubular fluid by cotransport or facilitated diffusion. If the concentrations of these nutrients rise in the tubular fluid, the rate of reabsorption increases until the carrier proteins are saturated. A concentration higher than the transport maximum exceeds the reabsorptive abilities of the nephron. In this case, some of the material will remain in the tubular fluid and appear in the urine. The transport maximum thus determines the **renal threshold**—the plasma concentration at which a specific substance or ion begins to appear in the urine.

The renal threshold varies with the substance involved. The renal threshold for glucose is approximately 180 mg/dL. When plasma glucose concentrations exceed 180 mg/dL, glucose appears in urine; this condition is called *glycosuria* or *glucosuria*. After you have eaten a meal rich in carbohydrates, your plasma glucose levels may exceed the T_m for a brief period. The liver quickly lowers circulating glucose levels, and very little glucose is lost in your urine. However, chronically increased plasma and urinary glucose concentrations are abnormal. (Glycosuria is one of the key signs of diabetes mellitus. ⤺ p. 664)

The renal threshold for amino acids is lower than that for glucose. Amino acids appear in urine when plasma concentrations exceed 65 mg/dL. Plasma amino acid levels commonly exceed the renal threshold after you have eaten a protein-rich meal, causing some amino acids to appear in your urine. This condition is termed **aminoaciduria** (am-i-nō-as-i-DYŪ-rē-uh).

T_m values for water-soluble vitamins are relatively low. As a result, you excrete excess quantities of these vitamins in urine. (This is typically the fate of water-soluble vitamins in daily supplements.) Cells of the renal tubule ignore a number of other substances in the tubular fluid. As water and other substances are removed, the concentrations of the ignored materials in the tubular fluid gradually rise. Table 26–3 lists some substances that are actively reabsorbed or secreted by the renal tubules, as well as several that are not transported at all.

Ways of Expressing Osmotic Concentration. The osmotic concentration, or *osmolarity*, of a solution is the total number of solute particles in each liter. ⤺ p. 118 We usually express osmolarity in **osmoles** per liter (Osm/L) or **milliosmoles** per liter (mOsm/L). If each liter of a fluid contains 1 mole of dissolved particles, the solute concentration is 1 Osm/L, or 1000 mOsm/L. Body fluids have an osmotic concentration of about 300 mOsm/L. For comparison, seawater is about 1000 mOsm/L and fresh water is about 5 mOsm/L. In a clinical setting, the osmolarity of urine is often reported as osmolality, expressed as milliosmoles per kilogram of water

Table 26–3	Tubular Reabsorption and Secretion	
Reabsorbed	**Secreted**	**No Transport Mechanism**
Ions Na^+, Cl^-, K^+ Ca_2^+, Mg^{2+}, SO_4^{2-}, HCO_3^-	**Ions** K^+, H^+, Ca^{2+}, PO_4^{3-}	Urea Water Urobilinogen Bilirubin
Metabolites Glucose Amino acids Proteins Vitamins	**Wastes** Creatinine Ammonia Organic acids and bases	
	Miscellaneous Neurotransmitters (ACh, NE, E, dopamine) Histamine Drugs (penicillin, atropine, morphine, many others)	

(mOsm/kg H_2O). When discussing osmotic concentrations in body fluids, these terms are often used interchangeably.

Ion concentrations are often reported in *milliequivalents* per liter (mEq/L). Milliequivalents indicate the number of positive or negative charges in solution, rather than the number of solute particles; multiply mmol/L by the charges on each ion to convert to mEq/L. For example, each sodium ion has one charge only, so for Na^+, 1 mmol/L = 1 mEq/L; for Ca^{2+}, each ion has two charges, so 1 mmol/L = 2 mEq/L. The concentrations of large organic molecules are usually reported in grams, milligrams, or micrograms per unit volume of solution (typically, per dL).

Cortical and Juxtamedullary Nephrons. In the next sections, we proceed along the nephron to consider the formation of filtrate and its subsequent changes in composition and concentration as it passes along the renal tubule. Most of what follows applies equally to cortical and juxtamedullary nephrons. The functions of the renal corpuscle and of the proximal and distal convoluted tubules are the same in all nephrons. The major difference between the two types of nephron is that the nephron loop of a cortical nephron is shorter. Note that it does not extend as far into the medulla as does the nephron loop of a juxtamedullary nephron (Figure 26–7a).

The long nephron loop in a juxtamedullary nephron extends deep into the renal pyramids. There it plays a key role in water conservation and the formation of concentrated urine. This process is vitally important, affecting the tubular fluid produced by every nephron in the kidney. For this reason, we use a juxtamedullary nephron as our example. Table 26–4 summarizes the functions of the various parts of the nephron.

Table 26–4	Renal Structures and Their Function		
Segment	**General Functions**	**Specific Functions**	**Mechanisms**
Renal corpuscle	*Filtration* of plasma; generates approximately 180 L/day of filtrate similar in composition to blood plasma without plasma proteins	*Filtration* of water, inorganic and organic solutes from plasma; retention of plasma proteins and blood cells	Glomerular hydrostatic (blood) pressure working across capillary endothelium, dense layer, and filtration slits
Proximal convoluted tubule (PCT)	*Reabsorption* of 60%–70% of the water (108–116 L/day), 99%–100% of the organic substrates, and 60%–70% of the sodium and chloride ions in the original filtrate	*Active reabsorption*: Glucose, other simple sugars, amino acids, vitamins, ions (including sodium, potassium, calcium, magnesium, phosphate, and bicarbonate)	Carrier-mediated transport, including facilitated transport (glucose, amino acids), cotransport (glucose, ions), and countertransport (with secretion of H^+)
		Passive reabsorption: Urea, chloride ions, lipid-soluble materials, water	Diffusion (solutes) or osmosis (water)
		Secretion: Hydrogen ions, ammonium ions, creatinine, drugs, and toxins (as at DCT)	Countertransport with sodium ions
Nephron loop	*Reabsorption* of 25% of the water (45 L/day) and 20%–25% of the sodium and chloride ions present in the original filtrate; creation of the concentration gradient in the medulla	*Reabsorption*: Sodium and chloride ions	Active transport by Na^+–K^+/2 Cl^- transporter
		Water	Osmosis
Distal convoluted tubule (DCT)	*Reabsorption* of a variable amount of water (usually 5%, or 9 L/day), under ADH stimulation, and a variable amount of sodium ions, under aldosterone stimulation	*Reabsorption*: Sodium and chloride ions	Cotransport
		Sodium ions (variable)	Countertransport with potassium ions; aldosterone-regulated
		Calcium ions (variable)	Carrier-mediated transport stimulated by parathyroid hormone and calcitriol
		Water (variable)	Osmosis; ADH regulated
		Secretion: Hydrogen ions, ammonium ions	Countertransport with sodium ions
		Creatinine, drugs, toxins	Carrier-mediated transport
Collecting system	*Reabsorption* of a variable amount of water (usually 9.3%, or 16.8 L/day) under ADH stimulation, and a variable amount of sodium ions, under aldosterone stimulation	*Reabsorption*: Sodium ions (variable)	Countertransport with potassium or hydrogen ions; aldosterone-regulated
		Bicarbonate ions (variable)	Diffusion, generated within tubular cells
		Water (variable)	Osmosis; ADH-regulated
		Urea (distal portions only)	Diffusion
		Secretion: Potassium and hydrogen ions (variable)	Carrier-mediated transport
Peritubular capillaries	*Redistribution* of water and solutes reabsorbed in the cortex	Return of water and solutes to the general circulation	Osmosis and diffusion
Vasa recta	*Redistribution* of water and solutes reabsorbed in the medulla and stabilization of the concentration gradient of the medulla	Return of water and solutes to the general circulation	Osmosis and diffusion

CHECKPOINT

7. Identify the three distinct processes that form urine in the kidney.

8. What occurs when the plasma concentration of a substance exceeds its tubular maximum?

See the blue Answers tab at the back of the book.

26-4 | Hydrostatic and colloid osmotic pressures influence glomerular filtration pressure, which in turn affects the glomerular filtration rate

Learning Outcome Describe the factors that influence glomerular filtration pressure and the rate of filtrate formation.

Filtration takes place in the renal corpuscle as fluids move across the wall of the glomerulus and into the capsular space. The process of **glomerular filtration** involves passage across a filtration membrane. Recall that this membrane has three components: (1) the capillary endothelium, (2) the dense layer, and (3) the filtration slits (**Figures 26–8b** and **26–10**).

Glomerular capillaries are fenestrated capillaries with pores ranging from 60 to 100 nm (0.06 to 0.1 μm) in diameter. These openings are small enough to prevent the passage of blood cells, but they are too large to restrict the diffusion of solutes, even those the size of plasma proteins. The dense layer is more selective: Only small plasma proteins, nutrients, and ions can cross it. The filtration slits are the finest filters of all. Their gaps are only 6–9 nm wide. These gaps are small enough to prevent the passage of most small plasma proteins. As a result, under

26

Figure 26–10 Glomerular Filtration.

Glomerulus

Efferent arteriole

Afferent arteriole

Podocyte

Dense layer

Capillary lumen

Filtration slit

Pedicels

Pore

Capsular space

Filtration membrane

a The glomerular filtration membrane

Factors Controlling Glomerular Filtration

The **glomerular hydrostatic pressure (GHP)** is the blood pressure in the glomerular capillaries. This pressure tends to push water and solute molecules out of the plasma and into the filtrate. The GHP, which averages 50 mm Hg, is significantly higher than capillary pressures elsewhere in the systemic circuit, because the efferent arteriole is smaller in diameter than the afferent arteriole.

Filtrate in capsular space

Plasma proteins

Solutes

50

25

15

10 mm Hg

The **blood colloid osmotic pressure (BCOP)** tends to draw water out of the filtrate and into the plasma; it thus opposes filtration. Over the entire length of the glomerular capillary bed, the BCOP averages about 25 mm Hg.

The **net filtration pressure (NFP)** is the net pressure acting across the glomerular capillaries. It represents the sum of the hydrostatic pressures and the colloid osmotic pressures. Under normal circumstances, the net filtration pressure is approximately 10 mm Hg. This is the average pressure forcing water and dissolved substances out of the glomerular capillaries and into the capsular space.

Capsular hydrostatic pressure (CsHP) opposes GHP. CsHP, which tends to push water and solutes out of the filtrate and into the plasma, results from the resistance of filtrate already present in the nephron that must be pushed toward the renal pelvis. The difference between GHP and CsHP is the *net hydrostatic pressure (NHP)*.

The **capsular colloid osmotic pressure** is usually zero because few, if any, plasma proteins enter the capsular space.

b Net filtration pressure

normal circumstances only a few plasma proteins—such as albumin molecules, with an average diameter of 7 nm—can cross the filtration membrane and enter the capsular space. However, plasma proteins are all that stay behind, so the filtrate contains dissolved ions and small organic molecules in roughly the same concentrations as in plasma.

Filtration Pressures

We discussed the major forces that act across capillary walls in Chapters 21 and 22. (You may find it helpful to review

Figures 21–10 and 21–11, pp. 765, 766, before you proceed.) The primary factor involved in glomerular filtration is basically the same one that regulates fluid and solute movement across capillaries throughout the body. This factor is the balance between *hydrostatic pressure*, or fluid pressure, and *colloid osmotic pressure*, or pressure due to materials in solution, on either side of the capillary walls.

Hydrostatic Pressure

Blood pressure is low in typical systemic capillaries. The reason is that capillary blood flows into the venous system, where

26

resistance is fairly low. However, at the glomerulus, blood leaving the glomerular capillaries flows into an efferent arteriole, whose diameter is *smaller* than that of the afferent arteriole. For this reason, the efferent arteriole offers considerable resistance. Relatively high pressures are needed to force blood into it. As a result, glomerular pressures are similar to those of small arteries. These pressures average about 50 mm Hg, instead of the 35 mm Hg typical of peripheral capillaries. Glomerular hydrostatic pressure (GHP), capsular hydrostatic pressure (CsHP), blood colloid osmotic pressure (BCOP), and net filtration pressure (NFP) are shown in **Figure 26–10**.

Capsular hydrostatic pressure (CsHP) opposes glomerular hydrostatic pressure. This pressure results from the resistance to flow along the nephron and the conducting system. (Before additional filtrate can enter the capsule, some of the filtrate already present must be forced into the PCT.) The CsHP averages about 15 mm Hg.

The *net hydrostatic pressure (NHP)* is the difference between the glomerular hydrostatic pressure, which tends to push water and solutes out of the bloodstream, and the capsular hydrostatic pressure, which tends to push water and solutes into the bloodstream. We can calculate net hydrostatic pressure as follows:

$$NHP = (GHP - CsHP) = (50 \text{ mm Hg} - 15 \text{ mm Hg})$$

$$= 35 \text{ mm Hg}$$

Colloid Osmotic Pressure

The *colloid osmotic pressure* of a solution is the osmotic pressure resulting from suspended proteins. Under normal conditions, very few plasma proteins enter the capsular space, so no opposing colloid osmotic pressure exists within the capsule. However, if the glomeruli are damaged by disease or injury, and plasma proteins begin passing into the capsular space, a *capsular colloid osmotic pressure* is created that promotes filtration and increases fluid losses in urine.

Net Filtration Pressure

The net filtration pressure (NFP) at the glomerulus is the difference between the net hydrostatic pressure and the blood colloid osmotic pressure acting across the glomerular capillaries. Under normal circumstances, we can summarize this relationship as

$$NFP = NHP - BCOP$$

or

$$NFP = 35 \text{ mm Hg} - 25 \text{ mm Hg} = 10 \text{ mm Hg}$$

This is the average pressure forcing water and dissolved materials out of the glomerular capillaries and into the capsular spaces (**Figure 26–10b**). Problems that affect the net filtration pressure can seriously disrupt kidney function and cause a variety of clinical signs and symptoms.

The Glomerular Filtration Rate

The **glomerular filtration rate (GFR)** is the amount of filtrate the kidneys produce each minute. Each kidney contains about 6 m² —some 64 square feet—of filtration surface, and the GFR averages an astounding *125 mL per minute*. This means that nearly 10 percent of the fluid delivered to the kidneys by the renal arteries leaves the bloodstream and enters the capsular spaces.

A *creatinine clearance test* is often used to estimate the GFR. Creatinine results from the breakdown of creatine phosphate in muscle tissue and is normally eliminated in urine. Creatinine enters the filtrate at the glomerulus and is not reabsorbed in significant amounts. By monitoring the creatinine concentrations in blood and the amount excreted in urine in a 24-hour period, a clinician can easily estimate the GFR.

Consider, for example, a person who eliminates 84 mg of creatinine each hour and has a plasma creatinine concentration of 1.4 mg/dL. The GFR is equal to the amount secreted divided by the plasma concentration, so this person's GFR is

$$\frac{84 \text{ mg/h}}{1.4 \text{ mg/dL}} = 60 \text{ dL/h} = 100 \text{ mL/min}$$

The GFR is usually reported in milliliters per minute.

The value 100 mL/min is only an approximation of the GFR. The reason is that up to 15 percent of creatinine in the urine enters by means of active tubular secretion. When necessary, a more accurate GFR can be obtained by using the complex carbohydrate *inulin*. This compound is not metabolized in the body and is neither reabsorbed nor secreted by the kidney tubules.

In the course of a single day, the glomeruli generate about 180 liters (48 gal) of filtrate, approximately 70 times the total plasma volume. But as filtrate passes through the renal tubules, about 99 percent of it is reabsorbed. You should now appreciate the significance of tubular reabsorption!

The glomerular filtration rate depends on the net filtration pressure across glomerular capillaries. Any factor that alters the net filtration pressure also alters the GFR and affects kidney function. One of the most significant factors is a decrease in renal blood pressure. If blood pressure at the glomeruli decreases by 20 percent (from 50 mm Hg to 40 mm Hg), kidney filtration ceases, because the net filtration pressure is 0 mm Hg. For this reason, the kidneys are sensitive to changes in blood pressure that have little or no effect on other organs. Hemorrhaging, shock, and dehydration are relatively common clinical conditions that can cause a dangerous decrease in the GFR and lead to acute renal failure (p. 1034).

Control of the GFR

Glomerular filtration is the vital first step for all other kidney functions. If filtration does not take place, waste products are not excreted, pH control is jeopardized, and an important mechanism for regulating blood volume is lost. It should be no surprise that a variety of regulatory mechanisms ensure that GFR remains within normal limits.

Filtration depends on adequate blood flow to the glomerulus and on the maintenance of normal filtration pressures. Three interacting levels of control stabilize GFR: (1) *autoregulation*, at the local level; (2) *hormonal regulation*, initiated by the kidneys; and (3) *autonomic regulation*, primarily by the sympathetic division of the autonomic nervous system.

Autoregulation of the GFR

Autoregulation (local blood flow regulation) maintains an adequate GFR despite changes in local blood pressure and blood flow. *Myogenic mechanisms*—how arteries and arterioles react to an increase or decrease in blood pressure—play a role in the autoregulation of blood flow. Changes to the diameters of afferent arterioles, efferent arterioles, and glomerular capillaries maintain GFR. The most important regulatory mechanisms stabilize the GFR when systemic blood pressure drops (**Figure 26–11**).

The GFR also remains relatively constant when systemic blood pressure increases. An increase in renal blood pressure stretches the walls of afferent arterioles, and the smooth muscle cells respond by contracting. The reduction in the diameter of afferent arterioles decreases glomerular blood flow and keeps the GFR within normal limits.

Hormonal Regulation of the GFR

The GFR is regulated by the hormones of the renin–angiotensin-aldosterone system and the natriuretic peptides (ANP and BNP). We introduced these hormones and their actions in Chapters 18 and 21. ⟳ pp. 665–667, 772–775 There are three triggers for the release of renin by the juxtaglomerular complex (JGC). They are (1) a decrease in blood pressure at the glomerulus as the result of a decrease in blood volume, a decrease in systemic pressures, or a blockage in the renal artery or its branches; (2) stimulation of juxtaglomerular cells by sympathetic innervation; or (3) a decrease in the osmotic concentration of the tubular fluid at the macula densa.

These triggers are often interrelated. For example, a decrease in systemic blood pressure reduces the glomerular filtration rate, while baroreceptor reflexes cause sympathetic activation. Meanwhile, a decrease in the GFR slows the movement of tubular fluid along the nephron. As a result, the tubular fluid is in the ascending limb of the nephron loop longer, and the concentration of sodium and chloride ions in the tubular fluid reaching the macula densa and DCT becomes abnormally low.

Figure 26–11 provides a general overview of the response of the renin–angiotensin-aldosterone system to a decrease in GFR. A decrease in GFR leads to the release of renin by the juxtaglomerular complex. Renin converts the inactive protein angiotensinogen to angiotensin I. Angiotensin I is also inactive but is then converted to angiotensin II by **angiotensin-converting enzyme (ACE)**. This conversion takes place primarily in the capillaries of the lungs. Angiotensin II acts at the nephron, adrenal glands, and in the CNS. In peripheral capillary beds, angiotensin II causes a brief but powerful vasoconstriction of arterioles and precapillary sphincters, increasing arterial pressures throughout the body. The combined effect is an increase in systemic blood volume and blood pressure and the restoration of normal GFR.

If blood volume increases, the GFR increases automatically. This increase promotes fluid losses that help return blood volume to normal levels. If the increase in blood volume is severe, hormonal factors further increase the GFR and speed up fluid losses in the urine. As noted in Chapter 18, the heart releases *natriuretic peptides* when increased blood volume or blood pressure stretches the walls of the heart. The atria release atrial natriuretic peptide (ANP), and the ventricles release brain natriuretic peptide (BNP). ⟳ pp. 666, 775 Among their other effects, these hormones trigger the dilation of afferent arterioles and the constriction of efferent arterioles. This mechanism increases glomerular pressures and increases the GFR. The natriuretic peptides also decrease sodium reabsorption at the renal tubules. The net result is increased urine production and decreased blood volume and pressure.

Autonomic Regulation of the GFR

Most of the autonomic innervation of the kidneys consists of sympathetic postganglionic fibers. (The role of the few parasympathetic fibers in regulating kidney function is not known.) Sympathetic activation has a direct effect on the GFR. It produces a powerful vasoconstriction of afferent arterioles, which decreases the GFR and slows the production of filtrate. In this way, the sympathetic activation triggered by an acute decrease in blood pressure or a heart attack overrides the local regulatory mechanisms that act to stabilize the GFR. As the crisis passes and sympathetic tone decreases, the filtration rate gradually returns to normal.

When the sympathetic division alters regional patterns of blood circulation, blood flow to the kidneys is often affected. For example, the dilation of superficial vessels in warm weather shunts blood away from the kidneys. As a result, glomerular filtration decreases temporarily. The effect becomes especially pronounced during strenuous exercise. As the blood flow to your skin and skeletal muscles increases, kidney perfusion gradually decreases. These changes may be opposed, with variable success, by autoregulation at the local level.

At maximal levels of exertion, renal blood flow may be less than 25 percent of normal resting levels. This reduction can create problems for endurance athletes. Metabolic wastes build up

Figure 26–11   **The Response to a Reduction in the GFR.**

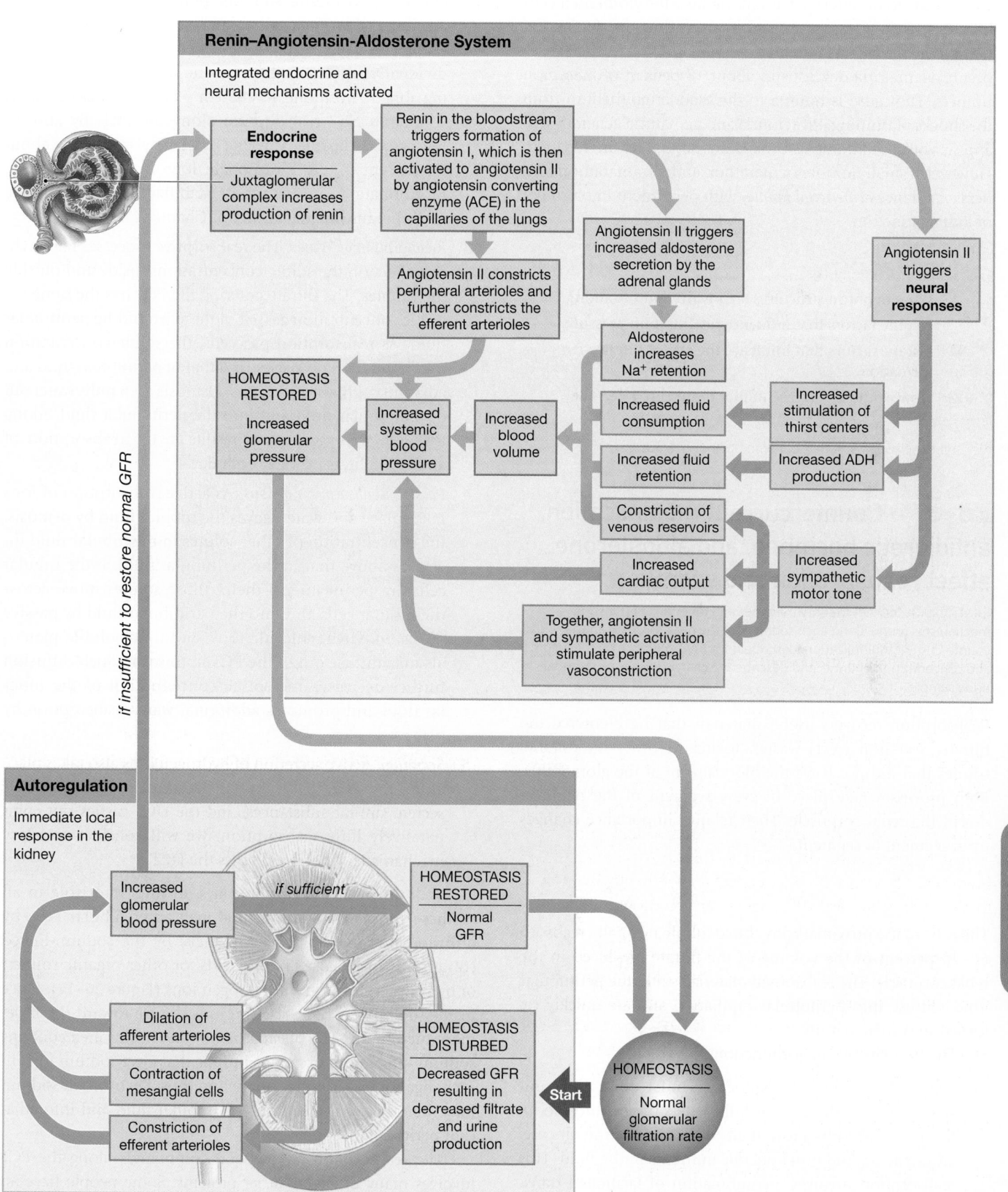

over the course of a long event. *Proteinuria* (protein in the urine) commonly occurs after such events because the glomerular cells have been injured by prolonged hypoxia (low oxygen levels). If the damage is substantial, *hematuria* (blood in the urine) occurs. Hematuria develops in about 18 percent of marathon runners. The cause is trauma to the bladder epithelium from the shocks of running. Proteinuria and hematuria generally disappear within 48 hours as the glomerular tissues are repaired. However, a small number of marathon and ultramarathon runners experience *acute renal failure*, with permanent impairment of kidney function.

CHECKPOINT

9. What nephron structures are involved in filtration?

10. List the factors that influence net filtration pressure.

11. List the factors that influence the rate of filtrate formation.

12. How would a decrease in blood pressure affect the GFR?

See the blue Answers tab at the back of the book.

26-5 ❙ Countercurrent multiplication, antidiuretic hormone, and aldosterone affect reabsorption and secretion

Learning Outcome Identify the types and functions of transport mechanisms found along each segment of the nephron, explain the role of countercurrent multiplication, describe hormonal influence on the volume and concentration of urine, and describe the characteristics of a normal urine sample.

Reabsorption recovers useful materials that have entered the filtrate. Secretion ejects wastes, toxins, or other undesirable solutes that did not leave the bloodstream at the glomerulus. Both processes take place in every segment of the nephron except the renal corpuscle. Their relative importance changes from segment to segment.

Reabsorption and Secretion at the PCT

The cells of the proximal convoluted tubule normally reabsorb 60–70 percent of the volume of the filtrate produced in the renal corpuscle. The reabsorbed materials enter the peritubular fluid, diffuse into peritubular capillaries, and are quickly returned to the circulation.

The PCT has five major functions:

1. *Reabsorption of Organic Nutrients.* Under normal circumstances, before the tubular fluid enters the nephron loop, the PCT reabsorbs more than 99 percent of the glucose, amino acids, and other organic nutrients in the fluid. This reabsorption involves a combination of facilitated transport and cotransport.

2. *Active Reabsorption of Ions.* The PCT actively transports several ions, including sodium, potassium, and bicarbonate ions, plus magnesium, phosphate, and sulfate ions (Figure 26–12). Although the ion pumps involved are individually regulated, they may be influenced by circulating ion or hormone levels. For example, angiotensin II stimulates Na^+ reabsorption along the PCT. By absorbing carbon dioxide, the PCT indirectly recaptures about 90 percent of the bicarbonate ions from tubular fluid. Bicarbonate is important in stabilizing blood pH. We examine this process further in Chapter 27.

3. *Reabsorption of Water.* The reabsorptive processes have a direct effect on the solute concentrations inside and outside the tubules. The filtrate entering the PCT has the same osmotic concentration as that of the surrounding peritubular fluid. As reabsorption proceeds, the solute concentration of tubular fluid decreases, and that of peritubular fluid and adjacent capillaries increases. Osmosis then pulls water out of the tubular fluid and into the peritubular fluid. Along the PCT, this mechanism results in the reabsorption of about 108 liters of water each day.

4. *Passive Reabsorption of Ions.* As active reabsorption of ions takes place and water leaves the tubular fluid by osmosis, the concentration of other solutes in the tubular fluid increases above that in the peritubular fluid. If the tubular cells are permeable to them, those solutes move across the tubular cells and into the peritubular fluid by passive diffusion. Urea, chloride ions, and lipid-soluble materials may diffuse out of the PCT in this way. Such diffusion further decreases the solute concentration of the tubular fluid and promotes additional water reabsorption by osmosis.

5. *Secretion.* Active secretion of hydrogen ions also takes place along the PCT (Table 26–4). Because the PCT and DCT secrete similar substances, and the DCT carries out comparatively little reabsorption, we will consider secretory mechanisms when we discuss the DCT.

Sodium ion reabsorption plays an important role in all of these processes. Sodium ions may enter tubular cells by diffusion through Na^+ leak channels; by the sodium-linked cotransport of glucose, amino acids, or other organic solutes; or by countertransport for hydrogen ions (Figure 26–12). Once inside the tubular cells, sodium ions diffuse toward the basement membrane. The plasma membrane in this area contains sodium–potassium exchange pumps that eject sodium ions in exchange for extracellular potassium ions. Reabsorbed sodium ions then diffuse through the peritubular fluid and into adjacent peritubular capillaries.

The reabsorption of ions and compounds along the PCT involves many different carrier proteins. Some people have an inherited inability to manufacture one or more of these carrier

26

proteins. For this reason, these individuals are unable to recover specific solutes from tubular fluid. In *renal glycosuria* (glī-kō-SYŪ-rē-uh), for example, a defective glucose carrier protein makes it impossible for the PCT to reabsorb glucose from tubular fluid.

The Nephron Loop and Countercurrent Multiplication

Approximately 60–70 percent of the volume of filtrate produced by the glomerulus has been reabsorbed before the tubular fluid

reaches the nephron loop. In the process, useful organic substrates and many mineral ions have been reclaimed. The nephron loop reabsorbs about half of the remaining water and two-thirds of the remaining sodium and chloride ions. This reabsorption takes place efficiently according to the principle of countercurrent exchange. We introduced this principle in Chapter 25 in our discussion of heat conservation mechanisms. ⤷ p. 991

The thin descending limb and the thick ascending limb of the nephron loop lie very close together. They are separated only by peritubular fluid. The exchange between these segments is

Figure 26–12 **Transport Activities at the PCT.** Sodium ions may enter a tubular cell from the filtrate by diffusion, cotransport, or countertransport. The sodium ions are then pumped into the peritubular fluid by the sodium–potassium exchange pump. Other reabsorbed solutes may be ejected into the peritubular fluid by separate active transport mechanisms. The absorption of bicarbonate is indirectly associated with the reabsorption of sodium ions and the secretion of hydrogen ions.

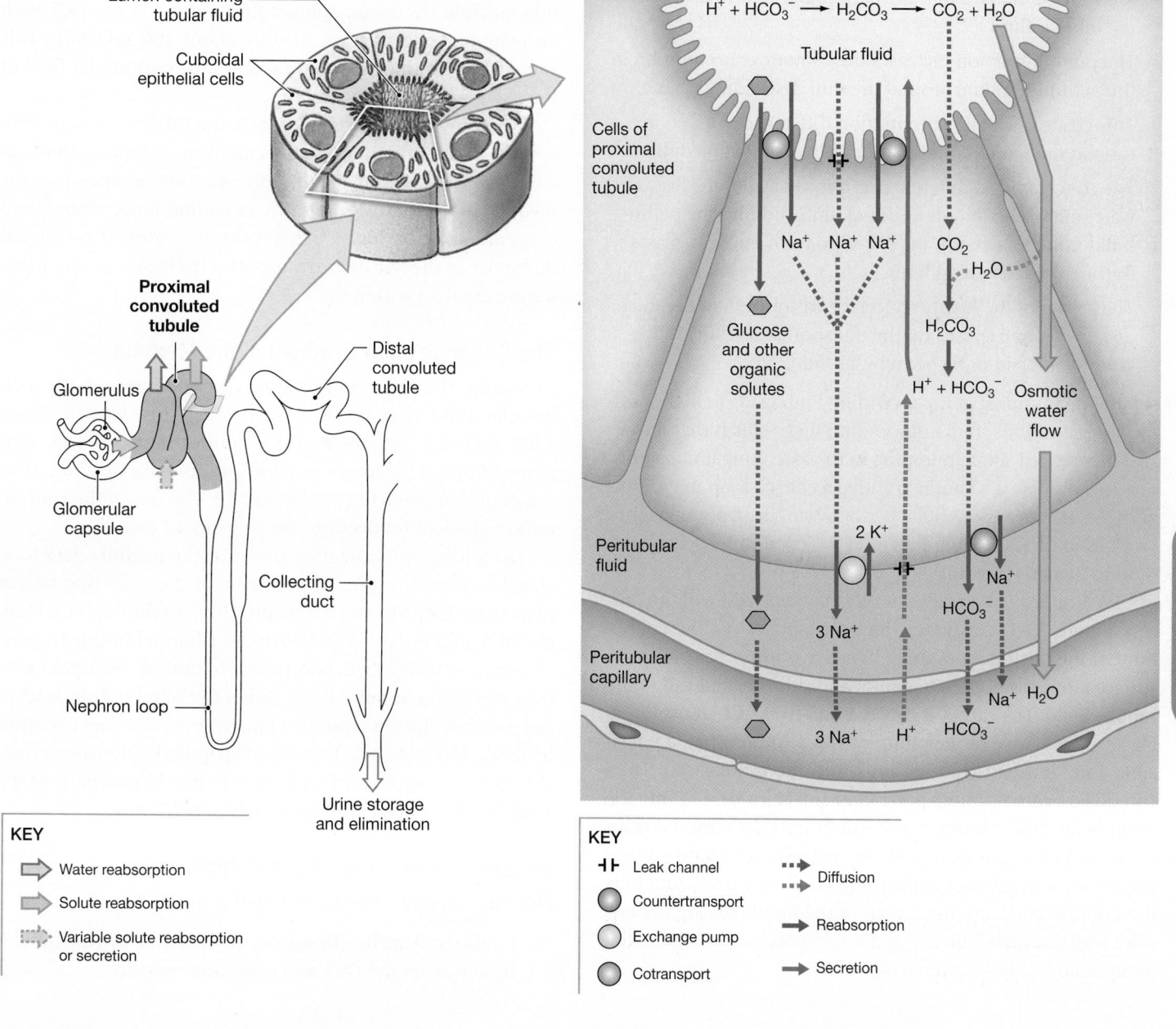

called **countercurrent multiplication**. *Countercurrent* refers to the fact that the exchange takes place between fluids moving in opposite directions: Tubular fluid in the descending limb flows toward the renal pelvis, while tubular fluid in the ascending limb flows toward the cortex. *Multiplication* refers to the fact that the effect of the exchange increases as movement of the fluid continues.

The two parallel limbs of the nephron loop have very different permeability characteristics. The thin descending limb is permeable to water but relatively impermeable to solutes. The thick ascending limb is relatively impermeable to both water and solutes, but it contains active transport mechanisms that pump sodium and chloride ions from the tubular fluid into the peritubular fluid of the medulla.

A quick overview of countercurrent multiplication will help you make sense of the details:

- Sodium and chloride ions are pumped out of the thick ascending limb and into the peritubular fluid. This dilutes the tubular fluid.

- The pumping action increases the osmotic concentration in the peritubular fluid around the thin descending limb.

- This creates a small concentration difference between the tubular fluid and peritubular fluid in the renal medulla.

- The concentration difference results in an osmotic flow of water out of the thin descending limb and into the peritubular fluid. As a result, the solute concentration increases in the thin descending limb.

- The arrival of the highly concentrated solution in the thick ascending limb speeds up the transport of sodium and chloride ions into the peritubular fluid.

Solute pumping at the ascending limb leads to higher solute concentrations in the descending limb, which then brings about increased solute pumping in the ascending limb. Notice that this process is a simple positive feedback loop that multiplies the concentration difference between the hypotonic tubular fluid in the ascending limb and the hypertonic peritubular fluid in the renal medulla.

We can now take a closer look at the mechanics of the process. **Figure 26–13a** diagrams ion transport across the epithelium of the thick ascending limb. Active transport at the apical surface moves sodium, potassium, and chloride ions out of the tubular fluid. The carrier is called a Na^+–K^+/2 Cl^- transporter, because each cycle of the pump carries a sodium ion, a potassium ion, and two chloride ions into the tubular cell. Then cotransport carriers pump potassium and chloride ions into the peritubular fluid. However, potassium ions are removed from the peritubular fluid as the sodium–potassium exchange pump pumps sodium ions out of the tubular cell. The potassium ions then diffuse back into the lumen of the tubule through potassium leak channels. The net result is that Na^+ and Cl^- enter the peritubular fluid of the renal medulla.

The removal of sodium and chloride ions from the tubular fluid in the ascending limb raises the osmotic concentration of the peritubular fluid around the thin descending limb (**Figure 26–13b**). Recall that the thin descending limb is permeable to water but not to solutes. As tubular fluid travels deeper into the medulla within the thin descending limb, osmosis moves water into the peritubular fluid. Solutes remain behind. As a result, the tubular fluid at the turn of the nephron loop has a higher osmotic concentration than it did at the start.

The pumping mechanism of the thick ascending limb is highly effective. Almost two-thirds of the sodium and chloride ions that enter it are pumped out of the tubular fluid before that fluid reaches the DCT. In other tissues, differences in solute concentration are quickly resolved by osmosis. However, osmosis cannot take place across the wall of the thick ascending limb, because the epithelium there is impermeable to water. So, as Na^+ and Cl^- are removed, the solute concentration in the tubular fluid decreases. Tubular fluid arrives at the DCT with an osmotic concentration of only about 100 mOsm/L. This value is one-third the concentration of the peritubular fluid of the renal cortex.

The rate of ion transport across the thick ascending limb is proportional to an ion's concentration in tubular fluid. As a result, more sodium and chloride ions are pumped into the medulla at the start of the thick ascending limb, where NaCl concentrations are highest, than near the cortex. This regional difference in the rate of ion transport is the basis of the concentration gradient within the medulla.

The Concentration Gradient of the Medulla

Normally, the maximum solute concentration of the peritubular fluid near the turn of the nephron loop is about 1200 mOsm/L (**Figure 26–13b**). Sodium and chloride ions pumped out of the loop's ascending limb make up about two-thirds of that gradient (750 mOsm/L). The rest of the concentration gradient results from the presence of urea.

To understand how urea arrives in the medulla, let's look ahead to events in the last segments of the collecting system (**Figure 26–13c**). The thick ascending limb of the nephron loop, the DCT, and the collecting ducts are all impermeable to urea. As water is reabsorbed, the concentration of urea gradually increases in the tubular fluid. When the tubular fluid reaches the papillary duct, it typically contains urea at a concentration of about 450 mOsm/L. Because the papillary ducts are permeable to urea, the urea concentration in the deepest parts of the medulla also averages 450 mOsm/L.

Benefits of Countercurrent Multiplication

The countercurrent mechanism performs two functions:

1. It efficiently reabsorbs solutes and water before the tubular fluid reaches the DCT and collecting system.

Figure 26–13   **Countercurrent Multiplication and Urine Concentration.**

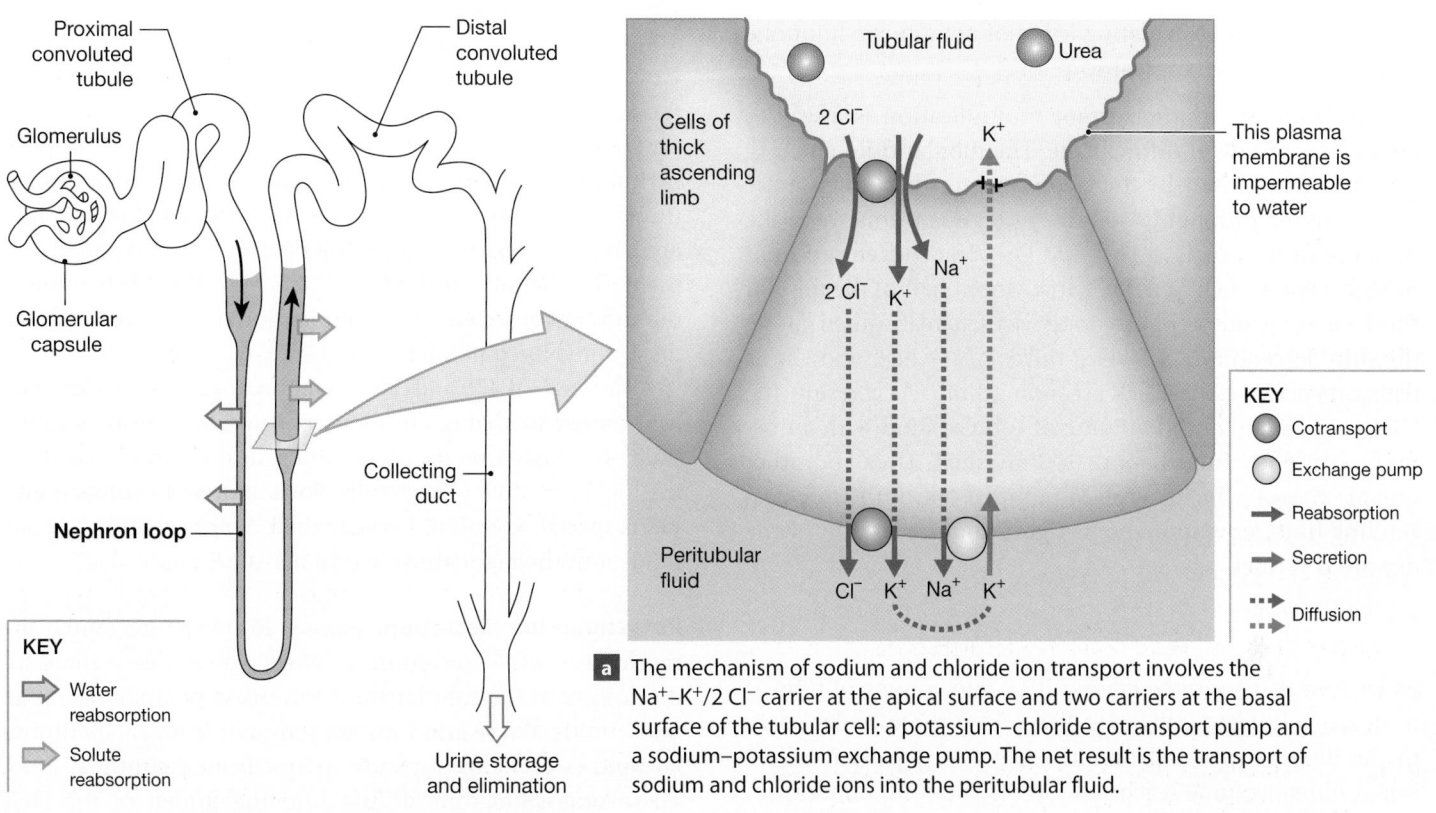

a The mechanism of sodium and chloride ion transport involves the Na⁺–K⁺/2 Cl⁻ carrier at the apical surface and two carriers at the basal surface of the tubular cell: a potassium–chloride cotransport pump and a sodium–potassium exchange pump. The net result is the transport of sodium and chloride ions into the peritubular fluid.

b Transport of NaCl along the ascending thick limb results in the movement of water from the descending limb.

c The permeability characteristics of both the loop and the collecting duct tend to concentrate urea in the tubular fluid and in the medulla. The nephron loop, DCT, and collecting duct are impermeable to urea. As water reabsorption occurs, the urea concentration increases. Papillary duct permeability to urea makes up nearly one-third of the solutes in the deepest portions of the medulla.

2. It establishes a concentration gradient that permits the passive reabsorption of water from the tubular fluid in the collecting system. Circulating levels of antidiuretic hormone (ADH) regulate this reabsorption.

In summary, countercurrent multiplication is a way to either concentrate or dilute urine. The tubular fluid entering the descending limb of the nephron loop has an osmotic concentration of roughly 300 mOsm/L, due primarily to the presence of ions such as Na^+ and Cl^-. The concentration of organic wastes, such as urea, is low. About half of the tubular fluid entering the nephron loop is then reabsorbed along the thin descending limb. Two-thirds of the Na^+ and Cl^- is reabsorbed along the thick ascending limb. As a result, the DCT receives a reduced volume of tubular fluid with an osmotic concentration of about 100 mOsm/L. Urea and other organic wastes, which were not pumped out of the thick ascending limb, now represent a significant proportion of the dissolved solutes.

Reabsorption and Secretion at the DCT

As we have just seen, the composition and volume of tubular fluid change dramatically as it flows from the capsular space to the distal convoluted tubule. Only 15–20 percent of the initial filtrate volume reaches the DCT. The concentrations of electrolytes and organic wastes in the arriving tubular fluid no longer resemble the concentrations in blood plasma. Selective reabsorption or secretion, primarily along the DCT, makes the final adjustments in the solute composition and volume of the tubular fluid.

Reabsorption at the DCT

Throughout most of the DCT, the tubular cells actively transport Na^+ and Cl^- out of the tubular fluid (**Figure 26–14a**). Tubular cells along the distal portions of the DCT also contain ion pumps that reabsorb tubular Na^+ in exchange for another cation (usually K^+) (**Figure 26–14b**). The hormone *aldosterone*, produced by the adrenal cortex, controls the Na^+ channels and the ion pump. This hormone stimulates the synthesis and incorporation of sodium channels and sodium ion pumps in plasma membranes along the DCT and collecting duct. The net result is a reduction in the number of sodium ions lost in urine.

However, sodium ion conservation is associated with potassium ion loss. Prolonged aldosterone stimulation can therefore produce *hypokalemia*, a dangerous reduction in the plasma K^+ concentration. The secretion of aldosterone and its actions on the DCT and collecting system are opposed by the natriuretic peptides (ANP and BNP).

The DCT is also the primary site of Ca^{2+} reabsorption. Circulating levels of parathyroid hormone and calcitriol regulate this process. ⊃ pp. 656–657

Secretion at the DCT

The blood entering the peritubular capillaries still contains a number of potentially undesirable substances that did not cross the filtration membrane at the glomerulus. In most cases, the concentrations of these substances are too low to cause physiological problems. However, any ions or compounds in peritubular capillaries will diffuse into the peritubular fluid. If those concentrations become too high, the tubular cells may absorb these substances from the peritubular fluid and secrete them into the tubular fluid. **Table 26–3** (p. 1012) lists some of the substances secreted into tubular fluid by the proximal and distal convoluted tubules.

The rate of K^+ and H^+ secretion increases or decreases in response to changes in their concentrations in peritubular fluid. The higher their concentration in the peritubular fluid, the higher the rate of secretion. Potassium and hydrogen ions merit special attention, because their concentrations in body fluids must be maintained within narrow limits.

Potassium Ion Secretion. **Figure 26–14a,b** diagrams the mechanism of K^+ secretion. In effect, tubular cells trade sodium ions in the tubular fluid for excess potassium ions in body fluids. Potassium ions are removed from the peritubular fluid in exchange for sodium ions from the tubular fluid. These potassium ions diffuse into the lumen of the DCT through potassium leak channels at the apical surfaces of the tubular cells.

Hydrogen Ion Secretion. Hydrogen ion secretion is also associated with the reabsorption of sodium. **Figure 26–14c** depicts two routes of secretion. Both involve the generation of carbonic acid by the enzyme *carbonic anhydrase.* ⊃ pp. 888, 926 Hydrogen ions generated by the dissociation of the carbonic acid are secreted by sodium-linked countertransport in exchange for Na^+ in the tubular fluid. The bicarbonate ions diffuse into the peritubular fluid and then into the bloodstream. There they help prevent changes in plasma pH.

Hydrogen ion secretion acidifies the tubular fluid while increasing the pH of the blood. Hydrogen ion secretion speeds up when the pH of the blood decreases. This can happen in *lactic acidosis*, which can develop after exhaustive muscle activity, or *ketoacidosis*, which can develop in starvation or diabetes mellitus. ⊃ p. 982 The combination of H^+ removal and HCO_3^- production by the kidneys plays an important role in the control of blood pH. Because one of the secretory pathways is aldosterone sensitive, aldosterone stimulates H^+ secretion. Prolonged aldosterone stimulation can cause *alkalosis*, or abnormally high blood pH.

In Chapter 25, we noted that the production of lactic acid and ketone bodies during the postabsorptive state could cause acidosis. Under these conditions, the PCT and DCT

Clinical Note

Diuretics *Diuresis* (dī-ū-RĒ-sis; *dia*, through + *ouresis*, urination) is the elimination of urine. *Urination* is an equivalent term in a general sense, but *diuresis* typically indicates the production of a large volume of urine. **Diuretics** (dī-ū-RET-iks) are drugs that promote water loss in urine. The usual goal in diuretic therapy is to decrease blood volume, blood pressure, extracellular fluid volume, or all three. The ability to control renal water losses with relatively safe and effective diuretics has saved the lives of many people, especially those with high blood pressure or congestive heart failure.

Diuretics have many mechanisms of action. However, all such drugs affect transport activities or water reabsorption along the nephron and collecting system. For example, consider the class of diuretics called *thiazides* (THĪ-uh-zīdz). These drugs increase water loss by decreasing sodium and chloride ion transport in the proximal and distal convoluted tubules.

Diuretic use for nonclinical reasons is on the rise. For example, some bodybuilders take large doses of diuretics to improve muscle definition temporarily. Some fashion models or horse jockeys do the same. Their goal is to reduce body weight for brief periods. This practice of "cosmetic dehydration" is extremely dangerous. It has caused several deaths due to electrolyte imbalance and consequent cardiac arrest.

deaminate amino acids in reactions that strip off the amino groups ($-NH_2$). The reaction sequence ties up H^+ and yields both **ammonium ions** (NH_4^+) and HCO_3^-. As indicated in **Figure 26–14c**, the ammonium ions are then pumped into the tubular fluid by sodium-linked countertransport, and the bicarbonate ions enter the bloodstream by way of the peritubular fluid.

Tubular deamination thus has two major benefits. It provides carbon chains suitable for catabolism. It also generates bicarbonate ions that add to the buffering capacity of plasma.

Reabsorption and Secretion along the Collecting System

The collecting ducts receive tubular fluid from many nephrons and carry it toward the renal sinus, through the concentration gradient in the medulla. The normal amount of water and solute loss in the collecting system is regulated in two ways:

- By aldosterone, which controls sodium ion pumps along most of the DCT and the proximal portion of the collecting system. As we have noted, these actions are opposed by the natriuretic peptides.

- By ADH, which controls the permeability of the DCT and collecting system to water. The secretion of ADH is suppressed by the natriuretic peptides, and this—combined with the effects of natriuretic peptide on aldosterone secretion and action—can dramatically increase urinary water losses.

The collecting system also has other reabsorptive and secretory functions. Many of them are important to the control of body fluid pH.

Reabsorption in the Collecting System

The collecting system reabsorbs sodium ions, bicarbonate ions, and urea as follows:

- *Sodium Ion Reabsorption.* The collecting system contains aldosterone-sensitive ion pumps that exchange Na^+ in tubular fluid for K^+ in peritubular fluid (**Figure 26–14b**).

- *Bicarbonate Reabsorption.* Bicarbonate ions are reabsorbed in exchange for chloride ions in the peritubular fluid (**Figure 26–14c**).

- *Urea Reabsorption.* The concentration of urea in the tubular fluid entering the collecting duct is relatively high. The fluid entering the papillary duct generally has the same osmotic concentration as that of interstitial fluid of the medulla—about 1200 mOsm/L—but contains a much higher concentration of urea. As a result, urea tends to diffuse out of the tubular fluid and into the peritubular fluid in the deepest portion of the medulla.

Secretion in the Collecting System

The collecting system is important in controlling the pH of body fluids through the secretion of hydrogen or bicarbonate ions. If the pH of the peritubular fluid decreases, carrier proteins pump hydrogen ions into the tubular fluid and reabsorb bicarbonate ions that help restore normal pH. If the pH of the peritubular fluid rises (a much less common event), the collecting system secretes bicarbonate ions and pumps hydrogen ions into the peritubular fluid. The net result is that the body eliminates a buffer and gains hydrogen ions that lower the pH. We examine these responses in more detail in Chapter 27, when we consider acid–base balance.

The Control of Urine Volume and Osmotic Concentration

Urine volume and osmotic concentration are regulated through the control of water reabsorption. Water is reabsorbed by osmosis along the proximal convoluted tubule and the descending limb of the nephron loop. The water permeabilities

26

Figure 26–14   **Tubular Secretion and Solute Reabsorption by the DCT.**

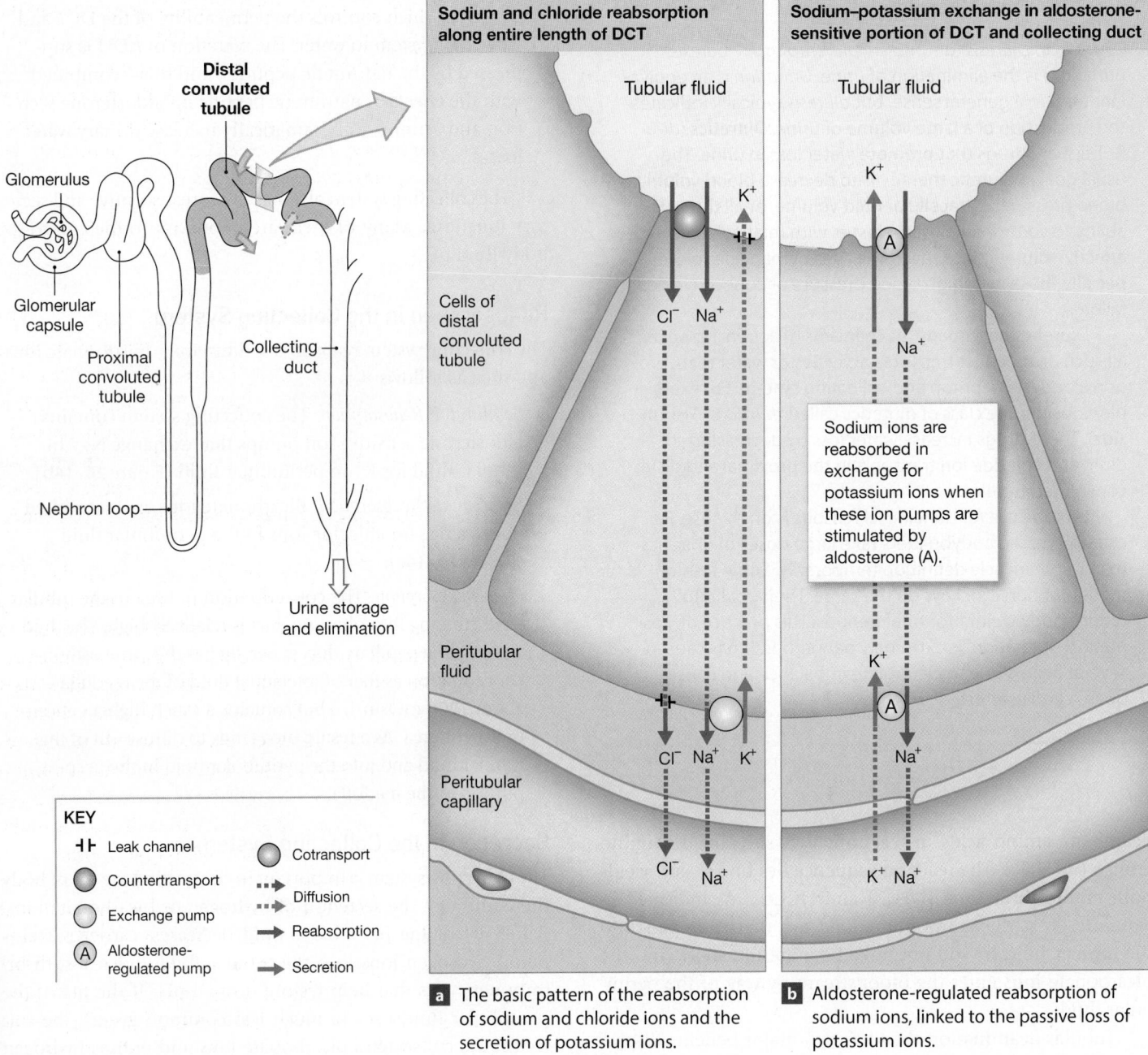

a The basic pattern of the reabsorption of sodium and chloride ions and the secretion of potassium ions.

b Aldosterone-regulated reabsorption of sodium ions, linked to the passive loss of potassium ions.

of these regions cannot be adjusted. As a result, water reabsorption takes place wherever the osmotic concentration of the peritubular fluid is greater than that of the tubular fluid. The ascending limb of the nephron loop is impermeable to water. In the distal convoluted tubule and collecting system, 1–2 percent of the volume of water in the original filtrate is recovered during sodium ion reabsorption. All these water movements represent *obligatory water reabsorption* because

they cannot be prevented. This reabsorption usually recovers 85 percent of the volume of filtrate.

The volume of water lost in urine depends on how much of the remaining water in the tubular fluid is reabsorbed along the DCT and collecting system. (This remaining water represents 15 percent of the filtrate volume, or approximately 27 liters per day.) The amount reabsorbed can be precisely controlled by a process called *facultative*

H⁺ secretion and HCO₃⁻ reabsorption along entire DCT and collecting duct

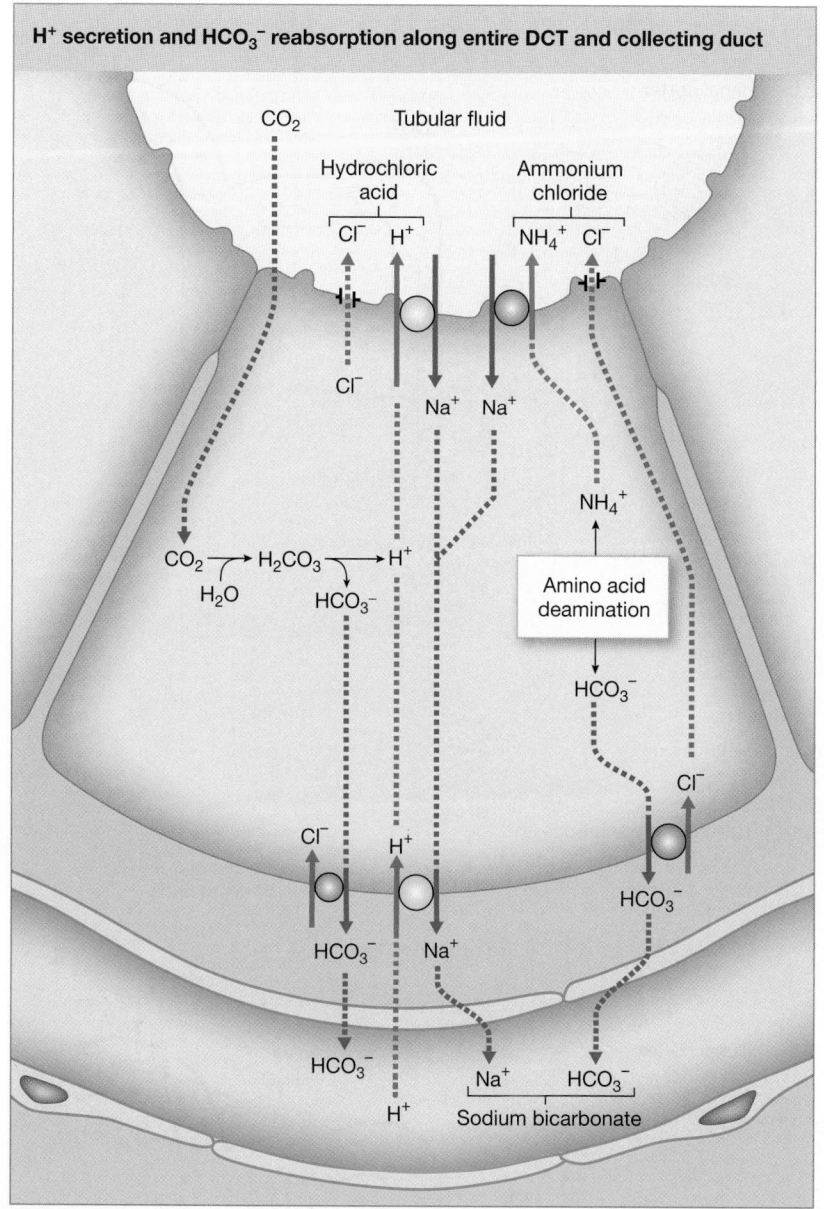

c Hydrogen ion secretion and the acidification of urine occur by two routes. The central theme is the exchange of hydrogen ions in the cytosol for sodium ions in the tubular fluid, and the reabsorption of the bicarbonate ions generated in the process.

water reabsorption. Precise control is possible because these segments are relatively impermeable to water except in the presence of ADH. This hormone causes special *water channels,* or *aquaporins,* to be inserted into the apical plasma membranes. These water channels dramatically enhance the rate of osmotic water movement. The higher the circulating levels of ADH, the greater the number of water channels, and the greater the water permeability of these segments.

As noted earlier in this chapter, the tubular fluid arriving at the DCT has an osmotic concentration of only about

100 mOsm/L. In the presence of ADH, osmosis takes place. Water moves out of the DCT until the osmotic concentration of the tubular fluid equals that of the surrounding cortex (roughly 300 mOsm/L).

The tubular fluid then flows along the collecting duct, which passes through the concentration gradient of the medulla. Additional water is then reabsorbed. The urine reaching the minor calyx has an osmotic concentration closer to 1200 mOsm/L. Just how closely the osmotic concentration approaches 1200 mOsm/L depends on how much ADH is present.

Figure 26–15 diagrams the effects of ADH on the DCT and collecting system. Without ADH (**Figure 26–15a**), water is not reabsorbed in these segments, so all the fluid reaching the DCT is lost in the urine. The person then produces large amounts of very dilute urine. That is just what happens in cases of *diabetes insipidus.* ⊃ p. 649 With this condition, urinary water losses may reach 24 liters (6.3 gal) per day and the urine osmotic concentration is 30–400 mOsm/L.

As ADH levels rise (**Figure 26–15b**), the DCT and collecting system become more permeable to water. As a result, the amount of water reabsorbed increases. At the same time, the osmotic concentration of the urine climbs. Under maximum ADH stimulation, the DCT and collecting system become so permeable to water that the osmotic concentration of the urine equals that of the deepest portion of the medulla. Note that the concentration of urine can never *exceed* that of the medulla, because the concentrating mechanism relies on osmosis.

The hypothalamus continuously secretes ADH at low levels. For this reason, the DCT and collecting system always have a significant degree of water permeability. At these low ADH levels, the DCT reabsorbs approximately 9 liters of water per day, or about 5 percent of the original volume of filtrate produced by the glomeruli. At normal ADH levels, the collecting system reabsorbs about 16.8 liters per day, or about 9.3 percent of the original volume of filtrate. A healthy adult typically produces 1200 mL of urine per day (about 0.6 percent of the filtrate volume). Its osmotic concentration range is 855–1335 mOsm/L.

The effects of ADH are opposed by those of the natriuretic peptides, ANP and BNP. These hormones stimulate the production of a large volume of relatively dilute urine. This water loss reduces plasma volume to normal.

26

Figure 26–15   **The Effects of ADH on the DCT and Collecting Duct.**

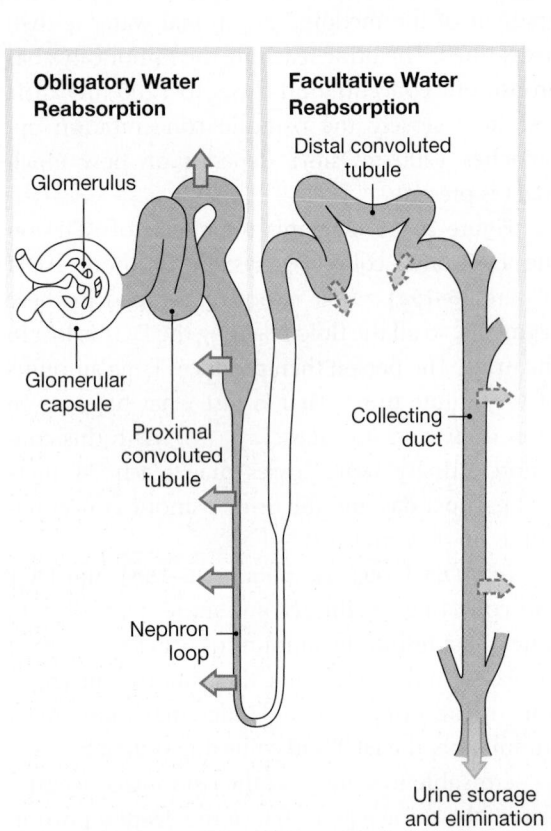

Obligatory Water Reabsorption

Glomerulus

Glomerular capsule

Proximal convoluted tubule

Nephron loop

Facultative Water Reabsorption

Distal convoluted tubule

Collecting duct

Urine storage and elimination

Renal cortex

PCT   DCT

Glomerulus

Na⁺Cl⁻   Na⁺Cl⁻

H_2O

H_2O Na⁺Cl⁻   Na⁺Cl⁻

H_2O Na⁺Cl⁻   Na⁺Cl⁻

H_2O Na⁺Cl⁻   Na⁺Cl⁻

H_2O

Solutes

Renal medulla   Collecting duct

a Tubule permeabilities and the osmotic concentration of urine without ADH

Large volume of dilute urine

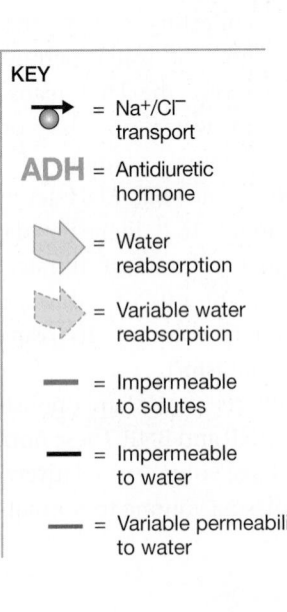

KEY

= Na⁺/Cl⁻ transport

ADH = Antidiuretic hormone

= Water reabsorption

= Variable water reabsorption

— = Impermeable to solutes

— = Impermeable to water

— = Variable permeability to water

Renal cortex

ADH H_2O

H_2O

Na⁺Cl⁻   Na⁺Cl⁻ H_2O

H_2O Na⁺Cl⁻   Na⁺Cl⁻ H_2O ←**ADH**

H_2O Na⁺Cl⁻   Na⁺Cl⁻ H_2O ←**ADH**

H_2O Na⁺Cl⁻   Na⁺Cl⁻ H_2O ←**ADH**

H_2O H_2O H_2O

H_2O H_2O

Renal medulla

b Tubule permeabilities and the osmotic concentration of urine with ADH

Small volume of concentrated urine

26

The Function of the Vasa Recta

The solutes and water reabsorbed in the renal medulla must be returned to the bloodstream without disrupting the concentration gradient. This return is the function of the vasa recta. Recall that the vasa recta are long, straight capillaries that parallel the long nephron loop of juxtamedullary nephrons.

Blood entering the vasa recta from the peritubular capillaries has an osmotic concentration of approximately 300 mOsm/L. As the blood descends into the medulla, it gradually increases in osmotic concentration as the solute concentration in the peritubular fluid increases. This increase in blood osmotic concentration involves both solute absorption and water loss. Solute absorption predominates, however, because the plasma proteins limit the osmotic flow of water out of the blood. ⟳ p. 765

Blood ascending toward the cortex gradually decreases in osmotic concentration as the solute concentration of the peritubular fluid decreases. Again, this decrease involves both solute diffusion and osmosis. In this case osmosis predominates, because the presence of plasma proteins does not oppose the osmotic flow of water into the blood.

The net results are that (1) some of the solutes absorbed in the descending portion of the vasa recta do not diffuse out in the ascending portion and (2) more water moves into the ascending portion of the vasa recta than moves out in the descending portion. Thus, the vasa recta carries both water and solutes out of the medulla. Under normal conditions, the removal of solutes and water by the vasa recta precisely balances the rates of solute reabsorption and osmosis in the medulla.

The Composition of Normal Urine

As we have seen, more than 99 percent of the 180 liters of filtrate produced each day by the glomeruli is reabsorbed. It never reaches the renal pelvis for elimination. General characteristics of the remaining filtrate—normal urine—are listed in Table 26–5. However, the composition of the urine produced each day varies with the metabolic and hormonal events under way.

The composition and concentration of urine are two related but distinct properties. The *composition* of urine reflects the filtration, reabsorption, and secretion activities of the nephrons. Some substances (such as urea) are neither actively excreted nor reabsorbed along the nephron. In contrast, organic nutrients are completely reabsorbed. Other substances, such as

Table 26–5	General Characteristics of Normal Urine
Characteristic	**Normal Range**
pH	4.5–8 (average: 6.0)
Specific gravity	1.003–1.030
Osmotic concentration (osmolarity)	855–1335 mOsm/L
Water content	93%–97%
Volume	700–2000 mL/day
Color	Clear yellow
Odor	Varies with composition
Bacterial content	None (sterile)

creatinine, are missed by filtration but are actively secreted into the tubular fluid.

Filtration, reabsorption, and secretion determine the kinds and amounts of substances excreted in urine. The *concentration* of these substances in a given urine sample depends on the osmotic movement of water across the walls of the tubules and collecting ducts. Because the composition and concentration of urine vary independently, you can produce a small volume of concentrated urine or a large volume of dilute urine and still excrete the same amount of dissolved substances. For this reason, physicians who are interested in a detailed assessment of renal function commonly analyze the urine produced over a 24-hour period rather than a single urine sample.

Urinalysis is the analysis of a urine sample. It is an important diagnostic tool, even in high-technology medicine. A standard urinalysis includes an assessment of the color and appearance of urine. These two characteristics can be determined without specialized equipment. In the 17th century, physicians classified the taste of the urine as sweet, salty, and so on, but quantitative analytical tests have long since replaced the taste-bud assay.

Normal urine is a clear, sterile solution. Its yellow color comes from the pigment urobilin. The kidneys generate this pigment from the urobilinogens produced by intestinal bacteria and absorbed in the colon (see **Figure 19–4**, p. 688). The characteristic odor of urine is due to the evaporation of small molecules, such as ammonia. Other substances not normally present, such as acetone or other ketone bodies, can also impart a distinctive smell.

Table 26–6 gives some typical values obtained from urinalysis. **Spotlight Figure 26–16** provides a summary of kidney function showing the major steps in the reabsorption of water and the production of concentrated urine.

26

1 The filtrate produced by the renal corpuscle has the same osmotic concentration as plasma—about 300 mOsm/L. It has the same composition as blood plasma but does not contain plasma proteins.

2 In the proximal convoluted tubule (PCT), the active removal of ions and organic nutrients produces a continuous osmotic flow of water out of the tubular fluid. This decreases the volume of filtrate but keeps the solutions inside and outside the tubule isotonic. Between 60 and 70 percent of the filtrate volume is absorbed here.

3 In the PCT and descending limb of the nephron loop, water moves into the surrounding peritubular fluids, leaving a small volume of highly concentrated tubular fluid. This reduction occurs by obligatory water reabsorption.

300 mOsm/L

H$_2$O

Nutrients

H$_2$O

Ions

RENAL CORTEX

RENAL MEDULLA

Vasa recta

Nephron loop

H$_2$O

KEY

= Water reabsorption

= Variable water reabsorption

= Na$^+$/Cl$^-$ transport

Ⓐ = Aldosterone-regulated pump

= Solutes

= Impermeable to solutes

= Impermeable to water

= Variable permeability to water

300 = Fluid osmotic concentration (mOsm/L)

4 The thick ascending limb is impermeable to water and solutes. The tubule cells actively transport Na^+ and Cl^- out of the tubule, thereby decreasing the osmotic concentration of the tubular fluid. Because only Na^+ and Cl^- are removed, urea makes up a higher proportion of the total osmotic concentration at the end of the nephron loop.

5 Further adjustments in the composition of the tubular fluid occur in the DCT and the collecting system. The osmotic concentration of the tubular fluid can be adjusted through active transport (reabsorption or secretion).

6 The final adjustments in the volume and osmotic concentration of the tubular fluid are made by controlling the water permeabilities of the distal portions of the DCT and the collecting system. ADH levels determine the final urine concentration.

7 The vasa recta absorbs the solutes and water reabsorbed by the nephron loop and the collecting ducts. By transporting these solutes and water into the bloodstream, the vasa recta maintain the concentration gradient of the renal medulla.

Tubular fluid from cortical nephrons

H_2O

K^+

Na^+

Na^+

K^+

100–300

300

600

900

1200

Na^+Cl^-

Na^+Cl^-

Na^+Cl^-

Na^+Cl^-

Na^+Cl^-

Na^+Cl^-

Nephron loop

H_2O

H_2O

H_2O

H_2O

600

900

1200

1200

ADH-regulated permeability

Vasa recta

1200

Table 26–6	Typical Values Obtained from Standard Urinalysis			
Compound	**Primary Source**	**Daily Elimination***	**Concentration**	**Remarks**
NITROGENOUS WASTES				
Urea	Deamination of amino acids by liver and kidneys	21 g	1800 mg/dL	Increases if negative nitrogen balance exists
Creatinine	Breakdown of creatine phosphate in skeletal muscle	1.8 g	150 mg/dL	Proportional to muscle mass; decreases during atrophy or muscle disease
Ammonia	Deamination by liver and kidney, absorption from intestinal tract	0.68 g	60 mg/dL	
Uric acid	Breakdown of purines	0.53 g	40 mg/dL	Increases in gout, liver diseases
Hippuric acid	Breakdown of dietary toxins	4.2 mg	350 μg/dL	
Urobilin	Urobilinogens absorbed at colon	1.5 mg	125 μg/dL	Gives urine its yellow color
Bilirubin	Hemoglobin breakdown product	0.3 mg	20 μg/dL	Increase may indicate problem with liver elimination or excess production; causes yellowing of skin and mucous membranes in jaundice
NUTRIENTS AND METABOLITES				
Carbohydrates		0.11 g	9 μg/dL	Primarily glucose; *glycosuria* develops if T_m is exceeded
Ketone bodies		0.21 g	17 μg/dL	Ketonuria may occur during postabsorptive state
Lipids		0.02 g	0.002 mg/dL	May increase in some kidney diseases
Amino acids		2.25 g	188 μg/dL	Note relatively high loss compared with other metabolites due to low T_m; excess (*aminoaciduria*) indicates T_m problem
IONS				
Sodium		4.0 g	40–220 mEq/L	Varies with diet, urine pH, hormones, etc.
Potassium		2.0 g	25–100 mEq/L	Varies with diet, urine pH, hormones, etc.
Chloride		6.4 g	110–250 mEq/L	
Calcium		0.2 g	17 mg/dL	Hormonally regulated (PTH/CT)
Magnesium		0.15 g	13 mg/dL	
BLOOD CELLS‡				
RBCs		130,000/day	100/mL	Excess (*hematuria*) indicates vascular damage in urinary system
WBCs		650,000/day	500/mL	Excess (*pyuria*) indicates renal infection or inflammation

*Representative values for a 70-kg (154-lb) male.
‡Usually estimated by counting the cells in a sample of sediment after urine centrifugation.

26

CHECKPOINT

13. What effect would increased amounts of aldosterone have on the K^+ concentration in urine?

14. What effect would a decrease in the Na^+ concentration of filtrate have on the pH of tubular fluid?

15. How would the lack of juxtamedullary nephrons affect the volume and osmotic concentration of urine?

16. Why does a decrease in the amount of Na^+ in the distal convoluted tubule lead to an increase in blood pressure?

See the blue Answers tab at the back of the book.

26-6 | Urine is transported by the ureters, stored in the bladder, and eliminated through the urethra, aided by the micturition reflex

Learning Outcome Describe the structures and functions of the ureters, urinary bladder, and urethra, discuss the voluntary and involuntary regulation of urination, and describe the micturition reflex.

Filtrate modification and urine production end when the fluid enters the renal pelvis. The urinary tract (the ureters, urinary bladder, and urethra) transports, stores, and eliminates urine. A **pyelogram** (PĪ-el-ō-gram) is an image of the urinary system (**Figure 26–17**). It is obtained by taking an x-ray of the kidneys after a radiopaque dye has been administered intravenously. Such an image provides an orientation to the relative sizes and positions of the main structures. Note that the sizes of the minor and major calyces, the renal pelvis, the ureters, the urinary bladder, and the proximal portion of the urethra are somewhat variable. These regions are lined by a *transitional epithelium* that can tolerate cycles of distension and contraction without damage. ⟲ p. 146

The Ureters

The ureters are a pair of muscular tubes that extend from the kidneys to the urinary bladder—a distance of about 30 cm (12 in.). Each ureter begins at the funnel-shaped renal pelvis (**Figure 26–4**). The ureters extend inferiorly and medially, passing over the anterior surfaces of the *psoas major muscles* (**Figure 26–3**).

Figure 26–17   A Pyelogram. This anterior-posterior view of urinary system structures is color-enhanced.   ATLAS: Plate 62b

11th and 12th ribs   Minor calyx   Major calyx

Ureter   Urinary bladder   Renal pelvis   Kidney

The ureters are retroperitoneal and are firmly attached to the posterior abdominal wall. The paths taken by the ureters in men and women are different, due to variations in the nature, size, and position of the reproductive organs. In males, the base of the urinary bladder lies between the rectum and the pubic symphysis (**Figure 26–18a**). In females, the base of the urinary bladder sits inferior to the uterus and anterior to the vagina (**Figure 26–18b**).

The ureters penetrate the posterior wall of the urinary bladder without entering the peritoneal cavity. They pass through the bladder wall at an oblique angle. The **ureteral openings** are slit-like rather than rounded (**Figure 26–18c**). This shape helps prevent the backflow of urine toward the ureter and kidneys when the urinary bladder contracts.

Histology of the Ureters

The wall of each ureter consists of three layers (**Figure 26–19a**): (1) an inner mucosa, made up of a transitional epithelium and the surrounding lamina propria; (2) a middle muscular layer made up of longitudinal and circular bands of smooth muscle; and (3) an outer connective tissue layer that is continuous with the fibrous capsule and peritoneum. About every 30 seconds, a peristaltic contraction begins at the renal pelvis. As it sweeps along the ureter, it forces urine toward the urinary bladder.

The Urinary Bladder

The urinary bladder is a hollow, muscular organ that serves as a temporary reservoir for urine (**Figure 26–18c**). The dimensions of the urinary bladder vary with its state of distension. A full urinary bladder can contain as much as a liter of urine.

A layer of peritoneum covers the superior surfaces of the urinary bladder. Several peritoneal folds assist in stabilizing its position. The **median umbilical ligament** extends from the anterior, superior border toward the umbilicus (navel). The **lateral umbilical ligaments** pass along the sides of the bladder to the umbilicus. These fibrous cords are the vestiges of the two *umbilical arteries*, which supplied blood to the placenta during embryonic and fetal development. ⟲ p. 798 The urinary bladder's posterior, inferior, and anterior surfaces lie outside the peritoneal cavity. In these areas, tough ligamentous bands anchor the urinary bladder to the pelvic and pubic bones.

In sectional view, the mucosa lining the urinary bladder is usually thrown into folds, or **rugae**, that disappear as the bladder fills. The triangular area bounded by the openings of the ureters and the entrance to the urethra makes up a region called the **trigone** (TRĪ-gōn) of the urinary bladder. There, the mucosa is smooth and very thick. The trigone acts as a funnel that channels urine into the urethra when the urinary bladder contracts.

The urethral entrance lies at the apex of the trigone, at the most inferior point in the urinary bladder. The region surrounding the urethral opening is known as the **neck** of the urinary bladder. It contains a muscular **internal urethral sphincter**.

26

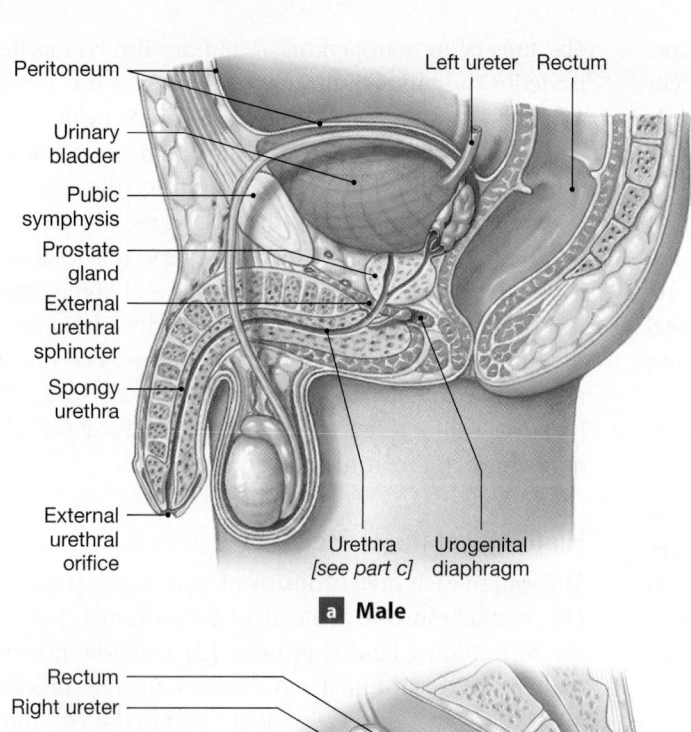

- Peritoneum
- Urinary bladder
- Pubic symphysis
- Prostate gland
- External urethral sphincter
- Spongy urethra
- External urethral orifice
- Left ureter
- Rectum
- Urethra [see part c]
- Urogenital diaphragm

a **Male**

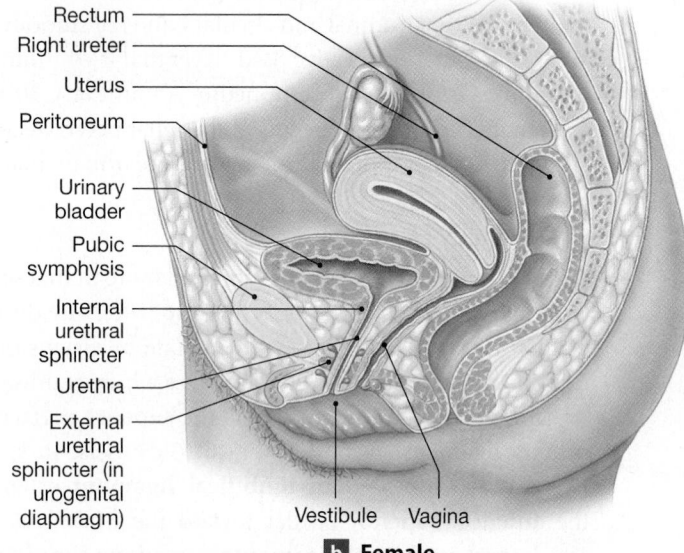

- Rectum
- Right ureter
- Uterus
- Peritoneum
- Urinary bladder
- Pubic symphysis
- Internal urethral sphincter
- Urethra
- External urethral sphincter (in urogenital diaphragm)
- Vestibule
- Vagina

b **Female**

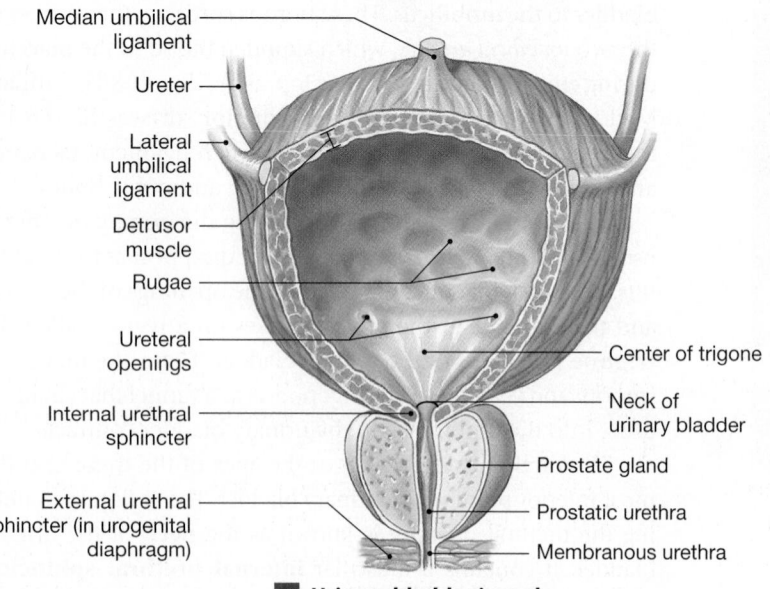

- Median umbilical ligament
- Ureter
- Lateral umbilical ligament
- Detrusor muscle
- Rugae
- Ureteral openings
- Internal urethral sphincter
- External urethral sphincter (in urogenital diaphragm)
- Center of trigone
- Neck of urinary bladder
- Prostate gland
- Prostatic urethra
- Membranous urethra

c **Urinary bladder in male**

Figure 26–18 **Organs for Conducting and Storing Urine.** ATLAS: Plates 62b; 64; 65

The smooth muscle fibers of this sphincter provide involuntary control over the discharge of urine from the bladder. The urinary bladder is innervated by postganglionic fibers from ganglia in the hypogastric plexus and by parasympathetic fibers from intramural ganglia that are controlled by branches of the pelvic nerves.

Histology of the Urinary Bladder

The wall of the urinary bladder contains mucosa, submucosa, and muscularis layers (**Figure 26–19b**). The muscularis layer consists of inner and outer layers of longitudinal smooth muscle, with a circular layer between the two. Together, these layers form the powerful **detrusor** (de-TRŪ-sor) **muscle** of the urinary bladder. When this muscle contracts, it compresses the urinary bladder and expels urine into the urethra.

The Urethra

The urethra extends from the neck of the urinary bladder and transports urine to the exterior of the body. The urethrae of males and females differ in length and in function. In males, the urethra extends from the neck of the urinary bladder to the tip of the penis (**Figure 26–18a,c**). This distance may be 18–20 cm (7–8 in.). We subdivide the male urethra into three segments: the prostatic urethra, the membranous urethra, and the spongy urethra. The **prostatic urethra** passes through the center of the prostate gland. The **membranous urethra** includes the short segment that penetrates the *urogenital diaphragm*, the muscular floor of the pelvic cavity. The **spongy urethra**, or *penile* (PĒ-nīl) *urethra*, extends from the distal border of the urogenital diaphragm to the external opening, or **external urethral orifice**, at the tip of the penis. In females, the urethra is very short. It extends 3–5 cm (1–2 in.) from the bladder to the vestibule (**Figure 26–18b**). The external urethral orifice is near the anterior wall of the vagina.

In both sexes, where the urethra passes through the urogenital diaphragm, a circular band of skeletal muscle forms the **external urethral sphincter**. This muscular band acts as a valve. The external urethral sphincter is under voluntary control through the perineal branch of the pudendal nerve. This sphincter has a resting muscle tone and must be voluntarily relaxed to permit micturition.

Figure 26–19   **The Histology of the Ureter, Urinary Bladder, and Urethra.**

Ureter   LM × 65

a   A transverse section through the ureter.

Transitional epithelium
Lamina propria ⎱ Mucosa

Smooth muscle

Outer connective tissue layer

Lumen of urethra

Smooth muscle

Stratified squamous epithelium of mucosa

Lamina propria containing mucous epithelial glands

Female urethra   LM × 50

c   A transverse section through the female urethra. A thick layer of smooth muscle surrounds the lumen.

Transitional epithelium
Lamina propria ⎱ Mucosa

Submucosa

Detrusor muscle

Visceral peritoneum

Urinary bladder   LM × 36

b   The wall of the urinary bladder.

TIPS & TRICKS

To remember that the urethra is the urinary tract's pipeline to the exterior, proclaim, "Eureka! Your urine! It's coming out the urethra!"

Histology of the Urethra

The urethral lining consists of a stratified epithelium that varies from transitional epithelium at the neck of the urinary bladder, to stratified columnar epithelium at the midpoint, to stratified squamous epithelium near the external urethral orifice. The lamina propria is thick and elastic. The mucosa is folded into longitudinal creases (**Figure 26–19c**). Mucin-secreting cells are located in the epithelial pockets. In males, the epithelial mucous glands may form tubules that extend into the lamina propria. Connective tissues of the lamina propria anchor the urethra to surrounding structures. In females, the lamina propria contains an extensive network of veins. Concentric layers of smooth muscle surround the entire complex.

The Micturition Reflex and Urination

As we have seen, urine reaches the urinary bladder by peristaltic contractions of the ureters. The **micturition reflex** coordinates the process of urination (**Figure 26–20**).

As the bladder fills with urine, stretch receptors in the bladder wall are stimulated. Afferent fibers in the pelvic nerves carry impulses to the sacral spinal cord. The increased level of activity in the fibers (1) facilitates parasympathetic motor neurons in

26

Figure 26–20   The Micturition Reflex.

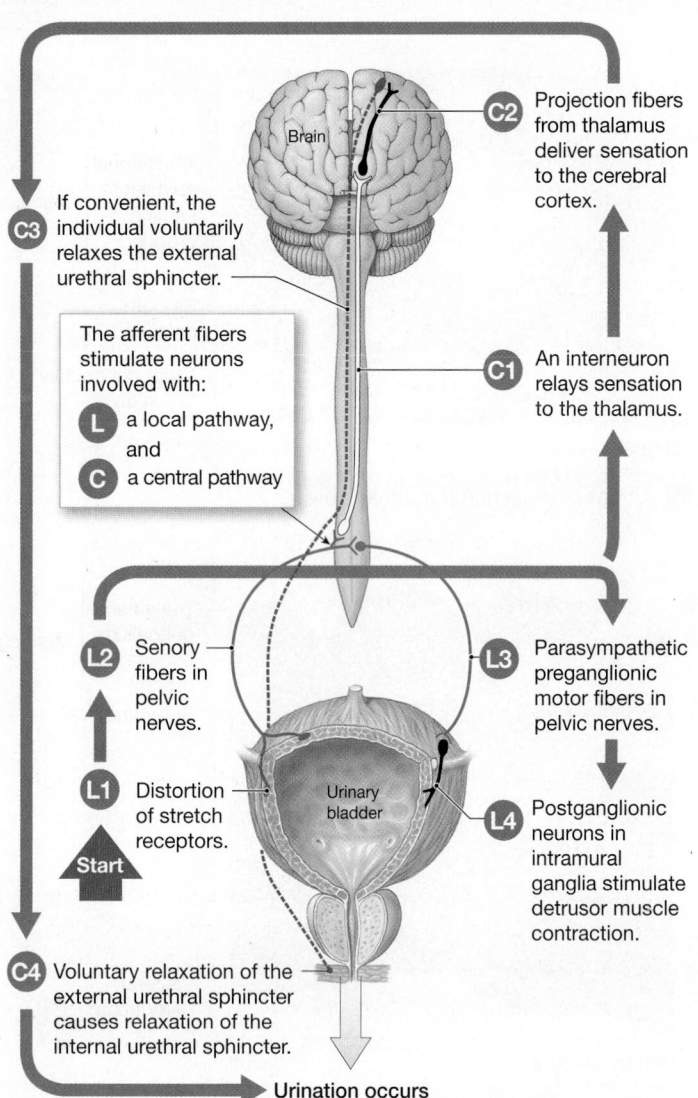

C2 Projection fibers from thalamus deliver sensation to the cerebral cortex.

C3 If convenient, the individual voluntarily relaxes the external urethral sphincter.

The afferent fibers stimulate neurons involved with:

L a local pathway, and

C a central pathway

C1 An interneuron relays sensation to the thalamus.

L2 Senory fibers in pelvic nerves.

L3 Parasympathetic preganglionic motor fibers in pelvic nerves.

L1 Distortion of stretch receptors.

Urinary bladder

Start

L4 Postganglionic neurons in intramural ganglia stimulate detrusor muscle contraction.

C4 Voluntary relaxation of the external urethral sphincter causes relaxation of the internal urethral sphincter.

→ Urination occurs

26

the sacral spinal cord and (2) stimulates interneurons that relay sensations to the thalamus and then, through projection fibers, to the cerebral cortex. As a result, you become aware of the fluid pressure in your urinary bladder.

The urge to urinate generally appears when your bladder contains about 200 mL of urine. As **Figure 26–20** shows, the micturition reflex begins when the stretch receptors provide adequate stimulation to parasympathetic preganglionic motor neurons. A further increase in bladder volume begins the cycle again, usually within an hour. Each increase in urinary volume leads to an increase in stretch receptor stimulation that makes the sensation more acute. Once the volume exceeds 500 mL, the bladder contractions triggered by the micturition reflex may generate enough pressure to force open the internal urethral sphincter. This opening leads to a reflexive relaxation of the external urethral sphincter. Urination takes place despite voluntary opposition or

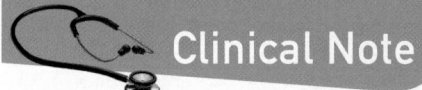

Clinical Note

Urinary Obstruction   Local blockages of the collecting ducts or ureters can result from *casts*—small blood clots, epithelial cells, lipids, or other materials that form in the collecting ducts. Casts are commonly eliminated in urine. They are visible in microscopic analyses of urine samples. **Renal calculi** (KAL-kū-lī), or *kidney stones*, form within the urinary tract from calcium deposits, magnesium salts, or crystals of uric acid. The condition is called *nephrolithiasis* (nef-rō-li-THĪ-uh-sis; *nephros*, kidney; *lithos*, stone). The blockage of the ureter by a stone or by other means (such as external compression) creates **urinary obstruction**. This problem is serious. In addition to causing pain, it decreases or prevents filtration in the affected kidney by increasing the capsular hydrostatic pressure.

Calculi are generally visible on an x-ray. If peristalsis and fluid pressures cannot dislodge them, they must be either surgically removed or destroyed. One nonsurgical procedure involves disintegrating the stones with a *lithotripter*, a device originally developed from machines used to de-ice airplane wings. Lithotripters focus sound waves on the stones, breaking them into smaller fragments that can be passed in the urine. Another nonsurgical approach is the insertion of a catheter armed with a laser that can shatter calculi with intense light beams.

potential inconvenience. At the end of a typical micturition, less than 10 mL of urine remains in the bladder.

Infants lack voluntary control over urination, because the necessary corticospinal connections have yet to be established. Accordingly, "toilet training" before age 2 often involves training the parent to anticipate the timing of the reflex rather than training the child to exert conscious control.

Incontinence (in-KON-ti-nens) is the inability to control urination voluntarily. Trauma to the internal or external urethral sphincter can contribute to incontinence in otherwise healthy adults. For example, some mothers develop *stress urinary incontinence* (SUI) if childbirth overstretches and damages the sphincter muscles. In this condition, increased intra-abdominal pressures—caused, for example, by a cough or sneeze—can overwhelm the sphincter muscles, causing urine to leak out. Incontinence can also develop in older people due to a general loss of muscle tone.

Damage to the central nervous system, the spinal cord, or the nerve supply to the urinary bladder or external urethral sphincter can also produce incontinence. For example, incontinence commonly accompanies Alzheimer's disease or spinal cord damage. In most cases, the affected individual develops an *automatic bladder*. The micturition reflex remains intact, but voluntary

Clinical Note

Renal Failure and Kidney Transplant **Renal failure** occurs when the kidneys cannot filter wastes from the blood and maintain homeostasis. When kidney filtration slows for any reason, urine production decreases. As the decrease continues, signs and symptoms of renal failure appear because water, ions, and metabolic wastes are retained. Virtually all systems in the body are affected. The person generally becomes hypertensive; anemia develops due to a decrease in erythropoietin production; and CNS problems can lead to sleeplessness, seizures, delirium, and even coma.

Acute renal failure occurs when filtration slows suddenly or stops due to renal ischemia, urinary obstruction, trauma, or nephrotoxic drugs. The reduction in kidney function takes place over a few days and may persist for weeks. Sensitized individuals can also develop acute renal failure after an allergic response to antibiotics or anesthetics.

In *chronic renal failure*, kidney function deteriorates gradually. The associated problems accumulate over years. This condition generally cannot be reversed. Its progression can only be slowed. Eventually, symptoms of *end-stage renal failure* develop. *Hemodialysis* is a treatment that "cleanses" the blood by serving as a substitute for normal kidney functioning. It can relieve the symptoms of end-stage and acute renal failure. However, this treatment is not a cure.

In terms of overall quality of life, probably the most satisfactory solution to the problem of end-stage renal failure is **kidney transplant**. This procedure involves implanting a new kidney from a living donor or a deceased donor. Of the 16,812 kidneys transplanted as of June 2013, 11,043 transplants came from deceased donors and 5,769 came from living donors. At the writing of this text, 96,645 people were awaiting a kidney transplant. The recipient's nonfunctioning kidney(s) may be removed, especially if an infection is present. The transplanted kidney and ureter are usually placed extraperitoneally in the pelvic cavity (within the iliac fossa). The ureter is connected to the recipient's urinary bladder. Patient survival is more than 90 percent at two years after the transplant.

---

control of the external urethral sphincter is lost, so the person cannot prevent the reflexive emptying of the urinary bladder. Damage to the pelvic nerves can abolish the micturition reflex entirely, because those nerves carry both afferent and efferent fibers of this reflex arc. The urinary bladder then becomes greatly distended with urine. It remains filled to capacity while the excess urine flows into the urethra in an uncontrolled stream. The insertion of a catheter is often needed to facilitate the discharge of urine.

CHECKPOINT

17. What effect would a high-protein diet have on the composition of urine?

18. Obstruction of a ureter by a kidney stone would interfere with the flow of urine between which two points?

19. The ability to control the micturition reflex depends on your ability to control which muscle?

See the blue Answers tab at the back of the book.

26-7 ▌ Age-related changes affect kidney function and the micturition reflex

Learning Outcome Describe the effects of aging on the urinary system.

In general, aging is associated with an increased incidence of kidney problems. One example—*nephrolithiasis*, the formation of calculi, or kidney stones—is described in the Urinary Obstruction Clinical Note (p. 1034). Other age-related changes in the urinary system include the following:

- *A Decrease in the Number of Functional Nephrons.* The total number of kidney nephrons decreases by 30–40 percent between ages 25 and 85.

- *A Reduction in the GFR.* This reduction results from fewer glomeruli, cumulative damage to the filtration apparatus in the remaining glomeruli, and reduced renal blood flow.

- *A Reduced Sensitivity to ADH.* With age, the distal portions of the nephron and collecting system become less responsive to ADH. Water and sodium ions are reabsorbed at a reduced rate, and more sodium ions are lost in urine.

- *Problems with the Micturition Reflex.* Three factors are involved in such problems: (1) The sphincter muscles lose muscle tone and become less effective at voluntarily retaining urine. This leads to incontinence, often involving a slow leakage of urine. (2) The ability to control micturition can be lost due to a stroke, Alzheimer's disease, or other CNS problems affecting the cerebral cortex or hypothalamus. (3) In males, *urinary retention* may develop if the prostate gland enlarges and compresses the urethra, restricting the flow of urine.

CHECKPOINT

20. List four age-related changes in the urinary system.

21. Define nephrolithiasis.

22. Describe how incontinence may develop in an elderly person.

See the blue Answers tab at the back of the book.

26

SYSTEM INTEGRATOR

Body System ——→ Urinary System

Urinary System ——→ Body System

Integumentary
Sweat glands assist in elimination of water and solutes, especially sodium and chloride ions; keratinized epidermis prevents excessive fluid loss through skin surface; epidermis produces vitamin D$_3$, important for the renal production of calcitriol

Skeletal
Axial skeleton provides some protection for kidneys and ureters; pelvis protects urinary bladder and proximal portion of urethra

Muscular
Sphincter controls urination by closing urethral opening; muscle layers of trunk provide some protection for urinary organs

Nervous
Adjusts renal blood pressure; monitors distension of urinary bladder and controls urination

Endocrine
Aldosterone and ADH adjust rates of fluid and electrolyte reabsorption by kidneys

Cardiovascular
Delivers blood to glomerular capillaries, where filtration occurs; accepts fluids and solutes reabsorbed during urine production

Lymphatic
Provides adaptive immunity against urinary tract infections

Respiratory
Assists in the regulation of pH by eliminating carbon dioxide

Digestive
Absorbs water needed to excrete wastes by kidneys; absorbs ions needed to maintain normal body fluid concentrations; liver removes bilirubin

Integumentary
Kidneys eliminate nitrogenous wastes; maintain fluid, electrolyte, and acid–base balance of blood that nourishes the skin
Page 200

Skeletal
Conserves calcium and phosphate needed for bone growth
Page 311

Muscular
Removes waste products of protein metabolism; assists in regulation of calcium and phosphate concentrations
Page 406

Nervous
Kidneys eliminate nitrogenous wastes; maintain fluid, electrolyte, and acid–base balance of blood, which is critical for neural function
Page 584

Endocrine
Kidney cells release renin when local blood pressure decreases and erythropoietin (EPO) when renal oxygen levels decrease
Page 673

Cardiovascular
Releases renin to increase blood pressure and erythropoietin to enhance red blood cell production
Page 802

Lymphatic
Eliminates toxins and wastes generated by cellular activities; acid pH of urine provides innate immunity against urinary tract infections
Page 850

Respiratory
Assists in the elimination of carbon dioxide; provides bicarbonate buffers that assist in pH regulation
Page 600

Digestive
Excretes toxins absorbed by the digestive epithelium; excretes bilirubin and nitrogenous wastes from the liver; calcitriol production by kidneys aids calcium and phosphate absorption along intestinal tract
Page 955

The URINARY System

Urinary
For all systems, the urinary system excretes wastes and maintains normal body fluid pH and ion composition.

Reproductive
Page 1116

Figure 26–21 diagrams the functional relationships between the urinary system and the other body systems we have studied so far.

1036

26-8 ┃ The urinary system is one of several body systems involved in waste excretion

Learning Outcome Give examples of interactions between the urinary system and other organ systems studied so far.

The urinary system excretes wastes produced by other body systems, but it is not the only organ system involved in excretion. Indeed, the urinary, integumentary, respiratory, and digestive systems are together regarded as an anatomically diverse **excretory system** whose components perform all the excretory functions that affect the composition of body fluids:

- *Integumentary System.* Water losses and electrolyte losses in sensible perspiration can affect the volume and composition of the plasma. The effects are most apparent when losses are extreme, such as during peak sweat production. Small amounts of metabolic wastes, including urea, also are eliminated in perspiration.

- *Respiratory System.* The lungs remove the carbon dioxide generated by cells. Small amounts of other compounds, such as acetone and water, evaporate into the alveoli and are eliminated when you exhale.

- *Digestive System.* The liver excretes small amounts of metabolic waste products in bile. You lose a variable amount of water in feces.

These excretory activities have an impact on the composition of body fluids. The respiratory system, for example, removes carbon dioxide from the body. Note that the excretory functions of these systems are not regulated as closely as are those of the kidneys. Under normal circumstances, the effects of integumentary and digestive excretory activities are minor compared with those of the urinary system.

Figure 26–21 summarizes the functional relationships between the urinary system and other systems. We explore many of these relationships further in Chapter 27 when we consider major aspects of fluid, pH, and electrolyte balance.

CHECKPOINT

23. Identify the role the urinary system plays for all other body systems.

24. Name the body systems that make up the excretory system.

See the blue Answers tab at the back of the book.

Related Clinical Terms

azotemia: The condition characterized by excessive urea or other nitrogen-containing compounds in the blood.

cystocele: Condition that occurs when the supportive tissue between a woman's bladder and vaginal wall weakens, stretches, and allows the bladder to bulge into the vagina.

cystoscopy: Diagnostic procedure using an optical instrument called a cystoscope that is inserted through the urethra to visually examine the bladder and lower urinary tract, to collect urine samples, or to view the prostate gland.

enuresis: Involuntary urination, especially that of a child while asleep.

nephroptosis: Condition in which the kidney is displaced downward from its usual and normal position; also called a floating kidney.

nephrotic syndrome: A kidney disorder that causes one to excrete excessive protein in the urine.

nephrotoxin: A toxin that has a specific harmful effect on the kidney.

polycystic kidney disease: An inherited abnormality that affects the development and structure of kidney tubules.

shock-wave lithotripsy: A noninvasive technique used to pulverize kidney stones by passing high-pressure shock waves through a water-filled tub in which the patient sits.

urologist: Physician who specializes in functions and disorders of the urinary system.

Chapter Review

Study Outline

MasteringA&P®

Access more chapter study tools online in the MasteringA&P Study Area:

- Chapter Quizzes, Chapter Practice Test, Art-labeling Activities, Animations, MP3 Tutor Sessions, and Clinical Case Studies
- Practice Anatomy Lab PAL™
- Interactive Physiology iP®
- A&P Flix *A&PFlix*™
- PhysioEx PhysioEx™

26-1 ┃ The kidneys, ureters, urinary bladder, and urethra make up the urinary system, which has three primary functions p. 999

1. The three major functions of the **urinary system** are *excretion*, the removal of organic waste products from body fluids; *elimination*, the discharge of these waste products into the environment; and homeostatic regulation of the volume and solute concentration of blood plasma. Other homeostatic functions include regulating blood volume and pressure by adjusting the volume of water lost and releasing hormones; regulating plasma concentrations of ions; helping to stabilize

blood pH; conserving nutrients; assisting the liver in detoxifying poisons; and, during starvation, deaminating amino acids so that they can be catabolized by other tissues.

2. The urinary system includes the **kidneys**, the **ureters**, the **urinary bladder**, and the **urethra**. The kidneys produce **urine**, which is a fluid that contains water, ions, and soluble compounds. During **urination (micturition)**, urine is forced out of the body. *(Figure 26–1)*

26-2 ▌ Kidneys are highly vascular organs containing functional units called nephrons, which filter, reabsorb, and secrete p. 1000

3. The left kidney extends superiorly slightly more than the right kidney. Both kidneys and the adrenal gland that overlies each are retroperitoneal. *(Figures 26–1, 26–2, 26–3)*

4. The **hilum**, a medial indentation, provides entry for the *renal artery* and *renal nerves* and exit for the *renal vein* and the ureter. *(Figures 26–3, 26–4)*

5. The superficial portion of the kidney, the cortex, surrounds the **medulla**. The ureter communicates with the **renal pelvis**, a chamber that branches into two **major calyces**. Each major calyx is connected to four or five **minor calyces**, which enclose the **renal papillae**. *(Figure 26–4)*

6. The blood supply to the kidneys includes the **renal**, **segmental**, **interlobar**, **arcuate**, and **cortical radiate arteries**. *(Figure 26–5)*

7. The **renal nerves**, which innervate the kidneys and ureters, are dominated by sympathetic postganglionic fibers.

8. The **nephron** is the basic functional unit in the kidney. It consists of the **renal corpuscle** and **renal tubule**. The renal tubule is long and narrow and divided into the *proximal convoluted tubule*, the *nephron loop*, and the *distal convoluted tubule*. **Filtrate** is produced at the renal corpuscle. The nephron empties **tubular fluid** into the **collecting system**, consisting of **collecting ducts** and **papillary ducts**. *(Figures 26–6, 26–7)*

9. Nephrons produce filtrate; reabsorb organic nutrients, water, and ions; and secrete into the tubular fluid various waste products. *(Table 26–1)*

10. Roughly 85 percent of the nephrons are **cortical nephrons**, located in the renal cortex. **Juxtamedullary nephrons** are closer to the renal medulla, with their *nephron loops* extending deep into the **renal pyramids**. *(Figure 26–7)*

11. Blood travels from the efferent arteriole to the **peritubular capillaries** and the **vasa recta**. *(Figure 26–7)*

12. The renal tubule begins at the renal corpuscle, which includes a knot of intertwined capillaries called the **glomerulus**, surrounded by the **glomerular capsule**. Blood arrives at the glomerulus by the **afferent arteriole** and departs in the **efferent arteriole**. *(Figure 26–8)*

13. At the glomerulus, **podocytes** cover the **dense layer** of the capillaries that project into the **capsular space**. The **pedicels** of the podocytes are separated by narrow **filtration slits**. **Mesangial cells** lie between adjacent capillaries and control capillary diameter. *(Figure 26–8)*

14. The **proximal convoluted tubule (PCT)** actively reabsorbs nutrients, plasma proteins, and ions from the filtrate. These substances are released into the **peritubular fluid**, which surrounds the nephron. *(Figure 26–6)*

15. The **nephron loop**, also called the **loop of Henle**, includes a **descending limb** and an **ascending limb**. Each limb contains a **thick segment** and a **thin segment**. The ascending limb delivers fluid to the **distal convoluted tubule (DCT)**, which actively secretes ions, toxins, and drugs, and reabsorbs sodium ions from the tubular fluid. *(Figures 26–6, 26–7)*

26-3 ▌ Different segments of the nephron form urine by filtration, reabsorption, and secretion p. 1009

16. Urine production maintains homeostasis by regulating blood volume and composition. In the process, organic waste products—notably urea, creatinine, and uric acid—are excreted.

17. Urine formation involves **filtration**, **reabsorption**, and **secretion**.

18. Four types of *carrier-mediated transport (facilitated diffusion, active transport, cotransport,* and *countertransport)* are involved in modifying filtrate. The saturation limit of a carrier protein is its **transport maximum**, which determines the **renal threshold**—the plasma concentration at which various compounds will appear in urine. *(Table 26–2)*

19. The transport maximum determines the renal threshold for the reabsorption of substances in tubular fluid. *(Table 26–3)*

20. Most regions of the nephron perform a combination of reabsorption and secretion. *(Figure 26–9; Table 26–4)*

26-4 ▌ Hydrostatic and colloid osmotic pressures influence glomerular filtration pressure, which in turn affects the glomerular filtration rate p. 1013

21. **Glomerular filtration** occurs as fluids move across the wall of the glomerulus into the capsular space in response to the **glomerular hydrostatic pressure (GHP)**—the hydrostatic (blood) pressure in the glomerular capillaries. This movement is opposed by the **capsular hydrostatic pressure (CsHP)** and by the **blood colloid osmotic pressure (BCOP)**. The **net filtration pressure (NFP)** at the glomerulus is the difference between the blood pressure and the opposing capsular and osmotic pressures. *(Figure 26–10)*

22. The **glomerular filtration rate (GFR)** is the amount of filtrate produced in the kidneys each minute. Any factor that alters the filtration pressure acting across the glomerular capillaries will change the GFR and affect kidney function.

23. A drop in filtration pressures stimulates the **juxtaglomerular complex (JGC)** to release renin and erythropoietin. *(Figure 26–11)*

24. Sympathetic activation (1) produces a powerful vasoconstriction of the afferent arterioles, decreasing the GFR and slowing the production of filtrate; (2) alters the GFR by changing the regional pattern of blood circulation; and (3) stimulates the release of renin by the juxtaglomerular complex. *(Figure 26–11)*

26-5 ▌ Countercurrent multiplication, antidiuretic hormone, and aldosterone affect reabsorption and secretion p. 1018

25. Glomerular filtration produces a filtrate with a composition similar to blood plasma, but with few, if any, plasma proteins.

26. The cells of the PCT normally reabsorb sodium and other ions, water, and almost all the organic nutrients that enter the filtrate. These cells also secrete various substances into the tubular fluid. *(Figure 26–12)*

27. Water and ions are reclaimed from tubular fluid by the nephron loop. A concentration gradient in the renal medulla encourages the osmotic flow of water out of the tubular fluid. The **countercurrent multiplication** between the ascending and descending limbs of the nephron loop helps create the osmotic gradient in the medulla. As water is lost by osmosis and the volume of tubular fluid decreases, the urea concentration increases. *(Figure 26–13)*

28. The DCT performs final adjustments by actively secreting or absorbing materials. Sodium ions are actively absorbed, in exchange for potassium or hydrogen ions discharged into

tubular fluid. Aldosterone secretion increases the rate of sodium reabsorption and potassium loss. *(Figure 26–14)*

29. The amount of water and solutes in the tubular fluid of the collecting ducts is further regulated by aldosterone and ADH secretions. *(Figure 26–14)*

30. Urine volume and osmotic concentration are regulated by controlling water reabsorption. Precise control over this occurs by *facultative water reabsorption. (Figure 26–15)*

31. Normally, the removal of solutes and water by the vasa recta precisely balances the rates of reabsorption and osmosis in the renal medulla.

32. More than 99 percent of the filtrate produced each day is reabsorbed before reaching the renal pelvis. *(Table 26–5)*

33. *Urinalysis* is the chemical and physical analysis of a urine sample. *(Table 26–6)*

34. Each segment of the nephron and collecting system contributes to the production of concentrated urine. *(Spotlight Figure 26–16; Table 26–4)*

26-6 ▌ Urine is transported by the ureters, stored in the bladder, and eliminated through the urethra, aided by the micturition reflex p. 1030

35. Urine production ends when tubular fluid enters the renal pelvis. The rest of the urinary system transports, stores, and eliminates urine. *(Figure 26–17)*

36. The ureters extend from the renal pelvis to the urinary bladder. Peristaltic contractions by smooth muscles move the urine along the tract. *(Figures 26–18, 26–19)*

37. The urinary bladder is stabilized by the **middle umbilical ligament** and the **lateral umbilical ligaments**. Internal features include the **trigone**, the **neck**, and the **internal**

urethral sphincter. The mucosal lining contains prominent **rugae** (folds). Contraction of the **detrusor muscle** compresses the urinary bladder and expels urine into the urethra. *(Figures 26–18, 26–19)*

38. In both sexes, where the urethra passes through the *urogenital diaphragm*, a circular band of skeletal muscles forms the **external urethral sphincter**, which is under voluntary control. *(Figure 26–18)*

39. Urination is coordinated by the **micturition reflex**, which is initiated by stretch receptors in the wall of the urinary bladder. Voluntary urination involves coupling this reflex with the voluntary relaxation of the external urethral sphincter, which allows the opening of the **internal urethral sphincter**. *(Figure 26–20)*

26-7 ▌ Age-related changes affect kidney function and the micturition reflex p. 1035

40. Aging is generally associated with increased kidney problems. Age-related changes in the urinary system include (1) declining numbers of functional nephrons, (2) reduced GFR, (3) reduced sensitivity to ADH, and (4) problems with the micturition reflex. (**Urinary retention** may develop in men whose prostate gland is enlarged.)

26-8 ▌ The urinary system is one of several body systems involved in waste excretion p. 1037

41. The urinary system is the major component of an anatomically diverse **excretory system** that includes the integumentary system, the respiratory system, and the digestive system. *(Figure 26–21)*

Review Questions

See the blue Answers tab at the back of the book.

LEVEL 1 Reviewing Facts and Terms

1. Identify the structures of the kidney in the following diagram.

(a) _____	(b) _____
(c) _____	(d) _____
(e) _____	(f) _____
(g) _____	(h) _____
(i) _____	(j) _____
(k) _____	(l) _____
(m) _____	

2. The glomeruli of the nephron are located in the **(a)** renal lobe, **(b)** renal pyramids, **(c)** cortex, **(d)** medulla.

3. The process of urine formation involves all of the following, *except* **(a)** filtration of plasma, **(b)** reabsorption of water, **(c)** reabsorption of certain solutes, **(d)** secretion of wastes, **(e)** secretion of excess lipoprotein and glucose molecules.

4. The glomerular filtration rate is regulated by all of the following, *except* **(a)** autoregulation, **(b)** sympathetic neural control, **(c)** cardiac output, **(d)** angiotensin II, **(e)** the hormone ADH.

5. The distal convoluted tubule is an important site for **(a)** active secretion of ions, **(b)** active secretion of acids and other materials, **(c)** selective reabsorption of sodium ions from the tubular fluid, **(d)** all of these.

6. Reabsorption in the proximal convoluted tubule uses **(a)** facilitated transport, **(b)** cotransport, **(c)** osmosis, **(d)** all of the above.

7. List three age-related changes that reduce kidney function.

8. List the structural components of the renal corpuscle.

9. Trace the pathway of the protein-free filtrate from where it is produced in the renal corpuscle until it drains into the renal pelvis in the form of urine. (Use arrows to indicate the direction of flow.)

10. Name the segments of the nephron distal to the renal corpuscle, and state the functions of each.

11. What is the function of the juxtaglomerular complex?

12. Using arrows, trace a drop of blood from its entry into the renal artery until its exit by a renal vein.

13. Name and define the three distinct processes involved in the production of urine.

26

14. Define what is the glomerular filtration rate. List three forces that determine its value.
15. Which structures of the urinary system are responsible for the transport, storage, and elimination of urine?

LEVEL 2 Reviewing Concepts

16. When the renal threshold for a substance exceeds its tubular maximum (a) more of the substance will be filtered, (b) more of the substance will be reabsorbed, (c) more of the substance will be secreted, (d) the amount of the substance that exceeds the tubular maximum will be found in the urine, (e) both a and d occur.
17. Sympathetic activation of nerve fibers in the nephron causes (a) the regulation of glomerular blood flow and pressure, (b) the stimulation of renin release from the juxtaglomerular complex, (c) the direct stimulation of water and Na$^+$ reabsorption, (d) all of these.
18. Sodium reabsorption in the DCT and in the cortical portion of the collecting system is accelerated by the secretion of (a) ADH, (b) renin, (c) aldosterone, (d) erythropoietin.
19. When ADH levels rise, (a) the amount of water reabsorbed increases, (b) the DCT becomes impermeable to water, (c) the amount of water reabsorbed decreases, (d) sodium ions are exchanged for potassium ions.
20. The control of blood pH by the kidneys during acidosis involves (a) the secretion of hydrogen ions and reabsorption of bicarbonate ions from the tubular fluid, (b) a decrease in the amount of water reabsorbed, (c) hydrogen ion reabsorption and bicarbonate ion loss, (d) potassium ion secretion.
21. Describe the role of antidiuretic hormone (ADH) in the formation of concentrated urine.
22. Describe how the urinary system supports cardiovascular function.

23. Describe the action of atrial and brain natriuretic peptides in the reduction of sodium reuptake in the distal convoluted tubule.
24. Describe two functions of countercurrent multiplication in the kidney.
25. Describe the micturition reflex.

LEVEL 3 Critical Thinking and Clinical Applications

26. In a normal kidney, which of the following conditions would cause an increase in the glomerular filtration rate (GFR)? (a) constriction of the afferent arteriole, (b) a decrease in the pressure of the glomerulus, (c) an increase in the capsular hydrostatic pressure, (d) a decrease in the concentration of plasma proteins in the blood, (e) a decrease in the net glomerular filtration process.
27. John and Terry are in arid mountainous terrain but have just consumed the remainder of their water. They calculate that it will take approximately seven days for them to reach the nearest supply. John thinks drinking their own urine will help them stay alive, whereas Tony thinks it may kill them. Who is right and why?
28. Why might people with hypertension (high blood pressure) be prescribed angiotensin-converting enzyme inhibitors (ACE inhibitors)?
29. Sylvia has kidney stones which have become lodged in the renal pelvis. Why might this lead to a reduction in GFR?
30. *Mannitol* is a sugar that is filtered, but not reabsorbed, by the kidneys. What effect would drinking a solution of mannitol have on the volume of urine produced?
31. The drug *Diamox* is sometimes used to treat mountain sickness. Diamox inhibits the action of carbonic anhydrase in the proximal convoluted tubule. Polyuria (the elimination of an unusually large volume of urine) is a side effect associated with the medication. Why does polyuria occur?

CLINICAL CASE Wrap-Up

A Case of "Hidden" Bleeding

Normally, when a kidney is injured, the urine appears very bloody (hematuria). Mike has made very little urine since his injury and massive blood loss, but the urine in his catheter is clear. This reduces the initial suspicion of a renal injury.

However, when the surgeon opens the left retroperitoneal space, he finds another large collection of free blood. When the surgeon clears this blood away, the source of Mike's continued bleeding becomes apparent. His left renal artery, renal vein, and ureter have been completely torn away from the kidney at the hilum. The renal artery is briskly pumping what circulatory blood volume is left into the retroperitoneal space, where it is walled off from the abdominal space by the parietal peritoneum. Because the left ureter is completely torn from the damaged kidney, there is no visible blood in the urine.

This kidney is too damaged to heal. The surgeon quickly performs a left nephrectomy (kidney removal). If Mike survives this trauma, he will have to live with a solitary kidney. With the bleeding stopped, including the "hidden" bleeding, the trauma team is finally able to sustain an adequate circulating blood volume.

1. How could you tell from Mike's x-rays that there is a high probability of renal injury?
2. What do you think will happen to the glomerular filtration rate (GFR) of Mike's remaining right kidney if he survives this trauma?

See the blue Answers tab at the back of the book.

27

Fluid, Electrolyte, and Acid–Base Balance

Learning Outcomes

These Learning Outcomes correspond by number to this chapter's sections and indicate what you should be able to do after completing the chapter.

27-1 ❙ Explain what is meant by the terms fluid balance, electrolyte balance, and acid–base balance, and discuss their importance for homeostasis.

27-2 ❙ Compare the composition of intracellular and extracellular fluids, explain the basic concepts involved in the regulation of fluids and electrolytes, and identify the hormones that play important roles in fluid and electrolyte regulation.

27-3 ❙ Describe the movement of fluid within the ECF, between the ECF and the ICF, and between the ECF and the environment.

27-4 ❙ Discuss the mechanisms by which sodium, potassium, calcium, magnesium, phosphate, and chloride ion concentrations are regulated to maintain electrolyte balance.

27-5 ❙ Explain the buffering systems that balance the pH of the intracellular and extracellular fluids, and describe the compensatory mechanisms involved in maintaining acid–base balance.

27-6 ❙ Identify the most frequent disturbances of acid–base balance, and explain how the body responds when the pH of body fluids varies outside normal limits.

27-7 ❙ Describe the effects of aging on fluid, electrolyte, and acid–base balance.

SPOTLIGHT
- The Diagnosis of Acid–Base Disorders p. 1070

CLINICAL CASE
- When Treatment Makes You Worse p. 1042

Clinical Notes
- Water and Weight Loss p. 1049
- Athletes and Salt Loss p. 1053

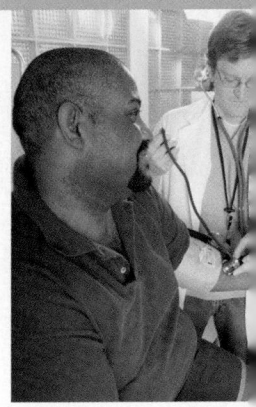

CLINICAL CASE

When Treatment Makes You Worse

Marquis is a 43-year-old office worker who is slightly overweight. At his annual checkup, his blood pressure measures 150/96, which is considered moderately high. Marquis feels fine, but his doctor tells him it is time to treat his hypertension. Marquis's doctor advises him to improve his diet and get more exercise. He also prescribes a new medication, hydrochlorothiazide, or HCTZ, which he instructs him to take every morning. HCTZ is a diuretic that is commonly used to lower blood pressure.

Marquis begins taking HCTZ and continues to feel well, although he notices the medication makes him need to urinate more frequently. A few weeks later he and a group of coworkers eat some bad potato salad at the office picnic and come down with food poisoning. After three days of diarrhea, Marquis feels absolutely horrible. He is so weak he is barely able to get out of bed. His muscles ache and his legs and feet cramp a lot. Worst of all, he feels as if there is a frog jumping around in his chest. He is often dizzy when this happens.

After a week in bed, Marquis feels even weaker. He goes to see his doctor, who sends him to the lab for some tests. "I know why you feel so terrible," says his doctor when he sees the lab results. **What is happening to Marquis? To find out, turn to Clinical Case Wrap-Up on p. 1075.**

An Introduction to Fluid, Electrolyte, and Acid–Base Balance

The next time you see a small pond, think about the fish in it. They live their entire lives totally dependent on the quality of that isolated environment. Severe water pollution will kill them, but even subtle changes can have equally grave effects. Changes in the volume of the pond, for example, can be quite important. If evaporation removes too much water, the fish become overcrowded. The oxygen and food supplies will run out, and the fish will suffocate or starve. The ionic concentration of the water is also crucial. Most of the fish in a freshwater pond will die if the water becomes too salty. Fish in a saltwater pond will die if their environment becomes too dilute. The pH of the pond water, too, is a vital factor. This is another reason that acid rain is such a problem.

Your cells live in a pond whose shores are the exposed surfaces of your skin. Most of your body weight is water. Water makes up about 99 percent of the volume of the fluid outside cells, and it is also an essential ingredient of cytoplasm. All of a cell's operations rely on water as a diffusion medium for the distribution of gases, nutrients, and wastes. If the water content of the body changes, cellular activities are jeopardized. For example, when the water content reaches very low levels, proteins denature, enzymes cease functioning, and cells die. In this chapter we discuss the homeostatic mechanisms that regulate ion concentrations, volume, and pH in the fluid surrounding cells.

27-1 | Fluid balance, electrolyte balance, and acid–base balance are interrelated and essential to homeostasis

Learning Outcome Explain what is meant by the terms fluid balance, electrolyte balance, and acid–base balance, and discuss their importance for homeostasis.

To survive, we must maintain a normal volume and composition of both the **extracellular fluid** or **ECF** (the interstitial fluid, plasma, and other body fluids) and the **intracellular fluid** or **ICF** (the cytosol). The ionic concentrations and pH (hydrogen ion concentration) of these fluids are as important as their absolute quantities. If concentrations of calcium or potassium ions in the ECF become too high, cardiac arrhythmias develop and death can result. A pH outside the normal range can also lead to a variety of serious problems. Low pH—acidity—is especially dangerous, because hydrogen ions break chemical bonds, change the shapes of complex molecules, disrupt plasma membranes, and impair tissue functions.

TIPS & TRICKS

The "p" in pH refers to power. Hence, pH refers to the **p**ower of **H**ydrogen.

In this chapter, we will explore the dynamics of exchange among the various body fluids, and between the body and the external environment. Stabilizing the volumes, solute concentrations, and pH of the ECF and the ICF involves three interrelated processes:

1. *Fluid Balance.* You are in **fluid balance** when the amount of water you gain each day is equal to the amount you lose to the environment. The maintenance of normal fluid balance involves regulating the content and distribution of body water in the ECF and the ICF. The digestive system is the main source of water gains. Metabolic activity generates a small amount of additional water. The urinary system is the primary route for water loss under normal conditions, but as we saw in Chapter 25, sweating can become important when body temperature is elevated. ⤴ p. 991 Although cells and tissues cannot transport water, they can transport ions and create concentration gradients that are then eliminated by osmosis.

2. *Electrolyte Balance.* **Electrolytes** are ions released when inorganic compounds dissociate. They are so named because they can conduct an electrical current in a solution. ⤴ p. 67 Each day, your body fluids gain electrolytes from the food and drink you consume. Your body fluids also lose electrolytes in urine, sweat, and feces. For each ion, daily gains must balance daily losses. For example, if you lose 500 mg of Na^+ in urine and sweat, you need to gain 500 mg of Na^+ from food and drink to remain in sodium balance. If the gains and losses for every electrolyte are in balance, you are said to be in **electrolyte balance**. Electrolyte balance primarily involves balancing the rates of absorption across the digestive tract with rates of loss by the kidneys, although losses at sweat glands and other sites can play a secondary role.

3. *Acid–Base Balance.* You are in **acid–base balance** when the production of hydrogen ions in your body is precisely offset by their loss. When acid–base balance exists, the pH of body fluids remains within normal limits. ⤴ p. 67 Preventing a decrease in pH is the primary problem, because your body generates a variety of acids during normal metabolic operations. The kidneys play a major role by secreting hydrogen ions into the urine and generating buffers that enter the bloodstream. Such secretion takes place primarily in the distal segments of the distal convoluted tubule (DCT) and along the collecting system. ⤴ p. 1022 The lungs also play a key role by eliminating carbon dioxide.

Much of the information in this chapter should be familiar to you from discussions in earlier chapters about aspects of fluid, electrolyte, or acid–base balance that affect specific systems. In this chapter we provide an overview that integrates those discussions to highlight important functional patterns.

This chapter has wide-ranging clinical importance: Steps to restore normal fluid, electrolyte, and acid–base balances must be part of the treatment of any serious illness affecting the nervous, cardiovascular, respiratory, urinary, or digestive system. Because this chapter builds on information in earlier chapters, we have included many references to relevant discussions and figures that can provide a quick review.

CHECKPOINT

1. Identify the three interrelated processes essential to stabilizing body fluid volumes.

2. List the components of extracellular fluid (ECF) and intracellular fluid (ICF).

See the blue Answers tab at the back of the book.

27-2 ▌ The ECF and ICF make up the fluid compartments, which also contain cations and anions

Learning Outcome Compare the composition of intracellular and extracellular fluids, explain the basic concepts involved in the regulation of fluids and electrolytes, and identify the hormones that play important roles in fluid and electrolyte regulation.

Figure 27–1a presents an overview of the body makeup of a 70-kg (154-lb) person with a minimum of body fat. The distribution is based on overall average values for males and females ages 18–40 years. Water makes up about 60 percent of the total body weight of an adult male, and 50 percent of that of an adult female (**Figure 27–1b**). This difference between the sexes reflects the proportionately larger mass of adipose tissue in adult females, and the greater average muscle mass in adult males. (Adipose tissue is only 10 percent water, whereas skeletal muscle is 75 percent water.)

In both sexes, intracellular fluid contains a greater proportion of total body water than does extracellular fluid. Exchange between the ICF and the ECF takes place across plasma membranes by osmosis, diffusion, and carrier-mediated transport. (To review the mechanisms involved, see **Spotlight Figure 3–22**, p. 124.)

The ECF and the ICF

The largest subdivisions of the ECF are the interstitial fluid of peripheral tissues and the plasma of circulating blood (**Figure 27–1a**). Minor components of the ECF include lymph, cerebrospinal fluid (CSF), synovial fluid, serous fluids (pleural, pericardial, and peritoneal fluids), aqueous humor, perilymph, and endolymph. More precise measurements of total body water provide additional information on differences between the sexes in the distribution of body water (**Figure 27–1b**). The greatest variation is in the ICF, as a result of differences in the intracellular water content of fat versus muscle. Less striking differences occur in the ECF values, due to variations in the interstitial fluid volume of various tissues and the larger blood volume in males versus females.

27

Figure 27–1   The Composition of the Human Body.

a The body composition (by weight, averaged for both sexes) and major body fluid compartments of a 70-kg (154-lb) person. For technical reasons, it is extremely difficult to determine the precise size of any of these compartments; estimates of their relative sizes vary widely.

b A comparison of the body compositions of adult males and females, ages 18–40 years.

In clinical situations, it is customary to estimate that two-thirds of the total body water is in the ICF and one-third in the ECF. This ratio underestimates the real volume of the ECF, but that underestimation is appropriate because portions of the ECF—including the water in bone, in many dense connective tissues, and in many of the minor ECF components—are relatively isolated. Exchange between these fluid volumes and the rest of the ECF takes place more slowly than does exchange between plasma and other interstitial fluids. For this reason, they can be safely ignored in many cases. Clinical attention usually focuses on the rapid fluid and solute movements that accompany the administration of blood, plasma, or saline solutions to counteract blood loss or dehydration.

Exchange among the subdivisions of the ECF occurs primarily across the endothelial lining of capillaries. Fluid may also travel from the interstitial spaces to plasma through lymphatic vessels that drain into the venous system. ⊃ p. 810 The kinds and quantities of dissolved electrolytes, proteins, nutrients, and wastes in the ECF vary regionally. (For a chemical analysis of the composition of ECF compartments, see the Appendix.) Still, the variations among the compartments of the ECF seem minor compared with the major differences between the ECF and the ICF.

The ECF and ICF are called **fluid compartments**, because they commonly behave as separate sections. The presence of a plasma membrane and active transport at the membrane surface

Figure 27–2   Cations and Anions in Body Fluids. Notice the differences in cation and anion concentrations in the various body fluid compartments.

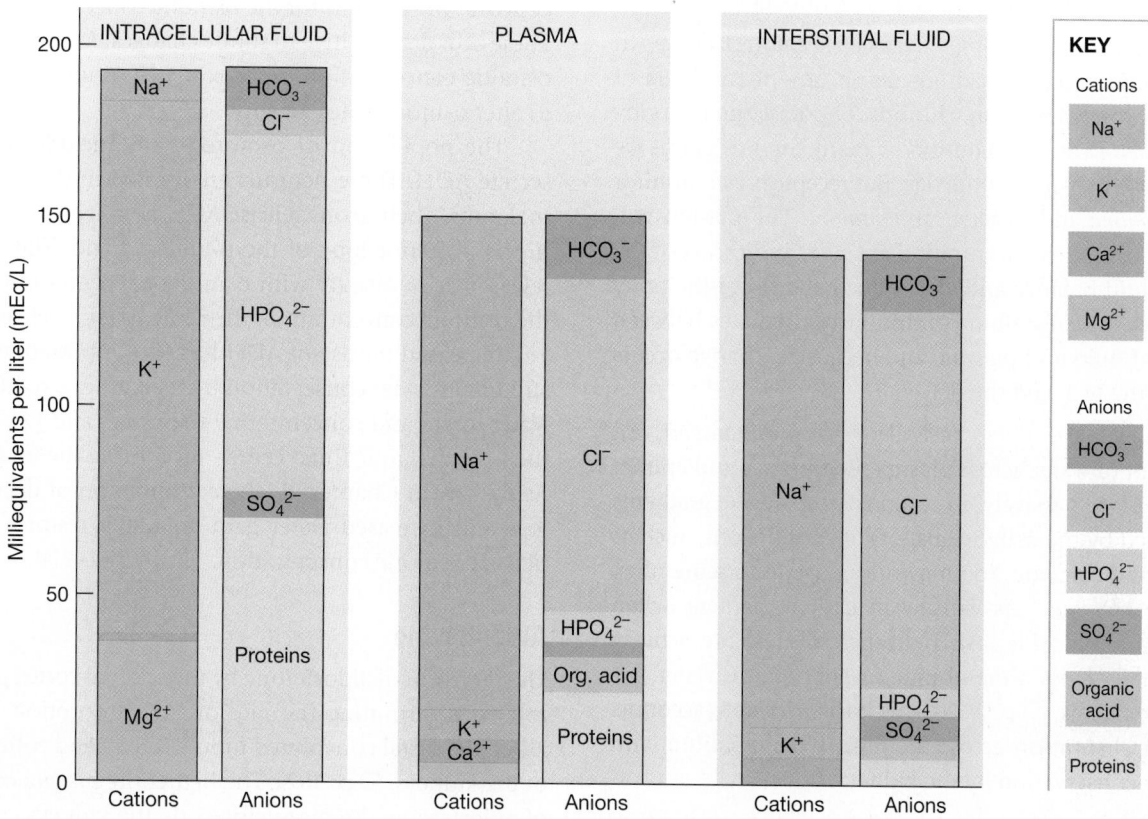

enable cells to maintain internal environments with a composition that differs from their surroundings. The principal ions in the ECF are sodium, chloride, and bicarbonate (HCO_3^-). The ICF contains an abundance of potassium, magnesium, and phosphate ions (HPO_4^{2-}), plus large amounts of negatively charged proteins. **Figure 27–2** compares the ICF with the two major subdivisions of the ECF (plasma and interstitial fluid).

If the plasma membrane were freely permeable, diffusion would continue until these ions were evenly distributed across the membrane. But it does not, because plasma membranes are selectively permeable: Ions can enter or leave the cell only by specific membrane channels. Also, carrier mechanisms move specific ions into or out of the cell.

Despite the differences in the concentration of specific substances, the osmotic concentrations of the ICF and ECF are identical. Osmosis eliminates minor differences in concentration almost at once, because most plasma membranes are freely permeable to water. (The only noteworthy exceptions are the apical surfaces of epithelial cells along the ascending limb of the nephron loop, the distal convoluted tubule, and the collecting system.) The regulation of fluid balance and electrolyte balance are tightly intertwined because changes in solute concentrations lead to immediate changes in water distribution.

Physiologists and clinicians pay particular attention to ionic distributions across membranes and to the electrolyte composition of body fluids. The Appendix at the end of the book reports normal values in the units most often used in clinical reports.

Basic Concepts in the Regulation of Fluids and Electrolytes

Four basic principles are key to understanding fluid balance and electrolyte balance:

1. *All the homeostatic mechanisms that monitor and adjust the composition of body fluids respond to changes in the ECF, not in the ICF.* Receptors monitoring the composition of two key components of the ECF—plasma and cerebrospinal fluid—detect significant changes in their composition or volume and trigger appropriate neural and endocrine responses. This arrangement makes functional sense, because a change in one ECF component will spread rapidly throughout the extracellular compartment and affect all the body's cells. In contrast, the ICF is contained within trillions of individual cells that are physically and chemically isolated from one another by their plasma membranes. For

this reason, changes in the ICF in one cell have no direct effect on the composition of the ICF in distant cells and tissues, unless those changes also affect the ECF.

2. *No receptors directly monitor fluid or electrolyte balance.* In other words, receptors cannot detect how many liters of water or grams of sodium, chloride, or potassium the body contains. Nor can they count how many liters or grams we gain or lose throughout the day. But receptors *can* monitor *plasma volume* and *osmotic concentration.* The plasma volume and osmotic concentration are good indicators of the state of fluid balance and electrolyte balance for the body as a whole, because fluid continuously circulates between interstitial fluid and plasma, and because exchange occurs between the ECF and the ICF.

3. *Cells cannot move water molecules by active transport.* All movement of water across plasma membranes and epithelia takes place passively, in response to osmotic gradients established by the active transport of specific ions, such as sodium and chloride. You may find it useful to remember, *"water follows salt."* As we saw in earlier chapters, when sodium and chloride ions (or other solutes) are actively transported across a membrane or epithelium, water follows by osmosis. ⤺ p. 1020 This basic principle accounts for water absorption across the digestive epithelium, and for water conservation by the kidneys.

4. *The body's content of water or electrolytes will increase if dietary gains exceed losses to the environment, and will decrease if losses exceed gains.* This basic rule is important when you consider the mechanics of fluid balance and electrolyte balance. Homeostatic adjustments affect the balance between urinary excretion and dietary absorption. As we saw in Chapter 26, circulating hormones regulate renal function. These hormones can also produce complementary changes in behavior. For example, the combination of angiotensin II and aldosterone can give you a sensation of thirst—which stimulates you to drink fluids—and a taste for heavily salted foods.

An Overview of the Primary Regulatory Hormones

Three hormones mediate physiological adjustments to fluid balance and electrolyte balance. They are (1) *antidiuretic hormone (ADH)*, (2) *aldosterone*, and (3) the *natriuretic peptides (ANP and BNP)*. We introduced and discussed these hormones in earlier chapters. We summarize their effects next. For a more detailed review, refer to the appropriate sections of Chapters 18, 21, and 26. The interactions among these hormones are illustrated in **Figures 18–19b, 21–15, 21–16,** and **26–11** (pp. 667, 774, 777, 1017).

Antidiuretic Hormone

The hypothalamus contains special cells known as **osmoreceptors**, which monitor the osmotic concentration of the ECF. These cells are sensitive to subtle changes: A 2 percent change in osmotic concentration (approximately 6 mOsm/L) is enough to alter osmoreceptor activity.

The population of osmoreceptors includes neurons that secrete ADH. These neurons are located in the anterior hypothalamus. Their axons release ADH near fenestrated capillaries in the posterior lobe of the pituitary gland. The rate of ADH release varies directly with osmotic concentration: The higher the osmotic concentration, the more ADH is released.

Increased release of ADH has two important effects: (1) It stimulates water conservation by the kidneys, reducing urinary water losses and concentrating the urine; and (2) it stimulates the hypothalamic thirst center, promoting the intake of fluids. As we saw in Chapter 21, the combination of decreased water loss and increased water gain gradually restores the normal plasma osmotic concentration. ⤺ pp. 772–774

Aldosterone

The secretion of aldosterone by the adrenal cortex plays a major role in determining the rate of Na^+ absorption and K^+ loss along the distal convoluted tubule (DCT) and collecting system of the kidneys. ⤺ p. 1022 The higher the plasma concentration of aldosterone, the more efficiently the kidneys conserve Na^+. Because "water follows salt," the conservation of Na^+ also increases water retention. As Na^+ is reabsorbed, Cl^- follows (see **Figure 26–14a,** p. 1024), and water follows by osmosis as sodium and chloride ions move out of the tubular fluid. Aldosterone also increases the sensitivity of salt receptors on the tongue. This effect may increase your consumption of salty foods.

Aldosterone is secreted in response to increasing K^+ or decreasing Na^+ levels in the blood reaching the adrenal cortex, or in response to the activation of the renin–angiotensin-aldosterone system. As we saw in earlier chapters, renin release occurs in response to any of three changes. They are (1) a decrease in plasma volume or blood pressure at the juxtaglomerular complex of the nephron; (2) a decrease in the osmotic concentration of tubular fluid at the DCT; or, as we will soon see, (3) decreasing Na^+ or increasing K^+ concentrations in the renal circulation.

Natriuretic Peptides

The natriuretic (*natrium,* sodium; *-uretic,* urine) peptides, atrial natriuretic peptide (ANP) and brain natriuretic peptide (BNP), are released by cardiac muscle cells in response to abnormal stretching of the heart walls. Increased blood pressure or increased blood volume causes this stretching. Among their other effects, these peptides reduce thirst and block the release of ADH and aldosterone that might otherwise lead to water and salt conservation. The resulting diuresis (fluid loss by the kidneys) decreases both blood pressure and plasma volume, eliminating the stretching.

The Interplay between Fluid Balance and Electrolyte Balance

At first glance, it can be very difficult to distinguish between water balance and electrolyte balance. For example, when you lose body water, plasma volume decreases and electrolyte concentrations increase. Conversely, when you gain or lose excess electrolytes, you also gain or lose water due to osmosis.

However, it is often useful to consider fluid balance and electrolyte balance separately because the regulatory mechanisms involved are quite different. This distinction is absolutely vital in a clinical setting, where problems with fluid balance and electrolyte balance must be identified and corrected promptly.

> **CHECKPOINT**
>
> 3. **Name three hormones that play a major role in adjusting fluid and electrolyte balance in the body.**
>
> 4. **What effect would drinking a half-gallon of distilled water have on ADH levels?**
>
> See the blue Answers tab at the back of the book.

27-3 ▌ Hydrostatic and osmotic pressures regulate the movement of water and electrolytes to maintain fluid balance

Learning Outcome Describe the movement of fluid within the ECF, between the ECF and the ICF, and between the ECF and the environment.

Water circulates freely within the ECF compartment. At capillary beds throughout the body, hydrostatic pressure forces water out of plasma and into interstitial spaces. Some of that water is then reabsorbed along the distal portion of the capillary bed, and the rest enters lymphatic vessels for transport to the venous circulation. Fluid also moves continuously among the minor components of the ECF:

1. Water moves back and forth across the mesothelial surfaces that line the peritoneal, pleural, and pericardial cavities, and through the synovial membranes that line joint capsules. The flow rate is significant. For example, about 7 liters (1.8 gal) of peritoneal fluid are produced and reabsorbed each day. However, the actual volume present at any time in the peritoneal cavity is very small—less than 35 mL (1.2 oz).

2. Water also moves between blood and cerebrospinal fluid (CSF), between the aqueous humor and vitreous humor of the eye, and between the perilymph and endolymph of the internal ear. The volumes involved in these water movements are very small, and the volume and composition of the fluids are closely regulated. For those reasons, we will largely ignore them in the discussion that follows.

Water movement can also take place between the ECF and the ICF, but under normal circumstances the two are in osmotic equilibrium. For this reason, no large-scale circulation occurs between the two compartments. (A small amount of water moves from the ICF to the ECF each day, as the result of mitochondrial water generation. We will consider this topic in a separate section.)

We cannot easily determine the body's water content. However, the concentration of Na^+, the most abundant ion in the ECF, provides useful clues to the state of water balance. When the body's water content increases, the Na^+ concentration of the ECF becomes abnormally low. And when the body's water content decreases, the Na^+ concentration becomes abnormally high.

Fluid Movement within the ECF

In our discussion of capillary dynamics in Chapter 21, we considered the basic principles that determine fluid movement among the divisions of the ECF. ⊃ p. 765 Recall that the exchange between plasma and interstitial fluid, by far the largest components of the ECF, is due to the relationship between the net hydrostatic pressure, which tends to push water out of the plasma and into the interstitial fluid, and the net colloid osmotic pressure, which tends to draw water out of the interstitial fluid and into the plasma. The interaction between these opposing forces is diagrammed in **Figure 21–11** (p. 766). It results in the continuous filtration of fluid from the capillaries into the interstitial fluid. This volume of fluid is then redistributed: After passing through the channels of the lymphatic system, the fluid returns to the venous system. At any moment, interstitial fluid and minor fluid compartments contain approximately 80 percent of the ECF volume, and plasma contains the other 20 percent.

Any factor that affects the net hydrostatic pressure or the net colloid osmotic pressure will alter the distribution of fluid within the ECF. The movement of abnormal amounts of water from plasma into interstitial fluid is called *edema*. Pulmonary edema, for example, can result from an increase in the blood pressure in pulmonary capillaries. Generalized edema can result from a decrease in blood colloid osmotic pressure, as in advanced starvation, when plasma protein concentrations decrease. Localized edema can result from damage to capillary walls (as in bruising), the constriction of regional venous circulation, or a blockage of the lymphatic drainage (as in *lymphedema*, introduced in Chapter 22). ⊃ p. 812

27

Fluid Gains and Losses

Figure 27–3 and **Table 27–1** indicate the major factors involved in fluid balance and highlight the routes of fluid exchange with the environment:

- *Water Losses.* You lose about 2500 mL of water each day through urine, feces, and *insensible perspiration*—the gradual movement of water across the epithelia of the skin and lungs. The losses due to *sensible perspiration*—the secretory activities of the sweat glands—vary with physical activity. Sensible perspiration can cause significant water deficits, with maximum perspiration rates reaching 4 liters per hour. ⊃ p. 990 Fever can also increase water losses. For each degree that body temperature rises above normal, daily insensible water losses increase by 200 mL. The advice "Drink plenty of fluids" for anyone who is sick has a definite physiological basis.

- *Water Gains.* A water gain of about 2500 mL/day is required to balance your average water losses. This value amounts to about 40 mL/kg of body weight per day. You obtain water through eating (1000 mL), drinking (1200 mL), and *metabolic generation* (300 mL).

Metabolic generation of water is the production of water within cells, primarily as a result of oxidative phosphorylation

Table 27–1	Water Balance
Source	**Daily Input (mL)**
Water content of food	1000
Water consumed as liquid	1200
Metabolic water produced during catabolism	300
Total	**2500**
Method of Elimination	**Daily Output (mL)**
Urination	1200
Evaporation at skin	750
Evaporation at lungs	400
Loss in feces	150
Total	**2500**

in mitochondria. (In Chapter 25 we described the synthesis of water at the end of the electron transport system. ⊃ p. 969) When cells break down 1 g of lipid, 1.7 mL of water is generated. Breaking down proteins or carbohydrates yields much lower amounts (0.41 mL/g and 0.55 mL/g, respectively).

A typical diet in the United States contains 46 percent carbohydrates, 40 percent lipids, and 14 percent protein. Such a diet produces approximately 300 mL of water per day, about 12 percent of your average daily requirement.

Fluid Shifts

A rapid water movement between the ECF and the ICF in response to an osmotic gradient is called a **fluid shift**. Fluid shifts occur quickly in response to changes in the osmotic concentration of the ECF and reach equilibrium within minutes to hours (**Figure 27–4**).

Figure 27–3 **Fluid Gains and Losses.** Fluid movements that maintain fluid balance in a normal person. The volumes are drawn to scale; the ICF is about twice as large as the ECF.

Water absorbed across digestive epithelium (2200 mL)

Water vapor lost at skin and lungs (1150 mL)

Water lost in feces (150 mL)

Water secreted by sweat glands (variable)

Metabolic water (300 mL)

ICF

ECF

Plasma membranes

Water lost in urine (1200 mL)

Figure 27–4 **Fluid Shifts between the ICF and ECF.**

Dehydration and drinking a glass of tap water are two examples that demonstrate the dynamic exchange of water between the ECF and ICF, as we will now see.

Allocation of Water Losses

Dehydration, or *water depletion*, develops when water losses outpace water gains. When you lose water but retain electrolytes, the osmotic concentration of the ECF increases. Osmosis then moves water out of the ICF and into the ECF until the two solutions are again isotonic. At that point, both the ECF and ICF are somewhat more concentrated than normal, and both volumes are lower than they were before the fluid loss. Because the ICF has about twice the functional volume of the ECF, the net change in the ICF is small. However, if the fluid imbalance continues unchecked, the loss of body water will produce severe thirst, dryness, and wrinkling of the skin. Eventually a significant decrease in plasma volume and blood pressure occurs, and shock may develop.

Conditions that cause severe water losses include excessive perspiration (as when you exercise in hot weather), inadequate water consumption, repeated vomiting, and diarrhea. These conditions promote water losses far in excess of electrolyte losses, so

- *If the osmotic concentration of the ECF increases, that fluid will become hypertonic with respect to the ICF.* Water will then move from the cells into the ECF until osmotic equilibrium is restored. The osmotic concentration of the ECF will increase if you lose water but retain electrolytes.

- *If the osmotic concentration of the ECF decreases, that fluid will become hypotonic with respect to the ICF.* Water will then move from the ECF into the cells, and the ICF volume will increase. The osmotic concentration of the ECF will decrease if you gain water but do not gain electrolytes.

To summarize, if the osmotic concentration of the ECF changes, a fluid shift between the ICF and ECF will tend to oppose the change. The ICF acts as a water reserve, because the volume of the ICF is much greater than that of the ECF. In effect, instead of a large change in the osmotic concentration of the ECF, smaller changes occur in both the ECF and ICF.

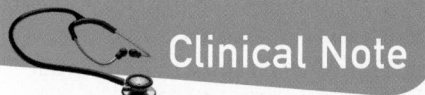

Clinical Note

Water and Weight Loss The safest way to lose weight is to reduce your intake of food while ensuring that all nutrient requirements are met. Water must be included on the list of nutrients, along with carbohydrates, fats, proteins, vitamins, and minerals. Nearly half of our normal water intake comes from food. For this reason, when you eat less, you become more dependent on drinking fluids and on whatever water is generated metabolically.

At the start of a diet, your body conserves water and catabolizes lipids. That is why the first week of dieting may seem rather unproductive. Over that week, the level of circulating ketone bodies gradually increases. In the weeks that follow, the rate of water loss by the kidneys increases in order to excrete wastes, such as the hydrogen ions released by ketone bodies and the urea and ammonia generated during protein catabolism. Your rate of fluid intake must also increase, or else you risk dehydration. While you are dieting, dehydration is especially serious, because as water is lost, the concentration of solute in the extracellular fluid (ECF) climbs, further increasing the concentration of wastes and acids generated during the catabolism of energy reserves. This soon becomes a positive feedback loop: These wastes enter the filtrate at the kidneys, and their excretion accelerates urinary water losses.

27

body fluids become increasingly concentrated, and sodium ion concentrations become abnormally high (a condition called *hypernatremia*).

Homeostatic responses include physiologic mechanisms (ADH and renin secretion) and behavioral changes (increasing fluid intake, preferably as soon as possible). Clinical therapies for acute dehydration include administering hypotonic fluids by mouth or intravenous infusion. These procedures rapidly increase ECF volume and promote the shift of water back into the ICF.

Distribution of Water Gains

When you drink a glass of tap water or when you are given hypotonic solutions intravenously, your body's water content increases without a corresponding increase in the concentration of electrolytes. As a result, the ECF increases in volume but becomes hypotonic with respect to the ICF. A fluid shift then occurs, and the volume of the ICF increases at the expense of the ECF. Once again, the larger volume of the ICF limits the amount of osmotic change. After the fluid shift, the ECF and ICF have slightly larger volumes and slightly lower osmotic concentrations than they did originally.

Normally, this situation will be promptly corrected. The decreased plasma osmotic concentration depresses the secretion of ADH, discouraging fluid intake and increasing water losses in urine. If the situation is *not* corrected, a variety of clinical problems will develop as water shifts into the intracellular fluid, distorting cells, changing the solute concentrations around enzymes, and disrupting normal cell functions. This condition is called **hyperhydration**, or *water excess*. It can be caused by (1) drinking a large volume of water or the infusion (injection into the bloodstream) of a hypotonic solution; (2) an inability to eliminate excess water in urine, due to chronic renal failure, heart failure, cirrhosis, or some other disorder; and (3) endocrine disorders, such as excessive ADH production.

The most obvious sign of overhydration is abnormally low sodium ion concentrations *(hyponatremia)*. This reduction in Na^+ concentrations in the ECF leads to a fluid shift into the ICF. The first signs are the effects on central nervous system function. The individual initially acts drunk. This condition, called *water intoxication*, may sound odd, but is extremely dangerous. Untreated cases can rapidly progress from confusion to hallucinations, convulsions, coma, and then death. To demonstrate the seriousness of this condition, in 2007 a 28-year-old woman died of water intoxication after entering a water-drinking contest, "Hold Your Wee for a Wii," sponsored by a Sacramento, California, radio station. Treatment of severe overhydration generally involves administering diuretics and infusing a concentrated salt solution that promotes a fluid shift from the ICF to the ECF and returns Na^+ concentrations to near-normal levels.

5. **Define edema.**
6. **Describe a fluid shift.**
7. **Define dehydration.**
8. **What effect would being in the desert without water for a day have on the osmotic concentration of your blood plasma?**

See the blue Answers tab at the back of the book.

27-4 | Sodium, potassium, calcium, magnesium, phosphate, and chloride balance is essential for maintaining homeostasis

Learning Outcome Discuss the mechanisms by which sodium, potassium, calcium, magnesium, phosphate, and chloride ion concentrations are regulated to maintain electrolyte balance.

You are in electrolyte balance when gains and losses are equal for each electrolyte in your body. Electrolyte balance is important because:

- *Total electrolyte concentrations directly affect water balance*, as previously described, and
- *The concentrations of individual electrolytes can affect cell functions*. We saw many examples in earlier chapters, including the effect of abnormal Na^+ concentrations on neuron activity and the effects of high or low Ca^{2+} and K^+ concentrations on cardiac muscle tissue.

Two cations, Na^+ and K^+, merit particular attention because (1) they are major contributors to the osmotic concentrations of the ECF and the ICF, respectively, and (2) they directly affect the normal functioning of all cells. Sodium is the main cation in the ECF. More than 90 percent of the osmotic concentration of the ECF results from sodium salts, mainly sodium chloride (NaCl) and sodium bicarbonate ($NaHCO_3$). Changes in the osmotic concentration of body fluids generally reflect changes in Na^+ concentration.

Electrolytes in body fluids are measured in terms of *equivalents (Eq)*. An **equivalent (Eq)** is the amount of a positive or negative ion that supplies one mole of electrical charge, and 1 equivalent = 1000 milliequivalents (mEq). For example, 1 mole of potassium ions (K^+) and 1 mole of chloride ions (Cl^-) are each 1 Eq (1000 mEq), but 1 mole of Ca^{2+} is 2 Eq (2000 mEq) because the calcium ion has a charge of +2. ↰ p. 1012

Normal Na^+ concentrations in the ECF average about 140 mEq/L, versus 10 mEq/L or less in the ICF. Potassium is the main cation in the ICF, where concentrations reach 160 mEq/L. Extracellular K^+ concentrations are generally very low, from 3.5 to 5.5 mEq/L (**Figure 27–2**).

Two general rules concerning sodium balance and potassium balance are worth noting:

1. *The most common problems with electrolyte balance are caused by an imbalance between gains and losses of sodium ions.*

2. *Problems with potassium balance are less common but significantly more dangerous than problems related to sodium balance.*

Sodium Balance

The total amount of sodium in the ECF represents a balance between two factors:

1. *Sodium Ion Uptake across the Digestive Epithelium.* Sodium ions enter the ECF by crossing the digestive epithelium through diffusion and carrier-mediated transport. The rate of absorption varies directly with the amount of sodium in the diet.

2. *Sodium Ion Excretion by the Kidneys and Other Sites.* Sodium losses take place primarily by excretion in urine and through perspiration. The kidneys are the most important sites of Na^+ regulation. We discussed the mechanisms for sodium reabsorption by the kidneys in Chapter 26. ⊃ pp. 1018, 1022

A person in sodium balance typically gains and loses 48–144 mEq (1.1–3.3 g) of Na^+ each day. When sodium gains exceed sodium losses, the total Na^+ content of the ECF increases. When losses exceed gains, the Na^+ content decreases. However, a change in the Na^+ content of the ECF does not produce a change in the Na^+ concentration. When sodium intake or output changes, a corresponding gain or loss of water tends to keep the Na^+ concentration constant. For example, if you eat a very salty meal, the osmotic concentration of the ECF will not increase. When sodium ions are pumped across the digestive epithelium, the solute concentration in that portion of the ECF increases, and that of the intestinal contents decreases. Osmosis then occurs. Additional water enters the ECF from the digestive tract, increasing the blood volume and blood pressure. For this reason, people with high blood pressure or a renal salt sensitivity are advised to restrict the amount of salt in their diets.

Sodium Balance and ECF Volume

The sodium regulatory mechanism is diagrammed in **Figure 27–5**. It changes the ECF volume but keeps the Na^+ concentration fairly stable. If you consume large amounts of salt *without* adequate fluid, as when you eat salty potato chips without taking a drink, the plasma Na^+ concentration increases temporarily. A change in ECF volume soon follows, however. Fluid will exit the ICF, increasing ECF volume and decreasing Na^+ concentrations somewhat. The secretion of ADH restricts water loss and stimulates thirst, promoting additional water consumption. Due to the inhibition of water receptors in the pharynx, ADH secretion begins even before Na^+ absorption

Figure 27–5 **The Homeostatic Regulation of Normal Sodium Ion Concentration in Body Fluids.**

occurs. The ADH secretion rate increases more after Na^+ absorption, due to osmoreceptor stimulation. ⊃ pp. 595, 649

When sodium losses exceed gains, the volume of the ECF decreases. This reduction occurs without a significant change in the osmotic concentration of the ECF. Thus, if you perspire heavily but consume only tap water, you will lose sodium, and the osmotic concentration of the ECF will decrease briefly. However, as soon as the osmotic concentration drops by 2 percent or more, ADH secretion decreases, so water losses by your kidneys increase. As water leaves the ECF, the osmotic concentration returns to normal.

Figure 27–6   **The Integration of Fluid Volume Regulation and Sodium Ion Concentration in Body Fluids.**

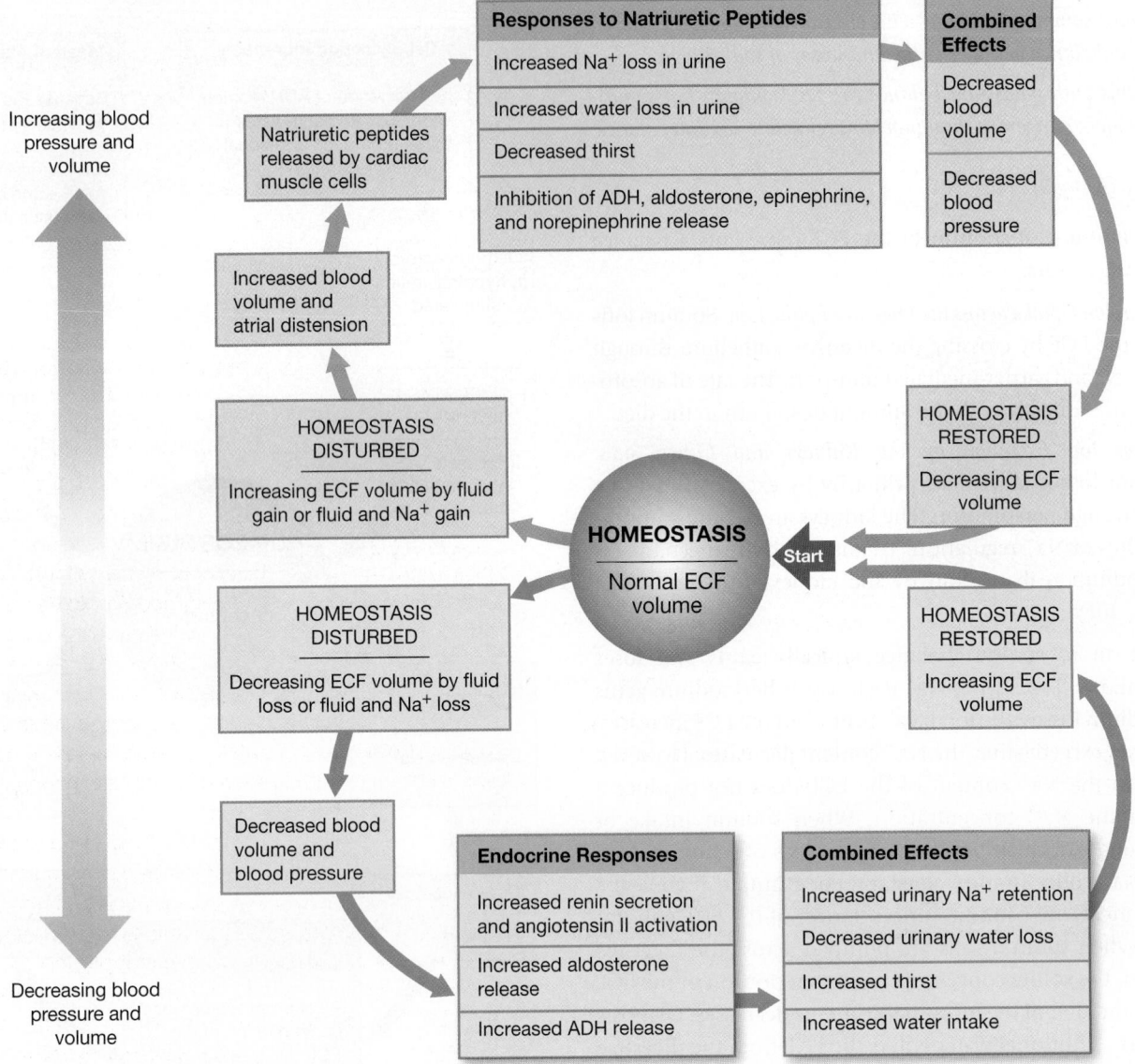

Minor changes in ECF volume do not matter, because they do not cause adverse physiological effects. If, however, regulation of Na⁺ concentrations results in a large change in ECF volume, the situation will be corrected by the same homeostatic mechanisms responsible for regulating blood volume and blood pressure. This is the case because when ECF volume changes, so does plasma volume and, in turn, blood volume. If ECF volume increases, blood volume increases, and if ECF volume decreases, blood volume decreases. As we saw in Chapter 21, blood volume has a direct effect on blood pressure. An increase in blood volume increases blood pressure, and a decrease in blood volume decreases blood pressure.

The net result is that homeostatic mechanisms can monitor ECF volume indirectly by monitoring blood pressure. The receptors involved are baroreceptors at the carotid sinus, the

aortic sinus, and the right atrium. The regulatory steps involved are reviewed in **Figure 27–6**.

Sustained abnormalities in the Na⁺ concentration in the ECF occur only when there are severe problems with fluid balance, such as dehydration or overhydration. When the body's water content increases enough to decrease the Na⁺ concentration of the ECF below 135 mEq/L, a state of **hyponatremia** (*natrium*, sodium) exists. When body water content decreases, the Na⁺ concentration increases. When that concentration exceeds 145 mEq/L, **hypernatremia** exists.

If the ECF volume is inadequate, both blood volume and blood pressure decrease, and the renin–angiotensin-aldosterone system is activated. In response, losses of water and Na⁺ are decreased, and gains of water and Na⁺ are increased. The net result is that ECF volume increases. Although the total amount of Na$^+$ in the ECF is increasing (gains exceed losses), the Na$^+$ concentration in the ECF remains unchanged, because osmotic water movement accompanies Na$^+$ absorption.

If the plasma volume becomes abnormally large, venous return increases, stretching the atrial and ventricular walls and stimulating the release of natriuretic peptides (ANP and BNP). This in turn decreases thirst and blocks the secretion of ADH and aldosterone, which together promote water or salt conservation. As a result, salt and water loss by the kidneys increases and the volume of the ECF decreases.

Potassium Balance

About 98 percent of the body's potassium is in the ICF. Cells expend energy to recover potassium ions as they diffuse across their plasma membranes and into the ECF. The K$^+$ concentration outside the cell is relatively low, and the concentration in the ECF at any moment represents a balance between (1) the rate of gain across the digestive epithelium and (2) the rate of loss into urine. Potassium loss in urine is regulated by controlling the activities of ion pumps along the distal portions of the nephron and collecting system. Whenever a sodium ion is reabsorbed from the tubular fluid, it generally is exchanged for a cation (typically K$^+$) from the peritubular fluid. ⟲ p. 1024

Urinary K$^+$ losses are usually limited to the amount gained by absorption across the digestive epithelium, typically 50–150 mEq (1.9–5.8 g) per day. (Potassium losses in feces and perspiration are negligible.) The K$^+$ concentration in the ECF is controlled by adjustments in the rate of active secretion along the distal convoluted tubule and collecting system of the nephron.

The rate of tubular secretion of K$^+$ into urine varies in response to three factors:

1. *Changes in the K$^+$ Concentration of the ECF.* In general, the higher the extracellular concentration of potassium, the higher the rate of secretion.

2. *Changes in pH.* When the pH of the ECF decreases, so does the pH of peritubular fluid. The rate of potassium secretion then decreases, because hydrogen ions, rather than potassium ions, are secreted in exchange for sodium ions in tubular fluid. The mechanisms for H$^+$ secretion were summarized in **Figure 26–14c** (p. 1025).

3. *Aldosterone Levels.* The hormone aldosterone strongly affects the rate at which K$^+$ is lost in urine. The ion pumps that are sensitive to this hormone reabsorb Na$^+$ from tubular fluid in exchange for K$^+$ from peritubular fluid. Aldosterone secretion is stimulated by angiotensin II as part of the regulation of blood volume. High plasma K$^+$ concentrations also stimulate aldosterone secretion directly. Either way, under the influence of aldosterone the amount of sodium conserved and the amount of potassium excreted in urine are directly related.

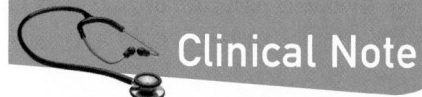

Clinical Note

Athletes and Salt Loss Unfounded notions and rumors about water and salt requirements during exercise are common. Sweat is a hypotonic solution that contains Na$^+$ in lower concentration than the ECF. As a result, a person who is sweating profusely loses more water than salt. This loss leads to an increase in the Na$^+$ concentration of the ECF. The water content of the ECF decreases as the water loss occurs, so blood volume decreases. Clinically, this condition is often called *volume depletion*. Because volume depletion takes place at the same time that blood is being shunted away from the kidneys, kidney function is impaired. As a result, wastes accumulate in the blood.

To prevent volume depletion, exercising athletes should drink liquids at regular intervals. The primary problem in volume depletion is water loss. Research has revealed no basis for the rumor that cramps will result if you drink while exercising. Salt pills and the various sports beverages that claim "faster absorption" and "better electrolyte balance" have no apparent benefits. Body reserves of electrolytes are sufficient to tolerate extended periods of strenuous activity. Problems with Na$^+$ balance are extremely unlikely, except during marathons or other activities that involve maximal exertion for more than six hours. However, both volume depletion (causing acute renal failure) and water intoxication (causing fatal hyponatremia) have occurred in marathon runners.

Some sports beverages contain sugars and vitamins as well as electrolytes. During endurance events (marathons, ultramarathons, and distance cycling), solutions containing less than 10 g/dL of glucose may improve performance if consumed late in the event, when metabolic reserves are exhausted. However, high sugar concentrations (above 10 g/dL) can cause cramps, diarrhea, and other problems. The benefit of "glucose polymers" (often cornstarch) in sports drinks has yet to be proved. Drinking beverages "fortified" with vitamins is actively discouraged: Vitamins are not lost during exercise, and the consumption of these beverages in large volumes could, over time, cause hypervitaminosis.

27

Two disorders of potassium imbalance are hypokalemia and hyperkalemia. **Hypokalemia** (*hypo-*, below + *kalium*, potassium) is a deficiency of potassium in the bloodstream. **Hyperkalemia** (*hyper-*, above) is an elevated level of potassium in the bloodstream (**Figure 27–7**).

TIPS & TRICKS

The chemical symbols for sodium (Na) and potassium (K) are derived from their Latin names, **Na**trium and **K**alium.

Figure 27–7 **Major Factors Involved in Disturbances of Potassium Balance.**

When the blood concentration of potassium falls below 2 mEq/L, extensive muscular weakness develops, followed by eventual paralysis. Severe hypokalemia is potentially lethal due to its effects on the heart.

Normal potassium levels in blood: (3.5–5.0 mEq/L)

High K$^+$ concentrations in the blood produce an equally dangerous condition known as hyperkalemia. Severe hyperkalemia occurs when the K$^+$ concentration exceeds 7 mEq/L, which results in cardiac arrhythmias.

Factors Promoting Hypokalemia

Several diuretics, including *Lasix*, can produce hypokalemia by increasing the volume of urine produced.	The endocrine disorder called aldosteronism, characterized by excessive aldosterone secretion, results in hypokalemia by overstimulating sodium retention and potassium loss.

Factors Promoting Hyperkalemia

Chronically low blood pH promotes hyperkalemia by interfering with K$^+$ excretion by the kidneys.	Kidney failure due to damage or disease will prevent normal K$^+$ secretion and thereby produce hyperkalemia.	Several drugs promote diuresis by blocking Na$^+$ reabsorption by the kidneys. When sodium reabsorption slows down, so does potassium secretion, and hyperkalemia can result.

Balance of Other Electrolytes

The ECF concentrations of other electrolytes are regulated as well. Here we will consider the most important ions involved: calcium, magnesium, phosphate, and chloride. Additional information about sodium, potassium, and these other ions is listed in Table 27–2, along with information about disorders that occur if we have too much or too little of them in the body.

Calcium Balance

Calcium is the most abundant mineral in the body. A typical person has 1–2 kg (2.2–4.4 lb) of this element. Ninety-nine percent of it is in the skeleton. In addition to forming the crystalline component of bone, calcium ions play key roles in controlling muscular and neural activities, in blood clotting, as cofactors for enzymatic reactions, and as second messengers.

As noted in Chapters 6 and 18, calcium homeostasis primarily reflects an interplay between the reserves in bone, the rate of absorption across the digestive tract, and the rate of loss by the kidneys. The hormones parathyroid hormone (PTH), calcitriol, and (to a lesser degree) calcitonin maintain calcium homeostasis in the ECF. Parathyroid hormone and calcitriol increase Ca^{2+} concentrations. Their actions are opposed by calcitonin. ⟳ pp. 655–657

A small amount of Ca^{2+} is lost in the bile. Under normal circumstances, very little Ca^{2+} escapes in urine or feces. To keep pace with these losses, an adult must absorb only 0.8–1.2 g/day of Ca^{2+}. That amount represents only about 0.03 percent of the calcium reserve in the skeleton. PTH from the parathyroid glands and calcitriol from the kidneys stimulate calcium absorption by the digestive tract and calcium reabsorption along the distal convoluted tubule.

Hypercalcemia exists when the Ca^{2+} concentration of the ECF exceeds 5.3 mEq/L. The primary cause of hypercalcemia

in adults is *hyperparathyroidism*, a condition resulting from PTH oversecretion. Less common causes include malignant cancers of the breast, lung, kidney, and bone marrow, and excessive use of calcium or vitamin D supplements. Severe hypercalcemia (12–13 mEq/L) causes such signs and symptoms as fatigue, confusion, cardiac arrhythmias, and calcification of the kidneys and soft tissues throughout the body.

Hypocalcemia (a Ca^{2+} concentration under 4.3 mEq/L) is much less common than hypercalcemia. *Hypoparathyroidism* (undersecretion of PTH), vitamin D deficiency, or chronic renal failure is typically responsible for hypocalcemia. Signs and symptoms include muscle spasms, sometimes with generalized convulsions, weak heartbeats, cardiac arrhythmias, and osteoporosis.

Magnesium Balance

The adult body contains about 29 g of magnesium. Almost 60 percent of it is in the skeleton. The magnesium in body fluids is found primarily in the ICF. There the concentration of Mg^{2+} averages about 26 mEq/L. Magnesium is required as a cofactor for several important enzymatic reactions, including the phosphorylation of glucose within cells and the use of ATP by contracting muscle fibers.

The typical range of Mg^{2+} concentrations in the ECF is 1.4–2.0 mEq/L, considerably lower than levels in the ICF. The proximal convoluted tubule reabsorbs magnesium very effectively. Keeping pace with the daily urinary loss requires a minimum dietary intake of only 24–32 mEq (0.3–0.4 g) per day.

Phosphate Balance

Phosphate ions (PO$_4^{3-}$) are required for bone mineralization. About 740 g are bound up in the mineral salts of the skeleton. In body fluids, the most important functions of PO$_4^{3-}$ involve

Table 27–2	Electrolyte Balance for Average Adult			
Ion and Normal ECF Range (mEq/L)	**Disorder (mEq/L)**	**Signs and Symptoms**	**Causes**	**Treatments**
Sodium (135–145)	Hypernatremia (>145)	Thirst, dryness and wrinkling of skin, reduced blood volume and pressure, eventual circulatory collapse	Dehydration; loss of hypotonic fluid	Ingestion of water or intravenous infusion of hypotonic solution
	Hyponatremia (<135)	Disturbed CNS function (water intoxication): confusion, hallucinations, convulsions, coma; death in severe cases	Infusion or ingestion of large volumes of hypotonic solution	Diuretic use and infusion of hypertonic salt solution
Potassium (3.5–5.0)	Hyperkalemia (>5.0)	Severe cardiac arrhythmias; muscle spasms	Renal failure; use of diuretics; chronic acidosis	Infusion of hypotonic solution; selection of different diuretics; infusion of buffers; dietary restrictions
	Hypokalemia (<3.5)	Muscular weakness and paralysis	Low-potassium diet; diuretics; hypersecretion of aldosterone; chronic alkalosis	Increase in dietary K^+ content; ingestion of K^+ tablets or solutions; infusion of potassium solution
Calcium (4.3–5.3)	Hypercalcemia (>5.3)	Confusion, muscle pain, cardiac arrhythmias, kidney stones, calcification of soft tissues	Hyperparathyroidism; cancer; vitamin D toxicity; calcium supplement overdose	Infusion of hypotonic fluid to lower Ca^{2+} levels; surgery to remove parathyroid gland; administration of calcitonin
	Hypocalcemia (<4.3)	Muscle spasms, convulsions, intestinal cramps, weak heartbeats, cardiac arrhythmias, osteoporosis	Poor diet; lack of vitamin D; renal failure; hypoparathyroidism; hypomagnesemia	Calcium supplements; administration of vitamin D
Magnesium (1.4–2.0)	Hypermagnesemia (>2.0)	Confusion, lethargy, respiratory depression, hypotension	Overdose of magnesium supplements or antacids (rare)	Infusion of hypotonic solution to lower plasma concentration
	Hypomagnesemia (<1.4)	Hypocalcemia, muscle weakness, cramps, cardiac arrhythmias, hypertension	Poor diet; alcoholism; severe diarrhea; kidney disease; malabsorption syndrome; ketoacidosis	Intravenous infusion of solution high in Mg^{2+}
Phosphate (1.8–3.0)	Hyperphosphatemia (>3.0)	No immediate symptoms; chronic elevation leads to calcification of soft tissues	High dietary phosphate intake; hypoparathyroidism	Dietary reduction; PTH supplementation
	Hypophosphatemia (<1.8)	Anorexia, dizziness, muscle weakness, cardiomyopathy, osteoporosis	Poor diet; kidney disease; malabsorption syndrome; hyperparathyroidism; vitamin D deficiency	Dietary improvement; vitamin D and/or calcitriol supplementation
Chloride (100–108)	Hyperchloremia (>108)	Acidosis, hyperkalemia	Dietary excess; increased chloride retention	Infusion of hypotonic solution to lower plasma concentration
	Hypochloremia (<100)	Alkalosis, anorexia, muscle cramps, apathy	Vomiting; hypokalemia	Diuretic use and infusion of hypertonic salt solution

the ICF. Inside cells the ions are required for the formation of high-energy compounds, the activation of enzymes, and the synthesis of nucleic acids.

The PO_4^{3-} concentration of the plasma is usually 1.8–3.0 mEq/L. Phosphate ions are reabsorbed from tubular fluid along the proximal convoluted tubule. Calcitriol stimulates phosphate ion reabsorption along the PCT. Urinary and fecal losses of PO_4^{3-} amount to 30–45 mEq (0.8–1.2 g) per day.

Chloride Balance

Chloride ions (Cl^-) are the most abundant anions in the ECF. The normal plasma concentration is 100–108 mEq/L. In the ICF, Cl^- concentrations are usually low (3 mEq/L). Chloride ions are absorbed across the digestive tract together with sodium

ions. Several carrier proteins along the renal tubules reabsorb Cl^- with Na^+. ⟳ pp. 1020, 1022 The amount of Cl^- loss is small. A gain of 48–146 mEq (1.7–5.1 g) per day will keep pace with losses in urine and perspiration.

CHECKPOINT

9. Identify four physiologically important cations and two important anions in the extracellular fluid.

10. Why does prolonged sweating increase plasma sodium ion levels?

11. Which is more dangerous, disturbances of sodium balance or disturbances of potassium balance? Explain your answer.

See the blue Answers tab at the back of the book.

27

27-5 ❙ In acid–base balance, regulation of hydrogen ions in body fluids involves buffer systems and respiratory and renal compensatory mechanisms

Learning Outcome Explain the buffering systems that balance the pH of the intracellular and extracellular fluids, and describe the compensatory mechanisms involved in maintaining acid–base balance.

We introduced the topic of pH and the chemical nature of acids, bases, and buffers in Chapter 2. Table 27–3 reviews key terms important to the discussion that follows. For a more detailed review, refer to the appropriate sections of Chapter 2 before you proceed. ⟳ pp. 65–69

The pH of body fluids can be altered by adding either acids or bases. In general, we can categorize acids and bases as either *strong* or *weak*.

- *Strong acids* and *strong bases* dissociate completely in solution. For example, hydrochloric acid (HCl), a strong acid, dissociates in solution by the reaction

$$HCl \longrightarrow H^+ + Cl^-$$

- When *weak acids* or *weak bases* enter a solution, a significant number of molecules remain intact, so dissociation is not complete. In other words, if you place molecules of a weak acid in one solution and the same number of molecules of a strong acid in another solution, the weak acid will release fewer hydrogen ions and have less effect on the pH of the solution than will the strong acid.

Table 27–3	A Review of Important Terms Relating to Acid–Base Balance
pH	The negative exponent (negative logarithm) of the hydrogen ion concentration $[H^+]$
Neutral	A solution with a pH of 7; the solution contains equal numbers of hydrogen ions and hydroxide ions
Acidic	A solution with a pH below 7; in this solution, hydrogen ions (H^+) predominate
Basic, or alkaline	A solution with a pH above 7; in this solution, hydroxide ions (OH^-) predominate
Acid	A substance that dissociates to release hydrogen ions, decreasing pH
Base	A substance that dissociates to release hydroxide ions or to tie up hydrogen ions, increasing pH
Salt	An ionic compound consisting of a cation other than hydrogen and an anion other than a hydroxide ion
Buffer	A substance that tends to oppose changes in the pH of a solution by removing or replacing hydrogen ions; in body fluids, buffers maintain blood pH within normal limits (7.35–7.45)

Carbonic acid is a weak acid. At the normal pH of the ECF, an equilibrium exists, and the reaction can be diagrammed as follows:

$$\underset{\text{Carbonic acid}}{H_2CO_3} \rightleftharpoons \underset{\text{hydrogen ion}}{H^+} + \underset{\text{bicarbonate ion}}{HCO_3^-}$$

The Importance of pH Control

The pH of body fluids reflects interactions among all the acids, bases, and salts in solution in the body. The pH of the ECF normally remains within narrow limits, usually 7.35–7.45. Any deviation from the normal range is extremely dangerous. Changes in H^+ concentrations disrupt the stability of plasma membranes, alter the structure of proteins, and change the activities of important enzymes. You could not survive for long with an ECF pH below 6.8 or above 7.7.

When the pH of blood falls below 7.35, *acidemia* exists. The physiological state that results is called **acidosis**. When the pH of blood rises above 7.45, *alkalemia* exists. The physiological state that results is called **alkalosis**.

Acidosis and alkalosis affect virtually all body systems, but the nervous and cardiovascular systems are particularly sensitive to pH fluctuations. For example, severe acidosis (pH below 7.0) can be deadly, because (1) central nervous system function deteriorates, and the person may become comatose; (2) cardiac contractions grow weak and irregular, and signs and symptoms of heart failure may develop; and (3) peripheral vasodilation produces a dramatic decrease in blood pressure, potentially producing circulatory collapse.

The control of pH is a homeostatic process of great physiological and clinical significance. Both acidosis and alkalosis are dangerous, but in practice, problems with acidosis are much more common. This is so because normal cellular activities generate several acids, including carbonic acid.

Types of Acids in the Body

There are three general categories of acids in the body. They are (1) **fixed acids**, which do not leave solution, (2) **organic acids**, which take part in or are produced by cellular metabolism, and (3) **volatile acids**, which can leave the body and enter the atmosphere through the lungs. **Figure 27–8** includes examples of each.

Carbonic acid (H_2CO_3) is an important volatile acid in body fluids. It forms spontaneously from carbon dioxide and water in body fluids, but this reaction proceeds much more rapidly in the presence of the enzyme *carbonic anhydrase*. This enzyme is found in the cytoplasm of red blood cells, liver and kidney cells, parietal cells of the stomach, and many other types of cells.

The partial pressure of carbon dioxide (P_{CO_2}) is the most important factor affecting the pH in body tissues. Most of the

Figure 27–8 **Three Classes of Acids That Can Threaten pH Balance.**

Classes of Acids

Fixed Acids

Fixed acids are acids that do not leave solution. Once produced, they remain in body fluids until they are eliminated by the kidneys. Sulfuric acid and phosphoric acid are the most important fixed acids in the body. They are generated in small amounts during the catabolism of amino acids and compounds that contain phosphate groups, including phospholipids and nucleic acids.

Organic Acids

Organic acids are acid participants in, or byproducts of, cellular metabolism. Important organic acids include lactic acid (produced by the anaerobic metabolism of pyruvate) and ketone bodies (synthesized from acetyl-CoA). Under normal conditions, most organic acids are metabolized rapidly, so significant accumulations do not occur.

Volatile Acids

Volatile acids can leave the body by entering the atmosphere at the lungs. Carbonic acid (H_2CO_3) is a volatile acid that forms through the interaction of water and carbon dioxide.

$$CO_2 + H_2O \rightleftharpoons H_2CO_3 \rightleftharpoons H^+ + HCO_3^-$$

Carbon dioxide   Water   Carbonic acid   Bicarbonate ion

carbon dioxide in solution is converted to carbonic acid, and most of the carbonic acid dissociates into hydrogen ions and bicarbonate ions. For this reason, the partial pressure of carbon dioxide and the pH are inversely related, as the seesaw in **Figure 27–9** illustrates. In other words, when carbon dioxide levels increase, additional hydrogen ions and bicarbonate ions are released, so the pH decreases. (Recall that the pH is a

negative exponent, so when the concentration of hydrogen ions goes up, the pH value goes down.)

In contrast, as carbon dioxide leaves the bloodstream and diffuses into the alveoli, the number of hydrogen ions and bicarbonate ions in the alveolar capillaries decreases, and blood pH increases. We will consider this process, which effectively removes hydrogen ions from solution, in more detail later in the chapter.

Figure 27–9 **The Basic Relationship between P_{CO_2} and Plasma pH.** The P_{CO_2} is inversely related to the pH.

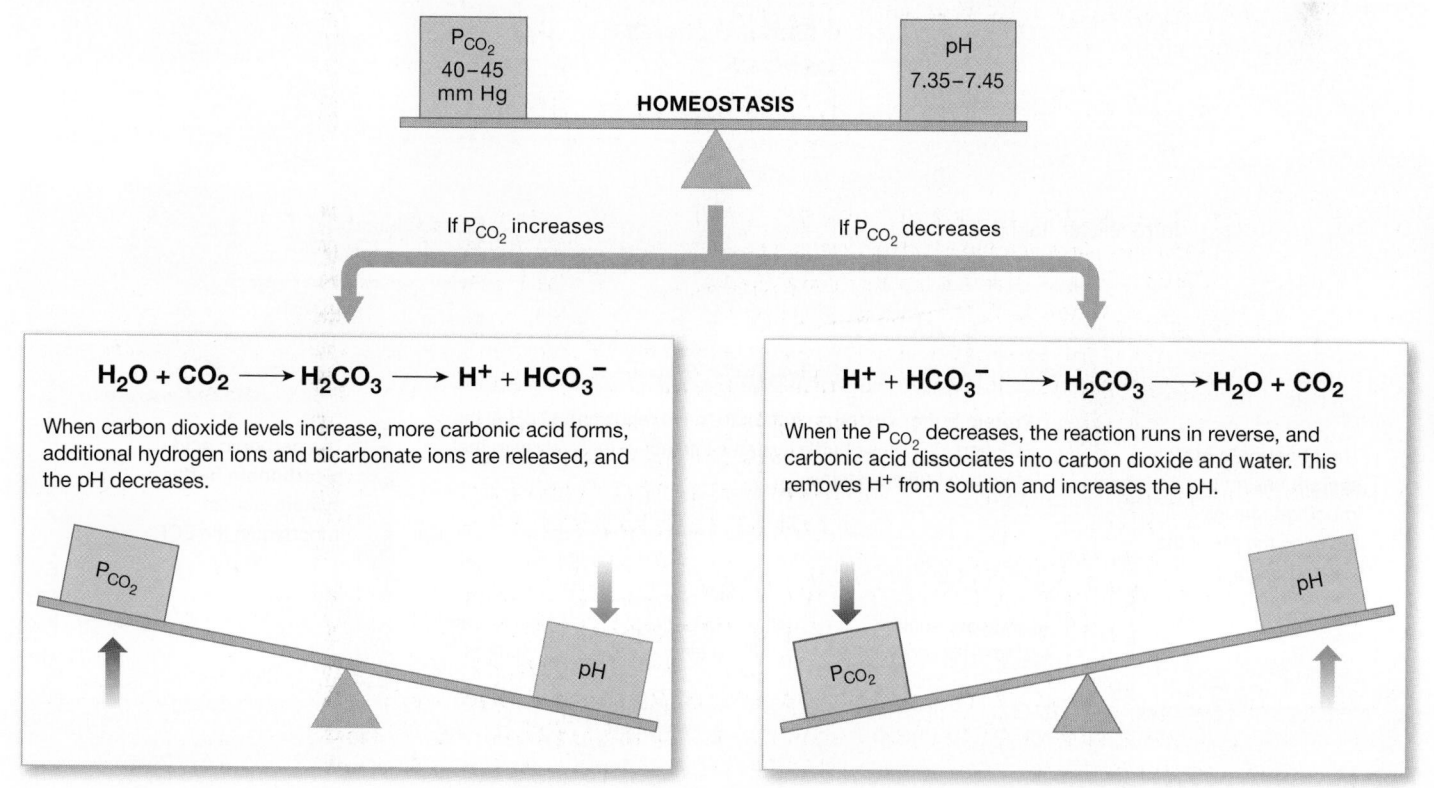

$$H_2O + CO_2 \longrightarrow H_2CO_3 \longrightarrow H^+ + HCO_3^-$$

When carbon dioxide levels increase, more carbonic acid forms, additional hydrogen ions and bicarbonate ions are released, and the pH decreases.

$$H^+ + HCO_3^- \longrightarrow H_2CO_3 \longrightarrow H_2O + CO_2$$

When the P_{CO_2} decreases, the reaction runs in reverse, and carbonic acid dissociates into carbon dioxide and water. This removes H^+ from solution and increases the pH.

27

Mechanisms of pH Control

To maintain acid–base balance over long periods of time, your body must balance gains and losses of hydrogen ions. Hydrogen ions are gained at the digestive tract and through metabolic activities within cells. Your kidneys eliminate these ions, by secreting H^+ into urine. The lungs eliminate them by forming water and carbon dioxide from H^+ and HCO_3^-. These sites of elimination are far removed from the sites of acid production. As the hydrogen ions travel through the body, they must be neutralized to avoid tissue damage.

Buffers and buffer systems in body fluids temporarily neutralize the acids produced by normal metabolic operations. *Buffers* are dissolved compounds that stabilize the pH of a solution by adding or removing H^+. Buffers include weak acids that can donate H^+, and weak bases that can absorb H^+. A **buffer system** in body fluids generally consists of a combination of a weak acid and the anion released by its dissociation. The anion functions as a weak base. In solution, molecules of the weak acid exist in equilibrium with its dissociation products. In chemical notation, this relationship is represented as

$$HY \rightleftharpoons H^+ + Y^-$$

Adding H^+ to the solution upsets the equilibrium. As a result, additional molecules of the weak acid form, removing some of the H^+ from the solution.

The body has three major buffer systems: phosphate buffer system, protein buffer system, and carbonic acid–bicarbonate buffer system. Each has slightly different characteristics and distributions (**Figure 27–10**).

As we look more closely at these buffer systems, remember that they provide only a temporary solution to an acid–base imbalance. To preserve homeostasis, the H^+ that these systems capture must ultimately be either permanently tied up in water molecules, through the removal of carbon dioxide by the lungs, or removed from body fluids, through secretion by the kidneys. For that, respiratory and renal mechanisms are needed, as we will see later in the chapter. The *combination* of buffer systems and these respiratory and renal mechanisms maintains body pH within narrow limits.

The Phosphate Buffer System

The **phosphate buffer system** consists of the anion $H_2PO_4^-$, which is a weak acid. The operation of the phosphate buffer system resembles that of the carbonic acid–bicarbonate buffer system, which is discussed below. The reversible reaction involved is

$$H_2PO_4^- \rightleftharpoons H^+ + PO_4^{2-}$$
$$\text{dihydrogen} \qquad\qquad \text{monohydrogen}$$
$$\text{phosphate} \qquad\qquad \text{phosphate}$$

Figure 27–10   **Buffer Systems in Body Fluids.**

The weak acid is *dihydrogen phosphate* ($H_2PO_4^-$), and the anion released is *monohydrogen phosphate* (HPO_4^{2-}). In the ECF, the phosphate buffer system plays only a supporting role in the regulation of pH, primarily because the concentration of HCO_3^- far exceeds that of HPO_4^{2-}. However, the phosphate buffer system is quite important in buffering the pH of the ICF. In addition, cells contain a *phosphate reserve* in the form of the weak base *sodium monohydrogen phosphate* (Na_2HPO_4). The phosphate buffer system is also important in stabilizing the pH of urine. The dissociation of Na_2HPO_4 provides additional HPO_4^{2-} for use by this buffer system:

$$2Na^+ \; + \; HPO_4^{2-} \; \rightleftharpoons \; Na_2HPO_4$$

monohydrogen   sodium
phosphate   monohydrogen
phosphate

Protein Buffer Systems

Protein buffer systems depend on the ability of amino acids to respond to pH changes by accepting or releasing H^+, that is, by acting as a base or an acid. Such molecules are said to be *amphoteric* (*ampho*, both). The underlying mechanism is shown in **Figure 27–11**:

- If pH increases, the carboxyl group ($-COOH$) of the amino acid can dissociate, acting as a weak acid and releasing a hydrogen ion. The carboxyl group then becomes a carboxylate ion ($-COO^-$). At the normal pH of body fluids (7.35–7.45), the carboxyl groups of most amino acids have already given up their hydrogen ions. (Proteins carry negative charges mainly for that reason.) However, some amino acids, notably *histidine* and *cysteine*, have R groups (side chains) that will donate hydrogen ions if the pH increases outside the normal range. Their buffering effects are very important in both the ECF and the ICF.

- In a neutral solution, an amino acid carries both + and – charges. This state is called a *zwitterion*. The term is derived from German meaning *hybrid ion*.

- If pH decreases, the carboxylate ion and the amino group ($-NH_2$) can act as weak bases and accept additional hydrogen ions, forming a carboxyl group ($-COOH$) and an amino ion ($-NH_3^+$), respectively. In free amino acids, both the main structural chain and the side chain can act as buffers. In a protein, most of the carboxyl and amino groups in the main chain are tied up in peptide bonds. Only the exposed amino group and carboxyl group at either end of a protein are available as buffers. So most of the buffering capacity of proteins is provided by the R groups of amino acids.

Plasma proteins contribute to the buffering capabilities of blood. Interstitial fluid contains extracellular protein fibers and dissolved amino acids that also assist in regulating pH. In the ICF of active cells, structural and other proteins provide an extensive buffering capability. Their buffering prevents destructive changes in pH when cellular metabolism produces organic acids, such as lactic acid or pyruvic acid.

The protein buffer system can also help stabilize the pH of the ECF, because exchange occurs between the ECF and the ICF. For example, when the pH of the ECF decreases, cells pump H^+ out of the ECF and into the ICF, where intracellular proteins can buffer them. When the pH of the ECF increases, pumps in plasma membranes move H^+ out of the ICF and into the ECF.

These mechanisms can assist in buffering the pH of the ECF, but the process is slow. Hydrogen ions must be individually transported across the plasma membrane. For this reason, the protein buffer system in most cells cannot make rapid, large-scale adjustments in the pH of the ECF.

Figure 27–11   **The Role of Amino Acids in Protein Buffer Systems.** Depending on the pH of their surroundings, amino acids either donate a hydrogen ion (as at left) or accept a hydrogen ion (as at right). An amino acid in solution is a zwitterion and has both a negative and positive charge. Additionally, several different R groups may release or absorb H^+.

The Hemoglobin Buffer System. The situation is somewhat different for red blood cells. These cells, which contain approximately 5.5 percent of the ICF, are normally suspended in the plasma. They are densely packed with hemoglobin, and their cytoplasm contains large amounts of carbonic anhydrase. Red blood cells have a significant effect on the pH of the ECF, because they absorb carbon dioxide from the plasma and convert it to carbonic acid. Carbon dioxide can diffuse across the RBC membrane very quickly, so no transport mechanism is needed. As the carbonic acid dissociates within an RBC, the bicarbonate ions diffuse into the plasma in exchange for chloride ions, a swap known as the *chloride shift*. ⊃ p. 890 Hemoglobin molecules buffer the hydrogen ions (**Figure 23–23**, p. 891). In the lungs, the entire reaction sequence diagrammed in **Figure 23–23** proceeds in reverse. This mechanism is known as the **hemoglobin buffer system**.

The hemoglobin buffer system is the only intracellular buffer system that can have an immediate effect on the pH of the ECF. The hemoglobin buffer system helps prevent drastic changes in pH when the plasma P_{CO_2} is increasing or decreasing.

The Carbonic Acid–Bicarbonate Buffer System

Body cells generate carbon dioxide virtually 24 hours a day. (Exceptions are red blood cells, some cancer cells, and tissues temporarily deprived of oxygen.) As we have seen, most of the carbon dioxide is converted to carbonic acid, which then

dissociates into a hydrogen ion and a bicarbonate ion. The carbonic acid and its dissociation products form the **carbonic acid–bicarbonate buffer system**. The primary role of this buffer system is to prevent changes in pH caused by organic acids and fixed acids in the ECF.

The carbonic acid–bicarbonate buffer system consists of the reaction introduced in our discussion of volatile acids (**Figure 27–12a**):

$$CO_2 + H_2O \rightleftharpoons H_2CO_3 \rightleftharpoons H^+ + HCO_3^-$$

carbon dioxide / water / carbonic acid / hydrogen ion / bicarbonate ion

Because the reaction is freely reversible, a change in the concentration of any one participant affects the concentrations of all others. For example, if hydrogen ions are added, most of them will be removed by interactions with HCO_3^-, forming H_2CO_3 (carbonic acid). In the process, the HCO_3^- acts as a weak base that buffers the excess H^+. The H_2CO_3 formed in this way in turn dissociates into CO_2 and H_2O (**Figure 27–12b**). The lungs can then excrete the extra CO_2. In effect, this reaction takes the H^+ released by a strong organic or fixed acid and generates a volatile acid that can easily be eliminated.

The carbonic acid–bicarbonate buffer system can also protect against increases in pH, although such changes are rare. If hydrogen ions are removed from the plasma, the reaction is

Figure 27–12   The Carbonic Acid–Bicarbonate Buffer System.

a Basic components of the carbonic acid–bicarbonate buffer system, and their relationships to carbon dioxide and the bicarbonate reserve

b The response of the carbonic acid–bicarbonate buffer system to hydrogen ions generated by fixed or organic acids in body fluids driven to the right: The P_{CO_2} decreases, and H_2CO_3 dissociates to replace the missing H^+.

The carbonic acid–bicarbonate buffer system has three important limitations:

1. *It cannot protect the ECF from changes in pH that result from increased or decreased levels of CO_2.* A buffer system cannot protect against changes in the concentration of its own weak acid. As **Figure 27–12a** indicates, an equilibrium exists among the components of this buffer system. Thus, in this system, the addition of excess H^+ from an outside source would drive the reaction to the left. But the addition of excess CO_2 would form H_2CO_3 and drive the reaction to the right. The dissociation of H_2CO_3 would release H^+ and HCO_3^-, decreasing the pH of the plasma.

2. *It can function only when the respiratory system and the respiratory control centers are working normally.* Normally, the increase in P_{CO_2} that occurs when fixed acids or organic acids are buffered stimulates an increase in the respiratory rate. This increase accelerates the removal of CO_2 by the lungs. The efficiency of the buffer system will be reduced if the respiratory passageways are blocked, or blood flow to the lungs is impaired, or the respiratory centers do not respond normally. This buffer system cannot eliminate H^+ and remove the threat to homeostasis unless the respiratory system is functioning normally.

3. *The ability to buffer acids is limited by the availability of bicarbonate ions.* Every time a hydrogen ion is removed from the plasma, a bicarbonate ion goes with it. When all the bicarbonate ions have been tied up, buffering capabilities are lost.

Problems due to a lack of bicarbonate ions are rare, for several reasons. First, body fluids contain a large reserve of HCO_3^-, mostly in the form of dissolved molecules of the weak base *sodium bicarbonate* ($NaHCO_3$). This readily available supply of HCO_3^- is known as the **bicarbonate reserve**. The reaction involved (**Figure 27–12a**) is

$$ Na^+ \; + \; HCO_3^- \; \rightleftharpoons \; NaHCO_3 $$

	bicarbonate	sodium
	ion	bicarbonate

When hydrogen ions enter the ECF, the bicarbonate ions tied up in H_2CO_3 molecules are replaced by HCO_3^- from the bicarbonate reserve (**Figure 27–12b**).

Second, additional HCO_3^- can be generated by the kidneys, through mechanisms described in Chapter 26 (**Figure 26–14c**, p. 1025). In the distal convoluted tubule and collecting system, carbonic anhydrase converts CO_2 within tubular cells into H_2CO_3, which then dissociates. The hydrogen ion is pumped into tubular fluid in exchange for a sodium ion, and the bicarbonate ion is transported into peritubular fluid in exchange for a chloride ion. In effect, tubular cells remove HCl from peritubular fluid in exchange for $NaHCO_3$.

Maintenance of Acid–Base Balance

Buffer systems can tie up excess H^+, but they provide only a temporary solution to an acid–base imbalance. The hydrogen ions are merely rendered harmless. They are not eliminated. To preserve homeostasis, the captured H^+ must ultimately be either permanently tied up in water molecules, through the removal of carbon dioxide by the lungs, or removed from body fluids, through secretion by the kidneys.

The underlying problem is that the body's supply of buffer molecules is limited. Suppose that a buffer molecule prevents a change in pH by binding a hydrogen ion that enters the ECF. That buffer molecule is then tied up, reducing the capacity of the ECF to cope with any additional H^+. Eventually, all the buffer molecules are bound to H^+, and further pH control becomes impossible.

The situation can be resolved only by removing H^+ from the ECF (thereby freeing the buffer molecules) or by replacing the buffer molecules. Similarly, if a buffer provides a hydrogen ion to maintain normal pH, homeostatic conditions will return only when either another hydrogen ion has been obtained or the buffer has been replaced.

The maintenance of acid–base balance thus includes balancing H^+ gains and losses. This "balancing act" involves coordinating the actions of buffer systems with respiratory mechanisms and renal mechanisms. These mechanisms support the buffer systems by (1) secreting or absorbing H^+, (2) controlling the excretion of acids and bases, and, when necessary, (3) generating additional buffers. As we have noted, it is the combination of buffer systems and these respiratory and renal mechanisms that maintains body pH within narrow limits.

Respiratory Compensation

Respiratory compensation is a change in the respiratory rate that helps stabilize the pH of the ECF. Respiratory compensation takes place whenever body pH strays outside normal limits. Such compensation works because respiratory activity has a direct effect on the carbonic acid–bicarbonate buffer system. Increasing or decreasing the rate of respiration alters pH by decreasing or increasing the P_{CO_2}. When the P_{CO_2} increases, the pH decreases, because the addition of CO_2 drives the carbonic acid–bicarbonate buffer system to the right. When the P_{CO_2} falls, the pH increases because the removal of CO_2 drives that buffer system to the left.

We described the mechanisms responsible for the control of respiratory rate in Chapter 23. Here we present only a brief summary. (For a review, see **Spotlight Figure 23–25** and **Figure 23–26**, pp. 894, 897.)

27

Chemoreceptors of the carotid and aortic bodies are sensitive to the P_{CO_2} of circulating blood. Other receptors, located on the ventrolateral surfaces of the medulla oblongata, monitor the P_{CO_2} of the CSF. A rise in P_{CO_2} stimulates the chemoreceptors, leading to an increase in the respiratory rate. As the rate of respiration increases, more CO_2 is lost at the lungs, so the P_{CO_2} returns to normal levels. Conversely, when the P_{CO_2} of the blood or CSF decreases, the chemoreceptors are inhibited. Respiratory activity becomes depressed and the breathing rate decreases, causing an elevation of the P_{CO_2} in the ECF.

Renal Compensation

Renal compensation is a change in the rates of H^+ and HCO_3^- secretion or reabsorption by the kidneys in response to changes in plasma pH. Under normal conditions, the body generates enough organic and fixed acids to add about 100 mEq of H^+ to the ECF each day. An equivalent number of hydrogen ions must be excreted in urine to maintain acid–base balance. In addition, the kidneys assist the lungs by eliminating any CO_2 that either enters the renal tubules during filtration or diffuses into the tubular fluid as it travels toward the renal pelvis.

Hydrogen ions are secreted into the tubular fluid along the proximal convoluted tubule (PCT), the distal convoluted tubule (DCT), and the collecting system. The basic mechanisms involved are shown in **Figures 26–12 and 26–14c** (pp. 1019, 1025). The ability to eliminate a large number of hydrogen ions in a normal volume of urine depends on the presence of buffers in the urine. The secretion of H^+ can continue only until the pH of the tubular fluid reaches 4.0–4.5. (At that point, the H^+ concentration gradient is so great that hydrogen ions leak out of the tubule as fast as they are pumped in.)

If the tubular fluid lacked buffers to absorb the hydrogen ions, the kidneys could secrete less than 1 percent of the acid produced each day before the pH reached this limit. To maintain acid balance under these conditions, the kidneys would have to produce about 1000 liters of urine each day just to keep pace with the generation of H^+ in the body! Buffers in tubular fluid are extremely important, because they keep the pH high enough for H^+ secretion to continue. Metabolic acids are being generated continuously. Without these buffering mechanisms, the kidneys could not maintain homeostasis.

Figure 27–13 diagrams the primary routes of H^+ secretion and the buffering mechanisms that stabilize the pH of tubular fluid. The three major buffers involved are the carbonic acid–bicarbonate buffer system, the phosphate buffer system, and the ammonia buffer system (**Figure 27–13a**). Glomerular filtration puts components of the carbonic acid–bicarbonate buffer system and the phosphate buffer system into the filtrate. Tubule cells (mainly those of the PCT) generate ammonia.

Figure 27–13a shows the secretion of H^+, which relies on carbonic anhydrase activity within tubular cells. The hydrogen ions generated may be pumped into the lumen in exchange for sodium ions, individually or together with chloride ions. The net result is the secretion of H^+, accompanied by the removal of CO_2 (from the tubular fluid, the tubule cells, and the ECF), and the release of sodium bicarbonate into the ECF.

Figure 27–13b shows the generation of ammonia within the tubules. As tubule cells use the enzyme *glutaminase* to break down the amino acid *glutamine*, amino groups are released as either ammonium ions (NH_4^+) or ammonia (NH_3). The ammonium ions are transported into the lumen in exchange for Na^+ in the tubular fluid. The NH_3, which is highly volatile and also toxic to cells, diffuses rapidly into the tubular fluid. There it reacts with a hydrogen ion, forming NH_4^+.

This reaction buffers the tubular fluid and removes a potentially dangerous compound from body fluids. The carbon chains of the glutamine molecules are ultimately converted to HCO_3^-, which is cotransported with Na^+ into the ECF. The generation of ammonia by tubule cells ties up H^+ in the tubular fluid and releases sodium bicarbonate into the ECF, where it contributes to the bicarbonate reserve. These mechanisms of H^+ secretion and buffering are always functioning, but their levels of activity vary widely with the pH of the ECF.

The Renal Responses to Acidosis and Alkalosis. Acidosis (low blood pH) develops when the normal plasma buffer mechanisms are stressed by excessive hydrogen ions. The kidney tubules do not distinguish among the various acids that may cause acidosis. Whether the decrease in pH results from the production of volatile, fixed, or organic acids, the renal contribution remains limited to (1) the secretion of H^+, (2) the activity of buffers in the tubular fluid, (3) the removal of CO_2, and (4) the reabsorption of $NaHCO_3$.

Tubule cells thus bolster the capabilities of the carbonic acid–bicarbonate buffer system. They do so by increasing the concentration of bicarbonate ions in the ECF, replacing those already used to remove hydrogen ions from solution. In a starving person, tubule cells break down amino acids, yielding ammonium ions that are pumped into the tubular fluid, bicarbonates to help buffer ketone bodies in the blood, and carbon chains for catabolism (**Figure 27–13b**).

When alkalosis (high blood pH) develops, (1) the rate of H^+ secretion by the kidneys declines, (2) tubule cells do not reclaim the bicarbonates in tubular fluid, and (3) the collecting system transports HCO_3^- into tubular fluid while releasing a strong acid (HCl) into peritubular fluid (**Figure 27–13c**). The concentration of HCO_3^- in blood decreases, promoting the dissociation of H_2CO_3 and the release of hydrogen ions. The additional H^+ generated by the kidneys helps return the pH to normal levels.

Figure 27–13 **Kidney Tubules and pH Regulation.**

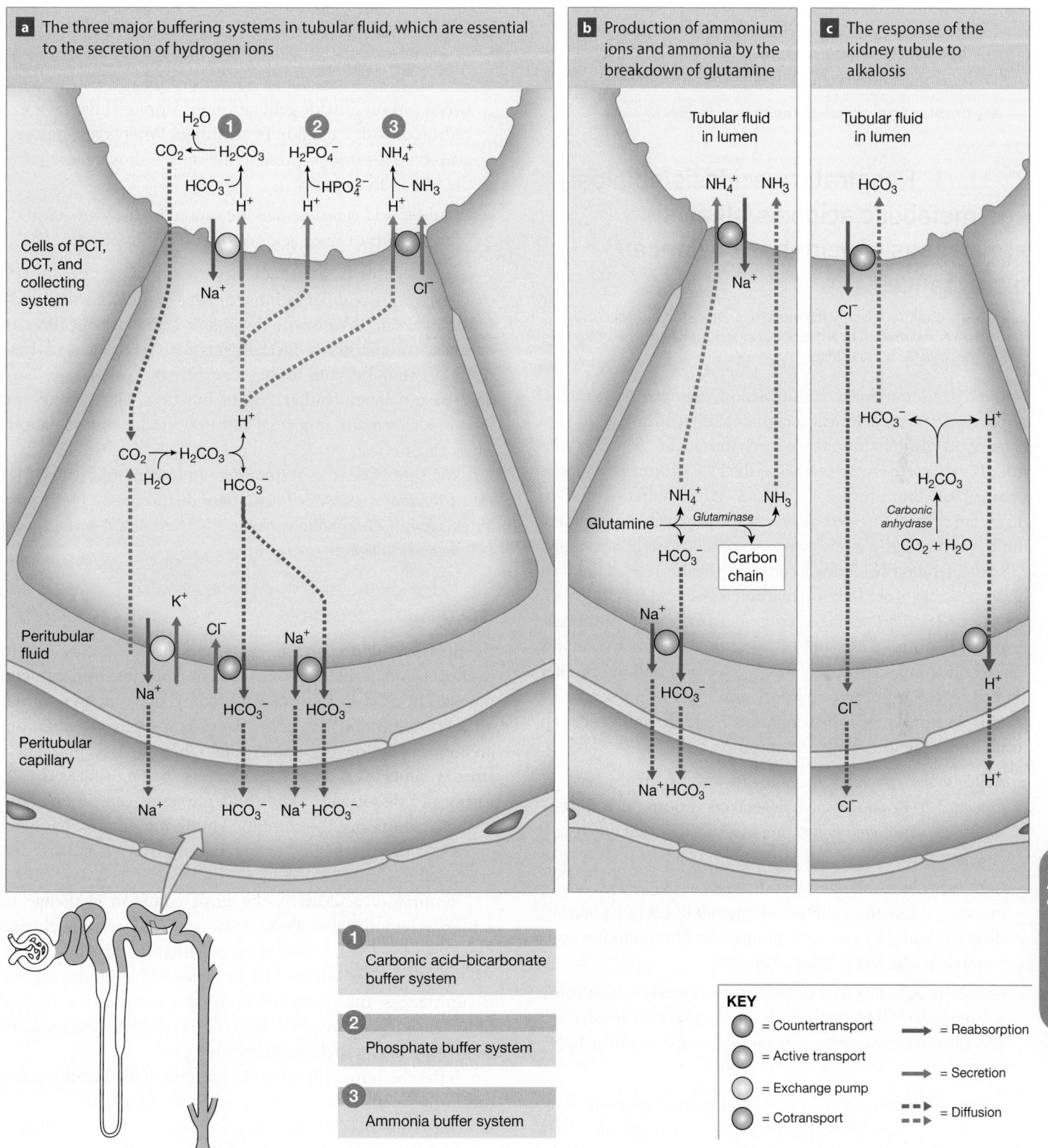

a The three major buffering systems in tubular fluid, which are essential to the secretion of hydrogen ions

b Production of ammonium ions and ammonia by the breakdown of glutamine

c The response of the kidney tubule to alkalosis

1 Carbonic acid–bicarbonate buffer system

2 Phosphate buffer system

3 Ammonia buffer system

KEY

= Countertransport

= Active transport

= Exchange pump

= Cotransport

→ = Reabsorption

→ = Secretion

‑‑► = Diffusion

27

CHECKPOINT

12. Identify the body's three major buffer systems.

13. What effect would a decrease in the pH of body fluids have on the respiratory rate?

14. Why must tubular fluid in nephrons be buffered?

See the blue Answers tab at the back of the book.

27-6 ▌ Respiratory acidosis/alkalosis and metabolic acidosis/alkalosis are classes of acid–base balance disturbances

Learning Outcome Identify the most frequent disturbances of acid–base balance, and explain how the body responds when the pH of body fluids varies outside normal limits.

Figure 27–14 summarizes the interactions among buffer systems, respiration, and renal function in maintaining normal acid–base balance. In combination, these mechanisms can generally control pH very precisely, so the pH of the ECF seldom varies more than 0.1 pH unit, from 7.35 to 7.45. When buffering mechanisms are severely stressed, however, the pH drifts outside these limits, producing signs and symptoms of alkalosis or acidosis.

For anyone considering a career in a health-related field, an understanding of acid–base dynamics is essential for clinical diagnosis and patient management under a variety of conditions. Temporary shifts in the pH of body fluids occur frequently. Rapid and complete recovery involves a combination of buffer system activity and the respiratory and renal responses. More serious and prolonged disturbances of acid–base balance can result under the following circumstances:

- *Any Disorder Affecting Circulating Buffers, Respiratory Performance, or Renal Function.* Several conditions, including *emphysema* and *renal failure*, are associated with dangerous changes in pH. ↪ pp. 901, 1034

- *Cardiovascular Conditions.* Conditions such as *heart failure* or *hypotension* can affect the pH of internal fluids by causing fluid shifts and by changing glomerular filtration rates and respiratory efficiency. ↪ pp. 736, 764

- *Conditions Affecting the Central Nervous System.* Neural damage or disease that affects the CNS can affect the respiratory and cardiovascular reflexes that are essential to normal pH regulation.

Serious abnormalities in acid–base balance generally have an initial *acute phase*, in which the pH moves rapidly out of the normal range. If the condition persists, physiological adjustments take place. The person then enters the *compensated phase*. Unless the underlying problem is corrected, compensation cannot be completed, and blood chemistry will remain abnormal. The pH typically remains outside normal limits even after compensation has occurred. Even if the pH is within the normal range, the P_{CO_2} or HCO_3^- concentrations can be abnormal.

The name given to the resulting condition usually indicates the primary source of the problem:

- *Respiratory acid–base disorders* result from a mismatch between carbon dioxide generation in peripheral tissues and carbon dioxide excretion by the lungs. When a respiratory acid–base disorder is present, the carbon dioxide level of the ECF is abnormal.

- *Metabolic acid–base disorders* are caused by the generation of organic acids or fixed acids or by conditions affecting the concentration of HCO_3^- in the ECF.

Respiratory compensation alone may restore normal acid–base balance in persons with respiratory acid–base disorders. In contrast, compensation mechanisms for metabolic acid–base disorders may be able to stabilize pH, but other aspects of acid–base balance (buffer system function, bicarbonate and P_{CO_2} levels) remain abnormal until the underlying metabolic cause is corrected.

We subdivide the respiratory and metabolic categories into four major classes of acid–base disturbances: (1) *respiratory acidosis*, (2) *respiratory alkalosis*, (3) *metabolic acidosis*, and (4) *metabolic alkalosis*.

Respiratory Acidosis

Respiratory acidosis develops when the respiratory system cannot eliminate all the carbon dioxide generated by peripheral tissues. The primary sign is low blood pH due to **hypercapnia**, an increased blood P_{CO_2}. The usual cause is hypoventilation, an abnormally low respiratory rate. When the P_{CO_2} in the ECF rises, H^+ and HCO_3^- concentrations also begin rising as H_2CO_3 forms and dissociates. Other buffer systems can tie up some of the H^+, but once the combined buffering capacity has been exceeded, the pH begins to fall rapidly. The effects are diagrammed in **Figure 27–15a**.

Respiratory acidosis is the most common challenge to acid–base equilibrium. Body tissues generate carbon dioxide rapidly. Even a few minutes of hypoventilation can cause acidosis, reducing the pH of the ECF to as low as 7.0. Under normal circumstances, the chemoreceptors that monitor the P_{CO_2} of blood and of cerebrospinal fluid (CSF) eliminate the problem by stimulating an increase in breathing rate.

If the chemoreceptors do not respond, if the breathing rate cannot be increased, or if the circulatory supply to the lungs is inadequate, pH will continue to decrease. If the decline is severe, **acute respiratory acidosis** develops. Acute respiratory acidosis is an immediate, life-threatening condition. It is especially dangerous in people whose tissues are generating large amounts of carbon dioxide, or in people who are incapable of normal respiratory activity. For this reason, the reversal of acute

Figure 27–14 **Interactions among the Carbonic Acid–Bicarbonate Buffer System and Compensatory Mechanisms in the Regulation of Plasma pH.** The central role of the carbonic acid–bicarbonate buffer system is highlighted.

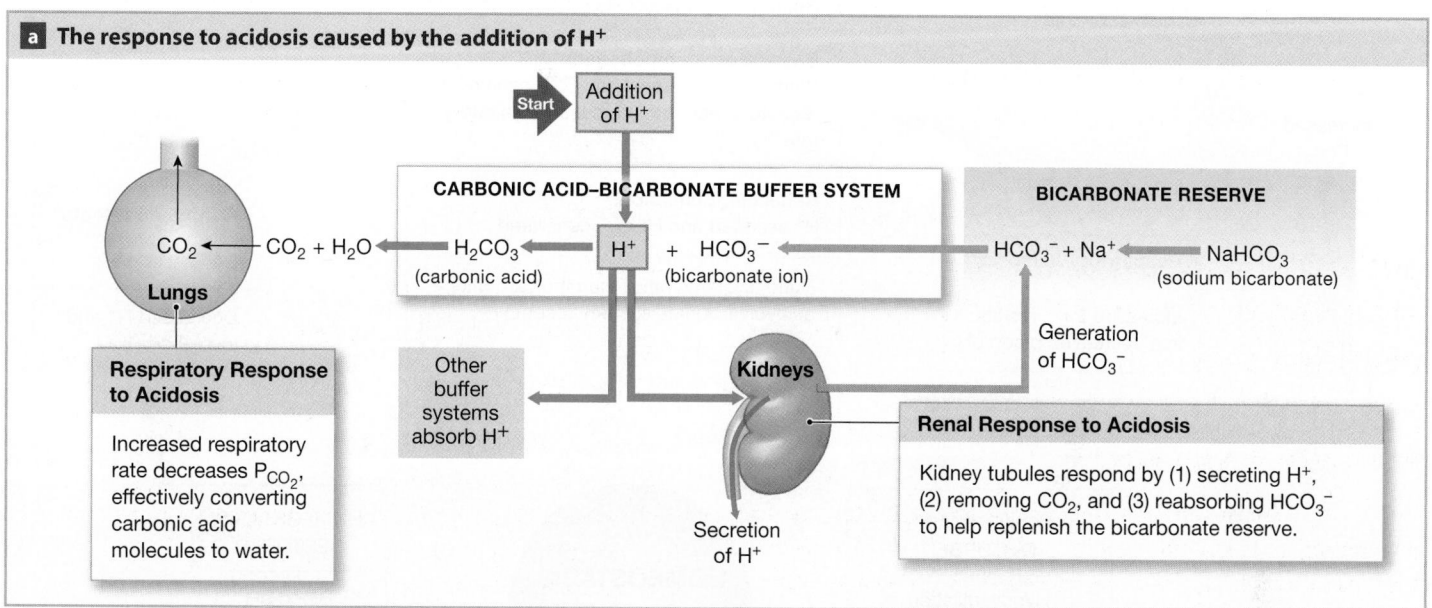

a The response to acidosis caused by the addition of H^+

b The response to alkalosis caused by the removal of H^+

respiratory acidosis is probably the major goal in the resuscitation of cardiac arrest or drowning victims. Thus, first-aid, CPR, and lifesaving courses always stress the "ABCs" of emergency care: *Airway, Breathing,* and *Circulation.*

Chronic respiratory acidosis develops when normal respiratory function has been compromised, but the compensatory mechanisms have not failed completely. For example, normal respiratory compensation may not occur in response to chemoreceptor stimulation in persons with CNS injuries and those whose respiratory centers have been desensitized by drugs

such as alcohol or barbiturates. As a result, these people are prone to developing acidosis due to chronic hypoventilation.

Even when respiratory centers are intact and functional, damage to some parts of the respiratory system can prevent increased pulmonary exchange. Examples of conditions fostering chronic respiratory acidosis include emphysema, congestive heart failure, and pneumonia (which typically cause alveolar damage or blockage). Pneumothorax and respiratory muscle paralysis have a similar effect, because they, too, limit adequate breathing rate.

27

Figure 27–15   **Respiratory Acid–Base Regulation.**

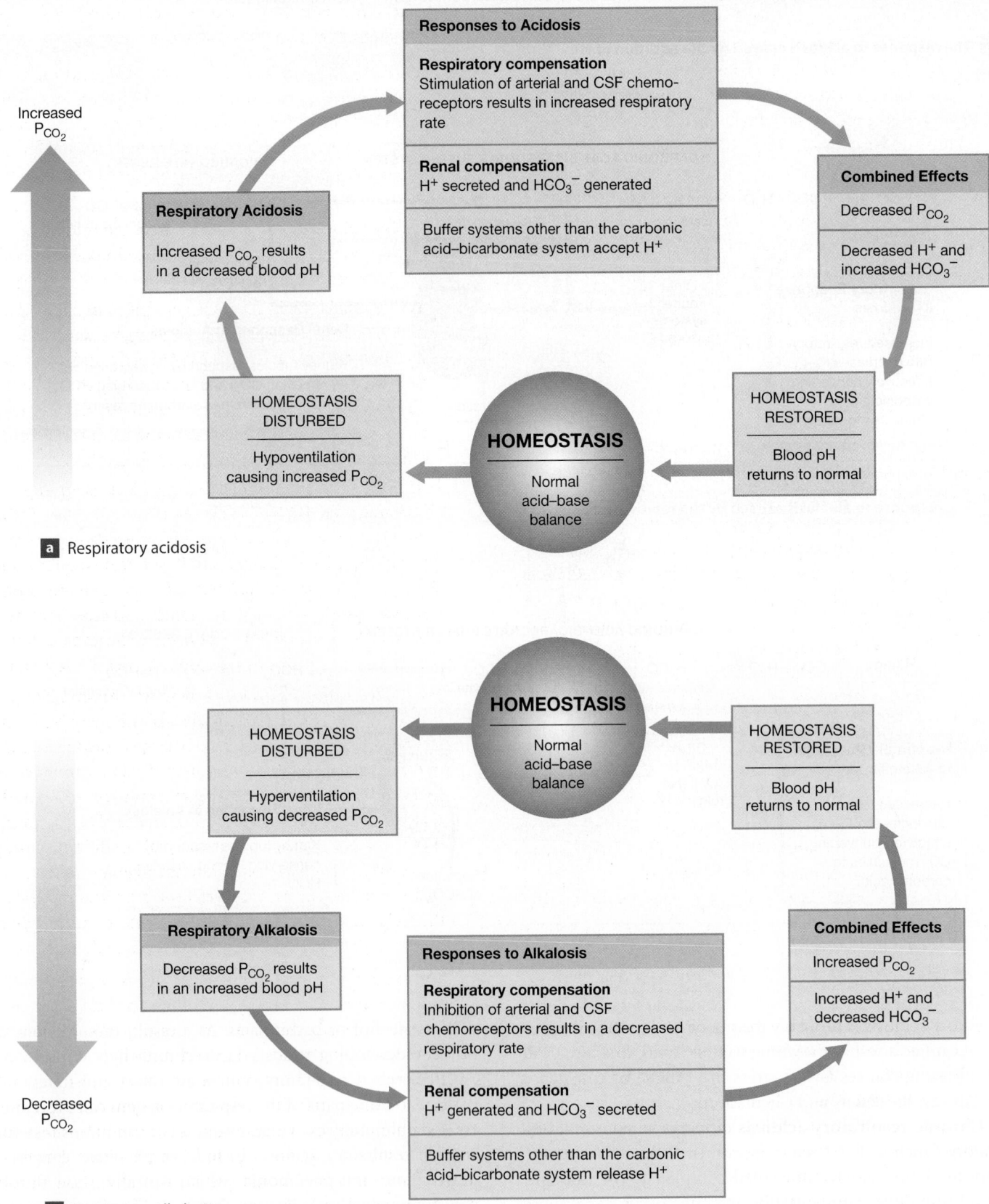

a Respiratory acidosis

b Respiratory alkalosis

When a normal pulmonary response does not take place, the kidneys respond by increasing the rate of H^+ secretion into tubular fluid. This response slows the rate of pH change. However, renal mechanisms alone cannot return the pH to normal until the underlying respiratory or circulatory problems are corrected.

The primary problem in respiratory acidosis is that the rate of pulmonary exchange is inadequate to keep the arterial P_{CO_2} within normal limits. Breathing efficiency can typically be improved temporarily by inducing bronchodilation or by using mechanical aids that provide air under positive pressure. If breathing has stopped, artificial respiration or a mechanical ventilator is required. These measures may restore normal pH if the respiratory acidosis was not severe or prolonged. Treatment of acute respiratory acidosis is complicated by the fact that, as we will soon see, it causes a complementary *metabolic acidosis* as oxygen-starved tissues generate lactic acid.

Respiratory Alkalosis

Problems resulting from **respiratory alkalosis** are fairly un-common (**Figure 27–15b**). Respiratory alkalosis develops when respiratory activity lowers blood P_{CO_2} to below-normal levels, a condition called **hypocapnia** (*-capnia*, presence of carbon di-oxide). A temporary hypocapnia can be produced by *hyperven-tilation* when increased respiratory activity leads to a reduction in the arterial P_{CO_2}. Continued hyperventilation can increase the pH to levels as high as 8.0. This condition generally cor-rects itself, because the reduction in P_{CO_2} halts the stimulation of the chemoreceptors, so the urge to breathe fades until carbon dioxide levels have returned to normal. Respiratory alkalosis caused by hyperventilation seldom lasts long enough to cause a clinical emergency.

Common causes of hyperventilation include physical stresses such as pain, or psychological stresses such as extreme anxiety. Hyperventilation gradually increases the pH of the cerebrospinal fluid, and central nervous system function is af-fected. The initial symptoms involve tingling sensations in the hands, feet, and lips. The person may also feel light-headed. If hyperventilation continues, the person may lose consciousness. Unconsciousness removes any contributing psychological stim-uli, and the breathing rate decreases. The P_{CO_2} then increases until pH returns to normal.

A simple treatment for respiratory alkalosis caused by hy-perventilation is to have the person rebreathe air exhaled into a small paper bag. As the P_{CO_2} in the bag increases, so do the person's alveolar and arterial CO_2 concentrations. This change eliminates the problem and restores the pH to normal levels. Other problems with respiratory alkalosis are rare and involve primarily (1) persons adapting to high altitudes, where the low P_{O_2} promotes hyperventilation; (2) patients on mechanical res-pirators; or (3) persons whose brain stem injuries render them incapable of responding to shifts in plasma CO_2 concentrations.

Metabolic Acidosis

Metabolic acidosis is the second most common type of acid–base imbalance. It has three major causes:

1. The most widespread cause of metabolic acidosis is the production of a large number of fixed acids or organic acids. The hydrogen ions released by these acids overload the carbonic acid–bicarbonate buffer system, so pH begins to decrease (**Figure 27–16a**). We considered two examples of metabolic acidosis earlier:

   - **Lactic acidosis** can develop after strenuous exercise or prolonged tissue hypoxia (oxygen starvation) as active cells rely on anaerobic respiration (see **Figure 10–20**, p. 345).

   - **Ketoacidosis** results from the generation of large quan-tities of ketone bodies during the postabsorptive state of metabolism. Ketoacidosis is a problem in starvation, and a potentially lethal complication of poorly controlled diabetes mellitus. In either case, peripheral tissues can-not obtain adequate glucose from the bloodstream and they begin metabolizing lipids and ketone bodies.

2. A less common cause of metabolic acidosis is an impaired ability to excrete H^+ by the kidneys (**Figure 27–16a**). For ex-ample, conditions marked by severe kidney damage, such as *glomerulonephritis*, typically result in severe metabolic acido-sis. ⟳ p. 1008 Diuretics that "turn off" the sodium–hydrogen transport system in the kidney tubules also cause metabolic acidosis. The secretion of H^+ is linked to the reabsorption of Na^+. When Na^+ reabsorption stops, so does H^+ secretion.

3. Metabolic acidosis occurs after severe bicarbonate loss (**Figure 27–16b**). The carbonic acid–bicarbonate buffer sys-tem relies on bicarbonate ions to balance hydrogen ions that threaten pH balance. A decrease in the HCO_3^- concentration in the ECF makes this buffer system less effective, and acido-sis soon develops. The most common cause of HCO_3^- deple-tion is chronic diarrhea. Under normal conditions, most of the bicarbonate ions secreted into the digestive tract in pan-creatic, hepatic, and mucous secretions are reabsorbed before the feces are eliminated. In diarrhea, these bicarbonates are lost, so the HCO_3^- concentration of the ECF decreases.

The nature of the problem must be understood before treatment can begin. Potential causes are so varied that clini-cians must piece together relevant clues to make a diagnosis. In some cases, the diagnosis is straightforward. For example, a patient with metabolic acidosis after a bicycle race probably has lactic acidosis. In other cases, clinicians must be detectives.

Compensation for metabolic acidosis generally involves a combination of respiratory and renal mechanisms. Hydrogen ions interacting with bicarbonate ions form carbon dioxide molecules that are eliminated by the lungs. The kidneys excrete additional hydrogen ions into the urine and generate bicarbon-ate ions that are released into the ECF.

27

Figure 27–16   **Responses to Metabolic Acidosis.**

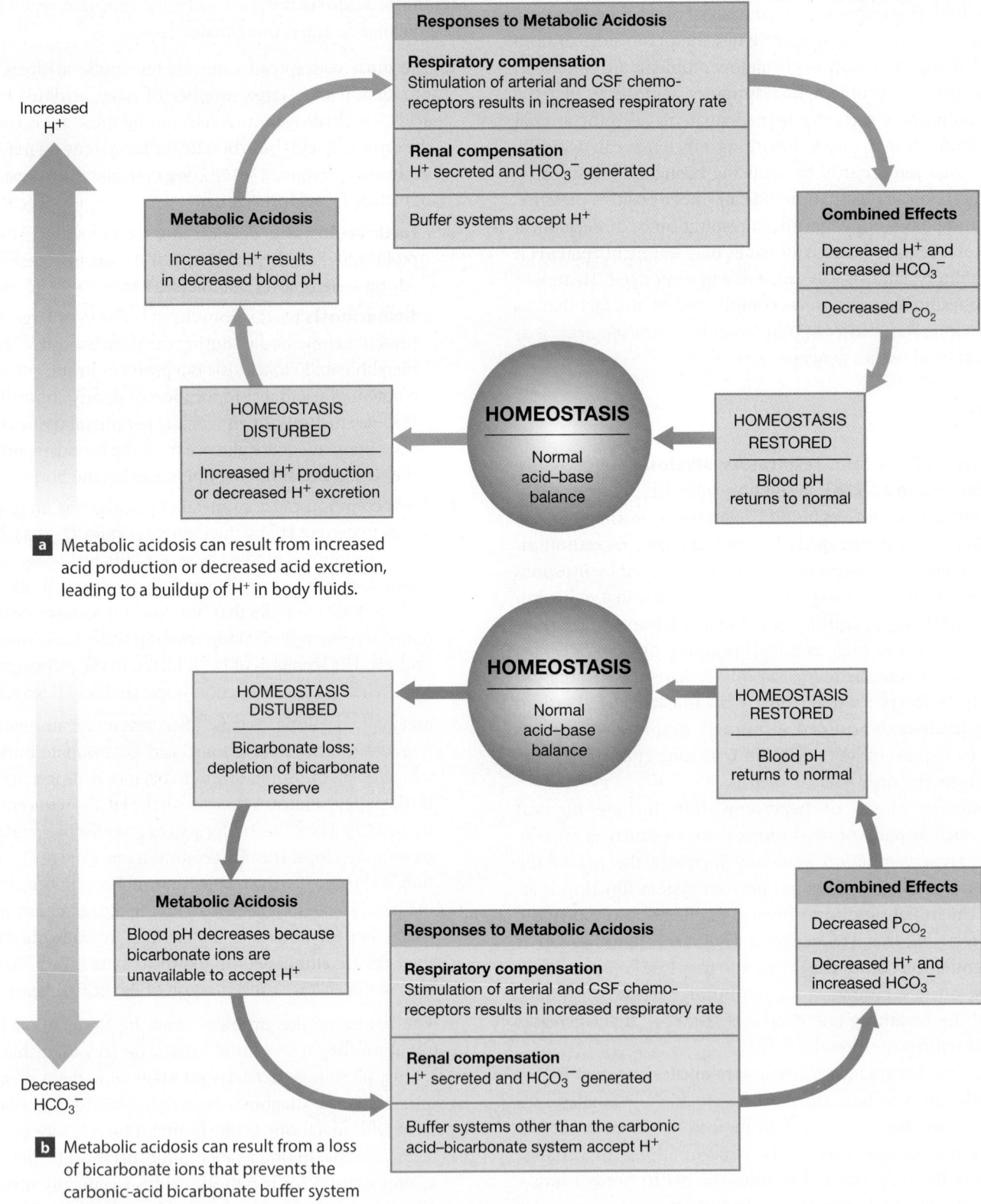

a Metabolic acidosis can result from increased acid production or decreased acid excretion, leading to a buildup of H⁺ in body fluids.

b Metabolic acidosis can result from a loss of bicarbonate ions that prevents the carbonic-acid bicarbonate buffer system from decreasing pH.

Combined Respiratory and Metabolic Acidosis

Respiratory acidosis and metabolic acidosis are typically linked, because oxygen-starved tissues generate large quantities of lactic acid, and because sustained hypoventilation leads to decreased arterial P_{O_2}. The problem can be especially serious in cases of near drowning, in which body fluids have high P_{CO_2}, low P_{O_2}, and large amounts of lactate generated by the muscles of the struggling person. (Lactate ions—and hydrogen ions—are released by the dissociation of lactic acid.) Prompt emergency treatment is essential. The usual procedure involves some form of artificial or mechanical respiratory assistance, coupled with intravenous infusion of an isotonic solution that contains sodium lactate, sodium gluconate, or sodium bicarbonate.

Metabolic Alkalosis

Metabolic alkalosis occurs when HCO_3^- concentrations are elevated (**Figure 27–17**). The bicarbonate ions then interact with hydrogen ions in solution, forming H_2CO_3. The resulting decrease in H^+ concentrations causes signs of alkalosis.

Metabolic alkalosis is rare, but we noted one interesting cause in Chapter 24. ⟲ p. 926 The phenomenon known as the *alkaline tide* is due to the influx into the ECF of large numbers of bicarbonate ions associated with the secretion of hydrochloric acid (HCl) by the gastric mucosa. The alkaline tide temporarily increases the HCO_3^- concentration in the ECF during meals. But bouts of repeated vomiting may bring on serious metabolic alkalosis, because the stomach continues to generate stomach acids to replace those that are lost. As a result, the HCO_3^- concentration of the ECF continues to increase.

Compensation for metabolic alkalosis involves a decrease in the breathing rate, coupled with an increased loss of HCO_3^- in urine. Treatment of mild cases typically addresses the primary cause—generally by controlling the vomiting—and may involve administering solutions that contain NaCl or KCl.

Treatment of acute cases of metabolic alkalosis may involve administering ammonium chloride (NH_4Cl). Metabolism of the ammonium ion in the liver liberates a hydrogen ion, so in effect the introduction of NH_4Cl leads to the internal generation of HCl, a strong acid. As the HCl diffuses into the bloodstream, pH falls toward normal levels.

The Detection of Acidosis and Alkalosis

Spotlight Figure 27–18 and Table 27–4 indicate the patterns that characterize the four major categories of acid–base disorders.

Figure 27–17 Metabolic Alkalosis. Metabolic alkalosis most commonly results from the loss of acids, especially stomach acid lost through vomiting. As replacement gastric acid is produced, the alkaline tide introduces many bicarbonate ions into the bloodstream, so pH increases.

ACID-BASE DISORDERS

Virtually anyone who has a problem that affects the cardiovascular, respiratory, urinary, digestive, or nervous system may develop potentially dangerous acid–base imbalances. For this reason, an arterial blood gas test is done to evaluate the body's acid–base balance. Standard tests measure arterial blood pH, partial pressure of carbon dioxide (P_{CO_2}), partial pressure of oxygen (P_{O_2}), and bicarbonate (HCO_3^-) levels. These measurements make diagnosing acid–base disorders relatively straightforward. The flowchart below shows the diagnostic steps used to determine whether a person has an acid–base disorder, and to further determine what type of acidosis or alkalosis exists.

Suspected Acid–Base Disorder

Check pH

Acidosis pH <7.35 (acidemia)

Check P_{CO_2}

Metabolic Acidosis

P_{CO_2} normal or decreased

Respiratory Acidosis

P_{CO_2} increased (>50 mm Hg)

Primary cause is hypoventilation

Check HCO_3^-

Acute Metabolic Acidosis

P_{CO_2} normal

Chronic (compensated) Metabolic Acidosis

P_{CO_2} decreased (<35 mm Hg)

Reduction due to respiratory compensation

Chronic (compensated) Respiratory Acidosis

HCO_3^- increased (>28 mEq/L)

Examples
• emphysema
• asthma

Acute Respiratory Acidosis

HCO_3^- normal

Examples
• respiratory failure
• CNS damage
• pneumothorax

*Check anion gap**

*The anion gap is defined as:
Na^+ concentration – (HCO_3^- concentration + Cl^- concentration)

Normal

Due to loss of HCO_3^- or to generation or ingestion of HCl

Example
• diarrhea

Increased

Due to generation or retention of organic or fixed acids

Examples
• lactic acidosis
• ketoacidosis
• chronic renal failure

Alkalosis pH >7.45 (alkalemia)

Check P_{CO_2}

Metabolic Alkalosis

P_{CO_2} increased (>45 mm Hg)

(HCO_3^- will be elevated)

Examples
• vomiting
• loss of gastric acid

Respiratory Alkalosis

P_{CO_2} decreased (<35 mm Hg)

Primary cause is hyperventilation

Check HCO_3^-

Acute Respiratory Alkalosis

Normal or slight decrease in HCO_3^-

Examples
• fever
• panic attacks

Chronic (compensated) Respiratory Alkalosis

Decreased HCO_3^- (<24 mEq/L)

Examples
• anemia
• CNS damage

Table 27–4	Changes in Blood Chemistry Associated with the Major Classes of Acid–Base Disorders				
Disorder	**pH (normal 7.35–7.45)**	**HCO_3^- (mEq/L) (normal = 21–28)**	**P_{CO_2} (mm Hg) (normal = 35–45)**	**Remarks**	**Treatments**
Respiratory acidosis	Decreased (below 7.35)	Acute: normal Compensated: increased (above 28)	Increased (above 45)	Generally caused by hypoventilation and CO_2 buildup in tissues and blood	Improve ventilation; in some cases, with bronchodilation and mechanical assistance
Metabolic acidosis	Decreased (below 7.35)	Decreased (below 24)	Acute: normal Compensated: decreased (below 35)	Caused by buildup of organic or fixed acid, impaired H^+ elimination at kidneys, or HCO_3^- loss in urine or feces	Administration of bicarbonate (gradual), with other steps as needed to correct primary cause
Respiratory alkalosis	Increased (above 7.45)	Acute: normal Compensated: decreased (below 24)	Decreased (below 35)	Generally caused by hyperventilation and reduction in plasma CO_2 levels	Reduce respiratory rate, allow rise in P_{CO_2}
Metabolic alkalosis	Increased (above 7.45)	Increased (above 28)	Increased (above 45)	Generally caused by prolonged vomiting and associated acid loss	pH below 7.55: no treatment; pH above 7.55: may require administration of NH_4Cl

Additional steps can help in identifying possible causes of the problem and in distinguishing compensated from uncompensated conditions. These steps may include determining the *anion gap*, plotting blood test results on a numerical graph called a *nomogram*, or using a diagnostic chart.

CHECKPOINT

15. How would a prolonged fast affect the body's pH?

16. Why can prolonged vomiting produce metabolic alkalosis?

See the blue Answers tab at the back of the book.

27-7 | Aging affects several aspects of fluid, electrolyte, and acid–base balance

Learning Outcome Describe the effects of aging on fluid, electrolyte, and acid–base balance.

Fetuses and infants have very different requirements for maintaining fluid, electrolyte, and acid–base balance than do adults. A fetus obtains the water, organic nutrients, and electrolytes it needs from the maternal bloodstream. Buffers in the fetal bloodstream provide short-term pH control, and the maternal kidneys eliminate the H^+ generated. A newborn's body water content is high: At birth, water makes up about 75 percent of body weight, compared with 50–60 percent in adults. Basic aspects of electrolyte balance are the same in newborns as in adults, but the effects of fluctuations in the diet are much more immediate in newborns because their reserves of minerals and energy sources are much smaller.

The descriptions of fluid, electrolyte, and acid–base balance in this chapter were based on the responses of normal, healthy adults under age 40. Aging affects many aspects of fluid, electrolyte, and acid–base balance, including the following:

- Total body water content gradually decreases with age. Between ages 40 and 60, average total body water content declines slightly, to 55 percent for males and 47 percent for females. After age 60, the values decline further, to about 50 percent for males and 45 percent for females. Among other effects, each decrease reduces the dilution of wastes, toxins, and any drugs that have been administered.

- A decrease in the glomerular filtration rate and in the number of functional nephrons reduces the body's ability to regulate pH through renal compensation.

- The body's ability to concentrate urine declines, so more water is lost in urine. In addition, the rate of insensible perspiration increases as the skin becomes thinner and more delicate. Maintaining fluid balance therefore requires a higher daily water intake. A reduction in ADH and aldosterone sensitivity makes older people less able than younger people to conserve body water when losses exceed gains.

- Many people over age 60 experience a net loss in body mineral content as muscle mass and skeletal mass decrease. This loss can be prevented, at least in part, by a combination of exercise and an increased dietary mineral supply.

- The reduction in vital capacity that accompanies aging reduces the body's ability to perform respiratory compensation, increasing the risk of respiratory acidosis. This

27

problem can be compounded by arthritis, which can reduce vital capacity by limiting rib movements, and by emphysema, another condition that, to some degree, develops with aging.

- Disorders affecting major systems become more common with increasing age. Most, if not all, of these disorders have some effect on fluid, electrolyte, and/or acid–base balance.

CHECKPOINT

17. As a person ages, the glomerular filtration rate and the number of functional nephrons decreases. What effect would these changes have on pH regulation?

18. After the age of 40, does the total body water content increase or decrease?

See the blue Answers tab at the back of the book.

Related Clinical Terms

antacid: A substance used to counteract or neutralize stomach acid.

bicarbonate loading: In sports, the act of ingesting bicarbonates prior to an athletic event to neutralize the lactic acid produced during strenuous physical activity.

enema: A procedure in which a liquid or gas is injected into the rectum, usually with the intention of having the rectum dispel its contents, but sometimes used as a way to introduce drugs or to permit x-ray imaging.

fluid replacement therapy: Procedure conducted with the intent to replace body fluids lost due to disease or restricted intake; or to maintain a higher-than-normal rate of fluid excretion to ensure the removal of toxins; or possibly to administer therapeutic or anesthetic agents slowly over time.

potassium adaptation: The tolerance to increasing amounts of potassium.

syndrome of inappropriate secretion of ADH (SIADH): A condition associated with excessive ADH secretion that results in the excretion of concentrated urine. It disturbs fluid and electrolyte balance, and causes nausea, vomiting, muscle cramps, confusion, and convulsions. This syndrome can occur with some forms of cancer, such as oat-cell lung cancer, pancreatic cancer, prostate cancer, Hodgkin's disease, and possibly due to a number of other disorders.

total body water: The sum of fluids within all compartments.

Chapter Review

Study Outline

MasteringA&P®

Access more chapter study tools online in the MasteringA&P Study Area:

- Chapter Quizzes, Chapter Practice Test, Art-labeling Activities, Animations, MP3 Tutor Sessions, and Clinical Case Studies
- Practice Anatomy Lab PAL™
- Interactive Physiology iP®
- A&P Flix A&PFlix
- PhysioEx PhysioEx

27-1 ▌ Fluid balance, electrolyte balance, and acid–base balance are interrelated and essential to homeostasis p. 1042

1. The maintenance of normal volume and composition of extracellular and intracellular fluids is vital to life. Three types of homeostasis are involved: **fluid balance**, **electrolyte balance**, and **acid–base balance**.

27-2 ▌ The ECF and ICF make up the fluid compartments, which also contain cations and anions p. 1043

2. The **intracellular fluid (ICF)** contains nearly two-thirds of the total body water. The **extracellular fluid (ECF)** contains the rest. Exchange occurs between the ICF and the ECF, but the

two **fluid compartments** retain their distinctive characteristics. *(Figures 27–1, 27–2)*

3. Homeostatic mechanisms that monitor and adjust the composition of body fluids respond to changes in the ECF.

4. No receptors directly monitor fluid or electrolyte balance. Receptors involved in fluid balance and in electrolyte balance respond to changes in plasma volume and osmotic concentration.

5. Body cells cannot move water molecules by active transport. All movements of water across plasma membranes and epithelia occur passively, in response to osmotic gradients.

6. The body's content of water or electrolytes will increase if intake exceeds outflow and will decrease if losses exceed gains.

7. ADH encourages water reabsorption by the kidneys and stimulates thirst. Aldosterone increases the rate of sodium reabsorption by the kidneys. Natriuretic peptides (ANP and BNP) oppose those actions and promote fluid and electrolyte losses in urine.

8. The regulatory mechanisms of fluid balance and electrolyte balance are quite different, and the distinction is clinically important.

27-3 ▌ Hydrostatic and osmotic pressures regulate the movement of water and electrolytes to maintain fluid balance p. 1047

9. Water circulates freely within the ECF compartment.

10. Water losses are normally balanced by gains through eating, drinking, and **metabolic generation**. *(Figure 27–3; Table 27–1)*

27

11. Water movement between the ECF and ICF is called a **fluid shift**. If the ECF becomes hypertonic relative to the ICF, water will move from the ICF into the ECF until osmotic equilibrium has been restored. If the ECF becomes hypotonic relative to the ICF, water will move from the ECF into the cells, and the volume of the ICF will increase. *(Figure 27–4)*

27-4 ▌ Sodium, potassium, calcium, magnesium, phosphate, and chloride balance is essential for maintaining homeostasis p. 1050

12. An **equivalent** is the amount of a positive ion or negative ion that supplies one mole of electrical charge, and 1 equivalent = 1000 milliequivalents (mEq).

13. Problems with electrolyte balance generally result from a mismatch between gains and losses of sodium. Problems with potassium balance are less common, but more dangerous.

14. The rate of sodium uptake across the digestive epithelium is directly proportional to the amount of sodium in the diet. Sodium losses occur mainly in urine and through perspiration. *(Figure 27–5)*

15. Shifts in sodium balance result in expansion or contraction of the ECF. Large variations in ECF volume are corrected by homeostatic mechanisms triggered by changes in blood volume. If the volume becomes too low, ADH and aldosterone are secreted; if the volume becomes too high, natriuretic peptides (ANP and BNP) are secreted. *(Figure 27–6)*

16. Potassium ion concentrations in the ECF are very low and not as closely regulated as are sodium ion concentrations. Potassium excretion increases as ECF concentrations rise, under aldosterone stimulation, and when the pH increases. Potassium retention occurs when the pH decreases. *(Figure 27–7)*

17. ECF concentrations of other electrolytes, such as calcium, magnesium, phosphate, and chloride, are also regulated. *(Table 27–2)*

27-5 ▌ In acid–base balance, regulation of hydrogen ions in body fluids involves buffer systems and respiratory and renal compensatory mechanisms p. 1056

18. Acids and bases are either **strong** or **weak**. *(Table 27–3)*

19. The pH of normal body fluids ranges from 7.35 to 7.45. Variations outside this relatively narrow range produce symptoms of **acidosis** or **alkalosis**.

20. **Volatile acids** can leave solution and enter the atmosphere. **Fixed acids** remain in body fluids until excreted by the kidneys. **Organic acids** are participants in, or are byproducts of, aerobic metabolism. *(Figure 27–8)*

21. Carbonic acid, a volatile acid, is the most important factor affecting the pH of the ECF. In solution, CO_2 reacts with water to form carbonic acid. An inverse relationship exists between pH and the concentration of CO_2. *(Figure 27–9)*

22. Sulfuric acid and phosphoric acid, the most important fixed acids, are generated during the catabolism of amino acids and compounds containing phosphate groups.

23. Organic acids include metabolic products such as lactic acid and ketone bodies.

24. A **buffer system** typically consists of a weak acid and the anion released by its dissociation. The anion functions as a weak base that can absorb H^+. The three major buffer systems are (1) the **phosphate buffer system** in the ICF and urine; (2) **protein buffer systems** in the ECF and ICF; and (3) the **carbonic acid–bicarbonate buffer system**, most important in the ECF. *(Figure 27–10)*

25. The phosphate buffer system plays a supporting role in regulating the pH of the ECF, but it is important in buffering the pH of the ICF and of urine.

26. In protein buffer systems, the initial, final, and R groups of component amino acids respond to changes in pH by accepting or releasing hydrogen ions. The **hemoglobin buffer system** is a protein buffer system that helps prevent drastic changes in pH when the P_{CO_2} is increasing or decreasing. *(Figure 27–11)*

27. The carbonic acid–bicarbonate buffer system prevents pH changes caused by organic acids and fixed acids in the ECF. The readily available supply of bicarbonate ions is the **bicarbonate reserve**. *(Figure 27–12)*

28. The lungs help regulate pH by affecting the carbonic acid–bicarbonate buffer system. A change in respiratory rate can raise or lower the P_{CO_2} of body fluids, affecting the body's buffering capacity. This process is called **respiratory compensation**.

29. In **renal compensation**, the kidneys vary their rates of hydrogen ion secretion and bicarbonate ion reabsorption, depending on the pH of the ECF. *(Figure 27–13)*

27-6 ▌ Respiratory acidosis/alkalosis and metabolic acidosis/alkalosis are classes of acid–base balance disturbances p. 1064

30. Interactions among buffer systems, respiration, and renal function normally maintain tight control of the pH of the ECF, generally within a range of 7.35–7.45. *(Figure 27–14)*

31. *Respiratory acid–base disorders* result when abnormal respiratory function causes an extreme increase or decrease in CO_2 levels in the ECF. *Metabolic acid–base disorders* are caused by the generation of organic or fixed acids or by conditions affecting the concentration of bicarbonate ions in the ECF.

32. **Respiratory acidosis** results from excessive levels of CO_2 in body fluids. *(Figure 27–15a)*

33. **Respiratory alkalosis** is a relatively rare condition associated with hyperventilation. *(Figure 27–15b)*

34. **Metabolic acidosis** results from the depletion of the bicarbonate reserve, caused by an inability to excrete hydrogen ions by the kidneys, the production of large numbers of fixed and organic acids, or bicarbonate loss that accompanies chronic diarrhea. *(Figure 27–16)*

35. **Metabolic alkalosis** results when bicarbonate ion concentrations become elevated, as occurs during extended periods of vomiting. *(Figure 27–17)*

36. An arterial blood gas test evaluates blood pH, P_{CO_2}, P_{O_2}, and HCO_3^- levels to recognize and classify acidosis and alkalosis as either respiratory or metabolic in nature. *(Spotlight Figure 27–18; Table 27–4)*

27-7 ▌ Aging affects several aspects of fluid, electrolyte, and acid–base balance p. 1071

37. Age-related changes that affect fluid, electrolyte, and acid–base balance include (1) reduced total body water content, (2) impaired ability to perform renal compensation, (3) increased water demands due to decreased ability to concentrate urine and reduced sensitivity to ADH and aldosterone, (4) a net loss of minerals, (5) reductions in respiratory efficiency that affect respiratory compensation, and (6) increased incidence of conditions that secondarily affect fluid, electrolyte, or acid–base balance.

27

Review Questions

See the blue Answers tab at the back of the book.

LEVEL 1 Reviewing Facts and Terms

1. The primary components of the extracellular fluid are (a) lymph and cerebrospinal fluid, (b) plasma and serous fluids, (c) interstitial fluid and plasma, (d) all of these.

2. The principal anions in the ICF are (a) phosphate and proteins (Pr^-), (b) phosphate and bicarbonate, (c) sodium and chloride, (d) sodium and potassium.

3. Osmoreceptors in the hypothalamus monitor the osmotic concentration of the ECF and secrete _____ in response to higher osmotic concentrations. (a) BNP, (b) ANP, (c) aldosterone, (d) ADH.

4. Write the missing names and molecular formulas for the following reactions between the carbonic acid–bicarbonate buffer system and the bicarbonate reserve.

(a) _____
(b) _____
(c) _____

5. Calcium homeostasis primarily reflects (a) a balance between absorption in the gut and excretion by the kidneys, (b) careful regulation of blood calcium levels by the kidneys, (c) an interplay between parathormone and aldosterone, (d) an interplay among reserves in the bones, the rate of absorption, and the rate of excretion, (e) hormonal control of calcium reserves in the bones.

6. Low blood pH results in (a) alkalosis and increased enzyme activity, (b) alkalosis and reduced enzyme activity, (c) acidosis and increased enzyme activity, (d) acidosis and reduced enzyme activity.

7. The four ways in which water is lost from the body are urine, _____, insensible perspiration (movement and loss of water through skin and lungs), and sensible perspiration (loss from the activity of sweat glands).

8. Respiratory acidosis develops when the blood pH is (a) increased due to a decreased blood P_{CO_2} level, (b) decreased due to an increased blood P_{CO_2} level, (c) increased due to an increased blood P_{CO_2} level, (d) decreased due to a decreased blood P_{CO_2} level.

9. Metabolic alkalosis occurs when (a) bicarbonate ion concentrations become elevated, (b) a severe bicarbonate loss occurs, (c) the kidneys fail to excrete hydrogen ions, (d) ketone bodies are generated in abnormally large quantities.

10. Identify four hormones that mediate major physiological adjustments affecting fluid and electrolyte balance. What are the primary effects of each hormone?

LEVEL 2 Reviewing Concepts

11. Drinking a solution hypotonic to the ECF causes the ECF to (a) increase in volume and become hypertonic to the ICF, (b) decrease in volume and become hypertonic to the ICF, (c) decrease in volume and become hypotonic to the ICF, (d) increase in volume and become hypotonic to the ICF.

12. The osmotic concentration of the ECF decreases if an individual gains water without a corresponding (a) gain of electrolytes, (b) loss of water, (c) fluid shift from the ECF to the ICF, (d) a, b, and c.

13. Acid–base compensation is reduced with age as there is (a) a reduction in aldosterone sensitivity, (b) a reduction in vital capacity and functional nephrons, (c) insufficient ADH release, (d) a decreased rate of insensible perspiration.

14. In a protein buffer system, if the pH increases, (a) the protein acquires a hydrogen ion from carbonic acid, (b) hydrogen ions are buffered by hemoglobin molecules, (c) a hydrogen ion is released and a carboxylate ion is formed, (d) a chloride shift occurs.

15. Differentiate among fluid balance, electrolyte balance, and acid–base balance, and explain why each is important to homeostasis.

16. What are fluid shifts? What is their function, and what factors can cause them?

17. How does the process of aging influence fluid balance?

18. Chronic respiratory acidosis can be compensated for by (a) increasing ventilation, (b) decreasing ventilation, (c) increasing HCO_3^- generation, (d) decreasing H^+ secretion.

19. Explain how ADH can be used in the homeostatic control of elevated sodium.

20. How do respiratory and renal mechanisms support the buffer systems?

21. Differentiate between respiratory compensation and renal compensation.

22. Distinguish between respiratory and metabolic disorders that disturb acid–base balance.

23. What is the difference between metabolic acidosis and respiratory acidosis? What can cause these conditions?

24. The most recent advice from medical and nutritional experts is to monitor one's intake of salt so that it does not exceed the amount needed to maintain a constant ECF volume. What effect does excessive salt ingestion have on blood pressure?

25. Exercise physiologists recommend that adequate amounts of fluid be ingested before, during, and after exercise. Why is fluid replacement during extensive sweating important?

LEVEL 3 Critical Thinking and Clinical Applications

26. After falling into an abandoned stone quarry filled with water and nearly drowning, a young boy is rescued. In assessing his condition, rescuers find that his body fluids have high P_{CO_2} and lactate levels, and low P_{O_2} levels. Identify the underlying problem and recommend the necessary treatment to restore homeostatic conditions.

27. Dan has been lost in the desert for two days with very little water. As a result of this exposure, you would expect to observe which of the following? (a) increased ADH levels, (b) decreased blood osmolarity, (c) normal urine production, (d) increased blood volume, (e) cells enlarged with fluid.

28. Mary, a nursing student, has been caring for burn patients. She notices that they consistently show elevated levels of potassium in their urine and wonders why. What would you tell her?

29. After several days of vomiting, Anthony is taken to the hospital. Blood tests show the following: pH – 7.50, P_{CO_2} – 43, HCO_3^- – 36. What is the acid–base balance status and what is the likely cause? Using evidence to support your answer, state whether compensation has occurred.

27

30. Anne was brought to the accident and emergency room after falling at home. Over a few days, her condition worsens, and she is moved to the intensive care unit after developing acute pneumonia. Her latest blood results show that she has lactate levels similar to that seen in athletes performing intensive exercise. What is the acid–base disorder Anne is experiencing? Explain the lactate observations.

31. Refer to the diagnostic flowchart in Spotlight Figure 27–18. Use information from the blood test results in the accompanying table to categorize the suspected acid–base disorders of the patients represented in the table.

Results	Patient 1	Patient 2	Patient 3	Patient 4
pH	7.5	7.2	7.0	7.7
P_{CO_2} (mmHg)	32	45	60	50
Na^- (mEq/L)	138	140	140	136
HCO_3^- (mEq/L)	22	20	28	34
Cl^- (mEq/L)	106	102	101	91
Anion gap* (mEq/L)	10	18	12	11

*Anion gap = Na^+ concentration – (HCO_3^- concentration + Cl^- concentration).

CLINICAL CASE Wrap-Up

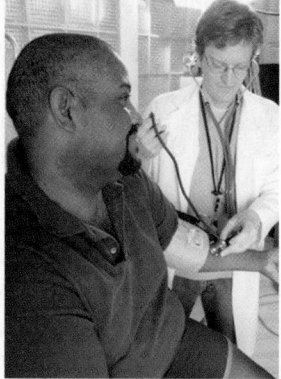

When Treatment Makes You Worse

"Your problem is a simple but critical electrolyte—potassium," the doctor tells Marquis.

Marquis had been in a state of electrolyte balance, in which the rates of gain and loss for each electrolyte are equal every day, but he had high blood pressure before beginning treatment. The hydrochlorothiazide diuretic he is taking works by inhibiting reabsorption of Na^+ and K^+ in the distal convoluted tubules. Recall that "water follows salt." The increased excretion of Na^+ and K^+ causes increased excretion of water, which reduces blood volume and therefore decreases pressure on the arterial walls. Ninety-eight percent of the body's potassium is in the ICF, leaving only a small amount in the ECF that is critically controlled. The fecal loss of K^+ is usually negligible, but diarrhea causes significant K^+ loss. Three days of diarrhea, on top of the new "potassium wasting" diuretic, have pushed Marquis into a state of hypokalemia. His blood K^+ measures only 2.5 mEq/L (normal 3.5–5.0 mEq/L).

Potassium is critical for a variety of cell functions, particularly in electrically active cells in the nervous system, skeletal muscle, and cardiac muscle. When blood K^+ levels are low, cells cannot repolarize and are unable to fire repeatedly. This causes muscle weakness, aching, and cramping. In cardiac muscle, hypokalemia increases the resting membrane potential and increases both the duration of the action potential and the duration of the refractory period, leading to cardiac arrhythmias.

1. How do you think Marquis's hypokalemia should be treated?

2. What is happening when Marquis complains of a sensation of a "frog jumping around" in his chest that is accompanied by dizziness?

See the blue Answers tab at the back of the book.

27

28

The Reproductive System

Learning Outcomes

These Learning Outcomes correspond by number to this chapter's sections and indicate what you should be able to do after completing the chapter.

28-1 ❙ List the basic structures of the human reproductive system, and summarize the functions of each.

28-2 ❙ Describe the structures of the male reproductive system and the roles played by the reproductive tract and accessory glands in producing spermatozoa; specify the composition of semen; and summarize the hormonal mechanisms that regulate male reproductive functions.

28-3 ❙ Describe the structures of the female reproductive system and the ovarian roles in oogenesis; explain the complete ovarian and uterine cycles; outline the histology, anatomy, and functions of the vagina; and summarize all aspects of the female reproductive cycle.

28-4 ❙ Discuss the physiology of sexual intercourse in males and females.

28-5 ❙ Describe the reproductive system changes that occur with aging.

28-6 ❙ Give examples of interactions between the reproductive system and each of the other organ systems.

SPOTLIGHTS

- Regulation of Male Reproduction p. 1093
- Regulation of Female Reproduction pp. 1110–1111

CLINICAL CASE

- A Post-Game Mystery p. 1077

Clinical Notes

- Dehydroepiandrosterone (DHEA) p. 1092

- Prostatic Hypertrophy and Prostate Cancer p. 1094
- Ovarian Cancer p. 1100
- Cervical Cancer p. 1101
- Breast Cancer p. 1109

A Post-Game Mystery

Dan and his high school football team have just beaten their archrivals in an exciting homecoming game, and they are looking forward to celebrating their victory at tonight's dance. Dan heads to the locker room. While showering, he starts to feel pain in the left side of his scrotum. He cannot recall getting hit there during the game. He does not see any bruising or swelling. However, his left testicle appears to be higher than his right, and is so tender to touch and movement that he can barely get dressed. Dan leaves the locker room hunched over in pain and feeling nauseous.

The coach notices Dan's odd posture and watches him stop to vomit in a trash can. He catches up to Dan and asks him what's going on. "I'm feeling a little sick to my stomach," says Dan. "And this is embarrassing, but my left testicle is killing me. I don't remember getting hit there, but it is so painful I can't touch it or stand up straight."

"Dan, I don't think you'll make it to the homecoming dance tonight," says Coach. "You're going straight to the emergency room." **Why is Coach so concerned about Dan's condition? To find out, turn to the Clinical Case Wrap-Up on p. 1120.**

An Introduction to the Reproductive System

Approximately 7.1 billion people currently live on the Earth. This is an astonishing number given that our reproductive system is the only system that is not essential to the life of an individual. Its activities do, however, affect other systems. In this chapter we discuss how the male and female reproductive organs produce and store specialized reproductive cells that combine to form new individuals. We also look at how various reproductive organs secrete hormones that play major roles in maintaining normal sexual function.

28-1 ┃ Basic reproductive system structures are gonads, ducts, accessory glands and organs, and external genitalia

Learning Outcome List the basic structures of the human reproductive system, and summarize the functions of each.

The human **reproductive system** ensures the continued existence of the human species—by producing, storing, nourishing, and transporting functional male and female reproductive cells. These cells are called **gametes** (GAM-ēts).

The reproductive system includes the following basic components:

- **Gonads** (GŌ-nadz; *gone*, seed, generation), or reproductive organs that produce gametes and hormones.
- Ducts that receive and transport the gametes.
- Accessory glands and organs that secrete fluids into the ducts of the reproductive system or into other excretory ducts.
- Perineal structures that are collectively known as the **external genitalia** (jen-i-TĀ-lē-uh).

In both males and females, the ducts are connected to chambers and passageways that open to the outside. The structures involved make up the *reproductive tract.* The male and female reproductive systems are functionally quite different, however.

In adult males, the **testes** (TES-tēz; singular, *testis*), or *testicles*, are male gonads that secrete sex hormones called *androgens.* The main androgen is *testosterone*, which was introduced in Chapter 18. ⟲ p. 667 The testes also produce the male gametes, called **spermatozoa** (sper-ma-tō-ZŌ-uh; singular, *spermatozoon*), or *sperm.* Males produce about one-half billion sperm each day. During *emission*, mature spermatozoa travel along a lengthy duct system, where they are mixed with the secretions of accessory glands. The mixture created is known as **semen** (SĒ-men). During *ejaculation*, semen is expelled from the body.

In adult females, the **ovaries**, or female gonads, release only one immature gamete, called an **oocyte**. Normally only one oocyte is released each month. This oocyte travels along one of two short *uterine tubes,* which end in the muscular organ called the *uterus* (Ū-ter-us). If a sperm reaches the oocyte and starts the process of *fertilization*, the oocyte matures into an **ovum** (plural, *ova*). A short passageway, the *vagina* (va-JĪ-nuh), connects the uterus with the exterior.

During *sexual intercourse*, ejaculation introduces semen into the vagina, and the spermatozoa then ascend the female reproductive tract. If fertilization takes place, the uterus will enclose and support a developing *embryo* as it grows into a *fetus* and prepares for birth.

Next we examine the anatomy of the male and female reproductive systems. We will consider the physiological and hormonal mechanisms that regulate reproductive function. Earlier chapters introduced the anatomical reference points used in the discussions that follow. You may find it helpful to review the figures on the pelvic girdle (**Figure 8–7** and **Figure 8–8**, pp. 277, 278), perineal musculature (**Figure 11–13**, p. 382),

28

pelvic innervation (**Figure 13–12**, p. 472), and regional blood supply (**Figures 21–24** and **21–28**, pp. 788, 794).

> **CHECKPOINT**
>
> 1. **Define gamete.**
> 2. **List the basic structures of the reproductive system.**
> 3. **Define gonads.**
>
> See the blue Answers tab at the back of the book.

28-2 ❙ Spermatogenesis occurs in the testes, and hormones from the hypothalamus, anterior lobe of the pituitary gland, and testes control male reproductive functions

Learning Outcome Describe the structures of the male reproductive system and the roles played by the reproductive tract and accessory glands in producing spermatozoa; specify the composition of semen; and summarize the hormonal mechanisms that regulate male reproductive functions.

Figure 28–1 shows the main structures of the male reproductive system. Starting from a testis, the spermatozoa travel within the *epididymis* (ep-i-DID-i-mus); the *ductus deferens* (DUK-tus

DEF-e-renz; plural, *ductus deferentia*), also called the *vas deferens;* the *ejaculatory duct;* and the *urethra* before leaving the body. Accessory organs—the *seminal* (SEM-i-nal) *glands* (seminal vesicles), the *prostate* (PROS-tāt) *gland,* and the *bulbourethral* (bul-bō-ū-RĒ-thral) *glands*—secrete various fluids into the ejaculatory ducts and urethra. The external genitalia consist of the *scrotum* (SKRŌ-tum), which encloses the testes, and the *penis* (PĒ-nis), an erectile organ. The distal portion of the urethra passes through the penis.

The Testes

Each testis is about 5 cm (2 in.) long, 3 cm (1.2 in.) wide, and 2.5 cm (1 in.) thick. Each weighs 10–15 g (0.35–0.53 oz). The testes hang within the **scrotum**, a fleshy pouch suspended inferior to the perineum. The scrotum is anterior to the anus, and posterior to the base of the penis (**Figure 28–1**). ATLAS: Embryology Summary 21: The Development of the Reproductive System

Descent of the Testes

During fetal development, the testes form inside the body cavity adjacent to the kidneys. A bundle of connective tissue fibers—called the **gubernaculum testis** (goo-bur-NAK-ū-lum TES-tis)—extends from each testis to the posterior wall of a small anterior and inferior pocket of the peritoneum (**Figure 28–2a**).

Figure 28–1   The Male Reproductive System. A sagittal section of the male reproductive structures.
ATLAS: Plate 64

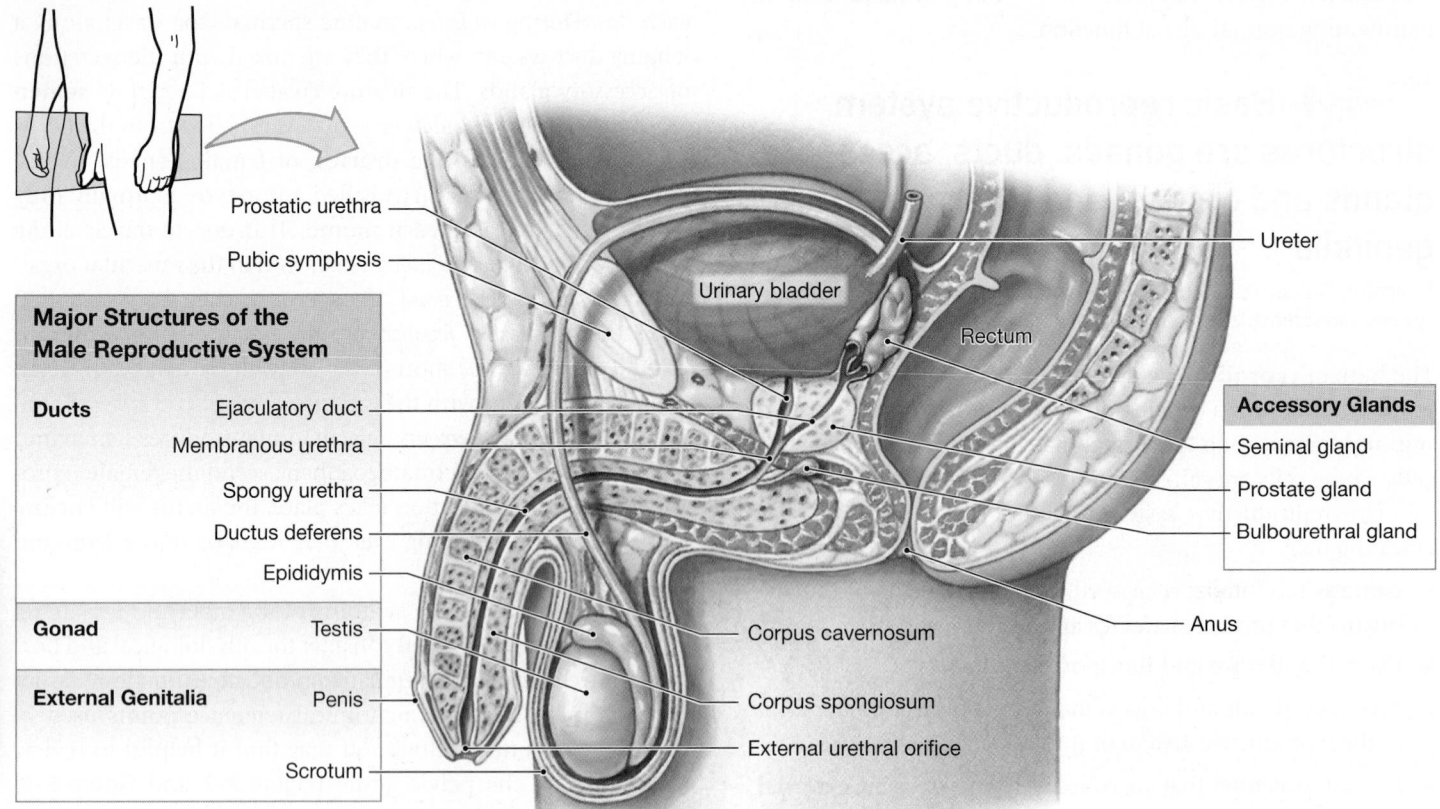

Major Structures of the Male Reproductive System	
Ducts	Ejaculatory duct
	Membranous urethra
	Spongy urethra
	Ductus deferens
	Epididymis
Gonad	Testis
External Genitalia	Penis
	Scrotum

Prostatic urethra
Pubic symphysis
Urinary bladder
Ureter
Rectum
Ejaculatory duct
Membranous urethra
Spongy urethra
Ductus deferens
Epididymis

Accessory Glands
Seminal gland
Prostate gland
Bulbourethral gland

Corpus cavernosum
Anus
Corpus spongiosum
External urethral orifice

28

Figure 28–2 The Descent of the Testes.

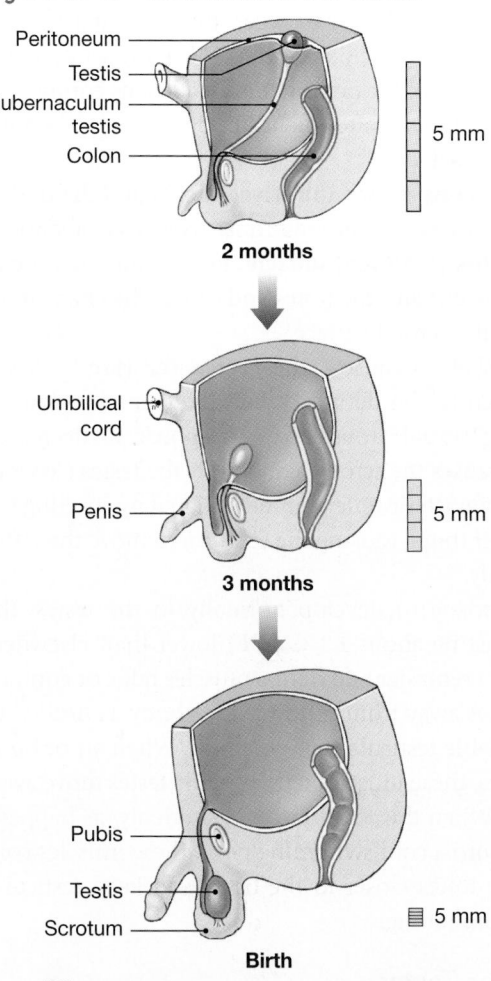

Peritoneum
Testis
Gubernaculum testis
Colon

5 mm

2 months

Umbilical cord

Penis

5 mm

3 months

Pubis

Testis

Scrotum

5 mm

Birth

a Sagittal sectional views of the positional changes involved in the descent of the right testis. Because the size of the gubernaculum testis remains constant (see the scale bar at the right) while the rest of the fetus grows, the relative position of the testis shifts.

Diaphragmatic ligament
Mesonephric duct

Developing kidneys
Gonads
Gubernaculum testis

2 months

Kidney
Diaphragmatic ligament
Epididymis
Testis

3 months

Epididymis
Testis
Urinary bladder

4 months

Ureter
Superficial inguinal ring
Scrotal cavity (opened) lined by tunica vaginalis

Testicular artery and vein
Vas deferens
Epididymis
Testis
Gubernaculum testis

7 months

Ureter
Spermatic cord

Testicular artery and vein
Urinary bladder
Scrotum (opened)

Birth

b Frontal views showing the descent of the testes and the formation of the spermatic cords.

As the fetus grows, the gubernacula do not get any longer, so they lock the testes in position. As a result, the position of each testis changes as the body enlarges. The testis gradually moves inferiorly and anteriorly toward the anterior abdominal wall. During the seventh developmental month, fetal growth continues rapidly, and circulating hormones stimulate a contraction of the gubernaculum testis. During this time, each testis moves through the abdominal musculature, along with small pockets of the peritoneal cavity. This process is called the **descent of the testes** into the scrotum (**Figure 28–2**).

In *cryptorchidism* (krip-TOR-ki-dizm; *crypto*, hidden + *orchis*, testis), one or both of the testes have not descended into the scrotum by the time of birth. Typically, the cryptorchid (abdominal) testes are lodged in the abdominal cavity or within the inguinal canal. Cryptorchidism occurs in about 3 percent

28

of full-term deliveries and in about 30 percent of premature births. In most instances, normal descent takes place a few weeks later.

The condition can be surgically corrected if it persists. Corrective measures should be taken before *puberty* (sexual maturation), because a cryptorchid testis will not produce spermatozoa. If both testes are cryptorchid, the male will be *sterile (infertile)* and unable to father children. If the testes cannot be moved into the scrotum, they will usually be removed. This surgical procedure to remove the testes is called an *orchiectomy* (or-kē-EK-tō-mē). About 10 percent of males with uncorrected cryptorchid testes eventually develop testicular cancer.

As each testis moves through the body wall, the ductus deferens and the testicular blood vessels, nerves, and lymphatic vessels accompany it. Together, these structures form the body of the spermatic cord, which we discuss next.

The Spermatic Cords

The **spermatic cords** are paired structures extending between the abdominopelvic cavity and the testes (**Figure 28–3**). Each spermatic cord begins at the entrance to the *inguinal canal* (a passageway through the abdominal musculature). After passing through the inguinal canal, the spermatic cord descends into the scrotum.

Each spermatic cord consists of layers of fascia and muscle enclosing the ductus deferens and the blood vessels, nerves, and lymphatic vessels that supply the testes. The blood vessels include the *deferential artery*, a *testicular artery*, and the **pampiniform** (pam-PIN-i-form; *pampinus*, tendril + *forma*, form) **plexus** of a testicular vein. Branches of the *genitofemoral nerve* from the lumbar plexus provide innervation.

The inguinal canals form during development as the testes descend into the scrotum. At that time, these canals link the scrotal cavities with the peritoneal cavity. In normal adult males, the inguinal canals are closed, but the presence of the spermatic cords creates weak points in the abdominal wall that remain throughout life. As a result, *inguinal hernias*—protrusions of visceral tissues or organs into the inguinal canal—are fairly common in males. The inguinal canals in females are very small, containing only the *ilioinguinal nerves* and the *round ligaments* of the uterus. The abdominal wall is nearly intact, so inguinal hernias in women are very rare.

The Scrotum and the Position of the Testes

The scrotum is divided internally into two chambers. A raised thickening in the scrotal surface known as the **raphe** (RĀ-fē) divides it in two (**Figure 28–3**). Each testis lies in a separate chamber, or **scrotal cavity**. Because the scrotal cavities are separated by a partition, infection or inflammation of one testis does not normally spread to the other.

A narrow space separates the inner surface of the scrotum from the outer surface of the testis. The **tunica vaginalis**

(TU-ni-ka vaj-i-NAL-is), a serous membrane, lines the scrotal cavity and reduces friction between the opposing parietal (scrotal) and visceral (testicular) surfaces. The tunica vaginalis is an isolated portion of the peritoneum that lost its connection with the peritoneal cavity after the testes descended, when the inguinal canal closed.

The scrotum consists of a thin layer of skin and the underlying superficial fascia. The dermis contains a layer of smooth muscle, the **dartos** (DAR-tōs) **muscle**. Resting muscle tone in the dartos muscle elevates the testes and causes the characteristic wrinkling of the scrotal surface.

A layer of skeletal muscle, the **cremaster** (krē-MAS-ter) **muscle**, lies deep to the dermis. When the cremaster muscle contracts during sexual arousal or in response to decreased temperature, it tenses the scrotum and pulls the testes closer to the body. The cremasteric reflex can be initiated by stroking the skin on the upper thigh, causing the scrotum to move the testes closer to the body.

For spermatozoa to develop normally in the testes, the temperature must be about 1.1°C (2°F) lower than elsewhere in the body. The cremaster and dartos muscles relax or contract to move the testes away from or toward the body as needed to maintain acceptable testicular temperatures. When air or body temperature rises, these muscles relax and the testes move away from the body. When the scrotum cools suddenly, as happens when jumping into a cold swimming pool, these muscles contract to pull the testes closer to the body and keep testicular temperatures from falling.

Structure of the Testes

Deep to the tunica vaginalis covering the testis is the **tunica albuginea** (al-bū-JIN-ē-uh), a dense layer of connective tissue rich in collagen fibers (**Figure 28–4a**). These fibers are continuous with those surrounding the adjacent epididymis and extend into the testis. There they form fibrous partitions, or *septa*, that converge toward the region nearest the entrance to the epididymis. The connective tissues in this region support the blood vessels and lymphatic vessels that supply and drain the testis, and the *efferent ductules*, which transport spermatozoa to the epididymis.

Histology of the Testes

The septa subdivide the testis into a series of **lobules** (**Figure 28–4a**). Distributed among the lobules are approximately 800 slender, tightly coiled **seminiferous** (sem-i-NIF-er-us) **tubules** (**Figures 28–4** and **28–5**). Each tubule averages about 80 cm (32 in.) in length. A typical testis contains nearly one-half mile of seminiferous tubules. Sperm production takes place within these tubules.

Each lobule contains several seminiferous tubules that merge into **straight tubules** that enter the mediastinum of the testis. Straight tubules are extensively interconnected, forming

Figure 28–3   **The Male Reproductive System in Anterior View.** In the anatomical position, the penis is erect, and the dorsal surface is closest to the navel. In a flaccid state, the dorsal surface is facing anteriorly.

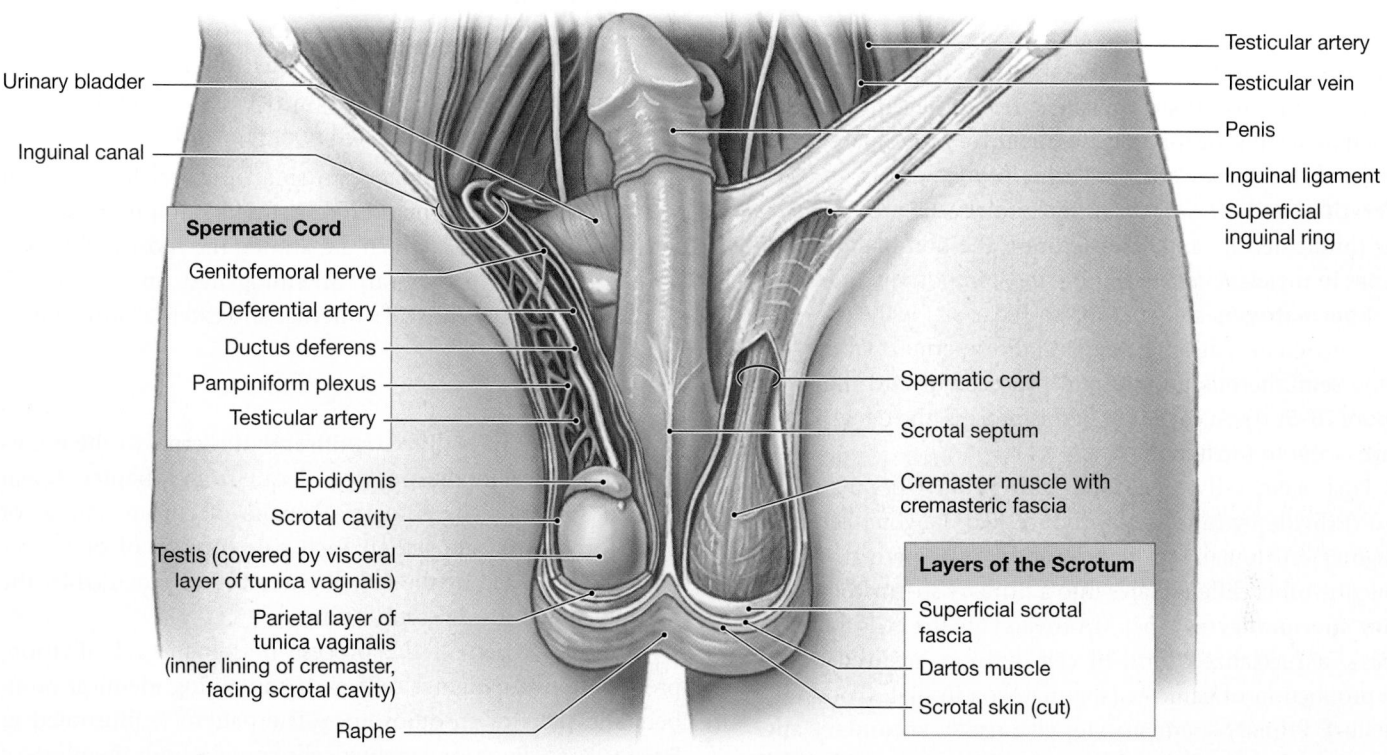

Figure 28–4   **The Structure of the Testes.**

a A transverse section of the scrotum and testes

b A section through a testis

Testis   LM × 26

28

a maze of passageways known as the **rete** (RĒ-tē; rete, a net) **testis** (**Figure 28–4**). Fifteen to 20 large **efferent ductules** connect the rete testis to the epididymis.

Because the seminiferous tubules are tightly coiled, most tissue slides show them in transverse section (**Figure 28–5a**). A delicate connective tissue capsule surrounds each tubule, and areolar tissue fills the spaces between the tubules (**Figure 28–5b**). Within those spaces are numerous blood vessels and large **interstitial cells** (*Leydig cells*). Interstitial cells produce *androgens* (androsterone and testosterone), the dominant sex hormones in males. *Testosterone* is the most important androgen.

Spermatogenesis (sper-ma-tō-JEN-e-sis) is the process of spermatozoa formation. It begins at the outermost layer of cells in the seminiferous tubules and proceeds toward the lumen (**Figure 28–5c,d**). At each step in this process, the daughter cells move closer to the lumen.

First, stem cells called **spermatogonia** (sper-ma-tō-GŌ-nē-uh) divide by mitosis to produce two daughter cells. One daughter cell remains at that location as a spermatogonium while the other differentiates into a primary spermatocyte. **Primary spermatocytes** (sper-MA-tō-sīts) are the cells that begin *meiosis*, a specialized form of cell division involved only in the production of gametes (spermatozoa in males, oocytes in females). Primary spermatocytes give rise to **secondary spermatocytes** that divide and differentiate into **spermatids** (SPER-ma-tidz)—immature gametes that subsequently differentiate into spermatozoa. The spermatozoa lose contact with the wall of the seminiferous tubule and enter the fluid in the lumen.

Each seminiferous tubule contains spermatogonia, spermatocytes at various stages of meiosis, spermatids, spermatozoa, and large **nurse cells.** Nurse cells are also known as *sustentacular* (sus-ten-TAK-ū-lar) cells or *Sertoli cells.* They provide a microenvironment that supports spermatogenesis (**Figure 28–5b,d**).

Spermatogenesis

Spermatogenesis begins at puberty and continues until relatively late in life (past age 70). It is a continuous process. All stages of meiosis are seen within the seminiferous tubules.

The complete process of spermatogenesis takes about 64 days. It involves three integrated processes:

1. *Mitosis.* Spermatogonia are stem cells that undergo cell divisions throughout adult life. (You can review the description of mitosis and cell division in Chapter 3. ⮌ pp. 130–131) One daughter cell from each division remains in place while the other is pushed toward the lumen (space) of the seminiferous tubule. The displaced cells differentiate into primary spermatocytes, which prepare to begin meiosis.

2. *Meiosis.* Meiosis (mī-Ō-sis) is a special form of cell division involved in gamete production. In humans, gametes contain 23 chromosomes, half the amount found in somatic cells. As a result, the fusion of the nuclei of a male gamete and a female gamete produces a cell that has the normal number of chromosomes (46), rather than twice that number. In the seminiferous tubules, meiotic divisions that begin with primary spermatocytes produce spermatids, the undifferentiated male gametes.

3. *Spermiogenesis.* Spermatids are small, unspecialized cells. In *spermiogenesis,* spermatids differentiate into physically mature spermatozoa, which are among the most highly specialized cells in the body. Spermiogenesis involves major changes in a spermatid's internal and external structures.

Mitosis and Meiosis

Mitosis and meiosis differ significantly in terms of the events that take place in the nucleus. Recall from Chapter 3 that human somatic cells contain 23 pairs of chromosomes, or 46 chromosomes altogether. Each pair consists of one chromosome provided by the father, and another provided by the mother, at the time of fertilization.

Mitosis is part of the process of somatic cell division, producing two daughter cells each containing identical numbers and pairs of chromosomes. The pattern is illustrated in **Figure 28–6a**. Because daughter cells contain both members of each chromosome pair (for a total of 46 chromosomes), they are called **diploid** (DIP-loyd; *diplo*, double) (2n) cells.

In contrast, meiosis follows the pattern in **Figure 28–6b**. It involves two cycles of cell division (*meiosis I* and *meiosis II*) and produces four cells, each containing 23 *individual* chromosomes (n=23). These cells are called **haploid** (HAP-loyd; *haplo*, single) (n) cells and contain only one member of each pair of chromosomes. The events in the nucleus shown in **Figure 28–6b** are the same for the formation of spermatozoa or oocytes.

As a cell prepares to begin meiosis, DNA replication occurs within the nucleus just as it does in a cell preparing for mitosis. The duplicated chromosomes condense and become visible with a light microscope. As in mitosis, each chromosome now consists of two duplicate *chromatids*.

At this point, the close similarities between meiosis and mitosis end. In meiosis, the corresponding maternal (inherited from the mother) and paternal (inherited from the father) chromosomes now come together in an event known as **synapsis** (si-NAP-sis). Synapsis involves 23 pairs of chromosomes, and each member of each pair consists of two chromatids. A matched set of four chromatids is called a **tetrad** (TET-rad; *tetras*, four) (**Figure 28–6b**). Some genetic material can be exchanged between the chromatids of a chromosome pair at this stage of meiosis. Such an exchange, called *crossing over*, increases genetic variation among offspring. We discuss genetics in Chapter 29.

Figure 28–5   **The Seminiferous Tubules.**

a A section through one or more seminiferous tubules.

b A cross section through a single tubule.

c Nurse cells surround the stem cells of the tubule and support the developing spermatocytes and spermatids.

d Stages in spermatogenesis in the wall of a seminiferous tubule.

28

Figure 28–6 **Chromosomes in Mitosis and Meiosis.**

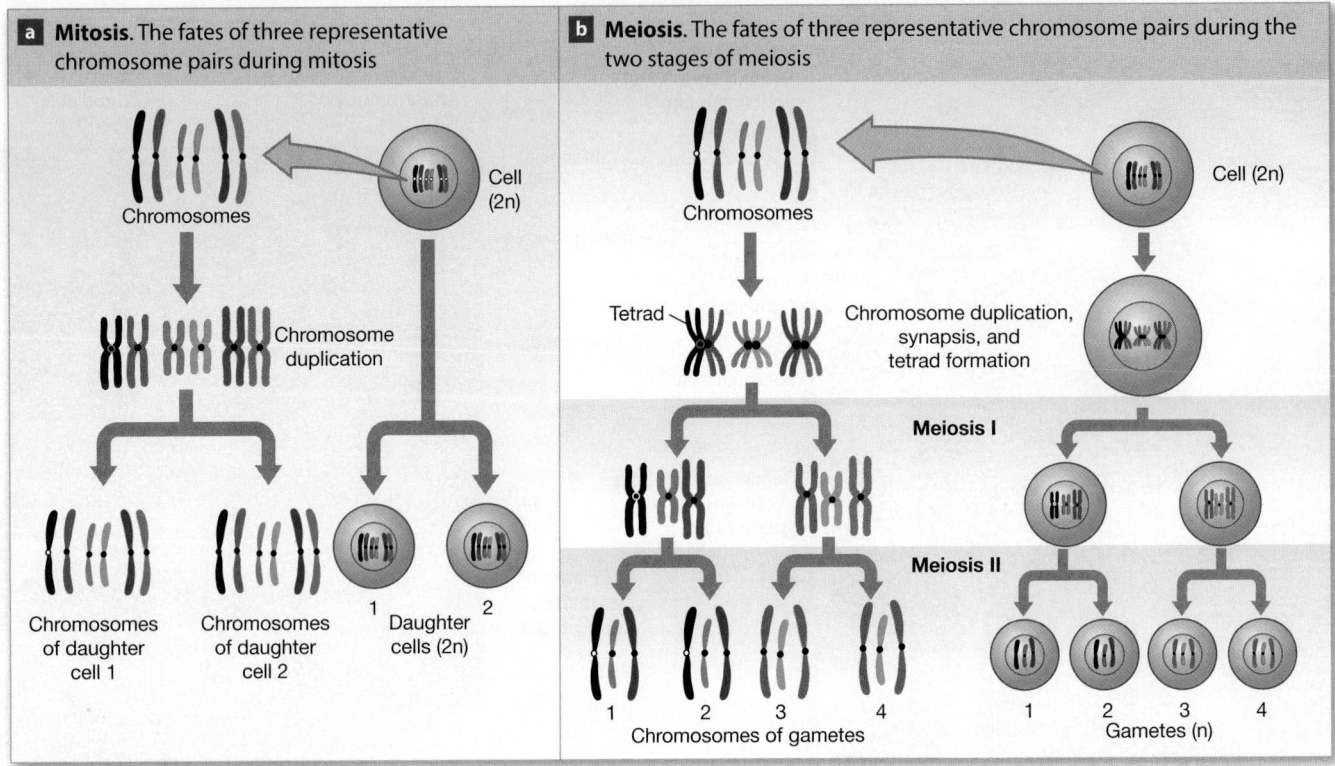

a **Mitosis**. The fates of three representative chromosome pairs during mitosis

Chromosomes

Cell (2n)

Chromosome duplication

Chromosomes of daughter cell 1

Chromosomes of daughter cell 2

Daughter cells (2n)
1   2

b **Meiosis**. The fates of three representative chromosome pairs during the two stages of meiosis

Chromosomes

Cell (2n)

Tetrad

Chromosome duplication, synapsis, and tetrad formation

Meiosis I

Meiosis II

Chromosomes of gametes
1   2   3   4

Gametes (n)
1   2   3   4

Meiosis includes two division cycles, referred to as **meiosis I** and **meiosis II**. We identify the stages within each cycle with I or II, for example as prophase I, metaphase II, and so on. The nuclear envelope disappears at the end of prophase I. As metaphase I begins, the tetrads line up along the metaphase plate. As anaphase I begins, the tetrads break up—the maternal and paternal chromosomes separate. This is a major difference between mitosis and meiosis: In mitosis, each daughter cell receives a copy of every chromosome, maternal and paternal, but in meiosis I, each daughter cell receives two copies of *either* the maternal chromosome *or* the paternal chromosome. (Compare the two parts of **Figure 28–6**.)

As anaphase I proceeds, the maternal and paternal components are randomly and independently distributed. That is, as each tetrad splits, we cannot predict which daughter cell will receive copies of the maternal chromosome, and which will receive copies of the paternal chromosome. As a result, telophase I ends with the formation of two daughter cells containing unique combinations of maternal and paternal chromosomes. Both cells contain 23 chromosomes.

The first meiotic division is called a **reductional division** because it reduces the number of chromosomes from 46 to 23. Each of these chromosomes still consists of two duplicate chromatids. The duplicates will separate during meiosis II.

The interphase between meiosis I and meiosis II is very brief. No DNA is replicated during that period. Each cell then proceeds through prophase II, metaphase II, and anaphase II. During anaphase II, the duplicate chromatids separate. Telophase II yields *four haploid cells*, each containing 23 chromosomes. Because the number of chromosomes has not changed, meiosis II is an **equational division**.

The chromosomes are evenly distributed among the four cells, but the cytoplasm may not be. In males, meiosis produces four immature gametes that are identical in size. Each will develop into a functional sperm. In females, meiosis produces one relatively huge oocyte and three tiny, nonfunctional polar bodies. (If fertilization takes place, the oocyte completes meiosis II, yielding an ovum.) We examine the details of spermatogenesis here (**Figure 28–7**). We will consider oogenesis in a later section.

Spermiogenesis

The four spermatids initially remain interconnected by bridges of cytoplasm because cytokinesis (cytoplasmic division) is not completed in meiosis I or meiosis II. These connections assist in transferring nutrients and hormones between the cells, helping ensure that the cells develop in synchrony. The bridges are not broken until the last stages of physical maturation.

In **spermiogenesis**, the last step of spermatogenesis, each spermatid matures into a single spermatozoon, or *sperm* (**Figure 28–7**). Developing spermatocytes undergoing meiosis, and spermatids undergoing spermiogenesis, are not free in

28

Figure 28–7   **Spermatogenesis.** The events depicted occur in the seminiferous tubules. The fates of three representative chromosome pairs are shown; for clarity, maternal and paternal chromatids are not identified.

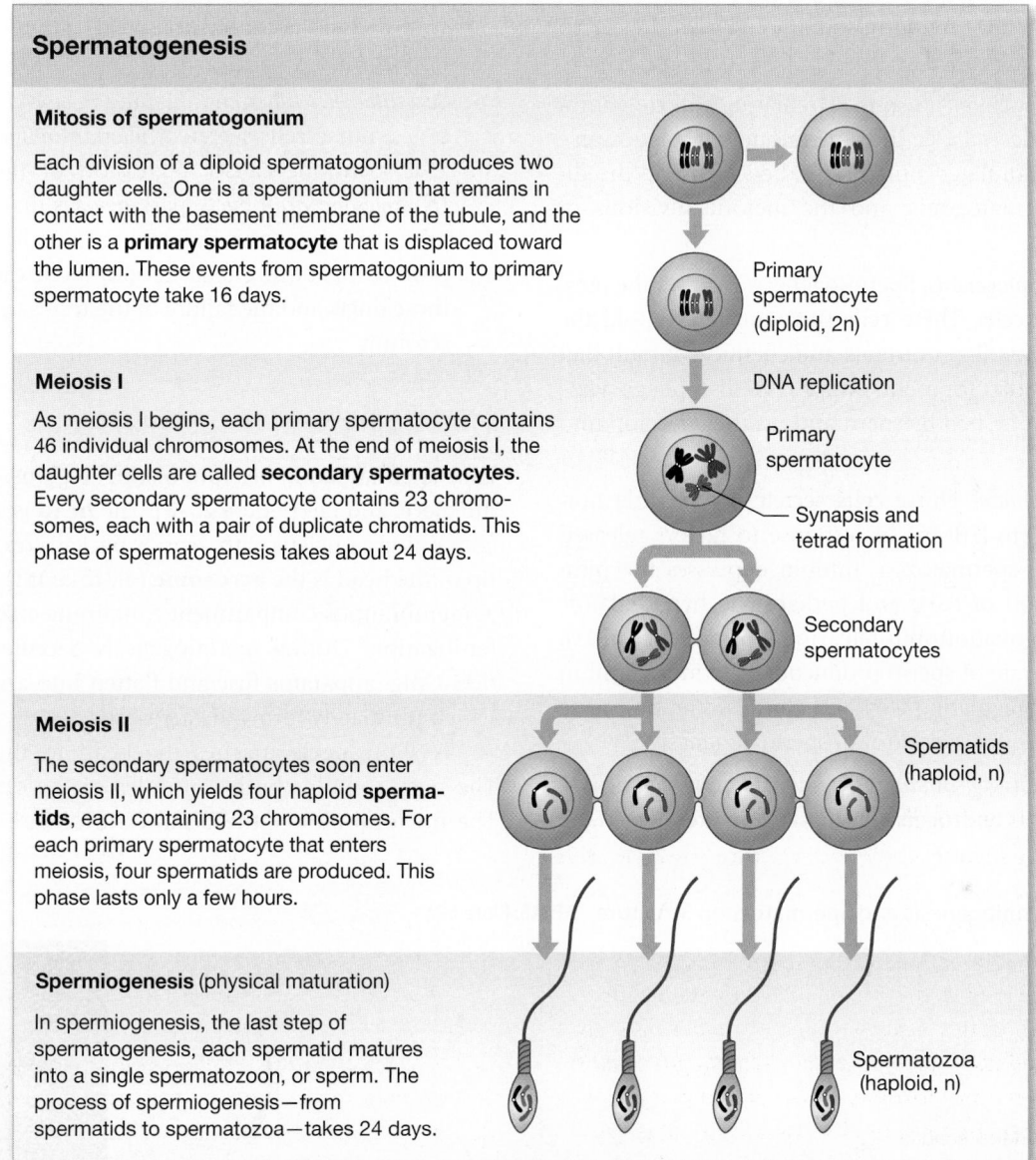

Spermatogenesis

Mitosis of spermatogonium

Each division of a diploid spermatogonium produces two daughter cells. One is a spermatogonium that remains in contact with the basement membrane of the tubule, and the other is a **primary spermatocyte** that is displaced toward the lumen. These events from spermatogonium to primary spermatocyte take 16 days.

Primary spermatocyte (diploid, 2n)

DNA replication

Meiosis I

As meiosis I begins, each primary spermatocyte contains 46 individual chromosomes. At the end of meiosis I, the daughter cells are called **secondary spermatocytes**. Every secondary spermatocyte contains 23 chromosomes, each with a pair of duplicate chromatids. This phase of spermatogenesis takes about 24 days.

Primary spermatocyte

Synapsis and tetrad formation

Secondary spermatocytes

Meiosis II

The secondary spermatocytes soon enter meiosis II, which yields four haploid **spermatids**, each containing 23 chromosomes. For each primary spermatocyte that enters meiosis, four spermatids are produced. This phase lasts only a few hours.

Spermatids (haploid, n)

Spermiogenesis (physical maturation)

In spermiogenesis, the last step of spermatogenesis, each spermatid matures into a single spermatozoon, or sperm. The process of spermiogenesis—from spermatids to spermatozoa—takes 24 days.

Spermatozoa (haploid, n)

the seminiferous tubules. Instead, they are surrounded by the cytoplasm of adjacent nurse cells. As spermiogenesis proceeds, the spermatids gradually develop into mature spermatozoa. At *spermiation,* a spermatozoon loses its attachment to the nurse cell and enters the lumen of the seminiferous tubule.

Nurse Cells. Nurse cells play a key role in spermatogenesis. These cells have six important functions that directly or indirectly affect mitosis, meiosis, and spermiogenesis within the seminiferous tubules:

1. *Maintenance of the Blood–Testis Barrier.* A **blood–testis barrier** isolates the seminiferous tubules from the general circulation. This barrier is comparable in function to the blood–brain barrier. ⊃ p. 495 Nurse cells are joined by

tight junctions. They form a layer that divides the seminiferous tubule into an outer basal compartment and an inner luminal compartment. The *basal compartment* contains the spermatogonia. Meiosis and spermiogenesis occur in the *luminal compartment* (**Figure 28–5d**). Transport across the nurse cells is tightly regulated, so conditions in the luminal compartment remain very stable.

The nurse cells produce the fluid in the lumen of a seminiferous tubule. They also regulate the fluid's composition. This luminal fluid is very different from the surrounding interstitial fluid. It is high in androgens, estrogens, potassium, and amino acids. The blood–testis barrier is essential to preserving the differences between the luminal fluid and the interstitial fluid.

28

In addition, this barrier prevents immune system cells from detecting and attacking the developing spermatozoa. The plasma membranes of spermatozoa contain sperm-specific antigens not found in somatic cell membranes, so they might be identified as "foreign."

2. *Support of Mitosis and Meiosis.* Circulating follicle-stimulating hormone (FSH) and testosterone stimulate nurse cells. These stimulated nurse cells then promote the division of spermatogonia and the meiotic divisions of spermatocytes.

3. *Support of Spermiogenesis.* Spermiogenesis requires the presence of nurse cells. These cells surround and enfold the spermatids, providing nutrients and chemical stimuli that promote their development. Nurse cells also phagocytize cytoplasm that is shed by spermatids as they develop into spermatozoa.

4. *Secretion of Inhibin.* Nurse cells secrete the peptide hormone *inhibin* (in-HIB-in) in response to factors released by developing spermatozoa. Inhibin depresses the pituitary production of FSH, and perhaps the hypothalamic secretion of gonadotropin-releasing hormone (GnRH). The faster the rate of sperm production, the more inhibin is secreted. By regulating FSH and GnRH secretion, nurse cells provide feedback control of spermatogenesis.

5. *Secretion of Androgen-Binding Protein. Androgen-binding protein (ABP)* binds androgens (primarily testosterone) in the luminal fluid of the seminiferous tubules. This protein is thought to be important in both elevating the concentration of androgens within the seminiferous tubules and stimulating spermiogenesis. FSH stimulates the production of ABP.

6. *Secretion of Müllerian-Inhibiting Factor.* In the developing testes, nurse cells secrete *Müllerian-inhibiting factor (MIF)*. This hormone causes regression of the fetal *Müllerian (paramesonephric) ducts,* passageways that form the uterine tubes and the uterus in females. In males, inadequate MIF production during fetal development causes retention of these ducts and the failure of the testes to descend into the scrotum.

The Anatomy of a Spermatozoon

Each spermatozoon has four distinct regions: head, neck, middle piece, and tail (**Figure 28–8**). The **head** is a flattened ellipse containing a nucleus with densely packed chromosomes. At the tip of the head is the **acrosome** (ak-rō-SŌM), or *acrosomal cap,* a membranous compartment containing enzymes essential to fertilization. During spermiogenesis, saccules of the spermatid's Golgi apparatus fuse and flatten into an *acrosomal vesicle,* which ultimately forms the acrosome.

A short **neck** attaches the head to the **middle piece**. The neck contains both centrioles of the original spermatid. The microtubules of the distal centriole are continuous with

Figure 28–8 **Spermiogenesis and Spermatozoon Structure.** ATLAS: Plate 60a

Spermatid

Mitochondria

Nucleus

Golgi apparatus

Acrosomal vesicle

Acrosome

Shed cytoplasm

Nucleus

Acrosome

Spermiogenesis. The differentiation of a spermatid into a spermatozoon. This process is completed in approximately 24 days.

Tail (55 μm)

Middle piece (5 μm)

Neck (1 μm)

Head (5 μm)

Fibrous sheath of flagellum

Mitochondrial spiral

Centrioles

Nucleus

Acrosome

Spermatozoon

Spermatozoa   SEM × 780

28

those of the middle piece and tail. Mitochondria in the middle piece are arranged in a spiral around the microtubules. Mitochondria provide the ATP required to move the tail.

The **tail** is the only flagellum in the human body. A *flagellum* is a whiplike organelle that moves a cell from one place to another. Cilia beat in a predictable, wavelike fashion, but the flagellum of a spermatozoon has a whiplike, corkscrew motion.

Unlike other, less specialized cells, a mature spermatozoon lacks an endoplasmic reticulum, a Golgi apparatus, lysosomes, peroxisomes, inclusions, and many other organelles. The loss of these organelles reduces the cell's size and mass. A sperm is essentially a mobile carrier for chromosomes, and extra weight would slow it down. Because a spermatozoon lacks glycogen or other energy reserves, it must absorb nutrients (primarily fructose) from the surrounding fluid.

The Male Reproductive Tract

The testes produce physically mature spermatozoa that are not yet capable of successfully fertilizing an oocyte. The other portions of the male reproductive system are responsible for the functional maturation, nourishment, storage, and transport of spermatozoa.

The Epididymis

Late in their development, spermatozoa detach from the nurse cells and lie within the lumen of the seminiferous tubule. They have most of the physical characteristics of mature spermatozoa, but are functionally immature and incapable of coordinated locomotion or fertilization. Fluid currents, created by cilia lining the efferent ductules, transport these immobile gametes into the epididymis (**Figure 28–4a**). The **epididymis** (ep-i-DID-i-mis; *epi*, on + *didymos*, twin) is the start of the male reproductive tract. It is a coiled tube bound to the posterior border of each testis.

The epididymides (ep-i-DID-i-mi-dēz) can be felt through the skin of the scrotum. Each epididymis is almost 7 m (23 ft) long, but it is coiled and twisted so it takes up very little space. It has a head, a body, and a tail (**Figure 28–9a,c**). The superior **head** is the portion of the epididymis proximal to the testis. The head receives spermatozoa from the efferent ductules.

The **body** begins distal to the last efferent ductule and extends inferiorly along the posterior margin of the testis. Near the inferior border of the testis, the number of coils decreases, marking the start of the **tail**. The tail re-curves and ascends to its connection with the ductus deferens. Spermatozoa are stored primarily within the tail of the epididymis.

Figure 28–9 **The Epididymis.** ATLAS: Plate 60a

Flagella of spermatozoa in lumen of epididymis

Pseudostratified columnar epithelium of epididymis

Stereocilia

Epididymis LM × 304

b Sectional view of the epididymis

Tunica vaginalis (reflected) Spermatic cord Ductus deferens

Spermatic cord

Ductus deferens

Efferent ductules

Straight tubule

Rete testis

Seminiferous tubule

Tunica albuginea

Testis

Scrotal cavity

Epididymis

Head of epididymis

Body of epididymis

Tail of epididymis

Testis

a A diagrammatic view of the head, body, and tail of an epididymis

c Appearance of the testis and epididymis on gross dissection

28

The epididymis has three functions:

1. *It monitors and adjusts the composition of the fluid produced by the seminiferous tubules.* The pseudostratified columnar epithelial lining of the epididymis has distinctive stereocilia (**Figure 28–9b**). These stereocilia increase the surface area available for absorption from, and secretion into, the fluid in the tubule.

2. *It acts as a recycling center for damaged spermatozoa.* Cellular debris and damaged spermatozoa are absorbed in the epididymis. The products of enzymatic breakdown are released into the surrounding interstitial fluids for pickup by the epididymal blood vessels.

3. *It stores and protects spermatozoa and facilitates their functional maturation.* It takes up to two weeks for a spermatozoon to pass through the epididymis and complete its functional maturation. During this time, spermatozoa exist in a sheltered environment that is precisely regulated by the surrounding epithelial cells.

Spermatozoa leaving the epididymis are mature but immobile. To become *motile* (actively swimming) and fully functional, they must undergo a process called **capacitation**. Capacitation normally takes place in two steps: (1) Spermatozoa become motile when they are mixed with secretions of the seminal glands, and (2) they become capable of successful fertilization when exposed to conditions in the female reproductive tract. The epididymis secretes a substance (as yet unidentified) that prevents premature capacitation.

Transport along the epididymis involves a combination of fluid movement and peristaltic contractions of smooth muscle in the epididymis. After passing along the tail of the epididymis, the spermatozoa enter the ductus deferens.

The Ductus Deferens

Each **ductus deferens**, or *vas deferens*, is 40–45 cm (16–18 in.) long. It begins at the tail of the epididymis (**Figure 28–9a,c**). As part of the spermatic cord, it ascends through the inguinal canal (**Figure 28–3**). Inside the abdominal cavity, the ductus deferens passes posteriorly, curving inferiorly along the lateral surface of the urinary bladder toward the superior and posterior margin of the prostate gland (**Figure 28–1**). Just before the ductus deferens reaches the prostate gland and seminal glands, its lumen enlarges. This expanded portion is known as the **ampulla** (am-PUL-luh) of the ductus deferens (**Figure 28–10a**).

The wall of the ductus deferens contains a thick layer of smooth muscle (**Figure 28–10b**). Peristaltic contractions in this layer propel spermatozoa and fluid along the duct, which is lined by a pseudostratified ciliated columnar epithelium. In addition to transporting spermatozoa, the ductus deferens can store spermatozoa for several months. During this time, the spermatozoa remain in a temporary state of inactivity with low metabolic rates.

The junction of the ampulla with the duct of the seminal gland marks the start of the **ejaculatory duct**. This short passageway is 2 cm, or less than 1 inch. It penetrates the muscular wall of the prostate gland and empties into the urethra (**Figures 28–1** and **28–10a**).

The Urethra

In males, the **urethra** is a passageway that extends 18–20 cm (7–8 in.) from the urinary bladder to the tip of the penis (**Figure 28–1**). It is divided into *prostatic, membranous,* and *spongy* regions. The male urethra is used by both the urinary and reproductive systems.

The Accessory Glands

The fluids secreted by the seminiferous tubules and the epididymis account for only about 5 percent of the volume of semen. The fluid component of semen is a mixture of secretions—each with distinctive biochemical characteristics—from many glands. Important glands include the *seminal glands,* the *prostate gland,* and the *bulbourethral glands.* All of these occur only in males.

Among the major functions of these glands are (1) activating spermatozoa; (2) providing the nutrients spermatozoa need for motility; (3) propelling spermatozoa and fluids along the reproductive tract, mainly by peristaltic contractions; and (4) producing buffers that counteract the acidity of the urethral and vaginal environments.

The Seminal Glands (Seminal Vesicles)

The ductus deferens on each side ends at the junction between the ampulla and the duct that drains the seminal gland (**Figure 28–10a**). The **seminal glands**, also called the **seminal vesicles**, are tubular glands embedded in connective tissue on either side of the midline, sandwiched between the posterior wall of the urinary bladder and the rectum. Each seminal gland is about 15 cm (6 in.). The body of the gland has many short side branches. The entire assemblage is coiled and folded into a compact, tapered mass roughly 5 cm × 2.5 cm (2 in. × 1 in.).

Seminal glands are very active secretory glands with an epithelial lining that contains extensive folds (**Figure 28–10c**). The seminal glands secrete about 60 percent of the volume of semen. The glandular fluid generally has the same osmotic concentration as that of blood plasma, but the compositions of the two fluids are quite different. In particular, the secretion of the seminal glands contains (1) higher concentrations of fructose, which is easily metabolized by spermatozoa; (2) prostaglandins, which can stimulate smooth muscle contractions along the male and female reproductive tracts; and (3) fibrinogen, which after ejaculation forms a temporary semen clot within the vagina. The secretions of the seminal glands are slightly alkaline, helping to neutralize acids in the

Figure 28–10   **The Ductus Deferens and Accessory Glands.**

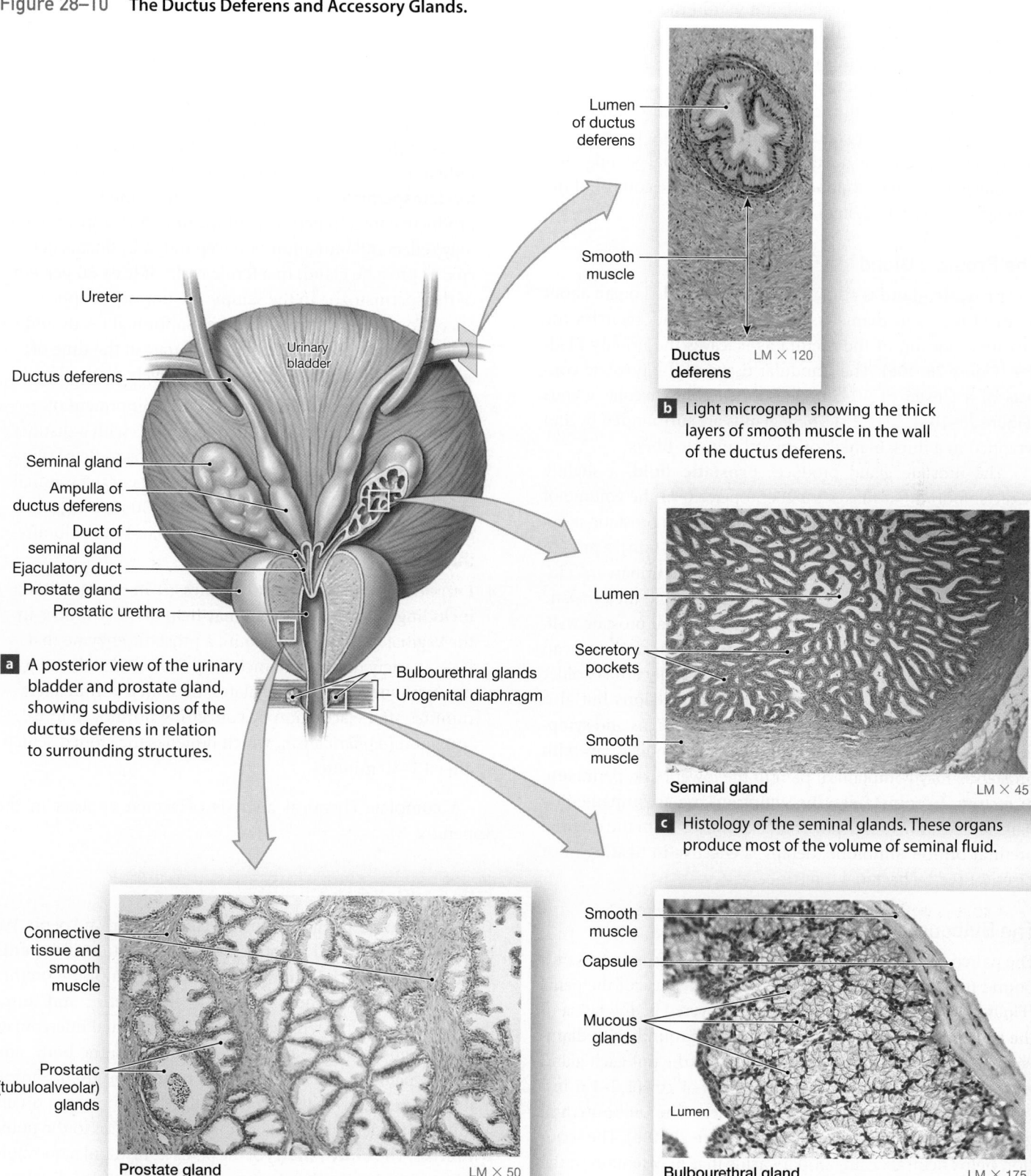

Lumen of ductus deferens

Smooth muscle

Ductus deferens   LM × 120

b Light micrograph showing the thick layers of smooth muscle in the wall of the ductus deferens.

Ureter

Urinary bladder

Ductus deferens

Seminal gland

Ampulla of ductus deferens

Duct of seminal gland

Ejaculatory duct

Prostate gland

Prostatic urethra

a A posterior view of the urinary bladder and prostate gland, showing subdivisions of the ductus deferens in relation to surrounding structures.

Bulbourethral glands

Urogenital diaphragm

Lumen

Secretory pockets

Smooth muscle

Seminal gland   LM × 45

c Histology of the seminal glands. These organs produce most of the volume of seminal fluid.

Connective tissue and smooth muscle

Prostatic (tubuloalveolar) glands

Prostate gland   LM × 50

d Histological detail of the glands of the prostate. The tissue between the individual glandular units consists largely of smooth muscle. Contractions of this muscle tissue help move the secretions into the ejaculatory duct and urethra.

Smooth muscle

Capsule

Mucous glands

Lumen

Bulbourethral gland   LM × 175

e Histology of the bulbourethral glands, which secrete a thick mucus into the spongy urethra.

28

secretions of the prostate gland and within the vagina. When mixed with the secretions of the seminal glands, spermatozoa undergo the first step in capacitation and begin beating their flagella, becoming highly motile.

The secretions of the seminal glands are discharged into the ejaculatory duct at *emission,* when peristaltic contractions are under way in the ductus deferens, seminal glands, and prostate gland. These contractions are under the control of the sympathetic nervous system.

The Prostate Gland

The **prostate gland** is a small, muscular, rounded organ about 4 cm (1.6 in.) in diameter. The prostate gland encircles the proximal portion of the urethra as it leaves the urinary bladder (**Figure 28–10a**). The glandular tissue of the prostate consists of a cluster of 30–50 compound tubuloalveolar glands (**Figure 28–10d**). ⟳ p. 151 These glands are surrounded by and wrapped in a thick blanket of smooth muscle fibers.

The prostate gland produces **prostatic fluid**, a slightly acidic solution that makes up 20–30 percent of the volume of semen. In addition to several other compounds, prostatic secretions contain **seminalplasmin** (sem-i-nal-PLAZ-min), a protein with antibiotic properties that may help prevent urinary tract infections in males. These secretions are ejected into the prostatic urethra by peristaltic contractions of the muscular prostate wall.

Prostatic inflammation, or **prostatitis** (pros-ta-TĪ-tis), can occur in males at any age, but it most commonly afflicts older men. Prostatitis can result from bacterial infections but also occurs in the apparent absence of pathogens. Signs and symptoms can resemble those of prostate cancer. Individuals with prostatitis may complain of pain in the lower back, perineum, or rectum. In some cases, the symptoms are accompanied by painful urination and the discharge of mucus from the external urethral orifice. Antibiotic therapy is effective in treating most cases caused by bacterial infection.

The Bulbourethral Glands

The paired **bulbourethral glands,** or *Cowper's glands,* are compound tubular mucous glands located at the base of the penis (**Figures 28–10a** and **28–11a**). They are covered by the fascia of the urogenital diaphragm. These glands are round, with diameters nearly 10 mm (less than 0.5 in.). The duct of each gland travels alongside the penile urethra for 3–4 cm (1.2–1.6 in.) before emptying into the urethral lumen. The bulbourethral glands secrete thick, alkaline mucus (**Figure 28–10e**). The secretion helps neutralize any urinary acids that may remain in the urethra, and it lubricates the *glans penis,* or tip of the penis.

Semen

A typical ejaculation releases 2–5 mL of semen. An abnormally low volume may indicate problems with the prostate gland or seminal glands. When sampled for analysis, semen is collected after a 36-hour period of sexual abstinence. The volume of fluid produced by an ejaculation, called the **ejaculate,** typically contains the following:

- *Spermatozoa.* The normal **sperm count** ranges from 20 million to 100 million spermatozoa per milliliter of semen. Most individuals with lower sperm counts are infertile, because too few spermatozoa survive the ascent of the female reproductive tract to perform fertilization. A low sperm count may reflect inflammation of the epididymis, ductus deferens, or prostate gland. In a fertile male, at least 60 percent of the spermatozoa in the sample are normal in appearance. Common abnormalities are malformed heads and "twin" spermatozoa that did not separate at the time of spermiation. Normal sperm will be actively swimming.

- *Seminal Fluid.* **Seminal fluid**, the fluid component of semen, is a mixture of glandular secretions with a distinct ionic and nutrient composition. A typical sample of seminal fluid contains the combined secretions of the seminal glands (60 percent), the prostate gland (30 percent), the nurse cells and epididymis (5 percent), and the bulbourethral glands (less than 5 percent).

- *Enzymes.* Several important enzymes are in seminal fluid, including (1) a protease that may help dissolve mucus in the vagina; (2) seminalplasmin, a prostatic enzyme that kills a variety of bacteria, including *Escherichia coli;* (3) a prostatic enzyme that coagulates the semen within a few minutes after ejaculation by converting fibrinogen to fibrin; and (4) *fibrinolysin,* which liquefies the clotted semen after 15–30 minutes.

A complete chemical analysis of semen appears in the Appendix.

The External Genitalia

The male external genitalia consist of the scrotum and penis. We have already described the structure of the scrotum. The **penis** is a tubular organ containing the distal portion of the urethra (**Figure 28–11a**). It conducts urine to the exterior and introduces semen into the female's vagina during sexual intercourse.

The penis has three main regions: the root, the body, and the glans penis (**Figure 28–11c**). The **root** is the fixed portion that attaches the penis to the body wall. This connection occurs within the urogenital triangle immediately inferior to the pubic symphysis. The **body (shaft)** of the penis is the tubular, movable portion of the organ. The **glans penis** is the expanded distal end that surrounds the external urethral orifice. The *neck* is the narrow portion of the penis between the shaft and the glans penis.

The skin overlying the penis resembles that of the scrotum. The dermis contains a layer of smooth muscle that is a continuation of the dartos muscle of the scrotum. The underlying

28

Figure 28–11   **The Penis.** ATLAS: Plate 60b

a A frontal section through the penis and associated organs

b A sectional view through the penis

c An anterior and lateral view of the penis, showing positions of the erectile tissues

areolar tissue allows the thin skin to move without distorting underlying structures. The subcutaneous layer also contains superficial arteries, veins, and lymphatic vessels.

A fold of skin called the **prepuce** (PRĒ-poos), or *foreskin*, surrounds the tip of the penis. The prepuce attaches to the relatively narrow neck of the penis and continues over the glans penis. **Preputial** (prē-PŪ-shē-al) **glands** in the skin of the neck and the inner surface of the prepuce secrete a waxy material known as **smegma** (SMEG-ma). Unfortunately, smegma can be an excellent nutrient source for bacteria. Mild inflammation and infections in this area are common, especially if the area is not washed thoroughly and frequently. One way to avoid such problems is **circumcision** (ser-kum-SIZH-un), the surgical removal of the prepuce. In Western societies (especially the United States), this procedure is generally performed shortly after birth. Circumcision lowers the risks of developing urinary tract infections, HIV infection, and penile cancer. Because it is a surgical procedure with risk of bleeding, infection, and other complications, the practice remains controversial.

Deep to the areolar tissue, a dense network of elastic fibers encircles the internal structures of the penis. Most of the body of the penis consists of three cylindrical columns of **erectile tissue (Figure 28–11b)**. Erectile tissue consists of a three-dimensional maze of vascular channels incompletely separated by partitions of elastic connective tissue and smooth muscle fibers. In the resting state, the arterial branches are constricted and the muscular partitions are tense. This combination restricts blood flow into the erectile tissue. During **erection** the penis stiffens and elevates to an upright position, but the flaccid (nonerect) penis hangs inferior to the pubic symphysis and anterior to the scrotum.

How does erection of the penis occur? The parasympathetic innervation of the penile arteries involves neurons that release nitric oxide at their axon terminals. The smooth muscles in the arterial walls relax when nitric oxide is released. At that

28

time, the vessels dilate, blood flow increases, and the vascular channels become engorged with blood.

The anterior surface of the flaccid penis covers two cylindrical masses of erectile tissue: the **corpora cavernosa** (KOR-por-uh ka-ver-NŌ-suh; singular, *corpus cavernosum*). The two are separated by a thin septum and encircled by a dense collagenous sheath (**Figure 28–11b**). The corpora cavernosa diverge at their bases, forming the **crura** (*crura*, legs; singular, *crus*) of the penis (**Figure 28–11a**). Each crus is bound to the ramus of the ischium and pubis by tough connective tissue ligaments. The corpora cavernosa extend along the length of the penis as far as its neck. The erectile tissue within each corpus cavernosum surrounds a central artery, or deep artery of the penis (**Figure 28–11b**).

The relatively slender **corpus spongiosum** (spon-jē-Ō-sum) surrounds the penile urethra (**Figure 28–11a,b**). This erectile body extends from the superficial fascia of the urogenital diaphragm to the tip of the penis, where it expands to form the glans penis. The sheath surrounding the corpus spongiosum contains more elastic fibers than does that of the corpora cavernosa, and the erectile tissue contains a pair of small arteries.

Hormones and Male Reproductive Function

The hormonal interactions that regulate male reproductive function are diagrammed in **Spotlight Figure 28–12**. We introduced major reproductive hormones in Chapter 18. ⟲ p. 667 The anterior lobe of the pituitary gland releases *follicle-stimulating hormone* (**FSH**) and *luteinizing hormone* (**LH**). The pituitary releases these hormones in response to *gonadotropin-releasing hormone* (**GnRH**). This peptide is synthesized in the hypothalamus and carried to the anterior lobe by the hypophyseal portal system.

The hormone GnRH is secreted in pulses rather than continuously. In adult males, small pulses occur at 60- to 90-minute intervals. As levels of GnRH change, so do the rates of secretion of FSH and LH (and testosterone, which is released in response to LH). The GnRH pulse frequency in adult males remains relatively steady from hour to hour, day to day, and year to year, unlike the situation in women, as we will see later in the chapter. As a result, plasma levels of FSH, LH, and testosterone remain within a relatively narrow range until relatively late in a man's life.

Testosterone plays a major role in maintaining male sexual function (**Spotlight Figure 28–12**). Testosterone functions like other steroid hormones, circulating in the bloodstream while bound to one of two types of transport proteins. These proteins are (1) *gonadal steroid-binding globulin* (GBG), which carries about two-thirds of the circulating testosterone, and (2) the albumins, which bind the remaining one-third. Testosterone diffuses across the plasma membrane of target cells and binds to an intracellular receptor. The hormone–receptor complex then binds to the DNA in the nucleus.

In many target tissues, some of the arriving testosterone is converted to **dihydrotestosterone (DHT)**. A small amount

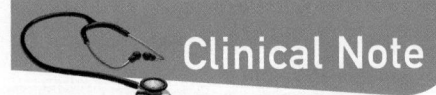

Clinical Note

Dehydroepiandrosterone (DHEA) *Dehydroepiandrosterone,* or **DHEA,** is the primary androgen secreted by the zona reticularis of the adrenal cortex. ⟲ p. 660 As noted in Chapter 18, these androgens, which are secreted in small amounts, are converted to testosterone (or estrogens) by other tissues. The significance of this small adrenal androgen secretion in both sexes remains unclear, but DHEA is being promoted as a wonder drug for increasing vitality, strength, and muscle mass. The effects of long-term high doses of DHEA remain largely unknown. However, high levels of DHEA in women have been linked to an increased risk of ovarian cancer as well as to masculinization, due to the conversion of DHEA to testosterone. The IOC (International Olympic Committee), NCAA, MLB, and NFL have banned the use of DHEA.

of DHT diffuses back out of the cell and into the bloodstream, and DHT levels are usually about 10 percent of circulating testosterone levels. Dihydrotestosterone can also enter peripheral cells and bind to the same hormone receptors targeted by testosterone. In addition, some tissues (notably those of the external genitalia) respond to DHT rather than to testosterone. Other tissues (including the prostate gland) are more sensitive to DHT than to testosterone.

Testosterone production begins around the seventh week of fetal development and reaches a prenatal peak after six months. Over this period, the developing nurse cells secrete Müllerian-inhibiting factor, which leads to the regression of the Müllerian ducts. The early surge in testosterone levels stimulates the differentiation of the male duct system and accessory organs and affects CNS development.

The best-known CNS effects occur in the developing hypothalamus. There, testosterone apparently programs the hypothalamic centers that are involved with (1) GnRH production and the regulation of pituitary FSH and LH secretion, (2) sexual behaviors, and (3) sexual drive. As a result of this prenatal exposure to testosterone, the hypothalamic centers will respond appropriately when the individual becomes sexually mature. The factors responsible for regulating the fetal production of testosterone are not known.

Testosterone levels are low at birth. Until puberty, background testosterone levels are relatively low, but they are higher in males than in females. Testosterone secretion accelerates markedly at puberty, initiating sexual maturation and the appearance of secondary sex characteristics. In adult males, negative feedback controls the level of testosterone production. Above-normal testosterone levels inhibit the release of GnRH by the hypothalamus, causing a reduction in LH secretion and lowering testosterone levels (**Spotlight Figure 28–12**).

Male reproductive function is regulated by the complex interaction of hormones from the hypothalamus, anterior lobe of the pituitary gland, and the testes. Negative feedback systems keep testosterone levels within a relatively narrow range until late in life.

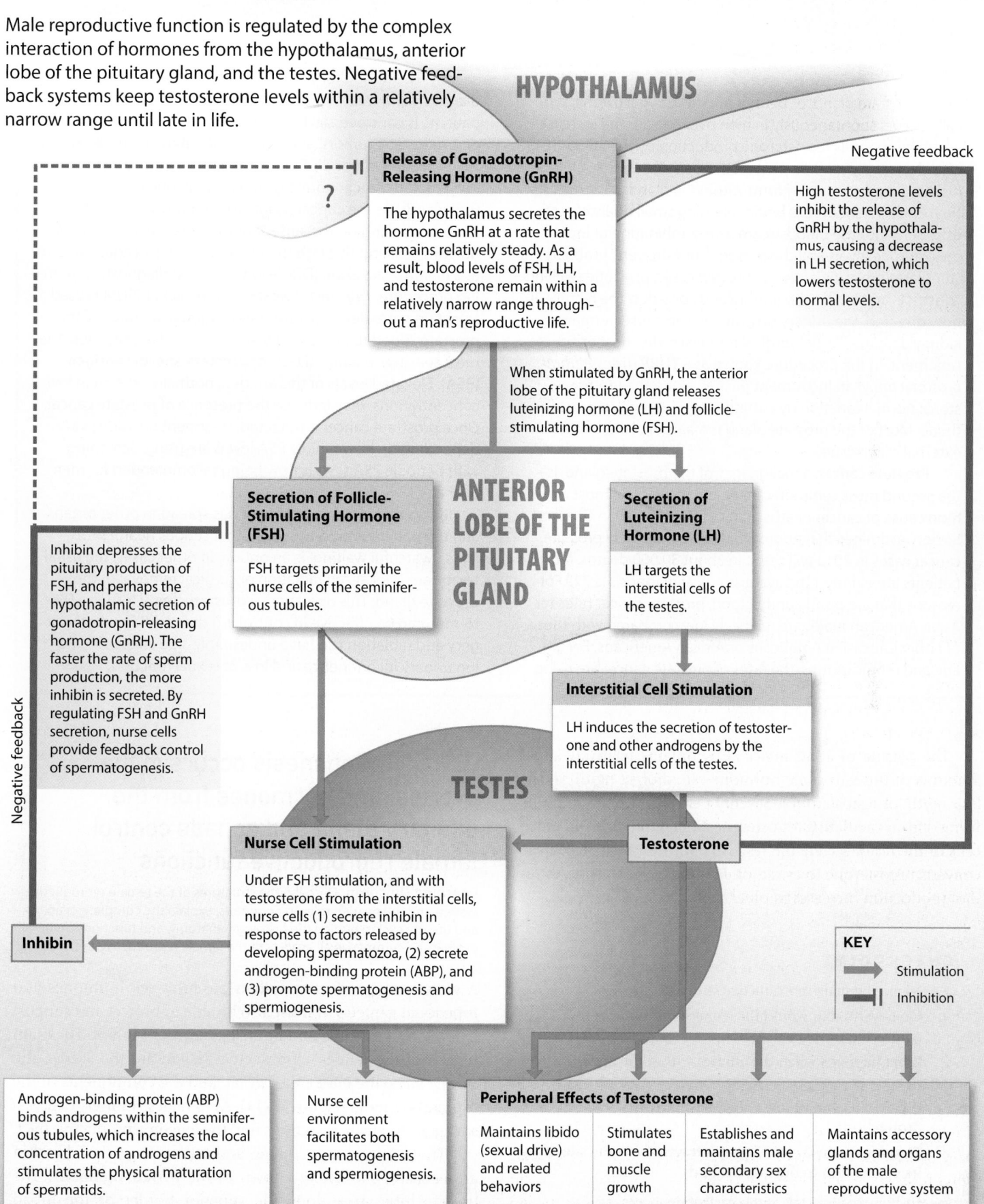

HYPOTHALAMUS

?

Release of Gonadotropin-Releasing Hormone (GnRH)

The hypothalamus secretes the hormone GnRH at a rate that remains relatively steady. As a result, blood levels of FSH, LH, and testosterone remain within a relatively narrow range throughout a man's reproductive life.

Negative feedback

High testosterone levels inhibit the release of GnRH by the hypothalamus, causing a decrease in LH secretion, which lowers testosterone to normal levels.

When stimulated by GnRH, the anterior lobe of the pituitary gland releases luteinizing hormone (LH) and follicle-stimulating hormone (FSH).

ANTERIOR LOBE OF THE PITUITARY GLAND

Secretion of Follicle-Stimulating Hormone (FSH)

FSH targets primarily the nurse cells of the seminiferous tubules.

Secretion of Luteinizing Hormone (LH)

LH targets the interstitial cells of the testes.

Inhibin depresses the pituitary production of FSH, and perhaps the hypothalamic secretion of gonadotropin-releasing hormone (GnRH). The faster the rate of sperm production, the more inhibin is secreted. By regulating FSH and GnRH secretion, nurse cells provide feedback control of spermatogenesis.

Interstitial Cell Stimulation

LH induces the secretion of testosterone and other androgens by the interstitial cells of the testes.

TESTES

Negative feedback

Testosterone

Nurse Cell Stimulation

Under FSH stimulation, and with testosterone from the interstitial cells, nurse cells (1) secrete inhibin in response to factors released by developing spermatozoa, (2) secrete androgen-binding protein (ABP), and (3) promote spermatogenesis and spermiogenesis.

Inhibin

KEY

→ Stimulation

⊣ Inhibition

Androgen-binding protein (ABP) binds androgens within the seminiferous tubules, which increases the local concentration of androgens and stimulates the physical maturation of spermatids.

Nurse cell environment facilitates both spermatogenesis and spermiogenesis.

Peripheral Effects of Testosterone

| Maintains libido (sexual drive) and related behaviors | Stimulates bone and muscle growth | Establishes and maintains male secondary sex characteristics | Maintains accessory glands and organs of the male reproductive system |

Clinical Note

Prostatic Hypertrophy and Prostate Cancer Enlargement of the prostate gland, or **benign prostatic hypertrophy,** typically occurs spontaneously in men over age 50. The increase in size happens as testosterone production by the interstitial cells decreases. For unknown reasons, small masses called *prostatic concretions* may form within the glands. At the same time, the interstitial cells begin releasing small quantities of estrogens into the bloodstream. The combination of lower testosterone levels and the presence of estrogens probably stimulates prostatic growth. In severe cases, prostatic swelling constricts and blocks the urethra and constricts the rectum. If not corrected, the urinary obstruction can cause permanent kidney damage. Partial surgical removal is the most effective treatment. In the procedure known as a **TURP** *(transurethral prostatectomy)* an instrument pushed along the urethra restores normal function by cutting away the swollen prostatic tissue. Most of the prostate gland remains in place, and no external scars result.

Prostate cancer, a malignancy of the prostate gland, is the second most common cancer and the second most common cause of cancer deaths in males. The American Cancer Society estimates that approximately 239,000 new prostate cancer cases in 2013 will result in about 30,000 deaths. Most patients are elderly. (The average age at diagnosis is 72.) For reasons that are poorly understood, prostate cancer rates for Asian American males are relatively low compared with those of either Caucasian Americans or African Americans. For all age and ethnic groups, the rates of prostate cancer are rising sharply. The reason for the increase is not known. Aggressive diagnosis and treatment of localized prostate cancer in elderly patients is controversial because many of these men have nonmetastatic tumors, and even if untreated are more likely to die of some other disease.

Prostate cancer normally originates in one of the secretory glands. As the cancer progresses, it produces a nodular lump or swelling on the surface of the prostate gland. Palpation of this gland through the rectal wall—a procedure known as a *digital rectal exam* (DRE)—is the easiest diagnostic screening procedure. *Transrectal prostatic ultrasound* (TRUS) is used to obtain more detailed information about the status of the prostate. Blood tests are also used for screening purposes. The most sensitive is a blood test for **prostate-specific antigen (PSA)**. Elevated levels of this antigen, normally present in low concentrations, may indicate the presence of prostate cancer. Once prostrate cancer is detected, treatment decisions vary depending on how rapidly PSA levels are rising. Screening with periodic PSA tests is now being recommended for men over age 50.

If cancer is detected before it has spread to other organs, and the patient is elderly or has other serious health problems, "watchful waiting" is an option. In other cases, the usual treatment is localized radiation or surgical removal of the prostate gland. This operation, a **prostatectomy** (pros-ta-TEK-tō-mē), can be effective in controlling the condition. Both surgery and radiation can have undesirable side effects, including urinary incontinence and loss of sexual function.

The plasma of adult males also contains relatively small amounts of the estrogenic hormone estradiol (2 ng/dL versus 525 ng/dL of testosterone). Seventy percent of the estradiol is formed from circulating testosterone. Interstitial cells and nurse cells of the testes secrete the rest. An enzyme called aromatase converts testosterone to estradiol. For unknown reasons, estradiol production increases in older men.

CHECKPOINT

4. Name the male reproductive structures.

5. On a warm day, would the cremaster muscle be contracted or relaxed? Why?

6. What happens when the arteries within the penis dilate?

7. What effect would low FSH levels have on sperm production?

8. Trace the pathway that a sperm travels from the site of its production to outside the body.

See the blue Answers tab at the back of the book.

28-3 ❘ Oogenesis occurs in the ovaries, and hormones from the pituitary gland and gonads control female reproductive functions

Learning Outcome Describe the structures of the female reproductive system and the ovarian roles in oogenesis; explain the complete ovarian and uterine cycles; outline the histology, anatomy, and functions of the vagina; and summarize all aspects of the female reproductive cycle.

A woman's reproductive system produces sex hormones and functional gametes. It must also be able to protect and support a developing embryo and nourish a newborn infant. The main organs of the female reproductive system are the *ovaries,* the *uterine tubes,* the *uterus,* the *vagina,* and the components of the external genitalia (**Figure 28–13**). As in males, a variety of accessory glands release secretions into the female reproductive tract.

The ovaries, uterine tubes, and uterus are enclosed within an extensive mesentery known as the **broad ligament.** The uterine tubes run along the superior border of the broad

Figure 28–13   **The Female Reproductive System.** A sagittal section of the female reproductive structures.   ATLAS: Plate 65

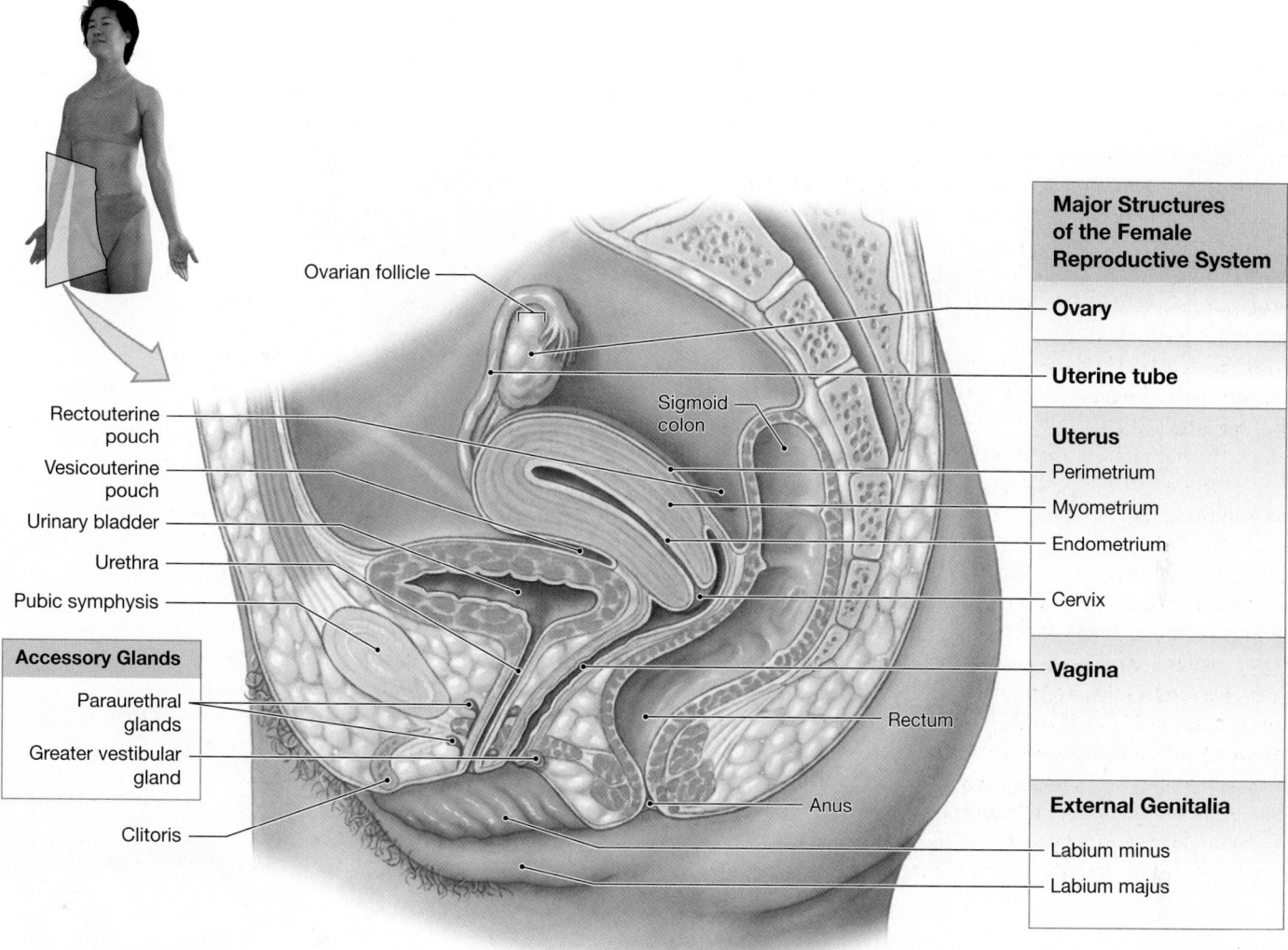

Ovarian follicle

Major Structures of the Female Reproductive System

Ovary

Uterine tube

Uterus
Perimetrium
Myometrium
Endometrium
Cervix

Vagina

External Genitalia
Labium minus
Labium majus

Rectouterine pouch
Vesicouterine pouch
Urinary bladder
Urethra
Pubic symphysis

Accessory Glands
Paraurethral glands
Greater vestibular gland
Clitoris

Sigmoid colon

Rectum

Anus

ligament and open into the pelvic cavity lateral to the ovaries. The **mesovarium** (mez-ō-VĀ-rē-um), a thickened fold of mesentery, supports and stabilizes the position of each ovary (**Figure 28–14a**). The broad ligament attaches to the sides and floor of the pelvic cavity, where it becomes continuous with the parietal peritoneum.

The broad ligament subdivides this part of the peritoneal cavity. The pocket formed between the posterior wall of the uterus and the anterior surface of the colon is the **rectouterine** (rek-tō-Ū-ter-in) **pouch** (**Figure 28–13**). The pocket formed between the uterus and the posterior wall of the bladder is the **vesicouterine** (ves-i-kō-Ū-ter-in) **pouch.** These subdivisions are easily seen in sagittal section.

Several other ligaments assist the broad ligament in supporting and stabilizing the uterus and associated reproductive organs. These ligaments lie within the mesentery sheet of the broad ligament and are connected to the ovaries or uterus. The broad ligament limits side-to-side movement and rotation, and

the other ligaments (described in our discussion of the ovaries and uterus) prevent superior–inferior movement.

The Ovaries

The paired ovaries are small, lumpy, almond-shaped organs near the lateral walls of the pelvic cavity (**Figure 28–14**). The ovaries perform three main functions. They (1) produce immature female gametes, or oocytes; (2) secrete female sex hormones, including estrogens and progesterone; and (3) secrete inhibin, involved in the feedback control of pituitary FSH production.

Each ovary is stabilized by the mesovarium and by a pair of supporting ligaments: the ovarian ligament and the suspensory (infundibulopelvic) ligament (**Figure 28–14a**). The **ovarian ligament** extends from the uterus, near the attachment of the uterine tube, to the medial surface of the ovary. The **suspensory ligament** extends from the lateral surface of the ovary

28

Figure 28–14 **The Ovaries and Their Relationships to the Uterine Tube and Uterus.** ATLAS: Plate 67

a A posterior view of the uterus, uterine tubes, and ovaries

b A sectional view of the ovary, uterine tube, and associated mesenteries

past the open end of the uterine tube to the pelvic wall. The suspensory ligament contains the major blood vessels of the ovary: the **ovarian artery** and **ovarian vein.** These vessels are connected to the ovary at the **ovarian hilum,** where the ovary attaches to the mesovarium (**Figure 28–14b**).

A typical ovary is about 5 cm long, 2.5 cm wide, and 8 mm thick (2 in. × 1 in. × 0.33 in.). It weighs 6–8 g (about 0.25 oz). An ovary is pink or yellowish and has a nodular consistency. The visceral peritoneum, or *germinal epithelium,* covers the surface of each ovary and consists of a layer of columnar epithelial cells that overlies a dense connective tissue layer called the **tunica albuginea** (**Figure 28–14b**). We divide the interior tissues, or **stroma,** of the ovary into a superficial *cortex* and a deeper *medulla.* Gametes are produced in the cortex.

Oogenesis

Ovum production, or **oogenesis** (ō-ō-JEN-e-sis; *oon,* egg), begins before a woman's birth, accelerates at puberty, and ends at

menopause. Between puberty and menopause, oogenesis occurs on a monthly basis as part of the *ovarian cycle.*

Oogenesis is summarized in **Figure 28–15**. Female reproductive stem cells produce *primary oocytes* by mitosis before birth. Then these cells proceed as far as prophase of meiosis I, and remain in that state until the individual reaches puberty.

Not all primary oocytes produced during fetal development survive until puberty. The ovaries have approximately 2 million *primordial follicles* at birth, each containing a primary oocyte. By puberty, the number has dropped to about 400,000. The rest of the primordial follicles degenerate in a process called **atresia** (a-TRĒ-zē-uh).

28

Figure 28–15 **Oogenesis.** For clarity, maternal and paternal chromatids are not identified.

Oogenesis

Mitosis of oogonium

Unlike spermatogonia, the **oogonia** (ō-ō-GŌ-nē-uh), or female reproductive stem cells, complete their mitotic divisions before birth.

Oogonium

Oogonium (stem cell)

Primary oocyte (diploid, 2n)

Meiosis I

Between the third and seventh months of fetal development, the daughter cells, or **primary oocytes** (Ō-ō-sīts), prepare to undergo meiosis. They proceed as far as the prophase of meiosis I, but then the process comes to a halt.

The primary oocytes remain in a state of suspended development until puberty, when rising levels of FSH trigger the start of the ovarian cycle. Each month after the ovarian cycle begins, some of the primary oocytes are stimulated to undergo further development. Meiosis I is then completed, yielding a secondary oocyte and a polar body.

DNA replication

Tetrad

Primary oocyte

First polar body

First polar body may not complete meiosis II

Secondary oocyte

Meiosis II

Each month after the ovarian cycle begins, one or more secondary oocytes leave the ovary suspended in metaphase of meiosis II.

At the time of fertilization, a second polar body forms and the fertilized oocyte is then called a mature **ovum**. (A cell in any of the preceding steps in oogenesis is sometimes called an immature ovum.)

Secondary oocyte released (ovulation) in metaphase of meiosis II

Second polar body

Nucleus of oocyte (n)

Sperm (n)

Fertilization *(see Figure 29–1)*

The nuclear events in the ovaries during meiosis are the same as those in the testes, but the process differs in two important details:

1. The cytoplasm of the primary oocyte is unevenly distributed during the two meiotic divisions. Oogenesis produces one secondary oocyte, which contains most of the original

cytoplasm, and two or three **polar bodies**, nonfunctional cells that later disintegrate.

2. The ovary releases a **secondary oocyte** rather than a mature ovum. The secondary oocyte is suspended in metaphase of meiosis II. Meiosis will not be completed unless fertilization occurs.

The Ovarian Cycle

Ovarian follicles are specialized structures in the cortex of the ovaries where both oocyte growth and meiosis I occur. We can divide the ovarian cycle into a **follicular phase**, or *preovulatory phase*, and a **luteal phase**, or *postovulatory phase*. Important steps in the ovarian cycle are summarized in **Figure 28–16**. (The drawing of the ovary illustrates the sequence and relative sizes of the various stages of the development, ovulation, and degeneration of an ovarian follicle. Follicles do not physically move around the cortex of the ovary.)

❶ *Primordial Follicles in Egg Nest.* Primary oocytes are located in the outer portion of the ovarian cortex, near the tunica albuginea, in clusters called *egg nests*. A single squamous layer of *follicle cells* surrounds each primary oocyte within an egg nest. The primary oocyte and its follicle cells form a **primordial follicle**. Beginning at puberty, primordial follicles are continuously activated to join other follicles already in development. The activating mechanism is unknown, but local hormones or growth factors within the ovary may be involved. The activated primordial follicle will either mature and be released as a secondary oocyte or degenerate (atresia). This monthly process is known as the **ovarian cycle**.

❷ *The Formation of Primary Follicles.* The preliminary steps in follicle development vary in length but may take almost

Figure 28–16   **The Ovarian Cycle.**

1   **Primordial follicles in egg nest**

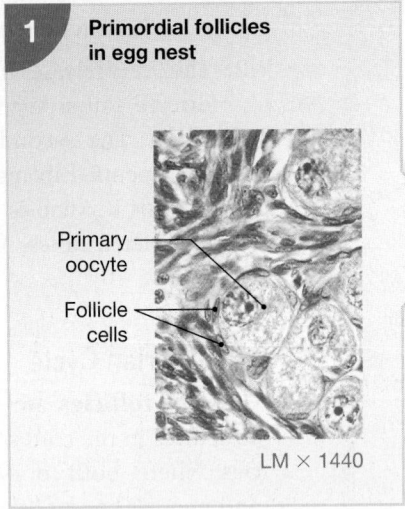

Primary oocyte

Follicle cells

LM × 1440

2   **Formation of primary follicle**

Granulosa cells

Primary oocytes

Zona pellucida

Thecal cells

LM × 1092

3   **Formation of secondary follicle**

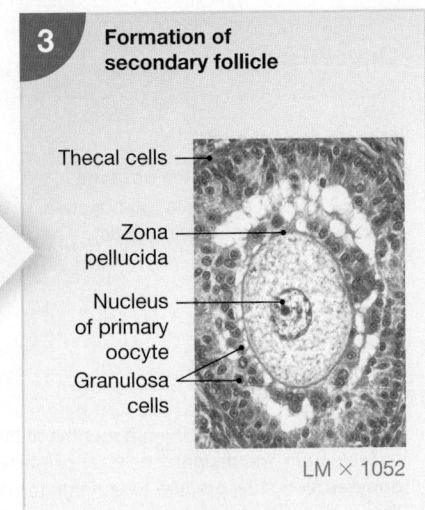

Thecal cells

Zona pellucida

Nucleus of primary oocyte

Granulosa cells

LM × 1052

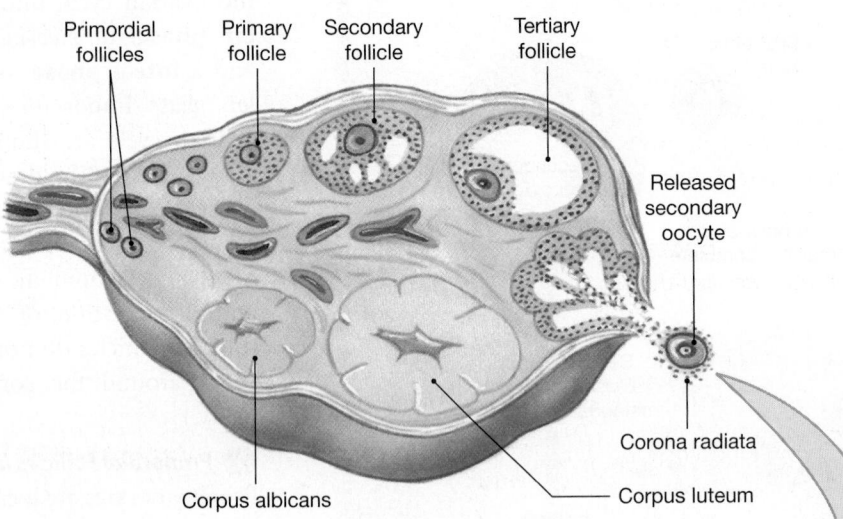

Primordial follicles

Primary follicle

Secondary follicle

Tertiary follicle

Released secondary oocyte

Corona radiata

Corpus albicans

Corpus luteum

4   **Formation of tertiary follicle**

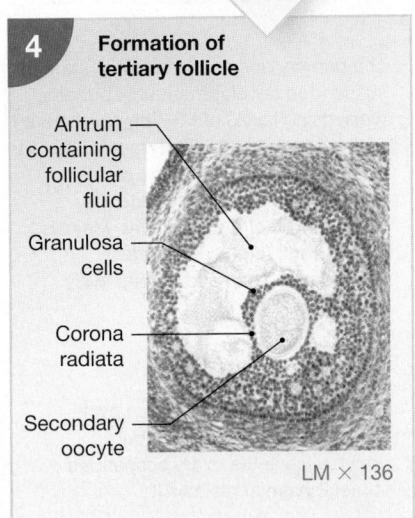

Antrum containing follicular fluid

Granulosa cells

Corona radiata

Secondary oocyte

LM × 136

7   **Formation of corpus albicans**

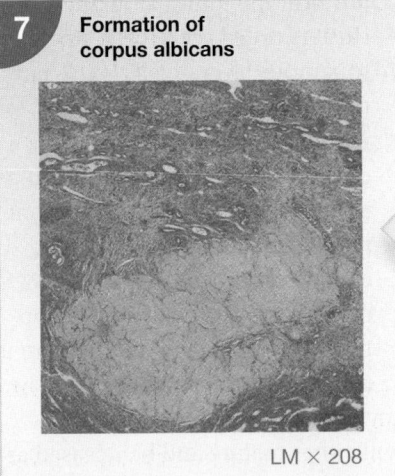

LM × 208

6   **Formation of corpus luteum**

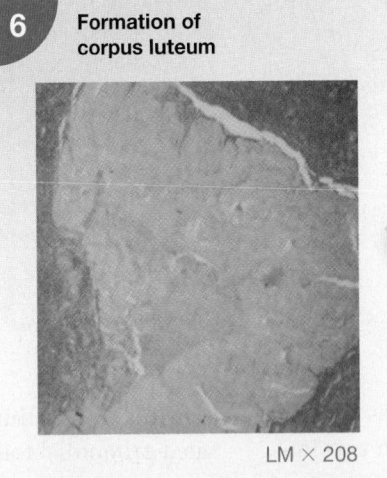

LM × 208

5   **Ovulation**

Follicular fluid

Secondary oocyte within corona radiata

Ruptured follicle wall

Outer surface of ovary

LM × 70

a year to complete. Follicle development begins with the activation of primordial follicles into **primary follicles**. The follicular cells enlarge, divide, and form several layers of cells around the growing primary oocyte. These follicle cells, which become rounded in appearance, are now called **granulosa cells**. Microvilli from the surrounding follicular cells intermingle with microvilli from the surface of the oocyte. This glycoprotein-rich region is called the **zona pellucida** (ZŌ-na pe-LŪ-sid-uh; *pellucidus,* translucent). The microvilli increase the surface area available for transferring materials from the follicular cells to the growing oocyte. As the granulosa cells enlarge and multiply, adjacent cells in the ovarian stroma form a layer of **thecal cells** (*theca,* a box) around the follicle. Thecal cells and granulosa cells work together to produce sex hormones called *estrogens.*

3 *The Formation of Secondary Follicles.* Many primordial follicles develop into primary follicles, but only a few of the primary follicles mature further. This process is apparently under the control of a growth factor produced by the oocyte. The transformation begins as the wall of the follicle thickens and the deeper follicular cells begin secreting small amounts of fluid. This **follicular fluid**, or *liquor folliculi,* accumulates in small pockets that gradually expand and separate the inner and outer layers of the follicle. At this stage, the complex is known as a **secondary follicle**. The primary oocyte continues to grow slowly, while the follicle as a whole enlarges rapidly as follicular fluid accumulates.

4 *The Formation of a Tertiary Follicle.* Eight to 10 days after the start of the ovarian cycle, the ovaries generally contain only a single secondary follicle destined for further development. By the 10th to the 14th day of the cycle, that follicle has become a **tertiary follicle**, or *mature graafian* (GRAH-fē-an) *follicle,* roughly 15 mm in diameter. This complex spans the entire width of the ovarian cortex and distorts the ovarian capsule, creating a prominent bulge in the surface of the ovary. The oocyte projects into the **antrum** (AN-trum), or expanded central chamber of the follicle. The antrum is surrounded by a mass of granulosa cells.

Until this time, the primary oocyte has been suspended in prophase of meiosis I. As the development of the tertiary follicle ends, LH levels begin rising, prompting the primary oocyte to complete meiosis I. Instead of producing two secondary oocytes, the first meiotic division yields a secondary oocyte and a small, nonfunctional polar body. The secondary oocyte then enters meiosis II, but stops once again at metaphase. Meiosis II will not be completed unless fertilization occurs.

Generally, on day 14 of a 28-day cycle, the secondary oocyte and the attached granulosa cells lose their connections with the follicular wall and drift free within the antrum. The granulosa cells still associated with the secondary oocyte form a protective layer known as the **corona radiata** (kō-RŌ-nuh rā-dē-AH-tuh).

5 *Ovulation.* At **ovulation**, the tertiary follicle releases the secondary oocyte. The distended follicular wall suddenly ruptures, ejecting the follicular contents, including the secondary oocyte and corona radiata, into the pelvic cavity. The sticky follicular fluid keeps the corona radiata (and the oocyte) attached to the surface of the ovary. The oocyte is then moved into the uterine tube by contact with the fimbriae that extend from the tube's funnel-like opening (**Figure 28–14a**), or by fluid currents produced by the cilia that line the tube. Ovulation marks the end of the **follicular phase** of the ovarian cycle and the start of the **luteal phase**.

6 *The Formation of the Corpus Luteum.* The empty tertiary follicle initially collapses, and ruptured vessels bleed into the antrum. The remaining granulosa cells then invade the area. They proliferate to create an endocrine structure known as the **corpus luteum** (LOO-tē-um; *lutea,* yellow), named for its yellow color. LH stimulates this process.

The cholesterol contained in the corpus luteum is used to manufacture the steroid hormone **progesterone** (prō-JES-ter-ōn). Progesterone is the main hormone in the luteal phase. Progesterone's primary function is to prepare the uterus for pregnancy by stimulating the maturation of the uterine lining and the secretions of uterine glands. The corpus luteum also secretes moderate amounts of estrogens, but levels are not as high as they were at ovulation.

TIPS & TRICKS

Progesterone literally means a steroid (*-one*) that favors (*pro-*) gestation (*-gest*).

7 *Formation of the Corpus Albicans.* The corpus luteum begins to degenerate about 12 days after ovulation (unless fertilization takes place). Progesterone and estrogen levels then fall markedly. Fibroblasts invade the nonfunctional corpus luteum, producing a knot of pale scar tissue called a **corpus albicans** (AL-bi-kanz). The disintegration, or *involution,* of the corpus luteum marks the end of the ovarian cycle. A new ovarian cycle then begins with the activation of another group of primordial follicles.

Age and Oogenesis

Generally only a single oocyte is released into the pelvic cavity at ovulation, although many primordial follicles may have developed into primary follicles, and several primary follicles may have been converted to secondary follicles. These undergo atresia. At puberty, each ovary contains about 200,000 primordial follicles. Forty years later, few if any follicles remain, although only about 500 secondary oocytes will have been ovulated.

28

The Uterine Tubes

Each **uterine tube** (*Fallopian tube* or *oviduct*) is a hollow, muscular cylinder measuring roughly 13 cm (5.2 in.) in length (**Figures 28–13** and **28–14**). We divide each uterine tube into three segments (**Figure 28–17a**):

1. *The Infundibulum.* The end closest to the ovary forms an expanded funnel, or **infundibulum**, with numerous fingerlike projections that extend into the pelvic cavity. The projections are called **fimbriae** (FIM-brē-ē). Fimbriae drape over the surface of the ovary, but there is no physical connection between the two structures. The inner surfaces of the infundibulum are lined with cilia that beat toward the middle segment of the uterine tube, called the *ampulla.*

2. *The Ampulla.* The **ampulla** is the middle region of the uterine tube. The thickness of its smooth muscle layers gradually increases as the tube approaches the uterus.

3. *The Isthmus.* The ampulla leads to the **isthmus** (IS-mus), a short segment connected to the uterine wall.

Histology of the Uterine Tube

The epithelium lining the uterine tube is composed of ciliated columnar epithelial cells with scattered mucin-secreting cells (**Figure 28–17c**). Concentric layers of smooth muscle surround the mucosa (**Figure 28–17b**). Oocytes are transported by a combination of ciliary movement and peristaltic contractions in the walls of the uterine tube. A few hours before ovulation, sympathetic and parasympathetic nerves from the hypogastric plexus "turn on" this beating pattern and initiate peristalsis. It normally takes three to four days for an oocyte to travel from the infundibulum to the *uterine cavity.* If fertilization is to occur, the secondary oocyte must encounter spermatozoa during the first 12–24 hours of its passage. Fertilization typically takes place near the boundary between the ampulla and isthmus of the uterine tube.

In addition to its transport function, the uterine tube provides a nutrient-rich environment that contains lipids and glycogen. This mixture nourishes both spermatozoa and a developing *pre-embryo* (the cell cluster produced by the initial cell

Figure 28–17 **The Uterine Tubes.**

a Regions of the uterine tubes, posterior view

b A sectional view of the isthmus

Isthmus LM × 122

c A colorized SEM of the ciliated lining of the uterine tube

Epithelial surface SEM × 4000

28

divisions following fertilization and before implantation in the uterine wall). Unfertilized oocytes degenerate in the terminal portions of the uterine tubes or within the uterus without completing meiosis.

In addition to ciliated cells, the epithelium lining the uterine tubes contains *peg cells* and scattered mucin-secreting cells. The peg cells project into the lumen of the uterine tube. They secrete a fluid that both completes the capacitation of spermatozoa and supplies nutrients to spermatozoa and the pre-embryo.

The Uterus

The **uterus** protects, nourishes, and removes wastes for the developing *embryo* (weeks 1–8) and *fetus* (week 9 through delivery). In addition, contractions of the muscular uterus are important in expelling the fetus at birth.

The uterus is a small, pear-shaped organ (**Figure 28–18a,b,c**). It is about 7.5 cm (3 in.) long with a maximum diameter of 5 cm (2 in.). It weighs 50–100 g (1.75–3.5 oz). In its normal position, the uterus bends anteriorly near its base, a condition known as *anteflexion* (an-tē-FLEK-shun) (**Figure 28–13**). In this position, the uterus covers the superior and posterior surfaces of the urinary bladder. If the uterus bends backward toward the sacrum, the condition is termed *retroflexion* (re-trō-FLEK-shun). Retroflexion occurs in about 20 percent of adult women and has no clinical significance. (A retroflexed uterus generally becomes anteflexed spontaneously during the third month of pregnancy.)

Suspensory Ligaments of the Uterus

In addition to the broad ligament, three pairs of suspensory ligaments stabilize the uterus and limit its range of movement (**Figure 28–18b**). The **uterosacral** (ū-te-rō-SĀ-krul) **ligaments** extend from the lateral surfaces of the uterus to the anterior face of the sacrum, keeping the body of the uterus from moving inferiorly and anteriorly. The **round ligaments** arise on the lateral margins of the uterus just posterior and inferior to the attachments of the uterine tubes. These ligaments extend through the inguinal canal and end in the connective tissues of the external genitalia. The round ligaments restrict posterior movement of the uterus. The **cardinal** *(lateral)* **ligaments** extend from the base of the uterus and vagina to the lateral walls of the pelvis. These ligaments tend to prevent inferior movement of the uterus. The muscles and fascia of the pelvic floor provide additional mechanical support.

Internal Anatomy of the Uterus

We divide the uterus into anatomical regions (**Figure 28–18a**). The uterine **body** is the largest portion of the uterus. The **fundus** is the rounded portion of the body superior to the attachment of the uterine tubes. The body ends at a constriction known as the **isthmus** of the uterus. The **cervix** (SER-viks) is the inferior portion of the uterus that extends from the isthmus to the vagina.

The tubular cervix projects about 1.25 cm (0.5 in.) into the vagina. Within the vagina, the distal end of the cervix forms a curving surface that surrounds the **external os** (*os,* an opening or mouth) of the uterus. The external os leads into the **cervical canal**, a constricted passageway that opens into the **uterine cavity** of the body at the **internal os**.

The uterus receives blood from branches of the **uterine arteries**. These vessels arise from branches of the *internal iliac arteries,* and from the *ovarian arteries,* which arise from the abdominal aorta inferior to the renal arteries. The arteries to the uterus are extensively interconnected, ensuring a reliable flow of blood to the organ despite changes in its position and shape during pregnancy. Numerous veins and lymphatic vessels also drain each portion of the uterus.

The organ is innervated by autonomic fibers from the hypogastric plexus (sympathetic) and from sacral segments S_3 and S_4 (parasympathetic). Sensory information reaches the central nervous system within the dorsal roots of spinal nerves T_{11} and T_{12}. The most delicate anesthetic procedures used during labor and delivery, known as *segmental blocks,* target only spinal nerves T_{10}–L_1.

The Uterine Wall

The dimensions of the uterus are highly variable. In women of reproductive age who have not given birth, the uterine wall is about 1.5 cm (0.6 in.) thick. The wall has a thick, middle, muscular **myometrium** (mī-ō-MĒ-trē-um; *myo-,* muscle + *metra,* uterus) and a thin, inner, glandular **endometrium** (en-dō-MĒ-trē-um) (**Figure 28–19**). The fundus and the posterior surface of the uterine body and isthmus are covered by an outer serous membrane that is continuous with the peritoneal lining. This incomplete serosa is called the **perimetrium**.

The endometrium makes up about 10 percent of the mass of the uterus. The glandular and vascular tissues of the

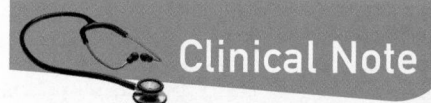

Clinical Note

Cervical Cancer **Cervical cancer** is the most common cancer of the reproductive system in women ages 15–34. Each year roughly 12,340 U.S. women are diagnosed with invasive cervical cancer, and approximately one-third of them eventually die from the condition. Another 35,000 women are diagnosed with a less aggressive form of cervical cancer. *Gardisil* is a vaccine that protects against four types of human papillomavirus (HPV) that cause most cervical cancers.

28

Figure 28–18 **The Uterus.** ATLAS: Plates 66; 67

Uterine Tube

Fimbriae Infundibulum Ampulla Isthmus

Fundus
of uterus

Uterine tube

Suspensory
ligament of ovary

Ovarian artery
and vein

Mesovarium

Body of
uterus

Ovary

Ovarian ligament

Uterine
cavity

Round ligament
of uterus

Perimetrium

Broad ligament

Myometrium

Endometrium

See Figure 28–19

Uterine artery
and vein

Internal os of uterus

Isthmus of uterus

Cervical canal

Cervix

Vaginal artery

External os
of uterus

a A posterior view with the
left portion of the uterus,
left uterine tube, and left
ovary shown in section.

Vaginal rugae

See Figure 28–21

Vagina

Uterosacral
ligament

POSTERIOR

Cardinal
ligaments
(under broad
ligament)

Sigmoid
colon

Suspensory
ligament
of ovary

Broad
ligament

Round
ligament
of uterus

Uterus

Urinary
bladder

ANTERIOR

b A superior view of the ligaments that stabilize
the position of the uterus in the pelvic cavity.

POSTERIOR

Rectouterine
pouch

Ovary

Ovarian
ligament

Uterine tube

Vesicouterine
pouch

Uterus

ANTERIOR

c Superior view of the female pelvic cavity showing supporting
ligaments of uterus and ovaries. In the photo, the urinary
bladder cannot be seen because it is covered by peritoneum.

28

Figure 28–19 **The Uterine Wall.**

Straight artery
Perimetrium
Myometrium
Endometrium
Uterine glands
Uterine cavity
Spiral artery
Arcuate arteries
Radial artery
Uterine artery

a A diagrammatic sectional view of the uterine wall, showing the endometrial regions and the blood supply to the endometrium

Endometrium

Simple columnar epithelium | Uterine glands | Functional zone | Basilar zone | Myometrium

Uterine cavity

Uterine wall LM × 32

b The basic histological structure of the uterine wall

endometrium support the physiological demands of the growing fetus. Vast numbers of uterine glands open onto the endometrial surface and extend deep into the lamina propria, almost to the myometrium. Under the influence of estrogens, the uterine glands, blood vessels, and epithelium change with the phases of the monthly *uterine cycle.*

The myometrium is the thickest portion of the uterine wall. It makes up almost 90 percent of the mass of the uterus. Smooth muscle in the myometrium is arranged into longitudinal, circular, and oblique layers. The smooth muscle tissue of the myometrium provides much of the force needed to move a fetus out of the uterus and into the vagina.

Histology of the Uterus. We divide the endometrium into a **functional zone**—the layer closest to the uterine cavity—and a **basilar zone**, adjacent to the myometrium (**Figure 28–19b**). The functional zone contains most of the **uterine glands** and contributes most of the endometrial thickness. This zone undergoes dramatic changes in thickness and structure during the menstrual cycle. The basilar zone attaches the endometrium to the myometrium and contains the terminal branches of the tubular uterine glands.

Within the myometrium, branches of the uterine arteries form **arcuate arteries**, which encircle the endometrium (**Figure 28–19a**). **Radial arteries** supply **straight arteries**,

which deliver blood to the basilar zone of the endometrium, and **spiral arteries**, which supply the functional zone.

The structure of the basilar zone remains fairly constant over time, but the functional zone undergoes cyclical changes in response to sex hormone levels. These cyclical changes produce the characteristic histological features of the uterine cycle.

The Uterine Cycle

The **uterine cycle**, or *menstrual* (MEN-strū-ul) *cycle,* is a repeating series of changes in the structure of the endometrium (**Figure 28–20**). The uterine cycle averages 28 days in length, but it can range from 21 to 35 days in healthy women of reproductive age.

We divide the uterine cycle into three phases: (1) *menses,* (2) the *proliferative phase,* and (3) the *secretory phase.* The phases take place in response to hormones associated with the regulation of the ovarian cycle. Menses and the proliferative phase occur during the follicular phase of the ovarian cycle. The secretory phase corresponds to the luteal phase of the ovarian cycle. We consider the regulatory mechanism involved in a later section.

Menses. The uterine cycle begins with the onset of **menses** (MEN-sēz), an interval marked by the degeneration of the functional zone of the endometrium (**Figure 28–20a**). This

28

Figure 28–20 **The Appearance of the Endometrium during the Uterine Cycle.**

a The endometrium at menses

b The endometrium during the proliferative phase

c The endometrium during the secretory phase of the uterine cycle

degeneration occurs in patches. It is caused by constriction of the spiral arteries, which reduces blood flow to areas of endometrium. Deprived of oxygen and nutrients, the secretory glands and other tissues in the functional zone begin to deteriorate. Eventually, the weakened arterial walls rupture, and blood pours into the connective tissues of the functional zone. Blood cells and degenerating tissues then break away and enter the uterine cavity, to be lost as they pass through the external os and into the vagina. Only the functional zone is affected, because the deeper, basilar layer receives blood from the straight arteries, which remain unconstricted.

The sloughing off, or shedding, of tissue is gradual. At each site repairs begin almost at once. Nevertheless, before menses has ended, the entire functional zone has been lost. The process of endometrial sloughing, called **menstruation** (men-strū-Ā-shun), generally lasts from one to seven days. During this time about 35 to 50 mL (1.2–1.7 oz) of blood are lost.

The process can be relatively painless. Painful menstruation, or **dysmenorrhea**, can result from uterine inflammation, myometrial contractions ("cramps"), or from conditions involving adjacent pelvic structures.

The Proliferative Phase. The basilar zone, including the basal parts of the uterine glands, survives menses intact. In the days after menses, the epithelial cells of the uterine glands multiply and spread across the endometrial surface, restoring the uterine epithelium (**Figure 28–20b**). Further growth and vascularization completely restore the functional zone.

During this reorganization, the endometrium is in the **proliferative phase**. Restoration takes place at the same time as the primary and secondary follicles enlarge in the ovary. Estrogens secreted by the developing ovarian follicles stimulate and sustain the proliferative phase.

By the time ovulation occurs, the functional zone is several millimeters thick. Prominent mucous glands extend to its border with the basilar zone. At this time, the uterine glands are manufacturing mucus that is rich in glycogen. This specialized mucus appears to be essential for the survival of the fertilized ovum through its earliest developmental stages. (We will consider these stages in Chapter 29.) The entire functional zone is highly vascularized, with small arteries spiraling toward the endometrial surface from larger arteries in the myometrium.

The Secretory Phase. During the **secretory phase** of the uterine cycle, the uterine glands enlarge, accelerating their rates of secretion. The arteries that supply the uterine wall elongate and spiral through the tissues of the functional zone (**Figure 28–20c**). This activity occurs under the combined stimulatory effects of progesterone and estrogens from the corpus luteum. The secretory phase begins at the time of ovulation and persists as long as the corpus luteum remains intact.

Secretory activities peak about 12 days after ovulation. Over the next day or two, the glands become less active. The uterine cycle ends as the corpus luteum stops producing stimulatory hormones. A new uterine cycle then begins with the onset of menses and the disintegration of the functional zone.

The secretory phase generally lasts 14 days. For this reason, you can identify the date of ovulation by counting backward 14 days from the first day of menses.

Menarche and Menopause. The uterine cycle begins at puberty. The first cycle, known as **menarche** (me-NAR-kē; *men,*

month + *arche,* beginning), typically occurs at age 11–12. The cycles continue until **menopause** (MEN-ō-pawz), the termination of the uterine cycle, at age 45–55. Over the interim, the regular appearance of uterine cycles is interrupted only by circumstances such as illness, stress, starvation, or pregnancy.

If menarche does not appear by age 16, or if the normal uterine cycle of an adult woman is interrupted for six months or more, the condition of **amenorrhea** (ā-men-ō-RĒ-uh) exists. *Primary amenorrhea* is the failure to initiate menses. This condition may indicate developmental abnormalities, such as nonfunctional ovaries, the absence of a uterus, or an endocrine or genetic disorder. It can also result from malnutrition: Puberty is delayed if leptin levels are too low. ⊃ p. 669

Severe physical or emotional stresses can cause transient *secondary amenorrhea.* In effect, the reproductive system gets "switched off." Factors associated with amenorrhea include drastic weight loss, anorexia nervosa, and severe depression or grief. Amenorrhea has also been observed in marathon runners and other women engaged in training programs that require sustained high levels of exertion, which severely reduce body lipid reserves.

The Vagina

The **vagina** is an elastic, muscular tube extending between the cervix and the exterior. It opens into the *vestibule,* a space bounded by the female external genitalia (**Figures 28–21** and **28–22**). The vagina is typically 7.5–9 cm (3–3.6 in.) long. Its diameter varies because it is highly distensible.

Figure 28–21 The Histology of the Vagina.

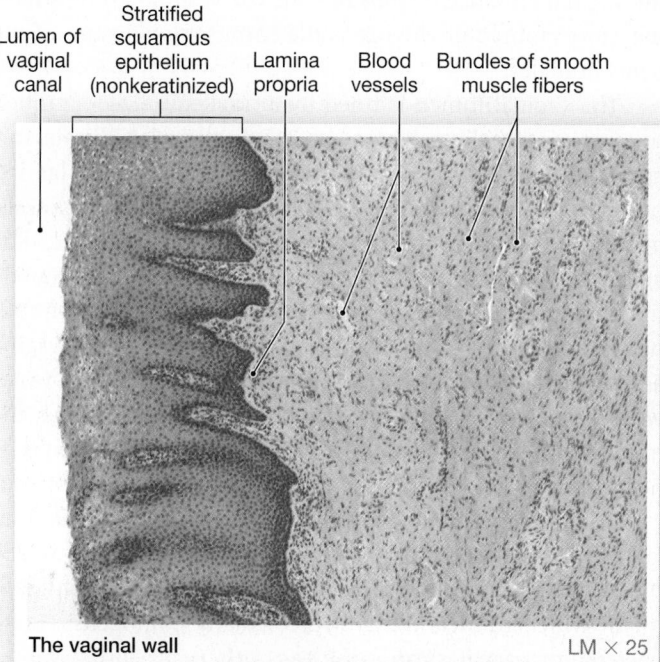

Lumen of vaginal canal | Stratified squamous epithelium (nonkeratinized) | Lamina propria | Blood vessels | Bundles of smooth muscle fibers

The vaginal wall   LM × 25

Fornix
Vaginal artery
Vaginal vein
Rugae
Vaginal canal
Hymen
Greater vestibular gland
Labia minora — Vestibule

At the proximal end of the vagina, the cervix projects into the **vaginal canal**. The shallow recess surrounding the cervical protrusion is known as the **fornix** (FOR-niks). The vagina lies parallel to the rectum, and the two are in close contact posteriorly. Anteriorly, the urethra extends along the superior wall of the vagina from the urinary bladder to the external urethral orifice, which opens into the vestibule.

The primary blood supply of the vagina is by the **vaginal branches** of the internal iliac (or uterine) arteries and veins. Innervation is from the hypogastric plexus, sacral nerves S_2–S_4, and branches of the pudendal nerve. ⤴ pp. 470, 789, 794

The vagina has three major functions. It (1) is a passageway for the elimination of menstrual fluids; (2) receives the penis during sexual intercourse, and holds spermatozoa prior to their passage into the uterus; and (3) forms the inferior portion of the *birth canal*, through which the fetus passes during delivery.

Anatomy and Histology of the Vagina

In sectional view, the lumen of the vagina appears constricted, forming a rough H-shape. The vaginal walls contain a network of blood vessels and layers of smooth muscle (**Figure 28–21**). The lining is moistened by secretions of the cervical glands and by the movement of water across the permeable epithelium. The **hymen** (HĪ-men) is an elastic epithelial fold of variable size that partially blocks the entrance to the vagina. An intact hymen is typically stretched or torn during first sexual intercourse, tampon use, pelvic examination, or physical activity.

The two *bulbospongiosus muscles* extend along either side of the vaginal entrance. Contractions of the bulbospongiosus muscles constrict the vagina. ⤴ p. 382 These muscles cover the **vestibular bulbs**, masses of erectile tissue on either side of the vaginal entrance (**Figure 28–22**). The vestibular bulbs have the same embryonic origins as the corpus spongiosum of the penis in males.

The vaginal lumen is lined by a nonkeratinized stratified squamous epithelium (**Figure 28–21**). In the relaxed state, this epithelium forms folds called *rugae*. The underlying lamina propria is thick and elastic, and it contains small blood vessels, nerves, and lymph nodes. The vaginal mucosa is surrounded by an elastic *muscularis* layer consisting of layers of smooth muscle fibers arranged in circular and longitudinal bundles continuous with the uterine myometrium. The portion of the vagina adjacent to the uterus has a serosal covering that is continuous with the pelvic peritoneum. Along the rest of the vagina, the muscularis layer is surrounded by an *adventitia* of fibrous connective tissue.

The vagina contains a population of resident bacteria, usually harmless, supported by nutrients in the cervical mucus. The metabolic activity of these bacteria creates an acidic environment, which restricts the growth of many pathogens. Fungi, bacteria, or parasites can cause **vaginitis** (vaj-i-NĪ-tis), an inflammation of the vaginal canal. In addition to any discomfort

that may result, the condition may affect the survival of spermatozoa and thereby reduce fertility. An acidic environment also inhibits the motility of sperm. For this reason, the buffers in semen are important to successful fertilization.

The hormonal changes associated with the ovarian cycle also affect the vaginal epithelium. By examining a *vaginal smear*—a sample of epithelial cells shed at the surface of the vagina—a clinician can estimate the corresponding stages in the ovarian and uterine cycles. This diagnostic procedure is an example of *exfoliative cytology*. ⤴ p. 146

The External Genitalia

The area containing the female external genitalia is the **vulva** (VUL-vuh), or *pudendum* (pū-DEN-dum; **Figure 28–22**). The vagina opens into the **vestibule**, a central space surrounded by small folds known as the **labia minora** (LĀ-bē-uh mi-NOR-uh; singular, *labium minus*). The labia minora are covered with a smooth, hairless skin. The urethra opens into the vestibule just anterior to the vaginal entrance. The **paraurethral glands**, or *Skene's glands*, discharge into the urethra near the external urethral opening.

Anterior to this opening, the **clitoris** (KLIT-ō-ris) projects into the vestibule. A small, rounded tissue projection, the clitoris is derived from the same embryonic structures as the penis in males. Internally, it contains erectile tissue comparable to the corpora cavernosa of the penis. A small erectile *glans* sits

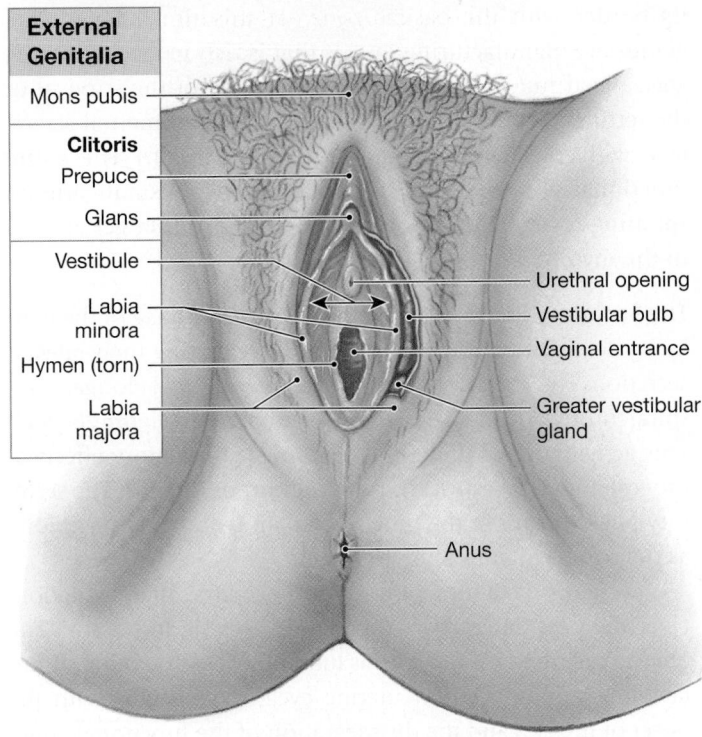

Figure 28–22 **The Female External Genitalia.**

External Genitalia	
Mons pubis	
Clitoris	Prepuce
	Glans
Vestibule	
Labia minora	
Hymen (torn)	
Labia majora	

Urethral opening
Vestibular bulb
Vaginal entrance
Greater vestibular gland
Anus

atop it. The vestibular bulbs along the sides of the vestibule are comparable to the corpus spongiosum. These erectile tissues engorge with blood during sexual arousal. Extensions of the labia minora encircle the body of the clitoris, forming its **prepuce**, or *hood*. ATLAS: Embryology Summary 21: The Development of the Reproductive System

A variable number of small **lesser vestibular glands** discharge their secretions onto the exposed surface of the vestibule between the orifices of the vagina and urethra. During sexual arousal, the **greater vestibular glands** (*Bartholin's glands*), located on either side of the distal portion of the vagina, secrete into the vestibule. These mucous glands keep the area moist and lubricated. The vestibular glands have the same embryonic origins as the bulbourethral glands in males.

The outer margins of the vulva are formed by the mons pubis and the labia majora. The **mons pubis** is a pad of adipose tissue covering the pubic symphysis. Adipose tissue also accumulates within the **labia majora** (singular, *labium majus*), prominent folds of skin that encircle and partially conceal the labia minora and adjacent structures. The outer margins of the labia majora and the mons pubis are covered with coarse hair, but the inner surfaces of the labia majora are hairless. Sebaceous glands and scattered apocrine sweat glands secrete onto the inner surface of the labia majora, moistening and lubricating them.

The Mammary Glands

A newborn infant cannot fend for itself, and several of its key systems have yet to complete development. Over the initial period of adjustment to an independent existence, the infant is nourished from the milk secreted by the maternal **mammary glands**. Milk production, or **lactation** (lak-TA-shun), takes place in these glands. In females, mammary glands are specialized organs of the integumentary system that are controlled mainly by hormones of the reproductive system and by the *placenta*, a temporary structure that provides the embryo and fetus with nutrients.

On each side, a mammary gland lies in the subcutaneous tissue of the **pectoral fat pad** deep to the skin of the chest (**Figure 28–23a**). Each breast has a **nipple**, a small conical projection where the ducts of the underlying mammary gland open onto the body surface. The reddish-brown skin around each nipple is the **areola** (a-RĒ-ō-luh). Large sebaceous glands deep to the areolar surface give it a grainy texture.

The glandular tissue of a mammary gland consists of separate lobes, each containing several secretory lobules. Ducts leaving the lobules converge, giving rise to a single **lactiferous** (lak-TIF-er-us) **duct** in each lobe. Near the nipple, that lactiferous duct enlarges, forming an expanded chamber called a **lactiferous sinus**. Typically, 15–20 lactiferous sinuses open onto the surface of each nipple. Dense connective tissue surrounds the duct system and forms partitions that extend between the lobes and the lobules. These bands of connective tissue, the *suspensory ligaments of the breast*, originate in the dermis of the overlying skin. A layer of areolar tissue separates the mammary gland complex from the underlying pectoralis muscles. Branches of the *internal thoracic artery* supply blood to each mammary gland (**Figure 21–20**, p. 783).

Figure 28–23b,c compares the histological organizations of inactive and active mammary glands. An inactive, or *resting*, mammary gland is dominated by a duct system rather than by active glandular cells. The size of the mammary glands in a nonpregnant woman reflects primarily the amount of adipose tissue present, not the amount of glandular tissue. The secretory apparatus normally does not complete its development unless pregnancy occurs. An active mammary gland is a tubuloalveolar gland, consisting of multiple glandular tubes that end in secretory alveoli. We will discuss the hormonal mechanisms involved in lactation in Chapter 29.

Hormones and Female Reproductive Function

The female reproductive tract is under hormonal control that involves interplay between secretions of both the pituitary gland and the gonads. Circulating hormones control the **female reproductive cycle**, coordinating the ovarian and uterine cycles for proper reproductive function. If the two cycles are not properly coordinated, infertility results. We do not yet know many of the biochemical details of the female reproductive cycle because the processes are complex and difficult to study, but the general patterns are reasonably clear.

As in males, GnRH from the hypothalamus regulates reproductive function in females. However, in females, the GnRH pulse frequency and amplitude (amount secreted per pulse) change throughout the ovarian cycle. If the hypothalamus were a radio station, the pulse frequency would correspond to the radio frequency it is transmitting on, and the amplitude would be the volume. We will consider changes in pulse frequency, as their effects are both dramatic and reasonably well understood. Circulating levels of estrogens and progesterone primarily control changes in GnRH pulse frequency. Estrogens increase the GnRH pulse frequency, and progesterone decreases it.

The endocrine cells of the anterior lobe of the pituitary gland respond as if each group of endocrine cells is monitoring different frequencies. As a result, each group of cells is sensitive to some GnRH pulse frequencies and insensitive to others. For example, consider the *gonadotropes*, the cells that produce FSH and LH. At one pulse frequency, the gonadotropes respond preferentially and secrete FSH, but at another frequency, these cells release primarily LH. FSH and LH production also occurs in pulses that follow the rhythm of GnRH pulses. If GnRH is absent or is supplied at a constant rate (without pulses), FSH and LH secretion stops in a matter of hours.

28

Figure 28–23   The Mammary Gland.   ATLAS: Plate 28

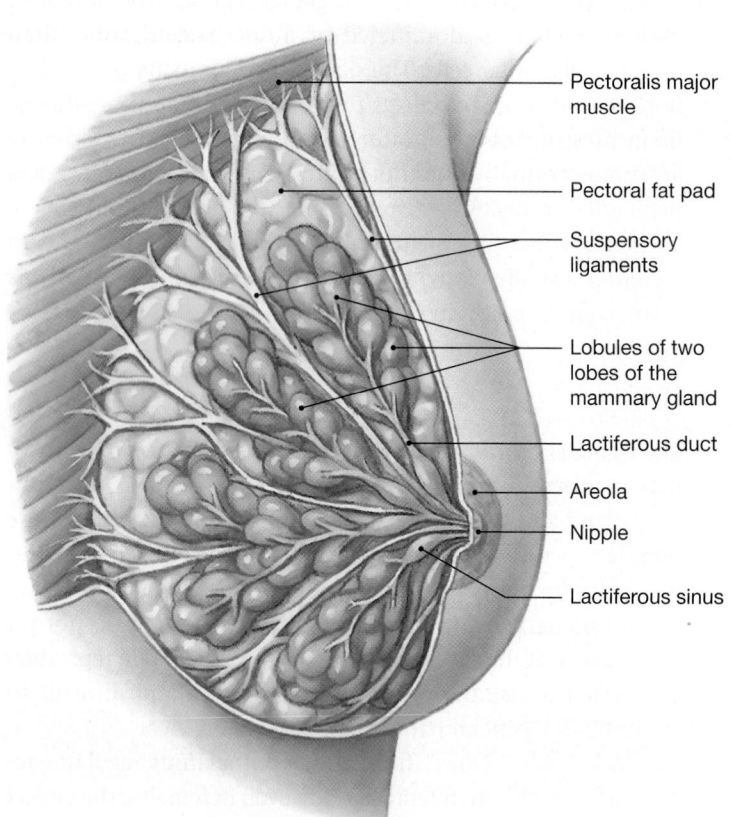

a   The mammary gland of the left breast

Labels for panel a:
- Pectoralis major muscle
- Pectoral fat pad
- Suspensory ligaments
- Lobules of two lobes of the mammary gland
- Lactiferous duct
- Areola
- Nipple
- Lactiferous sinus

Resting mammary gland   LM × 100

b   An inactive mammary gland of a nonpregnant woman

Labels:
- Secretory alveoli
- Lactiferous duct
- Connective tissue

Active mammary gland   LM × 131

c   An active mammary gland of a nursing woman

Labels:
- Secretory alveoli
- Lactiferous duct
- Milk

Spotlight Figure 28–24 shows the changes in circulating hormone levels that accompany the ovarian and uterine cycles. Early in the follicular phase of the ovarian cycle and prior to day 10, estrogen levels are low and the GnRH pulse frequency is 16–24 per day (one pulse every 60–90 minutes). At this frequency, FSH is the dominant hormone released by the anterior lobe of the pituitary gland. The estrogens released by developing follicles inhibit LH secretion. As secondary follicles develop, FSH levels decrease due to the negative feedback effects of inhibin. Follicular development and maturation continue, however, supported by the combination of estrogens, FSH, and LH.

As one or more tertiary follicles begin forming, the concentration of circulating estrogens rises steeply. As a result, the GnRH pulse frequency increases to about 36 per day (one pulse every 30–60 minutes). The increased pulse frequency stimulates LH secretion. In addition, around day 10 of the cycle, the effect of estrogen on LH secretion changes from inhibition to stimulation. The switchover takes place only after estrogen levels have risen above a specific threshold value for about 36 hours. (The threshold value and the time required vary among individuals.) High estrogen levels also increase gonadotrope sensitivity to GnRH. At about day 14, the estrogen level has peaked, the gonadotropes are at maximum sensitivity, and the GnRH pulses are arriving about every 30 minutes. The result is a massive release of LH from the anterior lobe of the pituitary gland. This sudden surge in LH concentration triggers (1) the completion of meiosis I by the primary oocyte, (2) the forceful rupture of the follicular wall, and (3) ovulation. Typically, ovulation occurs 34–38 hours after the LH surge begins, roughly 9 hours after the LH peak.

28

Clinical Note

Breast Cancer The uterine cycle also influences cyclical changes in the mammary glands. The effects usually go unnoticed, but occasional discomfort or even inflammation of mammary gland tissues can occur late in the cycle. If inflamed lobules become walled off by scar tissue, **cysts** are created. Clusters of cysts can be felt in the breast as discrete masses, a condition known as **fibrocystic disease**. Because its symptoms are similar to those of breast cancer, biopsies may be needed to distinguish between this benign condition and breast cancer.

Breast cancer is a malignant, metastasizing tumor of the mammary gland. It is the leading cause of death in women between ages 35 and 45, but it is most common in women over age 50. An estimated 12.4 percent of U.S. women will develop breast cancer at some point in their lifetime. Of the total of about 208,000 newly reported cases in 2010, about 1900 will occur in males. The incidence is highest among Caucasian Americans, somewhat lower in African Americans, and lowest in Asian Americans and American Indians. Notable risk factors include (1) a family history of breast cancer, (2) a first pregnancy after age 30, and (3) early menarche or late menopause.

Despite repeated studies, no links have been proven between breast cancer and oral contraceptive use, fat consumption, or alcohol use. It appears likely that multiple factors are involved. In some families an inherited genetic variation has been linked to higher-than-normal risk of developing the disease. However, most women never develop breast cancer—even women in families with a history of the disease. Mothers who breast-fed (nursed) their babies have a 20 percent lower incidence of breast cancer after menopause than do mothers who did not breast-feed. The reason is not known. Adding to the mystery, nursing does not appear to affect the incidence of premenopausal breast cancer.

Early detection of breast cancer is the key to saving lives. For this reason the use of clinical screening techniques has increased in recent years. **Mammography** involves the use of x-rays to examine breast tissues. The radiation dosage can be low, because only soft tissues must be penetrated.

Treatment is more successful if breast cancer is identified while it is still a relatively small, localized tumor. Once it has grown larger than 2 cm (0.78 in.), the chances of long-term survival worsen. A poor prognosis also follows if cancer cells have spread through the lymphatic system to the axillary lymph nodes. If the nodes are not yet involved, the chances of five-year survival are about 92 percent, but if four or more nodes are involved, the survival rate drops to 25 percent. More than 90 percent of breast cancers are now diagnosed at the local or regional stage.

Treatment of breast cancer includes surgical removal of the tumor. Because in many cases cancer cells begin to spread before the condition is diagnosed, part or all of the affected mammary gland is surgically removed and usually the axillary lymph nodes on that side are biopsied.

- In a *lumpectomy,* only a portion of the breast is removed.
- In a *total mastectomy,* the entire breast is removed.
- In *axillary lymph node dissection,* one or more of the axillary lymph nodes (called sentinel nodes) are removed to check for metastatic cells. If there are no cancer cells in the nodes checked, the other lymph nodes may be left in place.

A combination of chemotherapy, radiation treatments, and hormone treatments may be used to supplement the surgical procedures.

Hormones and the Follicular Phase of the Ovarian Cycle

Follicular development begins under FSH stimulation. Each month some of the primordial follicles begin to develop into primary follicles. As the follicles enlarge, thecal cells start producing *androstenedione,* a steroid hormone that is a key intermediate in the synthesis of estrogens and androgens. The granulosa cells absorb androstenedione and convert it to estrogens. In addition, interstitial cells scattered throughout the ovarian stroma secrete small quantities of estrogens. Circulating estrogens are bound primarily to albumins, with lesser amounts carried by gonadal steroid-binding globulin (GBG).

Of the three estrogens circulating in the bloodstream—estradiol, estrone, and estriol—**estradiol** (es-tra-DĪ-ol) is the most abundant and has the most pronounced effects on target tissues. It is the dominant hormone prior to ovulation. In estradiol synthesis, androstenedione is first converted to testosterone, which the enzyme aromatase converts to estradiol (**Figure 28–25**). The synthesis of both *estrone* and *estriol* proceeds directly from androstenedione.

Estrogens have multiple functions that affect the activities of many tissues and organs throughout the body. Among the important general functions of estrogens are (1) stimulating bone and muscle growth; (2) maintaining female secondary sex characteristics, such as body hair distribution and the location of adipose tissue deposits; (3) affecting central nervous system activity, especially in the hypothalamus, where estrogens increase the sexual drive; (4) maintaining functional accessory reproductive glands and organs; and (5) initiating the repair and growth of the endometrium.

28

The ovarian and uterine cycles must operate in synchrony to ensure proper reproductive function. If the two cycles are not properly coordinated, infertility results. A female who doesn't ovulate cannot conceive, even if her uterus is perfectly normal. A female who ovulates normally, but whose uterus is not ready to support an embryo, will also be infertile.

As in males, GnRH from the hypothalamus regulates reproductive function in females. However, in females, GnRH levels change throughout the course of the ovarian cycle.

HYPOTHALAMUS

① Release of Gonadotropin-Releasing Hormone (GnRH)

The cycle begins with the release of GnRH, which stimulates the production and secretion of FSH and the production—but not the secretion—of LH.

Release of GnRH

ANTERIOR LOBE OF PITUITARY GLAND

Production and secretion of FSH

Production of LH

Secretion of LH

② Follicular Phase of the Ovarian Cycle

The follicular phase begins when FSH stimulates some secondary follicles to develop into a tertiary follicle.

As secondary follicles develop, FSH levels decline due to the negative feedback effects of inhibin.

Developing follicles also secrete estrogens, especially estradiol, the dominant hormone prior to ovulation.

In low concentrations, estrogens inhibit LH secretion. This inhibition gradually decreases as estrogen levels increase.

③ Luteal Phase of the Ovarian Cycle

The combination of increased GnRH pulse frequency and elevated estrogen levels stimulates LH secretion.

On or around day 14, a massive surge in LH level triggers (1) the completion of meiosis I by the primary oocyte, (2) the forceful rupture of the follicular wall, (3) ovulation, roughly 9 hours after the LH peak, and (4) formation of the corpus luteum.

The corpus luteum secretes progesterone, which stimulates and sustains endometrial development.

After ovulation, progesterone levels rise and estrogen levels fall. This suppresses GnRH secretion. If pregnancy does not occur, the corpus luteum will degenerate after 12 days, and as progesterone levels decrease, GnRH secretion increases, and a new cycle begins.

Negative feedback

OVARY

Before day 10

After day 10

• Follicle development
• Secretion of inhibin
• Secretion of estrogens

• Meisois I completion
• Ovulation
• Corpus luteum formation

Secretion of progesterone

Effects on CNS

Stimulation of bone and muscle growth

Establishment and maintenance of female secondary sex characteristics

Maintenance of accessory glands and organs

Stimulation of endometrial growth and secretion

KEY

→ Stimulation

⊣ Inhibition

This illustration combines the key events in the ovarian and uterine cycles. The monthly hormonal fluctuations cause physiological changes that affect core body temperature. During the follicular phase—when estrogens are the dominant hormones—the **basal body temperature**, or the resting body temperature measured upon awakening in the morning, is about 0.3°C (0.5°F) lower than it is during the luteal phase, when progesterone dominates.

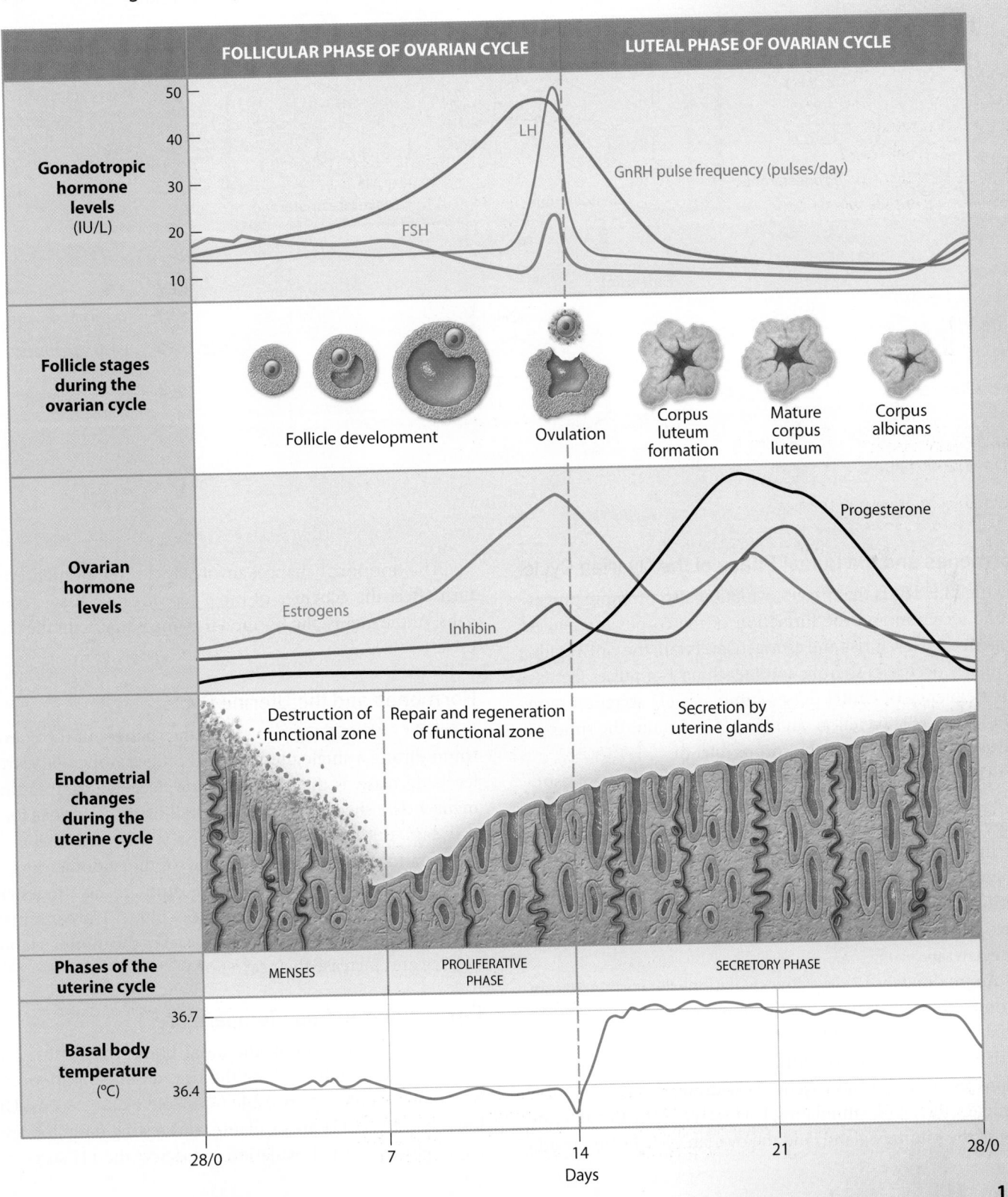

Figure 28–25   Pathways of Steroid Hormone Synthesis in Males and Females. All gonadal steroids are derived from cholesterol. In men, the pathway ends with the synthesis of testosterone, which may subsequently be converted to dihydrotestosterone. In women, an additional step after testosterone leads to estradiol synthesis. The synthesis of progesterone and estrogens other than estradiol involves alternative pathways.

Hormones and the Luteal Phase of the Ovarian Cycle

The high LH levels that trigger ovulation also promote progesterone secretion and the formation of the corpus luteum. As progesterone levels rise and estrogen levels fall, the GnRH pulse frequency decreases sharply, soon reaching 1–4 pulses per day. This frequency of GnRH pulses stimulates LH secretion more than it does FSH secretion, and the LH maintains the structure and secretory function of the corpus luteum.

The corpus luteum secretes moderate amounts of estrogens, but progesterone is the main hormone of the luteal phase. Its primary function is to continue the preparation of the uterus for pregnancy. Progesterone enhances the blood supply to the functional zone and stimulates the secretion of uterine glands. Progesterone levels remain high for the next week, but unless pregnancy occurs, the corpus luteum begins to degenerate.

Approximately 12 days after ovulation, the corpus luteum becomes nonfunctional, and progesterone and estrogen levels fall markedly. The blood supply to the functional zone is restricted, and the endometrial tissues begin to deteriorate. As progesterone and estrogen levels drop, the GnRH pulse frequency increases, stimulating FSH secretion by the anterior lobe of the pituitary gland, and the ovarian cycle begins again.

The hormonal changes involved with the ovarian cycle in turn affect the activities of other reproductive tissues and organs. At the uterus, the hormonal changes maintain the uterine cycle.

Hormones and the Uterine Cycle

Spotlight Figure 28–24 also shows the changes in the endometrium during a single uterine cycle. Progesterone and estrogen levels decrease as the corpus luteum degenerates, resulting in menses. The shedding of endometrial tissue continues for several days, until rising estrogen levels stimulate the repair and regeneration of the functional zone of the endometrium. The proliferative phase then continues until rising progesterone levels mark the arrival of the secretory phase. The combination of estrogen and progesterone then causes the uterine glands to enlarge and increase their secretions.

Hormones and Body Temperature

At the time of ovulation, the basal body temperature (BBT) decreases noticeably, making the rise in temperature over the next day even more noticeable (Spotlight Figure 28–24). Urine tests that detect LH are available, and testing daily for several days before expected ovulation can detect the LH surge more reliably than the BBT changes. This information can be important for individuals who wish to avoid or promote a pregnancy, because fertilization typically occurs within a day of ovulation. After that time, oocyte viability and the likelihood of successful fertilization decrease markedly.

CHECKPOINT

9. Name the structures of the female reproductive system.

10. What effect would blockage of both uterine tubes by scar tissue (resulting from an infection such as gonorrhea) have on a woman's ability to conceive?

11. What benefit does the acidic pH of the vagina provide?

12. Which layer of the uterus is sloughed off, or shed, during menstruation?

13. Would the blockage of a single lactiferous sinus interfere with the delivery of milk to the nipple? Explain.

14. What changes would you expect to observe in the ovarian cycle if the LH surge did not occur?

15. What effect would a blockage of progesterone receptors in the uterus have on the endometrium?

16. What event in the uterine cycle occurs when the levels of estrogens and progesterone decrease?

See the blue Answers tab at the back of the book.

28-4 | The autonomic nervous system influences male and female sexual function

Learning Outcome Discuss the physiology of sexual intercourse in males and females.

Sexual intercourse, also known as *coitus* (KŌ-i-tus) or *copulation*, introduces semen into the female reproductive tract. The process affects both the male and female reproductive systems.

Male Sexual Function

Complex neural reflexes coordinate sexual function in males. The reflex pathways utilize the sympathetic and parasympathetic divisions of the autonomic nervous system. During sexual **arousal**, erotic thoughts, the stimulation of sensory nerves in the genital region, or both lead to an increase in parasympathetic outflow over the pelvic nerves. This outflow in turn leads to **erection** of the penis (discussed on p. 1091). The skin covering the glans penis contains numerous sensory receptors, and erection tenses the skin and increases sensitivity. Subsequent stimulation can initiate the secretion of the bulbo-urethral glands, providing **lubrication** for the penile urethra and the surface of the glans penis.

During intercourse, the sensory receptors of the penis are rhythmically stimulated. This stimulation eventually results in

the coordinated processes of emission and ejaculation. **Emission** occurs under sympathetic stimulation. The process begins when the peristaltic contractions of the ampulla push seminal fluid and spermatozoa into the prostatic urethra. The seminal glands then begin contracting, and the contractions increase in force and duration over the next few seconds. Peristaltic contractions also appear in the walls of the prostate gland. The combination moves the seminal mixture into the membranous and penile portions of the urethra. While the contractions are proceeding, sympathetic commands also cause the contraction of the urinary bladder and the internal urethral sphincter. The combination of elevated pressure inside the bladder and the contraction of the sphincter effectively prevent semen from passing into the bladder.

Ejaculation occurs as powerful, rhythmic contractions appear in the *ischiocavernosus* and *bulbospongiosus muscles*, two superficial skeletal muscles of the pelvic floor. The ischiocavernosus muscles insert along the sides of the penis. Their contractions serve primarily to stiffen that organ. The bulbospongiosus muscle wraps around the base of the penis. The contraction of this muscle pushes semen toward the external urethral opening. The contractions of both muscles are controlled by somatic motor neurons in the inferior lumbar and superior sacral segments of the spinal cord. (The positions of these muscles are shown in Figure 11–13b, p. 382.) Contraction of the smooth muscle within the prostate acts to pinch off the urethra, preventing urine from passing through the erect penis.

Ejaculation is associated with intensely pleasurable sensations associated with perineal muscle contraction, an experience known as male **orgasm** (OR-gazm). Several other noteworthy physiological changes take place at this time, including pronounced but temporary increases in heart rate and blood pressure. After ejaculation, blood begins to leave the erectile tissue, and the erection begins to subside. This subsidence, called **detumescence** (dē-tū-MES-ens), is mediated by the sympathetic nervous system.

In sum, arousal, erection, emission, and ejaculation are controlled by a complex interplay between the sympathetic and parasympathetic divisions of the autonomic nervous system. Higher centers, including the cerebral cortex, can facilitate or inhibit many of the important reflexes, modifying sexual function. Any physical or psychological factor that affects a single component of the system can result in male sexual dysfunction, also called *impotence.*

Impotence, or **erectile dysfunction (ED)**, is defined as an inability to achieve or maintain an erection. Various physical causes may be responsible for impotence, because erection involves vascular changes as well as neural commands. For example, low blood pressure in the arteries supplying the penis, due to a circulatory blockage such as a plaque, will reduce the ability to attain an erection. Drugs, alcohol, trauma, or illnesses that affect the autonomic nervous system or the central nervous

28

system can have the same effect. But male sexual performance can also be strongly affected by the psychological state of the individual. Temporary periods of impotence are fairly common in healthy individuals who are experiencing severe stresses or emotional problems. Depression, anxiety, and fear of impotence are examples of emotional factors that can result in sexual dysfunction. The prescription drugs Viagra, Levitra, and Cialis, which enhance and prolong the effects of nitric oxide on the erectile tissue of the penis, have proven useful in treating many cases of impotence.

Female Sexual Function

The events in female sexual function are largely comparable to those of male sexual function. During sexual arousal, parasympathetic activation leads to engorgement of the erectile tissues of the clitoris and the vestibular bulbs, and increased secretion from cervical mucous glands and the greater vestibular glands. Clitoral erection increases the receptors' sensitivity to stimulation, and the cervical and vestibular glands lubricate the vaginal walls. A network of blood vessels in the vaginal walls becomes filled with blood at this time, and the vaginal surfaces are also moistened by fluid that moves across the epithelium from underlying connective tissues. (This process accelerates during intercourse as the result of mechanical stimulation.) Parasympathetic stimulation also causes contraction of subcutaneous smooth muscle of the nipples, making them more sensitive to touch and pressure.

During sexual intercourse, rhythmic contact of the penis with the clitoris and vaginal walls—reinforced by touch sensations from the breasts and other stimuli (visual, olfactory, and auditory)—provides stimulation that leads to orgasm. Female orgasm is accompanied by peristaltic contractions of the uterine and vaginal walls and, through impulses traveling over the pudendal nerves, rhythmic contractions of the bulbospongiosus and ischiocavernosus muscles. The latter contractions give rise to the intensely pleasurable sensations of orgasm.

Sexual activity carries with it the risk of infection with a variety of microorganisms. The consequences of such an infection may range from merely inconvenient to potentially lethal. **Sexually transmitted diseases (STDs)**, also called **sexually transmitted infections (STIs)**, are transferred from individual to individual, primarily or exclusively by sexual intercourse. At least two dozen bacterial, viral, and fungal infections are currently recognized as STDs. The bacterium *Chlamydia* can cause **pelvic inflammatory disease (PID)** and infertility. AIDS, caused by a virus, is deadly.

The incidence of STDs has been increasing in the United States since 1984. An estimated 20 million new cases occur each year, almost 50 percent in persons aged 15–24. Poverty, intravenous drug use, prostitution, and the appearance of drug-resistant pathogens all contribute to the problem.

CHECKPOINT

17. List the physiological events of sexual intercourse in both sexes, and indicate those that occur in males but not in females.

18. An inability to contract the ischiocavernosus and bulbospongiosus muscles would interfere with which part of the male sex act?

19. What changes occur in females during sexual arousal as the result of increased parasympathetic stimulation?

See the blue Answers tab at the back of the book.

28-5 | With age, decreasing levels of reproductive hormones cause functional changes

Learning Outcome Describe the reproductive system changes that occur with aging.

Sex hormones have widespread effects on the body. They affect brain development and behavioral drives, muscle mass, bone mass and density, body proportions, and the patterns of hair and body fat distribution. As aging occurs, reductions in sex hormone levels affect appearance, strength, and a variety of physiological functions. The aging process affects all body systems, including the reproductive systems of both men and women. As noted earlier in the chapter, these systems become fully functional at puberty. Later on, the most striking age-related changes in the female reproductive system occur at menopause. Comparable age-related changes in the male reproductive system occur more gradually and over a longer period of time.

Menopause

Menopause is usually defined as the time when ovulation and menstruation cease. Menopause typically occurs at age 45–55, but in the years immediately preceding it, the ovarian and uterine cycles become irregular. This interval is called *perimenopause*. A shortage of primordial follicles is the underlying cause of the irregular cycles. It has been estimated that almost 7 million potential oocytes are in fetal ovaries after five months of development, but the number decreases to about 2 million at birth, and to a few hundred thousand at puberty. With the arrival of perimenopause, the number of primordial follicles responding each month begins to decrease markedly. As their numbers decrease, estrogen levels decrease and may not increase enough to trigger ovulation. By age 50, there are often no primordial follicles left to respond to FSH. In **premature menopause**, this depletion occurs before age 40.

Menopause is accompanied by a decrease in circulating concentrations of estrogens and progesterone, and a sharp and sustained increase in the production of GnRH, FSH, and LH. The decrease in estrogen levels leads to reductions in the size of the uterus and breasts, accompanied by a thinning of the urethral and vaginal epithelia. The decreased estrogen concentrations have also been linked to the development of osteoporosis, presumably because bone deposition proceeds at a slower rate. A variety of neural effects are reported as well, including "hot flashes," anxiety, and depression. Hot flashes typically begin while estrogen levels are decreasing, and cease when estrogen levels reach minimal values. These intervals of increased body temperature are associated with surges in LH production. The hormonal mechanisms involved in other CNS effects of menopause are poorly understood. In addition, the risks of atherosclerosis and other forms of cardiovascular disease increase after menopause.

The majority of women experience only mild symptoms, but some individuals experience acutely unpleasant symptoms in perimenopause or during or after menopause. For most of those women, hormone replacement therapy (HRT) involving a combination of estrogens and progesterone can control the unpleasant neural and vascular changes associated with menopause. The hormones are administered as pills, by injection, or by transdermal "estrogen patches." However, recent studies suggest that taking estrogen replacement therapy for more than five years increases the risk of heart disease, breast cancer, Alzheimer's disease, blood clots, and stroke. HRT should be used only after a full discussion and assessment of the potential risks and benefits, and taken for as short a time as possible.

The Male Climacteric

Changes in the male reproductive system take place more gradually than do those in the female reproductive system. The period of declining reproductive function, which corresponds to perimenopause in women, is known as the **male climacteric** or *andropause*. Levels of circulating testosterone begin to decrease between the ages of 50 and 60, and levels of circulating FSH and LH increase. Although sperm production continues (men well into their 80s can father children), older men experience a gradual reduction in sexual activity. This decrease may be linked to declining testosterone levels. Some clinicians suggest the use of testosterone replacement therapy to enhance the libido (sexual drive) of elderly men, but this may increase the risk of prostate disease.

CHECKPOINT

20. **Define menopause.**

21. **Why does the level of FSH increase and remain high during menopause?**

22. **What is the male climacteric?**

See the blue Answers tab at the back of the book.

28-6 I The reproductive system secretes hormones affecting growth and metabolism of all body systems

Learning Outcome Give examples of interactions between the reproductive system and each of the other organ systems.

Normal human reproduction is a complex process that involves multiple systems. The hormones discussed in this chapter play a major role in coordinating reproductive events (Table 28–1).

Table 28–1	Hormones of the Reproductive System		
Hormone	**Source**	**Regulation of Secretion**	**Primary Effects**
Gonadotropin-releasing hormone (GnRH)	Hypothalamus	*Males:* inhibited by testosterone and possibly by inhibin *Females:* GnRH pulse frequency increased by estrogens, decreased by progesterone	Stimulates FSH secretion and LH synthesis in males Stimulates FSH secretion and LH synthesis in females
Follicle-stimulating hormone (FSH)	Anterior lobe of the pituitary gland	*Males:* stimulated by GnRH, inhibited by inhibin *Females:* stimulated by GnRH, inhibited by inhibin	*Males:* stimulates spermatogenesis and spermiogenesis through effects on nurse cells *Females:* stimulates follicle development, estrogen production, and oocyte maturation
Luteinizing hormone (LH)	Anterior lobe of the pituitary gland	*Males:* stimulated by GnRH *Females:* production stimulated by GnRH, secretion by the combination of high GnRH pulse frequencies and high estrogen levels	*Males:* stimulates interstitial cells to secrete testosterone *Females:* stimulates ovulation, formation of corpus luteum, and progesterone secretion
Androgens (primarily testosterone and dihydrotestosterone)	Interstitial cells of testes	Stimulated by LH	Establish and maintain male secondary sex characteristics and sexual behavior; promote maturation of spermatozoa; inhibit GnRH secretion
Estrogens (primarily estradiol)	Granulosa and thecal cells of developing follicles; corpus luteum	Stimulated by FSH	Stimulate LH secretion (at high levels); establish and maintain female secondary sex characteristics and sexual behavior; stimulate repair and growth of endometrium; increase frequency of GnRH pulses
Progesterone	Granulosa cells from midcycle through functional life of corpus luteum	Stimulated by LH	Stimulate endometrial growth and glandular secretion; reduce frequency of GnRH pulses
Inhibin	Nurse cells of testes and granulosa cells of ovaries	Stimulated by factors released by developing spermatozoa (male) and developing follicles (female)	Inhibits secretion of FSH (and possibly of GnRH)

28

SYSTEM INTEGRATOR

Body System ⟶ Reproductive System   Reproductive System ⟶ Body System

Integumentary
Covers external genitalia; provides sensations that stimulate sexual behaviors; mammary gland secretions nourish the newborn

Integumentary
Reproductive hormones affect distribution of body hair and subcutaneous fat deposits
Page 200

Skeletal
Pelvis protects reproductive organs of females, portion of ductus deferens and accessory glands in males

Skeletal
Sex hormones stimulate growth and maintenance of bones; sex hormones at puberty accelerate growth and closure of epiphyseal cartilages
Page 311

Muscular
Contractions of skeletal muscles eject semen from male reproductive tract; muscle contractions during sexual act produce pleasurable sensations in both sexes

Muscular
Reproductive hormones, especially testosterone, accelerate skeletal muscle growth
Page 406

Nervous
Controls sexual behaviors and sexual function

Nervous
Sex hormones affect CNS development and sexual behaviors
Page 584

Endocrine
Hypothalamic regulatory hormones and pituitary hormones regulate sexual development and function; oxytocin stimulates smooth muscle contractions in uterus and mammary glands

Endocrine
Steroid sex hormones and inhibin inhibit secretions from the hypothalamus and pituitary gland
Page 673

Cardiovascular
Distributes reproductive hormones; provides nutrients, oxygen, and waste removal for fetus; local blood pressure changes responsible for physical changes during sexual arousal

Cardiovascular
Estrogens may help maintain healthy vessels and slow development of atherosclerosis
Page 802

Lymphatic
Provides IgA for secretions by epithelial glands; assists in repairs and defense against infection

Lymphatic
Lysozymes and bactericidal chemicals in secretions provide innate immunity against reproductive tract infections
Page 850

Respiratory
Provides oxygen and removes carbon dioxide generated by tissues of reproductive system and (in pregnant women) by a growing embryo and fetus

The Male Reproductive System

Respiratory
Changes in respiratory rate and depth occur during sexual arousal, under control of the nervous system
Page 900

Digestive
Provides additional nutrients required to support gamete production and (in pregnant women) embryonic and fetal development

Digestive
In pregnant women, digestive organs are crowded by developing fetus, constipation is common, and appetite increases
Page 955

Urinary
Urethra in males carries semen to exterior; kidneys remove wastes generated by reproductive tissues and (in pregnant women) by a growing embryo and fetus

Urinary
Accessory organ secretions may have antibacterial action that helps prevent urethral infections in males
Page 1036

The REPRODUCTIVE System

Figure 28–26 diagrams the functional relationships between the reproductive system and the other body systems.

For all systems, the reproductive system secretes hormones with effects on growth and metabolism.

Physical factors also play a role. The man's sperm count must be adequate, the semen must have the correct pH and nutrients, and erection and ejaculation must occur in the proper sequence. The woman's ovarian and uterine cycles must be properly coordinated; ovulation and oocyte transport must occur normally. Her reproductive tract must provide a hospitable environment for the survival and movement of sperm, and for the subsequent fertilization of the oocyte. For these steps to occur, the reproductive, digestive, endocrine, nervous, cardiovascular, and urinary systems must all be functioning normally.

Even when all else is normal and fertilization occurs at the proper time and place, a healthy infant will not be produced unless the zygote—the union of an ovum with a sperm forming a single cell the size of a pinhead—manages to develop into a full-term fetus that typically weighs about 3 kg (6.6 lb). In Chapter 29 we will consider the process of development, focusing on the mechanisms that determine both the structure of the body and the distinctive characteristics of each individual.

Even though the reproductive system's primary function—producing offspring—doesn't play a role in maintaining homeostasis, reproduction depends on a variety of physical, physiological, and psychological factors. Many of these factors require intersystem cooperation. In addition, the hormones that control and coordinate sexual function have direct effects on the organs and tissues of other systems. For example, testosterone and estradiol affect both muscular development and bone density. **Figure 28–26** summarizes the relationships between the reproductive system and other body systems.

CHECKPOINT

23. **Describe the interaction between the reproductive system and the cardiovascular system.**

24. **Describe the interaction between the reproductive system and the skeletal system.**

See the blue Answers tab at the back of the book.

Related Clinical Terms

cervical dysplasia: Abnormal growth of epithelial cells in the uterine cervix; may progress to cancer.

endometriosis: The growth of endometrial tissue outside the uterus.

genital herpes: A sexually transmitted disease caused by a herpes virus and characterized by painful blisters in the genital area.

gonorrhea: A sexually transmitted bacterial disease caused by *Neisseria gonorrhoeae.* Commonly called "the clap."

gynecology: The branch of medicine that deals with the functions and diseases specific to women and girls affecting the reproductive system.

hydrocele: The accumulation of serous fluid in any body sac, but especially in the tunica vaginalis of the testis or along the spermatic cord.

hysterectomy: The surgical removal of the uterus.

menorrhagia: The condition of experiencing extremely heavy bleeding at menstruation.

oophorectomy: The surgical removal of one or both ovaries.

orchitis: Inflammation of one or both testicles.

ovarian cyst: A condition (usually harmless) in which fluid-filled sacs develop in or on the ovary.

premature ejaculation: A common complaint of ejaculating semen sooner than the man desires while achieving orgasm during intercourse. An estimated 30 percent of men regularly experience the problem.

polycystic ovary syndrome (PCOS): A condition in women that is characterized by irregular or no menstrual periods, acne, obesity, and excessive hair growth.

premenstrual dysphoric disorder (PMDD): A collection of physical and emotional symptoms that occur 5 to 11 days before a woman's period begins, and goes away once menstruation starts. Over 150 signs and symptoms have been associated with the condition.

premenstrual syndrome: A condition occurring in the last half of the menstrual cycle after ovulation that is a combination of physical and mood disturbances which normally end with the onset of the menstrual flow.

salpingitis: Inflammation of a uterine tube.

uterine fibroids (leiomyomas): Benign tumors of the uterus composed of smooth muscle tissue that grows in the wall of the uterus of some women. While not usually dangerous they can cause problems such as very heavy menstrual periods and pain.

uterine prolapse: Condition that occurs when a woman's pelvic floor muscles and ligaments stretch and weaken and provide inadequate support for the uterus, which then descends into the vaginal canal.

vasectomy: The surgical removal of a segment of each ductus (vas) deferens, making it impossible for spermatozoa to reach the distal portions of the male reproductive tract.

vulvovaginal candidiasis: A common female vaginal infection caused by the yeast *Candida,* usually *Candida albicans.*

28

Chapter Review

Study Outline

MasteringA&P®

Access more chapter study tools online in the MasteringA&P Study Area:

- Chapter Quizzes, Chapter Practice Test, Art-labeling Activities, Animations, MP3 Tutor Sessions, and Clinical Case Studies
- Practice Anatomy Lab   PAL™
- Interactive Physiology   iP®
- A&P Flix   A&PFlix
- PhysioEx   PhysioEx

28-1 ▍ Basic reproductive system structures are gonads, ducts, accessory glands and organs, and external genitalia p. 1077

1. The human **reproductive system** produces, stores, nourishes, and transports functional **gametes** (reproductive cells). **Fertilization** is the fusion of male and female gametes.

2. The reproductive system includes **gonads (testes or ovaries)**, ducts, accessory glands and organs, and the **external genitalia**.

3. In males, the testes produce **spermatozoa**, which are expelled from the body in **semen** during *ejaculation*. The ovaries of a sexually mature female produce **oocytes** (immature **ova**) that travel along *uterine tubes* toward the *uterus*. The *vagina* connects the uterus with the exterior of the body.

28-2 ▍ Spermatogenesis occurs in the testes, and hormones from the hypothalamus, anterior lobe of the pituitary gland, and testes control male reproductive functions p. 1078

4. Spermatozoa travel along the *epididymis*, the *ductus deferens*, the *ejaculatory duct*, and the *urethra* before leaving the body. Accessory organs (notably the *seminal glands, prostate gland,* and *bulbourethral glands*) secrete fluids into the ejaculatory ducts and the urethra. The *scrotum* encloses the testes, and the *penis* is an erectile organ. *(Figure 28–1)*

5. The **descent of the testes** through the *inguinal canals* occurs during fetal development. The testes remain connected to internal structures by the **spermatic cords**. The **raphe** marks the boundary between the two chambers in the **scrotum**. *(Figures 28–2, 28–3)*

6. The **dartos** muscle tightens the scrotum, giving it a wrinkled appearance as it elevates the testes. The **cremaster muscles** are more substantial muscles that pull the testes close to the body.

7. The **tunica albuginea** surrounds each testis. Septa extend from the tunica albuginea to the region of the testis closest to the entrance to the epididymis, creating a series of **lobules**. *(Figure 28–4)*

8. **Seminiferous tubules** within each lobule are the sites of sperm production. From there, spermatozoa pass through the **rete testis**. Seminiferous tubules connect to a **straight tubule**. **Efferent ductules** connect the rete testis to the

epididymis. Between the seminiferous tubules are **interstitial cells**, which secrete sex hormones. *(Figures 28–4, 28–5)*

9. Seminiferous tubules contain **spermatogonia**, stem cells involved in **spermatogenesis** (the production of spermatozoa), and **nurse** (sustentacular or Sertoli) **cells**, which sustain and promote the development of spermatozoa. *(Figures 28–6, 28–7)*

10. Each **spermatozoon** has a **head** capped by an **acrosome**, a **middle piece**, and a **tail**. *(Figure 28–8)*

11. From the testis, the spermatozoa enter the **epididymis**, an elongated tubule with **head, body**, and **tail** regions. The epididymis monitors and adjusts the composition of the fluid in the seminiferous tubules, serves as a recycling center for damaged spermatozoa, stores and protects spermatozoa, and facilitates their functional maturation. *(Figure 28–9)*

12. The **ductus deferens**, or *vas deferens*, begins at the epididymis and passes through the inguinal canal as part of the spermatic cord. Near the prostate gland, the ductus deferens enlarges to form the **ampulla**. The junction of the base of the seminal gland and the ampulla creates the **ejaculatory duct**, which empties into the urethra. *(Figures 28–9, 28–10)*

13. The **urethra** extends from the urinary bladder to the tip of the penis. The urethra can be divided into *prostatic, membranous,* and *spongy regions.*

14. Each **seminal gland (seminal vesicle)** is an active secretory gland that contributes about 60 percent of the volume of semen. Its secretions contain fructose (which is easily metabolized by spermatozoa), bicarbonate ions, prostaglandins, and fibrinogen. The **prostate gland** secretes slightly acidic **prostatic fluid**. Alkaline mucus secreted by the **bulbourethral glands** has lubricating properties. *(Figures 28–10, 28–11)*

15. A typical ejaculation releases 2–5 mL of semen **(ejaculate)**, which contains 20–100 million spermatozoa per milliliter. The fluid component of semen is **seminal fluid**.

16. The skin overlying the **penis** resembles that of the scrotum. Most of the **body** of the penis consists of three masses of **erectile tissue**. Beneath the superficial fascia are two **corpora cavernosa** and a single **corpus spongiosum**, which surrounds the urethra. Dilation of the blood vessels within the erectile tissue produces an **erection**. *(Figure 28–11)*

17. Important regulatory hormones include **FSH** *(follicle-stimulating hormone)*, **LH** *(luteinizing hormone)*, and **GnRH** *(gonadotropin-releasing hormone)*. *Testosterone* is the most important androgen. *(Spotlight Figure 28–12)*

28-3 ▍ Oogenesis occurs in the ovaries, and hormones from the pituitary gland and gonads control female reproductive functions p. 1094

18. Principal organs of the female reproductive system include the *ovaries, uterine tubes, uterus, vagina,* and external genitalia. *(Figure 28–13)*

19. The ovaries, uterine tubes, and uterus are enclosed within the **broad ligament**. The **mesovarium** supports and stabilizes each ovary. *(Figure 28–14)*

20. The ovaries are held in position by the **ovarian ligament** and the **suspensory ligament**. Major blood vessels enter the ovary at the **ovarian hilum**. Each ovary is covered by a **tunica albuginea**. *(Figure 28–14)*

21. **Oogenesis** (ovum production) occurs monthly in **ovarian follicles** as part of the **ovarian cycle**, which is divided into a **follicular** (*preovulatory*) **phase** and a **luteal** (*postovulatory*) **phase**. (*Figures 28–15, 28–16*)

22. As follicle development proceeds, **primary, secondary**, and **tertiary follicles** are produced in turn. At **ovulation**, a **secondary oocyte** and the attached follicular cells of the **corona radiata** are released through the ruptured ovarian wall. The follicular cells remaining within the ovary form the **corpus luteum**, which later degenerates into scar tissue called a **corpus albicans**. (*Figure 28–16*)

23. Each **uterine tube** has an **infundibulum** with **fimbriae** (fingerlike projections), an **ampulla**, and an **isthmus**. Each uterine tube opens into the *uterine cavity*. For fertilization to occur, a secondary oocyte must encounter spermatozoa during the first 12–24 hours of its passage from the infundibulum to the uterus. (*Figure 28–17*)

24. *Peg cells* lining the uterine tube secrete a fluid that completes the capacitation of spermatozoa.

25. The **uterus** provides mechanical protection, nutritional support, and waste removal for the developing embryo and fetus. Normally, the uterus bends anteriorly near its base (*anteflexion*). The **broad ligament**, **uterosacral ligaments**, **round ligaments**, and **lateral ligaments** stabilize the uterus. (*Figure 28–18*)

26. Major anatomical landmarks of the uterus include the **body, isthmus, cervix, external os** (*external orifice*), **uterine cavity, cervical canal**, and **internal os** (*internal orifice*). The uterine wall consists of an inner **endometrium**, a muscular **myometrium**, and a superficial **perimetrium** (an incomplete serous layer). (*Figures 28–18, 28–19*)

27. A typical 28-day **uterine**, or *menstrual*, **cycle** begins with the onset of **menses** and the destruction of the **functional zone** of the endometrium. This process of **menstruation** continues from one to seven days. (*Figure 28–20*)

28. After menses, the **proliferative phase** begins, and the functional zone thickens and undergoes repair. The proliferative phase is followed by the **secretory phase**, during which uterine glands enlarge. Menstrual activity begins at **menarche** and continues until **menopause**. (*Figure 28–20*)

29. The **vagina** is a muscular tube extending between the uterus and the external genitalia; it is lined by a nonkeratinized stratified squamous epithelium. A thin epithelial fold, the **hymen**, partially blocks the entrance to the vagina until physical distortion ruptures the membrane. (*Figures 28–21, 28–22*)

30. The components of the **vulva** are the **vestibule, labia minora, paraurethral glands, clitoris, labia majora**, and **lesser** and **greater vestibular glands**. (*Figure 28–22*)

31. A newborn infant is nourished from milk secreted by maternal **mammary glands**. (*Figure 28–23*)

32. Hormonal regulation of the **female reproductive cycle** involves the coordination of the ovarian and uterine cycles. (*Spotlight Figure 28–24*)

33. **Estradiol**, the most important *estrogen*, is the dominant hormone of the follicular phase. Ovulation occurs in response to a midcycle surge in LH. (*Spotlight Figure 28–24, Figure 28–25*)

34. The hypothalamic secretion of GnRH occurs in pulses that trigger the pituitary secretion of FSH and LH. FSH initiates follicular development, and activated follicles and ovarian interstitial cells produce estrogens. High estrogen levels stimulate LH secretion, increase pituitary sensitivity to GnRH, and increase the GnRH pulse frequency. **Progesterone** is the main hormone of the luteal phase. Changes in estrogen and progesterone levels are responsible for maintaining the uterine cycle. (*Spotlight Figure 28–24*)

28-4 ▌ The autonomic nervous system influences male and female sexual function p. 1113

35. During sexual **arousal** in males, erotic thoughts, sensory stimulation, or both lead to parasympathetic activity that produces erection. Stimuli accompanying **sexual intercourse** lead to **emission** and **ejaculation**. Contractions of the bulbospongiosus muscles are associated with **orgasm**.

36. The events of female sexual function resemble those of male sexual function, with parasympathetic arousal and skeletal muscle contractions associated with orgasm.

28-5 ▌ With age, decreasing levels of reproductive hormones cause functional changes p. 1114

37. Menopause (the time that ovulation and menstruation stop) typically occurs at ages 45–55. The production of GnRH, FSH, and LH rise, whereas circulating concentrations of estrogen and progesterone decrease.

38. During the **male climacteric**, at ages 50–60, circulating testosterone levels decrease, and FSH and LH levels increase.

28-6 ▌ The reproductive system secretes hormones affecting growth and metabolism of all body systems p. 1115

39. Normal human reproduction depends on a variety of physical, physiological, and psychological factors, many of which require intersystem cooperation. (*Figure 28–26; Table 28–1*)

Review Questions

See the blue Answers tab at the back of the book.

LEVEL 1 Reviewing Facts and Terms

1. Developing spermatozoa are nourished by (a) interstitial cells, (b) the seminal glands, (c) nurse cells, (d) Leydig cells, (e) the epididymis.
2. The ovaries are responsible for (a) the production of female gametes, (b) the secretion of female sex hormones, (c) the secretion of inhibin, (d) all of these.
3. Among the four cells produced by meiosis in females, one is the ovum. Which are the other three cells?
4. A sudden surge in LH secretion causes the (a) onset of menses, (b) rupture of the follicular wall and ovulation, (c) beginning of the proliferative phase, (d) end of the uterine cycle.
5. The main hormone of the postovulatory phase is (a) progesterone, (b) estradiol, (c) estrogen, (d) luteinizing hormone.
6. What is the key cellular organelle found in the middle piece of the sperm?
7. The _____ and _____ tissues of the endometrium support the physiological demands of the growing fetus.
8. Identify the three regions of the male urethra.
9. List the functions of testosterone in males.
10. List and summarize the important steps in the ovarian cycle.
11. How are oocytes transported along the fallopian tube?
12. What is the role of the clitoris in the female reproductive system?

28

13. Trace the path of milk flow from its site of production to outside the female.

14. Identify the main structures of the male reproductive system in the diagram below.

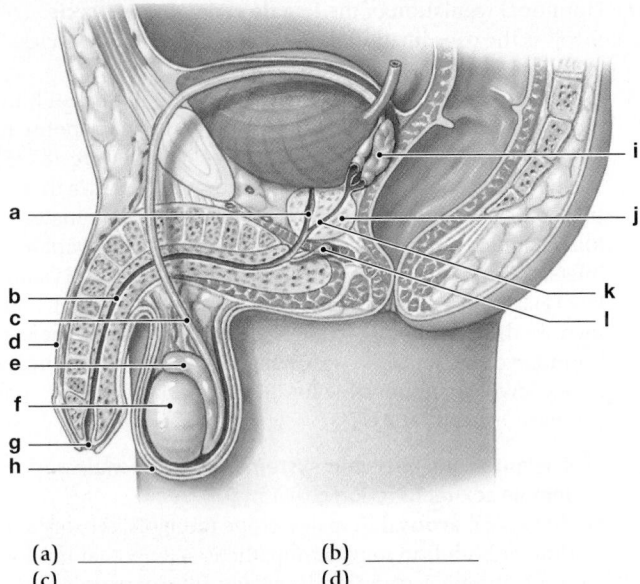

(a) _____	(b) _____
(c) _____	(d) _____
(e) _____	(f) _____
(g) _____	(h) _____
(i) _____	(j) _____
(k) _____	(l) _____

LEVEL 2 Reviewing Concepts

15. In the follicular phase of the ovarian cycle, the ovary (a) undergoes atresia, (b) forms a corpus luteum, (c) releases a mature ovum, (d) secretes progesterone, (e) matures a follicle.

16. Summarize the roles of the hormones in the ovarian and uterine cycles.

17. In the male reproductive tract, what is the role of the prostate?

18. Explain what happens to the follicle after the release of the oocytes.

19. In the penis, what is the function of the corpora cavernosa?

20. How does the aging process affect the female reproductive system?

21. What is the typical volume of an ejaculation?

22. What is the normal sperm count?

23. Where are the following hormones secreted? (a) GnRH, (b) progesterone, (c) luteinizing hormone.

LEVEL 3 Critical Thinking and Clinical Applications

24. Diane has peritonitis (an inflammation of the peritoneum), which her physician says resulted from a urinary tract infection. Why might this condition occur more readily in females than in males?

25. In a condition known as endometriosis, endometrial cells are believed to migrate from the body of the uterus into the uterine tubes or by way of the uterine tubes into the peritoneal cavity, where they become established. A major symptom of endometriosis is periodic pain. Why does such pain occur?

26. Contraceptive pills contain estradiol and progesterone, or progesterone alone, administered at programmed doses during the ovarian cycle to prevent follicle maturation and ovulation. Explain how such pills are effective.

27. Female bodybuilders and women with eating disorders such as anorexia nervosa commonly experience amenorrhea. What does this fact suggest about the relation between body fat and menstruation? What effect would amenorrhea have on achieving a successful pregnancy?

CLINICAL CASE   Wrap-Up

A Post-Game Mystery

The emergency room doctor examines Dan and notes how uncomfortable he is. Dan's left testis is extremely tender to the touch and is visibly higher than the right testis. It also appears swollen. The doctor further notes that the epididymis on the left testis is located anteromedially, near the scrotal raphe, instead of the normal posterior location. Dan's cremasteric reflex is intact on the right. But when the physician strokes Dan's inner left thigh, Dan's left testis does not retract toward his inguinal canal. His genitofemoral nerve is no longer working.

The doctor does a bedside ultrasonography, an imaging technique using high-frequency sound waves, and determines there is no blood supply to Dan's left testis. The doctor concludes that Dan is suffering from an acute testicular torsion. His left testis has twisted on its spermatic cord and cut off its blood supply,

including the deferential artery, the testicular artery, and the venous pampiniform plexus. This torsion has also twisted the genitofemoral nerve. Dan's left testis is infarcting, or dying from lack of blood.

A consulting urologist is quickly on the scene. "Sorry about ruining your homecoming night, Dan," she says. "But our best chance of saving your testis is to head straight to the operating room and untwist it."

1. What other structures in Dan's left spermatic cord are being twisted with this testicular torsion?

2. Epididymitis is a painful inflammation of the epididymis that can mimic the clinical signs and symptoms of testicular torsion. Do you think Dan's cremasteric reflex would be intact if he were suffering from epididymitis?

See the blue Answers tab at the back of the book.

28

29 Development and Inheritance

Learning Outcomes

These Learning Outcomes correspond by number to this chapter's sections and indicate what you should be able to do after completing the chapter.

29-1 ▍ Explain the relationship between differentiation and development, and specify the various stages of development.

29-2 ▍ Describe the process of fertilization, and explain how developmental processes are regulated.

29-3 ▍ List the three stages of prenatal development, and describe the major events of each.

29-4 ▍ Explain how the three germ layers are involved in forming the extraembryonic membranes, and discuss the importance of the placenta as an endocrine organ.

29-5 ▍ Describe the interplay between the maternal organ systems and the developing fetus, and discuss the structural and functional changes in the uterus during gestation.

29-6 ▍ List and discuss the events that occur during labor and delivery.

29-7 ▍ Identify the features and physiological changes of the postnatal stages of life.

29-8 ▍ Relate basic principles of genetics to the inheritance of human traits.

SPOTLIGHT
- Extraembryonic Membranes and Placenta Formation pp. 1130–1131

CLINICAL CASE
- The Twins That Looked Nothing Alike p. 1122

Clinical Notes
- Abortion p. 1140
- Chromosomal Abnormalities and Genetic Analysis p. 1156

The Twins That Looked Nothing Alike

Annie and her twin, Ruth, are in Twinsburg, Ohio, for the Twins Day Festival, the largest annual celebration of twins in the world. There are thousands of identical twins in attendance, and a healthy representation of fraternal twins as well. Annie and Ruth are amazed to see so many twins who not only look alike but also move alike, sound alike, and smile alike.

As they walk through the festival grounds, Annie and Ruth notice tents set up to recruit twins for everything from FBI face-recognition software development to fingerprint and handwriting analysis. The festival is a popular event for "nurture vs. nature"

researchers who are looking for twins to enlist in their studies.

Another centerpiece of the festival is the awarding of medals to twins in various categories, including the oldest twins, the youngest twins, the "most-alike" twins, and the "least-alike" twins. This year, the "least-alike" twins medals are awarded to two boys, one who is short and Caucasian, the other who is tall and of African descent. **Annie elbows Ruth and asks, "What do you think is the story with these twins?" To find out, turn to the Clinical Case Wrap-Up on p. 1160.**

An Introduction to Development and Inheritance

The physiological processes we have studied so far are relatively brief. Many last only a fraction of a second, and others may take hours. But some important processes are measured in months, years, or decades. A human being develops in the womb for nine months, grows to maturity in 15 to 20 years, and may live the better part of a century. During that time, he or she is always changing. Birth, growth, maturation, aging, and death are all parts of a single, continuous process. That process does not end with the individual, because humans can pass at least some of their characteristics on to their offspring. Each generation gives rise to a new generation that will repeat the cycle.

In this chapter, we explore how genetic programming, environmental factors, and various physiological processes affect the events following the union of male and female gametes—sperm and oocytes. The explanation begins with prenatal development and continues through childhood and adolescence and into maturity and senescence, or aging.

29-1 | Development, marked by various stages, is a continuous process that occurs from fertilization to maturity

Learning Outcome Explain the relationship between differentiation and development, and specify the various stages of development.

Time refuses to stand still. Today's infant will be tomorrow's adult. The gradual modification of anatomical structures and physiological characteristics during the period from fertilization to maturity is called **development**. The changes that

occur during development are truly remarkable. In a mere nine months, all the tissues, organs, and organ systems we have studied so far take shape and begin to function. What begins as a single cell slightly larger than the period at the end of this sentence becomes an individual like yourself whose body contains trillions of cells organized into a complex array of highly specialized structures. The formation of the different types of cells required in this process is called **differentiation**. Differentiation occurs through selective changes in genetic activity. As development proceeds, some genes are turned off and others are turned on. Exactly which genes are turned off and on varies from one type of cell to another, and the patterns change over time.

Development begins at **fertilization**, or **conception**, when the male and female gametes fuse. We divide development into stages characterized by specific anatomical changes. **Embryonic development** includes the events during the first two months after fertilization. The study of these events is called **embryology** (em-brē-OL-ō-jē). **Fetal development** begins at the start of the ninth week and continues until birth. Embryonic and fetal development are sometimes referred to collectively as **prenatal** (*natus,* birth) **development**, the primary focus of this chapter. **Postnatal development** begins at birth and continues to **maturity**, the state of full development or completed growth.

A basic understanding of human development gives important insights into anatomical structures. In addition, many of the mechanisms of development and growth are similar to those that repair injuries. In this chapter, we focus on major aspects of development. We consider highlights of the developmental process rather than examine the events in great detail. We also consider the regulatory mechanisms involved, and how developmental patterns can be modified—for good or harm. Few topics in the biological sciences are so fascinating, and fewer still confront investigators with so many scientific, technological, and ethical challenges. The ongoing debate over research with embryonic stem cells and fetal tissue has brought several ethical issues into the public eye. The information presented in this final chapter should help you form your opinions on many difficult moral, legal, and public policy questions.

All humans go through the same developmental stages, but differences in their genetic makeup produce distinctive individual characteristics. The term **inheritance** or **heredity** refers to the transfer of genetically determined characteristics from generation to generation. The study of the mechanisms responsible for inheritance is called **genetics**. In this chapter, we will consider basic genetics as it applies to inherited characteristics, such as sex, hair color, and various diseases.

CHECKPOINT

1. **Define differentiation.**
2. **What event marks the onset of development?**
3. **Define inheritance.**

See the blue Answers tab at the back of the book.

29-2 | Fertilization—the fusion of a secondary oocyte and a spermatozoon— forms a zygote

Learning Outcome Describe the process of fertilization, and explain how developmental processes are regulated.

Fertilization involves the fusion of two haploid gametes, each containing 23 chromosomes, producing a *zygote* that contains 46 chromosomes, the normal diploid number in a human somatic cell. The functional roles and contributions of the male and female gametes are very different. The spermatozoon delivers the paternal (father's) chromosomes to the site of fertilization. It must travel a relatively long distance and is small, efficient, and highly streamlined. In contrast, the female oocyte provides all the cellular organelles and inclusions, nourishment, and genetic programming necessary to support development of the embryo for nearly a week after conception. For this reason, the volume of this gamete is much greater than that of the spermatozoon. Recall from Chapter 28 that ovulation releases a secondary oocyte suspended in metaphase of meiosis II. At fertilization, the diameter of the secondary oocyte is more than twice the entire length of the spermatozoon (**Figure 29–1a**). The ratio of their volumes is even more striking—approximately 2000:1.

The spermatozoa deposited in the vagina are already motile, as a result of contact with secretions of the seminal

glands—the first step of *capacitation*. ↄ p. 1088 Capacitation involves the physical changes in the sperm coating and uncovering of receptors that allow it to penetrate and fertilize an oocyte. (An unidentified substance secreted by the epididymis appears to prevent premature capacitation.) The spermatozoa, however, cannot accomplish fertilization until they have been exposed to conditions in the female reproductive tract. Uterine tube peg cell secretions help with capacitation, but the exact mechanism responsible for capacitation remains unknown. Furthermore, how sperm "find" the oocyte remains a rich field of investigation. Fertilization typically takes place near the junction between the ampulla and isthmus of the uterine tube. It generally occurs within a day after ovulation. By this time, the secondary oocyte has traveled only a few centimeters, but spermatozoa must cover the distance between the vagina and the ampulla of the uterine tube. A spermatozoon can propel itself at speeds of only about 34 μm per second, or about 12.5 cm (5 in.) per hour. In theory it should take spermatozoa several hours to reach the upper portions of the uterine tubes. The actual time, however, ranges from two hours to as little as 30 minutes. Contractions of the uterine musculature and ciliary currents in the uterine tubes have been suggested as possibly helping the spermatozoa travel from the vagina to the site of fertilization.

Even with transport assistance, the passage is not easy. Of the nearly 200 million spermatozoa introduced into the vagina in a typical ejaculation, only about 10,000 enter the uterine tube, and fewer than 100 reach the isthmus. In general, a male with a sperm count below 20 million per milliliter is functionally sterile because too few spermatozoa survive to reach and fertilize an oocyte. While it is true that only one spermatozoon fertilizes an oocyte, dozens of spermatozoa are required for successful fertilization. The additional sperm are essential because one sperm does not contain enough acrosomal enzymes to disrupt the *corona radiata*, the layer of follicle cells that surrounds the oocyte.

The Oocyte at Ovulation

Ovulation takes place before the oocyte is completely mature. The secondary oocyte leaving the follicle is in metaphase of meiosis II. The cell's metabolic operations have been suspended as it awaits the stimulus for further development. If fertilization does not take place, the oocyte disintegrates without completing meiosis.

Fertilization is complicated by the fact that when the secondary oocyte leaves the ovary, it is surrounded by the corona radiata. Fertilization and the events that follow are diagrammed in **Figure 29–1b**. The cells of the corona radiata protect the secondary oocyte as it passes through the ruptured

29

Figure 29–1 **Fertilization.**

a A secondary oocyte and numerous sperm at the time of fertilization. Notice the difference in size between the gametes.

follicular wall, across the surface of the ovary, and into the infundibulum of the uterine tube. The physical process of fertilization requires that only a single spermatozoon contact the oocyte membrane, but that spermatozoon must first penetrate the corona radiata. The acrosome of each sperm contains several enzymes, including **hyaluronidase** (hī-uh-lū-RON-i-dās). Hyaluronidase breaks down the bonds between adjacent follicle cells. Dozens of spermatozoa must release hyaluronidase before the connections between the follicle cells break down enough to allow an intact spermatozoon to reach the oocyte.

No matter how many spermatozoa slip through the gap in the corona radiata, only a single spermatozoon fertilizes and activates the oocyte (**1**, **Figure 29–1b**). That spermatozoon must have an intact acrosome. The first step is the binding of the spermatozoon to *sperm receptors* in the zona pellucida, a thick envelope surrounding the oocyte. This binding triggers the rupture of the acrosome. The hyaluronidase and **acrosin**, another protein-digesting enzyme, then digest a path through the zona pellucida toward the surface of the oocyte. When the sperm contacts that surface, the sperm and oocyte membranes begin to fuse. This step triggers *oocyte activation*, a complex process we discuss in the next section.

Oocyte at Ovulation

Ovulation releases a secondary oocyte and the first polar body; both are surrounded by the corona radiata. The oocyte is suspended in metaphase of meiosis II.

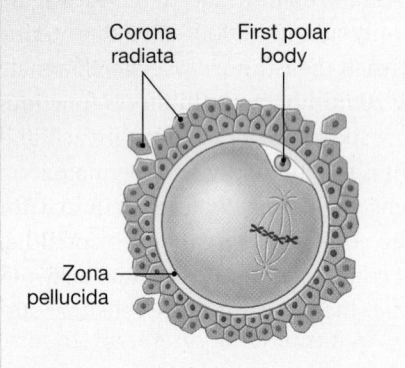

Corona radiata First polar body

Zona pellucida

---

1 **Fertilization and Oocyte Activation**

Acrosomal enzymes from multiple sperm create gaps in the corona radiata. A single sperm then makes contact with the oocyte membrane, and membrane fusion occurs, triggering oocyte activation and the completion of meiosis.

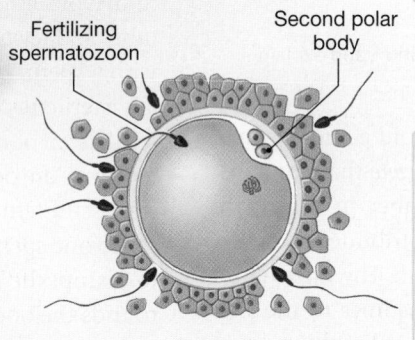

Fertilizing spermatozoon Second polar body

---

2 **Pronucleus Formation Begins**

The sperm is absorbed into the cytoplasm, and the female pronucleus develops.

Nucleus of fertilizing spermatozoon Female pronucleus

---

5 **Cleavage Begins**

The first cleavage division nears completion about 30 hours after fertilization.

Blastomeres

---

4 **Amphimixis Occurs and Cleavage Begins**

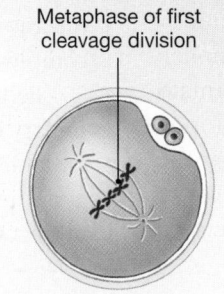

Metaphase of first cleavage division

---

3 **Spindle Formation and Cleavage Preparation**

The male pronucleus develops, and spindle fibers appear in preparation for the first cleavage division.

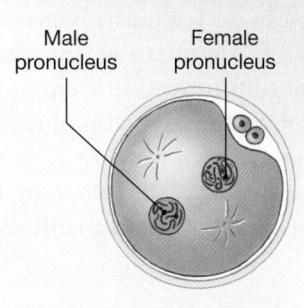

Male pronucleus Female pronucleus

---

b Fertilization and the preparations for cleavage.

29

Oocyte Activation

Oocyte activation involves a series of changes in the metabolic activity of the oocyte. The trigger for activation is contact and fusion of the plasma membranes of the sperm and oocyte. This process is accompanied by the depolarization of the oocyte membrane due to an increased permeability to sodium ions. The entry of sodium ions in turn causes the release of calcium ions from the smooth endoplasmic reticulum. The sudden rise in Ca^{2+} levels has important effects, including the following:

- *Exocytosis of Vesicles Adjacent to the Oocyte Membrane.* This process, called the *cortical reaction*, releases enzymes called zonal inhibiting proteins, or ZIPs, that both inactivate the sperm receptors and harden the zona pellucida. This combination prevents **polyspermy**, or fertilization by more than one sperm, which would create a zygote that is incapable of normal development. (Prior to completion of the cortical reaction, depolarization of the oocyte membrane probably prevents fertilization by any other sperm cells that penetrate the zona pellucida.)

- *Completion of Meiosis II and Formation of the Second Polar Body.* The sperm enters the oocyte and loses its plasma membrane. Meiosis II, which began in the tertiary follicle, can now be completed because fertilization occurred. The fertilized oocyte is now called an *ovum*.

- *Activation of Enzymes That Cause a Rapid Increase in the Cell's Metabolic Rate.* The cytoplasm contains a large number of mRNA strands that have been inactivated by special proteins. The mRNA strands are now activated, so protein synthesis accelerates quickly. Most of the proteins synthesized are required for development to proceed.

After oocyte activation and the completion of meiosis, the nuclear material remaining within the ovum reorganizes as the **female pronucleus** (**2**, **Figure 29–1b**). While these changes are under way, the nucleus of the spermatozoon swells. As it forms the **male pronucleus**, the rest of the sperm cell breaks down (**3**). The male pronucleus then migrates toward the center of the cell, and spindle fibers form. The male and female pronuclei then fuse in a process called *amphimixis* (am-fi-MIK-sis) (**4**). The cell is now a **zygote** that contains the normal diploid complement of 46 chromosomes, and fertilization is complete. This is the "moment of conception." Almost immediately, the chromosomes line up along a metaphase plate, and the cell prepares to divide. This is the start of the process of *cleavage*, a series of cell divisions that produce an ever-increasing number of smaller and smaller daughter cells. The first cleavage division is completed about 30 hours after fertilization. It yields two daughter cells, each one-half the size of the original zygote (**5**). These cells are called *blastomeres* (BLAS-tō-mērz).

TIPS & TRICKS

Amphimixis means "both mixed together."

CHECKPOINT

4. **Name two sperm enzymes important to secondary oocyte penetration.**

5. **How many chromosomes are contained within a human zygote?**

See the blue Answers tab at the back of the book.

29-3 | Gestation consists of three stages of prenatal development: the first, second, and third trimesters

Learning Outcome List the three stages of prenatal development, and describe the major events of each.

During prenatal development, a single cell ultimately forms a 3–4 kg (6.6–8.8 lb) infant. In postnatal development, the infant will grow through adolescence and maturity toward old age and eventual death. One of the most fascinating aspects of development is its apparent order. Continuity exists at all levels and at all times. Nothing "leaps" into existence without apparent precursors. Differentiation and increasing structural complexity take place hand in hand.

Differentiation involves changes in the genetic activity of some cells but not others. A continuous exchange of information takes place between the nucleus and the cytoplasm in a cell. Activity in the nucleus varies in response to chemical messages that arrive from the surrounding cytoplasm. In turn, ongoing nuclear activity alters conditions within the cytoplasm by directing the synthesis of specific proteins. In this way, the nucleus can affect enzyme activity, cell structure, and membrane properties.

In development, differences in the cytoplasmic composition of individual cells trigger changes in genetic activity. These changes in turn lead to further alterations in the cytoplasm, and the process continues in a sequence. But if all the cells of the embryo are derived from cell divisions of a zygote, how do the cytoplasmic differences originate? What sets the process in motion? The important first step occurs before fertilization, while the oocyte is in the ovary.

Before ovulation, the growing oocyte accepts amino acids, nucleotides, and glucose, as well as more complex materials such as phospholipids, mRNA molecules, and proteins, from

29

the surrounding granulosa cells. The contents of the cytoplasm are not evenly distributed because not all follicle cells manufacture and deliver the same nutrients and instructions to the oocyte. After fertilization, the zygote divides into ever-smaller cells that differ from one another in cytoplasmic composition. These differences alter the genetic activity of each cell, creating cell lines with increasingly diverse fates.

As development proceeds, some of the cells release chemical substances, including RNA molecules, polypeptides, and small proteins that affect the differentiation of other embryonic cells. This type of chemical interplay among developing cells is called *induction* (in-DUK-shun). It works over very short distances, such as when two types of cells are in direct contact. It may also operate over longer distances, with the inducing chemicals functioning as hormones.

This type of regulation, which involves an integrated series of interacting steps, can control highly complex processes. The mechanism is not always error free, however. The appearance of an abnormal or inappropriate inducer can change the course of development.

The time spent in prenatal development is known as **gestation** (jes-TĀ-shun). For convenience, we usually think of the gestation period as three integrated **trimesters**, each lasting three months:

1. The **first trimester** is the period of embryonic and early fetal development. During this time, the beginnings of all the major organ systems appear.

2. The **second trimester** is dominated by the development of organs and organ systems. This process nears completion by the end of the sixth month. During this time, body shape and proportions change. By the end of this trimester, the fetus looks distinctively human.

3. The **third trimester** is characterized by rapid fetal growth and deposition of adipose tissue. Early in the third trimester, most of the fetus's major organ systems become fully functional. An infant born one month or even two months prematurely has a reasonable chance of survival.

Martini's Atlas of the Human Body contains "Embryology Summaries" that introduce key steps in embryonic and fetal development. The summaries also trace the development of specific organ systems. We will refer to those summaries in the discussions that follow. As you proceed, reviewing the referenced material will help you understand the "big picture" as well as the specific details.

CHECKPOINT

6. Define gestation.

7. Characterize the key features of each trimester.

See the blue Answers tab at the back of the book.

29-4 | Cleavage, implantation, placentation, and embryogenesis are critical events of the first trimester

Learning Outcome Explain how the three germ layers are involved in forming the extraembryonic membranes, and discuss the importance of the placenta as an endocrine organ.

At the moment of conception, the fertilized ovum is a single cell about 0.135 mm (0.005 in.) in diameter. It weighs approximately 150 μg. By convention, pregnancies are clinically dated from the last menstrual period (LMP), which is usually two weeks before ovulation and conception. At the end of the first trimester (12 weeks from LMP, but only 10 developmental weeks), the fetus is almost 75 mm (3 in.) long and weighs perhaps 14 g (0.5 oz).

Many important and complex developmental events occur during the first trimester. Here we focus on four general processes: cleavage, implantation, placentation, and embryogenesis:

1. **Cleavage** (KLĒV-ij) is a sequence of cell divisions that begins immediately after fertilization (**Figure 29–1b**). During cleavage, the zygote becomes a **pre-embryo**, which develops into a multicellular complex known as a *blastocyst*. Cleavage ends when the blastocyst first contacts the uterine wall. Cleavage and blastocyst formation are introduced in the *Atlas*.

2. **Implantation** begins when the blastocyst attaches to the endometrium of the uterus. The process continues as the blastocyst invades maternal tissues. Important events during implantation set the stage for the formation of vital embryonic structures.

3. **Placentation** (plas-en-TĀ-shun) occurs as blood vessels form around the periphery of the blastocyst, and as the placenta develops. The **placenta** is a complex organ that permits exchange between maternal and embryonic blood. It supports the fetus from its formation early in the first trimester until just after birth, when the placenta stops functioning and is expelled from the uterus. From that point on, the newborn is physically independent of the mother.

4. **Embryogenesis** (em-brē-ō-JEN-e-sis) is the formation of a viable embryo. This process establishes the foundations for all major organ systems.

These processes are both complex and vital to the survival of the embryo. Perhaps because the events in the first trimester are so complex, it is the most dangerous prenatal stage. Only about 40 percent of conceptions produce embryos that survive the first trimester. For that reason, pregnant women are warned to avoid drugs, alcohol, tobacco, and other disruptive stresses during the first trimester, in the hope of preventing an error in the intricate processes that are under way.

Cleavage and Blastocyst Formation

Cleavage is a series of cell divisions that subdivides the cytoplasm of the zygote (**Figures 29–1b** and **29–2**). The first cleavage produces a pre-embryo consisting of two identical cells. As noted earlier, cleavage divisions produce identical cells called **blastomeres**. After the first division is completed approximately 30 hours after fertilization, subsequent divisions take place at intervals of 10–12 hours. During the initial divisions, all the blastomeres divide simultaneously. As the number of blastomeres increases, the timing becomes less predictable.

After three days of cleavage, the pre-embryo is a solid ball of cells resembling a mulberry. This stage is called the **morula** (MOR-ū-luh; *morus*, mulberry). The morula typically reaches the uterus on day 4. Over the next two days, the blastomeres form a **blastocyst**, a hollow ball with an inner cavity known as the **blastocoele** (BLAS-tō-sēl). The blastomeres are now no longer identical in size and shape. The outer layer of cells is called the **trophoblast** (TRŌ-fō-blast). As the word *trophoblast* implies (*trophos*, food + *blast*, precursor), cells in this layer provide nutrients to the developing embryo. A second group of cells, the **inner cell mass**, lies clustered at one end of the blastocyst. These cells are exposed only to the blastocoele and insulated from contact with the outside uterine environment by the trophoblast. In time, the inner cell mass will form the embryo.

Implantation

During blastocyst formation, enzymes released by the trophoblast erode an opening in the zona pellucida, which is then shed in a process known as *hatching* (**Figure 29–2**). The blastocyst is now freely exposed to the fluid contents of the uterine cavity. The uterine glands secrete this glycogen-rich fluid. Throughout the previous few days, the pre-embryo and early blastocyst had been absorbing fluid and nutrients from its surroundings. The process now accelerates, and the blastocyst enlarges. When fully formed, the blastocyst contacts the endometrium, and implantation takes place (**Figures 29–2** and **29–3**).

Implantation begins as the surface of the blastocyst closest to the inner cell mass adheres to the uterine lining (see day 7 in **Figure 29–3**). At the point of contact, the trophoblast cells divide rapidly, making the trophoblast several layers thick. The cells closest to the interior of the blastocyst remain intact, forming a layer of **cellular trophoblast**, or *cytotrophoblast*. Near the endometrial wall, however, the plasma membranes separating the trophoblast cells disappear, creating a layer of cytoplasm

Figure 29–2 Cleavage and Blastocyst Formation.

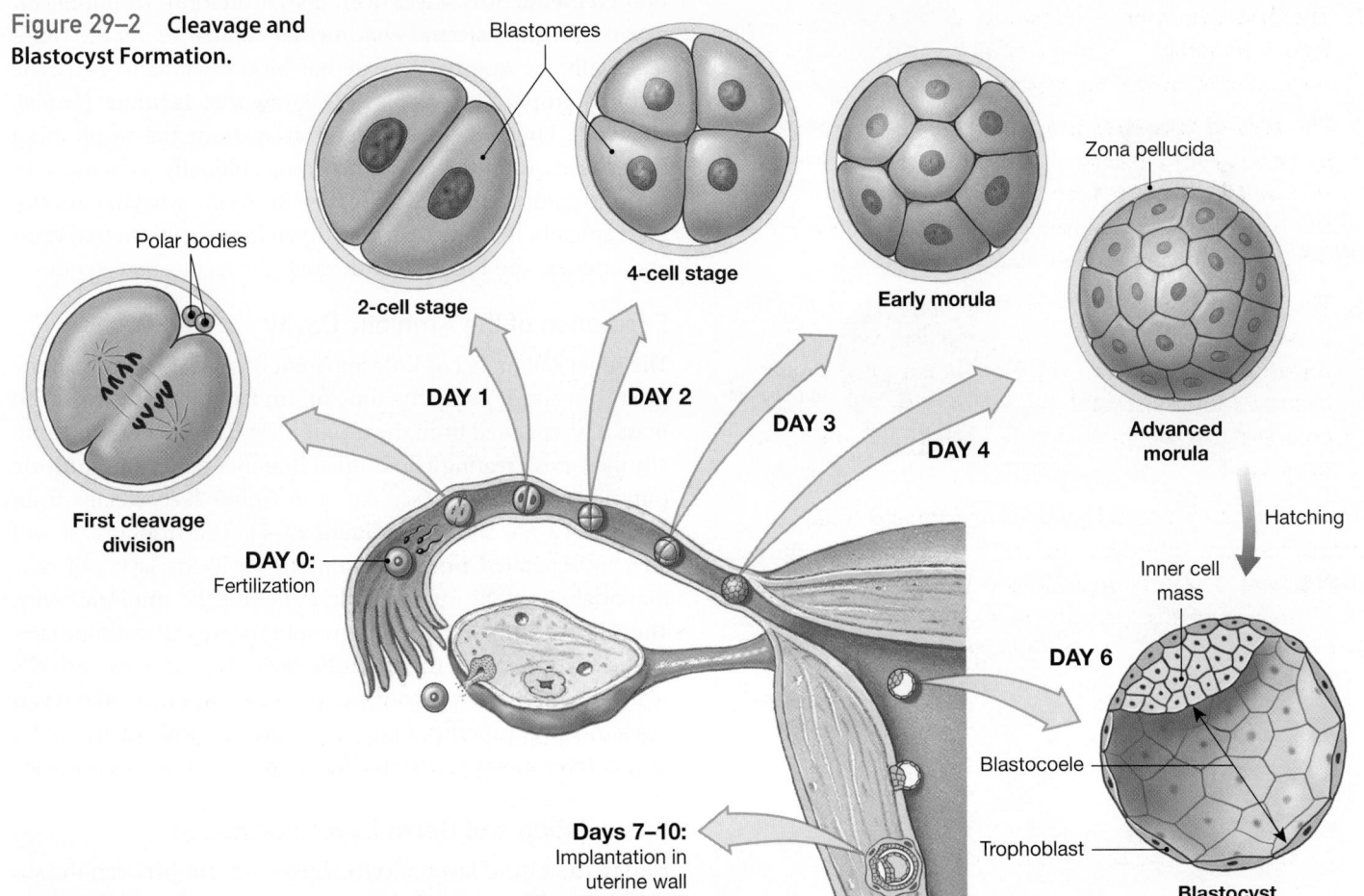

Figure 29–3 **Stages in Implantation.**

with multiple nuclei (day 8). This outer layer is called the **syncytial** (sin-SISH-ul) **trophoblast**, or *syncytiotrophoblast*.

The syncytial trophoblast erodes a path through the uterine epithelium by secreting hyaluronidase. This enzyme dissolves the proteoglycans between adjacent epithelial cells, just as hyaluronidase released by spermatozoa dissolved the connections between cells of the corona radiata. At first, the erosion creates a gap in the uterine lining, but maternal epithelial cells migrate and divide and soon repair the surface. By day 10 the repairs are complete, and the blastocyst has lost contact with the uterine cavity. Further development occurs entirely within the functional zone of the endometrium.

In most cases, implantation occurs in the fundus or in the body of the uterus. In an **ectopic pregnancy**, implantation occurs somewhere other than within the uterus, such as in one of the uterine tubes. Approximately 0.6 percent of pregnancies are ectopic pregnancies. They do not produce a viable embryo and can be life-threatening to the mother.

As implantation proceeds, the syncytial trophoblast continues to enlarge and spread into the surrounding endometrium (see day 9, **Figure 29–3**). The erosion of uterine glands releases nutrients that are absorbed by the syncytial trophoblast. They diffuse through the underlying cellular trophoblast to the inner cell mass. These nutrients provide the energy needed for the early stages of embryo formation. Trophoblastic extensions grow around endometrial capillaries. As the capillary walls are destroyed, maternal blood begins to percolate through trophoblastic channels known as **lacunae** (*lacuna*, singular). Fingerlike **villi** extend away from the trophoblast into the surrounding endometrium, gradually increasing in size and complexity until about day 21. As the syncytial trophoblast spreads, it begins breaking down larger endometrial veins and arteries, and blood flow through the lacunae accelerates.

Formation of the Amniotic Cavity

The inner cell mass has little apparent organization early in the blastocyst stage. Yet by the time of implantation, the inner cell mass has separated from the trophoblast. The separation gradually increases, creating a fluid-filled chamber called the **amniotic** (am-nē-OT-ik) **cavity** (see day 9 in **Figure 29–3**; details from days 10–12 are shown in **Figure 29–4**). The trophoblast will later be separated from the amniotic cavity by layers of cells that originate at the inner cell mass and line the amniotic cavity. These layers form the *amnion*, a membrane we will examine later in the chapter. When the amniotic cavity first appears, the cells of the inner cell mass are organized into an oval sheet that is two layers thick. A superficial layer faces the amniotic cavity, and a deeper layer is exposed to the fluid contents of the blastocoele.

Gastrulation and Germ Layer Formation

By day 12, a third layer of cells begins to form between the superficial and deep layers of the inner cell mass through a process

Figure 29–4   **The Inner Cell Mass and Gastrulation.**

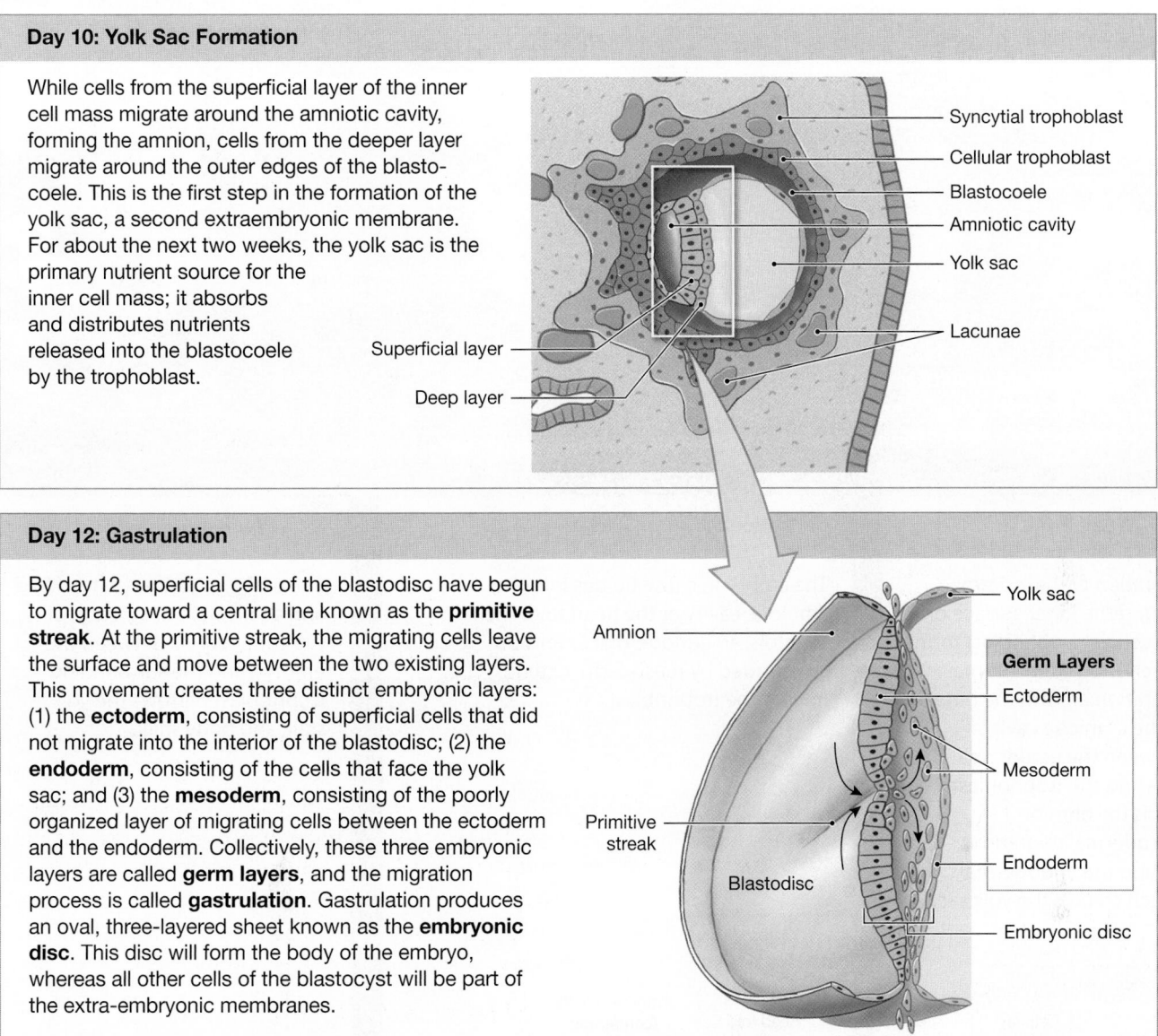

Day 10: Yolk Sac Formation

While cells from the superficial layer of the inner cell mass migrate around the amniotic cavity, forming the amnion, cells from the deeper layer migrate around the outer edges of the blastocoele. This is the first step in the formation of the yolk sac, a second extraembryonic membrane. For about the next two weeks, the yolk sac is the primary nutrient source for the inner cell mass; it absorbs and distributes nutrients released into the blastocoele by the trophoblast.

Superficial layer

Deep layer

Syncytial trophoblast

Cellular trophoblast

Blastocoele

Amniotic cavity

Yolk sac

Lacunae

Day 12: Gastrulation

By day 12, superficial cells of the blastodisc have begun to migrate toward a central line known as the **primitive streak**. At the primitive streak, the migrating cells leave the surface and move between the two existing layers. This movement creates three distinct embryonic layers: (1) the **ectoderm**, consisting of superficial cells that did not migrate into the interior of the blastodisc; (2) the **endoderm**, consisting of the cells that face the yolk sac; and (3) the **mesoderm**, consisting of the poorly organized layer of migrating cells between the ectoderm and the endoderm. Collectively, these three embryonic layers are called **germ layers**, and the migration process is called **gastrulation**. Gastrulation produces an oval, three-layered sheet known as the **embryonic disc**. This disc will form the body of the embryo, whereas all other cells of the blastocyst will be part of the extra-embryonic membranes.

Amnion

Primitive streak

Blastodisc

Yolk sac

Germ Layers

Ectoderm

Mesoderm

Endoderm

Embryonic disc

called **gastrulation** (gas-trū-LĀ-shun) (day 12, **Figure 29–4**). Together, the three layers of cells are called *germ layers*. The layers are known as the *ectoderm, mesoderm,* and *endoderm*. **Table 29–1** contains a listing of the contributions each germ layer makes to form the body systems described in earlier chapters. The formation of the mesoderm between the ectoderm and endoderm, and the developmental fates of the three germ layers are also summarized in the *Atlas.*

Formation of the Extraembryonic Membranes

The germ layers also form four **extraembryonic membranes**: the yolk sac, amnion, allantois, and chorion. **Spotlight Figure 29–5** illustrates and describes the processes involved in the development of these extraembryonic membranes.

We'll focus on the chorion here because of its early role in the maternal–embryo exchange of nutrients, gases, and wastes. When implantation first occurs, the nutrients absorbed by the trophoblast can easily reach the inner cell mass by simple diffusion. But as the embryo and the trophoblast enlarge, the distance between them increases, so diffusion alone can no longer keep pace with the demands of the developing embryo. Blood vessels now begin to develop within the mesoderm of the chorion, creating a rapid-transit system for nutrients that links the embryo with the trophoblast.

Chorionic villi continue to enlarge and branch, creating an intricate network within the endometrium. Embryonic blood vessels develop within each villus. Blood flow through those chorionic vessels begins early in the third week of development,

29

Figure 29–5
Extraembryonic Membranes and Placenta Formation

The germ layers introduced in Figure 29–4 also form four **extraembryonic membranes:** (1) The **yolk sac** (endoderm and mesoderm), (2) the **amnion** (ectoderm and mesoderm), (3) the **allantois** (endoderm and mesoderm), and (4) the **chorion** (mesoderm and trophoblast). These membranes support embryonic and fetal development, but few traces of their existence remain in adults.

Yolk sac

The yolk sac begins as a layer of cells spreads out around the outer edges of the blastocoele to form a complete pouch. It is the primary nutrient source for early embryonic development, and becomes an important site for blood cell formation.

Amnion

Ectodermal cells spread over the inner surface of the amniotic cavity, soon followed by mesodermal cells. Amniotic fluid is produced, which cushions the developing embryo.

---

1 **Week 2**

Migration of mesoderm around the inner surface of the cellular trophoblast forms the chorion. Mesodermal migration around the outside of the amniotic cavity, between the ectodermal cells and the trophoblast, forms the amnion. Mesodermal migration around the endodermal pouch creates the yolk sac.

Chorion

| Mesoderm | Cellular trophoblast | Amnion |

Blastocoele · Yolk sac · Syncytial trophoblast

---

2 **Week 3**

The embryonic disc bulges into the amniotic cavity at the head fold. The allantois, an endodermal extension surrounded by mesoderm, extends toward the trophoblast.

Head fold of embryo · Amniotic cavity (containing amniotic fluid)

Extraembryonic membranes
- Amnion
- Allantois
- Yolk sac
- Chorion

Chorionic villi of placenta · Syncytial trophoblast

---

3 **Week 4**

The embryo now has a head fold and a tail fold. Constriction of the connections between the embryo and the surrounding trophoblast narrows the yolk stalk and body stalk.

Tail fold · Body stalk · Yolk sac

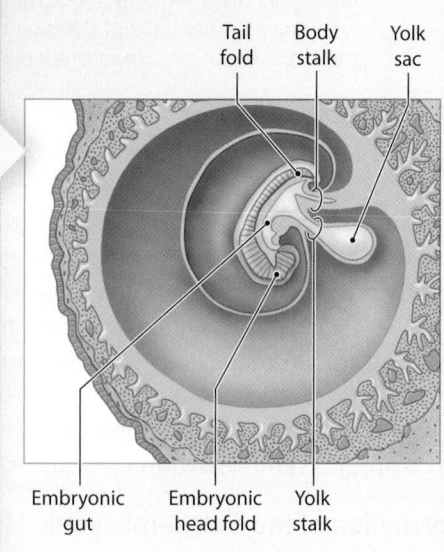

Embryonic gut · Embryonic head fold · Yolk stalk

Allantois

The allantois begins as an outpocket of the endoderm near the base of the yolk sac. The free endodermal tip then grows toward the wall of the blastocyst, surrounded by a mass of mesodermal cells. The base of the allantois eventually gives rise to the urinary bladder.

Chorion

The mesoderm associated with the allantois spreads around the entire blastocyst, separating the cellular trophoblast from the blastocoele. The appearance of blood vessels in the chorion is the first step in the creation of a functional placenta. By the third week of development, the mesoderm extends along the core of each trophoblastic villus, forming chorionic villi in contact with maternal tissues and blood vessels. These villi continue to enlarge and branch forming the placenta, the exchange platform between mother and fetus for nutrients, oxygen, and wastes.

4 **Week 5**

The developing embryo and extraembryonic membranes bulge into the uterine cavity. The trophoblast pushing out into the uterine cavity remains covered by endometrium but no longer participates in nutrient absorption and embryo support. The embryo moves away from the placenta, and the body stalk and yolk stalk fuse to form an umbilical stalk.

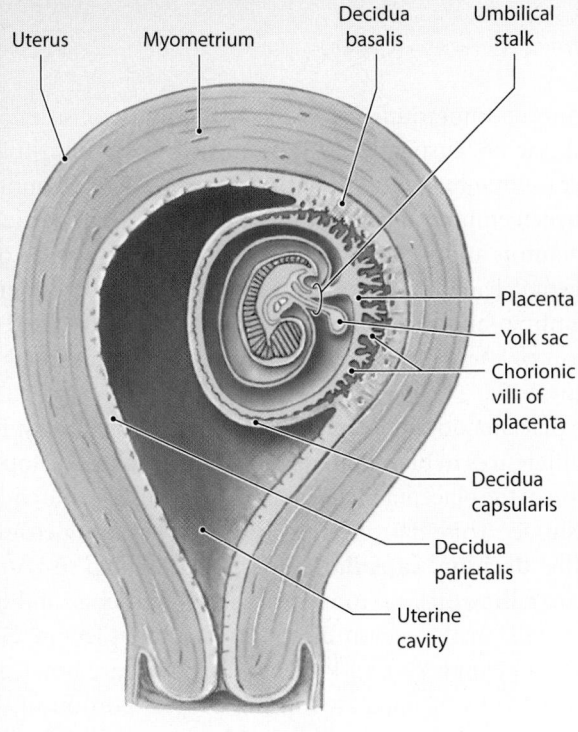

Uterus · Myometrium · Decidua basalis · Umbilical stalk · Placenta · Yolk sac · Chorionic villi of placenta · Decidua capsularis · Decidua parietalis · Uterine cavity

5 **Week 10**

The amnion has expanded greatly, filling the uterine cavity. The fetus is connected to the placenta by an elongated umbilical cord that contains a portion of the allantois, blood vessels, and the remnants of the yolk stalk.

Decidua parietalis · Umbilical cord · Decidua basalis · Placenta · Amnion · Amniotic cavity · Chorion · Decidua capsularis

Table 29–1	The Fates of the Germ Layers

ECTODERMAL CONTRIBUTIONS

Integumentary system: epidermis, hair follicles and hairs, nails, and glands communicating with the skin (sweat glands, mammary glands, and sebaceous glands)

Skeletal system: pharyngeal cartilages and their derivatives in adults (portion of sphenoid, the auditory ossicles, the styloid processes of the temporal bones, the cornu and superior rim of the hyoid bone)*

Nervous system: all neural tissue, including brain and spinal cord

Endocrine system: pituitary gland and adrenal medullae

Respiratory system: mucous epithelium of nasal passageways

Digestive system: mucous epithelium of mouth and anus, salivary glands

MESODERMAL CONTRIBUTIONS

Integumentary system: dermis and hypodermis

Skeletal system: all structures except some pharyngeal derivatives

Muscular system: all structures

Endocrine system: adrenal cortex, endocrine tissues of heart, kidneys, and gonads

Cardiovascular system: all structures

Lymphatic system: all structures

Urinary system: the kidneys, including the nephrons and the initial portions of the collecting system

Reproductive system: the gonads and the adjacent portions of the duct systems

Miscellaneous: the lining of the pleural, pericardial, and peritoneal cavities and the connective tissues that support all organ systems

ENDODERMAL CONTRIBUTIONS

Endocrine system: thymus, thyroid gland, and pancreas

Respiratory system: respiratory epithelium (except nasal passageways) and associated mucous glands

Digestive system: mucous epithelium (except mouth and anus), exocrine glands (except salivary glands), liver, and pancreas

Urinary system: urinary bladder and distal portions of the duct system

Reproductive system: distal portions of the duct system, stem cells that produce gametes

*The neural crest is derived from ectoderm and contributes to the formation of the skull and the skeletal derivatives of the embryonic pharyngeal arches.

when the embryonic heart starts beating. The blood supply to the chorionic villi arises from the allantoic arteries and veins.

As the chorionic villi enlarge, more maternal blood vessels are eroded. Maternal blood now moves slowly through complex lacunae lined by the syncytial trophoblast. Chorionic blood vessels pass close by, and gases and nutrients diffuse between the embryonic and maternal circulations across the layers of the trophoblast. Recall that fetal hemoglobin has a higher affinity for oxygen than does maternal hemoglobin, enabling fetal hemoglobin to strip oxygen from maternal hemoglobin. ⊃ p. 889 Maternal blood then reenters the venous system of the mother through the broken walls of small uterine veins. No mixing of maternal and fetal blood takes place, because layers of trophoblast always separate the two.

Placentation

At first, the entire blastocyst is surrounded by chorionic villi. The chorion continues to enlarge, expanding like a balloon within the endometrium. By week 4, the embryo, amnion, and yolk sac are suspended within an expansive, fluid-filled chamber (**Spotlight Figure 29–5**). The **body stalk**, the connection between embryo and chorion, contains the distal portions of the allantois and blood vessels that carry blood to and from the placenta. The narrow connection between the endoderm of the embryo and the yolk sac is called the **yolk stalk**. The formation of the yolk stalk and body stalk are illustrated in the *Atlas*.

The placenta does not continue to enlarge indefinitely. Regional differences in placental organization begin to develop as expansion of the placenta creates a prominent bulge in the endometrial surface. This relatively thin portion of the endometrium, called the **decidua capsularis** (dē-SID-ū-uh kap-sū-LA-ris; *deciduus*, a falling off), no longer exchanges nutrients, and the chorionic villi in the region disappear (**Spotlight Figure 29–5** [week 5] and **Figure 29–6a**). Placental functions are now concentrated in a disc-shaped area in the deepest portion of the endometrium, a region called the **decidua basalis** (bā-SĀ-lis).

29

The rest of the uterine endometrium, which has no contact with the chorion, is called the **decidua parietalis**.

As the end of the first trimester approaches, the fetus moves farther from the placenta (see weeks 5 and 10, Spotlight Figure 29–5). The fetus and placenta remain connected by the **umbilical cord**, which contains the allantois, the placental blood vessels, and the yolk stalk.

TIPS&TRICKS

Like a deep-sea diver's air hose or a space-walking astronaut's tether, the umbilical cord supplies the fetus with life-sustaining substances and removes wastes.

Placental Circulation

Figure 29–6a illustrates circulation at the placenta near the end of the first trimester. Blood flows from the fetus to the placenta through the paired **umbilical arteries** and returns in a single **umbilical vein**. ↻ p. 798 The chorionic villi provide the surface area for active and passive exchanges of gases, nutrients, and wastes between the fetal and maternal bloodstreams. The blood in the umbilical arteries is deoxygenated and contains wastes generated by fetal tissues. At the placenta, oxygen supplies are replenished, organic nutrients added, and carbon dioxide and other organic wastes removed.

The placenta places a considerable demand on the maternal cardiovascular system, and blood flow to the uterus and placenta is extensive. If the placenta is torn or otherwise damaged, the consequences may prove fatal to both fetus and mother.

The Endocrine Placenta

In addition to its role in the nutrition of the fetus, the placenta acts as an endocrine organ. Several hormones—including *human chorionic gonadotropin, human placental lactogen, placental prolactin, relaxin, progesterone,* and *estrogens*—are synthesized by the syncytial trophoblast and released into the maternal bloodstream.

Human Chorionic Gonadotropin. The hormone **human chorionic gonadotropin (hCG)** appears in the maternal bloodstream soon after implantation. The presence of hCG in blood or urine samples is a reliable indication of pregnancy. Home pregnancy tests detect the presence of hCG in the urine.

In function, hCG resembles luteinizing hormone (LH), because it maintains the integrity of the corpus luteum and promotes the continued secretion of progesterone. As a result, the endometrial lining remains perfectly functional, and menses does not normally occur. Without hCG, the pregnancy ends, because another uterine cycle begins and the functional zone of the endometrium disintegrates.

With hCG, the corpus luteum persists for three to four months before gradually decreasing in size and secretory function. The decrease in luteal function does not trigger the return of uterine cycles, because by the end of the first trimester, the placenta actively secretes both estrogens and progesterone.

Human Placental Lactogen and Placental Prolactin. **Human placental lactogen (hPL)** helps prepare the mammary glands for milk production. It also has a stimulatory effect on other maternal tissues comparable to that of growth hormone (GH), ensuring that glucose and protein are available for the fetus. At the mammary glands, the conversion from inactive to active status requires placental hormones (hPL, **placental prolactin**, estrogens, and progesterone) and several maternal hormones (GH, prolactin, and thyroid hormones). We consider the hormonal control of mammary gland function in a later section.

Relaxin. **Relaxin** is a peptide hormone secreted by the placenta and the corpus luteum during pregnancy. Relaxin (1) increases the flexibility of the pubic symphysis, permitting the pelvis to expand during delivery; (2) causes the cervix to dilate, making it easier for the fetus to enter the vaginal canal; and (3) suppresses the release of oxytocin by the hypothalamus and delays the onset of labor contractions.

Progesterone and Estrogens. After the first trimester, the placenta produces sufficient amounts of progesterone to maintain the endometrial lining and continue the pregnancy. As the end of the third trimester approaches, estrogen production by the placenta accelerates. As we will see in a later section, the increasing estrogen levels play a role in stimulating labor and delivery.

Embryogenesis

Shortly after gastrulation begins, the body of the embryo begins to separate itself from the rest of the embryonic disc. The body of the embryo and its internal organs now start to form. This process, called **embryogenesis**, begins as folding and differential growth of the embryonic disc produce a bulge that projects into the amniotic cavity (Spotlight Figure 29–5 **2**). This projection is known as the **head fold**. Similar movements lead to the formation of a **tail fold** (Spotlight Figure 29–5 **3**).

The embryo is now physically as well as developmentally distinct from the embryonic disc and the extraembryonic membranes. The definitive orientation of the embryo can now be seen, complete with dorsal and ventral surfaces and left and right sides.

29

Figure 29–6 **Views of Placental Structures.**

Syncytial trophoblast

Embryonic connective tissue

Area filled with maternal blood

Fetal blood vessels

Chorionic villus

LM × 280

b A cross section through a chorionic villus, showing the syncytial trophoblast exposed to the maternal blood space.

Decidua capsularis

Amnion

Umbilical cord (cut)

Placenta

Chorion

Yolk sac

Decidua basalis

Chorionic villi

Umbilical vein

Umbilical arteries

Area filled with maternal blood

Decidua parietalis

Myometrium

Uterine cavity

Cervical (mucous) plug in cervical canal

Cervix

Vagina

External os

Amnion

Trophoblast (cellular and syncytial layers)

Maternal blood vessels

a A view of the uterus after the fetus has been removed and the umbilical cord cut. Arrows in the enlarged view indicate the direction of blood flow. Blood flows into the placenta through ruptured maternal arteries and then flows around chorionic villi, which contain fetal blood vessels.

29

Figure 29–7 **The First 12 Weeks of Development.** ATLAS: Plate 90a

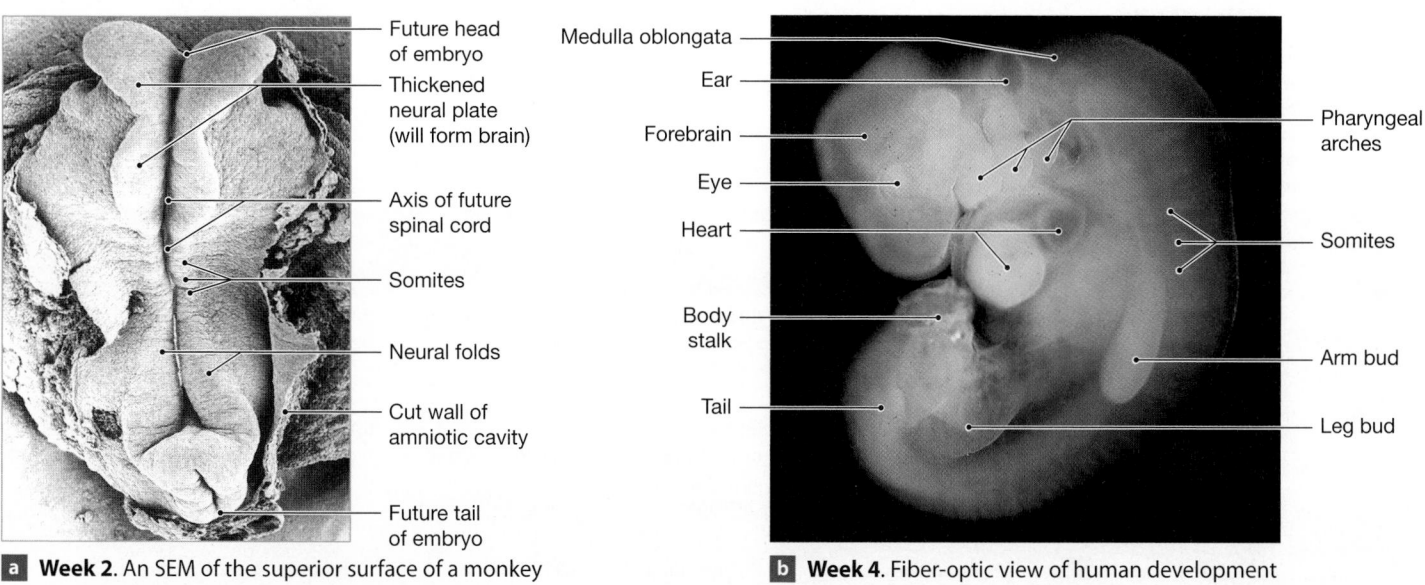

a **Week 2.** An SEM of the superior surface of a monkey embryo at 2 weeks of development. A human embryo at this stage would look essentially the same.

b **Week 4.** Fiber-optic view of human development at week 4 (about 5 mm in size).

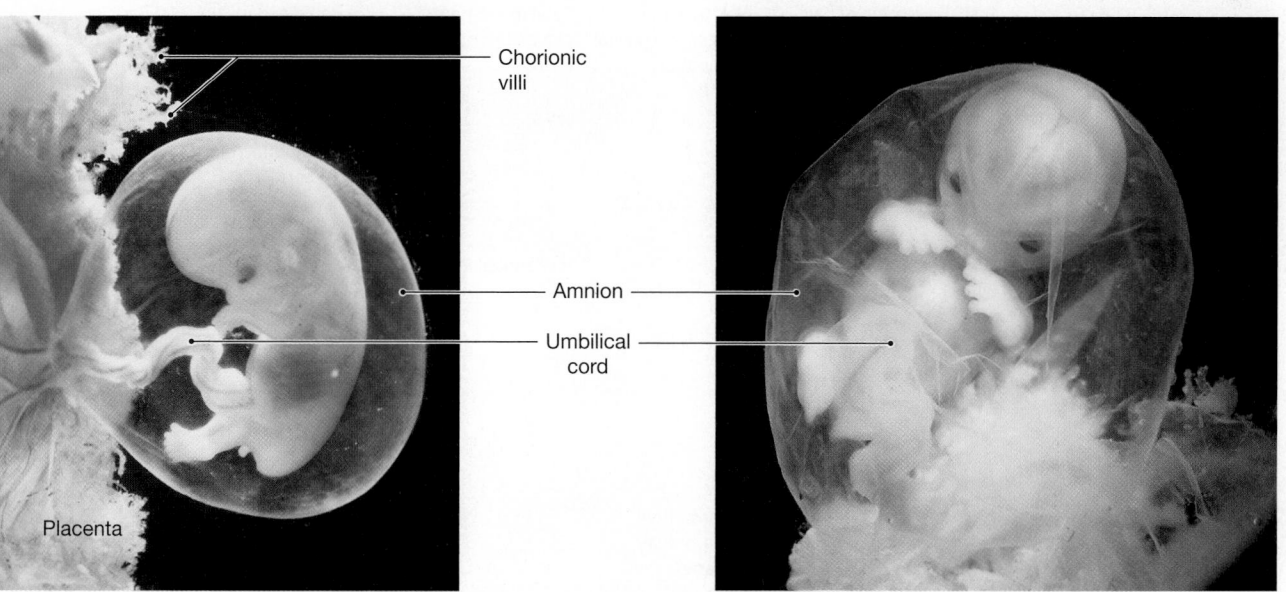

c **Week 8.** Fiber-optic view of human development at week 8 (about 1.6 cm in size).

d **Week 12.** Fiber-optic view of human development at week 12 (about 5.4 cm in size).

Table 29–2 provides an overview of the subsequent development of the major organs and body systems. The changes in proportions and appearance that occur between the second developmental week and the end of the first trimester are summarized in Figure 29–7.

The first trimester is a critical period for development, because events in the first 12 weeks establish the basis for **organogenesis**, the process of organ formation. The major features of organogenesis in each organ system are described in Embryology Summaries 4–21 in the *Atlas*. Important developmental milestones are indicated in Table 29–2.

CHECKPOINT

8. What is the developmental fate of the inner cell mass of the blastocyst?

9. Improper development of which of the extraembryonic membranes would affect the cardiovascular system?

10. Sue's pregnancy test indicates the presence of elevated levels of the hormone hCG (human chorionic gonadotropin). Explain whether she is pregnant or not.

11. What are two important functions of the placenta?

See the blue Answers tab at the back of the book.

29

Table 29–2 **An Overview of Prenatal Development.** From 8 weeks to about 20 weeks, the fetus is measured from crown to rump. From 20 weeks to birth, the baby is measured from crown to heel.

Background Material

ATLAS: Embryology Summaries 1–4:
The Formation of Tissues
The Development of Epithelia
The Origins of Connective Tissues
The Development of Organ Systems

Gestational Age (embryonic age/fetal age plus 2 weeks) (months)	Size and Weight	Integumentary System	Skeletal System	Muscular System	Nervous System	Special Sense Organs
1	5 mm (0.2 in.), 0.02 g (0.0007 oz.)		(b) Formation of somites	(b) Formation of somites	(b) Formation of neural tube	(b) Formation of eyes and ears
2	1.6 cm (.63 in.), 1.0 g (0.04 oz.)	(b) Formation of nail beds, hair follicles, sweat glands	(b) Formation of axial and appendicular cartilages	(c) Rudiments of axial musculature	(b) CNS, PNS organization, growth of cerebrum	(b) Formation of taste buds, olfactory epithelium
3	5.4 cm (2.13 in.), 14 g (0.5 oz.)	(b) Epidermal layers appear	(b) Spreading of ossification centers	(c) Rudiments of appendicular musculature	(c) Basic spinal cord and brain structure	
4	11.6 cm (4.6 in.), 100 g (3.53 oz.)	(b) Formation of hair, sebaceous glands (c) Sweat glands	(b) Articulations (c) Facial and palatal organization	Fetal movements can be felt by the mother	(b) Rapid expansion of cerebrum	(c) Basic eye and ear structure (b) Formation of peripheral receptors
5	16.4 cm (6.46 in.), 300 g (10.58 oz.)	(b) Keratin production, nail production			(b) Myelination of spinal cord	
6	30 cm (11.81 in.), 600 g (1.32 lb)			(c) Perineal muscles	(b) Formation of CNS tract (c) Layering of cortex	
7	37.6 cm (14.8 in.), 1.005 kg (2.22 lb)	(b) Keratinization, formation of nails, hair				(c) Eyelids open, retinas sensitive to light
8	42.4 cm (16.69 in.), 1.702 kg (3.75 lb)		(b) Formation of epiphyseal cartilages			(c) Taste receptors functional
9	47.4 cm (18.66 in.), 3.2 kg (7.05 lb)					
Early postnatal development		Hair changes in consistency and distribution	Formation and growth of epiphyseal cartilages continue	Muscle mass and control increase	Myelination, layering, CNS tract formation continue	
Location of relevant text and illustrations		ATLAS: Embryology Summary 5	Ch. 6: pp. 215–218 Ch. 7: pp. 250–252 ATLAS: Embryology Summaries 6, 7, 8	ATLAS: Embryology Summary 9	Ch. 14: p. 490 ATLAS: Embryology Summaries 10, 11, 12	ATLAS: Embryology Summary 13

Note: (b) = beginning; (c) = completion.

29

Gestational Age (embryonic age/fetal age plus 2 weeks) (months)	Endocrine System	Cardiovascular and Lymphatic Systems	Respiratory System	Digestive System	Urinary System	Reproductive System
1		(b) Heartbeat	(b) Formation of trachea and lungs	(b) Formation of intestinal tract, liver, pancreas (c) Yolk sac	(c) Allantois	
2	(b) Formation of thymus, thyroid, pituitary, adrenal glands	(c) Basic heart structure, major blood vessels, lymph nodes and ducts (b) Blood formation in liver	(b) Extensive bronchial branching into mediastinum (c) Diaphragm	(b) Formation of intestinal subdivisions, villi, salivary glands	(b) Formation of kidneys (metanephros)	(b) Formation of mammary glands
3	(c) Thymus, thyroid gland	(b) Tonsils, blood formation in bone marrow		(c) Gallbladder, pancreas		(b) Formation of gonads, ducts, genitalia; oogonia in female
4		(b) Migration of lymphocytes to lymphoid organs; blood formation in spleen			(b) Degeneration of embryonic kidneys (mesonephros)	
5		(c) Tonsils	(c) Nostrils open	(c) Intestinal subdivisions		
6	(c) Adrenal glands	(c) Spleen, liver, bone marrow	(b) Formation of alveoli	(c) Epithelial organization, glands		
7	(c) Pituitary gland			(c) Intestinal circular folds		(b) Descent of testes in male; primary oocytes in prophase I of meiosis in female
8			Complete pulmonary branching and alveolar structure		(c) Nephron formation	Descent of testes complete at or near time of birth
9						
Early postnatal development		Cardiovascular changes at birth; immune response gradually becomes fully operational				
Location of relevant text and illustrations	ATLAS: Embryology Summary 14	Ch. 19: pp. 687, 698 Ch. 21: pp. 798–800 ATLAS: Embryology Summaries 15, 16, 17	Ch. 23: p. 898 ATLAS: Embryology Summary 18	Ch. 24: pp. 909–911 ATLAS: Embryology Summary 19	Ch. 26: p. 1031 ATLAS: Embryology Summary 20	Ch. 28: pp. 1078–1080 ATLAS: Embryology Summary 21

Note: (b) = beginning; (c) = completion.

29-5 ▌ During the second and third trimesters, maternal organ systems support the developing fetus, and the uterus undergoes structural and functional changes

Learning Outcome Describe the interplay between the maternal organ systems and the developing fetus, and discuss the structural and functional changes in the uterus during gestation.

The rudiments of all the major organ systems have formed by the end of the first trimester (**Figure 29–7d**). Over the next three months, the fetus grows to a weight of about 0.6 kg (1.32 lb). Encircled by the amnion, the fetus grows faster than the surrounding placenta during this second trimester. When the mesoderm on the outer surface of the amnion contacts the mesoderm on the inner surface of the chorion, these layers fuse, creating a compound *amniochorionic membrane*. **Figure 29–8a** shows a four-month-old fetus. **Figure 29–8b** shows a six-month-old fetus.

During the third trimester, most of the organ systems become ready to perform their normal functions without maternal assistance. The rate of growth starts to slow, but in absolute terms this trimester sees the greatest weight gain. In the last three months of gestation, the fetus gains about 2.6 kg (5.7 lb), reaching a full-term weight of approximately 3.2 kg (7.05 lb). The Embryology Summaries in the *Atlas* illustrate organ system development in the second and third trimesters. Highlights are noted in **Table 29–2**.

At the end of gestation, a typical uterus has undergone a tremendous increase in size. **Figure 29–9a,b,d** shows the positions of the uterus, fetus, and placenta from 16 weeks to *full term* (nine months). When the pregnancy is at full term, the uterus and fetus push many of the maternal abdominal organs out of their normal positions (**Figure 29–9d**).

Pregnancy and Maternal Systems

The developing fetus is totally dependent on maternal organ systems for nourishment, respiration, and waste removal.

Figure 29–8   **The Second and Third Trimesters.**   ATLAS: Plate 90b

a A four-month-old fetus, seen through a fiber-optic endoscope (about 13.3 cm in size)

b Head of a six-month-old fetus, revealed through ultrasound (about 30 cm in size)

29

Figure 29–9 **Growth of the Uterus and Fetus.**

a Pregnancy at 16 weeks, showing the positions of the uterus, fetus, and placenta.

Placenta
Umbilical cord
Fetus at 16 weeks
Uterus
Amniotic fluid
Cervix
Vagina

b Pregnancy at three months to nine months (full term), showing the superior-most position of the uterus within the abdomen.

9 months
8 months
7 months
6 months
5 months
4 months
3 months
After dropping, in preparation to delivery

c A sectional view through the abdominopelvic cavity of a woman who is not pregnant.

Diaphragm
Liver
Stomach
Pancreas
Transverse colon
Small intestine
Umbilical cord
Placenta
Uterus
Urinary bladder
Pubic symphysis
Vagina
Urethra
Rectum

d Pregnancy at full term. Note the positions of the uterus and full-term fetus within the abdomen, and the displacement of abdominal organs.

Fundus of uterus
Aorta
Common iliac vein
Cervical (mucous) plug in cervical canal
External os

29

Maternal systems perform these functions in addition to their normal operations. For example, the mother must absorb enough oxygen, nutrients, and vitamins for herself *and* for her fetus, and she must eliminate all the wastes that are generated. This is not a burden over the initial weeks of gestation, but the demands placed on the mother become significant as the fetus grows. For the mother to survive under these conditions, maternal systems must compensate for changes introduced by the fetus. In practical terms, the mother must breathe, eat, and excrete for two. The major changes in maternal systems include the following:

- *Maternal respiratory rate increases and tidal volume increases.* As a result, the mother's lungs deliver the extra oxygen required, and remove the excess carbon dioxide generated by the fetus.

- *Maternal blood volume increases.* This increase occurs because blood flowing into the placenta decreases the volume in the rest of the systemic circuit, and because fetal metabolic activity both decreases blood P_{O_2} and increases P_{CO_2}. The latter combination stimulates the production of renin and erythropoietin, leading to an increase in maternal blood volume through mechanisms detailed in Chapter 21 (see **Figure 21–15,** p. 774). By the end of gestation, maternal blood volume has increased by almost 50 percent.

- *Maternal requirements for nutrients increase 10–30 percent.* Pregnant women tend to have increased appetites because they must nourish both themselves and their fetus.

- *Maternal glomerular filtration rate increases by approximately 50 percent.* This increase corresponds to the increase in blood volume and accelerates the excretion of metabolic wastes generated by the fetus. Pregnant women need to urinate frequently because the volume of urine produced increases and the weight of the uterus presses down on the urinary bladder.

- *The uterus undergoes a tremendous increase in size.* Structural and functional changes in the expanding uterus are so important that we will discuss them in a separate section.

- *The mammary glands increase in size, and secretory activity begins.* Mammary gland development requires a combination of hormones, including human placental lactogen and placental prolactin from the placenta, and prolactin (PRL), estrogens, progesterone, GH, and thyroxine from maternal endocrine organs. By the end of the sixth month of pregnancy, the mammary glands are fully developed and begin to produce clear secretions that are stored in the duct system and may be expressed from the nipple.

Structural and Functional Changes in the Uterus

At the end of gestation, a typical uterus has grown from 7.5 cm (3 in.) in length and 30–40 g (1–1.4 oz) in weight to

Clinical Note

Abortion **Abortion** is the termination of a pregnancy. Most references distinguish among spontaneous, therapeutic, and induced abortions. Most **spontaneous abortions**, or *miscarriages*, result from developmental problems (such as chromosomal defects in the embryo) or from hormonal problems, including abnormally low LH production by the maternal pituitary gland, reduced LH sensitivity at the corpus luteum, insufficient progesterone sensitivity in the endometrium, or insufficient placental production of hCG. Spontaneous abortions occur in at least 15 percent of recognized pregnancies. **Therapeutic abortions** are performed when continued pregnancy poses a threat to the life of the mother. **Induced abortions,** or *elective abortions*, are performed at the woman's request. Induced abortions remain the focus of considerable controversy.

30 cm (12 in.) in length and 1100 g (2.4 lb) in weight. The uterus may then contain 2 liters of fluid, plus fetus and placenta, for a total weight of about 6–7 kg (13–15.4 lb). This remarkable expansion occurs through the enlargement (hypertrophy) of existing cells, especially smooth muscle fibers, rather than by an increase in the total number of cells.

The tremendous stretching of the uterus is associated with a gradual increase in the rate of spontaneous smooth muscle contractions in the myometrium. In the early stages of pregnancy, the contractions are weak, painless, and brief. Evidence indicates that progesterone released by the placenta has an inhibitory effect on uterine smooth muscle, preventing more extensive and more powerful contractions. Placental and fetal factors are involved in the initiation of labor and delivery (**Figure 29–10**).

Late in pregnancy, some women experience occasional spasms in the uterine musculature, but these contractions are not regular or persistent. Such contractions are called **false labor**. **True labor** begins when biochemical and mechanical factors reach a point of no return. After nine months of gestation, multiple factors interact to initiate true labor. Once **labor contractions** have begun in the myometrium, positive feedback ensures that they will continue until delivery has been completed.

When labor begins, the fetal pituitary gland secretes oxytocin, which is released into the maternal bloodstream at the placenta. This may be the actual trigger for the onset of true labor, as it increases myometrial contractions and prostaglandin production, on top of the priming effects of estrogens and maternal oxytocin.

Figure 29–10 Factors Involved in the Initiation of Labor and Delivery.

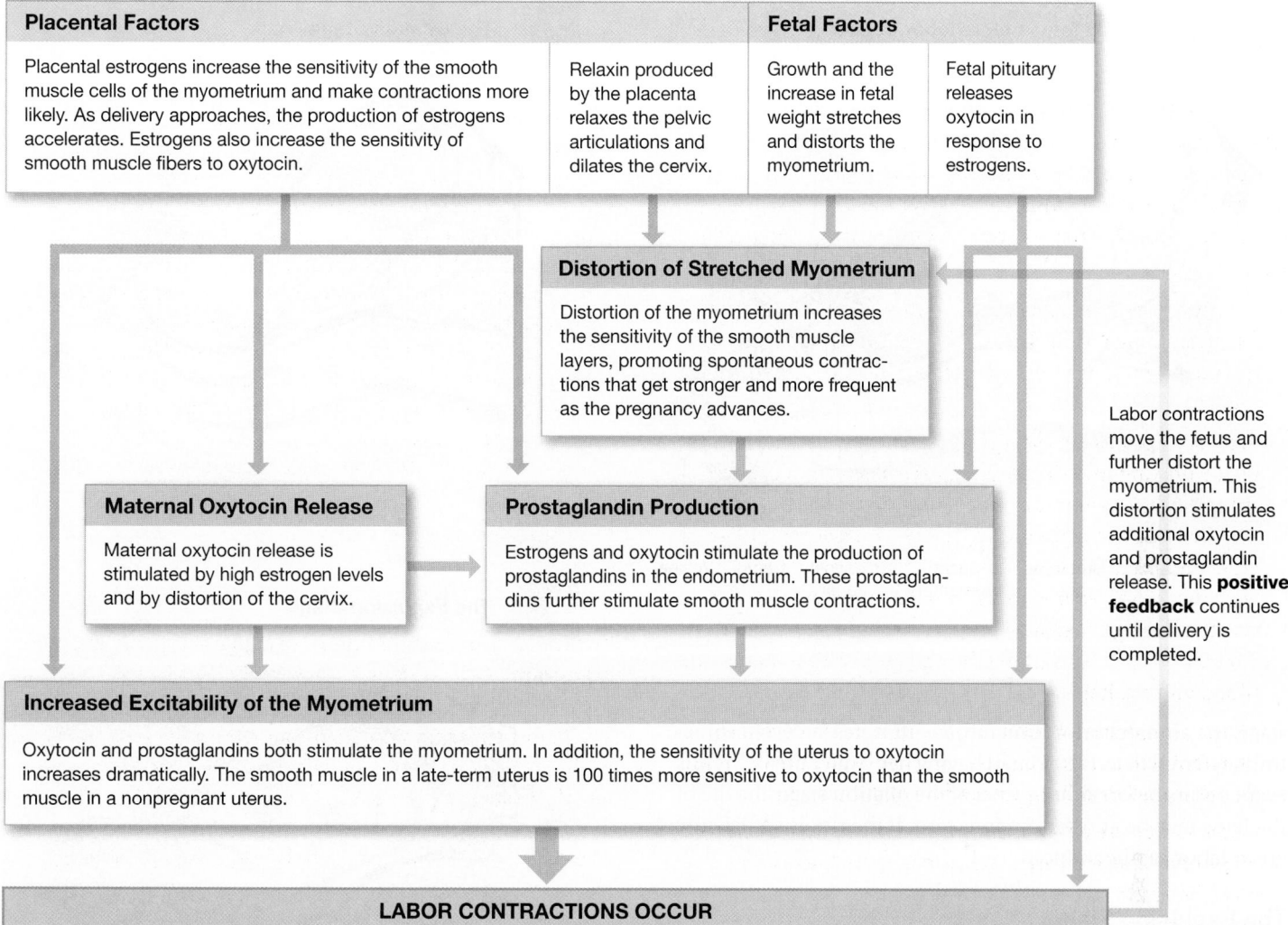

Placental Factors		Fetal Factors	
Placental estrogens increase the sensitivity of the smooth muscle cells of the myometrium and make contractions more likely. As delivery approaches, the production of estrogens accelerates. Estrogens also increase the sensitivity of smooth muscle fibers to oxytocin.	Relaxin produced by the placenta relaxes the pelvic articulations and dilates the cervix.	Growth and the increase in fetal weight stretches and distorts the myometrium.	Fetal pituitary releases oxytocin in response to estrogens.

Distortion of Stretched Myometrium

Distortion of the myometrium increases the sensitivity of the smooth muscle layers, promoting spontaneous contractions that get stronger and more frequent as the pregnancy advances.

Labor contractions move the fetus and further distort the myometrium. This distortion stimulates additional oxytocin and prostaglandin release. This **positive feedback** continues until delivery is completed.

Maternal Oxytocin Release

Maternal oxytocin release is stimulated by high estrogen levels and by distortion of the cervix.

Prostaglandin Production

Estrogens and oxytocin stimulate the production of prostaglandins in the endometrium. These prostaglandins further stimulate smooth muscle contractions.

Increased Excitability of the Myometrium

Oxytocin and prostaglandins both stimulate the myometrium. In addition, the sensitivity of the uterus to oxytocin increases dramatically. The smooth muscle in a late-term uterus is 100 times more sensitive to oxytocin than the smooth muscle in a nonpregnant uterus.

LABOR CONTRACTIONS OCCUR

CHECKPOINT

12. **Why do pregnant women experience breathing difficulty?**

13. **Identify three major factors opposing the calming action of progesterone on the uterus.**

14. **Why does a mother's blood volume increase during pregnancy?**

See the blue Answers tab at the back of the book.

29-6 ▌ Labor consists of the dilation, expulsion, and placental stages

Learning Outcome List and discuss the events that occur during labor and delivery.

The goal of labor is **parturition** (par-tū-RISH-un), or *childbirth*, the forcible expulsion of the fetus. During true labor, each contraction begins near the superior portion of the uterus and sweeps in a wave toward the cervix. The contractions are strong and occur at regular intervals. As parturition approaches, the contractions increase in force and frequency, changing the position of the fetus and moving it toward the cervical canal.

Stages of Labor

Labor has traditionally been divided into three stages: the *dilation stage*, the *expulsion stage*, and the *placental stage* (**Figure 29–11**).

The Dilation Stage

The **dilation stage** begins with the onset of true labor, as the cervix dilates and the fetus begins to shift toward the cervical canal, moved by gravity and uterine contractions (**Stage ❶** in **Figure 29–11**). This stage is highly variable in length but typically lasts eight or more hours. At the start of the dilation stage, labor contractions last up to half a minute and occur once every 10–30 minutes. Their frequency increases steadily. Late in this

29

Figure 29–11   **The Stages of Labor.**

Fully developed fetus before labor begins

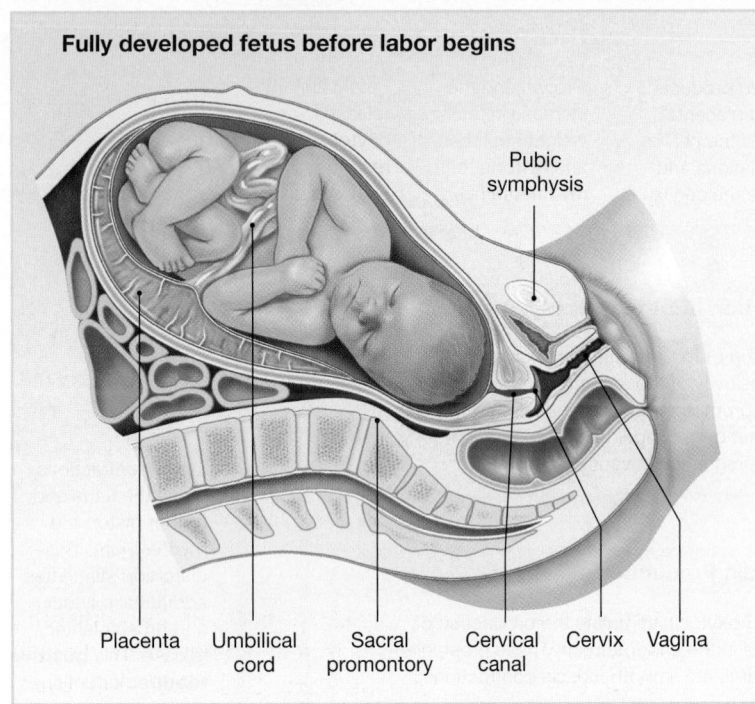

Pubic symphysis

Placenta   Umbilical cord   Sacral promontory   Cervical canal   Cervix   Vagina

1 **The Dilation Stage**

2 **The Expulsion Stage**

3 **The Placental Stage**

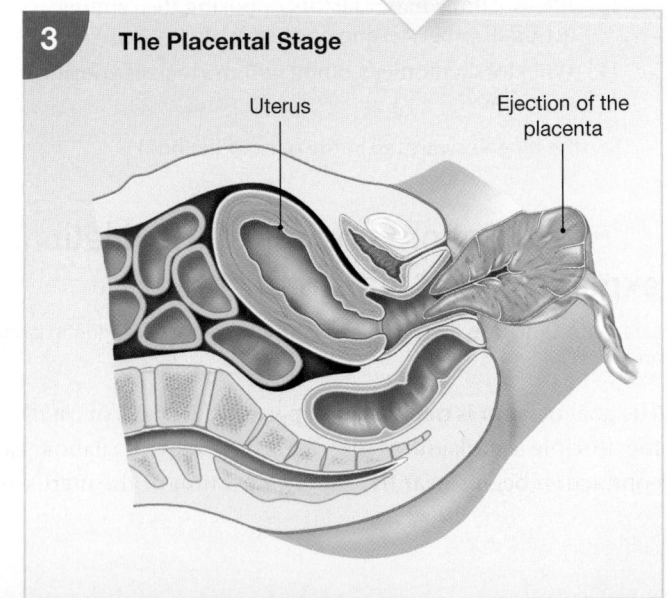

Uterus   Ejection of the placenta

stage, the amniochorionic membrane ruptures, an event sometimes referred to as the woman having her "water break." If this event occurs before other events of the dilation stage, the life of the fetus may be at risk from infection. If the risk is sufficiently great, labor can be induced.

The Expulsion Stage

The **expulsion stage** begins as the cervix, pushed open by the approaching fetus, completes its dilation to about 10 cm (4 in.) (**Stage 2** in **Figure 29–11**). In this stage, contractions reach maximum intensity. They occur at perhaps two- or three-minute intervals and last a full minute. Expulsion continues until the fetus has emerged from the vagina. In most cases, the expulsion stage lasts less than two hours. The arrival of the newborn infant into the outside world is **delivery**, or birth.

If the vaginal canal is too small to permit the passage of the fetus, posing acute danger of perineal tearing, a clinician may temporarily enlarge the passageway by performing an **episiotomy** (e-piz-ē-OT-ō-mē), an incision through the perineal musculature. After delivery, this surgical cut is repaired with sutures, a much simpler procedure than suturing the jagged edges associated with an extensive perineal tear.

If complications arise during the dilation or expulsion stage, the infant can be delivered by **cesarean section**, or "*C-section.*" In such cases, an incision is made through the abdominal wall, and the uterus is opened just enough to allow passage of the infant's head. The cesarean delivery rate had been

29

steadily increasing in the United States for the past 12 years. However, data released from the CDC's National Vital Statistics System showed that between 2009 and 2011 the cesarean delivery rate was unchanged and accounted for 31.3 percent of all live singleton births.

Immediately after birth, clinicians assess the newborn's health in five areas: heart rate, breathing, skin color, muscle tone, and reflex response. This assessment is called an **Apgar score**. Each criterion is scored from 0 to 2, with 8–10 indicating a healthy baby.

The Placental Stage

During the **placental stage** of labor, muscle tension builds in the walls of the partially empty uterus, which gradually decreases in size (**Stage ③** in **Figure 29–11**). This uterine contraction tears the connections between the endometrium and the placenta. In general, within an hour of delivery, the placental stage ends with the ejection of the placenta, or *afterbirth*. The disruption of the placenta is accompanied by a loss of blood, but associated uterine contraction compresses the uterine vessels and usually restricts this flow. Because maternal blood volume has increased greatly during pregnancy, the blood loss can normally be tolerated without difficulty.

Premature Labor

Premature labor occurs when true labor begins before the fetus has completed normal development. The newborn's chances of surviving are directly related to its body weight at delivery. Even with massive supportive efforts, newborns weighing less than 400 g (14 oz) at birth will not survive, primarily because their respiratory, cardiovascular, and urinary systems are unable to support life without aid from maternal systems. As a result, the dividing line between spontaneous abortion and **immature delivery** is usually set at 500 g (17.6 oz), the normal weight near the end of the second trimester.

Most fetuses born at 25–27 weeks of gestation (a birth weight under 600 g or 21.1 oz) die despite intensive neonatal care. Survivors have a high risk of developmental abnormalities. **Premature delivery** usually refers to birth at 28–36 weeks (a birth weight over 1 kg or 2.2 lb). With care, these newborns have a good chance of surviving and developing normally.

Difficult Deliveries

By the end of gestation in most pregnancies, the fetus has rotated within the uterus to transit the birth canal headfirst, facing the mother's sacrum. In about 6 percent of deliveries, the fetus faces the mother's pubis instead. These babies can be delivered normally, given enough time, but risks to infant and mother are reduced by a *forceps delivery*. Forceps resemble large, curved salad tongs that can be separated for insertion into the vaginal canal, one side at a time. Once in place, they are reunited and used to grasp the head of the fetus. An intermittent pull is applied, so that the forces on the head resemble those of normal delivery.

In 3–4 percent of deliveries, the legs or buttocks of the fetus enter the vaginal canal first. Such deliveries are **breech births**. Risks to the fetus are higher in breech births than in normal deliveries, because the umbilical cord can become constricted, cutting off placental blood flow.

The head is normally the widest part of the fetus. The mother's cervix may dilate enough to pass the baby's legs and body, but not the head. This entrapment compresses the umbilical cord, prolongs delivery, and subjects the fetus to severe distress and potential injury. If attempts to reposition the fetus or promote further dilation are unsuccessful over the short term, delivery by cesarean section may be required.

Multiple Births

Multiple births (twins, triplets, quadruplets, and so forth) can take place for several reasons. The ratio of twin births to single births in the U.S. population is nearly 1:89. "Fraternal twins," or **dizygotic twins** (dī-zī-GOT-ik), develop when two separate oocytes are ovulated and fertilized. Because chromosomes are shuffled during meiosis, the odds against any two zygotes from the same parents having identical genes exceed 1 in 8.4 million. Seventy percent of twins are dizygotic.

"Identical twins" or **monozygotic twins**, result either from the separation of blastomeres early in cleavage or from the splitting of the inner cell mass before gastrulation. In either event, the genetic makeup of the twins is identical because both formed from the same pair of gametes.

Triplets, quadruplets, and larger multiples can result from multiple ovulations, blastomere splitting, or some combination of the two. For unknown reasons, the rates of naturally occurring multiple births fall into a pattern: Twins occur in 1 of every 89 births, triplets in 1 of every 89^2 (or 7921) births, quadruplets in 1 of every 89^3 (704,969) births, and so forth. Taking fertility drugs that stimulate the maturation of abnormally high numbers of follicles can increase the incidence of multiple births.

Pregnancies with multiple fetuses pose special problems because the strains on the mother are multiplied. The chances of premature labor are increased, and the risks to the mother are higher than for single births. Increased risks also extend to the fetuses during gestation, and to the newborns, because even at full term such newborns have lower-than-average birth weights. They are also more likely to have problems during delivery. For example, in more than half of twin deliveries, one or both fetuses enter the vaginal canal in an abnormal position.

If the splitting of the blastomeres or of the embryonic disc is not complete, **conjoined** (*Siamese*) **twins** may develop.

29

These genetically identical twins typically share some skin, a portion of the liver, and perhaps other internal organs as well. When the fusion is minor, the infants can be surgically separated with some success. Most conjoined twins with more extensive fusions fail to survive delivery.

> **CHECKPOINT**
>
> 15. Name the three stages of labor.
> 16. Differentiate between immature delivery and premature delivery.
> 17. Give the biological terms for fraternal twins and identical twins.
>
> See the blue Answers tab at the back of the book.

29-7 | Postnatal stages are the neonatal period, infancy, childhood, adolescence, and maturity, followed by senescence

Learning Outcome Identify the features and physiological changes of the postnatal stages of life.

Developmental processes do not stop at delivery. Newborns have few of the anatomical, functional, or physiological characteristics of mature adults. The course of postnatal development typically includes five **life stages**: (1) the *neonatal period*, (2) *infancy*, (3) *childhood*, (4) *adolescence*, and (5) *maturity*. Each stage is typified by a distinctive combination of characteristics and abilities. These stages are familiar parts of human experience. Each stage has distinctive features, but the transitions between them are gradual, and the boundaries indistinct. At maturity, development ends and the process of aging, or *senescence*, begins.

The Neonatal Period, Infancy, and Childhood

The **neonatal period** extends from birth to one month of age. **Infancy**, or babyhood, then continues through the first year of life. **Childhood** lasts from infancy until **adolescence** (puberty), the period of sexual and physical maturation. Two major events are under way during these developmental stages:

1. The organ systems (except those associated with reproduction) become fully operational and gradually acquire the functional characteristics of adult structures.

2. The individual grows rapidly, and body proportions change significantly.

Pediatrics is the medical specialty that focuses on postnatal development from infancy through adolescence. Infants and young children cannot clearly describe the problems they are experiencing, so pediatricians and parents must be skilled observers. Standardized tests are used to assess developmental progress relative to average values.

The Neonatal Period

Physiological and anatomical changes take place as the fetus completes the transition to the status of newborn, or **neonate**. The neonatal period is the time from birth through 28 days. Before delivery, dissolved gases, nutrients, wastes, hormones, and antibodies were transferred across the placenta. At birth, the neonate must become relatively self-sufficient, performing respiration, digestion, and excretion using its own specialized organs and organ systems. The transition from fetus to neonate can be summarized as follows:

- At birth, the lungs are collapsed and filled with fluid. Filling them with air requires a massive and powerful inhalation. ⤺ p. 898

- When the lungs expand, the pattern of cardiovascular circulation changes due to alterations in blood pressure and flow rates. The ductus arteriosus closes, isolating the pulmonary and systemic trunks. Closure of the foramen ovale separates the atria of the heart, completing the separation of the pulmonary and systemic circuits. ⤺ p. 798

- The typical neonatal heart rate (120–140 beats per minute) and respiratory rate (30 breaths per minute) are considerably higher than in adults. In addition, the metabolic rate per unit of body weight in neonates is approximately twice that of adults.

- Before birth, the digestive system is relatively inactive, although it does accumulate a mixture of bile secretions, mucus, and epithelial cells. This collection of debris, called *meconium*, is excreted during the first few days of life. During that period, the newborn begins to nurse.

- As wastes build up in the arterial blood, the kidneys excrete them. Glomerular filtration is normal, but the neonate's kidneys cannot concentrate urine to any significant degree. As a result, urinary water losses are high, and neonatal fluid requirements are proportionally much greater than those of adults.

- The neonate has little ability to control its body temperature, particularly in the first few days after delivery. As the infant grows larger and its insulating subcutaneous adipose "blanket" gets thicker, its metabolic rate also increases. Daily and even hourly shifts in body temperature continue throughout childhood. ⤺ p. 993

Over the entire neonatal period, the newborn is dependent on nutrients contained in milk, typically breast milk secreted by the mother's mammary glands.

Lactation and the Mammary Glands. By the end of the sixth month of pregnancy, the mammary glands are fully developed, and the gland cells begin to secrete **colostrum** (kō-LOS-trum). Ingested by the infant during the first two or three days of life, colostrum contains more proteins and far less fat than breast milk. Many of the proteins are antibodies that may help the infant ward off infections until its own immune system becomes fully functional. In addition, the mucins present in both colostrum and milk can inhibit the replication of a family of viruses (*rotaviruses*) that can cause dangerous forms of gastroenteritis and diarrhea in infants.

As colostrum production drops, the mammary glands convert to milk production. Breast milk consists of water, proteins, amino acids, lipids, sugars, and salts. It also contains large quantities of *lysozyme*—an enzyme with antibiotic properties. In terms of energy, human milk provides about 750 kilocalories per liter. The secretory rate varies with the demand, but a 5–6-kg (11–13-lb) infant usually requires about 850 mL (3.6 cups) of milk per day. (Maternal milk production throughout this period is maintained through the combined actions of several hormones, as detailed in Chapter 18. つ p. 648)

Milk becomes available to infants through the **milk let-down (milk-ejection) reflex** (**Figure 29–12**). This reflex continues to function until *weaning*, withdrawing mother's milk, typically one to two years after birth. Milk production stops

soon after, and the mammary glands gradually return to a resting state. Earlier weaning is a common practice in the United States, where women take advantage of commercially prepared milk- or soy-based infant formulas that closely approximate the composition of natural breast milk. The major difference between such substitutes and natural milk is that the substitutes lack antibodies and lysozyme, which play important roles

Figure 29–12   **The Milk Let-Down Reflex.**

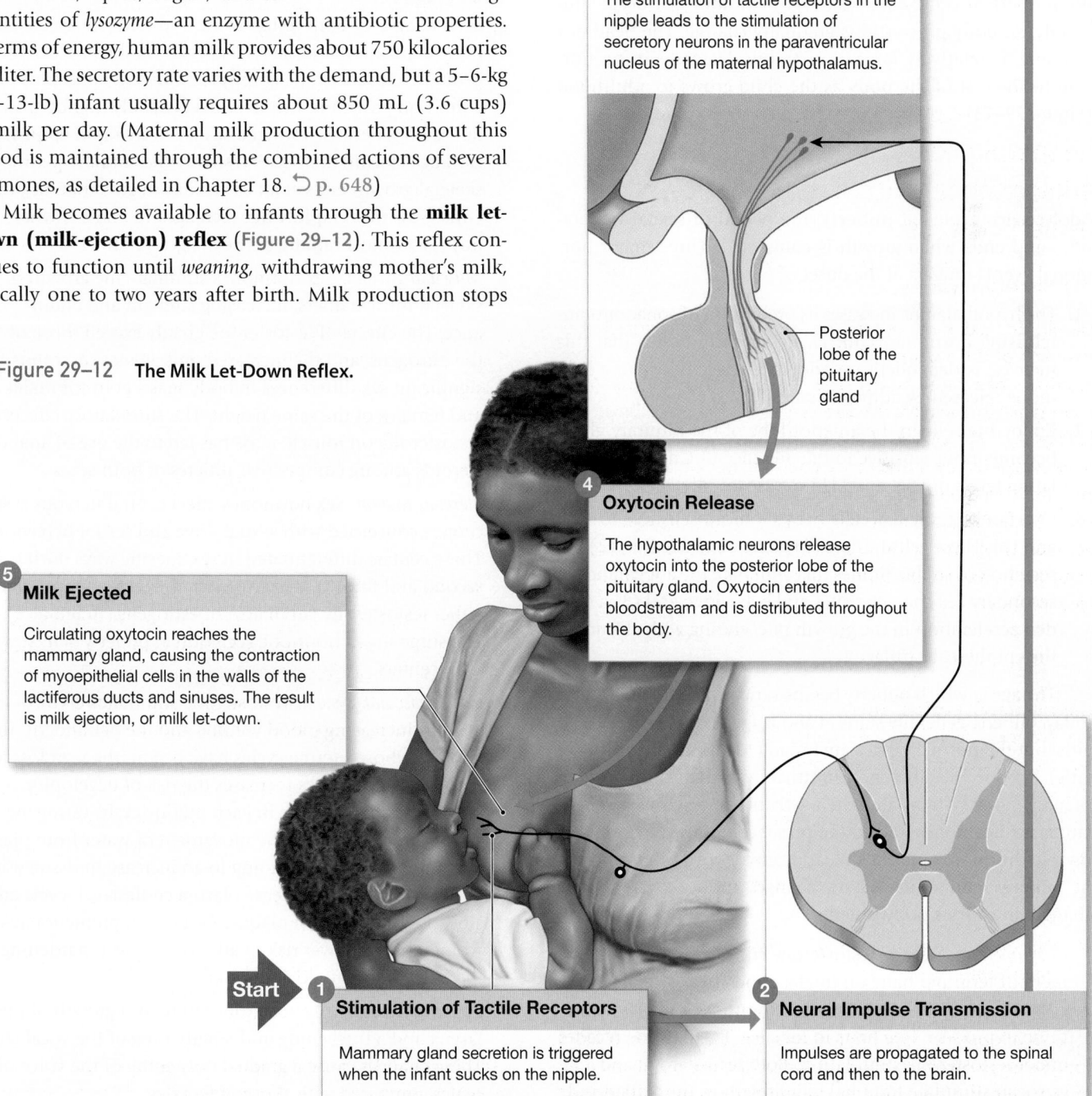

3 Oxytocin Secretion

The stimulation of tactile receptors in the nipple leads to the stimulation of secretory neurons in the paraventricular nucleus of the maternal hypothalamus.

Posterior lobe of the pituitary gland

4 Oxytocin Release

The hypothalamic neurons release oxytocin into the posterior lobe of the pituitary gland. Oxytocin enters the bloodstream and is distributed throughout the body.

5 Milk Ejected

Circulating oxytocin reaches the mammary gland, causing the contraction of myoepithelial cells in the walls of the lactiferous ducts and sinuses. The result is milk ejection, or milk let-down.

Start

1 Stimulation of Tactile Receptors

Mammary gland secretion is triggered when the infant sucks on the nipple.

2 Neural Impulse Transmission

Impulses are propagated to the spinal cord and then to the brain.

29

in maintaining the health of the infant. Consequently, early weaning is associated with an increased risk of infections and allergies in the infant.

Infancy and Childhood

The most rapid growth occurs during prenatal development, and the growth rate decreases after delivery. Growth during infancy and childhood occurs under the direction of circulating hormones, notably growth hormone, adrenal steroids, and thyroid hormones. These hormones affect each tissue and organ in specific ways, depending on the sensitivities of the individual cells. As a result, growth does not occur uniformly, so body proportions gradually change. The head, for example, is relatively large at birth but decreases in proportion to the rest of the body as the child grows to adulthood (**Figure 29–13**).

Adolescence and Maturity

Adolescence begins at **puberty**, the period of sexual maturation, and ends when growth is completed. Three major hormonal events interact at the onset of puberty:

1. The hypothalamus increases its production of gonadotropin-releasing hormone (GnRH). Evidence indicates that this increase is dependent on adequate levels of *leptin*, a hormone released by adipose tissues. ⊃ p. 669

2. Endocrine cells in the anterior lobe of the pituitary gland become more sensitive to the presence of GnRH, so circulating levels of FSH and LH increase rapidly.

3. Ovarian or testicular cells become more sensitive to FSH and LH. These cells initiate (1) gamete production; (2) the secretion of sex hormones that stimulate the appearance of secondary sex characteristics and behaviors; and (3) a sudden acceleration in the growth rate, ending with closure of the epiphyseal cartilages.

The age at which puberty begins varies. In the United States today, puberty generally starts at about age 12 in boys and 11 in girls, but the normal ranges are broad (10–15 in boys, 9–14 in girls). Sex-specific differences in structure and function develop as many body systems alter their activities in response to circulating sex hormones and to the presence of growth hormone, thyroid hormones, prolactin, and adrenocortical hormones. At puberty, endocrine system changes induce characteristic changes in various body systems:

- *Integumentary System.* Testosterone stimulates the development of terminal hairs on the face and chest of males. ⊃ p. 191 Under estrogen stimulation those follicles continue to produce fine hairs in females. The hairline recedes under testosterone stimulation. Both testosterone and estrogen stimulate terminal hair growth in the axillae and

in the genital area. Androgens, which are present in both sexes, also stimulate sebaceous gland secretion and may cause acne. Adipose tissues respond differently to testosterone than to estrogens, and this difference produces changes in the distribution of subcutaneous body fat. In women, the combination of estrogens, prolactin, growth hormone, and thyroid hormones promotes the initial development of the mammary glands. The duct system becomes more elaborate, but true secretory alveoli do not yet develop. Much of the growth of the breasts during this period reflects increased deposition of fat rather than glandular tissue.

- *Skeletal System.* Both testosterone and estrogen accelerate bone deposition and skeletal growth. In the process, they promote closure of the epiphyseal cartilages and thus place a limit on growth in height. Girls generally do not grow as tall as boys because estrogens cause more rapid epiphyseal cartilage closure than does testosterone, and the period of skeletal growth ends at an earlier age in girls than in boys. Girls grow most rapidly between ages 10 and 13, whereas boys grow most rapidly between ages 12 and 15.

- *Muscular System.* Sex hormones stimulate the growth of skeletal muscle fibers, increasing strength and endurance. The effects of testosterone greatly exceed those of the estrogens, and the increased muscle mass accounts for significant sex differences in body mass, even for males and females of the same height. The stimulatory effects of testosterone on muscle mass has led to the use of anabolic steroids among competitive athletes of both sexes.

- *Nervous System.* Sex hormones affect central nervous system centers concerned with sexual drive and sexual behaviors. These centers differentiated in sex-specific ways during the second and third trimesters, when the fetal gonads secrete either testosterone (in males) or estrogens (in females). The surge in sex hormone secretion at puberty activates the CNS centers.

- *Cardiovascular System.* Testosterone stimulates erythropoiesis, increasing blood volume and the hematocrit. In females whose uterine cycles have begun, the iron loss associated with menses increases the risk of developing iron-deficiency anemia. Late in each uterine cycle, estrogens and progesterone promote the movement of water from plasma into interstitial fluid, leading to an increase in tissue water content. Estrogens decrease plasma cholesterol levels and slow the formation of plaque. As a result, premenopausal women have a lower risk of atherosclerosis ("hardening of the arteries") than do adult men.

- *Respiratory System.* Testosterone stimulates growth of the larynx and a thickening and lengthening of the vocal cords. These changes cause a gradual deepening of the voice of males compared with those of females.

Figure 29–13   **Growth and Changes in Body Form and Proportion.** The views at 4, 8, and 16 weeks of gestation are presented at actual size. Notice the changes in body form and proportions as development proceeds. For example, the head, which contains the brain and sense organs, is proportionately large at birth.

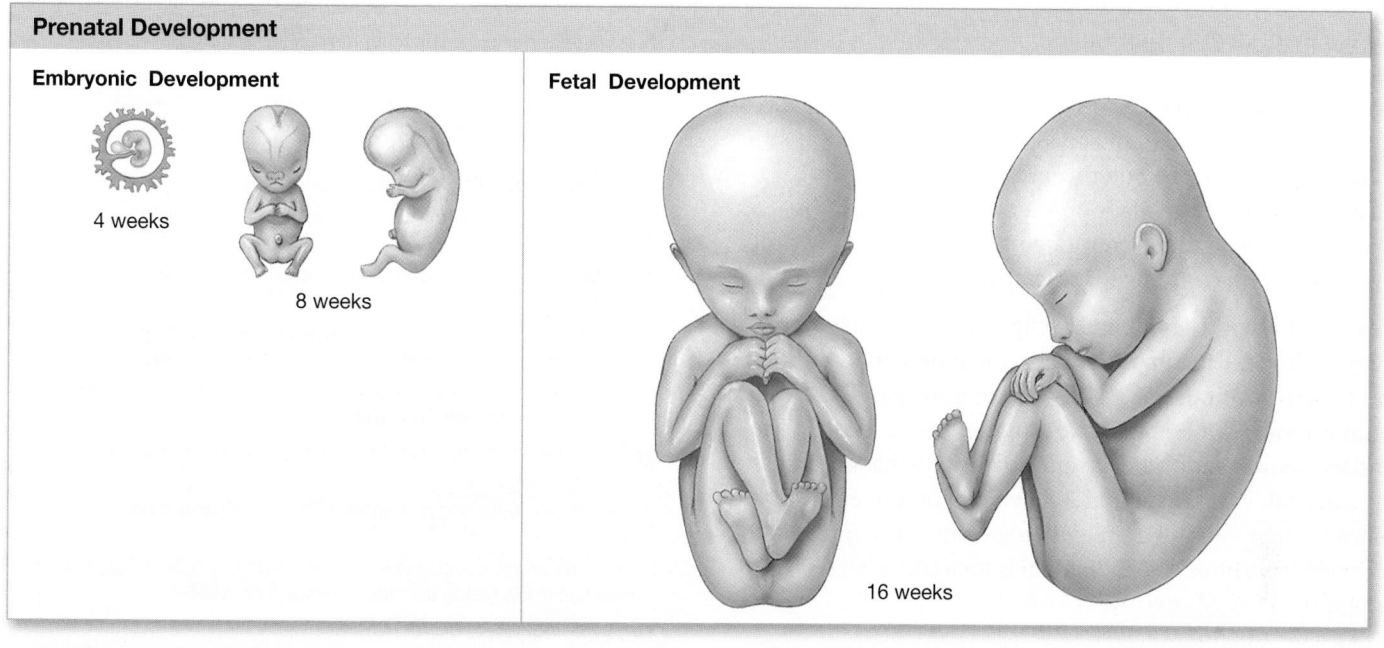

Prenatal Development

Embryonic Development

4 weeks

8 weeks

Fetal Development

16 weeks

Postnatal Development

Neonatal	Infancy	Childhood	Adolescence	Maturity

5 ft

4 ft

3 ft

2 ft

1 ft

0

1 month   1 year   Puberty   18 years
                                              (between 9–14 years)

■ *Reproductive System.* In males, testosterone stimulates the functional development of the accessory reproductive glands, such as the prostate gland and seminal glands, and helps promote spermatogenesis. In females, estrogens target the uterus, promoting a thickening of the myometrium, increasing blood flow to the endometrium, and stimulating cervical mucus production. Estrogens also promote the functional development of accessory reproductive organs in females. The first few uterine cycles may or may not be accompanied by ovulation. After the initial stage, the woman will be fertile, even though her growth and physical maturation will continue for several years.

After puberty, the continued background secretion of estrogens or androgens maintains sex-specific differences. In both sexes, growth continues at a slower pace until age 18–21. By that time most of the epiphyseal cartilages have closed. The boundary between adolescence and maturity is hazy, because it has physical, emotional, and behavioral components. Adolescence is often said to be over when growth stops, in the late teens or early twenties. The individual is then considered physically mature.

Senescence

Physical growth may stop at maturity, but physiological changes continue. The sex-specific differences produced at puberty remain, but further changes take place when sex hormone levels decline at menopause or the male climacteric. ⟲ p. 1114 All these changes are part of the process of **senescence** (*senesco*, to grow old), or aging, which reduces the functional capabilities of the person. Even without disease or injury, senescence-related changes at the molecular level ultimately lead to death.

Table 29–3 summarizes the age-related changes in physiological systems discussed in earlier chapters. Taken together, these changes both decrease the functional abilities of the individual and affect homeostatic mechanisms. As a result, elderly people are less able to make homeostatic adjustments in response to internal or environmental stresses. The risks of contracting a variety of infectious diseases are proportionately increased as immune function deteriorates. This deterioration leads to drastic physiological changes that affect all internal systems. Death ultimately occurs when the body's existing homeostatic mechanisms cannot counter some combination of stresses.

Physicians attempt to forestall death by adjusting homeostatic mechanisms or removing the sources of stress. **Geriatrics** (*geras*, old age) is the medical specialty that deals with the problems associated with aging. Physicians trained in geriatrics are known as **geriatricians**. Problems commonly encountered by geriatricians include infections, cancers, heart disease, strokes, arthritis, senile dementia, and anemia. These conditions are directly related to age-induced changes in vital systems.

Table 29–3	Effects of Aging on Organ Systems

The characteristic physical and functional changes that are part of the aging process affect all organ systems. Examples discussed in previous chapters include the following:

- A loss of elasticity in the skin that produces sagging and wrinkling. ⟲ pp. 198–199
- A decline in the rate of bone deposition, leading to weak bones, and degenerative changes in joints that make them less mobile. ⟲ pp. 225, 309
- Reductions in muscular strength and ability. ⟲ p. 403
- Impairment of coordination, memory, and intellectual function. ⟲ p. 583
- Reductions in the production of, and sensitivity to, circulating hormones. ⟲ p. 670
- Appearance of cardiovascular problems and a reduction in peripheral blood flow that can affect a variety of vital organs. ⟲ p. 802
- Reduced sensitivity and responsiveness of the immune system, leading to infection, cancer, or both. ⟲ p. 851
- Reduced elasticity in the lungs, leading to decreased respiratory function. ⟲ p. 899
- Decreased peristalsis and muscle tone along the digestive tract. ⟲ p. 954
- Decreased peristalsis and muscle tone in the urinary system, coupled with a reduction in the glomerular filtration rate. ⟲ p. 1035
- Functional impairment of the reproductive system, which eventually becomes inactive when menopause or the male climacteric occurs. ⟲ pp. 1114–1115

CHECKPOINT

18. Name the postnatal stages of development.

19. Describe the time frame for each of the following stages: neonatal period, infancy, and adolescence.

20. What is the difference between colostrum and breast milk?

21. Increases in the blood levels of GnRH, FSH, LH, and sex hormones mark the onset of which stage of development?

See the blue Answers tab at the back of the book.

29-8 | Genes and chromosomes determine patterns of inheritance

Learning Outcome Relate basic principles of genetics to the inheritance of human traits.

Recall that chromosomes contain DNA and proteins, and genes are functional segments of DNA. Each gene carries the information needed to direct the synthesis of a specific polypeptide. We introduced chromosome structure and the functions of genes in Chapter 3. ⟲ pp. 108–113 Every nucleated somatic cell in your body carries copies of the original 46 chromosomes present when you were a zygote. Those chromosomes and their genes make up your **genotype** (JĒN-ō-tīp; *geno-*, gene).

Through development and differentiation, the instructions contained in the genotype are expressed, or carried out, in many ways. No single cell or tissue uses all the information and instructions contained in the genotype. For example, in muscle fibers, the genes involved in the formation of excitable membranes and contractile proteins are active, but in cells of the pancreatic islets, a different set of genes is at work. Collectively, however, the instructions in your genotype determine the anatomical and physiological characteristics that make you a unique person. Those anatomical and physiological characteristics are your **phenotype** (FĒ-nō-tīp; *phaino*, to display). The phenotype results from the interaction between the person's genotype and the environment. In architectural terms, the genotype is a set of plans, and the phenotype is the finished building. Specific elements in your phenotype, such as hair and eye color, skin tone, and foot size, are called phenotypic *traits*, or *characters*.

Your genotype is derived from the genotypes of your parents. Yet you are not an exact copy of either parent. Nor are you an easily identifiable mixture of their characteristics. We begin our discussion of genetics with the basic patterns of inheritance and their implications. We then examine the mechanisms responsible for regulating the activities of the genotype during prenatal development.

Patterns of Inheritance

The 46 chromosomes carried by each human somatic cell occur in pairs. Every somatic cell contains 23 pairs of chromosomes. At amphimixis, the spermatozoon supplies one member of each pair, and the ovum supplies the other. The two members of each pair are known as **homologous** (huh-MOL-ō-gus) **chromosomes**. Twenty-two of those pairs are called **autosomal** (aw-tō-SŌ-mul) **chromosomes**. Most of the genes of the autosomal chromosomes affect somatic, or body, characteristics, such as hair color and skin pigmentation. The chromosomes of the 23rd pair are known as the **sex chromosomes**. One of their functions is to determine whether the individual is genetically male or female. **Figure 29–14** shows the **karyotype** (*karyon*, nucleus + *typos*, mark), or entire set of chromosomes, of a normal male. In the discussion that follows, we deal with the inheritance of traits carried on the autosomal chromosomes. We will examine the patterns of inheritance by the sex chromosomes in a later section.

The two chromosomes in a homologous autosomal pair have the same structure and carry genes that affect the same traits. For example, suppose that one member of the pair contains three genes in a row, with the first gene determining hair color, the second eye color, and the third skin pigmentation. The other chromosome, or *homolog*, carries genes that affect the same traits, and the genes are in the same sequence. The genes are also located at equivalent positions on their respective

Figure 29–14 **A Human Male Karyotype.**

chromosomes. A gene's position on a chromosome is called a **locus** (LŌ-kus; plural, *loci*).

The two chromosomes in a pair may not carry the same *form* of each gene, however. The various forms of a given gene are called **alleles** (uh-LĒLZ). These *alternative forms* determine the precise effect of the gene on your phenotype. If the two chromosomes of a homologous pair carry the same allele of a particular gene, you are **homozygous** (hō-mō-ZĪ-gus; *homos*, the same) for the trait affected by that gene. That allele will then indeed be expressed in your phenotype. For example, if you receive a gene for freckles from your father and a gene for freckles from your mother, you will be homozygous for freckles—and you will have freckles. About 80 percent of an individual's genome consists of homozygous alleles. In **simple inheritance**, the phenotype is determined by interactions between a single pair of alleles.

Interactions between Alleles

Because the chromosomes of a homologous pair have different origins, one paternal and the other maternal, they do not necessarily carry the same alleles. If you have two different alleles for the same gene, you are **heterozygous** (het-er-ō-ZĪ-gus; *heteros*, other) for the trait determined by that gene. The phenotype that results from a heterozygous genotype depends on the nature of the interaction between the corresponding alleles. The potential interactions, which include examples of normal and abnormal phenotypic traits, are diagrammed in **Figure 29–15**. For example, if you received a gene for curly hair from your

29

Figure 29–15 **Major Patterns of Inheritance.**

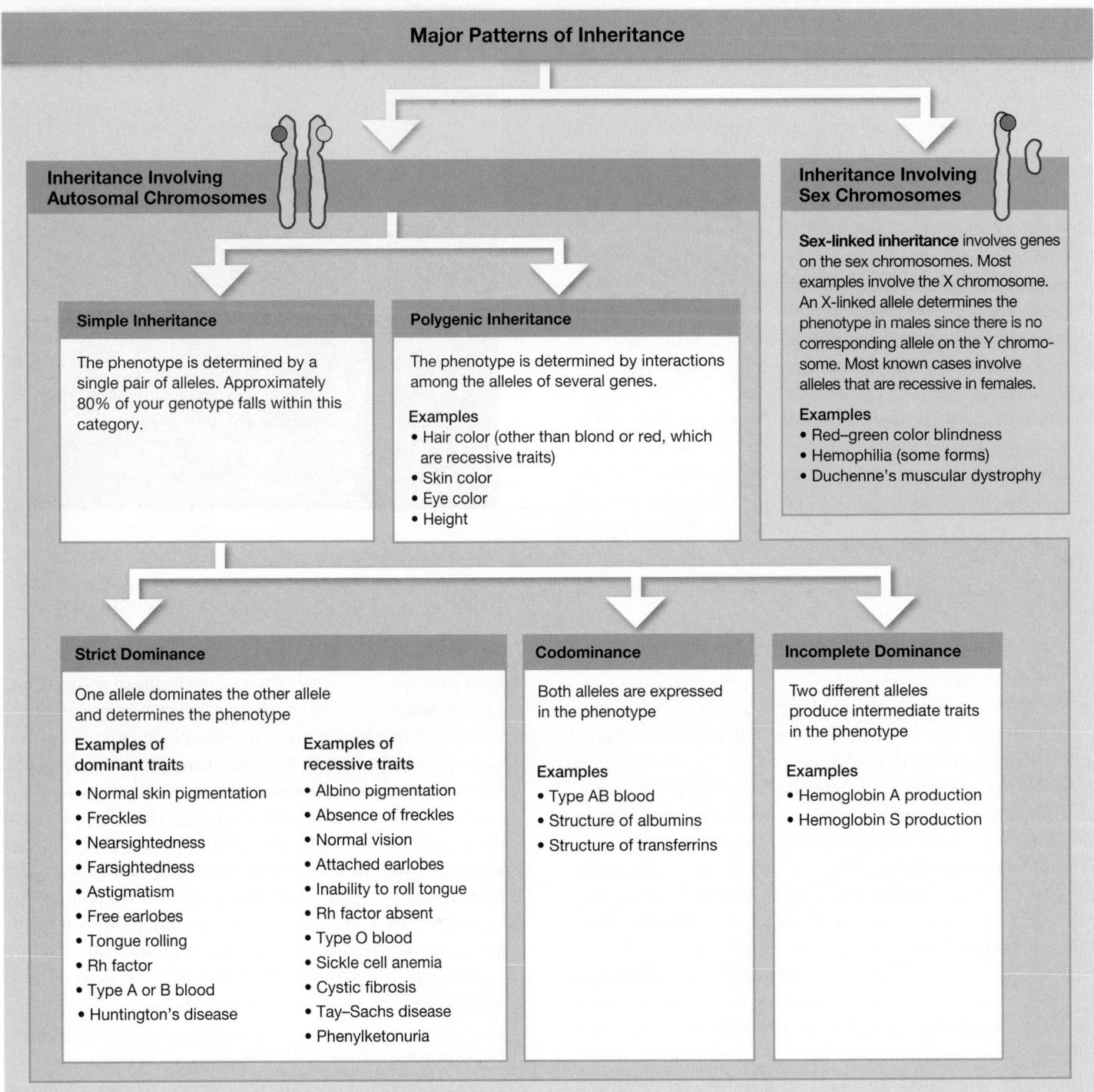

Major Patterns of Inheritance

Inheritance Involving Autosomal Chromosomes

Inheritance Involving Sex Chromosomes

Sex-linked inheritance involves genes on the sex chromosomes. Most examples involve the X chromosome. An X-linked allele determines the phenotype in males since there is no corresponding allele on the Y chromosome. Most known cases involve alleles that are recessive in females.

Examples
• Red–green color blindness
• Hemophilia (some forms)
• Duchenne's muscular dystrophy

Simple Inheritance

The phenotype is determined by a single pair of alleles. Approximately 80% of your genotype falls within this category.

Polygenic Inheritance

The phenotype is determined by interactions among the alleles of several genes.

Examples
• Hair color (other than blond or red, which are recessive traits)
• Skin color
• Eye color
• Height

Strict Dominance

One allele dominates the other allele and determines the phenotype

Examples of dominant traits
• Normal skin pigmentation
• Freckles
• Nearsightedness
• Farsightedness
• Astigmatism
• Free earlobes
• Tongue rolling
• Rh factor
• Type A or B blood
• Huntington's disease

Examples of recessive traits
• Albino pigmentation
• Absence of freckles
• Normal vision
• Attached earlobes
• Inability to roll tongue
• Rh factor absent
• Type O blood
• Sickle cell anemia
• Cystic fibrosis
• Tay–Sachs disease
• Phenylketonuria

Codominance

Both alleles are expressed in the phenotype

Examples
• Type AB blood
• Structure of albumins
• Structure of transferrins

Incomplete Dominance

Two different alleles produce intermediate traits in the phenotype

Examples
• Hemoglobin A production
• Hemoglobin S production

29

father, but a gene for straight hair from your mother, whether *you* will have curly hair, straight hair, or even wavy hair depends on the relationship between the alleles for those traits. The relationship can be one of *strict dominance, incomplete dominance,* or *codominance,* as follows:

• In **strict dominance**, an allele that is **dominant** will be expressed in the phenotype, *regardless of any conflicting*

instructions carried by the other allele. For instance, an individual with only one allele for freckles will have freckles, because that allele is dominant over the "nonfreckle" allele. An allele that is **recessive** will be expressed in the phenotype only if that same allele is present on *both chromosomes* of a homologous pair. For example, in Chapter 5 you learned that people with albinism cannot synthesize the yellow-brown pigment *melanin*. ⟲ p. 182 The presence of one allele that directs melanin production will result in normal color. Two recessive alleles must be present to produce an individual with albinism. A single gene can have many different alleles *in a population,* some dominant and others recessive. An individual can have a maximum of only two alleles—one from the mother and the other one from the father. If both parents have the same alleles, then an individual has only one kind of allele, but two copies of it, and is thus, homozygous.

■ In **incomplete dominance**, heterozygous alleles produce a phenotype that is intermediate (not completely dominant) to the phenotypes of individuals who are homozygous for one allele or the other. A good example is the sickling gene that affects the shape of red blood cells. Individuals with homozygous alleles that carry instructions for normal adult hemoglobin A have red blood cells of normal shape. Individuals with homozygous alleles for hemoglobin S, an abnormal form, have red blood cells that become sickle-shaped in peripheral capillaries when the P_{O_2} decreases. These individuals develop *sickle cell anemia.* ⟲ p. 686 Individuals who are heterozygous for this trait do not develop anemia but their red blood cells may sickle when tissue oxygen levels are extremely low. Incomplete dominance is relatively rare in humans.

■ In **codominance**, an individual who is heterozygous (has different alleles) for a given trait exhibits both phenotypes for that trait. Blood type in humans is determined by co-dominance. The alleles for type A and type B blood are dominant over the allele for type O blood, but a person with one type A allele and one type B allele has type AB blood, not A or B. Type AB blood has *both* type A antigens and type B antigens. The distinction between incomplete dominance and codominance is not always clear-cut. For example, a person who has alleles for hemoglobin A and hemoglobin S shows incomplete dominance for RBC shape, but codominance for hemoglobin. Each red blood cell contains a mixture of hemoglobin A and hemoglobin S.

Penetrance and Expressivity

Differences in genotype lead to distinct variations in phenotype, but the relationships are not always predictable. The presence of a particular pair of alleles does not affect the phenotype in the same way in every individual. **Penetrance** is the percentage of individuals with a particular genotype that show the "expected" phenotype. In other individuals, the activity of other genes or environmental factors may override the effects of that genotype. For example, *emphysema*, a respiratory disorder discussed in Chapter 23, has been linked to a specific abnormal genotype. ⟲ p. 901 However, about 20 percent of the individuals with this genotype do not develop emphysema, making the penetrance of this genotype approximately 80 percent. The effects of environmental factors are apparent: Most people who develop emphysema are cigarette smokers.

If a given genotype *does* affect the phenotype, it can do so to various degrees, again depending on the activity of other genes or environmental stimuli. For example, even though identical twins have the same genotype, they do not have exactly the same fingerprints. The extent to which a particular allele is expressed when it is present is termed its **expressivity**.

Environmental effects on genetic expression are particularly evident during embryonic and fetal development. Drugs, including certain antibiotics, alcohol, and nicotine in cigarette smoke, can disrupt fetal development. Factors that result in abnormal development are called **teratogens** (TER-uh-tō-jenz).

Predicting Inheritance

When an allele can be neatly characterized as dominant or recessive, you can predict the characteristics of individuals on the basis of their parents' alleles.

In such calculations, dominant alleles are traditionally indicated by capitalized abbreviations, and recessive alleles by lowercase abbreviations. For a given trait, the possibilities are indicated by *AA* (homozygous dominant), *Aa* (heterozygous), or *aa* (homozygous recessive). Each gamete involved in fertilization contributes a single allele for a given trait. That allele must be one of the two alleles contained by all cells in the parent's body. Consider, for example, the possible offspring of a mother with albinism (no skin pigmentation) and a father with normal skin pigmentation. The maternal alleles are abbreviated *aa* because albinism is a homozygous recessive trait. No matter which of her oocytes is fertilized, it will carry the recessive *a* allele. The father has normal pigmentation, a dominant trait. He is therefore either homozygous dominant *or* heterozygous for this trait, because both *AA* and *Aa* will produce the same phenotype: normal skin pigmentation. Every sperm produced by a homozygous dominant father will carry the *A* allele. In contrast, half the sperm produced by a heterozygous father will carry the dominant allele *A*, and the other half will carry the recessive allele *a*.

A simple box diagram known as a **Punnett square** helps us to predict the probabilities that children will have particular characteristics by showing the various combinations of parental alleles they can inherit. In the Punnett squares shown in **Figure 29–16**, the maternal alleles for skin pigmentation are listed along the horizontal axis, and the paternal ones along the vertical axis. The combinations of alleles appear in the small boxes. **Figure 29–16a** shows the possible offspring of an *aa* mother and an *AA* father. All the children will have the genotype *Aa*, so all will have normal skin pigmentation. Compare these results with those of **Figure 29–16b**, for a heterozygous father (*Aa*) and an *aa* mother. The heterozygous father produces two types of gametes, *A* and *a*. The mother's secondary oocyte,

29

Figure 29–16 **Predicting Phenotypic Characters by Using Punnett Squares.**

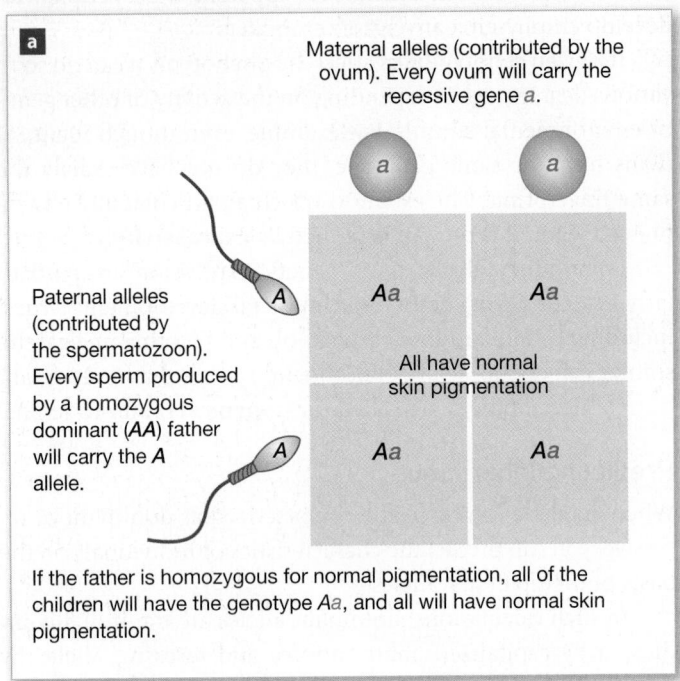

a

Maternal alleles (contributed by the ovum). Every ovum will carry the recessive gene *a*.

Paternal alleles (contributed by the spermatozoon). Every sperm produced by a homozygous dominant (*AA*) father will carry the *A* allele.

	a	*a*
A	*Aa*	*Aa*
A	*Aa*	*Aa*

All have normal skin pigmentation

If the father is homozygous for normal pigmentation, all of the children will have the genotype *Aa*, and all will have normal skin pigmentation.

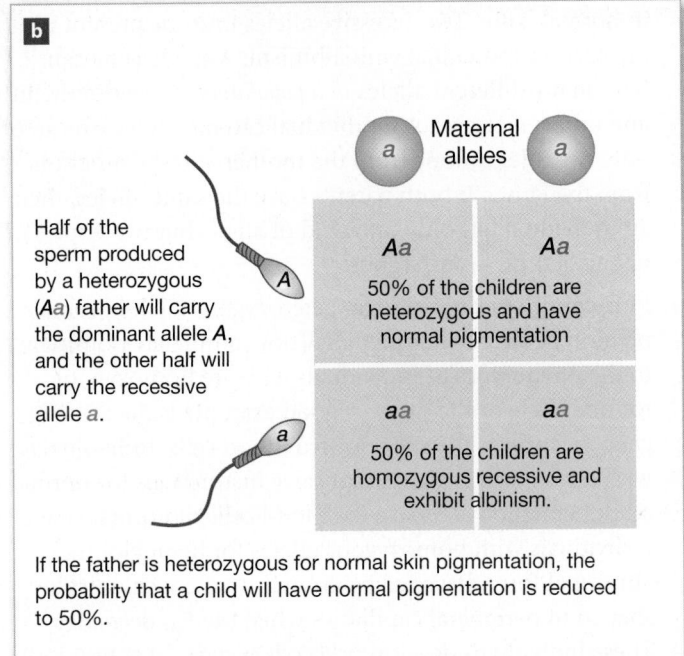

b

Maternal alleles

Half of the sperm produced by a heterozygous (*Aa*) father will carry the dominant allele *A*, and the other half will carry the recessive allele *a*.

	a	*a*
A	*Aa*	*Aa*
a	*aa*	*aa*

Aa *Aa* — 50% of the children are heterozygous and have normal pigmentation

aa *aa* — 50% of the children are homozygous recessive and exhibit albinism.

If the father is heterozygous for normal skin pigmentation, the probability that a child will have normal pigmentation is reduced to 50%.

carrying an *a* allele, may be fertilized by either one. As a result, the probability is 50 percent that a child of such a father will inherit the genotype *Aa* and so have normal skin pigmentation. The probability of inheriting the genotype *aa*, and thus having the albinism phenotype, is also 50 percent.

A Punnett square can also be used to draw conclusions about the identity and genotype of a parent. For example, in our scenario, a man with the genotype *AA* cannot be the father of an albino child (*aa*).

We can also predict the frequency of appearance of any inherited disorder that results from simple inheritance by using a Punnett square. Although they are rare in terms of overall numbers, more than 1200 inherited disorders have been identified that reflect the presence of one or two abnormal alleles for a single gene.

Some phenotypic traits are determined by interactions among several genes. Such interactions constitute **polygenic inheritance**. We cannot predict the presence or absence of these phenotypic traits using a simple Punnett square because the resulting phenotype depends not only on the nature of the alleles but also on how those alleles interact. In *suppression*, one gene suppresses the other. As a result, the second gene has no effect on the phenotype. In *complementary gene action*, dominant alleles on two genes interact to produce a phenotype different from that seen when one gene contains recessive alleles. The risks of developing several important adult disorders, including hypertension and coronary artery disease, are linked to polygenic inheritance.

Many of the developmental disorders responsible for fetal deaths and congenital malformations result from polygenic inheritance. In these cases, an individual's genetic composition does not by itself determine the onset of the disease. Instead, the conditions regulated by these genes establish a susceptibility to particular environmental influences. As a result, not every individual with the genetic tendency for a certain condition will develop that condition. For this reason, it is difficult to track polygenic conditions through successive generations. However, steps can be taken to prevent a crisis because many inherited polygenic conditions are *likely* (but not *guaranteed*) to occur. For example, you can reduce hypertension by controlling your diet and fluid volume, and you can prevent coronary artery disease by lowering your serum cholesterol levels.

Sources of Individual Variation

Just as you are not a copy of either of your parents, you also are not a mixture of their characteristics. We noted one reason for this in Chapter 28: During meiosis, maternal and paternal chromosomes are randomly distributed, so each gamete has a unique combination of maternal and paternal chromosomes. For this reason, you may have an allele for freckles from your father and an allele for nonfreckles from your mother, even though your sister received an allele for nonfreckles from each of your parents.

Only in very rare cases does an individual receive both alleles from one parent. The few documented cases appear to have resulted when duplicate maternal chromatids failed to

29

separate during meiosis II and the corresponding chromosome provided by the sperm did not participate in amphimixis. This condition, called *uniparental disomy,* generally remains undetected, because the individuals are phenotypically normal.

Genetic Recombination

During meiosis, various changes can occur in chromosome structure, producing gametes with chromosomes that differ from those of each parent. **Genetic recombination** greatly increases the genetic variation among gametes, and thus among the genotypes formed when those gametes combine in fertilization.

In one normal form of recombination, parts of homologous chromosomes become rearranged during synapsis of meiosis (**Figure 29–17**). When tetrads form, adjacent chromatids may overlap, an event called **crossing over**. The chromatids may then break, and the overlapping segments trade places. In general, the genetic exchange between homologous chromosomes is called crossing over, and between nonhomologous chromosomes it is called **translocation**. During recombination, portions of chromosomes may break away and be lost, or *deleted.* This event is abnormal, and it can have severe or even lethal effects on the zygote, depending on the nature of the lost genes.

Recombination that produces abnormal chromosome shapes or numbers is lethal for the zygote in almost all cases. Roughly 10 percent of zygotes have **chromosomal abnormalities**—that is, damaged, broken, missing, or extra copies of chromosomes—but only about 0.5 percent of newborns have such abnormalities. Few individuals with chromosomal abnormalities survive to full term. *Down's syndrome (trisomy 21),* which involves one of the smallest chromosomes in human beings, is an exception. In addition to contributing to prenatal mortality, chromosomal abnormalities produce a variety of serious clinical conditions. The high mortality rate and the severity of the problems reflect the fact that large numbers of genes have been added or deleted. Women who become pregnant later in life run a higher risk of birth defects and miscarriage due to chromosomal abnormalities in the oocyte. It seems that the longer the oocyte remains suspended in meiosis I, the more likely are recombination errors when meiosis is completed.

However, a lack of specific gene expression in a zygote is not necessarily due to chromosomal aberrations or deletions. Genes can be present, yet be prevented from being fully expressed. A genetic process called **genomic imprinting** is particularly important in the early embryo. Imprinting does not change the DNA itself. Instead, it results in specific (and usually reversible) chemical modifications, or markings, of DNA and its associated proteins. Because such chemical markings occur at a level *above* the DNA sequences, the study of such changes is called *epigenetics.* These changes then dictate whether the gene is expressed or not (silenced). As gametes mature, a proportion of their genes develop characteristic, maternal and paternal specific, imprinting markings. These markings persist after fertilization and can regulate whether or not the maternally or the paternally derived gene is transcribed during embryonic development. Many of the estimated 150 genes known to be affected by genomic imprinting regulate many aspects of early development, including rates of prenatal and postnatal growth, behavior, and language development. Imprinting is a normal process that acts as a "volume control" for parental genes since it results in one active copy of an imprinted gene.

Incorrectly imprinted genes can have the same effect as deletion of the same gene. This is exemplified by two human genetic disorders linked to genes in a specific portion of chromosome 15 and their differential imprinting during sperm or oocyte formation. *Angelman syndrome,* which results in hyperactivity, severe developmental disability, and seizures, occurs when the maternal genes are inactive. *Prader–Willi syndrome,* which results in short stature, reduced muscle tone and skin pigmentation, underdeveloped gonads, and some degree of developmental disability, occurs when paternal genes are inactive.

Mutations

Variations at the level of the individual gene can result from *mutations*—changes in the nucleotide sequence of an allele. **Spontaneous mutations** result from random errors in DNA replication. Such errors are relatively common, but in most cases the error is detected and repaired by enzymes in the nucleus. Those errors that go undetected and unrepaired have the potential to change the phenotype in some way.

Mutations occurring during meiosis can produce gametes that contain abnormal alleles. These alleles may be dominant or recessive, and they may occur on autosomal chromosomes or on sex chromosomes. The vast majority of mutations make the zygote incapable of completing normal development. Mutation, rather than chromosomal abnormalities, is probably the

Figure 29–17 Crossing Over and Recombination.

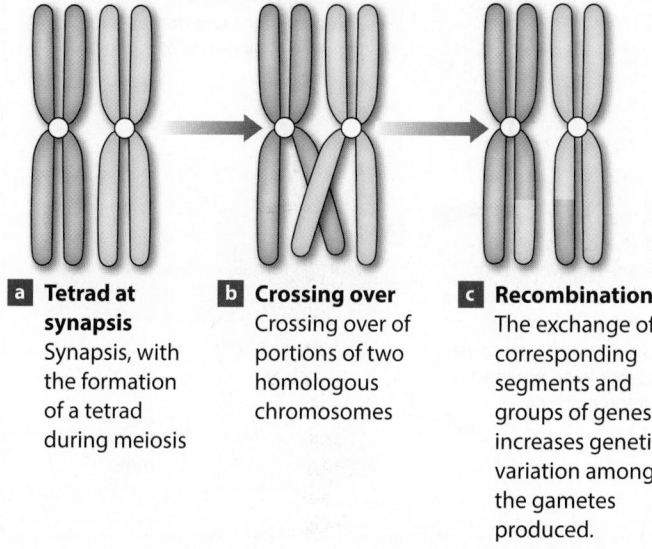

a **Tetrad at synapsis**
Synapsis, with the formation of a tetrad during meiosis

b **Crossing over**
Crossing over of portions of two homologous chromosomes

c **Recombination**
The exchange of corresponding segments and groups of genes increases genetic variation among the gametes produced.

29

primary cause of the high mortality rate among pre-embryos and embryos. (Approximately 50 percent of all zygotes fail to complete cleavage, and another 10 percent fail to reach the fifth month of gestation.)

If the abnormal allele is dominant but does not affect gestational survival, the individual's phenotype will show the effects of the mutation. If the abnormal allele is recessive and is on an autosomal chromosome, it will not affect the individual's phenotype as long as the zygote contains a normal allele contributed by the other parent at fertilization. Over generations, a recessive autosomal allele can spread through the population, remaining undetected until a fertilization occurs in which the two gametes contribute identical recessive alleles. This individual, who will be homozygous for the abnormal allele, will be the first to show the phenotypic effects of the original mutation. Individuals who are heterozygous for an abnormal allele but do not show the effects of the mutation are called **carriers**. Available genetic tests can determine whether an individual is a carrier for any of several autosomal recessive disorders, including Tay–Sachs disease. The information obtained from these tests can be useful in counseling prospective parents. For example, if both parents are carriers of the same disorder, they have a 25 percent probability of producing a child with the disease. This information may affect their decision to conceive.

Sex-Linked Inheritance

Sex-linked inheritance involves genes on the sex chromosomes. Unlike the other 22 chromosomal pairs, the sex chromosomes are not identical in appearance and gene content. There are two types of sex chromosomes: an **X chromosome** and a **Y chromosome**. X chromosomes are considerably larger and have more genes than do Y chromosomes. The smaller Y chromosome carries the *SRY gene* not found on the X chromosome. It specifies that an individual with that chromosome will be male. The normal pair of sex chromosomes in males is XY. Females do not have a Y chromosome. Their sex chromosome pair is XX.

All oocytes carry an X chromosome, because the only sex chromosomes females have are X chromosomes. But each sperm carries either an X or a Y chromosome, because males have one of each and can pass along either one during meiosis. As a Punnett square shows, the ratio of males to females in offspring should be 1:1. The birth statistics differ slightly from that prediction, with 106 males born for every 100 females. It has been suggested that more males are born because a sperm that carries the Y chromosome can reach the oocyte first, because that sperm does not have to carry the extra weight of the larger X chromosome.

The X chromosome also carries genes that affect somatic structures. Genes that are found on the X chromosome but not on the Y chromosome are called **X-linked**. The inheritance of characteristics, or traits, regulated by these X-linked genes does not follow the pattern of alleles on autosomal chromosomes.

(The SRY gene that determines "maleness" is an example of a Y-linked gene.)

The inheritance of color blindness shows the differences between sex-linked inheritance and autosomal inheritance. The presence of a dominant allele, *C*, on the X chromosome (X^C) results in normal color vision. A recessive allele, *c*, on the X chromosome (X^c) results in red–green color blindness. A woman, with her two (C) chromosomes, can be either homozygous dominant ($X^C X^C$) or heterozygous ($X^C X^c$) and still have normal color vision. She will be unable to distinguish reds from greens only if she carries two recessive alleles, $X^c X^c$. But a male has only one X chromosome, so whichever allele that chromosome carries determines whether he has normal color vision or has red–green color blindness. The Punnett square in **Figure 29–18** reveals that the sons produced by a father with normal vision and a heterozygous (carrier) mother have a 50 percent chance of being red–green color blind. Any daughters have normal color vision. Recessive alleles on X chromosomes produce genetic disorders in males at a higher frequency than in females.

A number of other clinical disorders noted earlier in the text are X-linked traits, including certain forms of hemophilia, diabetes insipidus, and muscular dystrophy. In several instances, advances in molecular genetics techniques have enabled geneticists to locate the specific genes on the X chromosome. These techniques provide a reasonably direct method of screening for the presence of a particular condition before any signs or symptoms appear, and even before birth.

Figure 29–18   Inheritance of an X-Linked Trait.

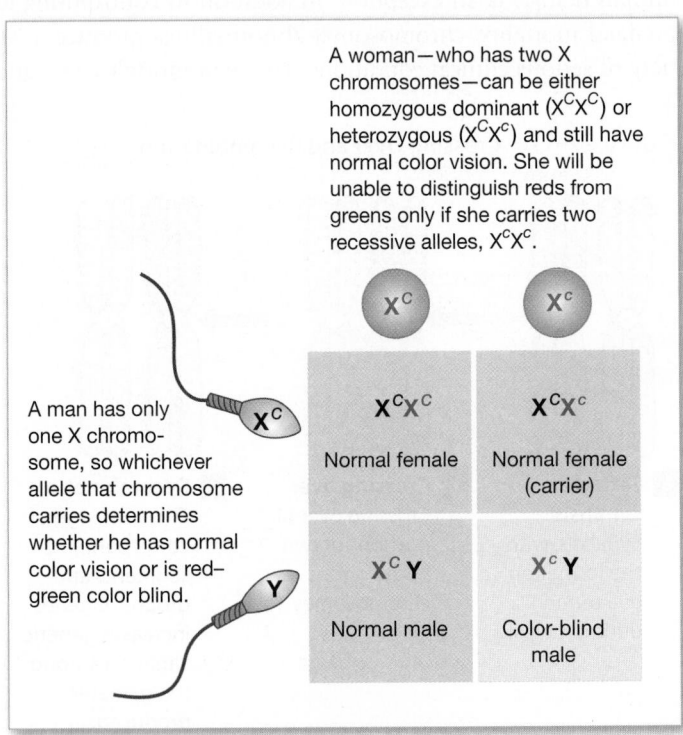

A woman—who has two X chromosomes—can be either homozygous dominant ($X^C X^C$) or heterozygous ($X^C X^c$) and still have normal color vision. She will be unable to distinguish reds from greens only if she carries two recessive alleles, $X^c X^c$.

A man has only one X chromosome, so whichever allele that chromosome carries determines whether he has normal color vision or is red–green color blind.

	X^C	X^c
X^C	$X^C X^C$ — Normal female	$X^C X^c$ — Normal female (carrier)
Y	$X^C Y$ — Normal male	$X^c Y$ — Color-blind male

The Human Genome Project and Beyond

The **Human Genome Project (HGP)** was completed about a decade ago. Its goal was to describe the **human genome**—the full set of genetic material (DNA) in our chromosomes.

One of its highlights was the expanded description of roughly 10,000 different single gene disorders. Most are very rare, but collectively they may affect 1 in every 200 births. Many of these disorders have been mapped to individual chromosomes. Several examples are included in **Figure 29–19**. Genetic screening and diagnostic tests for abnormal genes are now performed for many of these disorders.

The HGP sequenced the human genome's 3.2 billion base pairs nucleotide by nucleotide, and estimated that it contained 20,000–25,000 protein-coding genes. These genes accounted for only about 2 percent of the entire genome, which led to the remaining 98 percent of the genome to be assumed nonfunctional, or "junk."

It now turns out that our genome is not full of useless DNA sequences. Results of the recently completed, decade-long **Encyclopedia of DNA Elements (ENCODE)** project have revealed that some 80 percent of the genome is active in one way or another in gene regulation. For example, protein-coding genes are regulated by not only nearby DNA, but also by more distant sequences and by "non-coding" RNA. Non-coding RNA consists of short and long RNA strands that are not translated into protein. Some 18,400 non-coding RNA "genes" have so far been identified. In addition, ENCODE has documented a total

Figure 29–19 **A Map of Human Chromosomes.** The banding patterns of typical chromosomes in a male, and the locations of the genes associated with specific inherited disorders. The chromosomes are not drawn to scale.

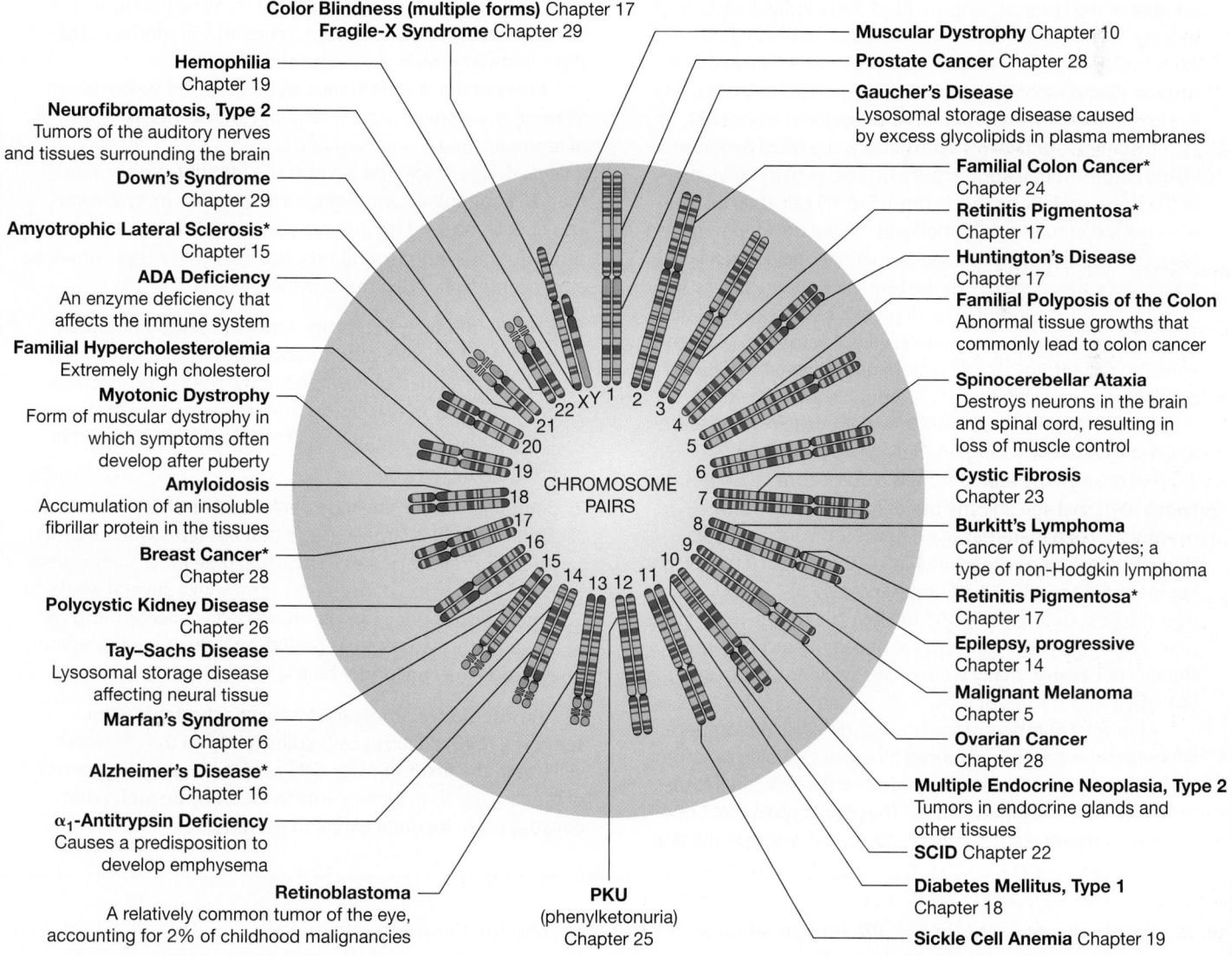

* One form of the disease

Clinical Note

Chromosomal Abnormalities and Genetic Analysis

Embryos that have abnormal autosomal chromosomes rarely survive. However, *translocation defects* and *trisomy* are two types of autosomal chromosome abnormalities that do not invariably result in prenatal death.

In a **translocation defect**, an exchange occurs between different (nonhomologous) chromosome pairs. For example, a piece of chromosome 8 may become attached to chromosome 14. The genes moved to their new position may function abnormally, becoming inactive or overactive. In a balanced translocation, where there is no net loss or gain of chromosomal material, embryos may survive.

In **trisomy**, a mistake occurs in meiosis. One of the gametes involved in fertilization carries an extra copy of one chromosome, so the zygote then has three copies of this chromosome rather than two. (The nature of the trisomy is indicated by the number of the chromosome involved. Thus, individuals with trisomy 13 have three copies of chromosome 13.) Zygotes with extra copies of chromosomes seldom survive. Individuals with trisomy 13 and trisomy 18 may survive until delivery but rarely live longer than a year. The notable exception is trisomy 21.

Trisomy 21, or **Down's syndrome**, is the most common viable chromosomal abnormality. Estimates of its incidence in the U.S. population range from 1.5 to 1.9 per 1000 births. Affected individuals exhibit delayed mental development and have characteristic physical malformations, including a facial appearance that gave rise to the term *mongolism*, once used to describe this condition. The degree of developmental disability ranges from moderate to severe. Anatomical problems affecting the cardiovascular system often prove fatal during childhood or early adulthood. Although some individuals survive to moderate old age, many develop Alzheimer's disease while still relatively young (before age 40).

For unknown reasons, there is a direct correlation between maternal age and the risk of having a child with trisomy 21. For a maternal age below 25, the incidence of Down's syndrome approaches 1 in 2000 births, or 0.05 percent. For maternal ages 30–34, the odds increase to 1 in 900, and over the next decade they go from 1 in 290 to 1 in 46, or more than 2 percent. These statistics are becoming increasingly significant because many women are delaying childbearing until their mid-thirties or later.

Abnormal numbers of sex chromosomes do not produce effects as severe as those induced by extra or missing autosomal chromosomes. In **Klinefelter's syndrome**, the individual carries the sex chromosome pattern XXY. The phenotype is male, but the extra X chromosome causes reduced androgen production.

As a result, the testes fail to mature so the individuals are sterile, and the breasts are slightly enlarged. The incidence of this condition among newborn males averages 1 in 750 births.

Individuals with **Turner's syndrome** have only a single, female sex chromosome; their sex chromosome complement is abbreviated XO. This kind of chromosomal deletion is known as **monosomy**. The incidence of this condition at delivery has been estimated as 1 in 10,000 live births. The condition may not be recognized at birth, because the phenotype is normal female. But maturational changes do not appear at puberty. The ovaries are nonfunctional, and estrogen production occurs at negligible levels.

Fragile-X syndrome causes developmental disability, abnormal facial development, and enlarged testes in affected males. The cause is an abnormal X chromosome that contains a *genetic stutter*, an abnormal repetition of a single nucleotide triplet. The presence of the stutter in some way disrupts the normal functioning of adjacent genes and so produces the signs and symptoms of the disorder.

Many of these conditions can be detected before birth through the analysis of fetal cells. In **amniocentesis**, a sample of amniotic fluid is removed and the fetal cells it contains are analyzed. This procedure permits the identification of more than 20 congenital conditions, including Down's syndrome. The needle inserted to obtain a fluid sample is guided into position during an ultrasound procedure. p. 47 Unfortunately, amniocentesis has two major drawbacks:

1. Because the sampling procedure represents a potential threat to the health of fetus and mother alike, amniocentesis is performed only when known risk factors are present. Examples of risk factors are a family history of specific conditions, or in the case of Down's syndrome, maternal age over 35.

2. Sampling cannot safely be performed until the volume of amniotic fluid is large enough that the fetus will not be injured during the process. The usual time for amniocentesis is at 14–15 weeks of gestation. It may take several weeks to obtain results once samples have been collected, and by the time the results are received, an induced or therapeutic abortion may no longer be a viable option.

An alternative procedure known as **chorionic villus sampling (CVS)** analyzes cells collected from the chorionic villi late in the first trimester. CVS carries a slightly higher risk of miscarriage than amniocentesis, but may be preferable because it can be done earlier in gestation.

---

of 20,687 protein-coding genes in the human genome, with perhaps another 50 or so to be determined.

Epigenetics is the study of inherited traits that are *not* due to changes in a person's genotype or DNA sequences. Similar

to genomic imprinting discussed on p. 1153, it acts at a level "above" the DNA by activating or inactivating specific genes. Such gene regulation occurs in a number of ways. One way is by chemically marking DNA and its histone proteins with methyl or acetyl groups, which permit or restrict transcription, respectively. Non-coding RNAs are another way gene expression can be modified.

Understanding the role of epigenetics in human clinical disorders, disease, and early development is an area of active study. Researchers are piecing together the human **epigenome**, that is, all the chemicals that mark our genome and affect its activities. Although the epigenetic modifiers are not part of the DNA, they can be passed on and inherited when cells divide.

The HGP and ENCODE have identified the normal genetic composition of a "typical" human and begun to reveal how our genes are regulated. Yet we all are variations on a basic theme. How do we decide what set of genes to accept as "normal"? Moreover, as we improve our abilities to alter our own genetic foundations, we will face many additional troubling ethical and legal dilemmas. Few people, for example, object to the insertion of a "correct" gene into somatic cells to cure a specific disease. But what if we could insert that modified gene into a gamete and change not only that individual, but all of his or her descendants as well? And what if the goal of manipulating the gene was not to correct or prevent any disorder, but instead to "improve" the individual by increasing his or her intelligence, height, or vision, or by altering some other phenotypic characteristic? Such difficult questions will not go away. In the years to come, we will have to find answers that are acceptable to us all.

CHECKPOINT

22. Describe the relationship between genotype and phenotype.

23. Define heterozygous.

24. Tongue rolling is an autosomal dominant trait. What would be the phenotype of a person who is heterozygous for this trait?

25. Why are children not identical copies of their parents?

See the blue Answers tab at the back of the book.

Related Clinical Terms

eclampsia: A condition in which one or more convulsions occur in a pregnant woman suffering from high blood pressure, often followed by coma and posing a threat to the health of mother and baby.

gamete intrafallopian transfer (GIFT): An assisted reproductive procedure in which a woman's eggs are removed, mixed with sperm, and replaced into the woman's uterine tube where the fertilization takes place, rather than in the laboratory.

infertility: The inability to achieve pregnancy after engaging in one year of appropriately timed intercourse.

in vitro fertilization: Fertilization outside the body, generally in a Petri dish.

neural tube defects (NTDs): Major birth defects caused by an abnormal development of the neural tube—the structure present during the embryonic stage that later becomes the central nervous system. These are very common birth defects that cause infant mortality and disability and include anencephaly and spina bifida.

placenta previa: Condition during pregnancy in which the placenta is abnormally placed so as to totally or partially cover the cervix.

placenta abruptio: Condition in which there is separation of the placenta from the uterine site of implantation before delivery of the baby.

preeclampsia: A condition in pregnancy characterized by sudden hypertension, albuminuria, and edema of the hands, feet, and face. It is the most common complication of pregnancy, affecting about 5 percent of pregnancies.

therapeutic cloning: A procedure that usually takes skin cells from a patient, and inserts a skin cell nucleus into a fertilized egg whose nucleus has been removed to create a new cell. That new cell divides repeatedly to form a blastocyst from which stem cells can be extracted to grow new tissue that is genetically matched to the patient.

Chapter Review

Study Outline

MasteringA&P®

Access more chapter study tools online in the MasteringA&P Study Area:

- Chapter Quizzes, Chapter Practice Test, Art-labeling Activities, Animations, MP3 Tutor Sessions, and Clinical Case Studies
- Practice Anatomy Lab   PAL™
- Interactive Physiology   iP®
- A&P Flix   *A&PFlix*
- PhysioEx   PhysioEx™

29-1 ▌ Development, marked by various stages, is a continuous process that occurs from fertilization to maturity p. 1122

1. **Development** is the gradual modification of anatomical structures and physiological characteristics from **conception** to maturity. The formation of different types of cells is **differentiation**.

2. **Prenatal development** occurs before birth; **postnatal development** begins at birth and continues to **maturity**, when aging begins. **Inheritance** is the transfer of genetically determined characteristics from generation to generation. **Genetics** is the study of the mechanisms of inheritance.

29

29-2 | Fertilization—the fusion of a secondary oocyte and a spermatozoon—forms a zygote p. 1123

3. **Fertilization**, or *conception*, normally occurs in the uterine tube within a day after ovulation. Spermatozoa cannot fertilize a secondary oocyte until they have undergone *capacitation*. (*Figure 29–1*)

4. The acrosomes of the spermatozoa release **hyaluronidase** and **acrosin**, enzymes required to penetrate the corona radiata and zona pellucida of the oocyte. When a single spermatozoon contacts the oocyte membrane, fertilization begins and **oocyte activation** follows. (*Figure 29–1*)

5. During activation, the oocyte completes meiosis II and thus becomes a functionally mature ovum. **Polyspermy** is prevented by membrane depolarization and the *cortical reaction*.

6. After activation, the **female pronucleus** and the **male pronucleus** fuse in a process called *amphimixis*. (*Figure 29–1*)

29-3 | Gestation consists of three stages of prenatal development: the first, second, and third trimesters p. 1125

7. During prenatal development, differences in the cytoplasmic composition of individual cells trigger changes in genetic activity. The chemical interplay among developing cells is **induction**.

8. The nine-month **gestation** period can be divided into three **trimesters**.

29-4 | Cleavage, implantation, placentation, and embryogenesis are critical events of the first trimester p. 1126

9. In the **first trimester**, **cleavage** subdivides the cytoplasm of the zygote in a series of mitotic divisions; the zygote becomes a **pre-embryo** and then a **blastocyst**. During **implantation**, the blastocyst becomes enclosed within the uterine endometrium. **Placentation** occurs as blood vessels form around the blastocyst and the **placenta** develops. **Embryogenesis** is the formation of a viable embryo.

10. The blastocyst consists of an outer **trophoblast** and an **inner cell mass**. (*Figure 29–2*)

11. Implantation occurs about seven days after fertilization as the blastocyst adheres to the uterine lining. (*Figure 29–3*)

12. As the trophoblast enlarges and spreads, maternal blood flows through open **lacunae**. After **gastrulation**, there is an **embryonic disc** composed of **endoderm**, **ectoderm**, and an intervening **mesoderm**. It is from these **germ layers** that the body systems differentiate. (*Figure 29–4; Table 29–1*)

13. Germ layers are involved in forming four **extraembryonic membranes**: the yolk sac, amnion, allantois, and chorion. (*Spotlight Figure 29–5*)

14. The **yolk sac** is an important site of blood cell formation. The **amnion** encloses fluid that surrounds and cushions the developing embryo. The base of the **allantois** later gives rise to the urinary bladder. Circulation within the vessels of the **chorion** provides a rapid-transit system that links the embryo with the trophoblast. (*Spotlight Figure 29–5, Figure 29–6*)

15. **Chorionic villi** extend outward into the maternal tissues, forming an intricate, branching network through which maternal blood flows. As development proceeds, the **umbilical cord** connects the fetus to the placenta. The syncytial trophoblast synthesizes **human chorionic gonadotropin (hCG)**, estrogens, progesterone, **human placental lactogen (hPL)**, **placental prolactin,** and **relaxin.** (*Figure 29–6*)

16. The first trimester is critical, because events in the first 12 weeks establish the basis for **organogenesis** (organ formation). (*Figure 29–7; Table 29–2*)

29-5 | During the second and third trimesters, maternal organ systems support the developing fetus, and the uterus undergoes structural and functional changes p. 1138

17. In the **second trimester**, the organ systems increase in complexity. During the **third trimester**, many of the organ systems become fully functional. (*Figure 29–8; Table 29–2*)

18. The fetus undergoes its largest weight gain in the third trimester. At the end of gestation, the fetus and the enlarged uterus displace many of the mother's abdominal organs. (*Figure 29–9*)

19. The developing fetus is totally dependent on maternal organs for nourishment, respiration, and waste removal. Maternal adaptations include increases in respiratory rate, tidal volume, blood volume, nutrient and vitamin intake, and glomerular filtration rate, as well as changes in the size of the uterus and mammary glands.

20. Progesterone produced by the placenta has an inhibitory effect on uterine muscles. Estrogens, oxytocin, and prostaglandins oppose its calming action. At some point, multiple factors interact to produce **labor contractions** in the uterine wall. (*Figure 29–10*)

29-6 | Labor consists of the dilation, expulsion, and placental stages p. 1141

21. The goal of **true labor** is **parturition**, the forcible expulsion of the fetus.

22. Labor can be divided into three stages: the **dilation stage**, the **expulsion stage**, and the **placental stage**. The Apgar score is used to assess the overall health of a newborn. (*Figure 29–11*)

23. **Premature labor** may result in **premature delivery**.

24. Difficult deliveries can include *forceps deliveries* and **breech births**—deliveries in which the legs or buttocks of the fetus, rather than the head, enter the vaginal canal first.

25. Twins are either **dizygotic** (fraternal) **twins** or **monozygotic** (identical) **twins**.

29-7 | Postnatal stages are the neonatal period, infancy, childhood, adolescence, and maturity, followed by senescence p. 1144

26. Postnatal development involves a series of five **life stages**: the neonatal period, infancy, childhood, adolescence, and maturity. *Senescence*, or aging, begins at maturity and ends in the death of the individual.

27. The **neonatal period** extends from birth to one month. In the transition from fetus to **neonate**, the respiratory, circulatory, digestive, and urinary systems of the infant begin functioning independently. The newborn must also begin thermoregulation.

28. Mammary gland cells produce protein-rich **colostrum** during the neonate's first few days of life and then convert to milk production. These secretions are released as a result of the **milk let-down (milk-ejection) reflex**. (*Figure 29–12*)

29. Body proportions gradually change during **infancy** (from age 1 month to 2 years) and during **childhood** (from age 2 years to puberty). (*Figure 29–13*)

30. **Adolescence** begins at **puberty**, when (1) the hypothalamus increases its production of GnRH, (2) circulating levels of FSH and LH rise rapidly, and (3) ovarian or testicular cells become more sensitive to FSH and LH. These changes

29

initiate gamete formation, the production of sex hormones, and a sudden increase in the growth rate. The hormonal changes at puberty, especially changes in sex hormone levels, produce sex-specific differences in the structure and function of many systems; these differences will be retained. Adolescence continues until growth is completed. Further changes occur when sex hormone levels decline at menopause or the male climacteric.

31. **Senescence** then begins, producing gradual reductions in the functional capabilities of all systems. *(Table 29–3)*

29-8 ▌ Genes and chromosomes determine patterns of inheritance p. 1148

32. Every somatic cell carries copies of the original 46 chromosomes in the zygote; these chromosomes and their component genes constitute the individual's **genotype**. The physical expression of the genotype is the individual's **phenotype.**

33. Every somatic human cell contains 23 pairs of chromosomes; each pair consists of **homologous chromosomes.** Twenty-two pairs are **autosomal chromosomes.** The chromosomes of the twenty-third pair are the **sex chromosomes**; they differ between the sexes. *(Figure 29–14)*

34. Chromosomes contain DNA, and genes are functional segments of DNA. The various forms of a given gene are called **alleles.** If both homologous chromosomes carry the same allele of a particular gene, the individual is **homozygous**; if they carry different alleles, the individual is **heterozygous.**

35. In **simple inheritance**, phenotypic traits are determined by interactions between a single pair of alleles. **Polygenic inheritance** involves interactions among alleles on several genes. *(Figure 29–15)*

36. Alleles are either **dominant** or **recessive**, depending on how their traits are expressed.

37. Combining maternal and paternal alleles in a **Punnett square** helps us predict the characteristics of offspring. *(Figure 29–16)*

38. **Genetic recombination**, the gene reshuffling in **crossing over** and **translocation** during meiosis, increases the genetic variation of male and female gametes. *(Figure 29–17)*

39. **Spontaneous mutations** are the result of random errors in DNA replication. Such mutations can cause the production of abnormal alleles.

40. The two types of sex chromosomes are an **X chromosome** and a **Y chromosome**. The normal sex chromosome complement of males is XY; that of females is XX. **Sex-linked inheritance** involves genes on the sex chromosomes. The X chromosome carries **X-linked genes**, which affect somatic structures but have no corresponding alleles on the Y chromosome. *(Figure 29–18)*

41. Together, the **HGP** and **ENCODE** project have revealed that the human genome contains protein-coding genes, DNA sequences for non-coding RNAs, and epigenetic markings affecting gene expression. Close to 22,000 protein-coding genes have been mapped, including some of those responsible for inherited disorders. *(Figure 29–19)*

Review Questions

See the blue Answers tab at the back of the book.

LEVEL 1 Reviewing Facts and Terms

1. The chorionic villi **(a)** form the umbilical cord, **(b)** form the umbilical vein, **(c)** form the umbilical arteries, **(d)** increase the surface area available for exchange between the placenta and maternal blood, **(e)** form the portion of the placenta called the decidua capsularis.

2. Identify the two extraembryonic membranes and the three different regions of the endometrium at week 10 of development in the following diagram.

(a) _____   (b) _____
(c) _____   (d) _____
(e) _____

3. The hormone that is the basis for a pregnancy test is **(a)** LH, **(b)** progesterone, **(c)** human chorionic gonadotropin (hCG), **(d)** human placental lactogen (hPL), **(e)** either c or d, depending on the type of test.

4. Recessive X-linked traits **(a)** are passed from fathers to their sons, **(b)** are more likely to be expressed in males, **(c)** always affect some aspect of the reproductive system, **(d)** are never expressed in females, **(e)** cannot be passed from mothers to daughters.

5. Describe the structure of the blastocyst, and state what the inner cell mass goes on to form.

6. The pre-embryo develops into a multicellular complex known as a _____. **(a)** blastocyst, **(b)** trophoblast, **(c)** lacuna, **(d)** blastomere.

7. The structure(s) that allow(s) active and passive exchange between the fetal and maternal bloodstreams is/are the **(a)** yolk stalk, **(b)** chorionic villi, **(c)** umbilical veins, **(d)** umbilical arteries.

8. If an allele must be present on both the maternal and paternal chromosomes to affect the phenotype, the allele is said to be **(a)** dominant, **(b)** recessive, **(c)** complementary, **(d)** heterozygous.

9. Describe the changes that occur in the oocyte immediately after fertilization.

10. What are the three germ layers in the embryo?

11. Identify the three stages of labor, and describe the events that characterize each stage.

12. List the factors involved in initiating labor contractions.

13. Identify the three life stages that occur between birth and approximately age 10. Describe the timing and characteristics of each stage.

14. What hormonal events are responsible for puberty? Which life stage does puberty initiate?

29

LEVEL 2 Reviewing Concepts

15. A normally pigmented woman whose father was an albino marries a normally pigmented man whose mother was an albino. What is the probability that they would have an albino child? **(a)** 50 percent, **(b)** 25 percent, **(c)** 12.5 percent, **(d)** 6.25 percent, **(e)** 100 percent.

16. If a sperm cell lacked sufficient quantities of hyaluronidase, it would *not* be able to **(a)** move its flagellum, **(b)** penetrate the corona radiata, **(c)** become capacitated, **(d)** survive the environment of the female reproductive tract, **(e)** metabolize fructose.

17. Problems involving the formation of the chorion would affect **(a)** the embryo's ability to produce blood cells, **(b)** the formation of limbs, **(c)** the embryo's ability to derive nutrition from the mother, **(d)** lung formation, **(e)** the urinary system.

18. What are the major physiological changes that happen to the mother's system during pregnancy?

19. Name the primary placental hormones and their functions.

20. During birth, which hormone stimulates uterine contraction?

21. During true labor, what physiological mechanisms ensure that uterine contractions continue until delivery has been completed?

22. How long after birth does the placenta get expelled?

23. Distinguish between the following paired terms: **(a)** genotype and phenotype, **(b)** heterozygous and homozygous, **(c)** simple inheritance and polygenic inheritance.

24. Indicate the type of inheritance involved in each of the following situations. **(a)** Children who exhibit the trait have at least one parent who also exhibits it, **(b)** Children exhibit the trait even though neither parent does, **(c)** The trait is expressed more commonly in sons than in daughters, **(d)** The trait is expressed equally in daughters and sons.

LEVEL 3 Critical Thinking and Clinical Applications

25. Hemophilia A, a condition in which blood does not clot properly, is a recessive trait located on the X chromosome (X^h). Suppose that a woman who is heterozygous for this trait (XX^h) has children with a normal male (XY). What is the probability that the couple will have hemophiliac daughters? What is the probability that the couple will have hemophiliac sons?

26. Joe and Jane desperately want to have children, and although they have tried for two years, they have not been successful. Finally, each of them consults a physician, and it turns out that Joe suffers from oligospermia (a low sperm count). He confides to you that he doesn't understand why this would interfere with his ability to have children since he remembers from biology class that it only takes one sperm to fertilize an egg. What would you tell him?

27. Cathy has just given birth to a little girl. When the nurses take the infant back to the nursery, she becomes cyanotic, a condition characterized by bluish discoloration of the skin. The episode passes, but when the infant is bathed, she becomes cyanotic again. Blood gas levels indicate that arterial blood is only 60 percent saturated. Physical examination reveals no structural deformities involving the respiratory or digestive system. What might be causing the problem?

28. Sally gives birth to a baby with a congenital deformity of the stomach. Sally believes that her baby's affliction is the result of a viral infection she suffered during her third trimester. Is this a possibility? Explain.

CLINICAL CASE Wrap-Up

The Twins That Looked Nothing Alike

The winners of the "least-alike" twins prize are named Hans and Willem. These charming boys traveled all the way from the Netherlands to win this award in Twinsburg.

Hans and Willem's parents used the services of a prestigious fertility clinic in the Netherlands in the 1990s. Their mother's ovaries were hormonally stimulated, and the secondary oocytes that were released were harvested laparoscopically (by abdominal surgery through small holes). These oocytes were then fertilized with sperm in the laboratory. On the sixth post-fertilization day, coinciding with the time a normally fertilized blastocyst would be finishing its journey through the uterine tube into the uterus, two blastocysts were transferred to the mother's uterus. Both had a successful implantation, placentation, embryogenesis, fetal development, and birth.

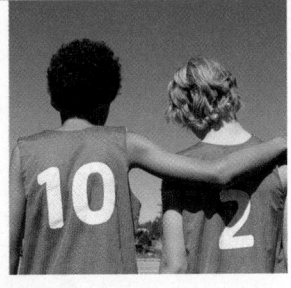

Why, then, do Hans and Willem look so different? As it turned out, the technician performing the in vitro fertilization mistakenly used a previously-used pipette to transfer the sperm. The couple that visited the clinic just before Hans and Willem's parents were of African descent, and the African man's sperm was still present in the pipette. As a result, one oocyte was fertilized by sperm from a Caucasian man, and the other was fertilized by sperm from an African man.

1. Why were the fertilized blastocysts implanted in Hans and Willem's mother on her sixth day after ovulation?

2. Are Hans and Willem monozygotic twins or dizygotic twins?

See the blue Answers tab at the back of the book.

29

Appendix

Normal Physiological Values

Tables 1 and 2 present normal averages or ranges for the chemical composition of body fluids. These values are approximations rather than absolute values, because test results vary from laboratory to laboratory due to differences in procedures, equipment, normal solutions, and so forth. Blanks in the tabular data appear where data are not available. The following locations in the text contain additional information about body fluid analysis:

Table 19–3 (p. 698) presents data on the formed elements of whole blood.

Table 26–2 (p. 1010) compares the normal values for solutes in plasma and urine.

Table 26–5 (p. 1027) gives the general characteristics of normal urine, and Table 26–2 (p. 1030) shows the typical values of substances in a standard urinalysis.

Table 1	The Composition of Minor Body Fluids						
	Normal Averages or Ranges						
Test	**Perilymph**	**Endolymph**	**Synovial Fluid**	**Sweat**	**Saliva**	**Semen**
pH			7.4	4–6.8	6.4*	7.19
Specific gravity			1.008–1.015	1.001–1.008	1.007	1.028
Electrolytes (mEq/L)						
Potassium	5.5–6.3	140–160	4.0	4.3–14.2	21	31.3
Sodium	143–150	12–16	136.1	0–104	14*	117
Calcium	1.3–1.6	0.05	2.3–4.7	0.2–6	3	12.4
Magnesium	1.7	0.02		0.03–4	0.6	11.5
Bicarbonate	17.8–18.6	20.4–21.4	19.3–30.6		6*	24
Chloride	121.5	107.1	107.1	34.3	17	42.8
Proteins (total) (mg/dL)	200	150	1.72 g/dL	7.7	386†	4.5 g/dL
Metabolites (mg/dL)						
Amino acids				47.6	40	1.26 g/dL
Glucose	104		70–110	3.0	11	224 (fructose)
Urea				26–122	20	72
Lipids (total)	12		20.9	‡	25–500§	188

*Increases under salivary stimulation.

†Primarily alpha-amylase, with some lysozymes.

‡Not present in merocrine (eccrine) secretions.

§Cholesterol.

Table 2 The Chemistry of Blood, Cerebrospinal Fluid (CSF), and Urine

Test	Blood*	CSF	Urine
		Normal Averages or Ranges	
pH	S: 7.35–7.45	7.31–7.34	4.5–8.0
Osmolarity (mOsm/L)	S: 280–295	292–297	855–1335
Electrolytes	|————————(mEq/L unless noted)————————|		(urinary loss, mEq per 24-hour period[†])
Bicarbonate	P: 20–28	20–24	0
Calcium	S: 8.5–10.3	2.1–3.0	6.5–16.5
Chloride	P: 97–107	100–108	110–250
Iron	S: 50–150 µg/L	23–52 µg/L	40–150 µg
Magnesium	S: 1.4–2.1	2–2.5	6.0–10.0
Phosphorus	S: 1.8–2.9	1.2–2.0	0.4–1.3 g
Potassium	P: 3.5–5.0	2.7–3.9	25–125
Sodium	P: 135–145	137–145	40–220
Sulfate	S: 0.2–1.3		1.07–1.3 g
Metabolites	|————————(mg/dL unless noted)————————|		(urinary loss, mg per 24-hour period[‡])
Amino acids	P/S: 2.3–5.0	10.0–14.7	41–133
Ammonia	P: 20–150 µg/dL	25–80 µg/dL	340–1200
Bilirubin	S: 0.5–1.0	<0.2	0
Creatinine	P/S: 0.6–1.5	0.5–1.9	770–1800
Glucose	P/S: 70–110	40–70	0
Ketone bodies	S: 0.3–2.0	1.3–1.6	10–100
Lactic acid	WB: 0.7–2.5 mEq/L[§]	10–20	100–600
Lipids (total)	P: 450–1000	0.8–1.7	0.002
Cholesterol (total)	S: 150–300	0.2–0.8	1.2–3.8
Triglycerides	S: 40–150	0–0.9	0
Urea	P: 8–25	12.0	1800
Uric acid	P: 2.0–6.0	0.2–1.5	250–750
Proteins	(g/dL)	(mg/dL)	(urinary loss, mg per 24-hour period[‡])
Total	P: 6.0–8.0	2.0–4.5	0–8
Albumin	S: 3.2–4.5	10.6–32.4	0–3.5
Globulins (total)	S: 2.3–3.5	2.8–15.5	7.3
Immunoglobulins	S: 1.0–2.2	1.1–1.7	3.1
Fibrinogen	P: 0.2–0.4	0.65	0

*S = serum, P = plasma, WB = whole blood.
[†]Because urinary output averages just over 1 liter per day, these electrolyte values are comparable to mEq/L.
[‡]Because urinary metabolite and protein data approximate mg/L or g/L, these data must be divided by 10 for comparison with CSF or blood concentrations.
[§]Venous blood sample.

The Periodic Table

The periodic table presents the known elements in order of their atomic weights. Each horizontal row represents a single electron shell. The number of elements in that row is determined by the maximum number of electrons that can be stored at that energy level. The element at the left end of each row contains a single electron in its outermost electron shell; the element at the right end of the row has a filled outer electron shell. Organizing the elements in this fashion highlights similarities that reflect the composition of the outer electron shell. These similarities are evident when you examine the vertical columns. All the gases of the right-most column—helium, neon, argon, krypton, xenon, and radon—have full electron shells; each is a gas at normal atmospheric temperature and pressure, and none reacts readily with other elements. These elements, highlighted in blue, are known as the noble, or inert, gases. In contrast, the elements of the left-most column below hydrogen—lithium, sodium, potassium, rubidium, caesium, and francium—are silvery, soft metals that are so highly reactive that pure forms cannot be found in nature. The fourth and fifth electron levels can hold up to 18 electrons. Table inserts are used for the so-called lanthanoids and actinoids to save space, as higher levels can store up to 32 electrons. Elements of particular importance to our discussion of human anatomy and physiology are highlighted in pink.

Legend:

Atomic number →	1	H	← Chemical symbol
		Hydrogen	← Element name
Atomic weight →		1.01	

Periodic Table (main body)

1	2	3	4	5	6	7	8	9	10	11	12	13	14	15	16	17	18
1 H Hydrogen 1.01																	2 He Helium 4.00
3 Li Lithium 6.94	4 Be Beryllium 9.01											5 B Boron 10.81	6 C Carbon 12.01	7 N Nitrogen 14.01	8 O Oxygen 16.00	9 F Fluorine 19.00	10 Ne Neon 20.18
11 Na Sodium 22.99	12 Mg Magnesium 24.31											13 Al Aluminum 26.98	14 Si Silicon 28.09	15 P Phosphorus 30.97	16 S Sulfur 32.07	17 Cl Chlorine 35.45	18 Ar Argon 39.95
19 K Potassium 39.10	20 Ca Calcium 40.08	21 Sc Scandium 44.96	22 Ti Titanium 47.87	23 V Vanadium 50.94	24 Cr Chromium 52.00	25 Mn Manganese 54.94	26 Fe Iron 55.85	27 Co Cobalt 58.93	28 Ni Nickel 58.69	29 Cu Copper 63.55	30 Zn Zinc 65.38	31 Ga Gallium 69.72	32 Ge Germanium 72.63	33 As Arsenic 74.92	34 Se Selenium 78.96	35 Br Bromine 79.90	36 Kr Krypton 83.80
37 Rb Rubidium 85.47	38 Sr Strontium 87.62	39 Y Yttrium 88.91	40 Zr Zirconium 91.22	41 Nb Niobium 92.91	42 Mo Molybdenum 95.96	43 Tc Technetium	44 Ru Ruthenium 101.1	45 Rh Rhodium 102.9	46 Pd Palladium 106.4	47 Ag Silver 107.9	48 Cd Cadmium 112.4	49 In Indium 114.8	50 Sn Tin 118.7	51 Sb Antimony 121.8	52 Te Tellurium 127.6	53 I Iodine 126.9	54 Xe Xenon 131.3
55 Cs Caesium 132.9	56 Ba Barium 137.3	57–71 Lanthanoids	72 Hf Hafnium 178.5	73 Ta Tantalum 180.9	74 W Tungsten 183.8	75 Re Rhenium 186.2	76 Os Osmium 190.2	77 Ir Iridium 192.2	78 Pt Platinum 195.1	79 Au Gold 197.0	80 Hg Mercury 200.6	81 Tl Thallium 204.3	82 Pb Lead 207.2	83 Bi Bismuth 209.0	84 Po Polonium	85 At Astatine	86 Rn Radon
87 Fr Francium	88 Ra Radium	89–103 Actinoids	104 Rf Rutherfordium	105 Db Dubnium	106 Sg Seaborgium	107 Bh Bohrium	108 Hs Hassium	109 Mt Meitnerium	110 Ds Darmstadtium	111 Rg Roentgenium	112 Cn Copernicium	114 Fl Flerovium	116 Lv Livermorium				

Lanthanoids (57–71)

57 La Lanthanum 138.9	58 Ce Cerium 140.1	59 Pr Praseodymium 140.9	60 Nd Neodymium 144.2	61 Pm Promethium	62 Sm Samarium 150.4	63 Eu Europium 152.0	64 Gd Gadolinium 157.3	65 Tb Terbium 158.9	66 Dy Dysprosium 162.5	67 Ho Holmium 164.9	68 Er Erbium 167.3	69 Tm Thulium 168.9	70 Yb Ytterbium 173.1	71 Lu Lutetium 175.0

Actinoids (89–103)

89 Ac Actinium	90 Th Thorium 232.0	91 Pa Protactinium 231.0	92 U Uranium 238.0	93 Np Neptunium	94 Pu Plutonium	95 Am Americium	96 Cm Curium	97 Bk Berkelium	98 Cf Californium	99 Es Einsteinium	100 Fm Fermium	101 Md Mendelevium	102 No Nobelium	103 Lr Lawrencium

Codon Chart

A codon is a sequence of three consecutive nucleotides in mRNA that codes for a particular amino acid or signals to stop protein synthesis. Because each mRNA codon consists of three nucleotides, the four nucleotides in mRNA (A, U, G, and C) can produce 64 different combinations. Of these, 61 codons correspond to amino acids and three act as stop signals of protein synthesis. There are only 20 different amino acids used to synthesize proteins, so most amino acids are represented by more than one codon. One codon, AUG, has a dual role. It codes for the amino acid methionine and also as the start signal for protein synthesis. It is always the first codon in a strand of mRNA.

Genetic Code (mRNA codons)

Second nucleotide

First nucleotide	U	C	A	G	Third nucleotide
U	UUU UUC Phenylalanine / UUA UUG Leucine	UCU UCC UCA UCG Serine	UAU UAC Tyrosine / UAA Stop / UAG Stop	UGU UGC Cysteine / UGA Stop / UGG Tryptophan	U C A G
C	CUU CUC CUA CUG Leucine	CCU CCC CCA CCG Proline	CAU CAC Histidine / CAA CAG Glutamine	CGU CGC CGA CGG Arginine	U C A G
A	AUU AUC Isoleucine / AUA / AUG Start or Met*	ACU ACC ACA ACG Threonine	AAU AAC Asparagine / AAA AAG Lysine	AGU AGC Serine / AGA AGG Arginine	U C A G
G	GUU GUC GUA GUG Valine	GCU GCC GCA GCG Alanine	GAU GAC Aspartic acid / GAA GAG Glutamic acid	GGU GGC GGA GGG Glycine	U C A G

*Abbreviation for methionine

Answers to Checkpoints, Review Questions, and Clinical Case Wrap-Ups

Chapter 1

Answers to Checkpoints

PAGE 29 **1.** Anatomy is the oldest medical science. **2.** Studying human anatomy and physiology is important because understanding normal physiology assists in recognizing when something abnormal occurs within the body. **PAGE 30** **3.** Anatomy is the study of internal and external body structures. **4.** Physiology is the study of how living organisms perform functions. **5.** Medical terminology is the use of prefixes, suffixes, word roots, and combining forms to construct anatomical, physiological, or medical terms. **6.** An eponym is a commemorative name for a structure or clinical condition that was originally named for a real or mythical person. **7.** The book used as the international standard for anatomical vocabulary is *International Anatomical Terminology (Terminologia Anatomica)*. **PAGE 32** **8.** Anatomy and physiology are closely related because all specific functions are performed by specific structures. **9.** Gross anatomy (often referred to as macroscopic anatomy) involves studying body structures that can be seen with the unaided eye. Microscopic anatomy is the study of body structures using a microscope to magnify the objects. **10.** Several specialties of physiology are cell physiology, organ physiology, systemic physiology, and pathological physiology. **11.** It is difficult to separate anatomy from physiology because the structures of body parts are so closely related to their functions; put another way, function follows form. **PAGE 33** **12.** The major levels of organization of the human body from the simplest to the most complex are the following: chemical level → cellular level → tissue level → organ level → organ system level → organism level. **13.** Major organ systems (and some structures of each) are the integumentary system (skin, hair, sweat glands, and nails), skeletal system (bones, cartilages, associated ligaments, and bone marrow), muscular system (skeletal muscles and associated tendons), nervous system (brain, spinal cord, peripheral nerves, and sense organs), endocrine system (pituitary gland, thyroid gland, pancreas, adrenal glands, gonads, and other endocrine tissues), cardiovascular system (heart, blood, and blood vessels), lymphatic system (spleen, thymus, lymphatic vessels, lymph nodes, and tonsils), respiratory system (nasal cavities, sinuses, larynx, trachea, bronchi, lungs, and alveoli), digestive system (teeth, tongue, pharynx, esophagus, stomach, small intestine, large intestine, liver, gallbladder, and pancreas), urinary system (kidneys, ureters, urinary bladder, and urethra), male reproductive system (testes, epididymides, ductus deferentia, seminal glands, prostate gland, penis, and scrotum), and female reproductive system (ovaries, uterine tubes, uterus, vagina, labia, clitoris, and mammary glands). **14.** A histologist investigates structures and properties at the tissue level of organization. **PAGE 36** **15.** Homeostasis refers to the existence of a stable internal environment. **16.** Extrinsic regulation is a type of homeostatic regulation resulting from activities of the nervous system or endocrine system. **17.** Physiological systems can function normally only under carefully controlled conditions. Homeostatic regulation prevents potentially disruptive changes in the body's internal environment. **PAGE 40** **18.** Negative feedback systems provide long-term control over the body's

internal conditions—that is, they maintain homeostasis—by counteracting the effects of a stimulus. **19.** When homeostasis fails, organ systems function less efficiently or even malfunction. The result is the state that we call disease. If the situation is not corrected, death can result. **20.** A positive feedback system amplifies or reinforces the effects of a stimulus. **21.** Positive feedback is useful in processes that must be completed quickly, such as blood clotting. In contrast, it is harmful in situations in which a stable condition must be maintained, because it tends to increase any departure from the desired condition. Positive feedback in the regulation of body temperature, for example, would cause a slight fever to spiral out of control, with fatal results. For this reason, physiological systems are typically regulated by negative feedback, which tends to oppose any departure from the norm. **22.** Equilibrium is a dynamic (constantly changing) state in which two opposing forces or processes are in balance. **23.** When the body continuously adapts, utilizing homeostatic systems, it is said to be in a state of dynamic equilibrium. **PAGE 44** **24.** The purpose of anatomical terms is to provide a standardized frame of reference for describing the human body. **25.** In the anatomical position, an anterior view is seen from the front and a posterior view is from the back. **PAGE 48** **26.** Body cavities protect internal organs and cushion them from thumps and bumps that occur while walking, running, or jumping. Body cavities also permit the organs that they surround to change in size and shape without disrupting the activities of nearby organs. **27.** The thoracic cavity includes the pleural and pericardial cavities, which enclose the lungs and heart respectively. The diaphragm forms the boundary between the superior thoracic cavity and the inferior abdominopelvic cavity. The abdominopelvic cavity contains the peritoneal cavity, which contains the superior abdominal cavity and the inferior pelvic cavity.

Answers to Review Questions

PAGE 50

Level 1 Reviewing Facts and Terms

1. (a) superior (b) inferior (c) posterior or dorsal (d) anterior or ventral (e) cranial (f) caudal (g) lateral (h) medial (i) proximal (j) distal **2.** g **3.** d **4.** a **5.** j **6.** b **7.** l **8.** n **9.** f **10.** h **11.** e **12.** c **13.** o **14.** k **15.** m **16.** i **17.** c **18.** d **19.** d **20.** a **21.** b **22.** (a) pericardial cavity (b) peritoneal cavity (c) pleural cavity (d) abdominal (or abdominopelvic) cavity **23.** A sign is objective whilst a symptom is subjective.

Level 2 Reviewing Concepts

24. (a) Anatomy is the study of internal and external structures and the physical relationships among body parts. (b) Physiology is the study of how organisms perform their vital functions. **25.** Specific functions are performed by specific structures, and the form of a structure relates to its function. **26.** Autoregulation occurs when the activities of a cell, tissue, organ, or organ system change automatically (that is, without neural or endocrine input) when faced with some environmental change. Extrinsic regulation results from the activities of the nervous or endocrine systems. It causes more extensive and potentially more effective adjustments in activities.   **27.** There is a lot of variation within a normal population.   **28.** The nervous system and endocrine system adjust heart rate and blood vessel diameter.   **29.** c

Level 3 Critical Thinking and Clinical Applications

30. Calcitonin is released when when blood calcium levels are elevated. This hormone should bring about a decrease in blood calcium levels, thus decreasing the stimulus for its own release. **31.** A lack of blood flow to the hypothalamus of the brain would cause brain cells there to die. Since this is where the body's thermoregulatory center lies, then control of body temperature would be affected.   **32.** The fluid in the pericardium compresses the heart and prevents filling, reducing output and pressure.

Answers to Clinical Case Wrap-Up

PAGE 51   1. In addition to the liver and most of the large intestine, the abdominal cavity contains the stomach, spleen, and small intestine.   **2.** A deep knife wound superior to the diaphragm would have entered the thoracic cavity, which contains the lungs and heart; associated organs of the respiratory, cardiovascular, and lymphatic systems; the inferior portions of the esophagus; and the thymus.

Chapter 2

Answers to Checkpoints

PAGE 57   1. An atom is the smallest stable unit of matter. **2.** Atoms of the same element that have the same atomic number but different numbers of neutrons are called isotopes.   **3.** Hydrogen has three isotopes: hydrogen-1, with a mass number of 1; deuterium, with a mass number of 2; and tritium, with a mass number of 3. The heavier sample must contain a higher proportion of one or both of the heavier isotopes.   **PAGE 62   4.** A chemical bond is an attractive force acting between two atoms that may be strong enough to hold them together in a molecule or compound. The strongest attractive forces result from the gain, loss, or sharing of electrons. Examples of such chemical bonds are ionic bonds and covalent bonds. In contrast, the weaker hydrogen bonds occur between molecules or compounds.   **5.** The atoms in a water molecule are held together by polar covalent bonds. Water molecules are attracted to one another by hydrogen bonds.   **6.** Atoms combine with each other to gain a complete set of electrons in their outer energy level. Oxygen atoms do not have a full outer energy level, so they readily react with many other elements to attain this stable arrangement. Neon already has a full outer energy level so it cannot combine with other elements.   **PAGE 64   7.** Using chemical notation, an ion is represented by one (or more) superscript plus or minus signs following the symbol of an element.   **8.** The molecular formula for glucose, a compound composed of 6 carbon atoms, 12 hydrogen atoms, and 6 oxygen atoms, is $C_6H_{12}O_6$. **9.** Three types of chemical reactions important to the study of human physiology include decomposition reactions, synthesis reactions, and exchange reactions. In a decomposition reaction, a chemical reaction breaks a molecule into smaller fragments. A synthesis reaction assembles smaller molecules into larger ones. In an exchange reaction, parts of the reacting molecules are shuffled around to produce new products.   **10.** This reaction involves a large molecule being broken down into two smaller ones, so it is a decomposition reaction.   **PAGE 65   11.** An enzyme is a protein that lowers the activation energy of a chemical reaction, which is the amount of energy required to start the reaction.   **12.** Without enzymes, chemical reactions would

proceed only under conditions that the cells could not tolerate (e.g., high temperatures). By lowering the activation energy, enzymes make it possible for chemical reactions to rapidly proceed under conditions compatible with life.   **13.** Organic compounds always contain carbon and hydrogen. Inorganic compounds do not contain carbon and hydrogen atoms as structural ingredients.   **PAGE 67   14.** Specific chemical properties of water that make life possible include its solubility (its strong polarity enables it to be used as an efficient solvent), its reactivity (it participates in many chemical reactions), its high heat capacity (it absorbs and releases heat slowly), and its ability to serve as a lubricant.   **PAGE 68   15.** The pH is a measure of the concentration of hydrogen ions in fluids. Acid and base concentrations are measured in pH, which is the negative logarithm of hydrogen ion concentration, expressed in moles per liter. On the pH scale, 7 represents neutrality; values below 7 indicate acidic solutions, and values above 7 indicate basic (alkaline) solutions.   **16.** If the body is to maintain homeostasis and thus health, the pH of different body fluids must remain within a fairly narrow range.   **PAGE 69   17.** An acid is a compound whose dissociation in solution releases a hydrogen ion and an anion; a base is a compound whose dissociation releases a hydroxide ion (OH^-) or removes a hydrogen ion (H^+) from the solution; a salt is an inorganic compound consisting of a cation other than H^+ and an anion other than OH^-.   **18.** Stomach discomfort is commonly the result of excessive stomach acidity ("acid indigestion"). Antacids contain a weak base that neutralizes the excess acid.   **PAGE 72   19.** A compound with a C:H:O ratio of 1:2:1 is a carbohydrate. The body uses carbohydrates mainly as an energy source.   **PAGE 77   20.** Lipids are a diverse group of compounds that include fatty acids, eicosanoids, glycerides, steroids, phospholipids, and glycolipids. They are organic compounds that contain carbon, hydrogen, and oxygen in a ratio that does not approximate 1:2:1.   **21.** Human plasma membranes contain mainly phospholipids, plus small amounts of cholesterol and glycolipids.   **PAGE 82   22.** Proteins are organic compounds formed from amino acids, which contain a central carbon atom, a hydrogen atom, an amino group ($-NH_2$), a carboxyl group ($-COOH_2$), and an R group or variable side chain. Proteins function in support, movement, transport, buffering, metabolic regulation, coordination and control, and defense.   **23.** The heat of boiling breaks bonds that maintain the protein's tertiary structure, quaternary structure, or both. The resulting structural change, known as denaturation, affects the ability of the protein molecule to perform its normal biological functions.   **PAGE 83   24.** A nucleic acid is a large organic molecule made of carbon, hydrogen, oxygen, nitrogen, and phosphorus. Nucleic acids regulate protein synthesis and make up the genetic material in cells.   **25.** Both DNA (deoxyribonucleic acid) and RNA (ribonucleic acid) contain nitrogenous bases and phosphate groups, but the nucleic acid described is RNA because it contains the sugar *ribose*.   **PAGE 85   26.** Adenosine triphosphate (ATP) is a high-energy compound consisting of adenosine to which three phosphate groups are attached. The second and third phosphate groups are each attached by a high-energy bond.   **27.** Phosphorylation of an ADP molecule yields a molecule of ATP.   **PAGE 86   28.** The biochemical building blocks that are components of cells include lipids (forming the plasma membrane), proteins (acting as enzymes), nucleic acids (directing the synthesis of cellular proteins), and carbohydrates (providing energy for cellular activities).   **29.** Metabolic turnover is the continuous breakdown and replacement of organic materials within cells.

Answers to Review Questions

PAGE 88
Level 1 Reviewing Facts and Terms
1. a.

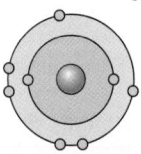

Oxygen atom

b. Two more electrons can fit into the outermost energy level of an oxygen atom. **2.** d **3.** c **4.** b **5.** a **6.** d **7.** b **8.** c **9.** d **10.** d **11.** b **12.** d **13.** a **14.** d **15.** b **16.** Solubility, reactivity, high heat capacity and lubrication **17.** carbohydrates, lipids, proteins, and nucleic acids **18.** Triglycerides (1) provide a significant energy reserve, (2) serve as insulation and thus act in heat conservation, and (3) protect organs by cushioning them. **19.** Primary, secondary, tertiary and quaternary **20.** (a) DNA: deoxyribose, phosphate, and nitrogenous bases (A, T, C, G); (b) RNA: ribose, phosphate, and nitrogenous bases (A, U, C, G) **21.** (1) adenosine, (2) phosphate groups, and (3) appropriate enzymes

Level 2 Reviewing Concepts
22. c **23.** d **24.** Saturation occurs when substrate levels allow all enzymes to cycle through the reaction sequence at maximum speed. Further increases in substrate level would not increase the rate of re-action, but increasing enzyme level will. **25.** High-energy bonds are covalent bonds whose breakdown releases energy. Phosphorylation of ADP using ATPase stores energy as ATP. **26.** Organic molecules are regularly replaced as part of metabolic turnover so that cells can change their chemical organization in response to internal and external stimuli. **27.** e **28.** c **29.** The molecule is a nucleic acid. Carbohydrates and lipids do not contain nitrogen. Although both proteins and nucleic acids contain nitrogen, only nucleic acids normally contain phosphorus.

Level 3 Critical Thinking and Clinical Applications
30. (a) number of electrons = 20; (b) atomic number = 20; (c) atomic weight = 40; (d) 2 electrons in first shell, 8 in second shell, 8 in third shell, and 2 in outer shell. **31.** The conversion of potential to kinetic energy is not a 100 percent efficient. For example, the breakdown of ATP to ADP and P during muscle contraction is an exothermic reac-tion (where energy is released), leading to the release of heat. **32.** If a person exhales large amounts of CO_2, the level of H^+ in the blood will decrease. A decrease in the amount of H^+ will cause the pH to increase.

Answers to Clinical Case Wrap-Up

PAGE 89 **1.** Baby Sean's stools contained undigested lipids and carbohydrates, which gave them a fatty or greasy appearance. **2.** Digestive enzymes aid in the decomposition reactions that break down lipids and carbohydrates in food, so they can be absorbed from the digestive tract. If these large molecules cannot be broken down, they cannot be used as energy sources or chemical building blocks.

Chapter 3

Answers to Checkpoints

PAGE 96 **1.** The general functions of the plasma membrane include physical isolation, regulation of exchange with the environment, sensitivity to the environment, and structural support. **2.** The com-ponents of the plasma membrane that allow it to perform its charac-teristic functions are membrane lipids, membrane proteins, and membrane carbohydrates. **3.** The phospholipid bilayer of the plasma membrane forms a physical barrier between the cell's inter-nal and external environments. **4.** Channel proteins are integral proteins that allow water and small ions to pass through the plasma membrane. PAGE 107 **5.** Cytoplasm is the material between the plasma membrane and the nuclear membrane. Cytosol is the fluid portion of the cytoplasm; cytosol is also called intracellular fluid. **6.** Cytosol has a higher concentration of potassium ions and suspended proteins, and a lower concentration of sodium ions, than the extracellular fluid. Cytosol also includes small quantities of car-bohydrates, and larger reserves of amino acids and lipids. **7.** The nonmembranous organelles and their functions include: (1) centri-ole = essential for movement of chromosomes during cell division; organization of microtubules in cytoskeleton; (2) cilia = movement of materials over cell surface (motile cilia), environmental sensor (primary cilium); (3) cytoskeleton = strength and support; move-ment of cellular structures and materials; cell movement; (4) micro-villi = increase surface area to facilitate absorption of extracellular materials; (5) proteasomes = breakdown and recycling of intracellu-lar proteins; (6) ribosomes = protein synthesis. **8.** The membra-nous organelles and their functions include: (1) endoplasmic reticulum = synthesis of secretory products; intracellular storage and transport; (2) rough ER = modification and packaging of newly syn-thesized proteins; (3) smooth ER = lipid and carbohydrate synthe-sis; (4) Golgi apparatus = storage, modification, and packaging of secretory products and lysosomal enzymes; (5) lysosomes = intra-cellular removal of damaged organelles or pathogens; (6) mitochon-dria = production of 95 percent of the ATP required by the cell; (7) peroxisomes = neutralization of toxic compounds. **9.** The SER synthesizes lipids, such as steroids. Ovaries and testes produce large amounts of steroid hormones, which are lipids, and thus need large amounts of SER. **10.** Mitochondria produce energy, in the form of ATP molecules, for the cell. A large number of mitochondria in a cell indicates a high demand for energy. PAGE 110 **11.** The nucleus is a cellular organelle that contains DNA, RNA, enzymes, and proteins. The nuclear envelope is a double membrane that surrounds the nu-cleus; the perinuclear space is the region between this double mem-brane. Nuclear pores enable chemical communication between the nucleus and the cytosol. **12.** A gene is a portion of a DNA strand that functions as a hereditary unit. Each gene is located at a particu-lar site on a specific chromosome and codes for a specific pro-tein. PAGE 113 **13.** Gene activation is the process of uncoiling the segment of DNA containing that gene, and temporarily remov-ing histones, so that the gene can affect the cell. **14.** Transcription is the synthesis of RNA from a DNA template. Translation uses the information from an mRNA strand to form a linear chain of amino acids, or protein. **15.** A cell that lacked the enzyme RNA polymerase would not be able to transcribe RNA from DNA. PAGE 119 **16.** A selectively permeable membrane allows the pas-sage of some substances and restricts the passage of others. It falls between two extremes: impermeable, which allows no substance to pass, and freely permeable, which permits the passage of any sub-stance. **17.** Diffusion is the passive molecular movement of a sub-stance from an area of higher concentration to an area of lower concentration. Diffusion proceeds until the concentration gradient is eliminated (equilibrium is reached). **18.** Five factors that influence the diffusion of substances in the body are (1) diffusion distance, (2) molecule size, (3) temperature, (4) concentration gradient, and (5) electrical forces. **19.** Diffusion is driven by a concentration gra-dient. The larger the concentration gradient, the faster the rate of dif-fusion, and the smaller the concentration gradient, the slower the rate of diffusion. If the concentration of oxygen in the lungs were to decrease, the concentration gradient between oxygen in the lungs and oxygen in the blood would decrease (as long as the oxygen level of the blood remained constant). Thus, oxygen would diffuse more slowly into the blood. **20.** Osmosis is the diffusion of water across a selectively permeable membrane from one solution to another solution that contains a higher solute concentration. **21.** The 10 percent salt solution is hypertonic with respect to the cells lining the nasal cavity, because it contains a higher salt (solute) concentration than do the cells. The hypertonic solution would draw water out of the cells, causing the cells to shrink and adding water to the mucus, thereby diluting it and relieving congestion. **PAGE 126 22.** In carrier-mediated transport, integral proteins bind specific ions or organic substrates and carry them across the plasma membrane. All forms of carrier-mediated transport are specific, have saturation limits, and are regulated (usually by hormones). **23.** An active transport process must be involved because it takes energy expenditure to move the hydrogen ions against their concentration gradient—that is, from a region where they are less concentrated (the cells lining the stomach) to a region where they are more concentrated (the interior of the stomach). **24.** Endocytosis is the movement of relatively large volumes of extracellular material into the cytoplasm by the formation of a membranous vesicle at the cell surface. Types of endocytosis are receptor-mediated endocytosis, the vesicle-mediated movement into the cytoplasm of specific target molecules (ligands); pinocytosis, the vesicle-mediated movement into the cytoplasm of extracellular fluid and its contents; and phagocytosis, the vesicle-mediated movement into the cytoplasm of extracellular solids, especially bacteria and debris. **25.** Exocytosis is the ejection of cytoplasmic materials by the fusion of a membranous vesicle with the plasma membrane. **26.** The process by which certain white blood cells engulf bacteria is called phagocytosis. **PAGE 127 27.** The membrane potential is the difference in electrical potential that results from the uneven distribution of positive and negative charges across the plasma membrane. It is expressed in millivolts (mV). **28.** If the plasma membrane were freely permeable to sodium ions, more of these positively charged ions would move into the cell, and the membrane potential would become less negative (move closer to zero). **PAGE 129 29.** The biological term for cellular reproduction is *cell division,* and the term for cell death is *apoptosis.* **30.** Important enzymes for DNA replication include helicases, DNA polymerase, and ligases. **31.** Interphase is a stage in a cell's life when it performs all its normal functions, and, if necessary, prepares for cell division. The stages of interphase include G_1, S, G_2, and G_0. A cell in G_0 is not preparing for cell division. **32.** This cell is likely in the G_1 phase of its life cycle. **33.** Mitosis is the essential step in cell division in which a single cell nucleus divides to produce two identical daughter cell nuclei. The four stages of mitosis are prophase, metaphase, anaphase, and telophase. **34.** If spindle fibers failed to form during mitosis, the cell would not be able to separate the chromosomes into two sets. If cytokinesis occurred, the result would be one cell with two sets of chromosomes and one cell with none. **PAGE 132 35.** A growth factor is a natural substance, such as a peptide or hormone that can stimulate the division of specific cell types. Representative growth factors include chalones, epidermal growth factor (EGF), erythropoietin, fibroblast growth factor (FGF), growth hormone, M-phase promoting factor (maturation-promoting factor), nerve growth factor (NGF), prolactin, and thymosins and related compounds. **PAGE 133 36.** An illness characterized by mutations that disrupt normal controls and produce potentially malignant cells is termed cancer. **37.** Metastasis is the spread of cancer cells to distant tissues and organs, leading to the establishment of secondary tumors. **PAGE 134 38.** Differentiation is the development of specific cellular characteristics and

functions that are different from the original cell. It results from genes in the cell being switched off, restricting the cell's potential functions.

Answers to Review Questions

PAGE 137
Level 1 Reviewing Facts and Terms
1. a. isotonic b. hypotonic c. hypertonic **2.** a **3.** a **4.** d **5.** a **6.** c **7.** a **8.** c **9.** Allows some materials to pass through it, but prevents others. **10.** b **11.** Aquaporins are special channel proteins that increase membrane permeability to water. Water passes through the channels diffusing from areas of high to low concentration across the plasma membrane. **12.** Four general functions of the plasma membrane are (1) physical isolation, (2) regulation of exchange with the environment, (3) sensitivity, and (4) structural support. **13.** Mitochondria, nucleus, Golgi apparatus, endoplasmic reticulum (smooth and rough), peroxisome, lysosome and vesicles. **14.** The major transport mechanisms are (1) diffusion, (2) carrier-mediated transport, and (3) vesicular transport. **15.** b **16.** Break down aging organelles, destroy bacteria, be involved in cell autolysis. Some of the vesicles produced by the Golgi apparatus become lysosomes.

Level 2 Reviewing Concepts
17. b **18.** d **19.** b **20.** c **21.** c **22.** d **23.** The concentration gradient of one substance provides the energy for the movement of another. For example, a sodium concentration gradient in the kidney tubule is used to cotransport other substances, such as glucose and amino acids during reabsorption. The same mechanism is used for cotransportation in the intestinal tract. **24.** G_0: normal cell functions; G_1: cell growth, duplication of organelles, and protein synthesis; S: DNA replication and synthesis of histones; G_2: protein synthesis **25.** Cells from the primary tumor break off and penetrate blood vessels circulating tissues, and escape into target organs that usually have large capillary beds. **26.** It is the development of new blood supply/vessels. Tumors stimulate angiogenesis to supply oxygen and nutrients, which allow/promote growth.

Level 3 Critical Thinking and Clinical Applications
27. This process is facilitated diffusion, which requires a carrier molecule but not cellular energy. The energy for this process is provided by the concentration gradient of the substance being transported. When all the carriers are actively involved in transport, the rate of transport reaches a saturation point. **28.** Solution A must have initially had more solutes than solution B. As a result, water moved by osmosis across the selectively permeable membrane from side B to side A, increasing the fluid level on side A. **29.** The concentration of solutes in blood will increase and water will move by osmosis from surrounding tissues, increasing blood volume. **30.** The isolation of the internal contents of membrane-bound organelles allows them to manufacture or store secretions, enzymes, or toxins that could adversely affect the cytoplasm in general. Another benefit is the increased efficiency of having specialized enzyme systems concentrated in one place. For example, the concentration of enzymes necessary for energy production in the mitochondrion increases the efficiency of cellular respiration.

Answers to Clinical Case Wrap-Up

PAGE 138 1. The other cell that might also be affected is the sperm cell, the only human cell with a tail-like flagellum, which has the same microtubule structure as motile cilia. **2.** Because Jackson's sperm cells will be affected, he is likely to experience difficulties in fathering children. Also, normal ciliary action is necessary to move sperm from the testes into the male reproductive tract.

Chapter 4

Answers to Checkpoints

PAGE 140 **1.** Histology is the study of tissues. **2.** The four major types of tissues that form all body structures are epithelial, connective, muscle, and neural tissue. **PAGE 144** **3.** Epithelial tissue is characterized by cellularity, polarity, attachment, avascularity, and regeneration. **4.** Epithelial tissue provides physical protection, controls permeability, provides sensation, and produces specialized secretions. **5.** An epithelium whose cells have many microvilli is probably involved in absorption or secretion. The microvilli greatly increase the surface area available for these processes. **6.** Epithelial cell junctions include tight junctions, gap junctions, spot desmosomes, and hemidesmosomes. **7.** Gap junctions allow small molecules and ions to pass from cell to cell. When connecting epithelial cells, they help coordinate such functions as the beating of cilia. In cardiac and smooth muscle tissues, they are essential for coordinating muscle cell contractions. **PAGE 152** **8.** The three cell shapes that are characteristic of epithelial cells are squamous (thin and flat), cuboidal (box-shaped), and columnar (tall and slender). **9.** A single layer of cells is termed simple, whereas multiple layers of cells are known as stratified. **10.** No, this is not a sample from the skin surface. A simple squamous epithelium does not provide enough protection against infection, abrasion, or dehydration. The skin surface has a stratified squamous epithelium. **11.** All these regions are subject to mechanical trauma and abrasion—by food (pharynx and esophagus), by feces (anus), and by intercourse or childbirth (vagina). **12.** The two primary types of glandular epithelia are endocrine glands and exocrine glands. **13.** Sebaceous glands exhibit holocrine secretion. **14.** This gland is an endocrine gland. **PAGE 159** **15.** Functions of connective tissues include: (1) defending the body from invading microorganisms; (2) establishing a structural framework for the body; (3) protecting delicate organs; (4) storing energy reserves; (5) supporting, surrounding, and interconnecting other types of tissue; and (6) transporting fluids and dissolved materials. **16.** The three categories of connective tissues are connective tissue proper, fluid connective tissues, and supporting connective tissues. **17.** Cells found in connective tissue proper are adipocytes, fibroblasts, fibrocytes, lymphocytes, macrophages, mast cells, melanocytes, mesenchymal cells, and microphages. **18.** The reduced collagen production resulting from a lack of vitamin C in the diet would cause connective tissue to be weak and prone to damage. **19.** Mast cells and basophils produce the molecule histamine, which antihistamines block. **20.** The type of connective tissue that contains triglycerides is adipose (fat) tissue. **21.** The two connective tissues that contain a fluid matrix are blood and lymph. **PAGE 163** **22.** The two types of supporting connective tissue are cartilage and bone. **23.** Unlike cartilage, bone has a direct blood supply, which is necessary for proper and rapid healing to occur. **24.** The type of cartilage damaged in a herniated intervertebral disc is fibrocartilage. **PAGE 165** **25.** The four types of membranes found in the body are the cutaneous membrane, mucous membranes, serous membranes, and synovial membranes. **26.** The pleural, peritoneal, and pericardial cavities are all lined by serous membranes. **27.** The lining of the nasal cavity is a mucous membrane. **PAGE 166** **28.** The three layers of fascia are superficial fascia (areolar and adipose tissue), deep fascia (dense irregular connective tissue), and subserous fascia (areolar tissue). **PAGE 168** **29.** The three types of muscle tissue in the body are skeletal muscle, cardiac muscle, and smooth muscle. **30.** Muscle tissue that lacks striations is smooth muscle. Both cardiac and skeletal muscles are striated. **31.** Skeletal muscle is repaired by the division of myosatellite cells, which are stem cells that persist in adult skeletal muscle tissue. **PAGE 169** **32.** The cells are most likely neurons. **PAGE 170** **33.** The two phases in the response to tissue injury are inflammation and regeneration. **34.** The four common signs of inflammation are swelling, redness, heat (warmth), and pain. **PAGE 172** **35.** With advancing age, the speed and effectiveness of tissue repair decrease, the rate of energy consumption in general declines, hormonal activity is altered, and other factors contribute to changes in structure and chemical composition. **36.** Longer life is associated with a longer period of potential exposure to chemicals and other environmental factors.

Answers to Review Questions

PAGE 174

Level 1 Reviewing Facts and Terms

1. (a) simple squamous epithelium, (b) simple cuboidal epithelium, (c) simple columnar epithelium, (d) stratified squamous epithelium, (e) stratified cuboidal epithelium, (f) stratified columnar epithelium. **2.** a **3.** b **4.** d **5.** c **6.** c **7.** b **8.** d **9.** c **10.** b **11.** c **12.** a **13.** b **14.** c **15.** a **16.** b **17.** Epithelial tissue (1) provides physical protection; (2) controls permeability; (3) provides sensations; and (4) produces specialized secretions. **18.** Endocrine glands secrete hormones onto the surface of the gland or directly into the surrounding fluid; exocrine glands secrete through ducts. **19.** Unlike in striated muscles, actin and myosin are not organized into repeated groups in smooth muscle. **20.** Connective tissues contain (1) specialized cells, (2) extracellular protein fibers, and (3) a fluid ground substance. **21.** Fibroblast, collagen **22.** Neural tissue contains (1) neurons, which transmit electrical impulses in the form of reversible changes in the membrane potential, and (2) neuroglia, which include several kinds of supporting cells and play a role in providing nutrients to neurons.

Level 2 Reviewing Concepts

23. Tight junctions form adhesion belts that prevent the passage of water and solutes isolating them to one side of the barrier. **24.** Tendons and ligaments tend to be subjected to forces in one direction whereas in muscle sheaths and the dermis, forces are from multiple angles. **25.** Fluid connective tissues have a liquid, watery matrix. They differ from supporting connective tissues in that they have many soluble proteins in the matrix, and they do not include insoluble fibers. **26.** The extensive connections between cells formed by cell junctions, proteoglycans, and physical interlocking hold skin cells together and can deny access to chemicals or pathogens that cover their free surfaces. If the skin is damaged and the connections are broken, infection can occur. **27.** b **28.** b **29.** Similarities: actin and myosin interactions produce contractions, calcium ions trigger and sustain contractions. Differences: skeletal muscles are relatively large, multinucleate, striated, and contract only under neural stimulation; cardiac muscles have 1–5 nuclei, are interconnected in a branching network, and contract in response to pacemaker cell activity; smooth muscles are small, spindle shaped, nonstriated, with only one nucleus, and are not under voluntary control.

Level 3 Critical Thinking and Clinical Applications

30. Because apocrine secretions are released by pinching off a portion of the secreting cell, you could test for the presence of plasma membranes, specifically for the phospholipids in plasma membranes. Merocrine secretions do not contain a portion of the secreting cell, so they would lack plasma membrane components. **31.** Skeletal muscle tissue would be made up of densely packed fibers running in the same direction, but since muscle fibers are composed of cells, they would have many nuclei and mitochondria.

Skeletal muscle also has an obvious banding pattern, or striations, due to the arrangement of the actin and myosin filaments within the cell. The student is probably looking at a slide of tendon (dense connective tissue). The small nuclei would be those of fibroblasts. **32.** Damage from infection causes the release of histamine and prostaglandins from mast cells. This increases blood flow, making the tissue appear red and increases the permeability of endothelial cells, allowing plasma into the surrounding connective tissue and causing swelling. This makes swallowing difficult. Sensory nerve fibers are sensitized by the inflammatory chemicals and stimulated by the swelling, which makes the throat feel sore.

Answers to Clinical Case Wrap-Up

PAGE 175 1. The collagen fibers within the matrix of connective tissue are defective in Ehlers-Danlos syndrome. **2.** Connective tissues throughout Anne Marie's body are affected by Ehlers-Danlos syndrome. This includes structures and organs composed of dense regular connective tissue (tendons, ligaments, aponeuroses), dense irregular connective tissue (skin, organ capsules of liver, kidneys, and spleen), and reticular tissue (spleen, liver, lymph nodes) as well.

Chapter 5

Answers to Checkpoints

PAGE 181 1. The layers of the epidermis are the stratum basale, stratum spinosum, stratum granulosum, stratum lucidum, and stratum corneum. **2.** Dandruff consists of cells from the stratum corneum. **3.** This splinter is lodged in the stratum granulosum. **4.** Wrinkles in the tips of the fingers (and toes) are due to the constriction of blood vessels in the underlying skin, which causes the skin to shrink. This effect is controlled by the autonomic nervous system, a division of the nervous system that acts outside our awareness. **5.** Sanding the tips of the fingers will not permanently remove fingerprints. The ridges of the fingerprints are formed in layers of the skin that are constantly regenerated, so these ridges will eventually reappear. The pattern of the ridges is determined by the arrangement of tissue in the dermis, which is not affected by sanding. **PAGE 183 6.** The two major pigments in the epidermis are carotene, an orange-yellow pigment, and melanin, which ranges in color from red-yellow (pheomelanin) to brown-black (eumelanin). **7.** When exposed to the ultraviolet radiation in sunlight, melanocytes in the epidermis and dermis synthesize the pigment melanin, darkening the skin. **8.** When skin gets warm, arriving oxygenated blood is diverted to the superficial dermis for the purpose of eliminating heat. The oxygenated blood gives a reddish coloration to the skin. **PAGE 185 9.** In the presence of ultraviolet radiation in sunlight, epidermal cells in the stratum spinosum and stratum basale convert a cholesterol-related steroid into cholecalciferol, or vitamin D_3. **10.** Cholecalciferol (vitamin D_3) is needed to form strong bones and teeth. When the body surface is covered, UV light cannot penetrate to the stratum basale in the skin to begin vitamin D_3 production, resulting in fragile bones. **PAGE 186 11.** Salivary glands and duodenal glands (glands of the duodenum) produce epidermal growth factor (EGF). **12.** Epidermal growth factor (EGF) promotes the divisions of basal cells in the stratum basale and stratum spinosum, accelerates the production of keratin in differentiating keratinocytes, stimulates epidermal development and epidermal repair after injury, and stimulates synthetic activity and secretion by epithelial glands. **PAGE 188 13.** The dermis (a connective tissue layer) lies between the epidermis and the hypodermis. **14.** The capillaries

and sensory neurons that supply the epidermis are located in the papillary layer of the dermis. **15.** The presence of elastic fibers and the flexibility and resilience of skin turgor allow the dermis to undergo repeated cycles of stretching and recoiling (returning to its original shape). **PAGE 189 16.** The tissue that connects the dermis to underlying tissues is the hypodermis or subcutaneous layer. **17.** The hypodermis is a layer of loose connective tissue (areolar tissue) and adipose tissue below (deep to) the dermis. It is also called the subcutaneous layer or superficial fascia. The hypodermis is not considered a part of the integument, but it is important in stabilizing the position of the skin in relation to underlying tissues. **18.** Subcutaneous fat provides insulation to help reduce heat loss, serves as an energy reserve, and acts as a shock absorber for the body. **PAGE 192 19.** A typical hair is a keratinous strand produced by epithelial cells of the hair follicle. **20.** The contraction of the arrector pili muscle pulls the hair follicle erect. The result is known as "goose bumps." **21.** Even though hair is a derivative of the epidermis, the hair follicles are in the dermis. Where the epidermis and deep dermis are destroyed, new hair will not grow. **PAGE 194 22.** Two types of exocrine glands found in the skin are sebaceous (oil) glands and sweat glands. **23.** Sebaceous secretions (called sebum) lubricate and protect the keratin of the hair shaft, lubricate and condition the surrounding skin, and inhibit the growth of bacteria. **24.** Deodorants are used to mask the odor of apocrine sweat gland secretions, which contain several kinds of organic compounds. Some of these compounds have an odor, and others produce an odor when metabolized by skin bacteria. **25.** Apocrine sweat glands enlarge and increase secretory activity in response to the increase in sex hormones that occurs at puberty. **PAGE 195 26.** Keratin is the substance that gives fingernails their strength. **27.** The area of thickened stratum corneum under the free edge of a nail is called the hyponychium. **28.** Nail growth occurs at the nail root, an epidermal fold that is not visible from the surface. **PAGE 198 29.** The combination of fibrin clots, fibroblasts, and the extensive network of capillaries in tissue that is healing is called granulation tissue. **30.** Skin can regenerate effectively even after considerable damage because stem cells persist in both the epithelial and connective tissue components of skin. When injury occurs, cells of the stratum basale replace epithelial cells, and mesenchymal cells replace cells lost from the dermis. **PAGE 199 31.** As a person ages, the blood supply to the dermis decreases and merocrine sweat glands become less active. These changes make it more difficult for older individuals to cool themselves in hot weather. **32.** With advancing age, melanocyte activity decreases, leading to gray or white hair.

Answers to Review Questions

PAGE 202
Level 1 Reviewing Facts and Terms
1. (a) epidermis **(b)** dermis **(c)** papillary layer **(d)** reticular layer **(e)** hypodermis (subcutaneous layer) **2.** a **3.** a **4.** b **5.** d **6.** a **7.** a **8.** d **9.** b **10.** a **11.** d **12.** d **13.** d **14.** a **15.** Epidermal cell division occurs in the stratum basale. **16.** These smooth muscles cause hairs to stand erect when stimulated. **17.** b **18.** c **19.** It is the blue colouration of skin as a result of low oxygen saturation of hemoglobin. It is easiest to see in areas of thinner skin such as the lips, tongue, and nail beds.

Level 2 Reviewing Concepts
20. Insensible perspiration is water loss by evaporation through the stratum corneum. Sensible perspiration is produced by active

sweat glands.　**21.** The integument is the first barrier to substances entering the body. This method of delivering drugs requires the substance to move across the epidermal plasma membranes. The lipid bilayer of the plasma membranes creates a barrier that prevents water from rapidly entering or leaving the cells. Fat-soluble (lipophilic) substances, such as oil containing a dissolved drug, can easily pass through this barrier to underlying connective tissue and enter the circulation. Water-soluble drugs are lipophobic and thus do not readily penetrate the integument.　**22.** A tan develops when the skin is exposed to the sun. The tan is a result of the synthesis of melanin in the skin. Melanin, in turn, helps prevent excessive skin damage by absorbing UV radiation before it reaches the deep layers of the epidermis and dermis. Within the epidermal cells, melanin concentrates around the nucleus, so it absorbs the UV light before it can damage nuclear DNA. A benefit of minimal sun exposure is that sunlight causes epidermal cells to convert a cholesterol-related steroid into cholecalciferol (Vitamin D_3), which is important to bone maintenance and growth.　**23.** Granulation tissue is composed of fibroblasts, new capillaries and clot material. Fibroblasts divide and invade the damaged tissue and stimulate the endothelial cells to replicate to produce new capillary networks which bring nutrients and oxygen.　**24.** c　**25.** c　**26.** d

Level 3 Critical Thinking and Clinical Applications

27. a　**28.** The puncture wound by the nail has a greater chance of becoming infected. Whereas the knife cut will bleed freely, washing many of the bacteria from the wound site, the nail can carry bacteria beneath the surface of the skin, where oxygen is limited (anaerobic), and past the skin's protective barriers.　**29. (a)** Ultraviolet radiation in sunlight converts a cholesterol-related steroid into vitamin D_3, or cholecalciferol. This compound is then converted to the hormone calcitriol, which is essential for absorbing the calcium and phosphorus by the small intestine that is necessary for normal bone maintenance and growth. **(b)** The child can drink more milk. Milk is routinely fortified with cholecalciferol, normally identified as "vitamin D," which is easily absorbed by the intestines.
30. Carotene is a/an orange/yellow pigment found in orange vegetables (carrots and oranges) which can accumulate in fatty tissue of the dermis and subcutaneous layers if eaten excessively. This gives the skin a pale orange appearance, especially in pale-skinned individuals.　**31.** The loss of epidermis and dermis means that there is no longer an effective barrier to fluid and electrolyte loss or bacterial colonisation. This means that dehydration and infection may become life threatening.

Answers to Clinical Case Wrap-Up

PAGE 203　1. The epidermis is affected by psoriasis.　**2.** The underlying problem in psoriasis is that cell division and maturation occur too rapidly. Instead of basal cells dividing at a rate to replace the keratinocytes that wear away at the surface of the skin, they divide so rapidly, and migrate through the epithelial layers so quickly, that keratinocytes pile up on the skin surface.

Chapter 6

Answers to Checkpoints

PAGE 205　1. The five primary functions of the skeletal system are support, storage of minerals and lipids, blood cell production, protection, and leverage.　**PAGE 208　2.** The six broad categories for classifying bones according to shape are sutural bones, irregular bones, short bones, flat bones, long bones, and sesamoid

bones.　**3.** A bone marking (surface feature) is an area on the surface of a bone structured for a specific function, such as joint formation, muscle attachment, or the passage of nerves and blood vessels.　**PAGE 210　4.** Mature bone cells are known as osteocytes, bone-building cells are called osteoblasts, and osteoclasts are bone-resorbing cells.　**5.** If the ratio of collagen to hydroxyapatite in a bone increased, the bone would become less strong (as well as more flexible).　**6.** Because osteoclasts break down or demineralize bone, the bone would have a decreased mineral content (less mass); as a result, it would also be weaker.　**PAGE 213　7.** Compact bone consists of osteons (Haversian systems) with little space between them. Compact bone lies over spongy bone and makes up most of the diaphysis. It functions to protect, support, and resist stress. Spongy bone consists of trabeculae with numerous red marrow–filled spaces. Spongy bone makes up most of the structure of short, flat, and irregular bones and is also found at the epiphyses of long bones. Spongy bone functions in storing marrow and providing some support.　**8.** The presence of lamellae that are not arranged in osteons is indicative of spongy bone, which is located in an epiphysis.　**PAGE 218　9.** In endochondral ossification, cells of the inner layer of the perichondrium differentiate into osteoblasts, and a cartilage model is gradually replaced by bone.
10. During intramembranous ossification, fibrous connective tissue is replaced by bone.　**11.** Long bones of the body, such as the femur, have an epiphyseal cartilage, a plate of cartilage that separates the epiphysis from the diaphysis so long as the bone is still growing lengthwise. An x-ray would indicate whether the epiphyseal cartilage is still present. If it is, growth is still occurring. If it is not, the bone has reached its adult length and thus, the person has reached full height.　**PAGE 220　12.** Bone remodeling refers to the process whereby old bone is continuously being destroyed by osteoclasts while new bone is being constructed by osteoblasts.　**13.** The biochemistry of some heavy-metal ions, such as strontium, cobalt, uranium, and plutonium, is very similar to that of calcium. Osteoblasts cannot differentiate these heavy-metal ions from normal calcium ions, so the heavy-metal ions become incorporated into the bone matrix. Over time, these dangerous ions can be released into the bloodstream during normal bone remodeling.　**PAGE 221　14.** The larger arm muscles of the weight lifter would apply more mechanical stress to the bones of the upper limbs. In response to that stress, the bones would grow thicker.　**15.** Growth continues throughout childhood. At puberty, a growth spurt takes place and is followed by the closure of the epiphyseal cartilages. The later puberty begins, the taller the child will be when the growth spurt begins, so the taller the person will be when growth is completed.　**16.** Increased levels of growth hormone prior to puberty will result in excessive bone growth, making the person taller.　**PAGE 223　17.** Parathyroid hormone (PTH) influences osteoclast activity, which causes a release of stored calcium ions from bone. Under the influence of calcitonin, osteoclast activity is inhibited, while osteoblasts continue to lock calcium ions in the bone matrix. In this way, PTH serves to increase blood calcium levels by causing its release from bone, and calcitonin decreases blood calcium levels by causing calcium to remain in bone.
18. The bones of children who have rickets are poorly mineralized and as a result are quite flexible. Under the weight of the body, the leg bones bend. The instability makes walking difficult and can lead to other problems of the legs and feet.　**19.** Parathyroid hormone (PTH) indirectly stimulates osteoclasts to release calcium ions from bone and enhances calcitriol's effect on the intestinal absorption of calcium. Increased PTH secretion would result in an increase in the

level of calcium ions in the blood. **20.** Calcitonin lowers blood calcium levels by inhibiting osteoclast activity and increasing the rate of calcium excretion by the kidneys. **PAGE 224** **21.** Immediately following a fracture, extensive bleeding occurs at the injury site. After several hours, a large blood clot called a fracture hematoma develops. Next, an internal callus forms as a network of spongy bone unites the inner edges, and an external callus of cartilage and bone stabilizes the outer edges. The cartilaginous external callus is eventually replaced by bone, and the struts of new spongy bone now unite the broken ends. With time, the swelling that initially marks the location of the fracture is remodeled, new compact bone is formed, and little evidence that a break occurred remains. **22.** An external callus forms early in the healing process, when cells from the endosteum and periosteum migrate to the area of the fracture. These cells form an enlarged collar (external callus) that encircles the bone in the area of the fracture. **PAGE 228** **23.** Osteopenia is inadequate ossification and is common during the aging process. It results from decreasing osteoblast activity accompanied by normal osteoclast activity. **24.** In women, the sex hormones known as estrogens play an important role in moving calcium into bones. After menopause, the level of these hormones decreases dramatically. As a result, older women have difficulty replacing the calcium in bones that is being lost due to normal aging. In men, the level of sex hormones (androgens) does not decrease until much later in life.

Answers to Review Questions

PAGE 230
Level 1 Reviewing Facts and Terms
1. b **2.** d **3.** a **4.** b **5.** b **6.** c **7.** (a) long bone (b) flat bone (c) sutural bone (d) irregular bone (e) short bones (f) sesamoid bone. **8.** a **9.** c **10.** c **11.** (1) support; (2) storage of minerals and lipids; (3) blood cell production; (4) protection; and (5) leverage **12.** (1) osteocytes; (2) osteoblasts; (3) osteoclasts; and (4) osteogenic cells **13.** (1) diaphysis (shaft); (2) epiphysis; (3) epiphyseal cartilage/line; (4) articular cartilage; (5) medullary canal; (6) periosteum; (7) endosteum **14.** d **15.** organic = collagen; inorganic = hydroxyapatite crystals **16.** (a) calcium salts, phosphate salts, and vitamins A, C, and D_3; (b) calcitriol, growth hormone, thyroxine, estrogens (in females) or androgens (in males), calcitonin, and parathyroid hormone (PTH) **17.** The three organs or tissues are the bones, the intestinal tract, and the kidneys. **18.** Parathyroid hormone stimulates osteoclast activity, increases the rate of intestinal absorption of calcium ions, and decreases the rate of excretion of calcium ions by the kidneys. These effects increase blood calcium levels.

Level 2 Reviewing Concepts
19. Nutrients reach the osteocytes in spongy bone by diffusing along canaliculi that open onto the surface of the trabeculae. **20.** The osteons are aligned parallel to the long axis of the shaft, which does not bend when forces are applied to either end. Stresses or impacts to the side of the shaft can lead to a fracture. **21.** The lack of physical stress during inactivity leads to the removal of calcium salts from bones. Up to one-third of the bone mass can be lost in this manner, causing the bones to become thin and brittle. **22.** The trabeculae of spongy bone are aligned along lines of stress. Forces from many directions can be transferred to adjoining compact bones. **23.** There are many long bones in the hand, each of which has an epiphyseal cartilage (plate). Measuring the width of these plates will provide clues to the hormonal control of growth in the child. **24.** Osteoclasts continuously remove older mineral deposits through

osteolysis whereas osteoblasts continuously add mineral deposits through osteogenesis. Spongy bone is most at risk as it is remodeled more frequently. **25.** b **26.** An internal callus forms whereby osteoblasts produce spongy bone that unites the broken ends of the bone. Osteoblasts replace the external cartilage callus which braces the fracture site. Osteoblasts and osteoclasts remodel, allowing the bone to be shaped by stress.

Level 3 Critical Thinking and Clinical Applications
27. The fracture might have damaged the epiphyseal cartilage in Sally's right leg. Even though the bone healed properly, the damaged leg did not produce as much cartilage as did the undamaged leg. The result would be a shorter bone on the side of the injury. **28.** After five weeks a simple fracture in a healthy patient will be in the hard callus stage moving towards the remodeling phase. Stimulating the osteoblasts and osteoclasts will maximize the functional stress of the bone by ensuring that remodeling occurs along lines of stress. **29.** a **30.** The matrix of bone will absorb traces of minerals from the diet. These minerals can be identified hundreds of years later. A diet rich in calcium and vitamin B will produce denser bones than will a diet lacking these. Cultural practices such as the binding of appendages or the wrapping of infant heads will manifest in misshapen bones. Heavy muscular activity will result in larger bone markings, indicating an athletic or physically-demanding lifestyle.

Answers to Clinical Case Wrap-Up

PAGE 231 **1.** Having multiple fractures in varying stages of healing, with no good explanation for how they occurred, is a hallmark sign of physical abuse in children. (Emergency personnel are obligated by law to report such signs.) **2.** Alex should avoid contact sports. As this incident shows, Alex is at much higher risk for fractures than any other soccer player on the field. He should be encouraged to pursue swimming, music, writing, art, or other stimulating, yet less physical, activities.

Chapter 7

Answers to Checkpoints

PAGE 233 **1.** The axial skeleton includes the skull (8 cranial bones and 14 facial bones), the bones associated with the skull (6 auditory ossicles and the hyoid bone), the vertebral column (24 vertebrae, the sacrum, and the coccyx), and the thoracic cage (the sternum and 24 ribs). **2.** The axial skeleton serves as a framework that supports and protects organs, provides an extensive surface area for muscle attachment, and it also stabilizes and positions parts of the appendicular skeleton that support the limbs. **PAGE 249** **3.** The foramen magnum is located in the base of the occipital bone. **4.** Tomás has fractured his right parietal bone. **5.** The sphenoid bone contains the sella turcica, which in turn contains the pituitary gland. **6.** The 14 facial bones include 2 inferior nasal conchae, 2 lacrimal bones, 1 mandible, 2 maxillae (maxillary bones), 2 nasal bones, 2 palatine bones, 1 vomer, and 2 zygomatic bones. **7.** The mental foramen is found in the mandible. Structures passing through this opening include the mental nerve, which carries sensory information to the brain, and mental vessels, which supply the chin and lips. **8.** The optic canal is found in the sphenoid bone. The optic nerve and ophthalmic artery pass through this structure. **9.** The foramina in the ethmoid bone are the olfactory foramina. **PAGE 250** **10.** The bones forming the orbital complex are the frontal, sphenoid, zygomatic, palatine, maxilla, lacrimal, and ethmoid. **11.** The bones forming the nasal complex are the frontal, ethmoid, nasal, maxilla, palatine, and sphenoid. **12.** The sphenoid, ethmoid, frontal, and paired palatine and maxillary bones contain the paranasal sinuses. **PAGE 251** **13.** A fontanelle is a relatively soft, flexible, fibrous region between two flat bones in the developing skull. The major fontanelles are the anterior fontanelle, occipital fontanelle, sphenoidal fontanelle, and mastoid fontanelle. **14.** Because they are not ossified at birth, fontanelles permit the skull to change shape during childbirth, and they allow for growth of the brain during infancy and early childhood. **PAGE 253** **15.** The secondary curves of the spine allow us to balance our body weight on our lower limbs with minimal muscular effort. Without the secondary curves, we would not be able to stand upright for extended periods. **16.** When you run your finger along a person's spine, you can feel the spinous processes of the vertebrae. **PAGE 261** **17.** The adult vertebral column has fewer vertebrae than a newborn because the five sacral vertebrae fuse to form a single sacrum, and the four coccygeal vertebrae fuse to form a single coccyx. **18.** The dens is part of the axis, or second cervical vertebra, which is located in the neck (cervical region). **19.** The presence of transverse foramina indicates that this vertebra is from the cervical region of the vertebral column. These foramina are passageways for important blood vessels that supply the brain. **20.** The lumbar vertebrae must support a great deal more weight than do vertebrae that are more superior in the spinal column. The large vertebral bodies allow the weight to be distributed over a larger area. **PAGE 263** **21.** True ribs (ribs 1–7) attach directly to the sternum by their own costal cartilage. False ribs (ribs 8–12) either attach by means of a common costal cartilage (as in the vertebrochondral ribs) or do not attach to the sternum (as in the floating ribs). **22.** Improper compression of the chest during CPR can—and commonly does—result in a fracture of the xiphoid process of the sternum or the ribs. **23.** Vertebrosternal ribs, or true ribs (ribs 1–7), attach directly to the sternum by separate costal cartilages; vertebrochondral ribs (ribs 8–10) have costal cartilages that merge with the cartilages of rib pair 7 before they reach the sternum.

Answers to Review Questions

PAGE 265
Level 1 Reviewing Facts and Terms
1. (a) occipital bone (b) parietal bone (c) frontal bone (d) temporal bone (e) sphenoid bone (f) ethmoid bone (g) vomer (h) mandible (i) lacrimal bone (j) nasal bone (k) zygomatic bone (l) maxilla **2.** a **3.** c **4.** d **5.** d **6.** a **7.** (a) cervical (b) thoracic (c) lumbar **8.** b **9.** c **10.** d **11.** At the junction of the pedicles and laminae. **12.** Manubrium, body, xiphoid process. **13.** The answers will include: lightening the skull; enhancing resonance of the voice; warming and humidifying air; assisting with regulation of intranasal pressure; increasing mucosal surface area; acting as shock absorbers and promoting facial growth. **14.** The fibrocartilage discs between adjacent vertebrae make the vertebral column more flexible.

Level 2 Reviewing Concepts
15. The petrous part of the temporal bone encloses the structures of the internal ear. The middle ear is located in the tympanic cavity within the petrous part. The external acoustic meatus (auditory canal) ends at the tympanic membrane, which leads into the middle ear. Mastoid air cells within the mastoid process are connected to the tympanic cavity. **16.** The ethmoid bone forms the superior surface of the nasal cavity. The olfactory foramina within the cribriform plate of the ethmoid bone allow neurons associated with the

sense of smell to extend into the nasal cavity. **17.** Protects organs and structures of the thoracic cavity. It serves as an attachment point for muscles involved in respiration, maintenance of the position of the vertebral column, and movements of the pectoral girdle (both clavicles and scapulae) and upper limbs. **18.** Keeping your back straight keeps the weight aligned along the axis of your vertebral column, where it can be transferred to your lower limbs. Bending your back would strain the muscles and ligaments of the back, increasing the risk of injury. **19.** Trudy. Protection. **20.** Superior and inferior costal facets (present on dorsolateral surface of vertebral body) articulate with the heads of two ribs (inferior and superior edges respectively). Transverse costal facet articulates with body of rib. **21.** a **22.** The facial nerve: CN VII. Motor control of facial muscles. Temporal bone, posterior to the base of the styloid process. **23.** e

Level 3 Critical Thinking and Clinical Applications
24. d **25.** Idiopathic scoliosis as it is common in adolescent girls during periods of growth and often leads to back pain. **26.** The atlas-axis complex is much smaller than the skull. The sudden movement Hayley experienced in the car could not be compensated for by the balancing muscles and causes partial or complete displacement of the cervical vertebrae, with injury to surrounding muscles and ligaments.

Answers to Clinical Case Wrap-Up

PAGE 266 **1.** The seven bones that make up the orbital complex are the frontal bone (forming the roof), the zygomatic bone (forming the lateral wall), the maxilla (forming most of the floor), the lacrimal bone, the lateral mass of the ethmoid bone, the sphenoid bone, and the palatine bone (forming the thin medial wall and floor). **2.** The four paranasal sinuses surrounding the orbital complex are the frontal, maxillary, and sphenoidal sinuses, and ethmoid air cells.

Chapter 8
Answers to Checkpoints

PAGE 271 **1.** Each of the two pectoral girdles consists of a clavicle (collarbone) and a scapula (shoulder blade). Each arm articulates with the trunk at the pectoral girdle. **2.** The clavicle attaches the scapula to the sternum, and restricts the scapula's range of motion. When the clavicle is broken, the scapula has a greater than normal range of motion and is less stable. **3.** The head of the humerus articulates with the scapula at the glenoid cavity. **PAGE 276** **4.** The bones of an upper limb are the humerus, ulna, radius, carpal bones (scaphoid, lunate, triquetrum, pisiform, trapezium, trapezoid, capitate, and hamate), 5 metacarpal bones, and 14 phalanges. **5.** The two rounded projections on either side of the elbow are the lateral and medial epicondyles of the humerus. **6.** The radius is lateral to the ulna while in the anatomical position; or the ulna is medial to the radius while in the anatomical position. **7.** The first distal phalanx is located at the tip of the thumb, so Bill's pollex is broken. **PAGE 280** **8.** The pelvic girdle is composed of the paired hip bones, also known as the coxal bones or pelvic bones. **9.** The three bones that make up a hip bone are the ilium, ischium, and pubis. **10.** The pelvis of females is adapted for supporting the weight of the developing fetus and enabling the newborn to pass through the pelvic outlet during delivery. Compared to males, such adaptations of the female pelvis include an enlarged pelvic outlet, a sacrum and coccyx with less curvature, a pelvic inlet that is wider and more circular, a relatively broad shape, ilia that project farther laterally, and an inferior angle between the pubic bones that is greater than 100 degrees (as opposed to 90 degrees or less for males).   **11.** When you are seated, your body weight is supported by the ischial tuberosities.   **PAGE 284**   **12.** The bones of the lower limb are the femur (thigh), patella (kneecap), tibia and fibula (leg), tarsal (ankle) bones (talus, calcaneus, cuboid, navicular, medial cuneiform, intermediate cuneiform, and lateral cuneiform), 5 metatarsal bones, and 14 phalanges.   **13.** The fibula is not part of the knee joint and does not bear weight, but it is an important point of attachment for many leg muscles. When the fibula is fractured, these muscles cannot function properly to move the leg, so walking is difficult and painful. The fibula also helps stabilize the ankle joint.   **14.** Joey has most likely fractured the calcaneus (heel bone).   **15.** The talus transmits the weight of the body from the tibia toward the toes.   **PAGE 285**   **16.** In general, the bones of males tend to be heavier than those of females, and bone markings are more prominent in males than in females.   **17.** Bones can reveal information about a person's sex, age, muscular development, nutritional state, handedness, and occupation, plus other information relative to the medical history.

Answers to Review Questions

PAGE 287
Level 1 Reviewing Facts and Terms
1. (a) anterior view; **(b)** lateral view; **(c)** posterior view; **(d)** acromion; **(e)** coracoid process; **(f)** spine; **(g)** glenoid cavity
2. d   **3. (a)** sacrum; **(b)** coccyx; **(c)** ilium; **(d)** pubis;
(e) ischium; **(f)** hip bone   **4.** a   **5.** b   **6.** d   **7.** Lunate, scaphoid, pisiform, triquetrum.   **8.** interosseus membrane   **9.** Articulation between the ilia of the hip bone and the sacrum.   **10.** Coccyx, ischial tuberosities, inferior border of pubic symphysis.

Level 2 Reviewing Concepts
11. d   **12.** a   **13.** e   **14.** d   **15.** d   **16.** d   **17.** c   **18.** The pelvic girdle consists of the two hip bones, or coxal bones. The pelvis is a composite structure; it consists of the two hip bones of the appendicular skeleton and the sacrum and coccyx of the axial skeleton.   **19.** d   **20.** The clavicles are small and fragile, so they are easy to break. Once this part of the pectoral girdle is broken, the assailant would no longer have efficient use of his arms.
21. e   **22.** The slender fibula parallels the tibia of the leg and provides an important site for muscle attachment. It does not help in transferring weight to the ankle and foot because it is excluded from the knee joint.   **23.** e

Level 3 Critical Thinking and Clinical Applications
24. Plantar flexion of the ankle, the talus, and calcaneus transfer weight to the metatarsal bones and phalanges through the tarsal bones.   **25.** Fred probably dislocated his shoulder, which is quite a common injury due to the weak nature of the glenohumeral joint.   **26.** Patellofemorol stress syndrome. Running on the road with a slight camber will force the patella to shift outside its normal track, shifting it laterally.

Answers to Clinical Case Wrap-Up

PAGE 288   **1.** The distal humeral fracture fragment is still attached to the radius and ulna at the elbow joint. The fracture site is above the medial and lateral humeral epicondyles, and above the joint itself.   **2.** In a severe fracture of the appendicular skeleton, massive swelling is likely to occur. A circular cast will not give, or expand, as swelling continues. This might literally squeeze off the circulation to the forearm and hand. To prevent this from happening, a posterior splint is used instead.

Chapter 9
Answers to Checkpoints
PAGE 292   **1.** The three types of joints as classified by their range of motion are the following: (1) an immovable joint or synarthrosis, (2) a slightly movable joint or amphiarthrosis, and (3) a freely movable joint or diarthrosis. A synarthrosis can be fibrous or cartilaginous, depending on the nature of the connection, or it can be a bony fusion, which develops over time. An amphiarthrosis is either fibrous or cartilaginous, depending on the nature of the connection. A diarthrosis is a synovial joint, which permits the greatest range of motion.   **2.** Both synarthrotic joints (except synostoses) and amphiarthrotic joints consist of bony regions separated by fibrocartilage or cartilaginous connective tissue.   **3.** These are fibrous joints known as sutures.   **PAGE 294**   **4.** Components of a synovial joint include a joint capsule (articular capsule), articular cartilages, synovial fluid, and various accessory structures (menisci, fat pads, ligaments, tendons, and bursae). Synovial joints are freely movable joints that have a joint (synovial) cavity, which is the space between articulating surfaces of two bones in a joint. Articular cartilages resembling hyaline cartilage cover the articulating bone surfaces. The fibrous joint capsule surrounds the joint, and a synovial membrane, which secretes synovial fluid, lines the walls of the joint cavity. Within the cavity, synovial fluid lubricates, distributes nutrients, and absorbs shocks. Menisci are articular discs made of fibrocartilage that allow for variation in the shapes of the articulating surfaces. Fat pads protect the cartilages. Ligaments are cords of fibrous tissue that support, strengthen, and reinforce the joint. Tendons passing across or around the joint limit the range of motion and provide mechanical support. Bursae are synovial fluid-filled pockets that reduce friction and absorb shocks.   **5.** Because articular cartilages are avascular (lack a blood supply), they rely on synovial fluid to supply nutrients and eliminate wastes. Impairing the circulation of synovial fluid would have the same effect as impairing a tissue's blood supply: Nutrients would not be delivered to meet the tissue's needs, and wastes would accumulate. Damage to, and ultimately the death of, the cells in the tissue would result.   **6.** A subluxation is a partial dislocation of a bone from its joint.   **PAGE 300**   **7.** Based on the shapes of the articulating surfaces, synovial joints are classified as gliding, hinge, pivot, condylar (ellipsoid), saddle, and ball-and-socket joints.   **8.** When doing jumping jacks, you move your lower limbs away from the body's midline; this movement is abduction. When you bring the lower limbs back together, the movement is adduction.   **9.** Flexion and extension are the movements associated with hinge joints.   **PAGE 302**   **10.** Movements possible across the intervertebral joints of the vertebral column are flexion (bending anteriorly), extension (bending posteriorly), lateral flexion (bending laterally), and rotation.   **11.** Intervertebral discs are not found between the first and second cervical vertebrae, between sacral vertebrae in the sacrum, or between coccygeal vertebrae in the coccyx. An intervertebral disc between the first and second cervical vertebrae would prohibit rotation. The vertebrae in the sacrum and coccyx are fused to provide a firm attachment for muscles and ligaments.
12. The vertebral movement involved in (a) bending forward is flexion. The movement involved in (b) bending to the side is lateral flexion. The movement involved in (c) moving the head to signify "no" is rotation.   **PAGE 305**   **13.** Ligaments and muscles provide most of the stability for the shoulder joint.   **14.** The subscapular bursa is located in the shoulder joint, so the tennis player would be more likely to develop inflammation of this structure (bursitis). The condition is associated with repetitive motion that occurs at

the shoulder, such as swinging a tennis racket. The jogger would be more at risk for injuries to the knee joint.   **15.** A shoulder separation is an injury involving partial or complete dislocation of the acromioclavicular joint.   **16.** Terry has most likely damaged his annular ligament.   **PAGE 309   17.** The bones making up the shoulder joint are the humerus and scapula. The bones involved with the knee joint are the femur, tibia, and patella. The fibula is not part of the knee joint.   **18.** The iliofemoral, pubofemoral, and ischiofemoral ligaments are at the hip joint.   **19.** Damage to the menisci of the knee joint decreases the joint's stability, so the person would have a difficult time locking the knee in place while standing and would have to use muscle contractions to stabilize the joint. If the person had to stand for a long time, the muscles would fatigue and the knee would "give out." It is also likely that the person would feel pain.   **20.** Like members of the clergy, carpet layers and roofers kneel a lot (and they also slide along on their knees), causing inflammation of the bursae in the knee joint.   **PAGE 310   21.** Rheumatism is a general term describing any painful condition of the bones, muscles, and joints (musculoskeletal system). One of several forms of rheumatism is arthritis.   **22.** Arthritis applies to all the rheumatic diseases that affect synovial joints. Of the many different types of arthritis, osteoarthritis is the most common.   **23.** Osteoblast activity must equal osteoclast activity or else the integrity of bone structure is compromised. For example, if osteoclast activity outpaces osteoblast activity, the result is thin, brittle bones.
24. The skeletal system provides structural support for all body systems and stores energy, calcium, and phosphate. The integumentary system synthesizes vitamin D_3, which is essential for calcium and phosphorus absorption. Calcium and phosphorus are needed for bone growth and maintenance.

Answers to Review Questions

PAGE 313
Level 1 Reviewing Facts and Terms
1. (a) joint (articular) capsule; **(b)** synovial membrane; **(c)** articular cartilage; **(d)** joint cavity (contains synovial fluid)   **2.** d   **3.** b   **4.** c   **5.** c   **6.** b   **7.** c   **8.** b   **9.** c   **10.** c   **11.** b   **12.** d   **13.** c   **14.** c   **15.** a   **16.** Protection, mineral storage, red blood cell production and providing energy reserves.   **17.** c   **18.** Lubrication, nutrient distribution, shock absorption.   **19.** Menisci: help to channel the flow of synovial fluid and allow variations in articular cartilage shape; Fat pads: protect the articular cartilage; Ligaments and tendons: support the synovial joint; Bursae: reduce friction of moving tendons and ligaments, and act as shock absorbers.

Level 2 Reviewing Concepts
20. b   **21.** b   **22.** A meniscus (artcular disc) subdivides a synovial cavity, channels the flow of synovial fluid, and allows variations in the shape of the articular surfaces. Menisci also act as cushions and shock absorbers.   **23.** a   **24.** Articular cartilages lack a perichondrium, and their matrix contains more water than does the matrix of other cartilages.   **25.** Osteoarthritis is a result of defective collagen formation in the articular cartilage, whereas rheumatoid arthritis is an inappropriate immune response to the articular cartilage.   **26.** Height decreases during adulthood in part as a result of osteoporosis in the vertebrae, and in part as a result of the decrease in water content of the nucleus pulposus region of intervertebral discs.   **27.** a) Protects the system and important for lung ventilation b) Provides structural support

Level 3 Critical Thinking and Clinical Applications
28. The problem is probably a sprained ankle. The ligaments have been damaged but not ruptured, and the joint remains

unaffected.   **29.** The synovial membrane which makes up part of the synovial joint is vascularised. Its capillaries allow the exchange of nutrients and waste products within the synovial fluid. During compression (e.g., flexion of the knee), the articular cartilage releases synovial fluid held within it. During extension, when compressional forces have dissipated, the ground substance of the cartilage retakes up synovial fluid, which will have been replenished from the synovial membrane.   **30.** Shoulder dislocations would occur more frequently than hip dislocations because the shoulder is a more mobile joint. Because the shoulder joint is not bound tightly by ligaments or other structures, it is easier to dislocate when excessive forces are applied. By contrast, the hip joint, although mobile, is less easily dislocated because it is stabilized by four heavy ligaments, and the bones fit together snugly in the joint. Additionally, the synovial capsule of the hip joint is larger than that of the shoulder, and its range of motion is decreased.

Answers to Clinical Case Wrap-Up
PAGE 314   1. Only diarthroses, or synovial joints, are affected by juvenile rheumatoid arthritis. This is a disease of the synovial lining, so it does not affect joints that have no synovial membrane.   **2.** Normal synovial membranes are thin, delicate, and smooth. The inflamed synovial membrane in a joint affected by rheumatoid arthritis would look thickened, red, excessively fluid-filled, swollen, and overgrown.

Chapter 10
Answers to Checkpoints
PAGE 317   1. The three types of muscle tissue are skeletal muscle, cardiac muscle, and smooth muscle.   **2.** Skeletal muscles produce skeletal movement, maintain posture and body position, support soft tissues, guard body entrances and exits, maintain body temperature, and store nutrient reserves.   **PAGE 318   3.** The epimysium is a dense layer of collagen fibers that surrounds the entire muscle; the perimysium divides the skeletal muscle into a series of compartments, each containing a bundle of muscle fibers called a fascicle; and the endomysium surrounds individual skeletal muscle cells (muscle fibers). The collagen fibers of the epimysium, perimysium, and endomysium come together to form either bundles known as tendons, or broad sheets called aponeuroses. Tendons and aponeuroses generally attach skeletal muscles to bones.   **4.** Because tendons attach muscles to bones, severing the tendon would disconnect the muscle from the bone, and so the muscle could not move a body part.   **PAGE 325   5.** Sarcomeres, the smallest contractile units of a striated muscle cell, are segments of myofibrils. Each sarcomere has dark A bands and light I bands. The A band contains the M line, the H band, and the zone of overlap. Each I band contains thin filaments, but not thick filaments. Z lines bisect the I bands and mark the boundaries between adjacent sarcomeres.   **6.** Skeletal muscle appears striated when viewed through a light microscope because the Z lines and thick filaments of the myofibrils within the muscle fibers are aligned.
7. You would expect the greatest concentration of calcium ions in resting skeletal muscle to be in the terminal cisternae of the sarcoplasmic reticulum.   **PAGE 334   8.** The neuromuscular junction, also known as the myoneural junction, is the synapse between a motor neuron and a skeletal muscle cell (fiber). This connection enables communication between the nervous system and a skeletal muscle fiber.   **9.** Acetylcholine (ACh) release is necessary for skeletal muscle contraction, because it serves as the first step in the

process that enables the formation of cross-bridges in the sarcomeres. A muscle's ability to contract depends on the formation of cross-bridges between the myosin heads and actin myofilaments. A drug that blocks ACh release would interfere with this cross-bridge formation and prevent muscle contraction. **10.** If the sarcolemma of a resting skeletal muscle suddenly became permeable to Ca^{2+}, the cytosolic concentration of Ca^{2+} would increase, and the muscle would contract. In addition, because the amount of calcium ions in the cytosol must decrease for relaxation to occur, the increased permeability of the sarcolemma to Ca^{2+} might prevent the muscle from relaxing completely. **11.** Without acetylcholinesterase (AChE), the motor end plate would be continuously stimulated by acetylcholine, locking the muscle in a state of contraction.

PAGE 342 **12.** A muscle's ability to contract depends on the formation of cross-bridges between the myosin and actin myofilaments in the muscle. In a muscle that is overstretched, the myofilaments would overlap very little, so very few cross-bridges between myosin and actin could form, and thus the contraction would be weak. If the myofilaments did not overlap at all, then no cross-bridges would form and the muscle could not contract. **13.** Yes, a skeletal muscle can contract without shortening. The muscle can shorten (isotonic, concentric contraction), elongate (isotonic, eccentric contraction), or remain the same length (isometric contraction), depending on the relationship between the load (resistance) and the tension produced by actin–myosin interactions.

PAGE 346 **14.** Muscle cells synthesize ATP continuously by utilizing creatine phosphate (CP) and metabolizing glycogen and fatty acids. Most cells generate ATP only through aerobic metabolism in the mitochondria and through glycolysis in the cytoplasm.

15. Muscle fatigue is a muscle's reduced ability to contract due to low pH (lactic acid buildup and its dissociation to lactate and a hydrogen ion), low ATP levels, or other problems. **16.** Oxygen debt is the amount of oxygen required to restore normal, pre-exertion conditions in muscle tissue. **PAGE 350** **17.** The three types of skeletal muscle fibers are (1) fast fibers (also called white muscle fibers, fast-twitch glycolytic fibers, Type II-B fibers, and fast fatigue fibers); (2) slow fibers (also called red muscle fibers, slow-twitch oxidative fibers, Type I fibers, and slow oxidative fibers); and (3) intermediate fibers (also called fast-twitch oxidative fibers, Type II-A fibers, and fast resistant fibers). **18.** A sprinter requires large amounts of energy for a short burst of activity. To supply this energy, the sprinter's muscles rely on anaerobic metabolism. Anaerobic metabolism is less efficient in producing energy than aerobic metabolism, and the process also produces acidic wastes; this combination contributes to muscle fatigue. Conversely, marathon runners derive most of their energy from aerobic metabolism, which is more efficient and produces fewer wastes than anaerobic metabolism does.

19. Activities that require short periods of strenuous activity produce a greater oxygen debt, because such activities rely heavily on energy production by anaerobic metabolism. Because lifting weights is more strenuous over the short term than swimming laps, which is an aerobic activity, weight lifting would likely produce a greater oxygen debt than would swimming laps. **20.** Individuals who excel at endurance activities have a higher than normal percentage of slow fibers. Slow fibers are physiologically better adapted to this type of activity than are fast fibers, which are less vascular and fatigue faster. **PAGE 351** **21.** Compared to skeletal muscle, cardiac muscle (1) has relatively small cells; (2) has cells with a centrally located nucleus (some may contain two or more nuclei); (3) has T tubules that are short and broad and do not form triads; (4) has an SR that lacks terminal cisternae and has tubules that contact

the plasma membrane as well as the T tubules; (5) has cells that are nearly totally dependent on aerobic metabolism as an energy source; and (6) contains intercalated discs that assist in stabilizing tissue structure and spreading action potentials. **22.** Cardiac muscle cells are joined by gap junctions, which allow ions and small molecules to flow directly between cells. As a result, action potentials generated in one cell spread rapidly to adjacent cells. Thus, all the cells contract simultaneously, as if they were a single unit (a syncytium). **PAGE 354** **23.** Smooth muscle cells lack sarcomeres, and thus smooth muscle tissue is nonstriated. Additionally, the thin filaments are anchored to dense bodies. **24.** Skeletal muscle contractions are least affected by changes in extracellular Ca^{2+} concentrations. In skeletal muscle, most of the calcium ions come from the sarcoplasmic reticulum. Most of the calcium ions that trigger a contraction in cardiac and smooth muscles come from the extracellular fluid. **25.** The looser organization of actin and myosin filaments in smooth muscle allows smooth muscle to contract over a wider range of resting lengths.

Answers to Review Questions

PAGE 356
Level 1 Reviewing Facts and Terms
1. (a) sarcolemma **(b)** sarcoplasm **(c)** mitochondria **(d)** myofibril **(e)** thin filament **(f)** thick filament **(g)** sarcoplasmic reticulum **(h)** T tubules **2.** d **3.** d **4.** c **5.** a **6.** d **7.** c **8.** d **9.** d **10.** (1) epimysium: surrounds entire muscle; (2) perimysium: surrounds muscle bundles (fascicles); and (3) endomysium: surrounds skeletal muscle fibers **11.** a **12.** The transverse (T) tubules propagate action potentials into the interior of the cell. **13.** (1) exposure of active sites; (2) attachment of cross-bridges; (3) pivoting of myosin heads (power stroke); (4) detachment of cross-bridges; and (5) activation of myosin heads (cocking) **14.** Both the frequency of motor unit stimulation and the number of motor units involved affect the amount of tension produced when a skeletal muscle contracts. **15.** ATP, creatine phosphate, glycogen **16.** Aerobic metabolism and glycolysis generate ATP from glucose in muscle cells. **17.** Depletion of metabolic substrates or enzymes. Decline in pH. Damage to the sacolemma or sarcoplasmic reticulum. Reduced blood flow and oxygen supply.

Level 2 Reviewing Concepts
18. In the inner range, neighboring thin filaments overlap each other at the sarcomere ends. In the outer range, the overlap of the thick and thin filaments almost disappears. Both situations reduce the potential number of crossbridges. **19.** Isotonic eccentric – the muscle lengthens during contraction. Isotonic concentric – the muscle shortens during contraction. Isometric – tension is generated but it does not exceed the load and so there is no movement of the muscle. **20.** a **21.** e **22.** They have a large capillary network and myoglobin stores which supply large quantities of oxygen to mitochondria for ATP production that use aerobic pathways. Crossbridges cycle slowly, so the potential of ATP demand outstripping supply is reduced. **23.** Calcium binds to troponin and weakens its bond with actin, which rolls the tropomyosin to expose the active sites. **24.** The timing of cardiac muscle contractions is determined by specialized cardiac muscle fibers called pacemaker cells; this property of cardiac muscle tissue is termed automaticity. **25.** If atracurium blocked the binding of ACh to ACh receptors at the motor end plates of neuromuscular junctions, the muscle's ability to contract would be inhibited. **26.** In rigor mortis, the membranes of the dead cells are no longer selectively permeable; the SR is no

longer able to retain calcium ions. As calcium ions enter the cytosol, a sustained contraction develops, making the body extremely stiff. Contraction persists because the dead muscle cells can no longer make the ATP required for cross-bridge detachment from the active sites. Rigor mortis begins a few hours after death and ends after one to six days, or when decomposition begins. Decomposition begins when the lysosomal enzymes released by autolysis break down the myofilaments.   **27.** The arrival of an action potential at the synaptic terminal stimulates secretory vesicles to release acetylcholine (ACh) into the synapse. ACh diffuses across the synaptic cleft and binds with the postsynaptic receptor, depolarising the postsynaptic membrane. The action potential is conducted along the sarcolemma into the muscle fiber via T-tubules and releases calcium ions from sarcoplasmic reticulum stores into the sacromere.

Level 3 Critical Thinking and Clinical Applications

28. Endurance athletes rely on aerobic metabolism for ATP production so increasing creatine phosphate levels would contribute little to improving performance. In contrast, sprinters and strength athletes use intense bursts of activity at peak levels, which rely on ATP, creatine phosphate stores and glycolysis to provide energy.   **29.** CK is the enzyme creatine kinase. It functions in the anaerobic reaction that transfers phosphate from creatine phosphate to ADP in muscle cells. The presence of cardiac troponin, a form found only in cardiac muscle cells, provides direct evidence that cardiac muscle cells have been severely damaged.   **30.** A skeletal muscle not regularly stimulated by a motor neuron will lose muscle tone and mass and become weak (it will atrophy). While his leg was immobilized, it did not receive sufficient stimulation to maintain proper muscle tone. It will take a while for Bill's muscles to build up enough to support his weight.

Answers to Clinical Case Wrap-Up

PAGE 357   1. A drug that inhibits acetylcholinesterase (AChE) is inhibiting the enzyme that breaks down the neurotransmitter acetylcholine (ACh). As a result, ACh stays around longer in the neuromuscular junction (NMJ). Because a patient with myasthenia gravis does not have enough functioning ACh receptors, a greater amount of ACh is better for strength of muscular response. In short, by inhibiting the breakdown of ACh, the drug strengthens the patient's muscular response.   **2.** Myasthenia gravis affects voluntary, skeletal muscle at the neuromuscular junction (NMJ). Cardiac muscle tissue contracts without neural stimulation, a property known as automaticity. There is no neural stimulation, no neuromuscular junction, no ACh neurotransmitter or receptor required for cardiac muscle contraction. For these reasons, myasthenia gravis would not affect cardiac muscle.

Chapter 11

Answers to Checkpoints

PAGE 361   1. Based on the patterns of fascicle organization, skeletal muscles can be classified as parallel muscles, convergent muscles, pennate muscles, or circular muscles.   **2.** Contraction of a pennate muscle generates more tension than would contraction of a parallel muscle of the same size because a pennate muscle contains more muscle fibers, and thus more myofibrils and sarcomeres, than does a parallel muscle of the same size.   **3.** The opening between the stomach and the small intestine would be guarded by a circular muscle, or sphincter. The concentric circles of muscle fibers found in sphincters are ideally suited for opening and closing

passageways and for acting as valves in the body.   **4.** A lever is a rigid structure—such as a board, a pry bar, or a bone—that moves on a fixed joint called the fulcrum. There are three classes of levers: In a first-class lever, the fulcrum lies between the applied force and load; in a second-class lever, the load lies between the applied force and fulcrum; and in a third-class lever, the applied force or pull exerted is between the fulcrum and the load. Third-class levers are the most common type in the body.   **5.** The joint between the occipital bone and the first cervical vertebra is part of a first-class lever system. The joint between the two bones (the fulcrum) lies between the skull (which provides the load) and the neck muscles (which provide the applied force).   **PAGE 365   6.** The origin of a muscle is the end that remains stationary during an action. Because the gracilis muscle moves the tibia, the origin of this muscle must be on the pelvis (pubis and ischium).   **7.** Muscles A and B are antagonists to each other, because they perform opposite actions.   **8.** A synergist is a muscle that helps a larger prime mover (or agonist—a muscle that is responsible for a specific movement) perform its actions more efficiently.   **PAGE 367   9.** Names of skeletal muscles are based on several factors, including location in the body, origin and insertion, fascicle organization, relative position, structural characteristics, and action. Names may also reflect the muscle shape, number of origins, and size.   **10.** The name *flexor carpi radialis longus* tells you that this muscle is a long muscle (longus) that lies next to the radius (radialis) and flexes (flexor) the wrist (carpi).   **PAGE 384   11.** Axial muscles arise on the axial skeleton. They position the head, neck, and vertebral column, move the rib cage, and form the perineum.   **12.** Contraction of the masseter muscle elevates the mandible, and relaxation of this muscle depresses the mandible. You would probably be chewing something.   **13.** You would expect the buccinator muscle, which positions the mouth for blowing, to be well developed in a trumpet player.   **14.** Swallowing involves contractions of the palatal muscles, which elevate the soft palate as well as portions of the superior pharyngeal wall. Elevation of the superior portion of the pharynx enlarges the opening to the auditory tube, permitting airflow to the middle ear and the inside of the eardrum. Making this opening larger by swallowing facilitates airflow into or out of the middle ear cavity.   **15.** Damage to the intercostal muscles would interfere with breathing.   **16.** A hit in the rectus abdominis muscle would cause that muscle to contract forcefully, resulting in flexion of the vertebral column. In other words, you would "double over."   **17.** The sore muscles are most likely the erector spinae muscles, especially the longissimus and the iliocostalis muscles of the lumbar region. These muscles would have to contract harder to counterbalance the increased anterior weight when carrying heavy boxes.

PAGE 401   18. When you shrug your shoulders, you are contracting your levator scapulae muscles.   **19.** The muscles involved in a rotator cuff injury are the supraspinatus, infraspinatus, teres minor, and subscapularis (SITS) muscles.   **20.** Injury to the flexor carpi ulnaris muscle would impair your ability to flex and adduct the wrist.   **21.** Injury to the obturator muscle would impair your ability to laterally rotate the hip.   **22.** A "pulled hamstring" refers to a strain affecting one or more of the three muscles that collectively flex the knee: the biceps femoris, semimembranosus, and semitendinosus muscles.   **23.** A torn calcaneal tendon would make plantar flexion difficult because this tendon attaches the soleus and gastrocnemius muscles to the calcaneus (heel bone).   **PAGE 405   24.** General age-related effects on skeletal muscles include decreased skeletal muscle fiber diameters, diminished muscle elasticity, decreased tolerance for exercise, and a decreased ability to recover from muscular injuries. **25.** Fibrosis is the development of increasing amounts of fibrous connective tissue. Fibrosis causes muscles to be less flexible, and the collagen fibers can restrict movement and circulation. **PAGE 407** **26.** The muscular system generates heat that maintains normal body temperature. **27.** Exercise affects the cardiovascular system by increasing heart rate and dilating blood vessels. With exercise, the rate and depth of breathing increases, and sweat gland secretion increases. The physiological effects that result from exercise are directed and coordinated by the nervous and endocrine systems.

Answers to Review Questions

PAGE 409
Level 1 Reviewing Facts and Terms
1. (a) deltoid muscle (multipennate); **(b)** extensor digitorum muscle (unipennate); **(c)** rectus femoris muscle (bipennate) **2. (a)** supraspinatus muscle; **(b)** infraspinatus muscle; **(c)** teres minor muscle **3.** The place where the fixed end attaches to a bone, cartilage, or connective tissue is called the origin of the muscle. The site where the movable end attaches to another structure is called the insertion of the muscle. **4.** a **5.** a **6.** a **7.** e **8.** c **9.** a **10.** Biceps brachii **11.** b **12.** The trapezius muscle **13.** a **14.** The sternocleidomastoid muscle extends from the clavicle and the sternum to the mastoid region of the skull. When it contracts the sternum is raised to increase lung inflation. **15.** It separates the abdominal and thoracic cavities or compartments. **16.** The axial musculature includes (1) muscles of the head and neck, (2) muscles of the vertebral column, (3) oblique and rectus muscles, and (4) muscles of the pelvic floor. **17.** The muscles of the pelvic floor (1) support the organs of the pelvic cavity, (2) flex joints of the sacrum and coccyx, and (3) control movement of materials through the urethra and anus. **18.** The tendons that cross the posterior and anterior surfaces of the wrist, pass through synovial tendon sheaths, elongated bursae that reduce friction. **19.** The functional muscle groups in the lower limbs are (1) muscles that move the thigh, (2) muscles that move the leg, and (3) muscles that move the foot and toes.

Level 2 Reviewing Concepts
20. It produces extension and lateral rotation at the hip joint. **21.** The spermatic cords pass through on their way from the testis. The condition is known as a hernia. **22.** The muscle tissue swells due to inflammation, and this puts pressure on the blood vessels which supply the tissue. The sheaths limit expansion and this leads to the blood vessels closing. The muscle becomes ischemic and may die. **23.** The Achilles tendon **24.** The hamstrings are connected to three posterior thigh muscles (semitendinosus, semimembranosus, and biceps femoris). **25.** In a convergent muscle, the direction of pull can be changed by stimulating only one group of muscle cells at any one time. When all the fibers contract at once, they do not pull as hard on the tendon as would a parallel muscle of the same size, because the muscle fibers on opposite sides of the tendon are pulling in different directions rather than working together. **26.** A pennate muscle contains more muscle fibers, and thus more myofibrils and sarcomeres, than does a parallel muscle of the same size, resulting in a contraction that generates more tension. **27.** Lifting heavy objects becomes easier as the elbow approaches a 90° angle. As you increase the angle at or near full extension, tension production decreases, so movement becomes more difficult. **28.** When the hamstrings are injured, flexion at the knee and extension at the hip are affected.

Level 3 Critical Thinking and Clinical Applications
29. Mary is not happy to see Jill. Contraction of the frontalis muscle would wrinkle Mary's brow, contraction of the procerus muscle would flare her nostrils, and contraction of the levator labii muscle on the right side would raise the right side of her lip, as in sneering. **30.** b **31.** Although the pectoralis muscle is located across the chest, it inserts on the greater tubercle of the humerus, the bone of the arm. When this muscle contracts, it contributes to flexion, adduction, and medial rotation of the humerus at the shoulder joint. All of these arm movements would be impaired if the muscle were damaged.

Answers to Clinical Case Wrap-Up
PAGE 410 **1.** In addition to the hamstrings (biceps femoris laterally, semitendinosus and semimembranosus medially) the sartorius muscle assists in knee flexion. The sartorius muscle is, however, too weak to do the job by itself. **2.** With the knee extended, and the avulsed hamstrings shortened to 70 percent of their resting length, the proximal tendons would not reach the ischial tuberosity. Flexing the knee moves the insertion closer to the origin, helping the tendons to reach their original attachment site.

Chapter 12
Answers to Checkpoints
PAGE 414 **1.** The two anatomical divisions of the nervous system are the central nervous system (CNS), consisting of the brain and spinal cord, and the peripheral nervous system (PNS), consisting of all neural tissue outside the CNS. **2.** The two functional divisions of the peripheral nervous system are the afferent division, which brings sensory information to the CNS from receptors in peripheral tissues and organs, and the efferent division, which carries motor commands from the CNS to muscles, glands, and adipose tissue. **3.** The two components of the efferent division of the PNS are the somatic nervous system (SNS) and the autonomic nervous system (ANS). **4.** Damage to the afferent division of the PNS, which is composed of nerves that carry sensory information to the brain and spinal cord, would interfere with a person's ability to experience a variety of sensory stimuli. **PAGE 418** **5.** Structural components of a typical neuron include a cell body or soma (which contains the nucleus and perikaryon), dendrites, an axon, telodendria, Nissl bodies, neurofilaments, intermediate neurotubules, neurofibrils, axoplasm, axolemma, initial segment, axon hillock, and collaterals. **6.** According to structure, neurons are classified as anaxonic, bipolar, unipolar, and multipolar. **7.** According to function, neurons are classified as sensory neurons, motor neurons, and interneurons. **8.** Because most sensory neurons of the PNS are unipolar, these neurons most likely function as sensory neurons. **PAGE 424** **9.** Central nervous system neuroglia include ependymal cells, astrocytes, oligodendrocytes, and microglia. **10.** Peripheral nervous system neuroglia include satellite cells and Schwann cells (neurilemma cells). **11.** The small phagocytic cells called microglia occur in increased numbers in infected (and damaged) areas of the CNS. **PAGE 432** **12.** The resting membrane potential is the membrane potential of a normal, unstimulated cell. **13.** If the voltage-gated sodium channels in the plasma membrane of a neuron could not open, sodium ions could not flood into the neuron and it would not be able to depolarize. **14.** If the extracellular concentration of potassium ions decreased, more potassium would leave the cell, and the membrane potential would become more negative. This condition is called hyperpolarization. **PAGE 438** **15.** An action potential is a propagated change

in the membrane potential of excitable cells, initiated by a change in the membrane permeability to sodium ions. **16.** The four steps involved in the generation of action potentials are (1) depolarization to threshold; (2) activation of voltage-gated sodium channels and rapid depolarization; (3) inactivation of voltage-gated sodium channels and activation of voltage-gated potassium channels; and (4) return to normal permeability. **PAGE 413** **17.** The presence of myelin greatly increases the propagation speed of action potentials. **18.** Action potentials travel along myelinated axons (by saltatory propagation) at much higher speeds than along unmyelinated axons. The axon with a propagation speed of 50 meters per second must be the myelinated axon. Typical propagation speeds for unmyelinated axons are about 1 m/sec. **PAGE 442** **19.** The major structural components of a synapse, the site where a neuron communicates with another cell, are a presynaptic cell and a postsynaptic cell, whose plasma membranes are separated by a narrow synaptic cleft. **20.** If a synapse involves direct physical contact between cells, it is termed electrical; if the synapse involves a neurotransmitter, it is termed chemical. **21.** If the voltage-gated calcium ion channels at a cholinergic synapse were blocked, Ca^{2+} could not enter the axon terminal and trigger the release of ACh into the synapse, so no communication would take place across the synapse. **22.** Because of synaptic delay, the pathway with fewer neurons (in this case, three) will transmit impulses more rapidly. **PAGE 447** **23.** Both neurotransmitters and neuromodulators are substances that are released by one neuron and that affect another neuron. A neurotransmitter alters the membrane potential of the other neuron, whereas a neuromodulator alters the other neuron's response to specific neurotransmitters. **24.** Neurotransmitters and neuromodulators are either (1) compounds that have a direct effect on membrane potential, (2) compounds that have an indirect effect on membrane potential, or (3) lipid-soluble gases that exert their effects inside the cell. **PAGE 451** **25.** No action potential will be generated. **26.** Yes, an action potential will be generated. **27.** Spatial summation would occur if the two EPSPs happened simultaneously.

Answers to Review Questions

PAGE 453
Level 1 Reviewing Facts and Terms
1. (a) dendrite; (b) nucleolus; (c) nucleus; (d) axon hillock; (e) initial segment; (f) axolemma; (g) axon; (h) telodendria; (i) axon terminals **2.** c **3.** a **4.** c **5.** b **6.** A shift from the resting membrane potential toward a more positive potential. For example, a move from the resting potential of -70mV to -45mV is a move from a negative to a more positive charge. **7.** a **8.** Voltage gated, mechanically gated, chemically gated. **9.** (a) the CNS: brain and spinal cord (b) the PNS: all other nerve fibers, divided between the efferent division (which consists of the somatic nervous system and the autonomic nervous system) and the afferent division (which consists of receptors and sensory neurons) **10.** Neuroglia in the PNS are (1) satellite cells and (2) Schwann cells. **11.** (1) sensory neurons: transmit impulses from the PNS to the CNS; (2) motor neurons: transmit impulses from the CNS to peripheral effectors; and (3) interneurons: analyze sensory inputs and coordinate motor outputs

Level 2 Reviewing Concepts
12. More Na^+ ions enter the cell, encouraging the cell to depolarize by the activation of more Na^+ channels. **13.** a **14.** Action potentials are changes in the transmembrane potential that propagate along the membrane. Graded potentials are changes in the transmembrane potential that cannot spread from the initial site. **15.** Chemically gated (ligand gated) channels open or close when they bind specific extracellular chemicals (ligands). Voltage-gated channels open or close in response to changes in the membrane potential. Mechanically gated channels open or close in response to physical distortion of the membrane surface. **16.** The all-or-none principle of action potentials states that any depolarization event sufficient to reach threshold will cause an action potential of the same strength, regardless of the amount of stimulation above threshold. **17.** The membrane depolarizes to threshold. Next, voltage-gated sodium channels are activated, and the membrane rapidly depolarizes. These sodium channels are then inactivated, and voltage-gated potassium channels are activated. Finally, normal permeability returns. The voltage-gated sodium channels become activated once the repolarization is complete; the voltage-gated potassium channels begin closing as the membrane potential reaches the normal resting membrane potential. **18.** In saltatory propagation, which occurs in myelinated axons, only the nodes along the axon can respond to a depolarizing stimulus. In continuous propagation, which occurs in unmyelinated axons, an action potential appears to move across the membrane surface in a series of tiny steps. **19.** Type A fibers are myelinated and carry action potentials very quickly (120 m/sec). Type B fibers are also myelinated, but carry action potentials more slowly due to their smaller diameter. Type C fibers are extremely slow due to their small diameter and lack of myelination. **20.** (1) The action potential arrives at the axon terminal, depolarizing it; (2) extracellular calcium enters the axon terminal, triggering the exocytosis of ACh; (3) ACh binds to the postsynaptic membrane and depolarizes the next neuron in the chain; (4) ACh is removed by AChE. **21.** Temporal summation is the addition of stimuli that arrive at a single synapse in rapid succession. Spatial summation occurs when simultaneous stimuli at multiple synapses have a cumulative effect on the membrane potential.

Level 3 Critical Thinking and Clinical Applications
22. Harry's kidney condition is causing the retention of potassium ions. As a result, the K^+ concentration of the extracellular fluid is higher than normal. Under these conditions, less potassium diffuses from heart muscle cells than normal, resulting in a resting membrane potential that is less negative (more positive). This change in the resting membrane potential moves the membrane potential closer to threshold, so it is easier to stimulate the muscle. The ease of stimulation accounts for the increased number of contractions evident in the rapid heart rate. **23.** To reach threshold, the postsynaptic membrane must receive enough neurotransmitter to produce an EPSP of +20 mV (+10 mV to reach threshold and +10 mV to cancel the IPSPs produced by the five inhibitory neurons). Each neuron releases enough neurotransmitter to produce a change of +2 mV, so at least 10 of the 15 excitatory neurons must be stimulated to produce this effect by spatial summation. **24.** Even if threshold is reached, action potentials do not occur as sodium ions are prevented from entering the axon and signals from pain receptors are not propagated. **25.** The absolute refractory period limits the number of action potentials that can travel along an axon in a given unit of time. During the absolute refractory period, the membrane cannot conduct an action potential, so a new depolarization event cannot occur until after the absolute refractory period has passed. If the absolute refractory period for a particular axon is 0.001 sec, then the maximum frequency of action potentials conducted by this axon would be 1000/sec.

Answers to Clinical Case Wrap-Up

PAGE 454 1. The polio virus attacks the motor neurons of the spinal cord gray matter. **2.** Guillain-Barré syndrome is an autoimmune disorder that attacks the myelin covering of peripheral nerves in adults. This causes both sensory and motor symptoms, usually in a symmetrical, ascending pattern over a few weeks. Involvement of the face, urinary bladder, and bowel is common.

Chapter 13

Answers to Checkpoints

PAGE 457 1. The central nervous system (CNS) is made up of the brain and spinal cord, while cranial nerves and spinal nerves make up the peripheral nervous system (PNS). **2.** A spinal reflex is a rapid, automatic response triggered by specific stimuli that is controlled in the spinal cord. **PAGE 461 3.** The three spinal meninges are the dura mater, arachnoid mater, and pia mater. **4.** Damage to the ventral root of a spinal nerve, which contains both visceral and somatic motor fibers, would interfere with motor function. **5.** The cerebrospinal fluid that surrounds the spinal cord is located in the subarachnoid space, which lies beneath the epithelium of the arachnoid mater and superficial to the pia mater. **PAGE 463 6.** Sensory nuclei receive and relay sensory information from peripheral receptors. Motor nuclei issue motor commands to peripheral effectors. **7.** The polio virus–infected neurons would be in the anterior gray horns of the spinal cord, where the cell bodies of somatic motor neurons are located. **8.** A disease that damages myelin sheaths would affect the columns in the white matter of the spinal cord, because the columns are composed of bundles of myelinated axons. **PAGE 473 9.** The major nerve plexuses are the cervical, brachial, lumbar, and sacral. **10.** An anesthetic that blocks the function of the dorsal rami of the cervical spinal nerves would affect the skin and muscles of the back of the neck and of the shoulders. **11.** Damage to the cervical plexus—or more specifically to the left and right phrenic nerves, which originate in this plexus and innervate the diaphragm—would greatly interfere with the ability to breathe and might even be fatal. **12.** Compression of the sciatic nerve produces the sensation that your leg has "fallen asleep." **PAGE 475 13.** A neuronal pool is a functional group of organized interneurons within the CNS. **14.** The five neuronal pool circuit patterns are divergence, convergence, serial processing, parallel processing, and reverberation. **PAGE 479 15.** A reflex is a rapid, automatic response to a specific stimulus. It is an important mechanism for maintaining homeostasis. **16.** The minimum number of neurons required for a reflex arc is two. One must be a sensory neuron that brings impulses to the CNS, and the other a motor neuron that transmits a response to the effector. **17.** The suckling reflex is an innate reflex. **PAGE 481 18.** All polysynaptic reflexes involve pools of interneurons; are intersegmental in distribution; and involve reciprocal inhibition, reverberating circuits, and the cooperation of several reflexes. **19.** When intrafusal fibers are stimulated by gamma motor neurons, the muscle spindles become more sensitive. As a result, little if any stretching stimulus would be needed to stimulate the contraction of the quadriceps muscles in the patellar reflex. For this reason, the reflex response would appear more quickly. **20.** This response is the tendon reflex. **21.** During a withdrawal reflex, the limb on the opposite side is extended. This response is called a crossed extensor reflex. **PAGE 482 22.** Reinforcement is an enhancement of spinal reflexes. It occurs when the postsynaptic neuron enters a state of generalized facilitation caused by chronically active excitatory synapses. **23.** A positive Babinski reflex is abnormal in adults. It indicates possible damage to descending tracts in the spinal cord.

Answers to Review Questions

PAGE 485
Level 1 Reviewing Facts and Terms
1. (a) white matter; **(b)** ventral root; **(c)** dorsal root; **(d)** pia mater; **(e)** arachnoid mater; **(f)** gray matter; **(g)** spinal nerve; **(h)** dorsal root ganglion; **(i)** dura mater **2.** d **3.** a **4.** d **5.** c **6.** c **7.** c **8.** Pia, arachnoid, dura. **9.** a **10.** d **11.** a **12.** Synapse **13.** c **14.** c **15. (a)** 1 **(b)** 7 **(c)** 3 **(d)** 5 **(e)** 4 **(f)** 6 **(g)** 2 **(h)** 8

Level 2 Reviewing Concepts
16. The vertebral column continues to grow, extending beyond the spinal cord. The end of the spinal cord is visible as the conus medullaris near L_1, and the cauda equina extends the remainder of the vertebral column. **17.** (1) arrival of stimulus and activation of receptor; (2) activation of sensory neuron; (3) information processing; (4) activation of a motor neuron; and (5) response by an effector (muscle, gland, or adipose tissue) **18.** d **19.** The first cervical nerve exits superior to vertebra C_1 (between the skull and vertebra); the last cervical nerve exits inferior to vertebra C_7 (between the last cervical vertebra and the first thoracic vertebra). So, there are 8 cervical nerves but only 7 cervical vertebrae. **20.** The axons are enclosed in myelin and the fatty material appears white. The cell bodies are free of myelin and they appear gray. **21.** Within the CNS, cerebrospinal fluid fills the central canal in the spinal cord, the ventricles of the brain, and the subarachnoid space between the arachnoid mater and the pia mater. CSF acts as a shock absorber and a diffusion medium for dissolved gases, nutrients, chemicals, and wastes. **22.** (1) involvement of pools of interneurons; (2) intersegmental distribution; (3) involvement of reciprocal innervation; (4) motor response prolonged by reverberating circuits; and (5) cooperation of reflexes to produce a coordinated, controlled response **23.** Transection of the spinal cord at segment C_7 would most likely result in paralysis from the neck down. Transection at segment T_{10} would produce paralysis and eliminate sensory input in the lower half of the body only. **24.** Convergence is the focus of several signal pathways directed to a single pathway, such as the various decision and modulating centers which operate the final motor pathways. Divergence is the routing of sensory signals along several pathways as the impulses reach the brain and which are conveyed to more than one brain center. **25.** b **26.** a **27.** The epineurium, or outermost layer, consists of a dense network of collagen fibers. The fibers of the perineurium, the middle layer, extend inward from the epineurium. These connective tissue partitions divide the nerve into a series of compartments that contain bundles of axons, or fascicles. Delicate connective tissue fibers of the endoneurium, the innermost layer, extend from the perineurium and surround individual axons.

Level 3 Critical Thinking and Clinical Applications
28. The median nerve is involved. **29.** The radial nerve is involved. **30.** The individual would still exhibit a defecation (bowel) and urination (urinary bladder) reflex because these spinal reflexes are processed at the level of the spinal cord. Efferent impulses from the organs would stimulate specific interneurons in the sacral region that synapse with the motor neurons controlling the sphincters, thus bringing about emptying when the organs began to fill. (This is the same situation that exists in a newborn infant who has not yet fully

developed the descending tracts required for conscious control.) However, an individual with the spinal cord transection at L₁ would lose voluntary control of the bowel and bladder because these functions rely on impulses carried by motor neurons in the brain that must travel down the cord and synapse with the interneurons and motor neurons involved in the reflex.   **31.** The anterior horn in the lumbar region of the spinal cord contains somatic motor neurons that direct the activity of skeletal muscles of the hip, lower limb, and foot. As a result of the injury, Karen would be expected to have poor control of most muscles of the lower limbs, causing difficulty walking (if she could walk at all) and problems maintaining balance (if she could stand).

Answers to Clinical Case Wrap-Up

PAGE 486   1. No recovery of motor activity or sensation is to be expected inferior to the level of Joe's complete injury at T₁₁. Central nervous tissue (brain and spinal cord) has very limited healing ability. **2.** Joe's Babinski reflexes would be positive, with toes going up, or dorsiflexed, which is abnormal in adults (but not in infants). The descending motor pathways to the feet have been completely disrupted at T₁₁.

Chapter 14

Answers to Checkpoints

PAGE 491   1. The six major regions of the brain are the cerebrum, cerebellum, diencephalon, midbrain, pons, and medulla oblongata.   **2.** The brain stem consists of the midbrain, pons, and medulla oblongata.   **3.** The rhombencephalon or hindbrain develops into the cerebellum, pons, and medulla oblongata.

PAGE 496   4. The layers of the cranial meninges are the outer dura mater, the middle arachnoid mater, and the inner pia mater.   **5.** If the normal circulation or resorption of cerebrospinal fluid (CSF) became blocked, CSF would continue to be produced at the choroid plexuses in each ventricle, but the fluid would remain there, causing the ventricles to swell—a condition known as hydrocephalus. **6.** If diffusion across the arachnoid granulations decreased, less CSF would re-enter the bloodstream, and CSF would accumulate in the ventricles. The increased pressure within the ventricles due to accumulated CSF could damage the brain.   **7.** Many water-soluble molecules are rare or absent in the extracellular fluid (ECF) of the brain because the blood–brain barrier regulates the movement of such molecules from the blood to the ECF of the brain.

PAGE 498   8. The nucleus gracilis and nucleus cuneatus are responsible for relaying somatic sensory information to the thalamus.   **9.** Damage to the medulla oblongata can be lethal because it contains many vital autonomic reflex centers, including those that control breathing and regulate heart rate and force of contraction, which affect blood pressure.   **PAGE 499   10.** The pons contains (1) sensory and motor nuclei of cranial nerves, (2) nuclei involved with the control of respiration, (3) nuclei and tracts that process and relay information heading to or from the cerebellum, and (4) ascending, descending, and transverse tracts.   **11.** Damage to the respiratory centers of the pons could result in loss of ability to modify the rhythmicity centers of the medulla oblongata, which set the basic pace for respiratory movements. **PAGE 475   12.** Components of the cerebellar gray matter include the cerebellar cortex and cerebellar nuclei.   **13.** The arbor vitae, which is part of the cerebellum, connects the cerebellar cortex and nuclei with cerebellar peduncles.   **PAGE 503   14.** Two pairs of sensory nuclei make up the corpora quadrigemina: the superior colliculi and inferior colliculi.   **15.** The superior colliculi of the

midbrain control reflexive movements of the eyes, head, and neck to visual stimuli, such as a bright light.   **PAGE 506   16.** The main components of the diencephalon are the epithalamus, thalamus, and hypothalamus.   **17.** Damage to the lateral geniculate nuclei would interfere with the sense of sight.   **18.** Changes in body temperature stimulate the preoptic area of the hypothalamus, a component of the diencephalon.   **PAGE 507   19.** The limbic system is responsible for processing memories and creating emotional states, drives, and associated behaviors.   **20.** Damage to the amygdaloid body would interfere with the sympathetic ("fight or flight") division of the autonomic nervous system (ANS).   **PAGE 517   21.** Projection fibers link the cerebral cortex to the spinal cord, passing through the diencephalon, brain stem, and cerebellum.   **22.** Damage to the basal nuclei would result in decreased muscle tone and the loss of coordination of learned movement patterns.   **23.** The primary motor cortex is located in the precentral gyrus of the frontal lobe of the cerebrum. **24.** Damage to the temporal lobes of the cerebrum would interfere with the processing of olfactory (smell) and auditory (sound) impulses.   **25.** The stroke has damaged the speech center, located in the frontal lobe.   **26.** The temporal lobe of the cerebrum is probably involved, specifically the hippocampus and the amygdaloid body. His problems may also involve other parts of the limbic system that act as a gate for loading and retrieving long-term memories.   **PAGE 529   27.** Cranial reflexes are automatic responses to stimuli that involve the sensory and motor fibers of cranial nerves. Both monosynaptic and polysynaptic reflex arcs occur. Cranial reflex testing is often used to assess damage to cranial nerves or to the associated processing centers in the brain.   **27.** Shining a light in a person's eye tests the direct light reflex and the consensual light reflex, which are visceral reflexes.

Answers to Review Questions

PAGE 532
Level 1 Reviewing Facts and Terms
1. (a) cerebrum; **(b)** diencephalon; **(c)** midbrain; **(d)** pons; **(e)** medulla oblongata; **(f)** cerebellum   **2. (a)** dura mater (periosteal layer); **(b)** dural sinus; **(c)** dura mater (meningeal layer); **(d)** subdural space; **(e)** arachnoid mater; **(f)** subarachnoid space; **(g)** pia mater   **3.** d   **4.** b   **5.** Vasomotor   **6.** d   **7.** b   **8.** Cardiovascular center   **9.** a   **10.** b   **11.** Fight or flight response **12.** a   **13.** a   **14.** It is a major fiber tract that connects the two cerebral hemispheres.   **15.** (1) portions of the hypothalamus where the capillary endothelium is extremely permeable; (2) capillaries in the pineal gland; and (3) capillaries at the choroid plexus   **16.** N I: olfactory nerve; N II: optic nerve; N III: oculomotor nerve; N IV: trochlear nerve; N V: trigeminal nerve; N VI: abducens nerve; N VII: facial nerve; N VIII: vestibulocochlear nerve; N IX: glossopharyngeal nerve; N X: vagus nerve; N XI: accessory nerve; and N XII: hypoglossal nerve

Level 2 Reviewing Concepts
17. The direct light reflex   **18.** Coordination   **19.** d   **20.** In Parkinson's disease, the substantia nigra is inhibited from secreting the neurotransmitter, dopamine, at the basal nuclei. This is generally due to the loss of dopamine-producing neurons in the substantia nigra.   **21.** Stroke   **22.** Alpha waves   **23.** This nucleus is the hippocampus, which is part of the limbic system.   **24.** The left cerebral hemisphere contains the general interpretive and speech centers and is responsible for analytical tasks, logical decision-making, and

language-based skills (reading, writing, and speaking). The right hemisphere analyzes sensory information and relates the body to the sensory environment. Interpretive centers in this hemisphere permit the identification of familiar objects by touch, smell, sight, taste, or feel. The right cerebral hemisphere is also important in understanding three-dimensional relationships and in analyzing the emotional context of a conversation.   **25.** d   **26.** c   **27.** Lesions in the general interpretive area (Wernicke's area, sensory) cause defective visual and auditory comprehension of language, repetition of spoken sentences, and defective naming of objects. Lesions in the speech center (Broca's area, motor) result in hesitant and distorted speech.

Level 3 Critical Thinking and Clinical Applications

28. Sensory innervation of the nasal mucosa occurs by the maxillary branch of the trigeminal nerve (N V). Irritation of the nasal lining by ammonia increases the frequency of action potentials along the maxillary branch of the trigeminal nerve through the semilunar ganglion to reach centers in the midbrain, which in turn excite the neurons of the reticular activating system (RAS). Increased activity by the RAS can bring the cerebrum back to consciousness.
29. The officer is testing the function of Bill's cerebellum. Many drugs, including alcohol, have pronounced effects on cerebellar function. A person who is under the influence of alcohol cannot properly anticipate the range and speed of limb movement, because processing and correction by the cerebellum are slow. Because of disturbance in muscular coordination (ataxia), Bill might have a difficult time walking a straight line or touching his finger to his nose.   **30.** Increasing pressure in the cranium could compress important blood vessels, leading to further brain damage in areas not directly affected by the hematoma. Pressure on the brain stem could disrupt vital respiratory, cardiovascular, and vasomotor functions and possibly cause death. Pressure on the motor nuclei of the cranial nerves would lead to drooping eyelids and dilated pupils. Pressure on descending motor tracts would impair muscle function and decrease muscle tone in the affected areas of the body.
31. In any inflamed tissue, edema occurs in the area of inflammation. The accumulation of fluid in the subarachnoid space can cause damage by pressing against neurons. If the intracranial pressure is excessive, brain damage can occur, and if the pressure involves vital autonomic reflex areas, death could occur.   **32.** Most of the functional problems observed in shaken baby syndrome are the result of trauma to the cerebral hemispheres due to contact between the brain and the inside of the skull. Damage to and distortion of the brain stem and medulla oblongata can cause death.

Answers to Clinical Case Wrap-Up

PAGE 533   1. The speech center, also called Broca's area or the motor speech area, is usually located in the left hemisphere. Dr. Taylor was unable to speak or understand speech after her stroke. The primary motor cortex that directs voluntary movements of the right side of the body is located in the left hemisphere, in the precentral gyrus of the cerebral cortex. Dr. Taylor was unable to use her right arm after her stroke. For these reasons, you would be able to determine that the left side of her brain was affected by the stroke.   **2.** Neuroplasticity is the ability of the nerve cells in the brain to make new connections to other nerve cells. Following trauma to the brain, as in Dr. Taylor's stroke, brain cells that are still alive but dormant can make new connections with other nerve cells. We can think of these changes as healing after a stroke or other trauma to the brain.

Chapter 15

Answers to Checkpoints

PAGE 535
1. The specialized cells that monitor specific conditions in the body or the external environment are sensory receptors.
2. Yes, it is possible for somatic motor commands to occur at the subconscious level. They also occur at the conscious level.
PAGE 539   3. Receptor A provides more precise sensory information because it has a smaller receptive field on the skin surface.
4. Adaptation is a decrease in receptor sensitivity or a decrease in perception after constant stimulation.   **PAGE 543   5.** The four types of general sensory receptors (and the stimuli that excite them) are nociceptors (pain), thermoreceptors (temperature), mechanoreceptors (physical distortion), and chemoreceptors (chemical concentration).   **6.** The three classes of mechanoreceptors are tactile receptors, baroreceptors, and proprioceptors.   **7.** If proprioceptors in your legs could not relay information about limb position and movement to the CNS (especially the cerebellum), your movements would be uncoordinated and you likely could not walk.
PAGE 549   8. The tract being compressed is the fasciculus gracilis in the posterior column of the spinal cord, which carries information about touch and pressure from the lower limbs to the brain.   **9.** The tracts that carry action potentials generated by nociceptors (pain receptors) are the lateral spinothalamic tracts.
10. The left cerebral hemisphere (specifically, the primary sensory cortex) receives impulses carried by the right fasciculus gracilis.   **PAGE 554   11.** The anatomical basis for opposite-side motor control is crossing-over (decussation) of axons, so the motor fibers of the corticospinal pathway innervate lower motor neurons on the opposite side of the body.   **12.** An injury involving the superior portion of the motor cortex would affect the ability to control the muscles in the upper limb and the proximal portion of the lower limb.   **13.** Increased stimulation of the motor neurons of the red nucleus would increase stimulation of the skeletal muscles in the upper limbs, thereby increasing their muscle tone.

Answers to Review Questions

PAGE 555
Level 1 Reviewing Facts and Terms
1. c   **2.** Phasic   **3.** Adaptation is a reduction in sensitivity in the presence of a constant stimulus.   **4.** d
5.

6. (1) free nerve endings: sensitive to touch and pressure; (2) root hair plexus: monitors distortions and movements across the body surface; (3) tactile discs: detect fine touch and pressure; (4) tactile corpuscles: detect fine touch and pressure; (5) lamellated corpuscles: sensitive to pulsing or vibrating stimuli (deep pressure); and (6) Ruffini corpuscles: sensitive to pressure and distortion of the skin **7.** (1) tactile receptors; (2) baroreceptors; and (3) proprioceptors **8.** (1) spinothalamic pathway: provides conscious sensations of poorly localized ("crude") touch, pressure, pain, and temperature; (2) posterior column pathway: provides conscious sensations of highly localized ("fine") touch, pressure, vibration, and proprioception; and (3) spinocerebellar pathway: carries proprioceptive information about the position of skeletal muscles, tendons, and joints to the cerebellum **9.** (1) corticobulbar tracts; (2) lateral corticospinal tracts; and (3) anterior corticospinal tracts **10.** (1) vestibulospinal tract; (2) tectospinal tract; and (3) reticulospinal tract **11.** (a) Carries sensation of fast pain or prickling pain, (b) carries sensation of slow pain, or burning and aching pain. **12.** a **13.** We use the term general senses to describe our sensitivity to temperature, pain, touch, pressure, vibration, and proprioception (body position). The special senses are olfaction (smell), vision (sight), gustation (taste), equilibrium (balance), and hearing.

Level 2 Reviewing Concepts
14. Muscles and glands **15.** A motor homunculus, a mapped-out area of the primary motor cortex, provides an indication of the degree of fine motor control available. A sensory homunculus indicates the relative number of peripheral sensory receptors. **16.** A sensory neuron that delivers sensations to the CNS is a first-order neuron. Within the CNS, the axon of the first-order neuron synapses on a second-order neuron, which is an interneuron located in the spinal cord or brain stem. The second-order neuron synapses on a third-order neuron in the thalamus. The axons of third-order neurons synapse on neurons of the primary sensory cortex of the cerebral hemispheres. **17.** Damage to the posterior spinocerebellar tract on the left side of the spinal cord at the L_1 level would interfere with the coordinated movement of the left leg. **18.** Injury to the primary motor cortex affects the ability to exert fine control over motor units. Gross movements are still possible, however, because they are controlled by the basal nuclei that use the reticulospinal or rubrospinal tracts. Thus, walking and other voluntary and involuntary movements can be performed with difficulty, and the movements are imprecise and awkward. **19.** Muscle tone is the continuous and passive partial contraction of the muscles when they are at rest. It helps maintain posture and is reduced during sleep. Muscle tone disappears if the nerve supply is severed and after death. **20.** Strong pain sensations arriving at a particular segment of the spinal cord can cause stimulation of the interneurons of the spinothalamic pathway. This stimulation is interpreted by the sensory cortex as originating in the region of the body surface associated with the origin of that same pathway.

Level 3 Critical Thinking and Clinical Applications
21. Kayla's tumor is most likely adjacent to the corticobulbar tracts. The axons of those tracts carry action potentials to motor nuclei of the cranial nerves, which control eye muscles and muscles of facial expression. **22.** Injuries to the motor cortex

eliminate the ability to produce fine control of motor units. However, as long as the cerebral nuclei are functional, gross movements would still be possible. Harry should still be able to walk, maintain his balance, and perform voluntary and involuntary movements using the rubrospinal and reticulospinal tracts in place of the corticospinal tracts. Although these movements may be awkward or difficult, they will still be possible. **23.** Denzel is experiencing phantom pain. Since pain perception occurs in the sensory cortex of the brain, he can still feel pain in his fingers if the brain projects feeling to that area. When he bumps the arm at the elbow, sensory receptors are stimulated to send impulses to the sensory cortex. The brain perceives a sensation from a general area, and projects that feeling to a body part. Since more sensory information reaches the brain from the hands and fingers, it is not unusual for the brain to project to this area.

Answers to Clinical Case Wrap-Up
PAGE 530 **1.** All the principal descending motor pathways carry information that would have been affected by the brain damage causing Michael's cerebral palsy. They include the corticospinal pathway, the medial pathway, and the lateral pathway. **2.** Poor motor control of the muscles used in swallowing can contribute to an increased risk of choking and poor infant nutrition, which can affect development and even be life-threatening.

Chapter 16
Answers to Checkpoints
PAGE 562 **1.** The two major divisions of the autonomic nervous system are the sympathetic division and the parasympathetic division. **2.** Two neurons are needed to carry an action potential from the spinal cord to smooth muscles in the intestine. The first neuron carries the action potential from the spinal cord to the autonomic ganglion, and a second neuron carries the action potential from the autonomic ganglion to the smooth muscle. **3.** The sympathetic division of the ANS is responsible for the physiological changes that occur in response to stress (confronting an angry dog) and increased activity (running). **4.** The sympathetic division of the autonomic nervous system includes preganglionic fibers from the lumbar and thoracic regions of the spinal cord, whereas the parasympathetic division includes preganglionic fibers from the cranial and sacral regions. **PAGE 566** **5.** The nerves that synapse in collateral ganglia originate in the inferior thoracic and superior lumbar regions of the spinal cord. The preganglionic fibers they contain pass through the sympathetic chain ganglia to the collateral ganglia. **PAGE 568** **6.** Because preganglionic fibers of the sympathetic nervous system release acetylcholine (ACh), a drug that stimulates ACh receptors would stimulate the postganglionic fibers of sympathetic nerves, resulting in increased sympathetic activity. **7.** Blocking the beta receptors on cells would decrease or prevent sympathetic stimulation of tissues containing those cells. Heart rate, force of contraction of cardiac muscle, and contraction of smooth muscle in the walls of blood vessels would decrease, lowering blood pressure. **PAGE 570** **8.** The vagus nerve (N X) carries preganglionic parasympathetic fibers that innervate the lungs, heart, stomach, liver, pancreas, and parts of the small and large intestines (as well as several other visceral organs). **9.** The parasympathetic division is sometimes referred to as the anabolic

system because parasympathetic stimulation leads to a general increase in the nutrient content of the blood. Cells throughout the body respond to the increase by absorbing the nutrients and using them to support growth and other anabolic activities. **PAGE 571   10.** Acetylcholine (ACh) is the neurotransmitter released by all parasympathetic neurons.   **11.** The two types of cholinergic receptors on the postsynaptic membranes of parasympathetic neurons are nicotinic receptors and muscarinic receptors.   **12.** Stimulation of muscarinic receptors, a type of cholinergic receptor located in postganglionic synapses of the parasympathetic nervous system, would cause K$^+$ channels to open, resulting in hyperpolarization of cardiac plasma membranes and a decreased heart rate.   **PAGE 575   13.** Most blood vessels receive sympathetic stimulation, so a loss of sympathetic tone would relax the smooth muscles lining the vessels; the resulting vasodilation would increase blood flow to the tissue. **14.** In anxious individuals, an increase in sympathetic stimulation would probably cause some or all of the following changes: a dry mouth; increased heart rate, blood pressure, and rate of breathing; cold sweats; an urge to urinate or defecate; a change in the motility of the digestive tract (that is, "butterflies in the stomach"); and dilated pupils.   **PAGE 577   15.** A visceral reflex is an autonomic reflex initiated in the viscera. It is an automatic motor response that can be modified, facilitated, or inhibited by higher centers, especially those of the hypothalamus. **16.** A brain tumor that interferes with hypothalamic function would also interfere with autonomic function, because centers in the hypothalamus are involved with autonomic (visceral) function.   **PAGE 582   17.** Higher-order functions require action by the cerebral cortex, involve both conscious and unconscious information processing, and are subject to modification and adjustment over time.   **18.** Test-taking involves short-term memory, although your instructor would like you to transfer this information to long-term memory.   **19.** The two general levels of sleep are deep sleep and rapid eye movement (REM) sleep.   **20.** If your RAS were suddenly stimulated, it would rouse the cerebrum to a state of consciousness and you would wake up.   **21.** A drug that increases the amount of serotonin released in the brain would produce a heightened perception of certain sensory stimuli (e.g., auditory or visual stimuli) and hallucinations.   **22.** Serotonin and norepinephrine are thought to be involved with sleep–wake cycles.   **23.** Amphetamines stimulate the secretion of dopamine.   **PAGE 583   24.** Some possible reasons for slower recall and for loss of memory in older people include a loss of neurons (possibly those involved in specific memories), changes in synaptic organization of the brain, changes in the neurons themselves, and decreased blood flow, which would affect the metabolic rate of neurons and perhaps slow the retrieval of information from memory.   **25.** Common age-related anatomical changes in the nervous system include a reduction in brain size and weight, a reduction in the number of neurons, a decrease in blood flow to the brain, changes in the synaptic organization of the brain, and intracellular and extracellular changes in CNS neurons. **26.** Alzheimer's disease is the most common form of senile dementia.   **27.** The nervous system controls the actions of the arrector pili muscles and sweat glands of the integumentary system. It also controls skeletal muscle contractions of the muscular system, which, in turn, affects the thickening of bones of the skeletal system. It also coordinates muscular activities associated with the respiratory and cardiovascular systems.

Answers to Review Questions

PAGE 587
Level 1 Reviewing Facts and Terms
1.

2. d   **3.** Sympathetic division – speeds up; parasympathetic division – slows down.   **4.** b   **5.** d   **6.** Amygdala and hippocampus   **7.** (1) ciliary ganglion; (2) pterygopalatine ganglion; (3) submandibular ganglion; and (4) otic ganglion   **8.** Visceral reflex arcs include a receptor, a sensory neuron, an interneuron (may or may not be present), and two visceral motor neurons. **9.** Increased neurotransmitter release, facilitation of synapses, and the formation of additional synaptic connections are thought to be involved in memory formation and storage.   **10.** During non-REM sleep, the entire body relaxes, and activity at the cerebral cortex is at a minimum; heart rate, blood pressure, respiratory rate, and energy utilization decrease. During REM sleep, active dreaming occurs, accompanied by alterations in blood pressure and respiratory rates; muscle tone decreases markedly, and response to outside stimuli decreases.   **11.** Commences arousal or wakening from sleep **12.** c   **13.** d   **14.** It is a secondary messenger; key enzymes **15.** Adrenaline and noradrenaline   **16.** The parasympathetic nervous system's role is of control and modulation. The sympathetic nervous system reaction in fight or flight is used in threat situations. There is no equivalent need to be calm.   **17.** The four pairs of cranial nerves are N III, N VII, N IX, and N X.   **18.** (1) cardiac plexus: heart rate increases (sympathetic)/decreases (parasympathetic); force of heart contraction increases (sympathetic)/decreases (parasympathetic); blood pressure increases (sympathetic)/decreases (parasympathetic); (2) pulmonary plexus: respiratory passageways dilate (sympathetic)/constrict (parasympathetic); (3) esophageal plexus: respiratory rate increases (sympathetic)/decreases (parasympathetic); (4) celiac plexus: digestion inhibited (sympathetic)/

stimulated (parasympathetic); (5) inferior mesenteric plexus: digestion inhibited (sympathetic)/stimulated (parasympathetic); and (6) hypogastric plexus: defecation inhibited (sympathetic)/stimulated (parasympathetic); urination inhibited (sympathetic)/stimulated (parasympathetic); sexual organs: stimulation of secretion (sympathetic)/erection (parasympathetic) **19.** Adrenaline or epinephrine

Level 2 Reviewing Concepts
20. a **21.** c **22.** The preganglionic fibers innervating the cervical ganglia originate in the ventral roots of the thoracic segments, which are undamaged. **23.** c **24.** b **25.** d **26.** Increased alertness by stimulation of the reticular activating system, causing the individual to feel "on edge"; A feeling of energy and euphoria often associated with a disregard for danger and a temporary insensitivity to painful stimuli; Increased activity in the cardiovascular and respiratory centers of the pons and medulla oblongata, which increases blood pressure, heart rate, breathing rate, and depth of respiration; A general elevation in muscle tone through stimulation of the medial and lateral pathways, so the individual looks tense and may begin to shiver; The mobilization of energy reserves, through the accelerated breakdown of glycogen in muscle and liver cells and the release of lipids by adipose tissues. **27.** If autonomic motor neurons maintain a background level of activity at all times, they can either increase or decrease their activity, providing a greater range of control options.
28. Cholinergic receptors are found in all the ganglia of the ANS, so nicotine would stimulate both sympathetic and parasympathetic responses in cardiovascular tissues. Although increased sympathetic stimulation increases heart rate and force of contraction, increased parasympathetic stimulation simultaneously decreases blood flow to the heart muscle. In addition to elevating heart rate and force of contraction, sympathetic stimulation also constricts peripheral blood vessels, all of which contribute to increased blood pressure.
29. A decreased vascular resistance results in hypotension, which along with bradycardia may result in the individual fainting.

Level 3 Critical Thinking and Clinical Applications
30. Epinephrine would be more effective, because it would reduce inflammation and relax the smooth muscle of the airways, making it easier for Phil to breathe. **31.** The molecule is probably mimicking NE and binding to alpha-1 receptors.

Answers to Clinical Case Wrap-Up
PAGE 588 **1.** Tim is less likely to hurt himself by falling if he is already lying down. Also, if he keeps his head on the same level as, or lower than, his heart, his brain will get more blood. Under parasympathetic stimulation the heart rate, force of contraction, and blood pressure decrease. The heart cannot push blood against gravity. If the head is level with or below the level of the heart, gravity need not be overcome. **2.** In addition to the reflex sympathetic activity at the heart and vascular system, the hypothalamus would call for sympathetic activation. This would stimulate the reticular activating system, increasing alertness. Stimulation of the respiratory centers of the pons and medulla oblongata would cause breathing rate and depth to increase. Stimulation of the medial and lateral pathways would contribute to a general increase in muscle tone.

Chapter 17

Answers to Checkpoints
PAGE 594 **1.** Olfaction is the sense of smell; it involves olfactory receptor cells in paired olfactory organs responding to chemical

stimuli. **2.** Axons from the olfactory epithelium collect into bundles that reach the olfactory bulb. In the olfactory pathway, axons leaving the olfactory bulb then travel along the olfactory tract to the olfactory cortex, hypothalamus, and portions of the limbic system. **3.** By the end of the lab period, central adaptation has occurred. Inhibition of synapses along the olfactory pathway reduces the amount of information reaching the olfactory cortex, even though the olfactory neurons remain active. **PAGE 596** **4.** Gustation is the sense of taste, provided by taste receptors responding to chemical stimuli. **5.** Taste receptors (in taste buds) are sensitive only to molecules and ions that are in solution. If you dry the surface of your tongue, the salt ions or sugar molecules have no moisture in which to dissolve, so they will not stimulate the taste receptors.
6. Your grandfather is experiencing the effects of several age-related changes. The number of taste buds decreases dramatically after age 50, and those that remain are not as sensitive as they once were. In addition, the loss of olfactory receptors contributes to the perception of fewer flavors in foods. **PAGE 607** **7.** The conjunctiva would be the first layer of the eye affected by inadequate tear production. Drying of the conjunctiva would produce an irritated, scratchy feeling. **8.** When the lens becomes more rounded, you are looking at an object that is close to you. **9.** Sue will likely be unable to see at all. The fovea (fovea centralis) contains only cones, which need high-intensity light to be stimulated. The dimly lit room contains light that is too weak to stimulate the cones. **10.** If the scleral venous sinus (canal of Schlemm) were blocked, the aqueous humor could not drain, producing an eye condition called glaucoma. Accumulation of this fluid increases the pressure within the eye, distorting soft tissues and interfering with vision. If untreated, the condition would ultimately cause blindness. **PAGE 616** **11.** If you were born without cones, you would still be able to see—so long as you had functioning rods—but you would lack color vision and see in black and white only. **12.** A vitamin A deficiency would reduce the quantity of retinal (retinene) the body could produce, thereby interfering with night vision (which operates at the body's threshold ability to respond to light). **13.** Vision would be impaired. Decreased phosphodiesterase activity would increase intracellular cGMP levels, which, by keeping gated sodium channels open, would prevent both hyperpolarization of the photoreceptor cell and a decrease in its neurotransmitter release. There would be no signal to the bipolar cell that a photon had been absorbed. **PAGE 629** **14.** If the round window could not move, the perilymph would not be moved by the vibration of the stapes at the oval window, reducing or eliminating the perception of sound. **15.** The loss of stereocilia (as a result of constant exposure to loud noises, for instance) would reduce hearing sensitivity and could lead to deafness. **16.** If the auditory tube were blocked, it would not be possible to equalize the pressure on both sides of the tympanic membrane. If external pressure then decreases, the pressure in the middle ear would be greater than that on the outside, forcing the tympanic membrane outward and producing pain.

Answers to Review Questions

PAGE 632
Level 1 Reviewing Facts and Terms
1. (a) vascular layer; **(b)** iris; **(c)** ciliary body; **(d)** choroid; **(e)** inner layer (retina); **(f)** neural part; **(g)** pigmented part; **(h)** fibrous layer; **(i)** cornea; **(j)** sclera **2.** d **3.** c **4.** c **5.** a **6.** c **7.** b **8.** d **9.** b **10. (a)** auricle; **(b)** external acoustic meatus; **(c)** tympanic membrane; **(d)** auditory ossicles; **(e)** semicircular canals; **(f)** vestibule; **(g)** auditory tube; **(h)** cochlea; **(i)** vestibulocochlear

nerve (N VIII)   **11.** d   **12.** d   **13.** c   **14.** (1) filiform papillae; (2) fungiform papillae; (3) foliate papillae; and (3) vallate papillae   **15.** The fibrous layer **(a)** is composed of the sclera and the cornea and **(b)** provides mechanical support and some physical protection, serves as an attachment site for the extrinsic eye muscles, and contains structures that assist in the focusing process.
16. The vascular layer consists of the iris, ciliary body, and choroid.   **17.** The malleus, incus, and stapes transmit a mechanical vibration (amplified along the way) from the tympanic membrane to the oval window.

Level 2 Reviewing Concepts

18. Axons leaving the olfactory epithelium collect into 20 or more bundles that penetrate the cribriform plate of the ethmoid bone to reach the olfactory bulbs of the cerebrum. Axons leaving the olfactory bulb travel along the olfactory tract to reach the olfactory cortex, hypothalamus, and portions of the limbic system.   **19.** The threshold for sour receptors is much lower than those that sense sweet chemicals.   **20.** An infected sebaceous gland of an eyelash or tarsal gland usually becomes a sty, a painful swelling.   **21.** a   **22.** c   **23.** f

Level 3 Critical Thinking and Clinical Applications

24. Your medial rectus muscles would contract, directing your gaze more medially. In addition, your pupils would constrict and the lenses would become more spherical.   **25.** Myopia is corrected by **(a)** concave lenses.   **26.** In removing the polyps, some of the olfactory epithelium was probably damaged or destroyed, decreasing the area available for the solution of odorants and reducing the intensity of the stimulus. As a result, after the surgery it would take a larger stimulus to provide the same level of smell.   **27.** The rapid descent in the elevator causes the otoliths of the saccular maculae to slide upward, producing the sensation of downward vertical motion. After the elevator abruptly stops, it takes a few seconds for the otoliths of the maculae to come to rest in the normal position. So long as the otoliths are displaced, you will perceive movement.   **28.** When Juan closes his eyes, visual cues are gone, and his brain must rely solely on proprioceptive information from the equilibrium centers of the internal ear to maintain normal posture. Because either the internal ear receptors or the sensory nerves are not functioning normally, he is unstable. The most likely reason for his drift to the left is that he is getting inappropriate sensations from equilibrium receptors, either at the maculae (affecting his ability to determine which way is "down") or one of the horizontal semicircular ducts (making him attempt to compensate for a perceived roll to the right).

Answers to Clinical Case Wrap-Up

PAGE 633   1. A visual acuity of 20/200 means Makena must be 20 feet away from an object to see it as clearly as someone with normal vision would see it at 200 feet. In the United States she would be considered legally blind.   **2.** If Makena's vision is corrected to a state of emmetropia, her vision would be corrected to normal.

Chapter 18

Answers to Checkpoints

PAGE 637   1. A hormone is a chemical messenger that is secreted by one cell and travels through the bloodstream to affect the activities of cells in other parts of the body.   **2.** Paracrine communication is the use of chemical messengers to transfer information from cell to cell within a single tissue.   **3.** The four mechanisms of intercellular communication are direct, paracrine, endocrine, and

synaptic.   **PAGE 645   4.** Neural responses occur within fractions of a second and are of short duration. Endocrine responses are slow to appear but last for minutes to days.   **5.** A substance that inhibits adenylate cyclase, the enzyme that converts ATP to cAMP, would block the action of any hormone that requires cAMP as a second messenger.   **6.** A cell's hormonal sensitivities are determined by the presence or absence of the receptor complex needed to bind a given hormone.   **PAGE 651   7.** The two lobes of the pituitary gland are the anterior lobe and the posterior lobe.   **8.** In dehydration, blood osmotic concentration is increased, which would stimulate the posterior lobe of the pituitary gland to release more ADH.
9. Somatomedins mediate the action of growth hormone. Elevated levels of growth hormone typically accompany elevated levels of somatomedins.   **10.** Elevated circulating levels of cortisol inhibit the endocrine cells that control the release of ACTH from the pituitary gland, so ACTH levels would decrease. This is an example of a negative feedback mechanism.   **PAGE 655   11.** Thyroxine (T_4), triiodothyronine (T_3), and calcitonin (CT) are hormones associated with the thyroid gland.   **12.** An individual whose diet lacks iodine would be unable to form the thyroid hormones thyroxine (T_4) and triiodothyronine (T_3). As a result, you would expect to see signs and symptoms associated with their deficiency: decreased metabolic rate, decreased body temperature, a poor response to physiological stress, and an increase in the size of the thyroid gland (goiter).
13. Most of the body's reserves of the thyroid hormones, T_4 and T_3, are bound to blood-borne proteins called thyroid-binding globulins. Because these compounds represent such a large reservoir of T_4 and T_3, it takes several days after removal of the thyroid gland for blood levels of thyroxine and T_3 to decline.   **PAGE 657   14.** The parathyroid glands are embedded in the posterior surface of the lateral lobes of the thyroid gland.   **15.** The hormone secreted by the parathyroid glands is parathyroid hormone (PTH).   **16.** The removal of the parathyroid glands would result in a decrease in the blood concentration of calcium ions.   **PAGE 660   17.** The two regions of the adrenal gland are the cortex and medulla. The cortex secretes mineralocorticoids (mainly aldosterone), glucocorticoids (mainly cortisol, hydrocortisone, and corticosterone), and androgens; the medulla secretes epinephrine and norepinephrine.   **18.** The three zones of the adrenal cortex from superficial to deep are the zona glomerulosa, zona fasciculata, and zona reticularis.   **19.** One function of cortisol is to decrease the cellular use of glucose while increasing both the available glucose (by promoting the breakdown of glycogen) and the conversion of amino acids to carbohydrates. Therefore, the net result of elevated cortisol levels would be an elevation of blood glucose.   **PAGE 661   20.** Pinealocytes are the special secretory cells in the pineal gland.
21. Increased amounts of light would inhibit the production (and release) of melatonin from the pineal gland, which receives neural input from the optic tracts. Melatonin secretion is influenced by day-night patterns.   **22.** Melatonin inhibits reproductive functions, protects against free radical damage, and influences circadian rhythms.   **PAGE 665   23.** The cells of the pancreatic islets (and their hormones) are alpha cells (glucagon), beta cells (insulin), delta cells (GH–IH), and F cells (pancreatic polypeptide).
24. A person with Type 1 or Type 2 diabetes has such high blood glucose levels that the kidneys cannot reabsorb all the glucose; some glucose is lost in urine. Because the urine contains high concentrations of glucose, less water can be reclaimed by osmosis, so the volume of urine production increases. The water losses reduce blood volume and elevate blood osmotic pressure, promoting thirst and triggering the secretion of ADH.   **25.** Increased levels

of glucagon stimulate the conversion of glycogen to glucose in the liver, which would in turn decrease the amount of glycogen in the liver. **PAGE 669 26.** Two hormones secreted by the kidneys are erythropoietin (EPO) and calcitriol. **27.** Leptin is a hormone released by adipose tissue. **28.** Once released into the bloodstream, renin functions as an enzyme, catalyzing the conversion of angiotensinogen to angiotensin I. **PAGE 672 29.** The type of hormonal interaction in which two hormones have opposite effects on their target tissues is called antagonism. **30.** A lack of GH, thyroid hormone, PTH, and the gonadal hormones would inhibit the formation and development of the skeletal system. **31.** During the resistance phase of the general adaptation syndrome, there is a high demand for glucose, especially by the nervous system. The hormones GH–RH and CRH increase the levels of GH and ACTH, respectively. Growth hormone (GH) mobilizes fat reserves and promotes the catabolism of protein; ACTH increases cortisol, which stimulates both the conversion of glycogen to glucose and the catabolism of fat and protein. Together, these metabolic processes conserve glucose for use by neural tissue. **32.** The endocrine system adjusts metabolic rates and substrate utilization, and regulates growth and development, in all other body systems.

33. Hormones of the endocrine system adjust muscle metabolism, energy production, and growth; hormones also regulate calcium and phosphate levels, which are critical to normal muscle functioning. Skeletal muscles protect some endocrine organs.

Answers to Review Questions

PAGE 676
Level 1 Reviewing Facts and Terms
1. b **2.** c **3.** d **4.** a **5.** d **6.** d **7.** d **8. (a)** hypothalamus; **(b)** pituitary gland; **(c)** thyroid gland; **(d)** thymus; **(e)** adrenal glands; **(f)** pineal gland; **(g)** parathyroid glands; **(h)** heart; **(i)** kidney; **(j)** adipose tissue; **(k)** digestive tract; **(l)** pancreatic islets (within pancreas); **(m)** gonads **9.** c **10.** c **11.** Amino acid derivatives, peptide hormones, and lipid derivatives.
12. Effects of thyroid hormones are (1) increased rate of energy consumption and utilization in cells; (2) accelerated production of sodium–potassium ATPase; (3) activation of genes coding for the synthesis of enzymes involved in glycolysis and energy production; (4) accelerated ATP production by mitochondria; and (5) in growing children, normal development of the skeletal, muscular, and nervous systems. **13.** Calcitonin decreases the concentration of calcium ions in body fluids; parathyroid hormone causes an increase in the concentration of calcium ions in body fluids. **14.** (1) zona glomerulosa: mineralocorticoids; (2) zona fasciculata: glucocorticoids; and (3) zona reticularis: androgens **15.** The kidneys release (1) erythropoietin, which stimulates the production of RBCs by the red bone marrow, and (2) calcitriol, which stimulates calcium and phosphate absorption along the digestive tract. **16.** Natriuretic peptides (1) promote the loss of sodium ions and water at the kidneys; (2) inhibit the secretion of water-conserving hormones, such as ADH and aldosterone; (3) suppress thirst; and (4) block the effects of angiotensin II and norepinephrine on arterioles. Angiotensin II opposes these actions by stimulating aldosterone secretion by the adrenal cortex and ADH by the posterior lobe of the pituitary gland, and further by retaining salt and water by the kidneys. Angiotensin II also stimulates thirst and elevates blood pressure. **17.** (1) alpha cells: glucagon; (2) beta cells: insulin; (3) delta cells: growth hormone–inhibiting hormone (GH–IH, or somatostatin); and (4) F cells: pancreatic polypeptide

Level 2 Reviewing Concepts
18. The primary difference involves speed and duration. In the nervous system, the source and destination of communication are quite specific, and the effects are extremely quick and short lived. In endocrine communication, the effects are slow to appear and commonly persist for days. A single hormone can alter the metabolic activities of multiple tissues and organs simultaneously. **19.** Hormones can (1) direct the synthesis of an enzyme (or other protein) not already present in the cytoplasm, (2) turn an existing enzyme "on" or "off," and (3) increase the rate of synthesis of a particular enzyme or other protein. **20.** Paracrine communication is from cell to cell in a single tissue, whereas endocrine communication modifies metabolic activities of target cells which make up many tissues and organs. **21.** Inactivation of phosphodiesterase (PDE), which converts cAMP to AMP, would prolong the effect of the hormone. **22.** The effect of each hormone is weaker than that produced by either hormone – antagonistic effect; The effect of the hormones is combined so that it is greater than the effect of each hormone alone – synergistic effect; The first hormone is required for the second to work – permissive effect; The effect each hormone has on specific tissues may be complementary – integrative effect. **23.** b **24.** a **25.** b

Level 3 Critical Thinking and Clinical Applications
26. Extreme thirst and frequent urination are characteristic of both diabetes insipidus and diabetes mellitus. To distinguish between the two, glucose levels in the blood and urine could be measured. A high glucose concentration would indicate diabetes mellitus. **27.** Phosphodiesterase (PDE) inhibition increases the level of cAMP, leading to increased kinase activity. Phosphorylation by the kinases increases SR calcium concentration, which is released during cardiac muscle contraction, increasing contractility. **28.** Sherry's signs and symptoms suggest hyperthyroidism. Blood tests could be performed to assay the levels of TSH, T_3, and T_4. From these tests, the physician could make a positive diagnosis (hormone levels would be elevated in hyperthyroidism) and also determine whether the condition is primary (a problem with the thyroid gland) or secondary (a problem with hypothalamo-pituitary control of the thyroid gland). **29.** One benefit of a portal system is that it ensures that the controlling hormones will be delivered directly to the target cells. Secondly, because the hormones go directly to their target cells without first passing through the general circulation, they are not diluted. The hypothalamus can control the cells of the anterior lobe of the pituitary gland with much smaller amounts of releasing and inhibiting hormones than would be necessary if the hormones had to first go through the circulatory pathway before reaching the pituitary. **30.** The natural effects of testosterone are to increase muscle mass, increase endurance, and enhance the "competitive spirit." Side effects in women include hirsutism (abnormal hair growth on the face or body), enlargement of the laryngeal cartilages, premature closure of the epiphyseal cartilages, and liver dysfunction.

Answers to Clinical Case Wrap-Up
PAGE 677 1. In addition to mobilizing Ca^{2+} from bone, PTH enhances the reabsorption of Ca^{2+} by the kidneys, decreasing urinary losses and keeping blood concentrations high. PTH also stimulates the secretion of calcitriol by the kidneys, which enhances absorption of Ca^{2+} in the digestive tract, which also increases blood concentrations of Ca^{2+}. **2.** Calcitonin, produced by the thyroid, decreases blood Ca^{2+} levels. Calcitonin does this by encouraging Ca^{2+} deposition in bone and by causing an increased excretion of Ca^{2+} by the kidneys.

Chapter 19

Answers to Checkpoints

PAGE 682 **1.** The five major functions of blood are transporting dissolved gases, nutrients, hormones, and metabolic wastes; regulating the pH and ion composition of interstitial fluids; restricting fluid losses at injury sites; defending against toxins and pathogens; and stabilizing body temperature. **2.** Red blood cells, white blood cells, and platelets are the formed elements of blood. **3.** Whole blood is composed of plasma and formed elements. **4.** Venipuncture is a common sampling technique because superficial veins are easy to locate, the walls of veins are thinner than those of arteries, and blood pressure in veins is relatively low, so the puncture wound seals quickly. **PAGE 683** **5.** The three major types of plasma proteins are albumins, globulins, and fibrinogen. **6.** A decrease in the amount of plasma proteins in the blood would lower plasma osmotic pressure, reduce the ability to fight infection, and decrease the transport and binding of some ions, hormones, and other molecules. **7.** During a viral infection, you would expect the level of immunoglobulins (antibodies) in the blood to be elevated. **PAGE 690** **8.** Hemoglobin is a protein composed of four globular subunits, each bound to a heme molecule, which gives red blood cells the ability to transport oxygen in the blood. **9.** After a significant loss of blood, the hematocrit—the amount of formed elements (mostly red blood cells) as a percentage of the total blood—would be reduced. **10.** Dave's hematocrit will increase, because reduced blood flow to the kidneys triggers the release of erythropoietin, which stimulates an increase in erythropoiesis (red blood cell formation). **11.** Bilirubin would accumulate in the blood, producing jaundice, because diseases that damage the liver, such as hepatitis or cirrhosis, impair the liver's ability to excrete bilirubin in the bile. **PAGE 693** **12.** Surface antigens on RBCs are glycoproteins or glycolipids in the plasma membrane; they determine blood type. **13.** Only type O blood can be safely transfused into a person whose blood type is O. **14.** If a person with type A blood receives a transfusion of type B blood, which contains anti-A antibodies, the red blood cells will agglutinate (clump), potentially blocking blood flow to various organs and tissues. **PAGE 700** **15.** The five types of white blood cells are neutrophils, eosinophils, basophils, monocytes, and lymphocytes. **16.** An infected cut would contain a large number of neutrophils, because these phagocytic white blood cells are the first to arrive at the site of an injury. **17.** The blood of a person fighting a viral infection would contain elevated numbers of lymphocytes, because B lymphocytes produce circulating antibodies. **18.** Basophils respond to an injury by releasing a variety of chemicals, including histamine and heparin. Histamine dilates blood vessels and heparin prevents blood clotting. Basophils also release other chemicals that attract eosinophils and other basophils to the area. **PAGE 701** **19.** Thrombocytopoiesis is the term for platelet production. **20.** Platelets are nonnucleated cell fragments in mammalian blood, whereas thrombocytes are nucleated platelets in nonmammalian vertebrate blood. **21.** Platelets release chemicals important to the clotting process, form a temporary patch in the walls of damaged blood vessels, and reduce the size of a break in the vessel wall. **PAGE 705** **22.** A decreased number of megakaryocytes would interfere with the blood's ability to clot properly, because fewer megakaryocytes would produce fewer platelets. **23.** Fruit juice and water do not contain fats, which are required for vitamin K absorption, leading to a vitamin K deficiency. This would lead to a decreased production of several clotting factors—most notably, prothrombin. As a result, clotting time would increase. **24.** The activation of Factor XII initiates the intrinsic pathway.

Answers to Review Questions

Level 1 Reviewing Facts and Terms

1. (a) neutrophil; **(b)** eosinophil; **(c)** basophil; **(d)** monocyte; **(e)** lymphocyte **2.** d **3.** f **4.** b **5.** d **6.** d **7.** b **8.** d **9.** a **10.** They are easy to locate, are low pressured, and have thin walls. **11.** Major types of plasma proteins are (1) albumins, which maintain the osmotic pressure of plasma and are important in the transport of fatty acids; (2) globulins, which (a) bind small ions, hormones, or compounds that might otherwise be filtered out of the blood at the kidneys or have very low solubility in water (transport globulins), or (b) attack foreign proteins and pathogens (immunoglobulins); and (3) fibrinogen, which functions in blood clotting. **12. (a)** anti-B antibodies; **(b)** anti-A antibodies; **(c)** neither anti-A nor anti-B antibodies; **(d)** both anti-A and anti-B antibodies **13.** WBCs exhibit (1) emigration (diapedesis), squeezing between adjacent endothelial cells in the capillary wall; (2) amoeboid movement, a gliding movement that transports the cell; (3) positive chemotaxis, the attraction to specific chemical stimuli, and (4) phagocytosis (engulfing particles for neutrophils, eosinophils, and monocytes). **14.** Flattened disc-shaped non-nucleated cytoplasmic fragments. Functions: transport and release of chemicals involved in the clotting process; Formation of a temporary platelet plug in damaged blood vessel walls; Contraction of platelet filaments reduce the size of the blood clot – this reduces the size of the break in the wall of the blood vessel. **15.** The primary lymphocytes are (1) T cells, which are responsible for cell-mediated immunity; (2) B cells, which are responsible for humoral immunity; and (3) NK cells, which are responsible for immune surveillance. **16.** Prothrombin is an inactive precursor that is converted to thrombin during coagulation. Thrombin is an enzyme that causes the clotting of blood by converting fibrinogen to fibrin. **17.** Erythropoietin is released (1) during anemia, (2) when blood flow to the kidneys declines, (3) when oxygen content of the air in the lungs declines, and (4) when the respiratory surfaces of the lungs are damaged. **18.** Initiation of the common pathway requires the activation of Factor X by the extrinsic and/or intrinsic pathways.

Level 2 Reviewing Concepts

19. a **20.** d **21.** c **22.** c **23.** Provides a large surface area to volume ratio for rapid gaseous exchange. Capable of forming rouleaux, which maximises smooth flow through narrow blood vessels. It allows the red blood cells to flex and squeeze through capillaries. **24.** White blood cells defend against toxins and pathogens. Neutrophils, eosinophils, and monocytes engulf and digest bacteria, protozoa, fungi, viruses, and cellular debris. Lymphocytes are specialized to attack and destroy specific foreign cells, proteins, and cancerous cells, directly or through the production of antibodies. **25.** Blood stabilizes and maintains body temperature by absorbing and redistributing the heat produced by active skeletal muscles. **26.** Red blood cells are engulfed by phagocytic cells in the liver. The globular proteins are broken down to amino acids and reused. The heme group is stripped of iron and converted to biliverdin, which is converted to blirubin and excreted in bile. The iron is bound to the plasma protein transferrin, and used in the production of RBC or stored bound to ferritin and hemosiderin in the liver and spleen. **27.** Lymphocytes release multi-CSF and GM-CSF, which stimulate the production of granulocytes and monocytes that are part of the nonspecific defenses.

Level 3 Critical Thinking and Clinical Applications

28. A prolonged prothrombin time and a normal partial thromboplastin time indicate a deficiency in the extrinsic system but

ANSWERS

not in the intrinsic system or common pathway. Factor VII would be deficient.   **29.** As the spleen enlarges, so does its capacity to store additional red blood cells, leading to fewer red blood cells in circulation, producing anemia. The decreased number of RBCs in circulation decreases the body's ability to deliver oxygen to the tissues, slowing their metabolism and producing the tired feeling and lack of energy. Because there are fewer RBCs than normal, the blood circulating through the skin is not as red, producing a pale complexion.   **30.** Taking a broad-spectrum antibiotic kills a wide range of bacteria, both pathogenic and nonpathogenic, including many of the normal flora of the intestine. Reducing the intestinal flora substantially decreases the amount of vitamin K they make available to the liver to produce prothrombin, a vital component of the common pathway. With decreased amounts of prothrombin in the blood, normal minor breaks in the vessels of the nasal passageways do not seal off as quickly, producing nosebleeds.   **31.** Low oxygen saturations release erythropoietin, increasing RBC production, increasing the hematocrit. Higher RBC counts can increase blood viscosity, slowing blood flow and increasing the risk of clot formation.

Answers to Clinical Case Wrap-Up

PAGE 709   **1.** Jasmine bleeds and bruises so easily because she has too few platelets to initiate the platelet phase of blood clotting. This also causes her prolonged menstrual periods.   **2.** Jasmine is short of breath climbing stairs because she has too few red blood cells carrying oxygen to her skeletal muscles.

Chapter 20

Answers to Checkpoints

PAGE 723   **1.** Damage to the semilunar valve of the right ventricle would affect blood flow to the pulmonary trunk.   **2.** Contraction of the papillary muscles (just before the rest of the ventricular myocardium contracts) pulls on the chordae tendineae, which prevent the AV valves from opening back into the atria.   **3.** The left ventricle is more muscular than the right ventricle because the left ventricle must generate enough force to propel blood throughout the body, except the lungs, whereas the right ventricle must generate only enough force to propel blood a few centimeters to the lungs.   **PAGE 732**   **4.** Automaticity, or autorhythmicity, is the ability of cardiac muscle tissue to contract without neural or hormonal stimulation.   **5.** The sinoatrial (SA) node is known as the cardiac pacemaker or the natural pacemaker.   **6.** If the cells of the SA node did not function, the heart would still continue to beat, but at a slower rate; and the AV node would act as the pacemaker.   **7.** If the impulses from the atria were not delayed at the AV node, they would be conducted through the ventricles so quickly by the bundle branches and Purkinje cells that the ventricles would begin contracting immediately, before the atria had finished their contraction. As a result, the ventricles would not be as full of blood as they could be, and the pumping of the heart would not be as efficient, especially during physical activity.   **PAGE 737**   **8.** The technical term for heart contraction is systole, and the term for heart relaxation is diastole.   **9.** The phases of the cardiac cycle are atrial systole, atrial diastole, ventricular systole, and ventricular diastole.   **10.** No. When pressure in the left ventricle first rises, the heart is contracting but no blood is leaving the heart. During this initial phase of contraction, both the AV valves and the semilunar valves are closed. The increase in pressure is the result of increased tension as the cardiac muscle contracts. When the pressure in the ventricle exceeds the pressure in the aorta, the aortic semilunar valves are forced open,

and blood is rapidly ejected from the ventricle.   **11.** One possible cause for an increase in the size of the QRS complex, which indicates a larger-than-normal amount of electrical activity during ventricular depolarization, is an enlarged heart. Because more cardiac muscle is depolarizing, the magnitude of the electrical event would be greater.   **PAGE 744**   **12.** Cardiac output is the amount of blood pumped by the left ventricle in one minute.   **13.** Caffeine acts directly on the conducting system and contractile cells of the heart, increasing the rate at which they depolarize. Drinking large amounts of caffeinated drinks would increase the heart rate.   **14.** Damage to the cardioinhibitory center of the medulla oblongata, which is part of the parasympathetic division of the autonomic nervous system, would reduce parasympathetic stimulation of the heart. The resulting sympathetic dominance would increase the heart rate.   **15.** A drug that increases the length of time required for the repolarization of pacemaker cells would decrease the heart rate, because the pacemaker cells would generate fewer action potentials per minute.
16. The heart pumps in proportion to the amount of blood that enters. A heart that beats too rapidly does not have sufficient time to fill completely between beats. Thus, when the heart beats too fast, very little blood leaves the ventricles and enters the circulation, so tissues suffer damage from inadequate blood supply.
17. Stimulating the acetylcholine receptors of the heart would slow the heart rate. Since cardiac output is the product of stroke volume and heart rate, a reduction in heart rate will lower the cardiac output (assuming that the stroke volume remains the same or doesn't increase).   **18.** The venous return fills the heart with blood, stretching the heart muscle. According to the Frank–Starling principle, the more the heart muscle is stretched, the more forcefully it will contract (to a point). The more forceful the contraction, the more blood the heart will eject with each beat (stroke volume). Therefore, increased venous return would increase the stroke volume (if all other factors are constant).   **19.** An increase in sympathetic stimulation of the heart would increase heart rate and force of contraction. The end-systolic volume (ESV) is the amount of blood that remains in a ventricle after a contraction (systole). The more forcefully the heart contracts, the more blood it ejects and the lower the ESV is. Therefore, increased sympathetic stimulation should result in a lower ESV.
20. SV = EDV − ESV, so SV = 125 mL − 40 mL = 85 mL

Answers to Review Questions

PAGE 747
Level 1 Reviewing Facts and Terms
1. (a) superior vena cava; **(b)** auricle of right atrium; **(c)** right ventricle; **(d)** left ventricle; **(e)** aortic arch; **(f)** left pulmonary artery; **(g)** pulmonary trunk; **(h)** auricle of left atrium   **2. (a)** ascending aorta; **(b)** opening of coronary sinus; **(c)** right atrium; **(d)** cusp of right AV (tricuspid) valve; **(e)** chordae tendineae; **(f)** right ventricle; **(g)** pulmonary valve; **(h)** left pulmonary veins; **(i)** left atrium; **(j)** aortic valve; **(k)** cusp of left AV (mitral) valve; **(l)** left ventricle; **(m)** interventricular septum   **3.** c   **4.** b   **5.** d   **6.** c   **7.** d   **8.** a and b   **9.** b   **10.** d   **11.** a   **12.** During ventricular contraction, tension in the papillary muscles pulls against the chordae tendineae, which keep the cusps of the AV valve from swinging into the atrium. This action prevents the backflow, or regurgitation, of blood into the atrium as the ventricle contracts.   **13.** Systole: contraction of atria and ventricles. At this time blood is pushed into an adjacent chamber. Diastole: relaxation of atria and ventricles. At this time the chamber fills with blood.   **14.** The right atrioventricular (AV) valve (the tricuspid valve) and the left AV valve (the bicuspid valve) prevent the backflow of blood from the ventricles into the atria. The pulmonary and aortic semilunar valves prevent the backflow

of blood from the pulmonary trunk and aorta into the right and left ventricles.   **15.** SA node → AV node → AV bundle → right and left bundle branches → Purkinje fibers (into the mass of ventricular muscle tissue)   **16.** The cardiac cycle comprises the events in a complete heartbeat, including a contraction/relaxation period for both atria and ventricles. The cycle begins with atrial systole as the atria contract and push blood into the relaxed ventricles. As the atria relax (atrial diastole), the ventricles contract (ventricular systole), forcing blood through the semilunar valves into the pulmonary trunk and aorta. The ventricles then relax (ventricular diastole). For the rest of the cardiac cycle, both the atria and ventricles are in diastole; passive filling occurs.   **17.** Stroke volume is the volume of blood ejected by the ventricle during one heart beat. Cardiac output is the volume of blood ejected by the left ventricle during one minute ($CO = SV \times HR$).

Level 2 Reviewing Concepts

18. b   **19.** a   **20.** d   **21.** The cells of atrioventricular node are smaller and have fewer gap junctions, which lead to a delay in impulse conduction. This allows the atria to empty before the ventricles contract.   **22.** The SA node, which is composed of cells that exhibit rapid prepotential, is the pacemaker of the heart. The AV node slows the impulse that signals contraction, because its cells are smaller than those of the conduction pathway.   **23.** The first sound ("lubb"), which marks the start of ventricular contraction, is produced as the AV valves close and the semilunar valves open. The second sound ("dupp") occurs when the semilunar valves close and the AV valves open, marking the start of ventricular diastole. The third heart sound is associated with blood flow into the ventricles, and the fourth sound is associated with atrial contraction. Listening to the heart sounds (auscultation) is a simple and effective diagnostic tool.   **24.** As sodium channels close at the end of the rapid depolarization phase, slow calcium channels open allowing the influx of positively charged calcium ions into the cell. This balances the loss of positively charged sodium ions, which are actively pumped out of the cell and lead to a plateau in the action potential. Skeletal muscle cells do not have slow calcium channels and so, lack the plateau phase.   **25.** Stroke volume and heart rate influence cardiac output.   **26.** Sympathetic activation increases the heart rate and the force of contractions; parasympathetic stimulation decreases the heart rate and the force of contractions.   **27.** All these hormones have positive inotropic effects, which means that they increase the strength of cardiac contraction.

Level 3 Critical Thinking and Clinical Applications

28. Venous return to the heart is reduced as gravitational forces increase on standing. This reduces cardiac output (CO), leading to reduced blood flow to the brain and dizziness. However, reduced CO also stimulates baroreceptors, and sympathetic stimulation is increased making the heart rate rise and causing vasoconstriction. These increase cardiac output and blood pressure [CO × Total Peripheral Resistance (TPR) = BP], which maintains blood flow despite increased gravitational forces.   **29.** Harvey probably has a regurgitating left atrioventricular (AV) valve. This valve is also called the mitral valve or the bicuspid valve. When an AV valve fails to close properly, blood flowing back into the atrium produces a murmur. A murmur at the beginning of systole implicates the AV valve because this is the period when the valve has just closed and the blood in the ventricle is under increasing pressure; thus the likelihood of backflow is the greatest. A sound heard at the end of systole or the beginning of diastole would implicate a regurgitating semilunar valve—in this case, the aortic semilunar valve.

30. Using $CO = HR \times SV$, person 1 has a cardiac output of 4500 mL, and person 2 has a cardiac output of 8550 mL. According to the Frank–Starling principle (Starling's law of the heart), in a normal heart the cardiac output is directly proportional to the venous return. Thus, person 2 has the greater venous return. Ventricular filling decreases with increased heart rate; person 1 has the lower heart rate and therefore the longer ventricular filling time.   **31.** Blood returning to the heart (venous return) is aided by the respiratory pump, which generates pressure gradients between the thoracic and abdominal cavity during the breathing cycle. The mechanical ventilator uses positive pressure, which reduces the effect of the respiratory pump. Venous return and the stretch of the heart (Starlings law) are reduced. As a consequence, stroke volume and cardiac output fall.

Answers to Clinical Case Wrap-Up

PAGE 748   1. The "lubb" heard when the AV valves close, and the "dupp" heard when the semilunar valves close, would be muffled and difficult to hear because the sounds were transmitted through a pericardial cavity filled with blood. This dampens the sounds when heard through a stethoscope.   **2.** The right atrium and right ventricle sit anteriorly, just deep to the sternum, and would likely be injured in this collision. Delicate structures that could also be injured include the anterior interventricular artery and great cardiac vein, running in the anterior interventricular sulcus. With a hard enough impact, the papillary muscles and chordae tendineae could rupture, leading to failure of the AV valves.

Chapter 21

Answers to Checkpoints

PAGE 760   1. The five general classes of blood vessels are arteries, arterioles, capillaries, venules, and veins.   **2.** The blood vessels are veins. Arteries and arterioles have a large amount of smooth muscle tissue in a thick, well-developed tunica media.   **3.** In the arterial system, pressures are high enough to keep the blood moving forward. In the venous system, blood pressure is too low to keep the blood moving on toward the heart. Valves in veins prevent blood from flowing backward whenever the venous pressure drops.   **4.** Fenestrated capillaries are located where fluids and small solutes move freely into and out of the blood, including endocrine glands, the choroid plexus of the brain, absorptive areas of the intestine, and filtration areas of the kidneys.   **PAGE 768   5.** The factors that contribute to total peripheral resistance are vascular resistance, vessel length, vessel diameter, blood viscosity, and turbulence.   **6.** In a healthy person, blood pressure is greater at the aorta than at the inferior vena cava. Blood, like other fluids, moves along a pressure gradient from areas of high pressure to areas of low pressure. If the pressure were higher in the inferior vena cava than in the aorta, the blood would flow backward.   **7.** While Sally was standing for a period of time, blood pooled in her lower limbs, which decreased venous return to her heart. In turn, cardiac output decreased, so less blood reached her brain, causing light-headedness and fainting. A hot day adds to this effect, because the loss of body water through sweating reduces blood volume.   **8.** Mike's mean arterial pressure is approximately 88.3 mm Hg; $70 + (125 - 70)/3 = 70 + 18.3 = 88.3$.   **PAGE 775   9.** Vasodilators promote the dilation of precapillary sphincters. Local vasodilators, such as decreased O_2 levels or increased CO_2 levels, act at the tissue level to accelerate blood flow through their tissue of origin.   **10.** Pressure on the common carotid artery would decrease blood pressure at the baroreceptors in the carotid sinus. This decrease would cause a decreased frequency of action potentials along the glossopharyngeal cranial nerve (IX) to the medulla oblongata, and more sympathetic impulses would be sent to the heart. The net result would be an increase in the heart rate. **11.** Vasoconstriction of the renal artery would decrease both blood flow and blood pressure at the kidney. In response, the kidney would increase the amount of renin it releases, which in turn would lead to an increase in the level of angiotensin II. The angiotensin II would bring about increased blood pressure and increased blood volume. **PAGE 779 12.** Blood pressure increases during exercise because (1) cardiac output increases and (2) resistance in visceral tissues increases. **13.** The immediate problem during hemorrhaging is maintaining adequate blood pressure and peripheral blood flow; the long-term problem is restoring normal blood volume. **14.** Both aldosterone and ADH promote fluid retention and reabsorption at the kidneys, preventing further reductions in blood volume. **PAGE 780 15.** The two circuits of the cardiovascular system are the pulmonary circuit and the systemic circuit. **16.** The three major patterns are the following: (1) The peripheral distributions of arteries and veins on the body's left and right sides are generally identical, except near the heart, where the largest vessels connect to the atria or ventricles; (2) a single vessel may have several names as it crosses specific anatomical boundaries, making accurate anatomical descriptions possible; and (3) tissues and organs are usually serviced by several arteries and veins. **PAGE 781 17.** The pulmonary arteries enter the lungs carrying deoxygenated blood, and the pulmonary veins leave the lungs carrying oxygenated blood. **18.** Right ventricle → pulmonary trunk → left and right pulmonary arteries → pulmonary arterioles → alveoli → pulmonary venules → pulmonary veins → left atrium **PAGE 798 19.** A blockage of the left subclavian artery would interfere with blood flow to the left arm. **20.** Compression of the common carotid arteries would decrease blood pressure at the carotid sinus and cause a rapid reduction in blood flow to the brain, resulting in a loss of consciousness. An immediate reflexive increase in heart rate and blood pressure would follow. **21.** Rupture of the celiac trunk would most directly affect the stomach, spleen, liver, and pancreas. **22.** The vein that is bulging in Tim's neck is the external jugular vein. **23.** A blockage of the popliteal vein would interfere with blood flow in the tibial and fibular veins (which form the popliteal vein) and the small saphenous vein (which joins the popliteal vein). **PAGE 799 24.** Two umbilical arteries supply blood to the placenta, and one umbilical vein returns blood from the placenta. The umbilical vein then drains into the ductus venosus within the fetal liver. (Remember, arteries carry blood away from the heart, and veins carry blood to the heart.) **25.** This blood sample was taken from the umbilical vein, which carries oxygenated, nutrient-rich blood from the placenta to the fetus. **26.** Structures specific to the fetal circulation include two umbilical arteries, an umbilical vein, the ductus venosus, the foramen ovale, and the ductus arteriosus. In the newborn, the foramen ovale closes and persists as the fossa ovalis, a shallow depression; the ductus arteriosus persists as the ligamentum arteriosum, a fibrous cord; and the umbilical vessels and ductus venosus persist throughout life as fibrous cords. **PAGE 801 27.** Components of the cardiovascular system affected by age include the blood, heart, and blood vessels. **28.** A thrombus is a stationary blood clot within the lumen of a blood vessel. **29.** An aneurysm is the ballooning out of a weakened arterial wall resulting from sudden pressure increases. **30.** The cardiovascular system provides other body systems with oxygen, hormones, nutrients, and white blood cells in

blood, while removing carbon dioxide and metabolic wastes; it also transfers heat to body tissues. **31.** The skeletal system provides calcium needed for normal cardiac muscle contraction, and it protects developing blood cells in the red bone marrow. The cardiovascular system provides calcium and phosphate for bone deposition, delivers erythropoietin to red bone marrow, and transports parathyroid hormone and calcitonin to osteoblasts and osteoclasts.

Answers to Review Questions

PAGE 805
Level 1 Reviewing Facts and Terms
1. (a) brachiocephalic trunk; (b) brachial; (c) radial; (d) external iliac; (e) anterior tibial; (f) right common carotid; (g) left subclavian; (h) common iliac; (i) femoral **2.** b **3.** e **4.** c **5.** b **6.** b **7.** b **8.** d **9.** (a) external jugular; (b) brachial; (c) median cubital; (d) radial; (e) great saphenous; (f) internal jugular; (g) superior vena cava; (h) left and right common iliac; (i) femoral **10.** b **11.** b **12.** c **13.** c **14.** a **15.** e **16.** i) Muscle compression near veins pushes blood towards the heart. ii) Respiratory pump reduces thoracic pressures of vessels during inhalation. This draws blood from higher pressure regions in the abdominal cavity and lower body. Exhalation increases thoracic pressures, which push blood into the right atrium. iii) Venous valves prevent the backflow of blood, and supports muscle compression and the respiratory pump. **17.** (a) Capillary hydrostatic pressure forces fluid out of the capillary at the arterial end. (b) Blood colloid osmotic pressure causes the movement of fluid back into the capillary at its venous end. **18.** Tunica intima – smooth endothelial cells help reduce resistance to blood flow. Tunica media – smooth muscle and elastic fibers allow the modification of vascular resistance. Tunica externa – fibrous connect tissue, supports and anchors the vessel.

Level 2 Reviewing Concepts
19. b **20.** c **21.** a **22.** Increasing tissue metabolism reduces local levels of oxygen and increases other metabolites, such as carbon dioxide, lactic acid and ADP. These local factors cause relaxation of precapillary sphincters and increase blood flow to match tissue oxygen and nutrient demand. **23.** Capillary walls are thin, so distances for diffusion are short. Continuous capillaries have small gaps between adjacent endothelial cells that permit the diffusion of water and small solutes into the surrounding interstitial fluid but prevent the loss of blood cells and plasma proteins. Fenestrated capillaries contain pores that permit very rapid exchange of fluids and solutes between interstitial fluid and plasma. The walls of arteries and veins are several cell layers thick and are not specialized for diffusion. **24.** Damaged cardiac cells are replaced by scar tissue; atherosclerosis restricts coronary blood flow, there is a reduction in the elasticity of the cardiac skeleton, and there are changes in nodal and conducting cells, all of which lead to reduction in maximum cardiac output. Atherosclerosis and calcium deposits reduce blood vessel wall elasticity, increasing vascular resistance, and the risk of aneurysm and thrombi formation. This can lead to myocardial infarction and stroke. Venous valves may leak, leading to blood pooling and reduced venous return. Blood hematocrit may decrease and peripheral veins may develop clots which become lodged in the lungs. **25.** Cardiac output and peripheral blood flow are directly proportional to blood pressure. Blood pressure is closely regulated by a combination of neural and hormonal mechanisms. The resistance of the vascular system opposes the movement of blood, so blood flow is inversely proportional to the resistance. Sources of peripheral resistance include vascular resistance, blood viscosity, and

turbulence. **26.** The brain receives arterial blood from four arteries that form anastomoses within the cranium. An interruption of any one vessel will not compromise the blood flow to the brain. **27.** The cardioacceleratory and vasomotor centers are stimulated when general sympathetic activation occurs. The result is an increase in cardiac output and blood pressure. When the parasympathetic division is activated, the cardioinhibitory centers are stimulated, reducing cardiac output.

Level 3 Critical Thinking and Clinical Applications

28. Fluid loss lowers blood volume, leading to sympathetic stimulation, which elevates blood pressure. So, Bob's blood pressure is high instead of low.
29. Following injury, endothelial cells of the tunica intima increase their uptake of fats and cholesterol from blood, much of which is dietary. Monocytes migrate to the tunica media and mature into macrophages, oxidize the excess fats and become fat-filled foam cells. Smooth muscle cells lay down collagen and elastin, and also take up fats to become foam cells. **30.** Angiotensin II is a potent vasoconstrictor and increases BP by increasing peripheral resistance (BP = CO × TPR). It also stimulates the release of aldosterone, which stimulates the kidney to increase the uptake of sodium and, therefore, water.

Answers to Clinical Case Wrap-Up

PAGE 806 1. Calcification follows the degeneration of smooth muscle in the tunica media of arteries. Calcification in arteries is related to atherosclerotic injury of the vessels. Calcification does not occur in normal, young, healthy blood vessel walls. **2.** Atherosclerosis is never seen in the venous system because veins are never subjected to the pressures of the arterial system. (In the average person the systemic arterial pressure is over 100 mm Hg.) This causes wear and tear in the arterial walls, leading to gradual degeneration of the smooth muscle in the tunica media, followed by deposition of calcium salts, and then by deposition of atherosclerotic plaques. The pressure within the venous system is always quite low (usually around 16 mm Hg). Damage to the endothelial surfaces does not get started.

Chapter 22

Answers to Checkpoints

PAGE 809 1. A pathogen is any disease-causing organism, such as a virus, bacterium, fungus, or parasite, that can survive and even thrive inside the body. **2.** Innate (nonspecific) defenses are anatomical barriers and defense mechanisms that either prevent or slow the entry of infectious organisms but do not distinguish one potential threat from another. Adaptive (specific) defenses involve an immune response against a specific type of threat. **PAGE 821 3.** The components of the lymphatic system are lymph, lymphatic vessels, primary lymphoid tissues and organs (red bone marrow and the thymus), and secondary lymphoid tissues and organs (lymph nodes, tonsils, MALT, appendix, and the spleen). **4.** A blockage of the thoracic duct would impair the drainage of lymph from inferior to the diaphragm and from the left side of the head and thorax, retarding the return of lymph to the venous blood and promoting the accumulation of fluid in the limbs (lymphedema). **5.** A lack of thymic hormones would drastically reduce the population of T lymphocytes by preventing their differentiation from lymphoid stem cells. **6.** Lymph nodes enlarge

during some infections because lymphocytes and phagocytes in the nodes multiply to defend against the infectious agent. **PAGE 828 7.** The body's innate (nonspecific) defenses include physical barriers, phagocytes, immune surveillance, interferons, complement, inflammation, and fever. **8.** A decrease in the number of monocyte-forming cells in red bone marrow would result in fewer macrophages of all types, including Kupffer cells of the liver, dendritic (Langerhans) cells in the skin and digestive tract, and alveolar macrophages. **9.** A rise in the level of interferon suggests a viral infection. Interferon does not help an infected cell, but "interferes" with the virus's ability to infect other cells. **10.** Pyrogens increase body temperature (produce a fever) by stimulating the temperature control area of the preoptic nucleus of the hypothalamus. **PAGE 831 11.** In cell-mediated (cellular) immunity, T cells defend against abnormal cells and pathogens inside cells. In antibody-mediated (humoral) immunity, B cells secrete antibodies that defend against antigens and pathogens in body fluids.
12. The two forms of active immunity are naturally acquired active immunity and artificially induced active immunity; the two forms of passive immunity are naturally acquired passive immunity and artificially induced passive immunity. **13.** The four general properties of adaptive immunity are specificity, versatility, memory, and tolerance. **PAGE 835 14.** The four major types of T cells are cytotoxic T (T_C) cells, memory T cells (both T_C and T_H), helper T (T_H) cells, and suppressor T (T_S) cells. **15.** Abnormal peptides in the cytoplasm of a cell can become attached to MHC (major histocompatibility complex) proteins and then displayed on the surface of the cell's plasma membrane. The recognition of such displayed peptides by T cells can initiate an immune response. **16.** A decrease in the number of cytotoxic T cells would affect cell-mediated immunity, reducing the effectiveness of T_C cells in killing foreign cells and virus-infected cells. **17.** Without helper T cells—which promote B cell division, the maturation of plasma cells, and antibody production by plasma cells—the antibody-mediated immune response would probably not occur. **PAGE 843 18.** Sensitization is the process by which a B cell becomes able to react with a specific antigen. **19.** Plasma cells produce and secrete antibodies, so observing an elevated number of plasma cells would lead us to expect increasing levels of antibodies in the blood. **20.** An antibody molecule consists of two parallel pairs of polypeptide chains: one pair of heavy chains and one pair of light chains. Each chain contains both constant segments and variable segments. **21.** Because production of a secondary response depends on the presence of memory B cells and memory T cells formed during a primary response, the secondary response would be more negatively affected by a lack of memory B cells for a particular antigen. **PAGE 849 22.** A developing fetus is protected primarily by natural passive immunity, through the maternal production of IgG antibodies that have crossed the placenta from the mother's bloodstream. **23.** An autoimmune disorder is a condition that results when the immune system's sensitivity to normal cells and tissues causes the production of autoantibodies. **24.** Stress can interfere with the immune response by depressing inflammation, reducing the number and activity of phagocytes, and inhibiting interleukin secretion. **PAGE 825 25.** Elderly people are more susceptible to viral and bacterial infections because the number of helper T cells declines with age and B cells are less responsive, so antibody levels rise more slowly after antigen exposure. **26.** As one ages, immune surveillance declines, so cancerous cells are not eliminated as effectively. **27.** Glucocorticoids released

by the endocrine system have anti-inflammatory effects; thymic hormones stimulate the development and maturation of lymphocytes; and many hormones affect immune function. The thymus gland secretes thymic hormones, and cytokines affect cells throughout the body. **28.** The lymphatic system provides defenses against infection; performs immune surveillance to eliminate cancer cells; and returns tissue fluid to the circulation.

Answers to Review Questions

PAGE 854
Level 1 Reviewing Facts and Terms
1. (a) tonsil; **(b)** cervical lymph nodes; **(c)** right lymphatic duct; **(d)** thymus; **(e)** cisterna chyli; **(f)** lumbar lymph nodes; **(g)** appendix; **(h)** lymphatics of lower limb; **(i)** lymphatics of upper limb; **(j)** axillary lymph nodes; **(k)** thoracic duct; **(l)** lymphatics of mammary gland; **(m)** spleen; **(n)** mucosa-associated lymphoid tissue (MALT); **(o)** pelvic lymph nodes; **(p)** inguinal lymph nodes; **(q)** red bone marrow **2.** b **3.** Lymphedema **4.** c **5.** c **6.** Kupffer cells are macrophages located in and around the liver sinusoids **7.** a **8.** c **9.** d **10.** d **11.** c **12.** (1) red bone marrow: maintain normal lymphocyte populations and other defense cells, such as monocytes and macrophages; (2) thymus: production of mature T cells and hormones that promote immune function; (3) spleen: filtration of blood, recycling of red blood cells, detection of blood-borne pathogens or toxins; (4) lymph nodes: filtration of lymph, detection of pathogens, initiation of immune response; (5) Lymphoid nodules, and MALT: defense of entrance and passageways of digestive tract and protection of epithelia lining the respiratory, urinary, and reproductive tracts against pathogens and foreign proteins or toxins; (6) lymphatics: movement of lymph from interstitial spaces to the venous system **13. (a)** lymphocytes responsible for cell-mediated immunity; **(b)** stimulate the activation and function of T cells and B cells; **(c)** inhibit the activation and function of both T cells and B cells; **(d)** produce and secrete antibodies; **(e)** recognize and destroy abnormal cells; **(f)** produce interleukin-7, which promotes the differentiation of B cells; **(g)** regulate T cell development and function.; **(h)** interfere with viral replication inside the cell and stimulate the activities of macrophages and NK cells; **(i)** reset the body's thermostat, causing a rise in body temperature (fever); **(j)** provide cell-mediated immunity, which defends against abnormal cells and pathogens inside cells; **(k)** provide humoral immunity, which defends against antigens and pathogens in the body (but not inside cells); **(l)** enhance innate (nonspecific) defenses and increase T cell sensitivity and stimulate B cell activity; **(m)** slow tumor growth and kill sensitive tumor cells; **(n)** stimulate the production of blood cells in the bone marrow and lymphocytes in lymphoid tissues and organs **14.** (1) T cells, derived from the thymus; (2) B cells, derived from red bone marrow; and (3) NK cells, derived from red bone marrow **15.** The acid secreted by the stomach kills bacteria, fungi, and viruses.

Level 2 Reviewing Concepts
16. c **17.** c **18.** b **19.** This is produced by the thymus, and promotes the maturation of lymphocytes to become T cells. **20.** It is the final stage which induces bacterial cell lysis. **21.** It is programmed cell death. This mechanism causes the target cell to implode without lysis due to sudden water loss from the cell. The T-cytotoxic cell initiates this process. **22.** On reaching the opposing plasma membrane, perforin molecules interact with one another and with the membrane to create a network of pores in it. These pores are large enough to allow the free passage of ions, proteins, and other intracellular materials. As a result, the target cell can no longer maintain its

internal environment, and it quickly disintegrates. **23.** (1) Free and fixed macrophages in connective tissues, (2) the Kupffer cells of the liver, and (3) the microglia in the central nervous system. **24.** The injections are timed to trigger the primary and secondary responses of the immune system. Upon first exposure to hepatitis antigens, B cells produce daughter cells that differentiate into plasma cells and memory B cells. The plasma cells begin producing antibodies, which represent the primary response to exposure. However, the primary response does not maintain elevated antibody levels for long periods, so the second and third injections are necessary to trigger secondary responses, when memory B cells differentiate into plasma cells and produce antibody concentrations that remain high much longer.

Level 3 Critical Thinking and Clinical Applications
25. Yes, the crime lab could determine whether the sample is blood plasma, which contains IgM, IgG, IgD, and IgE, or semen, which contains only IgA. **26.** Ted cannot yet know whether he'll come down with the measles. Ted's elevated blood IgM level and titer indicates that he is in the early stages of a primary response to the measles virus. If his immune response proves unable to control and then eliminate the virus, Ted will develop the measles. **27.** In a radical mastectomy, lymph nodes in the nearby axilla and surrounding region are removed along with the cancerous breast to try to prevent the spread of cancer cells by the lymphatic system. Lymphatic vessels from the limb on the affected side are tied off, and because there is no place for the lymph to drain, over time lymphedema causes swelling of the limb. **28.** A key characteristic of cancer cells is their ability to metastasize—to break free from a tumor and form new tumors in other tissues. The primary route of metastasis is the lymphatic system, including the lymph nodes. Examination of regional lymph nodes for the presence of cancer cells can help the physician determine if the cancer was caught in an early stage or whether it has started to spread to other tissues. **29.** Allergies occur when allergens bind to specific IgE antibodies that are bound to the surface of mast cells and basophils. A person becomes allergic when they develop IgE antibodies for a specific allergen. Theoretically, at least, a molecule that would bind to the specific IgE for ragweed allergen and prevent the allergen from binding should help to relieve the signs and symptoms of the allergy.

Answers to Clinical Case Wrap-Up
PAGE 855 **1.** Innate immunity is genetically determined and present at birth. For example, skin and mucous membranes provide an innate physical and chemical barrier to infecting organisms. All humans are susceptible to chickenpox, an infectious disease. The immunity developed after a chickenpox vaccination is adaptive immunity, developed to a specific antigen, the varicella-zoster virus (VZV). **2.** Baby Ruthie has developed life-long immunity to chickenpox as a result of having the disease. Her immune memory will provoke a secondary immune response if she is exposed to the varicella-zoster virus again. The recommended schedule (according to the Centers for Disease Control and Prevention, CDC) for varicella-zoster immunization is 12–15 months of age. However, there is no need for Ruthie to get a chickenpox vaccination after having had the disease.

Chapter 23

Answers to Checkpoints
PAGE 860 **1.** Functions of the respiratory system include providing an extensive surface area for gas exchange between blood and air, moving air to and from exchange surfaces, protecting exchange

surfaces from environmental variations and pathogens, producing sounds, detecting olfactory stimuli, and indirectly assisting in blood volume regulation and blood pressure through the conversion of angiotensin I to angiotensin II. **2.** The two anatomical divisions of the respiratory system are the upper respiratory system and the lower respiratory system. **3.** The respiratory mucosa lines the conducting portion of the respiratory tract. **PAGE 863 4.** The upper respiratory system consists of the nose, nasal cavity, paranasal sinuses, and pharynx. **5.** The rich vascularization to the nose delivers body heat to the nasal cavity, so inhaled air is warmed before it leaves the nasal cavity. The heat also evaporates moisture from the epithelium to humidify the incoming air. **6.** The lining of the nasopharynx, which receives air from only the nasal cavity, is the same as that of the nasal cavity: a pseudostratified ciliated columnar epithelium. Because the oropharynx and laryngopharynx receive air from the nasal cavity and potentially abrasive food from the oral cavity, they have a more highly protective lining: a stratified squamous epithelium. **PAGE 865 7.** The paired cartilages are the arytenoid cartilages, corniculate cartilages, and cuneiform cartilages. The unpaired laryngeal cartilages include the thyroid cartilage, cricoid cartilage, and epiglottis. **8.** The highly elastic vocal folds of the glottis are better known as the vocal cords. **9.** When vocal fold tension increases, the pitch of the voice is raised. **PAGE 867 10.** The trachea conveys air between the larynx and primary bronchi; cilia and the mucus produced by epithelial cells also protect the respiratory tree by trapping inhaled debris and sweeping it toward the pharynx, where it is removed through coughing or swallowing. **11.** The tracheal cartilages are C-shaped to allow space for expansion of the esophagus when food or liquid is swallowed. **12.** Objects are more likely to be lodged in the right bronchus because it is slightly larger and more vertical than the left bronchus. **PAGE 873 13.** Without surfactant, the alveoli would collapse as a result of surface tension in the thin layer of water that moistens the alveolar surfaces. **14.** Air passing through the glottis flows into the larynx and through the trachea. From there, the air flows into a primary bronchus, which supplies the lungs. In the lungs, the air passes to bronchi, bronchioles, a terminal bronchiole, a respiratory bronchiole, an alveolar duct, an alveolar sac, an alveolus, and ultimately to the respiratory membrane. **15.** The pulmonary arteries supply the gas exchange surfaces; the external carotid arteries, the thyrocervical trunks, and the bronchial arteries supply the conducting portions of the respiratory system. **16.** The pleura is a serous membrane that secretes pleural fluid, which lubricates the opposing parietal and visceral surfaces to prevent friction during breathing. **PAGE 874 17.** External respiration includes all the processes involved in the exchange of oxygen and carbon dioxide between the body's interstitial fluids and the external environment. Internal respiration is the absorption of oxygen and the release of carbon dioxide by the body's cells. **18.** The integrated steps involved in external respiration are pulmonary ventilation (breathing), gas diffusion, and the transport of oxygen and carbon dioxide. **PAGE 882 19.** Compliance is the ease with which the lungs expand. Factors affecting compliance include (a) the connective tissue structure of the lungs, (b) the level of surfactant production, and (c) the mobility of the thoracic cage. **20.** The pulmonary volumes are tidal volume (V_T), expiratory reserve volume (ERV), residual volume, and inspiratory reserve volume (IRV). **21.** When the rib penetrates Mark's chest wall, it also penetrates the pleural cavity, allowing atmospheric air to enter and increase the intrapleural pressure within the pleural cavity (a condition called pneumothorax). As a result, the natural elasticity of the lung may

cause the lung to collapse, a condition called atelectasis. **22.** Because the fluid produced in pneumonia takes up space that would normally be occupied by air, vital capacity would decrease. **PAGE 886 23.** Dalton's law states that each gas in a mixture exerts a pressure equal to its relative abundance. **24.** Henry's law states that at a constant temperature, the quantity of a particular gas that will dissolve in a liquid is directly proportional to the partial pressure of that gas. **PAGE 890 25.** Carbon dioxide is transported in the bloodstream as carbonic acid, bound to hemoglobin, or dissolved in the plasma. **26.** During exercise, active skeletal muscles generate heat and acidic wastes. A combination of increased body temperature and decreased pH causes hemoglobin to release more oxygen when you are exercising than when you are resting. **27.** Blockage of the trachea would interfere with the body's ability to gain oxygen and eliminate carbon dioxide. Because most carbon dioxide is transported in blood as bicarbonate ions formed from the dissociation of carbonic acid, an inability to eliminate carbon dioxide would result in an excess of hydrogen ions, which decreases blood pH. **PAGE 899 28.** Exciting the pneumotaxic centers would inhibit the inspiratory and apneustic centers, which would result in shorter and more rapid breaths. **29.** Peripheral chemoreceptors are more sensitive to carbon dioxide levels than to oxygen levels. When carbon dioxide dissolves, it forms carbonic acid, which dissociates and releases hydrogen ions, thereby decreasing pH and altering cell or tissue activity. **30.** Johnny's mother shouldn't worry. When Johnny holds his breath, blood carbon dioxide levels increase, causing increased stimulation of the inspiratory center and forcing Johnny to breathe again. **31.** Aging results in deterioration of elastic tissue (reducing compliance), arthritic changes in rib articulations, decreased flexibility at costal cartilages, decreased vital capacity, and some degree of emphysema. **PAGE 875 32.** The respiratory system provides oxygen and eliminates carbon dioxide for all body systems. **33.** The respiratory system provides alveolar phagocytes and respiratory defenses to trap pathogens and protect deeper tissues. Tonsils within the lymphatic system protect against respiratory infection, and lymph drainage from the lungs mobilizes defenses to ward off infection.

Answers to Review Questions

PAGE 904
Level 1 Reviewing Facts and Terms
1. b **2.** c **3.** a **4.** c **5.** c **6.** (a) nasal cavity; (b) pharynx; (c) right lung; (d) nose; (e) larynx; (f) trachea; (g) bronchus; (h) bronchioles; (i) left lung **7.** Since the air Brad is breathing is dry, large amounts of moisture are leaving the mucus in his respiratory tract to humidify inhaled air. Drying makes the mucus tacky and makes it difficult for the cilia to move, so mucus builds up, producing nasal congestion by morning. When Brad showers and drinks fluids, body water is replaced, so the mucus loosens up and can be moved along as usual. **8.** c **9.** The regions of the pharynx are the superior nasopharynx, where the nasal cavity opens into the pharynx; the middle oropharynx, posterior to the oral cavity; and the inferior laryngopharynx, which is posterior to the hyoid bone and glottis. **10.** External respiration is the exchange of blood gases with interstitial fluid and external environment. Internal respiration is the exchange of blood gases with interstitial fluid and cells. **11.** The steps of external respiration are **(a)** pulmonary ventilation (breathing); **(b)** gas diffusion across the respiratory membrane and between blood and interstitial fluids; and **(c)** the transport of oxygen and carbon dioxide between alveolar and peripheral capillaries. **12.** a **13.** Carbon dioxide is transported in the blood as carbonic acid, bound to hemoglobin, and dissolved in the plasma.

Level 2 Reviewing Concepts

14. c **15.** d **16.** c **17.** d **18.** Alveoli consist of simple squamous epithelium (type I pneumocytes), which offers a low diffusion distance for gas exchange. The surfactant released by type II pneumocytes allows lipid-soluble gases to dissolve. The shape of alveoli increases the surface area for gas exchange.
19. Contraction of the diaphragm increases thoracic volume, which decreases intrapleural and intrapulmonary pressure. Air moves down its pressure gradient from the atmosphere towards the lower intrapulmonary pressure. **20.** Type II pneumocytes (type II alveolar cells) produce surfactant, which reduces surface tension in the fluid coating the alveolar surface. Without surfactant, the surface tension would be so high that the delicate alveoli would collapse. **21.** Pulmonary ventilation, the physical movement of air into and out of the respiratory tract, maintains adequate alveolar ventilation. Alveolar ventilation, air movement into and out of the alveoli, prevents the buildup of carbon dioxide in the alveoli and ensures a continuous supply of oxygen that keeps pace with absorption by the bloodstream. **22. (a)** Boyle's law describes the inverse relationship between gas pressure and volume: If volume decreases, pressure rises; if volume increases, pressure falls. It is the basis for the direction of air movement in pulmonary ventilation. **(b)** Dalton's law states that each of the gases that make up a mixture of gases contributes to the total pressure in proportion to its relative abundance; that is, all the partial pressures added together equal the total pressure exerted by the gas mixture. This relationship is the basis for the calculation of the partial pressures of oxygen and carbon dioxide, and their exchange between blood and alveolar air. **(c)** Henry's law states that, at a given temperature, the amount of a particular gas that dissolves in a liquid is directly proportional to the partial pressure of that gas. Henry's law underlies the diffusion of gases between capillaries, alveoli, and interstitial fluid.
23. Reduction of elastic tissues changes lung compliance, which lowers vital capacity. Arthritic changes in the rib articulations restrict chest movement and reduce respiratory minute volume. The decline maybe accelerated by smoking. **24.** Pulmonary volumes are determined experimentally and include tidal volume (averaging 500 mL), expiratory reserve volume (approximately 1200 mL), residual volume (averaging 1200 mL), minimal volume (30–120 mL), and inspiratory reserve volume (approximately 3600 mL). Respiratory capacities include inspiratory capacity, functional residual capacity, vital capacity, and total lung capacity. The difference between the two measures is that respiratory capacities are the sums of various pulmonary volumes. **25.** The DRG is the inspiratory center that contains neurons that control lower motor neurons innervating the external intercostal muscles and the diaphragm. The DRG functions in every respiratory cycle, whether quiet or forced. The VRG functions only during forced respiration—active exhalation and maximal inhalation. The neurons involved with active exhalation are sometimes said to form an expiratory center.

Level 3 Critical Thinking and Clinical Applications

26. The partial pressure of oxygen at 550 mm Hg atmospheric pressure is 115.5 mm Hg (15.3 k Pa), an approximate 28% reduction. Hemoglobin oxygen saturation falls by about 5%. The same fall is not seen in hemoglobin as cooperative bonding of oxygen seen in the steep rising part of the oxygen–hemoglobin saturation curve means that hemoglobin maintains saturations over 90% even if alveolar oxygen partial pressures fall to 60 mm Hg. **27.** A person with chronic emphysema has constantly elevated blood levels of P_{CO_2} due to an inability to eliminate CO_2 efficiently because of

physical damage to the lungs. Over time, the brain ignores the stimulatory signals produced by the increased CO_2 and begins to rely on information from peripheral chemoreceptors to set the pace of breathing. (In other words, accommodation has occurred.) The peripheral chemoreceptors also accommodate to the elevated CO_2 and respond primarily to the level of O_2 in the blood, increasing breathing when O_2 levels are low and decreasing breathing when O_2 levels are high. When pure O_2 was administered, chemoreceptors responded with fewer action potentials to the medulla oblongata, so Mr. B. stopped breathing. **28.** Cary's hyperventilation resulted in abnormally low P_{CO_2}. This reduced his urge to breathe, so he stayed underwater longer, unaware that his P_{O_2} was dropping to the point of loss of consciousness. **29.** In anemia, the blood's ability to carry oxygen is decreased due to the lack of functional hemoglobin, red blood cells, or both. Anemia does not interfere with the exchange of carbon dioxide within the alveoli, nor with the amount of oxygen that will dissolve in the plasma. Because chemoreceptors respond to dissolved gases and pH, as long as the pH and the concentrations of dissolved carbon dioxide and oxygen are normal, ventilation patterns should not change significantly. **30.** d

Answers to Clinical Case Wrap-Up

PAGE 905 1. Andrea's cough is serving a good purpose. As she expels the irritating mucus, she is also expelling the virus. This is a good thing and should not be suppressed. She should, however, be very careful to use tissues to cover her cough and dispose of the infective mucus. **2.** Andrea's nose (runny), pharynx (sore throat), trachea, and bronchi (productive cough) seem to be infected by this virus. Her bronchioles are likely infected as well.

Chapter 24

Answers to Checkpoints

PAGE 915 1. Organs of the digestive system include the oral cavity, pharynx, esophagus, stomach, small intestine, large intestine, and accessory organs (teeth, tongue, gallbladder, salivary glands, liver, and pancreas). **2.** The six primary functions of the digestive system include the following: (1) ingestion = consciously eating food; (2) mechanical processing = crushing and shearing foodstuffs to make them more susceptible to enzymatic attack; (3) digestion = the chemical breakdown of food into smaller products for absorption; (4) secretion = the release of water, acids, and other substances by the epithelium of the digestive tract and by glandular organs; (5) absorption = movement of digested particles across the digestive epithelium and into the interstitial fluid of the digestive tract; and (6) excretion = the removal of waste products from the body. **3.** The mesenteries—sheets consisting of two layers of serous membrane connected by their loose connective tissue—support and stabilize the organs in the abdominopelvic cavity and provide a route for the associated blood vessels, nerves, and lymphatic vessels. **4.** The layers of the gastrointestinal tract, beginning at the surface of the lumen are the mucosa (adjacent to the lumen), submucosa, muscularis externa, and serosa. **5.** The waves of contractions responsible for peristalsis are more efficient in propelling intestinal contents than segmentation, which is basically a churning action that mixes intestinal contents with digestive fluids. **6.** A drug that blocks parasympathetic stimulation, which increases muscle tone and activity in the digestive tract, would slow peristalsis. **PAGE 919 7.** Structures associated with the oral cavity include the tongue, salivary glands, and teeth. **8.** Stratified squamous epithelium lines the oral cavity. It protects against friction or abrasion by food. **9.** Damage to the parotid salivary glands, which secrete the carbohydrate-digesting enzyme salivary amylase, would interfere with the digestion of complex carbohydrates. **10.** The incisors are the teeth most useful for chopping (or cutting or shearing) pieces of relatively rigid food, such as raw vegetables. **11.** The fauces are the dividing line between the oral cavity and the pharynx. **PAGE 921 12.** The pharynx is an anatomical space that receives a food bolus or liquids and passes them to the esophagus as part of the swallowing process. Air also passes through the pharynx. **13.** Muscles associated with the pharynx are pharyngeal constrictor muscles, the palatopharyngeus and stylopharyngeus muscles, and palatal muscles. **PAGE 923 14.** The structure connecting the pharynx to the stomach is the esophagus. **15.** The muscularis externa of the esophagus is an unusual segment of the digestive tract because it (1) contains skeletal muscle cells along most of the length of the esophagus and (2) is surrounded by an adventitia rather than a serosa. **16.** When the soft palate and larynx elevate and the glottis closes, swallowing (deglutition) is occurring. **PAGE 927 17.** The four regions of the stomach are the cardia, fundus, body, and pylorus. **18.** The low pH of the stomach creates an acidic environment that kills most microorganisms ingested with food, denatures proteins and inactivates most enzymes in food, helps break down plant cell walls and meat connective tissue, and activates pepsin. **19.** Large meals, especially meals with a high protein content, stimulate increased stomach acid secretion. When gastric glands are secreting, bicarbonate ions enter the bloodstream and increase the blood pH. This phenomenon is referred to as the alkaline tide. **20.** The vagus nerves contain parasympathetic motor fibers that can stimulate gastric secretions, even if food is not present in the stomach (the cephalic phase of gastric digestion). Cutting the branches of the vagus nerves that supply the stomach would prevent this type of secretion from occurring and thereby reduce the likelihood of ulcer formation. **PAGE 941 21.** The three regions of the small intestine are the duodenum, jejunum, and ileum. **22.** The small intestine has several adaptations that increase its surface area and thus its absorptive capacity for nutrients. The walls of the small intestine have folds called the circular folds (plicae circulares). The tissue that covers the circular folds forms fingerlike projections, the villi. The epithelial cells that cover the villi have an exposed surface covered by small fingerlike projections, the microvilli. In addition, the small intestine has a very rich supply of blood vessels and lymphatic vessels, which transport the nutrients that are absorbed. **23.** A high-fat meal would raise the cholecystokinin level in the blood. **24.** The hormone secretin, among other things, stimulates the pancreas to release fluid high in buffers to neutralize the chyme that enters the duodenum from the stomach. If the small intestine did not produce secretin, the pH of the intestinal contents would be lower than normal. **25.** Damage to the exocrine pancreas would most impair the digestion of fats (lipids), because it is the primary source of lipases. Even though such damage would also reduce carbohydrate and protein digestion, enzymes for digesting these nutrients are produced by other digestive system structures, including the salivary glands (carbohydrates), the small intestine (carbohydrates and proteins), and the stomach (proteins). **PAGE 948 26.** The four regions of the colon are the ascending colon, transverse colon, descending colon, and sigmoid colon. **27.** The large intestine is larger in diameter than the small intestine. The thin walls of the large intestine lack villi and have an abundance of mucous cells and intestinal glands. **28.** In haustral churning, segmentation movements mix the contents of nearby haustra. In mass movements, which occur a few times per day, strong peristaltic contractions move material from the transverse colon through the rest of the large intestine. **PAGE 954 29.** The nutrients the body requires are carbohydrates, lipids, proteins, vitamins, minerals, and water. **30.** Because chylomicrons are formed from the fats digested in a meal, fats increase the number of chylomicrons in the lacteals. **31.** Removal of the stomach would interfere with the absorption of vitamin B_{12}. Parietal cells of the stomach are the source of intrinsic factor, which is required for the vitamin's absorption. **32.** When someone with diarrhea loses fluid and electrolytes faster than they can be replaced, the resulting dehydration can be fatal. Although constipation can be quite uncomfortable, it does not interfere with any life-supporting processes; the few toxic wastes normally eliminated by the digestive system can move into the blood and be eliminated by the kidneys. **33.** General age-related digestive system changes include decreased secretory mechanisms, decreased gastric and intestinal motility, decreased mitotic activity of epithelial cells, and loss of muscle tone associated with tract sphincters; cumulative damage becomes more apparent, cancer rates increase, and dehydration occurs as a result of decreased osmoreceptor sensitivity. **34.** The digestive system absorbs the organic substrates, vitamins, ions, and water required by cells of all other body systems. **35.** The skeletal, muscular, nervous, endocrine, and cardiovascular systems may all be affected by inadequate absorption of calcium.

Answers to Review Questions

PAGE 958
Level 1 Reviewing Facts and Terms
1. d **2.** a **3.** c **4.** c **5.** (a) oral cavity (mouth); (b) liver; (c) gallbladder; (d) pancreas; (e) large intestine; (f) salivary glands; (g) pharynx; (h) esophagus; (i) stomach; (j) small intestine; (k) anus **6.** (a) mucosa; (b) submucosa; (c) muscularis externa; (d) serosa **7.** d **8.** a **9.** b **10.** (a) duodenum; (b) jejunum; (c) ileum **11.** c **12.** c **13.** a **14.** Thinner and weaker epithelium; reduced smooth muscle tone leading to weaker peristaltic contractions; increased rates of cancer; increased dehydration. **15.** d **16.** Activities of the digestive tract are regulated by local, hormonal, and neural mechanisms. **17.** The three phases of swallowing—the buccal, pharyngeal, and esophageal phases—are controlled by the swallowing center of the medulla oblongata by the trigeminal and glossopharyngeal cranial nerves. The motor commands originating at the swallowing center are distributed by cranial nerves V, IX, X, and XII. Along the esophagus, primary peristaltic contractions are coordinated by afferent and efferent fibers within the glossopharyngeal and vagus cranial nerves, but secondary peristaltic contractions occur in the absence of CNS instructions. **18.** The pancreas provides digestive enzymes, plus bicarbonate ions that elevate the pH of the chyme. The liver produces bile and is also the primary organ involved in regulating the composition of circulating blood. The gallbladder stores and releases bile, which contains additional buffers and bile salts that aid in the digestion and absorption of lipids. **19.** The hormones include the following: enterocrinin, which stimulates the submucosal glands of the duodenum; secretin, which stimulates the pancreas and liver to increase the secretion of water and bicarbonate ions; cholecystokinin (CCK), which causes an increase in the release of pancreatic secretions and bile into the duodenum, inhibits gastric activity, and appears to have CNS effects that reduce the sensation of hunger; gastric inhibitory peptide (GIP), which stimulates insulin release at pancreatic islets and the activity of the duodenal submucosal

glands; vasoactive intestinal peptide (VIP), which stimulates the secretion of intestinal glands, dilates regional capillaries, and inhibits acid production in the stomach; gastrin, which is secreted by G cells in the duodenum when they are exposed to large quantities of incompletely digested proteins; and, in small quantities, motilin, which stimulates intestinal contractions, villikinin, which promotes the movement of villi and associated lymph flow, and somatostatin, which inhibits gastric secretion. **20.** Respiratory muscle contraction increases abdominal pressures which aid defecation. The weight of the digestive organs aids the diaphragm during exhalation. **21.** Positive feedback loops in the defecation reflex involve (1) stretch receptors in the rectal walls, which promote a series of peristaltic contractions in the colon and rectum, moving feces toward the anus; and (2) the sacral parasympathetic system, also activated by the stretch receptors, which stimulates peristalsis by motor commands distributed by the pelvic nerves.

Level 2 Reviewing Concepts

22. e **23.** Emulsification of lipids to increase surface area for enzyme digestion. Chyme entering the duodenum stimulates enteroendocrine cells to release cholecystokinin into the blood stream. Cholecystokinin stimulates contraction of the gall bladder and hepatopancreatic sphincter relaxation allowing the release of bile. **24.** The pharynx functions as a common passageway for food, liquids, and air, and is subjected to abrasion, pathogen, and chemical attack. Squamous epithelium forms an effective protective layer. **25.** Circular muscle contracts behind the bolus of food whilst longitudinal muscle shortens in front. A wave of circular muscle contraction along with longitudinal muscle relaxation moves the bolus forward. **26.** Pepsin breaks down peptide bonds of proteins. The digestion of carbohydrates and lipids by salivary amylase and lingual lipase continues in the stomach until the pH falls sufficiently to cause their deactivation. **27.** (1) The cephalic phase of gastric secretion begins with the sight or thought of food. Directed by the CNS, this phase prepares the stomach to receive food. (2) The gastric phase begins with the arrival of food in the stomach; this phase is initiated by distension of the stomach, an increase in the pH of the gastric contents, and the presence of undigested materials in the stomach. (3) The intestinal phase begins when chyme starts to enter the small intestine. This phase controls the rate of gastric emptying and ensures that the secretory, digestive, and absorptive functions of the small intestine can proceed reasonably efficiently. **28.** After a heavy meal, bicarbonate ions pass from the parietal cells of the stomach into the interstitial fluid. The diffusion of bicarbonate ions from the interstitial fluid into the bloodstream increases blood pH. This sudden influx of bicarbonate ions into the bloodstream has been called the alkaline tide.

Level 3 Critical Thinking and Clinical Applications

29. If a gallstone is small enough, it can pass through the common bile duct and block the pancreatic duct. Enzymes from the pancreas then cannot reach the small intestine. As the enzymes accumulate, they irritate the duct and ultimately the exocrine pancreas, producing pancreatitis. **30.** Scarring from cirrhosis restricts blood flow through the liver, leading to an increase in pressure in the portal vein (portal hypertension). The portal veins connect with veins in the esophagus and the increase in pressure causes these to swell and rupture. **31.** If an individual cannot digest lactose, this sugar passes into the large intestine in an undigested form. The presence of extra sugar in the chyme increases its osmolarity, so less water is reabsorbed by the intestinal mucosa. The bacteria that inhabit the large intestine can metabolize the lactose, and in the process they

produce large amounts of carbon dioxide. This gas overstretches the intestine, which stimulates local reflexes that increase peristalsis. The combination of more-fluid contents and increased peristalsis causes diarrhea. The overexpansion of the intestine by gas, which is directly related to increased gas production by the bacteria, causes the severe pain and abdominal cramping. **32.** The primary effect of such surgeries would be a reduction in the volume of food (and thus in the amount of calories) consumed because the person feels full after eating a small amount. This can result in significant weight loss.

Answers to Clinical Case Wrap-Up

PAGE 960 **1.** If Mirasol's *Clostridium difficile* colitis is not successfully treated with a fecal transplant, she is likely to continue to experience bloody diarrhea. This may result in progressive dehydration that can eventually lead to her death. She may also continue to experience abdominal pain and blood loss. Eventually she may suffer from a lack of vitamin K (required for clotting factor production), biotin, and vitamin B_5. **2.** If you could look at the lining of Mirasol's large intestine with a colonoscope, the surface would look inflamed, swollen, reddened. You would likely see areas of raw, bleeding mucosa. You might even see areas of white cells, accumulated in pockets of pus, as Mirasol's immune system tries to fight this *C. difficile* infection.

Chapter 25

Answers to Checkpoints

PAGE 964 **1.** Metabolism is the sum of all biochemical processes under way within the human body; it includes anabolism and catabolism. **2.** Catabolism is the breakdown of organic molecules into simpler components, accompanied by the release of energy. **3.** Anabolism is the synthesis of new organic molecules from simpler components, and it requires energy. **PAGE 973** **4.** The primary role of the citric acid cycle in ATP production is to transfer electrons from substrates to coenzymes. These electrons provide energy for the production of ATP by the electron transport system. **5.** The NADH produced by glycolysis cannot enter the mitochondria, where the enzymes of the electron transport chain are located. However, an intermediary in the mitochondrial membrane can transfer the electrons from the NADH to a coenzyme within the mitochondria. In skeletal muscle cells, the intermediary transfers the electrons to FAD, whereas cardiac muscle cells use a different intermediary, which transfers the electrons to another NAD. In mitochondria, each NADH yields 3 molecules of ATP, whereas each $FADH_2$ yields just 2 molecules of ATP. The different intermediaries account for the difference in ATP yield. **6.** A decrease in cytosolic NAD would reduce ATP production in mitochondria. Decreased NAD would reduce the amount of pyruvate produced by glycolysis; less pyruvate means that the citric acid cycle could produce less ATP in mitochondria. **PAGE 978** **7.** Beta-oxidation is fatty acid catabolism that produces molecules of acetyl-CoA. **8.** The five major groups of lipoproteins are chylomicrons, very low density lipoproteins (VLDLs), intermediate-density lipoproteins (IDLs), low-density lipoproteins (LDLs), and high-density lipoproteins (HDLs). **9.** High-density lipoproteins (HDLs) are considered beneficial to health because they reduce the amount of fat (including cholesterol) in the bloodstream by transporting it to the liver for storage or excretion in the bile. **PAGE 982** **10.** Transamination is the transfer of an amino group from one molecule to another, especially from an amino acid to a keto acid. **11.** Deamination is the removal of an amino group from an amino acid.   **12.** A diet deficient in pyridoxine (vitamin B_6), an important coenzyme in deaminating and transaminating amino acids in cells, would interfere with the body's ability to metabolize proteins.

PAGE 983   13. Glycogenesis (the formation of glycogen) in the liver increases after a high-carbohydrate meal.   **14.** Blood levels of urea increase during the postabsorptive state because the deamination of many amino acids at that time yields ammonium ions, which are then converted to urea by the liver.   **15.** Accumulated acetyl-CoA is likely to be converted into ketone bodies.

PAGE 961   16. The two classes of vitamins are fat-soluble and water-soluble.   **17.** An athlete who is adding muscle mass through extensive training would try to maintain a positive nitrogen balance.   **18.** A decrease in the amount of bile salts, which are necessary for digesting and absorbing fats and fat-soluble vitamins (including vitamin A), would result in less vitamin A in the body, and perhaps a vitamin A deficiency.   **PAGE 994   19.** The BMR of a pregnant woman would be higher than her own BMR when she is not pregnant, due to both the increased metabolism associated with supporting the fetus and the contribution of fetal metabolism.   **20.** The vasoconstriction of peripheral vessels would decrease blood flow to the skin and thus the amount of heat the body can lose. As a result, body temperature would increase.
21. Infants have higher surface-to-volume ratios than do adults, and the body's temperature-regulating mechanisms are not fully functional at birth. As a result, infants must expend more energy to maintain body temperature, and they get cold more easily than do healthy adults.

Answers to Review Questions

PAGE 996
Level 1 Reviewing Facts and Terms

1. b   **2.** e   **3.** a   **4.** b   **5.** Secretion production   **6.** Determine osmotic potentials of body compartments; are important cofactors in many enzyme reactions; drive many physiological processes, such as muscle contraction, nerve conduction and blood clotting.
7. **(a)** fatty acids; **(b)** glucose; **(c)** amino acids; **(d)** two-carbon chains; **(e)** citric acid cycle; **(f)** CO_2; **(g)** ATP; **(h)** electron transport system; **(i)** H_2O **8.** It is the sum of all anabolic and catabolic reactions.   **9.** Increased blood flow to the skin, increased perspiration, increased respiratory drive.

Level 2 Reviewing Concepts

10. Oxidative phosphorylation is the generation of ATP within mitochondria, through a process called chemiosmosis that requires coenzymes, ATP synthase, and consumes oxygen. The electron transport system consists of a sequence of metal ion–containing proteins (metalloproteins) called cytochromes, embedded in the inner mitochondrial membrane. Energy from the stepwise passage of electrons (from H atoms) along the cytochrome molecules is used to pump hydrogen ions from the matrix into the intermembrane space. This forms a proton gradient across the inner mitochondrial membrane. The hydrogen ions diffuse back through ATP synthase and generate ATP. The electrons, hydrogen ions, and oxygen combine to produce water as a byproduct.   **11.** In lipid catabolism, a triglyceride is hydrolyzed, yielding glycerol and fatty acids. Glycerol is converted to pyruvate and enters the citric acid cycle. Fatty acids are broken into 2-carbon fragments by beta-oxidation inside mitochondria. The 2-carbon compounds then enter the citric acid cycle.
12. During the absorptive state, insulin prevents a large surge in blood glucose after a meal by causing the liver to remove glucose

from the hepatic portal circulation. During the postabsorptive state, blood glucose begins to decline, triggering the liver to release glucose produced through glycogenolysis and gluconeogenesis.
13. The catabolism is difficult and produces ammonium, which is toxic. They are also used as important structural components within the body, and significant loss can cause dysfunction.   **14.** i) Very low-density lipoproteins (LDL) are synthesised in the liver and secreted into the blood stream. Triglycerides are removed by the action of lipoprotein lipase to form intermediate-density lipoprotein (IDL). IDLs re-enter the liver where they are converted to LDL's, and re-enter the blood stream. LDL's enter peripheral tissue by vascular transport, and are absorbed into cells by endocytocysis and broken down by lysosomes. Cholesterol is released and used in the plasma membrane. ii) Cholesterol not used within the cell diffuses into the blood stream. Circulating high-density lipoproteins take up the cholesterol and return it to the liver where some is used in the production of bile.   **15.** Low-density lipids carry large amounts of cholesterol within the blood to peripheral tissue. This cholesterol can be used in the formation of arterial plaques.

Level 3 Critical Thinking and Clinical Applications

16. Cyanide binds with cytochrome a3, which inhibits the transfer of electrons to oxygen and prevents further electron transfer throughout the chain. This inhibits the TCA cycle. As oxygen cannot be used, it is not removed from arterial blood; hemoglobin in venous blood will remain fully saturated with oxygen, making the man's skin look red.   **17.** Based just on the information given, Charlie would appear to be in good health, at least relative to his diet and probable exercise. Problems are associated with elevated levels of LDLs, which carry cholesterol to peripheral tissues and make it available for the formation of atherosclerotic plaques in blood vessels. High levels of HDLs indicate that a considerable amount of cholesterol is being removed from the peripheral tissues and carried to the liver for disposal. You would encourage Charlie not to change, and keep up the good work.   **18.** It appears that Jill is suffering from ketoacidosis as a consequence of her anorexia. Because she is literally starving herself, her body is metabolizing large amounts of fatty acids and amino acids to provide energy, and in the process is producing large quantities of ketone bodies (normal metabolites from these catabolic processes). One of the ketones formed is acetone, which can be eliminated through the lungs. Acetone has a fruity aroma, so her breath would also smell fruity. The ketones are also converted into keto acids such as acetic acid. In large amounts, this lowers the body's pH. This is probably the cause of her arrhythmias.

Answers to Clinical Case Wrap-Up

PAGE 997   1. Diet and vitamin supplementation are the only known sources of vitamin A. Unlike some other vitamins that can be made by our skin (vitamin D_3) or by intestinal bacteria (vitamin K, Biotin [B_7], Pantothenic acid [B_5]), vitamin A is a true *essential* vitamin that must be consumed orally.   **2.** Justine's vitamin-deficient diet is likely to lead to deficiencies of all the fat-soluble and water-soluble vitamins, including D, E, K, all the B vitamins, and vitamin C.

Chapter 26
Answers to Checkpoints

PAGE 1000   1. The major functions of the urinary system include (1) the excretion of organic wastes from body fluids, (2) the elimination of body wastes to the exterior, and (3) the regulation of homeostasis by maintaining the volume and solute concentration of blood.   **2.** The urinary system consists of two kidneys, two ureters, a bladder, and a urethra.   **3.** Micturition, or urination, is the elimination of urine from the body.   **PAGE 1009   4.** The renal corpuscle, proximal convoluted tubule, distal convoluted tubule, and the proximal portions of the nephron loop and collecting duct are all in the renal cortex. (In a cortical nephron, most of the nephron loop is in the cortex; in a juxtamedullary nephron, most of the nephron loop is in the medulla.)   **5.** Plasma proteins are too large to pass through the pores of the glomerular capillaries; only the smallest plasma proteins can pass through the filtration slits of the podocytes.   **6.** Damage to the juxtaglomerular complex of a nephron would interfere with the hormonal control of blood pressure.   **PAGE 1013   7.** The three distinct processes that form urine in the kidney are filtration, reabsorption, and secretion.   **8.** When the plasma concentration of a substance exceeds its tubular maximum, the excess is not reabsorbed, so it is excreted in the urine.   **PAGE 1018   9.** The primary nephron structures involved in filtration are the glomerular capillaries, the dense layer, and the filtration slits of the podocytes.   **10.** The factors that influence net filtration pressure are net hydrostatic pressure and blood colloid osmotic pressure.   **11.** The factors that influence the rate of filtrate formation are the filtration pressure across glomerular capillaries, plus interactions among autoregulation, hormonal regulation, and autonomic regulation.   **12.** A decrease in blood pressure would decrease the GFR by decreasing the blood hydrostatic pressure within the glomerulus.   **PAGE 1030   13.** Increased amounts of aldosterone, which promotes Na^+ retention and K^+ secretion by the kidneys, would increase the K^+ concentration of urine.   **14.** If the concentration of Na^+ in the filtrate decreased, fewer hydrogen ions could be secreted by the countertransport mechanism involving these two ions. As a result, the pH of the tubular fluid would increase.   **15.** Without juxtamedullary nephrons, a large osmotic gradient could not exist in the medulla, and the kidneys would be unable to form concentrated urine.   **16.** When the amount of Na^+ in the tubular fluid passing through the distal convoluted tubule decreases, the cells of the macula densa are stimulated to release renin. The resulting activation of angiotensin II increases blood pressure.   **PAGE 1035   17.** Digestion of a high-protein diet would lead to increased production of urea, a nitrogenous waste formed from the metabolism of amino acids during the breakdown of proteins. As a result, the urine would contain more urea, and urine volume might also increase as a result of the need to flush the excess urea.   **18.** An obstruction of a ureter would interfere with the passage of urine between the renal pelvis and the urinary bladder.   **19.** The ability to control the micturition reflex depends on the ability to control the external urethral sphincter, a ring of skeletal muscle formed by the urogenital diaphragm, which acts as a valve.   **PAGE 1035   20.** Four general age-related changes in the urinary system are a decrease in the number of functional nephrons, a reduction in the GFR, a reduced sensitivity to ADH, and problems with the micturition reflex.   **21.** Nephrolithiasis is the formation of renal calculi, or kidney stones.   **22.** Incontinence may develop in elderly individuals as a result of either decreased muscle tone of sphincter muscles controlling micturition or disorders of the CNS affecting the ability to control urination.   **PAGE 1037   23.** For all systems, the urinary system excretes wastes collected from blood and maintains normal body fluid pH and ion composition.   **24.** Body systems that make up the body's excretory system include the urinary, integumentary, respiratory, and digestive systems.

Answers to Review Questions

PAGE 1039
Level 1 Reviewing Facts and Terms
1. (a) renal sinus; (b) renal pelvis; (c) hilum; (d) renal papilla; (e) ureter; (f) renal cortex; (g) renal medulla; (h) renal pyramid; (i) minor calyx; (j) major calyx; (k) kidney lobe; (l) renal columns; (m) fibrous capsule   **2.** c   **3.** e   **4.** c   **5.** d   **6.** d   **7.** Decline in functional nephron number, reduction in GFR, reduction in ADH sensitivity.   **8.** Glomerular capsule and the capillary network known as the glomerulus.   **9.** renal corpuscle (glomerulus/glomerular capsule) → proximal convoluted tubule → nephron loop → distal convoluted tubule → collecting duct → papillary duct → renal pelvis   **10.** proximal convoluted tubule: reabsorbs all the useful organic substrates from the filtrate; nephron loop: reabsorbs over 90 percent of the water in the filtrate; and distal convoluted tubule: secretes wastes into the tubular fluid that were not filtered or reabsorbed   **11.** The juxtaglomerular complex secretes the enzyme renin and the hormone erythropoietin.   **12.** renal artery → segmental arteries → interlobar arteries → arcuate arteries → cortical radiate arteries → afferent arterioles → glomerulus → venules → cortical radiate veins → arcuate veins → interlobar veins → renal vein   **13.** Processes in urine production are (1) filtration: the selective removal of large solutes and suspended materials from a solution on the basis of size; requires a filtration membrane and hydrostatic pressure, as provided by gravity or by blood pressure; (2) reabsorption: the removal of water and solute molecules from the filtrate after it enters the renal tubules; and (3) secretion: the transport of solutes from the peritubular fluid, across the tubular epithelium, and into the tubular fluid   **14.** Amount of filtrate produced by the kidney per minute. Glomerular hydrostatic pressure, capsular hydrostatic pressure, and blood colloid osmotic pressure.   **15.** Structures responsible for the transport, storage, and elimination of urine are the ureters, urinary bladder, and urethra.

Level 2 Reviewing Concepts
16. d   **17.** d   **18.** c   **19.** a   **20.** a   **21.** ADH stimulates the release of aquaporins (water channels) into the apical plasma membranes within the distal convoluted tubule and collecting duct. Water moves out of the filtrate and into the renal cortex by osmosis, which increases the osmotic concentration of urine.   **22.** Releases rennin, which, through the action of angiotensin II and aldosterone, helps regulate blood pressure. Releases erythropoietin, which increases red blood cell production.   **23.** Atria and ventricle stretch releases ANP and BNP, which inhibit the release of aldosterone, and the production of Na^+ ion pumps and channels in the distal convoluted tubule. This prevents the reuptake of sodium, which is lost along with water in the urine.   **24.** Countercurrent multiplication (1) is an efficient way to reabsorb solutes and water before the tubular fluid reaches the DCT and collecting system, and (2) establishes a concentration gradient that permits the passive reabsorption of water from urine in the collecting system.   **25.** The urge to urinate usually appears when the urinary bladder contains about 200 mL of urine. The micturition reflex begins to function when the stretch receptors have provided adequate stimulation to the parasympathetic motor neurons. The activity in the motor neurons generates action potentials that reach the smooth muscle in the wall of the urinary bladder. These efferent impulses travel over the pelvic nerves, producing a sustained contraction of the urinary bladder.

Level 3 Critical Thinking and Clinical Applications

26. d   **27.** Very dilute urine with osmotic concentrations approaching blood plasma (approximately 300 mOsm/L) could be produced if John and Tony were very well hydrated before they lost their supplies. Drinking urine that is produced at this stage would provide a net gain of water. However as osmotic concentrations of urine increase above blood plasma, water would be lost during the removal of excess sodium that was consumed from the urine. This would mean that consumption of more concentrated urine would actually add to water loss and quicken their demise.   **28.** ACE converts angiotensin I to II which stimulates (i) the reuptake of Na and water in the PCT, (ii) aldosterone release, increasing Na reuptake in the DCT and collecting duct (iii) ADH release, increasing the reuptake of water in DCT and collecting duct, and (iv) vasoconstriction. The combined effect is to increase blood volume and pressure. ACE inhibitors reduce this effect, which reduces hypertension.   **29.** The inability to drain urine into the bladder causes pressure build-up in the medulla and then cortex. Eventually the pressure opposes glomerular hydrostatic pressure, which reduces GFR.   **30.** Because mannitol is filtered but not reabsorbed, drinking a mannitol solution would lead to an increase in the osmolarity of the filtrate. Less water would be reabsorbed, and an increased volume of urine would be produced.   **31.** Carbonic anhydrase catalyzes the reaction that forms carbonic acid, a source of hydrogen ions that are excreted by the kidneys. Hydrogen ion excretion occurs by an antiport system in which sodium ions are exchanged for hydrogen ions. Fewer hydrogen ions would be available, so less sodium would be reabsorbed, causing an increased osmolarity of the filtrate. In turn, an increased volume of urine and more-frequent urination would result.

Answers to Clinical Case Wrap-Up

PAGE 1040   **1.** Mike's x-rays show fractures of the left 11th and 12th ribs, overlying the left kidney. They also show fractures of the transverse processes of the first three lumbar vertebrae on the left. The left renal hilum lies between the fractured ribs and the fractured lumbar vertebral transverse processes. These visible fractures are a strong indication that the soft tissues and organs in this area are also injured. **2.** If Mike survives this trauma, the GFR of his solitary right kidney is likely to increase to compensate for the loss of his left kidney. This is to be expected if there has been no trauma to his right kidney.

Chapter 27

Answers to Checkpoints

PAGE 1043

1. The three interrelated processes essential to stabilizing body fluid volumes are fluid balance, electrolyte balance, and acid–base balance.   **2.** The components of extracellular fluid (ECF) are interstitial fluid, plasma, and other body fluids; the intracellular fluid (ICF) is the cytosol of a cell.   **PAGE 1047**   **3.** Three hormones affecting fluid and electrolyte balance are antidiuretic hormone (ADH), aldosterone, and natriuretic peptides (ANP and BNP). **4.** Drinking a half-gallon of distilled water would temporarily decrease your blood osmolarity (osmotic concentration). This decrease in osmolarity would lead to a decrease in the level of ADH in your blood, because ADH release is triggered by increases in osmolarity.   **PAGE 1050**   **5.** Edema is the movement of abnormally large amounts of water from plasma into interstitial fluid.   **6.** A fluid shift is a rapid movement of water between the ECF and ICF in response to an osmotic gradient.   **7.** Dehydration is a decrease

in the water content of the body that develops when water losses outpace water gains, and this threatens homeostasis.   **8.** Being in the desert without water, you would lose fluid through perspiration, urine formation, and respiration. As a result, the osmotic concentration of your blood plasma (and other body fluids) would increase.   **PAGE 1055**   **9.** Important cations are sodium, potassium, calcium, and magnesium; important anions are phosphate and chloride.   **10.** Sweat is a hypotonic solution with lower sodium concentration than the ECF. Sweating causes a greater loss of water than of sodium, increasing plasma sodium ion levels.

11. Potassium ion imbalances are more dangerous than sodium ion imbalances because they can lead to extensive muscle weakness or even paralysis when blood concentrations are too low, and to cardiac arrhythmias when the levels are too high. Disturbances in sodium balance, by contrast, produce dehydration or tissue edema.   **PAGE 1064**   **12.** The body's three major buffer systems are the phosphate buffer system, the protein buffer systems, and the carbonic acid–bicarbonate buffer system.   **13.** A decrease in the pH of body fluids accompanies an increase in the partial pressure of carbon dioxide. Chemoreceptors sensitive to P_{CO_2} would stimulate the respiratory centers of the medulla oblongata, resulting in an increase in the respiratory rate.   **14.** Tubular fluid in nephrons must be buffered so that it can contain more H^+ without decreasing the pH below approximately 4.5, at which point H^+ secretion cannot continue because the H^+ concentration gradient becomes too steep. The hydrogen ions would leak out of the tubule as fast as they are pumped in.   **PAGE 1071**   **15.** In a prolonged fast, fatty acids are catabolized, producing large numbers of ketone bodies, which are organic acids that decrease the blood pH. (The decreased pH would eventually lead to ketoacidosis.)   **16.** In vomiting, large amounts of stomach hydrochloric acid are lost from the body. More acid is then formed by the parietal cells of the stomach. The H^+ that are released were produced by the dissociation of carbonic acid and exchange of bicarbonate ions with chloride ions from the blood. The alkaline tide of released bicarbonate ions increases the blood pH, leading to metabolic alkalosis.   **PAGE 1072**   **17.** A decrease in the glomerular filtration rate and the number of functional nephrons reduces the body's ability to regulate pH through renal compensation.   **18.** Total body water content gradually decreases with age.

Answers to Review Questions

PAGE 1074
Level 1 Reviewing Facts and Terms
1. c   **2.** a   **3.** d   **4.** (a) carbonic acid ($H_2CO_3^-$); (b) bicarbonate ion (HCO_3^-); (c) sodium bicarbonate ($NaHCO_3$)   **5.** d   **6.** d **7.** Feces   **8.** b   **9.** a   **10.** Four major hormones involved in fluid and electrolyte balance are (1) antidiuretic hormone (ADH): stimulates water conservation by the kidneys and stimulates the thirst center; (2) aldosterone: determines the rate of sodium reabsorption and potassium secretion along the DCT and collecting system of the kidney; and the natriuretic peptides, (3) ANP and (4) BNP: reduce thirst, promote the loss of Na^+ and water by the kidneys, and block the release of ADH and aldosterone.

Level 2 Reviewing Concepts
11. d   **12.** a   **13.** b   **14.** c   **15.** Fluid balance is a state in which the amount of water gained each day is equal to the amount lost to the environment. It is vital that the water content of the body remain stable, because water is an essential ingredient of cytoplasm and accounts for about 99 percent of ECF volume. Electrolyte balance exists when there is neither a net gain nor a net loss of any ion in body fluids. It is important that the ionic concentrations

in body water remain within normal limits; if levels of calcium or potassium become too high, for instance, cardiac arrhythmias can develop. Acid–base balance exists when the production of hydrogen ions precisely offsets their loss. The pH of body fluids must remain within a relatively narrow range; variations outside this range can be life threatening. **16.** Fluid shifts are rapid water movements between the ECF and the ICF that occur in response to increases or decreases in the osmotic concentration of the ECF. Such water movements dampen extreme shifts in electrolyte balance.
17. More water is lost in urine, and there is an increase in the rate of insensible perspiration as skin becomes thinner. Antidiuretic hormone and aldosterone sensitivity decline. This increases water and sodium loss and reduces fluid conservation. **18.** c **19.** Osmoreceptors would stimulate increased thirst and increase ADH release, resulting in water ingestion and reuptake in the kidney, diluting the ECF and reducing sodium concentrations. **20.** Respiratory and renal mechanisms support buffer systems by secreting or absorbing hydrogen ions, by controlling the excretion of acids and bases, and by generating additional buffers. **21.** Respiratory compensation is a change in the respiratory rate that helps stabilize the pH of the ECF. Increasing or decreasing the rate of respiration alters pH by decreasing or increasing the P_{CO_2}. When the P_{CO_2} decreases, the pH increases; when the P_{CO_2} increases, the pH decreases. Renal compensation is a change in the rates of hydrogen and bicarbonate ion secretion or reabsorption in response to changes in plasma pH. Tubular hydrogen ion secretion results in the diffusion of bicarbonate ions into the ECF. **22.** Respiratory disorders result from abnormal CO_2 levels in the ECF. An imbalance exists between the rate of CO_2 removal by the lungs and its generation in other tissues. Metabolic disorders are caused by the generation of organic or fixed acids or by conditions affecting the concentration of bicarbonate ions in the ECF. **23.** Respiratory acidosis, which results from an abnormally high level of carbon dioxide (hypercapnia), is usually caused by hypoventilation. Metabolic acidosis, which occurs when bicarbonate ion levels decrease, can result from overproduction of fixed or organic acids, impaired ability to secrete H^+ ions by the kidney, or severe bicarbonate loss. **24.** Excessive salt intake causes an increase in total blood volume and blood pressure due to an obligatory increase in water absorption across the intestinal lining and recall of fluid from the ICF. **25.** Since sweat is usually hypotonic, the loss of a large volume of sweat causes hypertonicity in body fluids. The loss of fluid volume is primarily from the interstitial space, which leads to a decrease in plasma volume and an increase in the hematocrit. Severe dehydration can cause the blood viscosity to increase substantially, resulting in an increased workload on the heart, ultimately increasing the probability of heart failure.

Level 3 Critical Thinking and Clinical Applications
26. The young boy has metabolic and respiratory acidosis. The metabolic acidosis resulted primarily from the large amounts of lactic acid generated by the boy's muscles as he struggled in the water. (The dissociation of lactic acid releases hydrogen ions and lactate ions.) Sustained hypoventilation during drowning contributed to both tissue hypoxia and respiratory acidosis. Respiratory acidosis developed as the P_{CO_2} increased in the ECF, increasing the production of carbonic acid and its dissociation into H^+ and HCO_3^-. Prompt emergency treatment is essential. The usual procedure involves some form of artificial or mechanical respiratory assistance (to increase the respiratory rate and decrease P_{CO_2} in the ECF) coupled with the intravenous infusion of a buffered isotonic solution containing sodium lactate, sodium gluconate, or sodium bicarbonate that would absorb the H^+ in the ECF and increase body fluid

pH. **27.** a **28.** When tissues are burned, cells are destroyed and their cytoplasmic contents leak into the interstitial fluid and then move into the plasma. Since potassium ions are normally found within the cell, damage to a large number of cells releases relatively large amounts of potassium ions into the blood. The elevated potassium level stimulates cells of the adrenal cortex to produce aldosterone and cells of the juxtaglomerular complex to produce renin. The renin activates the angiotensin-aldosterone system. Ultimately, angiotensin II stimulates increased aldosterone secretion, which promotes sodium retention and potassium secretion by the kidneys, thereby accounting for the elevated potassium levels in the patient's urine. **29.** Metabolic alkalosis. Replacing lost hydrochloric acid in stomach acid during vomiting causes an alkaline tide, whereby there is a large influx of HCO_3^- into the ECF. This is because H^+ is lost from the stomach and replaced from the blood, leaving excess HCO_3^-. The P_{CO_2}, although not outside of the normal range, is elevated, which suggests CO_2 retention. Retaining CO_2 increases H^+ and is the start of respiratory compensation. **30.** The acute pneumonia would reduce lung ventilation and increase CO_2 retention causing acidosis. As it is acute, the renal system would not have had time to compensate. So Anne would be in uncompensated respiratory acidosis. Failure to ventilate would also result in low blood oxygen levels, increasing anaerobic respiration in the tissues and the release of lactic acid. This would explain the elevated lactate levels and would result in a combined respiratory and metabolic acid–base status. **31.** Patient 1 has compensated respiratory alkalosis. Patient 2 has acute metabolic acidosis due to the generation or retention of organic or fixed acids. Patient 3 has acute respiratory acidosis. Patient 4 has metabolic alkalosis.

Answers to Clinical Case Wrap-Up
PAGE 1075 **1.** Marquis should be treated with potassium replacement. He can take K^+ orally, or it can be given intravenously (IV) for faster replacement. His doctor should consider changing his blood pressure medication to one that does not cause a loss of K^+. (These are often called potassium-sparing diuretics.) **2.** When Marquis complains of a "frog jumping around in his chest" with dizziness, he is likely describing a cardiac arrhythmia. This can be the most serious complication of hypokalemia and can lead to sudden cardiac death.

Chapter 28

Answers to Checkpoints
PAGE 1078 **1.** A gamete is a functional male or female reproductive cell. **2.** Basic structures of the reproductive system are gonads (reproductive organs), ducts (which receive and transport gametes), accessory glands (which secrete fluids), and external genitalia (perineal structures). **3.** Gonads are reproductive organs that produce gametes and hormones. **PAGE 1094** **4.** Male reproductive structures are the scrotum, testes, epididymides, ductus deferentia, ejaculatory duct, urethra, seminal glands, prostate gland, bulbourethral glands, and penis. **5.** On a warm day, the cremaster muscle (as well as the dartos muscle) would be relaxed so that the scrotal sac could descend away from the warmth of the body, thereby cooling the testes. **6.** When the arteries within the penis dilate, the increased blood flow causes the erectile tissues of the corpora cavernosa and corpus spongiosum to engorge with blood, producing an erection. **7.** Low FSH levels would lead to low levels of testosterone in the seminiferous tubules, decreasing both the sperm production rate and sperm count. **8.** The pathway of sperm is as

follows: seminiferous tubule → straight tubule → rete testis → efferent ductules → epididymis → ductus deferens → ejaculatory duct → urethra. **PAGE 1113** **9.** Structures of the female reproductive system include the ovaries, uterine tubes, uterus, vagina, and mammary glands. **10.** The blockage of both uterine tubes would cause sterility. **11.** The acidic pH of the vagina helps prevent bacterial, fungal, and parasitic infections in this region. **12.** The functional zone of the endometrium is sloughed off, or shed, during menstruation. **13.** Blockage of a single lactiferous sinus would not interfere with the delivery of milk to the nipple, because the mammary gland of each breast generally has 15–20 lactiferous sinuses. **14.** If the LH surge did not occur during an ovarian cycle, ovulation and corpus luteum formation would not occur. **15.** Blockage of progesterone receptors in the uterus would inhibit the development of the endometrium, making the uterus unprepared for pregnancy. **16.** A decrease in the levels of estrogens and progesterone signals the beginning of menses, the end of the uterine cycle. **PAGE 1114** **17.** The physiological events of sexual intercourse in both sexes are arousal, erection, lubrication, orgasm, and detumescence. Emission and ejaculation are additional phases that occur only in males. **18.** An inability to contract the ischiocavernosus and bulbospongiosus muscles would interfere with a male's ability to ejaculate and to experience orgasm. **19.** Parasympathetic stimulation in females during sexual arousal causes **(a)** engorgement of the erectile tissue of the clitoris and vestibular bulbs, **(b)** increased secretion of cervical and vaginal glands, **(c)** increased blood flow to the wall of the vagina, and **(d)** engorgement of the blood vessels in the nipples. **PAGE 1115** **20.** Menopause is the time when ovulation and menstruation cease, typically around ages 45–55. **21.** At menopause, circulating estrogen levels begin to decrease. Estrogen has an inhibitory effect on FSH (and on GnRH). As the level of estrogen decreases, the levels of FSH increase and remain high. **22.** The male climacteric, or andropause, is a period of declining reproductive function in men, typically between the ages of 50 and 60. **PAGE 1117** **23.** The cardiovascular system distributes reproductive hormones; provides nutrients, oxygen, and waste removal for the fetus; and produces local blood pressure changes responsible for the physical changes that occur during sexual intercourse. The reproductive system supplies estrogens that may help maintain healthy blood vessels and slow the development of atherosclerosis. **24.** Pelvic bones protect reproductive organs in females and portions of the ductus deferens and accessory glands in males; sex hormones stimulate growth and maintenance of bones, and accelerate growth and closure of epiphyseal cartilages at puberty.

Answers to Review Questions

PAGE 1119
Level 1 Reviewing Facts and Terms
1. (c) **2.** d **3.** Polar bodies **4.** b **5.** a **6.** Mitochondria
7. Glandular, vascular **8.** The three regions of the male urethra are the prostatic urethra, membranous urethra, and spongy urethra. **9.** In males, testosterone stimulates spermatogenesis and promotes the functional maturation of spermatozoa; maintains the male accessory reproductive organs; determines male secondary sex characteristics; stimulates metabolic pathways, especially those concerned with protein synthesis and muscle growth; and affects CNS function, by stimulating sexual drive and sexual behaviors.
10. Steps of the ovarian cycle are (1) the formation of primary follicles, (2) the formation of secondary follicles, (3) the formation of a tertiary follicle, (4) ovulation, and (5) the formation of the corpus luteum and its degeneration into the corpus albicans.
11. By a combination of ciliary movement and peristaltic contractions in the walls of the uterine tube. **12.** The clitoris—a part of the external genitalia of females that is derived from the same embryonic structures as the penis—contains erectile tissue that becomes engorged with blood during sexual arousal and provides pleasurable sensations. **13.** Path of milk flow: secretory lobules of glandular tissue (lobes) → ducts → lactiferous duct → lactiferous sinus, which opens onto the surface of the nipple. **14.** **(a)** prostatic urethra; **(b)** spongy urethra; **(c)** ductus deferens; **(d)** penis; **(e)** epididymis; **(f)** testis; **(g)** external urethral orifice; **(h)** scrotum; **(i)** seminal gland; **(j)** prostate gland; **(k)** ejaculatory duct; **(l)** bulbourethral gland.

Level 2 Reviewing Concepts
15. e **16.** The hypothalamic secretion of GnRH triggers the pituitary secretion of FSH and LH. FSH initiates follicular development, and activated follicles and ovarian interstitial cells produce estrogens. High estrogen levels stimulate LH secretion and increase anterior pituitary lobe sensitivity to GnRH, causing the release of LH. Progesterone is the main hormone of the luteal phase. Changes in estrogen and progesterone levels are responsible for maintaining the uterine cycle. **17.** The prostate gland produces prostatic fluid, a slightly acidic solution that makes up 20-30 percent of the volume of semen. In addition to several other compounds, prostatic secretions contain seminalplasmin, a protein with antibiotic properties that may help prevent urinary tract infections in males.
18. The empty tertiary follicle initially collapses, and ruptured vessels bleed into the antrum. The remaining granulosa cells then invade the area. They proliferate to create an endocrine structure known as the corpus luteum, named for its yellow colour. LH stimulates this process. Approximately 12 days after ovulation, the corpus luteum becomes nonfunctional, and progesterone and estrogen levels fall markedly. **19.** When flooded with blood, the penis becomes erect. **20.** (1) Menopause – ovulation and menstruation cease; (2) Thinning of vaginal epithelia; (3) The decrease in estrogen production can lead to osteoporosis. **21.** 2–5 mL **22.** 20 million to 100 million spermatozoa per milliliter of semen **23.** (1) Hypothalamus, (2) granulosa cells in the corpus luteum, (3) anterior lobe of the pituitary.

Level 3 Critical Thinking and Clinical Applications
24. Women more frequently experience peritonitis stemming from a urinary tract infection because infectious organisms exiting the urethral orifice can readily enter the nearby vagina. From there, they can then proceed to the uterus, into the uterine tubes, and finally into the peritoneal cavity. No such direct path of entry into the abdominopelvic cavity exists in men. **25.** Regardless of their location, endometrial cells have receptors that respond to estrogen and progesterone. Under the influence of estrogen at the beginning of the menstrual cycle, any endometrial cells in the peritoneal cavity proliferate and begin to develop glands and blood vessels, which then further develop under the control of progesterone. The dramatic increase in size of this tissue presses on neighboring abdominal tissues and organs, causing periodic painful sensations. **26.** Slightly elevated levels of estrogen and progesterone inhibit both GnRH release at the hypothalamus and the release of FSH and LH from the anterior lobe of the pituitary gland. Without FSH, primordial follicles do not begin to develop, and levels of estrogen remain low. An LH surge, triggered by the peaking of estradiol, is necessary for ovulation to occur. If the level of estradiol is not allowed to rise above the critical level, the LH surge will not occur,

and thus ovulation will not occur, even if a follicle managed to develop to a stage at which it could ovulate. Any mature follicles would ultimately degenerate, and no new follicles would mature to take their place. Although the ovarian cycle is interrupted, the level of hormones is still adequate to regulate a normal menstrual cycle.   **27.** These observations suggest that a certain amount of body fat is necessary for menstrual cycles to occur. The nervous system appears to respond to circulating levels of the adipose tissue hormone leptin; when leptin levels fall below a certain set point, menstruation ceases. Because a woman lacking adequate fat reserves might not be able to have a successful pregnancy, the body prevents pregnancy by shutting down the ovarian cycle, and thus the menstrual cycle. Once sufficient energy reserves become available, the cycles begin again.

Answers to Clinical Case Wrap-Up

PAGE 1120   1. In addition to the blood vessels and the genitofemoral nerve, the ductus deferens with its layers of fascia and muscle are twisted with testicular torsion.   **2.** If Dan were suffering from epididymitis, a painful inflammation of his epididymis, his cremasteric reflex would still be intact. His genitofemoral nerve would still be functioning and able to propagate this spinal reflex.

Chapter 29

Answers to Checkpoints

PAGE 1123   1. Differentiation is the formation of different types of cells during development.   **2.** Development begins at fertilization (conception), the union of a sperm and an oocyte, collectively known as gametes.   **3.** Inheritance refers to the transfer of genetically determined characteristics from one generation to the next.   **PAGE 1125   4.** Two sperm enzymes important to secondary oocyte penetration are hyaluronidase and acrosin.   **5.** A normal human zygote contains 46 chromosomes.   **PAGE 1126   6.** Gestation is the period of prenatal development. It consists of three trimesters.   **7.** The first trimester is the time of embryonic and early fetal development. The second trimester is a time of organ and organ system development. By the end of this stage, the fetus appears distinctly human. The third trimester is characterized by rapid fetal growth and the deposition of adipose tissue.   **PAGE 1135   8.** The inner cell mass of the blastocyst eventually develops into the embryo.   **9.** Improper development of the yolk sac—the mesoderm-derived structure that gives rise to blood vessels and is an important site of blood cell formation—would affect the development and function of the cardiovascular system.   **10.** Yes, Sue is pregnant. After fertilization, the developing trophoblast (and later, the placenta) produce and release the hormone hCG.   **11.** Placental functions include **(a)** supplying the developing fetus with a route for gas exchange, nutrient transfer, and waste elimination; and **(b)** producing hormones that affect maternal systems.   **PAGE 1141   12.** Pregnant women experience breathing difficulty because the pregnant uterus is pressing on the diaphragm and thus the lungs.   **13.** Three factors opposing the calming action of progesterone on the uterus are increasing estrogen levels, increasing oxytocin levels, and prostaglandin production.   **14.** A mother's blood volume increases during pregnancy to compensate for the reduction in maternal blood volume resulting from blood flow through the placenta.   **PAGE 1144   15.** The three stages of labor are the dilation stage, expulsion stage, and placental stage.   **16.** Immature delivery is the birth of a fetus weighing at least 500 g (17.6 oz), which is the normal weight near the end of the second trimester. Premature delivery

usually refers to birth at 28–36 weeks at a weight over 1 kg (2.2 lb).   **17.** Fraternal twins are dizygotic twins, and identical twins are monozygotic twins.   **PAGE 1148   18.** The postnatal stages of development are the neonatal period, infancy, childhood, adolescence, and maturity. These stages are followed by senescence, or aging.   **19.** The neonatal period is the time from birth to one month. Infancy continues from the neonatal period to age 1. Adolescence is the period after childhood in which sexual and physical maturity occur.   **20.** Colostrum is produced by the mammary glands from the end of the sixth month of pregnancy until a few days after birth. After that, the glands begin producing breast milk, which contains fewer proteins (including antibodies) and far more fat than colostrum.   **21.** Increases in the blood levels of GnRH, FSH, LH, and sex hormones mark the onset of puberty.   **PAGE 1157   22.** Genotype is a person's genetic makeup. Phenotype is a person's physical and physiological characteristics. It results from the interaction between the person's genotype and the environment.   **23.** Heterozygous refers to possessing two different alleles at corresponding sites of a chromosome pair. For a heterozygous trait, one or both of the alleles determines the individual's phenotype.   **24.** The phenotype of a person who is heterozygous for tongue rolling—who has one dominant allele and one recessive allele for that trait—would be "tongue roller."   **25.** One reason children are not identical copies of their parents is that during meiosis, parental chromosomes are randomly distributed such that each gamete has a unique set of chromosomes. Additionally, crossing-over and mutations during meiosis introduce new variations in gametes. Besides gene interactions, environmental factors can also affect the expression of genes and the resulting phenotype through epigenetic mechanisms.

Answers to Review Questions

PAGE 1159
Level 1 Reviewing Facts and Terms
1. d   **2.** (a) decidua parietalis; (b) decidua basalis; (c) amnion; (d) chorion; (e) decidua capsularis   **3.** c   **4.** b   **5.** The blastocyst is a hollow ball of cells around a central chamber called the blastocoele. The inner cell mass goes on to form the embryo.   **6.** a **7.** b   **8.** b   **9.** When a sperm contacts the secondary oocyte, their plasma membranes fuse. The oocyte is then activated: Its metabolic rate rises; it completes meiosis II; and the cortical reaction prevents additional sperm from entering. (Vesicles beneath the oocyte surface fuse with the plasma membrane and discharge their contents.) The male and female pronuclei fuse (amphimixis), and the zygote begins preparing for the first cleavage division.
10. The ectoderm, mesoderm and endoderm.   **11.** The dilation stage begins with the onset of true labor, as the cervix dilates and the fetus begins to move toward the cervical canal; late in this stage, the amniochorionic membrane ruptures. The expulsion stage begins as the cervix dilates completely and continues until the fetus has completely emerged from the vagina (delivery). In the placental stage, the uterus gradually contracts, tearing the connections between the endometrium and the placenta and ejecting the placenta.
12. The fetal pituitary secretes oxytocin, which is released into the maternal bloodstream at the placenta. This release may be the trigger for the onset of true labor.   **13.** (1) Neonatal period (birth to 1 month): The newborn becomes relatively self-sufficient and begins performing respiration (breathing), digestion, and excretion on its own. Heart rates and fluid requirements are higher than those of adults. Neonates have little ability to thermoregulate. (2) Infancy (1 month to 1 year): Major organ systems (other than those related to reproduction) become fully operational. (3) Childhood (1 year

to puberty): Growth continues; body proportions change significantly. **14.** Three events interact to promote increased hormone production and sexual maturation at puberty: (1) The hypothalamus increases its production of GnRH; (2) the anterior lobe of the pituitary gland becomes more sensitive to the presence of GnRH, and circulating levels of FSH and LH rise rapidly; and (3) ovarian or testicular cells become more sensitive to FSH and LH. Puberty initiates adolescence, which includes gametogenesis (gamete production) and the production of sex hormones that stimulate the appearance of secondary sexual characteristics and behaviors.

Level 2 Reviewing Concepts

15. b **16.** b **17.** c **18.** (a) Maternal respiratory rate and tidal volume increase, (b) increase in maternal blood volume, (c) increased nutrient requirement, (d) maternal glomerular filtration increases by about 50%, (e) increase in uterine size, (f) mammary glands increase in size and secretion begins. **19.** The placenta produces (1) human chorionic gonadotropin, which maintains the integrity of the corpus luteum and promotes the continued secretion of progesterone (keeping the endometrial lining functional); and (2) human placental lactogen and placental prolactin, which help prepare the mammary glands for milk production. Human placental lactogen also ensures adequate levels of glucose and protein for the developing fetus. (3) The placenta also produces relaxin, which increases the flexibility of the pubic symphysis, causes dilation of the cervix, and suppresses the release of oxytocin by the hypothalamus, delaying the onset of labor contractions. **20.** Oxytocin **21.** Positive feedback mechanisms between increasing levels of oxytocin and increased uterine distortion ensure that labor contractions continue until delivery has been completed. **22.** Usually within an hour. **23.** (a) Genotype refers to all of an individual's genes; phenotype refers to the individual's physical and physiological characteristics. (b) A heterozygous genotype carries different alleles for a given gene; a homozygous genotype carries identical alleles for that gene. (c) In simple inheritance, phenotypes are determined by interactions between a single pair of alleles; in polygenic inheritance, interactions occur among multiple genes. **24.** (a) dominant; (b) recessive; (c) X-linked; (d) autosomal

Level 3 Critical Thinking and Clinical Applications

25. The probability that this couple's daughters will have hemophilia is 0, because each daughter will receive a normal allele from her father. There is a 50 percent chance that a son will have hemophilia, because each son has a 50 percent chance of receiving the mother's normal allele and a 50 percent chance of receiving the mother's recessive allele. **26.** Although technically it only takes one sperm to fertilize an egg, the probability of this occurring if too few sperm are deposited is very low. Most sperm that enter the female reproductive tract are killed or disabled before they reach the uterus. Many of the sperm reaching the uterus are incapable of reaching the secondary oocyte, which is in a uterine tube. Once at the oocyte, the sperm must penetrate the corona radiata, which requires the combined acrosomal enzymes of dozens of sperm. If too few sperm are deposited in the vagina, the number reaching the uterine tube is too low to penetrate the corona radiata. **27.** The most obvious possibility is that the infant has a problem with the cardiovascular supply to the lungs. A good guess would be a patent ductus arteriosus (the ductus arteriosus has failed to close off completely). When the baby is not being stressed (bathing creates heat loss and thermal stress) or eating (less air is entering the lungs), the infant appears normal. Because some of the blood flow to the lungs is being shunted over to the aorta during stress and eating, too little blood is being oxygenated, so the infant becomes cyanotic. **28.** It is very unlikely that the baby's condition results from a viral infection contracted during the third trimester, because organ systems develop during the first trimester and are fully formed by the end of the second trimester.

Answers to Clinical Case Wrap-Up

PAGE 1160 **1.** Implanting the blastocysts on the sixth day after ovulation matches the fertility cycle in nature. This timing ensures the blastocysts are available for implantation in the uterine wall by days 7 through 10 after ovulation. **2.** Hans and Willem are not monozygotic twins because they did not result from the fertilization of a single oocyte with a single sperm, forming a single zygote that subsequently split. Neither are they technically dizygotic twins. They are the developmental result of two separate oocytes from the same mother, but unlike dizygotic twins, they have different *biological* fathers. In terms of biological inheritance, they are half brothers. Socially they are fraternal twins.

Glossary of Key Terms

A

abdomen: The region of the trunk between the inferior margin of the rib cage and the superior margin of the pelvis. (1)

abdominopelvic cavity: The term used to refer to the general region bounded by the abdominal wall and the pelvis; it contains the peritoneal cavity and visceral organs. (1)

abducens: Cranial nerve VI, which innervates the lateral rectus muscle of the eye. (14)

abduction: Movement away from the midline of the body, as viewed in the anatomical position. (9)

abortion: The premature loss or expulsion of an embryo or fetus. (29)

abscess: A localized collection of pus within a damaged tissue. (4, 22)

absorption: The active or passive uptake of gases, fluids, or solutes. (25)

accommodation: An alteration in the curvature of the lens of the eye to focus an image on the retina. (17)

acetabulum: The fossa on the lateral aspect of the pelvis that accommodates the head of the femur. (8)

acetylcholine (ACh): A chemical neurotransmitter in the brain and peripheral nervous system; the dominant neurotransmitter in the peripheral nervous system, released at neuromuscular junctions and synapses of the parasympathetic division. (10, 12, 16)

acetylcholinesterase (AChE): An enzyme found in the synaptic cleft, bound to the postsynaptic membrane, and in tissue fluids; breaks down and inactivates acetylcholine molecules; also called *cholinesterase*. (10, 12)

acetyl-CoA: An acetyl group bound to coenzyme A, a participant in the anabolic and catabolic pathways for carbohydrates, lipids, and many amino acids. (25)

acetyl group: —CH_3CO; an acetic acid molecule without the hydroxyl group. (25)

acid: A compound whose dissociation in solution releases a hydrogen ion and an anion; an acidic solution has a pH below 7.0 and has an excess of hydrogen ions. (2, 27)

acidosis: An abnormal physiological state characterized by blood pH below 7.35. (2, 25, 26, 27)

acinus/acini: A histological term referring to a blind pocket, pouch, or sac.

acoustic: Pertaining to sound or the sense of hearing. (17)

acquired immunodeficiency syndrome (AIDS): A disease caused by the human immunodeficiency virus (HIV); characterized by the destruction of helper T cells and a resulting severe impairment of the immune response. (22)

acromegaly: A condition caused by the overproduction of growth hormone in adults, characterized by a thickening of bones and an enlargement of cartilages and other soft tissues. (6, 18)

acromion: A continuation of the scapular spine that projects superior to the capsule of the glenohumeral joint. (8)

acrosome: A membranous sac at the tip of a spermatozoon that contains hyaluronidase; also called *acrosomal cap*. (28)

actin: The protein component of microfilaments that forms thin filaments in skeletal muscles and produces contractions of all muscles through interaction with thick (myosin) filaments; *see also* **sliding filament theory**. (3, 10)

action potential: A propagated change in the membrane potential of excitable cells, initiated by a change in the membrane permeability to sodium ions; *see also* **nerve impulse**. (10, 12)

active transport: The ATP-dependent absorption or secretion of solutes across a plasma membrane. (3, 26)

acute: Sudden in onset, typically severe in intensity, and brief in duration.

adaptation: A change in pupillary size in response to changes in light intensity (17); a decrease in receptor sensitivity or perception after chronic stimulation (15); physiological responses that produce acclimatization. (25)

Addison's disease: A condition resulting from the hyposecretion of glucocorticoids; characterized by lethargy, weakness, hypotension, and increased skin pigmentation. (5, 18)

adduction: Movement toward the axis or midline of the body, as viewed in the anatomical position. (9)

adenine: A purine; one of the nitrogenous bases in the nucleic acids RNA and DNA. (2)

adenohypophysis: The anterior lobe of the pituitary gland. (18)

adenoid: The pharyngeal tonsil. (22, 23)

adenosine: A compound consisting of adenine and ribose. (2)

adenosine diphosphate (ADP): A compound consisting of adenosine with two phosphate groups attached. (2, 25)

adenosine monophosphate (AMP): A nucleotide consisting of adenosine plus a phosphate group (PO_4^{3-}); also called *adenosine phosphate*.

adenosine triphosphate (ATP): A high-energy compound consisting of adenosine with three phosphate groups attached; the third is attached by a high-energy bond. (2, 10, 25)

adenylate cyclase: An enzyme bound to the inner surfaces of plasma membranes that can convert ATP to cyclic AMP; formerly called *adenylyl cyclase*. (12)

adhesion: The fusion of two mesenterial layers after damage or irritation of their opposing surfaces; this process restricts relative movement of the organs involved (4); the binding of a phagocyte to its target (22).

adipocyte: A fat cell. (4)

adipose tissue: Loose connective tissue dominated by adipocytes. (4, 18)

adrenal cortex: The superficial region of the adrenal gland that produces steroid hormones; also called the *suprarenal cortex*. (18)

adrenal gland: A small endocrine gland that secretes steroids and catecholamines and is located superior to each kidney; also called *suprarenal gland*. (18)

adrenal medulla: The core of the adrenal gland (18); a modified sympathetic ganglion that secretes catecholamines into the blood during sympathetic activation; also called *suprarenal medulla*. (16)

adrenergic: An axon terminal that, when stimulated, releases norepinephrine. (12)

adrenocortical hormone: Any steroid produced by the adrenal cortex. (18)

adrenocorticotropic hormone (ACTH): The hormone that stimulates the production and secretion of glucocorticoids by the zona fasciculata of the adrenal cortex; released by the anterior lobe of the pituitary gland (adenohypophysis) in response to corticotropin-releasing hormone. (18)

adventitia: The superficial layer of connective tissue surrounding an internal organ; fibers are continuous with those of surrounding tissues, providing support and stabilization. (24)

aerobic: Requiring oxygen.

aerobic metabolism: The complete breakdown of organic substrates into carbon dioxide and water, by pyruvate; a process that yields large amounts of ATP but requires mitochondria and oxygen. (3, 10, 25)

afferent: Conducting toward a center.

afferent arteriole: An arteriole that carries blood to a glomerulus of the kidney. (26)

afferent fiber: An axon that carries sensory information to the central nervous system. (12)

agglutination: The aggregation of red blood cells due to interactions between surface antigens and plasma antibodies. (19, 22)

agglutinins: Immunoglobulins (antibodies) in plasma that react with antigens on the surfaces of foreign red blood cells when donor and recipient differ in blood type. (19)

agglutinogens: Genetically determined surface antigens on red blood cells that stimulate the formation of specific agglutinins. (19)

aggregated lymphoid nodules: Lymphoid nodules beneath the epithelium of the small intestine; also called *Peyer's patches*. (22)

agonist: A muscle responsible for a specific movement; also called a *prime mover*. (11)

agranular: Without granules; *agranular leukocytes* are monocytes and lymphocytes. (19)

AIDS: *See* **acquired immunodeficiency syndrome**.

aldosterone: A mineralocorticoid produced by the zona glomerulosa of the adrenal cortex; stimulates the kidneys to conserve sodium and water; secreted in response to the presence of angiotensin II. (18, 26, 27)

alkalosis: The condition characterized by a blood pH greater than 7.45; associated with a relative deficiency of hydrogen ions or an excess of bicarbonate ions. (2, 27)

alpha receptors: Membrane receptors sensitive to norepinephrine or epinephrine; stimulation normally results in target cell excitation. (16)

alveolar sac: An air-filled chamber at the terminal end of the alveolar duct which gives rise to the alveoli in the lung. (23)

alveolus/alveoli: Blind pockets at the end of the respiratory tree, lined by a simple squamous epithelium and surrounded by a capillary network; sites of gas exchange with the blood (23); a bony socket that holds the root of a tooth. (24)

Alzheimer's disease: A disorder resulting from degenerative changes in populations of neurons in the cerebrum, causing dementia characterized by problems with attention, short-term memory, and emotions. (16)

amination: The attachment of an amino group to a carbon chain; performed by a variety of cells and important in the synthesis of amino acids. (25)

amino acids: Organic compounds made up of carbon, hydrogen, oxygen, and nitrogen; building block of protein; chemical structure can be summarized as R—CHNH₂—COOH. (2, 25)

amino group: —NH₂. (2)

amnion: One of the four extraembryonic membranes; surrounds the developing embryo or fetus. (29)

amniotic fluid: Fluid that fills the amniotic cavity; cushions and supports the embryo or fetus. (4, 29)

amphiarthrosis: A joint (articulation) that permits a small degree of independent movement; *see* **interosseous membrane** (8) and **pubic symphysis**. (9)

ampulla/ampullae: A localized dilation in the lumen of a canal or passageway. (17, 24, 28)

amygdaloid body: A basal nucleus that is a component of the limbic system and acts as an interface between the limbic system, the cerebrum, and sensory systems; also called *amygdala*. (14)

amylase: An enzyme that breaks down polysaccharides; produced by the salivary glands and pancreas. (24)

anabolism: The synthesis of complex organic compounds from simpler precursors. (2, 25)

anaerobic: Without oxygen.

analgesic: A substance that relieves pain. (15)

anal triangle: The posterior subdivision of the perineum. (11)

anaphase: The mitotic stage in which the paired chromatids separate and move toward opposite ends of the spindle apparatus. (3)

anaphylaxis: A hypersensitivity reaction due to the binding of antigens to immunoglobulins (IgE) on the surfaces of mast cells; the release of histamine, serotonin, and prostaglandins by mast cells then causes widespread inflammation; a sudden decrease in blood pressure may occur, producing anaphylactic shock. (22)

anastomosis: The joining of two tubes, usually referring to a connection between two peripheral vessels without an intervening capillary bed. (21)

anatomical position: An anatomical reference position; the body viewed from the anterior surface with the palms facing forward. (1)

anatomy: The study of the structure of the body. (1)

androgen: A steroid sex hormone mainly produced by the interstitial cells of the testis and manufactured in small quantities by the adrenal cortex in both sexes. (18, 28)

anemia: The condition marked by a decrease in the hematocrit, the hemoglobin content of the blood, or both. (19)

angiotensin I: The hormone produced by the activation of angiotensinogen by renin (18); angiotensin-converting enzyme (ACE) converts angiotensin I into angiotensin II in lung capillaries. (26)

angiotensin II: A hormone that causes an increase in systemic blood pressure, stimulates the secretion of aldosterone, promotes thirst, and causes the release of antidiuretic hormone (18); angiotensin-converting enzyme (ACE) in lung capillaries converts angiotensin I into angiotensin II. (21, 26)

angiotensinogen: The blood protein produced by the liver that is converted to angiotensin I by the enzyme renin. (18)

anion: An ion having a negative charge. (2, 27)

anoxia: Inadequate oxygen reaching body tissues. (23)

antagonist: A muscle that opposes the movement of an agonist. (10)

antebrachium: The forearm. (8)

anterior: On or near the front, or ventral surface, of the body. (1)

antibiotic: A chemical agent that selectively kills pathogens, primarily bacteria. (20)

antibody: A globular protein produced by plasma cells that will bind to specific antigens and promote their destruction or removal from the body. (19, 22)

antibody-mediated immunity: The form of immunity resulting from the presence of circulating antibodies produced by plasma cells; also called *humoral immunity*. (22)

anticodon: Three nitrogenous bases on a tRNA molecule that interact with a complementary codon on a strand of mRNA. (3)

antidiuretic hormone (ADH): A hormone synthesized in the hypothalamus and secreted at the posterior lobe of the pituitary gland (neurohypophysis); causes the kidneys to retain water and an increase in blood pressure. (18, 21, 26, 27)

antigen: A substance capable of inducing the production of antibodies. (22)

antigen–antibody complex: The combination of an antigen and a specific antibody. (22)

antigenic determinant site: A region of an antigen that can interact with an antibody molecule. (22)

antigen-presenting cell (APC): A cell that processes antigens and displays them, bound to Class II MHC proteins; essential to the initiation of a normal immune response. (22)

antihistamine: A chemical that blocks the action of histamine on peripheral tissues. (22)

antrum: A chamber or pocket. (28)

anulus: A cartilage or bone shaped like a ring; also spelled *annulus*. (9)

anus: The external opening of the anal canal. (24)

aorta: The large, elastic artery that carries blood away from the left ventricle and into the systemic circuit. (20)

apocrine secretion: A mode of secretion in which the glandular cell sheds portions of its cytoplasm. (4, 5)

aponeurosis/aponeuroses: Broad tendinous sheet(s) that may serve as the origin or insertion of a skeletal muscle. (4, 6, 10)

appendicular: Pertaining to the upper or lower limbs. (8)

appendix: A blind sac connected to the cecum of the large intestine. (24)

appositional growth: The enlargement of a cartilage or bone by the addition of cartilage or bony matrix at its surface. (4)

aqueous humor: A fluid similar to perilymph or cerebrospinal fluid that fills the anterior chamber of the eye. (17)

arachidonic acid: One of the essential fatty acids. (2, 18)

arachnoid granulations: Processes of the arachnoid mater that project into the superior sagittal sinus; sites where cerebrospinal fluid enters the venous circulation. (14)

arachnoid mater: The middle meninx (layer) that encloses cerebrospinal fluid and protects the central nervous system. (13, 14)

arbor vitae: The central, branching mass of white matter inside the cerebellum. (14)

arcuate: Curving.

areolar: Containing minute spaces, as in areolar tissue.

areolar tissue: Loose connective tissue with an open framework. (4)

arrector pili: Smooth muscles whose contractions force hairs to stand erect. (5)

arrhythmias: Abnormal patterns of cardiac contractions. (20)

arteriole: A small arterial branch that delivers blood to a capillary network. (21)

artery: A blood vessel that carries blood away from the heart and toward a peripheral capillary. (4, 20, 21)

articular: Pertaining to a joint.

articular capsule: The dense collagen fiber sleeve that surrounds a joint and provides protection and stabilization; also called *joint capsule*. (6, 9)

articular cartilage: The cartilage pad that covers the surface of a bone inside a joint cavity. (6, 9)

articulation: A joint. (9)

arytenoid cartilages: A pair of small cartilages in the larynx. (23)

ascending tract: A tract carrying information from the spinal cord to the brain. (13, 14)

association areas: Cortical areas of the cerebrum that are responsible for the integration of sensory inputs and/or motor commands. (14)

association neuron: See **interneuron**.

astrocyte: One of the four types of neuroglia in the central nervous system; responsible for maintaining the blood–brain barrier by the stimulation of endothelial cells. (12)

atherosclerosis: The formation of fatty plaques in the walls of arteries, restricting blood flow. (21)

atom: The smallest stable unit of matter. (2)

atomic number: The number of protons in the nucleus of an atom. (2)

atomic weight: The average of the different atomic masses and isotope proportions of a particular element. (2)

atria: Thin-walled chambers of the heart that receive venous blood from the pulmonary or systemic circuit.

atrial natriuretic peptide (ANP): See **natriuretic peptides**. (20)

atrial reflex: The reflexive increase in heart rate after an increase in venous return; due to mechanical and neural factors; also called *Bainbridge reflex*. (20, 21)

atrioventricular (AV) node: Specialized cardiocytes that relay the contractile stimulus to the atrioventricular bundle (bundle of His), the bundle branches, the Purkinje fibers, and the ventricular myocardium; located at the boundary between the atria and ventricles. (20)

atrioventricular (AV) valve: One of the valves that prevents backflow into the atria during ventricular systole (contraction). (20)

atrophy: The wasting away of tissues from a lack of use, ischemia, or nutritional abnormalities. (10)

auditory: Pertaining to the sense of hearing. (17)

auditory ossicles: The bones of the middle ear: malleus, incus, and stapes. (7, 17)

auditory tube: A passageway that connects the nasopharynx with the middle ear cavity; also called *Eustachian tube* or *pharyngotympanic tube*. (17)

auricle: A broad, flattened process that resembles the external ear; in the ear, the expanded, projecting portion that surrounds the external auditory meatus, also called *pinna* (17); in the heart, the externally visible flap formed by the collapse of the outer wall of a relaxed atrium (20).

autoantibodies: Antibodies that react with antigens on the surfaces of a person's own cells and tissues. (22)

autoimmunity: The immune system's sensitivity to one's own tissues, resulting in the production of autoantibodies. (22)

autolysis: The destruction of a cell due to the rupture of lysosomal membranes in its cytoplasm. (3)

automaticity: The spontaneous depolarization to threshold, characteristic of cardiac pacemaker cells. (10, 20)

autonomic ganglion: A collection of visceral motor neurons outside the central nervous system. (16)

autonomic nerve: A peripheral nerve consisting of preganglionic or postganglionic autonomic fibers. (16)

autonomic nervous system (ANS): Centers, nuclei, tracts, ganglia, and nerves involved in the unconscious regulation of visceral functions; includes components of the central nervous system and the peripheral nervous system. (12, 16)

autoregulation: Changes in activity that maintain homeostasis in direct response to changes in the local environment; does not require neural or endocrine control. (1, 21, 26)

autosome: A chromosome other than the X or Y sex chromosome. (29)

avascular: Without blood vessels. (4)

axilla: The armpit. (1, 8)

axolemma: The plasma membrane of an axon, continuous with the plasma membrane of the cell body and dendrites and distinct from any neuroglial coverings. (12)

axon: The elongated extension of a neuron that conducts an action potential. (4, 12)

axon hillock: In a multipolar neuron, the portion of the cell body adjacent to the initial segment. (12)

axoplasm: The cytoplasm within an axon. (12)

B

bacteria: Single-celled microorganisms, some pathogenic, that are common in the environment and in and on the body. (22)

Bainbridge reflex: *See* **atrial reflex**.

baroreception: The ability to detect changes in pressure. (15, 23)

baroreceptor reflex: A reflexive change in cardiac activity in response to changes in blood pressure. (21)

baroreceptors: The nerve endings that sense pressure changes. (15, 21)

basal nuclei: Nuclei of the cerebrum that are important in the subconscious control of skeletal muscle activity. (14, 15)

base: A compound whose dissociation releases a hydroxide ion (OH^-) or removes a hydrogen ion (H^+) from the solution. (2, 27)

basement membrane: A layer of filaments and fibers that attach an epithelium to the underlying connective tissue; also called *basal lamina*. (4)

basophils: Circulating granulocytes (white blood cells) similar in size and function to tissue mast cells. (19)

B cells: Lymphocytes capable of differentiating into plasma cells, which produce antibodies. (19, 22)

benign: Not malignant. (3)

beta cells: Cells of the pancreatic islets that secrete insulin in response to increased blood sugar concentrations. (18)

beta oxidation: Fatty acid catabolism that produces molecules of acetyl-CoA. (25)

beta receptors: Membrane receptors sensitive to epinephrine; stimulation may result in the excitation or inhibition of the target cell. (16)

bicarbonate ion: HCO_3^-; anion component of the carbonic acid–bicarbonate buffer system. (26, 27)

bicuspid: Having two cusps or points; refers to a premolar tooth, which has two roots, or to the left AV valve, which has two cusps. (24)

bicuspid valve: The left atrioventricular (AV) valve, also called *mitral valve*. (20)

bifurcate: To branch into two parts.

bile: The exocrine secretion of the liver; stored in the gallbladder and ejected into the duodenum. (24)

bile salts: Steroid derivatives in bile; responsible for the emulsification of ingested lipids. (2)

bilirubin: A pigment that is the by-product of hemoglobin catabolism. (19)

biopsy: The removal of a small sample of tissue for pathological analysis. (4, 13)

bladder: A muscular sac that distends as fluid is stored and whose contraction ejects the fluid at an appropriate time; used alone, the term usually refers to the urinary bladder. (26)

blastocyst: An early stage in the developing embryo, consisting of an outer trophoblast and an inner cell mass. (29)

blockers/blocking agents: Drugs that block membrane pores or prevent binding to membrane receptors. (16)

blood–brain barrier: The isolation of the central nervous system from the general circulation; primarily the result of astrocyte regulation of capillary permeabilities. (12, 14)

blood clot: *See* **clot**.

blood–CSF barrier: The isolation of the cerebrospinal fluid from the capillaries of the choroid plexus; primarily the result of specialized ependymal cells. (14)

blood pressure: A force exerted against vessel walls by the blood in the vessels, due to the push exerted by cardiac contraction and the elasticity of the vessel walls; usually measured along one of the muscular arteries, with systolic pressure measured during ventricular systole (contraction) and diastolic pressure during ventricular diastole (relaxation). (21)

blood–testis barrier: The isolation of the interior of the seminiferous tubules from the general circulation, due to the activities of the nurse (sustentacular) cells. (28)

Bohr effect: The increased oxygen release by hemoglobin in the presence of increased carbon dioxide levels. (23)

bolus: A compact mass; usually refers to compacted ingested material (food) on its way to the stomach. (23, 24)

bone: *See* **osseous tissue**.

bowel: The intestinal tract. (24)

Bowman's capsule: *See* **glomerular capsule**. (26)

brachial: Pertaining to the arm.

brachial plexus: A network formed by branches of spinal nerves C_5-T_1 en route to innervating the upper limb. (13)

brachium: The arm. (11)

bradycardia: An abnormally slow heart rate, usually below 50 bpm. (20)

brain natriuretic peptide (BNP): *See* **natriuretic peptides**.

brain stem: The brain minus the cerebrum, diencephalon, and cerebellum; midbrain, pons, and medulla oblongata. (14)

brevis: Short.

Broca's area: The speech center of the brain, normally located on the neural cortex of the left cerebral hemisphere. (14)

bronchial tree: The trachea, bronchi, and bronchioles. (23)

bronchodilation: The dilation of the bronchial passages; can be caused by sympathetic stimulation. (23)

bronchus/bronchi: Branch/branches of the bronchial tree between the trachea and bronchioles. (23)

buccal: Pertaining to the cheeks. (24)

buffer: A compound that stabilizes the pH of a solution by removing or releasing hydrogen ions. (2, 27)

buffer system: Interacting compounds that prevent increases or decreases in the pH of body fluids; includes the carbonic acid–bicarbonate buffer system, the phosphate buffer system, and the protein buffer systems. (27)

bulbar: Pertaining to the brain stem. (14)

bulbourethral glands: Mucous glands at the base of the penis that secrete into the penile urethra; the equivalent of the greater vestibular glands of females; also called *Cowper's glands*. (28)

bundle branches: Specialized conducting cells in the ventricles that carry the contractile stimulus from the atrioventricular bundle (bundle of His) to the Purkinje fibers. (20)

bundle of His: Specialized conducting cells in the interventricular septum that carry the contracting stimulus from the AV node to bundle branches and then to Purkinje fibers; also called the *atrioventricular bundle*. (20)

bursa: A small sac filled with synovial fluid that cushions adjacent structures and reduces friction. (9)

C

calcaneal tendon: The large tendon that inserts on the calcaneus; tension on this tendon produces extension (plantar flexion) of the foot; also called *Achilles tendon*. (8, 11)

calcaneus: The heel bone, the largest of the tarsal bones. (8)

calcification: The deposition of calcium salts within a tissue. (4, 6)

calcitonin: The hormone secreted by C cells of the thyroid when calcium ion concentrations are abnormally high; restores homeostasis by increasing the rate of bone deposition and the rate of calcium loss by the kidneys. (6, 18)

calculus/calculi: Solid mass/masses of insoluble materials that form within body fluids, especially the gallbladder, kidneys, or urinary bladder. (26)

callus: A localized thickening of the epidermis due to chronic mechanical stresses (5); a thickened area that forms at the site of a bone break as part of the repair process. (6)

calorigenic effect: The stimulation of energy production and heat loss by thyroid hormones. (18)

canaliculi: Microscopic passageways between cells; bile canaliculi carry bile to bile ducts in the liver (24); in bone, canaliculi permit the diffusion of nutrients and wastes to and from osteocytes. (4, 6)

cancellous bone: Spongy bone, composed of a network of bony struts. (6)

cancer: An illness caused by mutations leading to the uncontrolled growth and replication of the affected cells. (3)

cannula: A tube that can be inserted into the body; commonly placed in blood vessels prior to transfusion or dialysis. (19)

capacitation: The activation process that must occur before a spermatozoon can successfully fertilize an oocyte; occurs in the vagina after ejaculation. (28, 29)

capillary: A small blood vessel, located between an arteriole and a venule, whose thin wall permits the diffusion of gases, nutrients, and wastes between plasma and interstitial fluids. (4, 19, 20, 21)

capitulum: A general term for a small, elevated articular process; refers to the rounded distal surface of the humerus that articulates with the head of the radius. (8)

caput: The head. (7)

carbaminohemoglobin: Hemoglobin bound to carbon dioxide molecules. (19, 23)

carbohydrase: An enzyme that breaks down carbohydrate molecules. (24)

carbohydrate: An organic compound containing carbon, hydrogen, and oxygen in a ratio that approximates 1:2:1. (2, 25)

carbon dioxide: CO_2; a compound produced by the decarboxylation reactions of aerobic metabolism. (2, 23)

carbonic anhydrase: An enzyme that catalyzes the reaction $H_2O + CO_2 \rightarrow H_2CO_3$; important in carbon dioxide transport (23), gastric acid secretion (24), and renal pH regulation (26).

carcinogenic: Stimulating cancer formation in affected tissues. (3)

cardia: The region of the stomach surrounding its connection with the esophagus. (24)

cardiac: Pertaining to the heart. (10, 20)

cardiac cycle: One complete heartbeat, including atrial and ventricular systole and diastole. (20)

cardiac output: The amount of blood ejected by the left ventricle each minute; normally about 5 liters. (20)

cardiac reserve: The potential percentage increase in cardiac output above resting levels. (20)

cardiac tamponade: A compression of the heart due to fluid accumulation in the pericardial cavity. (20)

cardiocyte: A cardiac muscle cell. (4, 10, 20)

cardiovascular: Pertaining to the heart, blood, and blood vessels. (19, 20, 21)

cardiovascular centers: Poorly localized area in the reticular formation of the medulla oblongata of the brain; includes cardioaccelerary, cardioinhibitory, and vasomotor centers. (14, 21)

carotene: A yellow-orange pigment, found in carrots, sweet potatoes, and other vegetables, that the body can convert to vitamin A. (5)

carotid artery: The main artery of the neck, servicing cervical and cranial structures; one branch, the internal carotid, provides a major blood supply to the brain. (21)

carotid body: A group of receptors, adjacent to the carotid sinus, that are sensitive to changes in the carbon dioxide levels, pH, and oxygen concentrations of arterial blood. (15, 21)

carotid sinus: A dilated segment at the base of the internal carotid artery whose walls contain baroreceptors sensitive to changes in blood pressure. (21)

carotid sinus reflex: Reflexive changes in blood pressure that maintain homeostatic pressures at the carotid sinus, stabilizing blood flow to the brain. (21)

carpus/carpal: The wrist. (8, 11)

cartilage: A connective tissue with a gelatinous matrix that contains an abundance of fibers. (4)

catabolism: The breakdown of complex organic molecules into simpler components, accompanied by the release of energy. (2, 25)

catalyst: A substance that accelerates a specific chemical reaction but that is not altered by the reaction. (2)

catecholamines: Epinephrine, norepinephrine, dopamine, and related compounds. (18)

catheter: A tube surgically inserted into a body cavity or along a blood vessel or excretory passageway for the collection of body fluids, monitoring of blood pressure, or introduction of medications or radiographic dyes. (20)

cation: An ion having a positive charge. (2, 27)

cauda equina: Bundle of spinal nerve roots arising from the lumbosacral enlargement and medullary cone of the adult spinal cord; they extend caudally inside the vertebral canal en route to lumbar and sacral segments. (13)

caudal/caudally: Closest to or toward the tail (coccyx).

caudate nucleus: One of the basal nuclei involved with the subconscious control of skeletal muscular activity. (14)

cell: The smallest living structural unit in the human body. (3)

cell body: The body of a neuron; also called *soma*. (4, 12)

cell-mediated immunity: Resistance to disease by sensitized T cells that destroy antigen-bearing cells by direct contact or through the release of lymphotoxins; also called *cellular immunity*. (22)

center of ossification: The site in a connective tissue where bone formation begins. (6)

central canal: Longitudinal canal in the center of an osteon that contains blood vessels and nerves, also called *Haversian canal* (6); a passageway along the longitudinal axis of the spinal cord that contains cerebrospinal fluid. (13, 14)

central nervous system (CNS): The brain and spinal cord. (12)

centriole: A cylindrical intracellular organelle composed of nine groups of microtubules, three in each group (9 + 0 array); functions in mitosis or meiosis by organizing the microtubules of the spindle apparatus. (3)

centromere: The localized region where two chromatids remain connected after the chromosomes have replicated; site of spindle fiber attachment. (3)

centrosome: A region of cytoplasm that contains a pair of centrioles oriented at right angles to one another. (3)

cerebellum: The posterior portion of the metencephalon, containing the cerebellar hemispheres; includes the arbor vitae, cerebellar nuclei, and cerebellar cortex. (14, 15)

cerebral cortex: An extensive area of neural cortex covering the surfaces of the cerebral hemispheres. (14)

cerebral hemispheres: A pair of expanded portions of the cerebrum covered in neural cortex. (14)

cerebrospinal fluid (CSF): Fluid bathing the internal and external surfaces of the central nervous system; secreted by the choroid plexus. (12, 13, 14)

cerebrovascular accident (CVA): The occlusion of a blood vessel that supplies a portion of the brain, resulting in damage to the dependent neurons; also called *stroke*. (14)

cerebrum: The largest portion of the brain, composed of the cerebral hemispheres; includes the cerebral cortex, the basal nuclei, and the internal capsule. (14)

cerumen: The waxy secretion of the ceruminous glands along the external acoustic meatus. (5, 17)

ceruminous glands: Integumentary glands that secrete cerumen. (5, 17)

cervix: The inferior portion of the uterus. (28)

chemoreception: The detection of changes in the concentrations of dissolved compounds or gases (mainly CO_2 and O_2). (15, 17, 21, 23, 25)

chemotaxis: The attraction of phagocytic cells to the source of abnormal chemicals in tissue fluids. (22)

chemotherapy: The treatment of illness through the administration of specific chemicals.

chloride shift: The movement of plasma chloride ions into red blood cells in exchange for bicarbonate ions generated by the intracellular dissociation of carbonic acid. (23, 27)

cholecystokinin (CCK): A duodenal hormone that stimulates the contraction of the gallbladder and the secretion of enzymes by the exocrine pancreas. (24)

cholesterol: A steroid component of plasma membranes and a substrate for the synthesis of steroid hormones and bile salts. (2, 25)

choline: A breakdown product or precursor of acetylcholine. (12)

cholinergic synapse: A synapse where the presynaptic membrane releases acetylcholine on stimulation. (12, 16)

chondrocyte: A cartilage cell. (4)

chondroitin sulfate: The predominant proteoglycan in cartilage, responsible for the gelatinous consistency of the matrix. (4)

chordae tendineae: Fibrous cords that stabilize the position of the AV valves in the heart, preventing backflow during ventricular systole (contraction). (20)

chorion/chorionic: An extraembryonic membrane, consisting of the trophoblast and underlying mesoderm, that forms the placenta. (29)

choroid: The middle, vascular layer in the wall of the eye. (17)

choroid plexus: The vascular complex in the roof of the third and fourth ventricles of the brain, responsible for the production of cerebrospinal fluid. (14)

chromatid: One complete copy of a DNA strand and its associated nucleoproteins. (3, 28, 29)

chromatin: A histological term referring to the grainy material visible in cell nuclei during interphase; the appearance of the DNA content of the nucleus when the chromosomes are uncoiled. (3)

chromosomes: Dense structures, composed of tightly coiled DNA strands and associated histones, that become visible in the nucleus when a cell prepares to undergo mitosis or meiosis; normal human somatic cells each contain 46 chromosomes. (3, 28, 29)

chronic: Habitual or long term.

chylomicrons: Relatively large droplets that may contain triglycerides, phospholipids, and cholesterol in association with proteins; synthesized and released by intestinal cells and transported to the venous blood by the lymphatic system. (24, 25)

ciliary body: A thickened region of the choroid that encircles the lens of the eye; includes the ciliary muscle and the ciliary processes that support the suspensory ligaments of the lens. (17)

cilium/cilia: Slender organelle/organelles that extends above the free surface of an epithelial cell. Multiple *motile cilia* generally undergo cycles of movement and are composed of a basal body and microtubules in a 9 + 2 array; a nonmotile, solitary *primary cilium* acts as an environmental sensor, and has microtubules arranged in a 9 + 0 array. (3)

circulatory system: The network of blood vessels and lymphatic vessels that facilitate the distribution and circulation of extracellular fluid. (21, 22)

circumduction: A movement at a synovial joint in which the distal end of the bone moves in a circular direction, but the shaft does not rotate. (9)

cisterna: An expanded or flattened chamber derived from and associated with the endoplasmic reticulum. (3, 10)

citric acid cycle: The reaction sequence that occurs in the matrix of mitochondria; in the process, organic molecules are broken down, carbon dioxide molecules are released, and hydrogen molecules are transferred by coenzymes to the electron transport system. (3, 10, 25)

clot: A network of fibrin fibers and trapped blood cells; also called a *thrombus* if it occurs within blood vessels. (19)

clotting factors: Plasma proteins, synthesized by the liver, that are essential to the clotting response. (19)

clotting response: The series of events that result in the formation of a clot. (19)

coccygeal ligament: The fibrous extension of the dura mater and filum terminale; provides longitudinal stabilization to the spinal cord. (13)

coccyx: The terminal portion of the spinal column, consisting of relatively tiny, fused vertebrae. (7)

cochlea: The spiral portion of the bony labyrinth of the internal ear that surrounds the organ of hearing. (17)

cochlear duct: *See* **scala media**. (17)

codon: A sequence of three nitrogenous bases along an mRNA strand that will specify the location of a single amino acid in a peptide chain. (3)

coelom: The body cavity lined by a serous membrane and subdivided during fetal development into the pleural, pericardial, and peritoneal cavities.

coenzymes: Complex organic cofactors; most are structurally related to vitamins. (2, 25)

cofactor: Ions or molecules that must be attached to the active site before an enzyme can function; examples include mineral ions and several vitamins. (2)

collagen: A strong, insoluble protein fiber common in connective tissues. (4)

collateral ganglion: A sympathetic ganglion located anterior to the spinal column and separate from the sympathetic chain. (12, 16)

colliculus/colliculi: Little mound/mounds; in the brain, refers to one of the thickenings in the roof of the mesencephalon; the superior colliculi are associated with the visual system, and the inferior colliculi with the auditory system. (14, 15, 17)

colloid/colloidal suspension: A solution containing large organic molecules in suspension. (2, 26)

colon: The large intestine from the cecum to the rectum. (24)

coma: An unconscious state from which an individual cannot be aroused, even by strong stimuli. (16)

comminuted: Broken or crushed into small pieces.

commissure: A crossing over from one side to another.

common bile duct: The duct formed by the union of the cystic duct from the gallbladder and the bile ducts from the liver; terminates at the duodenal ampulla, where it meets the pancreatic duct. (24)

compact bone: Dense bone that contains parallel osteons. (6)

complement system: A group of distinct plasma proteins that interact in a chain reaction after exposure to activated antibodies or the surfaces of certain pathogens; complement proteins promote cell lysis, phagocytosis, and other defense mechanisms. (22)

compliance: Expandability; the ability of certain organs to tolerate changes in volume; indicates the presence of elastic fibers and smooth muscles. (23)

compound: A pure chemical substance made up of atoms of two or more different elements in a fixed proportion, regardless of the type of chemical bond joining them. (2)

concentration: The amount (in grams) or number of atoms, ions, or molecules (in moles) per unit volume. (2, 3, 25, 26)

concentration gradient: Regional differences in the concentration of a particular substance. (3, 25, 26)

conception: Fertilization. (29)

concha/conchae: Three pairs of thin, scroll-like bones that project into the nasal cavities; the superior and middle conchae are part of the ethmoid, and the inferior nasal conchae articulate with the ethmoid, lacrimal, maxilla, and palatine bones; also called *turbinate(s)*. (7)

condyle: A rounded articular projection on the surface of a bone. (8)

congenital: Present at birth.

congestive heart failure (CHF): The failure to maintain adequate cardiac output due to cardiovascular problems or myocardial damage. (23)

conjunctiva: A layer of stratified squamous epithelium that covers the inner surfaces of the eyelids and the anterior surface of the eye to the edges of the cornea. (17)

connective tissue: One of the four primary tissue types; provides a structural framework that stabilizes the relative positions of the other tissue types; includes connective tissue proper, cartilage, bone, and blood; contains cell products, cells, and ground substance. (4)

continuous propagation: The propagation of an action potential along an unmyelinated axon or a muscle plasma membrane, wherein the action potential affects every portion of the membrane surface. (12)

contractility: The ability to contract; possessed by skeletal, smooth, and cardiac muscle cells. (4, 20)

contralateral reflex: A reflex that affects the opposite side of the body from the stimulus. (13)

conus medullaris: The conical tip of the spinal cord that gives rise to the filum terminale. (13)

convergence: In the nervous system, the innervation of a single neuron by axons from several neurons; most common along motor pathways. (13)

coracoid process: A hook-shaped process of the scapula that projects above the anterior surface of the capsule of the shoulder joint. (8)

Cori cycle: The metabolic exchange of lactate from skeletal muscle for glucose from the liver; occurs during the recovery period after muscular exertion. (10)

cornea: The transparent portion of the fibrous layer of the anterior surface of the eye. (17)

corniculate cartilages: A pair of small laryngeal cartilages. (23)

cornu: Horn-shaped.

coronoid: Hooked or curved. (8)

corpora quadrigemina: The superior and inferior colliculi of the mesencephalic tectum (roof) in the brain. (14)

corpus/corpora: Body/bodies.

corpus callosum: A large bundle of axons that links centers in the left and right cerebral hemispheres. (14)

corpus luteum: The progesterone-secreting mass of follicle cells that develops in the ovary after ovulation. (18, 28)

cortex: The outer layer or region of an organ (5) or bone. (6)

corticobulbar tracts: Descending tracts that carry information or commands from the cerebral cortex to nuclei and centers in the brain stem. (15)

corticospinal tracts: Descending tracts that carry motor commands from the cerebral cortex to the anterior gray horns of the spinal cord. (15)

corticosteroid: A steroid hormone produced by the adrenal (suprarenal) cortex. (2, 18)

corticosterone: A corticosteroid secreted by the zona fasciculata of the adrenal (suprarenal) cortex; a glucocorticoid. (18)

corticotropin-releasing hormone (CRH): The releasing hormone, secreted by the hypothalamus, that stimulates secretion of adrenocorticotropic hormone (ACTH) by the anterior lobe of the pituitary gland (adenohypophysis). (18)

cortisol: A corticosteroid secreted by the zona fasciculata of the adrenal (suprarenal) cortex; a glucocorticoid. (18)

costa/costae: Rib/ribs. (7, 23)

cotransport: The membrane transport of a nutrient, such as glucose, in company with the movement of an ion, normally sodium; transport requires a carrier protein but does not involve direct ATP expenditure and can occur regardless of the concentration gradient for the nutrient. (3, 24, 26)

countercurrent exchange: The transfer of heat, water, or solutes between two fluids that travel in opposite directions. (25, 26)

countercurrent multiplication: Active transport between two limbs of a loop that contains a fluid moving in one direction in which the effect of the transport increases; responsible for the concentration of urine in the kidney tubules. (26)

covalent bond: A chemical bond between atoms that involves the sharing of electrons. (2)

coxal bone: Hip bone. (7, 8)

cranial: Pertaining to the head. (7)

cranial nerves: Peripheral nerves originating at the brain. (12)

craniosacral division: The parasympathetic division of the autonomic nervous system. (16)

cranium: The braincase; the skull bones that surround and protect the brain. (7)

creatine: A nitrogenous compound, synthesized in the body, that can form a high-energy bond by connecting to a phosphate group and that serves as an energy reserve. (10)

creatine phosphate: A high-energy compound in muscle cells; during muscle activity, the phosphate group is donated to ADP, regenerating ATP; also called *phosphorylcreatine*. (10)

creatinine: A breakdown product of creatine metabolism. (26)

crenation: Cellular shrinkage due to an osmotic movement of water out of the cytoplasm. (3)

cribriform plate: A portion of the ethmoid that contains the foramina used by the axons of olfactory receptors en route to the olfactory bulbs of the cerebrum. (7)

cricoid cartilage: A ring-shaped cartilage that forms the inferior margin of the larynx. (23)

crista/cristae: A ridge-shaped collection of hair cells in the ampulla of a semicircular duct; the crista and cupula form a receptor complex sensitive to movement along the plane of the semicircular canal. (17)

cross-bridge: The binding of a myosin head that projects from the surface of a thick filament at the active site of a thin filament in the presence of calcium ions. (10)

cuneiform cartilages: A pair of small cartilages in the larynx. (23)

cupula: A gelatinous mass that is located in the ampulla of a semicircular duct in the internal ear and whose movement stimulates the hair cells of the crista. (17)

Cushing's disease: A condition caused by the oversecretion of adrenal steroids. (18)

cutaneous membrane: The epidermis and papillary layer of the dermis. (4, 5)

cuticle: The layer of dead, keratinized cells that surrounds the shaft of a hair; for nails, *see* **eponychium**. (5)

cyanosis: An abnormal bluish coloration of the skin due to the presence of deoxygenated blood in vessels near the body surface. (5)

cystic duct: A duct that carries bile between the gallbladder and the common bile duct. (24)

cytochrome: A pigment component of the electron transport system; a structural relative of heme. (25)

cytokinesis: The cytoplasmic movement that separates two daughter cells at the completion of mitosis. (3)

cytology: The study of cells. (1, 3)

cytoplasm: The material between the plasma membrane and the nuclear membrane; cell contents. (3)

cytosine: A pyrimidine; one of the nitrogenous bases in the nucleic acids RNA and DNA. (2)

cytoskeleton: A network of microtubules and microfilaments in the cytoplasm. (3)

cytosol: The fluid portion of the cytoplasm; also called *intracellular fluid*. (3, 27)

cytotoxic: Poisonous to cells. (22)

cytotoxic T cells: Lymphocytes involved in cell-mediated immunity that kill target cells by direct contact or by the secretion of lymphotoxins; also called *killer T cells* and *T_C cells*. (22)

D

daughter cells: Genetically identical cells produced by somatic cell division. (3, 28)

deamination: The removal of an amino group from an amino acid. (25, 26)

decomposition reaction: A chemical reaction that breaks a molecule into smaller fragments. (2)

decussate: To cross over to the opposite side, usually referring to the crossover of the descending tracts of the corticospinal pathway on the ventral surface of the medulla oblongata. (15)

defecation: The elimination of fecal wastes. (24)

degradation: Breakdown, catabolism. (2, 25)

dehydration: A reduction in the water content of the body that threatens homeostasis. (27)

dehydration synthesis: The joining of two molecules associated with the removal of a water molecule. (2)

demyelination: The loss of the myelin sheath of an axon, normally due to chemical or physical damage to Schwann cells or oligodendrocytes. (12)

denaturation: A temporary or permanent change in the three-dimensional structure of a protein. (2)

dendrite: A sensory process of a neuron. (4, 12)

deoxyribonucleic acid (DNA): A nucleic acid consisting of a double chain of nucleotides that contains the sugar deoxyribose and the nitrogenous bases adenine, guanine, cytosine, and thymine. (2)

deoxyribose: A five-carbon sugar resembling ribose but lacking an oxygen atom. (2)

depolarization: A change in the membrane potential from a negative value toward 0 mV. (12, 20)

depression: Inferior (downward) movement of a body part.

dermatitis: An inflammation of the skin. (5)

dermatome: A sensory region monitored by the dorsal rami of a single spinal segment. (13)

dermis: The connective tissue layer beneath the epidermis of the skin. (5)

detrusor muscle: Collectively, the three layers of smooth muscle in the wall of the urinary bladder. (26)

detumescence: The loss of a penile erection. (28)

development: Growth and the acquisition of increasing structural and functional complexity; includes the period from conception to maturity.

diabetes insipidus: Disorder of the pituitary gland characterized by polyuria (excessive urination) and polydipsia (excessive thirst) that results from inadequate production of antidiuretic hormone (ADH). (18)

diabetes mellitus: Disorder characterized by polyuria (excessive urination) and glycosuria (glucose in the urine), most commonly due to the inadequate production or diminished sensitivity to insulin with a resulting increase of blood glucose levels. (18)

diapedesis: The movement of white blood cells through the walls of blood vessels by migration between adjacent endothelial cells. (19, 22)

diaphragm: Any muscular partition; the respiratory muscle that separates the thoracic cavity from the abdominopelvic cavity. (1, 11, 23)

diaphysis: The shaft of a long bone. (6)

diarthrosis: A synovial joint. (9)

diastolic pressure: Pressure measured in the walls of a muscular artery when the left ventricle is in diastole (relaxation). (20)

diencephalon: A division of the brain that includes the epithalamus, thalamus, and hypothalamus. (14)

differential count: An estimate of the number of each type of white blood cell on the basis of a random sampling of 100 white blood cells. (19)

differentiation: The gradual appearance of characteristic cellular specializations during development as the result of gene activation or repression. (3)

diffusion: Passive molecular movement from an area of higher concentration to an area of lower concentration. (3, 21, 23, 26)

digestion: The chemical breakdown of ingested food into simple molecules that can be absorbed by the cells of the digestive tract. (24)

digestive system: The digestive tract and associated glands. (24)

digestive tract: An internal passageway that begins at the mouth, ends at the anus, and is lined by a mucous membrane; also called *gastrointestinal tract*. (24)

dilate: To increase in diameter; to enlarge or expand.

disaccharide: A compound formed by the joining of two simple sugars by dehydration synthesis. (2)

dissociation: *See* ionization.

distal: A direction away from the point of attachment or origin; for a limb, away from its attachment to the trunk. (1, 8)

distal convoluted tubule (DCT): The segment of the nephron closest to the connecting tubules and collecting duct; an important site of active secretion. (26)

diuresis: Fluid loss by the kidneys; the production of unusually large volumes of urine. (26)

divergence: In neural tissue, the spread of information from one neuron to many neurons; an organizational pattern common along sensory pathways of the central nervous system. (13)

diverticulum: A sac or pouch in the wall of the colon or other organ. (24)

DNA molecule: Two DNA strands wound in a double helix and held together by hydrogen bonds between complementary nitrogenous base pairs. (3)

dopamine: An important neurotransmitter in the central nervous system. (12)

dorsal: Toward the back, posterior.

dorsal root ganglion: A peripheral nervous system ganglion containing the cell bodies of sensory neurons. (13)

dorsiflexion: Upward movement of the foot through flexion at the ankle. (9)

Down's syndrome: A genetic abnormality resulting from the presence of three copies of chromosome 21; people with this condition have characteristic physical and intellectual deficits; also called *trisomy 21*. (16)

duct: A passageway that delivers exocrine secretions to an epithelial surface. (4)

ductus arteriosus: A vascular connection between the pulmonary trunk and the aorta that functions throughout fetal life; normally closes at birth or shortly thereafter and persists as the ligamentum arteriosum. (21, 23)

ductus deferens: A passageway that carries spermatozoa from the epididymis to the ejaculatory duct; also called the *vas deferens*. (28)

duodenal ampulla: A chamber that receives bile from the common bile duct and pancreatic secretions from the pancreatic duct. (24)

duodenal papilla: A conical projection from the inner surface of the duodenum that contains the opening of the duodenal ampulla. (24)

duodenum: The proximal 25 cm (9.8 in.) of the small intestine that contains short villi and submucosal glands. (24)

dura mater: The outermost membrane (meninx) of the cranial and spinal meninges. (13, 14)

E

eccrine glands: Sweat glands of the skin that produce a watery secretion. (5)

ectoderm: One of the three primary germ layers; covers the surface of the embryo and gives rise to the nervous system, the epidermis and associated glands, and a variety of other structures. (29)

ectopic: Outside the normal location.

effector: A peripheral gland or muscle cell innervated by a motor neuron. (1, 12)

efferent: Conducting a fluid or nerve impulse away from an organ or structure.

efferent arteriole: An arteriole carrying blood away from a glomerulus of the kidney. (26)

efferent fiber: An axon that carries impulses away from the central nervous system. (12)

ejaculation: The ejection of semen from the penis as the result of muscular contractions of the bulbospongiosus and ischiocavernosus muscles. (28)

ejaculatory ducts: Short ducts that pass within the walls of the prostate gland and connect the ductus deferens with the prostatic urethra. (28)

elastase: A pancreatic enzyme that breaks down elastin fibers. (24)

elastin: Connective tissue fibers that stretch and recoil, providing elasticity to connective tissues. (4, 5)

electrical coupling: A connection between adjacent cells that permits the movement of ions and the transfer of graded or propagated changes in the membrane potential from cell to cell. (12)

electrocardiogram (ECG, EKG): A graphic record of the electrical activities of the heart, as monitored at specific locations on the body surface. (20)

electroencephalogram (EEG): A graphic record of the electrical activities of the brain. (14)

electrolytes: Soluble inorganic compounds whose ions will conduct an electrical current in solution. (2, 27)

electron: One of the three fundamental subatomic particles; has a negative charge and normally orbits the protons of the nucleus. (2, 25)

electron transport system (ETS): The cytochrome system responsible for aerobic energy production in cells; a complex bound to the inner mitochondrial membrane. (25)

element: All the atoms with the same atomic number. (2)

elevation: Movement in a superior, or upward, direction.

elimination: The ejection of wastes from the body through urination or defecation. (24, 26)

embolism: The obstruction of a vessel, typically by a blood clot. (19)

embolus: An air bubble, fat globule, or blood clot drifting in the bloodstream. (19)

embryo: The developmental stage beginning at fertilization and ending at the start of the third developmental month. (29)

embryology: The study of embryonic development, focusing on the first two months after fertilization. (1, 28)

endocardium: The simple squamous epithelium that lines the heart and is continuous with the endothelium of the great vessels. (20)

endochondral ossification: The replacement of a cartilaginous model with bone; the characteristic mode of formation for skeletal elements other than the bones of the cranium, the clavicles, and sesamoid bones. (6)

endocrine gland: A gland that secretes hormones into the blood. (4, 18)

endocrine system: The endocrine (ductless) glands and organs of the body. (18)

endocytosis: The movement of relatively large volumes of extracellular material into the cytoplasm by the formation of a membranous vesicle at the cell surface; includes pinocytosis and phagocytosis. (3)

endoderm: One of the three primary germ layers; the layer on the undersurface of the embryonic disc; gives rise to the epithelia and glands of the digestive system, the respiratory system, and portions of the urinary system. (29)

endogenous: Produced within the body.

endolymph: The fluid contents of the membranous labyrinth (the saccule, utricle, semicircular ducts, and cochlear duct) of the internal ear. (17)

endometrium: The mucous membrane lining the uterus. (28)

endomysium: A delicate network of connective tissue fibers that surrounds individual muscle cells. (10)

endoneurium: A delicate network of connective tissue fibers that surrounds individual nerve fibers. (13)

endoplasmic reticulum: A network of membranous channels in the cytoplasm of a cell that function in intracellular transport, synthesis, storage, packaging, and secretion. (3)

endorphins: Neuromodulators, produced in the central nervous system, that inhibit activity along pain pathways. (12)

endosteum: An incomplete cellular lining on the inner (medullary) surfaces of bones. (6)

endothelium: The simple squamous epithelial cells that line blood and lymphatic vessels. (4, 19, 21)

enkephalins: Neuromodulators, produced in the central nervous system, that inhibit activity along pain pathways. (12)

enterocrinin: A hormone secreted by the lining of the duodenum after exposure to chyme; stimulates the secretion of the submucosal glands. (24)

enteroendocrine cells: Endocrine cells scattered among the epithelial cells that line the digestive tract. (24)

enterogastric reflex: The reflexive inhibition of gastric secretion; initiated by the arrival of chyme in the small intestine. (24)

enterohepatic circulation: The excretion of bile salts by the liver, followed by the absorption of bile salts by intestinal cells for return to the liver by the hepatic portal vein. (24)

enteropeptidase: An enzyme in the lumen of the small intestine that activates the proenzymes secreted by the pancreas; formerly called enterokinase. (24)

enzyme: A protein that catalyzes a specific biochemical reaction. (2)

eosinophil: A type of white blood cell with a lobed nucleus and red-staining granules; participates in the immune response and is especially important during allergic reactions. (19)

ependyma: The layer of cells lining the ventricles and central canal of the central nervous system. (12)

epicardium: A serous membrane covering the outer surface of the heart; also called *visceral pericardium.* (20)

epidermis: The epithelium covering the surface of the skin. (5)

epididymis: A coiled duct that connects the rete testis to the ductus deferens; site of functional maturation of spermatozoa. (28)

epidural space: The space between the spinal dura mater and the walls of the vertebral foramen; contains blood vessels and adipose tissue; a common site of injection for regional anesthesia. (13)

epiglottis: A blade-shaped flap of tissue, reinforced by cartilage, that is attached to the posterior and superior surface of the thyroid cartilage; folds over the entrance to the larynx during swallowing. (23)

epimysium: A dense layer of collagen fibers that surrounds a skeletal muscle and is continuous with the tendons/aponeuroses of the muscle and with the perimysium. (10)

epineurium: A dense layer of collagen fibers that surrounds a peripheral nerve. (13)

epiphyseal cartilage: The cartilaginous region between the epiphysis and diaphysis of a growing bone; also called *epiphyseal plate.* (6)

epiphysis: The head of a long bone. (6)

epithelium: One of the four primary tissue types; a layer of cells that forms a superficial covering or an internal lining of a body cavity or vessel. (4, 24)

eponychium: A narrow zone of stratum corneum that extends across the surface of a nail at its exposed base; also called the *cuticle.* (5)

equilibrium: A dynamic state in which two opposing forces or processes are in balance. (1, 15, 17)

erection: The stiffening of the penis due to the erectile tissues of the corpora cavernosa and corpus spongiosum filling with blood. (28)

erythema: Redness and inflammation at the surface of the skin. (5, 22)

erythrocyte: A red blood cell (RBC); has no nucleus and contains large quantities of hemoglobin. (4, 19)

erythropoietin: A hormone released by most tissues, and especially by the kidneys, when oxygen levels decrease; stimulates erythropoiesis (red blood cell formation) in red bone marrow. (18, 19, 21)

Escherichia coli: A normal bacterial resident of the large intestine. (24)

esophagus: A muscular tube that connects the pharynx to the stomach. (24)

essential amino acids: Amino acids that cannot be synthesized in the body in adequate amounts and must be obtained from the diet. (25)

essential fatty acids: Fatty acids that cannot be synthesized in the body and must be obtained from the diet. (25)

estrogens: A class of steroid sex hormones that includes estradiol. (2, 18)

evaporation: A movement of molecules from the liquid state to the gaseous state.

eversion: A turning outward. (9)

excitable membranes: Membranes that propagate action potentials, a characteristic of muscle cells and nerve cells. (10, 12)

excitatory postsynaptic potential (EPSP): The depolarization of a postsynaptic membrane by a chemical neurotransmitter released by the presynaptic cell. (12)

excretion: The removal of wastes from the blood, tissues, or organs.

exocrine gland: A gland that secretes onto the body surface or into a passageway connected to the exterior. (4)

exocytosis: The ejection of cytoplasmic materials by the fusion of a membranous vesicle with the plasma membrane. (3)

expiration: Exhalation; breathing out.

extension: An increase in the angle between two articulating bones; the opposite of flexion. (8, 9)

external acoustic meatus: A passageway in the temporal bone that leads to the tympanic membrane of the middle ear. (17)

external ear: The auricle, external acoustic meatus, and tympanic membrane. (17)

external nares: The nostrils; the external openings into the nasal cavity. (23)

external respiration: The diffusion of gases between the alveolar air and the alveolar capillaries and between the systemic capillaries and peripheral tissues. (23)

exteroceptors: General sensory receptors in the skin, mucous membranes, and special sense organs that provide information about the external environment and about our position within it. (12)

extracellular fluid: All body fluids other than that contained within cells; includes plasma and interstitial fluid. (3, 27)

extraembryonic membranes: The yolk sac, amnion, chorion, and allantois. (29)

extrafusal fibers: Contractile muscle fibers (as opposed to the sensory intrafusal fibers, or muscle spindles). (13)

extrinsic pathway: A clotting pathway that begins with damage to blood vessels or surrounding tissues and ends with the formation of tissue factor complex. (19)

F

fabella: A sesamoid bone commonly located in the gastrocnemius muscle. (11)

facilitated: Brought closer to threshold, as in the depolarization of a nerve plasma membrane toward threshold; making the cell more sensitive to depolarizing stimuli. (12)

facilitated diffusion: The passive movement of a substance across a plasma membrane by means of a protein carrier. (3, 24, 26)

falciform ligament: A sheet of mesentery that contains the ligamentum teres, the fibrous remains of the umbilical vein of the fetus. (29)

falx: Sickle-shaped.

falx cerebri: The curving sheet of dura mater that extends between the two cerebral hemispheres; encloses the superior sagittal sinus. (7, 14)

fasciae: Connective tissue fibers, primarily collagen, that form sheets or bands beneath the skin to attach, stabilize, enclose, and separate muscles and other internal organs. (4)

fasciculus: A small bundle; usually refers to a collection of nerve axons or muscle fibers. (10, 15)

fatty acids: Hydrocarbon chains that end in a carboxyl group. (2)

fauces: The passage from the mouth to the pharynx, bounded by the palatal arches, the soft palate, and the uvula. (24)

febrile: Characterized by or pertaining to a fever. (22, 25)

feces: Waste products eliminated by the digestive tract at the anus; contains indigestible residue, bacteria, mucus, and epithelial cells. (24)

fenestra/fenestrae: Opening(s).

fertilization: The fusion of a secondary oocyte and a spermatozoon to form a zygote. (28)

fetus: The developmental stage lasting from the start of the third developmental month to delivery. (28, 29)

fibrin: Insoluble protein fiber that forms the basic framework of a blood clot. (19)

fibrinogen: A plasma protein that is the soluble precursor of the insoluble protein fibrin. (19)

fibroblasts: Cells of connective tissue proper that produce extracellular fibers and secrete the organic substances of the extracellular matrix. (4)

fibrocartilage: Cartilage containing an abundance of collagen fibers; located around the edges of joints, in the intervertebral discs, and the menisci of the knee; also referred to as *fibrous cartilage.* (4)

fibrocytes: Mature fibroblasts; maintain connective tissue fibers of connective tissue proper. (4)

fibrous layer: The outermost layer of the eye, composed of the sclera and cornea; also called *fibrous tunic.* (17)

fibula: The lateral, slender bone of the leg. (8)

filariasis: A condition resulting from infection by mosquito-borne parasites; can cause elephantiasis. (21, 22)

filiform papillae: Slender conical projections from the dorsal surface of the anterior two-thirds of the tongue. (17)

filtrate: The fluid produced by filtration at a glomerulus in the kidney. (26)

filtration: The movement of a fluid across a membrane whose pores restrict the passage of solutes on the basis of size. (21, 26)

filtration pressure: The hydrostatic pressure responsible for filtration. (21, 26)

filum terminale: A fibrous extension of the spinal cord, from the conus medullaris to the coccygeal ligament. (13)

fimbriae: Fringes; the fingerlike processes that surround the entrance to the uterine tube. (28)

fissure: An elongated groove or opening. (7, 14)

fistula: An abnormal passageway between two organs or from an internal organ or space to the body surface.

flaccid: Limp, soft, flabby; a muscle without muscle tone.

flagellum/flagella: An organelle that is structurally similar to a motile cilium but is used to propel a cell through a fluid; found on spermatozoa. (28)

flatus: Intestinal gas. (24)

flexion: A movement that decreases the angle between two articulating bones; the opposite of extension. (8, 9)

flexor: A muscle that produces flexion. (11)

flexor reflex: A reflex contraction of the flexor muscles of a limb in response to a painful stimulus. (13)

flexure: A bending.

folia: Leaflike folds; the slender folds in the surface of the cerebellar cortex. (14)

follicle: A small secretory sac or gland.

follicle-stimulating hormone (FSH): A hormone secreted by the adenohypophysis (anterior lobe of the pituitary gland); stimulates oogenesis (female) and spermatogenesis (male). (18, 28)

fontanelle: A relatively soft, flexible, fibrous region between two flat bones in the developing skull; also spelled *fontanel.* (7)

foramen/foramina: Opening/openings or passage/passages through a bone. (7, 20)

forearm: The distal portion of the upper limb between the elbow and wrist. (8)

forebrain: The cerebrum. (14)

fornix: An arch or the space bounded by an arch; in the brain, an arching tract that connects the hippocampus with the mammillary bodies (14); in the eye, a slender pocket located where the epithelium of the ocular conjunctiva folds back on itself as the palpebral conjunctiva (17); in the vagina, the shallow recess surrounding the protrusion of the cervix. (28)

fossa: A shallow depression or furrow in the surface of a bone. (8, 20)

fourth ventricle: An elongated ventricle of the metencephalon (pons and cerebellum) and the myelencephalon (medulla oblongata) of the brain; the roof contains a region of choroid plexus. (14)

fovea: The portion of the retina within the macula that provides the sharpest vision because it has the highest concentration of cones; also called *fovea centralis.* (17)

fracture: A break or crack in a bone. (6)

frenulum: A bridle; usually referring to a band of tissue that restricts movement, e.g., *lingual frenulum.* (24)

frontal plane: A sectional plane that divides the body into an anterior portion and a posterior portion; also called *coronal plane.* (1)

fructose: A hexose (six-carbon simple sugar) in foods and semen. (2, 28)

fundus: The base of an organ such as the stomach, uterus, or gallbladder.

G

gallbladder: The pear-shaped reservoir for bile after it is secreted by the liver. (24)

gametes: Reproductive cells (spermatozoa or oocytes) that contain half the normal chromosome complement. (28, 29)

gametogenesis: The formation of gametes. (28)

gamma aminobutyric acid (GABA): A neurotransmitter of the central nervous system whose effects are generally inhibitory. (12)

gamma motor neurons: Motor neurons that adjust the sensitivities of muscle spindles (intrafusal fibers). (13)

ganglion/ganglia: A collection of neuron cell bodies in the peripheral nervous system. (12, 16)

gangliosides: Glycolipids that are important components of plasma membranes in the central nervous system. (12)

gap junctions: Connections between cells that permit electrical coupling. (4)

gaster: The stomach (24); the body, or belly, of a skeletal muscle. (11)

gastric: Pertaining to the stomach. (24)

gastric glands: The tubular glands of the stomach whose cells produce acid, enzymes, intrinsic factor, and hormones. (24)

gastrointestinal (GI) tract: *See* **digestive tract**.

gene: A portion of a DNA strand that functions as a hereditary unit, is located at a particular site on a specific chromosome, and codes for a specific protein or polypeptide. (3, 29)

genetic engineering: Research and experiments involving the manipulation of the genetic makeup of an organism. (29)

genetics: The study of heredity and genetic variations. (29)

geniculates: The medial geniculate nuclei and the lateral geniculate nuclei are in the thalamus of the brain. (14)

genitalia: The reproductive organs. (28)

germinal centers: Pale regions in the interior of lymphoid tissues or lymphoid nodules, where cell divisions occur that produce additional lymphocytes. (22)

gestation: The period of intrauterine development; pregnancy. (29)

gland: Cells that produce exocrine or endocrine secretions. (4)

glenoid cavity: A rounded depression that forms the articular surface of the scapula at the shoulder joint. (8)

glial cells: *See* **neuroglia**.

globular proteins: Proteins whose tertiary structure makes them rounded and compact. (2)

glomerular capsule: The expanded initial portion of the nephron that surrounds the glomerulus. (26)

glomerular filtration rate (GFR): The rate of filtrate formation at the glomerulus. (26)

glomerulus: A knot of capillaries that projects into the enlarged, proximal end of a nephron; the site of filtration, the first step in the production of urine. (26)

glossopharyngeal nerve: Cranial nerve IX. (14)

glucagon: A hormone secreted by the alpha cells of the pancreatic islets; increases blood glucose concentrations. (18)

glucocorticoids: Hormones secreted by the zona fasciculata of the adrenal (suprarenal) cortex to modify glucose metabolism; cortisol and corticosterone are important examples. (18)

gluconeogenesis: The synthesis of glucose from noncarbohydrate precursors (e.g., lactate, glycerol, or amino acids). (25)

glucose: A six-carbon sugar, $C_6H_{12}O_6$; the preferred energy source for most cells and normally the only energy source for neurons. (2, 10, 18, 25)

glycerides: Lipids composed of glycerol bound to fatty acids. (2)

glycogen: A polysaccharide that is an important energy reserve; a polymer consisting of a long chain of glucose molecules. (2, 10)

glycogenesis: The synthesis of glycogen from glucose molecules. (25)

glycogenolysis: Glycogen breakdown and the liberation of glucose molecules. (25)

glycolipids: Compounds created by the combination of carbohydrate and lipid components. (2)

glycolysis: The anaerobic cytosolic breakdown of glucose into two 3-carbon molecules of pyruvate, with a net gain of two ATP molecules. (3, 10, 25)

glycoprotein: A compound containing a relatively small carbohydrate group attached to a large protein. (2, 18)

glycosuria: The presence of glucose in urine; also called *glucosuria.* (18, 26)

Golgi apparatus: A cellular organelle consisting of a series of membranous plates that give rise to lysosomes and secretory vesicles. (3)

gomphosis: A fibrous synarthrosis that binds a tooth to the bone of the jaw; *see* **periodontal ligament**. (24)

gonadotropin-releasing hormone (GnRH): A hypothalamic releasing hormone that causes the secretion of both follicle-stimulating hormone and luteinizing hormone by the anterior lobe of the pituitary gland (adenohypophysis). (18, 28)

gonadotropins: Follicle-stimulating hormone and luteinizing hormone, hormones that stimulate gamete development and sex hormone secretion. (18, 28)

gonads: Reproductive organs that produce gametes and sex hormones. (28)

granulocytes: White blood cells containing granules that are visible with the light microscope; includes eosinophils, basophils, and neutrophils; also called *granular leukocytes.* (19)

gray matter: Areas in the central nervous system that are dominated by neuron cell bodies, neuroglia, and unmyelinated axons. (12, 13, 14)

gray ramus: A bundle of postganglionic sympathetic nerve fibers distributed to effectors in the body wall, skin, and limbs by way of a spinal nerve. (13)

greater omentum: A large fold of the dorsal mesentery of the stomach; hangs anterior to the intestines. (24)

groin: The inguinal region. (11)

gross anatomy: The study of the structural features of the body without the aid of a microscope. (1)

growth hormone (GH): An adenohypophysis (anterior lobe of the pituitary gland) hormone that stimulates tissue growth and anabolism when nutrients are abundant and restricts tissue glucose dependence when nutrients are in short supply. (18)

growth hormone–inhibiting hormone (GH–IH): A hypothalamic regulatory hormone that inhibits growth hormone secretion by the anterior lobe of the pituitary gland (adenohypophysis); also called *somatostatin*. (18)

guanine: A purine; one of the nitrogenous bases in the nucleic acids RNA and DNA. (2)

gustation: Sense of taste. (15, 17)

gyrus: A prominent fold or ridge of neural cortex on the surfaces of the cerebral hemispheres. (14)

H

hair: A keratinous strand produced by epithelial cells of the hair follicle. (5)

hair cells: Sensory cells of the internal ear. (17)

hair follicle: An accessory structure of the integument; a tube lined by a stratified squamous epithelium that begins at the surface of the skin and ends at the hair papilla. (5)

hallux: The big toe; also called the *great toe*. (8)

haploid: Possessing half the normal number of chromosomes; a characteristic of gametes. (28, 29)

hard palate: The bony roof of the oral cavity, formed by the maxillae and palatine bones. (23, 24)

helper T cells: Lymphocytes whose secretions and other activities coordinate cell-mediated and antibody-mediated immunities; also called T_H *cells*. (22)

hematocrit: The percentage of formed elements in a sample of blood; also called *volume of packed red cells (VPRC)* or *packed cell volume (PCV)*. (19)

hematoma: A tumor or swelling filled with blood.

hematopoietic stem cells: Stem cells whose divisions produce each of the various populations of blood cells; also called *hemocytoblasts*. (19)

hematuria: The abnormal presence of red blood cells in urine. (19, 26)

heme: A porphyrin ring containing a central iron atom that can reversibly bind oxygen molecules; a component of the hemoglobin molecule. (19)

hemoglobin: A protein composed of four globular subunits, each bound to a heme molecule; gives red blood cells the ability to transport oxygen in the blood. (5, 19, 23, 27)

homozygous: Having identical alleles at corresponding sites on a chromosome pair. (29)

hemolysis: The breakdown of red blood cells. (3)

hemopoiesis: Blood cell formation and differentiation. (19)

hemorrhage: Blood loss; to bleed. (21)

hemostasis: The stoppage of bleeding. (19)

heparin: An anticoagulant released by activated basophils and mast cells. (4, 19)

hepatic duct: The duct that carries bile away from the liver lobes and toward the union with the cystic duct. (24)

hepatic portal vein: The vessel that carries blood from the intestinal capillaries to the sinusoids of the liver. (21)

hepatocyte: A liver cell. (24)

heterozygous: Possessing two different alleles at corresponding sites on a chromosome pair; a person's phenotype is determined by one or both of the alleles. (29)

hexose: A six-carbon simple sugar. (2)

hiatus: A gap, cleft, or opening.

high density lipoprotein (HDL): A lipoprotein with a relatively small lipid content; responsible for the movement of cholesterol from peripheral tissues to the liver. (25)

hilum: A localized region where blood vessels, lymphatic vessels, nerves, and/or other anatomical structures are attached to an organ. (22, 23, 26)

hippocampus: A region, deep to the floor of a lateral ventricle, involved in emotional states and the conversion of short-term to long-term memories. (12, 14)

histamine: The chemical released by stimulated mast cells or basophils to initiate or enhance inflammation. (4, 12)

histology: The study of tissues. (1, 4)

histones: Proteins associated with the DNA of the nucleus; the DNA strands are wound around them. (3)

holocrine: A form of exocrine secretion in which the secretory cell becomes swollen with vesicles and then ruptures. (4)

homeostasis: The maintenance of a relatively constant internal environment. (1)

hormone: A chemical that is secreted by one cell and travels through the bloodstream to affect the activities of cells in another part of the body. (2, 4, 6, 18, 21, 24, 28, 29)

human chorionic gonadotropin (hCG): The placental hormone that maintains the corpus luteum for the first three months of pregnancy. (29)

human immunodeficiency virus (HIV): The infectious agent that causes acquired immunodeficiency syndrome (AIDS). (22)

human leukocyte antigen (HLA): *See* **MHC protein**.

human placental lactogen (hPL): The placental hormone that stimulates the functional development of the mammary glands. (29)

humoral immunity: *See* **antibody-mediated immunity**.

hyaluronan: A carbohydrate component of proteoglycans in the matrix of many connective tissues. (4)

hyaluronidase: An enzyme that breaks down the bonds between adjacent follicle cells; produced by some bacteria and found in the acrosome of a spermatozoon. (29)

hydrogen bond: A weak interaction between the hydrogen atom on one molecule and a negatively charged portion of another molecule. (2)

hydrolysis: The breakage of a chemical bond through the addition of a water molecule; the reverse of dehydration synthesis. (2)

hydrophilic: Freely associating with water; readily entering into solution; water-loving. (2)

hydrophobic: Incapable of freely associating with water molecules; insoluble; water-fearing. (2)

hydrostatic pressure: Fluid pressure. (21, 26)

hydroxide ion: OH^-. (2)

hypercapnia: High blood carbon dioxide concentrations, commonly as a result of hypoventilation or inadequate tissue perfusion. (23, 27)

hyperplasia: An abnormal increase in the number of cells in a tissue or organ. (3)

hyperpolarization: The movement of the membrane potential away from the normal resting potential and farther from 0 mV. (12)

hypersecretion: The overactivity of glands that produce exocrine or endocrine secretions. (18)

hypertension: Abnormally high blood pressure. (21)

hypertonic: In comparing two solutions, the solution with the higher osmolarity. (3)

hypertrophy: An increase in tissue size without cell division. (10)

hyperventilation: A rate of respiration sufficient to decrease blood P_{CO_2} to levels below normal. (23, 27)

hypocapnia: An abnormally low blood P_{CO_2}; commonly results from hyperventilation. (27)

hypodermic needle: A needle inserted through the skin to introduce drugs into the subcutaneous layer. (5)

hypodermis: The layer of loose connective tissue below the dermis; also called *subcutaneous layer* or *superficial fascia*. (4, 5)

hypophyseal portal system: The network of vessels that carries blood from capillaries in the hypothalamus to capillaries in the anterior lobe of the pituitary gland (adenohypophysis). (18)

hypophysis: *See* **pituitary gland**. (18)

hyposecretion: Abnormally low exocrine or endocrine secretion. (18)

hypothalamus: The floor of the diencephalon; the region of the brain containing centers involved with the subconscious regulation of visceral functions, emotions, drives, and the coordination of neural and endocrine functions. (14)

hypothermia: An abnormally low body temperature. (25)

hypothesis: A prediction that can be subjected to scientific analysis and review.

hypotonic: In comparing two solutions, the solution with the lower osmolarity. (3)

hypoventilation: A respiratory rate that is insufficient to keep blood P_{CO_2} within normal levels. (23, 27)

hypoxia: A low tissue oxygen concentration. (19, 23)

I

ileum: The distal 2.5 m of the small intestine. (24)

ilium: The largest of the three bones whose fusion creates a coxal bone. (8)

immunity: Resistance to infection and disease caused by foreign substances, toxins, or pathogens. (22)

immunization: The production of immunity by the deliberate exposure to antigens under conditions that prevent the development of illness but stimulate the production of memory B cells. (22)

immunoglobulin (Ig): A circulating antibody. (19, 22)

implantation: The attachment of a blastocyst into the endometrium of the uterine wall. (29)

inclusions: Aggregations of insoluble pigments, nutrients, or other materials in cytoplasm. (3)

incus: The central auditory ossicle, located between the malleus and the stapes in the middle ear cavity. (17)

inducer: A stimulus that promotes the activity of a specific gene. (29)

inexcitable: Incapable of propagating an action potential. (12)

infarct: An area of dead cells that results from an interruption of blood flow. (19, 20)

infection: The invasion and colonization of body tissues by pathogens. (4)

inferior: Below, in reference to a particular structure, with the body in the anatomical position.

inferior vena cava: The vein that carries blood from the parts of the body inferior to the heart to the right atrium. (20, 21)

infertility: The inability to conceive; also called *sterility*. (28, 29)

inflammation: A nonspecific defense mechanism that operates at the tissue level; characterized by swelling, redness, heat (warmth), pain, and sometimes loss of function. (4, 22)

infundibulum: A tapering, funnel-shaped structure; in the brain, the connection between the pituitary gland and the hypothalamus (14, 18); in the uterine tube, the entrance bounded by fimbriae that receives the oocyte at ovulation. (28)

ingestion: The introduction of food into the digestive tract by way of the mouth; eating. (24)

inguinal canal: A passage through the abdominal wall that marks the path of testicular descent and that contains the testicular arteries, veins, and ductus deferens. (11, 28)

inguinal region: The area of the abdominal wall near the junction of the trunk and the thighs that contains the external genitalia; the groin. (28)

inhibin: A hormone, produced by nurse (sustentacular) cells of the testes and follicular cells of the ovaries, that inhibits the secretion of follicle-stimulating hormone by the adenohypophysis (anterior lobe of the pituitary gland). (18, 28)

inhibitory postsynaptic potential (IPSP): A hyperpolarization of the postsynaptic membrane after the arrival of a neurotransmitter. (12)

initial segment: The proximal portion of the axon where an action potential first appears. (12)

injection: The forcing of fluid into a body part or organ.

inner cell mass: Cells of the blastocyst that will form the body of the embryo. (29)

inner ear: *See* **internal ear**.

innervation: The distribution of sensory and motor nerves to a specific region or organ. (11, 16)

insensible perspiration: Evaporative water loss by diffusion across the epithelium of the skin or evaporation across the alveolar surfaces of the lungs. (5, 27)

insertion: A point of attachment of a muscle; the end that is easily movable. (11)

insoluble: Incapable of dissolving in solution. (2)

inspiration: Inhalation; the movement of air into the respiratory system. (23)

insulin: A hormone secreted by beta cells of the pancreatic islets; causes a decrease in blood glucose concentrations. (18)

integument: The skin. (5)

intercalated discs: Regions where adjacent cardiocytes interlock and where gap junctions permit electrical coupling between the cells. (4, 10, 20)

intercellular fluid: *See* **interstitial fluid**.

interferons: Peptides released by virus-infected cells, especially lymphocytes, that slow viral replication and make other cells more resistant to viral infection. (22)

interleukins: Peptides, released by activated monocytes and lymphocytes, that assist in the coordination of cell-mediated and antibody-mediated immunities. (22)

internal capsule: The collection of afferent and efferent fibers of the white matter of the cerebral

hemispheres, visible on gross dissection of the brain. (14)

internal ear: The membranous labyrinth that contains the organs of hearing and equilibrium. (17)

internal nares: The entrance to the nasopharynx from the nasal cavity. (23)

internal respiration: The diffusion of gases between interstitial fluid and cytoplasm. (23)

interneuron: An association neuron; central nervous system neurons that are between sensory and motor neurons. (12)

interoceptors: Sensory receptors monitoring the functions and status of internal organs and systems. (12)

interosseous membrane: The fibrous connective tissue membrane between the shafts of the tibia and fibula and between the radius and ulna; an example of a fibrous amphiarthrosis. (8)

interphase: The stage in the life cycle of a cell during which the chromosomes are uncoiled and all normal cellular functions except mitosis are under way. (29)

intersegmental reflex: A reflex that involves several segments of the spinal cord. (13)

interstitial fluid: The fluid in the tissues that fills the spaces between cells. (3)

interstitial growth: A form of cartilage growth in which the cartilage expands from within through the growth, mitosis, and secretion of chondrocytes in the matrix. (4)

interventricular foramen: The opening that permits fluid movement between the lateral and third ventricles of the brain. (14)

intervertebral disc: A fibrocartilage pad between the bodies of successive vertebrae that absorbs shocks. (7, 9)

intestinal crypt: A tubular epithelial pocket that is lined by secretory cells and opens into the lumen of the digestive tract; also called *intestinal gland*. (24)

intestine: The tubular organ of the digestive tract. (18, 24)

intracellular fluid: The cytosol. (27)

intrafusal fibers: Muscle spindle fibers. (13)

intramembranous ossification: The formation of bone within a connective tissue without the prior development of a cartilaginous model. (6)

intrinsic factor: A glycoprotein, secreted by the parietal cells of the stomach, that facilitates the intestinal absorption of vitamin B_{12}. (19, 24, 25)

intrinsic pathway: A pathway of the clotting system that begins with the activation of platelets and ends with the formation of factor X activator complex. (19)

inversion: A turning inward. (9)

in vitro: Outside the body, in an artificial environment.

in vivo: In the living body.

involuntary: Not under conscious control.

ion: An atom or molecule having a positive or negative charge due to the loss or gain, respectively, of one or more electrons. (2, 26, 27)

ionic bond: A molecular bond created by the attraction between ions with opposite charges. (2)

ionization: Dissociation; the breakdown of a molecule in solution to form ions. (2)

ipsilateral: A reflex response that affects the same side as the stimulus. (13)

iris: A contractile structure, made up of smooth muscle, that forms the colored portion of the eye. (17)

ischemia: An inadequate blood supply to a region of the body. (11)

ischium: One of the three bones whose fusion creates a coxal bone. (8)

islets of Langerhans: *See* **pancreatic islets**.

isotonic: A solution with an osmolarity that does not result in water movement across plasma membranes. (10)

isotopes: Forms of an element whose atoms contain the same number of protons but different numbers of neutrons (and thus differ in atomic mass). (2)

isthmus: A narrow band of tissue connecting two larger masses.

J

jejunum: The middle part of the small intestine. (24)

joint: An area where adjacent bones interact; also called *articulation*. (9)

juxtaglomerular cells: Modified smooth muscle cells between the walls of the distal convoluted tubule and the afferent and efferent arterioles adjacent to the glomerulus and the macula densa. (26)

juxtaglomerular complex: The macula densa and the juxtaglomerular cells; a complex responsible for the release of renin and erythropoietin. (26)

K

keratin: The tough, fibrous protein component of nails, hair, calluses, and the general integumentary surface. (5)

keto acid: The carbon chain that remains after the deamination or transamination of an amino acid. (25)

ketoacidosis: A condition characterized by a decrease in blood pH due to the presence of large numbers of ketone bodies. (25, 26, 27)

ketone bodies: Keto acids produced during the catabolism of lipids and ketogenic amino acids; specifically, acetone, acetoacetate, and beta-hydroxybutyrate. (25)

kidney: A component of the urinary system; an organ functioning in the regulation of blood composition, including the excretion of wastes and the maintenance of normal fluid and electrolyte balances. (18, 26)

killer T cells: *See* **cytotoxic T cells**.

Krebs cycle: *See* **citric acid cycle**.

Kupffer cells: Phagocytic cells of the liver sinusoids; also called *stellate reticuloendothelial cells*. (22, 24)

L

labium/labia: Lip/lips; the labia majora and labia minora are structures of the female external genitalia. (28)

labrum: A lip or rim.

labyrinth: A maze of passageways; the structures of the internal ear. (17)

lacrimal gland: A tear gland on the dorsolateral surface of the eye. (17)

lactase: An enzyme that breaks down the milk sugar, lactose. (24)

lactate: An anion released by the dissociation of lactic acid, produced from pyruvate under anaerobic conditions. (10)

lactation: The production of milk by the mammary glands. (28)

lacteal: A terminal lymphatic within an intestinal villus. (24)

lacuna: A small pit or cavity. (4, 6)

lambdoid suture: The synarthrosis between the parietal and occipital bones of the cranium. (7)

lamellae: Concentric layers; the concentric layers of bone within an osteon. (6)

lamellated corpuscle: A receptor sensitive to vibration. (15)

lamina: A thin sheet or layer.

lamina propria: The reticular tissue that underlies a mucous epithelium and forms part of a mucous membrane. (4, 23, 24)

Langerhans cells: Cells in the epithelium of the skin (15) and digestive tract (24) that participate in the immune response by presenting antigens to T cells; also called *dendritic cells.*

large intestine: The terminal portions of the intestinal tract, consisting of the colon, the rectum, and the anal canal. (24)

laryngopharynx: The division of the pharynx that is inferior to the epiglottis and superior to the esophagus. (23)

larynx: A complex cartilaginous structure that surrounds and protects the glottis and vocal cords; the superior margin is bound to the hyoid bone, and the inferior margin is bound to the trachea. (23)

latent period: The time between the stimulation of a muscle and the start of the contraction phase. (10)

lateral: Pertaining to the side.

lateral apertures: Openings in the roof of the fourth ventricle that permit the circulation of cerebrospinal fluid into the subarachnoid space. (14)

lateral ventricle: A fluid-filled chamber within a cerebral hemisphere. (14)

lens: The transparent refractive structure of the eye that is between the iris and the vitreous humor. (17)

lesion: A localized abnormality in tissue organization. (4)

lesser omentum: A double-layered peritoneal fold formed by the mesentery that connects the lesser curvature of the stomach to the liver. (24)

leukocyte: A white blood cell. (4, 19)

ligament: A dense band of connective tissue fibers that attaches one bone to another. (4, 9)

ligamentum arteriosum: The fibrous strand in adults that is the remnant of the ductus arteriosus of the fetal stage. (21)

ligamentum nuchae: An elastic ligament between the vertebra prominens and the occipital bone. (7)

ligamentum teres: The fibrous strand in the falciform ligament of adults that is the remnant of the umbilical vein; also called the *round ligament of the liver.* (24)

ligate: To tie off.

limbic system: The group of nuclei and centers in the cerebrum and diencephalon that are involved with emotional states, memories, and behavioral drives. (14)

lingual: Pertaining to the tongue. (17, 24)

lipid: An organic compound containing carbon, hydrogen, and oxygen in a ratio that does not approximate 1:2:1; includes fats, oils, and waxes. (2, 24, 25)

lipogenesis: The synthesis of lipids from nonlipid precursors. (25)

lipolysis: The catabolism of lipids as a source of energy. (25)

lipoprotein: A compound containing a relatively small lipid bound to a protein. (25)

liver: An organ of the digestive system that has varied and vital functions, including the production of plasma proteins, the excretion of bile, the storage of energy reserves, the detoxification of poisons, and the interconversion of nutrients. (24)

lobule: Histologically, the basic organizational unit of the liver. (24)

local hormone: *See* **prostaglandin.**

loop of Henle: *See* **nephron loop.** (26)

loose connective tissue: A loosely organized, easily distorted connective tissue that contains several fiber types, a varied population of cells, and a viscous ground substance. (4)

lumbar: Pertaining to the lower back. (7, 13)

lumen: The central space within a duct or other internal passageway. (4)

lungs: The paired organs of breathing enclosed by the pleural cavities. (23)

luteinizing hormone (LH): A hormone produced by the adenohypophysis (anterior lobe of the pituitary gland). In females, it assists FSH in follicle stimulation, triggers ovulation, and promotes the maintenance and secretion of endometrial glands. In males, it was formerly called *interstitial cell-stimulating hormone* because it stimulates testosterone secretion by the interstitial cells of the testes; also called *lutropin.* (18, 28)

lymph: The fluid contents of lymphatic vessels, similar in composition to interstitial fluid. (4, 22)

lymphatic vessels: The vessels of the lymphatic system; also called *lymphatics.* (4, 22)

lymph nodes: Lymphoid organs that monitor the composition of lymph. (22)

lymphocyte: A cell of the lymphatic system that plays a role in the immune response. (4, 19, 22)

lymphokines: Chemicals secreted by activated lymphocytes. (22)

lymphopoiesis: The production of lymphocytes from lymphoid stem cells. (19, 22)

lymphotoxin: A secretion of lymphocytes that kills the target cells. (22)

lysis: The destruction of a cell through the rupture of its plasma membrane. (3)

lysosome: An intracellular vesicle containing digestive enzymes. (3)

lysozyme: An enzyme, present in some exocrine secretions, that has antibiotic properties. (17)

M

macrophage: A phagocytic cell of the monocyte–macrophage system. (4, 22)

macula: A receptor complex, located in the saccule or utricle of the internal ear, that responds to linear acceleration or gravity. (17)

macula (of retina): The oval area in the retina whose center (the fovea) is the region of sharpest vision; also called *macula lutea.* (17)

macula densa: A group of specialized secretory cells that is located in a portion of the distal convoluted tubule, adjacent to the glomerulus and the juxtaglomerular cells; a component of the juxtaglomerular complex. (26)

major histocompatibility complex: *See* **MHC protein.**

malignant tumor: A form of cancer characterized by rapid cell growth and the spread of cancer cells throughout the body. (3)

malleus: An auditory ossicle, bound to the tympanic membrane and the incus. (17)

malnutrition: An unhealthy state produced by inadequate dietary intake or absorption of nutrients. (25)

mammillary bodies: Nuclei in the hypothalamus that affect eating reflexes and behaviors; a component of the limbic system. (14)

mammary glands: Milk-producing glands of the female breasts. (5, 28)

manus: The hand; manual. (8, 11)

marrow: A tissue that fills the internal cavities in bone; dominated by hemopoietic cells (red bone marrow) or by adipose tissue (yellow bone marrow). (6, 19)

mast cell: A connective tissue cell that, when stimulated, releases histamine, serotonin, and heparin, initiating the inflammatory response. (4)

mastication: Chewing. (11, 24)

mastoid sinus: Air-filled spaces in the mastoid process of the temporal bone. (7)

matrix: The extracellular fibers and ground substance of a connective tissue. (4)

maxillary sinus: One of the paranasal sinuses; an air-filled chamber lined by a respiratory epithelium that is located in a maxilla and opens into the nasal cavity. (7)

meatus: An opening or entrance into a passageway. (23, 26)

mechanoreception: The detection of mechanical stimuli, such as touch, pressure, or vibration. (15)

medial: Toward the midline of the body.

mediastinum: The central tissue mass that divides the thoracic cavity into two pleural cavities. (1, 20)

medulla: The inner layer or core of an organ.

medulla oblongata: The most caudal of the brain regions, also called the *myelencephalon.* (14)

medullary cavity: The space within a bone that contains the marrow. (6)

medullary rhythmicity center: The center in the medulla oblongata that sets the background pace of respiration; includes inspiratory and expiratory centers. (14)

megakaryocytes: Bone marrow cells responsible for the formation of platelets. (19)

meiosis: Cell division that produces gametes with half the normal somatic chromosome complement. (3, 28)

melanin: The yellow-brown pigment produced by the melanocytes of the skin. (5)

melanocyte: A specialized cell in the deeper layers of the stratified squamous epithelium of the skin; responsible for the production of melanin. (4, 5, 18)

melanocyte-stimulating hormone (MSH): A hormone, produced by the pars intermedia of the anterior lobe of the pituitary gland (adenohypophysis), that stimulates melanin production. (18)

melatonin: A hormone secreted by the pineal gland; inhibits secretion of MSH and GnRH. (14, 18)

membrane: Any sheet or partition; a layer consisting of an epithelium and the underlying connective tissue. (2)

membrane flow: The continuous movement and exchange of membrane segments. (3)

membrane potential: The potential difference, measured across a plasma membrane and expressed in millivolts, that results from the uneven distribution of positive and negative ions across the plasma membrane; also called *transmembrane potential.* (3, 10, 12)

membranous labyrinth: Endolymph-filled tubes that enclose the receptors of the internal ear. (17)

memory: The ability to recall information or sensations; can be divided into short-term and long-term memories. (16, 22)

meninges: Three membranes that surround the surfaces of the central nervous system; the dura mater, the pia mater, and the arachnoid. (13)

meniscus: A fibrocartilage pad between opposing surfaces in a joint. (9)

menses: The first part of the uterine cycle in which the endometrial lining sloughs away; menstrual period. (28)

merocrine: A method of secretion in which the cell ejects materials from secretory vesicles through exocytosis. (4, 5)

mesencephalon: The midbrain; the region between the diencephalon and pons. (14)

mesenchyme: Embryonic or fetal connective tissue. (4)

mesentery: A double layer of serous membrane that supports and stabilizes the position of an organ in the abdominopelvic cavity and provides a route for the associated blood vessels, nerves, and lymphatic vessels. (24)

mesoderm: The middle germ layer, between the ectoderm and endoderm of the embryo. (29)

mesothelium: A simple squamous epithelium that lines the body cavities enclosing the lungs, heart, and abdominal organs. (4)

messenger RNA (mRNA): RNA formed at transcription to direct protein synthesis in the cytoplasm. (2, 3)

metabolic turnover: The continuous breakdown and replacement of organic materials within cells. (2, 25)

metabolism: The sum of all biochemical processes under way within the human body at any moment; includes anabolism and catabolism. (25)

metabolites: Compounds produced in the body as a result of metabolic reactions. (2)

metacarpal bones: The five bones of the palm of the hand. (8)

metalloproteins: Proteins containing a metal ion cofactor; examples include plasma transport proteins, storage proteins, and enzymes. (19, 25)

metaphase: The stage of mitosis in which the chromosomes line up along the equatorial plane of the cell. (3)

metaphysis: The region of a long bone between the epiphysis and diaphysis, corresponding to the location of the epiphyseal cartilage of the developing bone. (6)

metarteriole: A vessel that connects an arteriole to a venule and that provides blood to a capillary plexus. (21)

metastasis: The spread of cancer cells from one organ to another, leading to the establishment of secondary tumors. (3)

metatarsal bone: One of the five bones of the foot that articulate with the tarsal bones (proximally) and the phalanges (distally). (8)

metencephalon: The pons and cerebellum of the brain. (14)

MHC protein: A surface antigen that is important to the recognition of foreign antigens and that plays a role in the coordination and activation of the immune response; also called *major histocompatibility complex (MHC) protein* or *human leukocyte antigen (HLA).* (22)

micelle: A droplet with hydrophilic portions on the outside; a spherical aggregation of bile salts, monoglycerides, and fatty acids in the lumen of the intestinal tract. (2, 24)

microfilaments: Fine protein filaments visible with the electron microscope; components of the cytoskeleton. (3)

microglia: Phagocytic neuroglia in the central nervous system. (12, 22)

microphages: Neutrophils and eosinophils. (4, 19, 22)

microtubules: Microscopic tubules that are part of the cytoskeleton and are a component in cilia, flagella, the centrioles, and spindle fibers. (3)

microvilli: Small, fingerlike extensions of the exposed plasma membrane of an epithelial cell. (3)

micturition: Urination. (26)

midbrain: The mesencephalon. (14)

middle ear: The space between the external and internal ears that contains auditory ossicles. (17)

midsagittal plane: A plane passing through the midline of the body that divides it into left and right halves. (1)

mineralocorticoids: Corticosteroids produced by the zona glomerulosa of the adrenal cortex; steroids such as aldosterone that affect mineral metabolism. (18)

mitochondrion: An intracellular organelle responsible for generating most of the ATP required for cellular operations. (3, 25)

mitosis: The division of a single cell nucleus that produces two identical daughter cell nuclei; an essential step in cell division. (3, 28)

mitral valve: *See* **bicuspid valve**.

mixed gland: A gland that contains exocrine and endocrine cells, or an exocrine gland that produces serous and mucous secretions. (4)

mixed nerve: A peripheral nerve that contains sensory and motor fibers. (13)

mole: A quantity of an element or compound having a mass in grams equal to the element's atomic weight or to the compound's molecular weight. (2)

molecular weight: The sum of the atomic weights of all the atoms in a molecule or compound. (3)

molecule: A chemical structure containing two or more atoms that are bonded together by shared electrons. (2)

monoclonal antibodies: Antibodies produced by genetically identical cells under laboratory conditions. (22)

monocytes: Phagocytic agranulocytes (white blood cells) in the circulating blood. (19)

monoglyceride: A lipid consisting of a single fatty acid bound to a molecule of glycerol. (2)

monokines: Secretions released by activated cells of the monocyte–macrophage system to coordinate various aspects of the immune response. (22)

monosaccharide: A simple sugar, such as glucose or ribose. (2, 24)

monosynaptic reflex: A reflex in which the sensory afferent neuron synapses directly on the motor efferent neuron. (13)

motor unit: All the muscle cells controlled by a single motor neuron. (10)

mucins: Proteoglycans responsible for the lubricating properties of mucus. (2, 24)

mucosa: A mucous membrane; the epithelium plus the lamina propria. (4, 24)

mucosa-associated lymphoid tissue (MALT): The extensive collection of lymphoid tissues linked with the epithelia of the digestive, respiratory, urinary, and reproductive tracts. (22)

mucous (adjective)**:** Indicating the presence or production of mucus.

mucous cell: A goblet-shaped, mucus-producing, unicellular gland in certain epithelia of the digestive and respiratory tracts; also called *goblet cells*. (4)

mucous membrane: *See* **mucosa**.

mucus (noun)**:** A lubricating fluid that is composed of water and mucins and is produced by unicellular and multicellular glands along the digestive, respiratory, urinary, and reproductive tracts. (2, 4)

multipolar neuron: A neuron with many dendrites and a single axon; the typical form of a motor neuron. (12)

multiunit smooth muscle: A smooth muscle tissue whose muscle cells are innervated in motor units. (10)

muscarinic receptors: Membrane receptors sensitive to acetylcholine and to muscarine, a toxin produced by certain mushrooms; located at all parasympathetic neuromuscular and neuroglandular junctions and at a few sympathetic neuromuscular and neuroglandular junctions. (16)

muscle: A contractile organ composed of muscle tissue, blood vessels, nerves, connective tissues, and lymphatic vessels. (10, 11)

muscle tissue: A tissue characterized by the presence of cells capable of contraction; includes skeletal, cardiac, and smooth muscle tissues. (4, 10)

muscularis externa: Concentric layers of smooth muscle responsible for peristalsis. (24)

muscularis mucosae: The layer of smooth muscle deep to the lamina propria; responsible for moving the mucosal surface. (24)

mutagens: Chemical agents that induce mutations and may be carcinogenic. (3)

mutation: A change in the nucleotide sequence of the DNA in a cell. (3)

myelencephalon: *See* **medulla oblongata**.

myelin: An insulating sheath around an axon; consists of multiple layers of neuroglial membrane; significantly increases the nerve impulse propagation rate along the axon. (12)

myelination: The formation of myelin. (12)

myenteric plexus: Parasympathetic motor neurons and sympathetic postganglionic fibers located between the circular and longitudinal layers of the muscularis externa. (24)

myocardial infarction: A heart attack; damage to the heart muscle due to an interruption of regional coronary circulation. (20)

myocardium: The cardiac muscle tissue of the heart. (20)

myofibril: Organized collections of myofilaments in skeletal and cardiac muscle cells. (10)

myofilaments: Fine protein filaments composed primarily of the proteins actin (thin filaments) and myosin (thick filaments). (10)

myoglobin: An oxygen-binding pigment that is especially common in slow skeletal muscle fibers and cardiac muscle cells. (2, 10)

myogram: A recording of the tension produced by muscle fibers on stimulation. (10)

myometrium: The thick layer of smooth muscle in the wall of the uterus. (28)

myosin: The protein component of thick filaments. (3, 10)

N

nail: A keratinous structure produced by epithelial cells of the nail root. (5)

nares, external: The entrance from the exterior to the nasal cavity. (23)

nares, internal: The entrance from the nasal cavity to the nasopharynx. (23)

nasal cavity: A chamber in the skull that is bounded by the internal and external nares. (7)

nasolacrimal duct: The passageway that transports tears from the nasolacrimal sac to the nasal cavity. (7, 17)

nasolacrimal sac: A chamber that receives tears from the lacrimal ducts. (17)

nasopharynx: A region that is posterior to the internal nares and superior to the soft palate and ends at the oropharynx. (23)

natriuretic peptides (NP): Hormones released by specialized cardiocytes when they are stretched by an abnormally large venous return; promotes fluid loss and reductions in blood pressure and in venous return. Includes atrial natriuretic peptide (ANP) and brain natriuretic peptide (BNP). (18, 21, 26, 27)

N compound: An organic compound containing nitrogen atoms. (25)

necrosis: The death of cells or tissues from disease or injury. (4, 22)

negative feedback: A corrective mechanism that opposes or negates a variation from normal limits. (18)

neonate: A newborn infant, or baby. (29)

neoplasm: A tumor, or mass of abnormal tissue. (3)

nephron: The basic functional unit of the kidney. (26)

nephron loop: The segment of the nephron that creates the concentration gradient in the renal medulla; also called *loop of Henle*. (26)

nerve impulse: An action potential in a neuron plasma membrane. (12)

neural cortex: An area of gray matter at the surface of the central nervous system. (13)

neurilemma: The outer surface of a Schwann cell that encircles an axon in the peripheral nervous system. (12)

neurofibrils: Microfibrils in the cytoplasm of a neuron. (12)

neurofilaments: Microfilaments in the cytoplasm of a neuron. (12)

neuroglandular junction: A cell junction at which a neuron controls or regulates the activity of a secretory (gland) cell. (12)

neuroglia: Cells of the central nervous system and peripheral nervous system that support and protect neurons; also called *glial cells*. (4, 12)

neurohypophysis: The posterior lobe of the pituitary gland, or pars nervosa; contains the axons of hypothalamic neurons, which release OXT and ADH. (18)

neuromodulator: A compound, released by a neuron, that adjusts the sensitivities of another neuron to specific neurotransmitters. (12)

neuromuscular junction: A synapse between a neuron and a muscle cell. (10, 12)

neuron: A cell in neural tissue that is specialized for intercellular communication through (1) changes in membrane potential and (2) synaptic connections. (4, 12, 15, 16)

neuroplasticity: The ability of nerve cells to make new connections. (14)

neurotransmitter: A chemical compound released by one neuron to affect the membrane potential of another. (12, 16)

neurotubules: Microtubules in the cytoplasm of a neuron. (12)

neutron: A fundamental particle that does not have a positive or a negative charge. (2)

neutrophil: A white blood cell that is very numerous and normally the first of the mobile phagocytic cells to arrive at an area of injury or infection. (19)

nicotinic receptors: Acetylcholine receptors on the surfaces of sympathetic and parasympathetic ganglion cells; respond to the compound nicotine. (16)

nipple: An elevated epithelial projection on the surface of the breast; contains the openings of the lactiferous sinuses. (28)

Nissl bodies: The ribosomes, Golgi apparatus, rough endoplasmic reticulum, and mitochondria of the perikaryon of a typical neuron. (12)

nitrogenous wastes: Organic waste products of metabolism that contain nitrogen, such as urea, uric acid, and creatinine. (25)

nociception: Pain perception. (15)

nodes: The areas between adjacent neuroglia where the myelin covering of an axon is incomplete; also called *nodes of Ranvier*. (12)

nodose ganglion: A sensory ganglion of cranial nerve X; also called *inferior ganglion*. (14)

noradrenaline: *See* **norepinephrine**.

norepinephrine (NE): A catecholamine neurotransmitter in the peripheral nervous system and central nervous system, released at most sympathetic neuromuscular and neuroglandular junctions, and a hormone secreted by the adrenal (suprarenal) medulla; also called *noradrenaline*. (12, 18)

nucleic acid: A polymer of nucleotides that contains a pentose sugar, a phosphate group, and one of four nitrogenous bases that regulate the synthesis of proteins and make up the genetic material in cells. (2)

nucleolus: The dense region in the nucleus that is the site of RNA synthesis. (3)

nucleoplasm: The fluid content of the nucleus. (3)

nucleoproteins: Proteins of the nucleus that are generally associated with DNA. (3)

nucleotide: A compound consisting of a nitrogenous base, a simple sugar, and a phosphate group. (2)

nucleus: A cellular organelle that contains DNA, RNA, and proteins; in the central nervous system, a mass of gray matter. (3)

nucleus pulposus: The gelatinous central region of an intervertebral disc. (9)

nurse cells: Supporting cells of the seminiferous tubules of the testis; responsible for the differentiation of spermatids, the maintenance of the blood–testis barrier, and the secretion of inhibin, androgen-binding protein, and Müllerian-inhibiting factor; also called *sustentacular cells*. (18, 28)

nutrient: An inorganic or organic substance that can be used by the body as a cofactor or to produce energy. (2, 25)

nystagmus: An unconscious, continuous movement of the eyes as if to adjust to constant motion. (17)

O

obesity: Body weight more than 20 percent above the ideal weight for a given individual. (25)

occlusal surface: The opposing surfaces of the teeth that come into contact when chewing food. (24)

ocular: Pertaining to the eye. (17)

oculomotor nerve: Cranial nerve III, which controls the extraocular muscles other than the superior oblique and the lateral rectus muscles. (14)

olecranon: The proximal end of the ulna that forms the prominent point of the elbow. (8)

olfaction: The sense of smell. (15, 17, 23)

olfactory bulb: The expanded ends of the olfactory tracts (17); the sites where the axons of the first cranial nerves (I) synapse on central nervous system interneurons that lie inferior to the frontal lobes of the cerebrum. (14)

oligodendrocytes: Central nervous system neuroglia that maintain cellular organization within gray matter and provide a myelin sheath in areas of white matter. (12)

oligopeptide: A short chain of amino acids. (2)

oocyte: A cell whose meiotic divisions will produce a single ovum and three polar bodies. (3, 28)

oogenesis: Formation and development of an oocyte. (28)

opsonization: An effect of coating an object with antibodies; the attraction and enhancement of phagocytosis. (22)

optic chiasm: The crossing point of the optic nerves. (14)

optic nerve: The second cranial nerve (II), which carries signals from the retina of the eye to the optic chiasm. (14)

optic tract: The tract over which nerve impulses from the retina are transmitted between the optic chiasm and the thalamus. (14)

orbit: The bony recess of the skull that contains the eyeball. (7)

organelle: An intracellular structure with a specific function or group of functions. (3)

organic compound: A compound containing carbon, hydrogen, and in most cases oxygen. (2)

organogenesis: The formation of organs during embryonic and fetal development. (29)

organs: Combinations of tissues that perform complex functions. (1)

origin: In a skeletal muscle, the point of attachment that does not change position when the muscle contracts; usually defined in terms of movements from the anatomical position. (11)

oropharynx: The middle portion of the pharynx, bounded superiorly by the nasopharynx, anteriorly by the oral cavity, and inferiorly by the laryngopharynx. (23)

osmolarity: The total concentration of dissolved materials in a solution, regardless of their specific identities, expressed in moles; also called *osmotic concentration*. (3, 26, 27)

osmoreceptor: A receptor sensitive to changes in the osmolarity of plasma. (27)

osmosis: The movement of water across a selectively permeable membrane from one solution to another solution that contains a higher solute concentration. (3, 21, 26, 27)

osmotic pressure: The force of osmotic water movement; the pressure that must be applied to prevent osmosis across a membrane. (3, 21, 26, 27)

osseous tissue: A strong connective tissue containing specialized cells and a mineralized matrix of crystalline calcium phosphate and calcium carbonate; also called *bone*. (4, 6)

ossicles: Small bones; middle ear bones. (17)

ossification: The formation of bone; osteogenesis. (6)

osteoblast: A cell that produces the fibers and matrix of bone. (6)

osteoclast: A cell that dissolves the fibers and matrix of bone. (6)

osteocyte: A bone cell responsible for the maintenance and turnover of the mineral content of the surrounding bone. (4, 6)

osteogenic layer: The inner, cellular layer of the periosteum that aids in bone growth and repair. (6)

osteolysis: The breakdown of the mineral matrix of bone. (6)

osteon: The basic histological unit of compact bone, consisting of osteocytes organized around a central canal and separated by concentric lamellae. (6)

otic: Pertaining to the ear. (17)

otoliths: Calcium carbonate crystals embedded in a gelatinous matrix; located on each macula of the vestibule. (17)

oval window: An opening in the bony labyrinth where the stapes attaches to the membranous wall of the scala vestibuli (vestibular duct). (17)

ovarian cycle: The monthly chain of events that leads to ovulation. (28)

ovary: The female reproductive organ that produces oocytes (gametes). (18, 28)

ovulation: The release of a secondary oocyte, surrounded by cells of the corona radiata, after the rupture of the wall of a tertiary follicle (29); in females, the periodic release of an oocyte from an ovary (28).

ovum/ova: The functional product of meiosis II, produced after the fertilization of a secondary oocyte. (28, 29)

oxytocin (OXT): A hormone produced by hypothalamic cells and secreted into capillaries at the posterior lobe of the pituitary gland (neurohypophysis); stimulates smooth muscle contractions of the uterus or mammary glands in females and the prostate gland in males. (18)

P

pacemaker cells: Cells of the sinoatrial node that set the pace of cardiac contraction. (4, 10, 20)

palate: The horizontal partition separating the oral cavity from the nasal cavity and nasopharynx;

divided into an anterior bony (hard) palate and a posterior fleshy (soft) palate. (7, 24)

palatine: Pertaining to the palate. (24)

palpate: To examine by touch.

palpebrae: Eyelids. (17)

pancreas: A digestive organ containing exocrine and endocrine tissues; the exocrine portion secretes pancreatic juice, and the endocrine portion secretes hormones, including insulin and glucagon. (18, 24)

pancreatic duct: A tubular duct that carries pancreatic juice from the pancreas to the duodenum. (18, 24)

pancreatic islets: Aggregations of endocrine cells in the pancreas; also called *islets of Langerhans*. (18, 24)

pancreatic juice: A mixture of buffers and digestive enzymes that is discharged into the duodenum under the stimulation of the enzymes secretin and cholecystokinin. (18, 24)

Papanicolaou (Pap) test: A test for the detection of malignancies based on the cytological appearance of epithelial cells, especially those of the cervix and uterus. (28)

papilla: A small, conical projection.

paralysis: The loss of voluntary motor control over a portion of the body. (13)

paranasal sinuses: Bony chambers, lined by respiratory epithelium, that open into the nasal cavity; the frontal, ethmoidal, sphenoidal, and maxillary sinuses. (7)

parasagittal: A section or plane that parallels the midsagittal plane but that does not pass along the midline. (1)

parasympathetic division: One of the two divisions of the autonomic nervous system; generally responsible for activities that conserve energy and lower the metabolic rate; the "rest and digest" division; also called *craniosacral division*. (16)

parathyroid glands: Four small glands embedded in the posterior surface of the thyroid gland that secrete parathyroid hormone. (6, 18)

parathyroid hormone (PTH): A hormone secreted by the parathyroid glands when blood calcium levels decrease below the normal range; causes increased osteoclast activity, increased intestinal calcium uptake, and decreased calcium ion loss by the kidneys. (6, 18)

parenchyma: The cells of a tissue or organ that are responsible for fulfilling its functional role; distinguished from the stroma of that tissue or organ. (4)

paresthesia: A sensory abnormality that produces a tingling sensation.

parietal: Relating to the parietal bone (7); referring to the wall of a cavity. (23)

parietal cells: Cells of the gastric glands that secrete hydrochloric acid and intrinsic factor. (24)

parotid salivary glands: Large salivary glands that secrete a saliva containing high concentrations of salivary (alpha) amylase. (24)

pars distalis: The large, anterior portion of the anterior lobe of the pituitary gland (adenohypophysis). (18)

pars intermedia: The portion of the anterior lobe of the pituitary gland (adenohypophysis) that is immediately adjacent to the posterior lobe of the pituitary gland (neurohypophysis) and the infundibulum. (18)

pars nervosa: The posterior lobe of the pituitary gland (neurohypophysis). (18)

pars tuberalis: The portion of the anterior lobe of the pituitary gland (adenohypophysis) that wraps around the infundibulum superior to the posterior lobe (neurohypophysis). (18)

patella: The sesamoid bone of the knee; also called the *kneecap*. (8)

pathogen: A disease-causing organism. (1, 22)

pathologist: A physician specializing in the identification of diseases on the basis of characteristic structural and functional changes in tissues and organs.

pelvic cavity: The inferior subdivision of the abdominopelvic cavity; encloses the urinary bladder, the sigmoid colon and rectum, and male or female reproductive organs. (1, 8)

pelvis: A bony complex created by the articulations among the coxal bones, the sacrum, and the coccyx. (8, 11)

penis: A component of the male external genitalia; a copulatory organ that surrounds the urethra and introduces semen into the female vagina; the developmental equivalent of the female clitoris. (28)

peptide: A chain of amino acids linked by peptide bonds. (2, 18)

peptide bond: A covalent bond between the amino group of one amino acid and the carboxyl group of another. (2)

pericardial cavity: The space between the parietal pericardium and the epicardium (visceral pericardium) that encloses the outer surface of the heart. (20)

pericardium: The fibrous sac that surrounds the heart; its inner, serous lining is continuous with the epicardium. (4, 20)

perichondrium: The layer that surrounds a cartilage, consisting of an outer fibrous region and an inner cellular region. (4)

perikaryon: The cytoplasm that surrounds the nucleus in the cell body of a neuron. (12)

perilymph: A fluid similar in composition to cerebrospinal fluid; located in the spaces between the bony labyrinth and the membranous labyrinth of the internal ear. (17)

perimysium: A connective tissue partition that separates adjacent fasciculi in a skeletal muscle. (10)

perineum: The pelvic floor and its associated structures. (11)

perineurium: A connective tissue partition that separates adjacent bundles of nerve fibers in a peripheral nerve. (13)

periodontal ligament: Collagen fibers that bind the cementum of a tooth to the periosteum of the surrounding alveolus. (24)

periosteum: The layer that surrounds a bone, consisting of an outer fibrous region and inner cellular region. (4, 6)

peripheral nervous system (PNS): All neural tissue outside the central nervous system. (12)

peripheral resistance: The resistance to blood flow; primarily caused by friction with the vessel walls. (21)

peristalsis: A wave of smooth muscle contractions that propels materials along the lumen of a tube such as the digestive tract (24), the ureters (26), or the ductus deferens. (28)

peritoneal cavity: The potential space within the abdominopelvic cavity lined by the peritoneum.

peritoneum: The serous membrane that lines the peritoneal cavity. (4, 28)

peritubular capillaries: A network of capillaries that surrounds the proximal and distal convoluted tubules of the kidneys. (26)

permeability: The ease with which dissolved materials can cross a membrane; if the membrane is freely permeable, any molecule can cross it; if impermeable, nothing can cross; most biological membranes are selectively permeable (allow some and restrict others). (3)

peroxisome: A membranous vesicle containing enzymes that break down hydrogen peroxide (H_2O_2). (3)

pes: The foot. (8, 11)

petrosal ganglion: A sensory ganglion of the glossopharyngeal nerve (N IX). (14, 15)

petrous: Stony; usually refers to the thickened portion of the temporal bone that encloses the internal ear. (17)

pH: The negative exponent (negative logarithm) of the hydrogen ion concentration, expressed in moles per liter. (2, 27)

phagocyte: A cell that performs phagocytosis. (22)

phagocytosis: The engulfing of extracellular materials or pathogens; the movement of extracellular materials into the cytoplasm by enclosure in a membranous vesicle. (3, 19, 22)

phalanx/phalanges: Bone(s) of the finger(s) or toe(s). (8)

pharmacology: The study of drugs, their physiological effects, and their clinical uses.

pharynx: The throat; a muscular passageway shared by the digestive and respiratory tracts. (11, 23, 24)

phasic response: A pattern of response to stimulation by sensory neurons that are normally inactive; stimulation causes a burst of neural activity that ends when the stimulus either stops or stops changing in intensity. (15)

phenotype: Physical characteristics that are genetically determined. (29)

phosphate group: PO_4^{3-}; a functional group that can be attached to an organic molecule; required for the formation of high-energy bonds. (2, 25, 27)

phospholipid: An important membrane lipid whose structure includes both hydrophilic and hydrophobic regions. (2, 3)

phosphorylation: The addition of a high-energy phosphate group to a molecule. (2, 25)

photoreception: Sensitivity to light. (17)

physiology: The study of function; deals with the ways organisms perform vital activities. (1)

pia mater: The innermost layer of the meninges bound to the underlying neural tissue. (13, 14)

pineal gland: Neural tissue in the posterior portion of the roof of the diencephalon; secretes melatonin. (14, 18)

pinna: *See* **auricle**.

pinocytosis: The introduction of fluids into the cytoplasm by enclosing them in membranous vesicles at the cell surface. (3)

pituitary gland: An endocrine organ that is located in the sella turcica of the sphenoid bone and is connected to the hypothalamus by the infundibulum; includes the posterior lobe (neurohypophysis) and the anterior lobe (adenohypophysis); also called the *hypophysis*. (14, 18)

placenta: A temporary structure in the uterine wall that permits diffusion between the fetal and maternal circulatory systems. (29)

plantar: Referring to the sole of the foot (1); muscles (11); plantar reflex. (13)

plantar flexion: Ankle extension; toe pointing. (8, 11)

plasma: The fluid ground substance of whole blood; what remains after the cells have been removed from a sample of whole blood. (4, 19)

plasma cell: An activated B cell that secretes antibodies. (4, 19, 22)

plasma membrane: A cell membrane; also called a *plasmalemma*. (3)

platelets: Small packets of cytoplasm that contain enzymes important in the clotting response; manufactured in bone marrow by megakaryocytes. (4, 19)

pleura: The serous membrane that lines the pleural (lung) cavities. (4, 23)

pleural cavities: Body cavities of the thoracic region that surround the lungs. (1, 23)

plexus: A network or braid.

polar body: A nonfunctional packet of cytoplasm that contains chromosomes eliminated from an oocyte during meiosis. (28, 29)

polar bond: A covalent bond in which electrons are shared unequally. (2)

polarized: Referring to cells that have regional differences in organelle distribution or cytoplasmic composition along a specific axis, such as between the basement membrane and free surface of an epithelial cell. (4)

pollex: The thumb. (8)

polymer: A large molecule consisting of a long chain of monomer subunits. (2)

polypeptide: A chain of amino acids strung together by peptide bonds; those containing more than 100 peptides are called *proteins*. (2)

polyribosome: Several ribosomes linked by their translation of a single mRNA strand. (3)

polysaccharide: A complex sugar, such as glycogen or a starch. (2)

polysynaptic reflex: A reflex in which interneurons are interposed between the sensory fiber and the motor neuron(s). (13)

polyunsaturated fats: Fatty acids containing carbon atoms that are linked by double bonds. (1, 2)

pons: The portion of the metencephalon that is anterior to the cerebellum. (14)

popliteal: Pertaining to the back of the knee. (9, 11, 21)

porphyrins: Ring-shaped molecules that form the basis of important respiratory and metabolic pigments, including heme and the cytochromes. (23)

positive feedback: A mechanism that increases a deviation from normal limits after an initial stimulus. (1)

postcentral gyrus: The primary sensory cortex, where touch, vibration, pain, temperature, and taste sensations arrive and are consciously perceived. (14)

posterior: Toward the back; dorsal.

postganglionic neuron: An autonomic neuron in a peripheral ganglion, whose activities control peripheral effectors. (16)

postsynaptic membrane: The portion of the plasma membrane of a postsynaptic cell that is part of a synapse. (12)

potential difference: The separation of opposite charges; requires a barrier that prevents ion migration. (3, 12)

precentral gyrus: The primary motor cortex of a cerebral hemisphere, located anterior to the central sulcus. (14)

prefrontal cortex: The anterior portion of each cerebral hemisphere; thought to be involved with higher intellectual functions, predictions, and calculations. (14)

preganglionic neuron: A visceral motor neuron in the central nervous system whose output controls one or more ganglionic motor neurons in the peripheral nervous system. (16)

premotor cortex: The motor association area between the precentral gyrus and the prefrontal area. (14)

preoptic nucleus: The hypothalamic nucleus that coordinates thermoregulatory activities. (14)

presynaptic membrane: The synaptic surface where neurotransmitter release occurs. (12)

prevertebral ganglion: *See* **collateral ganglion.**

prime mover: A muscle that performs a specific action. (11)

proenzyme: An inactive enzyme secreted by an epithelial cell. (19)

progesterone: The most important hormone secreted by the corpus luteum after ovulation. (18, 28)

prognosis: A prediction about the possible course or outcome from a specific disease.

projection fibers: Axons carrying information from the thalamus to the cerebral cortex. (14)

prolactin (PRL): The hormone that stimulates functional development of the mammary glands in females; a secretion of the anterior lobe of the pituitary gland (adenohypophysis). (18)

pronation: The rotation of the forearm that makes the palm face posteriorly. (9)

prone: Lying face down with the palms facing the floor. (1)

pronucleus: An enlarged ovum or spermatozoon nucleus that forms after fertilization but before amphimixis. (29)

prophase: The initial phase of mitosis; characterized by the appearance of chromosomes, the breakdown of the nuclear membrane, and the formation of the spindle apparatus. (3)

proprioception: The awareness of the positions of bones, joints, and muscles. (15)

prostaglandin: A fatty acid secreted by one cell that alters the metabolic activities or sensitivities of adjacent cells; also called *local hormone*. (2, 18)

prostate gland: An accessory gland of the male reproductive tract, contributing about one-third of the volume of semen. (28)

prosthesis: An artificial substitute for a body part.

protease: *See* **proteinase.**

protein: A large polypeptide with a complex structure. (2, 25)

proteinase: An enzyme that breaks down proteins into peptides and amino acids. (2, 3)

proteoglycan: A substance containing a large polysaccharide complex attached to a relatively small protein; examples include hyaluronan and chondroitin sulfate. (2)

proton: A fundamental subatomic particle having a positive charge. (2)

protraction: Movement anteriorly in the horizontal plane.

proximal: A direction toward the point of attachment or origin; for a limb, toward its attachment to the trunk. (1, 8)

proximal convoluted tubule (PCT): The segment of the nephron between the glomerular capsule (Bowman's capsule) and the nephron loop; the major site of active reabsorption from filtrate. (26)

pseudopodia: Temporary cytoplasmic extensions typical of mobile or phagocytic cells. (3)

pseudostratified epithelium: An epithelium that contains several layers of nuclei but whose cells are all in contact with the underlying basement membrane. (4)

puberty: A period of rapid growth, sexual maturation, and the appearance of secondary sexual characteristics; normally occurs at ages 10–15 years. (18, 28)

pubic symphysis: The fibrocartilaginous amphiarthrosis between the pubic bones of the hip. (8, 9)

pubis: The anterior, inferior component of the hip bone. (8)

pudendum: The external genitalia. (28)

pulmonary circuit: Blood vessels between the pulmonary semilunar valve of the right ventricle and the entrance to the left atrium; the blood flow through the lungs. (20)

pulmonary ventilation: The movement of air into and out of the lungs. (23)

pulvinar nucleus: The thalamic nucleus involved in the integration of sensory information prior to projection to the cerebral hemispheres. (14)

pupil: The opening in the center of the iris through which light enters the eye. (17)

purine: A nitrogen compound with a double ring-shaped structure; examples include adenine and guanine, two nitrogenous bases that are common in nucleic acids. (2, 12)

Purkinje cell: A large, branching neuron of the cerebellar cortex. (14)

Purkinje fibers: Specialized conducting cardiocytes in the ventricles of the heart. (20)

pus: An accumulation of debris, fluid, dead and dying cells, and necrotic tissue. (4, 20, 22)

pyloric sphincter: A ring of smooth muscle that regulates the passage of chyme from the stomach to the duodenum. (24)

pylorus: The gastric region between the body of the stomach and the duodenum; includes the pyloric sphincter. (24)

pyrimidine: A nitrogen compound with a single ring-shaped structure; examples include cytosine, thymine, and uracil, nitrogenous bases that are common in nucleic acids. (2)

pyruvate: The anion formed by the dissociation of pyruvic acid, a three-carbon compound produced by glycolysis. (25)

Q

quaternary structure: The three-dimensional protein structure produced by interactions between protein subunits. (2)

R

radiodensity: The relative resistance to the passage of x-rays. (1)

radiographic techniques: Methods of visualizing internal structures by using various forms of radiational energy. (1)

radiopaque: Having a high radiodensity. (1)

rami communicantes: Axon bundles that link the spinal nerves with the ganglia of the sympathetic chain. (8)

ramus/rami: Branch/branches.

raphe: A seam. (11, 28)

receptive field: The area monitored by a single sensory receptor. (15)

rectum: The inferior 15 cm (6 in.) of the digestive tract. (24)

rectus: Straight.

red blood cell (RBC): *See* **erythrocyte.**

reduction: The gain of hydrogen atoms or electrons or the loss of an oxygen molecule. (25)

reductional division: The first meiotic division, which reduces the chromosome number from 46 to 23. (28)

reflex: A rapid, automatic response to a stimulus. (12, 13, 16, 18, 21)

reflex arc: The receptor, sensory neuron, motor neuron, and effector involved in a particular reflex; interneurons may be present, depending on the reflex considered. (13)

refractory period: The time between the initiation of an action potential and the restoration of the normal resting membrane potential; during this period, the membrane will not respond normally to stimulation. (12)

relaxation phase: The period after a contraction when the tension in the muscle fiber returns to resting levels. (10)

relaxin: A hormone that loosens the pubic symphysis; secreted by the placenta. (29)

renal: Pertaining to the kidneys. (26)

renal corpuscle: The initial segment of the nephron, consisting of an expanded chamber that encloses the glomerulus. (26)

renin: The enzyme released by cells of the juxtaglomerular complex when renal blood flow decreases; converts angiotensinogen to angiotensin I. (18, 26)

rennin: A gastric enzyme that breaks down milk proteins. (24)

replication: Duplication. (29)

repolarization: The movement of the membrane potential away from a positive value and toward the resting potential. (12, 20)

respiration: The exchange of gases between cells and the environment; includes pulmonary ventilation, external respiration, and internal respiration. (23, 27)

respiratory minute volume ($\dot{V}_E$): The amount of air moved into and out of the respiratory system each minute. (23)

respiratory pump: A mechanism by which changes in the intrapleural pressures during the respiratory cycle assist the venous return to the heart; also called *thoracoabdominal pump*. (21, 23)

resting membrane potential: The membrane potential of a normal cell under homeostatic conditions. (3, 12)

rete: An interwoven network of blood vessels or passageways. (28)

reticular activating system (RAS): The mesencephalic (midbrain) portion of the reticular formation; responsible for arousal and the maintenance of consciousness. (16)

reticular formation: A diffuse network of gray matter that extends the entire length of the brain stem. (14)

reticulospinal tracts: Descending tracts of the medial pathway that carry involuntary motor commands issued by neurons of the reticular formation. (15)

retina: The innermost layer of the eye, lining the vitreous chamber; also called *inner layer*. (17)

retinal: A visual pigment derived from vitamin A. (17)

retraction: Movement posteriorly in the horizontal plane.

retroperitoneal: Behind or outside the peritoneal cavity. (1)

reverberation: A positive feedback along a chain of neurons such that they remain active once stimulated. (13)

rheumatism: A general term used to describe pain in muscles, tendons, bones, or joints. (9)

Rh factor: A surface antigen that may be present (Rh-positive) or absent (Rh-negative) from the surfaces of red blood cells. (19)

rhodopsin: The visual pigment in the membrane discs of the distal segments of rods. (17)

rhythmicity center: A medullary center responsible for the pace of respiration; includes inspiratory and expiratory centers. (23)

ribonucleic acid: A nucleic acid consisting of a chain of nucleotides that contain the sugar ribose and the nitrogenous bases adenine, guanine, cytosine, and uracil. (2, 3)

ribose: A five-carbon sugar that is a structural component of RNA. (2, 3)

ribosome: An organelle that contains rRNA and proteins and is essential to mRNA translation and protein synthesis. (2, 3)

rod: A photoreceptor responsible for vision in dim lighting. (17)

rough endoplasmic reticulum (RER): A membranous organelle that is a site of protein synthesis and storage. (3)

round window: An opening in the bony labyrinth of the internal ear that exposes the membranous wall of the tympanic duct to the air of the middle ear cavity. (17)

rubrospinal tracts: Descending tracts of the lateral pathway that carry involuntary motor commands issued by the red nucleus of the mesencephalon. (15)

rugae: Mucosal folds in the lining of the empty stomach that disappear as gastric distension occurs (24); folds in the urinary bladder (26).

S

saccule: A portion of the vestibular apparatus of the internal ear; contains a macula important for providing sensations of gravity and linear acceleration in a vertical dimension. (17)

sagittal plane: A sectional plane that divides the body into left and right portions. (1)

salt: An inorganic compound consisting of a cation other than H^+ and an anion other than OH^-. (2)

saltatory propagation: The relatively rapid propagation of an action potential between successive nodes of a myelinated axon. (12)

sarcolemma: The plasma membrane of a muscle cell. (10)

sarcomere: The smallest contractile unit of a striated muscle cell. (10)

sarcoplasm: The cytoplasm of a muscle cell. (10)

scala media: The central membranous tube within the cochlea that is filled with endolymph and contains the spiral organ (*organ of Corti*); also called *cochlear duct*. (17)

scala tympani: The perilymph-filled chamber of the internal ear, adjacent to the basilar membrane; pressure changes there distort the round window; also called *tympanic duct*. (17)

scala vestibuli: The perilymph-filled chamber of the internal ear, adjacent to the vestibular membrane; pressure waves are induced by movement of the stapes at the oval window; also called *vestibular duct*. (17)

scar tissue: The thick, collagenous tissue that forms at an injury site. (5)

Schwann cells: Neuroglia responsible for the neurilemma that surrounds axons in the peripheral nervous system; also called *neurilemma cells*. (12)

sciatic nerve: A nerve innervating the posteromedial portions of the thigh and leg. (13)

sclera: The fibrous, outer layer of the eye that forms the white area of the anterior surface; a portion of the fibrous layer of the eye. (17)

sclerosis: A hardening and thickening that commonly occurs secondary to tissue inflammation. (4)

scrotum: The loose-fitting, fleshy pouch that encloses the testes of the male. (28)

sebaceous glands: Glands that secrete sebum; normally associated with hair follicles. (5)

sebum: A waxy secretion that coats the surfaces of hairs. (5)

secondary sex characteristics: Physical characteristics that appear at puberty in response to sex hormones but are not involved in the production of gametes. (28)

secretin: A hormone, secreted by the duodenum, that stimulates the production of buffers by the pancreas and inhibits gastric activity. (24)

semen: The fluid ejaculate that contains spermatozoa and the secretions of accessory glands of the male reproductive tract. (28)

semicircular ducts: The tubular components of the membranous labyrinth of the internal ear; respond to rotational movements of the head. (17)

semilunar valve: A three-cusped valve guarding the exit from one of the cardiac ventricles; the pulmonary and aortic valves. (20)

seminal glands: Glands of the male reproductive tract that produce roughly 60 percent of the volume of semen; also called *seminal vesicles*. (28)

seminiferous tubules: Coiled tubules where spermatozoon production occurs in the testes. (28)

senescence: Aging.

sensible perspiration: Water loss due to secretion by sweat glands. (5, 27)

septa: Partitions that subdivide an organ. (20, 22)

serosa: *See* serous membrane.

serotonin: A neurotransmitter in the central nervous system; a substance that enhances inflammation and is released by activated mast cells and basophils. (12, 19)

serous cell: A cell that produces a serous secretion. (4)

serous membrane: A squamous epithelium and the underlying loose connective tissue; the lining of the pericardial, pleural, and peritoneal cavities; also called a *serosa*. (4, 24)

serous secretion: A watery secretion that contains high concentrations of enzymes; produced by serous cells. (4)

serum: The ground substance of blood plasma from which clotting agents have been removed. (19)

sesamoid bone: A bone that forms within a tendon. (6)

sigmoid colon: The S-shaped region of the colon between the descending colon and the rectum. (24)

sign: The visible, objective evidence of the presence of a disease.

simple epithelium: An epithelium containing a single layer of cells superficial to the basement membrane. (4)

sinoatrial (SA) node: The natural pacemaker of the heart; located in the wall of the right atrium. (20)

sinus: A chamber or hollow in a tissue; a large, dilated vein. (6, 7, 20)

sinusoid: An exchange vessel that is similar in general structure to a fenestrated capillary. The two differ in size (sinusoids are larger and more irregular in cross section), continuity (sinusoids have gaps between endothelial cells), and support (sinusoids have thin basement membranes, if present at all). (20)

skeletal muscle: A contractile organ of the muscular system. (10)

skeletal muscle tissue: A contractile tissue dominated by skeletal muscle fibers; characterized as striated, voluntary muscle. (4, 10)

sliding filament theory: The concept that a sarcomere shortens as the thick and thin filaments slide past one another. (10)

small intestine: The duodenum, jejunum, and ileum; the digestive tract between the stomach and the large intestine. (24)

smooth endoplasmic reticulum (SER): A membranous organelle in which lipid and carbohydrate synthesis and storage occur. (3)

smooth muscle tissue: Muscle tissue in the walls of many visceral organs; characterized as nonstriated, involuntary muscle. (4, 10, 24, 26)

soft palate: The fleshy posterior extension of the hard palate, separating the nasopharynx from the oral cavity. (24)

solute: Any materials dissolved in a solution. (2, 21, 26)

solution: A fluid containing dissolved materials. (2, 21)

solvent: The fluid component of a solution. (2, 21)

somatic: Pertaining to the body.

somatic nervous system (SNS): The efferent division of the nervous system that innervates skeletal muscles. (12, 15, 16)

somatomedins: Substances stimulating tissue growth; released by the liver after the secretion of growth hormone; also called *insulin-like growth factors*. (18)

somatotropin: Growth hormone (GH); produced by the anterior lobe of the pituitary gland (adenohypophysis) in response to growth hormone–releasing hormone (GH–RH). (18)

sperm: *See* **spermatozoon**.

spermatic cord: Collectively, the spermatic vessels, nerves, lymphatic vessels, and the ductus deferens, extending between the testes and the proximal end of the inguinal canal. (28)

spermatocyte: A cell of the seminiferous tubules that is engaged in meiosis. (28)

spermatogenesis: Spermatozoon production. (28)

spermatozoon/spermatozoa: Male gamete(s); also called *sperm*. (3, 28)

sphincter: A muscular ring that contracts to close the entrance or exit of an internal passageway. (10, 11, 26)

spinal nerve: One of 31 pairs of nerves that originate on the spinal cord from anterior and posterior roots. (12, 13)

spindle apparatus: Microtubule-based structure that distributes duplicated chromosomes to opposite ends of a dividing cell during mitosis. (3)

spinocerebellar tracts: Ascending tracts that carry sensory information to the cerebellum. (15)

spinothalamic tracts: Ascending tracts that carry poorly localized touch, pressure, pain, vibration, and temperature sensations to the thalamus. (15)

spinous process: The prominent posterior projection of a vertebra; formed by the fusion of two laminae. (7)

spiral organ: A receptor complex in the scala media of the cochlea that includes the inner and outer hair cells, supporting cells and structures, and the tectorial membrane; provides the sensation of hearing; also called the *organ of Corti*. (17)

spleen: A lymphoid organ important for the phagocytosis of red blood cells, the immune response, and lymphocyte production. (22)

squama: A broad, flat surface.

squamous: Flattened.

squamous epithelium: An epithelium whose superficial cells are flattened and platelike. (4)

stapes: The auditory ossicle attached to the tympanic membrane. (17)

stenosis: A constriction or narrowing of a passageway.

stereocilia: Elongated microvilli characteristic of the epithelium of the epididymis, sections of the ductus deferens (28), and the internal ear. (17)

steroid: A ring-shaped lipid structurally related to cholesterol. (2, 18)

stimulus: An environmental change that produces a change in cellular activities; often used to refer to events that alter the membrane potentials of excitable cells. (15)

stratified: Containing several layers. (4)

stratum: A layer; plural *strata*.

stretch receptors: Sensory receptors that respond to stretching of the surrounding tissues. (13)

stroma: The connective tissue framework of an organ; distinguished from the functional cells (parenchyma) of that organ.

subarachnoid space: A meningeal space containing cerebrospinal fluid; the area between the arachnoid membrane and the pia mater. (13)

subclavian: Pertaining to the region immediately posterior and inferior to the clavicle.

subcutaneous layer: *See* **hypodermis**.

submucosa: The region between the muscularis mucosae and the muscularis externa. (23, 24)

subserous fascia: The loose connective tissue layer deep to the serous membrane that lines body cavities. (4)

substrate: A participant (product or reactant) in an enzyme-catalyzed reaction. (2)

sulcus: A groove or furrow. (14)

summation: The temporal or spatial addition of contractile force or neural stimuli. (10, 12)

superficial fascia: *See* **hypodermis**.

superior: Above, in reference to a portion of the body in the anatomical position.

superior vena cava (SVC): The vein that carries blood to the right atrium from parts of the body that are superior to the heart. (20, 21)

supination: The rotation of the forearm such that the palm faces anteriorly. (9)

supine: Lying face up, with palms facing anteriorly. (1)

suppressor T cells: Lymphocytes that inhibit B cell activation and the secretion of antibodies by plasma cells. (22)

suprarenal cortex: *See* **adrenal cortex**. (18)

suprarenal gland: *See* **adrenal gland**. (18)

suprarenal medulla: *See* **adrenal medulla**. (16)

surfactant: A lipid secretion that coats the alveolar surfaces of the lungs and prevents their collapse. (23)

sustentacular cells: *See* **nurse cells**.

sutural bones: Irregular bones that form in fibrous tissue between the flat bones of the developing cranium; also called *Wormian bones*. (6)

suture: A fibrous joint between flat bones of the skull. (7, 9)

sympathetic division: The division of the autonomic nervous system that is responsible for "fight or flight" reactions; primarily concerned with the elevation of metabolic rate and increased alertness; also called the *thoracolumbar division*. (12, 16)

symphysis: A fibrous amphiarthrosis, such as that between adjacent vertebrae or between the pubic bones of the coxal bones. (9)

symptom: An abnormality of function as a result of disease; subjective experience of patient.

synapse: The site of communication between a nerve cell and some other cell; if the other cell is not a neuron, the term *neuromuscular junction* or *neuroglandular junction* is often used. (12, 16, 18)

synaptic delay: The period between the arrival of an impulse at the presynaptic membrane and the initiation of an action potential in the postsynaptic membrane. (12)

syncytium: A multinucleate mass of cytoplasm, produced by the fusion of cells or repeated mitoses without cytokinesis. (29)

syndrome: A discrete set of signs and symptoms that occur together.

synergist: A muscle that assists a prime mover in performing its primary action. (11)

synovial cavity: A fluid-filled chamber in a synovial joint. (4, 9)

synovial fluid: The substance secreted by synovial membranes that lubricates joints. (4, 9)

synovial joint: A freely movable joint where the opposing bone surfaces are separated by synovial fluid; a diarthrosis. (4, 9)

synovial membrane: An incomplete layer of fibroblasts confronting the synovial cavity, plus the underlying loose connective tissue. (4)

synthesis: Manufacture; anabolism. (23)

system: An interacting group of organs that performs one or more specific functions.

systemic circuit: The vessels between the aortic valve and the entrance to the right atrium; the system other than the vessels of the pulmonary circuit. (20)

systole: A period of contraction in a chamber of the heart, as part of the cardiac cycle. (20)

systolic pressure: The peak arterial pressure measured during ventricular systole. (20)

T

tactile: Pertaining to the sense of touch. (15)

tarsal bones: The bones of the ankle (the talus, calcaneus, navicular, and cuneiform bones). (8)

tarsus: The ankle. (8)

TCA (tricarboxylic acid) cycle: see **citric acid cycle**. (3, 10, 25)

T cells: Lymphocytes responsible for cell-mediated immunity and for the coordination and regulation of the immune response; includes regulatory T cells (helpers and suppressors) and cytotoxic (killer) T cells. (19, 22)

tectospinal tracts: Descending tracts of the medial pathway that carry involuntary motor commands issued by the colliculi. (15)

telodendria: Terminal axonal branches that end in axon terminals. (12)

telophase: The final stage of mitosis, characterized by the disappearance of the spindle apparatus, the reappearance of the nuclear membrane, the disappearance of the chromosomes, and the completion of cytokinesis. (3)

temporal: Pertaining to time (temporal summation) or to the temples (temporal bone). (7)

tendon: A collagenous band that connects a skeletal muscle to an element of the skeleton. (4, 10)

teres: Round.

terminal: Toward the end.

tertiary structure: The protein structure that results from interactions among distant portions of the same molecule; complex coiling and folding. (2)

testes: The male gonads, sites of gamete production and hormone secretion; also called *testicles*. (18, 28)

testosterone: The main androgen produced by the interstitial cells of the testes. (2, 18, 28)

tetraiodothyronine: T_4, or thyroxine, a thyroid hormone. (18)

thalamus: The walls of the diencephalon. (14)

therapy: The treatment of disease.

thermoreception: Sensitivity to temperature changes. (15)

thermoregulation: Homeostatic maintenance of body temperature. (1, 25)

thick filament: A cytoskeletal filament in a skeletal or cardiac muscle cell; composed of myosin, with a core of titin. (3, 10)

thin filament: A cytoskeletal filament in a skeletal or cardiac muscle cell; consists of actin, troponin, nebulin, and tropomyosin. (3, 10)

thoracolumbar division: The sympathetic division of the autonomic nervous system. (16)

thorax: The chest. (7)

threshold: The membrane potential at which an action potential begins. (12)

thrombin: The enzyme that converts soluble fibrinogen to insoluble fibrin. (19)

thymine: A pyrimidine; one of the nitrogenous bases in the nucleic acid DNA. (2)

thymosins: Thymic hormones essential to the development and differentiation of T cells. (18, 22)

thymus: A lymphoid organ, the site of T cell development and maturation. (18, 22)

thyroglobulin: A circulating transport globulin that binds thyroid hormones. (18)

thyroid gland: An endocrine gland whose lobes are lateral to the thyroid cartilage of the larynx. (18)

thyroid hormones: Thyroxine (T_4) and triiodothyronine (T_3), hormones of the thyroid gland; stimulate tissue metabolism, energy utilization, and growth. (18)

thyroid-stimulating hormone (TSH): The hormone, produced by the adenohypophysis (anterior lobe of the pituitary gland), that triggers the secretion of thyroid hormones by the thyroid gland. (18)

thyroxine: A thyroid hormone; also called T_4 or *tetraiodothyronine*. (18)

tidal volume: The volume of air moved into and out of the lungs during a normal quiet respiratory cycle. (23)

tissue: A collection of specialized cells and cell products that performs a specific function. (1, 4)

tonic response: An increase or decrease in the frequency of action potentials by sensory receptors that are chronically active. (15)

tonsil: A lymphoid nodule in the wall of the pharynx; the palatine, pharyngeal, and lingual tonsils. (22)

topical: Applied to the body surface.

toxic: Poisonous.

trabecula: A connective tissue partition that subdivides an organ. (22)

trachea: The windpipe; an airway extending from the larynx to the primary bronchi. (23)

tract: A bundle of axons in the central nervous system. (13, 14)

transcription: The encoding of genetic instructions on a strand of mRNA. (3)

transection: The severing or cutting of an object in the transverse plane.

translation: The process of peptide formation from the instructions carried by an mRNA strand. (3)

transmembrane potential: *See* **membrane potential**.

transudate: A fluid that diffuses across a serous membrane and lubricates opposing surfaces. (4)

transverse tubules: The transverse, tubular extensions of the sarcolemma that extend deep into the sarcoplasm, contacting cisternae of the sarcoplasmic reticulum; also called *T tubules*. (10)

tricuspid valve: The right atrioventricular valve, which prevents the backflow of blood into the right atrium during ventricular systole. (20)

trigeminal nerve: Cranial nerve V, which provides sensory information from the lower portions of the face (including the upper and lower jaws) and delivers motor commands to the muscles of mastication. (14)

triglyceride: A lipid that is composed of a molecule of glycerol attached to three fatty acids. (2, 25)

triiodothyronine: T_3, a thyroid hormone. (18)

trisomy: The abnormal possession of three copies of a chromosome; trisomy 21 is responsible for Down's syndrome. (29)

trochanter: Large process near the head of the femur. (8)

trochlea: A pulley; the spool-shaped medial portion of the condyle of the humerus. (8)

trochlear nerve: Cranial nerve IV, controlling the superior oblique muscle of the eye. (14)

trunk: The thoracic and abdominopelvic regions (1); a major arterial branch (21).

T tubules: *See* **transverse tubules**.

tuberculum: A small, localized elevation on a bony surface. (7)

tuberosity: A large, roughened elevation on a bony surface. (6)

tumor: A tissue mass formed by the abnormal growth and replication of cells. (3)

tunica: A layer or covering.

twitch: A single stimulus–contraction–relaxation cycle in a skeletal muscle. (10)

tympanic duct: *See* **scala tympani**.

tympanic membrane: The membrane that separates the external acoustic meatus from the middle ear; the membrane whose vibrations are transferred to the auditory ossicles and ultimately to the oval window; also called *eardrum* or *tympanum*. (17)

type A axons: Large myelinated axons. (12)

type B axons: Small myelinated axons. (12)

type C axons: Small unmyelinated axons. (12)

U

umbilical cord: The connecting stalk between the fetus and the placenta; contains the allantois, the umbilical arteries, and the umbilical vein. (21, 29)

umbilicus: The navel. (29)

unicellular gland: Mucous cells. (4)

unipolar neuron: A sensory neuron whose cell body is in a dorsal root ganglion or a sensory ganglion of a cranial nerve. (12)

unmyelinated axon: An axon whose neurilemma does not contain myelin and across which continuous propagation occurs. (12)

uracil: A pyrimidine; one of the nitrogenous bases in the nucleic acid RNA. (2)

ureters: Muscular tubes, lined by transitional epithelium, that carry urine from the renal pelvis to the urinary bladder. (26)

urethra: A muscular tube that carries urine from the urinary bladder to the exterior. (26)

urinary bladder: The muscular, distensible sac that stores urine prior to micturition. (26)

urination: The voiding of urine; also called *micturition*. (26)

uterus: The muscular organ of the female reproductive tract in which implantation, placenta formation, and fetal development occur. (28)

utricle: The largest chamber of the vestibular apparatus of the internal ear; contains a macula important for providing sensations of gravity and linear acceleration in a horizontal dimension. (17)

V

vagina: A muscular tube extending between the uterus and the vestibule. (28)

vallate papilla: One of the large, dome-shaped papillae on the superior surface of the tongue that forms a V, separating the body of the tongue from the root. (17)

vas deferens: *See* **ductus deferens**. (28)

vascular: Pertaining to blood vessels. (19)

vasoconstriction: A reduction in the diameter of arterioles due to the contraction of smooth muscles in the tunica media; increases peripheral resistance; may occur in response to local factors, through the action of hormones, or from the stimulation of the vasomotor center. (21)

vasodilation: An increase in the diameter of arterioles due to the relaxation of smooth muscles in the tunica media; decreases peripheral resistance; may occur in response to local factors, through the action of hormones, or after decreased stimulation of the vasomotor center. (21)

vasomotion: Rhythmic changes in the pattern of blood flow through a capillary bed due to the alternate contraction and relaxation of precapillary sphincters. (21)

vasomotor center: The center in the medulla oblongata whose stimulation produces vasoconstriction and an increase of peripheral resistance. (14)

vein: A blood vessel carrying blood from a capillary bed toward the heart. (20, 21)

vena cava: One of the major veins delivering systemic blood to the right atrium; superior and inferior venae cavae. (20, 21)

ventilation: Air movement into and out of the lungs. (23)

ventral: Pertaining to the anterior surface.

ventricle: A fluid-filled chamber; in the heart, one of the large chambers discharging blood into the pulmonary or systemic circuits (20); in the brain, one of four fluid-filled interior chambers. (14)

venule: Thin-walled veins that receive blood from capillaries. (21)

vermiform appendix: *See* **appendix**.

vertebral canal: The passageway that encloses the spinal cord; a tunnel bounded by the neural arches of adjacent vertebrae. (7)

vertebral column: The cervical, thoracic, and lumbar vertebrae, the sacrum, and the coccyx. (7, 11)

vesicle: A membranous sac in the cytoplasm of a cell. (3)

vestibular duct: *See* **scala vestibuli**.

vestibular nucleus: The processing center for sensations that arrive from the vestibular apparatus of the internal ear, located near the border between the pons and the medulla oblongata. (17)

vestibulospinal tracts: Descending tracts of the medial pathway that carry involuntary motor commands issued by the vestibular nucleus to stabilize the position of the head. (15)

villus/villi: Slender, finger-shaped projection(s) of a mucous membrane. (24)

virus: A noncellular pathogen. (22)

viscera: Internal organs of the thoracic and abdominopelvic cavities. (1)

visceral: Pertaining to viscera (internal organs) or their outer coverings. (1)

visceral smooth muscle: A smooth muscle tissue that forms sheets or layers in the walls of visceral organs; the cells may not be innervated, and the layers often show automaticity (rhythmic contractions). (10, 24)

viscosity: The resistance to flow that a fluid exhibits as a result of molecular interactions within the fluid. (21)

viscous: Thick, syrupy.

vitamin: An essential organic nutrient that functions as a coenzyme in vital enzymatic reactions. (25)

vitreous humor: The fluid component of the vitreous body, the gelatinous mass in the posterior cavity (vitreous chamber) of the eye. (17)

voluntary: Controlled by conscious thought processes.

W

white blood cells (WBCs): The granulocytes and agranulocytes of whole blood. (4, 19)

white matter: Regions in the central nervous system that are dominated by myelinated axons. (12, 13, 14)

white ramus: A nerve bundle containing the myelinated preganglionic axons of sympathetic motor neurons en route to the sympathetic chain or to a collateral ganglion. (13)

Wormian bones: *See* **sutural bones**.

X

xiphoid process: The slender, inferior extension of the sternum. (7)

Y

Y chromosome: The sex chromosome whose presence indicates that the individual is a genetic male. (29)

Z

zona fasciculata: The region of the adrenal cortex that secretes glucocorticoids. (18)

zona glomerulosa: The region of the adrenal cortex that secretes mineralocorticoids. (18)

zona reticularis: The region of the adrenal cortex that secretes androgens. (18)

zygote: The fertilized ovum, prior to the start of cleavage. (28)

Credits

Index

A

A band, 321–322
ABCD mnemonic, for melanoma, 184
Abdominal aorta, 787–789
Abdominal cavity, 48
Abdominal thrust, 902
Abdominopelvic cavity, 45, 48, 788, 794, 796–797
Abdominopelvic quadrants, 41–42
Abdominopelvic regions, 42
Abducens nerve (VI), 521, 528
Abduction, 295, 298, 364
Abductor (muscle name), 364, 366
Abductor digiti minimi muscle, 395, 402
Abductor hallucis muscle, 402
Abductor pollicis muscle, brevis/longus, 392, 395
ABO blood types, 690–691
Abortion, 1140
Abrasion, 196
Abscess, 170, 828
Absolute refractory period, 433
Absorption
 carbohydrates, 948–951
 definition, 909
 ions, 952–953
 in large intestine, 946–947
 lipids, 949, 951
 monosaccharides, 951
 proteins, 949, 951–952
 in stomach, 927, 949
 vitamins, 953
 water, 946
Absorptive state, 980, 982, 996
Accessory glands, male reproductive system, 1088–1090
Accessory ligaments, 293
Accessory nerve (XI), 527, 528
Accessory organs, digestive system, 907, 908
Accessory pancreatic duct, 933
Accessory respiratory muscles, 879
Accessory structures
 eye, 596–599
 glands. See Gland(s)
 hair, 189–192
 integument, 177–178
 nails, 195
 synovial joints, 293
Acclimatization, 992
Accommodation, vision, 605–606
Accommodation curves, 252
ACE (angiotensin-converting enzyme), 773, 872, 1016
ACE inhibitors, 666
Acetabular labrum, 306
Acetabular ligament, transverse, 306
Acetabular notch, 276, 306
Acetabulum, 276, 279
Acetate, 442
Acetoacetate, 982

Acetone, 982
Acetylcholine (ACh)
 in cholinergic synapses, 440–442, 567
 location, 444
 mechanism of action, 444
 muscle activity and, 326–327
 in parasympathetic division, 572
 structure, 444
 in sympathetic division, 567, 572
Acetylcholinesterase (AChE), 326–327, 442, 570
Acetyl-CoA, 965
Acetyl group, 965
Achilles (calcaneal) tendon, 283, 401–402
Achondroplasia, 228
Acid, 68, 85, 1056–1057
Acid–base balance
 age-related changes, 1070–1071
 definition, 1043
 maintenance, 1061–1063
 overview, 1043, 1073
 pH control in, 1056, 1058–1061
 renal compensation and, 1062–1063, 1066
 respiratory compensation and, 1061–1062, 1066
 terminology, 1056
Acid–base disorders
 acute phase, 1064
 blood chemistry in, 1071
 causes, 1064
 combined respiratory and metabolic acidosis, 1069
 compensated phase, 1064
 detection and diagnosis, 1069–1071
 metabolic acidosis, 1067–1069, 1071
 metabolic alkalosis, 1069, 1071
 respiratory acidosis, 1064–1067, 1071
 respiratory alkalosis, 1067, 1071
Acidemia, 1056
Acidic solution, 68, 1056
Acidophils. See Eosinophils
Acidosis
 combined respiratory and metabolic, 1069
 definition, 68, 1056
 detection and diagnosis, 1069–1071
 metabolic, 1067–1069, 1070, 1071
 renal responses, 1062, 1065
 respiratory, 1064–1067, 1070, 1071
 respiratory responses, 1065, 1068
Acinar (alveolar) glands, 151
Acoustic meatus, external/internal, 237, 238, 239, 242, 243
Acoustic nerve. See Vestibulocochlear nerve (VIII)
Acquired reflexes, 478
Acromegaly, 222, 668

Acromial end, clavicle, 270
Acromioclavicular joint, 270
Acromioclavicular ligament, 303
Acromion, 270
Acrosin, 1124
Acrosomal cap, 1086
Acrosomal vesicle, 1086
Acrosome, 1086
ACTH (adrenocorticotropic hormone), 183, 648, 651, 989
Actin, 97, 318
Actinic keratosis, 184
Actinins, 322
Action(s), muscle, 363–365
Action potentials
 all-or-none principle, 432
 axon diameter and propagation speed, 438–439
 cardiac muscle cells, 729, 731
 definition, 424, 432
 frequency, 449–450
 generation, 432–433, 434–435
 vs. graded potentials, 438
 overview, 452
 propagation, 433, 436–437, 438
 rate of generation, 449–450
 refractory period, 433
 skeletal muscle cells, 319, 326–327, 328
 sodium–potassium exchange pump and, 433
Activated channels, 428
Activation energy, 64
Active (gated) channels, 428–430
Active immunity, 829
Active site
 enzyme, 80
 G-actin molecule, 323
Active transport, 121–122, 125, 1011
Acute, 49
Acute epiglottitis, 865
Acute radiation syndrome, 55
Acute renal failure, 1018, 1035
Acute respiratory acidosis, 1064
ADA deficiency, 1155
Adam's apple, 863
Adaptation
 light and dark, 613
 sensory receptors, 538–539
Adaptive (specific) defenses, 808, 821, 853. See also Immune response
Adaptive (specific) immunity, 829, 830, 842
Addison's disease, 183, 668
Adduction, 295, 298, 364
Adductor (muscle name), 364
Adductor brevis muscle, 396, 398
Adductor hallucis muscle, 402
Adductor longus muscle, 396
Adductor magnus muscle, 364, 396, 398
Adductor pollicis muscle, 395
Adenine, 82
Adenitis, 851

Adenohypophysis, 645–649
Adenoid (pharyngeal tonsil), 816, 863
Adenosine, 445
Adenosine diphosphate (ADP), 84, 330–331, 702
Adenosine monophosphate (AMP), 84
Adenosine triphosphatase (ATPase), 84
Adenosine triphosphate. See ATP
Adenylate cyclase (adenylyl cyclase), 446, 640
ADH. See Antidiuretic hormone
ADHD (attention deficit hyperactivity disorder), 529
Adhesion, phagocyte, 824
Adhesion belt, 142, 143
Adipocytes (adipose cells, fat cells), 86, 153, 155–156
Adipose tissue, 152
 endocrine functions, 638, 666, 669
 liposuction, 157
 metabolic functions, 982
 structure and function, 155–157
 sympathetic effects, 567, 573
Adolescence, 1144, 1146–1148
ADP (adenosine diphosphate), 84, 330–331, 702
Adrenal artery, 788
Adrenal cortex, 648, 657–660
Adrenalectomy, 674
Adrenal gland, 638, 657, 659, 675
Adrenaline. See Epinephrine
Adrenal medulla, 562, 563, 564–565, 659–660
Adrenal vein, 796
Adrenergic receptors, 567, 571
Adrenergic synapses, 443, 566
Adrenocorticotropic hormone (ACTH), 183, 648, 651, 989
Adrenogenital syndrome, 668
Adventitia, 912, 1106
Adventitious bursa, 293
Aerobic endurance, 349
Aerobic metabolism (cellular respiration), 107, 343, 873, 965, 971
Afferent arterioles, nephron, 1002
Afferent division, PNS, 413
Afferent fibers, 416
Afferent lymphatics, 816
Afferent (sensory) neurons, 416
Afferent vessels. See Veins
Afterbirth. See Placenta
Afterload, 742
Age-related macular degeneration, 630
Ageusia, 630
Agglutination, 690, 839
Agglutinins, 690
Agglutinogens, 690
Aggregated lymphoid nodules, 816, 817, 932

Aging
cancer incidence and, 172
cardiovascular system, 801
cholesterol levels and, 977
digestive system, 954
female reproductive system, 1114–1115
fluid, electrolyte, and acid–base balance, 1070–1071
gustation, 596
hair and, 191, 192
hearing, 629
hormones, 670
immune response, 851
integumentary system, 198–199
joints, 309–310
male reproductive system, 1115
muscular system, 403–405
nervous system, 583
olfaction, 594
oogenesis and, 1099
respiratory system, 899
skeletal system, 284–285
summary, 1148
telomerase in, 133
tissue structure and, 170
urinary system, 1035
vision, 606
Agonist (muscle), 363
Agranular leukocytes (agranulocytes), 693
AIDS (acquired immune deficiency syndrome), 845, 848, 872
Air, gas concentrations, 883
Ala, sacrum, 261
Alarm phase, general adaptation syndrome, 671
Albinism, 182
Albumins, 652, 681, 682
Alcohol, 927, 988
Alcoholism, 988
Aldosterone, 658, 1022, 1046, 1053
Aldosteronism, 668
Alimentary canal. *See* **Digestive tract**
Alkalemia, 1056
Alkaline, 68
Alkaline solution, 1056
Alkaline tide, 926, 1069
Alkalosis
definition, 68, 1022, 1056
detection and diagnosis, 1069–1071
metabolic, 1069–1071
renal responses, 1062, 1065, 1066, 1069
respiratory, 1066, 1067, 1071
respiratory responses, 1065, 1066, 1069
Allantois, 1130, 1131
Alleles, 1149
Allergens, 697, 848
Allergic rhinitis, 251, 848
Allergy, 844, 848–849
Allograft, 851
All-or-none principle, action potential, 432
Alpha$_1$-antitrypsin deficiency, 1155
Alpha$_1$-receptor agonists, 585
Alpha-2-macroglobulin, 705
Alpha$_2$-receptor agonists, 585

Alpha-amylase (salivary amylase), 916, 917, 950
Alpha cells, 662, 665
Alpha (α) chains, 685
Alpha helix, 78, 79
Alpha (α) receptors, 567, 571
Alpha waves, 516
Alternative pathway, complement activation, 826–827
Alveolar air, 884
Alveolar ducts, 869, 871
Alveolar epithelium, 860
Alveolar (acinar) glands, 151
Alveolar macrophages, 823, 869
Alveolar process, 246, 248
Alveolar sacs, 869
Alveolar ventilation, 874, 880, 891
Alveolus (lung), 778, 858, 870
Alveolus (tooth socket), 918
Alzheimer's disease, 583, 585, 1155
Amacrine cells, retina, 602
Amanita muscaria, 570
Amenorrhea, 1105
Amination, 979
Amino acid(s)
catabolism, 978
essential, 979
neurotransmitters, 444
nonessential, 979
in plasma, 1010
in protein buffer systems, 1059
structure, 76–77
in urine, 1010, 1030
Amino acid derivatives (biogenic amines), 444, 637, 639
Aminoaciduria, 1012
Amino group, 70, 77
Ammonia, 1010, 1030
Ammonium ions, 946, 1023
Amnesia, 579
Amniocentesis, 146, 1156
Amniochorionic membrane, 1138
Amnion, 1128, 1130
Amniotic cavity, 1128
Amniotic fluid, 146, 1130
Amoeboid movement, 693
AMP (adenosine monophosphate), 84
Amphetamines, 582
Amphiarthrosis, 290, 291
Amphimixis, 1124, 1125
Amphoteric, 1059
Amplification, 640
Amplitude, sound wave, 625
Ampulla
ductus deferens, 1088
duodenal, 933
semicircular duct, 621
uterine tube, 1100
Ampulla of Vater, 933
Amygdaloid body, 506, 511, 578
Amyloid β protein, 583
Amyloidosis, 1155
Amyotrophic lateral sclerosis (ALS), 552, 1155
Anabolic steroids, 672. *See also* **Androgen(s)**
Anabolic system, 568. *See also* **Parasympathetic division, ANS**
Anabolism, 63, 963–964
Anaerobic endurance, 348

Anaerobic process, 343
Anal canal, 945–946
Anal columns, 945–946
Analgesia, 554
Anal orifice, 945
Anal sphincter, internal/external, 383, 945–946
Anal triangle, 382–383
Anamnestic response, 851
Anandamide, 445
Anaphase, mitosis, 131
Anaphylactic shock, 849
Anaphylaxis, 848–849
Anaplasia, 134
Anastomosis, 757
Anatomical directions, 42, 43
Anatomical landmarks, 40–41
Anatomical neck, humerus, 271
Anatomical position, 40, 41
Anatomical regions, 41–42
Anatomic dead space, 880
Anatomy
clinical, 31
definition, 29, 49
developmental, 31
gross, 30–31, 49
history, 28–29
microscopic, 31
vs. physiology, 30
regional, 30
sectional, 42, 44
superficial, 40
surface, 30
systemic, 30–31
Anaxonic neurons, 416, 417
Anchoring proteins, 95
Anconeus muscle, 389, 391
Androgen(s)
abuse, 672
effects, 221, 228, 660, 980, 1115
in normal growth, 670
production, 648, 660, 1077, 1082, 1092, 1112
regulation of secretion, 1115
underproduction/overproduction, 668
Androgen-binding protein (ABP), 1086
Andropause, 1114
Androstenedione, 672, 1109
Anemia
aplastic, 679, 709
definition, 686
diagnosis, 690
effects, 686
iron-deficiency, 687
pernicious, 689
Anesthesia, 461, 543
Anesthetics, 116, 451
Aneurysm, 755, 801
Angelman syndrome, 1153
Angina pectoris, 724
Angiogenesis, 133, 758
Angiogram, 801
Angiography, 724
Angiotensin-converting enzyme (ACE), 773, 872, 1016
Angiotensin I, 666
Angiotensin II, 666, 773, 774, 1016
Angiotensinogen, 666
Angle, rib, 261

Angular motion, 295, 296, 297
Animal starch. *See* **Glycogen**
Anion gap, 1070, 1071
Anions, 57
Anisotropic, 321
Ankle (tarsus), 283, 308, 401–405, 473
Ankyloglossia, 916
Ankylosing spondylitis, 312
Annular ligament, 304
Annulus fibrosus, 300
Anorexia, 994
Anorexia nervosa, 994
Anosmia, 630
Anoxia, 874
Ansa cervicalis nerve, superior/inferior, 469
Antabuse (disulfiram), 988
Antacid, 1072
Antagonist (muscle), 363
Antagonistic effects, hormones, 669
Antebrachial cutaneous nerve, medial, 471
Antebrachial vein, median, 793
Antebrachium, 273
Anteflexion, uterus, 1101
Antegrade flow, axon, 416
Anterior, 40, 41, 42
Anterior cavity, eye, 599
Anterior chamber, eye, 599
Anterior commissure, 508, 510
Anterior cruciate ligament (ACL), 306, 309
Anterior fontanelle, 250
Anterior lobe
cerebellum, 501
pituitary gland, 645–649
Anterior margin, tibial tuberosity, 282
Anterior median fissure, 457
Anterior nuclei, thalamus, 503, 506
Anterior white commissure, 463
Anterograde amnesia, 579
Antiangiogenesis factor, 159
Antibody(ies). *See also* **Immunoglobulin(s)**
actions, 153, 679, 681, 838–839
antigen–antibody complex, 837–838
classes, 838–839
structure, 837
Antibody-mediated (humoral) immunity, 697, 814, 815, 828, 842. *See also* **Immune response**
Antibody titer, 840
Anticholinesterase drug, 451
Anticoagulants, 705
Anticodon, 112, 114
Antidiuretic hormone (ADH)
cardiovascular effects, 773
for diabetes insipidus, 649
in fluid balance, 1046
functions, 650, 651
renal effects, 1025, 1026
secretion, 649
in sodium balance, 1051–1052
underproduction/overproduction, 668
Antigen(s), 690, 814, 830

Antigen–antibody complex,
    837–838
Antigen binding sites, 837
Antigenic determinant sites, 838
Antigen presentation, 817,
    831–832
Antigen-presenting cells (APCs),
    832, 841
Antigen processing, 832
Antigen recognition, 832–834
Antihistamines, 849
Anti-inflammatory effects, 658
Antioxidant, 661
Antiport (countertransport), 121
Anti-Rh antibodies, 694–695
Antithrombin-III, 705
Antiviral proteins, 825
Antrum, follicle, 1099
Anus, 945–946
Aorta
    abdominal, 787–789
    ascending, 718, 781
    descending, 719, 786–789
    terminal segment, 787
    thoracic, 786–787
Aortic arch, 719, 781
Aortic bodies, 543, 772
Aortic reflex, 770
Aortic sinus, 542, 721
Aortic sinuses, 770
Aortic valve, 718, 720
APCs (antigen-presenting cells),
    832, 841
Apex
    heart, 712
    nose, 861
    patella, 280
    sacrum, 261
Apgar score, 1143
Aphasia, 515
Apical foramen, 918
Apical surface, epithelium, 141
Aplastic anemia, 679, 709
Apnea, 897
Apneustic center, 499, 893, 894
Apocrine secretion, 149, 150
Apocrine sweat glands, 193
Apolipoproteins, 682
Aponeuroses, 359
Aponeurosis, 157, 318
Apoptosis, 127, 834, 845
Appendicitis, 943
Appendicular muscles
    overview, 367, 368–369, 384, 408
    pelvis and lower limbs, 395–403
    shoulders and upper limbs,
        384–395
    trunk, 385–386
Appendicular skeleton
    joints, 308
    lower limbs, 280–284
    movements, 308
    overview, 233, 268, 269
    pectoral girdle, 268–271
    pelvic girdle, 276–278
    upper limbs, 271–275
Appendix, 816, 943
Applied force, 361
Appositional growth, 160, 215, 216
Aquaporins, 118, 1025
Aqueous humor, 599, 603–604

Aqueous solutions, 66–67
Arachidonic acid, 73
Arachnoid granulations, 494
Arachnoid mater, 460, 491
Arachnoid villi, 494
Arbor vitae, 500, 501
Arch
    arcuate (dorsal), 789
    atlas, 257
    foot, 284
    palmar, 793
    plantar, 789, 793
Arcuate arteries, 1002, 1103
Arcuate fibers, 508, 510
Arcuate line, 276
Arcuate veins, 1003
Areflexia, 483
Areola, 1107
Areolar tissue, 155, 156
Arginine, 952, 979
Aromatase, 1094
Arousal
    sexual, 576, 1113, 1114
    from sleep, 581
Arrector pili muscle, 189
Arresting solution, 991
Arrhythmias, 728–729, 730
Arterial anastomoses, 722, 757
Arterial puncture (stick), 682, 706
Arteries
    abdominopelvic organs, 788
    brain, 784–786
    chest and upper limb, 781, 783
    elastic, 752, 753
    functions, 157, 711, 750
    head and neck, 784
    lower limb, 789, 790
    lower limbs, 789
    muscular, 753, 754
    overview, 750, 782, 803
    pelvis, 789
    pulmonary circuit, 780–781
    structure, 751–752
    systemic circuit, 781–789
    trunk, 786–789
    types, 752–754
    vs. veins, 752, 753
Arteries listed
    adrenal, 788
    aorta. See Aorta
    arcuate, 1002, 1103
    axillary, 781, 783
    basilar, 785
    brachial, 682, 754, 783, 784
    bronchial, 787
    carotid, 495, 752, 770, 781,
        784–785
    celiac trunk, 564
    central retinal, 603
    cerebral, 785, 786
    circumflex, 721
    colic, 788
    communicating, 786
    coronary, 721
    cortical radiate, 1002
    cystic, 788
    deferential, 1080
    digital, 784
    dorsalis pedis, 789
    esophageal, 787
    femoral, 754, 780, 789

    fibular, 789
    gastric, 788
    gastroduodenal, 788
    gastroepiploic, 788
    genicular, 789
    gluteal, 789
    gonadal, 789
    hepatic, 788
    hypophyseal, 646
    iliac, 752, 780, 787, 789, 1101, 1106
    inferior mesenteric, 564
    intercostal, 787
    interlobar, 1002
    internal thoracic, 781, 783, 1107
    interventricular, 721, 757
    intestinal, 788
    lumbar, 789
    mediastinal, 787
    mesenteric, 754, 788
    obturator, 789
    ophthalmic, 785
    ovarian, 789, 1096, 1101
    palmar arches, 783, 784
    pancreatic, 788
    pancreaticoduodenal, 788
    pericardial, 787
    phrenic, 787, 788
    plantar, 789
    popliteal, 789
    pudendal, 789
    pulmonary, 718, 752, 781
    radial, 682, 783, 784, 1103
    rectal, 788
    renal, 788, 1001, 1002
    sacral, 789
    segmental, 1002
    sigmoid, 788
    spiral, 1103
    splenic, 788, 820
    straight, 1103
    subclavian, 752, 781
    superior mesenteric, 564
    testicular, 789, 1080
    tibial, 789
    trabecular, 820
    ulnar, 783, 784
    umbilical, 798, 1031, 1133
    uterine, 1101
    vertebral, 495, 781
Arterioles, 750, 753
Arteriosclerosis, 583, 754–755
Arteriovenous anastomoses, 758
Arthritis, 309, 313
Arthroplasty, 312
Arthroscopy, 312
Articular capsule, 292
Articular cartilage, 160, 217, 292
Articular disc (meniscus), 273, 293
Articular facet, vertebrae, 253
Articular processes, superior/
    inferior, vertebrae, 253
Articulation, in speech, 864
Articulations, 165, 290. See also
    Joint(s)
Artificially induced active
    immunity, 829
Artificially induced passive
    immunity, 830
Artificial pacemaker, 744
Artificial sweeteners, 71, 86
Aryepiglottic folds, 863

Arytenoid cartilages, 863
Asbestosis, 901
Ascending aorta, 718, 781
Ascending colon, 943
Ascending limb, nephron
    loop, 1008
Ascending pathways. See Sensory
    pathways
Ascending tracts, 463, 498, 499
Ascites, 909
Ascorbic acid (vitamin C), 987
Aspartate, 444, 446
Aspect, 40
Asphyxia, 901
Aspirate, 901
Aspirin, 639, 705, 927
Association areas, 513
Association fibers, 508, 510
Association neurons
    (interneurons), 418
Asthma, 869
Astigmatism, 606
Astral rays, 131
Astrocytes, 418–419
Asynchronous motor unit
    summation, 339
Asystole, 744
Ataxia, 501
Atelectasis, 878
Atenolol, 742
Atherectomy, 725
Atherosclerosis, 750, 754–755, 806
Atherosclerotic plaque, 724,
    754–755
Athletes
    amenorrhea in, 1105
    blood doping, 689
    carbohydrate depletion/
        loading, 973
    cardiac reserve, 744
    cardiovascular fitness, 775–776
    DHEA use, 1092
    hormones to improve
        performance, 672
    knee injuries, 309
    proteinuria and hematuria, 1018
    salt loss in, 1053
    training at high altitude, 901
Atlas, 256, 257, 302
Atmosphere (atm), 875
Atmospheric pressure, 874
Atomic mass, 55
Atomic mass unit (amu), 55
Atomic number, 53
Atomic weight, 55
Atoms, 32, 53–54, 58, 86
ATP (adenosine triphosphate)
    in active transport, 121–122
    demand for, in skeletal muscle,
        44–45
    in endocytosis, 122
    locations, 445
    mechanism of action, 445
    mitochondrial production,
        965–970
    in muscle contraction, 330–331,
        342–346, 355
    production, 106–107, 343
    production in glycolysis, 965
    production in lipid
        catabolism, 973

ATP (adenosine triphosphate)
(Continued)
production in oxidative
phosphorylation, 969–970
production in protein
catabolism, 978
structure, 84
ATPase (adenosine
triphosphatase), 84
ATP synthase, 969
Atresia, 1096
Atria, 711, 715, 718
Atrial appendage, 712
Atrial baroreceptors, 772
Atrial fibrillation (AF), 730
Atrial natriuretic peptide (ANP)
in cardiovascular regulation,
774, 775
effects, 666
in fluid and electrolyte
balance, 1047
in glomerular filtration rate
regulation, 1016
targets, 666
Atrial reflex, 740, 772
Atrial systole, 734
Atrioventricular (AV) bundle, 727
Atrioventricular (AV) node, 721,
723, 726–727
Atrioventricular (AV) septal
defect, 800
Atrioventricular (AV) valves, 715,
718, 719, 720
Atrophy, muscle, 348
Atropine, 451
Attention deficit hyperactivity
disorder (ADHD), 529
Auditory association area, 513
Auditory cortex, 513
Auditory nerve. See
Vestibulocochlear nerve
Auditory ossicles, 242, 617, 618
Auditory pathways, 628
Auditory reflexes, 529
Auditory sensitivity, 628–629
Auditory tube, 160, 242, 617
Auricle (atrial appendage), 712
Auricle (ear), 160, 617
Auricular surface
ilium, 276
sacrum, 259
Auscultation, 40, 736
Autism, 529
Autoantibodies, 848
Autograft, 851
Autoimmune disease/disorders,
309, 844, 848
Autolysis, 103
Automated external defibrillator
(AED), 744
Automatic bladder, 1034
Automatic implantable
cardioverted defibrillator
(AICD), 744
Automaticity (autorhythmicity),
351, 723
Autonomic ganglia, 417, 559, 574
Autonomic nervous system (ANS)
autonomic tone, 572, 574, 740
dual innervation, 571–572
heart contractility and, 742

heart rate and, 739–740
organization, 558–559
overview, 413, 456, 558, 585–586
parasympathetic division. See
Parasympathetic division, ANS
vs. somatic nervous system, 558,
559, 577
sympathetic division. See
Sympathetic division, ANS
Autonomic plexuses, 571, 574
Autonomic (visceral) reflexes,
478, 529, 575–576
Autonomic tone, 572, 574, 740
Autopsy, 31, 172
Autoregulation
blood flow, 768–769
capillary, 758
definition, 33, 49
glomerular filtration rate, 1016
Autorhythmicity (automaticity),
351, 723
Autosomal chromosomes, 1149
Avascularity, epithelium, 141
Avogadro's number, 56
Axial muscles
head and neck, 370–377
oblique and rectus, 377, 380–381
overview, 367, 368–369, 408
pelvic floor, 381–383
trunk, 385–386
vertebral column, 377–381
Axial skeleton
joints, 302
overview, 233, 234
skull, 233–252
thoracic cage, 261–263
vertebral column, 252–261
Axillary artery, 781, 783
Axillary border, scapula, 270
Axillary lymph node dissection,
1109
Axillary nerve, 469, 471
Axillary vein, 793
Axis, 257–258, 302
Axoaxonic synapse, 449
Axolemma, 415
Axon(s)
classification, 439
diameter and propagation speed,
438–439
structure, 169, 414, 415
Axon hillock, 415
Axon terminal, 328, 415
Axoplasm, 415
Axoplasmic transport, 416
Azotemia, 1037
Azygos vein, 793

B

Babinski reflex/sign, 482
Baby teeth, 919
Back pain, 252–253
Bacteria, 841, 843, 946–947, 986
Bainbridge reflex, 740
Balance. See Equilibrium
Balanced diet, 983
Ball-and-socket joints, 297
Balloon angioplasty, 725
Band cells, 699, 700
Barium contrast x-ray, 46

Baroreceptor(s), 540, 542
Baroreceptor reflexes, 576,
770–771, 896
Bartholin's glands, 1107
Basal body, 99
Basal body temperature,
1111, 1112
Basal cell(s). See Stem cells
Basal cell carcinoma, 184
Basal compartment, seminiferous
tubule, 1085
Basal ganglia, 510
Basal lamina (basement
membrane), 141
Basal metabolic rate (BMR), 989
Basal nuclei, 508, 510–512,
552–553
Basal surface, epithelium, 141
Base
heart, 48, 712
patella, 280
sacrum, 261
Bases, 69, 85, 1056
Basic food groups, 983
Basic solution, 68, 1056
Basilar artery, 785
Basilar membrane, 623
Basilar zone, endometrium, 1103
Basilic vein, 793
Basolateral surface, epithelium, 141
Basophils
abundance, 698
functions, 153, 159, 697, 698, 841
histology, 693, 696, 697
B cells
functions, 697, 813–814, 841, 853
life span and circulation, 814
primary/secondary response to
antigens, 840–841
sensitization and activation, 815,
836–837
bcl-2 gene, 127
Bedsore (decubitus ulcer), 187
Behavior
brain chemistry and, 582
hormones and, 670
Behavioral drives, 506
Bell's palsy, 523
Belly, muscle, 359
Benadryl (diphenhydramine
hydrochloride), 849
Bends, the, 884
Benign prostatic hypertrophy
(BPH), 1094
Benign tumor, 132
Beta-adrenergic blockers (beta
blockers), 585, 742
Beta cells, 662, 665
Beta (β) chains, 685
Betahydroxybutyrate, 982
Beta-oxidation, 973, 974
Beta (β) receptors, 567, 571
Beta sheet, 78, 79
Beta waves, 516
Biaxial joint, 296
Biaxial movement, 291
Bicarbonate ions
absorption, 953
blood, 890
formation, 69
in plasma, 1010

reabsorption at proximal
convoluted tubule, 1018, 1019
reabsorption in collecting
system, 1023
secretion in collecting system,
1023, 1025
stomach, 926
in urine, 1010
Bicarbonate loading, 1072
Bicarbonate reserve, 1060,
1061, 1065
Biceps brachii muscle
actions, 304, 342, 364, 389–391
anatomy, 304
classification, 359, 360
innervation, 391
insertion, 273, 389–390
origin, 391
Biceps femoris muscle, 398
Bicuspids, 919
Bicuspid (mitral) valve, 718
Bifid, 256
Bile, 655, 936, 939–940
Bile canaliculi, 936
Bile duct, common, 933, 936, 937
Bile duct system, 936–937
Bile ductules, 936
Bile salts, 75, 939
Bilirubin, 687, 939, 1030
Biliverdin, 687
Binge-purge syndrome, 994
Biogenic amines (amino acid
derivatives), 444, 637, 639
Biopsy, 172
Biotin (vitamin B$_7$), 221, 987
Bipennate muscle, 360
Bipolar cells, retina, 602
Bipolar neurons, 416, 417
Birth canal, 1106
2,3-Bisphosphoglycerate (BPG),
889–890
Bitter receptors, 593
Bladder, urinary, 999, 1031–1032,
1034
Blast cells, 699, 700
Blastocoele, 1127
Blastocyst, 1127
Blastomeres, 1125, 1127
Bleaching, 611, 612, 613
Bleeding time, 704
Blepharitis, 630
Blindness, 607
Blind spot, 603
Blisters, 181
Blood. See also specific components
age-related changes, 801
coagulation. See Coagulation
collection of sample, 682
composition, 157–159, 680–681
distribution, 759
formed elements, 680–681,
698–699
functions, 679, 706
hemostasis. See Hemostasis
pH, 682
physical characteristics, 682
plasma. See Plasma
viscosity, 761
Blood bank, 706
Blood–brain barrier, 419, 490,
495–496

Blood clot, 703
Blood colloid osmotic pressure (BCOP), 766, 767, 1014
Blood–CSF barrier, 495
Blood doping, 689
Blood flow, 762
  autoregulation, 768–769
  during exercise, 775–776
  hormones and, 772–775
  neural mechanisms of control, 769–772
  renal, 1003
  to special regions, 778–779
Blood gas analysis, 886
Blood groups, 690–691
Blood pressure (BP)
  arterial, 763–764
  definition, 760, 762
  hormonal control, 772–775
  reflex control, 770–772
  regulation, 39
  renal, 1015
  response to hemorrhage, 777–778
  in small arteries and arterioles, 764
  venous, 764
Blood reservoir, 188
Blood smear, 682
Blood–testis barrier, 1085
Blood–thymus barrier, 814
Blood types
  compatibility testing, 692–693
  cross-reactions, 691, 692
  distribution in human population, 691
  overview, 690
Blood vessels. See also Arteries; Veins
  age-related changes, 801
  blood distribution in, 759
  capacitance, 759
  comparison of diameter, cross-sectional area, pressure, and velocity, 762–763
  pressure gradient, 760
  pulmonary circuit, 711, 780–781, 885
  resistance, 760–762
  systemic circuit. See Systemic circuit
  turbulence, 761–762
  wall structure, 751–752
"Blue baby," 774
Blue cones, 607
Body
  epididymis, 1087
  gallbladder, 940
  hyoid, 248
  mandible, 248
  muscle, 359
  pancreas, 933
  penis, 1090
  rib, 261
  scapula, 270
  sphenoid, 243
  sternum, 263
  stomach, 923
  tongue, 916
  uterus, 1101
  vertebra, 253

Bodybuilders, 672, 1023
Body cavities, 44–45
Body composition, 1043, 1044
Body fluids
  balance. See Fluid balance
  buffer systems. See Buffer systems
  composition, 1045
  electrolytes in, 67, 1045. See also Electrolyte balance
  pH, 67–68, 1058–1061
  regulation, 39
Body position, 339
Body stalk, 1132
Body temperature
  basal, 1111
  cardiac output and, 734
  control. See Thermoregulation
  uterine cycle and, 1112–1113
  variability in normal range, 37–38
Bohr effect, 888
Bolus, 865, 913
Bone
  age-related changes, 225, 228
  blood and nerve supplies, 218–220
  calcium and, 222–223
  vs. cartilage, 162–163
  cells, 209–210
  compact, 210–213, 227
  composition, 208–210, 222–223, 229
  as connective tissue, 152
  effects of exercise on, 220
  effects of hormones on, 220–221, 228
  fracture. See Fracture(s)
  growth and development, 220–221
  heterotopic, 214
  markings, 207
  matrix, 208–209
  metabolism, 163
  overview, 173, 229
  shapes, 206–207
  spongy, 208, 212–213, 227
  structure, 161, 208
Bone graft, 286
Bone markings, 207
Bone marrow, 212–213
Bone marrow biopsy, 706
Bone marrow transplant, 228
Bone mineral density (BMD) test, 228
Bone scan, 228
Bones listed
  auditory ossicles, 242, 617
  calcaneus, 283
  capitate, 274
  carpal, 273–274, 308
  clavicle, 262, 268, 270, 308
  coccyx, 252, 260, 261, 279
  coxal, 275, 308
  cuboid, 283
  cuneiform, 283
  ethmoid, 237, 238–239, 244–245
  femur, 208, 213, 280, 308
  fibula, 282–283, 308
  frontal, 237, 238, 241
  hamate, 274
  humerus, 271, 308
  hyoid, 248, 285
  ilium, 276, 279

  ischium, 275
  lacrimal, 247
  lunate, 274
  mandible, 237, 247–248, 279, 285
  maxillae, 237, 245–246
  metacarpal, 274–275, 308
  metatarsal, 284, 308
  nasal, 246
  nasal conchae, 244, 245, 247
  navicular, 283
  occipital, 237, 238–239, 285
  palatine, 246
  parietal, 208, 238
  patella, 206, 280
  phalanges, 275, 284, 308
  pisiform, 274
  pubis, 275
  radius, 273, 308
  ribs, 261–263
  sacrum, 252, 258–261, 279, 308
  scaphoid, 273
  scapula, 268, 270–271, 308
  sphenoid, 237, 238–239, 243–244
  sternum, 262–263, 308
  talus, 283, 308
  tarsal, 283, 308
  temporal, 237, 238–239, 241–243
  tibia, 282, 308
  trapezium, 274
  trapezoid, 274
  triquetrum, 274
  ulna, 273, 308
  vertebrae, 254–261
  vomer, 246–247
  zygomatic, 247
Bony labyrinth, 617, 619
Borborygmus, 956
Botulism, 354
Bouchard's nodes, 312
Bowman's (glomerular) capsule, 1004, 1007–1008
Boyle's law, 875
BPG (2,3-bisphosphoglycerate), 889–890
Brachial artery, 682, 754, 783, 784
Brachialis muscle, 365, 389, 391
Brachial plexus, 469–471
Brachial vein, 793
Brachiocephalic trunk, 781
Brachiocephalic vein, 793
Brachioradialis muscle, 389, 391
Brachium, 271
Bradycardia, 727
Brain
  activity monitoring, 515–518
  age-related changes, 583
  blood supply, 495, 778, 784–786
  landmarks, 488–490
  lateral view, 509
  midsagittal section, 509
  overview, 488, 530
  prenatal development, 490
  protection and support, 492–493
  regions, 488–490
  respiratory centers, 892–895
  tumors, 418
  veins, 789, 792, 793
  ventricles, 418, 490–491
Braincase (cranium), 233, 279
Brain natriuretic peptide (BNP)

  in cardiovascular regulation, 774, 775
  effects, 666
  in fluid and electrolyte balance, 1046–1047
  in glomerular filtration rate regulation, 1016
  targets, 666
Brain stem, 489, 497, 615
Brain vesicles, primary/secondary, 490
Brain waves, 516–517
Branched glands, 151
Breast, 1107, 1108
Breastbone (sternum), 262–263
Breast cancer, 1109, 1155
Breathing
  at birth, 898–899
  mechanics, 876, 878–880
  modes, 878, 880
  pressure changes during, 875–878
  protective reflexes, 898
  voluntary control, 898
Breech birth, 1143
Brevis (muscle name), 366
Broad ligament, 1094
Broca's area, 514
Brodmann, Korbinian, 515
Brodmann areas, 512, 515
Bronchi, 865, 866, 870
Bronchial artery, 787
Bronchial tree, 867
Bronchioles, 858, 867
Bronchitis, 867
Bronchoconstriction, 867
Bronchodilation, 867, 892
Bronchography, 901
Bronchomediastinal trunk, left/right, 812
Bronchopulmonary carcinoma, 901
Bronchopulmonary segment, 867
Bronchoscope, 901
Brown fat, 155, 993
Brown-Sequard syndrome, 483
Bruise (contusion), 187
Bruit, 762
Brunner's glands, 932
Brush border, 931
Brush border enzymes, 932, 950–951
Bruxism, 248
Buccal cavity. See Oral cavity
Buccal phase, swallowing, 922
Buccinator muscle, 367, 370, 372
Buffer(s), 69, 1056, 1058
Buffer systems
  carbonic acid–bicarbonate, 1058, 1060–1061
  hemoglobin, 1060
  interactions among, 1063, 1064, 1065
  overview, 69
  phosphate, 1058–1059
  protein, 1058, 1059–1060
Buffy coat, 684
Bulbar conjunctiva, 596
Bulbospongiosus muscle, 383, 1106, 1113
Bulbourethral glands, 1078, 1088, 1090

Bulging disc, 301
Bulimia, 994
Bulk transport, 122
Bundle branches, left/right, 727
Bundle of His (AV bundle), 727
Bunion, 294
Burkitt's lymphoma, 851, 1155
Burning/aching pain, 539
Burns, 197
Bursae, 293, 304
Bursitis, 294
Butyrate, 685

C

Cabot, Richard, 31
Cachexia, 172
Cadaver dissection, 31
Caffeine, 734
Calcaneal (Achilles) tendon, 283, 401
Calcaneus, 283
Calcification, 152, 215
Calcitonin (CT)
  bone remodeling and, 221
  calcium balance and, 223, 225, 1054
  functions, 655, 658
  regulatory control, 658
  secretion, 655
Calcitriol
  in bone growth and maintenance, 221
  in calcium homeostasis, 1054
  functions, 75, 665–666
  in growth, 670
  production, 665
  synthesis, 184
Calcium (Ca²⁺)
  absorption, 953
  in blood clotting, 705
  bone physiology and, 222–223, 230
  in cardiac contractions, 732
  control of blood levels, 222–225
  functions in human body, 54, 655, 985
  G proteins and, 641
  homeostatic regulation, 656–657, 1054
  imbalances, 1054, 1055
  in muscle contraction, 320, 328, 333–334
  normal ECF range, 1055
  reabsorption at distal convoluted tubule, 1022
  recommended daily allowance, 985
  as second messenger, 640, 641
  in urine, 1030
Calcium channel(s), slow/fast, 731
Calcium channel blockers, 725, 742
Calcium hydroxide, 209
Calcium phosphate, 209
Calcium reserve, skeleton as, 222
Calcium salts, 209
Calculi, 1034
Callus, 226
Calmodulin, 353, 641, 642
Calorie, 989
calorie, 989

Calorigenic effect, 654
Calorimeter, 989
Calsequestrin, 320
Calvaria, 235
cAMP (cyclic-AMP), 446, 567, 640–641
Canal (bone marking), 207
Canaliculi, 162, 209, 210
Canal of Schlemm (scleral venous sinus), 603
Cancellous (trabecular) bone, 208
Cancer
  aging and, 172
  breast, 1109, 1155
  cell division in, 132–133
  cervical, 1101
  CNS, 418
  colorectal, 943, 1155
  definition, 132
  lung, 901
  lymphatic system and, 814
  osteoporosis and, 228
  ovarian, 1100, 1155
  pancreatic, 956
  prostate, 1094, 1155
  skin, 184
  testicular, 1080
Canines, 919
Capacitance, blood vessel, 759
Capacitance vessels, 759
Capacitation, 1088, 1123
Capillary(ies)
  anatomy, 756
  autoregulation, 758
  continuous, 753, 756
  diffusion in, 765
  fenestrated, 646, 753, 756–757
  filtration in, 765
  lymphatic, 810
  overview, 157, 711, 750, 755–756
  reabsorption in, 765–766
  sinusoidal, 756
  structure, 753, 756
  vasomotion, 758
Capillary bed, 757–758
Capillary exchange, 760, 765–766, 803
Capillary hydrostatic pressure (CHP), 760, 765, 766
Capillary plexus, 757–758
Capillary pressure, 760, 765–768
Capitate, 274
Capitulum. See Head (bone part)
Capsular colloid osmotic pressure, 1014, 1015
Capsular epithelium, 1007
Capsular hydrostatic pressure (CsHP), 1014, 1015
Capsular ligaments, 293
Capsular space, 1005, 1007
Capsule, 157, 248
Carbaminohemoglobin, 686, 890
Carbohydrases, 934, 948, 950
Carbohydrate(s)
  breakdown and synthesis, 972
  definition, 70
  digestion and absorption, 948–951
  in human body, 1044
  metabolism, 965–973, 980–981, 995

overview, 69–72, 85, 87
  in plasma expanders, 683
  plasma membrane, 96
  in urine, 1030
Carbohydrate depletion/ loading, 973
Carbohydrate loading, 973
Carbon (C), 54
Carbon dioxide (CO₂)
  binding to hemoglobin, 890
  Bohr effect, 888
  cerebrospinal fluid, 772–773
  partial pressures, 883
  structure, 60
  transport, 889–890
Carbonic acid (H₂CO₃), 69, 890, 1056
Carbonic acid–bicarbonate buffer system, 1060–1061, 1063, 1065
Carbonic anhydrase, 888, 1022, 1056
Carbon monoxide (CO), 443, 445, 446, 889
Carboxyl group, 70, 77
Carboxyl ion, 78
Carboxypeptidase, 935, 950
Carbuncle, 201
Carcinogen, 132
Carcinoma, 172. See also Cancer
Cardia, 923
Cardiac arrest, 730, 744
Cardiac arrhythmias, 728–729, 730
Cardiac centers, 496, 739, 769
Cardiac conduction system. See Conducting system, heart
Cardiac cycle
  heart sounds, 736–737
  overview, 732–733
  phases, 733–734
  pressure and volume changes, 734–736
Cardiac muscle cells (cardiac myocytes; cardiocytes), 168, 350, 715, 716, 729, 731
Cardiac muscle tissue, 356
  function, 167, 351
  locations, 167
  overview, 166
  vs. skeletal muscle tissue, 353, 715, 716
  vs. smooth muscle tissue, 353
  structure, 350–351, 715, 716
Cardiac notch, 867
Cardiac output (CO)
  abnormal conditions affecting, 734
  calculation, 737
  control, 743–744
  exercise and, 775–776
  factors affecting, 738–742
Cardiac pacemaker. See Sinoatrial (SA) node
Cardiac plexus, 572, 739
Cardiac reflexes, 739
Cardiac reserve, 744
Cardiac skeleton, 721
Cardiac sphincter, 921
Cardiac tamponade, 712, 748
Cardiac troponin I, 725
Cardiac troponin T, 725

Cardiac veins, 715, 722
Cardinal ligaments, uterus, 1101
Cardinal signs and symptoms, inflammation, 827
Cardioacceleratory center, 738, 769, 770
Cardioacceleratory reflex, 576
Cardiocytes. See Cardiac muscle cells
Cardiodynamics, 737, 746. See also Cardiac output; Heart rate; Stroke volume
Cardioinhibitory center, 738, 770
Cardiology, 744
Cardiomegaly, 744
Cardiomyoplasty, 745
Cardiopulmonary resuscitation (CPR), 901
Cardiovascular centers, 496, 498, 770, 772
Cardiovascular fitness, 775–776
Cardiovascular physiology
  arterial blood pressure, 763–764
  capillary exchange, 765–768
  elastic rebound, 764
  key terms and relationships, 762–763
  overview, 760
  pressures in small arteries and arteriole, 764
  pressures in systemic circuit, 760
  total peripheral resistance, 760–762
  venous pressure and venous return, 764–765
Cardiovascular system. See also Blood; Blood vessels; Heart
  age-related changes, 801
  autoregulation of blood flow, 768–769
  blood distribution in, 759
  blood supply to special regions, 778–779
  changes at birth, 799
  circulation patterns, 779–780
  congenital problems, 800
  fetal, 798–799
  hormones and, 772–775
  integration with other systems, 802
  neural mechanisms of control, 769–772
  overview, 34, 679, 711, 745
  parasympathetic control of, 574
  prenatal development, 1137
  at puberty, 1146
  pulmonary circuit, 711, 780–781
  regulatory mechanisms, 768–775, 804
  respiratory system and, 899, 901
  response to exercise, 407, 775–776
  response to hemorrhage, 777–778
  smooth muscle in, 351
  sympathetic control of, 574–575
  systemic circuit, 711, 781–798
Carditis, 721
Carina, 866
Carotene, 182, 612
Carotid artery, 752, 770, 781, 784–785

Carotid bodies, 543, 772
Carotid canal, 237, 239, 242
Carotid sinus, 542, 770, 784
Carotid sinus massage, 801
Carpal bones, proximal/distal, 273, 308
Carpal tunnel syndrome, 275, 392
Carpus (wrist), 273
Carrier-mediated transport, 120–122, 136, 1010–1011
Carrier proteins, 95, 1011
Carriers, 1154
Cartilage
vs. bone, 162–163
growth, 159–160
metabolism, 163
overview, 173
structure, 152, 159, 163
types, 160–162
Casts, 1034
Catabolism, 63, 999
Catalysts, 64
Cataract, 604
Catecholamines, 637, 639
Catechol-O-methyltransferase (COMT), 567
Cathartics, 956
Catheter, cardiac, 725
Cations, 57
Cauda equina, 459
Caudal, 43
Caudate lobe, liver, 935
Caudate nucleus, 510, 511
Cavernous sinus, 793
C cells (parafollicular cells), 223
C (clear) cells, thyroid gland, 655
CD3 receptor complex, 833
CD4 markers, 833
CD4 T cells, 833, 835
CD8 markers, 833
CD8 T cells, 833–835
Cdc2 protein, 129
CD markers, 832
Cecum, 930, 943
Celiac ganglion, 564
Celiac plexus, 572
Celiac trunk, 564, 788
Cell(s)
abnormal growth and division, 132–133
cytoplasm and organelles, 92–93, 96–107
differentiation, 134
division, 127–129
life cycle, 127–132, 136–137
model, anatomy, 92–93
nucleus. See Nucleus
overview, 31, 32, 85–86, 91, 135
plasma membrane. See Plasma membrane
Cell adhesion molecules (CAMs), 142
Cell biology, 91
Cell body, neuron (soma), 91, 169, 414–415
Cell junctions, 141, 143
Cell-mediated (cellular) immunity, 697, 813, 815, 828, 842. See also Immune response
Cell membrane. See Plasma membrane

Cell physiology, 31
Cell theory, 91
Cellular level of organization, 32, 34, 91. See also Cell(s)
Cellular metabolism, 962–964
Cellular respiration (aerobic metabolism), 107, 343, 873, 965, 971
Cellular trophoblast, 1127
Cellulose, 71
Cementum, 918
Centimeters of water (cm H₂O), 876
Central adaptation, 538
Central adiposity, 189
Central canal, spinal cord, 418, 420, 463
Central (Haversian) canal, bone, 211
Central chemoreceptors, 896
Central nervous system (CNS). See also Brain; Spinal cord
age-related changes, 583
cardiovascular centers and, 772
neuroglia, 418–423
overview, 412, 456–457
primary tumors, 418
secondary tumors, 418
Central pattern generator, 892
Central retinal artery, 603
Central retinal vein, 603
Central sulcus, 508
Central vein, 935
Centrioles, 92, 99, 100
Centromere, 131
Centrosome, 92, 99
Cephalic (cranial), 43
Cephalic phase, gastric activity, 928
Cephalic vein, 793
Cerebellar cortex, 488, 500
Cerebellar hemispheres, 498, 501
Cerebellar nuclei, 500, 501
Cerebellar peduncles, superior/middle/inferior, 497, 500, 501
Cerebellum, 488, 489, 499–501, 530, 553
Cerebral aqueduct, 490
Cerebral arterial circle, 786
Cerebral artery, anterior/middle/posterior, 785, 786
Cerebral cortex, 488, 508, 578. See also Cerebrum
Cerebral hemispheres, 488
Cerebral meningitis, 459
Cerebral palsy, 535, 553, 556
Cerebral peduncles, 502, 550
Cerebral surface, 241
Cerebral vein, superficial/inferior, 789, 793
Cerebrospinal fluid (CSF)
carbon dioxide, 772–773
formation and circulation, 493–494
functions, 418, 460, 493, 494
pH, 772–773
Cerebrovascular accident (stroke), 488, 495, 533, 583, 778, 786
Cerebrovascular diseases, 495

Cerebrum
basal nuclei, 508, 510–512
cortex, 488, 508
hemispheric lateralization, 515
motor and sensory areas, 512–515
overview, 488, 489, 531
white matter, 508, 509
Cerumen, 194, 617
Ceruminous glands, 194, 617
Cervical canal, 1101
Cervical cancer, 1101
Cervical curve, 252
Cervical dysplasia, 1117
Cervical enlargement, 457
Cervical plexus, 468–469
Cervical vertebrae, 254, 255–257
Cervix, 146, 1101
Cesarean section, 1142
cGMP (cyclic guanosine monophosphate), 610, 611, 640
Chalazion, 596
Chalones, 132
Channel-mediated diffusion, 118
Channels, plasma membrane, 95
Characters, phenotypic, 1149
Charley horse, 407
Cheekbone, 242
Cheeks, 915
Chemical bonds, 57–61, 86
Chemical gradients, 117, 426
Chemical level of organization
atoms and molecules, 53–56
chemical bonds, 57–62
chemical reactions, 62–65
chemicals and cells, 85–86
inorganic compounds, 65–69
organic compounds, 69–85
overview, 32, 34, 86–88
Chemically gated (ligand-gated) channels, 428, 429
Chemical notation, 58
Chemical reactions, 58, 62, 63–64, 87
Chemical synapses, 440
Chemiosmosis, 969
Chemiosmotic phosphorylation, 969
Chemistry, 53
Chemoreceptor(s), 539, 543
Chemoreceptor reflexes, 772, 896–897
Chemosensitive area, 896
Chemotaxis, 693, 824
Chemotherapy, 49
Chenodeoxycholate, 939
Chest, 783, 793, 794
Cheyne-Stokes breathing, 901
Chickenpox, 465
Chief cells
gastric gland, 925
parathyroid, 656
Childbirth. See Delivery; Labor
Childhood, 1144, 1146, 1147
Chlamydia, 1114
Chlamydia pneumoniae, 755
Chloride (Cl⁻)
absorption, 953
functions in human body, 54, 985
homeostatic regulation of, 1055
imbalances, 1055

ion formation, 59
normal ECF range, 1055
in plasma, 1010
reabsorption at distal convoluted tubule, 1023
reabsorption at nephron loop, 1020
recommended daily allowance, 985
in urine, 1010, 1030
Chloride shift, 890, 1060
Cholate, 939
Cholecalciferol (vitamin D₃), 184–185, 665
Cholecystitis, 940
Cholecystokinin (CCK), 929, 940, 941, 942, 989
Cholelithiasis, 956
Cholera, 956
Cholesterol
dietary fats and, 977
functions, 74–75
serum levels, 977
sources, 75
steroid hormone synthesis from, 1112
structure, 75
transport and distribution, 975–976
Choline, 442
Cholinergic receptors, 571
Cholinergic synapses, 440–442, 566
Cholinesterase. See Acetylcholinesterase
Chondrocytes, 159, 170
Chondroitin sulfates, 159
Chondromalacia, 312
Chordae tendineae, 718
Chorion, 1129, 1130, 1131
Chorionic villi, 1129, 1134
Chorionic villus sampling, 1156
Choroid, 601
Choroid plexus, 493, 494, 495, 756
Chromatid, 131, 1082
Chromatin, 109
Chromium, 985
Chromosomal abnormalities, 1153, 1156
Chromosomal microtubule, 131
Chromosomes, 109, 1155
Chronic, 49
Chronic fatigue syndrome, 851
Chronic obstructive pulmonary disease (COPD), 902
Chronic renal failure, 1035
Chronic respiratory acidosis, 1065
Chylomicrons, 932, 951, 975
Chyme, 923
Chymosin (rennin), 926, 950
Chymotrypsin, 935, 950
Chymotrypsinogen, 935, 950
Cialis, 1114
Ciliary body, 601
Ciliary dyskinesia, 138
Ciliary ganglion, 520, 522, 568
Ciliary muscle, 601
Ciliary processes, 601
Ciliary zonule, 601
Ciliated epithelium, 141–142
Ciliopathies, 99
Cilium (cilia), 93, 99, 100, 141

C₁ inactivator, 705
Cingulate gyrus, 506
Circadian rhythm, 615, 661
Circle of Willis, 786
Circular folds, small intestine, 930
Circular muscle (sphincter), 351, 361, 912
Circulation. *See* Blood flow
Circulatory pressure, 760, 762
Circumcision, 1091
Circumduction, 295, 296, 298
Circumferential lamellae, 211
Circumflex artery, 721
Circumventricular organs, 495
Cirrhosis, 935
Cis face, Golgi apparatus, 105
Cisterna(e), 93, 101
Cisterna chyli, 811
Citric acid cycle, 106, 343, 965–968, 971
CK-MB, 725
Classical pathway, complement activation, 826–827
Class II MHC proteins, 832
Class I MHC proteins, 832
Claustrum, 511
Clavicle, 262, 268, 270, 308
Clear layer, 43
Cleavage, 1124, 1125, 1126, 1127
Cleavage furrow, 131
Cleavage (tension) lines, skin, 187
Cleft palate, 246
Clinical anatomy, 31
Clitoris, 1106
Clone, 830
Closed fracture, 226
Closed reduction, 228
Clostridium difficile, 907, 960
Clostridium tetani, 332
Clot retraction, 703, 705
Clotting factors, 702, 703–704
Clotting system, 679, 700. *See also* Coagulation; Hemostasis
Clubfoot (talipes equinovarus), 284
Club hair, 191
Coagulation, 703
  calcium ions and, 705
  clot retraction, 703, 705
  clotting factors, 701, 702, 703–704
  common pathway, 703, 704
  extrinsic pathway, 703, 704
  feedback control, 38, 704–705
  interaction among pathways, 704
  intrinsic pathway, 703, 704
  vitamin K and, 705
Coagulation time, 704
Coated vesicles, 123
Cobalamin. *See* Vitamin B₁₂
Cobalt, 985
Cocaine, 443
Coccygeal cornua, 261
Coccygeal ligament, 457, 460
Coccygeus muscle, 383
Coccyx, 252, 260, 261, 279, 302
Cochlea, 618, 624
Cochlear branch, vestibulocochlear nerve, 524, 628
Cochlear duct (scala media), 618, 623

Cochlear nuclei, dorsal/ventral, 499, 524, 628
Coding strand, 111
Codominance, 1150, 1151
Codon, 111
Coenzyme(s), 81
Coenzyme A (CoA), 442, 963
Coenzyme Q, 969
Cofactor, 81
Coitus (sexual intercourse), 1077, 1113–1114
Cold sore, 201
Colic artery, right/middle/left, 788
Colic flexure, right/left, 943
Colic vein, right/middle/left, 797, 798
Colitis, 956
Collagen, 80
Collagen fibers, 154, 186
Collagenous (dense connective) tissues, 157
Collateral ganglia, 562, 563, 564
Collateral ligament
  fibular, 306
  radial, 304
  tibial, 306
  ulnar, 304
Collaterals
  arteries, 757
  axons, 415
Collecting duct, 1006, 1009, 1026
Collecting system
  anatomy, 1006, 1009
  effects of antidiuretic hormone, 1025, 1026
  functions, 1009, 1013
  histology, 1006
  mechanisms, 1013
  reabsorption and secretion along, 1023
Colles fracture, 227
Colliculus, superior/inferior, 501, 502, 552
Colloid osmotic pressure, 1014, 1015
Colloids, 67, 652
Colon, 943–945
Colonoscopy, 943
Colony-stimulating factors (CSFs), 700, 847
Color blindness, 613, 1154, 1155
Colorectal cancer, 943, 1155
Color vision, 613
Colostrum, 1145
Columnar epithelium, 144, 146, 148, 149
Columns (funiculi), white matter, 463
Coma, 580
Combining forms, 29
Comedo, 201
Comminuted fracture, 227
Commissural fibers, 508
Common bile duct, 933, 936, 937
Common pathway, coagulation, 703, 704
Commotio cordis, 745
Communicating artery, anterior/posterior, 786
Compact (dense) bone, 208, 210–213, 227

Compartment syndrome, 407
Compatibility, blood type, 692
Compazine (prochlorperazine), 619
Compensation curves, 252
Complementary base pairs, 82
Complementary gene action, 1152
Complementary strands, 82
Complement (C) proteins, 696, 825
Complement system, 823, 825–827
Complete antigen, 838
Complete proteins, 983
Complete tetanus, 337, 338
Compliance, lungs, 875
Compound, 57
Compound fracture, 226
Compound glands, 151
Compression, 325
Compression fracture, 226
COMT (catechol-O-methyltransferase), 567
Concentration, 67
Concentration gradient, 116, 117, 765
Concentric contractions, 340, 341
Concentric lamellae, 211
Conception (fertilization), 1077, 1123–1125
Concussion, 529
Condensation reaction, 63
Conducting arteries, 752
Conducting cells, 723
Conducting portion, respiratory tract, 858, 872
Conducting system, heart
  AV bundle, 723, 727
  AV node, 723, 726–727
  bundle branches, 723, 727
  calcium ions and, 732
  conduction deficits, 727
  contractile cells, 729–732
  electrocardiogram, 728–729
  energy for, 732
  overview, 718, 723, 726, 746
  physiology, 723
  Purkinje fibers, 723, 727–728
  SA node, 723, 726
Conduction, 433, 990
Conduction deficits, 728
Conductive deafness, 630
Condylar joint, 297
Condylar process, 248
Condyles
  definition, 207
  femur, medial/lateral, 280
  humerus, 273
  tibia, medial/lateral, 282
Cones, 601, 607, 609
Congenital heart disease, 800
Congenital talipes equinovarus (clubfoot), 284
Congenital thymic aplasia, 851
Congestive heart failure, 745
Conjoined twins, 1143
Conjunctiva, 596
Conjunctivitis, 597
Connective tissue(s)
  fluid, 152, 157–159. *See also* Blood
  heart, 721
  proper. *See* Connective tissue proper

supporting, 152, 159–163. *See also* Bone; Cartilage
Connective tissue proper
  cell populations, 153
  classification, 152
  components, 152
  dense, 157
  embryonic, 154–155
  fascia, 165–166
  fibers, 154
  functions, 152
  ground substance, 154
  loose, 155–157
  mucous, 155
  overview, 152, 173
Connexons, 142, 439
Consciousness, 580–582
Consensual light reflex, 529, 575, 576, 600
Constant segments, antibody, 837
Continuous capillary, 756
Continuous passive motion (CPM), 310
Continuous propagation, 433
Contractile cells, 723, 729–732
Contractile proteins, 77
Contractility, 742
Contraction, skeletal muscle
  control, 328
  cycle, 328, 330–332
  electrical impulses in, 328
  energy sources, 342–346
  energy uses, 344–345
  excitation–contraction coupling, 328–329
  fatigue and, 344
  initiation, 333
  isometric, 340
  isotonic, 339–340
  load and speed, 340, 342
  muscle tone and, 339
  opposing, 342
  recovery period, 344–346
  relaxation after, 333, 334, 342
  sarcomere appearance during, 325
  shortening during, 334
  sliding filament theory, 324–325
  tension production, 338
Contraction, smooth muscle, 353–354
Contraction cycle, 328, 330–332
Contraction phase, twitch, 337
Contralateral reflex arcs, 477, 481
Control center, 33, 36
Contusion (bruise), 187
Conus arteriosus, 718
Conus medullaris, 457
Convection, 990
Convergence
  neural circuits, 474
  vision, 614
Convergent muscle, 359–360
Coombs test, 851
Copper, 985
Copulation (sexual intercourse), 1077, 1113–1114
Coracoacromial ligament, 303
Coracobrachialis muscle, 386, 389
Coracoclavicular ligament, 303
Coracohumeral ligament, 303
Coracoid process, 270

Cords, spinal nerves, 469
Cori cycle, 344
Cornea, 597, 599
Corneal epithelium, 597
Corneal limbus, 599
Corneal reflex, 529
Corneal replacement, 599
Corniculate cartilages, 863
Cornification (keratinization), 181, 191
Coronal (frontal) plane/section, 44, 235
Coronal suture, 235, 236
Corona radiata, 1099, 1123
Coronary artery, left/right, 721
Coronary artery bypass graft (CABG), 725
Coronary artery disease (CAD), 724, 754
Coronary circulation, 721–722
Coronary ischemia, 724
Coronary sinus, 715
Coronary spasms, 778
Coronary sulcus, 712
Coronary thrombosis, 725
Coronoid fossa, 273
Coronoid process
   mandible, 248
   ulna, 273
Corpora cavernosa, 1092
Corpora quadrigemina, 497, 501
Cor pulmonale, 745
Corpus albicans, 1099
Corpus callosum, 508, 510
Corpus luteum, 667, 1099
Corpus spongiosum, 1092
Corpus striatum, 510
Corrugator supercilii muscle, 372
Cortex
   adrenal, 657–660
   bone, 208
   cerebellar, 488, 500. See also Cerebellum
   cerebral, 488, 508, 578. See also Cerebrum
   hair, 191
   ovary, 1096
   renal, 1002
   thymus, 819
Cortical bone, 208
Cortical nephron, 1007, 1012
Cortical radiate artery, 1002
Cortical radiate vein, 1003
Cortical reaction, 1125
Corticobulbar tracts, 550–551
Corticospinal pathway, 550–551
Corticospinal tracts, anterior/lateral, 550, 551
Corticosteroids, 75, 657
Corticosterone, 658
Corticotropin (adrenocorticotropic hormone), 183, 648, 651, 989
Corticotropin-releasing hormone (CRH), 648
Cortisol, 658, 659
Cortisone, 658
Costae (ribs), 261–263
Costal breathing, 878
Costal cartilages, 261
Costal facets, vertebrae, 258

Costal groove, 261
Costal processes, vertebrae, 256
Costimulation, 834
Cotransport (symport), 120–121, 1011
Cough, 857, 905
Coughing reflex, 576, 865, 898
Countercurrent exchange, 991, 992
Countercurrent multiplication, 1019–1022, 1038
Countertransport, 121, 1011
Coupling factors, 969
Covalent bonds, 60–61
Cowper's (bulbo-urethral) glands, 1089, 1090
Coxal bone, 276, 308
Cradle cap, 193
Cranial (cephalic), 43
Cranial bones, 233, 235, 240–245
Cranial cavity, 233, 238
Cranial meninges, 459, 491–493
Cranial nerves
   abducens (VI), 521, 528
   accessory (XI), 527, 528
   classification, 518–519
   facial (VII), 242, 523, 528
   glossopharyngeal (IX), 525, 528, 896
   hypoglossal (XII), 240, 527–528
   oculomotor (III), 520, 528
   olfactory (I), 519, 528
   optic (II), 520, 528
   origins, 518
   overview, 412, 518–519, 528, 531
   trigeminal (V), 522, 528
   trochlear (IV), 521, 528
   vagus (X), 525–527, 528, 896
   vestibulocochlear (VIII), 524, 528, 621, 628
Cranial reflexes, 456, 479, 529
Cranial root, accessory nerve, 527
Craniosacral division, ANS. See Parasympathetic division, ANS
Craniostenosis, 252
Craniotomy, 263
Cranium, 233, 279
Creatine, 342, 984
Creatine kinase (CK), 343
Creatine phosphate (CP), 342, 1009
Creatinine, 1009, 1010, 1030
Creatinine clearance test, 1015
Cremasteric reflex, 1080
Cremaster muscle, 1080
Crenation, 119
Crest (bone marking), 207
Creutzfeldt-Jakob disease (CJD), 529
Crib death, 893
Cribriform plate, 238, 244
Cricoid cartilage, 863
Crista ampullaris, 621
Cristae, mitochondria, 103
Crista galli, 238, 244
Crohn's disease, 956
Cross-bridges, 324, 330
Crossed extensor reflexes, 477, 481
Crossing over, 1082, 1153
Cross-match testing, 693
Cross-reactions, in transfusions, 691, 692

Cross (transverse) section, 44
Cross-training, 349
Crown, tooth, 918
Crude touch and pressure receptors, 540, 546
Crura, penis, 1092
Cryptorchidism, 1079–1080
Crypts of Lieberkühn, 932
Crystal arthritis, 309
Crystallins, 604
CT (computed tomography) scans, 46
Cubital vein, median, 793
Cuboid, 283
Cuboidal epithelium, 144, 146, 147
Cuneiform bone, medial/intermediate/lateral, 283
Cuneiform cartilages, 160, 863
Cupula, 621
Current, 425
Cushing's disease, 668
Cuspids, 919
Cusps, 717–718
Cutaneous membrane, 164, 165, 177, 178. See also Skin
Cutaneous plexus, 187
Cuticle
   hair, 191
   nail, 195
Cyanide ion, 969
Cyanosis, 183, 800
Cycles, sound, 624
Cyclic-AMP (cAMP), 446, 567, 640–641
Cyclic-GMP (cyclic guanosine monophosphate; cGMP), 610, 611, 640
Cyclin, 129
Cyclosporin A (CsA), 832
Cyst
   breast, 1109
   ovarian, 1117
Cysteine, 1059
Cystic artery, 788
Cystic duct, 937
Cystic fibrosis, 53, 89, 860, 1155
Cystic vein, 798
Cystocele, 1037
Cystoscopy, 1037
Cytochromes, 969
Cytokines, 846–847, 814, 825, 844
Cytokinesis, 131
Cytology
   definition, 31, 91
   exfoliative, 146, 1106
   overview, 49
Cytoplasm, 92, 96
Cytosine, 82
Cytoskeleton, 92, 97, 98
Cytosol. See Intracellular fluid
Cytotoxic reactions, 848
Cytotoxic T (T_C) cells, 813, 815, 831, 834, 841
Cytotrophoblast (cellular trophoblast), 1127

D

DAG (diacylglycerol), 642
Dalton, 55
Dalton's law, 882

Dandruff, 193
Dark-adapted state, 613
Dark current, 610, 611
Dartos muscle, 1080
"Date rape" drug, 672
Daughter cells, 127
Daughter chromosomes, 131
D cells, 926
Deamination, 937–939, 978, 979
Decarboxylation, 965
Decibels, 625
Decidua basalis, 1132
Decidua capsularis, 1132
Decidua parietalis, 1133
Deciduous teeth, 919
Decomposition reaction, 63
Decompression sickness, 884
Decubitus ulcer (bedsore), 187
Decussation, 496, 548
Decussation of pyramids, 496
Deep, 43
Deep breathing, 878
Deep cortex, lymph node, 816
Deep fasciae, 165, 166
Deep femoral artery, 789
Deep femoral vein, 794
Deep lymphatics, 811
Deep sleep, 580
Deep tendon (stretch) reflexes, 475, 476, 478, 479
Deep venous thrombosis (DVT), 801
Defecation, 576, 909, 948
Defecation reflex, 947–948
Defensins, 696
Deferential artery, 1080
Deflation reflex, 897
Degenerative arthritis, 309
Degenerative joint disease, 309
Deglutition (swallowing), 863, 866, 922
Degranulation, 696
Dehydration (water depletion), 1049
Dehydration synthesis, 63, 71, 82
Dehydroepiandrosterone (DHEA), 1092
Delayed hypersensitivity, 848
Delayed-onset muscle soreness (DOMS), 349
Delirium, 529
Delivery, 1142, 1143. See also Labor
Delta cells, 662, 665
Delta waves, 516
Deltoid muscle, 273, 361, 365, 386, 389
Deltoid tuberosity, humerus, 271
Dementia, 529
Demyelination, 421
Denaturation, 81
Dendrites, 169, 414, 415
Dendritic (Langerhans) cells, 179, 816, 832
Dendritic spines, 415
Dens, 257
Dense area, 143
Dense bodies, 352
Dense (compact) bone, 208, 210–213
Dense connective tissues, 152, 157, 158

Dense irregular connective tissue, 157, 158
Dense layer
    basement membrane, 144
    glomerular capillaries, 1007
Dense regular connective tissue, 157, 158
Dental calculus, 919
Dental caries, 954
Dental plaque, 919
Dental succession, 919
Dentate gyrus, 506
Denticulate ligaments, 461
Dentin, 918
Deoxyhemoglobin, 685
Deoxyribonucleic acid. See DNA
Depolarization, 328, 430, 432
Depression (movement), 299
Depressor anguli oris muscle, 372
Depressor labii inferioris muscle, 372
Depth perception, 615
Dermal bones, 218, 219
Dermal (intramembranous) ossification, 218, 219
Dermal papillae, 179
Dermatitis, 186
Dermatology, 201
Dermatome, 464, 465
Dermatomyositis, 851
Dermicidin, 194
Dermis, 177, 202
    blood supply, 187
    innervation, 187–188
    lines of cleavage, 187
    organization, 186
    properties, 186
Descending aorta, 719, 786–789
Descending colon, 943
Descending limb, nephron loop, 1008, 1028–1029
Descending (motor) pathways, 549–554
Descending tracts, 463, 498, 499
Descent, testes, 1078–1079
Desmin, 352
Desmopressin, 649
Desmosome, 142
Detached retina, 600
Detoxification, 101
Detrusor muscle, 1032
Detumescence, 1113
Deuterium, 54, 55
Development, 1122, 1157. See also Postnatal development; Prenatal development
Developmental anatomy, 31
Developmental dyslexia, 515
Deviated nasal septum, 251, 263
Dextran, 119
DHEA (dehydroepiandrosterone), 1092
Diabetes insipidus, 649, 668, 1025
Diabetes mellitus
    genes associated with, 1155
    insulin-dependent (type 1), 663, 664, 848
    non-insulin-dependent (type 2), 663, 664
    signs and symptoms, 668

Diabetic nephropathy, 664
Diabetic neuropathy, 664
Diabetic retinopathy, 600, 664
Diabetogenic effect, 649
Diacylglycerol (DAG), 642
Diapedesis, 693, 824
Diaphragm
    actions, 381, 878
    gross anatomy, 45
    innervation, 318, 381
    insertion, 263, 381
    origin, 381
Diaphragmatic breathing, 878
Diaphragmatic hernia, 400
Diaphysis, 208, 268. See also Shaft
Diarthrosis, 290, 291
Diastole, 732
Diastolic pressure, 763
Diazepam (Valium), 579
Dicrotic notch, 736
Diencephalon, 488, 489, 497, 503–506, 531
Diet
    balanced, 983
    disease and, 987
    energy gains and losses, 988–989
    fats and cholesterol, 977
    nutrition and, 983–987
Differential count, WBCs, 698
Differentiation, 134, 1122
Difficult deliveries, 1143
Diffusion
    across plasma membranes, 117–119
    in capillary exchange, 765
    channel-mediated, 118
    facilitated, 121, 124
    factors influencing rate, 117
    liquid-gas, 883–884
    osmosis, 118–119, 124
    overview, 113–115, 124, 136
    principles, 115–117
    respiratory function and, 884–886
    simple, 117–118
Digastric muscle, 375, 376, 377
Digestion
    carbohydrates, 948–951
    chemical events, 949, 950
    ion absorption, 952–953
    lipids, 949, 951
    overview, 908–909, 956, 958
    proteins, 951–952
    vitamin absorption, 953
    water absorption, 952
Digestive epithelium, 911–912
Digestive system
    age-related changes, 954
    disorders, 946
    enzymes, 950
    esophagus, 921–922
    functions, 908–909
    gallbladder, 940
    hormonal control, 914, 940
    integration with other systems, 954–955
    intestinal absorption, 941
    large intestine, 942–948
    liver, 935–940
    neural mechanisms, 914, 940
    oral cavity, 915–919
    overview, 35, 907–908
    pancreas, 933–935

parasympathetic effects, 573
peritoneum and, 909–911
prenatal development, 909, 910, 1137
regulation of activities, 913–914
small intestine, 927, 930–932
smooth muscle in, 351
stomach, 923–927
sympathetic effects, 573
Digestive tract
    definition, 907
    endocrine activity, 638, 665, 666
    epithelial stem cells, 912
    histological organization, 911–912
    movement of materials, 912–913
Digital artery, 784
Digitalis, 742
Digital rectal exam, 1094
Digital subtraction angiography (DSA), 47, 724
Digital vein, 793
Diglyceride, 74
Dihydrogen phosphate, 1059
Dihydrotestosterone (DHT), 1092
Dihydroxyacetone phosphate, 975
Dilation stage, labor, 1141–1142
Dimenhydrinate (Dramamine), 619
Dipeptidases, 950, 952
Dipeptide, 78
Dipeptides, 952
Diphenhydramine hydrochloride (Benadryl), 849
Diphtheria, 421
Diploë, 208
Diploid cells, 1082
Direct communication, intercellular, 635–637
Directional terms, 42, 43
Direct light reflex, 529, 576
Disaccharides, 71, 72, 950
Disconnection syndrome, 514
Discs, photoreceptor, 607
Disease, 39
Dislocation (luxation), 294
Displaced fracture, 226
Disseminated intravascular coagulation, 706
Dissociation, 66
Distal, 43
Distal convoluted tubule (DCT)
    anatomy, 1005–1007, 1008–1009
    effects of antidiuretic hormone, 1025, 1026
    functions, 1008, 1013, 1028–1029
    mechanisms, 1013
    reabsorption and secretion at, 1022–1024, 1025
Distensible, 759
Distribution (muscular) arteries, 753, 754
Disulfide bond, 79
Disulfiram (Antabuse), 988
Diuresis, 1023
Diuretics, 1023
Divergence, neural circuits, 474
Diverticulitis, 956
Diverticulosis, 956
Dizygotic twins, 1143
DNA (deoxyribonucleic acid)
    cell control by, 113

coding strand, 111
functions, 82, 136
mutations, 112
organization in nucleus, 109–110
in protein synthesis, 110–113
replication, 128
vs. RNA, 82–84
structure, 82, 83
template strand, 111
transcription of mRNA, 110–111, 114–115, 136
triplet codes, 109
DNA fingerprinting, 110
DNA polymerase, 127, 128
Dobutamine, 742
Dominant hemisphere, 515
Dominant inheritance, 1150–1151
Dopamine, 443, 444, 512, 742
Dorsal (posterior), 43
Dorsal (arcuate) arch, 789
Dorsal interosseus muscles, 395, 402
Dorsalis pedis artery, 789
Dorsal ramus, 466–467
Dorsal respiratory group (DRG), 892, 894
Dorsal root(s), 459
Dorsal root ganglia, 459
Dorsal venous arch, 793
Dorsiflexion, 284, 299
Dorsum nasi, 861
Double covalent bond, 60
Down-regulation, 640
Down's syndrome (trisomy 21)
    Alzheimer's disease and, 585
    chromosomal abnormalities in, 1153, 1155, 1156
    diagnosis, 146
    heart defects and, 800
Dramamine (dimenhydrinate), 619
Drives, 506
Drugs
    absorption in stomach, 927
    hypodermal injection, 370
    immunosuppressive, 832, 848
    inactivation in liver, 939
    intradermal injection, 370
    intramuscular injection, 370
    plasma membrane and, 116
    subcutaneous injection, 188
d-tubocurarine, 451
Dual energy x-ray absorptiometry (DEXA), 228
Dual innervation, 571–572
Duchenne muscular dystrophy, 354
Duct(s), 149
Ductless glands, 149
Duct of Santorini (accessory pancreatic duct), 933
Duct of Wirsung (pancreatic duct), 933
Ductus arteriosus, 798, 898
Ductus deferens (vas deferens), 651, 1078, 1088, 1089
Ductus venosus, 798
Duodenal ampulla, 933, 940
Duodenal glands, 932
Duodenal papilla, 940
Duodenum, 923, 930, 941
Dural folds, 492
Dural sinuses, 492

Dura mater, 459–460, 491
Dust cells (alveolar macrophages), 823, 869
Dynamic equilibrium, 39
Dynein, 99, 416
Dynorphins, 443, 445
Dyscrasia, 706
Dyslexia, 515
Dysmenorrhea, 1104
Dysphagia, 956
Dysplasia, 134
Dyspnea, 902
Dysthymia, 451

E

Ear, 372, 616–619. *See also* Hearing
Eardrum (tympanic membrane), 242, 617, 629
Earwax (cerumen), 194, 617
Eating disorders, 994
Eccentric contraction, 340, 341
Ecchymosis, 706
Eccrine (merocrine) sweat glands, 193–194
Echocardiography, 745
Echogram, 47
Eclampsia, 1157
Ectoderm, 1129, 1132
Ectopic pacemaker, 728, 730
Ectopic pregnancy, 1128
Eczema, 201, 851
Edema, 767, 768, 1047
Effector(s), 33, 36, 413, 475
Efferent arteriole, glomerulus, 1005
Efferent division, PNS, 413
Efferent ductules, 1082
Efferent fibers, 417
Efferent lymphatics, 816
Efferent (motor) neurons, 417
Efferent vessels. *See* Arteries
Ehlers-Danlos syndrome, 140, 175
Eicosanoids, 73, 77, 637–638, 639
Ejaculate, 1090
Ejaculation, 576, 1077, 1090, 1113
Ejaculatory duct, 1078, 1088
Ejection fraction, 735, 737
Elastase, 935, 950
Elastic arteries, 752, 753
Elastic cartilage, 160–161
Elastic fibers, 154, 186
Elasticity, muscle relaxation and, 342
Elastic ligaments, 154
Elastic membrane, internal/external, 751
Elastic rebound, 721, 764, 880
Elastic tissue, 157, 158
Elastin, 154
Elbow joint, 273, 304, 389–391
Elective abortion, 1140
Electrical force, 53
Electrical gradients, 425, 426
Electrical synapses, 439–440
Electrocardiogram (ECG, EKG), 723, 728–729
Electrocardiography, 723
Electrochemical gradient, 117, 425–428
Electroencephalogram (EEG), 515–518

Electrolyte(s), 67
Electrolyte balance
    age-related changes, 1071–1072
    in burn patient, 197
    calcium, 1054
    chloride, 1055
    definition, 1043
    disorders, 1055
    in extracellular fluid, 1044–1045
    fluid balance and, 1047
    hormonal control, 1046–1047
    importance, 1050
    in intracellular fluid, 1044–1045
    magnesium, 1055
    normal ECF ranges, 1055
    overview, 1043, 1050–1051, 1073
    phosphate, 1054–1055
    potassium, 1053
    regulation, 1045–1046
    sodium, 1051–1053
Electron(s), 53, 57
Electron acceptor, 59
Electron cloud, 53
Electron donor, 59
Electron shell, 53
Electron transport system, 343, 968, 969–970, 971
Elements, 54–55
Elevation, 299
Ellipsoid (condylar) joints, 297
Embolism, 706
Embryo, 1101
Embryogenesis, 1126, 1133, 1134
Embryology, 31, 1122
Embryonic connective tissue, 154–155
Embryonic development, 1122
Embryonic disc, 1129
Embryonic stem cells, 134
Emesis, 929
Emigration, 693, 824
Emission, 651, 1077, 1090, 1113
Emmetropia, 608
Emphysema, 899, 901, 1151
Empty sella syndrome, 674
Emulsification, 939
Emulsion droplets, 940
Enamel, tooth, 918
Encyclopedia of DNA Elements (ENCODE), 1155
End-diastolic volume (EDV), 734, 737, 741–742
Endergonic reactions, 65
Endocarditis, 745
Endocardium, 714
Endochondral ossification, 215–217
Endocrine communication, 636
Endocrine glands, 149
Endocrine pancreas, 662. *See also* Pancreas
Endocrine reflexes, 642–644
Endocrine secretions, 149
Endocrine system. *See also specific glands and organs*
    control by endocrine reflexes, 642–644
    disorders, 668
    exercise and, 407
    hormones. *See* Hormones
    immune response and, 851

    integration with other systems, 673
    intercellular communication, 635–637
    overview, 34, 635, 637, 638, 674
    parasympathetic effects, 573
    prenatal development, 1137
    sympathetic effects, 573
Endocytosis, 103, 122–123, 126
Endoderm, 1129, 1132
Endogenous pyrogen (interleukin-1), 828, 846
Endolymph, 617
Endolymphatic duct, 621
Endolymphatic sac, 621
Endometriosis, 1117
Endometrium, 1101, 1104
Endomorphins, 443, 445
Endomysium, 318
Endoneurium, 464
Endoplasmic reticulum (ER), 93, 101–102
Endorphins, 443, 445, 540
Endosomes, 103, 122
Endosteum, 210, 213, 214
Endothelins, 701
Endothelium, 146
Endotracheal tube, 902
End-stage renal failure, 1035
End-systolic volume (ESV), 736, 737, 742
Endurance, 346, 348, 349
Enema, 1072
Energetics, 962, 995
Energy, 62–63, 428
Energy levels, electrons, 56
Energy reserves
    cardiac muscle, 351
    skeletal muscle, 342–343
Enkephalins, 443, 445, 540
Enlargements, spinal cord, 457
Enteric nervous system, 562, 576
Enterocrinin, 932, 941
Enteroendocrine cells, 911, 914
Enterogastric reflex, 929
Enterohepatic circulation, bile, 940
Enteropeptidase, 932, 950
Enuresis, 1037
Enzymes
    activation energy and, 64
    cofactors, 81
    pH changes and, 81
    plasma membrane, 95
    regulation, 81
    saturation limit, 81
    seminal fluid, 1090
    specificity, 80
    structure and function, 80–81, 87
    temperature changes and, 81–82
Eosin, 696
Eosinophils (acidophils), 153, 159, 693, 696, 698, 841
Ependyma, 418, 490
Ependymal cells, 418
Ependymocytes, 418
Epicardium, 712
Epicondyles, medial/lateral
    femur, 280
    humerus, 273
Epicranial aponeurosis, 370
Epicranium, 370, 372

Epidemiology, 49
Epidermal derivatives, 189
Epidermal growth factor (EGF), 132, 185
Epidermal ridges, 179
Epidermis
    overview, 178, 201–202
    pigmentation, 182–183
    structure, 179–181
    tumors, 184
    vitamin D_3 production in, 184–185
Epididymis, 1078, 1087–1088
Epidural block, 460
Epidural hemorrhage, 493
Epidural space, 460
Epigenetics, 1153, 1156
Epigenome, 1157
Epiglottis, 864
Epiglottitis, acute, 865
Epilepsies, 517, 1155
Epimysium, 318
Epinephrine
    effects, 660, 981
    heart rate and, 741
    muscle metabolism and, 346
    as neurotransmitter, 444
    overproduction, 668
    production in adrenal medulla, 565, 660
    regulatory control, 660
    in stress response, 671
    stroke volume and, 742
    sympathetic stimulation and release of, 567
Epineurium, 463
Epiphyseal cartilage (plate), 215, 217, 285
Epiphyseal closure, 215, 217
Epiphyseal fracture, 227
Epiphyseal line, 215, 217
Epiphysis, 208
Epiploic appendages, 943
Episiotomy, 1142
Epistaxis, 861
Epithalamus, 503
Epithelia (epithelium)
    characteristics, 140
    definition, 140
    digestive, 911–912
    glandular. *See* Gland(s)
    renewal and repair, 912
    respiratory, 859–860
Epithelial tissue
    attachment to basement membrane, 143–144
    classification, 144–145, 173
    columnar, 146
    cuboidal, 146
    functions, 141
    as innate defense, 822
    intercellular connections, 142–143
    maintenance and repair, 144
    overview, 140–141
    specializations, 141–142
    squamous, 145–146
    transitional, 146
Epitopes, 838
Eponychium, 195
Eponyms, 29
Equational division, 1084

Equilibrium, 39, 64
Equilibrium (balance)
  muscle tone and, 339
  overview, 536, 631
  pathways, 621–622, 623
  semicircular ducts, 619–622
  utricle and saccule, 621, 622
Equilibrium potential, 425
Equinovarus, 483
Equivalents (Eq), 1050
Erb's palsy (Erb-Duchenne
  palsy), 483
Erectile dysfunction (ED),
  1113–1114
Erectile tissue, 1091, 1106
Erection, 1091–1092, 1113
Erector spinae muscles, 377–379
Eruption, teeth, 919
ERV (expiratory reserve
  volume), 881
Erythema, 197
Erythroblast, 687
Erythroblastosis fetalis, 695
Erythrocytes. See Red blood cells
Erythrocytosis, 690
Erythropoiesis, 687, 689–690
Erythropoietin (EPO;
  erythropoiesis-stimulating
  hormone)
  abuse, 672
  cardiovascular regulation
    and, 775
  effects and targets, 132, 666,
    684, 689
  secretion, 666, 1009
Esophageal artery, 787
Esophageal hiatus, 400, 921
Esophageal phase,
  swallowing, 922
Esophageal plexus, 572
Esophageal sphincter, upper/
  lower, 921
Esophageal varices, 956
Esophageal vein, 793
Esophagus, 921–922, 957
Essential amino acids, 979
Essential fatty acids, 975
Estradiol. See also Estrogen(s)
  effects, 1109, 1115
  in males, 1094
  regulation of secretion, 1115
  source, 669, 1115
  structure, 1112
Estriol, 1109
Estrogen(s)
  effects, 221, 228, 669, 1109, 1112,
    1115, 1146
  in labor, 1141
  in normal growth, 670
  in pregnancy, 1133
  production, 669, 1099, 1112
  regulation of secretion, 648, 669,
    1110, 1115
  structure, 75, 639
  underproduction/
    overproduction, 668
Estrone, 1109
Ethmoid, 237, 238–239, 244–245
Ethmoidal air cells, 244
Ethmoidal labyrinth, 244
Etiology, 49

Eupnea, 878
Eustachian (auditory) tube, 160,
  242, 617
Evaporation, 990
Eversion, 299
Excess postexercise oxygen
  consumption (EPOC), 344
Exchange pump, 121, 125
Exchange reaction, 63
Exchange vessels, 711. See also
  Capillary(ies)
Excitable membranes, 328, 428–429
Excitation–contraction coupling,
  328–329, 353
Excitatory interneurons, 475
Excitatory neurotransmitters, 440
Excitatory postsynaptic potential
  (EPSP), 447, 449
Excitotoxicity, 451
Excretion, 909
Excretory system, 1037
Exercise
  body system response, 407
  bone remodeling and, 220
  cardiovascular disease and, 776
  cardiovascular system response,
    775–776
  salt loss during, 1053
Exergonic reactions, 65
Exfoliative cytology, 146, 1106
Exhalation (expiration), 875, 876,
  877, 879
Exhaustion phase, general
  adaptation syndrome, 671
Exocrine glands, 149–151
Exocrine pancreas, 662. See also
  Pancreas
Exocrine secretions, 149
Exocytosis, 125, 126
Exons, 112
Exophthalmos, 674
Expiration (exhalation), 875, 876,
  877, 879
Expiratory center, 892
Expiratory reserve volume
  (ERV), 881
Expressivity, 1151
Expulsion stage, labor, 1142–1143
Extension, 273, 284, 295, 364
Extensor (muscle name), 364, 366
Extensor carpi radialis muscle,
  brevis/longus, 366, 391
Extensor carpi ulnaris muscle, 391
Extensor digiti minimi muscle, 392
Extensor digitorum muscle,
  360, 392
Extensor digitorum muscle,
  brevis/longus, 401–403
Extensor hallucis longus muscle,
  401–403
Extensor indicis muscle, 392
Extensor pollicis muscle, brevis/
  longus, 392
Extensor retinacula, 392, 401–402
External acoustic meatus, 617
External branch, accessory
  nerve, 527
External callus, 226
External ear, 617
External genitalia, 1077,
  1090–1092, 1106–1107

External oblique muscle, 381
External occipital crest, 240
External os, uterus, 1101
External respiration, 873, 885
External root sheath, hair, 190
Externus (muscle name), 365
Exteroceptors, 417, 539
Extracapsular ligaments, 293
Extracellular fluid (ECF)
  composition, 1044–1045
  fluid and electrolyte regulation in,
    1045–1046
  fluid movement in, 1047
  fluid shifts, 1048–1049
  overview, 92, 1042, 1072
  sodium balance and, 1051–1053
Extracorporeal liver assist device
  (ELAD), 935
Extraembryonic membranes, 1129,
  1130–1131
Extrafusal muscle fibers, 479
Extrapulmonary bronchi, 867
Extrapyramidal system (EPS), 551
Extrinsic (muscle name), 365
Extrinsic ligaments, 293
Extrinsic pathway, coagulation,
  703, 704
Extrinsic regulation, 33
Eye. See also Vision
  accessory structures, 596–599
  age-related changes, 606
  chambers, 603–604
  fibrous layer, 599
  lens, 604–607
  muscles, 370, 373
  overview, 596, 631
  parasympathetic effects, 573
  retina. See Retina
  sectional anatomy, 598
  sympathetic effects, 573
  vascular layer, 599–601
Eyelashes, 596
Eyelids, 596

F

Fabella, 401
Facet (bone marking), 207
Facial bones, 235, 245–248
Facial expression, muscles, 370,
  371, 372
Facial nerve (VII), 242, 523, 528
Facial vein, 793
Facilitated diffusion, 121, 124,
  951, 1010
Facilitation, 448
F-actin (filamentous actin), 323
Fact memories, 578
Factor I, 704
Factor II, 704
Factor III, 704
Factor IV, 704
Factor V, 704
Factor VII, 704
Factor VIII, 704
Factor IX, 704
Factor X, 704
Factor XI, 704
Factor XII, 704
Factor XIII, 704
Factor B, 826

Factor D, 826
Factor P, 826
Facultative water reabsorption, 1024
Fainting, 558, 588, 764
Falciform ligament, 909, 935
Fallopian (uterine) tubes, 1077,
  1100–1101, 1123
False labor, 1140
False (greater) pelvis, 277
False ribs, 261
Falx cerebelli, 492
Falx cerebri, 244, 492
Familial colon cancer, 1155
Familial hypercholesterolemia,
  994, 1155
Familial polyposis of the
  colon, 1155
Farsightedness, 608
Fascia(e), 165–166
Fascicles
  muscle, 317, 318, 323, 359,
    365, 407
  spinal nerve, 464
Fasciculation, 338
Fasciculus cuneatus, 545, 548
Fasciculus gracilis, 545, 548
Fast-adapting receptors, 538
Fast fibers, 346, 347, 348
Fast pain, 539
Fast sodium channels, 729
Fat cells (adipocytes), 85–86, 153,
  155–156
Fatigue, muscle, 344
Fat pads, 293
Fat-soluble vitamins, 953,
  986–987
Fat substitutes, 72
Fatty acids
  dietary, 977
  essential, 975
  free, 975
  monounsaturated, 73
  omega-3, 74
  polyunsaturated, 73, 74
  saturated, 73, 74
  structure and function, 72–73, 77
  unsaturated, 73, 74
Fatty appendices, 943
Fauces, 916
F cells, 662, 665
Fecal occult blood test, 956
Fecal transplant, 907, 960
Feces, 909
Feeding center, 506
Female pronucleus, 1124, 1125
Female reproductive cycle,
  1107–1113
Female reproductive system
  age-related changes, 1099,
    1114–1115
  endocrine functions, 667, 669
  external genitalia, 1106–1107
  hormonal regulation,
    1107–1113, 1115
  integration with other systems,
    1115–1117
  mammary glands, 1107, 1108
  oogenesis, 1096–1099
  ovarian cycle. See Ovarian cycle
  ovaries, 1095–1096
  overview, 35, 1077, 1094–1095

parasympathetic effects, 573
prenatal development, 1137
at puberty, 1146
smooth muscle in, 352
sympathetic effects, 573
uterine cycle, 1103–1105, 1112
uterine tubes, 1100–1101
uterus, 1101–1103
vagina, 1105–1106
Female sexual function, 1114
Femoral artery, 754, 780, 789
Femoral circumflex artery, 789
Femoral circumflex vein, 794
**Femoral cutaneous nerve, lateral/
posterior, 470, 472**
Femoral head, 280
Femoral nerve, 470, 472
Femoral vein, 793, 794
Femur, 208, 213, 280, 308
**Fenestrated capillaries, 646, 753,
756–757**
Ferritin, 687
**Fertilization, 1077, 1123–1125,
1158**
Fetal alcohol syndrome, 988
Fetal circulation, 798–799
Fetal hemoglobin, 685, 889
Fetus
cardiovascular changes at birth,
799
circulation in heart and great
vessels, 798–799
definition, 1101
descent of testes, 1078–1079
development, 1122. *See also*
Prenatal development
development of immunological
competence, 844
fluid, electrolyte, and acid–base
balance, 1071
placental blood supply, 798
respiratory changes at birth, 898
Fever, 823, 828, 994
Fibrillation, 745
Fibrillin, 155
Fibrin, 196, 681, 683
Fibrinogen, 681, 683
Fibrinolysin, 1090
Fibrinolysis, 705
Fibroblast growth factor (FGF), 132
Fibroblasts, 153
Fibrocartilage, 161–162
Fibrocystic disease, 1109
Fibrocytes, 153
**Fibrodysplasia ossificans
progressiva, 214**
Fibromyalgia, 354
Fibromyositis, 407
Fibrosis, 170, 403
Fibrous capsule, 292
Fibrous capsule, kidney, 1001
Fibrous layer, eye, 599
**Fibrous pericardium (pericardial
sac), 712**
Fibrous proteins, 80
Fibrous skeleton, heart, 721
Fibula, 282–283, 308
**Fibularis muscle, brevis/longus,
401–403**
Fibular nerve, 471

Fibular vein, 793
Field of vision, 615
**"Fight or flight" response, 506,
560, 671**
Filamentous actin (F-actin), 323
Filgrastim (Neupogen), 700
Filiform papillae, 594
Filling time, 741
Filtrate, 1005
Filtration
capillary, 765–768
pressures, 1014–1015
in urine formation, 1009, 1010
Filtration membrane, 1008
Filtration slits, 1008
Filum terminale, 457
Fimbria, 1100
**Fine touch and pressure
receptors, 540, 547**
Fingerprints, 179
First breath, 898
First-class lever, 361, 362
First-degree burn, 197
First heart sound, 737
First messengers, 446, 640, 641
First-order neuron, 544
First trimester
blastocyst formation, 1127
cleavage, 1127
embryogenesis, 1133–1134
implantation, 1127–1129, 1132
overview, 1126, 1136–1137
placentation, 1132–1135
Fissure
bone marking, 207
cerebellar, 501
cerebral, 488
lung, 867
Five-year survival rate, 184
Fixator, 365
Fixed acids, 1056, 1057
Fixed macrophages, 153, 822
Fixed ribosomes, 100
Flaccid paralysis, 554
Flagellum (flagella), 99, 1086
Flat bones, 206
Flatfeet, 284
Flatus, 947
**Flavin adenine dinucleotide
(FAD), 966**
Flavin mononucleotide (FMN), 969
Flavor, 594–595
Flexion, 273, 284, 295, 299, 364
Flexor (muscle name), 366
**Flexor carpi muscle, radialis/
ulnaris, 391**
**Flexor digiti minimi brevis
muscle, 395, 402**
Flexor digitorum muscle
brevis/longus, 303, 402
superficialis/profundus, 392
**Flexor hallucis muscle, brevis/
longus, 402–403, 405**
**Flexor pollicis muscle, brevis/
longus, 392, 395**
Flexor reflex, 477, 480–481
Flexor retinaculum, 275, 392
Floaters, 607
**Floating kidney (nephroptosis),
1001, 1037**
Floating ribs, 261

Flocculonodular lobe, 501
Fluid balance
age-related changes, 1071–1072
in burn patient, 197
definition, 1043
electrolyte balance and, 1047
in extracellular fluid, 1044–1045
fluid gains and losses, 1048
fluid movement within ECF, 1047
fluid shifts, 1048–1049
hormonal control, 1046–1047
in intracellular fluid, 1044–1045
overview, 1043
regulation, 1045–1046
Fluid compartments, 1044
**Fluid connective tissues, 152,
157–159.** *See also* Blood
Fluid replacement therapy, 1072
Fluid shifts, 1048–1049
Fluoxetine (Prozac), 443, 582
Foam cells, 754
Focal calcification, 754
Focal distance, 605
Focal point, 605
Folate (folic acid; vitamin B$_9$), 987
Folia, 501
Foliate papillae, 594, 595
Follicle(s)
hair, 189
ovarian, 667, 1097
thyroid, 652–654
Follicle cavity, 652
Follicle cells, 1097
**Follicle-stimulating hormone
(FSH)**
effects, 648, 651, 1115
in female reproductive cycle, 1110,
1115
regulation of secretion, 648,
651, 1115
in spermatogenesis, 1092,
1093, 1115
Follicular fluid, 1099
**Follicular phase, ovarian cycle,
1097, 1109, 1110–1111**
Follitropin. *See* Follicle-
stimulating hormone
Fontanelles, 250–251, 285
Food groups, 983, 984
Foot
arch, 284
arteries, 790
bones, 283–284
in diabetes, 664
muscles, 401–405
nerves, 473
Foramen (foramina)
apical, 918
definition, 207
infraorbital, 237, 246
jugular, 237, 240, 242
mandibular, 248, 249
mental, 237, 248
obturator, 276, 279
stylomastoid, 237, 242, 243
supraorbital, 237, 241
vertebral, 253
zygomaticofacial, 247
Foramen lacerum, 237, 242
**Foramen magnum, 237, 238,
239, 240**

**Foramen of Monro (interventricular
foramen), 490**
**Foramen ovale, 237, 239, 244, 718,
798, 898**
Foramen rotundum, 239, 244
Foramen spinosum, 239, 244
Force, 346
**Forced breathing (hyperpnea),
878, 879, 880, 892–893**
Forceps delivery, 1143
Foreskin, 1091
**Formed elements, blood, 157–159,
680–681, 699**
Fornix
eye, 597
limbic system, 506
vagina, 1106
Fossa (bone marking), 207
Fossa ovalis, 718, 799
Fourth heart sound, 737
Fourth ventricle, 490
Fovea, 601
Fovea capitis, 280–281
Fovea centralis, 601
Fractionated blood, 682
Fracture(s)
definition, 224
distal humerus, 268, 288
hip, 280, 306
repair process, 226–227
rib, 262
stress, 284
types, 226–227
Fracture hematoma, 226
Fragile-X syndrome, 1155, 1156
Frank, Otto, 742
Frank–Starling principle, 742
**FRC (functional residual
capacity), 881, 882**
Freckles, 182
Free edge, nail, 195
Free fatty acids, 975
Freely permeable membrane, 114
Free macrophages, 153, 823
Free nerve endings, 537, 540, 541
Free radical, 60
Free ribosomes, 100
Freeze fracture (freeze-etching), 108
Frenulum, lower lip, 915
Frequency, sound, 623, 624, 628
Frons, 241
Frontal bone, 237, 238, 241
Frontal eye field, 513
Frontal lobe, 504, 508, 512, 513
**Frontal (coronal) plane/section,
44, 235**
Frontal sinuses, 238, 241
**Frontal (metopic) suture,
241, 285**
Frostbite, 201
Fructose, 70, 950
Fulcrum, 361
Full term, 1138
Full-thickness burn, 197
Full-thickness graft, 197
Functional groups, 69
**Functional residual capacity
(FRC), 881, 882**
Functional syncytium, 351
**Functional zone, endometrium,
1103**

Fundus
    gallbladder, 940
    stomach, 923
    uterus, 1101
Fungiform papillae, 594, 595
Funiculi (columns), white
    matter, 463
Funny bone, 273
Furuncle, 201

G

G$_0$ phase, interphase, 130
G$_1$ phase, interphase, 130
G$_2$ phase, interphase, 130
G-actin, 323
Galactorrhea, 674
Galactose, 950
Galen, Claudius, 31
Gallbladder, 940
Gallstones, 940
Gamete intrafallopian transfer
    (GIFT), 1157
Gametes, 1077
Gamma-aminobutyric acid
    (GABA), 443, 444
Gamma efferents, 480
Gamma-hydroxybutyrate
    (GHB), 672
Gamma motor neurons, 480
Ganglion (ganglia), 416, 423,
    560, 561
Ganglion cells, retina, 602
Ganglionic neurons, 559, 562
Gangrene, 201
Gap junction, 142, 143
Gas(es)
    definition, 61
    dissolved, 85
    partial pressures, 882–883
    pressure measurement, 874
    pressure/volume relationship,
        874–875
Gas exchange, 882–886
Gas laws, 882–884
Gastrectomy, 956
Gastric artery, left/right, 788
Gastric glands, 925–926
Gastric inhibitory peptide (GIP),
    929, 941, 942
Gastric juice, 925
Gastric lipase, 926
Gastric phase, gastric
    activity, 928
Gastric pits, 925
Gastric reflex, 576
Gastric secretion, 928, 929
Gastric ulcer, 923
Gastric vein, 798
Gastrin, 926, 941, 942
Gastritis, 923
Gastrocnemius muscle, 363,
    400–401
Gastroduodenal artery, 788
Gastroenteric reflex, 929
Gastroepiploic artery, left/right, 788
Gastroepiploic vein, left/right,
    797, 798
Gastroesophageal reflux disease
    (GERD), 400, 956
Gastroileal reflex, 929

Gastrointestinal system. *See*
    Digestive system
Gastrointestinal tract. *See*
    Digestive tract
Gastroscope, 956
Gastrosplenic ligament, 820
Gastrulation, 1128–1129
Gated (active) channels, 428–430
Gaucher's disease, 1155
G cells, 926
G-CSF, 700
GDP (guanosine diphosphate),
    968
Gemelli muscle, superior/
    inferior, 396
Gender differences. *See* Sex
    differences
Gene, 110
Gene activation, 110
General adaptation syndrome,
    670, 671
General interpretive area, 514
General senses
    adaptation, 538–539
    chemoreceptors, 543
    definition, 536
    interpretation of sensory
        information, 537–538
    mechanoreceptors, 540–543
    nociceptors, 539–540
    somatic pathways, 544–549
    stimuli detection, 536–537
    thermoreceptors, 540
    visceral pathways, 549
Generator potential, 537, 592
Genetics, 1123
Genetic analysis, 1156
Genetic code, 109
Genetic engineering, 134
Genetic recombination, 1153
Genetic stutter, 1156
Genicular artery, 789
Geniculate ganglia, 523
Genioglossus muscle, 365, 375
Geniohyoid muscle, 375, 377
Genital herpes, 1117
Genitofemoral nerve, 470,
    472, 1080
Genome, 1155
Genomic imprinting, 1153
Genotype, 1148
Genu valgum, 286
Geriatricians, 1148
Geriatrics, 1148
Germ cells (sex cells; reproductive
    cells), 91
Germinal center, 816
Germinal epithelium, 1096
Germinative cells, 144, 179. *See
    also* Stem cells
Germ layers, 1129, 1132
Gestation, 1125–1126, 1158.
    See also Prenatal
    development
Ghrelin, 926–927, 990
Giemsa stain, 693
Gigantism, 222, 668
Gingivae, 915
Gingival sulcus, 918
Gingivitis, 918
Glabella, 241

Gland(s). *See also specific glands*
    ceruminous, 194
    characteristics, 140
    classification, 149
    control of, 194
    endocrine, 149
    modes of secretion, 149
    sebaceous, 191–192
    structure, 150–151
    sweat, 193–194
    types of secretions, 149
Gland cells, 141
Glandular epithelium, 141
Glans
    clitoris, 1106–1107
    penis, 1090
Glasgow coma scale, 529
Glassy membrane, hair, 190
Glaucoma, 603
Glenohumeral (shoulder) joint,
    270, 302–304, 313
Glenohumeral ligament, 303
Glenoid cavity, 270
Glenoid labrum, 303
Glial cells. *See* Neuroglia
Gliding joints, 295, 297
Gliding movement, 295
Global aphasia, 515
Globular proteins, 80
Globulins, 682–683
Globus pallidus, 510
Glomerular (Bowman's) capsule,
    1004, 1007–1008
Glomerular filtration, 1013–1014
Glomerular filtration rate (GFR),
    1015, 1016–1018, 1140
Glomerular hydrostatic pressure
    (GHP), 1014, 1015
Glomerulonephritis, 1008, 1067
Glomerulus, 1005, 1007, 1028–1029
Glossopharyngeal nerve (IX), 525,
    528, 896
Glottis, 863, 865
Glucagon, 663, 742
Glucocorticoids, 648, 658, 660, 668
Gluconeogenesis, 663, 939, 972
Glucose
    in blood, 662–663
    gluconeogenesis, 663, 972
    glycolysis. *See* Glycolysis
    movement across plasma
        membrane, 121
    in plasma, 1010
    reabsorption at proximal
        convoluted tubule, 1018
    in solution, 66
    in stress response, 671
    structure, 70–71
    in urine, 664, 1010, 1012
Glucose-6-phosphate, 965
Glucose-sparing effect, 649
Glucosuria (glycosuria), 664, 1012
Glutamate, 444, 539, 580
Glutaminase, 1062
Glutamine, 1062
Gluteal artery, 789
Gluteal lines, 276
Gluteal muscles, 395–397
Gluteal nerves, superior/
    inferior, 472
Gluteal vein, 794

Gluteus muscle, minimus/medius/
    maximus, 364, 395–397
Glycerides, 74, 77
Glycerol, 74
Glycine, 444, 446
Glycocalyx, 96
Glycogen, 71–72, 649, 972
Glycogenesis, 972
Glycogenolysis, 973
Glycolipids, 75–77
Glycolysis, 106, 343, 965, 966,
    971–972
Glycoproteins, 82, 639
Glycosaminoglycans, 142
Glycosuria (glucosuria), 664, 1012
GM-CSF, 700
Golgi apparatus (Golgi complex),
    93, 102, 104–105
Golgi tendon organs, 543
Gomphosis, 291, 918
Gonad(s), 648, 667, 1077
Gonadal artery, 789
Gonadal steroid-binding globulin
    (GBG), 1092
Gonadal vein, 796
Gonadotropes, 1107
Gonadotropin-releasing hormone
    (GnRH), 1092, 1093, 1110, 1115
Gonadotropins, 648
Gonorrhea, 1117
Goose bumps, 189, 559
Gout, 310
Gouty arthritis, 309
G protein-coupled receptors, 567
G proteins, 446, 640–641
Gracilis muscle, 396, 398
Graded (local) potentials, 328,
    424, 430–432, 438
Graft rejection, 832
Granular leukocytes
    (granulocytes), 693
Granulation tissue, 196
Granulosa cells, 1099
Graves disease, 668
Gravity, muscle contraction and, 342
Gray commissures, 463
Gray horns, posterior/anterior/
    lateral, 463
Gray matter
    cerebellum, 500
    characteristics, 415, 421
    medulla oblongata, 498
    midbrain, 502
    pons, 499
    spinal cord, 461–463, 483–484
Gray ramus, 564
Gray ramus communicans, 467
Great auricular nerve, 469
Great cardiac vein, 722
Great cerebral vein, 793
Greater curvature, stomach, 923
Greater horns, hyoid, 248
Greater omentum, 909
Greater sciatic notch, 276
Greater wings, sphenoid, 244
Great saphenous vein, 793
Green cones, 607
Greenstick fracture, 227
Groin pull, 398, 407
Gross anatomy, 30–31, 49
Ground substance, 152, 154

Growth factors, 129, 132
Growth hormone (GH)
    bone growth and, 221, 222
    effects, 132, 160, 648–649, 651
    muscle metabolism and, 346
    regulation, 651
    sources, 132
    targets, 132
    underproduction/
      overproduction, 668
Growth hormone–inhibiting
    hormone (GH-IH;
    somatostatin), 649, 926, 941
Growth hormone–releasing
    hormone (GH-RH;
    somatocrinin), 649
Guanine, 82
Guanosine diphosphate (GDP), 968
Guanosine triphosphate (GTP),
    445, 610, 968
Gubernaculum testis, 1078
Guillain–Barré syndrome, 412,
    421, 454
Gums, 915
Gustation
    age-related changes, 596
    discrimination, 595–596
    overview, 536, 630–631
    pathways, 594–595
    receptors, 592–593, 594, 595
Gustatory cortex, 513
Gustatory receptor cells, 594
Gustducins, 593
Gynecology, 1117
Gynecomastia (adrenogenital
    syndrome), 668
Gyri, 488

H

H₂CO₃ (carbonic acid), 69, 890, 1056
Hair
    age-related changes, 199
    color, 191–192
    functions, 189
    growth cycle, 191
    overview, 189, 202
    production, 191
    structure, 189–191
    types, 191
Hair bulb, 191
Hair cells, 619, 627
Hair follicles, 189
Hair matrix, 191
Hair papilla, 191
Hair root, 191
Hair shaft, 191
Halcion (triazolam), 579
Half-life, 55
Halitosis, 956
Hallux, 284
Hamate, 274
Hamstrings, 359, 401, 410
Hand, 274–275, 389–395, 470
Haploid cells, 1082
Haptens, 838
Hard keratin, 190
Hard palate, 246, 861, 916
Hashimoto's disease, 674
Hassall's (thymic) corpuscles,
    819, 820

Hatching, 1127
Haustra, 943
Haustral churning, 947
Haversian canal, 211
Haversian system (osteon), 211
Hawking, Stephen, 552
H band, 322
HDL (high-density
    lipoprotein), 975
Head
    epididymis, 1087
    pancreas, 933
    spermatozoon, 1085
    thick filament, 324
Head (body part), 370–377, 784,
    789, 792, 793
Head (bone part)
    definition, 207, 268
    femur, 280
    humerus, 271, 273
    radius, 273
    rib, 261
    scapula, 271
    ulna, 273
Head fold, 1133
Hearing
    age-related changes, 629
    auditory pathways, 628
    auditory sensitivity, 628–629
    cochlear duct, 623
    overview, 536, 622–623,
      631–632
    process, 626–627
    sound characteristics, 623–626
Heart. See also cardiac headings
    age-related changes, 801
    autonomic innervation and
      regulation, 739–741
    blood flow to, 778
    blood supply, 721–722
    cardiac cycle, 732–737
    cardiac skeleton, 721
    cardiodynamics, 737–743
    conducting system. See
      Conducting system, heart
    congenital problems, 800
    connective tissues, 721
    endocrine functions, 638, 666
    gross anatomy, 712, 745
    internal anatomy and
      organization, 715–721
    left atrium, 718
    left ventricle, 718–719
    left vs. right ventricles, 719
    location in thoracic cavity, 712,
      713, 714
    pericardium, 48, 163, 712, 713
    relationship to lung, 867, 869
    right atrium, 711, 715, 718
    right ventricle, 718
    sectional anatomy, 717
    superficial anatomy, 712, 714
    valves, 719–721
    wall, 712, 715
Heart attack (myocardial
    infarction), 725
Heart block, 745
Heart disease
    congenital, 800
    coronary, 724–725, 754
    valvular, 721

Heart failure, 736
Heart murmur, 737
Heart rate, 738–741, 1144
Heart sounds, 736–737
Heat, 63
Heat capacity, 66
Heat conservation, 991
Heat cramps, 994
Heat exhaustion, 993
Heat gain, 991
Heat-gain center, 991
Heat generation, 992–993
Heat loss, 346, 990–991
Heat-loss center, 991
Heat production, 346
Heat prostration, 993
Heat stroke, 993
Heat transfer, 990–991. See also
    Thermoregulation
Heavy chains, antibody, 837
Heavy metals, 86, 218, 421
Heberden's nodes, 312
Heimlich maneuver, 902
Helicases, 127, 128
Helium (He), 56
Helper T (T_H) cells, 813, 831, 841
Hematocrit (Hct), 680, 683, 690
Hematologists, 687
Hematology, 706
Hematoma, fracture, 226
Hematopoiesis (hemopoiesis), 680
Hematopoietic stem cells
    (HSCs), 687
Hematuria, 687, 1018, 1030
Heme, 685
Hemiazygos vein, 793
Hemidesmosomes, 143
Hemiparesis, 483
Hemispheric lateralization,
    515, 516
Hemochromatosis, 706
Hemocytoblasts, 687, 699
Hemodialysis, 1035, 1037
Hemoglobin (Hb; Hgb)
    abnormal, 686, 690
    BPG and, 888–889
    carbon dioxide binding, 890
    carbon monoxide binding, 889
    conservation and recycling,
      686–687, 688
    fetal, 685, 889
    functions, 685–686
    normal values, 690
    overview, 685
    in oxygen transport, 886–889
    partial pressure of oxygen and,
      886–887
    pH and, 887–888
    skin color and, 183
    structure, 80, 685
    temperature and, 888
Hemoglobin buffer system, 1060
Hemoglobin saturation, 886
Hemoglobinuria, 687
Hemolysis, 119, 687
Hemolytic disease of the
    newborn, 694–695, 839
Hemolyze/hemolyzed, 687, 688
Hemophilia, 706, 1155
Hemopoiesis, 680
Hemopoietic growth factor, 706

Hemoptysis, 902
Hemorrhoids, 759, 945
Hemosiderin, 687
Hemosiderosis, 706
Hemostasis, 701–705. See also
    Coagulation
Hemothorax, 902
Henry's law, 883–884
Heparin, 153, 697, 705
Hepatic artery, 788
Hepatic duct, right/left/
    common, 937
Hepatic flexure, 943
Hepatic portal system, 797–798
Hepatic portal vein, 797
Hepatic vein, 796
Hepatitis, 935, 936
Hepatocytes, 935
Hepatopancreatic sphincter, 940
Heptose, 70
Heredity, 1123
Hering–Breuer reflexes, 896–897
Hernia, 400
Herniated disc, 263, 301
Hertz (Hz), 624
Heterotopic bones, 214
Heterozygous, 1149
Hexose, 70
Hgb. See Hemoglobin
High-density lipoprotein
    (HDL), 975
High-energy bonds, 84, 106
High-energy compound, 84, 85
Higher-order functions
    arousal, 581
    characteristics, 578, 587
    memory, 578–580
    reticular activating system, 502, 581
    sleep, 580–581
    states of consciousness, 580–581
High-frequency sound, 624
Hilum
    kidney, 1001
    lung, 866
    lymph node, 816
    ovary, 1096
    spleen, 820
Hinge joints, 297
Hip bones, 276
Hip fracture, 280, 310
Hip joint, 305–306, 313
Hippocampus, 415, 506, 580
Hippuric acid, 1030
Hirsutism, 674
Histamine, 153, 444, 697
Histidine, 979, 1059
Histiocytes (fixed macrophages),
    153, 822
Histology, 31, 49, 140
Histones, 109
Hives, 849
Hodgkin's lymphoma, 851
Holocrine secretion, 149, 150
Homeostasis
    definition, 28, 33, 49
    negative feedback in, 36–38
    overview, 33, 36, 671
    positive feedback in, 38–39
    systems integration and, 39
Homeostatic regulation, 33
Homologous chromosomes, 1149

Homozygous, 1149
Hooke, Robert, 91
Horizontal cells, retina, 602
Horizontal plate, palatine bones, 246
Hormonal stimuli, 642
Hormone-binding proteins, 682
Hormone replacement therapy, 1115
Hormones. *See also specific hormones*
  age-related changes, 670
  athletic performance and, 672
  behavior and, 670
  bone remodeling and, 221, 228
  calcium balance and, 222–225
  cardiac contractility and, 742
  in cardiovascular regulation, 772–775
  classification, 637–638, 639
  in digestive system control, 914, 940, 941, 942
  in female reproductive cycle, 1107–1113
  glomerular filtration rate regulation, 1016
  heart rate control, 741
  interactions, 669
  intestinal, 914, 940
  intracellular receptors and, 642, 643
  in male reproductive function, 1092–1094
  mechanisms of action, 640–642
  in metabolism, 980–981
  muscle metabolism and, 346
  as neurotransmitters, 445
  in normal growth, 669–670
  overview, 635–636, 676
  plasma membrane receptors and, 640–642
  regulatory, 33, 643
  secretion and distribution, 149, 638, 640
  in stress response, 670, 671
Horns, spinal cord, 463
Host versus graft disease, 851
"Hot flashes," 1089
Housemaid's knee, 294, 312
Howship's lacunae, 213
Human chorionic gonadotropin (hCG), 1133
Human Genome Project (HGP), 1155–1157
Human immunodeficiency virus (HIV), 845, 848
Human leukocyte antigens (HLAs), 831–832
Human physiology, 31
Human placental lactogen (hPL), 1133
Human tetanus immune globulin, 332
Humeroradial joint, 304
Humeroulnar joint, 304
Humerus, 268, 271–273, 288, 308
Humoral (antibody-mediated) immunity, 697, 814, 815, 828, 842. *See also* Immune response
Humoral stimuli, 642
Hunter's response, 992

Huntington's disease, 582, 1155
Hyaline cartilage, 160, 161, 216
Hyaluronan (hyaluronic acid), 142
Hyaluronidase, 1124
Hybridoma, 851
Hydration sphere, 66
Hydrocele, 1047
Hydrocephalus, 493
Hydrochloric acid (HCl), 69, 926
Hydrocortisone (cortisol), 658, 659
Hydrogen (H_2)
  functions in human body, 54
  secretion at distal convoluted tubule, 1022–1023, 1025
  secretion in the collection system, 1023, 1025
  structure, 53–54, 55, 56, 58
Hydrogen bonds, 61, 62
Hydrogen cyanide, 969
Hydrogen peroxide, 696
Hydrolysis, 63, 948
Hydrophilic molecules, 27
Hydrophobic molecules, 27
Hydrostatic pressure, 118, 762, 1014
Hydroxide ion, 68
Hydroxyapatite, 209
Hydroxyl group, 70
Hydroxyurea, 685
Hymen, 1106
Hyoglossus muscle, 374, 375
Hyoid bone, 248, 285
Hyperalgesia, 554
Hyperbaric oxygenation, 902
Hypercalcemia, 1054, 1055
Hypercapnia, 896, 1064
Hyperchloremia, 1055
Hypercholesterolemia, 75, 755, 1155
Hyperchromic anemia, 690
Hyperextension, 298
Hyperglycemia, 664
Hyperhydration, 1050
Hyperkalemia, 1053, 1054, 1055
Hypermagnesemia, 1055
Hypernatremia, 1050, 1052, 1055
Hyperopia, 608
Hyperparathyroidism, 635, 668, 677, 1054
Hyperphosphatemia, 1055
Hyperplasia, 134
Hyperpnea (forced breathing), 878, 879, 880, 892–893
Hyperpolarization, 328, 430, 432
Hypersensitivity, immediate, 848
Hypertension, 764
Hyperthyroidism, 668
Hypertonic solution, 119, 120
Hypertrophy, 134, 347
Hyperuricemia, 994
Hyperventilation, 896, 1067
Hypervitaminosis, 986, 987
Hypervolemic, 706
Hypesthesia, 543
Hypoaldosteronism, 668
Hypocalcemia, 1054, 1055
Hypocalcemic tetany, 674
Hypocapnia, 896, 1067
Hypochloremia, 1055
Hypochromic anemia, 690
Hypodermal injection, 370

Hypodermic needle, 188
Hypodermis (subcutaneous layer; superficial fascia), 165, 166, 178, 188–189
Hypogastric plexus, 572
Hypoglossal canal, 238, 239, 240
Hypoglossal nerve (XII), 240, 527–528
Hypogonadism, 648, 668
Hypokalemia, 1022, 1042, 1053, 1054, 1055, 1075
Hypomagnesemia, 1055
Hyponatremia, 1050, 1052, 1055
Hyponychium, 195
Hypoparathyroidism, 668, 1054
Hypophosphatemia, 1055
Hypophyseal artery, superior/inferior, 646
Hypophyseal fossa, 238, 243
Hypophyseal portal system, 646
Hypophysectomy, 674
Hypophysis. *See* Pituitary gland
Hyposomia, 630
Hypotension, 764
Hypothalamus
  control of anterior pituitary, 646–647, 650
  control of endocrine function, 642–644
  in fluid balance regulation, 1046
  functions, 504–506
  gross anatomy, 504, 505
  hormones, 638
  overview, 488, 489, 495
  in thermoregulation, 36–37
Hypothermia, 991, 993
Hypothyroidism, 668
Hypotonic solution, 119, 120
Hypoventilation, 896
Hypovitaminosis, 986
Hypovolemic, 706
Hypoxia, 689, 874
Hysterectomy, 1117

I bands, 321, 322–323
Idiopathic, 49
IDLs (intermediate-density lipoproteins), 975
Ileocecal valve, 929, 930, 943
Ileocolic artery, 788
Ileocolic vein, 798
Ileum, 930
Iliac artery, 752, 787, 789, 1101, 1106
Iliac crest, 276
Iliac fossa, 276, 279
Iliac spines, 276
Iliac tuberosity, 276
Iliacus muscle, 396, 398
Iliac vein, common/external/internal, 794, 796
Iliococcygeus muscle, 383
Iliocostalis muscle, cervicis/lumborum/thoracis, 377–379
Iliofemoral ligament, 306
Iliohypogastric nerve, 472
Ilioinguinal nerve, 472, 1080
Iliopsoas muscle, 398
Iliotibial tract, 395

Ilium, 276, 279
Image reversal, 606–607
Immature delivery, 1143
Immediate hypersensitivity, 848
Immune complex, 839
Immune complex disorder, 848, 1008
Immune disorders, 844, 848–849
Immune response
  age-related changes, 851
  antibodies in, 837–839
  antigens in, 690, 840–841
  B cells in, 836–837
  cytokines in, 846–847
  endocrine system influences on, 851
  nervous system influences on, 851
  overview, 808, 830–831, 841–843
  primary, 840
  secondary, 840–841
  stress and, 849
  T cells in. *See* T cells
Immune surveillance, 697, 815, 823, 824
Immune system, 808. *See also* Lymphatic system
Immunity. *See also* Immune response
  active, 829
  adaptive, 829, 830
  antibody-mediated (humoral), 697, 814, 815, 828, 842
  cell-mediated, 697, 813, 815, 828, 842
  innate (nonspecific), 829, 842, 846
  overview, 808
  passive, 829–830
Immunization, 332, 829, 855, 916
Immunodeficiency disease, 844, 848
Immunoglobulin(s), 681, 814, 838–839. *See also* Antibody(ies)
Immunoglobulin A (IgA), 838, 839
Immunoglobulin D (IgD), 838, 839
Immunoglobulin E (IgE), 838, 839
Immunoglobulin G (IgG), 838, 839, 840
Immunoglobulin M (IgM), 838, 839, 840
Immunological competence, 844
Immunological escape, 824
Immunology, 851
Immunosuppression, 832
Immunosuppressive drugs, 832, 848
Immunotherapy, 172
Impacted teeth, 919
Impermeable, 113
Impetigo, 201
Impingement syndrome, 407
Implantation, 1126, 1127–1128
Impotence, 1113–1114
Inactivated channels, 428
Incision, 196
Incisors, 919
Inclusions, cell, 97
Incomplete dominance, 1150, 1151
Incomplete proteins, 983
Incomplete tetanus, 337, 338
Incontinence, 1034
Incus, 617
Indole, 946

Induced abortion, 1140
Induced hypothermia, 991
Induced pluripotent stem (iPS) cells, 44
Induction, 1126
Inert, 57
Inert gases, 57
Infancy, 1144, 1146, 1147
Infants. *See also* Neonates
  fluid, electrolyte, and acid–base balance, 1071
  skull, 250–252
  thermoregulatory problems, 993
Infection, 171
Infectious mononucleosis, 851
Inferior, 43
Inferior colliculus, 501, 502, 552
Inferior ganglion, 525, 526
Inferior meatus, 599
Inferior mesenteric artery, 564
Inferior mesenteric ganglion, 564
Inferior mesenteric plexus, 572
Inferior mesenteric vein, 797
Inferior oblique muscle, 370, 373, 380–381
Inferior rectus muscle, 370, 373
Inferior vena cava, 715, 793–797
Infertility, 1080, 1090, 1110, 1123, 1157
Inflammation (inflammatory response), 170, 171, 823, 827–828
Inflammatory phase, injury repair, 196
Inflammatory T cells, 813
Inflation reflex, 897
Information processing, CNS
  overview, 424, 447, 475
  postsynaptic potentials, 447–449
  presynaptic inhibition and facilitation, 449, 450
  rate of action potential generation and, 449–450
Infraorbital foramen, 237, 246
Infraperitoneal, 48
Infraspinatus muscle, 303, 386, 389
Infraspinous fossa, 271
Infundibulopelvic (suspensory) ligament, 1095
Infundibulum, 488, 504, 645, 1100
Ingestion, 908
Inguinal canal, 400, 1080
Inguinal hernia, 400, 1080
Inhalation (inspiration), 875, 876, 877, 879
Inheritance
  definition, 1123
  interactions between alleles, 1149–1151
  overview, 1149
  patterns, 1149, 1150
  penetrance and expressivity, 1151
  prediction, 1151–1152
  sex-linked, 1154
  sources of individual variation, 1152–1154
Inherited disorders, mitochondrial, 106
Inhibin, 648, 669, 1086, 1093, 1115
Inhibiting hormone (IH), 646–647

Inhibition
  neuron, 447, 481
  presynaptic, 449, 450
Inhibitory interneurons, 475
Inhibitory neurotransmitters, 440
Inhibitory postsynaptic potential (IPSP), 447, 449
Initial segment, axon, 415
Injury(ies)
  cartilage, 160
  chest wall, 878
  cornea, 599
  elbow, 304
  head, 493, 579
  immune response, 827–828
  integumentary system, 195–198
  knee, 306, 309
  neural response, 423, 424
  peripheral nerves, 464
  rotator cuff, 303, 389
  shoulder, 303
  spleen, 820
  tissue response, 170, 171
Innate (nonspecific) defenses
  complement system, 825–827
  fever, 828
  immune surveillance, 824–825
  inflammation, 827–828
  interferons, 825
  overview, 696, 808, 820–821, 822, 823, 853
  phagocytes, 822–824
  physical barriers, 822
Innate (nonspecific) immunity, 829, 842, 846. *See also* Immune response
Innate reflexes, 478
Inner cell mass, 1127
Inner layer, eye. *See* Retina
Inner membrane, mitochondria, 965
Innervation, 367, 415
Innominate artery (brachiocephalic trunk), 781
Innominate vein (brachiocephalic vein), 793
Inorganic compounds, 65–69, 85, 87
Inositol triphosphate (IP_3), 642
Inotropic effects, 446, 742
Insensible perspiration, 181, 1048
Insensible water loss, 990
Insertion, muscle, 332, 363, 365, 408
Insoluble fiber, 956
Inspiration (inhalation), 875, 876, 877, 879
Inspiratory capacity, 881, 882
Inspiratory center, 892
Inspiratory reserve volume (IRV), 881, 882
Insula, 508
Insulin
  effects, 644, 663, 980–981
  in normal growth, 670
  regulation of, 663
  underproduction/overproduction, 668. *See also* Diabetes mellitus
Insulin dependent cells, 663
Insulin dependent (type 1) diabetes, 663, 664, 848
Insulin independent cells, 663

Insulin-like growth factors (somatomedins), 648
Integral proteins, 95
Integrative centers, 513–514
Integrative effects, hormones, 669
Integument, 177
Integumentary system
  age-related changes, 198–199
  dermis, 186–188
  epidermis, 179–186
  exercise and, 407
  hair, 189–192
  hypodermis, 188–189
  injury repair, 195–198, 202
  innervation, 187–188
  integration with other systems, 200
  nails, 195
  overview, 34, 177–178, 201–202
  prenatal development, 1136
  at puberty, 1146
  smooth muscle in, 351
Intensity, sound, 625, 626
Interatrial opening (foramen ovale), 237, 239, 244, 718, 798, 898
Interatrial septum, 715
Intercalated discs, 168, 351, 715, 716
Intercellular communication, 635–637
Intercondylar eminence, 282
Intercondylar fossa, 280
Intercostal artery, 787
Intercostal muscle, 261, 380–381, 878
Intercostal vein, 793
Interferons, 823, 825, 847
Interferon alpha (α), 825, 847
Interferon beta (β), 825, 847
Interferon gamma (γ), 825, 847
Interleukin(s), 846
Interleukin-1 (IL-1), 828, 846
Interleukin-2 (IL-2), 846
Interleukin-6 (IL-6), 701
Interleukin-7 (IL-7), 814
Interlobar arteries, 1002
Interlobar veins, 1003
Interlobular arteries, 1002
Interlobular septa
  liver, 935
  lung, 869
Interlobular veins, 1003
Intermediate-density lipoproteins (IDLs), 975
Intermediate fibers, 347
Intermediate filaments, 97
Intermembrane space, mitochondria, 965
Intermittent claudication, 801
Intermuscular septa, 363
Internal branch, accessory nerve, 527
Internal callus, 226
Internal capsule, 508
Internal carotid artery, 495
Internal ear, 617–619
Internal oblique muscle, 381
Internal occipital crest, 492
Internal os, uterus, 1101
Internal respiration, 873, 885
Internal root sheath, hair, 190

Internal thoracic artery, 781, 783, 1107
Internal thoracic vein, 793
International Anatomical Terminology, 29–30
Interneurons, 418, 473–474
Internodal pathways, 723
Internodes, 421
Internus (muscle name), 365
Interoceptors, 417, 539
Interosseous membrane, 273, 282
Interphase, 127, 130
Intersegmental reflex arcs, 479
Interspinales muscle, 377–379
Interspinous ligament, 301
Interstitial cell(s), 648, 667, 1082, 1093
Interstitial cell–stimulating hormone, 648
Interstitial fluid, 92, 680, 1045
Interstitial fluid colloid osmotic pressure (ICOP), 767
Interstitial fluid hydrostatic pressure (IHP), 766
Interstitial growth, 160, 216
Interstitial lamellae, 211
Interthalamic adhesion, 503
Intertransversarii muscle, 377–379
Intertrochanteric crest, 280
Intertrochanteric line, 280
Intertubercular groove (sulcus), humerus, 271
Intervals, electrocardiogram, 728
Interventricular artery, anterior/posterior, 721, 757
Interventricular foramen (foramen of Monro), 490
Interventricular septum, 715
Interventricular sulcus, anterior/posterior, 712
Intervertebral discs, 253, 300–301, 313
Intervertebral foramina, 253
Intervertebral ligaments, 301
Intestinal artery, 788
Intestinal glands, 932
Intestinal juice, 932
Intestinal phase, gastric activity, 929
Intestinal reflex, 576
Intestinal trunk, 812
Intestinal vein, 798
Intestinal villi, 930–931
Intestinal wall, 931
Intestines. *See* Large intestine; Small intestine
Intra-alveolar pressure, 875
Intracapsular ligaments, 293
Intracellular fluid (ICF; cytosol)
  composition, 96–97, 1044–1045
  fluid and electrolyte regulation in, 1045–1046
  osmotic concentrations, 1045
  overview, 92, 1042, 1072
Intradermal injections, 370
Intrafusal muscle fibers, 479
Intramembranous ossification, 215, 219
Intramural ganglion, 568
Intramuscular (IM) injections, 370
Intraocular pressure, 603
Intrapleural pressure, 877

Intrapulmonary bronchi, 867
Intrapulmonary pressure, 875
Intrinsic (muscle name), 365
Intrinsic factor, 689, 925, 953, 987
Intrinsic ligaments, 293
Intrinsic pathway, coagulation, 703, 704
Intrinsic regulation (autoregulation), 35
Introns, 112
Inuit, heart disease in, 74
Inulin, 1015
Invasion, 132
Inversion, 299
In vitro fertilization, 1157
Involution
  corpus luteum, 1099
  thymus, 819
Iodine (I), 54, 655, 953
Ion(s). *See also specific ions*
  absorption, 953
  chemical notation, 58
  overview, 57
  in urine vs. plasma, 1010
Ionic bonds, 57, 59
Ionic compound, 59
Ionization, 66
Ion pumps, 121, 320
Ipsilateral reflex arcs, 481
Iris, 600
Iron (Fe)
  absorption, 953
  functions in human body, 54, 985
  recommended daily allowance, 985
  recycling and storage, 687
  reserves, 986
Iron-deficiency anemia, 687
Iron overload, 706
Irregular bones, 206
Irritable bowel syndrome (IBS), 956
Ischemia, 407, 724
Ischial ramus, 276
Ischial spine, 276
Ischial tuberosity, 276
Ischiocavernosus muscle, 383, 1113
Ischiofemoral ligament, 306
Ischium, 276
Islets of Langerhans (pancreatic islets), 662, 665, 933
Isoleucine, 979
Isomers, 70
Isometric contractions, 332, 340, 341
Isoproterenol, 742
Isotonic contractions, 339–340
Isotonic solution, 118, 120
Isotopes, 55
Isotropic, 321
Isovolumetric contraction, 734
Isovolumetric relaxation, 736
Isozymes, 80
Isthmus
  thyroid gland, 652
  uterine tube, 1100
  uterus, 1101
Itch sensation, 542

J

Jaundice, 183, 198, 687
Jejunum, 930

Joint(s)
  age-related changes, 309–310
  appendicular skeleton, 308
  axial skeleton, 302
  classification, 290, 291, 312
  overview, 290
  synovial. *See* Synovial joint
Joint capsule, 292
Joint cavity, 165
Joint mice, 312
Jugular foramen, 237, 239, 240, 242
Jugular notch, 262
Jugular trunk, left/right, 812
Jugular vein, internal/external, 491, 495, 793
Jugular venous pulse (JVP), 793
Juvenile rheumatoid arthritis, 290, 314
Juxtaglomerular cells, 666, 773, 1009
Juxtaglomerular complex, 1009
Juxtamedullary nephrons, 1007, 1012

K

Kaposi sarcoma, 847
Karyotype, 1149
Keloid, 198
Keratin, 80, 146, 181, 190
Keratinization, 181, 191
Keratinized epithelium, 146
Keratinocytes, 179
Keratohyalin, 181
Kernig's sign, 483
Keto acid, 978
Ketoacidosis, 982, 1022, 1067
Ketone bodies, 981, 982, 1030
Ketonemia, 982
Ketonuria, 994
Ketosis, 982
Kidney(s). *See also renal headings*
  in acid–base balance regulation, 1062–1063, 1065, 1066, 1068
  age-related changes, 1035
  blood supply, 1002–1003
  connective tissue layers, 1001
  endocrine functions, 638, 665–666
  gross anatomy, 999–1001, 1038
  injury, 999, 1040
  innervation, 1004
  nephrons, 1004–1009
  overview, 1028–1029, 1038. *See also* Renal physiology
  red blood cell recycling in, 688
  sectional anatomy, 1001–1002
Kidney dialysis (hemodialysis), 1035, 1037
Kidney lobe, 1002
Kidney stones, 1034
Kidney transplantation, 1035
Kilocalorie, 989
Kinase, 640
Kinesin, 99, 416
Kinetic energy, 62–63
Kinetochore, 131
Kinocilium, 621
Klinefelter's syndrome, 1156
Knee-jerk (patellar) reflex, 476, 478, 479

Knee joint, 306–309, 313, 398, 401
Korotkoff sounds, 801
Krebs cycle. *See* Citric acid cycle
Kupffer cells, 822, 935
Kwashiorkor, 979, 994
Kyphosis, 255

L

Labeled line, 537
Labetalol, 742
Labia, 915
Labia majora, 1107
Labia minora, 1106
Labor
  dilation stage, 1141–1142
  expulsion stage, 1142–1143
  initiation, 1141
  placental stage, 1142, 1143
  premature, 1143
  true vs. false, 1140
Labor contractions, 1140
Lacrimal apparatus, 597, 599
Lacrimal bone, 247
Lacrimal canaliculi, 597, 599
Lacrimal caruncle, 596
Lacrimal fossa, 241
Lacrimal gland, 241, 597
Lacrimal lake, 599
Lacrimal puncta, 599
Lacrimal sac, 246, 597, 599
Lacrimal sulcus, 247
Lactase, 950
Lactation, 1107, 1144–1146
Lacteals, 810, 932
Lactic acid, 69, 345
Lactic acidosis, 1022, 1067
Lactiferous duct, 1107
Lactiferous sinus, 1107
Lactose, 950
Lactose intolerant, 951
Lacunae
  bone, 162, 209
  cartilage, 159
  in implantation, 1128
Lambdoid suture, 235, 236
Lamella(e), 210, 211
Lamellated (pacinian) corpuscles, 188, 541, 542
Lamina(e), vertebral, 253
Lamina lucida, 143
Lamina propria, 163, 859, 909, 912
Laminectomy, 263
Langerhans (dendritic) cells, 179, 816, 832
Lanugo, 191
Large bowel. *See* Large intestine
Large granular lymphocytes. *See* NK (natural killer) cells
Large intestine
  absorption in, 946–947
  cecum, 943
  colon, 943–945
  gross anatomy, 942–944
  histology, 945–946
  movements, 947–948
  overview, 942, 958
  rectum, 945
  red blood cell recycling in, 688

Large ribosomal unit, 100, 113
Large veins, 753, 758
Laryngeal elevator muscles, 374, 376
Laryngeal prominence, 863
Laryngeal spasms, 898
Laryngitis, 865
Laryngopharynx, 863
Larynx, 370, 863–865
LASIK (laser-assisted in-situ keratomileusis), 608
Latent period, twitch, 336–337
Lateral, 43
Lateral border, scapula, 270
Lateral canthus, 596
Lateral excursion, 370
Lateral flexion, 299
Lateral geniculate nucleus, 504
Lateral masses, ethmoid, 244
Lateral nail folds, 195
Lateral nail grooves, 195
Lateral pathway, 550, 551–552
Lateral rectus muscle, 370, 373
Lateral rotation, 298
Lateral rotators, 364, 396, 397
Lateral sulcus, 508
Lateral ventricle, 490
Latex allergy, 851
Latissimus dorsi muscle, 365, 386
Lauric acid, 73
LDL (low-density lipoprotein), 975
Leads, electrocardiogram, 728
Leak (passive) channels, 118, 425, 428
Lecithin, 76
Left atrioventricular (mitral) valve, 718
Left atrium, 711
Left-handedness, 515
Left-to-right shunt, 800
Left ventricle, 711, 718–719
Leg. *See* Lower limb
Leiomyomas (uterine fibroids), 1117
Length–tension relationships, muscle, 335–336, 353
Lens, 604–607
Lens fibers, 604
Lentiform nucleus, 510, 511
Lentigos, 182
Leptin, 666, 669, 927, 1146
Lesion, 172
Lesser curvature, stomach, 923
Lesser horns, hyoid, 248
Lesser occipital nerve, 469
Lesser omentum, 909
Lesser sciatic notch, 276
Lesser wings, sphenoid, 244
Leucine, 979
Leukemia, 697
Leukocytes. *See* White blood cells
Leukocytosis, 697
Leukopenia, 697
Leukotrienes, 73, 639
Levator anguli oris muscle, 372
Levator ani muscle, 383
Levator labii superioris muscle, 372
Levator palpebrae superioris muscle, 372, 596
Levator scapulae muscle, 384, 385

Levator veli palatini muscle, 374, 376
Levels of organization, 32–35
Levers, 361–362, 407–408
Levitra, 1114
Lewis Wave, 992
Leydig (interstitial) cells, 648, 667, 1082, 1093
Lidocaine, 116
Life stages, 1144. *See also* Postnatal development
Ligament(s)
  characteristics, 154, 157
  denticulate, 154
  elbow joint, 304
  femoral head, 306
  intervertebral, 301
  knee joint, 306–308
  patellar, 280
  shoulder joint, 303
  synovial joints, 293
Ligament teres, 306
Ligamentum arteriosum, 719, 799
Ligamentum flavum, 301
Ligamentum nuchae, 258, 301
Ligand, 95
Ligand-gated (chemically gated) channels, 428
Ligases, 128
Light-adapted state, 613
Light chains, antibody, 837
Limbic lobe, 506
Limbic system, 503, 506–507
Line (bone marking), 207
Linea alba, 381
Linea aspera, 280
Linear motion, 296
Linea terminalis, 277
Lines of action, 364
Lingual frenulum, 916
Lingual lipase, 916, 950
Lingual papillae, 594, 595, 916
Lingual tonsils, 816
Linoleic acid, 975
Linolenic acid, 975
Lipases, 948, 950
Lipid(s)
  catabolism, 973–975, 995
  digestion and absorption, 949, 951, 980–981
  in human body, 1044
  overview, 72–77, 85
  in plasma, 1010
  plasma membrane, 94–95
  synthesis, 975
  transport and distribution, 975–976
  in urine, 1010, 1030
Lipid derivatives, 637, 639
Lipofuscin, 583
Lipogenesis, 975
Lipolysis, 567, 973–975
Lipoprotein(s), 682, 754, 975, 976
Lipoprotein lipase, 976
Liposome, 134
Liposuction (lipoplasty), 157, 189
Liquids, 61
Liquor folliculi, 1099

Lithium (Li), 56
Lithotripter, 1034
Liver
  bile duct system, 936–937
  gross anatomy, 935
  hematological regulation, 939
  histology, 935–936
  lobules, 935
  metabolic functions, 982
  metabolic regulation, 937–939
  physiology, 937–940
  plasma protein processing and, 682–683
  red blood cell recycling in, 688
  turnover times, 85–86
Liver spots (senile lentigos), 182
Liver transplant, 935
Load (resistance), 325
Lobar bronchi, 867
Lobes
  cerebellum, 501
  kidney, 1002
  liver, 935
  lung, 867
  thymus, 819
  thyroid gland, 652
Lobules
  liver, 935
  pulmonary, 869, 870
  testes, 1080
  thymus, 819
Local anesthetics, 116
Local current, 430
Local hormones, 73, 635
Local (graded) potentials, 424, 430–432, 438
Local vasoconstrictors, 768
Local vasodilators, 768
Lockjaw, 332
Locus, 1149
Long bones, 206
Longissimus (muscle name), 366
Longissimus muscle, cervicis/ thoracis, 377–379
Longitudinal arch, foot, 284
Longitudinal fasciculi, 508, 510
Longitudinal fissure, cerebrum, 508
Longitudinal layer, digestive tract, 912
Longitudinal ligament, anterior/ posterior, 301
Long reflexes, 575, 914
Long-term memories, 578
Longus (muscle name), 366
Longus muscle, 377–379
Loop of Henle. *See* Nephron loop
Loose connective tissues, 152, 155–157
Lordosis, 255
Lou Gehrig's disease, 552
Low-density lipoproteins (LDLs), 975
Lower limb
  arteries, 789, 790
  bones, 280–284, 286
  joints, 308
  movements, 308
  muscles, 395–405
  veins, 793–794, 796

Lower motor neuron, 549
Lower respiratory system
  larynx, 863–865
  lungs. *See* Lung(s)
  overview, 858, 902
  primary bronchi, 866
  trachea, 865–866
Low-frequency sound, 624
LSD (lysergic acid diethylamide), 582
Lubrication
  ingested material, 915
  sexual arousal and, 1113, 1114
  by synovial fluid, 293
Lumbar artery, 789
Lumbar curve, 252
Lumbar enlargement, spinal cord, 457
Lumbar plexus, 469–473
Lumbar puncture (spinal tap), 460
Lumbar trunks, 812
Lumbar vein, 796
Lumbar vertebrae, 254, 257, 258, 260
Lumbrical muscles, 394–395, 402–403
Lumen, 142
Luminal compartment, seminiferous tubule, 1085
Lumpectomy, 1109
Lunate, 274
Lunate surface, acetabulum, 276
Lung(s)
  in acid–base balance regulation, 1061–1062, 1065, 1066, 1068
  alveolar ducts and alveoli, 869–871
  blood flow to, 778–779
  blood supply, 872
  bronchi, 867, 870
  bronchioles, 867, 869
  compliance, 875
  gross anatomy, 867, 868
  lobes, 867
  lobules, 869, 870
  pleural cavities and pleural membranes, 872–873
  pressure and airflow to, 875
  relationship to heart, 867, 869
  root, 866
  surfaces, 866, 867, 898
Lung cancer, 901
Lung perfusion, 891
Lunula, 195
Luteal phase, ovarian cycle, 1097, 1110–1111, 1112
Luteinizing hormone (LH), 648, 651, 1092, 1093, 1115
Lutropin. *See* Luteinizing hormone
Luxation (dislocation), 294
Lyme disease, 312, 846
Lymph, 157, 809
Lymphadenopathy, 818
Lymphatic capillaries, 810
Lymphatic duct, left/right, 811, 812
Lymphatic nodule, 816
Lymphatics. *See* Lymphatic vessels
Lymphatic system
  body defenses and, 820–821
  cancer and, 814

functions, 810
integration with other systems, 850
organs, 816–820
overview, 35, 808, 809, 852
prenatal development, 1137
tissues, 700, 816
vessels, 157, 809, 810–813
Lymphatic tissues, 700
Lymphatic trunks, 811
Lymphatic valves, 811
Lymphatic vessels, 157, 809, 810–813, 816
Lymphedema, 812–813, 1047
Lymph glands, 818
Lymph nodes, 814, 816–818
Lymphocytes. *See also* B cells; Natural killer (NK) cells; T cells
  abundance, 813
  appearance, 693
  functions, 153, 698, 808, 841
  life span and circulation, 814
  location, 159
  production, 814–815
  types, 697, 813–814, 841
Lymphocytosis, 698
Lymphoid nodules, 816
Lymphoid organs, 809, 816–820
Lymphoid stem cells, 687, 699, 814
Lymphoid tissues, 809, 816
Lymphomas, 814
Lymphopenia, 698
Lymphopoiesis, 700, 814, 815
Lymphotoxin, 834
Lysergic acid diethylamide (LSD), 582
Lysine, 952, 979
Lysosomal storage diseases, 103
Lysosome, 93, 102–103, 106
Lysozyme, 597, 822, 917, 1145

M

Macrocytic anemia, 690
Macrophages, 126, 153, 688, 822–823
Macula, 601
Macula densa, 1009
Maculae, 621
Macular degeneration, age-related, 630
Magnesium (Mg)
  absorption, 953
  functions in human body, 54, 985
  homeostatic regulation of, 1054
  imbalances, 1055
  normal ECF range, 1055
  recommended daily allowance, 985
  in urine, 1030
Magnetic resonance imaging (MRI), 47
Magnus (muscle name), 366
Major (muscle name), 366
Major calyx, kidney, 1002
Major histocompatibility complex (MHC), 831
Male climacteric, 1115
Male pattern baldness, 191
Male pronucleus, 1124, 1125

Male reproductive system
accessory glands, 1088–1090
age-related changes, 1115
bulbo-urethral glands, 1089, 1090
disorders, 1094
ductus deferens, 1088
endocrine functions, 667
epididymis, 1087–1088
external genitalia, 1090–1092
hormonal regulation, 1092–1094, 1115
integration with other systems, 1115–1117
overview, 35, 1077–1078, 1081
parasympathetic effects, 573
prenatal development, 1137
prostate gland, 1089, 1090
at puberty, 1146
reproductive tract, 1087–1090
scrotum, 1078, 1080
semen, 1077, 1090
seminal glands (vesicles), 1088–1090
smooth muscle, 352
spermatic cords, 1080
spermatogenesis. *See* Spermatogenesis
sympathetic effects, 573
testes, 1078–1082
urethra, 1078, 1088
Male sexual function, 1113–1114
Malignant melanoma, 183, 184, 1155
Malignant tumor, 132
Malleolus, medial/lateral, 282
Malleus, 617
Malnutrition, 983
MALT (mucosa-associated lymphoid tissue), 816
Maltase, 950
Maltose, 950
Mamillary bodies, 504, 505
Mammary glands
anatomy, 1107, 1108
development, 194, 648
histology, 1107, 1108
lactation and, 1144–1146
in pregnancy, 1140
Mammography, 1109
Mandible, 237, 247–248, 279, 285, 302
Mandibular angle, 248
Mandibular arcade, 919
Mandibular canal, 248
Mandibular foramen, 248
Mandibular fossa, 242
Mandibular groove, 917
Mandibular notch, 248
Manganese, 985
Manubrium, 262
Marasmus, 979, 994
Marfan's syndrome, 155, 222, 1155
Marginal arteries, coronary, 721
Margination, 693
Marrow (medullary) cavity, 208
Mass, 53
Masseter muscle, 370, 374
Mass movements, large intestine, 947
Mass number, 54
Mast cell activation, 171

Mast cells, 153, 841
Mastectomy, 1109
Mastication, 370, 373, 918, 919
Mastoid cells, 242
Mastoid fontanelle, 250
Mastoiditis, 242
Mastoid process, 242
Matrix
connective tissue, 152
hair, 191
mitochondria, 103
Matter, 53, 61–62
Maturation promoting factor, 129, 132
Mature graafian follicle, 1099
Maturity, 1122, 1146–1147
Maxillae, 237, 245–246, 302
Maxillary arcade, 919
Maxillary sinuses, 245, 246
Maxillary vein, 793
Maximus (muscle name), 366
M cells, 614
M-CSF, 700
Mean arterial pressure (MAP), 763, 764
Mean corpuscular hemoglobin concentration (MCHC), 690
Mean corpuscular volume (MCV), 690
Meatus, 207
Meatus, inferior/middle/ superior, 861
Mechanically gated channels, 429
Mechanical processing, digestion, 908
Mechanoreceptors, 539, 540–543
Meconium, 1144
Medial, 43
Medial antebrachial cutaneous nerve, 471
Medial border, scapula, 270
Medial canthus, 596
Medial geniculate nucleus, 504
Medial lemniscus, 548
Medial nuclei, thalamus, 504
Medial pathway, 550, 551–552
Medial rectus muscle, 370, 373
Medial rotator, 364
Median cubital vein, 682
Median eminence, 646
Median nerve, 275, 392, 469, 470
Mediastinal artery, 787
Mediastinum, 45, 48, 712, 713
Medical terminology, 29, 49
Medium-sized (muscular) arteries, 753, 754
Medium-sized veins, 753, 758
Medulla
adrenal. *See* Adrenal medulla
hair, 191
lymph node, 816
ovary, 1096
renal, 1002
thymus, 819
Medulla oblongata, 489–490, 496–498, 530, 892, 894–895
Medullary (marrow) cavity, 208
Medullary cords, 816
Megakaryocytes, 701
Meibomian (tarsal) glands, 596

Meiosis
definition, 127
in oogenesis, 1096–1097
overview, 1082, 1084
in spermatogenesis, 1082, 1084–1085
Meiosis I, 1082, 1084, 1085, 1097
Meiosis II, 1082, 1084, 1085, 1097
Meissner's (tactile) corpuscles, 188, 541, 542
Melanin, 97, 154, 182
Melanocyte(s), 154, 179, 182
Melanocyte-stimulating hormone (MSH; melanotropin), 183, 649, 651
Melanosomes, 182
Melatonin, 503, 639, 661
Membrane
composition, 76
cutaneous, 164, 165. *See also* Skin
mucous, 163, 164
plasma. *See* Plasma membrane
serous, 44, 163–165
synovial, 164, 165
types, 163, 164
Membrane attack complex (MAC), 826
Membrane flow, 103
Membrane potential
changes in, 428–430
chemical gradients and, 425
electrical gradients and, 425, 426
electrochemical gradients and, 425–428
overview, 126–127, 424–425, 426, 452
resting, 126, 424, 426, 428
sodium–potassium exchange pump, 428
Membrane renewal vesicles, 105
Membranous labyrinth, 617
Membranous organelles, 92–93, 97. *See also specific organelles*
Membranous urethra, 1032
Memory, 578–580
Memory, immunologic, 830
Memory B cells, 836, 841
Memory consolidation, 578
Memory engram, 580
Memory helper (T$_H$) cells, 835, 841
Memory T$_C$ cells, 835, 841
Memory T cells, 813, 831
Menarche, 1105
Meningeal layers, dura mater, 491
Meninges
cranial, 491–493
spinal, 459, 460
Meningitis, 459
Meniscus (articular disc), 273, 293
Menopause, 1096, 1105, 1114–1115
Menorrhagia, 1117
Menses, 1103–1104, 1112
Menstrual cycle. *See* Uterine cycle
Menstruation, 1104
Mental foramen, 237, 248
Mentalis muscle, 372
Mental protuberance, 248
Mercaptan, 594
Merkel cells, 179, 541
Merkel (tactile) disc, 179, 188, 540, 541

Merocrine secretion, 149, 150
Merocrine (eccrine) sweat glands, 193–194
Mesangial cells, 1008
Mesencephalon (midbrain), 489, 490, 497, 501–502, 530
Mesenchymal cells, 153, 154
Mesenchyme, 154
Mesenteric artery, inferior/ superior, 564, 754, 788
Mesenteric vein, inferior/ superior, 798
Mesenteries, dorsal/ventral, 909, 910
Mesentery proper, 911
Mesoappendix, 943
Mesocolon, 910
Mesoderm, 1129, 1132
Mesothelium, 146
Mesovarium, 1095
Messenger RNA (mRNA), 100, 110–111, 114–115
Metabolic acidosis, 1067–1069, 1070, 1071
Metabolic alkalosis, 1069, 1070, 1071
Metabolic generation, of water, 1048
Metabolic pathway, 64
Metabolic rate, 988–990, 996
Metabolic turnover, 85–86, 962, 964
Metabolism
absorptive state, 980, 982
aerobic, 106, 343, 965, 971
amino acids, 978
carbohydrates, 965–973
cellular, 962–964
definition, 62, 962, 995
energy gains and losses, 988–990
hormonal effects, 980–981
lipids, 973–978
postabsorptive state, 981, 982
proteins, 978
Metabolites, 65
Metabotropic effects, 446
Metacarpal bones, 274–275, 308
Metalloproteins, 682
Metaphase, 131
Metaphase plate, 131
Metaphyseal vessels, 218
Metaphysis, 208
Metaplasia, 172
Metarteriole, 757
Metastasis, 132
Metatarsal bones, 284, 308
Metatarsus, 284
Metencephalon, 490
Methionine, 979
Metopic (frontal) suture, 241, 285
Metoprolol, 725, 742
MHC proteins, 831–833
Micelles, 75–76, 951
Microcephaly, 252
Microcytic anemia, 690
Microfilaments, 97, 98
Microglia, 423, 822
Micromole (µmol), 56
Microphages, 153, 693, 822
Microscopic anatomy, 31
Microtubules, 97–99, 131
Microvilli, 92, 99, 141

Micturition. *See* Urination
Micturition reflex, 1033–1035
Midbrain (mesencephalon), 489, 490, 497, 501–502, 530
Middle ear (tympanic cavity), 242, 617, 618
Middle piece, spermatozoon, 1086
Midsagittal plane, 44
Midsagittal section, 44, 235
Migraine, 530
Migratory phase, injury repair, 196
Milk let-down (milk-ejection) reflex, 145, 650
Milk teeth, 919
Milliequivalents, 1012
Millimeters of mercury (mm Hg), 876
Millimole (mmol), 56
Milliosmoles, 1012
Millivolt (mV), 126
Mineralocorticoids, 658, 659, 668
Minerals, 985–986, 1044. *See also specific minerals*
Minimal volume, 881, 882
Minimus (muscle name), 366
Minor (muscle name), 366
Minor calyx, kidney, 1002
Minoxidil (Rogaine), 191
Miscarriage, 1140
Mitochondria, 93, 103, 106–107, 965–970
Mitosis, 127, 131, 1082, 1084, 1085
Mitotic rate, 129
Mitral valve, 718
Mitral valve prolapse, 745
Mixed exocrine glands, 149
Mixed nerves, 459
Mixing waves, 928
M line, 322
Modality, 537
Moderator band, 718
Molars, 919
Mole (mol), 55, 57
Molecular biology, 29
Molecular motors, 99
Molecular weight, 57
Molecule, 57, 58, 62
Monaxial joint, 296
Monaxial movement, 291
Mongolism. *See* Down's syndrome
Monoamine oxidase (MAO), 567
Monocyte–macrophage system, 822
Monocytes, 159, 693, 697, 698
Monoglyceride, 74
Monohydrogen phosphate, 1059
Monosaccharides, 70–71, 72, 950–951
Monosomy, 1156
Monosynaptic reflexes, 476, 479–480
Monounsaturated fatty acid, 73
Monozygotic twins, 1143
Mons pubis, 1107
Morbidity, 49
Morgagni, Giovanni, 31
Morphine, 443
Morula, 1126, 1127
Motilin, 940
Motility
    gastric, 928
    spermatozoa, 1088

Motion sickness, 619
Motor end plate, 328
Motor homunculus, 551
Motor neurons, 417
Motor nuclei, 463
Motor (descending) pathways, 536, 549–554
Motor speech area, 514
Motor unit, 338
Mouth, 372
Movement(s)
    describing, 295–299
    lower limb, 308
    pectoral girdle, 308
    pelvic girdle, 308
    skull, 302
    synovial joints, 295–299
    thoracic cage, 302
    upper limb, 308
    vertebral, 301–302
M-phase promoting factor (MPF), 129, 132
Mucin(s), 82, 149
Mucins, 917
Mucosa
    digestive tract, 911–912
    nasal, 861
    oral, 915
    respiratory, 859
Mucosa-associated lymphoid tissue (MALT), 816
Mucous (goblet) cells, 149
Mucous connective tissue, 154
Mucous glands, 149
Mucous membranes (mucosae), 163, 164
Mucus, 82, 149, 822
Mucus escalator, 860
Müllerian (paramesonephric) ducts, 1086
Müllerian-inhibiting factor (MIF), 1086
Multicellular glands, 150, 151
Multi-CSF, 700, 701
Multifidus muscle, 377–379
Multinucleate, 166, 318
Multipennate muscle, 361
Multiple births, 1143
Multiple endocrine neoplasia, 1155
Multiple sclerosis (MS), 421, 848
Multipolar neurons, 416, 417
Multiunit smooth muscle cells, 353
Mumps, 916
Muscarine, 570
Muscarinic receptors, 570, 571
Muscle contraction. *See* Contraction
Muscle fibers
    characteristics, 317, 318, 323
    intrafusal/extrafusal, 479
    types, 167, 317. *See also* Cardiac muscle tissue; Skeletal muscle fibers; Smooth muscle tissue
Muscles listed
    abductor digiti minimi, 395, 402
    abductor hallucis, 402
    abductor pollicis, 392, 395
    adductor brevis, 396, 398
    adductor hallucis, 402
    adductor longus, 396

adductor magnus, 364, 396, 398
adductor pollicis, 395
anal sphincter, 383
anconeus, 389, 391
arrector pili, 189
biceps brachii. *See* Biceps brachii muscle
biceps femoris, 398
brachialis, 365, 389, 391
brachioradialis, 389, 391
buccinator, 367, 370, 372
bulbospongiosus, 383, 1106, 1113
ciliary, 601
coccygeus, 383
coracobrachialis, 386, 389
corrugator supercilii, 372
cremaster, 1080
dartos, 1080
deltoid, 273, 361, 365, 386, 389
depressor anguli oris, 372
depressor labii inferioris, 372
detrusor, 1032
diaphragm. *See* Diaphragm
digastric, 375, 377
dorsal interosseus, 395, 402
erector spinae, 377–379
extensor carpi radialis, 366, 391
extensor carpi ulnaris, 391
extensor digiti minimi, 392
extensor digitorum, 360, 392
extensor digitorum brevis, 402–403
extensor digitorum longus, 401, 402
extensor hallucis longus, 401–402
extensor indicis, 392
extensor pollicis, brevis/longus, 392
external oblique, 381
fibularis, brevis/longus, 401–403
flexor carpi, 391
flexor digiti minimi brevis, 395, 402
flexor digitorum, 392, 402
flexor hallucis, 402–403, 405
flexor pollicis, 392, 395
gastrocnemius, 363, 400–401
gemelli, 396
genioglossus, 365, 374, 375
geniohyoid, 375, 377
gluteus, 364, 395–397
gracilis, 396, 398
hyoglossus, 374, 375
iliacus, 396, 398
iliococcygeus, 383
iliocostalis, 377–379
iliopsoas, 398
inferior oblique, 370, 373, 380–381
inferior rectus, 370, 373
infraspinatus, 303, 386, 389
intercostal, 261, 380–381, 878
internal oblique, 381
interspinales, 377–379
intertransversarii, 377–379
ischiocavernosus, 383, 1113
laryngeal elevators, 374–375
lateral rectus, 370, 373
latissimus dorsi, 365, 386, 389
levator anguli oris, 372
levator ani, 383
levator labii superioris, 372

levator palpebrae superioris, 372, 596
levator scapulae, 384, 385
levator veli palatini, 374, 376
longissimus, 377–379
longus, 377–379
lumbricals, 394–395, 402–403
masseter, 370, 374
medial rectus, 370, 373
mentalis muscle, 372
multifidus, 377–379
mylohyoid, 248, 375, 377
nasalis, 372
oblique, 380–381
obturators, 396
occipitofrontalis, 370, 372
omohyoid, 377
opponens digiti minimi, 394–395
opponens pollicis, 394–395
orbicularis oculi, 372, 596
orbicularis oris, 361, 365, 370, 372
palatal, 920
palatoglossus, 374, 375
palatopharyngeus, 920
palmar interosseus, 395
palmaris brevis/longus, 391, 395
papillary, 718
pectinate, 718
pectineus, 396, 398
pectoralis, 359, 384, 386, 389
pharyngeal constrictor, 374, 376, 920
piriformis, 396
plantar interosseous, 402
plantaris, 400
platysma, 370, 372
popliteus, 398, 401
procerus, 372
pronator quadratus, 391
pronator teres, 391
psoas major, 396, 1031
pterygoid, 370, 374
pubococcygeus, 383
pupillary, 600
quadratus femoris, 396
quadratus lumborum, 377–379
quadratus plantae, 402
quadriceps femoris, 280, 401
rectus abdominis, 263, 360, 365, 381
rectus femoris, 360, 365, 397, 400
rhomboid, 365, 384, 385
risorius muscle, 367, 372
rotatores, 377–379
sartorius, 366, 398
scalene, 380–381
semimembranosus, 398
semispinalis, 377–379
semitendinosus, 398
serratus anterior, 384, 385
serratus posterior, 380–381
soleus, 401, 403
spinalis cervicis, 377–379
spinalis thoracis, 377–379
splenius capitis, 377–379
splenius cervicis, 377–379
stapedius, 617
sternocleidomastoid, 377
sternohyoid, 377
sternothyroid, 377

Muscles listed (Continued)
styloglossus, 374, 375
stylohyoid, 377
stylopharyngeus, 920
subclavius, 384, 385
subscapularis, 303, 364, 386, 389
superior oblique, 370, 373
superior rectus muscle, 370, 373
supinator, 391
supraspinatus, 303, 386, 389
temporalis, 240, 365, 370, 374
temporoparietalis, 370, 372
tensor fasciae latae, 395–396
tensor tympani, 617
tensor veli palatini, 374, 376
teres major, 365, 386, 389
teres minor, 303, 364, 386, 389
thyrohyoid, 377
tibialis, anterior/posterior,
    400–403
tongue, 916
trachealis, 866
transverse perineal, 383
transversus abdominis, 380–381
transversus thoracis, 380–381
trapezius, 365, 384
triceps brachii, 342, 363, 364,
    389–391
vastus intermedius, 398
vastus lateralis, 370, 398
vastus medialis, 398
zygomaticus major, 372
zygomaticus minor, 372
Muscle spindles, 479–480
Muscle tissue
cardiac. See Cardiac muscle tissue
cell turnover times, 85–86
characteristics, 166–168, 174,
    317–318
skeletal. See Skeletal muscle tissue
smooth. See Smooth muscle tissue
Muscle tone, 339
Muscular (medium-sized) arteries,
    753, 754
Muscular compression, 764
Muscular dystrophies, 354, 1155
Muscularis externa, 912
Muscularis layer, vagina, 1106
Muscularis mucosae, 912
Muscular system
age-related changes, 403–405
appendicular. See Appendicular
    muscles
axial. See Axial muscles
integration with other systems, 406
muscle types, 359–361
overview, 34, 359, 410
prenatal development, 1136
at puberty, 1146
Musculocutaneous nerve, 469, 470
Mutagens, 132
Mutations, 112, 1153–1154
Myasthenia gravis, 316, 354, 357
Mycobacterium tuberculosis, **860**
Mydriasis, 630
Myelencephalon, 490
Myelin, 421
Myelinated axon, 421, 437
Myelin sheath, 421, 422
Myelocytes, 699, 700
Myelography, 483

Myeloid stem cells, 687, 699
Myeloid tissue (red bone
    marrow), 205, 212–213,
    687–689, 814, 815
Myeloproliferative disorder, 706
Myenteric plexus, 912
Myenteric (short) reflexes, 914
Mylohyoid line, 248
Mylohyoid muscle, 248, 375, 377
Myoblasts, 318
Myocardial infarction (MI), 725
Myocarditis, 745
Myocardium, 714
Myoclonus, 530
Myoepithelial cells, 193
Myofibrils, 319–320, 323
Myofilaments, 319
Myogenic mechanisms, 1016
Myoglobin, 80, 346
Myogram, 336
Myometrium, 1101
Myoneural junction
    (neuromuscular junction),
    326–327, 328
Myopathy, 354
Myopia (nearsightedness), 590,
    608, 633
Myosatellite cells (satellite cells),
    168, 318, 423
Myosin, 99, 331
Myosin light chain kinase, 353
Myotatic (stretch) reflexes, 475,
    476, 478, 479
Myotonic dystrophy, 1155
MyPlate Plan, 984
Myxedema, 668

N

Nail(s), 195, 202
Nail bed, 195
Nail body, 195
Nail root, 195
Na^+–K^+/2 Cl^- transporter, 1020
Naltrexone (ReVia; Depade;
    Vivitrol), 987
Nares, internal/external, 246, 861
Nasal bones, 246
Nasal cavity, 861
Nasal complex, 249–250
Nasal conchae, inferior/middle/
    superior, 244, 245, 247,
    861, 862
Nasalis muscle, 372
Nasal mucosa, 861
Nasal polyps, 902
Nasal septum, 233, 237, 251,
    861, 862
Nasal spine, anterior, 245
Nasal vestibule, 861
Nasolacrimal canal, 246, 599
Nasolacrimal duct, 246, 597, 599
Nasopharyngeal meatus, 863
Nasopharynx, 617, 863
Natriuretic peptides, 666, 1016. See
    ***also* Atrial natriuretic peptide;**
    Brain natriuretic peptide
Natural killer (NK) cells, 697, 813,
    814, 815, 824–825, 841
Naturally acquired active
    immunity, 829

Naturally acquired passive
    immunity, 830
Natural pacemaker. *See* Sinoatrial
    (SA) node
Navicular, 283
N compounds, 983–984
Near point of vision, 606
Nearsightedness (myopia), 590,
    608, 633
Nebulin, 323
Neck
gallbladder, 940
spermatozoon, 1086
tooth, 918
urinary bladder, 1031
Neck (body part), 375–377, 784,
    789, 792
Neck (bone part)
definition, 207, 268
femur, 280
humerus, anatomical/surgical, 271
rib, 261
Necrosis, 134, 170, 187, 828
Negative Babinski reflex, 482
Negative exponent, 1057
Negative feedback, 36–38, 50,
    475, 644
Negative nitrogen balance, 984
Neon (Ne), 56
Neonates
cardiovascular changes at birth, 799
definition, 1144
fluid, electrolyte, and acid–base
    balance, 1071
neonatal period, 1144–1145
respiratory changes at birth,
    898–899, 1144
thermoregulatory problems,
    993, 1144
Neoplasm, 132, 418. *See also* **Cancer**
Nephrolithiasis, 1034
Nephron(s)
age-related changes, 1035
anatomy, 1004–1009
functions, 1005, 1038
organization, 1004–1006
types, 1007
urine formation in, 1009–1012
Nephron loop (loop of Henle),
    1005–1007, 1008, 1013,
    1019–1022
Nephropathy, diabetic, 664
Nephroptosis (floating kidney),
    1001, 1037
Nephrotic syndrome, 1037
Nephrotoxin, 1037
Nerve(s), 412. *See also specific*
    body regions and organs
cranial. *See* Cranial nerves
peripheral. *See* Peripheral nerves
spinal. *See* Spinal nerves
Nerve conduction study, 483
Nerve deafness, 630
Nerve fibers, 169, 318. *See also*
    Axon(s)
Nerve growth factor (NGF), 132, 483
Nerve impulses, 432. *See also*
    Action potentials
Nerve palsy, peripheral
    (peripheral neuropathy),
    464, 664

Nerve plexus(es), 464–473
Nerves listed. *See also* **Cranial**
    nerves
ansa cervicalis, 469
antebrachial cutaneous,
    medial, 471
axillary, 469, 471
cervical, 469
dorsal scapular, 471
femoral, 470, 472
femoral cutaneous, lateral/
    posterior, 470, 472
fibular, 471
genitofemoral, 470, 472, 1080
gluteal, 472
great auricular, 469
iliohypogastric, 472
ilioinguinal, 472, 1080
lesser occipital, 469
median, 275, 392, 469, 470
musculocutaneous, 469, 470
obturator, 472
pectoral, 471
pelvic, 568
phrenic, 469
pudendal, 470, 472
radial, 273, 304, 469, 470
renal, 1001, 1004
sciatic, 276, 470
splanchnic, 564
subscapular, 471
supraclavicular, 469
suprascapular, 471
sural, 471
thoracodorsal, 471
tibial, 471
transverse cervical, 469
ulnar, 273, 469, 470
Nerve tissue. *See* **Neural tissue**
Nervous system
age-related changes, 583
anatomical divisions, 412, 451
autonomic. *See* Autonomic
    nervous system
disorders, categorization, 581
exercise and, 407
functional divisions, 413, 451
immune response and, 851
integration with other
    systems, 584
overview, 34
prenatal development, 1136
at puberty, 1146
somatic. *See* Somatic nervous
    system
Net capillary colloid osmotic
    pressure, 766
Net diffusion, 116
Net filtration pressure (NFP), 767,
    1014, 1015
Net hydrostatic pressure
    (NHP), 1015
Neupogen (filgrastim), 700
Neural circuits, 474–475
Neural integration
autonomic nervous system. *See*
    Autonomic nervous system
higher-order functions. *See*
    Higher-order functions
overview, 535
sensory pathways, 544–549

sensory receptors, 536–543
somatic nervous system, 549–554.
See also Somatic nervous system
Neural stimuli, 642
Neural tissue. *See also* **Nervous**
system
information processing, 447–451
metabolic functions, 982
neuroglia. *See* Neuroglia
neurons. *See* Neurons
overview, 168–169
response to injury, 423, 424
synaptic activity, 439. *See also*
Synapse
Neural tube, 490
Neural tube defects (NTDs), 1157
Neurilemma, 423
Neurilemma (Schwann) cells,
422, 423
Neuroblastoma, 451
Neurocoel, 490
Neuroendocrine cells, 564
Neuroendocrine reflexes, 644
Neurofibrillary tangles, 583
Neurofibrils, 414
Neurofibromatosis, type 2, 1155
Neurofilaments, 414
Neuroglandular junction, 415
Neuroglia
characteristics, 169, 412, 452
CNS, 418–423
PNS, 418, 423
Neurohypophysis, 649–651
Neurological examinations, 581
Neuromodulators, 443–446
Neuromuscular junction (NMJ),
326–327, 328, 355, 415
Neuronal pools, 473–474, 484
Neurons
action potentials. *See* Action
potentials
characteristics, 169, 412, 452
functional classification, 416–418
information processing by,
447–451
membrane potential. *See*
Membrane potential
response to injury, 423, 424
structural classification, 416
structure, 414–416
turnover times, 85–86
Neuropathy, 451, 464, 664
Neuropeptides, 443, 445
Neuropeptide Y, 445
Neurotoxin, 451
Neurotransmitters. *See also*
specific neurotransmitters
activities, 442–443
of adrenal medulla, 564–565
brain function and, 582
characteristics, 326, 415, 440
mechanisms of action, 446
pain perception and, 539–540
representative, 442–443, 444–445
sympathetic, 566–567
Neurotubules, 414
Neutral fats. *See* **Triglycerides**
Neutralization, antigen, 839
Neutral solution, 68, 1056
Neutrons, 53
Neutrophilic band cell, 700

Neutrophilic myelocyte, 700
Neutrophils
abundance, 698
characteristics, 693, 696, 698
functions, 153, 159, 698, 841
life span, 696
in response to injury, 828
Nevus, 201
Newborn. *See* **Neonates**
Niacin (vitamin B$_3$), 987
Nicotinamide adenine
dinucleotide (NAD), 965
Nicotine, 448, 570, 734
Nicotinic receptors, 570, 571
Nifedipine, 742
Night blindness, 612
9 + 0 array, 99
9 + 2 array, 99
Nipple, 1107
Nissl bodies, 415
Nitrate, absorption, 953
Nitric oxide (NO), 60, 443, 445,
446, 567
Nitrogen (N), 54, 60, 883
Nitrogen balance, 983–985
Nitrogenous base, 82
Nitroglycerin, 725, 915
Nitroxidergic synapse, 567
NK (natural killer) cells, 697, 813,
814, 815, 824–825, 841
NMDA (N-methyl D-aspartate)
receptors, 580
Nociceptors, 538, 539–540
Nodal system. *See* **Conducting**
system, heart
Nodes, 421
Nodes of Ranvier, 421
Nomogram, 1071
Nondisplaced fracture, 226
Nonessential amino acids, 979
Non-insulin dependent (type 2)
diabetes, 663, 664
Nonkeratinized epithelium, 146
Nonmembranous organelles,
92–93, 97. *See also specific*
organelles
Nonpolar covalent bonds, 60–61
Nonself antigens, 830
Nonshivering thermogenesis, 992
Nonspecific defenses. *See* **Innate**
(nonspecific) defenses
Nonspecific (innate)
immunity, 829
Nonspecific resistance, 821
Nonstriated involuntary
muscle, 168
Nonstriated muscle, 352
Noradrenaline. *See*
Norepinephrine
Norepinephrine, 442–443, 565,
660, 741, 742
Normal saline, 119
Normoblast, 689
Normochromic, 690
Normocytic, 690
Normotensive, 801
Normovolemic, 706
Nose, 246, 372, 590–591, 861–863.
See also **Olfaction**
Nostrils (external nares),
246, 861

Nuchal lines, inferior/superior, 240
Nuclear envelope, 93, 108
Nuclear matrix, 109
Nuclear pore, 93, 108–109
Nucleases, 934, 950
Nuclei, nervous system
basal, 508, 510–512, 552–553
hypothalamus, 506
limbic system, 506–507
medulla oblongata, 498
pons, 499
spinal cord, 463
thalamus, 503–504
Nucleic acids, 82, 83, 85. *See also*
DNA; RNA
Nucleolus (nucleoli), 93, 109
Nucleoplasm, 109
Nucleosome, 109
Nucleotides, 82, 83
Nucleus, atom, 53
Nucleus, cell, 93, 108–110, 113, 136
Nucleus basalis, 578
Nucleus cuneatus, 496, 498
Nucleus gracilis, 496, 498
Nucleus pulposus, 300
Nurse (sustentacular) cells, 648,
669, 1082, 1083, 1085–1086,
1093
Nursemaid's elbow, 304
Nutrient(s), 65
Nutrient artery, 218
Nutrient foramina, 218
Nutrient pool, 963
Nutrient vein, 218
Nutrition. *See also* **Diet**
definition, 983, 996
disease and, 987
food groups, 983, 984
MyPlate Plan, 984
nitrogen balance, 983–985
vitamins, 986–987
Nyctalopia (night blindness), 612
Nystagmus, 622

O

Obesity, 989–990
Obestatin, 927
Obligatory water reabsorption,
1024
Oblique (muscle name), 365
Oblique layer, stomach, 925
Oblique muscle, 380–381
Obturator artery, 789
Obturator foramen, 276, 279
Obturator muscle, externus/
internus, 396
Obturator nerve, 472
Obturator vein, 794
Occipital bone, 237, 238–239, 240,
285, 302
Occipital condyles, 240
Occipital crest, external/internal,
240, 492
Occipital fontanelle, 250
Occipital lobe, 504, 508, 512, 513
Occipital protuberance, external, 240
Occipital sinus, 789
Occipitofrontalis muscle, 370, 372
Occlusal surfaces, 919
Ocular conjunctiva, 596

Oculomotor muscles, 370
Oculomotor nerve (III), 520, 528
Odorants, 590, 593
Off-center neurons, 614
Oil (sebaceous) glands, 192–193
Olecranon, 273
Olecranon fossa, 273
Olfaction
age-related changes, 594
discrimination, 591, 594
organs, 590–591
overview, 536, 590, 630
pathways, 591
receptors, 590–593
Olfactory bulbs, 519, 591
Olfactory cortex, 513, 591
Olfactory epithelium, 590
Olfactory foramina, 239, 245
Olfactory glands, 590
Olfactory nerve (I), 519, 528
Olfactory organs, 590
Olfactory receptor(s), 245, 590–591
Olfactory receptor cells, 590
Olfactory region, 861
Olfactory tracts, 519
Oligodendrocytes, 421
Olivary nuclei, 496, 497, 628
Olives, 496
Omega-3 fatty acids, 74
Omohyoid muscle, 376, 377
On-center neurons, 614
Oncogenes, 132–133
Oncologist, 134
Oncotic pressure, 766
Onycholysis, 201
Oocyte
activation, 1125
characteristics, 1077
at ovulation, 1123–1124
primary, 1096, 1097
secondary, 1097
Oogenesis, 1096–1099
Oogonia, 1097
Oophorectomy, 1117
Open fracture, 226
Open reduction, 228
Ophthalmic artery, 785
Ophthalmologist, 630
Opioids, 443, 445
Opponens digiti minimi
muscle, 394–395
Opponens pollicis muscle, 394–395
Opportunistic infections, 845
Opposition, 299
Opsin, 607, 610–611, 612
Opsonization, 826, 839
Optic canal, 237, 239, 244
Optic chiasm, 520
Optic disc, 603
Optic nerve (II), 520, 528
Optic radiation, 615
Optic tracts, 504, 520
Optometrist, 630
Oral cavity
anatomy, 915–916
functions, 915–916
overview, 915–916, 956–957
pharynx, 374–375, 862, 863, 920
salivary glands, 248, 523, 916–918
teeth. *See* Teeth
tongue, 916

Oral mucosa, 915
Ora serrata, 601
Orbicularis oculi muscle, 372, 596
Orbicularis oris muscle, 361, 365, 370, 372
Orbit(s), 241, 249
Orbital complex, 249
Orbital fat, 599
Orbital fissure, inferior/superior, 237, 244, 246
Orbital fracture, 233, 266
Orbital process, 246
Orbital rim, 245
Orchiectomy, 1080
Orchitis, 1117
Orexigenic, 994
Organ(s), 31, 32, 49
Organelles, 92–93, 96–107, 135–136. *See also specific organelles*
Organic acids, 1056, 1057
Organic compounds, 65, 69–85, 87–88
Organism, 32
Organism level of organization, 32, 35
Organizational levels, 32–35
Organ level of organization, 32, 35
Organ of Corti (spiral organ), 625
Organogenesis, 1135
Organ physiology, 31
Organ system level of organization, 32, 35
Organ systems, 30–31, 32, 34–35, 39, 49. *See also specific systems*
Organ transplantation, 832, 1035
Orgasm, 1113, 1114
Origin, muscle, 332, 363, 365, 408
Oropharynx, 863
Orthopedics, 229
Orthopnea, 902
Orthostatic hypertension, 801
Osmolarity, 118, 1012
Osmoles, 1012
Osmoreceptors, 649, 1046
Osmosis, 118–119, 124, 765
Osmotic concentration, 118, 1012, 1046
Osmotic pressure (OP), 118, 766
Osseous tissue. *See* Bone
Ossification (osteogenesis)
   endochondral, 215–217
   intramembranous, 215, 218
   overview, 210, 215
Osteoarthritis, 309
Osteoblasts, 209, 210
Osteoclast(s), 209, 210
Osteoclast-activating factor, 228
Osteoclastic crypts, 213
Osteocytes, 162, 209–210
Osteogenesis. *See* Ossification
Osteogenesis imperfecta (OI), 205, 231
Osteogenic (osteoprogenitor) cells, 209, 210
Osteoid, 210
Osteolysis, 210
Osteomalacia, 223
Osteomyelitis, 229
Osteon (Haversian system), 211
Osteopenia, 225

Osteopetrosis, 229
Osteoporosis, 170, 225, 228
Osteosarcoma, 229
Ostia, 251
Otalgia, 630
Otic ganglion, 522, 525, 568
Otitis media, 617
Otolith, 621
Otorhinolaryngology, 902
Outer cortex, lymph node, 816
Outer membrane, mitochondria, 965
Oval window, 618
Ovarian artery, 789, 1096, 1101
Ovarian cancer, 1100, 1155
Ovarian cycle, 1097–1099, 1109, 1110–1112
Ovarian cyst, 1117
Ovarian follicles, 1097
Ovarian hilum, 1096
Ovarian ligament, 1095
Ovarian vein, 796, 1096
Ovary(ies), 638, 667, 1077, 1095–1096
Oviduct (uterine tube), 1077, 1100–1101, 1123
Ovulation, 1099, 1123–1124
Ovum, 1077, 1097
Oxidation, 968
Oxidative phosphorylation, 968, 969–970
Oxygen (O_2)
   blood, 772–773
   CSF, 772–773
   delivery to tissues, 886–889, 891
   functions in human body, 54
   hemoglobin and, 685, 886–887
   partial pressures, 883
   structure, 58
Oxygen debt, 344
Oxygen–hemoglobin saturation (dissociation) curve, 887
Oxyhemoglobin (HbO_2), 685, 886
Oxytocin (OXT), 650, 651, 1141, 1145
Ozone hole, 184

P

p53 gene, 129
Pacemaker cells, 168, 351, 726
Pacemaker potential, 726
Pacesetter cells, 354, 912
Pacinian (lamellated) corpuscles, 188, 541, 542
Packed cell volume (PCV), 680
Paget's disease, 229
Pain pathways, 539
Pain threshold, 554
Pain tolerance, 554
Palatal muscles, 374, 376, 920
Palatine bones, 246
Palatine processes, 246
Palatine tonsils, 816
Palatoglossal arch, 916
Palatoglossus muscle, 375
Palatopharyngeal arch, 916
Palatopharyngeus muscle, 920
Pallidectomy, 530
Pallor, 201
Palmar arches, superficial/ deep, 784

Palmar interosseus muscle, 395
Palmaris muscle, brevis/longus, 391, 395
Palmar vein, superficial/deep, 793
Palmar venous arches, 793
Palpebrae, 596
Palpebral conjunctiva, 596
Palpebral fissure, 596
Palpitation, 745
Pampiniform plexus, 1080
Pancreas
   anatomy, 661
   endocrine functions, 662, 663, 675
   exocrine functions, 662, 675
   gross anatomy, 933
   histology, 933–934
   islets, 638, 662, 665
   physiology, 934–935
Pancreatic acini, 662, 933
Pancreatic alpha-amylase, 934, 950
Pancreatic artery, 788
Pancreatic cancer, 956
Pancreatic duct, 933
Pancreatic islets, 638, 662, 665, 933
Pancreatic juice, 933
Pancreatic lipase, 934, 950
Pancreaticoduodenal artery, superior/inferior, 788
Pancreaticoduodenal vein, 798
Pancreatic polypeptide (PP), 662
Pancreatic vein, 797
Pancreatitis, 933
Pannus, 312
Pantothenic acid (vitamin B_5), 221, 987
Papanicolaou, George, 146
Papilla(e)
   dermal, 179
   duodenal, 940
   hair, 191
   lingual, 594, 595
   renal, 1002
Papillary duct, 1006, 1009
Papillary layer, dermis, 186
Papillary muscle, 718
Pap test, 146
Para-aminohippuric acid (PAH), 1003
Paracrine communication, 635, 636
Paracrine factors, 635
Parafollicular cells (C cells), 223
Parahippocampal gyrus, 506
Parallel muscle, 359, 360
Parallel processing, neural circuits, 474
Paralysis agitans. *See* Parkinson's disease
Paramesonephric ducts, 1086
Paranasal sinuses, 249–250, 861
Paraplegia, 483
Parasagittal section, 44, 235
Parasympathetic blocking agents, 585
Parasympathetic division, ANS
   activation, 562, 568–569
   comparison to sympathetic division, 572
   effects, 569–570
   functional effects, by system, 573

   interaction with sympathetic division, 571–575
   membrane receptors and responses, 570
   neurotransmitter release, 570
   organization and anatomy, 568
   overview, 413, 561, 562, 570–571, 586
Parasympathetic reflexes, 576
Parasympathomimetic drugs, 585
Parathyroid (chief) cells, 656
Parathyroid glands, 222–223, 638, 656, 675
Parathyroid hormone (PTH)
   bone remodeling and, 221
   calcium balance and, 222–224, 656–657, 1054
   effects, 656–657, 658
   in normal growth, 670
   regulatory control, 658
   underproduction/ overproduction, 668
Paraurethral glands, 1106
Paraventricular nucleus, 504, 505, 649
Paravertebral ganglia, 562
Parenchyma, 154
Paresthesia, 543
Parietal bones, 208, 238, 240
Parietal cells, 925
Parietal lobe, 504, 508, 512, 513
Parietal pericardium, 48, 712
Parietal peritoneum, 48, 909
Parietal pleura, 48, 872
Parietal portion, serous membrane, 164
Parietal serosa, 48
Parieto-occipital sulcus, 508
Parkinson's disease, 134, 443, 501, 512, 553
Parotid duct, 917
Parotid salivary glands, 916–917
Paroxetine (Paxil), 443
Paroxysmal atrial tachycardia (PAT), 730
Pars distalis, 645
Pars intermedia, 645
Pars tuberalis, 645
Partial antigens, 838
Partial pressure, of gases, 882–883
Partial-thickness burns, 197
Parturition, 1141
Passive (leak) channels, 118, 425, 428
Passive immunity, 829–830
Patella(e), 206, 280, 309
Patellar ligament, 280, 306
Patellar reflex, 476, 478, 479
Patellar retinacula, 306
Patellar surface, 280
Patellofemoral stress syndrome, 280
Patent ductus arteriosus, 800
Patent foramen ovale, 800
Pathogens, 808
Pathological anatomy, 31
Pathological physiology, 31
Pathologist, 172
Pathophysiology, 49
Paxil (paroxetine), 443

P cells, 614
P/D1 cells, 927
Pectinate muscles, 718
Pectineal line, 276
Pectineus muscle, 396, 398
Pectoral fat pad, 1107
Pectoral girdle
    bones, 268–271, 286
    joints, 308
    movements, 308
    muscles, 384–395
Pectoralis muscle, major/minor,
    359, 384, 386, 389
Pectoral nerves, medial/
    lateral, 471
Pediatrics, 1144
Pedicels, 1008
Pedicles, vertebral arch, 253
Peg cells, 1101
Pelvic bones, 276–278
Pelvic brim, 277
Pelvic cavity, 48, 277
Pelvic diaphragm, 382–383
Pelvic floor, 381–383
Pelvic girdle, 276–278, 285,
    286, 308
Pelvic inflammatory disease
    (PID), 1114
Pelvic inlet, 277, 279
Pelvic nerves, 568
Pelvic outlet, 277–279
Pelvimetry, 286
Pelvis
    arteries, 789
    bones, 276–278, 279
    muscles, 395–398
    sex differences, 277, 279
    veins, 794, 796
Penetrance, 1151
Penicillin, 838
Penile urethra, 1032
Penis, 1078, 1090–1091
Pennate muscle, 360
Pentose, 70
Pepsin, 81, 926, 950
Pepsinogen, 926, 950
Peptic ulcer, 923, 946
Peptidases, 935, 950
Peptide, 78
Peptide bond, 78
Peptide hormones, 637, 639
Perception, 536
Percutaneous transluminal
    coronary angiography
    (PTCA), 745
Perforating (Volkmann's)
    canals, 211
Perforating (Sharpey's) fibers, 213
Perforins, 824
Pericardial artery, 787
Pericardial cavity, 48
Pericardial fluid, 712
Pericardial sac, 712
Pericardiocentesis, 711, 748
Pericarditis, 712
Pericardium, 48, 163, 712, 713
Perichondrium, 157, 159
Perikaryon, 414
Perilymph, 617
Perimenopause, 1114
Perimetrium, 1101

Perimysium, 318
Perinephric fat, 1001
Perineum, 277, 383
Perineurium, 463
Perinuclear space, 108
Periodontal disease, 956
Periodontal ligament, 918
Periosteal layers, dura mater, 491
Periosteum, 157, 162, 209,
    213, 214
Peripheral adaptation, 538
Peripheral chemoreceptors, 896
Peripheral nerves, 464–473
Peripheral nervous system (PNS),
    412, 456–457
Peripheral neuropathy, 464, 664
Peripheral proteins, 95
Peripheral resistance (PR), 762
Peristalsis, 913
Peristaltic wave, primary/
    secondary, 922
Peritoneal cavity, 48
Peritoneum, 48, 164, 909–911
Peritonitis, 909
Peritubular capillaries, 1005,
    1007, 1013
Peritubular fluid, 1008
Permanent dentition, 919–920
Permeability, 113
Permissive effects, hormones, 669
Pernicious anemia, 689
Peroneal (fibular) artery, 789
Peroneal (fibular) nerve, 471
Peroneus muscles (fibularis
    brevis/longus), 401, 403
Peroxisome(s), 93, 103
Perpendicular plate
    ethmoid bone, 245
    palatine bone, 246
Perspiration, 181, 194, 990, 1048
Petrosal sinus, 789
Petrous part, temporal bone, 242
Peyer's patches (aggregated
    lymphoid nodules), 816,
    817, 932
PF-3, 704
pH
    blood, 68, 679, 887–888
    body fluid, 67–68, 1057
    cerebrospinal fluid, 772–773
    control mechanisms, 1058–1061
    definition, 68
    enzyme function and, 81
    gastric contents, 928–929
    hydrogen ion concentration
        and, 68
    sweat, 193
    urine, 1027
Phagocyte-activating
    chemicals, 847
Phagocytes, 822–824
Phagocytic dust cells (alveolar
    macrophages), 823, 869
Phagocytosis, 125, 126, 824
Phagosomes, 122
Phalanges, 275, 284
Phalanx, 308
Phantom limb pain, 545
Pharyngeal arches, 916
Pharyngeal constrictor muscles,
    374, 376, 920

Pharyngeal phase, swallowing, 922
Pharyngeal tonsil (adenoid), 816,
    817, 863
Pharyngotympanic (auditory)
    tube, 160, 242, 617
Pharynx, 374–375, 862, 863, 920
Phasic receptors, 538
Phenotype, 1149
Phenylalanine, 952, 978, 979
Phenylalanine hydroxylase, 978
Phenylketonuria (PKU), 978, 1155
Phenylthiocarbamide (PTC), 596
Pheochromocytoma, 668
Phlebitis, 801
Phlebotomist, 706
Phonation, 864
Phosphate balance, 1054–1055
Phosphate buffer system, 1058–
    1059, 1063
Phosphate group, 70, 75
Phosphate ion, absorption, 953
Phosphate reserve, 1059
Phosphodiesterase (PDE), 611, 642
Phospholipase C (PLC), 642
Phospholipid(s), 75–77
Phospholipid bilayer, 94
Phosphorus (P), 54, 985
Phosphorylation, 84, 640, 642
Photophobia, 630
Photorefractive keratectomy
    (PRK), 608
Photoreception, 607, 610–611
Photoreceptors, 601–602, 607,
    610–611
Phrenic artery, superior/inferior,
    787, 788
Phrenic nerve, 469
Phrenic vein, 797
Physical conditioning, 348–349
Physical therapist, 407
Physiology
    vs. anatomy, 30
    cell, 31
    definition, 29, 49
    human, 31
    organ, 31
    pathological, 31
    systemic, 31
Pia mater, 460–461, 492
Pica, 994
Pineal gland, 495, 503, 615, 638,
    660–661
Pinealocytes, 660
Pinkeye, 597
Pinna, 160, 617
Pinocytosis, 125, 126
Piriformis muscle, 396
Pisiform, 274
Pitch, sound, 624
Pitcher's arm, 391
Pituitary gland
    anatomy and orientation, 243,
        488, 645
    anterior lobe, 645–649
    hormone release, 495
    hormones, 638, 647–649, 650, 651
    hypothalamic control, 646–647
    overview, 645, 674–675
    portal system and blood
        supply, 646
    posterior lobe, 649–651

Pituitary growth failure (pituitary
    dwarfism), 222, 668
Pivot joints, 297
Placenta
    blood supply, 798
    circulation, 1133
    endocrine functions, 1133
    formation, 1130–1131
    functions, 1126
    structure, 1132–1133, 1134
Placenta abruptio, 1157
Placental prolactin, 1133
Placental stage, labor, 1142, 1143
Placenta previa, 1157
Placentation, 1126, 1132–1133
Plane(s)
    of movement, 291
    sectional, 44
Plantar arch, 789
Plantar artery, medial/lateral, 789
Plantar fasciitis, 407
Plantar flexion, 284, 299
Plantar interosseous muscles, 402
Plantaris muscle, 400
Plantar reflex, 482
Plantar vein, 793
Plantar venous arch, 793
Plaque
    atherosclerotic, 724, 754–755
    central nervous system, 583
    definition, 706
    dental, 919
Plasma
    cations and ions, 1045
    composition, 680–681,
        682–683, 707
    as connective tissue, 157
    vs. interstitial fluid, 680
    solute concentrations, 1010
Plasma cells, 153, 697,
    813–814, 815
Plasma expanders, 683
Plasma membrane
    carbohydrates, 96
    drugs affecting, 116
    functions, 92, 94, 135
    lipids, 94–95
    movement across, 113–126
    overview, 91, 94, 135
    proteins, 95–96
    receptors, 640
    structure, 92, 94, 135
Plasmapheresis, 706
Plasma proteins, 680–681,
    682–683
Plasmin, 705
Plasminogen, 705
Plasticity, 353
Plateau, action potential, 731
Platelet(s), 159, 680–681, 698,
    700–701, 707
Platelet adhesion, 702
Platelet aggregation, 702
Platelet-derived growth factor
    (PDGF), 702
Platelet phase, hemostasis,
    702–703
Platelet plug, 702
Platysma muscle, 370, 372
Pleura, 48, 163, 872
Pleural cavity, 48, 872

Pleural fluid, 873
Pleurisy, 873
Plexus of Auerbach (myenteric plexus), 912
Plexus of Meissner (submucosal plexus), 912
Plicae, 912
Plicae circulares, 930
Pneumocystis jiroveci, 872
Pneumocytes type I, 869
Pneumocytes type II, 869
Pneumonia, 872
Pneumotaxic center, 499, 893, 894
Pneumothorax, 878
Podocytes, 1008
Point mutation, 112
Poisoning
    carbon monoxide, 889
    heavy-metal, 421
Polar body, 1097
Polar covalent bond, 61
Polarity, epithelium, 141, 142
Polarized cells, 328
Polar molecule, 61, 66
Pollex, 275
Polycystic kidney disease, 1037, 1155
Polycystic ovary syndrome (PCOS), 1117
Polycythemia, 690
Polygenic inheritance, 1150, 1152
Polymorphonuclear leukocytes. *See* Neutrophils
Polymyositis, 851
Polypeptides, 78
Polyps, 956
Polysaccharides, 71, 72
Polyspermy, 1125
Polysynaptic reflexes, 477, 480–481
Polyunsaturated fatty acid, 73
Polyuria, 664
Pons, 489, 498–499, 530, 893
Popliteal artery, 789
Popliteal ligaments, 306
Popliteal vein, 794
Popliteus muscle, 398, 401
Porphyria, 201
Porphyrins, 984
Porta hepatis, 935
Portal areas, liver, 936
Portal hypertension, 936
Portal system, 646
Portal vessels, 646
Positive Babinski reflex, 482
Positive chemotaxis, 693
Positive feedback, 38–39, 50
Positive feedback loop, 38
Positive nitrogen balance, 984
Positron emission tomography (PET), 47
Postabsorptive state, 981, 982, 996
Postcentral gyrus, 513
Posterior (dorsal), 40, 41, 42
Posterior cavity, eye, 599
Posterior chamber, eye, 599
Posterior column pathway, 547, 548–549
Posterior cruciate ligament (PCL), 306
Posterior femoral cutaneous nerve, 472

Posterior lobe
    cerebellum, 501
    pituitary gland, 649–651
Posterior median sulcus, 457
Postganglionic fibers, 418, 559, 572
Postherpetic neuralgia, 465
Postnatal development
    adolescence and maturity, 1146–1148
    definition, 1122, 1158–1159
    early, 1136–1137
    infancy and childhood, 1146
    neonatal period, 1144–1146
    senescence, 1148. *See also* Aging
Postsynaptic cell, 415
Postsynaptic membrane, 416
Postsynaptic neuron, 439
Postsynaptic potentials, 447
Post-traumatic amnesia, 579
Posttraumatic stress disorder (PTSD), 674
Postural reflexes, 477, 480
Potassium (K)
    absorption, 953
    electrochemical gradient, 425, 427
    functions in human body, 54, 985
    homeostatic regulation of, 1053, 1054
    imbalances, 1053, 1054, 1055
    normal ECF range, 1055
    in plasma, 1010
    reabsorption at distal convoluted tubule, 1024
    recommended daily allowance, 985
    secretion at distal convoluted tubule, 1022
    in urine, 1010, 1030
Potassium adaptation, 1072
Potassium channels, 435, 731
Potential difference, 126, 426
Potential energy, 62–63, 127, 428
Potential spaces, 48
Pott's fracture, 227, 283
Pounds per square inch (psi), 876
Power stroke, cilium, 99, 100
Prader–Willi syndrome, 1153
Pre-Bötzinger complex, 892
Precapillary arteriole, 757
Precapillary sphincter, 757
Precentral gyrus, 513
Precipitation, of antigens, 839
Precocious puberty, 668
Prednisone, 832
Preeclampsia, 1157
Pre-embryo, 1100–1101, 1126
Prefixes, 29
Prefrontal cortex, 514
Prefrontal lobotomy, 514–515
Preganglionic fibers, 417, 559, 561, 572
Preganglionic neurons, 559, 560
Pregnancy. *See also* Labor; Prenatal development
    back pain in, 253
    ectopic, 1128
    first trimester. *See* First trimester
    maternal systems changes during, 1138–1140
    multiple, 1143

Rh factors in, 692, 694–695
    second trimester. *See* Second trimester
    third trimester. *See* Third trimester
    uterine growth during, 1139, 1140
Pre-hypertension, 764
Preload, 741
Premature atrial contractions (PACs), 730
Premature delivery, 1143
Premature ejaculation, 1047
Premature labor, 1143
Premature menopause, 1114
Premature ventricular contractions (PVCs), 730
Premenstrual syndrome, 1117
Premolars, 919
Premotor cortex, 513
Pre-mRNA, 112
Prenatal development
    brain, 490
    definition, 1122, 1158
    first trimester. *See* First trimester
    overview, 1125–1126, 1135, 1136, 1147
    second trimester. *See* Second trimester
    third trimester. *See* Third trimester
Preoptic area, hypothalamus, 505, 506
Prepatellar bursitis, 312
Prepotential, 726
Prepuce
    clitoris, 1107
    penis, 1091
Preputial glands, 1091
Presbyopia, 608
Pressure (P), 760
Pressure gradient, blood vessel, 760
Pressure points, 754
Pressure wave, sound, 623, 626
Presynaptic cell, 415
Presynaptic facilitation, 449, 450
Presynaptic inhibition, 449, 450
Presynaptic membrane, 415
Presynaptic neuron, 439
Prevertebral ganglia, 562
Prickling pain, 539
Primary amenorrhea, 1105
Primary brain vesicles, 490
Primary bronchi, left/right, 866
Primary cilium, 99
Primary curves, 252
Primary dentition, 919–920
Primary fissure, cerebellum, 501
Primary follicles, 1097–1099
Primary motor cortex, 513
Primary oocytes, 1096, 1097
Primary ossification center, 216–217
Primary respiratory muscles, 879
Primary response, to antigen, 839
Primary sensory cortex, 513
Primary spermatocytes, 1082
Primary structure, protein, 78, 79
Primary taste sensations, 595
Primary tumor, 132
Prime mover, 363
Primitive streak, 1129
Primordial follicles, 1096, 1097, 1098
P–R interval, 728
Prion, 134
Procaine, 116

Procarboxypeptidase, 935, 950
Procerus muscle, 372
Process (bone marking), 207
Prochlorperazine (Compazine), 619
Procoagulants (clotting factors), 702, 703–704
Products, of chemical reaction, 62
Proelastase, 935, 950
Proenzymes, 703
Proerythroblasts, 687
Profundus (muscle name), 365
Progenitor cells, 698, 699
Progesterone, 648
    functions, 669, 1099, 1112, 1115
    in pregnancy, 1133
    production, 1112
    regulatory control, 669, 1115
Prohormones, 639
Projection fibers, 508
Prolactin (PRL), 132, 647, 648, 651
Prolactin-inhibiting hormone (PIH), 647, 648
Prolactinoma, 674
Prolactin-releasing factors (PRF), 647, 648
Proliferation phase, injury repair, 196
Proliferative phase, uterine cycle, 1104–1105, 1111
Promethazine, 619
Pronation, 298
Pronator (muscle name), 366
Pronator quadratus muscle, 391
Pronator teres muscle, 391
Prone, 41
Pronucleus, male/female, 1124, 1125
Propagation, 433
Properdin, 826
Prophase, mitosis, 131
Propranolol, 725, 742
Proprioceptors, 417, 539, 540, 542–543
Prosencephalon, 490
Prosopagnosia, 530
Prostacyclin, 703
Prostacyclins, 639
Prostaglandins, 73, 639, 705, 1141
Prostate cancer, 1094, 1155
Prostatectomy, 1094
Prostate gland, 1089, 1090
Prostate-specific antigen (PSA), 1094
Prostatic concretions, 1094
Prostatic fluid, 1090
Prostatic urethra, 1032
Prostatitis, 1090
Prosthesis, 312
Proteases, 101, 935, 948, 950
Proteasomes, 93, 101
Protectin, 824
Protein(s)
    catabolism, 978, 995
    complete/incomplete, 983
    digestion and absorption, 949, 951–952, 980–981
    fibrous, 80
    functions, 77, 80, 85, 87
    globular, 80
    in human body, 77, 1044
    in plasma, 1010

plasma membrane, 95–96
shape, 78–80
sources, 85
structure, 77–78, 80, 85, 87
synthesis, 104–105, 110–113, 978–979
in urine, 1010
Protein buffer systems, 1058, 1059–1060
Protein C, 705
Protein-calorie malnutrition, 994
Protein deficiency diseases, 979
Protein kinase C (PKC), 642
Proteinuria, 1018
Proteoglycans, 82
Proteolytic, 926
Proteolytic enzymes, 935
Prothrombin activator, 704
Proton(s), 53
Proton acceptor, 69
Proton donor, 69
Protraction, 299
Proximal, 43
Proximal convoluted tubule (PCT)
anatomy, 1005–1007, 1008
functions, 1008, 1013, 1028–1029
mechanisms, 1013
reabsorption and secretion at, 1018–1019
Prozac (fluoxetine), 443, 582
Pseudopodia, 126
Pseudostratified columnar epithelium, 148, 149
Pseudounipolar (unipolar) neurons, 416, 417
Psoas major muscle, 396, 398, 1031
Psoriasis, 177, 195, 203, 851
Psychosis, 530
Psychosocial dwarfism, 674
Pterygoid muscle, 370, 374
Pterygoid plate, 244
Pterygoid processes, 244
Pterygopalatine ganglion, 523, 568
Ptyalin (salivary amylase), 916, 917, 950
Puberty
apocrine sweat glands, 193
bone growth, 217, 221
female, 191, 1146
hair changes, 191
male, 191, 660, 865, 1080, 1146
precocious, 668
testosterone levels in, 1092
Pubic symphysis, 276, 278
Pubic tubercle, 276
Pubis, 276
Pubococcygeus muscle, 383
Pubofemoral ligament, 306
Pudendal artery, 789
Pudendal nerve, 470, 472
Pudendal vein, 794
Pudendum, 1106
Pulled groin, 398, 407
Pulled hamstring, 401
Pulmonary arterioles, 781
Pulmonary artery, left/right, 718, 752, 781
Pulmonary circuit, 711, 780–781, 804, 872, 885
Pulmonary embolism, 801, 872
Pulmonary lobules, 869

Pulmonary plexus, 572
Pulmonary trunk, 718
Pulmonary valve, 718
Pulmonary vein, left/right, 718, 781
Pulmonary ventilation
air movement, 874–875
carbon dioxide transport, 889–890
definition, 873
gas exchange, 882–886
mechanics of breathing, 876, 878–880
oxygen transport, 886–889, 891
pressure changes during, 875–878
respiratory rates and volumes, 880–882
Pulp, spleen, 820
Pulp cavity, 918
Pulpitis, 956
Pulse (heartbeat), 763
Pulse (hormone release), 644
Pulse pressure, 763
Pulvinar nucleus, 504
Punnett square, 1151, 1152
Pupil, 600
Pupillary constrictor muscles, 600
Pupillary constrictor reflex, 613
Pupillary dilator muscles, 600
Pupillary reflex, 575, 576
Purines, 82, 83, 445
Purkinje cells, 501, 549
Purkinje fibers, 723, 727–728
Pus, 170, 696, 828
Putamen, 510
P wave, 728
Pyelogram, 1031
Pyloric antrum, 923
Pyloric canal, 923
Pyloric glands, 926–927
Pyloric sphincter, 923
Pyloric stenosis, 956
Pylorospasm, 956
Pylorus, 923
Pyramidal cells, 513, 550
Pyramidal system, 550–551
Pyramids, 551
Pyrexia (fever), 994
Pyridoxine (vitamin B$_6$), 987
Pyrimidines, 82, 83
Pyrogens, 828, 994
Pyruvate, 106, 965
Pyruvic acid, 965
Pyuria, 1030

Q

QRS complex, 728
Q–T interval, 728
Quadrate lobe, liver, 935
Quadratus femoris muscle, 396
Quadratus lumborum muscle, 377–379
Quadratus plantae muscle, 402
Quadriceps femoris, 280, 401
Quadriplegia, 483
Quadruplets, 1143
Quaternary structure, protein, 79–80
Quiet breathing, 878, 879, 892–893

R

Rabies, 416
Radial artery, 682, 783, 784, 1103
Radial collateral ligament, 304
Radial fossa, humerus, 273
Radial groove, humerus, 273
Radial head, 273
Radial nerve, 273, 304, 469, 470
Radial notch, 273
Radial tuberosity, 273
Radial vein, 793
Radiation, as heat transfer mechanism, 990
Radiation sickness, 55
Radioactive decay, 55
Radiographic anatomy, 31
Radioisotopes, 55
Radiopacity, 46
Radiopaque, 46
Radioulnar joint, distal/ proximal, 273
Radius, 273, 308
Rafts, 96
Rales, 902
Rami communicantes, 467
Ramus (bone marking)
definition, 207
mandible, 248
pubis, ischial/inferior/superior, 276
Range of motion, 290
Raphe, 359, 916, 1080
Rapid eye movement (REM) sleep, 580
Raynaud's phenomenon, 801
Reabsorption
in capillary exchange, 765–768
in collecting system, 1023
in distal convoluted tubule, 1022
in proximal convoluted tubule, 1018–1019
in urine formation, 1009, 1010–1012
Reactant, 62
Reactive atoms, 57
Recall of fluids, 768
Receptive field, 537
Receptive speech area, 514
Receptor(s), 33, 36, 413, 475. See also specific receptors
Receptor cascade, 640
Receptor-mediated endocytosis, 122–123, 125, 126
Receptor potential, 537
Receptor proteins, 95
Receptor site, 121
Receptor specificity, 537
Recessive inheritance, 1150–1151
Reciprocal inhibition, 481
Recognition proteins, 95
Recombination, 1153
Recommended daily allowance (RDA), 985, 986, 987
Recovery period, muscle, 344–346
Recruitment, motor units, 338
Rectal artery, 788
Rectal vein, 797
Rectouterine pouch, 1095
Rectum, 945–946
Rectus (muscle name), 365
Rectus abdominis muscle, 263, 360, 365, 381

Rectus femoris muscle, 360, 365, 397, 400
Red blood cell count, 683, 690
Red blood cells (RBCs)
abundance, 159, 683–684, 698
crenation, 119, 120
formation, 686–687, 699
functions, 159, 698
hemoglobin. See Hemoglobin
hemolysis, 119, 120
histology, 698
overview, 680–681, 707
oxygen transport, 886–889
production, 687–690
recycling of components, 686–687, 688
structure, 684–685
tests, 690, 692–693
turnover, 686–687, 698
in urine, 1030
Red bone marrow, 205, 212–213, 687–689, 814, 815
Red cones, 607
Red–green color blindness, 613, 1154
Red muscles, 347
Red nucleus, 501, 502
Red pulp, spleen, 820
Reduction, 968
Reductional division, 1084
Referred pain, 548
Reflex(es). See also Spinal reflexes; specific reflexes
acquired, 478
classification, 478–479
cranial, 456, 479, 529
definition, 413, 471
innate, 478
overview, 475, 484
somatic, 478, 529
visceral, 478, 529
Reflex arc, 475, 476
Reflex centers, 496, 498
Refracted, 605
Refraction, 604–605, 608
Refractory period
of action potential, 731
of membrane, 328, 433, 435
Regeneration, 170, 171
Regional anatomy, 30
Regulation
extrinsic, 33
homeostatic, 35
Regulatory hormones, 643, 644
Regulatory T cells. See Helper T (T$_H$) cells; Suppressor T (T$_S$) cells
Regurgitation, cardiac, 719
Reinforcement, 482
Relative refractory period, 433
Relaxation, muscle, 334, 342
Relaxation phase, twitch, 337
Relaxin, 277, 1133, 1141
Releasing hormone (RH), 646, 647
Remission, 172
Remodeling, bone, 218
Renal artery, 788, 1001, 1002
Renal calculi, 1034
Renal columns, 1002
Renal compensation, 1062–1063, 1065, 1066

Renal corpuscle, 1004, 1006, 1007–1008, 1013
Renal cortex, 1002
Renal failure, 1035
Renal fascia, 1001
Renal glycosuria, 1019
Renal medulla, 1002
Renal nerves, 1001, 1004
Renal papilla, 1002
Renal pelvis, 1002
Renal physiology
  filtration pressures, 1013–1015
  glomerular filtration rate, 1015–1018
  nephron loop and countercurrent multiplication, 1019–1022
  overview, 1009–1012
  reabsorption and secretion along the collecting system, 1023, 1026
  reabsorption and secretion at the distal convoluted tubule, 1022–1023, 1026
  reabsorption and secretion at the proximal convoluted tubule, 1018–1019
  summary, 1028–1029
  urine formation, 1009–1012
  urine volume and osmotic concentration, 1023–1025
  vasa recta, 1026, 1027
Renal pyramids, 1002
Renal sinus, 1001
Renal threshold, 1012
Renal tubule, 1005, 1006
Renal vein, 796, 1001, 1003
Renin, 666, 1009, 1016
Renin-angiotensin-aldosterone system, 666, 667, 1016, 1017
Rennin (chymosin), 926, 950
Replication, DNA, 127
Repolarization, 328, 430, 432
Reposition, 299
Repressor genes, 129
Reproductive cells (sex cells; germ cells), 91
Reproductive system
  basic structures, 1077
  female. See Female reproductive system
  hormones, 1115
  integration with other systems, 1115–1117
  male. See Male reproductive system
  prenatal development, 1137
  at puberty, 1148
Residual volume, 881, 882
Resistance
  blood flow, 755, 760, 762
  membrane, 425
  muscle contraction and, 325
Resistance phase, general adaptation syndrome, 671
Resistance vessels, 755
Resonance, 625
Resorption (osteolysis), 210
Respiration, 873–874. See also Respiratory physiology
Respiratory acidosis, 1064–1067, 1070, 1071
Respiratory alkalosis, 1066, 1067, 1070, 1071

Respiratory bronchioles, 69
Respiratory burst, 696, 828
Respiratory centers, 892–895
Respiratory chain (electron transport system), 343, 968, 969–970
Respiratory compensation, 1061–1062, 1065, 1066
Respiratory cycle, 877
Respiratory defense system, 860
Respiratory distress syndrome, 872
Respiratory epithelium, 859–860
Respiratory infection, 857, 905
Respiratory membrane, 872, 884–885
Respiratory minute volume, 880
Respiratory mucosa, 859
Respiratory muscles, 878, 879
Respiratory physiology
  control of respiration, 890–898
  gas exchange, 882–886
  gas transport, 886–890
  pulmonary ventilation, 874–882
Respiratory portion, respiratory tract, 858, 872
Respiratory pump, 765, 877
Respiratory rate, 880, 1140, 1144
Respiratory reflexes, 892, 896–898
Respiratory rhythmicity centers, 496, 498, 892, 894–895
Respiratory system
  age-related changes, 899
  cardiovascular system and, 899, 901
  changes at birth, 898
  exercise and, 407
  functions, 857
  integration with other systems, 899–901
  lower. See Lower respiratory system
  organization, 858–860
  overview, 35, 857
  parasympathetic effects, 573
  prenatal development, 1137
  at puberty, 1146
  pulmonary ventilation. See Pulmonary ventilation
  smooth muscle in, 351
  structure, 858
  sympathetic effects, 573
  upper, 858, 861–863
Respiratory tract, 858
"Rest and digest" response, 561
Resting membrane potential, 126, 424, 426, 428. See also Membrane potential
Rete testis, 1082
Reticular activating system, 502, 581–582
Reticular fibers, 154
Reticular formation, 496, 498, 499, 502, 552
Reticular layer, dermis, 186
Reticular tissue, 156, 157
Reticulocyte, 689
Reticulocyte count, 690
Reticulocytosis, 690
Reticuloendothelial system, 822

Reticulospinal tracts, 550, 552
Retina, 600, 601–603, 614
Retinal (retinene), 607, 612
Retinitis pigmentosa, 613, 1155
Retinoblastoma, 1155
Retinopathy, 600
Retraction, 299
Retroflexion, uterus, 1101
Retrograde amnesia, 579
Retrograde flow, axon, 416
Retroperitoneal, 48, 911
Retrovirus, 845
Return stroke, cilium, 99, 100
Reverberation, neural circuits, 474–475
Reverse transcriptase, 845
Reversible reaction, 63–64
Rheumatic fever, 721
Rheumatism, 309, 313
Rheumatoid arthritis, 309, 848
Rh factors, 690, 692
Rhinitis, 902
Rhinoplasty, 902
Rhinovirus, 857
Rh negative (Rh⁻), 690
Rhodopsin, 607, 609, 612
RhoGAM, 695
Rhombencephalon, 490
Rhomboid muscle, major/minor, 365, 384, 385
Rh positive (Rh⁺), 690
Rib cage, 261
Riboflavin (vitamin B₂), 987
Ribosomal RNA (rRNA), 100
Ribosomes, 93, 99–100
Ribs (costae), 261–263, 302
RICE (rest, ice, compression, and elevation), 354
Rickets, 185, 223
Right atrioventricular (tricuspid) valve, 718
Right atrium, 711, 715, 718
Right-to-left shunt, 800
Right ventricle, 711, 718, 719
Rigor mortis, 334
Ringer's solution, 683
Risorius muscle, 367, 372
RNA
  vs. DNA, 82–84
  messenger. See Messenger RNA
  ribosomal, 100
  structure, 82, 83
  transfer, 112
  types, 82
RNA polymerase, 110
RNA processing, 112
Rods, 601, 607, 609
Root
  hair, 191
  lung, 866
  nail, 195
  penis, 1090
  tongue, 916
  tooth, 918
Root canal, 918, 956
Root hair plexus, 189, 540, 541
Root sheath, hair follicle, 190
Rosacea, 201
Rotation, 296, 298–299, 364
Rotator cuff, 303, 389
Rotatores muscle, 377–379

Rotaviruses, 1145
Rough endoplasmic reticulum (RER), 93, 102
Rouleaux, 684
Round ligaments
  liver, 935
  uterus, 1080, 1101
Round window, 618
Rubrospinal tracts, 550
Ruffini corpuscles, 541, 542
Rugae
  stomach, 923
  urinary bladder, 1031
  vagina, 1106
Rule of nines, 197
Runner's knee, 280
R wave, 728

S

Saccule, 618, 621, 622
Sacral artery, middle/lateral, 789
Sacral canal, 259
Sacral cornua, 259
Sacral crest, median/lateral, 259
Sacral curve, 252
Sacral foramina, 259
Sacral hiatus, 259
Sacral plexus, 469–473
Sacral promontory, 261
Sacral tuberosity, 261
Sacral vein, 794, 796
Sacroiliac joint, 259, 276
Sacrum, 252, 258–261, 279, 302, 308
Saddle joints, 297
SAD PUCKER mnemonic, for retroperitoneal organs, 1001
Sagittal plane, 44
Sagittal section, 44
Sagittal sinus, inferior/superior, 492, 789, 793
Sagittal suture, 235, 236
Saliva, 917
Salivary amylase, 916, 917, 950
Salivary glands, 248, 523, 916–918
Salivatory nuclei, 917–918
Salpingitis, 1117
Saltatory propagation, 437, 438
Salts, 69, 85, 1056
Salty receptors, 593
Sarcolemma, 318–319
Sarcoma, 172
Sarcomere
  contracted, 325, 331, 334–335, 355
  length–tension relationships, 335–336
  resting, 325, 330
  structure, 321–325
Sarcoplasm, 318–319
Sarcoplasmic reticulum, 320
Sartorius muscle, 366, 398
Satellite cells (myosatellite cells), 168, 318, 423
Satiety center, 506
Saturated enzyme, 81, 1011
Saturated fatty acid, 73
Saturation limit, enzyme, 81
Saturation limits, 120
Scab, 196
Scala media (cochlear duct), 618, 623

Scala tympani (tympanic duct), 623, 626
Scala vestibuli (vestibular duct), 623, 626
Scalene muscles, 380–381
Scalp (epicranium), 370, 372
Scanning electron micrograph (SEM), 98, 135
Scanning techniques, 46
Scaphoid, 273
Scapula, 268, 270–271
Scapular nerve, dorsal, 471
Scapular spine, 271
Scarring phase, injury repair, 196
Scar tissue, 171, 196
Schilling test, 706
Schizophrenia, 582
Schwann cells, 422, 423
Sciatic nerve, 276, 470, 472
Sciatic notch, greater/lesser, 276
Scientific method, 32
Sclera, 599
Scleral venous sinus, 603
Scleroderma, 201
Sclerotherapy, 801
Scoliosis, 255
Scopolamine (Transderm Scop), 619
Scotoma, 607
Scrotal cavity, 1080
Scrotum, 1078, 1081
Scurvy, 221
Seasonal affective disorder (SAD), 661
Sebaceous follicles, 192
Sebaceous (oil) glands, 192–193, 202
Seborrheic dermatitis, 193
Sebum, 192
Secondary active transport, 122, 123
Secondary amenorrhea, 1105
Secondary brain vesicles, 490
Secondary bronchi, 867
Secondary curves, 252
Secondary dentition, 919–920
Secondary follicles, 1098, 1099
Secondary memories, 578
Secondary oocyte, 1097
Secondary ossification center, 216–217
Secondary response, to antigen, 840–841
Secondary spermatocytes, 1082
Secondary structure, protein, 78, 79
Secondary tumor, 132
Second-class lever, 361, 362
Second-degree burn, 197
Second heart sound, 737
Second messengers, 446, 567, 640, 641
Second-order neuron, 544
Second trimester
    changes in maternal systems, 1138–1140
    changes in uterus, 1139, 1140
    fetal development, 1138
    overview, 1126, 1136–1137
Secretin, 941, 942
Secretion
    in collecting system, 1023
    digestive, 909
    at distal convoluted tubule, 1022

modes, 149
    at proximal convoluted tubule, 1018–1019
    in small intestine, 932–933
    types, 149
    in urine formation, 1010–1012
Secretory phase, uterine cycle, 1105, 1111
Secretory piece, immunoglobulin, 839
Secretory sheet, 151, 925
Secretory vesicles, 105
Section, 42
Sectional anatomy, 42, 44
Sectional planes, 42, 44
Segmental blocks, 1101
Segmental bronchi, 867
Segmentation, 913
Segments, electrocardiogram, 728
Seizure, 517
Selectively permeable membrane, 114
Selective serotonin reuptake inhibitors (SSRIs), 443, 582
Selenium, 985
Self-antigens, 830
Sellar diaphragm, 645
Sella turcica, 243, 645
Semen, 1077, 1090
Semicircular canals, 618
Semicircular ducts, 618, 619–622
Semilunar ganglion, 522
Semilunar valves, 721
Semimembranosus muscle, 398
Seminal fluid, 1090
Seminal glands (vesicles), 1078, 1088–1090
Seminalplasmin, 1090
Seminiferous tubules, 1080, 1083
Semispinalis muscle, capitis/cervicis/thoracis, 377–379
Semitendinosus muscle, 398
Senescence, 1148. See also Aging
Senile cataracts, 604
Senile dementia, 583
Senile lentigos (liver spots), 182
Senility, 583
Sensation, 536
Sense organs, 536
Sensible perspiration, 181, 990, 1048
Sensitization
    B cells, 836–837
    Rh factor, 692
Sensory association areas, 513
Sensory coding, 538
Sensory ganglia, 416
Sensory homunculus, 546, 548
Sensory neurons, 416, 475
Sensory nuclei, 463
Sensory pathways, 535–536, 544–549
Sensory processing centers, 535
Sensory receptors. See also Special senses
    adaptation, 538–539
    classification, 539–543
    information interpretation, 537–538
    overview, 535–536, 554–555
    stimuli detection, 536–537

Sentinel node, 851
Sepsis, 197
Septa
    lung, 869
    testes, 1080
    thymus, 819
Septicemia, 706
Septum pellucidum, 490
Serial processing, neural circuits, 474
Serosa, 164, 912
Serotonin, 443, 444, 449, 660, 702
Serous fluid, 164
Serous glands, 149
Serous membranes (serosa), 44, 163–165
Serratus anterior muscle, 384, 385
Serratus posterior muscle, 380–381
Sertoli cells. See Nurse (sustentacular) cells
Sertraline (Zoloft), 443
Serum, 683
Set point, 33
Severe acute respiratory syndrome (SARS), 902
Severe combined immunodeficiency disease (SCID), 848, 1155
Sex cells (germ cells; reproductive cells), 91
Sex chromosomes, 1149
Sex differences
    blood volume, 682
    body composition, 1043, 1044
    hematocrit, 680
    pelvis, 277, 279
    pulmonary volume and capacity, 881
    skeletal system, 279
Sex hormones, 221, 228, 670. See also Androgen(s); Estrogen(s)
Sex-linked inheritance, 1150, 1154
Sexual arousal, 576, 1113, 1114
Sexual function, 1113–1114
Sexual intercourse, 1077, 1113–1114
Sexually transmitted diseases (STDs), 1114
Sexually transmitted infections (STIs), 1114
Shaft (diaphysis)
    definition, 208, 268
    femur, 280
    rib, 261
Shaft, penis, 1090
Shallow breathing, 878
Sharpey's (perforating) fibers, 213
Shingles, 465
Shin-splints, 407
Shiver, 991
Shivering thermogenesis, 991
Shock, 778
Shock-wave lithotripsy, 1037
Short bones, 206
Short chain polypeptides, 639
Short reflexes, 575
Short (myenteric) reflexes, 914

Short-term memories, 578
Shoulder girdle. See Pectoral girdle
Shoulder (glenohumeral) joint, 270, 302–304, 313
Shoulder separation, 303
Siamese (conjoined) twins, 1143
Sickle cell anemia, 685, 686, 1151, 1155
Sick sinus syndrome, 745
Sigmoid artery, 788
Sigmoid colon, 943
Sigmoid mesocolon, 911
Sigmoid sinus, 793
Signs, 32
Simple columnar epithelium, 144, 148, 149
Simple cuboidal epithelium, 144, 146, 147
Simple diffusion, 117–118
Simple epithelium, 144, 145
Simple fracture, 226
Simple gland, 151
Simple inheritance, 1149, 1150
Simple squamous epithelium, 144, 145, 146
Simple sugars, 70
Single covalent bond, 60
Sinoatrial (SA) node, 721, 723, 726, 740
Sinus(es), 207, 233
Sinusitis, 251
Sinusoidal capillaries, 756, 757
Sinusoids, 756, 757, 935
Situs inversus, 91, 138
Skatole, 946
Skeletal muscle(s)
    actions, 363–365, 366–367
    age-related changes, 403
    atrophy, 347
    blood supply, 318
    connective tissue organization, 317–318
    functions, 316, 355
    hormones and, 346
    hypertrophy, 348
    insertions, 332, 363, 365, 408
    names, 365–367, 408
    nerves, 318
    origins, 332, 363, 365, 408
    overview, 168, 367, 368–369
    parasympathetic effects, 573
    physical conditioning and, 348–349
    sympathetic effects, 573
    tension production, 338
    tone, 339
Skeletal muscle fibers
    atrophy, 348
    contraction. See Contraction, skeletal muscle
    control, 328
    development, 319
    distribution, 347
    hypertrophy, 347
    innervation, 318
    muscle performance and, 347
    overview, 318
    sliding filament theory, 324–325
    structure, 318–325, 355
    tension production, 334–337
    types, 346–347

Skeletal muscle tissue
vs. cardiac muscle tissue, 353, 715, 716
connective tissue organization, 317–318
contraction. *See* Contraction, skeletal muscle
functional organization, 323
functions, 167
locations, 167
metabolic functions, 982
metabolism, 342–346
overview, 166
vs. smooth muscle tissue, 353
tension production, 338
Skeletal system. *See also* **Bone**
age-related changes, 285
axial skeleton, 233–263
functions, 205, 310
integration with other systems, 310–311
overview, 34, 205, 229
prenatal development, 1136
at puberty, 1146
sex differences, 279
Skene's glands, 1106
Skill memories, 578
Skin. *See also* **Integumentary system**
abnormalities, 198
age-related changes, 198–199
cleavage lines, 187
dermis. *See* Dermis
epidermis. *See* Epidermis
functions, 177–178
hypodermis, 165, 166, 178, 188–189
injury repair, 195–198
innervation, 187–188
parasympathetic effects, 573
structure, 177–178
sympathetic effects, 573
thick, 179, 180
thin, 179
water immersion, 181
Skin cancers, 184
Skin-fold test, 994
Skin graft, 197
Skin tests, 844
Skin turgor, 186
Skull
age-related changes, 285
anterior view, 237
cranial bones, 233, 235, 240–245
facial bones, 235, 245–248
fontanelles, 250–251
foramina and fissures, 237
horizontal section, 239
infant, 250–252
inferior view, 237
joints, 302
lateral view, 236
movements, 302
nasal complex, 249–250
orbital complex, 249
overview, 233
posterior view, 236
sagittal section, 238
sex differences, 279
superior view, 236
sutures, 235, 236

Sleep, 580–581
Sleep disorders, 581
Sliding filament theory, 324–325
Slow-adapting receptors, 538
Slow calcium channels, 731
Slow fibers, 346–347, 348
Slow pain, 539
Slow potassium channels, 731
Small intestine
absorption in, 941, 949
glands, 932
gross anatomy, 927, 930
histology, 930–932
hormones, 940, 941, 942
movements, 933
overview, 927, 957
regional specializations, 932
secretions, 932–933, 949
Small proteins, 639
Small ribosomal unit, 100, 113
Small saphenous vein, 793
Smegma, 1091
Smoking, 899
Smooth endoplasmic reticulum (SER), 93, 101
Smooth muscle tissue
vs. cardiac muscle tissue, 353
control of contractions, 353–354
excitation–contraction coupling, 353
functions, 167, 351–352
length–tension relationships, 353
locations, 167
overview, 166, 356
vs. skeletal muscle tissue, 353
structure, 352–353
tone, 354
Sneezing, 898
Snellen chart, 630
Sodium (Na$^+$)
absorption, 953
electrochemical gradient, 425, 427
functions in human body, 54, 985
homeostatic regulation, 1051–1053
imbalances, 1050, 1052–1053, 1055
ionic bond formation, 59
normal ECF range, 1055
in plasma, 1010
reabsorption at distal convoluted tubule, 1022, 1024
reabsorption at proximal convoluted tubule, 1018, 1019
reabsorption in collecting system, 1023
recommended daily allowance, 985
in urine, 1010, 1030
Sodium bicarbonate, 1061
Sodium channels
in action potential generation, 434–435
fast/slow, 731
in photoreception, 610–611
voltage-gated, 328
Sodium chloride (NaCl), 59, 66
Sodium hydroxide (NaOH), 69
Sodium monohydrogen phosphate, 1059

Sodium–potassium ATPase, 122, 428, 433
Sodium–potassium exchange pump, 122, 426–428, 433
Soft keratin, 190
Soft palate, 861, 916
Solar plexus, 572
Soleus muscle, 401, 403
Solids, 61
Solitary nucleus, 496, 549
Solubility, 65
Soluble fiber, 956
Solute, 65
Solution, 65
Solvent, 65
Soma, 91, 169, 414–415
Somatic cells, 91. *See also* **Cell(s)**
Somatic motor association area, 513
Somatic motor neurons, 417
Somatic motor pathways, 535
Somatic motor system. *See* **Somatic nervous system**
Somatic nervous system (SNS)
basal nuclei, 552–553
cerebellum, 553
comparison to autonomic nervous system, 558, 559, 577
corticospinal pathway, 550–551
lateral pathway, 552
levels of processing and motor control, 553–554
medial pathway, 552
motor homunculus, 551
overview, 413, 549–550, 555
Somatic reflexes, 478, 529
Somatic sensory association area, 513
Somatic sensory neurons, 416
Somatic sensory pathways, 544–549
Somatocrinin (growth hormone–releasing hormone), 649
Somatomedins (insulin-like growth factors), 648
Somatostatin (growth hormone–inhibiting hormone), 649, 926, 940
Somatotropin. *See* **Growth hormone**
Sound, 623–626
Sounds of Korotkoff, 801
Sour receptors, 593
Spastic paralysis, 554
Spatial summation, 447, 448
Special circulation, 778–779
Special senses
gustation. *See* Gustation
hearing. *See* Hearing
olfaction. *See* Olfaction
overview, 536, 590
prenatal development, 1136
vision. *See* Vision
Specific (adaptive) defenses, 696, 821, 830. *See also* **Immune response**
Specific (adaptive) immunity, 118, 829
Specificity
of adaptive immunity, 830
enzyme, 80

Specific resistance, 821
Speech center, 514
Spermatic cords, 400, 1080, 1081
Spermatids, 1082
Spermatocytes, primary/secondary, 1082
Spermatogenesis, 1082–1085, 1092, 1093
Spermatogonia, 1082
Spermatozoa, 1077, 1086–1087, 1090, 1123
Sperm count, 1090
Spermiation, 1085
Spermiogenesis, 1082, 1084–1085
Sperm receptors, 1124
S phase, interphase, 130
Sphenoid, 237, 238–239, 243–244
Sphenoidal fontanelle, 250
Sphenoidal sinus, 238, 243, 244
Sphenoidal spine, 244
Sphenopalatine ganglia, 522
Sphincter(s), 351, 361
Sphincter of Oddi, 940
Sphygmomanometer, 801
Spicules, 219
Spina bifida, 263
Spinal accessory nerve (accessory nerve), 527, 528
Spinal anesthesia, 461
Spinal cord, 418, 420, 457–459, 462, 483
Spinal curves, 252
Spinal fusion, 263
Spinalis cervicis muscle, 377–379
Spinalis thoracis muscle, 377–379
Spinal meninges, 459, 460
Spinal meningitis, 459
Spinal nerves
anatomy, 463–464
overview, 412, 459
peripheral distribution, 464–468
plexuses, 464–473, 484
Spinal reflexes
brain's effect on, 481–482
monosynaptic, 476, 479–480
overview, 456, 476–477, 484
polysynaptic, 477, 480–481
reinforcement and inhibition, 482
voluntary movement and, 482
Spinal root, accessory nerve, 527
Spinal shock, 483
Spinal tap (lumbar puncture), 460
Spindle apparatus, 100
Spindle fibers, 131
Spine. *See* **Vertebral column**
Spine (bone marking), 207
Spinoaccessory nerve (accessory nerve), 527, 528
Spinocerebellar ataxia, 1155
Spinocerebellar pathway, 547, 549
Spinocerebellar tracts, anterior/posterior, 545, 547, 549
Spinothalamic pathway, 544–545, 546, 548
Spinothalamic tracts, anterior/lateral, 544, 545, 546
Spinous process, vertebrae, 253
Spiral arteries, 1103
Spiral CT scan, 47
Spiral fracture, 226
Spiral ganglion, 627

Spiral organ, 623, 625
Spirometer, 881
Spirulina, 983
Splanchnic nerves, 609
Spleen, 820
Splenectomy, 820
Splenic artery, 788, 820
Splenic flexure, 943
Splenic vein, 797, 820
Splenius capitis muscle, 377–379
Splenius cervicis muscle, 377–379
Splenomegaly, 851
Split-thickness graft, 197
Spongy bone, 212–213, 227
Spongy urethra, 1032
Spontaneous abortion, 1140
Spontaneous mutation, 1153
Sports drinks, 349, 1053
Sports injuries, 309. *See also*
    Athletes
Spot desmosomes, 143
Sprain, 293
Sputum, 902
Squamous cell carcinoma, 184
Squamous epithelium, 144,
    145–146
Squamous part, temporal
    bone, 241
Squamous suture, 235, 236
Stapedius muscle, 617
Stapes, 617
Starches, 71
Starling, Ernest H., 742
Starling's law of the heart, 742
Start codon, 114
State of equilibrium, 39
States of consciousness, 580–582
States of matter, 61–62
Stato-acoustic nerve. *See*
    Vestibulocochlear nerve
Stellate reticuloendothelial
    (Kupffer) cells, 935
Stem cells (basal cells)
    epidermal, 179
    epithelial, 144
    gustatory, 594
    lymphoid, 687, 699
    myeloid, 687, 699
    olfactory, 590
    overview, 129
    therapeutic uses, 134
Stent, cardiac, 725
Stercobilinogens, 687, 946
Stercobilins, 687, 946
Stereocilia, 621
Sterile, 1080
Sternal end, clavicle, 270
Sternoclavicular joint, 270
Sternocleidomastoid muscle, 375,
    376, 377
Sternohyoid muscle, 377
Sternothyroid muscle, 377
Sternum, 262–263, 302, 308
Steroid-binding proteins, 683
"Steroid creams," 632
Steroid hormones, 637–638,
    639, 642, 643, 1112. *See also*
    specific hormones
Steroids, 74–75, 77
Stethoscope, 736
Stimulus, 33

Stimulus frequency, 338
Stomach
    digestion and absorption in,
        927, 949
    functions, 923, 957
    gastric glands, 925–926
    gross anatomy, 923–925
    histology, 925
    pyloric glands, 926–927
    regulation of activity, 927
Stop codon, 115
Strabismus, 630
Straight artery, 1103
Straight sinus, 789, 793
Straight tubule, 1080
Stratified columnar epithelium,
    144, 148, 149
Stratified cuboidal epithelium,
    144, 146, 147
Stratified epithelium, 144, 145
Stratified squamous epithelium,
    144, 145, 146
Stratum, 179
Stratum basale (stratum
    germinativum), 179, 180
Stratum corneum, 180
Stratum granulosum, 180
Stratum lucidum, 180
Stratum spinosum, 179–180
Stress, 670, 849
Stress fractures, 284
Stress response (general
    adaptation syndrome),
    670, 671
Stress urinary incontinence, 1034
Stretch marks, 186
Stretch reflexes, 475, 476,
    478, 479
Striated involuntary muscle, 168
Striated voluntary muscle, 168
Striations, 323
Strict dominance, 1150
Stridor, 902
Stroke. *See* Cerebrovascular
    accident
Stroke volume (SV), 735, 737,
    738, 741–743
Stroma
    ovarian, 1096
    reticular tissue, 154
Stromal cells, 814
Strong acid, 69, 1056
Strong base, 69, 1056
Structural formula, 70
Structural lipids, 76
Structural proteins, 77
Student's elbow, 294
Stupor, 530
Stuttering, 902
Sty, 596
Styloglossus muscle, 375
Stylohyoid ligaments, 248
Stylohyoid muscle, 375, 377
Styloid process
    radius, 273
    skull, 238, 242, 285
    ulna, 273
Stylomastoid foramen, 237,
    242, 243
Stylopharyngeus muscle, 920
Subacromial bursa, 304

Subarachnoid space, 460, 491
Subatomic particles, 53
Subcapsular space, 816
Subclavian artery, 752, 781, 783
Subclavian trunk, left/right, 812
Subclavian vein, 793
Subclavius muscle, 384, 385
Subcoracoid bursa, 304
Subcutaneous injection, 188
Subcutaneous layer (hypodermis;
    superficial fascia), 165, 166,
    178, 188–189
Subdeltoid bursa, 304
Subdural hemorrhage, 493
Subdural space, 460
Sublingual ducts, 917
Sublingual salivary glands, 917
Subluxation, 294
Submandibular ducts, 917
Submandibular ganglion, 522,
    523, 568
Submandibular salivary glands,
    248, 917
Submucosa
    digestive tract, 912
    trachea, 866
Submucosal glands, 932
Submucosal plexus, 912
Subpapillary plexus, 187
Subscapular bursa, 304
Subscapular fossa, 270
Subscapularis muscle, 303, 364,
    386, 389
Subscapular nerve, 471
Subserous fasciae, 166
Substance P, 443, 445, 539
Substantia nigra, 501, 502
Substrate-level
    phosphorylation, 968
Substrates, 80
Sucrase, 950
Sucrose, 71, 950
Sudden infant death syndrome
    (SIDS), 893
Sudoriferous glands, 193
Suffixes, 29
Sulci, cerebral cortex, 488
Sulcus (bone marking), 207
Sulfadiazine, 496
Sulfate ion, absorption, 953
Sulfisoxazole, 496
Sulfur (S), 54
Summation, 447
Sunburn, 197
Sun protection factor (SPF), 184
Sunscreen, 184
Superficial, 43
Superficial anatomy, 40
Superficial fascia (hypodermis;
    subcutaneous layer), 165,
    166, 178, 188–189
Superficialis (muscle name), 365
Superficial lymphatics, 811
Superficial reflexes, 478
Superior, 43
Superior border, scapula, 270
Superior colliculus, 501, 502, 552
Superior ganglion, 525, 526
Superior mesenteric artery, 564
Superior mesenteric ganglion, 564
Superior mesenteric vein, 798

Superior oblique muscle, 370, 373
Superior rectus muscle, 370, 373
Superior vena cava, 715, 789,
    793, 795
Superoxide anions, 696
Supination, 299
Supinator muscle, 391
Supine, 41
Supporting connective tissues,
    152, 159–163. *See also* Bone;
    Cartilage
Suppression (genetic), 1152
Suppression factors, 835
Suppressor T (T_S) cells, 813, 831,
    835, 841
Suprachiasmatic nucleus, 505, 506
Supraclavicular nerve, 469
Supraoptic nucleus, 504, 505, 649
Supraorbital foramen, 237, 241
Supraorbital margin, 241
Supraorbital notch, 241
Suprarenal gland. *See* Adrenal gland
Suprascapular nerve, 471
Supraspinatus muscle, 303,
    386, 389
Supraspinous fossa, 271
Supraspinous ligament, 301
Sural nerve, 471
Surface anatomy, 30
Surface antigens, 690
Surface features, bone, 207
Surface tension, 61, 871
Surfactant, 869
Surgical anatomy, 31
Surgical neck, humerus, 271
Suspensions, 67
Suspensory ligaments
    breast, 1107
    lens, 601
    ovary, 1095–1096
    uterus, 1101
Sustentacular (nurse) cells,
    648, 667, 1082, 1083,
    1085–1086, 1093
Sutural (Wormian) bone, 206, 235
Suture, 291
Sutures, skull, 235, 236
Swallowing (deglutition), 863,
    866, 922
Swallowing center, 922
Swallowing reflex, 576, 922
Sweat, 193–194, 1053. *See also*
    Perspiration
Sweet receptors, 593
Sympathetic activation, 560, 566
Sympathetic blocking agents, 585
Sympathetic chain ganglia, 562,
    563–564
Sympathetic division, ANS
    activation, 562, 566
    comparison to parasympathetic
        division, 572
    functional effects, by system, 573
    interaction with parasympathetic
        division, 571–575
    neurotransmitter release and,
        566–568
    organization and anatomy,
        563–565, 572
    overview, 413, 560, 562–563,
        568, 586

Sympathetic nerves, 467, 564
Sympathetic reflexes, 576
Sympathomimetic drugs, 585
Symphysis, 291
Symport (cotransport), 120–121, 1011
Symptoms, 32
Synapse
  chemical, 440
  cholinergic, 440–442
  electrical, 439–440
  functions, 328, 415–416, 439, 452
  nitroxidergic, 567
  structure, 415–416
Synapsis, 1082
Synaptic cleft, 326, 328, 415
Synaptic communication, 636, 637
Synaptic delay, 442
Synaptic fatigue, 442
Synaptic terminal, 328, 415
Synaptic vesicles, 415
Synarthrosis, 290, 291
Synchondrosis, 291
Syncope, 801
Syncytial trophoblast
  (syncytiotrophoblast), 1127
Syncytium, functional, 351
Syndesmosis, 291
Syndrome, 49
Syndrome of inappropriate
  secretion of ADH (SIADH),
  668, 1072
Syneresis (clot retraction), 703, 705
Synergist (muscle), 363
Synergistic effect, hormones, 669
Synesthesia, 630
Synostosis, 291
Synovial fluid, 165, 293
Synovial joint
  accessory structures, 293
  movements, 294–299, 312–313
  stabilization, 294–295
  structure, 292–294, 312
  types, 297
Synovial membrane, 164, 165, 292
Synovial tendon sheath, 293, 392
Synovitis, 312
Synthesis reaction, 63
Syphilis, 554
Systemic anatomy, 30–31
Systemic arteries, 711, 781–789.
  See also Arteries
Systemic circuit
  arteries, 711, 781–789. See also
    Arteries
  overview, 711, 804
  partial pressures in, 885–886
  pressures within, 760, 762, 763
  veins, 711, 789–798. See also Veins
  velocity within, 762
Systemic lupus erythematosus
  (SLE), 851
Systemic physiology, 31
Systemic veins, 711, 789–798. See
  also Veins
Systems integrators
  cardiovascular system, 802
  digestive system, 955
  endocrine system, 673
  integumentary system, 200
  lymphatic system, 850

muscular system, 406
nervous system, 584
reproductive system, 1116
respiratory system, 900
skeletal system, 311
urinary system, 1036
Systole, 732, 734–736
Systolic pressure, 763

T

T₄ (thyroxine), 221, 639, 652, 658,
  668, 989
Tabes dorsalis, 483
Tachycardia, 727
Tachypnea, 902
Tactile corpuscles, 188, 541, 542
Tactile (Merkel) disc, 179, 188,
  540, 541
Tactile receptors, 540–542
Taeniae coli, 943
Tail
  epididymis, 1087
  pancreas, 933
  spermatozoon, 1087
  thick filament, 324
Tail fold, 1133
Talipes equinovarus
  (clubfoot), 284
Talus, 283, 308
Tanning bed, 184
Tanycytes, 418
Target cells, 636
Tarsal bones, 283, 308
Tarsal glands, 596
Tartar, 918
Taste buds, 594, 595
Taste hairs, 594
Taste pore, 594
Taste receptors, 592–593, 594
Taste sensations, 593, 595–596.
  See also Gustation
Tattoos, 198
Tay–Sachs disease, 451, 1154, 1155
T cells, 697, 813, 815, 853
  activation, 832, 834–836
  antigen presentation, 831–832
  antigen recognition, 832–834
  CD4, 835
  CD8, 834–835
  life span and circulation, 814
  types, 813, 831
Tear (lacrimal) gland, 241, 597
Tears, 597, 599
Tectorial membrane, 623
Tectospinal tracts, 550, 552
Tectum, 501, 552
Teeth
  age-related changes, 285
  decay, 918–919
  dental succession, 919
  sex differences, 279
  structure, 918–919
  types, 919
Teeth grinding, 248
Tegmentum, 501
Telencephalon, 490
Telodendria, 414, 415
Telomerase, 133
Telomeres, 129
Telophase, mitosis, 131

Temperature
  body. See Body temperature
  effect on diffusion rate, 117
  effect on enzymes, 81–82
  effect on hemoglobin, 888
  regulation. See Thermoregulation
Template strand, DNA, 111
Temporal bone, 237, 238–239,
  241–243
Temporalis muscle, 240, 365,
  370, 374
Temporal lines, inferior/superior,
  240, 241
Temporal lobe, 504, 508, 513
Temporal processes, 241, 247
Temporal summation, 447, 448
Temporal vein, 793
Temporomandibular joint (TMJ)
  syndrome, 248
Temporoparietalis muscle,
  370, 372
Tendinous inscriptions, 363, 381
Tendon(s), 152, 154, 157, 293, 318
Tendon reflex, 480
Tenosynovitis, 407
Tension, 325
  motor units and production of,
    338–339
  production by muscle fibers,
    334–338
  production in skeletal
    muscles, 338
Tensor fasciae latae muscle,
  395–396
Tensor tympani muscle, 617
Tensor veli palatini muscle,
  374, 376
Tentorium cerebelli, 492
Teratogens, 1151
Teres (muscle name), 366
Teres major muscle, 365, 386, 389
Teres minor muscle, 303, 364,
  386, 389
Terminal bronchioles, 867
Terminal cisternae, 320
Terminal ganglion, 568
Terminal hairs, 191
Terminal lymphatics (lymphatic
  capillaries), 810
Terminal segment, aorta, 787
Terminal web, 97
Terminologia Anatomica (TA), 29–30
Tertiary bronchi, 867
Tertiary follicle, 1099
Tertiary memories, 578
Tertiary structure, protein, 78–79
Testes, 638, 667, 1077–1082
Testicular artery, 789, 1080
Testicular cancer, 1080
Testicular torsion, 1077, 1120
Testicular vein, 796
Testosterone
  effects, 667, 1092, 1093,
    1115, 1146
  integumentary system and, 1146
  muscle metabolism and, 346
  production and secretion, 648,
    1077, 1082, 1092, 1115
  structure, 75
Testosterone-binding globulin
  (TeBG), 683

Tetanus, 332, 337, 338
Tetracycline, 496
Tetrad, 1082
Tetraiodothyronine. See Thyroxine
Tetralogy of Fallot, 800
Tetrose, 70
Thalamic nuclei, 503–504
Thalamus, 488, 489, 503–504, 506
Thalassemia, 685, 686
Thecal cells, 1099
Therapeutic abortion, 1140
Therapeutic cloning, 1157
Thermal inertia, 66
Thermoreceptors, 538, 539, 540
Thermoregulation
  blood in, 679
  in burn patient, 197
  disorders, 993
  heat conservation, 991
  heat gain/loss regulation,
    991–992
  heat generation, 991–992
  heat transfer mechanisms,
    990–991
  hypothalamus in, 36–37
  individual variations, 992–994
  integumentary system in, 194
  muscle activity and, 346
  negative feedback in, 36–38
  in neonates, 993, 1144
  organ systems involved, 39
  overview, 990, 996
  skeletal muscle and, 316
  sweat glands and, 194
Theta waves, 516
Thiamine (vitamin B₁), 987
Thiazides, 1023
Thick filaments, 97, 99, 319,
  323–325
Thick segment, nephron loop, 1008
Thick skin, 179, 180
Thigh. See Lower limb
Thin filaments, 319, 323–325
Thin segment, nephron loop, 1008
Thin skin, 179, 180
Third-class lever, 361, 362
Third-degree burn, 197
Third heart sound, 737
Third-order neuron, 544
Third trimester
  changes in maternal systems, 1140
  changes in uterus, 1139, 1140
  fetal development, 1138
  overview, 1126, 1136–1137
Third ventricle, 490
Thirst center, 506
Thoracentesis, 873
Thoracic aorta, 786–787
Thoracic cage, 261–263, 302
Thoracic cavity, 45, 48
Thoracic curve, 252
Thoracic duct, 811, 812
Thoracic vein, internal, 793
Thoracic vertebrae, 254, 257,
  258, 259
Thoracodorsal nerve, 471
Thoracolumbar division, ANS. See
  Sympathetic division, ANS
Thoroughfare channel, 757
Threonine, 979
Threshold, 432

Threshold potential, 328
Thrill, 801
Thrombin, 704
Thrombocyte(s). *See* Platelet(s)
Thrombocyte-stimulating
   factor, 701
Thrombocytopenia, 701
Thrombocytopoiesis, 701
Thrombocytosis, 701
Thrombolytic, 706
Thrombomodulin, 705
Thrombophlebitis, 801
Thrombopoietin, 701
Thromboxane A_2, 702
Thromboxanes, 639
Thrombus, 706, 801
Thymic (Hassall's) corpuscles,
   819, 820
Thymic epithelial cells (TECs),
   819–820
Thymic lobes, 819
Thymine, 82
Thymosin(s), 132, 666, 820
Thymus, 638, 666, 814, 815,
   818–820
Thyrocervical trunk, 781, 783
Thyroglobulin, 652
Thyrohyoid muscle, 377
Thyroid-binding globulins
   (TBGs), 652, 682
Thyroid-binding prealbumin
   (TBPA), 652
Thyroid cartilage, 652, 863
Thyroidectomy, 674
Thyroid follicles, 652–654
Thyroid function tests, 674
Thyroid gland
   anatomy, 652, 653, 675
   C cells and calcitonin, 655
   follicles, 652–654
   hormones. *See* Thyroid
      hormone(s)
Thyroid hormone(s). *See also*
   specific hormones
   functions, 654–655, 658
   heart rate and, 741
   intracellular binding, 642, 643
   iodine and, 655
   muscle metabolism and, 346
   in normal growth, 670
   overview, 638
   regulatory control, 658
   structure, 639
Thyroiditis, 848
Thyroid peroxidase, 652
Thyroid-stimulating hormone
   (TSH; thyrotropin), 648,
   651, 654, 992
Thyrotoxicosis, 674
Thyrotropin-releasing hormone
   (TRH), 648, 992
Thyroxine (tetraiodothyronine;
   T_4), 221, 639, 652, 658,
   668, 989
Tibia, 282, 308
Tibial artery, posterior/
   anterior, 789
Tibial collateral ligament, 306
Tibialis muscle, anterior/
   posterior, 400–403
Tibial nerve, 471

Tibial tuberosity, 282
Tibial vein, anterior/
   posterior, 793
Tic douloureux, 522
Tickle sensation, 542
Tidal volume (V_T), 877, 881
Tight junction, 142, 143
Timolol, 742
Tinea, 201
Tinnitus, 630
Tissue(s)
   age-related changes, 170
   connective. *See* Connective
      tissue(s)
   definition, 31, 32, 49, 140
   epithelial. *See* Epithelia
   inflammation, 170, 171
   membrane. *See* Membrane
   muscle, 166–168. *See also* Cardiac
      muscle tissue; Skeletal muscle
      tissue; Smooth muscle tissue
   neural. *See* Neural tissue
   regeneration, 170
   response to injury, 170, 171
   types, 140–141
Tissue cholinesterase, 570
Tissue engineering, 172
Tissue factor (TF; factor III), 704
Tissue level of organization, 32,
   35, 140. *See also* Tissue(s)
Tissue perfusion, 768
Tissue plasminogen activator
   (t-PA), 705
Tissue rejection, 172
Tissue repair, 827
Tissue transplantation, 172
Titin, 319, 322
Tolerance, of immune system, 830
Tongue, 374, 375, 594–596, 916
Tongue muscles, extrinsic/
   intrinsic, 916
Tonicity, 118
Tonic receptors, 538
Tonsillitis, 816
Tonsils, 816, 817
Tooth. *See* Teeth
Tophi, 312
Torr, 876
Torticollis, 407
Total body water, 1072
Total lung capacity, 881, 882
Total mastectomy, 1109
Total peripheral resistance,
   760–762
Trabeculae
   lymph nodes, 816
   pulmonary, 869
   spongy bone, 212
Trabeculae carneae, 718
Trabecular artery, 820
Trabecular vein, 820
Trace elements, 54
Trachea, 865–866
Tracheal cartilages, 866
Trachealis muscle, 866
Tracheostomy, 864
Traction, 229
Tracts, CNS, 463, 498, 499
Traits, 1149
Transamination, 978, 979
Transcortins, 657, 682

Transcription, 110–111
Transderm Scop
   (scopolamine), 619
Transducin, 611
Transduction, 536, 537
Trans face, Golgi apparatus, 105
Trans fatty acids, 74
Transferrin, 682, 687
Transfer RNA (tRNA), 112
Transfusion(s), 686, 691–693
Transfusion reaction, 692
Transient ischemic attack
   (TIA), 530
Transitional epithelium, 146,
   147, 1031
Translation, 112–113, 114–115
Translocation, 1153
Translocation defect, 1156
Transmembrane potential. *See*
   Membrane potential
Transmission electron
   micrograph (TEM), 135
Transplantation, 172, 832, 1035
Transport globulins, 681
Transport maximum (T_m),
   1011–1012
Transport proteins, 77
Transport vesicles, 102
Transposition of the great
   vessels, 800
Transrectal prostatic ultrasound,
   1094
Transthyretin, 652, 682
Transudate, 165
Transurethral prostatectomy
   (TURP), 1094
Transverse acetabular
   ligament, 306
Transverse arch, foot, 284
Transverse cervical nerve, 469
Transverse colon, 943
Transverse costal facets,
   vertebrae, 258
Transverse fibers, 499, 500
Transverse foramina,
   vertebrae, 256
Transverse fracture, 226
Transverse lines, sacrum, 261
Transverse mesocolon, 911
Transverse perineal muscle, 383
Transverse (horizontal) plane, 44
Transverse processes,
   vertebrae, 253
Transverse (cross) section, 44
Transverse sinus, 492, 789
Transverse tubules (T tubules), 319
Transversus (muscle name), 365
Transversus abdominis muscle,
   380–381
Transversus thoracis muscle,
   380–381
Trapezium, 274
Trapezius muscle, 365, 384
Trapezoid, 274
Trauma. *See* Injury(ies)
Treppe, 337
Tretinoin (Retin-A), 186
Triacylglycerols. *See* Triglycerides
Triad, 320
Triaxial joint, 296
Triazolam (Halcion), 579

Tricarboxylic acid cycle. *See* Citric
   acid cycle
Triceps brachii muscle, 342, 363,
   364, 389–391
Trick knee, 309
Tricuspid valve, 718
Trigeminal nerve (V), 522, 528
Trigeminal neuralgia (tic
   douloureux), 522
Triglycerides
   blood levels, 977
   catabolism, 567, 649, 916, 951,
      973–975
   formation, 75, 663
   storage, 74
   structure and function, 74
Trigone, 1031
Triiodothyronine (T_3), 652,
   658, 668
Trimesters, 1126
Triose, 70
Tripeptide, 78
Triple covalent bond, 60
Triplet code, 109
Triplets, 1143
Triquetrum, 274
Trisaccharides, 950
Trisomy, 1156
Trisomy 21. *See* Down's syndrome
Tritium, 54, 55
Trochanter, 207, 280
Trochlea
   definition, 207
   humerus, 273
   talus, 283
Trochlear nerve (IV), 521, 528
Trochlear notch, 273
Trophoblast, 1127
Tropic (trophic) hormones, 648
Tropomyosin, 323
Troponin, 323
True labor, 1140
True (lesser) pelvis, 277
True ribs, 261
Trunk(s)
   arteries, 786–789
   lymphatic, 811
   spinal nerves, 469
   veins, 793
Trypsin, 81, 935, 950
Trypsinogen, 935, 950
Tryptophan, 637, 639, 979
T (transverse) tubules, 319
Tuberal area, hypothalamus,
   504, 505
Tubercle
   definition, 207
   humerus, greater/lesser, 271
   pubic, 276
   rib, 261
Tuberculin skin test, 844
Tuberculosis, 860
Tuberosity
   definition, 207
   humerus, deltoid, 271
   iliac, 276
   ischial, 276
   radial, 273
   tibial, 280
Tubuloacinar glands, 151
Tubuloalveolar glands, 151

d-Tubocurarine, 451
Tubular fluid, 1005
Tubular glands, 151
Tubular maximum, 1011–1012
Tubulin, 97
Tumor(s), 132, 418. *See also* Cancer
Tumor grading, 172
Tumor necrosis factor (TNF), 822, 828, 847
Tumor-specific antigens, 824
Tumor staging, 172
Tunica albuginea, 1080, 1096
Tunica externa (tunica adventitia), 751
Tunica intima (tunica interna), 751
Tunica media, 751
Tunica vaginalis, 1080
Turbulence, 761–762
Turner's syndrome, 1156
Turnover time, 86
Tussis, 902
T wave, 728
Twins, 1143
Twitch, 336–337
Two-point discrimination test, 543
Tympanic cavity (middle ear), 242, 617, 618
Tympanic duct (scala tympani), 623, 626
Tympanic membrane, 242, 617, 618, 629
Tympanic reflex, 529
Type 1 (insulin dependent) diabetes, 663, 664, 848
Type 2 (non-insulin dependent) diabetes, 663, 664
Type AB blood, 690, 691
Type A blood, 690, 691
Type A fibers, axon, 439, 539
Type B blood, 690, 691
Type B fibers, axon, 439
Type C fibers, axon, 439, 539
Type O blood, 690, 691
Tyrosine, 182, 637, 639, 652, 952

U

Ubiquinone, 969
Ubiquitin, 101
Ulcer, 187
Ulna, 273, 308
Ulnar artery, 783, 784
Ulnar collateral ligament, 304
Ulnar head, 273
Ulnar nerve, 273, 469, 470
Ulnar notch, 273
Ulnar vein, 793
Ultrasound, 47
Ultraviolet (UV) radiation, 182–183
Umami, 593, 595
Umbilical artery, 798, 1031, 1133
Umbilical cord (stalk), 1133
Umbilical ligaments, middle/lateral, 1031
Umbilical vein, 798, 1133
Unconscious, 580
Unicellular exocrine glands, 150
Unicellular glands, 150
Uniparental disomy, 1153
Unipennate muscle, 360

Unipolar neurons, 416, 417
Universal donors, 692
Universal recipients, 692
Unmyelinated axon, 421, 422, 436
Unsaturated fatty acid, 73
Upper limbs
    arteries, 783
    bones, 271–275, 286
    joints, 308
    movements, 308
    muscles, 386–395
    veins, 793, 794
Upper motor neuron, 549
Upper respiratory system, 858, 861–863
Up-regulation, 640
Uracil, 82
Urea, 939, 978, 1009, 1010, 1023, 1030
Urea cycle, 978
Ureteral openings, 1031
Ureters, 999, 1031, 1033
Urethra, 999, 1032–1033, 1078, 1088
Urethral orifice, external, 1032
Urethral sphincter, internal/external, 383, 1031, 1032
Uric acid, 1009, 1010, 1030
Uridine triphosphate (UTP), 973
Urinalysis, 1027, 1030
Urinary bladder, 999, 1032, 1033–1034
Urinary obstruction, 1034
Urinary system
    age-related changes, 1035
    anatomy, 999–1000
    functions, 999–1000
    integration with other systems, 1036, 1037
    kidney. *See* Kidney(s)
    overview, 35, 999, 1037–1038
    parasympathetic effects, 573
    prenatal development, 1137
    smooth muscle in, 351–352
    sympathetic effects, 573
    urine transport, storage, and elimination, 1030–1035
Urinary tract, 999
Urination, 576, 999, 1033–1035
Urine
    characteristics, 999, 1027
    color, 1027
    composition, 1027, 1030
    formation, 1009–1012
    osmotic concentration, 1023–1025
    solutes in, 1010
    transport, storage, and elimination, 1030–1035, 1039
    volume regulation, 1023–1025
Urobilin, 1027
Urobilinogens, 687, 946
Urobilins, 687, 946
Urogastrone, 932
Urogenital diaphragm, 382–383, 1032
Urogenital triangle, 382–383
Urologist, 1037
Urticaria, 201
Uterine artery, 1101
Uterine cavity, 1100, 1101
Uterine cycle
    hormonal regulation, 1112

menses, 1103–1104, 1111
    overview, 1103
    proliferative phase, 1104–1105, 1111
    secretory phase, 1105, 1111
Uterine fibroids (leiomyomas), 1117
Uterine glands, 1103
Uterine prolapse, 1117
Uterine tubes, 1077, 1100–1101, 1123
Uterine wall, 1101–1103
Uterosacral ligaments, 1101
Uterus
    anatomy, 1101–1103
    changes in pregnancy, 1139, 1140
    histology, 1103–1104
    overview, 1077
    suspensory ligaments, 1101
Utricle, 618, 621, 622
Uvea (vascular layer), eye, 599–601
Uvula, 916

V

Vaccine, 332, 465, 829, 916
Vagina, 1077, 1105–1106
Vaginal branches, blood vessels, 1106
Vaginal canal, 1106
Vaginal smear, 1106
Vaginitis, 1106
Vagus nerve (X), 525–527, 528, 896
Valence shell, 56
Valine, 979
Valium (diazepam), 579
Vallate papillae, 594, 595
Valsalva maneuver, 772
Valves
    cardiac, 719–721
    lymphatic, 811
    venous, 752, 758–759
Valvular heart disease, 721
Variable segments, antibody, 837
Varicella-zoster virus, 465
Varicose veins, 759
Varicosity, sympathetic, 566
Vasa recta, 1007, 1013, 1026, 1027
Vasa vasorum, 752
Vascular endothelial growth factor (VEGF), 758
Vascular layer (uvea), eye, 599–601
Vascular murmur, 801
Vascular phase, hemostasis, 701–702
Vascular resistance, 760–761, 762
Vascular spasm, 701
Vas deferens (ductus deferens), 651, 1078, 1088, 1089
Vasectomy, 1117
Vasoactive intestinal peptide (VIP), 941, 942
Vasoconstriction, 567, 650, 752
Vasoconstrictors, 769
Vasodilation, 567, 752
Vasodilators, 768, 769
Vasomotion, 758
Vasomotor center, 496, 759, 769
Vasomotor reflex, 576
Vasomotor tone, 770

Vasopressin. *See* Antidiuretic hormone
Vasovagal, 588
Vastus intermedius muscle, 398
Vastus lateralis muscle, 370, 398
Vastus medialis muscle, 398
Veins
    abdomen, 794, 796–797
    vs. arteries, 752, 753
    brain, 789, 792, 793
    characteristics, 157, 750
    chest, 793, 794
    head and neck, 789, 792, 793
    hepatic portal system, 797–798
    lower limbs, 793–794, 796
    pelvis, 794, 796
    structure, 751–752, 753
    systemic, 789–798
    types, 758
    upper limbs, 793, 794
    valves, 758–759
Veins listed
    adrenal, 796
    antebrachial, 793
    arcuate, 1003
    axillary, 793
    azygos, 793
    basilic, 793
    brachial, 793
    brachiocephalic, 793
    cardiac, 715, 722
    central retinal, 603
    cephalic, 793
    cerebral, 789, 793
    colic, 797, 798
    cortical radiate, 1003
    cystic, 798
    digital, 793
    esophageal, 793
    facial, 793
    femoral, 793, 794
    fibular, 793
    gastric, 798
    gastroepiploic, 797, 798
    gluteal, 794
    gonadal, 796
    great cerebral, 793
    great saphenous, 793
    hemiazygos, 793
    hepatic, 796
    hepatic portal, 797
    ileocolic, 798
    iliac, 794, 796
    inferior mesenteric, 797
    inferior vena cava, 715, 793–797
    intercostal, 793
    interlobar, 1003
    intestinal, 798
    jugular, 491, 495, 793
    lumbar, 796
    maxillary, 793
    median cubital, 682
    obturator, 794
    ovarian, 796, 1096
    palmar, 793
    pancreatic, 797
    pancreaticoduodenal, 798
    phrenic, 797
    plantar, 793

popliteal, 794
pudendal, 794
pulmonary, 718, 781
radial, 790
rectal, 797
renal, 796, 1001, 1003
sacral, 794
small saphenous, 793
splenic, 797, 820
subclavian, 781, 793
superior mesenteric, 798
superior vena cava, 715, 789, 793, 795
temporal, 793
testicular, 796
thoracic, 793
tibial, 793
trabecular, 820
ulnar, 793
umbilical, 798, 1133
vertebral, 793
Vellus hairs, 191
Venipuncture, 682
Venoconstriction, 759
Venous pressure, 760, 762, 764–765
Venous reserve, 759
Venous return, 740, 741, 758, 764–765, 775
Venous valves, 752, 758–759
Ventral, 42
Ventral nuclei, thalamus, 504
Ventral ramus, 466–467, 471
Ventral respiratory group (VRG), 892, 894
Ventral root(s), 459
Ventricles
brain, 418, 490–491
heart, 711, 718–719
Ventricular diastole, 736
Ventricular ejection, 734
Ventricular septal defect, 800
Ventricular systole, 734–736
Ventricular tachycardia (VT), 730
Venules, 750, 753, 758
Verapamil, 742
Vermiform appendix, 816, 943
Vermis, 501
Vernix caseosa, 192
Versatility, of adaptive immunity, 830
Vertebrae
age-related changes, 285
anatomy, 253–254
cervical, 254, 255–257
coccyx, 261
functions, 252
lumbar, 254, 258, 260
regional differences, 257
sacrum, 258–261
thoracic, 254
Vertebral arch, 253
Vertebral artery, 495, 781
Vertebral body, 253
Vertebral border, scapula, 270
Vertebral canal, 253
Vertebral column
joints, 302
movements, 301–302
muscles, 377–381
regions, 254–261

spinal curvature, 252–253
vertebral anatomy, 253–254
Vertebral end, rib, 261
Vertebral end plates, 300
Vertebral foramen, 253
Vertebral ribs, 261
Vertebral vein, 793
Vertebra prominens, 258
Vertebrates, 28
Vertebrochondral ribs, 261
Vertebrosternal ribs, 261
Vertigo, 630
Very low-density lipoproteins (VLDLs), 975
Vesicles
brain, 490
cell, 122
Vesicouterine pouch, 1095
Vesicular transport, 122, 126, 136
Vessels
blood. *See* Blood vessels
lymphatic, 809
Vestibular branch, vestibulocochlear nerve, 524, 621
Vestibular bulbs, 1106
Vestibular complex, 618
Vestibular duct (scala vestibuli), 623, 626
Vestibular folds, 864
Vestibular ganglia, 621
Vestibular glands, greater/lesser, 1107
Vestibular ligaments, 864
Vestibular nuclei, 499, 524, 552, 621
Vestibule
female external genitalia, 1105, 1106
internal ear, 618
nasal, 861
oral cavity, 370, 915
Vestibulocochlear nerve (VIII), 524, 528, 621, 628
Vestibulo-ocular reflexes, 529
Vestibulospinal tracts, 550, 552, 622
Viagra, 1114
Villi, 1128
Villikinin, 940
VIP (vasoactive intestinal peptide), 941, 942
Virilism, 674
Virus(es), 843
Viscera, 45
Visceral epithelium, 1007–1008
Visceral motor neurons, 417
Visceral motor system. *See* **Autonomic nervous system**
Visceral nervous system. *See* **Autonomic nervous system**
Visceral pericardium, 48, 712
Visceral peritoneum, 48, 909, 1096
Visceral pleura, 48, 869, 872
Visceral portion, serous membrane, 164
Visceral reflex arc, 575
Visceral reflexes, 478, 529, 575–577, 586
Visceral sensory neurons, 416
Visceral sensory pathways, 549

Visceral smooth muscle cells, 353
Visceral smooth muscle tissue, 912
Viscera serosa, 45
Viscosity, 761, 762
Vision
accommodation, 605–606
accommodation problems, 606
acuity, 607
age-related changes, 606
image reversal, 606–607
light and dark adaptation, 613
overview, 536, 631
physiology, 607–615
refraction, 604–605
Visual acuity, 607
Visual association area, 513
Visual axis, eye, 601
Visual cortex, 513
Visual pathways, 613–615
Visual pigments, 607, 609
Vital capacity, 881, 882
Vitamin(s)
absorption, 946, 953
definition, 185
fat-soluble, 953, 986–987
water-soluble, 953, 987
Vitamin A
bone growth and, 221
deficiency/excess, 962, 986, 997
functions in human body, 986
recommended daily allowance, 986
in visual physiology, 607, 612
Vitamin B_1 (thiamine), 987
Vitamin B_2 (riboflavin), 987
Vitamin B_3 (niacin), 987
Vitamin B_5 (pantothenic acid), 221, 987
Vitamin B_6 (pyridoxine), 987
Vitamin B_7 (biotin), 945, 987
Vitamin B_9 (folic acid), 987
Vitamin B_{12} (cobalamin)
absorption, 689, 925
bone growth and, 221
deficiency/excess, 987
functions in human body, 987
recommended daily allowance, 987
Vitamin C (ascorbic acid), 987
Vitamin D
deficiency/excess, 986
functions in human body, 986
recommended daily allowance, 986
Vitamin D_3 (cholecalciferol), 184–185, 221
Vitamin deficiency disease, 986
Vitamin E, 986
Vitamin K
absorption in large intestine, 946
in blood clotting, 705
bone growth and, 221
deficiency, 705
deficiency/excess, 946, 986
functions in human body, 986
recommended daily allowance, 986
Vitamin toxicity, 986
Vitiligo, 183
Vitrectomy, 600

Vitreous body, 603, 604
Vitreous humor, 604
Vocal cords, 864
Vocal folds, 864
Vocal ligaments, 864
Volatile acids, 1056, 1057
Volkmann's canals, 211
Volt (V), 126
Voltage-gated channels, 328, 428–429
Volume depletion, 1053
Vomer, 246–247
Vomiting (emesis), 929
V_T (tidal volume), 877, 881
VT; V-tach (ventricular tachycardia), 730
Vulva, 1106
Vulvovaginal candidiasis, 1117

W
Wadlow, Robert, 222
Wallerian degeneration, 424
Wandering (free) macrophages, 153, 823
Waste products
concentration, 39
processing in large intestine, 946–947
Water (H_2O)
absorption, 952
allocation of losses, 1049–1050
balance. *See* Fluid balance
digestive secretion and reabsorption, 952
distribution of gains, 1050
facultative reabsorption, 1024–1025
functions, 85
in human body, 65, 1044
metabolic generation, 1048
movement, 1045–1046
obligatory reabsorption, 1024
properties, 65–66
reabsorption at proximal convoluted tubule, 1018
reabsorption in collecting system, 1025, 1026
reabsorption in large intestine, 946
in solutions, 66–67
sources, 85
structure, 58, 61, 85
weight loss and, 1049
Water channels (aquaporins), 118, 1024
Water depletion (dehydration), 1049
Water excess, 1050
"Water follows salt," 1020
Water gains, 1048
Water intoxication, 1050
Water receptors, 595–596
Water-soluble vitamins, 953, 987
Water vapor, 883
Wavelength, 624, 626
Wave summation, 337, 338
Weak acid, 69, 1056
Weak base, 69, 1056
Weaning, 1145
Weight loss, water and, 1049
Wernicke's area, 514
Wharton's jelly, 154

INDEX

Wheeze, 902
Whiplash, 257
White blood cells (WBCs), 159,
    679, 680–681, 693, 707
  abundance, 693, 698
  circulation and movement, 693
  differential count, 697–698
  production, 698–700
  types, 681, 693, 696–697, 698
  in urine, 1030
White coat hypertension, 801
White column, posterior/anterior/
    lateral, 463
White commissure, anterior, 463
White fat, 155
White matter
  cerebellum, 500
  cerebrum, 508, 510

characteristics, 421
  medulla oblongata, 498
  midbrain, 502
  pons, 499
  spinal cord, 463
White muscles, 347
White pulp, spleen, 820
White ramus, 563
White ramus communicans, 467
Whole blood, 680–681. *See also*
    Blood
Wisdom teeth, 919
Withdrawal reflex, 477, 480–481
Word roots, 29
Work, 62
Wormian (sutural) bone, 206, 235
Wright's stain, 693
Wrist (carpus), 273

X

X chromosome, 1154
Xenograft, 851
Xenotransplant, 172
Xerosis, 181
Xiphoid process, 263
X-linked inheritance, 1154
X-linked traits, 1154
X-rays, 46

Y

Y chromosome, 1154
Yellow bone marrow, 205,
    213, 687
Yolk sac, 1129, 1130
Yolk stalk, 1132

Z

Zinc, 985
Z line, 322–323
Zoloft (sertraline), 443
Zona fasciculata, 658–660
Zona glomerulosa, 658, 659
Zona pellucida, 1099
Zona reticularis, 659, 660
Zone of overlap, 322
Zostavax, 465
Zwitterions, 78, 1059
Zygomatic arch, 241, 242
Zygomatic bones, 247
Zygomaticofacial foramen, 247
Zygomatic process, 241
Zygomaticus muscle, major/
    minor, 372
Zygote, 1125

COMMON ABBREVIATIONS USED
IN SCIENCE

ACh	acetylcholine
AChE	acetylcholinesterase
ACTH	adrenocorticotropic hormone
ADH	antidiuretic hormone
ADP	adenosine diphosphate
AIDS	acquired immunodeficiency syndrome
ALS	amyotrophic lateral sclerosis
AMP	adenosine monophosphate
ANP	atrial natriuretic peptide
ANS	autonomic nervous system
AP	arterial pressure
ARDS	adult respiratory distress syndrome
atm	atmospheric pressure
ATP	adenosine triphosphate
ATPase	adenosine triphosphatase
AV	atrioventricular
AVP	arginine vasopressin
BCOP	blood colloid osmotic pressure
BMR	basal metabolic rate
BPG	bisphosphoglycerate
bpm	beats per minute
BUN	blood urea nitrogen
C	large calorie; Celsius
CABG	coronary artery bypass graft
CAD	coronary artery disease
cAMP	cyclic-AMP
CAPD	continuous ambulatory peritoneal dialysis
CCK	cholecystokinin
CD	cluster of differentiation
CF	cystic fibrosis
CHF	congestive heart failure
CHP	capillary hydrostatic pressure
CsHP	capsular hydrostatic pressure
CNS	central nervous system
CO	cardiac output; carbon monoxide
CoA	coenzyme A
COMT	catechol-O-methyltransferase
COPD	chronic obstructive pulmonary disease
CP	creatine phosphate
CPK, CK	creatine phosphokinase
CPM	continuous passive motion
CPR	cardiopulmonary resuscitation
CRF	chronic renal failure
CRH	corticotropin-releasing hormone
CSF	cerebrospinal fluid; colony-stimulating factors
CT	computerized tomography; calcitonin
CVA	cerebrovascular accident
CVS	cardiovascular system
DAG	diacylglycerol
DC	Doctor of Chiropractic
DCT	distal convoluted tubule
DDST	Denver Developmental Screening Test
DIC	disseminated intravascular coagulation
DJD	degenerative joint disease
DMD	Duchenne's muscular dystrophy
DNA	deoxyribonucleic acid
DO	Doctor of Osteopathy
DPM	Doctor of Podiatric Medicine
DSA	digital subtraction angiography
E	epinephrine
ECF	extracellular fluid
ECG	electrocardiogram
EDV	end-diastolic volume
EEG	electroencephalogram
EKG	electrocardiogram
ELISA	enzyme-linked immunosorbent assay
EPSP	excitatory postsynaptic potential
ERV	expiratory reserve volume
ESV	end-systolic volume
ETS	electron transport system
FAD	flavin adenine dinucleotide
FAS	fetal alcohol syndrome
FES	functional electrical stimulation
FMN	flavin mononucleotide

FRC	functional residual capacity
FSH	follicle-stimulating hormone
GABA	gamma aminobutyric acid
GAS	general adaptation syndrome
GC	glucocorticoid
GFR	glomerular filtration rate
GH	growth hormone
GH–IH	growth hormone–inhibiting hormone
GHP	glomerular hydrostatic pressure
GH–RH	growth hormone–releasing hormone
GIP	gastric inhibitory peptide
GnRH	gonadotropin-releasing hormone
GTP	guanosine triphosphate
Hb	hemoglobin
hCG	human chorionic gonadotropin
HCl	hydrochloric acid
HDL	high-density lipoprotein
HDN	hemolytic disease of the newborn
hGH	human growth hormone
HIV	human immunodeficiency virus
HLA	human leukocyte antigen
HMD	hyaline membrane disease
HP	hydrostatic pressure
hPL	human placental lactogen
HR	heart rate
Hz	Hertz
ICF	intracellular fluid
ICOP	interstitial fluid colloid osmotic pressure
IGF	insulin-like growth factor
IH	inhibiting hormone
IM	intramuscular
IP3	inositol triphosphate
IPSP	inhibitory postsynaptic potential
IRV	inspiratory reserve volume
ISF	interstitial fluid
IUD	intrauterine device
IVC	inferior vena cava
IVF	in vitro fertilization
kcal	kilocalorie
LDH	lactate dehydrogenase
LDL	low density lipoprotein
L-DOPA	levodopa
LH	luteinizing hormone
LLQ	left lower quadrant
LM	light micrograph
LSD	lysergic acid diethylamide
LUQ	left upper quadrant
MAO	monoamine oxidase
MAP	mean arterial pressure
MC	mineralocorticoid
MD	Doctor of Medicine
mEq	millequivalent; (10^{-3})
MHC	major histocompatibility complex
MI	myocardial infarction
mm Hg	millimeters of mercury
mmol	millimole
mOsm	milliosmole
MRI	magnetic resonance imaging
mRNA	messenger RNA
MS	multiple sclerosis
MSH	melanocyte-stimulating hormone
MSH–IH	melanocyte-stimulating hormone–inhibiting hormone
NAD	nicotinamide adenine dinucleotide
NE	norepinephrine
NFP	net filtration pressure
NHP	net hydrostatic pressure
NO	nitric oxide
NRDS	neonatal respiratory distress syndrome
OP	osmotic pressure
Osm	osmoles
OXT	oxytocin
PAC	premature atrial contraction
PAT	paroxysmal atrial tachycardia
PCT	proximal convoluted tubule

PCV	packed cell volume
PEEP	positive end-expiratory pressure
PET	positron emission tomography
PFC	perfluorochemical emulsion
PG	prostaglandin
PID	pelvic inflammatory disease
PIH	prolactin-inhibiting hormone
PIP	phosphatidylinositol
PKC	protein kinase C
PKU	phenylketonuria
PLC	phospholipase C
PMN	polymorphonuclear leukocyte
PNS	peripheral nervous system
PR	peripheral resistance
PRF	prolactin-releasing factor
PRL	prolactin
psi	pounds per square inch
PT	prothrombin time
PTA	post-traumatic amnesia; plasma thromboplastin antecedent
PTC	phenylthiocarbamide
PTH	parathyroid hormone
PVC	premature ventricular contraction
RAS	reticular activating system
RBC	red blood cell
RDA	recommended daily allowance
RDS	respiratory distress syndrome
REM	rapid eye movement
RER	rough endoplasmic reticulum
RH	releasing hormone
RHD	rheumatic heart disease
RLQ	right lower quadrant
RNA	ribonucleic acid
rRNA	ribosomal RNA
RUQ	right upper quadrant
SA	sinoatrial
SCA	sickle cell anemia
SCID	severe combined immunodeficiency disease
SEM	scanning electron micrograph
SER	smooth endoplasmic reticulum
SGOT	serum glutamic oxaloacetic transaminase
SIADH	syndrome of inappropriate ADH secretion
SIDS	sudden infant death syndrome
SLE	systemic lupus erythematosus
SNS	somatic nervous system
STD	sexually transmitted disease
SV	stroke volume
SVC	superior vena cava
T_3	triiodothyronine
T_4	tetraiodothyronine, or thyroxine
TB	tuberculosis
TBG	thyroid-binding globulin
TEM	transmission electron micrograph
TIA	transient ischemic attack
T_m	transport (tubular) maximum
TMJ	temporomandibular joint
t-PA	tissue plasminogen activator
TRH	thyrotropin-releasing hormone
tRNA	transfer RNA
TSH	thyroid-stimulating hormone
TSS	toxic shock syndrome
U.S.	United States
UTI	urinary tract infection
UTP	uridine triphosphate
UV	ultraviolet
$\dot{V}_A$	alveolar ventilation
V_D	anatomic dead space
$\dot{V}_E$	respiratory minute volume
V_T	tidal volume
VF	ventricular fibrillation
VLDL	very low-density lipoprotein
VPRC	volume of packed red cells
VT	ventricular tachycardia
WBC	white blood cell

FOREIGN WORD ROOTS, PREFIXES, SUFFIXES, AND COMBINING FORMS

Each entry starts with the commonly used form or forms of the prefix, suffix, or combining form followed by the word root (shown in italics) and its English translation. One example is also given to illustrate the use of each entry.

a-, *a-,* without: avascular
ab-, *ab,* from: abduct
-ac, *-akos,* pertaining to: cardiac
acr-, *akron,* extremity: acromegaly
ad-, *ad,* to, toward: adduct
aden-, adeno-, *adenos,* gland: adenoid
adip-, *adipos,* fat: adipocytes
aer-, *aeros,* air: aerobic metabolism
-al, *-alis,* pertaining to: brachial
alb-, *albicans,* white: albino
-algia, *algos,* pain: neuralgia
allo-, *allos,* other: allograft
ana-, *ana,* up, back: anaphase
andro-, *andros,* male: androgen
angio-, *angeion,* vessel: angiogram
ante-, *ante,* before: antebrachial
anti-, ant-, *anti,* against: antibiotic
apo-, *apo,* from: apocrine
arachn-, *arachne,* spider: arachnoid
arter-, *arteria,* artery: arterial
arthro-, *arthros,* joint: arthroscopy
astro-, *aster,* star: astrocyte
atel-, *ateles,* imperfect: atelectasis
aur-, *auris,* ear: auricle
auto-, *auto,* self: autonomic
baro-, *baros,* pressure: baroreceptor
bi-, *bi-* two: bifurcate
bio-, *bios,* life: biology
-blast, *blastos,* precursor: osteoblast
brachi-, *brachium,* arm: brachiocephalic
brachy-, *brachys,* short: brachydactyly
brady-, *bradys,* slow: bradycardia
bronch-, *bronchus,* windpipe, airway: bronchial
carcin-, *karkinos,* cancer: carcinoma
cardi-, cardio-, *kardia,* heart: cardiac
-cele, *kele,* tumor, hernia, or swelling: blastocele
-centesis, *kentesis,* puncture: thoracocentesis
cephal-, *cephalos,* head: brachiocephalic
cerebr-, *cerebrum,* brain: cerebral hemispheres
cerebro-, *cerebros,* brain: cerebrospinal fluid
cervic-, *cervicis,* neck: cervical vertebrae
chole-, *chole,* bile: cholecystitis
-chondrion, *chondrion,* granule: mitochondrion
chondro-, *chondros,* cartilage: chondrocyte
chrom-, chromo-, *chroma,* color: chromatin
circum-, *circum,* around: circumduction
-clast, *klastos,* broken: osteoclast
coel-, -coel, *koila,* cavity: coelom
colo-, *kolon,* colon: colonoscopy
contra-, *contra,* against: contralateral
corp-, *corpus,* body: corpuscle
cortic-, *cortex,* rind or bark: corticospinal
cost-, *costa,* rib: costal
cranio-, *cranium,* skull: craniosacral
cribr-, *cribrum,* sieve: cribriform
-crine, *krinein,* to separate: endocrine
cut-, *cutis,* skin: cutaneous
cyan-, *kyanos,* blue: cyanosis
cyst-, -cyst, *kystis,* sac: blastocyst
cyt-, cyto-, *kytos,* a hollow cell: cytology
de-, *de,* from, away: deactivation

dendr-, *dendron,* tree: dendrite
dent-, *dentes,* teeth: dentition
derm-, *derma,* skin: dermatome
desmo-, *desmos,* band: desmosome
di-, *dis,* twice: disaccharide
dia-, *dia,* through: diameter
digit-, *digit,* a finger or toe: digital
dipl-, *diploos,* double: diploid
dis-, *dis,* apart, away from: disability
diure-, *diourein,* to urinate: diuresis
dys-, *dys,* painful: dysmenorrhea
-ectasis, *ektasis,* expansion: atelectasis
ecto-, *ektos,* outside: ectoderm
-ectomy, *ektome,* excision: appendectomy
ef-, *ex,* away from: efferent
emmetro-, *emmetros,* in proper measure: emmetropia
encephalo-, *enkephalos,* brain: encephalitis
end-, endo-, *endon,* within: endometrium
entero-, *enteron,* intestine: enteric
epi-, *epi,* upon: epimysium
erythema-, *erythema,* flushed (skin): erythematosis
erythro-, *erythros,* red: erythrocyte
ex-, *ex,* out of, away from: exocytosis
extra-, *exter,* outside of, beyond, in addition: extracellular
ferr-, *ferrum,* iron: transferrin
fil-, *filum,* thread: filament
-form, *-formis,* shape: fusiform
gastr-, *gaster,* stomach: gastrointestinal
-gen, -genic, *gennan,* to produce: mutagen
genicula-, *geniculum,* kneelike structure: geniculates
genio-, *geneion,* chin: geniohyoid
gest-, *gesto,* to bear: gestation
glosso-, -glossus, *glossus,* tongue: hypoglossal
glyco-, *glykys,* sugar: glycogen
-gram, *gramma,* record: myogram
gran-, *granulum,* grain: granulocyte
-graph, -graphia, *graphein,* to write, record: electroencephalograph
gyne-, gyno-, *gynaikos,* woman: gynecologist
hem-, hemo-, *haima,* blood: hemopoiesis
hemi-, *hemi-,* one half: hemisphere
hepato-, *hepaticus,* liver: hepatocyte
hetero-, *heteros,* other: heterozygous
histo-, *histos,* tissue: histology
holo-, *holos,* entire: holocrine
homeo-, *homoios,* similar: homeostasis
homo-, *homos,* same: homozygous
hyal-, hyalo-, *hyalos,* glass: hyaline
hydro-, *hydros,* water: hydrolysis
hyo-, *hyoeides,* U-shaped: hyoid bone
hyper-, *hyper,* above: hyperpolarization
hypo-, *hypo,* under: hypothyroid
hyster-, *hystera,* uterus: hysterectomy
-ia, *-ia,* state or condition: insomnia
idi-, *idios,* one's own: idiopathic
in-, *in-,* in, within, or denoting negative effect: inactivate
infra-, *infra,* beneath: infraorbital
inter-, *inter,* between: interventricular
intra-, *intra,* within: intracapsular
ipsi-, *ipse,* itself: ipsilateral
iso-, *isos,* equal: isotonic
-itis, *-itis,* inflammation: dermatitis

karyo-, *karyon,* body: megakaryocyte
kerato-, *keros,* horn: keratin
kino-, -kinin, *kinein,* to move: bradykinin
lact-, lacto-, -lactin, *lac,* milk: prolactin
lapar-, *lapara,* flank or loins: laparoscopy
-lemma, *lemma,* husk: plasmalemma
leuk-, leuko-, *leukos,* white: leukemia, leukocyte
liga-, *ligare,* to bind together: ligase
lip-, lipo-, *lipos,* fat: lipoid
lith-, *lithos,* stone: cholelithiasis
lys- lyso-, *lysis,* a loosening: hydrolysis
macr-, *makros,* large: macrophage
mal-, *mal,* abnormal: malabsorption
mammilla-, *mammilla,* nipple: mammillary
mast-, masto-, *mastos,* breast: mastoid
mega-, *megas,* big: megakaryocyte
melan-, *melas,* black: melanocyte
men-, *men,* month: menstrual
mero-, *meros,* part: merocrine
meso-, *mesos,* middle: mesoderm
meta-, *meta,* after, beyond: metaphase
micr-, *mikros,* small: microscope
mono-, *monos,* single: monocyte
morph-, morpho-, *morphe,* form: morphology
multi-, *multus,* much, many: multicellular
-mural, *murus,* wall: intramural
myelo-, *myelos,* marrow: myeloblast
myo-, *mys,* muscle: myofilament
narc-, *narkoun,* to numb or deaden: narcotics
nas-, *nasus,* nose: nasolacrimal duct
natri-, *natrium,* sodium: natriuretic
necr-, *nekros,* corpse: necrosis
nephr-, *nephros,* kidney: nephron
neur-, neuri-, neuro-, *neuron,* nerve: neuromuscular
oculo-, *oculus,* eye: oculomotor
odont-, *odontos,* tooth: odontoid process
-oid, *eidos,* form, resemblance: odontoid process
oligo-, *oligos,* little, few: oligopeptide
-ology, *logos,* the study of: physiology
-oma, *-oma,* swelling: carcinoma
onco-, *onkos,* mass, tumor: oncology
oo-, *oon,* egg: oocyte
ophthalm-, *ophthalmos,* eye: ophthalmic nerve
-opia, *ops,* eye: myopia
orb-, *orbita,* a circle: orbicularis oris
orchi-, *orchis,* testis: orchiectomy
orth-, *orthos,* correct, straight: orthopedist
-osis, *-osis,* state, condition: neurosis
ost-, oste-, osteo-, *osteon,* bone: osteocyte
oto-, *otikos,* ear: otolith
para-, *para,* beyond: paraplegia
path-, -pathy, patho-, *pathos,* disease:pathology
pedia-, *paidos,* child: pediatrician
per-, *per,* through, throughout: percutaneous
peri-, *peri,* around: perineurium
phago-, *phago,* to eat: phagocyte
-phasia, *phasis,* speech: aphasia
-phil, -philia, *philos,* love: hydrophilic
phleb-, *phleps,* a vein: phlebitis
-phobia, *phobos,* fear: hydrophobic
phot-, photo-, *phos,* light: photoreceptor

-phylaxis, *phylax,* a guard: prophylaxis
physio-, *physis,* nature: physiology
-plasia, *plasis,* formation: dysplasia
platy-, *platys,* flat: platysma
-plegia, *plege,* a blow, paralysis: paraplegia
-plexy, *plessein,* to strike: apoplexy
pneum-, *pneuma,* air: pneumotaxic center
pod-, podo-, *podos,* foot: podocyte
-poiesis, *poiesis,* making: hemopoiesis
poly-, *polys,* many: polysaccharide
post-, *post,* after: postanal
pre-, *prae,* before: precapillary sphincter
presby-, *presbys,* old: presbyopia
pro-, *pro,* before: prophase
proct-, *proktos,* anus: proctology
pterygo-, *pteryx,* wing: pterygoid
pulmo-, *pulmo,* lung: pulmonary
pulp-, *pulpa,* flesh: pulpitis
pyel-, *pyelos,* trough or pelvis: pyelitis
quadr-, *quadrans,* one quarter: quadriplegia
re-, *re-,* back, again: reinfection
retro-, *retro,* backward: retroperitoneal
rhin-, *rhis,* nose: rhinitis
-rrhage, *rhegnymi,* to burst forth: hemorrhage
-rrhea, *rhein,* flow, discharge: amenorrhea
sarco-, *sarkos,* flesh: sarcomere
scler-, sclero-, *skleros,* hard: sclera
-scope, *skopeo,* to view: colonoscope
-sect, *sectio,* to cut: transect
semi-, *semis,* half: semitendinosus
-septic, *septikos,* putrid: antiseptic
-sis, *-sis,* state or condition: metastasis
som-, -some, *soma,* body: somatic
spino-, *spina,* spine, vertebral column:
   spinothalamic pathway
stalsis, *staltikos,* contractile: peristalsis
sten-, *stenos,* a narrowing: stenosis
stomy, *stoma,* mouth, opening: colostomy
stylo-, *stylus,* stake, pole: styloid process
sub-, *sub,* below: subcutaneous
super-, *super,* above or beyond: superficial
supra-, *supra,* on the upper side: supraspinous fossa
syn-, *syn,* together: synthesis
tachy-, *tachys,* swift: tachycardia
telo-, *telos,* end: telophase
tetra-, *tettares,* four: tetralogy of Fallot
therm-, thermo-, *therme,* heat: thermoregulation
thorac-, *thorax,* chest: thoracentesis
thromb-, *thrombos,* clot: thrombocyte
-tomy, *tome,* to cut: appendectomy
tox-, *toxikon,* poison: toxemia
trans-, *trans,* through: transudate
tri-, *tres,* three: trimester
tropho-, *trophe,* nutrition: trophoblast
trophy, *trophikos,* nourishment: atrophy
-tropic, *trope,* turning: adrenocorticotropic
tropo-, *tropikos,* turning: troponin
uni-, *unus,* one: unicellular
uro-, -uria, *ouron,* urine: glycosuria
vas-, *vas,* vessel: vascular
zyg-, *zygotos,* yoke: zygote